Function Notation (2-3, 2-5, 2-6)

$f(x)$	Value of f at x
$(f \circ g)(x) = f[g(x)]$	Composite function
$f^{-1}(x)$	Value of inverse of f at x

Linear Equations and Functions (1-1, 2-2, 2-4)

$y = mx + b$	Slope–intercept form
$(y - y_1) = m(x - x_1)$	Point–slope form
$f(x) = mx + b$	Linear function
$y = b$	Horizontal line
$x = a$	Vertical line

Polynomial and Rational Forms (A-2, A-3, A-4, 1-8, 2-4, 3-1–3-5)

$f(x) = ax^2 + bx + c$	Quadratic function
$f(x) = a_n x^n + a_{n-1}x^{n-1} + \cdots + a_1 x + a_0$, $a_n \neq 0$, n a nonnegative integer	Polynomial function
$f(x) = \dfrac{p(x)}{q(x)}$, p and q polynomial functions, $q(x) \neq 0$	Rational function

Exponential and Logarithmic Functions (4-1–4-5)

$f(x) = b^x, \quad b > 0, b \neq 1$	Exponential function
$f(x) = \log_b x, \quad b > 0, b \neq 1$	Logarithmic function

$y = \log_b x$ if and only if $x = b^y, \quad b > 0, b \neq 1$

Matrices and Determinants (8-1, 9-1–9-6)

$\begin{bmatrix} a & b & c \\ d & e & f \end{bmatrix}$ Matrix

$\begin{vmatrix} a & b & c \\ d & e & f \\ g & h & i \end{vmatrix}$ Determinant

Arithmetic Sequence (

$a_1, a_2, \ldots, a_n, \ldots$

$a_n - a_{n-1} = d$	Common difference
$a_n = a_1 + (n - 1)d$	nth-term formula
$S_n = a_1 + \cdots + a_n = \dfrac{n}{2}[2a_1 + (n - 1)d]$	Sum of n terms
$S_n = \dfrac{n}{2}(a_1 + a_n)$	

Geometric Sequence (10-3)

$a_1, a_2, \ldots, a_n, \ldots$

$\dfrac{a_n}{a_{n-1}} = r$	Common ratio
$a_n = a_1 r^{n-1}$	nth-term formula
$S_n = a_1 + \cdots + a_n = \dfrac{a_1 - a_1 r^n}{1 - r}, \quad r \neq 1$	Sum of n terms
$S_n = \dfrac{a_1 - r a_n}{1 - r}, \quad r \neq 1$	
$S_\infty = a_1 + a_2 + \cdots = \dfrac{a_1}{1 - r}, \quad \lvert r \rvert < 1$	Sum of infinitely many terms

Factorial and Binomial Formulas (10-4)

$n! = n(n - 1) \cdots 2 \cdot 1, \quad n \in N$	n factorial
$0! = 1$	
$\dbinom{n}{r} = \dfrac{n!}{r!(n - r)!}, \quad 0 \leq r \leq n$	
$(a + b)^n = \displaystyle\sum_{k=0}^{n} \binom{n}{k} a^{n-k} b^k, \quad n \geq 1$	Binomial formula

Circle (2-1)

$(x - h)^2 + (y - k)^2 = r^2$	Center at (h, k); radius r
$x^2 + y^2 = r^2$	Center at $(0, 0)$; radius r

Parabola (2-4, 11-1)

$y^2 = 4ax$, $a > 0$, opens right; $a < 0$, opens left
Focus: $(a, 0)$; Directrix: $x = -a$;
Axis: x axis

$x^2 = 4ay$, $a > 0$, opens up; $a < 0$, opens down
Focus: $(0, a)$; Directrix: $y = -a$;
Axis: y axis

(Continued on back endpaper)

FIFTH EDITION

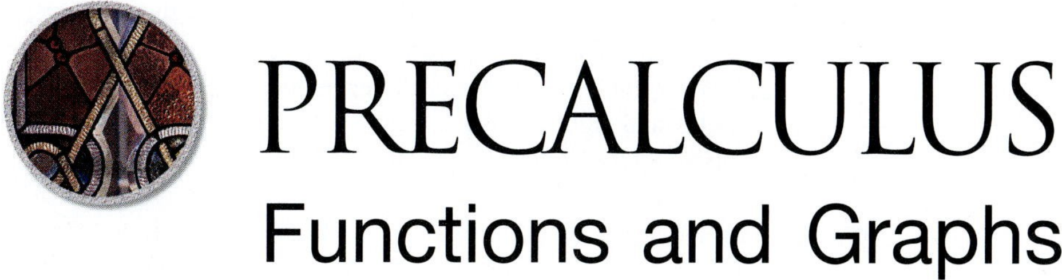

PRECALCULUS

Functions and Graphs

Barnett, Ziegler & Byleen's Precalculus Series

College Algebra, Seventh Edition
This book is the same as *College Algebra with Trigonometry* without the three chapters on trigonometry.
ISBN 0-07-236868-3

College Algebra with Trigonometry, Seventh Edition
This book is the same as *College Algebra* with three chapters on trigonometry added. Comparing *College Algebra with Trigonometry* with *Precalculus*, *College Algebra with Trigonometry* has more intermediate algebra review and starts trigonometry with angles and right triangles.
ISBN 0-07-236869-1

Precalculus: Functions and Graphs, Fifth Edition
This book differs from *College Algebra with Trigonometry* in that *Precalculus* starts at a higher level, placing intermediate algebra review in the appendix; and it starts trigonometry with the unit circle and circular functions.
ISBN 0-07-236871-3

College Algebra: A Graphing Approach
This book is the same as *Precalculus: A Graphing Approach* without the three chapters on trigonometry. This text assumes the use of a graphing utility.
ISBN 0-07-005710-9

Precalculus: A Graphing Approach
This book is the same as *College Algebra: A Graphing Approach* with three additional chapters on trigonometry. This text assumes the use of a graphing utility.
ISBN 0-07-005717-6

FIFTH EDITION

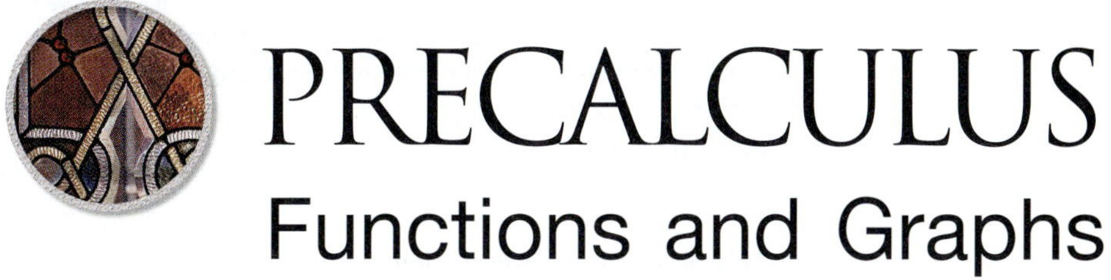

PRECALCULUS
Functions and Graphs

RAYMOND A. BARNETT
Merritt College

MICHAEL R. ZIEGLER
Marquette University

KARL E. BYLEEN
Marquette University

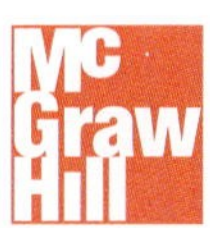

Boston Burr Ridge, IL Dubuque, IA Madison, WI New York San Francisco St. Louis
Bangkok Bogotá Caracas Lisbon London Madrid
Mexico City Milan New Delhi Seoul Singapore Sydney Taipei Toronto

McGraw-Hill Higher Education

A Division of The **McGraw-Hill** *Companies*

PRECALCULUS: FUNCTIONS AND GRAPHS, FIFTH EDITION

This book is printed on acid-free paper.

3 4 5 6 7 8 9 0 VNH/VNH 0 9 8 7 6 5 4 3 2 1

ISBN 0–07–236871–3

Vice president and editor-in-chief: *Kevin T. Kane*
Publisher: *JP Lenney*
Senior sponsoring editor: *Maggie Rogers*
Developmental editor: *Michelle Munn*
Editorial assistant: *Allyndreth Cassidy*
Marketing manager: *Mary K. Kittell*
Project manager: *Sheila M. Frank*
Lead producer: *Steve Metz*
Senior production supervisor: *Sandra Hahn*
Designer: *K. Wayne Harms*
Cover/interior designer: *Rokusek Design*
Senior supplement coordinator: *David A. Welsh*
Compositor: *GTS Graphics, Inc.*
Typeface: *10/12 Times Roman*
Printer: *Von Hoffmann Press, Inc.*

Library of Congress Cataloging-in-Publication Data

Barnett, Raymond A.
Precalculus: functions and graphs / Raymond A. Barnett, Michael R. Ziegler, Karl E. Byleen. — 5th ed.
p cm. — (Barnett, Ziegler & Byleen's precalculus series)
Includes indexes.
ISBN 0-07-236871-3 (acid-free paper)
1. Functions. 2. Algebra—Graphic methods. I. Ziegler, Michael R. II. Byleen, Karl E. III. Title.

QA331.3 .B38 2001
512.1—dc21 00-035141
CIP

www.mhhe.com

ABOUT THE AUTHORS

Raymond A. Barnett, a native of and educated in California, received his B.A. in mathematical statistics from the University of California at Berkeley and his M.A. in mathematics from the University of Southern California. He has been a member of the Merritt College Mathematics Department and was chairman of the department for four years. Associated with four different publishers, Raymond Barnett has authored or co-authored eighteen textbooks in mathematics, most of which are still in use. In addition to international English editions, a number of books have been translated into Spanish. Co-authors include Michael Ziegler, Marquette University; Thomas Kearns, Northern Kentucky University; Charles Burke, City College of San Francisco; John Fujii, Merritt College; and Karl Byleen, Marquette University.

Michael R. Ziegler received his B.S. from Shippensburg State College and his M.S. and Ph.D. from the University of Delaware. After completing postdoctoral work at the University of Kentucky, he was appointed to the faculty of Marquette University where he currently holds the rank of Professor in the Department of Mathematics, Statistics, and Computer Science. Dr. Ziegler has published over a dozen research articles in complex analysis and has co-authored over a dozen undergraduate mathematics textbooks with Raymond Barnett and Karl Byleen.

Karl E. Byleen received his B.S., M.A., and Ph.D. degrees in mathematics from the University of Nebraska. He is currently an Associate Professor in the Department of Mathematics, Statistics, and Computer Science of Marquette University. He has published a dozen research articles on the algebraic theory of semigroups, and has coauthored several undergraduate mathematics textbooks with Raymond Barnett and Michael Ziegler.

CONTENTS

PREFACE

Precalculus: Functions and Graphs is one of five books in the authors' precalculus series (see page ii for a brief comparison of all five books). The primary goal of this text is student comprehension. Great care has been taken to write a book that is mathematically correct and accessible to all students, regardless of their level of understanding. Emphasis is placed on problem solving, ideas, and computational skills rather than mathematical theory. Unless their inclusion adds significant insight into a particular concept, most derivations and proofs are omitted. The text presents concrete examples first, followed by a discussion of related general concepts and results.

Improvements in this edition evolved from the generous response of a large number of users of the last and previous editions, as well as survey results from instructors, mathematics departments, course outlines, and college catalogs. Fundamental to a book's growth and effectiveness is classroom use and feedback. Now in its fifth edition, *Precalculus: Functions and Graphs* has had the benefit of having a substantial amount of both. The most significant change in this fifth edition is the extensive updating of end-of-section and end-of-chapter problem sets. Continuing in our goal of addressing all levels of students, these new problem sets are appropriately balanced across the routine, easy mechanical problems and the more difficult mechanics with some theory. (See the *Features* section within this preface for further explanation.) Additional Explore-Discuss boxes have also been included where appropriate, further emphasizing the emerging use of technology in mathematics.

Also new to this edition, a free Student tutorial CD-ROM accompanies every copy of this text. This text-specific SMART-CD includes an extended real-world application video to begin each chapter. At the section level, this tutorial CD-ROM uses video, descriptive discussion, interactive diagrams, and extended matched problems to reinforce a student's understanding and further exploration of key concepts in the text. Algorithmically generated diagnostic Pre-Tests, Practice, and Post-Tests are also included within each chapter for further reinforcement.

Due to extremely positive feedback from our users, we have continued our emphasis on learning tools that enable a student to easily understand the material being introduced. These tools include Examples with Matched Problems, Explore-Discuss boxes, important student aids such as Think and Caution boxes, and extensive Application problem sets for real-world understanding of concepts. Our Group Activities, Chapter Reviews, and Cumulative Review Exercises continue to reinforce the understanding of critical concepts throughout the text. (See the *Features* section within this preface for further discussion of these tools.)

Technology In the Text

The generic term "graphing utility" is used to refer to any of the various graphing calculators or computer software packages that might be available to a student using this book. TI-83 screen shots are used throughout the text to illustrate graphing principles, but the discussion and explanations lend themselves to any graphing utility being used. Although use of a graphing utility is optional, it is likely that many students and instructors will want to make use of one of these devices. To assist these students and instructors, optional graphing utility activities are included throughout the book beginning in Chapter 2. These activities include brief discussions in the text, examples or portions of examples solved on a graphing utility, and problems for the student to solve. All of the graphing utility material is optional and is clearly identified by the following symbol and can be omitted without loss of continuity, if desired.

Technology Support

The Barnett precalculus series provides extensive technology-based support to assist in student comprehension. Each of the following supplements has been created specifically for this series, clarifying and exploring the concepts introduced within the text.

SMART-CD Student Tutorial

This CD-ROM tutorial is included with every copy of the text. In recognition of the visual, auditory, and interactive aspects of individual learning styles, this tutorial makes use of video, descriptive discussion, interactive diagrams, and extended matched problems. Each chapter on the CD-ROM begins with an extended real-world application video designed to convince even the most skeptical of students that mathematics is important. Algorithmically generated diagnostic Pre-Test, Practice, and Post-Test sections are included to direct students to particular sections where further practice is needed. Key concepts from each section have been identified in the tutorial portion of the CD-ROM, and interactive Matched Problems are included for concept mastery. This tutorial portion also includes forty-five Interactive Diagrams developed by Jere Confrey and Alan Maloney of Quest Math and Science Multimedia, Inc., designed specifically for use with the Barnett series. Each Interactive Diagram (ID) is a separate Java Applet that contains an illustration that can be manipulated by the user for further conceptual understanding of the topic presented. For each section of the text where an ID has been created, an icon has been placed in the margin.

The SMART CD-ROM is available in both Windows and Macintosh platforms.

Website and Online Learning Center

In addition to the SMART-CD, an extensive text-specific website has been created to further enhance the technological support for this series. A visit to the website at *www.mhhe.com/barnett* will prove our commitment to supporting the ever-changing technology needs in mathematical instruction and learning. The following features are only a few of the many technological supplements available to students that can be accessed through this site:

- **NetTutor-Your Online Tutorial Service:** McGraw-Hill is proud to offer this web-based "homework hotline" via an Internet whiteboard. NetTutor provides live "office hours" with mathematics tutors every weekday, as well as a 24-hour response posting service for students accessing the service during off hours.
- **Graphing Calculator Workshops:** Keystroke guides have been created for each of the main calculators in use today. We will continue to update these guides as new calculators are introduced into the market.
- **Explore the Web Exercises, Additional Applications and Answers,** and **Practice Quizzes** are also available for further practice of the mathematical concepts being introduced.

The Barnett series website continues in its support of instructors using this text by providing PageOut for course-specific website creation, access to the course solutions consultants for the series, and extended instructor materials such as PowerPoint presentations and solutions to all even-numbered problems online.

FEATURES

The following pages serve as an introduction to the specific features included within this text.

Examples and Matched Problems

Integrated throughout the text, completely worked examples and practice problems are used to introduce concepts and demonstrate problem-solving techniques. Each Example is followed by a similar Matched Problem for the student to work through while reading the material. Answers to the matched problems are located at the end of each section, for easy reference. This active involvement in the learning process helps students develop a more thorough understanding of algebraic concepts and processes.

Technology

The generic term "graphing utility" is used to refer to any of the various graphing calculators or computer software packages that might be available to students using this book. The use of a graphing utility is optional within this text. To assist those that choose to use a graphing utility, optional activities are included throughout the book beginning in Chapter 2. These include brief discussions in the text, examples or portions of examples solved on a graphing utility, and problems for the student to solve. All optional graphing material is clearly identified by the following symbol and can be omitted without loss of continuity, if desired.

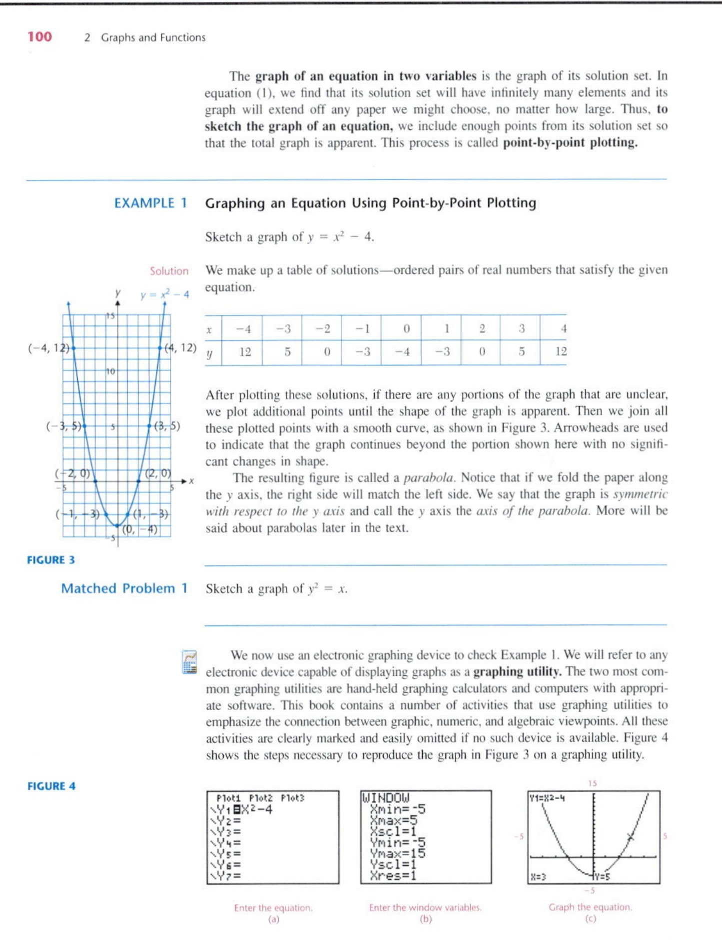

100 2 Graphs and Functions

The **graph of an equation in two variables** is the graph of its solution set. In equation (1), we find that its solution set will have infinitely many elements and its graph will extend off any paper we might choose, no matter how large. Thus, **to sketch the graph of an equation,** we include enough points from its solution set so that the total graph is apparent. This process is called **point-by-point plotting.**

EXAMPLE 1 **Graphing an Equation Using Point-by-Point Plotting**

Sketch a graph of $y = x^2 - 4$.

Solution We make up a table of solutions—ordered pairs of real numbers that satisfy the given equation.

x	−4	−3	−2	−1	0	1	2	3	4
y	12	5	0	−3	−4	−3	0	5	12

After plotting these solutions, if there are any portions of the graph that are unclear, we plot additional points until the shape of the graph is apparent. Then we join all these plotted points with a smooth curve, as shown in Figure 3. Arrowheads are used to indicate that the graph continues beyond the portion shown here with no significant changes in shape.

The resulting figure is called a *parabola*. Notice that if we fold the paper along the y axis, the right side will match the left side. We say that the graph is *symmetric with respect to the y axis* and call the y axis the *axis of the parabola*. More will be said about parabolas later in the text.

FIGURE 3

Matched Problem 1 Sketch a graph of $y^2 = x$.

We now use an electronic graphing device to check Example 1. We will refer to any electronic device capable of displaying graphs as a **graphing utility.** The two most common graphing utilities are hand-held graphing calculators and computers with appropriate software. This book contains a number of activities that use graphing utilities to emphasize the connection between graphic, numeric, and algebraic viewpoints. All these activities are clearly marked and easily omitted if no such device is available. Figure 4 shows the steps necessary to reproduce the graph in Figure 3 on a graphing utility.

FIGURE 4

Enter the equation. (a)

Enter the window variables. (b)

Graph the equation. (c)

Exploration and Discussion

Interspersed at appropriate places in every section, Explore-Discuss boxes encourage students to think critically about mathematics and to explore key concepts in more detail. Verbalization of mathematical concepts, results, and processes is encouraged in these Explore-Discuss boxes, as well as in some matched problems, and in particular problems in almost every exercise set. Explore-Discuss material can be used in class or as an out-of-class activity.

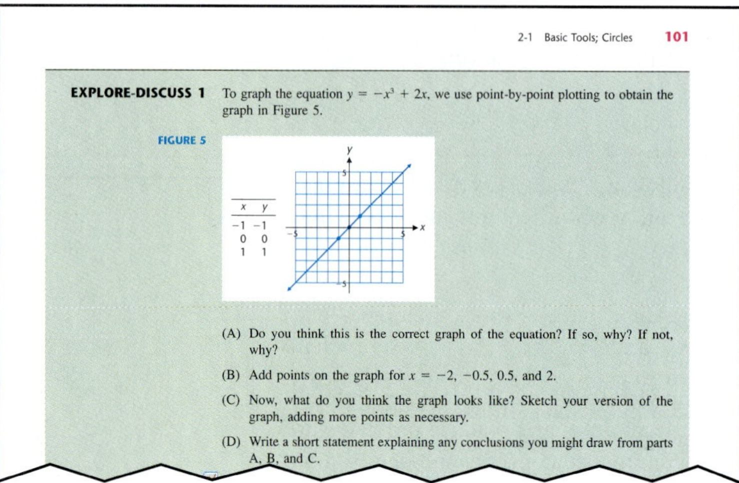

2-1 Basic Tools; Circles 101

EXPLORE-DISCUSS 1 To graph the equation $y = -x^3 + 2x$, we use point-by-point plotting to obtain the graph in Figure 5.

FIGURE 5

x	y
−1	−1
0	0
1	1

(A) Do you think this is the correct graph of the equation? If so, why? If not, why?

(B) Add points on the graph for $x = -2, -0.5, 0.5$, and 2.

(C) Now, what do you think the graph looks like? Sketch your version of the graph, adding more points as necessary.

(D) Write a short statement explaining any conclusions you might draw from parts A, B, and C.

Balanced Exercise Sets

Precalculus: Functions and Graphs contains over 5900 problems. Each Exercise Set is designed so that an average or below-average student will experience success and a very capable student will be challenged. Exercise Sets are found at the end of each section in the text, and are divided into A (routine, easy mechanics), B (more difficult mechanics), and C (difficult mechanics and some theory) levels of difficulty so that students at all levels can be challenged. Problem numbers that appear in blue indicate exercises that require the students to apply their reasoning and writing skills to the solution of the problem.

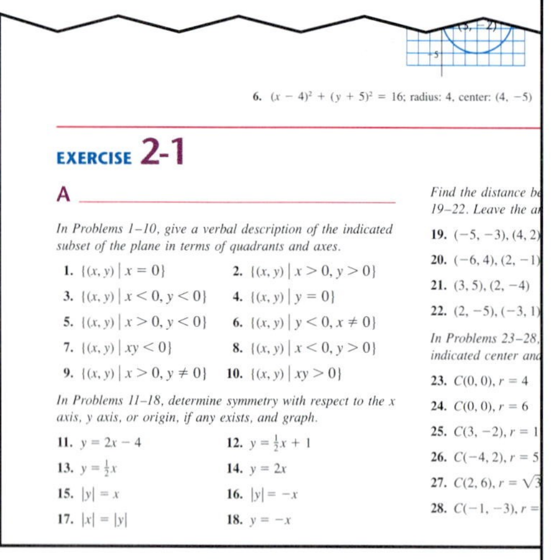

6. $(x - 4)^2 + (y + 5)^2 = 16$; radius: 4, center: $(4, -5)$

EXERCISE 2-1

A

In Problems 1–10, give a verbal description of the indicated subset of the plane in terms of quadrants and axes.

1. $\{(x, y) \mid x = 0\}$
2. $\{(x, y) \mid x > 0, y > 0\}$
3. $\{(x, y) \mid x < 0, y < 0\}$
4. $\{(x, y) \mid y = 0\}$
5. $\{(x, y) \mid x > 0, y < 0\}$
6. $\{(x, y) \mid y < 0, x \neq 0\}$
7. $\{(x, y) \mid xy < 0\}$
8. $\{(x, y) \mid x < 0, y > 0\}$
9. $\{(x, y) \mid x > 0, y \neq 0\}$
10. $\{(x, y) \mid xy > 0\}$

In Problems 11–18, determine symmetry with respect to the x axis, y axis, or origin, if any exists, and graph.

11. $y = 2x - 4$
12. $y = \frac{1}{2}x + 1$
13. $y = \frac{1}{2}x$
14. $y = 2x$
15. $|y| = x$
16. $|y| = -x$
17. $|x| = |y|$
18. $y = -x$

Find the distance be… 19–22. Leave the a…

19. $(-5, -3), (4, 2)$
20. $(-6, 4), (2, -1)$
21. $(3, 5), (2, -4)$
22. $(2, -5), (-3, 1)$

In Problems 23–28, … indicated center and…

23. $C(0, 0), r = 4$
24. $C(0, 0), r = 6$
25. $C(3, -2), r = 1$
26. $C(-4, 2), r = 5$
27. $C(2, 6), r = \sqrt{3}$
28. $C(-1, -3), r =$

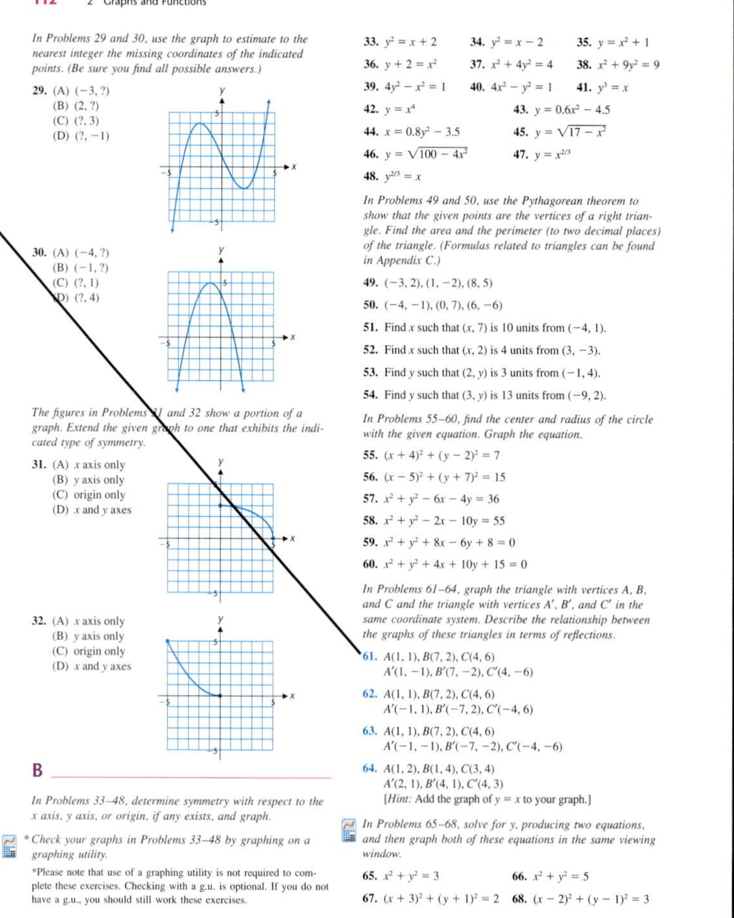

112 2 Graphs and Functions

In Problems 29 and 30, use the graph to estimate to the nearest integer the missing coordinates of the indicated points. (Be sure you find all possible answers.)

29. (A) $(-3, ?)$ (B) $(2, ?)$ (C) $(?, 3)$ (D) $(?, -1)$

30. (A) $(-4, ?)$ (B) $(-1, ?)$ (C) $(?, 1)$ (D) $(?, 4)$

The figures in Problems 31 and 32 show a portion of a graph. Extend the given graph to one that exhibits the indicated type of symmetry.

31. (A) x axis only (B) y axis only (C) origin only (D) x and y axes

32. (A) x axis only (B) y axis only (C) origin only (D) x and y axes

B

In Problems 33–48, determine symmetry with respect to the x axis, y axis, or origin, if any exists, and graph.

* *Check your graphs in Problems 33–48 by graphing on a graphing utility.*

*Please note that use of a graphing utility is not required to complete these exercises. Checking with a g.u. is optional. If you do not have a g.u., you should still work these exercises.

33. $y^2 = x + 2$
34. $y^2 = x - 2$
35. $y = x^2 + 1$
36. $y + 2 = x^2$
37. $x^2 + 4y^2 = 4$
38. $x^2 + 9y^2 = 9$
39. $4y^2 - x^2 = 1$
40. $4x^2 - y^2 = 1$
41. $y^3 = x$
42. $y = x^4$
43. $y = 0.6x^2 - 4.5$
44. $x = 0.8y^2 - 3.5$
45. $y = \sqrt{17 - x^2}$
46. $y = \sqrt{100 - 4x^2}$
47. $y = x^{2/3}$
48. $y^{2/3} = x$

In Problems 49 and 50, use the Pythagorean theorem to show that the given points are the vertices of a right triangle. Find the area and the perimeter (to two decimal places) of the triangle. (Formulas related to triangles can be found in Appendix C.)

49. $(-3, 2), (1, -2), (8, 5)$
50. $(-4, -1), (0, 7), (6, -6)$
51. Find x such that $(x, 7)$ is 10 units from $(-4, 1)$.
52. Find x such that $(x, 2)$ is 4 units from $(3, -3)$.
53. Find y such that $(2, y)$ is 3 units from $(-1, 4)$.
54. Find y such that $(3, y)$ is 13 units from $(-9, 2)$.

In Problems 55–60, find the center and radius of the circle with the given equation. Graph the equation.

55. $(x + 4)^2 + (y - 2)^2 = 7$
56. $(x - 5)^2 + (y + 7)^2 = 15$
57. $x^2 + y^2 - 6x - 4y = 36$
58. $x^2 + y^2 - 2x - 10y = 55$
59. $x^2 + y^2 + 8x - 6y + 8 = 0$
60. $x^2 + y^2 + 4x + 10y + 15 = 0$

In Problems 61–64, graph the triangle with vertices A, B, and C and the triangle with vertices A′, B′, and C′ in the same coordinate system. Describe the relationship between the graphs of these triangles in terms of reflections.

61. $A(1, 1), B(7, 2), C(4, 6)$ $A'(1, -1), B'(7, -2), C'(4, -6)$
62. $A(1, 1), B(7, 2), C(4, 6)$ $A'(-1, 1), B'(-7, 2), C'(-4, 6)$
63. $A(1, 1), B(7, 2), C(4, 6)$ $A'(-1, -1), B'(-7, -2), C'(-4, -6)$
64. $A(1, 2), B(1, 4), C(3, 4)$ $A'(2, 1), B'(4, 1), C'(4, 3)$ [*Hint:* Add the graph of $y = x$ to your graph.]

In Problems 65–68, solve for y, producing two equations, and then graph both of these equations in the same viewing window.

65. $x^2 + y^2 = 3$
66. $x^2 + y^2 = 5$
67. $(x + 3)^2 + (y + 1)^2 = 2$
68. $(x - 2)^2 + (y - 1)^2 = 3$

Applications

One of the primary objectives of this book is to give the student substantial experience in modeling and solving real-world problems. Over 600 application exercises help convince even the most skeptical student that mathematics is relevant to everyday life. The most difficult application problems are marked with two stars (★★), the moderately difficult application problems with one star (★), and the easier application problems are not marked. An **Applications Index** is included in the back of this textbook to help locate particular applications.

89. Physics—Rate. The distance in feet that an object falls in a vacuum is given by $s(t) = 16t^2$, where t is time in seconds. Find:
(A) $s(0), s(1), s(2), s(3)$
(B) $\dfrac{s(2 + h) - s(2)}{h}$
(C) What happens in part B when h tends to 0? Interpret physically.

90. Physics—Rate. An automobile starts from rest and travels along a straight and level road. The distance in feet traveled by the automobile is given by $s(t) = 10t^2$, where t is time in seconds. Find:
(A) $s(8), s(9), s(10), s(11)$
(B) $\dfrac{s(11 + h) - s(11)}{h}$
(C) What happens in part B as h tends to 0? Interpret physically.

91. Manufacturing. A candy box is to be made out of a piece of cardboard that measures 8 by 12 inches. Squares, x inches on a side, will be cut from each corner, and then the ends and sides will be folded down (see the figure). Find a formula for the volume of the box $V(x)$ in terms of x. From practical considerations, what is the domain of the function V?

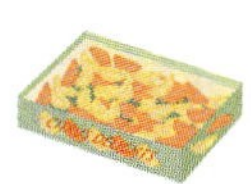

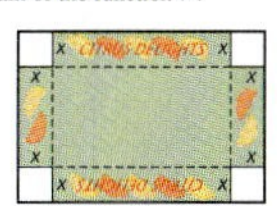

92. Construction. A rancher has 20 miles of fencing to fence a rectangular piece of grazing land along a straight river. If no fence is required along the river and the sides perpendicular to the river are x miles long, find a formula for the area $A(x)$ of the rectangle in terms of x. From practical considerations, what is the domain of the function A?

★ **93. Construction.** The manager of an animal clinic wants to construct a kennel with four individual pens, as indicated in the figure. State law requires that each pen have a gate 3 feet wide and an area of 50 square feet. If x is the width of one pen, express the total amount of fencing $F(x)$ (excluding the gates) required for the construction of the kennel as a function of x. Complete the following table [round values of $F(x)$ to one decimal place]:

x	4	5	6	7
$F(x)$				

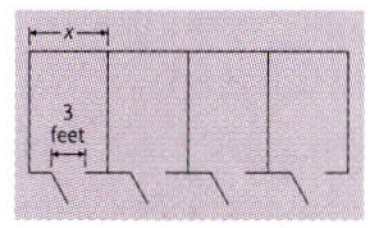

Figure for 93

★ **94. Architecture.** An architect wants to design a window with an area of 24 square feet in the shape of a rectangle surmounted by a semicircle, as indicated in the figure. If x is the width of the window, express the perimeter $P(x)$ of the window as a function of x. Complete the table below [round each value of $P(x)$ to one decimal place]:

x	4	5	6	7
$P(x)$				

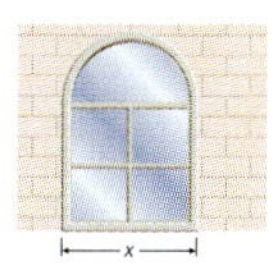

★ **95. Construction.** A freshwater pipeline is to be run from a source on the edge of a lake to a small resort community on an island 8 miles offshore, as indicated in the figure. It costs \$10,000 per mile to lay the pipe on land and \$15,000 per mile to lay the pipe in the lake. Express the total cost $C(x)$ of constructing the pipeline as a function of x. From practical considerations, what is the domain of the function C?

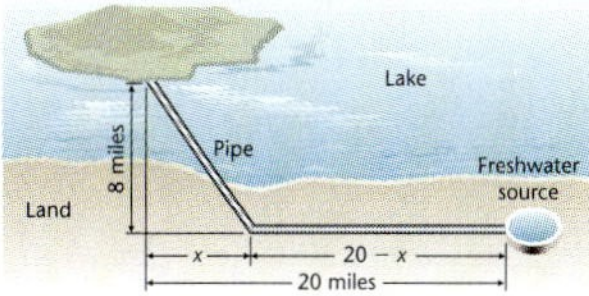

★ **96. Weather.** An observation balloon is released at a point 10 miles from the station that receives its signal and rises vertically, as indicated in the figure. Express the distance $d(h)$ between the balloon and the receiving station as a function of the altitude h of the balloon.

Foundation for Calculus

As many students will use this book to prepare for a calculus course, examples and exercises that are especially pertinent to calculus are marked with an icon. ∫

B

In Problems 25–28, find the equation of the line passing through the given point with the given slope. Write the final answer in the slope-intercept form $y = mx + b$.

25. $(0, 3); m = -2$ **26.** $(4, 0); m = 3$

27. $(-5, 4); m = \frac{3}{2}$ **28.** $(2, -3); m = -\frac{4}{5}$

In Problems 29–34, find the equation of the line passing through the two given points. Write the final answer in the slope-intercept form $y = mx + b$ or in the form $x = c$.

29. $(2, 5); (4, -3)$ **30.** $(-1, 4); (3, 2)$

31. $(-3, 2); (-3, 5)$ **32.** $(0, 5); (2, 5)$

33. $(-4, 2); (0, 2)$ **34.** $(5, -4); (5, 6)$

In Problems 35–46, write an equation of the line that contains the indicated point and meets the indicated condition(s). Write the final answer in the standard form $Ax + By = C$, $A \ge 0$.

35. $(2, -1)$; parallel to $y = -3x + 7$

36. $(-3, 2)$; parallel to $y = 4x - 5$

37. $(0, -4)$; parallel to $2x + 3y = 9$

∫ *Problems 53–58 are calculus-related. Recall that a line tangent to a circle at a point is perpendicular to the radius drawn to that point (see the figure). Find the equation of the line tangent to the circle at the indicated point. Write the final answer in the standard form $Ax + By = C$, $A \ge 0$. Graph the circle and the tangent line on the same coordinate system.*

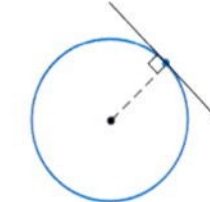

53. $x^2 + y^2 = 25, (3, 4)$

54. $x^2 + y^2 = 100, (-8, 6)$

55. $x^2 + y^2 = 50, (5, -5)$

56. $x^2 + y^2 = 80, (-4, -8)$

57. $(x - 3)^2 + (y + 4)^2 = 169, (8, -16)$

58. $(x + 5)^2 + (y - 9)^2 = 289, (-13, -6)$

Group Activities

A Group Activity is located at the end of each chapter and involves many of the concepts discussed in that chapter. These activities strongly encourage the verbalization of mathematical concepts, results, and processes. All of these special activities are highlighted to emphasize their importance.

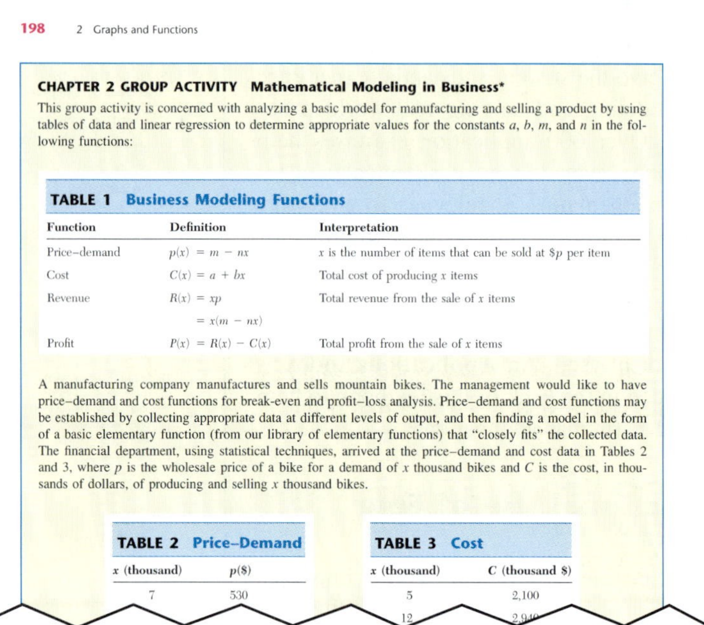

198 2 Graphs and Functions

CHAPTER 2 GROUP ACTIVITY Mathematical Modeling in Business*

This group activity is concerned with analyzing a basic model for manufacturing and selling a product by using tables of data and linear regression to determine appropriate values for the constants a, b, m, and n in the following functions:

TABLE 1 Business Modeling Functions

Function	Definition	Interpretation
Price–demand	$p(x) = m - nx$	x is the number of items that can be sold at \$$p$ per item
Cost	$C(x) = a + bx$	Total cost of producing x items
Revenue	$R(x) = xp$	Total revenue from the sale of x items
	$= x(m - nx)$	
Profit	$P(x) = R(x) - C(x)$	Total profit from the sale of x items

A manufacturing company manufactures and sells mountain bikes. The management would like to have price–demand and cost functions for break-even and profit–loss analysis. Price–demand and cost functions may be established by collecting appropriate data at different levels of output, and then finding a model in the form of a basic elementary function (from our library of elementary functions) that "closely fits" the collected data. The financial department, using statistical techniques, arrived at the price–demand and cost data in Tables 2 and 3, where p is the wholesale price of a bike for a demand of x thousand bikes and C is the cost, in thousands of dollars, of producing and selling x thousand bikes.

TABLE 2 Price–Demand

x (thousand)	p(\$)
7	530

TABLE 3 Cost

x (thousand)	C (thousand \$)
5	2,100
12	2,940

Graphs and Illustrations

All graphs in this text are computer generated to ensure mathematical accuracy. Graphing utility screens displayed in the text are actual output from a graphing calculator.

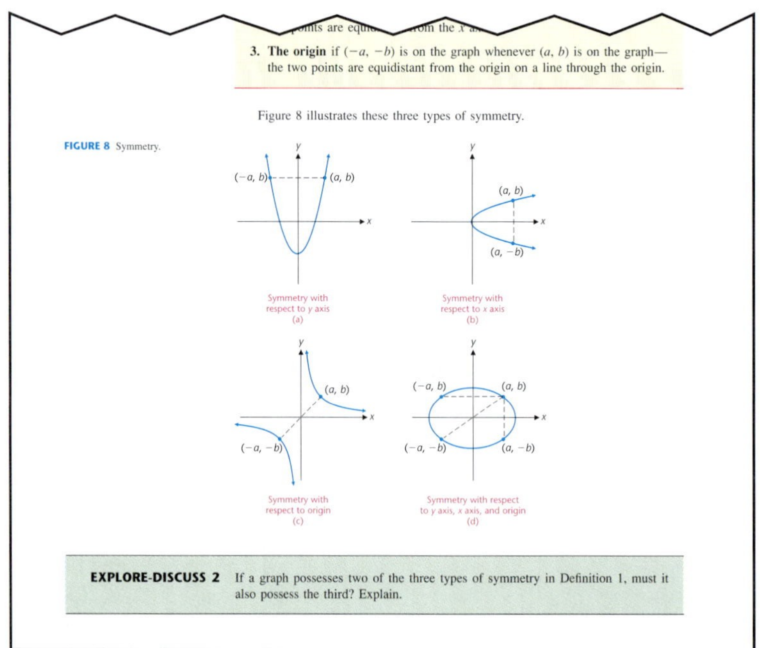

3. The origin if $(-a, -b)$ is on the graph whenever (a, b) is on the graph—the two points are equidistant from the origin on a line through the origin.

Figure 8 illustrates these three types of symmetry.

FIGURE 8 Symmetry.

EXPLORE-DISCUSS 2 If a graph possesses two of the three types of symmetry in Definition 1, must it also possess the third? Explain.

Boldface Type

Boldface type is used to introduce new terms and highlight important comments.

Think Boxes

Think boxes are dashed boxes used to enclose steps that are usually performed mentally.

Caution Boxes

Caution boxes appear throughout the text to indicate where student errors often occur.

Annotation

Annotation of examples and developments, in small colored type, is found throughout the text to help students through critical stages.

2-3 Functions **141**

In addition to evaluating functions at specific numbers, it is important to be able to evaluate functions at expressions that involve one or more variables. For example, the **difference quotient**

$$\frac{f(x+h)-f(x)}{h} \qquad x \text{ and } x+h \text{ in the domain of } f,\ h \neq 0$$

is studied extensively in a calculus course.

EXAMPLE 6 **Evaluating and Simplifying a Difference Quotient**

For $f(x) = x^2 + 4x + 5$, find and simplify:

(A) $f(x+h)$ (B) $\frac{f(x+h)-f(x)}{h},\ h \neq 0$

Solution (A) To find $f(x+h)$, we replace x with $x+h$ everywhere it appears in the equation that defines f and simplify:

$$f(x+h) = (x+h)^2 + 4(x+h) + 5$$
$$= x^2 + 2xh + h^2 + 4x + 4h + 5$$

(B) Using the result of part A, we get

$$\frac{f(x+h)-f(x)}{h} = \frac{x^2 + 2xh + h^2 + 4x + 4h + 5 - (x^2 + 4x + 5)}{h}$$
$$= \frac{x^2 + 2xh + h^2 + 4x + 4h + 5 - x^2 - 4x - 5}{h}$$
$$= \frac{2xh + h^2 + 4h}{h} = \frac{h(2x + h + 4)}{h} = 2x + h + 4$$

Matched Problem 6 Repeat Example 6 for $f(x) = x^2 + 3x + 7$.

CAUTION

1. If f is a function, then the symbol $f(x+h)$ represents the value of f at the number $x+h$ and must be evaluated by replacing the independent variable in the equation that defines f with the expression $x+h$, as we did in Example 6. Do not confuse this notation with the familiar algebraic notation for multiplication:

$f(x+h) \neq fx + fh$ $f(x+h)$ is function notation.

$4(x+h) = 4x + 4h$ $4(x+h)$ is algebraic multiplication notation.

Screened Boxes

Screened boxes are used to highlight important definitions, theorems, results, and step-by-step processes.

Properties of a Quadratic Function and Its Graph

Given a quadratic function and the form obtained by completing the square

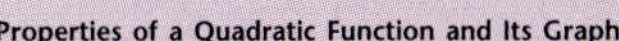

$$f(x) = ax^2 + bx + c = a(x-h)^2 + k \qquad a \neq 0$$

we summarize general properties as follows:

1. The graph of f is a parabola:

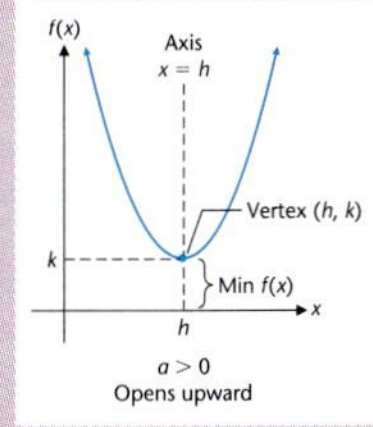

a > 0
Opens upward

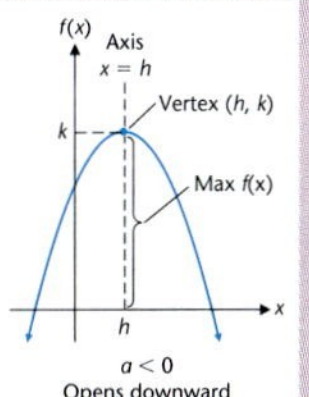

a < 0
Opens downward

2. Vertex: (h, k) (Parabola rises on one side of the vertex and falls on the other.)
3. Axis (of symmetry): $x = h$ (Parallel to y axis)

Chapter Review

Chapter Review sections are provided at the end of each chapter and include a thorough review of all the important terms and symbols. This recap is followed by a comprehensive set of review exercises.

5. Find a and b for $r = 0.1\overline{9}$ and then find a terminating decimal expansion for r.

Appendix A Review

A-1 ALGEBRA AND REAL NUMBERS

A **set** is a collection of objects called **elements** or **members** of the set. Sets are usually described by **listing** the elements or by stating a **rule** that determines the elements. A set may be **finite** or **infinite.** A set with no elements is called the **empty set** or the **null set** and is denoted $\varnothing$. A **variable** is a symbol that represents unspecified elements from a **replacement set.** A **constant** is a symbol for a single object. If each element of set A is also in set B, we say A is a **subset** of B and write $A \subset B$.

Real numbers:

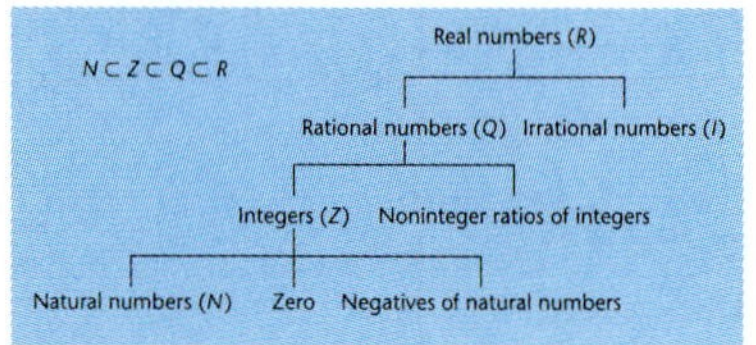

Real number line:

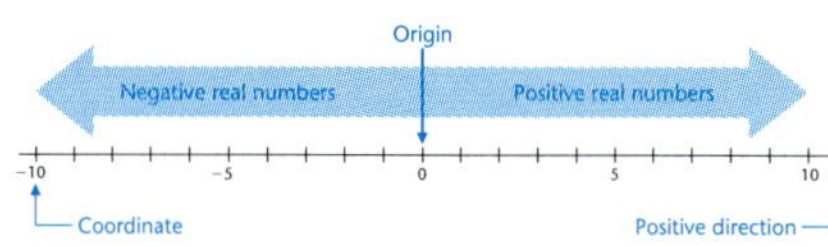

Cumulative Review Exercise

A Cumulative Review Exercise is provided after every second or third chapter, for additional reinforcement.

Cumulative Review Exercise Chapters 1 and 2 **207**

Cumulative Review Exercise Chapters 1 and 2

Work through all the problems in this cumulative review and check answers in the back of the book. Answers to all review problems are there, and following each answer is a number in italics indicating the section in which that type of problem is discussed. Where weaknesses show up, review appropriate sections in the text.

A

1. Solve for x: $\frac{7x}{5} - \frac{3+2x}{2} = \frac{x-10}{3} + 2$

2. Solve for x and y: $2x - 3y = 8$
$4x + y = 2$

Solve and graph Problems 3–5.

3. $2(3 - y) + 4 \le 5 - y$

4. $|x - 2| < 7$ **5.** $x^2 + 3x \ge 10$

6. Perform the indicated operations and write the answer in standard form:
(A) $(2 - 3i) - (-5 + 7i)$ (B) $(1 + 4i)(3 - 5i)$
(C) $\frac{5+i}{2+3i}$

Solve Problems 7–10.

7. $3x^2 = -12x$ **8.** $4x^2 - 20 = 0$

9. $x^2 - 6x + 2 = 0$ **10.** $x - \sqrt{12 - x} = 0$

11. For what values of x does $\sqrt{2 + 3x}$ represent a real number?

Given the poin… and $B(5, 6)$ …

17. How are the graphs of the following related to the graph of $y = |x|$?
(A) $y = 2|x|$ (B) $y = |x - 2|$ (C) $y = |x| - 2$

18. Sketch a graph of each of the functions in parts A and B using the graph of function f in the figure.
(A) $y = -f(x + 1)$ (B) $y = 2f(x) - 2$

f(x) 5 -5 5 x -5

B

Solve Problems 19–22.

19. $\frac{x+3}{2x+2} + \frac{5x+2}{3x+3} = \frac{5}{6}$ **20.** $\frac{3}{x} = \frac{6}{x+1} - \frac{1}{x-1}$

21. $2x + 1 = 3\sqrt{2x - 1}$

22. $2x - 3y = 9$
$4x + 2y = 23$

Solve and graph Problems 23–25.

23. $|4x - 9| > 3$ **24.** $\sqrt{(3m-4)^2} \le 2$

SUPPLEMENTS

A comprehensive set of ancillary materials for both the student and the instructor is available for use with this text.

Student's Solutions Manual

This supplement is available for sale to the student, and includes detailed solutions to all odd-numbered problems and most review exercises.

Instructor's Solutions Manual

This manual provides solutions to even-numbered problems and answers to all problems in the text.

Instructor's Resource Manual

This supplement provides transparency masters and sample tests for each chapter in the text.

Print and Computerized Testbanks

A Computerized Testbank is available that provides a variety of formats to enable the instructor to create tests using both algorithmically generated test questions and those from a static testbank. This testing system enables the instructor to choose questions either manually or randomly by section, question types, difficulty level, and other criteria. This testing software supports online testing and is available for PC and Macintosh computers. A softcover print version of the testbank includes the static questions found in the computerized version.

Barnett/Ziegler/Byleen Video Series

Course videotapes, created specifically for this series, provide students with additional reinforcement of the topics presented in the book. These videos are keyed to the text and feature an effective combination of learning techniques, including personal instruction, state-of-the-art graphics, and real-world applications.

Course Solutions

Fully integrated multimedia, a full-scale Online Learning Center, and a Course Integration Guide are designed specifically to help you with your individual precalculus course needs. Assembled by an expert in your field of study, this printed manual fully integrates the numerous products available to accompany *Precalculus: Functions and Graphs*. The Course Integration Guide will also contain detailed solutions for the Group Activities found in the text, a description of each Interactive Diagram, and detailed solutions to the exploratory questions that accompany each Interactive Diagram.

SMART-CD Student Tutorial

This CD-ROM tutorial is included with every copy of the text. In recognition of the visual, auditory, and interactive aspects of individual learning styles, this tutorial makes use of video, descriptive discussion, interactive diagrams, and extended matched problems. Each chapter on the CD-ROM begins with an extended real-world application video designed to convince even the most skeptical of students that mathematics is important. Algorithmically generated diagnostic Pre-Test, Practice, and Post-Test sections are included to direct students to particular sections where further practice is needed. Key concepts from each section have been identified in the tutorial portion of the CD-ROM, and interactive Matched Problems are included for concept mastery. This tutorial portion also includes forty-five Interactive Diagrams developed by Jere Confrey and Alan Maloney of Quest Math and Science Multimedia, Inc., designed specifically for use with the Barnett series. Each Interactive Diagram (ID) is a separate Java Applet that contains an illustration that can be manipulated by the user for further conceptual understanding of the topic presented. For each section of the text where an ID has been created, an icon has been placed in the margin. The SMART-CD is available in both Windows and Macintosh platforms.

Website and Online Learning Center

An extensive text-specific website has been created to further enhance the technological support for this series. Visit the site at **www.mhhe.com/barnett** to explore the numerous features available to both students and instructors. Among the many tools available to students is NetTutor-Your Online Tutorial Service, a "homework hotline" via an Internet whiteboard. NetTutor provides live "office hours" with mathematics tutors every weekday, as well as a 24-hour response posting service for students accessing the service during off hours. The Graphing Calculator Workshop features keystroke guides for each of the main calculators in use today. Instructors have access to PageOut for course-specific website creation, extended instructor materials such as PowerPoint presentations, and solutions to all even-numbered problems online, and access to the course solutions consultants for the series.

For further information about these or any supplements, please contact your local McGraw-Hill sales representative, or visit our website at **www.mhhe.com/barnett**.

Accuracy

Because of the careful checking and proofing by a number of mathematics instructors (acting independently), the authors and publisher believe this book to be substantially error free. For any errors remaining, the authors would be grateful if they were sent to: Karl E. Byleen, 9322 W. Garden Court, Hales Corners, WI 53130; or by email, to *byleen@execpc.com*.

Acknowledgments

In addition to the authors, many others are involved in the successful publication of a book. We wish to thank personally:

Hossein Hamedani and Caroline Woods for providing a careful and thorough check of all the mathematical calculations in the book, the student solutions manual, and the answer manual (a tedious but extremely important job); all of the supplements authors for developing the supplemental manuals that are so important to the success of a text; Jeanne Wallace for accurately and efficiently producing many of the manuals that supplement the text; George Morris and his staff at Scientific Illustrators for their effective illustrations and accurate graphs; Maggie Rogers, Sheila Frank, Michelle Munn, and all the other people at McGraw-Hill who contributed their efforts to the production of this book. Producing this new edition with the help of all these extremely competent people has been a most satisfying experience.

We also wish to thank the following manuscript reviewers: Yungchen Cheng, S.W. Missouri State University; Allan Cochran, University of Arkansas; Wayne Ehler, Anne Arundel Community College; Abdulla Elusta, Broward Community College; Betsy Farber, Bucks County Community College; Jeffrey Graham, Western Carolina University; Selwyn Hollis, Armstrong Atlantic State University; Randal Hoppens, Blinn College; Linda Horner, Broward Community College–N. Campus; Beverly Reed, Kent State University; Mike Rosenthal, Florida International University; Robert Woodside, East Carolina State University; Mary Wright, Southern Illinois University. Chapter reviewers: Diane Abbott, Bowling Green State University; Victor Akatsa, Chicago State University; Judith Beauford, University of the Incarnate Word; George Bradley, Duquesne University; Ronald Brent, University of Massachusetts-Lowell; Jeff Brown, University of North Carolina at Wilmington; Kimberly Brown, Tarrant County Junior College; Ed Buman, Creighton University; Roxanne Byrne, University of Colorado at Denver; Magdalena Caproiu, Antelope Valley College; Harold Carda, South Dakota School of Mines and Technology; Allan Cochran, University of Arkansas; Barbara Cortzen, DePaul University; Patrick Crowe, Johnson City Community College; Elizabeth Chu, Sussex County Community College; Marilyn Danchanko, Camria County Area Community College; Karin Deck, University of North Carolina at Wilmington; Joe Di Costanzo, Johnson County Community College; Michelle Diehl, University of New Mexico; Jeff Dinitz, University of Vermont; Charles Douglas, South Georgia College; Mary Ehlers, Seattle University; Laura Fernelius, University of Wisconsin at Oshkosh; Larry Friesen, Butler County Community College; Doris Fuller, Virginia State University; Dan Gardner, Elgin Community College; Rebecca Gehrke-Griswold, Jefferson Community College-SW; Sheryl Griffith, Iowa Central Community College; Vernon Gwaltney, John Tyler Community College; Rutger Hangelbroek, Western Illinois University; Keith Howell, Southwestern Michigan College; Brian Jackson, Connor's State College; Juan Carlos Jimenez, Springfield Technical Community College; Nancy Johnson, Manatee Community College; Klaus Kaiser, University of Houston; Rahim Karimpour, Southern Illinois University at Edwardsville; Karla Karstens, University of Vermont; Thomas Keller, Southwest Texas State University; Warren Koepp, Texas A&M at Commerce; Kent Kromarek, University of Wisconsin-Waukesha; Sonja Kung, University of Wisconsin at Stevens Point; Mark Lesperance, Kansas State University; Carol Lucas, University of Kansas; Richard Mason, Indian Hills Community College; Marilyn McCollum, North Carolina State University; Gael Mericle, Minnesota State University-Mankato; Michael Montano, Riverside Community College; Patricia Moreland, Cowley County Community College; Robert Moyer, Fort Valley State University; Arumudam Muhundan, Manatee Community College; Milt Myers, Delaware County Community College; Richard Nadel, Florida International University; Roxie Novak, Radford University; Vicki Partin, Lexington Community College; Gloria Phoenix, North Carolina Agricultural & Technical State University; Donna Russo, Quincy College; Mohammed Saleem, San Jose State University; Donna Salinardi, University of Connecticut; Jean Sanders, University of Wisconsin at Platteville; Radha Sankaran, Passaic County Community College; Ellen Scheiber, Drexel University; Patricia Schmidt, University of Pittsburgh at Greensburg; Samkar Sethuraman, Augusta State University; James Shockley, Virginia Polytechnic Institute; Debbye Stapleton, Georgia Southern University; Lora Stewart, Cosumnes River College; Joseph Sukta, Moraine Valley Community College; Ralph Svetic, Winona State University; Hussain Talibi, Tuskegee University; Stuart Thomas, University of Oregon; Richard Vinson, University of South Alabama; James Ward, Pensacola Junior College; Lyndon Weberg, University of Wisconsin; Chad Wheatley, Delaware Technical & Community College; Edward White, Frostburg State University; Clifton Whyburn, University of Houston; Tom Williams, Rowan-Cabarrus Community College; Terry Wilson, San Jacinto College; Zhanbo Yang, University of the Incarnate Word; George Zazi, Chicago State University; Cathleen Zucco-Teveloff, Rutgers University.

R. A. Barnett
M. R. Ziegler
K. E. Byleen

CHAPTER DEPENDENCIES

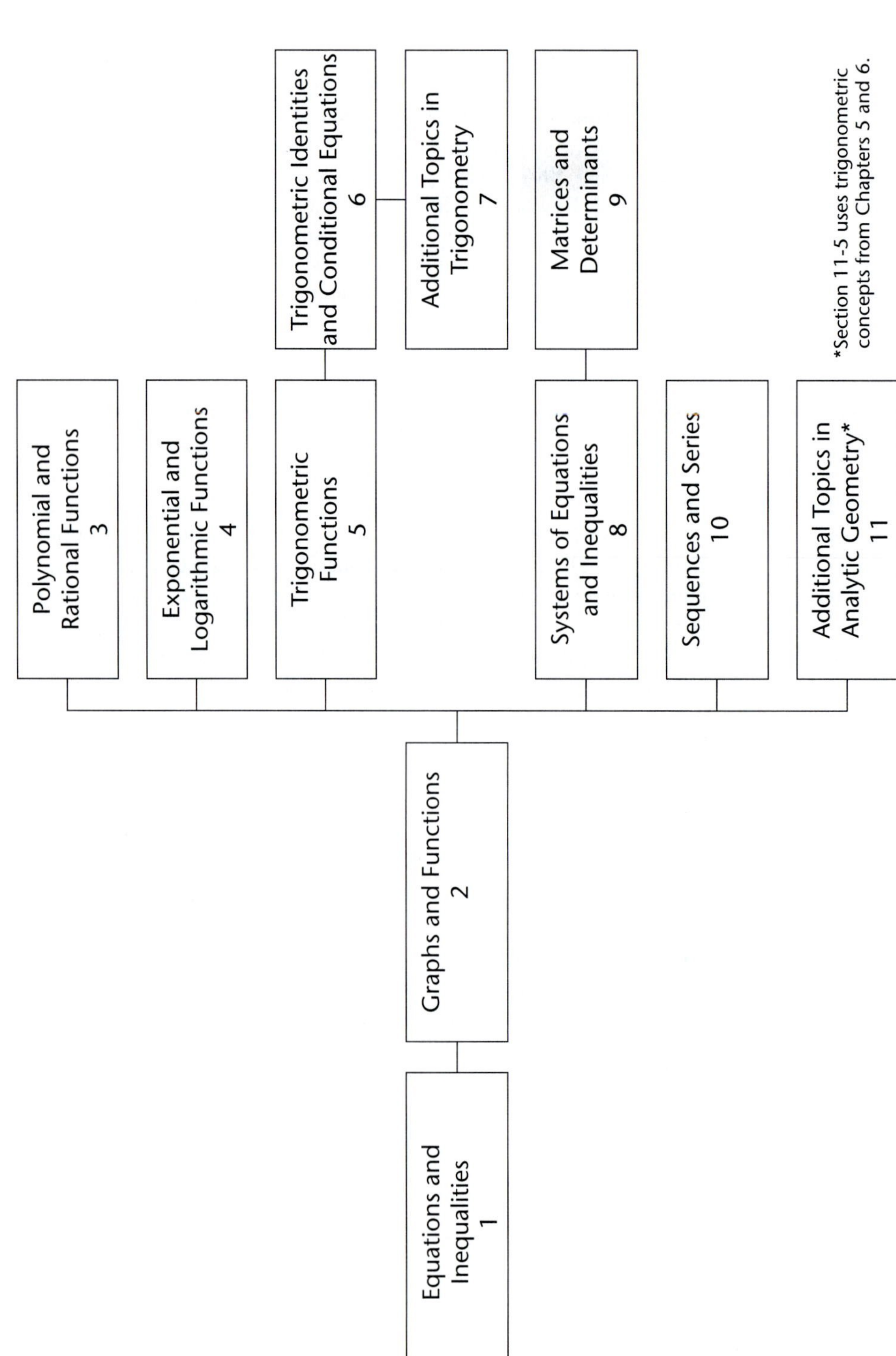

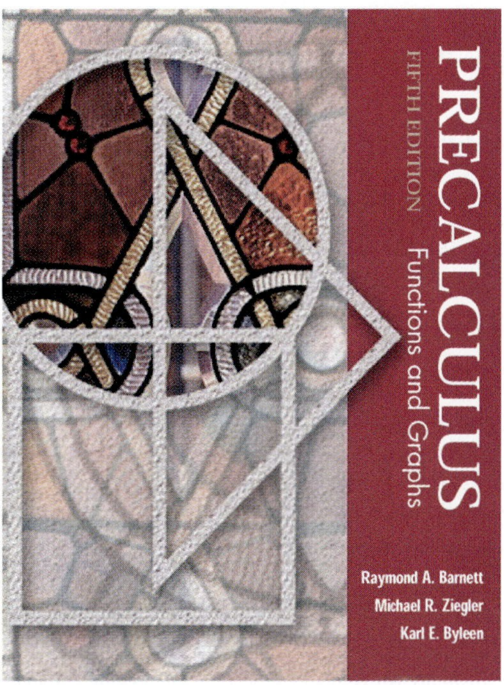

McGraw-Hill is proud to offer an exciting new suite of multimedia products and services called Course Solutions.

Designed specifically to help you with your individual course needs, **Course Solutions** will assist you in integrating your syllabus with our premier titles and state-of-the-art new media tools that support them.

At the heart of Course Solutions you'll find:

- Fully integrated multimedia
- A full-scale Online Learning Center
- A Course Integration Guide

As well as these unparalleled services:

- McGraw-Hill Learning Architecture
- McGraw-Hill Course Consultant Service
- Visual Resource Library (VRL)
- Image Licensing
- McGraw-Hill Student Tutorial Service
- McGraw-Hill Instructor Syllabus Service
- PageOut Lite
- PageOut: The Course Website Development Center
- Other Delivery Options

Course Solutions truly has the solutions to your every teaching need. Read on to learn how we can specifically help you with your classroom challenges.

SPECIAL ATTENTION

to your specific needs.

These "perks" are all part of the extra service delivered through McGraw-Hill's **Course Solutions:**

McGraw-Hill Learning Architecture

Each McGraw-Hill *Online Learning Center* is ready to be ported into our *McGraw-Hill Learning Architecture*—a full course management software system for *Local Area Networks* and *Distance Learning Classes*. Developed in conjunction with Top Class software, *McGraw-Hill Learning Architecture* is a powerful course management system available upon special request.

McGraw-Hill Course Consultant Service

In addition to the *Course Integration Guide*, instructors using **Course Solutions** textbooks can access a special curriculum-based *Course Consultant Service,* via a web-based threaded discussion list within each *Online Learning Center.* A **McGraw-Hill Course Solutions Consultant** will personally help you—as a text adopter—integrate this text and media into your course to fit your specific needs. This content-based service is offered in addition to our usual software support services.

McGraw-Hill Instructor Syllabus Service

For *new* adopters of **Course Solutions** textbooks, McGraw-Hill will help correlate all text, supplement, and appropriate materials and services to your course syllabus. Simply call your McGraw-Hill sales representative for assistance.

PageOut Lite

Free to **Course Solutions** textbook adopters, *PageOut Lite* is perfect for instructors who want to create their own website. In just a few minutes, even novices can turn their syllabus into a website using *PageOut Lite.*

PageOut: The Course Website Development Center

For those who want the benefits of *PageOut Lite's* no-hassle approach to site development, but with even more features, we offer *PageOut: The Course Website Development Center.*

PageOut shares many of *PageOut Lite's* features, but its syllabus page lets you create links that will take your students to your original material, other website addresses, and to *McGraw-Hill Online Learning Center* content. This means you can assign *Online Learning Center* content within your syllabus-based website. *PageOut*'s grade book function will tell you when each student has taken a quiz or worked through an exercise, automatically recording his or her scores for you. *PageOut* also features a discussion board list where you and your students can exchange questions and post announcements, as well as an area for students to build personal web pages. Contact your McGraw-Hill sales representative for assistance.

Other Delivery Options

Online Learning Centers are also compatible with a number of full-service online course delivery systems or outside educational service providers. For a current list of compatible delivery systems, contact your McGraw-Hill sales representative.

And for your students . . .

McGraw-Hill Student Tutorial Service

Within each *Online Learning Center* resides a **FREE** *Student Tutorial Service.* This web-based "homework hotline"—available via a threaded discussion list—features guaranteed, 24-hour response time on weekdays. Students can also receive immediate tutorial assistance via an Internet whiteboard during regularly scheduled NetTutor office hours.

www.mhhe.com/barnett

TO THE STUDENT

The following suggestions are made to help you get the most out of this book and your efforts.

As you study the text we suggest a five step-process. For each section:

1. Read a mathematical development.
2. Work through the illustrative examples.
3. Work the matched problem.

} Repeat the 1-2-3 cycle until the section is finished.

4. Review the main ideas in the section.
5. Work the assigned exercises at the end of the section.

All of this should be done with plenty of paper, pencils, a calculator, and a wastebasket at hand. In fact, no mathematics text should be read without pencil and paper in hand; mathematics is not a spectator sport. Just as you cannot learn to swim by watching someone else swim, you cannot learn mathematics by simply reading worked examples—you must work problems, lots of them.

If you have a graphing calculator or access to a computer with mathematical software, such as Maple or Mathematica, you should pay particular attention to the remarks, explore-discuss boxes, and exercises marked with . This is optional material that we have included to help you learn to make effective use of technology as part of the problem-solving process. If you do not have one of these devices, please omit material so marked, since it may involve calculations that cannot be done by hand.

If you have difficulty with the course, then, in addition to doing the regular assignments, spend more time on the examples and matched problems and work more A exercises, even if they are not assigned. If you find the course too easy, then work more C exercises and applied problems, even if they are not assigned.

Raymond A. Barnett
Michael R. Ziegler
Karl E. Byleen

Equations and Inequalities

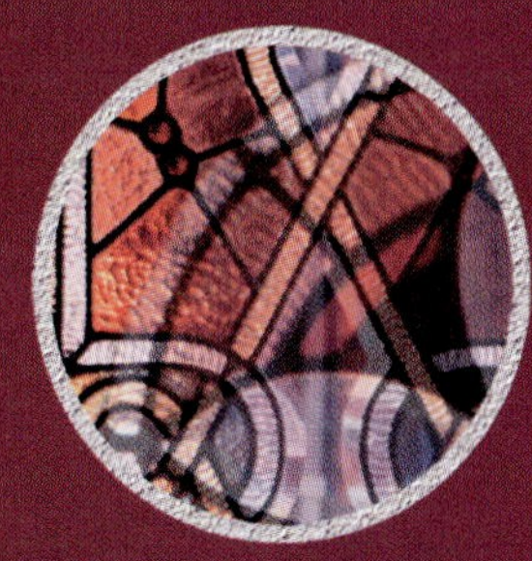

CHAPTER 1

One of the important uses of algebra is the solving of equations and inequalities. In this chapter we look at techniques for solving linear and nonlinear equations and inequalities. In addition, we consider a number of applications that can be solved using these techniques. Additional techniques for solving polynomial equations will be discussed in Chapter 3.

SECTION 1-1 Linear Equations and Applications

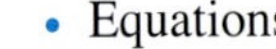
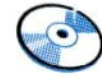

- Equations
- Solving Linear Equations
- A Strategy for Solving Word Problems
- Number and Geometric Problems
- Rate–Time Problems
- Mixture Problems
- Some Final Observations on Linear Equations

• Equations

An **algebraic equation** is a mathematical statement that relates two algebraic expressions involving at least one variable. Some examples of equations with x as the variable are

$$3x - 2 = 7 \qquad \frac{1}{1 + x} = \frac{x}{x - 2}$$

$$2x^2 - 3x + 5 = 0 \qquad \sqrt{x + 4} = x - 1$$

The **replacement set,** or **domain,** for a variable is defined to be the set of numbers that are permitted to replace the variable.

Assumption

On Domains of Variables

Unless stated to the contrary, we assume that the domain for a variable is the set of those real numbers for which the algebraic expressions involving the variable are real numbers.

For example, the domain for the variable x in the expression

$$2x - 4$$

is R, the set of all real numbers, since $2x - 4$ represents a real number for all replacements of x by real numbers. The domain of x in the equation

$$\frac{1}{x} = \frac{2}{x - 3}$$

is the set of all real numbers except 0 and 3. These values are excluded because the left member is not defined for $x = 0$ and the right member is not defined for $x = 3$.

The left and right members represent real numbers for all other replacements of x by real numbers.

The **solution set** for an equation is defined to be the set of elements in the domain of the variable that makes the equation true. Each element of the solution set is called a **solution,** or **root,** of the equation. To **solve an equation** is to find the solution set for the equation.

An equation is called an **identity** if the equation is true for all elements from the domain of the variable. An equation is called a **conditional equation** if it is true for certain domain values and false for others. For example,

$$2x - 4 = 2(x - 2) \qquad \text{and} \qquad \frac{5}{x^2 - 3x} = \frac{5}{x(x - 3)}$$

are identities, since both equations are true for all elements from the respective domains of their variables. On the other hand, the equations

$$3x - 2 = 5 \qquad \text{and} \qquad \frac{2}{x - 1} = \frac{1}{x}$$

are conditional equations, since, for example, neither equation is true for the domain value 2.

Knowing what we mean by the solution set of an equation is one thing; finding it is another. To this end we introduce the idea of equivalent equations. Two equations are said to be **equivalent** if they both have the same solution set for a given replacement set. A basic technique for solving equations is to perform operations on equations that produce simpler equivalent equations, and to continue the process until an equation is reached whose solution is obvious.

Application of any of the properties of equality given in Theorem 1 will produce equivalent equations.

Theorem 1 **Properties of Equality**

For a, b, and c any real numbers:

1. If $a = b$, then $a + c = b + c$. **Addition Property**
2. If $a = b$, then $a - c = b - c$. **Subtraction Property**
3. If $a = b$, then $ca = cb$, $c \neq 0$. **Multiplication Property**
4. If $a = b$, then $\frac{a}{c} = \frac{b}{c}$, $c \neq 0$. **Division Property**
5. If $a = b$, then either may replace the other in any statement without changing the truth or falsity of the statement. **Substitution Property**

• Solving Linear Equations

We now turn our attention to methods of solving *first-degree,* or *linear, equations* in one variable.

DEFINITION 1 **Linear Equation in One Variable**

Any equation that can be written in the form

$$ax + b = 0 \qquad a \neq 0 \qquad \textbf{Standard Form}$$

where a and b are real constants and x is a variable, is called a **linear,** or **first-degree, equation** in one variable.

$5x - 1 = 2(x + 3)$ is a linear equation, since it can be written in the standard form $3x - 7 = 0$.

EXAMPLE 1 Solving a Linear Equation

Solve $5x - 9 = 3x + 7$ and check.

Solution We use the properties of equality to transform the given equation into an equivalent equation whose solution is obvious.

$5x - 9 = 3x + 7$	Original equation
$5x - 9 + 9 = 3x + 7 + 9$	Add 9 to both sides.
$5x = 3x + 16$	Combine like terms.
$5x - 3x = 3x + 16 - 3x$	Subtract $3x$ from both sides.
$2x = 16$	Combine like terms.
$\frac{2x}{2} = \frac{16}{2}$	Divide both sides by 2.
$x = 8$	Simplify.

The solution set for this last equation is obvious:

$$\text{Solution set: } \{8\}$$

And since the equation $x = 8$ is equivalent to all the preceding equations in our solution, $\{8\}$ is also the solution set for all these equations, including the original equation. [*Note:* If an equation has only one element in its solution set, we generally use the last equation (in this case, $x = 8$) rather than set notation to represent the solution.]

Check

$5x - 9 = 3x + 7$	Original equation
$5(8) - 9 \stackrel{?}{=} 3(8) + 7$	Substitute $x = 8$.
$40 - 9 \stackrel{?}{=} 24 + 7$	Simplify each side.
$31 \stackrel{\checkmark}{=} 31$	A true statement

Matched Problem 1 Solve and check: $7x - 10 = 4x + 5$

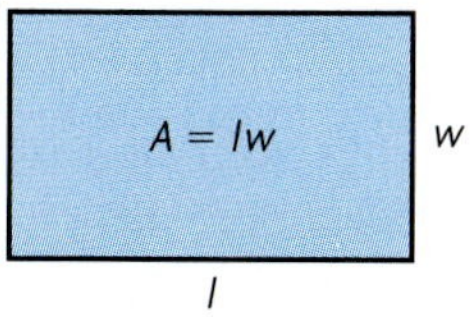

$A = lw$ w

l

FIGURE 1 Area of a rectangle.

We frequently encounter equations involving more than one variable. For example, if l and w are the length and width of a rectangle, respectively, the area of the rectangle is given by (see Fig. 1).

$$A = lw$$

Depending on the situation, we may want to solve this equation for l or w. To solve for w, we simply consider A and l to be constants and w to be a variable. Then the equation $A = lw$ becomes a linear equation in w which can be solved easily by dividing both sides by l:

$$w = \frac{A}{l} \qquad l \neq 0$$

EXAMPLE 2 Solving an Equation with More Than One Variable

Solve for P in terms of the other variables: $A = P + Prt$

Solution

$$A = P + Prt \qquad \text{Think of } A, r, \text{ and } t \text{ as constants.}$$

$$A = P(1 + rt) \qquad \text{Factor to isolate } P.$$

$$\frac{A}{1 + rt} = P \qquad \text{Divide both sides by } 1 + rt.$$

$$P = \frac{A}{1 + rt} \qquad \text{Restriction: } 1 + rt \neq 0$$

Matched Problem 2 Solve for F in terms of C: $C = \frac{5}{9}(F - 32)$

• A Strategy for Solving Word Problems

A great many practical problems can be solved using algebraic techniques—so many, in fact, that there is no one method of attack that will work for all. However, we can formulate a strategy that will help you organize your approach.

Strategy for Solving Word Problems

1. Read the problem carefully—several times if necessary—that is, until you understand the problem, know what is to be found, and know what is given.

2. Let one of the unknown quantities be represented by a variable, say x, and try to represent all other unknown quantities in terms of x. This is an important step and must be done carefully.
3. If appropriate, draw figures or diagrams and label known and unknown parts.
4. Look for formulas connecting the known quantities to the unknown quantities.
5. Form an equation relating the unknown quantities to the known quantities.
6. Solve the equation and write answers to *all* questions asked in the problem.
7. Check and interpret all solutions in terms of the original problem—not just the equation found in step 5—since a mistake may have been made in setting up the equation in step 5.

The remaining examples in this section contain solutions to a variety of word problems illustrating both the process of setting up word problems and the techniques used to solve the resulting equations. It is suggested that you cover up a solution, try solving the problem yourself, and uncover just enough of a solution to get you going again in case you get stuck. After successfully completing an example, try the matched problem. After completing the section in this way, you will be ready to attempt a fairly large variety of applications.

• Number and Geometric Problems

The first examples introduce the process of setting up and solving word problems in a simple mathematical context. Following these, the examples are of a more substantive nature.

EXAMPLE 3 Setting Up and Solving a Word Problem

Find four consecutive even integers such that the sum of the first three exceeds the fourth by 8.

Solution Let x = the first even integer, then

$$x \qquad x + 2 \qquad x + 4 \qquad \text{and} \qquad x + 6$$

represent four consecutive even integers starting with the even integer x. (Remember, even integers increase by 2.) The phrase "the sum of the first three exceeds the fourth by 8" translates into an equation:

$$\text{Sum of the first three} = \text{Fourth} + \text{Excess}$$

$$x + (x + 2) + (x + 4) = (x + 6) + 8$$

$$3x + 6 = x + 14$$

$$2x = 8$$

$$x = 4$$

The four consecutive integers are 4, 6, 8, and 10.

Check

$$\begin{array}{rl} 4 + 6 + 8 = 18 & \text{Sum of first three} \\ -8 & \text{Excess} \\ \hline 10 & \text{Fourth} \end{array}$$

Matched Problem 3 Find three consecutive odd integers such that 3 times their sum is 5 more than 8 times the middle one.

EXPLORE-DISCUSS 1 According to property 1 of Theorem 1, multiplying both sides of an equation by a nonzero number always produces an equivalent equation. By what number would you choose to multiply both sides of the following equation to eliminate all the fractions?

$$\frac{x+1}{3} - \frac{x}{4} = \frac{1}{2}$$

If you did not choose 12, the LCD of all the fractions in this equation, you could still solve the resulting equation, but with more effort. (For a discussion of LCDs and how to find them, see Section A-4.)

EXAMPLE 4 Using a Diagram in the Solution of a Word Problem

If one side of a triangle is one-third the perimeter, the second side is one-fifth the perimeter, and the third side is 7 meters, what is the perimeter of the triangle?

Solution Let p = the perimeter. Draw a triangle and label the sides, as shown in Figure 2. Then

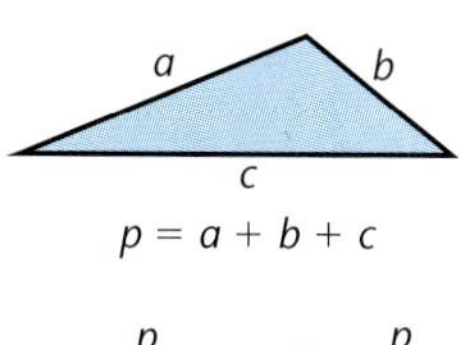

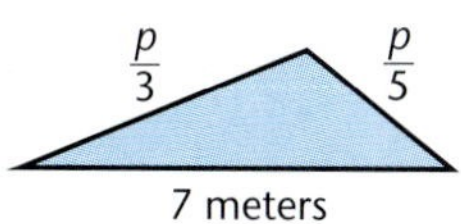

FIGURE 2

$$p = a + b + c$$

$$p = \frac{p}{3} + \frac{p}{5} + 7$$

$$\mathbf{15} \cdot p = \mathbf{15}\left(\frac{p}{3} + \frac{p}{5} + 7\right)$$

Multiply both sides by 15, the LCD. This and the next step usually can be done mentally.

$$15p = 15 \cdot \frac{p}{3} + 15 \cdot \frac{p}{5} + 15 \cdot 7$$

$$15p = 5p + 3p + 105$$

$$7p = 105$$

$$p = 15$$

The perimeter is 15 meters.

Check

$$\frac{p}{3} = \frac{15}{3} = 5 \quad \text{Side 1}$$

$$\frac{p}{5} = \frac{15}{5} = 3 \quad \text{Side 2}$$

$$7 \quad \text{Side 3}$$

$$\overline{15 \text{ meters}} \quad \text{Perimeter}$$

Matched Problem 4 If one side of a triangle is one-fourth the perimeter, the second side is 7 centimeters, and the third side is two-fifths the perimeter, what is the perimeter?

CAUTION A very common error occurs about now—students tend to confuse *algebraic expressions* involving fractions with *algebraic equations* involving fractions.

Consider these two problems:

(A) Solve: $\frac{x}{2} + \frac{x}{3} = 10$ (B) Add: $\frac{x}{2} + \frac{x}{3} + 10$

The problems look very much alike but are actually very different. To solve the equation in (A) we multiply both sides by 6 (the LCD) to clear the fractions. This works so well for equations that students want to do the same thing for problems like (B). The only catch is that (B) is not an equation, and the multiplication property of equality does not apply. If we multiply (B) by 6, we simply obtain an expression 6 times as large as the original! Compare the following:

(A)

$$\frac{x}{2} + \frac{x}{3} = 10$$

$$6 \cdot \frac{x}{2} + 6 \cdot \frac{x}{3} = 6 \cdot 10$$

$$3x + 2x = 60$$

$$5x = 60$$

$$x = 12$$

(B)

$$\frac{x}{2} + \frac{x}{3} + 10$$

$$= \frac{3 \cdot x}{3 \cdot 2} + \frac{2 \cdot x}{2 \cdot 3} + \frac{6 \cdot 10}{6 \cdot 1}$$

$$= \frac{3x}{6} + \frac{2x}{6} + \frac{60}{6}$$

$$= \frac{5x + 60}{6}$$

• Rate–Time Problems

There are many types of quantity–rate–time problems and distance–rate–time problems. In general, if Q is the quantity of something produced (kilometers, words, parts, and so on) in T units of time (hours, years, minutes, seconds, and so on), then the formulas given in the box are relevant.

Quantity–Rate–Time Formulas

$$R = \frac{Q}{T} \qquad \text{Rate} = \frac{\text{Quantity}}{\text{Time}}$$

$$Q = RT \qquad \text{Quantity} = (\text{Rate})(\text{Time})$$

$$T = \frac{Q}{R} \qquad \text{Time} = \frac{\text{Quantity}}{\text{Rate}}$$

If Q is distance D, then

$$R = \frac{D}{T} \qquad D = RT \qquad T = \frac{D}{R}$$

[*Note:* R is an average or uniform rate.]

EXAMPLE 5 A Distance–Rate–Time Problem

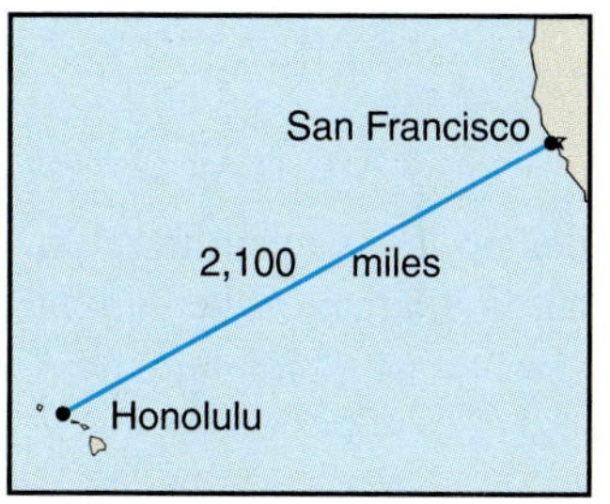

The distance along a shipping route between San Francisco and Honolulu is 2,100 nautical miles. If one ship leaves San Francisco at the same time another leaves Honolulu, and if the former travels at 15 knots* and the latter at 20 knots, how long will it take the two ships to rendezvous? How far will they be from Honolulu and San Francisco at that time?

Solution Let T = number of hours until both ships meet. Draw a diagram and label known and unknown parts. Both ships will have traveled the same amount of time when they meet.

$D_1 = 20T$ — 20 knots → | $D_2 = 15T$ — 15 knots ←

H ——— Meeting ——— SF

$$\begin{pmatrix}\text{Distance ship 1}\\ \text{from Honolulu}\\ \text{travels to}\\ \text{meeting point}\end{pmatrix} + \begin{pmatrix}\text{Distance ship 2}\\ \text{from San Francisco}\\ \text{travels to}\\ \text{meeting point}\end{pmatrix} = \begin{pmatrix}\text{Total distance}\\ \text{from Honolulu}\\ \text{to San Francisco}\end{pmatrix}$$

$$D_1 + D_2 = 2{,}100$$

$$20T + 15T = 2{,}100$$

$$35T = 2{,}100$$

$$T = 60$$

Therefore, it takes 60 hours, or 2.5 days, for the ships to meet.

*15 knots means 15 nautical miles per hour. There are 6,076.1 feet in 1 nautical mile.

$$\text{Distance from Honolulu} = 20 \cdot 60 = 1{,}200 \text{ nautical miles}$$

$$\text{Distance from San Francisco} = 15 \cdot 60 = 900 \text{ nautical miles}$$

Check

$$1{,}200 + 900 = 2{,}100 \text{ nautical miles}$$

Matched Problem 5 An old piece of equipment can print, stuff, and label 38 mailing pieces per minute. A newer model can handle 82 per minute. How long will it take for both pieces of equipment to prepare a mailing of 6,000 pieces? [*Note:* The mathematical form is the same as in Example 5.]

Some equations involving variables in a denominator can be transformed into linear equations. We may proceed in essentially the same way as in the preceding example; however, we must exclude any value of the variable that will make a denominator 0. With these values excluded, we may multiply through by the LCD even though it contains a variable, and, according to Theorem 1, the new equation will be equivalent to the old.

EXAMPLE 6 A Distance–Rate–Time Problem

An excursion boat takes 1.5 times as long to go 360 miles up a river than to return. If the boat cruises at 15 miles per hour in still water, what is the rate of the current?

Solution Let

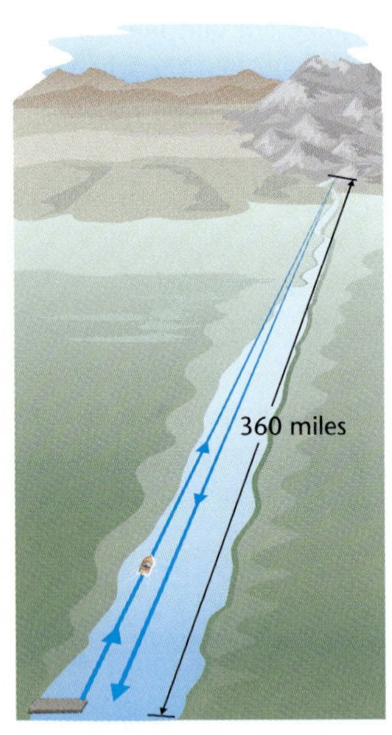

$$x = \text{Rate of current (in miles per hour)}$$

$$15 - x = \text{Rate of boat upstream}$$

$$15 + x = \text{Rate of boat downstream}$$

$$\text{Time upstream} = (1.5)(\text{Time downstream})$$

$$\frac{\text{Distance upstream}}{\text{Rate upstream}} = (1.5)\frac{\text{Distance downstream}}{\text{Rate downstream}} \qquad \text{Recall: } T = \frac{D}{R}$$

$$\frac{360}{15 - x} = (1.5)\frac{360}{15 + x} \qquad x \neq 15,\ x \neq -15$$

$$\frac{360}{15 - x} = \frac{540}{15 + x}$$

$$360(15 + x) = 540(15 - x) \qquad \text{Multiply both sides by } (15 - x)(15 + x).$$

$$5{,}400 + 360x = 8{,}100 - 540x$$

$$900x = 2{,}700$$

$$x = 3$$

Therefore, the rate of the current is 3 miles per hour. The check is left to the reader.

Matched Problem 6 A jetliner takes 1.2 times as long to fly from Paris to New York (3,600 miles) as to return. If the jet cruises at 550 miles per hour in still air, what is the average rate of the wind blowing in the direction of Paris from New York?

EXPLORE-DISCUSS 2 Consider the following solution:

$$\frac{x}{x-2} + 2 = \frac{2x-2}{x-2}$$

$$x + 2x - 4 = 2x - 2$$

$$x = 2$$

Is $x = 2$ a root of the original equation? If not, why? Discuss the importance of excluding values that make a denominator 0 when solving equations.

EXAMPLE 7 **A Quantity–Rate–Time Problem**

An advertising firm has an old computer that can prepare a whole mailing in 6 hours. With the help of a newer model the job is complete in 2 hours. How long would it take the newer model to do the job alone?

Solution Let x = time (in hours) for the newer model to do the whole job alone.

$$\begin{pmatrix}\text{Part of job completed}\\ \text{in a given length of time}\end{pmatrix} = (\text{Rate})(\text{Time})$$

$$\text{Rate of old model} = \frac{1}{6}\text{ job per hour}$$

$$\text{Rate of new model} = \frac{1}{x}\text{ job per hour}$$

$$\begin{pmatrix}\text{Part of job completed}\\ \text{by old model}\\ \text{in 2 hours}\end{pmatrix} + \begin{pmatrix}\text{Part of job completed}\\ \text{by new model}\\ \text{in 2 hours}\end{pmatrix} = 1 \text{ whole job}$$

$$\begin{pmatrix}\text{Rate of}\\ \text{old model}\end{pmatrix}\begin{pmatrix}\text{Time of}\\ \text{old model}\end{pmatrix} + \begin{pmatrix}\text{Rate of}\\ \text{new model}\end{pmatrix}\begin{pmatrix}\text{Time of}\\ \text{new model}\end{pmatrix} = 1 \qquad \text{Recall: } Q = RT$$

$$\frac{1}{6}(2) + \frac{1}{x}(2) = 1 \qquad x \neq 0$$

$$\frac{1}{3} + \frac{2}{x} = 1$$
$$x + 6 = 3x$$
$$-2x = -6$$
$$x = 3$$

Therefore, the new computer could do the job alone in 3 hours.

Check

$$\begin{aligned} \text{Part of job completed by old model in 2 hours} &= 2(\tfrac{1}{6}) = \tfrac{1}{3} \\ \text{Part of job completed by new model in 2 hours} &= 2(\tfrac{1}{3}) = \underline{\tfrac{2}{3}} \\ \text{Part of job completed by both models in 2 hours} &= 1 \end{aligned}$$

Matched Problem 7 Two pumps are used to fill a water storage tank at a resort. One pump can fill the tank by itself in 9 hours, and the other can fill it in 6 hours. How long will it take both pumps operating together to fill the tank?

• Mixture Problems

A variety of applications can be classified as mixture problems. Even though the problems come from different areas, their mathematical treatment is essentially the same.

EXAMPLE 8 A Mixture Problem

How many liters of a mixture containing 80% alcohol should be added to 5 liters of a 20% solution to yield a 30% solution?

Solution Let x = amount of 80% solution used.

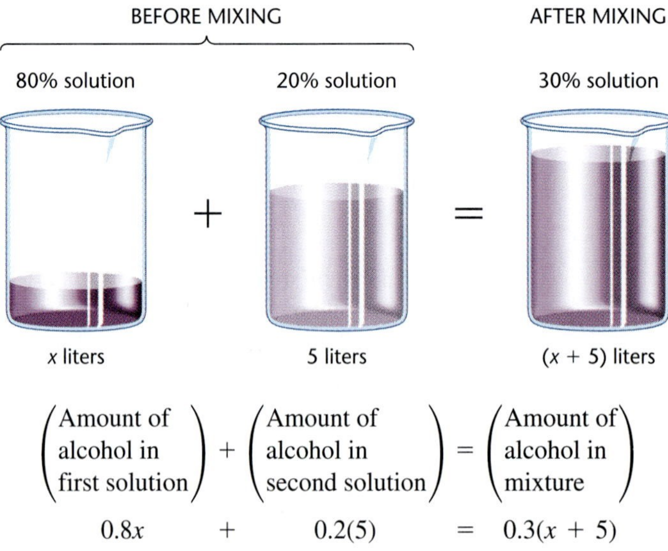

$$\begin{pmatrix}\text{Amount of}\\ \text{alcohol in}\\ \text{first solution}\end{pmatrix} + \begin{pmatrix}\text{Amount of}\\ \text{alcohol in}\\ \text{second solution}\end{pmatrix} = \begin{pmatrix}\text{Amount of}\\ \text{alcohol in}\\ \text{mixture}\end{pmatrix}$$

$$0.8x \quad + \quad 0.2(5) \quad = \quad 0.3(x + 5)$$

$$0.8x + 1 = 0.3x + 1.5$$
$$0.5x = 0.5$$
$$x = 1$$

Add 1 liter of the 80% solution.

Check

	Liters of solution	Liters of alcohol	Percent alcohol
First solution	1	$0.8(1) = 0.8$	80
Second solution	5	$0.2(5) = 1$	20
Mixture	6	1.8	$1.8/6 = 0.3$, or 30%

Matched Problem 8 A chemical storeroom has a 90% acid solution and a 40% acid solution. How many centiliters of the 90% solution should be added to 50 centiliters of the 40% solution to yield a 50% solution?

• Some Final Observations on Linear Equations

It can be shown that any equation that can be written in the form

$$ax + b = 0 \qquad a \neq 0 \tag{1}$$

with no restrictions on x, has exactly one solution, and the solution can be found as follows:

$$ax + b = 0$$
$$ax = -b \qquad \text{Subtraction property of equality}$$
$$x = \frac{-b}{a} \qquad \text{Division property of equality}$$

Requiring $a \neq 0$ in equation (1) is an important restriction, because without it we are able to write equations with first-degree members that have no solutions or infinitely many solutions. For example,

$$2x - 3 = 2x + 5$$

has no solution, and

$$3x - 4 = 5 + 3(x - 3)$$

has infinitely many solutions. Try to solve each equation to see what happens.

Answers to Matched Problems

1. $x = 5$ **2.** $F = \frac{9}{5}C + 32$ **3.** 3, 5, 7 **4.** 20 centimeters
5. 50 minutes **6.** 50 miles per hour **7.** 3.6 hours **8.** 12.5 centiliters

EXERCISE 1-1

A

In Problems 1–16, solve each equation.

1. $4(x + 5) = 6(x - 2)$ **2.** $3(y - 4) + 2y = 18$

3. $5 + 4(w - 1) = 2w + 2(w + 4)$

4. $4 - 3(t + 2) + t = 5(t - 1) - 7t$

5. $5 - \frac{3a - 4}{5} = \frac{7 - 2a}{2}$ **6.** $\frac{3b}{7} + \frac{2b - 5}{2} = -4$

7. $\frac{x}{2} + \frac{2x - 1}{3} = \frac{3x + 4}{4}$ **8.** $\frac{x}{5} + \frac{3x - 1}{2} = \frac{6x + 5}{4}$

9. $0.1(t + 0.5) + 0.2t = 0.3(t - 0.4)$

10. $0.1(w + 0.5) + 0.2w = 0.2(w - 0.4)$

11. $0.35(s + 0.34) + 0.15s = 0.2s - 1.66$

12. $0.35(u + 0.34) - 0.15u = 0.2u - 1.66$

13. $\frac{2}{y} + \frac{5}{2} = 4 - \frac{2}{3y}$ **14.** $\frac{3 + w}{6w} = \frac{1}{2w} + \frac{4}{3}$

15. $\frac{z}{z - 1} = \frac{1}{z - 1} + 2$ **16.** $\frac{t}{t - 1} = \frac{2}{t - 1} + 2$

B

In Problems 17–24, solve each equation.

17. $\frac{2m}{5} + \frac{m - 4}{6} = \frac{4m + 1}{4} - 2$

18. $\frac{3(n - 2)}{5} + \frac{2n + 3}{6} = \frac{4n + 1}{9} + 2$

19. $1 - \frac{x - 3}{x - 2} = \frac{2x - 3}{x - 2}$ **20.** $\frac{2x - 3}{x + 1} = 2 - \frac{3x - 1}{x + 1}$

21. $\frac{6}{y + 4} + 1 = \frac{5}{2y + 8}$ **22.** $\frac{4y}{y - 3} + 5 = \frac{12}{y - 3}$

23. $\frac{3a - 1}{a^2 + 4a + 4} - \frac{3}{a^2 + 2a} = \frac{3}{a}$

24. $\frac{1}{b - 5} - \frac{10}{b^2 - 5b + 25} = \frac{1}{b + 5}$

In Problems 25–28, use a calculator to solve each equation to 3 significant digits.

25. $3.142x - 0.4835(x - 4) = 6.795$

26. $0.0512x + 0.125(x - 2) = 0.725x$

27. $\frac{2.32x}{x - 2} - \frac{3.76}{x} = 2.32$ **28.** $\frac{6.08}{x} + 4.49 = \frac{4.49x}{x + 3}$

In Problems 29–36, solve for the indicated variable in terms of the other variables.

29. $a_n = a_1 + (n - 1)d$ for d (arithmetic progressions)

30. $F = \frac{9}{5}C + 32$ for C (temperature scale)

31. $\frac{1}{f} = \frac{1}{d_1} + \frac{1}{d_2}$ for f (simple lens formula)

32. $\frac{1}{R} = \frac{1}{R_1} + \frac{1}{R_2}$ for R_1 (electric circuit)

33. $A = 2ab + 2ac + 2bc$ for a (surface area of a rectangular solid)

34. $A = 2ab + 2ac + 2bc$ for c

35. $y = \frac{2x - 3}{3x + 5}$ for x **36.** $x = \frac{3y + 2}{y - 3}$ for y

In Problems 37 and 38, imagine that the indicated "solutions" were given to you by a student whom you were tutoring in this class. Is the solution right or wrong? If the solution is wrong, explain what is wrong and show a correct solution.

37. $\frac{x}{x - 3} + 4 = \frac{2x - 3}{x - 3}$

$x + 4x - 12 = 2x - 3$

$x = 3$

38. $\frac{x^2 + 1}{x - 1} = \frac{x^2 + 4x - 3}{x - 1}$

$x^2 + 1 = x^2 + 4x - 3$

$x = 1$

C

In Problems 39–41, solve for x.

39. $\frac{x - \frac{1}{x}}{1 + \frac{1}{x}} = 3$ **40.** $\frac{x - \frac{1}{x}}{x + 1 - \frac{2}{x}} = 1$

41. $\frac{x + 1 - \frac{2}{x}}{1 - \frac{1}{x}} = x + 2$

42. Solve for y in terms of x: $\frac{y}{1-y} = \left(\frac{x}{1-x}\right)^3$

43. Solve for x in terms of y: $y = \cfrac{a}{1 + \cfrac{b}{x+c}}$

44. Let m and n be real numbers with m larger than n. Then there exists a positive real number p such that $m = n + p$. Find the fallacy in the following argument:

$$\begin{aligned} m &= n + p \\ (m-n)m &= (m-n)(n+p) \\ m^2 - mn &= mn + mp - n^2 - np \\ m^2 - mn - mp &= mn - n^2 - np \\ m(m-n-p) &= n(m-n-p) \\ m &= n \end{aligned}$$

APPLICATIONS

These problems are grouped according to subject area. As before, the most difficult problems are marked with two stars (★★), the moderately difficult problems are marked with one star (★), and the easier problems are not marked.

Numbers

45. Find a number such that 10 less than two-thirds the number is one-fourth the number.

46. Find a number such that 6 more than one-half the number is two-thirds the number.

47. Find four consecutive even integers so that the sum of the first three is 2 more than twice the fourth.

48. Find three consecutive even integers so that the first plus twice the second is twice the third.

Geometry

49. Find the dimensions of a rectangle with a perimeter of 54 meters, if its length is 3 meters less than twice its width.

50. A rectangle 24 meters long has the same area as a square that is 12 meters on a side. What are the dimensions of the rectangle?

51. Find the perimeter of a triangle if one side is 16 feet, another side is two-sevenths the perimeter, and the third side is one-third the perimeter.

52. Find the perimeter of a triangle if one side is 11 centimeters, another side is one-fifth the perimeter, and the third side is one-fourth the perimeter.

Business and Economics

53. The sale price on a camera after a 20% discount is \$72. What was the price before the discount?

54. A stereo store marks up each item it sells 60% above wholesale price. What is the wholesale price on a cassette player that retails at \$144?

55. One employee of a computer store is paid a base salary of \$2,150 a month plus an 8% commission on all sales over \$7,000 during the month. How much must the employee sell in 1 month to earn a total of \$3,170 for the month?

56. A second employee of the computer store in Problem 55 is paid a base salary of \$1,175 a month plus a 5% commission on all sales during the month.

(A) How much must this employee sell in 1 month to earn a total of \$3,170 for the month?

(B) Determine the sales level where both employees receive the same monthly income. If employees can select either of these payment methods, how would you advise an employee to make this selection?

Earth Science

★ **57.** In 1984, the Soviets led the world in drilling the deepest hole in the Earth's crust—more than 12 kilometers deep. They found that below 3 kilometers the temperature T increased 2.5°C for each additional 100 meters of depth.

(A) If the temperature at 3 kilometers is 30°C and x is the depth of the hole in kilometers, write an equation using x that will give the temperature T in the hole at any depth beyond 3 kilometers.

(B) What would the temperature be at 15 kilometers? (The temperature limit for their drilling equipment was about 300°C.)

(C) At what depth (in kilometers) would they reach a temperature of 280°C?

★ **58.** Because air is not as dense at high altitudes, planes require a higher ground speed to become airborne. A rule of thumb is 3% more ground speed per 1,000 feet of elevation, assuming no wind and no change in air temperature. (Compute numerical answers to 3 significant digits.)

(A) Let

V_S = Takeoff ground speed at sea level for a particular plane (in miles per hour)

A = Altitude above sea level (in thousands of feet)

V = Takeoff ground speed at altitude A for the same plane (in miles per hour)

Write a formula relating these three quantities.

(B) What takeoff ground speed would be required at Lake Tahoe airport (6,400 feet), if takeoff ground speed at San Francisco airport (sea level) is 120 miles per hour?

(C) If a landing strip at a Colorado Rockies hunting lodge (8,500 feet) requires a takeoff ground speed of 125 miles per hour, what would be the takeoff ground speed in Los Angeles (sea level)?

(D) If the takeoff ground speed at sea level is 135 miles per hour and the takeoff ground speed at a mountain resort is 155 miles per hour, what is the altitude of the mountain resort in thousands of feet?

★★ **59.** An earthquake emits a primary wave and a secondary wave. Near the surface of the Earth the primary wave travels at about 5 miles per second, and the secondary wave travels at about 3 miles per second. From the time lag between the two waves arriving at a given seismic station, it is possible to estimate the distance to the quake. Suppose a station measures a time difference of 12 seconds between the arrival of the two waves. How far is the earthquake from the station? (The *epicenter* can be located by obtaining distance bearings at three or more stations.)

★★ **60.** A ship using sound-sensing devices above and below water recorded a surface explosion 39 seconds sooner on its underwater device than on its above-water device. If sound travels in air at about 1,100 feet per second and in water at about 5,000 feet per second, how far away was the explosion?

Life Science

61. A naturalist for a fish and game department estimated the total number of trout in a certain lake using the popular capture–mark–recapture technique. She netted, marked, and released 200 trout. A week later, allowing for thorough mixing, she again netted 200 trout and found 8 marked ones among them. Assuming that the ratio of marked trout to the total number in the second sample is the same as the ratio of all marked fish in the first sample to the total trout population in the lake, estimate the total number of fish in the lake.

62. Repeat Problem 61 with a first (marked) sample of 300 and a second sample of 180 with only 6 marked trout.

Chemistry

★ **63.** How many gallons of distilled water must be mixed with 50 gallons of 30% alcohol solution to obtain a 25% solution?

★ **64.** How many gallons of hydrochloric acid must be added to 12 gallons of a 30% solution to obtain a 40% solution?

★ **65.** A chemist mixes distilled water with a 90% solution of sulfuric acid to produce a 50% solution. If 5 liters of distilled water is used, how much 50% solution is produced?

★ **66.** A fuel oil distributor has 120,000 gallons of fuel with 0.9% sulfur content, which exceeds pollution control standards of 0.8% sulfur content. How many gallons of fuel oil with a 0.3% sulfur content must be added to the 120,000 gallons to obtain fuel oil that will comply with the pollution control standards?

Rate–Time

★ **67.** An old computer can do the weekly payroll in 5 hours. A newer computer can do the same payroll in 3 hours. The old computer starts on the payroll, and after 1 hour the newer computer is brought on-line to work with the older computer until the job is finished. How long will it take both computers working together to finish the job? (Assume the computers operate independently.)

★ **68.** One pump can fill a gasoline storage tank in 8 hours. With a second pump working simultaneously, the tank can be filled in 3 hours. How long would it take the second pump to fill the tank operating alone?

★★ **69.** The cruising speed of an airplane is 150 miles per hour (relative to the ground). You wish to hire the plane for a 3-hour sightseeing trip. You instruct the pilot to fly north as far as he can and still return to the airport at the end of the allotted time.

(A) How far north should the pilot fly if the wind is blowing from the north at 30 miles per hour?

(B) How far north should the pilot fly if there is no wind?

★ **70.** Suppose you are at a river resort and rent a motor boat for 5 hours starting at 7 A.M. You are told that the boat will travel at 8 miles per hour upstream and 12 miles per hour returning. You decide that you would like to go as far up the river as you can and still be back at noon. At what time should you turn back, and how far from the resort will you be at that time?

Music

★ **71.** A major chord in music is composed of notes whose frequencies are in the ratio 4:5:6. If the first note of a chord has a frequency of 264 hertz (middle C on the piano), find the frequencies of the other two notes. [*Hint:* Set up two proportions using 4:5 and 4:6.]

★ **72.** A minor chord is composed of notes whose frequencies are in the ratio 10:12:15. If the first note of a minor chord is A, with a frequency of 220 hertz, what are the frequencies of the other two notes?

Psychology

73. In an experiment on motivation, Professor Brown trained a group of rats to run down a narrow passage in a cage to receive food in a goal box. He then put a harness on each rat and connected it to an overhead wire attached to a scale. In this way he could place the rat different distances from the food and measure the pull (in grams) of the rat toward the food. He found that the relationship between motivation (pull) and position was given approximately by the equation

$$p = -\tfrac{1}{5}d + 70 \qquad 30 \le d \le 170$$

where pull p is measured in grams and distance d in centimeters. When the pull registered was 40 grams, how far was the rat from the goal box?

74. Professor Brown performed the same kind of experiment as described in Problem 73, except that he replaced the food in the goal box with a mild electric shock. With the same kind of apparatus, he was able to measure the avoidance strength relative to the distance from the object to be avoided. He found that the avoidance strength a (measured in grams) was related to the distance d that the rat was from the shock (measured in centimeters) approximately by the equation

$$a = -\tfrac{4}{3}d + 230 \qquad 30 \le d \le 170$$

If the same rat were trained as described in this problem and in Problem 73, at what distance (to one decimal place) from the goal box would the approach and avoidance strengths be the same? (What do you think the rat would do at this point?)

Puzzle

75. An oil-drilling rig in the Gulf of Mexico stands so that one-fifth of it is in sand, 20 feet of it is in water, and two-thirds of it is in the air. What is the total height of the rig?

76. During a camping trip in the North Woods in Canada, a couple went one-third of the way by boat, 10 miles by foot, and one-sixth of the way by horse. How long was the trip?

★★ **77.** After exactly 12 o'clock noon, what time will the hands of a clock be together again?

SECTION 1-2 Systems of Linear Equations and Applications

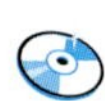

- Systems of Equations
- Substitution
- Applications

• Systems of Equations

In the preceding section, we solved word problems by introducing a single variable representing one of the unknown quantities and then tried to represent all other unknown quantities in terms of this variable. In certain word problems, it is more convenient to introduce several variables, find equations relating these variables, and then solve the resulting system of equations. For example, if a 12-foot board is cut into two pieces so that one piece is 2 feet longer than the other, then letting

$$x = \text{Length of the longer piece}$$
$$y = \text{Length of the shorter piece}$$

we see that x and y must satisfy both the following equations:

$$x + y = 12$$
$$x - y = 2$$

We now have a system of two linear equations in two variables. Thus, we can solve this problem by finding all pairs of numbers x and y that satisfy both equations.

In general, we are interested in solving linear systems of the type:

$$\begin{aligned} ax + by &= h \\ cx + dy &= k \end{aligned} \qquad \text{System of two linear equations in two variables}$$

where x and y are variables and a, b, c, d, h, and k are real constants. A pair of numbers $x = x_0$ and $y = y_0$ is a **solution** of this system if each equation is satisfied by the pair. The set of all such pairs of numbers is called the **solution set** for the system. To **solve** a system is to find its solution set. In this section, we will restrict our discussion to simple solution techniques for systems of two linear equations in two variables. Larger systems and more sophisticated solution methods will be discussed in Chapter 8.

• Substitution

To solve a system by **substitution,** we first choose one of the two equations in a system and solve for one variable in terms of the other. (We make a choice that avoids fractions, if possible.) Then we substitute the result in the other equation and solve the resulting linear equation in one variable. Finally, we substitute this result back into the expression obtained in the first step to find the second variable. We illustrate this process by returning to the board problem stated at the beginning of the section.

EXAMPLE 1 Solving a System by Substitution

Solve the board problem by solving the system

$$x + y = 12$$
$$x - y = 2$$

Solution Solve either equation for one variable and substitute into the remaining equation. We choose to solve the first equation for y in terms of x:

$$x + y = 12 \qquad \text{Solve the first equation for } y \text{ in terms of } x.$$
$$y = 12 - x \qquad \text{Substitute into the second equation.}$$

$$x - y = 2$$
$$x - (\mathbf{12 - x}) = 2$$
$$x - 12 + x = 2$$
$$2x = 14$$
$$\mathbf{x = 7}$$

Now, replace x with 7 in $y = 12 - x$:

$$y = 12 - x$$
$$y = 12 - \mathbf{7}$$
$$\mathbf{y = 5}$$

Thus, the longer board measures 7 feet and the shorter board measures 5 feet.

Check

$$x + y = 12 \qquad x - y = 2$$
$$7 + 5 \stackrel{?}{=} 12 \qquad 7 - 5 \stackrel{?}{=} 2$$
$$12 \stackrel{\checkmark}{=} 12 \qquad 2 \stackrel{\checkmark}{=} 2$$

Matched Problem 1 Solve by substitution and check:

$$\begin{aligned} x - y &= 3 \\ x + 2y &= -3 \end{aligned}$$

EXAMPLE 2 Solving a System by Substitution

Solve by substitution and check:

$$\begin{aligned} 2x - 3y &= 7 \\ 3x - y &= 7 \end{aligned}$$

Solution To avoid fractions, we choose to solve the second equation for y:

$$\begin{aligned}
3x - y &= 7 && \text{Solve for } y \text{ in terms of } x. \\
-y &= -3x + 7 \\
y &= 3x - 7 && \text{Substitute into first equation.} \\
2x - 3y &= 7 && \text{First equation} \\
2x - 3(\mathbf{3x - 7}) &= 7 && \text{Solve for } x. \\
2x - 9x + 21 &= 7 \\
-7x &= -14 \\
\mathbf{x} &= \mathbf{2} && \text{Substitute } x = 2 \text{ in } y = 3x - 7. \\
y &= 3x - 7 \\
y &= 3(\mathbf{2}) - 7 \\
\mathbf{y} &= \mathbf{-1}
\end{aligned}$$

Thus, the solution is $x = 2$ and $y = -1$.

Check

$$2x - 3y = 7 \qquad 3x - y = 7$$
$$2(2) - 3(-1) \stackrel{?}{=} 7 \qquad 3(2) - (-1) \stackrel{?}{=} 7$$
$$7 \stackrel{\checkmark}{=} 7 \qquad 7 \stackrel{\checkmark}{=} 7$$

Matched Problem 2 Solve by substitution and check:

$$\begin{aligned} 3x - 4y &= 18 \\ 2x + y &= 1 \end{aligned}$$

EXPLORE-DISCUSS 1 Use substitution to solve each of the following systems. Discuss the nature of the solution sets you obtain.

$$x + 3y = 4 \qquad x + 3y = 4$$
$$2x + 6y = 7 \qquad 2x + 6y = 8$$

• Applications

The following examples illustrate the advantages of using systems of equations and substitution in solving word problems.

EXAMPLE 3 Diet

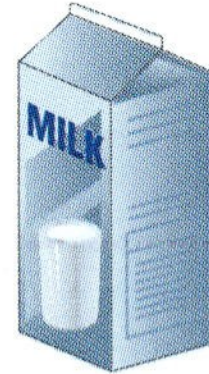

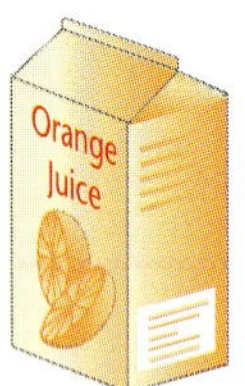

An individual wants to use milk and orange juice to increase the amount of calcium and vitamin A in her daily diet. An ounce of milk contains 38 milligrams of calcium and 56 micrograms* of vitamin A. An ounce of orange juice contains 5 milligrams of calcium and 60 micrograms of vitamin A. How many ounces of milk and orange juice should she drink each day to provide exactly 550 milligrams of calcium and 1,200 micrograms of vitamin A?

Solution First we define the relevant variables:

$$x = \text{Number of ounces of milk}$$
$$y = \text{Number of ounces of orange juice}$$

Next we summarize the given information in a table. It is convenient to organize the tables so that the quantities represented by the variables correspond to columns in the table (rather than to rows), as shown.

	Milk	Orange Juice	Total Needed
Calcium	38	5	550
Vitamin A	56	60	1,200

Now we use the information in the table to form equations involving x and y:

$$\begin{pmatrix}\text{Calcium in } x \text{ oz}\\ \text{of milk}\end{pmatrix} + \begin{pmatrix}\text{Calcium in } y \text{ oz}\\ \text{of orange juice}\end{pmatrix} = \begin{pmatrix}\text{Total calcium}\\ \text{needed (mg)}\end{pmatrix}$$
$$38x + 5y = 550$$
$$\begin{pmatrix}\text{Vitamin A in } x \text{ oz}\\ \text{of milk}\end{pmatrix} + \begin{pmatrix}\text{Vitamin A in } y \text{ oz}\\ \text{of orange juice}\end{pmatrix} = \begin{pmatrix}\text{Total vitamin A}\\ \text{needed } (\mu\text{g})\end{pmatrix}$$
$$56x + 60y = 1{,}200$$

*A microgram (μg) is one-millionth (10^{-6}) of a gram.

$$5y = 550 - 38x \qquad \text{Solve first equation for } y.$$

$$y = 110 - 7.6x \qquad (1)$$

$$56x + 60(110 - 7.6x) = 1{,}200 \qquad \text{Substitute for } y \text{ in second equation.}$$

$$56x + 6{,}600 - 456x = 1{,}200$$

$$-400x = -5{,}400$$

$$\mathbf{x = 13.5} \qquad \text{Substitute in (1).}$$

$$y = 110 - 7.6(13.5)$$

$$\mathbf{y = 7.4}$$

Drinking 13.5 ounces of milk and 7.4 ounces of orange juice each day will provide the required amounts of calcium and vitamin A.

Check

$$38x + 5y = 550 \qquad 56x + 60y = 1{,}000$$

$$38(13.5) + 5(7.4) \stackrel{?}{=} 500 \qquad 56(13.5) + 60(7.4) \stackrel{?}{=} 1{,}200$$

$$500 \stackrel{\checkmark}{=} 500 \qquad 1{,}200 \stackrel{\checkmark}{=} 1{,}200$$

Matched Problem 3 An individual wants to use cottage cheese and yogurt to increase the amount of protein and calcium in his daily diet. An ounce of cottage cheese contains 3 grams of protein and 12 milligrams of calcium. An ounce of yogurt contains 1 gram of protein and 44 milligrams of calcium. How many ounces of cottage cheese and yogurt should he eat each day to provide exactly 57 grams of protein and 840 milligrams of calcium?

EXAMPLE 4 Airspeed

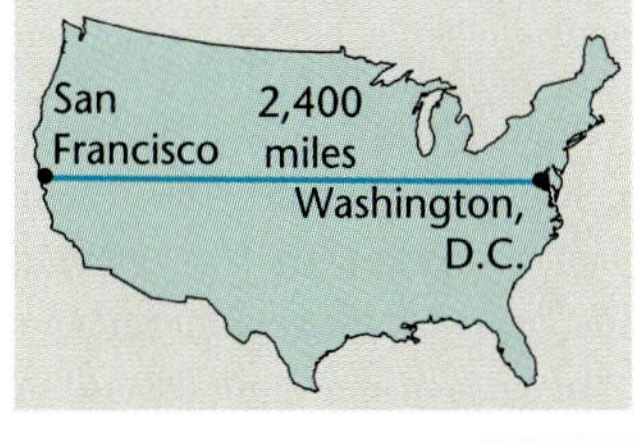

An airplane makes the 2,400-mile trip from Washington, D.C., to San Francisco in 7.5 hours and makes the return trip in 6 hours. Assuming that the plane travels at a constant airspeed and that the wind blows at a constant rate from west to east, find the plane's airspeed and the wind rate.

Solution Let x represent the airspeed of the plane and let y represent the rate at which the wind is blowing (both in miles per hour). The ground speed of the plane is determined by combining these two rates; that is,

$$x - y = \text{Ground speed flying east to west (headwind)}$$

$$x + y = \text{Ground speed flying west to east (tailwind)}$$

Applying the familiar formula $D = RT$ to each leg of the trip leads to the following system of equations:

$$2{,}400 = 7.5(x - y) \qquad \text{From Washington to San Francisco}$$

$$2{,}400 = 6(x + y) \qquad \text{From San Francisco to Washington}$$

After simplification, we have

$$x - y = 320$$
$$x + y = 400$$

Solve using substitution:

$$x = y + 320 \quad \text{Solve first equation for } x.$$
$$y + 320 + y = 400 \quad \text{Substitute in second equation.}$$
$$2y = 80$$
$$\mathbf{y = 40 \text{ mph}} \quad \text{Wind rate}$$
$$x = 40 + 320$$
$$\mathbf{x = 360 \text{ mph}} \quad \text{Airspeed}$$

Check

$$2{,}400 = 7.5(x - y) \qquad 2{,}400 = 6(x + y)$$
$$2{,}400 \stackrel{?}{=} 7.5(360 - 40) \qquad 2{,}400 \stackrel{?}{=} 6(360 + 40)$$
$$2{,}400 \stackrel{\checkmark}{=} 2{,}400 \qquad 2{,}400 \stackrel{\checkmark}{=} 2{,}400$$

Matched Problem 4 A boat takes 8 hours to travel 80 miles upstream and 5 hours to return to its starting point. Find the speed of the boat in still water and the speed of the current.

EXAMPLE 5 Supply and Demand

The quantity of a product that people are willing to buy during some period of time depends on its price. Generally, the higher the price, the less the demand; the lower the price, the greater the demand. Similarly, the quantity of a product that a supplier is willing to sell during some period of time also depends on the price. Generally, a supplier will be willing to supply more of a product at higher prices and less of a product at lower prices. The simplest supply and demand model is a linear model.

Suppose we are interested in analyzing the sale of cherries each day in a particular city. Using special analytic techniques (regression analysis) and collected data, an analyst arrives at the following price–demand and price–supply equations:

$$p = -0.3q + 5 \quad \text{Demand equation (consumer)}$$
$$p = 0.06q + 0.68 \quad \text{Supply equation (supplier)}$$

where q represents the quantity in thousands of pounds and p represents the price in dollars. For example, we see that consumers will purchase 11 thousand pounds ($q = 11$) when the price is $p = -0.3(11) + 5 = \$1.70$ per pound. On the other hand, suppliers will be willing to supply 17 thousand pounds of cherries at \$1.70 per pound (solve $1.7 = 0.06q + 0.68$ for q). Thus, at \$1.70 per pound the suppliers are willing to supply more cherries than the consumers are willing to purchase. The supply exceeds the demand at that price, and the price will come down. At what price will cherries stabilize for the day? That is, at what price will supply equal demand? This price, if it exists, is called the **equilibrium price,** and the quantity sold at that price is called the **equilibrium quantity.** How do we find these quantities? We solve the linear system

$$p = -0.3q + 5 \quad \text{Demand equation}$$
$$p = 0.06q + 0.68 \quad \text{Supply equation}$$

using substitution (substituting $p = -0.3q + 5$ into the second equation).

$$-0.3q + 5 = 0.06q + 0.68$$
$$-0.36q = -4.32$$
$$\mathbf{q = 12 \text{ thousand pounds}} \qquad \textbf{(equilibrium quantity)}$$

Now substitute $q = 12$ back into either of the original equations in the system and solve for p (we choose the second equation):

$$p = 0.06(12) + 0.68$$
$$\mathbf{p = \$1.40 \text{ per pound}} \qquad \textbf{(equilibrium price)}$$

Matched Problem 5 The price–demand and price–supply equations for strawberries in a certain city are

$$p = -0.2q + 4 \quad \text{Demand equation}$$
$$p = 0.04q + 1.84 \quad \text{Supply equation}$$

where q represents the quantity in thousands of pounds and p represents the price in dollars.

Find the equilibrium quantity and the equilibrium price.

EXAMPLE 6 Cost and Revenue

A publisher is planning to produce a new textbook. The **fixed costs** (reviewing, editing, typesetting, and so on) are \$320,000, and the **variable costs** (printing, sales commissions, and so on) are \$31.25 per book. The wholesale price (the amount received by the publisher) will be \$43.75 per book. How many books must the publisher sell to **break even;** that is, so that costs will equal revenues?

Solution If x represents the number of books printed and sold, then the cost and revenue equations for the publisher are

$$y = 320{,}000 + 31.25x \quad \text{Cost equation}$$
$$y = 43.75x \quad \text{Revenue equation}$$

The publisher breaks even when costs equal revenues. We can find when this occurs by solving this system. Using the second equation to substitute for y in the first equation, we have

$$43.75x = 320{,}000 + 31.25x$$
$$12.5x = 320{,}000$$
$$x = 25{,}600$$

Thus, the publisher will break even when 25,600 books are printed and sold.

Matched Problem 6 A computer software company is planning to market a new word processor. The fixed costs (design, programming, and so on) are \$720,000, and the variable costs (disk duplication, manual production, and so on) are \$25.40 per copy. The wholesale price of the word processor will be \$44.60 per copy. How many copies of the word processor must the company manufacture and sell to break even?

Answers to Matched Problems

1. $x = 1, y = -2$ **2.** $x = 2, y = -3$
3. 13.9 oz cottage cheese, 15.3 oz of yogurt
4. Boat: 13 mph; current: 3 mph
5. Equilibrium quantity = 9 thousand pounds; Equilibrium price = \$2.20 per pound
6. 37,500 copies

EXERCISE 1-2

A

Solve Problems 1–6 by substitution.

1. $y = 3x + 5$, $y = 4x + 7$

2. $y = 5x - 8$, $y = 2x + 4$

3. $4x + 5y = -8$, $3x + y = 5$

4. $2x + 7y = 4$, $x + 3y = 1$

5. $6x + 11y = 16$, $2x - 3y = -8$

6. $9x + 7y = 8$, $-4x + 3y = 27$

B

Solve Problems 7–16 by substitution.

7. $3s - 5t = -30$, $7s + 11t = 32$

8. $2s - 9t = 22$, $8s - 11t = 48$

9. $13m + 3n = 10$, $7m - 3n = -10$

10. $12m - 11n = 2$, $18m + 7n = 3$

11. $y = 5.46x$, $y = 6{,}300 + 4.86x$

12. $y = 7.15x$, $y = 13{,}860 + 3.85x$

13. $0.7u - 0.2v = 0.5$, $0.3u - 0.5v = 0.09$

14. $0.4u - 0.9v = 0.71$, $0.8u - 0.2v = 0.3$

15. $\frac{6}{5}a + \frac{3}{2}b = -4$, $\frac{1}{3}a - \frac{5}{4}b = 1$

16. $\frac{3}{8}a - \frac{1}{2}b = 2$, $\frac{9}{10}a + \frac{2}{5}b = 4$

17. In the process of solving a system by substitution, suppose you encounter a contradiction, such as $0 = 1$. How would you describe the solutions to such a system? Illustrate your ideas with the system

$$x - 2y = -3$$
$$-2x + 4y = 7$$

18. In the process of solving a system by substitution, suppose you encounter an identity, such as $0 = 0$. How would you describe the solutions to such a system? Illustrate your ideas with the system

$$x - 2y = -3$$
$$-2x + 4y = 6$$

C

In Problems 19 and 20, solve each system for p and q in terms of x and y. Explain how you could check your solution and then perform the check.

19. $x = 2 + p - 2q$
$y = 3 - p + 3q$

20. $x = -1 + 2p - q$
$y = 4 - p + q$

Problems 21 and 22 refer to the system

$$ax + by = h$$
$$cx + dy = k$$

where x and y are variables and a, b, c, d, h, and k are real constants.

21. Solve the system for x and y in terms of the constants a, b, c, d, h, and k. Clearly state any assumptions you must make about the constants during the solution process.

22. Discuss the nature of solutions to systems that do not satisfy the assumptions you made in Problem 21.

APPLICATIONS 

23. **Airspeed.** It takes a private airplane 8.75 hours to make the 2,100-mile flight from Atlanta to Los Angeles and 5 hours to make the return trip. Assuming that the wind blows at a constant rate from Los Angeles to Atlanta, find the airspeed of the plane and the wind rate.

24. **Airspeed.** A plane carries enough fuel for 20 hours of flight at an airspeed of 150 miles per hour. How far can it fly into a 30 mph headwind and still have enough fuel to return to its starting point? (This distance is called the *point of no return.*)

25. **Rate–Time.** A crew of eight can row 20 kilometers per hour in still water. The crew rows upstream and then returns to its starting point in 15 minutes. If the river is flowing at 2 km/h, how far upstream did the crew row?

26. **Rate–Time.** It takes a boat 2 hours to travel 20 miles down a river and 3 hours to return upstream to its starting point. What is the rate of the current in the river?

27. **Chemistry.** A chemist has two solutions of hydrochloric acid in stock: a 50% solution and an 80% solution. How much of each should be used to obtain 100 milliliters of a 68% solution?

28. **Business.** A jeweler has two bars of gold alloy in stock, one of 12 carats and the other of 18 carats (24-carat gold is pure gold, 12-carat is $\frac{12}{24}$ pure, 18-carat gold is $\frac{18}{24}$ pure, and so on). How many grams of each alloy must be mixed to obtain 10 grams of 14-carat gold?

29. **Break-Even Analysis.** It costs a small recording company $17,680 to prepare a compact disc. This is a one-time fixed cost that covers recording, package design, and so on. Variable costs, including such things as manufacturing, marketing, and royalties, are $4.60 per CD. If the CD is sold to music shops for $8 each, how many must be sold for the company to break even?

30. **Break-Even Analysis.** A videocassette manufacturer has determined that its weekly cost equation is $C = 3{,}000 + 10x$, where x is the number of cassettes produced and sold each week. If cassettes are sold for $15 each to distributors, how many must be sold each week for the manufacturer to break even? (Refer to Problem 29.)

31. **Finance.** Suppose you have $12,000 to invest. If part is invested at 10% and the rest at 15%, how much should be invested at each rate to yield 12% on the total amount invested?

32. **Finance.** An investor has $20,000 to invest. If part is invested at 8% and the rest at 12%, how much should be invested at each rate to yield 11% on the total amount invested?

33. **Production.** A supplier for the electronics industry manufactures keyboards and screens for graphing calculators at plants in Mexico and Taiwan. The hourly production rates at each plant are given in the table. How many hours should each plant be operated to exactly fill an order for 4,000 keyboards and screens?

Plant	Keyboards	Screens
Mexico	40	32
Taiwan	20	32

34. **Production.** A company produces Italian sausages and bratwursts at plants in Green Bay and Sheboygan. The hourly production rates at each plant are given in the table. How many hours should each plant be operated to exactly fill an order for 62,250 Italian sausages and 76,500 bratwursts?

Plant	Italian sausage	Bratwurst
Green Bay	800	800
Sheboygan	500	1,000

35. Nutrition. Animals in an experiment are to be kept on a strict diet. Each animal is to receive, among other things, 20 grams of protein and 6 grams of fat. The laboratory technician is able to purchase two food mixes of the following compositions: Mix A has 10% protein and 6% fat; mix B has 20% protein and 2% fat. How many grams of each mix should be used to obtain the right diet for a single animal?

36. Nutrition. A fruit grower can use two types of fertilizer in an orange grove, brand A and brand B. Each bag of brand A contains 8 pounds of nitrogen and 4 pounds of phosphoric acid. Each bag of brand B contains 7 pounds of nitrogen and 7 pounds of phosphoric acid. Tests indicate that the grove needs 720 pounds of nitrogen and 500 pounds of phosphoric acid. How many bags of each brand should be used to provide the required amounts of nitrogen and phosphoric acid?

★ **37. Supply and Demand.** At \$0.60 per bushel, the daily supply for wheat is 450 bushels and the daily demand is 645 bushels. When the price is raised to \$0.90 per bushel, the daily supply increases to 750 bushels and the daily demand decreases to 495 bushels. Assume that the supply and demand equations are linear.

(A) Find the supply equation. [*Hint:* Write the supply equation in the form $p = aq + b$ and solve for a and b.]

(B) Find the demand equation.

(C) Find the equilibrium price and quantity.

★ **38. Supply and Demand.** At \$1.40 per bushel, the daily supply for soybeans is 1,075 bushels and the daily demand is 580 bushels. When the price falls to \$1.20 per bushel, the daily supply decreases to 575 bushels and the daily demand increases to 980 bushels. Assume that the supply and demand equations are linear.

(A) Find the supply equation. [See the hint in Problem 37.]

(B) Find the demand equation.

(C) Find the equilibrium price and quantity.

★ **39. Physics.** An object dropped off the top of a tall building falls vertically with constant acceleration. If s is the distance of the object above the ground (in feet) t seconds after its release, then s and t are related by an equation of the form

$$s = a + bt^2$$

where a and b are constants. Suppose the object is 180 feet above the ground 1 second after its release and 132 feet above the ground 2 seconds after its release.

(A) Find the constants a and b.

(B) How high is the building?

(C) How long does the object fall?

★ **40. Physics.** Repeat Problem 39 if the object is 240 feet above the ground after 1 second and 192 feet above the ground after 2 seconds.

★ **41. Earth Science.** An earthquake emits a primary wave and a secondary wave. Near the surface of the Earth the primary wave travels at about 5 miles per second and the secondary wave at about 3 miles per second. From the time lag between the two waves arriving at a given receiving station, it is possible to estimate the distance to the quake. (The *epicenter* can be located by obtaining distance bearings at three or more stations.) Suppose a station measured a time difference of 16 seconds between the arrival of the two waves. How long did each wave travel, and how far was the earthquake from the station?

★ **42. Earth Science.** A ship using sound-sensing devices above and below water recorded a surface explosion 6 seconds sooner by its underwater device than its above-water device. Sound travels in air at about 1,100 feet per second and in seawater at about 5,000 feet per second.

(A) How long did it take each sound wave to reach the ship?

(B) How far was the explosion from the ship?

SECTION 1-3 Linear Inequalities

- Inequality Relations and Interval Notation
- Solving Linear Inequalities
- Applications

We now turn to the problem of solving linear inequalities in one variable, such as

$$3(x - 5) \geq 5(x + 7) - 10 \qquad \text{and} \qquad -4 \leq 3 - 2x < 7$$

• Inequality Relations and Interval Notation

The above mathematical forms involve the **inequality,** or **order, relation**—that is, "less than" and "greater than" relations. Just as we use = to replace the words "is equal to," we use the **inequality symbols** $<$ and $>$ to represent "is less than" and "is greater than," respectively.

While it probably seems obvious to you that

$$2 < 4 \qquad 5 > 0 \qquad 25{,}000 > 1$$

are true, it may not seem as obvious that

$$-4 < -2 \qquad 0 > -5 \qquad -25{,}000 < -1$$

To make the inequality relation precise so that we can interpret it relative to all real numbers, we need a precise definition of the concept.

DEFINITION 1 **$a < b$ and $b > a$**

For a and b real numbers, we say that **a is less than b** or **b is greater than a** and write

$$a < b \qquad \text{or} \qquad b > a$$

if there exists a positive real number p such that $a + p = b$ (or equivalently, $b - a = p$).

We certainly expect that if a positive number is added to *any* real number, the sum is larger than the original. That is essentially what the definition states.

When we write

$$a \le b$$

we mean $a < b$ or $a = b$ and say **a is less than or equal to b.** When we write

$$a \ge b$$

we mean $a > b$ or $a = b$ and say **a is greater than or equal to b.**

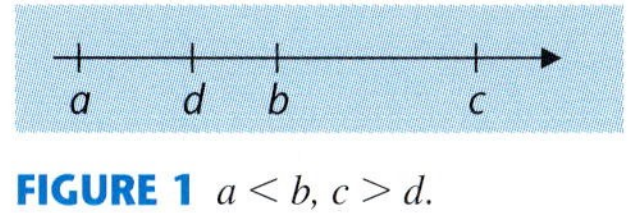

FIGURE 1 $a < b, c > d$.

The inequality symbols $<$ and $>$ have a very clear geometric interpretation on the real number line. If $a < b$, then a is to the left of b; if $c > d$, then c is to the right of d (Fig.1).

It is an interesting and useful fact that for any two real numbers a and b, either $a < b$, or $a > b$, or $a = b$. This is called the **trichotomy property** of real numbers.

The double inequality $a < x \le b$ means that $x > a$ *and* $x \le b$; that is, x is between a and b, including b but not including a. The set of all real numbers x satisfying the inequality $a < x \le b$ is called an **interval** and is represented by $(a, b]$. Thus,

$$(a, b] = \{x \mid a < x \le b\}^*$$

*In general, $\{x \mid P(x)\}$ represents the set of all x such that statement $P(x)$ is true. To express this set verbally, just read the vertical bar as "such that."

The number a is called the **left endpoint** of the interval, and the symbol "(" indicates that a is not included in the interval. The number b is called the **right endpoint** of the interval, and the symbol "]" indicates that b is included in the interval. Other types of intervals of real numbers are shown in Table 1.

TABLE 1 Interval Notation

Interval notation	Inequality notation	Line graph	Type
$[a, b]$	$a \le x \le b$	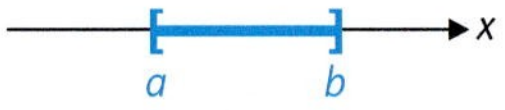	Closed
$[a, b)$	$a \le x < b$	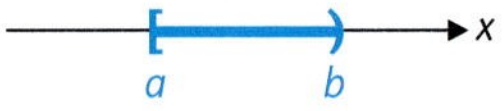	Half-open
$(a, b]$	$a < x \le b$	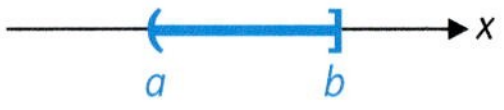	Half-open
(a, b)	$a < x < b$	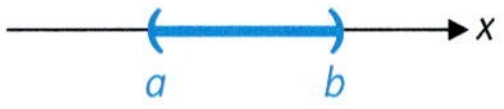	Open
$[b, \infty)$	$x \ge b$	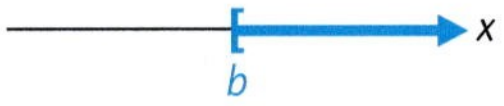	Closed
(b, ∞)	$x > b$	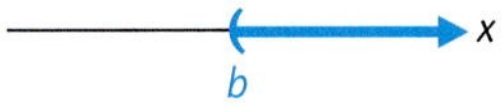	Open
$(-\infty, a]$	$x \le a$	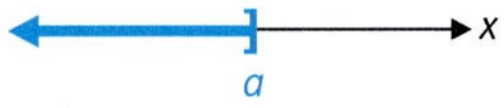	Closed
$(-\infty, a)$	$x < a$	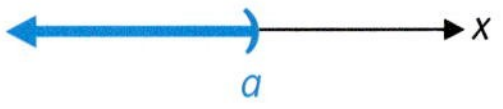	Open

Note that the symbol "∞," read "infinity," used in Table 1 is not a numeral. When we write $[b, \infty)$, we are simply referring to the interval starting at b and continuing indefinitely to the right. We would never write $[b, \infty]$ or $b \le x \le \infty$, since ∞ cannot be used as an endpoint of an interval. The interval $(-\infty, \infty)$ represents the set of real numbers R, since its graph is the entire real number line.

CAUTION

It is important to note that

$$5 > x \ge -3 \qquad \text{is equivalent to } [-3, 5) \text{ and not to } (5, -3]$$

In interval notation, the smaller number is always written to the left. Thus, it may be useful to rewrite the inequality as $-3 \le x < 5$ before rewriting it in interval notation.

EXAMPLE 1 Graphing Intervals and Inequalities

Write each of the following in inequality notation and graph on a real number line:

(A) $[-2, 3)$ (B) $(-4, 2)$ (C) $[-2, \infty)$ (D) $(-\infty, 3)$

Solutions (A) $-2 \le x < 3$

-5 -2 0 3 5 x

(B) $-4 < x < 2$

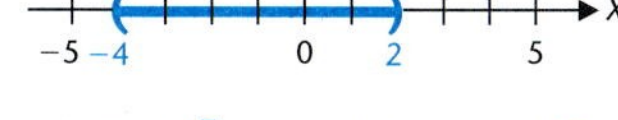

(C) $x \ge -2$

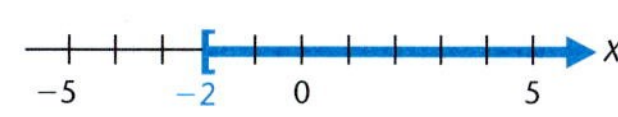

(D) $x < 3$

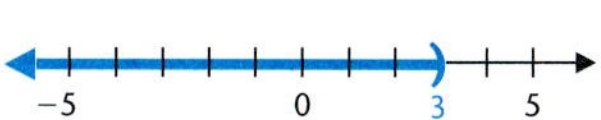

Matched Problem 1 Write each of the following in interval notation and graph on a real number line:

(A) $-3 < x \le 3$ (B) $2 \ge x \ge -1$ (C) $x > 1$ (D) $x \le 2$

EXPLORE-DISCUSS 1 Example 1C shows the graph of the inequality $x \ge -2$. What is the graph of $x < -2$? What is the corresponding interval? Describe the relationship between these sets.

Since intervals are sets of real numbers, the set operations of *union* and *intersection* are often useful when working with intervals. The **union** of sets A and B, denoted by $A \cup B$, is the set formed by combining all the elements of A and all the elements of B. The **intersection** of sets A and B, denoted by $A \cap B$, is the set of elements of A that are also in B. Symbolically:

DEFINITION 2

Union and Intersection

Union: $A \cup B = \{x \mid x \text{ is in } A \textbf{ or } x \text{ is in } B\}$

$\{1, 2, 3\} \cup \{2, 3, 4, 5\} = \{1, 2, 3, 4, 5\}$

Intersection: $A \cap B = \{x \mid x \text{ is in } A \textbf{ and } x \text{ is in } B\}$

$\{1, 2, 3\} \cap \{2, 3, 4, 5\} = \{2, 3\}$

EXAMPLE 2 **Graphing Unions and Intersections of Intervals**

If $A = [-2, 3]$, $B = (1, 6)$, and $C = (4, \infty)$, graph the indicated sets and write as a single interval, if possible:

(A) $A \cup B$ and $A \cap B$ (B) $A \cup C$ and $A \cap C$

Solution (A)

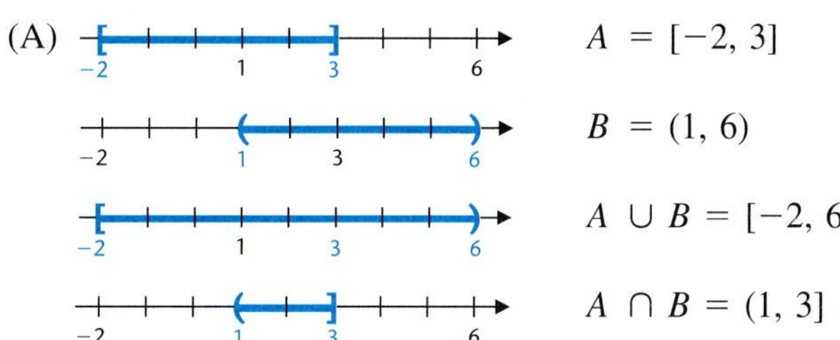

$A = [-2, 3]$

$B = (1, 6)$

$A \cup B = [-2, 6)$

$A \cap B = (1, 3]$

(B)

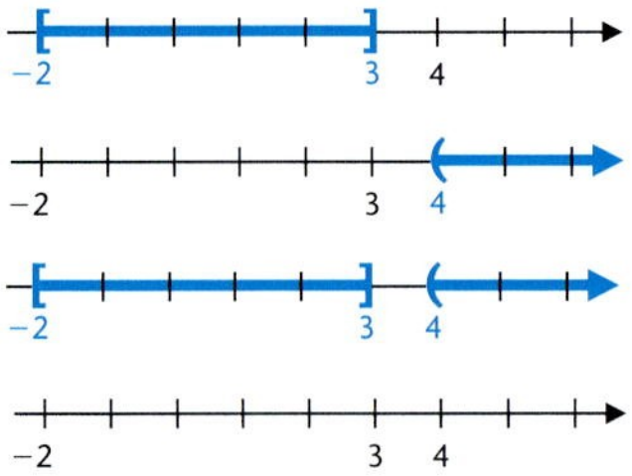

$A = [-2, 3]$

$C = (4, \infty)$

$A \cup C = [-2, 3] \cup (4, \infty)$

$A \cap C = \varnothing$

Matched Problem 2 If $D = [-4, 1)$, $E = (-1, 3]$, and $F = [2, \infty)$, graph the indicated sets and write as a single interval, if possible:

(A) $D \cup E$ (B) $D \cap E$ (C) $E \cup F$ (D) $E \cap F$

EXPLORE-DISCUSS 2 Replace ? with < or > in each of the following.

(A) $-1 \ ? \ 3$ and $2(-1) \ ? \ 2(3)$

(B) $-1 \ ? \ 3$ and $-2(-1) \ ? \ -2(3)$

(C) $12 \ ? \ -8$ and $\dfrac{12}{4} \ ? \ \dfrac{-8}{4}$

(D) $12 \ ? \ -8$ and $\dfrac{12}{-4} \ ? \ \dfrac{-8}{-4}$

Based on these examples, describe verbally the effect of multiplying both sides of an inequality by a number.

• Solving Linear Inequalities

We now turn to the problem of solving linear inequalities in one variable, such as

$$2(2x + 3) < 6(x - 2) + 10 \qquad \text{and} \qquad -3 < 2x + 3 \le 9$$

The **solution set** for an inequality is the set of all values of the variable that make the inequality a true statement. Each element of the solution set is called a **solution** of the inequality. To **solve an inequality** is to find its solution set. Two inequalities are **equivalent** if they have the same solution set for a given replacement set. Just as with equations, we perform operations on inequalities that produce simpler equivalent inequalities, and continue the process until an inequality is reached whose solution is obvious. The properties of inequalities given in Theorem 1 can be used to produce equivalent inequalities.

Theorem 1

Inequality Properties

For a, b, and c any real numbers:

1. If $a < b$ and $b < c$, then $a < c$. **Transitive Property**
2. If $a < b$, then $a + c < b + c$. **Addition Property**
   $-2 < 4$ $\quad -2 + 3 < 4 + 3$
3. If $a < b$, then $a - c < b - c$. **Subtraction Property**
   $-2 < 4$ $\quad -2 - 3 < 4 - 3$
4. If $a < b$ and c is positive, then $ca < cb$.
   $-2 < 4$ $\quad 3(-2) < 3(4)$
5. If $a < b$ and c is negative, then $ca > cb$.
   $-2 < 4$ $\quad (-3)(-2) > (-3)(4)$

   Multiplication Property (Note difference between 4 and 5.)
6. If $a < b$ and c is positive, then $\frac{a}{c} < \frac{b}{c}$.
   $-2 < 4$ $\quad \frac{-2}{2} < \frac{4}{2}$
7. If $a < b$ and c is negative, then $\frac{a}{c} > \frac{b}{c}$.
   $-2 < 4$ $\quad \frac{-2}{-2} > \frac{4}{-2}$

   Division Property (Note difference between 6 and 7.)

Similar properties hold if each inequality sign is reversed, or if $<$ is replaced with $\leq$ and $>$ is replaced with $\geq$. Thus, we find that we can perform essentially the same operations on inequalities that we perform on equations. When working with inequalities, however, we have to be particularly careful of the use of the multiplication and division properties.

The order of the inequality reverses if we multiply or divide both sides of an inequality statement by a negative number.

EXPLORE-DISCUSS 3 Properties of equality are easily summarized: We can add, subtract, multiply, or divide both sides of an equation by any nonzero real number to produce an equivalent equation. Write a similar summary for the properties of inequalities.

Now let's see how the inequality properties are used to solve linear inequalities. Several examples will illustrate the process.

EXAMPLE 3 Solving a Linear Inequality

Solve and graph: $2(2x + 3) - 10 < 6(x - 2)$

Solution

$$2(2x + 3) - 10 < 6(x - 2)$$

$$4x + 6 - 10 < 6x - 12 \quad \text{Simplify left and right sides.}$$

$$4x - 4 < 6x - 12$$

$$4x - 4 + 4 < 6x - 12 + 4 \quad \text{Addition property}$$

$$4x < 6x - 8$$

$$4x - 6x < 6x - 8 - 6x \quad \text{Subtraction property}$$

$$-2x < -8$$

$$\frac{-2x}{-2} > \frac{-8}{-2}$$

Division property—note that order reverses since −2 is negative

$$x > 4 \quad \text{or} \quad (4, \infty)$$ Solution set

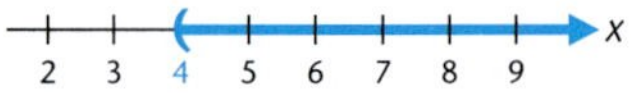

Graph of solution set

Matched Problem 3 Solve and graph: $3(x - 1) \geq 5(x + 2) - 5$

EXAMPLE 4 Solving a Linear Inequality Involving Fractions

Solve and graph: $\dfrac{2x - 3}{4} + 6 \geq 2 + \dfrac{4x}{3}$

Solution

$$\frac{2x - 3}{4} + 6 \geq 2 + \frac{4x}{3}$$

$$12 \cdot \frac{2x - 3}{4} + 12 \cdot 6 \geq 12 \cdot 2 + 12 \cdot \frac{4x}{3}$$

Multiply both sides by 12, the LCD.

$$3(2x - 3) + 72 \geq 24 + 4(4x)$$

$$6x - 9 + 72 \geq 24 + 16x$$

$$6x + 63 \geq 24 + 16x$$

$$-10x \geq -39$$

$$x \leq 3.9 \quad \text{or} \quad (-\infty, 3.9]$$

Order reverses since both sides are divided by −10, a negative number.

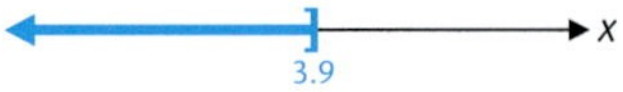

Matched Problem 4 Solve and graph: $\frac{4x-3}{3} + 8 < 6 + \frac{3x}{2}$

EXAMPLE 5 **Solving a Double Inequality**

Solve and graph: $-3 \le 4 - 7x < 18$

Solution We proceed as before, except we try to isolate x in the middle with a coefficient of 1.

$$-3 \le 4 - 7x < 18$$

$$-3 - 4 \le 4 - 7x - 4 < 18 - 4 \qquad \text{Subtract 4 from each member.}$$

$$-7 \le -7x < 14$$

$$\frac{-7}{-7} \ge \frac{-7x}{-7} > \frac{14}{-7} \qquad \text{Divide each member by } -7 \text{ and reverse each inequality.}$$

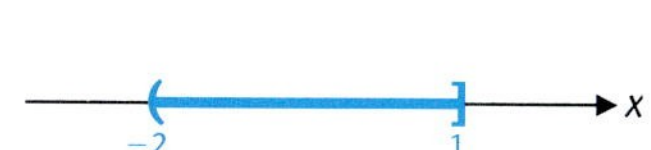

$$1 \ge x > -2 \quad \text{or} \quad -2 < x \le 1 \quad \text{or} \quad (-2, 1]$$

Matched Problem 5 Solve and graph: $-3 < 7 - 2x \le 7$

• Applications

EXAMPLE 6 **Chemistry**

In a chemistry experiment, a solution of hydrochloric acid is to be kept between 30°C and 35°C—that is, $30 \le C \le 35$. What is the range in temperature in degrees Fahrenheit if the Celsius/Fahrenheit conversion formula is $C = \frac{5}{9}(F - 32)$?

Solution

$$30 \le C \le 35$$

$$30 \le \frac{5}{9}(F - 32) \le 35 \qquad \text{Replace } C \text{ with } \frac{5}{9}(F - 32).$$

$$\frac{9}{5} \cdot 30 \le \frac{9}{5} \cdot \frac{5}{9}(F - 32) \le \frac{9}{5} \cdot 35 \qquad \text{Multiply each member by } \frac{9}{5}.$$

$$54 \le F - 32 \le 63$$

$$54 + 32 \le F - 32 + 32 \le 63 + 32$$ Add 32 to each member.

$$86 \le F \le 95$$

The range of the temperature is from 86°F to 95°F, inclusive.

Matched Problem 6 A film developer is to be kept between 68°F and 77°F—that is, $68 \le F \le 77$. What is the range in temperature in degrees Celsius if the Celsius/Fahrenheit conversion formula is $F = \frac{9}{5}C + 32$?

Answers to Matched Problems

1. (A) $(-3, 3]$
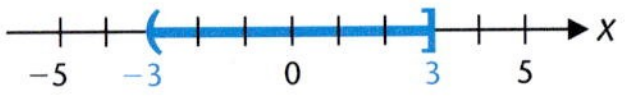

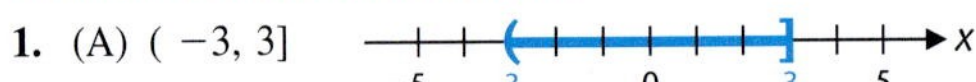

(B) $[-1, 2]$
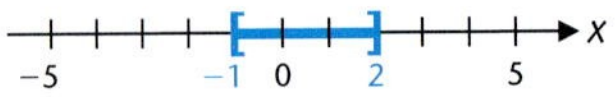

(C) $(1, \infty)$
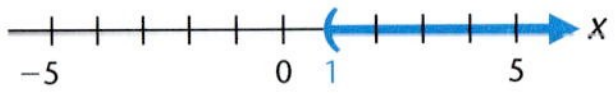

(D) $(-\infty, 2]$
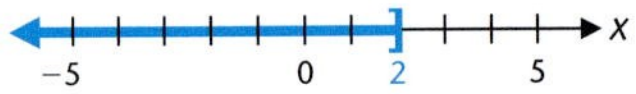

2. (A)
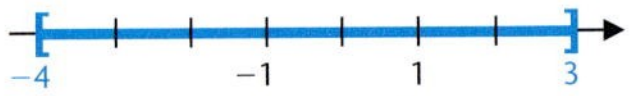

$D \cup E = [-4, 3]$

(B)
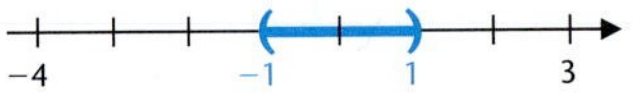

$D \cap E = (-1, 1)$

(C)
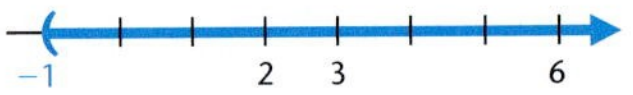

$E \cup F = (-1, \infty)$

(D)
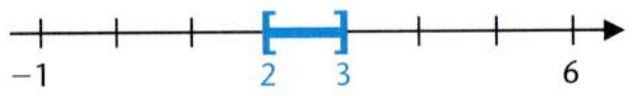

$E \cap F = [2, 3]$

3. $x \le -4$ or $(-\infty, -4]$
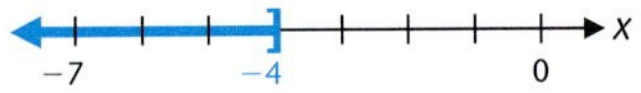

4. $x > 6$ or $(6, \infty)$
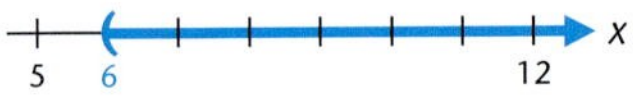

5. $5 > x \ge 0$ or $0 \le x < 5$ or $[0, 5)$
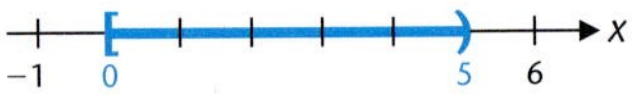

6. $20 \le C \le 25$: the range in temperature is from 20°C to 25°C

EXERCISE 1-3

A

In Problems 1–6, rewrite in inequality notation and graph on a real number line.

1. $[-8, 7]$ **2.** $(-4, 8)$ **3.** $[-6, 6)$

4. $(-3, 3]$ **5.** $[-6, \infty)$ **6.** $(-\infty, 7)$

In Problems 7–12, rewrite in interval notation and graph on a real number line.

7. $-2 < x \le 6$ **8.** $-5 \le x \le 5$ **9.** $-7 < x < 8$

10. $-4 \le x < 5$ **11.** $x \le -2$ **12.** $x > 3$

In Problems 13–16, write in interval and inequality notation.

13.
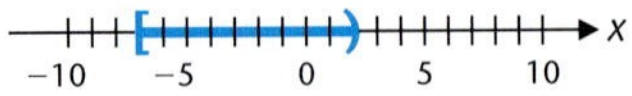

14.
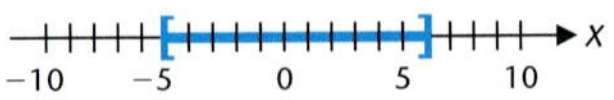

15.
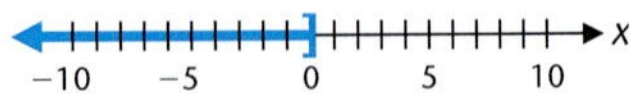

16.
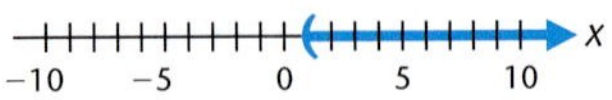

In Problems 17–30, solve and graph.

17. $7x - 8 < 4x + 7$ **18.** $4x + 8 \ge x - 1$

19. $3 - x \geq 5(3 - x)$

20. $2(x - 3) + 5 < 5 - x$

21. $\frac{N}{-2} > 4$

22. $\frac{M}{-3} \leq -2$

23. $-5t < -10$

24. $-7n \geq 21$

25. $3 - m < 4(m - 3)$

26. $2(1 - u) \geq 5u$

27. $-2 - \frac{B}{4} \leq \frac{1 + B}{3}$

28. $\frac{y - 3}{4} - 1 > \frac{y}{2}$

29. $-4 < 5t + 6 \leq 21$

30. $2 \leq 3m - 7 < 14$

B

In Problems 31–42, graph the indicated set and write as a single interval, if possible.

31. $(-5, 5) \cup [4, 7]$

32. $(-5, 5) \cap [4, 7]$

33. $[-1, 4) \cap (2, 6]$

34. $[-1, 4) \cup (2, 6]$

35. $(-\infty, 1) \cup (-2, \infty)$

36. $(-\infty, 1) \cap (2, \infty)$

37. $(-\infty, -1) \cup [3, 7)$

38. $(1, 6] \cup [9, \infty)$

39. $[2, 3] \cup (1, 5)$

40. $[2, 3] \cap (1, 5)$

41. $(-\infty, 4) \cup (-1, 6]$

42. $(-3, 2) \cup [0, \infty)$

In Problems 43–54, solve and graph.

43. $\frac{q}{7} - 3 > \frac{q - 4}{3} + 1$

44. $\frac{p}{3} - \frac{p - 2}{2} \leq \frac{p}{4} - 4$

45. $\frac{2x}{5} - \frac{1}{2}(x - 3) \leq \frac{2x}{3} - \frac{3}{10}(x + 2)$

46. $\frac{2}{3}(x + 7) - \frac{x}{4} > \frac{1}{2}(3 - x) + \frac{x}{6}$

47. $-4 \leq \frac{9}{5}x + 32 \leq 68$

48. $-1 \leq \frac{2}{3}A + 5 \leq 11$

49. $-12 < \frac{3}{4}(2 - x) \leq 24$

50. $24 \leq \frac{2}{3}(x - 5) < 36$

51. $16 < 7 - 3x \leq 31$

52. $-1 \leq 9 - 2x < 5$

53. $-6 < -\frac{2}{5}(1 - x) \leq 4$

54. $15 \leq 7 - \frac{2}{5}x \leq 21$

Use a calculator to solve each of the inequalities in Problems 55–58. Write answers using inequality notation.

55. $5.23(x - 0.172) \leq 6.02x - 0.427$

56. $72.3x - 4.07 > 9.02(11.7x - 8.22)$

57. $-0.703 < 0.122 - 2.28x < 0.703$

58. $-4.26 < 3.88 - 6.07x < 5.66$

Problems 59–64 are calculus-related. For what real number(s) x does each expression represent a real number?

59. $\sqrt{1 - x}$

60. $\sqrt{x + 5}$

61. $\sqrt{3x + 5}$

62. $\sqrt{7 - 2x}$

63. $\frac{1}{\sqrt[4]{2x + 3}}$

64. $\frac{1}{\sqrt[4]{5 - 6x}}$

65. What can be said about the signs of the numbers a and b in each case?
(A) $ab > 0$ (B) $ab < 0$
(C) $\frac{a}{b} > 0$ (D) $\frac{a}{b} < 0$

66. What can be said about the signs of the numbers a, b, and c in each case?
(A) $abc > 0$ (B) $\frac{ab}{c} < 0$
(C) $\frac{a}{bc} > 0$ (D) $\frac{a^2}{bc} < 0$

67. Replace each question mark with $<$ or $>$, as appropriate:
(A) If $a - b = 1$, then a ? b.
(B) If $u - v = -2$, then u ? v.

68. For what p and q is $p + q < p - q$?

C

69. If both a and b are negative numbers and b/a is greater than 1, then is $a - b$ positive or negative?

70. If both a and b are positive numbers and b/a is greater than 1, then is $a - b$ positive or negative?

71. Indicate true (T) or false (F):
(A) If $p > q$ and $m > 0$, then $mp < mq$.
(B) If $p < q$ and $m < 0$, then $mp > mq$.
(C) If $p > 0$ and $q < 0$, then $p + q > q$.

72. Assume that $m > n > 0$; then

$$mn > n^2$$
$$mn - m^2 > n^2 - m^2$$
$$m(n - m) > (n + m)(n - m)$$
$$m > n + m$$
$$0 > n$$

But it was assumed that $n > 0$. Find the error.

Prove each inequality property in Problems 73–76, given a, b, and c are arbitrary real numbers.

73. If $a < b$, then $a + c < b + c$.

74. If $a < b$, then $a - c < b - c$.

75. (A) If $a < b$ and c is positive, then $ca < cb$.
(B) If $a < b$ and c is negative, then $ca > cb$.

76. (A) If $a < b$ and c is positive, then $\frac{a}{c} < \frac{b}{c}$.
(B) If $a < b$ and c is negative, then $\frac{a}{c} > \frac{b}{c}$.

APPLICATIONS 

Write all answers using inequality notation.

77. Earth Science. The Soviets, in drilling the world's deepest hole in 1984, found that the temperature x kilometers below the surface of the Earth was given by

$$T = 30 + 25(x - 3) \qquad 3 \le x \le 15$$

where T is temperature in degrees Celsius. At what depth will the temperature be between 200°C and 300°C, inclusive?

78. Earth Science. As dry air moves upward it expands, and in so doing it cools at a rate of about 5.5°F for each 1,000-foot rise up to about 40,000 feet. If the ground temperature is 70°F, then the temperature T at height h is given approximately by $T = 70 - 0.0055h$. For what range in altitude will the temperature be between 26°F and −40°F, inclusive?

79. Business and Economics. An electronics firm is planning to market a new graphing calculator. The fixed costs are $650,000 and the variable costs are $47 per calculator. The wholesale price of the calculator will be $63. For the company to make a profit, it is clear that revenues must be greater than costs.

(A) How many calculators must be sold for the company to make a profit?

(B) How many calculators must be sold for the company to break even?

(C) Discuss the relationship between the results in parts A and B.

80. Business and Economics. A video game manufacturer is planning to market a 64-bit version of its game machine. The fixed costs are $550,000 and the variable costs are $120 per machine. The wholesale price of the machine will be $140.

(A) How many game machines must be sold for the company to make a profit?

(B) How many game machines must be sold for the company to break even?

(C) Discuss the relationship between the results in parts A and B.

81. Business and Economics. The electronics firm in Problem 79 finds that rising prices for parts increases the variable costs to $50.50 per calculator.

(A) Discuss possible strategies the company might use to deal with this increase in costs.

(B) If the company continues to sell the calculators for $63, how many must they sell now to make a profit?

(C) If the company wants to start making a profit at the same production level as before the cost increase, how much should they increase the wholesale price?

82. Business and Economics. The video game manufacturer in Problem 80 finds that unexpected programming problems increases the fixed costs to $660,000.

(A) Discuss possible strategies the company might use to deal with this increase in costs.

(B) If the company continues to sell the game machines for $140, how many must they sell now to make a profit?

(C) If the company wants to start making a profit at the same production level as before the cost increase, how much should they increase the wholesale price?

83. Energy. If the power demands in a 110-volt electric circuit in a home vary between 220 and 2,750 watts, what is the range of current flowing through the circuit? ($W = EI$, where W = Power in watts, E = Pressure in volts, and I = Current in amperes.)

84. Psychology. A person's IQ is given by the formula

$$\text{IQ} = \frac{\text{MA}}{\text{CA}}100$$

where MA is mental age and CA is chronological age. If

$$80 \le \text{IQ} \le 140$$

for a group of 12-year-old children, find the range of their mental ages.

★ **85. Finance.** If an individual aged 65–69 continues to work after Social Security benefits start, benefits will be reduced when earnings exceed an earnings limitation. In 1989, benefits were reduced by $1 for every $2 earned in excess of $8,880. Find the range of benefit reductions for individuals earning between $13,000 and $16,000.

★ **86. Finance.** Refer to Problem 85. In 1990 the law was changed so that benefits were reduced by $1 for every $3 earned in excess of $8,880. Find the range of benefit reductions for individuals earning between $13,000 and $16,000.

SECTION 1-4 Absolute Value in Equations and Inequalities

- Absolute Value and Distance
- Absolute Value in Equations and Inequalities
- Absolute Value and Radicals

This section discusses solving absolute value equations and inequalities.

• Absolute Value and Distance

We start with a geometric definition of absolute value. If a is the coordinate of a point on a real number line, then the distance from the origin to a is represented by $|a|$ and is referred to as the **absolute value** of a. Thus, $|5| = 5$, since the point with coordinate 5 is five units from the origin, and $|-6| = 6$, since the point with coordinate -6 is six units from the origin (Fig.1).

FIGURE 1 Absolute value.

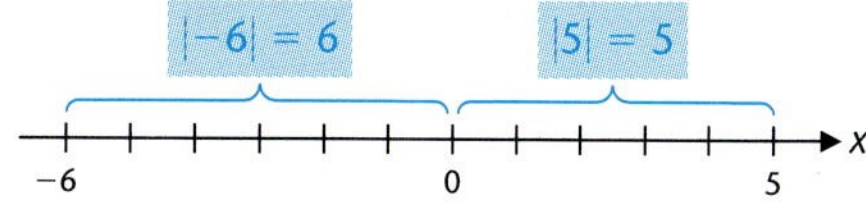

Symbolically, and more formally, we define absolute value as follows:

DEFINITION 1

Absolute Value

$$|x| = \begin{cases} x & \text{if } x \geq 0 \\ -x & \text{if } x < 0 \end{cases}$$

$|4| = 4$

$|-3| = -(-3) = 3$

[*Note:* $-x$ is positive if x is negative.]

Both the geometric and nongeometric definitions of absolute value are useful, as will be seen in the material that follows. Remember:

The absolute value of a number is never negative.

EXAMPLE 1 **Absolute Value of a Real Number**

(A) $|\pi - 3| = \pi - 3$ Since $\pi \approx 3.14$, $\pi - 3$ is positive.
(B) $|3 - \pi| = -(3 - \pi) = \pi - 3$ Since $3 - \pi$ is negative

Matched Problem 1 Write without the absolute value sign:

(A) $|8|$ (B) $|\sqrt[3]{9} - 2|$ (C) $|-\sqrt{2}|$ (D) $|2 - \sqrt[3]{9}|$

Following the same reasoning used in Example 1, the next theorem can be proved (see Problem 79 in Exercise 1-4).

Theorem 1

For all real numbers a and b,

$$|b - a| = |a - b|$$

We use this result in defining the distance between two points on a real number line.

DEFINITION 2

Distance between Points A and B

Let A and B be two points on a real number line with coordinates a and b, respectively. The **distance between A and B** is given by

$$d(A, B) = |b - a|$$

This distance is also called the **length of the line segment** joining A and B.

EXAMPLE 2 **Distance between Points on a Number Line**

Find the distance between points A and B with coordinates a and b, respectively, as given.

(A) $a = 4,\ b = 9$ (B) $a = 9,\ b = 4$ (C) $a = 0,\ b = 6$

Solutions

(A)

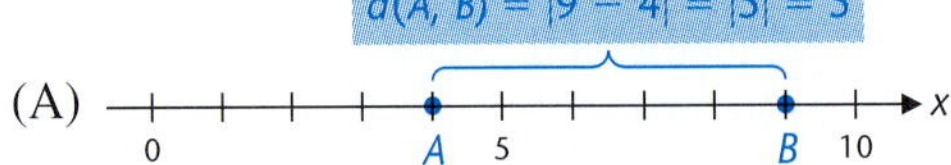

(B) $d(A, B) = |4 - 9| = |-5| = 5$

0 B 5 A 10 x

(C) $d(A, B) = |6 - 0| = |6| = 6$

0 A 5 B 10 x

It should be clear, since $|b - a| = |a - b|$, that

$$d(A, B) = d(B, A)$$

Hence, in computing the distance between two points on a real number line, it does not matter how the two points are labeled—point A can be to the left or to the right of point B. Note also that if A is at the origin O, then

$$d(O, B) = |b - 0| = |b|$$

Matched Problem 2 Use the number line below to find the indicated distances.

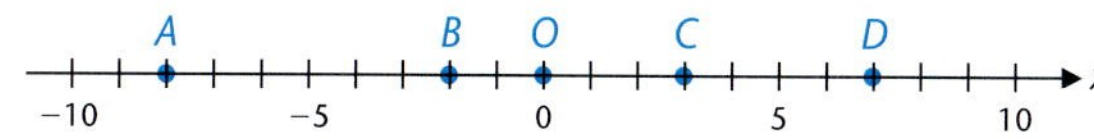

(A) $d(C, D)$ (B) $d(D, C)$ (C) $d(A, B)$
(D) $d(A, C)$ (E) $d(O, A)$ (F) $d(D, A)$

• Absolute Value in Equations and Inequalities

The interplay between algebra and geometry is an important tool when working with equations and inequalities involving absolute value. For example, the algebraic statement

$$|x - 1| = 2$$

can be interpreted geometrically as stating that the distance from x to 1 is 2.

EXPLORE-DISCUSS 1 Write geometric interpretations of the following algebraic statements:

(A) $|x - 1| < 2$ (B) $0 < |x - 1| < 2$ (C) $|x - 1| > 2$

EXAMPLE 3 **Solving Absolute Value Problems Geometrically**

Interpret geometrically, solve, and graph. Write solutions in both inequality and interval notation, where appropriate.

(A) $|x - 3| = 5$ (B) $|x - 3| < 5$
(C) $0 < |x - 3| < 5$ (D) $|x - 3| > 5$

Solutions (A) Geometrically, $|x - 3|$ represents the distance between x and 3. Thus, in $|x - 3| = 5$, x is a number whose distance from 3 is 5. That is,

$$x = 3 \pm 5 = -2 \quad \text{or} \quad 8$$

The solution set is $\{-2, 8\}$. This *is not* interval notation.

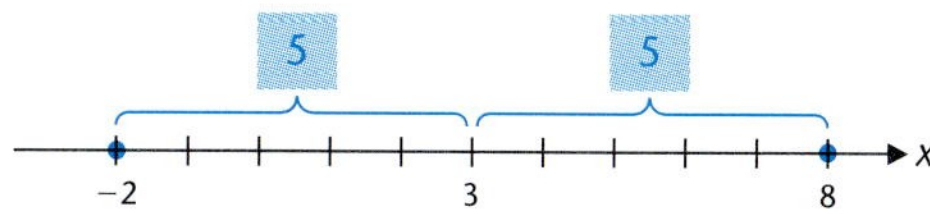

(B) Geometrically, in $|x - 3| < 5$, x is a number whose distance from 3 is less than 5; that is,

$$-2 < x < 8$$

The solution set is $(-2, 8)$. This *is* interval notation.

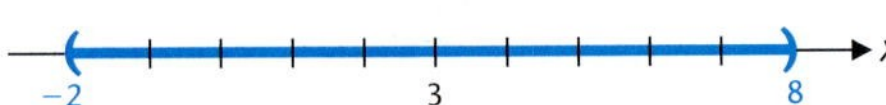

(C) The form $0 < |x - 3| < 5$ is frequently encountered in calculus and more advanced mathematics. Geometrically, x is a number whose distance from 3 is less than 5, but x cannot equal 3. Thus,

$$-2 < x < 8 \qquad x \neq 3 \qquad \text{or} \qquad (-2, 3) \cup (3, 8)$$

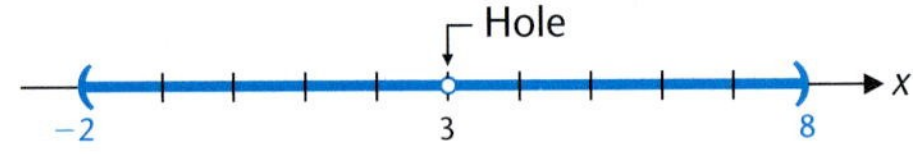

(D) Geometrically, in $|x - 3| > 5$, x is a number whose distance from 3 is greater than 5; that is,

$$x < -2 \qquad \text{or} \qquad x > 8 \qquad \text{or} \qquad (-\infty, -2) \cup (8, \infty)$$

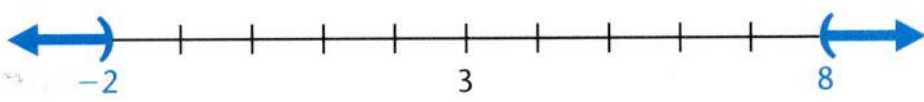

CAUTION

Do not confuse solutions like

$$-2 < x \qquad \text{and} \qquad x < 8$$

which can also be written as

$$-2 < x < 8 \qquad \text{or} \qquad (-2, 8)$$

with solutions like

$$x < -2 \qquad \text{or} \qquad x > 8$$

which cannot be written as a double inequality or as a single interval.

We summarize the preceding results in Table 1.

TABLE 1 Geometric Interpretation of Absolute Value Equations and Inequalities

Form ($d > 0$)	Geometric interpretation	Solution	Graph
$\lvert x - c \rvert = d$	Distance between x and c is equal to d.	$\{c - d, c + d\}$	d d; $c - d$ c $c + d$ x
$\lvert x - c \rvert < d$	Distance between x and c is less than d.	$(c - d, c + d)$	$c - d$ c $c + d$ x
$0 < \lvert x - c \rvert < d$	Distance between x and c is less than d, but $x \neq c$.	$(c - d, c) \cup (c, c + d)$	$c - d$ c $c + d$ x
$\lvert x - c \rvert > d$	Distance between x and c is greater than d.	$(-\infty, c - d) \cup (c + d, \infty)$	$c - d$ c $c + d$ x

Matched Problem 3 Interpret geometrically, solve, and graph. Write solutions in both inequality and interval notation, where appropriate.

(A) $|x + 2| = 6$ (B) $|x + 2| < 6$
(C) $0 < |x + 2| < 6$ (D) $|x + 2| > 6$

[*Hint:* $|x + 2| = |x - (-2)|$.]

EXPLORE-DISCUSS 2

Describe the set of numbers that satisfies each of the following:

(A) $2 > x > 1$ (B) $2 > x < 1$
(C) $2 < x > 1$ (D) $2 < x < 1$

Explain why it is never necessary to use double inequalities with inequality symbols pointing in different directions. Standard mathematical notation requires that all inequality symbols in an expression must point in the same direction.

Reasoning geometrically as before (noting that $|x| = |x - 0|$) leads to Theorem 2.

Theorem 2 **Properties of Equations and Inequalities Involving $|x|$**

For $p > 0$:

1. $|x| = p$ is equivalent to $x = p$ or $x = -p$. 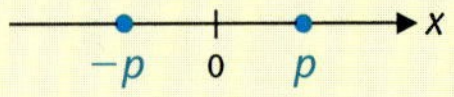
2. $|x| < p$ is equivalent to $-p < x < p$.
3. $|x| > p$ is equivalent to $x < -p$ or $x > p$.

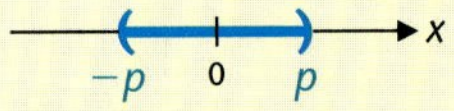

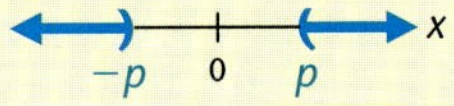

If we replace x in Theorem 2 with $ax + b$, we obtain the more general Theorem 3.

Theorem 3 **Properties of Equations and Inequalities Involving $|ax + b|$**

For $p > 0$:

1. $|ax + b| = p$ is equivalent to $ax + b = p$ or $ax + b = -p$.
2. $|ax + b| < p$ is equivalent to $-p < ax + b < p$.
3. $|ax + b| > p$ is equivalent to $ax + b < -p$ or $ax + b > p$.

EXAMPLE 4 Solving Absolute Value Problems

Solve, and write solutions in both inequality and interval notation, where appropriate.

(A) $|3x + 5| = 4$ (B) $|x| < 5$ (C) $|2x - 1| < 3$ (D) $|7 - 3x| \leq 2$

Solutions (A) $|3x + 5| = 4$

$$3x + 5 = \pm 4$$

$$3x = -5 \pm 4$$

$$x = \frac{-5 \pm 4}{3}$$

$$x = -3, -\tfrac{1}{3}$$

or $\{-3, -\tfrac{1}{3}\}$

(B) $|x| < 5$

$$-5 < x < 5$$

or $(-5, 5)$

(C) $|2x - 1| < 3$

$$-3 < 2x - 1 < 3$$
$$-2 < 2x < 4$$
$$-1 < x < 2$$
or $(-1, 2)$

(D) $|7 - 3x| \le 2$

$$-2 \le 7 - 3x \le 2$$
$$-9 \le -3x \le -5$$
$$3 \ge x \ge \tfrac{5}{3}$$
$$\tfrac{5}{3} \le x \le 3$$
or $[\tfrac{5}{3}, 3]$

Matched Problem 4 Solve, and write solutions in both inequality and interval notation, where appropriate.

(A) $|2x - 1| = 8$ (B) $|x| \le 7$ (C) $|3x + 3| \le 9$ (D) $|5 - 2x| < 9$

EXAMPLE 5 Solving Absolute Value Inequalities

Solve, and write solutions in both inequality and interval notation.

(A) $|x| > 3$ (B) $|2x - 1| \ge 3$ (C) $|7 - 3x| > 2$

Solutions

(A) $|x| > 3$

$x < -3$ or $x > 3$ Inequality notation

$(-\infty, -3) \cup (3, \infty)$ Interval notation

(B) $|2x - 1| \ge 3$

$2x - 1 \le -3$ or $2x - 1 \ge 3$

$2x \le -2$ or $2x \ge 4$

$x \le -1$ or $x \ge 2$ Inequality notation

$(-\infty, -1] \cup [2, \infty)$ Interval notation

(C) $|7 - 3x| > 2$

$7 - 3x < -2$ or $7 - 3x > 2$

$-3x < -9$ or $-3x > -5$

$x > 3$ or $x < \tfrac{5}{3}$ Inequality notation

$(-\infty, \tfrac{5}{3}) \cup (3, \infty)$ Interval notation

Matched Problem 5 Solve, and write solutions in both inequality and interval notation.

(A) $|x| \ge 5$ (B) $|4x - 3| > 5$ (C) $|6 - 5x| > 16$

EXAMPLE 6 An Absolute Value Problem with Two Cases

Solve: $|x + 4| = 3x - 8$

Solution Theorem 3 does not apply directly, since $3x - 8$ is not known to be positive. However, we can use the definition of absolute value and two cases: $x + 4 \geq 0$ and $x + 4 < 0$.

Case 1. $x + 4 \geq 0$ (that is, $x \geq -4$)
For this case, the possible values of x are in the set $\{x \mid x \geq -4\}$.

$$|x + 4| = 3x - 8$$

$$x + 4 = 3x - 8 \qquad |a| = a \text{ for } a \geq 0$$

$$-2x = -12$$

$$x = 6 \qquad \text{A solution, since 6 is among the possible values of } x$$

The check is left to the reader.

Case 2. $x + 4 < 0$ (that is, $x < -4$)
In this case, the possible values of x are in the set $\{x \mid x < -4\}$.

$$|x + 4| = 3x - 8$$

$$-(x + 4) = 3x - 8 \qquad |a| = -a \text{ for } a < 0$$

$$-x - 4 = 3x - 8$$

$$-4x = -4$$

$$x = 1 \qquad \text{Not a solution, since 1 is not among the possible values of } x$$

Combining both cases, we see that the only solution is $x = 6$.

Check As a final check, we substitute $x = 6$ and $x = 1$ in the original equation.

$$|x + 4| = 3x - 8 \qquad |x + 4| = 3x - 8$$

$$|6 + 4| \stackrel{?}{=} 3(6) - 8 \qquad |1 + 4| \stackrel{?}{=} 3(1) - 8$$

$$10 \stackrel{\checkmark}{=} 10 \qquad 5 \neq -5$$

Matched Problem 6 Solve: $|3x - 4| = x + 5$

• Absolute Value and Radicals

In Section A-7, we show that if x is positive or 0, then

$$\sqrt{x^2} = x$$

If x is negative, however, we must write

$$\sqrt{x^2} = -x \qquad \sqrt{(-2)^2} = -(-2) = 2$$

Thus, for x any real number,

$$\sqrt{x^2} = \begin{cases} x & \text{if } x \geq 0 \\ -x & \text{if } x < 0 \end{cases}$$

But this is exactly how we defined $|x|$ at the beginning of this section (see Definition 1). Thus, for x any real number,

$$\sqrt{x^2} = |x|$$

Answers to Matched Problems

1. (A) 8 (B) $\sqrt[3]{9} - 2$ (C) $\sqrt{2}$ (D) $\sqrt[3]{9} - 2$

2. (A) 4 (B) 4 (C) 6 (D) 11 (E) 8 (F) 15

3. (A) x is a number whose distance from -2 is 6.
$x = -8, 4$ or $\{-8, 4\}$

(B) x is a number whose distance from -2 is less than 6.
$-8 < x < 4$ or $(-8, 4)$

(C) x is a number whose distance from -2 is less than 6, but x cannot equal -2.
$-8 < x < 4, x \neq -2$, or $(-8, -2) \cup (-2, 4)$

(D) x is a number whose distance from -2 is greater than 6.
$x < -8$ or $x > 4$, or $(-\infty, -8) \cup (4, \infty)$

4. (A) $x = -\frac{7}{2}, \frac{9}{2}$ or $\{-\frac{7}{2}, \frac{9}{2}\}$ (B) $-7 \leq x \leq 7$ or $[-7, 7]$ (C) $-4 \leq x \leq 2$ or $[-4, 2]$
(D) $-2 < x < 7$ or $(-2, 7)$

5. (A) $x \leq -5$ or $x \geq 5$, or $(-\infty, -5] \cup [5, \infty)$ (B) $x < -\frac{1}{2}$ or $x > 2$, or $(-\infty, -\frac{1}{2}) \cup (2, \infty)$
(C) $x < -2$ or $x > \frac{22}{5}$, or $(-\infty, -2) \cup (\frac{22}{5}, \infty)$

6. $x = -\frac{1}{4}, \frac{9}{2}$ or $\{-\frac{1}{4}, \frac{9}{2}\}$

EXERCISE 1-4

A

In Problems 1–8, simplify, and write without absolute value signs. Do not replace radicals with decimal approximations.

1. $|\sqrt{5}|$ **2.** $|-\frac{3}{4}|$

3. $|(-6) - (-2)|$ **4.** $|(-2) - (-6)|$

5. $|5 - \sqrt{5}|$ **6.** $|\sqrt{7} - 2|$

7. $|\sqrt{5} - 5|$ **8.** $|2 - \sqrt{7}|$

In Problems 9–12, find the distance between points A and B with coordinates a and b respectively, as given.

9. $a = -7,\ b = 5$ **10.** $a = 3,\ b = 12$

11. $a = 5,\ b = -7$ **12.** $a = -9,\ b = -17$

In Problems 13–18, use the number line below to find the indicated distances.

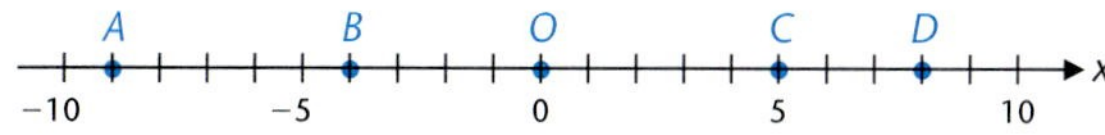

13. $d(B, O)$ **14.** $d(A, B)$ **15.** $d(O, B)$

16. $d(B, A)$ **17.** $d(B, C)$ **18.** $d(D, C)$

Write each of the statements in Problems 19–28 as an absolute value equation or inequality.

19. x is 4 units from 3.

20. y is 3 units from 1.

21. m is 5 units from -2.

22. n is 7 units from -5.

23. x is less than 5 units from 3.

24. z is less than 8 units from -2.

25. p is more than 6 units from -2.

26. c is no greater than 7 units from -3.

27. q is no less than 2 units from 1.

28. d is no more than 4 units from 5.

B

In Problems 29–44, solve, interpret geometrically, and graph. When applicable, write answers using both inequality notation and interval notation.

29. $|x| \le 7$ **30.** $|t| \le 5$ **31.** $|x| \ge 7$

32. $|x| \ge 5$ **33.** $|y - 5| = 3$ **34.** $|t - 3| = 4$

35. $|y - 5| < 3$ **36.** $|t - 3| < 4$ **37.** $|y - 5| > 3$

38. $|t - 3| > 4$ **39.** $|u + 8| = 3$ **40.** $|x + 1| = 5$

41. $|u + 8| \le 3$ **42.** $|x + 1| \le 5$ **43.** $|u + 8| \ge 3$

44. $|x + 1| \ge 5$

In Problems 45–62, solve each equation or inequality. When applicable, write answers using both inequality notation and interval notation.

45. $|3x - 7| \le 4$ **46.** $|5y + 2| \ge 8$

47. $|4 - 2t| > 6$ **48.** $|10 + 4s| < 6$

49. $|7m + 11| = 3$ **50.** $|4 - 5n| \le 8$

51. $|\frac{1}{2}w - \frac{3}{4}| < 2$ **52.** $|\frac{1}{3}z + \frac{5}{6}| = 1$

53. $|0.2u + 1.7| \ge 0.5$ **54.** $|0.5v - 2.5| > 1.6$

55. $|\frac{9}{5}C + 32| < 31$ **56.** $|\frac{5}{9}(F - 32)| < 40$

57. $\sqrt{x^2} < 2$ **58.** $\sqrt{m^2} > 3$

59. $\sqrt{(1 - 3t)^2} \le 2$ **60.** $\sqrt{(3 - 2x)^2} < 5$

61. $\sqrt{(2t - 3)^2} > 3$ **62.** $\sqrt{(3m + 5)^2} \ge 4$

C

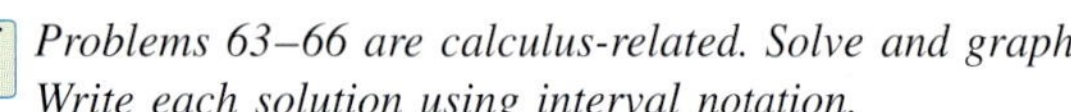

Problems 63–66 are calculus-related. Solve and graph. Write each solution using interval notation.

63. $0 < |x - 3| < 0.1$ **64.** $0 < |x - 5| < 0.01$

65. $0 < |x - c| < d$ **66.** $0 < |x - 4| < d$

In Problems 67–76, for what values of x does each hold?

67. $|x - 2| = x - 2$ **68.** $|x + 4| = -(x + 4)$

69. $|2x - 3| = 3 - 2x$ **70.** $|3x - 9| = 3x - 9$

71. $|3x + 5| = 2x + 6$ **72.** $|7 - 2x| = 5 - x$

73. $|x| + |x + 3| = 3$ **74.** $|x| - |x - 5| = 5$

75. $|2x + 7| - |6 - 3x| = 8$ **76.** $|3x + 1| + |3 - 2x| = 11$

77. What are the possible values of $\frac{x}{|x|}$?

78. What are the possible values of $\frac{|x - 1|}{x - 1}$?

79. Prove that $|b - a| = |a - b|$ for all real numbers a and b.

80. Prove that $|x|^2 = x^2$ for all real numbers x.

81. Prove that the average of two numbers is between the two numbers; that is, if $m < n$, then

$$m < \frac{m + n}{2} < n$$

82. Prove that for $m < n$,

$$d\left(m, \frac{m + n}{2}\right) = d\left(\frac{m + n}{2}, n\right)$$

83. Prove that $|-m| = |m|$.

84. Prove that $|m| = |n|$ if and only if $m = n$ or $m = -n$.

85. Prove that for $n \neq 0$,

$$\left|\frac{m}{n}\right| = \frac{|m|}{|n|}$$

86. Prove that $|mn| = |m||n|$.

87. Prove that $-|m| \le m \le |m|$.

88. Prove the **triangle inequality:**

$$|m + n| \le |m| + |n|$$

Hint: Use Problem 87 to show that

$$-|m| - |n| \le m + n \le |m| + |n|$$

89. If a and b are real numbers, prove that the maximum of a and b is given by

$$\max(a, b) = \tfrac{1}{2}[a + b + |a - b|]$$

90. Prove that the minimum of a and b is given by

$$\min(a, b) = \tfrac{1}{2}[a + b - |a - b|]$$

APPLICATIONS

91. Statistics. Inequalities of the form

$$\left|\frac{x - m}{s}\right| < n$$

occur frequently in statistics. If $m = 45.4$, $s = 3.2$, and $n = 1$, solve for x.

92. Statistics. Repeat Problem 91 for $m = 28.6$, $s = 6.5$, and $n = 2$.

★ **93. Business.** The daily production P in an automobile assembly plant is within 20 units of 500 units. Express the daily production as an absolute value inequality.

★ **94. Chemistry.** In a chemical process, the temperature T is to be kept within 10°C of 200°C. Express this restriction as an absolute value inequality.

95. Approximation. The area A of a region is approximately equal to 12.436. The error in this approximation is less than 0.001. Describe the possible values of this area both with an absolute value inequality and with interval notation.

96. Approximation. The volume V of a solid is approximately equal to 6.94. The error in this approximation is less than 0.02. Describe the possible values of this volume both with an absolute value inequality and with interval notation.

★ **97. Significant Digits.** If $N = 2.37$ represents a measurement, then we assume an accuracy of 2.37 ± 0.005. Express the accuracy assumption using an absolute value inequality.

★ **98. Significant Digits.** If $N = 3.65 \times 10^{-3}$ is a number from a measurement, then we assume an accuracy of $3.65 \times 10^{-3} \pm 5 \times 10^{-6}$. Express the accuracy assumption using an absolute value inequality.

SECTION 1-5 Complex Numbers

- Introductory Remarks
- The Complex Number System
- Complex Numbers and Radicals

• Introductory Remarks

The Pythagoreans (500–275 B.C.) found that the simple equation

$$x^2 = 2 \tag{1}$$

had no rational number solutions. If equation (1) were to have a solution, then a new kind of number had to be invented—an irrational number. The irrational numbers $\sqrt{2}$ and $-\sqrt{2}$ are both solutions to equation (1). Irrational numbers were not put on a firm mathematical foundation until the nineteenth century. The rational and irrational numbers together constitute the real number system.

Is there any need to consider another number system? Yes, if we want the simple equation

$$x^2 = -1$$

to have a solution. If x is any real number, then $x^2 \geq 0$. Thus, $x^2 = -1$ cannot have any real number solutions. Once again a new type of number must be invented, a number whose square can be negative. These new numbers are among the numbers called *complex numbers*. The complex numbers evolved over a long period of time, but, like the real numbers, it was not until the nineteenth century that they were placed on a firm mathematical foundation.

• The Complex Number System

We start the development of the complex number system by defining a complex number and several special types of complex numbers. We then define equality, addition, and multiplication in this system, and from these definitions the important special properties and operational rules for addition, subtraction, multiplication, and division will follow.

TABLE 1 Brief History of Complex Numbers

Approximate date	Person	Event
50	Heron of Alexandria	First recorded encounter of a square root of a negative number
850	Mahavira of India	Said that a negative has no square root, since it is not a square
1545	Cardano of Italy	Solutions to cubic equations involved square roots of negative numbers.
1637	Descartes of France	Introduced the terms *real* and *imaginary*
1748	Euler of Switzerland	Used i for $\sqrt{-1}$
1832	Gauss of Germany	Introduced the term *complex number*

DEFINITION 1 Complex Number

A **complex number** is a number of the form

$$a + bi \qquad \textbf{Standard Form}$$

where a and b are real numbers and i is called the **imaginary unit.**

The imaginary unit i introduced in Definition 1 is not a real number. It is a special symbol used in the representation of the elements in this new complex number system.

Some examples of complex numbers are

$$3 - 2i \qquad \tfrac{1}{2} + 5i \qquad 2 - \tfrac{1}{3}i$$
$$0 + 3i \qquad 5 + 0i \qquad 0 + 0i$$

Particular kinds of complex numbers are given special names as follows:

DEFINITION 2 Names for Particular Kinds of Complex Numbers

Imaginary Unit:	i	
Complex Number:	$a + bi$	a and b real numbers
Imaginary Number:	$a + bi$	$b \neq 0$
Pure Imaginary Number:	$0 + bi = bi$	$b \neq 0$
Real Number:	$a + 0i = a$	
Zero:	$0 + 0i = 0$	
Conjugate of $a + bi$:	$a - bi$	

EXAMPLE 1 Special Types of Complex Numbers

Given the list of complex numbers:

$$3 - 2i \qquad \tfrac{1}{2} + 5i \qquad 2 - \tfrac{1}{3}i$$

$$0 + 3i = 3i \qquad 5 + 0i = 5 \qquad 0 + 0i = 0$$

(A) List all the imaginary numbers, pure imaginary numbers, real numbers, and zero.

(B) Write the conjugate of each.

Solutions

(A) Imaginary numbers: $3 - 2i$, $\tfrac{1}{2} + 5i$, $2 - \tfrac{1}{3}i$, $3i$
Pure imaginary numbers: $0 + 3i = 3i$
Real numbers: $5 + 0i = 5$, $0 + 0i = 0$
Zero: $0 + 0i = 0$

(B) $3 + 2i \qquad \tfrac{1}{2} - 5i \qquad 2 + \tfrac{1}{3}i$
$0 - 3i = -3i \qquad 5 - 0i = 5 \qquad 0 - 0i = 0$

Matched Problem 1

Given the list of complex numbers:

$$6 + 7i \qquad \sqrt{2} - \tfrac{1}{3}i \qquad 0 - i = -i$$

$$0 + \tfrac{2}{3}i = \tfrac{2}{3}i \qquad -\sqrt{3} + 0i = -\sqrt{3} \qquad 0 - 0i = 0$$

(A) List all the imaginary numbers, pure imaginary numbers, real numbers, and zero.

(B) Write the conjugate of each.

In Definition 2, notice that we identify a complex number of the form $a + 0i$ with the real number a, a complex number of the form $0 + bi$, $b \neq 0$, with the **pure imaginary number** bi, and the complex number $0 + 0i$ with the real number 0. Thus, a real number is also a complex number, just as a rational number is also a real number. Any complex number that is not a real number is called an **imaginary number.** If we combine the set of all real numbers with the set of all imaginary numbers, we obtain ***C*, the set of complex numbers.** The relationship of the complex number system to the other number systems we have studied is shown in Figure 1.

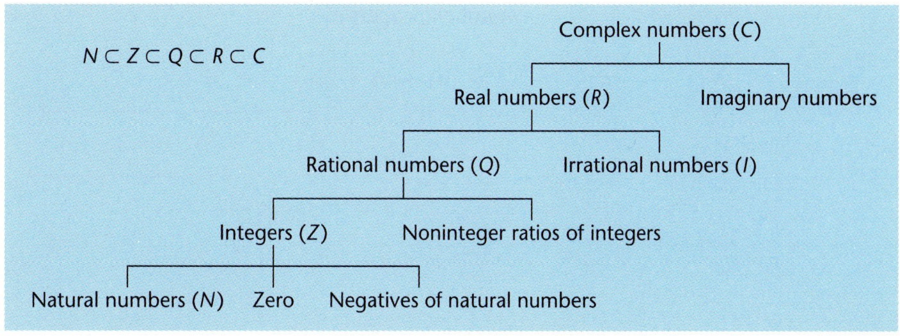

FIGURE 1 Complex numbers and important subsets

To use complex numbers, we must know how to add, subtract, multiply, and divide them. We start by defining equality, addition, and multiplication.

DEFINITION 3

Equality and Basic Operations

1. **Equality:** $a + bi = c + di$ if and only if $a = c$ and $b = d$
2. **Addition:** $(a + bi) + (c + di) = (a + c) + (b + d)i$
3. **Multiplication:** $(a + bi)(c + di) = (ac - bd) + (ad + bc)i$

In Section A-1 we list the basic properties of the real number system. Using Definition 3, it can be shown that the complex number system possesses the same properties. That is:

1. Addition and multiplication of complex numbers are commutative and associative operations.
2. There is an additive identity and a multiplicative identity for complex numbers.
3. Every complex number has an additive inverse or negative.
4. Every nonzero complex number has a multiplicative inverse or reciprocal.
5. Multiplication distributes over addition.

As a consequence of these properties, we can manipulate complex number symbols of the form $a + bi$ just like we manipulate binomials of the form $a + bx$, as long as we remember that i is a special symbol for the imaginary unit, not for a real number. Thus, you will not have to memorize the definitions of addition and multiplication of complex numbers. We now discuss these operations and some of their properties. Others will be considered in Exercise 1-5.

EXAMPLE 2

Addition of Complex Numbers

Carry out each operation and express the answer in standard form:

(A) $(2 - 3i) + (6 + 2i)$ (B) $(-5 + 4i) + (0 + 0i)$

Solution (A) We could apply the definition of addition directly, but it is easier to use complex number properties.

$$(2 - 3i) + (6 + 2i) = 2 - 3i + 6 + 2i \quad \text{Remove parentheses.}$$

$$= (2 + 6) + (-3 + 2)i \quad \text{Combine like terms.}$$

$$= 8 - i$$

(B) $(-5 + 4i) + (0 + 0i) = -5 + 4i + 0 + 0i$
$= -5 + 4i$

Matched Problem 2 Carry out each operation and express the answer in standard form:

(A) $(3 + 2i) + (6 - 4i)$ (B) $(0 + 0i) + (7 - 5i)$

Example 2B and Matched Problem 2B illustrate the following general result: For any complex number $a + bi$,

$$(a + bi) + (0 + 0i) = (0 + 0i) + (a + bi) = a + bi$$

Thus, $0 + 0i$ is the **additive identity** or **zero** for the complex numbers. We anticipated this result in Definition 1 when we identified the complex number $0 + 0i$ with the real number 0.

We could define negatives and subtraction in terms of the additive inverse of a complex number, as we did for real numbers, but once again it is much easier to use the properties of complex numbers.

EXAMPLE 3 Negation and Subtraction

Carry out each operation and express the answer in standard form:

(A) $-(4 - 5i)$ (B) $(7 - 3i) - (6 + 2i)$ (C) $(-2 + 7i) + (2 - 7i)$

Solution (A) $-(4 - 5i) = (-1)(4 - 5i) = -4 + 5i$

(B) $(7 - 3i) - (6 + 2i) = 7 - 3i - 6 - 2i$
$= 1 - 5i$

(C) $(-2 + 7i) + (2 - 7i) = -2 + 7i + 2 - 7i = 0$

Matched Problem 3 Carry out each operation and express the answer in standard form:

(A) $-(-3 + 2i)$ (B) $(3 - 5i) - (1 - 3i)$ (C) $(-4 + 9i) + (4 - 9i)$

In general, the **additive inverse** or **negative** of $a + bi$ is $-a - bi$ since

$$(a + bi) + (-a - bi) = (-a - bi) + (a + bi) = 0$$

(see Example 3C and Matched Problem 3C).

Now we turn our attention to multiplication. First, we use the definition of multiplication to see what happens to the complex unit i when it is squared:

$$\begin{aligned} i^2 &= (\overset{a}{0} + \overset{b}{1}i)(\overset{c}{0} + \overset{d}{1}i) \\ &= (\overset{a}{0} \cdot \overset{c}{0} - \overset{b}{1} \cdot \overset{d}{1}) + (\overset{a}{0} \cdot \overset{d}{1} + \overset{b}{1} \cdot \overset{c}{0})i \\ &= -1 + 0i \\ &= -1 \end{aligned}$$

Thus, we have proved that

$$\mathbf{i^2 = -1}$$

We now have a number whose square is negative and a solution to $x^2 = -1$. Since $i^2 = -1$, we define $\sqrt{-1}$ to be the imaginary unit i. Thus,

$$i = \sqrt{-1} \quad \textbf{and} \quad -i = -\sqrt{-1}$$

Just as was the case with addition and subtraction, multiplication of complex numbers can be carried out by using the properties of complex numbers rather than the definition of multiplication. We just replace i^2 with -1 each time it occurs.

EXAMPLE 4 Multiplying Complex Numbers

Carry out each operation and express the answer in standard form:

(A) $(2 - 3i)(6 + 2i)$ (B) $1(3 - 5i)$ (C) $i(1 + i)$ (D) $(3 + 4i)(3 - 4i)$

Solution

(A) $(2 - 3i)(6 + 2i) = 2(6 + 2i) - 3i(6 + 2i)$

$= 12 + 4i - 18i - 6i^2$

$= 12 - 14i - 6(-1)$ Replace i^2 with -1.

$= 18 - 14i$

(B) $1(3 - 5i) = 1 \cdot 3 - 1 \cdot 5i = 3 - 5i$

(C) $i(1 + i) = i + i^2 = i - 1 = -1 + i$

(D) $(3 + 4i)(3 - 4i) = 9 - 12i + 12i - 16i^2$

$= 9 + 16 = 25$

Matched Problem 4 Carry out each operation and express the answer in standard form:

(A) $(5 + 2i)(4 - 3i)$ (B) $3(-2 + 6i)$ (C) $i(2 - 3i)$ (D) $(2 + 3i)(2 - 3i)$

For any complex number $a + bi$,

$$1(a + bi) = (a + bi)1 = a + bi$$

(see Example 4B). Thus, 1 is the **multiplicative identity** for complex numbers, just as it is for real numbers.

Earlier we stated that every nonzero complex number has a multiplicative inverse or reciprocal. We will denote this as a fraction, just as we do with real numbers. Thus.

$$\frac{1}{a + bi} \quad \text{is the } \textbf{reciprocal} \text{ of} \quad a + bi \qquad a + bi \neq 0$$

The following important property of the conjugate of a complex number is used to express reciprocals and quotients in standard form.

Theorem 1 **Product of a Complex Number and Its Conjugate**

$$(a + bi)(a - bi) = a^2 + b^2 \quad \text{A real number}$$

EXAMPLE 5 **Reciprocals and Quotients**

Carry out each operation and express the answer in standard form:

(A) $\dfrac{1}{2 + 3i}$ (B) $\dfrac{7 - 3i}{1 + i}$

Solution (A) Multiply numerator and denominator by the conjugate of the denominator:

$$\frac{1}{2 + 3i} = \frac{1}{2 + 3i} \cdot \frac{2 - 3i}{2 - 3i} = \frac{2 - 3i}{4 - 9i^2} = \frac{2 - 3i}{4 + 9}$$

$$= \frac{2 - 3i}{13} = \frac{2}{13} - \frac{3}{13}i$$

This answer can be checked by multiplication:

Check

$$(2 + 3i)\left(\frac{2}{13} - \frac{3}{13}i\right) = \frac{4}{13} - \frac{6}{13}i + \frac{6}{13}i - \frac{9}{13}i^2$$

$$= \frac{4}{13} + \frac{9}{13} = 1$$

$$\text{(B)}\ \frac{7-3i}{1+i} = \frac{7-3i}{1+i} \cdot \frac{1-i}{1-i} = \frac{7-7i-3i+3i^2}{1-i^2}$$

$$= \frac{4-10i}{2} = 2-5i$$

Check $(1+i)(2-5i) = 2-5i+2i-5i^2 = 7-3i$

Matched Problem 5 Carry out each operation and express the answer in standard form:

(A) $\dfrac{1}{4+2i}$ (B) $\dfrac{6+7i}{2-i}$

EXAMPLE 6 Combined Operations

Carry out the indicated operations and write each answer in standard form:

(A) $(3-2i)^2 - 6(3-2i) + 13$ (B) $\dfrac{2-3i}{2i}$

Solutions (A)
$$\begin{aligned}(3-2i)^2 - 6(3-2i) + 13 &= 9 - 12i + 4i^2 - 18 + 12i + 13\\ &= 9 - 12i - 4 - 18 + 12i + 13\\ &= 0\end{aligned}$$

(B) If a complex number is divided by a pure imaginary number, we can make the denominator real by multiplying numerator and denominator by i.

$$\frac{2-3i}{2i} \cdot \frac{i}{i} = \frac{2i-3i^2}{2i^2} = \frac{2i+3}{-2} = -\frac{3}{2} - i$$

Matched Problem 6 Carry out the indicated operations and write each answer in standard form:

(A) $(3+2i)^2 - 6(3+2i) + 13$ (B) $\dfrac{4-i}{3i}$

EXPLORE-DISCUSS 1 Natural number powers of i take on particularly simple forms:

i	$i^5 = i^4 \cdot i = (1)i = i$
$i^2 = -1$	$i^6 = i^4 \cdot i^2 = 1(-1) = -1$
$i^3 = i^2 \cdot i = (-1)i = -i$	$i^7 = i^4 \cdot i^3 = 1(-i) = -i$
$i^4 = i^2 \cdot i^2 = (-1)(-1) = 1$	$i^8 = i^4 \cdot i^4 = 1 \cdot 1 = 1$

In general, what are the possible values for i^n, n a natural number? Explain how you could easily evaluate i^n for any natural number n. Then evaluate each of the following:

(A) i^{17} (B) i^{24} (C) i^{38} (D) i^{47}

• Complex Numbers and Radicals

Recall that we say that a is a square root of b if $a^2 = b$. If x is a positive real number, then x has two square roots, the principal square root, denoted by $\sqrt{x}$, and its negative, $-\sqrt{x}$ (Section A-7). If x is a negative real number, then x still has two square roots, but now these square roots are imaginary numbers.

DEFINITION 4 **Principal Square Root of a Negative Real Number**

The **principal square root of a negative real number,** denoted by $\sqrt{-a}$ where a is positive, is defined by

$$\sqrt{-a} = i\sqrt{a} \qquad \sqrt{-3} = i\sqrt{3} \qquad \sqrt{-9} = i\sqrt{9} = 3i$$

The other square root of $-a$, $a > 0$, is $-\sqrt{-a} = -i\sqrt{a}$.

Note in Definition 4 that we wrote $i\sqrt{a}$ and $i\sqrt{3}$ in place of the standard forms $\sqrt{a}i$ and $\sqrt{3}i$. We follow this convention whenever it appears that i might accidentally slip under a radical sign ($\sqrt{a}i \neq \sqrt{ai}$, but $\sqrt{a}i = i\sqrt{a}$). Definition 4 is motivated by the fact that

$$(i\sqrt{a})^2 = i^2a = -a$$

EXAMPLE 7 **Complex Numbers and Radicals**

Write in standard form:

(A) $\sqrt{-4}$ (B) $4 + \sqrt{-5}$ (C) $\dfrac{-3 - \sqrt{-5}}{2}$ (D) $\dfrac{1}{1 - \sqrt{-9}}$

Solutions (A) $\sqrt{-4} = i\sqrt{4} = 2i$ (B) $4 + \sqrt{-5} = 4 + i\sqrt{5}$

(C) $\dfrac{-3 - \sqrt{-5}}{2} = \dfrac{-3 - i\sqrt{5}}{2} = -\dfrac{3}{2} - \dfrac{\sqrt{5}}{2}i$

(D) $\dfrac{1}{1 - \sqrt{-9}} = \dfrac{1}{1 - 3i} = \dfrac{1 \cdot (1 + 3i)}{(1 - 3i) \cdot (1 + 3i)}$

$$= \frac{1 + 3i}{1 - 9i^2} = \frac{1 + 3i}{10} = \frac{1}{10} + \frac{3}{10}i$$

Matched Problem 7 Write in standard form:

(A) $\sqrt{-16}$ (B) $5 + \sqrt{-7}$ (C) $\dfrac{-5 - \sqrt{-2}}{2}$ (D) $\dfrac{1}{3 - \sqrt{-4}}$

EXPLORE-DISCUSS 2 From Theorem 1 in Section 1-7, we know that if a and b are positive real numbers, then

$$\sqrt{a}\sqrt{b} = \sqrt{ab} \tag{1}$$

Thus, we can evaluate expressions like $\sqrt{9}\sqrt{4}$ two ways:

$$\sqrt{9}\sqrt{4} = \sqrt{(9)(4)} = \sqrt{36} = 6 \quad \text{and} \quad \sqrt{9}\sqrt{4} = (3)(2) = 6$$

Evaluate each of the following two ways. Is (1) a valid property to use in all cases?

(A) $\sqrt{9}\sqrt{-4}$ (B) $\sqrt{-9}\sqrt{4}$ (C) $\sqrt{-9}\sqrt{-4}$

CAUTION Note that in Example 7D, we wrote $1 - \sqrt{-9} = 1 - 3i$ before proceeding with the simplification. This is a necessary step because some of the properties of radicals that are true for real numbers turn out not to be true for complex numbers. In particular, for positive real numbers a and b,

$$\sqrt{a}\sqrt{b} = \sqrt{ab} \quad \text{but} \quad \sqrt{-a}\sqrt{-b} \neq \sqrt{(-a)(-b)}$$

(See Explore-Discuss 2.)

Early resistance to these new numbers is suggested by the words used to name them: *complex* and *imaginary*. In spite of this early resistance, complex numbers have come into widespread use in both pure and applied mathematics. They are used extensively, for example, in electrical engineering, physics, chemistry, statistics, and aeronautical engineering. Our first use of them will be in connection with solutions of second-degree equations in the next section.

Answers to Matched Problems

1. (A) Imaginary numbers: $6 + 7i$, $\sqrt{2} - \frac{1}{3}i$, $0 - i = -i$, $0 + \frac{2}{3}i = \frac{2}{3}i$
Pure imaginary numbers: $0 - i = -i$, $0 + \frac{2}{3}i = \frac{2}{3}i$
Real numbers: $-\sqrt{3} + 0i = -\sqrt{3}$, $0 - 0i = 0$
Zero: $0 - 0i = 0$
(B) $6 - 7i$, $\sqrt{2} + \frac{1}{3}i$, $0 + i = i$, $0 - \frac{2}{3}i = -\frac{2}{3}i$, $-\sqrt{3} - 0i = -\sqrt{3}$, $0 + 0i = 0$
2. (A) $9 - 2i$ (B) $7 - 5i$
3. (A) $3 - 2i$ (B) $2 - 2i$ (C) 0
4. (A) $26 - 7i$ (B) $-6 + 18i$ (C) $3 + 2i$ (D) 13
5. (A) $\frac{1}{5} - \frac{1}{10}i$ (B) $1 + 4i$
6. (A) 0 (B) $-\frac{1}{3} - \frac{4}{3}i$
7. (A) $4i$ (B) $5 + i\sqrt{7}$ (C) $-\frac{5}{2} - (\sqrt{2}/2)i$ (D) $\frac{3}{13} + \frac{2}{13}i$

EXERCISE 1-5

A

In Problems 1–26, perform the indicated operations and write each answer in standard form.

1. $(3 + 5i) + (2 + 4i)$
2. $(4 + i) + (5 + 3i)$
3. $(8 - 3i) + (-5 + 6i)$
4. $(-1 + 2i) + (4 - 7i)$
5. $(9 + 5i) - (6 + 2i)$
6. $(3 + 7i) - (2 + 5i)$
7. $(3 - 4i) - (-5 + 6i)$
8. $(-4 - 2i) - (1 + i)$
9. $2 + (3i + 5)$
10. $(2i + 7) - 4i$
11. $(2i)(4i)$
12. $(3i)(5i)$
13. $-2i(4 - 6i)$
14. $(-4i)(2 - 3i)$
15. $(1 + 2i)(3 - 4i)$
16. $(2 - i)(-5 + 6i)$
17. $(3 - i)(4 + i)$
18. $(5 + 2i)(4 - 3i)$
19. $(2 + 9i)(2 - 9i)$
20. $(3 + 8i)(3 - 8i)$
21. $\dfrac{1}{2 + 4i}$
22. $\dfrac{i}{3 + i}$
23. $\dfrac{4 + 3i}{1 + 2i}$
24. $\dfrac{3 - 5i}{2 - i}$
25. $\dfrac{7 + i}{2 + i}$
26. $\dfrac{-5 + 10i}{3 + 4i}$

B

In Problems 27–36, convert imaginary numbers to standard form, perform the indicated operations, and express answers in standard form.

27. $(2 - \sqrt{-4}) + (5 - \sqrt{-9})$
28. $(3 - \sqrt{-4}) + (-8 + \sqrt{-25})$
29. $(9 - \sqrt{-9}) - (12 - \sqrt{-25})$
30. $(-2 - \sqrt{-36}) - (4 + \sqrt{-49})$
31. $(3 - \sqrt{-4})(-2 + \sqrt{-49})$
32. $(2 - \sqrt{-1})(5 + \sqrt{-9})$
33. $\dfrac{5 - \sqrt{-4}}{7}$
34. $\dfrac{6 - \sqrt{-64}}{2}$
35. $\dfrac{1}{2 - \sqrt{-9}}$
36. $\dfrac{1}{3 - \sqrt{-16}}$

Write Problems 37–42 in standard form.

37. $\dfrac{2}{5i}$
38. $\dfrac{1}{3i}$
39. $\dfrac{1 + 3i}{2i}$
40. $\dfrac{2 - i}{3i}$
41. $(2 - 3i)^2 - 2(2 - 3i) + 9$
42. $(2 - i)^2 + 3(2 - i) - 5$
43. Evaluate $x^2 - 2x + 2$ for $x = 1 - i$.
44. Evaluate $x^2 - 2x + 2$ for $x = 1 + i$.
45. Simplify: i^{18}, i^{32}, and i^{67}
46. Simplify: i^{21}, i^{43}, and i^{52}
47. For what real values of x and y will the following equation be a true statement?

$$(2x - 1) + (3y + 2)i = 5 - 4i$$

48. For what real values of x and y will the following equation be a true statement?

$$3x + (y - 2)i = (5 - 2x) + (3y - 8)i$$

In Problems 49–52, for what real values of x does each expression represent an imaginary number?

49. $\sqrt{3 - x}$
50. $\sqrt{5 + x}$
51. $\sqrt{2 - 3x}$
52. $\sqrt{3 + 2x}$

Use a calculator to compute Problems 53–56. Write in standard form $a + bi$, where a and b are computed to two decimal places.

53. $(3.17 - 4.08i)(7.14 + 2.76i)$
54. $(6.12 + 4.92i)(1.82 - 5.05i)$
55. $\dfrac{8.14 + 2.63i}{3.04 + 6.27i}$
56. $\dfrac{7.66 + 3.33i}{4.72 - 2.68i}$

C

In Problems 57–62, perform the indicated operations, and write each answer in standard form.

57. $(a + bi) + (c + di)$
58. $(a + bi) - (c + di)$
59. $(a + bi)(a - bi)$
60. $(u - vi)(u + vi)$
61. $(a + bi)(c + di)$
62. $\dfrac{a + bi}{c + di}$
63. Show that $i^{4k} = 1$, k a natural number
64. Show that $i^{4k+1} = i$, k a natural number

Supply the reasons in the proofs for the theorems stated in Problems 65 and 66.

65. *Theorem:* The complex numbers are commutative under addition.

Proof: Let $a + bi$ and $c + di$ be two arbitrary complex numbers; then:

Statement

1. $(a + bi) + (c + di) = (a + c) + (b + d)i$

2. $= (c + a) + (d + b)i$

3. $= (c + di) + (a + bi)$

Reason

1.

2.

3.

66. *Theorem:* The complex numbers are commutative under multiplication.

Proof: Let $a + bi$ and $c + di$ be two arbitrary complex numbers; then:

Statement

1. $(a + bi)\cdot(c + di) = (ac - bd) + (ad + bc)i$

2. $= (ca - db) + (da + cb)i$

3. $= (c + di)(a + bi)$

Reason

1.

2.

3.

Letters z and w are often used as complex variables, where $z = x + yi$, $w = u + vi$, and x, y, u, v are real numbers. The conjugates of z and w, denoted by $\bar{z}$ and $\bar{w}$, respectively, are given by $\bar{z} = x - yi$ and $\bar{w} = u - vi$. In Problems 67–74, express each property of conjugates verbally and then prove the property.

67. $z\bar{z}$ is a real number.

68. $z + \bar{z}$ is a real number.

69. $\bar{z} = z$ if and only if z is real.

70. $\bar{\bar{z}} = z$

71. $\overline{z + w} = \bar{z} + \bar{w}$

72. $\overline{z - w} = \bar{z} - \bar{w}$

73. $\overline{zw} = \bar{z} \cdot \bar{w}$

74. $\overline{z/w} = \bar{z}/\bar{w}$

SECTION 1-6 Quadratic Equations and Applications

- Solution by Factoring
- Solution by Square Root
- Solution by Completing the Square
- Solution by Quadratic Formula
- Applications

The next class of equations we consider are the second-degree polynomial equations in one variable, called *quadratic equations.*

DEFINITION 1 **Quadratic Equation**

A **quadratic equation** in one variable is any equation that can be written in the form

$$ax^2 + bx + c = 0 \qquad a \neq 0 \qquad \textbf{Standard Form}$$

where x is a variable and a, b, and c are constants.

Now that we have discussed the complex number system, we will use complex numbers when solving equations. Recall that a solution of an equation is also called a *root* of the equation. A real number solution of an equation is called a **real root,** and an imaginary number solution is called an **imaginary root.** In this section we develop methods for finding all real and imaginary roots of a quadratic equation.

• Solution by Factoring

If $ax^2 + bx + c$ can be written as the product of two first-degree factors, then the quadratic equation can be quickly and easily solved. The method of solution by factoring rests on the zero property of complex numbers, which is a generalization of the zero property of real numbers reviewed in Section A-1.

> **Zero Property**
>
> If m and n are complex numbers, then
>
> $$m \cdot n = 0 \qquad \text{if and only if} \qquad m = 0 \text{ or } n = 0 \text{ (or both)}$$

EXAMPLE 1 Solving Quadratic Equations by Factoring

Solve by factoring:

(A) $6x^2 - 19x - 7 = 0$ (B) $x^2 - 6x + 5 = -4$ (C) $2x^2 = 3x$

Solutions

(A) $6x^2 - 19x - 7 = 0$

$(2x - 7)(3x + 1) = 0$ Factor left side.

$2x - 7 = 0$ or $3x + 1 = 0$

$x = \frac{7}{2}$ $\qquad x = -\frac{1}{3}$

The solution set is $\{-\frac{1}{3}, \frac{7}{2}\}$.

(B) $x^2 - 6x + 5 = -4$

$x^2 - 6x + 9 = 0$ Write in standard form.

$(x - 3)^2 = 0$ Factor left side.

$x = 3$

The solution set is $\{3\}$. The equation has one root, 3. But since it came from two factors, we call 3 a **double root.**

(C) $2x^2 = 3x$

$2x^2 - 3x = 0$

$x(2x - 3) = 0$

$x = 0$ or $2x - 3 = 0$

$x = \frac{3}{2}$

Solution set: $\{0, \frac{3}{2}\}$

Matched Problem 1 Solve by factoring:

(A) $3x^2 + 7x - 20 = 0$ (B) $4x^2 + 12x + 9 = 0$ (C) $4x^2 = 5x$

CAUTION

1. One side of an equation must be 0 before the zero property can be applied. Thus

$$x^2 - 6x + 5 = -4$$
$$(x - 1)(x - 5) = -4$$

does not imply that $x - 1 = -4$ or $x - 5 = -4$. See Example 1B for the correct solution of this equation.

2. The equations

$$2x^2 = 3x \qquad \text{and} \qquad 2x = 3$$

are not equivalent. The first has solution set $\{0, \frac{3}{2}\}$, while the second has solution set $\{\frac{3}{2}\}$. The root $x = 0$ is lost when each member of the first equation is divided by the variable x. See Example 1C for the correct solution of this equation.

Do not divide both members of an equation by an expression containing the variable for which you are solving. You may be dividing by 0.

• Solution by Square Root

We now turn our attention to quadratic equations that do not have the first-degree term—that is, equations of the special form

$$ax^2 + c = 0 \qquad a \neq 0$$

The method of solution of this special form makes direct use of the square root property:

Square Root Property

If $A^2 = C$, then $A = \pm\sqrt{C}$.

EXPLORE-DISCUSS 1 Determine if each of the following pairs of equations is equivalent or not. Explain your answer.

(A) $x^2 = 4$ and $x = |2|$
(B) $x^2 = 4$ and $x = -2$
(C) $x = \sqrt{4}$ and $x = 2$
(D) $x = \sqrt{4}$ and $x = -2$

The use of the square root property is illustrated in the next example.

Note: It is common practice to represent solutions of quadratic equations informally by the last equation rather than by writing a solution set using set notation. From now on, we will follow this practice unless a particular emphasis is desired.

EXAMPLE 2 Using the Square Root Property

Solve using the square root property:

(A) $2x^2 - 3 = 0$ (B) $3x^2 + 27 = 0$ (C) $(x + \frac{1}{2})^2 = \frac{5}{4}$

Solutions

(A)
$$2x^2 - 3 = 0$$
$$x^2 = \frac{3}{2}$$
$$x = \pm\sqrt{\frac{3}{2}} \quad \text{or} \quad \pm\frac{\sqrt{6}}{2}$$
Solution set: $\left\{\frac{-\sqrt{6}}{2}, \frac{\sqrt{6}}{2}\right\}$

(B)
$$3x^2 + 27 = 0$$
$$x^2 = -9$$
$$x = \pm\sqrt{-9} \quad \text{or} \quad \pm 3i$$
Solution set: $\{-3i, 3i\}$

(C)
$$(x + \tfrac{1}{2})^2 = \tfrac{5}{4}$$
$$x + \tfrac{1}{2} = \pm\sqrt{\tfrac{5}{4}}$$
$$x = -\frac{1}{2} \pm \frac{\sqrt{5}}{2}$$
$$= \frac{-1 \pm \sqrt{5}}{2}$$

Matched Problem 2 Solve using the square root property:

(A) $3x^2 - 5 = 0$ (B) $2x^2 + 8 = 0$ (C) $(x + \frac{1}{3})^2 = \frac{2}{9}$

EXPLORE-DISCUSS 2 Replace ? in each of the following with a number that makes the equation valid.

(A) $(x + 1)^2 = x^2 + 2x + ?$ (B) $(x + 2)^2 = x^2 + 4x + ?$
(C) $(x + 3)^2 = x^2 + 6x + ?$ (D) $(x + 4)^2 = x^2 + 8x + ?$

Replace ? in each of the following with a number that makes the trinomial a perfect square.

(E) $x^2 + 10x + ?$ (F) $x^2 + 12x + ?$ (G) $x^2 + bx + ?$

• Solution by Completing the Square

The methods of square root and factoring are generally fast when they apply; however, there are equations, such as $x^2 + 6x - 2 = 0$ (see Example 4A), that cannot be solved directly by these methods. A more general procedure must be developed to take care of this type of equation—for example, the method of completing the square. This method is based on the process of transforming the standard quadratic equation

$$ax^2 + bx + c = 0$$

into the form

$$(x + A)^2 = B$$

where A and B are constants. The last equation can easily be solved by using the square root property. But how do we transform the first equation into the second? The following brief discussion provides the key to the process.

What number must be added to $x^2 + bx$ so that the result is the square of a first-degree polynomial? There is a simple mechanical rule for finding this number, based on the square of the following binomials:

$$(x + m)^2 = x^2 + 2mx + m^2$$

$$(x - m)^2 = x^2 - 2mx + m^2$$

In either case, we see that the third term on the right is the square of one-half of the coefficient of x in the second term on the right. This observation leads directly to the rule for completing the square.

Completing the Square

To complete the square of a quadratic of the form $x^2 + bx$, add the square of one-half the coefficient of x; that is, add $(b/2)^2$. Thus,

$$x^2 + bx \qquad\qquad x^2 + 5x$$

$$x^2 + bx + \left(\frac{b}{2}\right)^2 = \left(x + \frac{b}{2}\right)^2 \qquad x^2 + 5x + \left(\frac{5}{2}\right)^2 = \left(x + \frac{5}{2}\right)^2$$

EXAMPLE 3 **Completing the Square**

Complete the square for each of the following:

(A) $x^2 - 3x$ (B) $x^2 - bx$

Solutions (A) $x^2 - 3x$

$$x^2 - 3x + \tfrac{9}{4} = (x - \tfrac{3}{2})^2$$ Add $\left(\frac{-3}{2}\right)^2$; that is, $\frac{9}{4}$.

(B) $x^2 - bx$

$$x^2 - bx + \frac{b^2}{4} = \left(x - \frac{b}{2}\right)^2$$ Add $\left(\frac{-b}{2}\right)^2$; that is, $\frac{b^2}{4}$.

Matched Problem 3 Complete the square for each of the following:

(A) $x^2 - 5x$ (B) $x^2 + mx$

It is important to note that the rule for completing the square applies only to quadratic forms in which the coefficient of the second-degree term is 1. This causes little trouble, however, as you will see. We now solve two equations by the method of completing the square.

EXAMPLE 4 **Solution by Completing the Square**

Solve by completing the square:

(A) $x^2 + 6x - 2 = 0$ (B) $2x^2 - 4x + 3 = 0$

Solutions (A) $x^2 + 6x - 2 = 0$

$$x^2 + 6x = 2$$

$$x^2 + 6x + 9 = 2 + 9$$ Complete the square on the left side, and add the same number to the right side.

$$(x + 3)^2 = 11$$

$$x + 3 = \pm\sqrt{11}$$

$$x = -3 \pm \sqrt{11}$$

(B) $2x^2 - 4x + 3 = 0$

$$x^2 - 2x + \tfrac{3}{2} = 0$$ Make the leading coefficient 1 by dividing by 2.

$$x^2 - 2x = -\tfrac{3}{2}$$

$$x^2 - 2x + 1 = -\tfrac{3}{2} + 1$$ Complete the square on the left side and add the same number to the right side.

$$(x-1)^2 = -\tfrac{1}{2} \qquad \text{Factor the left side.}$$

$$x - 1 = \pm\sqrt{-\tfrac{1}{2}}$$

$$x = 1 \pm i\sqrt{\tfrac{1}{2}}$$

$$= 1 \pm \frac{\sqrt{2}}{2}i \qquad \text{Answer in } a + bi \text{ form.}$$

Matched Problem 4 Solve by completing the square:

(A) $x^2 + 8x - 3 = 0$ (B) $3x^2 - 12x + 13 = 0$

• Solution by Quadratic Formula

Now consider the general quadratic equation with unspecified coefficients:

$$ax^2 + bx + c = 0 \qquad a \neq 0$$

We can solve it by completing the square exactly as we did in Example 4B. To make the leading coefficient 1, we must multiply both sides of the equation by $1/a$. Thus,

$$x^2 + \frac{b}{a}x + \frac{c}{a} = 0$$

Adding $-c/a$ to both sides of the equation and then completing the square of the left side, we have

$$x^2 + \frac{b}{a}x + \frac{b^2}{4a^2} = \frac{b^2}{4a^2} - \frac{c}{a}$$

We now factor the left side and solve using the square root property:

$$\left(x + \frac{b}{2a}\right)^2 = \frac{b^2 - 4ac}{4a^2}$$

$$x + \frac{b}{2a} = \pm\sqrt{\frac{b^2 - 4ac}{4a^2}}$$

$$x = -\frac{b}{2a} \pm \frac{\sqrt{b^2 - 4ac}}{2a} \qquad \text{See Problem 77 in Exercise 2-6.}$$

$$= \frac{-b \pm \sqrt{b^2 - 4ac}}{2a}$$

We have thus derived the well-known and widely used **quadratic formula:**

Theorem 1 **Quadratic Formula**

If $ax^2 + bx + c = 0$, $a \neq 0$, then

$$x = \frac{-b \pm \sqrt{b^2 - 4ac}}{2a}$$

The quadratic formula should be memorized and used to solve quadratic equations when other methods fail or are more difficult to apply.

EXAMPLE 5 Using the Quadratic Formula

Solve $2x + \frac{3}{2} = x^2$ by use of the quadratic formula. Leave the answer in simplest radical form.

Solution

$$2x + \tfrac{3}{2} = x^2$$

$$4x + 3 = 2x^2 \quad \text{Multiply both sides by 2.}$$

$$2x^2 - 4x - 3 = 0 \quad \text{Write in standard form.}$$

$$x = \frac{-b \pm \sqrt{b^2 - 4ac}}{2a} \quad a = 2,\ b = -4,\ c = -3$$

$$= \frac{-(-4) \pm \sqrt{(-4)^2 - 4(2)(-3)}}{2(2)}$$

$$= \frac{4 \pm \sqrt{40}}{4} = \frac{4 \pm 2\sqrt{10}}{4} = \frac{2 \pm \sqrt{10}}{2}$$

CAUTION

1. $-4^2 \neq (-4)^2$ $\quad -4^2 = -16$ and $(-4)^2 = 16$
2. $2 + \dfrac{\sqrt{10}}{2} \neq \dfrac{2 + \sqrt{10}}{2}$ $\quad 2 + \dfrac{\sqrt{10}}{2} = \dfrac{4 + \sqrt{10}}{2}$
3. $\dfrac{\not{4} \pm 2\sqrt{10}}{\not{4}} \neq \pm 2\sqrt{10}$ $\quad \dfrac{4 \pm 2\sqrt{10}}{4} = \dfrac{2(2 \pm \sqrt{10})}{4} = \dfrac{2 \pm \sqrt{10}}{2}$

Matched Problem 5 Solve $x^2 - \frac{5}{2} = -3x$ by use of the quadratic formula. Leave the answer in simplest radical form.

EXAMPLE 6 Using the Quadratic Formula with a Calculator

Solve $5.37x^2 - 6.03x + 1.17 = 0$ to two decimal places using a calculator.

Solution

$$5.37x^2 - 6.03x + 1.17 = 0$$

$$x = \frac{6.03 \pm \sqrt{(-6.03)^2 - 4(5.37)(1.17)}}{2(5.37)}$$

$$= 0.25, 0.87$$

Matched Problem 6 Solve $2.79x^2 + 5.07x - 7.69 = 0$ to two decimal places using a calculator.

We conclude this part of the discussion by noting that $b^2 - 4ac$ in the quadratic formula is called the **discriminant** and gives us useful information about the corresponding roots as shown in Table 1.

TABLE 1 Discriminant and Roots

Discriminant $b^2 - 4ac$	**Roots of** $ax^2 + bx + c = 0$ a, b, **and** c **real numbers,** $a \neq 0$
Positive	Two distinct real roots
0	One real root (a double root)
Negative	Two imaginary roots, one the conjugate of the other

For example:

(A) $2x^2 - 3x - 4 = 0$ has two real roots, since

$$b^2 - 4ac = (-3)^2 - 4(2)(-4) = 41 > 0$$

(B) $4x^2 - 4x + 1 = 0$ has one real (double) root, since

$$b^2 - 4ac = (-4)^2 - 4(4)(1) = 0$$

(C) $2x^2 - 3x + 4 = 0$ has two imaginary roots, since

$$b^2 - 4ac = (-3)^2 - 4(2)(4) = -23 < 0$$

• Applications

We now consider several applications that make use of quadratic equations. First, the strategy for solving word problems, presented earlier in Section 1-1, is repeated below.

Strategy for Solving Word Problems

1. Read the problem carefully—several times if necessary—that is, until you understand the problem, know what is to be found, and know what is given.

2. Let one of the unknown quantities be represented by a variable, say x, and try to represent all other unknown quantities in terms of x. This is an important step and must be done carefully.
3. If appropriate, draw figures or diagrams and label known and unknown parts.
4. Look for formulas connecting the known quantities to the unknown quantities.
5. Form an equation relating the unknown quantities to the known quantities.
6. Solve the equation and write answers to *all* questions asked in the problem.
7. Check and interpret all solutions in terms of the original problem—not just the equation found in step 5—since a mistake may have been made in setting up the equation in step 5.

EXAMPLE 7 Setting Up and Solving a Word Problem

The sum of a number and its reciprocal is $\frac{13}{6}$. Find all such numbers.

Solution Let x = the number; then:

$$x + \frac{1}{x} = \frac{13}{6}$$

$$(6x)x + (6x)\frac{1}{x} = (6x)\frac{13}{6}$$ Multiply both sides by $6x$. [*Note:* $x \neq 0$.]

$$6x^2 + 6 = 13x$$ A quadratic equation

$$6x^2 - 13x + 6 = 0$$

$$(2x - 3)(3x - 2) = 0$$

$$2x - 3 = 0 \quad \text{or} \quad 3x - 2 = 0$$

$$x = \tfrac{3}{2} \qquad\qquad x = \tfrac{2}{3}$$

Thus, two such numbers are $\frac{3}{2}$ and $\frac{2}{3}$.

Check $\frac{3}{2} + \frac{2}{3} = \frac{13}{6}$ $\quad \frac{2}{3} + \frac{3}{2} = \frac{13}{6}$

Matched Problem 7 The sum of two numbers is 23 and their product is 132. Find the two numbers. [*Hint:* If one number is x, then the other number is $23 - x$.]

EXAMPLE 8 A Distance–Rate–Time Problem

An excursion boat takes 1.6 hours longer to go 36 miles up a river than to return. If the rate of the current is 4 miles per hour, what is the rate of the boat in still water?

Solution Let

$$x = \text{Rate of boat in still water}$$
$$x + 4 = \text{Rate downstream}$$
$$x - 4 = \text{Rate upstream}$$

$$\begin{pmatrix}\text{Time}\\ \text{upstream}\end{pmatrix} - \begin{pmatrix}\text{Time}\\ \text{downstream}\end{pmatrix} = 1.6$$

$$\frac{36}{x-4} - \frac{36}{x+4} = 1.6 \qquad T = \frac{D}{R},\ x \neq 4,\ x \neq -4$$

$$36(x+4) - 36(x-4) = 1.6(x-4)(x+4)$$
$$36x + 144 - 36x + 144 = 1.6x^2 - 25.6$$
$$1.6x^2 = 313.6$$
$$x^2 = 196$$
$$x = \sqrt{196} = 14$$

The rate in still water is 14 miles per hour.

[*Note:* $-\sqrt{196} = -14$ must be discarded, since it doesn't make sense in the problem to have a negative rate.]

Check

$$\text{Time upstream} = \frac{D}{R} = \frac{36}{14-4} = 3.6$$
$$\text{Time downstream} = \frac{D}{R} = \frac{36}{14+4} = 2$$
$$1.6 \quad \text{Difference of times}$$

Matched Problem 8 Two boats travel at right angles to each other after leaving a dock at the same time. One hour later they are 25 miles apart. If one boat travels 5 miles per hour faster than the other, what is the rate of each? [*Hint:* Use the Pythagorean theorem,* remembering that distance equals rate times time.]

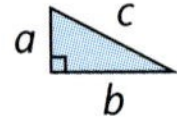

**Pythagorean theorem:* A triangle is a right triangle if and only if the square of the length of the longest side is equal to the sum of the squares of the lengths of the two shorter sides: $c^2 = a^2 + b^2$.

EXAMPLE 9 A Quantity–Rate–Time Problem

A payroll can be completed in 4 hours by two computers working simultaneously. How many hours are required for each computer to complete the payroll alone if the older model requires 3 hours longer than the newer model? Compute answers to two decimal places.

Solution Let

$$x = \text{Time for new model to complete the payroll alone}$$

$$x + 3 = \text{Time for old model to complete the payroll alone}$$

$$4 = \text{Time for both computers to complete the payroll together}$$

Then,

$$\frac{1}{x} = \text{Rate for new model} \qquad \text{Completes } \frac{1}{x} \text{ of the payroll per hour}$$

$$\frac{1}{x+3} = \text{Rate for old model} \qquad \text{Completes } \frac{1}{x+3} \text{ of the payroll per hour}$$

$$\begin{pmatrix}\text{Part of job}\\ \text{completed by}\\ \text{new model in}\\ \text{4 hours}\end{pmatrix} + \begin{pmatrix}\text{Part of job}\\ \text{completed by}\\ \text{old model in}\\ \text{4 hours}\end{pmatrix} = 1 \text{ whole job}$$

$$\frac{1}{x}(4) + \frac{1}{x+3}(4) = 1 \qquad x \neq 0,\ x \neq -3$$

$$\frac{4}{x} + \frac{4}{x+3} = 1$$

$$4(x + 3) + 4x = x(x + 3) \qquad \text{Multiply both sides by } x(x + 3).$$

$$4x + 12 + 4x = x^2 + 3x$$

$$x^2 - 5x - 12 = 0$$

$$x = \frac{5 \pm \sqrt{73}}{2}$$

$$x = \frac{5 + \sqrt{73}}{2} \approx 6.77 \qquad \frac{5 - \sqrt{73}}{2} \approx -1.77 \text{ is discarded since } x \text{ cannot be negative.}$$

$$x + 3 = 9.77$$

The new model would complete the payroll in 6.77 hours working alone, and the old model would complete the payroll in 9.77 hours working alone.

Check

$$\frac{1}{6.77}(4) + \frac{1}{9.77}(4) \stackrel{?}{=} 1$$

$$1.000\,259 \stackrel{\checkmark}{\approx} 1$$

Note: We do not expect the check to be exact, since we rounded the answers to two decimal places. An exact check would be produced by using $x = (5 + \sqrt{73})/2$. The latter is left to the reader.

Matched Problem 9 Two technicians can complete a mailing in 3 hours when working together. Alone, one can complete the mailing 2 hours faster than the other. How long will it take each person to complete the mailing alone? Compute the answers to two decimal places.

Answers to Matched Problems

1. (A) $x = -4, \frac{5}{3}$ (B) $x = -\frac{3}{2}$ (a double root) (C) $x = 0, \frac{5}{4}$
2. (A) $x = \pm\sqrt{\frac{5}{3}}$ or $\pm\sqrt{15}/3$ (B) $x = \pm 2i$ (C) $x = (-1 \pm \sqrt{2})/3$
3. (A) $x^2 - 5x + \frac{25}{4} = (x - \frac{5}{2})^2$ (B) $x^2 + mx + (m^2/4) = [x + (m/2)]^2$
4. (A) $x = -4 \pm \sqrt{19}$ (B) $x = (6 \pm i\sqrt{3})/3$ or $2 \pm (\sqrt{3}/3)i$ **5.** $x = (-3 \pm \sqrt{19})/2$
6. $x = -2.80, 0.98$ **7.** 11 and 12 **8.** 15 and 20 miles per hour **9.** 5.16 and 7.16 hours

EXERCISE 1-6

Leave all answers involving radicals in simplified radical form unless otherwise stated.

A

In Problems 1–6, solve by factoring.

1. $2x^2 = 8x$ **2.** $2y^2 + 5y = 3$

3. $4t^2 + 9 = 12t$ **4.** $3s^2 = -6s$

5. $3w^2 + 13w = 10$ **6.** $16x^2 + 9 = 24x$

In Problems 7–18, solve by using the square root property.

7. $m^2 - 25 = 0$ **8.** $n^2 + 16 = 0$

9. $c^2 + 9 = 0$ **10.** $d^2 - 36 = 0$

11. $4y^2 + 9 = 0$ **12.** $9x^2 - 25 = 0$

13. $25z^2 - 32 = 0$ **14.** $16w^2 + 27 = 0$

15. $(s + 1)^2 = 5$ **16.** $(t - 2)^2 = -3$

17. $(n - 3)^2 = -4$ **18.** $(m + 4)^2 = 1$

In Problems 19–26, solve using the quadratic formula.

19. $x^2 - 2x - 1 = 0$ **20.** $y^2 - 4y + 7 = 0$

21. $x^2 - 2x + 3 = 0$ **22.** $y^2 - 4y + 1 = 0$

23. $2t^2 + 8 = 6t$ **24.** $9s^2 + 2 = 12s$

25. $2t^2 + 1 = 6t$ **26.** $9s^2 + 7 = 12s$

B

In Problems 27–34, solve by completing the square.

27. $x^2 - 4x - 1 = 0$ **28.** $y^2 + 4y - 3 = 0$

29. $2r^2 + 10r + 11 = 0$ **30.** $2s^2 - 6s + 7 = 0$

31. $4u^2 + 8u + 15 = 0$ **32.** $4v^2 + 16v + 23 = 0$

33. $3w^2 + 4w + 3 = 0$ **34.** $3z^2 - 8z + 1 = 0$

In Problems 35–52, solve by any method.

35. $12x^2 + 7x = 10$ **36.** $9x^2 + 9x = 4$

37. $(2y - 3)^2 = 5$ **38.** $(3m + 2)^2 = -4$

39. $x^2 = 3x + 1$ **40.** $x^2 + 2x = 2$

41. $7n^2 = -4n$ **42.** $8u^2 + 3u = 0$

43. $1 + \frac{8}{x^2} = \frac{4}{x}$ **44.** $\frac{2}{u} = \frac{3}{u^2} + 1$

45. $\frac{24}{10 + m} + 1 = \frac{24}{10 - m}$ **46.** $\frac{1.2}{y - 1} + \frac{1.2}{y} = 1$

47. $\frac{2}{x - 2} = \frac{4}{x - 3} - \frac{1}{x + 1}$ **48.** $\frac{3}{x - 1} - \frac{2}{x + 3} = \frac{4}{x - 2}$

49. $\frac{x+2}{x+3} - \frac{x^2}{x^2-9} = 1 - \frac{x-1}{3-x}$

50. $\frac{11}{x^2-4} + \frac{x+3}{2-x} = \frac{2x-3}{x+2}$

51. $|3u - 2| = u^2$

52. $|12 + 7x| = x^2$

In Problems 53–56, solve for the indicated variable in terms of the other variables. Use positive square roots only.

53. $s = \frac{1}{2}gt^2$ for t

54. $a^2 + b^2 = c^2$ for a

55. $P = EI - RI^2$ for I

56. $A = P(1 + r)^2$ for r

Solve Problems 57–60 to two decimal places using a calculator.

57. $2.07x^2 - 3.79x + 1.34 = 0$

58. $0.61x^2 - 4.28x + 2.93 = 0$

59. $4.83x^2 + 2.04x - 3.18 = 0$

60. $5.13x^2 + 7.27x - 4.32 = 0$

61. Consider the quadratic equation

$$x^2 + 4x + c = 0$$

where c is a real number. Discuss the relationship between the values of c and the three types of roots listed in Table 1.

62. Consider the quadratic equation

$$x^2 - 2x + c = 0$$

where c is a real number. Discuss the relationship between the values of c and the three types of roots listed in Table 1.

Use the discriminant to determine whether the equations in Problems 63–66 have real solutions.

63. $0.0134x^2 + 0.0414x + 0.0304 = 0$

64. $0.543x^2 - 0.182x + 0.003\ 12 = 0$

65. $0.0134x^2 + 0.0214x + 0.0304 = 0$

66. $0.543x^2 - 0.182x + 0.0312 = 0$

C

Solve Problems 67–70 and leave answers in simplified radical form (i is the imaginary unit).

67. $\sqrt{3}x^2 = 8\sqrt{2}x - 4\sqrt{3}$

68. $2\sqrt{2}x + \sqrt{3} = \sqrt{3}x^2$

69. $x^2 + 2ix = 3$

70. $x^2 = 2ix - 3$

In Problems 71 and 72, find all solutions.

71. $x^3 - 1 = 0$

72. $x^4 - 1 = 0$

73. Can a quadratic equation with rational coefficients have one rational root and one irrational root? Explain.

74. Can a quadratic equation with real coefficients have one real root and one imaginary root? Explain.

75. Show that if r_1 and r_2 are the two roots of $ax^2 + bx + c = 0$, then $r_1r_2 = c/a$.

76. For r_1 and r_2 in Problem 75, show that $r_1 + r_2 = -b/a$.

77. In one stage of the derivation of the quadratic formula, we replaced the expression

$$\pm\sqrt{(b^2 - 4ac)/4a^2}$$

with

$$\pm\sqrt{b^2 - 4ac}/2a$$

What justifies using $2a$ in place of $|2a|$?

78. Find the error in the following "proof" that two arbitrary numbers are equal to each other: Let a and b be arbitrary numbers such that $a \neq b$. Then

$$(a - b)^2 = a^2 - 2ab + b^2 = b^2 - 2ab + a^2$$
$$(a - b)^2 = (b - a)^2$$
$$a - b = b - a$$
$$2a = 2b$$
$$a = b$$

APPLICATIONS 

79. Numbers. Find two numbers such that their sum is 21 and their product is 104.

80. Numbers. Find all numbers with the property that when the number is added to itself the sum is the same as when the number is multiplied by itself.

81. Numbers. Find two consecutive positive even integers whose product is 168.

82. Numbers. The sum of a number and its reciprocal is $\frac{10}{3}$. Find the number.

83. Geometry. If the length and width of a 4- by 2-inch rectangle are each increased by the same amount, the area of the new rectangle will be twice that of the original. What are the dimensions of the new rectangle (to two decimal places)?

84. Geometry. Find the base b and height h of a triangle with an area of 2 square feet if its base is 3 feet longer than its height and the formula for area is $A = \frac{1}{2}bh$.

85. Business. If \$$P$ are invested at an interest rate r compounded annually, at the end of 2 years the amount will be $A = P(1 + r)^2$. At what interest rate will \$1,000 increase to \$1,440 in 2 years? [*Note:* $A =$ \$1,440 and $P =$ \$1,000.]

★ **86. Economics.** In a certain city, the price–demand and price–supply equations for CDs are

$$p = \frac{75{,}000}{q} \qquad \text{Demand equation}$$

$$p = 0.0005q + 12.5 \qquad \text{Supply equation}$$

where q represents quantity and p represents the price in dollars. Find the equilibrium price.

87. Puzzle. Two planes travel at right angles to each other after leaving the same airport at the same time. One hour later they are 260 miles apart. If one travels 140 miles per hour faster than the other, what is the rate of each?

88. Navigation. A speedboat takes 1 hour longer to go 24 miles up a river than to return. If the boat cruises at 10 miles per hour in still water, what is the rate of the current?

★ **89. Engineering.** One pipe can fill a tank in 5 hours less than another. Together they can fill the tank in 5 hours. How long would it take each alone to fill the tank? Compute the answer to two decimal places.

★★ **90. Engineering.** Two gears rotate so that one completes 1 more revolution per minute than the other. If it takes the smaller gear 1 second less than the larger gear to complete $\frac{1}{5}$ revolution, how many revolutions does each gear make in 1 minute?

★ **91. Physics—Engineering.** For a car traveling at a speed of v miles per hour, under the best possible conditions the shortest distance d necessary to stop it (including reaction time) is given by the empirical formula $d = 0.044v^2 + 1.1v$, where d is measured in feet. Estimate the speed of a car that requires 165 feet to stop in an emergency.

★ **92. Physics—Engineering.** If a projectile is shot vertically into the air (from the ground) with an initial velocity of 176 feet per second, its distance y (in feet) above the ground t seconds after it is shot is given by $y = 176t - 16t^2$ (neglecting air resistance).

(A) Find the times when y is 0, and interpret the results physically.

(B) Find the times when the projectile is 16 feet off the ground. Compute answers to two decimal places.

★ **93. Construction.** A developer wants to erect a rectangular building on a triangular-shaped piece of property that is 200 feet wide and 400 feet long (see the figure). Find the dimensions of the building if its cross-sectional area is 15,000 square feet. [*Hint:* Use Euclid's theorem* to find a relationship between the length and width of the building.]

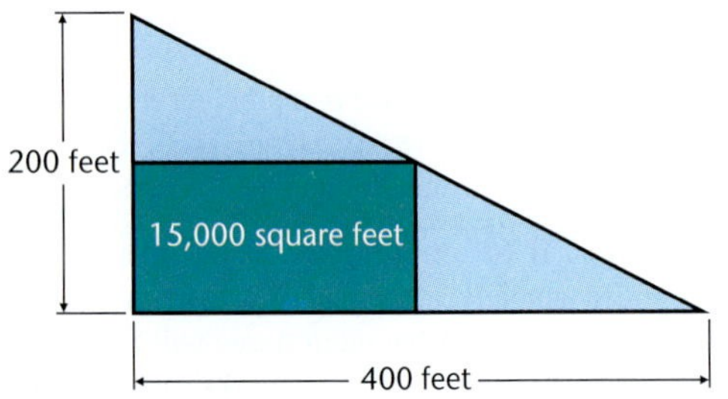

★ **94. Architecture.** An architect is designing a small A-frame cottage for a resort area. A cross section of the cottage is an isosceles triangle with an area of 98 square feet. The front wall of the cottage must accommodate a sliding door that is 6 feet wide and 8 feet high (see the figure). Find the width and height of the cross section of the cottage. [*Recall:* The area of a triangle with base b and altitude h is $bh/2$.]

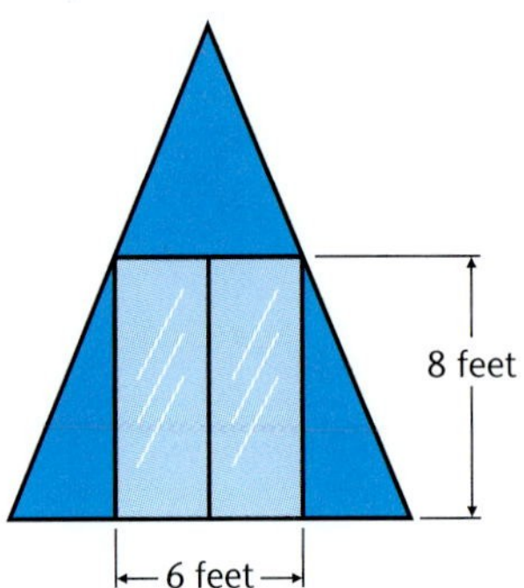

95. Transportation. A delivery truck leaves a warehouse and travels north to factory A. From factory A the truck travels east to factory B and then returns directly to the warehouse (see the figure). The driver recorded the truck's odometer reading at the warehouse at both the beginning and the end of the trip and also at factory B, but forgot to record it at factory A (see the table). The driver does recall that it was further from the warehouse to factory A than it was from factory A to factory B. Since delivery charges are based on distance from the warehouse, the driver needs to know how far factory A is from the warehouse. Find this distance.

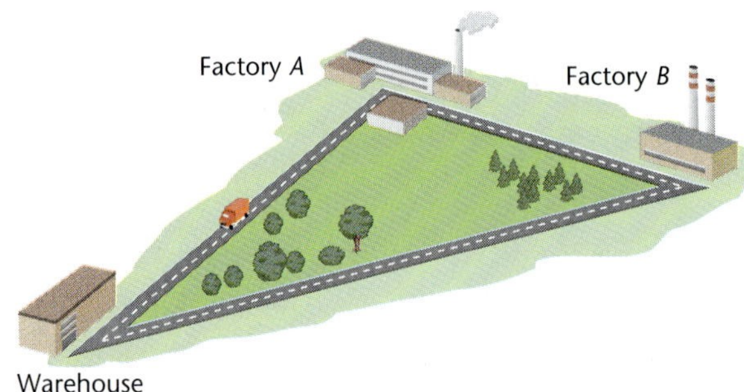

**Euclid's theorem:* If two triangles are similar, their corresponding sides are proportional:

$$\frac{a}{a'} = \frac{b}{b'} = \frac{c}{c'}$$

Odometer Readings	
Warehouse	5 2 8 4 6
Factory A	5 2 ? ? ?
Factory B	5 2 9 3 7
Warehouse	5 3 0 0 2

★★ **96. Construction.** A $\frac{1}{4}$-mile track for racing stock cars consists of two semicircles connected by parallel straightaways (see the figure). In order to provide sufficient room for pit crews, emergency vehicles, and spectator parking, the track must enclose an area of 100,000 square feet. Find the length of the straightaways and the diameter of the semicircles to the nearest foot. [*Recall:* The area A and circumference C of a circle of diameter d are given by $A = \pi d^2/4$ and $C = \pi d$.]

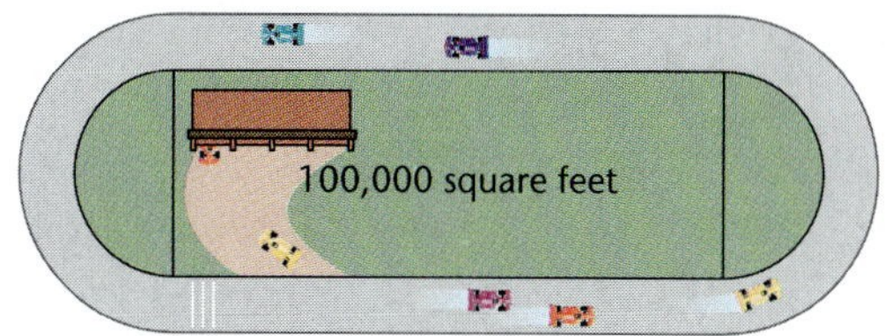

SECTION 1-7 Equations Reducible to Quadratic Form

- Equations Involving Radicals
- Equations Involving Rational Exponents

• Equations Involving Radicals

In solving an equation involving a radical like

$$x = \sqrt{x + 2}$$

it appears that we can remove the radical by squaring each side and then proceed to solve the resulting quadratic equation. Thus,

$$\begin{aligned} x^2 &= (\sqrt{x + 2})^2 \\ x^2 &= x + 2 \\ x^2 - x - 2 &= 0 \\ (x - 2)(x + 1) &= 0 \\ x &= 2, -1 \end{aligned}$$

Now we check these results in the original equation.

Check: $x = 2$	Check: $x = -1$
$x = \sqrt{x + 2}$	$x = \sqrt{x + 2}$
$\mathbf{2} \stackrel{?}{=} \sqrt{\mathbf{2} + 2}$	$\mathbf{-1} \stackrel{?}{=} \sqrt{\mathbf{-1} + 2}$
$2 \stackrel{?}{=} \sqrt{4}$	$-1 \stackrel{?}{=} \sqrt{1}$
$2 \stackrel{\checkmark}{=} 2$	$-1 \neq 1$

Thus, 2 is a solution, but -1 is not. These results are a special case of Theorem 1.

Theorem 1 **Squaring Operation on Equations**

If both sides of an equation are squared, then the solution set of the original equation is a subset of the solution set of the new equation.

Equation	Solution Set
$x = 3$	$\{3\}$
$x^2 = 9$	$\{-3, 3\}$

This theorem provides us with a method of solving some equations involving radicals. It is important to remember that any new equation obtained by raising both members of an equation to the same power may have solutions, called **extraneous solutions,** that are not solutions of the original equation. On the other hand, any solution of the original equation must be among those of the new equation.

Every solution of the new equation must be checked in the original equation to eliminate extraneous solutions.

EXPLORE-DISCUSS 1 Squaring both sides of the equations $x = \sqrt{x}$ and $x = -\sqrt{x}$ produces the new equation $x^2 = x$. Find the solutions to the new equation and then check for extraneous solutions in each of the original equations.

CAUTION Remember that $\sqrt{9}$ represents the *positive* square root of 9 and $-\sqrt{9}$ represents the *negative* square root of 9. It is correct to use the symbol $\pm$ to combine these two roots when solving an equation:

$$x^2 = 9 \quad \text{implies} \quad x = \pm\sqrt{9} = \pm 3$$

But it is incorrect to use $\pm$ when evaluating the positive square root of a number:

$$\sqrt{9} \neq \pm 3 \qquad \sqrt{9} = 3$$

EXAMPLE 1 **Solving Equations Involving Radicals**

Solve:

(A) $x + \sqrt{x - 4} = 4$ (B) $\sqrt{2x + 3} - \sqrt{x - 2} = 2$

Solutions (A) $x + \sqrt{x - 4} = 4$

$\sqrt{x - 4} = 4 - x$ Isolate radical on one side.

$x - 4 = 16 - 8x + x^2$ Square both sides.

$$x^2 - 9x + 20 = 0$$
$$(x - 5)(x - 4) = 0$$
$$x = 5, 4$$

Check

$x = 5$	$x = 4$
$x + \sqrt{x - 4} = 4$	$x + \sqrt{x - 4} = 4$
$\mathbf{5} + \sqrt{\mathbf{5} - 4} \stackrel{?}{=} 4$	$\mathbf{4} + \sqrt{\mathbf{4} - 4} \stackrel{?}{=} 4$
$6 \neq 4$	$4 \stackrel{\checkmark}{=} 4$

This shows that 4 is a solution to the original equation and 5 is extraneous. Thus,

$$x = 4$$ Only one solution

(B) To solve an equation that contains more than one radical, isolate one radical at a time and square both sides to eliminate the isolated radical. Repeat this process until all the radicals are eliminated.

$$\sqrt{2x + 3} - \sqrt{x - 2} = 2$$
$$\sqrt{2x + 3} = \sqrt{x - 2} + 2$$ Isolate one of the radicals.
$$2x + 3 = x - 2 + 4\sqrt{x - 2} + 4$$ Square both sides.
$$x + 1 = 4\sqrt{x - 2}$$ Isolate the remaining radical.
$$x^2 + 2x + 1 = 16(x - 2)$$ Square both sides.
$$x^2 - 14x + 33 = 0$$
$$(x - 3)(x - 11) = 0$$
$$x = 3, 11$$

Check

$x = 3$	$x = 11$
$\sqrt{2x + 3} - \sqrt{x - 2} = 2$	$\sqrt{2x + 3} - \sqrt{x - 2} = 2$
$\sqrt{2(\mathbf{3}) + 3} - \sqrt{\mathbf{3} - 2} \stackrel{?}{=} 2$	$\sqrt{2(\mathbf{11}) + 3} - \sqrt{\mathbf{11} - 2} \stackrel{?}{=} 2$
$2 \stackrel{\checkmark}{=} 2$	$2 \stackrel{\checkmark}{=} 2$

Both solutions check. Thus,

$$x = 3, 11$$ Two solutions

Matched Problem 1 Solve:

(A) $x - 5 = \sqrt{x - 3}$ (B) $\sqrt{2x + 5} + \sqrt{x + 2} = 5$

CAUTION

When squaring an expression like $\sqrt{x-2}+2$, be certain to correctly apply the formula for squaring the sum of two terms (see Section A-2):

$$\begin{aligned}(u+v)^2 &= u^2 + 2uv + v^2\\ (\sqrt{x-2}+2)^2 &= (\sqrt{x-2})^2 + 2(\sqrt{x-2})(2) + (2)^2\\ &= x-2+4\sqrt{x-2}+4\end{aligned}$$

Do not omit the middle term in this product:

$$(\sqrt{x-2}+2)^2 \neq x-2+4$$

• Equations Involving Rational Exponents

To solve the equation

$$x^{2/3} - x^{1/3} - 6 = 0$$

write it in the form

$$(x^{1/3})^2 - x^{1/3} - 6 = 0$$

You can now recognize that the equation is quadratic in $x^{1/3}$. So, we solve for $x^{1/3}$ first, and then solve for x. We can solve the equation directly or make the substitution $u = x^{1/3}$, solve for u, and then solve for x. Both methods of solution are shown below.

Method I. Direct solution:

$$(x^{1/3})^2 - x^{1/3} - 6 = 0$$

$$(x^{1/3} - 3)(x^{1/3} + 2) = 0 \qquad \text{Factor left side.}$$

$$x^{1/3} = 3 \quad \text{or} \quad x^{1/3} = -2$$

$$(x^{1/3})^3 = 3^3 \qquad (x^{1/3})^3 = (-2)^3 \qquad \text{Cube both sides.}$$

$$x = 27 \qquad x = -8$$

Solution set: $\{-8, 27\}$

Method II. Using substitution:
Let $u = x^{1/3}$, solve for u, and then solve for x.

$$u^2 - u - 6 = 0$$

$$(u-3)(u+2) = 0$$

$$u = 3, -2$$

Replacing u with $x^{1/3}$, we obtain

$$x^{1/3} = 3 \quad \text{or} \quad x^{1/3} = -2$$

$$x = 27 \qquad x = -8$$

Solution set: $\{-8, 27\}$

In general, if an equation that is not quadratic can be transformed to the form

$$au^2 + bu + c = 0$$

where u is an expression in some other variable, then the equation is called an **equation of quadratic type.** Once recognized as an equation of quadratic type, an equation often can be solved using quadratic methods.

EXPLORE-DISCUSS 2 Which of the following can be transformed into an equation of quadratic type by making a substitution of the form $u = x^n$?

(A) $3x^{-4} + 2x^{-2} + 7$ (B) $7x^5 - 3x^2 + 3$
(C) $2x^5 + 4x^2\sqrt{x} - 6$ (D) $8x^{-2}\sqrt{x} - 5x^{-1}\sqrt{x} - 2$

In general, if a, b, c, m, and n are nonzero real numbers, when can an expression of the form $ax^m + bx^n + c$ be transformed into an equation of quadratic type?

EXAMPLE 2 Solving Equations of Quadratic Type

Solve:

(A) $x^4 - 3x^2 - 4 = 0$
(B) $3x^{-2/5} - 6x^{-1/5} + 2 = 0$

Solutions (A) The equation is quadratic in x^2. We solve for x^2 and then for x:

$$(x^2)^2 - 3x^2 - 4 = 0 \quad \text{Quadratic in } x^2$$

$$(x^2 - 4)(x^2 + 1) = 0 \quad \text{Factor the left side.}$$

$$x^2 = 4 \quad \text{or} \quad x^2 = -1$$

$$x = \pm 2 \quad \text{or} \quad x = \pm i$$

Solution set: $\{-2, 2, -i, i\}$

Since we did not raise each side of the equation to a natural number power, we do not have to check for extraneous solutions. (You should still check the accuracy of the solutions.)

(B) The equation $3x^{-2/5} - 6x^{-1/5} + 2 = 0$ is quadratic in $x^{-1/5}$. We substitute $u = x^{-1/5}$ and solve for u:

$$3u^2 - 6u + 2 = 0$$

$$u = \frac{6 \pm \sqrt{12}}{6} = \frac{3 \pm \sqrt{3}}{3} \quad \text{Use the quadratic formula.}$$

$$x = u^{-5}$$

$$= \left(\frac{3 \pm \sqrt{3}}{3}\right)^{-5}$$

Thus, the two solutions are

$$x = \left(\frac{3}{3 \pm \sqrt{3}}\right)^5 \quad \text{Two real solutions}$$

Matched Problem 2 Solve:

(A) $x^4 + 3x^2 - 4 = 0$ (B) $3x^{-2/5} - x^{-1/5} - 2 = 0$

EXAMPLE 3 Setting Up and Solving a Word Problem

The diagonal of a rectangle is 10 inches, and the area is 45 square inches. Find the dimensions of the rectangle correct to one decimal place.

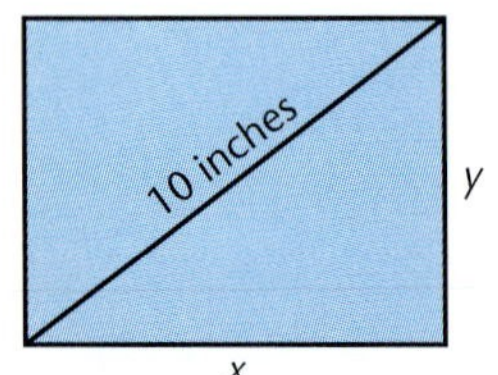

FIGURE 1

Solution Draw a rectangle and label the dimensions as shown in Figure 1. From the Pythagorean theorem,

$$x^2 + y^2 = 10^2$$

Thus,

$$y = \sqrt{100 - x^2}$$

Since the area of the rectangle is given by xy, we have

$$x\sqrt{100 - x^2} = 45 \quad \text{Area of the rectangle}$$

$$x^2(100 - x^2) = 2{,}025 \quad \text{Square both sides.}$$

$$100x^2 - x^4 = 2{,}025$$

$$(x^2)^2 - 100x^2 + 2{,}025 = 0 \quad \text{Quadratic in } x^2$$

$$x^2 = \frac{100 \pm \sqrt{100^2 - 4(1)(2{,}025)}}{2}$$

$$x^2 = 50 \pm 5\sqrt{19}$$

$$x = \sqrt{50 \pm 5\sqrt{19}} \quad \text{Discard the negative solutions since } x > 0.$$

If $x = \sqrt{50 + 5\sqrt{19}} \approx 8.5$, then

$$\begin{aligned} y &= \sqrt{100 - x^2} \\ &= \sqrt{100 - (50 + 5\sqrt{19})} \\ &= \sqrt{50 - 5\sqrt{19}} \approx 5.3 \end{aligned}$$

Thus, the dimensions of the rectangle to one decimal place are 8.5 inches by 5.3 inches. Notice that if $x = \sqrt{50 - 5\sqrt{19}}$, then $y = \sqrt{50 + 5\sqrt{19}}$, and the dimensions are still 8.5 inches by 5.3 inches.

Check Area: $(8.5)(5.3) = 45.05 \approx 45$

Diagonal: $\sqrt{8.5^2 + 5.3^2} = \sqrt{100.34} \approx 10$

Note: An exact check can be obtained by using $\sqrt{50 - 5\sqrt{19}}$ and $\sqrt{50 + 5\sqrt{19}}$ in place of these decimal approximations. This is left to the reader.

Matched Problem 3 If the area of a right triangle is 24 square inches and the hypotenuse is 12 inches, find the lengths of the legs of the triangle correct to one decimal place.

Answers to Matched Problems

1. (A) $x = 7$ (B) $x = 2$

2. (A) $x = \pm 1, \pm 2i$ (B) $x = 1, -\frac{243}{32}$

3. 11.2 inches by 4.3 inches

EXERCISE 1-7

In Problems 1–6, determine the validity of each statement. If a statement is false, explain why.

A

1. If $x^2 = 5$, then $x = \pm\sqrt{5}$.

2. $\sqrt{25} = \pm 5$

3. $(\sqrt{x-1} + 1)^2 = x$

4. $(\sqrt{x-1})^2 + 1 = x$

5. If $x^3 = 2$, then $x = 8$.

6. If $x^{1/3} = 8$, then $x = 2$.

Solve:

7. $\sqrt{x+2} = 4$

8. $\sqrt{x-4} = 2$

9. $\sqrt{3y-2} = y - 2$

10. $\sqrt{4y+1} = 5 - y$

11. $\sqrt{5w+6} - w = 2$

12. $\sqrt{2w-3} + w = 1$

13. $\sqrt{3t-2} = 1 - 2\sqrt{t}$

14. $\sqrt{5t+4} - 2\sqrt{t} = 1$

15. $m^4 + 2m^2 - 15 = 0$

16. $m^4 + 4m^2 - 12 = 0$

17. $3x = \sqrt{x^2 - 2}$

18. $x = \sqrt{5x^2 + 9}$

19. $2y^{2/3} + 5y^{1/3} - 12 = 0$

20. $3y^{2/3} + 2y^{1/3} - 8 = 0$

21. $(m^2 - 2m)^2 + 2(m^2 - 2m) = 15$

22. $(m^2 + 2m)^2 - 6(m^2 + 2m) = 16$

B

Solve:

23. $\sqrt{2t+3} + 2 = \sqrt{t-2}$

24. $\sqrt{2x-1} - \sqrt{x-5} = 3$

25. $\sqrt{w+3} + \sqrt{2-w} = 3$

26. $\sqrt{w+7} = 2 + \sqrt{3-w}$

27. $\sqrt{8-z} = 1 + \sqrt{z+5}$

28. $\sqrt{3z+1} + 2 = \sqrt{z-1}$

29. $\sqrt{4x^2 + 12x + 1} - 6x = 9$

30. $6x - \sqrt{4x^2 - 20x + 17} = 15$

31. $y^{-2} - 2y^{-1} + 3 = 0$

32. $y^{-2} - 3y^{-1} + 4 = 0$

33. $2t^{-4} - 5t^{-2} + 2 = 0$

34. $15t^{-4} - 23t^{-2} + 4 = 0$

35. $3z^{-1} - 3z^{-1/2} + 1 = 0$

36. $2z^{-1} - 3z^{-1/2} + 2 = 0$

Solve Problems 37–40 two ways: by squaring and by substitution.

37. $4m + 8\sqrt{m} - 5 = 0$

38. $4m + 8\sqrt{m} - 21 = 0$

39. $2w + 3\sqrt{w} = 14$

40. $3w + 5\sqrt{w} = 12$

C

Solve:

41. $\sqrt{7-2x} - \sqrt{x+2} = \sqrt{x+5}$

42. $\sqrt{1+3x} - \sqrt{2x-1} = \sqrt{x+2}$

43. $3 + x^{-4} = 5x^{-2}$

44. $2 + 4x^{-4} = 7x^{-2}$

45. $2\sqrt{x+5} = 0.01x + 2.04$

46. $3\sqrt{x-1} = 0.05x + 2.9$

47. $2x^{-2/5} - 5x^{-1/5} + 1 = 0$

48. $x^{-2/5} - 3x^{-1/5} + 1 = 0$

APPLICATIONS

49. Manufacturing. A lumber mill cuts rectangular beams from circular logs (see the figure). If the diameter of the log is 16 inches and the cross-sectional area of the beam is 120 square inches, find the dimensions of the cross section of the beam correct to one decimal place.

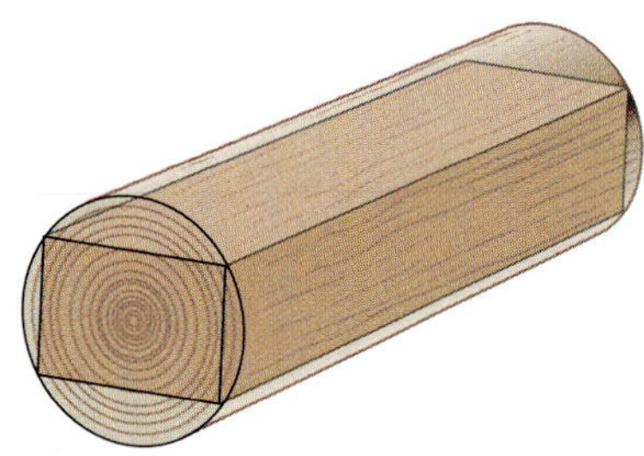

50. Design. A food-processing company packages an assortment of their products in circular metal tins 12 inches in diameter. Four identically sized rectangular boxes are used to divide the tin into eight compartments (see the figure). If the cross-sectional area of each box is 15 square inches, find the dimensions of the boxes correct to one decimal place.

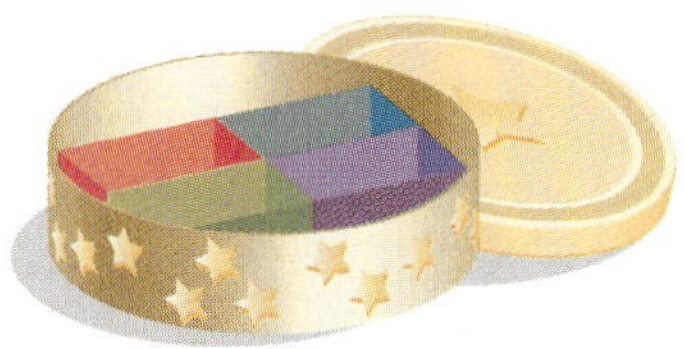

★ **51. Construction.** A water trough is constructed by bending a 4- by 6-foot rectangular sheet of metal down the middle and attaching triangular ends (see the figure). If the volume of the trough is 9 cubic feet, find the width correct to two decimal places.

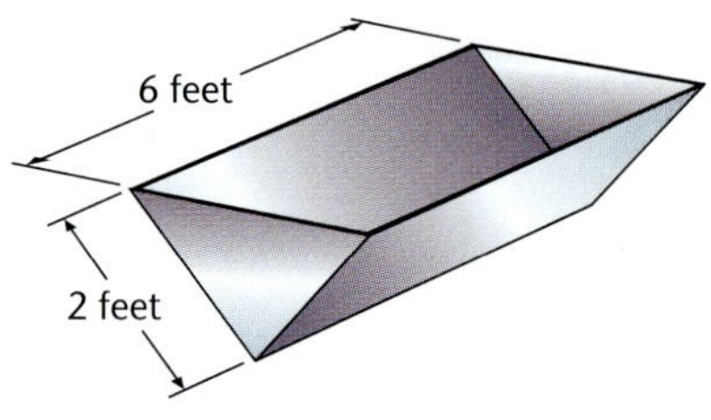

★ **52. Design.** A paper drinking cup in the shape of a right circular cone is constructed from 125 square centimeters of paper (see the figure). If the height of the cone is 10 centimeters, find the radius correct to two decimal places.

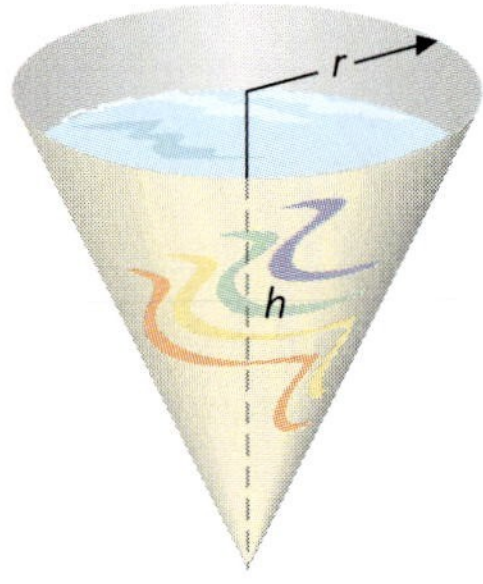

Lateral surface area:

$S = \pi r\sqrt{r^2 + h^2}$

53. Supply and Demand. The weekly supply and demand equations for a certain brand of telephones are given by

$p = 14 + 0.01q$ Supply equation

$p = 50 - 0.5\sqrt{q}$ Demand equation

where q is the number of telephones and $\$p$ is the price. Find the equilibrium price and the equilibrium quantity.

SECTION 1-8 Polynomial and Rational Inequalities

- Polynomial Inequalities
- Rational Inequalities

In this section we solve fairly simple polynomial and rational inequalities of the form

$$2x^2 - 3x - 4 < 0 \qquad \text{and} \qquad \frac{x - 2}{x^2 - x - 3} \geq 0$$

Even though we limit the discussion to quadratic inequalities and rational inequalities with numerators and denominators of degree 2 or less (you will see why below), the theory presented applies to polynomial and rational inequalities in general. In Chapter 3, with additional theory, we will be able to use the methods developed here to solve polynomial and rational inequalities of a more general nature. Also, the process—with only slight modification of key theorems—applies to other forms encountered in calculus.

Why so much interest in solving inequalities? Most significant applications of mathematics involve more use of inequalities than equalities. In the real world few things are exact.

• Polynomial Inequalities

We know how to solve linear inequalities such as

$$3x - 7 \geq 5(x - 2) + 3$$

But how do we solve quadratic (or higher-degree polynomial) inequalities such as the one given below?

$$x^2 + 2x < 8 \tag{1}$$

We first write the inequality in **standard form;** that is, we transfer all nonzero terms to the left side, leaving only 0 on the right side:

$$x^2 + 2x - 8 < 0 \quad \text{Standard form} \tag{2}$$

In this example, we are looking for values of x that will make the quadratic on the left side less than 0—that is, negative.

The following theorem provides the basis for an effective way of solving this problem. Theorem 1 makes direct use of the *real zeros* of the polynomial on the left side of inequality (2). **Real zeros** of a polynomial are those real numbers that make the polynomial equal to 0—that is, the real roots of the corresponding polynomial equation. If a polynomial has one or more real zeros, then plotting these zeros on a real number line divides the line into two or more intervals.

Theorem 1 **Sign of a Polynomial over a Real Number Line**

A nonzero polynomial will have a constant sign (either always positive or always negative) within each interval determined by its real zeros plotted on a number line. If a polynomial has no real zeros, then the polynomial is either positive over the whole real number line or negative over the whole real number line.

We now complete the solution of inequality (1) using Theorem 1. After writing (1) in standard form, as we did in inequality (2), we find the real zeros of the polynomial on the left side by solving the corresponding polynomial equation:

$$x^2 + 2x - 8 = 0 \qquad \text{Can be solved by factoring}$$

$$(x - 2)(x + 4) = 0$$

$$x = -4, 2 \qquad \text{Real zeros of the polynomial } x^2 + 2x - 8$$

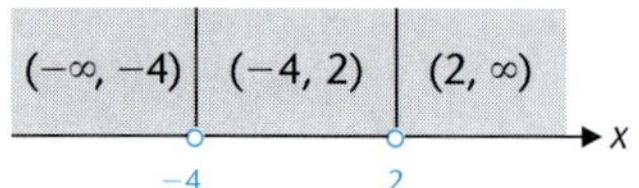

FIGURE 1 Real zeros of $x^2 + 2x - 8$.

Next, we plot the real zeros, -4 and 2, on a number line (Fig. 1) and note that three intervals are determined: $(-\infty, -4)$, $(-4, 2)$, and $(2, \infty)$.

From Theorem 1 we know that the polynomial has constant sign on each of these three intervals. If we select a **test number** in each interval and evaluate the polynomial at that number, then the sign of the polynomial at this test number must be the sign for the whole interval. Since any number within an interval can be used as a test number, we generally choose test numbers that result in easy computations. In this example, we choose -5, 0, and 3. Table 1 shows the computations.

TABLE 1 **Polynomial: $x^2 + 2x - 8 = (x - 2)(x + 4)$**

Test number	-5	0	3
Value of polynomial for test number	7	-8	7
Sign of polynomial in interval containing test number	+	−	+
Interval containing test number	$(-\infty, -4)$	$(-4, 2)$	$(2, \infty)$

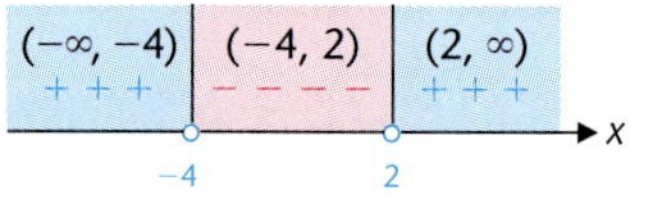

FIGURE 2 Sign chart for $x^2 + 2x - 8$.

Using the information in Table 1, we construct a **sign chart** for the polynomial, as shown in Figure 2.

Thus, $x^2 + 2x - 8$ is negative within the interval $(-4, 2)$, and we have solved the inequality. The solution and graph are given below and in Figure 3:

$$-4 < x < 2 \qquad \text{Inequality notation}$$

$$(-4, 2) \qquad \text{Interval notation}$$

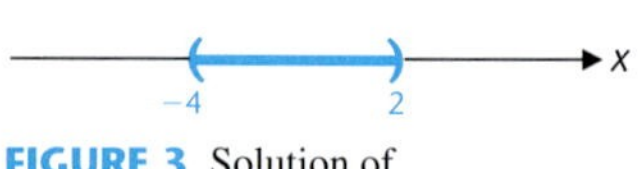

FIGURE 3 Solution of $x^2 + 2x < 8$.

Note: If $<$ in the original problem had been $\leq$ instead, then we would have included the zeros of the polynomial in the solution set.

The steps in the above process are summarized in the following box:

Key Steps in Solving Polynomial Inequalities

Step 1. Write the polynomial inequality in standard form (a form where the right-hand side is 0).

Step 2. Find all real zeros of the polynomial (the left side of the standard form).

Step 3. Plot the real zeros on a number line, dividing the number line into intervals.

Step 4. Choose a test number (that is easy to compute with) in each interval, and evaluate the polynomial for each number (a small table is useful).

Step 5. Use the results of step 4 to construct a sign chart, showing the sign of the polynomial in each interval.

Step 6. From the sign chart, write down the solution of the original polynomial inequality (and draw the graph, if required).

With a little experience, many of the above steps can be combined and the process streamlined to two or three key operational steps. The critical part of the method is step 2, finding all real zeros of the polynomial. At this point we can find all real zeros of any quadratic polynomial (see Section 1-6). Finding real zeros of higher-degree polynomials is more difficult, and the process is considered in detail in Chapter 3.

EXPLORE-DISCUSS 1 We can solve a quadratic equation by factoring the quadratic polynomial and setting each factor equal to 0, as we did in the preceding example. Can we solve quadratic inequalities the same way? That is, can we solve

$$(x - 2)(x + 4) < 0$$

by considering linear inequalities involving the factors $x - 2$ and $x + 4$? Discuss how you could arrive at the correct solution, $-4 < x < 2$, by considering various combinations of

$$x - 2 < 0 \qquad x - 2 > 0 \qquad x + 4 < 0 \qquad x + 4 > 0$$

We now turn to a significant application that involves a polynomial inequality.

EXAMPLE 1 Profit and Loss Analysis

A company manufactures and sells flashlights. For a particular model, the marketing research and financial departments estimate that at a price of $\$p$ per unit, the weekly cost C and revenue R (in thousands of dollars) will be given by the equations

$$C = 7 - p \qquad \text{Cost equation}$$

$$R = 5p - p^2 \qquad \text{Revenue equation}$$

Find prices (including a graph) for which the company will realize:

(A) A profit (B) A loss

Solutions (A) A profit will result if cost is less than revenue, that is, if

$$C < R$$
$$7 - p < 5p - p^2$$

We solve this inequality following the steps outlined above.

Step 1. Write the polynomial inequality in standard form.

$$p^2 - 6p + 7 < 0 \quad \text{Standard form}$$

Step 2. Find all real zeros of the polynomial.

$$p^2 - 6p + 7 = 0$$
$$p = \frac{6 \pm \sqrt{36 - 28}}{2} \quad \text{Solve, using the quadratic formula.}$$
$$= 3 \pm \sqrt{2}$$
$$\approx \$1.59, \$4.41 \quad \text{Real zeros of the polynomial rounded to the nearest cent.}$$

Step 3. Plot the real zeros on a number line.

The two real zeros determine three intervals: $(-\infty, 1.59)$, $(1.59, 4.41)$, and $(4.41, \infty)$.

(−∞, 1.59) | (1.59, 4.41) | (4.41, ∞) → p

$1.59 $4.41

Step 4. Choose a test number in each interval, and construct a table.

Polynomial: $p^2 - 6p + 7$			
Test number	1	2	5
Value of polynomial for test number	2	−1	2
Sign of polynomial in interval containing test number	+	−	+
Interval containing test number	$(-\infty, 1.59)$	$(1.59, 4.41)$	$(4.41, \infty)$

Step 5. Construct a sign chart.

(−∞, 1.59) + + + | (1.59, 4.41) − − − − | (4.41, ∞) + + + → p

$1.59 $4.41

Sign chart for $p^2 - 6p - 7$

Step 6. Write the solution and draw the graph.

Referring to the sign chart for the polynomial $p^2 - 6p + 7$ in step 5, we see that $p^2 - 6p + 7 < 0$, and a profit will occur ($C < R$), for

$\$1.59 < p < \$4.41$

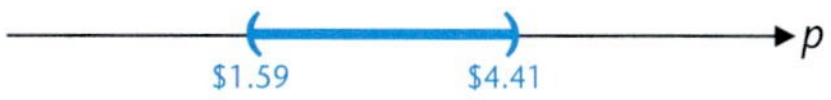

Profit

(B) A loss will result if cost is greater than revenue; that is, if

$$C > R$$
$$7 - p > 5p - p^2$$

Writing this polynomial inequality in standard form, we obtain the same inequality that was obtained in step 1 of part A, except the order of the inequality is reversed:

$$p^2 - 6p + 7 > 0 \quad \text{Standard form}$$

Referring to the sign chart for the polynomial $p^2 - 6p + 7$ in step 5 of part A, we see that $p^2 - 6p + 7 > 0$, and a loss will occur ($C > R$), for

$p < \$1.59$ or $p > \$4.41$

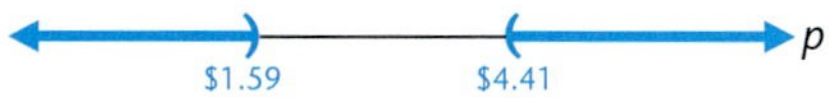

Since a negative price doesn't make sense, we must modify this result by deleting any number to the left of 0. Thus, a loss will occur for the following prices:

$\$0 \le p < \$1.59$ or $p > \$4.41$

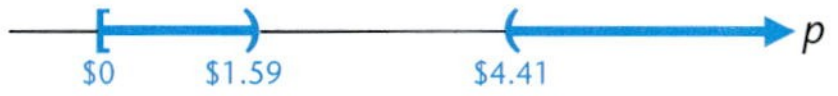

The real zeros are not included, because they are the values for which $R = C$, the **break-even** values for the company.

Matched Problem 1 A company manufactures and sells computer printer ribbons. For a particular ribbon, the marketing research and financial departments estimate that at a price of $\$p$ per unit, the weekly cost C and revenue R (in thousands of dollars) will be given by the equations

$$C = 13 - p \quad \text{Cost equation}$$
$$R = 7p - p^2 \quad \text{Revenue equation}$$

Find prices (including a graph) for which the company will realize:

(A) A profit (B) A loss

• Rational Inequalities

The steps for solving polynomial inequalities can, with slight modification, be used to solve rational inequalities such as

$$\frac{x-3}{x+5} > 0 \qquad \text{and} \qquad \frac{x^2+5x-6}{5-x} \le 3$$

If, after suitable operations on an inequality, the right side is 0 and the left side is of the form P/Q, where P and Q are nonzero polynomials, then the inequality is said to be a **rational inequality in standard form.** When the real zeros (if they exist) of the polynomials P and Q are plotted on a number line, they divide the line into two or more intervals. The following theorem, which includes Theorem 1 as a special case, provides a basis for solving rational inequalities in standard form.

Theorem 2 **Sign of a Rational Expression over a Real Number Line**

The rational expression P/Q, where P and Q are nonzero polynomials, will have a constant sign (either always positive or always negative) within each interval determined by the real zeros of P and Q plotted on a number line. If neither P nor Q have real zeros, then the rational expression P/Q is either positive over the whole real number line or negative over the whole real number line.

We will illustrate the use of Theorem 2 through an example.

EXAMPLE 2 Solving a Rational Inequality

Solve and graph: $\dfrac{x^2-3x-10}{1-x} \ge 2$

Solution We might be tempted to start by multiplying both sides by $1 - x$ (as we would do if the inequality were an equation). However, since we don't know whether $1 - x$ is positive or negative, we don't know whether the order of the inequality is to be changed. We proceed instead as follows (modifying the steps for solving polynomial inequalities as needed):

Step 1. Write the inequality in standard form.

$$\frac{x^2-3x-10}{1-x} \ge 2$$

$$\frac{x^2-3x-10}{1-x} - 2 \ge 0 \qquad \text{Subtract 2 from both sides.}$$

$$\frac{x^2-3x-10-2(1-x)}{1-x} \ge 0 \qquad \text{Combine left side into a single fraction.}$$

$$\frac{x^2-x-12}{1-x} \ge 0 \qquad \text{Standard form: } \frac{P}{Q} \ge 0$$

The left side of the last inequality is a rational expression of the form P/Q, where $P = x^2 - x - 12$ and $Q = 1 - x$. Our problem now is to find all values of x so that $P/Q \geq 0$; that is, so that P/Q is positive or 0.

Step 2. Find all real zeros for polynomials P and Q.

$$x^2 - x - 12 = 0$$

$$(x + 3)(x - 4) = 0$$

$$x = -3, 4 \quad \text{Real zeros for } P$$

$$1 - x = 0$$

$$x = 1 \quad \text{Real zero for } Q$$

Note: The real zeros for P make P/Q equal to 0; thus, the equality part of the original inequality is satisfied for these zeros and they must be included in the final solution set. On the other hand, since division by 0 is never allowed, P/Q is not defined at the zeros of Q. Thus, the real zeros of Q must *not* be included in the solution set.

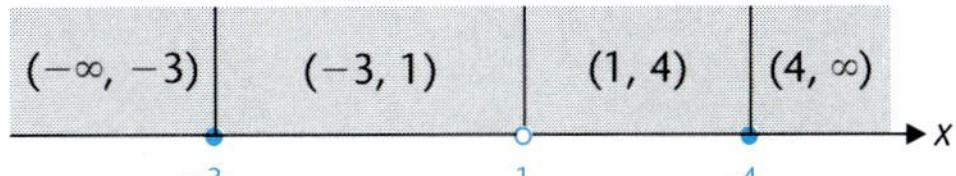

Step 3. Plot the real zeros for P and Q on a number line.

The three zeros of P and Q determine four intervals: $(-\infty, -3)$, $(-3, 1)$, $(1, 4)$, and $(4, \infty)$. Note that we use solid dots at -3 and 4 to indicate that these zeros of P are part of the solution set. However, we use an open dot at 1 to indicate that this zero of Q is not part of the solution set. Remember, P/Q is not defined at the zeros of Q.

Step 4. Choose a test number in each interval, and construct a table.

Rational expression: $\dfrac{x^2 - x - 12}{1 - x} = \dfrac{(x + 3)(x - 4)}{1 - x}$

Test number	-4	0	2	5
Value of P/Q	$\frac{8}{5}$	-12	10	-2
Sign of P/Q	+	−	+	−
Interval	$(-\infty, -3)$	$(-3, 1)$	$(1, 4)$	$(4, \infty)$

Step 5. Construct a sign chart.

(−∞, −3) + + + + | (−3, 1) − − − − − − | (1, 4) + + + + + | (4, ∞) − − −

−3 1 4 x

Sign chart for $\dfrac{x^2 - x - 12}{1 - x}$

Step 6. Write the solution and draw the graph.

From the sign chart, we see that

$$\frac{x^2 - x - 12}{1 - x} \geq 0 \qquad \text{and} \qquad \frac{x^2 - 3x - 10}{1 - x} \geq 2$$

for

$x \leq -3$ or $1 < x \leq 4$ Inequality notation

$(-\infty, -3] \cup (1, 4]$ Interval notation

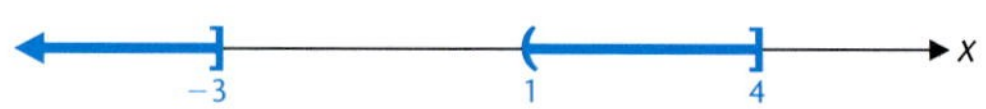

Matched Problem 2 Solve and graph: $\dfrac{3}{2 - x} \leq \dfrac{1}{x + 4}$

Answers to Matched Problems

1. (A) Profit: $\$2.27 < p < \$5.73$ (B) Loss: $\$0 \leq p < \$2.27$ or $p > \$5.73$

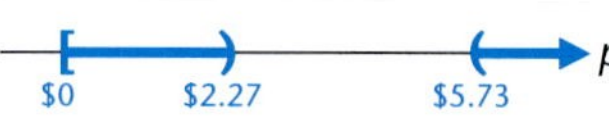

2. $-4 < x \leq -\frac{5}{2}$ or $x > 2$
$(-4, -\frac{5}{2}] \cup (2, \infty)$

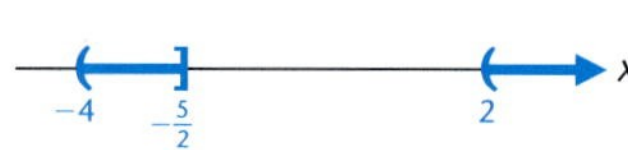

EXERCISE 1-8

A

In Problems 1–14, solve and graph. Express answers in both inequality and interval notation.

1. $x^2 < 10 - 3x$

2. $x^2 + x < 12$

3. $x^2 + 21 > 10x$

4. $x^2 + 7x + 10 > 0$

5. $x^2 \leq 8x$

6. $x^2 + 6x \geq 0$

7. $x^2 + 5x \leq 0$

8. $x^2 \leq 4x$

9. $x^2 > 4$

10. $x^2 \leq 9$

11. $\dfrac{x - 2}{x + 4} \leq 0$

12. $\dfrac{x + 3}{x - 1} \geq 0$

13. $\dfrac{x + 4}{1 - x} \leq 0$

14. $\dfrac{3 - x}{x + 5} \leq 0$

B

In Problems 15–26, solve and graph. Express answers in both inequality and interval notation.

15. $\dfrac{x^2 + 5x}{x - 3} \geq 0$

16. $\dfrac{x - 4}{x^2 + 2x} \leq 0$

17. $\dfrac{(x + 1)^2}{x^2 + 2x - 3} \leq 0$

18. $\dfrac{x^2 - x - 12}{x^2 + 4} \leq 0$

19. $\dfrac{1}{x} < 4$

20. $\dfrac{5}{x} > 3$

21. $\dfrac{3x + 1}{x + 4} \leq 1$

22. $\dfrac{5x - 8}{x - 5} \geq 2$

23. $\dfrac{2}{x + 1} \geq \dfrac{1}{x - 2}$

24. $\dfrac{3}{x - 3} \leq \dfrac{2}{x + 2}$

25. $x^3 + 2x^2 \le 8x$

26. $2x^3 + x^2 > 6x$

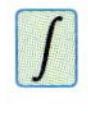

For what real values of x will each expression in Problems 27–32 represent a real number? Write answers using inequality notation.

27. $\sqrt{x^2 - 9}$

28. $\sqrt{4 - x^2}$

29. $\sqrt{2x^2 + x - 6}$

30. $\sqrt{3x^2 - 7x - 6}$

31. $\sqrt{\dfrac{x + 7}{3 - x}}$

32. $\sqrt{\dfrac{x - 1}{x + 3}}$

If a, b, and c are real numbers, the quadratic equation $ax^2 + bx + c = 0$ must have either two distinct real roots, one double real root, or two conjugate imaginary roots. In Problems 33–36, use the given information concerning the roots to describe the possible solution sets for the indicated inequality. Illustrate your conclusions with specific examples.

33. $ax^2 + bx + c > 0$, given distinct real roots r_1 and r_2 with $r_1 < r_2$.

34. $ax^2 + bx + c \le 0$, given distinct real roots r_1 and r_2 with $r_1 < r_2$.

35. $ax^2 + bx + c \ge 0$, given one (double) real root r.

36. $ax^2 + bx + c < 0$, given one (double) real root r.

37. Give an example of a quadratic inequality whose solution set is the entire real line.

38. Give an example of a quadratic inequality whose solution set is the empty set.

C

In Problems 39–50, solve and graph. Express answers in both inequality and interval notation.

39. $x^2 + 1 < 2x$

40. $x^2 + 25 < 10x$

41. $x^2 < 3x - 3$

42. $x^2 + 3 > 2x$

43. $x^2 - 1 \ge 4x$

44. $2x + 2 > x^2$

45. $x^3 > 2x^2 + x$

46. $x^3 \le 4x^2 + 3x$

47. $4x^4 + 4 \le 17x^2$

48. $x^4 + 36 \ge 13x^2$

49. $|x^2 - 1| \le 3$

50. $\left|\dfrac{x + 1}{x}\right| > 2$

APPLICATIONS

51. Profit and Loss Analysis. At a price of $\$p$ per unit, the marketing department in a company estimates that the weekly cost C and revenue R (in thousands of dollars) will be given by the equations

$$C = 28 - 2p \quad \text{Cost equation}$$
$$R = 9p - p^2 \quad \text{Revenue equation}$$

Find the prices for which the company has
(A) A profit (B) A loss

52. Profit and Loss Analysis. At a price of $\$p$ per unit, the marketing department in a company estimates that the weekly cost C and revenue R (in thousands of dollars) will be given by the equations

$$C = 27 - 2p \quad \text{Cost equation}$$
$$R = 10p - p^2 \quad \text{Revenue equation}$$

Find the prices for which the company has
(A) A profit (B) A loss

53. Physics. If an object is shot straight up from the ground with an initial velocity of 112 feet per second, its distance d (in feet) above the ground at the end of t seconds (neglecting air resistance) is given by $d = 112t - 16t^2$. Find the interval of time for which the object is 160 feet above the ground or higher.

54. Physics. In Problem 53, find the interval of time for which the object is above the ground.

★ **55. Safety Research.** It is of considerable importance to know the shortest distance d (in feet) in which a car can be stopped, including reaction time of the driver, at various speeds v (in miles per hour). Safety research has produced the formula $d = 0.044v^2 + 1.1v$ for a given car. At what speeds will it take the car more than 330 feet to stop?

★ **56. Safety Research.** Using the information in Problem 55, at what speeds will it take a car less than 220 feet to stop?

★★ **57. Marketing.** When successful new software is first introduced, the weekly sales generally increase rapidly for a period of time and then begin to decrease. Suppose that the weekly sales S (in thousands of units) t weeks after the software is introduced are given by

$$S = \frac{200t}{t^2 + 100}$$

When will sales be 8 thousand units per week or more?

★★ **58. Medicine.** A drug is injected into the bloodstream of a patient through her right arm. The concentration (in milligrams per milliliter) of the drug in the bloodstream of the left arm t hours after the injection is given approximately by

$$C = \frac{0.12t}{t^2 + 2}$$

When will the concentration of the drug in the left arm be 0.04 milligram per milliliter or greater?

CHAPTER 1 GROUP ACTIVITY Rates of Change

1. **Average Rate.** If you score 90 on your first math exam and 100 on the second exam, then your average exam score for the two exams is $\frac{1}{2}(90 + 100) = 95$. The number 95 is called the **arithmetic average** of 90 and 100. Now suppose you walk uphill at a rate of 3 mph for 5 hours and then turn around and return to your starting point by walking downhill at 6 mph for 2.5 hours. The arithmetic average of the rates for each leg of the trip is $\frac{1}{2}(3 + 6) = 4.5$ mph. On the other hand, you walked a total distance of 30 miles in 7.5 hours so that the rate for the round-trip is $30/7.5 = 4$ mph. Which is your *average rate?* The basic formula $D = RT$ is valid whenever an object travels a distance D at a *constant* rate R for a fixed time T. If the rate is not constant, then this formula can still be used but must be interpreted differently. To be precise, for objects moving at nonconstant rates, **average rate is total distance divided by total time.** Thus, your average rate for the total trip up and down the hill is 4 mph, not 4.5 mph. The formula $R = D/T$ now has two interpretations: $R = D/T$ is the *rate* for an object moving at a constant rate and the *average rate* for an object whose rate is not always the same.

(A) If r is the rate for one leg of a round-trip and s is the rate for the return trip, express the average rate for the round-trip in terms of r and s.

(B) A boat can travel 10 mph in still water. The boat travels 60 miles up a river with a 5 mph current and then returns to its starting point. Find the average rate for the round-trip using the definition of average rate and then check with the formula you found in part A.

(C) Referring to the hill-climbing example discussed earlier, if you walk up the hill at 3 mph, how fast must you walk downhill to average 6 mph for the round-trip? (This is similar to a famous problem communicated to Albert Einstein by Max Wertheimer. See Abraham S. Luchins and Edith H. Luchins, The Einstein–Wertheimer Correspondence on Geometric Proofs and Mathematical Puzzles, *Mathematical Intelligencer 2,* Spring 1990, pp. 40–41. For a discussion of this and other interesting rate–time problems, see Lawrence S. Braden, My Favorite Rate–Time Problems, *Mathematics Teacher,* November 1991, pp. 635–638.)

2. **Instantaneous Rate.** One of the fundamental concepts of calculus is the *instantaneous rate* of a moving object, which is closely related to the average rate discussed above. To introduce this concept, consider the following problem.

A small steel ball dropped from a tower will fall a distance of y feet in x seconds, as given approximately by the formula (from physics)

$$y = 16x^2$$

Figure 1 shows the position of the ball on a number line (positive direction down) at the end of 0, 1, 2, and 3 seconds. Clearly, the ball is not falling at a constant rate.

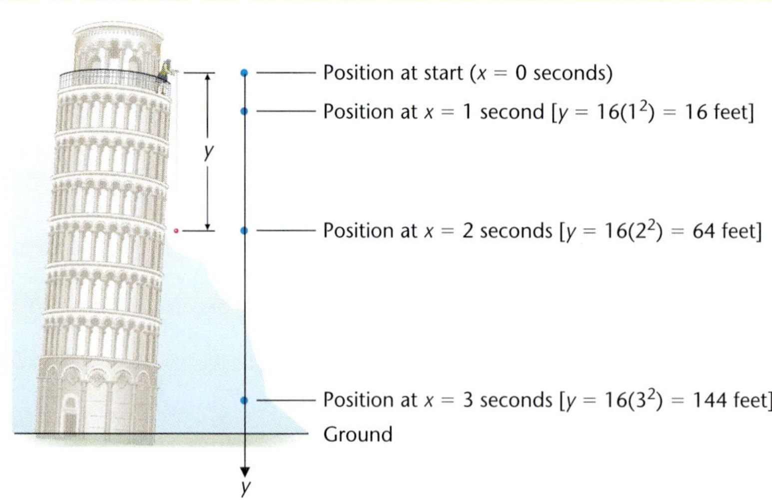

FIGURE 1 Position of a falling object [*Note:* Positive direction is down.]

(A) What is the average rate that the ball falls during the first second (from $x = 0$ to $x = 1$ second)? During the second second? During the third second?

By definition, average rate involves the distance an object travels over an *interval* of time, as in part A. How can we determine the rate of an object at a *given instant* of time? For example, how fast is the ball falling at exactly 2 seconds after it was released? We will approach this problem from two directions, numerically and algebraically.

(B) Complete the following table of average rates. What number do these average rates appear to approach?

Time interval	[1.9, 2]	[1.99, 2]	[1.999, 2]	[1.9999, 2]
Distance fallen				
Average rate				

(C) Show that the average rate over the time interval $[t, 2]$ is $\dfrac{64 - 16t^2}{2 - t}$. Simplify this algebraic expression and discuss its values for t very close to 2.

(D) Based on the results of parts B and C, how fast do you think the ball is falling at 2 seconds?

Chapter 1 Review

1-1 LINEAR EQUATIONS AND APPLICATIONS

A **solution** or **root** of an equation is a number in the **domain** or **replacement set** of the variable that when substituted for the variable makes the equation a true statement. An equation is an **identity** if it is true for all values from the domain of the variable and a **conditional equation** if it is true for some domain values and false for others. Two equations are **equivalent** if they have the same **solution set.** The **properties of equality** are used to solve equations:

1. If $a = b$, then $a + c = b + c$. Addition Property
2. If $a = b$, then $a - c = b - c$. Subtraction Property
3. If $a = b$, then $ca = cb$, $c \neq 0$. Multiplication Property
4. If $a = b$, then $\dfrac{a}{c} = \dfrac{b}{c}$, $c \neq 0$. Division Property
5. If $a = b$, then either may replace the other in any statement without changing the truth or falsity of statement. Substitution Property

An equation that can be written in the **standard form** $ax + b = 0$, $a \neq 0$, is a **linear** or **first-degree equation.**

Strategy for Solving Word Problems

1. Read the problem carefully—several times if necessary—that is, until you understand the problem, know what is to be found, and know what is given.
2. Let one of the unknown quantities be represented by a variable, say x, and try to represent all other unknown quantities in terms of x. This is an important step and must be done carefully.
3. If appropriate, draw figures or diagrams and label known and unknown parts.
4. Look for formulas connecting the known quantities to the unknown quantities.
5. Form an equation relating the unknown quantities to the known quantities.
6. Solve the equation and write answers to *all* questions asked in the problem.
7. Check and interpret all solutions in terms of the original problem—not just the equation found in step 5—since a mistake may have been made in setting up the equation in step 5.

If Q is **quantity** produced or **distance** traveled at an average or uniform **rate** R in T units of **time,** then the **quantity–rate–time formulas** are

$$R = \frac{Q}{T} \qquad Q = RT \qquad T = \frac{Q}{R}$$

1-2 SYSTEMS OF LINEAR EQUATIONS AND APPLICATIONS

A system of two linear equations with two variables is a system of the form:

$$\begin{aligned} ax + by &= h \\ cx + dy &= k \end{aligned} \qquad (1)$$

where x and y are the two variables. The ordered pair of numbers (x_0, y_0) is a **solution** to system (1) if each equation is satisfied by the pair. The set of all such ordered pairs of numbers is called the **solution set** for the system. To **solve** a system is to find its solution set. To solve a system by **substitution,** solve either equation for either variable, substitute in the other equation, solve the resulting linear equation in one variable, and then substitute this value into the expression obtained in the first step to find the other variable.

If one equation in a system is a **demand equation** and the other is a **supply equation,** then the solution produces the **equilibrium price** and the **equilibrium quantity.** If one equation in a system is a **cost equation** (often formed by using **fixed costs** and **variable costs**) and the other is a **revenue equation,** then the solution produces the number of units that must be manufactured to **break even.**

1-3 LINEAR INEQUALITIES

The inequality symbols $<, >, \leq, \geq$ are used to express **inequality relations. Line graphs, interval notation,** and the set operations of **union** and **intersection** are used to describe inequality relations. A **solution** of a linear inequality in one variable is a value of the variable that makes the inequality a true statement. Two inequalities are **equivalent** if they have the same **solution set. Inequality properties** are used to solve inequalities:

1. If $a < b$ and $b < c$, then $a < c$. — Transitive Property
2. If $a < b$, then $a + c < b + c$. — Addition Property
3. If $a < b$, then $a - c < b - c$. — Subtraction Property
4. If $a < b$ and $c > 0$, then $ca < cb$. — Multiplication Property
5. If $a < b$ and $c < 0$, then $ca > cb$. — Multiplication Property
6. If $a < b$ and $c > 0$, then $\frac{a}{c} < \frac{b}{c}$. — Division Property
7. If $a < b$ and $c < 0$, then $\frac{a}{c} > \frac{b}{c}$. — Division Property

The order of an inequality reverses if we multiply or divide both sides of an inequality statement by a negative number.

1-4 ABSOLUTE VALUE IN EQUATIONS AND INEQUALITIES

The **absolute value** of a number x is the distance on a real number line from the origin to the point with coordinate x and is given by

$$|x| = \begin{cases} x & \text{if } x \geq 0 \\ -x & \text{if } x < 0 \end{cases}$$

The **distance between points A and B** with coordinates a and b, respectively, is $d(A, B) = |b - a|$, which has the following **geometric interpretations:**

$\lvert x - c\rvert = d$	Distance between x and c is equal to d.
$\lvert x - c\rvert < d$	Distance between x and c is less than d.
$0 < \lvert x - c\rvert < d$	Distance between x and c is less than d, but $x \neq c$.
$\lvert x - c\rvert > d$	Distance between x and c is greater than d.

Equations and inequalities involving absolute values are solved using the following relationships for $p > 0$:

1. $|x| = p$ is equivalent to $x = p$ or $x = -p$.
2. $|x| < p$ is equivalent to $-p < x < p$.
3. $|x| > p$ is equivalent to $x < -p$ or $x > p$.

These relationships also hold if x is replaced with $ax + b$. For x any real number, $\sqrt{x^2} = |x|$.

1-5 COMPLEX NUMBERS

A **complex number** in **standard form** is a number in the form $a + bi$, where a and b are real numbers and i is the **imaginary unit.** If $b \neq 0$, then $a + bi$ is also called an **imaginary number.** If $a = 0$, then $0 + bi = bi$ is also called a **pure imaginary number.** If $b = 0$, then $a + 0i = a$ is a **real number.** The complex **zero** is $0 + 0i = 0$. The **conjugate** of $a + bi$ is $a - bi$. **Equality, addition,** and **multiplication** are defined as follows:

1. $a + bi = c + di$ if and only if $a = c$ and $b = d$

2. $(a + bi) + (c + di) = (a + c) + (b + d)i$

3. $(a + bi)(c + di) = (ac - bd) + (ad + bc)i$

Since complex numbers obey the same commutative, associative, and distributive properties as real numbers, most operations with complex numbers are performed by using these properties and the fact that $\mathbf{i^2 = -1}$. The **property of conjugates,**

$$(a + bi)(a - bi) = a^2 + b^2$$

can be used to find **reciprocals** and **quotients.** If $a > 0$, then the **principal square root of the negative real number** $-a$ is $\sqrt{-a} = i\sqrt{a}$.

1-6 QUADRATIC EQUATIONS AND APPLICATIONS

A **quadratic equation** in **standard form** is an equation that can be written in the form

$$ax^2 + bx + c = 0 \qquad a \neq 0$$

where x is a variable and a, b, and c are constants. Methods of solution include:

1. **Factoring** and using the **zero property:**

$$m \cdot n = 0 \quad \text{if and only if} \quad m = 0 \text{ or } n = 0 \text{ (or both)}$$

2. Using the **square root property:**

$$\text{If } A^2 = C, \text{ then } A = \pm\sqrt{C}$$

3. **Completing the square:**

$$x^2 + bx + \left(\frac{b}{2}\right)^2 = \left(x + \frac{b}{2}\right)^2$$

4. Using the **quadratic formula:**

$$x = \frac{-b \pm \sqrt{b^2 - 4ac}}{2a}$$

If the **discriminant** $\mathbf{b^2 - 4ac}$ is positive, the equation has two distinct **real roots;** if the discriminant is 0, the equation has one real **double root;** and if the discriminant is negative, the equation has two **imaginary roots,** each the conjugate of the other.

1-7 EQUATIONS REDUCIBLE TO QUADRATIC FORM

A **square root radical** can be eliminated from an equation by isolating the radical on one side of the equation and squaring both sides of the equation. The new equation formed by squaring both sides may have **extraneous solutions.** Consequently, **every solution of the new equation must be checked in the original equation to eliminate extraneous solutions.** If an equation contains more than one radical, then the process of isolating a radical and squaring both sides can be repeated until all radicals are eliminated. If a substitution transforms an equation into the form $au^2 + bu + c = 0$, where u is an expression in some other variable, then the equation is an **equation of quadratic type** that can be solved by quadratic methods.

1-8 POLYNOMIAL AND RATIONAL INEQUALITIES

An inequality is in **standard form** if the right side is 0. If the left side is a **polynomial,** then the **real zeros** of this polynomial divide the real number line into intervals with the property that the polynomial has constant sign over each interval. Selecting a **test number** in each interval and constructing a **sign chart** produces the solution to the inequality. If the left side of an inequality is a **rational expression** of the form P/Q, where P and Q are polynomials, then the real zeros of both polynomials are used to divide the real number line into intervals over which P/Q has constant sign. Since **division by zero is never allowed,** the real zeros of Q must always be excluded from the solution set.

Chapter 1 Review Exercise

Work through all the problems in this chapter review and check answers in the back of the book. Answers to all review problems are there, and following each answer is a number in italics indicating the section in which that type of problem is discussed. Where weaknesses show up, review appropriate sections in the text.

A

Solve Problems 1–3.

1. $0.05x + 0.25(30 - x) = 3.3$

2. $\dfrac{5x}{3} - \dfrac{4 + x}{2} = \dfrac{x - 2}{4} + 1$

3. $y = 4x - 9$
$y = -x + 6$

Solve and graph Problems 4–8.

4. $3(2 - x) - 2 \leq 2x - 1$

5. $|y + 9| < 5$

6. $|3 - 2x| \leq 5$

7. $x^2 + x < 20$

8. $x^2 \geq 4x + 21$

9. Perform the indicated operations and write the answers in standard form:
(A) $(-3 + 2i) + (6 - 8i)$
(B) $(3 - 3i)(2 + 3i)$
(C) $\dfrac{13 - i}{5 - 3i}$

Solve Problems 10–16.

10. $2x^2 - 7 = 0$

11. $2x^2 = 4x$

12. $2x^2 = 7x - 3$

13. $m^2 + m + 1 = 0$

14. $y^2 = \frac{3}{2}(y + 1)$

15. $\sqrt{5x - 6} - x = 0$

16. $3x + 2y = 5$
$4x - y = 14$

17. For what values of x does $\sqrt{3 - 5x}$ represent a real number?

B

Solve Problems 18–20.

18. $\dfrac{7}{2 - x} = \dfrac{10 - 4x}{x^2 + 3x - 10}$

19. $\dfrac{u - 3}{2u - 2} = \dfrac{1}{6} - \dfrac{1 - u}{3u - 3}$

20. $5m + 6n = 2$
$4m - 9n = 20$

Solve and graph Problems 21–25.

21. $\dfrac{x + 3}{8} \le 5 - \dfrac{2 - x}{3}$

22. $|3x - 8| > 2$

23. $\dfrac{1}{x} < 2$

24. $\dfrac{3}{x - 4} \le \dfrac{2}{x - 3}$

25. $\sqrt{(1 - 2m)^2} \le 3$

26. For what real values of x does the expression below represent a real number?

$$\sqrt{\frac{x + 4}{2 - x}}$$

27. If the coordinates of A and B on a real number line are -8 and -2, respectively, find:
(A) $d(A, B)$ (B) $d(B, A)$

28. Perform the indicated operations and write the final answers in standard form:
(A) $(3 + i)^2 - 2(3 + i) + 3$ (B) i^{27}

29. Convert to $a + bi$ forms, perform the indicated operations, and write the final answers in standard form:
(A) $(2 - \sqrt{-4}) - (3 - \sqrt{-9})$
(B) $\dfrac{2 - \sqrt{-1}}{3 + \sqrt{-4}}$ (C) $\dfrac{4 + \sqrt{-25}}{\sqrt{-4}}$

Solve Problems 30–35.

30. $\left(u + \dfrac{5}{2}\right)^2 = \dfrac{5}{4}$

31. $1 + \dfrac{3}{u^2} = \dfrac{2}{u}$

32. $\dfrac{x}{x^2 - x - 6} - \dfrac{2}{x - 3} = 3$

33. $2x^{2/3} - 5x^{1/3} - 12 = 0$

34. $m^4 + 5m^2 - 36 = 0$

35. $\sqrt{y - 2} - \sqrt{5y + 1} = -3$

Use a calculator to solve Problems 36–40, and compute to two decimal places.

36. $2.15x - 3.73(x - 0.93) = 6.11x$

37. $-1.52 \le 0.77 - 2.04x \le 5.33$

38. $\dfrac{3.77 - 8.47i}{6.82 - 7.06i}$

39. $6.09x^2 + 4.57x - 8.86 = 0$

40. $15.2x + 5.6y = 20$
$2.5x + 7.5y = 10$

Solve Problems 41–43 for the indicated variable in terms of the other variables.

41. $P = M - Mdt$ for M (mathematics of finance)

42. $P = EI - RI^2$ for I (electrical engineering)

43. $x = \dfrac{4y + 5}{2y + 1}$ for y

44. Find the error in the following "solution" and then find the correct solution.

$$\frac{4}{x^2 - 4x + 3} = \frac{3}{x^2 - 3x + 2}$$

$$4x^2 - 12x + 8 = 3x^2 - 12x + 9$$

$$x^2 = 1$$

$$x = -1 \quad \text{or} \quad x = 1$$

45. Consider the quadratic equation

$$x^2 - 6x + c = 0$$

where c is a real number. Discuss the relationship between the values of c and the three types of roots listed in Table 1 in Section 2-6.

C

46. For what values of a and b is the inequality $a + b < b - a$ true?

47. If a and b are negative numbers and $a > b$, then is a/b greater than 1 or less than 1?

48. Solve for x in terms of y: $y = \dfrac{1}{1 - \dfrac{1}{1 - x}}$

49. Solve and graph: $0 < |x - 6| < d$

Solve Problems 50 and 51.

50. $2x^2 = \sqrt{3}x - \frac{1}{2}$

51. $4 = 8x^{-2} - x^{-4}$

52. Evaluate: $(a + bi)\left(\dfrac{a}{a^2 + b^2} - \dfrac{b}{a^2 + b^2}i\right)$, $a, b \neq 0$

Solve Problems 53–55.

53. $2x > \dfrac{x^2}{5} + 5$

54. $\dfrac{x^2}{4} + 4 \geq 2x$

55. $\left|x - \dfrac{8}{x}\right| \geq 2$

56. Solve for u and v in terms of x and y and check.

$$x = 2 + 3u + 7v$$
$$y = -3 + 2u + 5v$$

57. Discuss the nature of the solution sets for each of the following systems:
(A) $2x - y = -5$, $-6x + 3y = 15$
(B) $2x - y = -5$, $-6x + 3y = 10$

APPLICATIONS 

58. **Numbers.** Find a number such that subtracting its reciprocal from the number gives $\frac{16}{15}$.

59. **Sports Medicine.** The following quotation was found in a sports medicine handout: "The idea is to raise and sustain your heart rate to 70% of its maximum safe rate for your age. One way to determine this is to subtract your age from 220 and multiply by 0.7."
(A) If H is the maximum safe sustained heart rate (in beats per minute) for a person of age A (in years), write a formula relating H and A.
(B) What is the maximum safe sustained heart rate for a 20-year-old?
(C) If the maximum safe sustained heart rate for a person is 126 beats per minute, how old is the person?

★ 60. **Chemistry.** A chemical storeroom has an 80% alcohol solution and a 30% alcohol solution. How many milliliters of each should be used to obtain 50 milliliters of a 60% solution?

★ 61. **Rate–Time.** An excursion boat takes 2 hours longer to go 45 miles up a river than to return. If the boat's speed in still water is 12 miles per hour, what is the rate of the current?

★ 62. **Rate–Time.** A crew of four practices by rowing up a river for a fixed distance and then returning to their starting point. The river has a current of 3 km/h.
(A) Currently the crew can row 15 km/h in still water. If it takes them 25 minutes to make the round-trip, how far upstream did they row?
(B) After some additional practice the crew cuts the round-trip time to 23 minutes. What is their still-water speed now? Round answers to one decimal place.
(C) If the crew wants to increase their still-water speed to 18 km/h, how fast must they make the round-trip? Express answer in minutes rounded to one decimal place.

63. **Nutrition.** A fruit grower can use two types of fertilizer in an orange grove, brand A and brand B. Each bag of brand A contains 8 pounds of nitrogen and 9 pounds of phosphoric acid. Each bag of brand B contains 4 pounds of nitrogen and 7 pounds of phosphoric acid. Tests indicate that the grove needs 860 pounds of nitrogen and 1,080 pounds of phosphoric acid. How many bags of each brand should be used to provide the required amounts of nitrogen and phosphoric acid?

64. **Cost Analysis.** Cost equations for manufacturing companies are often quadratic in nature. If the cost equation for manufacturing inexpensive calculators is $C = x^2 - 10x + 31$, where C is the cost of manufacturing x units per week (both in thousands), find:
(A) The output for a \$15 thousand weekly cost
(B) The output for a \$6 thousand weekly cost

65. **Break-Even Analysis.** The manufacturing company in Problem 64 sells its calculators to wholesalers for \$3 each. Thus, its revenue equation is $R = 3x$, where R is revenue and x is the number of units sold per week (both in thousands). Find the break-even point(s) for the company—that is, the output at which revenue equals cost.

★ 66. **Profit Analysis.** Referring to Problems 64 and 65, find all output levels for which a profit will result—that is, for which $R > C$.

★ 67. **Chemistry.** If the temperature T of a solution must be kept within 5°C of 110°C, express this restriction as an absolute value inequality.

★ 68. **Design.** The pages of a textbook have uniform margins of 2 centimeters on all four sides (see the figure on the next page). If the area of the entire page is 480 square centimeters and the area of the printed portion is 320 square centimeters, find the dimensions of the page.

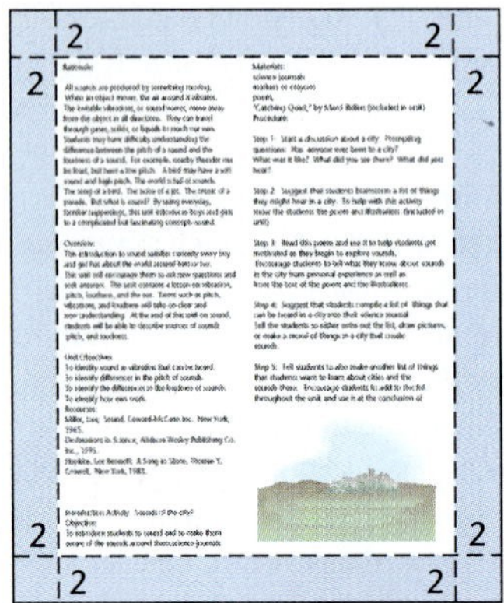

Figure for 68

★ **69. Design.** A landscape designer uses 8-foot timbers to form a pattern of isosceles triangles along the wall of a building (see the figure). If the area of each triangle is 24 square feet, find the base correct to two decimal places.

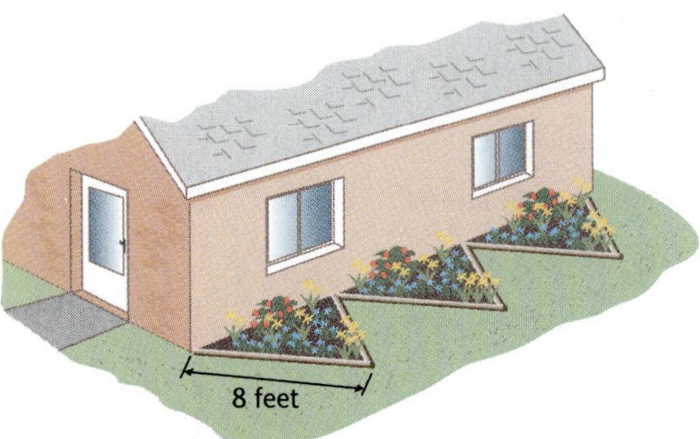

Graphs and Functions

CHAPTER 2

The function concept is one of the most important ideas in mathematics. The study of mathematics beyond the most elementary level requires a firm understanding of a basic list of elementary functions, their properties, and their graphs. In the first two sections of this chapter we consider some basic geometric concepts, including the graphs of circles and straight lines. In the remaining sections we introduce the important concept of a function, discuss basic properties, and consider operations that can be performed with functions. As we progress through this and subsequent chapters of the book, we will encounter a number of different types of elementary functions. A thorough understanding of the definitions, graphs, and properties of these elementary functions will provide you with a set of tools that should become a part of your mathematical toolbox for use in this and most future courses or activities that involve mathematics.

SECTION 2-1 Basic Tools; Circles

- Cartesian Coordinate System
- Graphing: Point by Point
- Symmetry
- Distance between Two Points
- Circles

In this section we develop some of the basic tools used in analytic geometry and apply these tools to the graphing of equations and to the derivation of the equation of a circle.

• Cartesian Coordinate System

Just as a real number line is formed by establishing a one-to-one correspondence between the points on a line and the elements in the set of real numbers, we can form a **real plane** by establishing a one-to-one correspondence between the points in a plane and elements in the set of all ordered pairs of real numbers. This can be done by means of a Cartesian coordinate system.

Recall that to form a **Cartesian** or **rectangular coordinate system,** we select two real number lines, one horizontal and one vertical, and let them cross through their origins as indicated in Figure 1. Up and to the right are the usual choices for the positive directions. These two number lines are called the **horizontal axis** and the **vertical axis,** or, together, the **coordinate axes.** The horizontal axis is usually referred to as the ***x* axis** and the vertical axis as the ***y* axis,** and each is labeled accordingly. Other labels may be used in certain situations. The coordinate axes divide the plane into four parts called **quadrants,** which are numbered counterclockwise from I to IV (see Fig. 1).

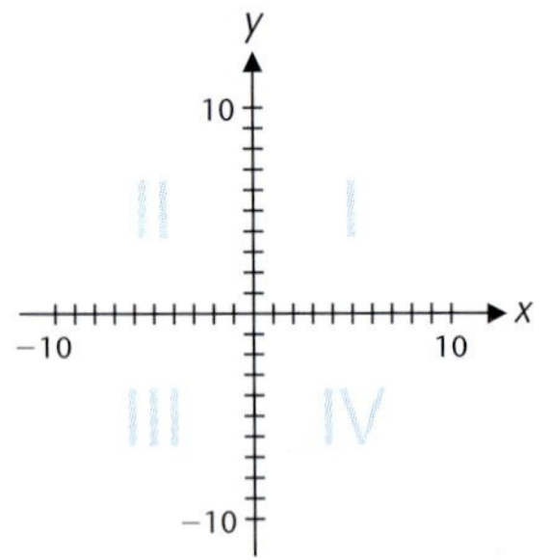

FIGURE 1 Cartesian coordinate system.

Now we want to assign *coordinates* to each point in the plane. Given an arbitrary point P in the plane, pass horizontal and vertical lines through the point (Fig. 2). The vertical line will intersect the horizontal axis at a point with coordinate a, and the horizontal line will intersect the vertical axis at a point with coordinate b. These two

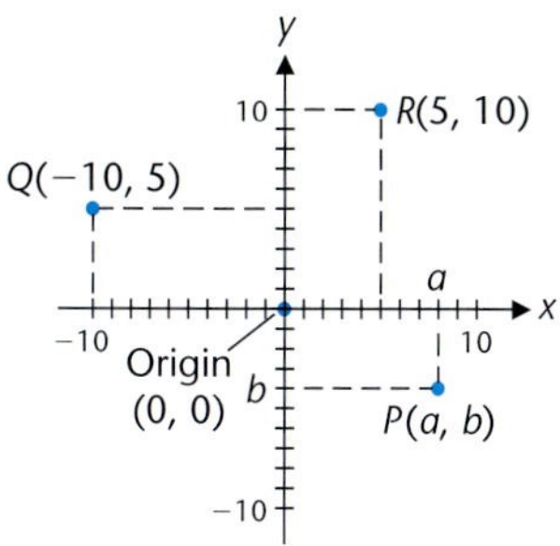

FIGURE 2 Coordinates in a plane.

numbers written as the ordered pair* (a, b) form the **coordinates** of the point P. The first coordinate a is called the **abscissa** of P; the second coordinate b is called the **ordinate** of P. The abscissa of Q in Figure 2 is -10, and the ordinate of Q is 5. The coordinates of a point can also be referenced in terms of the axis labels. The ***x* coordinate** of R in Figure 2 is 5, and the ***y* coordinate** of R is 10. The point with coordinates $(0, 0)$ is called the **origin.**

The procedure we have just described assigns to each point P in the plane a unique pair of real numbers (a, b). Conversely, if we are given an ordered pair of real numbers (a, b), then, reversing this procedure, we can determine a unique point P in the plane. Thus:

There is a one-to-one correspondence between the points in a plane and the elements in the set of all ordered pairs of real numbers.

This is often referred to as the **fundamental theorem of analytic geometry.**

• Graphing: Point by Point

The fundamental theorem of analytic geometry allows us to look at algebraic forms geometrically and to look at geometric forms algebraically. We begin by considering an algebraic form, an equation in two variables:

$$y = x^2 - 4 \tag{1}$$

A **solution** to equation (1) is an ordered pair of real numbers (a, b) such that

$$b = a^2 - 4$$

The **solution set** of equation (1) is the set of all its solutions. More formally,

$$\text{Solution set of equation (1): } \{(x, y) \mid y = x^2 - 4\}$$

To find a solution to equation (1) we simply replace one of the variables with a number and solve for the other variable. For example, if $x = 2$, then $y = 2^2 - 4 = 0$, and the ordered pair $(2, 0)$ is a solution. Similarly, if $y = 5$, then $5 = x^2 - 4$, $x^2 = 9$, $x = \pm 3$, and the ordered pairs $(3, 5)$ and $(-3, 5)$ are solutions.

Sometimes replacing one variable with a number and solving for the other variable will introduce imaginary numbers. For example, if $y = -5$ in equation (1), then

$$\begin{aligned} -5 &= x^2 - 4 \\ x^2 &= -1 \\ x &= \pm\sqrt{-1} = \pm i \end{aligned}$$

Thus, $(-i, 5)$ and $(i, 5)$ are solutions to $y = x^2 - 4$. However, the coordinates of a point in a rectangular coordinate system must be real numbers. Therefore, **when graphing an equation, we only consider those values of the variables that produce real solutions to the equation.**

*An **ordered pair** of real numbers is a pair of numbers in which the order is specified. We now use (a, b) as the coordinates of a point in a plane. In Chapter 1 we used (a, b) to represent an interval on a real number line. These concepts are not the same. You must always interpret the symbol (a, b) in terms of the context in which it is used.

The **graph of an equation in two variables** is the graph of its solution set. In equation (1), we find that its solution set will have infinitely many elements and its graph will extend off any paper we might choose, no matter how large. Thus, **to sketch the graph of an equation,** we include enough points from its solution set so that the total graph is apparent. This process is called **point-by-point plotting.**

EXAMPLE 1 Graphing an Equation Using Point-by-Point Plotting

Sketch a graph of $y = x^2 - 4$.

Solution We make up a table of solutions—ordered pairs of real numbers that satisfy the given equation.

x	-4	-3	-2	-1	0	1	2	3	4
y	12	5	0	-3	-4	-3	0	5	12

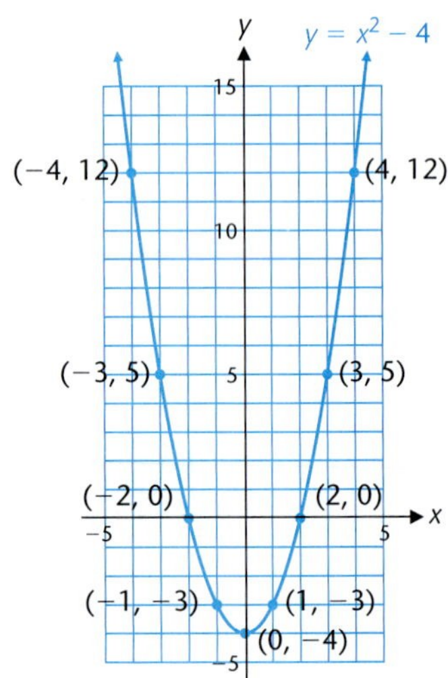

FIGURE 3

After plotting these solutions, if there are any portions of the graph that are unclear, we plot additional points until the shape of the graph is apparent. Then we join all these plotted points with a smooth curve, as shown in Figure 3. Arrowheads are used to indicate that the graph continues beyond the portion shown here with no significant changes in shape.

The resulting figure is called a *parabola.* Notice that if we fold the paper along the y axis, the right side will match the left side. We say that the graph is *symmetric with respect to the y axis* and call the y axis the *axis of the parabola.* More will be said about parabolas later in the text.

Matched Problem 1 Sketch a graph of $y^2 = x$.

We now use an electronic graphing device to check Example 1. We will refer to any electronic device capable of displaying graphs as a **graphing utility.** The two most common graphing utilities are hand-held graphing calculators and computers with appropriate software. This book contains a number of activities that use graphing utilities to emphasize the connection between graphic, numeric, and algebraic viewpoints. All these activities are clearly marked and easily omitted if no such device is available. Figure 4 shows the steps necessary to reproduce the graph in Figure 3 on a graphing utility.

FIGURE 4

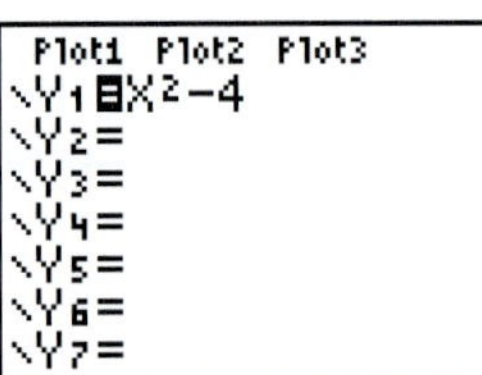

Enter the equation.
(a)

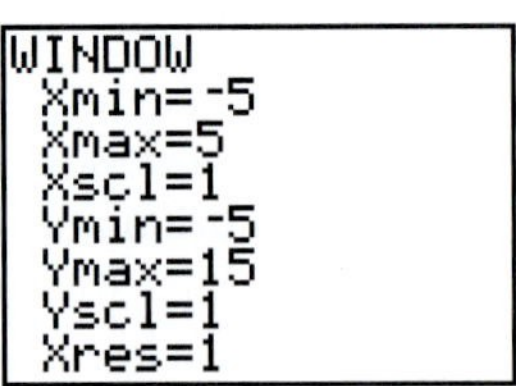

Enter the window variables.
(b)

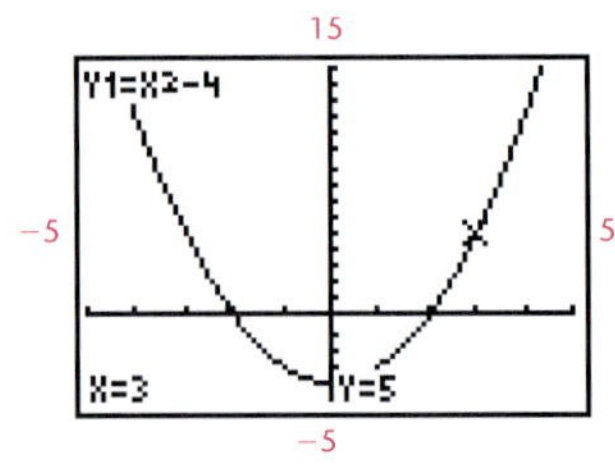

Graph the equation.
(c)

EXPLORE-DISCUSS 1 To graph the equation $y = -x^3 + 2x$, we use point-by-point plotting to obtain the graph in Figure 5.

FIGURE 5

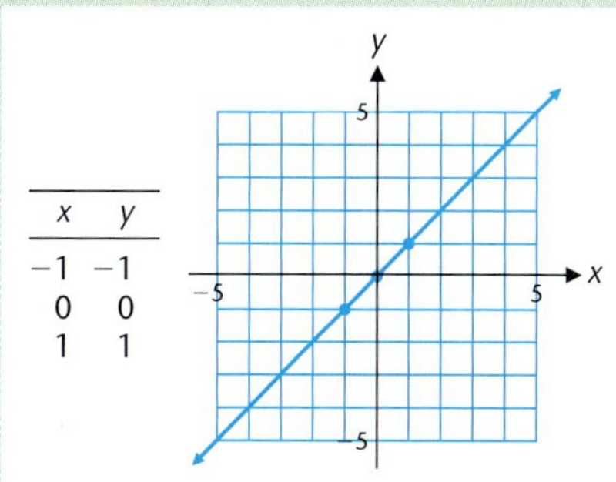

(A) Do you think this is the correct graph of the equation? If so, why? If not, why?

(B) Add points on the graph for $x = -2, -0.5, 0.5$, and 2.

(C) Now, what do you think the graph looks like? Sketch your version of the graph, adding more points as necessary.

(D) Write a short statement explaining any conclusions you might draw from parts A, B, and C.

 (E) Compare your version of the graph with one produced on a graphing utility.

The use of graphs to illustrate relationships between quantities is commonplace. Estimating the coordinates of points on a graph provides specific examples of this relationship, even if no equation for the graph is available. The next example illustrates this process.

EXAMPLE 2 Ozone Levels

The ozone level is measured in parts per billion (ppb). The ozone level during a 12-hour period in a suburb of Milwaukee, Wisconsin, on a particular summer day is given in Figure 6 on the next page (*source:* Wisconsin Department of Natural Resources). Use this graph to estimate the following ozone levels to the nearest integer and times to the nearest quarter hour.

(A) The ozone level at 6 P.M.
(B) The highest ozone level and the time when it occurs.
(C) The time(s) when the ozone level is 90 ppb.

FIGURE 6 Ozone level.

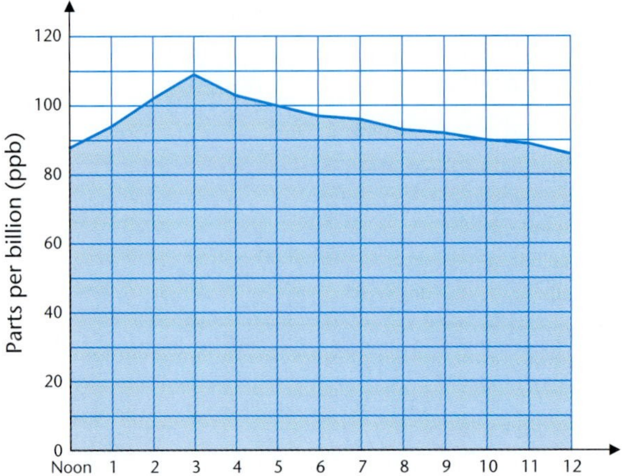

Solution (A) The ordinate of the point on the graph with abscissa 6 is 97 ppb (see Fig. 7).

(B) The highest ozone level is 109 ppb at 3 P.M.

(C) The ozone level is 90 ppb at about 12:30 P.M. and again at 10 P.M.

FIGURE 7

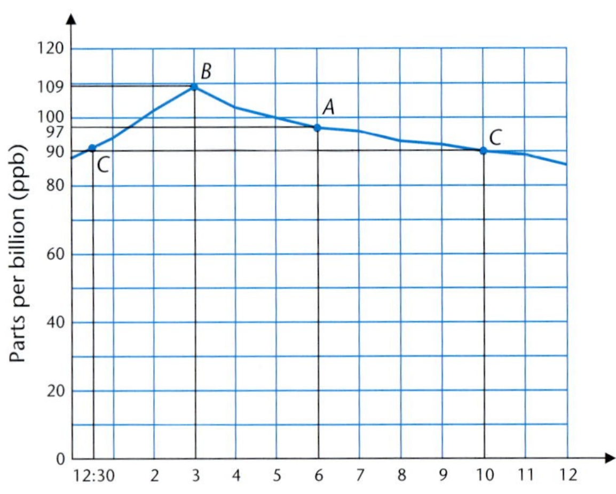

Matched Problem 2 Use Figure 6 to estimate the following ozone levels to the nearest integer and times to the nearest quarter hour.

(A) The ozone level at 7 P.M.

(B) The time(s) when the ozone level is 100 ppb.

An important aspect of this course, and later in calculus, is the development of tools that can be used to analyze graphs, whether using point-by-point plotting or a graphing utility. A particularly useful tool is *symmetry,* which we now discuss.

• Symmetry

We noticed that the graph of $y = x^2 - 4$ in Example 1 is *symmetric with respect to the y axis;* that is, the two parts of the graph coincide if the paper is folded along the

y axis. Similarly, we say that a graph is *symmetric with respect to the x axis* if the parts above and below the x axis coincide when the paper is folded along the x axis. In general, we define symmetry with respect to the y axis, x axis, and origin as follows:

DEFINITION 1

Symmetry

A graph is **symmetric with respect to:**

1. **The y axis** if $(-a, b)$ is on the graph whenever (a, b) is on the graph—the two points are equidistant from the y axis.
2. **The x axis** if $(a, -b)$ is on the graph whenever (a, b) is on the graph—the two points are equidistant from the x axis.
3. **The origin** if $(-a, -b)$ is on the graph whenever (a, b) is on the graph—the two points are equidistant from the origin on a line through the origin.

Figure 8 illustrates these three types of symmetry.

FIGURE 8 Symmetry.

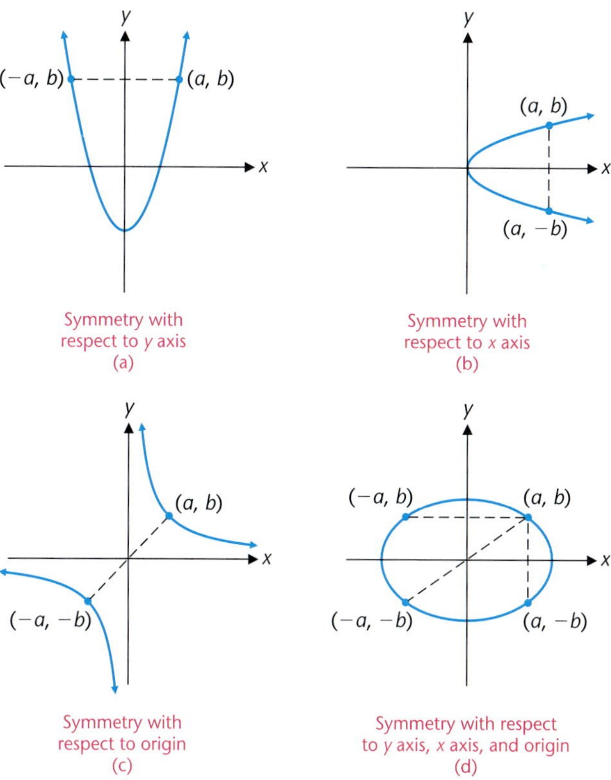

EXPLORE-DISCUSS 2 If a graph possesses two of the three types of symmetry in Definition 1, must it also possess the third? Explain.

Given an equation, if we can determine the symmetry properties of its graph ahead of time, we can save a lot of time and energy in sketching the graph. For example, we know that the graph of $y = x^2 - 4$ in Example 1 is symmetric with respect to the y axis, so we can carefully sketch only the right side of the graph; then reflect the result across the y axis to obtain the whole sketch—the point-by-point plotting is cut in half!

The tests for symmetry are given in Theorem 1. These tests are easily applied and are very helpful aids to graphing. Recall, two equations are equivalent if they have the same solution set.

Theorem 1 **Tests for Symmetry**

Symmetry with respect to the:	Equation is equivalent when:
y axis	x is replaced with $-x$
x axis	y is replaced with $-y$
Origin	x and y are replaced with $-x$ and $-y$

EXAMPLE 3 Using Symmetry as an Aid to Graphing

Test for symmetry and graph:

(A) $y = x^3$ (B) $y = |x|$ (C) $x^2 + 4y^2 = 36$

Solution (A) **Symmetry tests for $y = x^3$.**

Test y Axis
Replace x with $-x$:

$$y = (-x)^3$$
$$y = -x^3$$

Test x Axis
Replace y with $-y$:

$$-y = x^3$$
$$y = -x^3$$

Test Origin
Replace x with $-x$ and y with $-y$:

$$-y = (-x)^3$$
$$-y = -x^3$$
$$y = x^3$$

The only test that produces an equivalent equation is replacing x with $-x$ and y with $-y$. Thus, the only symmetry property for the graph of $y = x^3$ is symmetry with respect to the origin.

Graph. Note that positive values of x produce positive values for y and negative values of x produce negative values for y. Therefore, the graph occurs in the first and third quadrants. We make a careful sketch in the first quadrant; then reflect these points through the origin to obtain the complete sketch shown in Figure 9.

FIGURE 9

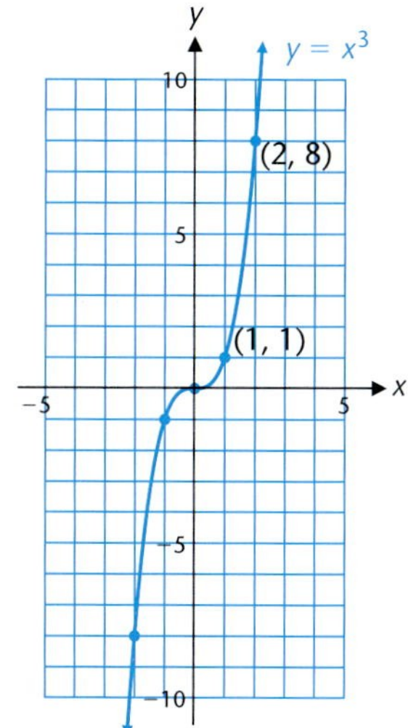

At a glance, the graph shows us how y varies as x varies. A graph is a visual aid and should be constructed to impart the maximum amount of information using the least amount of effort on the part of the observer. Label coordinate axes and indicate scales on both axes.

x	0	1	2
y	0	1	8

(B) **Symmetry tests for $y = |x|$.**

Test y Axis	**Test x Axis**	**Test Origin**
Replace x with $-x$:	Replace y with $-y$:	Replace x with $-x$ and y with $-y$:
$y = |-x|$	$-y = |x|$	$-y = |-x|$
$y = |x|$	$y = -|x|$	$-y = |x|$
		$y = -|x|$

Thus, the only symmetry property for the graph of $y = |x|$ is symmetry with respect to the y axis.

Graph. Since $|x|$ is never negative, this graph occurs in the first and second quadrants. We make a careful sketch in the first quadrant; then reflect this graph across the y axis to obtain the complete sketch shown in Figure 10.

FIGURE 10

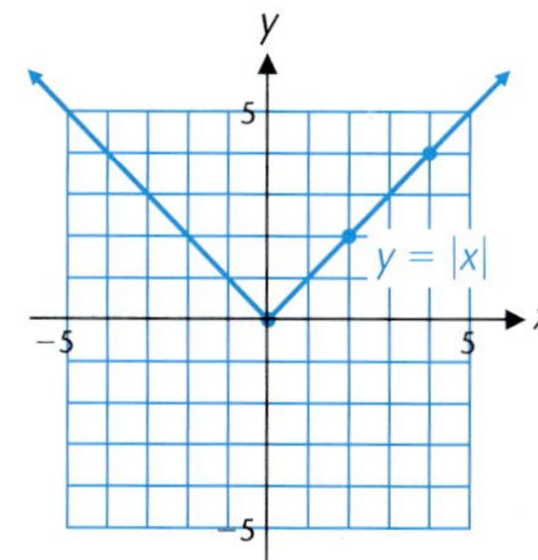

x	0	2	4
y	0	2	4

(C) Since both x and y occur to only even powers in $x^2 + 4y^2 = 36$, the equation will remain equivalent if x is replaced with $-x$ or if y is replaced with $-y$. Consequently, the graph is symmetric with respect to the y axis, x axis, and origin. We need to make a careful sketch in only the first quadrant, reflect this graph across the y axis, and then reflect everything across the x axis. To find quadrant I solutions, we solve the equation for either y in terms of x or x in terms of y. We choose the latter because the result is simpler to work with.

$$x^2 + 4y^2 = 36$$
$$x^2 = 36 - 4y^2$$
$$x = \pm\sqrt{36 - 4y^2}$$

To obtain the quadrant I portion of the graph, we sketch $x = \sqrt{36 - 4y^2}$ for $0 \le y \le 3$. Note that $36 - 4y^2 < 0$ for $y > 3$, so there are no real solutions for $y > 3$. The final graph is shown in Figure 11.

x	6	$\sqrt{32} \approx 5.7$	$\sqrt{20} \approx 4.5$	0
y	0	1	2	3

Choose values for y and solve for x.

FIGURE 11

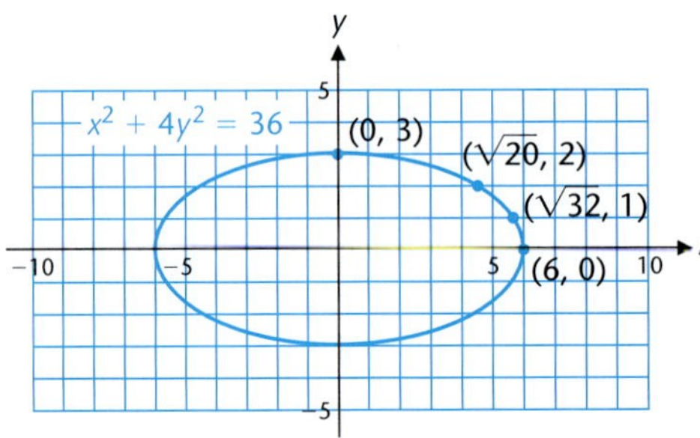

This figure is called an *ellipse*.

Parts A and B of Example 3 are easily checked on a graphing utility, and you should do so if you have one of these devices. Checking part C is more involved. Since most graphing utilities only graph equations of the form $y =$ (expression in x), we must solve the equation $x^2 + 4y^2 = 36$ for y (details omitted) and graph both solutions as shown in Figure 12.

FIGURE 12

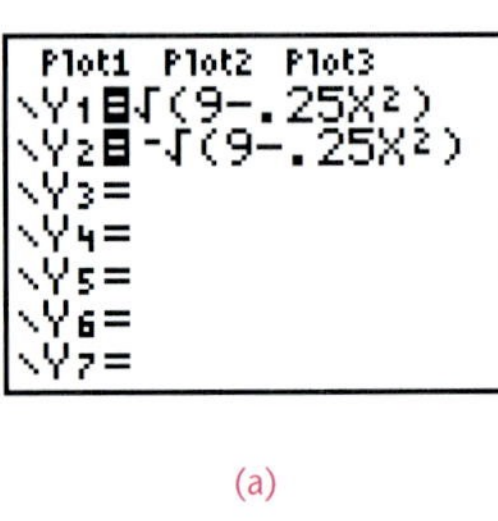

(a)

5

−10 10

−5

(b)

Matched Problem 3 Test for symmetry and graph:

(A) $y = x$ (B) $y = -|x|$ (C) $9x^2 + y^2 = 36$

• Distance between Two Points

Analytic geometry is concerned with two basic problems:

1. Given an equation, find its graph.

2. Given a figure (line, circle, parabola, ellipse, etc.) in a coordinate system, find its equation.

So far we have concentrated on the first problem. We now introduce a basic tool that is used extensively in solving the second problem. This basic tool is the *distance-between-two-points formula,* which is easily derived using the Pythagorean theorem (see footnote, Section 2-6). Let $P_1(x_1, y_1)$ and $P_2(x_2, y_2)$ be two points in a rectangular coordinate system (the scale on each axis is assumed to be the same). Then referring to Figure 13, we see that

$$[d(P_1, P_2)]^2 = |x_2 - x_1|^2 + |y_2 - y_1|^2$$
$$= (x_2 - x_1)^2 + (y_2 - y_1)^2 \quad \text{Since } |N|^2 = N^2.$$

FIGURE 13 Distance between two points.

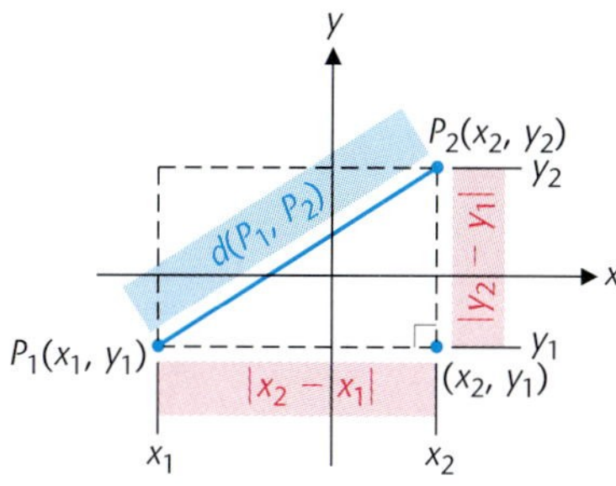

Thus:

Theorem 2 **Distance between $P_1(x_1, y_1)$ and $P_2(x_2, y_2)$**

$$d(P_1, P_2) = \sqrt{(x_2 - x_1)^2 + (y_2 - y_1)^2}$$

EXAMPLE 4 Using the Distance-between-Two-Points Formula

Find the distance between the points $(-3, 5)$ and $(-2, -8)$.*

Solution It doesn't matter which point we designate as P_1 or P_2 because of the squaring in the formula. Let $(x_1, y_1) = (-3, 5)$ and $(x_2, y_2) = (-2, -8)$. Then

$$d = \sqrt{[(-2) - (-3)]^2 + [(-8) - 5]^2}$$
$$= \sqrt{(-2 + 3)^2 + (-8 - 5)^2} = \sqrt{1^2 + (-13)^2} = \sqrt{1 + 169} = \sqrt{170}$$

Matched Problem 4 Find the distance between the points $(6, -3)$ and $(-7, -5)$.

*We often speak of the point (a, b) when we are referring to the point with coordinates (a, b). This shorthand, though not accurate, causes little trouble, and we will continue the practice.

• Circles

The distance-between-two-points formula would still be helpful if its only use were to find actual distances between points, such as in Example 4. However, its more important use is in finding equations of figures in a rectangular coordinate system. We will use it to derive the standard equation of a circle. We start with a coordinate-free definition of a circle.

DEFINITION 2 **Circle**

A **circle** is the set of all points in a plane equidistant from a fixed point. The fixed distance is called the **radius,** and the fixed point is called the **center.**

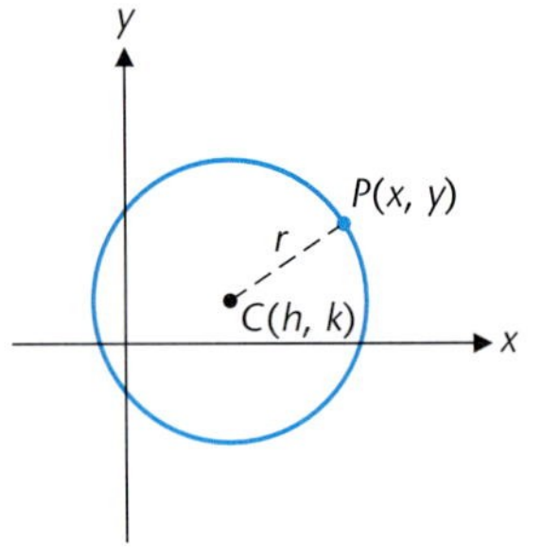

FIGURE 14 Circle.

Let's find the equation of a circle with radius r ($r > 0$) and center C at (h, k) in a rectangular coordinate system (Fig. 14). The point $P(x, y)$ is on the circle if and only if $d(P, C) = r$; that is, if and only if

$$\sqrt{(x-h)^2 + (y-k)^2} = r \qquad r > 0$$

or, equivalently,

$$(x-h)^2 + (y-k)^2 = r^2 \qquad r > 0$$

Theorem 3 **Standard Equation of a Circle**

1. Circle with radius r and center at (h, k):

$$(x-h)^2 + (y-k)^2 = r^2 \qquad r > 0$$

2. Circle with radius r and center at $(0, 0)$:

$$x^2 + y^2 = r^2 \qquad r > 0$$

EXPLORE-DISCUSS 3 Describe geometrically the set of all points (x, y) that are equidistant from the points $(-1, 0)$ and $(1, 0)$, and then use the distance formula to verify your result algebraically.

EXAMPLE 5 Equations and Graphs of Circles

Find the equation of a circle with radius 4 and center at:

(A) $(-3, 6)$ (B) $(0, 0)$

Graph each equation.

Solutions

(A) $(h, k) = (-3, 6)$ and $r = 4$:

$$(x - h)^2 + (y - k)^2 = r^2$$
$$[x - (-3)]^2 + (y - 6)^2 = 4^2$$
$$(x + 3)^2 + (y - 6)^2 = 16$$

To graph the equation, locate the center $C(-3, 6)$ and draw a circle of radius 4 (Fig. 15).

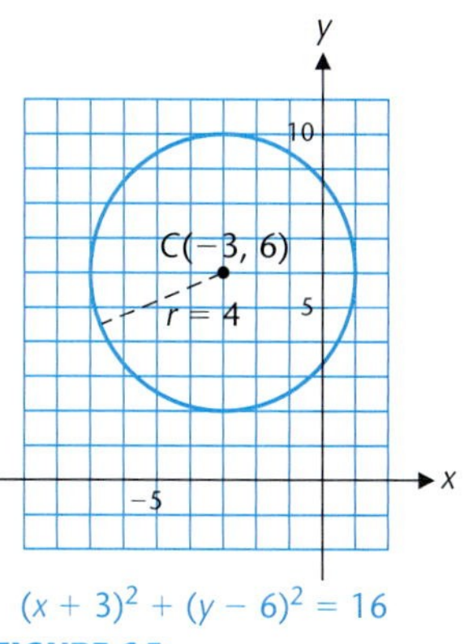

$(x + 3)^2 + (y - 6)^2 = 16$

FIGURE 15

(B) $(h, k) = (0, 0)$ and $r = 4$:

$$x^2 + y^2 = r^2$$
$$x^2 + y^2 = 4^2$$
$$x^2 + y^2 = 16$$

To graph the equation, locate the center at the origin and draw a circle of radius 4 (Fig. 16).

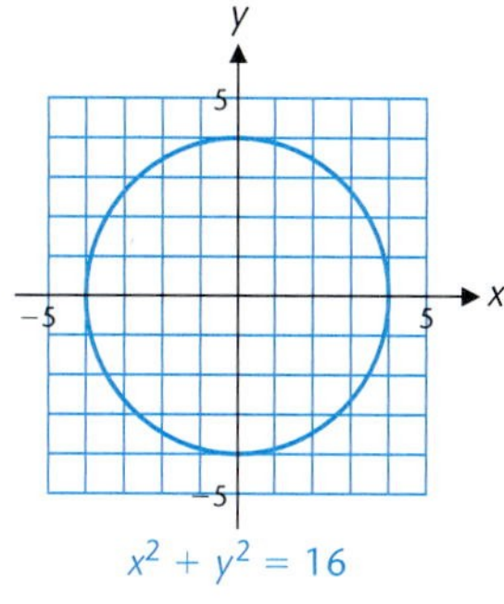

$x^2 + y^2 = 16$

FIGURE 16

Matched Problem 5 Find the equation of a circle with radius 3 and center at:

(A) $(3, -2)$ (B) $(0, 0)$

Graph each equation.

To graph the circle in Figure 16 on a graphing utility, we must graph two equations: $y_1 = \sqrt{16 - x^2}$ and $y_2 = -\sqrt{16 - x^2}$ [Fig. 17(a)]. Notice that the graph does not look circular, because the units on the x axis are physically longer than the units on the y axis in the rectangular viewing window. To rectify this, we must choose the window variables so that a length of one unit on the x axis equals a length of one unit on the y axis. The resulting window, called a **squared window,** displays circles that look circular [Fig. 17(b)]. Most graphing utilities have a routine, usually denoted by Zoom Square or something similar, that will adjust the window variables automatically to produce a squared window. Consult your manual for details.

FIGURE 17

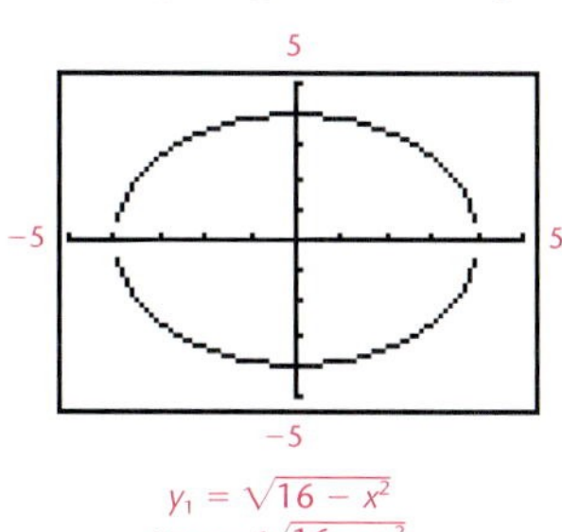

$y_1 = \sqrt{16 - x^2}$
$y_2 = -\sqrt{16 - x^2}$
(a)

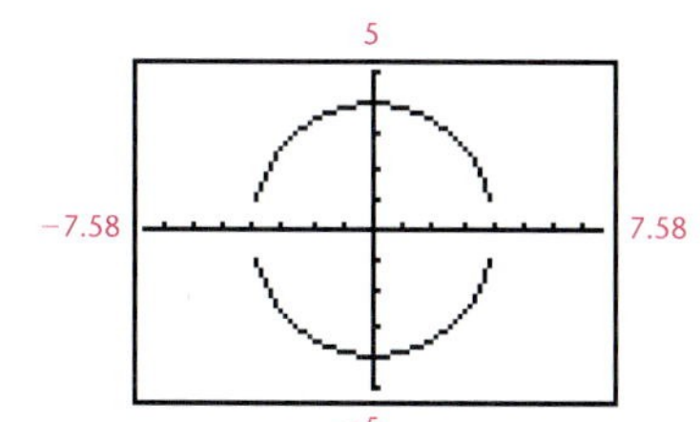

A squared window
(b)

Also notice that the graphs in Figure 17 have gaps near $x = -4$ and $x = 4$ due to the discrepancy between the actual coordinates of a point and the screen coordinates of the pixel containing that point. (Try tracing along the graph of $y_1 = \sqrt{16 - x^2}$ and observe what happens when x gets close to -4 or 4.) It is important to remember that graphing utilities are low-resolution devices that produce only rough approximations to graphs. It is up to you to visualize the correct appearance of a graph and fill in any missing gaps.

EXAMPLE 6 Finding the Center and Radius of a Circle

Find the center and radius of the circle with equation $x^2 + y^2 + 6x - 4y = 23$.

Solution We transform the equation into the form $(x - h)^2 + (y - k)^2 = r^2$ by completing the square relative to x and relative to y (see Section 1-6). From this standard form we can determine the center and radius.

$$x^2 + y^2 + 6x - 4y = 23$$

$$(x^2 + 6x \quad\quad) + (y^2 - 4y \quad\quad) = 23$$

$$(x^2 + 6x + 9) + (y^2 - 4y + 4) = 23 + 9 + 4 \qquad \text{Complete the squares.}$$

$$(x + 3)^2 + (y - 2)^2 = 36$$

$$[x - (-3)]^2 + (y - 2)^2 = 6^2$$

Center: $C(h, k) = C(-3, 2)$

Radius: $r = \sqrt{36} = 6$

Matched Problem 6 Find the center and radius of the circle with equation $x^2 + y^2 - 8x + 10y = -25$.

Answers to Matched Problems

1.

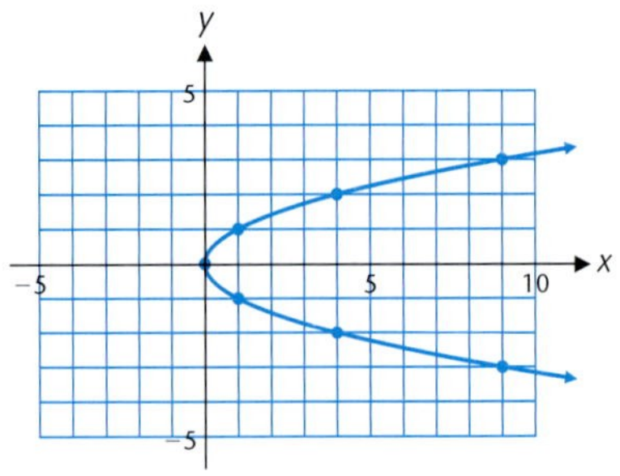

2. (A) 96 ppb (B) 1:45 P.M. and 5 P.M.

3. (A) Symmetric with respect to the origin

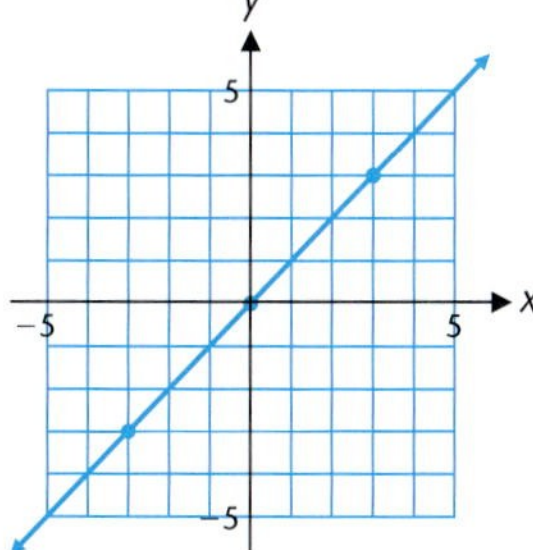

(B) Symmetric with respect to the y axis

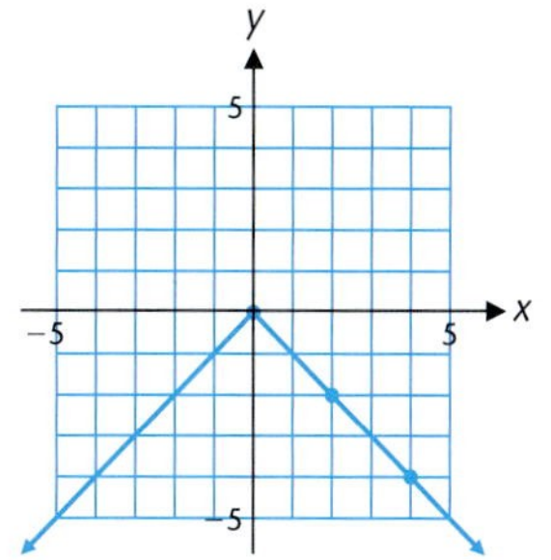

(C) Symmetric with respect to the x axis, y axis, and origin

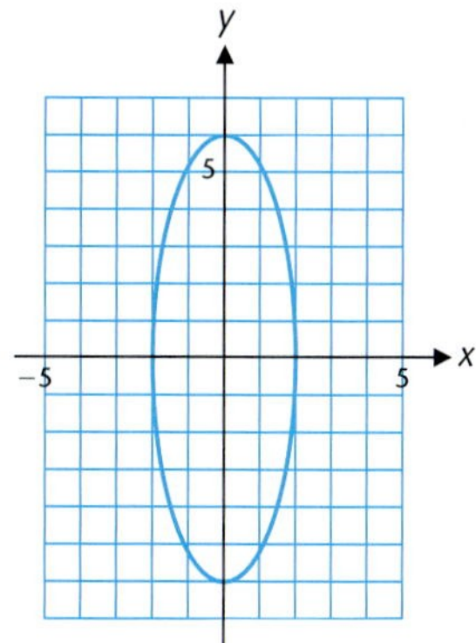

4. $d = \sqrt{173}$

5. (A) $(x - 3)^2 + (y + 2)^2 = 9$

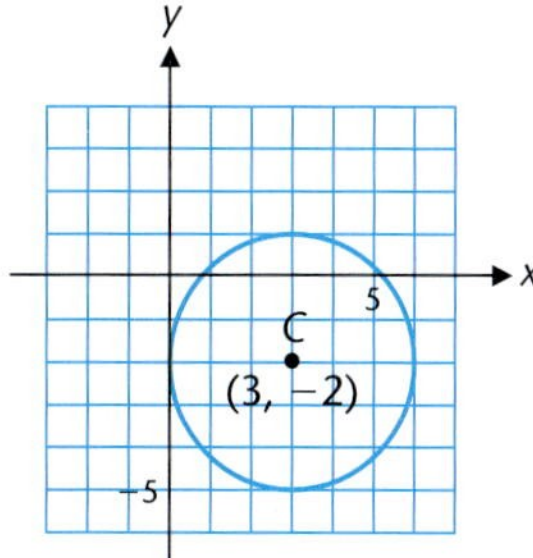

(B) $x^2 + y^2 = 9$

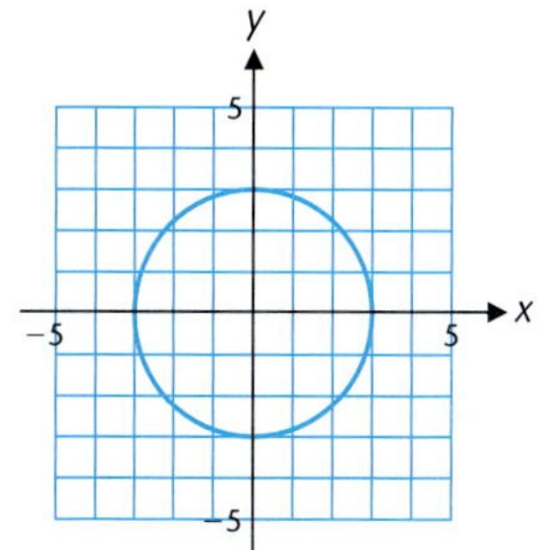

6. $(x - 4)^2 + (y + 5)^2 = 16$; radius: 4, center: $(4, -5)$

EXERCISE 2-1

A

In Problems 1–10, give a verbal description of the indicated subset of the plane in terms of quadrants and axes.

1. $\{(x, y) \mid x = 0\}$

2. $\{(x, y) \mid x > 0, y > 0\}$

3. $\{(x, y) \mid x < 0, y < 0\}$

4. $\{(x, y) \mid y = 0\}$

5. $\{(x, y) \mid x > 0, y < 0\}$

6. $\{(x, y) \mid y < 0, x \neq 0\}$

7. $\{(x, y) \mid xy < 0\}$

8. $\{(x, y) \mid x < 0, y > 0\}$

9. $\{(x, y) \mid x > 0, y \neq 0\}$

10. $\{(x, y) \mid xy > 0\}$

In Problems 11–18, determine symmetry with respect to the x axis, y axis, or origin, if any exists, and graph.

11. $y = 2x - 4$

12. $y = \frac{1}{2}x + 1$

13. $y = \frac{1}{2}x$

14. $y = 2x$

15. $|y| = x$

16. $|y| = -x$

17. $|x| = |y|$

18. $y = -x$

Find the distance between the indicated points in Problems 19–22. Leave the answer in radical form.

19. $(-5, -3), (4, 2)$

20. $(-6, 4), (2, -1)$

21. $(3, 5), (2, -4)$

22. $(2, -5), (-3, 1)$

In Problems 23–28, write the equation of a circle with the indicated center and radius.

23. $C(0, 0), r = 4$

24. $C(0, 0), r = 6$

25. $C(3, -2), r = 1$

26. $C(-4, 2), r = 5$

27. $C(2, 6), r = \sqrt{3}$

28. $C(-1, -3), r = \sqrt{5}$

In Problems 29 and 30, use the graph to estimate to the nearest integer the missing coordinates of the indicated points. (Be sure you find all possible answers.)

29. (A) $(-3, ?)$
(B) $(2, ?)$
(C) $(?, 3)$
(D) $(?, -1)$

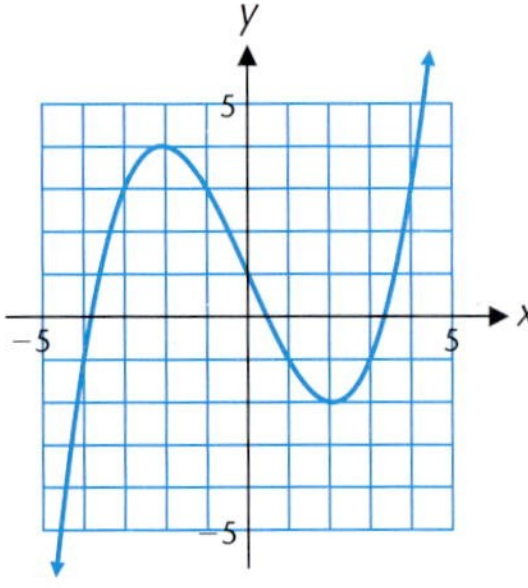

30. (A) $(-4, ?)$
(B) $(-1, ?)$
(C) $(?, 1)$
(D) $(?, 4)$

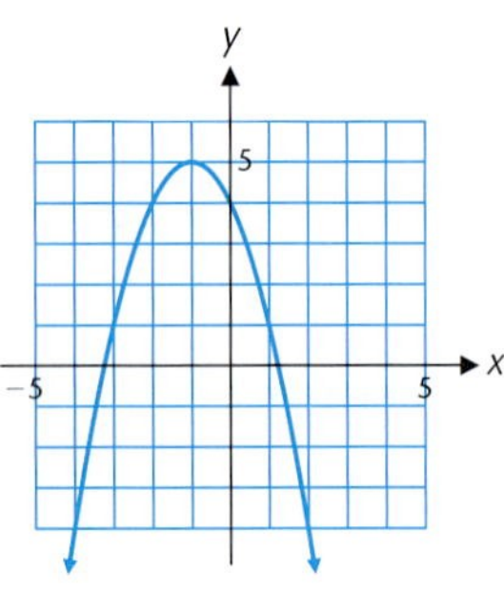

The figures in Problems 31 and 32 show a portion of a graph. Extend the given graph to one that exhibits the indicated type of symmetry.

31. (A) x axis only
(B) y axis only
(C) origin only
(D) x and y axes

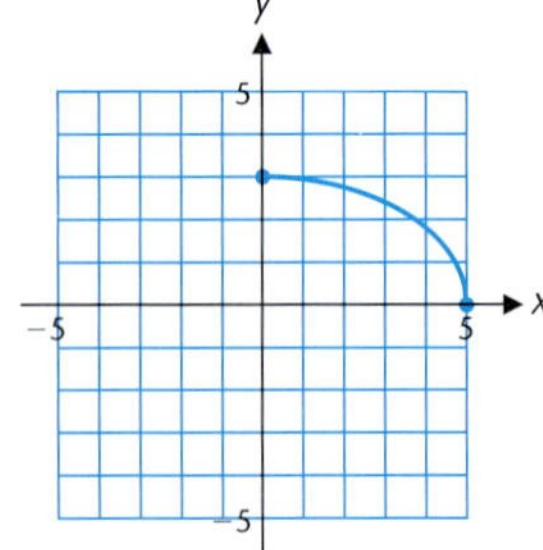

32. (A) x axis only
(B) y axis only
(C) origin only
(D) x and y axes

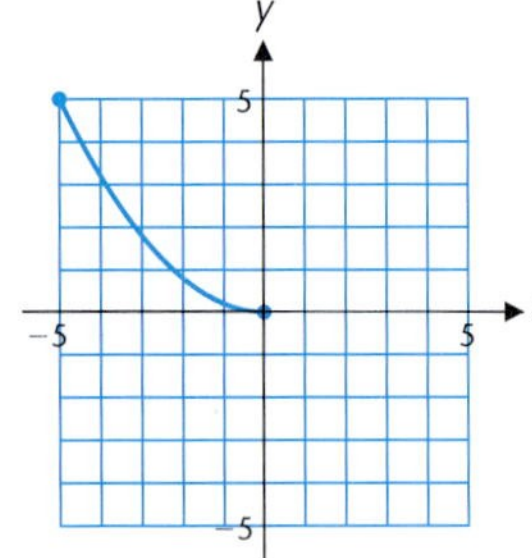

B

In Problems 33–48, determine symmetry with respect to the x axis, y axis, or origin, if any exists, and graph.

* *Check your graphs in Problems 33–48 by graphing on a graphing utility.*

*Please note that use of a graphing utility is not required to complete these exercises. Checking with a g.u. is optional. If you do not have a g.u., you should still work these exercises.

33. $y^2 = x + 2$ **34.** $y^2 = x - 2$ **35.** $y = x^2 + 1$

36. $y + 2 = x^2$ **37.** $x^2 + 4y^2 = 4$ **38.** $x^2 + 9y^2 = 9$

39. $4y^2 - x^2 = 1$ **40.** $4x^2 - y^2 = 1$ **41.** $y^3 = x$

42. $y = x^4$ **43.** $y = 0.6x^2 - 4.5$

44. $x = 0.8y^2 - 3.5$ **45.** $y = \sqrt{17 - x^2}$

46. $y = \sqrt{100 - 4x^2}$ **47.** $y = x^{2/3}$

48. $y^{2/3} = x$

In Problems 49 and 50, use the Pythagorean theorem to show that the given points are the vertices of a right triangle. Find the area and the perimeter (to two decimal places) of the triangle. (Formulas related to triangles can be found in Appendix C.)

49. $(-3, 2), (1, -2), (8, 5)$

50. $(-4, -1), (0, 7), (6, -6)$

51. Find x such that $(x, 7)$ is 10 units from $(-4, 1)$.

52. Find x such that $(x, 2)$ is 4 units from $(3, -3)$.

53. Find y such that $(2, y)$ is 3 units from $(-1, 4)$.

54. Find y such that $(3, y)$ is 13 units from $(-9, 2)$.

In Problems 55–60, find the center and radius of the circle with the given equation. Graph the equation.

55. $(x + 4)^2 + (y - 2)^2 = 7$

56. $(x - 5)^2 + (y + 7)^2 = 15$

57. $x^2 + y^2 - 6x - 4y = 36$

58. $x^2 + y^2 - 2x - 10y = 55$

59. $x^2 + y^2 + 8x - 6y + 8 = 0$

60. $x^2 + y^2 + 4x + 10y + 15 = 0$

In Problems 61–64, graph the triangle with vertices A, B, and C and the triangle with vertices A′, B′, and C′ in the same coordinate system. Describe the relationship between the graphs of these triangles in terms of reflections.

61. $A(1, 1), B(7, 2), C(4, 6)$
$A'(1, -1), B'(7, -2), C'(4, -6)$

62. $A(1, 1), B(7, 2), C(4, 6)$
$A'(-1, 1), B'(-7, 2), C'(-4, 6)$

63. $A(1, 1), B(7, 2), C(4, 6)$
$A'(-1, -1), B'(-7, -2), C'(-4, -6)$

64. $A(1, 2), B(1, 4), C(3, 4)$
$A'(2, 1), B'(4, 1), C'(4, 3)$
[*Hint:* Add the graph of $y = x$ to your graph.]

In Problems 65–68, solve for y, producing two equations, and then graph both of these equations in the same viewing window.

65. $x^2 + y^2 = 3$ **66.** $x^2 + y^2 = 5$

67. $(x + 3)^2 + (y + 1)^2 = 2$ **68.** $(x - 2)^2 + (y - 1)^2 = 3$

In Problems 69–72, graph each pair of equations in the same viewing window for the indicated values of x. Find the center and radius of the resulting circle by examining the graph, and find the equation of the circle. Explain how you could check your work, and then check it.

69. $y = \sqrt{2x - x^2}, y = -\sqrt{2x - x^2}, 0 \le x \le 2$

70. $y = 1 + \sqrt{1 - x^2}, y = 1 - \sqrt{1 - x^2}, -1 \le x \le 1$

71. $y = 1 + \sqrt{5 + 4x - x^2}, y = 1 - \sqrt{5 + 4x - x^2}, -1 \le x \le 5$

72. $y = -1 + \sqrt{4x - x^2}, y = -1 - \sqrt{4x - x^2}, 0 \le x \le 4$

C

In Problems 73–78, determine symmetry with respect to the x axis, y axis, or origin, if any exists, and graph.

Check your graphs in Problems 73–78 by graphing on a graphing utility.

73. $y^3 = |x|$ **74.** $|y| = x^3$ **75.** $xy = 1$

76. $xy = -1$ **77.** $y = 6x - x^2$ **78.** $y = x^2 - 6x$

79. Find the equation of the perpendicular bisector of the line segment joining $(-6, -2)$ and $(4, 4)$ by using the distance-between-two-points formula.

80. Use the distance-between-two-points formula to show that the point

$$\left(\frac{x_1 + x_2}{2}, \frac{y_1 + y_2}{2}\right)$$

is the **midpoint** of the line segment joining (x_1, y_1) and (x_2, y_2).

Find the equation of a circle that has a diameter with the end-points given in Problems 81 and 82. [Hint: See Problem 80.]

81. $(7, -3), (1, 7)$ **82.** $(-3, 2), (7, -4)$

83. Find the equation of a circle with center $(2, 2)$ whose graph passes through the point $(3, -5)$.

84. Find the equation of a circle with center $(-5, 4)$ whose graph passes through the point $(2, -3)$.

85. If a graph is symmetric with respect to the x axis and to the origin, must it be symmetric with respect to the y axis? Explain your answer.

86. If a graph is symmetric with respect to the y axis and to the origin, must it be symmetric with respect to the x axis? Explain your answer.

APPLICATIONS

87. Price and Demand. The quantity of a product that consumers are willing to buy during some period of time depends on its price. The price p and corresponding weekly demand q for a particular brand of diet soda in a city are shown in the figure. Use this graph to estimate the following demands to the nearest 100 cases.

(A) What is the demand when the price is \$6.00 per case?

(B) Does the demand increase or decrease if the price is increased from \$6.00 to \$6.30 per case? By how much?

(C) Does the demand increase or decrease if the price is decreased from \$6.00 to \$5.70? By how much?

(D) Write a brief description of the relationship between price and demand illustrated by this graph.

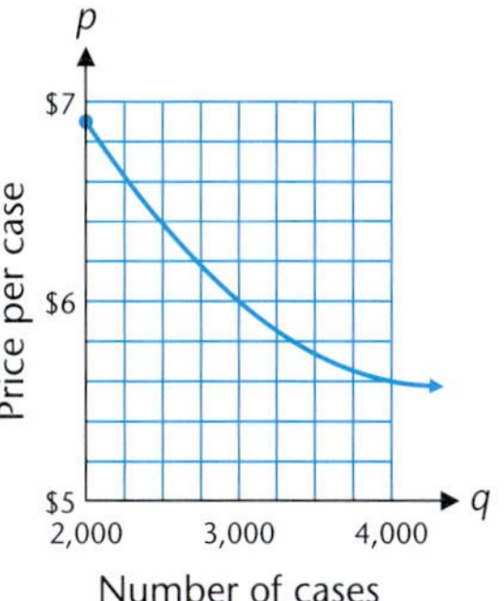

88. Price and Supply. The quantity of a product that suppliers are willing to sell during some period of time depends on its price. The price p and corresponding weekly supply q for a particular brand of diet soda in a city are shown in the figure. Use this graph to estimate the following supplies to the nearest 100 cases.

(A) What is the supply when the price is \$5.60 per case?

(B) Does the supply increase or decrease if the price is increased from \$5.60 to \$5.80 per case? By how much?

(C) Does the supply increase or decrease if the price is decreased from \$5.60 to \$5.40 per case? By how much?

(D) Write a brief description of the relationship between price and supply illustrated by this graph.

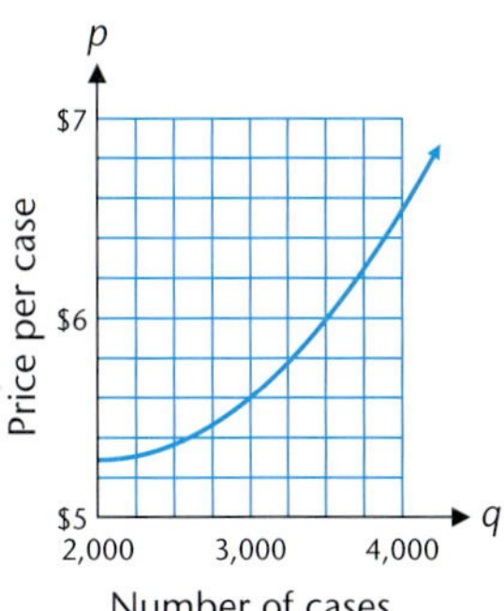

89. Temperature. The temperature during a spring day in the Midwest is given in the figure. Use this graph to estimate the following temperatures to the nearest degree and times to the nearest hour.

(A) The temperature at 9:00 A.M.

(B) The highest temperature and the time when it occurs.

(C) The time(s) when the temperature is 49°F.

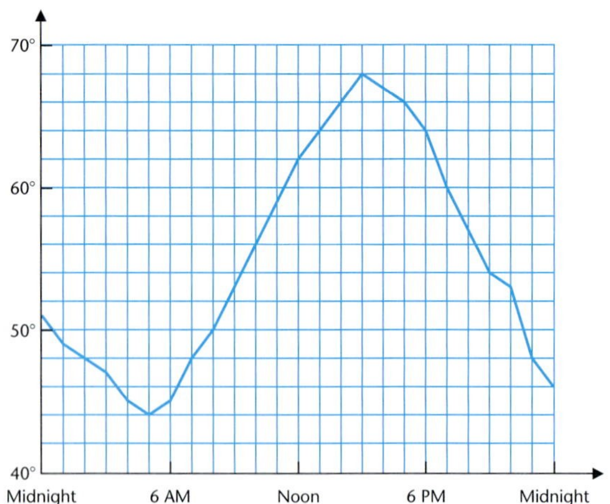

90. Temperature. Use the graph in Problem 89 to estimate the following temperatures to the nearest degree and times to the nearest half hour.

(A) The temperature at 7:00 P.M.

(B) The lowest temperature and the time when it occurs.

(C) The time(s) when the temperature is 52°F.

91. Physics. The speed (in meters per second) of a ball swinging at the end of a pendulum is given by

$$v = 0.5\sqrt{2 - x}$$

where x is the vertical displacement (in centimeters) of the ball from its position at rest (see the figure).

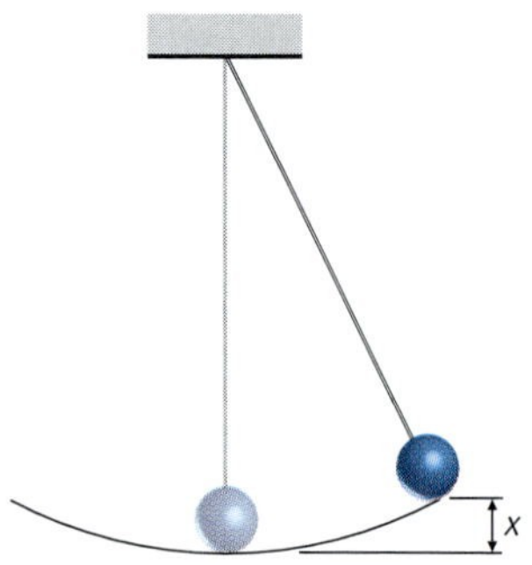

(A) Graph v for $0 \leq x \leq 2$.

(B) Describe the relationship between this graph and the physical behavior of the ball as it swings back and forth.

92. Physics. The speed (in meters per second) of a ball oscillating at the end of a spring is given by

$$v = 4\sqrt{25 - x^2}$$

where x is the vertical displacement (in centimeters) of the ball from its position at rest (positive displacement measured downwards—see the figure).

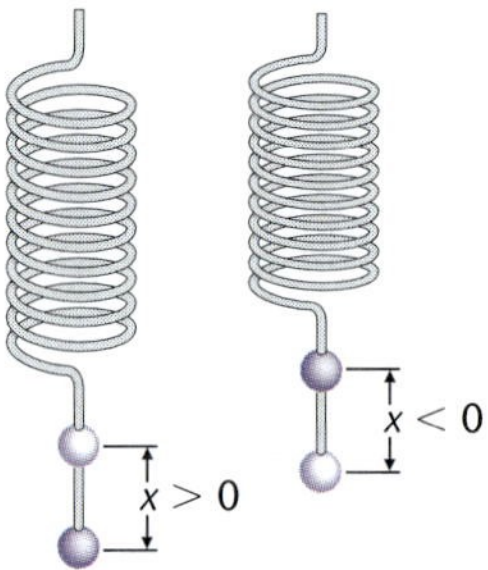

(A) Graph v for $-5 \leq x \leq 5$.

(B) Describe the relationship between this graph and the physical behavior of the ball as it oscillates up and down.

93. Architecture. An arched doorway is formed by placing a circular arc on top of a rectangle (see the figure). If the doorway is 4 feet wide and the height of the arc above its ends is 1 foot, what is the radius of the circle containing the arc? [*Hint:* Note that $(2, r - 1)$ must satisfy $x^2 + y^2 = r^2$.]

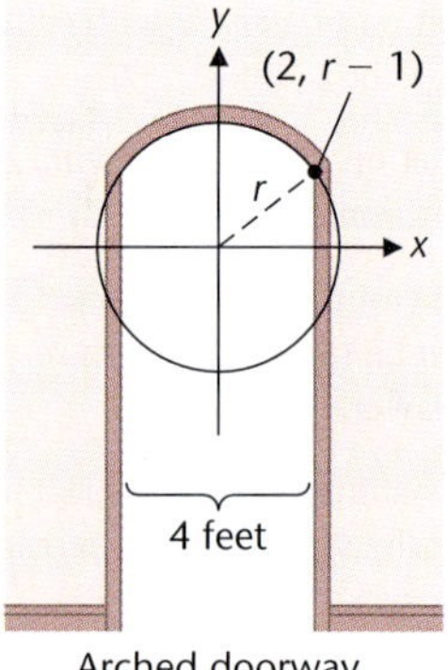

Arched doorway

94. Engineering. The cross section of a rivet has a top that is an arc of a circle (see the figure). If the ends of the arc are 12 millimeters apart and the top is 4 millimeters above the ends, what is the radius of the circle containing the arc?

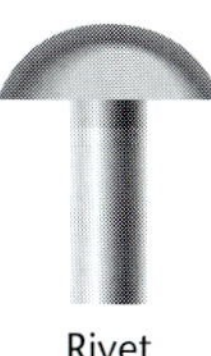

Rivet

★ **95. Construction.** Town B is located 36 miles east and 15 miles north of town A (see the figure). A local telephone company wants to position a relay tower so that the distance from the tower to town B is twice the distance from the tower to town A.

(A) Show that the tower must lie on a circle, find the center and radius of this circle, and graph.

(B) If the company decides to position the tower on this circle at a point directly east of town A, how far from town A should they place the tower? Compute answer to one decimal place.

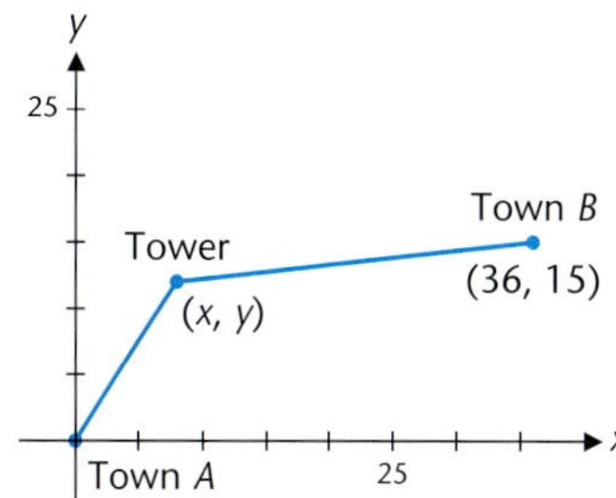

★ **96. Construction.** Repeat Problem 95 if the distance from the tower to town A is twice the distance from the tower to town B.

SECTION 2-2 Straight Lines

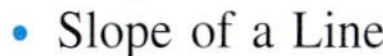

- Graphs of First-Degree Equations in Two Variables
- Slope of a Line
- Equations of a Line—Special Forms
- Parallel and Perpendicular Lines

In this section we will consider one of the most basic geometric figures—a straight line. We will learn how to graph straight lines, given various standard equations, and how to find the equation of a straight line, given information about the line. Adding these important tools to our mathematical toolbox will enable us to use straight lines as an effective problem-solving tool, as evidenced by the application exercises at the end of this section.

• Graphs of First-Degree Equations in Two Variables

With your past experience in graphing equations in two variables, you probably remember that first-degree equations in two variables, such as

$$y = -3x + 5 \qquad 3x - 4y = 9 \qquad y = -\tfrac{2}{3}x$$

have graphs that are straight lines. This fact is stated in Theorem 1. For a partial proof of this theorem, see Problem 80 of the exercises at the end of this section.

Theorem 1 **The Equation of a Straight Line**

If A, B, and C are constants, with A and B not both 0, and x and y are variables, then the graph of the equation

$$Ax + By = C \qquad \textbf{Standard Form} \qquad (1)$$

is a straight line. Any straight line in a rectangular coordinate system has an equation of this form.

Also, the graph of any equation of the form

$$y = mx + b \tag{2}$$

where m and b are constants, is a straight line. Form (2), which we will discuss in detail later, is simply a special case of form (1) for $B \neq 0$. This can be seen by solving form (1) for y in terms of x:

$$y = -\frac{A}{B}x + \frac{C}{B} \qquad B \neq 0$$

To graph either equation (1) or (2), we plot any two points from the solution set and use a straightedge to draw a line through these two points. The points where the line crosses the axes are convenient to use and easy to find. The **y intercept*** is the ordinate of the point where the graph crosses the y axis, and the **x intercept** is the abscissa of the point where the graph crosses the x axis. To find the y intercept, let $x = 0$ and solve for y; to find the x intercept, let $y = 0$ and solve for x. It is often advisable to find a third point as a checkpoint. All three points must lie on the same straight line or a mistake has been made.

EXAMPLE 1 Using Intercepts to Graph a Straight Line

Graph the equation $3x - 4y = 12$.

Solution Find intercepts, a third checkpoint (optional), and draw a line through the two (three) points (Fig. 1).

FIGURE 1

x	0	4	8
y	-3	0	3

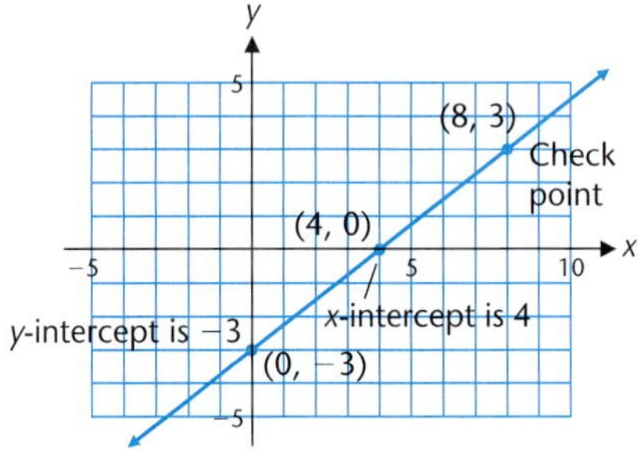

Matched Problem 1 Graph the equation $4x + 3y = 12$.

*If the x intercept is a and the y intercept is b, then the graph of the line passes through the points $(a, 0)$ and $(0, b)$. It is common practice to refer to both the numbers a and b and the points $(a, 0)$ and $(0, b)$ as the x and y intercepts of the line.

To check the answer to Example 1 on a graphing utility, we first solve the equation for y and then graph (Fig. 2):

$$3x - 4y = 12$$
$$-4y = -3x + 12$$
$$y = 0.75x - 3$$

FIGURE 2

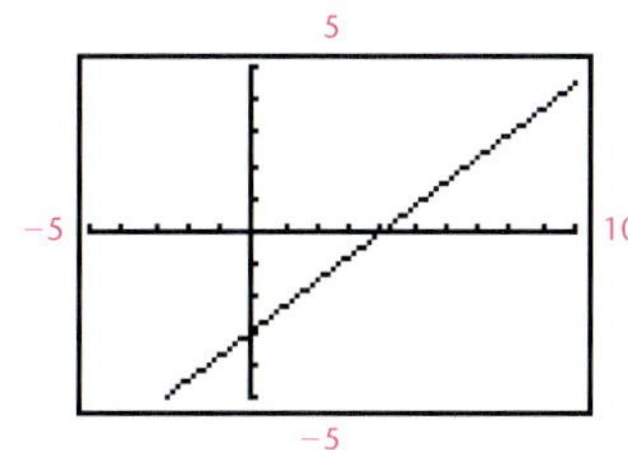

• Slope of a Line

If we take two points $P_1(x_1, y_1)$ and $P_2(x_2, y_2)$ on a line, then the ratio of the change in y to the change in x as we move from point P_1 to point P_2 is called the **slope** of the line. Roughly speaking, slope is a measure of the "steepness" of a line. Sometimes the change in x is called the **run** and the change in y is called the **rise.**

DEFINITION 1 **Slope of a Line**

If a line passes through two distinct points $P_1(x_1, y_1)$ and $P_2(x_2, y_2)$, then its slope m is given by the formula

$$m = \frac{y_2 - y_1}{x_2 - x_1} \qquad x_1 \neq x_2$$

$$= \frac{\text{Vertical change (rise)}}{\text{Horizontal change (run)}}$$

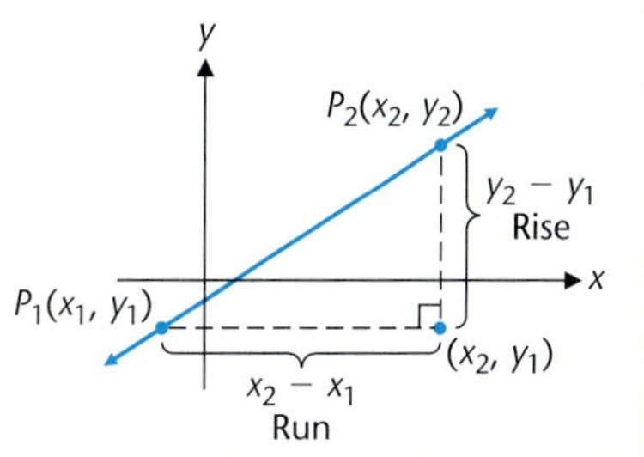

For a horizontal line, y doesn't change as x changes; hence, its slope is 0. On the other hand, for a vertical line, x doesn't change as y changes; hence, $x_1 = x_2$ and its slope is not defined:

$$\frac{y_2 - y_1}{x_2 - x_1} = \frac{y_2 - y_1}{0}$$ For a vertical line, slope is not defined.

In general, the slope of a line may be positive, negative, 0, or not defined. Each of these cases is interpreted geometrically as shown in Table 1.

TABLE 1 Geometric Interpretation of Slope

Line	Slope	Example
Rising as x moves from left to right	Positive	
Falling as x moves from left to right	Negative	
Horizontal	0	
Vertical	Not defined	

In using the formula to find the slope of the line through two points, it doesn't matter which point is labeled P_1 or P_2, since changing the labeling will change the sign in both the numerator and denominator of the slope formula:

$$\frac{y_2 - y_1}{x_2 - x_1} = \frac{y_1 - y_2}{x_1 - x_2} \qquad \text{For example: } \frac{5-2}{7-3} = \frac{2-5}{3-7}$$

In addition, it is important to note that the definition of slope doesn't depend on the two points chosen on the line as long as they are distinct. This follows from the fact that the ratios of corresponding sides of similar triangles are equal.

EXAMPLE 2 Finding Slopes

Sketch a line through each pair of points and find the slope of each line.

(A) $(-3, -4), (3, 2)$ (B) $(-2, 3), (1, -3)$
(C) $(-4, 2), (3, 2)$ (D) $(2, 4), (2, -3)$

Solutions (A)

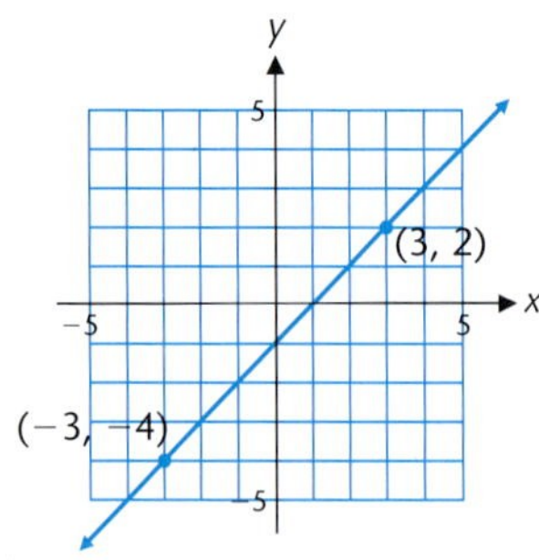

$$m = \frac{2-(-4)}{3-(-3)} = \frac{6}{6} = 1$$

(B)

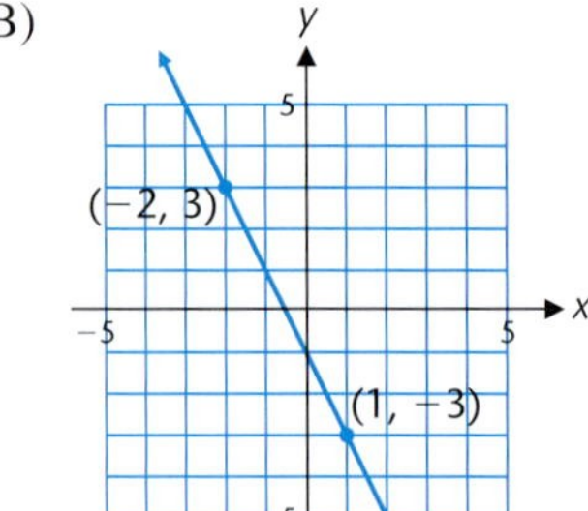

$$m = \frac{-3-3}{1-(-2)} = \frac{-6}{3} = -2$$

(C)

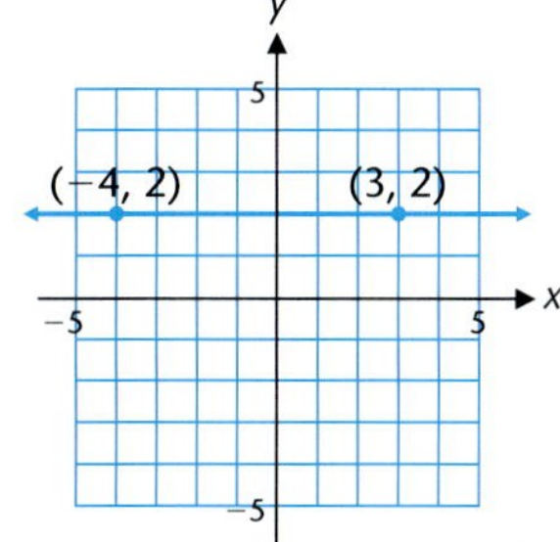

$$m = \frac{2 - 2}{3 - (-4)} = \frac{0}{7} = 0$$

(D)

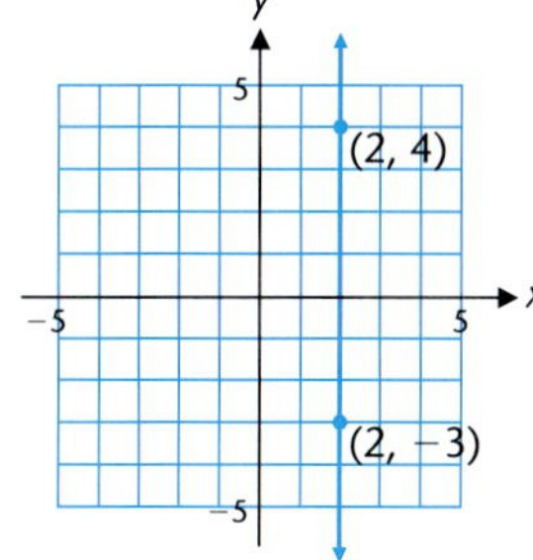

$$m = \frac{-3 - 4}{2 - 2} = \frac{-7}{0};$$

slope is not defined

Matched Problem 2 Find the slope of the line through each pair of points. Do not graph.

(A) $(-3, -3)$, $(2, -3)$ (B) $(-2, -1)$, $(1, 2)$
(C) $(0, 4)$, $(2, -4)$ (D) $(-3, 2)$, $(-3, -1)$

• Equations of a Line—Special Forms

Let us start by investigating why $y = mx + b$ is called the *slope–intercept form* for a line.

EXPLORE-DISCUSS 1

(A) Graph $y = x + b$ for $b = -5, -3, 0, 3,$ and 5 simultaneously in the same coordinate system. Verbally describe the geometric significance of b.

(B) Graph $y = mx - 1$ for $m = -2, -1, 0, 1,$ and 2 simultaneously in the same coordinate system. Verbally describe the geometric significance of m.

(C) Using a graphing utility, explore the graph of $y = mx + b$ for different values of m and b.

As you can see from the above exploration, the constants m and b in

$$y = mx + b \tag{3}$$

have special geometric significance, which we now explicitly state.

If we let $x = 0$, then $y = b$, and we observe that the graph of equation (3) crosses the y axis at $(0, b)$. The constant b is the y *intercept.* For example, the y intercept of the graph of $y = -3x - 2$ is -2.

To determine the geometric significance of m, we proceed as follows: If $y = mx + b$, then by setting $x = 0$ and $x = 1$, we conclude that both $(0, b)$ and $(1, m + b)$ lie on the graph, which is a line. Hence, the slope of this line is given by

$$\text{Slope} = \frac{y_2 - y_1}{x_2 - x_1} = \frac{(m + b) - b}{1 - 0} = m$$

Thus, m is the slope of the line given by $y = mx + b$. Now we know why equation (3) is called the **slope–intercept form** of an equation of a line.

Theorem 2 **Slope–Intercept Form**

$$y = mx + b$$

$$m = \frac{\text{Rise}}{\text{Run}} = \text{Slope}$$

$$b = y \text{ intercept}$$

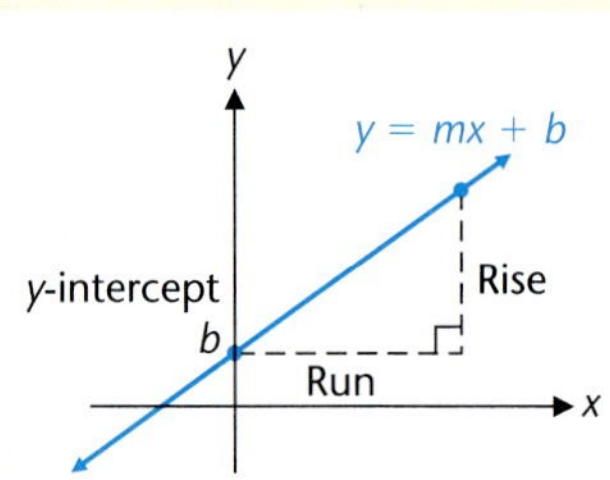

EXAMPLE 3 **Using the Slope–Intercept Form**

(A) Write the slope–intercept equation of a line with slope $\frac{2}{3}$ and y intercept -5.

(B) Find the slope and y intercept, and graph $y = -\frac{3}{4}x - 2$.

Solutions (A) Substitute $m = \frac{2}{3}$ and $b = -5$ in $y = mx + b$ to obtain

$$y = \tfrac{2}{3}x - 5$$

(B) The y intercept of $y = -\frac{3}{4}x - 2$ is -2, so the point $(0, -2)$ is on the graph. The slope of the line is $-\frac{3}{4}$, so when the x coordinate of $(0, -2)$ increases (runs) by 4 units, the y coordinate changes (rises) by -3. The resulting point $(4, -5)$ is easily plotted, and the two points yield the graph of the line. In short, we start at the y intercept -2, and move 4 units to the right and 3 units down to obtain a second point. We then draw a line through the two points, as shown in Figure 3.

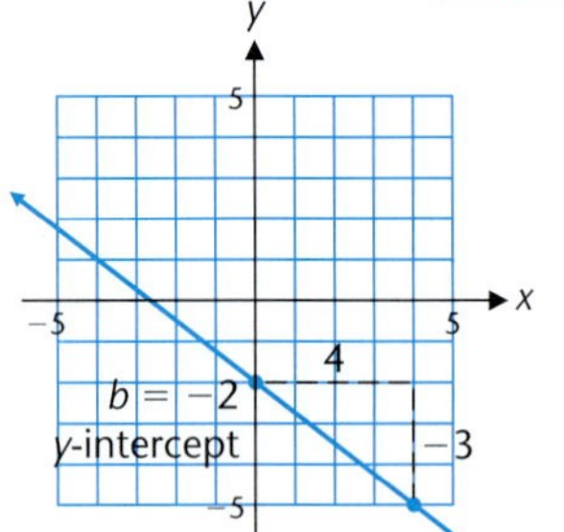

FIGURE 3

Matched Problem 3 Write the slope–intercept equation of the line with slope $\frac{2}{3}$ and y intercept 1. Graph the equation.

In Example 3 we found the equation of a line with a given slope and y intercept. It is also possible to find the equation of a line passing through a given point with a given slope or to find the equation of a line containing two given points.

Suppose a line has slope m and passes through a fixed point (x_1, y_1). If the point (x, y) is any other point on the line, then

$$\frac{y - y_1}{x - x_1} = m \qquad x \neq x_1$$

that is,

$$y - y_1 = m(x - x_1) \tag{4}$$

We now observe that (x_1, y_1) also satisfies equation (4) and conclude that (4) is an equation of a line with slope m that passes through (x_1, y_1).

We have just obtained the **point–slope form** of the equation of a line.

Theorem 3 **Point–Slope Form**

An equation of a line through a point $P_1(x_1, y_1)$ with slope m is

$$y - y_1 = m(x - x_1)$$

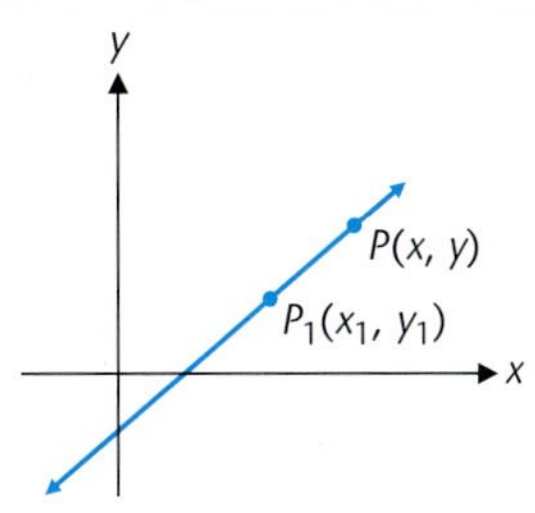

Remember that $P(x, y)$ is a variable point and $P_1(x_1, y_1)$ is fixed.

The point–slope form is extremely useful, since it enables us to find an equation for a line if we know its slope and the coordinates of a point on the line or if we know the coordinates of two points on the line. In the latter case, we find the slope first using the coordinates of the two points; then we use the point–slope form with either of the two given points.

EXAMPLE 4 **Using the Point–Slope Form**

(A) Find an equation for the line that has slope $\frac{2}{3}$ and passes through the point $(-2, 1)$. Write the final answer in the standard form $Ax + By = C$.

(B) Find an equation for the line that passes through the two points $(4, -1)$ and $(-8, 5)$. Write the final answer in the slope–intercept form $y = mx + b$.

Solutions (A) Let $m = \frac{2}{3}$ and $(x_1, y_1) = (-2, 1)$. Then

$$y - y_1 = m(x - x_1)$$

$$y - 1 = \tfrac{2}{3}[x - (-2)]$$

$$y - 1 = \tfrac{2}{3}(x + 2)$$

$$3y - 3 = 2x + 4$$

$$-2x + 3y = 7 \qquad \text{or} \qquad 2x - 3y = -7$$

(B) First, find the slope of the line by using the slope formula:

$$m = \frac{y_2 - y_1}{x_2 - x_1} = \frac{5 - (-1)}{-8 - 4} = \frac{6}{-12} = -\frac{1}{2}$$

Now let (x_1, y_1) be either of the two given points and proceed as in part A—we choose $(x_1, y_1) = (4, -1)$:

$$\begin{aligned} y - y_1 &= m(x - x_1) \\ y - (-1) &= -\tfrac{1}{2}(x - 4) \\ y + 1 &= -\tfrac{1}{2}(x - 4) \\ y + 1 &= -\tfrac{1}{2}x + 2 \\ y &= -\tfrac{1}{2}x + 1 \end{aligned}$$

You should verify that using $(-8, 5)$, the other given point, produces the same equation.

Matched Problem 4 (A) Find an equation for the line that has slope $-\frac{2}{5}$ and passes through the point $(3, -2)$. Write the final answer in the standard form $Ax + By = C$.

(B) Find an equation for the line that passes through the two points $(-3, 1)$ and $(7, -3)$. Write the final answer in the slope–intercept form $y = mx + b$.

EXAMPLE 5 Business Markup Policy

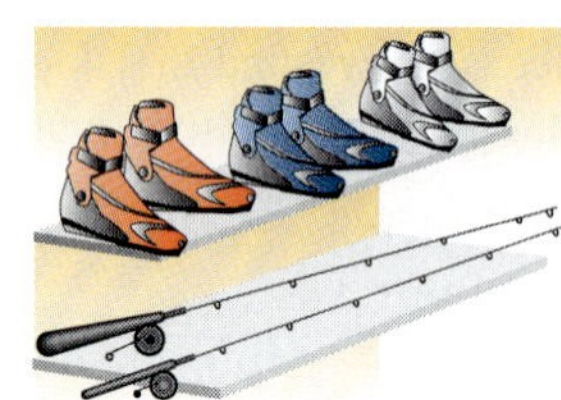

A sporting goods store sells a fishing rod that cost \$60 for \$82 and a pair of cross-country ski boots that cost \$80 for \$106.

(A) If the markup policy of the store for items that cost more than \$30 is assumed to be linear and is reflected in the pricing of these two items, write an equation that relates retail price R to cost C.

(B) Use the equation to find the retail price for a pair of running shoes that cost \$40.

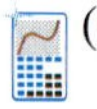

(C) Check with a graphing utility.

Solutions (A) If the retail price R is assumed to be linearly related to cost C, then we are looking for an equation whose graph passes through $(C_1, R_1) = (60, 82)$ and $(C_2, R_2) = (80, 106)$. We find the slope, and then use the point–slope form to find the equation.

$$m = \boxed{\frac{R_2 - R_1}{C_2 - C_1}} = \frac{106 - 82}{80 - 60} = \frac{24}{20} = 1.2$$

$$\begin{aligned} R - R_1 &= m(C - C_1) \\ R - 82 &= 1.2(C - 60) \\ R - 82 &= 1.2C - 72 \\ R &= 1.2C + 10 \end{aligned}$$

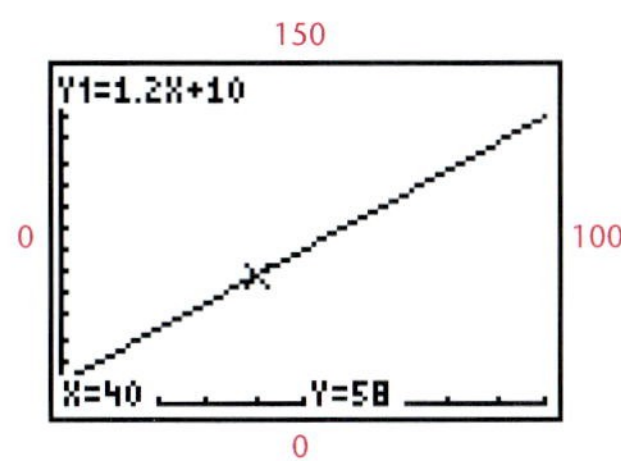

FIGURE 4

(B) $R = 1.2(40) + 10 = \$58$

(C) The check is shown in Figure 4.

Matched Problem 5 The management of a company that manufactures ballpoint pens estimates costs for running the company to be \$200 per day at zero output and \$700 per day at an output of 1,000 pens.

(A) Assuming total cost per day C is linearly related to total output per day x, write an equation relating these two quantities.
(B) What is the total cost per day for an output of 5,000 pens?

The simplest equations of lines are those for horizontal and vertical lines. Consider the following two equations:

$$x + 0y = a \quad \text{or} \quad x = a \tag{5}$$

$$0x + y = b \quad \text{or} \quad y = b \tag{6}$$

In equation (5), y can be any number as long as $x = a$. Thus, the graph of $x = a$ is a vertical line crossing the x axis at $(a, 0)$. In equation (6), x can be any number as long as $y = b$. Thus, the graph of $y = b$ is a horizontal line crossing the y axis at $(0, b)$. We summarize these results as follows:

Theorem 4

Vertical and Horizontal Lines

Equation		Graph
$x = a$	(short for $x + 0y = a$)	Vertical line through $(a, 0)$ (Slope is undefined.)
$y = b$	(short for $0x + y = b$)	Horizontal line through $(0, b)$ (Slope is 0.)

EXAMPLE 6 Graphing Horizontal and Vertical Lines

Graph the line $x = -2$ and the line $y = 3$.

Solution

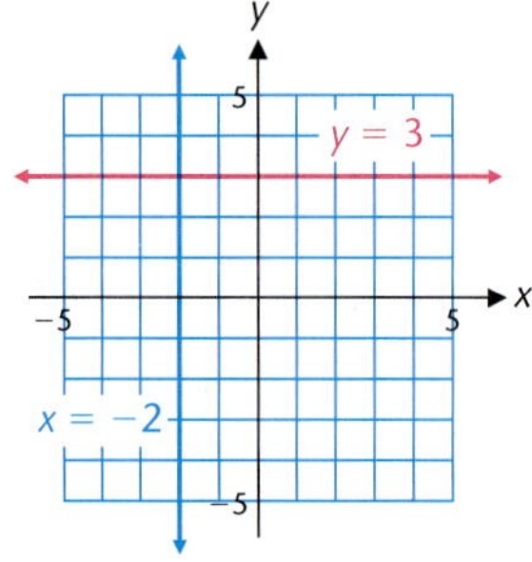

Matched Problem 6 Graph the line $x = 4$ and the line $y = -2$.

The various forms of the equation of a line that we have discussed are summarized in Table 2 for convenient reference.

TABLE 2 Equations of a Line

Standard form	$Ax + By = C$	A and B not both 0
Slope–intercept form	$y = mx + b$	Slope: m; y intercept: b
Point–slope form	$y - y_1 = m(x - x_1)$	Slope: m; Point: (x_1, y_1)
Horizontal line	$y = b$	Slope: 0
Vertical line	$x = a$	Slope: Undefined

EXPLORE-DISCUSS 2 Determine conditions on A, B, and C so that the linear equation $Ax + By = C$ can be written in each of the following forms, and discuss the possible number of x and y intercepts in each case.

1. $y = mx + b$, $m \neq 0$
2. $y = b$
3. $x = a$

• Parallel and Perpendicular Lines

From geometry, we know that two vertical lines are parallel to each other and that a horizontal line and a vertical line are perpendicular to each other. How can we tell when two nonvertical lines are parallel or perpendicular to each other? Theorem 5, which we state without proof, provides a convenient test.

Theorem 5 **Parallel and Perpendicular Lines**

Given two nonvertical lines L_1 and L_2 with slopes m_1 and m_2, respectively, then

$$L_1 \parallel L_2 \quad \text{if and only if} \quad m_1 = m_2$$

$$L_1 \perp L_2 \quad \text{if and only if} \quad m_1 m_2 = -1$$

The symbols $\parallel$ and $\perp$ mean, respectively, "is parallel to" and "is perpendicular to." In the case of perpendicularity, the condition $m_1 m_2 = -1$ also can be written as

$$m_2 = -\frac{1}{m_1} \quad \text{or} \quad m_1 = -\frac{1}{m_2}$$

Thus:

Two nonvertical lines are perpendicular if and only if their slopes are the negative reciprocals of each other.

EXAMPLE 7 **Parallel and Perpendicular Lines**

Given the line: L: $3x - 2y = 5$ and the point $P(-3, 5)$, find an equation of a line through P that is:

(A) Parallel to L (B) Perpendicular to L

Write the final answers in the slope–intercept form $y = mx + b$.

Solutions First, find the slope of L by writing $3x - 2y = 5$ in the equivalent slope–intercept form $y = mx + b$:

$$3x - 2y = 5$$

$$-2y = -3x + 5$$

$$y = \tfrac{3}{2}x - \tfrac{5}{2}$$

Thus, the slope of L is $\frac{3}{2}$. The slope of a line parallel to L is the same, $\frac{3}{2}$, and the slope of a line perpendicular to L is $-\frac{2}{3}$. We now can find the equations of the two lines in parts A and B using the point–slope form.

(A) Parallel ($m = \frac{3}{2}$):

$$y - y_1 = m(x - x_1)$$
$$y - 5 = \tfrac{3}{2}(x + 3)$$
$$y - 5 = \tfrac{3}{2}x + \tfrac{9}{2}$$
$$y = \tfrac{3}{2}x + \tfrac{19}{2}$$

(B) Perpendicular ($m = -\frac{2}{3}$):

$$y - y_1 = m(x - x_1)$$
$$y - 5 = -\tfrac{2}{3}(x + 3)$$
$$y - 5 = -\tfrac{2}{3}x - 2$$
$$y = -\tfrac{2}{3}x + 3$$

Matched Problem 7 Given the Line L: $4x + 2y = 3$ and the point $P(2, -3)$, find an equation of a line through P that is:

(A) Parallel to L (B) Perpendicular to L

Write the final answers in the slope–intercept form $y = mx + b$.

Answers to Matched Problems

1.

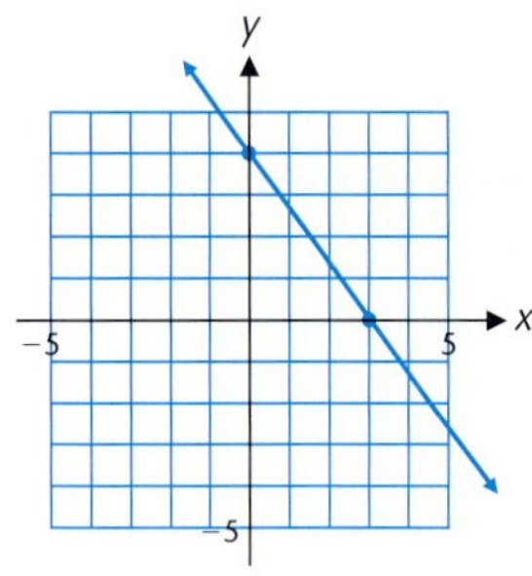

2. (A) $m = 0$ (B) $m = 1$
(C) $m = -4$ (D) m is not defined

3. $y = \frac{2}{3}x + 1$

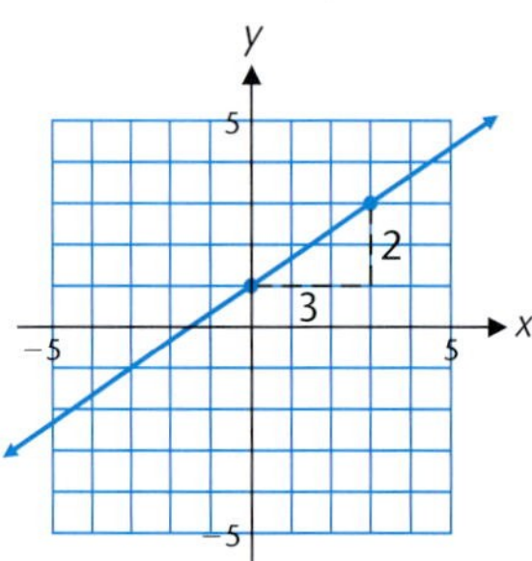

4. (A) $2x + 5y = -4$ (B) $y = -\frac{2}{5}x - \frac{1}{5}$

5. (A) $C = 0.5x + 200$ (B) $2,700

6.

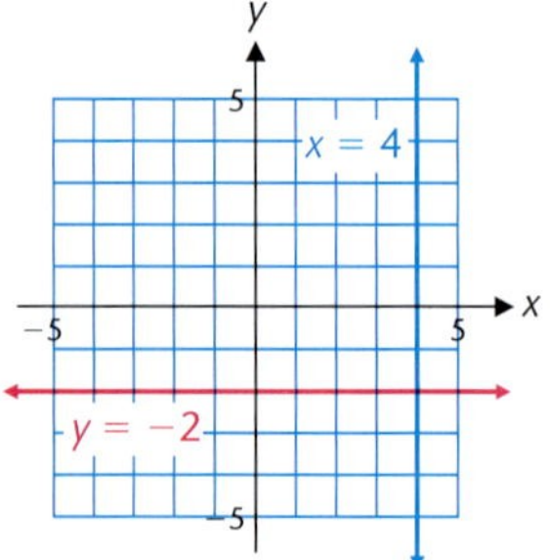

7. (A) $y = -2x + 1$ (B) $y = \frac{1}{2}x - 4$

EXERCISE 2-2

A

In Problems 1–6, use the graph of each line to find the x intercept, y intercept, and slope. Write the slope-intercept form of the equation of the line.

1.

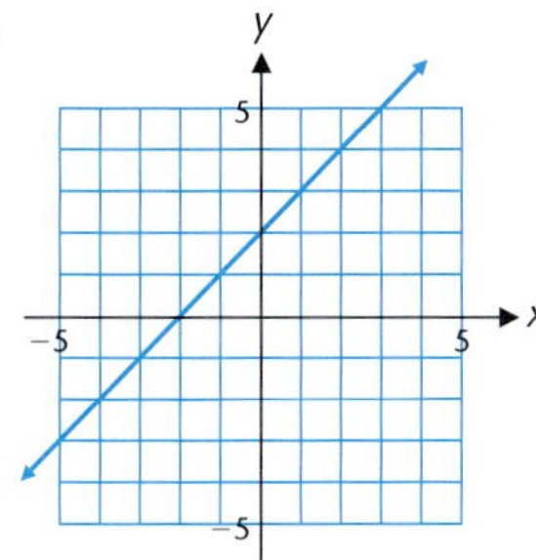

2.

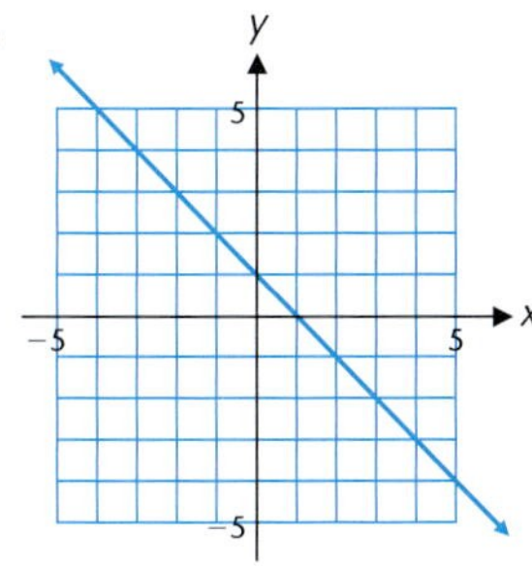

3.

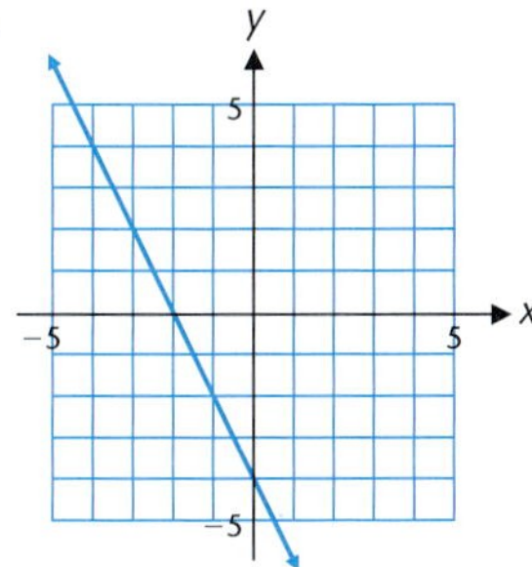

4.

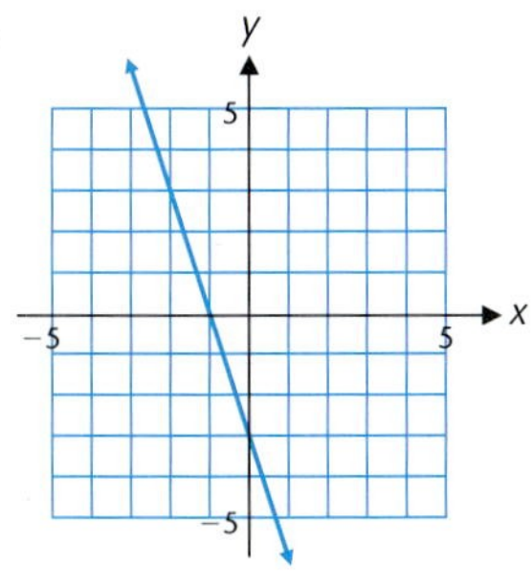

5.

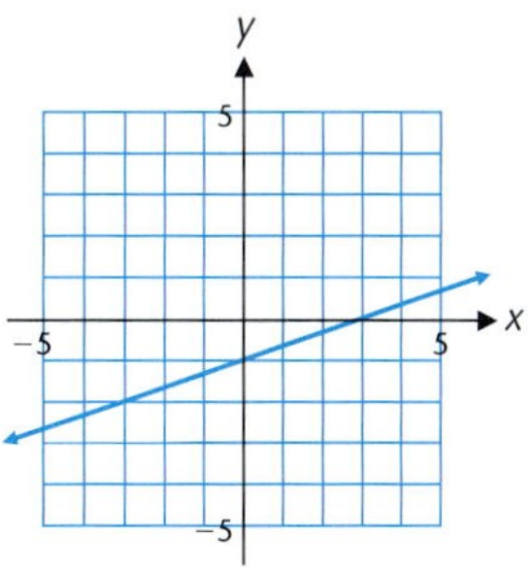

6.

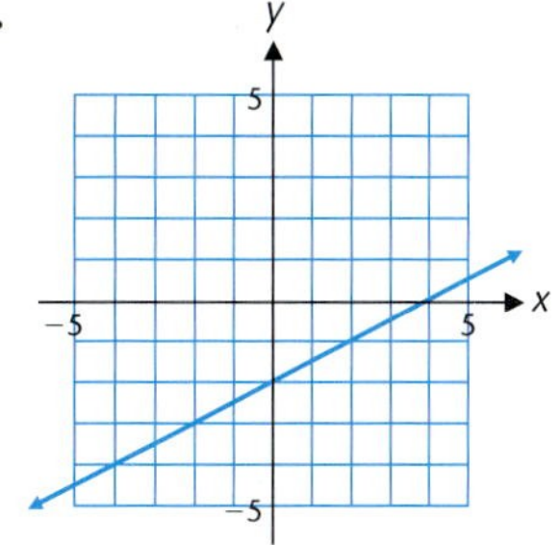

Graph each equation in Problems 7–20, and indicate the slope, if it exists.

Check your graphs in Problems 7–20 by graphing each on a graphing utility.

7. $y = -\frac{3}{5}x + 4$

8. $y = -\frac{3}{2}x + 6$

9. $y = -\frac{3}{4}x$

10. $y = \frac{2}{3}x - 3$

11. $2x - 3y = 15$

12. $4x + 3y = 24$

13. $4x - 5y = -24$

14. $6x - 7y = -49$

15. $\frac{y}{8} - \frac{x}{4} = 1$

16. $\frac{y}{6} - \frac{x}{5} = 1$

17. $x = -3$

18. $y = -2$

19. $y = 3.5$

20. $x = 2.5$

In Problems 21–24, find the equation of the line with the indicated slope and y intercept. Write the final answer in the standard form $Ax + By = C$, $A \geq 0$.

21. Slope $= 1$; y intercept $= 0$

22. Slope $= -1$; y intercept $= 7$

23. Slope $= -\frac{2}{3}$; y intercept $= -4$

24. Slope $= \frac{5}{3}$; y intercept $= 6$

B

In Problems 25–28, find the equation of the line passing through the given point with the given slope. Write the final answer in the slope-intercept form $y = mx + b$.

25. $(0, 3)$; $m = -2$

26. $(4, 0)$; $m = 3$

27. $(-5, 4)$; $m = \frac{3}{2}$

28. $(2, -3)$; $m = -\frac{4}{5}$

In Problems 29–34, find the equation of the line passing through the two given points. Write the final answer in the slope-intercept form $y = mx + b$ or in the form $x = c$.

29. $(2, 5)$; $(4, -3)$

30. $(-1, 4)$; $(3, 2)$

31. $(-3, 2)$; $(-3, 5)$

32. $(0, 5)$; $(2, 5)$

33. $(-4, 2)$; $(0, 2)$

34. $(5, -4)$; $(5, 6)$

In Problems 35–46, write an equation of the line that contains the indicated point and meets the indicated condition(s). Write the final answer in the standard form $Ax + By = C$, $A \geq 0$.

35. $(2, -1)$; parallel to $y = -3x + 7$

36. $(-3, 2)$; parallel to $y = 4x - 5$

37. $(0, -4)$; parallel to $2x + 3y = 9$

38. $(-2, 0)$; parallel to $-3x + 4y = 10$

39. $(3, 3)$; parallel to x axis

40. $(-2, -1)$; parallel to y axis

41. $(4, 5)$; perpendicular to $y = \frac{3}{2}x - 4$

42. $(-1, 3)$; perpendicular to $y = -\frac{3}{5}x + 2$

43. $(5, 0)$; perpendicular to $-5x + 2y = 1$

44. $(0, 3)$; perpendicular to $2x + y = 1$

45. $(-2, -3)$; perpendicular to x axis

46. $(1, -7)$; perpendicular to y axis

In Problems 47–50, classify the quadrilateral ABCD with the indicated vertices as a trapezoid, a parallelogram, a rectangle, or none of these.

47. $A(-2, 2)$; $B(8, 7)$; $C(10, 1)$; $D(-4, -6)$

48. $A(-5, -2)$; $B(-3, 4)$; $C(6, 10)$; $D(4, 4)$

49. $A(0, 2)$; $B(4, -1)$; $C(1, -5)$; $D(-3, -2)$

50. $A(-6, 3)$; $B(3, 7)$; $C(2, 4)$; $D(-4, -1)$

51. Find the equation of the perpendicular bisector of the line segment joining $(-4, -3)$ and $(2, 4)$ by using the point-slope form of the equation of a line.

52. Solve Problem 51 by using the distance between two points formula, and compare the results.

$\int$ *Problems 53–58 are calculus-related. Recall that a line tangent to a circle at a point is perpendicular to the radius drawn to that point (see the figure). Find the equation of the line tangent to the circle at the indicated point. Write the final answer in the standard form $Ax + By = C$, $A \geq 0$. Graph the circle and the tangent line on the same coordinate system.*

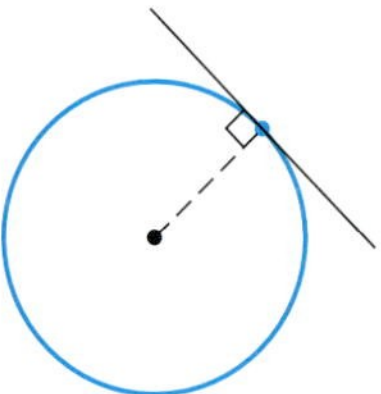

53. $x^2 + y^2 = 25$, $(3, 4)$

54. $x^2 + y^2 = 100$, $(-8, 6)$

55. $x^2 + y^2 = 50$, $(5, -5)$

56. $x^2 + y^2 = 80$, $(-4, -8)$

57. $(x - 3)^2 + (y + 4)^2 = 169$, $(8, -16)$

58. $(x + 5)^2 + (y - 9)^2 = 289$, $(-13, -6)$

C

59. (A) Graph the following equations in the same coordinate system:

$$3x + 2y = 6 \qquad 3x + 2y = 3$$
$$3x + 2y = -6 \qquad 3x + 2y = -3$$

(B) From your observations in part A, describe the family of lines obtained by varying C in $Ax + By = C$ while holding A and B fixed.

(C) Verify your conclusions in part B with a proof.

60. (A) Graph the following two equations in the same coordinate system:

$$3x + 4y = 12 \qquad 4x - 3y = 12$$

(B) Graph the following two equations in the same coordinate system:

$$2x + 3y = 12 \qquad 3x - 2y = 12$$

(C) From your observations in parts A and B, describe the apparent relationship of the graphs of $Ax + By = C$ and $Bx - Ay = C$.

(D) Verify your conclusions in part C with a proof.

Sketch the graphs of the equations in Problems 61–66.

61. $y = |\frac{1}{2}x|$

62. $y = |x + 2|$

63. $y = 2|x| - 4$

64. $y = -\frac{1}{2}|x| + 1$

65. $x^2 - y^2 = 0$

66. $4y^2 - 9x^2 = 0$

67. Describe the relationship between the graphs of $y = mx + b$ and $y = |mx + b|$. (See Problems 61 and 62.)

68. Describe the relationship between the graphs of $y = mx + b$ and $y = m|x| + b$. (See Problems 63 and 64.)

69. Prove that if a line L has x intercept $(a, 0)$ and y intercept $(0, b)$, then the equation of L can be written in the **intercept form**

$$\frac{x}{a} + \frac{y}{b} = 1 \qquad a, b \neq 0$$

In Problems 70 and 71, write the equation of the line with the indicated intercepts in the standard form $Ax + By = C$, $A \geq 0$.

70. (3, 0) and (0, 5)

71. (−2, 0) and (0, 7)

72. Let

$$P_1(x_1, y_1) = P_1(x_1, mx_1 + b)$$
$$P_2(x_2, y_2) = P_2(x_2, mx_2 + b)$$
$$P_3(x_3, y_3) = P_3(x_3, mx_3 + b)$$

be three arbitrary points that satisfy $y = mx + b$ with $x_1 < x_2 < x_3$. Show that P_1, P_2, and P_3 are **collinear;** that is, they lie on the same line. [*Hint:* Use the distance formula and show that $d(P_1, P_2) + d(P_2, P_3) = d(P_1, P_3)$.] This proves that the graph of $y = mx + b$ is a straight line.

APPLICATIONS

73. Boiling Point of Water. At sea level, water boils when it reaches a temperature of 212°F. At higher altitudes, the atmospheric pressure is lower and so is the temperature at which water boils. The boiling point B in degrees Fahrenheit at an altitude of x feet is given approximately by

$$B = 212 - 0.0018x$$

(A) Complete Table 1.

TABLE 1

x	0	5,000	10,000	15,000	20,000	25,000	30,000
B							

(B) Based on the information in the table, write a brief verbal description of the relationship between altitude and the boiling point of water.

74. Air Temperature. As dry air moves upward, it expands and cools. The air temperature A in degrees Celsius at an altitude of x kilometers is given approximately by

$$A = 25 - 9x$$

(A) Complete Table 2.

TABLE 2

x	0	1	2	3	4	5
A						

(B) Based on the information in the table, write a brief verbal description of the relationship between altitude and air temperature.

75. Car Rental. A car rental agency computes daily rental charges for compact cars with the equation

$$c = 25 + 0.25x$$

where c is the daily charge in dollars and x is the daily mileage. Translate this algebraic statement into a verbal statement that can be used to explain the daily charges to a customer.

76. Installation Charges. A telephone store computes charges for phone installation with the equation

$$c = 15 + 0.7x$$

where c is the installation charge in dollars and x is the time in minutes spent performing the installation. Translate this algebraic statement into a verbal statement that can be used to explain the installation charges to a customer.

Merck & Co., Inc., is the world's largest pharmaceutical company. Problems 77 and 78 refer to the data in Table 3, taken from the company's 1993 annual report.

TABLE 3 Selected Financial Data (billion \$) for Merck & Co., Inc.

	1988	1989	1990	1991	1992
Sales	\$5.9	\$6.5	\$7.7	\$8.6	\$9.7
Net income	\$1.2	\$1.5	\$1.8	\$2.1	\$2.4

77. Sales Analysis. A mathematical model for Merck's sales is given by

$$y = 5.74 + 0.97x$$

where $x = 0$ corresponds to 1988.

(A) Complete Table 4. Round values of y to one decimal place.

TABLE 4

x	0	1	2	3	4
Sales	5.9	6.5	7.7	8.6	9.7
y					

(B) Sketch the graph of y and the sales data on the same axes.

(C) Use the modeling equation to estimate the sales in 1993. In 2000.

(D) Write a brief verbal description of the company's sales from 1988 to 1992.

78. Income Analysis. A mathematical model for Merck's income is given by

$$y = 1.2 + 0.3x$$

where $x = 0$ corresponds to 1988.

(A) Complete Table 5. Round values of y to one decimal place.

TABLE 5

x	0	1	2	3	4
Net income	1.2	1.5	1.8	2.1	2.4
y					

(B) Sketch the graph of the modeling equation and the income data on the same axes.

(C) Use the modeling equation to estimate the income in 1993. In 2000.

(D) Write a brief verbal description of the company's income from 1988 to 1992.

79. Physics. The two temperature scales Fahrenheit (F) and Celsius (C) are linearly related. It is known that water freezes at 32°F or 0°C and boils at 212°F or 100°C.

(A) Find a linear equation that expresses F in terms of C.

(B) If a European family sets its house thermostat at 20°C, what is the setting in degrees Fahrenheit? If the outside temperature in Milwaukee is 86°F, what is the temperature in degrees Celsius?

(C) What is the slope of the graph of the linear equation found in part A? (The slope indicates the change in Fahrenheit degrees per unit change in Celsius degrees.)

80. Physics. Hooke's law states that the relationship between the stretch s of a spring and the weight w causing the stretch is linear (a principle upon which all spring scales are constructed). For a particular spring, a 5-pound weight causes a stretch of 2 inches, while with no weight the stretch of the spring is 0.

(A) Find a linear equation that expresses s in terms of w.

(B) What weight will cause a stretch of 3.6 inches?

(C) What is the slope of the graph of the equation? (The slope indicates the amount of stretch per pound increase in weight.)

81. Business—Depreciation. A copy machine was purchased by a law firm for $8,000 and is assumed to have a depreciated value of $0 after 5 years. The firm takes straight-line depreciation over the 5-year period.

(A) Find a linear equation that expresses value V in dollars in terms of time t in years.

(B) What is the depreciated value after 3 years?

(C) What is the slope of the graph of the equation found in part A? Interpret verbally.

82. Business—Markup Policy. A clothing store sells a shirt costing $20 for $33 and a jacket costing $60 for $93.

(A) If the markup policy of the store for items costing over $10 is assumed to be linear, write an equation that expresses retail price R in terms of cost C (wholesale price).

(B) What does a store pay for a suit that retails for $240?

(C) What is the slope of the graph of the equation found in part A? Interpret verbally.

83. Flight Conditions. In stable air, the air temperature drops about 5°F for each 1,000-foot rise in altitude.

(A) If the temperature at sea level is 70°F and a commercial pilot reports a temperature of −20°F at 18,000 feet, write a linear equation that expresses temperature T in terms of altitude A (in thousands of feet).

(B) How high is the aircraft if the temperature is 0°F?

(C) What is the slope of the graph of the equation found in part A? Interpret verbally.

★ **84. Flight Navigation.** An airspeed indicator on some aircraft is affected by the changes in atmospheric pressure at different altitudes. A pilot can estimate the true airspeed by observing the indicated airspeed and adding to it about 2% for every 1,000 feet of altitude.

(A) If a pilot maintains a constant reading of 200 miles per hour on the airspeed indicator as the aircraft climbs from sea level to an altitude of 10,000 feet, write a linear equation that expresses true airspeed T (miles per hour) in terms of altitude A (thousands of feet).

(B) What would be the true airspeed of the aircraft at 6,500 feet?

(C) What is the slope of the graph of the equation found in part A? Interpret verbally.

★ **85. Oceanography.** After about 9 hours of a steady wind, the height of waves in the ocean is approximately linearly related to the duration of time the wind has been blowing. During a storm with 50-knot winds, the wave height after 9 hours was found to be 23 feet, and after 24 hours it was 40 feet.

(A) If t is time after the 50-knot wind started to blow and h is the wave height in feet, write a linear equation that expresses height h in terms of time t.

(B) How long will the wind have been blowing for the waves to be 50 feet high?

Express all calculated quantities to three significant digits.

86. Oceanography. As a diver descends into the ocean, pressure increases linearly with depth. The pressure is 15 pounds per square inch on the surface and 30 pounds per square inch 33 feet below the surface.

(A) If p is the pressure in pounds per square inch and d is the depth below the surface in feet, write an equation that expresses p in terms of d.

(B) How deep can a scuba diver go if the safe pressure for his equipment and experience is 40 pounds per square inch?

★ **87. Medicine.** Cardiovascular research has shown that above the 210 cholesterol level, each 1% increase in cholesterol level increases coronary risk 2%. For a particular age group, the coronary risk at a 210 cholesterol level is found to be 0.160 and at a level of 231 the risk is found to be 0.192.

(A) Find a linear equation that expresses risk R in terms of cholesterol level C.

(B) What is the risk for a cholesterol level of 260?

(C) What is the slope of the graph of the equation found in part A? Interpret verbally.

Express all calculated quantities to three significant digits.

★ **88. Demographics.** The average number of persons per household in the United States has been shrinking steadily for as long as statistics have been kept and is approximately linear with respect to time. In 1900, there were about 4.76 persons per household and in 1990, about 2.5.

(A) If N represents the average number of persons per household and t represents the number of years since 1900, write a linear equation that expresses N in terms of t.

(B) What is the predicted household size in the year 2000?

Express all calculated quantities to three significant digits.

SECTION 2-3 Functions

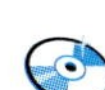

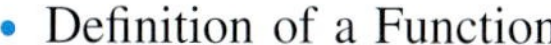

- Definition of a Function

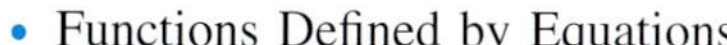

- Functions Defined by Equations
- Function Notation
- Application
- A Brief History of the Function Concept

The idea of correspondence plays a central role in the formulation of the function concept. You have already had experiences with correspondences in everyday life. For example:

To each person there corresponds an age.

To each item in a store there corresponds a price.

To each automobile there corresponds a license number.

To each circle there corresponds an area.

To each number there corresponds its cube.

One of the most important aspects of any science (managerial, life, social, physical, computer, etc.) is the establishment of correspondences among various types of phenomena. Once a correspondence is known, predictions can be made. A chemist can use a gas law to predict the pressure of an enclosed gas, given its temperature. An engineer can use a formula to predict the deflections of a beam subject to different loads. A computer scientist can use formulas to compare the efficiency of algorithms for sorting data stored in a computer. An economist would like to be able to predict interest rates, given the rate of change of the money supply. And so on.

• Definition of a Function

What do all the preceding examples have in common? Each describes the matching of elements from one set with the elements in a second set. Consider Tables 1–3, which list values for the cube, square, and square root, respectively.

TABLE 1

Domain (number)	Range (cube)
−2 →	−8
−1 →	−1
0 →	0
1 →	1
2 →	8

TABLE 2

Domain (number)	Range (square)
−2 →	4
−1 →	1
0 →	0
1 →	1
2 →	4

TABLE 3

Domain (number)	Range (square root)
0 →	0
1 →	1
1 →	−1
4 →	2
4 →	−2
9 →	3
9 →	−3

Tables 1 and 2 specify functions, but Table 3 does not. Why not? The definition of the term *function* will explain.

DEFINITION 1

Rule Form of the Definition of a Function

A **function** is a rule that produces a correspondence between two sets of elements such that to each element in the first set there corresponds *one and only one* element in the second set.

The first set is called the **domain,** and the set of all corresponding elements in the second set is called the **range.**

Tables 1 and 2 specify functions, since to each domain value there corresponds exactly one range value (for example, the cube of -2 is -8 and no other number). On the other hand, Table 3 does not specify a function, since to at least one domain value there corresponds more than one range value (for example, to the domain value 9 there corresponds -3 and 3, both square roots of 9).

EXPLORE-DISCUSS 1

Consider the set of students enrolled in a college and the set of faculty members of that college. Define a correspondence between the two sets by saying that a student corresponds to a faculty member if the student is currently enrolled in a course taught by the faculty member. Is this correspondence a function? Discuss.

Since a function is a rule that pairs each element in the domain with a corresponding element in the range, this correspondence can be illustrated by using ordered pairs of elements, where the first component represents a domain element and the second component represents the corresponding range element. Thus, the functions defined in Tables 1 and 2 can be written as follows:

$$\text{Function 1} = \{(-2, -8), (-1, -1), (0, 0), (1, 1), (2, 8)\}$$

$$\text{Function 2} = \{(-2, 4), (-1, 1), (0, 0), (1, 1), (2, 4)\}$$

In both cases, notice that no two ordered pairs have the same first component and different second components. On the other hand, if we list the set A of ordered pairs determined by Table 3, we have

$$A = \{(0, 0), (1, 1), (1, -1), (4, 2), (4, -2), (9, 3), (9, -3)\}$$

In this case, there are ordered pairs with the same first component and different second components; for example, $(1, 1)$ and $(1, -1)$ both belong to the set A. Once again, we see that Table 3 does not define a function.

This suggests an alternative but equivalent way of defining functions that produces additional insight into this concept.

DEFINITION 2 **Set Form of the Definition of a Function**

A **function** is a set of ordered pairs with the property that no two ordered pairs have the same first component and different second components.

The set of all first components in a function is called the **domain** of the function, and the set of all second components is called the **range.**

EXAMPLE 1 **Functions Defined as Sets of Ordered Pairs**

(A) The set $S = \{(1, 4), (2, 3), (3, 2), (4, 3), (5, 4)\}$ defines a function since no two ordered pairs have the same first component and different second components. The domain and range are

$$\text{Domain} = \{1, 2, 3, 4, 5\} \quad \text{Set of first components}$$

$$\text{Range} = \{2, 3, 4\} \quad \text{Set of second components}$$

(B) The set $T = \{(1, 4), (2, 3), (3, 2), (2, 4), (1, 5)\}$ does not define a function since there are ordered pairs with the same first component and different second components [for example, $(1, 4)$ and $(1, 5)$].

Matched Problem 1 Determine whether each set defines a function. If it does, then state the domain and range.

(A) $S = \{(-2, 1), (-1, 2), (0, 0), (-1, 1), (-2, 2)\}$
(B) $T = \{(-2, 1), (-1, 2), (0, 0), (1, 2), (2, 1)\}$

• Functions Defined by Equations

Both versions of the definition of a function are quite general, with no restrictions on the type of elements that make up the domain or range. Points in the plane and complex numbers are two examples of domain and range elements that are used in more advanced courses. In this text, unless otherwise indicated, **the domain and range of a function will be sets of real numbers.**

Defining a function by displaying the rule of correspondence in a table or listing all the ordered pairs in the function only works if the domain and range are finite sets. Functions with finite domains and ranges are used extensively in certain specialized areas, such as computer science, but most applications of functions involve infinite domains and ranges. If the domain and range of a function are infinite sets, then the rule of correspondence cannot be displayed in a table, and it is not possible to actually list all the ordered pairs belonging to the function. For most functions, we use an equation in two variables to specify both the rule of correspondence and the set of ordered pairs.

Consider the equation

$$y = x^2 + 2x \qquad x \text{ any real number} \tag{1}$$

This equation assigns to each domain value x exactly one range value y. For example,

$$\text{If } x = 4 \qquad \text{then} \qquad y = (4)^2 + 2(4) = 24$$

$$\text{If } x = -\tfrac{1}{3} \qquad \text{then} \qquad y = (-\tfrac{1}{3})^2 + 2(-\tfrac{1}{3}) = -\tfrac{5}{9}$$

Thus, we can view equation (1) as a function with rule of correspondence

$$y = x^2 + 2x \qquad x^2 + 2x \text{ corresponds to } x$$

or, equivalently, as a function with set of ordered pairs

$$\{(x, y) \mid y = x^2 + 2x, \ x \text{ a real number}\}$$

The variable x is called an *independent variable,* indicating that values can be assigned "independently" to x from the domain. The variable y is called a *dependent variable,* indicating that the value of y "depends" on the value assigned to x and on the given equation. In general, any variable used as a placeholder for domain values is called an **independent variable;** any variable that is used as a placeholder for range values is called a **dependent variable.**

Which equations can be used to define functions?

Functions Defined by Equations

In an equation in two variables, if to each value of the independent variable there corresponds exactly one value of the dependent variable, then the equation defines a function.

If there is any value of the independent variable to which there corresponds more than one value of the dependent variable, then the equation does not define a function.

EXAMPLE 2 **Determining if an Equation Defines a Function**

Determine which of the following equations define functions with independent variable x and domain all real numbers:

(A) $y^3 - x = 1$ (B) $y^2 - x^2 = 9$

Solutions (A) Solving for the dependent variable y, we have

$$y^3 - x = 1 \qquad (2)$$
$$y^3 = 1 + x$$
$$y = \sqrt[3]{1 + x}$$

Since $1 + x$ is a real number for each real number x and since each real number has exactly one real cube root, equation (2) assigns exactly one value of the dependent variable, $y = \sqrt[3]{1 + x}$, to each value of the independent variable x. Thus, equation (2) defines a function.

(B) Solving for the dependent variable y, we have

$$y^2 - x^2 = 9 \qquad (3)$$
$$y^2 = 9 + x^2$$
$$y = \pm\sqrt{9 + x^2}$$

Since $9 + x^2$ is always a positive real number and since each positive real number has two real square roots, each value of the independent variable x corresponds to two values of the dependent variable, $y = -\sqrt{9 + x^2}$ and $y = \sqrt{9 + x^2}$. Thus, equation (3) does not define a function.

Matched Problem 2 Determine which of the following equations define functions with independent variable x and domain all real numbers:

(A) $y^2 + x^4 = 4$ (B) $y^3 - x^3 = 3$

Notice that we have used the phrase "an equation defines a function" rather than "an equation is a function." This is a somewhat technical distinction, but it is employed consistently in mathematical literature and we will adhere to it in this text.

It is very easy to determine whether an equation defines a function if you have the graph of the equation. The two equations we considered in Example 2 are graphed in Figure 1 on the next page.

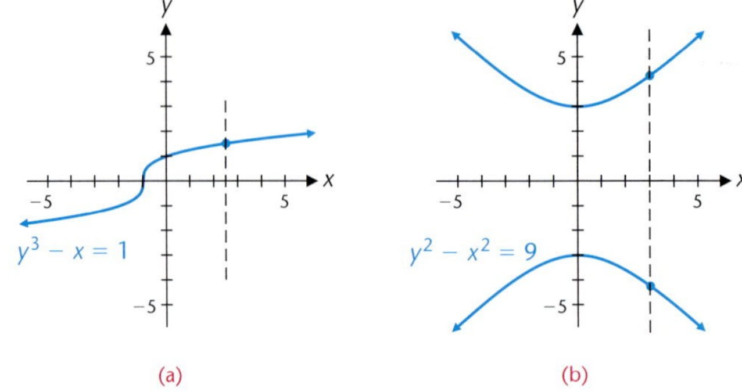

FIGURE 1 Graphs of equations and the vertical line test.

In Figure 1(a), each vertical line intersects the graph of the equation $y^3 - x = 1$ in exactly one point. This shows that each value of the independent variable x corresponds to exactly one value of the dependent variable y and confirms our conclusion that this equation defines a function. On the other hand, Figure 1(b) shows that there exist vertical lines that intersect the graph of $y^2 - x^2 = 9$ in two points. This indicates that there exist values of the independent variable x that correspond to two different values of the dependent variable y, which confirms our conclusion that this equation does not define a function. These observations are generalized in Theorem 1.

Theorem 1 **Vertical Line Test for a Function**

An equation defines a function if each vertical line in the rectangular coordinate system passes through at most one point on the graph of the equation.

If any vertical line passes through two or more points on the graph of an equation, then the equation does not define a function.

EXPLORE-DISCUSS 2 The definition of a function specifies that to each element in the domain there corresponds one and only one element in the range.

(A) Give an example of a function such that to each element of the range there correspond exactly two elements of the domain.

(B) Give an example of a function such that to each element of the range there corresponds exactly one element of the domain.

In Example 2, the domain was given in the statement of the problem. In many cases, this will not be done. Unless stated to the contrary, we will adhere to the following convention regarding domains and ranges for functions defined by equations:

Agreement on Domains and Ranges

If a function is defined by an equation and the domain is not indicated, then we assume that the domain is the set of all real number replacements of the independent variable that produce *real values* for the dependent variable. The range is the set of all values of the dependent variable corresponding to these domain values.

EXAMPLE 3 **Finding the Domain of a Function**

Find the domain of the function defined by the equation $y = \sqrt{x - 3}$, assuming x is the independent variable.

Solution For y to be real, $x - 3$ must be greater than or equal to 0. That is,

$$x - 3 \geq 0 \qquad \text{or} \qquad x \geq 3$$

Thus,

$$\text{Domain: } \{x \mid x \geq 3\} \text{ or } [3, \infty)$$

Note that in many cases we will dispense with the use of set notation and simply write $x \geq 3$ instead of $\{x \mid x \geq 3\}$.

Matched Problem 3 Find the domain of the function defined by the equation $y = \sqrt{x + 5}$, assuming x is the independent variable.

• Function Notation

We will use letters to name functions and to provide a very important and convenient notation for defining functions. For example, if f is the name of the function defined by the equation $y = 2x + 1$, then instead of the more formal representations

$$f\colon y = 2x + 1 \qquad \text{Rule of correspondence}$$

or

$$f\colon \{(x, y) \mid y = 2x + 1\} \qquad \text{Set of ordered pairs}$$

we simply write

$$f(x) = 2x + 1 \qquad \text{Function notation}$$

The symbol $f(x)$ is read "f of x," "f at x," or "the value of f at x" and represents the number in the range of the function f to which the domain value x is paired. Thus, $f(3)$ is the range value for the function f associated with the domain value 3. We find this range value by replacing x with 3 wherever x occurs in the function definition

$$f(x) = 2x + 1$$

and evaluating the right side,

$$\begin{aligned} f(3) &= 2 \cdot 3 + 1 \\ &= 6 + 1 \\ &= 7 \end{aligned}$$

The statement $f(3) = 7$ indicates in a concise way that the function f assigns the range value 7 to the domain value 3 or, equivalently, that the ordered pair (3, 7) belongs to f.

The symbol $f{:}x \to f(x)$, read "f maps x into $f(x)$," is also used to denote the relationship between the domain value x and the range value $f(x)$ (see Fig. 2). Whenever we write $y = f(x)$, we assume that x is an independent variable and that y and $f(x)$ both represent the dependent variable.

FIGURE 2 Function notation.

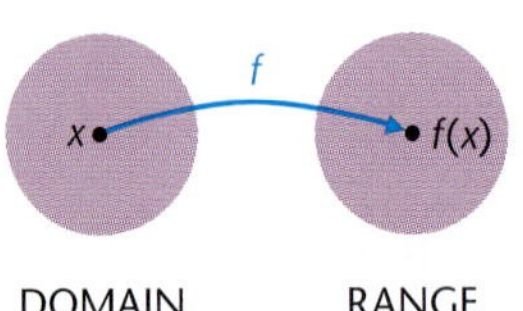

The function f "maps" the domain value x into the range value $f(x)$.

Letters other than f and x can be used to represent functions and independent variables. For example,

$$g(t) = t^2 - 3t + 7$$

defines g as a function of the independent variable t. To find $g(-2)$, we replace t by -2 wherever t occurs in

$$g(t) = t^2 - 3t + 7$$

and evaluate the right side:

$$\begin{aligned} g(-2) &= (-2)^2 - 3(-2) + 7 \\ &= 4 + 6 + 7 \\ &= 17 \end{aligned}$$

Thus, the function g assigns the range value 17 to the domain value -2; the ordered pair $(-2, 17)$ belongs to g.

It is important to understand and remember the definition of the symbol $f(x)$:

DEFINITION 3 **The Symbol $f(x)$**

The symbol $f(x)$ represents the real number in the range of the function f corresponding to the domain value x. Symbolically, $f: x \to f(x)$. The ordered pair $(x, f(x))$ belongs to the function f. If x is a real number that is not in the domain of f, then f is **not defined** at x and $f(x)$ **does not exist.**

EXAMPLE 4 Evaluating Functions

For

$$f(x) = \frac{15}{x-3} \qquad g(x) = 16 + 3x - x^2 \qquad h(x) = \sqrt{25 - x^2}$$

find:

(A) $f(6)$ (B) $g(-7)$ (C) $h(10)$ (D) $f(0) + g(4) - h(-3)$

Solution (A) $f(6) = \dfrac{15}{6-3} = \dfrac{15}{3} = 5$

(B) $g(-7) = 16 + 3(-7) - (-7)^2 = 16 - 21 - 49 = -54$

(C) $h(10) = \sqrt{25 - 10^2} = \sqrt{25 - 100} = \sqrt{-75}$

But $\sqrt{-75}$ is not a real number. Since we have agreed to restrict the domain of a function to values of x that produce real values for the function, 10 is not in the domain of h and $h(10)$ is not defined.

(D) $f(0) + g(4) - h(-3)$

$$= \frac{15}{0-3} + [16 + 3(4) - 4^2] - \sqrt{25 - (-3)^2}$$

$$= \frac{15}{-3} + 12 - \sqrt{16}$$

$$= -5 + 12 - 4 = 3$$

Matched Problem 4 Use the functions in Example 4 to find:

(A) $f(-2)$ (B) $g(6)$ (C) $h(-8)$ (D) $\dfrac{f(8)}{h(4)}$

EXAMPLE 5 Finding Domains of Functions

Find the domains of functions f, g, and h:

$$f(x) = \frac{15}{x - 3} \qquad g(x) = 16 + 3x - x^2 \qquad h(x) = \sqrt{25 - x^2}$$

Solutions

Domain of f

The fraction $15/(x - 3)$ represents a real number for all replacements of x by real numbers except $x = 3$, since division by 0 is not defined. Thus, $f(3)$ does not exist, and the domain of f is the set of all real numbers except 3. We often indicate this by writing

$$f(x) = \frac{15}{x - 3} \qquad x \neq 3$$

Domain of g

The domain is R, the set of all real numbers, since $16 + 3x - x^2$ represents a real number for all replacements of x by real numbers.

Domain of h

The domain is the set of all real numbers x such that $\sqrt{25 - x^2}$ is a real number—that is, such that $25 - x^2 \geq 0$. Solving $25 - x^2 = (5 - x)(5 + x) \geq 0$ by methods discussed in Section 2-8, we find

$$\text{Domain: } -5 \leq x \leq 5 \qquad \text{or} \qquad [-5, 5]$$

Matched Problem 5 Find the domains of functions F, G, and H:

$$F(x) = x^2 + 5x - 2 \qquad G(x) = \sqrt{\frac{x - 2}{x + 3}} \qquad H(x) = \frac{4}{x - 2}$$

EXPLORE-DISCUSS 3 Let x and h be real numbers.

(A) If $f(x) = 4x + 3$, which of the following is true:
(1) $f(x + h) = 4x + 3 + h$
(2) $f(x + h) = 4x + 4h + 3$
(3) $f(x + h) = 4x + 4h + 6$

(B) If $g(x) = x^2$, which of the following is true:
(1) $g(x + h) = x^2 + h$
(2) $g(x + h) = x^2 + h^2$
(3) $g(x + h) = x^2 + 2hx + h^2$

(C) If $M(x) = x^2 + 4x + 3$, describe the operations that must be performed to evaluate $M(x + h)$.

In addition to evaluating functions at specific numbers, it is important to be able to evaluate functions at expressions that involve one or more variables. For example, the **difference quotient**

$$\frac{f(x+h)-f(x)}{h} \qquad x \text{ and } x+h \text{ in the domain of } f,\ h \neq 0$$

is studied extensively in a calculus course.

EXAMPLE 6 Evaluating and Simplifying a Difference Quotient

For $f(x) = x^2 + 4x + 5$, find and simplify:

(A) $f(x+h)$ (B) $\dfrac{f(x+h)-f(x)}{h}, h \neq 0$

Solution (A) To find $f(x+h)$, we replace x with $x+h$ everywhere it appears in the equation that defines f and simplify:

$$\begin{aligned} f(x+h) &= (x+h)^2 + 4(x+h) + 5 \\ &= x^2 + 2xh + h^2 + 4x + 4h + 5 \end{aligned}$$

(B) Using the result of part A, we get

$$\begin{aligned} \frac{f(x+h)-f(x)}{h} &= \frac{x^2 + 2xh + h^2 + 4x + 4h + 5 - (x^2 + 4x + 5)}{h} \\ &= \frac{x^2 + 2xh + h^2 + 4x + 4h + 5 - x^2 - 4x - 5}{h} \\ &= \frac{2xh + h^2 + 4h}{h} = \frac{h(2x + h + 4)}{h} = 2x + h + 4 \end{aligned}$$

Matched Problem 6 Repeat Example 6 for $f(x) = x^2 + 3x + 7$.

CAUTION

1. If f is a function, then the symbol $f(x+h)$ represents the value of f at the number $x+h$ and must be evaluated by replacing the independent variable in the equation that defines f with the expression $x+h$, as we did in Example 6. Do not confuse this notation with the familiar algebraic notation for multiplication:

$$f(x+h) \neq fx + fh \qquad f(x+h) \text{ is function notation.}$$

$$4(x+h) = 4x + 4h \qquad 4(x+h) \text{ is algebraic multiplication notation.}$$

2. There is another common incorrect interpretation of the symbol $f(x + h)$. If f is an arbitrary function, then

$$f(x + h) \neq f(x) + f(h)$$

It is possible to find some particular functions for which $f(x + h) = f(x) + f(h)$ is a true statement, but in general these two expressions are not equal.

• Application

EXAMPLE 7 **Construction**

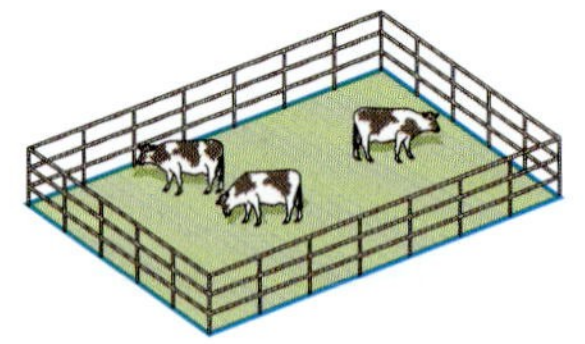

A rectangular feeding pen for cattle is to be made with 100 meters of fencing.

(A) If x represents the width of the pen, express its area $A(x)$ in terms of x.
(B) What is the domain of the function A (determined by the physical restrictions)?

Solutions (A) Draw a figure and label the sides.

x (Width)

$50 - x$ (Length)

Perimeter = 100 meters of fencing.
Half the perimeter = 50.
If x = Width, then $50 - x$ = Length.

$$A(x) = \text{(Width)(Length)} = x(50 - x)$$

(B) To have a pen, x must be positive, but x must also be less than 50 (or the length will not exist). Thus,

Domain: $0 < x < 50$ Inequality notation

$(0, 50)$ Interval notation

Matched Problem 7 Rework Example 7 with the added assumption that a large barn is to be used as one side of the pen.

• A Brief History of the Function Concept

The history of the use of functions in mathematics illustrates the tendency of mathematicians to extend and generalize each concept. The word "function" appears to have been first used by Leibniz in 1694 to stand for any quantity associated with a curve. By 1718, Johann Bernoulli considered a function any expression made up of constants and a variable. Later in the same century, Euler came to regard a function as any equation made up of constants and variables. Euler made extensive use of the extremely important notation $f(x)$, although its origin is generally attributed to Clairaut (1734).

The form of the definition of function that has been used until well into the twentieth century (many texts still contain this definition) was formulated by Dirichlet (1805–1859). He stated that, if two variables x and y are so related that for each value of x there corresponds exactly one value of y, then y is said to be a (single-valued) function of x. He called x, the variable to which values are assigned at will, the independent variable, and y, the variable whose values depend on the values assigned to x, the dependent variable. He called the values assumed by x the domain of the function, and the corresponding values assumed by y the range of the function.

Now, since set concepts permeate almost all mathematics, we have the more general definition of function presented in this section in terms of sets of ordered pairs of elements.

Answers to Matched Problems

1. (A) S does not define a function
(B) T defines a function with domain $\{-2, -1, 0, 1, 2\}$ and range $\{0, 1, 2\}$

2. (A) Does not define a function (B) Defines a function

3. $x \geq -5$ Inequality notation
$[-5, \infty)$ Interval notation

4. (A) -3 (B) -2 (C) Does not exist (D) 1

5. Domain of F: all real numbers
Domain of G: $x < -3$ or $x \geq 2$ Inequality notation
$(-\infty, -3) \cup [2, \infty)$ Interval notation
Domain of H: All real numbers except 2

6. (A) $x^2 + 2xh + h^2 + 3x + 3h + 7$ (B) $2x + h + 3$

7. (A) $A(x) = x(100 - 2x)$ (B) Domain: $0 < x < 50$ Inequality notation
$(0, 50)$ Interval notation

EXERCISE 2-3

A

Indicate whether each table in Problems 1–6 defines a function.

1.

Domain	Range
−1 → 3	1
0 → 2	2
1 → 1	3

2.

Domain	Range
2 → 1	1
4 → 3	3
6 → 5	5

3. Domain Range

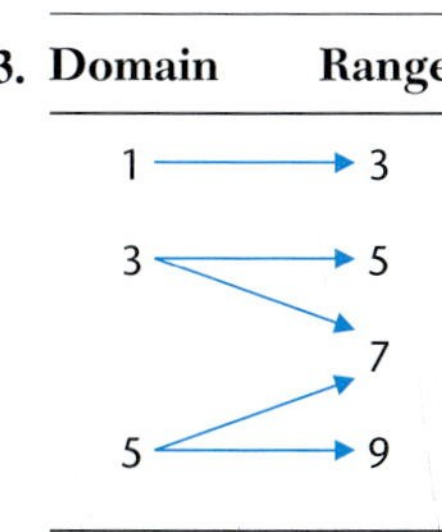

4. Domain Range

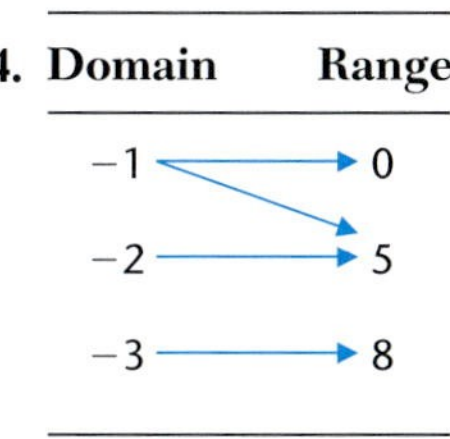

5. Domain Range

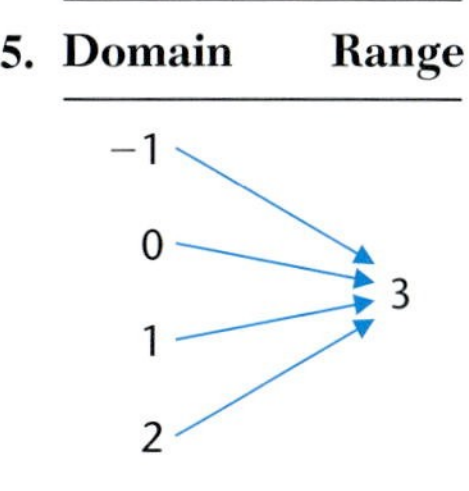

6.

Domain	Range
2 → 8	8
3 → 8	
4 → 9	9
5 → 9	

Indicate whether each set in Problems 7–12 defines a function. Find the domain and range of each function.

7. $\{(2, 4), (3, 6), (4, 8), (5, 10)\}$

8. $\{(-1, 4), (0, 3), (1, 2), (2, 1)\}$

9. $\{(10, -10), (5, -5), (0, 0), (5, 5), (10, 10)\}$

10. $\{(-10, 10), (-5, 5), (0, 0), (5, 5), (10, 10)\}$

11. $\{(0, 1), (1, 1), (2, 1), (3, 2), (4, 2), (5, 2)\}$

12. $\{(1, 1), (2, 1), (3, 1), (1, 2), (2, 2), (3, 2)\}$

Indicate whether each graph in Problems 13–18 is the graph of a function.

13.

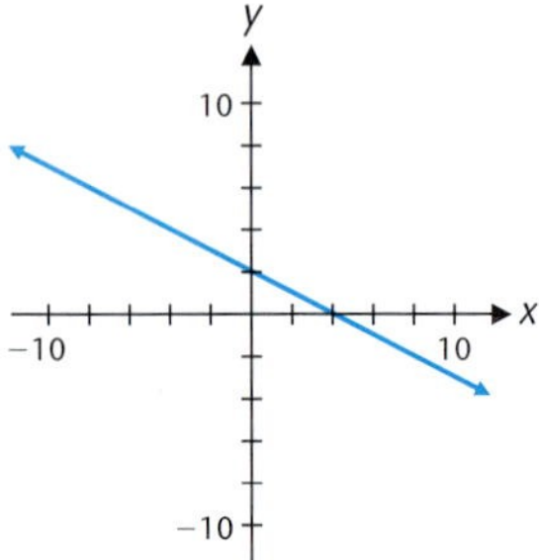

14.

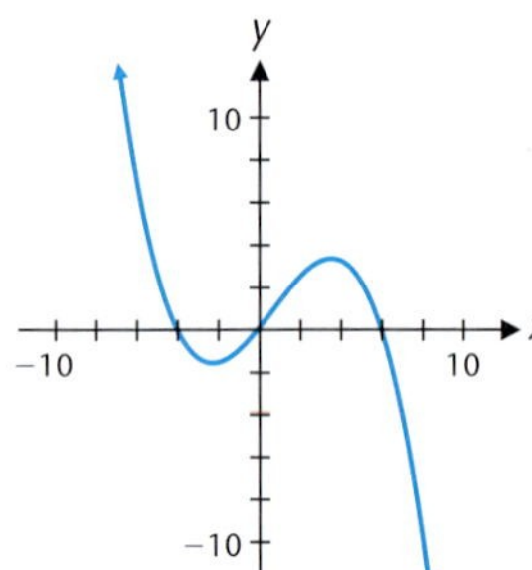

15.

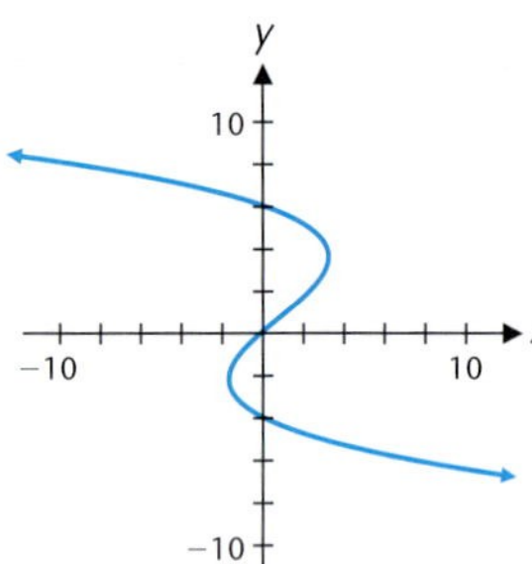

16.

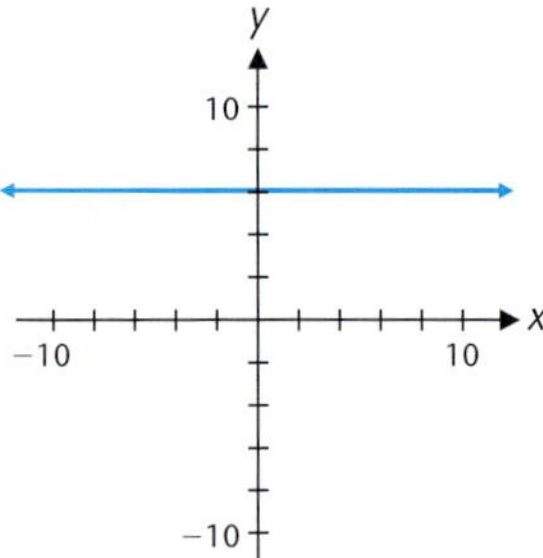

17.

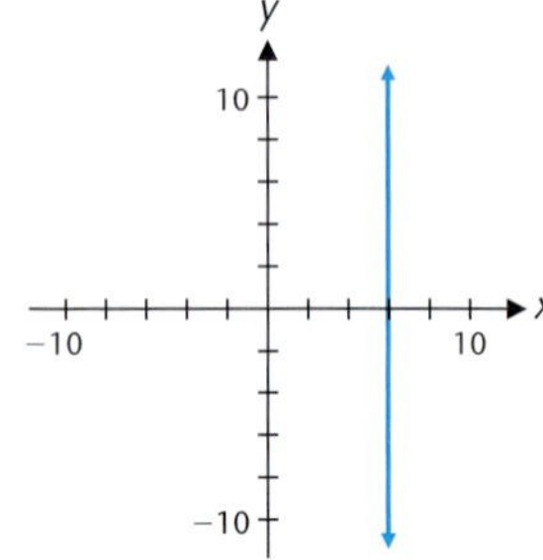

18.

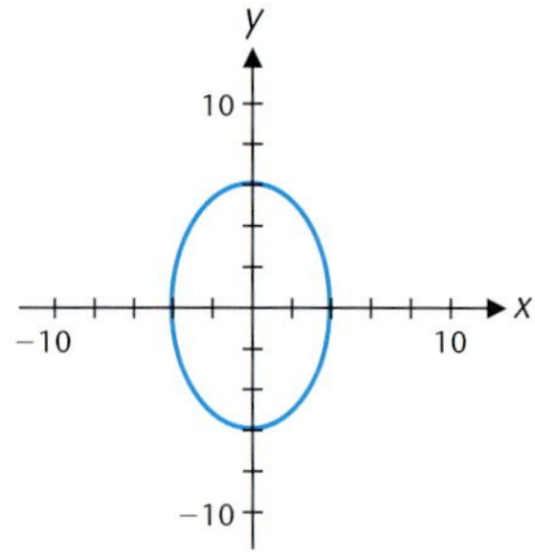

Problems 19–28 refer to the functions

$$f(x) = 2x + 6 \qquad g(t) = 3 - 2t$$
$$F(m) = 2m^2 + 3m - 1 \qquad G(u) = u^2 + u - 2$$

Evaluate as indicated.

19. $f(-1)$

20. $g(6)$

21. $G(-2)$

22. $F(-3)$

23. $F(-1) + f(3)$

24. $G(2) - g(-3)$

25. $2F(-2) - G(-1)$

26. $3G(-2) + 2F(-1)$

27. $\dfrac{f(0) \cdot g(-2)}{F(-3)}$

28. $\dfrac{g(4) \cdot f(2)}{G(1)}$

In Problems 29–32, use the following graph of a function f to determine x or y to the nearest integer, as indicated. Some problems may have more than one answer.

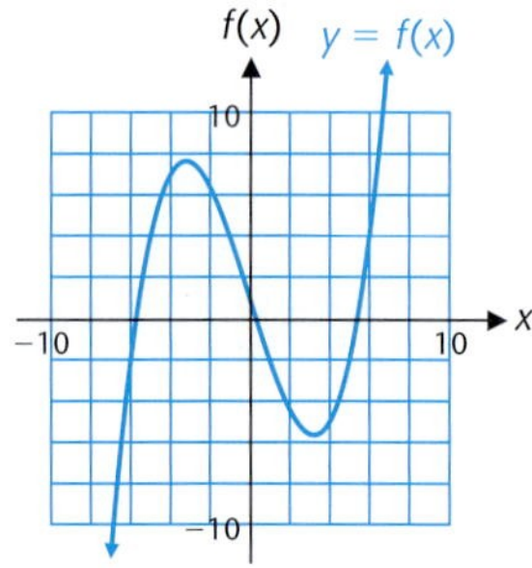

29. $y = f(-4)$

30. $y = f(4)$

31. $4 = f(x)$

32. $-2 = f(x)$

B

Determine which of the equations in Problems 33–42 define a function with independent variable x. For those that do, find the domain. For those that do not, find a value of x to which there corresponds more than one value of y.

33. $y - x^2 = 1$

34. $y^2 - x = 1$

35. $2x^3 + y^2 = 4$

36. $3x^2 + y^3 = 8$

37. $x^3 - y = 2$

38. $x^3 + |y| = 6$

39. $xy = 0$

40. $xy = 1$

41. $x^2 + xy = 1$

42. $x^2 + xy = 0$

In Problems 43–56, find the domain of the indicated function.

43. $f(x) = \sqrt{x - 2}$

44. $g(t) = \sqrt[3]{5 - t}$

45. $h(u) = \sqrt[3]{u + 9}$

46. $k(s) = \sqrt{-1 - s}$

47. $F(w) = \dfrac{2w + 3}{w - 4}$

48. $G(m) = \dfrac{1 - 2m}{m^2 + 3}$

49. $H(n) = \dfrac{3n + 7}{n^2 + n - 2}$

50. $K(v) = \dfrac{2v^2 - 9}{v^2 - v - 6}$

51. $I(y) = \sqrt{y^2 - 2y - 3}$

52. $J(z) = \sqrt{z^2 - 2z + 5}$

53. $f(x) = \sqrt{\dfrac{x^2 - 4x + 5}{x^2 + 2x + 3}}$

54. $g(t) = \sqrt{\dfrac{t - 1}{5 - t}}$

55. $F(y) = \sqrt{\dfrac{y - 3}{y + 2}}$

56. $G(z) = \sqrt{\dfrac{z^2 + 4z + 6}{z^2 + 4z + 3}}$

The verbal statement "function f multiplies the square of the domain element by 3 and then subtracts 7 from the result" and the algebraic statement "$f(x) = 3x^2 - 7$" define the same function. In Problems 57–60, translate each verbal definition of a function into an algebraic definition.

57. Function g subtracts 5 from twice the cube of the domain element.

58. Function f multiplies the domain element by -3 and adds 4 to the result.

59. Function G multiplies the square root of the domain element by 2 and subtracts the square of the domain element from the result.

60. Function F multiplies the cube of the domain element by -8 and adds three times the square root of 3 to the result.

In Problems 61–64, translate each algebraic definition of the function into a verbal definition.

61. $f(x) = 2x - 3$

62. $g(x) = -2x + 7$

63. $F(x) = 3x^3 - 2\sqrt{x}$

64. $G(x) = 4\sqrt{x} - x^2$

65. If $F(s) = 3s + 15$, find: $\dfrac{F(2 + h) - F(2)}{h}$

66. If $K(r) = 7 - 4r$, find: $\dfrac{K(1 + h) - K(1)}{h}$

67. If $g(x) = 2 - x^2$, find: $\dfrac{g(3 + h) - g(3)}{h}$

68. If $P(m) = 2m^2 + 3$, find: $\dfrac{P(2 + h) - P(2)}{h}$

69. If $L(w) = -2w^2 + 3w - 1$, find: $\dfrac{L(-2 + h) - L(-2)}{h}$

70. If $D(p) = -3p^2 - 4p + 9$, find: $\dfrac{D(-1 + h) - D(-1)}{h}$

71. Find $f(x)$, given that

$$f(x + h) = 3(x + h)^2 - 5(x + h) + 9$$

72. Find $g(w)$, given that

$$g(w + h) = -4(w + h)^3 + 7(w + h) - 5$$

73. Find $m(t)$, given that

$$m(t + h) = -2(t + h)^2 - 5\sqrt{t + h} - 2$$

74. Find $s(z)$, given that

$$s(z + h) = 3(z + h) + 9\sqrt{z + h} + 1$$

C

In Problems 75–82, find and simplify:

(A) $\dfrac{f(x + h) - f(x)}{h}$ (B) $\dfrac{f(x) - f(a)}{x - a}$

75. $f(x) = 4x - 7$

76. $f(x) = -5x + 2$

77. $f(x) = 2x^2 - 4$

78. $f(x) = 5 - 3x^2$

79. $f(x) = -4x^2 + 3x - 2$

80. $f(x) = 3x^2 - 5x - 9$

81. $f(x) = x^3 - 2x$

82. $f(x) = x^2 - x^3$

83. The area of a rectangle is 64 square inches. Express the perimeter $P(w)$ as a function of the width w and state the domain.

84. The perimeter of a rectangle is 50 inches. Express the area $A(w)$ as a function of the width w and state the domain.

85. The altitude of a right triangle is 5 meters. Express the hypotenuse $h(b)$ as a function of the base b and state the domain.

86. The altitude of a right triangle is 4 meters. Express the base $b(h)$ as a function of the hypotenuse h and state the domain.

APPLICATIONS

Most of the applications in this section are calculus-related. That is, similar problems will appear in a calculus course, but additional analysis of the functions will be required.

87. Cost Function. The fixed costs per day for a doughnut shop are \$300, and the variable costs are \$1.75 per dozen doughnuts produced. If x dozen doughnuts are produced daily, express the daily cost $C(x)$ as a function of x.

88. Cost Function. The fixed costs per day for a ski manufacturer are \$3,750, and the variable costs are \$68 per pair of skis produced. If x pairs of skis are produced daily, express the daily cost $C(x)$ as a function of x.

89. Physics—Rate. The distance in feet that an object falls in a vacuum is given by $s(t) = 16t^2$, where t is time in seconds. Find:

(A) $s(0), s(1), s(2), s(3)$

(B) $\dfrac{s(2 + h) - s(2)}{h}$

(C) What happens in part B when h tends to 0? Interpret physically.

90. Physics—Rate. An automobile starts from rest and travels along a straight and level road. The distance in feet traveled by the automobile is given by $s(t) = 10t^2$, where t is time in seconds. Find:

(A) $s(8), s(9), s(10), s(11)$

(B) $\dfrac{s(11 + h) - s(11)}{h}$

(C) What happens in part B as h tends to 0? Interpret physically.

91. Manufacturing. A candy box is to be made out of a piece of cardboard that measures 8 by 12 inches. Squares, x inches on a side, will be cut from each corner, and then the ends and sides will be folded down (see the figure). Find a formula for the volume of the box $V(x)$ in terms of x. From practical considerations, what is the domain of the function V?

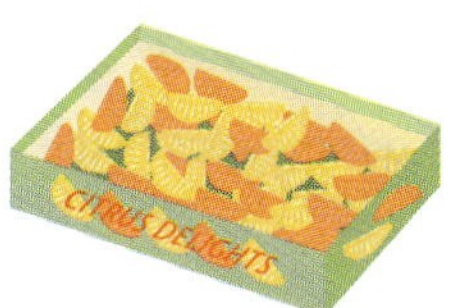

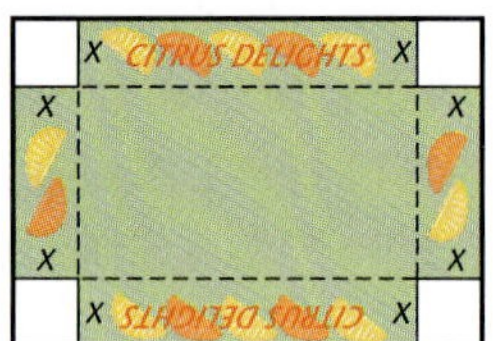

92. Construction. A rancher has 20 miles of fencing to fence a rectangular piece of grazing land along a straight river. If no fence is required along the river and the sides perpendicular to the river are x miles long, find a formula for the area $A(x)$ of the rectangle in terms of x. From practical considerations, what is the domain of the function A?

★ **93. Construction.** The manager of an animal clinic wants to construct a kennel with four individual pens, as indicated in the figure. State law requires that each pen have a gate 3 feet wide and an area of 50 square feet. If x is the width of one pen, express the total amount of fencing $F(x)$ (excluding the gates) required for the construction of the kennel as a function of x. Complete the following table [round values of $F(x)$ to one decimal place]:

x	4	5	6	7
$F(x)$				

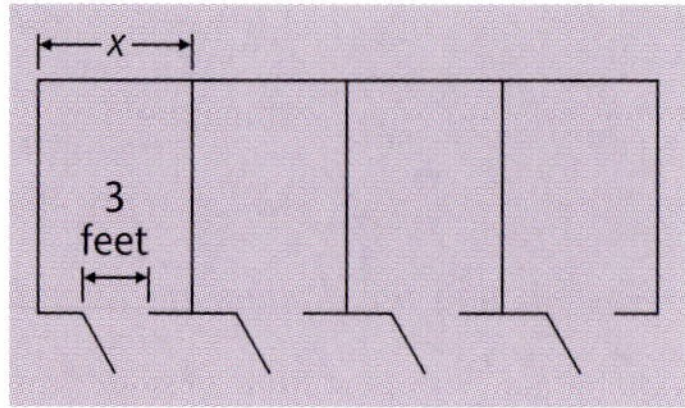

Figure for 93

★ **94. Architecture.** An architect wants to design a window with an area of 24 square feet in the shape of a rectangle surmounted by a semicircle, as indicated in the figure. If x is the width of the window, express the perimeter $P(x)$ of the window as a function of x. Complete the table below [round each value of $P(x)$ to one decimal place]:

x	4	5	6	7
$P(x)$				

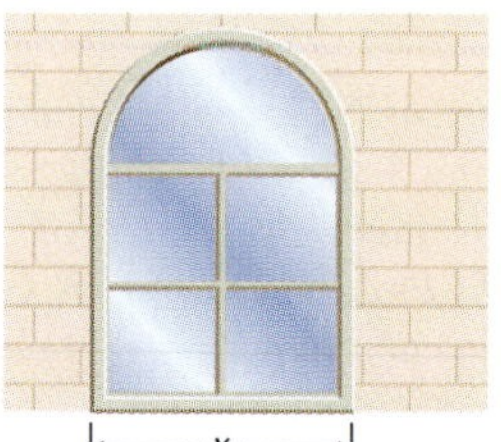

★ **95. Construction.** A freshwater pipeline is to be run from a source on the edge of a lake to a small resort community on an island 8 miles offshore, as indicated in the figure. It costs $10,000 per mile to lay the pipe on land and $15,000 per mile to lay the pipe in the lake. Express the total cost $C(x)$ of constructing the pipeline as a function of x. From practical considerations, what is the domain of the function C?

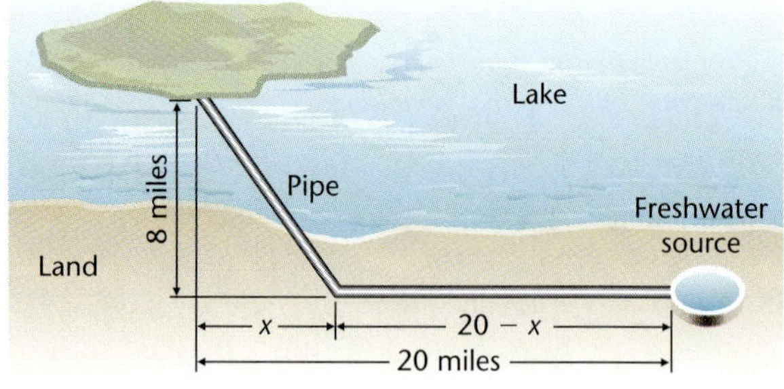

★ **96. Weather.** An observation balloon is released at a point 10 miles from the station that receives its signal and rises vertically, as indicated in the figure. Express the distance $d(h)$ between the balloon and the receiving station as a function of the altitude h of the balloon.

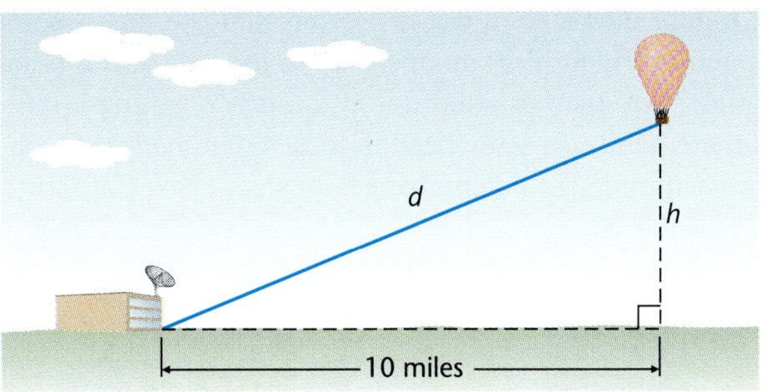

Figure for 96

★★ **97. Operational Costs.** The cost per hour for fuel for running a train is $v^2/5$ dollars, where v is the speed in miles per hour. (Note that cost goes up as the square of the speed.) Other costs, including labor, are $400 per hour. Express the total cost of a 500-mile trip as a function of the speed v.

★★ **98. Operational Costs.** Refer to Problem 97. If it takes t hours for the train to complete the 500-mile trip, express the total cost as a function of t.

SECTION 2-4 Graphing Functions

- Basic Concepts
- Linear Functions
- Quadratic Functions
- Piecewise-Defined Functions
- The Greatest Integer Function

In this section we take another look at the graphs of linear equations, this time using the function concepts introduced in the preceding section. We also develop procedures for graphing functions defined by quadratic equations and functions formed by "piecing" together two or more other functions. We begin by discussing some general concepts related to the graphs of functions.

• Basic Concepts

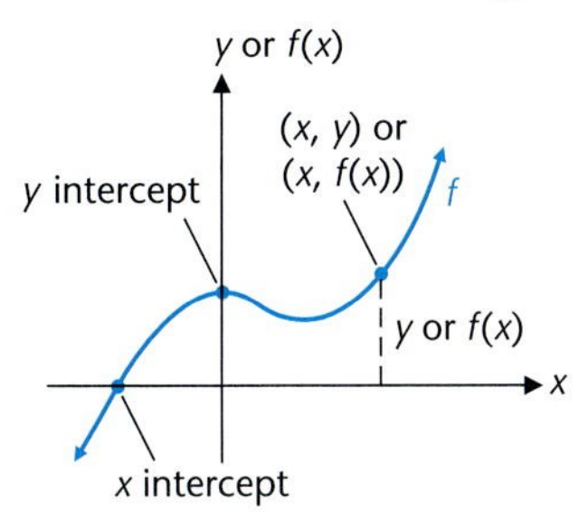

FIGURE 1 Graph of a function.

Each function that has a real number domain and range has a graph—the graph of the ordered pairs of real numbers that constitute the function. When functions are graphed, domain values usually are associated with the horizontal axis and range values with the vertical axis. Thus, the **graph of a function *f*** is the same as the graph of the equation

$$y = f(x)$$

where x is the independent variable and the abscissa of a point on the graph of f. The variables y and $f(x)$ are dependent variables, and either is the ordinate of a point on the graph of f (see Fig. 1)

The abscissa of a point where the graph of a function intersects the x axis is called an ***x* intercept** or **zero** of the function. The x intercept is also a real solution or **root** of the equation $f(x) = 0$. The ordinate of a point where the graph of a function crosses the y axis is called the ***y* intercept** of the function. The y intercept is given by $f(0)$, provided 0 is in the domain of f. Note that a function can have more than one x intercept but can never have more than one y intercept—a consequence of the vertical line test discussed in the preceding section.

The domain of a function is the set of all the x coordinates of points on the graph of the function, and the range is the set of all the y coordinates. It is instructive to view the domain and range as subsets of the coordinate axes as in Figure 2 on the next page. Note the effective use of interval notation in describing the domain and range of the functions in this figure. In Figure 2(a) a solid dot is used to indicate that a point is on the graph of the function and in Figure 2(b) an open dot to indicate that a point is not on the graph of the function. An open or solid dot at the end of a graph indicates that the graph terminates there, while an arrowhead indicates that the graph continues beyond the portion shown with no significant changes in shape [see Fig. 2(b)].

FIGURE 2 Domain and range.

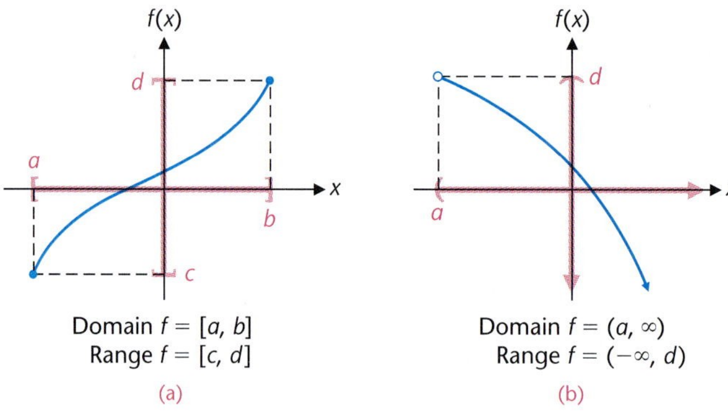

EXAMPLE 1 Finding the Domain and Range from a Graph

Find the domain and range of the function f in Figure 3:

Solution The dots at each end of the graph of f indicate that the graph terminates at these points. Thus, the x coordinates of the points on the graph are between -3 and 6. The open dot at $(-3, 4)$ indicates that -3 is not in the domain of f, while the closed dot at $(6, -3)$ indicates that 6 is in the domain of f. That is,

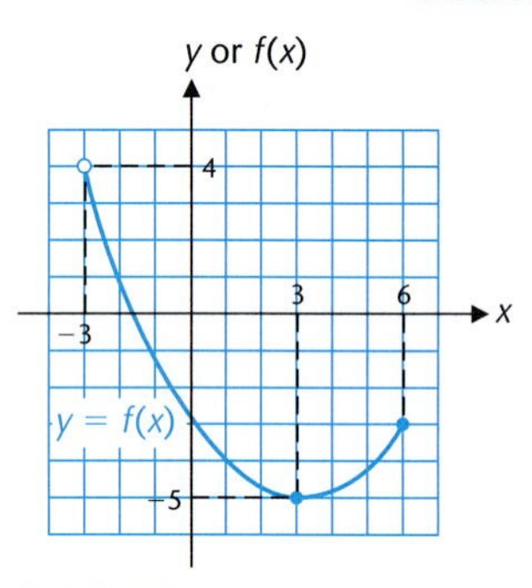

FIGURE 3

$$\text{Domain: } -3 < x \leq 6 \quad \text{or} \quad (-3, 6]$$

The y coordinates are between -5 and 4, and, as before, the open dot at $(-3, 4)$ indicates that 4 is not in the range of f and the closed dot at $(3, -5)$ indicates that -5 is in the range of f. Thus,

$$\text{Range: } -5 \leq y < 4 \quad \text{or} \quad [-5, 4)$$

Matched Problem 1 Find the domain and range of the function f given by the graph in Figure 4.

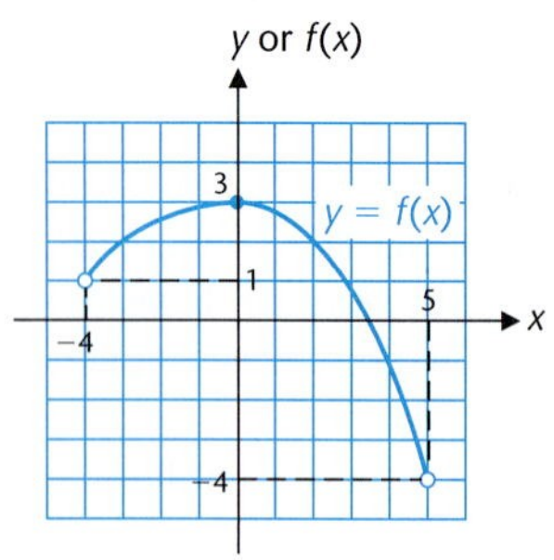

FIGURE 4

In general, you will not be expected to be able to determine the range of every function defined by an equation you will encounter in this course, but for certain basic functions collected in our library of elementary functions, it is important to know both the domain and range of each, and these will be discussed as these elementary functions are introduced. One of the primary goals of this course is to provide you with a library of basic mathematical functions, including their graphs and other important properties. These can then be used to analyze graphs and properties of a wide variety of more complex functions that arise quite naturally in important applications. In this section we begin this process by introducing some of the language commonly used to describe the behavior of a graph.

We now take a look at increasing and decreasing properties of functions. Intuitively, a function is increasing over an interval I in its domain if its graph rises as the independent variable increases over I. A function is decreasing over I if its graph falls as the independent variable increases over I (Fig. 5).

FIGURE 5 Increasing, decreasing, and constant functions.

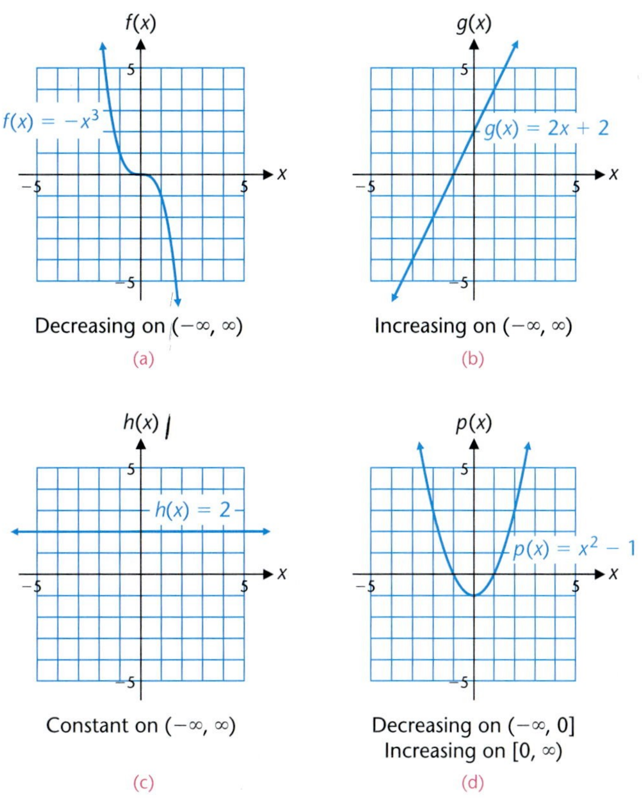

More formally, we define increasing, decreasing, and constant functions as follows:

DEFINITION 1 **Increasing, Decreasing, and Constant Functions**

Let I be an interval in the domain of a function f. Then:

1. **f is increasing** on I if $f(b) > f(a)$ whenever $b > a$ in I.
2. **f is decreasing** on I if $f(b) < f(a)$ whenever $b > a$ in I.
3. **f is constant** on I if $f(a) = f(b)$ for all a and b in I.

• Linear Functions

We now apply the general concepts discussed above to a specific class of functions known as *linear functions.*

DEFINITION 2 **Linear Function**

A function f is a **linear function** if

$$f(x) = mx + b \qquad m \neq 0$$

where m and b are real numbers.

Graphing a linear function is equivalent to graphing the equation

$$y = mx + b$$

which we recognize as the equation of a line with slope m and y intercept b. Since the expression $mx + b$ represents a real number for all real number replacements of x, the domain of a linear function is the set of all real numbers. The restriction $m \neq 0$ in the definition of a linear function implies that the graph is not a horizontal line. Hence, the range of a linear function is also the set of all real numbers.

Graph of $f(x) = mx + b$, $m \neq 0$

The graph of a linear function f is a nonvertical and nonhorizontal straight line with slope m and y intercept b.

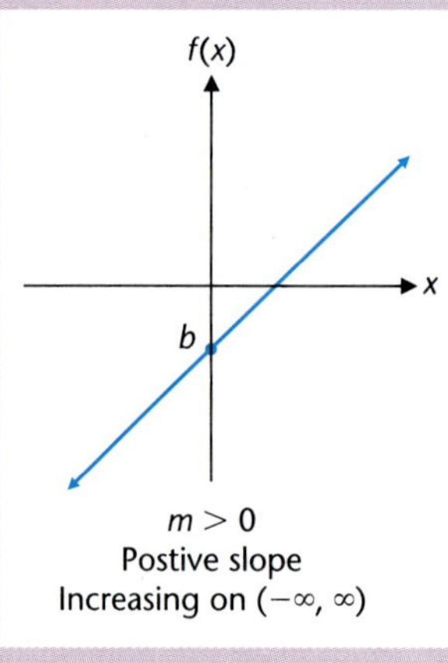

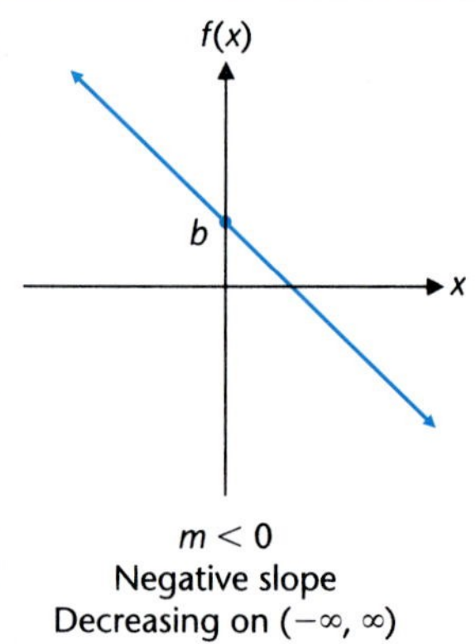

Domain: All real numbers Range: All real numbers

Notice that two types of lines are not the graphs of linear functions. A vertical line with equation $x = a$ does not pass the vertical line test and cannot define a function. A horizontal line with equation $y = b$ does pass the vertical line test and does define a function. However, a function of the form

$$f(x) = b \qquad \text{Constant function}$$

is called a **constant function,** not a linear function.

EXPLORE-DISCUSS 1

(A) Is it possible for a linear function to have two x intercepts? No x intercepts? If either of your answers is yes, give an example.
(B) Is it possible for a linear function to have two y intercepts? No y intercept? If either of your answers is yes, give an example.
(C) Discuss the possible number of x and y intercepts for a constant function.

EXAMPLE 2 Graphing a Linear Function

Find the slope and intercepts, and then sketch the graph of the linear function defined by

$$f(x) = -\tfrac{2}{3}x + 4$$

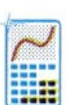
Check with a graphing utility.

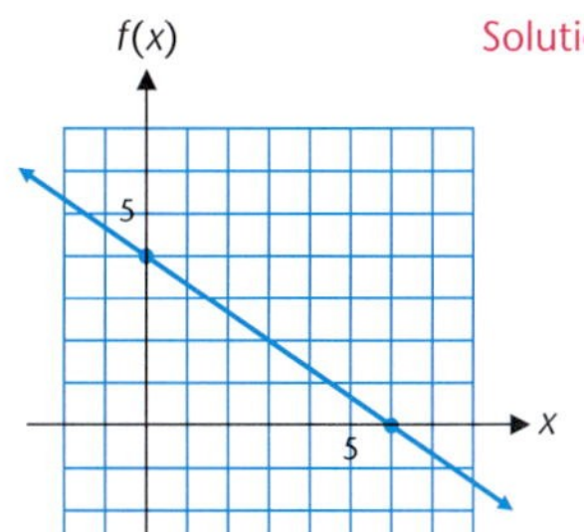

FIGURE 6

Solution The y intercept is $f(0) = 4$, and the slope is $-\frac{2}{3}$. To find the x intercept, we solve the equation $f(x) = 0$ for x:

$$\begin{aligned} f(x) &= 0 \\ -\tfrac{2}{3}x + 4 &= 0 \\ -\tfrac{2}{3}x &= -4 \\ x &= (-\tfrac{3}{2})(-4) = 6 \quad \text{x intercept} \end{aligned}$$

The graph of f is shown in Figure 6.

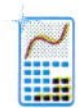
To find the y intercept with a graphing utility, simply evaluate the function at $x = 0$ [Fig. 7(a)]. Most graphing utilities have a built-in procedure for approximating x intercepts, usually called *root* or *zero* [Fig. 7(b)].

FIGURE 7

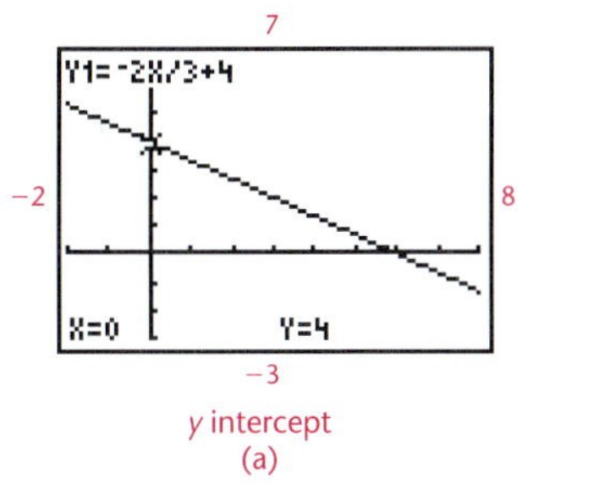

y intercept
(a)

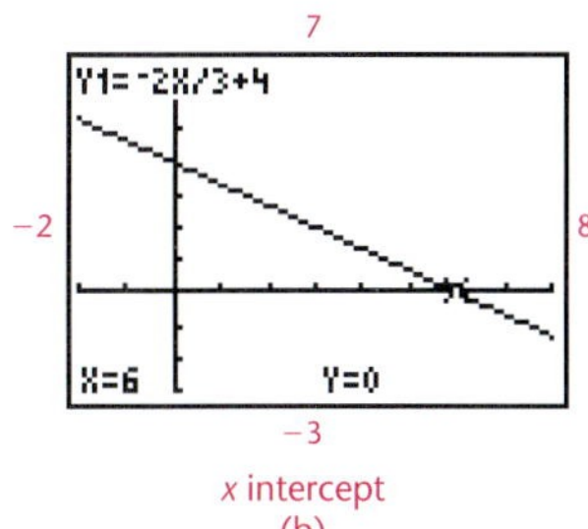

x intercept
(b)

Matched Problem 2 Find the slope and intercepts, and then sketch the graph of the linear function defined by

$$f(x) = \tfrac{3}{2}x - 6$$

• Quadratic Functions

Just as we used the first-degree polynomial $mx + b$, $m \neq 0$, to define a linear function, we use the second-degree polynomial $ax^2 + bx + c$, $a \neq 0$, to define a *quadratic function.*

DEFINITION 3 **Quadratic Function**

A function f is a **quadratic function** if

$$f(x) = ax^2 + bx + c \qquad a \neq 0 \tag{1}$$

where a, b, and c are real numbers.

The graphs of three quadratic functions are shown in Figure 8. The graph of a quadratic function is called a **parabola.**

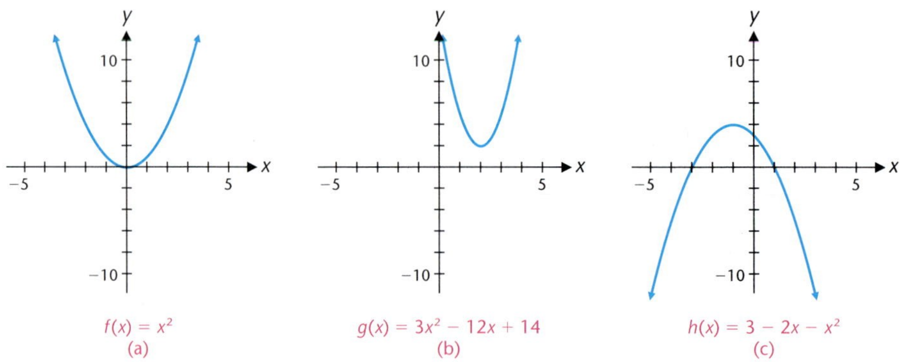

FIGURE 8 Graphs of quadratic functions.

Since the expression $ax^2 + bx + c$ represents a real number for all real number replacements of x:

The domain of a quadratic function is the set of all real numbers.

The range of a quadratic function and many important features of its graph can be determined by first transforming equation (1) by completing the square into the form

$$f(x) = a(x - h)^2 + k \tag{2}$$

A brief review of completing the square, which we discussed in Section 1-6, might prove helpful at this point. We illustrate this method through an example and then generalize the results.

Consider the quadratic function given by

$$f(x) = 2x^2 - 8x + 4 \tag{3}$$

We start by transforming equation (3) into form (2) by completing the square as follows:

$$
\begin{aligned}
f(x) &= 2x^2 - 8x + 4 \\
&= 2(x^2 - 4x) + 4 && \text{Factor the coefficient of } x^2 \text{ out of the first two terms.} \\
&= 2(x^2 - 4x + ?) + 4 \\
&= 2(x^2 - 4x + 4) + 4 - 8 && \text{We add 4 to complete the square inside the parentheses. But because of the 2 outside the parentheses, we have actually added 8, so we must subtract 8.} \\
&= 2(x - 2)^2 - 4 && \text{The transformation is complete.}
\end{aligned}
$$

Thus,

$$f(x) = 2(x - 2)^2 - 4 \tag{4}$$

If $x = 2$, then $2(x - 2)^2 = 0$ and $f(2) = -4$. For any other value of x, the positive number $2(x - 2)^2$ is added to -4, thus making $f(x)$ larger. Therefore,

$$f(2) = -4$$

is the *minimum value* of $f(x)$ for all x—a very important result! Furthermore, if we choose any two x values that are equidistant from the vertical line $x = 2$, we will obtain the same value for the function. For example, $x = 1$ and $x = 3$ are each one unit from $x = 2$, and the corresponding functional values are

$$f(1) = 2(-1)^2 - 4 = -2$$
$$f(3) = 2(1)^2 - 4 = -2$$

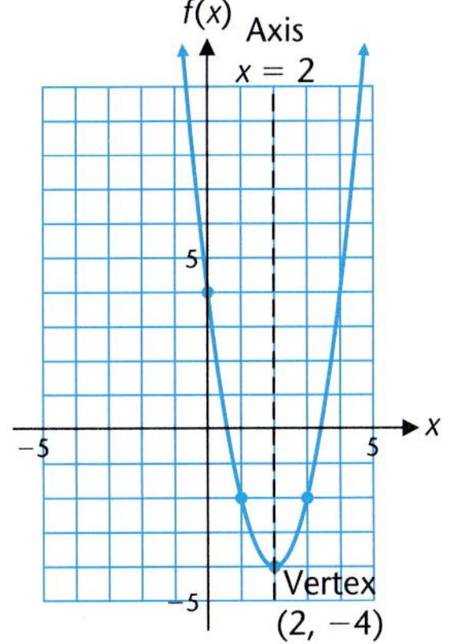

FIGURE 9 Graph of $f(x) = 2(x - 2)^2 - 4$.

Thus, the vertical line $x = 2$ is a line of symmetry. That is, if the graph is drawn on a piece of paper and the paper is folded along the line $x = 2$, then the two sides of the parabola will match exactly. All these results are illustrated by graphing equations (3) or (4) and the line $x = 2$ in the same coordinate system. (Fig. 9).

From the above discussion, we see that as x moves from left to right, $f(x)$ is decreasing on $(-\infty, 2]$ and increasing on $[2, \infty)$. Furthermore, $f(x)$ can assume any values greater than or equal to -4, but no values less than -4. Thus,

$$\text{Range of } f\text{: } y \geq -4 \quad \text{or} \quad [-4, \infty)$$

In general, the graph of a quadratic function is a parabola with line of symmetry parallel to the vertical axis. The lowest or highest point on the parabola, whichever exists, is called the **vertex.** The maximum or minimum value of a quadratic function always occurs at the vertex of the parabola. The line of symmetry through the vertex is called the **axis** of the parabola. In the above example, $x = 2$ is the axis of the parabola and $(2, -4)$ is its vertex.

Note the important results we have obtained by transforming equation (3) into equation (4):

The vertex of the parabola

The axis of the parabola

The minimum value of $f(x)$

The range of the function f

Now, let us explore the effect of changing the constants a, h, and k on the graph of $y = a(x - h)^2 + k$.

EXPLORE-DISCUSS 2 Explore the effect of changing the constants a, h, and k on the graph of $f(x) = a(x - h)^2 + k$.

(A) Let $a = 1$ and $h = 5$. Graph function f for $k = -4$, 0, and 3 simultaneously in the same coordinate system. Explain the effect of changing k on the graph of f.
(B) Let $a = 1$ and $k = 2$. Graph function f for $h = -4$, 0, 5 simultaneously in the same coordinate system. Explain the effect of changing h on the graph of f.
(C) Let $h = 5$ and $k = -2$. Graph function f for $a = 0.25$, 1, and 3 simultaneously in the same coordinate system. Graph function f for $a = 1$, -1, and -0.25 simultaneously in the same coordinate system.

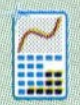

(D) Discuss parts A–C using a graphing utility and a standard viewing window.

The above discussion is generalized for all quadratic functions in the following box:

Properties of a Quadratic Function and Its Graph

Given a quadratic function and the form obtained by completing the square

$$f(x) = ax^2 + bx + c = a(x - h)^2 + k \qquad a \neq 0$$

we summarize general properties as follows:

1. The graph of f is a parabola:

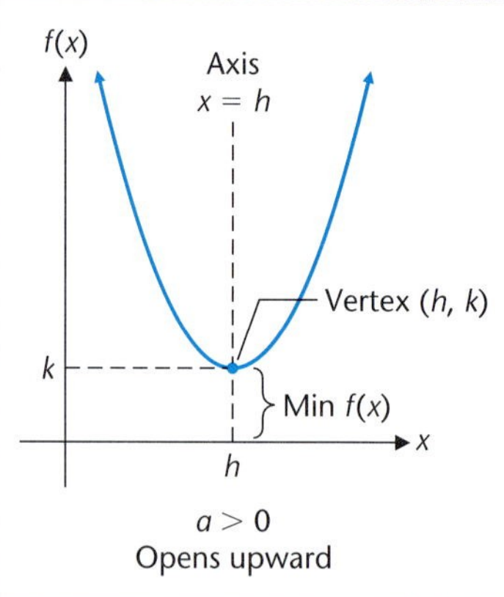

$a > 0$
Opens upward

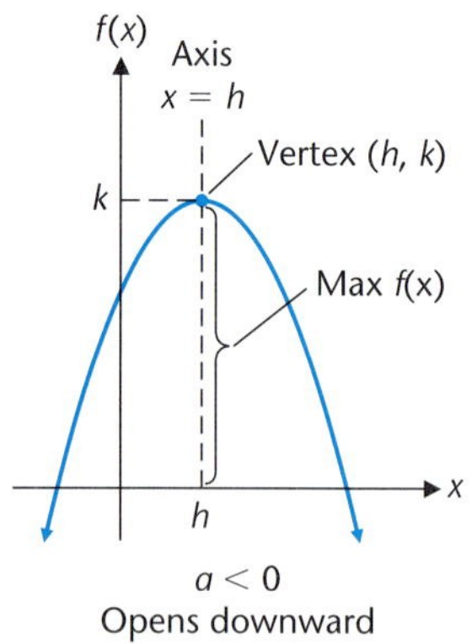

$a < 0$
Opens downward

2. Vertex: (h, k) (Parabola rises on one side of the vertex and falls on the other.)

3. Axis (of symmetry): $x = h$ (Parallel to y axis)

4. $f(h) = k$ is the minimum if $a > 0$ and the maximum if $a < 0$.

5. Domain: All real numbers

   Range: $(-\infty, k]$ if $a < 0$ or $[k, \infty)$ if $a > 0$

EXAMPLE 3 Graph of a Quadratic Function

Graph, finding the vertex, axis, maximum or minimum of $f(x)$, intervals where f is increasing or decreasing, and range.

$$f(x) = -0.5x^2 - x + 2$$

Solution Complete the square:

$$\begin{aligned} f(x) &= -0.5x^2 - x + 2 \\ &= -0.5(x^2 + 2x + ?) + 2 \\ &= -0.5(x^2 + 2x + 1) + 2 + 0.5 \\ &= -0.5(x + 1)^2 + 2.5 \end{aligned}$$

From this last form we see that $h = -1$ and $k = 2.5$. Thus, the vertex is $(-1, 2.5)$, the axis of symmetry is $x = -1$, and the maximum value is $f(-1) = 2.5$. To graph f, locate the axis and vertex; then plot several points on either side of the axis (Fig. 10).

FIGURE 10

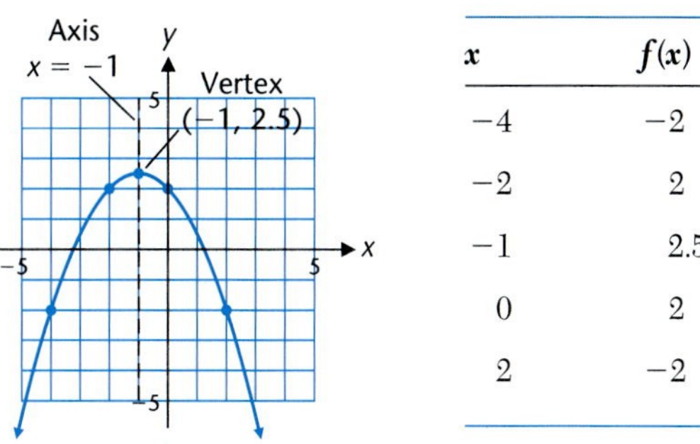

x	$f(x)$
−4	−2
−2	2
−1	2.5
0	2
2	−2

From the graph we see that f is increasing on $(-\infty, -1]$ and decreasing on $[-1, \infty)$. Also, $y = f(x)$ can be any number less than or equal to 2.5. Thus, the range of f is $y \le 2.5$ or $(-\infty, 2.5]$.

Matched Problem 3 Graph, finding the vertex, axis, maximum or minimum of $f(x)$, intervals where f is increasing or decreasing, and range.

$$f(x) = -x^2 + 4x - 4$$

• Piecewise-Defined Functions

The **absolute value function** can be defined using the definition of absolute value from Section 1-4:

$$f(x) = |x| = \begin{cases} -x & \text{if } x < 0 \\ x & \text{if } x \geq 0 \end{cases}$$

Notice that this function is defined by different formulas for different parts of its domain. Functions whose definitions involve more than one formula are called **piecewise-defined functions.** As the next example illustrates, piecewise-defined functions occur naturally in many applications.

EXAMPLE 4 **Rental Charges**

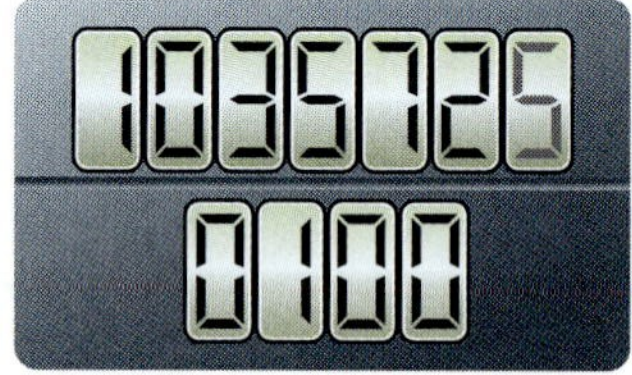

A car rental agency charges \$0.25 per mile if the total mileage does not exceed 100. If the total mileage exceeds 100, the agency charges \$0.25 per mile for the first 100 miles plus \$0.15 per mile for the additional mileage. If x represents the number of miles a rented vehicle is driven, express the mileage charge $C(x)$ as a function of x. Find $C(50)$ and $C(150)$, and graph C.

Solution If $0 \leq x \leq 100$, then

$$C(x) = 0.25x$$

If $x > 100$, then

Charge for the first 100 miles — Charge for the additional mileage

$$\begin{aligned} C(x) &= 0.25(100) + 0.15(x - 100) \\ &= 25 + 0.15x - 15 \\ &= 10 + 0.15x \end{aligned}$$

Thus, we see that C is a piecewise-defined function:

$$C(x) = \begin{cases} 0.25x & \text{if } 0 \leq x \leq 100 \\ 10 + 0.15x & \text{if } x > 100 \end{cases}$$

Piecewise-defined functions are evaluated by first determining which rule applies and then using the appropriate rule to find the value of the function. For example, to evaluate $C(50)$, we use the first rule and obtain

$$C(50) = 0.25(50) = \$12.50 \qquad x = 50 \text{ satisfies } 0 \leq x \leq 100$$

To evaluate $C(150)$, we use the second rule and obtain

$$C(150) = 10 + 0.15(150) = \$32.50 \qquad x = 150 \text{ satisfies } x > 100$$

To graph C, we graph each rule in the definition for the indicated values of x (Fig. 11).

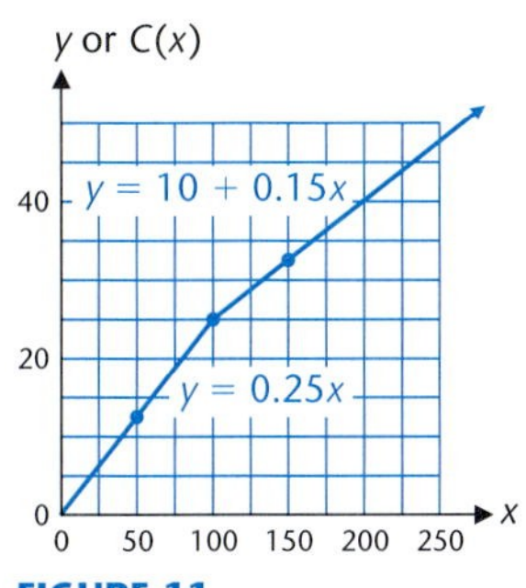

FIGURE 11

x	$y = 0.25x$
50	12.5
100	25

x	$y = 10 + 0.15x$
100	25
150	32.5

Notice that the two formulas produce the same value at $x = 100$ and that the graph of C contains no breaks. Informally, a graph (or portion of a graph) is said to be **continuous** if it contains no breaks or gaps. (A formal presentation of continuity may be found in calculus texts.)

Matched Problem 4 Refer to Example 4. Find $C(x)$ if the agency charges \$0.30 per mile when the total mileage does not exceed 75, and \$0.30 per mile for the first 75 miles plus \$0.20 per mile for additional mileage when the total mileage exceeds 75. Find $C(50)$ and $C(100)$, and graph C.

EXAMPLE 5 Graphing a Function Involving Absolute Value

Graph the function f given by

$$f(x) = x + \frac{x}{|x|}$$

and find its domain and range.

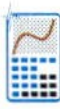
Check with a graphing utility.

Solution We use the piecewise definition of $|x|$ to find a piecewise definition of f that does not involve $|x|$.

If $x < 0$, then $|x| = -x$ and

$$f(x) = x + \frac{x}{|x|} = x + \frac{x}{-x} = x - 1$$

If $x = 0$, then f is not defined, since division by 0 is not permissible.

If $x > 0$, then $|x| = x$ and

$$f(x) = x + \frac{x}{|x|} = x + \frac{x}{x} = x + 1$$

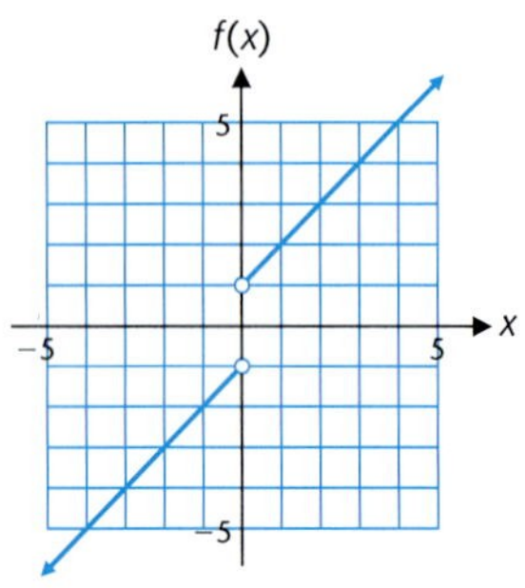

FIGURE 12

Thus, a piecewise definition for f is

$$f(x) = \begin{cases} x - 1 & \text{if } x < 0 \\ x + 1 & \text{if } x > 0 \end{cases}$$

$$\text{Domain: } x \neq 0 \quad \text{or} \quad (-\infty, 0) \cup (0, \infty)$$

We use this definition to graph f as shown in Figure 12. Examining this graph, we see that $y = f(x)$ can be any number less than -1 or any number greater than 1. Thus,

$$\text{Range: } y < -1 \quad \text{or} \quad y > 1 \quad \text{or} \quad (-\infty, -1) \cup (1, \infty)$$

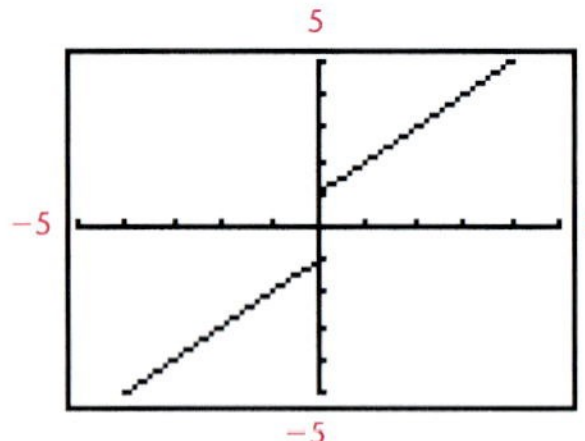

FIGURE 13

Notice that we used open dots in the figure at $(0, -1)$ and $(0, 1)$ to indicate that these points do not belong to the graph of f. Because of the break in the graph at $x = 0$, we say that f is **discontinuous** at $x = 0$.

The check is shown in Figure 13. Most graphing utilities denote the absolute value function by abs(x)—check your manual. 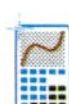

Matched Problem 5 Graph the function f given by

$$f(x) = -\frac{2x}{|x|} - x$$

and find its domain and range.

• The Greatest Integer Function

We conclude this section with a discussion of an interesting and useful function called the *greatest integer function.*

The **greatest integer** of a real number x, denoted by $[\![x]\!]$, is the integer n such that $n \le x < n + 1$; that is, $[\![x]\!]$ is the largest integer less than or equal to x. For example,

$$[\![3.45]\!] = 3 \qquad [\![-2.13]\!] = -3 \quad \text{Not } -2$$

$$[\![7]\!] = 7 \qquad [\![-8]\!] = -8$$

$$[\![0]\!] = 0$$

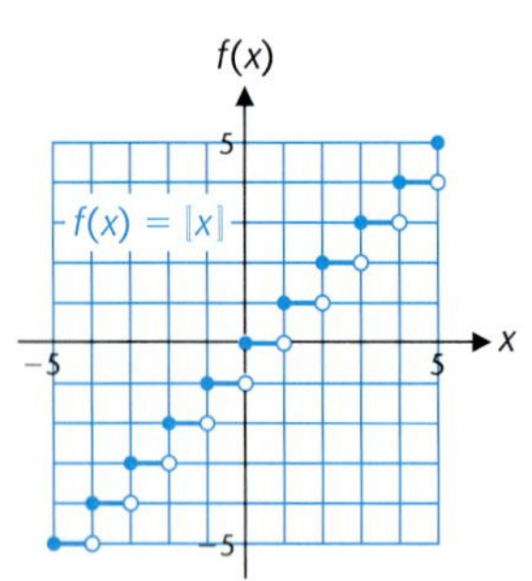

FIGURE 14 Greatest integer function.

The **greatest integer function** f is defined by the equation $f(x) = [\![x]\!]$. A piecewise definition of f for $-2 \le x < 3$ is shown at the top of the next page and a sketch of the graph of f for $-5 \le x \le 5$ is shown in Figure 14. Since the domain of f is all real numbers, the piecewise definition continues indefinitely in both directions, as does the stairstep pattern in the figure. Thus, the range of f is the set of all integers. The greatest integer function is an example of a more general class of functions called **step functions.**

$$f(x) = [\![x]\!] = \begin{cases} \vdots & \\ -2 & \text{if } -2 \le x < -1 \\ -1 & \text{if } -1 \le x < 0 \\ 0 & \text{if } 0 \le x < 1 \\ 1 & \text{if } 1 \le x < 2 \\ 2 & \text{if } 2 \le x < 3 \\ \vdots & \end{cases}$$

Notice in Figure 14 that at each integer value of x there is a break in the graph, and between integer values of x there is no break. Thus, the greatest integer function is discontinuous at each integer n and continuous on each interval of the form $[n, n + 1)$.

EXPLORE-DISCUSS 3 Most graphing utilities denote the greatest integer function as int (x), although not all define it the same way we have here. Graph $y = \text{int}(x)$ for $-5 \le x \le 5$ and $-5 \le y \le 5$ and discuss any differences between your graph and Figure 14. If your graphing utility supports both a connected mode and a dot mode for graphing functions (consult your manual), which mode is preferable for this graph?

EXAMPLE 6 Computer Science

Let

$$f(x) = \frac{[\![10x + 0.5]\!]}{10}$$

Find:

(A) $f(6)$ (B) $f(1.8)$ (C) $f(3.24)$ (D) $f(4.582)$ (E) $f(-2.68)$

What operation does this function perform?

Solutions

(A) $f(6) = \dfrac{[\![60.5]\!]}{10} = \dfrac{60}{10} = 6$

(B) $f(1.8) = \dfrac{[\![18.5]\!]}{10} = \dfrac{18}{10} = 1.8$

(C) $f(3.24) = \dfrac{[\![32.9]\!]}{10} = \dfrac{32}{10} = 3.2$

(D) $f(4.582) = \dfrac{[\![46.32]\!]}{10} = \dfrac{46}{10} = 4.6$

(E) $f(-2.68) = \dfrac{[\![-26.3]\!]}{10} = \dfrac{-27}{10} = -2.7$

Comparing the values of x and $f(x)$ in Table 1 in the margin, we conclude that this function rounds decimal fractions to the nearest tenth.

Matched Problem 6 Let $f(x) = [\![x + 0.5]\!]$. Find:

(A) $f(6)$ (B) $f(1.8)$ (C) $f(3.24)$ (D) $f(-4.3)$ (E) $f(-2.69)$

What operation does this function perform?

TABLE 1

x	$f(x)$
6	6
1.8	1.8
3.24	3.2
4.582	4.6
−2.68	−2.7

Answers to Matched Problems

1. Domain: $-4 < x < 5$ or $(-4, 5)$
   Range: $-4 < y \le 3$ or $(-4, 3]$

2. y intercept: $f(0) = -6$
   x intercept: 4
   Slope: $\frac{3}{2}$

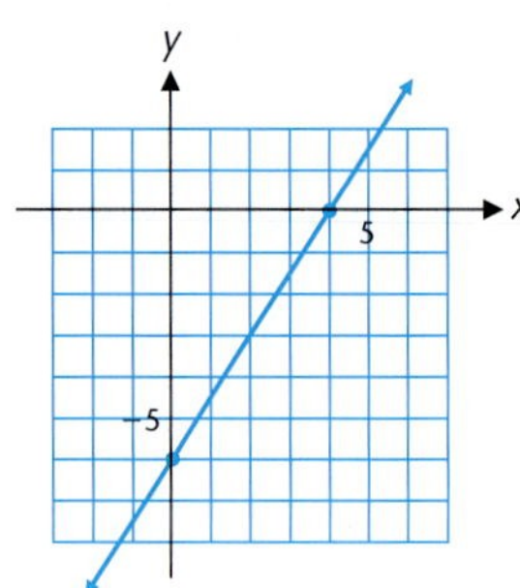

3. Axis: $x = 2$
   Vertex: $(2, f(2)) = (2, 0)$
   Max $f(x)$: $f(2) = 0$
   Increasing: $(-\infty, 2]$
   Decreasing: $[2, \infty)$
   Range: $(-\infty, f(2)] = (-\infty, 0]$

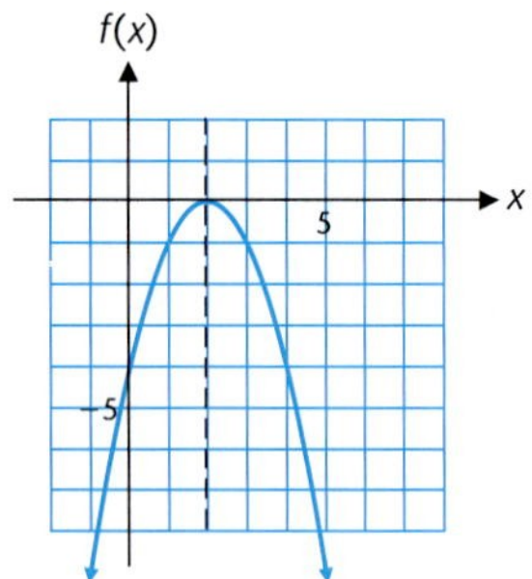

4. $C(x) = \begin{cases} 0.3x & \text{if } 0 \le x \le 75 \\ 7.5 + 0.2x & \text{if } x > 75 \end{cases}$
   $C(50) = \$15$; $C(100) = \$27.50$

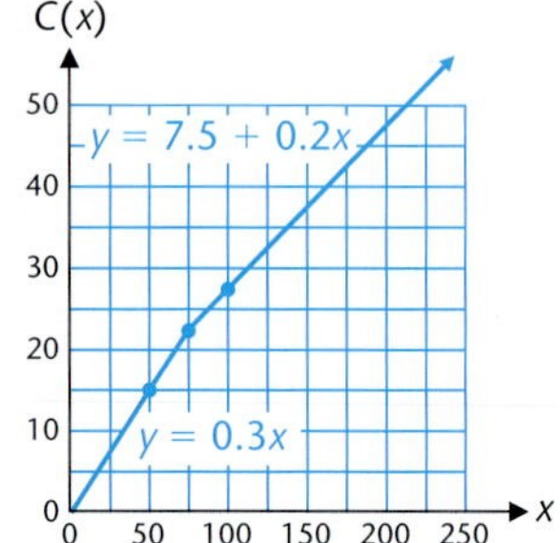

5. $f(x) = \begin{cases} 2 - x & \text{if } x < 0 \\ -2 - x & \text{if } x > 0 \end{cases}$
   Domain: $x \ne 0$ or $(-\infty, 0) \cup (0, \infty)$
   Range: $(-\infty, -2) \cup (2, \infty)$

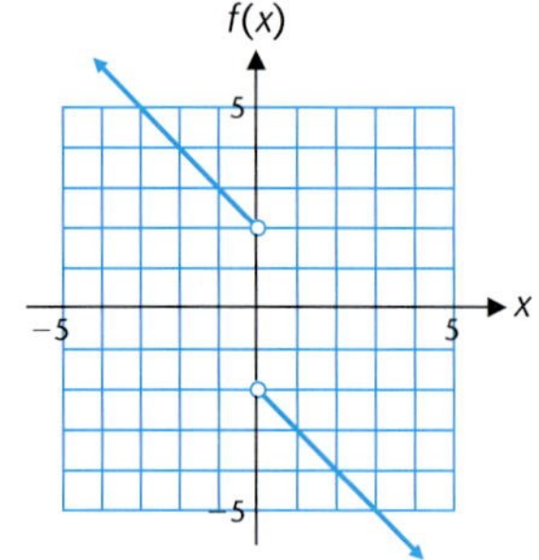

6. (A) 6 (B) 2 (C) 3 (D) −4 (E) −3;
   f rounds decimal fractions to the nearest integer.

EXERCISE 2-4

A

Problems 1–6 refer to functions f, g, h, k, p, and q given by the following graphs. (Assume the graphs continue as indicated beyond the parts shown.)

1. For the function f, find:
(A) Domain (B) Range
(C) x intercepts (D) y intercept
(E) Intervals over which f is increasing
(F) Intervals over which f is decreasing
(G) Intervals over which f is constant
(H) Any points of discontinuity

2. Repeat Problem 1 for the function g.

3. Repeat Problem 1 for the function h.

4. Repeat Problem 1 for the function k.

5. Repeat Problem 1 for the function p.

6. Repeat Problem 1 for the function q.

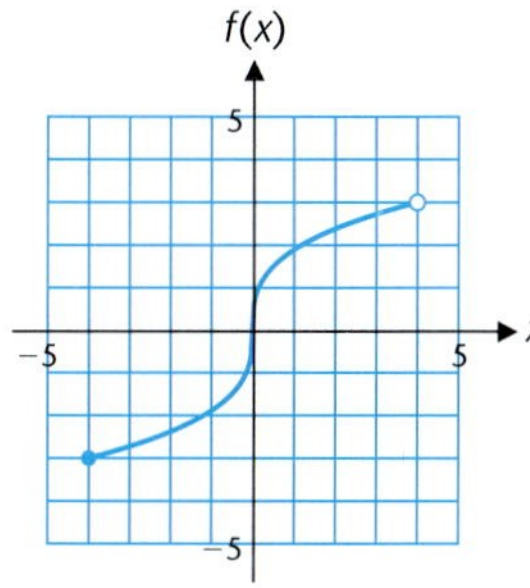

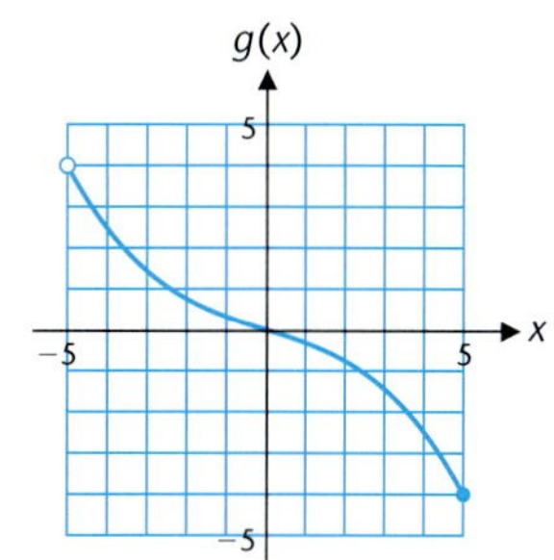

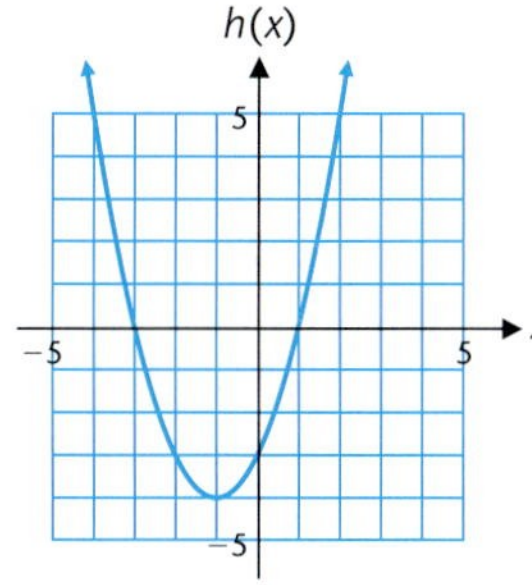

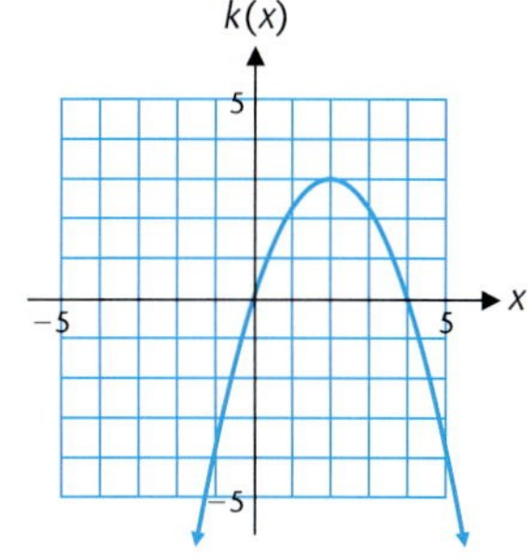

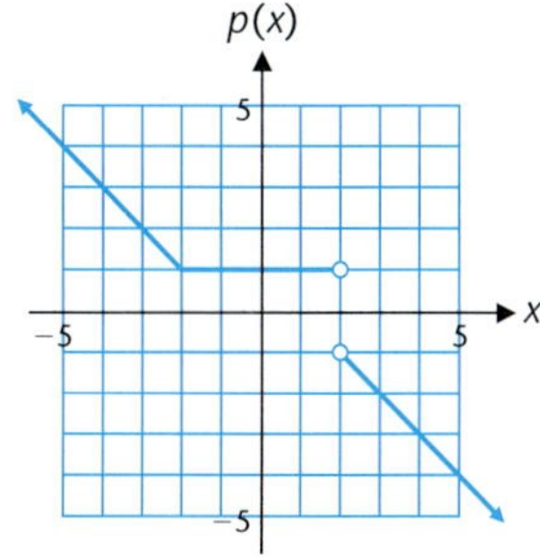

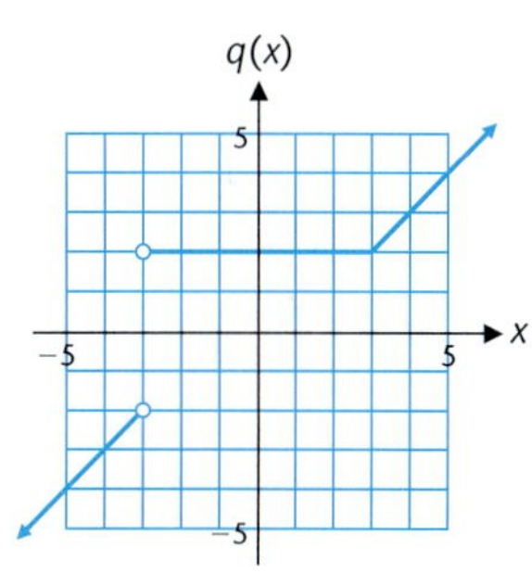

Problems 7–12 describe the graph of a continuous function f over the interval [−5, 5]. Sketch the graph of a function that is consistent with the given information.

7. The function f is increasing on $[-5, -2]$, constant on $[-2, 2]$, and decreasing on $[2, 5]$.

8. The function f is decreasing on $[-5, -2]$, constant on $[-2, 2]$, and increasing on $[2, 5]$.

9. The function f is decreasing on $[-5, -2]$, constant on $[-2, 2]$, and decreasing on $[2, 5]$.

10. The function f is increasing on $[-5, -2]$, constant on $[-2, 2]$, and increasing on $[2, 5]$.

11. The function f is decreasing on $[-5, -2]$, increasing on $[-2, 2]$, and decreasing on $[2, 5]$.

12. The function f is increasing on $[-5, -2]$, decreasing on $[-2, 2]$, and increasing on $[2, 5]$.

In Problems 13–16, find the slope and intercepts, and then sketch the graph.

13. $f(x) = 2x + 4$

14. $f(x) = 3x - 3$

15. $f(x) = -\frac{1}{2}x - \frac{5}{3}$

16. $f(x) = -\frac{3}{4}x + \frac{6}{5}$

In Problems 17 and 18, find a linear function f satisfying the given conditions.

17. $f(-2) = 7$ and $f(4) = -2$

18. $f(-3) = -2$ and $f(5) = 4$

B

In Problems 19–22, graph, finding the axis, vertex, maximum or minimum, and range.

19. $f(x) = (x - 3)^2 + 2$

20. $f(x) = \frac{1}{2}(x + 2)^2 - 4$

21. $f(x) = -(x + 3)^2 - 2$

22. $f(x) = -(x - 2)^2 + 4$

In Problems 23–26, graph, finding the axis, vertex, x intercepts, and y intercept.

23. $f(x) = x^2 - 4x - 5$

24. $f(x) = x^2 - 6x + 5$

25. $f(x) = -x^2 + 6x$

26. $f(x) = -x^2 + 2x + 8$

In Problems 27–30, graph, finding the axis, vertex, intervals over which f is increasing, and intervals over which f is decreasing.

27. $f(x) = x^2 + 6x + 11$

28. $f(x) = x^2 - 8x + 14$

29. $f(x) = -x^2 + 6x - 6$

30. $f(x) = -x^2 - 10x - 24$

In Problems 31–38, graph, finding the domain, range, and any points of discontinuity.

31. $f(x) = \begin{cases} x + 1 & \text{if } -1 \le x < 0 \\ -x + 1 & \text{if } 0 \le x \le 1 \end{cases}$

32. $f(x) = \begin{cases} x & \text{if } -2 \le x < 1 \\ -x + 2 & \text{if } 1 \le x \le 2 \end{cases}$

33. $f(x) = \begin{cases} -2 & \text{if } -3 \le x < -1 \\ 4 & \text{if } -1 < x \le 2 \end{cases}$

34. $f(x) = \begin{cases} 1 & \text{if } -2 \le x < 2 \\ -3 & \text{if } 2 < x \le 5 \end{cases}$

35. $f(x) = \begin{cases} x + 2 & \text{if } x < -1 \\ x - 2 & \text{if } x \ge -1 \end{cases}$

36. $f(x) = \begin{cases} -1 - x & \text{if } x \le 2 \\ 5 - x & \text{if } x > 2 \end{cases}$

37. $g(x) = \begin{cases} x^2 + 1 & \text{if } x < 0 \\ -x^2 - 1 & \text{if } x > 0 \end{cases}$

38. $h(x) = \begin{cases} -x^2 - 2 & \text{if } x < 0 \\ x^2 + 2 & \text{if } x > 0 \end{cases}$

C

In Problems 39–44, graph, finding the axis, vertex, maximum or minimum of f(x), range, intercepts, intervals over which f is increasing, and intervals over which f is decreasing.

39. $f(x) = \frac{1}{2}x^2 + 2x + 3$

40. $f(x) = 2x^2 - 12x + 14$

41. $f(x) = 4x^2 - 12x + 9$

42. $f(x) = -\frac{1}{2}x^2 + 4x - 10$

43. $f(x) = -2x^2 - 8x - 2$

44. $f(x) = -4x^2 - 4x - 1$

In Problems 45–50, find a piecewise definition of f that does not involve the absolute value function (see Example 5). Sketch the graph, and find the domain, range, and any points of discontinuity.

 *Check your graphs in Problems 45–50 by graphing the given definition of f on a graphing utility.*

45. $f(x) = \dfrac{|x|}{x}$

46. $f(x) = x|x|$

47. $f(x) = x + \dfrac{|x - 1|}{x - 1}$

48. $f(x) = x + 2\dfrac{|x + 1|}{x + 1}$

49. $f(x) = |x| + |x - 2|$

50. $f(x) = |x| - |x - 3|$

In Problems 51–56, write a piecewise definition for f (see the discussion of Fig. 14 in this section) and sketch the graph of f. Include sufficient intervals to clearly illustrate both the definition and the graph. Find the domain, range, and any points of discontinuity.

 *Check your graphs in Problems 51–56 by graphing the given definition of f on a graphing utility.*

51. $f(x) = [\![x/2]\!]$

52. $f(x) = [\![x/3]\!]$

53. $f(x) = [\![3x]\!]$

54. $f(x) = [\![2x]\!]$

55. $f(x) = x - [\![x]\!]$

56. $f(x) = [\![x]\!] - x$

57. Given that f is a quadratic function with min $f(x) = f(2) = 4$, find the axis, vertex, range, and x intercepts.

58. Given that f is a quadratic function with max $f(x) = f(-3) = -5$, find the axis, vertex, range, and x intercepts.

59. The function f is continuous and increasing on the interval $[1, 9]$ with $f(1) = -5$ and $f(9) = 4$.
(A) Sketch a graph of f that is consistent with the given information.
(B) How many times does your graph cross the x axis? Could the graph cross more times? Fewer times? Support your conclusions with additional sketches and/or verbal arguments.

60. Repeat Problem 59 if the function does not have to be continuous.

61. The function f is continuous on the interval $[-5, 5]$ with $f(-5) = -4$, $f(1) = 3$, and $f(5) = -2$.
(A) Sketch a graph of f that is consistent with the given information.
(B) How many times does your graph cross the x axis? Could the graph cross more times? Fewer times? Support your conclusions with additional sketches and/or verbal arguments.

62. Repeat Problem 61 if f is continuous on $[-8, 8]$ with $f(-8) = -6$, $f(-4) = 3$, $f(3) = -2$, and $f(8) = 5$.

Problems 63–66 are calculus-related. In geometry, a line that intersects a circle in two distinct points is called a secant line, as shown in figure (a). In calculus, the line through the points $(x_1, f(x_1))$ and $(x_2, f(x_2))$ is called a ***secant line*** *for the graph of the function f, as shown in figure (b).*

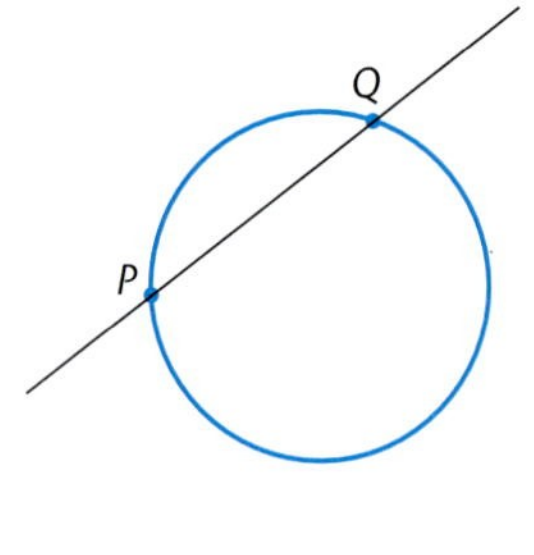

Secant line for a circle
(a)

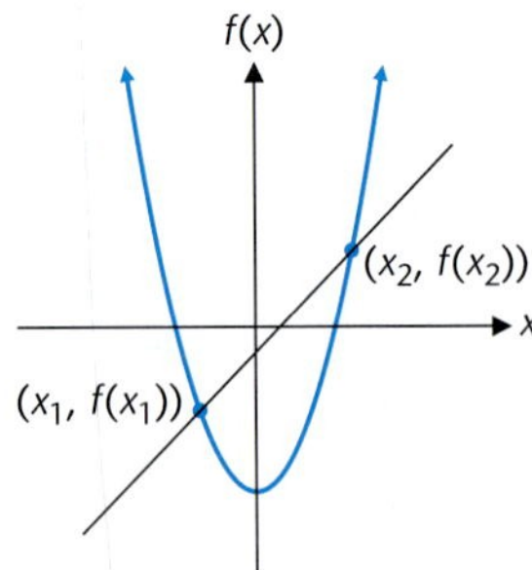

Secant line for the graph of a function
(b)

In Problems 63 and 64, find the equation of the secant line through the indicated points on the graph of f. Graph f and the secant line on the same coordinate system.

63. $f(x) = x^2 - 4$; $(-1, -3)$, $(3, 5)$

64. $f(x) = 9 - x^2$; $(-2, 5)$, $(4, -7)$

65. Let $f(x) = x^2 - 3x + 5$. If h is a nonzero real number, then $(2, f(2))$ and $(2 + h, f(2 + h))$ are two distinct points on the graph of f.

(A) Find the slope of the secant line through these two points.

(B) Evaluate the slope of the secant line for $h = 1$, $h = 0.1$, $h = 0.01$, and $h = 0.001$. What value does the slope seem to be approaching?

66. Repeat Problem 65 for $f(x) = x^2 + 2x - 6$.

Problems 67–74 require the use of a graphing utility.

In Problems 67–72, first graph functions f and g in the same viewing window, then graph m(x) and n(x) in their own viewing windows:

$$m(x) = 0.5[f(x) + g(x) + |f(x) - g(x)|]$$

$$n(x) = 0.5[f(x) + g(x) - |f(x) - g(x)|]$$

67. $f(x) = -2x$, $g(x) = 0.5x$

68. $f(x) = 3x + 1$, $g(x) = -0.5x - 4$

69. $f(x) = 5 - 0.2x^2$, $g(x) = 0.3x^2 - 4$

70. $f(x) = 0.15x^2 - 5$, $g(x) = 5 - 1.5|x|$

71. $f(x) = 0.2x^2 - 0.4x - 5$, $g(x) = 0.3x - 3$

72. $f(x) = 8 + 1.5x - 0.4x^2$, $g(x) = -0.2x + 5$

73. How would you characterize the relationship between f, g, and m in Problems 67–72? [*Hint:* See Problem 89 in Exercise 1-4.]

74. How would you characterize the relationship between f, g, and n in Problems 67–72? [*Hint:* See Problem 90 in Exercise 1-4.]

APPLICATIONS

75. Tire Mileage. An automobile tire manufacturer collected the data in Table 1 relating tire pressure x, in pounds per square inch (lb/in.^{2}) and mileage, in thousands of miles.

TABLE 1

x	28	30	32	34	36
Mileage	45	52	55	51	47

A mathematical model for this data is given by

$$f(x) = -0.518x^2 + 33.3x - 481$$

(A) Complete Table 2. Round values of $f(x)$ to one decimal place.

TABLE 2

x	28	30	32	34	36
Mileage	45	52	55	51	47
$f(x)$					

(B) Sketch the graph of f and the mileage data on the same axes.

(C) Use values of the modeling function rounded to two decimal places to estimate the mileage for a tire pressure of 31 lb/in^2. For 35 lb/in^2.

(D) Write a brief description of the relationship between tire pressure and mileage.

76. Automobile Production. Table 3 lists General Motors' total U.S. vehicle production in millions of units from 1989 to 1993.

TABLE 3

Year	89	90	91	92	93
Production	4.7	4.1	3.5	3.7	5.0

A mathematical model for GM's production data is given by

$$f(x) = 0.33x^2 - 1.3x + 4.8$$

where $x = 0$ corresponds to 1989.

(A) Complete Table 4. Round values of x to one decimal place.

TABLE 4

x	0	1	2	3	4
Production	4.7	4.1	3.5	3.7	5.0
$f(x)$					

(B) Sketch the graph of f and the production data on the same axes.

(C) Use values of the modeling function f rounded to two decimal places to estimate the production in 1994. In 1995.

(D) Write a brief verbal description of GM's production from 1989 to 1993.

77. Physics—Spring Stretch. Hooke's law states that the relationship between the stretch s of a spring and the weight w causing the stretch is linear (a principle upon which all spring scales are constructed). A 10-pound weight stretches a spring 1 inch, while with no weight the stretch of the spring is 0.

(A) Find a linear function f: $s = f(w) = mw + b$ that represents this relationship. [*Hint:* Both points (10, 1) and (0, 0) are on the graph of f.]

(B) Find $f(15)$ and $f(30)$—that is, the stretch of the spring for 15-pound and 30-pound weights.

(C) What is the slope of the graph of f? (The slope indicates the increase in stretch for each pound increase in weight.)

(D) Graph f for $0 \le w \le 40$.

78. Business—Depreciation. An electronic computer was purchased by a company for \$20,000 and is assumed to have a salvage value of \$2,000 after 10 years. Its value is depreciated linearly from \$20,000 to \$2,000.

(A) Find the linear function f: $V = f(t)$ that relates value V in dollars to time t in years.

(B) Find $f(4)$ and $f(8)$, the values of the computer after 4 and 8 years, respectively.

(C) Find the slope of the graph of f. (The slope indicates the decrease in value per year.)

(D) Graph f for $0 \le t \le 10$.

★ **79. Sales Commissions.** An appliance salesperson receives a base salary of \$200 a week and a commission of 4% on all sales over \$3,000 during the week. In addition, if the weekly sales are \$8,000 or more, the salesperson receives a \$100 bonus. If x represents weekly sales (in dollars), express the weekly earnings $E(x)$ as a function of x, and sketch its graph. Identify any points of discontinuity. Find $E(5{,}750)$ and $E(9{,}200)$.

★ **80. Service Charges.** On weekends and holidays, an emergency plumbing repair service charges \$2.00 per minute for the first 30 minutes of a service call and \$1.00 per minute for each additional minute. If x represents the duration of a service call in minutes, express the total service charge $S(x)$ as a function of x, and sketch its graph. Identify any points of discontinuity. Find $S(25)$ and $S(45)$.

81. Construction. A rectangular dog pen is to be made with 100 feet of fence wire.

(A) If x represents the width of the pen, express its area $A(x)$ in terms of x.

(B) Considering the physical limitations, what is the domain of the function A?

(C) Graph the function for this domain.

(D) Determine the dimensions of the rectangle that will make the area maximum.

82. Construction. Rework Problem 81 with the added assumption that an existing property fence will be used for one side of the pen. (Let x = Width; see the figure.)

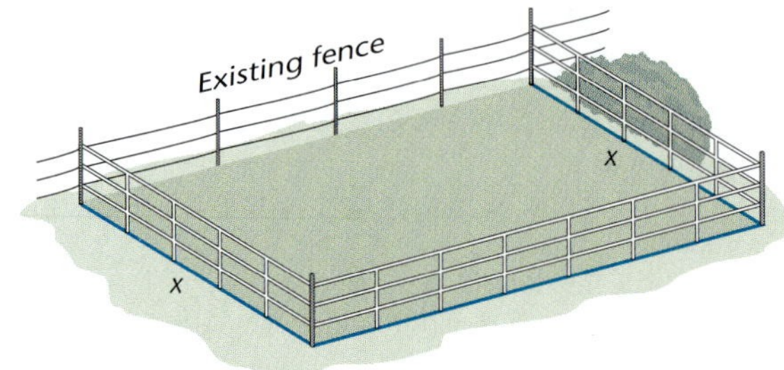

83. Computer Science. Let $f(x) = 10[\![0.5 + x/10]\!]$. Evaluate f at 4, −4, 6, −6, 24, 25, 247, −243, −245, and −246. What operation does this function perform?

84. Computer Science. Let $f(x) = 100[\![0.5 + x/100]\!]$. Evaluate f at 40, −40, 60, −60, 740, 750, 7,551, −601, −649, and −651. What operation does this function perform?

★ **85. Computer Science.** Use the greatest integer function to define a function f that rounds real numbers to the nearest hundredth.

★ **86. Computer Science.** Use the greatest integer function to define a function f that rounds real numbers to the nearest thousandth.

87. Delivery Charges. A nationwide package delivery service charges \$15 for overnight delivery of packages weighing 1 pound or less. Each additional pound (or fraction thereof) costs an additional \$3. Let $C(x)$ be the charge for overnight delivery of a package weighing x pounds.

(A) Write a piecewise definition of C for $0 < x \le 6$, and sketch the graph of C by hand.

(B) Can the function f defined by $f(x) = 15 + 3[\![x]\!]$ be used to compute the delivery charges for all x, $0 < x \le 6$? Justify your answer.

88. Telephone Charges. Calls to 900 numbers are charged to the caller. A 900 number hot line for tips and hints for video games charges \$4 for the first minute of the call and \$2 for each additional minute (or fraction thereof). Let $C(x)$ be the charge for a call lasting x minutes.

(A) Write a piecewise definition of C for $0 < x \le 6$, and sketch the graph of C by hand.

(B) Can the function f defined by $f(x) = 4 + 2[\![x]\!]$ be used to compute the charges for all x, $0 < x \le 6$? Justify your answer.

★★ **89. Car Rental.** A car rental agency rents 300 cars a day at a rate of \$40 per day. For each \$1 increase in rate, five fewer cars are rented. At what rate should the cars be rented to produce the maximum income? What is the maximum income?

★★ **90. Rental Income.** A 400-room hotel in Las Vegas is filled to capacity every night at \$70 a room. For each \$1 increase in rent, four fewer rooms are rented. If each rented room costs \$10 to service per day, how much should the management charge for each room to maximize profit? What is the maximum profit?

★★ **91. Physics.** A stunt driver is planning to jump a motorcycle from one ramp to another as illustrated in the figure. The ramps are 10 feet high, and the distance between the ramps is 80 feet. The trajectory of the cycle through the air is given by the graph of

$$f(x) = \frac{1}{4}x - \left(\frac{16}{v^2}\right)x^2$$

where v is the velocity of the cycle in feet per second as it leaves the ramp.

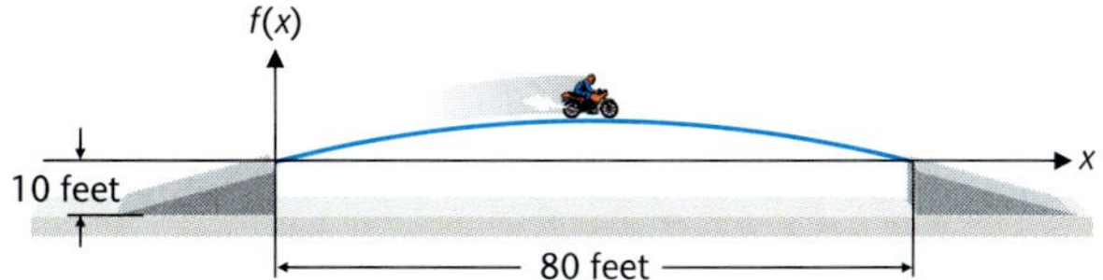

(A) How fast must the cycle be traveling when it leaves the ramp in order to follow the trajectory illustrated in the figure?

(B) What is the maximum height of the cycle above the ground as it follows this trajectory?

★★ **92. Physics.** The trajectory of a circus performer shot from a cannon is given by the graph of the function

$$f(x) = x - \frac{1}{100}x^2$$

Both the cannon and the net are 10 feet high (see the figure).

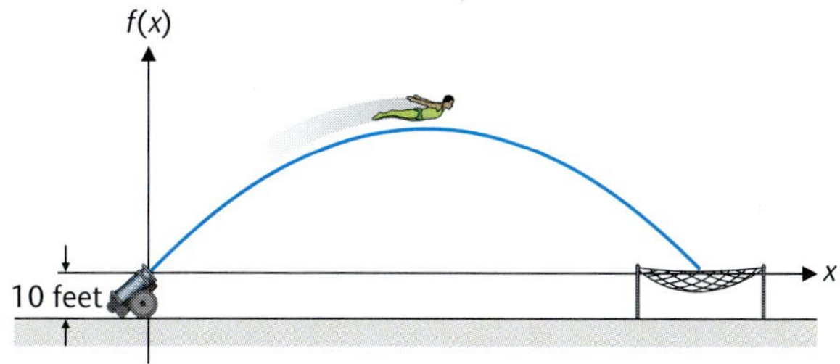

(A) How far from the muzzle of the cannon should the center of the net be placed so that the performer lands in the center of the net?

(B) What is the maximum height of the performer above the ground?

SECTION 2-5 Combining Functions

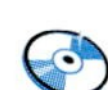

- Operations on Functions
- Composition
- Elementary Functions
- Vertical and Horizontal Shifts
- Reflections, Expansions, and Contractions

If two functions f and g are both defined at a real number x, and if $f(x)$ and $g(x)$ are both real numbers, then it is possible to perform real number operations such as addition, subtraction, multiplication, or division with $f(x)$ and $g(x)$. Furthermore, if $g(x)$ is a number in the domain of f, then it is also possible to evaluate f at $g(x)$. In this section we see how operations on the values of functions can be used to define operations on the functions themselves. We also investigate the graphic implications of some of these operations.

• Operations on Functions

The functions f and g given by

$$f(x) = 2x + 3 \qquad \text{and} \qquad g(x) = x^2 - 4$$

are defined for all real numbers. Thus, for any real x we can perform the following operations:

$$f(x) + g(x) = 2x + 3 + x^2 - 4 = x^2 + 2x - 1$$
$$f(x) - g(x) = 2x + 3 - (x^2 - 4) = -x^2 + 2x + 7$$
$$f(x)g(x) = (2x + 3)(x^2 - 4) = 2x^3 + 3x^2 - 8x - 12$$

For $x \neq \pm 2$ we can also form the quotient

$$\frac{f(x)}{g(x)} = \frac{2x + 3}{x^2 - 4} \qquad x \neq \pm 2$$

Notice that the result of each operation is a new function. Thus, we have

$$(f + g)(x) = f(x) + g(x) = x^2 + 2x - 1 \qquad \text{Sum}$$

$$(f - g)(x) = f(x) - g(x) = -x^2 + 2x + 7 \qquad \text{Difference}$$

$$(fg)(x) = f(x)g(x) = 2x^3 + 3x^2 - 8x - 12 \qquad \text{Product}$$

$$\left(\frac{f}{g}\right)(x) = \frac{f(x)}{g(x)} = \frac{2x + 3}{x^2 - 4} \qquad x \neq \pm 2 \qquad \text{Quotient}$$

Notice that the sum, difference, and product functions are defined for all values of x, as were f and g, but the domain of the quotient function must be restricted to exclude those values where $g(x) = 0$.

DEFINITION 1

Operations on Functions

The **sum, difference, product,** and **quotient** of the functions f and g are the functions defined by

$$(f + g)(x) = f(x) + g(x) \qquad \text{Sum function}$$

$$(f - g)(x) = f(x) - g(x) \qquad \text{Difference function}$$

$$(fg)(x) = f(x)g(x) \qquad \text{Product function}$$

$$\left(\frac{f}{g}\right)(x) = \frac{f(x)}{g(x)} \qquad g(x) \neq 0 \qquad \text{Quotient function}$$

Each function is defined on the intersection of the domains of f and g, with the exception that the values of x where $g(x) = 0$ must be excluded from the domain of the quotient function.

EXAMPLE 1 **Finding the Sum, Difference, Product, and Quotient Functions**

Let $f(x) = \sqrt{4 - x}$ and $g(x) = \sqrt{3 + x}$. Find the functions $f + g$, $f - g$, fg, and f/g, and find their domains.

Solution

$$(f + g)(x) = f(x) + g(x) = \sqrt{4 - x} + \sqrt{3 + x}$$

$$(f - g)(x) = f(x) - g(x) = \sqrt{4 - x} - \sqrt{3 + x}$$

$$\begin{aligned}(fg)(x) = f(x)g(x) &= \sqrt{4 - x}\sqrt{3 + x} \\ &= \sqrt{(4 - x)(3 + x)} \\ &= \sqrt{12 + x - x^2}\end{aligned}$$

$$\left(\frac{f}{g}\right)(x) = \frac{f(x)}{g(x)} = \frac{\sqrt{4 - x}}{\sqrt{3 + x}} = \sqrt{\frac{4 - x}{3 + x}}$$

Domain of f

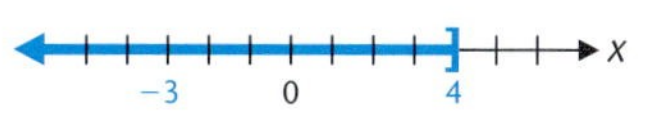

Domain of g

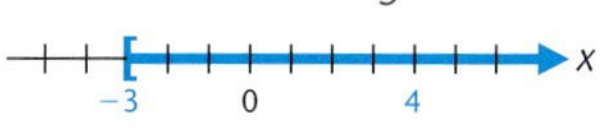

Domain of $f + g$, $f - g$, and fg

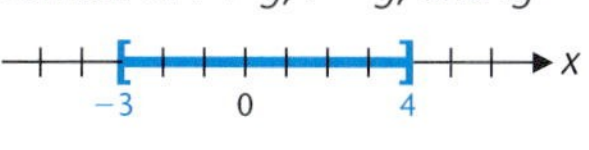

Domain of $\frac{f}{g}$

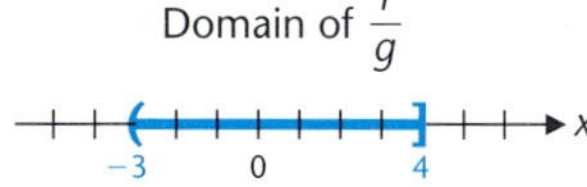

The domains of f and g are

$$\text{Domain of } f\text{: } x \le 4 \text{ or } (-\infty, 4]$$

$$\text{Domain of } g\text{: } x \ge -3 \text{ or } [-3, \infty)$$

The intersection of these domains is

$$(-\infty, 4] \cap [-3, \infty) = [-3, 4]$$

This is the domain of the functions $f + g$, $f - g$, and fg. Since $g(-3) = 0$, $x = -3$ must be excluded from the domain of the quotient function. Thus,

$$\text{Domain of } \frac{f}{g}\text{: } (-3, 4]$$

Matched Problem 1 Let $f(x) = \sqrt{x}$ and $g(x) = \sqrt{10 - x}$. Find the functions $f + g$, $f - g$, fg, and f/g, and find their domains.

• Composition

Consider the function h given by the equation

$$h(x) = \sqrt{2x + 1}$$

Inside the radical is a first-degree polynomial that defines a linear function. So the function h is really a combination of a square root function and a linear function. We can see this more clearly as follows. Let

$$u = 2x + 1 = g(x)$$

$$y = \sqrt{u} = f(u)$$

Then

$$h(x) = f[g(x)]$$

The function h is said to be the *composite* of the two functions f and g. (Loosely speaking, we can think of h as a function of a function.) What can we say about the domain of h given the domains of f and g? In forming the composite $h(x) = f[g(x)]$: **x must be restricted so that x is in the domain of g and $g(x)$ is in the domain of f.** Since the domain of f, where $f(u) = \sqrt{u}$, is the set of nonnegative real numbers, we see that $g(x)$ must be nonnegative; that is,

$$g(x) \ge 0$$

$$2x + 1 \ge 0$$

$$x \ge -\tfrac{1}{2}$$

Thus, the domain of h is this restricted domain of g.

A special function symbol is often used to represent the *composite of two functions,* which we define in general terms below.

DEFINITION 2

Composite Functions

Given functions f and g, then $f \circ g$ is called their **composite** and is defined by the equation

$$(f \circ g)(x) = f[g(x)]$$

The domain of $f \circ g$ is the set of all real numbers x in the domain of g where $g(x)$ is in the domain of f.

As an immediate consequence of Definition 2, we have (see Fig. 1):

The domain of $f \circ g$ is always a subset of the domain of g, and the range of $f \circ g$ is always a subset of the range of f.

FIGURE 1 Composite functions.

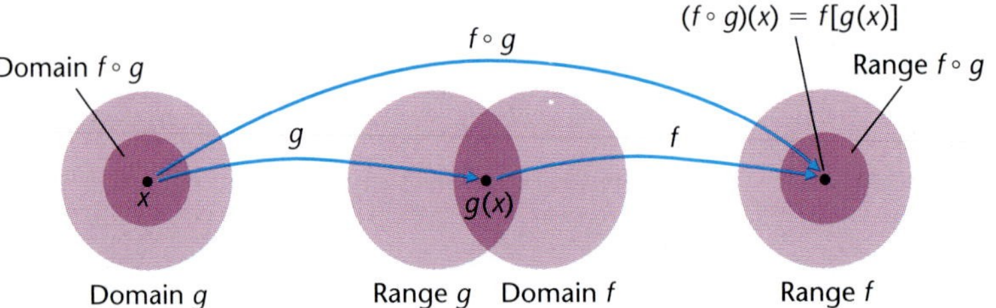

EXAMPLE 2 Finding the Composition of Two Functions

Find $(f \circ g)(x)$ and $(g \circ f)(x)$ and their domains for $f(x) = x^{10}$ and $g(x) = 3x^4 - 1$.

Solution

$$(f \circ g)(x) = f[g(x)] = f(3x^4 - 1) = (3x^4 - 1)^{10}$$
$$(g \circ f)(x) = g[f(x)] = g(x^{10}) = 3(x^{10})^4 - 1 = 3x^{40} - 1$$

The functions f and g are both defined for all real numbers. If x is any real number, then x is in the domain of g, $g(x)$ is in the domain of f, and, consequently, x is in the domain of $f \circ g$. Thus, the domain of $f \circ g$ is the set of all real numbers. Using similar reasoning, the domain of $g \circ f$ also is the set of all real numbers.

Matched Problem 2 Find $(f \circ g)(x)$ and $(g \circ f)(x)$ and their domains for $f(x) = 2x + 1$ and $g(x) = (x - 1)/2$.

If two functions are both defined for all real numbers, then so is their composition.

EXPLORE-DISCUSS 1 Verify that if $f(x) = 1/(1 - 2x)$ and $g(x) = 1/x$, then $(f \circ g)(x) = x/(x - 2)$. Clearly, $f \circ g$ is not defined at $x = 2$. Are there any other values of x where $f \circ g$ is not defined? Explain.

If either function in a composition is not defined for some real numbers, then, as Example 3 illustrates, the domain of the composition may not be what you first think it should be.

EXAMPLE 3 **Finding the Composition of Two Functions**

Find $(f \circ g)(x)$ and its domain for $f(x) = \sqrt{4 - x^2}$ and $g(x) = \sqrt{3 - x}$.

Solution We begin by stating the domains of f and g, a good practice in any composition problem:

$$\text{Domain } f\text{: } -2 \le x \le 2 \quad \text{or} \quad [-2, 2]$$

$$\text{Domain } g\text{: } x \le 3 \quad \text{or} \quad (-\infty, 3]$$

Next we find the composition:

$$\begin{aligned}(f \circ g)(x) &= f[g(x)] = f(\sqrt{3 - x}) \\ &= \sqrt{4 - (\sqrt{3 - x})^2} \\ &= \sqrt{4 - (3 - x)} \qquad (\sqrt{t})^2 = t,\ t \ge 0 \\ &= \sqrt{1 + x}\end{aligned}$$

Even though $\sqrt{1 + x}$ is defined for all $x \ge -1$, we must restrict the domain of $f \circ g$ to those values that also are in the domain of g. Thus,

$$\text{Domain } f \circ g\text{: } x \ge -1 \text{ and } x \le 3 \quad \text{or} \quad [-1, 3]$$

Matched Problem 3 Find $(f \circ g)(x)$ and its domain for $f(x) = \sqrt{9 - x^2}$ and $g(x) = \sqrt{x - 1}$.

CAUTION The domain of $f \circ g$ cannot always be determined simply by examining the final form of $(f \circ g)(x)$. Any numbers that are excluded from the domain of g must also be excluded from the domain of $f \circ g$.

In calculus, it is not only important to be able to find the composition of two functions, but also to recognize when a given function is the composition of two simpler functions.

EXAMPLE 4 Recognizing Composition Forms

Express h as a composition of two simpler functions for

$$h(x) = (3x + 5)^5$$

Solution If we let $f(x) = x^5$ and $g(x) = 3x + 5$, then

$$h(x) = (3x + 5)^5 = f(3x + 5) = f[g(x)] = (f \circ g)(x)$$

and we have expressed h as the composition of f and g.

Matched Problem 4 Express h as a composition of the square root function and a linear function for $h(x) = \sqrt{4x - 7}$.

• Elementary Functions

The functions

$$g(x) = x^2 - 4 \qquad h(x) = (x - 4)^2 \qquad k(x) = -4x^2$$

can all be obtained from the function $f(x) = x^2$ by performing simple operations on f:

$$g(x) = f(x) - 4 \qquad h(x) = f(x - 4) \qquad k(x) = -4f(x)$$

It follows that the graphs of functions g, h, and k are closely related to the graph of function f. Before exploring relationships of this type, we want to identify some elementary functions, summarize their basic properties, and include them in our library of elementary functions. Figure 2 shows six basic functions that you will encounter frequently. You should know the definition, domain, and range of each and be able to sketch their graphs.

FIGURE 2 Some basic functions and their graphs. [*Note:* Letters used to designate these functions may vary from context to context; R is the set of all real numbers.]

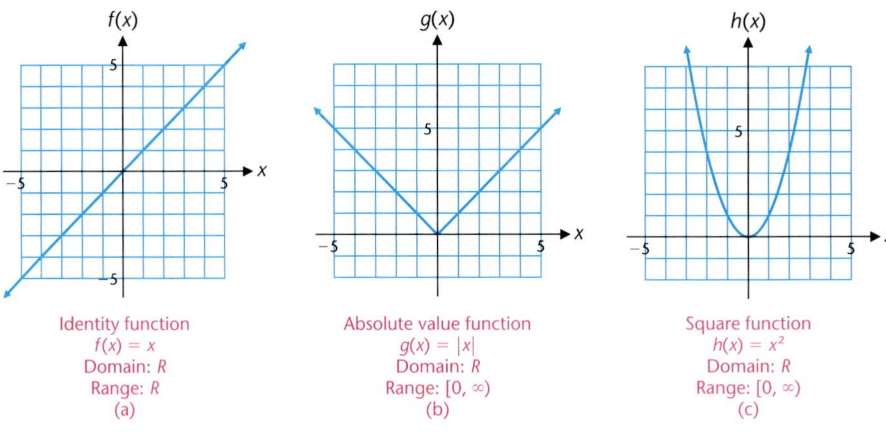

Identity function
$f(x) = x$
Domain: R
Range: R
(a)

Absolute value function
$g(x) = |x|$
Domain: R
Range: $[0, \infty)$
(b)

Square function
$h(x) = x^2$
Domain: R
Range: $[0, \infty)$
(c)

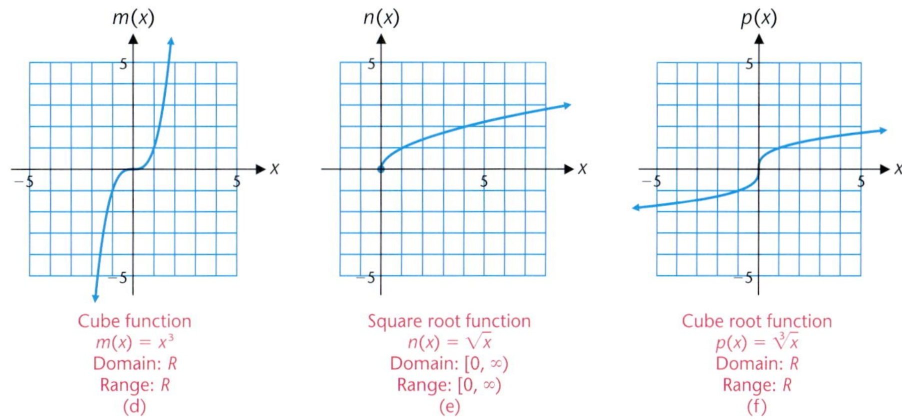

Cube function
$m(x) = x^3$
Domain: R
Range: R
(d)

Square root function
$n(x) = \sqrt{x}$
Domain: $[0, \infty)$
Range: $[0, \infty)$
(e)

Cube root function
$p(x) = \sqrt[3]{x}$
Domain: R
Range: R
(f)

• Vertical and Horizontal Shifts

How are the graphs of $y = (f + g)(x)$, $y = (fg)(x)$, and $y = (f \circ g)(x)$ related to the graphs of $y = f(x)$ and $y = g(x)$? In general, this is a difficult question to answer. However, if g is chosen to be a very simple function, such as $g(x) = k$ or $g(x) = x + h$, then we can establish some very useful relationships between the graph of $y = f(x)$ and the graphs of $y = f(x) + k$, $y = kf(x)$, and $y = f(x + h)$. We refer to the graph obtained by performing one of these operations on a function f as a **transformation** of the graph of $y = f(x)$.

EXPLORE-DISCUSS 2 Let $f(x) = |x|$.

(A) Graph $y = f(x) + k$ for $k = -2$, 0, and 1 simultaneously in the same coordinate system. Describe the relationship between the graph of $y = f(x)$ and the graph of $y = f(x) + k$ for k, any real number.

(B) Graph $y = f(x + h)$ for $h = -2$, 0, and 1 simultaneously in the same coordinate system. Describe the relationship between the graph of $y = f(x)$ and the graph of $y = f(x + h)$ for h, any real number.

EXAMPLE 5 Vertical and Horizontal Shifts

(A) How are the graphs of $y = x^2 + 2$ and $y = x^2 - 3$ related to the graph of $y = x^2$? Confirm your answer by graphing all three functions simultaneously in the same coordinate system.

(B) How are the graphs of $y = (x + 2)^2$ and $y = (x - 3)^2$ related to the graph of $y = x^2$? Confirm your answer by graphing all three functions simultaneously in the same coordinate system.

Solutions (A) The graph of $y = x^2 + 2$ is the same as the graph of $y = x^2$ shifted upward 2 units, and the graph of $y = x^2 - 3$ is the same as the graph of $y = x^2$ shifted downward 3 units. Figure 3 on the next page confirms these conclusions. [It appears that the graph of $y = f(x) + k$ is the graph of $y = f(x)$ shifted up if k is positive and down if k is negative.]

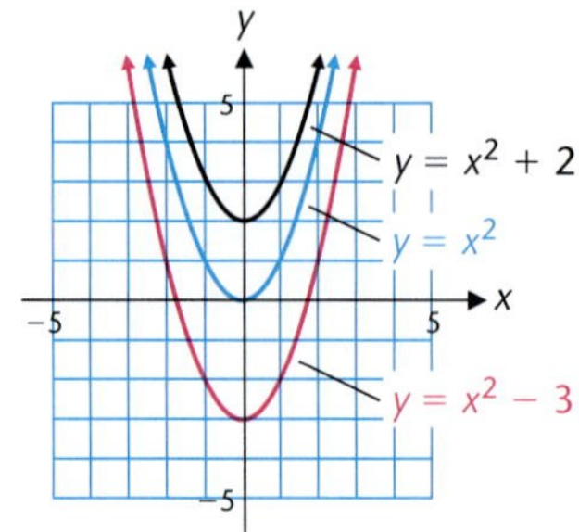

FIGURE 3 Vertical shifts.

(B) The graph of $y = (x + 2)^2$ is the same as the graph of $y = x^2$ shifted to the left 2 units, and the graph of $y = (x - 3)^2$ is the same as the graph of $y = x^2$ shifted to the right 3 units. Figure 4 confirms these conclusions. [It appears that the graph of $y = f(x + h)$ is the graph of $y = f(x)$ shifted right if h is negative and left if h is positive—the opposite of what you might expect.]

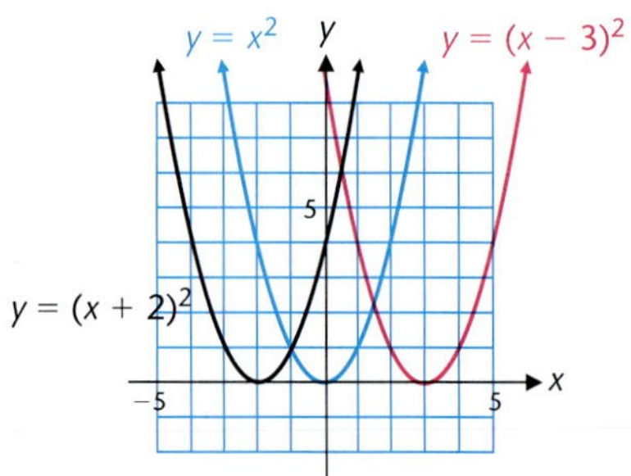

FIGURE 4 Horizontal shifts.

Matched Problem 5

(A) How are the graphs of $y = \sqrt{x} + 3$ and $y = \sqrt{x} - 1$ related to the graph of $y = \sqrt{x}$? Confirm your answer by graphing all three functions simultaneously in the same coordinate system.

(B) How are the graphs of $y = \sqrt{x + 3}$ and $y = \sqrt{x - 1}$ related to the graph of $y = \sqrt{x}$? Confirm your answer by graphing all three functions simultaneously in the same coordinate system.

Comparing the graph of $y = f(x) + k$ with the graph of $y = f(x)$, we see that the graph of $y = f(x) + k$ can be obtained from the graph of $y = f(x)$ by **vertically translating** (shifting) the graph of the latter upward k units if k is positive and downward $|k|$ units if k is negative. Comparing the graph of $y = f(x + h)$ with the graph of $y = f(x)$, we see that the graph of $y = f(x + h)$ can be obtained from the graph of $y = f(x)$ by **horizontally translating** (shifting) the graph of the latter h units to the left if h is positive and $|h|$ units to the right if h is negative.

EXAMPLE 6 **Vertical and Horizontal Translations (Shifts)**

The graphs in Figure 5 are either horizontal or vertical shifts of the graph of $f(x) = |x|$. Write appropriate equations for functions H, G, M, and N in terms of f.

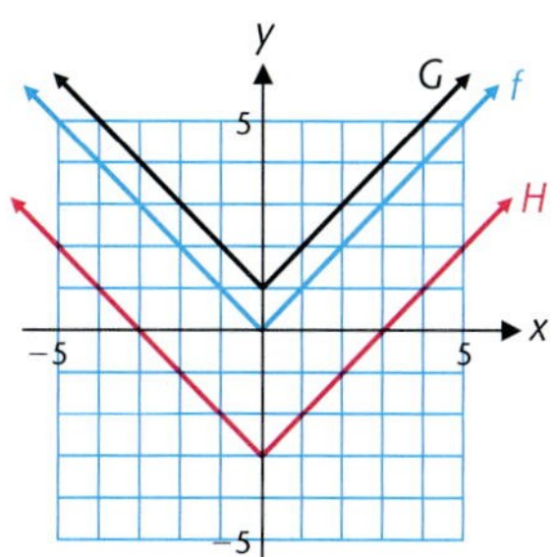

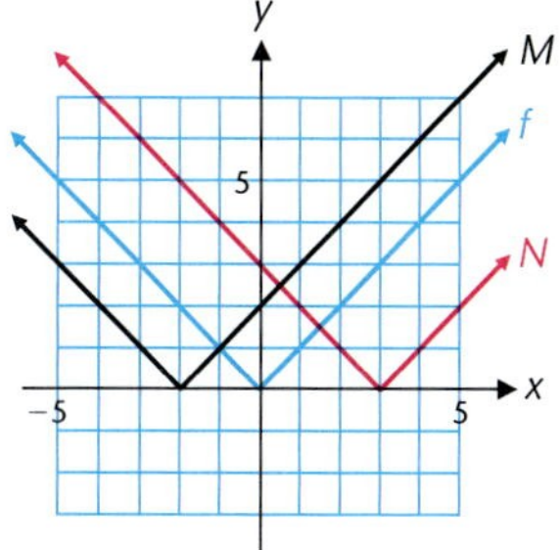

FIGURE 5 Vertical and horizontal shifts.

Solution Functions H and G are vertical shifts given by

$$H(x) = |x| - 3$$
$$G(x) = |x| + 1$$

Functions M and N are horizontal shifts given by

$$M(x) = |x + 2|$$
$$N(x) = |x - 3|$$

Matched Problem 6 The graphs in Figure 6 are either horizontal or vertical shifts of the graph of $f(x) = x^3$. Write appropriate equations for functions H, G, M, and N in terms of f.

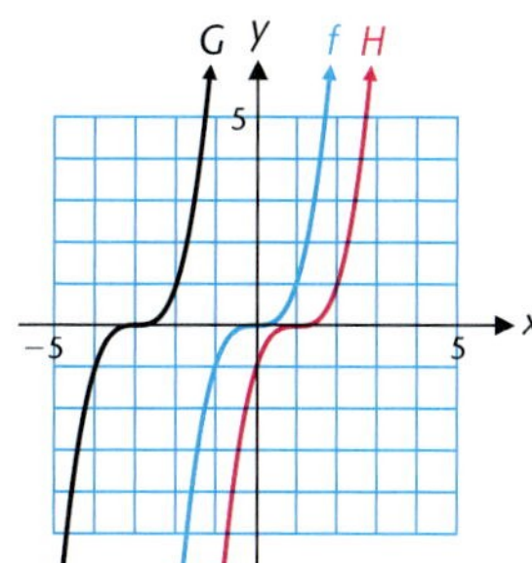

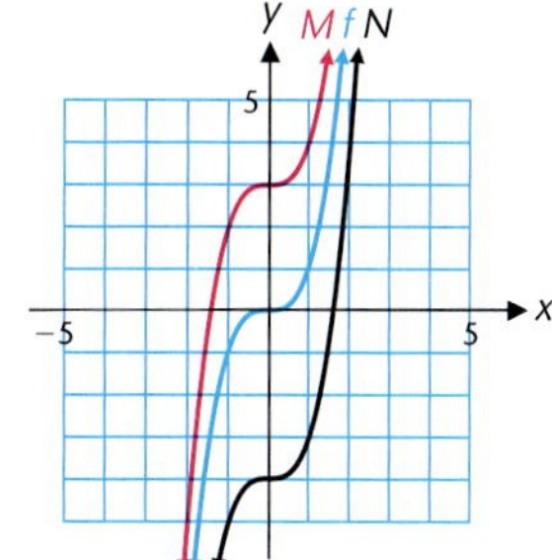

FIGURE 6 Vertical and horizontal shifts.

• Reflections, Expansions, and Contractions

We now investigate how the graph of $y = Af(x)$ is related to the graph of $y = f(x)$ for different real numbers A.

EXPLORE-DISCUSS 3

(A) Graph $y = A\sqrt{x}$ for $A = 1, 2$, and $\frac{1}{2}$ simultaneously in the same coordinate system.

(B) Graph $y = A\sqrt{x}$ for $A = -1, -2$, and $-\frac{1}{2}$ simultaneously in the same coordinate system.

(C) Describe the relationship between the graph of $h(x) = \sqrt{x}$ and the graph of $G(x) = A\sqrt{x}$ for A any real number.

Comparing the graph of $y = Af(x)$ with the graph of $y = f(x)$, we see that the graph of $y = Af(x)$ can be obtained from the graph of $y = f(x)$ by multiplying each ordinate value of the latter by A. The result is a **vertical expansion** of the graph of $y = f(x)$ if $A > 1$, a **vertical contraction** of the graph of $y = f(x)$ if $0 < A < 1$, and a **reflection in the x axis** if $A = -1$.

EXAMPLE 7 Reflections, Expansions, and Contractions

(A) How are the graphs of $y = 2\sqrt[3]{x}$ and $y = 0.5\sqrt[3]{x}$ related to the graph of $y = \sqrt[3]{x}$? Confirm your answer by graphing all three functions simultaneously in the same coordinate system.

(B) How is the graph of $y = -2\sqrt[3]{x}$ related to the graph of $y = \sqrt[3]{x}$? Confirm your answer by graphing both functions simultaneously in the same coordinate system.

Solution (A) The graph of $y = 2\sqrt[3]{x}$ is a vertical expansion of the graph of $y = \sqrt[3]{x}$ by a factor of 2, and the graph of $y = 0.5\sqrt[3]{x}$ is a vertical contraction of the graph of $y = \sqrt[3]{x}$ by a factor of 0.5. Figure 7 confirms this conclusion.

FIGURE 7 Vertical expansion and contraction.

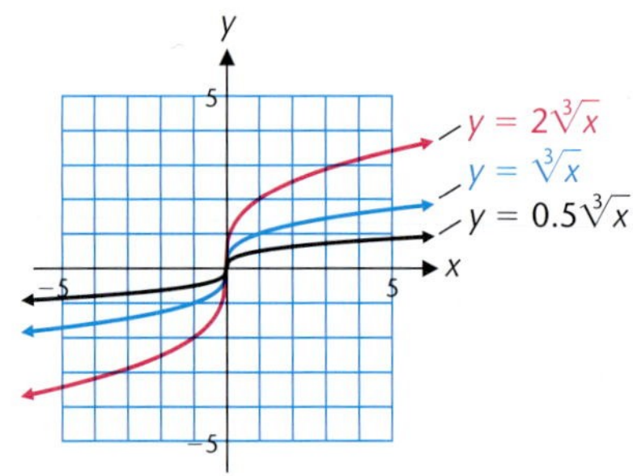

(B) The graph of $y = -2\sqrt[3]{x}$ is a reflection in the x axis and a vertical expansion of the graph of $y = \sqrt[3]{x}$. Figure 8 confirms this conclusion.

FIGURE 8 Reflection and vertical expansion.

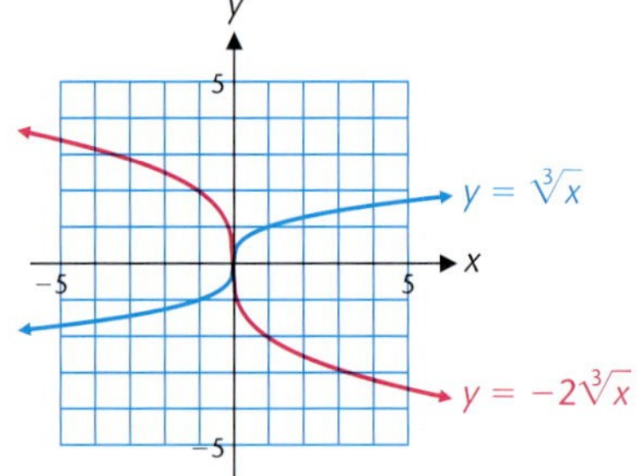

Matched Problem 7 (A) How are the graphs of $y = 2x$ and $y = 0.5x$ related to the graph of $y = x$? Confirm your answer by graphing all three functions simultaneously in the same coordinate system.

(B) How is the graph of $y = -0.5x$ related to the graph of $y = x$? Confirm your answer by graphing both functions in the same coordinate system.

The various transformations considered above are summarized in the following box for easy reference:

Graph Transformations (Summary)

Vertical Translation [see Fig. 9(a)]:

$$y = f(x) + k \quad \begin{cases} k > 0 & \text{Shift graph of } y = f(x) \text{ up } k \text{ units} \\ k < 0 & \text{Shift graph of } y = f(x) \text{ down } |k| \text{ units} \end{cases}$$

Horizontal Translation [see Fig. 9(b)]:

$$y = f(x + h) \quad \begin{cases} h > 0 & \text{Shift graph of } y = f(x) \text{ left } h \text{ units} \\ h < 0 & \text{Shift graph of } y = f(x) \text{ right } |h| \text{ units} \end{cases}$$

Reflection [see Fig. 9(c)]:

$$y = -f(x) \quad \text{Reflect graph of } y = f(x) \text{ in the } x \text{ axis}$$

Vertical Expansion and Contraction [see Fig. 9(d)]:

$$y = Af(x) \quad \begin{cases} A > 1 & \text{Vertically expand graph of } y = f(x) \text{ by multiplying each ordinate value by } A \\ 0 < A < 1 & \text{Vertically contract graph of } y = f(x) \text{ by multiplying each ordinate value by } A \end{cases}$$

FIGURE 9 Graph transformations.

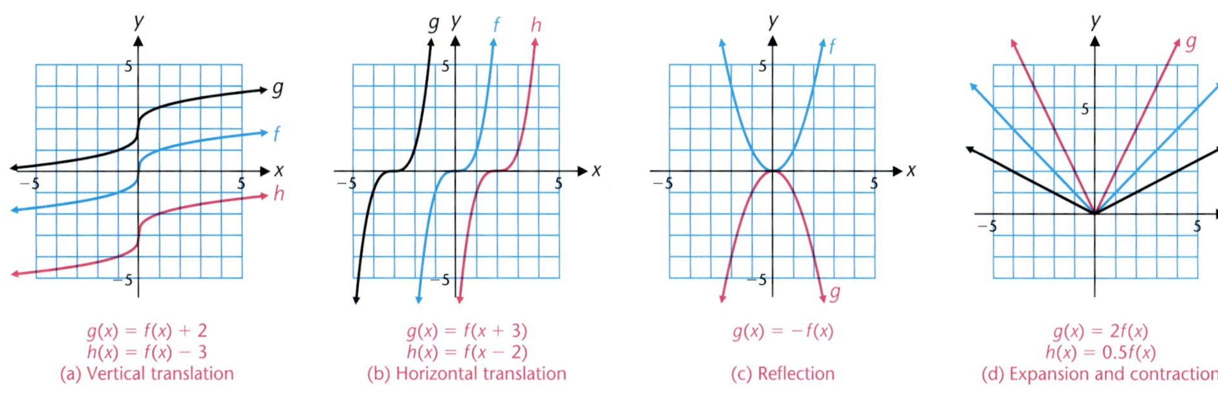

EXPLORE-DISCUSS 4 Use a graphing utility to explore the graph of $y = A(x + h)^2 + k$ for various values of the constants A, h, and k. Discuss how the graph of $y = A(x + h)^2 + k$ is related to the graph of $y = x^2$.

EXAMPLE 8 Combining Graph Transformations

The graph of $y = g(x)$ in Figure 10 is a transformation of the graph of $y = x^2$. Find an equation for the function g.

FIGURE 10

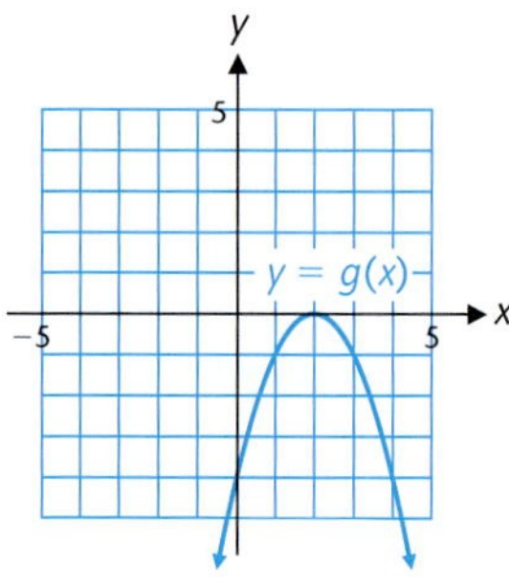

Solution To transform the graph of $y = x^2$ [Fig. 11(a)] into the graph of $y = g(x)$, we first reflect the graph of $y = x^2$ in the x axis [Fig. 11(b)], then shift it to the right two units [Fig. 11(c)]. Thus, an equation for the function g is

$$g(x) = -(x - 2)^2$$

FIGURE 11

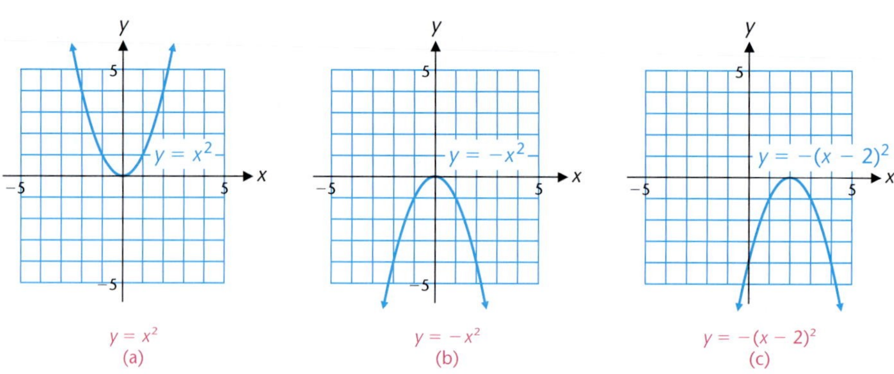

$y = x^2$
(a)

$y = -x^2$
(b)

$y = -(x - 2)^2$
(c)

Matched Problem 8

The graph of $y = h(x)$ in Figure 12 is a transformation of the graph of $y = x^3$. Find an equation for the function h.

FIGURE 12

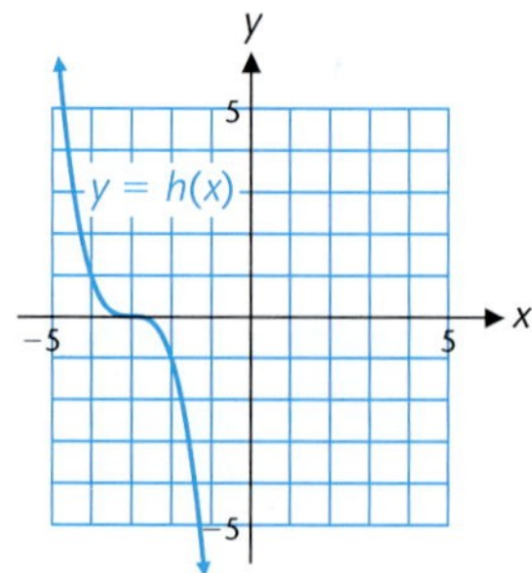

Answers to Matched Problems

1. $(f + g)(x) = \sqrt{x} + \sqrt{10 - x}$, $(f - g)(x) = \sqrt{x} - \sqrt{10 - x}$, $(fg)(x) = \sqrt{10x - x^2}$, $(f/g)(x) = \sqrt{x/(10 - x)}$; the functions $f + g, f - g$, and fg have domain $[0, 10]$, the domain of f/g is $[0, 10)$

2. $(f \circ g)(x) = x$, domain $= (-\infty, \infty)$
$(g \circ f)(x) = x$, domain $= (-\infty, \infty)$

3. $(f \circ g)(x) = \sqrt{10 - x}$; domain: $x \geq 1$ and $x \leq 10$ or $[1, 10]$

4. $h(x) = (f \circ g)(x)$, where $f(x) = \sqrt{x}$ and $g(x) = 4x - 7$

5. (A) The graph of $y = \sqrt{x} + 3$ is the same as the graph of $y = \sqrt{x}$ shifted upward 3 units, and the graph of $y = \sqrt{x} - 1$ is the same as the graph of $y = \sqrt{x}$ shifted downward 1 unit. The figure confirms these conclusions.

(B) The graph of $y = \sqrt{x + 3}$ is the same as the graph of $y = \sqrt{x}$ shifted to the left 3 units, and the graph of $y = \sqrt{x - 1}$ is the same as the graph of $y = \sqrt{x}$ shifted to the right 1 unit. The figure confirms these conclusions.

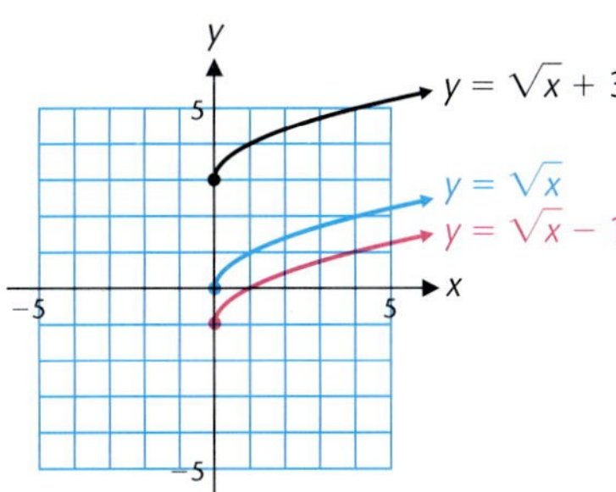

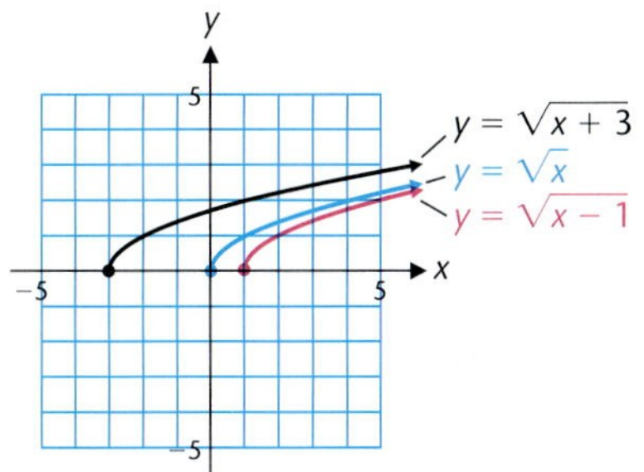

6. $G(x) = (x + 3)^3$, $H(x) = (x - 1)^3$, $M(x) = x^3 + 3$, $N(x) = x^3 - 4$

7. (A) The graph of $y = 2x$ is a vertical expansion of the graph of $y = x$, and the graph of $y = 0.5x$ is a vertical contraction of the graph of $y = x$. The figure confirms these conclusions.

(B) The graph of $y = -0.5x$ is a vertical contraction and a reflection in the x axis of the graph of $y = x$. The figure confirms this conclusion.

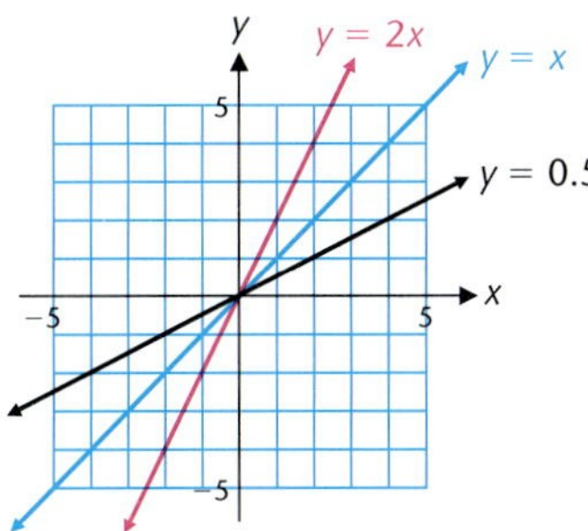

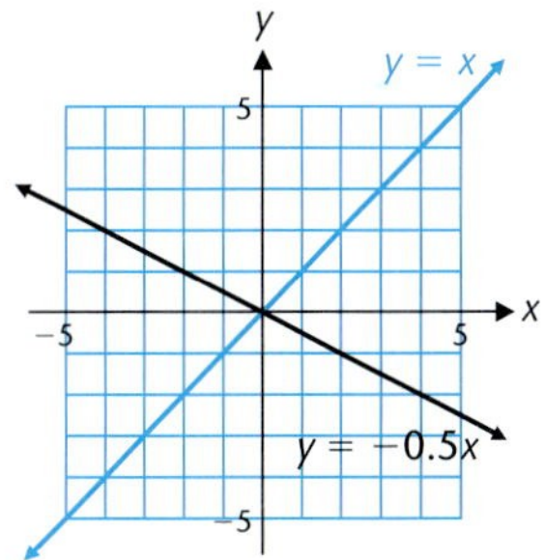

8. The graph of function h is a reflection in the x axis and a horizontal translation of 3 units to the left of the graph of $y = x^3$. An equation for h is $h(x) = -(x + 3)^3$.

EXERCISE 2-5

A

Without looking back in the text, indicate the domain and range of each of the following functions. (Making rough sketches on scratch paper may help.)

1. $h(x) = -\sqrt{x}$ **2.** $m(x) = -\sqrt[3]{x}$ **3.** $g(x) = -2x^2$

4. $f(x) = -0.5|x|$ **5.** $F(x) = -0.5x^3$ **6.** $G(x) = 4x^3$

In Problems 7–10, for the indicated functions f and g, find the functions $f + g$, $f - g$, fg, and f/g, and find their domains.

7. $f(x) = 4x$; $g(x) = x + 1$

8. $f(x) = 3x$; $g(x) = x - 2$

9. $f(x) = 2x^2$; $g(x) = x^2 + 1$

10. $f(x) = 3x; \quad g(x) = x^2 + 4$

In Problems 11–14, for the indicated functions f and g, find the functions $f \circ g$ and $g \circ f$, and find their domains.

11. $f(x) = x^2 + 3; \quad g(x) = x^2 - 4x$

12. $f(x) = x^2 - 5x; \quad g(x) = x^2 + 1$

13. $f(x) = 2x^{2/3}; \quad g(x) = x^3 - 1$

14. $f(x) = 4 - x^3; \quad g(x) = 3x^{1/3}$

Problems 15–22 refer to the functions f and g given by the graphs below (the domain of each function is $[-2, 2]$).

Use the graph of f or g, as required, to graph each given function.

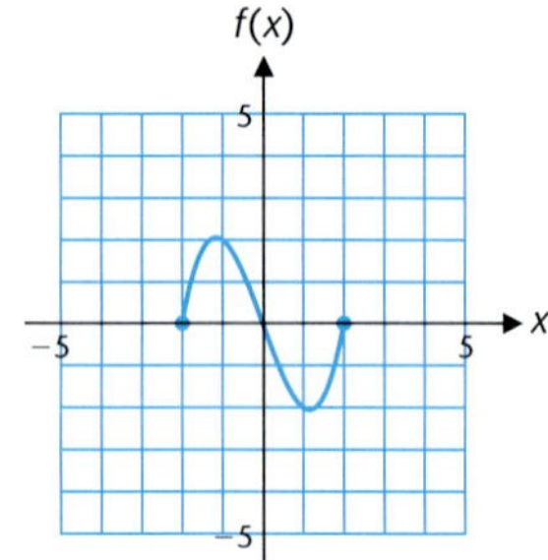

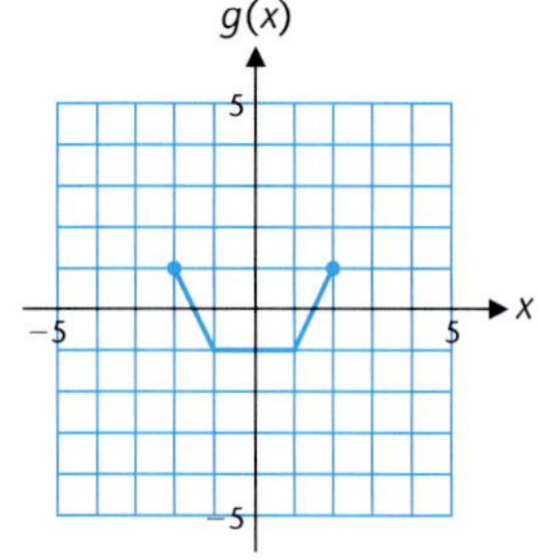

15. $f(x) + 2$ **16.** $g(x) - 1$ **17.** $g(x + 2)$

18. $f(x - 1)$ **19.** $-f(x)$ **20.** $-g(x)$

21. $2g(x)$ **22.** $\frac{1}{2}f(x)$

B

In Problems 23–28, indicate how the graph of each function is related to the graph of one of the six basic functions in Figure 2. Sketch a graph of each function.

Check your descriptions and graphs in Problems 23–28 by graphing each function on a graphing utility.

23. $g(x) = -|x + 2|$ **24.** $h(x) = -|x - 4|$

25. $f(x) = (x - 2)^2 - 4$ **26.** $m(x) = (x + 1)^2 + 3$

27. $f(x) = 4 - 2\sqrt{x}$ **28.** $g(x) = -2 + 3\sqrt[3]{x}$

In Problems 29–34, for the indicated functions f and g, find the functions $f + g$, $f - g$, fg, and f/g, and find their domains.

29. $f(x) = \sqrt{x + 2}; \quad g(x) = \sqrt{4 - x}$

30. $f(x) = \sqrt{5 - x}; \quad g(x) = \sqrt{x + 1}$

31. $f(x) = 5 - 2\sqrt{x}; \quad g(x) = 3 - \sqrt{x}$

32. $f(x) = 3\sqrt{x} + 6; \quad g(x) = \sqrt{x} - 1$

33. $f(x) = \sqrt{x^2 + x - 2}; \quad g(x) = \sqrt{24 + 2x - x^2}$

34. $f(x) = \sqrt{x^2 + 3x - 10}; \quad g(x) = \sqrt{x^2 - x - 12}$

In Problems 35–40, for the indicated functions f and g, find the functions $f \circ g$ and $g \circ f$, and find their domains.

35. $f(x) = x + 2; \quad g(x) = \sqrt{4 - x}$

36. $f(x) = \sqrt{x + 1}; \quad g(x) = x - 2$

37. $f(x) = x + 3; \quad g(x) = \dfrac{1}{x - 2}$

38. $f(x) = \dfrac{x}{x + 4}; \quad g(x) = 2 - x$

39. $f(x) = |x + 2|; \quad g(x) = \dfrac{x}{x - 3}$

40. $f(x) = \dfrac{x}{x - 4}; \quad g(x) = |x + 3|$

Each graph in Problems 41–46 is the result of applying a sequence of transformations to the graph of one of the six basic functions in Figure 2. Identify the basic function and describe the transformation verbally. Write an equation for the given graph.

Check your equations in Problems 41–46 by graphing each on a graphing utility.

41.

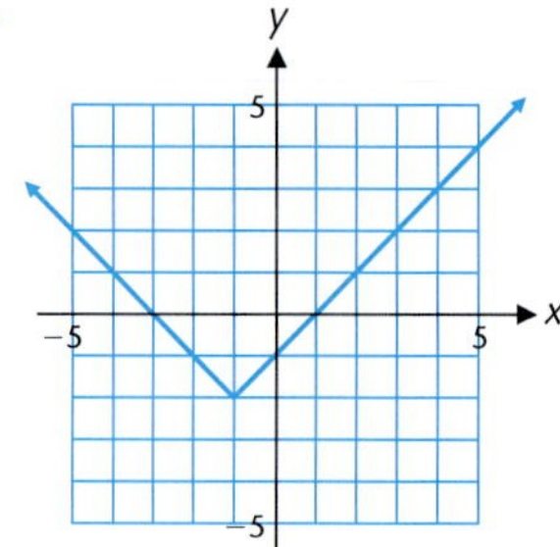

42.

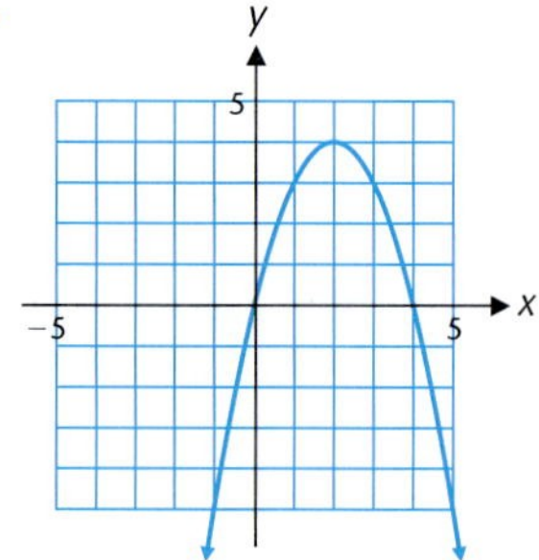

43.

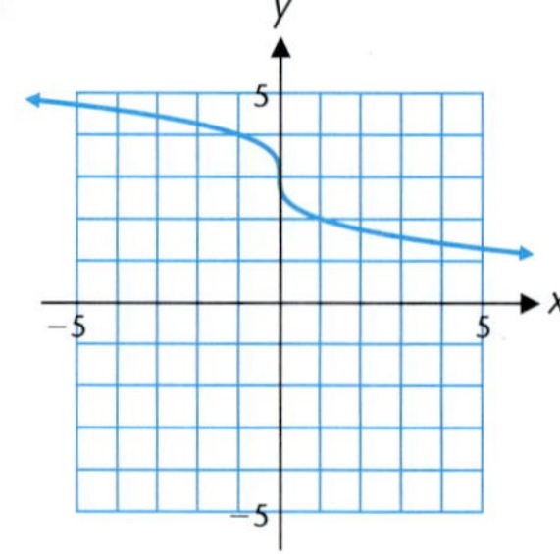

44.

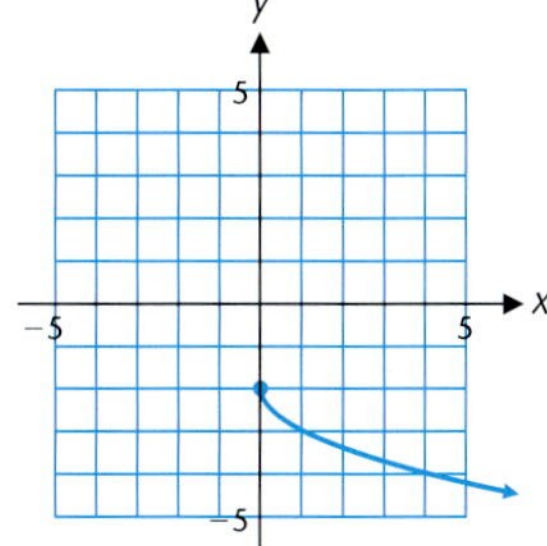

45.

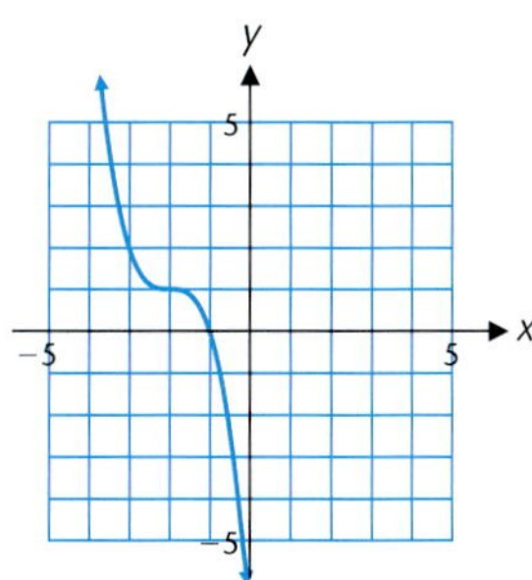

46.

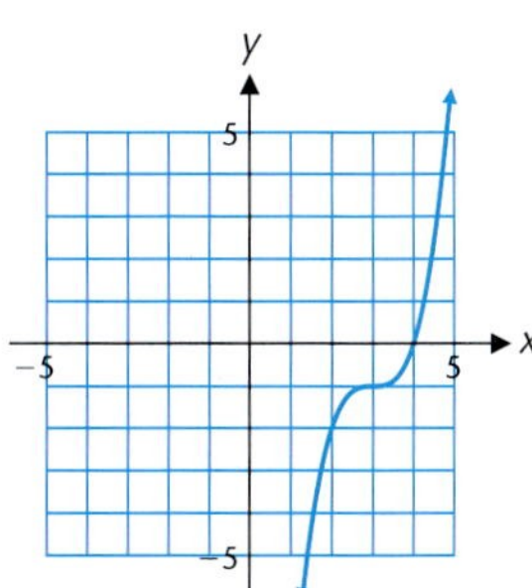

In Problems 47–52, the graph of the function g is formed by applying the indicated sequence of transformations to the given function f. Find an equation for the function g and graph g using $-5 \le x \le 5$ and $-5 \le y \le 5$.

47. The graph of $f(x) = \sqrt{x}$ is shifted 2 units to the left and 3 units up.

48. The graph of $f(x) = \sqrt[3]{x}$ is shifted 3 units to the right and 2 units down.

49. The graph of $f(x) = |x|$ is reflected in the x axis and shifted to the right 3 units.

50. The graph of $f(x) = |x|$ is reflected in the x axis and shifted to the left 1 unit.

51. The graph of $f(x) = x^3$ is reflected in the x axis and shifted to the left 2 units and up 1 unit.

52. The graph of $f(x) = x^2$ is reflected in the x axis and shifted to the right 2 units and down 4 units.

Changing the order in a sequence of transformations may change the final result. Investigate each pair of transformations in Problems 53–58 to determine whether reversing their order can produce a different result. Support your conclusions with specific examples and/or mathematical arguments.

53. Vertical shift, horizontal shift.

54. Vertical shift, reflection in y axis.

55. Vertical shift, reflection in x axis.

56. Vertical shift, vertical expansion.

57. Horizontal shift, reflection in x axis.

58. Horizontal shift, vertical contraction.

In Problems 59–66, express h as a composition of two simpler functions f and g of the form $f(x) = x^n$ and $g(x) = ax + b$, where n is a rational number and a and b are integers.

59. $h(x) = (2x - 7)^4$

60. $h(x) = (3 - 5x)^7$

61. $h(x) = \sqrt{4 + 2x}$

62. $h(x) = \sqrt{3x - 11}$

63. $h(x) = 3x^7 - 5$

64. $h(x) = 5x^6 + 3$

65. $h(x) = \dfrac{4}{\sqrt{x}} + 3$

66. $h(x) = -\dfrac{2}{\sqrt{x}} + 1$

C

Each of the following graphs involves a reflection in the x axis and/or a vertical expansion or contraction of one of the basic functions in Figure 2. Identify the basic function and describe the transformation verbally. Write an equation for the given graph.

Check your equations in Problems 67–70 by graphing each on a graphing utility.

67.

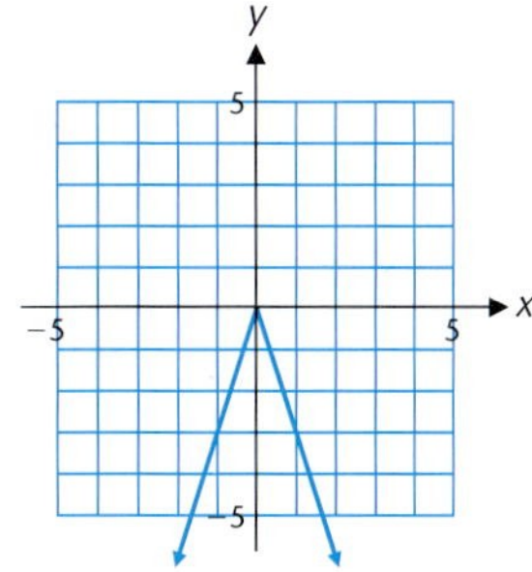

68.

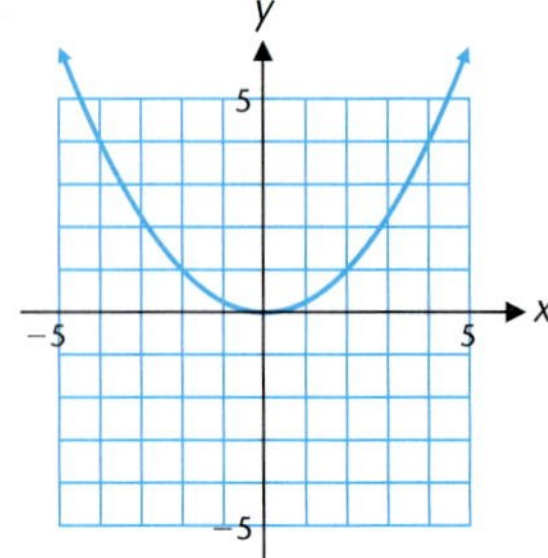

69.

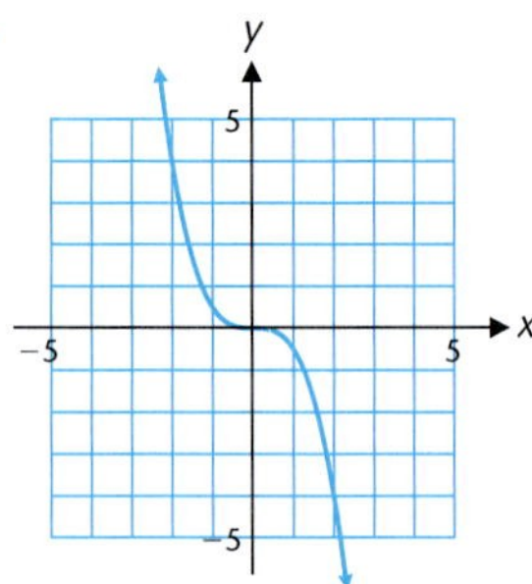

70.

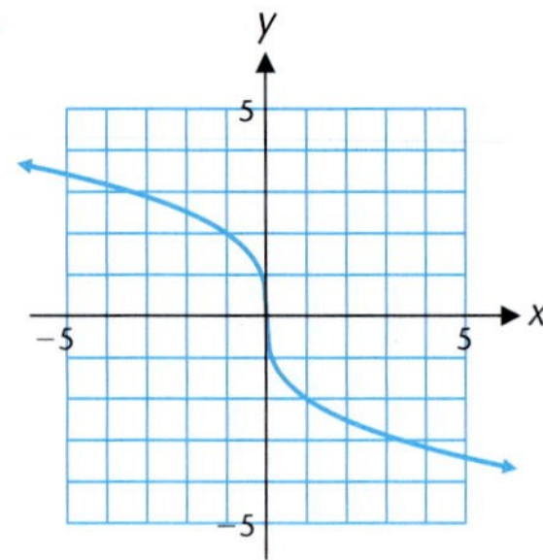

71. Are the functions fg and gf identical? Justify your answer.

72. Are the functions $f \circ g$ and $g \circ f$ identical? Justify your answer.

73. Is there a function g that satisfies $f \circ g = g \circ f = f$ for all functions f? If so, what is it?

74. Is there a function g that satisfies $fg = gf = f$ for all functions f? If so, what is it?

In Problems 75–78, for the indicated functions f and g, find the functions $f + g$, $f - g$, fg, and f/g, and find their domains.

75. $f(x) = x + \dfrac{1}{x}$; $g(x) = x - \dfrac{1}{x}$

76. $f(x) = x - 1$; $g(x) = x - \dfrac{6}{x - 1}$

77. $f(x) = 1 - \dfrac{x}{|x|}$; $g(x) = 1 + \dfrac{x}{|x|}$

78. $f(x) = x + |x|$; $g(x) = x - |x|$

In Problems 79–84, for the indicated functions f and g, find the functions $f \circ g$ and $g \circ f$, and find their domains.

79. $f(x) = x^2$; $g(x) = \sqrt{9 - x}$

80. $f(x) = \sqrt{x - 16}$; $g(x) = x^2$

81. $f(x) = \dfrac{2x + 1}{x - 2}$; $g(x) = \dfrac{x + 2}{x - 3}$

82. $f(x) = \dfrac{x + 2}{x - 4}$; $g(x) = \dfrac{x - 5}{x + 1}$

83. $f(x) = \sqrt{x^2 + 5}$; $g(x) = \sqrt{x^2 - 4}$

84. $f(x) = \sqrt{x^2 + 8}$; $g(x) = \sqrt{x^2 - 9}$

APPLICATIONS

85. Market Research. The demand x and the price p (in dollars) for a certain product are related by

$$x = f(p) = 4{,}000 - 200p$$

The revenue (in dollars) from the sale of x units is given by

$$R(x) = 20x - \frac{1}{200}x^2$$

and the cost (in dollars) of producing x units is given by

$$C(x) = 10x + 30{,}000$$

Express the profit as a function of the price p.

86. Market Research. The demand x and the price p (in dollars) for a certain product are related by

$$x = f(p) = 5{,}000 - 100p$$

The revenue (in dollars) from the sale of x units and the cost (in dollars) of producing x units are given, respectively, by

$$R(x) = 50x - \frac{1}{100}x^2 \quad \text{and} \quad C(x) = 20x + 40{,}000$$

Express the profit as a function of the price p.

87. Family of Curves. In calculus, solutions to certain types of problems often involve an unspecified constant. For example, consider the equation

$$y = \frac{1}{C}x^2 - C$$

where C is a positive constant. The collection of graphs of this equation for all permissible values of C is called a **family of curves.** On the same axes, graph the members of this family corresponding to $C = 1, 2, 3$, and 4.

88. Family of Curves. A family of curves is defined by the equation

$$y = 2C - \frac{2}{C}x^2$$

where C is a positive constant. On the same axes, graph the members of this family corresponding to $C = 1, 2, 3$, and 4.

89. Fluid Flow. A cubic tank is 4 feet on a side and is initially full of water. Water flows out an opening in the bottom of the tank at a rate proportional to the square root of the depth (see the figure). Using advanced concepts from mathematics and physics, it can be shown that the volume of the water in the tank t minutes after the water begins to flow is given by

$$V(t) = \frac{64}{C^2}(C - t)^2 \qquad 0 \le t \le C$$

where C is a constant that depends on the size of the opening. Graph $V(t)$ for $C = 1$, $C = 2$, $C = 4$, and $C = 8$.

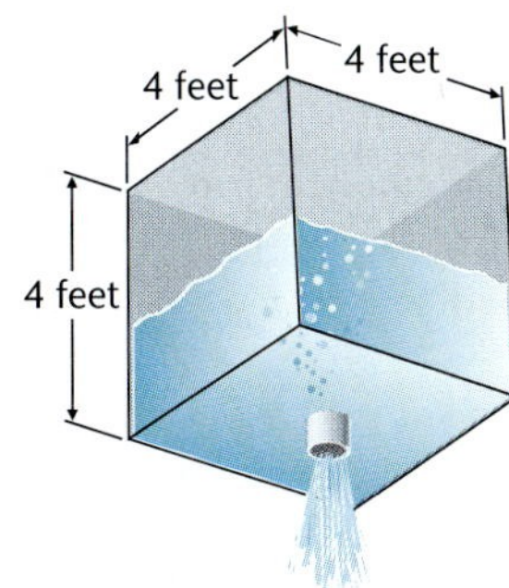

90. Evaporation. A water trough with triangular ends is 9 feet long, 4 feet wide, and 2 feet deep (see the figure). Initially, the trough is full of water, but due to evaporation, the volume of the water in the trough decreases at a rate proportional to the square root of the volume. Using advanced concepts from mathematics and physics, it can be shown that the volume after t hours is given by

$$V(t) = \frac{1}{C^2}(t + 6C)^2 \qquad 0 \le t \le 6|C|$$

where C is a constant. Graph $V(t)$ for $C = -4$, $C = -5$, and $C = -6$.

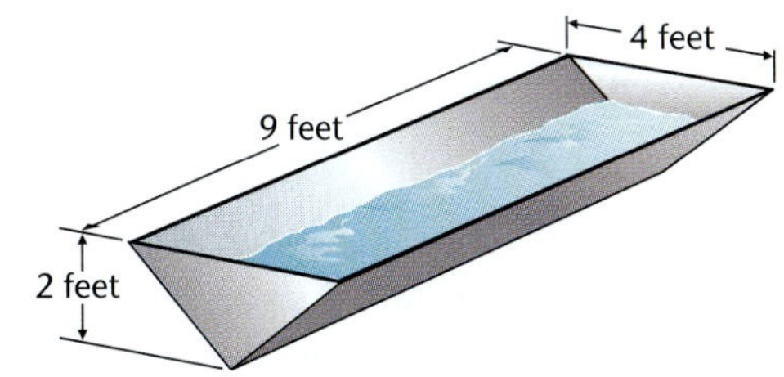

★ **91. Fluid Flow.** A conical paper cup with diameter 4 inches and height 4 inches is initially full of water. A small hole is made in the bottom of the cup and the water begins to flow out of the cup. Let h and r be the height and radius, respectively, of the water in the cup t minutes after the water begins to flow.

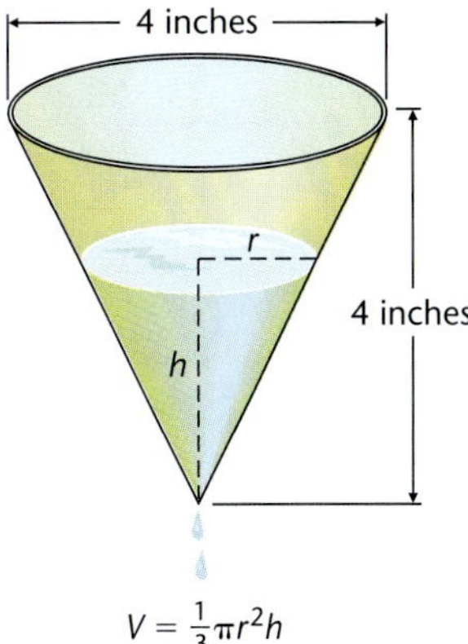

(A) Express r as a function of h.
(B) Express the volume V as a function of h.
(C) If the height of the water after t minutes is given by

$$h(t) = 0.5\sqrt{t}$$

express V as a function of t.

★ **92. Evaporation.** A water trough with triangular ends is 6 feet long, 4 feet wide, and 2 feet deep. Initially, the trough is full of water, but due to evaporation, the volume of the water is decreasing. Let h and w be the height and width, respectively, of the water in the tank t hours after it began to evaporate.

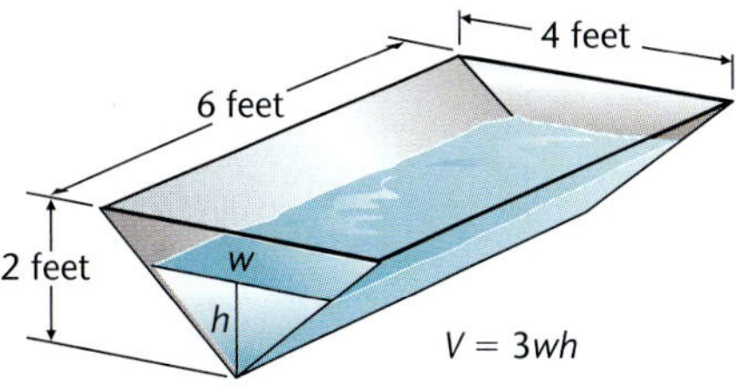

(A) Express w as a function of h.
(B) Express V as a function of h.
(C) If the height of the water after t hours is given by

$$h(t) = 2 - 0.2\sqrt{t}$$

express V as a function of t.

SECTION 2-6 Inverse Functions

- One-to-One Functions
- Inverse Functions

Many important mathematical relationships can be expressed in terms of functions. For example,

$C = \pi d = f(d)$ The circumference of a circle is a function of the diameter d.

$V = s^3 = g(s)$ The volume of a cube is a function of the edge s.

$d = 1{,}000 - 100p = h(p)$ The demand for a product is a function of the price p.

$F = \frac{9}{5}C + 32$ Temperature measured in °F is a function of temperature in °C.

In many cases, we are interested in *reversing* the correspondence determined by a function. Thus,

$d = \frac{C}{\pi} = m(C)$ The diameter of a circle is a function of the circumference C.

$s = \sqrt[3]{V} = n(V)$ The edge of a cube is a function of the volume V.

$p = 10 - \frac{1}{100}d = r(d)$ The price of a product is a function of the demand d.

$C = \frac{5}{9}(F - 32)$ Temperature measured in °C is a function of temperature in °F.

As these examples illustrate, reversing the relationship between two quantities often produces a new function. This new function is called the *inverse* of the original function. Later in this text we will see that many important functions (for example, logarithmic functions) are actually defined as the inverses of other functions.

In this section, we develop techniques for determining whether the inverse function exists, some general properties of inverse functions, and methods for finding the rule of correspondence that defines the inverse function. A review of Section 2-3 will prove very helpful at this point.

• One-to-One Functions

Recall the set form of the definition of a function:

A function is a set of ordered pairs with the property that no two ordered pairs have the same first component and different second components.

However, it is possible that two ordered pairs in a function could have the same second component and different first components. If this does not happen, then we call the function a *one-to-one function.* It turns out that one-to-one functions are the only functions that have inverse functions.

DEFINITION 1 **One-to-One Function**

A function is **one-to-one** if no two ordered pairs in the function have the same second component and different first components.

To illustrate this concept, consider the following three sets of ordered pairs:

$$f = \{(0, 3), (0, 5), (4, 7)\}$$
$$g = \{(0, 3), (2, 3), (4, 7)\}$$
$$h = \{(0, 3), (2, 5), (4, 7)\}$$

Set f is not a function because the ordered pairs (0, 3) and (0, 5) have the same first component and different second components. Set g is a function, but it is not a one-to-one function because the ordered pairs (0, 3) and (2, 3) have the same second component and different first components. But set h is a function, and it is one-to-one. Representing these three sets of ordered pairs as rules of correspondence provides some additional insight into this concept.

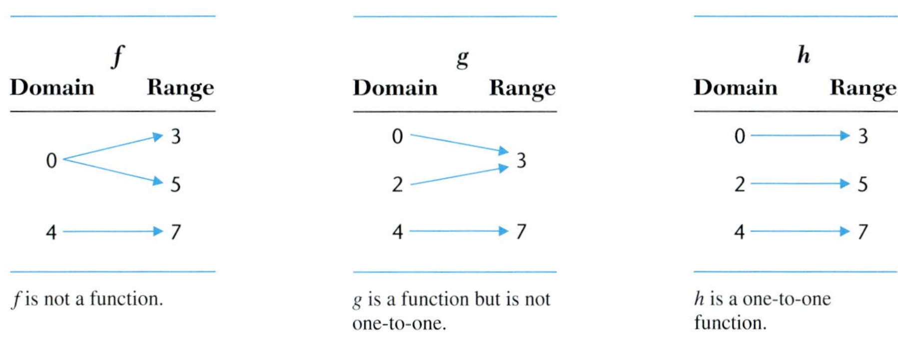

f is not a function.

g is a function but is not one-to-one.

h is a one-to-one function.

EXAMPLE 1 Determining Whether a Function Is One-to-One

Determine whether f is a one-to-one function for:

(A) $f(x) = x^2$ (B) $f(x) = 2x - 1$

Solutions

(A) To show that a function is not one-to-one, all we have to do is find two different ordered pairs in the function with the same second component and different first components. Since

$$f(2) = 2^2 = 4 \quad \text{and} \quad f(-2) = (-2)^2 = 4$$

the ordered pairs (2, 4) and (−2, 4) both belong to f and f is not one-to-one.

(B) To show that a function is one-to-one, we have to show that no two ordered pairs have the same second component and different first components. To do this, we assume there are two ordered pairs $(a, f(a))$ and $(b, f(b))$ in f with the same second components and then show that the first components must also be the same. That is, we show that $f(a) = f(b)$ implies $a = b$. We proceed as follows:

$f(a) = f(b)$	Assume second components are equal.
$2a - 1 = 2b - 1$	Evaluate $f(a)$ and $f(b)$.
$2a = 2b$	Simplify.
$a = b$	Conclusion: f is one-to-one

Thus, by Definition 1, f is a one-to-one function.

Matched Problem 1 Determine whether f is a one-to-one function for:

(A) $f(x) = 4 - x^2$ (B) $f(x) = 4 - 2x$

The methods used in the solution of Example 1 can be stated as a theorem.

Theorem 1

One-to-One Functions

1. If $f(a) = f(b)$ for at least one pair of domain values a and b, $a \neq b$, then f is not one-to-one.
2. If the assumption $f(a) = f(b)$ always implies that the domain values a and b are equal, then f is one-to-one.

Applying Theorem 1 is not always easy—try testing $f(x) = x^3 + 2x + 3$, for example. However, if we are given the graph of a function, then there is a simple graphic procedure for determining if the function is one-to-one. If a horizontal line intersects the graph of a function in more than one point, then the function is not one-to-one, as shown in Figure 1(a). However, if each horizontal line intersects the graph in one point, or not at all, then the function is one-to-one, as shown in Figure 1(b). These observations form the basis for the *horizontal line test.*

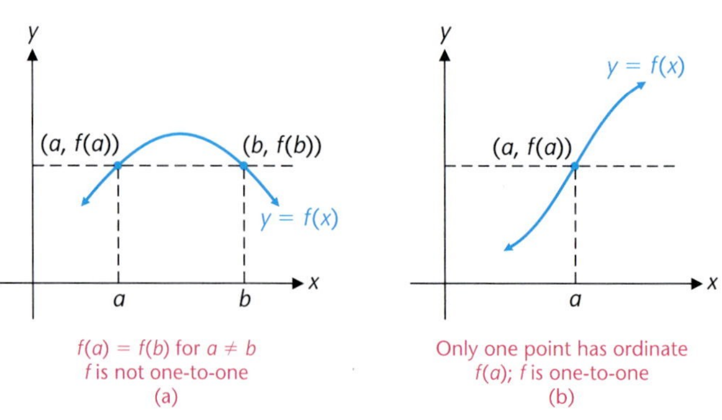

FIGURE 1 Intersections of graphs and horizontal lines.

Theorem 2 **Horizontal Line Test**

A function is one-to-one if and only if each horizontal line intersects the graph of the function in at most one point.

The graphs of the functions considered in Example 1 are shown in Figure 2. Applying the horizontal line test to each graph confirms the results we obtained in Example 1.

FIGURE 2 Applying the horizontal line test.

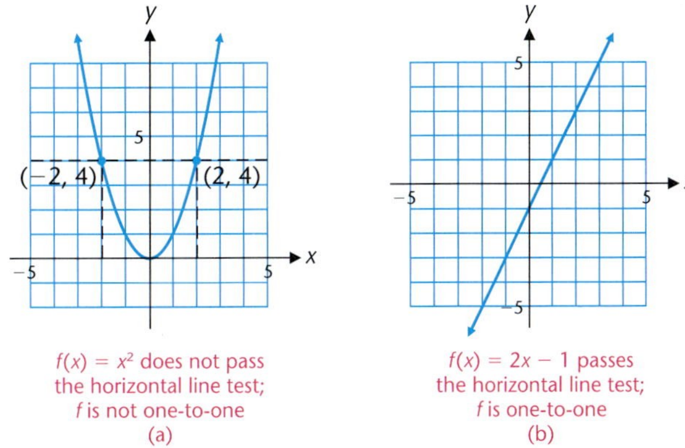

A function that is increasing throughout its domain or decreasing throughout its domain will always pass the horizontal line test [see Figs. 3(a) and 3(b)]. Thus, we have the following theorem.

Theorem 3 **Increasing and Decreasing Functions**

If a function f is increasing throughout its domain or decreasing throughout its domain, then f is a one-to-one function.

FIGURE 3 Increasing, decreasing, and one-to-one functions.

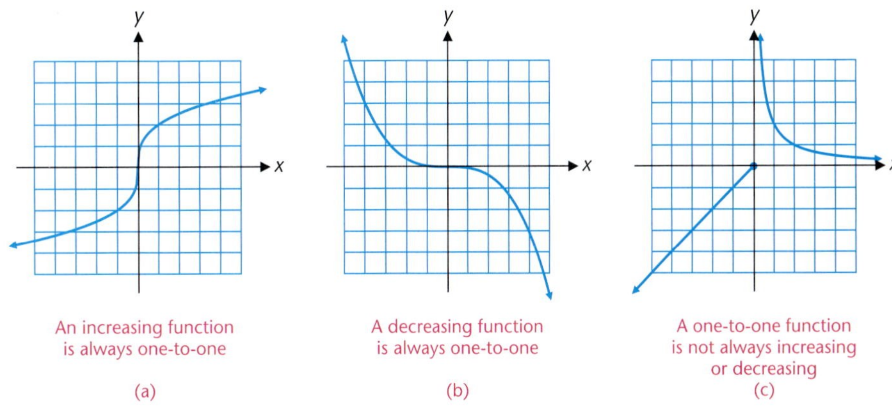

The converse of Theorem 3 is false. To see this, consider the function graphed in Figure 3(c). This function is increasing on $(-\infty, 0]$ and decreasing on $(0, \infty)$, yet the graph passes the horizontal line test. Thus, this is a one-to-one function that is neither an increasing function nor a decreasing function.

• Inverse Functions

Now we want to see how we can form a new function by reversing the correspondence determined by a given function. Let g be the function defined as follows:

$$g = \{(-3, 9), (0, 0), (3, 9)\} \quad \text{g is not one-to-one.}$$

Notice that g is not one-to-one because the domain elements -3 and 3 both correspond to the range element 9. We can reverse the correspondence determined by function g simply by reversing the components in each ordered pair in g, producing the following set:

$$G = \{(9, -3), (0, 0), (9, 3)\} \quad \text{G is not a function.}$$

But the result is not a function because the domain element 9 corresponds to two different range elements, -3 and 3. On the other hand, if we reverse the ordered pairs in the function

$$f = \{(1, 2), (2, 4), (3, 9)\} \quad \text{f is one-to-one.}$$

we obtain

$$F = \{(2, 1), (4, 2), (9, 3)\} \quad \text{F is a function.}$$

This time f is a one-to-one function, and the set F turns out to be a function also. This new function F, formed by reversing all the ordered pairs in f, is called the *inverse* of f and is usually denoted* by f^{-1}. Thus,

$$f^{-1} = \{(2, 1), (4, 2), (9, 3)\} \quad \text{The inverse of f}$$

Notice that f^{-1} is also a one-to-one function and that the following relationships hold:

$$\text{Domain of } f^{-1} = \{2, 4, 9\} = \text{Range of } f$$
$$\text{Range of } f^{-1} = \{1, 2, 3\} = \text{Domain of } f$$

Thus, reversing all the ordered pairs in a one-to-one function forms a new one-to-one function and reverses the domain and range in the process. We are now ready to present a formal definition of the inverse of a function.

*f^{-1}, read "f inverse," is a special symbol used here to represent the inverse of the function f. It does *not* mean $1/f$.

DEFINITION 2 **Inverse of a Function**

If f is a one-to-one function, then the **inverse** of f, denoted f^{-1}, is the function formed by reversing all the ordered pairs in f. Thus,

$$f^{-1} = \{(y, x) \mid (x, y) \text{ is in } f\}$$

If f is not one-to-one, then f **does not have an inverse** and f^{-1} **does not exist.**

The following properties of inverse functions follow directly from the definition.

Theorem 4 **Properties of Inverse Functions**

If f^{-1} exists, then

1. f^{-1} is a one-to-one function.

2. Domain of f^{-1} = Range of f

3. Range of f^{-1} = Domain of f

EXPLORE-DISCUSS 1 Most graphing utilities have a routine, usually denoted by Draw Inverse (or an abbreviation of this phrase—consult your manual), that will draw the graph formed by reversing the ordered pairs of all the points on the graph of a function. For example, Figure 4(a) shows the graph of $f(x) = 2x - 1$ along with the graph obtained by using the Draw Inverse routine. Figure 4(b) does the same for $f(x) = x^2$.

(A) Is the graph produced by Draw Inverse in Figure 4(a) the graph of a function? Does f^{-1} exist? Explain.

(B) Is the graph produced by Draw Inverse in Figure 4(b) the graph of a function? Does f^{-1} exist? Explain.

(C) If you have a graphing utility with a Draw Inverse routine, apply it to the graphs of $y = \sqrt{x - 1}$ and $y = 4x - x^2$ to determine if the result is the graph of a function and if the inverse of the original function exists.

FIGURE 4

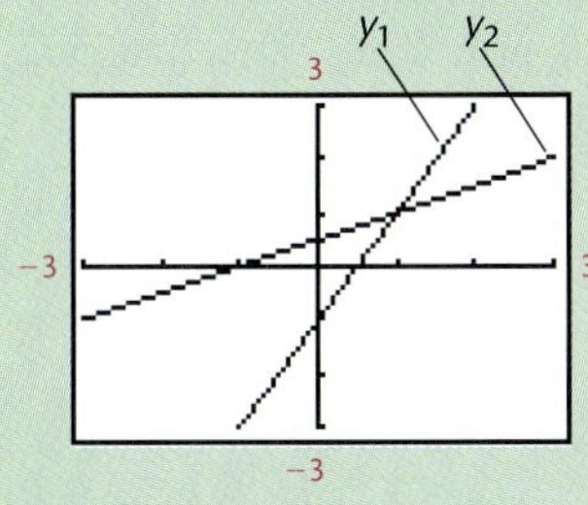

$y_1 = 2x - 1$
y_2 = Draw Inverse y_1
(a)

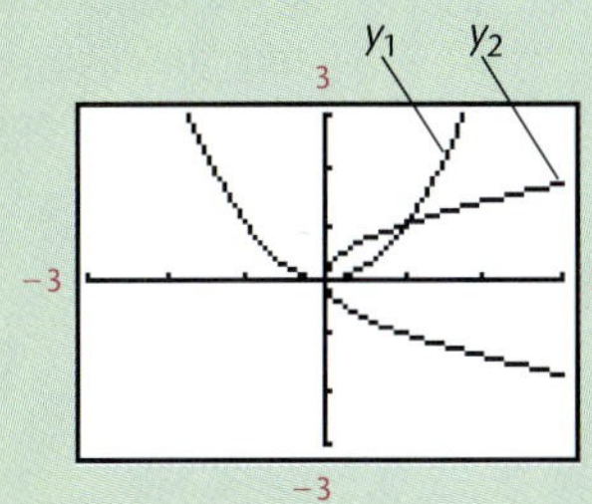

$y_1 = x^2$
y_2 = Draw Inverse y_1
(b)

Finding the inverse of a function defined by a finite set of ordered pairs is easy; just reverse each ordered pair. But how do we find the inverse of a function defined by an equation? Consider the one-to-one function f defined by

$$f(x) = 2x - 1$$

To find f^{-1}, we let $y = f(x)$ and solve for x:

$$\begin{aligned} y &= 2x - 1 \\ y + 1 &= 2x \\ \tfrac{1}{2}y + \tfrac{1}{2} &= x \end{aligned}$$

Since the ordered pair (x, y) is in f if and only if the reversed ordered pair (y, x) is in f^{-1}, this last equation defines f^{-1}:

$$x = f^{-1}(y) = \tfrac{1}{2}y + \tfrac{1}{2} \tag{1}$$

Something interesting happens if we form the composition* of f and f^{-1} in either of the two possible orders

$$f^{-1}[f(x)] = f^{-1}[2x - 1] = \tfrac{1}{2}(2x - 1) + \tfrac{1}{2} = x - \tfrac{1}{2} + \tfrac{1}{2} = x$$

and

$$f[f^{-1}(y)] = f(\tfrac{1}{2}y + \tfrac{1}{2}) = 2(\tfrac{1}{2}y + \tfrac{1}{2}) - 1 = y + 1 - 1 = y$$

These compositions indicate that if f maps x into y, then f^{-1} maps y back into x and if f^{-1} maps y into x, then f maps x back into y. This is interpreted schematically in Figure 5.

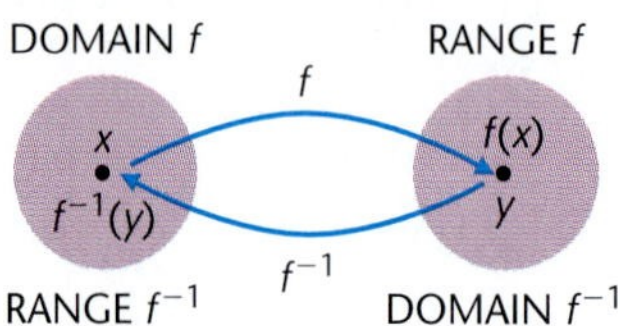

FIGURE 5 Composition of f and f^{-1}.

Finally, we note that we usually use x to represent the independent variable and y the dependent variable in an equation that defines a function. It is customary to do this for inverse functions also. Thus, interchanging the variables x and y in equation (1), we can state that the inverse of

$$y = f(x) = 2x - 1$$

is

$$y = f^{-1}(x) = \tfrac{1}{2}x + \tfrac{1}{2}$$

*When working with inverse functions, it is customary to write compositions as $f[g(x)]$ rather than as $(f \circ g)(x)$.

In general, we have the following result:

Theorem 5 **Relationship between f and f^{-1}**

If f^{-1} exists, then

1. $x = f^{-1}(y)$ if and only if $y = f(x)$.
2. $f^{-1}[f(x)] = x$ for all x in the domain of f.
3. $f[f^{-1}(y)] = y$ for all y in the domain of f^{-1} or, if x and y have been interchanged, $f[f^{-1}(x)] = x$ for all x in the domain of f^{-1}.

If f and g are one-to-one functions satisfying

$$f[g(x)] = x \qquad \text{for all } x \text{ in the domain of } g$$

$$g[f(x)] = x \qquad \text{for all } x \text{ in the domain of } f$$

then it can be shown that $g = f^{-1}$ and $f = g^{-1}$. Thus, the inverse function is the only function that satisfies both these compositions. We can use this fact to check that we have found the inverse correctly.

EXPLORE-DISCUSS 2 Find $f[g(x)]$ and $g[f(x)]$ for

$$f(x) = (x - 1)^3 + 2 \qquad \text{and} \qquad g(x) = (x - 2)^{1/3} + 1$$

How are f and g related?

The procedure for finding the inverse of a function defined by an equation is given in the next box. This procedure can be applied whenever it is possible to solve $y = f(x)$ for x in terms of y.

Finding the Inverse of a Function f

Step 1. Find the domain of f and verify that f is one-to-one. If f is not one-to-one, then stop, since f^{-1} does not exist.

Step 2. Solve the equation $y = f(x)$ for x. The result is an equation of the form $x = f^{-1}(y)$.

Step 3. Interchange x and y in the equation found in step 2. This expresses f^{-1} as a function of x.

Step 4. Find the domain of f^{-1}. Remember, the domain of f^{-1} must be the same as the range of f.

Check your work by verifying that

$$f^{-1}[f(x)] = x \qquad \text{for all } x \text{ in the domain of } f$$

and

$$f[f^{-1}(x)] = x \qquad \text{for all } x \text{ in the domain of } f^{-1}$$

EXAMPLE 2 **Finding the Inverse of a Function**

Find f^{-1} for $f(x) = \sqrt{x - 1}$

Solution *Step 1.* Find the domain of f and verify that f is one-to-one. The domain of f is $[1, \infty)$. The graph of f in Figure 6 shows that f is one-to-one, hence f^{-1} exists.

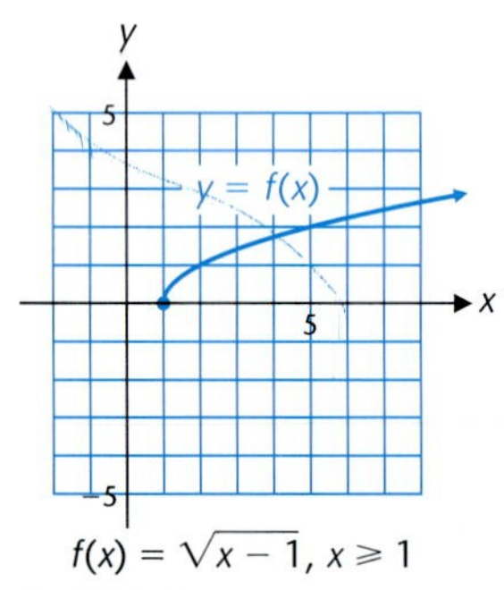

$f(x) = \sqrt{x - 1},\ x \geq 1$

FIGURE 6

Step 2. Solve the equation $y = f(x)$ for x.

$$y = \sqrt{x - 1}$$
$$y^2 = x - 1$$
$$x = y^2 + 1$$

Thus,

$$x = f^{-1}(y) = y^2 + 1$$

Step 3. Interchange x and y.

$$y = f^{-1}(x) = x^2 + 1$$

Step 4. Find the domain of f^{-1}. The equation $f^{-1}(x) = x^2 + 1$ is defined for all values of x, but this does not tell us what the domain of f^{-1} is. Remember, the domain of f^{-1} must equal the range of f. From the graph of f, we see that the range of f is $[0, \infty)$. Thus, the domain of f^{-1} is also $[0, \infty)$. That is,

$$f^{-1}(x) = x^2 + 1 \qquad x \geq 0$$

Check For x in $[1, \infty)$, the domain of f, we have

$$\begin{aligned} f^{-1}[f(x)] &= f^{-1}(\sqrt{x - 1}) \\ &= (\sqrt{x - 1})^2 + 1 \\ &= x - 1 + 1 \\ &\stackrel{\checkmark}{=} x \end{aligned}$$

For x in $[0, \infty)$, the domain of f^{-1}, we have

$$\begin{aligned} f[f^{-1}(x)] &= f(x^2 + 1) \\ &= \sqrt{(x^2 + 1) - 1} \\ &= \sqrt{x^2} \\ &= |x| && \sqrt{x^2} = |x| \text{ for any real number } x. \\ &\overset{\checkmark}{=} x && |x| = x \text{ for } x \geq 0. \end{aligned}$$

Matched Problem 2 Find f^{-1} for $f(x) = \sqrt{x + 2}$.

EXPLORE-DISCUSS 3 Most basic arithmetic operations can be reversed by performing a second operation: subtraction reverses addition, division reverses multiplication, squaring reverses taking the square root, etc. Viewing a function as a sequence of reversible operations gives additional insight into the inverse function concept. For example, the function $f(x) = 2x - 1$ can be described verbally as a function that multiplies each domain element by 2 and then subtracts 1. Reversing this sequence describes a function g that adds 1 to each domain element and then divides by 2, or $g(x) = (x + 1)/2$, which is the inverse of the function f. For each of the following functions, write a verbal description of the function, reverse your description, and write the resulting algebraic equation. Verify that the result is the inverse of the original function.

(A) $f(x) = 3x + 5$ (B) $f(x) = \sqrt{x - 1}$ (C) $f(x) = \dfrac{1}{x + 1}$

There is an important relationship between the graph of any function and its inverse that is based on the following observation: In a rectangular coordinate system, the points (a, b) and (b, a) are symmetric with respect to the line $y = x$ [see Fig. 7(a)]. Theorem 6 is an immediate consequence of this observation.

FIGURE 7 Symmetry with respect to the line $y = x$.

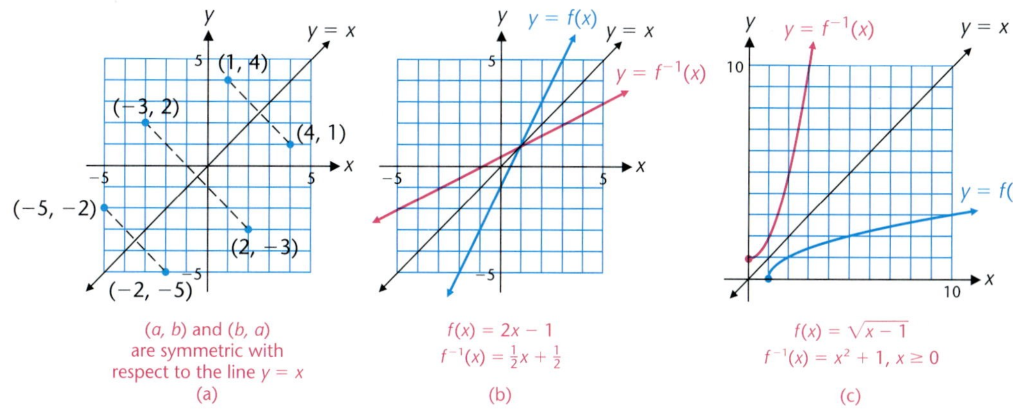

Theorem 6 **Symmetry Property for the Graphs of f and f^{-1}**

The graphs of $y = f(x)$ and $y = f^{-1}(x)$ are symmetric with respect to the line $y = x$.

Knowledge of this symmetry property makes it easy to graph f^{-1} if the graph of f is known, and vice versa. Figures 7(b) and 7(c) illustrate this property for the two inverse functions we found earlier in this section.

If a function is not one-to-one, we usually can restrict the domain of the function to produce a new function that is one-to-one. Then we can find an inverse for the restricted function. Suppose we start with $f(x) = x^2 - 4$. Since f is not one-to-one, f^{-1} does not exist [Fig. 8(a)]. But there are many ways the domain of f can be restricted to obtain a one-to-one function. Figures 8(b) and 8(c) illustrate two such restrictions.

FIGURE 8 Restricting the domain of a function.

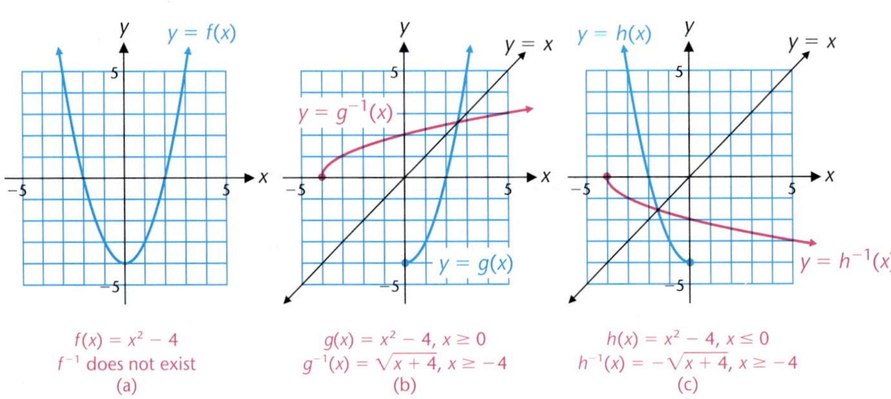

EXPLORE-DISCUSS 4 To graph the function

$$g(x) = x^2 - 4, \qquad x \geq 0$$

on a graphing utility, enter

$$y_1 = (x^2 - 4)/(x \geq 0)$$

(A) The Boolean expression $(x \geq 0)$ is assigned the value 1 if the inequality is true and 0 if it is false. How does this result in restricting the graph of $x^2 - 4$ to just those values of x satisfying $x \geq 0$?

(B) Use this concept to reproduce Figures 8(b) and 8(c) on a graphing utility.

(C) Do your graphs appear to be symmetric with respect to the line $y = x$? What happens if you use a squared window for your graph?

Recall from Theorem 2 that increasing and decreasing functions are always one-to-one. This provides the basis for a convenient and popular method of restricting the domain of a function:

If the domain of a function f is restricted to an interval on the x axis over which f is increasing (or decreasing), then the new function determined by this restriction is one-to-one and has an inverse.

We used this method to form the functions g and h in Figure 8.

EXAMPLE 3 Finding the Inverse of a Function

Find the inverse of $f(x) = 4x - x^2$, $x \leq 2$. Graph f, f^{-1}, and $y = x$ in the same coordinate system.

Solution *Step 1.* Find the domain of f and verify that f is one-to-one. The graph of $y = 4x - x^2$ is the parabola shown in Figure 9(a). Restricting the domain of f to $x \leq 2$ restricts the graph of f to the left side of this parabola [Fig. 9(b)]. Thus, f is a one-to-one function.

FIGURE 9

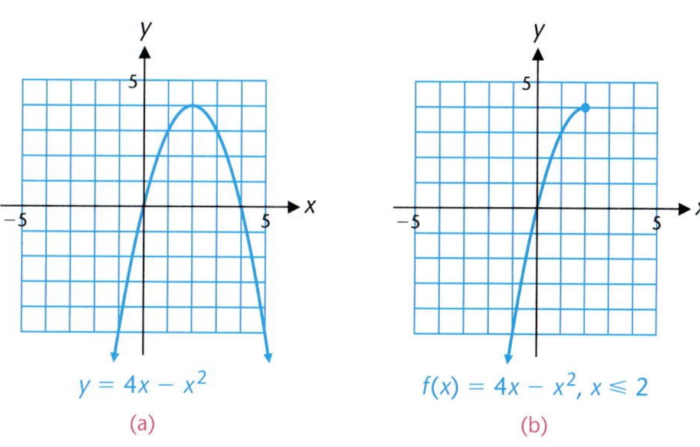

$y = 4x - x^2$
(a)

$f(x) = 4x - x^2, x \leq 2$
(b)

Step 2. Solve the equation $y = f(x)$ for x.

$$y = 4x - x^2$$

$$x^2 - 4x = -y \quad \text{Rearrange terms.}$$

$$x^2 - 4x + 4 = -y + 4 \quad \text{Add 4 to complete the square on the left side.}$$

$$(x - 2)^2 = 4 - y$$

Taking the square root of both sides of this last equation, we obtain two possible solutions:

$$x - 2 = \pm\sqrt{4 - y}$$

The restricted domain of f tells us which solution to use. Since $x \leq 2$ implies $x - 2 \leq 0$, we must choose the negative square root. Thus,

$$x - 2 = -\sqrt{4 - y}$$
$$x = 2 - \sqrt{4 - y}$$

and we have found

$$x = f^{-1}(y) = 2 - \sqrt{4 - y}$$

Step 3. Interchange x and y.

$$y = f^{-1}(x) = 2 - \sqrt{4 - x}$$

Step 4. Find the domain of f^{-1}. The equation $f^{-1}(x) = 2 - \sqrt{4 - x}$ is defined for $x \leq 4$. From the graph in Figure 9(b), the range of f also is $(-\infty, 4]$. Thus,

$$f^{-1}(x) = 2 - \sqrt{4 - x} \qquad x \leq 4$$

The check is left for the reader.

The graphs of f, f^{-1}, and $y = x$ are shown in Figure 10. Sometimes it is difficult to visualize the reflection of the graph of f in the line $y = x$. Choosing some points on the graph of f and plotting their reflections first makes it easier to sketch the graph of f^{-1}. Figure 11 shows a check on a graphing utility.

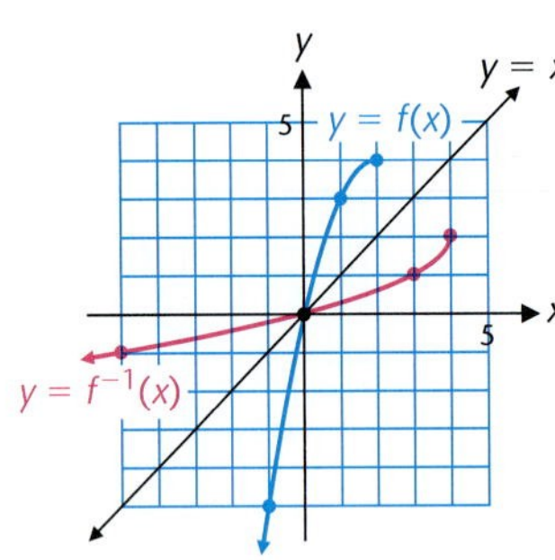

FIGURE 10

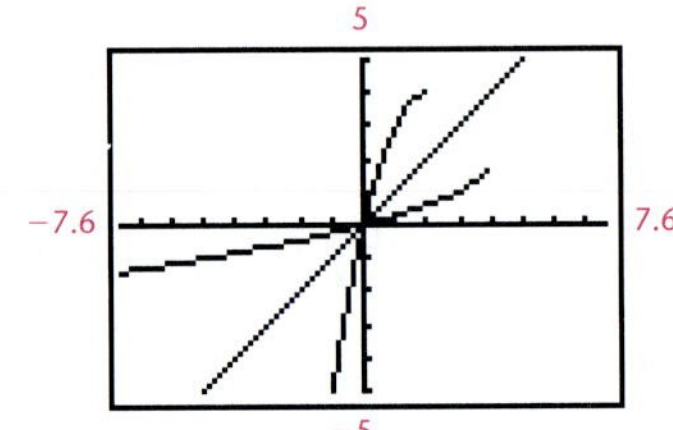

FIGURE 11

Matched Problem 3 Find the inverse of $f(x) = 4x - x^2$, $x \geq 2$. Graph f, f^{-1}, and $y = x$ in the same coordinate system.

Answers to Matched Problems

1. (A) Not one-to-one (B) One-to-one
2. $f^{-1}(x) = x^2 - 2$, $x \geq 0$
3. $f^{-1}(x) = 2 + \sqrt{4 - x}$, $x \leq 4$

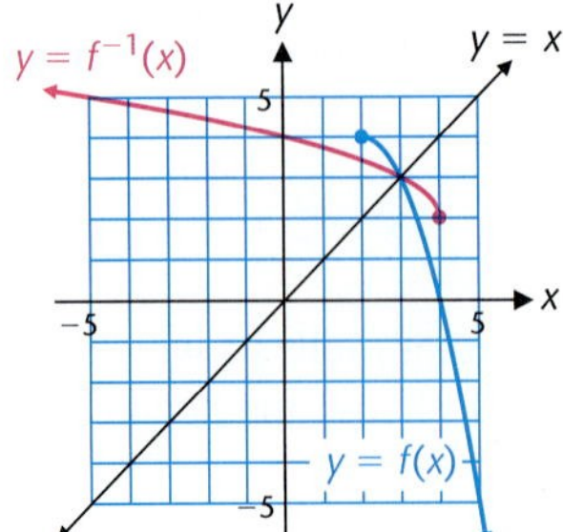

EXERCISE 2-6

A

Which of the functions in Problems 1–16 are one-to-one?

1. $\{(1, 2), (2, 1), (3, 4), (4, 3)\}$
2. $\{(-1, 0), (0, 1), (1, -1), (2, 1)\}$
3. $\{(5, 4), (4, 3), (3, 3), (2, 4)\}$
4. $\{(5, 4), (4, 3), (3, 2), (2, 1)\}$

5. 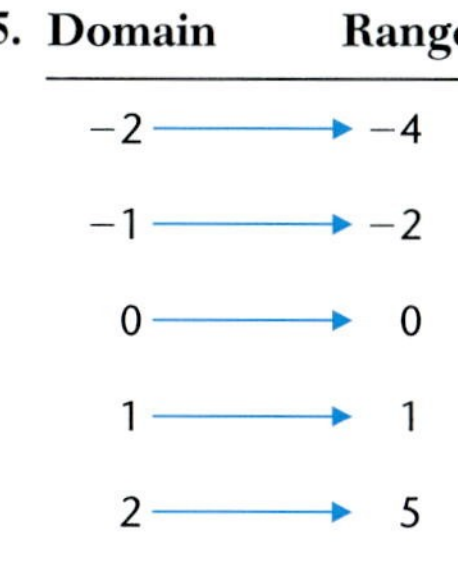

6.

Domain	Range
−2 →	−3
−1 →	−3
0 →	7
1 →	9
2 →	9

7. 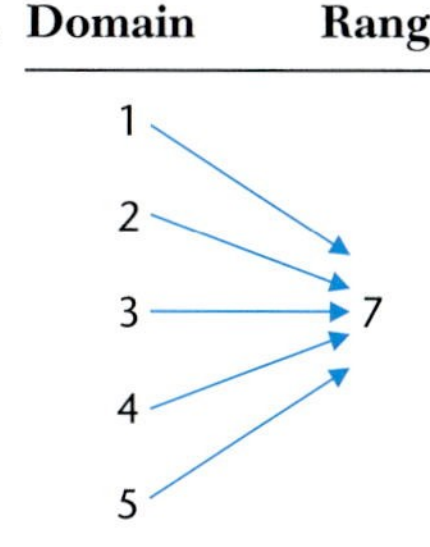

8. 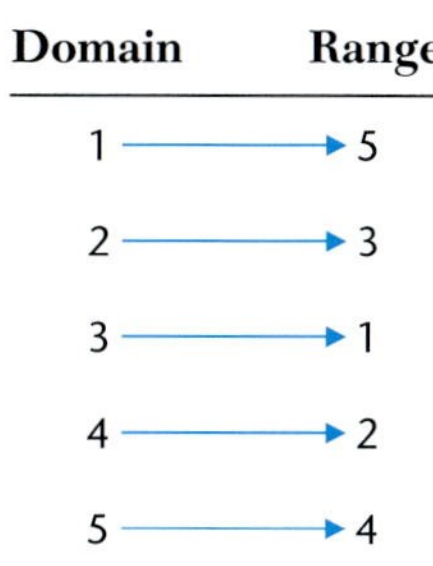

9. 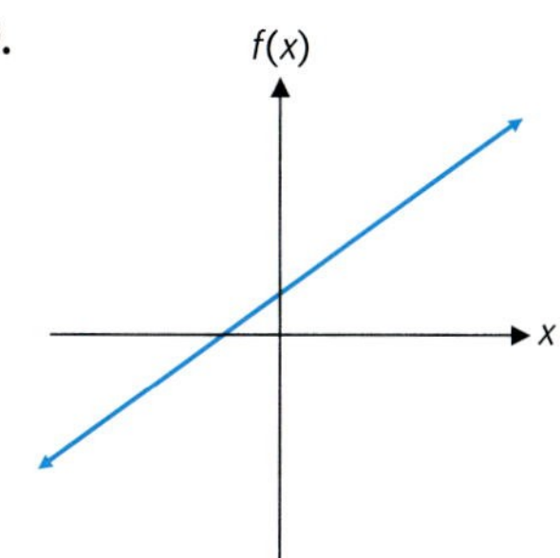

10. 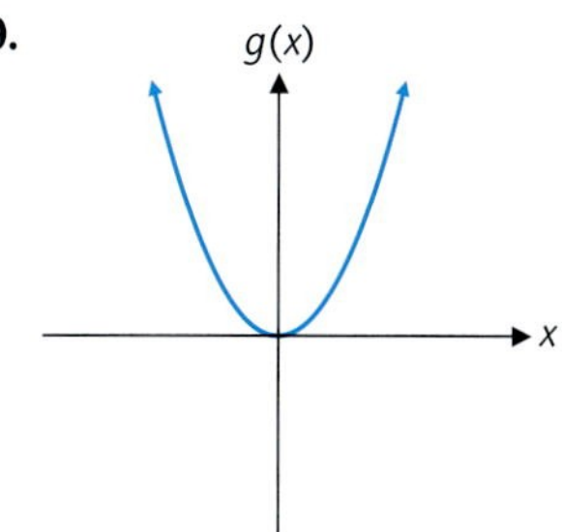

11. 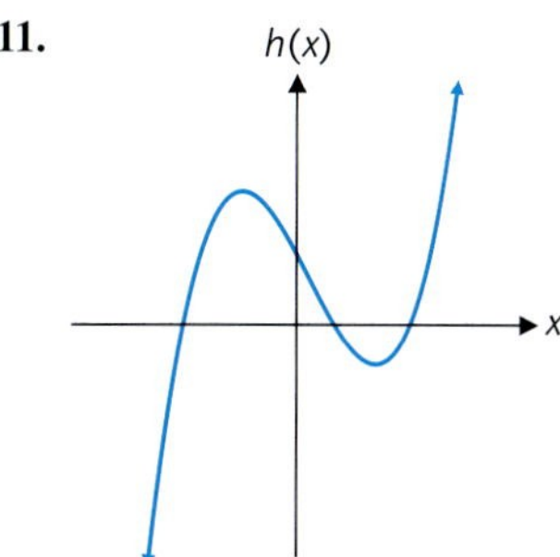

12. 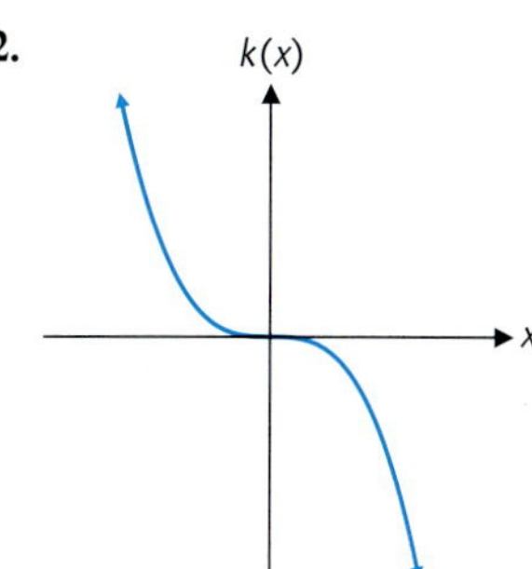

13. 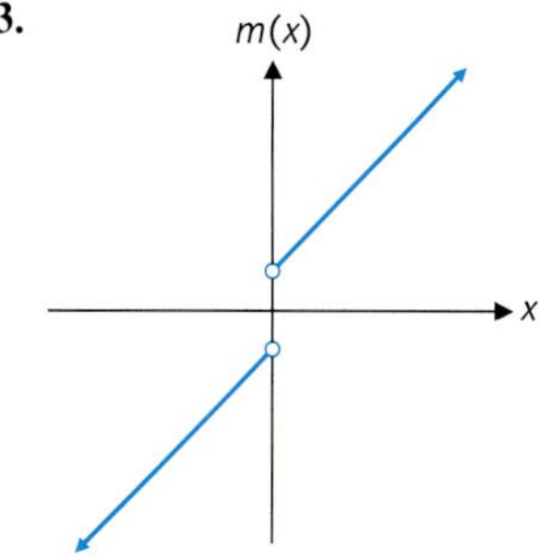

14.

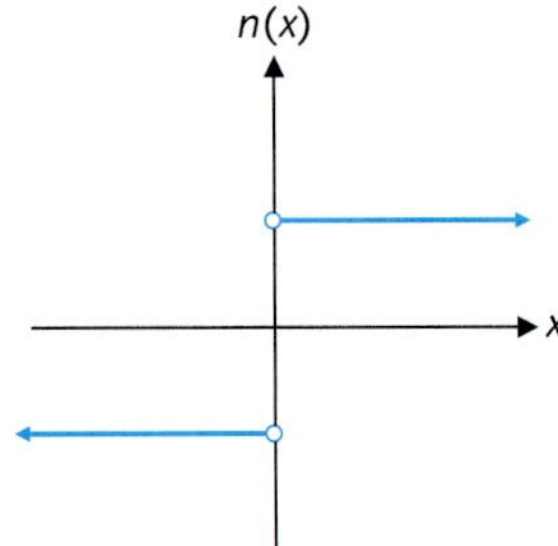

15.

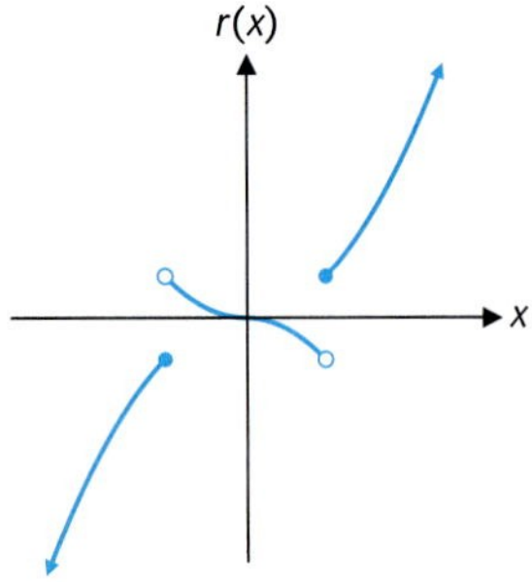

16.

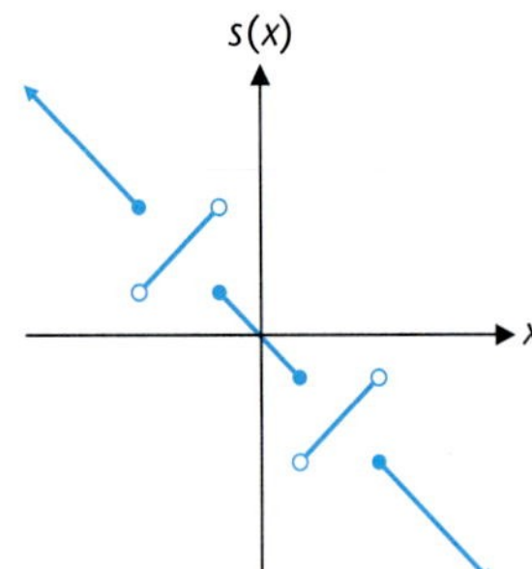

B

Which of the functions in Problems 17–22 are one-to-one?

17. $F(x) = \frac{1}{2}x + 2$ **18.** $G(x) = -\frac{1}{3}x + 1$

19. $H(x) = 4 - x^2$ **20.** $K(x) = \sqrt{4 - x}$

21. $M(x) = \sqrt{x + 1}$ **22.** $N(x) = x^2 - 1$

Problems 23–30 require the use of a graphing utility. Graph each function, and use the graph to determine if the function is one-to-one.

23. $f(x) = \dfrac{x^2 + |x|}{x}$ **24.** $f(x) = \dfrac{x^2 - |x|}{x}$

25. $f(x) = \dfrac{x^3 + |x|}{x}$ **26.** $f(x) = \dfrac{|x|^3 + |x|}{x}$

27. $f(x) = \dfrac{x^2 - 4}{|x - 2|}$ **28.** $f(x) = \dfrac{1 - x^2}{|x + 1|}$

29. $f(x) = \dfrac{x^3 - 9x}{|x^2 - 9|}$ **30.** $f(x) = \dfrac{4x - x^3}{|x^2 - 4|}$

In Problems 31–34, use the graph of the one-to-one function f to sketch the graph of f^{-1}. State the domain and range of f^{-1}.

31.

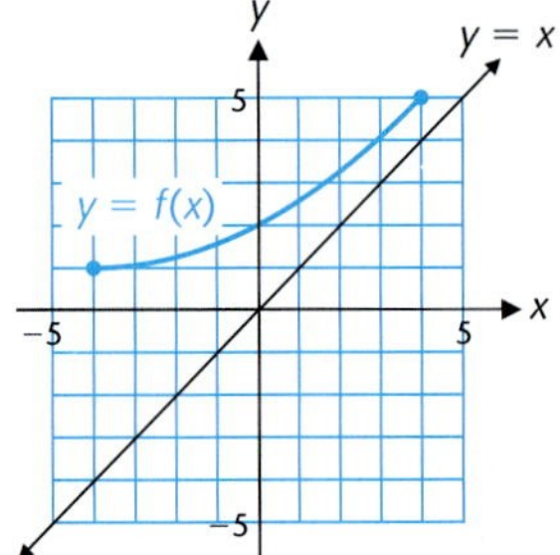

32.

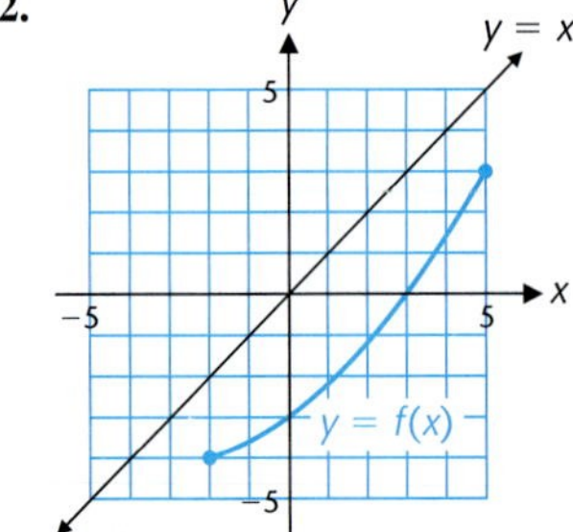

33.

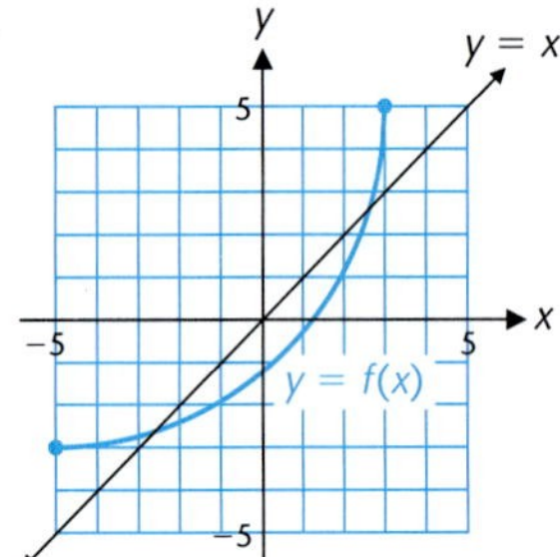

34.

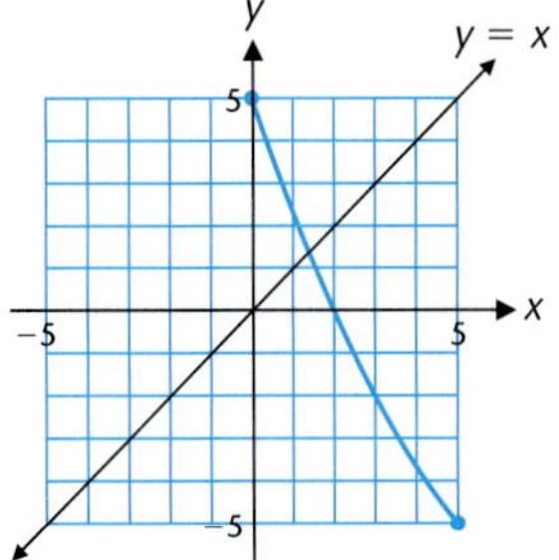

In Problems 35–40, verify that g is the inverse of the one-to-one function f by showing that $g[f(x)] = x$ and $f[g(x)] = x$. Sketch the graphs of f, g, and the line $y = x$ in the same coordinate system.

Check your graphs in Problems 35–40 by graphing f, g, and the line y = x in a squared viewing window on a graphing utility.

35. $f(x) = 3x + 6; g(x) = \frac{1}{3}x - 2$

36. $f(x) = -\frac{1}{2}x + 2; g(x) = -2x + 4$

37. $f(x) = 4 + x^2, x \geq 0; g(x) = \sqrt{x - 4}$

38. $f(x) = \sqrt{x + 2}; g(x) = x^2 - 2, x \geq 0$

39. $f(x) = -\sqrt{x - 2}; g(x) = x^2 + 2, x \leq 0$

40. $f(x) = 6 - x^2, x \leq 0; g(x) = -\sqrt{6 - x}$

The functions in Problems 41–60 are one-to-one. Find f^{-1}.

41. $f(x) = \frac{1}{5}x$

42. $f(x) = 4x$

43. $f(x) = 2x + 7$

44. $f(x) = 0.25x + 2.25$

45. $f(x) = 0.2x + 0.4$

46. $f(x) = 7 - 8x$

47. $f(x) = 3 - \dfrac{2}{x}$

48. $f(x) = 5 + \dfrac{4}{x}$

49. $f(x) = \dfrac{2x}{x + 1}$

50. $f(x) = \dfrac{4x}{2 - x}$

51. $f(x) = \dfrac{0.2x - 0.4}{0.1x + 0.5}$

52. $f(x) = \dfrac{x - 0.2}{x + 0.5}$

53. $f(x) = 8x^3 - 5$

54. $f(x) = 2x^5 + 9$

55. $f(x) = 2 + \sqrt[5]{3x - 7}$

56. $f(x) = -1 + \sqrt[3]{4 - 5x}$

57. $f(x) = 2\sqrt{9 - x}$

58. $f(x) = 3\sqrt{x - 4}$

59. $f(x) = 2 + \sqrt{3 - x}$

60. $f(x) = 4 - \sqrt{x + 5}$

61. How are the x and y intercepts of a function and its inverse related?

62. Does a constant function have an inverse? Explain.

C

The functions in Problems 63–66 are one-to-one. Find f^{-1}.

63. $f(x) = (x - 1)^2 + 2, x \geq 1$

64. $f(x) = 3 - (x - 5)^2, x \leq 5$

65. $f(x) = x^2 + 2x - 2, x \leq -1$

66. $f(x) = x^2 + 8x + 7, x \geq -4$

The graph of each function in Problems 67–70 is one-quarter of the graph of the circle with radius 3 and center (0, 0). Find f^{-1}, find the domain and range of f^{-1}, and sketch the graphs of f and f^{-1} in the same coordinate system.

67. $f(x) = -\sqrt{9 - x^2}, 0 \leq x \leq 3$

68. $f(x) = \sqrt{9 - x^2}, 0 \leq x \leq 3$

69. $f(x) = \sqrt{9 - x^2}, -3 \leq x \leq 0$

70. $f(x) = -\sqrt{9 - x^2}, -3 \leq x \leq 0$

The graph of each function in Problems 71–74 is one-quarter of the graph of the circle with radius 1 and center (0, 1). Find f^{-1}, find the domain and range of f^{-1}, and sketch the graphs of f and f^{-1} in the same coordinate system.

71. $f(x) = 1 + \sqrt{1 - x^2}, 0 \leq x \leq 1$

72. $f(x) = 1 - \sqrt{1 - x^2}, 0 \leq x \leq 1$

73. $f(x) = 1 - \sqrt{1 - x^2}, -1 \leq x \leq 0$

74. $f(x) = 1 + \sqrt{1 - x^2}, -1 \leq x \leq 0$

75. Find $f^{-1}(x)$ for $f(x) = ax + b, a \neq 0$.

76. Find $f^{-1}(x)$ for $f(x) = \sqrt{a^2 - x^2}, a > 0, 0 \leq x \leq a$.

77. Refer to Problem 75. For which a and b is f its own inverse?

78. How could you recognize the graph of a function that is its own inverse?

79. Show that the line through the points (a, b) and (b, a), $a \neq b$, is perpendicular to the line $y = x$ (see the figure).

80. Show that the point $((a + b)/2, (a + b)/2)$ bisects the line segment from (a, b) to (b, a), $a \neq b$ (see the figure).

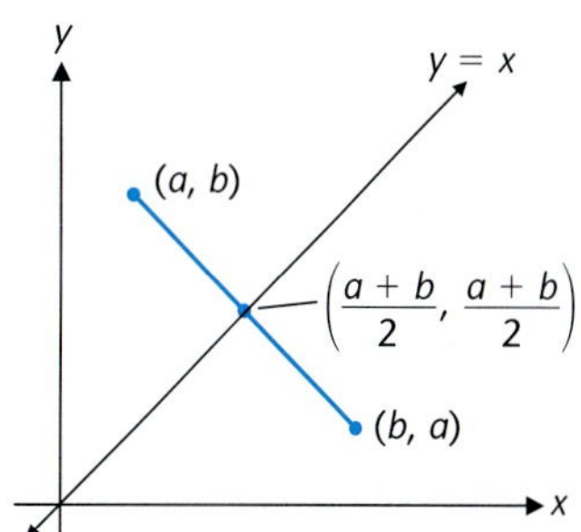

In Problems 81–84, the function f is not one-to-one. Find the inverses of the functions formed by restricting the domain of f as indicated.

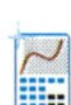

Check Problems 81–84 by graphing f, g, and the line y = x in a squared viewing window on a graphing utility. [Hint: To restrict the graph of y = f(x) to an interval of the form $a \leq x \leq b$, enter y = f(x)/((a ≤ x)(x ≤ b)).]*

81. $f(x) = (2 - x)^2$:
(A) $x \leq 2$ (B) $x \geq 2$

82. $f(x) = (1 + x)^2$:
(A) $x \leq -1$ (B) $x \geq -1$

83. $f(x) = \sqrt{4x - x^2}$:
(A) $0 \leq x \leq 2$ (B) $2 \leq x \leq 4$

84. $f(x) = \sqrt{6x - x^2}$:
(A) $0 \leq x \leq 3$ (B) $3 \leq x \leq 6$

CHAPTER 2 GROUP ACTIVITY Mathematical Modeling in Business*

This group activity is concerned with analyzing a basic model for manufacturing and selling a product by using tables of data and linear regression to determine appropriate values for the constants a, b, m, and n in the following functions:

TABLE 1 Business Modeling Functions

Function	Definition	Interpretation
Price–demand	$p(x) = m - nx$	x is the number of items that can be sold at $\$p$ per item
Cost	$C(x) = a + bx$	Total cost of producing x items
Revenue	$R(x) = xp$	Total revenue from the sale of x items
	$= x(m - nx)$	
Profit	$P(x) = R(x) - C(x)$	Total profit from the sale of x items

A manufacturing company manufactures and sells mountain bikes. The management would like to have price–demand and cost functions for break-even and profit–loss analysis. Price–demand and cost functions may be established by collecting appropriate data at different levels of output, and then finding a model in the form of a basic elementary function (from our library of elementary functions) that "closely fits" the collected data. The financial department, using statistical techniques, arrived at the price–demand and cost data in Tables 2 and 3, where p is the wholesale price of a bike for a demand of x thousand bikes and C is the cost, in thousands of dollars, of producing and selling x thousand bikes.

TABLE 2 Price–Demand

x (thousand)	p($)
7	530
13	360
19	270
25	130

TABLE 3 Cost

x (thousand)	C (thousand $)
5	2,100
12	2,940
19	3,500
26	3,920

(A) Building a Mathematical Model for Price–Demand. Plot the data in Table 2 and observe that the relationship between p and x is almost linear. After observing a relationship between variables, analysts often try to model the relationship in terms of a basic function, from a portfolio of elementary functions, which "best fits" the data.

1. **Linear regression lines** are frequently used to model linear phenomena. This is a process of fitting to a set of data a straight line that minimizes the sum of the squares of the distances of all the points in the graph of the data to the line by using the **method of least squares.** Many graphing utilities have this routine built in. Read your user's manual for your particular graphing utility, and discuss among the members of the

*This group project may be done without the use of a graphing utility, but significant additional insight into mathematical modeling will be gained if one is available.

group how this is done. After obtaining the linear regression line for the data in Table 2, graph the line and the data in the same viewing window.

2. The linear regression line found in part 1 is a mathematical model for the price–demand function and is given by

$$p(x) = 666.5 - 21.5x \quad \text{Price–demand function}$$

Graph the data points from Table 2 and the price–demand function in the same rectangular coordinate system.

3. The linear regression line defines a linear price–demand function. Interpret the slope of the function. Discuss its domain and range. Using the mathematical model, determine the price for a demand of 10,000 bikes. For a demand of 20,000 bikes.

(B) Building a Mathematical Model for Cost. Plot the data in Table 3 in a rectangular coordinate system. Which type of function appears to best fit the data?

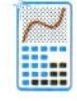

1. Fit a linear regression line to the data in Table 3. Then plot the data points and the line in the same viewing window.

2. The linear regression line found in part 1 is a mathematical model for the cost function and is given by

$$C(x) = 86x + 1{,}782 \quad \text{Cost function}$$

Graph the data points from Table 3 and the cost function in the same rectangular coordinate system.

3. Interpret the slope and the y intercept of the cost function. Discuss its domain and range. Using the mathematical model, determine the cost for an output and sales of 10,000 bikes. For an output and sales of 20,000 bikes.

(C) Break-Even and Profit–Loss Analysis. Write an equation for the revenue function and state its domain. Write the equation for the profit function and state its domain.

1. Graph the revenue function and the cost function simultaneously in the same rectangular coordinate system. Algebraically determine at what outputs (to the nearest unit) the company breaks even. Determine where costs exceed revenues and revenues exceed costs.

2. Graph the revenue function and the cost function simultaneously in the same viewing window. Graphically determine at what outputs (to the nearest unit) the company breaks even and where costs exceed revenues and revenues exceed costs.

3. Graph the profit function in a rectangular coordinate system. Algebraically determine at what outputs (to the nearest unit) the company breaks even. Determine where profits occur and where losses occur. At what output and price will a maximum profit occur? Do the maximum revenue and maximum profit occur for the same output? Discuss.

4. Graph the profit function in a graphing utility. Graphically determine at what outputs (to the nearest unit) the company breaks even and where losses occur and profits occur. At what output and price will a maximum profit occur? Do the maximum revenue and maximum profit occur for the same output? Discuss.

Chapter 2 Review

2-1 BASIC TOOLS; CIRCLES

A **Cartesian** or **rectangular coordinate system** is formed by the intersection of a horizontal real number line and a vertical real number line at their origins. These lines are called the **coordinate axes.** The **horizontal axis** is often referred to as the ***x* axis** and the **vertical axis** as the ***y* axis.** These axes divide the plane into four **quadrants.** Each point in the plane corresponds to its **coordinates**—an ordered pair (a, b) determined by passing horizontal and vertical lines through the point. The **abscissa** or ***x* coordinate** a is the coordinate of the intersection of the vertical line with the horizontal axis, and the **ordinate** or ***y* coordinate** b is the coordinate of the intersection of the horizontal line with the vertical axis. The point $(0, 0)$ is called the **origin.** The **solution set** of an equation in two variables is the set of all ordered pairs of real numbers that make the equation a true statement. The **graph of an equation in two variables** is the graph of its solution set formed using **point-by-point plotting** or with the aid of a **graphing utility.**

The **distance between the two points** $P_1(x_1, y_1)$ and $P_2(x_2, y_2)$ is

$$d(P_1, P_2) = \sqrt{(x_2 - x_1)^2 + (y_2 - y_1)^2}$$

The **standard equations for a circle** are

$$(x - h)^2 + (y - k)^2 = r^2 \qquad \text{Radius: } r > 0, \quad \text{Center: } (h, k)$$

$$x^2 + y^2 = r^2 \qquad \text{Radius: } r > 0, \quad \text{Center: } (0, 0)$$

Using a **squared window** will improve the appearance of a circle on a graphing utility.

2-2 STRAIGHT LINES

The **standard form** for the equation of a line is $Ax + By = C$, where A, B, and C are constants, A and B not both 0. The ***y* intercept** is the ordinate of the point where the graph crosses the y axis, and the ***x* intercept** is the abscissa of the point where the graph crosses the x axis. The **slope** of the line through the points (x_1, y_1) and (x_2, y_2) is

$$m = \frac{y_2 - y_1}{x_2 - x_1} \qquad \text{if } x_1 \neq x_2$$

The slope is not defined for a vertical line where $x_1 = x_2$. Two lines with slopes m_1 and m_2 are **parallel** if and only if $m_1 = m_2$ and **perpendicular** if and only if $m_1 m_2 = -1$.

Equations of a Line

Standard form	$Ax + By = C$	A and B not both 0
Slope–intercept form	$y = mx + b$	Slope: m; y intercept: b
Point–slope form	$y - y_1 = m(x - x_1)$	Slope: m; Point: (x_1, y_1)
Horizontal line	$y = b$	Slope: 0
Vertical line	$x = a$	Slope: Undefined

2-3 FUNCTIONS

A **function** is a **rule** that produces a correspondence between two sets of elements such that to each element in the first set there corresponds one and only one element in the second set. The first set is called the **domain** and the set of all corresponding elements in the second set is called the **range.** Equivalently, a **function** is a **set of ordered pairs** with the property that no two ordered pairs have the same first component and different second components. The **domain** is the set of all first components, and the **range** is the set of all second components. An **equation** in two variables **defines a function** if to each value of the **independent variable,** the placeholder for domain values, there corresponds exactly one value of the **dependent variable,** the placeholder for range values. A **vertical line** will intersect the graph of a function in at most one point. Unless otherwise specified, the **domain of a function defined by an equation** is assumed to be the set of all real number replacements for the independent variable that produce real values for the dependent variable. The symbol $\boldsymbol{f(x)}$ represents the real number in the range of the function f corresponding to the domain value x. Equivalently, the ordered pair $(x, f(x))$ belongs to the function f.

2-4 GRAPHING FUNCTIONS

The **graph of a function** $\boldsymbol{f}$ is the graph of the equation $y = f(x)$. The abscissa of any point where the graph of a function f crosses the x axis is called an ***x* intercept** of f. The ordinate of a point where the graph crosses the y axis is called the ***y* intercept.**

Let I be an interval in the domain of a function f. Then:

1. **f is increasing** on I if $f(b) > f(a)$ whenever $b > a$ in I.
2. **f is decreasing** on I if $f(b) < f(a)$ whenever $b > a$ in I.
3. **f is constant** on I if $f(a) = f(b)$ for all a and b in I.

A function f is a **linear function** if $f(x) = mx + b$, $m \neq 0$, and a **quadratic function** if $f(x) = ax^2 + bx + c$, $a \neq 0$. The graph of a linear function is a straight line that is neither horizontal nor vertical.

Properties of $f(x) = ax^2 + bx + c = a(x - h)^2 + k$, $a \neq 0$, and its graph:

1. The graph of f is a parabola:

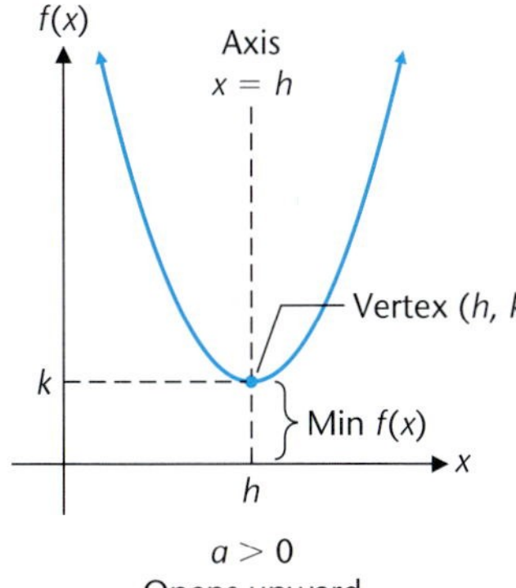

$a > 0$
Opens upward

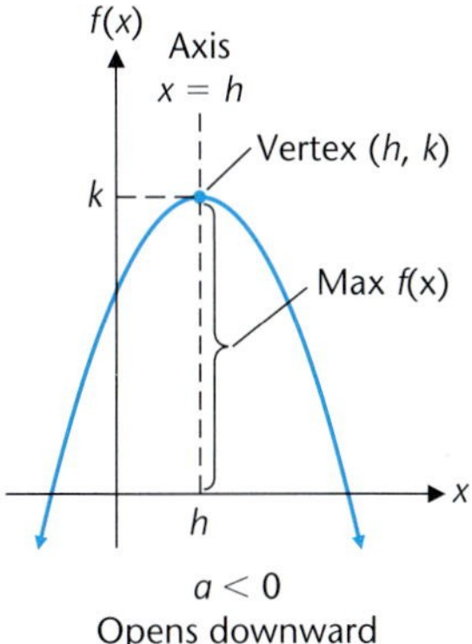

$a < 0$
Opens downward

2. Vertex: (h, k) (Parabola increases on one side of the vertex and decreases on the other.)
3. Axis (of symmetry): $x = h$ (Parallel to y axis)
4. $f(h) = k$ is the minimum if $a > 0$ and the maximum if $a < 0$.
5. Domain: All real numbers

   Range: $(-\infty, k]$ if $a < 0$ or $[k, \infty)$ if $a > 0$

A **piecewise-defined function** is a function whose definition involves more than one formula. The graph of a function is **continuous** if it has no holes or breaks and **discontinuous** at any point where it has a hole or break. The **greatest integer function** is defined by

$$f(x) = [\![x]\!] = n \qquad \text{where } n \text{ is an integer, } n \leq x < n + 1$$

2-5 COMBINING FUNCTIONS

The **sum, difference, product,** and **quotient** of the functions f and g are defined by

$$(f + g)(x) = f(x) + g(x) \qquad (f - g)(x) = f(x) - g(x)$$

$$(fg)(x) = f(x)g(x) \qquad \left(\frac{f}{g}\right)(x) = \frac{f(x)}{g(x)} \qquad g(x) \neq 0$$

The **domain** of each function is the intersection of the domains of f and g, with the exception that values of x where $g(x) = 0$ must be excluded from the domain of f/g.

The **composition** of functions f and g is defined by $(f \circ g)(x) = f[g(x)]$. The **domain** of $f \circ g$ is the set of all real numbers x in the domain of g where $g(x)$ is in the domain of f. The domain of $f \circ g$ is always a subset of the domain of g.

Vertical Translation:

$y = f(x) + k$, $k > 0$	Shift the graph of $y = f(x)$ upward k units.
$y = f(x) - k$, $k > 0$	Shift the graph of $y = f(x)$ downward k units.

Horizontal Translation:

$y = f(x - h)$, $h > 0$	Shift the graph of $y = f(x)$ to the right h units.
$y = f(x + h)$, $h > 0$	Shift the graph of $y = f(x)$ to the left h units.

Reflection:

$y = -f(x)$	Reflect the graph of $y = f(x)$ in the x axis.

Expansion and Contraction:

$y = Cf(x)$, $C > 1$	Expand the graph of $y = f(x)$ by multiplying each ordinate value by C.
$y = Cf(x)$, $0 < C < 1$	Contract the graph of $y = f(x)$ by multiplying each ordinate value by C.

2-6 INVERSE FUNCTIONS

A function is **one-to-one** if no two ordered pairs in the function have the same second component and different first components. A **horizontal line** will intersect the graph of a one-to-one function in at most one point. A function that is increasing (decreasing) throughout its domain is one-to-one. The **inverse** of the one-to-one function f is the function f^{-1} formed by reversing all the ordered pairs in f. If f is not one-to-one, then f^{-1} **does not exist.**

Assuming that f^{-1} exists, then:

1. f^{-1} is one-to-one.
2. Domain of f^{-1} = Range of f.

3. Range of f^{-1} = Domain of f.
4. $x = f^{-1}(y)$ if and only if $y = f(x)$.
5. $f^{-1}[f(x)] = x$ for all x in the domain of f.
6. $f[f^{-1}(x)] = x$ for all x in the domain of f^{-1}.
7. To find f^{-1}, solve the equation $y = f(x)$ for x and then interchange x and y.
8. The graphs of $y = f(x)$ and $y = f^{-1}(x)$ are symmetric with respect to the line $y = x$.

Chapter 2 Review Exercise

Work through all the problems in this chapter review and check answers in the back of the book. Answers to all review problems are there, and following each answer is a number in italics indicating the section in which that type of problem is discussed. Where weaknesses show up, review appropriate sections in the text.

A

1. Given the points $A(-2, 3)$ and $B(4, 0)$, find:
(A) Distance between A and B
(B) Slope of the line through A and B
(C) Slope of a line perpendicular to the line through A and B

2. Write the equation of a circle with radius $\sqrt{7}$ and center:
(A) $(0, 0)$ (B) $(3, -2)$

3. Find the center and radius of the circle given by

$$(x + 3)^2 + (y - 2)^2 = 5$$

4. Graph $3x + 2y = 9$ and indicate its slope.

5. Write an equation of a line with x intercept 6 and y intercept 4. Write the final answer in the standard form $Ax + By = C$, where A, B, and C are integers.

6. Write the slope–intercept form of the equation of the line with slope $-\frac{2}{3}$ and y intercept 2.

7. Write the equations of the vertical and horizontal lines passing through the point $(-3, 4)$. What is the slope of each?

8. Indicate whether each set defines a function. Find the domain and range of each function.
(A) $\{(1, 1), (2, 4), (3, 9)\}$
(B) $\{(1, 1), (1, -1), (2, 2), (2, -2)\}$
(C) $\{(-2, 2), (-1, 2), (0, 2), (1, 2), (2, 2)\}$

9. Indicate whether each graph specifies a function:

(A)
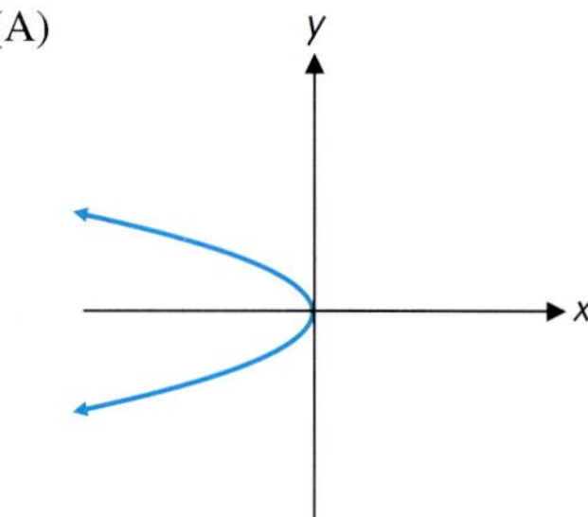

(B)
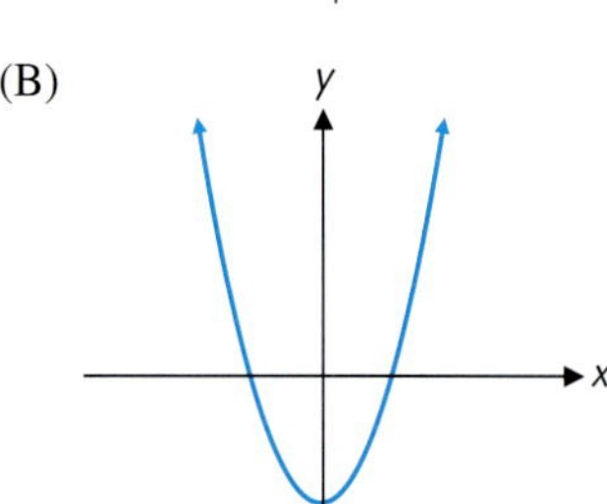

(C)
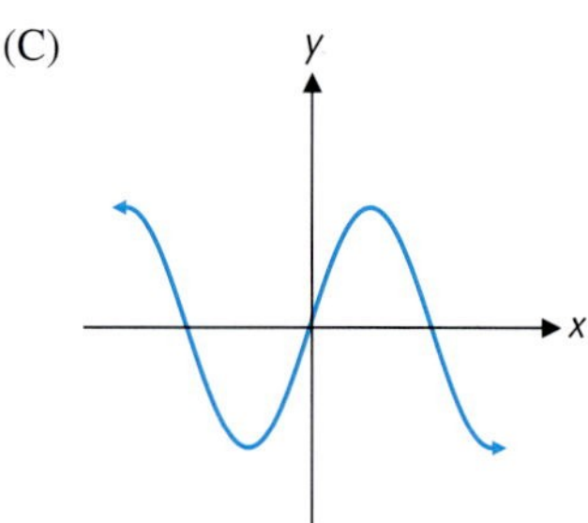

(D)
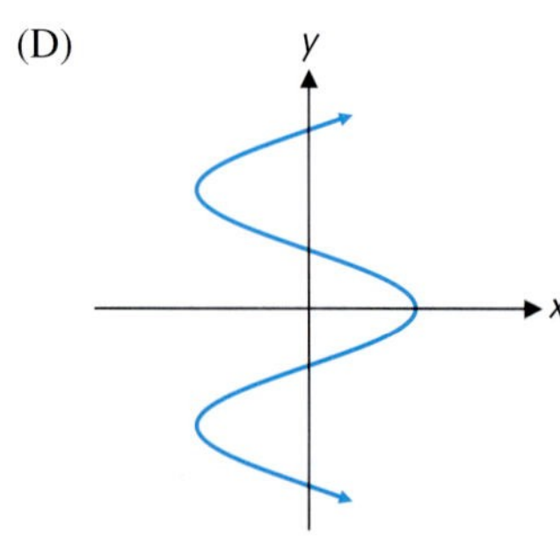

10. Which of the following equations define functions?
(A) $y = x$ (B) $y^2 = x$
(C) $y^3 = x$ (D) $|y| = x$

Problems 11–20 refer to the functions f, g, k, and m given by:

$$f(x) = 3x + 5 \qquad g(x) = 4 - x^2$$
$$k(x) = 5$$
$$m(x) = 2|x| - 1$$

Find the indicated quantities or expressions.

11. $f(2) + g(-2) + k(0)$
12. $\dfrac{m(-2) + 1}{g(2) + 4}$

13. $\dfrac{f(2 + h) - f(2)}{h}$
14. $\dfrac{g(a + h) - g(a)}{h}$

15. $(f + g)(x)$
16. $(f - g)(x)$

17. $(fg)(x)$
18. $\left(\dfrac{f}{g}\right)(x)$

19. $(f \circ g)(x)$
20. $(g \circ f)(x)$

21. Sketch a graph of each of the functions in parts A–D using the graph of function f in the figure.
(A) $y = -f(x)$ (B) $y = f(x) + 4$
(C) $y = f(x - 2)$ (D) $y = -f(x + 3) - 3$

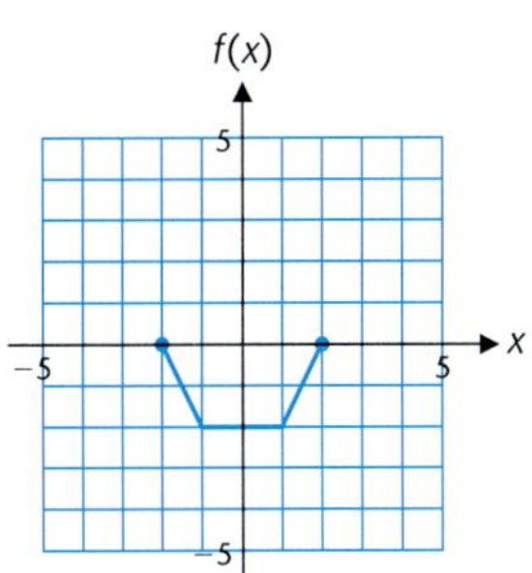

22. Match each equation with a graph of one of the functions f, g, m, or n in the figure. Each graph is a graph of one of the equations and is assumed to continue without bound beyond the portion shown.
(A) $y = (x - 2)^2 - 4$ (B) $y = -(x + 2)^2 + 4$
(C) $y = -(x - 2)^2 + 4$ (D) $y = (x + 2)^2 - 4$

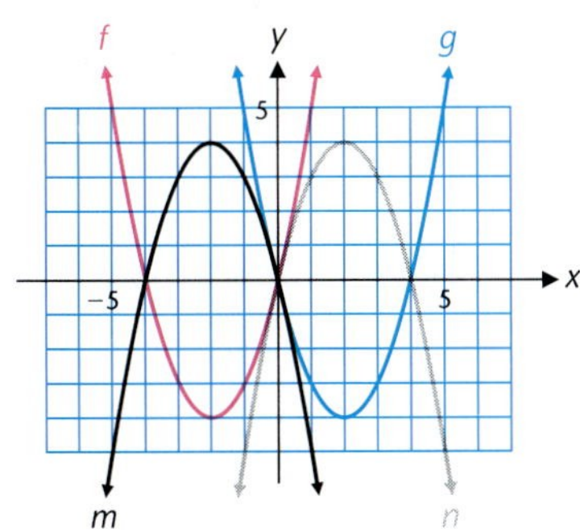

23. Referring to the graph of function f in the figure for Problem 22 and using known properties of quadratic functions, find each of the following to the nearest integer:
(A) Intercepts (B) Vertex
(C) Maximum or minimum (D) Range
(E) Increasing interval (F) Decreasing interval

24. Find the maximum or minimum value of $f(x) = x^2 - 6x + 11$ without graphing. What are the coordinates of the vertex of the graph?

25. How are the graphs of the following related to the graph of $y = x^2$?
(A) $y = -x^2$ (B) $y = x^2 - 3$
(C) $y = (x + 3)^2$

B

Problems 26–32 refer to the function q given by the following graph. (Assume the graph continues as indicated beyond the part shown.)

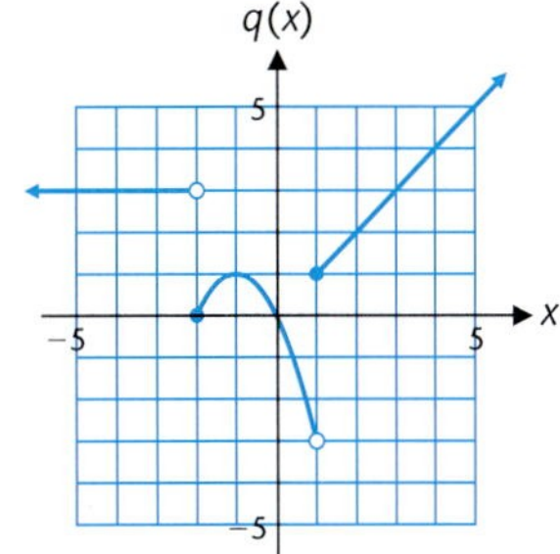

26. Find y to the nearest integer:
(A) $y = q(0)$ (B) $y = q(1)$
(C) $y = q(2)$ (D) $y = q(-2)$

27. Find x to the nearest integer:
(A) $q(x) = 0$ (B) $q(x) = 1$
(C) $q(x) = -3$ (D) $q(x) = 3$

28. Find the domain and range of q.

29. Find the intervals over which q is increasing.

30. Find the intervals over which q is decreasing.

31. Find the intervals over which q is constant.

32. Identify any points of discontinuity.

33. The function f multiplies the cube of the domain element by 4 and then subtracts the square root of the domain element. Write an algebraic definition of f.

34. Write a verbal description of the function $f(x) = 3x^2 + 4x - 6$.

35. (A) Find an equation of the line through $P(-4, 3)$ and $Q(0, -3)$. Write the final answer in the standard form $Ax + By = C$, where A, B, and C are integers with $A > 0$.
(B) Find $d(P, Q)$.

36. Write equations of the lines
(A) Parallel to (B) Perpendicular to
the line $6x + 3y = 5$ and passing through the point $(-2, 1)$. Write the final answers in the slope–intercept form $y = mx + b$.

37. Discuss the graph of $4x^2 + 9y^2 = 36$ relative to symmetry with respect to the x axis, y axis and origin.

38. Find the domain of $g(x) = 1/\sqrt{3 - x}$

39. Graph $f(x) = x^2 - 6x + 5$. Show the axis of symmetry and vertex, and find the range, intercepts, and maximum or minimum value of $f(x)$.

40. Find the domain of $h(x) = 1/(4 - \sqrt{x})$.

41. Given $f(x) = \sqrt{x} - 8$ and $g(x) = |x|$:
(A) Find $f \circ g$ and $g \circ f$.
(B) Find the domains of $f \circ g$ and $g \circ f$.

42. Which of the following functions are one-to-one?
(A) $f(x) = x^3$
(B) $g(x) = (x - 2)^2$
(C) $h(x) = 2x - 3$
(D) $F(x) = (x + 3)^2, x \geq -3$

43. Given $f(x) = 3x - 7$:
(A) Find $f^{-1}(x)$.
(B) Find $f^{-1}(5)$.
(C) Find $f^{-1}[f(x)]$.
(D) Is f increasing, decreasing, or constant on $(-\infty, \infty)$?

 Check by graphing f, f^{-1}, and $y = x$ in a squared viewing window on a graphing utility.

44. Graph, finding the domain, range, and any points of discontinuity:

$$f(x) = \begin{cases} 2 - x & \text{if } -1 \leq x < 0 \\ x^2 & \text{if } 0 \leq x \leq 1 \end{cases}$$

45. The following graph is the result of applying a sequence of transformations to the graph of $y = x^2$. Describe the transformations verbally and write an equation for the given graph.

 Check by graphing your equation on a graphing utility.

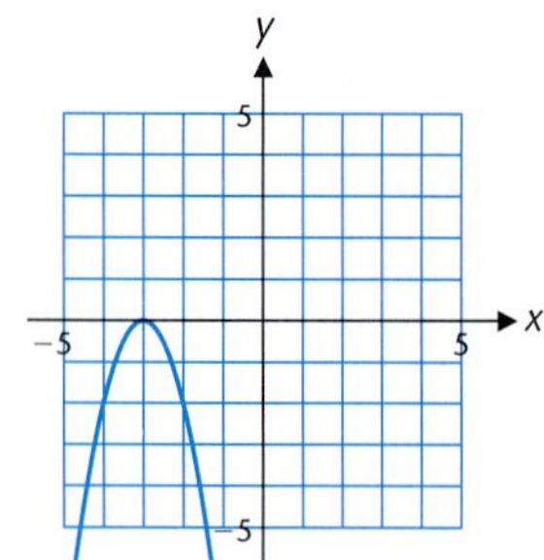

46. The graph of $f(x) = |x|$ is expanded by a factor of 3, reflected in the x axis, and shifted 2 units to the right and 5 units up to form the graph of the function g. Find an equation for the function g and graph g.

47. Write an equation for the following graph in the form $y = a(x - h)^2 + k$, where a is either -1 or $+1$ and h and k are integers.

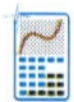 Check by graphing your equation on a graphing utility.

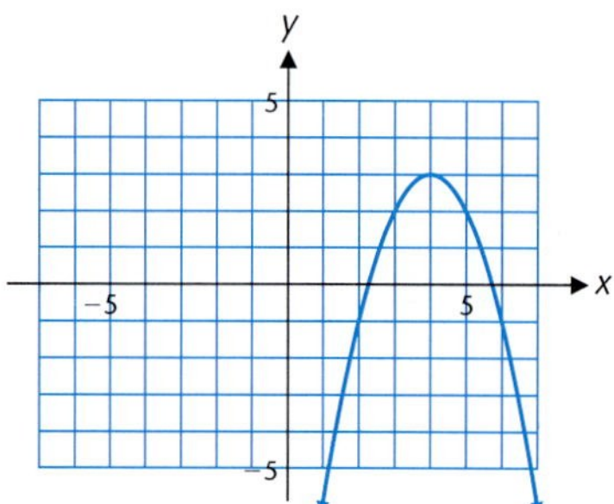

48. Graph:
(A) $y = |x| - 2$ (B) $y = |x + 1|$
(C) $y = \frac{1}{2}|x|$

49. Given $f(x) = \sqrt{x - 1}$:
(A) Find $f^{-1}(x)$.
(B) Find the domain and range of f and f^{-1}.
(C) Graph f, f^{-1}, and $y = x$ on the same coordinate system.

 Check by graphing f, f^{-1}, and $y = x$ in a squared window on a graphing utility.

50. Find the equation of a circle that passes through the point $(-1, 4)$ with center at $(3, 0)$.

51. Find the center and radius of the circle given by $x^2 + y^2 + 4x - 6y = 3$.

52. Determine symmetry with respect to the x axis, y axis, and origin and graph $xy = 4$.

53. If the slope of a line is negative, is the linear function represented by the graph of the line increasing, decreasing, or constant on $(-\infty, \infty)$?

54. Given $f(x) = x^2 - 1, x \geq 0$:
(A) Find the domain and range of f and f^{-1}.
(B) Find $f^{-1}(x)$.
(C) Find $f^{-1}(3)$.
(D) Find $f^{-1}[f(4)]$.
(E) Find $f^{-1}[f(x)]$.

 Check by graphing f, f^{-1}, and $y = x$ in a squared window on a graphing utility.

C

55. The graph at the top of next page is the result of applying a sequence of transformations to the graph of $y = \sqrt[3]{x}$. Describe the transformations verbally, and write an equation for the given graph.

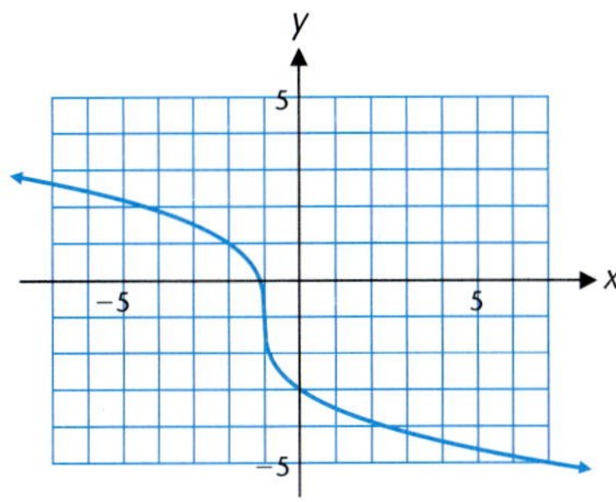

Check by graphing your equation on a graphing utility.

56. How is the graph of $f(x) = -(x - 2)^2 - 1$ related to the graph of $g(x) = x^2$?

57. Graph: $f(x) = -|x + 1| - 1$

58. Find the domain of $f(x) = \sqrt{25 - x^2}$.

59. Given $f(x) = x^2$ and $g(x) = \sqrt{1 - x}$, find each function and its domain.
(A) fg (B) f/g (C) $f \circ g$ (D) $g \circ f$

60. For the one-to-one function f given by

$$f(x) = \frac{x + 2}{x - 3}$$

(A) Find $f^{-1}(x)$.
(B) Find $f^{-1}(3)$.
(C) Find $f^{-1}[f(x)]$.

61. Find a piecewise definition of $f(x) = |x + 1| - |x - 1|$ that does not involve the absolute value function. Find the domain and range of f.

62. Find the equation of the set of points equidistant from (3, 3) and (6, 0). What is the name of the geometric figure formed by this set?

63. Prove that two nonvertical lines are parallel if and only if their slopes are the same.

64. Prove that the lines $mx - y = b$ and $x + my = c$ are perpendicular.

65. Graph:

(A) $f(x) = [\![|x|]\!]$ (B) $g(x) = |[\![x]\!]|$

66. Graph in a standard viewing window:

$$f(x) = 0.1(x - 2)^2 + \frac{|3x - 6|}{x - 2}$$

Assuming the graph continues as indicated beyond the part shown in this viewing window, find the domain, range, and any points of discontinuity. (Use the dot mode on your graphing utility, if it has one.)

67. A partial graph of the function f is shown in the figure. Complete the graph of f over the interval [0, 5] given that:
(A) f is symmetric with respect to the y axis.
(B) f is symmetric with respect to the origin.

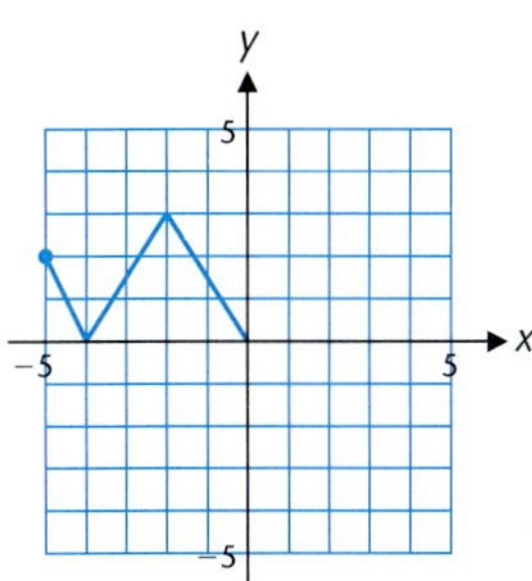

68. The function f is decreasing on $[-5, 5]$ with $f(-5) = 4$ and $f(5) = -3$.
(A) If f is continuous on $[-5, 5]$, how many times can the graph of f cross the x axis? Support your conclusion with examples and/or verbal arguments.
(B) Repeat part A if the function does not have to be continuous.

APPLICATIONS

69. Linear Depreciation. A computer system was purchased by a small company for \$12,000 and is assumed to have a depreciated value of \$2,000 after 8 years. If the value is depreciated linearly from \$12,000 to \$2,000:
(A) Find the linear equation that relates value V (in dollars) to time t (in years).
(B) What would be the depreciated value of the system after 5 years?

70. Business—Pricing. A sporting goods store sells tennis shorts that cost \$30 for \$48 and sunglasses that cost \$20 for \$32.
(A) If the markup policy of the store for items that cost over \$10 is assumed to be linear and is reflected in the pricing of these two items, write an equation that expresses retail price R as a function of cost C.
(B) What should be the retail price of a pair of skis that cost \$105?

★ **71. Income.** A salesperson receives a base salary of \$200 per week and a commission of 10% on all sales over \$3,000 during the week. If x represents the salesperson's weekly sales, express the total weekly earnings $E(x)$ as a function of x. Find $E(2{,}000)$ and $E(5{,}000)$.

72. Demand. Egg consumption has been decreasing for some time, presumably because of increasing awareness of the high cholesterol in egg yolks. Table 1 lists the annual per capita consumption of eggs in the United States (*source:* Department of Agriculture).

TABLE 1

Year	1970	1975	1980	1985	1990
Consumption	309	276	271	255	233

A mathematical model for this data is given by

$$f(x) = 303.4 - 3.46x$$

where $x = 0$ corresponds to 1970.

(A) Complete Table 2. Round values of $f(x)$ to the nearest integer.

TABLE 2

x	0	5	10	15	20
Consumption	309	276	271	255	233
$f(x)$					

(B) Graph $y = f(x)$ and the data in the table on the same set of axes.

(C) Use the modeling function f to estimate the per capita egg consumption in 1995. In 2000.

(D) Based on the information in the table, write a brief verbal description of egg consumption from 1970 to 1990.

73. Pricing. An office supply store sells ballpoint pens for \$0.49 each. For an order of 3 dozen or more pens, the price per pen for all pens ordered is reduced to \$0.44, and for an order of 6 dozen or more, the price per pen for all pens ordered is reduced to \$0.39.

(A) If $C(x)$ is the total cost in dollars for an order of x pens, write a piecewise definition for C.

(B) Graph $y = C(x)$ for $0 \le x \le 108$, and identify any points of discontinuity.

74. Break-Even Analysis. A video production company is planning to produce an instructional videotape. The producer estimates that it will cost \$84,000 to shoot the video and \$15 per unit to copy and distribute the tape. The wholesale price of the tape is \$50 per unit.

(A) Write the cost equation and the revenue equation, and graph both simultaneously in the rectangular coordinate system.

(B) Determine when $R = C$; then, with the aid of part A, determine when $R < C$ and $R > C$.

★ **75. Market Research.** If x units of a product are produced each week and sold for a price of \$$p$ per unit, then the weekly demand, revenue, and cost equations are, respectively,

$$x = 500 - 10p$$
$$R(x) = 50x - \tfrac{1}{10}x^2$$
$$C(x) = 20x + 4{,}000$$

Express the weekly profit as a function of the price p.

★ **76. Architecture.** A circular arc forms the top of an entryway with 6-foot vertical sides 8 feet apart. If the top of the arc is 2 feet above the ends, what is the radius of the arc?

77. Construction. A farmer has 120 feet of fencing to be used in the construction of two identical rectangular pens sharing a common side (see the figure).

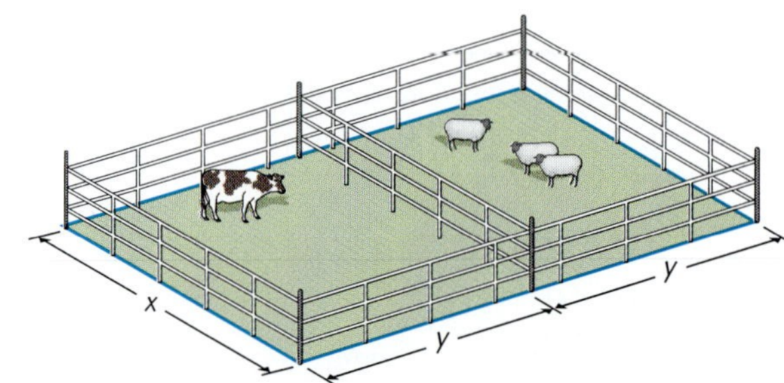

(A) Express the total area $A(x)$ enclosed by both pens as a function of the width x.

(B) From physical considerations, what is the domain of the function A?

(C) Find the dimensions of the pens that will make the total enclosed area maximum.

78. Computer Science. In computer programming, it is often necessary to check numbers for certain properties (even, odd, perfect square, etc.). The greatest integer function provides a convenient method for determining some of these properties. For example, the following function can be used to determine whether a number is the square of an integer:

$$f(x) = x - ([\![\sqrt{x}]\!])^2$$

(A) Find $f(1)$. (B) Find $f(2)$. (C) Find $f(3)$.
(D) Find $f(4)$. (E) Find $f(5)$.
(F) Find $f(n^2)$, where n is a positive integer.

Cumulative Review Exercise Chapters 1 and 2

Work through all the problems in this cumulative review and check answers in the back of the book. Answers to all review problems are there, and following each answer is a number in italics indicating the section in which that type of problem is discussed. Where weaknesses show up, review appropriate sections in the text.

A

1. Solve for x: $\frac{7x}{5} - \frac{3 + 2x}{2} = \frac{x - 10}{3} + 2$

2. Solve for x and y: $2x - 3y = 8$
$4x + y = 2$

Solve and graph Problems 3–5.

3. $2(3 - y) + 4 \le 5 - y$

4. $|x - 2| < 7$

5. $x^2 + 3x \ge 10$

6. Perform the indicated operations and write the answer in standard form:
(A) $(2 - 3i) - (-5 + 7i)$ (B) $(1 + 4i)(3 - 5i)$
(C) $\frac{5 + i}{2 + 3i}$

Solve Problems 7–10.

7. $3x^2 = -12x$

8. $4x^2 - 20 = 0$

9. $x^2 - 6x + 2 = 0$

10. $x - \sqrt{12 - x} = 0$

11. For what values of x does $\sqrt{2 + 3x}$ represent a real number?

12. Given the points $A(3, 2)$ and $B(5, 6)$, find:
(A) Distance between A and B
(B) Slope of the line through A and B
(C) Slope of a line perpendicular to the line through A and B

13. Find the equation of the circle with radius $\sqrt{2}$ and center:
(A) $(0, 0)$ (B) $(-3, 1)$

14. Graph $2x - 3y = 6$ and indicate its slope and intercepts.

15. Indicate whether each set defines a function. Find the domain and range of each function.
(A) $\{(1, 1), (2, 1), (3, 1)\}$
(B) $\{(1, 1), (1, 2), (1, 3)\}$
(C) $\{(-2, 2), (-1, -1), (0, 0), (1, -1), (2, 2)\}$

16. For $f(x) = x^2 - 2x + 5$ and $g(x) = 3x - 2$, find:
(A) $f(-2) + g(3)$ (B) $(f + g)(x)$
(C) $(f \circ g)(x)$
(D) $\frac{f(a + h) - f(a)}{h}$

17. How are the graphs of the following related to the graph of $y = |x|$?
(A) $y = 2|x|$ (B) $y = |x - 2|$ (C) $y = |x| - 2$

18. Sketch a graph of each of the functions in parts A and B using the graph of function f in the figure.
(A) $y = -f(x + 1)$ (B) $y = 2f(x) - 2$

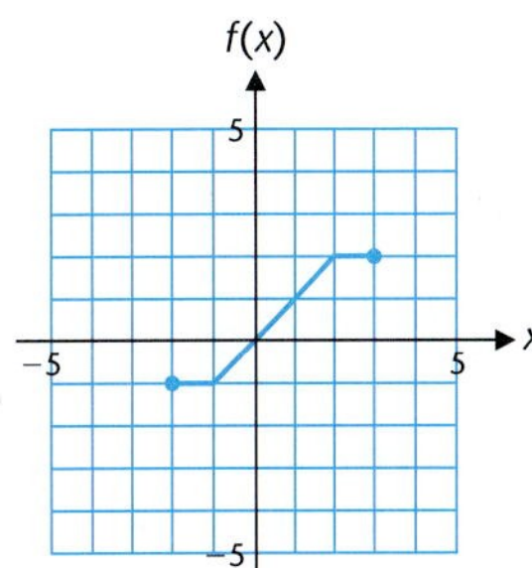

B

Solve Problems 19–22.

19. $\frac{x + 3}{2x + 2} + \frac{5x + 2}{3x + 3} = \frac{5}{6}$

20. $\frac{3}{x} = \frac{6}{x + 1} - \frac{1}{x - 1}$

21. $2x + 1 = 3\sqrt{2x - 1}$

22. $2x - 3y = 9$
$4x + 2y = 23$

Solve and graph Problems 23–25.

23. $|4x - 9| > 3$

24. $\sqrt{(3m - 4)^2} \le 2$

25. $\frac{2}{x + 1} \ge \frac{1}{x - 2}$

26. For what real values of x does the expression below represent a real number?

$$\frac{\sqrt{x - 2}}{x - 4}$$

27. Perform the indicated operations and write the final answers in standard form:
(A) $(2 - 3i)^2 - (4 - 5i)(2 - 3i) - (2 + 10i)$
(B) $\frac{3}{5} + \frac{4}{5}i + \frac{1}{\frac{3}{5} + \frac{4}{5}i}$ (C) i^{35}

28. Convert to $a + bi$ forms, perform the indicated operations, and write the final answers in standard form:
(A) $(5 + 2\sqrt{-9}) - (2 - 3\sqrt{-16})$
(B) $\frac{2 + 7\sqrt{-25}}{3 - \sqrt{-1}}$ (C) $\frac{12 - \sqrt{-64}}{\sqrt{-4}}$

29. Find each of the following for the function f given by the graph shown below.

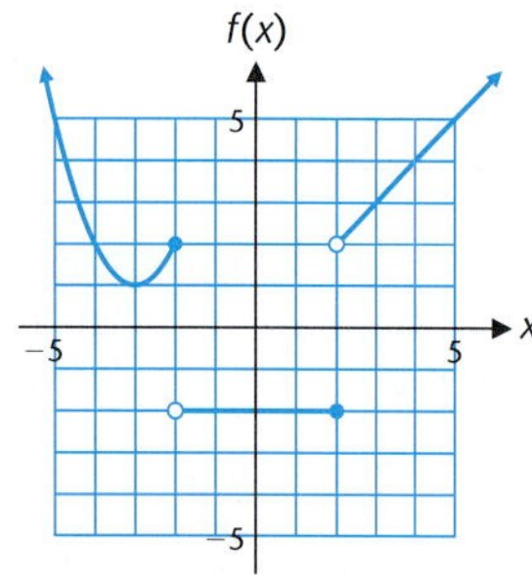

(A) The domain of f (B) The range of f

(C) $f(-3) + f(-2) + f(2)$

(D) The intervals over which f is increasing.

(E) The x coordinates of any points of discontinuity.

30. Write equations of the lines

(A) Parallel to (B) Perpendicular to

the line $3x + 2y = 12$ and passing through the point $(-6, 1)$. Write the final answers in the slope–intercept form $y = mx + b$.

31. Find the domain of $g(x) = \sqrt{x + 4}$.

32. Graph $f(x) = x^2 - 2x - 8$. Show the axis of symmetry and vertex, and find the range, intercepts, and maximum or minimum value of $f(x)$.

33. Given $f(x) = 1/(x - 2)$ and $g(x) = (x + 3)/x$, find $f \circ g$. What is the domain of $f \circ g$?

34. Find $f^{-1}(x)$ for $f(x) = 2x + 5$.

35. Graph, finding the domain, range, and any points of discontinuity:

$$f(x) = \begin{cases} x - 1 & \text{if } x < 0 \\ x^2 + 1 & \text{if } x \geq 0 \end{cases}$$

36. Graph:

(A) $y = 2\sqrt{x} + 1$ (B) $y = -\sqrt{x + 1}$

37. The graph in the figure is the result of applying a sequence of transformations to the graph of $y = |x|$. Describe the transformations verbally, and write an equation for the graph in the figure.

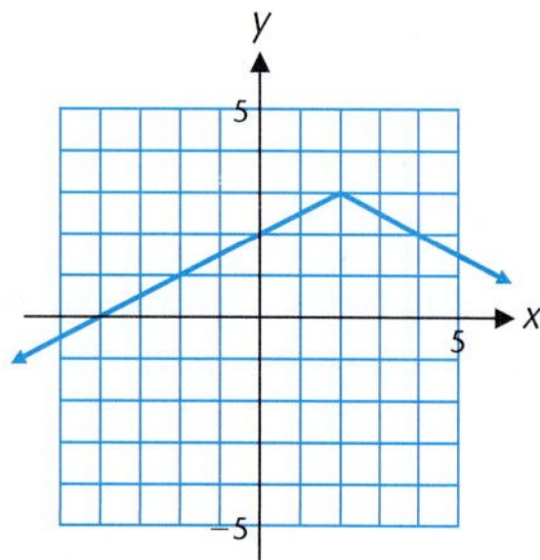

Check by graphing your equation on a graphing utility.

38. Let $f(x) = \sqrt{x + 4}$

(A) Find $f^{-1}(x)$.

(B) Find the domain and range of f and f^{-1}.

(C) Graph f, f^{-1}, and $y = x$ on the same coordinate system. Check by graphing f, f^{-1}, and $y = x$ in a squared window on a graphing utility.

39. Find the center and radius of the circle given by $x^2 - 6x + y^2 + 2y = 0$. Graph the circle and show the center and the radius.

40. Discuss symmetry with respect to the x axis, y axis, and the origin for the equation

$$xy + |xy| = 5$$

41. Write an equation for the graph in the figure in the form $y = a(x - h)^2 + k$, where a is either -1 or $+1$ and h and k are integers.

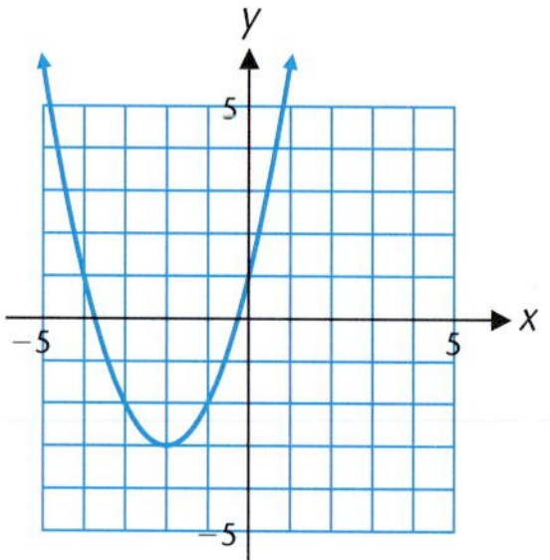

Check by graphing your equation on a graphing utility.

Solve Problems 42–45.

42. $1 + \dfrac{14}{y^2} = \dfrac{6}{y}$

43. $4x^{2/3} - 4x^{1/3} - 3 = 0$

44. $u^4 + u^2 - 12 = 0$

45. $\sqrt{8t - 2} - 2\sqrt{t} = 1$

Use a calculator to solve Problems 46–48. Compute answers to two decimal places.

46. $-3.45 < 1.86 - 0.33x \leq 7.92$

47. $2.35x^2 + 10.44x - 16.47 = 0$

48. $12.5x + 2.5y = 20$
$3.5x + 8.7y = 10$

49. Solve for y in terms of x:

$$\frac{x - 2}{x + 1} = \frac{2y + 1}{y - 2}$$

50. Solve for s and t in terms of x and y and check:

$$x = -1 + 5s + 2t$$
$$y = 2 + 2s + t$$

51. Graph $y = -2\sqrt{x + 1} + 3$.

C

52. Evaluate $x^2 - x + 2$ for $x = \frac{1}{2} - \frac{i}{2}\sqrt{7}$.

53. For what values of a and b is the inequality $a - b < b - a$ true?

54. Solve for y in terms of x:

$$\frac{x + y}{y - \frac{x + y}{x - y}} = 1$$

55. Find all roots: $3x^2 = 2\sqrt{2}x - 1$

56. Consider the quadratic equation

$$x^2 + bx + 1 = 0$$

where b is a real number. Discuss the relationship between the values of b and the three types of roots listed in Table 1 in Section 1-6.

57. Find all real roots: $1 = 6x^{-2} + 9x^{-4}$

58. Write in standard form: $\frac{a + bi}{a - bi}$, $a, b \neq 0$

59. Solve: $\left|\frac{x + 4}{x}\right| < 3$

60. Find a piecewise definition of $f(x) = |x + 2| + |x - 2|$ that does not involve the absolute value function. Graph f and find the domain and range.

61. Given $f(x) = x^2$ and $g(x) = \sqrt{4 - x^2}$, find:
(A) Domain of g (B) f/g and its domain
(C) $f \circ g$ and its domain

62. Let $f(x) = x^2 - 2x - 3$, $x \geq 1$.
(A) Find $f^{-1}(x)$.
(B) Find the domain and range of f^{-1}.
(C) Graph f, f^{-1}, and $y = x$ on the same coordinate system. Check by graphing f, f^{-1}, and $y = x$ in a squared window on a graphing utility.

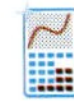

63. Write a piecewise definition for $f(x) = 2x - [\![2x]\!]$ and sketch the graph of f. Include sufficient intervals to clearly illustrate both the definition and the graph. Find the domain, range, and any points of discontinuity.

APPLICATIONS

64. Numbers. Find a number such that the number exceeds its reciprocal by $\frac{3}{2}$.

65. Rate–Time. An older conveyor belt working alone can fill a railroad car with ore in 14 minutes. The older belt and a newer belt operating together can fill the car in 6 minutes. How long will it take the newer belt working alone to fill the car with ore?

★**66. Rate–Time.** A boat travels upstream for 35 miles and then returns to its starting point. If the round trip took 4.8 hours and the boat's speed in still water is 15 miles per hour, what is the speed of the current?

★**67. Chemistry.** How many gallons of distilled water must be mixed with 24 gallons of a 90% sulfuric acid solution to obtain a 60% solution?

68. Break-Even Analysis. The publisher's fixed costs for the production of a new cookbook are \$41,800. Variable costs are \$4.90 per book. If the book is sold to bookstores for \$9.65, how many must be sold for the publisher to break even?

69. Finance. An investor instructs a broker to purchase a certain stock whenever the price per share p of the stock is within \$10 of \$200. Express this instruction as an absolute value inequality.

70. Supply and Demand. Suppose the supply and demand equations for Styrofoam "cheese heads" in Green Bay for a particular week are

$p = 5.5 + 0.002q$ Supply equation

$p = 22 - 0.001q$ Demand equation

where p is the price in dollars and q is the number of cheese heads. Find the equilibrium price and quantity.

71. Profit and Loss Analysis. At a price of \$$p$ per unit, the marketing department in a company estimates that the weekly cost C and the weekly revenue R, in thousands of dollars, will be given by the equations

$C = 88 - 12p$ Cost equation

$R = 15p - 2p^2$ Revenue equation

Find the prices for which the company has:
(A) A profit (B) A loss

★**72. Shipping.** A ship leaves port A, sails east to port B, and then north to port C, a total distance of 115 miles. The next day the ship sails directly from port C back to port A, a distance of 85 miles. Find the distance between ports A and B and between ports B and C.

73. Price and Demand. The weekly demand for mouthwash in a chain of drug stores is 1,160 bottles at a price of \$3.79 each. If the price is lowered to \$3.59, the weekly demand increases to 1,340 bottles. Assuming the relationship between the weekly demand x and the price per bottle p is linear, express x as a function of p. How many bottles would the store sell each week if the price were lowered to \$3.29?

74. Business–Pricing. A telephone company begins a new pricing plan that charges customers for local calls as follows: The first 60 calls each month are 6 cents each, the next 90 are 5 cents each, the next 150 are 4 cents each, and any additional calls are 3 cents each. If C is the cost, in dollars, of placing x calls per month, write a piecewise definition of C as a function of x and graph.

75. Construction. A homeowner has 80 feet of chain-link fencing to be used to construct a dog pen adjacent to a house (see the figure).

(A) Express the area $A(x)$ enclosed by the pen as a function of the width x.

(B) From physical considerations, what is the domain of the function A?

(C) Graph A and determine the dimensions of the pen that will make the area maximum.

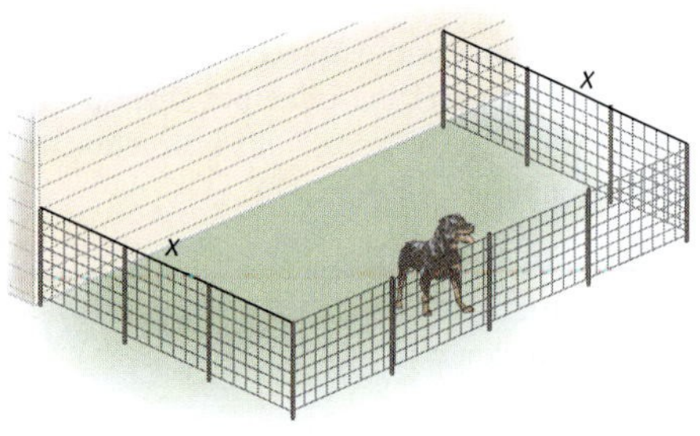

76. Computer Science. Let $f(x) = x - 2[\![x/2]\!]$. This function can be used to determine if an integer is odd or even.

(A) Find $f(1), f(2), f(3), f(4)$.

(B) Find $f(n)$ for any integer n. [*Hint:* Consider two cases, $n = 2k$ and $n = 2k + 1$, k an integer.]

★**77. Physics.** The distance s above the ground (in feet) of an object dropped from a hot-air balloon t seconds after it is released is given by

$$s = a + bt^2$$

where a and b are constants. Suppose the object is 2,100 feet above the ground 5 seconds after its release and 900 feet above the ground 10 seconds after its release.

(A) Find the constants a and b.

(B) How high is the balloon?

(C) How long does the object fall?

POLYNOMIAL AND RATIONAL FUNCTIONS

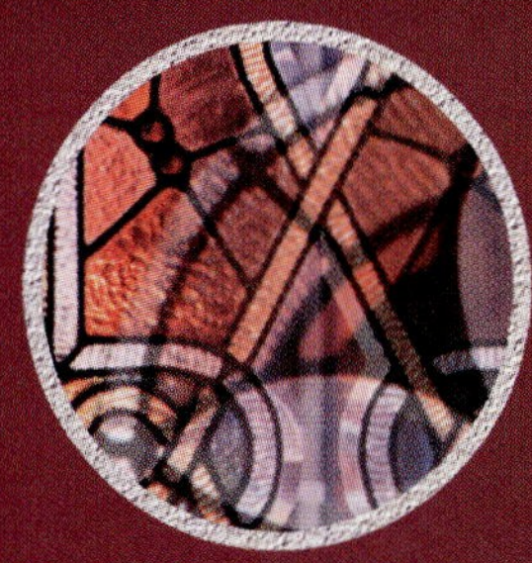

CHAPTER 3

Recall that the zeros of a function f are the solutions or roots of the equation $f(x) = 0$, if any exist. We know how to find all real and imaginary zeros of linear and quadratic functions (see Table 1).

TABLE 1 Zeros of Linear and Quadratic Functions

Function	Form	Equation	Zeros/roots
Linear	$f(x) = ax + b, a \neq 0$	$ax + b = 0$	$x = -\frac{b}{a}$
Quadratic	$f(x) = ax^2 + bx + c, a \neq 0$	$ax^2 + bx + c = 0$	$x = \frac{-b \pm \sqrt{b^2 - 4ac}}{2a}$

Linear and quadratic functions are also called first- and second-degree *polynomial functions,* respectively. Thus, Table 1 contains formulas for the zeros of any first- or second-degree polynomial function. What about higher-degree polynomial functions such as

$$p(x) = 4x^3 - 2x^2 + 3x + 5 \quad \text{Third degree}$$
$$q(x) = -2x^4 + 5x^2 - 6 \quad \text{Fourth degree}$$
$$r(x) = x^5 - x^4 + x^3 - 10 \quad \text{Fifth degree}$$

It turns out that there are direct, though complicated, methods for finding all zeros for any third- or fourth-degree polynomial function. However, the Frenchman Évariste Galois (1811–1832) proved at the age of 20 that for polynomial functions of degree greater than 4 there is no finite step-by-step process that will always yield all zeros.* This does not mean that we give up looking for zeros of higher-degree polynomial functions. It just means that we will have to use a variety of specialized methods, and sometimes we will have to approximate the zeros. The development of these methods is one of the primary objectives of this chapter.

We begin in Section 3-1 by discussing the graphic properties of polynomial functions. In Section 3-2 we develop tools for finding all the rational zeros of a polynomial equation with rational coefficients. In Section 3-3 we discuss methods for locating the real zeros of a polynomial with real coefficients. Once located, the real zeros are easily approximated with a graphing utility. Section 3-4 deals with rational functions and their graphs, and Section 3-5 discusses the decomposition of rational functions into simpler forms, an important tool for calculus.

*Galois' contribution, using the new concept of "group," was of the highest mathematical significance and originality. However, his contemporaries hardly read his papers, dismissing them as "almost unintelligible." At the age of 21, involved in political agitation, Galois met an untimely death in a duel. A short but fascinating account of Galois' tragic life can be found in E. T. Bell's *Men of Mathematics* (New York: Simon & Schuster, 1937), pp. 362–377.

SECTION 3-1 Polynomial Functions and Graphs

- Polynomial Functions
- Polynomial Division
- Division Algorithm
- Remainder Theorem
- Graphing Polynomial Functions

• Polynomial Functions

In Chapter 2 you were introduced to the basic functions

$$f(x) = b \qquad \text{Constant function}$$

$$f(x) = ax + b, \qquad a \neq 0 \qquad \text{Linear function}$$

$$f(x) = ax^2 + bx + c, \qquad a \neq 0 \qquad \text{Quadratic function}$$

as well as some special cases of more complicated functions such as

$$f(x) = ax^3 + bx^2 + cx + d, \qquad a \neq 0 \qquad \text{Cubic function}$$

Notice the evolving pattern going from the constant function to the cubic function—the terms in each equation are of the form ax^n, where n is a nonnegative integer and a is a real number. All these functions are special cases of the general class of functions called *polynomial functions.* The function

$$P(x) = a_nx^n + a_{n-1}x^{n-1} + \cdots + a_1x + a_0 \qquad a_n \neq 0$$

is called an ***n*th-degree polynomial function.** We will also refer to $P(x)$ as a **polynomial of degree *n*** or, more simply, as a **polynomial.** The numbers $a_n, a_{n-1}, \ldots, a_1, a_0$ are called the **coefficients of the function.** A nonzero constant function is a zero-degree polynomial, a linear function is a first-degree polynomial, and a quadratic function is a second-degree polynomial. The zero function $Q(x) = 0$ is also considered to be a polynomial but is not assigned a degree. The coefficients of a polynomial function may be complex numbers, or may be restricted to real numbers, rational numbers, or integers, depending on our interests. The domain of a polynomial function can be the set of complex numbers, the set of real numbers, or appropriate subsets of either, depending on our interests. In general, the context will dictate the choice of coefficients and domain.

The number r is said to be a **zero of the function *P*,** or a **zero of the polynomial *P(x)*,** or a **solution or root of the equation *P(x)* = 0,** if

$$P(r) = 0$$

A zero of a polynomial may or may not be the number 0. A zero of a polynomial is any number that makes the value of the polynomial 0. If the coefficients of a polynomial $P(x)$ are real numbers, then a real zero is simply an x intercept for the graph of $y = P(x)$. Consider the polynomial

$$P(x) = x^2 - 4x + 3$$

The graph of P is shown in Figure 1.

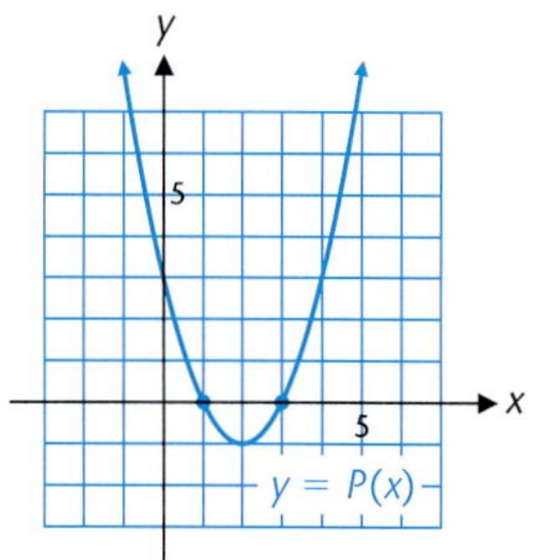

FIGURE 1 Zeros, roots, and x intercepts.

The x intercepts 1 and 3 are zeros of $P(x) = x^2 - 4x + 3$, since $P(1) = 0$ and $P(3) = 0$. The x intercepts 1 and 3 are also solutions or roots for the equation $x^2 - 4x + 3 = 0$.

In general:

> **Zeros and Roots**
>
> If the coefficients of a polynomial $P(x)$ are real, then the x intercepts of the graph of $y = P(x)$ are real **zeros** of P and $P(x)$ and real **solutions,** or **roots,** for the equation $P(x) = 0$.

• Polynomial Division

We can find quotients of polynomials by a long-division process similar to that used in arithmetic. An example will illustrate the process.

EXAMPLE 1 Algebraic Long Division

Divide $5 + 4x^3 - 3x$ by $2x - 3$.

Solution

$$\begin{array}{r}
2x^2 + 3x + 3 \\
2x - 3\,\overline{)\,4x^3 + 0x^2 - 3x + 5} \\
\underline{4x^3 - 6x^2} \\
6x^2 - 3x \\
\underline{6x^2 - 9x} \\
6x + 5 \\
\underline{6x - 9} \\
14 = R \\
\text{Remainder}
\end{array}$$

Arrange the dividend and the divisor in descending powers of the variable. Insert, with 0 coefficients, any missing terms of degree less than 3.
Divide the first term of the divisor into the first term of the dividend.
Multiply the divisor by $2x^2$, line up like terms, subtract as in arithmetic, and bring down $-3x$. Repeat the process until the degree of the remainder is less than that of the divisor.

Thus,

$$\frac{4x^3 - 3x + 5}{2x - 3} = 2x^2 + 3x + 3 + \frac{14}{2x - 3}$$

Check

$$(2x - 3)\left[(2x^2 + 3x + 3) + \frac{14}{2x - 3}\right] = (2x - 3)(2x^2 + 3x + 3) + 14$$

$$= 4x^3 - 3x + 5$$

Matched Problem 1 Divide $6x^2 - 30 + 9x^3$ by $3x - 4$.

Being able to divide a polynomial $P(x)$ by a linear polynomial of the form $x - r$ quickly and accurately will be of great help in the search for zeros of higher-degree polynomial functions. This kind of division can be carried out more efficiently by a method called **synthetic division.** The method is most easily understood through an example. Let's start by dividing $P(x) = 2x^4 + 3x^3 - x - 5$ by $x + 2$, using ordinary long division. The critical parts of the process are indicated in color.

$$
\begin{array}{lrl}
 & 2x^3 - 1x^2 + 2x - 5 & \text{Quotient} \\
\text{Divisor} \quad x + 2 & \overline{)2x^4 + 3x^3 + 0x^2 - 1x - 5} & \text{Dividend} \\
 & \underline{2x^4 + 4x^3} \qquad\qquad\qquad\quad & \\
 & -1x^3 + 0x^2 \qquad\qquad\quad & \\
 & \underline{-1x^3 - 2x^2} \qquad\qquad\quad & \\
 & 2x^2 - 1x \qquad\; & \\
 & \underline{2x^2 + 4x} \qquad\; & \\
 & -5x - 5 & \\
 & \underline{-5x - 10} & \\
 & 5 & \text{Remainder}
\end{array}
$$

The numerals printed in color, which represent the essential part of the division process, are arranged more conveniently as follows:

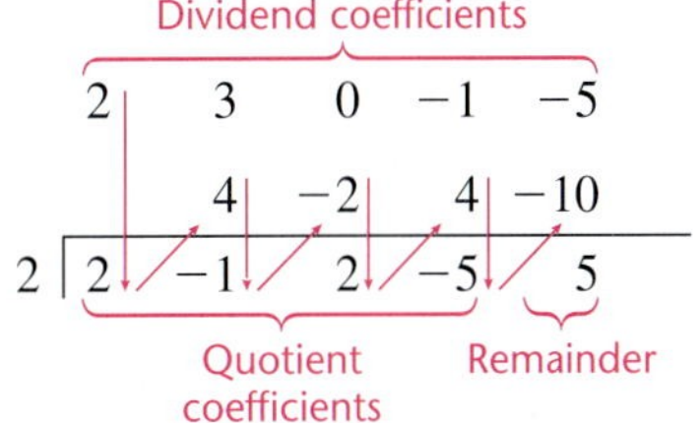

Mechanically, we see that the second and third rows of numerals are generated as follows. The first coefficient, 2, of the dividend is brought down and multiplied by 2 from the divisor; and the product, 4, is placed under the second dividend coefficient, 3, and subtracted. The difference, -1, is again multiplied by the 2 from the divisor; and the product is placed under the third coefficient from the dividend and subtracted. This process is repeated until the remainder is reached. The process can be made a little faster, and less prone to sign errors, by changing $+2$ from the divisor to -2 and adding instead of subtracting. Thus

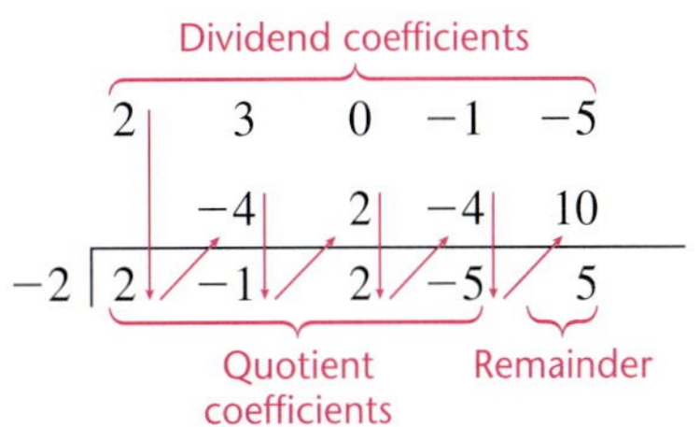

Key Steps in the Synthetic Division Process

To divide the polynomial $P(x)$ by $x - r$:

Step 1. Arrange the coefficients of $P(x)$ in order of descending powers of x. Write 0 as the coefficient for each missing power.

Step 2. After writing the divisor in the form $x - r$, use r to generate the second and third rows of numbers as follows. Bring down the first coefficient of the dividend and multiply it by r; then add the product to the second coefficient of the dividend. Multiply this sum by r, and add the product to the third coefficient of the dividend. Repeat the process until a product is added to the constant term of $P(x)$.

Step 3. The last number to the right in the third row of numbers is the remainder. The other numbers in the third row are the coefficients of the quotient, which is of degree 1 less than $P(x)$.

EXAMPLE 2 Synthetic Division

Use synthetic division to find the quotient and remainder resulting from dividing $P(x) = 4x^5 - 30x^3 - 50x - 2$ by $x + 3$. Write the answer in the form $Q(x) + R/(x - r)$, where R is a constant.

Solution Since $x + 3 = x - (-3)$, we have $r = -3$, and

$$\begin{array}{r|rrrrrr} & 4 & 0 & -30 & 0 & -50 & -2 \\ & & -12 & 36 & -18 & 54 & -12 \\ \hline -3 & 4 & -12 & 6 & -18 & 4 & -14 \end{array}$$

The quotient is $4x^4 - 12x^3 + 6x^2 - 18x + 4$ with a remainder of -14. Thus,

$$\frac{P(x)}{x + 3} = 4x^4 - 12x^3 + 6x^2 - 18x + 4 + \frac{-14}{x + 3}$$

Matched Problem 2 Repeat Example 2 with $P(x) = 3x^4 - 11x^3 - 18x + 8$ and divisor $x - 4$.

A calculator is a convenient tool for performing synthetic division. Any type of calculator can be used, although one with a memory will save some keystrokes. The flowchart in Figure 2 shows the repetitive steps in the synthetic division process, and Figure 3 illustrates the results of applying this process to Example 2 on a graphing calculator.

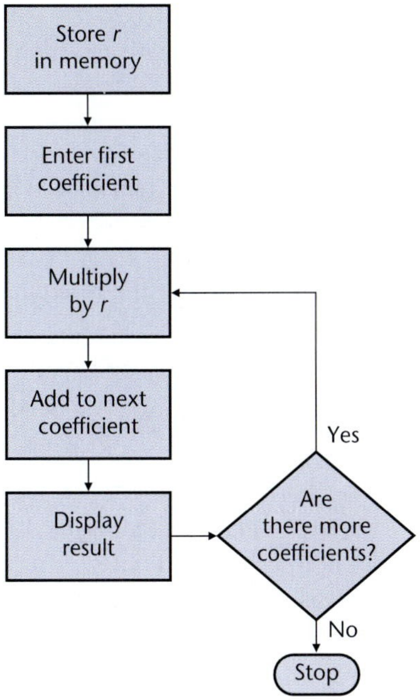

FIGURE 2 Synthetic division.

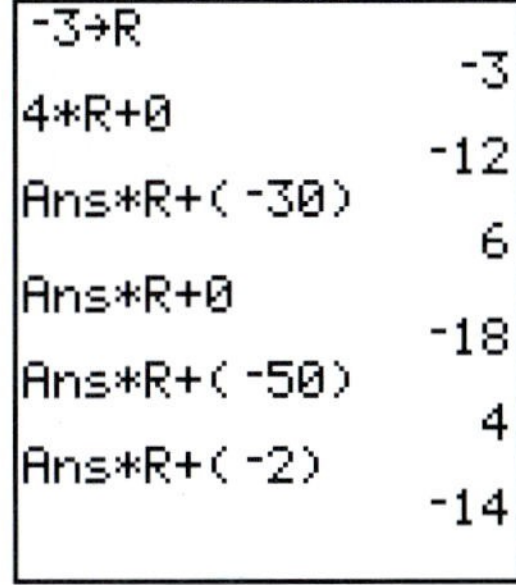

FIGURE 3

• Division Algorithm

If we divide $P(x) = 2x^4 - 5x^3 - 4x^2 + 13$ by $x - 3$, we obtain

$$\frac{2x^4 - 5x^3 - 4x^2 + 13}{x - 3} = 2x^3 + x^2 - x - 3 + \frac{4}{x - 3} \qquad x \neq 3$$

If we multiply both sides of this equation by $x - 3$, then we get

$$2x^4 - 5x^3 - 4x^2 + 13 = (x - 3)(2x^3 + x^2 - x - 3) + 4$$

This last equation is an identity in that the left side is equal to the right side for *all* replacements of x by real or imaginary numbers, including $x = 3$. This example suggests the important **division algorithm,** which we state as Theorem 1 without proof.

Theorem 1

Division Algorithm

For each polynomial $P(x)$ of degree greater than 0 and each number r, there exists a unique polynomial $Q(x)$ of degree 1 less than $P(x)$ and a unique number R such that

$$P(x) = (x - r)Q(x) + R$$

The polynomial $Q(x)$ is called the **quotient,** $x - r$ is the **divisor,** and R is the **remainder.** Note that R may be 0.

EXPLORE-DISCUSS 1 Let $P(x) = x^3 - 3x^2 - 2x + 8$.

(A) Evaluate $P(x)$ for
(i) $x = -2$ (ii) $x = 1$ (iii) $x = 3$
(B) Use synthetic division to find the remainder when $P(x)$ is divided by
(i) $x + 2$ (ii) $x - 1$ (iii) $x - 3$

What conclusion does a comparison of the results in parts A and B suggest?

• Remainder Theorem

We now use the division algorithm in Theorem 1 to prove the *remainder theorem*.

The equation in Theorem 1,

$$P(x) = (x - r)Q(x) + R$$

is an identity; that is, it is true for all real or imaginary replacements for x. In particular, if we let $x = r$, then we observe a very interesting and useful relationship:

$$\begin{aligned} P(r) &= (r - r)Q(r) + R \\ &= 0 \cdot Q(r) + R \\ &= 0 + R \\ &= R \end{aligned}$$

In words, the value of a polynomial $P(x)$ at $x = r$ is the same as the remainder R obtained when we divide $P(x)$ by $x - r$. We have proved the well-known remainder theorem:

Theorem 2 **Remainder Theorem**

If R is the remainder after dividing the polynomial $P(x)$ by $x - r$, then

$$P(r) = R$$

EXAMPLE 3 **Two Methods for Evaluating Polynomials**

If $P(x) = 4x^4 + 10x^3 + 19x + 5$, find $P(-3)$ by:

(A) Using the remainder theorem and synthetic division
(B) Evaluating $P(-3)$ directly

Solutions (A) Use synthetic division to divide $P(x)$ by $x - (-3)$.

$$\begin{array}{r|rrrrr} & 4 & 10 & 0 & 19 & 5 \\ & & -12 & 6 & -18 & -3 \\ \hline -3 & 4 & -2 & 6 & 1 & 2 = R = P(-3) \end{array}$$

(B) $P(-3) = 4(-3)^4 + 10(-3)^3 + 19(-3) + 5$

$= 2$

Matched Problem 3 Repeat Example 3 for $P(x) = 3x^4 - 16x^2 - 3x + 7$ and $x = -2$.

• Graphing Polynomial Functions

The shape of the graph of a polynomial function is connected to the degree of the polynomial. The shapes of odd-degree polynomial functions have something in common, and the shapes of even-degree polynomial functions have something in common. Figure 4 shows graphs of representative polynomial functions from degrees 1 to 6 and suggests some general properties of graphs of polynomial functions.

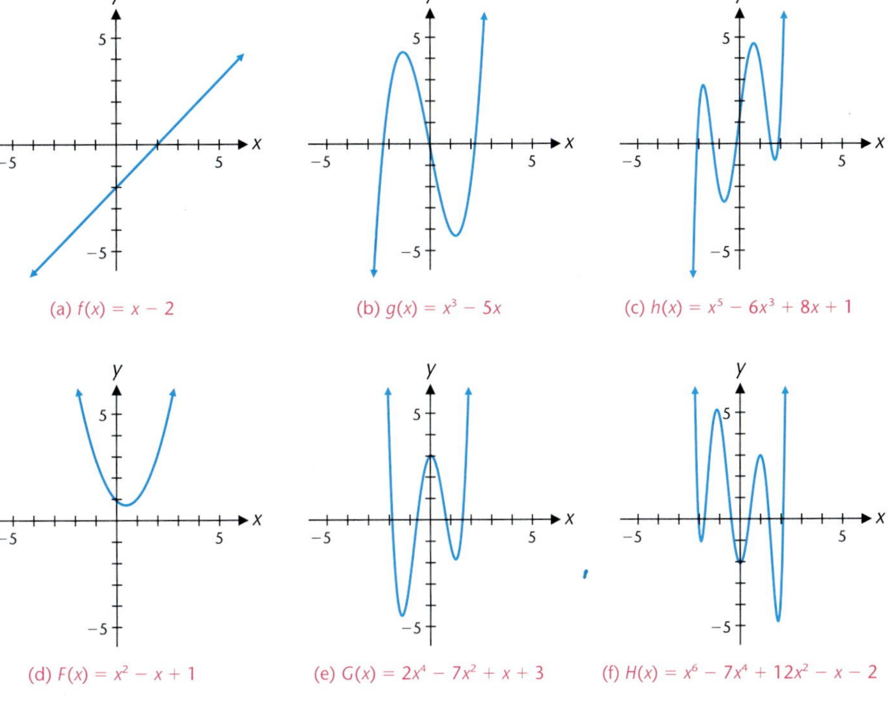

FIGURE 4 Graphs of polynomial functions.

Notice that the odd-degree polynomial graphs start negative, end positive, and cross the x axis at least once. The even-degree polynomial graphs start positive, end positive, and may not cross the x axis at all. In all cases in Figure 4, the coefficient of the highest-degree term was chosen positive. If any leading coefficient had been chosen negative, then we would have a similar graph but reflected in the x axis.

The shape of the graph of a polynomial is also related to the shape of the graph of the highest-degree or **leading term** of the polynomial. Figure 5 compares the graph of one of the polynomials from Figure 4 with the graph of its leading term. Although quite dissimilar for points close to the origin, as we "zoom out" to points distant from the origin, the graphs become quite similar. The leading term in the polynomial dominates all other terms combined.

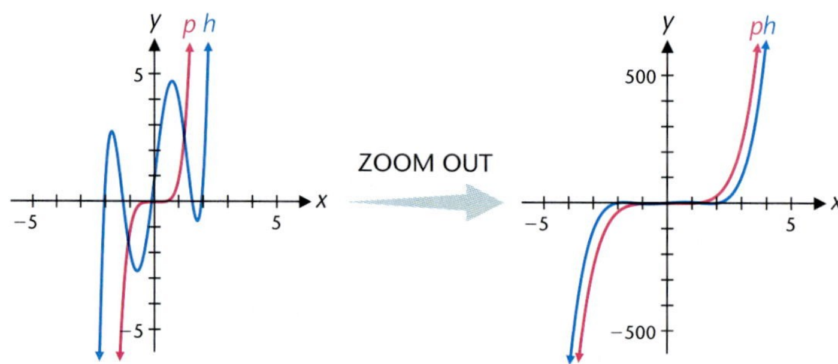

FIGURE 5 $p(x) = x^5$, $h(x) = x^5 - 6x^3 + 8x + 1$.

In general, the behavior of the graph of a polynomial function as x decreases without bound to the left or as x increases without bound to the right is determined by its leading term. We often use the symbols $-\infty$ and ∞ to help describe this left and right behavior.* The various possibilities are summarized in Theorem 3.

Theorem 3

Left and Right Behavior of a Polynomial

$$P(x) = a_n x^n + a_{n-1}x^{n-1} + \cdots + a_1 x + a_0, \qquad a_n \neq 0$$

1. $a_n > 0$ and n even
Graph of $P(x)$ increases without bound as x decreases to the left and as x increases to the right.

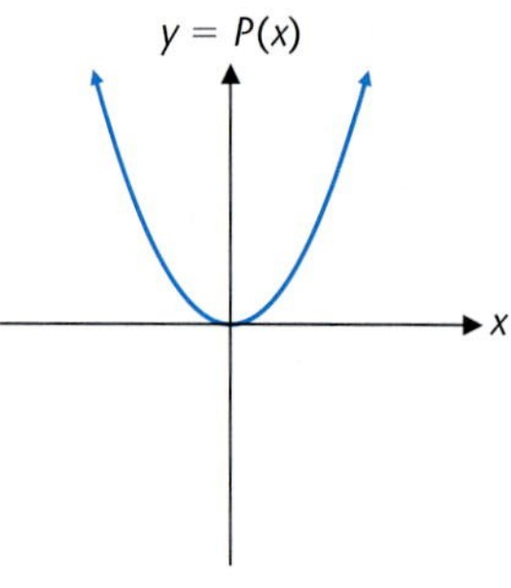

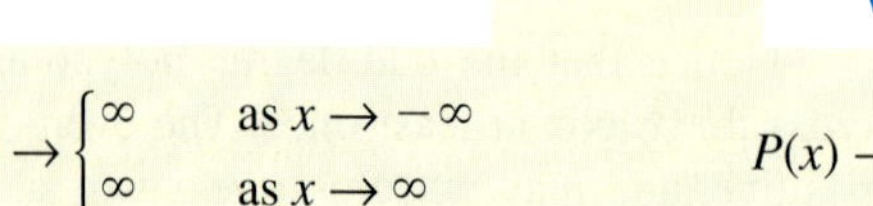

$$P(x) \to \begin{cases} \infty & \text{as } x \to -\infty \\ \infty & \text{as } x \to \infty \end{cases}$$

2. $a_n > 0$ and n odd
Graph of $P(x)$ decreases without bound as x decreases to the left and increases without bound as x increases to the right.

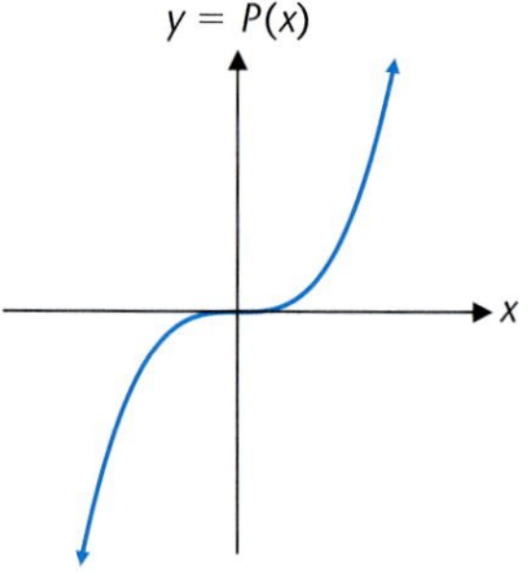

$$P(x) \to \begin{cases} -\infty & \text{as } x \to -\infty \\ \infty & \text{as } x \to \infty \end{cases}$$

*Remember, the symbol ∞ does not represent a real number. Earlier, we used ∞ to denote unbounded intervals, such as $[0, \infty)$. Now we are using it to describe quantities that are growing with no upper limit on their size.

3. $a_n < 0$ and n even
Graph of $P(x)$ decreases without bound as x decreases to the left and as x increases to the right.

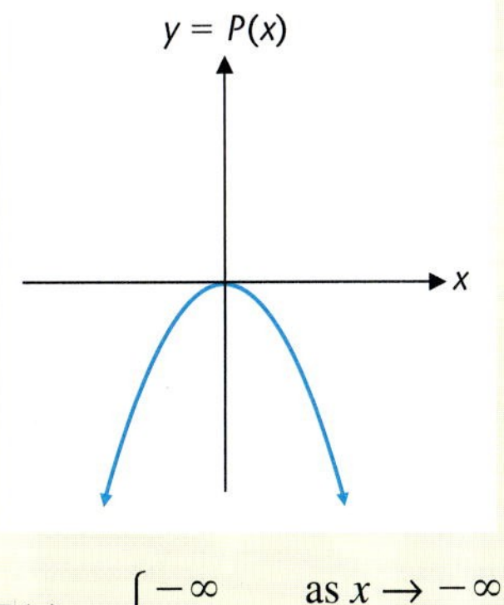

$$P(x) \to \begin{cases} -\infty & \text{as } x \to -\infty \\ -\infty & \text{as } x \to \infty \end{cases}$$

4. $a_n < 0$ and n odd
Graph of $P(x)$ increases without bound as x decreases to the left and decreases without bound as x increases to the right.

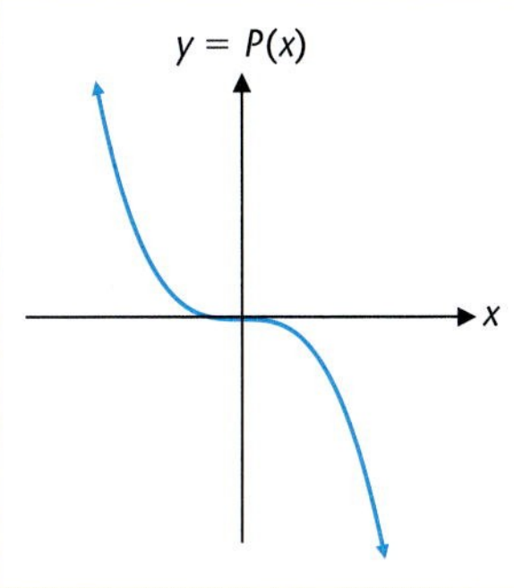

$$P(x) \to \begin{cases} \infty & \text{as } x \to -\infty \\ -\infty & \text{as } x \to \infty \end{cases}$$

Figure 4 gives examples of polynomial functions with graphs containing the maximum number of *turning points* possible for a polynomial of that degree. A **turning point** on a continuous graph is a point that separates an increasing portion from a decreasing portion. Listed in Theorem 4 below are useful properties of polynomial functions we accept without proof. Property 3 is discussed in detail later in this chapter. The other properties are established in calculus.

Theorem 4 **Graph Properties of Polynomial Functions**

Let P be an nth-degree polynomial function with real coefficients.

1. P is continuous for all real numbers.

2. The graph of P is a smooth curve.

3. The graph of P has at most n x intercepts.

4. P has at most $n - 1$ turning points.

EXPLORE-DISCUSS 2

(A) What is the least number of turning points an odd-degree polynomial function can have? An even-degree polynomial function?

(B) What is the maximum number of x intercepts the graph of a polynomial function of degree n can have?

(C) What is the maximum number of real solutions an nth-degree polynomial equation can have?

(D) What is the least number of x intercepts the graph of a polynomial function of odd degree can have? Of even degree?

(E) What is the least number of real solutions a polynomial equation of odd degree can have? Of even degree?

EXAMPLE 4 Graphing a Polynomial

Graph $P(x) = x^3 - 12x + 2$, $-4 \leq x \leq 4$. Find points by using synthetic division and the remainder theorem. How many x intercepts does the graph have? How many turning points? Describe the left and right behavior of $P(x)$.

Solution We evaluate $P(x)$ from $x = -4$ to $x = 4$ for integer values of x. The process is speeded up by forming a synthetic division table. To simplify the form of the table, we dispense with writing the product of r with each coefficient in the quotient and perform the calculations mentally or with a calculator. A calculator becomes increasingly useful as the coefficients become more numerous or complicated. The table also provides other important information, as will be seen in subsequent sections.

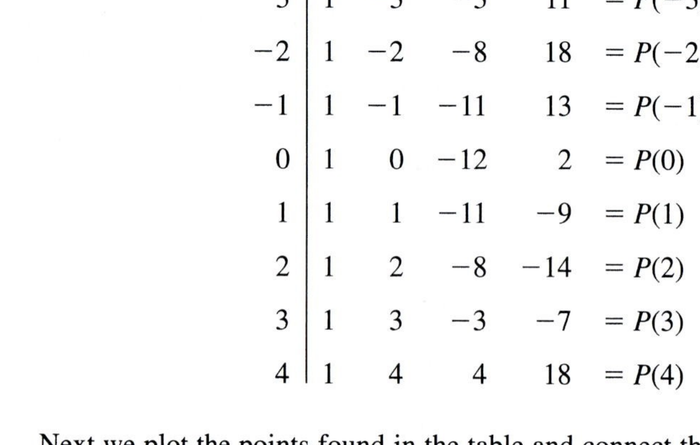

	1	0	−12	2	
−4	1	−4	4	−14	$= P(-4)$
−3	1	−3	−3	11	$= P(-3)$
−2	1	−2	−8	18	$= P(-2)$
−1	1	−1	−11	13	$= P(-1)$
0	1	0	−12	2	$= P(0)$
1	1	1	−11	−9	$= P(1)$
2	1	2	−8	−14	$= P(2)$
3	1	3	−3	−7	$= P(3)$
4	1	4	4	18	$= P(4)$

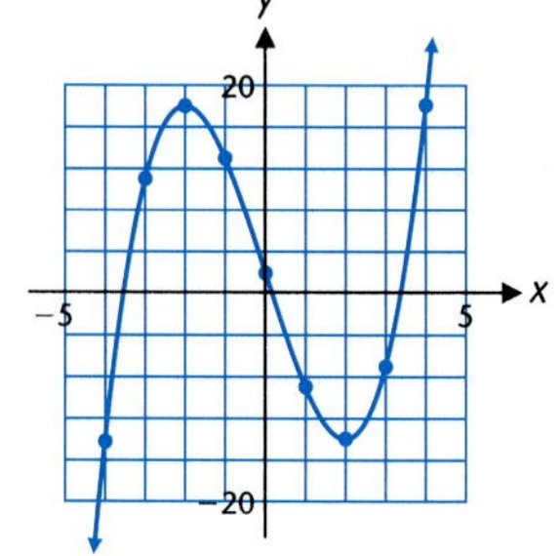

FIGURE 6 $P(x) = x^3 - 12x + 2$.

Next we plot the points found in the table and connect them with a smooth curve (Fig. 6). As we draw this curve, we notice that the graph crosses the x axis three times and changes direction twice. The next two sections will address the question of determining precisely where a graph crosses the x axis. Precise determination of the location of turning points requires calculus techniques. Lacking this precise information, we simply change direction at $x = -2$ and $x = 2$.

The leading term of $P(x)$ is x^3. From case 2 in Theorem 3 we see that $P(x) \to -\infty$ as $x \to -\infty$ and $P(x) \to \infty$ as $x \to \infty$.

Matched Problem 4 Graph $P(x) = x^3 - 4x^2 - 4x + 16$, $-3 \leq x \leq 5$. Find points by using synthetic division and the remainder theorem. How many x intercepts does the graph have? How many turning points? Describe the left and right behavior of $P(x)$.

Remark. A graphing utility can quickly produce a table of values, without using synthetic division, and can graph a polynomial just as quickly. In Section 3-3 we will find that a synthetic division table is a valuable tool when used in conjunction with a graphing utility. Thus, students with graphing utilities should also learn to construct synthetic division tables. (See Table 2 in Section 3-3 for a more efficient way to construct a synthetic division table on a graphing utility.)

Answers to Matched Problems

1. $3x^2 + 6x + 8 + \dfrac{2}{3x - 4}$

2. $\dfrac{P(x)}{x - 4} = 3x^3 + x^2 + 4x - 2 + \dfrac{0}{x - 4} = 3x^3 + x^2 + 4x - 2$

3. $P(-2) = -3$ for both parts, as it should.

4.

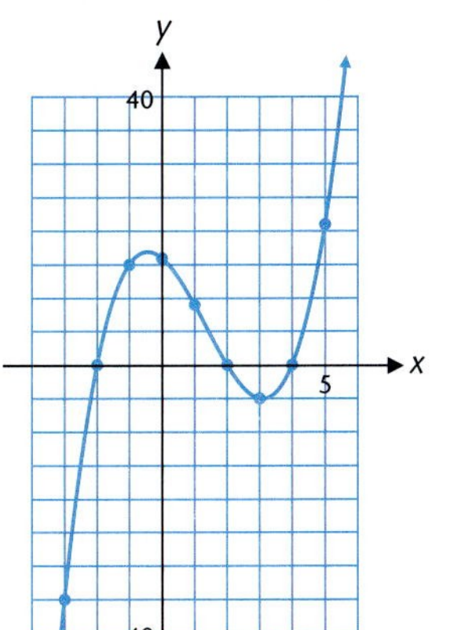

x	$P(x)$
−3	−35
−2	0
−1	15
0	16
1	9
2	0
3	−5
4	0
5	21

Three x intercepts and two turning points; $P(x) \to -\infty$ as $x \to -\infty$; $P(x) \to \infty$ as $x \to \infty$.

EXERCISE 3-1

A

In Problems 1–4, a is a positive real number. Match each function with one of graphs (a)–(d).

1. $f(x) = ax^3$

2. $g(x) = -ax^4$

3. $h(x) = ax^6$

4. $k(x) = -ax^5$

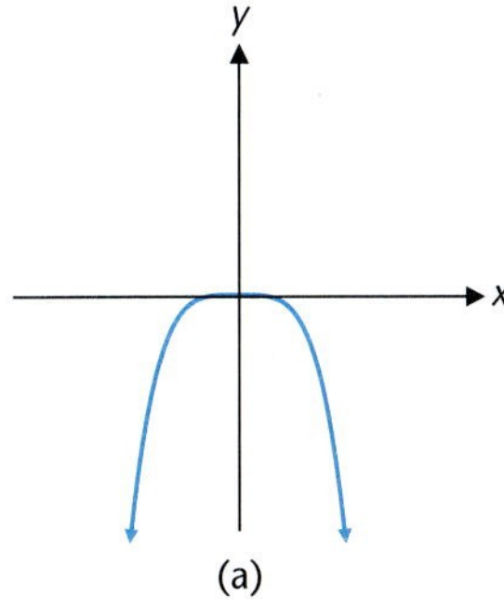

(a)

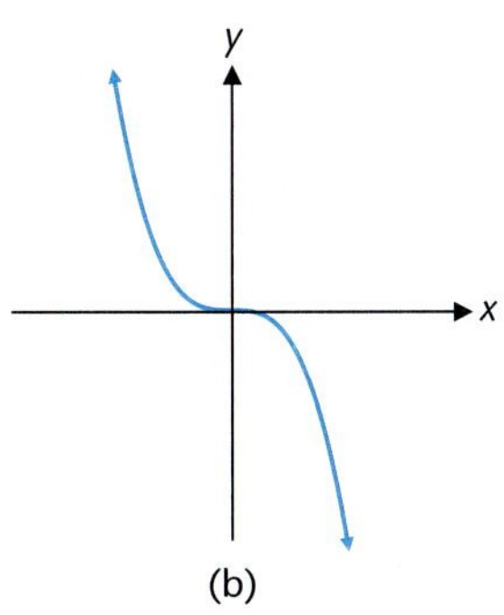

(b)

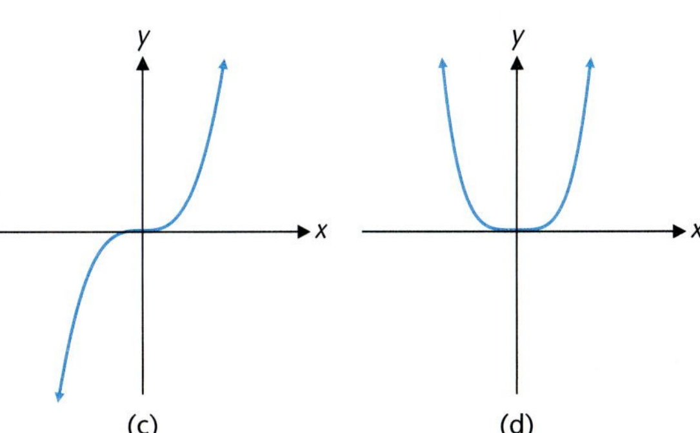

(c) (d)

Problems 5–8 refer to the graphs of functions f, g, h, and k shown below.

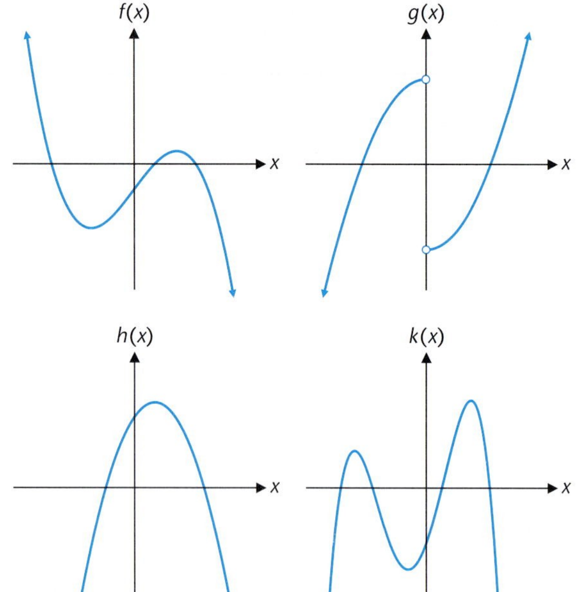

5. Which of these functions could be a second-degree polynomial?

6. Which of these functions could be a third-degree polynomial?

7. Which of these functions could be a fourth-degree polynomial?

8. Which of these functions is not a polynomial?

In Problems 9–16, divide, using algebraic long division. Write the quotient, and indicate the remainder.

9. $(a^2 - 4) \div (a + 2)$ **10.** $(a^2 + 4) \div (a + 2)$

11. $(b^2 - 6b - 9) \div (b - 3)$

12. $(b^2 - 6b + 9) \div (b - 3)$

13. $(3x + 2 + x^3 - x^2) \div (x - 1)$

14. $(2x^2 + 4x - x^3 - 8) \div (x - 2)$

15. $(1 + 8y^3 - 4y^2 - 2y) \div (2y + 1)$

16. $(3 + 8y^3 - 6y^2 - 7y) \div (2y - 3)$

In Problems 17–22, use synthetic division to write the quotient $P(x) \div (x - r)$ in the form $P(x)/(x - r) = Q(x) + R/(x - r)$, where R is a constant.

17. $(x^2 + 4x - 15) \div (x - 3)$

18. $(x^2 - 2x - 1) \div (x - 4)$ **19.** $(3x^2 - x - 7) \div (x + 2)$

20. $(4x^2 + 18x + 4) \div (x + 5)$

21. $(2x^3 + 3x^2 - 8x + 1) \div (x + 3)$

22. $(3x^3 - 4x^2 - 7x + 9) \div (x - 2)$

B

Use synthetic division and the remainder theorem in Problems 23–28.

23. Find $P(-5)$, given $P(x) = 2x^2 + 8x - 6$.

24. Find $P(2)$, given $P(x) = 3x^2 - 4x + 2$.

25. Find $P(-4)$, given $P(x) = 4x^3 + 12x^2 - 8x + 25$.

26. Find $P(-3)$, given $P(x) = 5x^3 + 14x^2 + 3x + 10$.

27. Find $P(3)$, given $P(x) = 2x^4 - 5x^3 + 2x^2 - 11x - 14$.

28. Find $P(-6)$, given $P(x) = 3x^4 + 18x^3 + x^2 + 4x - 7$.

In Problems 29–44, divide, using synthetic division. Write the quotient, and indicate the remainder. As coefficients get more involved, a calculator should prove helpful. Do not round off—all quantities are exact.

29. $(2x^5 - 5x^2 + 3) \div (x - 1)$

30. $(3x^4 - 2x - 5) \div (x + 1)$

31. $(x^4 - 16) \div (x + 4)$ **32.** $(x^5 - 32) \div (x - 2)$

33. $(4x^4 - 9x^3 - 8x^2 - 2x - 7) \div (x - 3)$

34. $(2x^4 + 6x^3 - 4x^2 - 5x + 7) \div (x + 3)$

35. $(x^6 + 7x^5 + 10x^4 - x^2 - 5x) \div (x + 5)$

36. $(x^6 + 6x^5 + 2x^4 + 12x^3 - 3x - 18) \div (x + 6)$

37. $(2x^4 + 9x^3 + 5x^2 - 4x + 3) \div (x + \frac{3}{2})$

38. $(2x^4 + 5x^3 + 5x + 8) \div (x + \frac{1}{2})$

39. $(3x^4 + 5x^3 - 5x^2 + 10x - 1) \div (x - \frac{1}{3})$

40. $(4x^4 - 11x^3 + 18x^2 - 5x + 4) \div (x - \frac{3}{4})$

41. $(5x^4 - 4x^3 + 2x - 5) \div (x - 0.2)$

42. $(3x^4 - 4x^2 + 5x + 8) \div (x + 0.8)$

43. $(5x^5 + 2x^4 + 4x^3 + 6x^2 - 6) \div (x + 0.6)$

44. $(10x^5 - 4x^4 + 2x^2 + 4x - 1) \div (x - 0.4)$

In Problems 45–52, graph each polynomial function using synthetic division and the remainder theorem. Then describe each graph verbally, including the number of x intercepts, the number of turning points, and the left and right behavior.

**Check your work in Problems 45–52 by graphing on a graphing utility.*

45. $P(x) = x^3 - 5x^2 + 2x + 8, -2 \le x \le 5$

46. $P(x) = x^3 + 2x^2 - 5x - 6, -4 \le x \le 3$

47. $P(x) = x^3 + 4x^2 - x - 4, -5 \le x \le 2$

48. $P(x) = x^3 - 2x^2 - 5x + 6, -3 \le x \le 4$

49. $P(x) = -x^3 + 2x^2 - 3, -2 \le x \le 3$

*Please note that use of a graphing utility is not required to complete these exercises. Checking them with a g.u. is optional.

50. $P(x) = -x^3 - x + 4, -2 \le x \le 2$

51. $P(x) = -x^3 + 3x^2 - 3x + 2, -1 \le x \le 3$

52. $P(x) = -x^3 + x^2 + 4x + 6, -3 \le x \le 4$

In Problems 53–56, either give an example of a polynomial with real coefficients that satisfies the given conditions or explain why such a polynomial cannot exist.

53. $P(x)$ is a third-degree polynomial with one x intercept.

54. $P(x)$ is a fourth-degree polynomial with no x intercepts.

55. $P(x)$ is a third-degree polynomial with no x intercepts.

56. $P(x)$ is a fourth-degree polynomial with no turning points.

C

In Problems 57–60, divide, using algebraic long division. Write the quotient, and indicate the remainder.

57. $(x^4 + x^3 + x^2 - x - 2) \div (x^2 - 1)$

58. $(x^4 + 2x^3 - 3x^2 - 8x - 4) \div (x^2 - 4)$

59. $(x^4 + x^3 - 7x^2 + 8x + 1) \div (x^2 + 3x - 2)$

60. $(x^4 - 7x^3 + 15x^2 - 9x + 1) \div (x^2 - 4x + 1)$

In Problems 61 and 62, divide, using synthetic division. Do not use a calculator.

61. $(x^4 + 2x^3 - 2x^2 + 2x - 3) \div (x - i)$

62. $(x^4 + 2x^3 - 2x^2 + 2x - 3) \div (x + i)$

63. Let $P(x) = x^2 + 2ix - 10$. Find:
(A) $P(2 - i)$ (B) $P(5 - 5i)$
(C) $P(3 - i)$ (D) $P(-3 - i)$

64. Let $P(x) = x^2 - 4ix - 13$. Find:
(A) $P(5 + 6i)$ (B) $P(1 + 2i)$
(C) $P(3 + 2i)$ (D) $P(-3 + 2i)$

In Problems 65–72, graph each polynomial function using synthetic division and the remainder theorem. Then describe each graph verbally, including the number of x intercepts, the number of turning points, and the left and right behavior.

Check your work in Problems 65–72 by graphing on a graphing utility.

65. $P(x) = x^4 - 2x^3 - 2x^2 + 8x - 8$

66. $P(x) = x^4 + x^3 - 3x^2 + 7x - 6$

67. $P(x) = x^4 + 4x^3 - x^2 - 10x - 8$

68. $P(x) = x^4 - 8x^2 - 4x + 10$

69. $P(x) = -x^4 + 2x^3 + 10x^2 - 10x - 9$

70. $P(x) = -x^4 - 5x^3 + x^2 + 20x + 5$

71. $P(x) = x^5 - 6x^4 + 4x^3 + 17x^2 - 5x - 7$

72. $P(x) = x^5 - 9x^3 + 4x^2 + 15x - 10$

73. (A) Divide $P(x) = a_2x^2 + a_1x + a_0$ by $x - r$, using both synthetic division and the long-division process, and compare the coefficients of the quotient and the remainder produced by each method.
(B) Expand the expression representing the remainder. What do you observe?

74. Repeat Problem 73 for

$$P(x) = a_3x^3 + a_2x^2 + a_1x + a_0$$

75. Polynomials also can be evaluated conveniently using a "nested factoring" scheme. For example, the polynomial $P(x) = 2x^4 - 3x^3 + 2x^2 - 5x + 7$ can be written in a nested factored form as follows:

$$\begin{aligned} P(x) &= 2x^4 - 3x^3 + 2x^2 - 5x + 7 \\ &= (2x - 3)x^3 + 2x^2 - 5x + 7 \\ &= [(2x - 3)x + 2]x^2 - 5x + 7 \\ &= \{[(2x - 3)x + 2]x - 5\}x + 7 \end{aligned}$$

Use the nested factored form to find $P(-2)$ and $P(1.7)$. [*Hint:* To evaluate $P(-2)$, store -2 in your calculator's memory and proceed from left to right recalling -2 as needed.]

76. Find $P(-2)$ and $P(1.3)$ for $P(x) = 3x^4 + x^3 - 10x^2 + 5x - 2$ using the nested factoring scheme presented in Problem 75.

SECTION 3-2 Finding Rational Zeros of Polynomials

- Factor Theorem
- Fundamental Theorem of Algebra
- Imaginary Zeros
- Rational Zeros

In this section we develop some important properties of polynomials with arbitrary coefficients. Then we consider the problem of finding all the rational zeros of a polynomial with rational coefficients. In some cases, this process will also enable us to find irrational or imaginary zeros.

• Factor Theorem

The division algorithm (Theorem 1 in Section 3-1)

$$P(x) = (x - r)Q(x) + R$$

may, because of the remainder theorem (Theorem 2 in Section 3-1), be written in a form where R is replaced by $P(r)$:

$$P(x) = (x - r)Q(x) + P(r)$$

It is easy to see that $x - r$ is a factor of $P(x)$ if and only if $P(r) = 0$; that is, if and only if r is a zero of the polynomial $P(x)$. This result is known as the **factor theorem:**

Theorem 1 **Factor Theorem**

If r is a zero of the polynomial $P(x)$, then $x - r$ is a factor of $P(x)$. Conversely, if $x - r$ is a factor of $P(x)$, then r is a zero of $P(x)$.

The relationship between zeros, roots, factors, and x intercepts is fundamental to the study of polynomials. With the addition of the factor theorem, we now know that the following statements are equivalent for any polynomial $P(x)$:

1. r is a root of the equation $P(x) = 0$.
2. r is a zero of $P(x)$.
3. $x - r$ is a factor of $P(x)$.

If, in addition, the coefficients of $P(x)$ are real numbers and r is a real number, then we can add a fourth statement to this list:

4. r is an x intercept of the graph of $P(x)$.

EXAMPLE 1 Factors, Zeros, Roots, and Intercepts

(A) Use the factor theorem to show that $x + 1$ is a factor of $P(x) = x^{25} + 1$.
(B) What are the zeros of $P(x) = 3(x - 5)(x + 2)(x - 3)$?
(C) What are the roots of $x^4 - 1 = 0$?
(D) What are the x intercepts of the graph of $P(x) = x^4 - 1$?

Solutions (A) Since $x + 1 = x - (-1)$, we have $r = -1$ and

$$P(r) = P(-1) = (-1)^{25} + 1 = -1 + 1 = 0$$

Hence, -1 is a zero of $P(x) = x^{25} + 1$. By the factor theorem, $x - (-1) = x + 1$ is a factor of $x^{25} + 1$.

(B) Since $(x - 5)$, $(x + 2)$, and $(x - 3)$ are all factors of $P(x)$, 5, -2, and 3 are zeros of $P(x)$.

(C) Factoring the left side, we have

$$x^4 - 1 = 0$$

$$(x^2 - 1)(x^2 + 1) = 0$$

$$(x - 1)(x + 1)(x - i)(x + i) = 0$$

Thus, the roots of $x^4 - 1 = 0$ are 1, -1, i, and $-i$.

(D) From part C, the zeros of $P(x)$ are 1, -1, i, and $-i$. However, x intercepts must be real numbers. Thus the x intercepts of the graph of $P(x) = x^4 - 1$ are 1 and -1 (see Fig. 1).

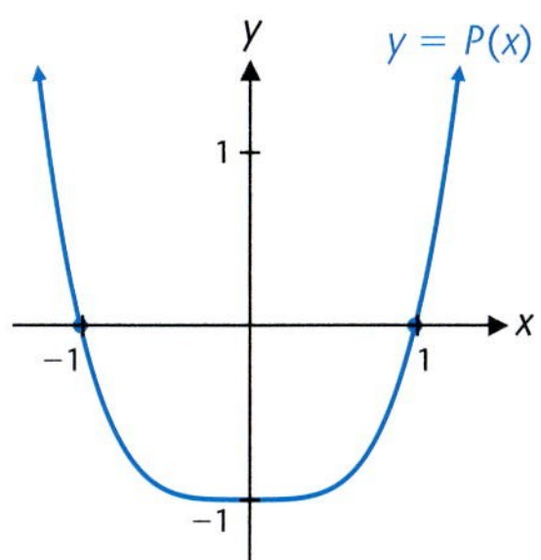

FIGURE 1 x intercepts of $P(x) = x^4 - 1$.

Matched Problem 1

(A) Use the factor theorem to show that $x - 1$ is a factor of $P(x) = x^{54} - 1$.

(B) What are the zeros of

$$P(x) = 2(x + 3)(x + 7)(x - 8)(x + 1)?$$

(C) What are the roots of $x^2 + 4 = 0$?

(D) What are the x intercepts of $P(x) = x^2 + 4$?

• Fundamental Theorem of Algebra

Theorem 2, often referred to as the **fundamental theorem of algebra,** requires verification that is beyond the scope of this book, so we state it without proof.

Theorem 2 **Fundamental Theorem of Algebra**

Every polynomial $P(x)$ of degree $n > 0$ has at least one zero.

If $P(x) = a_nx^n + a_{n-1}x^{n-1} + \cdots + a_1x + a_0$ is a polynomial of degree $n > 0$ with complex coefficients, then, according to Theorem 2, it has at least one zero, say r_1. According to the factor theorem, $x - r_1$ is a factor of $P(x)$. Thus,

$$P(x) = (x - r_1)Q_1(x)$$

where $Q_1(x)$ is a polynomial of degree $n - 1$. If $n - 1 = 0$, then $Q_1(x) = a_n$. If $n - 1 > 0$, then, by Theorem 2, $Q_1(x)$ has at least one zero, say r_2. And

$$Q_1(x) = (x - r_2)Q_2(x)$$

where $Q_2(x)$ is a polynomial of degree $n - 2$. Thus,

$$P(x) = (x - r_1)(x - r_2)Q_2(x)$$

If $n - 2 = 0$, then $Q_2(x) = a_n$. If $n - 2 > 0$, then $Q_2(x)$ has at least one zero, say r_3. And

$$Q_2(x) = (x - r_3)Q_3(x)$$

where $Q_3(x)$ is a polynomial of degree $n - 3$.

We continue in this way until $Q_k(x)$ is of degree zero—that is, until $k = n$. At this point, $Q_n(x) = a_n$, and we have

$$P(x) = (x - r_1)(x - r_2) \cdot \cdots \cdot (x - r_n)a_n$$

Thus, $r_1, r_2, \ldots, r_n$ are n zeros, not necessarily distinct, of $P(x)$. Is it possible for $P(x)$ to have more than these n zeros? Let's assume that r is a number different from the zeros above. Then

$$P(r) = a_n(r - r_1)(r - r_2) \cdot \cdots \cdot (r - r_n) \neq 0$$

since r is not equal to any of the zeros. Hence, r is not a zero, and we conclude that $r_1, r_2, \ldots, r_n$ are the only zeros of $P(x)$. We have just sketched a proof of Theorem 3.

Theorem 3 ***n* Zeros Theorem**

Every polynomial $P(x)$ of degree $n > 0$ can be expressed as the product of n linear factors. Hence, $P(x)$ has exactly n zeros—not necessarily distinct.

Theorems 2 and 3 were first proved in 1797 by Carl Friedrich Gauss, one of the greatest mathematicians of all time, at the age of 20.

If $P(x)$ is represented as the product of linear factors and $x - r$ occurs m times, then r is called a **zero of multiplicity *m*.** For example, if

$$P(x) = 4(x - 5)^3(x + 1)^2(x - i)(x + i)$$

then this seventh-degree polynomial has seven zeros, not all distinct. That is, 5 is a zero of multiplicity 3, or a triple zero; -1 is a zero of multiplicity 2, or a double zero; and i and $-i$ are zeros of multiplicity 1, or simple zeros. Thus, this seventh-degree polynomial has exactly seven zeros if we count 5 and -1 with their respective multiplicities.

EXPLORE-DISCUSS 1 If r is a real zero of a polynomial $P(x)$ with real coefficients, then r is also an x intercept for the graph of $P(x)$. Discuss the difference between the graph of $P(x)$ at a real zero of odd multiplicity and at a real zero of even multiplicity.

EXAMPLE 2 **Factoring a Polynomial**

If -2 is a double zero of $P(x) = x^4 - 7x^2 + 4x + 20$, write $P(x)$ as a product of first-degree factors.

Solution Since -2 is a double zero of $P(x)$, we can write

$$\begin{aligned} P(x) &= (x + 2)^2 Q(x) \\ &= (x^2 + 4x + 4)Q(x) \end{aligned}$$

and find $Q(x)$ by dividing $P(x)$ by $x^2 + 4x + 4$. Carrying out the algebraic long division, we obtain

$$Q(x) = x^2 - 4x + 5$$

The zeros of $Q(x)$ are found, using the quadratic formula, to be $2 - i$ and $2 + i$. Thus, $P(x)$ written as a product of linear factors is

$$P(x) = (x + 2)^2[x - (2 - i)][x - (2 + i)]$$

[*Note:* Any time $Q(x)$ is a quadratic polynomial, its zeros can be found using the quadratic formula.]

Matched Problem 2 If 3 is a double zero of $P(x) = x^4 - 12x^3 + 55x^2 - 114x + 90$, write $P(x)$ as a product of first-degree factors.

• Imaginary Zeros

Something interesting happens if we restrict the coefficients of a polynomial to real numbers. Let's use the quadratic formula to find the zeros of the polynomial

$$P(x) = x^2 - 6x + 13$$

To find the zeros of $P(x)$, we solve $P(x) = 0$:

$$\begin{aligned} x^2 - 6x + 13 &= 0 \\ x &= \frac{6 \pm \sqrt{36 - 52}}{2} \\ &= \frac{6 \pm \sqrt{-16}}{2} = \frac{6 \pm 4i}{2} = 3 \pm 2i \end{aligned}$$

The zeros of $P(x)$ are $3 - 2i$ and $3 + 2i$, conjugate imaginary numbers (see Section 1-5). Also observe that the imaginary zeros in Example 2 are the conjugate imaginary numbers $2 - i$ and $2 + i$.

This is generalized in the following theorem:

Theorem 4 **Imaginary Zeros Theorem**

Imaginary zeros of polynomials with real coefficients, if they exist, occur in conjugate pairs.

As a consequence of Theorems 3 and 4, we also know (think this through) the following:

Theorem 5 **Real Zeros and Odd-Degree Polynomials**

A polynomial of odd degree with real coefficients always has at least one real zero.

EXPLORE-DISCUSS 2

(A) Let $P(x)$ be a third-degree polynomial with real coefficients. Indicate which of the following statements are true and which are false. Justify your conclusions.
 (i) $P(x)$ has at least one real zero.
 (ii) $P(x)$ has three zeros.
 (iii) $P(x)$ can have two real zeros and one imaginary zero.

(B) Let $P(x)$ be a fourth-degree polynomial with real coefficients. Indicate which of the following statements are true and which are false. Justify your conclusions.
 (i) $P(x)$ has four zeros.
 (ii) $P(x)$ has at least two real zeros.
 (iii) If we know $P(x)$ has three real zeros, then the fourth zero must be real.

• Rational Zeros

First note that a polynomial with rational coefficients can always be written as a constant times a polynomial with integer coefficients. For example,

$$P(x) = \tfrac{1}{2}x^3 - \tfrac{2}{3}x^2 + \tfrac{7}{4}x + 5$$
$$= \tfrac{1}{12}(6x^3 - 8x^2 + 21x + 60)$$

Thus, it is sufficient to confine our attention to polynomials with integer coefficients.

We introduce the rational zero theorem by examining the following quadratic polynomial whose zeros can be found easily by factoring:

$$P(x) = 6x^2 - 13x - 5 = (2x - 5)(3x + 1)$$

Zeros of $P(x)$: $\quad \dfrac{5}{2} \quad$ and $\quad -\dfrac{1}{3} = \dfrac{-1}{3}$

Notice that the numerators, 5 and -1, of the zeros are both integer factors of -5, the constant term in $P(x)$. The denominators 2 and 3 of the zeros are both integer factors of 6, the coefficient of the highest-degree term in $P(x)$. These observations are generalized in Theorem 6.

Theorem 6

Rational Zero Theorem

If the rational number b/c, in lowest terms, is a zero of the polynomial

$$P(x) = a_nx^n + a_{n-1}x^{n-1} + \cdots + a_1x + a_0 \qquad a_n \neq 0$$

with integer coefficients, then b must be an integer factor of a_0 and c must be an integer factor of a_n.

$$P(x) = a_nx^n + a_{n-1}x^{n-1} + \cdots + a_1x + a_0$$

$\frac{b}{c}$: b must be a factor of a_0; c must be a factor of a_n

The proof of Theorem 6 is not difficult and is instructive, so we sketch it here.

Proof Since b/c is a zero of $P(x)$,

$$a_n\left(\frac{b}{c}\right)^n + a_{n-1}\left(\frac{b}{c}\right)^{n-1} + \cdots + a_1\left(\frac{b}{c}\right) + a_0 = 0 \tag{1}$$

If we multiply both sides of equation (1) by c^n, we obtain

$$a_nb^n + a_{n-1}b^{n-1}c + \cdots + a_1bc^{n-1} + a_0c^n = 0 \tag{2}$$

which can be written in the form

$$a_nb^n = c(-a_{n-1}b^{n-1} - \cdots - a_0c^{n-1}) \tag{3}$$

Since both sides of equation (3) are integers, c must be a factor of a_nb^n. And since the rational number b/c is given to be in lowest terms, b and c can have no common factors other than ± 1. That is, b and c are **relatively prime.** This implies that b^n and c also are relatively prime. Hence, c must be a factor of a_n.

Now, if we solve equation (2) for a_0c^n and factor b out of the right side, we have

$$a_0c^n = b(-a_nb^{n-1} - \cdots - a_1c^{n-1})$$

We see that b is a factor of a_0c^n and, hence, a factor of a_0, since b and c are relatively prime.

EXPLORE-DISCUSS 3 Let $P(x) = a_3x^3 + a_2x^2 + a_1x + a_0$, where a_3, a_2, a_1, and a_0 are integers.

1. If $P(2) = 0$, there is one coefficient that must be an even integer. Identify this coefficient and explain why it must be even.
2. If $P(\frac{1}{2}) = 0$, there is one coefficient that must be an even integer. Identify this coefficient and explain why it must be even.
3. If $a_3 = a_0 = 1$, $P(-1) \neq 0$, and $P(1) \neq 0$, does $P(x)$ have any rational zeros? Support your conclusion with verbal arguments and/or examples.

It is important to understand that Theorem 6 does not say that a polynomial $P(x)$ with integer coefficients must have rational zeros.

It simply states that if $P(x)$ does have a rational zero, then the numerator of the zero must be an integer factor of a_0 and the denominator of the zero must be an integer factor of a_n. Since every integer has a finite number of integer factors, Theorem 6 enables us to construct a finite list of possible rational zeros. Finding any rational zeros then becomes a routine, although sometimes tedious, process of elimination.

EXAMPLE 3 Finding Rational Zeros

Find all the rational zeros for $P(x) = 2x^3 - 9x^2 + 7x + 6$.

Solution If b/c in lowest terms is a rational zero of $P(x)$, then b must be a factor of 6 and c must be a factor of 2.

Possible values of b are the integer factors of 6: $\pm 1, \pm 2, \pm 3, \pm 6$ (4)

Possible values of c are the integer factors of 2: $\pm 1, \pm 2$ (5)

Writing all possible fractions b/c where b is from (4) and c is from (5), we have

Possible rational zeros for $P(x)$: $\pm 1, \pm 2, \pm 3, \pm 6, \pm \frac{1}{2}, \pm \frac{3}{2}$ (6)

[*Note:* All fractions are in lowest terms, and duplicates like $\pm 6/\pm 2 = \pm 3$ are not repeated.] If $P(x)$ has any rational zeros, they must be in list (6). We use a synthetic division table to test the numbers in this list until we find a zero. If we complete the list without finding a zero, we can conclude that $P(x)$ does not have any rational zeros.

	2	−9	7	6	
$\frac{1}{2}$	2	−8	3	$\frac{15}{2}$	
1	2	−7	0	6	
$\frac{3}{2}$	2	−6	−2	3	
2	2	−5	−3	0	$P(2) = 0$

We have found a zero! We could continue to test the remaining eight numbers in list (6) to see if there are more rational zeros. However, it is generally more efficient to factor the original polynomial at this point, producing a polynomial of lower degree that is referred to as the **reduced polynomial** for $P(x)$. Using the last line in the synthetic division table, we have

$$P(x) = 2x^3 - 9x^2 + 7x + 6 = (x - 2)(2x^2 - 5x - 3)$$

The reduced polynomial $Q(x) = 2x^2 - 5x - 3$ is a second-degree polynomial whose remaining zeros can be found by factoring or, if necessary, the quadratic formula. Thus,

$$P(x) = (x - 2)(2x^2 - 5x - 3) = (x - 2)(x - 3)(2x + 1)$$

and the rational zeros of $P(x)$ are 2, 3, and $-\frac{1}{2}$. Since $P(x)$ is a cubic polynomial, we can conclude that we have found all the zeros of $P(x)$, without testing the remaining numbers in list (6).

Matched Problem 3 Find all rational zeros for $P(x) = 2x^3 + x^2 - 11x - 10$.

Strategy for Finding Rational Zeros

Assume that $P(x)$ is a polynomial with integer coefficients and is of degree greater than 2.

Step 1. List the possible rational zeros of $P(x)$ using the rational zero theorem (Theorem 6).

Step 2. Construct a synthetic division table. If a rational zero r is found, stop, write

$$P(x) = (x - r)Q(x)$$

and immediately proceed to find the rational zeros for $Q(x)$, the reduced polynomial relative to $P(x)$. If the degree of $Q(x)$ is greater than 2, return to step 1, using $Q(x)$ in place of $P(x)$. If $Q(x)$ is quadratic, find all its zeros, using standard methods for solving quadratic equations.

EXAMPLE 4 Finding Rational and Irrational Zeros

Find all zeros exactly for $P(x) = 2x^3 - 7x^2 + 4x + 3$.

Solution *Step 1.* $\pm 1, \pm 3, \pm\frac{1}{2}, \pm\frac{3}{2}$ Possible rational zeros

Step 2.

	2	−7	4	3
$\frac{1}{2}$	2	−6	1	$\frac{7}{2}$
1	2	−5	−1	2
$\frac{3}{2}$	2	−4	−2	0

We see that $x = \frac{3}{2}$ is a zero. Thus,

$$P(x) = (x - \tfrac{3}{2})(2x^2 - 4x - 2)$$

The reduced polynomial $Q(x) = 2x^2 - 4x - 2$ is quadratic, so its zeros can be found by standard methods. This time $Q(x)$ does not factor by inspection, so we use the quadratic formula:

$$2x^2 - 4x - 2 = 0$$
$$x^2 - 2x - 1 = 0$$
$$x = \frac{2 \pm \sqrt{4 - 4(1)(-1)}}{2}$$
$$= \frac{2 \pm 2\sqrt{2}}{2} = 1 \pm \sqrt{2}$$

The exact zeros of $P(x)$ are $\frac{3}{2}$ and $1 \pm \sqrt{2}$.

Matched Problem 4 Find all zeros exactly for $P(x) = 3x^3 - 10x^2 + 5x + 4$.

Remark. A graphing utility can speed up the process of looking for rational zeros. Figure 2 shows the graph of the polynomial $P(x)$ discussed in Example 4. A glance at the graph shows that we need not have tested $x = \frac{1}{2}$ or $x = 1$. A graphing utility can also be used to evaluate the polynomial in order to test possible roots (see Fig. 2); however, synthetic division is necessary to factor $P(x)$. In the next section we will discuss using graphing utilities to approximate irrational zeros, such as $1 \pm \sqrt{2}$. And we will see that a synthetic division table is a useful tool when searching for the zeros of a polynomial.

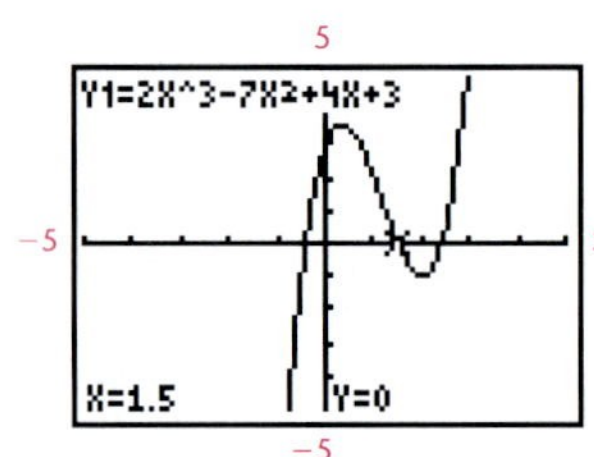

FIGURE 2

EXAMPLE 5 Finding Rational and Imaginary Zeros

Find all zeros exactly for $P(x) = x^4 - 6x^3 + 14x^2 - 14x + 5$.

Solution *Step 1.* ± 1 and ± 5 Possible rational zeros

Step 2.

$$\begin{array}{r|rrrrr} & 1 & -6 & 14 & -14 & 5 \\ \hline 1 & 1 & -5 & 9 & -5 & 0 \end{array}$$

Thus, 1 is a zero of $P(x)$, and we can write

$$P(x) = (x - 1)(x^3 - 5x^2 + 9x - 5)$$

This time the reduced polynomial is a cubic, so we repeat steps 1 and 2 using $Q(x)$.

$$Q(x) = x^3 - 5x^2 + 9x - 5 \quad \text{Reduced polynomial}$$

Step 1. ± 1 and ± 5 Possible rational zeros

Step 2.

$$\begin{array}{r|rrrr} & 1 & -5 & 9 & -5 \\ \hline 1 & 1 & -4 & 5 & 0 \end{array}$$

$$Q(x) = (x - 1)(x^2 - 4x + 5)$$

We find the zeros of the reduced quadratic polynomial $Q_1(x) = (x^2 - 4x + 5)$ by using the quadratic formula:

$$x^2 - 4x + 5 = 0$$

$$x = \frac{4 \pm \sqrt{16 - 4(1)(5)}}{2}$$

$$= \frac{4 \pm \sqrt{-4}}{2} = 2 \pm i$$

The exact zeros of $P(x)$ are 1 (multiplicity 2), $2 - i$, and $2 + i$.

Matched Problem 5 Find all zeros exactly for $P(x) = x^4 + 4x^3 + 10x^2 + 12x + 5$.

Remark. Example 5 illustrates the importance of using the reduced polynomial whenever a zero is encountered. Testing possible rational zeros in the original polynomial will never reveal any multiple zeros.

Answers to Matched Problems

1. (A) $P(1) = 1^{54} - 1 = 0$ implies that $x - 1$ is a factor of $P(x)$.
(B) $-3, -7, 8, -1$ (C) $-2i, 2i$ (D) No x intercepts

2. $p(x) = (x - 3)^2[x - (3 - i)][x - (3 + i)]$

3. $-2, -1, \frac{5}{2}$

4. $\frac{4}{3}, 1 - \sqrt{2}, 1 + \sqrt{2}$

5. -1 (multiplicity 2), $-1 - 2i, -1 + 2i$

EXERCISE 3-2

A

Write the zeros of each polynomial in Problems 1–4, and indicate the multiplicity of each if over 1. What is the degree of each polynomial?

1. $P(x) = (x + 2)^5(x - 3)^4$

2. $P(x) = (x + 6)^2(x - 5)^3$

3. $P(x) = 2(x + 1)(x - 1)^4(x - 7)^3$

4. $P(x) = 2(x + 4)^3(x - 3)^2(x - 4)$

In Problems 5–10, find a polynomial P(x) of lowest degree, with leading coefficient 1, that has the indicated set of zeros. Leave the answer in a factored form. Indicate the degree of the polynomial.

5. 2 (multiplicity 3), -1

6. -4 (multiplicity 2), 0, 2 (multiplicity 3)

7. $0, 2, -1 + \sqrt{3}, -1 - \sqrt{3}$

8. 1 (multiplicity 3), $-3 + \sqrt{2}, -3 - \sqrt{2}$

9. -2 (multiplicity 2), 2 (multiplicity 2), $3 - i, 3 + i$

10. $2i$ (multiplicity 2), $-2i$ (multiplicity 2), 2 (multiplicity 3)

In Problems 11–16, find a polynomial of lowest degree, with leading coefficient 1, that has the indicated graph. Assume all zeros are integers. Leave the answer in a factored form. Indicate the degree of each polynomial.

11.

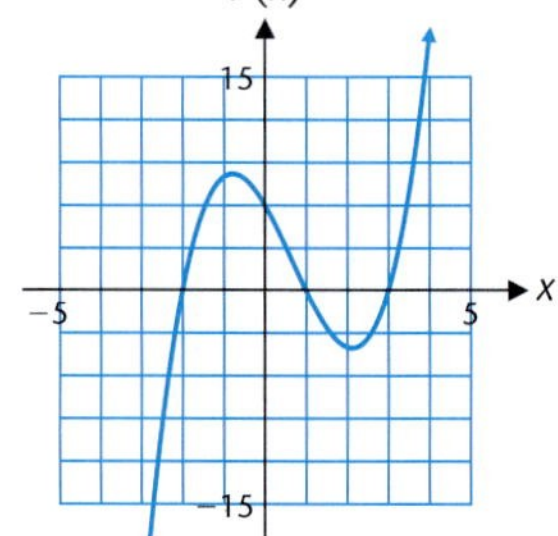

12.

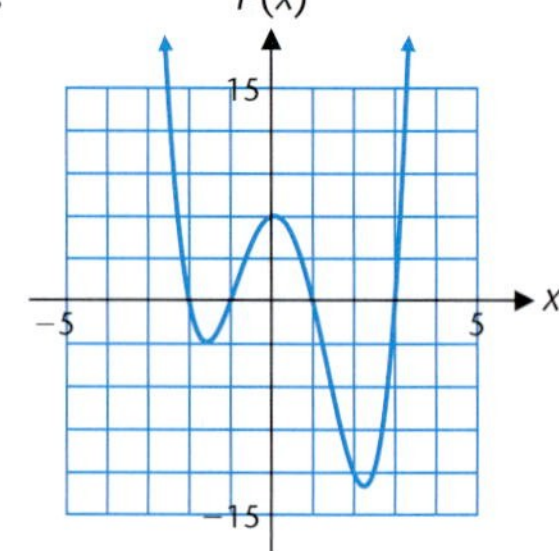

13.

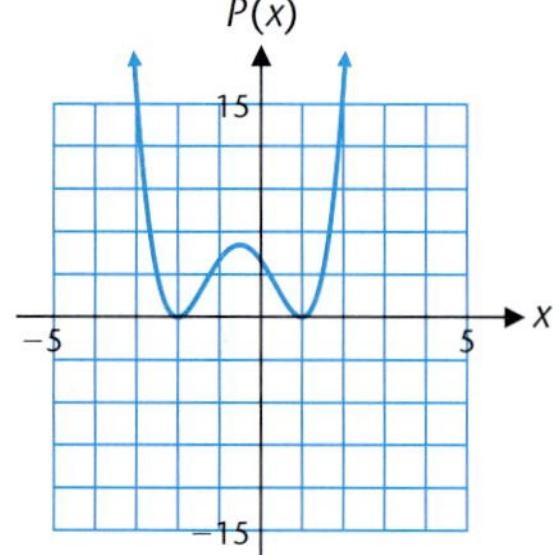

14.

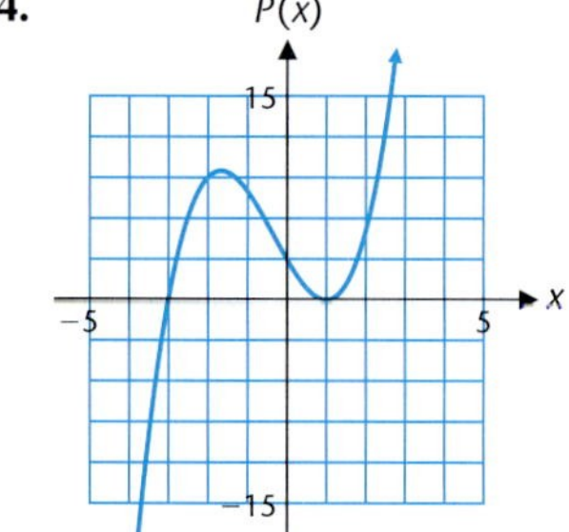

15.

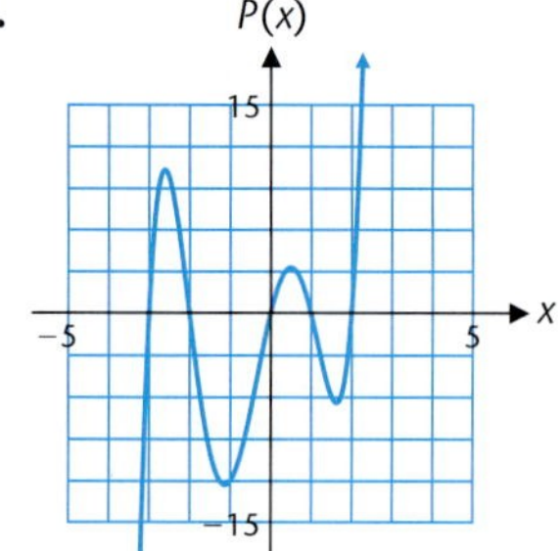

16.

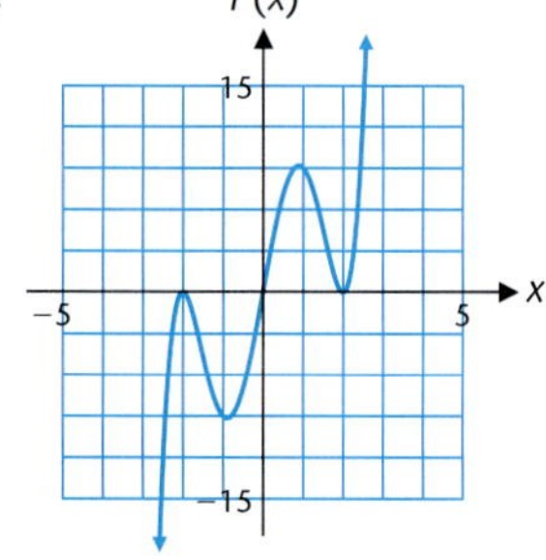

In Problems 17–20, determine whether the second polynomial is a factor of the first polynomial without dividing or using synthetic division. [Hint: Evaluate directly and use the factor theorem.]

17. $x^{23} - 1; x - 1$

18. $x^{23} - 1; x + 1$

19. $3x^3 - 5x^2 - 4x + 6; x + 1$

20. $3x^3 - 5x^2 - 4x + 6; x - 1$

B

For each polynomial in Problems 21–26, list all possible rational zeros (Theorem 6).

21. $P(x) = x^3 - 3x^2 + 2x - 10$

22. $P(x) = x^3 + 5x^2 - 8x + 14$

23. $P(x) = 2x^3 + 9x^2 - 6x + 5$

24. $P(x) = 3x^3 - 2x^2 - 4x + 2$

25. $P(x) = 6x^3 + 5x^2 + 2x - 25$

26. $P(x) = 10x^3 + 2x^2 - 7x - 9$

In Problems 27–32, write P(x) as a product of linear terms.

27. $P(x) = x^3 + 9x^2 + 24x + 16$; -1 is a zero

28. $P(x) = x^3 - 4x^2 - 3x + 18$; 3 is a double zero

29. $P(x) = x^4 - 1$; 1 and -1 are zeros

30. $P(x) = x^4 + 2x^2 + 1$; i is a double zero

31. $P(x) = 2x^3 - 17x^2 + 90x - 41$; $\frac{1}{2}$ is a zero

32. $P(x) = 3x^3 - 10x^2 + 31x + 26$; $-\frac{2}{3}$ is a zero

In Problems 33–40, find all roots exactly (rational, irrational, and imaginary) for each polynomial equation.

33. $2x^3 - 7x^2 + 2x + 6 = 0$

34. $2x^3 - 7x^2 - 6x - 1 = 0$

35. $x^4 + 2x^3 - 2x^2 - 6x - 3 = 0$

36. $x^4 - 11x^2 + 12x + 4 = 0$

37. $x^4 + 2x^3 - 10x^2 - 18x + 9 = 0$

38. $x^4 - 2x^3 + 9x^2 + 2x - 10 = 0$

39. $2x^5 - x^4 - 5x^3 + 10x^2 - 2x - 4 = 0$

40. $3x^5 + 10x^4 + 4x^3 - 20x^2 - 7x + 10 = 0$

In Problems 41–48, find all zeros exactly (rational, irrational, and imaginary) for each polynomial.

41. $P(x) = x^3 + 5x^2 - 2x - 24$

42. $P(x) = x^3 - 4x^2 - 9x + 36$

43. $P(x) = x^4 - 3.3x^3 + 2.3x^2 + 0.6x$

44. $P(x) = x^4 - 4.1x^3 + 0.1x^2 + 1.2x$

45. $P(x) = x^4 - 2x^3 - 14x^2 + 30x + 9$

46. $P(x) = x^4 + 9x^3 + 23x^2 + 8x - 16$

47. $P(x) = 3x^5 - 2x^4 + 6x^3 + 20x^2 - x - 10$

48. $P(x) = 4x^5 - 18x^4 + 24x^3 - 7x^2 - 4x + 4$

In Problems 49–54, write each polynomial as a product of linear factors.

49. $P(x) = 6x^3 + 19x^2 + 11x - 6$

50. $P(x) = 6x^3 - 11x^2 - 4x + 4$

51. $P(x) = x^3 - x^2 - 13x - 3$

52. $P(x) = x^3 - 4x^2 + 2x + 4$

53. $P(x) = 4x^4 - 4x^3 - 19x^2 + 16x + 12$

54. $P(x) = 4x^4 + 16x^3 + 7x^2 - 18x - 9$

In Problems 55–60, solve each inequality (see Section 1-8).

55. $x^2 \le 4x - 1$

56. $x^2 > 2x + 1$

57. $x^3 + 3 \le 3x^2 + x$

58. $9x + 9 \le x^3 + x^2$

59. $2x^3 + 6 \ge 13x - x^2$

60. $5x^3 - 3x^2 < 10x - 6$

In Problems 61–64, multiply.

61. $[x - (4 - 5i)][x - (4 + 5i)]$

62. $[x - (5 + 2i)][x - (5 - 2i)]$

63. $[x - (a + bi)][x - (a - bi)]$

64. $(x - bi)(x + bi)$

C

In Problems 65–70, find all other zeros of P(x), given the indicated zero.

65. $P(x) = x^3 + x + 10$; $1 + 2i$ is one zero

66. $P(x) = x^3 + 2x^2 - 3x - 10$, $-2 + i$ is one zero

67. $P(x) = x^3 + 4x^2 + 9x + 36$; $-3i$ is one zero

68. $P(x) = x^3 - 5x^2 + 4x - 20$; $2i$ is one zero

69. $P(x) = x^4 - 8x^3 + 24x^2 - 20x - 13$; $3 - 2i$ is one zero

70. $P(x) = x^4 - 6x^3 + 19x^2 - 42x + 10$; $1 - 3i$ is one zero

In Problems 71–74, solve each inequality (see Section 1-8).

71. $\dfrac{4}{2x^3 + 5x^2 - 2x - 5} \ge 0$

72. $\dfrac{7}{2x^3 - x^2 - 8x + 4} \le 0$

73. $\dfrac{x^2 - 3x - 10}{x^3 - 4x^2 + x + 6} \le 0$

74. $\dfrac{x^2 + 4x - 21}{x^3 + 7x^2 + 7x - 15} \ge 0$

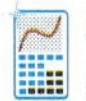

Problems 75–80 require the use of a graphing utility. Graph the polynomial and use the graph to help locate the real zeros. Then find all zeros (rational, irrational, and imaginary) exactly.

75. $P(x) = 3x^3 - 37x^2 + 84x - 24$

76. $P(x) = 2x^3 - 9x^2 - 2x + 30$

77. $P(x) = 4x^4 + 4x^3 + 49x^2 + 64x - 240$

78. $P(x) = 6x^4 + 35x^3 + 2x^2 - 233x - 360$

79. $P(x) = 4x^4 - 44x^3 + 145x^2 - 192x + 90$

80. $P(x) = x^5 - 6x^4 + 6x^3 + 28x^2 - 72x + 48$

81. The solutions to the equation $x^3 - 1 = 0$ are all the cube roots of 1.
(A) How many cube roots of 1 are there?
(B) 1 is obviously a cube root of 1; find all others.

82. The solutions to the equation $x^3 - 8 = 0$ are all the cube roots of 8.
(A) How many cube roots of 8 are there?
(B) 2 is obviously a cube root of 8; find all others.

83. If P is a polynomial function with real coefficients of degree n, with n odd, then what is the maximum number of times the graph of $y = P(x)$ can cross the x axis? What is the minimum number of times?

84. Answer the questions in Problem 83 for n even.

85. Given $P(x) = x^2 + 2ix - 5$ with $2 - i$ a zero, show that $2 + i$ is not a zero of $P(x)$. Does this contradict Theorem 4? Explain.

86. If $P(x)$ and $Q(x)$ are two polynomials of degree n, and if $P(x) = Q(x)$ for more than n values of x, then how are $P(x)$ and $Q(x)$ related?

APPLICATIONS 

Find all rational solutions exactly, and find irrational solutions to two decimal places.

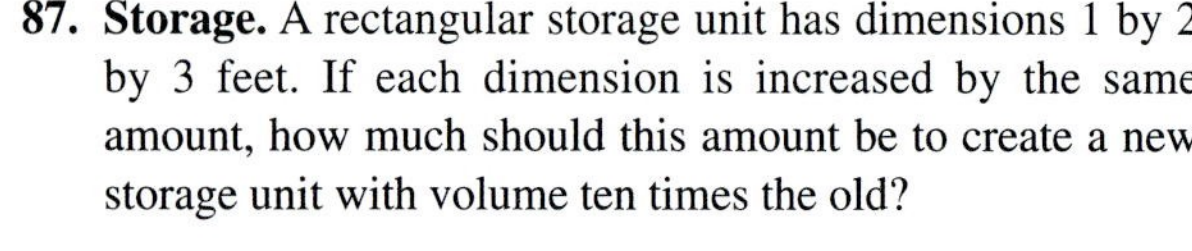

87. Storage. A rectangular storage unit has dimensions 1 by 2 by 3 feet. If each dimension is increased by the same amount, how much should this amount be to create a new storage unit with volume ten times the old?

88. Construction. A rectangular box has dimensions 1 by 1 by 2 feet. If each dimension is increased by the same amount, how much should this amount be to create a new box with volume six times the old?

★ **89. Packaging.** An open box is to be made from a rectangular piece of cardboard that measures 8 by 5 inches, by cutting out squares of the same size from each corner and bending up the sides (see the figure). If the volume of the box is to be 14 cubic inches, how large a square should be cut from each corner? [*Hint:* Determine the domain of x from physical considerations before starting.]

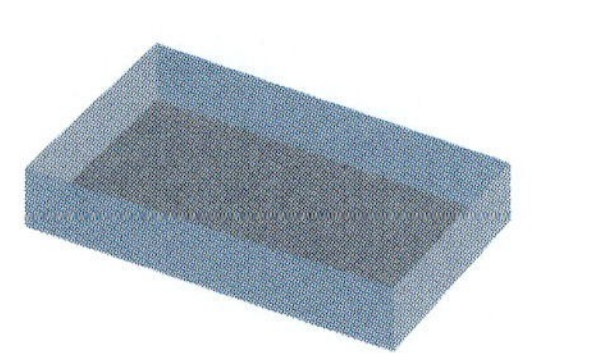

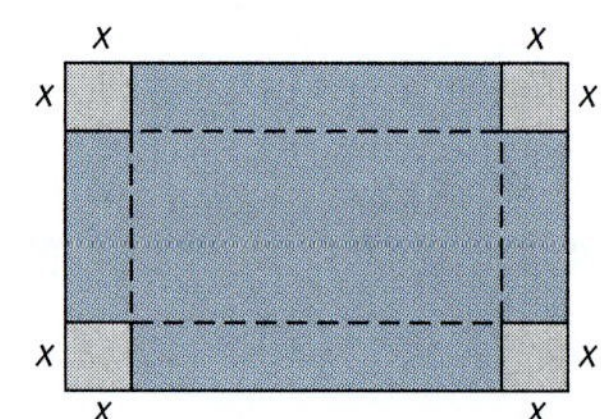

★ **90. Fabrication.** An open metal chemical tank is to be made from a rectangular piece of stainless steel that measures 10 by 8 feet, by cutting out squares of the same size from each corner and bending up the sides (see the figure). If the volume of the tank is to be 48 cubic feet, how large a square should be cut from each corner?

SECTION 3-3 Approximating Real Zeros of Polynomials

- Locating Real Zeros
- The Bisection Method
-  Approximating Real Zeros Using a Graphing Utility
- Application

The strategy for finding zeros discussed in the preceding section is designed to find as many exact real and imaginary zeros as possible. But there are zeros that cannot be found by using the strategy. For example, the polynomial

$$P(x) = x^5 + x - 1$$

must have at least one real zero (Theorem 5 in Section 3-2). Since the only possible rational zeros are ± 1 and neither of these turns out to be a zero, $P(x)$ must have at least one irrational zero. We cannot find the exact value of this zero, but it can be approximated using various well-known methods.

In this section we will develop two important tools for locating real zeros, the *location theorem* and the *upper and lower bound theorem.* Next we will discuss how the location theorem forms the basis for the *method of bisection,* a popular method that is used by most graphing utilities to approximate real zeros. Finally, we will see how the upper and lower bound theorem can aid in approximating real zeros with a graphing utility. We will restrict our attention to the real zeros of polynomials with real coefficients.

• Locating Real Zeros

Let us return to the polynomial function

$$P(x) = x^5 + x - 1$$

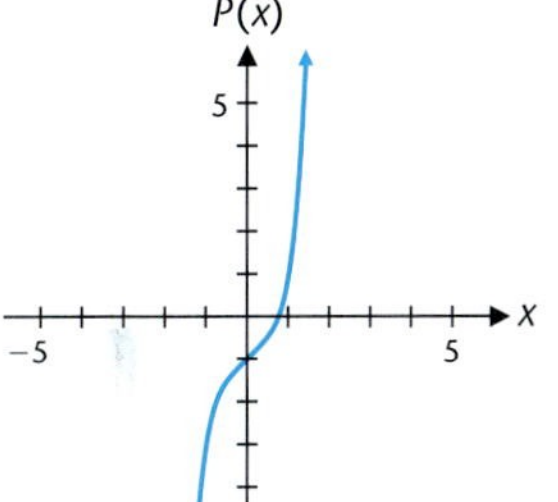

FIGURE 1 $P(x) = x^5 + x - 1.$

As we found above, $P(x)$ has no rational zeros and at least one irrational zero. The graph of $P(x)$ is shown in Figure 1.

Note that $P(0) = -1$ and $P(1) = 1$. Since the graph of a polynomial function is continuous, the graph of $P(x)$ must cross the x axis at least once between $x = 0$ and $x = 1$. This observation is the basis for Theorem 1 and leads to an effective method for locating zeros.

Theorem 1 **Location Theorem**

If f is continuous on an interval I, a and b are two numbers in I, and $f(a)$ and $f(b)$ are of opposite sign, then there is at least one x intercept between a and b.

We will find Theorem 1 very useful when we are searching for real zeros, hence the name *location theorem.* It is important to remember that "at least" in Theorem 1 means "one or more." Notice in Figure 2(a) that $f(-3) = -15 < 0$, $f(3) = 15 > 0$, and f has one zero between -3 and 3. In Figure 2(b), $f(-3) = -15$ and $f(3) = 15$, but this time there are three zeros between -3 and 3.

FIGURE 2 The location theorem.

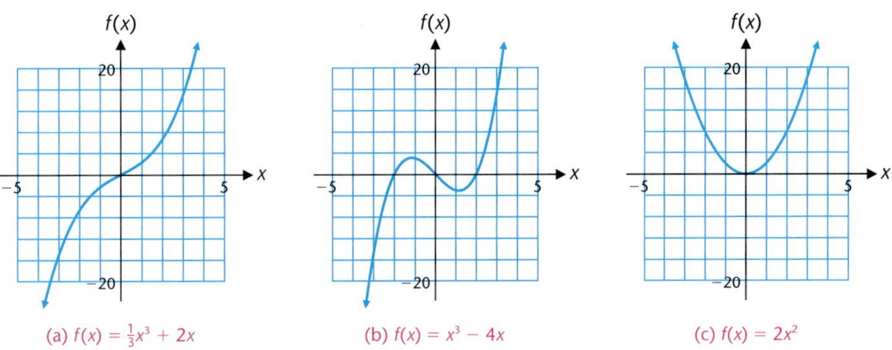

The converse to the location theorem (Theorem 1) is false; that is, if c is a zero of f, then f may or may not change sign at c. Compare Figure 2(a) and (c). Both functions have a zero at $x = 0$, but the first changes sign at 0 and the second does not.

EXAMPLE 1 Locating Real Zeros

Let $P(x) = x^3 - 6x^2 + 9x - 3$. Use a synthetic division table to locate the zeros of $P(x)$ between successive integers.

Solution We construct a synthetic division table and look for sign changes.

	1	−6	9	−3	
0	1	−6	9	−3	
					Sign change
1	1	−5	4	1	
					Sign change
2	1	−4	1	−1	
3	1	−3	0	−3	
					Sign change
4	1	−2	1	1	

According to Theorem 1, $P(x)$ must have a real zero in each of the intervals (0, 1), (1, 2), and (3, 4). Since $P(x)$ is a cubic polynomial, we have located all its zeros.

Matched Problem 1 Let $P(x) = x^3 - 8x^2 + 15x - 2$. Use a synthetic division table to locate the zeros of $P(x)$ between successive integers.

In the solution to Example 1, we located three zeros in a relatively few number of steps and could stop searching because we knew that a cubic polynomial could not have more than three zeros. But what if we had not found three zeros? Some cubic polynomials have only one real zero. How can we tell when we have searched far enough? The next theorem tells us how to find *upper* and *lower bounds* for the real zeros of a polynomial. Any number that is greater than or equal to the largest zero of a polynomial is called an **upper bound** of the zeros of the polynomial. Similarly, any number that is less than or equal to the smallest zero of the polynomial is called a **lower bound** of the zeros of the polynomial. Theorem 2, based on the synthetic division process, enables us to determine upper and lower bounds of the real zeros of a polynomial with real coefficients.

Theorem 2 **Upper and Lower Bounds of Real Zeros**

Given an nth-degree polynomial $P(x)$ with real coefficients, $n > 0$, $a_n > 0$, and $P(x)$ divided by $x - r$ using synthetic division:

1. **Upper Bound.** If $r > 0$ and all numbers in the quotient row of the synthetic division, including the remainder, are nonnegative, then r is an upper bound of the real zeros of $P(x)$.
2. **Lower Bound.** If $r < 0$ and all numbers in the quotient row of the synthetic division, including the remainder, alternate in sign, then r is a lower bound of the real zeros of $P(x)$.

[*Note:* In the lower-bound test, if 0 appears in one or more places in the quotient row, including the remainder, the sign in front of it can be considered either positive or negative, but not both. For example, the numbers 1, 0, 1 can be considered to alternate in sign, while 1, 0, −1 cannot.]

We sketch a proof of part 1 of Theorem 2. The proof of part 2 is similar, only a little more difficult.

Proof If all the numbers in the quotient row of the synthetic division are nonnegative after dividing $P(x)$ by $x - r$, then

$$P(x) = (x - r)Q(x) + R$$

where the coefficients of $Q(x)$ are nonnegative and R is nonnegative. If $x > r > 0$, then $x - r > 0$ and $Q(x) > 0$; hence,

$$P(x) = (x - r)Q(x) + R > 0$$

Thus, $P(x)$ cannot be 0 for any x greater than r, and r is an upper bound for the real zeros of $P(x)$.

EXAMPLE 2 Bounding Real Zeros

Let $P(x) = x^4 - 2x^3 - 10x^2 + 40x - 90$. Find the smallest positive integer and the largest negative integer that, by Theorem 2, are upper and lower bounds, respectively, for the real zeros of $P(x)$. Also note the location of any zeros discovered in the process of building the synthetic division table.

Solution An easy way to locate the upper and lower bounds is to test $r = 1, 2, 3, \ldots$ until the quotient row turns nonnegative; then test $r = -1, -2, -3, \ldots$ until the quotient row alternates in sign. It is also useful to include $r = 0$ in the table to detect any sign changes between $r = 0$ and $r = \pm 1$.

		1	−2	−10	40	−90	
	0	1	−2	−10	40	−90	
	1	1	−1	−11	29	−61	
	2	1	0	−10	20	−50	
	3	1	1	−7	19	−33	
	4	1	2	−2	32	38	
UB	5	1	3	5	65	235	← This quotient row is nonnegative; hence, 5 is an upper bound (UB).
	−1	1	−3	−7	47	−137	
	−2	1	−4	−2	44	−178	
	−3	1	−5	5	25	−165	
	−4	1	−6	14	−16	−26	
LB	−5	1	−7	25	−85	335	← This quotient row alternates in sign; hence, −5 is a lower bound (LB).

Because of Theorem 2, we conclude that all real zeros of $P(x) = x^4 - 2x^3 - 10x^2 + 40x - 90$ must lie between -5 and 5. We also note that there must be at least one zero in (3, 4) and at least one in $(-5, -4)$.

Matched Problem 2 Let $P(x) = x^4 - 5x^3 - x^2 + 40x - 70$. Find the smallest positive integer and the largest negative integer that, by Theorem 2, are upper and lower bounds, respectively, for the real zeros of $P(x)$. Also note the location of any zeros discovered in the process of building the synthetic division table.

• The Bisection Method

Now that we know how to locate real zeros of a polynomial, we turn to the problem of actually approximating a real zero. Explore-Discuss 1 provides an introduction to the repeated systematic application of the location theorem (Theorem 1) called the *bisection method.* This is the method for approximating real zeros that is programmed into many graphing utilities.

EXPLORE-DISCUSS 1 Let $P(x) = x^3 + x - 1$. Since $P(0) = -1$ and $P(1) = 1$, the location theorem implies that $P(x)$ must have at least one zero in (0, 1).

•

(A) Is $P(0.5)$ positive or negative? Is there a zero in (0, 0.5) or in (0.5, 1)?
(B) Let m be the midpoint of the interval from part A that contains a zero. Is $P(m)$ positive or negative? What does this tell you about the location of the zero?
(C) Explain how this process could be used repeatedly to approximate a zero to any desired accuracy.

The **bisection method** used to approximate real zeros is straightforward: Let $P(x)$ be a polynomial with real coefficients. If $P(x)$ has opposite signs at the endpoints of the interval (a, b), then a real zero r lies in this interval. We bisect this interval [find the midpoint $m = (a + b)/2$], check the sign of $P(m)$, and choose the interval (a, m) or (m, b) on which $P(x)$ has opposite signs at the endpoints. We repeat this bisecting process (producing a set of "nested" intervals, each half the size of the preceding one and each containing the real zero r) until we get the desired decimal accuracy for the zero approximation. At any point in the process if $P(m) = 0$, we stop, since m is a real zero. An example will help clarify the process.

EXAMPLE 3 Approximating Real Zeros by Bisection

For the polynomial $P(x) = x^4 - 2x^3 - 10x^2 + 40x - 90$ in Example 2, we found that all the real zeros lie between -5 and 5 and that each of the intervals $(-5, -4)$ and (3, 4) contained at least one zero. Use bisection to approximate a real zero on the interval (3, 4) to one decimal place accuracy.

Solution We start the process with a synthetic division table:

	1	−2	−10	40	−90	
3	1	1	−7	19	−33	$= P(3)$
4	1	2	−2	32	38	$= P(4)$

TABLE 1 Bisection Approximation

Sign change interval (a, b)	Midpoint m	Sign of P: $P(a)$	$P(m)$	$P(b)$
(3, 4)	3.5	−	−	+
(3.5, 4)	3.75	−	+	+
(3.5, 3.75)	3.625	−	+	+
(3.5, 3.625)	3.563	−	−	+
(3.563, 3.625)	We stop here	−		+

Since the sign of $P(x)$ changes at the endpoints of the interval (3.563, 3.625), we conclude that a real zero lies on this interval and is given by $r = 3.6$ to one decimal place accuracy (each endpoint rounds to 3.6).

Figure 3 illustrates the nested intervals produced by the bisection method in Table 1. Match each step in Table 1 with an interval in Figure 3. Note how each interval that contains a zero gets smaller and smaller and is contained in the preceding interval that contained the zero.

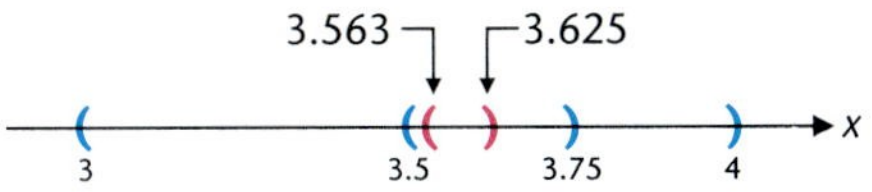

FIGURE 3 Nested intervals produced by the bisection method in Table 1.

If we had wanted two decimal place accuracy, we would use four decimal places for the values of x and continue the process in Table 1 until the endpoints of a sign change interval rounded to the same two decimal place number.

Matched Problem 3 Use the bisection method to approximate to one decimal place accuracy a zero on the interval $(-5, -4)$ for the polynomial in Example 3.

• Approximating Real Zeros Using a Graphing Utility

The bisection method is easy to understand but tedious to carry out, especially if the approximation must be accurate to more than two decimal places. Fortunately, this is the type of repetitive calculation that a graphing utility can be programmed to carry out. In fact, we have been using a graphing utility for some time now to find the zeros of a function (see Section 2-4). Now we will see how the upper and lower bound theorem can be used in conjunction with the zero approximation routine on a graphing utility to approximate all the real zeros of a polynomial.

EXAMPLE 4 Approximating Real Zeros Using a Graphing Utility

Given the polynomial $P(x) = x^5 + x - 1$:

(A) Form a synthetic division table to find upper and lower bounds for any real zeros, and locate real zeros between successive integers.

(B) Graph $P(x)$ in a graphing utility, and approximate any real zeros to four decimal places using a built-in zero approximation routine.

Solution (A) Form a synthetic division table:

		1	0	0	0	1	−1	
	0	1	0	0	0	1	−1	Real zero
UB	1	1	1	1	1	2	1	
LB	−1	1	−1	1	−1	2	−3	

From the table we see that all real zeros of $P(x)$ are between -1 and 1, and a real zero lies on the interval (0, 1).

(B) Enter $P(x)$ in a graphing utility, and set the window dimensions with the synthetic division table in part A as a guide. Figure 4(a) shows the graph of $P(x)$, and Figure 4(b) shows the zero approximation using a built-in routine.

FIGURE 4

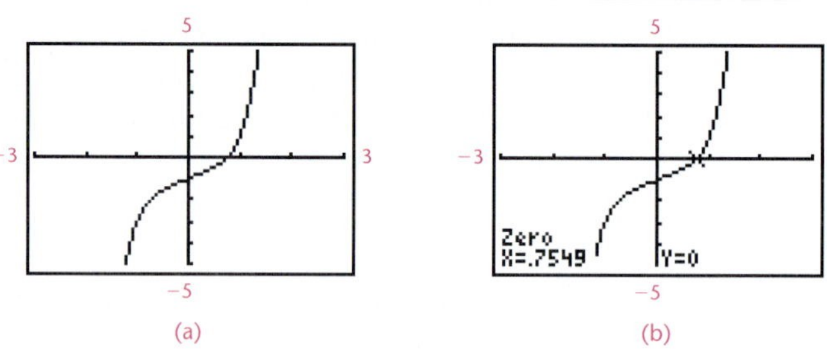

(a) (b)

It is clear from the graph and the upper and lower bounds of the zeros found in part A that $P(x)$ has only one real zero, which is, to four decimal places, $x = 0.7549$.

Matched Problem 4

Given the polynomial $P(x) = x^5 - x^2 + 1$:

(A) Form a synthetic division table to find upper and lower bounds for any real zeros, and locate real zeros between successive integers.

(B) Graph $P(x)$ in a graphing utility and approximate any real zeros to four decimal places using a built-in zero approximation routine.

Earlier in this section and in Section 3-1 we saw that a calculator is a useful tool for constructing a synthetic division table. A graphing utility that can store and execute programs is even more useful. Table 2 shows a simple program for a graphing calculator that will perform synthetic division. The table also shows the output generated when we use this program to construct the synthetic division table in Example 2.

TABLE 2 Synthetic Division on a Graphing Utility

Program SYNDIV*

TI-82/TI-83	TI-85/TI-86
Lbl A	Lbl A
Prompt R	Prompt R
2→I	2→I
dim(L1)→N	dimL L1→N
{0}→L2	{0}→L2
N→dim(L2)	N→dimL L2
L1(1)→L2(1)	L1(1)→L2(1)
Lbl B	Lbl B
L2(I-1)*R+L1(I)→L2(I)	L2(I-1)*R+L1(I)→L2(I)
1+I→I	1+I→I
If I≤N	If I≤N
Goto B	Goto B
Pause L2	Pause L2
Goto A	Goto A

Output

```
{1,-2,-10,40,-90}→L1
        {1 -2 -10 40 -90}
SYNDIV
R=?1
       {1 -1 -11 29 -61}
R=?2
        {1 0 -10 20 -50}
R=?3
         {1 1 -7 19 -33}
R=?4
          {1 2 -2 32 38}
R=?5
          {1 3 5 65 235}
R=?-1
       {1 -3 -7 47 -137}
R=?-2
       {1 -4 -2 44 -178}
R=?-3
        {1 -5 5 25 -165}
R=?-4
       {1 -6 14 -16 -26}
R=?-5
       {1 -7 25 -85 335}
R=?
```

*Available for download at *www.mhhe.com/barnett.*

 **EXPLORE-DISCUSS 2** If you have a TI-82, TI-83, TI-85, or TI-86 graphing calculator, enter the appropriate version of SYNDIV in your calculator exactly as shown in Table 2. To use the program, store the coefficients of the polynomial in L1 (see the first line of output in Table 2) and execute the program. Press ENTER to continue after each line is displayed. Press QUIT at the "R=?" prompt to terminate the program.

If you have some other graphing utility that can store and execute programs, consult your manual and modify the statements in SYNDIV so that the program works on your graphing utility.

 **EXAMPLE 5 Approximating Real Zeros with a Graphing Utility**

Let $P(x) = x^3 - 30x^2 + 275x - 720$:

(A) Find the smallest positive integer multiple of 10 and the largest negative integer multiple of 10 that, by Theorem 2, are upper and lower bounds, respectively, for the real zeros of $P(x)$.

(B) Use a graphing utility to approximate the real zeros of $P(x)$ to two decimal places.

Solution (A) We construct a synthetic division table to search for bounds for the zeros of $P(x)$. The size of the coefficients in $P(x)$ indicates that we can speed up this search by choosing larger increments between test values.

		1	−30	275	−720
	10	1	−20	75	30
	20	1	−10	75	780
UB	30	1	0	275	7,530
LB	−10	1	−40	675	−7,470

Thus, all real zeros of $P(x) = x^3 - 30x^2 + 275x - 720$ must lie between -10 and 30.

(B) Graphing $P(x)$ for $-10 \le x \le 30$ (Fig. 5) shows that $P(x)$ has three zeros. The approximate values of these zeros (details omitted) are 4.48, 11.28, and 14.23.

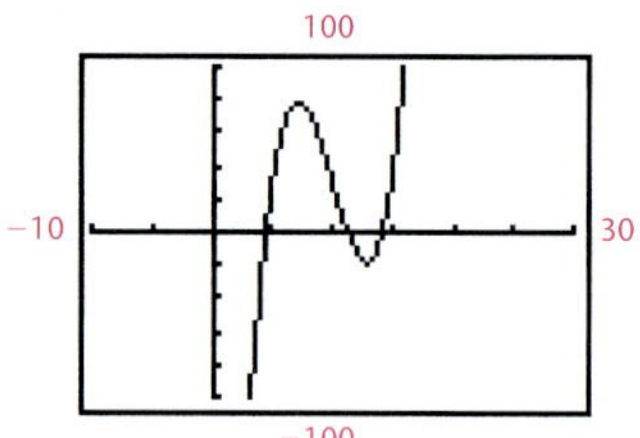

FIGURE 5 $P(x) = x^3 - 30x^2 + 275x - 720$.

Matched Problem 5 Let $P(x) = x^3 - 25x^2 + 170x - 170$.

(A) Find the smallest positive integer multiple of 10 and the largest negative integer multiple of 10 that, by Theorem 2, are upper and lower bounds, respectively, for the real zeros of $P(x)$.

(B) Use a graphing utility to approximate the real zeros of $P(x)$ to two decimal places.

Remark: One of the most frequently asked questions concerning graphing utilities is how to determine the correct viewing window. The upper and lower bound theorem provides an answer to this question for polynomial functions. As Example 5 illustrates, the upper and lower bound theorem and the zero approximation routine on a graphing utility are two important mathematical tools that work very well together.

• Application

EXAMPLE 6 **Construction**

An oil tank is in the shape of a right circular cylinder with a hemisphere at each end (see Fig. 6). The cylinder is 55 inches long, and the volume of the tank is $11{,}000\pi$ cubic inches (approximately 20 cubic feet). Let x denote the common radius of the hemispheres and the cylinder.

(A) Find a polynomial equation that x must satisfy.
(B) Approximate x to one decimal place.

FIGURE 6

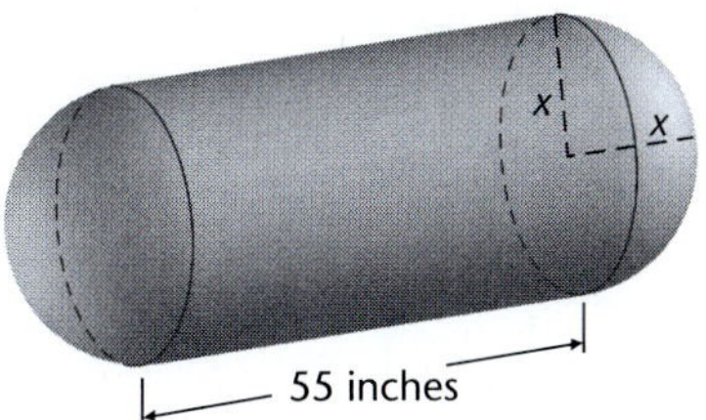

Solution (A) If x is the common radius of the hemispheres and the cylinder in inches, then

$$\begin{pmatrix}\text{Volume}\\ \text{of}\\ \text{tank}\end{pmatrix} = \begin{pmatrix}\text{Volume}\\ \text{of two}\\ \text{hemispheres}\end{pmatrix} + \begin{pmatrix}\text{Volume}\\ \text{of}\\ \text{cylinder}\end{pmatrix}$$

$$11{,}000\pi = \tfrac{4}{3}\pi x^3 + 55\pi x^2 \quad \text{Multiply by } 3/\pi.$$

$$33{,}000 = 4x^3 + 165x^2$$

$$0 = 4x^3 + 165x^2 - 33{,}000$$

Thus, x must be a positive zero of

$$P(x) = 4x^3 + 165x^2 - 33{,}000$$

(B) Since the coefficients of $P(x)$ are large, we use larger increments in the synthetic division table:

		4	165	0	−33,000
	10	4	205	2,050	−12,500
UB	20	4	245	4,900	65,000

Graphing $y = P(x)$ for $0 \le x \le 20$ (Fig. 7), we see that $x = 12.4$ inches (to one decimal place). [If you do not have a graphing utility, construct a table like Table 1 to approximate the zero of $P(x)$.]

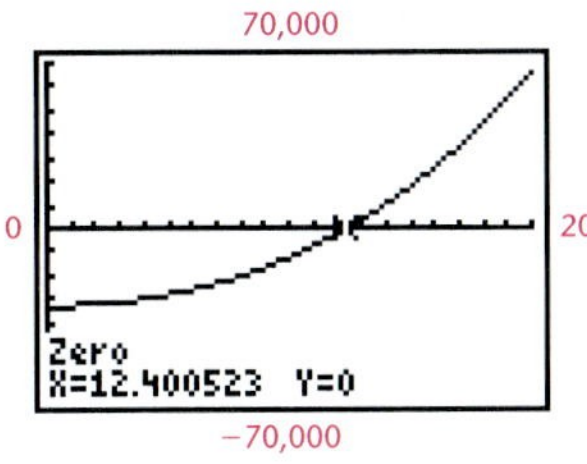

FIGURE 7 $P(x) = 4x^3 + 165x^2 - 33{,}000$.

Matched Problem 6 Repeat Example 6 if the volume of the tank is $44{,}000\pi$ cubic inches.

Answers to Matched Problems

1. Intervals containing zeros: (0, 1), (2, 3), (5, 6)
2. Lower bound: −3; Upper bound: 6
   Intervals containing zeros: (−3, −2), (3, 4)
3. $x = -4.1$
4. (A) Lower bound: −1; Upper bound: 1
   Intervals containing zeros: (−1, 0)
   (B) Real zero: $x = -0.8087$

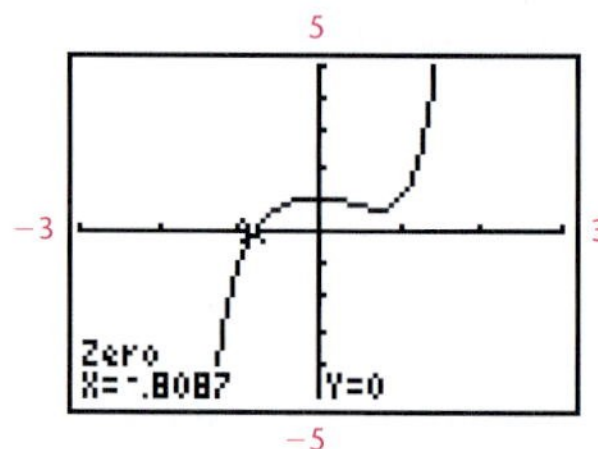

5. (A) Lower bound: −10; Upper bound: 30
   (B) Real zeros: 1.20, 11.46, 12.34
6. (A) $P(x) = 4x^3 + 165x^2 - 132{,}000 = 0$ (B) 22.7 in.

EXERCISE 3-3

A

In Problems 1–4, use the table of values for the polynomial function P to discuss the possible locations of the x intercepts of the graph of $y = P(x)$.

1.

x	−7	−5	−1	3	5	8
$P(x)$	9	4	−3	6	4	−2

2.

x	−8	−2	0	2	4	9
$P(x)$	−3	4	5	2	−5	6

3.

x	−6	−4	0	2	4	7
$P(x)$	−5	3	−4	−6	3	−5

4.

x	−5	−3	−1	0	2	5
$P(x)$	7	4	2	−1	3	−6

In Problems 5–8, use a synthetic division table and Theorem 1 to locate each real zero between successive integers.

5. $P(x) = x^3 - 9x^2 + 23x - 14$

6. $P(x) = x^3 - 12x^2 + 44x - 49$

7. $P(x) = x^3 + 3x^2 - x - 5$

8. $P(x) = x^3 + x^2 - 4x - 3$

Find the smallest positive integer and largest negative integer that, by Theorem 2, are upper and lower bounds, respectively, for the real zeros of each of the polynomials given in Problems 9–14.

9. $P(x) = x^3 - 3x + 1$

10. $P(x) = x^3 - 4x^2 + 4$

11. $P(x) = x^4 - 3x^3 + 4x^2 + 2x - 9$

12. $P(x) = x^4 - 4x^3 + 6x^2 - 4x - 7$

13. $P(x) = x^5 - 3x^3 + 3x^2 + 2x - 2$

14. $P(x) = x^5 - 3x^4 + 3x^2 + 2x - 1$

B

In Problems 15–22:

(A) Find the smallest positive integer and largest negative integer that, by Theorem 2, are upper and lower bounds, respectively, for the real zeros of P(x). Also note the location of any zeros between successive integers.

(B) Approximate to one decimal place the largest real zero of P(x) using the bisection method.

15. $P(x) = x^3 - 2x^2 - 5x + 4$

16. $P(x) = x^3 + x^2 - 4x - 1$

17. $P(x) = x^3 - 2x^2 - x + 5$

18. $P(x) = x^3 - 3x^2 - x - 2$

19. $P(x) = x^4 - 2x^3 - 7x^2 + 9x + 7$

20. $P(x) = x^4 - x^3 - 9x^2 + 9x + 4$

21. $P(x) = x^4 - x^3 - 4x^2 + 4x + 3$

22. $P(x) = x^4 - 3x^3 - x^2 + 3x + 3$

In Problems 23–30:

(A) Find the smallest positive integer and largest negative integer that, by Theorem 2, are upper and lower bounds, respectively, for the real zeros of P(x).

(B) Approximate the real zeros of each polynomial to two decimal places.

23. $P(x) = x^3 - 2x^2 + 3x - 8$

24. $P(x) = x^3 + 3x^2 + 4x + 5$

25. $P(x) = x^4 + x^3 - 5x^2 + 7x - 22$

26. $P(x) = x^4 - x^3 - 8x^2 - 12x - 25$

27. $P(x) = x^5 - 3x^3 - 4x + 4$

28. $P(x) = x^5 - x^4 - 2x^2 - 4x - 5$

29. $P(x) = x^5 + x^4 + 3x^3 + x^2 + 2x - 5$

30. $P(x) = x^5 - 2x^4 - 6x^2 - 9x + 10$

C

In Problems 31–34:

(A) Find the smallest positive integer and largest negative integer that, by Theorem 2, are upper and lower bounds, respectively, for the real zeros of P(x). Also note the location of any zeros between successive integers.

(B) Approximate to two decimal places the largest real zero of P(x) using the bisection method.

31. $P(x) = x^5 - 5x^4 + 5x^3 + 5x^2 - 10x + 5$

32. $P(x) = x^5 - 2x^4 - 7x^3 + 8x^2 + 12x - 5$

33. $P(x) = x^5 - 10x^3 + 9x + 10$

34. $P(x) = x^5 - 9x^3 + 4x^2 + 12x - 15$

In Problems 35–44:

(A) Find the smallest positive integer multiple of 10 and largest negative integer multiple of 10 that, by Theorem 2, are upper and lower bounds, respectively, for the real zeros of each polynomial.

(B) Approximate the real zeros of each polynomial to two decimal places.

35. $P(x) = x^3 - 24x^2 - 25x + 10$

36. $P(x) = x^3 - 37x^2 + 70x - 20$

37. $P(x) = x^4 + 12x^3 - 900x^2 + 5{,}000$

38. $P(x) = x^4 - 12x^3 - 425x^2 + 7{,}000$

39. $P(x) = x^4 - 100x^2 - 1{,}000x - 5{,}000$

40. $P(x) = x^4 - 5x^3 - 50x^2 - 500x + 7{,}000$

41. $P(x) = 4x^4 - 40x^3 - 1{,}475x^2 + 7{,}875x - 10{,}000$

42. $P(x) = 9x^4 + 120x^3 - 3{,}083x^2 - 25{,}674x - 48{,}400$

43. $P(x) = 0.01x^5 - 0.1x^4 - 12x^3 + 9{,}000$

44. $P(x) = 0.1x^5 + 0.7x^4 - 18.775x^3 - 340x^2 - 1{,}645x - 2{,}450$

APPLICATIONS

Express the solutions to Problems 45–50 as the roots of a polynomial equation of the form $P(x) = 0$ and approximate these solutions to one decimal place. Use a graphing utility, if available; otherwise, use the bisection method.

★ **45. Geometry.** Find all points on the graph of $y = x^2$ that are 1 unit away from the point (1, 2). [*Hint:* Use the distance-between-two-points formula from Section 2-1.]

★ **46. Geometry.** Find all points on the graph of $y = x^2$ that are 1 unit away from the point (2, 1).

★ **47. Manufacturing.** A box is to be made out of a piece of cardboard that measures 18 by 24 inches. Squares, x inches on a side, will be cut from each corner, and then the ends and sides will be folded up (see the figure). Find the value of x that would result in a box with a volume of 600 cubic inches.

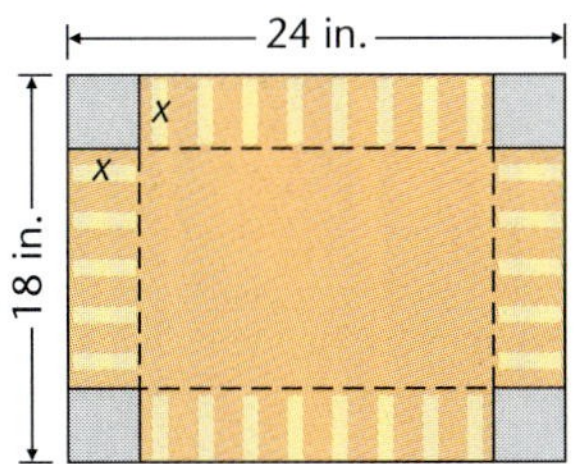

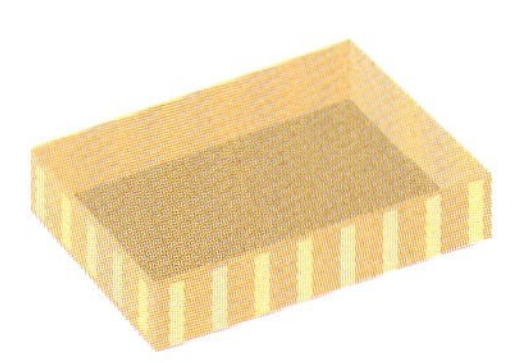

★ **48. Manufacturing.** A box with a hinged lid is to be made out of a piece of cardboard that measures 20 by 40 inches. Six squares, x inches on a side, will be cut from each corner and the middle, and then the ends and sides will be folded up to form the box and its lid (see the figure). Find the value of x that would result in a box with a volume of 500 cubic inches.

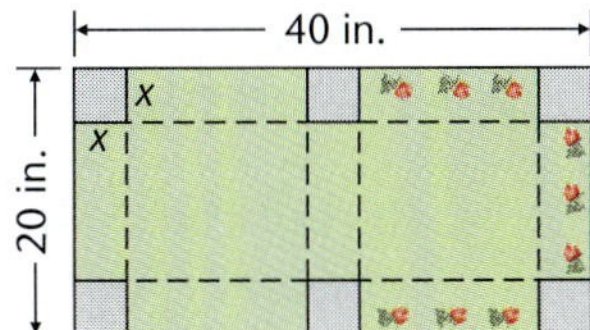

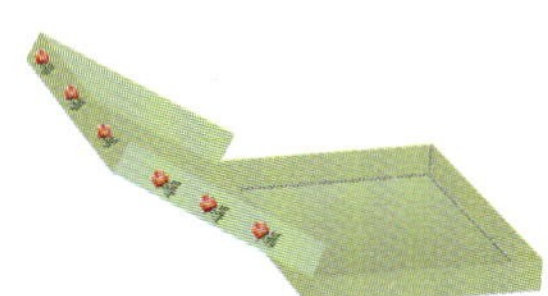

★ **49. Construction.** A propane gas tank is in the shape of a right circular cylinder with a hemisphere at each end (see the figure). If the overall length of the tank is 10 feet and the volume is 20π cubic feet, find the common radius of the hemispheres and the cylinder.

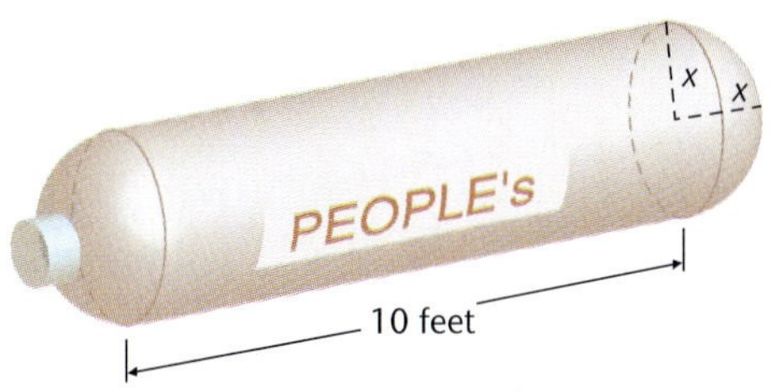

★ **50. Shipping.** A shipping box is reinforced with steel bands in all three directions (see the figure). A total of 20.5 feet of steel tape is to be used, with 6 inches of waste because of a 2-inch overlap in each direction. If the box has a square base and a volume of 2 cubic feet, find its dimensions.

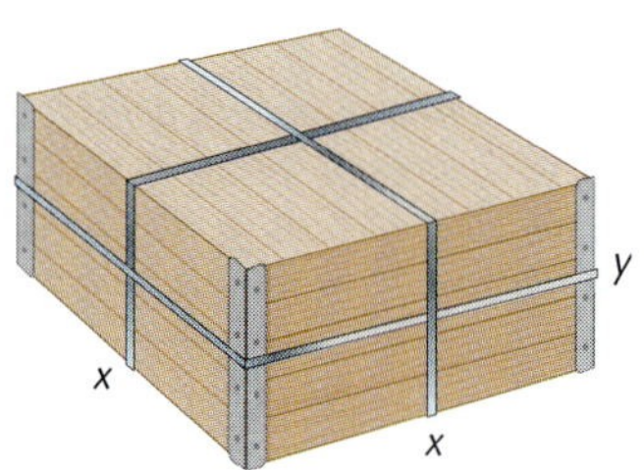

SECTION 3-4 Rational Functions

- Rational Functions
- Vertical and Horizontal Asymptotes
- Graphing Rational Functions

• Rational Functions

Just as rational numbers are defined in terms of quotients of integers, rational functions are defined in terms of quotients of polynomials. The following equations define rational functions:

$$f(x) = \frac{x - 1}{x^2 - x - 6} \qquad g(x) = \frac{1}{x} \qquad h(x) = \frac{x^3 - 1}{x}$$

$$p(x) = 2x^2 - 3 \qquad q(x) = 3 \qquad r(x) = 0$$

In general, a function f is a **rational function** if

$$f(x) = \frac{n(x)}{d(x)} \qquad d(x) \neq 0$$

where $n(x)$ and $d(x)$ are polynomials. The **domain of f** is the set of all real numbers x such that $d(x) \neq 0$.

If $x = a$ and $d(a) = 0$, then f is not defined at $x = a$ and there can be no point on the graph of f with abscissa $x = a$. Remember, division by 0 is never allowed. It can be shown that:

If $f(x) = n(x)/d(x)$ and $d(a) = 0$, then f is discontinuous at $x = a$ and the graph of f has a hole or break at $x = a$.

If $x = a$ is in the domain of $f(x)$ and $n(a) = 0$, then the graph of f crosses the x axis at $x = a$. Thus:

If $f(x) = n(x)/d(x)$, $n(a) = 0$, and $d(a) \neq 0$, then $x = a$ is an x intercept for the graph of f.

What happens if both $n(a) = 0$ and $d(a) = 0$? In this case, we know that $x - a$ is a factor of both $n(x)$ and $d(x)$, and thus, $f(x)$ is not in lowest terms (see Section A-4).

Unless specifically stated to the contrary, we assume that all the rational functions we consider are reduced to lowest terms.

EXAMPLE 1 **Finding the Domain and x Intercepts for a Rational Function**

Find the domain and x intercepts for $f(x) = \dfrac{2x^2 - 2x - 4}{x^2 - 9}$

Solution

$$f(x) = \frac{n(x)}{d(x)} = \frac{2x^2 - 2x - 4}{x^2 - 9} = \frac{2(x - 2)(x + 1)}{(x - 3)(x + 3)}$$

Since $d(3) = 0$ and $d(-3) = 0$, the domain of f is

$$x \neq \pm 3 \qquad \text{or} \qquad (-\infty, -3) \cup (-3, 3) \cup (3, \infty)$$

Since $n(2) = 0$ and $n(-1) = 0$, the graph of f crosses the x axis at $x = 2$ and $x = -1$.

Matched Problem 1 Find the domain and x intercepts for: $f(x) = \dfrac{3x^2 - 12}{x^2 + 2x - 3}$

• Vertical and Horizontal Asymptotes

Even though a rational function f may be discontinuous at $x = a$ (no graph for $x = a$), it is still useful to know what happens to the graph of f when x is close to a. For example, consider the very simple rational function f defined by

$$f(x) = \frac{1}{x}$$

It is clear that the function f is discontinuous at $x = 0$. But what happens to $f(x)$ when x approaches 0 from either side of 0? A numerical approach will give us an idea of what happens to $f(x)$ when x gets close to 0. From Table 1, we see that as x approaches 0 from the right, $1/x$ gets larger and larger—that is, $1/x$ increases without bound. We write this symbolically* as

$$\frac{1}{x} \to \infty \qquad \text{as} \qquad x \to 0^+$$

TABLE 1 **Behavior of $1/x$ as $x \to 0^+$**

x	1	0.1	0.01	0.001	0.0001	0.000 01	0.000 001	. . . x approaches 0 from the right ($x \to 0^+$)
$1/x$	1	10	100	1,000	10,000	100,000	1,000,000	. . . $1/x$ increases without bound ($1/x \to \infty$)

*Remember, the symbol ∞ does not represent a real number. In this context, ∞ is used to indicate that the values of $1/x$ increase without bound. That is, $1/x$ exceeds any given number N no matter how large N is chosen.

If x approaches 0 from the left, then x and $1/x$ are both negative and the values of $1/x$ decrease without bound (see Table 2). This is denoted as

$$\frac{1}{x} \to -\infty \qquad \text{as} \qquad x \to 0^-$$

TABLE 2 Behavior of $1/x$ as $x \to 0^-$

x	-1	-0.1	-0.01	-0.001	-0.0001	$-0.000\ 01$	$-0.000\ 001$	...	x approaches 0 from the left ($x \to 0^-$)
$1/x$	-1	-10	-100	$-1{,}000$	$-10{,}000$	$-100{,}000$	$-1{,}000{,}000$	...	$1/x$ decreases without bound ($1/x \to -\infty$)

The graph of $f(x) = 1/x$ for $-1 \le x \le 1$, $x \ne 0$, is shown in Figure 1. The behavior of f as x approaches 0 from the right is illustrated on the graph by drawing a curve that becomes almost vertical and placing an arrow on the curve to indicate that the values of $1/x$ continue to increase without bound as x approaches 0 from the right. The behavior as x approaches 0 from the left is illustrated in a similar manner.

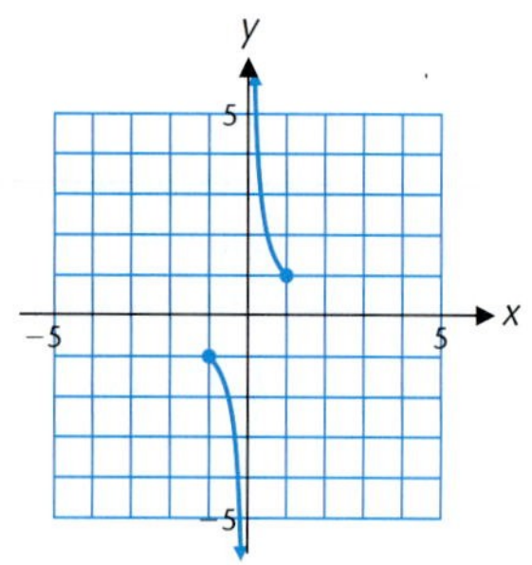

FIGURE 1 $f(x) = \dfrac{1}{x}$ near $x = 0$.

EXPLORE-DISCUSS 1 Construct tables similar to Tables 1 and 2 for $g(x) = 1/x^2$, and discuss the behavior of the graph of $g(x)$ near $x = 0$.

The preceding discussion suggests that vertical asymptotes are associated with the zeros of the denominator of a rational function. Using the same kind of reasoning, we state the following general method of locating vertical asymptotes for rational functions.

Theorem 1 **Vertical Asymptotes and Rational Functions**

Let f be a rational function defined by

$$f(x) = \frac{n(x)}{d(x)}$$

where $n(x)$ and $d(x)$ are polynomials. If a is a real number such that $d(a) = 0$ and $n(a) \neq 0$, then the line $x = a$ is a vertical asymptote of the graph of $y = f(x)$.

Now we look at the behavior of $f(x) = 1/x$ as $|x|$ gets very large—that is, as $x \to \infty$ and as $x \to -\infty$. Consider Tables 3 and 4. As x increases without bound, $1/x$ is positive and approaches 0 from above. As x decreases without bound, $1/x$ is negative and approaches 0 from below. For our purposes, it is not necessary to distinguish between $1/x$ approaching 0 from above and from below. Thus, we will describe this behavior by writing

$$\frac{1}{x} \to 0 \qquad \text{as} \qquad x \to \infty \text{ and as } x \to -\infty$$

TABLE 3 Behavior of $1/x$ as $x \to \infty$

x	1	10	100	1,000	10,000	100,000	1,000,000	. . .	x increases without bound ($x \to \infty$)
$1/x$	1	0.1	0.01	0.001	0.0001	0.000 01	0.000 001	. . .	$1/x$ approaches 0 ($1/x \to 0$)

TABLE 4 Behavior of $1/x$ as $x \to -\infty$

x	−1	−10	−100	−1,000	−10,000	−100,000	−1,000,000	. . .	x decreases without bound ($x \to -\infty$)
$1/x$	−1	−0.1	−0.01	−0.001	−0.0001	−0.000 01	−0.000 001	. . .	$1/x$ approaches 0 ($1/x \to 0$)

The completed graph of $f(x) = 1/x$ is shown in Figure 2. Notice that the behavior as $x \to \infty$ and as $x \to -\infty$ is illustrated by drawing a curve that is almost horizontal and adding arrows at the ends. The curve in Figure 2 is an example of a plane curve called a **hyperbola,** and the coordinate axes are called **asymptotes** for this curve. The y axis is a *vertical asymptote* for $1/x$, and the x axis is a *horizontal asymptote* for $1/x$.

FIGURE 2 $f(x) = \frac{1}{x}, x \neq 0.$

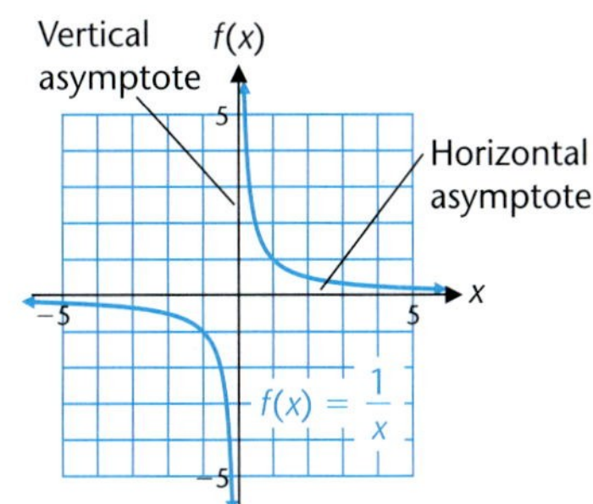

DEFINITION 1 **Horizontal and Vertical Asymptotes**

The line $x = a$ is a **vertical asymptote** for the graph of $y = f(x)$ if $f(x)$ either increases or decreases without bound as x approaches a from the right or from the left. Symbolically,

$$f(x) \to \infty \quad \text{or} \quad f(x) \to -\infty \quad \text{as} \quad x \to a^+ \quad \text{or} \quad x \to a^-$$

The line $y = b$ is a **horizontal asymptote** for the graph of $y = f(x)$ if $f(x)$ approaches b as x increases without bound or as x decreases without bound. Symbolically,

$$f(x) \to b \quad \text{as} \quad x \to \infty \quad \text{or} \quad x \to -\infty$$

EXPLORE-DISCUSS 2 Construct tables similar to Tables 3 and 4 for each of the following functions and discuss the behavior of each as $x \to \infty$ and as $x \to -\infty$:

(A) $f(x) = \dfrac{3x}{x^2 + 1}$ (B) $g(x) = \dfrac{3x^2}{x^2 + 1}$ (C) $h(x) = \dfrac{3x^3}{x^2 + 1}$

In Section 3-1, we saw that the behavior of a polynomial

$$P(x) = a_n x^n + \ldots + a_1 x + a_0$$

as $x \to \pm\infty$ is determined by its leading term, $a_n x^n$. In a similar manner, the behavior of a rational function is determined by the ratio of the leading terms of its numerator and denominator; that is, the graphs of

$$f(x) = \frac{a_m x^m + \cdots + a_1 x + a_0}{b_n x^n + \cdots + b_1 x + b_0} \quad \text{and} \quad g(x) = \frac{a_m x^m}{b_n x^n}$$

exhibit the same behavior as $x \to \infty$ and as $x \to -\infty$ (see Fig. 3).

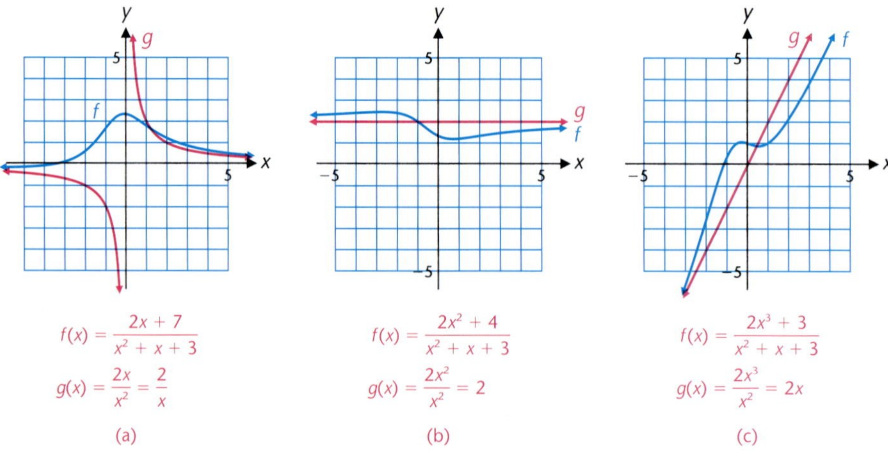

FIGURE 3 Graphs of rational functions as $x \to \pm\infty$.

In Figure 3(a), the degree of the numerator is less than the degree of the denominator and the x axis is a horizontal asymptote. In Figure 3(b), the degree of the numerator equals the degree of the denominator and the line $y = 2$ is a horizontal asymptote. In Figure 3(c), the degree of the numerator is greater than the degree of the denominator and there are no horizontal asymptotes. These ideas are generalized in Theorem 2 to provide a simple way to locate the horizontal asymptotes of any rational function.

Theorem 2 **Horizontal Asymptotes and Rational Functions**

Let f be a rational function defined by the quotient of two polynomials as follows:

$$f(x) = \frac{a_m x^m + \cdots + a_1 x + a_0}{b_n x^n + \cdots + b_1 x + b_0}$$

1. For $m < n$, the line $y = 0$ (the x axis) is a horizontal asymptote.
2. For $m = n$, the line $y = a_m/b_n$ is a horizontal asymptote.
3. For $m > n$, the graph will increase or decrease without bound, depending on m, n, a_m, and b_n, and there are no horizontal asymptotes.

EXAMPLE 2 Finding Vertical and Horizontal Asymptotes for a Rational Function

Find all vertical and horizontal asymptotes for

$$f(x) = \frac{n(x)}{d(x)} = \frac{2x^2 - 2x - 4}{x^2 - 9}$$

Solution Since $d(x) = x^2 - 9 = (x - 3)(x + 3)$, the graph of $f(x)$ has vertical asymptotes at $x = 3$ and $x = -3$ (Theorem 1). Since $n(x)$ and $d(x)$ have the same degree, the line

$$y = \frac{a_2}{b_2} = \frac{2}{1} = 2 \qquad a_2 = 2,\ b_2 = 1$$

is a horizontal asymptote (Theorem 2, part 2).

Matched Problem 2 Find all vertical and horizontal asymptotes for

$$f(x) = \frac{3x^2 - 12}{x^2 + 2x - 3}$$

• Graphing Rational Functions

We now use the techniques for locating asymptotes, along with other graphing aids discussed in the text, to graph several rational functions. First, we outline a systematic approach to the problem of graphing rational functions:

> **Graphing a Rational Function: $f(x) = n(x)/d(x)$**
>
> *Step 1.* *Intercepts.* Find the real solutions of the equation $n(x) = 0$ and use these solutions to plot any x intercepts of the graph of f. Evaluate $f(0)$, if it exists, and plot the y intercept.
>
> *Step 2.* *Vertical Asymptotes.* Find the real solutions of the equation $d(x) = 0$ and use these solutions to determine the domain of f, the points of discontinuity, and the vertical asymptotes. Sketch any vertical asymptotes as dashed lines.
>
> *Step 3.* *Sign Chart.* Construct a sign chart for f and use it to determine the behavior of the graph near each vertical asymptote.
>
> *Step 4.* *Horizontal Asymptotes.* Determine whether there is a horizontal asymptote and if so, sketch it as a dashed line.
>
> *Step 5.* *Complete the Sketch.* Complete the sketch of the graph by plotting additional points and joining these points with a smooth continuous curve over each interval in the domain of f. Do not cross any points of discontinuity.

EXAMPLE 3 Graphing a Rational Function

Graph: $y = f(x) = \dfrac{2x}{x - 3}$

Solution

$$f(x) = \frac{2x}{x - 3} = \frac{n(x)}{d(x)}$$

Step 1. *Intercepts.* Find real zeros of $n(x) = 2x$ and find $f(0)$:

$$2x = 0$$

$$x = 0 \quad x \text{ intercept}$$

$$f(0) = 0 \quad y \text{ intercept}$$

The graph crosses the coordinate axes only at the origin. Plot this intercept, as shown in Figure 4.

FIGURE 4

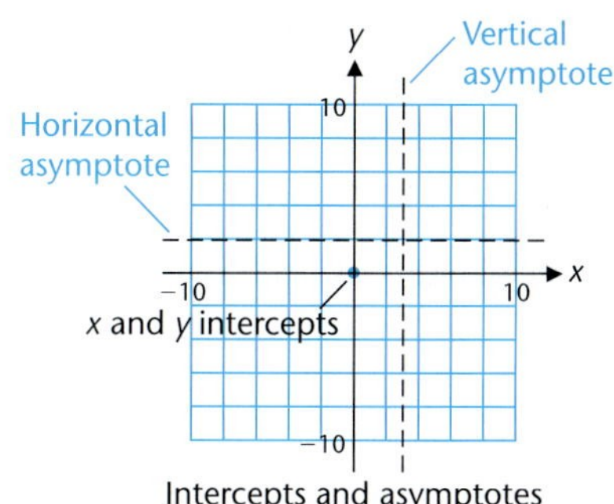

Intercepts and asymptotes

Step 2. *Vertical Asymptotes.* Find real zeros of $d(x) = x - 3$:

$$x - 3 = 0$$
$$x = 3$$

The domain of f is $(-\infty, 3) \cup (3, \infty)$, f is discontinuous at $x = 3$, and the graph has a vertical asymptote at $x = 3$. Sketch this asymptote, as shown in Figure 4.

Test number	-1	1	4
Value of f	$\frac{1}{2}$	-1	8
Sign of f	+	−	+

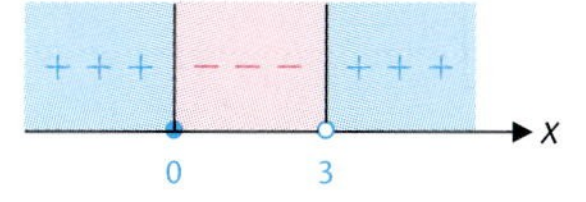

Step 3. *Sign Chart.* Construct a sign chart for $f(x)$ (review Section 2-8), as shown in the margin. Since $x = 3$ is a vertical asymptote and $f(x) < 0$ for $0 < x < 3$,

$$f(x) \to -\infty \quad \text{as} \quad x \to 3^-$$

Since $x = 3$ is a vertical asymptote and $f(x) > 0$ for $x > 3$,

$$f(x) \to \infty \quad \text{as} \quad x \to 3^+$$

Notice how much information is contained in the sign chart for f. The solid dot determines the x intercept; the open dot determines the domain, the point of discontinuity, and the vertical asymptote; and the signs of $f(x)$ determine the behavior of the graph at the vertical asymptote.

Step 4. *Horizontal Asymptote.* Since $n(x)$ and $d(x)$ have the same degree, the line $y = 2$ is a horizontal asymptote. Sketch this asymptote, as shown in Figure 4.

Step 5. *Complete the Sketch.* By plotting a few additional points, we obtain the graph in Figure 5. Notice that the graph is a smooth continuous curve over the interval $(-\infty, 3)$ and over the interval $(3, \infty)$. As expected, there is a break in the graph at $x = 3$.

FIGURE 5

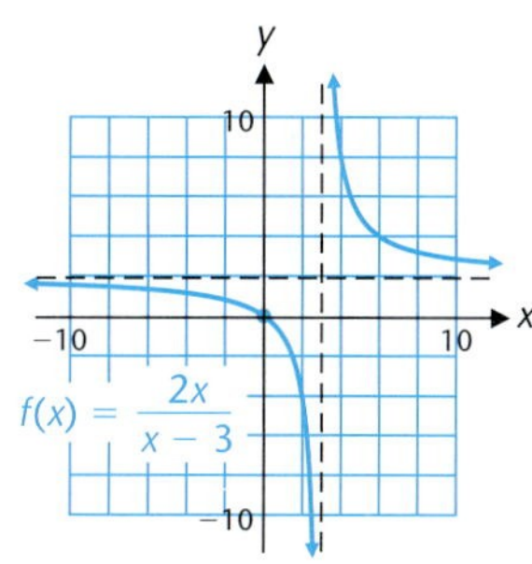

As you gain experience in graphing, many of the steps in Example 3 can be done mentally (or on scratch paper) and the process can be speeded up considerably.

Matched Problem 3 Proceed as in Example 3 and graph: $y = f(x) = \dfrac{3x}{x + 2}$

Remark: Refer to Example 3. When we graph $f(x) = 2x/(x - 3)$ on a graphing utility [Fig. 6(a)], it appears that the graphing utility has also drawn the vertical asymptote, but this is not the case. Most graphing utilities, when set in *connected mode,* calculate points on a graph and connect these points with line segments. The last point plotted to the left of the asymptote and the first plotted to the right of the asymptote will usually have very large y coordinates. If these y coordinates have opposite sign, then the graphing utility may connect the two points with a nearly vertical line segment, which gives the appearance of an asymptote. If you wish, you can set the calculator in *dot mode* to plot the points without the connecting line segments (Fig. 6(b)].

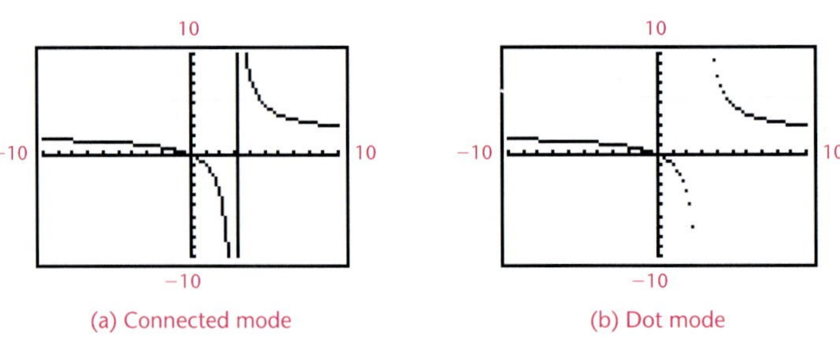

FIGURE 6 Graphing utility graphs of $f(x) = \dfrac{2x}{x - 3}$.

In the remaining examples we will just list the results of each step in the graphing strategy and omit the computational details.

EXAMPLE 4 Graphing a Rational Function

Graph: $y = f(x) = \dfrac{x^2 - 6x + 9}{x^2 + x - 2}$

Solution

$$f(x) = \frac{x^2 - 6x + 9}{x^2 + x - 2} = \frac{(x - 3)^2}{(x + 2)(x - 1)}$$

x intercept: $x = 3$

y intercept: $y = f(0) = -\frac{9}{2} = -4.5$

Domain: $(-\infty, -2) \cup (-2, 1) \cup (1, \infty)$

Points of discontinuity: $x = -2$ and $x = 1$

Vertical asymptotes: $x = -2$ and $x = 1$

Test number	-3	0	2	4
Value of f	9	$-\frac{9}{2}$	$\frac{1}{4}$	$\frac{1}{18}$
Sign of f	$+$	$-$	$+$	$+$

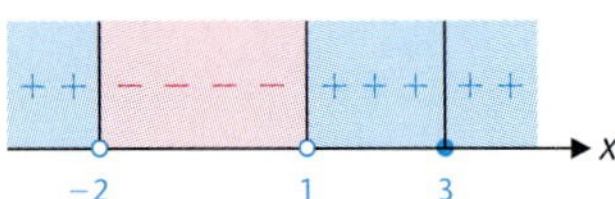

$$f(x) \to \infty \quad \text{as} \quad x \to -2^-$$
$$f(x) \to -\infty \quad \text{as} \quad x \to -2^+$$
$$f(x) \to -\infty \quad \text{as} \quad x \to 1^-$$
$$f(x) \to \infty \quad \text{as} \quad x \to 1^+$$

Horizontal asymptote: $y = 1$

Sketch in the intercepts and asymptotes (Fig. 7), then sketch the graph of f (Fig. 8).

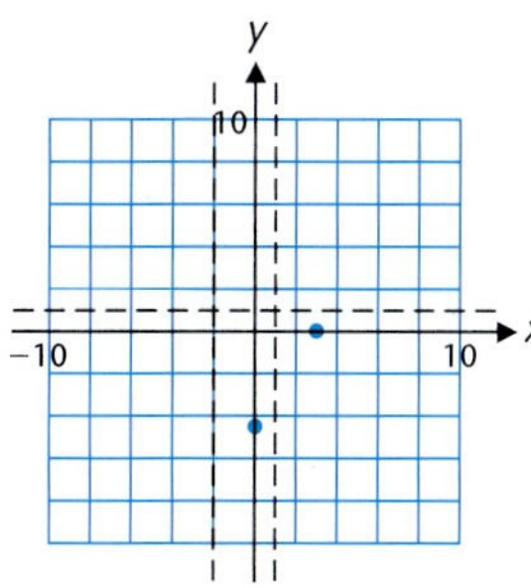

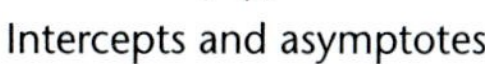
Intercepts and asymptotes

FIGURE 7

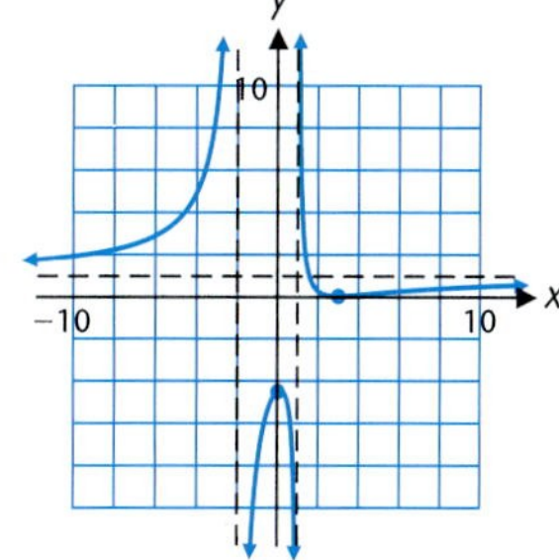

FIGURE 8

Matched Problem 4 Graph: $y = f(x) = \dfrac{x^2}{x^2 - 7x + 10}$

CAUTION The graph of a function cannot cross a vertical asymptote, but the same statement is not true for horizontal asymptotes. The graph in Example 4 clearly shows that **the graph of a function can cross a horizontal asymptote.** The definition of a horizontal asymptote requires $f(x)$ to approach b as x increases or decreases without bound, but it does not preclude the possibility that $f(x) = b$ for one or more values of x. In fact, using the cosine function from trigonometry, it is possible to construct a function whose graph crosses a horizontal asymptote an infinite number of times (see Fig. 9).

FIGURE 9 Multiple intersections of a graph and a horizontal asymptote.

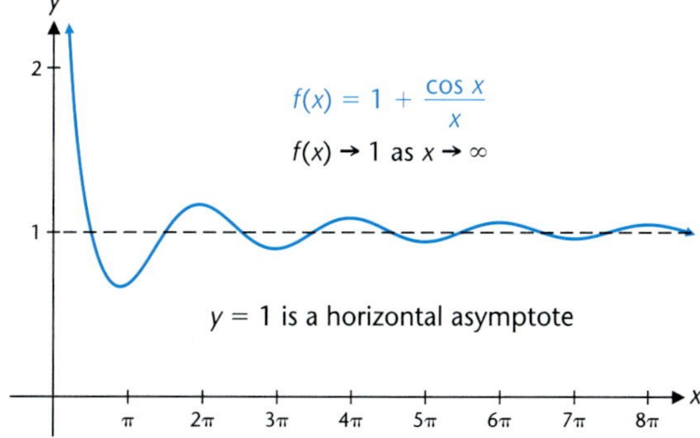

EXAMPLE 5 Graphing a Rational Function

Graph: $y = f(x) = \dfrac{x^2 - 3x - 4}{x - 2}$

Solution

$$f(x) = \frac{x^2 - 3x - 4}{x - 2} = \frac{(x + 1)(x - 4)}{x - 2}$$

Test number	-2	0	3	5
Value of f	$-\frac{3}{2}$	2	-4	2
Sign of f	$-$	$+$	$-$	$+$

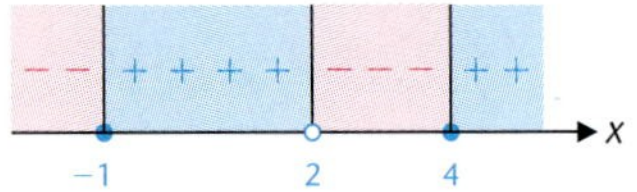

x intercepts: $x = -1$ and $x = 4$

y intercept: $y = f(0) = 2$

Domain: $(-\infty, 2) \cup (2, \infty)$

Points of discontinuity: $x = 2$

Vertical asymptote: $x = 2$

$f(x) \to \infty$ as $x \to 2^-$

$f(x) \to -\infty$ as $x \to 2^+$

No horizontal asymptote

Even though the graph of f does not have a horizontal asymptote, we can still gain some useful information about the behavior of the graph as $x \to -\infty$ and as $x \to \infty$ if we first perform a long division:

$$\begin{array}{r} x - 1 \\ x - 2\overline{)x^2 - 3x - 4} \\ \underline{x^2 - 2x} \\ -x - 4 \\ \underline{-x + 2} \\ -6 \end{array}$$

Quotient: $x - 1$; Remainder: -6

Thus,

$$f(x) = \frac{x^2 - 3x - 4}{x - 2} = x - 1 - \frac{6}{x - 2}$$

As $x \to -\infty$ or $x \to \infty$, $6/(x - 2) \to 0$ and the graph of f approaches the line $y = x - 1$. This line is called an **oblique asymptote** for the graph of f. The asymptotes and intercepts are sketched in Figure 10, and the graph of f is sketched in Figure 11.

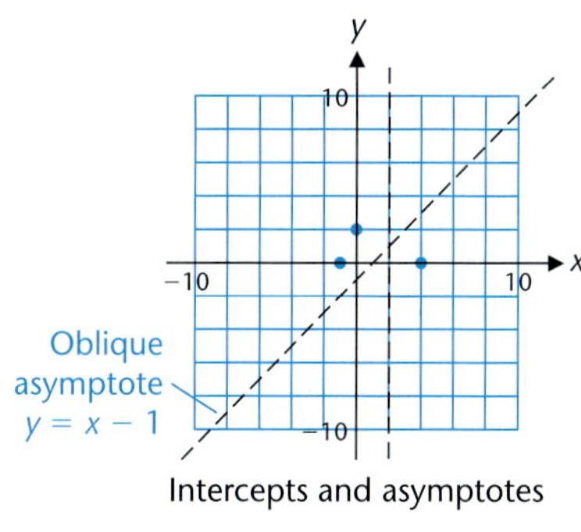

FIGURE 10

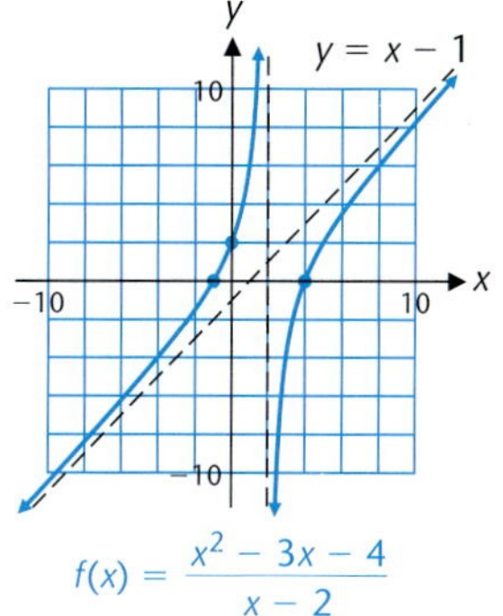

FIGURE 11

Generalizing the results of Example 5, we have Theorem 3.

Theorem 3 **Oblique Asymptotes and Rational Functions**

If $f(x) = n(x)/d(x)$, where $n(x)$ and $d(x)$ are polynomials and the degree of $n(x)$ is 1 more than the degree of $d(x)$, then $f(x)$ can be expressed in the form

$$f(x) = mx + b + \frac{r(x)}{d(x)}$$

where the degree of $r(x)$ is less than the degree of $d(x)$. The line

$$y = mx + b$$

is an oblique asymptote for the graph of f. That is,

$$[f(x) - (mx + b)] \to 0 \quad \text{as} \quad x \to -\infty \quad \text{or} \quad x \to \infty$$

Matched Problem 5 Graph, including any oblique asymptotes: $y = f(x) = \dfrac{x^2 + 5}{x + 1}$

Answers to Matched Problems

1. Domain: $(-\infty, -3) \cup (-3, 1) \cup (1, \infty)$; x intercepts: $x = -2$, $x = 2$
2. Vertical asymptotes: $x = -3$, $x = 1$; horizontal asymptote: $y = 3$
3.

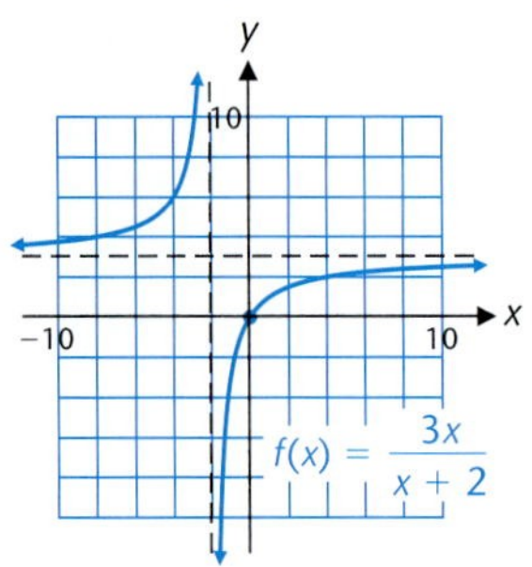

4.

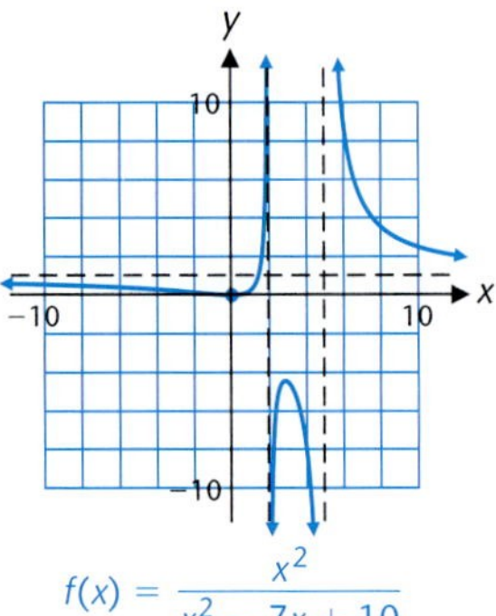

5.

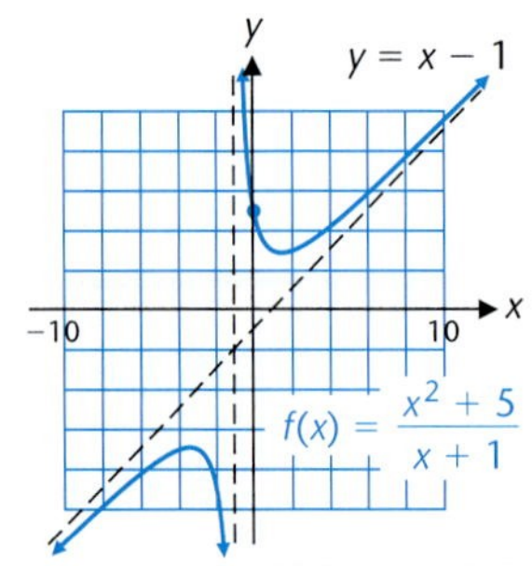

EXERCISE 3-4

A

In Problems 1–4, match each graph with one of the following functions:

$$f(x) = \frac{2x - 4}{x + 2} \qquad g(x) = \frac{2x + 4}{2 - x}$$

$$h(x) = \frac{2x + 4}{x - 2} \qquad k(x) = \frac{4 - 2x}{x + 2}$$

1.

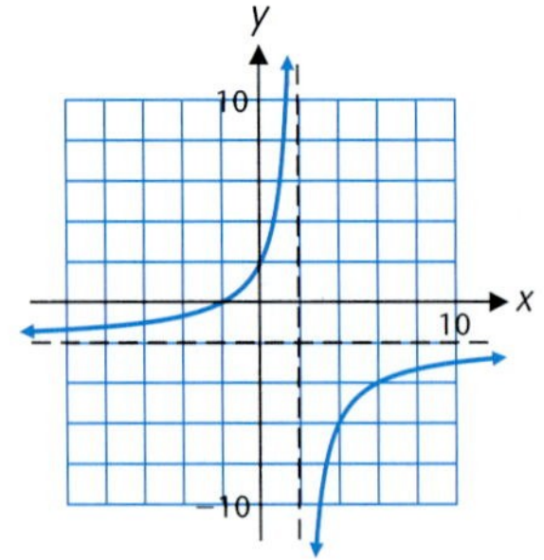

2.

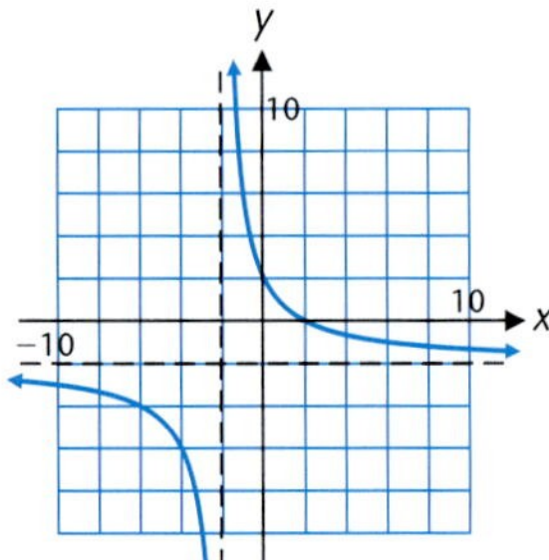

3.

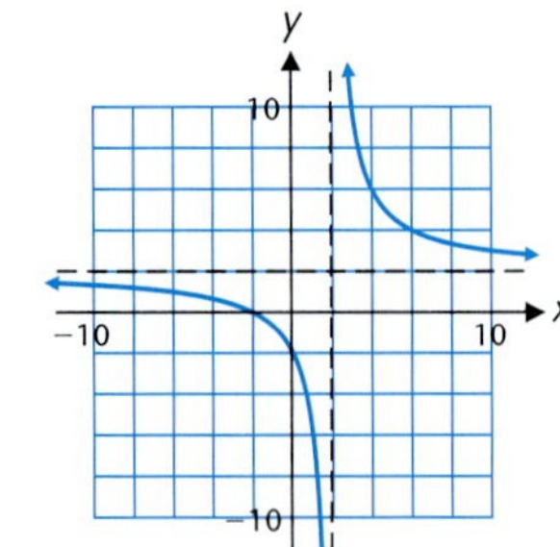

4.

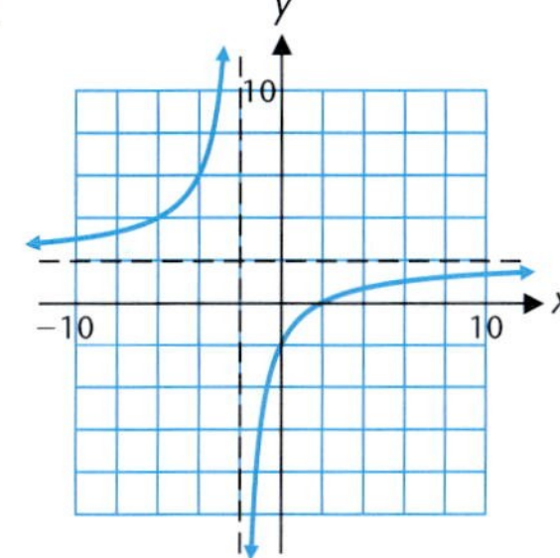

In Problems 5–12, find the domain and x intercepts. Do not graph.

5. $f(x) = \dfrac{2x - 4}{x + 1}$ **6.** $g(x) = \dfrac{3x + 6}{x - 1}$

7. $h(x) = \dfrac{x^2 - 1}{x^2 - 16}$ **8.** $k(x) = \dfrac{x^2 - 36}{x^2 - 25}$

9. $r(x) = \dfrac{x^2 - x - 6}{x^2 - x - 12}$ **10.** $s(x) = \dfrac{x^2 + x - 12}{x^2 + x - 6}$

11. $F(x) = \dfrac{x}{x^2 + 4}$ **12.** $G(x) = \dfrac{x^2}{x^2 + 16}$

In Problems 13–20, find all vertical and horizontal asymptotes. Do not graph.

13. $f(x) = \dfrac{2x}{x - 4}$ **14.** $h(x) = \dfrac{3x}{x + 5}$

15. $s(x) = \dfrac{2x^2 + 3x}{3x^2 - 48}$ **16.** $r(x) = \dfrac{5x^2 - 7x}{2x^2 - 50}$

17. $p(x) = \dfrac{2x}{x^4 + 1}$ **18.** $q(x) = \dfrac{5x^4}{2x^2 + 3x - 2}$

19. $t(x) = \dfrac{6x^4}{3x^2 - 2x - 5}$ **20.** $g(x) = \dfrac{3x}{x^4 + 2x^2 + 1}$

B

In Problems 21–40, use the graphing strategy outlined in the text to sketch the graph of each function.

Check Problems 21–40 on a graphing utility.

21. $f(x) = \dfrac{1}{x - 4}$ **22.** $g(x) = \dfrac{1}{x + 3}$

23. $f(x) = \dfrac{x}{x + 1}$ **24.** $f(x) = \dfrac{3x}{x - 3}$

25. $h(x) = \dfrac{x}{2x - 2}$ **26.** $p(x) = \dfrac{3x}{4x + 4}$

27. $f(x) = \dfrac{2x - 4}{x + 3}$ **28.** $f(x) = \dfrac{3x + 3}{2 - x}$

29. $g(x) = \dfrac{1 - x^2}{x^2}$ **30.** $f(x) = \dfrac{x^2 + 1}{x^2}$

31. $f(x) = \dfrac{9}{x^2 - 9}$ **32.** $g(x) = \dfrac{6}{x^2 - x - 6}$

33. $f(x) = \dfrac{x}{x^2 - 1}$ **34.** $p(x) = \dfrac{x}{1 - x^2}$

35. $g(x) = \dfrac{2}{x^2 + 1}$ **36.** $f(x) = \dfrac{x}{x^2 + 1}$

37. $f(x) = \dfrac{12x^2}{(3x + 5)^2}$ **38.** $f(x) = \dfrac{7x^2}{(2x - 3)^2}$

39. $f(x) = \dfrac{x^2 - 1}{x^2 + 7x + 10}$ **40.** $f(x) = \dfrac{x^2 + 6x + 8}{x^2 - x - 2}$

41. If $f(x) = n(x)/d(x)$, where $n(x)$ and $d(x)$ are quadratic functions, what is the maximum number of x intercepts $f(x)$ can have? What is the minimum number? Illustrate both cases with examples.

42. If $f(x) = n(x)/d(x)$, where $n(x)$ and $d(x)$ are quadratic functions, what is the maximum number of vertical asymptotes $f(x)$ can have? What is the minimum number? Illustrate both cases with examples.

In Problems 43–48, find all vertical, horizontal, and oblique asymptotes. Do not graph.

43. $f(x) = \dfrac{2x^2}{x - 1}$ **44.** $g(x) = \dfrac{3x^2}{x + 2}$

45. $p(x) = \dfrac{x^3}{x^2 + 1}$ **46.** $q(x) = \dfrac{x^5}{x^3 - 8}$

47. $r(x) = \dfrac{2x^2 - 3x + 5}{x}$ **48.** $s(x) = \dfrac{-3x^2 + 5x + 9}{x}$

In Problems 49–52, use a graphing utility to investigate the behavior of each function as $x \to \infty$ and as $x \to -\infty$, and to find any horizontal asymptotes.

49. $f(x) = \dfrac{5x}{\sqrt{x^2 + 1}}$ **50.** $f(x) = \dfrac{2x}{\sqrt{x^2 - 1}}$

51. $f(x) = \dfrac{4\sqrt{x^2 - 4}}{x}$ **52.** $f(x) = \dfrac{3\sqrt{x^2 + 1}}{x - 1}$

C

In Problems 53–58, use the graphing strategy outlined in the text to sketch the graph of each function. Include any oblique asymptotes.

Check Problems 53–58 with a graphing utility.

53. $f(x) = \dfrac{x^2 + 1}{x}$ **54.** $g(x) = \dfrac{x^2 - 1}{x}$

55. $k(x) = \dfrac{x^2 - 4x + 3}{2x - 4}$

56. $h(x) = \dfrac{x^2 + x - 2}{2x - 4}$

57. $F(x) = \dfrac{8 - x^3}{4x^2}$

58. $G(x) = \dfrac{x^4 + 1}{x^3}$

If $f(x) = n(x)/d(x)$, where the degree of $n(x)$ is greater than the degree of $d(x)$, then long division can be used to write $f(x) = p(x) + q(x)/d(x)$, where $p(x)$ and $q(x)$ are polynomials with the degree of $q(x)$ less than the degree of $d(x)$. In Problems 59–62, perform the long division and discuss the relationship between the graphs of $f(x)$ and $p(x)$ as $x \to \infty$ and as $x \to -\infty$.

59. $f(x) = \dfrac{x^4}{x^2 + 1}$

60. $f(x) = \dfrac{x^5}{x^2 + 1}$

61. $f(x) = \dfrac{x^5}{x^2 - 1}$

62. $f(x) = \dfrac{x^5}{x^3 - 1}$

In calculus, it is often necessary to consider rational functions that are not in lowest terms, such as the functions given in Problems 63–66. For each function, state the domain, reduce the function to lowest terms, and sketch its graph. Remember to exclude from the graph any points with x values that are not in the domain.

63. $f(x) = \dfrac{x^2 - 4}{x - 2}$

64. $g(x) = \dfrac{x^2 - 1}{x + 1}$

65. $r(x) = \dfrac{x + 2}{x^2 - 4}$

66. $s(x) = \dfrac{x - 1}{x^2 - 1}$

APPLICATIONS

67. Employee Training. A company producing electronic components used in television sets has established that on the average, a new employee can assemble $N(t)$ components per day after t days of on-the-job training, as given by

$$N(t) = \frac{50t}{t + 4} \qquad t \geq 0$$

Sketch the graph of N, including any vertical or horizontal asymptotes. What does N approach as $t \to \infty$?

68. Physiology. In a study on the speed of muscle contraction in frogs under various loads, researchers W. O. Fems and J. Marsh found that the speed of contraction decreases with increasing loads. More precisely, they found that the relationship between speed of contraction S (in centimeters per second) and load w (in grams) is given approximately by

$$S(w) = \frac{26 + 0.06w}{w} \qquad w \geq 5$$

Sketch the graph of S, including any vertical or horizontal asymptotes. What does S approach as $w \to \infty$?

69. Retention. An experiment on retention is conducted in a psychology class. Each student in the class is given 1 day to memorize the same list of 40 special characters. The lists are turned in at the end of the day, and for each succeeding day for 20 days each student is asked to turn in a list of as many of the symbols as can be recalled. Averages are taken, and it is found that a good approximation of the average number of symbols, $N(t)$, retained after t days is given by

$$N(t) = \frac{5t + 30}{t} \qquad t \geq 1$$

Sketch the graph of N, including any vertical or horizontal asymptotes. What does N approach as $t \to \infty$?

70. Learning Theory. In 1917, L. L. Thurstone, a pioneer in quantitative learning theory, proposed the function

$$f(x) = \frac{a(x + c)}{(x + c) + b}$$

to describe the number of successful acts per unit time that a person could accomplish after x practice sessions. Suppose that for a particular person enrolling in a typing class,

$$f(x) = \frac{50(x + 1)}{x + 5} \qquad x \geq 0$$

where $f(x)$ is the number of words per minute the person is able to type after x weeks of lessons. Sketch the graph of f, including any vertical or horizontal asymptotes. What does f approach as $x \to \infty$?

Using calculus techniques, it can be shown that the minimum value of a function of the form

$$g(x) = ax + b + \frac{c}{x} \qquad a > 0, c > 0, x > 0$$

is min $g(x) = g(\sqrt{c/a})$. Use this fact in Problems 71–74.

★ **71. Replacement Time.** A desktop office copier has an initial price of \$2,500. A maintenance/service contract costs \$200 for the first year and increases \$50 per year thereafter. It can be shown that the total cost of the copier after n years is given by

$$C(n) = 2{,}500 + 175n + 25n^2$$

The average cost per year for n years is $\overline{C}(n) = C(n)/n$.

(A) Find the rational function $\overline{C}$.

(B) When is the average cost per year minimum? (This is frequently referred to as the *replacement time* for this piece of equipment.)

(C) Sketch the graph of $\overline{C}$, including any asymptotes.

★ **72. Average Cost.** The total cost of producing x units of a certain product is given by

$$C(x) = \tfrac{1}{5}x^2 + 2x + 2{,}000$$

The average cost per unit for producing x units is $\overline{C}(x) = C(x)/x$.

(A) Find the rational function $\overline{C}$.

(B) At what production level will the average cost per unit be minimal?

(C) Sketch the graph of $\overline{C}$, including any asymptotes.

★ **73. Construction.** A rectangular dog pen is to be made to enclose an area of 225 square feet.

(A) If x represents the width of the pen, express the total length $L(x)$ of the fencing material required for the pen in terms of x.

(B) Considering the physical limitations, what is the domain of the function L?

(C) Find the dimensions of the pen that will require the least amount of fencing material.

(D) Graph the function L, including any asymptotes.

★ **74. Construction.** Rework Problem 73 with the added assumption that the pen is to be divided into two sections, as shown in the figure.

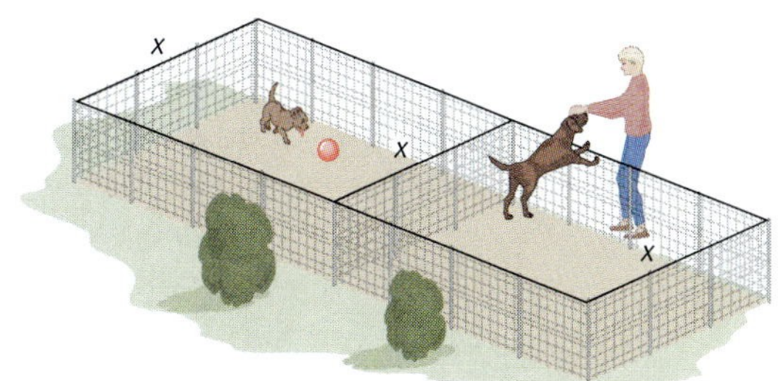

SECTION 3-5 Partial Fractions

- Basic Theorems
- Partial Fraction Decomposition

You have now had considerable experience combining two or more rational expressions into a single rational expression. For example, problems such as

$$\frac{2}{x+5} + \frac{3}{x-4} = \frac{2(x-4) + 3(x+5)}{(x+5)(x-4)} = \frac{5x+7}{(x+5)(x-4)}$$

should seem routine. Frequently in more advanced courses, particularly in calculus, it is advantageous to be able to reverse this process—that is, to be able to express a rational expression as the sum of two or more simpler rational expressions called **partial fractions.** As is often the case with reverse processes, the process of decomposing a rational expression into partial fractions is more difficult than combining rational expressions. Basic to the process is the factoring of polynomials, so the topics discussed earlier in this chapter can be put to effective use.

We confine our attention to rational expressions of the form $P(x)/D(x)$, where $P(x)$ and $D(x)$ are polynomials with real coefficients and no common factors. In addition, we assume that the degree of $P(x)$ is less than the degree of $D(x)$. If the degree of $P(x)$ is greater than or equal to that of $D(x)$, we have only to divide $P(x)$ by $D(x)$ to obtain

$$\frac{P(x)}{D(x)} = Q(x) + \frac{R(x)}{D(x)}$$

where the degree of $R(x)$ is less than that of $D(x)$. For example,

$$\frac{x^4 - 3x^3 + 2x^2 - 5x + 1}{x^2 - 2x + 1} = x^2 - x - 1 + \frac{-6x + 2}{x^2 - 2x + 1}$$

If the degree of $P(x)$ is less than that of $D(x)$, then $P(x)/D(x)$ is called a **proper fraction.**

• Basic Theorems

Our task now is to establish a systematic way to decompose a proper fraction into the sum of two or more partial fractions. The following three theorems take care of the problem completely. Theorems 1 and 3 are stated without proof.

Theorem 1 **Equal Polynomials**

Two polynomials are equal to each other if and only if the coefficients of terms of like degree are equal.

For example, if

Equate the constant terms.

$$(A + 2B)x + B = 5x - 3$$

Equate the coefficients of x.

then

$$B = -3$$ Substitute $B = -3$ into the second equation to solve for A.

$$A + 2B = 5$$

$$A + 2(-3) = 5$$

$$A = 11$$

EXPLORE-DISCUSS 1 If

$$x + 5 = A(x + 1) + B(x - 3) \tag{1}$$

is a polynomial identity (that is, both sides represent the same polynomial), then equating coefficients produces the system

$1 = A + B$ Equating coefficients of x

$5 = A - 3B$ Equating constant terms

(A) Solve this system (see Section 1-2).
(B) For an alternate method of solution, substitute $x = 3$ in (1) to find A and then substitute $x = -1$ in (1) to find B. Explain why method B is valid.

Theorem 2 **Linear and Quadratic Factor Theorem**

For a polynomial with real coefficients, there always exists a complete factoring involving only linear and/or quadratic factors with real coefficients where the linear and quadratic factors are prime relative to the real numbers.

That Theorem 2 is true can be seen as follows: From earlier theorems in this chapter, we know that an nth-degree polynomial $P(x)$ has n zeros and n linear factors. The real zeros of $P(x)$ correspond to linear factors of the form $(x - r)$, where r is a real number. Since $P(x)$ has real coefficients, the imaginary zeros occur in conjugate pairs. Thus, the imaginary zeros correspond to pairs of factors of the form $[x - (a + bi)]$ and $[x - (a - bi)]$, where a and b are real numbers. Multiplying these two imaginary factors, we have

$$[x - (a + bi)][x - (a - bi)] = x^2 - 2ax + a^2 + b^2$$

This quadratic polynomial with real coefficients is a factor of $P(x)$. Thus, $P(x)$ can be factored into a product of linear factors and quadratic factors, all with real coefficients.

• Partial Fraction Decomposition

We are now ready to state Theorem 3, which forms the basis for partial fraction decomposition.

Theorem 3 **Partial Fraction Decomposition**

Any proper fraction $P(x)/D(x)$ reduced to lowest terms can be decomposed into the sum of partial fractions as follows:

1. If $D(x)$ has a nonrepeating linear factor of the form $ax + b$, then the partial fraction decomposition of $P(x)/D(x)$ contains a term of the form

$$\frac{A}{ax + b} \qquad A \text{ a constant}$$

2. If $D(x)$ has a k-repeating linear factor of the form $(ax + b)^k$, then the partial fraction decomposition of $P(x)/D(x)$ contains k terms of the form

$$\frac{A_1}{ax + b} + \frac{A_2}{(ax + b)^2} + \cdots + \frac{A_k}{(ax + b)^k} \qquad A_1, A_2, \ldots, A_k \text{ constants}$$

3. If $D(x)$ has a nonrepeating quadratic factor of the form $ax^2 + bx + c$, which is prime relative to the real numbers, then the partial fraction decomposition of $P(x)/D(x)$ contains a term of the form

$$\frac{Ax + B}{ax^2 + bx + c} \qquad A, B \text{ constants}$$

4. If $D(x)$ has a k-repeating quadratic factor of the form $(ax^2 + bx + c)^k$, where $ax^2 + bx + c$ is prime relative to the real numbers, then the partial fraction decomposition of $P(x)/D(x)$ contains k terms of the form

$$\frac{A_1x + B_1}{ax^2 + bx + c} + \frac{A_2x + B_2}{(ax^2 + bx + c)^2} + \cdots + \frac{A_kx + B_k}{(ax^2 + bx + c)^k}$$

$$A_1, \ldots, A_k, \quad B_1, \ldots, B_k \text{ constants}$$

Let's see how the theorem is used to obtain partial fraction decompositions in several examples.

EXAMPLE 1 Nonrepeating Linear Factors

Decompose into partial fractions: $\dfrac{5x + 7}{x^2 + 2x - 3}$

Solution We first try to factor the denominator. If it can't be factored in the real numbers, then we can't go any further. In this example, the denominator factors, so we apply part 1 from Theorem 3:

$$\frac{5x + 7}{(x - 1)(x + 3)} = \frac{A}{x - 1} + \frac{B}{x + 3} \tag{2}$$

To find the constants A and B, we combine the fractions on the right side of equation (2) to obtain

$$\frac{5x + 7}{(x - 1)(x + 3)} = \frac{A(x + 3) + B(x - 1)}{(x - 1)(x + 3)}$$

Since these fractions have the same denominator, their numerators must be equal. Thus,

$$5x + 7 = A(x + 3) + B(x - 1) \tag{3}$$

We could multiply the right side and find A and B by using Theorem 1, but in this case it is easier to take advantage of the fact that equation (3) is an identity—that is, it must hold for all values of x. In particular, we note that if we let $x = 1$, then the second term of the right side drops out and we can solve for A:

$$5 \cdot 1 + 7 = A(1 + 3) + B(1 - 1)$$
$$12 = 4A$$
$$A = 3$$

Similarly, if we let $x = -3$, the first term drops out and we find

$$-8 = -4B$$
$$B = 2$$

Hence,

$$\frac{5x+7}{x^2+2x-3} = \frac{3}{x-1} + \frac{2}{x+3} \tag{4}$$

as can easily be checked by adding the two fractions on the right.

Matched Problem 1 Decompose into partial fractions: $\dfrac{7x+6}{x^2+x-6}$

EXPLORE-DISCUSS 2 A graphing utility can also be used to check a partial fraction decomposition. To check Example 1, we graph the left and right sides of (4) in a graphing utility (Fig. 1). Discuss how the Trace feature on the graphing utility can be used to check that the graphing utility is displaying two identical graphs.

FIGURE 1

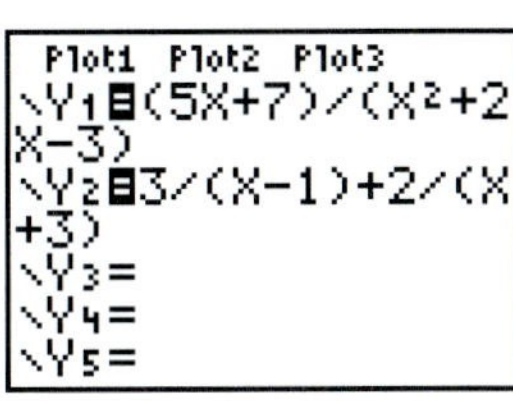

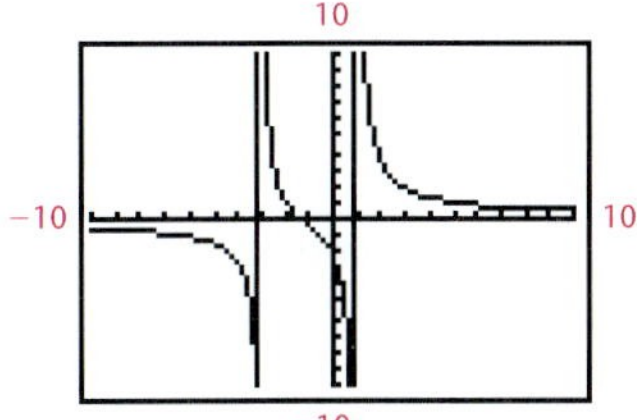

EXAMPLE 2 **Repeating Linear Factors**

Decompose into partial fractions: $\dfrac{6x^2-14x-27}{(x+2)(x-3)^2}$

Solution Using parts 1 and 2 from Theorem 3, we write

$$\frac{6x^2-14x-27}{(x+2)(x-3)^2} = \frac{A}{x+2} + \frac{B}{x-3} + \frac{C}{(x-3)^2}$$
$$= \frac{A(x-3)^2 + B(x+2)(x-3) + C(x+2)}{(x+2)(x-3)^2}$$

Thus, for all x,

$$6x^2 - 14x - 27 = A(x-3)^2 + B(x+2)(x-3) + C(x+2)$$

If $x = 3$, then If $x = -2$, then

$$-15 = 5C \qquad 25 = 25A$$

$$C = -3 \qquad A = 1$$

There are no other values of x that will cause terms on the right to drop out. Since any value of x can be substituted to produce an equation relating A, B, and C, we let $x = 0$ and obtain

$$-27 = 9A - 6B + 2C \qquad \text{Substitute } A = 1 \text{ and } C = -3.$$

$$-27 = 9 - 6B - 6$$

$$B = 5$$

Thus,

$$\frac{6x^2 - 14x - 27}{(x+2)(x-3)^2} = \frac{1}{x+2} + \frac{5}{x-3} - \frac{3}{(x-3)^2}$$

Matched Problem 2 Decompose into partial fractions: $\dfrac{x^2 + 11x + 15}{(x-1)(x+2)^2}$

EXAMPLE 3 Nonrepeating Linear and Quadratic Factors

Decompose into partial fractions: $\dfrac{5x^2 - 8x + 5}{(x-2)(x^2 - x + 1)}$

Solution First, we see that the quadratic in the denominator can't be factored further in the real numbers. Then, we use parts 1 and 3 from Theorem 3 to write

$$\frac{5x^2 - 8x + 5}{(x-2)(x^2 - x + 1)} = \frac{A}{x-2} + \frac{Bx + C}{x^2 - x + 1}$$

$$= \frac{A(x^2 - x + 1) + (Bx + C)(x - 2)}{(x-2)(x^2 - x + 1)}$$

Thus, for all x,

$$5x^2 - 8x + 5 = A(x^2 - x + 1) + (Bx + C)(x - 2)$$

If $x = 2$, then

$$9 = 3A$$

$$A = 3$$

If $x = 0$, then, using $A = 3$, we have

$$5 = 3 - 2C$$
$$C = -1$$

If $x = 1$, then, using $A = 3$ and $C = -1$, we have

$$2 = 3 + (B - 1)(-1)$$
$$B = 2$$

Hence,

$$\frac{5x^2 - 8x + 5}{(x - 2)(x^2 - x + 1)} = \frac{3}{x - 2} + \frac{2x - 1}{x^2 - x + 1}$$

Matched Problem 3 Decompose into partial fractions: $\dfrac{7x^2 - 11x + 6}{(x - 1)(2x^2 - 3x + 2)}$

EXAMPLE 4 Repeating Quadratic Factors

Decompose into partial fractions: $\dfrac{x^3 - 4x^2 + 9x - 5}{(x^2 - 2x + 3)^2}$

Solution Since $x^2 - 2x + 3$ can't be factored further in the real numbers, we proceed to use part 4 from Theorem 3 to write

$$\frac{x^3 - 4x^2 + 9x - 5}{(x^2 - 2x + 3)^2} = \frac{Ax + B}{x^2 - 2x + 3} + \frac{Cx + D}{(x^2 - 2x + 3)^2}$$
$$= \frac{(Ax + B)(x^2 - 2x + 3) + Cx + D}{(x^2 - 2x + 3)^2}$$

Thus, for all x,

$$x^3 - 4x^2 + 9x - 5 = (Ax + B)(x^2 - 2x + 3) + Cx + D$$

Since the substitution of carefully chosen values of x doesn't lead to the immediate determination of A, B, C, or D, we multiply and rearrange the right side to obtain

$$x^3 - 4x^2 + 9x - 5 = Ax^3 + (B - 2A)x^2 + (3A - 2B + C)x + (3B + D)$$

Now we use Theorem 1 to equate coefficients of terms of like degree:

$$A = 1$$
$$B - 2A = -4$$
$$3A - 2B + C = 9$$
$$3B + D = -5$$

From these equations we easily find that $A = 1$, $B = -2$, $C = 2$, and $D = 1$. Now we can write

$$\frac{x^3 - 4x^2 + 9x - 5}{(x^2 - 2x + 3)^2} = \frac{x - 2}{x^2 - 2x + 3} + \frac{2x + 1}{(x^2 - 2x + 3)^2}$$

Matched Problem 4 Decompose into partial fractions: $\dfrac{3x^3 - 6x^2 + 7x - 2}{(x^2 - 2x + 2)^2}$

Answers to Matched Problems

1. $\dfrac{4}{x - 2} + \dfrac{3}{x + 3}$ **2.** $\dfrac{3}{x - 1} - \dfrac{2}{x + 2} + \dfrac{1}{(x + 2)^2}$ **3.** $\dfrac{2}{x - 1} + \dfrac{3x - 2}{2x^2 - 3x + 2}$

4. $\dfrac{3x}{x^2 - 2x + 2} + \dfrac{x - 2}{(x^2 - 2x + 2)^2}$

EXERCISE 3-5

A

In Problems 1–4, find constants A and B so that the indicated equation is a polynomial identity.

1. $5x - 10 = A(x + 4) + B(x - 6)$

2. $3x + 18 = A(x - 1) + B(x + 2)$

3. $-9x + 5 = A(2x + 5) + B(3x + 2)$

4. $-32x - 13 = A(4x - 7) + B(2x + 8)$

In Problems 5–10, find constants A, B, C, and D so that the right side is equal to the left.

5. $\dfrac{5x^2 + 7x + 6}{(x - 1)(x + 2)^2} = \dfrac{A}{x - 1} + \dfrac{B}{x + 2} + \dfrac{C}{(x + 2)^2}$

6. $\dfrac{x^2 + 5x - 12}{(x + 1)(x - 3)^2} = \dfrac{A}{x + 1} + \dfrac{B}{x - 3} + \dfrac{C}{(x - 3)^2}$

7. $\dfrac{x^2 + 5x + 3}{x(x^2 + 1)} = \dfrac{A}{x} + \dfrac{Bx + C}{x^2 + 1}$

8. $\dfrac{3x^2 - 2x - 16}{x(x^2 + 4)} = \dfrac{A}{x} + \dfrac{Bx + C}{x^2 + 4}$

9. $\dfrac{2x^3 + x^2 - 2}{(x^2 + x + 2)^2} = \dfrac{Ax + B}{x^2 + x + 2} + \dfrac{Cx + D}{(x^2 + x + 2)^2}$

10. $\dfrac{4x^2 - 2x + 13}{(x^2 - x + 4)^2} = \dfrac{Ax + B}{x^2 - x + 4} + \dfrac{Cx + D}{(x^2 - x + 4)^2}$

B

In Problems 11–22, decompose into partial fractions.

11. $\dfrac{3x - 40}{x^2 - x - 12}$

12. $\dfrac{4x - 43}{x^2 + 3x - 10}$

13. $\dfrac{8x - 22}{6x^2 + 17x - 14}$

14. $\dfrac{x - 27}{10x^2 - 13x - 3}$

15. $\dfrac{6x^2 - 14x + 4}{4x^3 - 4x^2 + x}$

16. $\dfrac{7x - 8}{9x^3 - 12x^2 + 4x}$

17. $\dfrac{10x^2 + 4x + 3}{2x^3 + x^2 + x}$

18. $\dfrac{9x^2 + 14}{x^3 + x^2 + 2x}$

19. $\dfrac{4x^3 - 5x^2 + 6x - 5}{x^4 + 2x^2 + 1}$

20. $\dfrac{2x^3 - x^2 + 11x}{x^4 + 8x^2 + 16}$

21. $\dfrac{4x^2 - 5x + 3}{x^3 + x + 2}$

22. $\dfrac{5x^2 - 8x + 12}{x^3 - x^2 - 4}$

C

In Problems 23–26, decompose into a sum of a polynomial and partial fractions.

23. $\dfrac{x^5 - 4x^2 - 16x + 40}{x^4 - 2x^3 - 8x + 16}$

24. $\dfrac{2x^5 + 2x^4 - 6x^3}{x^4 + 4x + 3}$

25. $\dfrac{2x^5 - 7x^2 + 20x + 21}{x^4 - 4x^2 - 12x - 9}$

26. $\dfrac{x^5 + 2x^4 + 3x^2 - 14x}{x^4 - x^2 - 4x - 4}$

In Problems 27–30, a and b are real constants. Decompose into partial fractions.

27. $\frac{x}{(x+a)^2}$, $a \neq 0$

28. $\frac{1}{x(x+a)^2}$, $a \neq 0$

29. $\frac{1}{(x-a)(x-b)}$, $a \neq b$

30. $\frac{x}{(x-a)(x-b)}$, $a \neq b$

CHAPTER 3 GROUP ACTIVITY Interpolating Polynomials

Given two points in the plane, we can use the point-slope form of the equation of a line to find a polynomial whose graph passes through these two points. How can we proceed if we are given more than two points? For example, how can we find the equation of a polynomial $P(x)$ whose graph passes through the points listed in Table 1 and graphed in Figure 1?

TABLE 1

x	1	2	3	4
$P(x)$	1	3	−3	1

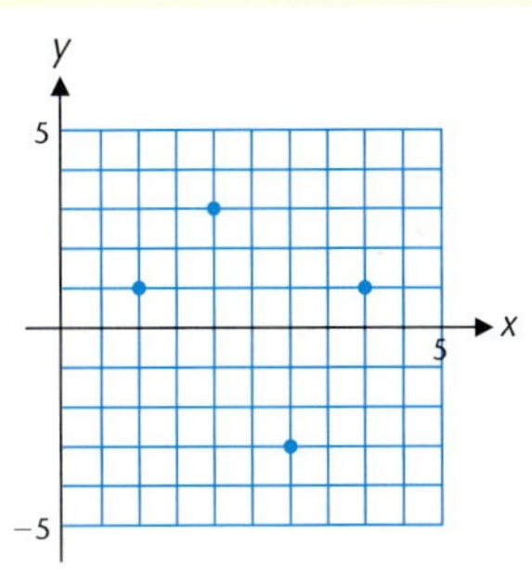

FIGURE 1

The key to solving this problem is to write the unknown polynomial $P(x)$ in the following special form:

$$P(x) = a_0 + a_1(x-1) + a_2(x-1)(x-2) + a_3(x-1)(x-2)(x-3) \quad (1)$$

Since the graph of $P(x)$ is to pass through each point in Table 1, we can substitute each value of x in (1) to determine the coefficients a_0, a_1, a_2, and a_3. First we evaluate (1) at $x = 1$ to determine a_0:

$$1 = P(1)$$

$$= a_0 \qquad \text{All other terms in (1) are 0 when } x = 1.$$

Using this value for a_0 in (1) and evaluating at $x = 2$, we have

$$3 = P(2) = 1 + a_1(1) \qquad \text{All other terms are 0.}$$

$$2 = a_1$$

Continuing in this manner, we have

$$-3 = P(3) = 1 + 2(2) + a_2(2)(1)$$

$$-8 = 2a_2$$

$$-4 = a_2$$

$$1 = P(4) = 1 + 2(3) - 4(3)(2) + a_3(3)(2)(1)$$
$$18 = 6a_3$$
$$3 = a_3$$

We have now evaluated all the coefficients in (1) and can write

$$P(x) = 1 + 2(x - 1) - 4(x - 1)(x - 2) + 3(x - 1)(x - 2)(x - 3) \qquad (2)$$

If we expand the products in (2) and collect like terms, we can express $P(x)$ in the more conventional form (verify this)

$$P(x) = 3x^3 - 22x^2 + 47x - 27$$

(A) To check these calculations, evaluate $P(x)$ at $x = 1, 2, 3,$ and 4 and compare the results with Table 1. Then add the graph of $P(x)$ to Figure 1.
(B) Write a verbal description of the special form of $P(x)$ in (1).

In general, given a set of $n + 1$ points:

x	x_0	x_1	...	x_n
y	y_0	y_1	...	y_n

the **interpolating polynomial** for these points is the polynomial $P(x)$ of degree less than or equal to n that satisfies $P(x_k) = y_k$ for $k = 0, 1, \ldots, n$. The **general form** of the interpolating polynomial is

$$P(x) = a_0 + a_1(x - x_0) + a_2(x - x_0)(x - x_1) + \cdots + a_n(x - x_0)(x - x_1) \cdot \cdots \cdot (x - x_{n-1})$$

(C) Summarize the procedure for using the points in the table to find the coefficients in the general form.
(D) Give an example to show that the interpolating polynomial can have degree strictly less than n.
(E) Could there be two different polynomials of degree less than or equal to n whose graph passes through the given $n + 1$ points? Justify your answer.
(F) Find the interpolating polynomial for each of Tables 2 and 3. Check your answers by evaluating the polynomial, and illustrate by graphing the points in the table and the polynomial on the same axes.

TABLE 2

x	-1	0	1	2
y	5	3	3	11

TABLE 3

x	-2	-1	0	1	2
y	-3	0	5	0	-3

A surprisingly short program on a graphing utility can be used to calculate the coefficients in the general form of an interpolating polynomial. Table 4 shows such a program for a Texas Instruments graphing calculator and the output generated when we use the program to find the coefficients of the interpolating polynomial for Table 1.

TABLE 4 Interpolating Polynomial Coefficients on a Graphing Utility

Program INTERP*	Output

L2→L3
dimL L3→M
For(I,2,M,1)
For(J,M,I,-1)
(L3(J)-L3(J-1))/(L1(J)-L1(J-I+1))→L3(J)
End
End
Disp L3
	{1,2,3,4}→L1
	{1 2 3 4}
	{1,3,-3,1}→L2
	{1 3 -3 1}
	INTERP
	{1 2 -4 3}
	Done

(G) If you have a TI-85 or TI-86 graphing calculator, enter INTERP in your calculator exactly as shown in Table 4. To use the program, enter the x values in L1 and the corresponding y values in L2 (see the output in Table 4) and then execute the program. If you have some other graphing utility that can store and execute programs, consult your manual and modify the statements in INTERP so that the program works on your graphing utility. Use INTERP to check your answers to part F.

*Available for download at *www.mhhe.com/barnett.*

Chapter 3 Review

In this chapter, unless indicated otherwise, the coefficients of the ***n*th-degree polynomial function** $P(x) = a_nx^n + a_{n-1}x^{n-1} + \cdots + a_1x + a_0$ are complex numbers and the domain is the set of complex numbers. The number r is said to be a **zero of the function *P*,** or a **zero of the polynomial *P*(*x*),** or a **solution or root of the equation *P*(*x*) = 0,** if $P(r) = 0$. If the coefficients of $P(x)$ are real numbers, then the x intercepts of the graph of $y = P(x)$ are real **zeros** of P and $P(x)$ and real **solutions** or **roots** for the equation $P(x) = 0$.

3-1 POLYNOMIAL FUNCTIONS AND GRAPHS

Synthetic division is an efficient method for dividing polynomials by linear terms of the form $x - r$ that is well-suited to calculator use.

Let $P(x)$ be a polynomial of degree greater than 0 and let r be a real number. Then we have the following important theorems:

Division Algorithm. $P(x) = (x - r)Q(x) + R$, where $x - r$ is the **divisor;** $Q(x)$, a unique polynomial of degree 1 less than $P(x)$, is the **quotient;** and R, a unique real number, is the **remainder.**

Remainder Theorem. $P(r) = R$.

The left and right behavior of an nth-degree polynomial $P(x)$ with real coefficients is determined by its highest degree or **leading term.** As $x \to \pm\infty$, a_nx^n and $P(x)$ both approach $\pm\infty$, depending on n and the sign of a_n. A **turning point** on a continuous graph is a point that separates an increasing portion from a decreasing portion. Important graph properties are:

1. P is continuous for all real numbers.
2. The graph of P is a smooth curve.
3. The graph of P has at most n x intercepts.
4. P has at most $n - 1$ turning points.

3-2 FINDING RATIONAL ZEROS OF POLYNOMIALS

If $P(x)$ is a polynomial of degree $n > 0$, then we have the following important theorems:

Factor Theorem. The number r is a zero of $P(x)$ if and only if $(x - r)$ is a factor of $P(x)$.

Fundamental Theorem of Algebra. $P(x)$ has at least one zero.

***n* Zeros Theorem.** $P(x)$ can be expressed as a product of n linear factors and has n zeros, not necessarily distinct.

If $P(x)$ is represented as the product of linear factors and $x - r$ occurs m times, then r is called a **zero of multiplicity *m*.**

Imaginary Zeros Theorem. If $P(x)$ has real coefficients, then imaginary zeros of $P(x)$, if they exist, must occur in conjugate pairs.

Real Zeros and Odd-Degree Polynomials. If $P(x)$ has real coefficients and is of odd degree, then $P(x)$ always has at least one real zero.

Rational Zero Theorem. If the rational number b/c, in lowest terms, is a zero of the polynomial

$$P(x) = a_nx^n + a_{n-1}x^{n-1} + \ldots + a_1x + a_0,$$

$a_n \neq 0$ with integer coefficients, then b must be an integer factor of a_0 and c must be an integer factor of a_n.

Strategy for Finding Rational Zeros

Assume that $P(x)$ is a polynomial with integer coefficients and is of degree greater than 2.

Step 1. List the possible rational zeros of $P(x)$ using the rational zero theorem (Theorem 6).

Step 2. Construct a synthetic division table. If a rational zero r is found, stop, write

$$P(x) = (x - r)Q(x)$$

and immediately proceed to find the rational zeros for $Q(x)$, the **reduced polynomial** relative to $P(x)$. If the degree of $Q(x)$ is greater than 2, return to step 1 using $Q(x)$ in place of $P(x)$. If $Q(x)$ is quadratic, find all its zeros using standard methods for solving quadratic equations.

3-3 APPROXIMATING REAL ZEROS OF POLYNOMIALS

The following theorems are important tools for locating the real zeros of a polynomial with real coefficients. Once located, a graphing utility can be used to approximate the zeros.

Location Theorem. If f is continuous on an interval I, a and b are two numbers in I, and $f(a)$ and $f(b)$ are of opposite sign, then there is at least one x intercept between a and b.

Upper and Lower Bounds of Real Zeros. If $a_n > 0$ and $P(x)$ is divided by $x - r$ using synthetic division:

1. If $r > 0$ and all numbers in the quotient row of the synthetic division, including the remainder, are nonnegative, then r is greater than or equal to the largest zero of $P(x)$ and is called an **upper bound** of the real zeros of $P(x)$.

2. If $r < 0$ and all numbers in the quotient row of the synthetic division, including the remainder, alternate in sign, then r is less than or equal to the smallest zero of $P(x)$ and is called a **lower bound** of the real zeros of $P(x)$.

3-4 RATIONAL FUNCTIONS

A function of the form $f(x) = n(x)/d(x)$, where $n(x)$ and $d(x)$ are polynomials, is a **rational function.** The line $x = a$ is a **vertical asymptote** for the graph of $y = f(x)$ if $f(x) \to \infty$ or $f(x) \to -\infty$ as $x \to a^+$ or $x \to a^-$. If $d(a) = 0$ and $n(a) \neq 0$, then the line $x = a$ is a vertical asymptote. The line $y = b$ is a **horizontal asymptote** for the graph of $y = f(x)$ if $f(x) \to b$ as $x \to \infty$ or $x \to -\infty$. The line $y = mx + b$ is an **oblique asymptote** if $[f(x) - (mx + b)] \to 0$ as $x \to \infty$ or $x \to -\infty$.

Let

$$f(x) = \frac{a_mx^m + \cdots + a_1x + a_0}{b_nx^n + \cdots + b_1x + b_0}, \qquad a_m, b_n \neq 0$$

The behavior of the graph of f as $x \to \infty$ or $x \to -\infty$ is determined by the ratio of the leading terms of the numerator and denominator, a_mx^m/b_nx^n

1. If $m < n$, then the x axis is a horizontal asymptote.
2. If $m = n$, then the line $y = a_m/b_n$ is a horizontal asymptote.
3. If $m > n$, then there are no horizontal asymptotes.

Graphing a Rational Function: $f(x) = n(x)/d(x)$

Step 1. Intercepts. Find the real solutions of the equation $n(x) = 0$ and use these solutions to plot any x intercepts of the graph of f. Evaluate $f(0)$, if it exists, and plot the y intercept.

Step 2. Vertical Asymptotes. Find the real solutions of the equation $d(x) = 0$ and use these solutions to determine the domain of f, the points of discontinuity, and the vertical asymptotes. Sketch any vertical asymptotes as dashed lines.

Step 3. Sign Chart. Construct a sign chart for f and use it to determine the behavior of the graph near each vertical asymptote.

Step 4. Horizontal Asymptotes. Determine whether there is a horizontal asymptote and if so, sketch it as a dashed line.

Step 5. Complete the Sketch. Complete the sketch of the graph by plotting additional points and joining these points with a smooth continuous curve over each interval in the domain of f. (Do not cross any points of discontinuity.)

3-5 PARTIAL FRACTIONS

A rational function $P(x)/D(x)$ often can be decomposed into a sum of simpler rational functions called **partial fractions.** If the degree of $P(x)$ is less than the degree of $D(x)$, then $P(x)/D(x)$ is called a **proper fraction.** We have the following important theorems:

Equal Polynomials. Two polynomials are equal to each other if and only if the coefficients of terms of like degree are equal.

Linear and Quadratic Factor Theorem. A polynomial with real coefficients can be factored into a product of linear and/or quadratic factors with real coefficients where the linear and quadratic factors are prime relative to the real numbers.

Partial Fraction Decomposition. Any proper fraction $P(x)/D(x)$ reduced to lowest terms can be decomposed into the sum of partial fractions as follows:

1. If $D(x)$ has a nonrepeating linear factor of the form $ax + b$, then the partial fraction decomposition of $P(x)/D(x)$ contains a term of the form

$$\frac{A}{ax + b} \qquad A \text{ a constant}$$

2. If $D(x)$ has a k-repeating linear factor of the form $(ax + b)^k$, then the partial fraction decomposition of $P(x)/D(x)$ contains k terms of the form

$$\frac{A_1}{ax + b} + \frac{A_2}{(ax + b)^2} + \cdots + \frac{A_k}{(ax + b)^k}$$

$$A_1, A_2, \ldots, A_k \text{ constants}$$

3. If $D(x)$ has a nonrepeating quadratic factor of the form $ax^2 + bx + c$, which is prime relative to the real numbers, then the partial fraction decomposition of $P(x)/D(x)$ contains a term of the form

$$\frac{Ax + B}{ax^2 + bx + c} \qquad A, B \text{ constants}$$

4. If $D(x)$ has a k-repeating quadratic factor of the form $(ax^2 + bx + c)^k$, where $ax^2 + bx + c$ is prime relative to the real numbers, then the partial fraction decomposition of $P(x)/D(x)$ contains k terms of the form

$$\frac{A_1x + B_1}{ax^2 + bx + c} + \frac{A_2x + B_2}{(ax^2 + bx + c)^2} + \cdots + \frac{A_kx + B_k}{(ax^2 + bx + c)^k}$$

$$A_1, \ldots, A_k, \quad B_1, \ldots, B_k \text{ constants}$$

Chapter 3 Review Exercise

Work through all the problems in this chapter review, and check answers in the back of the book. Answers to all review problems are there, and following each answer is a number in italics indicating the section in which that type of problem is discussed. Where weaknesses show up, review appropriate sections in the text.

A

1. Use synthetic division to divide $P(x) = 2x^3 + 3x^2 - 1$ by $D(x) = x + 2$, and write the answer in the form $P(x) = D(x)Q(x) + R$.

2. If $P(x) = x^5 - 4x^4 + 9x^2 - 8$, find $P(3)$ using the remainder theorem and synthetic division.

3. What are the zeros of $P(x) = 3(x - 2)(x + 4)(x + 1)$?

4. If $P(x) = x^2 - 2x + 2$ and $P(1 + i) = 0$, find another zero of $P(x)$.

5. Let $P(x)$ be the polynomial whose graph is shown in the figure.
   (A) Assuming that $P(x)$ has integer zeros and leading coefficient 1, find the lowest-degree equation that could produce this graph.
   (B) Describe the left and right behavior of $P(x)$.

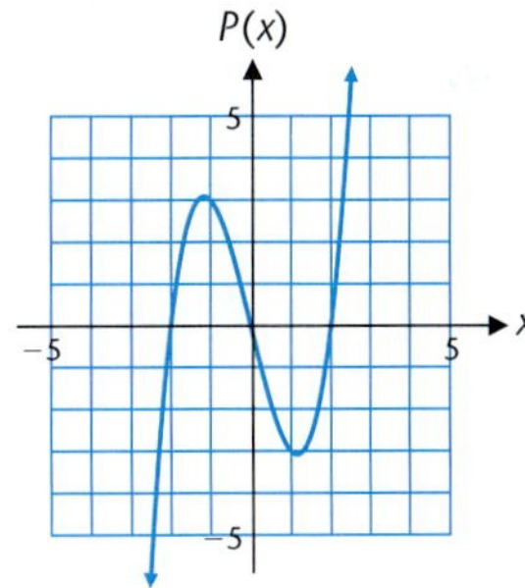

6. According to the upper and lower bound theorem, which of the following are upper or lower bounds of the zeros of $P(x) = x^3 - 4x^2 + 2$?

$$-2, -1, 3, 4$$

7. How do you know that $P(x) = 2x^3 - 3x^2 + x - 5$ has at least one real zero between 1 and 2?

8. Write the possible rational zeros for $P(x) = x^3 - 4x^2 + x + 6$.

9. Find all rational zeros for $P(x) = x^3 - 4x^2 + x + 6$.

10. Find the domain and x intercept(s) for:

(A) $f(x) = \dfrac{2x - 3}{x + 4}$ (B) $g(x) = \dfrac{3x}{x^2 - x - 6}$

11. Find the horizontal and vertical asymptotes for the functions in Problem 10.

12. Decompose into partial fractions: $\dfrac{7x - 11}{(x - 3)(x + 2)}$

B

13. Let $P(x) = x^3 - 3x^2 - 3x + 4$.

(A) Graph $P(x)$ and describe the graph verbally, including the number of x intercepts, the number of turning points, and the left and right behavior.

(B) Use the bisection method to approximate the largest x intercept to one decimal place.

14. If $P(x) = 8x^4 - 14x^3 - 13x^2 - 4x + 7$, find $Q(x)$ and R such that $P(x) = (x - \frac{1}{4})Q(x) + R$. What is $P(\frac{1}{4})$?

15. If $P(x) = 4x^3 - 8x^2 - 3x - 3$, find $P(-\frac{1}{2})$ using the remainder theorem and synthetic division.

16. Use the quadratic formula and the factor theorem to factor $P(x) = x^2 - 2x - 1$.

17. Is $x + 1$ a factor of $P(x) = 9x^{26} - 11x^{17} + 8x^{11} - 5x^4 - 7$? Explain, without dividing or using synthetic division.

18. Determine all rational zeros of $P(x) = 2x^3 - 3x^2 - 18x - 8$.

19. Factor the polynomial in Problem 18 into linear factors.

20. Find all rational zeros of $P(x) = x^3 - 3x^2 + 5$.

21. Find all zeros (rational, irrational, and imaginary) exactly for $P(x) = 2x^4 - x^3 + 2x - 1$.

22. Factor the polynomial in Problem 21 into linear factors.

23. Solve $2x^3 + 3x^2 \le 11x + 6$. Write the answer in inequality and interval notation.

24. Let $P(x) = x^4 - 2x^3 - 30x^2 - 25$.

(A) Find the smallest positive and the largest negative integers that, by Theorem 2 in Section 3-3, are upper and lower bounds, respectively, for the real zeros of $P(x)$.

(B) Use the bisection method to approximate the largest real zero of $P(x)$ to two decimal places.

(C) Use a graphing utility to approximate the real zeros of $P(x)$ to two decimal places.

25. Let $f(x) = \dfrac{x - 1}{2x + 2}$

(A) Find the domain and the intercepts for f.

(B) Find the vertical and horizontal asymptotes for f.

(C) Sketch a graph of f. Draw vertical and horizontal asymptotes with dashed lines.

26. Decompose into partial fractions: $\dfrac{-x^2 + 3x + 4}{x(x - 2)^2}$

27. Decompose into partial fractions: $\dfrac{8x^2 - 10x + 9}{2x^3 - 3x^2 + 3x}$

C

28. Use synthetic division to divide $P(x) = x^3 + 3x + 2$ by $[x - (1 + i)]$, and write the answer in the form $P(x) = D(x)Q(x) + R$.

29. Find a polynomial of lowest degree with leading coefficient 1 that has zeros $-\frac{1}{2}$ (multiplicity 2), -3, and 1 (multiplicity 3). (Leave the answer in factored form.) What is the degree of the polynomial?

30. Repeat Problem 29 for a polynomial $P(x)$ with zeros -5, $2 - 3i$, and $2 + 3i$.

31. Find all zeros (rational, irrational, and imaginary) exactly for $P(x) = 2x^5 - 5x^4 - 8x^3 + 21x^2 - 4$.

32. Factor the polynomial in Problem 31 into linear factors.

33. Solve

$$\frac{4x^2 + 4x - 3}{2x^3 + 3x^2 - 11x - 6} \ge 0$$

Write the answer in both inequality and interval notation.

34. What is the minimal degree of a polynomial $P(x)$, given that $P(-1) = -4$, $P(0) = 2$, $P(1) = -5$, and $P(2) = 3$? Justify your conclusion.

35. If $P(x)$ is a cubic polynomial with integer coefficients and if $1 + 2i$ is a zero of $P(x)$, can $P(x)$ have an irrational zero? Explain.

36. The solutions to the equation $x^3 - 27 = 0$ are the cube roots of 27.

(A) How many cube roots of 27 are there?

(B) 3 is obviously a cube root of 27; find all others.

37. Let $P(x) = x^4 + 2x^3 - 500x^2 - 4{,}000$.

(A) Find the smallest positive integer multiple of 10 and the largest negative integer multiple of 10 that, by Theorem 2 in Section 4-3, are upper and lower bounds, respectively, for the real zeros of $P(x)$.

(B) Approximate the real zeros of $P(x)$ to two decimal places.

38. Graph

$$f(x) = \frac{x^2 + 2x + 3}{x + 1}$$

Indicate any vertical, horizontal, or oblique asymptotes with dashed lines.

39. Use a graphing utility to find any horizontal asymptotes for

$$f(x) = \frac{2x}{\sqrt{x^2 + 3x + 4}}$$

40. Decompose into partial fractions:

$$\frac{5x^2 + 2x + 9}{x^4 - 3x^3 + x^2 - 3x}$$

APPLICATIONS

Express the solutions to each problem as the roots of a polynomial equation of the form $P(x) = 0$. Find rational solutions exactly and irrational solutions to one decimal place. Use a graphing utility or bisection only if necessary.

41. Architecture. An entryway is formed by placing a rectangular door inside an arch in the shape of the parabola with graph $y = 16 - x^2$, x and y in feet (see the figure). If the area of the door is 48 square feet, find the dimensions of the door.

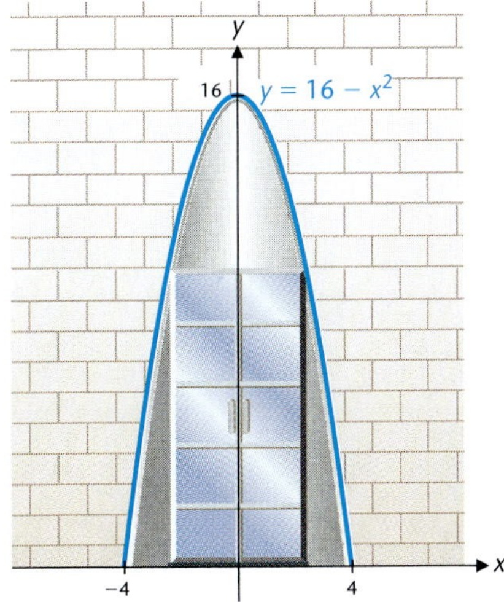

42. Construction. A grain silo is formed by attaching a hemisphere to the top of a right circular cylinder (see the figure). If the cylinder is 18 feet high and the volume of the silo is 486π cubic feet, find the common radius of the cylinder and the hemisphere.

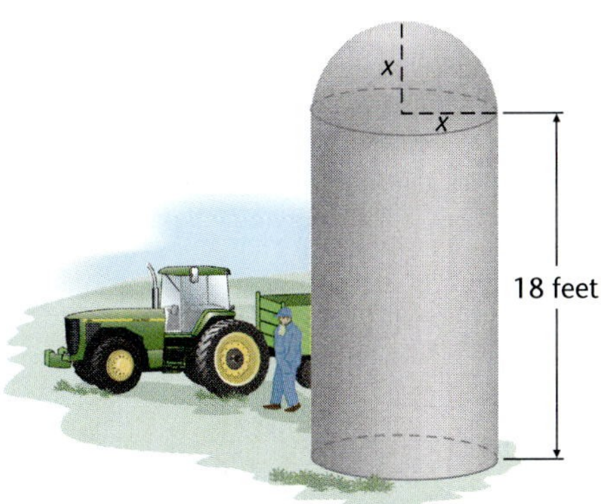

★ **43. Manufacturing.** A box is to be made out of a piece of cardboard that measures 15 by 20 inches. Squares, x inches on a side, will be cut from each corner, and then the ends and sides will be folded up (see the figure). Find the value of x that would result in a box with a volume of 300 cubic inches.

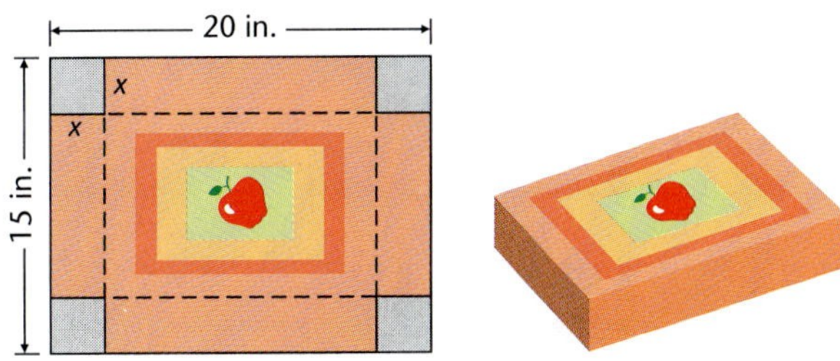

★ **44. Geometry.** Find all points on the graph of $y = x^2$ that are 3 units from the point (1, 4).

EXPONENTIAL AND LOGARITHMIC FUNCTIONS

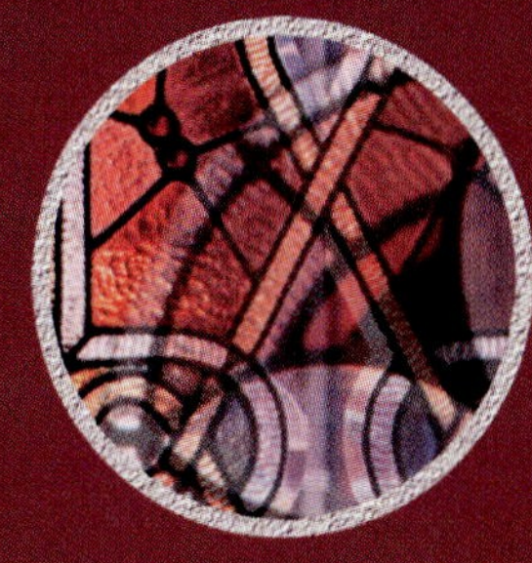

CHAPTER 4

Most of the functions we have considered so far have been polynomial and rational functions, with a few others involving roots or powers of polynomial or rational functions. The general class of functions defined by means of the algebraic operations of addition, subtraction, multiplication, division, and the taking of powers and roots on variables and constants are called *algebraic functions.*

In this chapter we define and investigate the properties of two new and important types of functions called *exponential functions* and *logarithmic functions.* These functions are not algebraic, but are members of another class of functions called *transcendental functions.* The exponential functions and logarithmic functions are used in describing and solving a wide variety of real-world problems, including growth of populations of people, animals, and bacteria; radioactive decay; growth of money at compound interest; absorption of light as it passes through air, water, or glass; and magnitudes of sounds and earthquakes. We consider applications in these areas plus many more in the sections that follow.

SECTION 4-1 Exponential Functions

- Exponential Functions
- Basic Exponential Graphs
- Additional Exponential Properties
- Applications

In this section we define exponential functions, look at some of their important properties—including graphs—and consider several significant applications.

• Exponential Functions

Let's start by noting that the functions f and g given by

$$f(x) = 2^x \qquad \text{and} \qquad g(x) = x^2$$

are not the same function. Whether a variable appears as an exponent with a constant base or as a base with a constant exponent makes a big difference. The function g is a quadratic function, which we have already discussed. The function f is a new type of function called an *exponential function.*

Many students, if asked to graph an exponential function such as $f(x) = 2^x$, would not hesitate at all. They would likely make up a table by assigning integers to x, plot the resulting points, and then join these points with a smooth curve, as shown in Figure 1.

FIGURE 1 $f(x) = 2^x$.

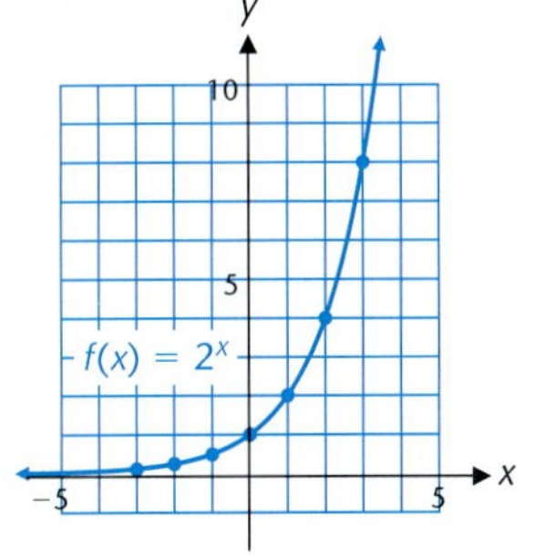

x	$f(x)$
−3	$\frac{1}{8}$
−2	$\frac{1}{4}$
−1	$\frac{1}{2}$
0	1
1	2
2	4
3	8

The only catch is that we have not defined 2^x for all real numbers x. We know what 2^5, 2^{-3}, $2^{2/3}$, $2^{-3/5}$, $2^{1.4}$, and $2^{-3.15}$ mean because we have defined 2^p for any rational number p, but what does

$$2^{\sqrt{2}}$$

mean? The question is not easy to answer at this time. In fact, a precise definition of $2^{\sqrt{2}}$ must wait for more advanced courses, where we can show that, if b is a positive real number and x is any real number, then

$$b^x$$

names a real number, and the graph of $f(x) = 2^x$ is as indicated in Figure 1. We also can show that for x irrational, b^x can be approximated as closely as we like by using rational number approximations for x. Since $\sqrt{2} = 1.414213\ldots$, for example, the sequence

$$2^{1.4}, 2^{1.41}, 2^{1.414}, \ldots$$

approximates $2^{\sqrt{2}}$, and as we use more decimal places, the approximation improves.

DEFINITION 1 **Exponential Function**

The equation

$$f(x) = b^x \qquad b > 0, b \neq 1$$

defines an **exponential function** for each different constant b, called the **base.** The independent variable x may assume any real value.

Thus, the **domain of f** is the set of all real numbers, and it can be shown that the **range of f** is the set of all positive real numbers. We require the base b to be positive to avoid imaginary numbers such as $(-2)^{1/2}$.

• Basic Exponential Graphs

EXPLORE-DISCUSS 1 Compare the graphs of $f(x) = 3^x$ and $g(x) = 2^x$ by plotting both functions on the same coordinate system. Find all points of intersection of the graphs. For which values of x is the graph of f above the graph of g? Below the graph of g? Are the graphs of f and g close together as $x \to \infty$? As $x \to -\infty$? Discuss.

It is useful to compare the graphs of $y = 2^x$ and $y = (\frac{1}{2})^x = 2^{-x}$ by plotting both on the same coordinate system, as shown in Figure 2(a). The graph of

$$f(x) = b^x \qquad b > 1 \qquad \text{[Fig. 2(b)]}$$

looks very much like the graph of the particular case $y = 2^x$, and the graph of

$$f(x) = b^x \qquad 0 < b < 1 \qquad \text{[Fig. 2(b)]}$$

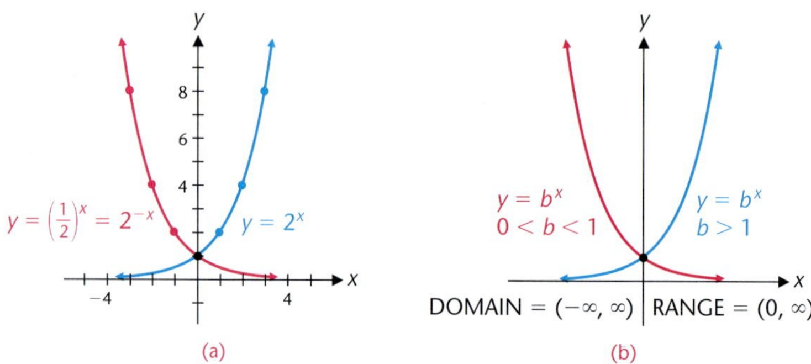

FIGURE 2 Basic exponential graphs.

looks very much like the graph of $y = (\frac{1}{2})^x$. Note in both cases that the x axis is a *horizontal asymptote* for the graph.

The graphs in Figure 2 suggest the following important general properties of exponential functions, which we state without proof:

Basic Properties of the Graph of $f(x) = b^x$, $b > 0$, $b \neq 1$

1. All graphs pass through the point (0, 1). $b^0 = 1$ for any permissible base b.
2. All graphs are continuous, with no holes or jumps.
3. The x axis is a horizontal asymptote.
4. If $b > 1$, then b^x increases as x increases.
5. If $0 < b < 1$, then b^x decreases as x increases.
6. The function f is one-to-one.

Property 6 implies that an exponential function has an inverse, called a *logarithmic function,* which we will discuss in Section 4-3.

A calculator may be used to create an accurate table of values from which the graph of an exponential function is drawn. Example 1 illustrates the process. (Of course, we may bypass the creation of a table of values with a graphing utility, which graphs the function directly.)

EXAMPLE 1 Graphing Multiples of Exponential Functions

Use integer values of x from -3 to 3 to construct a table of values for $y = \frac{1}{2}(4^x)$, and then graph this function.

Solution Use a calculator to create the table of values shown below. Then plot the points, and join these points with a smooth curve (see Fig. 3).

FIGURE 3 $y = \frac{1}{2}(4^x)$.

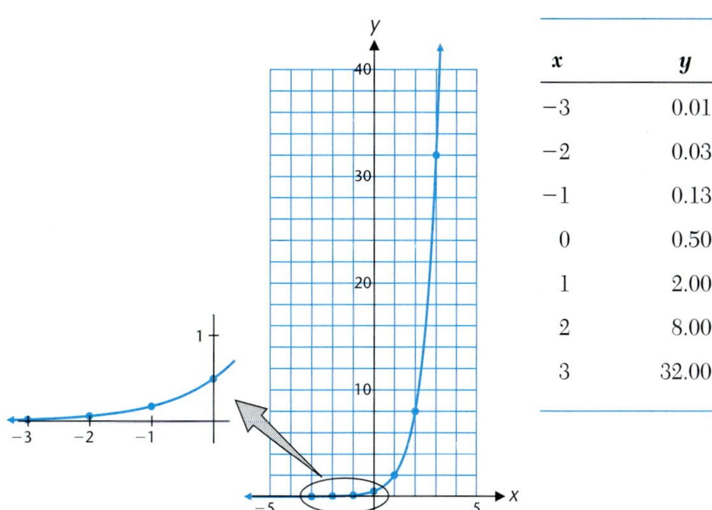

x	y
−3	0.01
−2	0.03
−1	0.13
0	0.50
1	2.00
2	8.00
3	32.00

Matched Problem 1 Repeat Example 1 for $y = \frac{1}{2}(\frac{1}{4})^x = \frac{1}{2}(4^{-x})$.

• Additional Exponential Properties

Exponential functions, whose domains include irrational numbers, obey the familiar laws of exponents we discussed earlier for rational exponents. We summarize these exponent laws here and add two other important and useful properties.

Exponential Function Properties

For a and b positive, $a \neq 1$, $b \neq 1$, and x and y real:

1. Exponent laws:

$$a^x a^y = a^{x+y} \qquad (a^x)^y = a^{xy} \qquad (ab)^x = a^x b^x$$

$$\left(\frac{a}{b}\right)^x = \frac{a^x}{b^x} \qquad \frac{a^x}{a^y} = a^{x-y} \qquad \frac{2^{5x}}{2^{7x}} \boxed{= 2^{5x-7x}} = 2^{-2x}$$

2. $a^x = a^y$ if and only if $x = y$. If $6^{4x} = 6^{2x+4}$, then $4x = 2x + 4$, and $x = 2$.

3. For $x \neq 0$, then $a^x = b^x$ if and only if $a = b$. If $a^4 = 3^4$, then $a = 3$.

EXAMPLE 2 **Using Exponential Function Properties**

Solve $4^{x-3} = 8$ for x.

Solution Express both sides in terms of the same base, and use property 2 to equate exponents.

$$4^{x-3} = 8$$

$$(2^2)^{x-3} = 2^3 \quad \text{Express 4 and 8 as powers of 2.}$$

$$2^{2x-6} = 2^3 \quad (a^x)^y = a^{xy}$$

$$2x - 6 = 3 \quad \text{Property 2}$$

$$2x = 9$$

$$x = \tfrac{9}{2}$$

Check $$4^{(9/2)-3} = 4^{3/2} = (\sqrt{4})^3 = 2^3 \stackrel{\checkmark}{=} 8$$

Matched Problem 2 Solve $27^{x+1} = 9$ for x.

• Applications

We now consider three applications that utilize exponential functions in their analysis: population growth, radioactive decay, and compound interest. Population growth and compound interest are examples of exponential growth, while radioactive decay is an example of negative exponential growth.

Our first example involves the growth of populations, such as people, animals, insects, and bacteria. Populations tend to grow exponentially and at different rates. A convenient and easily understood measure of growth rate is the **doubling time**—that is, the time it takes for a population to double. Over short periods of time the **doubling time growth model** is often used to model population growth:

$$P = P_0 2^{t/d}$$

where P = Population at time t
P_0 = Population at time $t = 0$
d = Doubling time

Note that when $t = d$,

$$P = P_0 2^{d/d} = P_0 2$$

and the population is double the original, as it should be. We use this model to solve a population growth problem in Example 3.

EXAMPLE 3 Population Growth

Mexico has a population of around 100 million people, and it is estimated that the population will double in 21 years. If population growth continues at the same rate, what will be the population:

(A) 15 years from now? (B) 30 years from now?

Calculate answers to 3 significant digits.

Solutions We use the doubling time growth model:

$$P = P_0 2^{t/d}$$

Substituting $P_0 = 100$ and $d = 21$, we obtain

$$P = 100(2^{t/21})$$ See Figure 4.

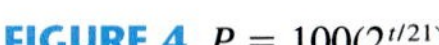

FIGURE 4 $P = 100(2^{t/21})$

(A) Find P when $t = 15$ years:

$$P = 100(2^{15/21}) \approx 164 \text{ million people}$$ Use a calculator.

(B) Find P when $t = 30$ years:

$$P = 100(2^{30/21}) \approx 269 \text{ million people}$$ Use a calculator.

Matched Problem 3 The bacterium *Escherichia coli* (*E. coli*) is found naturally in the intestines of many mammals. In a particular laboratory experiment, the doubling time for *E. coli* is found to be 25 minutes. If the experiment starts with a population of 1,000 *E. coli* and there is no change in the doubling time, how many bacteria will be present:

(A) In 10 minutes? (B) In 5 hours?

Write answers to 3 significant digits.

EXPLORE-DISCUSS 2 The doubling time growth model would *not* be expected to give accurate results over long periods of time. According to the doubling time growth model of Example 3, what was the population of Mexico 500 years ago at the height of Aztec civilization? What will the population of Mexico be 200 years from now? Explain why these results are unrealistic. Discuss factors that affect human populations that are not taken into account by the doubling time growth model.

Our second application involves radioactive decay, which is often referred to as negative growth. Radioactive materials are used extensively in medical diagnosis and therapy, as power sources in satellites, and as power sources in many countries. If we start with an amount A_0 of a particular radioactive isotope, the amount declines exponentially in time. The rate of decay varies from isotope to isotope. A convenient and easily understood measure of the rate of decay is the **half-life** of the isotope—that is, the time it takes for half of a particular material to decay. In this section we use the following **half-life decay model:**

$$\begin{aligned} A &= A_0(\tfrac{1}{2})^{t/h} \\ &= A_0 2^{-t/h} \end{aligned}$$

where A = Amount at time t

A_0 = Amount at time $t = 0$

h = Half-life

Note that when $t = h$,

$$A = A_0 2^{-h/h} = A_0 2^{-1} = \frac{A_0}{2}$$

and the amount of isotope is half the original amount, as it should be.

EXAMPLE 4 Radioactive Decay

The radioactive isotope gallium 67 (^{67}Ga), used in the diagnosis of malignant tumors, has a biological half-life of 46.5 hours. If we start with 100 milligrams of the isotope, how many milligrams will be left after:

(A) 24 hours? (B) 1 week?

Compute answers to 3 significant digits.

Solutions We use the half-life decay model:

$$A = A_0(\tfrac{1}{2})^{t/h} = A_0 2^{-t/h}$$

Using $A_0 = 100$ and $h = 46.5$, we obtain

$$A = 100(2^{-t/46.5})$$ See Figure 5.

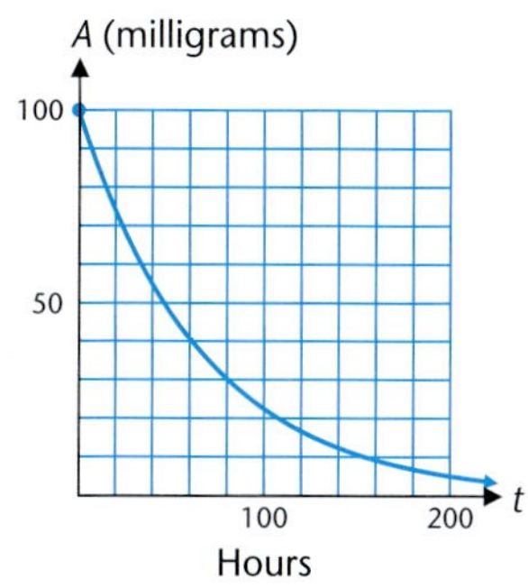

FIGURE 5 $A = 100(2^{-t/46.5})$.

(A) Find A when $t = 24$ hours:

$$A = 100(2^{-24/46.5})$$
$$= 69.9 \text{ milligrams} \quad \text{Use a calculator.}$$

(B) Find A when $t = 168$ hours (1 week = 168 hours):

$$A = 100(2^{-168/46.5})$$
$$= 8.17 \text{ milligrams} \quad \text{Use a calculator.}$$

Matched Problem 4 Radioactive gold 198 (^{198}Au), used in imaging the structure of the liver, has a half-life of 2.67 days. If we start with 50 milligrams of the isotope, how many milligrams will be left after:

(A) $\frac{1}{2}$ day? (B) 1 week?

Compute answers to 3 significant digits.

Our third application deals with the growth of money at compound interest. This topic is important to most people and is fundamental to many topics in the mathematics of finance.

The fee paid to use another's money is called **interest.** It is usually computed as a percent, called the **interest rate,** of the principal over a given period of time. If, at the end of a payment period, the interest due is reinvested at the same rate, then the interest earned as well as the principal will earn interest during the next payment period. Interest paid on interest reinvested is called **compound interest.**

Suppose you deposit \$1,000 in a savings and loan that pays 8% compounded semiannually. How much will the savings and loan owe you at the end of 2 years? Compounded semiannually means that interest is paid to your account at the end of each 6-month period, and the interest will in turn earn interest. The **interest rate per period** is the annual rate, 8% = 0.08, divided by the number of compounding periods per year, 2. If we let A_1, A_2, A_3, and A_4 represent the new amounts due at the end of the first, second, third, and fourth periods, respectively, then

$$A_1 = \$1{,}000 + \$1{,}000\left(\frac{0.08}{2}\right)$$
$$= \$1{,}000(1 + 0.04) \qquad P\left(1 + \frac{r}{n}\right)$$
$$A_2 = A_1(1 + 0.04)$$
$$= [\$1{,}000(1 + 0.04)](1 + 0.04)$$
$$= \$1{,}000(1 + 0.04)^2 \qquad P\left(1 + \frac{r}{n}\right)^2$$
$$A_3 = A_2(1 + 0.04)$$
$$= [\$1{,}000(1 + 0.04)^2](1 + 0.04)$$
$$= \$1{,}000(1 + 0.04)^3 \qquad P\left(1 + \frac{r}{n}\right)^3$$

$$\begin{aligned} A_4 &= A_3(1 + 0.04) \\ &= [\$1{,}000(1 + 0.04)^3](1 + 0.04) \\ &= \$1{,}000(1 + 0.04)^4 \qquad P\left(1 + \frac{r}{n}\right)^4 \end{aligned}$$

What do you think the savings and loan will owe you at the end of 6 years? If you guessed

$$A = \$1{,}000(1 + 0.04)^{12}$$

you have observed a pattern that is generalized in the following compound interest formula:

Compound Interest

If a **principal** P is invested at an annual **rate** r compounded n times a year, then the **amount** A in the account at the end of t years is given by

$$A = P\left(1 + \frac{r}{n}\right)^{nt}$$

The annual rate r is expressed in decimal form.

Since the principal P represents the initial amount in the account and A represents the amount in the account t years later, we also call P the **present value** of the account and A the **future value** of the account.

EXAMPLE 5 **Compound Interest**

If you deposit \$5,000 in an account paying 9% compounded daily, how much will you have in the account in 5 years? Compute the answer to the nearest cent.

Solution We use the compound interest formula as follows:

$$\begin{aligned} A &= P\left(1 + \frac{r}{n}\right)^{nt} \\ &= 5{,}000\left(1 + \frac{0.09}{365}\right)^{(365)(5)} \\ &= \$7{,}841.12 \qquad \text{Use a calculator.} \end{aligned}$$

The graph of

$$A = 5{,}000\left(1 + \frac{0.09}{365}\right)^{365t}$$

is shown in Figure 6.

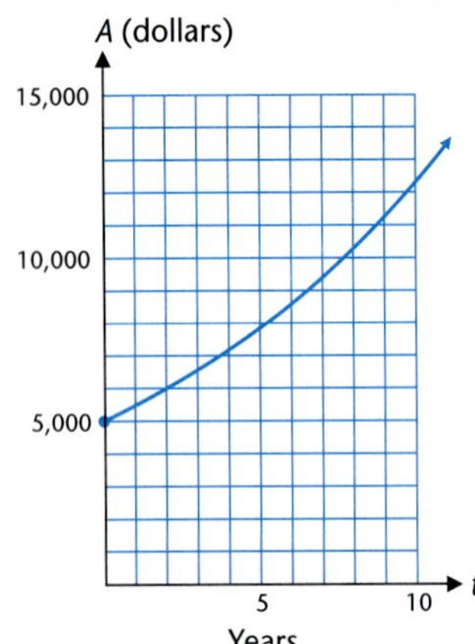

FIGURE 6

Matched Problem 5 If \$1,000 is invested in an account paying 10% compounded monthly, how much will be in the account at the end of 10 years? Compute the answer to the nearest cent.

EXAMPLE 6 Visualizing Investments with a Graphing Utility

Use a graphing utility to compare the growth of an investment of \$1,000 at 10% compounded monthly with an investment of \$2,000 at 5% compounded monthly. When do the two investments have the same value?

Solution We use the compound interest formula to express the future value y_1 of the first investment by $y_1 = 1{,}000(1 + 0.10/12)^{12x}$, and the future value y_2 of the second investment by $y_2 = 2{,}000(1 + 0.05/12)^{12x}$, where x is time in years. We graph both functions and use the intersection routine of the graphing utility to conclude that the investments have the same value when $x \approx 14$ years, as shown in Figure 7. After that time the \$1,000 investment has the greater value.

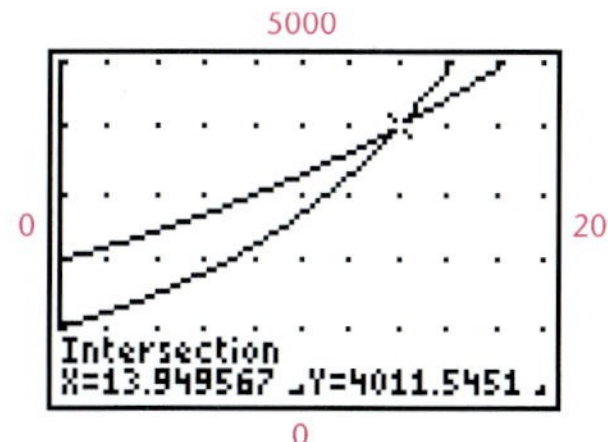

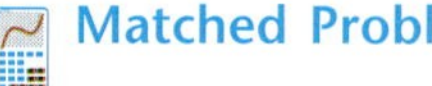
FIGURE 7

Matched Problem 6 Use a graphing utility to determine when an investment of \$5,000 at 6% compounded quarterly has the same value as an investment of \$4,000 at 10% compounded daily.

Answers to Matched Problems

1. $y = \frac{1}{2}(4^{-x})$

x	y
−3	32.00
−2	8.00
−1	2.00
0	0.50
1	0.13
2	0.03
3	0.01

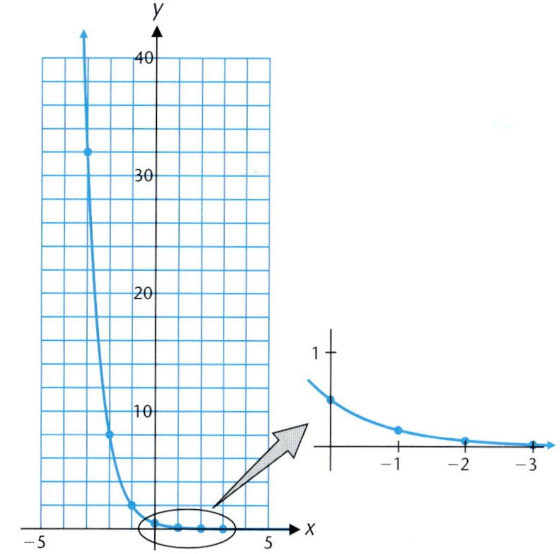

2. $x = -\frac{1}{3}$
3. (A) 1,320 (B) $4{,}100{,}000 = 4.10 \times 10^6$
4. (A) 43.9 mg (B) 8.12 mg
5. \$2,707.04
6. 5 years, 6 months

EXERCISE 4-1

A

In Problems 1–10, construct a table of values for integer values of x over the indicated interval and then graph the function.

1. $y = 3^x; [-3, 3]$ **2.** $y = 5^x; [-2, 2]$

3. $y = (\frac{1}{3})^x = 3^{-x}; [-3, 3]$ **4.** $y = (\frac{1}{5})^x = 5^{-x}; [-2, 2]$

5. $g(x) = -3^{-x}; [-3, 3]$ **6.** $f(x) = -5^x; [-2, 2]$

7. $h(x) = 5(3^x); [-3, 3]$ **8.** $f(x) = 4(5^x); [-2, 2]$

9. $y = 3^{x+3} - 5; [-6, 0]$ **10.** $y = 5^{x+2} + 4; [-4, 0]$

In Problems 11–22, simplify.

11. $2^{5x+1}2^{3-2x}$ **12.** $5^{2x-3}5^{6-4x}$ **13.** $\dfrac{3^{2y-x}}{3^{3x-5y}}$

14. $\dfrac{7^{-3y-x}}{7^{-4x-6y}}$ **15.** $(4^x)^3$ **16.** $(x^3)^4$

17. $(10^2)^x(10^x)^2$ **18.** $2^{x^3}2^{x^2}$ **19.** $\left(\dfrac{5^x}{4^y}\right)^3$

20. $\left(\dfrac{3^{2x}}{7^{3y}}\right)^2$ **21.** $\left(\dfrac{a^{-2}b^3c^{-1}}{a^{-3}b^2c^{-3}}\right)^2$ **22.** $\left(\dfrac{a^{-4}b^3c^2}{a^{-6}b^2c^{-1}}\right)^3$

B

In Problems 23–34, solve for x.

23. $3^{2x-5} = 3^{4x+2}$ **24.** $4^{4x+1} = 4^{2x-2}$

25. $10^{x^2+2} = 10^{2x+2}$ **26.** $5^{x^2-5} = 5^{3x+5}$

27. $(2x + 1)^3 = 8$ **28.** $(2x - 1)^5 = -32$

29. $5^{3x} = 25^{x+3}$ **30.** $4^{5x+1} = 16^{2x-1}$

31. $4^{2x+2} = 8^{x+2}$ **32.** $100^{2x+4} = 1{,}000^{x+4}$

33. $100^{x^2} = 10^{5x-3}$ **34.** $3^{x^2} = 9^{x+4}$

35. Find all real numbers a such that $a^2 = a^{-2}$. Explain why this does not violate the second exponential function property in the box on page 286.

36. Find real numbers a and b such that $a \neq b$ but $a^4 = b^4$. Explain why this does not violate the third exponential function property in the box on page 286.

Graph each function in Problems 37–46 by constructing a table of values.

* *Check Problems 37–46 with a graphing utility.*

37. $G(t) = 3^{t/100}$ **38.** $f(t) = 2^{t/10}$ **39.** $y = 11(3^{-x/2})$

40. $y = 7(2^{-2x})$ **41.** $g(x) = 2^{-|x|}$ **42.** $f(x) = 2^{|x|}$

43. $y = 1{,}000(1.08)^x$ **44.** $y = 100(1.03)^x$

45. $y = 2^{-x^2}$ **46.** $y = 3^{-x^2}$

C

In Problems 47–50, simplify.

47. $(6^x + 6^{-x})(6^x - 6^{-x})$ **48.** $(3^x - 3^{-x})(3^x + 3^{-x})$

49. $(6^x + 6^{-x})^2 - (6^x - 6^{-x})^2$ **50.** $(3^x - 3^{-x})^2 + (3^x + 3^{-x})^2$

Graph each function in Problems 51–54 by constructing a table of values.

Check Problems 51–54 with a graphing utility.

51. $m(x) = x(3^{-x})$ **52.** $h(x) = x(2^x)$

53. $f(x) = \dfrac{2^x + 2^{-x}}{2}$ **54.** $g(x) = \dfrac{3^x + 3^{-x}}{2}$

In Problems 55–58:

(A) Approximate the real zeros of each function to two decimal places.

(B) Investigate the behavior of each function as $x \to \infty$ and $x \to -\infty$ and find any horizontal asymptotes.

55. $f(x) = 3^x - 5$ **56.** $f(x) = 4 + 2^{-x}$

57. $f(x) = 1 + x + 10^x$ **58.** $f(x) = 8 - x^2 + 2^{-x}$

APPLICATIONS 

59. Gaming. A person bets on red and black on a roulette wheel using a *Martingale strategy.* That is, a \$2 bet is placed on red, and the bet is doubled each time until a win occurs. The process is then repeated. If black occurs n times in a row, then $L = 2^n$ dollars is lost on the nth bet. Graph this function for $1 \leq n \leq 10$. Even though the function is defined only for positive integers, points on this type of graph are usually jointed with a smooth curve as a visual aid.

60. Bacterial Growth. If bacteria in a certain culture double every $\frac{1}{2}$ hour, write an equation that gives the number of bacteria N in the culture after t hours, assuming the culture has 100 bacteria at the start. Graph the equation for $0 \leq t \leq 5$.

61. Population Growth. Because of its short life span and frequent breeding, the fruit fly *Drosophila* is used in some genetic studies. Raymond Pearl of Johns Hopkins University,

*Please note that use of graphing utility is not required to complete these exercises. Checking them with a g.u. is optional.

for example, studied 300 successive generations of descendants of a single pair of *Drosophila* flies. In a laboratory situation with ample food supply and space, the doubling time for a particular population is 2.4 days. If we start with 5 male and 5 female flies, how many flies should we expect to have in:
(A) 1 week? (B) 2 weeks?

62. Population Growth. If Kenya has a population of about 30,000,000 people and a doubling time of 19 years and if the growth continues at the same rate, find the population in:
(A) 10 years (B) 30 years
Compute answers to 2 significant digits.

63. Insecticides. The use of the insecticide DDT is no longer allowed in many countries because of its long-term adverse effects. If a farmer uses 25 pounds of active DDT, assuming its half-life is 12 years, how much will still be active after:
(A) 5 years? (B) 20 years?
Compute answers to 2 significant digits.

64. Radioactive Tracers. The radioactive isotope technetium 99m (^{99m}Tc) is used in imaging the brain. The isotope has a half-life of 6 hours. If 12 milligrams are used, how much will be present after:
(A) 3 hours? (B) 24 hours?
Compute answers to 3 significant digits.

65. Finance. Suppose \$4,000 is invested at 11% compounded weekly. How much money will be in the account in:
(A) $\frac{1}{2}$ year? (B) 10 years?
Compute answers to the nearest cent.

66. Finance. Suppose \$2,500 is invested at 7% compounded quarterly. How much money will be in the account in:
(A) $\frac{3}{4}$ year? (B) 15 years?
Compute answers to the nearest cent.

★ **67. Finance.** A couple just had a new child. How much should they invest now at 8.25% compounded daily in order to have \$40,000 for the child's education 17 years from now? Compute the answer to the nearest dollar.

★ **68. Finance.** A person wishes to have \$15,000 cash for a new car 5 years from now. How much should be placed in an account now if the account pays 9.75% compounded weekly? Compute the answer to the nearest dollar.

★ **69. Finance.** Will an investment of \$10,000 at 8.9% compounded daily ever be worth more at the end of a quarter than an investment of \$10,000 at 9% compounded quarterly? Explain.

★ **70. Finance.** A sum of \$5,000 is invested at 13% compounded semiannually. Suppose that a second investment of \$5,000 is made at interest rate r compounded daily. For which values of r, to the nearest tenth of a percent, is the second investment better than the first? Discuss.

SECTION 4-2 The Exponential Function with Base e

- Base e Exponential Function
- Growth and Decay Applications Revisited
- Continuous Compound Interest
- A Comparison of Exponential Growth Phenomena

Until now the number π has probably been the most important irrational number you have encountered. In this section we will introduce another irrational number, e, that is just as important in mathematics and its applications.

• Base e Exponential Function

The following expression is important to the study of calculus and, as we will see later in this section, also is closely related to the compound interest formula discussed in the preceding section:

$$\left(1 + \frac{1}{m}\right)^m$$

EXPLORE-DISCUSS 1 (A) Calculate the value of $[1 + (1/m)]^m$ for $m = 1,2,3,4,5$. Are the values increasing or decreasing as m gets larger?

(B) What is the smallest positive integer m such that $[1 + (1/m)]^m$ is greater than 2.5? Greater than 2.7? Greater than 2.9?

TABLE 1

m	$\left(1+\frac{1}{m}\right)^m$
1	2
10	2.59374 . . .
100	2.70481 . . .
1,000	2.71692 . . .
10,000	2.71814 . . .
100,000	2.71827 . . .
1,000,000	2.71828 . . .
⋮	⋮

Interestingly, by calculating the value of the expression for larger and larger values of m (see Table 1), it appears that $[1 + (1/m)]^m$ approaches a number close to 2.7183. In a calculus course we can show that as m increases without bound, the value of $[1 + (1/m)]^m$ approaches an irrational number that we call $\boldsymbol{e}$. Just as irrational numbers such as π and $\sqrt{2}$ have unending, nonrepeating decimal representations (see Section A-1), e also has an unending, nonrepeating decimal representation. To 12 decimal places,

$$e = 2.718\ 281\ 828\ 459$$

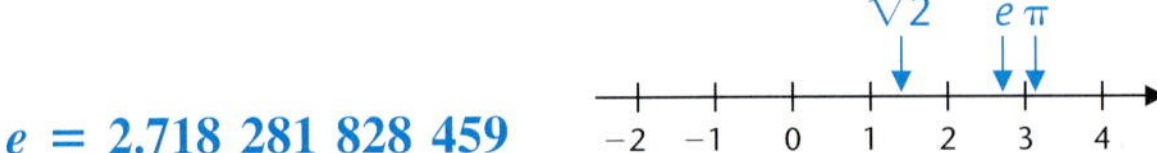

Exactly who discovered e is still being debated. It is named after the great Swiss mathematician Leonhard Euler (1707–1783), who computed e to 23 decimal places using $[1 + (1/m)]^m$.

The constant e turns out to be an ideal base for an exponential function because in calculus and higher mathematics many operations take on their simplest form using this base. This is why you will see e used extensively in expressions and formulas that model real-world phenomena.

DEFINITION 1 **Exponential Function with Base *e***

For x a real number, the equation

$$f(x) = e^x$$

defines the **exponential function with base *e*.**

The exponential function with base e is used so frequently that it is often referred to as *the* exponential function. The graphs of $y = e^x$ and $y = e^{-x}$ are shown in Figure 1.

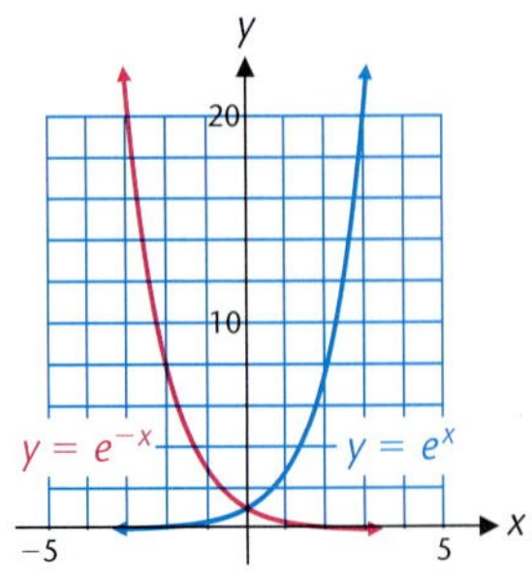

FIGURE 1 Exponential functions with base e.

EXPLORE-DISCUSS 2 A graphing utility was used to graph the functions $f(x) = 3^x$, $g(x) = 2^x$, and $h(x) = e^x$ in Figure 2. Where do the graphs intersect? Which graph lies between the others? Which graph is above the others when $x > 0$? When $x < 0$? Discuss the behavior of the three functions as $x \to \infty$ and as $x \to -\infty$.

FIGURE 2

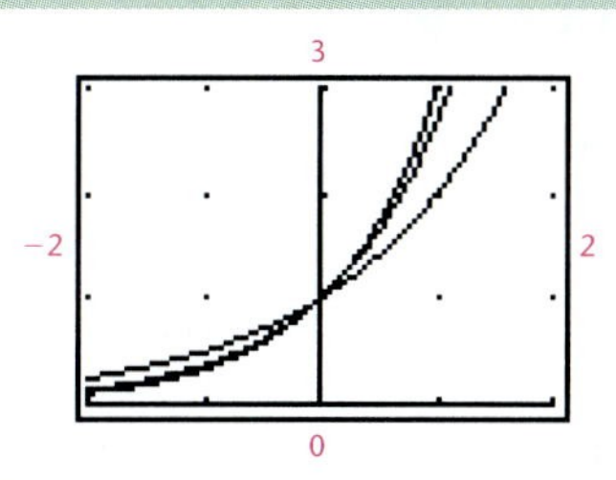

EXAMPLE 1 Graphing Exponential Functions

Graph $y = 4 - e^{x/2}$

Solution Use a calculator to construct a table of values for integer values of x. Then plot the points and join these points with a smooth curve (see Fig. 3).

FIGURE 3 $y = 4 - e^{x/2}$

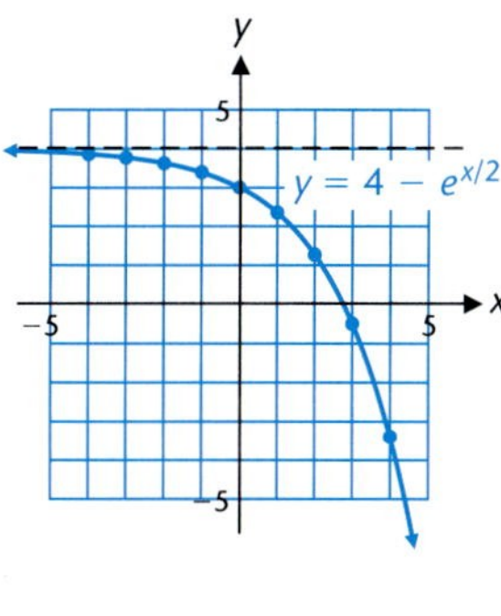

x	y
−4	3.86
−3	3.78
−2	3.63
−1	3.39
0	3
1	2.35
2	1.28
3	−0.48
4	−3.39

Notice that as x approaches $-\infty$, the values of $e^{x/2}$ approach 0 and the values of $4 - e^{x/2}$ approach 4. The line $y = 4$ is a horizontal asymptote for the graph.

Matched Problem 1 Graph $y = 2e^{x/2} - 5$.

• Growth and Decay Applications Revisited

Most exponential growth and decay problems are modeled using base e exponential functions. We present two applications here and many more in Exercise 4-2.

EXAMPLE 2 **Medicine—Bacteria Growth**

Cholera, an intestinal disease, is caused by a cholera bacterium that multiplies exponentially by cell division as modeled by

$$N = N_0 e^{1.386t}$$

where N is the number of bacteria present after t hours and N_0 is the number of bacteria present at $t = 0$. If we start with 1 bacterium, how many bacteria will be present in:

(A) 5 hours? (B) 12 hours?

Compute the answers to 3 significant digits.

Solutions (A) Use $N_0 = 1$ and $t = 5$:

$$\begin{aligned} N &= N_0 e^{1.386t} \\ &= e^{1.386(5)} \\ &= 1{,}020 \end{aligned}$$

(B) Use $N_0 = 1$ and $t = 12$:

$$\begin{aligned} N &= N_0 e^{1.386t} \\ &= e^{1.386(12)} \\ &= 16{,}700{,}000 \end{aligned}$$

Matched Problem 2 Graph the exponential growth model for cholera bacteria over the indicated interval:

$$N = e^{1.386t} \qquad 0 \le t \le 5$$

EXAMPLE 3 **Carbon 14 Dating**

Cosmic-ray bombardment of the atmosphere produces neutrons, which in turn react with nitrogen to produce radioactive carbon 14. Radioactive carbon 14 enters all living tissues through carbon dioxide, which is first absorbed by plants. As long as a plant or animal is alive, carbon 14 is maintained in the living organism at a constant level. Once the organism dies, however, carbon 14 decays according to the equation.

$$A = A_0 e^{-0.000124t}$$

where A is the amount of carbon 14 present after t years and A_0 is the amount present at time $t = 0$. If 1,000 milligrams of carbon 14 are present at the start, how many milligrams will be present in:

(A) 10,000 years? (B) 50,000 years?

Compute answers to 3 significant digits.

Solutions Substituting $A_0 = 1{,}000$ in the decay equation, we have

$$A = 1{,}000e^{-0.000124t} \quad \text{See Figure 4.}$$

(A) Solve for A when $t = 10{,}000$:

$$A = 1{,}000e^{-0.000124(10{,}000)}$$
$$= 289 \text{ milligrams}$$

(B) Solve for A when $t = 50{,}000$:

$$A = 1{,}000e^{-0.000124(50{,}000)}$$
$$= 2.03 \text{ milligrams}$$

FIGURE 4

More will be said about carbon 14 dating in Exercise 4-5, where we will be interested in solving for t after being given information about A and A_0.

Matched Problem 3 Referring to Example 3, how many milligrams of carbon 14 would have to be present at the beginning in order to have 10 milligrams present after 20,000 years? Compute the answer to 4 significant digits.

• Continuous Compound Interest

The constant e occurs naturally in the study of compound interest. Returning to the compound interest formula discussed in Section 4-1,

$$A = P\left(1 + \frac{r}{n}\right)^{nt} \quad \text{Compound interest}$$

recall that P is the principal invested at an annual rate r compounded n times a year and A is the amount in the account after t years. Suppose P, r, and t are held fixed and n is increased without bound. Will the amount A increase without bound or will it tend to some limiting value? Let's perform a calculator experiment before we attack the general problem. If $P = \$100$, $r = 0.08$, and $t = 2$ years, then

$$A = 100\left(1 + \frac{0.08}{n}\right)^{2n}$$

The amount A is computed for several values of n in Table 2. Notice that the largest gain appears in going from annually to semiannually. Then, the gains slow down as n increases. In fact, it appears that A might be tending to something close to \$117.35 as n gets larger and larger.

TABLE 2 Effect of Compounding Frequency

Compounding frequency	n	$A = 100\left(1 + \frac{0.08}{n}\right)^{2n}$
Annually	1	\$116.6400
Semiannually	2	116.9859
Quarterly	4	117.1659
Weekly	52	117.3367
Daily	365	117.3490
Hourly	8,760	117.3501

We now return to the general problem to see if we can determine what happens to $A = P[1 + (r/n)]^{nt}$ as n increases without bound. A little algebraic manipulation of the compound interest formula will lead to an answer and a significant result in the mathematics of finance.

$$A = P\left(1 + \frac{r}{n}\right)^{nt}$$

$$= P\left(1 + \frac{1}{n/r}\right)^{(n/r)rt} \qquad \text{Change algebraically.}$$

$$= P\left[\left(1 + \frac{1}{m}\right)^{m}\right]^{rt} \qquad \text{Let } m = n/r.$$

The expression within the square brackets should look familiar. Recall from the first part of this section that

$$\left(1 + \frac{1}{m}\right)^{m} \text{ approaches } e \qquad \text{as} \qquad m \text{ approaches } \infty$$

Since r is fixed, $m = n/r$ approaches ∞ as n approaches ∞. Thus,

$$P\left(1 + \frac{r}{n}\right)^{nt} \text{ approaches } Pe^{rt} \qquad \text{as} \qquad n \text{ approaches } \infty$$

and we have arrived at the **continuous compound interest formula,** a very important and widely used formula in business, banking, and economics.

Continuous Compound Interest Formula

If a principal P is invested at an annual rate r compounded continuously, then the amount A in the account at the end of t years is given by

$$A = Pe^{rt}$$

The annual rate r is expressed as a decimal.

EXAMPLE 4 Continuous Compound Interest

If \$100 is invested at an annual rate of 8% compounded continuously, what amount, to the nearest cent, will be in the account after 2 years?

Solution Use the continuous compound interest formula to find A when $P = \$100$, $r = 0.08$, and $t = 2$:

$$\begin{aligned} A &= Pe^{rt} \\ &= \$100e^{(0.08)(2)} \quad \text{8\% is equivalent to } r = 0.08. \\ &= \$117.35 \end{aligned}$$

Compare this result with the values calculated in Table 2.

Matched Problem 4 What amount will an account have after 5 years if \$100 is invested at an annual rate of 12% compounded annually? Quarterly? Continuously? Compute answers to the nearest cent.

The continuous compound interest formula may also be used to model short-term population growth. If a population P is assumed to grow continuously at an annual rate r, then the population A at the end of t years is given by $A = Pe^{rt}$.

• A Comparison of Exponential Growth Phenomena

The equations and graphs given in Table 3 compare several widely used growth models. These are divided basically into two groups: unlimited growth and limited growth. Following each equation and graph is a short, incomplete list of areas in which the models are used. We have only touched on a subject that has been extensively developed and which you are likely to study in greater depth in the future.

TABLE 3 Exponential Growth and Decay

Description	Equation	Graph	Uses
Unlimited growth	$y = ce^{kt}$ $c, k > 0$		Short-term population growth (people, bacteria, etc.); growth of money at continuous compound interest
Exponential decay	$y = ce^{-kt}$ $c, k > 0$		Radioactive decay; light absorption in water, glass, etc.; atmospheric pressure; electric circuits
Limited growth	$y = c(1 - e^{-kt})$ $c, k > 0$		Learning skills; sales fads; company growth; electric circuits
Logistic growth	$y = \dfrac{M}{1 + ce^{-kt}}$ $c, k, M > 0$		Long-term population growth; epidemics; sales of new products; company growth

Answers to Matched Problems

1.

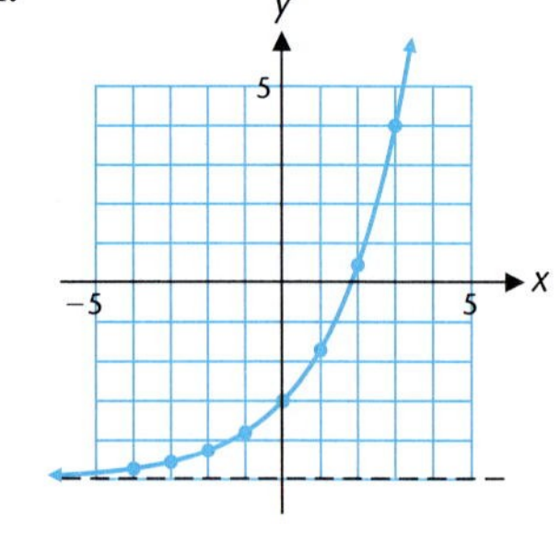

x	y
−4	−4.73
−3	−4.55
−2	−4.26
−1	−3.79
0	−3
1	−1.7
2	0.44
3	3.96

2.

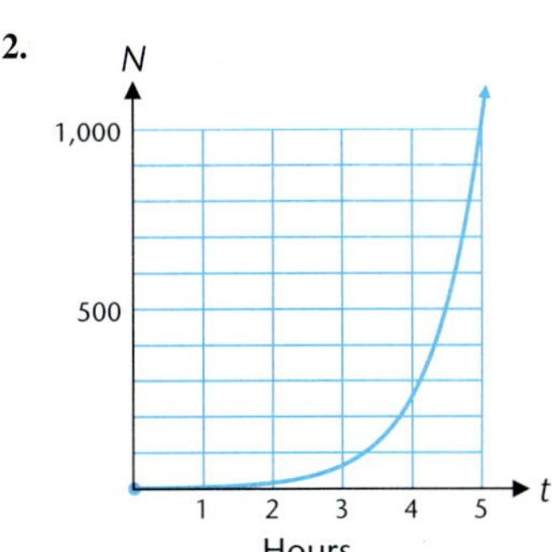

3. 119.4 mg

4. Annually: \$176.23; quarterly: \$180.61; continuously: \$182.21

EXERCISE 4-2

A

In Problems 1–6, construct a table of values for integer values of x over the indicated interval and then graph the function.

Check Problems 1–6 with a graphing utility.

1. $y = -e^x$; $[-3, 3]$
2. $y = -e^{-x}$; $[-3, 3]$
3. $y = 10e^{0.2x}$; $[-5, 5]$
4. $y = 100e^{0.1x}$; $[-5, 5]$
5. $f(t) = 100e^{-0.1t}$; $[-5, 5]$
6. $g(t) = 10e^{-0.2t}$; $[-5, 5]$

In Problems 7–12, simplify.

7. $e^{2x}e^{-3x}$
8. $(e^{-x})^4$
9. $(e^x)^3$
10. $e^{-4x}e^{6x}$
11. $\dfrac{e^{5x}}{e^{2x+1}}$
12. $\dfrac{e^{4-3x}}{e^{2-5x}}$

13. (A) Explain what is wrong with the following reasoning about the expression $[1 + (1/m)]^m$: As m gets large, $1 + (1/m)$ approaches 1 because $1/m$ approaches 0, and 1 raised to any power is 1, so $[1 + (1/m)]^m$ approaches 1.
(B) Which number does $[1 + (1/m)]^m$ approach as $m \to \infty$?

14. (A) Explain what is wrong with the following reasoning about the expression $[1 + (1/m)]^m$: If $b > 1$, then the exponential function $b^x \to \infty$ as $x \to \infty$, and $1+(1/m)$ is greater than 1, so $[1 + (1/m)]^m$ approaches infinity as $m \to \infty$.
(B) Which number does $[1 + (1/m)]^m$ approach as $m \to \infty$?

B

In Problems 15–22, graph each function by constructing a table of values.

Check Problems 15–22 with a graphing utility.

15. $y = 2 + e^{x-2}$
16. $y = -3 + e^{1+x}$
17. $y = e^{-|x|}$
18. $y = e^{|x|}$
19. $M(x) = e^{x/2} + e^{-x/2}$
20. $C(x) = \dfrac{e^x + e^{-x}}{2}$
21. $N = \dfrac{200}{1 + 3e^{-t}}$
22. $N = \dfrac{100}{1 + e^{-t}}$

In Problems 23–28, simplify.

23. $\dfrac{-2x^3e^{-2x} - 3x^2e^{-2x}}{x^6}$
24. $\dfrac{5x^4e^{5x} - 4x^3e^{5x}}{x^8}$
25. $(e^x + e^{-x})^2 + (e^x - e^{-x})^2$
26. $e^x(e^{-x} + 1) - e^{-x}(e^x + 1)$
27. $\dfrac{e^{-x}(e^x - e^{-x}) + e^{-x}(e^x + e^{-x})}{e^{-2x}}$
28. $\dfrac{e^x(e^x + e^{-x}) - (e^x - e^{-x})e^x}{e^{2x}}$

In Problems 29–32, solve each equation. [Remember: $e^{-x} \neq 0$.]

29. $2xe^{-x} = 0$
30. $(x - 3)e^x = 0$
31. $x^2e^x - 5xe^x = 0$
32. $3xe^{-x} + x^2e^{-x} = 0$

C

One of the most important functions in statistics is the ***normal probability density function***

$$f(x) = \frac{1}{\sigma\sqrt{2\pi}}e^{-(x-\mu)^2/2\sigma^2}$$

where μ is the ***mean*** *and σ is the* ***standard deviation.*** *The graph of this function is the "bell-shaped" curve that instructors refer to when they say that they are grading on a curve.*

Graph the related functions given in Problems 33 and 34.

33. $f(x) = e^{-x^2}$
34. $g(x) = \dfrac{1}{\sqrt{\pi}}e^{-x^2/2}$

35. Given $f(s) = (1 + s)^{1/s}$, $s \neq 0$:
(A) Complete the tables below to four decimal places.
(B) What does $(1 + s)^{1/s}$ seem to tend to as s approaches 0?

s	$f(s)$
−0.5	4.0000
−0.2	3.0518
−0.1	
−0.01	
−0.001	
−0.0001	

s	$f(s)$
0.5	2.2500
0.2	2.4883
0.1	
0.01	
0.001	
0.0001	

36. Refer to Problem 35. Graph $f(s) = (1 + s)^{1/s}$ for s in $[-0.5, 0) \cup (0, 0.5]$.

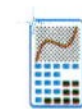

Problems 37–40 require the use of a graphing utility.

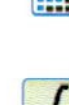

It is common practice in many applications of mathematics to approximate nonpolynomial functions with appropriately selected polynomials. For example, the polynomials in Problems 37–40, called ***Taylor polynomials,*** *can be used to approximate the exponential function $f(x) = e^x$. To illustrate this approximation graphically, in each problem graph $f(x) = e^x$ and the indicated polynomial in the same viewing window, $-4 \leq x \leq 4$ and $-5 \leq y \leq 50$.*

37. $P_1(x) = 1 + x + \frac{1}{2}x^2$

38. $P_2(x) = 1 + x + \frac{1}{2}x^2 + \frac{1}{6}x^3$

39. $P_3(x) = 1 + x + \frac{1}{2}x^2 + \frac{1}{6}x^3 + \frac{1}{24}x^4$

40. $P_4(x) = 1 + x + \frac{1}{2}x^2 + \frac{1}{6}x^3 + \frac{1}{24}x^4 + \frac{1}{120}x^5$

41. Investigate the behavior of the functions $f_1(x) = x/e^x$, $f_2(x) = x^2/e^x$, and $f_3(x) = x^3/e^x$ as $x \to \infty$ and as $x \to -\infty$, and find any horizontal asymptotes. Generalize to functions of the form $f_n(x) = x^n/e^x$, where n is any positive integer.

42. Investigate the behavior of the functions $g_1(x) = xe^x$, $g_2(x) = x^2e^x$, and $g_3(x) = x^3e^x$ as $x \to \infty$ and as $x \to -\infty$, and find any horizontal asymptotes. Generalize to functions of the form $g_n(x) = x^ne^x$, where n is any positive integer.

APPLICATIONS

43. **Population Growth.** If the world population is about 6 billion people now and if the population grows continuously at an annual rate of 1.7%, what will the population be in 10 years? Compute the answer to 2 significant digits.

44. **Population Growth.** If the population in Mexico is around 100 million people now and if the population grows continuously at an annual rate of 2.3%, what will the population be in 8 years? Compute the answer to 2 significant digits.

45. **Population Growth.** In 1996 the population of Russia was 148 million and the population of Nigeria was 104 million. If the populations of Russia and Nigeria grow continuously at annual rates of −0.62% and 3.0%, respectively, when will Nigeria have a greater population than Russia?

46. **Population Growth.** In 1996 the population of Germany was 84 million and the population of Egypt was 64 million. If the populations of Germany and Egypt grow continuously at annual rates of −0.15% and 1.9%, respectively, when will Egypt have a greater population than Germany?

47. **Space Science.** Radioactive isotopes, as well as solar cells, are used to supply power to space vehicles. The isotopes gradually lose power because of radioactive decay. On a particular space vehicle the nuclear energy source has a power output of P watts after t days of use as given by

$$P = 75e^{-0.0035t}$$

Graph this function for $0 \leq t \leq 100$.

48. **Earth Science.** The atmospheric pressure P, in pounds per square inch, decreases exponentially with altitude h, in miles above sea level, as given by

$$P = 14.7e^{-0.21h}$$

Graph this function for $0 \leq h \leq 10$.

49. **Marine Biology.** Marine life is dependent upon the microscopic plant life that exists in the *photic zone,* a zone that goes to a depth where about 1% of the surface light still remains. Light intensity I relative to depth d, in feet, for one of the clearest bodies of water in the world, the Sargasso Sea in the West Indies, can be approximated by

$$I = I_0e^{-0.00942d}$$

where I_0 is the intensity of light at the surface. What percentage of the surface light will reach a depth of:
(A) 50 feet? (B) 100 feet?

50. **Marine Biology.** Refer to Problem 49. In some waters with a great deal of sediment, the photic zone may go down only 15 to 20 feet. In some murky harbors, the intensity of light d feet below the surface is given approximately by

$$I = I_0e^{-0.23d}$$

What percentage of the surface light will reach a depth of:
(A) 10 feet? (B) 20 feet?

51. **Money Growth.** If you invest \$5,250 in an account paying 11.38% compounded continuously, how much money will be in the account at the end of:
(A) 6.25 years? (B) 17 years?

52. **Money Growth.** If you invest \$7,500 in an account paying

8.35% compounded continuously, how much money will be in the account at the end of:
(A) 5.5 years? (B) 12 years?

53. Money Growth. *Barron's,* a national business and financial weekly, published the following "Top Savings Deposit Yields" for $2\frac{1}{2}$-year certificate of deposit accounts:

Gill Saving	8.30% (CC)
Richardson Savings and Loan	8.40% (CQ)
USA Savings	8.25% (CD)

where CC represents compounded continuously, CQ compounded quarterly, and CD compounded daily. Compute the value of \$1,000 invested in each account at the end of $2\frac{1}{2}$ years.

54. Money Growth. Refer to Problem 53. In another issue of *Barron's,* 1-year certificate of deposit accounts included:

Alamo Savings	8.25% (CQ)
Lamar Savings	8.05% (CC)

Compute the value of \$10,000 invested in each account at the end of 1 year.

★ **55. Present Value.** A promissory note will pay \$30,000 at maturity 10 years from now. How much should you be willing to pay for the note now if the note gains value at a rate of 9% compounded continuously?

★ **56. Present Value.** A promissory note will pay \$50,000 at maturity $5\frac{1}{2}$ years from now. How much should you be willing to pay for the note now if the note gains value at a rate of 10% compounded continuously?

57. AIDS Epidemic. In June of 1996 the World Health Organization estimated that 7.7 million cases of AIDS (acquired immunodeficiency syndrome) had occurred worldwide since the beginning of the epidemic. Assuming that the disease spreads continuously at an annual rate of 17%, estimate the total number of AIDS cases which will have occurred by June of the year:
(A) 2000 (B) 2004

58. AIDS Epidemic. In June 1996 the World Health Organization estimated that 28 million people worldwide had been infected with HIV (human immunodeficiency virus) since the beginning of the AIDS epidemic. Assuming that HIV infection spreads continuously at an annual rate of 19%, estimate the total number of people who will have been infected with HIV by June of the year:
(A) 2000 (B) 2004

★ **59. Learning Curve.** People assigned to assemble circuit boards for a computer manufacturing company undergo on-the-job training. From past experience it was found that the learning curve for the average employee is given by

$$N = 40(1 - e^{-0.12t})$$

where N is the number of boards assembled per day after t days of training. Graph this function for $0 \le t \le 30$. What is the maximum number of boards an average employee can be expected to produce in 1 day?

★ **60. Advertising.** A company is trying to expose a new product to as many people as possible through television advertising in a large metropolitan area with 2 million possible viewers. A model for the number of people N, in millions, who are aware of the product after t days of advertising was found to be

$$N = 2(1 - e^{-0.037t})$$

Graph this function for $0 \le t \le 50$. What value does N tend to as t increases without bound?

61. Newton's Law of Cooling. This law states that the rate at which an object cools is proportional to the difference in temperature between the object and its surrounding medium. The temperature T of the object t hours later is given by

$$T = T_m + (T_0 - T_m)e^{-kt}$$

where T_m is the temperature of the surrounding medium and T_0 is the temperature of the object at $t = 0$. Suppose a bottle of wine at a room temperature of 72°F is placed in the refrigerator to cool before a dinner party. If the temperature in the refrigerator is kept at 40°F and $k = 0.4$, find the temperature of the wine, to the nearest degree, after 3 hours. (In Exercise 4-5 we will find out how to determine k.)

62. Newton's Law of Cooling. Refer to Problem 61. What is the temperature, to the nearest degree, of the wine after 5 hours in the refrigerator?

★ **63. Photography.** An electronic flash unit for a camera is activated when a capacitor is discharged through a filament of wire. After the flash is triggered, and the capacitor is discharged, the circuit (see the figure) is connected and the battery pack generates a current to recharge the capacitor. The time it takes for the capacitor to recharge is called the *recycle time.* For a particular flash unit using a 12-volt battery pack, the charge q, in coulombs, on the capacitor t seconds after recharging has started is given by

$$q = 0.0009(1 - e^{-0.2t})$$

Graph this function for $0 \le t \le 10$. Estimate the maximum charge on the capacitor.

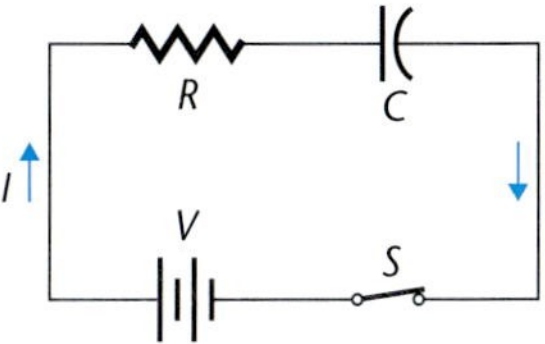

★ **64. Medicine.** An electronic heart pacemaker utilizes the same type of circuit as the flash unit in Problem 63, but it is designed so that the capacitor discharges 72 times a minute. For a particular pacemaker, the charge on the capacitor t seconds after it starts recharging is given by

$$q = 0.000\,008(1 - e^{-2t})$$

Graph this function for $0 \le t \le \frac{5}{6}$. Estimate the maximum charge on the capacitor.

★ **65. Wildlife Management.** A herd of 20 white-tailed deer is introduced to a coastal island where there had been no deer before. Their population is predicted to increase according to the logistic curve

$$N = \frac{100}{1 + 4e^{-0.14t}}$$

where N is the number of deer expected in the herd after t years. Graph the equation for $0 \le t \le 30$. Estimate the herd size that the island can support.

★ **66. Training.** A trainee is hired by a computer manufacturing company to learn to test a particular model of a personal computer after it comes off the assembly line. The learning curve for an average trainee is given by

$$N = \frac{200}{4 + 21e^{-0.1t}}$$

where N is the number of computers tested per day after t days on the job. Graph the equation for $0 \le t \le 50$. What is the maximum number of computers that an average tester can be expected to test per day after training?

★ **67. Catenary.** A free-hanging power line between two supporting towers looks something like a parabola, but it is actually a curve called a **catenary.** A catenary has an equation of the form

$$y = \frac{e^{mx} + e^{-mx}}{2m}$$

where m is a constant. Graph the equation for $m = 0.25$.

★ **68. Catenary.** Graph the equation in Problem 67 for $m = 0.4$.

SECTION 4-3 Logarithmic Functions

- Definition of Logarithmic Function
- From Logarithmic Form to Exponential Form, and Vice Versa
- Properties of Logarithmic Functions

• Definition of Logarithmic Function

We now define a new class of functions, called **logarithmic functions,** as inverses of exponential functions. Since exponential functions are one-to-one, their inverses exist. Here you will see why we placed special emphasis on the general concept of inverse functions in Section 2-6. If you know quite a bit about a function, then, based on a knowledge of inverses in general, you will automatically know quite a bit about its inverse. For example, the graph of f^{-1} is the graph of f reflected across the line $y = x$, and the domain and range of f^{-1} are, respectively, the range and domain of f.

If we start with the exponential function,

$$f\colon \quad y = 2^x$$

and interchange the variables x and y, we obtain the inverse of f:

$$f^{-1}\colon \quad x = 2^y$$

The graphs of f, f^{-1}, and the line $y = x$ are shown in Figure 1. This new function is given the name **logarithmic function with base 2.** Since we cannot solve the equation $x = 2^y$ for y using the algebra properties discussed so far, we introduce a new symbol to represent this inverse function:

$$y = \log_2 x \quad \text{Read "log to the base 2 of } x\text{."}$$

Thus,

$$y = \log_2 x \qquad \text{is equivalent to} \qquad x = 2^y$$

that is, $\log_2 x$ is the exponent to which 2 must be raised to obtain x. Symbolically, $x = 2^y = 2^{\log_2 x}$.

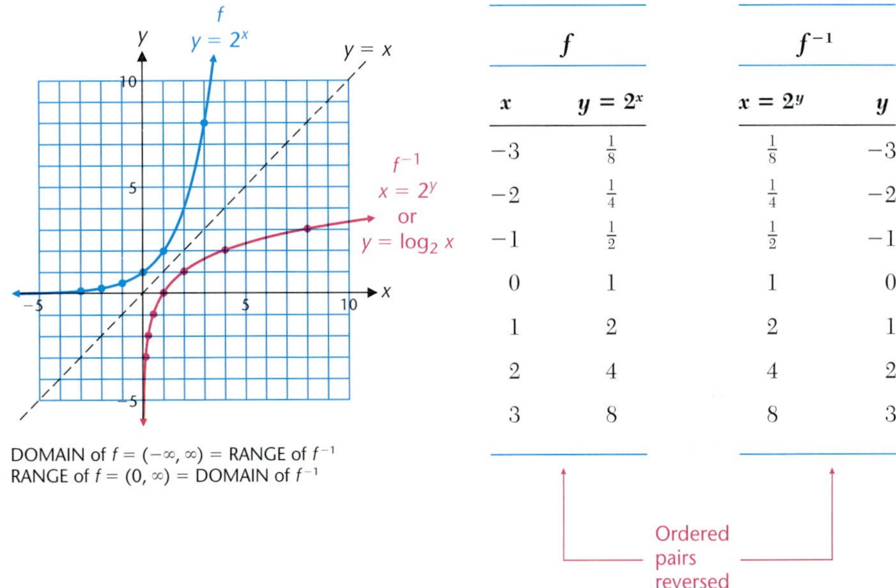

f		f^{-1}	
x	$y = 2^x$	$x = 2^y$	y
-3	$\frac{1}{8}$	$\frac{1}{8}$	-3
-2	$\frac{1}{4}$	$\frac{1}{4}$	-2
-1	$\frac{1}{2}$	$\frac{1}{2}$	-1
0	1	1	0
1	2	2	1
2	4	4	2
3	8	8	3

FIGURE 1 Logarithmic function with base 2.

In general, we define the **logarithmic function with base *b*** to be the inverse of the exponential function with base b ($b > 0$, $b \neq 1$).

DEFINITION 1

Definition of Logarithmic Function

For $b > 0$ and $b \neq 1$,

Logarithmic form		Exponential form
$y = \log_b x$	is equivalent to	$x = b^y$

The log to the base b of x is the exponent to which b must be raised to obtain x.

$$y = \log_{10} x \qquad \text{is equivalent to} \qquad x = 10^y$$

$$y = \log_e x \qquad \text{is equivalent to} \qquad x = e^y$$

Remember: A logarithm is an exponent.

It is very important to remember that $y = \log_b x$ and $x = b^y$ define the same function, and as such can be used interchangeably.

Since the domain of an exponential function includes all real numbers and its range is the set of positive real numbers, the **domain** of a logarithmic function is the set of all positive real numbers and its **range** is the set of all real numbers. Thus, $\log_{10} 3$ is defined, but $\log_{10} 0$ and $\log_{10}(-5)$ are not defined. That is, 3 is a logarithmic domain value, but 0 and -5 are not. Typical logarithmic curves are shown in Figure 2.

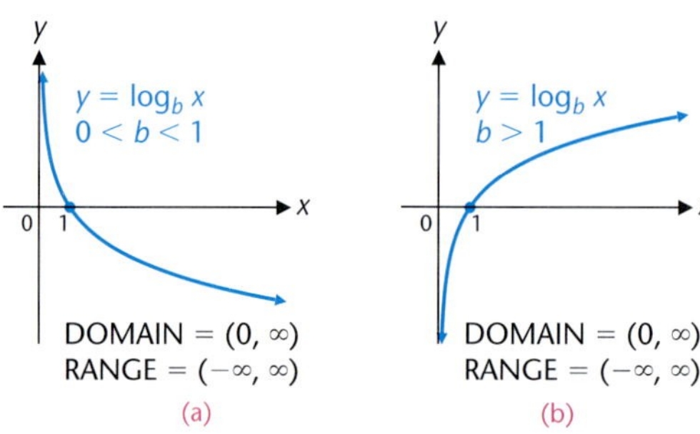

FIGURE 2 Typical logarithmic graphs.

EXPLORE-DISCUSS 1 For the exponential function $f = \{(x, y) \mid y = (\frac{2}{3})^x\}$, graph f, f^{-1}, and $y = x$ on the same coordinate system. Discuss the domains and ranges of f and its inverse. By what other name is f^{-1} known?

• From Logarithmic Form to Exponential Form, and Vice Versa

We now look into the matter of converting logarithmic forms to equivalent exponential forms, and vice versa.

EXAMPLE 1 Logarithmic–Exponential Conversions

Change each logarithmic form to an equivalent exponential form:

(A) $\log_2 8 = 3$ (B) $\log_{25} 5 = \frac{1}{2}$ (C) $\log_2 (\frac{1}{4}) = -2$

Solutions

(A) $\log_2 8 = 3$ is equivalent to $8 = 2^3$

(B) $\log_{25} 5 = \frac{1}{2}$ is equivalent to $5 = 25^{1/2}$

(C) $\log_2 (\frac{1}{4}) = -2$ is equivalent to $\frac{1}{4} = 2^{-2}$

Matched Problem 1 Change each logarithmic form to an equivalent exponential form:

(A) $\log_3 27 = 3$ (B) $\log_{36} 6 = \frac{1}{2}$ (C) $\log_3 (\frac{1}{9}) = -2$

EXAMPLE 2 Logarithmic–Exponential Conversions

Change each exponential form to an equivalent logarithmic form:

(A) $49 = 7^2$ (B) $3 = \sqrt{9}$ (C) $\frac{1}{5} = 5^{-1}$

Solutions (A) $49 = 7^2$ is equivalent to $\log_7 49 = 2$
(B) $3 = \sqrt{9}$ is equivalent to $\log_9 3 = \frac{1}{2}$
(C) $\frac{1}{5} = 5^{-1}$ is equivalent to $\log_5 (\frac{1}{5}) = -1$

Matched Problem 2 Change each exponential form to an equivalent logarithmic form:

(A) $64 = 4^3$ (B) $2 = \sqrt[3]{8}$ (C) $\frac{1}{16} = 4^{-2}$

To gain a little deeper understanding of logarithmic functions and their relationship to the exponential functions, we consider a few problems where we want to find x, b, or y in $y = \log_b x$, given the other two values. All values were chosen so that the problems can be solved without tables or a calculator.

EXAMPLE 3 Solutions of the Equation $y = \log_b x$

Find x, b, or y as indicated:

(A) Find y: $y = \log_4 8$ (B) Find x: $\log_3 x = -2$
(C) Find b: $\log_b 1{,}000 = 3$

Solutions (A) Write $y = \log_4 8$ in equivalent exponential form:

$$8 = 4^y$$

$$2^3 = 2^{2y} \quad \text{Write each number to the same base 2.}$$

$$2y = 3 \quad \text{Recall that } b^m = b^n \text{ if and only if } m = n.$$

$$y = \tfrac{3}{2}$$

Thus, $\frac{3}{2} = \log_4 8$.

(B) Write $\log_3 x = -2$ in equivalent exponential form:

$$x = 3^{-2}$$

$$= \frac{1}{3^2} = \frac{1}{9}$$

Thus, $\log_3 (\frac{1}{9}) = -2$.

(C) Write $\log_b 1{,}000 = 3$ in equivalent exponential form:

$$1{,}000 = b^3$$

$$10^3 = b^3 \quad \text{Write 1,000 as a third power.}$$

$$b = 10$$

Thus, $\log_{10} 1{,}000 = 3$.

Matched Problem 3 Find x, b, or y as indicated.

(A) Find y: $y = \log_9 27$ (B) Find x: $\log_2 x = -3$

(C) Find b: $\log_b 100 = 2$

• Properties of Logarithmic Functions

The familiar properties of exponential functions imply corresponding properties of logarithmic functions.

EXPLORE-DISCUSS 2 Discuss the connection between the exponential equation and the logarithmic equation, and explain why each equation is valid.

(A) $2^4 2^7 = 2^{11}$; $\log_2 2^4 + \log_2 2^7 = \log_2 2^{11}$

(B) $2^{13}/2^5 = 2^8$; $\log_2 2^{13} - \log_2 2^5 = \log_2 2^8$

(C) $(2^6)^9 = 2^{54}$; $9 \log_2 2^6 = \log_2 2^{54}$

Several of the powerful and useful properties of logarithmic functions are listed in Theorem 1.

Theorem 1 **Properties of Logarithmic Functions**

If b, M, and N are positive real numbers, $b \neq 1$, and p and x are real numbers, then:

1. $\log_b 1 = 0$

2. $\log_b b = 1$

3. $\log_b b^x = x$

4. $b^{\log_b x} = x, x > 0$

5. $\log_b MN = \log_b M + \log_b N$

6. $\log_b \dfrac{M}{N} = \log_b M - \log_b N$

7. $\log_b M^p = p \log_b M$

8. $\log_b M = \log_b N$ if and only if $M = N$

The first two properties in Theorem 1 follow directly from the definition of a logarithmic function:

$$\log_b 1 = 0 \quad \text{since} \quad b^0 = 1$$

$$\log_b b = 1 \quad \text{since} \quad b^1 = b$$

The third and fourth properties look more complicated than they are. They follow directly from the fact that exponential and logarithmic functions are inverses of each other. Recall from Section 1-8 that if f is one-to-one, then f^{-1} is a one-to-one function satisfying

$$f^{-1}[f(x)] = x \qquad \text{and} \qquad f[f^{-1}(x)] = x$$

Applying these general properties to $f(x) = b^x$ and $f^{-1}(x) = \log_b x$, we see that

$$\begin{aligned} f^{-1}[f(x)] &= x & f[f^{-1}(x)] &= x \\ \log_b [f(x)] &= x & b^{f^{-1}(x)} &= x \\ \log_b b^x &= x & b^{\log_b x} &= x \end{aligned}$$

Properties 5 to 7 enable us to convert multiplication into addition, division into subtraction, and power and root problems into multiplication. The proofs of these properties are based on properties of exponents. A sketch of a proof of the fifth property follows: To bring exponents into the proof, we let

$$u = \log_b M \qquad \text{and} \qquad v = \log_b N$$

and convert these to the equivalent exponential forms

$$M = b^u \qquad \text{and} \qquad N = b^v$$

Now, see if you can provide the reasons for each of the following steps:

$$\log_b MN = \log_b b^u b^v = \log_b b^{u+v} = u + v = \log_b M + \log_b N$$

The other properties are established in a similar manner (see Problems 111 and 112 in Exercise 4-3.)

Finally, the eighth property follows from the fact that logarithmic functions are one-to-one.

We now illustrate the use of these properties in several examples.

EXAMPLE 4 **Using Logarithmic Properties**

Simplify using the properties in Theorem 1:

(A) $\log_e 1$ (B) $\log_{10} 10$ (C) $\log_e e^{2x+1}$
(D) $\log_{10} 0.01$ (E) $10^{\log_{10} 7}$ (F) $e^{\log_e x^2}$

Solutions
(A) $\log_e 1 = 0$
(B) $\log_{10} 10 = 1$
(C) $\log_e e^{2x+1} = 2x + 1$
(D) $\log_{10} 0.01 = \log_{10} 10^{-2} = -2$
(E) $10^{\log_{10} 7} = 7$
(F) $e^{\log_e x^2} = x^2$

Matched Problem 4 Simplify using the properties in Theorem 1:

(A) $\log_{10} 10^{-5}$ (B) $\log_5 25$ (C) $\log_{10} 1$
(D) $\log_e e^{m+n}$ (E) $10^{\log_{10} 4}$ (F) $e^{\log_e (x^4+1)}$

EXAMPLE 5 Using Logarithmic Properties

Write in terms of simpler logarithmic forms:

(A) $\log_b 3x$ (B) $\log_b \frac{x}{5}$ (C) $\log_b x^7$

(D) $\log_b \frac{mn}{pq}$ (E) $\log_b (mn)^{2/3}$ (F) $\log_b \frac{x^8}{y^{1/5}}$

Solutions

(A) $\log_b 3x = \log_b 3 + \log_b x$ $\quad \log_b MN = \log_b M + \log_b N$

(B) $\log_b \frac{x}{5} = \log_b x - \log_b 5$ $\quad \log_b \frac{M}{N} = \log_b M - \log_b N$

(C) $\log_b x^7 = 7 \log_b x$ $\quad \log_b M^p = p \log_b M$

(D) $\log_b \frac{mn}{pq} = \log_b mn - \log_b pq$ $\quad \log_b \frac{M}{N} = \log_b M - \log_b N$

$= \log_b m + \log_b n - (\log_b p + \log_b q)$ $\quad \log_b MN = \log_b M + \log_b N$

$= \log_b m + \log_b n - \log_b p - \log_b q$

(E) $\log_b (mn)^{2/3} = \frac{2}{3} \log_b mn$ $\quad \log_b M^p = p \log_b M$

$= \frac{2}{3}(\log_b m + \log_b n)$ $\quad \log_b MN = \log_b M + \log_b N$

(F) $\log_b \frac{x^8}{y^{1/5}} = \log_b x^8 - \log_b y^{1/5}$ $\quad \log_b \frac{M}{N} = \log_b M - \log_b N$

$= 8 \log_b x - \frac{1}{5} \log_b y$ $\quad \log_b M^p = p \log_b M$

Matched Problem 5

Write in terms of simpler logarithmic forms, as in Example 5.

(A) $\log_b \frac{r}{uv}$ (B) $\log_b \left(\frac{m}{n}\right)^{3/5}$ (C) $\log_b \frac{u^{1/3}}{v^5}$

EXAMPLE 6 Using Logarithmic Properties

If $\log_e 3 = 1.10$ and $\log_e 7 = 1.95$, find:

(A) $\log_e (\frac{7}{3})$ (B) $\log_e \sqrt[3]{21}$

Solutions

(A) $\log_e (\frac{7}{3}) = \log_e 7 - \log_e 3 = 1.95 - 1.10 = 0.85$

(B) $\log_e \sqrt[3]{21} = \log_e (21)^{1/3} = \frac{1}{3} \log_e (3 \cdot 7) = \frac{1}{3}(\log_e 3 + \log_e 7)$

$= \frac{1}{3}(1.10 + 1.95) = 1.02$

Matched Problem 6 If $\log_e 5 = 1.609$ and $\log_e 8 = 2.079$, find:

(A) $\log_e \frac{5^{10}}{8}$ (B) $\log_e \sqrt[4]{\frac{8}{5}}$

The following example and problem, though somewhat artificial, will give you additional practice in using the properties in Theorem 1.

EXAMPLE 7 Using Logarithmic Properties

Find x so that $\log_b x = \frac{2}{3} \log_b 27 + 2 \log_b 2 - \log_b 3$ without using a calculator or table.

Solution

First we use properties from Theorem 1 to express the right side as the logarithm of a single number.

$$\begin{aligned}
\log_b x &= \tfrac{2}{3} \log_b 27 + 2 \log_b 2 - \log_b 3 \\
&= \log_b 27^{2/3} + \log_b 2^2 - \log_b 3 \\
&= \log_b 9 + \log_b 4 - \log_b 3 && 27^{2/3} = 9;\ 2^2 = 4 \\
&= \log_b \frac{9 \cdot 4}{3} && \text{Properties 5 and 6 of Theorem 1} \\
&= \log_b 12 \\
x &= 12 && \text{Property 8 of Theorem 1}
\end{aligned}$$

Matched Problem 7 Find x so that $\log_b x = \frac{2}{3} \log_b 8 + \frac{1}{2} \log_b 9 - \log_b 6$ without using a calculator or table.

CAUTION

We conclude this section by noting three common errors:

1. $\dfrac{\log_b M}{\log_b N} \neq \log_b M - \log_b N$ — $\log_b M - \log_b N = \log_b \dfrac{M}{N}$; $\dfrac{\log_b M}{\log_b N}$ cannot be simplified.
2. $\log_b (M + N) \neq \log_b M + \log_b N$ — $\log_b M + \log_b N = \log_b MN$; $\log_b (M + N)$ cannot be simplified.
3. $(\log_b M)^p \neq p \log_b M$ — $p \log_b M = \log_b M^p$; $(\log_b M)^p$ cannot be simplified.

Answers to Matched Problems

1. (A) $27 = 3^3$ (B) $6 = 36^{1/2}$ (C) $\frac{1}{9} = 3^{-2}$
2. (A) $\log_4 64 = 3$ (B) $\log_8 2 = \frac{1}{3}$ (C) $\log_4 (\frac{1}{16}) = -2$
3. (A) $y = \frac{3}{2}$ (B) $x = \frac{1}{8}$ (C) $b = 10$
4. (A) -5 (B) 2 (C) 0 (D) $m + n$ (E) 4 (F) $x^4 + 1$
5. (A) $\log_b r - \log_b u - \log_b v$ (B) $\frac{3}{5}(\log_b m - \log_b n)$ (C) $\frac{1}{3}\log_b u - 5\log_b v$
6. (A) 14.01 (to 4 significant digits) (B) 0.1175 (to 4 significant digits)
7. $x = 2$

EXERCISE 4-3

A

Rewrite Problems 1–8 in equivalent exponential form.

1. $\log_2 64 = 6$
2. $\log_6 216 = 3$
3. $\log_{10} 100{,}000 = 5$
4. $\log_{10} 0.000001 = -6$
5. $\log_9 3 = \frac{1}{2}$
6. $\log_{16} 2 = \frac{1}{4}$
7. $\log_{1/4} 64 = -3$
8. $\log_{1/5} 25 = -2$

Rewrite Problems 9–16 in equivalent logarithmic form.

9. $0.001 = 10^{-3}$
10. $10{,}000{,}000 = 10^7$
11. $4 = 8^{2/3}$
12. $64 = 16^{3/2}$
13. $81^{-1/4} = \frac{1}{3}$
14. $32^{-2/5} = \frac{1}{4}$
15. $5 = \sqrt[3]{125}$
16. $11 = \sqrt{121}$

In Problems 17–30, simplify each expression using Theorem 1.

17. $\log_8 8$
18. $\log_5 1$
19. $\log_{0.5} 1$
20. $\log_{0.2} 0.2$
21. $\log_{10} 1{,}000$
22. $\log_2 64$
23. $\log_e \sqrt{e}$
24. $\log_5 \sqrt[3]{5}$
25. $\log_{10} 0.001$
26. $\log_{10} 0.0001$
27. $10^{\log_{10} x^2}$
28. $10^{\log_{10}\sqrt{x}}$
29. $e^{3\log_e(x-1)}$
30. $e^{-2\log_e(x+1)}$

B

Find x, y, or b, as indicated in Problems 31–44.

31. $\log_2 x = 3$
32. $\log_3 x = 2$
33. $\log_5 x = -2$
34. $\log_4 x = -3$
35. $\log_3 81 = y$
36. $\log_4 64 = y$
37. $\log_{16} x = \frac{3}{2}$
38. $\log_8 x = \frac{2}{3}$
39. $\log_{1/2} 8 = y$
40. $\log_{1/3} 81 = y$
41. $\log_b 32 = 2.5$
42. $\log_b 8 = 0.5$
43. $\log_b 0.0001 = -2$
44. $\log_b 1{,}000 = -3$

Write Problems 45–58 in terms of simpler logarithmic forms (see Example 5).

45. $\log_b x^6y^9$
46. $\log_b x^{2/3}y^{4/3}$
47. $\log_b \dfrac{u^{3/2}}{v^{5/3}}$
48. $\log_b \dfrac{v^7}{u^8}$
49. $\log_b \dfrac{mn}{pq}$
50. $\log_b \dfrac{mnp}{q}$
51. $\log_b \dfrac{1}{a^4}$
52. $\log_b \dfrac{1}{a^3}$
53. $\log_b \sqrt{c^2 + d^2}$
54. $\log_b \sqrt[4]{c^4 + d^4}$
55. $\log_b \dfrac{\sqrt{u}}{vw^2}$
56. $\log_b \dfrac{u^3v^4}{\sqrt{w}}$
57. $\log_b \sqrt[3]{\dfrac{x^2\sqrt{y}}{z^3}}$
58. $\log_b \sqrt[4]{\left(\dfrac{\sqrt{x}}{y^2z^3}\right)^3}$

In Problems 59–68, write each expression in terms of a single logarithm with a coefficient of 1. Example: $\log_b u^2 - \log_b v = \log_b (u^2/v)$.

59. $2\log_b x - \log_b y$
60. $\log_b m - \frac{1}{2}\log_b n$
61. $\log_b w - \log_b x - \log_b y$
62. $\log_b w + \log_b x - \log_b y$
63. $3\log_b x + 2\log_b y - \frac{1}{4}\log_b z$
64. $\frac{1}{3}\log_b w - 3\log_b x - 5\log_b y$
65. $5(\frac{1}{2}\log_b u - 2\log_b v)$
66. $7(4\log_b m + \frac{1}{3}\log_b n)$
67. $\frac{1}{5}(2\log_b x + 3\log_b y)$
68. $\frac{1}{3}(4\log_b x - 2\log_b y)$

In Problems 69–76, write each expression in terms of logarithms of first-degree polynomials. Example:

$$\log_b \frac{(2x+1)^3}{(3x-5)^4} = 3\log_b (2x+1) - 4\log_b (3x-5)$$

69. $\log_b [(x+3)^5(2x-7)^2]$
70. $\log_b [(5x-4)^3(3x+2)^4]$
71. $\log_b \dfrac{(x+10)^7}{(1+10x)^2}$
72. $\log_b \dfrac{(x-3)^5}{(5+x)^3}$

73. $\log_b \dfrac{x^2}{\sqrt{x+1}}$ **74.** $\log_b \dfrac{\sqrt{x-1}}{x^3}$

75. $\log_b (x^4 + x^3 - 20x^2)$ **76.** $\log_b (x^5 + 5x^4 - 14x^3)$

In Problems 77–86, solve for x without using a calculator or table.

77. $\log_2 (x + 5) = 2 \log_2 3$

78. $\log_{10} (5 - x) = 3 \log_{10} 2$

79. $2 \log_5 x = \log_5 (x^2 - 6x + 2)$

80. $\log_{10} (x^2 - 2x - 2) = 2 \log_{10} (x - 2)$

81. $\log_e (x + 8) - \log_e x = 3 \log_e 2$

82. $\log_7 4x - \log_7 (x + 1) = \frac{1}{2} \log_7 4$

83. $2 \log_3 x = \log_3 2 + \log_3 (4 - x)$

84. $\log_4 x + \log_4 (x + 2) = \frac{1}{2} \log_4 9$

85. $3 \log_b 2 + \frac{1}{2} \log_b 25 - \log_b 20 = \log_b x$

86. $\frac{3}{2} \log_b 4 - \frac{2}{3} \log_b 8 + 2 \log_b 2 = \log_b x$

If $\log_b 2 = 0.69$, $\log_b 3 = 1.10$, and $\log_b 5 = 1.61$, find the value of each expression in Problems 87–96.

87. $\log_b 30$ **88.** $\log_b 12$ **89.** $\log_b \frac{2}{5}$

90. $\log_b \frac{5}{3}$ **91.** $\log_b 27$ **92.** $\log_b 16$

93. $\log_b \sqrt[3]{2}$ **94.** $\log_b \sqrt{3}$ **95.** $\log_b \sqrt{0.9}$

96. $\log_b \sqrt[3]{1.5}$

C

Graph Problems 97–100.

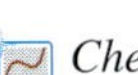
Check Problems 97–100 with a graphing utility by graphing the inverse of each function.

97. $y = \log_2 (x - 2)$

98. $y = \log_2 (x + 3)$

99. $y = \log_2 x - 2$

100. $y = \log_2 x + 3$

101. (A) For $f = \{(x, y) \mid y = (\frac{1}{2})^x = 2^{-x}\}$, graph f, f^{-1}, and $y = x$ on the same coordinate system.
(B) Indicate the domain and range of f and f^{-1}.
(C) What other name can you use for the inverse of f?

102. (A) For $f = \{(x, y) \mid y = (\frac{1}{3})^x = 3^{-x}\}$, graph f, f^{-1}, and $y = x$ on the same coordinate system.
(B) Indicate the domain and range of f and f^{-1}.
(C) What other name can you use for the inverse of f?

Find the inverse of each function in Problems 103–106.

103. $f(x) = 5^{3x-1} + 4$

104. $g(x) = 3^{2x-3} - 2$

105. $g(x) = 3 \log_e (5x - 2)$

106. $f(x) = 2 + \log_e (5x - 3)$

107. Explain why the graph of the reflection of the function $y = 3^{x^2}$ in the line $y = x$ is not the graph of a function.

108. Explain why the graph of the reflection of the function $y = 2^{|x|}$ in the line $y = x$ is not the graph of a function.

109. Write $\log_e x - \log_e 100 = -0.08t$ in an exponential form that is free of logarithms.

110. Write $\log_e x - \log_e C + kt = 0$ in an exponential form that is free of logarithms.

111. Prove that $\log_b (M/N) = \log_b M - \log_b N$ under the hypotheses of Theorem 1.

112. Prove that $\log_b M^p = p \log_b M$ under the hypotheses of Theorem 1.

SECTION 4-4 Common and Natural Logarithms

- Common and Natural Logarithms—Definition and Evaluation
- Applications

John Napier (1550–1617) is credited with the invention of logarithms, which evolved out of an interest in reducing the computational strain in research in astronomy. This new computational tool was immediately accepted by the scientific world. Now, with the availability of inexpensive calculators, logarithms have lost most of their importance as a computational device. However, the logarithmic concept has been greatly generalized since its conception, and logarithmic functions are used widely in both theoretical and applied sciences.

Of all possible logarithmic bases, the base e and the base 10 are used almost exclusively. Before we can use logarithms in certain practical problems, we need to be able to approximate the logarithm of any positive number to either base 10 or base e. And conversely, if we are given the logarithm of a number to base 10 or base e, we need to be able to approximate the number. Historically, tables were used for this purpose, but now calculators are used since they are faster and can find far more values than any table can possibly include.

• Common and Natural Logarithms—Definition and Evaluation

Common logarithms, also called **Briggsian logarithms,** are logarithms with base 10. **Natural logarithms,** also called **Napierian logarithms,** are logarithms with base e. Most calculators have a function key labeled "log" and a function key labeled "ln." The former represents a common logarithm and the latter a natural logarithm. In fact, "log" and "ln" are both used extensively in mathematical literature, and whenever you see either used in this book without a base indicated, they should be interpreted as follows:

Logarithmic Notation

$\log x = \log_{10} x$ **Common logarithm**

$\ln x = \log_e x$ **Natural logarithm**

EXAMPLE 1 Calculator Evaluation of Logarithms

Use a calculator to evaluate each to six decimal places:

(A) log 3,184 (B) ln 0.000 349 (C) log (−3.24)

Solutions

(A) $\log 3{,}184 = 3.502973$
(B) $\ln 0.000349 = -7.960439$
(C) $\log(-3.24) =$ Error

Why is an error indicated in part C? Because −3.24 is not in the domain of the log function. [*Note:* Calculators display error messages in various ways. Some calculators use a more advanced definition of logarithmic functions that involves complex numbers. They will display an ordered pair, representing a complex number, as the value of log (−3.24), rather than an error message. You should interpret such a display as indicating that the number entered is not in the domain of the logarithmic function as we have defined it.]

Matched Problem 1

Use a calculator to evaluate each to six decimal places:

(A) log 0.013 529 (B) ln 28.693 28 (C) ln (−0.438)

When working with common and natural logarithms, we follow the common practice of using the equal sign "=" where it might be more appropriate to use the approximately equal sign "≈." No harm is done as long as we keep in mind that in a statement such as log 3.184 = 0.503, the number on the right is only assumed accurate to three decimal places and is not exact.

EXPLORE-DISCUSS 1 Graphs of the functions $f(x) = \log x$ and $g(x) = \ln x$ are shown in the graphing utility display of Figure 1. Which graph belongs to which function? It appears from the display that one of the functions may be a constant multiple of the other. Is that true? Find and discuss the evidence for your answer.

FIGURE 1

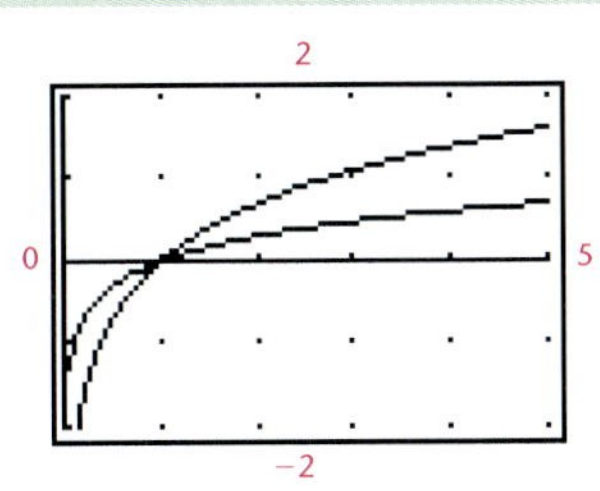

EXAMPLE 2 **Calculator Evaluation of Logarithms**

Use a calculator to evaluate each expression to three decimal places:

(A) $\dfrac{\log 2}{\log 1.1}$ (B) $\log \dfrac{2}{1.1}$ (C) $\log 2 - \log 1.1$

Solutions

(A) $\dfrac{\log 2}{\log 1.1} = 7.273$

(B) $\log \dfrac{2}{1.1} = 0.260$

(C) $\log 2 - \log 1.1 = 0.260$ Note that $\dfrac{\log 2}{\log 1.1} \neq \log 2 - \log 1.1$, but $\log \dfrac{2}{1.1} = \log 2 - \log 1.1$ (see Theorem 1, Section 4-3).

Matched Problem 2 Use a calculator to evaluate each to three decimal places:

(A) $\dfrac{\ln 3}{\ln 1.08}$ (B) $\ln \dfrac{3}{1.08}$ (C) $\ln 3 - \ln 1.08$

We now turn to the second problem: Given the logarithm of a number, find the number. To solve this problem, we make direct use of the logarithmic–exponential relationships discussed in Section 4-3.

Logarithmic–Exponential Relationships

$\log x = y$ is equivalent to $x = 10^y$

$\ln x = y$ is equivalent to $x = e^y$

EXAMPLE 3 **Solving $\log_b x = y$ for x**

Find x to 3 significant digits, given the indicated logarithms.

(A) $\log x = -9.315$ (B) $\ln x = 2.386$

Solutions (A) $\log x = -9.315$

$$x = 10^{-9.315} \quad \text{Change to equivalent exponential form.}$$

$$= 4.84 \times 10^{-10}$$

Notice that the answer is displayed in scientific notation in the calculator.

(B) $\ln x = 2.386$

$$x = e^{2.386} \quad \text{Change to equivalent exponential form.}$$

$$= 10.9$$

Matched Problem 3 Find x to 4 significant digits, given the indicated logarithms.

(A) $\ln x = -5.062$ (B) $\log x = 12.0821$

• Applications

We now consider three applications that are solved using common and natural logarithms. The first application concerns sound intensity; the second, earthquake intensity; and the third, rocket flight theory.

Sound Intensity

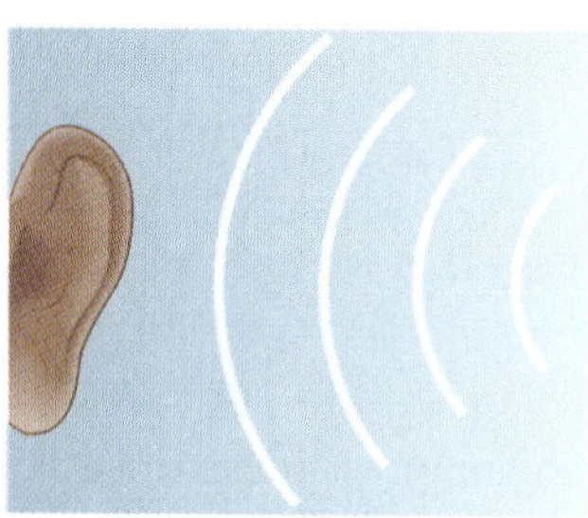

The human ear is able to hear sound over an incredible range of intensities. The loudest sound a healthy person can hear without damage to the eardrum has an intensity 1 trillion (1,000,000,000,000) times that of the softest sound a person can hear. Working directly with numbers over such a wide range is very cumbersome. Since the logarithm, with base greater than 1, of a number increases much more slowly than the number itself, logarithms are often used to create more convenient compressed scales. The decibel scale for sound intensity is an example of such a scale. The **decibel,** named after the inventor of the telephone, Alexander Graham Bell (1847–1922), is defined as follows:

$$D = 10 \log \frac{I}{I_0} \quad \text{Decibel scale} \qquad (1)$$

where D is the **decibel level** of the sound, I is the **intensity** of the sound measured in watts per square meter (W/m^2), and I_0 is the intensity of the least audible sound that an average healthy young person can hear. The latter is standardized to be $I_0 = 10^{-12}$ watt per square meter. Table 1 lists some typical sound intensities from familiar sources.

TABLE 1 Typical Sound Intensities

Sound intensity, W/m^2	Sound
1.0×10^{-12}	Threshold of hearing
5.2×10^{-10}	Whisper
3.2×10^{-6}	Normal conversation
8.5×10^{-4}	Heavy traffic
3.2×10^{-3}	Jackhammer
1.0×10^{0}	Threshold of pain
8.3×10^{2}	Jet plane with afterburner

EXAMPLE 4 **Sound Intensity**

Find the number of decibels from a whisper with sound intensity 5.20×10^{-10} watt per square meter. Compute the answer to two decimal places.

Solution We use the decibel formula (1):

$$\begin{aligned} D &= 10 \log \frac{I}{I_0} \\ &= 10 \log \frac{5.2 \times 10^{-10}}{10^{-12}} \\ &= 10 \log (5.2 \times 10^{2}) \\ &= 10 \log 520 \\ &= 27.16 \text{ decibels} \end{aligned}$$

Matched Problem 4 Find the number of decibels from a jackhammer with sound intensity 3.2×10^{-3} watt per square meter. Compute the answer to two decimal places.

EXPLORE-DISCUSS 2 Imagine using a large sheet of graph paper, ruled with horizontal and vertical lines $\frac{1}{8}$ inch apart, to plot the sound intensities of Table 1 on the x axis and the corresponding decibel levels on the y axis. Suppose that each $\frac{1}{8}$-inch unit on the x axis represents the intensity of the least audible sound (10^{-12} W/m^2), and each $\frac{1}{8}$-inch unit on the y axis represents 1 decibel. If the point corresponding to a jet plane with afterburner is plotted on the graph paper, how far is it from the x axis? From the y axis? (Give the first answer in inches and the second in miles!) Discuss.

Earthquake Intensity

The energy released by the largest earthquake recorded, measured in joules, is about 100 billion (100,000,000,000) times the energy released by a small earthquake that is barely felt. Over the past 150 years several people from various countries have devised different types of measures of earthquake magnitudes so that their severity could be easily compared. In 1935 the California seismologist Charles Richter devised a logarithmic scale that bears his name and is still widely used in the United States. The **magnitude** M on the **Richter scale*** is given as follows:

$$M = \frac{2}{3}\log\frac{E}{E_0} \qquad \text{Richter scale} \qquad (2)$$

where E is the energy released by the earthquake, measured in joules, and E_0 is the energy released by a very small reference earthquake which has been standardized to be

$$E_0 = 10^{4.40} \text{ joules}$$

The destructive power of earthquakes relative to magnitudes on the Richter scale is indicated in Table 2.

TABLE 2
The Richter Scale

Magnitude on Richter scale	Destructive power
$M < 4.5$	Small
$4.5 < M < 5.5$	Moderate
$5.5 < M < 6.5$	Large
$6.5 < M < 7.5$	Major
$7.5 < M$	Greatest

EXAMPLE 5 Earthquake Intensity

The 1906 San Francisco earthquake released approximately 5.96×10^{16} joules of energy. What was its magnitude on the Richter scale? Compute the answer to two decimal places.

Solution We use the magnitude formula (2):

$$M = \frac{2}{3}\log\frac{E}{E_0}$$

$$= \frac{2}{3}\log\frac{5.96 \times 10^{16}}{10^{4.40}}$$

*Originally, Richter defined the magnitude of an earthquake in terms of logarithms of the maximum seismic wave amplitude, in thousandths of a millimeter, measured on a standard seismograph. Formula (2) gives essentially the same magnitude that Richter obtained for a given earthquake but in terms of logarithms of the energy released by the earthquake.

$$= \frac{2}{3} \log (5.96 \times 10^{11.6})$$

$$= 8.25$$

Matched Problem 5 The 1985 earthquake in central Chile released approximately 1.26×10^{16} joules of energy. What was its magnitude on the Richter scale? Compute the answer to two decimal places.

EXAMPLE 6 Earthquake Intensity

If the energy release of one earthquake is 1,000 times that of another, how much larger is the Richter scale reading of the larger than the smaller?

Solution Let

$$M_1 = \frac{2}{3} \log \frac{E_1}{E_0}$$

and

$$M_2 = \frac{2}{3} \log \frac{E_2}{E_0}$$

be the Richter equations for the smaller and larger earthquakes, respectively. Substituting $E_2 = 1{,}000E_1$ into the second equation, we obtain

$$\begin{aligned} M_2 &= \frac{2}{3} \log \frac{1{,}000E_1}{E_0} \\ &= \frac{2}{3}\left(\log 10^3 + \log \frac{E_1}{E_0}\right) \\ &= \frac{2}{3}(3) + \frac{2}{3} \log \frac{E_1}{E_0} \\ &= 2 + M_1 \end{aligned}$$

Thus, an earthquake with 1,000 times the energy of another has a Richter scale reading of 2 more than the other.

Matched Problem 6 If the energy release of one earthquake is 10,000 times that of another, how much larger is the Richter scale reading of the larger than the smaller?

Rocket Flight Theory

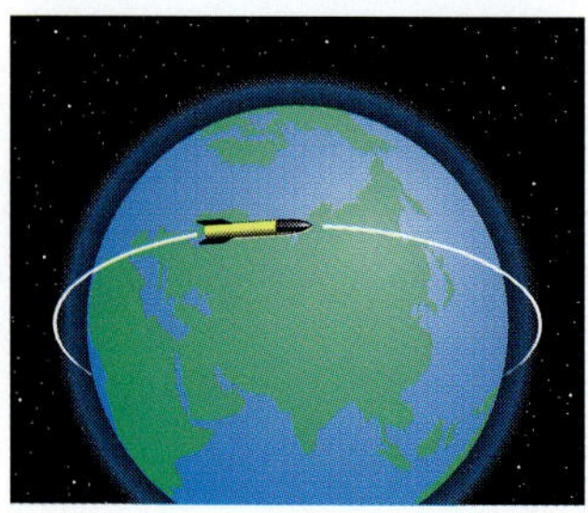

The theory of rocket flight uses advanced mathematics and physics to show that the velocity v of a rocket at burnout (depletion of fuel supply) is given by

$$v = c \ln \frac{W_t}{W_b} \quad \text{Rocket equation} \qquad (3)$$

where c is the exhaust velocity of the rocket engine, W_t is the takeoff weight (fuel, structure, and payload), and W_b is the burnout weight (structure and payload).

Because of the Earth's atmospheric resistance, a launch vehicle velocity of at least 9.0 kilometers per second is required in order to achieve the minimum altitude needed for a stable orbit. It is clear that to increase velocity v, either the weight ratio W_t/W_b must be increased or the exhaust velocity c must be increased. The weight ratio can be increased by the use of solid fuels, and the exhaust velocity can be increased by improving the fuels, solid or liquid.

EXAMPLE 7 Rocket Flight Theory

A typical single-stage, solid-fuel rocket may have a weight ratio $W_t/W_b = 18.7$ and an exhaust velocity $c = 2.38$ kilometers per second. Would this rocket reach a launch velocity of 9.0 kilometers per second?

Solution We use the rocket equation (3):

$$\begin{aligned} v &= c \ln \frac{W_t}{W_b} \\ &= 2.38 \ln 18.7 \\ &= 6.97 \text{ kilometers per second} \end{aligned}$$

The velocity of the launch vehicle is far short of the 9.0 kilometers per second required to achieve orbit. This is why multiple-stage launchers are used—the deadweight from a preceding stage can be jettisoned into the ocean when the next stage takes over.

Matched Problem 7

A launch vehicle using liquid fuel, such as a mixture of liquid hydrogen and liquid oxygen, can produce an exhaust velocity of $c = 4.7$ kilometers per second. However, the weight ratio W_t/W_b must be low—around 5.5 for some vehicles—because of the increased structural weight to accommodate the liquid fuel. How much more or less than the 9.0 kilometers per second required to reach orbit will be achieved by this vehicle?

Answers to Matched Problems

1. (A) $-1.868\ 734$ (B) $3.356\ 663$ (C) Not possible
2. (A) 14.275 (B) 1.022 (C) 1.022
3. (A) $x = 0.006\ 333$ (B) $x = 1.21 \times 10^{12}$ **4.** 95.05 decibels **5.** 7.80 **6.** 2.67
7. 1 km/s less

EXERCISE 4-4

A

In Problems 1–8, evaluate to four decimal places.

1. $\log 27{,}593$ **2.** $\log 0.000\,539$

3. $\log 0.004\,321$ **4.** $\log 120{,}564$

5. $\ln 0.000\,654$ **6.** $\ln 0.023\,198$

7. $\ln 456.76$ **8.** $\ln 132.43$

In Problems 9–16, evaluate x to 4 significant digits, given:

9. $\ln x = 3.4797$ **10.** $\ln x = -0.2985$

11. $\ln x = -2.2643$ **12.** $\ln x = 6.8236$

13. $\log x = -1.2543$ **14.** $\log x = -2.6123$

15. $\log x = 3.5324$ **16.** $\log x = 2.5017$

B

In Problems 17–24, evaluate to three decimal places.

17. $\dfrac{\log 3.215}{\log 2.569}$ **18.** $\log \dfrac{3.215}{2.569}$

19. $\ln \dfrac{0.5545}{0.0545}$ **20.** $\dfrac{\ln 0.5545}{\ln 0.0545}$

21. $\dfrac{\ln 0.6}{0.01}$ **22.** $\dfrac{\log 0.7}{0.005}$

23. $\dfrac{\log 300}{\log 3}$ **24.** $\dfrac{\ln 300}{\ln 3}$

In Problems 25–32, evaluate x to 5 significant digits.

25. $x = \ln (3.4562 \times 10^{15})$

26. $x = \ln (4.3931 \times 10^{-11})$

27. $x = \log (6.7744 \times 10^{-13})$

28. $x = \log (5.1212 \times 10^{14})$

29. $\log x = 21.667\,503$ **30.** $\ln x = 14.561\,094$

31. $\ln x = -11.112\,445$ **32.** $\log x = -15.599\,943$

Graph each function in Problems 33–40.

Check Problems 33–40 with a graphing utility.

33. $y = \ln x$ **34.** $y = -\ln x$

35. $y = |\ln x|$ **36.** $y = \ln |x|$

37. $y = 2 \ln (x + 2)$ **38.** $y = 2 \ln x + 2$

39. $y = 4 \ln x - 3$ **40.** $y = 4 \ln (x - 3)$

C

41. Find the fallacy:

$$1 < 3$$
$$\tfrac{1}{27} < \tfrac{3}{27} \qquad \text{Divide both sides by 27.}$$
$$\tfrac{1}{27} < \tfrac{1}{9}$$
$$(\tfrac{1}{3})^3 < (\tfrac{1}{3})^2$$
$$\log (\tfrac{1}{3})^3 < \log (\tfrac{1}{3})^2$$
$$3 \log \tfrac{1}{3} < 2 \log \tfrac{1}{3}$$
$$3 < 2 \qquad \text{Divide both sides by } \log \tfrac{1}{3}.$$

42. Find the fallacy:

$$3 > 2$$
$$3 \log \tfrac{1}{2} > 2 \log \tfrac{1}{2} \qquad \text{Multiply both sides by } \log \tfrac{1}{2}.$$
$$\log (\tfrac{1}{2})^3 > \log (\tfrac{1}{2})^2$$
$$(\tfrac{1}{2})^3 > (\tfrac{1}{2})^2$$
$$\tfrac{1}{8} > \tfrac{1}{4}$$
$$1 > 2 \qquad \text{Multiply both sides by 8.}$$

43. The function $f(x) = \log x$ increases extremely slowly as $x \to \infty$, but the composite function $g(x) = \log(\log x)$ increases still more slowly.
(A) Illustrate this fact by computing the values of both functions for several large values of x.
(B) Determine the domain and range of the function g.
(C) Discuss the graphs of both functions.

44. The function $f(x) = \ln x$ increases extremely slowly as $x \to \infty$, but the composite function $g(x) = \ln(\ln x)$ increases still more slowly.
(A) Illustrate this fact by computing the values of both functions for several large values of x.
(B) Determine the domain and range of the function g.
(C) Discuss the graphs of both functions.

In Problems 45–48, use a graphing utility to find the coordinates of all points of intersection to two decimal places.

45. $f(x) = \ln x$, $g(x) = 0.1x - 0.2$

46. $f(x) = \log x$, $g(x) = 4 - x^2$

47. $f(x) = \ln x$, $g(x) = x^{1/3}$

48. $f(x) = 3 \ln(x-2)$, $g(x) = 4e^{-x}$

Problems 49–52 require the use of a graphing utility.

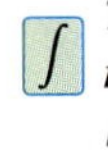
*The polynomials in Problems 49–52, called **Taylor polynomials,** can be used to approximate the function $g(x) = \ln(1 + x)$. To illustrate this approximation graphically, in each problem, graph $g(x) = \ln(1 + x)$ and the indicated polynomial in the same viewing window, $-1 \le x \le 3$ and $-2 \le y \le 2$.*

49. $P_1(x) = x - \frac{1}{2}x^2$

50. $P_2(x) = x - \frac{1}{2}x^2 + \frac{1}{3}x^3$

51. $P_3(x) = x - \frac{1}{2}x^2 + \frac{1}{3}x^3 - \frac{1}{4}x^4$

52. $P_4(x) = x - \frac{1}{2}x^2 + \frac{1}{3}x^3 - \frac{1}{4}x^4 + \frac{1}{5}x^5$

APPLICATIONS 

53. Sound. What is the decibel level of:
(A) The threshold of hearing, 1.0×10^{-12} watt per square meter?
(B) The threshold of pain, 1.0 watt per square meter?
Compute answers to 2 significant digits.

54. Sound. What is the decibel level of:
(A) A normal conversation, 3.2×10^{-6} watt per square meter?
(B) A jet plane with an afterburner, 8.3×10^2 watts per square meter?
Compute answers to 2 significant digits.

55. Sound. If the intensity of a sound from one source is 1,000 times that of another, how much more is the decibel level of the louder sound than the quieter one?

56. Sound. If the intensity of a sound from one source is 10,000 times that of another, how much more is the decibel level of the louder sound than the quieter one?

57. Earthquakes. The largest recorded earthquake to date was in Colombia in 1906, with an energy release of 1.99×10^{17} joules. What was its magnitude on the Richter scale? Compute the answer to one decimal place.

58. Earthquakes. Anchorage, Alaska, had a major earthquake in 1964 that released 7.08×10^{16} joules of energy. What was its magnitude on the Richter scale? Compute the answer to one decimal place.

★★ **59. Earthquakes.** The 1933 Long Beach, California, earthquake had a Richter scale reading of 6.3, and the 1964 Anchorage, Alaska, earthquake had a Richter scale reading of 8.3. How many times more powerful was the Anchorage earthquake than the Long Beach earthquake?

★★ **60. Earthquakes.** Generally, an earthquake requires a magnitude of over 5.6 on the Richter scale to inflict serious damage. How many times more powerful than this was the great 1906 Colombia earthquake, which registered a magnitude of 8.6 on the Richter scale?

61. Space Vehicles. A new solid-fuel rocket has a weight ratio $W_t/W_b = 19.8$ and an exhaust velocity $c = 2.57$ kilometers per second. What is its velocity at burnout? Compute the answer to two decimal places.

62. Space Vehicles. A liquid-fuel rocket has a weight ratio $W_t/W_b = 6.2$ and an exhaust velocity $c = 5.2$ kilometers per second. What is its velocity at burnout? Compute the answer to two decimal places.

63. Chemistry. The hydrogen ion concentration of a substance is related to its acidity and basicity. Because hydrogen ion concentrations vary over a very wide range, logarithms are used to create a compressed **pH scale,** which is defined as follows:

$$\text{pH} = -\log\,[\text{H}^+]$$

where $[\text{H}^+]$ is the hydrogen ion concentration, in moles per liter. Pure water has a pH of 7, which means it is neutral. Substances with a pH less than 7 are acidic, and those with a pH greater than 7 are basic. Compute the pH of each substance listed, given the indicated hydrogen ion concentration.
(A) Seawater, 4.63×10^{-9}
(B) Vinegar, 9.32×10^{-4}
Also, indicate whether it is acidic or basic. Compute answers to one decimal place.

64. Chemistry. Refer to Problem 63. Compute the pH of each substance below, given the indicated hydrogen ion concentration. Also, indicate whether it is acidic or basic. Compute answers to one decimal place.
(A) Milk, 2.83×10^{-7}
(B) Garden mulch, 3.78×10^{-6}

★ **65. Ecology.** Refer to Problem 63. Many lakes in Canada and the United States will no longer sustain some forms of wildlife because of the increase in acidity of the water from acid rain and snow caused by sulfur dioxide emissions from industry. If the pH of a sample of rainwater is 5.2, what is its hydrogen ion concentration in moles per liter? Compute the answer to two significant digits.

★ **66. Ecology.** Refer to Problem 63. If normal rainwater has a pH of 5.7, what is its hydrogen ion concentration in moles per liter? Compute the answer to two significant digits.

SECTION 4-5 Exponential and Logarithmic Equations

- Exponential Equations
- Logarithmic Equations
- Change of Base

Equations involving exponential and logarithmic functions, such as

$$2^{3x-2} = 5 \qquad \text{and} \qquad \log (x + 3) + \log x = 1$$

are called **exponential** and **logarithmic equations,** respectively. Logarithmic properties play a central role in their solution.

• Exponential Equations

The following examples illustrate the use of logarithmic properties in solving exponential equations.

EXAMPLE 1 Solving an Exponential Equation

Solve $2^{3x-2} = 5$ for x to four decimal places.

Solution How can we get x out of the exponent? Use logs! Since the logarithm function is one-to-one, if two positive quantities are equal, their logs are equal. See Theorem 1 in Section 4-3.

$$2^{3x-2} = 5$$

$$\log 2^{3x-2} = \log 5 \qquad \text{Take the common or natural log of both sides.}$$

$$(3x - 2) \log 2 = \log 5 \qquad \text{Use } \log_b N^p = p \log_b N \text{ to get } 3x - 2 \text{ out of the exponent position.}$$

$$3x - 2 = \frac{\log 5}{\log 2}$$

$$x = \frac{1}{3}\left(2 + \frac{\log 5}{\log 2}\right) \qquad \textit{Remember: } \frac{\log 5}{\log 2} \neq \log 5 - \log 2.$$

$$= 1.4406 \qquad \text{To four decimal places}$$

A graphing utility provides an alternative approach. We graph the functions $y_1 = 2^{3x-2}$ and $y_2 = 5$ and use the intersection routine, as indicated in Figure 1.

FIGURE 1

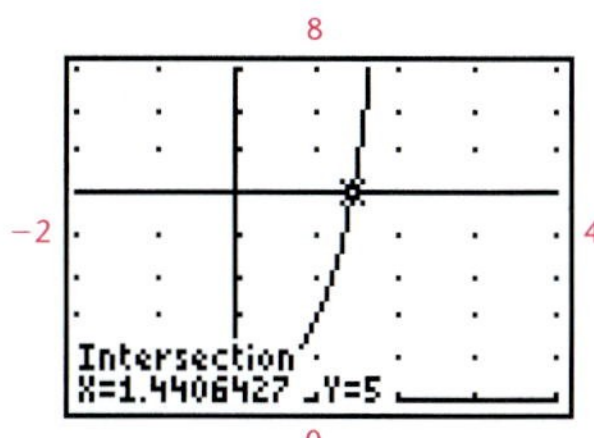

Matched Problem 1 Solve $35^{1-2x} = 7$ for x to four decimal places.

EXAMPLE 2 Compound Interest

A certain amount of money P (principal) is invested at an annual rate r compounded annually. The amount of money A in the account after t years, assuming no withdrawals, is given by

$$A = P\left(1 + \frac{r}{n}\right)^{nt} = P(1 + r)^t \qquad n = 1 \text{ for annual compounding}$$

How many years to the nearest year will it take the money to double if it is invested at 6% compounded annually?

Solution To find the doubling time, we replace A in $A = P(1.06)^t$ with $2P$ and solve for t.

$$2P = P(1.06)^t$$

$$2 = 1.06^t \qquad \text{Divide both sides by } P.$$

$$\log 2 = \log 1.06^t \qquad \text{Take the common or natural log of both sides.}$$

$$= t \log 1.06 \qquad \text{Note how log properties are used to get } t \text{ out of the exponent position.}$$

$$t = \frac{\log 2}{\log 1.06}$$

$$= 12 \text{ years} \qquad \text{To the nearest year}$$

Matched Problem 2 Repeat Example 2, changing the interest rate to 9% compounded annually.

EXAMPLE 3 Atmospheric Pressure

The atmospheric pressure P, in pounds per square inch, at x miles above sea level is given approximately by

$$P = 14.7e^{-0.21x}$$

At what height will the atmospheric pressure be half the sea-level pressure? Compute the answer to 2 significant digits.

Solution Sea-level pressure is the pressure at $x = 0$. Thus,

$$P = 14.7e^0 = 14.7$$

One-half of sea-level pressure is $14.7/2 = 7.35$. Now our problem is to find x so that $P = 7.35$; that is, we solve $7.35 = 14.7e^{-0.21x}$ for x:

$$
\begin{aligned}
7.35 &= 14.7e^{-0.21x} \\
0.5 &= e^{-0.21x} && \text{Divide both sides by 14.7 to simplify.} \\
\ln 0.5 &= \ln e^{-0.21x} && \text{Since the base is } e\text{, take the natural log of both sides.} \\
&= -0.21x && \ln e = 1 \\
x &= \frac{\ln 0.5}{-0.21} \\
&= 3.3 \text{ miles} && \text{To 2 significant digits}
\end{aligned}
$$

Matched Problem 3 Using the formula in Example 3, find the altitude in miles so that the atmospheric pressure will be one-eighth that at sea level. Compute the answer to 2 significant digits.

The graph of

$$y = \frac{e^x + e^{-x}}{2} \tag{1}$$

is a curve called a **catenary** (Fig. 2). A uniform cable suspended between two fixed points is a physical example of such a curve.

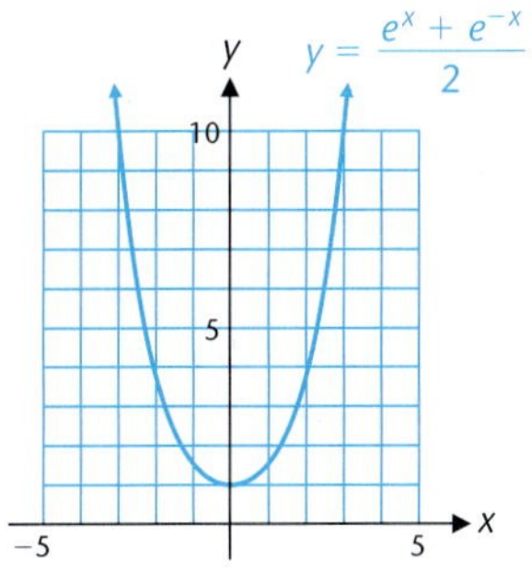

FIGURE 2 Catenary.

EXAMPLE 4 Solving an Exponential Equation

Given equation (1), find x for $y = 2.5$. Compute the answer to four decimal places.

Solution

$$
\begin{aligned}
y &= \frac{e^x + e^{-x}}{2} \\
2.5 &= \frac{e^x + e^{-x}}{2} \\
5 &= e^x + e^{-x} \\
5e^x &= e^{2x} + 1 && \text{Multiply both sides by } e^x. \\
e^{2x} - 5e^x + 1 &= 0 && \text{This is a quadratic in } e^x.
\end{aligned}
$$

Let $u = e^x$; then

$$u^2 - 5u + 1 = 0$$

$$u = \frac{5 \pm \sqrt{25 - 4(1)(1)}}{2}$$

$$= \frac{5 \pm \sqrt{21}}{2}$$

$$e^x = \frac{5 \pm \sqrt{21}}{2}$$ Replace u with e^x and solve for x.

$$\ln e^x = \ln \frac{5 \pm \sqrt{21}}{2}$$ Take the natural log of both sides (both values on the right are positive).

$$x = \ln \frac{5 \pm \sqrt{21}}{2}$$ $\log_b b^x = x$.

$$= -1.5668, 1.5668$$

Matched Problem 4  Given $y = (e^x - e^{-x})/2$, find x for $y = 1.5$. Compute the answer to three decimal places.

EXPLORE-DISCUSS 1 Let $y = e^{2x} + 3e^x + e^{-x}$

(A) Try to find x when $y = 7$ using the method of Example 4. Explain the difficulty that arises.

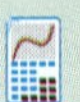 (B) Use a graphing utility to find x when $y = 7$.

• Logarithmic Equations

We now illustrate the solution of several types of logarithmic equations.

EXAMPLE 5 Solving a Logarithmic Equation

Solve $\log(x + 3) + \log x = 1$, and check.

Solution First use properties of logarithms to express the left side as a single logarithm, then convert to exponential form and solve for x.

$$\log(x + 3) + \log x = 1$$

$$\log[x(x + 3)] = 1$$ Combine left side using $\log M + \log N = \log MN$.

$$x(x + 3) = 10^1$$ Change to equivalent exponential form.

$$x^2 + 3x - 10 = 0 \quad \text{Write in } ax^2 + bx + c = 0 \text{ form and solve.}$$
$$(x + 5)(x - 2) = 0$$
$$x = -5, 2$$

Check $x = -5$: $\log(-5 + 3) + \log(-5)$ is not defined because the domain of the log function is $(0, \infty)$.

$x = 2$: $\log(2 + 3) + \log 2 = \log 5 + \log 2$
$$= \log(5 \cdot 2) = \log 10 \stackrel{\checkmark}{=} 1$$

Thus, the only solution to the original equation is $x = 2$. Remember, answers should be checked in the original equation to see whether any should be discarded.

Matched Problem 5 Solve $\log(x - 15) = 2 - \log x$, and check.

EXAMPLE 6 Solving a Logarithmic Equation

Solve $(\ln x)^2 = \ln x^2$.

Solution There are no logarithmic properties for simplifying $(\ln x)^2$. However, we can simplify $\ln x^2$, obtaining an equation involving $\ln x$ and $(\ln x)^2$.

$$(\ln x)^2 = \ln x^2$$
$$= 2 \ln x \quad \text{This is a quadratic equation in } \ln x. \text{ Move all nonzero terms to the left and factor.}$$
$$(\ln x)^2 - 2 \ln x = 0$$
$$(\ln x)(\ln x - 2) = 0$$
$$\ln x = 0 \quad \text{or} \quad \ln x - 2 = 0$$
$$x = e^0 \qquad \ln x = 2$$
$$= 1 \qquad x = e^2$$

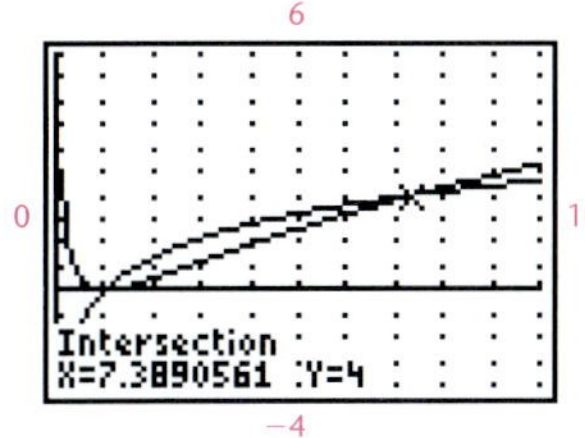

FIGURE 3

Checking that both $x = 1$ and $x = e^2$ are solutions to the original equation is left to you.

A graphing utility provides an alternative approach. We graph the functions $y_1 = (\ln x)^2$ and $y_2 = \ln x^2$ and use the intersection routine as indicated in Figure 3. Note that $x=7.389$ is an approximation to the value of e^2, which is one of the solutions of the original equation. A graphing utility may be used in this manner to solve many complicated equations involving exponential and logarithmic functions that cannot be solved by the algebraic methods of Examples 1–6. See Exercise 4-5.

Matched Problem 6 Solve $\log x^2 = (\log x)^2$.

CAUTION Note that

$$(\log_b x)^2 \neq \log_b x^2 \qquad \begin{aligned}(\log_b x)^2 &= (\log_b x)(\log_b x)\\ \log_b x^2 &= 2 \log_b x\end{aligned}$$

EXAMPLE 7 Earthquake Intensity

Recall from Section 4-4 that the magnitude of an earthquake on the Richter scale is given by

$$M = \frac{2}{3} \log \frac{E}{E_0}$$

Solve for E in terms of the other symbols.

Solution

$$M = \frac{2}{3} \log \frac{E}{E_0}$$

$$\log \frac{E}{E_0} = \frac{3M}{2} \qquad \text{Multiply both sides by } \tfrac{3}{2}.$$

$$\frac{E}{E_0} = 10^{3M/2} \qquad \text{Change to exponential form.}$$

$$E = E_0 10^{3M/2}$$

Matched Problem 7 Solve the rocket equation from Section 4-4 for W_b in terms of the other symbols:

$$v = c \ln \frac{W_t}{W_b}$$

• Change of Base

How would you find the logarithm of a positive number to a base other than 10 or e? For example, how would you find $\log_3 5.2$? In Example 8 we evaluate this logarithm using a direct process. Then we develop a change-of-base formula to find such logarithms in general. You may find it easier to remember the process than the formula.

EXAMPLE 8 Evaluating a Base 3 Logarithm

Evaluate $\log_3 5.2$ to four decimal places.

Solution Let $y = \log_3 5.2$ and proceed as follows:

$$\log_3 5.2 = y$$

$$5.2 = 3^y \qquad \text{Change to exponential form.}$$

$$\ln 5.2 = \ln 3^y \qquad \text{Take the natural log (or common log) of each side.}$$

$$= y \ln 3 \qquad \log_b M^p = p \log_b M$$

$$y = \frac{\ln 5.2}{\ln 3} \qquad \text{Solve for } y.$$

Replace y with $\log_3 5.2$ from the first step, and use a calculator to evaluate the right side:

$$\log_3 5.2 = \frac{\ln 5.2}{\ln 3} = 1.5007$$

Matched Problem 8 Evaluate $\log_{0.5} 0.0372$ to four decimal places.

To develop a change-of-base formula for arbitrary positive bases, with neither base equal to 1, we proceed as above. Let $y = \log_b N$, where N and b are positive and $b \neq 1$. Then

$$\log_b N = y$$

$$N = b^y \qquad \text{Write in exponential form.}$$

$$\log_a N = \log_a b^y \qquad \text{Take the log of each side to another positive base } a, a \neq 1.$$

$$= y \log_a b \qquad \log_b M^p = p \log_b M$$

$$y = \frac{\log_a N}{\log_a b} \qquad \text{Solve for } y.$$

Replacing y with $\log_b N$ from the first step, we obtain the **change-of-base formula:**

$$\log_b N = \frac{\log_a N}{\log_a b}$$

In words, this formula states that the logarithm of a number to a given base is the logarithm of that number to a new base divided by the logarithm of the old base to the new base. In practice, we usually choose either e or 10 for the new base so that a calculator can be used to evaluate the necessary logarithms (see Example 8).

EXPLORE-DISCUSS 2 If b is any positive real number different from 1, the change-of-base formula implies that the function $y = \log_b x$ is a constant multiple of the natural logarithmic function; that is, $\log_b x = k \ln x$ for some k.

(A) Graph the functions $y = \ln x$, $y = 2 \ln x$, $y = 0.5 \ln x$, and $y = -3 \ln x$.

(B) Write each function of part A in the form $y = \log_b x$ by finding the base b to two decimal places.

(C) Is every exponential function $y = b^x$ a constant multiple of $y = e^x$? Explain.

Answers to Matched Problems

1. $x = 0.2263$ **2.** More than double in 9 years, but not quite double in 8 years
3. 9.9 miles **4.** $x = 1.195$ **5.** $x = 20$ **6.** $x = 1{,}100$ **7.** $W_b = W_t e^{-v/c}$
8. 4.7486

EXERCISE 4-5

A

Solve Problems 1–12 to 3 significant digits.

1. $10^x = 27.5$ **2.** $e^x = 9.62$ **3.** $e^{-x} = 0.0028$

4. $10^{-x} = 1.25$ **5.** $10^{2x+5} = 43.7$ **6.** $e^{1-3x} = 9.62$

7. $e^{4-2x} = 45$ **8.** $10^{4x-1} = 5{,}000$ **9.** $3^x = 35$

10. $2^x = 0.0525$ **11.** $5^{-x} = 250$ **12.** $4^{-x} = 0.0001$

Solve Problems 13–18 exactly.

13. $\log x + \log 4 = 1$ **14.** $\ln x + \ln 4 = 1$

15. $\ln 8 - \ln x = 2$

16. $\log 8 - \log x = 2$

17. $\log (x + 10) + \log (x - 5) = 2$

18. $\log (x + 5) + \log (x - 10) = 3$

B

Solve Problems 19–26 to 3 significant digits.

19. $2 = 1.002^{4x}$ **20.** $3 = 1.001^{12x}$

21. $e^{-0.005x} = 100$ **22.** $e^{25x} = 1.25$

23. $1{,}000 = 75e^{0.5x}$ **24.** $1{,}000 = 46e^{-0.4x}$

25. $e^{-0.1x^2} = 0.2$ **26.** $e^{-0.2x^2} = 0.5$

Solve Problems 27–38 exactly.

27. $\log x - \log 5 = \log 2 - \log (x - 3)$

28. $\log (6x + 5) - \log 3 = \log 2 - \log x$

29. $\ln x = \ln (2x - 1) - \ln (x - 2)$

30. $\ln (x + 1) = \ln (3x + 1) - \ln x$

31. $\log (2x + 1) = 1 - \log (x - 1)$

32. $1 - \log (x - 2) = \log (3x + 1)$

33. $(\ln x)^3 = \ln x^4$ **34.** $(\log x)^3 = \log x^4$

35. $\ln (\ln x) = 1$ **36.** $\log (\log x) = 1$

37. $x^{\log x} = 100x$ **38.** $3^{\log x} = 3x$

In Problems 39–40:
(A) Explain the difficulty in solving the equation exactly.
(B) Determine the number of solutions by graphing the functions on each side of the equation.

39. $e^{x/2} = 5 \ln x$ **40.** $\ln(\ln x) + \ln x = 2$

In Problems 41–42:
(A) Explain the difficulty in solving the equation exactly.
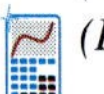
(B) Use a graphing utility to find all solutions to three decimal places.

41. $3^x + 2 = 7 + x - e^{-x}$ **42.** $e^{x/4} = 5 \log x + 4 \ln x$

Evaluate Problems 43–48 to four decimal places.

43. $\log_5 372$ **44.** $\log_4 23$ **45.** $\log_8 0.0352$

46. $\log_2 0.005\,439$ **47.** $\log_3 0.1483$ **48.** $\log_{12} 435.62$

C

Solve Problems 49–56 for the indicated variable in terms of the remaining symbols. Use the natural log for solving exponential equations.

49. $A = Pe^{rt}$ for r (finance)

50. $A = P\left(1 + \frac{r}{n}\right)^{nt}$ for t (finance)

51. $D = 10 \log \frac{I}{I_0}$ for I (sound)

52. $t = \frac{-1}{k}(\ln A - \ln A_0)$ for A (decay)

53. $M = 6 - 2.5 \log \frac{I}{I_0}$ for I (astronomy)

54. $L = 8.8 + 5.1 \log D$ for D (astronomy)

55. $I = \dfrac{E}{R}(1 - e^{-Rt/L})$ for t (circuitry)

56. $S = R\dfrac{(1 + i)^n - 1}{i}$ for n (annuity)

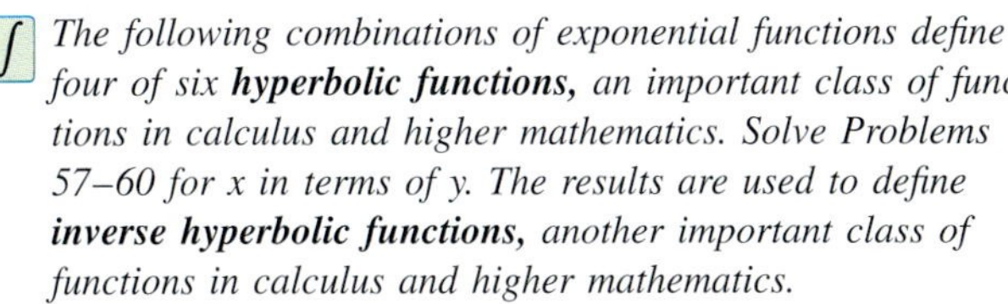

*The following combinations of exponential functions define four of six **hyperbolic functions,** an important class of functions in calculus and higher mathematics. Solve Problems 57–60 for x in terms of y. The results are used to define **inverse hyperbolic functions,** another important class of functions in calculus and higher mathematics.*

57. $y = \dfrac{e^x + e^{-x}}{2}$

58. $y = \dfrac{e^x - e^{-x}}{2}$

59. $y = \dfrac{e^x - e^{-x}}{e^x + e^{-x}}$

60. $y = \dfrac{e^x + e^{-x}}{e^x - e^{-x}}$

Problems 61–76 require the use of a graphing utility.

In Problems 61–64, use a graphing utility to graph each function. [Hint: Use the change-of-base formula first.]

61. $y = 3 + \log_2 (2 - x)$

62. $y = \log_3 (4 + x) - 5$

63. $y = \log_3 x - \log_2 x$

64. $y = \log_3 x + \log_2 x$

In Problems 65–76, use a graphing utility to approximate to two decimal places any solutions of the equation in the interval $0 \le x \le 1$. None of these equations can be solved exactly using any step-by-step algebraic process.

65. $2^{-x} - 2x = 0$

66. $3^{-x} - 3x = 0$

67. $x3^x - 1 = 0$

68. $x2^x - 1 = 0$

69. $e^{-x} - x = 0$

70. $xe^{2x} - 1 = 0$

71. $xe^x - 2 = 0$

72. $e^{-x} - 2x = 0$

73. $\ln x + 2x = 0$

74. $\ln x + x^2 = 0$

75. $\ln x + e^x = 0$

76. $\ln x + x = 0$

APPLICATIONS

77. Compound Interest. How many years, to the nearest year, will it take a sum of money to double if it is invested at 15% compounded annually? Use the formula $A = P[1 + (r/n)]^{nt}$.

78. Compound Interest. How many years, to the nearest year, will it take money to quadruple if it is invested at 20% compounded annually? Use the formula $A = P[1 + (r/n)]^{nt}$.

79. Compound Interest. At what annual rate compounded continuously will \$1,000 have to be invested to amount to \$2,500 in 10 years? Use the formula $A = Pe^{rt}$. Compute the answer to 3 significant digits.

80. Compound Interest. How many years will it take \$5,000 to amount to \$8,000 if it is invested at an annual rate of 9% compounded continuously? Use the formula $A = Pe^{rt}$. Compute the answer to 3 significant digits.

★★ **81. Astronomy.** The brightness of stars is expressed in terms of magnitudes on a numerical scale that increases as the brightness decreases. The magnitude m is given by the formula

$$m = 6 - 2.5 \log \frac{L}{L_0}$$

where L is the light flux of the star and L_0 is the light flux of the dimmest stars visible to the naked eye.

(A) What is the magnitude of the dimmest stars visible to the naked eye?

(B) How many times brighter is a star of magnitude 1 than a star of magnitude 6?

82. Astronomy. An optical instrument is required to observe stars beyond the sixth magnitude, the limit of ordinary vision. However, even optical instruments have their limitations. The limiting magnitude L of any optical telescope with lens diameter D, in inches, is given by

$$L = 8.8 + 5.1 \log D$$

(A) Find the limiting magnitude for a homemade 6-inch reflecting telescope.

(B) Find the diameter of a lens that would have a limiting magnitude of 20.6.

Compute answers to 3 significant digits.

83. World Population. A mathematical model for world population growth over short periods of time is given by

$$P = P_0 e^{rt}$$

where P is the population after t years, P_0 is the population at $t = 0$, and the population is assumed to grow continuously at the annual rate r. How many years, to the nearest year, will it take the world population to double if it grows continuously at an annual rate of 2%?

★ **84. World Population.** Refer to Problem 83. Starting with a world population of 4 billion people and assuming that the population grows continuously at an annual rate of 2%, how many years, to the nearest year, will it be before there is only 1 square yard of land per person? Earth contains approximately 1.7×10^{14} square yards of land.

★ **85. Archaeology—Carbon 14 Dating.** As long as a plant or animal is alive, carbon 14 is maintained in a constant amount in its tissues. Once dead, however, the plant or animal ceases taking in carbon, and carbon 14 diminishes by radioactive decay according to the equation

$$A = A_0 e^{-0.000124t}$$

where A is the amount after t years and A_0 is the amount when $t = 0$. Estimate the age of a skull uncovered in an archaeological site if 10% of the original amount of carbon 14 is still present. Compute the answer to 3 significant digits.

★ **86. Archaeology—Carbon 14 Dating.** Refer to Problem 85. What is the half-life of carbon 14? That is, how long will it take for half of a sample of carbon 14 to decay? Compute the answer to 3 significant digits.

★ **87. Photography.** An electronic flash unit for a camera is activated when a capacitor is discharged through a filament of wire. After the flash is triggered and the capacitor is discharged, the circuit (see the figure) is connected and the battery pack generates a current to recharge the capacitor. The time it takes for the capacitor to recharge is called the *recycle time.* For a particular flash unit using a 12-volt battery pack, the charge q, in coulombs, on the capacitor t seconds after recharging has started is given by

$$q = 0.0009(1 - e^{-0.2t})$$

How many seconds will it take the capacitor to reach a charge of 0.0007 coulomb? Compute the answer to 3 significant digits.

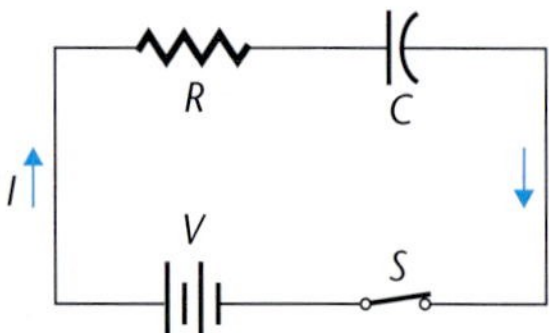

★ **88. Advertising.** A company is trying to expose a new product to as many people as possible through television advertising in a large metropolitan area with 2 million possible viewers. A model for the number of people N, in millions, who are aware of the product after t days of advertising was found to be

$$N = 2(1 - e^{-0.037t})$$

How many days, to the nearest day, will the advertising campaign have to last so that 80% of the possible viewers will be aware of the product?

★★ **89. Newton's Law of Cooling.** This law states that the rate at which an object cools is proportional to the difference in temperature between the object and its surrounding medium. The temperature T of the object t hours later is given by

$$T = T_m + (T_0 - T_m)e^{-kt}$$

where T_m is the temperature of the surrounding medium and T_0 is the temperature of the object at $t = 0$. Suppose a bottle of wine at a room temperature of 72°F is placed in a refrigerator at 40°F to cool before a dinner party. After an hour the temperature of the wine is found to be 61.5°F. Find the constant k, to two decimal places, and the time, to one decimal place, it will take the wine to cool from 72 to 50°F.

★ **90. Marine Biology.** Marine life is dependent upon the microscopic plant life that exists in the *photic zone,* a zone that goes to a depth where about 1% of the surface light still remains. Light intensity is reduced according to the exponential function

$$I = I_0e^{-kd}$$

where I is the intensity d feet below the surface and I_0 is the intensity at the surface. The constant k is called the *coefficient of extinction.* At Crystal Lake in Wisconsin it was found that half the surface light remained at a depth of 14.3 feet. Find k, and find the depth of the photic zone. Compute answers to 3 significant digits.

★ **91. Wildlife Management.** A herd of 20 white-tailed deer is introduced to a coastal island where there had been no deer before. Their population is predicted to increase according to the logistic curve

$$N = \frac{100}{1 + 4e^{-0.14t}}$$

where N is the number of deer expected in the herd after t years. In how many years, to the nearest year, will the herd number 50?

★ **92. Training.** A trainee is hired by a computer manufacturing company to learn to test a particular model of a personal computer after it comes off the assembly line. The learning curve for an average trainee is given by

$$N = \frac{200}{4 + 21e^{-0.1t}}$$

where N is the number of computers tested per day after t days on the job. How many days, to the nearest day, will it take an average trainee to achieve 40 tested computers per day?

CHAPTER 4 GROUP ACTIVITY Growth of Increasing Functions

The exponential function 2^x and the logarithmic function $\log_2 x$ are both increasing functions: their graphs rise as x increases. Moreover, they increase without bound. But they grow in strikingly different ways. The exponential function increases extremely rapidly, the logarithmic function extremely slowly. By computing values of functions and studying their graphs, we can investigate the manner in which various increasing functions grow.

(A) The functions $f(x) = 2^x$ and $g(x) = x^3$ are both increasing, but for large values of x, f increases much faster than g does. Illustrate this fact by computing values of both functions at several large values of x. Graph both functions and determine the number of intersection points.

(B) The functions $h(x) = \ln x$ and $k(x) = x^{1/4}$ are both increasing for $x>0$, but for large values of x, h increases much more slowly than k does. Illustrate this fact by computing values of both functions at several large values of x. Graph both functions and determine the number of intersection points.

(C) The following 12 functions all increase without bound for $x>1$. Arrange them in order, starting with the function that increases most slowly and ending with the function that increases most rapidly: $\sqrt{x}$, 2^x, $\log x$, e^{e^x}, x^6, $x \ln x$, e^x, $x^{0.1}$, $\ln(\ln x)$, e^{x^2}, $2x$, $\ln x$.

Chapter 4 Review

4-1 EXPONENTIAL FUNCTIONS

The equation $f(x) = b^x$, $b > 0$, $b \neq 1$, defines an **exponential function** with **base** b. The domain of f is $(-\infty, \infty)$ and the range is $(0, \infty)$. The **graph** of an exponential function is a continuous curve that always passes through the point $(0, 1)$ and has the x axis as a horizontal asymptote. If $b > 1$, then b^x increases as x increases, and if $0 < b < 1$, then b^x decreases as x increases. The function f is one-to-one and has an inverse. We have the following **exponential function properties:**

1. $a^x a^y = a^{x+y}$ $\quad (a^x)^y = a^{xy}$ $\quad (ab)^x = a^x b^x$

   $\left(\dfrac{a}{b}\right)^x = \dfrac{a^x}{b^x}$ $\quad \dfrac{a^x}{a^y} = a^{x-y}$
2. $a^x = a^y$ if and only if $x = y$.
3. For $x \neq 0$, then $a^x = b^x$ if and only if $a = b$.

Exponential functions are used to describe various types of **growth.**

1. **Population growth** can be modeled by using the **doubling time growth model** $P = P_0 2^{t/d}$, where P is population at time t, P_0 is the population at time $t = 0$, and d is the **doubling time**—the time it takes for the population to double.
2. **Radioactive decay** can be modeled by using the **half-life decay model** $A = A_0(\frac{1}{2})^{t/h} = A_0 2^{-t/h}$, where A is the amount at time t, A_0 is the amount at time $t = 0$, and h is the **half-life**—the time it takes for half the material to decay.
3. The growth of money in an account paying **compound interest** is described by $A = P(1 + r/n)^{nt}$, where P is the **principal,** r is the annual **rate,** n is the number of compounding periods in one year, and A is the **amount** in the account after t years. We also call P the **present value** and A the **future value** of the account.

4-2 THE EXPONENTIAL FUNCTION WITH BASE *e*

As m approaches ∞, the expression $(1 + 1/m)^m$ approaches the irrational number $e \approx 2.718\,281\,828\,459$. The function $f(x) = e^x$ is called the **exponential function with base *e*.** Exponential functions with base e are used to model a variety of different types of exponential growth and decay, including growth of money in accounts that pay **continuous compound interest.** If a principal P is invested at an annual rate r compounded continuously, then the amount A in the account after t years is given by $A = Pe^{rt}$.

4-3 LOGARITHMIC FUNCTIONS

The **logarithmic function with base *b*** is defined to be the inverse of the exponential function with base b and is denoted by $y = \log_b x$. Thus, $y = \log_b x$ if and only if $x = b^y$, $b > 0$, $b \neq 1$. The domain of a logarithmic function is $(0, \infty)$ and the range is $(-\infty, \infty)$. The graph of a logarithmic function is a continuous curve that always passes through the point $(1, 0)$ and has the y

axis as a vertical asymptote. We have the following **properties of logarithmic functions:**

1. $\log_b 1 = 0$
2. $\log_b b = 1$
3. $\log_b b^x = x$
4. $b^{\log_b x} = x, x > 0$
5. $\log_b MN = \log_b M + \log_b N$
6. $\log_b \frac{M}{N} = \log_b M - \log_b N$
7. $\log_b M^p = p \log_b M$
8. $\log_b M = \log_b N$ if and only if $M = N$

4-4 COMMON AND NATURAL LOGARITHMS

Logarithms to the base 10 are called **common logarithms** and are denoted by $\log x$. Logarithms to the base e are called **natural logarithms** and are denoted by $\ln x$. Thus, $\log x = y$ is equivalent to $x = 10^y$, and $\ln x = y$ is equivalent to $x = e^y$.

The following applications involve logarithms:

1. The **decibel** is defined by $D = 10 \log (I/I_0)$, where D is the **decibel level** of the sound, I is the **intensity** of the sound, and $I_0 = 10^{-12}$ watt per square meter is a standardized sound level.
2. The **magnitude** M of an earthquake on the **Richter scale** is given by $M = \frac{2}{3} \log (E/E_0)$, where E is the energy released by the earthquake and $E_0 = 10^{4.40}$ joules is a standardized energy level.
3. The **velocity** v of a rocket at burnout is given by the **rocket equation** $v = c \ln (W_t/W_b)$, where c is the exhaust velocity, W_t is the takeoff weight, and W_b is the burnout weight.

4-5 EXPONENTIAL AND LOGARITHMIC EQUATIONS

Various techniques for solving **exponential equations,** such as $2^{3x-2} = 5$, and **logarithmic equations,** such as $\log (x + 3) + \log x = 1$, are illustrated by examples. The **change-of-base formula,** $\log_b N = (\log_a N)/(\log_a b)$, relates logarithms to two different bases and can be used, along with a calculator, to evaluate logarithms to bases other than e or 10.

Chapter 4 Review Exercise

Work through all the problems in this chapter review and check answers in the back of the book. Answers to all review problems are there, and following each answer is a number in italics indicating the section in which that type of problem is discussed. Where weaknesses show up, review appropriate sections in the text.

A

1. Write in logarithmic form using base 10: $m = 10^n$
2. Write in logarithmic form using base e: $x = e^y$

Write Problems 3 and 4 in exponential form.

3. $\log x = y$
4. $\ln y = x$

In Problems 5 and 6, simplify.

5. $\frac{7^{x+2}}{7^{2-x}}$
6. $\left(\frac{e^x}{e^{-x}}\right)^x$

Solve Problems 7–9 for x exactly. Do not use a calculator or table.

7. $\log_2 x = 3$
8. $\log_x 25 = 2$
9. $\log_3 27 = x$

Solve Problems 10–13 for x to 3 significant digits.

10. $10^x = 17.5$
11. $e^x = 143{,}000$
12. $\ln x = -0.015\,73$
13. $\log x = 2.013$

B

Solve Problems 14–24 for x exactly. Do not use a calculator or table.

14. $\ln (2x - 1) = \ln (x + 3)$
15. $\log (x^2 - 3) = 2 \log (x - 1)$
16. $e^{x^2-3} = e^{2x}$
17. $4^{x-1} = 2^{1-x}$
18. $2x^2e^{-x} = 18e^{-x}$
19. $\log_{1/4} 16 = x$
20. $\log_x 9 = -2$
21. $\log_{16} x = \frac{3}{2}$
22. $\log_x e^5 = 5$
23. $10^{\log_{10} x} = 33$
24. $\ln x = 0$

Evaluate Problems 25–28 to 4 significant digits using a calculator.

25. $\ln \pi$
26. $\log (-e)$
27. $\pi^{\ln 2}$
28. $\frac{e^{\pi} + e^{-\pi}}{2}$

Solve Problems 29–38 for x to 3 significant digits.

29. $x = 2(10^{1.32})$
30. $x = \log_5 23$
31. $\ln x = -3.218$
32. $x = \log (2.156 \times 10^{-7})$
33. $x = \frac{\ln 4}{\ln 2.31}$
34. $25 = 5(2^x)$

35. $4{,}000 = 2{,}500(e^{0.12x})$

36. $0.01 = e^{-0.05x}$

37. $5^{2x-3} = 7.08$

38. $\dfrac{e^x - e^{-x}}{2} = 1$

Solve Problems 39–44 for x exactly. Do not use a calculator or table.

39. $\log 3x^2 - \log 9x = 2$

40. $\log x - \log 3 = \log 4 - \log (x + 4)$

41. $\ln (x + 3) - \ln x = 2 \ln 2$

42. $\ln (2x + 1) - \ln (x - 1) = \ln x$

43. $(\log x)^3 = \log x^9$

44. $\ln (\log x) = 1$

In Problems 45 and 46, simplify.

45. $(e^x + 1)(e^{-x} - 1) - e^x(e^{-x} - 1)$

46. $(e^x + e^{-x})(e^x - e^{-x}) - (e^x - e^{-x})^2$

Graph each function in Problems 47–50.

Check Problems 47–50 with a graphing utility.

47. $y = 2^{x-1}$

48. $f(t) = 10e^{-0.08t}$

49. $y = \ln (x + 1)$

50. $N = \dfrac{100}{1 + 3e^{-t}}$

51. If the graph of $y = e^x$ is reflected in the line $y=x$, the graph of the function $y = \ln x$ is obtained. Discuss the functions that are obtained by reflecting the graph of $y = e^x$ in the x axis and in the y axis.

52. (A) Explain why the equation $e^{-x/3} = 4 \ln (x+1)$ has exactly one solution.

(B) Find the solution of the equation to three decimal places.

53. Approximate all real zeros of $f(x) = 4 - x^2 + \ln x$ to three decimal places.

54. Find the coordinates of the points of intersection of $f(x) = 10^{x-3}$ and $g(x) = 8 \log x$ to three decimal places.

C

Solve Problems 55–58 for the indicated variable in terms of the remaining symbols.

55. $D = 10 \log \dfrac{I}{I_0}$ for I (sound intensity)

56. $y = \dfrac{1}{\sqrt{2\pi}} e^{-x^2/2}$ for x (probability)

57. $x = -\dfrac{1}{k} \ln \dfrac{I}{I_0}$ for I (X-ray intensity)

58. $r = P\dfrac{i}{1 - (1 + i)^{-n}}$ for n (finance)

Find the inverse of each function in Problems 59 and 60.

59. $f(x) = 2 \ln (x - 1)$

60. $f(x) = \dfrac{e^x - e^{-x}}{2}$

61. Write $\ln y = -5t + \ln c$ in an exponential form free of logarithms; then solve for y in terms of the remaining symbols.

62. For $f = \{(x, y) \mid y = \log_2 x\}$, graph f and f^{-1} on the same coordinate system. What are the domains and ranges for f and f^{-1}?

63. Explain why 1 cannot be used as a logarithmic base.

64. Prove that $\log_b (M/N) = \log_b M - \log_b N$.

APPLICATIONS

65. **Population Growth.** Many countries have a population growth rate of 3% (or more) per year. At this rate, how many years will it take a population to double? Use the annual compounding growth model $P = P_0(1 + r)^t$. Compute the answer to 3 significant digits.

66. **Population Growth.** Repeat Problem 65 using the continuous compounding growth model $P = P_0e^{rt}$.

67. **Carbon 14 Dating.** How many years will it take for carbon 14 to diminish to 1% of the original amount after the death of a plant or animal? Use the formula $A = A_0e^{-0.000124t}$. Compute the answer to 3 significant digits.

★ 68. **Medicine.** One leukemic cell injected into a healthy mouse will divide into two cells in about $\frac{1}{2}$ day. At the end of the day these two cells will divide into four. This doubling continues until 1 billion cells are formed; then the animal dies with leukemic cells in every part of the body.

(A) Write an equation that will give the number N of leukemic cells at the end of t days.

(B) When, to the nearest day, will the mouse die?

69. **Money Growth.** Assume $1 had been invested at an annual rate of 3% compounded continuously at the birth of Christ. What would be the value of the account in the year 2000? Compute the answer to 2 significant digits.

70. **Present Value.** Solving $A = Pe^{rt}$ for P, we obtain $P = Ae^{-rt}$, which is the **present value** of the amount A due in t years if money is invested at a rate r compounded continuously.

(A) Graph $P = 1{,}000(e^{-0.08t})$, $0 \le t \le 30$.

(B) What does it appear that P tends to as t tends to infinity? [*Conclusion:* The longer the time until the amount A is due, the smaller its present value, as we would expect.]

71. **Earthquakes.** The 1971 San Fernando, California, earthquake released 1.99×10^{14} joules of energy. Compute its magnitude on the Richter scale using the formula

$M = \frac{2}{3} \log (E/E_0)$, where $E_0 = 10^{4.40}$ joules. Compute the answer to one decimal place.

72. Earthquakes. Refer to Problem 71. If the 1906 San Francisco earthquake had a magnitude of 8.3 on the Richter scale, how much energy was released? Compute the answer to 3 significant digits.

★ **73. Sound.** If the intensity of a sound from one source is 100,000 times that of another, how much more is the decibel level of the louder sound than the softer one? Use the formula $D = 10 \log (I/I_0)$.

★★ **74. Marine Biology.** The intensity of light entering water is reduced according to the exponential function

$$I = I_0 e^{-kd}$$

where I is the intensity d feet below the surface, I_0 is the intensity at the surface, and k is the coefficient of extinction. Measurements in the Sargasso Sea in the West Indies have indicated that half the surface light reaches a depth of 73.6 feet. Find k, and find the depth at which 1% of the surface light remains. Compute answers to 3 significant digits.

★ **75. Wildlife Management.** A lake formed by a newly constructed dam is stocked with 1,000 fish. Their population is expected to increase according to the logistic curve

$$N = \frac{30}{1 + 29e^{-1.35t}}$$

where N is the number of fish, in thousands, expected after t years. The lake will be open to fishing when the number of fish reaches 20,000. How many years, to the nearest year, will this take?

Cumulative Review Exercise Chapters 3 and 4

Work through all the problems in this cumulative review and check answers in the back of the book. Answers to all review problems are there, and following each answer is a number in italics indicating the section in which that type of problem is discussed. Where weaknesses show up, review appropriate sections in the text.

A

1. Let $P(x)$ be the polynomial whose graph is shown in the figure.
(A) Assuming that $P(x)$ has integer zeros and leading coefficient 1, find the lowest-degree equation that could produce this graph.
(B) Describe the left and right behavior of $P(x)$.

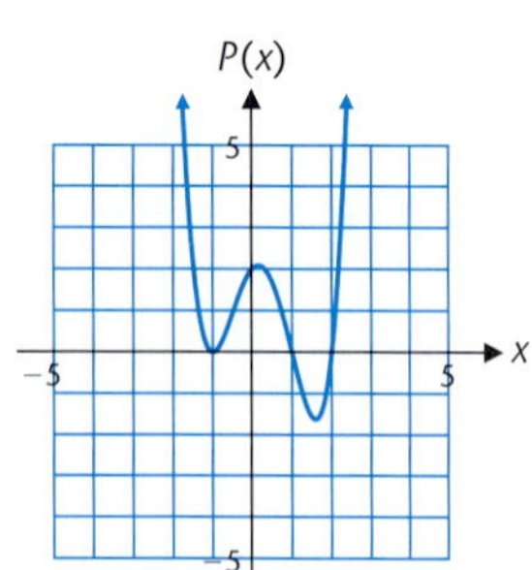

2. For $P(x) = 3x^3 + 5x^2 - 18x - 3$ and $D(x) = x + 3$, use synthetic division to divide $P(x)$ by $D(x)$, and write the answer in the form $P(x) = D(x)Q(x) + R$.

3. Let $P(x) = 2(x + 2)(x - 3)(x - 5)$. What are the zeros of $P(x)$?

4. Let $P(x) = 4x^3 - 5x^2 - 3x - 1$. How do you know that $P(x)$ has at least one real zero between 1 and 2?

5. Let $P(x) = x^3 + x^2 - 10x + 8$. Find all rational zeros for $P(x)$.

6. Decompose into partial fractions:

$$\frac{5x - 4}{(x - 2)(x + 1)}$$

7. Solve for x:
(A) $y = 10^x$ (B) $y = \ln x$

8. Simplify:
(A) $(2e^x)^3$ (B) $\dfrac{e^{3x}}{e^{-2x}}$

9. Solve for x exactly. Do not use a calculator.
(A) $\log_3 x = 2$ (B) $\log_3 81 = x$
(C) $\log_x 4 = -2$

10. Solve for x to 3 significant digits.
(A) $10^x = 2.35$ (B) $e^x = 87{,}500$
(C) $\log x = -1.25$ (D) $\ln x = 2.75$

B

11. The function f subtracts the square root of the domain element from 3 times the natural log of the domain element. Write an algebraic definition of f.

12. Write a verbal description of the function $f(x) = 100e^{0.5x} - 50$.

13. If $P(x) = 2x^3 - 5x^2 + 3x + 2$, find $P(\frac{1}{2})$ using the remainder theorem and synthetic division.

14. Which of the following is a factor of

$$P(x) = x^{25} - x^{20} + x^{15} + x^{10} - x^5 + 1$$

(A) $x - 1$ (B) $x + 1$

15. Let $P(x) = x^4 - 8x^2 + 3$.
(A) Graph $P(x)$ and describe the graph verbally, including the number of x intercepts, the number of turning points, and the left and right behavior.
(B) Use the bisection method to approximate the largest x intercept to one decimal place.

16. Let $P(x) = x^4 + 2x^3 - 20x^2 - 30$.
(A) Find the smallest positive and the largest negative integers that, by Theorem 2 in Section 3-3, are upper and lower bounds, respectively, for the real zeros of $P(x)$.
(B) Use the bisection method to approximate the largest real zero of $P(x)$ to two decimal places.

(C) Use a graphing utility to approximate the real zeros of $P(x)$ to two decimal places.

17. Let $P(x) = 2x^4 - 9x^3 + 10x^2 + x - 4$. Find $Q(x)$ and R such that $P(x) = (x - 2)\,Q(x) + R$. What is $P(2)$?

18. Find all zeros (rational, irrational, and imaginary) exactly for $P(x) = 4x^3 - 20x^2 + 29x - 15$.

19. Find all zeros (rational, irrational, and imaginary) exactly for $P(x) = x^4 + 5x^3 + x^2 - 15x - 12$, and factor $P(x)$ into linear factors.

20. Solve $x^3 + 36 \leq 7x^2$. Write answers in inequality and interval notation.

21. Approximate all real solutions to two decimal places: $x^3 + 4x - 20 = 0$

22. Decompose into partial fractions:

$$\frac{3x^2 - x + 1}{x(x + 1)^2}$$

23. Decompose into partial fractions:

$$\frac{x^2 + x - 2}{x^3 - x^2 + x}$$

24. Let $f(x) = \dfrac{2x + 8}{x + 2}$.
(A) Find the domain and the intercepts for f.
(B) Find the vertical and horizontal asymptotes for f.
(C) Sketch the graph of f. Draw vertical and horizontal asymptotes with dashed lines.

Solve Problems 25–34 for x exactly. Do not use a calculator.

25. $2^{x^2} = 4^{x+4}$

26. $2x^2e^{-x} + xe^{-x} = e^{-x}$

27. $e^{\ln x} = 2.5$

28. $\log_x 10^4 = 4$

29. $\log_9 x = -\frac{3}{2}$

30. $\ln(x + 4) - \ln(x - 4) = 2\ln 3$

31. $\ln(2x^2 + 2) = 2\ln(2x - 4)$

32. $\log x + \log(x + 15) = 2$

33. $\log(\ln x) = -1$

34. $4(\ln x)^2 = \ln x^2$

Solve Problems 35–39 for x to 3 significant digits.

35. $x = \log_3 41$

36. $\ln x = 1.45$

37. $4(2^x) = 20$

38. $10e^{-0.5x} = 1.6$

39. $\dfrac{e^x - e^{-x}}{e^x + e^{-x}} = \dfrac{1}{2}$

Graph each function in Problems 40–43.

Check Problems 40–43 with a graphing utility.

40. $y = 3^{1-x}$

41. $f(x) = \ln(2 - x)$

42. $A(t) = 100e^{-0.3t}$

43. $y = -2e^{-x} + 3$

44. If the graph of $y = \ln x$ is reflected in the line $y = x$, the graph of the function $y = e^x$ is obtained. Discuss the functions that are obtained by reflecting the graph of $y = \ln x$ in the x axis and in the y axis.

45. (A) Explain why the equation $e^{-x} = \ln x$ has exactly one solution.

(B) Use a graphing utility to approximate the solution of the equation to two decimal places.

C

46. If $P(x)$ is a fourth-degree polynomial with integer coefficients and if i is a zero of $P(x)$, can $P(x)$ have any irrational zeros? Explain.

47. Let $P(x) = x^4 + 9x^3 - 500x^2 + 20{,}000$.
(A) Find the smallest positive integer multiple of 10 and the largest negative integer multiple of 10 that, by Theorem 2 in Section 3-3, are upper and lower bounds, respectively, for the real zeros of $P(x)$.
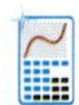
(B) Approximate the real zeros of $P(x)$ to two decimal places.

48. Graph f and indicate any horizontal, vertical, or oblique asymptotes with dashed lines:

$$f(x) = \frac{x^2 + 4x + 8}{x + 2}$$

49. Find a polynomial of lowest degree with leading coefficient 1 that has zeros -1 (multiplicity 2), 0 (multiplicity 3), $3 + 5i$, and $3 - 5i$. Leave the answer in factored form. What is the degree of the polynomial?

50. Find all zeros (rational, irrational, and imaginary) exactly for

$$P(x) = x^5 - 4x^4 + 3x^3 + 10x^2 - 10x - 12$$

and factor $P(x)$ into linear factors.

51. Find rational roots exactly and irrational roots to two decimal places for

$$P(x) = x^5 + 4x^4 + x^3 - 11x^2 - 8x + 4$$

52. Decompose into partial fractions:

$$\frac{x^2 - 4x + 11}{x^4 - x^3 + x^2 - 3x + 2}$$

53. Let $f(x) = 3 \ln (x - 2)$.
(A) Find $f^{-1}(x)$.
(B) Find the domain and range of f and f^{-1}.
(C) Graph f, f^{-1}, and $y = x$ on the same coordinate system.

Check by graphing f, f^{-1}, and $y = x$ in a squared window on a graphing utility.

54. Use natural logarithms to solve for n:

$$A = P\frac{(1 + i)^n - 1}{i}$$

55. Solve $\ln y = 5x + \ln A$ for y. Express the answer in a form that is free of logarithms.

56. Solve $y = \dfrac{e^x - 2e^{-x}}{2}$ for x:

APPLICATIONS

57. Shipping. A mailing service provides customers with rectangular shipping containers. The length plus the girth of one of these containers is 10 feet (see the figure). If the end of the container is square and the volume is 8 cubic feet, find the dimensions. Find rational solutions exactly and irrational solutions to one decimal place.

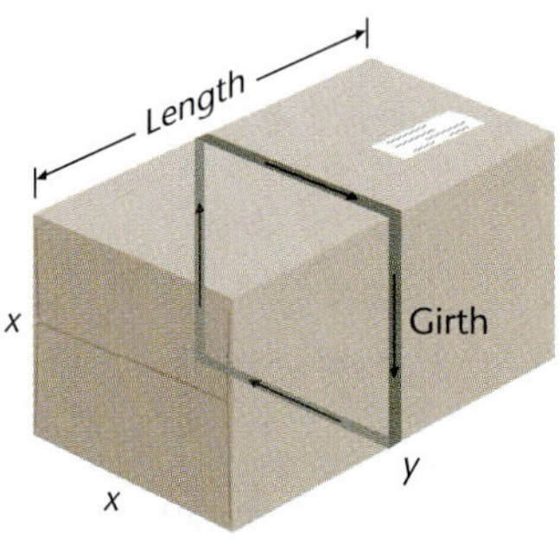

58. Geometry. The diagonal of a rectangle is 2 feet longer than one of the sides, and the area of the rectangle is 6 square feet. Find the dimensions of the rectangle. Find rational solutions exactly and irrational solutions to one decimal place.

59. Population Growth. If the Democratic Republic of Congo has a population of about 40 million people and a doubling time of 22 years, find the population in:
(A) 5 years (B) 30 years
Compute answers to 3 significant digits.

60. Compound Interest. How long will it take money invested in an account earning 7% compounded annually to double? Use the annual compounding growth model $P = P_0(1 + r)^t$, and compute the answer to 3 significant digits.

61. Compound Interest. Repeat Problem 60 using the continuous compound interest model $P = P_0e^{rt}$.

62. Earthquakes. If the 1906, and 1989 San Francisco earthquakes registered 8.3 and 7.1, respectively, on the Richter scale, how many times more powerful was the 1906 earthquake than the 1989 earthquake? Use the formula $M = \frac{2}{3} \log (E/E_0)$, where $E_0 = 10^{4.40}$ joules, and compute the answer to one decimal place.

63. Sound. If the decibel level at a rock concert is 88, find the intensity of the sound at the concert. Use the formula $D = 10 \log (I/I_0)$, where $I_0 = 10^{-12}$ watt per square meter, and compute the answer to 2 significant digits.

TRIGONOMETRIC FUNCTIONS

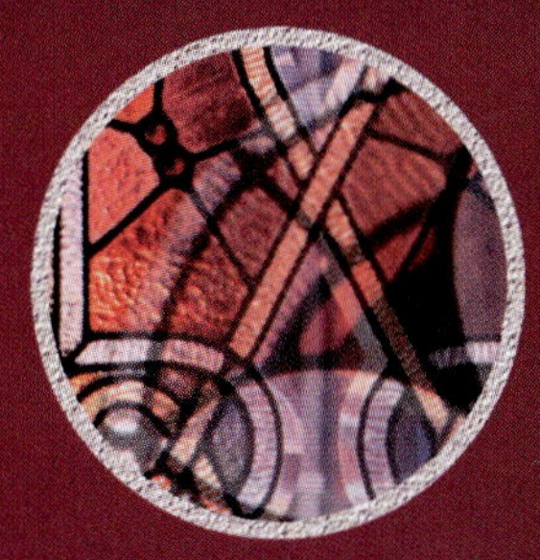

CHAPTER 5

Trigonometric functions seem to have had their origins with the Greeks' investigation of the indirect measurement of distances and angles in the "celestial sphere." (The ancient Egyptians had used some elementary geometry to build the pyramids and remeasure lands flooded by the Nile, but neither they nor the ancient Babylonians had developed the concept of angle measure.) The word "trigonometry," based on the Greek words for *triangle measurement,* was first used as the title for a text by the German mathematician Pitiscus in A.D. 1600.

Originally, the trigonometric functions were restricted to angles and their applications to the indirect measurement of angles and distances. These functions gradually broke free of these restrictions, and modern applications range over many types of problems that have little to do with angles—applications involving sound, light, and electrical waves; business cycles; and planetary motion.

Our approach to the subject does not follow historical lines in that we first introduce trigonometric functions (circular functions) with real number domains; then we define trigonometric functions with angle domains.

SECTION 5-1 The Wrapping Function

- Definition of the Wrapping Function
- Exact Values for Particular Real Numbers
- The Wrapping Function Is Not One-to-One

• Definition of the Wrapping Function

The important *circular function* definitions presented in the next section are based on a function W, called the *wrapping function,* whose domain is the set of real numbers and whose range is the set of points on the *unit circle.* By the **unit circle** we mean the circle of radius 1 with center at the origin of a rectangular coordinate system. Its equation is $u^2 + v^2 = 1$ (see Fig. 1).*

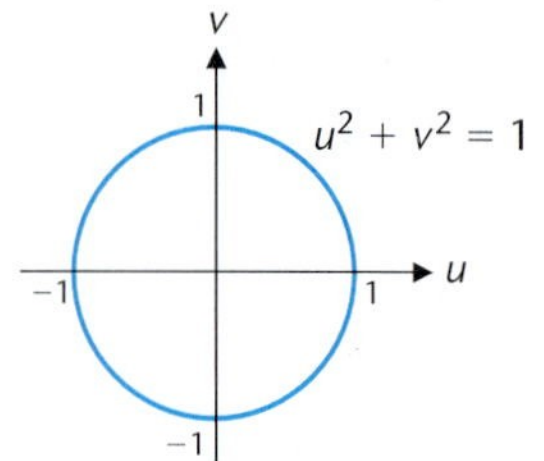

FIGURE 1 Unit circle.

In defining the **wrapping function,** we "wrap" a real number line with origin at (1, 0) around the unit circle—the positive real axis is wrapped counterclockwise, and the negative real axis is wrapped clockwise. In this way, each real number on the real line is paired with a unique point, called a **circular point,** on the unit circle, as shown in Figure 2.

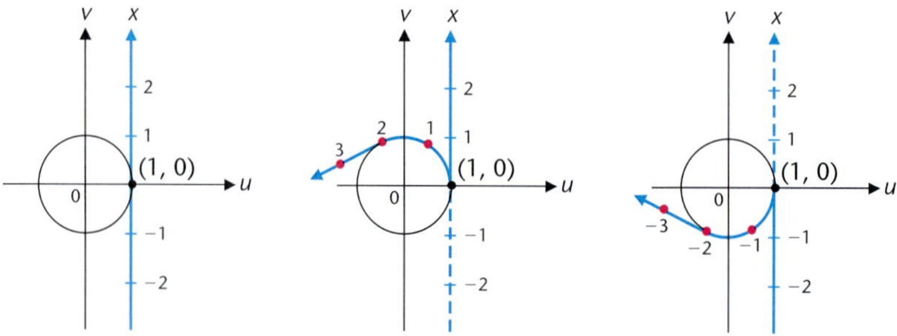

FIGURE 2 The wrapping function.

*We use the variables u and v instead of x and y so that x can be used without ambiguity as an independent variable in defining the wrapping and circular functions.

To locate the circular point associated with a number such as 37 or -105, the number line is wrapped many times around the circle.

An equivalent way of pairing real numbers with points on the unit circle is to think in terms of *arc length*. To find the circular point P associated with the real number x, we start at $A(1, 0)$ and move $|x|$ units along the unit circle, counterclockwise if x is positive and clockwise if x is negative. The length of arc AP is $|x|$ (see Fig. 3).

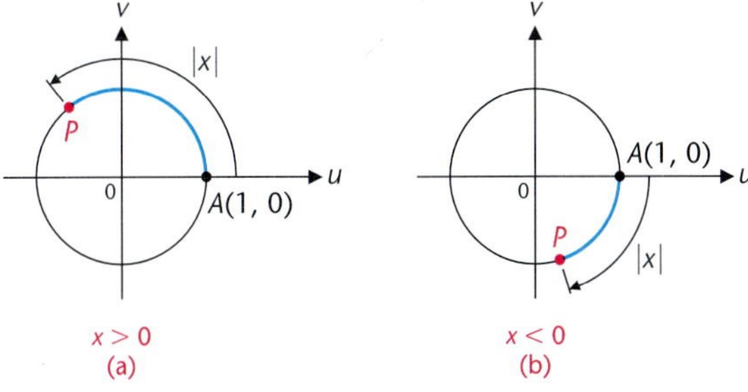

FIGURE 3 The wrapping function and arc length.

It is important to be able to find the coordinates (a, b) of the circular point P associated with a given real number x so that we can write $W(x) = (a, b)$. In general, this is difficult and requires the use of a calculator. However, for certain real numbers, integer multiples of $\pi/6$, $\pi/4$, $\pi/3$, and $\pi/2$, we can find the exact coordinates of the corresponding circular points using simple geometric properties of a circle.

• Exact Values for Particular Real Numbers

We start our investigation by finding the circumference of the unit circle. Since radius $r = 1$, the circumference is

$$2\pi r = 2\pi(1) = 2\pi \quad \text{Circumference of the unit circle}$$

One-fourth, one-half, and three-fourths of the circumference are, respectively, $\pi/2$, π, and $3\pi/2$. The circular points corresponding to these real numbers are on the coordinate axes, and hence, their coordinates are easily determined (see Fig. 4).

$$W(0) = (1, 0)$$

$$W\left(\frac{\pi}{2}\right) = (0, 1)$$

$$W(\pi) = (-1, 0)$$

$$W\left(\frac{3\pi}{2}\right) = (0, -1)$$

$$W(2\pi) = (1, 0)$$

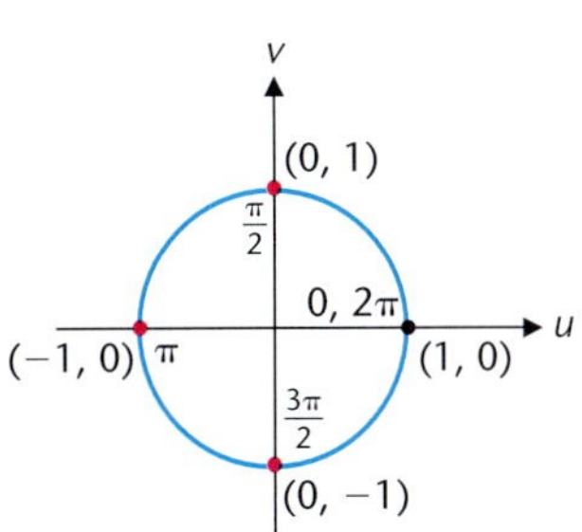

FIGURE 4 Circular points on the coordinate axes.

Following the same procedure, we can find the coordinates of *any* circular point on a coordinate axis—that is, for any circular point corresponding to a real number that is an integer multiple of $\pi/2$.

EXAMPLE 1 **Finding the Coordinates of Circular Points**

Find the coordinates of the circular points:

(A) $W(-\pi/2)$ (B) $W(5\pi/2)$

Solution (A) Starting at (1, 0), we go one-fourth the way around the unit circle in a clockwise direction (see Fig. 4). Thus,

$$W\left(\frac{-\pi}{2}\right) = (0, -1)$$

(B) Starting at (1, 0) and proceeding counterclockwise, we count quarter-circle steps, $\pi/2$, $2\pi/2$, $3\pi/2$, $4\pi/2$, and ending at $5\pi/2$. Thus, the circular point is on the positive vertical axis, and we have

$$W\left(\frac{5\pi}{2}\right) = (0, 1)$$

Matched Problem 1 Find the coordinates of the circular points:

(A) $W(-\pi)$ (B) $W(3\pi)$

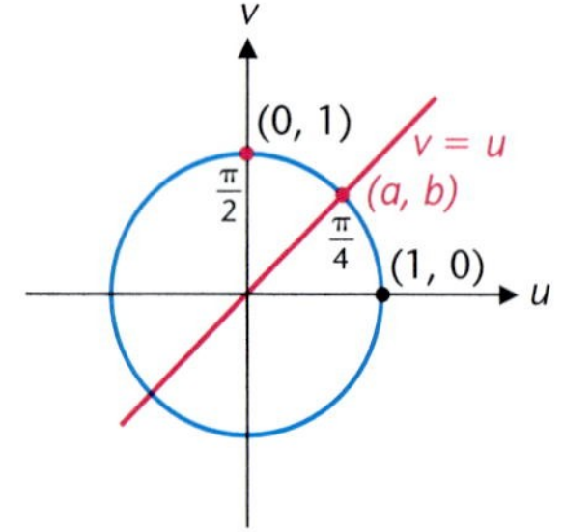

FIGURE 5 Circular point $W(\pi/4)$.

We now find the coordinates of the circular point $W(\pi/4)$. Since $\pi/4$ is one-half the arc joining (1, 0) and (0, 1), the circular point $W(\pi/4)$ must lie on the line $v = u$, as shown in Figure 5. Since $W(\pi/4)$ is on the line $v = u$ and on the circle $u^2 + v^2 = 1$, its coordinates (a, b) must satisfy both equations. That is,

$$a = b \qquad \text{and} \qquad a^2 + b^2 = 1$$

Substituting a for b in the second equation, we have

$$a^2 + a^2 = 1$$

$$2a^2 = 1$$

$$a^2 = \frac{1}{2}$$

$$a = \pm\frac{1}{\sqrt{2}}$$

$$a = \frac{1}{\sqrt{2}}$$ $a = -1/\sqrt{2}$ must be discarded, since $W(\pi/4)$ is in the first quadrant.

Using the first equation, we see that

$$b = a = \frac{1}{\sqrt{2}}$$

Therefore,

$$W\left(\frac{\pi}{4}\right) = \left(\frac{1}{\sqrt{2}}, \frac{1}{\sqrt{2}}\right)$$

Using symmetry properties of a circle—the unit circle is symmetric with respect to both axes and the origin—we can easily find the coordinates of any circular point that is reflected across the vertical axis, horizontal axis, or origin from $W(\pi/4)$.

EXAMPLE 2 Finding the Coordinates of Circular Points

Find the coordinates of the circular points:

(A) $W(5\pi/4)$ (B) $W(-\pi/4)$

Solutions (A) Starting at (1, 0) and counting in one-eighth circle steps counterclockwise ($\pi/4$, $2\pi/4$, $3\pi/4$, $4\pi/4$, $5\pi/4$), we find ourselves in the third quadrant on the circle halfway between $(-1, 0)$ and $(0, -1)$, as indicated in Figure 6. Using symmetry with respect to the origin, we have

$$W\left(\frac{5\pi}{4}\right) = \left(-\frac{1}{\sqrt{2}}, -\frac{1}{\sqrt{2}}\right)$$

FIGURE 6

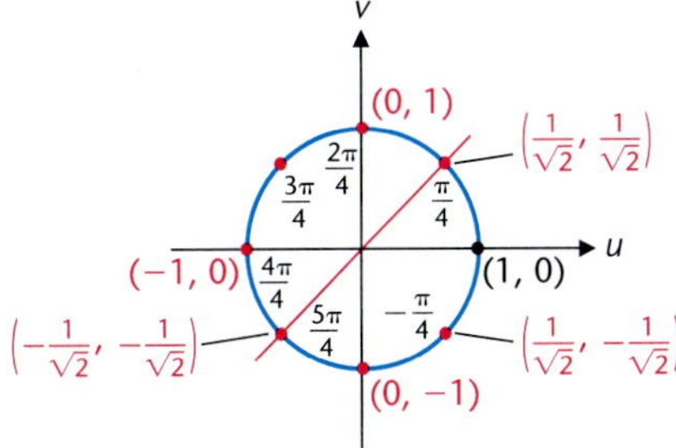

(B) Starting at (1, 0), we proceed one-eighth the way around the unit circle in a clockwise direction and end up in the fourth quadrant on the circle halfway between $(0, -1)$ and $(1, 0)$, as indicated in Figure 6. Using symmetry with respect to the horizontal axis, we see that

$$W\left(-\frac{\pi}{4}\right) = \left(\frac{1}{\sqrt{2}}, -\frac{1}{\sqrt{2}}\right)$$

Matched Problem 2

Find the coordinates of the circular points:

(A) $W(3\pi/4)$ (B) $W(-7\pi/4)$

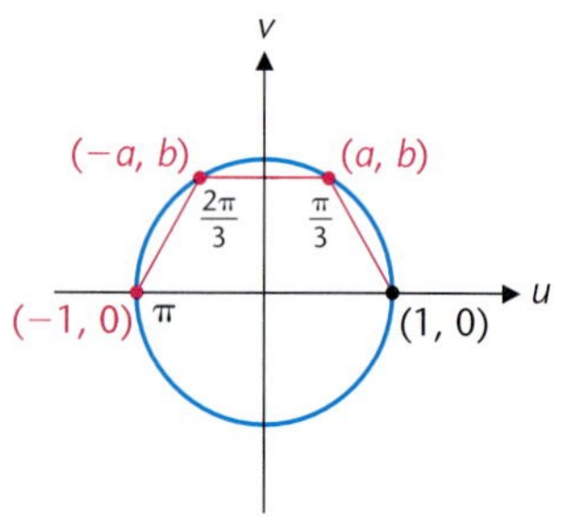

FIGURE 7 Circular point $W(\pi/3)$.

We continue our investigation by finding the coordinates of the circular point $W(\pi/3)$. Referring to Figure 7, we divide the upper semicircle from $(1, 0)$ to $(-1, 0)$ into thirds. The circular points $W(\pi/3)$ and $W(2\pi/3)$ are symmetric with respect to the v axis; hence, if $W(\pi/3)$ is given coordinates (a, b), then $W(2\pi/3)$ must have coordinates $(-a, b)$. The chord joining $W(2\pi/3)$ and $W(\pi/3)$ is thus $2a$ units long. Using the distance formula (see Section 2-1), we find the length of the chord joining $W(0)$ and $W(\pi/3)$ to be given by $\sqrt{(a-1)^2 + b^2}$. The two chords are equal in length, since congruent arcs are opposite congruent chords on the same circle. Thus,

$$\sqrt{(a-1)^2 + b^2} = 2a$$

Squaring both sides, we obtain

$$(a-1)^2 + b^2 = 4a^2$$

$$a^2 - 2a + 1 + b^2 = 4a^2$$

$$a^2 + b^2 - 2a + 1 = 4a^2$$

$$1 - 2a + 1 = 4a^2 \qquad a^2 + b^2 = 1 \text{ (Why?)}$$

$$4a^2 + 2a - 2 = 0$$

$$2a^2 + a - 1 = 0$$

$$(2a - 1)(a + 1) = 0$$

$$a = \tfrac{1}{2} \qquad \text{or} \qquad a = -1$$

$$a = \tfrac{1}{2} \qquad a = -1 \text{ must be discarded. (Why?)}$$

Substitute $a = \frac{1}{2}$ into $a^2 + b^2 = 1$ and solve for b:

$$\left(\frac{1}{2}\right)^2 + b^2 = 1$$

$$b^2 = \frac{3}{4}$$

$$b = \pm\frac{\sqrt{3}}{2}$$

$$b = \frac{\sqrt{3}}{2} \qquad b = -\sqrt{3}/2 \text{ must be discarded. (Why?)}$$

Thus,

$$W\left(\frac{\pi}{3}\right) = \left(\frac{1}{2}, \frac{\sqrt{3}}{2}\right)$$

Proceeding in a similar manner, or using symmetry with respect to the line $v = u$, we can obtain

$$W\left(\frac{\pi}{6}\right) = \left(\frac{\sqrt{3}}{2}, \frac{1}{2}\right)$$

The key results from the above discussion for the first quadrant are summarized in Figure 8.

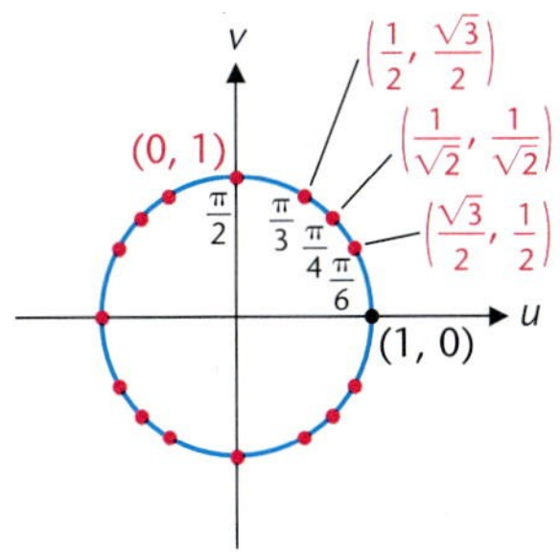

FIGURE 8 Coordinates of Key Circular Points

It is important that you memorize these first-quadrant relationships.

EXPLORE-DISCUSS 1 An effective **memory aid** for recalling the coordinates of the key circular points in Figure 8 can be created by writing the coordinates of the circular points $W(0)$, $W(\pi/6)$, $W(\pi/4)$, $W(\pi/3)$, and $W(\pi/2)$, keeping this order, in a form where each numerator is the square root of an appropriate number and each denominator is 2. For example, $W(0) = (1, 0) = (\sqrt{4}/2, \sqrt{0}/2)$. Describe the pattern that results.

The reason for memorizing the coordinates of key circular points in the first quadrant is that by using these, along with the symmetry of the unit circle, we can find the coordinates of *any* circular point that corresponds to *any* integer multiple of $\pi/6$, $\pi/4$, $\pi/3$, and $\pi/2$.

EXAMPLE 3 Finding Coordinates of Circular Points

Find the coordinates of the circular points:

(A) $W(5\pi/6)$ (B) $W(-2\pi/3)$

Solutions (A) Note that $5\pi/6$ is $\pi/6$ less than $\pi = 6\pi/6$. Locate $5\pi/6$ in the second quadrant and use Figure 8 and symmetry with respect to the vertical axis to find $W(5\pi/6)$. See Figure 9.

$$W\left(\frac{5\pi}{6}\right) = \left(-\frac{\sqrt{3}}{2}, \frac{1}{2}\right)$$

FIGURE 9

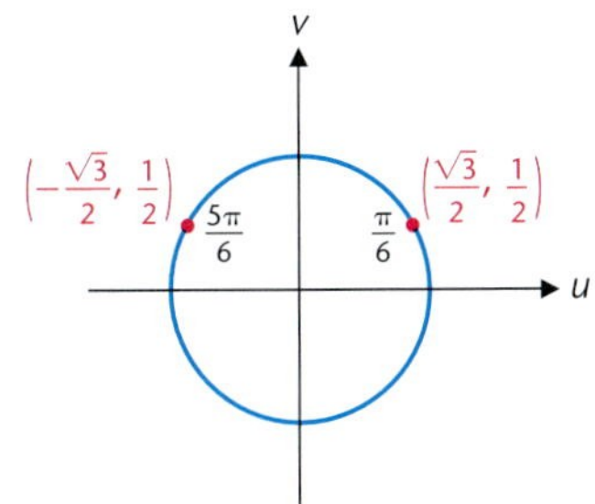

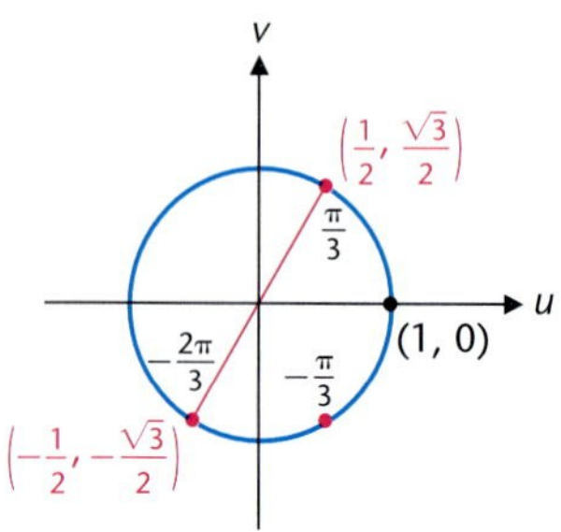

FIGURE 10

(B) Locate $-2\pi/3$ in the third quadrant and use Figure 8 and symmetry with respect to the origin to find $W(-2\pi/3)$. See Figure 10.

$$W\left(-\frac{2\pi}{3}\right) = \left(-\frac{1}{2}, -\frac{\sqrt{3}}{2}\right)$$

Matched Problem 3 Find the coordinates of the circular points:

(A) $W(5\pi/3)$ (B) $W(-7\pi/6)$

• The Wrapping Function Is Not One-to-One

It is easy to see that the wrapping function is not a one-to-one function. Each domain value, a real number, corresponds to exactly one range value, a point on the unit circle. However, each range value, a point on the unit circle, corresponds to infinitely many domain values, real numbers. For example, we see that

$$W\left(\frac{\pi}{2}\right) = (0, 1)$$

That is, exactly one range value corresponds to the domain value $\pi/2$. But how many domain values correspond to the range value (0, 1)? Every time we go around the unit circle 2π units in either direction from (0, 1), we return to the same point. Thus, if we are asked to solve

$$W(x) = (0, 1)$$

we have to write

$$x = \frac{\pi}{2} + 2k\pi \qquad k \text{ any integer}$$

and there are infinitely many domain values of W that correspond to the range value (0, 1). In general:

Theorem 1 **A Wrapping Function Property**

For all real numbers x,

$$W(x) = W(x + 2k\pi) \qquad k \text{ any integer*}$$

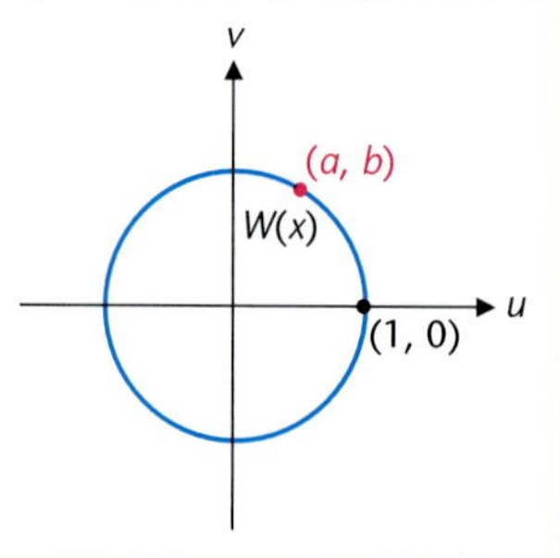

*Think of a point P moving around the unit circle in either direction. Every time P covers a distance of 2π, the circumference of the circle, it is back at the point where it started.

We will have more to say about the implications of this important property of the wrapping function in subsequent sections.

EXPLORE-DISCUSS 2 (A) Solve the circular point equation $W(x) = (0, -1)$, $-2\pi \le x \le 2\pi$.

(B) Describe an expression that would represent all solutions to $W(x) = (0, -1)$.

Answers to Matched Problems

1. (A) $(-1, 0)$ (B) $(-1, 0)$
2. (A) $(-1/\sqrt{2}, 1/\sqrt{2})$ (B) $(1/\sqrt{2}, 1/\sqrt{2})$
3. (A) $(1/2, -\sqrt{3}/2)$ (B) $(-\sqrt{3}/2, 1/2)$

EXERCISE 5-1

A

In Problems 1–12, find the coordinates for each circular point. Sketch your own figures—don't look back in the text.

1. $W(\pi)$ **2.** $W(0)$ **3.** $W(6\pi)$

4. $W(3\pi)$ **5.** $W(-\pi)$ **6.** $W(-5\pi)$

7. $W(3\pi/2)$ **8.** $W(\pi/2)$ **9.** $W(-\pi/2)$

10. $W(-3\pi/2)$ **11.** $W(11\pi/2)$ **12.** $W(-15\pi/2)$

B

In Problems 13–24, find the coordinates for each circular point. Sketch your own figures—don't look back in the text.

13. $W(\pi/3)$ **14.** $W(2\pi/3)$ **15.** $W(-\pi/4)$

16. $W(-5\pi/4)$ **17.** $W(-5\pi/6)$ **18.** $W(\pi/4)$

19. $W(13\pi/4)$ **20.** $W(7\pi/6)$ **21.** $W(19\pi/2)$

22. $W(-7\pi)$ **23.** $W(-13\pi/6)$ **24.** $W(-11\pi/4)$

Determine the signs of a and b for the coordinates (a, b) of each circular point indicated in Problems 25–34. First determine the quadrant in which each circular point lies. [Note: $\pi/2 \approx 1.57$, $\pi \approx 3.14$, $3\pi/2 \approx 4.71$, and $2\pi \approx 6.28$.]

25. $W(3.2)$ **26.** $W(-0.9)$ **27.** $W(-5.7)$

28. $W(1.3)$ **29.** $W(\sqrt{2})$ **30.** $W(\sqrt{3})$

31. $W(12.5)$ **32.** $W(-37)$ **33.** $W(-27)$

34. $W(-7.8)$

In Problems 35–38, for each equation find all solutions for $0 \le x < 2\pi$, then write an expression that represents all solutions for the equation without any restrictions on x.

35. $W(x) = (1, 0)$ **36.** $W(x) = (-1, 0)$

37. $W(x) = (-1/\sqrt{2}, 1/\sqrt{2})$ **38.** $W(x) = (1/\sqrt{2}, -1/\sqrt{2})$

39. Describe in words why $W(x) = W(x + 4\pi)$ for every real number x.

40. Describe in words why $W(x) = W(x - 6\pi)$ for every real number x.

C

If $W(x) = (a, b)$, indicate whether the statements in Problems 41–46 are true (T) or false (F). Sketching figures should help you decide.

41. $W(x + \pi) = (-a, -b)$ **42.** $W(x + \pi) = (a, b)$

43. $W(-x) = (-a, b)$ **44.** $W(-x) = (a, -b)$

45. $W(x + 2\pi) = (a, b)$ **46.** $W(x + 2\pi) = (-a, -b)$

In Problems 47–52, find all solutions x, $-2\pi \le x \le 2\pi$, such that:

47. $W(x) = \left(\frac{1}{\sqrt{2}}, \frac{1}{\sqrt{2}}\right)$ **48.** $W(x) = \left(\frac{\sqrt{3}}{2}, \frac{1}{2}\right)$

49. $W(x) = \left(-\frac{1}{2}, \frac{\sqrt{3}}{2}\right)$ **50.** $W(x) = \left(\frac{1}{2}, -\frac{\sqrt{3}}{2}\right)$

51. $W(x) = \left(-\frac{\sqrt{3}}{2}, -\frac{1}{2}\right)$ **52.** $W(x) = \left(-\frac{1}{\sqrt{2}}, \frac{1}{\sqrt{2}}\right)$

Find all solutions to each equation in Problems 53 and 54.

53. $W(x) = W(\pi/4)$ **54.** $W(x) = W(2\pi/3)$

SECTION 5-2 Circular Functions

- Definition of the Circular Functions
- Exact Values for Particular Real Numbers
- Sign Properties
- Basic Identities
- Calculator Evaluation

• Definition of the Circular Functions

In Section 5-1 we saw that the wrapping function W pairs each real number x with an ordered pair of real numbers (a, b), the coordinates of the circular point $W(x)$. We use this association to construct the six **circular functions,** also called **trigonometric functions*: sine, cosine, tangent, cotangent, secant,** and **cosecant.** The values of these functions for a real number x are denoted by **sin *x*, cos *x*, tan *x*, cot *x*, sec *x*,** and **csc *x*,** respectively. These values are expressed in terms of the coordinates of the circular point $W(x) = (a, b)$ as indicated in Definition 1.

DEFINITION 1

Circular Functions

If x is a real number and (a, b) are the coordinates of the circular point $W(x)$, then

$$\sin x = b \qquad \csc x = \frac{1}{b} \quad b \neq 0$$

$$\cos x = a \qquad \sec x = \frac{1}{a} \quad a \neq 0$$

$$\tan x = \frac{b}{a} \quad a \neq 0 \qquad \cot x = \frac{a}{b} \quad b \neq 0$$

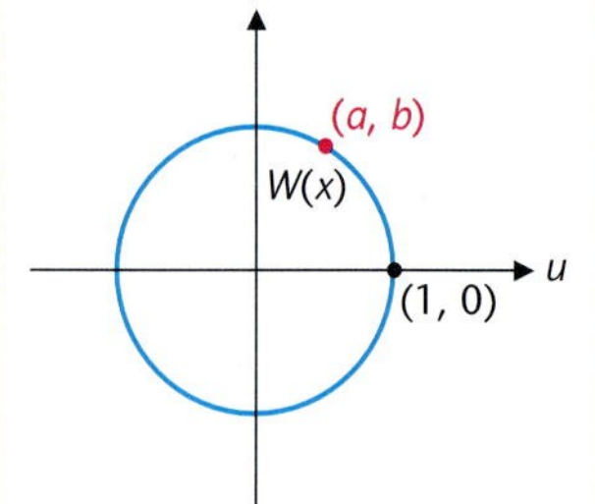

The domain of both the sine and cosine functions is the set of real numbers R. The range of both the sine and cosine functions is $[-1, 1]$. This is the set of numbers assumed by b, for sine, and a, for cosine, as the circular point (a, b) moves around the unit circle. The domain of cosecant is the set of real numbers x such that b in $W(x) = (a, b)$ is not 0. Similar restrictions are made on the domains of the other three circular functions. We will have more to say about the domains and ranges of all six circular functions in subsequent sections.

• Exact Values for Particular Real Numbers

Using the results in Section 5-1, we can evaluate any one of the six circular functions exactly, when it exists, for integer multiples of the real numbers $\pi/6$, $\pi/4$, $\pi/3$, and $\pi/2$. Figure 8 in Section 5-1, which you should have memorized, and symmetry properties of the unit circle are central to the process. Later in this section we will show

*Strictly speaking, the term *trigonometric* is used when we are dealing with angle domains and *circular* is used when we are dealing with real number domains. We will not insist on this distinction and often, as is the convention, use *trigonometric* for both.

how a calculator can be used to evaluate the circular functions to 8 or more significant digits for arbitrary real numbers. You might ask why we don't go directly to the calculator. The answer is that there are many situations in which it is more desirable to work with exact forms, if available, than the corresponding decimal approximations that are produced by a calculator.

EXAMPLE 1 **Exact Evaluation of Circular Functions**

Evaluate each circular function exactly for $x = \pi/3$.

Solution From Section 5-1 we know that

$$W\left(\frac{\pi}{3}\right) = \left(\frac{1}{2}, \frac{\sqrt{3}}{2}\right) \quad \text{[See Fig. 1]}$$

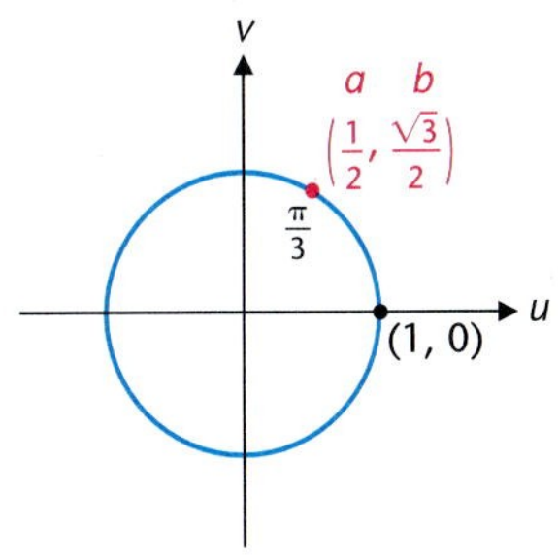

FIGURE 1

Thus,

$$\sin\frac{\pi}{3} = b = \frac{\sqrt{3}}{2} \qquad \csc\frac{\pi}{3} = \frac{1}{b} = \frac{1}{\sqrt{3}/2} = \frac{2}{\sqrt{3}}$$

$$\cos\frac{\pi}{3} = a = \frac{1}{2} \qquad \sec\frac{\pi}{3} = \frac{1}{a} = \frac{1}{\frac{1}{2}} = 2$$

$$\tan\frac{\pi}{3} = \frac{b}{a} = \frac{\sqrt{3}/2}{\frac{1}{2}} = \sqrt{3} \qquad \cot\frac{\pi}{3} = \frac{a}{b} = \frac{\frac{1}{2}}{\sqrt{3}/2} = \frac{1}{\sqrt{3}}$$

Matched Problem 1 Evaluate each circular function exactly for $x = \pi/6$.

EXAMPLE 2 **Exact Evaluation of Circular Functions**

Evaluate exactly:

(A) $\sin(5\pi/6)$ (B) $\cot(-\pi)$ (C) $\sec(-2\pi/3)$ (D) $\tan(7\pi/4)$

Solutions Sketch a figure for each part, then use Figure 8 in Section 5-1 and symmetry properties of the unit circle.

(A)

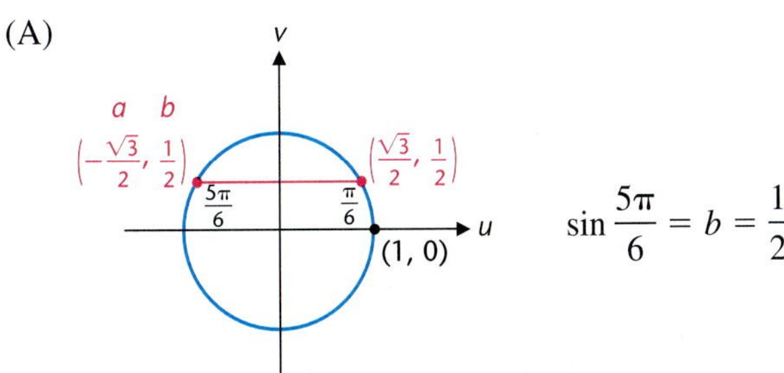

$$\sin\frac{5\pi}{6} = b = \frac{1}{2}$$

(B)

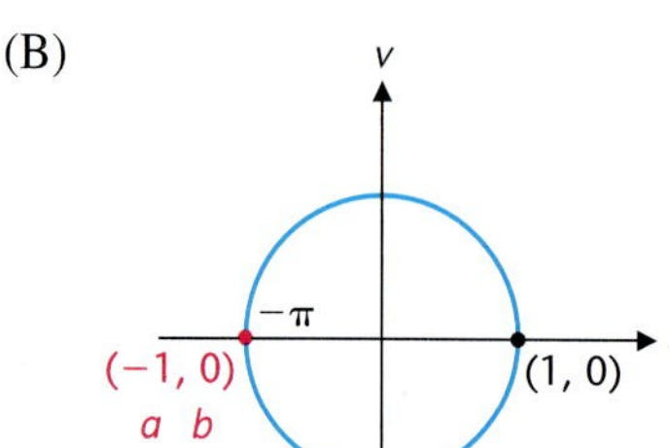

$$\cot(-\pi) = \frac{a}{b} = \frac{-1}{0}$$

Not defined

(C)

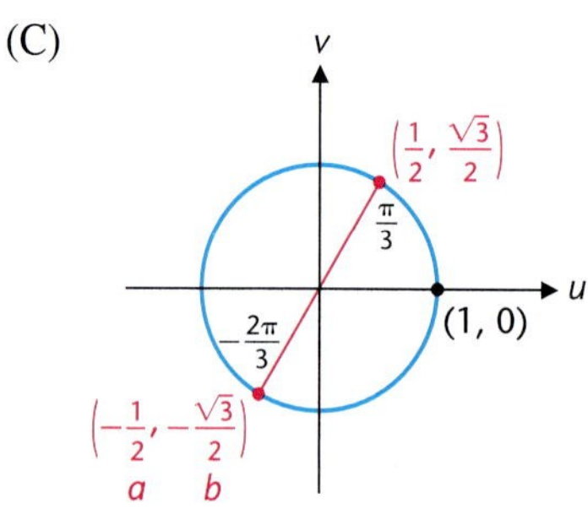

$$\sec\left(-\frac{2\pi}{3}\right) = \frac{1}{a} = \frac{1}{-\frac{1}{2}} = -2$$

(D)

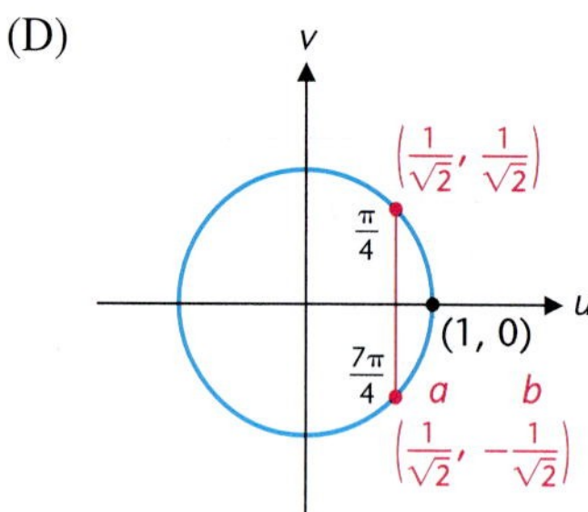

$$\tan\frac{7\pi}{4} = \frac{b}{a} = \frac{-1/\sqrt{2}}{1/\sqrt{2}} = -1$$

Matched Problem 2 Evaluate exactly:

(A) $\cos(5\pi/6)$ (B) $\sin(-3\pi/4)$ (C) $\csc 3\pi$ (D) $\tan(-\pi/3)$

• Sign Properties

As a circular point $W(x)$ moves from quadrant to quadrant, its coordinates (a, b) undergo sign changes. Hence, the circular functions also undergo sign changes. It is important to know the sign of each circular function in each quadrant. Table 1 shows the sign behavior for each function. It is not necessary to memorize Table 1, since the sign of each function for each quadrant is easily determined from its definition (which *should* be memorized).

TABLE 1 Sign Properties

Circular function	Sign in quadrant I	II	III	IV
$\sin x = b$	+	+	−	−
$\csc x = 1/b$	+	+	−	−
$\cos x = a$	+	−	−	+
$\sec x = 1/a$	+	−	−	+
$\tan x = b/a$	+	−	+	−
$\cot x = a/b$	+	−	+	−

II: a b (−, +) I: a b (+, +)

III: a b (−, −) IV: a b (+, −)

EXPLORE-DISCUSS 1

(A) Determine the quadrant in which both $\tan x < 0$ and $\sin x > 0$. Draw diagrams, and explain your reasoning.

(B) Determine the quadrant in which both $\cos x > 0$ and $\cot x < 0$. Draw diagrams, and explain your reasoning.

• Basic Identities

Returning to the definitions of the circular functions and noting that

$$\sin x = b \qquad \text{and} \qquad \cos x = a$$

we can obtain the following useful relationships among the six circular functions:

$$\csc x = \frac{1}{b} = \frac{1}{\sin x} \tag{1}$$

$$\sec x = \frac{1}{a} = \frac{1}{\cos x} \tag{2}$$

$$\cot x = \frac{a}{b} = \frac{1}{b/a} = \frac{1}{\tan x} \tag{3}$$

$$\tan x = \frac{b}{a} = \frac{\sin x}{\cos x} \tag{4}$$

$$\cot x = \frac{a}{b} = \frac{\cos x}{\sin x} \tag{5}$$

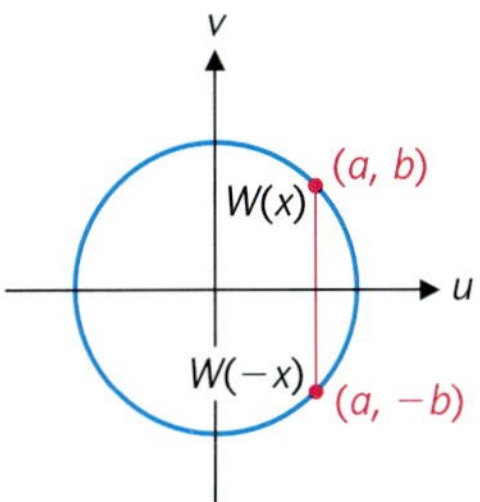

FIGURE 2 Symmetry property.

Because the circular points $W(x)$ and $W(-x)$ are symmetric with respect to the horizontal axis (Fig. 2), we have the following sign properties:

$$\sin(-x) = -b = -\sin x \tag{6}$$

$$\cos(-x) = a = \cos x \tag{7}$$

$$\tan(-x) = \frac{-b}{a} = -\frac{b}{a} = -\tan x \tag{8}$$

Finally, because $(a, b) = (\cos x, \sin x)$ is on the unit circle $u^2 + v^2 = 1$, it follows that

$$(\cos x)^2 + (\sin x)^2 = 1$$

which is usually written as

$$\sin^2 x + \cos^2 x = 1 \tag{9}$$

where $\sin^2 x$ and $\cos^2 x$ are concise ways of writing $(\sin x)^2$ and $(\cos x)^2$, respectively.

CAUTION

$$(\sin x)^2 \neq \sin x^2$$

$$(\cos x)^2 \neq \cos x^2$$

Equations (1)–(9) are called **basic identities.** They hold true for all replacements of x by real numbers for which both sides of an equation are defined. These basic identities must be memorized along with the definitions of the six circular functions, since the material is used extensively in developments that follow. Note that most of Chapter 6 is devoted to trigonometric identities.

We summarize the basic identities for convenient reference in Theorem 1.

Theorem 1 **Basic Trigonometric Identities**

For x any real number (in all cases restricted so that both sides of an equation are defined):

Reciprocal Identities

(1) $\csc x = \dfrac{1}{\sin x}$ (2) $\sec x = \dfrac{1}{\cos x}$ (3) $\cot x = \dfrac{1}{\tan x}$

Quotient Identities

(4) $\tan x = \dfrac{\sin x}{\cos x}$ (5) $\cot x = \dfrac{\cos x}{\sin x}$

Identities for Negatives

(6) $\sin(-x) = -\sin x$ (7) $\cos(-x) = \cos x$ (8) $\tan(-x) = -\tan x$

Pythagorean Identity

(9) $\sin^2 x + \cos^2 x = 1$

EXAMPLE 3 Using Basic Identities

Use the basic identities to find the values of the other five circular functions given $\sin x = -\frac{1}{2}$ and $\tan x > 0$.

Solution We first note that the circular point $W(x)$ is in quadrant III, since that is the only quadrant in which $\sin x < 0$ and $\tan x > 0$. We next find $\cos x$ using identity (9):

$$\sin^2 x + \cos^2 x = 1 \qquad \text{Pythagorean identity (9)}$$
$$(-\tfrac{1}{2})^2 + \cos^2 x = 1$$
$$\cos^2 x = \tfrac{3}{4}$$
$$\cos x = -\frac{\sqrt{3}}{2} \qquad \text{Since } W(x) \text{ is in quadrant III}$$

Now, since we have values for $\sin x$ and $\cos x$, we can find values for the other four circular functions using identities (1), (2), (4), and (5):

$$\csc x = \frac{1}{\sin x} = \frac{1}{-\frac{1}{2}} = -2 \qquad \text{Reciprocal identity (1)}$$
$$\sec x = \frac{1}{\cos x} = \frac{1}{-\sqrt{3}/2} = -\frac{2}{\sqrt{3}} \qquad \text{Reciprocal identity (2)}$$
$$\tan x = \frac{\sin x}{\cos x} = \frac{-\frac{1}{2}}{-\sqrt{3}/2} = \frac{1}{\sqrt{3}} \qquad \text{Quotient identity (4)}$$
$$\cot x = \frac{\cos x}{\sin x} = \frac{-\sqrt{3}/2}{-\frac{1}{2}} = \sqrt{3} \qquad \text{Quotient identity (5)}$$

[*Note:* We could also use identity (3).]

In Example 3 it is important to note that we were able to find the values of the other five circular functions without finding x.

Matched Problem 3 Use the basic identities to find the values of the other five circular functions given $\cos x = 1/\sqrt{2}$ and $\cot x < 0$.

EXPLORE-DISCUSS 2 Given the conditions on x in Example 3: $\sin x = -\frac{1}{2}$ and $\tan x > 0$. Find, using basic identities and the results in Example 3, each of the following:

(A) $\sin(-x)$ (B) $\sec(-x)$ (C) $\tan(-x)$

Verbally justify each step in your solution process.

• Calculator Evaluation

Evaluating circular functions for real numbers other than integer multiples of $\pi/6$, $\pi/4$, $\pi/3$, and $\pi/2$ is difficult without the use of a calculator. Using advanced mathematics, calculators are internally programmed to evaluate these functions automatically to an accuracy of 8 or more significant digits.

If you look at the function keys on your calculator, you will find three keys labeled

SIN COS TAN

These keys are used to evaluate the sine, cosine, and tangent functions directly. A careful look at the function keys on your calculator also will reveal that there are no keys for cosecant, secant, and cotangent. Why is it not necessary to have these additional keys? Because of the reciprocal identities (1)–(3), we can use the function keys for sine, cosine, and tangent, along with the reciprocal function key

$\boxed{x^{-1}}$ or $\boxed{1/x}$

to obtain csc x, sec x, and cot x. Don't use the keys marked $\sin^{-1}$, $\cos^{-1}$, and $\tan^{-1}$ to evaluate csc, sec, and cot, respectively. You will see why in Section 5-9. Some examples should make the process of calculator evaluation of the circular functions clear.

CAUTION

Setting Calculator Mode: Before commencing with the examples and exercises, read the instruction book accompanying your calculator to determine how to put it in radian (rad) mode. It is in this mode that we can evaluate the circular functions for real numbers. (This process is justified in Section 5-4 when we discuss trigonometric functions with angle domains.) Forgetting to set the calculator in the correct mode before starting calculations involving circular or trigonometric functions is a frequent cause of error when using a calculator.

EXAMPLE 4 Calculator Evaluation

Evaluate to 4 significant digits using a calculator:

(A) sin 2 (B) tan (−1.612) (C) csc 3.2

Solutions

(A) sin 2 = 0.9093

(B) tan (−1.612) = 24.26

(C) csc 3.2 = −17.13

Matched Problem 4 Evaluate to 4 significant digits using a calculator:

(A) cos 4 (B) sec 1.605 (C) cot (−3.133)

EXPLORE-DISCUSS 3 Use a calculator to evaluate each of the following, and explain the results shown on the calculator:

(A) tan ($\pi/2$) (B) cot 0 (C) sec ($-\pi/2$)

Answers to Matched Problems

1. $\sin(\pi/6) = \frac{1}{2}$, $\cos(\pi/6) = \sqrt{3}/2$, $\tan(\pi/6) = 1/\sqrt{3}$, $\csc(\pi/6) = 2$, $\sec(\pi/6) = 2/\sqrt{3}$, $\cot(\pi/6) = \sqrt{3}$
2. (A) $-\sqrt{3}/2$ (B) $-1/\sqrt{2}$ (C) Not defined (D) $-\sqrt{3}$
3. $\sin x = -1/\sqrt{2}$, $\csc x = -\sqrt{2}$, $\sec x = \sqrt{2}$, $\tan x = -1$, $\cot x = -1$
4. (A) -0.6536 (B) -29.24 (C) 116.4

EXERCISE 5-2

Figure 8 in Section 5-1, the definition of the circular functions, and the basic identities should now be memorized. Work the problems in this exercise without looking back in the text. Draw lots of pictures, if necessary.

A

1. Write the value of each circular function in terms of the coordinates (a, b) of the circular point $W(x)$.
   (A) $\cos x$ (B) $\csc x$ (C) $\cot x$
   (D) $\sec x$ (E) $\tan x$ (F) $\sin x$

2. Given $W(x) = (a, b)$, identify each quantity using one of the circular function values $\sin x$, $\cos x$, and so on.
   (A) b (B) $1/a$ (C) b/a
   (D) $1/b$ (E) a (F) a/b

In Problems 3–20, find the exact value of each expression (if it exists) without the use of a calculator.

3. $\cos 0$ 4. $\sin 0$ 5. $\sin(\pi/6)$
6. $\cos(\pi/6)$ 7. $\sin(\pi/2)$ 8. $\cos(\pi/2)$
9. $\tan(\pi/3)$ 10. $\cos(\pi/3)$ 11. $\tan(\pi/2)$
12. $\cot 0$ 13. $\sec 0$ 14. $\cot(\pi/4)$
15. $\sec(\pi/4)$ 16. $\csc(\pi/3)$ 17. $\tan(\pi/4)$
18. $\tan 0$ 19. $\csc 0$ 20. $\cot(\pi/6)$

In Problems 21–26, in which quadrants must $W(x)$ lie so that:

21. $\cos x < 0$ 22. $\tan x > 0$ 23. $\sin x > 0$
24. $\sec x > 0$ 25. $\cot x < 0$ 26. $\csc x < 0$

Evaluate Problems 27–32 to 4 significant digits using a calculator.

27. $\tan 4.728$ 28. $\cos 3.167$ 29. $\sec(-1.489)$
30. $\csc(-13.25)$ 31. $\sin(-9.841)$ 32. $\cot 6.386$

B

In Problems 33–48, find the exact value of each expression (if it exists) without the use of a calculator.

33. $\tan(-\pi/2)$ 34. $\sec(3\pi/4)$ 35. $\csc(3\pi/2)$
36. $\sin(\pi/3)$ 37. $\cos(5\pi/3)$ 38. $\cot(-3\pi)$
39. $\sec(-\pi/4)$ 40. $\tan(7\pi/4)$ 41. $\sin(7\pi/6)$
42. $\cos(-5\pi/3)$ 43. $\cot(3\pi/4)$ 44. $\csc(-\pi/3)$
45. $\tan(-2\pi/3)$ 46. $\sec(-5\pi/2)$ 47. $\csc 5\pi$
48. $\cot(-13\pi/6)$

In Problems 49–52, find the value of each to one significant digit. Use only the accompanying figure, Definition 1, and a calculator as necessary for multiplication and division. Check your results by evaluating each directly on a calculator.

49. (A) $\sin 0.4$ (B) $\cos 0.4$ (C) $\tan 0.4$
50. (A) $\sin 0.8$ (B) $\cos 0.8$ (C) $\cot 0.8$
51. (A) $\sec 2.2$ (B) $\tan 5.9$ (C) $\cot 3.8$
52. (A) $\csc 2.5$ (B) $\cot 5.6$ (C) $\tan 4.3$

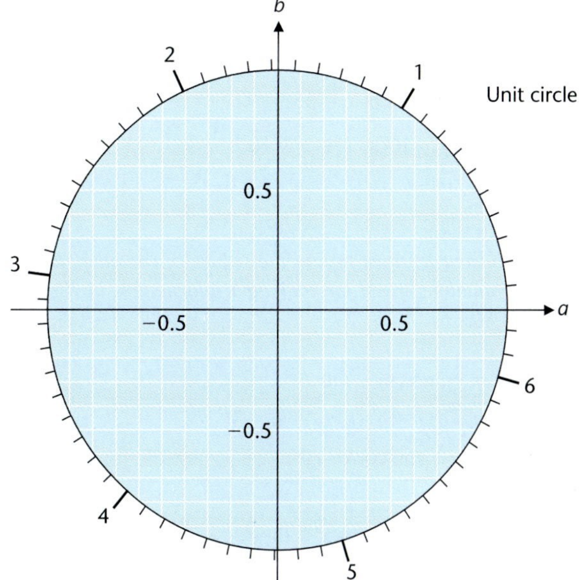

In Problems 53–56, in which quadrants are the statements true and why?

53. $\sin x < 0$ and $\cot x < 0$

54. $\cos x > 0$ and $\tan x < 0$

55. $\cos x < 0$ and $\sec x > 0$

56. $\sin x > 0$ and $\csc x < 0$

For which values of x, $0 \le x \le 2\pi$, is each of Problems 57–62 not defined?

57. $\cos x$ **58.** $\sin x$ **59.** $\tan x$

60. $\cot x$ **61.** $\sec x$ **62.** $\csc x$

How does the functional value indicated in Problems 63 and 64 vary as x varies over the indicated intervals? [Hint: Draw a unit circle and note that $W(x) = (a, b) = (\cos x, \sin x)$.]

63. $\sin x$:
(A) $[0, \pi/2]$ (B) $[\pi/2, \pi]$
(C) $[\pi, 3\pi/2]$ (D) $[3\pi/2, 2\pi]$

64. $\cos x$:
(A) $[0, \pi/2]$ (B) $[\pi/2, \pi]$
(C) $[\pi, 3\pi/2]$ (D) $[3\pi/2, 2\pi]$

Complete Problems 65–68 to 4 significant digits using a calculator.

65. $\sin(\cos 0.3157)$ **66.** $\cos(\tan 5.183)$

67. $\cos[\csc(-1.408)]$ **68.** $\sec[\cot(-3.566)]$

Use appropriate identities to solve Problems 69–74.

69. Find $\sin(-x)$ if $\sin x = -\frac{1}{3}$.

70. Find $\cos(-x)$ if $\cos x = -\frac{1}{3}$.

71. Find $\tan(-x)$ if $\tan x = -\sqrt{5}$.

72. Find $\sec(-x)$ if $\sec x = 50$.

73. Find $\cot(-x)$ if $\cot x = 25$.

74. Find $\csc(-x)$ if $\csc x = -\sqrt{10}$.

C

Use the basic identities to find the values of the other five circular functions given the indicated information in Problems 75–80.

75. $\cos x = \dfrac{1}{2}$ and $\tan x < 0$

76. $\sin x = \dfrac{\sqrt{3}}{2}$ and $\cot x < 0$

77. $\sin x = -\dfrac{1}{\sqrt{2}}$ and $\cos x < 0$

78. $\sec x = 2$ and $\sin x < 0$

79. $\tan x = \sqrt{3}$ and $\sin x < 0$

80. $\cot x = -1$ and $\sin x > 0$

In Problems 81–86, find the smallest positive x (in terms of π) for which:

81. $\cos x = -1$

82. $\sin x = -\dfrac{\sqrt{3}}{2}$

83. $\cot x = -\sqrt{3}$

84. $\tan x = -1$

85. $\sec x = -\dfrac{2}{\sqrt{3}}$

86. $\csc x = -\sqrt{2}$

In Problems 87 and 88, fill in the blanks citing the appropriate identity (1)–(9).

87.

Statement	Reason
$\cot^2 x + 1 = \left(\dfrac{\cos x}{\sin x}\right)^2 + 1$	(A) ________
$= \dfrac{\cos^2 x}{\sin^2 x} + 1$	Algebra
$= \dfrac{\cos^2 x + \sin^2 x}{\sin^2 x}$	Algebra
$= \dfrac{1}{\sin^2 x}$	(B) ________
$= \left(\dfrac{1}{\sin x}\right)^2$	Algebra
$= \csc^2 x$	(C) ________

88.

Statement	Reason
$\tan^2 x + 1 = \left(\dfrac{\sin x}{\cos x}\right)^2 + 1$	(A) ________
$= \dfrac{\sin^2 x}{\cos^2 x} + 1$	Algebra
$= \dfrac{\sin^2 x + \cos^2 x}{\cos^2 x}$	Algebra
$= \dfrac{1}{\cos^2 x}$	(B) ________
$= \left(\dfrac{1}{\cos x}\right)^2$	Algebra
$= \sec^2 x$	(C) ________

APPLICATIONS 

If an n-sided regular polygon is inscribed in a circle of radius r, then it can be shown that the area of the polygon is given by

$$A = \frac{1}{2}nr^2 \sin \frac{2\pi}{n}$$

Compute each area exactly and then to 4 significant digits using a calculator if the area is not an integer.

89. $n = 12, r = 5$ meters

90. $n = 4, r = 3$ inches

91. $n = 3, r = 4$ inches

92. $n = 8, r = 10$ centimeters

Approximating π. *Problems 93 and 94 refer to a sequence of numbers generated as follows:*

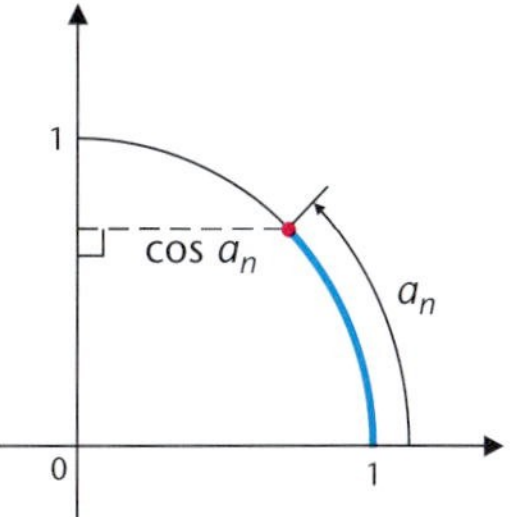

$$\begin{aligned} &a_1 \\ &a_2 = a_1 + \cos a_1 \\ &a_3 = a_2 + \cos a_2 \\ &\vdots \\ &a_{n+1} = a_n + \cos a_n \end{aligned}$$

93. Let $a_1 = 0.5$, and compute the first five terms of the sequence to six decimal places and compare the fifth term with $\pi/2$ computed to six decimal places.

94. Repeat Problem 93, starting with $a_1 = 1$.

SECTION 5-3 Angles and Their Measure

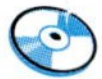

- Angles
- Degree and Radian Measure
- From Degrees to Radians, and Vice Versa

In this section, we will introduce the idea of angle and two measures of angles, *degree* and *radian.*

• Angles

The study of trigonometry depends on the concept of angle. An **angle** is formed by rotating (in a plane) a ray m, called the **initial side** of the angle, around its endpoint until it coincides with a ray n, called the **terminal side** of the angle. The common endpoint V of m and n is called the **vertex** (see Fig. 1). A counterclockwise rotation produces a **positive angle,** and a clockwise rotation produces a **negative angle,** as shown in Figure 2(a) and (b). The amount of rotation in either direction is not restricted.

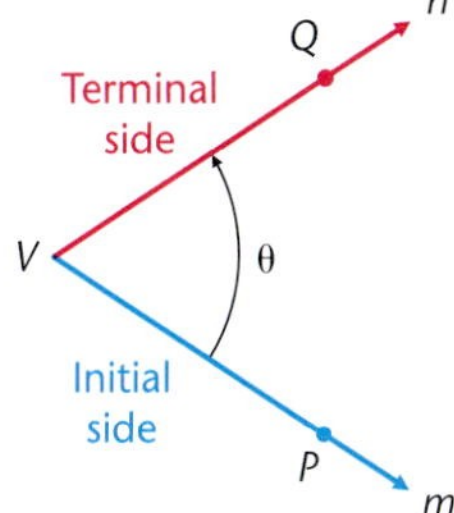

FIGURE 1 Angle θ or angle PVQ or $\angle V$.

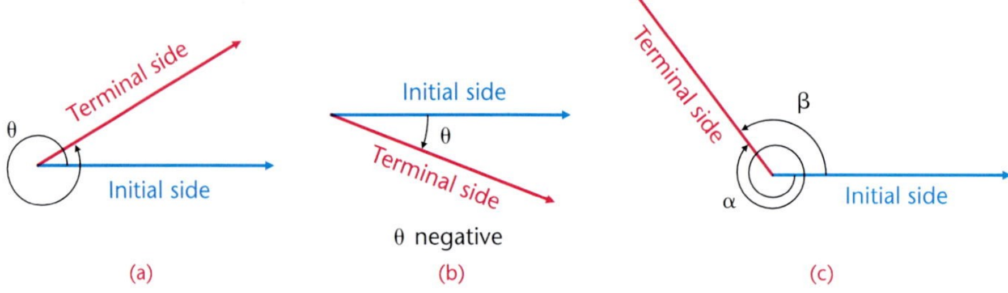

FIGURE 2 Angles and rotation.

Two different angles may have the same initial and terminal sides, as shown in Figure 2(c). Such angles are said to be **coterminal.**

An angle in a rectangular coordinate system is said to be in **standard position** if its vertex is at the origin and the initial side is along the positive x axis. If the terminal side of an angle in standard position lies along a coordinate axis, the angle is said to be a **quadrantal angle.** If the terminal side does not lie along a coordinate axis, then the angle is often referred to in terms of the quadrant in which the terminal side lies (Fig. 3).

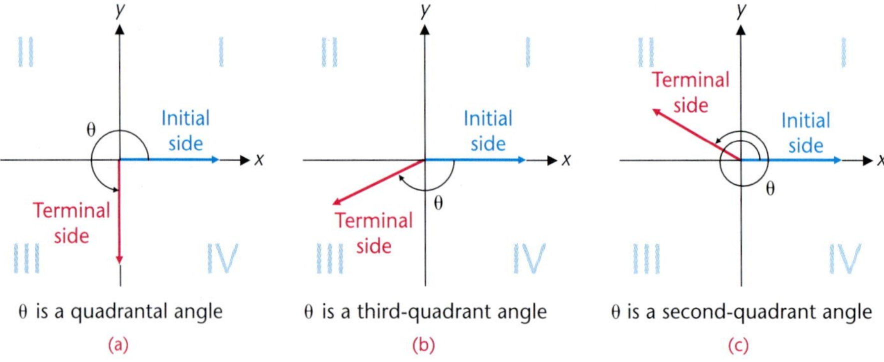

FIGURE 3 Angles in standard positions.

• Degree and Radian Measure

Just as line segments are measured in centimeters, meters, inches, or miles, angles are measured in different units. The two most commonly used units for angle measure are *degree* and *radian.*

DEFINITION 1

Degree Measure

An angle formed by one complete rotation is said to have a measure of 360 degrees (360°). An angle formed by $\frac{1}{360}$ of a complete rotation is said to have a measure of **1 degree** (1°). The symbol ° denotes degrees.

Certain angles have special names. Figure 4 shows a **straight angle,** a **right angle,** an **acute angle,** and an **obtuse angle.**

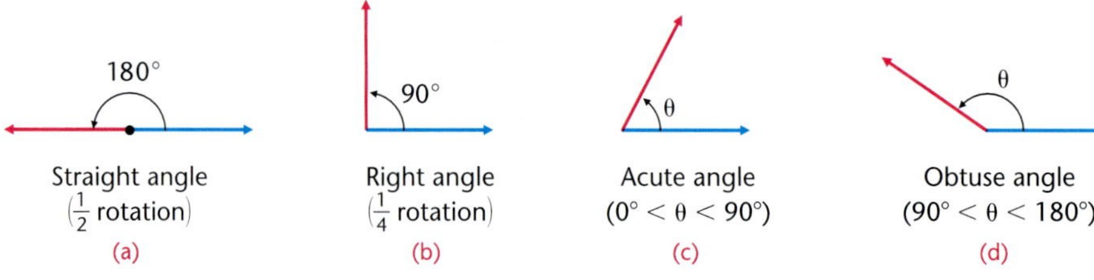

FIGURE 4 Types of angles.

Two positive angles are **complementary** if their sum is 90°; they are **supplementary** if their sum is 180°.

A degree can be divided further using decimal notation. For example, 42.75° represents an angle of degree measure 42 plus three-quarters of 1 degree. A degree can also be divided further using minutes and seconds just as an hour is divided into minutes and seconds. Each degree is divided into 60 equal parts called **minutes,** and each minute is divided into 60 equal parts called **seconds.** Symbolically, minutes are represented by ′ and seconds by ″. Thus,

$$12°23'14''$$

is a concise way of writing 12 degrees, 23 minutes, and 14 seconds.

Decimal degrees (DD) are useful in some instances and degrees–minutes–seconds (DMS) are useful in others. You should be able to go from one form to the other, as demonstrated in Example 1.

Conversion Accuracy

If an angle is measured to the nearest second, the converted decimal form should not go beyond three decimal places, and vice versa.

EXAMPLE 1 From DMS to DD and Back

(A) Convert 21°47′12″ to decimal degrees.
(B) Convert 105.183° to degree–minute–second form.

Solution

(A) $21°47'12'' = \left(21 + \frac{47}{60} + \frac{12}{3600}\right)^{\circ} = 21.787°$

(B) $$\begin{aligned} 105.183° &= 105°(0.183\cdot 60)' \\ &= 105°10.98' \\ &= 105°10'(0.98\cdot 60)'' \\ &= 105°10'59'' \end{aligned}$$

Matched Problem 1

(A) Convert 193°17′34″ to DD form.
(B) Convert 237.615° to DMS form.

Some scientific and some graphing calculators can convert the DD and DMS forms automatically, but the process differs significantly among the various types of calculators. Check your owner's manual for your particular calculator. The conversion methods outlined in Example 1 show you the reasoning behind the process and are sometimes easier to use than the "automatic" methods for some calculators.

Degree measure of angles is used extensively in engineering, surveying, and navigation. Another unit of angle measure, called the *radian,* is better suited for certain mathematical developments, scientific work, and engineering applications.

DEFINITION 2 **Radian Measure**

If the vertex of an angle θ is placed at the center of a circle with radius $r > 0$, and the length of the arc opposite θ on the circumference is s, then the **radian measure of θ** is given by

$$\theta = \frac{s}{r} \text{ radians}$$

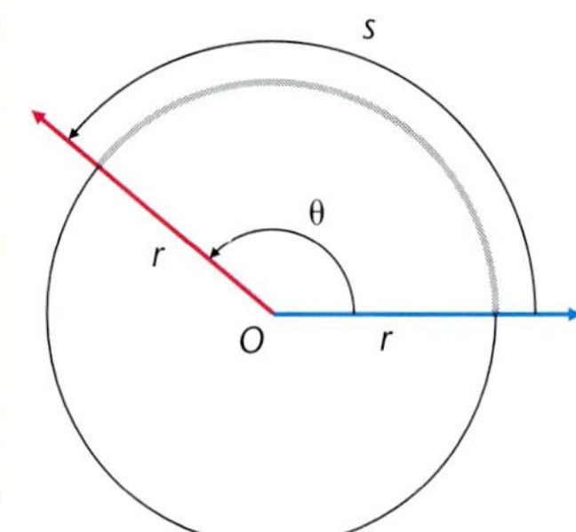

Also,

$$s = r\theta$$

If $s = r$, then

$$\theta = \frac{r}{r} = 1 \text{ radian}$$

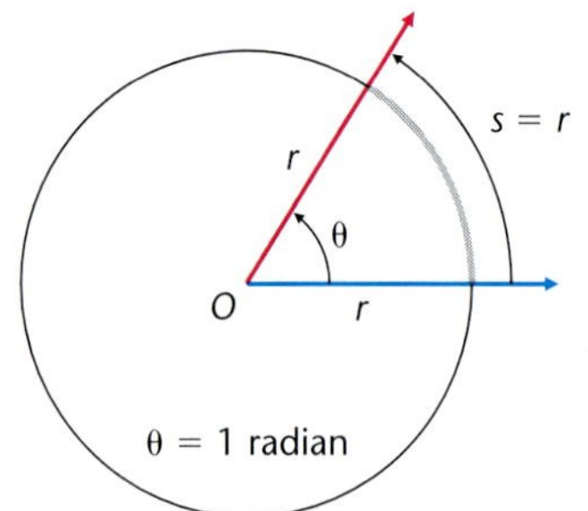

Thus, **one radian** is the size of the central angle of a circle that intercepts an arc the same length as the radius of the circle. [*Note:* s and r must be measured in the same units. Also note that θ is being used in two ways: as the name of an angle and as the measure of the angle. The context determines the choice. Thus, when we write $\theta = s/r$, we mean the radian measure of angle θ is s/r.]

EXAMPLE 2 Computing Radian Measure

What is the radian measure of a central angle θ opposite an arc of 24 meters in a circle of radius 6 meters?

Solution

$$\theta = \frac{s}{r} = \frac{24 \text{ meters}}{6 \text{ meters}} = 4 \text{ radians}$$

Matched Problem 2 What is the radian measure of a central angle θ opposite an arc of 60 feet in a circle of radius 12 feet?

Remark. *Radian measure is a unitless number.* The units in which the arc length and radius are measured cancel; hence, we are left with a "unitless," or pure, number. For this reason, the word "radian" is often omitted when we are dealing with the radian measure of angles unless a special emphasis is desired.

EXPLORE-DISCUSS 1 Discuss why the radian measure of an angle is independent of the size of the circle having the angle as a central angle.

• From Degrees to Radians, and Vice Versa

What is the radian measure of an angle of 180°? A central angle of 180° is subtended by an arc one-half the circumference of a circle. Thus, if C is the circumference of a circle, then one-half the circumference is given by

$$s = \frac{C}{2} = \frac{2\pi r}{2} = \pi r \qquad \text{and} \qquad \theta = \frac{s}{r} = \frac{\pi r}{r} = \pi \text{ rad}$$

Hence, 180° corresponds to π* rad. This is important to remember, since the radian measures of many special angles can be obtained from this correspondence. For example, 90° is 180°/2; therefore, 90° corresponds to $\pi/2$ rad.

EXPLORE-DISCUSS 2 Complete Table 1:

TABLE 1

Radians	$\pi/6$	$\pi/4$	$\pi/3$	$\pi/2$	π	$3\pi/2$	2π
Degrees				90	180		

*The constant π has a long and interesting history; a few important dates are listed below:

1650 B.C.	Rhind Papyrus	$\pi \approx \frac{256}{81} = 3.16049\ldots$
240 B.C.	Archimedes	$3\frac{10}{71} < \pi < 3\frac{1}{7}$ $(3.1408\ldots < \pi < 3.1428\ldots)$
A.D. 264	Liu Hui	$\pi \approx 3.14159$
A.D. 470	Tsu Ch'ung-chih	$\pi \approx \frac{355}{113} = 3.1415929\ldots$
A.D. 1674	Leibniz	$\pi = 4(1 - \frac{1}{3} + \frac{1}{5} - \frac{1}{7} + \frac{1}{9} - \frac{1}{11} + \ldots)$ $\approx 3.1415926535897932384626$ (This and other series can be used to compute π to any decimal accuracy desired.)
A.D. 1761	Johann Lambert	Showed π to be irrational (π as a decimal is nonrepeating and nonterminating)

The results in Table 1 are summarized in Figure 5 for easy reference. These correspondences and multiples of them will be used extensively in work that follows.

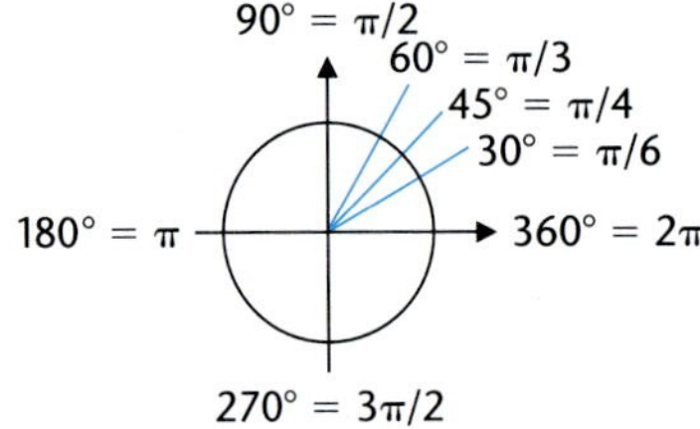

FIGURE 5 Radian–degree correspondences.

In general, the following proportion can be used to convert degree measure to radian measure, and vice versa.

Radian–Degree Conversion Formulas

$$\frac{\theta_{\text{deg}}}{180°} = \frac{\theta_{\text{rad}}}{\pi \text{ rad}} \qquad \text{Basic proportion}$$

$$\theta_{\text{deg}} = \frac{180°}{\pi \text{ rad}}\theta_{\text{rad}} \qquad \text{Radians to degrees}$$

$$\theta_{\text{rad}} = \frac{\pi \text{ rad}}{180°}\theta_{\text{deg}} \qquad \text{Degrees to radians}$$

[*Note:* The basic proportion is usually easier to remember. Also we will omit units in calculations until the final answer. If your calculator does not have a key labeled π, use $\pi \approx 3.14159$.]

Some scientific and graphing calculators can automatically convert radian measure to degree measure, and vice versa. Check your owner's manual for your particular calculator.

EXAMPLE 3 **Radian–Degree Conversions**

(A) Find the radian measure, exact and to 3 significant digits, of an angle of 75°.
(B) Find the degree measure, exact and to 4 significant digits, of an angle of 5 radians.
(C) Find the radian measure to two decimal places of an angle of 41°12′.

Solution (A)

$$\theta_{\text{rad}} = \frac{\pi \text{ rad}}{180°}\theta_{\text{deg}} = \frac{\pi}{180}(75) = \overset{\text{exact}}{\frac{5\pi}{12}} = \overset{\text{3 sig. dig.}}{1.31}$$

(B)

$$\theta_{\text{deg}} = \frac{180°}{\pi \text{ rad}}\theta_{\text{rad}} = \frac{180}{\pi}(5) = \overset{\text{exact}}{\frac{900}{\pi}} = \overset{\text{4 sig. dig.}}{286.5°}$$

```
75°
              1.31
5ʳ
             286.5
41°12'
             41.20
```

FIGURE 6 Automatic conversion.

(C) $41°12' = \left(41 + \dfrac{12}{60}\right)^\circ = 41.2°$ Change 41°12′ to DD first.

$$\theta_{\text{rad}} = \frac{\pi \text{ rad}}{180°}\,\theta_{\text{deg}} = \frac{\pi}{180}(41.2) = 0.72$$ To 2 dec. places

Figure 6 shows the three conversions above done automatically on a graphing calculator.

Matched Problem 3 (A) Find the radian measure, exact and to 3 significant digits, of an angle of 240°.
(B) Find the degree measure, exact and to 3 significant digits, of an angle of 1 radian.
(C) Find the radian measure to three decimal places of an angle of 125°23′.

Remark. We will write θ in place of θ_{deg} and θ_{rad} when it is clear from the context whether we are dealing with degree or radian measure.

EXAMPLE 4 Engineering

A belt connects a pulley of 2-inch radius with a pulley of 5-inch radius. If the larger pulley turns through 10 radians, through how many radians will the smaller pulley turn?

Solution We first sketch Figure 7.

FIGURE 7

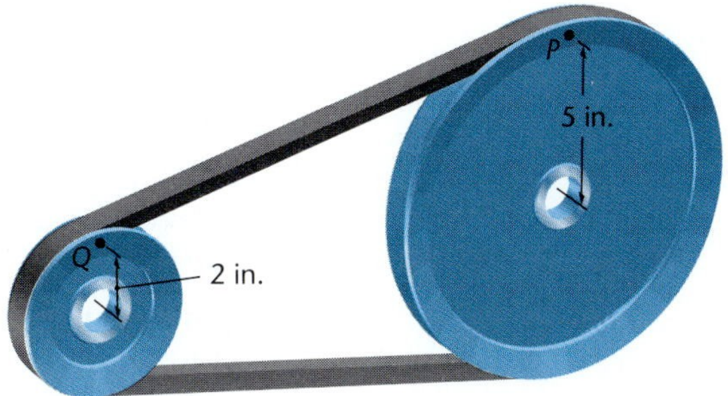

When the larger pulley turns through 10 radians, the point P on its circumference will travel the same distance (arc length) that point Q on the smaller circle travels. For the larger pulley:

$$\theta = \frac{s}{r}$$

$$s = r\theta = (5)(10) = 50 \text{ inches}$$

For the smaller pulley:

$$\theta = \frac{s}{r} = \frac{50}{2} = 25 \text{ radians}$$

Matched Problem 4 In Example 4, through how many radians will the larger pulley turn if the smaller pulley turns through 4 radians?

Answers to Matched Problems

1. (A) 193.293° (B) 237°36′54″ **2.** 5 rad

3. (A) $\frac{4\pi}{3} = 4.19$ (B) $\frac{180}{\pi} = 57.3°$ (C) 2.188 **4.** 1.6 rad

EXERCISE 5-3

In all problems, if angle measure is expressed by a number that is not in degrees, it is assumed to be in radians.

A

Find the degree measure of each of the angles in Problems 1–4, keeping in mind that an angle of one complete rotation corresponds to 360°.

1. $\frac{1}{9}$ rotation **2.** $\frac{1}{5}$ rotation

3. $\frac{3}{4}$ rotation **4.** $\frac{7}{6}$ rotations

Find the radian measure of a central angle θ opposite an arc s in a circle of radius r, where r and s are as given in Problems 5–8.

5. $r = 4$ centimeters, $s = 24$ centimeters

6. $r = 8$ inches, $s = 16$ inches

7. $r = 12$ feet, $s = 30$ feet

8. $r = 18$ meters, $s = 27$ meters

Find the radian measure of each angle in Problems 9–12, keeping in mind that an angle of one complete rotation corresponds to 2π radians.

9. $\frac{1}{8}$ rotation **10.** $\frac{1}{6}$ rotation

11. $\frac{3}{4}$ rotation **12.** $\frac{11}{8}$ rotations

B

Find the exact radian measure, in terms of π, of each angle in Problems 13–16.

13. 30°, 60°, 90°, 120°, 150°, 180°

14. 60°, 120°, 180°, 240°, 300°, 360°

15. −45°, −90°, −135°, −180°

16. −90°, −180°, −270°, −360°

Find the exact degree measure of each angle in Problems 17–20.

17. $\frac{\pi}{3}, \frac{2\pi}{3}, \pi, \frac{4\pi}{3}, \frac{5\pi}{3}, 2\pi$ **18.** $\frac{\pi}{6}, \frac{\pi}{3}, \frac{\pi}{2}, \frac{2\pi}{3}, \frac{5\pi}{6}, \pi$

19. $-\frac{\pi}{2}, -\pi, -\frac{3\pi}{2}, -2\pi$ **20.** $-\frac{\pi}{4}, -\frac{\pi}{2}, -\frac{3\pi}{4}, -\pi$

In Problems 21–26, determine whether the statement is true or false. If true, explain why. If false, give a counterexample.

21. If two positive angles are supplementary, then one is obtuse and the other is acute.

22. If two positive angles are complementary, then both are acute.

23. If two angles in standard position are coterminal, then they have the same measure.

24. If two angles in standard position have the same measure, then they are coterminal.

25. If an angle is a right angle, then it is a quadrantal angle.

26. If an angle is a quadrantal angle, then it is a right angle.

Convert each angle in Problems 27–30 to decimal degrees to three decimal places.

27. 5°51′33″ **28.** 14°18′37″

29. 354°8′29″ **30.** 184°31′7″

Convert each angle in Problems 31–34 to degree–minute–second form.

31. 3.042° **32.** 49.715°

33. 403.223° **34.** 156.808°

Find the radian measure to three decimal places for each angle in Problems 35–38.

35. 18° **36.** 79°

37. 23°45′32″ **38.** 48°55′12″

Find the degree measure to two decimal places for each angle in Problems 39–42.

39. 1.52 **40.** 0.64

41. −0.83 **42.** −2.65

Indicate whether each angle in Problems 43–54 is a I, II, III, or IV quadrant angle or a quadrantal angle. All angles are in standard position in a rectangular coordinate system. (A sketch may be of help in some problems.)

43. 130° **44.** 97° **45.** −1.34

46. −4.75 **47.** $\frac{7\pi}{4}$ **48.** $\frac{8\pi}{3}$

49. $\frac{-5\pi}{2}$ **50.** $\frac{11\pi}{5}$ **51.** −835°

52. −630° **53.** 9.73 **54.** 24.14

C

Which angles in Problems 55–60 are coterminal with 120° if all angles are placed in standard position in a rectangular coordinate system?

55. −600° **56.** 240° **57.** 960°

58. 840° **59.** $\frac{2\pi}{3}$ **60.** $-\frac{8\pi}{3}$

Which angles in Problems 61–66 are coterminal with $-\frac{3\pi}{4}$ *if all angles are placed in standard position in a rectangular coordinate system?*

61. $\frac{11\pi}{4}$ **62.** $\frac{5\pi}{4}$ **63.** $\frac{13\pi}{4}$

64. $-\frac{19\pi}{4}$ **65.** −495° **66.** 855°

APPLICATIONS

67. Circumference of the Earth. The early Greeks used the proportion $s/C = \theta°/360°$, where s is an arc length on a circle, $\theta°$ is the degree measure of the corresponding central angle, and C is the circumference of the circle ($C = 2\pi r$). Eratosthenes (240 B.C.), in his famous calculation of the circumference of the Earth, reasoned as follows: He knew that at Syene (now Aswan) during the summer solstice the noon sun was directly overhead and shined on the water straight down a deep well. On the same day at the same time, 5,000 stadia (approx. 500 miles) due north in Alexandria, sun rays crossed a vertical pole at an angle of 7.5°, as indicated in the figure. Carry out Eratosthenes' calculation for the circumference of the Earth to the nearest thousand miles. (The current calculation for the equatorial circumference is 24,902 miles).

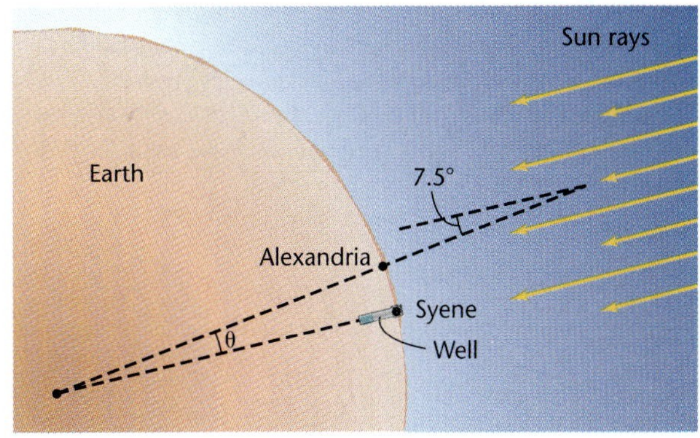

68. Circumference of the Earth. Repeat Problem 67 with the sun crossing the vertical pole in Alexandria at 7°12′.

69. Circumference of the Earth. In Problem 67, verbally explain how θ in the figure was determined.

70. Circumference of the Earth. Verbally explain how the radius, surface area, and volume of the Earth can be determined from the result of Problem 67.

71. Radian Measure. What is the radian measure of the larger angle made by the hands of a clock at 4:30? Express the answer exactly in terms of π.

72. Radian Measure. What is the radian measure of the smaller angle made by the hands of a clock at 1:30? Express the answer exactly in terms of π.

73. Engineering. Through how many radians does a pulley of 10-centimeter diameter turn when 10 meters of rope is pulled through it without slippage?

74. Engineering. Through how many radians does a pulley of 6-inch diameter turn when 4 feet of rope is pulled through it without slippage?

75. Astronomy. A line from the sun to the Earth sweeps out an angle of how many radians in 1 week? Assume the Earth's orbit is circular and there are 52 weeks in a year. Express the answer in terms of π and as a decimal to two decimal places.

76. Astronomy. A line from the center of the Earth to the equator sweeps out an angle of how many radians in 9 hours? Express the answer in terms of π and as a decimal to two decimal places.

★ **77. Engineering.** A trail bike has a front wheel with a diameter of 40 centimeters and a back wheel of diameter 60 centimeters. Through what angle in radians does the front wheel turn if the back wheel turns through 8 radians?

★ **78. Engineering.** In Problem 77, through what angle in radians will the back wheel turn if the front wheel turns through 15 radians?

The arc length on a circle is easy to compute if the corresponding central angle is given in radians and the radius of the circle is known ($s = r\theta$). If the radius of a circle is large and a central angle is small, then an arc length is often used to approximate the length of the corresponding chord, as shown in the figure. If an angle is given in degree measure, converting to radian measure first may be helpful in certain problems. This information will be useful in Problems 79–82.

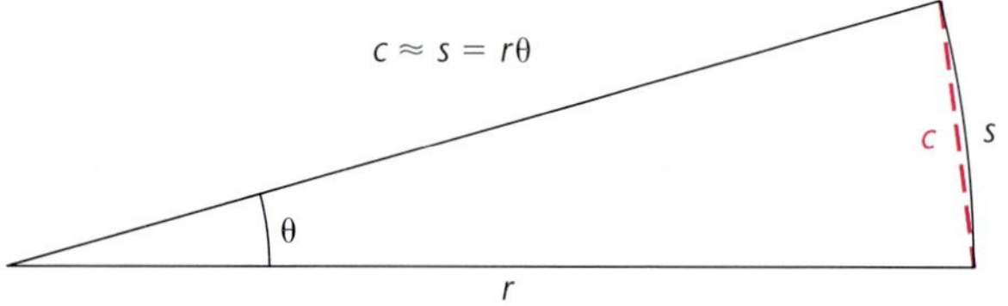

79. Astronomy. The sun is about 9.3×10^7 mi from the Earth. If the angle subtended by the diameter of the sun on the surface of the Earth is 9.3×10^{-3} rad, approximately what is the diameter of the sun to the nearest thousand miles in standard decimal notation?

80. Astronomy. The moon is about 381,000 kilometers from the Earth. If the angle subtended by the diameter of the moon on the surface of the Earth is 0.0092 rad, approximately what is the diameter of the moon to the nearest hundred kilometers?

81. Photography. The angle of view of a 1000-mm telephoto lens is 2.5°. At 750 ft, what is the width of the field of view to the nearest foot?

82. Photography. The angle of view of a 300-mm lens is 8°. At 500 ft, what is the width of the field of view to the nearest foot?

SECTION 5-4 Trigonometric Functions

- Definition of the Trigonometric Functions
- Calculator Evaluation of Trigonometric Functions
- Definition of the Trigonometric Functions—Alternate Form
- Exact Values for Special Angles and Real Numbers
- Summary of Special Angle Values

In this section we define trigonometric functions with angle domains, where angles can have either degree or radian measure. We also show how circular functions are related to trigonometric functions so that you will be able to move easily from one to the other, as needed.

• Definition of the Trigonometric Functions

We are now ready to define trigonometric functions with angle domains. Since we have already defined the circular functions with real number domains, we can take advantage of these results and define the trigonometric functions with angle domains in terms of the circular functions. To each of the six circular functions we associate a trigonometric function of the same name. If θ is an angle, in either radian or degree measure, we assign values to $\sin \theta$, $\cos \theta$, $\tan \theta$, $\csc \theta$, $\sec \theta$, and $\cot \theta$ as given in Definition 1.

DEFINITION 1 **Trigonometric Functions with Angle Domains**

If θ is an angle with radian measure x, then the value of each **trigonometric function** at θ is given by its value at the real number x.

Trigonometric Function		**Circular Function**
$\sin \theta$	=	$\sin x$
$\cos \theta$	=	$\cos x$
$\tan \theta$	=	$\tan x$
$\csc \theta$	=	$\csc x$
$\sec \theta$	=	$\sec x$
$\cot \theta$	=	$\cot x$

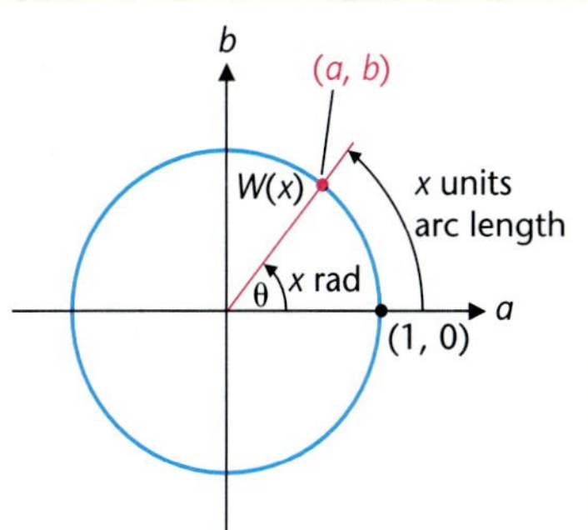

If θ is an angle in degree measure, convert to radian measure and proceed as above.

[*Note:* To reduce the number of different symbols in certain figures, the u and v axes we started with will often be labeled as the a and b axes, respectively. Also, an expression such as $\sin 30°$ denotes the sine of the angle whose measure is 30°.]

The figure in Definition 1 makes use of the important fact that in a unit circle the arc length s opposite an angle of x radians is x units long, and vice versa:

$$s = r\theta = 1 \cdot x = x$$

EXAMPLE 1 **Exact Evaluation for Special Angles**

Evaluate exactly without a calculator:

(A) $\sin\left(\frac{\pi}{6} \text{ radians}\right)$ (B) $\tan\left(\frac{3\pi}{4} \text{ radians}\right)$ (C) $\cos 180°$ (D) $\csc(-150°)$

Solution

(A) $\sin\left(\frac{\pi}{6} \text{ radians}\right) = \sin \frac{\pi}{6} = \frac{1}{2}$

(B) $\tan\left(\frac{3\pi}{4} \text{ radians}\right) = \tan \frac{3\pi}{4} = -1$

(C) $\cos 180° \quad \boxed{= \cos(\pi \text{ radians})} \quad = \cos \pi = -1$

(D) $\csc(-150°) \quad \boxed{= \csc\left(-\frac{5\pi}{6} \text{ radians}\right)} \quad = \csc\left(-\frac{5\pi}{6}\right) = -2$

Matched Problem 1 Evaluate exactly without a calculator:

(A) $\tan(-\pi/4 \text{ radians})$ (B) $\cos(2\pi/3 \text{ radians})$
(C) $\sin 90^\circ$ (D) $\sec(-120^\circ)$

• Calculator Evaluation of Trigonometric Functions

How do we evaluate trigonometric functions for arbitrary angles? Just as a calculator can be used to approximate circular functions for arbitrary real numbers, a calculator can be used to approximate trigonometric functions for arbitrary angles.

Most calculators have a choice of three trigonometric modes: degree (decimal), radian, or grad.

$$\text{The measure of a right angle} = 90^\circ = \frac{\pi}{2} \text{ radians} = 100 \text{ grads}$$

The **grad unit** is used in certain engineering applications and will not be used in this book. We repeat a caution stated earlier:

CAUTION Read the instruction book accompanying your calculator to determine how to put your calculator in degree or radian mode. Forgetting to set the correct mode before starting calculations involving trigonometric functions is a frequent cause of error when using a calculator.

Using a calculator with degree and radian modes, we can evaluate trigonometric functions directly for angles in either degree or radian measure without having to convert degree measure to radian measure first. (Some calculators work only with decimal degrees, and others work with either decimal degrees or degree–minute–second forms. Consult your manual.)

We generalize the reciprocal identities (stated first in Theorem 1, Section 5-2) to evaluate cosecant, secant, and cotangent.

Theorem 1 **Reciprocal Identities**

For x any real number or angle in degree or radian measure:

$$\csc x = \frac{1}{\sin x} \qquad \sin x \neq 0$$

$$\sec x = \frac{1}{\cos x} \qquad \cos x \neq 0$$

$$\cot x = \frac{1}{\tan x} \qquad \tan x \neq 0$$

EXAMPLE 2 Calculator Evaluation

Evaluate to 4 significant digits using a calculator:

(A) $\cos 173.42°$ (B) $\sin$ (3 radians) (C) $\tan 7.183$
(D) $\cot(-102°51')$ (E) $\sec$ (-12.59 radians) (F) [illegible]

Solutions

(A) $\cos 173.42° = -0.9934$ — Degree mode
(B) $\sin$ (3 radians) $= 0.1411$ — Radian mode
(C) $\tan 7.183 = 1.260$ — Radian mode
(D) $\cot(-102°51') = \cot(-102.85°)$ — Degree mode (S[illegible] ators require decimal degrees.)
$= 0.2281$
(E) $\sec$ (-12.59 radians) $= 1.000$ — Radian mode
(F) $\csc(-206.3) = 1.156$ — Radian mode

Matched Problem 2

Evaluate to 4 significant digits using a calculator:

(A) $\sin 239.12°$ (B) $\cos$ (7 radians) (C) $\cot 1$[illegible]
(D) $\tan(-212°33')$ (E) $\sec$ (-8.09 radians) (F) $\csc$ ([illegible]4.5)

• Definition of the Trigonometric Functions—Alternate Form

For many applications involving the use of trigonometric functions, including triangle applications, it is useful to write Definition 1 in an alternate form—a form that utilizes the coordinates of an arbitrary point $(a, b) \neq (0, 0)$ on the terminal side of an angle θ (see Fig. 1).

This alternate form of Definition 1 is easily found by inserting a unit circle in Figure 1, drawing perpendiculars from points P and Q to the horizontal axis (Fig. 2), and utilizing the fact that ratios of corresponding sides of similar triangles are proportional.

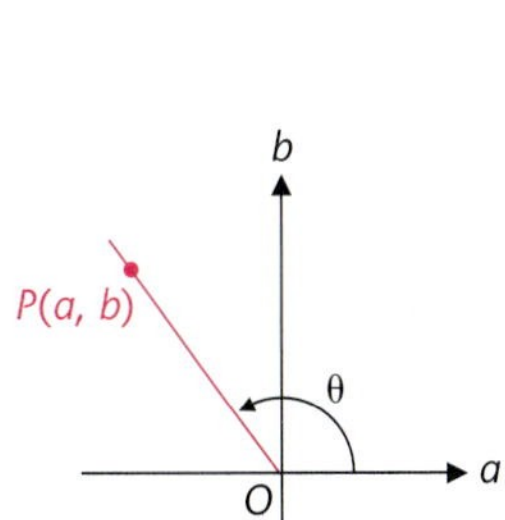

FIGURE 1 Angle θ.

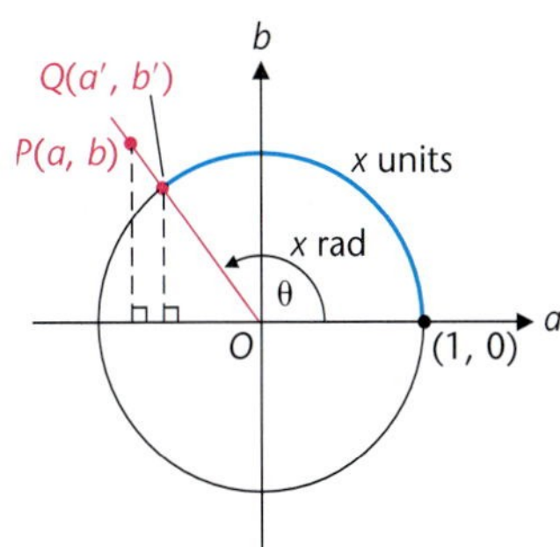

FIGURE 2 Similar triangles.

Letting $r = d(O, P)$ and noting that $d(O, Q) = 1$, we have

$$\sin\theta = \sin x = b' = \frac{b'}{1} = \frac{b}{r} \qquad b \text{ and } b' \text{ always have the same sign.}$$

$$\cos\theta = \cos x = a' = \frac{a'}{1} = \frac{a}{r} \qquad a \text{ and } a' \text{ always have the same sign.}$$

The values of the other four trigonometric functions can be obtained using basic identities. For example,

$$\tan\theta = \frac{\sin\theta}{\cos\theta} = \frac{b/r}{a/r} = \frac{b}{a}$$

We now have the very useful alternate form of Definition 1 given below.

DEFINITION 1 (ALTERNATE FORM)

Trigonometric Functions with Angle Domains

If θ is an arbitrary angle in standard position in a rectangular coordinate system and $P(a, b)$ is a point r units from the origin on the terminal side of θ, then:

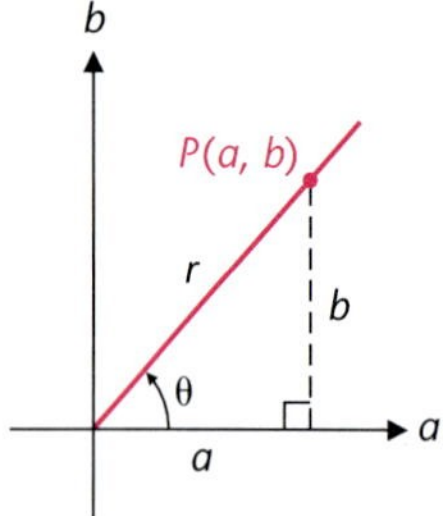

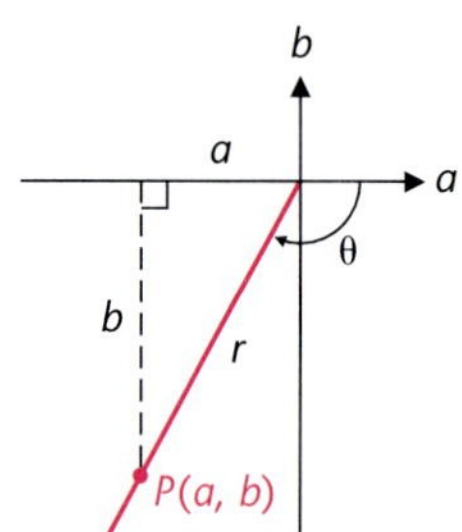

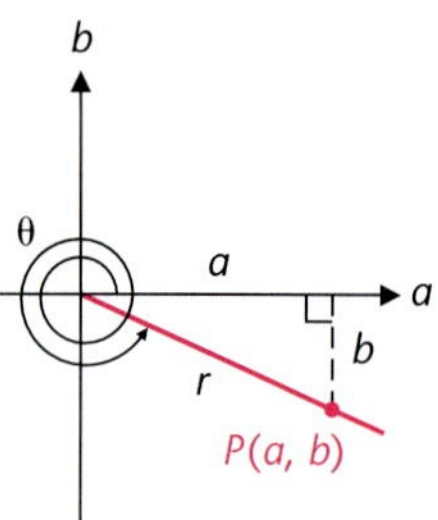

$\sin\theta = \dfrac{b}{r}$ $\qquad$ $\csc\theta = \dfrac{r}{b},\ b \neq 0$ $\qquad$ $r = \sqrt{a^2 + b^2} > 0;$

$\cos\theta = \dfrac{a}{r}$ $\qquad$ $\sec\theta = \dfrac{r}{a},\ a \neq 0$

$\tan\theta = \dfrac{b}{a},\ a \neq 0$ $\qquad$ $\cot\theta = \dfrac{a}{b},\ b \neq 0$

Domains: Sets of all possible angles for which the ratios are defined

Ranges: Subsets of the set of real numbers

(Domains and ranges will be stated more precisely in Section 5-6.)

[*Note:* The right triangle formed by drawing a perpendicular from $P(a, b)$ to the horizontal axis is called the **reference triangle** associated with the angle θ. We will often refer to this triangle.]

EXPLORE-DISCUSS 1 Discuss why, for a given angle θ, the ratios in Definition 1 are independent of the choice of $P(a, b)$ on the terminal side of θ as long as $(a, b) \neq (0, 0)$.

The alternate form of Definition 1 should be memorized. As a memory aid, note that when $r = 1$, then $P(a, b)$ is on the unit circle, and all function values correspond to the values obtained using Definition 1 for circular functions in Section 5-2. In fact, using the alternate form of Definition 1 in conjunction with the original statement of Definition 1 in this section, we have an alternate way of evaluating circular functions:

Circular Functions and Trigonometric Functions

For x any real number:

$$\begin{aligned} \sin x &= \sin (x \text{ radians}) & \cos x &= \cos (x \text{ radians}) \\ \sec x &= \sec (x \text{ radians}) & \csc x &= \csc (x \text{ radians}) \\ \tan x &= \tan (x \text{ radians}) & \cot x &= \cot (x \text{ radians}) \end{aligned} \tag{1}$$

Thus, we are now free to evaluate circular functions in terms of trigonometric functions, using reference triangles where appropriate, or in terms of circular points and the wrapping function discussed earlier. Each approach has certain advantages in particular situations, and you should become familiar with the uses of both approaches.

It is because of equations (1) that we are able to evaluate circular functions using a calculator set in radian mode (see Section 5-2). Generally, unless a certain emphasis is desired, we will not use "rad" after a real number. That is, we will interpret expressions such as "sin 5.73" as the "circular function value sin 5.73" or the "trigonometric function value sin (5.73 rad)" by the context in which the expression occurs or the form we wish to emphasize. We will remain flexible and often switch back and forth between circular function emphasis and trigonometric function emphasis, depending on which approach provides the most enlightenment for a given situation.

EXAMPLE 3 **Evaluating Trigonometric Functions**

Find the value of each of the six trigonometric functions for the illustrated angle θ with terminal side that contains $P(-3, -4)$. See Figure 3.

FIGURE 3

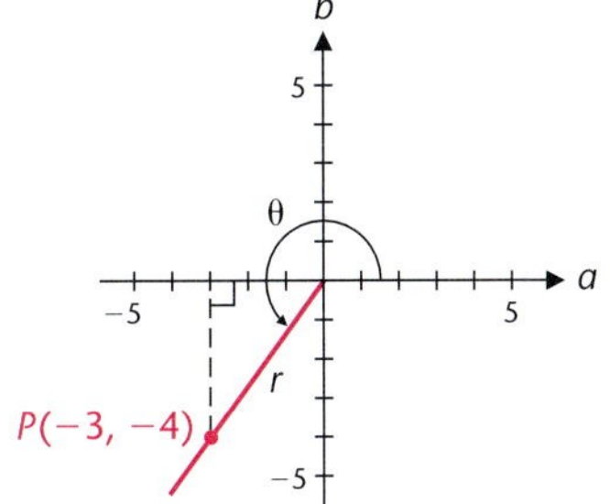

Solution

$$(a, b) = (-3, -4)$$

$$r = \sqrt{a^2 + b^2} = \sqrt{(-3)^2 + (-4)^2} = \sqrt{25} = 5$$

$$\sin\theta = \frac{b}{r} = \frac{-4}{5} = -\frac{4}{5} \qquad \csc\theta = \frac{r}{b} = \frac{5}{-4} = -\frac{5}{4}$$

$$\cos\theta = \frac{a}{r} = \frac{-3}{5} = -\frac{3}{5} \qquad \sec\theta = \frac{r}{a} = \frac{5}{-3} = -\frac{5}{3}$$

$$\tan\theta = \frac{b}{a} = \frac{-4}{-3} = \frac{4}{3} \qquad \cot\theta = \frac{a}{b} = \frac{-3}{-4} = \frac{3}{4}$$

Matched Problem 3 Find the value of each of the six trigonometric functions if the terminal side of θ contains the point $(-6, -8)$. [*Note:* This point lies on the terminal side of the angle in Example 3; hence, the final results should be the same as those obtained in Example 3.]

EXAMPLE 4 Evaluating Trigonometric Functions

Find the value of each of the other five trigonometric functions for an angle θ (without finding θ) given that θ is a IV quadrant angle and $\sin\theta = -\frac{4}{5}$.

Solution The information given is sufficient for us to locate a reference triangle in quadrant IV for θ, even though we don't know what θ is. We sketch a reference triangle, label what we know (Fig. 4), and then complete the problem as indicated.

FIGURE 4

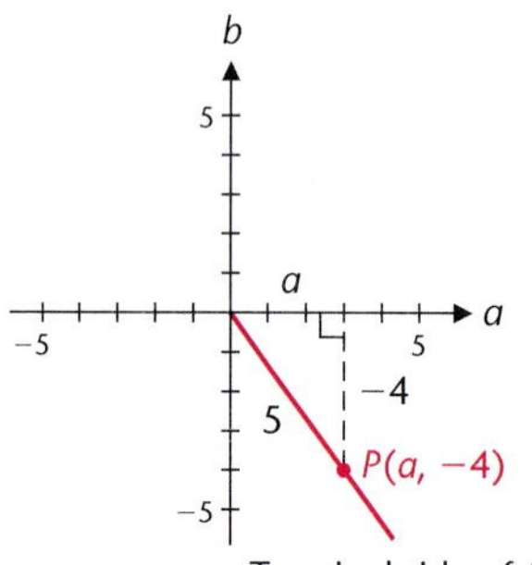

Since $\sin\theta = b/r = -\frac{4}{5}$, we can let $b = -4$ and $r = 5$ (r is never negative). If we can find a, then we can determine the values of the other five functions.

Use the Pythagorean theorem to find a:

$$a^2 + (-4)^2 = 5^2$$

$$a^2 = 9$$

$$a = \pm 3$$

$$= 3$$ a cannot be negative because θ is a IV quadrant angle.

Using $(a, b) = (3, -4)$ and $r = 5$, we have

$$\cos\theta = \frac{a}{r} = \frac{3}{5} \qquad \sec\theta = \frac{r}{a} = \frac{5}{3} \qquad \csc\theta = \frac{r}{b} = \frac{5}{-4} = -\frac{5}{4}$$

$$\tan\theta = \frac{b}{a} = \frac{-4}{3} = -\frac{4}{3} \qquad \cot\theta = \frac{a}{b} = \frac{3}{-4} = -\frac{3}{4}$$

Matched Problem 4 Find the value of each of the other five trigonometric functions for an angle θ (without finding θ) given that θ is a II quadrant angle and $\tan\theta = -\frac{3}{4}$.

• Exact Values for Special Angles and Real Numbers

Assuming a trigonometric function is defined, it can be evaluated exactly without the use of a calculator or table (which is different from finding approximate values using a calculator or table) for any integer multiple of 30°, 45°, 60°, 90°, $\pi/6$, $\pi/4$, $\pi/3$, or $\pi/2$. With a little practice you will be able to determine these values mentally. Working with exact values has advantages over working with approximate values in many situations.

The easiest angles to deal with are quadrantal angles since these angles are integer multiples of 90° or $\pi/2$. It is easy to find the coordinates of a point on a coordinate axis. Since any nonorigin point will do, we shall for convenience choose points 1 unit from the origin, as shown in Figure 5.

FIGURE 5 Quadrantal angles.

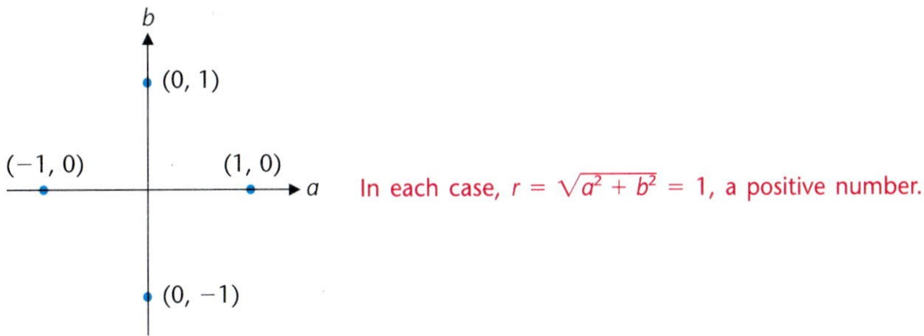

EXAMPLE 5 Trig Functions of Quadrantal Angles

Find:

(A) $\sin 90°$ (B) $\cos\pi$ (C) $\tan(-2\pi)$ (D) $\cot(-180°)$

Solutions For each, visualize the location of the terminal side of the angle relative to Figure 3. With a little practice, you should be able to do most of the following mentally.

(A) $\sin 90° = \frac{b}{r} = \frac{1}{1} = 1$ $(a, b) = (0, 1);\ r = 1$

(B) $\cos\pi = \frac{a}{r} = \frac{-1}{1} = -1$ $(a, b) = (-1, 0);\ r = 1$

(C) $\tan(-2\pi) = \frac{b}{a} = \frac{0}{1} = 0$ $(a, b) = (1, 0); r = 1$

(D) $\cot(-180°) = \frac{a}{b} = \frac{-1}{0}$ $(a, b) = (-1, 0); r = 1$

Not defined

Matched Problem 5 Find:

(A) $\sin(3\pi/2)$ (B) $\sec(-\pi)$ (C) $\tan 90°$ (D) $\cot(-270°)$

EXPLORE-DISCUSS 2 Notice in Example 5D that $\cot(-180°)$ is not defined. Discuss other angles in degree measure for which the cotangent is not defined. For what angles in degree measure is the cosecant function not defined?

Because the concept of *reference triangle* introduced in Definition 1 (alternate form) plays an important role in much of the material that follows, we restate its definition here and define the related concept of *reference angle.*

Reference Triangle and Reference Angle

1. To form a **reference triangle** for θ, draw a perpendicular from a point $P(a, b)$ on the terminal side of θ to the horizontal axis.
2. The **reference angle** α is the acute angle (always taken positive) between the terminal side of θ and the horizontal axis.

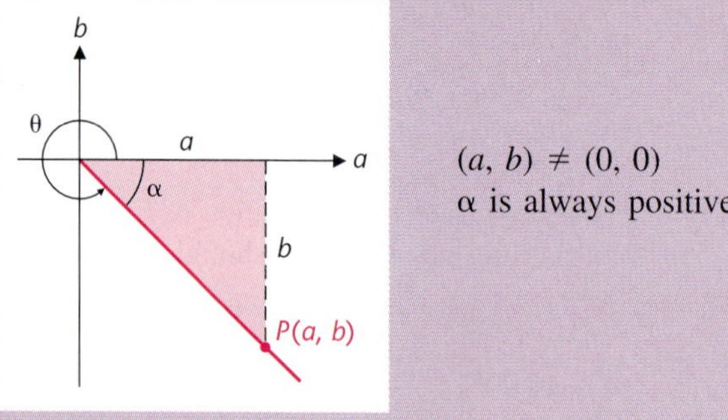

$(a, b) \neq (0, 0)$
α is always positive

Figure 6 shows several reference triangles and reference angles corresponding to particular angles.

FIGURE 6 Reference triangles and reference angles.

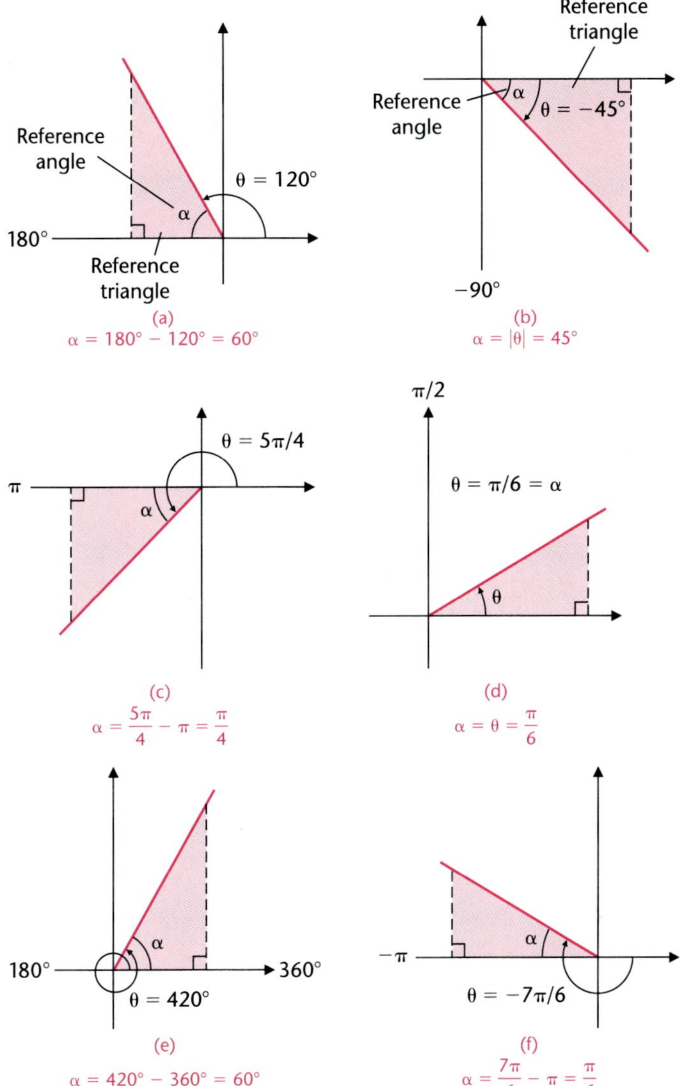

If a reference triangle of a given angle is a 30°–60° right triangle or a 45° right triangle, then we can find exact coordinates, other than (0, 0), on the terminal side of the given angle. To this end, we first note that a 30°–60° triangle forms half of an equilateral triangle, as indicated in Figure 7. Because all sides are equal in an equilateral triangle, we can apply the Pythagorean theorem to obtain a useful relationship among the three sides of the original triangle:

FIGURE 7 30°–60° right triangle.

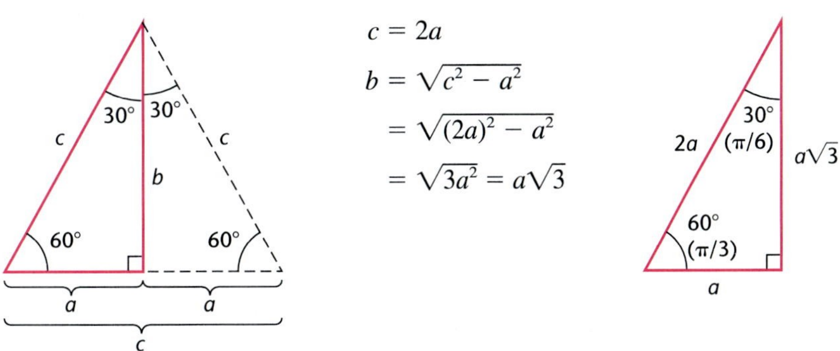

Similarly, using the Pythagorean theorem on a 45° right triangle, we obtain the result shown in Figure 8.

FIGURE 8 45° right triangle.

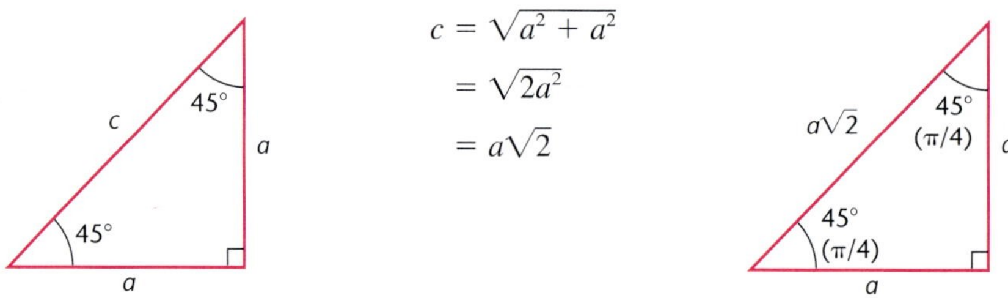

Figure 9 illustrates the results shown in Figures 7 and 8 for the case $a = 1$. This case is the easiest to remember. All other cases can be obtained from this special case by multiplying or dividing the length of each side of a triangle in Figure 9 by the same nonzero quantity. For example, if we wanted the hypotenuse of a special 45° right triangle to be 1, we would simply divide each side of the 45° triangle in Figure 9 by $\sqrt{2}$.

FIGURE 9

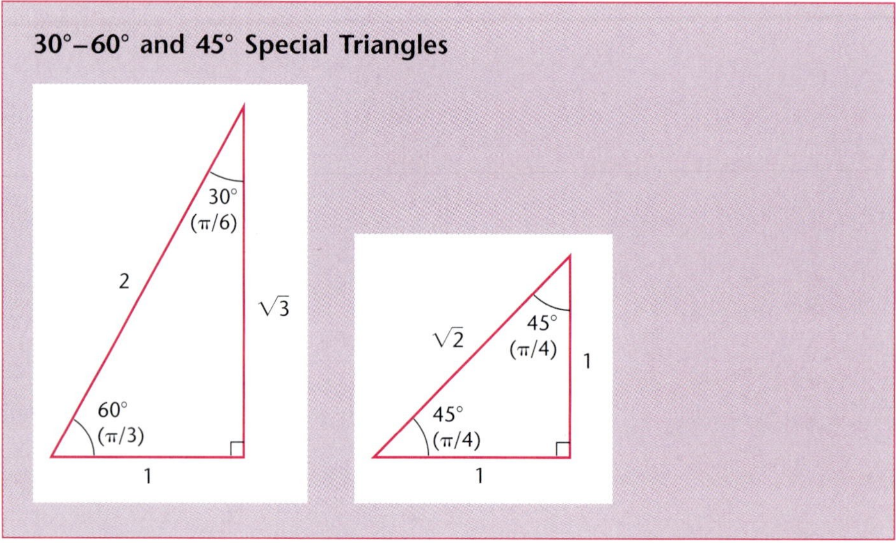

If an angle or a real number has a 30°–60° or a 45° reference triangle, then we can use Figure 9 to find exact coordinates of a nonorigin point on the terminal side of the angle. Using the definition of the trigonometric functions, Definition 1 alternate form, we will then be able to find the exact value of any of the six functions for the indicated angle or real number.

EXAMPLE 6 Exact Evaluation

Evaluate exactly using appropriate reference triangles:

(A) cos 60°, sin (π/3), tan (π/3) (B) sin 45°, cot (π/4), sec (π/4)

Solutions (A) Use the special 30°–60° triangle with sides 1, 2, and $\sqrt{3}$ as the reference triangle, and use 60° or π/3 as the reference angle (Fig. 10). Use the sides of the reference triangle to determine $P(a, b)$ and r; then use the appropriate definitions.

FIGURE 10

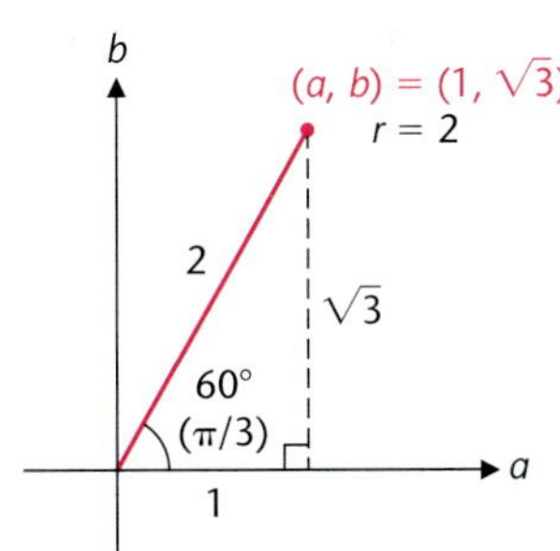

$$\cos 60° = \frac{a}{r} = \frac{1}{2}$$

$$\sin\frac{\pi}{3} = \frac{b}{r} = \frac{\sqrt{3}}{2}$$

$$\tan\frac{\pi}{3} = \frac{b}{a} = \frac{\sqrt{3}}{1} = \sqrt{3}$$

(B) Use the special 45° triangle with sides 1, 1, and $\sqrt{2}$ as the reference triangle, and use 45° or $\pi/4$ as the reference angle (Fig. 11). Use the sides of the reference triangle to determine $P(a, b)$ and r; then use the appropriate definitions.

FIGURE 11

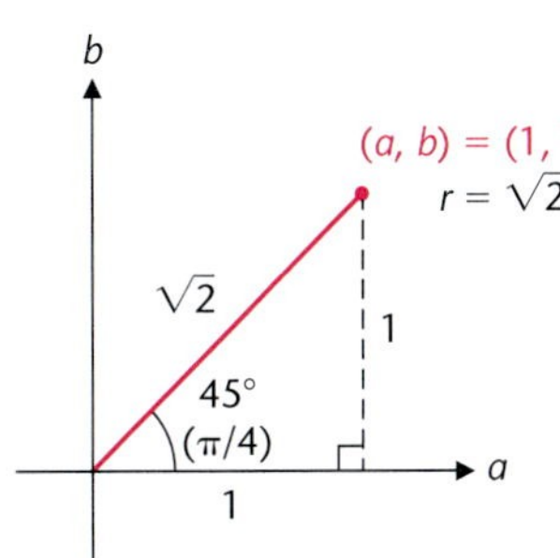

$$\sin 45° = \frac{b}{r} = \frac{1}{\sqrt{2}} \text{ or } \frac{\sqrt{2}}{2}$$

$$\cot\frac{\pi}{4} = \frac{a}{b} = \frac{1}{1} = 1$$

$$\sec\frac{\pi}{4} = \frac{r}{a} = \frac{\sqrt{2}}{1} = \sqrt{2}$$

Matched Problem 6 Evaluate exactly using appropriate reference triangles:

(A) cos 45°, tan ($\pi/4$), csc ($\pi/4$) (B) sin 30°, cos ($\pi/6$), cot ($\pi/6$)

Before proceeding, it is useful to observe from a geometric point of view multiples of $\pi/3$ (60°), $\pi/6$ (30°), and $\pi/4$ (45°). These are illustrated in Figure 12.

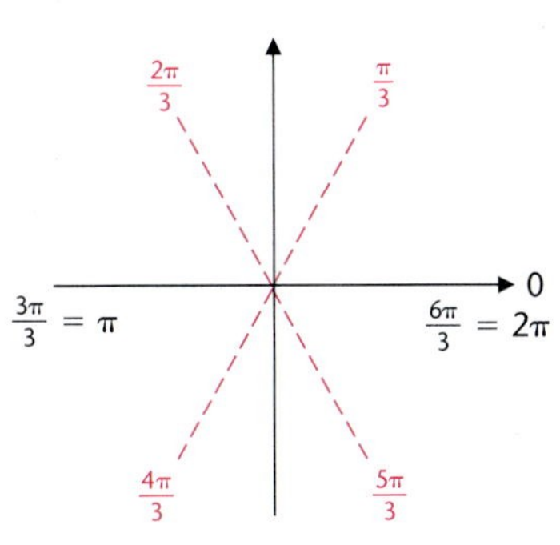

Multiples of $\pi/3$ (60°)

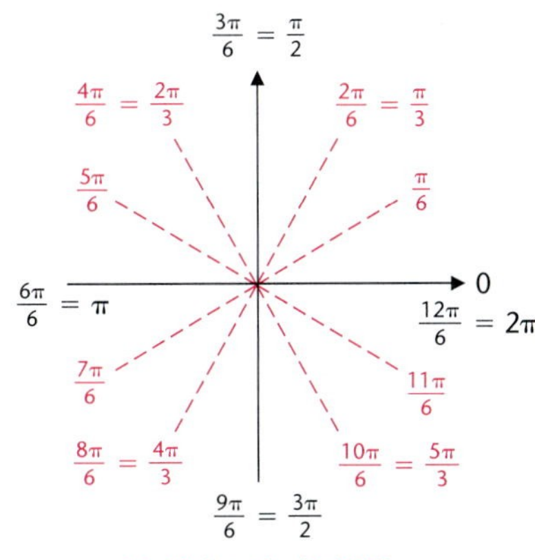

Multiples of $\pi/6$ (30°)

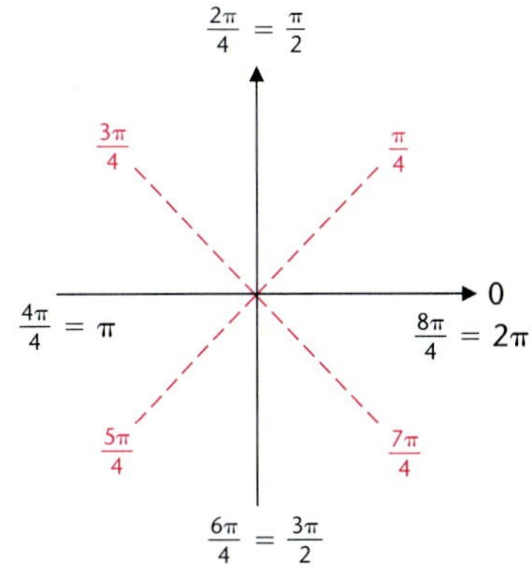

Multiples of $\pi/4$ (45°)

FIGURE 12 Multiples of special angles.

EXAMPLE 7 **Exact Evaluation**

Evaluate exactly using appropriate reference triangles:

(A) $\cos(7\pi/4)$ (B) $\sin(2\pi/3)$ (C) $\tan 210°$ (D) $\sec(-240°)$

Solutions Each angle (or real number) has a 30°–60° or 45° reference triangle. Locate it, determine (a, b) and r as in Example 6, and then evaluate.

(A) $\cos \dfrac{7\pi}{4} = \dfrac{1}{\sqrt{2}}$ or $\dfrac{\sqrt{2}}{2}$ (B) $\sin \dfrac{2\pi}{3} = \dfrac{\sqrt{3}}{2}$

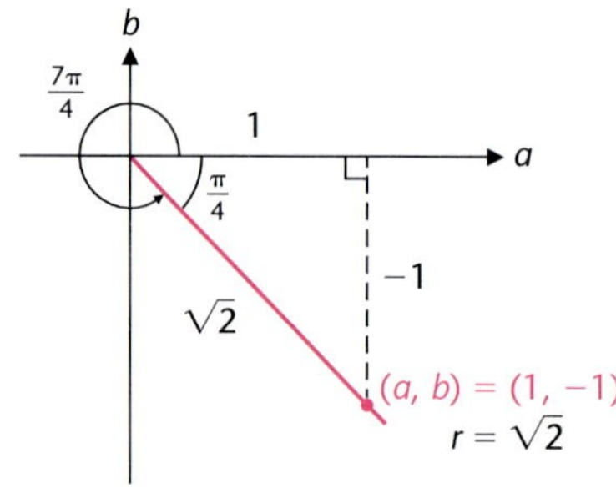

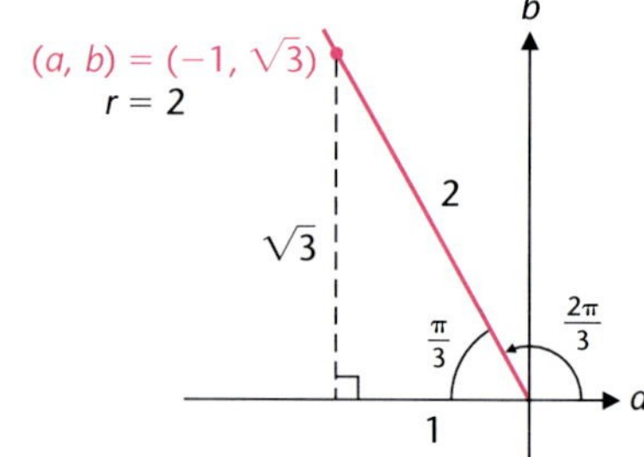

(C) $\tan 210° = \dfrac{-1}{-\sqrt{3}} = \dfrac{1}{\sqrt{3}}$ or $\dfrac{\sqrt{3}}{3}$

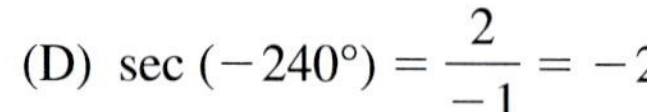

(D) $\sec(-240°) = \dfrac{2}{-1} = -2$

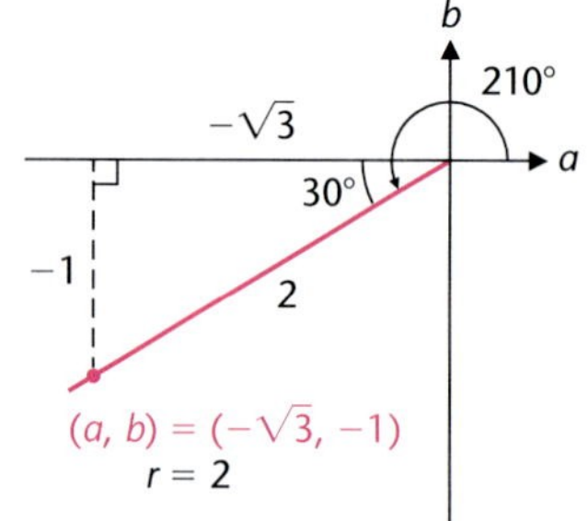

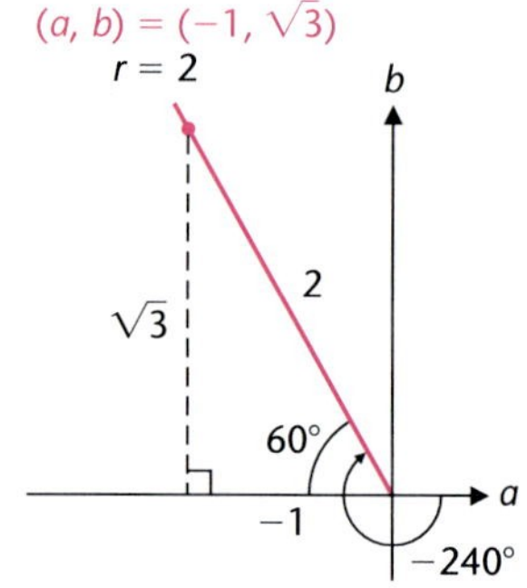

Matched Problem 7 Evaluate exactly using appropriate triangles:

(A) $\tan(-\pi/4)$ (B) $\sin 210°$ (C) $\cos(2\pi/3)$ (D) $\csc(-240°)$

Now the problem is reversed; that is, let the exact value of one of the six trigonometric functions be given and assume this value corresponds to one of the special reference triangles. Can a smallest positive θ be found for which the trigonometric function has that value? Example 8 shows how.

EXAMPLE 8 Finding Special Angles

Find the smallest positive θ in degree and radian measure for which each is true.

(A) $\tan \theta = 1/\sqrt{3}$ (B) $\sec \theta = -\sqrt{2}$

Solutions (A)

$$\tan \theta = \frac{b}{a} = \frac{1}{\sqrt{3}}$$

We can let $(a, b) = (\sqrt{3}, 1)$ or $(-\sqrt{3}, -1)$. The smallest positive θ for which this is true is a quadrant I angle with reference triangle as drawn in Figure 13.

FIGURE 13

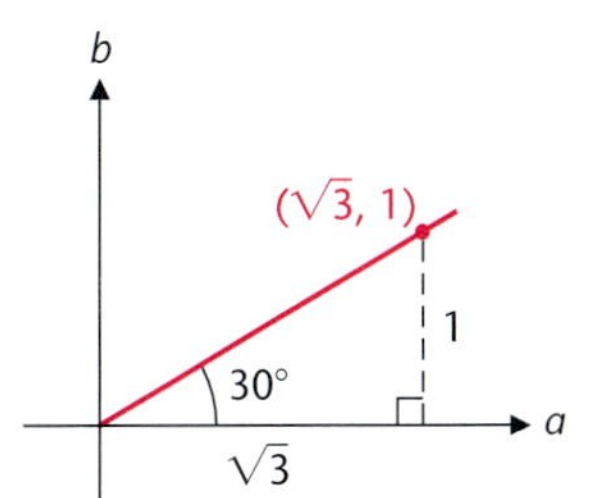

$$\theta = 30° \text{ or } \frac{\pi}{6}$$

(B)

$$\sec \theta = \frac{r}{a} = \frac{\sqrt{2}}{-1} \quad \text{Because } r > 0$$

a is negative in quadrants II and III. The smallest positive θ is associated with a 45° reference triangle in quadrant II as drawn in Figure 14.

FIGURE 14

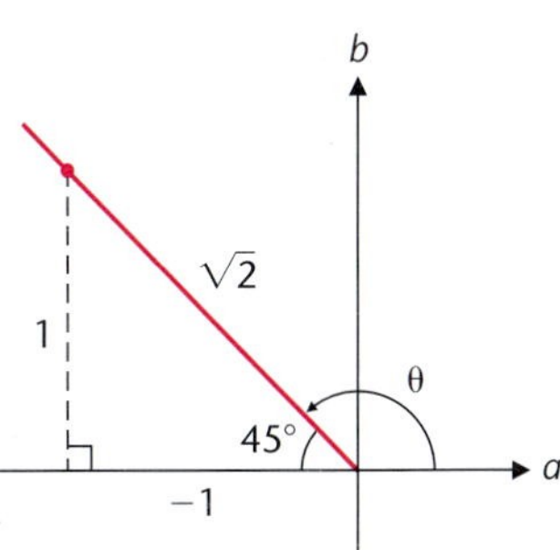

$$\theta = 135° \text{ or } \frac{3\pi}{4}$$

Matched Problem 8

Find the smallest positive θ in degree and radian measure for which each is true.

(A) $\sin \theta = \sqrt{3}/2$ (B) $\cos \theta = -1/\sqrt{2}$

Remark. After quite a bit of practice, the reference triangle figures in Examples 7 and 8 can be visualized mentally; however, when in doubt, draw a figure.

• Summary of Special Angle Values

Table 1 includes a summary of the exact values of the sine, cosine, and tangent for the special angle values from 0° to 90°. Some people like to memorize these values, while others prefer to memorize the triangles in Figure 9. Do whichever is easier for you.

TABLE 1 Special Angle Values

θ	$\sin\theta$	$\cos\theta$	$\tan\theta$
0°	0	1	0
30°	$\frac{1}{2}$	$\sqrt{3}/2$	$1/\sqrt{3}$ or $\sqrt{3}/3$
45°	$1/\sqrt{2}$ or $\sqrt{2}/2$	$1/\sqrt{2}$ or $\sqrt{2}/2$	1
60°	$\sqrt{3}/2$	$\frac{1}{2}$	$\sqrt{3}$
90°	1	0	Not defined

These special angle values are easily remembered for sine and cosine if you note the unexpected pattern after completing Table 2 in Explore-Discuss 3.

EXPLORE-DISCUSS 3 Fill in the cosine column in Table 2 with a pattern of values that is similar to those in the sine column. Discuss how the two columns of values are related.

TABLE 2 Special Angle Values—Memory Aid

θ	$\sin\theta$	$\cos\theta$
0°	$\sqrt{0}/2 = 0$	
30°	$\sqrt{1}/2 = \frac{1}{2}$	
45°	$\sqrt{2}/2$	
60°	$\sqrt{3}/2$	
90°	$\sqrt{4}/2 = 1$	

Cosecant, secant, and cotangent can be found for these special angles by using the values in Tables 1 or 2 and the reciprocal identities from Theorem 1.

Answers to Matched Problems

1. (A) -1 (B) $-\frac{1}{2}$ (C) 1 (D) -2
2. (A) -0.8582 (B) 0.7539 (C) 1.542 (D) -0.6383 (E) -4.277 (F) 1.137
3. $\sin\theta = -\frac{4}{5}$, $\cos\theta = -\frac{3}{5}$, $\tan\theta = \frac{4}{3}$, $\csc\theta = -\frac{5}{4}$, $\sec\theta = -\frac{5}{3}$, $\cot\theta = \frac{3}{4}$
4. $\sin\theta = \frac{3}{5}$, $\cos\theta = -\frac{4}{5}$, $\csc\theta = \frac{5}{3}$, $\sec\theta = -\frac{5}{4}$, $\cot\theta = -\frac{4}{3}$
5. (A) -1 (B) -1 (C) Not defined (D) 0
6. (A) $\cos 45° = 1/\sqrt{2}$, $\tan(\pi/4) = 1$, $\csc(\pi/4) = \sqrt{2}$
(B) $\sin 30° = \frac{1}{2}$, $\cos(\pi/6) = \sqrt{3}/2$, $\cot(\pi/6) = \sqrt{3}$
7. (A) -1 (B) $-\frac{1}{2}$ (C) $-\frac{1}{2}$ (D) $2/\sqrt{3}$ (E) $\sqrt{3}$ (F) 2
8. (A) 60° or $\pi/3$ (B) 135° or $3\pi/4$

EXERCISE 5-4

A

Find the value of each of the six trigonometric functions for an angle θ that has a terminal side containing the point indicated in Problems 1–4.

1. $(6, 8)$ **2.** $(-3, 4)$

3. $(-1, \sqrt{3})$ **4.** $(\sqrt{3}, 1)$

Evaluate Problems 5–14 to 4 significant digits using a calculator. Make sure your calculator is in the correct mode (degree or radian) for each problem.

5. $\sin 68°$ **6.** $\tan 21°$

7. $\cot 5$ **8.** $\csc(-11)$

9. $\cos 78.24°$ **10.** $\sin 45.01°$

11. $\csc 365°52'48''$ **12.** $\sec 88°27'15''$

13. $\tan(-1.58)$ **14.** $\cot 25.1$

In Problems 15–26, evaluate exactly, using reference triangles where appropriate, without using a calculator.

15. $\sin 0°$ **16.** $\cos 0°$ **17.** $\tan 60°$

18. $\cos 30°$ **19.** $\sin 45°$ **20.** $\csc 60°$

21. $\sec 45°$ **22.** $\cot 45°$ **23.** $\cot 0°$

24. $\cot 90°$ **25.** $\tan 90°$ **26.** $\sec 0°$

Find the reference angle α for each angle θ in Problems 27–32.

27. $\theta = 300°$ **28.** $\theta = 135°$ **29.** $\theta = \dfrac{7\pi}{6}$

30. $\theta = \dfrac{\pi}{4}$ **31.** $\theta = -\dfrac{5\pi}{3}$ **32.** $\theta = -\dfrac{5\pi}{4}$

B

In Problems 33–48, evaluate exactly, using reference angles where appropriate, without using a calculator.

33. $\tan(3\pi/4)$ **34.** $\cos(7\pi/6)$

35. $\sin(-30°)$ **36.** $\cot 120°$

37. $\sec(5\pi/6)$ **38.** $\csc(-5\pi/4)$

39. $\cot 315°$ **40.** $\sin 240°$

41. $\csc(-150°)$ **42.** $\tan(-135°)$

43. $\cos(13\pi/6)$ **44.** $\sec(13\pi/4)$

45. $\tan(-7\pi/3)$ **46.** $\cos(2\pi/3)$

47. $\sec(23\pi/4)$ **48.** $\csc(-17\pi/6)$

For which values of θ, $0° \le \theta < 360°$, is each of Problems 49–54 not defined? Explain why.

49. $\cos\theta$ **50.** $\sec\theta$ **51.** $\tan\theta$

52. $\cot\theta$ **53.** $\csc\theta$ **54.** $\sin\theta$

In Problems 55–60, find the smallest positive θ in degree and radian measure for which:

55. $\cos\theta = \dfrac{-1}{2}$ **56.** $\sin\theta = \dfrac{-\sqrt{3}}{2}$

57. $\sin\theta = \dfrac{-1}{2}$ **58.** $\tan\theta = -\sqrt{3}$

59. $\csc\theta = \dfrac{-2}{\sqrt{3}}$ **60.** $\sec\theta = -\sqrt{2}$

Find the value of each of the other five trigonometric functions for an angle θ, without finding θ, given the information indicated in Problems 61–64. Sketching a reference triangle should be helpful.

61. $\sin\theta = \frac{3}{5}$ and $\cos\theta < 0$

62. $\tan\theta = -\frac{4}{3}$ and $\sin\theta < 0$

63. $\cos\theta = -\sqrt{5}/3$ and $\cot\theta > 0$

64. $\cos\theta = -\sqrt{5}/3$ and $\tan\theta > 0$

65. Which trigonometric functions are not defined when the terminal side of an angle lies along the vertical axis? Why?

66. Which trigonometric functions are not defined when the terminal side of an angle lies along the horizontal axis? Why?

67. Find exactly, all θ, $0° \le \theta < 360°$, for which $\cos\theta = -\sqrt{3}/2$.

68. Find exactly, all θ, $0° \le \theta < 360°$, for which $\cot\theta = -1/\sqrt{3}$.

69. Find exactly, all θ, $0 \le \theta < 2\pi$, for which $\tan\theta = 1$.

70. Find exactly, all θ, $0 \le \theta < 2\pi$, for which $\sec\theta = -\sqrt{2}$.

C

For Problems 71 and 72, refer to the following figure.

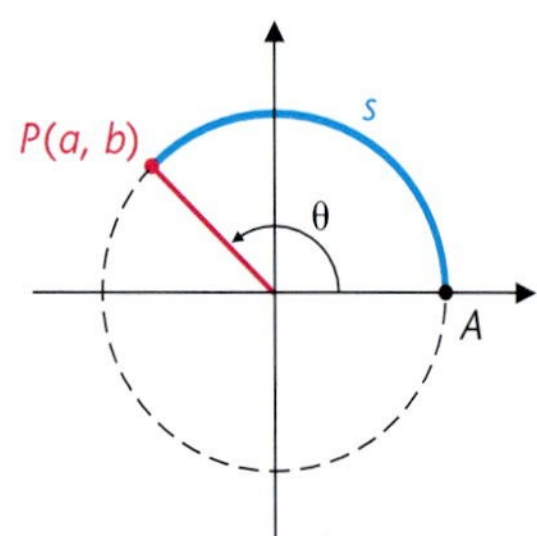

71. If the coordinates of A are (4, 0) and arc length s is 7 units, find:

(A) The exact radian measure of θ

(B) The coordinates of P to three decimal places

72. If the coordinates of A are (2, 0) and arc length s is 8 units, find:

(A) The exact radian measure of θ

(B) The coordinates of P to three decimal places

73. In a rectangular coordinate system, a circle with its center at the origin passes through the point $(6\sqrt{3}, 6)$. What is the length of the arc on the circle in quadrant I between the positive horizontal axis and the point $(6\sqrt{3}, 6)$?

74. In a rectangular coordinate system, a circle with its center at the origin passes through the point $(2, 2\sqrt{3})$. What is the length of the arc on the circle in quadrant I between the positive horizontal axis and the point $(2, 2\sqrt{3})$?

APPLICATIONS 

75. Solar Energy. The intensity of light I on a solar cell changes with the angle of the sun and is given by the formula $I = k \cos \theta$, where k is a constant (see the figure).

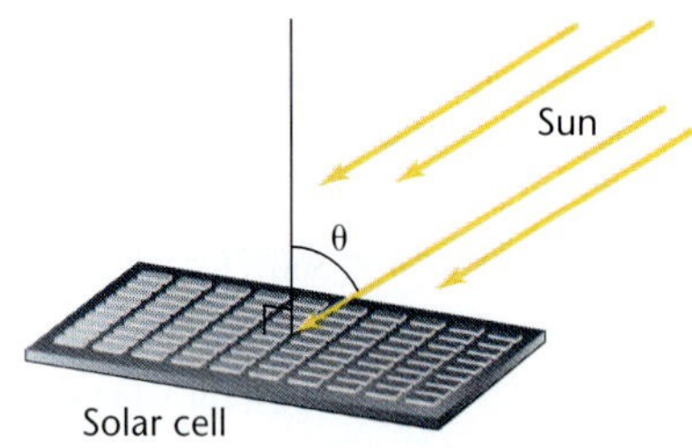

Find light intensity I in terms of k for $\theta = 0°$, $\theta = 30°$, and $\theta = 60°$.

76. Solar Energy. Refer to Problem 75.
Find light intensity I in terms of k for $\theta = 20°$, $\theta = 50°$, and $\theta = 90°$.

77. Physics—Engineering. The figure illustrates a piston connected to a wheel that turns 3 revolutions per second; hence, the angle θ is being generated at $3(2\pi) = 6\pi$ radians per second, or $\theta = 6\pi t$, where t is time in seconds. If P is at (1, 0) when $t = 0$, show that

$$\begin{aligned} y &= b + \sqrt{4^2 - a^2} \\ &= \sin 6\pi t + \sqrt{16 - (\cos 6\pi t)^2} \end{aligned}$$

for $t \geq 0$.

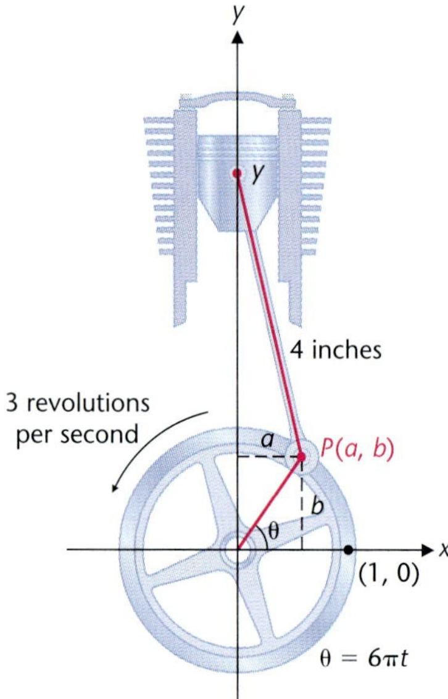

78. Physics—Engineering. In Problem 77, find the position of the piston y when $t = 0.2$ second (to 3 significant digits).

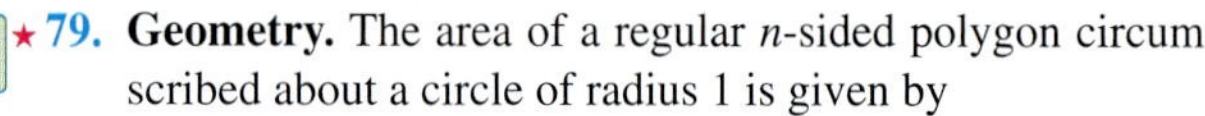

★ **79. Geometry.** The area of a regular n-sided polygon circumscribed about a circle of radius 1 is given by

$$A = n \tan \frac{180°}{n}$$

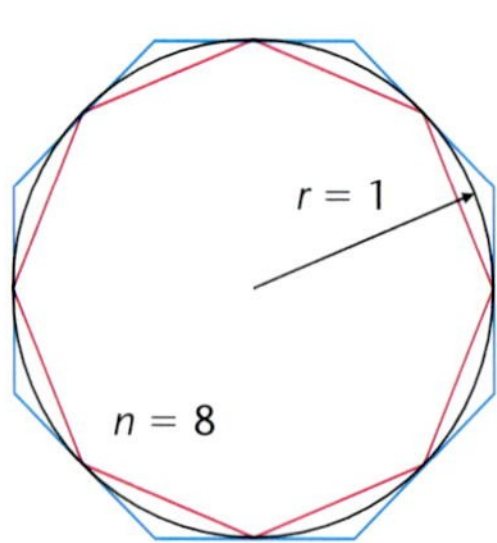

(A) Find A for $n = 8$, $n = 100$, $n = 1{,}000$, and $n = 10{,}000$. Compute each to five decimal places.

(B) What number does A seem to approach as $n \to \infty$? (What is the area of a circle with radius 1?)

★80. **Geometry.** The area of a regular n-sided polygon inscribed in a circle of radius 1 is given by

$$A = \frac{n}{2} \sin \frac{360°}{n}$$

(A) Find A for $n = 8$, $n = 100$, $n = 1{,}000$, and $n = 10{,}000$. Compute each to five decimal places.

(B) What number does A seem to approach as $n \to \infty$? (What is the area of a circle with radius 1?)

81. **Angle of Inclination.** Recall (Section 2-2) that the **slope** of a nonvertical line passing through points $P_1(x_1, y_1)$ and $P_2(x_2, y_2)$ is given by Slope $= m = (y_2 - y_1)/(x_2 - x_1)$. The angle θ that the line L makes with the x axis, $0° \le \theta < 180°$, is called the **angle of inclination** of the line L (see figure). Thus,

$$\text{Slope} = m = \tan \theta \qquad 0° \le \theta < 180°$$

(A) Compute the slopes to two decimal places of the lines with angles of inclination 88.7° and 162.3°.

(B) Find the equation of a line passing through $(-4, 5)$ with an angle of inclination 137°. Write the answer in the form $y = mx + b$, with m and b to two decimal places.

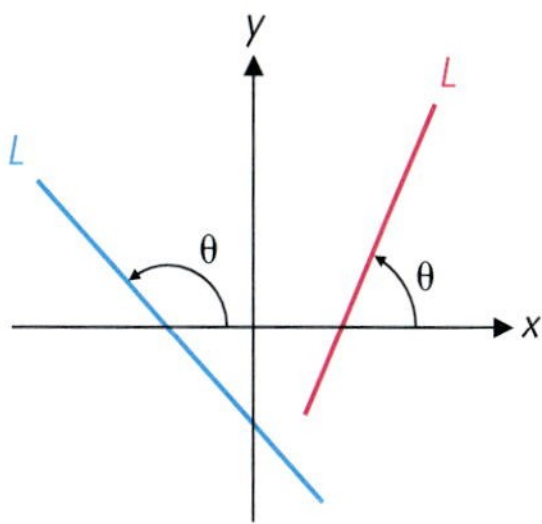

82. **Angle of Inclination.** Refer to Problem 81.

(A) Compute the slopes to two decimal places of the lines with angles of inclination 5.34° and 92.4°.

(B) Find the equation of a line passing through $(6, -4)$ with an angle of inclination 106°. Write the answer in the form $y = mx + b$, with m and b to two decimal places.

SECTION 5-5 Solving Right Triangles*

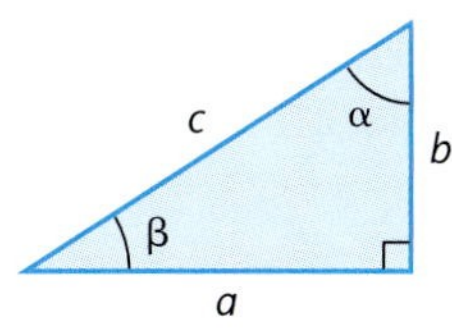

FIGURE 1

In the previous sections we applied trigonometric and circular functions in the solutions of a variety of significant problems. In this section we are interested in the particular class of problems involving right triangles. Referring to Figure 1, our objective is to find all unknown parts of a right triangle, given the measure of two sides or the measure of one acute angle and a side. This is called **solving a right triangle.** Trigonometric functions play a central role in this process.

To start, we locate a right triangle in the first quadrant of a rectangular coordinate system and observe, from the definition of the trigonometric functions, six trigonometric ratios involving the sides of the triangle. (Note that the right triangle is the reference triangle for the angle θ.)

Trigonometric Ratios

(a, b)
c
b
θ
a
$0° < \theta < 90°$

$$\sin \theta = \frac{b}{c} \qquad \csc \theta = \frac{c}{b}$$

$$\cos \theta = \frac{a}{c} \qquad \sec \theta = \frac{c}{a}$$

$$\tan \theta = \frac{b}{a} \qquad \cot \theta = \frac{a}{b}$$

*This section provides a significant application of trigonometric functions to real-world problems. However, it may be postponed or omitted without loss of continuity, if desired. Some may want to cover the section just before Sections 7-1 and 7-2.

Side b is often referred to as the **side opposite** angle θ, a as the **side adjacent** to angle θ, and c as the **hypotenuse.** Using these designations for an arbitrary right triangle removed from a coordinate system, we have the following:

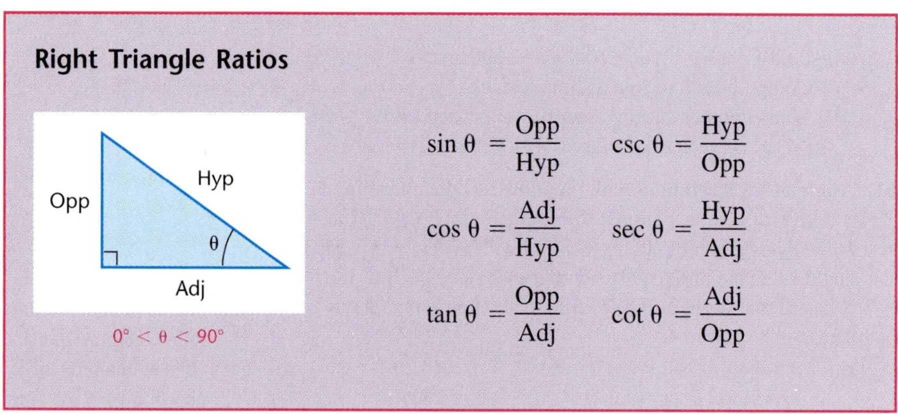

EXPLORE-DISCUSS 1 For a given value θ, $0 < \theta < 90°$, explain why the value of each of the six trigonometric functions is independent of the size of the right triangle that contains θ.

TABLE 1

Angle to nearest	Significant digits for side measure
1°	2
10′ or 0.1°	3
1′ or 0.01°	4
10″ or 0.001°	5

The use of the trigonometric ratios for right triangles is made clear in the following examples. Regarding computational accuracy, we use Table 1 as a guide. (The table is also printed inside the back cover of this book for easy reference.) We will use $=$ rather than $\approx$ in many places, realizing the accuracy indicated in Table 1 is all that is assumed. Another word of caution: when using your calculator be sure it set in degree mode.

EXAMPLE 1 **Right Triangle Solution**

Solve the right triangle with $c = 6.25$ feet and $\beta = 32.2°$.

Solution First draw a figure and label the parts (Fig. 2):

FIGURE 2

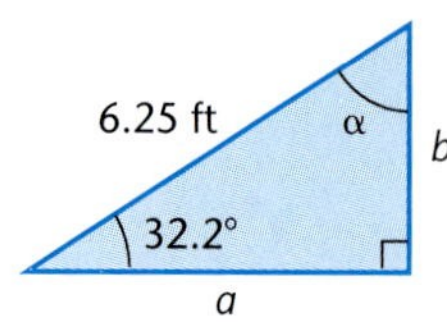

Solve for α $\qquad \alpha = 90° - 32.2° = 57.8°$ $\quad \alpha$ and β are complementary.

Solve for b

$$\sin\beta = \frac{b}{c} \qquad \text{Or use } \csc\beta = \frac{c}{b}.$$

$$\sin 32.2^\circ = \frac{b}{6.25}$$

$$b = 6.25 \sin 32.2^\circ$$

$$= 3.33 \text{ feet}$$

Solve for a

$$\cos\beta = \frac{a}{c} \qquad \text{Or use } \sec\beta = \frac{c}{a}.$$

$$\cos 32.2^\circ = \frac{a}{6.25}$$

$$a = 6.25 \cos 32.2^\circ$$

$$= 5.29 \text{ feet}$$

Matched Problem 1 Solve the right triangle with $c = 27.3$ meters and $\alpha = 47.8^\circ$.

In the next example we are confronted with a problem of the following type: Find θ given

$$\sin\theta = 0.4196$$

We know how to find (or approximate) $\sin\theta$ given θ, but how do we reverse the process? How do we find θ given $\sin\theta$? First, we note that the solution to the problem can be written symbolically as either

$$\theta = \arcsin 0.4196$$

"arcsin" and "$\sin^{-1}$" both represent the same thing.

or

$$\theta = \sin^{-1} 0.4196$$

Both expressions are read "θ is the angle whose sine is 0.4196".

CAUTION

It is important to note that $\sin^{-1} 0.4196$ does not mean $1/(\sin 0.4196)$. The superscript $^{-1}$ is part of a function symbol, and **$\sin^{-1}$ represents the inverse sine function.** Inverse trigonometric functions are developed in detail in Section 5-9.

Fortunately, we can find θ directly using a calculator. Most calculators of the type used in this book have the function keys [$\sin^{-1}$], [$\cos^{-1}$], [$\tan^{-1}$] or their equivalents (check your manual). These function keys take us from a trigonometric ratio back to the corresponding acute angle in degree measure when the calculator is in degree mode. Thus, if $\sin \theta = 0.4196$, then we can write either $\theta = \arcsin 0.4196$ or $\theta = \sin^{-1} 0.4196$. We choose the latter and proceed as follows:

$$\theta = \sin^{-1} 0.4196$$

$$= 24.81° \qquad \text{To the nearest hundredth degree}$$

$$\text{or } 24°49' \qquad \text{To the nearest minute}$$

Check

$$\sin 24.81° = 0.4196$$

EXPLORE-DISCUSS 2 Solve each of the following for θ to the nearest hundredth of a degree using a calculator. Explain why an error message occurs in one of the problems.

(A) $\cos \theta = 0.2044$ (B) $\tan \theta = 1.4138$ (C) $\sin \theta = 1.4138$

EXAMPLE 2 Right Triangle Solution

Solve the right triangle with $a = 4.32$ centimeters and $b = 2.62$ centimeters. Compute the angle measures to the nearest $10'$.

Solution Draw a figure and label the known parts (Fig. 3):

FIGURE 3

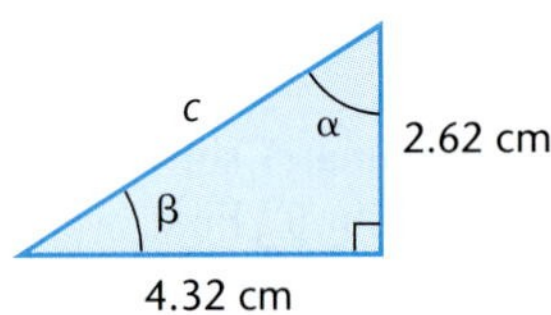

Solve for β

$$\tan \beta = \frac{2.62}{4.32}$$

$$\beta = \tan^{-1} \frac{2.62}{4.32}$$

$$= 31.2° \text{ or } 31°10' \qquad 0.2° = [(0.2)(60)]' = 12' \approx 10' \text{ to nearest } 10'$$

Solve for α

$$\alpha = 90° - 31°10' \quad \boxed{= 89°60' - 31°10'} \quad = 58°50'$$

Solve for c

$$\sin \beta = \frac{2.62}{c} \qquad \text{Or use } \csc \beta = \frac{c}{2.62}.$$

$$c = \frac{2.62}{\sin 31.2°} = 5.06 \text{ centimeters}$$

or, using the Pythagorean theorem,

$$c = \sqrt{4.32^2 + 2.62^2} = 5.05 \text{ centimeters}$$

Note the slight difference in the values obtained for c (5.05 versus 5.06). This was caused by rounding β to the nearest $10'$ in the first calculation for c.

Matched Problem 2 Solve the right triangle with $a = 1.38$ kilometers and $b = 6.73$ kilometers.

EXAMPLE 3 Geometry

If a pentagon (a five-sided regular polygon) is inscribed in a circle of radius 5.35 centimeters, find the length of one side of the pentagon.

Solution Sketch a figure and insert triangle ACB with C at the center (Fig. 4). Add the auxiliary line CD as indicated. We will find AD and double it to find the length of the side wanted.

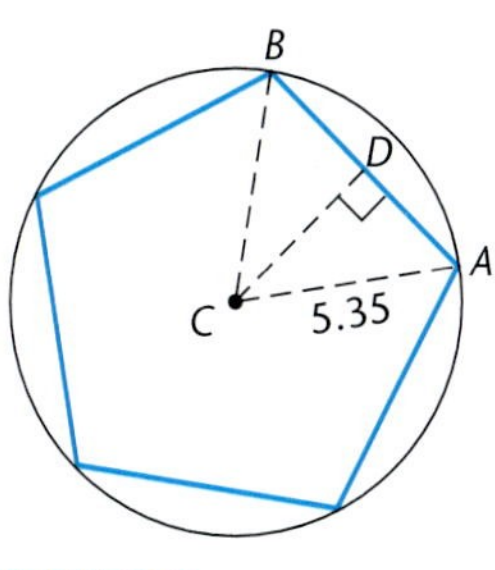

FIGURE 4

$$\text{Angle } ACB = \frac{360°}{5} = 72° \quad \text{Exact}$$

$$\text{Angle } ACD = \frac{72°}{2} = 36° \quad \text{Exact}$$

$$\sin(\text{angle } ACD) = \frac{AD}{AC}$$

$$\begin{aligned} AD &= AC \sin(\text{angle } ACD) \\ &= 5.35 \sin 36° \\ &= 3.14 \text{ centimeters} \\ AB &= 2AD = 6.28 \text{ centimeters} \end{aligned}$$

Matched Problem 3 If a square of side 43.6 meters is inscribed in a circle, what is the radius of the circle?

EXAMPLE 4 Architecture

In designing a house an architect wishes to determine the amount of overhang of a roof so that it shades the entire south wall at noon during the summer solstice (Fig. 5). Minimally, how much overhang should be provided for this purpose?

FIGURE 5

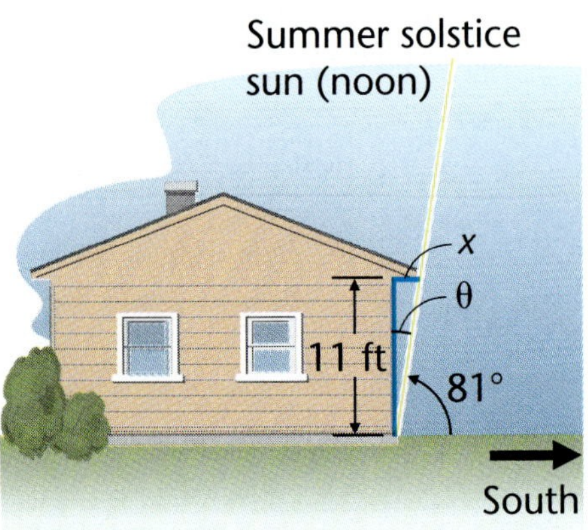

Solution From Figure 5 we solve for x:

$$\theta = 90° - 81° = 9°$$

$$\tan \theta = \frac{x}{11}$$

$$x = 11 \tan 9° = 1.7 \text{ feet}$$

Matched Problem 4 With the overhang found in Example 4, how far will the shadow of the overhang come down the south wall at noon during the winter solstice, when the angle of elevation of the sun is 34°?

EXPLORE-DISCUSS 3 At what latitude is the home of Example 4 located?

Answers to Matched Problems

1. $\beta = 42.2°$, $a = 20.2$ m, $b = 18.3$ m **2.** $\alpha = 11°40'$, $\beta = 78°20'$, $c = 6.87$ km
3. 30.8 m **4.** 1.1 ft

EXERCISE 5-5

A

For the triangle in the figure for Problems 1–12, write the ratios of sides corresponding to each trigonometric function given in Problems 1–6. Do not look back at the definitions.

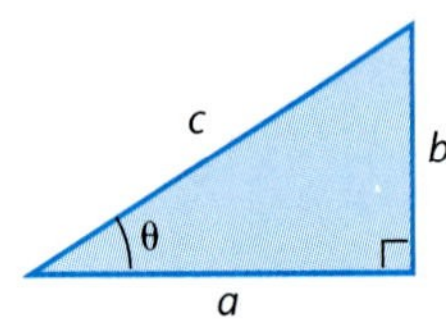

1. $\sin \theta$ **2.** $\cot \theta$ **3.** $\csc \theta$

4. $\cos \theta$ **5.** $\tan \theta$ **6.** $\sec \theta$

Each ratio in Problems 7–12 defines a trigonometric function of the complement of θ (refer to the figure). Indicate which function without looking back at the definitions.

7. b/a **8.** c/b **9.** a/c

10. a/b **11.** b/c **12.** c/a

In Problems 13–18, find each acute angle θ in degree measure to two decimal places using a calculator.

13. $\sin\theta = 0.9243$

14. $\cos\theta = 0.5277$

15. $\theta = \tan^{-1} 9.533$

16. $\theta = \sin^{-1} 0.0317$

17. $\theta = \cos^{-1} 0.7425$

18. $\theta = \tan^{-1} 4.296$

B

Solve each triangle in Problems 19–30 using the information given and the triangle labeling in the figure.

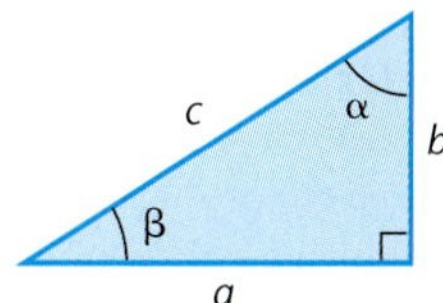

19. $\beta = 17.8°, c = 3.45$

20. $\beta = 33.7°, b = 22.4$

21. $\beta = 43°20', a = 123$

22. $\beta = 62°30', c = 42.5$

23. $\alpha = 23°0', a = 54.0$

24. $\alpha = 54°, c = 4.3$

25. $\alpha = 53.21°, b = 23.82$

26. $\alpha = 35.73°, b = 6.482$

27. $a = 6.00, b = 8.46$

28. $a = 22.0, b = 46.2$

29. $b = 10.0, c = 12.6$

30. $b = 50.0, c = 165$

Problems 31–36 give a geometric interpretation of the ***trigonometric ratios.*** *Refer to the figure, where O is the center of a circle of radius 1, θ is the acute angle AOD, D is the intersection point of the terminal side of angle θ with the circle, and EC is tangent to the circle at D.*

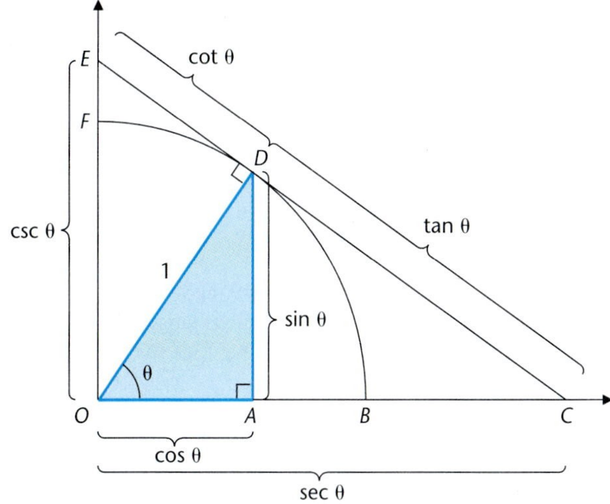

31. Explain why:
(A) $\cos\theta = OA$ (B) $\cot\theta = DE$ (C) $\sec\theta = OC$

32. Explain why:
(A) $\sin\theta = AD$ (B) $\tan\theta = DC$ (C) $\csc\theta = OE$

33. Explain what happens to each of the following as the acute angle θ approaches 90°:
(A) $\cos\theta$ (B) $\cot\theta$ (C) $\sec\theta$

34. Explain what happens to each of the following as the acute angle θ approaches 90°:
(A) $\sin\theta$ (B) $\tan\theta$ (C) $\csc\theta$

35. Explain what happens to each of the following as the acute angle θ approaches 0°:
(A) $\sin\theta$ (B) $\tan\theta$ (C) $\csc\theta$

36. Explain what happens to each of the following as the acute angle θ approaches 0°:
(A) $\cos\theta$ (B) $\cot\theta$ (C) $\sec\theta$

C

37. Show that (see figure): $h = \dfrac{d}{\cot\alpha - \cot\beta}$

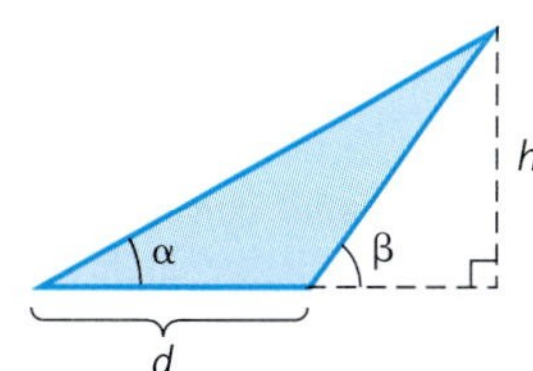

38. Show that (see figure): $h = \dfrac{d}{\cot\alpha + \cot\beta}$

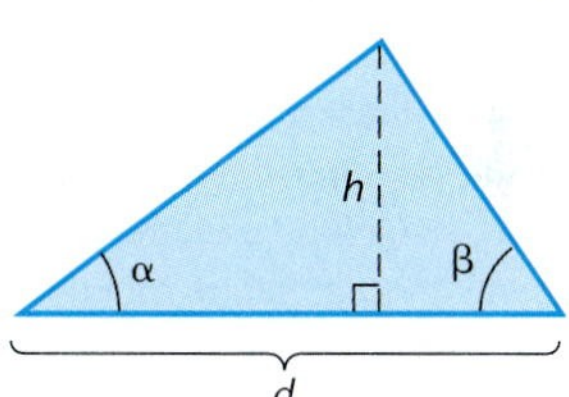

APPLICATIONS 

39. Surveying. Find the height of a tree (growing on level ground) if at a point 105 feet from the base of the tree the angle to its top relative to the horizontal is found to be 65.3°.

40. Air Safety. To measure the height of a cloud ceiling over an airport, a searchlight is directed straight upward to produce a lighted spot on the clouds. Five hundred meters away an observer reports the angle of the spot relative to the horizontal to be 32.2°. How high (to the nearest meter) are the clouds above the airport?

41. Engineering. If a train climbs at a constant angle of 1°23′, how many vertical feet has it climbed after going 1 mile? (1 mile = 5,280 feet)

42. Air Safety. If a jet airliner climbs at an angle of 15°30′ with a constant speed of 315 miles per hour, how long will it take (to the nearest minute) to reach an altitude of 8.00 miles? Assume there is no wind.

★**43. Astronomy.** Find the diameter of the moon (to the nearest mile) if at 239,000 miles from Earth it produces an angle of 32′ relative to an observer on Earth.

★**44. Astronomy.** If the sun is 93,000,000 miles from Earth and its diameter is opposite an angle of 32′ relative to an observer on Earth, what is the diameter of the sun (to 2 significant digits)?

★**45. Geometry.** If a circle of radius 4 centimeters has a chord of length 3 centimeters, find the central angle that is opposite this chord (to the nearest degree).

★**46. Geometry.** Find the length of one side of a nine-sided regular polygon inscribed in a circle of radius 4.06 inches.

47. Physics. In a course in physics it is shown that the velocity v of a ball rolling down an inclined plane (neglecting air resistance and friction) is given by

$$v = gt \sin \theta$$

where g is a gravitational constant (acceleration due to gravity), t is time, and θ is the angle of inclination of the plane (see the figure). Galileo (1564–1642) used this equation in the form

$$g = \frac{v}{t \sin \theta}$$

to estimate g after measuring v experimentally. (At that time, no timing devices existed to measure the velocity of a free-falling body, so Galileo used the inclined plane to slow the motion down.) A steel ball is rolled down a glass plane inclined at 8.0°. Approximate g to one decimal place if at the end of 3.0 seconds the ball has a measured velocity of 4.1 meters per second.

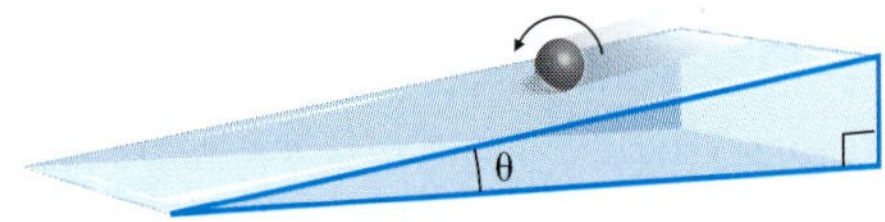

48. Physics. Refer to Problem 47. A steel ball is rolled down a glass plane inclined at 4.0°. Approximate g to one decimal place if at the end of 4.0 seconds the ball has a measured velocity of 9.0 feet per second.

★**49. Engineering—Cost Analysis.** A cable television company wishes to run a cable from a city to a resort island 3 miles offshore. The cable is to go along the shore, then underwater to the island, as indicated in the accompanying figure. The cost of running the cable along the shore is \$15,000 per mile and underwater, \$25,000 per mile.

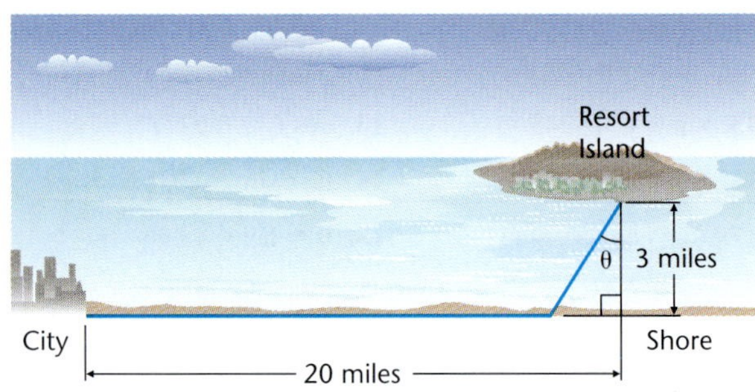

(A) Referring to the figure, show that the cost in terms of θ is given by

$$C(\theta) = 75{,}000 \sec \theta - 45{,}000 \tan \theta + 300{,}000$$

(B) Calculate a table of costs, each cost to the nearest dollar, for the following values of θ: 10°, 20°, 30°, 40°, and 50°. (Notice how the costs vary with θ. In a course in calculus, students are asked to find θ so that the cost is minimized.)

★**50. Engineering—Cost Analysis.** Refer to Problem 49. Suppose the island is 4 miles offshore and the cost of running the cable along the shore is \$20,000 per mile and underwater, \$30,000 per mile.

(A) Referring to the figure for Problem 49 with appropriate changes, show that the cost in terms of θ is given by

$$C(\theta) = 120{,}000 \sec \theta - 80{,}000 \tan \theta + 400{,}000$$

(B) Calculate a table of costs, each cost to the nearest dollar, for the following values of θ: 10°, 20°, 30°, 40°, and 50°.

★★**51. Geometry.** Find r in the accompanying figure (to 2 significant digits) so that the circle is tangent to all three sides of the isosceles triangle. [*Hint:* The radius of a circle is perpendicular to a tangent line at the point of tangency.]

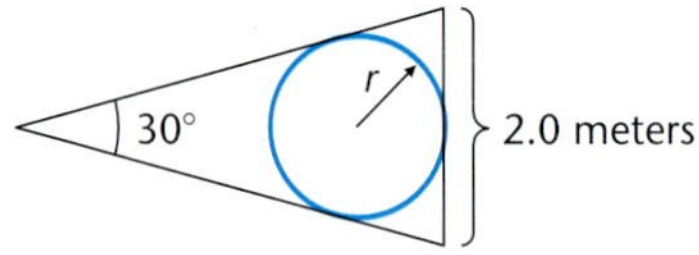

★★**52. Geometry.** Find r in the accompanying figure (to 2 significant digits) so that the smaller circle is tangent to the larger circle and the two sides of the angle. [See the hint in Problem 51.]

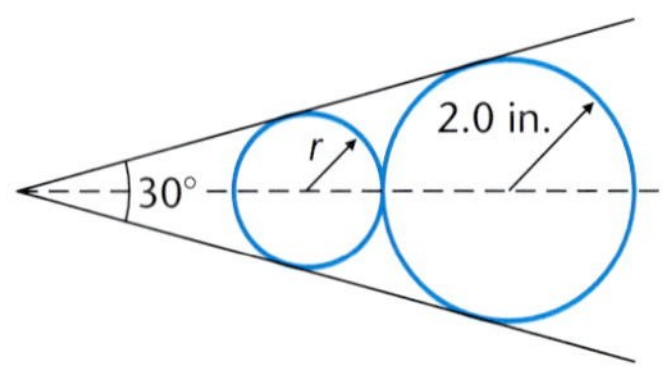

SECTION 5-6 Graphing Basic Trigonometric Functions

- Periodic Functions
- Graphs of $y = \sin x$ and $y = \cos x$
- Graphs of $y = \tan x$ and $y = \cot x$
- Graphs of $y = \csc x$ and $y = \sec x$
- Graphs on a Graphing Utility

Consider the graphs of sunrise times and sound waves shown in Figure 1. What is a common feature of the two graphs? Both represent repetitive phenomena; that is, both appear to be periodic. Trigonometric functions are particularly suited to describe periodic phenomena.

FIGURE 1 Periodic phenomena.

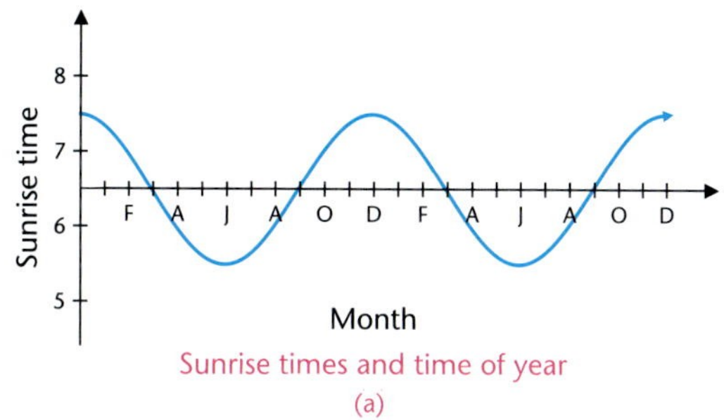

Sunrise times and time of year
(a)

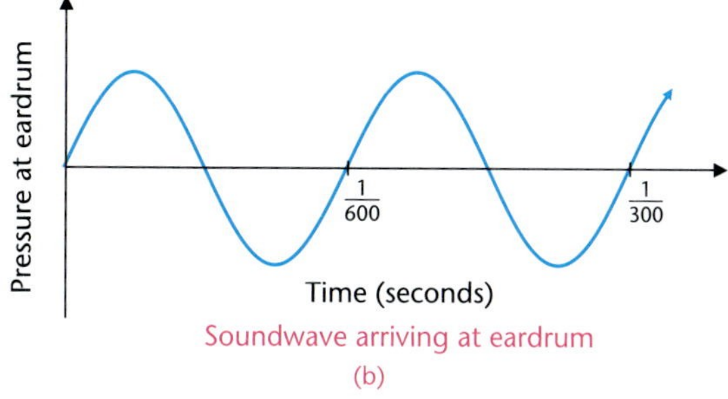

Soundwave arriving at eardrum
(b)

This section discusses the graphs of the six trigonometric functions with the real number domains introduced earlier. Also discussed are the domains, ranges, and periodic properties of these functions. The circular functions introduced in Section 5-2 will prove particularly useful in this regard.

It appears there is a lot to remember in this section. However, you only need to be familiar with the graphs and properties of the sine, cosine, and tangent functions. The reciprocal relationships discussed earlier will enable you to determine the graphs and properties of the other three trigonometric functions from the graphs and properties of the sine, cosine, and tangent functions.

• Periodic Functions

Let's return to circular points and wrapping functions discussed in Sections 5-1 and 5-2. Because the unit circle has a circumference of 2π, we find that for a given value of x (see Fig. 2 on the following page) we will return to the circular point $W(x) = (a, b)$ if we add any integer multiple of 2π to x. Think of a point P moving around the unit circle in either direction. Every time P covers a distance of 2π, the circumference of the circle, it is back at the point where it started. Thus, for x any real number,

$$\sin(x + 2k\pi) = \sin x \qquad k \text{ any integer}$$
$$\cos(x + 2k\pi) = \cos x \qquad k \text{ any integer}$$

FIGURE 2

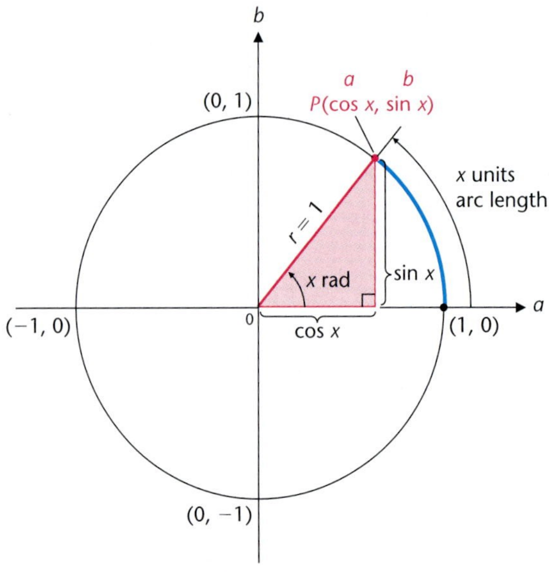

Functions with this kind of repetitive behavior are called **periodic functions.** In general, we have Definition 1.

DEFINITION 1 **Periodic Functions**

A function f is **periodic** if there exists a positive real number p such that

$$f(x + p) = f(x)$$

for all x in the domain of f. The smallest such positive p, if it exists, is called the **fundamental period of f** (or often just the **period of f**).

Both the sine and cosine functions are periodic with period 2π.

• Graphs of $y = \sin x$ and $y = \cos x$

We start by graphing

$$y = \sin x \qquad x \text{ a real number} \tag{1}$$

The graph of the sine function is the graph of the set of all ordered pairs of real numbers (x, y) that satisfy equation (1). Obtaining the graphs using point-by-point plotting is tedious and tends to obscure many important properties. We gain significantly more insight into the nature of these functions by observing how $y = \sin x = b$ varies as the circular point $P(a, b)$ moves around the unit circle.

We now know that the domain of the sine function is the set of all real numbers R, the range is $[-1, 1]$, and the period is 2π. Because the sine function has a period of 2π, we will concentrate on the graph over one period, from 0 to 2π. Once we have the graph for one period, we can complete as much of the rest of the graph as is needed by repeating the graph to the left or to the right.

Figure 3 illustrates how $y = \sin x = b$ varies as x increases from 0 to 2π and $P(a, b)$ moves around the unit circle.

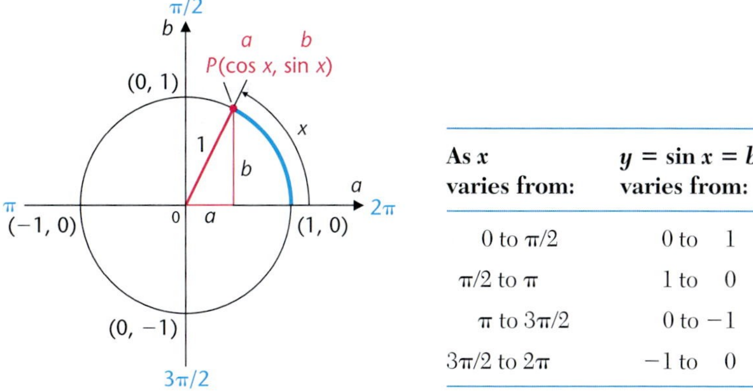

As x varies from:	$y = \sin x = b$ varies from:
0 to $\pi/2$	0 to 1
$\pi/2$ to π	1 to 0
π to $3\pi/2$	0 to -1
$3\pi/2$ to 2π	-1 to 0

FIGURE 3 Variation in $\sin x$.

To sketch a graph of $y = \sin x$ over the interval $[0, 2\pi]$, we divide the interval into four equal parts corresponding to the quadrants through which x varies and y behaves uniformly. We choose as our basic unit on the x axis $2\pi/4 = \pi/2$. Of course, all other real numbers are on the x axis, but for clarity we choose only to mark the multiples of $\pi/2$. To complete the sketch, we use the results in Figure 3 supplemented where necessary by special real values (integer multiples of $\pi/6$ or $\pi/4$) or calculator values. The final graph is shown in Figure 4. The circle on the left—which is used to define the sine function—is usually referred to mentally and is not part of the graph of $y = \sin x$.

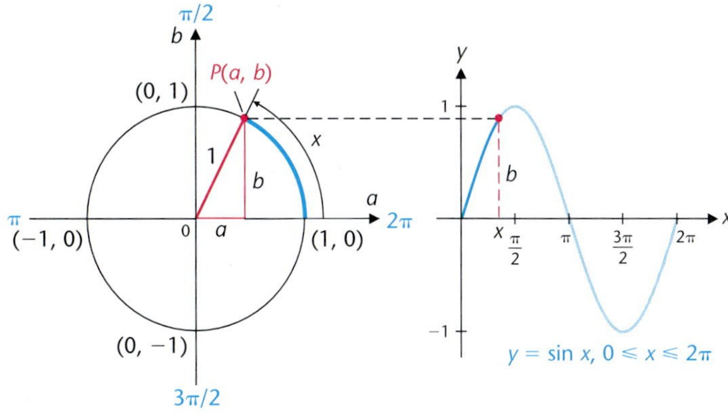

FIGURE 4 Graphing $y = \sin x$.

Figure 5 summarizes the above results and shows the sine graph over several periods.

FIGURE 5

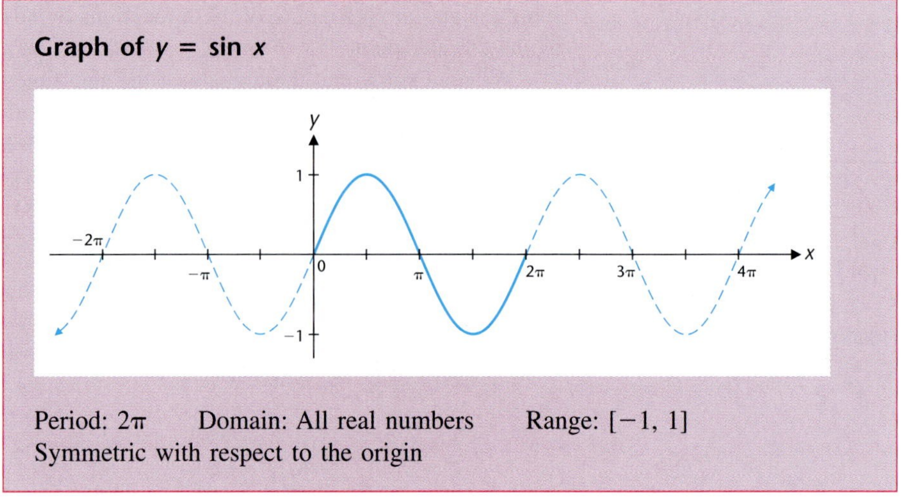

If we proceed in the same way for the cosine function, we can obtain its graph. Figure 6 shows how $\cos x = a = y$ varies as $P(a, b)$ moves around the unit circle.

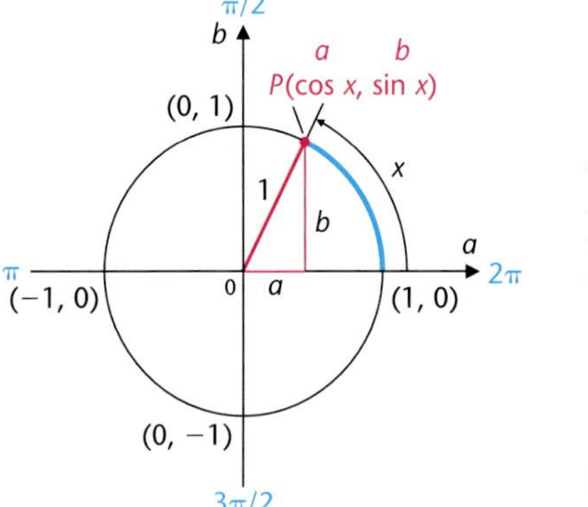

FIGURE 6 Variation in cos *x*.

As x varies from:	$y = \cos x = a$ varies from:
0 to $\pi/2$	1 to 0
$\pi/2$ to π	0 to -1
π to $3\pi/2$	-1 to 0
$3\pi/2$ to 2π	0 to 1

We can use the results in Figure 6, the fact that the cosine function is periodic with period 2π, and special or calculator values where necessary to obtain Figure 7.

FIGURE 7

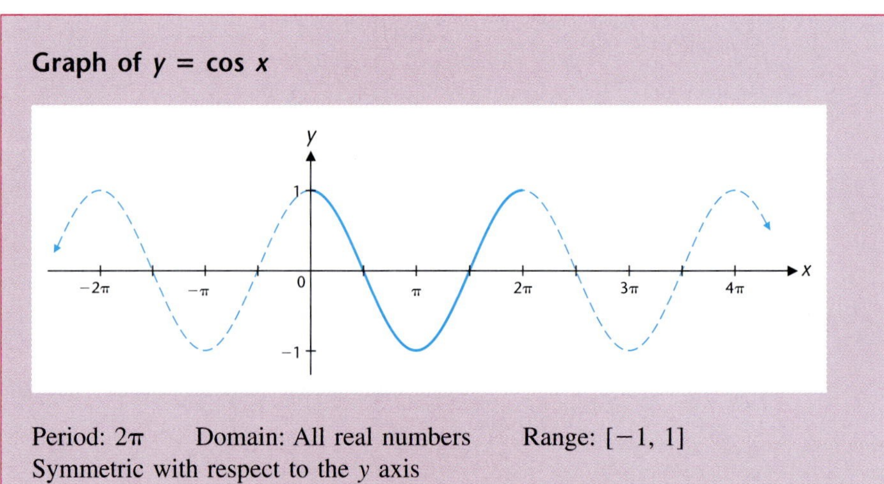

The basic characteristics of the sine and cosine graphs should be learned so that the curves can be sketched quickly. In particular, you should be able to answer the following questions:

1. What is the period of each function (how often does the graph repeat)?
2. Where are the x intercepts?
3. Where are the y intercepts?
4. How far does each curve deviate from the x axis?
5. Where do the high and low points occur?
6. What are the symmetry properties?

EXPLORE-DISCUSS 1

(A) Discuss how the graphs of the sine and cosine functions are related.
(B) How would you shift and/or reflect the sine graph to obtain the cosine graph?
(C) Is the graph of either $y = \sin(x - \pi/2)$ or $y = \sin(x + \pi/2)$ the same as the graph of $y = \cos x$? Explain in terms of shifts and/or reflections.

• Graphs of $y = \tan x$ and $y = \cot x$

We first discuss the graph of $y = \tan x$. Then from this graph, because $\cot x = 1/(\tan x)$, we will be able to get the graph of $y = \cot x$ using reciprocals of ordinates.

Figure 8 shows that whenever the circular point $P(a, b)$ is on the horizontal axis (that is, whenever $x = k\pi$, k an integer), then $(a, b) = (\pm 1, 0)$, and $\tan x = b/a = 0/(\pm 1) = 0$. Whenever $P(a, b)$ is on the vertical axis [that is, when $x = (\pi/2) + k\pi$, k an integer], then $(a, b) = (0, \pm 1)$, and $\tan x = b/a = (\pm 1)/0$ is not defined (the tangent function is discontinuous).

$\tan x = \dfrac{b}{a}$

FIGURE 8 The unit circle and tan x.

The values of x such that $P(a, b)$ is on the horizontal axis in Figure 8 are the zeros for tan x, or the x intercepts for the graph of $y = \tan x$. Thus, we can write

$$x \text{ intercepts: } k\pi \qquad k \text{ an integer}$$

As a first step in graphing $y = \tan x$, we locate the x intercepts on the x axis as illustrated in Figure 9.

FIGURE 9 Intercepts and asymptotes for $y = \tan x$.

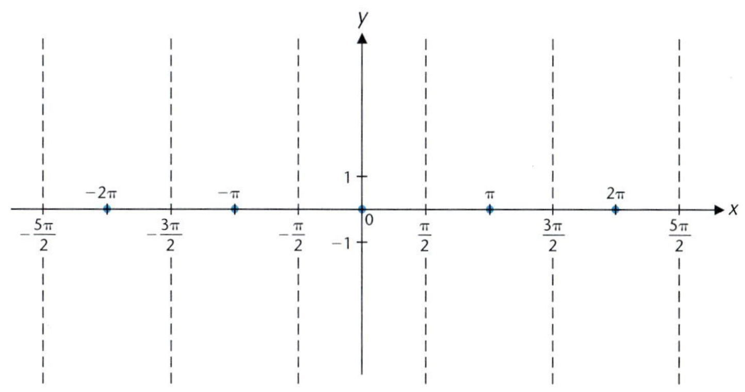

The values of x such that $P(a, b)$ is on a vertical axis in Figure 8 are points of discontinuity. As a second step in graphing $y = \tan x$, we draw dashed vertical lines through these points of discontinuity as illustrated in Figure 9—the graph cannot cross these lines. These dashed vertical lines, called *asymptotes,* are convenient guidelines for sketching the graph of $y = \tan x$. The line $x = a$ is a **vertical asymptote** for the graph of $y = f(x)$ if $f(x)$ either increases or decreases without bound as x approaches a from the left or from the right. Thus, we write

$$\text{Vertical asymptotes:} \quad x = \frac{\pi}{2} + k\pi \qquad k \text{ an integer}$$

We next investigate the behavior of the graph of $y = \tan x$ in more detail between the two asymptotes nearest the origin, that is, over the interval $(-\pi/2, \pi/2)$. Since $\tan(-x) = -\tan x$ (Section 5-2), we only need to develop the graph for the interval $[0, \pi/2)$, then we can reflect this graph through the origin to obtain the graph for the entire interval $(-\pi/2, \pi/2)$.

Two points on the graph for the interval $[0, \pi/2)$ are easy to compute: $\tan 0 = 0$ and $\tan(\pi/4) = 1$. What happens to $\tan x$ as x approaches $\pi/2$ from the left? If x approaches $\pi/2$ from the left, the circular point $P(a, b)$ in Figure 9 stays in the first quadrant, and a approaches 0 through positive values and b approaches 1. What happens to $y = \tan x$ in the process? The calculator experiment in Example 1 will help determine an answer.

EXAMPLE 1 Calculator Experiment

Form a table of values for $y = \tan x$ approaching $\pi/2 \approx 1.570\ 796$ through values less than $\pi/2$, starting at 0. Conclusion?

Solution A table is created as follows:

x	0	0.5	1	1.57	1.5707	1.570 796
$\tan x$	0	0.5	1.6	1,256	10,381	3,060,022

As x approaches $\pi/2$ from the left, $\tan x$ appears to increase without bound.

Matched Problem 1 Form a table of values for $y = \tan x$ with x approaching $-\pi/2 \approx -1.570\ 796$ through values greater than $-\pi/2$, starting at 0. That is, use the negative of the x values used in Example 1. Conclusion?

Figure 10(a) shows the graph resulting from the Example 1 analysis. The graph can be completed for the interval $(-\pi/2, \pi/2)$ by reflecting the graph in Figure 10(a) through the origin. Figure 10(b) shows the result.

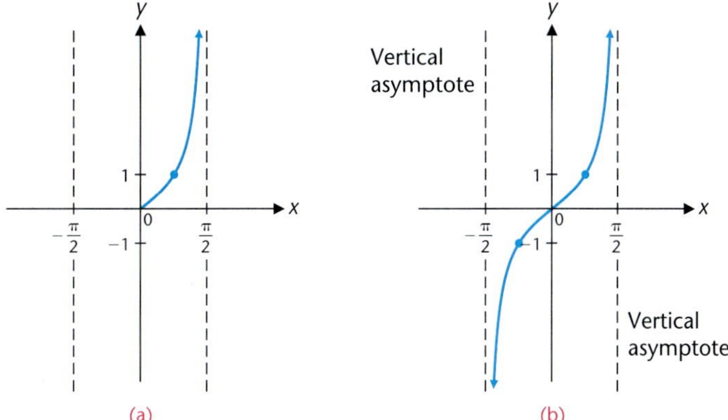

FIGURE 10 Graph of $y = \tan x$, $-\pi/2 < x < \pi/2$.

Proceeding in the same way for the other intervals between asymptotes, it appears that the tangent function is periodic with period π. To verify this, return again to Figure 8. If (a, b) are the coordinates of the circular point associated with x, then, using symmetry of the unit circle and congruent reference triangles, $(-a, -b)$ are the coordinates of the circular point associated with $x + \pi$. Hence,

$$\tan (x + \pi) = \frac{-b}{-a} = \frac{b}{a} = \tan x$$

and we conclude that the tangent function is periodic with period π. In general,

$$\tan (x + k\pi) = \tan x \qquad k \text{ an integer}$$

for all values of x for which both sides of the equation are defined.

To complete the graph of $y = \tan x$ we need only to repeat the graph in Figure 10 to the left and right over intervals of π to produce as much of the general graph as we need (see Fig. 11). The main characteristics of the graph of the tangent function should be learned so that the graph can be sketched quickly.

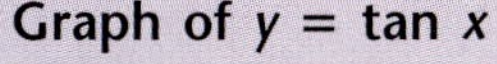

Graph of $y = \tan x$

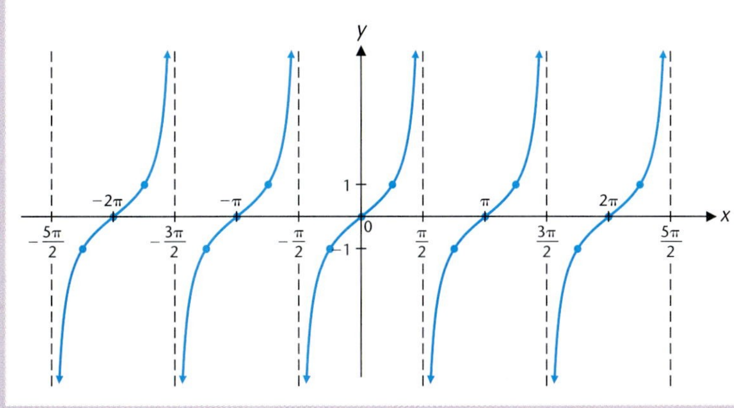

Period: π

Domain: All real numbers except $\pi/2 + k\pi$, k an integer

Range: All real numbers

Symmetric with respect to the origin

Increasing function between consecutive asymptotes

Discontinuous at $x = \pi/2 + k\pi$, k an integer

FIGURE 11

We now turn to the cotangent function. Since $\cot x = 1/\tan x$, we can graph $y = \cot x$ by taking reciprocals of the y values in the graph of $y = \tan x$ in Figure 11. Note that the x intercepts and the vertical asymptotes are interchanged. The graph of $y = \cot x$ is shown in Figure 12. As with the tangent function, its main characteristics should be learned so that its graph can be sketched quickly.

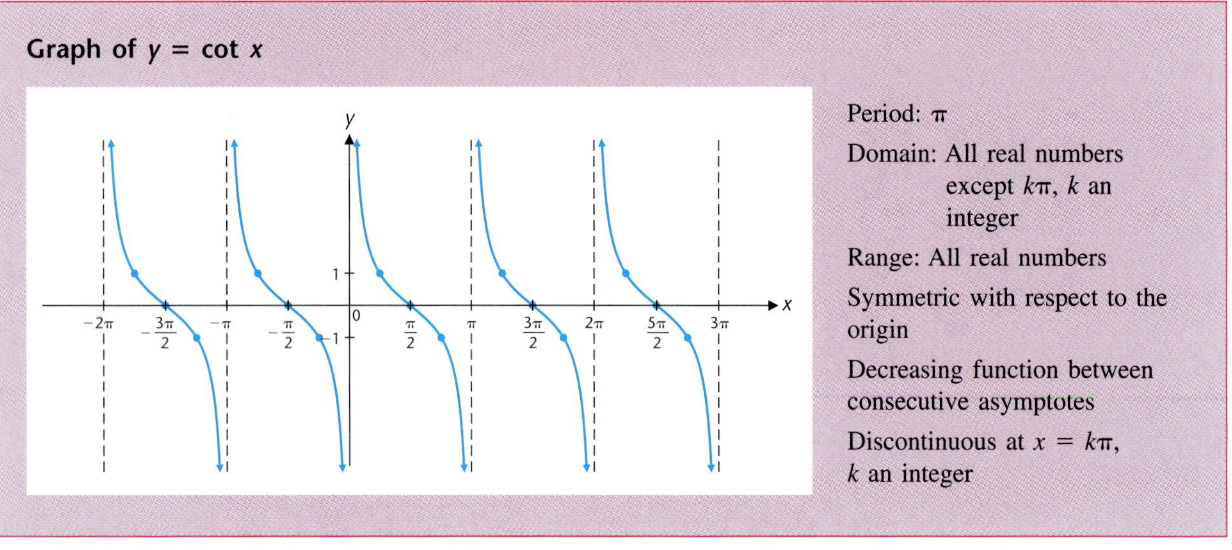

FIGURE 12

EXPLORE-DISCUSS 2

(A) Discuss how the graphs of the tangent and cotangent functions are related.

(B) How would you shift and/or reflect the tangent graph to obtain the cotangent graph?

(C) Is the graph of either $y = \tan(x - \pi/2)$ or $y = -\tan(x - \pi/2)$ the same as the graph of $y = \cot x$? Explain in terms of shifts and/or reflections.

• Graphs of $y = \csc x$ and $y = \sec x$

Just as we obtained the graph of $y = \cot x$ by taking reciprocals of the y values in the graph of $y = \tan x$, since

$$\csc x = \frac{1}{\sin x} \qquad \text{and} \qquad \sec x = \frac{1}{\cos x}$$

we can obtain the graphs of $y = \csc x$ and $y = \sec x$ by taking reciprocals of the y values in the graphs of $y = \sin x$ and $y = \cos x$, respectively. Vertical asymptotes occur at the x intercepts of either $\sin x$ or $\cos x$.

The graphs of $y = \csc x$ and $y = \sec x$ are shown in Figures 13 and 14, respectively. As a graphing aid, we sketch in broken lines of $y = \sin x$ and $y = \cos x$ first and then draw vertical asymptotes through the x intercepts. Check a few points on the graphs with a calculator.

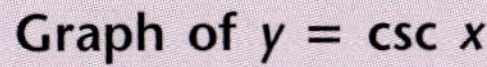

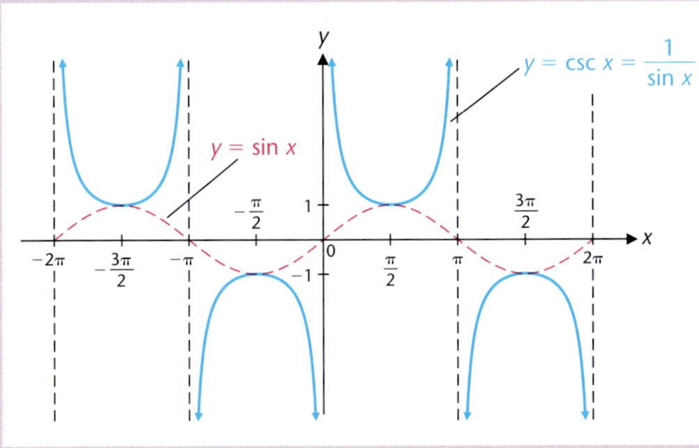

Period: 2π

Domain: All real numbers except $k\pi$, k an integer

Range: All real numbers y such that $y \leq -1$ or $y \geq 1$

Symmetric with respect to the origin

Discontinuous at $x = k\pi$, k an integer

FIGURE 13

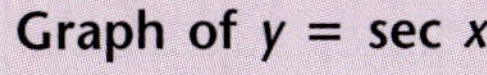

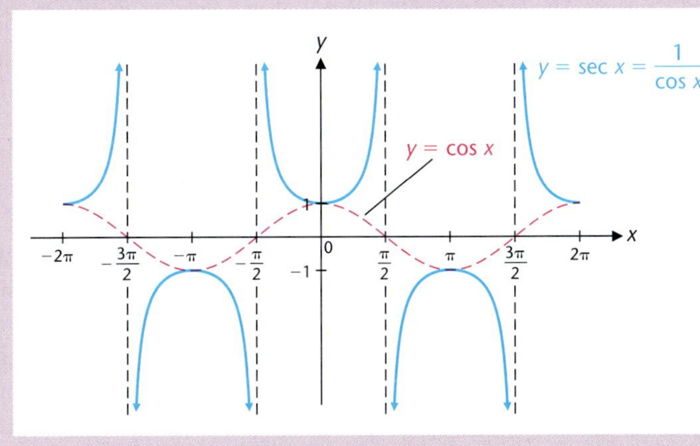

Period: 2π

Domain: All real numbers except $\pi/2 + k\pi$, k an integer

Range: All real numbers y such that $y \leq -1$ or $y \geq 1$

Symmetric with respect to the y axis

Discontinuous at $x = \pi/2 + k\pi$, k an integer

FIGURE 14

We have completed the discussion of the graphs of the six basic trigonometric functions and their fundamental properties. In all cases we proceeded from basic definitions and properties of these functions. You should be able to sketch any of these graphs and describe their fundamental attributes. Keeping the unit circle in mind should prove helpful.

• Graphs on a Graphing Utility

Now that you know how the graphs of the six trigonometric functions are generated from basic definitions and properties, we turn to their graphs on a graphing utility, which can produce these graphs almost instantaneously.

EXAMPLE 2 Trigonometric Graphs on a Graphing Utility

Use a graphing utility to graph the functions

$$y = \sin x \qquad y = \tan x \qquad y = \sec x$$

for $-2\pi \le x \le 2\pi$, $-5 \le y \le 5$. Display each graph in a separate viewing window using a "connected" mode.

Solution First set the graphing utility in radian and connected modes. Next enter the following window parameters, using 6.3 as an approximation for 2π:

$$\text{Xmin} = -6.3 \qquad \text{Xmax} = 6.3 \qquad \text{Xscl} = 1$$

$$\text{Ymin} = -5 \qquad \text{Ymax} = 5 \qquad \text{Yscl} = 1$$

Now enter each function and produce its graph as indicated in Figure 15.

FIGURE 15 Graphing utility graphs in "connected" mode.

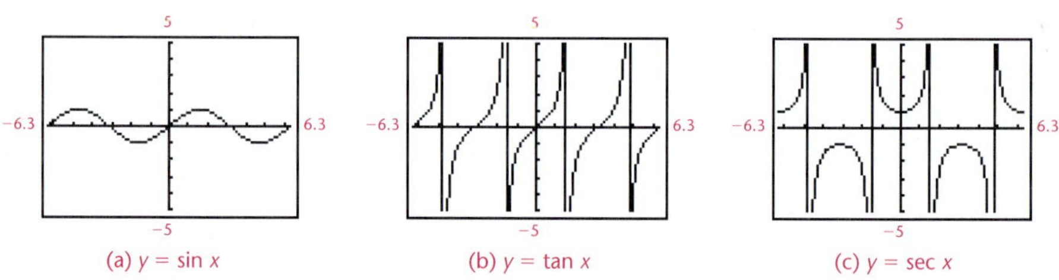

In Figure 15(b) and (c), it appears that the graphing utility has also drawn the vertical asymptotes for these functions. This is not the case. Most graphing utilities calculate points on a graph and connect these points with line segments. The last point plotted to the left of an asymptote and the first point plotted to the right of the asymptote will usually have very large y coordinates. If these y coordinates have opposite sign, then the utility will connect the two points with a line that is nearly vertical, and the line has the appearance of an asymptote. The utility is not performing any asymptote analysis, it is simply connecting points with straight lines. No harm is done as long as you recognize this, and the visual effect is close to that produced with the asymptotes drawn in. You can set a utility to plot points without straight-line connections ("dot" mode), as shown in Figure 16. Unless stated to the contrary, we will graph in the connected mode.

FIGURE 16 Graphing utility graphs in "dot" mode.

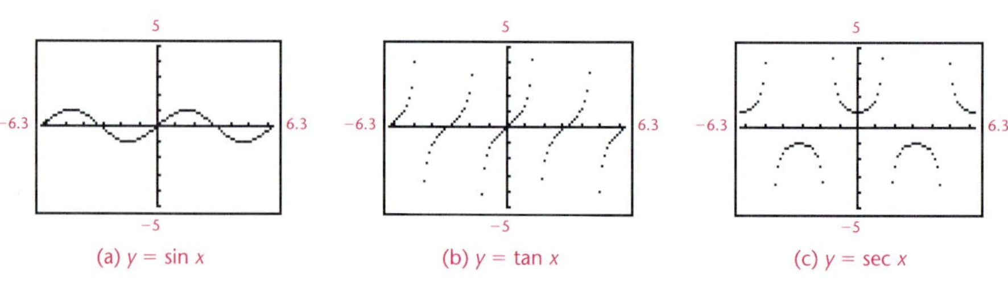

Matched Problem 2

Repeat Example 2 for (A) $y = \cos x$, (B) $y = \cot x$, and (C) $y = \csc x$. (Use connected mode.)

Answers to Matched Problems

1.

x	0	-0.5	-1	-1.57	-1.5707	$-1.570\ 796$
$\tan x$	0	-0.5	-1.6	$-1{,}256$	$-10{,}381$	$-3{,}060{,}022$

As x approaches $-\pi/2$ from the right, $\tan x$ appears to decrease without bound.

2. (A) $y = \cos x$ (B) $y = \cot x$ (C) $y = \csc x$

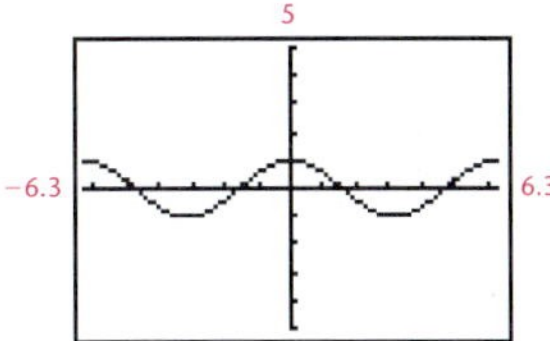

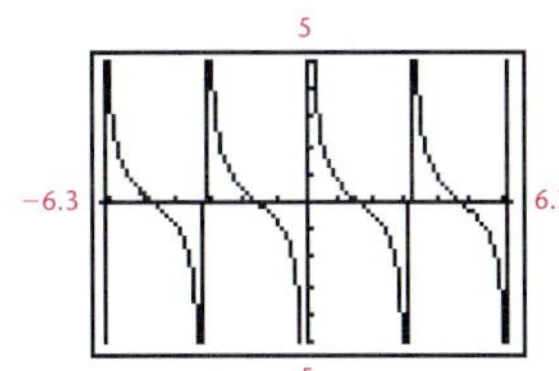

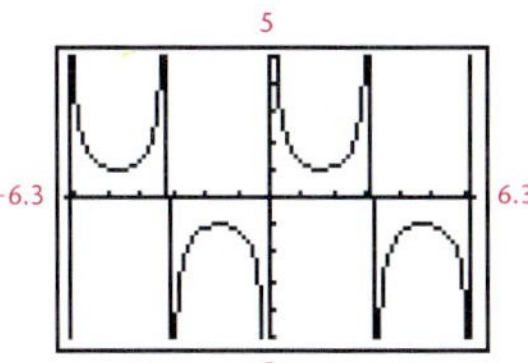

EXERCISE 5-6

The figure below will be useful in many of the problems in this exercise.

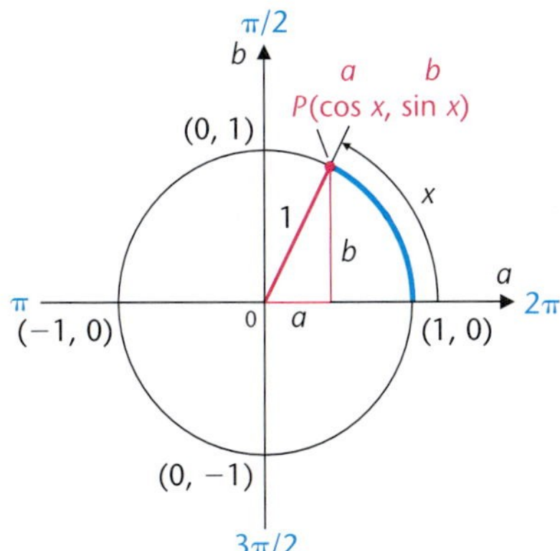

A

Answer Problems 1–12 without looking back in the text or using a calculator. You can refer to the above figure.

1. What are the periods of the sine, cotangent, and cosecant functions?

2. What are the periods of the cosine, tangent, and secant functions?

3. How far does the graph of each function deviate from the x axis?
(A) $y = \cos x$ (B) $y = \tan x$ (C) $y = \csc x$

4. How far does the graph of each function deviate from the x axis?
(A) $y = \sin x$ (B) $y = \cot x$ (C) $y = \sec x$

5. What are the x intercepts for the graph of each function over the interval $-2\pi \le x \le 2\pi$?
(A) $y = \sin x$ (B) $y = \cot x$ (C) $y = \csc x$

6. What are the x intercepts for the graph of each function over the interval $-2\pi \le x \le 2\pi$?
(A) $y = \cos x$ (B) $y = \tan x$ (C) $y = \sec x$

7. Which of the six basic trigonometric functions are undefined at $x = 0$?

8. Which of the six basic trigonometric functions are undefined at $x = \pi/2$?

B

9. At what values of x, $-2\pi \le x \le 2\pi$, do the vertical asymptotes for the following functions cross the x axis?
(A) $y = \cos x$ (B) $y = \tan x$ (C) $y = \csc x$

10. At what values of x, $-2\pi \le x \le 2\pi$, do the vertical asymptotes for the following functions cross the x axis?
(A) $y = \sin x$ (B) $y = \cot x$ (C) $y = \sec x$

11. Sketch graphs of each of the following functions over the interval $[-2\pi, 2\pi]$. Indicate the scale on the x axis in terms of π, and draw in vertical asymptotes using dashed lines where appropriate.
(A) $y = \cos x$ (B) $y = \tan x$ (C) $y = \csc x$

12. Sketch graphs of each of the following functions over the interval $[-2\pi, 2\pi]$. Indicate the scale on the x axis in terms of π, and draw in vertical asymptotes using dashed lines where appropriate.
(A) $y = \sin x$ (B) $y = \cot x$ (C) $y = \sec x$

13. (A) Describe a shift and/or reflection that will transform the graph of $y = \csc x$ into the graph of $y = \sec x$.
(B) Is either the graph of $y = -\csc(x + \pi/2)$ or $y = -\csc(x - \pi/2)$ the same as the graph of $y = \sec x$? Explain in terms of shifts and/or reflections.

14. (A) Describe a shift and/or reflection that will transform the graph of $y = \sec x$ into the graph of $y = \csc x$.

(B) Is either the graph of $y = -\sec(x - \pi/2)$ or $y = -\sec(x + \pi/2)$ the same as the graph of $y = \csc x$? Explain in terms of shifts and/or reflections.

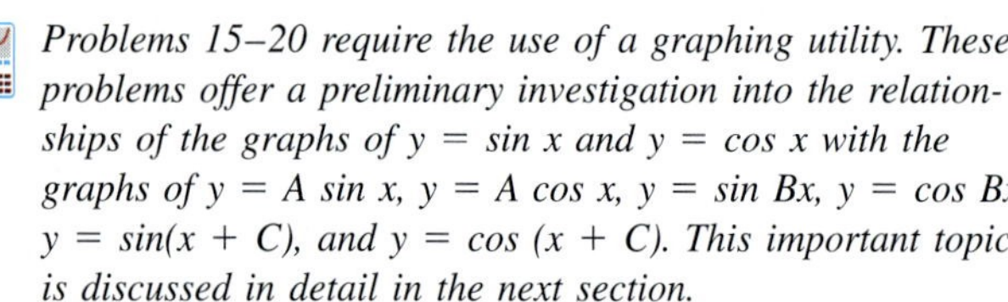

Problems 15–20 require the use of a graphing utility. These problems offer a preliminary investigation into the relationships of the graphs of $y = \sin x$ and $y = \cos x$ with the graphs of $y = A \sin x$, $y = A \cos x$, $y = \sin Bx$, $y = \cos Bx$, $y = \sin(x + C)$, and $y = \cos(x + C)$. This important topic is discussed in detail in the next section.

15. (A) Graph $y = A \cos x$ $(-2\pi \le x \le 2\pi, -3 \le y \le 3)$ for $A = 1, 2$, and -3, all in the same viewing window.
(B) Do the x intercepts change? If so, where?
(C) How far does each graph deviate from the x axis? (Experiment with additional values of A.)
(D) Describe how the graph of $y = \cos x$ is changed by changing the values of A in $y = A \cos x$.

16. Repeat Problem 15, replacing each occurrence of cosine (cos) by sine (sin).

17. (A) Graph $y = \sin Bx$ $(-\pi \le x \le \pi, -2 \le y \le 2)$ for $B = 1, 2$, and 3, all in the same viewing window.
(B) How many periods of each graph appear in this viewing window? (Experiment with additional positive integer values of B.)
(C) Based on the observations in part B, how many periods of the graph of $y = \sin nx$, n a positive integer, would appear in this viewing window?

18. Repeat Problem 17, replacing each occurrence of sine (sin) by cosine (cos).

19. (A) Graph $y = \cos(x + C)$ $(-2\pi \le x \le 2\pi, -1.5 \le y \le 1.5)$ for $C = 0, -\pi/2$, and $\pi/2$, all in the same viewing window. (Experiment with additional values of C.)
(B) Describe how the graph of $y = \cos x$ is changed by changing the values of C in $y = \cos(x + C)$.

20. Repeat Problem 19, replacing each occurrence of cosine (cos) by sine (sin).

C

21. Try to calculate each of the following on your calculator. Explain the results.
(A) $\sec(\pi/2)$ (B) $\tan(-\pi/2)$ (C) $\cot(-\pi)$

22. Try to calculate each of the following on your calculator. Explain the results.
(A) $\csc \pi$ (B) $\tan(\pi/2)$ (C) $\cot 0$

23. Explain why a function that is increasing throughout its domain cannot be periodic.

24. Explain why a nonconstant polynomial function cannot be periodic.

Problems 25 and 26 require the use of a graphing utility.

25. Graph $f(x) = \sin x$ and $g(x) = x$ in the same viewing window $(-1 \le x \le 1, -1 \le y \le 1)$.
(A) What do you observe about the two graphs when x is close to 0, say $-0.5 \le x \le 0.5$?
(B) Complete the table to three decimal places (use the table feature on your graphing utility if it has one):

x	-0.3	-0.2	-0.1	0.0	0.1	0.2	0.3
$\sin x$							

(In applied mathematics certain derivations, formulas and calculations are simplified by replacing $\sin x$ with x for small $|x|$.)

26. Graph $h(x) = \tan x$ and $g(x) = x$ in the same viewing window $(-1 \le x \le 1, -1 \le y \le 1)$.
(A) What do you observe about the two graphs when x is close to 0, say $-0.5 \le x \le 0.5$?
(B) Complete the table to three decimal places (use the table feature on your graphing utility if it has one):

x	-0.3	-0.2	-0.1	0.0	0.1	0.2	0.3
$\tan x$							

(In applied mathematics certain derivations, formulas and calculations are simplified by replacing $\tan x$ with x for small $|x|$.)

In Problems 27–34, determine whether the statement is true or false. If true, explain why. If false, give a counterexample.

27. If functions f and g are periodic with the same period, then the sum $f + g$ and difference $f - g$ are also periodic.

28. If functions f and g are periodic with the same period, then the product fg and quotient f/g are also periodic.

29. If f is any function and g is periodic, then the composite $f \circ g$ is periodic.

30. If f is periodic and g is any function, then the composite $f \circ g$ is periodic.

31. If function f is periodic, then the function g defined by $g(x) = 5f(x)$ is periodic with the same period as f.

32. If function f is periodic, then the function h defined by $h(x) = f(5 + x)$ is periodic with the same period as f.

33. If function f is periodic, then the function j defined by $j(x) = f(5x)$ is periodic with the same period as f.

34. If function f is periodic, then the function k defined by $k(x) = 5 - f(x)$ is periodic with the same period as f.

SECTION 5-7 Graphing $y = k + A \sin (Bx + C)$ and $y = k + A \cos (Bx + C)$

- $y = A \sin x$ and $y = A \cos x$
- $y = \sin Bx$ and $y = \cos Bx$
- $y = k + A \sin Bx$ and $y = k + A \cos Bx$
- $y = k + A \sin (Bx + C)$ and $y = k + A \cos (Bx + C)$
- Finding an Equation of the Graph of a Simple Harmonic

Imagine a weight suspended from the ceiling by a spring. If the weight were pulled downward and released, then, assuming no air resistance or friction, it would move up and down with the same frequency and amplitude forever. This idealized motion is an example of **simple harmonic motion.** Simple harmonic motion can be described by functions of the form $y = A \sin (Bx + C)$ or $y = A \cos (Bx + C)$, called **simple harmonics.**

Simple harmonics are extremely important in both pure and applied mathematics. In applied mathematics they are used in the analysis of sound waves, radio waves, X rays, gamma rays, visible light, infrared radiation, ultraviolet radiation, seismic waves, ocean waves, electric circuits, electric generators, vibrations, bridge and building construction, spring-mass systems, bow waves of boats, sonic booms, and so on. Analysis involving simple harmonics is called **harmonic analysis.**

The graphs of simple harmonics, and of the more general forms $y = k + A \sin (Bx + C)$ and $y = k + A \cos (Bx + C)$, can be investigated by studying the effects that A, B, C, and k have on the basic graphs of $y = \sin x$ and $y = \cos x$ discussed in Section 5-6. A brief review of Section 2-5 should also prove helpful.

• $y = A \sin x$ and $y = A \cos x$

We first investigate the effect of A by comparing

$$y = \sin x \qquad \text{and} \qquad y = A \sin x$$

The graph of $y = A \sin x$ can be obtained from the graph of $y = \sin x$ by multiplying each y value of $y = \sin x$ by A. The graph of $y = A \sin x$ still crosses the x axis where the graph of $y = \sin x$ crosses the x axis, because $A \cdot 0 = 0$. Since the maximum value of $\sin x$ is 1, the maximum value of $A \sin x$ is $|A| \cdot 1 = |A|$. The constant $|A|$ is called the **amplitude** of the graph of $y = A \sin x$ and indicates the maximum deviation of the graph of $y = A \sin x$ from the x axis. Finally, the period of $y = A \sin x$ is also 2π, since $A \sin (x + 2\pi) = A \sin x$.

EXAMPLE 1 Comparing Graphs with Different Amplitudes

Compare the amplitudes of $y = \frac{1}{2} \sin x$ and $y = -2 \sin x$ with the amplitude of $y = \sin x$ and sketch a graph of each on the same coordinate system for $0 \le x \le 2\pi$.

Solution The amplitude of the graph of $y = \frac{1}{2} \sin x$ is $|\frac{1}{2}| = \frac{1}{2}$, the amplitude of the graph of $y = -2 \sin x$ is $|-2| = 2$, and the amplitude of the graph of $y = \sin x$ is $|1| = 1$. The negative sign in $y = -2 \sin x$ reflects the graph of $y = 2 \sin x$ across the x axis (turns it upside-down). The graphs of all three equations are shown in Figure 1:

FIGURE 1 Change in amplitude.

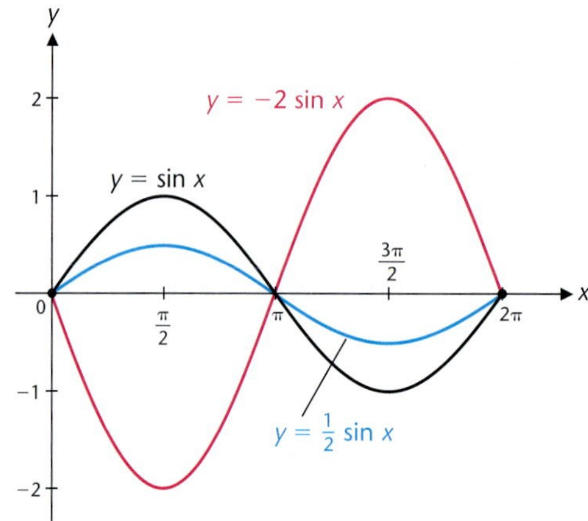

Matched Problem 1

Compare the graphs of $y = \frac{1}{3} \cos x$ and $y = -3 \cos x$ with the graph of $y = \cos x$ by graphing each on the same coordinate system for $-\pi/2 \le x \le 3\pi/2$.

• $y = \sin Bx$ and $y = \cos Bx$

We now investigate the effect of B by comparing

$$y = \sin x \qquad \text{and} \qquad y = \sin Bx \qquad B > 0$$

Both have the same amplitude, 1, but how do their periods compare? Since $\sin x$ has a period of 2π, it follows that $\sin Bx$ completes one cycle as Bx varies from

$$Bx = 0 \qquad \text{to} \qquad Bx = 2\pi$$

or as x varies from

$$x = 0 \qquad \text{to} \qquad x = \frac{2\pi}{B}$$

We conclude that the period of $\sin Bx$ is $2\pi/B$, as can be checked as follows: If $f(x) = \sin Bx$, then

$$f\left(x + \frac{2\pi}{B}\right) = \sin \left[B\left(x + \frac{2\pi}{B}\right)\right]$$
$$= \sin (Bx + 2\pi) = \sin Bx = f(x)$$

EXAMPLE 2 Comparing Graphs with Different Periods

Compare the periods of $y = \sin 2x$ and $y = \sin (x/2)$ with the period of $y = \sin x$ and sketch a graph of each on the same coordinate system for one period starting at the origin.

Solution What is the period of $y = \sin 2x$?

$$\text{Period} = \frac{2\pi}{B} = \frac{2\pi}{2} = \pi \qquad \text{Half the period for } \sin x.$$

What is the period of $y = \sin(x/2)$?

$$\text{Period} = \frac{2\pi}{B} = \frac{2\pi}{\frac{1}{2}} = 4\pi \qquad \text{Double the period for } \sin x.$$

The graphs of all three equations are shown in Figure 2.

FIGURE 2 Change in period.

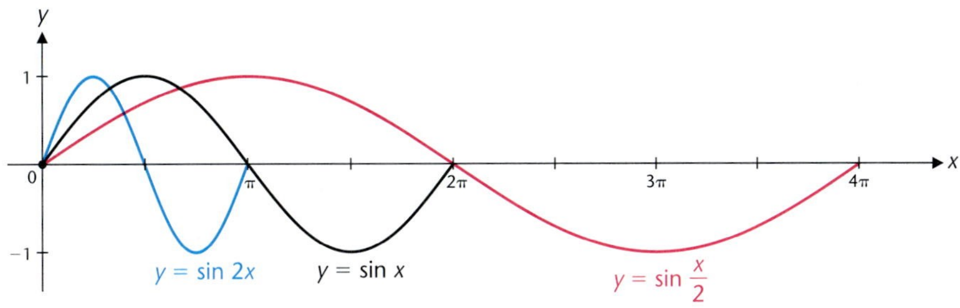

It is clear from Example 2 that the effect of B is to compress or stretch the basic sine curve by changing the period of the function. A similar analysis applies to $y = \cos Bx$, for $B > 0$, where it can be shown that its period is also $2\pi/B$.

Matched Problem 2 Compare the graphs of $y = \cos 2x$ and $y = \cos (x/2)$ with the graph of $y = \cos x$ by graphing each on the same coordinate system for one period starting at the origin.

• $y = k + A \sin Bx$ and $y = k + A \cos Bx$

Combining the discussions on amplitude and period, we summarize the results as follows:

> For $y = A \sin Bx$ or $y = A \cos Bx$, $B > 0$:
>
> $$\text{Amplitude} = |A| \qquad \text{Period} = \frac{2\pi}{B}$$
>
> If $0 < B < 1$, the basic sine or cosine curve is stretched.
>
> If $B > 1$, the basic sine or cosine curve is compressed.

You can either memorize the formula for the period, $2\pi/B$, or use the reasoning we used in deriving the formula. Recall, sin Bx or cos Bx completes one cycle as Bx varies from

$$Bx = 0 \qquad \text{to} \qquad Bx = 2\pi$$

that is, as x varies from

$$x = 0 \qquad \text{to} \qquad x = \frac{2\pi}{B}$$

Some prefer to memorize a formula, others a process.

We now consider some examples where we show how graphs of $y = A \sin Bx$ and $y = A \cos Bx$ can be sketched rather quickly.

EXAMPLE 3 **Graphing an Equation of the Form $y = A \sin Bx$**

State the amplitude and period for $y = 2 \sin 2x$, and graph the equation for the interval $-\pi \le x \le 2\pi$.

Solution

$$\text{Amplitude} = |2| = 2 \qquad \text{Period} = \frac{2\pi}{2} = \pi$$

To sketch the graph, divide the interval of one period $[0, \pi]$ into four equal parts, locate the high and low points and x intercepts, sketch in one period, and then extend this sketch to cover the desired interval. Scale the x axis using $\pi/4$ (the period divided by 4) as the basic unit, and adjust the scale on the y axis to accommodate the amplitude 2 (Fig. 3). *The scales on both axes do not have to be the same.*

Matched Problem 3 State the amplitude and period for $y = -\frac{1}{2} \sin (x/2)$, and graph the equation for $-4\pi \le x \le 4\pi$.

FIGURE 3 $y = 2 \sin 2x$.

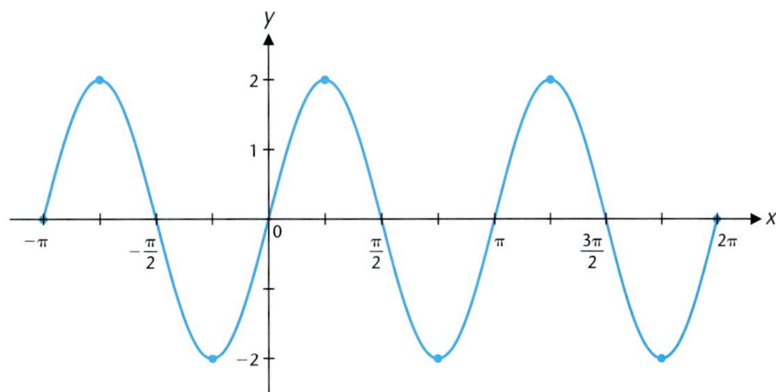

EXAMPLE 4 Graphing an Equation of the Form $y = A \cos Bx$

State the amplitude and period for $y = -3 \cos (\pi x/2)$, and graph the equation for $-4 \leq x \leq 4$.

Solution

$$\text{Amplitude} = |-3| = 3 \qquad \text{Period} = \frac{2\pi}{\pi/2} = 4$$

The graph of $y = -3 \cos (\pi x/2)$ is the same as the graph of $y = 3 \cos (\pi x/2)$ reflected across the x axis. Divide the interval of one period [0, 4] into four equal parts, locate the high and low points and x intercepts, sketch in one period, and then extend the sketch to cover the desired interval. Scale the x axis using $\frac{4}{4} = 1$ as the basic unit (the period divided by 4), and adjust the vertical axis to accommodate the amplitude 3 (Fig. 4).

FIGURE 4 $y = -3 \cos \dfrac{\pi x}{2}$.

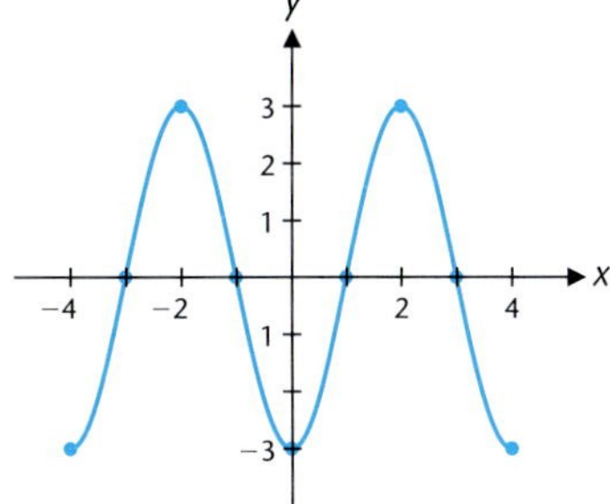

Matched Problem 4 State the amplitude and period for $y = \frac{1}{4} \cos 2\pi x$, and graph the equation for the interval $-1 \leq x \leq 1$.

EXPLORE-DISCUSS 1 Find an equation of the form $y = A \cos Bx$ that produces the following graph:

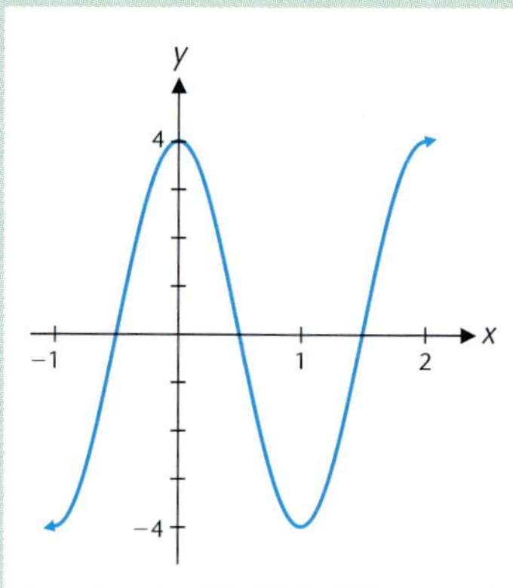

Is it possible for an equation of the form $y = A \sin Bx$ to produce the same graph? Explain.

How would you graph a function such as

$$y = k + A \sin Bx \qquad \text{or} \qquad y = k + A \cos Bx?$$

Applying the methods of Section 2-5, you would graph $y = A \sin Bx$ or $y = A \cos Bx$ and translate the curve vertically—up k units if k is positive and down $|k|$ units if k is negative.

EXAMPLE 5 Graphing an Equation of the Form $y = k + A \sin Bx$

Graph $y = -2 - 3 \cos (\pi x/2)$, $-4 \leq x \leq 4$.

Solution We first graph $y = -3 \cos (\pi x/2)$, as we did in Example 4, then translate the graph $|-2| = 2$ units downward (Fig. 5).

FIGURE 5 $y = -2 - 3 \cos \frac{\pi x}{2}$.

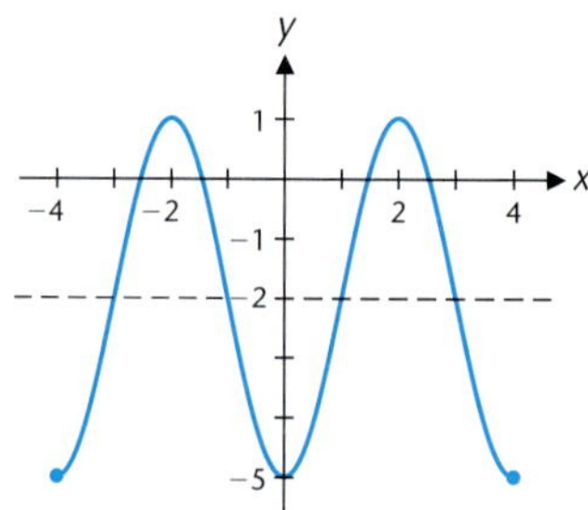

Many find it easier to first draw the horizontal dashed line shown in the figure (which represents a vertical translation 2 units below the x axis), then draw the graph of $y = -3 \cos (\pi x/2)$ as if this horizontal line is the original x axis.

Matched Problem 5 Graph $y = 1 + \frac{1}{4} \cos 2\pi x, \ -1 \le x \le 1$.

• $y = k + A \sin (Bx + C)$ and $y = k + A \cos (Bx + C)$

We now consider graphs of simple harmonics, that is, functions of the form

$$y = A \sin (Bx + C) \qquad \text{and} \qquad y = A \cos (Bx + C)$$

We will find that the graphs of these functions are simply the graphs of

$$y = A \sin Bx \qquad \text{or} \qquad y = A \cos Bx$$

translated horizontally to the left or to the right. We can see this as follows: Since $A \sin x$ has a period of 2π, it follows that $A \sin (Bx + C)$ completes one cycle as $Bx + C$ varies from

$$Bx + C = 0 \qquad \text{to} \qquad Bx + C = 2\pi$$

or (solving for x in each equation) as x varies from

Phase shift — Period

$$x = -\frac{C}{B} \qquad \text{to} \qquad x = -\frac{C}{B} + \frac{2\pi}{B}$$

We conclude that $y = A \sin (Bx + C)$ has a period of $2\pi/B$, and its graph is translated $|-C/B|$ units to the right if $-C/B$ is positive and $|-C/B|$ units to the left if $-C/B$ is negative. The number $-C/B$ is also referred to as the **phase shift.**

What are the period and phase shift for $y = \sin (x + \pi/2)$? To find an answer, use the formulas above for period and phase shift or follow the process used in deriving the formulas. Most will probably find that the process is easier to remember: Set $x + \pi/2$ equal to 0 and to 2π, then solve each equation for x:

$$x + \frac{\pi}{2} = 0 \qquad x + \frac{\pi}{2} = 2\pi$$

$$x = -\frac{\pi}{2} \qquad x = -\frac{\pi}{2} + 2\pi$$

Thus, $-\pi/2$ is the phase shift and 2π is the period. The graph of $y = \sin x$ is translated horizontally $|-\pi/2| = \pi/2$ units to the left. Figure 6(a) on the next page shows the graphs of $y = \sin x$ and $y = \sin (x + \pi/2)$.

Going through a similar process, we graph $y = \sin (x - \pi/2)$, as shown in Figure 6(b). Here, the phase shift is $\pi/2$, and the graph of $y = \sin x$ is translated horizontally $\pi/2$ units to the right.

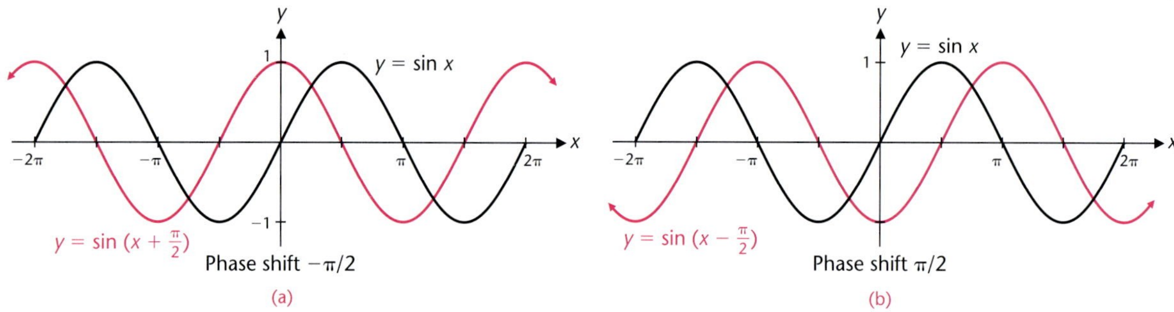

FIGURE 6

A similar analysis applies to $y = A \cos (Bx + C)$. The results are summarized in the following boxes.

> **Properties of $y = A \sin (Bx + C)$ and $y = A \cos (Bx + C)$**
>
> For $B > 0$:
>
> $$\text{Amplitude} = |A| \qquad \text{Period} = \frac{2\pi}{B} \qquad \text{Phase shift} = -\frac{C}{B}$$

It is not necessary to memorize the formulas for period and phase shift unless you wish to. They are easily derived, as illustrated in the following steps for graphing.

> **Graphing $y = A \sin (Bx + C)$ and $y = A \cos (Bx + C)$**
>
> ***Step 1.*** Find the amplitude $|A|$.
>
> ***Step 2.*** Solve $Bx + C = 0$ and $Bx + C = 2\pi$:
>
> $$Bx + C = 0 \qquad \text{and} \qquad Bx + C = 2\pi$$
>
> $$x = \underbrace{-\frac{C}{B}}_{\text{Phase shift}} \qquad\qquad x = \underbrace{-\frac{C}{B}}_{\text{Phase shift}} + \underbrace{\frac{2\pi}{B}}_{\text{Period}}$$
>
> $$\text{Phase shift} = -\frac{C}{B} \qquad \text{Period} = \frac{2\pi}{B}$$
>
> The graph completes one full cycle as $Bx + C$ varies from 0 to 2π—that is, as x varies over the interval
>
> $$\left[-\frac{C}{B}, -\frac{C}{B} + \frac{2\pi}{B}\right]$$
>
> ***Step 3.*** Graph one cycle over the interval $[-C/B, -C/B + 2\pi/B]$.
>
> ***Step 4.*** Extend the graph in step 3 to the left or right as desired.

EXAMPLE 6 Graphing an Equation of the Form $y = A \cos(Bx + C)$

Find the amplitude, period, and phase shift for $y = \frac{1}{2}\cos(4x - \pi)$, then sketch its graph for $-\pi \le x \le \pi$.

Solution *Step 1. Find the amplitude:*

$$\text{Amplitude} = |A| = |\tfrac{1}{2}| = \tfrac{1}{2}$$

Step 2. Solve $Bx + C = 0$ and $Bx + C = 2\pi$:

$$4x - \pi = 0 \qquad 4x - \pi = 2\pi$$

$$x = \frac{\pi}{4} \qquad x = \frac{\pi}{4} + \frac{\pi}{2} = \frac{3\pi}{4}$$

Phase shift Period

$$\text{Phase shift} = \frac{\pi}{4} \qquad \text{Period} = \frac{\pi}{2}$$

The graph completes one cycle as x varies over the interval $[\pi/4, 3\pi/4]$.

Step 3. Graph one cycle over the interval $[\pi/4, 3\pi/4]$. Divide the interval into four equal parts and sketch in one cycle (Fig. 7). (Scale the x axis in units of $\pi/8$.)

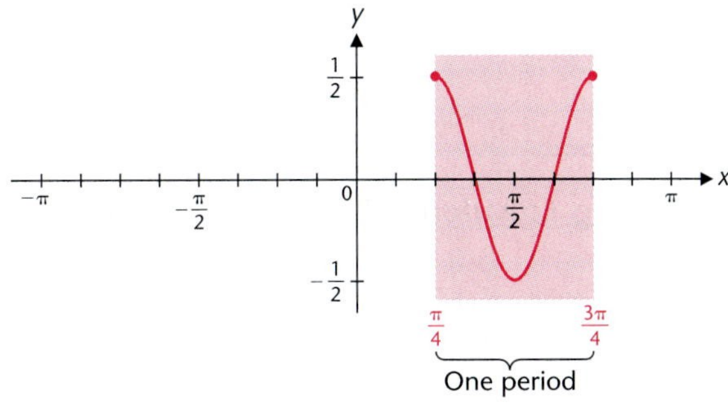

FIGURE 7 $y = \frac{1}{2}\cos(4x - \pi)$, $\frac{\pi}{4} \le x \le \frac{3\pi}{4}$.

Step 4. Extend the graph in step 3 to cover the interval $[-\pi, \pi]$ as shown in Figure 8.

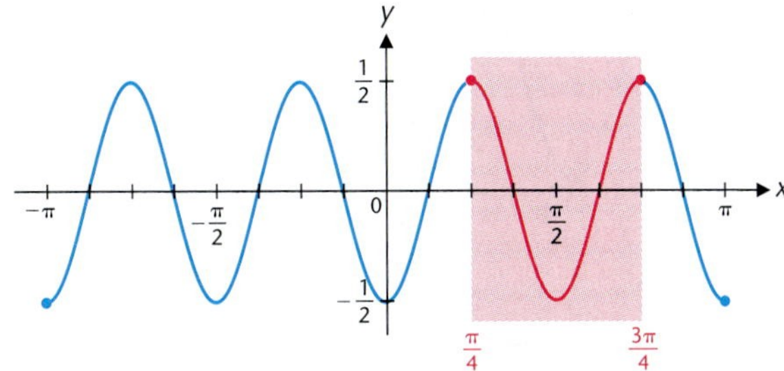

FIGURE 8 $y = \frac{1}{2}\cos(4x - \pi)$, $-\pi \le x \le \pi$.

Matched Problem 6 Graph $y = \frac{3}{4} \sin (2x + \pi)$, $-\pi \leq x \leq \pi$. State the amplitude, period, and phase shift.

EXPLORE-DISCUSS 2 Find an equation of the form $y = A \sin (Bx + C)$ that produces the following graph:

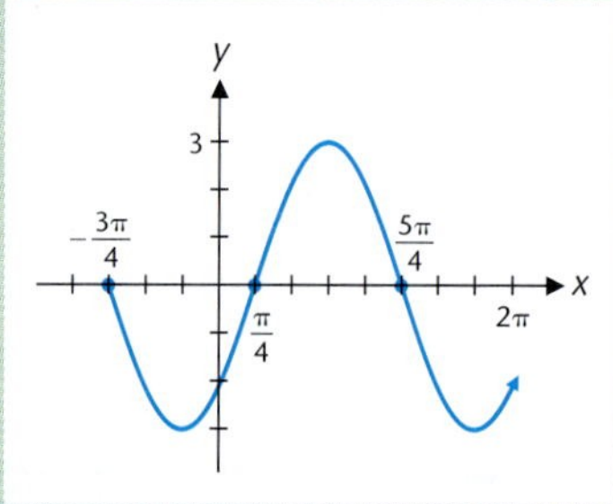

Is it possible for an equation of the form $y = A \cos (Bx + C)$ to produce the same graph? Explain.

Finally, graphing equations of the form $y = k + A \sin (Bx + C)$ or $y = k + A \cos (Bx + C)$ involves both horizontal translations (phase shifts) and vertical translations.

EXAMPLE 7 Graphing Equations of the Form $y = k + A \cos (Bx + C)$

Graph $y = \frac{3}{2} + \frac{1}{2} \cos (4x - \pi)$, $-\pi \leq x \leq \pi$.

Solution Graph $y = \frac{1}{2} \cos (4x - \pi)$ as in Example 6, then vertically translate the graph up $\frac{3}{2}$ units (Fig. 9):

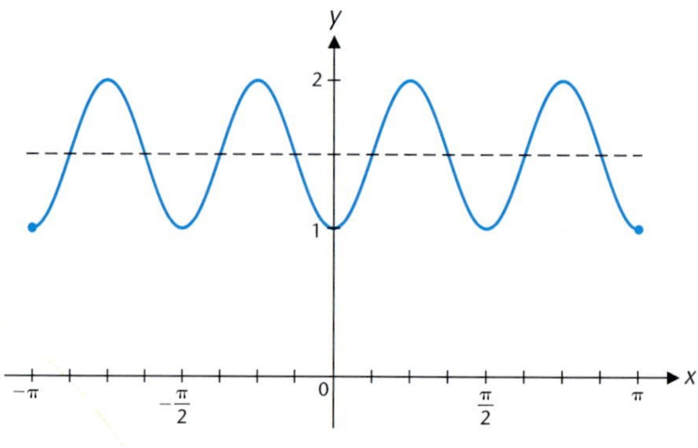

FIGURE 9 $y = \frac{3}{2} + \frac{1}{2} \cos (4x - \pi)$.

Matched Problem 7 Graph $y = -\frac{1}{2} + \frac{3}{4} \sin (2x + \pi)$, $-\pi \le x \le \pi$.

• Finding an Equation of the Graph of a Simple Harmonic

Given the graph of a simple harmonic, we wish to find an equation of the form

$$y = A \sin (Bx + C) \qquad \text{or} \qquad y = A \cos (Bx + C)$$

that produces the graph. An example will illustrate the process.

EXAMPLE 8 Finding an Equation of the Graph of a Simple Harmonic

Graph $y_1 = 3 \sin x + 4 \cos x$ using a graphing utility, and find an equation of the form $y_2 = A \sin (Bx + C)$ that has the same graph as y_1. Find A and B exactly and C to three decimal places.

Solution The graph of y_1 is shown in Figure 10.

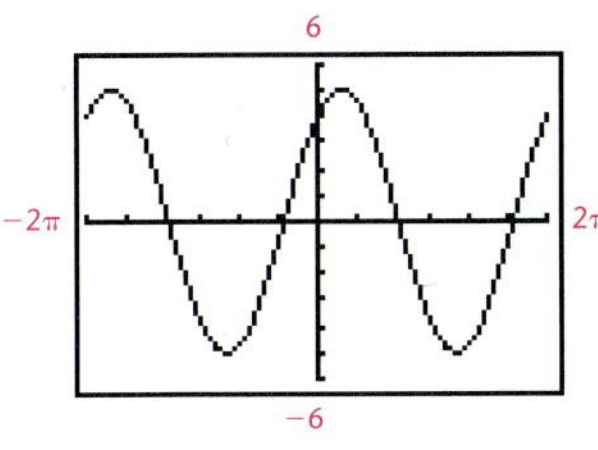

FIGURE 10 $y_1 = 3 \sin x + 4 \cos x$.

The graph appears to be a sine curve shifted to the left. The amplitude and period appear to be 5 and 2π, respectively. (We will assume this for now and check it at the end.) Thus, $A = 5$, and since $P = 2\pi/B$, then $B = 2\pi/P = 2\pi/2\pi = 1$. Using a graphing utility, we find that the x intercept closest to the origin, to three decimal places, is -0.927. To find C, substitute $B = 1$ and $x = -0.927$ into the phase-shift formula $x = -C/B$ and solve for C:

$$x = -\frac{C}{B}$$

$$-0.927 = -\frac{C}{1}$$

$$C = 0.927$$

We now have the equation we are looking for:

$$y_2 = 5 \sin (x + 0.927)$$

Check Graph y_1 and y_2 in the same viewing window. If the graphs are the same, it appears that only one graph is drawn—the second graph is drawn over the first. To check further that the graphs are the same, use TRACE and switch back and forth between y_1 and y_2 at different values of x. Figure 11 shows a comparison at $x = 0$ (both graphs appear in the same viewing window).

FIGURE 11

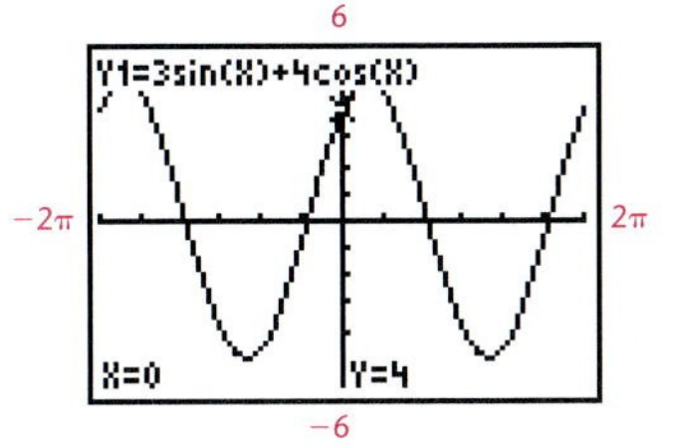

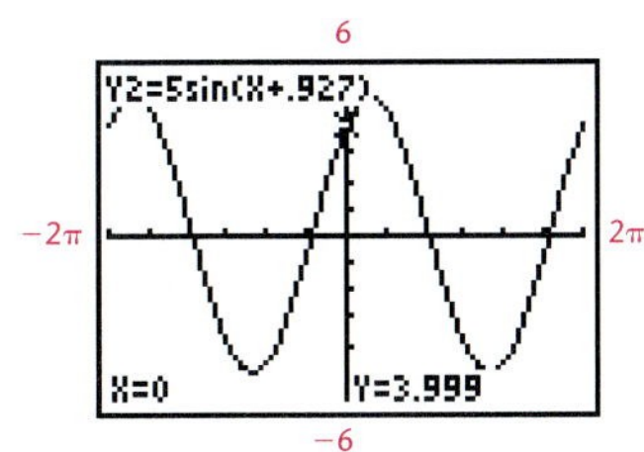

Matched Problem 8 Graph $y_1 = 4 \sin x - 3 \cos x$ using a graphing utility, and find an equation of the form $y_2 = A \sin (Bx + C)$ that has the same graph as y_1. (Find the x intercept closest to the origin to three decimal places.)

Answers to Matched Problems

1.

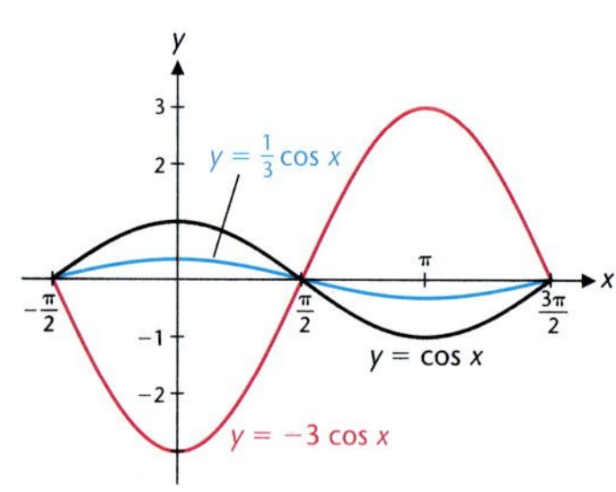

2.

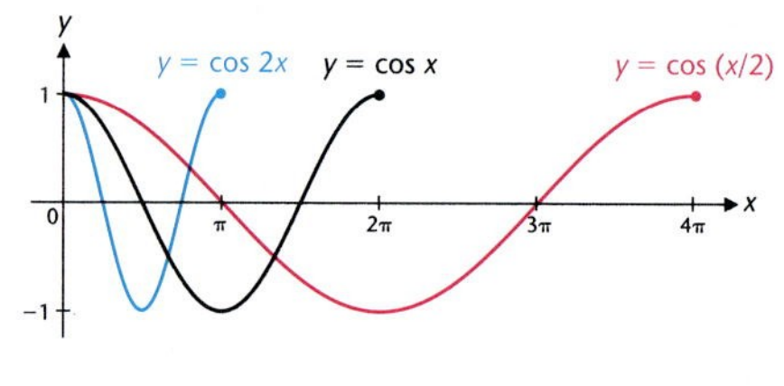

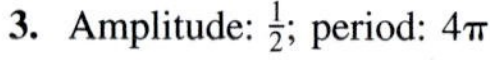

3. Amplitude: $\frac{1}{2}$; period: 4π

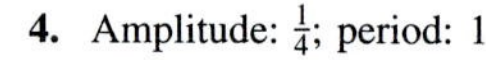

4. Amplitude: $\frac{1}{4}$; period: 1

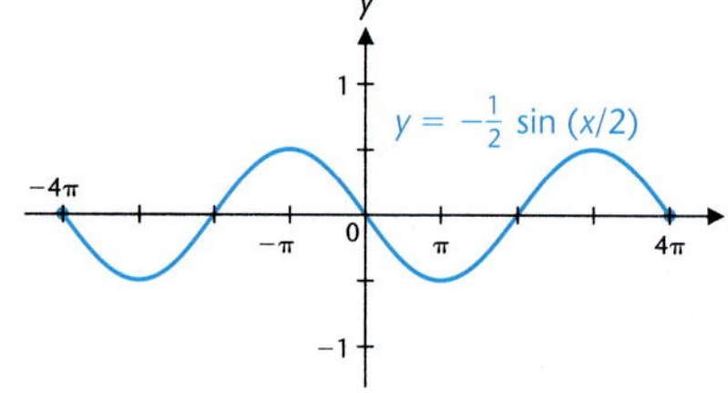

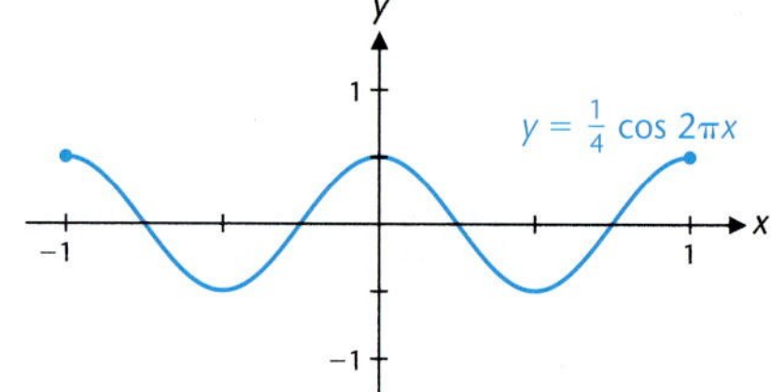

5.

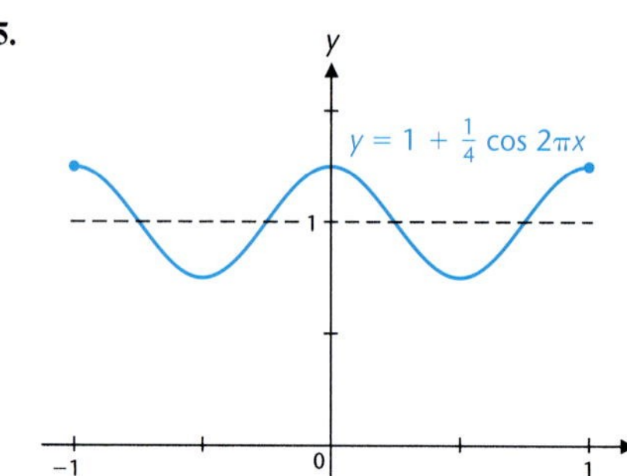

6. Amplitude: $\frac{3}{4}$; period: π; phase shift: $-\pi/2$

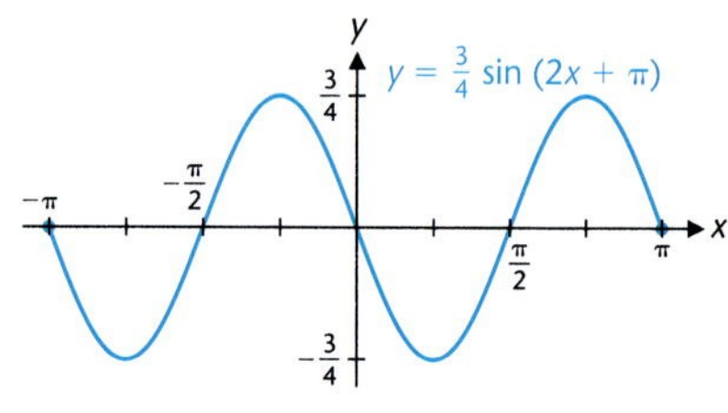

7.

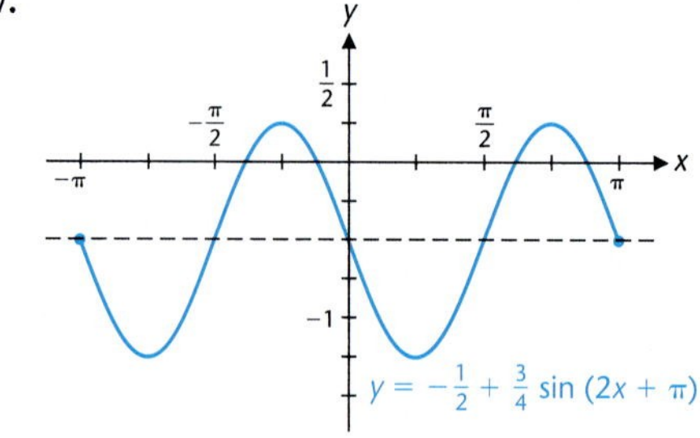

8. $y_2 = 5 \sin (x - 0.644)$

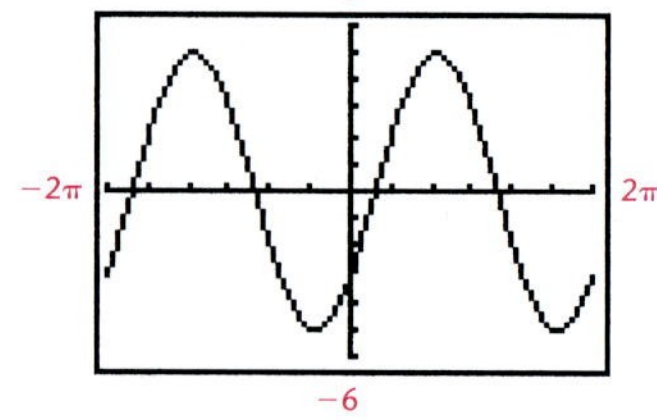

EXERCISE 5-7

A

State the amplitude A and period P of each function in Problems 1–12, and graph the function over the indicated interval.

1. $y = 3 \sin x, \ -2\pi \le x \le 2\pi$

2. $y = \frac{1}{4} \cos x, \ -2\pi \le x \le 2\pi$

3. $y = -\frac{1}{2} \cos x, \ -2\pi \le x \le 2\pi$

4. $y = -2 \sin x, \ -2\pi \le x \le 2\pi$

5. $y = \sin 3x, \ -\pi \le x \le 2\pi$

6. $y = \cos 2x, \ -\pi \le x \le \pi$

7. $y = \cos (x/2), \ -4\pi \le x \le 4\pi$

8. $y = \sin (x/3), \ -6\pi \le x \le 6\pi$

9. $y = \sin \pi x, \ -2 \le x \le 2$

10. $y = \cos \pi x, \ -2 \le x \le 2$

11. $y = 3 \cos 2x, \ -\pi \le x \le \pi$

12. $y = 2 \sin 4x, \ -\pi \le x \le \pi$

B

State the amplitude A and period P of each function in Problems 13–20, and graph the function over the indicated interval.

13. $y = -\frac{1}{2} \sin 2\pi x, \ -2 \le x \le 2$

14. $y = -\frac{1}{3} \cos 2\pi x, \ -2 \le x \le 2$

15. $y = -3 \cos (x/2), \ -4\pi \le x \le 4\pi$

16. $y = -\frac{1}{4} \sin (x/2), \ -4\pi \le x \le 4\pi$

17. $y = 2 + 2 \sin (\pi x/2), \ -4 \le x \le 4$

18. $y = 3 + 3 \cos (\pi x/2), \ -4 \le x \le 4$

19. $y = 4 - 2 \cos (x/2), \ -4\pi \le x \le 4\pi$

20. $y = 3 - 2 \sin (x/2), \ -4\pi \le x \le 4\pi$

In Problems 21–24, find the equation of the form $y = A \sin Bx$ that produces the graph shown.

21.

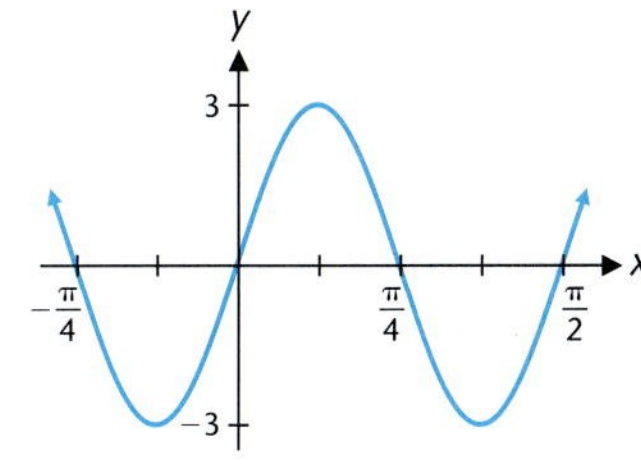

22.

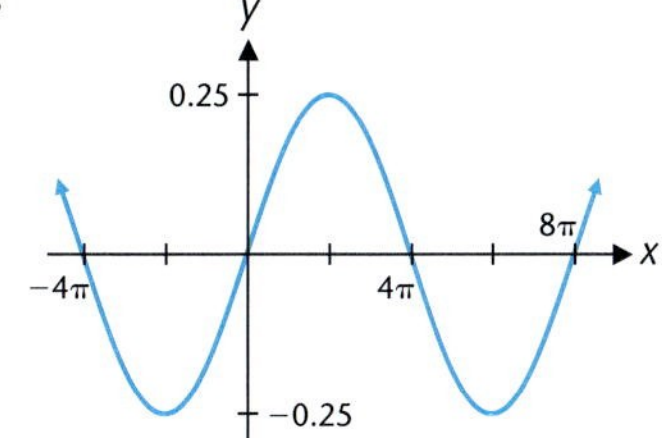

23.

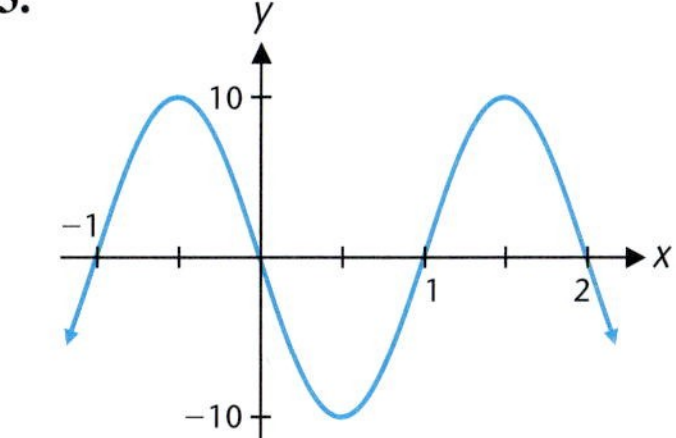

24.

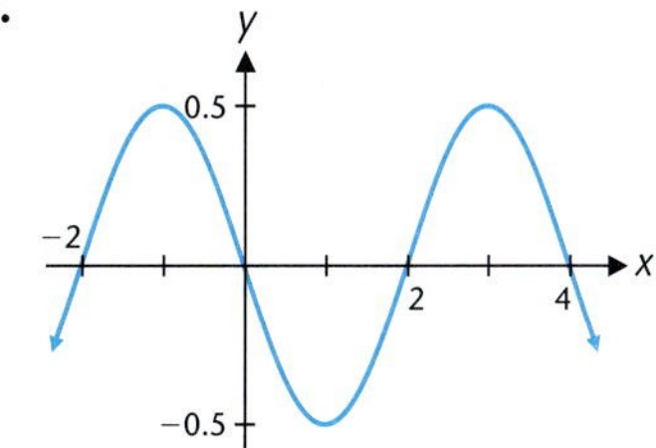

In Problems 25–28, find the equation of the form $y = A \cos Bx$ that produces the graph shown.

25.

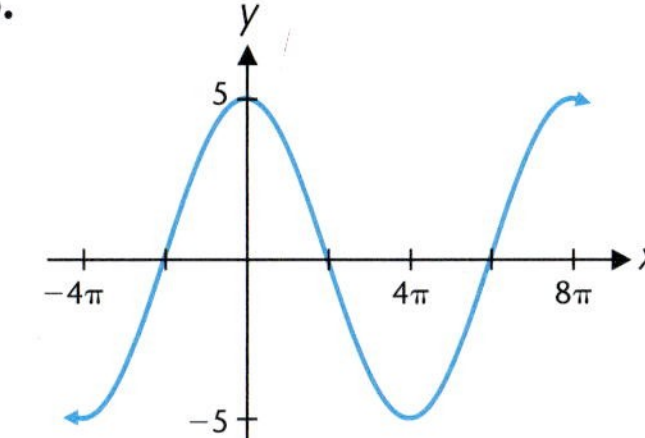

26.

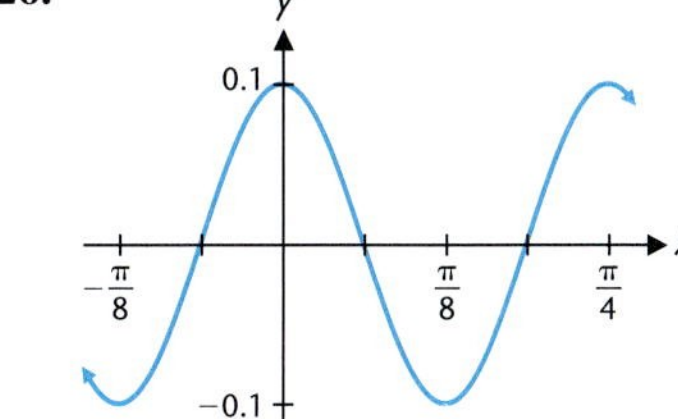

27.

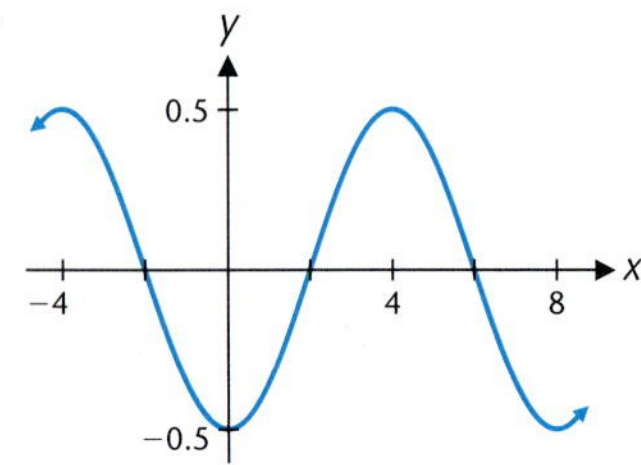

28.

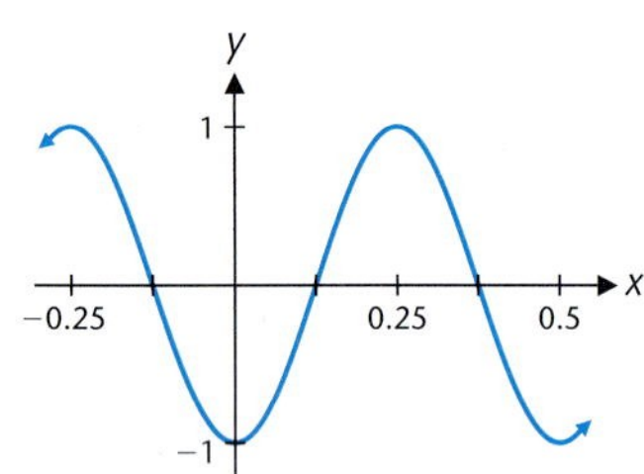

Graph each function in Problems 29–32 in a graphing utility. (Select the dimensions of each viewing window so that at least two periods are visible.) Find an equation of the form $y = k + A \sin Bx$ or $y = k + A \cos Bx$ that has the same graph as the given equation. (These problems suggest the existence of further identities in addition to the basic identities discussed in Section 5-2.)

29. $y = \cos^2 x - \sin^2 x$

30. $y = \sin x \cos x$

31. $y = 2 \sin^2 x$

32. $y = 2 \cos^2 x$

State the amplitude A, period P, and phase shift of each function in Problems 33–40, and graph the function over the indicated interval.

33. $y = \sin (x + \pi),\ -\pi \le x \le 3\pi$

34. $y = \cos (x - \pi),\ -\pi \le x \le 3\pi$

35. $y = \frac{1}{2} \cos (x - \pi/4),\ -\pi \le x \le 3\pi$

36. $y = 2 \sin (x + \pi/4),\ -2\pi \le x \le 2\pi$

37. $y = \sin [\pi(x - 1)],\ -2 \le x \le 3$

38. $y = \cos [2\pi(x - \frac{1}{2})],\ -1 \le x \le 2$

39. $y = 3 \cos (\pi x + \pi/2),\ -2 \le x \le 2$

40. $y = 2 \sin (\pi x - \pi/4),\ -1 \le x \le 3$

In Problems 41–44, graph each equation over the indicated interval.

41. $y = -1 + \sin (x + \pi),\ -\pi \le x \le 3\pi$

42. $y = 1 + \cos (x - \pi),\ -\pi \le x \le 3\pi$

43. $y = 2 - 4 \cos (2x - \pi),\ -\pi \le x \le 3\pi$

44. $y = -1 - 2 \cos (4x + \pi),\ -\pi \le x \le \pi$

C

Problems 45 and 46 refer to the following graph:

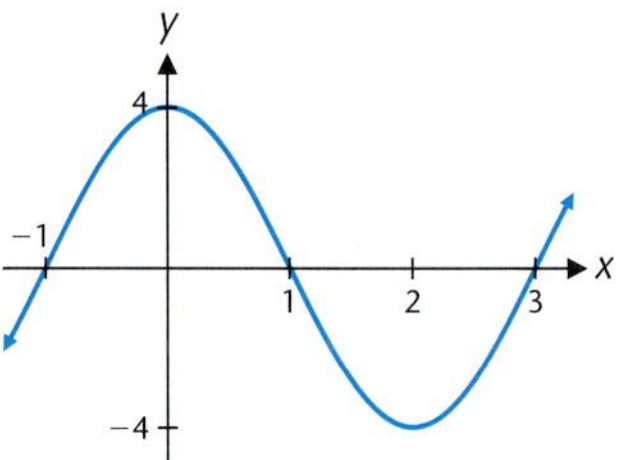

45. If the graph is a graph of an equation of the form $y = A \sin (Bx + C)$, $0 < -C/B < 2$, find the equation.

46. If the graph is a graph of an equation of the form $y = A \sin (Bx + C)$, $-2 < -C/B < 0$, find the equation.

Problems 47 and 48 refer to the following graph:

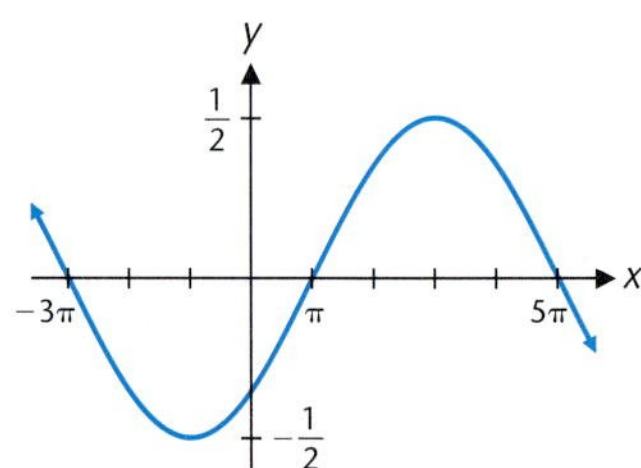

47. If the graph is a graph of an equation of the form $y = A \cos (Bx + C)$, $0 < -C/B < 4\pi$, find the equation.

48. If the graph is a graph of an equation of the form $y = A \cos (Bx + C)$, $-2\pi < -C/B < 0$, find the equation.

49. How many functions of the form $y = A \sin (Bx + C)$ have amplitude 4, period π, and phase shift $-\pi/3$? Explain. Find their equations.

50. How many functions of the form $y = A \cos (Bx + C)$ have amplitude 8, period 5, and phase shift 2? Explain. Find their equations.

Problems 51–66 require the use of a graphing utility.

In Problems 51–54, state the amplitude, period, and phase shift of each function and sketch a graph of the function with the aid of a graphing utility.

51. $y = 3.5 \sin \left[\frac{\pi}{2}(t + 0.5)\right],\ 0 \le t \le 10$

52. $y = 5.4 \sin \left[\frac{\pi}{2.5}(t - 1)\right],\ 0 \le t \le 6$

53. $y = 50 \cos [2\pi(t - 0.25)],\ 0 \le t \le 2$

54. $y = 25 \cos [5\pi(t - 0.1)],\ 0 \le t \le 2$

In Problems 55–60, graph each equation in a graphing utility. (Select the dimensions of each viewing window so that at least two periods are visible.) Find an equation of the form $y = A \sin (Bx + C)$ that has the same graph as the given equation. Find A and B exactly and C to three decimal places. Use the x intercept closest to the origin as the phase shift.

55. $y = \sqrt{2} \sin x + \sqrt{2} \cos x$

56. $y = \sqrt{2} \sin x - \sqrt{2} \cos x$

57. $y = \sqrt{3} \sin x - \cos x$

58. $y = \sin x + \sqrt{3} \cos x$

59. $y = 4.8 \sin 2x - 1.4 \cos 2x$

60. $y = 1.4 \sin 2x + 4.8 \cos 2x$

In Problems 61–64, determine whether the statement is true or false. If true, explain why. If false, give a counterexample.

61. If f is a simple harmonic, then the function g defined by $g(x) = f(3x)$ is a simple harmonic.

62. If f is a simple harmonic, then the function h defined by $h(x) = f(x + 3)$ is a simple harmonic.

63. If f is a simple harmonic, then the function j defined by $j(x) = 3f(x)$ is a simple harmonic.

64. If f is a simple harmonic, then the function k defined by $k(x) = 3 - f(x)$ is a simple harmonic.

Problems 65–70 illustrate combinations of functions that occur in harmonic analysis applications. Graph parts A, B, and C of each problem in the same viewing window. In Problems 65–68, what is happening to the amplitude of the function in part C? Give an example of a physical phenomenon that might be modeled by a similar function.

65. $0 \le x \le 16$

(A) $y = \frac{1}{x}$ (B) $y = -\frac{1}{x}$ (C) $y = \frac{1}{x} \sin \frac{\pi}{2} x$

66. $0 \le x \le 10$

(A) $y = \frac{2}{x}$ (B) $y = -\frac{2}{x}$ (C) $y = \frac{2}{x} \cos \pi x$

67. $0 \le x \le 10$

(A) $y = x$ (B) $y = -x$ (C) $y = x \sin \frac{\pi}{2} x$

68. $0 \le x \le 10$

(A) $y = \frac{x}{2}$ (B) $y = -\frac{x}{2}$ (C) $y = \frac{x}{2} \cos \pi x$

69. $0 \le x \le 2\pi$

(A) $y = \sin x$

(B) $y = \sin x + \frac{\sin 3x}{3}$

(C) $y = \sin x + \frac{\sin 3x}{3} + \frac{\sin 5x}{5}$

70. $0 \le x \le 4$

(A) $y = \sin \pi x$

(B) $y = \sin \pi x + \frac{\sin 2\pi x}{2}$

(C) $y = \sin \pi x + \frac{\sin 2\pi x}{2} + \frac{\sin 3\pi x}{3}$

APPLICATIONS 

71. Spring-Mass System. A 6-pound weight hanging from the end of a spring is pulled $\frac{1}{3}$ foot below the equilibrium position and then released (see figure). If air resistance and friction are neglected, the distance x that the weight is from the equilibrium position relative to time t (in seconds) is given by

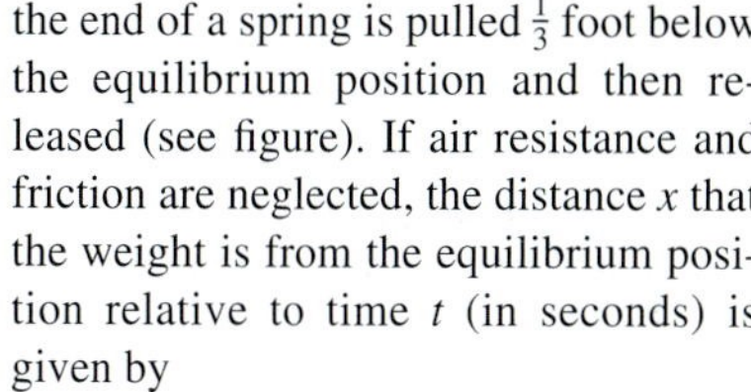

$$x = \frac{1}{3} \cos 8t$$

State the period P and amplitude A of this function, and graph it for $0 \le t \le \pi$.

72. Electric Circuit. An alternating current generator generates a current given by

$$I = 30 \sin 120t$$

where t is time in seconds. What are the amplitude A and period P of this function? What is the frequency of the current; that is, how many cycles (periods) will be completed in 1 second?

★ **73. Spring-Mass System.** Assume the motion of the weight in Problem 71 has an amplitude of 8 inches and a period of 0.5 second, and that its position when $t = 0$ is 8 inches below its position at rest (displacement above rest position is positive and below is negative). Find an equation of the form $y = A \cos Bt$ that describes the motion at any time $t \ge 0$. (Neglect any damping forces—that is, friction and air resistance.)

★ **74. Electric Circuit.** If the voltage E in an electric circuit has an amplitude of 110 volts and a period of $\frac{1}{60}$ second, and if $E = 110$ volts when $t = 0$ seconds, find an equation of the form $E = A \cos Bt$ that gives the voltage at any time $t \ge 0$.

75. Pollution. The amount of sulfur dioxide pollutant from heating fuels released in the atmosphere in a city varies seasonally. Suppose the number of tons of pollutant released into the atmosphere during the nth week after January 1 for a particular city is given by

$$A(n) = 1.5 + \cos\frac{n\pi}{26} \qquad 0 \le n \le 104$$

Graph the function over the indicated interval and describe what the graph shows.

76. Medicine. A seated normal adult breathes in and exhales about 0.82 liter of air every 4.00 seconds. The volume of air in the lungs t seconds after exhaling is approximately

$$V(t) = 0.45 - 0.37\cos\frac{\pi t}{2} \qquad 0 \le t \le 8$$

Graph the function over the indicated interval and describe what the graph shows.

77. Electric Circuit. The current in an electric circuit is given by $I = 15\cos(120\pi t + \pi/2)$, $0 \le t \le \frac{2}{60}$, where I is measured in amperes. State the amplitude A, period P, and phase shift. Graph the equation.

78. Electric Circuit. The current in an electric circuit is given by $I = 30\cos(120\pi t - \pi)$, $0 \le t \le \frac{3}{60}$, where I is measured in amperes. State the amplitude A, period P, and phase shift. Graph the equation.

79. Physics—Engineering. The thin, plastic disk shown in the figure is rotated at 3 revolutions per second, starting at $\theta = 0$ (thus at the end of t seconds, $\theta = 6\pi t$—why?). If the disk has a radius of 3, show that the position of the shadow on the y scale from the small steel ball B is given by

$$y = 3\sin 6\pi t$$

Graph this equation for $0 \le t \le 1$.

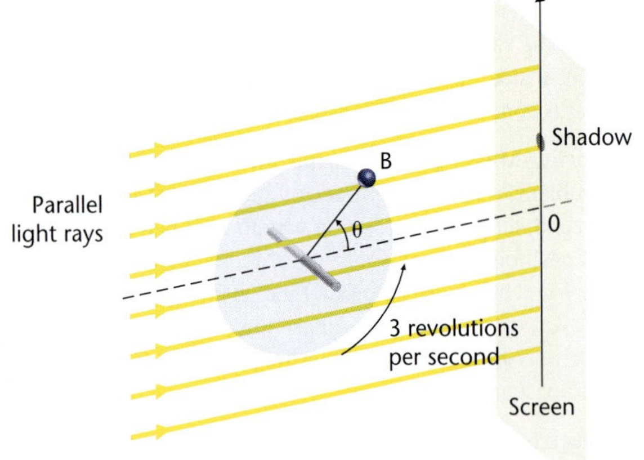

80. Physics—Engineering. If in Problem 79 the disk started rotating at $\theta = \pi/2$, show that the position of the shadow at time t (in seconds) is given by

$$y = 3\sin\left(6\pi t + \frac{\pi}{2}\right)$$

Graph this equation for $0 \le t \le 1$.

81. Modeling Sunset Times. Sunset times for the fifth of each month over a period of 1 year were taken from a tide booklet for the San Francisco Bay to form Table 1. Daylight saving time was ignored, and the times are for a 24-hour clock starting at midnight.

(A) Using 1 month as the basic unit of time, enter the data for a 2-year period in your graphing utility and produce a scatter plot in the viewing window. Before entering Table 1 data into your graphing utility, convert sunset times from hours and minutes to decimal hours rounded to two decimal places. Choose $15 \le y \le 20$ for the viewing window.

(B) It appears that a sine curve of the form

$$y = k + A\sin(Bx + C)$$

will closely model these data. The constants k, A, and B are easily determined from Table 1 as follows: $A = (\text{Max } y - \text{Min } y)/2$, $B = 2\pi/\text{Period}$, $k = \text{Min } y + A$. To estimate C, visually estimate to one decimal place the smallest positive phase shift from the plot in part A. After determining A, B, k, and C, write the resulting equation. (Your value of C may differ slightly from the answer in the back of the book.)

(C) Plot the results of parts A and B in the same viewing window. (An improved fit may result by adjusting your value of C slightly.)

TABLE 1

x (mos.)	1	2	3	4	5	6
y (sunset)*	17:05	17:38	18:07	18:36	19:04	19:29
x (mos.)	7	8	9	10	11	12
y (sunset)*	19:35	19:15	18:34	17:47	17:07	16:51

*Time on a 24-hr clock, starting at midnight.

82. Modeling Temperature Variation. The 30-year average monthly temperature, F°, for each month of the year for Washington, D.C., is given in Table 2 (*World Almanac*).

(A) Using 1 month as the basic unit of time, enter the data for a 2-year period in your graphing utility and produce a scatter plot in the viewing window. Choose $0 \le y \le 80$ for the viewing window.

(B) It appears that a sine curve of the form

$$y = k + A \sin (Bx + C)$$

will closely model this data. The constants k, A, and B are easily determined from Table 2 as follows: $A = (\text{Max } y - \text{Min } y)/2$, $B = 2\pi/\text{Period}$, $k = \text{Min } y + A$. To estimate C, visually estimate to one decimal place the smallest positive phase shift from the plot in part A. After determining A, B, k, and C, write the resulting equation.

(C) Plot the results of parts A and B in the same viewing window. (An improved fit may result by adjusting your value of C slightly.)

TABLE 2

x (mos.)	1	2	3	4	5	6	7	8	9	10	11	12
y (temp.)	31	34	43	53	62	71	76	74	67	55	45	35

SECTION 5-8 Graphing More General Tangent, Cotangent, Secant, and Cosecant Functions

- Graphing $y = A \tan (Bx + C)$ and $y = A \cot (Bx + C)$
- Graphing $y = A \sec (Bx + C)$ and $y = A \csc (Bx + C)$

In this section the graphing of the more general forms of the tangent, cotangent, secant, and cosecant functions is discussed. Essentially, we follow the same process we developed for graphing $y = A \sin (Bx + C)$ and $y = A \cos (Bx + C)$. The process is not difficult if you have a clear understanding of the basic graphs and periodic properties for each of these functions.

• Graphing $y = A \tan (Bx + C)$ and $y = A \cot (Bx + C)$

For convenient reference, we repeat the graphs that were shown for $y = \tan x$ and $y = \cot x$ in Section 5-6 (see Figs. 1 and 2).

Graph of $y = \tan x$

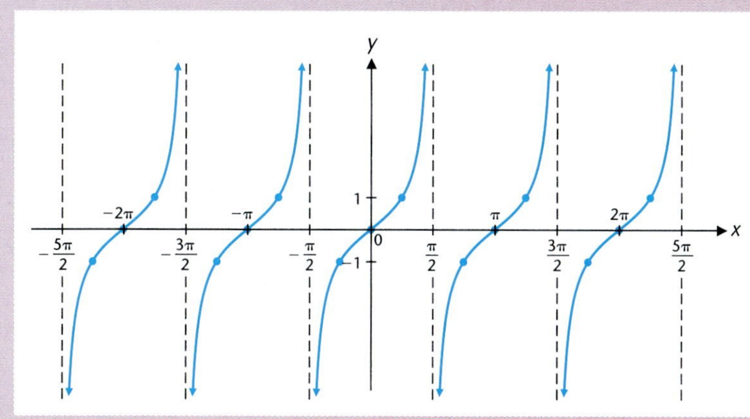

Period: π

Domain: All real numbers except $\pi/2 + k\pi$, k an integer

Range: All real numbers

Symmetric with respect to the origin

Increasing function between consecutive asymptotes

Discontinuous at $x = \pi/2 + k\pi$, k an integer

FIGURE 1

Graph of $y = \cot x$

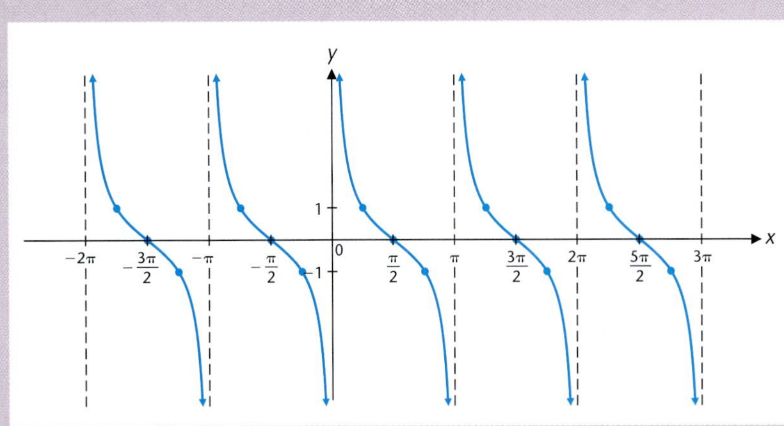

Period: π

Domain: All real numbers except $k\pi$, k an integer

Range: All real numbers

Symmetric with respect to the origin

Decreasing function between consecutive asymptotes

Discontinuous at $x = k\pi$, k an integer

FIGURE 2

EXPLORE-DISCUSS 1 (A) Match each function to its graph and discuss how the graph compares to the graph of $y = \tan x$ or $y = \cot x$.

(1) $y = 4 \tan x$ (2) $y = \tan 2x$ (3) $y = \cot (x - \pi/2)$

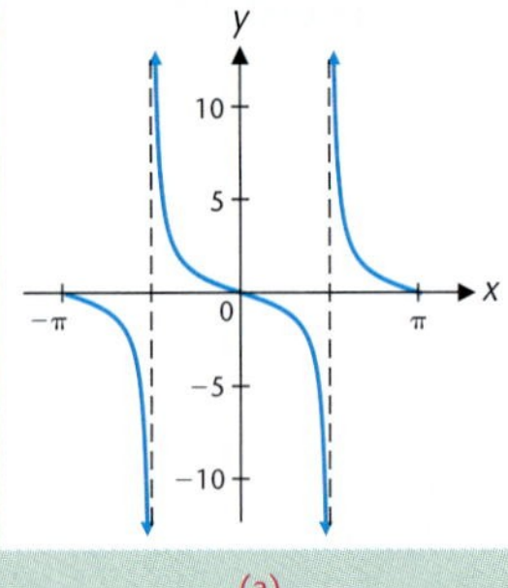

(a)

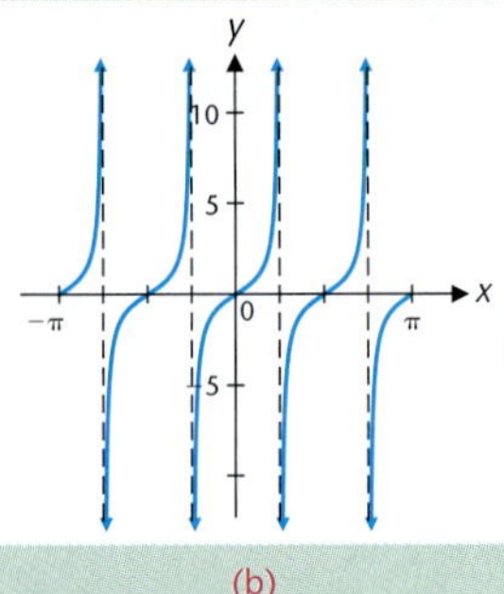

(b)

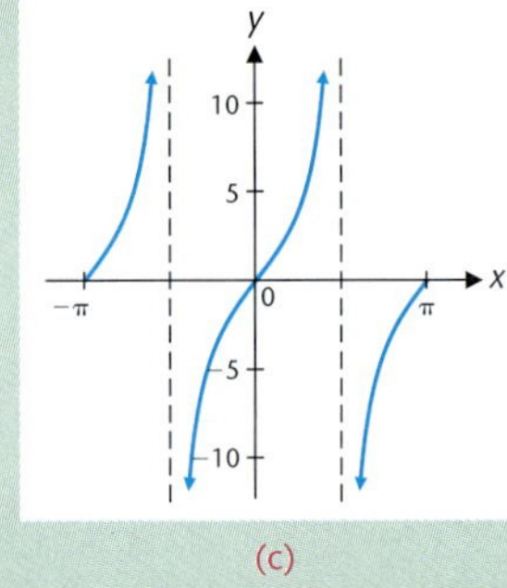

(c)

(B) Use a graphing utility to explore the nature of the changes in the graphs of the following functions when the values of A, B, and C are changed. Discuss what happens in each case.

$y = A \tan x$ and $y = A \cot x$ for different values of A

$y = \tan Bx$ and $y = \cot Bx$ for different values of B

$y = \tan (x + C)$ and $y = \cot (x + C)$ for different values of C

To quickly sketch the graphs of equations of the form $y = A \tan (Bx + C)$, you need to know how the constants A, B, and C affect the basic graphs of $y = \tan x$ and $y = \cot x$, respectively.

First note that **amplitude is not defined for the tangent and cotangent functions.** The graphs of both deviate without end from the x axis. The effect of A is to make the graph steeper if $|A| > 1$ or to make the graph less steep if $|A| < 1$. If A is negative, the graph is reflected across the x axis.

Just as with the sine and cosine functions, the constants B and C, respectively, effect a change in period and phase shift. Since $A \tan x$ and $A \cot x$ each has a period of π, it follows that $A \tan (Bx + C)$ and $A \cot (Bx + C)$ each completes one cycle as $Bx + C$ varies from

$$Bx + C = 0 \qquad \text{to} \qquad Bx + C = \pi$$

or (solving for x) as x varies from

Phase shift — Period

$$x = -\frac{C}{B} \qquad \text{to} \qquad x = -\frac{C}{B} + \frac{\pi}{B}$$

Thus, $y = A \tan (Bx + C)$ and $y = A \cot (Bx + C)$ each has a **period of π/B** and a **phase shift of $-C/B$.** The basic graph is shifted to the right if $-C/B$ is positive and to the left if $-C/B$ is negative.

As before, you do not need to memorize the formulas for period and phase shift. You only need to remember the process used to obtain the formulas.

EXAMPLE 1 Graphing an Equation of the Form $y = A \cot (Bx + C)$

Find the period and phase shift for $y = 2 \cot (x/2)$, then sketch its graph for $-2\pi < x < 2\pi$.

Solution One cycle of $y = 2 \cot (x/2)$ is completed as $x/2$ varies from 0 to π. Solve each equation for x:

$$\frac{x}{2} = 0 \qquad \frac{x}{2} = \pi$$

$$x = 0 \qquad x = 0 + 2\pi$$

$$\text{Phase shift} = 0 \qquad \text{Period} = 2\pi$$

In general, if $C = 0$, there is no phase shift. The graph is sketched for one period, $(0, 2\pi)$, then extended over the interval $(-2\pi, 2\pi)$ as in Figure 3.

FIGURE 3

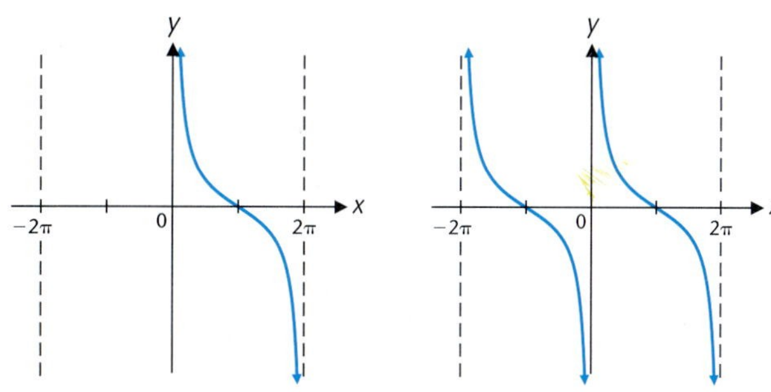

Matched Problem 1 Find the period and phase shift for $y = 3 \tan(\pi x/2)$, then sketch its graph for $-3 < x < 3$.

EXAMPLE 2 Graphing an Equation of the Form $y = A \cot(Bx + C)$

Find the period and phase shift for $y = \cot(2x + \pi/2)$, then sketch the graph for $-\pi/2 < x < \pi$.

Solution *Step 1.* Find the period and phase shift by solving $Bx + C = 0$ and $Bx + C = \pi$ for x:

$$2x + \frac{\pi}{2} = 0 \qquad 2x + \frac{\pi}{2} = \pi$$

$$2x = -\frac{\pi}{2} \qquad 2x = -\frac{\pi}{2} + \pi$$

$$x = -\frac{\pi}{4} \qquad x = -\frac{\pi}{4} + \frac{\pi}{2}$$

$$\text{Phase shift} = -\frac{\pi}{4} \qquad \text{Period} = \frac{\pi}{2}$$

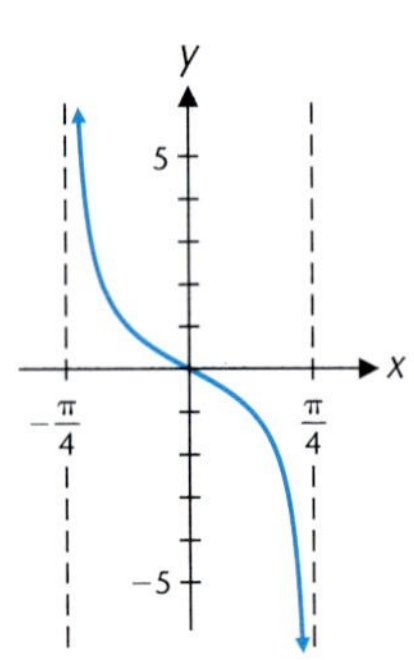

FIGURE 4

Step 2. Sketch one period of the graph starting at $x = -\pi/4$ (the phase shift) and ending at $x = -\pi/4 + \pi/2$ (the phase shift plus one period)—Figure 4.

Step 3. Extend the graph over the interval $(-\pi/2, \pi)$—Figure 5.

FIGURE 5

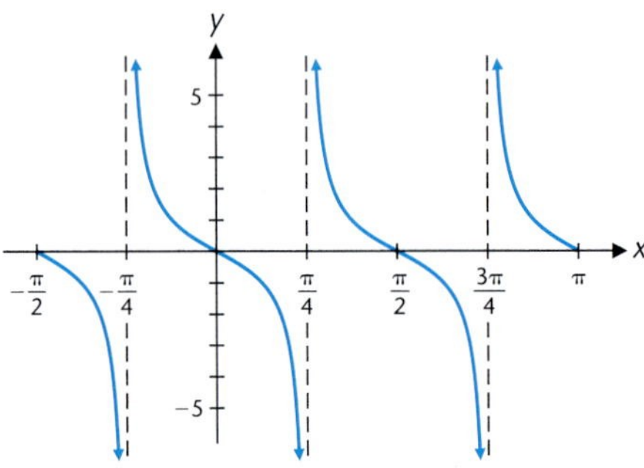

Matched Problem 2 Find the period and phase shift for $y = \tan(\pi x/2 + \pi/4)$, then sketch the graph for $-3 \le x \le 3$.

• Graphing $y = A \sec(Bx + C)$ and $y = A \csc(Bx + C)$

For convenient reference, we repeat the graphs that were shown for $y = \csc x$ and $y = \sec x$ in Section 5-6 (see Figs. 6 and 7).

Graph of $y = \csc x$

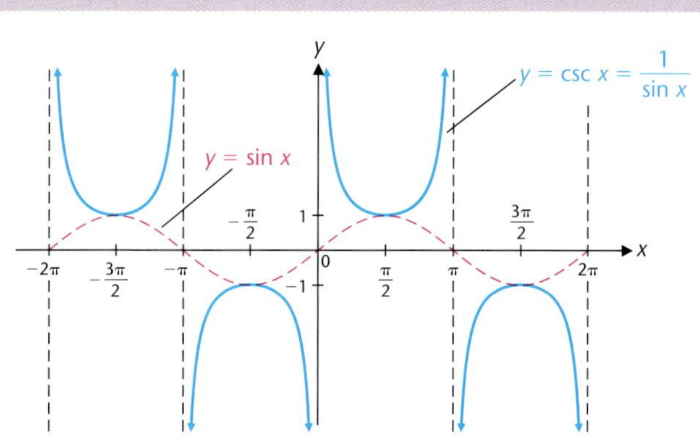

Period: 2π

Domain: All real numbers except $k\pi$, k an integer

Range: All real numbers y such that $y \le -1$ or $y \ge 1$

Symmetric with respect to the origin

Discontinuous at $x = k\pi$, k an integer

FIGURE 6

Graph of $y = \sec x$

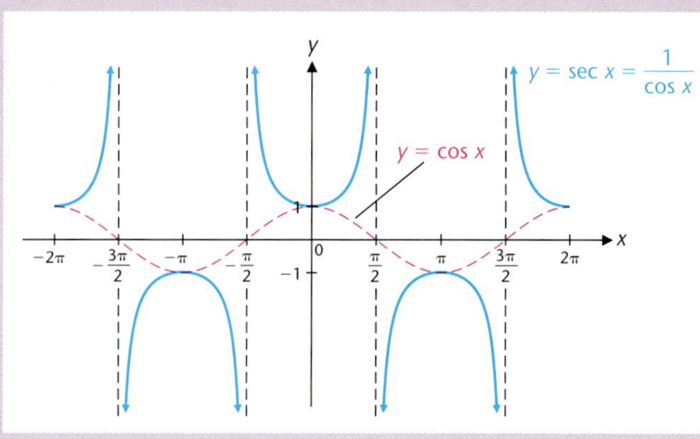

Period: 2π

Domain: All real numbers except $\pi/2 + k\pi$, k an integer

Range: All real numbers y such that $y \le -1$ or $y \ge 1$

Symmetric with respect to the y axis

Discontinuous at $x = \pi/2 + k\pi$, k an integer

FIGURE 7

EXPLORE-DISCUSS 2 (A) Match each function to its graph, and discuss how the graph compares to the graph of $y = \csc x$ or $y = \sec x$.

(1) $y = \frac{1}{2} \csc x$ (2) $y = \sec \pi x$ (3) $y = \csc (x - \pi/2)$

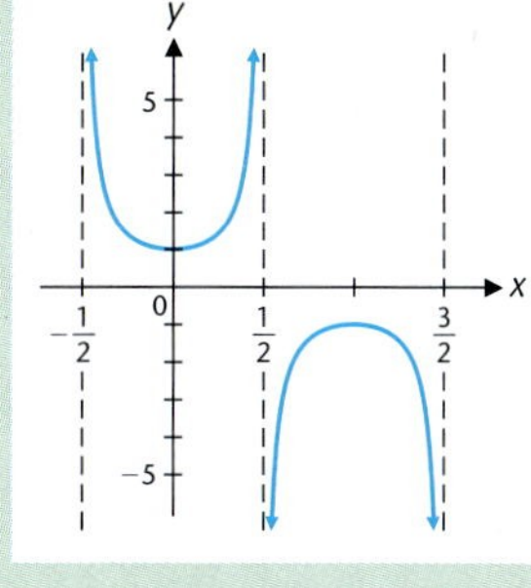

(a)

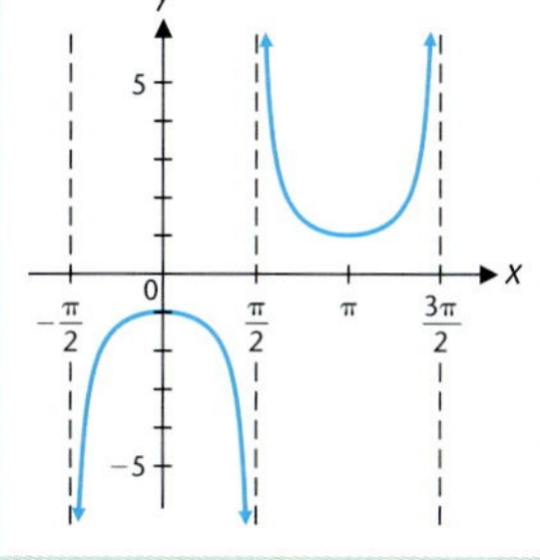

(b)

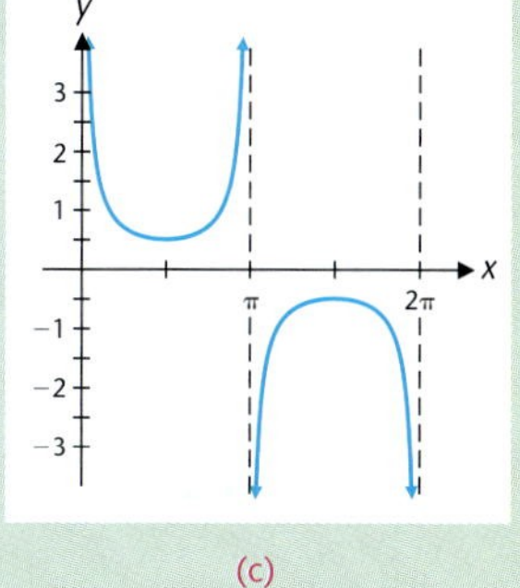

(c)

(B) Use a graphing utility to explore the nature of the changes in the graphs of the following functions when the values of A, B, and C are changed. Discuss what happens in each case.

$y = A \sec x$ and $y = A \csc x$ for different values of A

$y = \sec Bx$ and $y = \csc Bx$ for different values of B

$y = \sec (x + C)$ and $y = \csc (x + C)$ for different values of C

As with the tangent and cotangent functions, **amplitude is not defined for either the secant or the cosecant functions.** Since both functions have a period of 2π, we find the period and phase shift for each by solving $Bx + C = 0$ and $Bx + C = 2\pi$.

To graph either $y = A \sec (Bx + C)$ or $y = A \csc (Bx + C)$, you will probably find it easier to graph $y = (1/A) \cos (Bx + C)$ or $y = (1/A) \sin (Bx + C)$ with a dashed curve, then take reciprocals. An example should help to make the process clear.

EXAMPLE 3 Graphing an Equation of the Form $y = A \sec (Bx + C)$

Find the period and phase shift for $y = \frac{1}{2} \sec (2x + \pi)$, then sketch the graph for $-3\pi/4 < x < 3\pi/4$.

Solution *Step 1.* Find the period and phase shift by solving $Bx + C = 0$ and $Bx + C = 2\pi$ for x:

$$\begin{aligned} 2x + \pi &= 0 & 2x + \pi &= 2\pi \\ 2x &= -\pi & 2x &= -\pi + 2\pi \\ x &= -\frac{\pi}{2} & x &= -\frac{\pi}{2} + \pi \end{aligned}$$

$$\text{Phase shift} = -\frac{\pi}{2} \qquad \text{Period} = \pi$$

Step 2. Since

$$\frac{1}{2} \sec (2x + \pi) = \frac{1}{2 \cos (2x + \pi)}$$

we graph

$$y = 2 \cos (2x + \pi)$$

for one cycle from $-\pi/2$ to $-\pi/2 + \pi$, and then take reciprocals. Notice that we also place vertical asymptotes through the x intercepts of the cosine graph to guide us when we sketch the secant function—Figure 8.

FIGURE 8

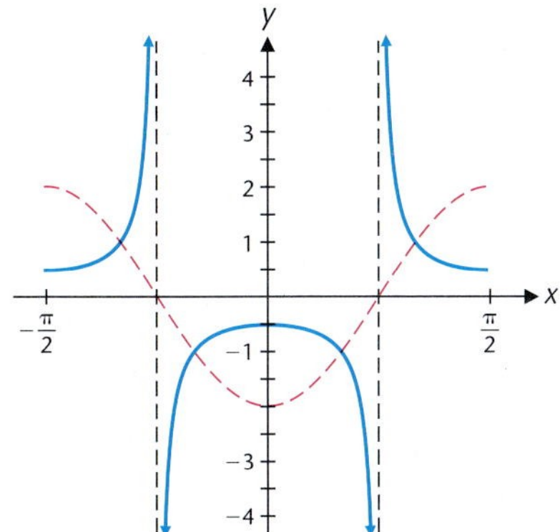

Step 3. Extend the graph over the required interval $(-3\pi/4, 3\pi/4)$—Figure 9.

FIGURE 9

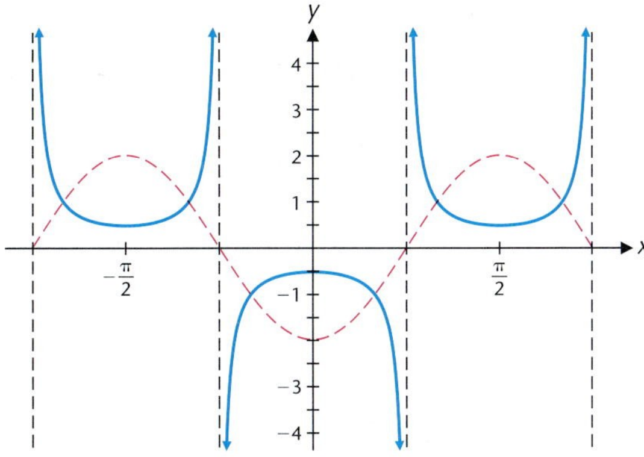

Matched Problem 3 Find the period and phase shift for $y = 2 \csc (\pi x/2 - \pi)$, then sketch the graph for $-2 \le x \le 10$.

Answers to Matched Problems

1. Period 2, phase shift 0

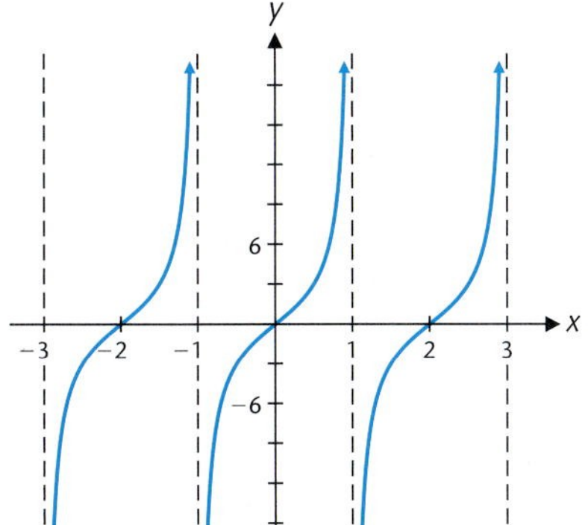

2. Period 2, phase shift $-\frac{1}{2}$

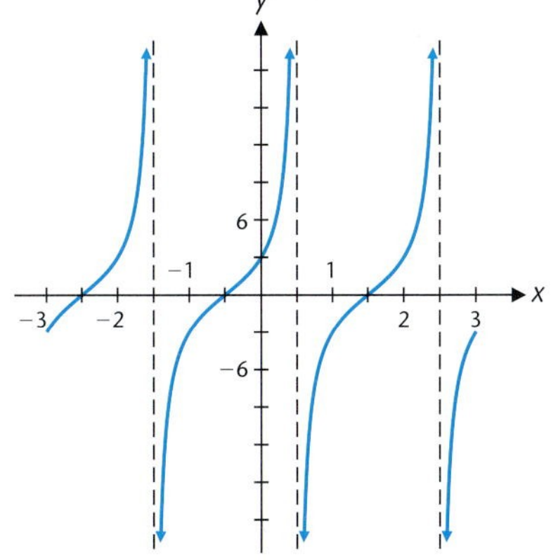

3. Period 4, phase shift 2

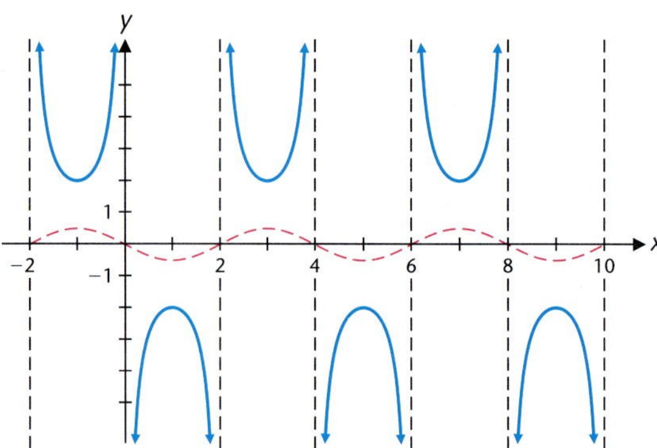

EXERCISE 5-8

A

In Problems 1–8, find the period of each function, and graph the function for the indicated interval.

1. $y = 2 \cot 4x,\ 0 < x < \pi/2$

2. $y = 3 \tan 2x,\ -\pi \le x \le \pi$

3. $y = -\frac{1}{4} \tan 8\pi x,\ 0 < x < \frac{1}{2}$

4. $y = -\frac{1}{2} \cot 2\pi x,\ 0 < x < 1$

5. $y = \csc (x/2),\ -3\pi \le x \le 3\pi$

6. $y = \sec \pi x,\ -1.5 \le x \le 3.5$

7. $y = 2 \sec \pi x,\ -1 \le x \le 3$

8. $y = 2 \csc (x/2),\ 0 < x < 8\pi$

B

In Problems 9–14, find the period and phase shift, then graph each function.

9. $y = \cot \left(x + \dfrac{\pi}{2}\right),\ -\dfrac{\pi}{2} < x < \dfrac{3\pi}{2}$

10. $y = \tan \left(x - \dfrac{\pi}{2}\right),\ -\pi < x < \pi$

11. $y = \tan (2x + \pi),\ -\dfrac{3\pi}{4} < x < \dfrac{3\pi}{4}$

12. $y = \cot (2x - \pi),\ -\dfrac{\pi}{2} \le x \le \dfrac{\pi}{2}$

13. $y = \sec \left(\pi x + \dfrac{\pi}{2}\right),\ -1 < x < 1$

14. $y = \csc \left(\pi x - \dfrac{\pi}{2}\right),\ -1 < x < 1$

In Problems 15–18, determine whether the statement is true or false. If true, explain why. If false, give a counterexample.

15. The graphs of $y = \cos (\pi x)$ and $y = \csc (\pi x)$ have infinitely many intersection points.

16. The graphs of $y = \sin (\pi x)$ and $y = \csc (\pi x)$ have infinitely many intersection points.

17. Every horizontal line intersects the graph of $y = 0.1 \sec (5x + 1)$ infinitely many times.

18. The maximum deviation of the graph of $y = 7 \tan (3\pi x + 2)$ from the x axis is 7.

In Problems 19–22, graph at least two cycles of the given equation in a graphing utility, then find an equation of the form $y = A \tan Bx$, $y = A \cot Bx$, $y = A \sec Bx$, or $y = A \csc Bx$ that has the same graph. (These problems suggest additional identities beyond those discussed in Section 5-2. Additional identities are discussed in detail in Chapter 6.)

19. $y = \cot x - \tan x$

20. $y = \cot x + \tan x$

21. $y = \csc x + \cot x$

22. $y = \csc x - \cot x$

C

In Problems 23–26, find the period and phase shift, then graph each function.

23. $y = -2 \tan \left(\dfrac{\pi}{4}x - \dfrac{\pi}{4}\right),\ -1 < x < 7$

24. $y = -3 \cot (\pi x - \pi),\ -2 < x < 2$

25. $y = 3 \csc \left(\dfrac{\pi}{2}x + \dfrac{\pi}{2}\right),\ -1 < x < 3$

26. $y = 2 \sec \left(\pi x - \dfrac{\pi}{2}\right),\ -1 < x < 3$

In Problems 27–30, graph at least two cycles of the given equation in a graphing utility, then find an equation of the form $y = A \tan Bx$, $y = A \cot Bx$, $y = A \sec Bx$, or $y = A \csc Bx$ that has the same graph. (These problems suggest additional identities beyond those discussed in Section 5-2. Additional identities are discussed in detail in Chapter 6.)

27. $y = \sin 3x + \cos 3x \cot 3x$

28. $y = \cos 2x + \sin 2x \tan 2x$

29. $y = \dfrac{\sin 4x}{1 + \cos 4x}$

30. $y = \dfrac{\sin 6x}{1 - \cos 6x}$

APPLICATIONS

★ **31. Motion.** A beacon light 20 ft from a wall rotates clockwise at the rate of 1/4 rps (see figure); thus, $\theta = \pi t/2$.

(A) Start counting time in seconds when the light spot is at N and write an equation for the length c of the light beam in terms of t.

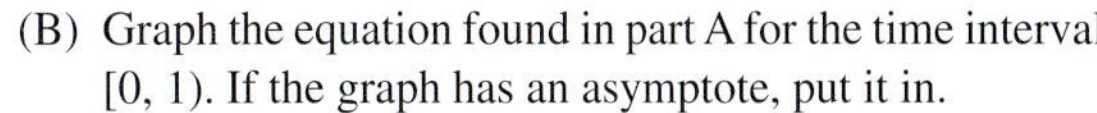

(B) Graph the equation found in part A for the time interval [0, 1). If the graph has an asymptote, put it in.

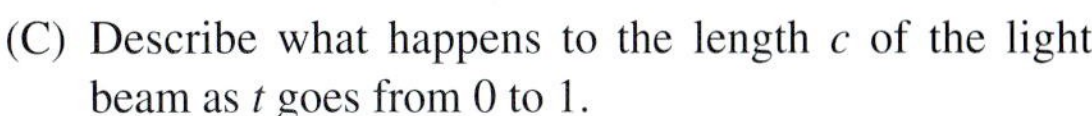

(C) Describe what happens to the length c of the light beam as t goes from 0 to 1.

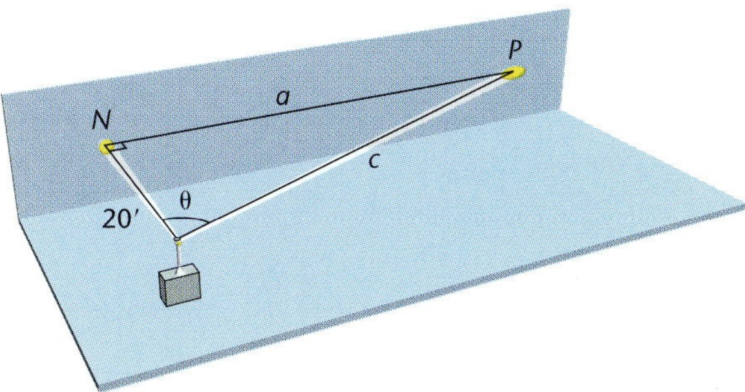

★ **32. Motion.** Refer to Problem 31.

(A) Write an equation for the distance a the light spot travels along the wall in terms of time t.

(B) Graph the equation found in part A for the time interval [0, 1). If the graph has an asymptote, put it in.

(C) Describe what happens to the distance a along the wall as t goes from 0 to 1.

SECTION 5-9 Inverse Trigonometric Functions

- Inverse Sine Function
- Inverse Cosine Function
- Inverse Tangent Function
- Summary
- Inverse Cotangent, Secant, and Cosecant Functions (Optional)

A brief review of the general concept of inverse functions discussed in Section 2-6 should prove helpful before proceeding with this section. In the following box we restate a few important facts about inverse functions from that section.

Facts about Inverse Functions

For f a one-to-one function and f^{-1} its inverse:

1. If (a, b) is an element of f, then (b, a) is an element of f^{-1}, and conversely.

2. Range of f = Domain of f^{-1}

Domain of f = Range of f^{-1}

3.

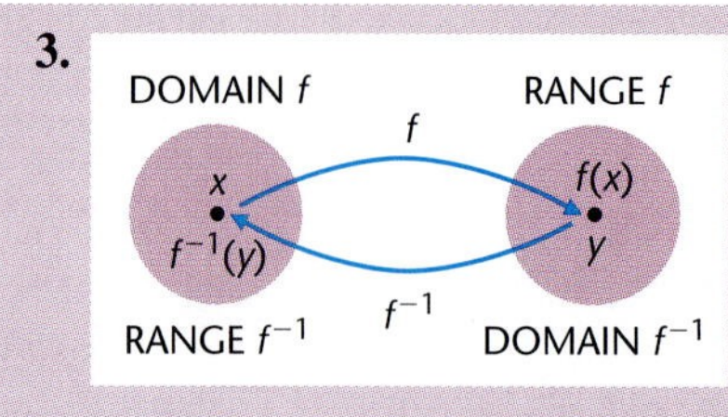

4. If $x = f^{-1}(y)$, then $y = f(x)$ for y in the domain of f^{-1} and x in the domain of f, and conversely.

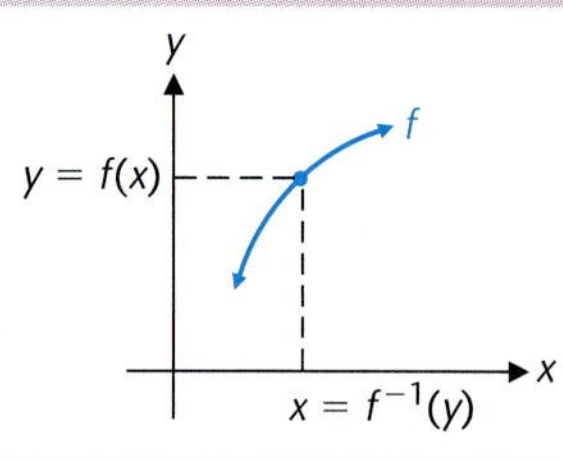

5. $f[f^{-1}(y)] = y$ for y in the domain of f^{-1}
$f^{-1}[f(x)] = x$ for x in the domain of f

All trigonometric functions are periodic; hence, each range value can be associated with infinitely many domain values (Fig. 1). As a result, no trigonometric function is one-to-one. Without restrictions, no trigonometric function has an inverse function. To resolve this problem, we restrict the domain of each function so that it is one-to-one over the restricted domain. Thus, for this restricted domain, an inverse function is guaranteed.

FIGURE 1 $y = \sin x$ is not one-to-one over $(-\infty, \infty)$.

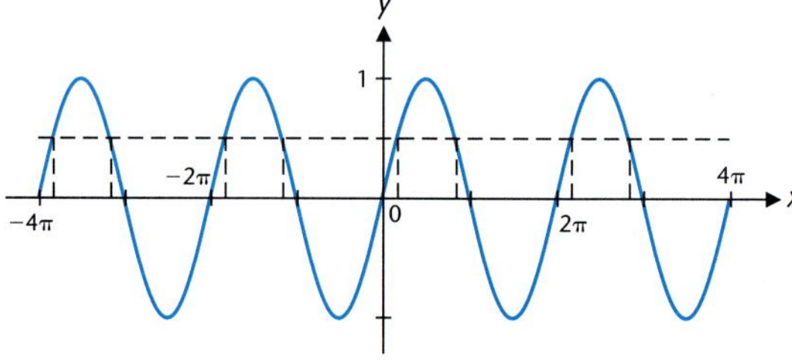

Inverse trigonometric functions represent another group of basic functions that are added to our library of elementary functions. These functions are used in many applications and mathematical developments, and they will be particularly useful to us when we solve trigonometric equations in Section 6-5.

• Inverse Sine Function

How can the domain of the sine function be restricted so that it is one-to-one? This can be done in infinitely many ways. A fairly natural and generally accepted way is illustrated in Figure 2.

FIGURE 2 $y = \sin x$ is one-to-one over $[-\pi/2, \pi/2]$.

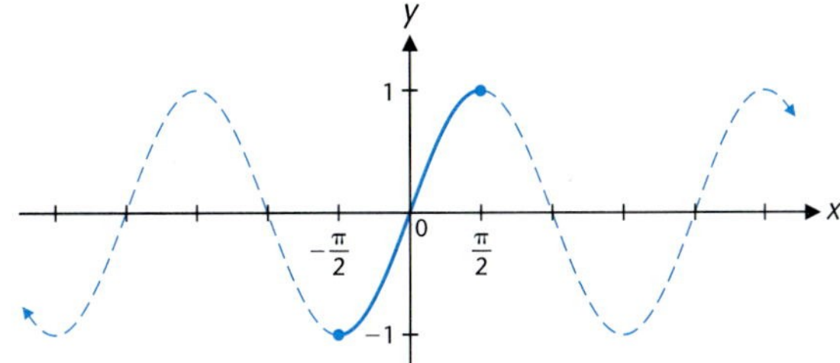

If the domain of the sine function is restricted to the interval $[-\pi/2, \pi/2]$, we see that the restricted function passes the horizontal line test (Section 1-8) and thus is one-to-one. Note that each range value from -1 to 1 is assumed exactly once as x moves from $-\pi/2$ to $\pi/2$. We use this restricted sine function to define the *inverse sine function.*

DEFINITION 1

Inverse Sine Function

The **inverse sine function,** denoted by $\mathbf{sin^{-1}}$ or **arcsin,** is defined as the inverse of the restricted sine function $y = \sin x$, $-\pi/2 \le x \le \pi/2$. Thus,

$$y = \sin^{-1} x \qquad \text{and} \qquad y = \arcsin x$$

are equivalent to

$$\sin y = x \qquad \text{where } -\pi/2 \le y \le \pi/2,\ -1 \le x \le 1$$

In words, the inverse sine of x, or the arcsine of x, is the number or angle y, $-\pi/2 \le y \le \pi/2$, whose sine is x.

To graph $y = \sin^{-1} x$, take each point on the graph of the restricted sine function and reverse the order of the coordinates. For example, since $(-\pi/2, -1)$, $(0, 0)$, and $(\pi/2, 1)$ are on the graph of the restricted sine function (Fig. 3(a)) then $(-1, -\pi/2)$, $(0, 0)$, and $(1, \pi/2)$ are on the graph of the inverse sine function, as shown in Figure 3(b). Using these three points provides us with a quick way of sketching the graph of the inverse sine function. A more accurate graph can be obtained by using a calculator.

FIGURE 3 Inverse sine function.

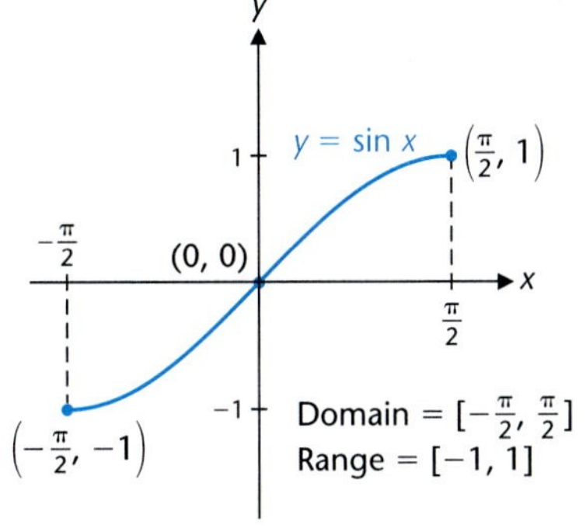

Restricted sine function

(a)

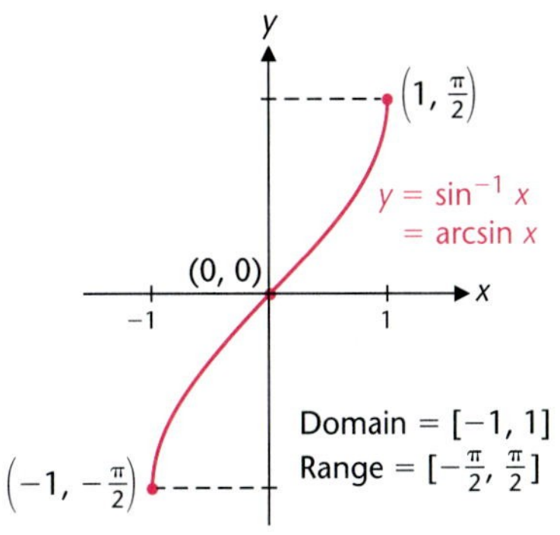

Inverse sine function

(b)

We state the important sine–inverse sine identities which follow from the general properties of inverse functions given in the box at the beginning of this section.

Sine–Inverse Sine Identities

$\sin(\sin^{-1} x) = x \qquad -1 \le x \le 1 \qquad f[f^{-1}(x)] = x$

$\sin^{-1}(\sin x) = x \qquad -\pi/2 \le x \le \pi/2 \qquad f^{-1}[f(x)] = x$

$\sin(\sin^{-1} 0.7) = 0.7 \qquad \sin(\sin^{-1} 1.3) \ne 1.3$

$\sin^{-1}[\sin(-1.2)] = -1.2 \qquad \sin^{-1}[\sin(-2)] \ne -2$

[*Note:* The number 1.3 is not in the domain of the inverse sine function, and −2 is not in the restricted domain of the sine function. Try calculating all these examples with your calculator and see what happens!]

EXAMPLE 1 Exact Values

Find exact values without using a calculator:

(A) $\arcsin(-\frac{1}{2})$ (B) $\sin^{-1}(\sin 1.2)$ (C) $\cos(\sin^{-1}\frac{2}{3})$

Solution (A) $y = \arcsin(-\frac{1}{2})$ is equivalent to

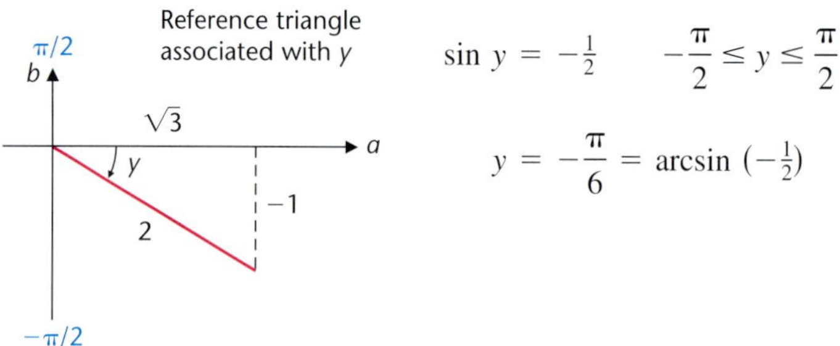

$$\sin y = -\tfrac{1}{2} \qquad -\frac{\pi}{2} \le y \le \frac{\pi}{2}$$

$$y = -\frac{\pi}{6} = \arcsin(-\tfrac{1}{2})$$

[*Note:* $y \ne 11\pi/6$, even though $\sin(11\pi/6) = -\frac{1}{2}$. y must be between $-\pi/2$ and $\pi/2$, inclusive.]

(B) $\sin^{-1}(\sin 1.2) = 1.2$ Sine–inverse sine identity, since $-\pi/2 \le 1.2 \le \pi/2$

(C) Let $y = \sin^{-1}\frac{2}{3}$; then $\sin y = \frac{2}{3}$, $-\pi/2 \le y \le \pi/2$. Draw the reference triangle associated with y. Then $\cos y = \cos(\sin^{-1}\frac{2}{3})$ can be determined directly from the triangle (after finding the third side) without actually finding y.

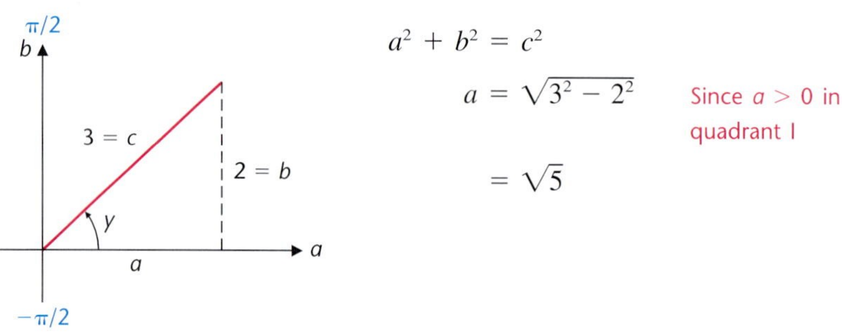

$$a^2 + b^2 = c^2$$

$$a = \sqrt{3^2 - 2^2} \qquad \text{Since } a > 0 \text{ in quadrant I}$$

$$= \sqrt{5}$$

Thus, $\cos(\sin^{-1}\frac{2}{3}) = \cos y = \sqrt{5}/3$.

Matched Problem 1 Find exact values without using a calculator:

(A) arcsin $(\sqrt{2}/2)$
(B) sin $[\sin^{-1}(-0.4)]$
(C) tan $[\sin^{-1}(-1/\sqrt{5})]$

EXAMPLE 2 **Calculator Values**

Find to 4 significant digits using a calculator:

(A) arcsin (-0.3042)
(B) $\sin^{-1} 1.357$
(C) cot $[\sin^{-1}(-0.1087)]$

Solution The function keys used to represent inverse trigonometric functions vary among different brands of calculators, so read the user's manual for your calculator. Set your calculator in radian mode and follow your manual for key sequencing.

(A) arcsin $(-0.3042) = -0.3091$
(B) $\sin^{-1} 1.357 =$ Error 1.357 is not in the domain of $\sin^{-1}$
(C) cot $[\sin^{-1}(-0.1087)] = -9.145$

Matched Problem 2 Find to 4 significant digits using a calculator:

(A) $\sin^{-1} 0.2903$
(B) arcsin (-2.305)
(C) cot $[\sin^{-1}(-0.3446)]$

• Inverse Cosine Function

To restrict the cosine function so that it becomes one-to-one, we choose the interval $[0, \pi]$. Over this interval the restricted function passes the horizontal line test, and each range value is assumed exactly once as x moves from 0 to π (Fig. 4). We use this restricted cosine function to define the *inverse cosine function.*

FIGURE 4 $y = \cos x$ is one-to-one over $[0, \pi]$.

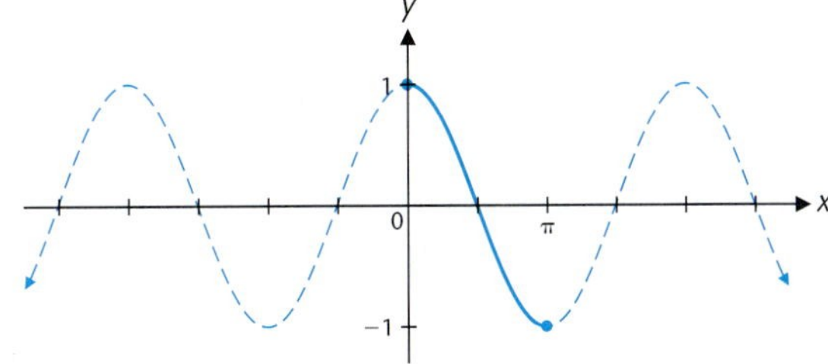

DEFINITION 2

Inverse Cosine Function

The **inverse cosine function,** denoted by $\cos^{-1}$ or **arccos,** is defined as the inverse of the restricted cosine function $y = \cos x$, $0 \le x \le \pi$. Thus,

$$y = \cos^{-1} x \qquad \text{and} \qquad y = \arccos x$$

are equivalent to

$$\cos y = x \qquad \text{where } 0 \le y \le \pi,\ -1 \le x \le 1$$

In words, the inverse cosine of x, or the arccosine of x, is the number or angle y, $0 \le y \le \pi$, whose cosine is x.

Figure 5 compares the graphs of the restricted cosine function and its inverse. Notice that $(0, 1)$, $(\pi/2, 0)$, and $(\pi, -1)$ are on the restricted cosine graph. Reversing the coordinates gives us three points on the graph of the inverse cosine function.

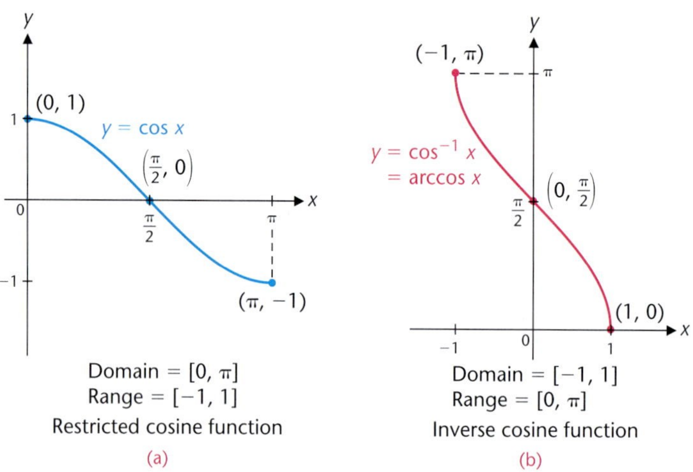

FIGURE 5 Inverse cosine function.

We complete the discussion by giving the cosine–inverse cosine identities:

Cosine–Inverse Cosine Identities

$\cos(\cos^{-1} x) = x$	$-1 \le x \le 1$	$f[f^{-1}(x)] = x$
$\cos^{-1}(\cos x) = x$	$0 \le x \le \pi$	$f^{-1}[f(x)] = x$

EXPLORE-DISCUSS 1

Evaluate each of the following with a calculator. Which illustrate a cosine–inverse cosine identity and which do not? Discuss why.

(A) $\cos(\cos^{-1} 0.2)$ (C) $\cos^{-1}(\cos 2)$

(B) $\cos[\cos^{-1}(-2)]$ (D) $\cos^{-1}[\cos(-3)]$

EXAMPLE 3 **Exact Values**

Find exact values without using a calculator:

(A) $\arccos(-\sqrt{3}/2)$ (B) $\cos(\cos^{-1} 0.7)$ (C) $\sin[\cos^{-1}(-\frac{1}{3})]$

Solutions (A) $y = \arccos(-\sqrt{3}/2)$ is equivalent to

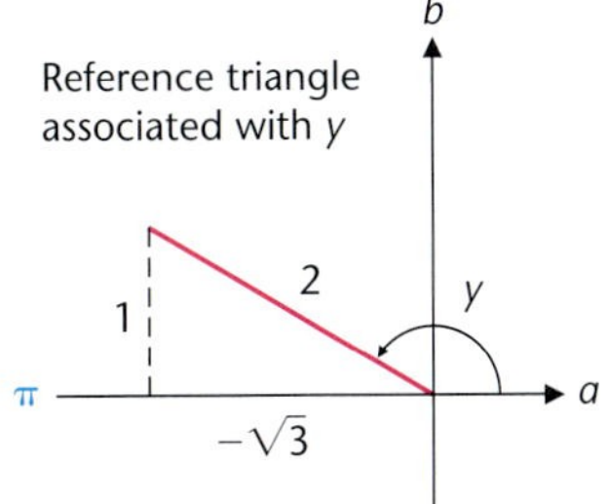

$$\cos y = -\frac{\sqrt{3}}{2} \qquad 0 \le y \le \pi$$

$$y = \frac{5\pi}{6} = \arccos\left(-\frac{\sqrt{3}}{2}\right)$$

[*Note:* $y \neq -5\pi/6$, even though $\cos(-5\pi/6) = -\sqrt{3}/2$. y must be between 0 and π, inclusive.]

(B) $\cos(\cos^{-1} 0.7) = 0.7$ Cosine–inverse cosine identity, since $-1 \le 0.7 \le 1$

(C) Let $y = \cos^{-1}(-\frac{1}{3})$; then $\cos y = -\frac{1}{3}$, $0 \le y \le \pi$. Draw a reference triangle associated with y. Then $\sin y = \sin[\cos^{-1}(-\frac{1}{3})]$ can be determined directly from the triangle (after finding the third side) without actually finding y.

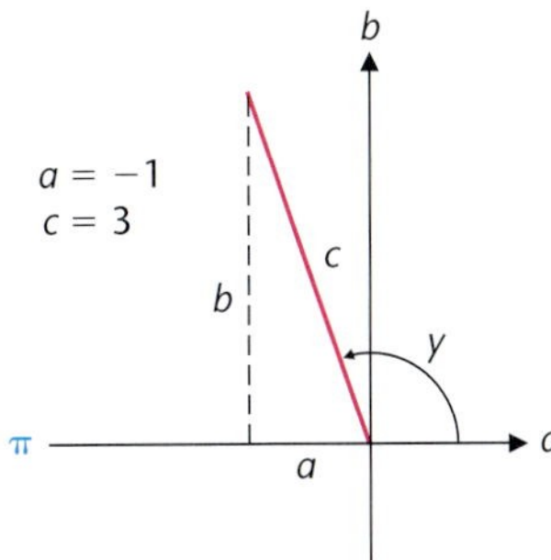

$$a^2 + b^2 = c^2$$

$$b = \sqrt{3^2 - (-1)^2} \qquad \text{Since } b > 0 \text{ in quadrant II}$$

$$= \sqrt{8} = 2\sqrt{2}$$

Thus, $\sin[\cos^{-1}(-\frac{1}{3})] = \sin y = 2\sqrt{2}/3$.

Matched Problem 3 Find exact values without using a calculator:

(A) $\arccos(\sqrt{2}/2)$ (B) $\cos^{-1}(\cos 3.05)$ (C) $\cot[\cos^{-1}(-1/\sqrt{5})]$

EXAMPLE 4 **Calculator Values**

Find to 4 significant digits using a calculator:

(A) $\arccos 0.4325$ (B) $\cos^{-1} 2.137$ (C) $\csc[\cos^{-1}(-0.0349)]$

Solution Set your calculator in radian mode.

(A) $\arccos 0.4325 = 1.124$
(B) $\cos^{-1} 2.137 = \text{Error}$ 2.137 is not in the domain of $\cos^{-1}$
(C) $\csc [\cos^{-1} (-0.0349)] = 1.001$

Matched Problem 4 Find to 4 significant digits using a calculator:

(A) $\cos^{-1} 0.6773$ (B) $\arccos (-1.003)$ (C) $\cot [\cos^{-1} (-0.5036)]$

• Inverse Tangent Function

To restrict the tangent function so that it becomes one-to-one, we choose the interval $(-\pi/2, \pi/2)$. Over this interval the restricted function passes the horizontal line test, and each range value is assumed exactly once as x moves across this restricted domain (Fig. 6). We use this restricted tangent function to define the *inverse tangent function.*

FIGURE 6 $y = \tan x$ is one-to-one over $(-\pi/2, \pi/2)$.

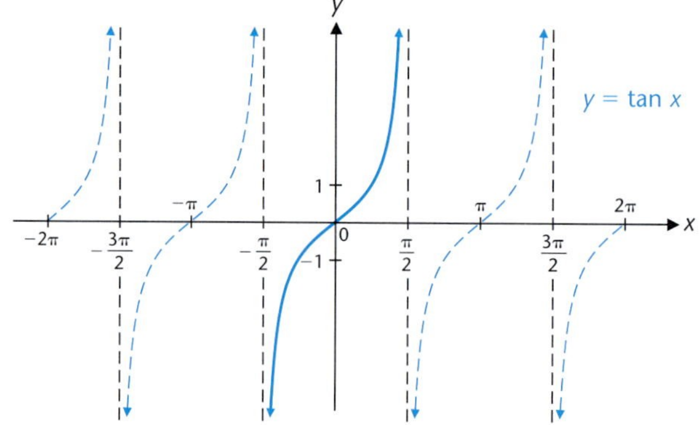

DEFINITION 3 **Inverse Tangent Function**

The **inverse tangent function,** denoted by $\mathbf{tan^{-1}}$ or **arctan,** is defined as the inverse of the restricted tangent function $y = \tan x$, $-\pi/2 < x < \pi/2$. Thus,

$$y = \tan^{-1} x \quad \text{and} \quad y = \arctan x$$

are equivalent to

$$\tan y = x \quad \text{where } -\pi/2 < y < \pi/2 \text{ and } x \text{ is a real number}$$

In words, the inverse tangent of x, or the arctangent of x, is the number or angle y, $-\pi/2 < y < \pi/2$, whose tangent is x.

Figure 7 compares the graphs of the restricted tangent function and its inverse. Notice that $(-\pi/4, -1)$, $(0, 0)$, and $(\pi/4, 1)$ are on the restricted tangent graph. Reversing the coordinates gives us three points on the graph of the inverse tangent function. Also note that the vertical asymptotes become horizontal asymptotes for the inverse function.

FIGURE 7 Inverse tangent function.

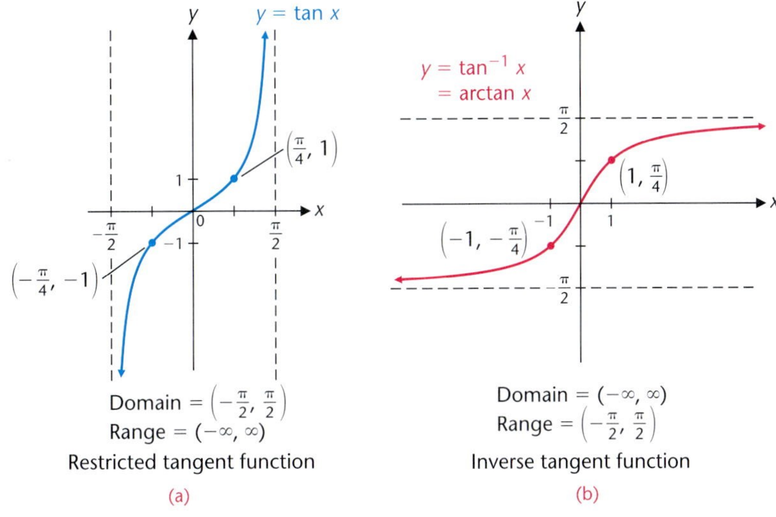

We now state the tangent–inverse tangent identities.

Tangent–Inverse Tangent Identities

$\tan(\tan^{-1} x) = x$	$-\infty < x < \infty$	$f[f^{-1}(x)] = x$
$\tan^{-1}(\tan x) = x$	$-\pi/2 < x < \pi/2$	$f^{-1}[f(x)] = x$

EXPLORE-DISCUSS 2 Evaluate each of the following with a calculator. Which illustrate a tangent–inverse tangent identity and which do not? Discuss why.

(A) $\tan(\tan^{-1} 30)$ (C) $\tan^{-1}(\tan 1.4)$
(B) $\tan[\tan^{-1}(-455)]$ (D) $\tan^{-1}[\tan(-3)]$

EXAMPLE 5 **Exact Values**

Find exact values without using a calculator:

(A) $\tan^{-1}(-1/\sqrt{3})$ (B) $\tan^{-1}(\tan 0.63)$

Solutions (A) $y = \tan^{-1}(-1/\sqrt{3})$ is equivalent to

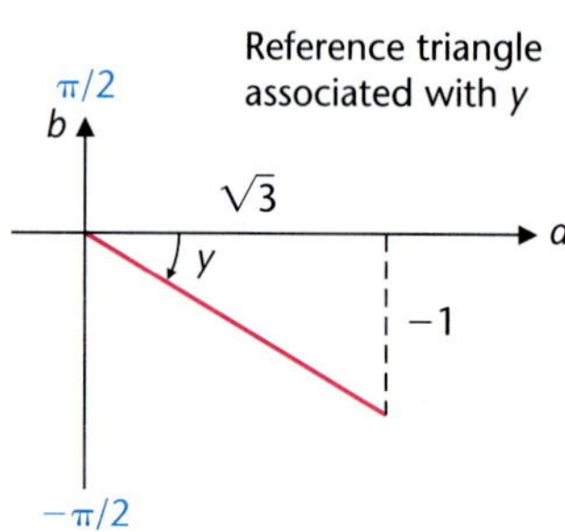

$$\tan y = -\frac{1}{\sqrt{3}} \qquad -\frac{\pi}{2} < y < \frac{\pi}{2}$$

$$y = -\frac{\pi}{6} = \tan^{-1}\left(-\frac{1}{\sqrt{3}}\right)$$

[*Note:* y cannot be $11\pi/6$. y must be between $-\pi/2$ and $\pi/2$.]

(B) $\tan^{-1}(\tan 0.63) = 0.63$ Tangent–inverse tangent identity, since $-\pi/2 < 0.63 < \pi/2$

Matched Problem 5 Find exact values without using a calculator:

(A) $\arctan(-\sqrt{3})$ (B) $\tan(\tan^{-1} 43)$

• Summary

We summarize the definitions and graphs of the inverse trigonometric functions discussed so far for convenient reference.

Summary of $\sin^{-1}$, $\cos^{-1}$, and $\tan^{-1}$

$y = \sin^{-1} x$	is equivalent to	$x = \sin y$	$-1 \le x \le 1,\ -\pi/2 \le y \le \pi/2$
$y = \cos^{-1} x$	is equivalent to	$x = \cos y$	$-1 \le x \le 1,\ 0 \le y \le \pi$
$y = \tan^{-1} x$	is equivalent to	$x = \tan y$	$-\infty < x < \infty,\ -\pi/2 < y < \pi/2$

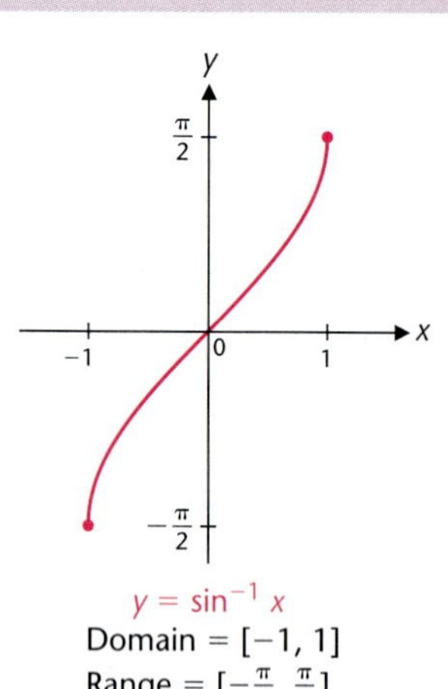

$y = \sin^{-1} x$
Domain = $[-1, 1]$
Range = $[-\frac{\pi}{2}, \frac{\pi}{2}]$

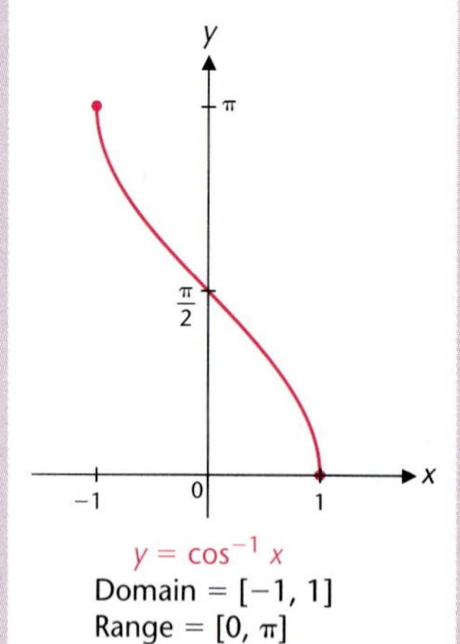

$y = \cos^{-1} x$
Domain = $[-1, 1]$
Range = $[0, \pi]$

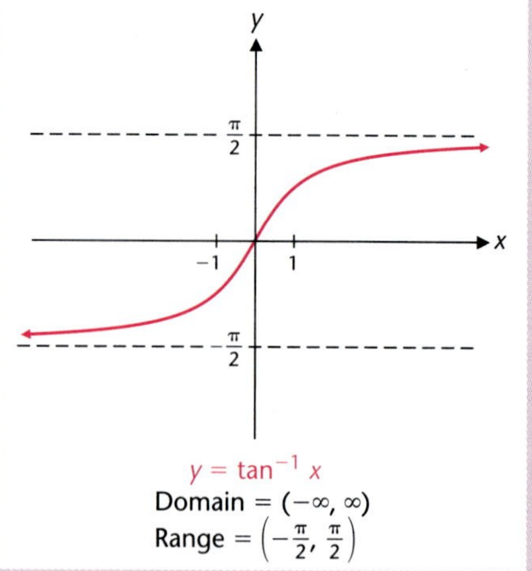

$y = \tan^{-1} x$
Domain = $(-\infty, \infty)$
Range = $\left(-\frac{\pi}{2}, \frac{\pi}{2}\right)$

• Inverse Cotangent, Secant, and Cosecant Functions (Optional)

For completeness, we include the definitions and graphs of the inverse cotangent, secant, and cosecant functions.

DEFINITION 4 Inverse Cotangent, Secant, and Cosecant Functions

$y = \cot^{-1} x$	is equivalent to	$x = \cot y$	where $0 < y < \pi$, $-\infty < x < \infty$
$y = \sec^{-1} x$	is equivalent to	$x = \sec y$	where $0 \le y \le \pi$, $y \ne \pi/2$, $\lvert x\rvert \ge 1$
$y = \csc^{-1} x$	is equivalent to	$x = \csc y$	where $-\pi/2 \le y \le \pi/2$, $y \ne 0$, $\lvert x\rvert \ge 1$

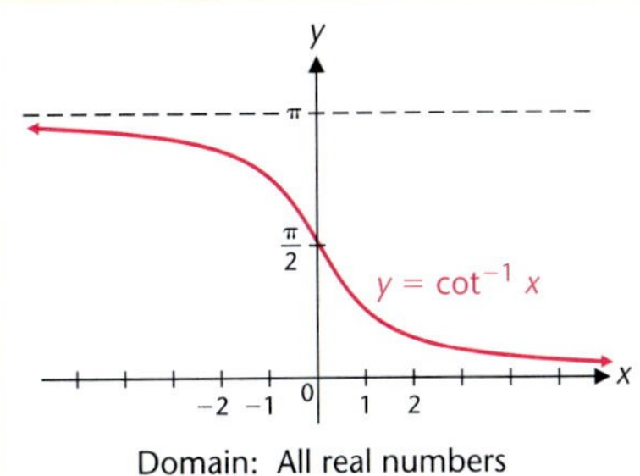

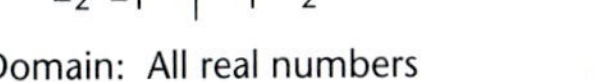
Domain: All real numbers
Range: $0 < y < \pi$

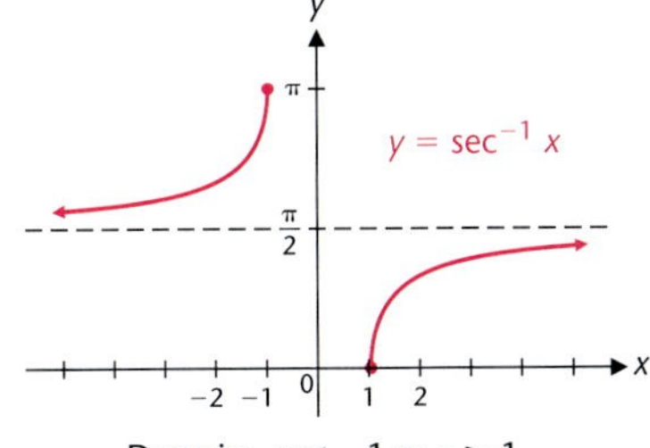

Domain: $x \le -1$ or $x \ge 1$
Range: $0 \le y \le \pi$, $y \ne \pi/2$

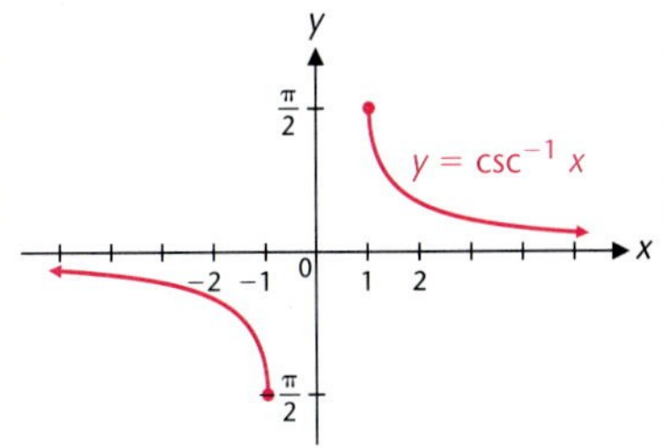

Domain: $x \le -1$ or $x \ge 1$
Range: $-\pi/2 \le y \le \pi/2$, $y \ne 0$

[*Note:* The definitions of $\sec^{-1}$ and $\csc^{-1}$ are not universally agreed upon.]

Answers to Matched Problems

1. (A) $\pi/4$ (B) -0.4 (C) $-1/2$
2. (A) 0.2945 (B) Not defined (C) -2.724
3. (A) $\pi/4$ (B) 3.05 (C) $-1/2$
4. (A) 0.8267 (B) Not defined (C) -0.5829
5. (A) $-\pi/3$ (B) 43

EXERCISE 5-9

Unless stated to the contrary, the inverse trigonometric functions are assumed to have real number ranges (use radian mode in calculator problems). A few problems involve ranges with angles in degree measure, and these are clearly indicated (use degree mode in calculator problems).

A

In Problems 1–12, find exact values without using a calculator.

1. $\cos^{-1} 0$
2. $\sin^{-1} 0$
3. $\arcsin (\sqrt{3}/2)$
4. $\arccos (\sqrt{3}/2)$
5. $\arctan \sqrt{3}$
6. $\tan^{-1} 1$
7. $\sin^{-1} (\sqrt{2}/2)$
8. $\cos^{-1} \frac{1}{2}$
9. $\arccos 1$
10. $\arctan (1/\sqrt{3})$
11. $\sin^{-1} \frac{1}{2}$
12. $\tan^{-1} 0$

In Problems 13–18, evaluate to 4 significant digits using a calculator.

13. $\cos^{-1} 0.8217$
14. $\arcsin 0.5625$
15. $\arctan 133.3$
16. $\arccos 0.0127$
17. $\sin^{-1} 2.153$
18. $\tan^{-1} 8.529$

B

In Problems 19–28, find exact values without using a calculator.

19. $\arccos (-\sqrt{3}/2)$
20. $\arcsin (-\sqrt{3})$
21. $\arctan (-1)$
22. $\cos^{-1} (-\sqrt{2}/2)$
23. $\sin^{-1} (-\frac{1}{2})$
24. $\tan^{-1} (-1/\sqrt{3})$
25. $\tan (\tan^{-1} 25)$
26. $\sin [\sin^{-1} (-0.6)]$

27. $\sin(\cos^{-1}\sqrt{3}/2)$ **28.** $\tan(\cos^{-1} 1/2)$

In Problems 29–32, evaluate to 4 significant digits using a calculator.

29. $\cot[\cos^{-1}(-0.7003)]$ **30.** $\sec[\sin^{-1}(-0.0399)]$

31. $\sqrt{5 + \cos^{-1}(1 - \sqrt{2})}$ **32.** $\sqrt{2} + \tan^{-1}\sqrt[3]{5}$

In Problems 33–38, find the exact degree measure of each without the use of a calculator.

33. $\sin^{-1}(-\sqrt{2}/2)$ **34.** $\cos^{-1}(-1/2)$ **35.** $\arctan(-\sqrt{3})$

36. $\arctan(-1)$ **37.** $\cos^{-1}(-1)$ **38.** $\sin^{-1}(-1)$

In Problems 39–42, find the degree measure of each to two decimal places using a calculator set in degree mode.

39. $\cos^{-1} 0.7253$ **40.** $\tan^{-1} 12.4304$

41. $\arcsin(-0.3662)$ **42.** $\arccos(-0.9206)$

43. Evaluate $\sin^{-1}(\sin 2)$ with a calculator set in radian mode, and explain why this does or does not illustrate the inverse sine–sine identity.

44. Evaluate $\cos^{-1}[\cos(-0.5)]$ with a calculator set in radian mode, and explain why this does or does not illustrate the inverse cosine–cosine identity.

Problems 45–54 require the use of a graphing utility.

In Problems 45–52, graph each function in a graphing utility over the indicated interval.

45. $y = \sin^{-1} x$, $-1 \le x \le 1$

46. $y = \cos^{-1} x$, $-1 \le x \le 1$

47. $y = \cos^{-1}(x/3)$, $-3 \le x \le 3$

48. $y = \sin^{-1}(x/2)$, $-2 \le x \le 2$

49. $y = \sin^{-1}(x - 2)$, $1 \le x \le 3$

50. $y = \cos^{-1}(x + 1)$, $-2 \le x \le 0$

51. $y = \tan^{-1}(2x - 4)$, $-2 \le x \le 6$

52. $y = \tan^{-1}(2x + 3)$, $-5 \le x \le 2$

53. The identity $\cos(\cos^{-1} x) = x$ is valid for $-1 \le x \le 1$.
(A) Graph $y = \cos(\cos^{-1} x)$ for $-1 \le x \le 1$.
(B) What happens if you graph $y = \cos(\cos^{-1} x)$ over a wider interval, say, $-2 \le x \le 2$? Explain.

54. The identity $\sin(\sin^{-1} x) = x$ is valid for $-1 \le x \le 1$.
(A) Graph $y = \sin(\sin^{-1} x)$ for $-1 \le x \le 1$.
(B) What happens if you graph $y = \sin(\sin^{-1} x)$ over a wider interval, say, $-2 \le x \le 2$? Explain.

C

In Problems 55–58, find the exact solutions to the equation. Explain your reasoning.

55. $\sin^{-1} x = \cos^{-1} x$

56. $\sin^{-1} x = \tan^{-1} x$

57. $\sin^{-1}(\sqrt{3}x) = \cos^{-1} x$

58. $\sin^{-1} x = \sin^{-1}(1/x)$

In Problems 59–62, write each expression as an algebraic expression in x free of trigonometric or inverse trigonometric functions.

59. $\cos(\sin^{-1} x)$ **60.** $\sin(\cos^{-1} x)$

61. $\cos(\arctan x)$ **62.** $\tan(\arcsin x)$

In Problems 63 and 64, find $f^{-1}(x)$. How must x be restricted in $f^{-1}(x)$?

63. $f(x) = 4 + 2\cos(x - 3)$, $3 \le x \le (3 + \pi)$

64. $f(x) = 3 + 5\sin(x - 1)$, $(1 - \pi/2) \le x \le (1 + \pi/2)$

Problems 65–66 require the use of a graphing utility.

65. The identity $\cos^{-1}(\cos x) = x$ is valid for $0 \le x \le \pi$.
(A) Graph $y = \cos^{-1}(\cos x)$ for $0 \le x \le \pi$.
(B) What happens if you graph $y = \cos^{-1}(\cos x)$ over a larger interval, say, $-2\pi \le x \le 2\pi$? Explain.

66. The identity $\sin^{-1}(\sin x) = x$ is valid for $-\pi/2 \le x \le \pi/2$.
(A) Graph $y = \sin^{-1}(\sin x)$ for $-\pi/2 \le x \le \pi/2$.
(B) What happens if you graph $y = \sin^{-1}(\sin x)$ over a larger interval, say, $-2\pi \le x \le 2\pi$? Explain.

APPLICATIONS 

67. Photography. The viewing angle changes with the focal length of a camera lens: A 28-mm wide-angle lens has a wide viewing angle and a 300-mm telephoto lens has a narrow viewing angle. For a 35-mm-format camera the viewing angle θ, in degrees, is given by

$$\theta = 2\tan^{-1}\frac{21.634}{x}$$

where x is the focal length of the lens being used. What is the viewing angle (in decimal degrees to two decimal places) of a 28-mm lens? Of a 100-mm lens?

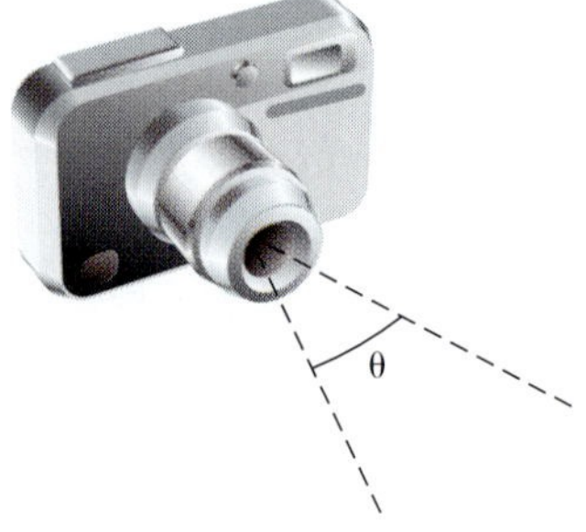

68. Photography. Referring to Problem 67, what is the viewing angle (in decimal degrees to two decimal places) of a 17-mm lens? Of a 70-mm lens?

69. (A) Graph the function in Problem 67 in a graphing utility using degree mode. The graph should cover lenses with focal lengths from 10 mm to 100 mm.

(B) What focal-length lens, to two decimal places, would have a viewing angle of 40°? Solve by graphing $\theta = 40$ and $\theta = 2\tan^{-1}(21.634/x)$ in the same viewing window and finding the point of intersection using an approximation routine.

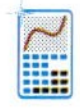

70. (A) Graph the function in Problem 67 in a graphing utility, in degree mode, with the graph covering lenses with focal lengths from 100 mm to 1000 mm.

(B) What focal-length lens, to two decimal places, would have a viewing angle of 10°? Solve by graphing $\theta = 10$ and $\theta = 2\tan^{-1}(21.634/x)$ in the same viewing window and finding the point of intersection using an approximation routine.

★ **71. Engineering.** The length of the belt around the two pulleys in the figure is given by

$$L = \pi D + (d - D)\theta + 2C\sin\theta$$

where θ (in radians) is given by

$$\theta = \cos^{-1}\frac{D - d}{2C}$$

Verify these formulas, and find the length of the belt to two decimal places if $D = 4$ inches, $d = 2$ inches, and $C = 6$ inches.

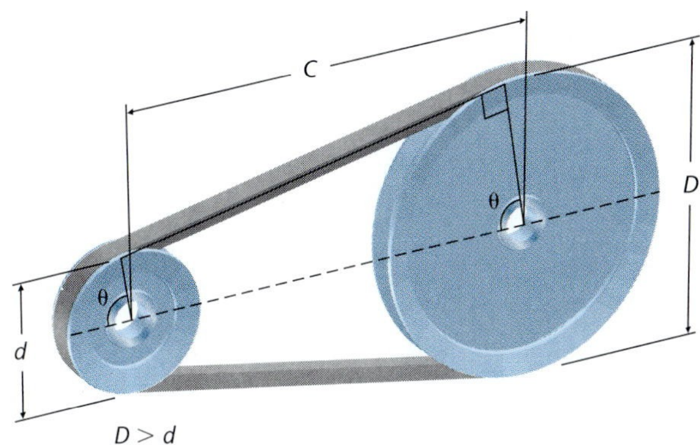

★ **72. Engineering.** For Problem 71, find the length of the belt if $D = 6$ inches, $d = 4$ inches, and $C = 10$ inches.

73. Engineering. The function

$$y_1 = 4\pi - 2\cos^{-1}\frac{1}{x} + 2x\sin\left(\cos^{-1}\frac{1}{x}\right)$$

represents the length of the belt around the two pulleys in Problem 71 when the centers of the pulleys are x inches apart.

(A) Graph y_1 in a graphing utility (in radian mode), with the graph covering pulleys with their centers from 3 to 10 inches apart.

(B) How far, to two decimal places, should the centers of the two pulleys be placed to use a belt 24 inches long? Solve by graphing y_1 and $y_2 = 24$ in the same viewing window and finding the point of intersection using an approximation routine.

74. Engineering. The function

$$y_1 = 6\pi - 2\cos^{-1}\frac{1}{x} + 2x\sin\left(\cos^{-1}\frac{1}{x}\right)$$

represents the length of the belt around the two pulleys in Problem 72 when the centers of the pulleys are x inches apart.

(A) Graph y_1 in a graphing utility (in radian mode), with the graph covering pulleys with their centers from 3 to 20 inches apart.

(B) How far, to two decimal places, should the centers of the two pulleys be placed to use a belt 36 inches long? Solve by graphing y_1 and $y_2 = 36$ in the same viewing window and finding the point of intersection using an approximation routine.

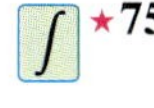

★ **75. Motion.** The figure represents a circular courtyard surrounded by a high stone wall. A floodlight located at E shines into the courtyard.

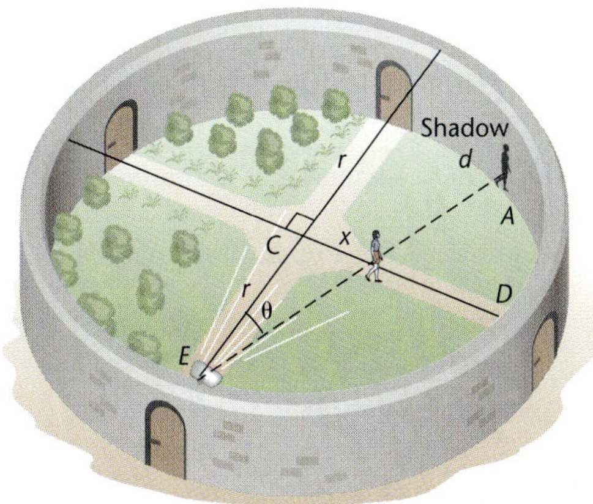

(A) If a person walks x feet away from the center along DC, show that the person's shadow will move a distance given by

$$d = 2r\theta = 2r\tan^{-1}\frac{x}{r}$$

where θ is in radians. [*Hint:* Draw a line from A to C.]

(B) Find d to two decimal places if $r = 100$ feet and $x = 40$ feet.

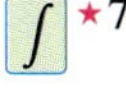

★ **76. Motion.** In Problem 75, find d for $r = 50$ feet and $x = 25$ feet.

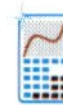

CHAPTER 5 GROUP ACTIVITY A Predator–Prey Analysis Involving Mountain Lions and Deer

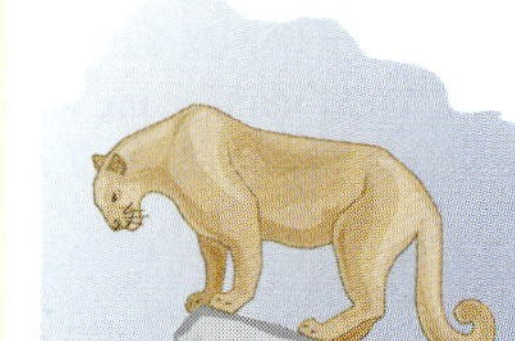

In some western state wilderness areas, deer and mountain lion populations are interrelated, since the mountain lions rely on the deer as a food source. The population of each species goes up and down in cycles, but out of phase with each other. A wildlife management research team estimated the respective populations in a particular region every 2 years over a 16-year period, with the results shown in Table 1:

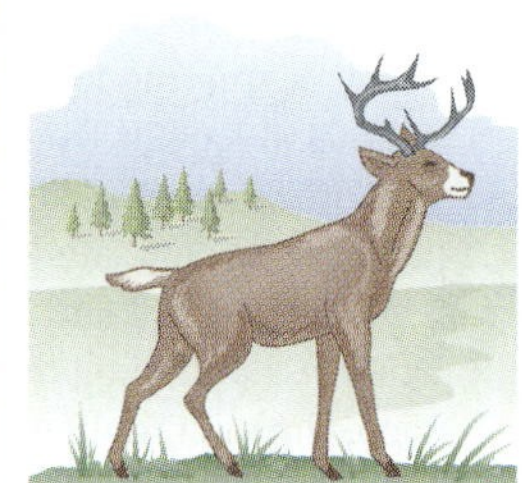

TABLE 1 Mountain Lion–Deer Populations

Years	0	2	4	6	8	10	12	14	16
Deer	1272	1523	1152	891	1284	1543	1128	917	1185
Mtn. Lions	39	47	63	54	37	48	60	46	40

(A) Deer Population Analysis

1. Enter the data for the deer population for the time interval [0, 16] in a graphing utility and produce a scatter plot of the data.
2. A function of the form $y = k + A \sin (Bx + C)$ can be used to model this data. Use the data in Table 1 to determine k, A, and B. Use the graph in part 1 to visually estimate C to one decimal place.
3. Plot the data from part 1 and the equation from part 2 in the same viewing window. If necessary, adjust the value of C for a better fit.
4. Write a summary of the results, describing fluctuations and cycles of the deer population.

(B) Mountain Lion Population Analysis

1. Enter the data for the mountain lion population for the time interval [0, 16] in a graphing utility and produce a scatter plot of the data.
2. A function of the form $y = k + A \sin (Bx + C)$ can be used to model this data. Use the data in Table 1 to determine k, A, and B. Use the graph in part 1 to visually estimate C to one decimal place.
3. Plot the data from part 1 and the equation from part 2 in the same viewing window. If necessary, adjust the value of C for a better fit.
4. Write a summary of the results, describing fluctuations and cycles of the mountain lion population.

(C) Interrelationship of the Two Populations

1. Discuss the relationship of the maximum predator populations to the maximum prey populations relative to time.
2. Discuss the relationship of the minimum predator populations to the minimum prey populations relative to time.
3. Discuss the dynamics of the fluctuations of the two interdependent populations. What causes the two populations to rise and fall, and why are they out of phase from one another?

Chapter 5 Review

5-1 THE WRAPPING FUNCTION

The **unit circle** is a circle of radius 1 with center at the origin of a rectangular coordinate system. The **wrapping function** wraps a real number line with the origin at (1, 0) around the unit circle—the positive real axis is wrapped counterclockwise, and the negative real axis is wrapped clockwise. Thus, each real number on the real line is paired with a unique point, called a **circular point,** on the unit circle.

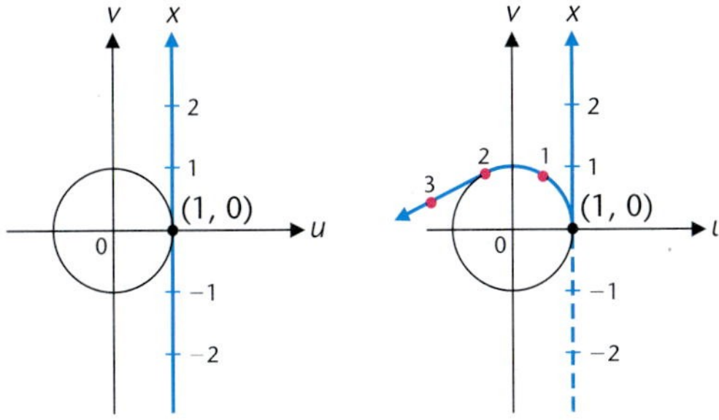

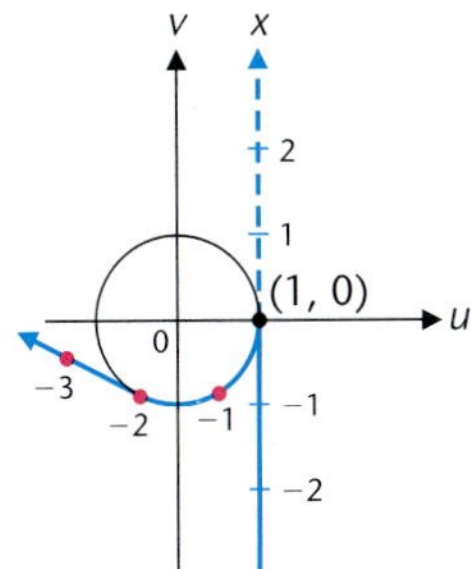

An equivalent way of pairing real numbers with points on the unit circle is to think in terms of *arc length.* To find the circular point P associated with the real number x, we start at $A(1, 0)$ and move $|x|$ units along the unit circle, counterclockwise if x is positive and clockwise if x is negative. The length of arc AP is $|x|$ (see the figure).

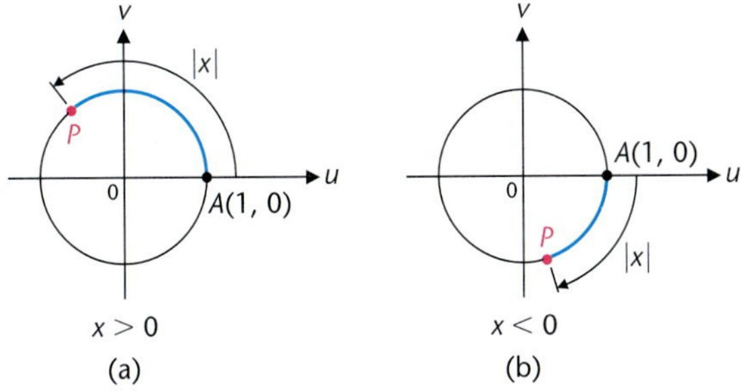

$x > 0$
(a)

$x < 0$
(b)

Coordinates of Key Circular Points

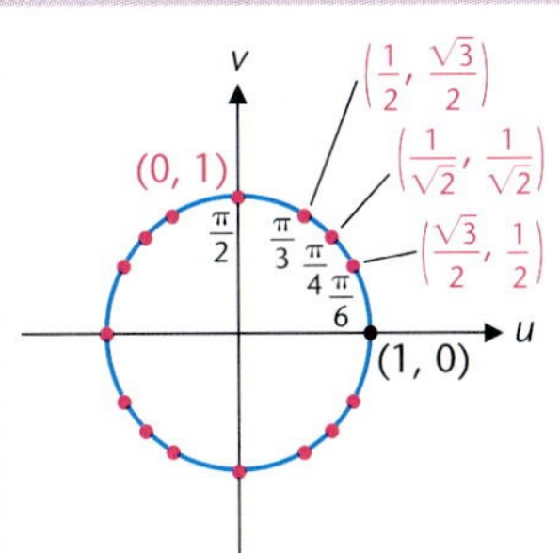

A Memory Aid

$$W\left(\frac{\pi}{2}\right) = (0, 1) = \left(\frac{\sqrt{0}}{2}, \frac{\sqrt{4}}{2}\right)$$

$$W\left(\frac{\pi}{3}\right) = \left(\frac{1}{2}, \frac{\sqrt{3}}{2}\right) = \left(\frac{\sqrt{1}}{2}, \frac{\sqrt{3}}{2}\right)$$

$$W\left(\frac{\pi}{4}\right) = \left(\frac{1}{\sqrt{2}}, \frac{1}{\sqrt{2}}\right) = \left(\frac{\sqrt{2}}{2}, \frac{\sqrt{2}}{2}\right)$$

$$W\left(\frac{\pi}{6}\right) = \left(\frac{\sqrt{3}}{2}, \frac{1}{2}\right) = \left(\frac{\sqrt{3}}{2}, \frac{\sqrt{1}}{2}\right)$$

$$W(0) = (1,0) = \left(\frac{\sqrt{4}}{2}, \frac{\sqrt{0}}{2}\right)$$

Note patterns

The following is an **important wrapping function property:** For all real numbers x,

$W(x) = W(x + 2k\pi)$
k any integer

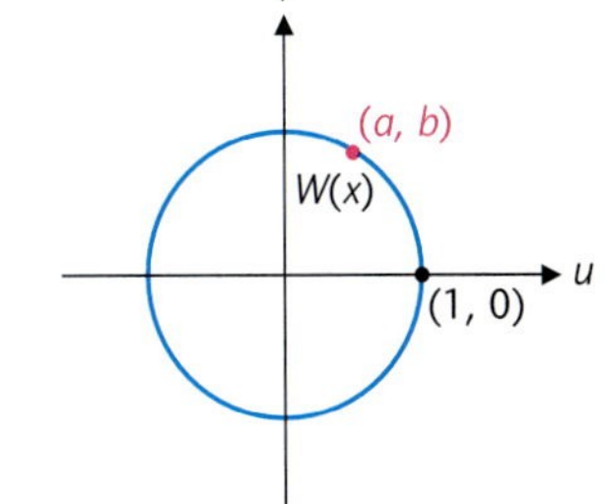

5-2 CIRCULAR FUNCTIONS

The wrapping function W pairs each real number x with an ordered pair of real numbers (a, b), the coordinates of the circular point $W(x)$. This association is used in the following definition of the six **circular functions:** If x is a real number and (a, b) are the coordinates of the circular point $W(x)$, then

$$\sin x = b \qquad \csc x = \frac{1}{b} \quad b \neq 0$$

$$\cos x = a \qquad \sec x = \frac{1}{a} \quad a \neq 0$$

$$\tan x = \frac{b}{a} \quad a \neq 0 \qquad \cot x = \frac{a}{b} \quad b \neq 0$$

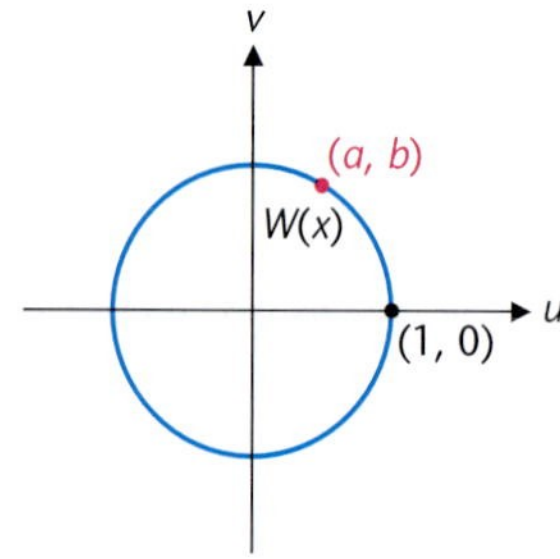

Using the results in Section 5-1, we can evaluate any one of the six circular functions exactly, when it exists, for integer multiples of the real numbers $\pi/6$, $\pi/4$, $\pi/3$, and $\pi/2$. A calculator can be used to evaluate the circular functions for arbitrary real numbers.

The following **basic trigonometric identities** hold true for all replacements of x by real numbers for which both sides of an equation are defined:

Reciprocal Identities

$$\csc x = \frac{1}{\sin x} \qquad \sec x = \frac{1}{\cos x} \qquad \cot x = \frac{1}{\tan x}$$

Quotient Identities

$$\tan x = \frac{\sin x}{\cos x} \qquad \cot x = \frac{\cos x}{\sin x}$$

Identities for Negatives

$$\sin(-x) = -\sin x \qquad \cos(-x) = \cos x \qquad \tan(-x) = -\tan x$$

Pythagorean Identity

$$\sin^2 x + \cos^2 x = 1$$

5-3 ANGLES AND THEIR MEASURE

An **angle** has two sides and a common point called a **vertex.** An **angle** can be formed by starting with the **initial side** in a fixed position and rotating the **terminal side** from that fixed position to its final position—counterclockwise, **positive;** clockwise, **negative.** An angle is in **standard position** in a rectangular coordinate system if its vertex is at the origin and its initial side is along the positive x axis. **Quadrantal angles** have their terminal sides on a coordinate axis. An angle of **1 degree** is $\frac{1}{360}$ of a complete rotation. An angle of **1 radian** is a central angle of a circle subtended by an arc having the same length as the radius.

$$\text{Radian measure: } \theta = \frac{s}{r}$$

$$\text{Radian–degree conversion: } \frac{\theta_{\text{deg}}}{180°} = \frac{\theta_{\text{rad}}}{\pi \text{ rad}}$$

5-4 TRIGONOMETRIC FUNCTIONS

To each of the six circular functions we associate a **trigonometric function** of the same name. If θ is an angle with radian measure x, then the value of each trigonometric function at θ is given by its value at the real number x:

Trigonometric Function		Circular Function
$\sin\theta$	=	$\sin x$
$\cos\theta$	=	$\cos x$
$\tan\theta$	=	$\tan x$
$\csc\theta$	=	$\csc x$
$\sec\theta$	=	$\sec x$
$\cot\theta$	=	$\cot x$

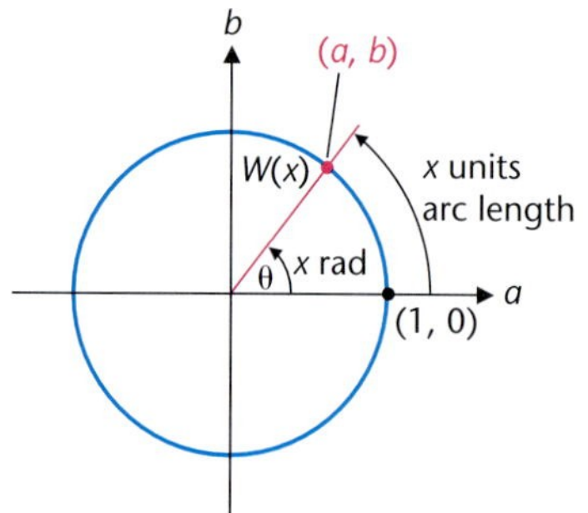

For many applications involving the use of trigonometric functions, including triangle applications, it is useful to have an **alternate definition of a trigonometric function** that utilizes the coordinates of an arbitrary point $(a, b) \neq (0, 0)$ on the terminal side of an angle θ: If θ is an arbitrary angle in standard position in a rectangular coordinate system and $P(a, b)$ is a point r units from the origin on the terminal side of θ, then:

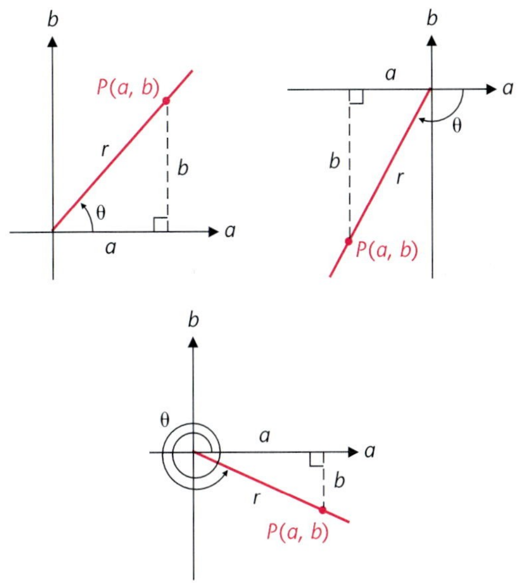

$\sin\theta = \dfrac{b}{r}$ $\qquad$ $\csc\theta = \dfrac{r}{b} \quad b \neq 0$

$\cos\theta = \dfrac{a}{r}$ $\qquad$ $\sec\theta = \dfrac{r}{a} \quad a \neq 0$

$\tan\theta = \dfrac{b}{a}, \quad a \neq 0$ $\qquad$ $\cot\theta = \dfrac{a}{b} \quad b \neq 0$

$r = \sqrt{a^2 + b^2} > 0$; $P(a, b)$ is an arbitrary point on the terminal side of θ, $(a, b) \neq (0, 0)$

Domains: Sets of all possible angles for which the ratios are defined

Ranges: Subsets of the set of real numbers

(Domains and ranges are stated more precisely in the review for Section 5-6.)

Associated with each angle that does not terminate on a coordinate axis is a **reference triangle** for θ. The reference triangle is formed by drawing a perpendicular from point $P(a, b)$ on the terminal side of θ to the horizontal axis. The **reference angle** α is the acute angle, always taken positive, between the terminal side of θ and the horizontal axis as indicated in the following figure.

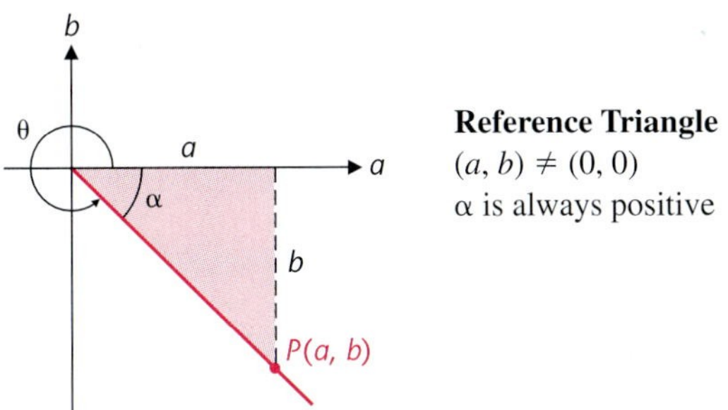

Reference Triangle
$(a, b) \neq (0, 0)$
α is always positive

If a reference triangle of a given angle is a 30°–60° right triangle or a 45° right triangle, then we can find exact coordinates, other than (0, 0), on the terminal side of the given angle. The following 30°–60° and 45° right triangle relationships are useful in this regard:

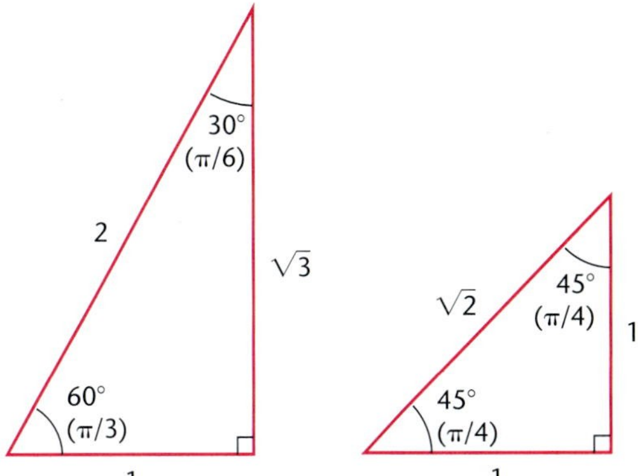

Some special angle values are summarized in the following table:

TABLE 1 Special Angle Values

θ	sin θ	cos θ	tan θ
0°	0	1	0
30°	$\frac{1}{2}$	$\sqrt{3}/2$	$1/\sqrt{3}$ or $\sqrt{3}/$
45°	$1/\sqrt{2}$ or $\sqrt{2}/2$	$1/\sqrt{2}$ or $\sqrt{2}/2$	1
60°	$\sqrt{3}/2$	$\frac{1}{2}$	$\sqrt{3}$
90°	1	0	Not defined

The circular functions are related to the trigonometric functions as follows: For x any real number,

$\sin x = \sin (x \text{ radians})$ $\qquad$ $\cos x = \cos (x \text{ radians})$

$\sec x = \sec (x \text{ radians})$ $\qquad$ $\csc x = \csc (x \text{ radians})$

$\tan x = \tan (x \text{ radians})$ $\qquad$ $\cot x = \cot (x \text{ radians})$

You are now free to evaluate circular functions in terms of trigonometric functions, using reference triangles where appropriate, or in terms of circular points and the wrapping function. Each approach has certain advantages in particular situations.

5-5 SOLVING RIGHT TRIANGLES

A **right triangle** is a triangle with one 90° angle. To **solve a right triangle** is to find all unknown angles and sides, given the measures of two sides or the measures of one side and an acute angle.

Trigonometric Functions of Acute Angles

$\sin\theta = \dfrac{\text{Opp}}{\text{Hyp}}$ $\csc\theta = \dfrac{\text{Hyp}}{\text{Opp}}$

$\cos\theta = \dfrac{\text{Adj}}{\text{Hyp}}$ $\sec\theta = \dfrac{\text{Hyp}}{\text{Adj}}$

$\tan\theta = \dfrac{\text{Opp}}{\text{Adj}}$ $\cot\theta = \dfrac{\text{Adj}}{\text{Opp}}$

Opp Hyp θ Adj

Computational Accuracy

Angle to nearest	Significant digits for side measure
1°	2
10′ or 0.1°	3
1′ or 0.01°	4
10″ or 0.001°	5

5-6 GRAPHING BASIC TRIGONOMETRIC FUNCTIONS

A function f is **periodic** if there exists a positive real number p such that

$$f(x + p) = f(x)$$

for all x in the domain of f. The smallest such positive p, if it exists, is called the **fundamental period of *f*,** or often just the **period of *f*.** All the circular and trigonometric functions are periodic.

Graph of $y = \sin x$:

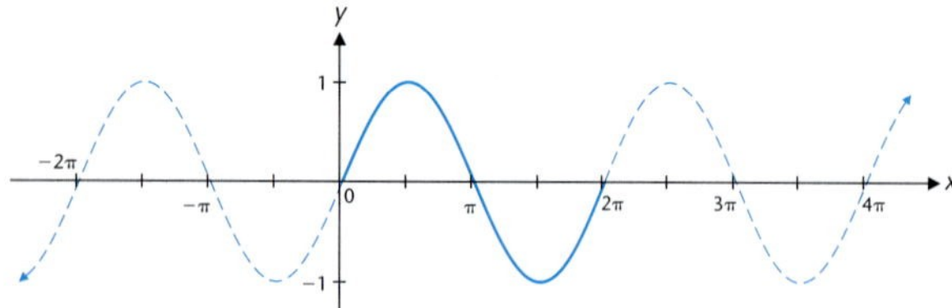

Period: 2π
Domain: All real numbers
Range: $[-1, 1]$

Graph of $y = \cos x$:

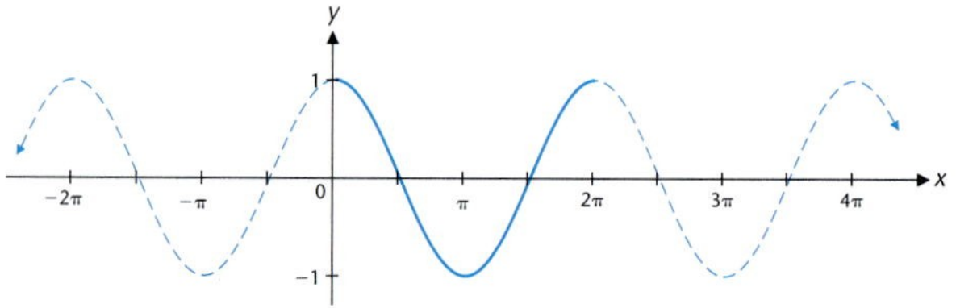

Period: 2π
Domain: All real numbers
Range: $[-1, 1]$

Graph of $y = \tan x$:

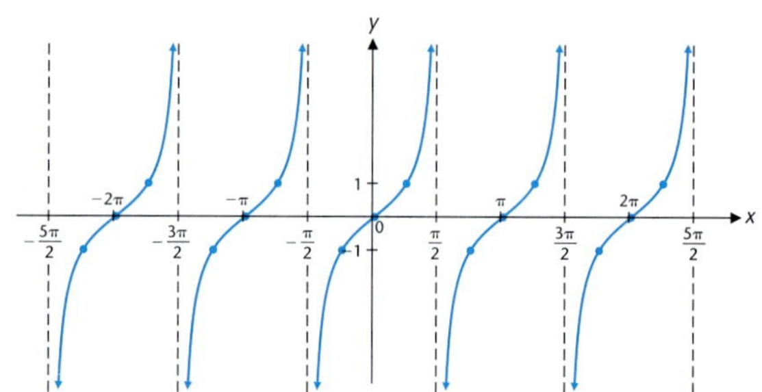

Period: π
Domain: All real numbers except $\pi/2 + k\pi$, k an integer
Range: All real numbers

Graph of $y = \cot x$:

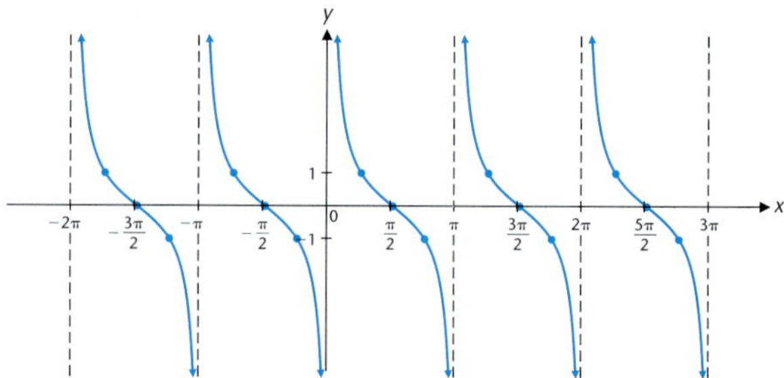

Period: π
Domain: All real numbers except $k\pi$, k an integer
Range: All real numbers

Graph of $y = \csc x$:

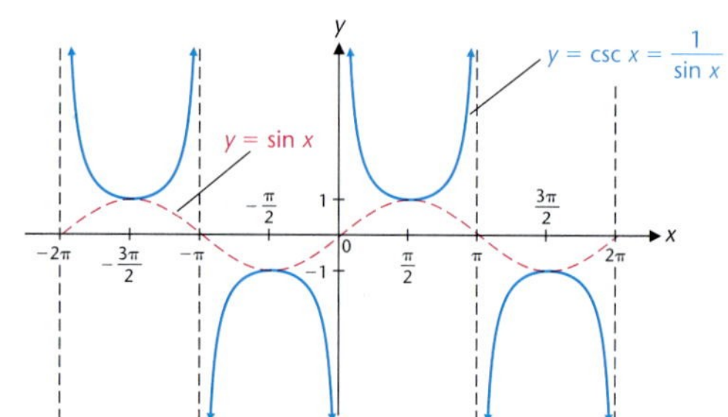

Period: 2π
Domain: All real numbers except $k\pi$, k an integer
Range: All real numbers y such that $y \le -1$ or $y \ge 1$

Graph of $y = \sec x$:

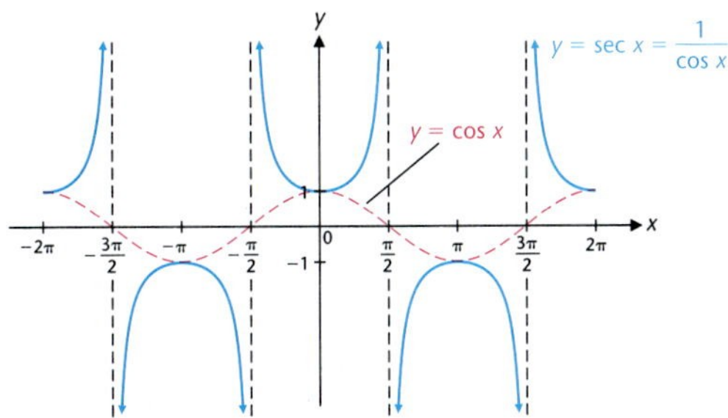

Period: 2π
Domain: All real numbers except $\pi/2 + k\pi$, k an integer
Range: All real numbers y such that $y \le -1$ or $y \ge 1$

5-7 GRAPHING $y = k + A \sin (Bx + C)$ and $y = k + A \cos (Bx + C)$

$$|A| = \text{Amplitude}$$

To find the period and phase shift, solve $(Bx + C) = 0$ and $(Bx + C) = 2\pi$:

$$x = -\frac{C}{B} \quad \text{to} \quad x = -\frac{C}{B} + \frac{2\pi}{B}$$

(Phase shift: $-\frac{C}{B}$; Period: $\frac{2\pi}{B}$)

Period is $2\pi/B$. Phase shift is a horizontal translation to the right if $-C/B$ is positive and to the left if $-C/B$ is negative.

$|k|$ is vertical translation: up if k is positive and down if k is negative.

5-8 GRAPHING MORE GENERAL TANGENT, COTANGENT, SECANT, AND COSECANT FUNCTIONS

Amplitude is not defined for these functions.

To find the period and phase shift for $y = A \tan (Bx + C)$ or $y = A \cot (Bx + C)$, solve $Bx + C = 0$ and $Bx + C = \pi$:

$$x = -\frac{C}{B} \quad \text{to} \quad x = -\frac{C}{B} + \frac{\pi}{B}$$

(Phase shift: $-\frac{C}{B}$; Period: $\frac{\pi}{B}$)

The period is π/B. The phase shift is a horizontal translation to the right if $-C/B$ is positive and to the left if $-C/B$ is negative.

To find the period and phase shift for $y = A \sec (Bx + C)$ or $y = A \csc (Bx + C)$, solve $Bx + C = 0$ and $Bx + C = 2\pi$:

$$x = -\frac{C}{B} \quad \text{to} \quad x = -\frac{C}{B} + \frac{2\pi}{B}$$

(Phase shift: $-\frac{C}{B}$; Period: $\frac{2\pi}{B}$)

Period is $2\pi/B$. Phase shift is a horizontal translation to the right if $-C/B$ is positive and to the left if $-C/B$ is negative.

5-9 INVERSE TRIGONOMETRIC FUNCTIONS

$y = \sin^{-1} x = \arcsin x$ if and only if $\sin y = x$, $-\pi/2 \le y \le \pi/2$ and $-1 \le x \le 1$.

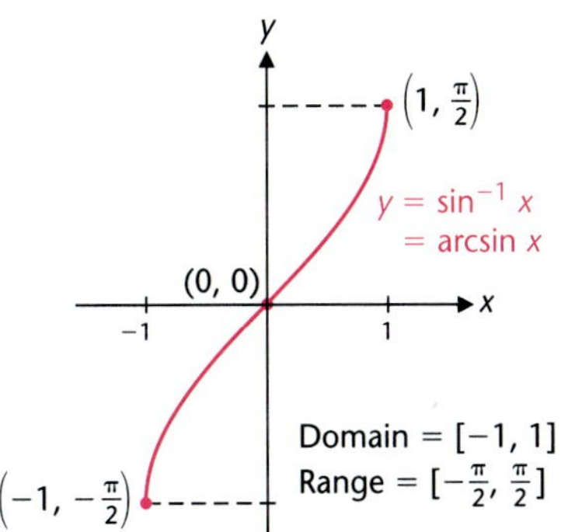

Inverse sine function

$y = \cos^{-1} x = \arccos x$ if and only if $\cos y = x$, $0 \le y \le \pi$ and $-1 \le x \le 1$.

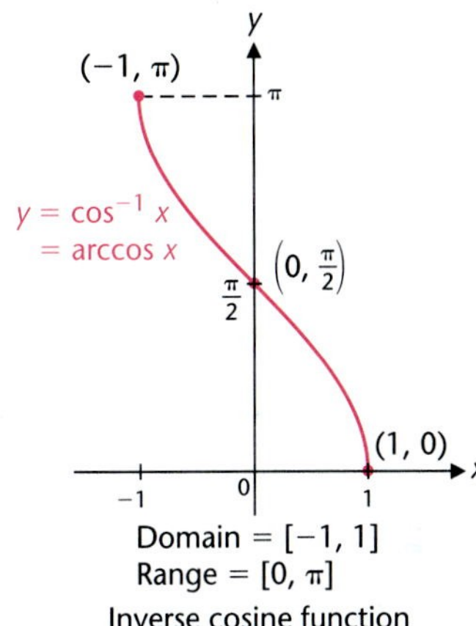

Domain = [−1, 1]
Range = [0, π]
Inverse cosine function

$y = \tan^{-1} x = \arctan x$ if and only if $\tan y = x$, $-\pi/2 < y < \pi/2$ and x is any real number.

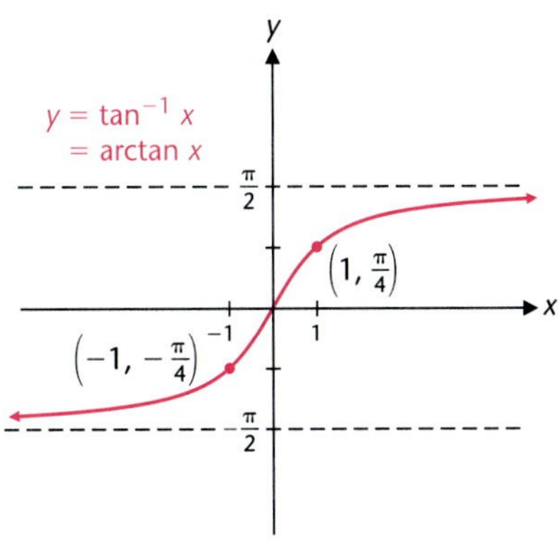

Domain = $(-\infty, \infty)$
Range = $\left(-\frac{\pi}{2}, \frac{\pi}{2}\right)$
Inverse tangent function

Chapter 5 Review Exercise

Work through all the problems in this chapter review and check answers in the back of the book. Answers to all review problems are there, and following each answer is a number in italics indicating the section in which that type of problem is discussed. Where weaknesses show up, review appropriate sections in the text.

A

1. Find the radian measure of a central angle opposite an arc 15 centimeters long on a circle of radius 6 centimeters.

2. In a circle of radius 3 centimeters, find the length of an arc opposite an angle of 2.5 radians.

3. Solve the triangle:

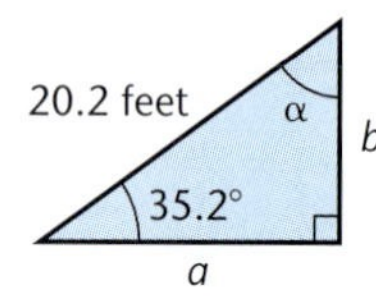

4. Find the reference angle associated with each angle θ:
(A) $\theta = \pi/3$ (B) $\theta = -120°$
(C) $\theta = -13\pi/6$ (D) $\theta = 210°$

5. In which quadrants is each negative?
(A) $\sin\theta$ (B) $\cos\theta$ (C) $\tan\theta$

6. If $(4, -3)$ is on the terminal side of angle θ, find:
(A) $\sin\theta$ (B) $\sec\theta$ (C) $\cot\theta$

7. Complete Table 1 using exact values. Do not use a calculator.

TABLE 1

θ°	θ rad	sin θ	cos θ	tan θ	csc θ	sec θ	cot θ
0°					ND*		
30°							
45°	$\pi/4$		$1/\sqrt{2}$				
60°							
90°							
180°							
270°							
360°							

*ND = Not Defined

8. What is the period of each of the following?
(A) $y = \cos x$ (B) $y = \csc x$ (C) $y = \tan x$

9. Indicate the domain and range of each.
(A) $y = \sin x$ (B) $y = \tan x$

10. Sketch a graph of $y = \sin x$, $-2\pi \le x \le 2\pi$.

11. Sketch a graph of $y = \cot x$, $-\pi < x < \pi$.

12. Verbally describe the meaning of a central angle in a circle with radian measure 0.5.

13. Describe the smallest shift of the graph of $y = \sin x$ that produces the graph of $y = \cos x$.

B

14. Change 1.37 radians to decimal degrees to two decimal places.

15. Solve the triangle:

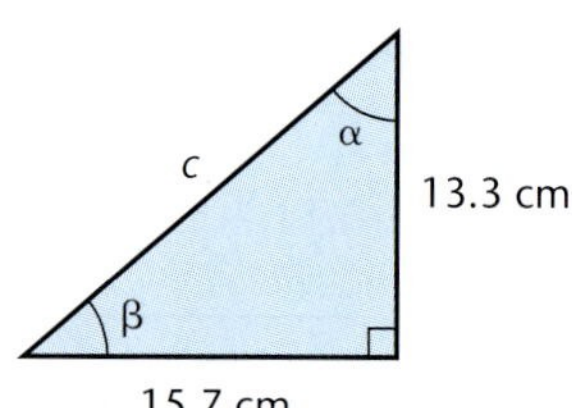

16. Indicate whether the angle is a I, II, III, or IV quadrant angle or a quadrantal angle.
(A) $-210°$ (B) $5\pi/2$ (C) 4.2 radians

17. Which of the following angles are coterminal with 120°?
(A) $-240°$ (B) $-7\pi/6$ (C) 840°

18. Which of the following have the same value as cos 3?
(A) cos 3° (B) cos (3 radians) (C) $\cos(3 + 2\pi)$

19. For which values of x, $0 \le x < 2\pi$, is each of the following not defined?
(A) $\tan x$ (B) $\cot x$ (C) $\csc x$

20. A circular point $P(a, b)$ moves clockwise around the circumference of a unit circle starting at (1, 0) and stops after covering a distance of 8.305 units. Explain how you would find the coordinates of point P at its final position and how you would determine the quadrant in which P is located. Find the coordinates of P to three decimal places and the quadrant for the final position of P.

In Problems 21–36, evaluate exactly without the use of a calculator.

21. $\sin(-30°)$ **22.** $\cos(-45°)$ **23.** $\tan\dfrac{3\pi}{4}$

24. $\cot\dfrac{10\pi}{3}$ **25.** arccos 0 **26.** $\sin^{-1}(-1)$

27. arctan 1

28. $\tan^{-1}(-\sqrt{3})$

29. $\cos^{-1}\sqrt{2}$

30. $\arccos\left(-\frac{1}{2}\right)$

31. $\arcsin\left(\frac{\sqrt{3}}{2}\right)$

32. $\sin^{-1}(-\sqrt{3})$

33. $\sin\left(\sin^{-1}\frac{2}{3}\right)$

34. $\cos^{-1}\left(\cos\frac{7\pi}{6}\right)$

35. sec (arctan 2)

36. $\sin\left(\arccos\frac{3}{4}\right)$

Evaluate Problems 37–44 to 4 significant digits using a calculator.

37. cos 423.7°

38. tan 93°46′17″

39. sec (−2.073)

40. $\sin^{-1}(-0.8277)$

41. arccos (−1.3281)

42. $\tan^{-1} 75.14$

43. $\csc[\cos^{-1}(-0.4081)]$

44. $\sin^{-1}(\tan 1.345)$

45. Find the exact degree measure of each without a calculator:
(A) $\theta = \sin^{-1}(-1/2)$ (B) $\theta = \arccos(-1/2)$

46. Find the degree measure of each to two decimal places using a calculator:
(A) $\theta = \cos^{-1}(-0.8763)$ (B) $\theta = \arctan 7.3771$

47. Evaluate $\cos^{-1}[\cos(-2)]$ with a calculator set in radian mode, and explain why this does or does not illustrate the inverse cosine–cosine identity.

48. Sketch a graph of $y = -2\cos\pi x$, $-1 \le x \le 3$. Indicate amplitude A and period P.

49. Sketch a graph of $y = -2 + 3\sin(x/2)$, $-4\pi \le x \le 4\pi$.

50. Find the equation of the form $y = A\cos Bx$ that has the graph

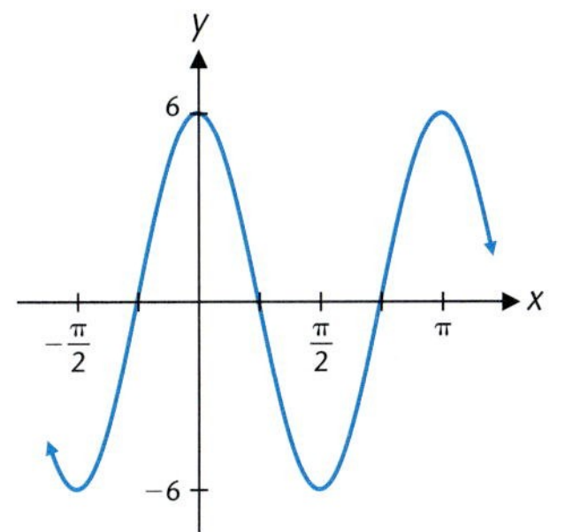

51. Find the equation of the form $y = A\sin Bx$ that has the graph

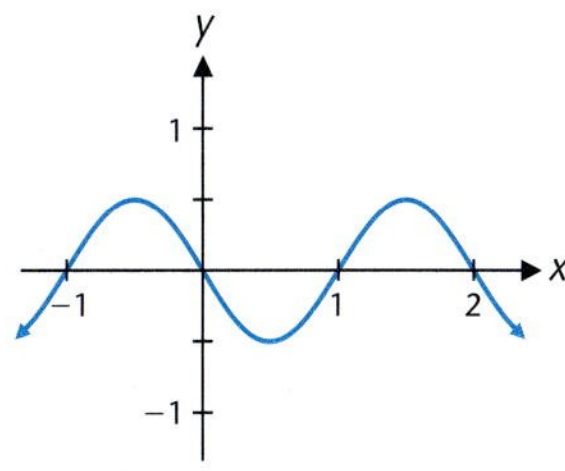

52. Describe the smallest shift and/or reflection that transforms the graph of $y = \tan x$ into the graph of $y = \cot x$.

53. Simplify each of the following using appropriate basic identities:

(A) $\sin(-x)\cot(-x)$ (B) $\dfrac{\sin^2 x}{1 - \sin^2 x}$

54. Sketch a graph of $y = 3\sin[(x/2) + (\pi/2)]$ over the interval $-4\pi \le x \le 4\pi$.

55. Indicate the amplitude A, period P, and phase shift for the graph of $y = -2\cos[(\pi/2)x - (\pi/4)]$. Do not graph.

56. Sketch a graph of $y = \cos^{-1} x$, and indicate the domain and range.

57. Graph $y = 1/(1 + \tan^2 x)$ in a graphing utility that displays at least two full periods of the graph. Find an equation of the form $y = k + A\sin Bx$ or $y = k + A\cos Bx$ that has the same graph.

58. Graph each equation in a graphing utility and find an equation of the form $y = A\tan Bx$ or $y = A\cot Bx$ that has the same graph as the given equation. Select the dimensions of the viewing window so that at least two periods are visible.

(A) $y = \dfrac{2\sin^2 x}{\sin 2x}$ (B) $y = \dfrac{2\cos^2 x}{\sin 2x}$

C

59. If in the figure the coordinates of A are (8, 0) and arc length s is 20 units, find:
(A) The exact radian measure of θ
(B) The coordinates of P to 3 significant digits

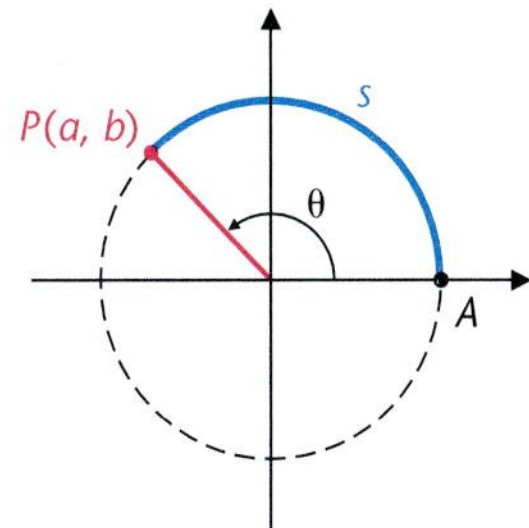

60. Find exactly the least positive real number for which:
(A) $\cos x = -\frac{1}{2}$ (B) $\csc x = -\sqrt{2}$

61. Sketch a graph of $y = \sec x$, $-\pi/2 < x < 3\pi/2$.

62. Sketch a graph of $y = \tan^{-1} x$, and indicate the domain and range.

63. Indicate the period P and phase shift for the graph of $y = -5 \tan (\pi x + \pi/2)$. Do not graph.

64. Indicate the period and phase shift for the graph of $y = 3 \csc (x/2 - \pi/4)$. Do not graph.

65. Indicate whether each is symmetric with respect to the x axis, y axis, or origin.
(A) Sine (B) Cosine (C) Tangent

66. Write as an algebraic expression in x free of trigonometric or inverse trigonometric functions:

$$\sec (\sin^{-1} x)$$

67. Try to calculate each of the following on your calculator. Explain the results.
(A) $\csc (-\pi)$ (B) $\tan (-3\pi/2)$ (C) $\sin^{-1} 2$

68. The accompanying graph is a graph of an equation of the form $y = A \sin (Bx + C)$, $-1 < -C/B < 0$. Find the equation.

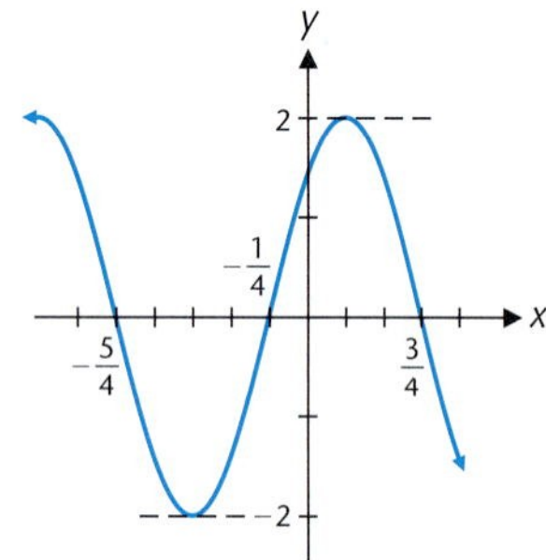

In Problems 69–76, determine whether the statement is true or false. If true, explain why. If false, give a counterexample.

69. If positive angles α and β are complementary, then 2α and 2β are supplementary.

70. If α and β are two of the angles of a triangle and $\alpha < \beta$, then $\sin \alpha < \sin \beta$.

71. If α and β are angles such that $0 < \alpha < \beta < \pi$, then $\sin \alpha < \sin \beta$.

72. If f and g are periodic functions with fundamental period 2π, then f/g is periodic with fundamental period 2π.

73. Each of the six basic trigonometric functions is periodic.

74. Each of the inverses of the six basic trigonometric functions is periodic.

75. If $f(x) = A \sin (Bx + C)$, then the function g defined by $g(x) = f(x - \frac{C}{B})$ has the same graph as f.

76. If $f(x) = A \sin (Bx + C)$, then the function g defined by $g(x) = f(x - \frac{2\pi}{B})$ has the same graph as f.

77. Is the wrapping function a periodic function? Explain.

78. If $c = 1$, explain why the area of the triangle in the figure is given by $A = \frac{1}{2} \cos \theta \sin \theta$.

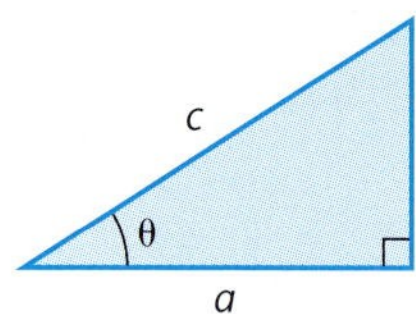

Figure for 78 and 79

79. If $a = 1$, explain why the area of the triangle in the figure is given by $A = \frac{1}{2} \tan \theta$

80. Explain why the area of a regular n-sided polygon circumscribed about a circle of radius 1 is given by

$$A = n \tan \frac{180°}{n}$$

81. Explain why the area of a regular n-sided polygon inscribed in a circle of radius 1 is given by

$$A = n \cos \frac{180°}{n} \sin \frac{180°}{n}$$

82. Explain why the area of the inscribed polygon of Problem 81 is also given by the formula

$$A = \frac{n}{2} \sin \frac{360°}{n}$$

83. Graph $y = 1.2 \sin 2x + 1.6 \cos 2x$ in a graphing utility. (Select the dimensions of the viewing window so that at least two periods are visible.) Find an equation of the form $y = A \sin (Bx + C)$ that has the same graph as the given equation. Find A and B exactly and C to three decimal places. Use the x intercept closest to the origin as the phase shift.

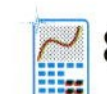
84. A particular wave form is approximated by the first six terms of a Fourier series:

$$y = \frac{4}{\pi}\left(\sin x + \frac{\sin 3x}{3} + \frac{\sin 5x}{5} + \frac{\sin 7x}{7} + \frac{\sin 9x}{9} + \frac{\sin 11x}{11}\right)$$

(A) Graph this equation in a graphing utility for $-3\pi \le x \le 3\pi$ and $-2 \le y \le 2$.

(B) The graph in part A approximates a wave form that is made up entirely of straight-line segments. Sketch by hand the waveform that the Fourier series approximates.

This wave form is called a **pulse wave** or a **square wave,** and is used, for example, to test distortion and to synchronize operations in computers.

APPLICATIONS

85. Astronomy. A line from the sun to the Earth sweeps out an angle of how many radians in 73 days? Express the answer in terms of π.

★ **86. Geometry.** Find the perimeter of a square inscribed in a circle of radius 5.00 centimeters.

★ **87. Alternating Current.** The current I in an alternating electric current has an amplitude of 30 amperes and a period of $\frac{1}{60}$ second. If $I = 30$ amperes when $t = 0$, find an equation of the form $I = A \cos Bt$ that gives the current at any time $t \geq 0$.

88. Restricted Access. A 10-ft-wide canal makes a right turn into a 15-ft-wide canal. Long narrow logs are to be floated through the canal around the right angle turn (see figure). We are interested in finding the longest log that will go around the corner, ignoring the log's diameter.

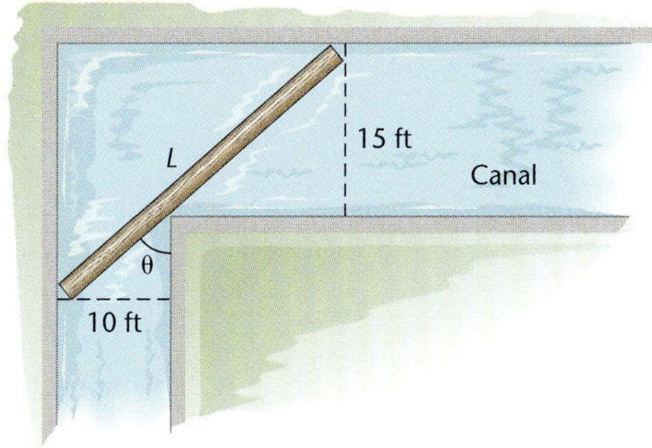

(A) Express the length of the line L that touches the two outer sides of the canal and the inside corner in terms of θ.

(B) Complete Table 2, each to one decimal place, and estimate from the table the longest log to the nearest foot that can make it around the corner. (The longest log is the shortest distance L.)

TABLE 2

θ (rads)	0.4	0.5	0.6	0.7	0.8	0.9	1.0
L (ft)	42.0						

(C) Graph the function in part A in a graphing utility and use an approximation method to find the shortest distance L to one decimal place, hence, the length of the longest log that can make it around the corner.

(D) Explain what happens to the length L as θ approaches 0 or $\pi/2$.

89. Modeling Seasonal Business Cycles. A soft drink company has revenues from sales over a 2-year period as shown by the accompanying graph, where $R(t)$ is revenue (in millions of dollars) for a month of sales t months after February 1.

(A) Find an equation of the form $R(t) = k + A \cos Bt$ that produces this graph.

(B) Verbally interpret the graph

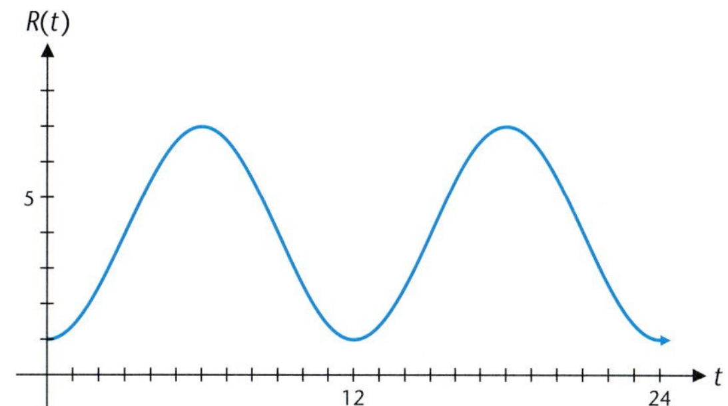

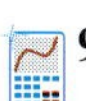

90. Modeling Temperature Variation. The 30-year average monthly temperature, °F, for each month of the year for Los Angeles is given in Table 3 (*World Almanac*).

(A) Using 1 month as the basic unit of time, enter the data for a 2-year period in your graphing utility and produce a scatter plot in the viewing window. Choose $40 \leq y \leq 90$ for the viewing window.

(B) It appears that a sine curve of the form

$$y = k + A \sin (Bx + C)$$

will closely model this data. The constants k, A, and B are easily determined from Table 3. To estimate C, visually estimate to one decimal place the smallest positive phase shift from the plot in part A. After determining A, B, k, and C, write the resulting equation. (Your value of C may differ slightly from the answer book.)

(C) Plot the results of parts A and B in the same viewing window. (An improved fit may result by adjusting your value of C slightly.)

TABLE 3

x (mos.)	1	2	3	4	5	6	7	8	9	10	11	12
y (temp.)	58	60	61	63	66	70	74	75	74	70	63	58

Trigonometric Identities and Conditional Equations

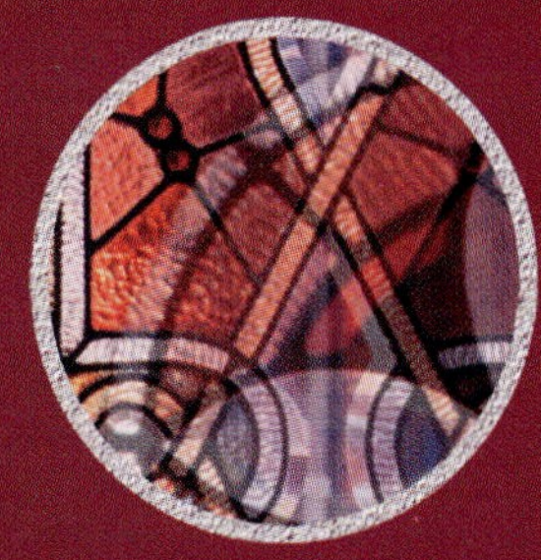

Chapter 6

Trigonometric functions are widely used in solving real-world problems as well as in the development of mathematics. Whatever their use, it is often of value to be able to change a trigonometric expression from one form to an equivalent more useful form. This involves the use of identities. Recall that an equation in one or more variables is said to be an *identity* if the left side is equal to the right side for all replacements of the variables for which both sides are defined. For example, the equation

$$x^2 - 2x - 8 = (x - 4)(x + 2)$$

is an identity, but

$$x^2 - 2x - 8 = 0$$

is not. The latter is called a *conditional equation,* since it holds only for certain values of x and not for all values for which both sides are defined. The first four sections of the chapter deal with trigonometric identities and the last section with conditional trigonometric equations.

SECTION 6-1 Basic Identities and Their Use

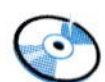

- Basic Identities
- Establishing Other Identities

In this section we review the basic identities introduced in Section 5-2 and show how they are used to establish other identities.

• Basic Identities

In the following box we list for convenient reference the basic identities introduced in Section 5-2. These identities will be used very frequently in the work that follows and should be memorized.

Basic Trigonometric Identities

Reciprocal Identities

$$\csc x = \frac{1}{\sin x} \qquad \sec x = \frac{1}{\cos x} \qquad \cot x = \frac{1}{\tan x}$$

Quotient Identities

$$\tan x = \frac{\sin x}{\cos x} \qquad \cot x = \frac{\cos x}{\sin x}$$

Identities for Negatives

$$\sin(-x) = -\sin x \qquad \cos(-x) = \cos x \qquad \tan(-x) = -\tan x$$

Pythagorean Identities

$$\sin^2 x + \cos^2 x = 1 \qquad \tan^2 x + 1 = \sec^2 x \qquad 1 + \cot^2 x = \csc^2 x$$

All these identities were established in Section 5-2 (the second and third Pythagorean identities were established in Problems 87 and 88 in Exercise 5-2).

EXPLORE-DISCUSS 1 Discuss an easy way to recall the second and third Pythagorean identities from the first. [*Hint:* Divide through the first Pythagorean identity by appropriate expressions.]

• Establishing Other Identities

Identities are established in order to convert one form to an equivalent form that may be more useful. To *verify an identity* means to prove that both sides of an equation are equal for all replacements of the variables for which both sides are defined. Such a proof might use basic identities or other verified identities and algebraic operations such as multiplication, factoring, combining and reducing fractions, and so on. The following examples illustrate some of the techniques used to verify certain identities. The steps illustrated are not necessarily unique—often, there is more than one path to a desired goal. To become proficient in the use of identities, it is important that you work out many problems on your own.

EXAMPLE 1 Identity Verification

Verify the identity: $\cos x \tan x = \sin x$

Verification

Generally, we proceed by starting with the more complicated of the two sides, and transform that side into the other side in one or more steps using basic identities, algebra, or other established identities. Thus,

$$\cos x \tan x = \cos x \frac{\sin x}{\cos x} \qquad \text{Quotient identity}$$

$$= \sin x \qquad \text{Algebra}$$

Matched Problem 1 Verify the identity: $\sin x \cot x = \cos x$

EXAMPLE 2 Identity Verification

Verify the identity: $\sec(-x) = \sec x$

Verification

$$\begin{aligned} \sec(-x) &= \frac{1}{\cos(-x)} && \text{Reciprocal identity} \\ &= \frac{1}{\cos x} && \text{Identity for negatives} \\ &= \sec x && \text{Reciprocal identity} \end{aligned}$$

Matched Problem 2 Verify the identity: $\csc(-x) = -\csc x$

EXAMPLE 3 Identity Verification

Verify the identity: $\cot x \cos x + \sin x = \csc x$

Verification

$$\begin{aligned} \cot x \cos x + \sin x &= \frac{\cos x}{\sin x}\cos x + \sin x && \text{Quotient identity} \\ &= \frac{\cos^2 x}{\sin x} + \sin x && \text{Algebra} \\ &= \frac{\cos^2 x + \sin^2 x}{\sin x} && \text{Algebra} \\ &= \frac{1}{\sin x} && \text{Pythagorean identity} \\ &= \csc x && \text{Reciprocal identity} \end{aligned}$$

Key Algebraic Steps in Example 3

$$\frac{a}{b}a + b = \frac{a^2}{b} + b = \frac{a^2 + b^2}{b}$$

Matched Problem 3 Verify the identity: $\tan x \sin x + \cos x = \sec x$

To verify an identity, proceed from one side to the other, or both sides to the middle, making sure all steps are reversible. Do not use properties of equality to perform the same operation on both sides of the equation. Even though no fixed method of verification works for all identities, certain steps help in many cases.

Suggested Steps in Verifying Identities

1. Start with the more complicated side of the identity, and transform it into the simpler side.

2. Try algebraic operations such as multiplying, factoring, combining fractions, and splitting fractions.

3. If other steps fail, express each function in terms of sine and cosine functions, and then perform appropriate algebraic operations.

4. At each step, keep the other side of the identity in mind. This often reveals what you should do in order to get there.

EXAMPLE 4 **Identity Verification**

Verify the identity: $\dfrac{1 + \sin x}{\cos x} + \dfrac{\cos x}{1 + \sin x} = 2 \sec x$

Verification

$$\begin{aligned}
\frac{1 + \sin x}{\cos x} + \frac{\cos x}{1 + \sin x} &= \frac{(1 + \sin x)^2 + \cos^2 x}{\cos x\,(1 + \sin x)} && \text{Algebra}\\
&= \frac{1 + 2\sin x + \sin^2 x + \cos^2 x}{\cos x\,(1 + \sin x)} && \text{Algebra}\\
&= \frac{1 + 2\sin x + 1}{\cos x\,(1 + \sin x)} && \text{Pythagorean identity}\\
&= \frac{2 + 2\sin x}{\cos x\,(1 + \sin x)} && \text{Algebra}\\
&= \frac{2(1 + \sin x)}{\cos x\,(1 + \sin x)} && \text{Algebra}\\
&= \frac{2}{\cos x} && \text{Algebra}\\
&= 2 \sec x && \text{Reciprocal identity}
\end{aligned}$$

Key Algebraic Steps in Example 4

$$\frac{a}{b} + \frac{b}{a} = \frac{a^2 + b^2}{ba} \qquad (1 + c)^2 = 1 + 2c + c^2 \qquad \frac{m(a + b)}{n(a + b)} = \frac{m}{n}$$

Matched Problem 4 Verify the identity: $\dfrac{1 + \cos x}{\sin x} + \dfrac{\sin x}{1 + \cos x} = 2 \csc x$

EXAMPLE 5 **Identity Verification**

Verify the identity: $\dfrac{\sin^2 x + 2\sin x + 1}{\cos^2 x} = \dfrac{1 + \sin x}{1 - \sin x}$

Verification

$$\frac{\sin^2 x + 2\sin x + 1}{\cos^2 x} = \frac{(\sin x + 1)^2}{\cos^2 x} \qquad \text{Algebra}$$

$$= \frac{(\sin x + 1)^2}{1 - \sin^2 x} \qquad \text{Pythagorean identity}$$

$$= \frac{(1 + \sin x)^2}{(1 - \sin x)(1 + \sin x)} \qquad \text{Algebra}$$

$$= \frac{1 + \sin x}{1 - \sin x} \qquad \text{Algebra}$$

Key Algebraic Steps in Example 5

$$a^2 + 2a + 1 = (a + 1)^2 \qquad 1 - b^2 = (1 - b)(1 + b)$$

Matched Problem 5 Verify the identity: $\sec^4 x - 2\sec^2 x \tan^2 x + \tan^4 x = 1$

EXAMPLE 6 Identity Verification

Verify the identity: $\dfrac{\tan x - \cot x}{\tan x + \cot x} = 1 - 2\cos^2 x$

Verification

$$\frac{\tan x - \cot x}{\tan x + \cot x} = \frac{\dfrac{\sin x}{\cos x} - \dfrac{\cos x}{\sin x}}{\dfrac{\sin x}{\cos x} + \dfrac{\cos x}{\sin x}}$$

Change to sines and cosines (quotient identities).

$$= \frac{(\sin x)(\cos x)\left(\dfrac{\sin x}{\cos x} - \dfrac{\cos x}{\sin x}\right)}{(\sin x)(\cos x)\left(\dfrac{\sin x}{\cos x} + \dfrac{\cos x}{\sin x}\right)}$$

Multiply numerator and denominator by $(\sin x)(\cos x)$, and use algebra to transform the compound fraction into a simple fraction.

$$= \frac{\sin^2 x - \cos^2 x}{\sin^2 x + \cos^2 x}$$

$$= \frac{1 - \cos^2 x - \cos^2 x}{1} \qquad \text{Pythagorean identity}$$

$$= 1 - 2\cos^2 x \qquad \text{Algebra}$$

Key Algebraic Steps in Example 6

$$\frac{\dfrac{a}{b} - \dfrac{b}{a}}{\dfrac{a}{b} + \dfrac{b}{a}} = \frac{ab\left(\dfrac{a}{b} - \dfrac{b}{a}\right)}{ab\left(\dfrac{a}{b} + \dfrac{b}{a}\right)} = \frac{a^2 - b^2}{a^2 + b^2}$$

Matched Problem 6 Verify the identity: $\cot x - \tan x = \dfrac{2\cos^2 x - 1}{\sin x \cos x}$

Just observing how others verify identities won't make *you* good at it. You must verify a large number on your own. With practice the process will seem less complicated.

EXAMPLE 7 Testing Identities Using a Graphing Utility

Use a graphing utility to test whether each of the following is an identity. If an equation appears to be an identity, verify it. If the equation does not appear to be an identity, find a value of x for which both sides are defined but are not equal.

(A) $\tan x + 1 = (\sec x)(\sin x - \cos x)$
(B) $\tan x - 1 = (\sec x)(\sin x - \cos x)$

Solution (A) Graph each side of the equation in the same viewing window (Fig. 1).

FIGURE 1

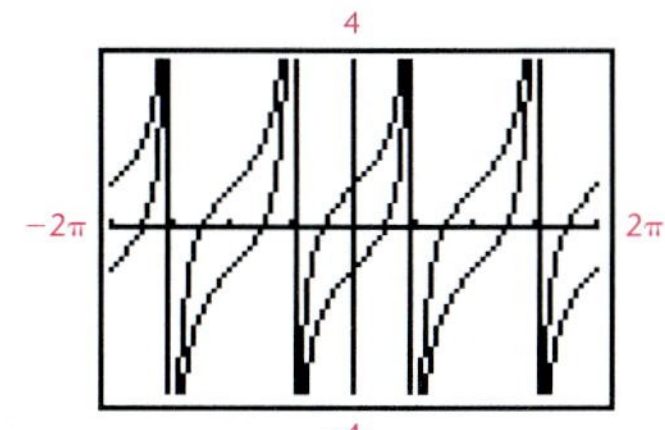

Not an identity, since the graphs do not match. Try $x = 0$

$$\text{Left side:} \quad \tan 0 + 1 = 1$$

$$\text{Right side:} \quad (\sec 0)(\sin 0 - \cos 0) = -1$$

Finding one value of x for which both sides are defined, but are not equal, is enough to verify that the equation is not an identity.

(B) Graph each side of the equation in the same viewing window (Fig. 2).

FIGURE 2

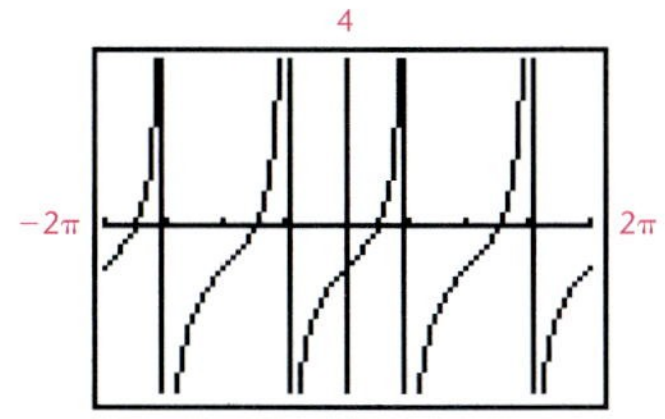

The equation appears to be an identity, which we now verify:

$$(\sec x)(\sin x - \cos x) = \left(\frac{1}{\cos x}\right)(\sin x - \cos x)$$
$$= \frac{\sin x}{\cos x} - \frac{\cos x}{\cos x}$$
$$= \tan x - 1$$

Matched Problem 7 Use a graphing utility to test whether each of the following is an identity. If an equation appears to be an identity, verify it. If the equation does not appear to be an identity, find a value of x for which both sides are defined but are not equal.

(A) $\dfrac{\sin x}{1 - \cos^2 x} = \csc x$ (B) $\dfrac{\sin x}{1 - \cos^2 x} = \sec x$

Answers to Matched Problems

In the following identity verifications, other correct sequences of steps are possible—the process is not unique.

1. $\sin x \cot x = \sin x \dfrac{\cos x}{\sin x} = \cos x$
2. $\csc(-x) = \dfrac{1}{\sin(-x)} = \dfrac{1}{-\sin x} = -\csc x$
3. $\tan x \sin x + \cos x = \dfrac{\sin^2 x}{\cos x} + \cos x = \dfrac{\sin^2 x + \cos^2 x}{\cos x} = \dfrac{1}{\cos x} = \sec x$
4. $\dfrac{1 + \cos x}{\sin x} + \dfrac{\sin x}{1 + \cos x} = \dfrac{(1 + \cos x)^2 + \sin^2 x}{\sin x(1 + \cos x)} = \dfrac{1 + 2\cos x + \cos^2 x + \sin^2 x}{\sin x(1 + \cos x)}$
$= \dfrac{2(1 + \cos x)}{\sin x(1 + \cos x)} = 2\csc x$
5. $\sec^4 x - 2\sec^2 x \tan^2 x + \tan^4 x = (\sec^2 x - \tan^2 x)^2 = 1^2 = 1$
6. $\cot x - \tan x = \dfrac{\cos x}{\sin x} - \dfrac{\sin x}{\cos x} = \dfrac{\cos^2 x - \sin^2 x}{\sin x \cos x} = \dfrac{\cos^2 x - (1 - \cos^2 x)}{\sin x \cos x} = \dfrac{2\cos^2 x - 1}{\sin x \cos x}$
7. (A) An identity:

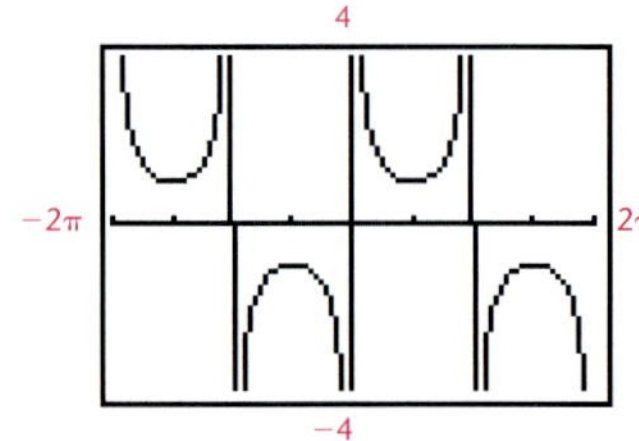

$$\frac{\sin x}{1 - \cos^2 x} = \frac{\sin x}{\sin^2 x} = \frac{1}{\sin x} = \csc x$$

(B) Not an identity; left side is not equal to the right side for $x = 1$, for example.

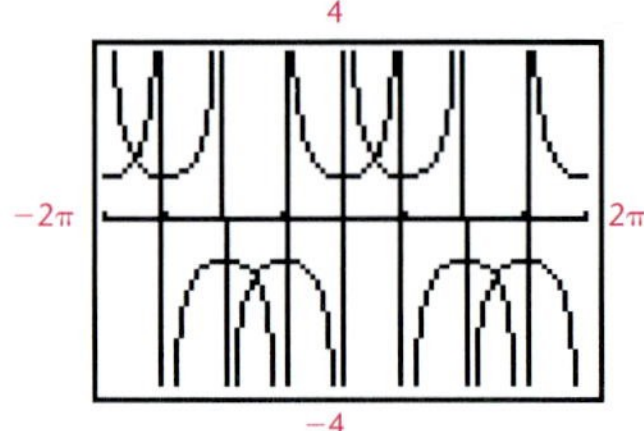

EXERCISE 6-1

A

Verify that Problems 1–24 are identities.

1. $\sin\theta\sec\theta = \tan\theta$
2. $\cos\theta\csc\theta = \cot\theta$
3. $\cot u\sec u\sin u = 1$
4. $\tan\theta\csc\theta\cos\theta = 1$
5. $\dfrac{\sin(-x)}{\cos(-x)} = -\tan x$
6. $\cot(-x)\tan x = -1$
7. $\sin\alpha = \dfrac{\tan\alpha\cot\alpha}{\csc\alpha}$
8. $\tan\alpha = \dfrac{\cos\alpha\sec\alpha}{\cot\alpha}$
9. $\cot u + 1 = (\csc u)(\cos u + \sin u)$
10. $\tan u + 1 = (\sec u)(\sin u + \cos u)$
11. $\dfrac{\cos x - \sin x}{\sin x\cos x} = \csc x - \sec x$
12. $\dfrac{\cos^2 x - \sin^2 x}{\sin x\cos x} = \cot x - \tan x$
13. $\dfrac{\sin^2 t}{\cos t} + \cos t = \sec t$
14. $\dfrac{\cos^2 t}{\sin t} + \sin t = \csc t$
15. $\dfrac{\cos x}{1 - \sin^2 x} = \sec x$
16. $\dfrac{\sin u}{1 - \cos^2 u} = \csc u$
17. $(1 - \cos u)(1 + \cos u) = \sin^2 u$
18. $(1 - \sin t)(1 + \sin t) = \cos^2 t$
19. $\cos^2 x - \sin^2 x = 1 - 2\sin^2 x$
20. $(\sin x + \cos x)^2 = 1 + 2\sin x\cos x$
21. $(\sec t + 1)(\sec t - 1) = \tan^2 t$
22. $(\csc t - 1)(\csc t + 1) = \cot^2 t$
23. $\csc^2 x - \cot^2 x = 1$
24. $\sec^2 u - \tan^2 u = 1$

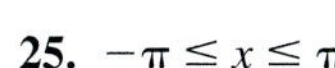
In Problems 25–28, graph all parts of each problem in the same viewing window in a graphing utility.

25. $-\pi \le x \le \pi$
 (A) $y = \sin^2 x$ (B) $y = \cos^2 x$
 (C) $y = \sin^2 x + \cos^2 x$
26. $-\pi \le x \le \pi$
 (A) $y = \sec^2 x$ (B) $y = \tan^2 x$
 (C) $y = \sec^2 x - \tan^2 x$
27. $-\pi \le x \le \pi$
 (A) $y = \dfrac{\cos x}{\cot x\sin x}$ (B) $y = 1$
28. $-\pi \le x \le \pi$
 (A) $y = \dfrac{\sin x}{\cos x\tan x}$ (B) $y = 1$

B

In Problems 29–36, is the equation an identity? Explain.

29. $\dfrac{x}{|x|} = 1$
30. $\dfrac{x^2 - 4}{x + 2} = x - 2$
31. $\dfrac{1}{\sqrt{x - 1}} = \dfrac{\sqrt{x - 1}}{|x - 1|}$
32. $\sqrt{x^2 + 2x + 1} = x + 1$
33. $\sin x + \cos x = 1$
34. $\sin^2 x - \cos^2 x = 1$
35. $\sin^4 x + \cos^4 x = 1$
36. $\sin^3 x + \cos^3 x = 1$

Verify that Problems 37–64 are identities.

37. $\dfrac{1 - (\sin x - \cos x)^2}{\sin x} = 2\cos x$
38. $\dfrac{1 - \cos^2 y}{(1 - \sin y)(1 + \sin y)} = \tan^2 y$
39. $\cos\theta + \sin\theta = \dfrac{\cot\theta + 1}{\csc\theta}$
40. $\sin\theta + \cos\theta = \dfrac{\tan\theta + 1}{\sec\theta}$
41. $\dfrac{1 + \cos y}{1 - \cos y} = \dfrac{\sin^2 y}{(1 - \cos y)^2}$
42. $1 - \sin y = \dfrac{\cos^2 y}{1 + \sin y}$
43. $\tan^2 x - \sin^2 x = \tan^2 x\sin^2 x$
44. $\sec^2 x + \csc^2 x = \sec^2 x\csc^2 x$
45. $\dfrac{\csc\theta}{\cot\theta + \tan\theta} = \cos\theta$
46. $\dfrac{1 + \sec\theta}{\sin\theta + \tan\theta} = \csc\theta$
47. $\ln(\tan x) = \ln(\sin x) - \ln(\cos x)$
48. $\ln(\cot x) = \ln(\cos x) - \ln(\sin x)$
49. $\ln(\cot x) = -\ln(\tan x)$
50. $\ln(\csc x) = -\ln(\sin x)$
51. $\dfrac{1 - \cos A}{1 + \cos A} = \dfrac{\sec A - 1}{\sec A + 1}$
52. $\dfrac{1 - \csc y}{1 + \csc y} = \dfrac{\sin y - 1}{\sin y + 1}$
53. $\sin^4 w - \cos^4 w = 1 - 2\cos^2 w$
54. $\sin^4 x + 2\sin^2 x\cos^2 x + \cos^4 x = 1$
55. $\dfrac{\cos^2 z - 3\cos z + 2}{\sin^2 z} = \dfrac{2 - \cos z}{1 + \cos z}$

56. $\dfrac{\sin^2 t + 4\sin t + 3}{\cos^2 t} = \dfrac{3 + \sin t}{1 - \sin t}$

57. $\dfrac{\cos^3\theta - \sin^3\theta}{\cos\theta - \sin\theta} = 1 + \sin\theta\cos\theta$

58. $\dfrac{\cos^3 u + \sin^3 u}{\cos u + \sin u} = 1 - \sin u \cos u$

59. $(\sec x - \tan x)^2 = \dfrac{1 - \sin x}{1 + \sin x}$

60. $(\cot u - \csc u)^2 = \dfrac{1 - \cos u}{1 + \cos u}$

61. $\dfrac{\csc^4 x - 1}{\cot^2 x} = 2 + \cot^2 x$

62. $\dfrac{\sec^4 x - 1}{\tan^2 x} = 2 + \tan^2 x$

63. $\dfrac{1 + \sin v}{\cos v} = \dfrac{\cos v}{1 - \sin v}$

64. $\dfrac{\sin x}{1 - \cos x} = \dfrac{1 + \cos x}{\sin x}$

Use a graphing utility to test whether each of Problems 65–76 is an identity. If an equation appears to be an identity, verify it. If the equation does not appear to be an identity, find a value of x for which both sides are defined but are not equal.

65. $\dfrac{\sin(-x)}{\cos(-x)\tan(-x)} = -1$

66. $\dfrac{\cos(-x)}{\sin x \cot(-x)} = 1$

67. $\dfrac{\sin x}{\cos x \tan(-x)} = -1$

68. $\dfrac{\cos x}{\sin(-x)\cot(-x)} = 1$

69. $\sin x + \dfrac{\cos^2 x}{\sin x} = \sec x$

70. $\dfrac{1 - \tan^2 x}{1 - \cot^2 x} = \tan^2 x$

71. $\sin x + \dfrac{\cos^2 x}{\sin x} = \csc x$

72. $\dfrac{\tan^2 x - 1}{1 - \cot^2 x} = \tan^2 x$

73. $\dfrac{\tan x}{\sin x - 2\tan x} = \dfrac{1}{\cos x - 2}$

74. $\dfrac{\cos x}{1 - \sin x} + \dfrac{\cos x}{1 + \sin x} = 2\sec x$

75. $\dfrac{\tan x}{\sin x + 2\tan x} = \dfrac{1}{\cos x - 2}$

76. $\dfrac{\cos x}{\sin x + 1} - \dfrac{\cos x}{\sin x - 1} = 2\csc x$

C

Verify that Problems 77–82 are identities.

77. $\dfrac{2\sin^2 x + 3\cos x - 3}{\sin^2 x} = \dfrac{2\cos x - 1}{1 + \cos x}$

78. $\dfrac{3\cos^2 z + 5\sin z - 5}{\cos^2 z} = \dfrac{3\sin z - 2}{1 + \sin z}$

79. $\dfrac{\tan u + \sin u}{\tan u - \sin u} - \dfrac{\sec u + 1}{\sec u - 1} = 0$

80. $\dfrac{\sin x \cos y + \cos x \sin y}{\cos x \cos y - \sin x \sin y} = \dfrac{\tan x + \tan y}{1 - \tan x \tan y}$

81. $\tan\alpha + \cot\beta = \dfrac{\tan\beta + \cot\alpha}{\tan\beta\cot\alpha}$

82. $\dfrac{\cot\alpha + \cot\beta}{\cot\alpha\cot\beta - 1} = \dfrac{\tan\alpha + \tan\beta}{1 - \tan\alpha\tan\beta}$

Problems 83–88 require the use of a graphing utility. From the graph of $y1 = f(x)$, find a simpler function of the form $g(x) = k + AT(x)$, where $T(x)$ is one of the six trigonometric functions, that has the same graph as $y1 = f(x)$. Verify the identity $f(x) = g(x)$.

83. $f(x) = \dfrac{1 - \sin^2 x}{\tan x} + \sin x \cos x$

84. $f(x) = \dfrac{1 + \sin x}{2\cos x} - \dfrac{\cos x}{2 + 2\sin x}$

85. $f(x) = \dfrac{\cos^2 x}{1 + \sin x - \cos^2 x}$

86. $f(x) = \dfrac{\tan x \sin x}{1 - \cos x}$

87. $f(x) = \dfrac{1 + \cos x - 2\cos^2 x}{1 - \cos x} - \dfrac{\sin^2 x}{1 + \cos x}$

88. $f(x) = \dfrac{3\sin x - 2\sin x\cos x}{1 - \cos x} - \dfrac{1 + \cos x}{\sin x}$

Each of the equations in Problems 89–96 is an identity in certain quadrants associated with x. Indicate which quadrants.

89. $\sqrt{1 - \cos^2 x} = -\sin x$

90. $\sqrt{1 - \sin^2 x} = \cos x$

91. $\sqrt{1 - \cos^2 x} = \sin x$

92. $\sqrt{1 - \sin^2 x} = -\cos x$

93. $\sqrt{1 - \sin^2 x} = |\cos x|$

94. $\sqrt{1 - \cos^2 x} = |\sin x|$

95. $\dfrac{\sin x}{\sqrt{1 - \sin^2 x}} = \tan x$

96. $\dfrac{\sin x}{\sqrt{1 - \sin^2 x}} = -\tan x$

In calculus, trigonometric substitutions provide an effective way to rationalize the radical forms $\sqrt{a^2 - u^2}$ and $\sqrt{a^2 + u^2}$, which in turn leads to the solution to an important class of problems. Problems 97–100 involve such transformations. [Recall: $\sqrt{x^2} = |x|$ for all real numbers x.]

97. In the radical form $\sqrt{a^2 - u^2}$, $a > 0$, let $u = a\sin x$, $-\pi/2 < x < \pi/2$. Simplify, using a basic identity, and write the final form free of radicals.

98. In the radical form $\sqrt{a^2 - u^2}$, $a > 0$, let $u = a \cos x$, $0 < x < \pi$. Simplify, using a basic identity, and write the final form free of radicals.

99. In the radical form $\sqrt{a^2 + u^2}$, $a > 0$, let $u = a \tan x$, $0 < x < \pi/2$. Simplify, using a basic identity, and write the final form free of radicals.

100. In the radical form $\sqrt{a^2 + u^2}$, $a > 0$, let $u = a \cot x$, $0 < x < \pi/2$. Simplify, using a basic identity, and write the final form free of radicals.

SECTION 6-2 Sum, Difference, and Cofunction Identities

- Sum and Difference Identities for Cosine
- Cofunction Identities
- Sum and Difference Identities for Sine and Tangent
- Summary and Use

The basic identities discussed in Section 6-1 involved only one variable. In this section we consider identities that involve two variables.

• Sum and Difference Identities for Cosine

We start with the important **difference identity for cosine:**

$$\cos(x - y) = \cos x \cos y + \sin x \sin y \tag{1}$$

Many other useful identities can be readily verified from this particular one.

Here, we sketch a proof of equation (1) assuming x and y are in the interval $(0, 2\pi)$ and $x > y > 0$. It then follows easily, by periodicity and basic identities, that (1) holds for all real numbers and angles in radian or degree measure.

First, associate x and y with arcs and angles on the unit circle as indicated in Figure 1(a). Using the definitions of the circular functions given in Section 5-2, label the terminal points of x and y as shown in Figure 1(a).

FIGURE 1 Difference identity.

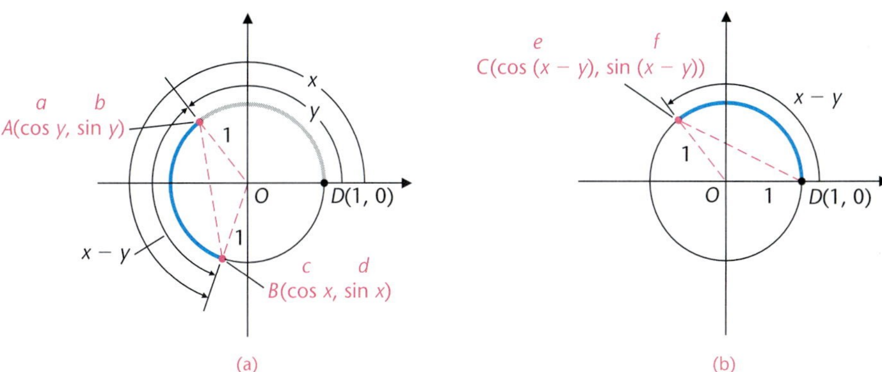

Now if you rotate the triangle AOB clockwise about the origin until the terminal point A coincides with $D(1, 0)$, then terminal point B will be at C, as shown in Figure 1(b). Thus, since rotation preserves lengths,

$$d(A, B) = d(C, D)$$

$$\sqrt{(c - a)^2 + (d - b)^2} = \sqrt{(1 - e)^2 + (0 - f)^2}$$

$$(c - a)^2 + (d - b)^2 = (1 - e)^2 + f^2$$

$$c^2 - 2ac + a^2 + d^2 - 2db + b^2 = 1 - 2e + e^2 + f^2$$

$$(c^2 + d^2) + (a^2 + b^2) - 2ac - 2db = 1 - 2e + (e^2 + f^2) \tag{2}$$

Since points A, B, and C are on unit circles, $c^2 + d^2 = 1$, $a^2 + b^2 = 1$, and $e^2 + f^2 = 1$, and equation (2) simplifies to

$$e = ac + bd \tag{3}$$

Replacing e, a, c, b, and d with $\cos(x - y)$, $\cos y$, $\cos x$, $\sin y$, and $\sin x$, respectively (see Fig. 1), we obtain

$$\begin{aligned}\cos(x - y) &= \cos y \cos x + \sin y \sin x \\ &= \cos x \cos y + \sin x \sin y \qquad (4)\end{aligned}$$

We have thus established the difference identity for cosine.

If we replace y with $-y$ in equation (4) and use the identities for negatives (a good exercise for you), we obtain

$$\cos(x + y) = \cos x \cos y - \sin x \sin y \tag{5}$$

This is the **sum identity for cosine.**

EXPLORE-DISCUSS 1 Discuss how you would show that, in general,

$$\cos(x - y) \neq \cos x - \cos y$$

and

$$\cos(x + y) \neq \cos x + \cos y$$

• Cofunction Identities

To obtain sum and difference identities for the sine and tangent functions, we first derive *cofunction identities* directly from equation (1), the difference identity for cosine:

$$\begin{aligned}\cos(x - y) &= \cos x \cos y + \sin x \sin y \\ \cos\left(\frac{\pi}{2} - y\right) &= \cos\frac{\pi}{2}\cos y + \sin\frac{\pi}{2}\sin y \\ &= (0)(\cos y) + (1)(\sin y) \\ &= \sin y\end{aligned}$$

Thus, we have the **cofunction identity for cosine:**

$$\cos\left(\frac{\pi}{2} - y\right) = \sin y \tag{6}$$

for y any real number or angle in radian measure. If y is in degree measure, replace $\pi/2$ with $90°$.

Now, if we let $y = \pi/2 - x$ in equation (6), we have

$$\cos\left[\frac{\pi}{2} - \left(\frac{\pi}{2} - x\right)\right] = \sin\left(\frac{\pi}{2} - x\right)$$

$$\cos x = \sin\left(\frac{\pi}{2} - x\right)$$

This is the **cofunction identity for sine;** that is,

$$\sin\left(\frac{\pi}{2} - x\right) = \cos x \qquad (7)$$

where x is any real number or angle in radian measure. If x is in degree measure, replace $\pi/2$ with 90°.

Finally, we state the **cofunction identity for tangent** (and leave its derivation to Problem 12 in Exercise 6-2):

$$\tan\left(\frac{\pi}{2} - x\right) = \cot x \qquad (8)$$

for x any real number or angle in radian measure. If x is in degree measure, replace $\pi/2$ with 90°.

Remark. If $0 < x < 90°$, then x and $90° - x$ are complementary angles. Originally, "cosine," "cotangent," and "cosecant" meant, respectively, "complements sine," "complements tangent," and "complements secant." Now we simply refer to cosine, cotangent, and cosecant as **cofunctions** of sine, tangent, and secant, respectively.

• Sum and Difference Identities for Sine and Tangent

To derive a difference identity for sine, we use equations (1), (6), and (7) as follows:

$$\sin(x - y) = \cos\left[\frac{\pi}{2} - (x - y)\right]$$ Use equation (6).

$$= \cos\left[\left(\frac{\pi}{2} - x\right) - (-y)\right]$$ Algebra

$$= \cos\left(\frac{\pi}{2} - x\right)\cos(-y) + \sin\left(\frac{\pi}{2} - x\right)\sin(-y)$$ Use equation (1).

$$= \sin x \cos y - \cos x \sin y$$ Use equations (6) and (7) and identities for negatives.

The same result is obtained by replacing $\pi/2$ with 90°. Thus,

$$\sin(x - y) = \sin x \cos y - \cos x \sin y \qquad (9)$$

is the **difference identity for sine.**

Now, if we replace y in equation (9) with $-y$ (a good exercise for you), we obtain

$$\sin(x + y) = \sin x \cos y + \cos x \sin y \qquad (10)$$

the **sum identity for sine.**

It is not difficult to derive sum and difference identities for the tangent function. See if you can supply the reason for each step:

$$\tan(x - y) = \frac{\sin(x - y)}{\cos(x - y)}$$

$$= \frac{\sin x \cos y - \cos x \sin y}{\cos x \cos y + \sin x \sin y}$$

$$= \frac{\dfrac{\sin x \cos y}{\cos x \cos y} - \dfrac{\cos x \sin y}{\cos x \cos y}}{\dfrac{\cos x \cos y}{\cos x \cos y} + \dfrac{\sin x \sin y}{\cos x \cos y}}$$

Divide the numerator and denominator by $\cos x \cos y$.

$$= \frac{\dfrac{\sin x}{\cos x} - \dfrac{\sin y}{\cos y}}{1 + \dfrac{\sin x \sin y}{\cos x \cos y}}$$

$$= \frac{\tan x - \tan y}{1 + \tan x \tan y}$$

Thus, for all angles or real numbers x and y,

$$\tan(x - y) = \frac{\tan x - \tan y}{1 + \tan x \tan y} \qquad (11)$$

is the **difference identity for tangent.**

If we replace y in equation (11) with $-y$ (another good exercise for you), we obtain

$$\tan(x + y) = \frac{\tan x + \tan y}{1 - \tan x \tan y} \qquad (12)$$

the **sum identity for tangent.**

EXPLORE-DISCUSS 2 Discuss how you would show that, in general,

$$\tan(x - y) \neq \tan x - \tan y$$

and

$$\tan(x + y) \neq \tan x + \tan y$$

• Summary and Use

Before proceeding with examples illustrating the use of these new identities, review the list given in the following box.

Summary of Identities

Sum Identities

$$\sin(x + y) = \sin x \cos y + \cos x \sin y$$

$$\cos(x + y) = \cos x \cos y - \sin x \sin y$$

$$\tan(x + y) = \frac{\tan x + \tan y}{1 - \tan x \tan y}$$

Difference Identities

$$\sin(x - y) = \sin x \cos y - \cos x \sin y$$

$$\cos(x - y) = \cos x \cos y + \sin x \sin y$$

$$\tan(x - y) = \frac{\tan x - \tan y}{1 + \tan x \tan y}$$

Cofunction Identities
(Replace $\pi/2$ with 90° if x is in degrees.)

$$\cos\left(\frac{\pi}{2} - x\right) = \sin x \qquad \sin\left(\frac{\pi}{2} - x\right) = \cos x \qquad \tan\left(\frac{\pi}{2} - x\right) = \cot x$$

EXAMPLE 1 **Using the Difference Identity**

Simplify $\cos(x - \pi)$ using the difference identity.

Solution

$$\begin{aligned}\cos(x - y) &= \cos x \cos y + \sin x \sin y\\ \cos(x - \pi) &= \cos x \cos \pi + \sin x \sin \pi\\ &= (\cos x)(-1) + (\sin x)(0)\\ &= -\cos x\end{aligned}$$

Matched Problem 1 Simplify $\sin(x + 3\pi/2)$ using a sum identity.

EXAMPLE 2 **Checking the Use of Sum and Difference Identities on a Graphing Utility**

Simplify $\sin(x - \pi)$ using a difference identity. Enter the original form as $y1$ and the converted form as $y2$ in a graphing utility, then graph both in the same viewing window.

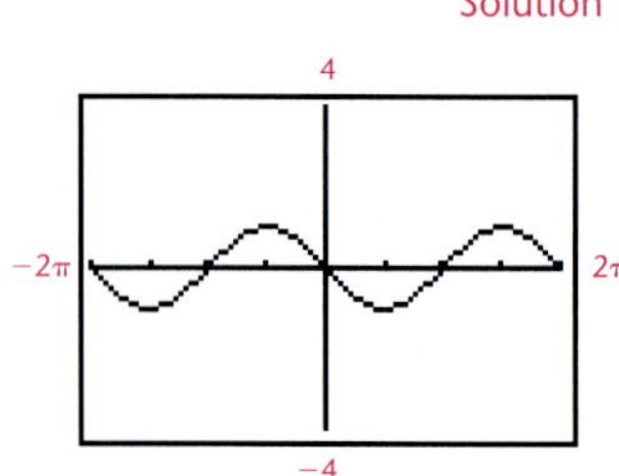

FIGURE 2

Solution

$$\sin(x - y) = \sin x \cos y - \cos x \sin y$$
$$\begin{aligned}\sin(x - \pi) &= \sin x \cos \pi - \cos x \sin \pi \\ &= (\sin x)(-1) - (\cos x)(0) \\ &= -\sin x\end{aligned}$$

Graph $y1 = \sin(x - \pi)$ and $y2 = -\sin x$ in the same viewing window (Fig. 2). Use TRACE and move back and forth between $y1$ and $y2$ for different values of x to see that the corresponding y values are the same, or nearly the same.

Matched Problem 2 Simplify $\cos(x + 3\pi/2)$ using a sum identity. Enter the original form as $y1$ and the converted form as $y2$ in a graphing utility, then graph both in the same viewing window.

EXAMPLE 3 Finding Exact Values

Find the value of tan 75° in exact radical form.

Solution Since we can write 75° = 45° + 30°, the sum of two special angles, we can use the sum identity for tangent with $x = 45°$ and $y = 30°$:

$$\tan(x + y) = \frac{\tan x + \tan y}{1 - \tan x \tan y}$$

$$\begin{aligned}\tan(45° + 30°) &= \frac{\tan 45° + \tan 30°}{1 - \tan 45° \tan 30°} && \text{Sum identity} \\ &= \frac{1 + (1/\sqrt{3})}{1 - 1(1/\sqrt{3})} && \text{Evaluate functions exactly.} \\ &= \frac{\sqrt{3} + 1}{\sqrt{3} - 1} && \text{Multiply numerator and denominator by } \sqrt{3} \text{ and simplify.} \\ &= 2 + \sqrt{3} && \text{Rationalize denominator and simplify.}\end{aligned}$$

Matched Problem 3 Find the value of cos 15° in exact radical form.

EXAMPLE 4 Finding Exact Values

Find the exact value of $\cos(x + y)$, given $\sin x = \frac{3}{5}$, $\cos y = \frac{4}{5}$, x is an angle in quadrant II, and y is an angle in quadrant I. Do not use a calculator or table.

Solution We start with the sum identity for cosine,

$$\cos(x + y) = \cos x \cos y - \sin x \sin y$$

We know $\sin x$ and $\cos y$ but not $\cos x$ and $\sin y$. We find the latter two using two different methods as follows (use the method that is easiest for you).

Given $\sin x = \frac{3}{5}$ and x is an angle in quadrant II, find $\cos x$:

Method I. Use a reference triangle:

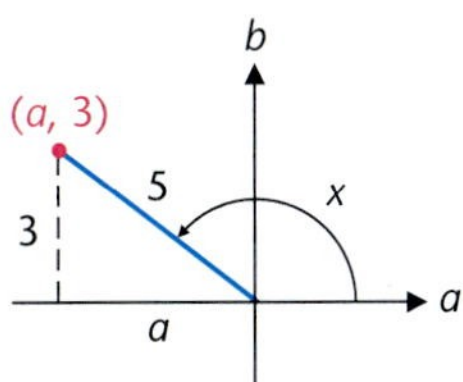

$$\cos x = \frac{a}{5}$$
$$a^2 + 3^2 = 5^2$$
$$a^2 = 16$$
$$a = \pm 4$$

In quadrant II, $a = -4$

Therefore, $\cos x = -\frac{4}{5}$

Method II. Use a unit circle:

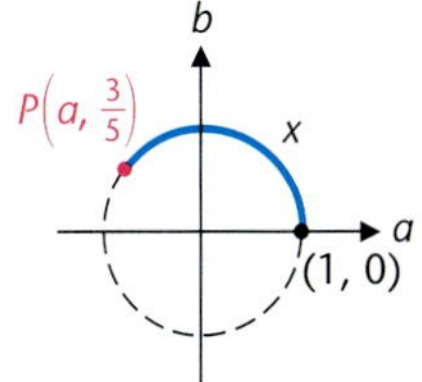

$$\cos x = a$$
$$a^2 + (\tfrac{3}{5})^2 = 1$$
$$a^2 = \tfrac{16}{25}$$
$$a = \pm\tfrac{4}{5}$$

In quadrant II, $a = -\frac{4}{5}$

Therefore, $\cos x = -\frac{4}{5}$

Given $\cos y = \frac{4}{5}$ and y is an angle in quadrant I, find $\sin y$:

Method I. Use a reference triangle:

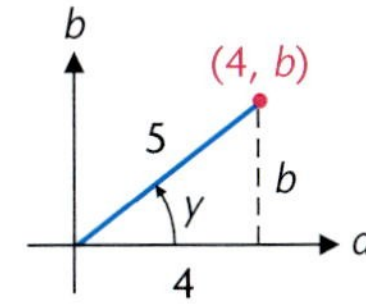

$$\sin y = \frac{b}{5}$$
$$4^2 + b^2 = 5^2$$
$$b^2 = 9$$
$$b = \pm 3$$

In quadrant I, $b = 3$

Therefore, $\sin y = \frac{3}{5}$

Method II. Use a unit circle:

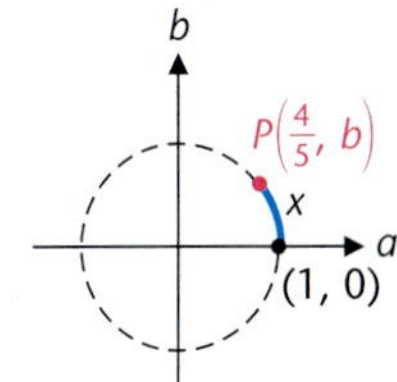

$$\sin y = b$$
$$(\tfrac{4}{5})^2 + b^2 = 1$$
$$b^2 = \tfrac{9}{25}$$
$$b = \pm\tfrac{3}{5}$$

In quadrant I, $b = \frac{3}{5}$

Therefore, $\sin y = \frac{3}{5}$

We can now evaluate $\cos(x + y)$ without knowing x and y:

$$\begin{aligned}\cos(x + y) &= \cos x \cos y - \sin x \sin y \\ &= (-\tfrac{4}{5})(\tfrac{4}{5}) - (\tfrac{3}{5})(\tfrac{3}{5}) = -\tfrac{25}{25} = -1\end{aligned}$$

Matched Problem 4 Find the exact value of $\sin(x - y)$, given $\sin x = -\frac{2}{3}$, $\cos y = \sqrt{5}/3$, x is an angle in quadrant III, and y is an angle in quadrant IV. Do not use a calculator or table.

EXAMPLE 5 Identity Verification

Verify the identity: $\tan x + \cot y = \dfrac{\cos(x - y)}{\cos x \sin y}$

Verification

$$\frac{\cos(x-y)}{\cos x \sin y} = \frac{\cos x \cos y + \sin x \sin y}{\cos x \sin y} \quad \text{Difference identity for cosine}$$

$$= \frac{\cancel{\cos x} \cos y}{\cancel{\cos x} \sin y} + \frac{\sin x \cancel{\sin y}}{\cos x \cancel{\sin y}} \quad \text{Algebra}$$

$$= \cot y + \tan x \quad \text{Quotient identities}$$

$$= \tan x + \cot y$$

Matched Problem 5 Verify the identity: $\cot y - \cot x = \dfrac{\sin(x - y)}{\sin x \sin y}$

Answers to Matched Problems

1. $-\cos x$ **2.** $y1 = \cos(x + 3\pi/2)$, $y2 = \sin x$

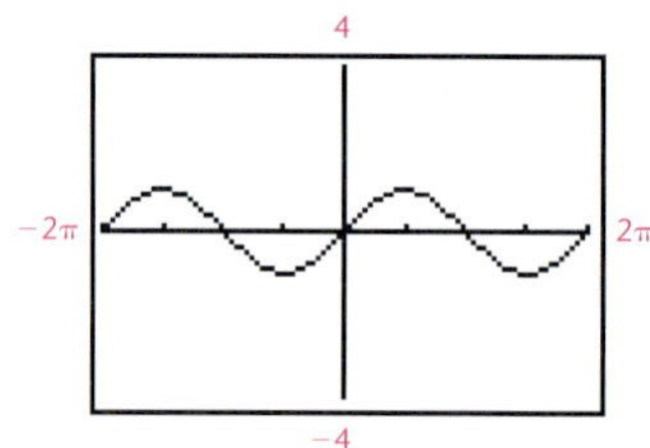

3. $(1 + \sqrt{3})/2\sqrt{2}$ or $(\sqrt{6} + \sqrt{2})/4$ **4.** $-4\sqrt{5}/9$

5. $\dfrac{\sin(x-y)}{\sin x \sin y} = \dfrac{\sin x \cos y - \cos x \sin y}{\sin x \sin y} = \dfrac{\cancel{\sin x} \cos y}{\cancel{\sin x} \sin y} - \dfrac{\cos x \cancel{\sin y}}{\sin x \cancel{\sin y}} = \cot y - \cot x$

EXERCISE 6-2

A

In Problems 1–10, is the equation an identity? Explain, making use of sum or difference identities.

1. $\sin(x + 2\pi) = \sin x$ **2.** $\cos(x - 2\pi) = \cos x$

3. $\cos(x + \pi) = \cos x$ **4.** $\sin(x - \pi) = -\sin x$

5. $\tan(x + \pi) = \tan x$ **6.** $\tan(\pi - x) = \tan x$

7. $\csc(2\pi - x) = \csc x$ **8.** $\sec(2\pi - x) = \sec x$

9. $\sin(x + 2k\pi) = \sin x$, k an integer

10. $\tan(x + k\pi) = \tan x$, k an integer

Verify each identity in Problems 11–14 using cofunction identities for sine and cosine and basic identities discussed in Section 6-1.

11. $\cot\left(\dfrac{\pi}{2} - x\right) = \tan x$ **12.** $\tan\left(\dfrac{\pi}{2} - x\right) = \cot x$

13. $\csc\left(\dfrac{\pi}{2} - x\right) = \sec x$ **14.** $\sec\left(\dfrac{\pi}{2} - x\right) = \csc x$

Convert Problems 15–20 to forms involving sin x, cos x, and/or tan x using sum or difference identities.

15. $\cos(x + 45°)$ **16.** $\sin(x + 30°)$

17. $\tan\left(\frac{\pi}{3} + x\right)$

18. $\cos(\pi - x)$

19. $\sin(x - 90°)$

20. $\tan(x - 45°)$

B

Use appropriate identities to find exact values for Problems 21–28. Do not use a calculator.

21. $\cos 20° \cos 25° - \sin 20° \sin 25°$

22. $\sin 75° \cos 15° - \cos 75° \sin 15°$

23. $\dfrac{\tan 50° - \tan 20°}{1 + \tan 50° \tan 20°}$

24. $\dfrac{\tan 35° + \tan 25°}{1 - \tan 35° \tan 25°}$

25. $\sin 15°$

26. $\cos 15°$

27. $\cos \frac{11\pi}{12}\left[\textit{Hint: } \frac{11\pi}{12} = \frac{2\pi}{3} + \frac{\pi}{4}\right]$

28. $\sin\left(-\frac{\pi}{12}\right)\left[\textit{Hint: } -\frac{\pi}{12} = \frac{\pi}{6} - \frac{\pi}{4}\right]$

Find $\sin(x - y)$ and $\tan(x + y)$ exactly without a calculator using the information given in Problems 29–32.

29. $\sin x = -\frac{3}{5}$, $\sin y = \sqrt{8}/3$, x is a quadrant IV angle, y is a quadrant I angle.

30. $\sin x = \frac{2}{3}$, $\cos y = -\frac{1}{4}$, x is a quadrant II angle, y is a quadrant III angle.

31. $\tan x = \frac{3}{4}$, $\tan y = -\frac{1}{2}$, x is a quadrant III angle, y is a quadrant IV angle.

32. $\cos x = -\frac{1}{3}$, $\tan y = \frac{1}{2}$, x is a quadrant II angle, y is a quadrant III angle.

Verify each identity in Problems 33–46.

33. $\cos 2x = \cos^2 x - \sin^2 x$

34. $\sin 2x = 2 \sin x \cos x$

35. $\cot(x + y) = \dfrac{\cot x \cot y - 1}{\cot x + \cot y}$

36. $\cot(x - y) = \dfrac{\cot x \cot y + 1}{\cot y - \cot x}$

37. $\tan 2x = \dfrac{2 \tan x}{1 - \tan^2 x}$

38. $\cot 2x = \dfrac{\cot^2 x - 1}{2 \cot x}$

39. $\dfrac{\sin(v + u)}{\sin(v - u)} = \dfrac{\cot u + \cot v}{\cot u - \cot v}$

40. $\dfrac{\sin(u + v)}{\sin(u - v)} = \dfrac{\tan u + \tan v}{\tan u - \tan v}$

41. $\cot x - \tan y = \dfrac{\cos(x + y)}{\sin x \cos y}$

42. $\tan x - \tan y = \dfrac{\sin(x - y)}{\cos x \cos y}$

43. $\tan(x - y) = \dfrac{\cot y - \cot x}{\cot x \cot y + 1}$

44. $\tan(x + y) = \dfrac{\cot x + \cot y}{\cot x \cot y - 1}$

45. $\dfrac{\cos(x + h) - \cos x}{h} = \cos x\left(\dfrac{\cos h - 1}{h}\right) - \sin x\left(\dfrac{\sin h}{h}\right)$

46. $\dfrac{\sin(x + h) - \sin x}{h} = \sin x\left(\dfrac{\cos h - 1}{h}\right) + \cos x\left(\dfrac{\sin h}{h}\right)$

Evaluate both sides of the difference identity for sine and the sum identity for tangent for the values of x and y indicated in Problems 47–50. Evaluate to 4 significant digits using a calculator.

47. $x = 5.288$, $y = 1.769$

48. $x = 3.042$, $y = 2.384$

49. $x = 42.08°$, $y = 68.37°$

50. $x = 128.3°$, $y = 25.62°$

51. Explain how you would show that, in general,

$$\sec(x - y) \neq \sec x - \sec y$$

52. Explain how you would show that, in general,

$$\csc(x + y) \neq \csc x + \csc y$$

In Problems 53–58, use sum or difference identities to convert each equation to a form involving sin x, cos x, and/or tan x. Enter the original equation in a graphing utility as y1 and the converted form as y2, then graph y1 and y2 in the same viewing window. Use TRACE to compare the two graphs.

53. $y = \sin(x + \pi/6)$

54. $y = \sin(x - \pi/3)$

55. $y = \cos(x - 3\pi/4)$

56. $y = \cos(x + 5\pi/6)$

57. $y = \tan(x + 2\pi/3)$

58. $y = \tan(x - \pi/4)$

C

In Problems 59–62, evaluate exactly as real numbers without the use of a calculator.

59. $\sin[\cos^{-1}(-\frac{4}{5}) + \sin^{-1}(-\frac{3}{5})]$

60. $\cos[\sin^{-1}(-\frac{3}{5}) + \cos^{-1}(\frac{4}{5})]$

61. $\sin[\arccos \frac{1}{2} + \arcsin(-1)]$

62. $\cos[\arccos(-\sqrt{3}/2) - \arcsin(-\frac{1}{2})]$

63. Express $\sin(\sin^{-1} x + \cos^{-1} y)$ in an equivalent form free of trigonometric and inverse trigonometric functions.

64. Express $\cos(\sin^{-1} x - \cos^{-1} y)$ in an equivalent form free of trigonometric and inverse trigonometric functions.

Verify the identities in Problems 65 and 66.

65. $\cos(x + y + z) = \cos x \cos y \cos z - \sin x \sin y \cos z - \sin x \cos y \sin z - \cos x \sin y \sin z$

66. $\sin(x + y + z) = \sin x \cos y \cos z + \cos x \sin y \cos z + \cos x \cos y \sin z - \sin x \sin y \sin z$

In Problems 67–68, write each equation in terms of a single trigonometric function. Enter the original equation in a graphing utility as y1 and the converted form as y2, then graph y1 and y2 in the same viewing window. Use TRACE to compare the two graphs.

67. $y = \cos 1.2x \cos 0.8x - \sin 1.2x \sin 0.8x$

68. $y = \sin 0.8x \cos 0.3x - \cos 0.8x \sin 0.3x$

APPLICATIONS

69. Analytic Geometry. Use the information in the figure to show that

$$\tan(\theta_2 - \theta_1) = \frac{m_2 - m_1}{1 + m_1 m_2}$$

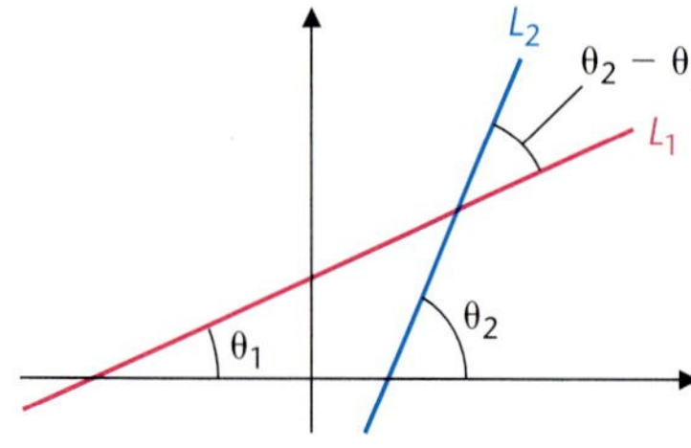

$$\tan\theta_1 = \text{Slope of } L_1 = m_1$$

$$\tan\theta_2 = \text{Slope of } L_2 = m_2$$

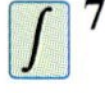

70. Analytic Geometry. Find the acute angle of intersection between the two lines $y = 3x + 1$ and $y = \frac{1}{2}x - 1$. (Use the results of Problem 69.)

★★ **71. Light Refraction.** Light rays passing through a plate glass window are refracted when they enter the glass and again when they leave to continue on a path parallel to the entering rays (see the figure). If the plate glass is M inches thick, the parallel displacement of the light rays is N inches, the angle of incidence is α, and the angle of refraction is β, show that

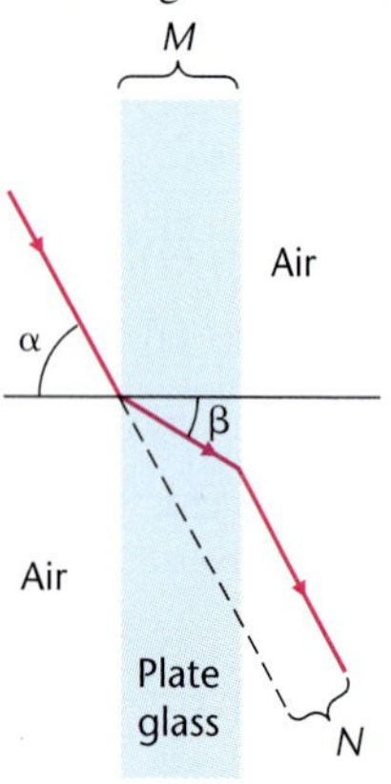

$$\tan\beta = \tan\alpha - \frac{N}{M}\sec\alpha$$

[*Hint:* First use geometric relationships to obtain

$$\frac{M}{\sin(90° - \beta)} = \frac{N}{\sin(\alpha - \beta)}$$

then use difference identities and fundamental identities to complete the derivation.]

72. Light Refraction. Use the result of Problem 71 to find β to the nearest degree if $\alpha = 43°$, $M = 0.25$ inch, and $N = 0.11$ inch.

★ **73. Surveying.** El Capitan is a large monolithic granite peak that rises straight up from the floor of Yosemite Valley in Yosemite National Park. It attracts rock climbers worldwide. At certain times the reflection of the peak can be seen in the Merced River that runs along the valley floor. How can the height H of El Capitan above the river be determined by using only a sextant h feet high to measure the angle of elevation β to the top of the peak, and the angle of depression α of the reflected peak top in the river? (See accompanying figure, which is not to scale.)

(A) Using right triangle relationships, show that

$$H = h\frac{1 + \tan\beta\cot\alpha}{1 - \tan\beta\cot\alpha}$$

(B) Using sum or difference identities, show that the result in part A can be written in the form

$$H = h\frac{\sin(\alpha + \beta)}{\sin(\alpha - \beta)}$$

(C) If a sextant of height 4.90 feet measures α to be 46.23° and β to be 46.15°, compute the height H of El Capitan above the Merced River to 3 significant digits.

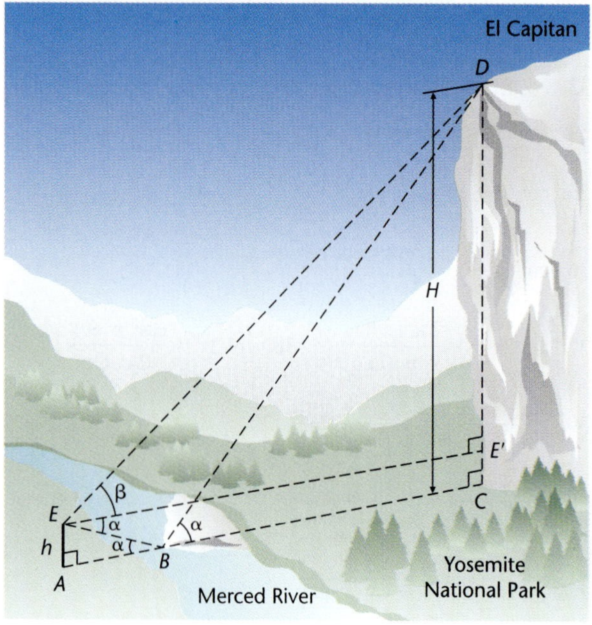

SECTION 6-3 Double-Angle and Half-Angle Identities

- Double-Angle Identities
- Half-Angle Identities

This section develops another important set of identities called *double-angle* and *half-angle identities.* We can derive these identities directly from the sum and difference identities given in Section 6-2. Even though the names use the word "angle," the new identities hold for real numbers as well.

• Double-Angle Identities

Start with the sum identity for sine,

$$\sin (x + y) = \sin x \cos y + \cos x \sin y$$

and replace y with x to obtain

$$\sin (x + x) = \sin x \cos x + \cos x \sin x$$

On simplification, this gives

$$\sin 2x = 2 \sin x \cos x \qquad \text{Double-angle identity for sine} \tag{1}$$

If we start with the sum identity for cosine,

$$\cos (x + y) = \cos x \cos y - \sin x \sin y$$

and replace y with x, we obtain

$$\cos (x + x) = \cos x \cos x - \sin x \sin x$$

On simplification, this gives

$$\cos 2x = \cos^2 x - \sin^2 x \qquad \text{First double-angle identity for cosine} \tag{2}$$

Now, using the Pythagorean identity

$$\sin^2 x + \cos^2 x = 1 \tag{3}$$

in the form

$$\cos^2 x = 1 - \sin^2 x \tag{4}$$

and substituting it into equation (2), we get

$$\cos 2x = 1 - \sin^2 x - \sin^2 x$$

On simplification, this gives

$$\cos 2x = 1 - 2 \sin^2 x \qquad \text{Second double-angle identity for cosine} \tag{5}$$

Or, if we use equation (3) in the form

$$\sin^2 x = 1 - \cos^2 x$$

and substitute it into equation (2), we get

$$\cos 2x = \cos^2 x - (1 - \cos^2 x)$$

On simplification, this gives

$$\cos 2x = 2\cos^2 x - 1 \qquad \text{Third double-angle identity for cosine} \qquad (6)$$

Double-angle identities can be established for the tangent function in the same way by starting with the sum formula for tangent (a good exercise for you).

We list the double-angle identities below for convenient reference.

Double-Angle Identities

$$\sin 2x = 2 \sin x \cos x$$

$$\cos 2x = \cos^2 x - \sin^2 x = 1 - 2\sin^2 x = 2\cos^2 x - 1$$

$$\tan 2x = \frac{2 \tan x}{1 - \tan^2 x} = \frac{2 \cot x}{\cot^2 x - 1} = \frac{2}{\cot x - \tan x}$$

The identities in the second row are used to a good advantage in calculus in the form

$$\sin^2 x = \frac{1 - \cos 2x}{2} \qquad \cos^2 x = \frac{1 + \cos 2x}{2}$$

to transform a power form to a nonpower form.

EXPLORE-DISCUSS 1 (A) Discuss how you would show that, in general,

$$\sin 2x \neq 2 \sin x \qquad \cos 2x \neq 2 \cos x \qquad \tan 2x \neq 2 \tan x$$

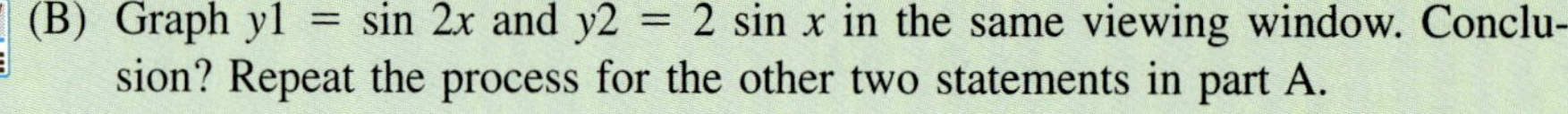

(B) Graph $y1 = \sin 2x$ and $y2 = 2 \sin x$ in the same viewing window. Conclusion? Repeat the process for the other two statements in part A.

EXAMPLE 1 Identity Verification

Verify the identity: $\cos 2x = \dfrac{1 - \tan^2 x}{1 + \tan^2 x}$

Verification We start with the right side:

$$\frac{1 - \tan^2 x}{1 + \tan^2 x} = \frac{1 - \dfrac{\sin^2 x}{\cos^2 x}}{1 + \dfrac{\sin^2 x}{\cos^2 x}} \qquad \text{Quotient identities}$$

$$= \frac{\cos^2 x - \sin^2 x}{\cos^2 x + \sin^2 x} \qquad \text{Algebra}$$

$$= \cos^2 x - \sin^2 x \qquad \text{Pythagorean identity}$$

$$= \cos 2x \qquad \text{Double-angle identity}$$

Key Algebraic Steps in Example 1

$$\frac{1 - \dfrac{a^2}{b^2}}{1 + \dfrac{a^2}{b^2}} = \frac{b^2\left(1 - \dfrac{a^2}{b^2}\right)}{b^2\left(1 + \dfrac{a^2}{b^2}\right)} = \frac{b^2 - a^2}{b^2 + a^2}$$

Matched Problem 1 Verify the identity: $\sin 2x = \dfrac{2 \tan x}{1 + \tan^2 x}$

EXAMPLE 2 Finding Exact Values

Find the exact values, without using a calculator, of $\sin 2x$ and $\cos 2x$ if $\tan x = -\frac{3}{4}$ and x is a quadrant IV angle.

Solution First draw the reference triangle for x and find any unknown sides:

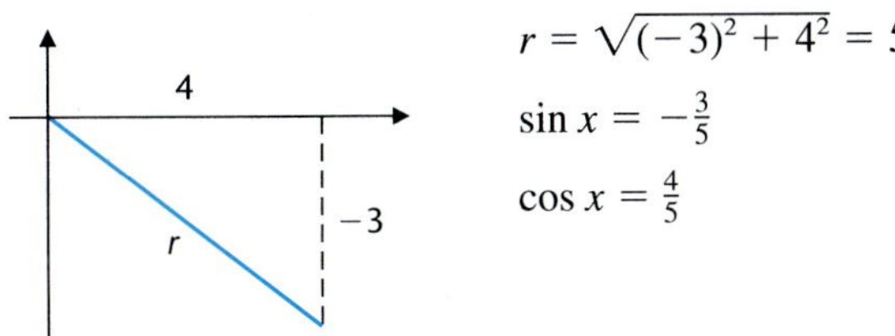

$$r = \sqrt{(-3)^2 + 4^2} = 5$$

$$\sin x = -\tfrac{3}{5}$$

$$\cos x = \tfrac{4}{5}$$

Now use double-angle identities for sine and cosine:

$$\sin 2x = 2 \sin x \cos x = 2(-\tfrac{3}{5})(\tfrac{4}{5}) = -\tfrac{24}{25}$$

$$\cos 2x = 2 \cos^2 x - 1 = 2(\tfrac{4}{5})^2 - 1 = \tfrac{7}{25}$$

Matched Problem 2 Find the exact values, without using a calculator, of $\cos 2x$ and $\tan 2x$ if $\sin x = \frac{4}{5}$ and x is a quadrant II angle.

• Half-Angle Identities

Half-angle identities are simply double-angle identities stated in an alternate form. Let's start with the double-angle identity for cosine in the form

$$\cos 2m = 1 - 2\sin^2 m$$

Now replace m with $x/2$ and solve for $\sin(x/2)$ [if $2m$ is twice m, then m is half of $2m$—think about this]:

$$\cos x = 1 - 2\sin^2 \frac{x}{2}$$

$$\sin^2 \frac{x}{2} = \frac{1 - \cos x}{2}$$

$$\sin \frac{x}{2} = \pm\sqrt{\frac{1 - \cos x}{2}} \qquad \textbf{Half-angle identity for sine} \qquad (7)$$

where the choice of the sign is determined by the quadrant in which $x/2$ lies.

To obtain a half-angle identity for cosine, start with the double-angle identity for cosine in the form

$$\cos 2m = 2\cos^2 m - 1$$

and let $m = x/2$ to obtain

$$\cos \frac{x}{2} = \pm\sqrt{\frac{1 + \cos x}{2}} \qquad \textbf{Half-angle identity for cosine} \qquad (8)$$

where the sign is determined by the quadrant in which $x/2$ lies.

To obtain a *half-angle identity for tangent,* use the quotient identity and the half-angle formulas for sine and cosine:

$$\tan \frac{x}{2} = \frac{\sin \frac{x}{2}}{\cos \frac{x}{2}} = \frac{\pm\sqrt{\frac{1 - \cos x}{2}}}{\pm\sqrt{\frac{1 + \cos x}{2}}} = \pm\sqrt{\frac{1 - \cos x}{1 + \cos x}}$$

Thus,

$$\tan \frac{x}{2} = \pm\sqrt{\frac{1 - \cos x}{1 + \cos x}} \qquad \textbf{Half-angle identity for tangent} \qquad (9)$$

where the sign is determined by the quadrant in which $x/2$ lies.

Simpler versions of equation (9) can be obtained as follows:

$$\left|\tan \frac{x}{2}\right| = \sqrt{\frac{1 - \cos x}{1 + \cos x}} \qquad (10)$$

$$= \sqrt{\frac{1 - \cos x}{1 + \cos x} \cdot \frac{1 + \cos x}{1 + \cos x}}$$

$$= \sqrt{\frac{1 - \cos^2 x}{(1 + \cos x)^2}}$$

$$= \sqrt{\frac{\sin^2 x}{(1 + \cos x)^2}}$$

$$= \frac{\sqrt{\sin^2 x}}{\sqrt{(1 + \cos x)^2}}$$

$$= \frac{|\sin x|}{1 + \cos x}$$

$\sqrt{\sin^2 x} = |\sin x|$ and
$\sqrt{(1 + \cos x)^2} = 1 + \cos x$, since $1 + \cos x$ is never negative.

All absolute value signs can be dropped, since it can be shown that $\tan(x/2)$ and $\sin x$ always have the same sign (a good exercise for you). Thus,

$$\tan \frac{x}{2} = \frac{\sin x}{1 + \cos x} \qquad \textbf{Half-angle identity for tangent} \qquad (11)$$

By multiplying the numerator and the denominator in the radicand in equation (10) by $1 - \cos x$ and reasoning as before, we also can obtain

$$\tan \frac{x}{2} = \frac{1 - \cos x}{\sin x} \qquad \textbf{Half-angle identity for tangent} \qquad (12)$$

We now list all the half-angle identities for convenient reference.

Half-Angle Identities

$$\sin \frac{x}{2} = \pm\sqrt{\frac{1 - \cos x}{2}}$$

$$\cos \frac{x}{2} = \pm\sqrt{\frac{1 + \cos x}{2}}$$

$$\tan \frac{x}{2} = \pm\sqrt{\frac{1 - \cos x}{1 + \cos x}} = \frac{\sin x}{1 + \cos x} = \frac{1 - \cos x}{\sin x}$$

where the sign is determined by the quadrant in which $x/2$ lies.

EXPLORE-DISCUSS 2 (A) Discuss how you would show that, in general,

$$\sin \frac{x}{2} \neq \tfrac{1}{2} \sin x \qquad \cos \frac{x}{2} \neq \tfrac{1}{2} \cos x \qquad \tan \frac{x}{2} \neq \tfrac{1}{2} \tan x$$

(B) Graph $y1 = \sin(x/2)$ and $y2 = \frac{1}{2} \sin x$ in the same viewing window. Conclusion? Repeat the process for the other two statements in part A.

EXAMPLE 3 **Finding Exact Values**

Compute the exact value of sin 165° without a calculator using a half-angle identity.

Solution

$$\sin 165° = \sin \frac{330°}{2}$$

$$= \sqrt{\frac{1 - \cos 330°}{2}}$$ Use half-angle identity for sine with a positive radical, since sin 165° is positive.

$$= \sqrt{\frac{1 - (\sqrt{3}/2)}{2}}$$

$$= \frac{\sqrt{2 - \sqrt{3}}}{2}$$

Matched Problem 3 Compute the exact value of tan 105° without a calculator using a half-angle identity.

EXAMPLE 4 **Finding Exact Values**

Find the exact values of $\cos(x/2)$ and $\cot(x/2)$ without using a calculator if $\sin x = -\frac{3}{5}$, $\pi < x < 3\pi/2$.

Solution Draw a reference triangle in the third quadrant, and find $\cos x$. Then use appropriate half-angle identities.

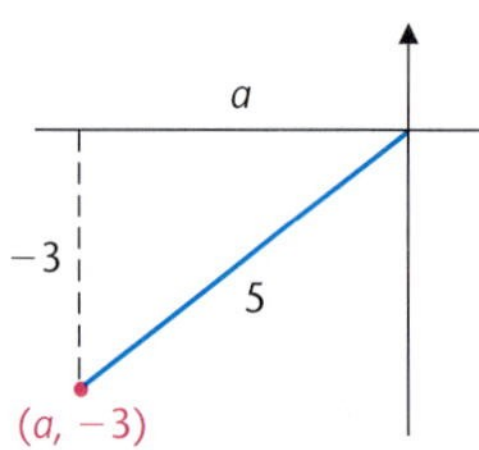

$$a = -\sqrt{5^2 - (-3)^2} = -4$$

$$\cos x = -\tfrac{4}{5}$$

If $\pi < x < 3\pi/2$, then

$$\frac{\pi}{2} < \frac{x}{2} < \frac{3\pi}{4}$$ Divide each member of $\pi < x < 3\pi/2$ by 2.

Thus, $x/2$ is an angle in the second quadrant where cosine and cotangent are negative, and

$$\cos \frac{x}{2} = -\sqrt{\frac{1 + \cos x}{2}}$$

$$= -\sqrt{\frac{1 + (-\frac{4}{5})}{2}}$$

$$= -\sqrt{\frac{1}{10}} \text{ or } \frac{-\sqrt{10}}{10}$$

$$\cot \frac{x}{2} = \frac{1}{\tan(x/2)} = \frac{\sin x}{1 - \cos x}$$

$$= \frac{-\frac{3}{5}}{1 - (-\frac{4}{5})} = -\frac{1}{3}$$

Matched Problem 4 Find the exact values of sin $(x/2)$ and tan $(x/2)$ without using a calculator if $\cot x = -\frac{4}{3}$, $\pi/2 < x < \pi$.

EXAMPLE 5 Identity Verification

Verify the identity: $\sin^2 \frac{x}{2} = \frac{\tan x - \sin x}{2 \tan x}$

Verification

$$\sin \frac{x}{2} = \pm\sqrt{\frac{1 - \cos x}{2}} \qquad \text{Half-angle identity for sine}$$

$$\sin^2 \frac{x}{2} = \frac{1 - \cos x}{2} \qquad \text{Square both sides.}$$

$$= \frac{\tan x}{\tan x} \cdot \frac{1 - \cos x}{2} \qquad \text{Algebra}$$

$$= \frac{\tan x - \tan x \cos x}{2 \tan x} \qquad \text{Algebra}$$

$$= \frac{\tan x - \sin x}{2 \tan x} \qquad \text{Quotient identity}$$

Matched Problem 5 Verify the identity: $\cos^2 \frac{x}{2} = \frac{\tan x + \sin x}{2 \tan x}$

Answers to Matched Problems

1. $\frac{2 \tan x}{1 + \tan^2 x} = \frac{2\left(\frac{\sin x}{\cos x}\right)}{1 + \frac{\sin^2 x}{\cos^2 x}} = \frac{\cos^2 x\left[2\left(\frac{\sin x}{\cos x}\right)\right]}{\cos^2 x\left(1 + \frac{\sin^2 x}{\cos^2 x}\right)} = \frac{2 \sin x \cos x}{\cos^2 x + \sin^2 x} = 2 \sin x \cos x = \sin 2x$

2. $\cos 2x = -\frac{7}{25}$, $\tan 2x = \frac{24}{7}$ 3. $-\sqrt{3} - 2$ 4. $\sin (x/2) = 3\sqrt{10}/10$, $\tan (x/2) = 3$

5. $\cos^2 \frac{x}{2} = \frac{1 + \cos x}{2} = \frac{\tan x}{\tan x} \cdot \frac{1 + \cos x}{2} = \frac{\tan x + \tan x \cos x}{2 \tan x} = \frac{\tan x + \sin x}{2 \tan x}$

EXERCISE 6-3

A

In Problems 1–6, verify each identity for the values indicated.

1. $\cos 2x = \cos^2 x - \sin^2 x$, $x = 30°$

2. $\sin 2x = 2 \sin x \cos x$, $x = 45°$

3. $\tan 2x = \frac{2}{\cot x - \tan x}$, $x = \frac{\pi}{3}$

4. $\tan 2x = \frac{2 \tan x}{1 - \tan^2 x}$, $x = \frac{\pi}{6}$

5. $\sin \frac{x}{2} = \pm\sqrt{\frac{1 - \cos x}{2}}$, $x = \pi$

(Choose the correct sign.)

6. $\cos \frac{x}{2} = \pm\sqrt{\frac{1 + \cos x}{2}}$, $x = \frac{\pi}{2}$

(Choose the correct sign.)

In Problems 7–10, find the exact value without a calculator using half-angle identities.

7. $\tan 15°$ **8.** $\sin 165°$

9. $\cos 112.5°$ **10.** $\tan 157.5°$

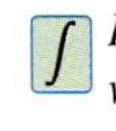

In Problems 11–14, graph y1 and y2 in the same viewing window for $-2\pi \le x \le 2\pi$. Use TRACE to compare the two graphs.

11. $y1 = \cos 2x,\ y2 = \cos^2 x - \sin^2 x$

12. $y1 = \sin 2x,\ y2 = 2 \sin x \cos x$

13. $y1 = \tan \frac{x}{2},\ y2 = \frac{\sin x}{1 + \cos x}$

14. $y1 = \tan 2x,\ y2 = \frac{2 \tan x}{1 - \tan^2 x}$

B

Verify the identities in Problems 15–28.

15. $(\sin x + \cos x)^2 = 1 + \sin 2x$

16. $\sin 2x = (\tan x)(1 + \cos 2x)$

17. $\sin^2 x = \frac{1}{2}(1 - \cos 2x)$

18. $\cos^2 x = \frac{1}{2}(\cos 2x + 1)$

19. $1 - \cos 2x = \tan x \sin 2x$

20. $1 + \sin 2t = (\sin t + \cos t)^2$

21. $\sin^2 \frac{x}{2} = \frac{1 - \cos x}{2}$ **22.** $\cos^2 \frac{x}{2} = \frac{1 + \cos x}{2}$

23. $\cot \frac{\theta}{2} = \frac{\sin \theta}{1 - \cos \theta}$ **24.** $\cot \frac{\theta}{2} = \frac{1 + \cos \theta}{\sin \theta}$

25. $\cos 2u = \frac{1 - \tan^2 u}{1 + \tan^2 u}$ **26.** $\frac{\cos 2u}{1 - \sin 2u} = \frac{1 + \tan u}{1 - \tan u}$

27. $2 \csc 2x = \frac{1 + \tan^2 x}{\tan x}$ **28.** $\sec 2x = \frac{\sec^2 x}{2 - \sec^2 x}$

In Problems 29–34, is the equation an identity? Explain.

29. $\cos 2x = 2 \sin x \cos x$ **30.** $\sin 4x = 4 \sin x \cos x$

31. $\tan 2x = \frac{-2 \tan x}{\tan^2 x - 1}$ **32.** $\tan 6x = \frac{6 \tan x}{1 - \tan^2 x}$

33. $\cot 2x = \frac{2 \cot x}{1 - \cot^2 x}$ **34.** $2 \csc 2x = \sec x \csc x$

Compute the exact values of sin 2x, cos 2x, and tan 2x using the information given in Problems 35–38 and appropriate identities. Do not use a calculator.

35. $\sin x = \frac{3}{5},\ \pi/2 < x < \pi$ **36.** $\cos x = -\frac{4}{5},\ \pi/2 < x < \pi$

37. $\tan x = -\frac{5}{12},\ -\pi/2 < x < 0$

38. $\cot x = -\frac{5}{12},\ -\pi/2 < x < 0$

In Problems 39–42, compute the exact values of sin (x/2), cos (x/2), and tan (x/2) using the information given and appropriate identities. Do not use a calculator.

39. $\sin x = -\frac{1}{3},\ \pi < x < 3\pi/2$

40. $\cos x = -\frac{1}{4},\ \pi < x < 3\pi/2$

41. $\cot x = \frac{3}{4},\ -\pi < x < -\pi/2$

42. $\tan x = \frac{3}{4},\ -\pi < x < -\pi/2$

Suppose you are tutoring a student who is having difficulties in finding the exact values of sin θ and cos θ from the information given in Problems 43 and 44. Assuming that you have worked through each problem and have identified the key steps in the solution process, proceed with your tutoring by guiding the student through the solution process using the following questions. Record the expected correct responses from the student.

(A) The angle 2θ is in what quadrant and how do you know?

(B) How can you find sin 2θ and cos 2θ? Find each.

(C) What identities relate sin θ and cos θ with either sin 2θ or cos 2θ?

(D) How would you use the identities in part C to find sin θ and cos θ exactly, including the correct sign?

(E) What are the exact values for sin θ and cos θ?

43. Find the exact values of $\sin \theta$ and $\cos \theta$, given $\tan 2\theta = -\frac{4}{3}$, $0° < \theta < 90°$.

44. Find the exact values of $\sin \theta$ and $\cos \theta$, given $\sec 2\theta = -\frac{5}{4}$, $0° < \theta < 90°$.

Verify each of the following identities for the value of x indicated in Problems 45–48. Compute values to 5 significant digits using a calculator.

(A) $\tan 2x = \frac{2 \tan x}{1 - \tan^2 x}$

(B) $\cos \frac{x}{2} = \pm\sqrt{\frac{1 + \cos x}{2}}$

(Choose the correct sign.)

45. $x = 252.06°$ **46.** $x = 72.358°$

47. $x = 0.934\ 57$ **48.** $x = 4$

In Problems 49–52, graph y1 and y2 in the same viewing window for $-2\pi \le x \le 2\pi$, and state the intervals for which the equation y1 = y2 is an identity.

49. $y1 = \cos (x/2),\ y2 = \sqrt{\frac{1 + \cos x}{2}}$

50. $y1 = \cos (x/2),\ y2 = -\sqrt{\frac{1 + \cos x}{2}}$

51. $y1 = \sin (x/2),\ y2 = -\sqrt{\frac{1 - \cos x}{2}}$

52. $y1 = \sin (x/2),\ y2 = \sqrt{\frac{1 - \cos x}{2}}$

C

Verify the identities in Problems 53–56.

53. $\cos 3x = 4\cos^3 x - 3\cos x$

54. $\sin 3x = 3\sin x - 4\sin^3 x$

55. $\cos 4x = 8\cos^4 x - 8\cos^2 x + 1$

56. $\sin 4x = (\cos x)(4\sin x - 8\sin^3 x)$

In Problems 57–62, find the exact value of each without using a calculator.

57. $\cos(2\cos^{-1}\frac{3}{5})$

58. $\sin(2\cos^{-1}\frac{3}{5})$

59. $\tan[2\cos^{-1}(-\frac{4}{5})]$

60. $\tan[2\tan^{-1}(-\frac{3}{4})]$

61. $\cos[\frac{1}{2}\cos^{-1}(-\frac{3}{5})]$

62. $\sin[\frac{1}{2}\tan^{-1}(-\frac{4}{3})]$

In Problems 63–68, graph f(x) in a graphing utility, find a simpler function g(x) that has the same graph as f(x), and verify the identity f(x) = g(x). [Assume g(x) = k + AT(Bx), where T(x) is one of the six trigonometric functions.]

63. $f(x) = \csc x - \cot x$

64. $f(x) = \csc x + \cot x$

65. $f(x) = \dfrac{1 - 2\cos 2x}{2\sin x - 1}$

66. $f(x) = \dfrac{1 + 2\cos 2x}{1 + 2\cos x}$

67. $f(x) = \dfrac{1}{\cot x \sin 2x - 1}$

68. $f(x) = \dfrac{\cot x}{1 + \cos 2x}$

APPLICATIONS

★**69. Indirect Measurement.** Find the exact value of x in the figure; then find x and θ to three decimal places. [*Hint:* Use $\cos 2\theta = 2\cos^2\theta - 1$.]

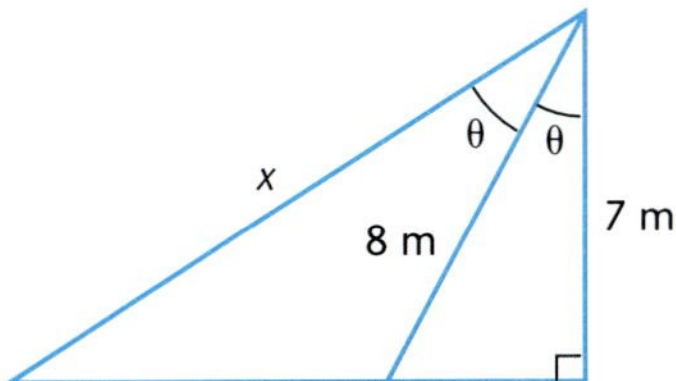

★**70. Indirect Measurement.** Find the exact value of x in the figure; then find x and θ to three decimal places. [*Hint:* Use $\tan 2\theta = (2\tan\theta)/(1 - \tan^2\theta)$.]

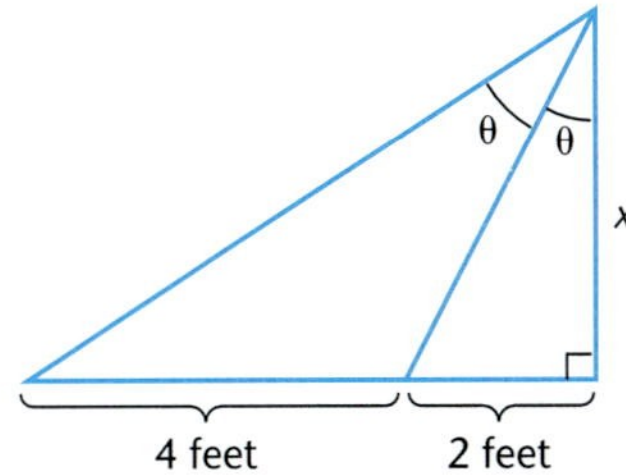

★**71. Sports—Physics.** The theoretical distance d that a shot-putter, discus thrower, or javelin thrower can achieve on a given throw is found in physics to be given approximately by

$$d = \frac{2v_0^2 \sin\theta\cos\theta}{32 \text{ feet per second per second}}$$

where v_0 is the initial speed of the object thrown (in feet per second) and θ is the angle above the horizontal at which the object leaves the hand (see the figure).

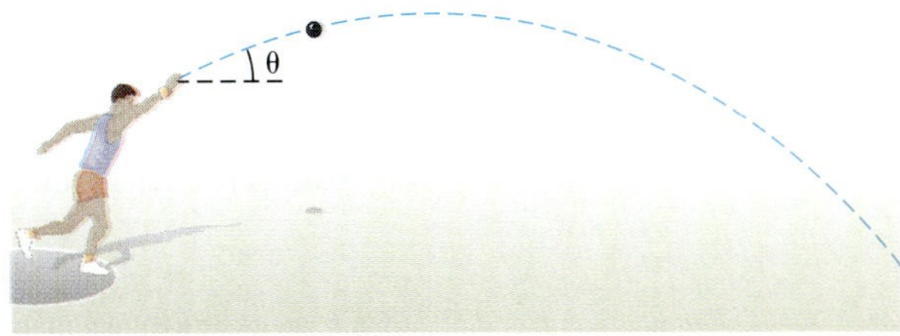

(A) Write the formula in terms of $\sin 2\theta$ by using a suitable identity.

(B) Using the resulting equation in part A, determine the angle θ that will produce the maximum distance d for a given initial speed v_0. This result is an important consideration for shot-putters, javelin throwers, and discus throwers.

72. Geometry. In part (a) of the figure, M and N are the midpoints of the sides of a square. Find the exact value of $\cos\theta$. [*Hint:* The solution uses the Pythagorean theorem, the definition of sine and cosine, a half-angle identity, and some auxiliary lines as drawn in part (b) of the figure.]

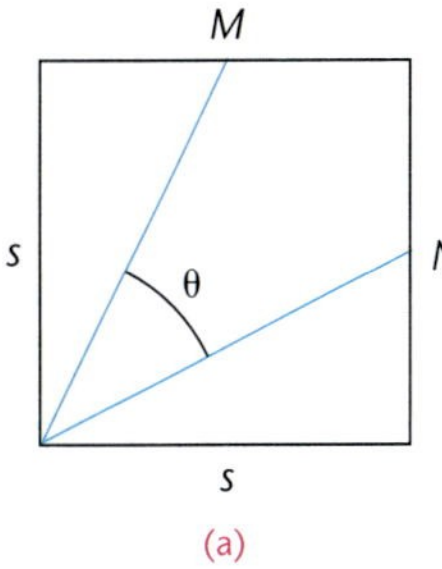

(a)

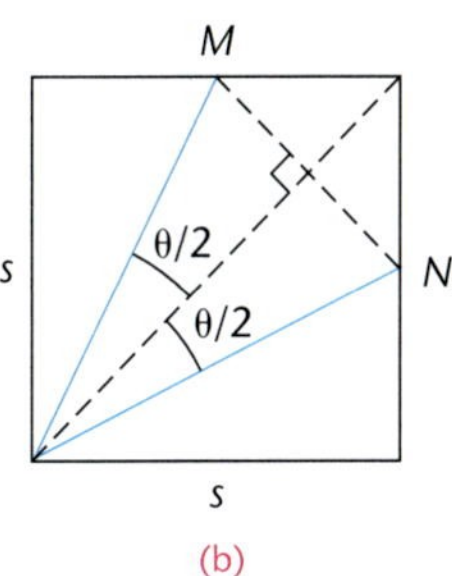

(b)

73. Area. An n-sided regular polygon is inscribed in a circle of radius R.

(A) Show that the area of the n-sided polygon is given by

$$A_n = \tfrac{1}{2}nR^2 \sin\frac{2\pi}{n}$$

[*Hint:* Area of a triangle $= \frac{1}{2}$ (base)(altitude). Also, a double-angle identity is useful.]

(B) For a circle of radius 1, complete Table 1, to five decimal places, using the formula in part A.

TABLE 1

n	10	100	1,000	10,000
A_n				

(C) What number does A_n seem to approach as n increases without bound? (What is the area of a circle of radius 1?)

(D) Will A_n exactly equal the area of the circumscribed circle for some sufficiently large n? How close can A_n be made to get to the area of the circumscribed circle. [In calculus, the area of the circumscribed circle is called the *limit* of A_n as n increases without bound. In symbols, for a circle of radius 1, we would write $\lim_{n\to\infty} A_n = \pi$. The limit concept is the cornerstone on which calculus is constructed.]

SECTION 6-4 Product–Sum and Sum–Product Identities

- Product–Sum Identities
- Sum–Product Identities

Our work with identities is concluded by developing the *product–sum* and *sum–product identities,* which are easily derived from the sum and difference identities developed in Section 6-2. These identities are used in calculus to convert product forms to more convenient sum forms. They also are used in the study of sound waves in music to convert sum forms to more convenient product forms.

• Product–Sum Identities

First, add, left side to left side and right side to right side, the sum and difference identities for sine:

$$\sin(x+y) = \sin x \cos y + \cos x \sin y$$
$$\sin(x-y) = \sin x \cos y - \cos x \sin y$$
$$\sin(x+y) + \sin(x-y) = 2\sin x \cos y$$

or

$$\sin x \cos y = \tfrac{1}{2}[\sin(x+y) + \sin(x-y)]$$

Similarly, by adding or subtracting the appropriate sum and difference identities, we can obtain three other **product–sum identities.** These are listed below for convenient reference.

Product–Sum Identities

$$\sin x \cos y = \tfrac{1}{2}[\sin(x+y) + \sin(x-y)]$$
$$\cos x \sin y = \tfrac{1}{2}[\sin(x+y) - \sin(x-y)]$$
$$\sin x \sin y = \tfrac{1}{2}[\cos(x-y) - \cos(x+y)]$$
$$\cos x \cos y = \tfrac{1}{2}[\cos(x+y) + \cos(x-y)]$$

EXAMPLE 1 A Product as a Difference

Write the product $\cos 3t \sin t$ as a sum or difference.

Solution

$$\cos x \sin y = \tfrac{1}{2}[\sin (x + y) - \sin (x - y)] \qquad \text{Let } x = 3t \text{ and } y = t.$$

$$\cos 3t \sin t = \tfrac{1}{2}[\sin (3t + t) - \sin (3t - t)]$$

$$= \tfrac{1}{2}\sin 4t - \tfrac{1}{2}\sin 2t$$

Matched Problem 1 Write the product $\cos 5\theta \cos 2\theta$ as a sum or difference.

EXAMPLE 2 Finding Exact Values

Evaluate $\sin 105° \sin 15°$ exactly using an appropriate product–sum identity.

Solution

$$\sin x \sin y = \tfrac{1}{2}[\cos (x - y) - \cos (x + y)]$$

$$\sin 105° \sin 15° = \tfrac{1}{2}[\cos (105° - 15°) - \cos (105° + 15°)]$$

$$= \tfrac{1}{2}[\cos 90° - \cos 120°]$$

$$= \tfrac{1}{2}[0 - (-\tfrac{1}{2})] = \tfrac{1}{4}, \text{ or } 0.25$$

Matched Problem 2 Evaluate $\cos 165° \sin 75°$ exactly using an appropriate product–sum identity.

• Sum–Product Identities

The product–sum identities can be transformed into equivalent forms called **sum–product identities.** These identities are used to express sums and differences involving sines and cosines as products involving sines and cosines. We illustrate the transformation for one identity. The other three identities can be obtained by following similar procedures.

We start with a product–sum identity:

$$\sin \alpha \cos \beta = \tfrac{1}{2}[\sin (\alpha + \beta) + \sin (\alpha - \beta)] \qquad (1)$$

We would like

$$\alpha + \beta = x$$

$$\alpha - \beta = y$$

Solving this system, we have

$$\alpha = \frac{x + y}{2} \qquad \beta = \frac{x - y}{2} \qquad (2)$$

Substituting (2) into equation (1) and simplifying, we obtain

$$\sin x + \sin y = 2 \sin \frac{x + y}{2} \cos \frac{x - y}{2}$$

All four sum–product identities are listed below for convenient reference.

Sum–Product Identities

$$\sin x + \sin y = 2 \sin \frac{x+y}{2} \cos \frac{x-y}{2}$$

$$\sin x - \sin y = 2 \cos \frac{x+y}{2} \sin \frac{x-y}{2}$$

$$\cos x + \cos y = 2 \cos \frac{x+y}{2} \cos \frac{x-y}{2}$$

$$\cos x - \cos y = -2 \sin \frac{x+y}{2} \sin \frac{x-y}{2}$$

EXAMPLE 3 A Difference as a Product

Write the difference $\sin 7\theta - \sin 3\theta$ as a product.

Solution

$$\sin x - \sin y = 2 \cos \frac{x+y}{2} \sin \frac{x-y}{2}$$

$$\sin 7\theta - \sin 3\theta = 2 \cos \frac{7\theta + 3\theta}{2} \sin \frac{7\theta - 3\theta}{2}$$

$$= 2 \cos 5\theta \sin 2\theta$$

Matched Problem 3 Write the sum $\cos 3t + \cos t$ as a product.

EXAMPLE 4 Finding Exact Values

Find the exact value of $\sin 105° - \sin 15°$ using an appropriate sum–product identity.

Solution

$$\sin x - \sin y = 2 \cos \frac{x+y}{2} \sin \frac{x-y}{2}$$

$$\sin 105° - \sin 15° = 2 \cos \frac{105° + 15°}{2} \sin \frac{105° - 15°}{2}$$

$$= 2 \cos 60° \sin 45°$$

$$= 2\left(\frac{1}{2}\right)\left(\frac{\sqrt{2}}{2}\right) = \frac{\sqrt{2}}{2}$$

Matched Problem 4 Find the exact value of $\cos 165° - \cos 75°$ using an appropriate sum–product identity.

EXPLORE-DISCUSS 1 The following "proof without words" of two of the sum–product identities is based on a similar "proof" by Sidney H. Kung, Jacksonville University, that was printed in the October 1996 issue of *Mathematics Magazine.* Discuss how the relationships below the figure are verified from the figure.

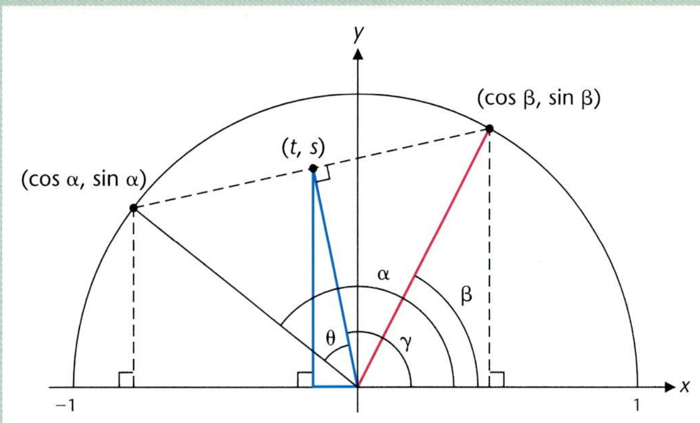

$$\theta = \frac{\alpha - \beta}{2} \qquad \gamma = \frac{\alpha + \beta}{2}$$

$$\frac{\sin\alpha + \sin\beta}{2} = s = \cos\frac{\alpha - \beta}{2}\sin\frac{\alpha + \beta}{2}$$

$$\frac{\cos\alpha + \cos\beta}{2} = t = \cos\frac{\alpha - \beta}{2}\cos\frac{\alpha + \beta}{2}$$

Answers to Matched Problems

1. $\frac{1}{2}\cos 7\theta + \frac{1}{2}\cos 3\theta$ **2.** $(-\sqrt{3} - 2)/4$ **3.** $2\cos 2t\cos t$ **4.** $-\sqrt{6}/2$

EXERCISE 6-4

A

In Problems 1–4, write each product as a sum or difference involving sine and cosine.

1. $\sin 3m \cos m$ **2.** $\cos 7A \cos 5A$

3. $\sin u \sin 3u$ **4.** $\cos 2\theta \sin 3\theta$

In Problems 5–8, write each difference or sum as a product involving sines and cosines.

5. $\sin 3t + \sin t$ **6.** $\cos 7\theta + \cos 5\theta$

7. $\cos 5w - \cos 9w$ **8.** $\sin u - \sin 5u$

B

Evaluate Problems 9–12 exactly using an appropriate identity.

9. $\sin 75° \sin 15°$ **10.** $\cos 105° \sin 165°$

11. $\cos 7.5° \cos 52.5°$ **12.** $\sin 82.5° \cos 37.5°$

Evaluate Problems 13–16 exactly using an appropriate identity.

13. $\sin 195° - \sin 105°$

14. $\cos 75° + \cos 15°$

15. $\cos 165° - \cos 105°$

16. $\sin 285° + \sin 195°$

Use sum and difference identities to verify the identities in Problems 17 and 18.

17. $\cos x \cos y = \frac{1}{2}[\cos(x + y) + \cos(x - y)]$

18. $\sin x \sin y = \frac{1}{2}[\cos(x - y) - \cos(x + y)]$

19. Explain how you can transform the product–sum identity

$$\sin u \sin v = \tfrac{1}{2}[\cos(u - v) - \cos(u + v)]$$

into the sum–product identity

$$\cos x - \cos y = -2 \sin \frac{x + y}{2} \sin \frac{x - y}{2}$$

by a suitable substitution.

20. Explain how you can transform the product–sum identity

$$\cos u \cos v = \tfrac{1}{2}[\cos(u + v) + \cos(u - v)]$$

into the sum–product identity

$$\cos x + \cos y = 2 \cos \frac{x + y}{2} \cos \frac{x - y}{2}$$

by a suitable substitution.

Verify each identity in Problems 21–24.

21. $\dfrac{\sin 2t + \sin 4t}{\cos 2t - \cos 4t} = \cot t$

22. $\dfrac{\cos t - \cos 3t}{\sin t + \sin 3t} = \tan t$

23. $\dfrac{\sin x - \sin y}{\cos x - \cos y} = -\cot \dfrac{x + y}{2}$

24. $\dfrac{\sin x + \sin y}{\cos x + \cos y} = \tan \dfrac{x + y}{2}$

In Problems 25–30, is the equation an identity? Explain.

25. $2 \sin 2x \sin x = \cos x - \cos 3x$

26. $(\sin 2x + \sin 2y) \tan(x - y) = \cos 2x - \cos 2y$

27. $\cos 2x = \sin(x + y)\cos(x - y) - \cos(x + y)\sin(x - y)$

28. $\sin 2x = \sin(x + y)\cos(x - y) + \cos(x + y)\sin(x - y)$

29. $\sec(2x + y)\csc(2x - y)[\sin 4x + \sin 2y] = 2$

30. $\sec x \sec y [\cos(x + y) + \cos(x - y)] = 2$

Verify each of the following identities for the values of x and y indicated in Problems 31–34. Evaluate each side to 5 significant digits.

(A) $\cos x \sin y = \frac{1}{2}[\sin(x + y) - \sin(x - y)]$

(B) $\cos x + \cos y = 2 \cos \dfrac{x + y}{2} \cos \dfrac{x - y}{2}$

31. $x = 172.63°$, $y = 20.177°$

32. $x = 50.137°$, $y = 18.044°$

33. $x = 1.1255$, $y = 3.6014$

34. $x = 0.039\,17$, $y = 0.610\,52$

In Problems 35–42, write each as a product if y is a sum or difference, or as a sum or difference if y is a product. Enter the original equation in a graphing utility as y1, the converted form as y2, and graph y1 and y2 in the same viewing window. Use TRACE to compare the two graphs.

35. $y = \sin 2x + \sin x$

36. $y = \cos 3x + \cos x$

37. $y = \cos 1.7x - \cos 0.3x$

38. $y = \sin 2.1x - \sin 0.5x$

39. $y = \sin 3x \cos x$

40. $y = \cos 5x \cos 3x$

41. $y = \sin 2.3x \sin 0.7x$

42. $y = \cos 1.9x \sin 0.5x$

C

Verify each identity in Problems 43 and 44.

43. $\cos x \cos y \cos z = \frac{1}{4}[\cos(x + y - z) + \cos(y + z - x) + \cos(z + x - y) + \cos(x + y + z)]$

44. $\sin x \sin y \sin z = \frac{1}{4}[\sin(x + y - z) + \sin(y + z - x) + \sin(z + x - y) - \sin(x + y + z)]$

In Problems 45–48:

(A) Graph y1, y2, and y3 in a graphing utility for $0 \le x \le 1$ and $-2 \le y \le 2$.

(B) Convert y1 to a sum or difference and repeat part A.

45. $y1 = 2\cos(28\pi x)\cos(2\pi x)$
$y2 = 2\cos(2\pi x)$
$y3 = -2\cos(2\pi x)$

46. $y1 = 2\sin(24\pi x)\sin(2\pi x)$
$y2 = 2\sin(2\pi x)$
$y3 = -2\sin(2\pi x)$

47. $y1 = 2\sin(20\pi x)\cos(2\pi x)$
$y2 = 2\cos(2\pi x)$
$y3 = -2\cos(2\pi x)$

48. $y1 = 2\cos(16\pi x)\sin(2\pi x)$
$y2 = 2\sin(2\pi x)$
$y3 = -2\sin(2\pi x)$

APPLICATIONS

Problems 49 and 50 involve the phenomena of sound called beats. If two tones having the same loudness and close together in pitch (frequency) are sounded, one following the other, most people would have difficulty in differentiating the

two tones. However, if the tones are sounded simultaneously, they will interact with each other, producing a low warbling sound called a **beat.** *Musicians, when tuning an instrument with other instruments or a tuning fork, listen for these lower-beat frequencies and try to eliminate them by adjusting their instruments. Problems 49 and 50 provide a visual illustration of the beat phenomena.*

49. Music—Beat Frequencies. Equations $y = 0.5 \cos 128\pi t$ and $y = -0.5 \cos 144\pi t$ are equations of sound waves with frequencies 64 and 72 hertz, respectively. If both sounds are emitted simultaneously, a *beat* frequency results.

(A) Show that

$$0.5 \cos 128\pi t - 0.5 \cos 144\pi t = \sin 8\pi t \sin 136\pi t$$

(The product form is more useful to sound engineers.)

(B) Graph each equation in a different viewing window for $0 \le t \le 0.25$:

$$y = 0.5 \cos 128\pi t$$

$$y = -0.5 \cos 144\pi t$$

$$y = 0.5 \cos 128\pi t - 0.5 \cos 144\pi t$$

$$y = \sin 8\pi t \sin 136\pi t$$

50. Music—Beat Frequencies. $y = 0.25 \cos 256\pi t$ and $y = -0.25 \cos 288\pi t$ are equations of sound waves with frequencies 128 and 144 hertz, respectively. If both sounds are emitted simultaneously, a *beat* frequency results.

(A) Show that

$$0.25 \cos 256\pi t - 0.25 \cos 288\pi t = 0.5 \sin 16\pi t \sin 272\pi t$$

(The product form is more useful to sound engineers.)

(B) Graph each equation in a different viewing window for $0 \le t \le 0.125$:

$$y = 0.25 \cos 256\pi t$$

$$y = -0.25 \cos 288\pi t$$

$$y = 0.25 \cos 256\pi t - 0.25 \cos 288\pi t$$

$$y = 0.5 \sin 16\pi t \sin 272\pi t$$

SECTION 6-5 Trigonometric Equations

- Solving Trigonometric Equations Using an Algebraic Approach
- Solving Trigonometric Equations Using a Graphing Utility

The first four sections of this chapter consider trigonometric equations called *identities.* These are equations that are true for all replacements of the variable(s) for which both sides are defined. We now consider another class of trigonometric equations, called **conditional equations,** which may be true for some replacements of the variable but false for others. For example,

$$\cos x = \sin x$$

is a conditional equation, since it is true for some values, for example, $x = \pi/4$, and false for others, such as $x = 0$. (Check both values.)

This section considers two approaches for solving conditional trigonometric equations: an algebraic approach and a graphing utility approach. Solving trigonometric equations using an algebraic approach often requires the use of algebraic manipulation, identities, and ingenuity. In some cases algebraic methods lead to exact solutions, which are very useful in certain contexts. Graphing utility methods can be used to approximate solutions to a greater variety of trigonometric equations but usually do not produce exact solutions. Each method has its strengths.

EXPLORE-DISCUSS 1 We are interested in solutions to the equation

$$\cos x = 0.5$$

The figure below shows a partial graph of the left and right sides of the equation.

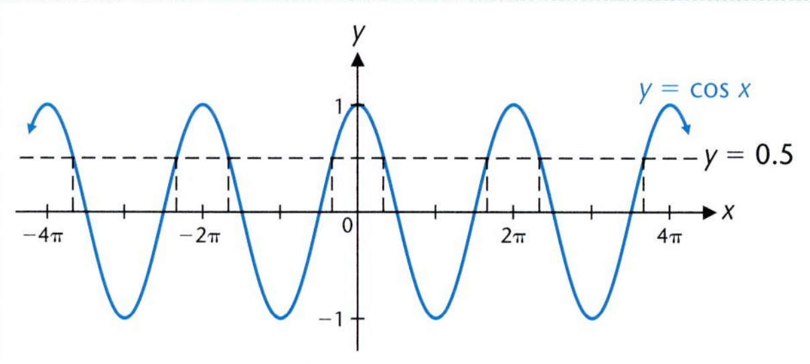

(A) How many solutions does the equation have on the interval $[0, 2\pi)$? What are they?

(B) How many solutions does the equation have on the interval $(-\infty, \infty)$? Discuss a method of writing all solutions to the equation.

• Solving Trigonometric Equations Using an Algebraic Approach

You might find the following suggestions for solving trigonometric equations using an algebraic approach useful:

Suggestions for Solving Trigonometric Equations Algebraically

1. Regard one particular trigonometric function as a variable, and solve for it.

(a) Consider using algebraic manipulation such as factoring, combining or separating fractions, and so on.

(b) Consider using identities.

2. After solving for a trigonometric function, solve for the variable.

A number of examples should help make the algebraic approach clear.

EXAMPLE 1 **Exact Solutions Using Factoring**

Find all solutions exactly for $2\cos^2 x - \cos x = 0$.

Solution ***Step 1.*** *Solve for* $\cos x$*:*

$$2\cos^2 x - \cos x = 0 \quad \text{Factor out } \cos x.$$

$$\cos x\,(2\cos x - 1) = 0 \quad ab = 0 \text{ only if } a = 0 \text{ or } b = 0.$$

$$\cos x = 0 \quad \text{or} \quad 2\cos x - 1 = 0$$

$$\cos x = \tfrac{1}{2}$$

Step 2. *Solve each equation over one period [0, 2π):* Sketch a graph of $y = \cos x$, $y = 0$, and $y = \frac{1}{2}$ in the same coordinate system to provide an aid to writing all solutions over one period (Fig. 1).

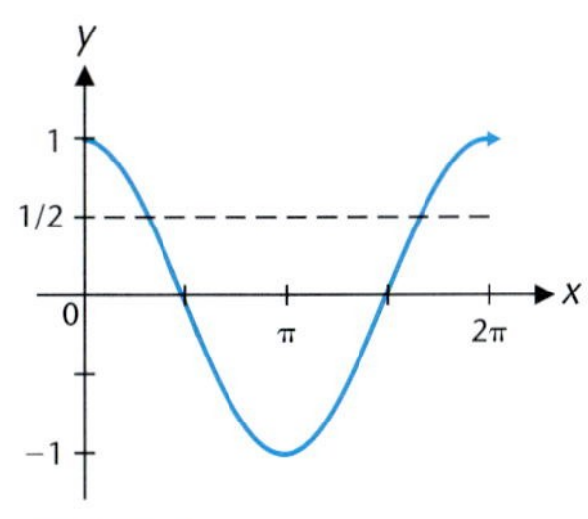

FIGURE 1

$$\cos x = 0 \qquad \cos x = \tfrac{1}{2}$$

$$x = \frac{\pi}{2}, \frac{3\pi}{2} \qquad x = \frac{\pi}{3}, \frac{5\pi}{3}$$

Step 3. *Write an expression for all solutions.* Since the cosine function is periodic with period 2π, all solutions are given by

$$x = \begin{cases} \pi/3 + 2k\pi \\ \pi/2 + 2k\pi \\ 3\pi/2 + 2k\pi \\ 5\pi/3 + 2k\pi \end{cases} \quad k \text{ any integer}$$

Matched Problem 1 Find all solutions exactly for $2\sin^2 x + \sin x = 0$.

EXAMPLE 2 Approximate Solutions Using Identities and Factoring

Find all real solutions for $3\cos^2 x + 8\sin x = 7$. Compute all inverse functions to four decimal places.

Solution

Step 1. *Solve for sin x and/or cos x.* Move all nonzero terms to the left of the equal sign and express the left side in terms of $\sin x$:

$$3\cos^2 x + 8\sin x = 7$$

$$3\cos^2 x + 8\sin x - 7 = 0 \quad \cos^2 x = 1 - \sin^2 x$$

$$3(1 - \sin^2 x) + 8\sin x - 7 = 0$$

$$3\sin^2 x - 8\sin x + 4 = 0 \quad 3u^2 - 8u + 4 = (u - 2)(3u - 2)$$

$$(\sin x - 2)(3\sin x - 2) = 0 \quad ab = 0 \text{ only if } a = 0 \text{ or } b = 0$$

$$\sin x - 2 = 0 \quad \text{or} \quad 3\sin x - 2 = 0$$

$$\sin x = 2 \qquad \sin x = \tfrac{2}{3}$$

Step 2. *Solve each equation over one period [0, 2π).* Sketch a graph of

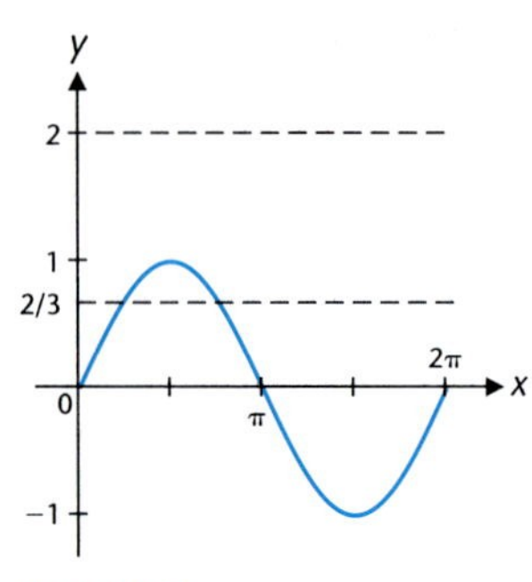

FIGURE 2

$y = \sin x$, $y = 2$, and $y = \frac{2}{3}$ in the same coordinate system to provide an aid to writing all solutions over one period (Fig. 2).

Solve the first equation:

$$\sin x = 2 \quad \text{No solution, since } -1 \le \sin x \le 1.$$

Solve the second equation:

$$\sin x = \tfrac{2}{3}$$ From the graph we see there are solutions in the first and second quadrants.

$$x = \sin^{-1}\tfrac{2}{3} = 0.7297$$ First-quadrant solution

$$x = \pi - 0.7297 = 2.4119$$ Second-quadrant solution

Check $\sin 0.7297 = 0.6667$; $\sin 2.4119 = 0.6666$ (Checks may not be exact because of round-off errors.)

Step 3. Write an expression for all solutions. Since the sine function is periodic with period 2π, all solutions are given by

$$x = \begin{cases} 0.7297 + 2k\pi \\ 2.4119 + 2k\pi \end{cases} \quad k \text{ any integer}$$

Matched Problem 2 Find all real solutions to $8 \sin^2 x = 5 - 10 \cos x$. Compute all inverse functions to four decimal places.

EXAMPLE 3 Approximate Solutions Using Substitution

Find θ in degree measure to three decimal places so that

$$5 \sin (2\theta - 5) = -3.045 \qquad 0° \le 2\theta - 5 \le 360°$$

Solution *Step 1. Make a substitution.* Let $u = 2\theta - 5$ to obtain

$$5 \sin u = -3.045 \qquad 0° \le u \le 360°$$

Step 2. Solve for sin u.

$$\sin u = \frac{-3.045}{5} = -0.609$$

Step 3. Solve for u over $0° \le u \le 360°$. Sketch a graph of $y = \sin u$ and $y = -0.609$ in the same coordinate system to provide an aid to writing all solutions over $0° \le u \le 360°$ (Fig. 3).

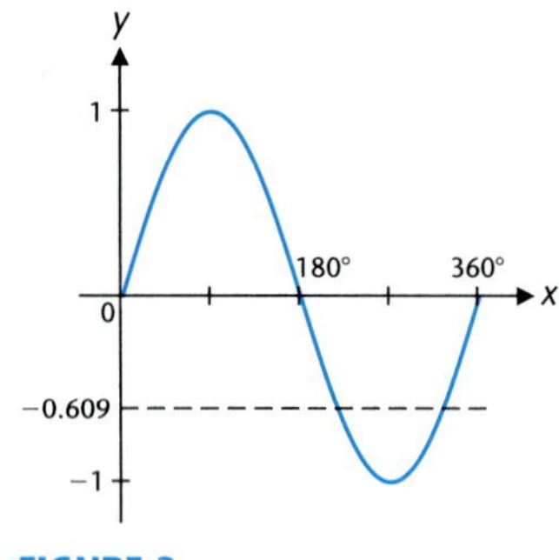

FIGURE 3

Solutions are in the third and fourth quadrants. If the reference angle is α, then $u = 180° + \alpha$ or $u = 360° - \alpha$.

$$\alpha = \sin^{-1}0.609 = 37.517° \quad \text{Reference angle.}$$
$$u = 180° + 37.517°$$
$$= 217.517° \quad \text{Third-quadrant solution}$$
$$u = 360° - 37.517°$$
$$= 322.483° \quad \text{Fourth-quadrant solution}$$

Check $\sin 217.517° = -0.609 \qquad \sin 322.483° = -0.609$

Step 4. Now solve for θ:

$$u = 217.517° \qquad u = 322.483°$$
$$2\theta - 5 = 217.517° \qquad 2\theta - 5 = 322.483°$$
$$\boldsymbol{\theta = 111.259°} \qquad \boldsymbol{\theta = 163.742°}$$

A final check in the original equation is left to the reader.

Matched Problem 3 Find θ in degree measure to three decimal places so that

$$8 \tan (6\theta + 15) = -64.328 \qquad -90° < 6\theta + 15 < 90°$$

EXAMPLE 4 Exact Solutions Using Identities and Factoring

Find exact solutions for $\sin^2 x = \frac{1}{2} \sin 2x$, $0 \le x < 2\pi$.

Solution The following solution includes only the key steps. Sketch graphs as appropriate on scratch paper.

$$\sin^2 x = \tfrac{1}{2} \sin 2x \quad \text{Use double-angle identity.}$$
$$= \tfrac{1}{2}(2 \sin x \cos x)$$
$$\sin^2 x - \sin x \cos x = 0 \quad a^2 - ab = a(a - b)$$
$$\sin x(\sin x - \cos x) = 0 \quad a(a - b) = 0 \text{ only if } a = 0 \text{ or } a - b = 0$$
$$\sin x = 0 \quad \text{or} \quad \sin x - \cos x = 0$$
$$x = 0, \pi \qquad \sin x = \cos x$$
$$\frac{\sin x}{\cos x} = 1$$
$$\tan x = 1$$
$$x = \frac{\pi}{4}, \frac{5\pi}{4}$$

Combining the solutions from both equations, we have the complete set of solutions:

$$x = 0, \frac{\pi}{4}, \pi, \frac{5\pi}{4}$$

Matched Problem 4 Find exact solutions for $\sin 2x = \sin x$, $0 \le x < 2\pi$.

EXAMPLE 5 Approximate Solutions Using Identities and the Quadratic Formula

Solve $\cos 2x = 4 \cos x - 2$ for all real x. Compute inverse functions to four decimal places.

Solution *Step 1. Solve for* $\cos x$.

$$\cos 2x = 4 \cos x - 2 \qquad \text{Use double-angle identity.}$$

$$2 \cos^2 x - 1 = 4 \cos x - 2$$

$$2 \cos^2 x - 4 \cos x + 1 = 0$$

Quadratic in $\cos x$. Left side does not factor using integer coefficients. Solve using quadratic formula.

$$\cos x = \frac{4 \pm \sqrt{16 - 4(2)(1)}}{2(2)}$$

$$= 1.707107 \text{ or } 0.292893$$

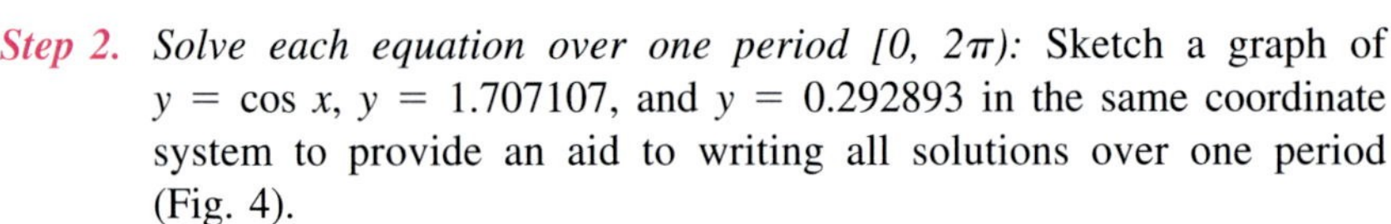

Step 2. Solve each equation over one period $[0, 2\pi)$: Sketch a graph of $y = \cos x$, $y = 1.707107$, and $y = 0.292893$ in the same coordinate system to provide an aid to writing all solutions over one period (Fig. 4).

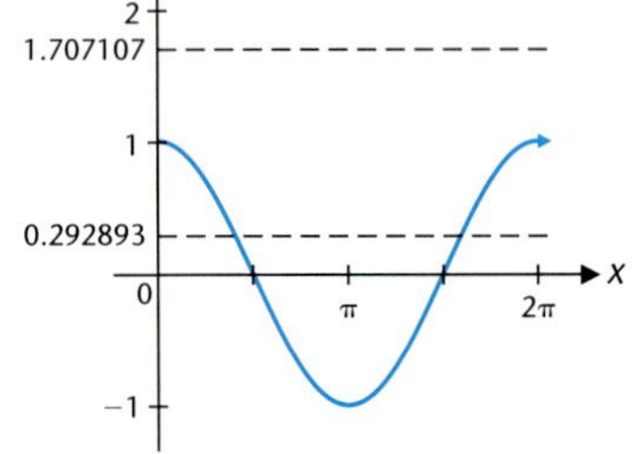

FIGURE 4

Solve the first equation:

$$\cos x = 1.707107 \qquad \text{No solution, since } -1 \le \cos x \le 1$$

Solve the second equation:

$$\cos x = 0.292893$$

Figure 4 indicates a first-quadrant solution and a fourth-quadrant solution. If the reference angle is α, then $x = \alpha$ or $x = 2\pi - \alpha$.

$$\alpha = \cos^{-1} 0.292893 = 1.2735$$

$$2\pi - \alpha = 2\pi - 1.2735 = 5.0096$$

Check $\cos 1.2735 = 0.292936 \qquad \cos 5.0096 = 0.292854$

Step 3. *Write an expression for all solutions.* Since the cosine function is periodic with period 2π, all solutions are given by

$$x = \begin{cases} 1.2735 + 2k\pi \\ 5.0096 + 2k\pi \end{cases} \qquad k \text{ any integer}$$

Matched Problem 5 Solve $\cos 2x = 2(\sin x - 1)$ for all real x. Compute inverse functions to four decimal places.

• Solving Trigonometric Equations Using a Graphing Utility

All the trigonometric equations that were solved earlier with algebraic methods can also be solved, though usually not exactly, with graphing utility methods. In addition, there are many trigonometric equations that can be solved (to any decimal accuracy desired) using graphing utility methods but cannot be solved in a finite sequence of steps using algebraic methods. Examples 6–8 are such examples.

EXAMPLE 6 Solution Using a Graphing Utility

Find all real solutions to four decimal places for $2 \cos 2x = 1.35x - 2$.

Solution This relatively simple trigonometric equation cannot be solved using a finite number of algebraic steps (try it!). However, it can be solved rather easily to the accuracy desired using a graphing utility. Graph $y1 = 2 \cos 2x$ and $y2 = 1.35x - 2$ in the same viewing window, and find any points of intersection using a built-in routine. The first point of intersection is shown in Figure 5. It appears there may be more than one point of intersection, but zooming in on the portion of the graph in question shows that the two graphs do not intersect in that region (Fig. 6). The only solution is

$$x = 0.9639$$

Check Left side: $2 \cos 2(0.9639) = -0.6989$
Right side: $1.35(0.9639) - 2 = -0.6987$

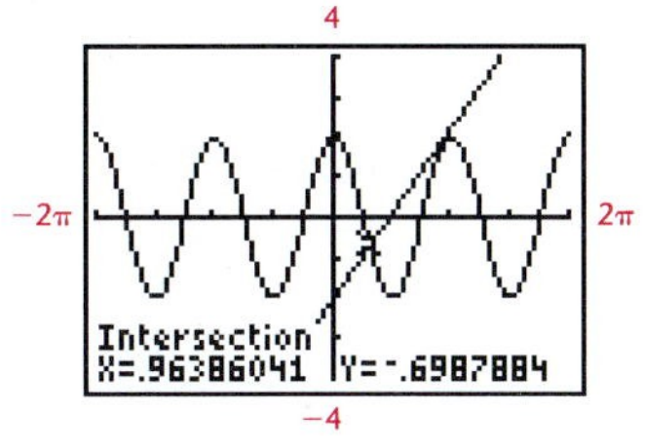

FIGURE 5

3
π/2
3π/2
1

FIGURE 6

Matched Problem 6 Find all real solutions to four decimal places for $\sin x/2 = 0.2x - 0.5$.

EXAMPLE 7 Geometric Application

A 10-centimeter arc on a circle has an 8-centimeter chord. What is the radius of the circle to four decimal places? What is the radian measure of the central angle, to four decimal places, subtended by the arc?

Solution Sketch a figure with auxiliary lines (Fig. 7). From the figure, θ in radians is

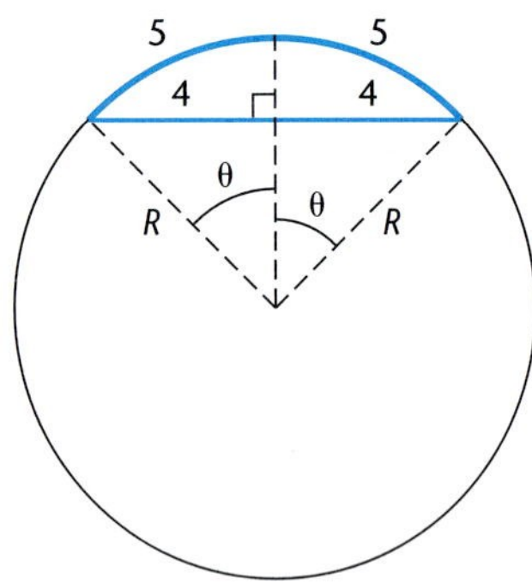

FIGURE 7

$$\theta = \frac{5}{R} \qquad \text{and} \qquad \sin\theta = \frac{4}{R}$$

Thus,

$$\sin\frac{5}{R} = \frac{4}{R}$$

and our problem is to solve this trigonometric equation for R. Algebraic methods will not isolate R, so turn to the use of a graphing utility. Start by graphing $y1 = \sin(5/x)$ and $y2 = 4/x$ in the same viewing window for $1 \le x \le 10$ and $-2 \le y \le 2$ (Fig. 8). It appears that the graphs intersect for x between 4 and 5. To get a clearer look at the intersection point, we change the window dimensions to $4 \le x \le 5$ and $0.5 \le y \le 1.5$ and use a built-in routine to find the point of intersection (Fig. 9).

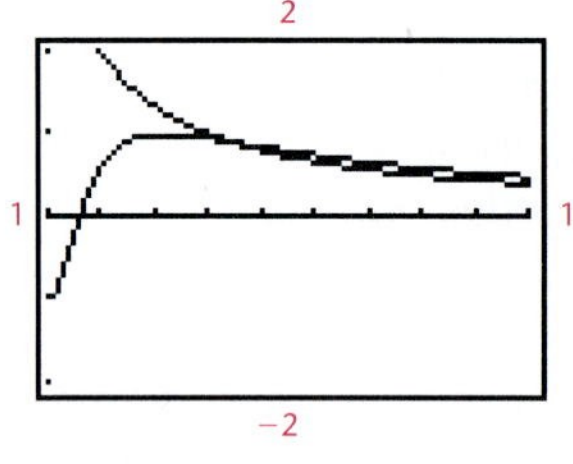

FIGURE 8

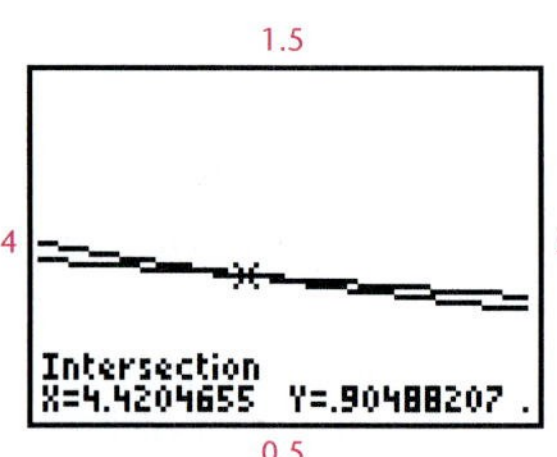

FIGURE 9

From Figure 9, we see that

$$R = 4.4205 \text{ centimeters}$$

Check
$$\sin\frac{5}{R} = \sin\frac{5}{4.4205} = 0.9049 \qquad \frac{4}{R} = \frac{4}{4.4205} = 0.9049$$

Having R, we can compute the radian measure of the central angle subtended by the 10-centimeter arc:

$$\text{Central angle} = \frac{10}{R} = \frac{10}{4.4205} = 2.2622 \text{ radians}$$

Matched Problem 7 An 8.2456-inch arc on a circle has a 6.0344-inch chord. What is the radius of the circle to four decimal places? What is the measure of the central angle, to four decimal places, subtended by the arc?

EXAMPLE 8 **Solution Using a Graphing Utility**

Find all real solutions, to four decimal places, for $\tan(x/2) = 1/x$, $-\pi < x \leq 3\pi$.

Solution Graph $y1 = \tan(x/2)$ and $y2 = 1/x$ in the same viewing window for $-\pi < x < 3\pi$ (Fig. 10). Solutions are at points of intersection.

FIGURE 10

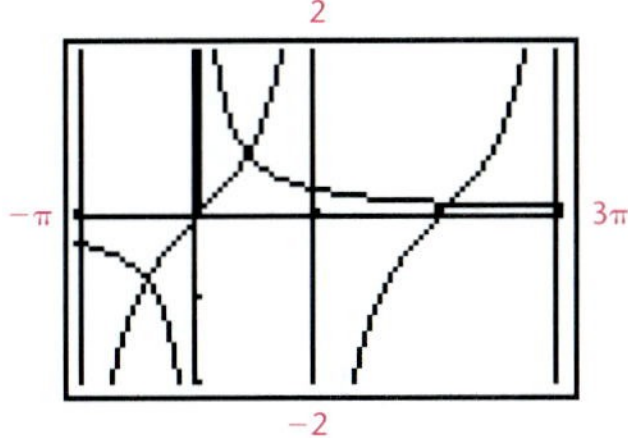

Using a built-in routine, the three solutions are found to be

$$x = -1.3065, 1.3065, 6.5846$$

Checking these solutions is left to the reader.

Matched Problem 8 Find all real solutions, to four decimal places, for $0.25 \tan(x/2) = \ln x$, $0 < x < 4\pi$.

Solving trigonometric inequalities using a graphing utility is as easy as solving trigonometric equations using a graphing utility. Example 9 illustrates the process.

EXAMPLE 9 **Solving a Trigonometric Inequality**

Solve $\sin x - \cos x < 0.25x - 0.5$ using two decimal place accuracy.

Solution Graph $y1 = \sin x - \cos x$ and $y2 = 0.25x - 0.5$ in the same viewing window (Fig. 11).

FIGURE 11

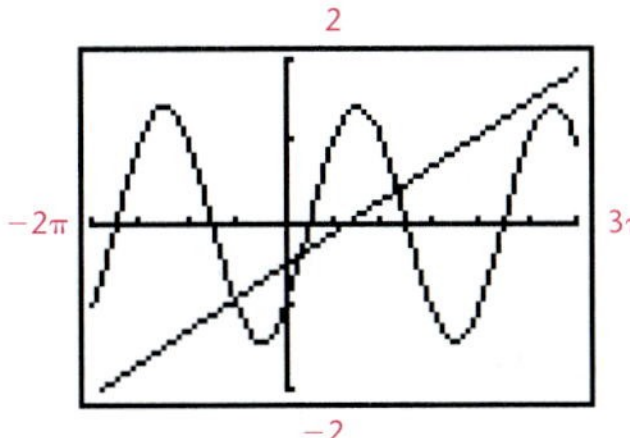

Finding the three points of intersection by a built-in routine, we see that the graph of $y1$ is below the graph of $y2$ on the following two intervals: $(-1.65, 0.52)$ and $(3.63, \infty)$. Thus, the solution set to the inequality is $(-1.65, 0.52) \cup (3.63, \infty)$.

Matched Problem 9 Solve $\cos x - \sin x > 0.4 - 0.3x$ using two decimal place accuracy.

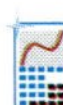

EXPLORE-DISCUSS 2 How many solutions does the following equation have?

$$\sin \frac{1}{x} = 0 \qquad (1)$$

Graph $y1 = \sin (1/x)$ and $y2 = 0$ for each of the indicated intervals in parts A–G. From each graph estimate the number of solutions that equation (1) appears to have. What final conjecture would you be willing to make regarding the number of solutions to equation (1)? Explain.

(A) $[-20, 20]$; can 0 be a solution? Explain.

(B) $[-2, 2]$ (C) $[-1, 1]$ (D) $[-0.1, 0.1]$ (E) $[-0.01, 0.01]$

(F) $[-0.001, 0.001]$ (G) $[-0.0001, 0.0001]$

Answers to Matched Problems

1. $x = \begin{cases} 0 + 2k\pi \\ \pi + 2k\pi \\ 7\pi/6 + 2k\pi \\ 11\pi/6 + 2k\pi \end{cases}$ k any integer

2. $x = \begin{cases} 1.8235 + 2k\pi \\ 4.4597 + 2k\pi \end{cases}$ k any integer

3. $-16.318°$ **4.** $x = 0, \pi/3, \pi, 5\pi/3$

5. $x = \begin{cases} 0.9665 + 2k\pi \\ 2.1751 + 2k\pi \end{cases}$ k any integer

6. $x = 5.1609$ **7.** $R = 3.1103$ in.; central angle $= 2.6511$ rad

8. $x = 1.1828, 2.6369, 9.2004$ **9.** $(-1.67, 0.64) \cup (3.46, \infty)$

EXERCISE 6-5

A

In Problems 1–10, find exact solutions over the indicated intervals, x a real number, θ in degrees.

1. $2 \sin x + 1 = 0, 0 \le x < 2\pi$

2. $2 \cos x + 1 = 0, 0 \le x < 2\pi$

3. $2 \sin x + 1 = 0$, all real x

4. $2 \cos x + 1 = 0$, all real x

5. $\tan x + \sqrt{3} = 0, 0 \le x < \pi$

6. $\sqrt{3} \tan x + 1 = 0, 0 \le x < \pi$

7. $\tan x + \sqrt{3} = 0$, all real x

8. $\sqrt{3} \tan x + 1 = 0$, all real x

9. $2 \cos \theta - \sqrt{3} = 0, 0° \le \theta < 360°$

10. $\sqrt{2}\sin\theta - 1 = 0, 0° \le \theta < 360°$

Solve Problems 11–16 to four decimal places (θ in degrees, x real).

11. $7\cos x - 3 = 0, 0 \le x < 2\pi$

12. $5\cos x - 2 = 0, 0 \le x < 2\pi$

13. $2\tan\theta - 7 = 0, 0° \le \theta < 180°$

14. $4\tan\theta + 15 = 0, 0° \le \theta < 180°$

15. $1.3224\sin x + 0.4732 = 0$, all real x

16. $5.0118\sin x - 3.1105 = 0$, all real x

Solve Problems 17–20 to four decimal places using a graphing utility.

17. $1 - x = 2\sin x$, all real x

18. $2x - \cos x = 0$, all real x

19. $\tan(x/2) = 8 - x, 0 \le x < \pi$

20. $\tan 2x = 1 + 3x, 0 \le x < \pi/4$

B

In Problems 21–30, find exact solutions for x real and θ in degrees.

21. $2\sin^2\theta + \sin 2\theta = 0$, all θ

22. $\cos^2\theta = \frac{1}{2}\sin 2\theta$, all θ

23. $\tan x = -2\sin x, 0 \le x < 2\pi$

24. $\cos x = \cot x, 0 \le x < 2\pi$

25. $2\cos^2\theta + 3\sin\theta = 0, 0° \le \theta < 360°$

26. $\sin^2\theta + 2\cos\theta = -2, 0° \le \theta < 360°$

27. $\cos 2\theta + \cos\theta = 0, 0° \le \theta < 360°$

28. $\cos 2\theta + \sin^2\theta = 0, 0° \le \theta < 360°$

29. $2\sin^2(x/2) - 3\sin(x/2) + 1 = 0, 0 \le x \le 2\pi$

30. $4\cos^2 2x - 4\cos 2x + 1 = 0, 0 \le x \le 2\pi$

Solve Problems 31–36, for x real and θ in degrees. Compute inverse functions to 4 significant digits.

31. $6\sin^2\theta + 5\sin\theta = 6, 0° \le \theta \le 90°$

32. $4\cos^2\theta = 7\cos\theta + 2, 0° \le \theta \le 180°$

33. $3\cos^2 x - 8\cos x = 3, 0 \le x \le \pi$

34. $8\sin^2 x + 10\sin x = 3, 0 \le x \le \pi/2$

35. $2\sin x = \cos 2x, 0 \le x < 2\pi$

36. $\cos 2x + 10\cos x = 5, 0 \le x < 2\pi$

In Problems 37–42, how many solutions does the equation have? Explain.

37. $\sin^2 x = (1 - \cos x)(1 + \cos x)$

38. $2\sec^2 x = \tan^2 x + 1$

39. $3\sin x = x - 3$

40. $\pi\cos x + 2x = \pi$

41. $\sin 2x = 3\sin x\cos x$

42. $(\cot x - \tan x)\tan 2x = 2$

Solve Problems 43–52 to four decimal places using a graphing utility.

43. $2\sin x = \cos 2x, 0 \le x < 2\pi$

44. $\cos 2x + 10\cos x = 5, 0 \le x < 2\pi$

45. $2\sin^2 x = 1 - 2\sin x$, all real x

46. $\cos^2 x = 3 - 5\cos x$, all real x

47. $\cos 2x > x^2 - 2$, all real x

48. $2\sin(x - 2) < 3 - x^2$, all real x

49. $\cos(2x + 1) \le 0.5x - 2$, all real x

50. $\sin(3 - 2x) \ge 1 - 0.4x$, all real x

51. $e^{\sin x} = 2x - 1$, all real x

52. $e^{-\sin x} = 3 - x$, all real x

53. Explain the difference between evaluating $\tan^{-1}(-5.377)$ and solving the equation $\tan x = -5.377$.

54. Explain the difference between evaluating $\cos^{-1}(-0.7334)$ and solving the equation $\cos x = -0.7334$.

C

Find exact solutions to Problems 55–58. [Hint: Square both sides at an appropriate point, solve, then eliminate extraneous solutions at the end.]

55. $\cos x - \sin x = 1, 0 \le x < 2\pi$

56. $\sin x + \cos x = 1, 0 \le x < 2\pi$

57. $\tan x - \sec x = 1, 0 \le x < 2\pi$

58. $\sec x + \tan x = 1, 0 \le x < 2\pi$

Solve Problems 59–60 to 4 significant digits using a graphing utility.

59. $\sin(1/x) = 1.5 - 5x, 0.04 \le x \le 0.2$

60. $2\cos(1/x) = 950x - 4, 0.006 < x < 0.007$

61. We are interested in the zeros of the function $f(x) = \sin(1/x)$ for $x > 0$.

(A) Explore the graph of f over different intervals $[0.1, b]$ for various values of b, $b > 0.1$. Does the function f have a largest zero? If so, what is it (to four decimal places)? Explain what happens to the graph of f as x increases without bound. Does the graph have an asymptote? If so, what is its equation?

(B) Explore the graph of f over different intervals $(0, b]$ for various values of b, $0 < b \le 0.1$. How many zeros exist between 0 and b, for any $b > 0$, however small? Explain why this happens. Does f have a smallest positive zero? Explain.

62. We are interested in the zeros of the function $g(x) = \cos(1/x)$ for $x > 0$.

(A) Explore the graph of g over different intervals $[0.1, b]$ for various values of b, $b > 0.1$. Does the function g have a largest zero? If so, what is it (to four decimal places)? Explain what happens to the graph of g as x increases without bound. Does the graph have an asymptote? If so, what is its equation?

(B) Explore the graph of g over different intervals $(0, b]$ for various values of b, $0 < b \le 0.1$. How many zeros exist between 0 and b, for any $b > 0$, however small? Explain why this happens. Does g have a smallest positive zero? Explain.

APPLICATIONS

63. Electric Current. An alternating current generator produces a current given by the equation

$$I = 30 \sin 120\pi t$$

where t is time in seconds and I is current in amperes. Find the smallest positive t (to 4 significant digits) such that $I = -10$ amperes.

64. Electric Current. Refer to Problem 63. Find the smallest positive t (to 4 significant digits) such that $I = 25$ amperes.

65. Optics. A polarizing filter for a camera contains two parallel plates of polarizing glass, one fixed and the other able to rotate. If θ is the angle of rotation from the position of maximum light transmission, then the intensity of light leaving the filter is $\cos^2 \theta$ times the intensity I of light entering the filter (see the figure).

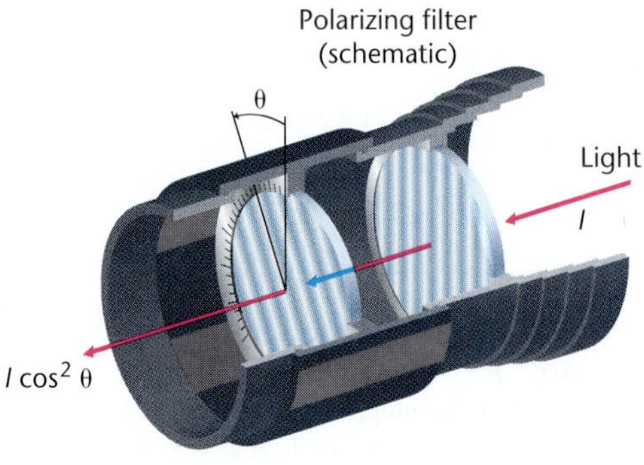

Find the smallest positive θ (in decimal degrees to two decimal places) so that the intensity of light leaving the filter is 40% of that entering.

66. Optics. Refer to Problem 65. Find the smallest positive θ so that the light leaving the filter is 70% of that entering.

67. Astronomy. The planet Mercury travels around the sun in an elliptical orbit given approximately by

$$r = \frac{3.44 \times 10^7}{1 - 0.206 \cos \theta}$$

(see the figure). Find the smallest positive θ (in decimal degrees to 3 significant digits) such that Mercury is 3.09×10^7 miles from the sun.

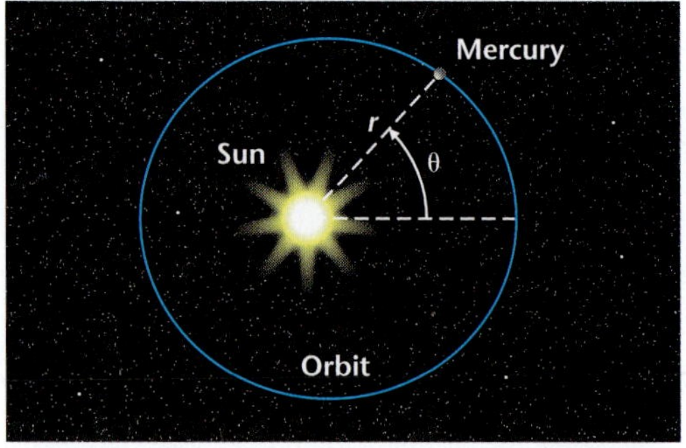

68. Astronomy. Refer to Problem 67. Find the smallest positive θ (in decimal degrees to 3 significant digits) such that Mercury is 3.78×10^7 miles from the sun.

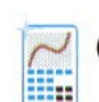

69. Geometry. The area of the segment of a circle in the figure is given by

$$A = \tfrac{1}{2}R^2(\theta - \sin\theta)$$

where θ is in radian measure. Use a graphing utility to find the radian measure, to three decimal places, of angle θ if the radius is 8 inches and the area of the segment is 48 square inches.

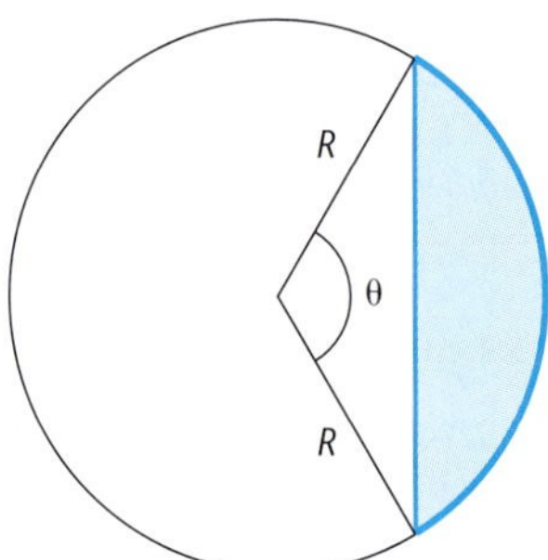

70. Geometry. Repeat Problem 69 if the radius is 10 centimeters and the area of the segment is 40 square centimeters.

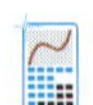

71. Eye Surgery. A surgical technique for correcting an astigmatism involves removing small pieces of tissue in order to change the curvature of the cornea.* In the cross section of

*Based on the article "The Surgical Correction of Astigmatism," Sheldon Rothman and Helen Strassberg, *UMAP Journal*, vol V, no. 2, 1984.

a cornea shown in the figure, the circular arc, with radius R and central angle 2θ, represents a cross section of the surface of the cornea.

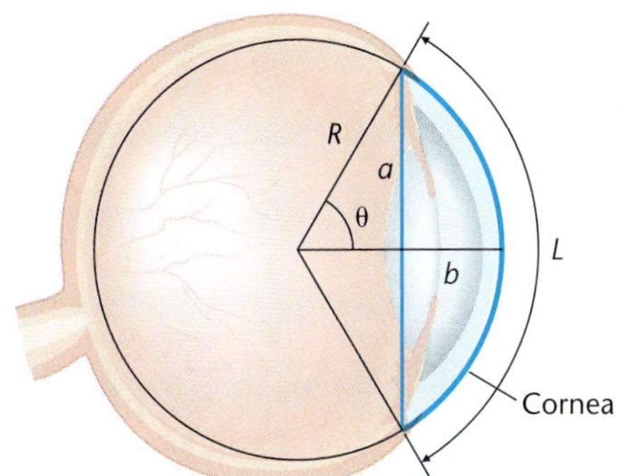

(A) If $a = 5.5$ millimeters and $b = 2.5$ millimeters, find L correct to four decimal places.

(B) Reducing the chord length $2a$ without changing the length L of the arc has the effect of pushing the cornea outward and giving it a rounder, yet still circular, shape. With the aid of a graphing utility in part of the solution, approximate b to four decimal places if a is reduced to 5.4 millimeters and L remains the same as it was in part A.

 **72. Eye Surgery.** Refer to Problem 71.

(A) If in the figure $a = 5.4$ millimeters and $b = 2.4$ millimeters, find L correct to four decimal places.

(B) Increasing the chord length without changing the arc length L has the effect of pulling the cornea inward and giving it a flatter, yet still circular, shape. With the aid of a graphing utility in part of the solution, approximate b to four decimal places if a is increased to 5.5 millimeters and L remains the same as it was in part A.

 **Analytic Geometry.** *Find simultaneous solutions for each system of equations in Problems 73 and 74 ($0° \le \theta \le 360°$). These are polar equations, which will be discussed in the next chapter.*

★ **73.** $r = 2 \sin \theta$
$r = \sin 2\theta$

★ **74.** $r = 2 \sin \theta$
$r = 2(1 - \sin \theta)$

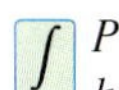 *Problems 75 and 76 are related to rotation of axes in analytic geometry.*

★★ **75. Analytic Geometry.** Give the equation $2xy = 1$, replace x and y with

$$x = u \cos \theta - v \sin \theta$$
$$y = u \sin \theta + v \cos \theta$$

and simplify the left side of the resulting equation. Find the smallest positive θ in degree measure so that the coefficient of the uv term is 0.

★★ **76. Analytic Geometry.** Repeat Problem 75 for $xy = -2$.

CHAPTER 6 GROUP ACTIVITY From $M \sin Bt + N \cos Bt$ to $A \sin (Bt + C)$, a Harmonic Analysis Tool

In solving certain kinds of more advanced applied mathematical problems—problems dealing with electric circuits, spring–mass systems, heat flow, and so on—the solution process leads naturally to a function of the form

$$y = M \sin Bt + N \cos Bt \qquad (1)$$

The following investigation will show that phenomena leading to this type of equation are simple harmonic and can be represented by an equation of the form

$$y = A \sin (Bt + C)$$

 **(A) Graphing Utility Exploration.** Use a graphing utility to explore the nature of the graph of equation (1) for various values of M, N, and B. Does the graph appear to be simple harmonic; that is, does it appear to be a graph of an equation of the form $y = A \sin (Bt + C)$?

The graph of $y = 2 \sin(\pi t) - 3 \cos(\pi t)$, which is typical of the various graphs from equation (1), is shown in Figure 1. It turns out that the graph in Figure 1 can also be obtained from an equation of the form

$$y = A \sin(Bt + C) \qquad (2)$$

for suitable values of A, B and C.

FIGURE 1 $y = 2 \sin(\pi t) - 3 \cos(\pi t)$.

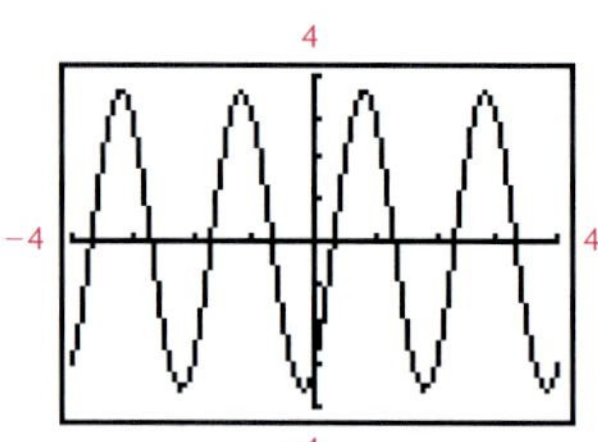

The problem now is: Given M, N, and B in equation (1), find A, B, and C in equation (2) so that equation (2) produces the same graph as equation (1). The latter is often preferred over the former, because from (2) one can easily read amplitude, period, and phase shift, and recognize a phenomenon as simple harmonic.

The process of finding A, B, and C, given M, N, and B requires a little ingenuity and the use of the sum identity

$$\sin(x + y) = \sin x \cos y + \cos x \sin y \qquad (3)$$

How do we proceed? We start by trying to get the right side of equation (1) to look like the right side of identity (3). Then we use (3), from right to left, to obtain (2).

(B) Establishing a Transformation Identity. Show that

$$y = M \sin Bt + N \cos Bt = \sqrt{M^2 + N^2} \sin(Bt + C) \qquad (4)$$

where C is any angle (in radians if t is real) having $P(M, N)$ on its terminal side. *Hint:* A first step is the following:

$$M \sin Bt + N \cos Bt = \frac{\sqrt{M^2 + N^2}}{\sqrt{M^2 + N^2}} (M \sin Bt + N \cos Bt)$$

(C) Use of Transformation Identity. Use equation (4) to transform

$$y1 = -4 \sin \frac{t}{2} + 3 \cos \frac{t}{2}$$

into the form $y2 = A \sin(Bt + C)$, where C is chosen so that $|C|$ is minimum. Compute C to three decimal places. From the new equation determine the amplitude, period, and phase shift.

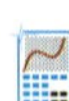

(D) Graphing Utility Visualization and Verification. Graph $y1$ and $y2$ from part C in the same viewing window.

(E) Physics Application. A weight suspended from a spring, with spring constant 64, is pulled 4 centimeters below its equilibrium position and is then given a downward thrust to produce an initial downward velocity of 24 centimeters per second. In more advanced mathematics (differential equations) the equation of motion (neglecting air resistance and friction) is found to be given approximately by

$$y1 = -3 \sin 8t - 4 \cos 8t$$

where $y1$ is the position of the bottom of the weight on the scale in Figure 2 at time t (y is in centimeters and t is in seconds). Transform the equation into the form

$$y2 = A \sin (Bt + C)$$

and indicate the amplitude, period, and phase shift of the motion. Choose the least positive C and keep A positive.

FIGURE 2 Spring–mass system.

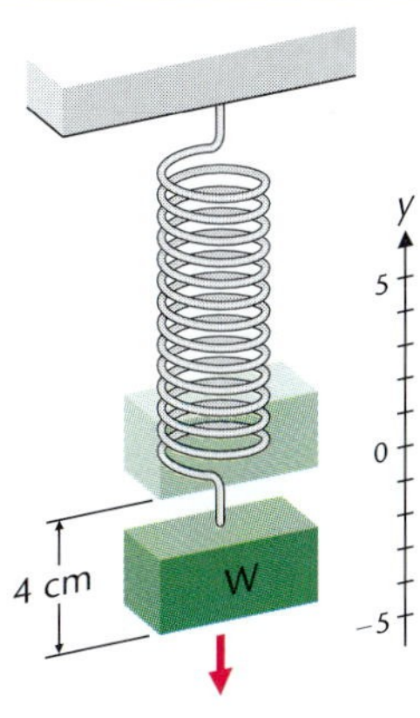

(F) Graphing Utility Visualization and Verification. Graph $y1$ and $y2$ from part E in the same viewing window of a graphing utility, $0 \le t \le 6$. How many times will the bottom of the weight pass $y = 2$ in the first 6 seconds?

(G) Solving a Trigonometric Equation. How long, to three decimal places, will it take the bottom of the weight to reach $y = 2$ for the first time?

Chapter 6 Review

6-1 BASIC IDENTITIES AND THEIR USE

The following 11 identities are basic to the process of changing trigonometric expressions to equivalent but more useful forms:

Reciprocal Identities

$$\csc x = \frac{1}{\sin x} \qquad \sec x = \frac{1}{\cos x} \qquad \cot x = \frac{1}{\tan x}$$

Quotient Identities

$$\tan x = \frac{\sin x}{\cos x} \qquad \cot x = \frac{\cos x}{\sin x}$$

Identities for Negatives

$$\sin (-x) = -\sin x \qquad \cos (-x) = \cos x$$

$$\tan (-x) = -\tan x$$

Pythagorean Identities

$$\sin^2 x + \cos^2 x = 1 \qquad \tan^2 x + 1 = \sec^2 x$$

$$1 + \cot^2 x = \csc^2 x$$

Even though there is no fixed method of verification that works for all identities, the following suggested steps are helpful in many cases.

Suggested Steps in Verifying Identities

1. Start with the more complicated side of the identity, and transform it into the simpler side.
2. Try algebraic operations such as multiplying, factoring, combining fractions, and splitting fractions.
3. If other steps fail, express each function in terms of sine and cosine functions, and then perform appropriate algebraic operations.
4. At each step, keep the other side of the identity in mind. This often reveals what you should do in order to get there.

6-2 SUM, DIFFERENCE, AND COFUNCTION IDENTITIES

Sum Identities

$$\sin(x + y) = \sin x \cos y + \cos x \sin y$$

$$\cos(x + y) = \cos x \cos y - \sin x \sin y$$

$$\tan(x + y) = \frac{\tan x + \tan y}{1 - \tan x \tan y}$$

Difference Identities

$$\sin(x - y) = \sin x \cos y - \cos x \sin y$$

$$\cos(x - y) = \cos x \cos y + \sin x \sin y$$

$$\tan(x - y) = \frac{\tan x - \tan y}{1 + \tan x \tan y}$$

Cofunction Identities
(Replace $\pi/2$ with 90° if x is in degrees.)

$$\cos\left(\frac{\pi}{2} - x\right) = \sin x \qquad \sin\left(\frac{\pi}{2} - x\right) = \cos x$$

$$\tan\left(\frac{\pi}{2} - x\right) = \cot x$$

6-3 DOUBLE-ANGLE AND HALF-ANGLE IDENTITIES

Double-Angle Identities

$$\sin 2x = 2 \sin x \cos x$$

$$\cos 2x = \cos^2 x - \sin^2 x = 1 - 2\sin^2 x = 2\cos^2 x - 1$$

$$\tan 2x = \frac{2 \tan x}{1 - \tan^2 x} = \frac{2 \cot x}{\cot^2 x - 1} = \frac{2}{\cot x - \tan x}$$

Half-Angle Identities

$$\sin\frac{x}{2} = \pm\sqrt{\frac{1 - \cos x}{2}}$$

$$\cos\frac{x}{2} = \pm\sqrt{\frac{1 + \cos x}{2}}$$

$$\tan\frac{x}{2} = \pm\sqrt{\frac{1 - \cos x}{1 + \cos x}} = \frac{\sin x}{1 + \cos x} = \frac{1 - \cos x}{\sin x}$$

6-4 PRODUCT–SUM AND SUM–PRODUCT IDENTITIES

Product–Sum Identities

$$\sin x \cos y = \tfrac{1}{2}[\sin(x + y) + \sin(x - y)]$$

$$\cos x \sin y = \tfrac{1}{2}[\sin(x + y) - \sin(x - y)]$$

$$\sin x \sin y = \tfrac{1}{2}[\cos(x - y) - \cos(x + y)]$$

$$\cos x \cos y = \tfrac{1}{2}[\cos(x + y) + \cos(x - y)]$$

Sum–Product Identities

$$\sin x + \sin y = 2 \sin\frac{x + y}{2} \cos\frac{x - y}{2}$$

$$\sin x - \sin y = 2 \cos\frac{x + y}{2} \sin\frac{x - y}{2}$$

$$\cos x + \cos y = 2 \cos\frac{x + y}{2} \cos\frac{x - y}{2}$$

$$\cos x - \cos y = -2 \sin\frac{x + y}{2} \sin\frac{x - y}{2}$$

6-5 TRIGONOMETRIC EQUATIONS

The first four sections of the chapter considered trigonometric equations called **identities.** Identities are true for all replacements of the variable(s) for which both sides are defined. This section considers **conditional equations.** Conditional equations may be true for some variable replacements but are false for other variable replacements for which both sides are defined. The equation $\sin x = \cos x$ is a conditional equation.

In **solving a trigonometric equation using an algebraic approach,** no particular rule will always lead to all solutions of every trigonometric equation you are likely to encounter. Solving trigonometric equations algebraically often requires the use of algebraic manipulation, identities, and ingenuity.

Suggestions for Solving Trigonometric Equations Algebraically

1. Regard one particular trigonometric function as a variable, and solve for it.
   (a) Consider using algebraic manipulation such as factoring, combining or separating fractions, and so on.
   (b) Consider using identities.
2. After solving for a trigonometric function, solve for the variable.

In **solving a trigonometric equation using a graphing utility approach** one can solve a larger variety of problems than with the algebraic approach. The solutions are generally approximations (to whatever decimal accuracy desired). In some cases exact solutions can be found using an algebraic approach.

Chapter 6 Review Exercise

Work through all the problems in this chapter review and check answers in the back of the book. Answers to all review problems except verifications are there, and following each answer is a number in italics indicating the section in which that type of problem is discussed. Where weaknesses show up, review appropriate sections in the text.

A

Verify each identity in Problems 1–4.

1. $\tan x + \cot x = \sec x \csc x$

2. $\sec^4 x - 2\sec^2 x \tan^2 x + \tan^4 x = 1$

3. $\dfrac{1}{1 - \sin x} + \dfrac{1}{1 + \sin x} = 2\sec^2 x$

4. $\cos\left(x - \dfrac{3\pi}{2}\right) = -\sin x$

5. Write as a sum: $\cos 2\alpha \cos 3\alpha$

6. Write as a product: $\sin 9x - \sin 5x$

7. Simplify: $\sin\left(x + \dfrac{9\pi}{2}\right)$

Solve Problems 8 and 9 exactly (θ in degrees, x real).

8. $\sqrt{2}\cos\theta + 1 = 0$, all θ

9. $\sin x \tan x - \sin x = 0$, all real x

Solve Problems 10–13 to four decimal places (θ in degrees and x real).

10. $\cos\theta = 0.8215$, all θ

11. $\cot x = -154.3$, $-\pi/2 < x < \pi/2$

12. $\csc x = 1.786$, all real x

13. $3\tan(11 - 3x) = 23.46$, $-\pi/2 < 11 - 3x < \pi/2$

14. Use a graphing utility to test whether each of the following is an identity. If an equation appears to be an identity, verify it. If the equation does not appear to be an identity, find a value of x for which both sides are defined but are not equal.
(A) $(\sin x + \cos x)^2 = 1 - 2\sin x \cos x$
(B) $\cos^2 x - \sin^2 x = 1 - 2\sin^2 x$

B

Verify each identity in Problems 15–23.

15. $\dfrac{1 - 2\cos x - 3\cos^2 x}{\sin^2 x} = \dfrac{1 - 3\cos x}{1 - \cos x}$

16. $(1 - \cos x)(\csc x + \cot x) = \sin x$

17. $\dfrac{1 + \sin x}{\cos x} = \dfrac{\cos x}{1 - \sin x}$

18. $\cos 2x = \dfrac{1 - \tan^2 x}{1 + \tan^2 x}$

19. $\cot \dfrac{x}{2} = \dfrac{\sin x}{1 - \cos x}$

20. $\cot x - \tan x = \dfrac{4\cos^2 x - 2}{\sin 2x}$

21. $\left(\dfrac{1 - \cot x}{\csc x}\right)^2 = 1 - \sin 2x$

22. $\tan m + \tan n = \dfrac{\sin(m + n)}{\cos m \cos n}$

23. $\tan(x + y) = \dfrac{\cot x + \cot y}{\cot x \cot y - 1}$

Evaluate Problems 24 and 25 exactly using appropriate sum–product or product–sum identities.

24. $\cos 195° \sin 75°$

25. $\cos 195° + \cos 105°$

In Problems 26–29, is the equation an identity? Explain.

26. $\tan^2 x = \sec^2 x + 1$

27. $\sin 4x = 4 \sin x \cos x \cos 2x$

28. $\cos(x - \pi/2) = \sin x$

29. $\sin(x - \pi/2) = \cos x$

Solve Problems 30–34 exactly (θ in degrees, x real).

30. $4 \sin^2 x - 3 = 0, 0 \le x < 2\pi$

31. $2 \sin^2 \theta + \cos \theta = 1, 0° \le \theta \le 180°$

32. $2 \sin^2 x - \sin x = 0$, all real x

33. $\sin 2x = \sqrt{3} \sin x$, all real x

34. $2 \sin^2 \theta + 5 \cos \theta + 1 = 0$, all θ

Solve Problems 35–37 to 4 significant digits (θ in degrees, x real).

35. $\tan \theta = 0.2557$, all θ

36. $\sin^2 x + 2 = 4 \sin x$, all real x

37. $\tan^2 x = 2 \tan x + 1, 0 \le x < \pi$

In Problems 38–41, how many solutions does the equation have? Explain.

38. $\cos x = x^2 - 1$

39. $e^x = 1 + \sin x$

40. $\sin x = \sec x$

41. $\tan x = 7x - 9$

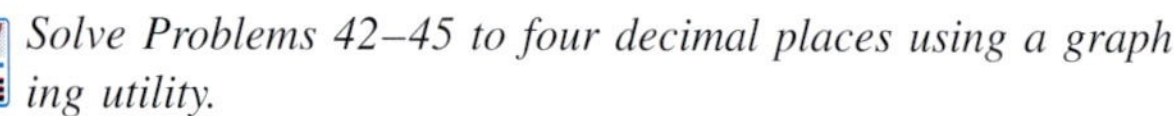

Solve Problems 42–45 to four decimal places using a graphing utility.

42. $3 \sin 2x = 2x - 2.5$, all real x

43. $3 \sin 2x > 2x - 2.5$, all real x

44. $2 \sin^2 x - \cos 2x = 1 - x^2$, all real x

45. $2 \sin^2 x - \cos 2x \le 1 - x^2$, all real x

46. Given the equation $\tan(x + y) = \tan x + \tan y$.
(A) Is $x = 0$ and $y = \pi/4$ a solution?
(B) Is the equation an identity or a conditional equation? Explain.

47. Explain the difference in evaluating $\sin^{-1} 0.3351$ and solving the equation $\sin x = 0.3351$.

48. Use a graphing utility to test whether each of the following is an identity. If an equation appears to be an identity, verify it. If the equation does not appear to be an identity, find a value of x for which both sides are defined but are not equal.

(A) $\dfrac{\tan x}{\sin x + 2 \tan x} = \dfrac{1}{\cos x - 2}$

(B) $\dfrac{\tan x}{\sin x - 2 \tan x} = \dfrac{1}{\cos x - 2}$

49. Use a sum or difference identity to convert $y = \cos(x - \pi/3)$ to a form involving $\sin x$ and/or $\cos x$. Enter the original equation in a graphing utility as $y1$, the converted form as $y2$, and graph $y1$ and $y2$ in the same viewing window. Use TRACE to compare the two graphs.

50. (A) Solve $\tan(x/2) = 2 \sin x$ exactly, $0 \le x < 2\pi$, using algebraic methods.

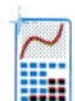

(B) Solve $\tan(x/2) = 2 \sin x$, $0 \le x < 2\pi$, to four decimal places using a graphing utility.

51. Solve $3 \cos(x - 1) = 2 - x^2$, for all real x, to three decimal places using a graphing utility.

C

Solve Problems 52–54 exactly without the use of a calculator.

52. Given $\tan x = -\frac{3}{4}$, $\pi/2 \le x \le \pi$, find:
(A) $\sin(x/2)$ (B) $\cos 2x$

53. $\sin[2 \tan^{-1}(-\frac{3}{4})]$

54. $\sin(\sin^{-1} \frac{3}{5} + \cos^{-1} \frac{4}{5})$

55. (A) Solve $\cos^2 2x = \cos 2x + \sin^2 2x$, $0 \le x < \pi$, exactly using algebraic methods.

(B) Solve $\cos^2 2x = \cos 2x + \sin^2 2x$, $0 \le x < \pi$, to four decimal places using a graphing utility.

56. We are interested in the zeros of

$$f(x) = \sin \frac{1}{x - 1} \qquad \text{for } x > 0$$

(A) Explore the graph of f over different intervals $[a, b]$ for various values of a and b, $0 < a < b$. Does the function f have a smallest zero? If so, what is it (to four decimal places)? Does the function have a largest zero? If so, what is it (to four decimal places)?
(B) Explain what happens to the graph as x increases without bound. Does the graph have an asymptote? If so, what is its equation?
(C) Explore the graph of f over smaller and smaller intervals containing $x = 1$. How many zeros exist on any interval containing $x = 1$? Is $x = 1$ a zero? Explain.

APPLICATIONS 

57. Indirect Measurement. Find the exact value of x in the figure, then find x and θ to three decimal places. [*Hint:* Use a suitable identity involving $\tan 2\theta$.]

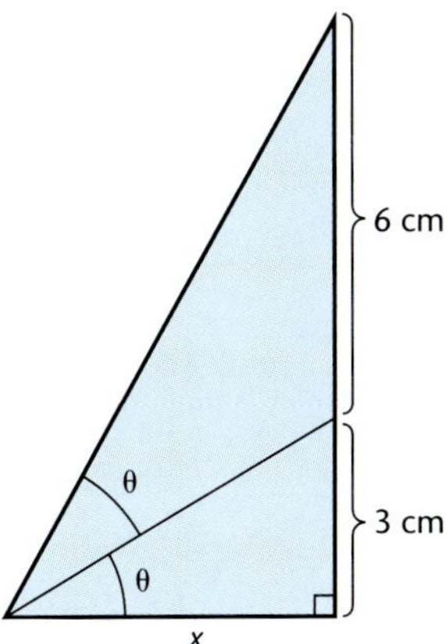

58. Electric Current. An alternating current generator produces a current given by the equation

$$I = 50 \sin 120\pi (t - 0.001)$$

where t is time in seconds and I is current in amperes. Find the smallest positive t, to 3 significant digits, such that $I = 40$ amperes.

59. Music—Beat Frequencies. $y = 0.6 \cos 184\pi t$ and $y = -0.6 \cos 208\pi t$ are equations of sound waves with frequencies 92 and 104 hertz, respectively. If both sounds are emitted simultaneously, a beat frequency results.

(A) Show that

$$0.6 \cos 184\pi t - 0.6 \cos 208\pi t = 1.2 \sin 12\pi t \sin 196\pi t$$

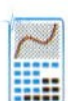

(B) Graph each of the following equations in a different viewing window for $0 \le t \le 0.2$.

$$y = 0.6 \cos 184\pi t$$

$$y = -0.6 \cos 208\pi t$$

$$y = 0.6 \cos 184\pi t - 0.6 \cos 208\pi t$$

$$y = 1.2 \sin 12\pi t \sin 196\pi t$$

60. Engineering. A bridge with a circular arch with an arc length of 36 feet spans a 32-foot canal (see figure). Determine the height of the circular arc above the water at the center of the bridge, and the radius of the circular arc, both to three decimal places. Start by drawing auxiliary lines in the figure, labeling appropriate parts, then explain how the following trigonometric equation

$$\sin \theta = \tfrac{8}{9}\theta$$

is related to the problem. After solving the trigonometric equation for θ, the radius is easy to find and the height of the arch above the water can be found with a little ingenuity.

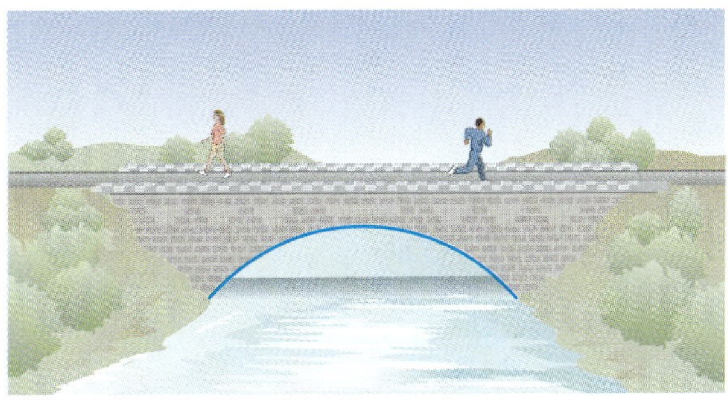

Additional Topics in Trigonometry

Chapter 7

In this chapter a number of additional topics involving trigonometry are considered. First, the problem of solving triangles is returned to—not just right triangles but any triangle. Then some of these ideas are used to develop the important concept of a vector. With knowledge of trigonometry, the *polar coordinate system* is introduced, probably the most important coordinate system after the rectangular coordinate system. After considering polar equations and their graphs, complex numbers are represented in *polar form.* Once a complex number is in polar form, it will be possible to find nth powers and nth roots of the number using an ingenious theorem due to De Moivre.

SECTION 7-1 Law of Sines

- Law of Sines Derivation
- Solving the ASA and AAS Cases
- Solving the SSA Case—Including the Ambiguous Case

The law of sines (developed in this section) and the law of cosines (developed in the next section) play fundamental roles in solving **oblique triangles**—triangles without a right angle. Every oblique triangle is either **acute,** all angles between 0° and 90°, or **obtuse,** one angle between 90° and 180°. Figure 1 illustrates both types of triangles.

FIGURE 1 Oblique triangles.

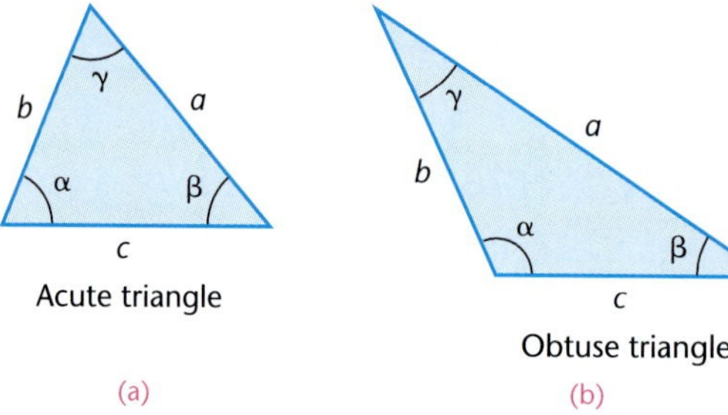

TABLE 1 Triangles and Significant Digits

Angle to nearest	Significant digits for side measure
1°	2
10′ or 0.1°	3
1′ or 0.01°	4
10″ or 0.001°	5

Note how the sides and angles of the oblique triangles in Figure 1 have been labeled: Side a is opposite angle α, side b is opposite angle β, and side c is opposite angle γ. Also note that the largest side of a triangle is opposite the largest angle. Given any three of the six quantities indicated in Figure 1, we are interested in finding the remaining three, if they exist. This process is called **solving the triangle.**

The *law of sines* is developed in this section, and the *law of cosines* is developed in the next section. These two laws provide the basic tools for solving oblique triangles. If the given quantities include a side and the opposite angle, the law of sines should be used; otherwise, start with the law of cosines.

Before proceeding with specific examples, it is important to recall the rules in Table 1 regarding accuracy of angle and side measure. Table 1 is also repeated inside the back cover of the text for easy reference.

> **Calculator Calculations**
>
> When solving for a particular side or angle, carry out all operations within the calculator and then round to the appropriate number of significant digits (as specified in Table 1) at the end of the calculation. Your answer may still differ slightly from those in the book, depending on the order in which you solve for the sides and angles.

• Law of Sines Derivation

The law of sines is relatively easy to prove using the right triangle properties studied in Section 5-5. We will also use the fact that

$$\sin (180° - x) = \sin x$$

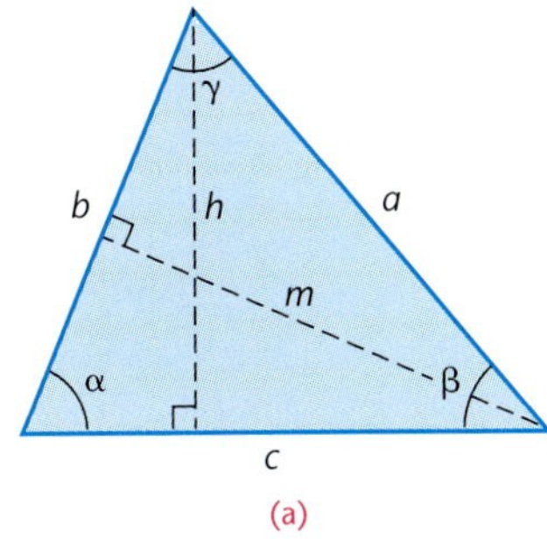

(a)

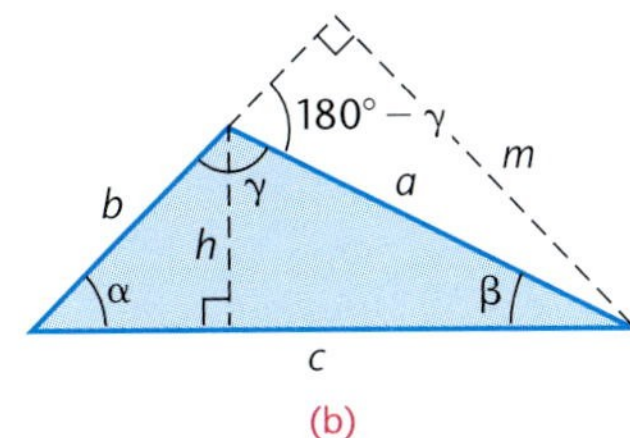

(b)

FIGURE 2

which is readily obtained using a difference identity (a good exercise for you). Referring to the triangles in Figure 2, we proceed as follows: For each triangle,

$$\sin \alpha = \frac{h}{b} \qquad \text{and} \qquad \sin \beta = \frac{h}{a}$$

Solving each equation for h, we obtain

$$h = b \sin \alpha \qquad \text{and} \qquad h = a \sin \beta$$

Thus,

$$b \sin \alpha = a \sin \beta$$

$$\frac{\sin \alpha}{a} = \frac{\sin \beta}{b} \tag{1}$$

Similarly, for each triangle in Figure 2,

$$\sin \alpha = \frac{m}{c} \qquad \text{and} \qquad \sin \gamma = \sin (180° - \gamma) = \frac{m}{a}$$

Solving each equation for m, we obtain

$$m = c \sin \alpha \qquad \text{and} \qquad m = a \sin \gamma$$

Thus,

$$c \sin \alpha = a \sin \gamma$$

$$\frac{\sin \alpha}{a} = \frac{\sin \gamma}{c} \tag{2}$$

If we combine equations (1) and (2), we obtain the **law of sines.**

Theorem 1 **Law of Sines**

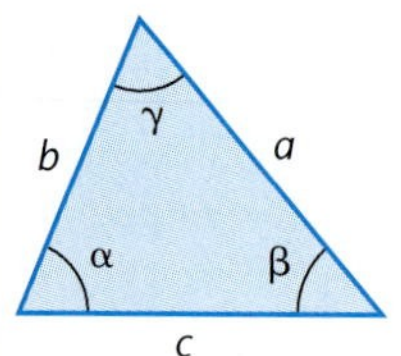

$$\frac{\sin \alpha}{a} = \frac{\sin \beta}{b} = \frac{\sin \gamma}{c}$$

In words, the ratio of the sine of an angle to its opposite side is the same as the ratio of the sine of either of the other angles to its opposite side.

Suppose that an angle of a triangle and its opposite side are known. Then the ratio of Theorem 1 can be calculated. So if one additional part of the triangle, either of the other angles or either of the other sides, is known, then the law of sines can be used to solve the triangle.

Thus, the law of sines is used to solve triangles, given:

1. Two sides and an angle opposite one of them (SSA)
2. Two angles and any side (ASA or AAS)

If the given information for a triangle consists of two sides and the included angle (SAS) or three sides (SSS), then the law of sines cannot be applied. The key to handling these two cases, the law of cosines, is developed in Section 7-2.

We will apply the law of sines to the easier ASA and AAS cases first, then we will turn to the more challenging SSA case.

• Solving the ASA and AAS Cases

EXAMPLE 1 **Solving the ASA Case**

Solve the triangle in Figure 3.

FIGURE 3

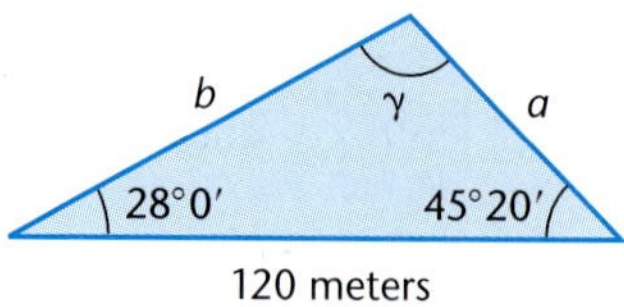

Solution We are given two angles and the included side, which is the ASA case. Find the third angle, then solve for the other two sides using the law of sines.

Solve for γ

$$\alpha + \beta + \gamma = 180°$$
$$\gamma = 180° - (\alpha + \beta)$$
$$= 180° - (28°0' + 45°20')$$
$$= 106°40'$$

Solve for a

$$\frac{\sin \alpha}{a} = \frac{\sin \gamma}{c}$$
$$a = \frac{c \sin \alpha}{\sin \gamma}$$
$$= \frac{120 \sin 28°0'}{\sin 106°40'}$$
$$= 58.8 \text{ meters}$$

Solve for b

$$\frac{\sin \beta}{b} = \frac{\sin \gamma}{c}$$
$$b = \frac{c \sin \beta}{\sin \gamma}$$
$$= \frac{120 \sin 45°20'}{\sin 106°40'}$$
$$= 89.1 \text{ meters}$$

Matched Problem 1 Solve the triangle in Figure 4.

FIGURE 4

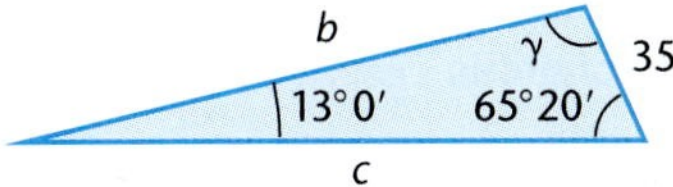

Note that the **AAS case** can always be converted to the ASA case by first solving for the third angle. For the ASA or AAS case to determine a unique triangle, the sum of the two angles must be between 0° and 180°, since the sum of all three angles in a triangle is 180° and no angle can be zero or negative.

• Solving the SSA Case—Including the Ambiguous Case

We now look at the case where we are given two sides and an angle opposite one of the sides–the SSA case. This case has several possible outcomes, depending on the measures of the two sides and the angle. Table 2 illustrates the various possibilities.

TABLE 2 SSA Variations

α	a ($h = b \sin \alpha$)	Number of triangles	Figure	Case
Acute	$0 < a < h$	0	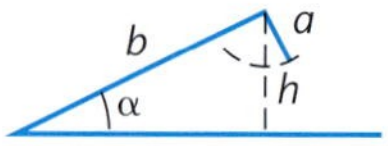	(a)
Acute	$a = h$	1	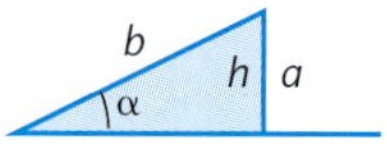	(b)
Acute	$h < a < b$	2	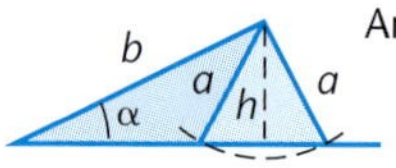 Ambiguous case	(c)
Acute	$a \geq b$	1	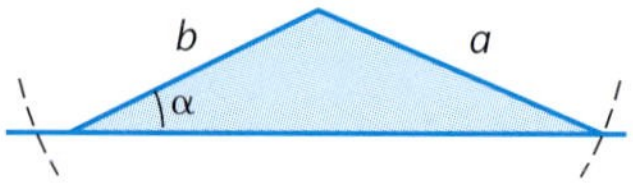	(d)
Obtuse	$0 < a \leq b$	0	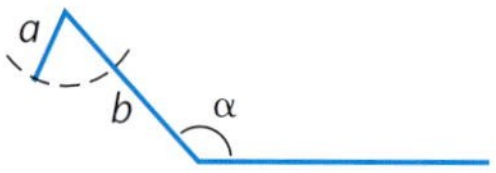	(e)
Obtuse	$a > b$	1	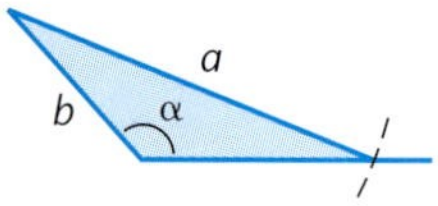	(f)

Table 2 need not be committed to memory. Usually a rough sketch of a particular situation will indicate which of the variations applies. The case where $h < a < b$ is referred to as the **ambiguous case,** because two triangles, one acute and the other obtuse, are always possible.

EXPLORE-DISCUSS 1 Discuss which cases in Table 2 apply and why if in the solution process of solving an SSA triangle with α acute it is found that:

(1) $\sin \beta > 1$

(2) $\sin \beta = 1$

(3) $0 < \sin \beta < 1$

EXAMPLE 2 Solving the SSA Case

Solve the triangle(s) with $\alpha = 123°$, $b = 23$ centimeters, and $a = 47$ centimeters.

Solution From a rough sketch (Fig. 5), we see that there is only one triangle:

FIGURE 5

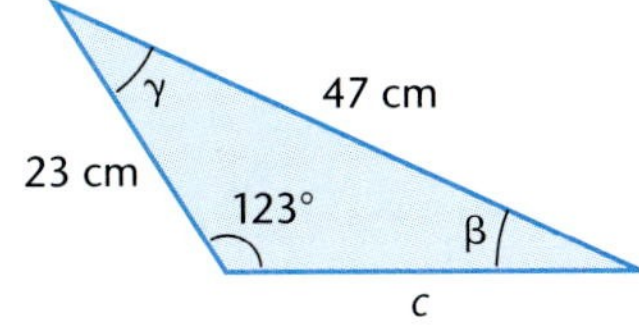

Solve for β

$$\frac{\sin \beta}{b} = \frac{\sin \alpha}{a}$$

$$\sin \beta = \frac{b \sin \alpha}{a} = \frac{23 \sin 123°}{47}$$

$$\beta = \sin^{-1}\left(\frac{23 \sin 123°}{47}\right) = 24°$$

Solve for γ

$$\alpha + \beta + \gamma = 180°$$

$$\gamma = 180° - 123° - 24° = 33°$$

Solve for c

$$\frac{\sin \alpha}{a} = \frac{\sin \gamma}{c}$$

$$c = \frac{a \sin \gamma}{\sin \alpha} = \frac{47 \sin 33°}{\sin 123°} = 31 \text{ centimeters}$$

Matched Problem 2 Solve the triangle(s) with $\beta = 98°$, $a = 62$ meters, and $b = 88$ meters.

EXAMPLE 3 Solving the SSA (Ambiguous) Case

Solve the triangle(s) with $\alpha = 26°$, $a = 1.0$ meter, and $b = 1.8$ meters.

Solution If we try to draw a triangle with the indicated sides and angle, we find that two triangles, I and II, are possible (Fig. 6). This is verified by the fact that $h < a < b$, where $h = b \sin \alpha = 0.79$ meters, $a = 1.0$ meter, and $b = 1.8$ meters.

FIGURE 6

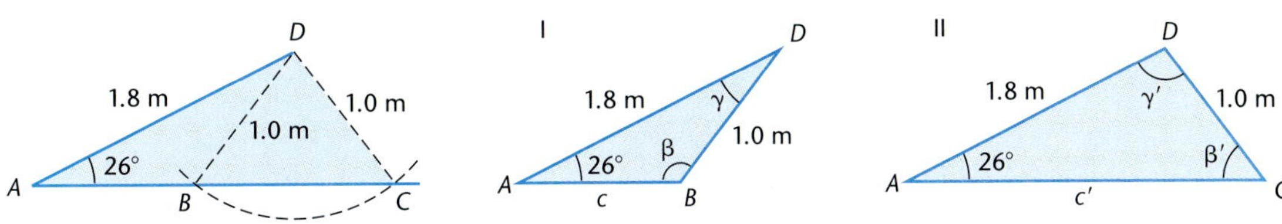

Solve for β and β′ We start by finding β and β′ using the law of sines:

$$\frac{\sin \beta}{b} = \frac{\sin \alpha}{a}$$

$$\sin \beta = \frac{b \sin \alpha}{a} = \frac{1.8 \sin 26°}{1.0} = 0.7891$$

Angle β can be either obtuse or acute:

$$\beta = 180° - \sin^{-1} 0.7891 \qquad \text{or} \qquad \beta' = \sin^{-1} 0.7891$$
$$= 180° - 52° = 128° \qquad\qquad = 52°$$

Solve for γ and γ'

We next find γ and γ':

$$\gamma = 180° - (26° + 128°) = 26°$$
$$\gamma' = 180° - (26° + 52°) = 102°$$

Solve for c and c'

Finally, we solve for c and c':

$$\frac{\sin\alpha}{a} = \frac{\sin\gamma}{c} \qquad\qquad \frac{\sin\alpha}{a} = \frac{\sin\gamma'}{c'}$$
$$c = \frac{a\sin\gamma}{\sin\alpha} \qquad\qquad c' = \frac{a\sin\gamma'}{\sin\alpha}$$
$$= \frac{1.0\sin 26°}{\sin 26°} \qquad\qquad = \frac{1.0\sin 102°}{\sin 26°}$$
$$= 1.0 \text{ meter} \qquad\qquad = 2.2 \text{ meters}$$

In summary:

Triangle I:	$\beta = 128°$	$\gamma = 26°$	$c = 1.0$ meter
Triangle II:	$\beta' = 52°$	$\gamma' = 102°$	$c' = 2.2$ meters

Matched Problem 3 Solve the triangle(s) with $a = 8$ kilometers, $b = 10$ kilometers, and $\alpha = 35°$.

The law of sines is useful in many applications, as can be seen in Example 4 and the applications in Exercise 7-1.

EXAMPLE 4 Surveying

To measure the length d of a lake (see Fig. 7), a baseline AB is established and measured to be 125 meters. Angles A and B are measured to be 41.6° and 124.3°, respectively. How long is the lake?

FIGURE 7

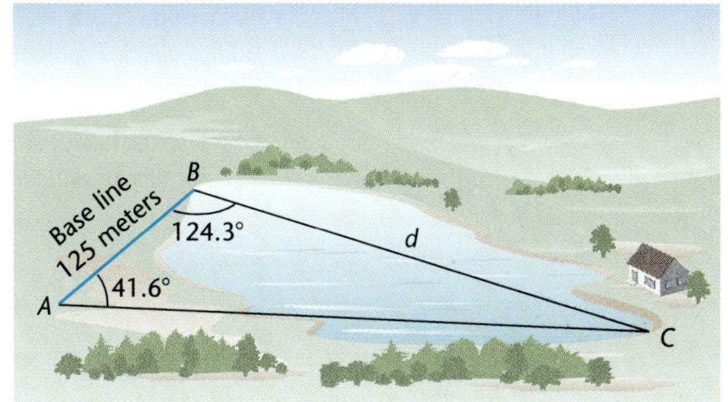

Solution Find angle C and use the law of sines.

$$\text{Angle } C = 180° - (124.3° + 41.6°) = 14.1°$$

$$\frac{\sin 14.1°}{125} = \frac{\sin 41.6°}{d}$$

$$d = 125\left(\frac{\sin 41.6°}{\sin 14.1°}\right) = 341 \text{ meters}$$

Matched Problem 4 In Example 4, find the distance AC.

Answers to Matched Problems

1. $\gamma = 101°40'$, $b = 141$, $c = 152$ 2. $\alpha = 44°$, $\gamma = 38°$, $c = 55$ m
3. $\beta = 134°$, $\beta' = 46°$, $\gamma = 11°$, $\gamma' = 99°$, $c = 2.7$ km, $c' = 14$ km 4. 424 m

EXERCISE 7-1

The labeling in the figure below is the convention we will follow in this exercise set. Your answers to some problems may differ slightly from those in the book, depending on the order in which you solve for the sides and angles of a given triangle.

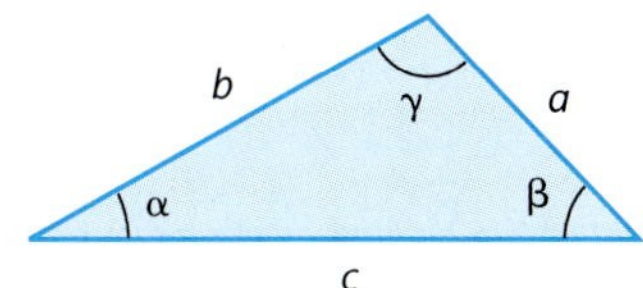

A

Solve each triangle in Problems 1–8.

1. $\alpha = 73°$, $\beta = 28°$, $c = 42$ feet

2. $\alpha = 41°$, $\beta = 33°$, $c = 21$ centimeters

3. $\alpha = 122°$, $\gamma = 18°$, $b = 12$ kilometers

4. $\beta = 43°$, $\gamma = 36°$, $a = 92$ millimeters

5. $\beta = 112°$, $\gamma = 19°$, $c = 23$ yards

6. $\alpha = 52°$, $\gamma = 105°$, $c = 47$ meters

7. $\alpha = 52°$, $\gamma = 47°$, $a = 13$ centimeters

8. $\beta = 83°$, $\gamma = 77°$, $c = 25$ miles

In Problems 9–20, are there zero, one, or two triangles with the given measurements? (Do not solve the triangle.) Which case, if any, of Table 2 applies? Explain.

9. $a = 5$ inches, $b = 6$ inches, $\alpha = 30°$

10. $a = 6$ inches, $b = 5$ inches, $\alpha = 30°$

11. $a = 4$ feet, $b = 8$ feet, $\alpha = 30°$

12. $a = 2$ feet, $b = 5$ feet, $\alpha = 30°$

13. $a = 5$ inches, $b = 8$ inches, $\alpha = 45°$

14. $a = 3$ inches, $b = 4$ inches, $\alpha = 45°$

15. $a = 4$ feet, $b = 5$ feet, $\alpha = 120°$

16. $a = 6$ feet, $b = 4$ feet, $\alpha = 120°$

17. $a = 8$ inches, $b = 6$ inches, $\alpha = 60°$

18. $a = 7$ inches, $b = 8$ inches, $\alpha = 60°$

19. $a = 13$ feet, $b = 12$ feet, $\alpha = 90°$

20. $a = 12$ feet, $b = 13$ feet, $\alpha = 90°$

B

Solve each triangle in Problems 21–32. If a problem has no solution, say so.

21. $\alpha = 118.3°$, $\gamma = 12.2°$, $b = 17.3$ feet

22. $\beta = 27.5°$, $\gamma = 54.5°$, $a = 9.27$ inches

23. $\alpha = 67.7°$, $\beta = 54.2°$, $b = 123$ meters

24. $\alpha = 122.7°$, $\beta = 34.4°$, $b = 18.3$ kilometers

25. $\alpha = 46.5°$, $a = 7.9$ millimeters, $b = 13.1$ millimeters

26. $\alpha = 26.3°$, $a = 14.7$ inches, $b = 35.2$ inches

27. $\alpha = 38.9°$, $a = 30.0$ inches, $b = 42.7$ inches

28. $\alpha = 27.3°$, $a = 135$ centimeters, $b = 244$ centimeters

29. $\alpha = 123.2°$, $a = 101$ yards, $b = 152$ yards

30. $\alpha = 137.3°$, $a = 13.9$ meters, $b = 19.1$ meters

31. $\beta = 29°30'$, $a = 43.2$ millimeters, $b = 56.5$ millimeters

32. $\beta = 33°50'$, $a = 673$ meters, $b = 1{,}240$ meters

C

33. Let $\alpha = 42.3°$ and $b = 25.2$ centimeters. Determine a value k so that if $0 < a < k$, there is no solution; if $a = k$, there is one solution; and if $k < a < b$, there are two solutions.

34. Let $\alpha = 37.3°$ and $b = 42.8$ centimeters. Determine a value k so that if $0 < a < k$, there is no solution; if $a = k$, there is one solution; and if $k < a < b$, there are two solutions.

35. Mollweide's equation,

$$(a - b)\cos\frac{\gamma}{2} = c\sin\frac{\alpha - \beta}{2}$$

is often used to check the final solution of a triangle, since all six parts of a triangle are involved in the equation. If the left side does not equal the right side after substitution, then an error has been made in solving a triangle. Use this equation to check Problem 1. (Because of rounding errors, both sides may not be exactly the same.)

36. (A) Use the law of sines and suitable identities to show that for any triangle

$$\frac{a - b}{a + b} = \frac{\tan\dfrac{\alpha - \beta}{2}}{\tan\dfrac{\alpha + \beta}{2}}$$

(B) Verify the formula with values from Problem 1.

APPLICATIONS 

37. Coast Guard. Two lookout posts, A and B (10.0 miles apart), are established along a coast to watch for illegal ships coming within the 3-mile limit. If post A reports a ship S at angle $BAS = 37°30'$ and post B reports the same ship at angle $ABS = 20°0'$, how far is the ship from post A? How far is the ship from the shore (assuming the shore is along the line joining the two observation posts)?

38. Fire Lookout. A fire at F is spotted from two fire lookout stations, A and B, which are 10.0 miles apart. If station B reports the fire at angle $ABF = 53°0'$ and station A reports the fire at angle $BAF = 28°30'$, how far is the fire from station A? From station B?

★ **39. Natural Science.** The tallest trees in the world grow in Redwood National Park in California; they are taller than a football field is long. Find the height of one of these trees, given the information in the figure. (The 100-foot measurement is accurate to 3 significant digits.)

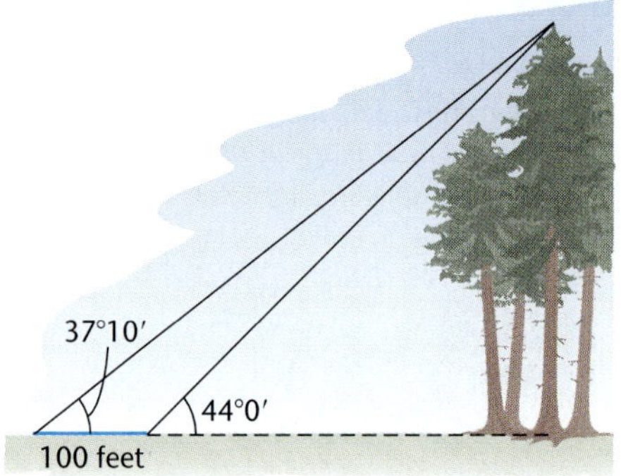

★ **40. Surveying.** To measure the height of Mt. Whitney in California, surveyors used a scheme like the one shown in the figure in Problem 39. They set up a horizontal baseline 2,000 feet long at the foot of the mountain and found the angle nearest the mountain to be 43°5′; the angle farthest from the mountain was found to be 38°0′. If the baseline was 5,000 feet above sea level, how high is Mt. Whitney above sea level?

41. **Engineering.** A 4.5-inch piston rod joins a piston to a 1.5-inch crankshaft (see figure). How far is the base of the piston from the center of the crankshaft (distance d) when the rod makes an angle of 9° with the centerline? There are two answers to the problem.

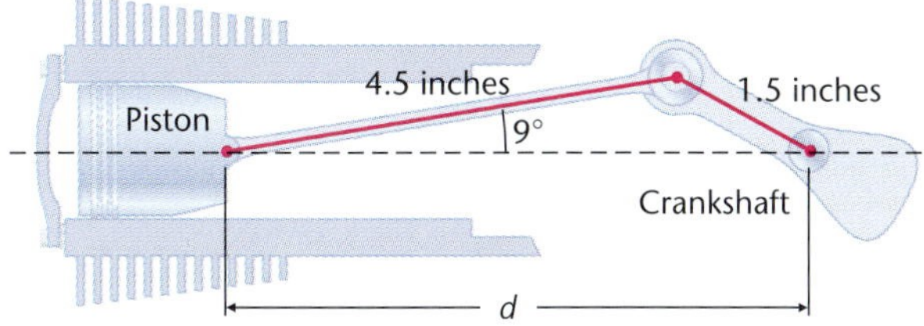

42. **Engineering.** Repeat Problem 41 if the piston rod is 6.3 inches, the crankshaft is 1.7 inches, and the angle is 11°.

43. **Astronomy.** The orbits of the Earth and Venus are approximately circular, with the sun at the center. A sighting of Venus is made from Earth, and the angle α is found to be 18°40′. If the radius of the orbit of the Earth is 1.495×10^8 kilometers and the radius of the orbit of Venus is 1.085×10^8 kilometers, what are the possible distances from the Earth to Venus (see figure)?

44. **Astronomy.** In Problem 43, find the maximum angle α. [*Hint:* The angle is maximum when a straight line joining the Earth and Venus is tangent to Venus's orbit.]

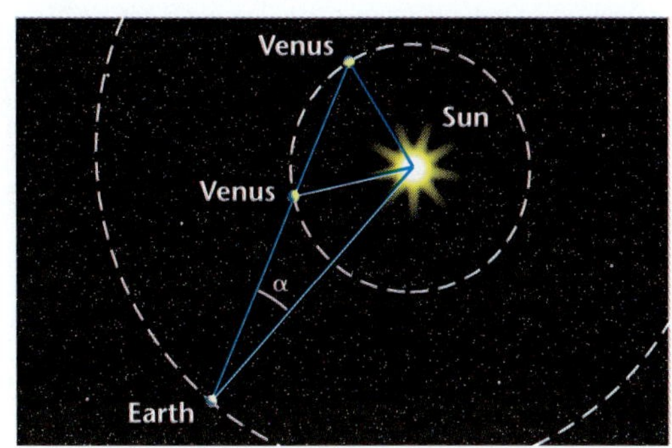

★ 45. **Surveying.** A tree growing on a hillside casts a 102-foot shadow straight down the hill (see figure). Find the vertical height of the tree if, relative to the horizontal, the hill slopes 15.0° and the angle of elevation of the sun is 62.0°.

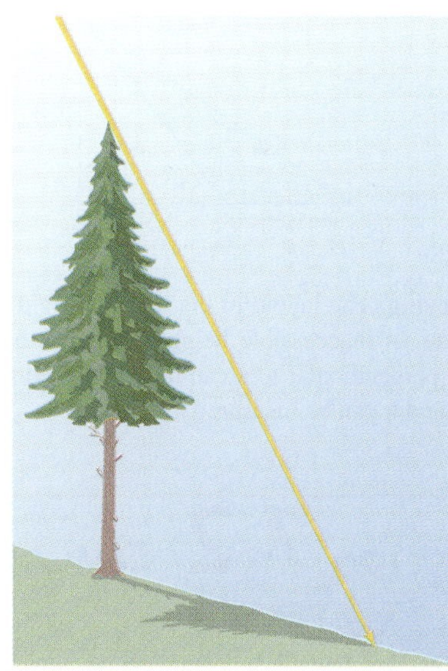

★ 46. **Surveying.** Find the height of the tree in Problem 45 if the shadow length is 157 feet and, relative to the horizontal, the hill slopes 11.0° and the angle of elevation of the sun is 42.0°.

★ 47. **Life Science.** A cross section of the cornea of an eye, a circular arc, is shown in the figure. Find the arc radius R and the arc length s, given the chord length $C = 11.8$ millimeters and the central angle $\theta = 98.9°$.

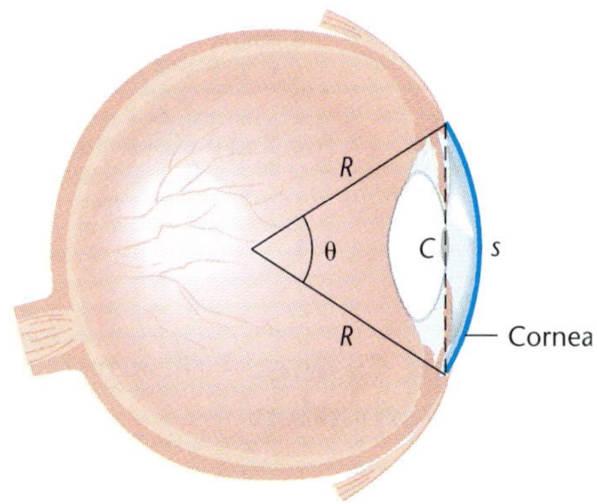

★ 48. **Life Science.** Referring to the figure, find the arc radius R and the arc length s, given the chord length $C = 10.2$ millimeters and the central angle $\theta = 63.2°$.

★ 49. **Surveying.** The procedure illustrated in Problems 39 and 40 is used to determine an inaccessible height h when a baseline d on a line perpendicular to h can be established (see the figure on the next page) and the angles α and β can be measured. Show that

$$h = d\left[\frac{\sin \alpha \sin \beta}{\sin (\beta - \alpha)}\right]$$

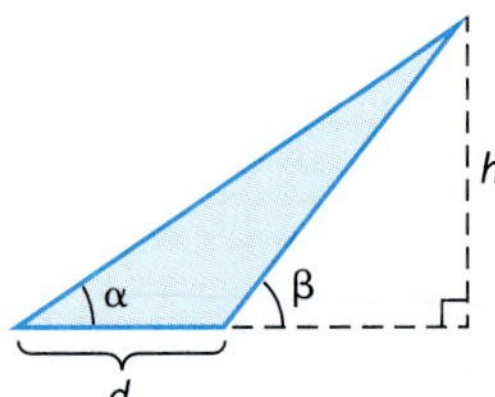

★★ **50. Surveying.** The layout in the figure at right is used to determine an inaccessible height h when a baseline d in a plane perpendicular to h can be established and the angles α, β, and γ can be measured. Show that

$$h = d \sin \alpha \csc (\alpha + \beta) \tan \gamma$$

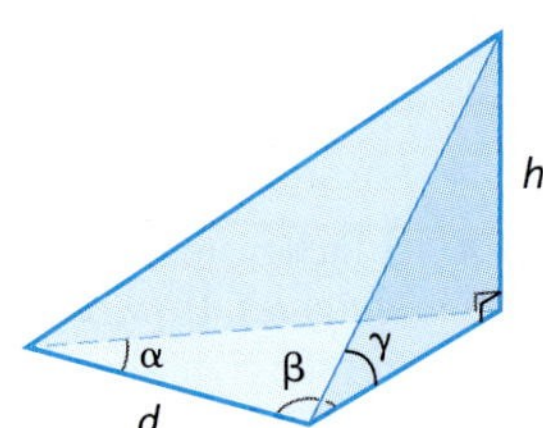

SECTION 7-2 Law of Cosines

- Law of Cosines Derivation
- Solving the SAS Case
- Solving the SSS Case

If in a triangle two sides and the included angle are given (SAS) or three sides are given (SSS), the law of sines cannot be used to solve the triangle—neither case involves an angle and its opposite side (Fig. 1). Both cases can be solved starting with the *law of cosines,* which is the subject matter for this section.

FIGURE 1

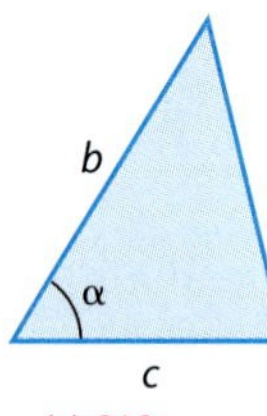

(a) SAS case

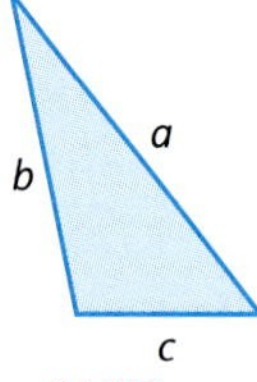

(b) SSS case

• Law of Cosines Derivation

Theorem 1 states the *law of cosines.*

Theorem 1 **Law of Cosines**

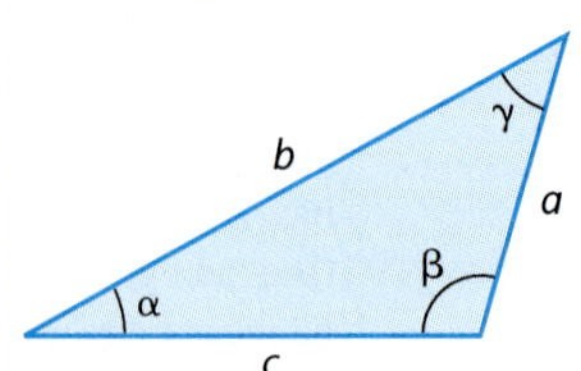

$$a^2 = b^2 + c^2 - 2bc \cos \alpha$$
$$b^2 = a^2 + c^2 - 2ac \cos \beta$$
$$c^2 = a^2 + b^2 - 2ab \cos \gamma$$

All three equations say essentially the same thing.

The law of cosines is used to solve triangles, given:

1. Two sides and the included angle (SAS)
2. Three sides (SSS)

We will establish the first equation in Theorem 1. The other two equations then can be obtained from this one simply by relabeling the figure. We start by locating a triangle in a rectangular coordinate system. Figure 2 shows three typical triangles.

For an arbitrary triangle located as in Figure 2, the distance-between-two-points formula is used to obtain

$$a = \sqrt{(h-c)^2 + (k-0)^2}$$

$$a^2 = (h-c)^2 + k^2 \qquad \text{Square both sides.}$$

$$= h^2 - 2hc + c^2 + k^2 \tag{1}$$

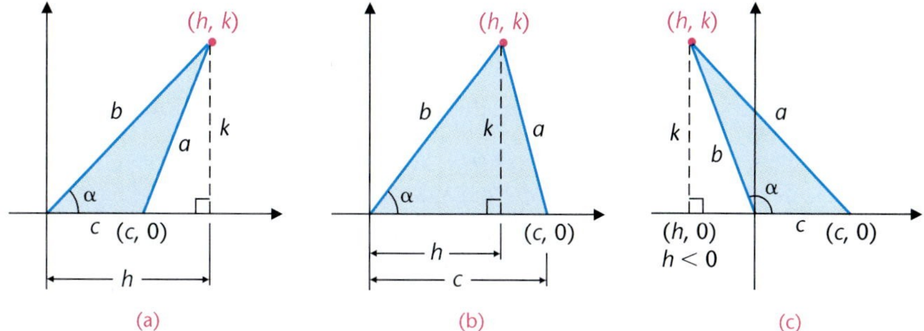

FIGURE 2 Three representative triangles.

From Figure 2, we note that

$$b^2 = h^2 + k^2$$

Substituting b^2 for $h^2 + k^2$ in equation (1), we obtain

$$a^2 = b^2 + c^2 - 2hc \tag{2}$$

But

$$\cos \alpha = \frac{h}{b}$$

$$h = b \cos \alpha$$

Thus, by replacing h in equation (2) with $b \cos \alpha$, we reach our objective:

$$a^2 = b^2 + c^2 - 2bc \cos \alpha$$

[*Note:* If α is acute, then $\cos \alpha$ is positive; if α is obtuse, then $\cos \alpha$ is negative.]

• Solving the SAS Case

For the SAS case, start by using the law of cosines to find the side opposite the given angle. Then use either the law of cosines or the law of sines to find a second angle. Because of the simpler computations, the law of sines will generally be used to find the second angle.

EXPLORE-DISCUSS 1 After using the law of cosines to find the side opposite the angle for a SAS case, the law of sines is used to find a second angle. Figure 2 (a) shows that there are two choices for a second angle.

(A) If the given angle is obtuse, can either of the remaining angles be obtuse? Explain.

(B) If the given angle is acute, then one of the remaining angles may or may not be obtuse. Explain why choosing the angle opposite the shorter side guarantees the selection of an acute angle.

(C) Starting with $(\sin \alpha)/a = (\sin \beta)/b$, show that

$$\alpha = \sin^{-1}\left(\frac{a \sin \beta}{b}\right) \tag{1}$$

(D) Explain why equation (1) gives us the correct angle α only if α is acute.

The above discussion leads to the following strategy for solving the SAS case:

Strategy for Solving the SAS Case

Step	Find	Method
1.	Side opposite given angle	Law of cosines
2.	Second angle (Find the angle opposite the shorter of the two given sides—this angle will always be acute.)	Law of sines
3.	Third angle	Subtract the sum of the measures of the given angle and the angle found in step 2 from 180°.

EXAMPLE 1 Solving the SAS Case

Solve the triangle in Figure 3.

FIGURE 3

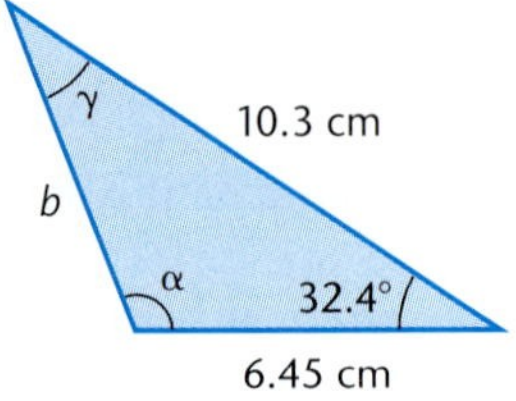

Solution

Solve for b Use the law of cosines:

$$b^2 = a^2 + c^2 - 2ac \cos \beta \quad \text{Solve for } b.$$

$$b = \sqrt{a^2 + c^2 - 2ac \cos \beta}$$

$$= \sqrt{(10.3)^2 + (6.45)^2 - 2(10.3)(6.45) \cos 32.4°}$$

$$= 5.96 \text{ cm}$$

Solve for γ Since side c is shorter than side a, γ must be acute, and the law of sines is used to solve for γ.

$$\frac{\sin \gamma}{c} = \frac{\sin \beta}{b} \quad \text{Solve for } \sin \gamma.$$

$$\sin \gamma = \frac{c \sin \beta}{b} \quad \text{Solve for } \gamma.$$

$$\gamma = \sin^{-1}\left(\frac{c \sin \beta}{b}\right)$$

Since γ is acute, the inverse sine function gives us γ directly.

$$= \sin^{-1}\left(\frac{6.45 \sin 32.4°}{5.96}\right)$$

$$= 35.4°$$

Solve for α

$$\alpha = 180° - (\beta + \gamma)$$

$$= 180° - (32.4° + 35.4°) = 112.2°$$

Matched Problem 1 Solve the triangle with $\alpha = 77.5°$, $b = 10.4$ feet, and $c = 17.7$ feet.

• Solving the SSS Case

Starting with three sides of a triangle, the problem is to find the three angles. Subsequent calculations are simplified if we solve for the obtuse angle first, if present. The law of cosines is used for this purpose. A second angle, which must be acute, can be found using either law, although computations are usually simpler with the law of sines.

EXPLORE-DISCUSS 2 (A) Starting with $a^2 = b^2 + c^2 - 2bc \cos \alpha$, show that

$$\alpha = \cos^{-1}\left(\frac{a^2 - b^2 - c^2}{-2bc}\right) \qquad (2)$$

(B) Does equation (2) give us the correct angle α irrespective of whether α is acute or obtuse? Explain.

The above discussion leads to the following strategy for solving the SSS case.

Strategy for Solving the SSS Case

Step	Find	Method
1.	Angle opposite longest side—this will take care of an obtuse angle, if present.	Law of cosines
2.	Either of the remaining angles, which will be acute (why?)	Law of sines
3.	Third angle	Subtract the sum of the measures of the angles found in steps 1 and 2 from 180°.

EXAMPLE 2 Solving the SSS Case

Solve the triangle with $a = 27.3$ meters, $b = 17.8$ meters, and $c = 35.2$ meters.

Solution Three sides of the triangle are given and we are to find the three angles. This is the SSS case.

Sketch the triangle (Fig. 4) and use the law of cosines to find the largest angle, then use the law of sines to find one of the two remaining acute angles.

FIGURE 4

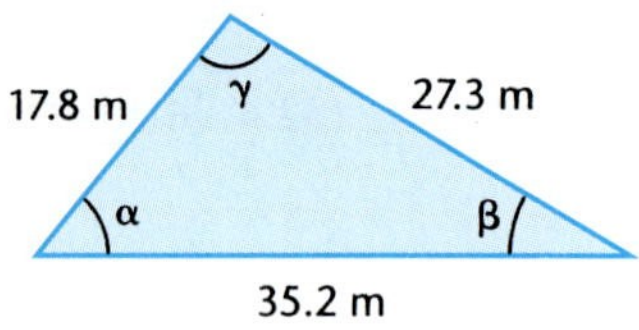

Since γ is the largest angle, we solve for it first using the law of cosines.

Solve for γ

$$c^2 = a^2 + b^2 - 2ab\cos\gamma$$

$$\cos\gamma = \frac{a^2 + b^2 - c^2}{2ab} \qquad \text{Solve for } \cos\gamma.$$

$$\gamma = \cos^{-1}\left(\frac{a^2 + b^2 - c^2}{2ab}\right) \qquad \text{Solve for } \gamma.$$

$$= \cos^{-1}\left[\frac{(27.3)^2 + (17.8)^2 - (35.2)^2}{2(27.3)(17.8)}\right]$$

$$= 100.5°$$

Solve for α We now solve for either α or β, using the law of sines. We choose α.

$$\frac{\sin\alpha}{a} = \frac{\sin\gamma}{c}$$

$$\sin \alpha = \frac{a \sin \gamma}{c} = \frac{27.3 \sin 100.5}{35.2} \qquad \text{Solve for } \sin \alpha.$$

$$\alpha = \sin^{-1}\left(\frac{27.3 \sin 100.5}{35.2}\right) \qquad \text{Solve for } \alpha.$$

$$= 49.7° \qquad \alpha \text{ is acute.}$$

Solve for β

$$\alpha + \beta + \gamma = 180°$$

$$\beta = 180° - (\alpha + \gamma)$$

$$= 180° - (49.7° + 100.5°)$$

$$= 29.8°$$

Matched Problem 2 Solve the triangle with $a = 1.25$ yards, $b = 2.05$ yards, and $c = 1.52$ yards.

EXAMPLE 3 Finding the Side of a Regular Polygon

If a seven-sided regular polygon is inscribed in a circle of radius 22.8 centimeters, find the length of one side of the polygon.

Solution Sketch a figure (Fig. 5) and use the law of cosines:

FIGURE 5

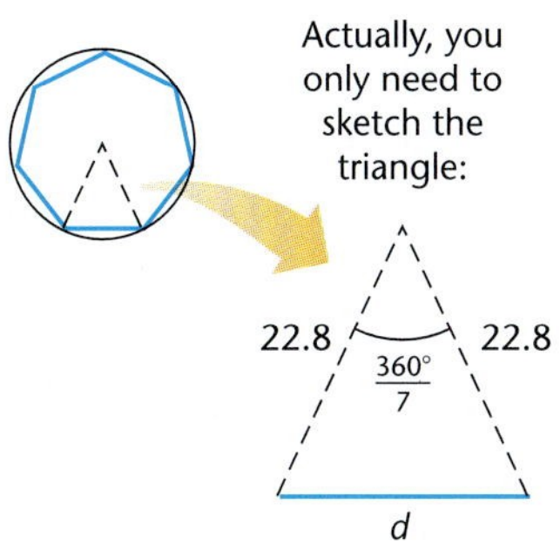

$$d^2 = 22.8^2 + 22.8^2 - 2(22.8)(22.8) \cos \frac{360°}{7}$$

$$d = \sqrt{2(22.8)^2 - 2(22.8)^2 \cos \frac{360°}{7}}$$

$$= 19.8 \text{ centimeters}$$

Matched Problem 3 If an 11-sided regular polygon is inscribed in a circle with radius 4.63 inches, find the length of one side of the polygon.

Answers to Matched Problems

1. $a = 18.5$ ft, $\beta = 33.3°$, $\gamma = 69.2°$
2. $\alpha = 37.4°$, $\beta = 95.0°$, $\gamma = 47.6°$ **3.** 2.61 in.

EXERCISE 7-2

The labeling in the figure below is the convention we will follow in this exercise set. Your answers to some problems may differ slightly from those in the book, depending on the order in which you solve for the sides and angles of a given triangle.

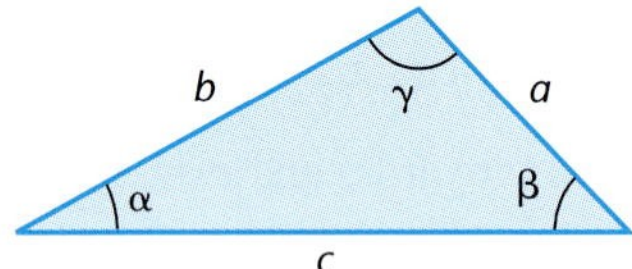

A

1. Referring to the figure above, if $\alpha = 47.3°$, $b = 11.7$ centimeters, and $c = 6.04$ centimeters, which of the two angles, β or γ, can you say for certain is acute and why?

2. Referring to the figure above, if $\alpha = 93.5°$, $b = 5.34$ inches, and $c = 8.77$ inches, which of the two angles, β or γ, can you say for certain is acute and why?

Solve each triangle in Problems 3–6.

3. $\alpha = 71.2°$, $b = 5.32$ yards, $c = 5.03$ yards

4. $\beta = 57.3°$, $a = 6.08$ centimeters, $c = 5.25$ centimeters

5. $\gamma = 120°20'$, $a = 5.73$ millimeters, $b = 10.2$ millimeters

6. $\alpha = 135°50'$, $b = 8.44$ inches, $c = 20.3$ inches

B

7. Referring to the figure at the beginning of the exercise, if $a = 13.5$ feet, $b = 20.8$ feet, and $c = 8.09$ feet, then if the triangle has an obtuse angle, which angle must it be and why?

8. Suppose you are told that a triangle has sides $a = 12.5$ centimeters, $b = 25.3$ centimeters, and $c = 10.7$ centimeters. Explain why the triangle has no solution.

Solve each triangle in Problems 9–12 if the triangle has a solution. Use decimal degrees for angle measure.

9. $a = 4.00$ meters, $b = 10.2$ meters, $c = 9.05$ meters

10. $a = 10.5$ miles, $b = 20.7$ miles, $c = 12.2$ miles

11. $a = 6.00$ kilometers, $b = 5.30$ kilometers, $c = 5.52$ kilometers

12. $a = 31.5$ meters, $b = 29.4$ meters, $c = 33.7$ meters

Problems 13–26 represent a variety of problems involving both the law of sines and the law of cosines. Solve each triangle. If a problem does not have a solution, say so.

13. $\alpha = 92.6°$, $\beta = 88.9°$, $a = 15.2$ centimeters

14. $\alpha = 79.4°$, $\gamma = 102.3°$, $a = 6.4$ millimeters

15. $\beta = 126.2°$, $a = 13.8$ inches, $c = 12.5$ inches

16. $\gamma = 19.1°$, $a = 16.4$ yards, $b = 28.2$ yards

17. $a = 23.4$ meters, $b = 6.9$ meters, $c = 31.3$ meters

18. $a = 86$ inches, $b = 32$ inches, $c = 53$ inches

19. $\beta = 38.4°$, $a = 11.5$ inches, $b = 14.0$ inches

20. $\gamma = 66.4°$, $b = 25.5$ meters, $c = 25.5$ meters

21. $a = 32.9$ meters, $b = 42.4$ meters, $c = 20.4$ meters

22. $a = 10.5$ centimeters, $b = 5.23$ centimeters, $c = 8.66$ centimeters

23. $\gamma = 58.4°$, $b = 7.23$ meters, $c = 6.54$ meters

24. $\alpha = 46.7°$, $a = 18.1$ meters, $b = 22.6$ meters

25. $\beta = 39.8°$, $a = 12.5$ inches, $b = 7.31$ inches

26. $\gamma = 47.9°$, $b = 35.2$ inches, $c = 25.5$ inches

C

27. Show, using the law of cosines, that if $\gamma = 90°$, then $c^2 = a^2 + b^2$ (the Pythagorean theorem).

28. Show, using the law of cosines, that if $c^2 = a^2 + b^2$, then $\gamma = 90°$.

29. Show that for any triangle,

$$\frac{a^2 + b^2 + c^2}{2abc} = \frac{\cos \alpha}{a} + \frac{\cos \beta}{b} + \frac{\cos \gamma}{c}$$

30. Show that for any triangle,

$$a = b \cos \gamma + c \cos \beta$$

31. Give a solution to Example 3 that does not use the law of cosines by showing that $\frac{d}{2} = 22.8 \sin \frac{360°}{14}$.

32. Show that the length d of one side of an n-sided regular polygon, inscribed in a circle of radius r, is given by $d = 2r \sin \frac{180^\circ}{n}$.

APPLICATIONS

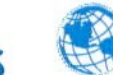

33. **Surveying.** To find the length AB of a small lake, a surveyor measured angle ACB to be 96°, AC to be 91 yards, and BC to be 71 yards. What is the approximate length of the lake?

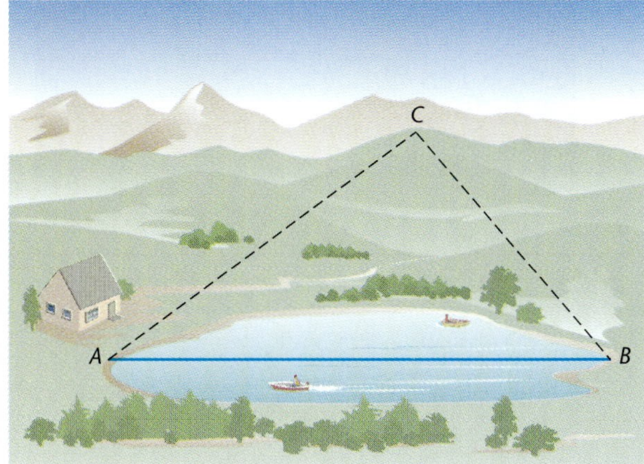

34. **Surveying.** Suppose the figure for this problem represents the base of a large rock outcropping on a farmer's land. If a surveyor finds $\angle ACB = 110°$, $AC = 85$ meters, and $BC = 73$ meters, what is the approximate length (to one decimal place) of the outcropping?

35. **Geometry.** Two adjacent sides of a parallelogram meet at an angle of 35°10′ and have lengths of 3 and 8 feet. What is the length of the shorter diagonal of the parallelogram (to 3 significant digits)?

36. **Geometry.** What is the length of the longer diagonal of the parallelogram in Problem 35 (to 3 significant digits)?

37. **Navigation.** Los Angeles and Las Vegas are approximately 200 miles apart. A pilot 80 miles from Los Angeles finds that she is 6°20′ off course relative to her start in Los Angeles. How far is she from Las Vegas at this time? (Compute the answer to 3 significant digits.)

38. **Search and Rescue.** At noon, two search planes set out from San Francisco to find a downed plane in the ocean. Plane A travels due west at 400 miles per hour, and plane B flies northwest at 500 miles per hour. At 2 P.M. plane A spots the survivors of the downed plane and radios plane B to come and assist in the rescue. How far is plane B from plane A at this time (to 3 significant digits)?

39. **Geometry.** Find the perimeter of a pentagon inscribed in a circle of radius 12.6 meters.

40. **Geometry.** Find the perimeter of a nine-sided regular polygon inscribed in a circle of radius 7.09 centimeters.

★ 41. **Analytic Geometry.** If point A in the figure has coordinates (3, 4) and point B has coordinates (4, 3), find the radian measure of angle θ to three decimal places.

★ 42. **Analytic Geometry.** If point A in the figure has coordinates (4, 3) and point B has coordinates (5, 1), find the radian measure of angle θ to three decimal places.

★ 43. **Engineering.** Three circles of radius 2.03, 5.00, and 8.20 centimeters are tangent to one another (see figure). Find the three angles formed by the lines joining their centers (to the nearest 10′).

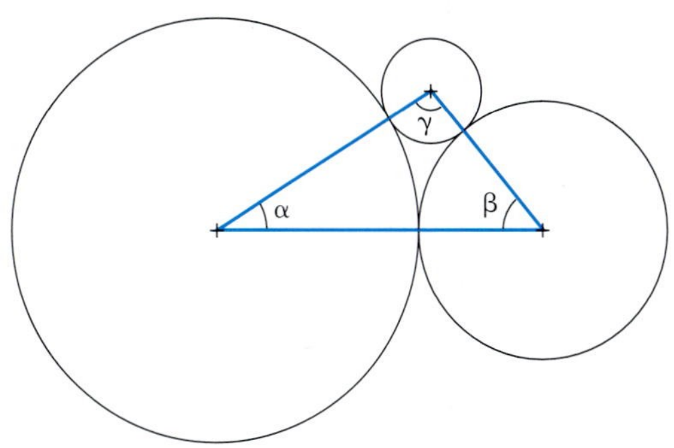

★ 44. **Engineering.** Three circles of radius 2.00, 5.00, and 8.00 inches are tangent to each other (see figure). Find the three angles formed by the lines joining their centers (to the nearest 10′).

45. **Geometry.** A rectangular solid has sides as indicated in the figure. Find $\angle CAB$ to the nearest degree.

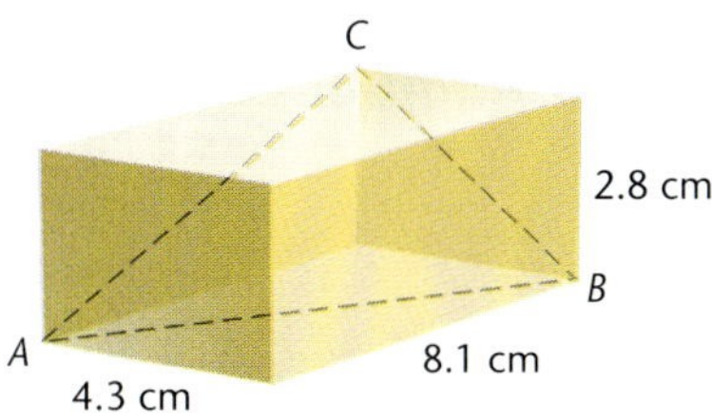

46. **Geometry.** Referring to the figure, find $\angle ACB$ to the nearest degree.

★ 47. **Space Science.** For communications between a space shuttle and the White Sands tracking station in southern New Mexico, two satellites are placed in geostationary orbit, 130° apart relative to the center of the Earth and 22,300 miles above the surface of the Earth (see figure). (A satellite in **geostationary** orbit remains stationary above a fixed point on the surface of the Earth.) Radio signals are sent from the tracking station by way of the satellites to the shuttle, and vice versa. This system allows the tracking station to be in contact with the shuttle over most of the Earth's surface. How far to the nearest 100 miles is one of the geostationary satellites from the White Sands tracking station W? The radius of the Earth is 3,964 miles.

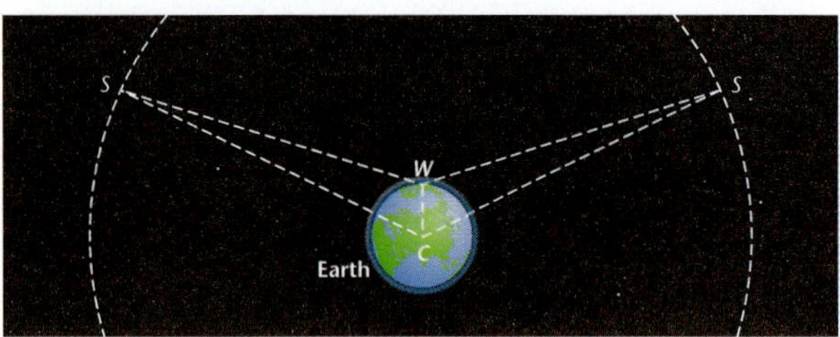

★ **48. Space Science.** A satellite S, in circular orbit around the Earth, is sighted by a tracking station T (see figure). The distance TS is determined by radar to be 1,034 miles, and the angle of elevation above the horizon is 32.4°. How high is the satellite above the Earth at the time of the sighting? The radius of the Earth is 3,964 miles.

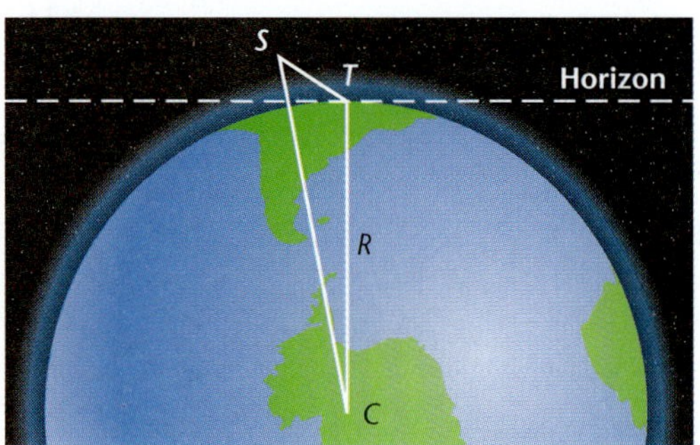

SECTION 7-3 Geometric Vectors

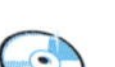

- Geometric Vectors and Vector Addition
- Velocity Vectors
- Force Vectors
- Resolution of Vectors into Vector Components

Many physical quantities, such as length, area, or volume, can be completely specified by a single real number. Other quantities, such as directed distances, velocities, and forces, require for their complete specification both a magnitude and a direction. The former are often called **scalar quantities,** and the latter are called **vector quantities.**

In this section we limit our discussion to the intuitive idea of geometric vectors in a plane. In Section 7-4 we introduce algebraic vectors, a first step in the generalization of a concept that has far-reaching consequences. Vectors are widely used in many areas of science and engineering.

• Geometric Vectors and Vector Addition

A line segment to which a direction has been assigned is called a **directed line segment.** A **geometric vector** is a directed line segment and is represented by an arrow (see Fig. 1). A vector with an *initial point* O and a **terminal point** P (the end with the arrowhead) is denoted by $\overrightarrow{OP}$. Vectors are also denoted by a boldface letter, such as **v**. Since it is difficult to write boldface on paper, we suggest that you use an arrow over a single letter, such as $\vec{v}$, when you want the letter to denote a vector.

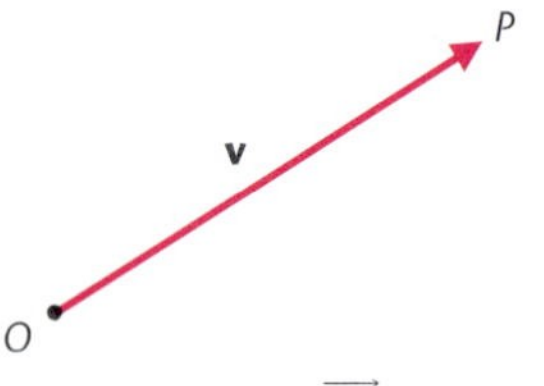

FIGURE 1 Vector $\overrightarrow{OP}$, or **v**.

The **magnitude** of the vector $\overrightarrow{OP}$, denoted by $|\overrightarrow{OP}|$, $|\vec{v}|$ or $|\mathbf{v}|$, is the length of the directed line segment. Two vectors have the **same direction** if they are parallel and point in the same direction. Two vectors have **opposite direction** if they are parallel and point in opposite directions. The **zero vector,** denoted by $\vec{0}$ or **0**, has a magni-

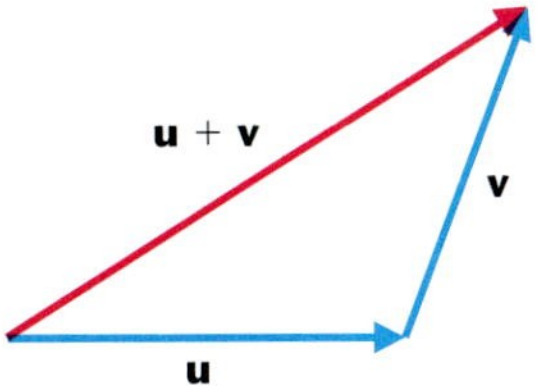

FIGURE 2 Vector addition: tail-to-tip rule.

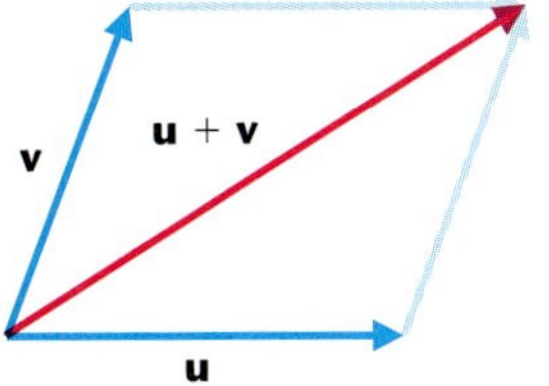

FIGURE 3 Vector addition: parallelogram rule.

tude of zero and an arbitrary direction. Two vectors are **equal** if they have the same magnitude and direction. Thus, a vector may be **translated** from one location to another as long as the magnitude and direction do not change.

The **sum of two vectors u and v** can be defined using the **tail-to-tip rule:** Translate **v** so that its tail end (initial point) is at the tip end (terminal point) of **u**. Then, the vector from the tail end of **u** to the tip end of **v** is the sum, denoted by $\mathbf{u} + \mathbf{v}$, of the vectors **u** and **v** (see Fig. 2).

The sum of two nonparallel vectors also can be defined using the **parallelogram rule:** The **sum of two nonparallel vectors u and v** is the diagonal of the parallelogram formed using **u** and **v** as adjacent sides (see Fig. 3). If **u** and **v** are parallel, use the tail-to-tip rule.

Both rules give the same sum. The choice of which rule to use depends on the situation and what seems most natural.

The vector $\mathbf{u} + \mathbf{v}$ is also called the **resultant** of the two vectors **u** and **v**, and **u** and **v** are called **vector components** of $\mathbf{u} + \mathbf{v}$. It is useful to observe that vector addition is **commutative** and **associative**. That is, $\mathbf{u} + \mathbf{v} = \mathbf{v} + \mathbf{u}$ and $\mathbf{u} + (\mathbf{v} + \mathbf{w}) = (\mathbf{u} + \mathbf{v}) + \mathbf{w}$.

EXPLORE-DISCUSS 1 If **a**, **b** and **c** represent three arbitrary geometric vectors, illustrate using either definition of vector addition that:

1. $\mathbf{a} + \mathbf{b} = \mathbf{b} + \mathbf{a}$
2. $\mathbf{a} + (\mathbf{b} + \mathbf{c}) = (\mathbf{a} + \mathbf{b}) + \mathbf{c}$

• Velocity Vectors

A vector that represents the direction and speed of an object in motion is called a **velocity vector.** Problems involving objects in motion often can be analyzed using vector methods. Many of these problems involve the use of a **navigational compass,** which is marked clockwise in degrees starting at north as indicated in Figure 4.

FIGURE 4

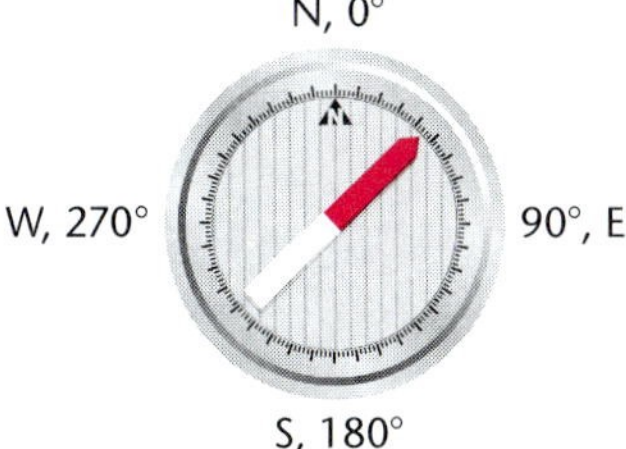

EXAMPLE 1 Apparent and Actual Velocity

An airplane has a compass heading (the direction the plane is pointing) of 85° and an airspeed (relative to the air) of 140 miles per hour. The wind is blowing from north to south at 66 miles per hour. The velocity of a plane relative to the air is called **apparent velocity,** and the velocity relative to the ground is called **resultant,** or **actual velocity.** The resultant velocity is the vector sum of the apparent velocity and the wind velocity. Find the resultant velocity; that is, find the actual speed and direction of the airplane relative to the ground. Directions are given to the nearest degree and magnitudes to 2 significant digits.

Solution Geometric vectors [Fig. 5 (a)] are used to represent the apparent velocity vector and the wind velocity vector. Add the two vectors using the tail-to-tip method of addition of vectors to obtain the resultant (actual) velocity vector [Fig. 5 (b)]. From the vector diagram [Fig. 5 (b)], we obtain the triangle in Figure 6 and solve for γ, c, and α.

FIGURE 5

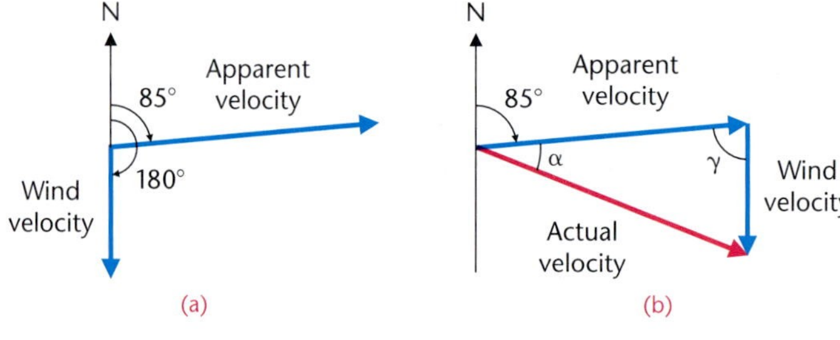

FIGURE 6

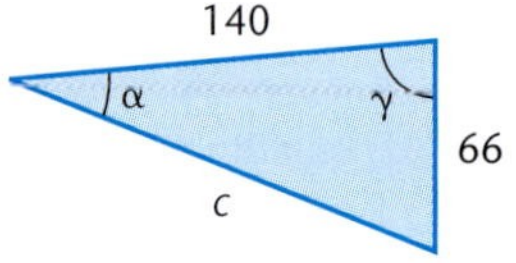

Solve for γ Since the wind velocity vector is parallel to the north–south line, $\gamma = 85°$ [alternate interior angles of two parallel lines cut by a transversal are equal—see Fig. 5 (b)].

Solve for c Use the law of cosines:

$$\begin{aligned} c^2 &= a^2 + b^2 - 2ab\cos\gamma \\ c &= \sqrt{a^2 + b^2 - 2ab\cos\gamma} \\ &= \sqrt{66^2 + 140^2 - 2(66)(140)\cos 85°} \\ &= 150 \text{ miles per hour} \end{aligned}$$

Speed relative to the ground.

Solve for α Use the law of sines:

$$\frac{\sin\alpha}{a} = \frac{\sin\gamma}{c}$$

$$\begin{aligned} \alpha &= \sin^{-1}\left(\frac{a\sin\gamma}{b}\right) \\ &= \sin^{-1}\left(\frac{66\sin 85}{150}\right) = 26° \end{aligned}$$

$$\text{Actual heading} = 85° + \alpha = 85° + 26° = 111°$$

Thus, the magnitude and direction of the resultant velocity vector are 150 miles per hour and 111°, respectively. That is, the plane, relative to the ground, is traveling at 150 miles per hour in a direction of 111°.

Matched Problem 1 A river is flowing southwest (225°) at 3.0 miles per hour. A boat crosses the river with a compass heading of 90°. If the speedometer on the boat reads 5.0 miles per hour (the boat's speed relative to the water), what is the resultant velocity? That is, what is the boat's actual speed and direction relative to the ground? Directions are to the nearest degree, and magnitudes are to 2 significant digits.

• Force Vectors

A vector that represents the direction and magnitude of an applied force is called a **force vector.** If an object is subjected to two forces, then the sum of these two forces, the **resultant force,** is a single force. If the resultant force replaced the original two forces, it would act on the object in the same way as the two original forces taken together. In physics it is shown that the resultant force vector can be obtained using vector addition to add the two individual force vectors. It seems natural to use the parallelogram rule for adding force vectors, as is illustrated in the next example.

EXAMPLE 2 Finding the Resultant Force

Two forces of 30 and 70 pounds act on a point in a plane. If the angle between the force vectors is 40°, what are the magnitude and direction (relative to the 70-pound force) of the resultant force? The magnitudes of the forces are to 2 significant digits and the angles to the nearest degree.

Solution We start with a diagram (Fig. 7), letting geometric vectors represent the various forces:

FIGURE 7

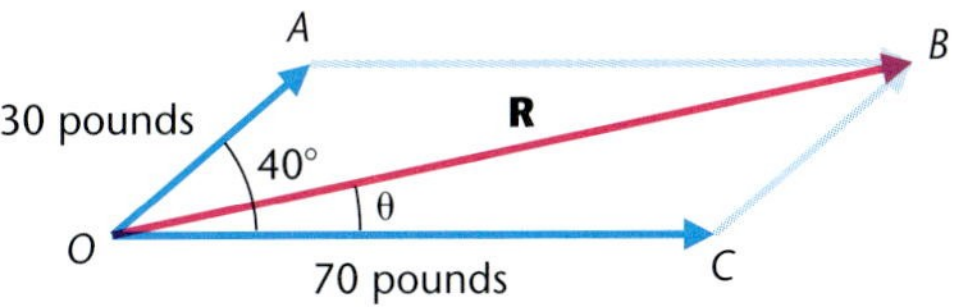

Because adjacent angles in a parallelogram are supplementary, the measure of angle $OCB = 180° - 40° = 140°$. We can now find the magnitude of the resultant vector **R** using the law of cosines (see Fig. 8):

$$|\mathbf{R}|^2 = 30^2 + 70^2 - 2(30)(70)\cos 140°$$

$$|\mathbf{R}| = \sqrt{30^2 + 70^2 - 2(30)(70)\cos 140°}$$

$$= 95 \text{ pounds}$$

FIGURE 8

To find θ, the direction of **R**, we use the law of sines (see Fig. 9):

$$\frac{\sin\theta}{30} = \frac{\sin 140°}{95}$$

$$\sin\theta = \frac{30\sin 140°}{95}$$

FIGURE 9

$$\theta = \sin^{-1}\left(\frac{30 \sin 140^\circ}{95}\right) = 12^\circ$$

Thus, the two given forces are equivalent to a single force of 95 pounds in the direction of 12° (relative to the 70-pound force).

Matched Problem 2 Repeat Example 2 using an angle of 100° between the two forces.

• Resolution of Vectors into Vector Components

Instead of adding vectors, many problems require the resolution of vectors into components. As we indicated earlier, whenever a vector is expressed as the sum or resultant of two vectors, the two vectors are called **vector components** of the given vector. Example 3 illustrates an application of the process of resolving a vector into vector components.

EXAMPLE 3 Resolving a Force Vector into Components

A car weighing 3,210 pounds is on a driveway inclined 20.2° to the horizontal. Neglecting friction, find the magnitude of the force parallel to the driveway that will keep the car from rolling down the hill.

Solution We start by drawing a vector diagram (Fig. 10):

FIGURE 10

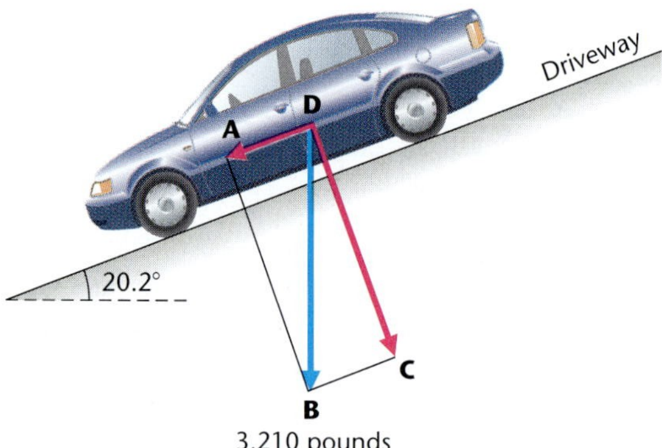

The force vector $\overrightarrow{DB}$ acts in a downward direction and represents the weight of the car. Note that $\overrightarrow{DB} = \overrightarrow{DC} + \overrightarrow{DA}$, where $\overrightarrow{DC}$ is the perpendicular component of $\overrightarrow{DB}$ relative to the driveway and $\overrightarrow{DA}$ is the parallel component of $\overrightarrow{DB}$ relative to the driveway.

To keep the car at D from rolling down the hill, we need a force with the magnitude of $\overrightarrow{DA}$ but oppositely directed. To find $|\overrightarrow{DA}|$, we first observe that $\angle ABD = 20.2^\circ$. This is true because $\angle ABD$ and the driveway angle have the same complement, $\angle ADB$.

$$\sin 20.2° = \frac{|\overrightarrow{DA}|}{3{,}210}$$

$$|\overrightarrow{DA}| = 3{,}210 \sin 20.2°$$

$$= 1{,}110 \text{ pounds}$$

Matched Problem 3 Find the magnitude of the perpendicular component of $\overrightarrow{DB}$ in Example 3.

Answers to Matched Problems

1. Resultant velocity: magnitude = 3.6 mph, direction = 126°
2. $|\mathbf{R}| = 71$ lb, $\theta = 25°$
3. $|\overrightarrow{DC}| = 3{,}010$ lb

EXERCISE 7-3

Express all angle measures in decimal degrees.

A

Problems 1–8 refer to figures (a) and (b) showing vector addition for vectors **u** *and* **v** *at right angles to each other.*

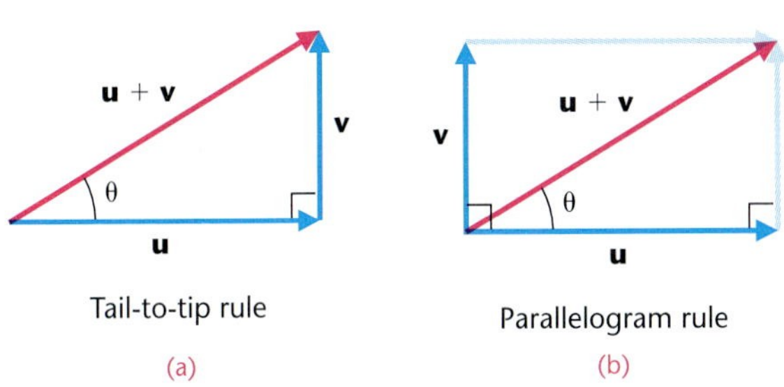

Tail-to-tip rule (a)

Parallelogram rule (b)

In Problems 1–4, find $|\mathbf{u} + \mathbf{v}|$ *and* θ*, given* $|\mathbf{u}|$ *and* $|\mathbf{v}|$ *in figures (a) and (b).*

1. $|\mathbf{u}| = 30$ miles per hour, $|\mathbf{v}| = 72$ miles per hour

2. $|\mathbf{u}| = 216$ miles per hour, $|\mathbf{v}| = 63$ miles per hour

3. $|\mathbf{u}| = 29$ kilograms, $|\mathbf{v}| = 29$ kilograms

4. $|\mathbf{u}| = 78$ kilograms, $|\mathbf{v}| = 45$ kilograms

In Problems 5–8, find $|\mathbf{u}|$ *and* $|\mathbf{v}|$*, the magnitudes of the horizontal and vertical components of* $\mathbf{u} + \mathbf{v}$*, given* $|\mathbf{u} + \mathbf{v}|$ *and* θ *in figures (a) and (b).*

5. $|\mathbf{u} + \mathbf{v}| = 24$ pounds, $\theta = 60°$

6. $|\mathbf{u} + \mathbf{v}| = 48$ pounds, $\theta = 45°$

7. $|\mathbf{u} + \mathbf{v}| = 390$ miles per hour, $\theta = 6°$

8. $|\mathbf{u} + \mathbf{v}| = 75$ miles per hour, $\theta = 81°$

B

Problems 9–16 refer to figures (c) and (d) showing vector addition for vectors **u** *and* **v**.

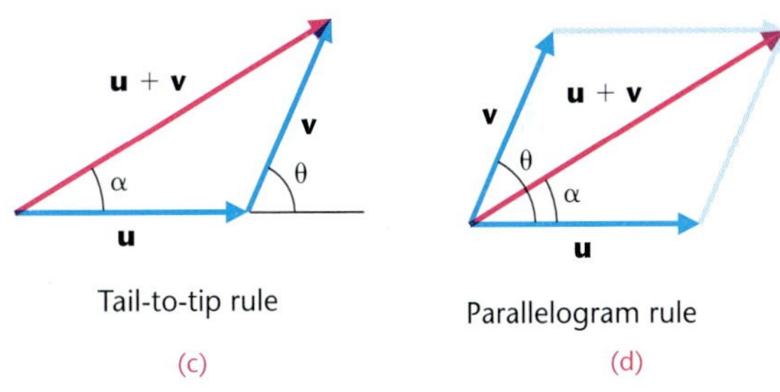

Tail-to-tip rule (c)

Parallelogram rule (d)

In Problems 9–12, find $|\mathbf{u} + \mathbf{v}|$ *and* α*, given* $|\mathbf{u}|$*,* $|\mathbf{v}|$*, and* θ *in figures (c) and (d).*

9. $|\mathbf{u}| = 66$ grams, $|\mathbf{v}| = 22$ grams, $\theta = 68°$

10. $|\mathbf{u}| = 120$ grams, $|\mathbf{v}| = 84$ grams, $\theta = 44°$

11. $|\mathbf{u}| = 21$ knots, $|\mathbf{v}| = 3.2$ knots, $\theta = 53°$

12. $|\mathbf{u}| = 8.0$ knots, $|\mathbf{v}| = 2.0$ knots, $\theta = 64°$

In Problems 13–16, find $|\mathbf{u}|$ *and* $|\mathbf{v}|$*, given* $|\mathbf{u} + \mathbf{v}|$*,* α *and* θ *in figures (c) and (d).*

13. $|\mathbf{u} + \mathbf{v}| = 14$ kilograms, $\alpha = 25°$, $\theta = 79°$

14. $|\mathbf{u} + \mathbf{v}| = 33$ kilograms, $\alpha = 17°$, $\theta = 43°$

15. $|\mathbf{u} + \mathbf{v}| = 223$ miles per hour, $\alpha = 42.3°$, $\theta = 69.4°$

16. $|\mathbf{u} + \mathbf{v}| = 437$ miles per hour, $\alpha = 17.8°$, $\theta = 50.5°$

C

In Problems 17–24, determine whether the statement is true or false. If true, explain why. If false, give a counterexample.

17. The zero vector is perpendicular to every vector.

18. The zero vector is parallel to every vector.

19. Vectors having the same magnitude are equal.

20. Equal vectors have the same magnitude.

21. Equal vectors have the same direction.

22. Perpendicular vectors have the opposite direction.

23. The magnitude of every vector is positive.

24. The magnitude of $\mathbf{u} + \mathbf{v}$ is greater than $|\mathbf{u}|$.

APPLICATIONS

In navigation problems, refer to the figure of a navigational compass below:

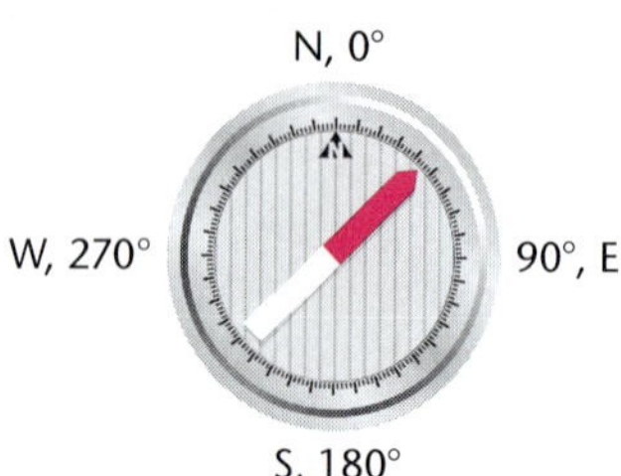

Navigational compass

In Problems 25–28, assume the north, east, south, and west directions are exact.

25. Navigation. An airplane is flying with a compass heading of 285° and an airspeed of 230 miles per hour. A steady wind of 35 miles per hour is blowing in the direction of 260°. What is the plane's actual velocity; that is, what is its speed and direction relative to the ground?

26. Navigation. A powerboat crossing a wide river has a compass heading of 25° and speed relative to the water of 15 miles per hour. The river is flowing in the direction of 135° at 3.9 miles per hour. What is the boat's actual velocity; that is, what is its speed and direction relative to the ground?

★ **27. Navigation.** Two docks are directly opposite each other on a southward-flowing river. A boat pilot wishes to go in a straight line from the east dock to the west dock in a ferryboat with a cruising speed in still water of 8.0 knots. If the river's current is 2.5 knots, what compass heading should be maintained while crossing the river? What is the actual speed of the boat relative to the land?

★ **28. Navigation.** An airplane can cruise at 255 miles per hour in still air. If a steady wind of 46.0 miles per hour is blowing from the west, what compass heading should the pilot fly in order for the course of the plane relative to the ground to be north (0°)? Compute the ground speed for this course.

★ **29. Resultant Force.** A large ship has gone aground in a harbor and two tugs, with cables attached, attempt to pull it free. If one tug pulls with a compass course of 52° and a force of 2,300 pounds and a second tug pulls with a compass course of 97° and a force of 1,900 pounds, what is the compass direction and the magnitude of the resultant force?

★ **30. Resultant Force.** Repeat Problem 29 if one tug pulls with a compass direction of 161° and a force of 2,900 kilograms and a second tug pulls with a compass direction of 192° and a force of 3,600 kilograms.

31. Resolution of Forces. An automobile weighing 4,050 pounds is standing on a driveway inclined 5.5° with the horizontal.

(A) Find the magnitude of the force parallel to the driveway necessary to keep the car from rolling down the hill.

(B) Find the magnitude of the force perpendicular to the driveway.

32. Resolution of Forces. Repeat Problem 31 for a car weighing 2,500 pounds parked on a hill inclined at 15° to the horizontal.

★ **33. Resolution of Forces.** If two weights are fastened together and placed on inclined planes as shown in the figure, neglecting friction, which way will they slide?

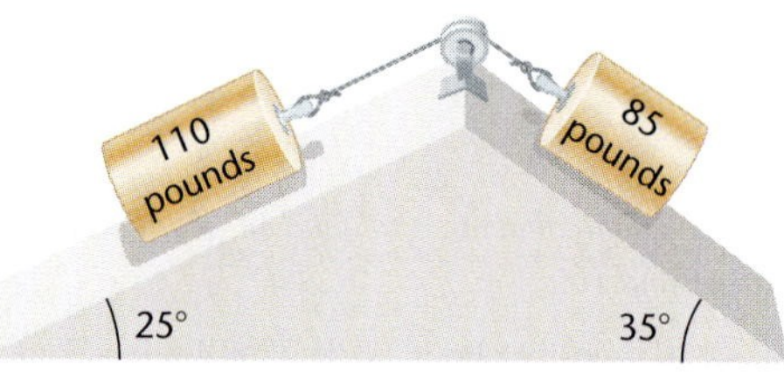

★ **34. Resolution of Forces.** If two weights are fastened together and placed on inclined planes as indicated in the figure, neglecting friction, which way will they slide?

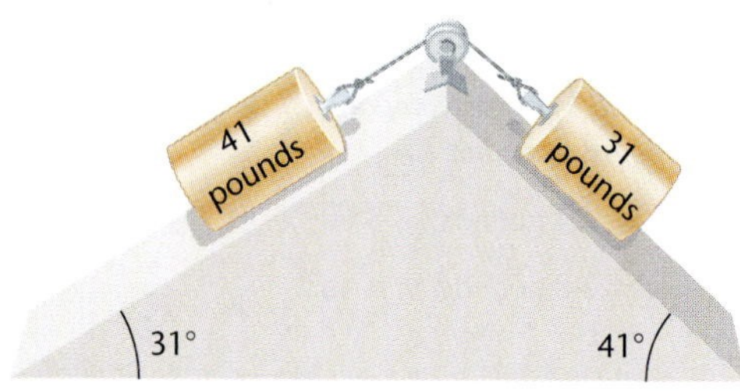

SECTION 7-4 Algebraic Vectors

- From Geometric Vectors to Algebraic Vectors
- Vector Addition and Scalar Multiplication
- Unit Vectors
- Algebraic Properties
- Static Equilibrium

Geometric vectors in a plane are readily generalized to three-dimensional space. However, to generalize vectors further to higher-dimensional abstract spaces, it is essential to define the vector concept algebraically. This is done in such a way that the geometric vectors become special cases of the more general algebraic vectors. Algebraic vectors have many advantages over geometric vectors. One advantage will become apparent when we consider static equilibrium problems at the end of the section.

The development of algebraic vectors in this book is introductory in nature and is restricted to the plane. Further study of vectors in three- and higher-dimensional spaces is reserved for more advanced mathematical courses.

• From Geometric Vectors to Algebraic Vectors

The transition from geometric vectors to algebraic vectors is begun by placing geometric vectors in a rectangular coordinate system. A geometric vector $\overrightarrow{AB}$ in a rectangular coordinate system translated so that its initial point is at the origin is said to be in **standard position.** The vector $\overrightarrow{OP}$ such that $\overrightarrow{OP} = \overrightarrow{AB}$ is said to be the **standard vector** for $\overrightarrow{AB}$ (see Fig. 1).

Note that the vector $\overrightarrow{OP}$ in Figure 1 is the standard vector for infinitely many vectors—all vectors with the same magnitude and direction as $\overrightarrow{OP}$.

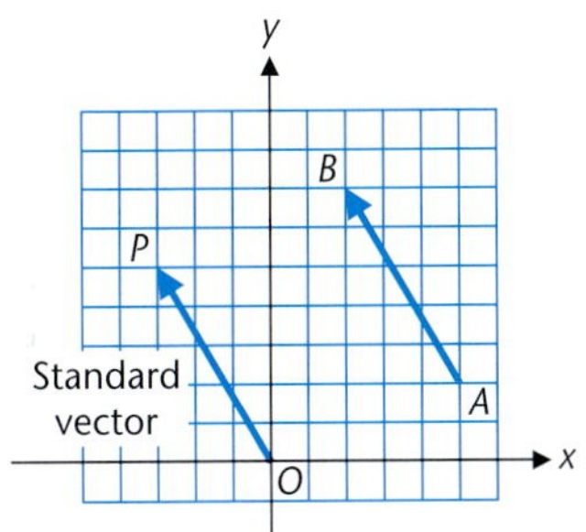

FIGURE 1 $\overrightarrow{OP}$ is the standard vector for $\overrightarrow{AB}$.

EXPLORE-DISCUSS 1

(A) In a copy of Figure 1, draw in three other vectors having $\overrightarrow{OP}$ as their standard vector.

(B) If the tail of a vector is at point $A(-3, 2)$ and its tip is at $B(6, 4)$, discuss how you would find the coordinates of P so that $\overrightarrow{OP}$ is the standard vector for $\overrightarrow{AB}$.

Given the coordinates of the endpoints of a geometric vector in a rectangular coordinate system, how do we find its corresponding standard vector? The process is not difficult. The coordinates of the initial point, O, of $\overrightarrow{OP}$ are always (0, 0). Thus, we have only to find the coordinates of P, the terminal point of $\overrightarrow{OP}$. The coordinates of P are given by

$$(x_p, y_p) = (x_b - x_a, y_b - y_a) \tag{1}$$

where the coordinates of A are (x_a, y_a) and the coordinates of B are (x_b, y_b). Example 1 illustrates the use of equation (1).

EXAMPLE 1 Finding a Standard Vector for a Given Vector

Given the geometric vector $\overrightarrow{AB}$ with initial point $A(3, 4)$ and terminal point $B(7, -1)$, find the standard vector $\overrightarrow{OP}$ for $\overrightarrow{AB}$. That is, find the coordinates of the point P such that $\overrightarrow{OP} = \overrightarrow{AB}$.

Solution The coordinates of P are given by

$$\begin{aligned}(x_p, y_p) &= (x_b - x_a, y_b - y_a)\\ &= (7 - 3, -1 - 4)\\ &= (4, -5)\end{aligned}$$

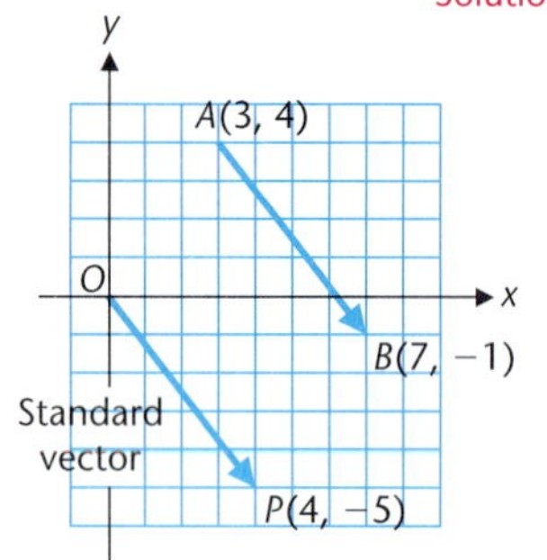

FIGURE 2

Note in Figure 2 that if we start at A, then move to the right 4 units and down 5 units, we will be at B. If we start at the origin, then move to the right 4 units and down 5 units, we will be at P.

Matched Problem 1

Given the geometric vector $\overrightarrow{AB}$ with initial point $A(8, -3)$ and terminal point $B(4, 5)$, find the standard vector $\overrightarrow{OP}$ for $\overrightarrow{AB}$.

The preceding discussion suggests another way of looking at vectors. Since, given any geometric vector $\overrightarrow{AB}$ in a rectangular coordinate system, there always exists a point $P(x_p, y_p)$ such that $\overrightarrow{OP} = \overrightarrow{AB}$, the point $P(x_p, y_p)$ completely specifies the vector $\overrightarrow{AB}$, except for its position. And we are not concerned about its position because we are free to translate $\overrightarrow{AB}$ anywhere we please. Conversely, given any point $P(x_p, y_p)$ in a rectangular coordinate system, the directed line segment joining O to P forms the geometric vector $\overrightarrow{OP}$.

This leads us to define an **algebraic vector** as an ordered pair of real numbers. To avoid confusing a point (a, b) with a vector (a, b), **we use $\langle a, b\rangle$ to represent an algebraic vector.** Geometrically, the algebraic vector $\langle a, b\rangle$ corresponds to the standard (geometric) vector $\overrightarrow{OP}$ with terminal point $P(a, b)$ and initial point $O(0, 0)$, as illustrated in Figure 3.

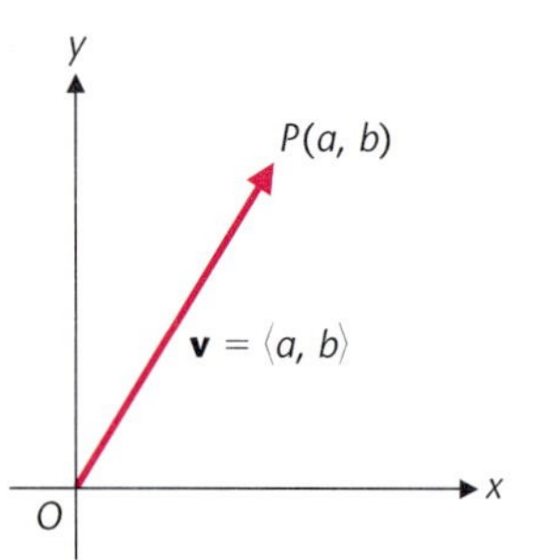

FIGURE 3 Algebraic vector $\langle a, b\rangle$ associated with a geometric vector $\overrightarrow{OP}$.

The real numbers a and b are **scalar components** of the vector $\langle a, b\rangle$. The word **scalar** means real number and is often used in the context of vectors where one refers to "scalar quantities" as opposed to "vector quantities." Thus, we talk about "scalar components" and "vector components" of a given vector. The words "scalar" and "vector" are often dropped if the meaning of component is clear from the context.

Two vectors $\mathbf{u} = \langle a, b\rangle$ and $\mathbf{v} = \langle c, d\rangle$ are said to be **equal** if their corresponding components are equal, that is, if $a = c$ and $b = d$. The **zero vector** is denoted by $\mathbf{0} = \langle 0, 0\rangle$.

Geometric vectors are limited to spaces we can visualize, that is, to two- and three-dimensional spaces. Algebraic vectors do not have these restrictions. The following are algebraic vectors from two-, three-, four-, and five-dimensional spaces:

$$\langle -2, 5\rangle \quad \langle 3, 0, -8\rangle \quad \langle 5, 1, 1, -2\rangle \quad \langle -1, 0, 1, 3, 4\rangle$$

As we said earlier, the discussion in this book is limited to algebraic vectors in a two-dimensional space, which represents a plane.

We now define the *magnitude* of an algebraic vector:

DEFINITION 1

Magnitude of v = $\langle a, b\rangle$

The **magnitude,** or **norm,** of a vector $\mathbf{v} = \langle a, b\rangle$ is denoted by $|\mathbf{v}|$ and is given by

$$|\mathbf{v}| = \sqrt{a^2 + b^2}$$

FIGURE 4 Magnitude of vector $\langle a, b\rangle$ geometrically interpreted.

Geometrically, $\sqrt{a^2 + b^2}$ is the length of the standard geometric vector $\overrightarrow{OP}$ associated with the algebraic vector $\langle a, b\rangle$ (see Fig. 4).

The definition of magnitude is readily generalized to higher-dimensional vector spaces. For example, if $|\mathbf{v}| = \langle a, b, c, d\rangle$, then the magnitude, or norm, is given by $\sqrt{a^2 + b^2 + c^2 + d^2}$. But now we are not able to interpret the result in terms of geometric vectors.

EXAMPLE 2 **Finding the Magnitude of a Vector**

Find the magnitude of the vector $\mathbf{v} = \langle 3, -5\rangle$.

Solution

$$|\mathbf{v}| = \sqrt{3^2 + (-5)^2} = \sqrt{34}$$

Matched Problem 2 Find the magnitude of the vector $\mathbf{v} = \langle -2, 4\rangle$.

• Vector Addition and Scalar Multiplication

To add two algebraic vectors, add the corresponding components as indicated in the following definition of addition:

DEFINITION 2

Vector Addition

If $\mathbf{u} = \langle a, b\rangle$ and $\mathbf{v} = \langle c, d\rangle$, then

$$\mathbf{u} + \mathbf{v} = \langle a + c, b + d\rangle$$

The definition of addition of algebraic vectors is consistent with the parallelogram and tail-to-tip definitions for adding geometric vectors given in Section 7-3 (see Explore-Discuss 2).

EXPLORE-DISCUSS 2 If $\mathbf{u} = \langle -3, 2\rangle$, $\mathbf{v} = \langle 7, 3\rangle$, then $\mathbf{u} + \mathbf{v} = \langle -3 + 7, 2 + 3\rangle = \langle 4, 5\rangle$. Locate $\mathbf{u}$, $\mathbf{v}$, and $\mathbf{u} + \mathbf{v}$ in a rectangular coordinate system and interpret geometrically in terms of the parallelogram and tail-to-tip rules discussed in the last section.

To multiply a vector by a scalar (a real number) multiply each component by the scalar:

DEFINITION 3

Scalar Multiplication

If $\mathbf{u} = \langle a, b\rangle$ and k is a scalar, then

$$k\mathbf{u} = k\langle a, b\rangle = \langle ka, kb\rangle$$

Geometrically, if a vector $\mathbf{v}$ is multiplied by a scalar k, the magnitude of the vector $\mathbf{v}$ is multiplied by $|k|$. If k is positive, then $k\mathbf{v}$ has the same direction as $\mathbf{v}$. If k is negative, then $k\mathbf{v}$ has the opposite direction as $\mathbf{v}$. These relationships are illustrated in Figure 5.

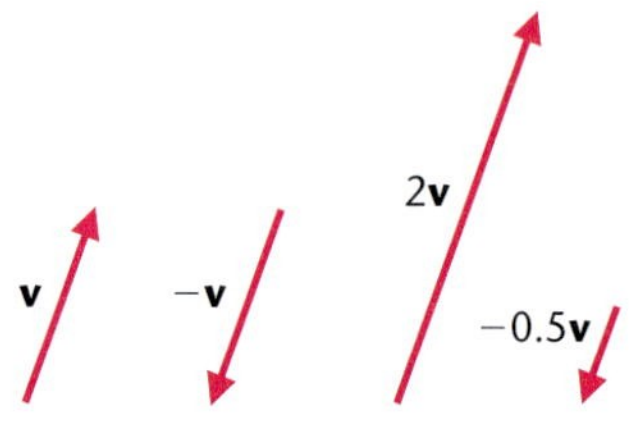

FIGURE 5 Scalar multiplication geometrically interpreted.

EXAMPLE 3 Vector Addition and Scalar Multiplication

Let $\mathbf{u} = \langle 4, -3\rangle$, $\mathbf{v} = \langle 2, 3\rangle$, and $\mathbf{w} = \langle 0, -5\rangle$, find:

(A) $\mathbf{u} + \mathbf{v}$ (B) $-2\mathbf{u}$ (C) $2\mathbf{u} - 3\mathbf{v}$ (D) $3\mathbf{u} + 2\mathbf{v} - \mathbf{w}$

Solutions

(A) $\mathbf{u} + \mathbf{v} = \langle 4, -3\rangle + \langle 2, 3\rangle = \langle 6, 0\rangle$

(B) $-2\mathbf{u} = -2\langle 4, -3\rangle = \langle -8, 6\rangle$

(C) $2\mathbf{u} - 3\mathbf{v} = 2\langle 4, -3\rangle - 3\langle 2, 3\rangle$
$= \langle 8, -6\rangle + \langle -6, -9\rangle = \langle 2, -15\rangle$

(D) $3\mathbf{u} + 2\mathbf{v} - \mathbf{w} = 3\langle 4, -3\rangle + 2\langle 2, 3\rangle - \langle 0, -5\rangle$
$= \langle 12, -9\rangle + \langle 4, 6\rangle + \langle 0, 5\rangle$
$= \langle 16, 2\rangle$

Matched Problem 3 Let $\mathbf{u} = \langle -5, 3\rangle$, $\mathbf{v} = \langle 4, -6\rangle$, and $\mathbf{w} = \langle -2, 0\rangle$, find:

(A) $\mathbf{u} + \mathbf{v}$ (B) $-3\mathbf{u}$ (C) $3\mathbf{u} - 2\mathbf{v}$ (D) $2\mathbf{u} - \mathbf{v} + 3\mathbf{w}$

• Unit Vectors

If $|\mathbf{v}| = 1$, then $\mathbf{v}$ is called a **unit vector.** A unit vector can be formed from an arbitrary nonzero vector as follows:

A Unit Vector with the Same Direction as v

If $\mathbf{v}$ is a nonzero vector, then

$$\mathbf{u} = \frac{1}{|\mathbf{v}|}\mathbf{v}$$

is a unit vector with the same direction as $\mathbf{v}$.

EXAMPLE 4 **Finding a Unit Vector with the Same Direction as a Given Vector**

Given a vector $\mathbf{v} = \langle 1, -2 \rangle$, find a unit vector $\mathbf{u}$ with the same direction as $\mathbf{v}$.

Solution

$$|\mathbf{v}| = \sqrt{1^2 + (-2)^2} = \sqrt{5}$$

$$\mathbf{u} = \frac{1}{|\mathbf{v}|}\mathbf{v} = \frac{1}{\sqrt{5}}\langle 1, -2 \rangle$$

$$= \left\langle \frac{1}{\sqrt{5}}, \frac{-2}{\sqrt{5}} \right\rangle$$

Check

$$|\mathbf{u}| = \sqrt{\left(\frac{1}{\sqrt{5}}\right)^2 + \left(\frac{-2}{\sqrt{5}}\right)^2} = \sqrt{\frac{1}{5} + \frac{4}{5}} = \sqrt{1} = 1$$

And we see that $\mathbf{u}$ is a unit vector with the same direction as $\mathbf{v}$.

Matched Problem 4 Given a vector $\mathbf{v} = \langle 3, 1 \rangle$, find a unit vector $\mathbf{u}$ with the same direction as $\mathbf{v}$.

We now define two very important unit vectors, the $\mathbf{i}$ and $\mathbf{j}$ unit vectors.

The i and j Unit Vectors

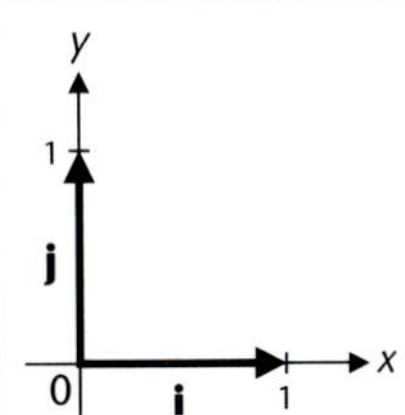

$$\mathbf{i} = \langle 1, 0 \rangle$$

$$\mathbf{j} = \langle 0, 1 \rangle$$

Why are the **i** and **j** unit vectors so important? One of the reasons is that any vector $\mathbf{v} = \langle a, b \rangle$ can be expressed as a linear combination of those two vectors; that is, as $a\mathbf{i} + b\mathbf{j}$.

$$\begin{aligned} \mathbf{v} = \langle a, b \rangle &= \langle a, 0 \rangle + \langle 0, b \rangle \\ &= a\langle 1, 0 \rangle + b\langle 0, 1 \rangle = a\mathbf{i} + b\mathbf{j} \end{aligned}$$

EXAMPLE 5 Expressing a Vector in Terms of the i and j Vectors

Express each vector as a linear combination of the **i** and **j** unit vectors.

(A) $\langle -2, 4 \rangle$ (B) $\langle 2, 0 \rangle$ (C) $\langle 0, -7 \rangle$

Solutions
(A) $\langle -2, 4 \rangle = -2\mathbf{i} + 4\mathbf{j}$
(B) $\langle 2, 0 \rangle = 2\mathbf{i} + 0\mathbf{j} = 2\mathbf{i}$
(C) $\langle 0, -7 \rangle = 0\mathbf{i} - 7\mathbf{j} = -7\mathbf{j}$

Matched Problem 5 Express each vector as a linear combination of the **i** and **j** unit vectors.

(A) $\langle 5, -3 \rangle$ (B) $\langle -9, 0 \rangle$ (C) $\langle 0, 6 \rangle$

• Algebraic Properties

Vector addition and scalar multiplication possess algebraic properties similar to the real numbers. These properties enable us to manipulate symbols representing vectors and scalars in much the same way we manipulate symbols that represent real numbers in algebra. These properties are listed below for convenient reference.

Algebraic Properties of Vectors

A. Addition Properties. For all vectors **u**, **v**, and **w**:

1. $\mathbf{u} + \mathbf{v} = \mathbf{v} + \mathbf{u}$ **Commutative Property**
2. $\mathbf{u} + (\mathbf{v} + \mathbf{w}) = (\mathbf{u} + \mathbf{v}) + \mathbf{w}$ **Associative Property**
3. $\mathbf{u} + \mathbf{0} = \mathbf{0} + \mathbf{u} = \mathbf{u}$ **Additive Identity**
4. $\mathbf{u} + (-\mathbf{u}) = (-\mathbf{u}) + \mathbf{u} = \mathbf{0}$ **Additive Inverse**

B. Scalar Multiplication Properties. For all vectors **u** and **v** and all scalars m and n:

1. $m(n\mathbf{u}) = (mn)\mathbf{u}$ **Associative Property**
2. $m(\mathbf{u} + \mathbf{v}) = m\mathbf{u} + m\mathbf{v}$ **Distributive Property**
3. $(m + n)\mathbf{u} = m\mathbf{u} + n\mathbf{u}$ **Distributive Property**
4. $1\mathbf{u} = \mathbf{u}$ **Multiplicative Identity**

EXAMPLE 6 Algebraic Operations on Vectors Expressed in Terms of the i and j Vectors

For $\mathbf{u} = \mathbf{i} - 2\mathbf{j}$ and $\mathbf{v} = 5\mathbf{i} + 2\mathbf{j}$, compute each of the following:

(A) $\mathbf{u} + \mathbf{v}$ (B) $\mathbf{u} - \mathbf{v}$ (C) $2\mathbf{u} + 3\mathbf{v}$

Solutions

(A) $\mathbf{u} + \mathbf{v} = (\mathbf{i} - 2\mathbf{j}) + (5\mathbf{i} + 2\mathbf{j})$
$= \mathbf{i} - 2\mathbf{j} + 5\mathbf{i} + 2\mathbf{j} = 6\mathbf{i} + 0\mathbf{j} = 6\mathbf{i}$

(B) $\mathbf{u} - \mathbf{v} = (\mathbf{i} - 2\mathbf{j}) - (5\mathbf{i} + 2\mathbf{j})$
$= \mathbf{i} - 2\mathbf{j} - 5\mathbf{i} - 2\mathbf{j} = -4\mathbf{i} - 4\mathbf{j}$

(C) $2\mathbf{u} + 3\mathbf{v} = 2(\mathbf{i} - 2\mathbf{j}) + 3(5\mathbf{i} + 2\mathbf{j})$
$= 2\mathbf{i} - 4\mathbf{j} + 15\mathbf{i} + 6\mathbf{j} = 17\mathbf{i} + 2\mathbf{j}$

Matched Problem 6

For $\mathbf{u} = 2\mathbf{i} - \mathbf{j}$ and $\mathbf{v} = 4\mathbf{i} + 5\mathbf{j}$, compute each of the following:

(A) $\mathbf{u} + \mathbf{v}$ (B) $\mathbf{u} - \mathbf{v}$ (C) $3\mathbf{u} - 2\mathbf{v}$

• Static Equilibrium

Algebraic vectors can be used to solve many types of problems in physics and engineering. We complete this section by considering a few problems involving *static equilibrium.* Fundamental to our approach are two basic principles regarding forces and objects subject to these forces:

Conditions for Static Equilibrium

1. An object at rest is said to be in **static equilibrium.**
2. For an object located at the origin in a rectangular coordinate system to remain in static equilibrium, at rest, it is necessary that the sum of all the force vectors acting on the object be the zero vector.

Example 7 shows how some important physics/engineering problems can be solved using algebraic vectors and the conditions for static equilibrium. It is assumed that you know how to solve a system of two equations with two variables. In case you need a reminder, procedures are reviewed in Section 1-2.

EXAMPLE 7 Tension in Cables

A cable car, used to ferry people and supplies across a river, weighs 2,500 pounds fully loaded. The car stops when partway across and deflects the cable relative to the horizontal, as indicated in Figure 6. What is the tension in each part of the cable running to each tower?

FIGURE 6

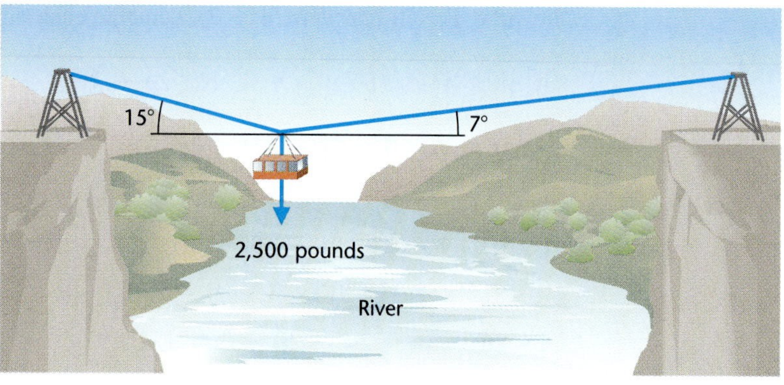

Solution

Step 1. Draw a force diagram with all force vectors in standard position at the origin (Fig. 7). The objective is to find $|\mathbf{u}|$ and $|\mathbf{v}|$.

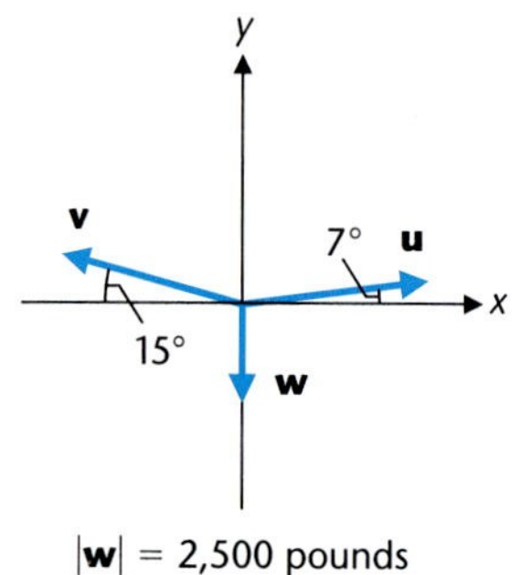

$|\mathbf{w}| = 2{,}500$ pounds

FIGURE 7

Step 2. Write each force vector in terms of the **i** and **j** unit vectors:

$$\mathbf{u} = |\mathbf{u}|(\cos 7°)\mathbf{i} + |\mathbf{u}|(\sin 7°)\mathbf{j}$$
$$\mathbf{v} = |\mathbf{v}|(-\cos 15°)\mathbf{i} + |\mathbf{v}|(\sin 15°)\mathbf{j}$$
$$\mathbf{w} = -2{,}500\mathbf{j}$$

Step 3. For the system to be in static equilibrium, the sum of the force vectors must be the zero vector. That is,

$$\mathbf{u} + \mathbf{v} + \mathbf{w} = \mathbf{0}$$

Replacing vectors **u**, **v**, and **w** from step 2, we obtain

$$[|\mathbf{u}|(\cos 7°)\mathbf{i} + |\mathbf{u}|(\sin 7°)\mathbf{j}] + [|\mathbf{v}|(-\cos 15°)\mathbf{i} + |\mathbf{v}|(\sin 15°)\mathbf{j}] - 2{,}500\mathbf{j} = 0\mathbf{i} + 0\mathbf{j}$$

which on combining **i** and **j** vectors becomes

$$[|\mathbf{u}|(\cos 7°) + |\mathbf{v}|(-\cos 15°)]\mathbf{i} + [|\mathbf{u}|(\sin 7°) + |\mathbf{v}|(\sin 15°) - 2{,}500]\mathbf{j} = 0\mathbf{i} + 0\mathbf{j}$$

Since two vectors are equal if and only if their corresponding components are equal, we are led to the following system of two equations in the two variables $|\mathbf{u}|$ and $|\mathbf{v}|$:

$$\begin{aligned}(\cos 7°)|\mathbf{u}| + (-\cos 15°)|\mathbf{v}| \phantom{- 2{,}500} &= 0\\ (\sin 7°)|\mathbf{u}| + (\sin 15°)|\mathbf{v}| - 2{,}500 &= 0\end{aligned}$$

Solving this system by standard methods, we find that

$$|\mathbf{u}| = 6{,}400 \text{ pounds} \qquad \text{and} \qquad |\mathbf{v}| = 6{,}600 \text{ pounds}$$

Did you expect that the tension in each part of the cable is more than the weight hanging from the cable?

Matched Problem 7 Repeat Example 7 with 15° replaced with 13°, 7° replaced with 9°, and the 2,500 pounds replaced with 1,900 pounds.

Answers to Matched Problems

1. $P(-4, 8)$ 2. $2\sqrt{5}$
3. (A) $\langle -1, -3\rangle$ (B) $\langle 15, -9\rangle$ (C) $\langle -23, 21\rangle$ (D) $\langle -20, 12\rangle$
4. $\mathbf{u} = \langle 3/\sqrt{10}, 1/\sqrt{10}\rangle$
5. (A) $5\mathbf{i} - 3\mathbf{j}$ (B) $-9\mathbf{i}$ (C) $6\mathbf{j}$
6. (A) $6\mathbf{i} + 4\mathbf{j}$ (B) $-2\mathbf{i} - 6\mathbf{j}$ (C) $-2\mathbf{i} - 13\mathbf{j}$
7. $|\mathbf{u}| = 4{,}900$ lb, $|\mathbf{v}| = 5{,}000$ lb

EXERCISE 7-4

A

In Problems 1–6, represent each geometric vector $\overrightarrow{AB}$, with endpoints as indicated, as an algebraic vector in the form $\langle a, b\rangle$.

1. $A(0, 0), B(7, 2)$
2. $A(5, 3), B(0, 0)$
3. $A(4, 0), B(0, 8)$
4. $A(0, -5), B(6, 0)$
5. $A(9, -4), B(7, 5)$
6. $A(-6, -3), B(9, 1)$

In Problems 7–12, find the magnitude of each vector.

7. $\langle -15, 0\rangle$
8. $\langle 0, 32\rangle$
9. $\langle -21, 72\rangle$
10. $\langle -48, -20\rangle$
11. $\langle -155, 468\rangle$
12. $\langle 836, 123\rangle$

B

In Problems 13–16, find:

(A) $\mathbf{u} + \mathbf{v}$ (B) $\mathbf{u} - \mathbf{v}$ (C) $2\mathbf{u} - \mathbf{v} + 3\mathbf{w}$

13. $\mathbf{u} = \langle 2, 1\rangle, \mathbf{v} = \langle -1, 3\rangle, \mathbf{w} = \langle 3, 0\rangle$
14. $\mathbf{u} = \langle -1, 2\rangle, \mathbf{v} = \langle 3, -2\rangle, \mathbf{w} = \langle 0, -2\rangle$
15. $\mathbf{u} = \langle -4, -1\rangle, \mathbf{v} = \langle 2, 2\rangle, \mathbf{w} = \langle 0, 1\rangle$
16. $\mathbf{u} = \langle -3, 2\rangle, \mathbf{v} = \langle -2, 2\rangle, \mathbf{w} = \langle -3, 0\rangle$

In Problems 17–22, express $\mathbf{v}$ *in terms of the* $\mathbf{i}$ *and* $\mathbf{j}$ *unit vectors.*

17. $\langle -8, 0\rangle$
18. $\langle 0, 14\rangle$
19. $\langle 6, -12\rangle$
20. $\langle -5, -18\rangle$
21. $\mathbf{v} = \overrightarrow{AB}$, where $A = (2, 3)$ and $B = (-3, 1)$
22. $\mathbf{v} = \overrightarrow{AB}$, where $A = (-2, -1)$ and $B = (0, 2)$

In Problems 23–28, let $\mathbf{u} = 3\mathbf{i} - 2\mathbf{j}$, $\mathbf{v} = 2\mathbf{i} + 4\mathbf{j}$, *and* $\mathbf{w} = 2\mathbf{i}$, *and perform the indicated operations.*

23. $\mathbf{u} + \mathbf{v}$
24. $\mathbf{u} - \mathbf{v}$
25. $2\mathbf{u} - 3\mathbf{v}$
26. $3\mathbf{u} + 2\mathbf{v}$
27. $2\mathbf{u} - \mathbf{v} - 2\mathbf{w}$
28. $\mathbf{u} - 3\mathbf{v} + 2\mathbf{w}$

In Problems 29–32, find a unit vector $\mathbf{u}$ *with the same direction as* $\mathbf{v}$.

29. $\mathbf{v} = \langle -1, 1\rangle$
30. $\mathbf{v} = \langle 2, 1\rangle$
31. $\mathbf{v} = \langle -12, 5\rangle$
32. $\mathbf{v} = \langle -7, -24\rangle$

In Problems 33–36, determine whether the statement is true or false. If true, explain why. If false, give a counterexample.

33. If $\mathbf{u}$ is a scalar multiple of $\mathbf{v}$, then $\mathbf{u}$ and $\mathbf{v}$ have the same direction.
34. If $\mathbf{u}$ and $\mathbf{v}$ are nonzero vectors that have the same direction, then $\mathbf{u}$ is a scalar multiple of $\mathbf{v}$.
35. The sum of two unit vectors is a unit vector.
36. If $\mathbf{u}$ is a unit vector and k is a scalar, then the magnitude of $k\mathbf{u}$ is k.

C

In Problems 37–44, let $\mathbf{u} = \langle a, b\rangle$, $\mathbf{v} = \langle c, d\rangle$, *and* $\mathbf{w} = \langle e, f\rangle$ *be vectors and* m *and* n *be scalars. Prove each of the following vector properties using appropriate properties of real numbers and the definitions of vector addition and scalar multiplication.*

37. $\mathbf{u} + (\mathbf{v} + \mathbf{w}) = (\mathbf{u} + \mathbf{v}) + \mathbf{w}$
38. $\mathbf{u} + \mathbf{v} = \mathbf{v} + \mathbf{u}$
39. $\mathbf{u} + \mathbf{0} = \mathbf{u}$
40. $\mathbf{u} + (-\mathbf{u}) = \mathbf{0}$
41. $(m + n)\mathbf{u} = m\mathbf{u} + n\mathbf{u}$
42. $m(\mathbf{u} + \mathbf{v}) = m\mathbf{u} + m\mathbf{v}$
43. $m(n\mathbf{u}) = (mn)\mathbf{u}$
44. $1\mathbf{u} = \mathbf{u}$

APPLICATIONS 

In Problems 45–52, compute all answers to 3 significant digits.

45. Static Equilibrium. A unicyclist at a certain point on a tightrope deflects the rope as indicated in the figure. If the total weight of the cyclist and the unicycle is 155 pounds, how much tension is in each part of the cable?

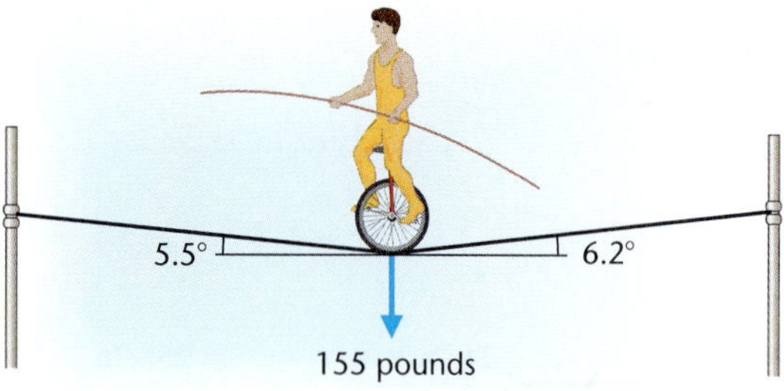

46. Static Equilibrium. Repeat Problem 45 with the left angle 4.2°, the right angle 5.3°, and the total weight 112 pounds.

47. Static Equilibrium. A weight of 1,000 pounds is suspended from two cables as shown in the figure. What is the tension in each cable?

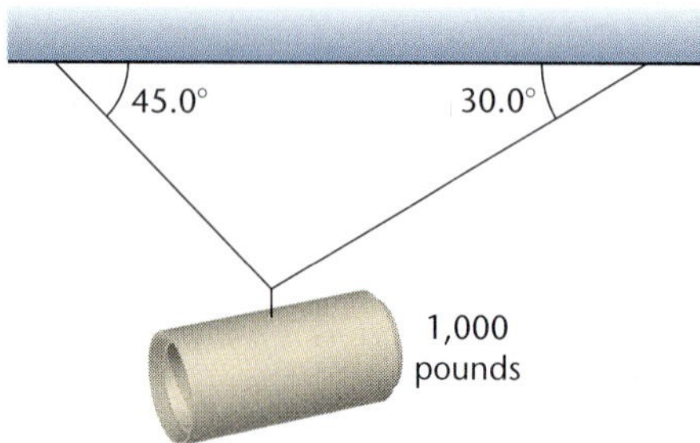

48. Static Equilibrium. A weight of 500 pounds is supported by two cables as illustrated. What is the tension in each cable?

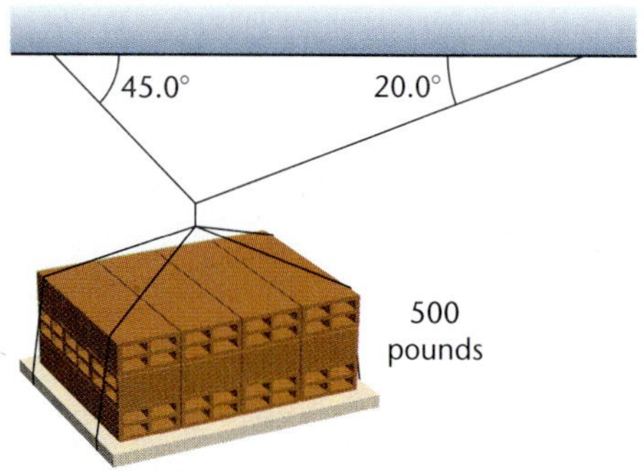

49. Static Equilibrium. A 400-pound sign is suspended as shown in figure (a). The corresponding force diagram (b) is formed by observing the following: Member AB is "pushing" at B and is under compression. This "pushing" force also can be thought of as the force vector **a** "pulling" to the right at B. The force vector **b** reflects the fact that member CB is under tension—that is, it is "pulling" at B. The force vector **c** corresponds to the weight of the sign "pulling" down at B. Find the magnitudes of the forces in the rigid supporting members; that is, find $|\mathbf{a}|$ and $|\mathbf{b}|$ in the force diagram (b).

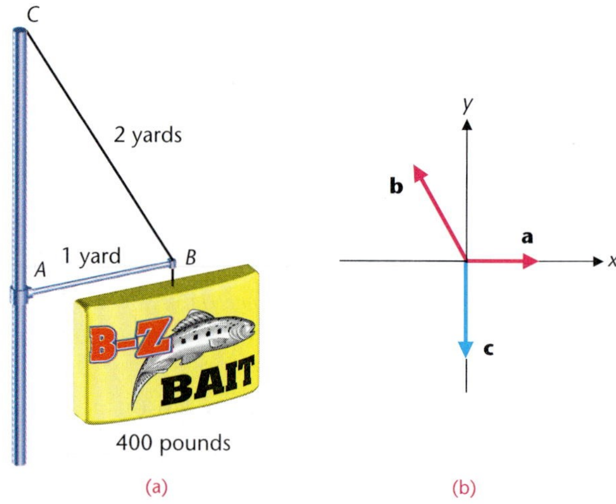

(a) (b)

50. Static Equilibrium. A weight of 1,000 kilograms is supported as shown in the figure. What are the magnitudes of the forces on the members AB and BC?

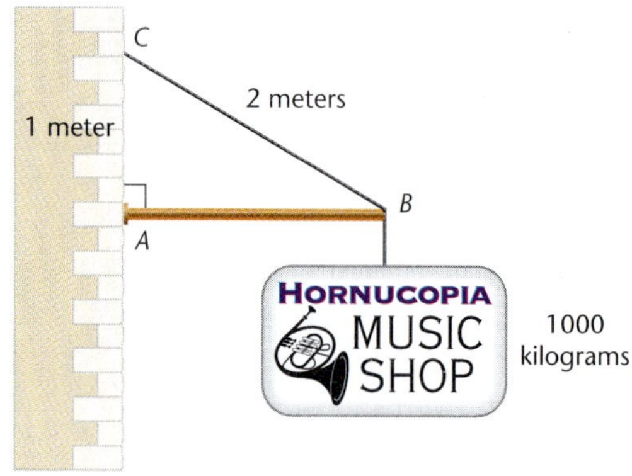

51. Static Equilibrium. A 1,250-pound weight is hanging from a hoist as indicated in the figure on the next page. What are the magnitudes of the forces on the members AB and BC?

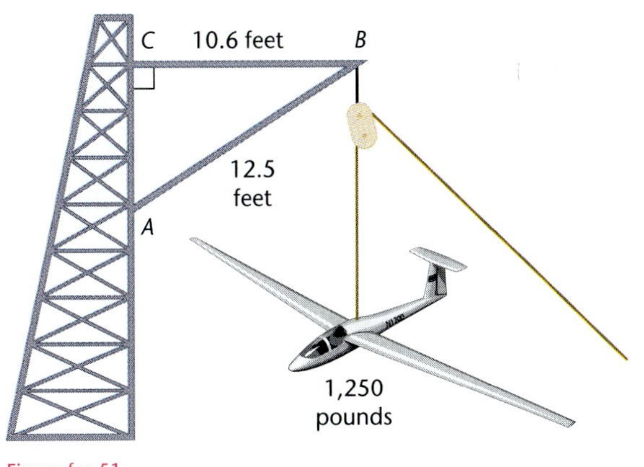

Figure for 51

52. Static Equilibrium. A weight of 5,000 kilograms is supported as shown in the figure. What are the magnitudes of the forces on the members AB and BC?

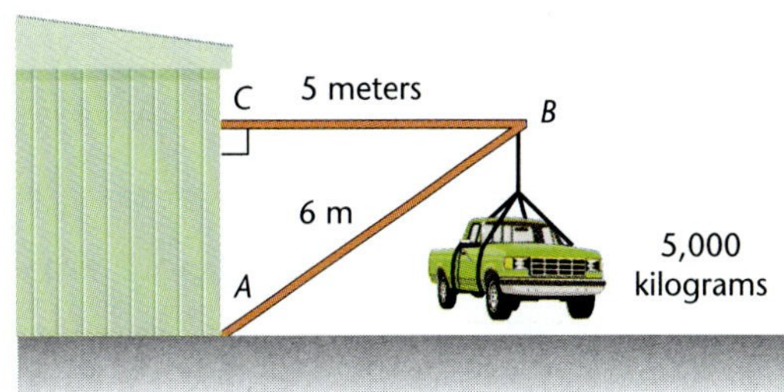

Figure for 52

SECTION 7-5 Polar Coordinates and Graphs

- Polar Coordinate System
- Converting from Polar to Rectangular Form, and Vice Versa
- Graphing Polar Equations
- Some Standard Polar Curves
- Application

Up until now we have used only the rectangular coordinate system. Other coordinate systems have particular advantages in certain situations. Of the many that are possible, the *polar coordinate system* ranks second in importance to the rectangular coordinate system and forms the subject matter for this section.

• Polar Coordinate System

To form a **polar coordinate system** in a plane (see Fig. 1), start with a fixed point O and call it the **pole,** or **origin.** From this point draw a half line, or ray (usually horizontal and to the right), and call this line the **polar axis.**

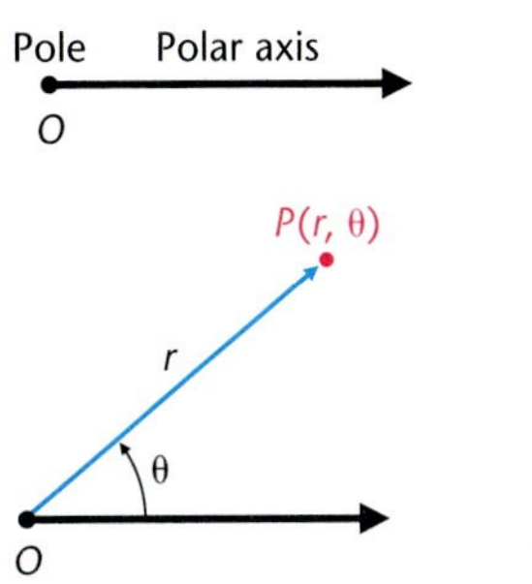

FIGURE 1 Polar coordinate system.

If P is an arbitrary point in a plane, then associate polar coordinates (r, θ) with it as follows: Starting with the polar axis as the initial side of an angle, rotate the terminal side until it, or the extension of it through the pole, passes through the point. The θ coordinate in (r, θ) is this angle, in degree or radian measure. The angle θ is positive if the rotation is counterclockwise and negative if the rotation is clockwise. The r coordinate in (r, θ) is the directed distance from the pole to the point P. It is positive if measured from the pole along the terminal side of θ and negative if measured along the terminal side extended through the pole.

Figure 2 illustrates a point P with three different sets of polar coordinates. Study this figure carefully. The pole has polar coordinates $(0, \theta)$ for arbitrary θ. For example, $(0, 0°)$, $(0, \pi/3)$, and $(0, -371°)$ are all coordinates of the pole.

FIGURE 2 Polar coordinates of a point.

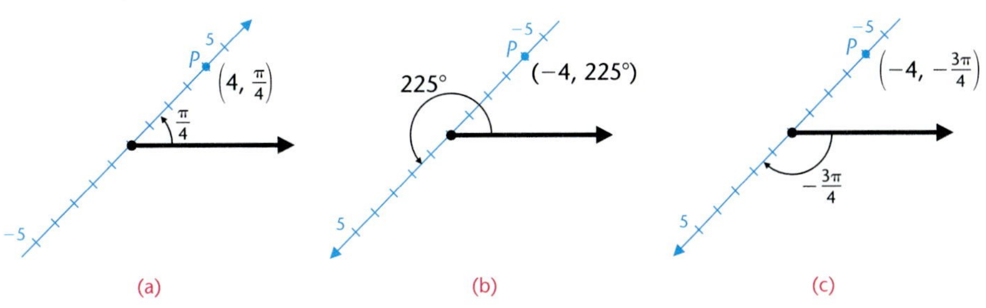

We now see a distinct difference between rectangular and polar coordinates for the given point. For a given point in a rectangular coordinate system, there exists exactly one set of rectangular coordinates. On the other hand, in a polar coordinate system, a point has infinitely many sets of polar coordinates.

Just as graph paper with a rectangular grid is readily available for plotting rectangular coordinates, polar graph paper is available for plotting polar coordinates.

EXAMPLE 1 **Plotting Points in a Polar Coordinate System**

Plot the following points in a polar coordinate system:

(A) $A(3, 30°)$, $B(-8, 180°)$, $C(5, -135°)$, $D(-10, -45°)$
(B) $A(5, \pi/3)$, $B(-6, 5\pi/6)$, $C(7, -\pi/2)$, $D(-4, -\pi/6)$

Solutions

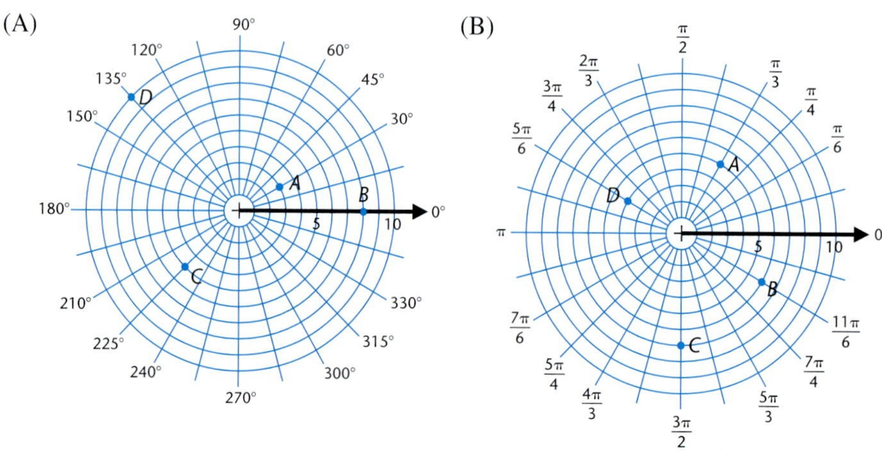

Matched Problem 1 Plot the following points in a polar coordinate system:

(A) $A(8, 45°)$, $B(-5, 150°)$, $C(4, -210°)$, $D(-6, -90°)$
(B) $A(9, \pi/6)$, $B(-3, -\pi)$, $C(-7, 7\pi/4)$, $D(5, -5\pi/6)$

EXPLORE-DISCUSS 1 A point in a polar coordinate system has coordinates $(5, 30°)$. How many other polar coordinates does the point have for θ restricted to $-360° \le \theta \le 360°$? Find the other coordinates of the point, and explain how they are found.

• Converting from Polar to Rectangular Form, and Vice Versa

Often, it is necessary to transform coordinates or equations in rectangular form to polar form, or vice versa. The following polar–rectangular relationships are useful in this regard:

Polar–Rectangular Relationships

We have the following relationships between rectangular coordinates (x, y) and polar coordinates (r, θ):

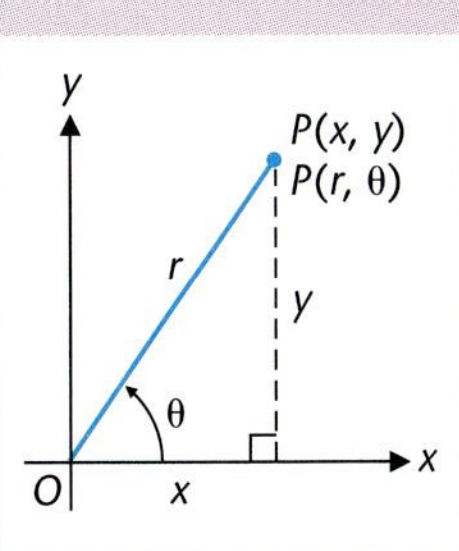

$$r^2 = x^2 + y^2$$

$$\sin\theta = \frac{y}{r} \quad \text{or} \quad y = r\sin\theta$$

$$\cos\theta = \frac{x}{r} \quad \text{or} \quad x = r\cos\theta$$

$$\tan\theta = \frac{y}{x}$$

[*Note:* The signs of x and y determine the quadrant for θ. The angle θ is chosen so that $-\pi < \theta \le \pi$ or $-180° < \theta \le 180°$, unless directed otherwise.]

Many calculators can automatically convert rectangular coordinates to polar form, and vice versa. (Read your manual for your particular calculator.) Example 2 illustrates calculator conversions in both directions.

EXAMPLE 2 **Converting from Polar to Rectangular Form, and Vice Versa**

(A) Convert the polar coordinates $(-4, 1.077)$ to rectangular coordinates to three decimal places.

(B) Convert the rectangular coordinates $(-3.207, -5.719)$ to polar coordinates with θ in degree measure, $-180° < \theta \le 180°$, and $r \ge 0$.

Solution (A) Use a calculator set in radian mode.

$$(r, \theta) = (-4, 1.077)$$

$$x = r\cos\theta = (-4)\cos 1.077 = -1.896$$

$$y = r\sin\theta = (-4)\sin 1.077 = -3.522$$

Rectangular coordinates are $(-1.\ 896, -3.522)$

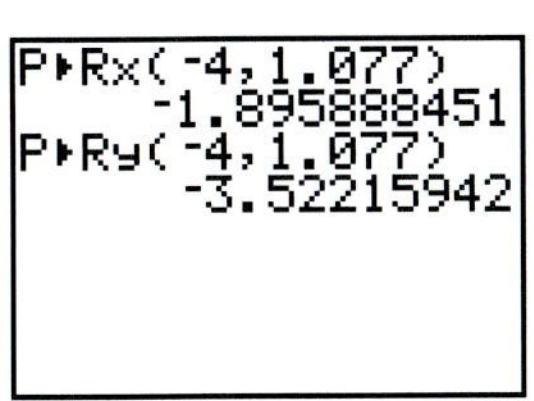

FIGURE 3

Figure 3 shows the same conversion done in a graphing utility with a built-in conversion routine.

(B) Use a calculator set in degree mode.

$$(x, y) = (-3.207, -5.719)$$

$$r = \sqrt{x^2 + y^2} = \sqrt{(-3.207)^2 + (-5.719)^2} = 6.557$$

$$\tan\theta = \frac{y}{x} = \frac{-5.719}{-3.207}$$

θ is a third-quadrant angle and is to be chosen so that $-180° < \theta \le 180°$.

$$\theta = -180° + \tan^{-1}\frac{-5.719}{-3.207} = -119.28°$$

Polar coordinates are $(6.557, -119.28°)$.

```
R▸Pr(-3.207,-5.7
19)
      6.556814013
R▸Pθ(-3.207,-5.7
19)
     -119.2820682
```

FIGURE 4

Figure 4 shows the same conversion done in a graphing utility with a built-in conversion routine.

Matched Problem 2 (A) Convert the polar coordinates $(8.677, -1.385)$ to rectangular coordinates to three decimal places.
(B) Convert the rectangular coordinates $(-6.434, 4.023)$ to polar coordinates with θ in degree measure, $-180° < \theta \le 180°$, and $r \ge 0$.

Generally, a more important use of the polar–rectangular relationships is in the conversion of equations in rectangular form to polar form, and vice versa.

EXAMPLE 3 Converting an Equation from Rectangular Form to Polar Form

Change $x^2 + y^2 - 4y = 0$ to polar form.

Solution Use $r^2 = x^2 + y^2$ and $y = r \sin \theta$.

$$x^2 + y^2 - 4y = 0$$
$$r^2 - 4r \sin \theta = 0$$
$$r(r - 4 \sin \theta) = 0$$
$$r = 0 \quad \text{or} \quad r - 4 \sin \theta = 0$$

The graph of $r = 0$ is the pole. Because the pole is included in the graph of $r - 4 \sin \theta = 0$ (let $\theta = 0$), we can discard $r = 0$ and keep only

$$r - 4 \sin \theta = 0$$

or

$$r = 4 \sin \theta \quad \text{The polar form of } x^2 + y^2 - 4y = 0$$

Matched Problem 3 Change $x^2 + y^2 - 6x = 0$ to polar form.

EXAMPLE 4 Converting an Equation from Polar Form to Rectangular Form

Change $r = -3\cos\theta$ to rectangular form.

Solution The transformation of this equation as it stands into rectangular form is fairly difficult. With a little trick, however, it becomes easy. We multiply both sides by r, which simply adds the pole to the graph. But the pole is already part of the graph of $r = -3\cos\theta$ (let $\theta = \pi/2$), so we haven't actually changed anything.

$$r = -3\cos\theta$$

$$r^2 = -3r\cos\theta \quad \text{Multiply both sides by } r.$$

$$x^2 + y^2 = -3x \quad r^2 = x^2 + y^2 \text{ and } r\cos\theta = x$$

$$x^2 + y^2 + 3x = 0$$

Matched Problem 4 Change $r + 2\sin\theta = 0$ to rectangular form.

• Graphing Polar Equations

We now turn to graphing polar equations. The **graph** of a polar equation, such as $r = 3\theta$ or $r = 6\cos\theta$, in a polar coordinate system is the set of all points having coordinates that satisfy the polar equation. Certain curves have simpler representations in polar coordinates, and other curves have simpler representations in rectangular coordinates.

To establish fundamentals in graphing polar equations, we start with a point-by-point graph. We then consider a more rapid way of making rough sketches of certain polar curves. And, finally, we show how polar curves are graphed in a graphing utility.

To plot a polar equation using **point-by-point plotting,** just as in rectangular coordinates, make a table of values that satisfy the equation, plot these points, then join them with a smooth curve. Example 5 illustrates the process.

EXAMPLE 5 Point-by-Point Plotting

(A) Graph $r = 8\cos\theta$ with θ in radians.
(B) Convert the polar equation in part A to rectangular form, and identify the graph.

Solution (A) We construct a table using multiples of $\pi/6$, plot these points, then join the points with a smooth curve (Fig. 5):

θ	r
0	8.0
$\pi/6$	6.9
$\pi/3$	4.0
$\pi/2$	0.0
$2\pi/3$	−4.0
$5\pi/6$	−6.9
π	8.0
Graph Repeats	

FIGURE 5

(B)

$$r = 8 \cos \theta$$

$$r^2 = 8r \cos \theta \quad \text{Multiply both sides by } r.$$

$$x^2 + y^2 = 8x \quad \text{Change to rectangular form.}$$

$$x^2 - 8x + y^2 = 0$$

$$x^2 - 8x + 16 + y^2 = 16 \quad \text{Complete the square on the left side.}$$

$$(x - 4)^2 + y^2 = 4^2 \quad \text{Standard equation of a circle}$$

The graph in part A is a circle with center at (4, 0) and radius 4 (see Section 2-1).

Matched Problem 5

(A) Graph $r = 8 \sin \theta$ with θ in degrees.
(B) Convert the polar equation in part A to rectangular form, and identify the graph.

If only a rough sketch of a polar equation involving $\sin \theta$ or $\cos \theta$ is desired, you can speed up the point-by-point graphing process by taking advantage of the uniform variation of $\sin \theta$ and $\cos \theta$ as θ moves around a unit circle. This process is referred to as **rapid polar sketching.** It is convenient to visualize Figure 6 in the process. With a little practice most of the table work in rapid sketching can be done mentally, and a rough sketch can be made directly from the equation.

FIGURE 6

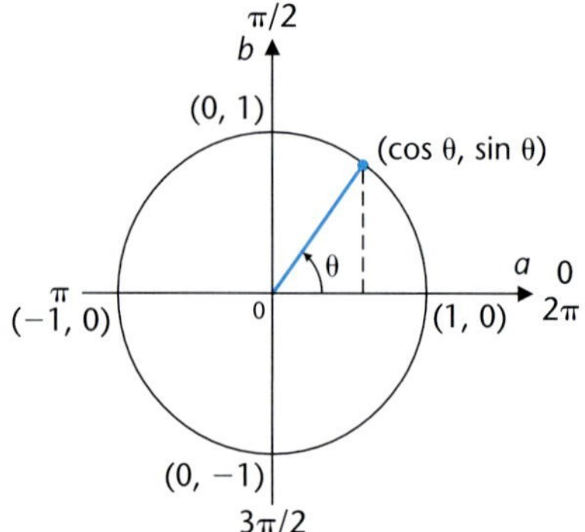

EXAMPLE 6 Rapid Polar Sketching

Sketch $r = 4 + 4 \cos \theta$ using rapid sketching techniques with θ in radians.

Solution We set up a table that indicates how r varies as we let θ vary through each set of quadrant values:

θ varies from	$\cos \theta$ varies from	$4 \cos \theta$ varies from	$r = 4 + 4 \cos \theta$ varies from
0 to $\pi/2$	1 to 0	4 to 0	8 to 4
$\pi/2$ to π	0 to -1	0 to -4	4 to 0
π to $3\pi/2$	-1 to 0	-4 to 0	0 to 4
$3\pi/2$ to 2π	0 to 1	0 to 4	4 to 8

Notice that as θ increases from 0 to $\pi/2$, $\cos \theta$ decreases from 1 to 0, $4 \cos \theta$ decreases from 4 to 0, and $r = 4 + 4 \cos \theta$ decreases from 8 to 4, and so on. Sketching these values, we obtain the graph in Figure 7, called a **cardioid.**

FIGURE 7

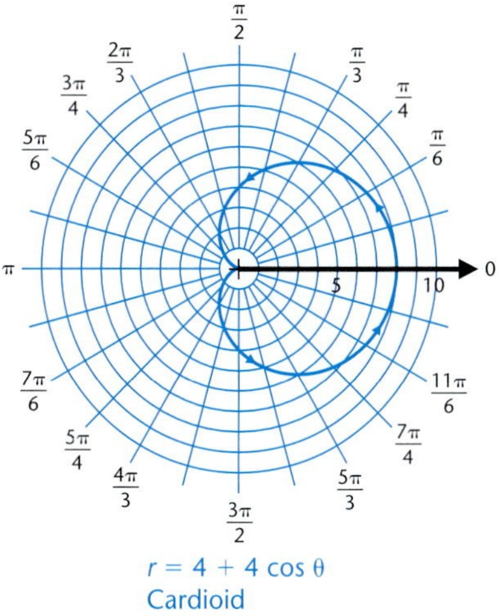

$r = 4 + 4 \cos \theta$
Cardioid

Matched Problem 6 Sketch $r = 5 + 5 \sin \theta$ using rapid sketching techniques with θ in radians.

EXAMPLE 7 Rapid Polar Sketching

Sketch $r = 8 \cos 2\theta$ with θ in radians.

Solution Start by letting 2θ (instead of θ) range through each set of quadrant values. That is, start with values for 2θ in the second column of the table, fill in the table at the top of the next page, and then fill in the first column for θ.

Start with the second column

θ varies from	2θ varies from	$\cos 2\theta$ varies from	$r = 8 \cos 2\theta$ varies from
0 to $\pi/4$	0 to $\pi/2$	1 to 0	8 to 0
$\pi/4$ to $\pi/2$	$\pi/2$ to π	0 to -1	0 to -8
$\pi/2$ to $3\pi/4$	π to $3\pi/2$	-1 to 0	-8 to 0
$3\pi/4$ to π	$3\pi/2$ to 2π	0 to 1	0 to 8
π to $5\pi/4$	2π to $5\pi/2$	1 to 0	8 to 0
$5\pi/4$ to $3\pi/2$	$5\pi/2$ to 3π	0 to -1	0 to -8
$3\pi/2$ to $7\pi/4$	3π to $7\pi/2$	-1 to 0	-8 to 0
$7\pi/4$ to 2π	$7\pi/2$ to 4π	0 to 1	0 to 8

As 2θ increases from 0 to $\pi/2$, θ increases from 0 to $\pi/4$, and r decreases from 8 to 0. As 2θ increases from $\pi/2$ to π, θ increases from $\pi/4$ to $\pi/2$, and r decreases from 0 to -8, and so on. Continue until the graph starts to repeat. Plotting the values, we obtain the graph in Figure 8, called a **four-leafed rose:**

FIGURE 8

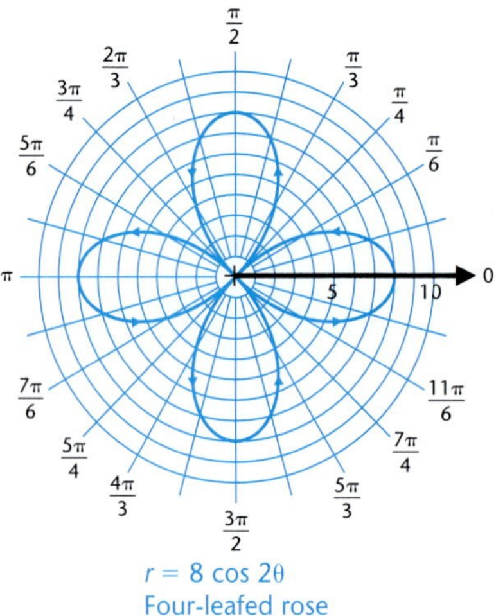

$r = 8 \cos 2\theta$
Four-leafed rose

Matched Problem 7 Sketch $r = 6 \sin 2\theta$ with θ in radians.

We now turn to **graphing polar equations in a graphing utility.** Example 8 illustrates the process.

EXAMPLE 8 Graphing in a Graphing Utility

Graph each of the following polar equations in a graphing utility (parts B and C are from Examples 6 and 7).

(A) $r = 3\theta$, $0 \le \theta \le 3\pi/2$ (Archimedes' spiral)
(B) $r = 4 + 4 \cos \theta$ (cardioid)
(C) $r = 8 \cos 2\theta$ (four-leafed rose)

Solution Set the graphing utility in polar mode, and select polar coordinates and radian measure. Adjust window values to accommodate the whole graph. A squared graph is often desirable in showing the true shape of the curve and is used here. Many graphing utilities, including the one used here, do not show a polar grid. When using TRACE, many graphing utilities offer a choice between polar coordinates and rectangular coordinates for points on the polar curve. The graphs of the polar equations above are shown in Figure 9.

(A) $r = 3\theta$, $0 \le \theta \le 3\pi/2$

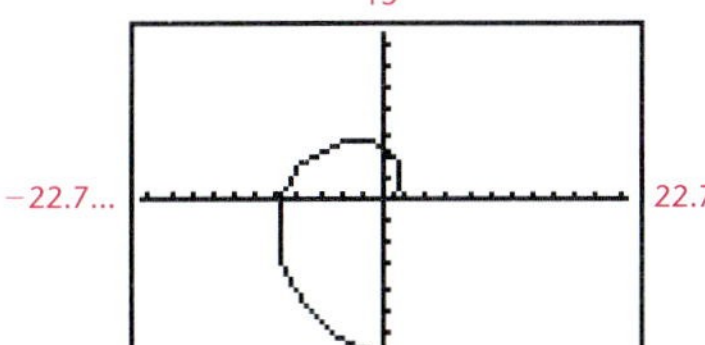

(B) $r = 4 + 4 \cos \theta$

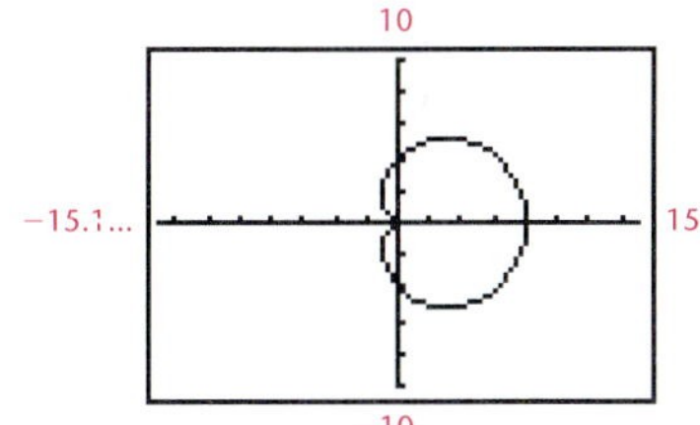

(C) $r = 8 \cos 2\theta$

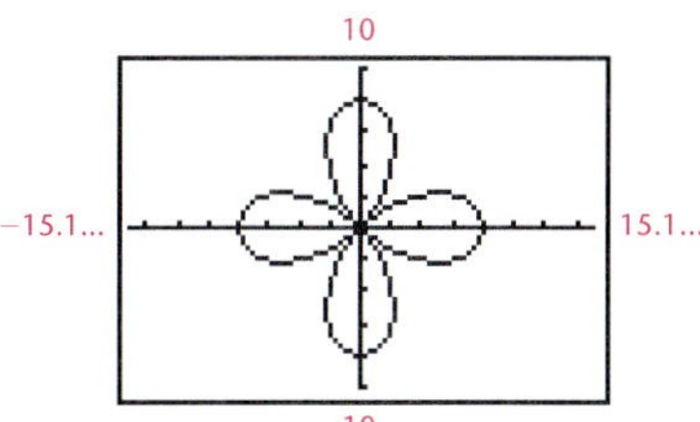

FIGURE 9

Matched Problem 8

Graph each of the following polar equations in a graphing utility.

(A) $r = 2\theta$, $0 \le \theta \le 2\pi$
(B) $r = 5 + 5 \sin \theta$ (C) $r = 6 \sin 2\theta$

EXPLORE-DISCUSS 2

(A) Graph $r1 = 10 \sin \theta$ and $r2 = 10 \cos \theta$ in the same viewing window. Use TRACE on $r1$, and estimate the polar coordinates where the two graphs intersect. Do the same thing for $r2$. Which intersection point appears to have the same polar coordinates on each curve and consequently represents a simultaneous solution to both equations? Which intersection point appears to have different polar coordinates on each curve and consequently does not represent a simultaneous solution? Solve the system for r and θ.

(B) Explain how rectangular coordinate systems differ from polar coordinate systems relative to intersection points and simultaneous solutions of systems of equations in the respective systems.

• Some Standard Polar Curves

In a rectangular coordinate system the simplest types of equations to graph are found by setting the rectangular variables x and y equal to constants:

$$x = a \qquad \text{and} \qquad y = b$$

The graphs are straight lines: The graph of $x = a$ is a vertical line, and the graph of $y = b$ is a horizontal line. A glance at Table 1 on the next page shows that horizontal and vertical lines do not have simple equations in polar coordinates.

Two of the simplest types of polar equations to graph in a polar coordinate system are found by setting the polar variables r and θ equal to constants:

$$r = a \qquad \text{and} \qquad \theta = b$$

Figure 10 illustrates the graphs of $\theta = \pi/4$ and $r = 5$.

FIGURE 10

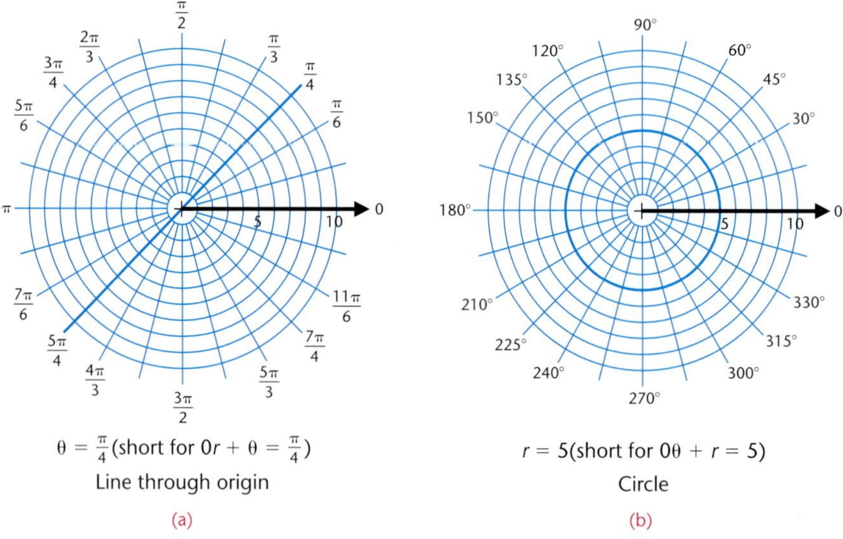

$\theta = \frac{\pi}{4}$ (short for $0r + \theta = \frac{\pi}{4}$)
Line through origin
(a)

$r = 5$ (short for $0\theta + r = 5$)
Circle
(b)

Table 1 illustrates a number of standard polar graphs and their equations. Polar graphing is often made easier if you have some idea of the final form.

• Application

Serious sailboat racers make polar plots of boat speed at various angles to the wind with various sail combinations at different wind speeds. With many polar plots for different sizes and types of sails at different wind speeds, they are able to accurately choose a sail for the optimum performance for different points of sail relative to any given wind strength. Figure 11 illustrates one such polar plot, where the maximum speed appears to be about 7.5 knots at 105° off the wind (with spinnaker sail set).

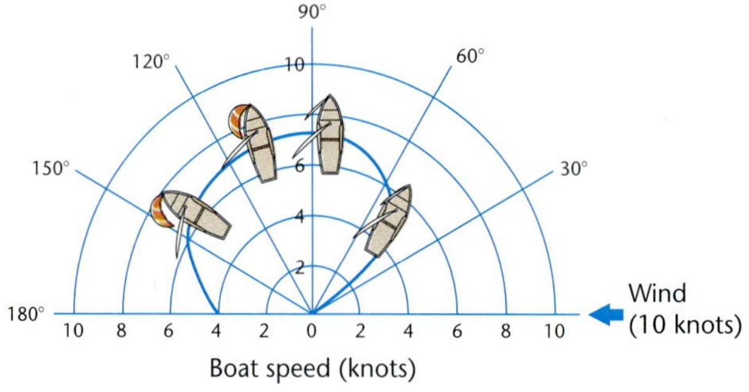

FIGURE 11 Polar diagram showing optimum sailing speed at different sailing angles to the wind.

TABLE 1 Standard Polar Graphs

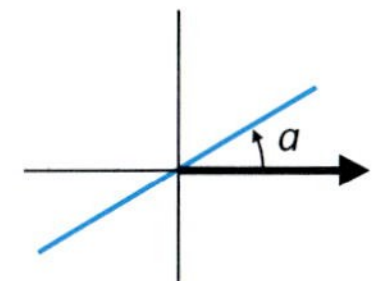

Line through origin:
$\theta = a$

(a)

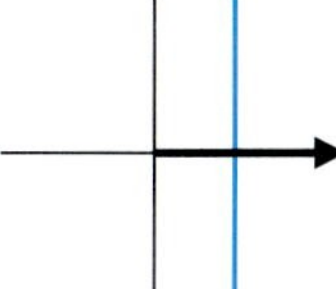

Vertical line:
$r = a/\cos\theta$
$= a \sec\theta$

(b)

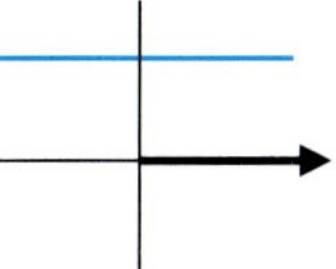

Horizontal line:
$r = a/\sin\theta$
$= a \csc\theta$

(c)

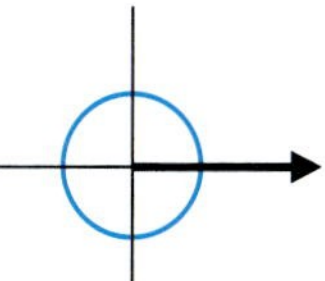

Circle:
$r = a$

(d)

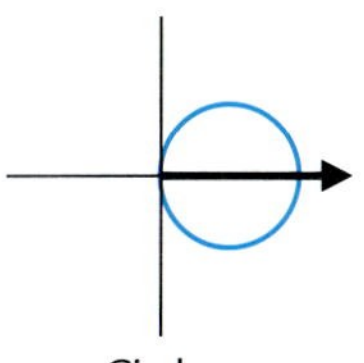

Circle:
$r = a\cos\theta$

(e)

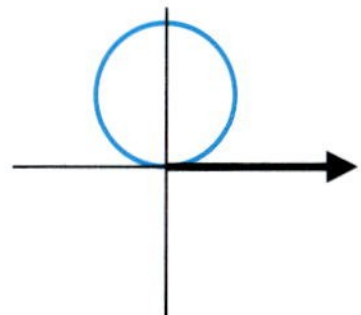

Circle:
$r = a\sin\theta$

(f)

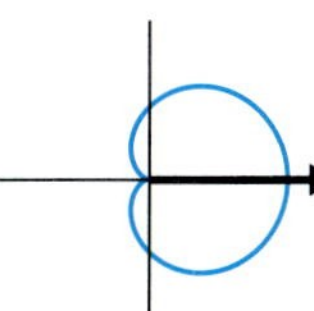

Cardioid:
$r = a + a\cos\theta$

(g)

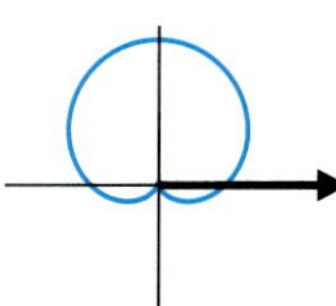

Cardioid:
$r = a + a\sin\theta$

(h)

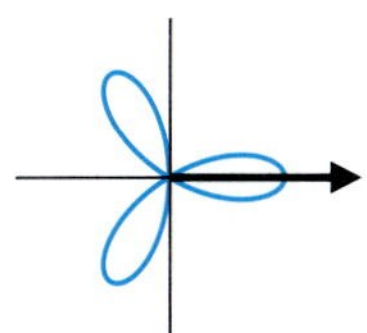

Three-leafed rose
$r = a\cos 3\theta$

(i)

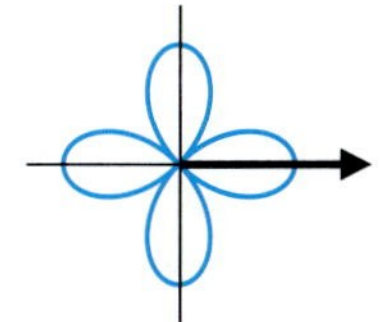

Four-leafed rose
$r = a\cos 2\theta$

(j)

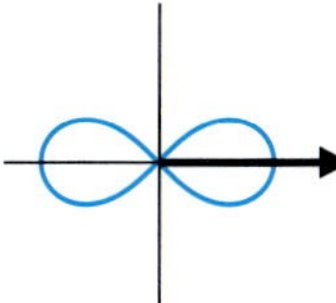

Lemniscate:
$r^2 = a^2\cos 2\theta$

(k)

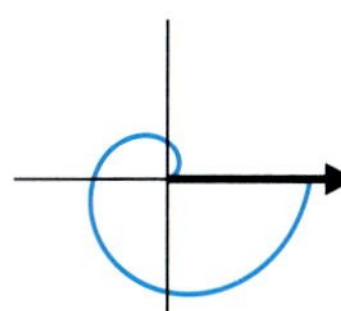

Archimedes' spiral:
$r = a\theta$, $a > 0$

(l)

Answers to Matched Problems

1. (A)

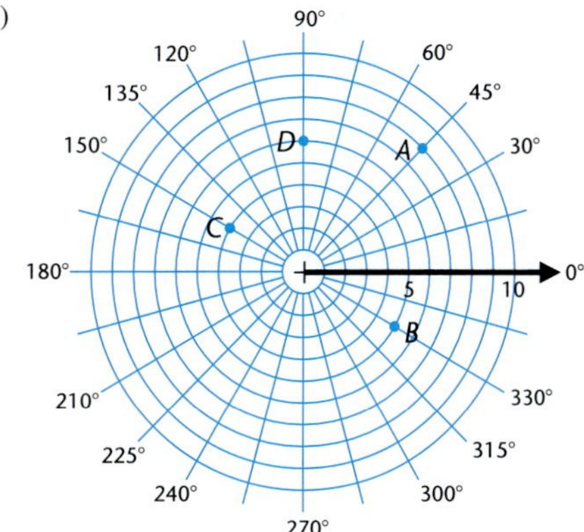

(B)

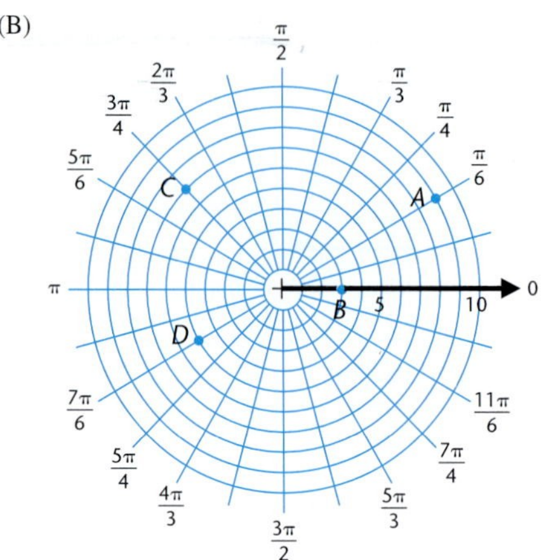

2. (A) (1.603, −8.528) (B) (7.588, 147.98°)

3. $r = 6 \cos \theta$ **4.** $x^2 + y^2 + 2y = 0$

5. (A)

θ	r
0°	0.0
30°	4.0
60°	6.9
90°	8.0
120°	6.9
150°	4.0
180°	0.0
Graph Repeats	

Circle:
$r = 8 \sin \theta$

(B) $x^2 + (y - 4)^2 = 4^2$

A circle with center at (0, 4) and radius 4.

6. $r = 5 + 5 \sin \theta$

Cardioid

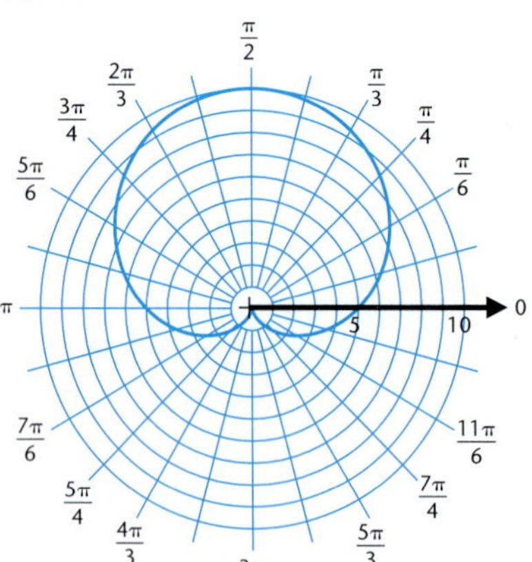

7. $r = 6 \sin 2\theta$

Four-leafed rose

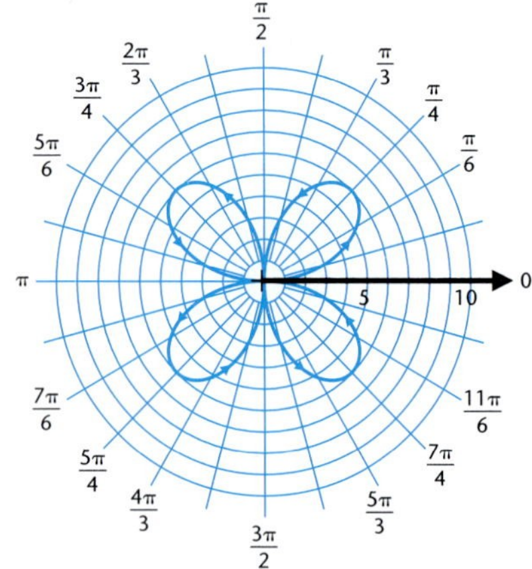

8. (A) $r = 2\theta,\ 0 \le \theta \le 2\pi$

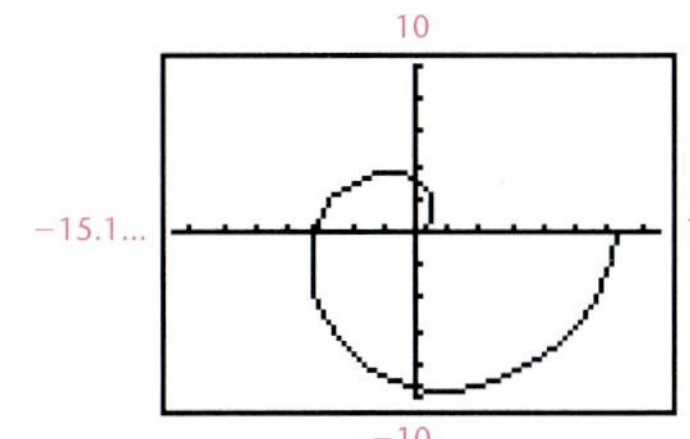

(B) $r = 5 + 5\sin\theta$

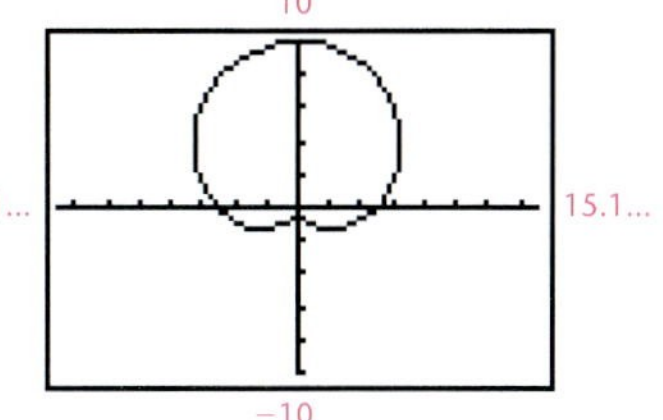

(C) $r = 6\sin 2\theta$

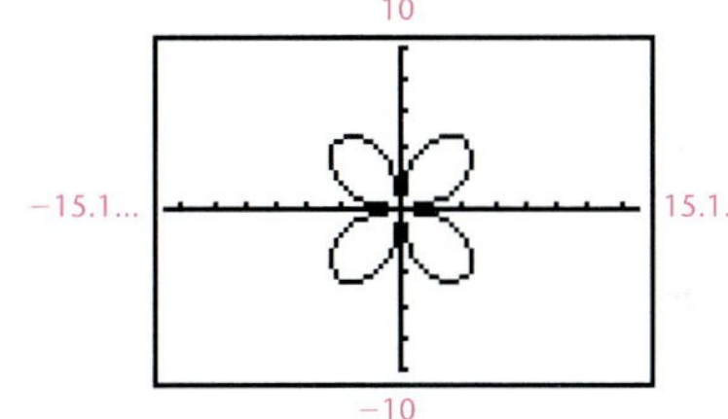

EXERCISE 7-5

A

Plot A, B, and C in Problems 1–8 in a polar coordinate system.

1. $A(4, 0°), B(7, 180°), C(9, 45°)$
2. $A(8, 0°), B(5, 90°), C(6, 30°)$
3. $A(-4, 0°), B(-7, 180°), C(-9, 45°)$
4. $A(-8, 0°), B(-5, 90°), C(-6, 30°)$
5. $A(8, -\pi/3), B(4, -\pi/4), C(10, -\pi/6)$
6. $A(6, -\pi/6), B(5, -\pi/2), C(8, -\pi/4)$
7. $A(-6, -\pi/6), B(-5, -\pi/2), C(-8, -\pi/4)$
8. $A(-6, -\pi/2), B(-5, -\pi/3), C(-8, -\pi/4)$

9. A point in a polar coordinate system has coordinates $(-5, 3\pi/4)$. Find all other polar coordinates for the point, $-2\pi \le \theta \le 2\pi$, and verbally describe how the coordinates are associated with the point.

10. A point in a polar coordinate system has coordinates $(6, -30°)$. Find all other polar coordinates for the point, $-360° \le \theta \le 360°$, and verbally describe how the coordinates are associated with the point.

Graph Problems 11 and 12 in a polar coordinate system using point-by-point plotting and the special values 0, $\pi/6$, $\pi/4$, $\pi/3$, $\pi/2$, $2\pi/3$, $3\pi/4$, $5\pi/6$, and π for θ.

Verify the graphs of Problems 11 and 12 on a graphing utility.

11. $r = 10\sin\theta$

12. $r = 10\cos\theta$

Graph Problems 13–16 in a polar coordinate system.

13. $r = 8$

14. $r = 5$

15. $\theta = \pi/3$

16. $\theta = \pi/6$

In Problems 17–22, convert the polar coordinates to rectangular coordinates to three decimal places.

17. $(6, \pi/6)$

18. $(7, 2\pi/3)$

19. $(-2, 7\pi/8)$

20. $(3, -3\pi/7)$

21. $(-4.233, -2.084)$

22. $(-9.028, -0.663)$

B

In Problems 23–28, convert the rectangular coordinates to polar coordinates with θ in degree measure, $-180° < \theta \le 180°$, and $R \ge 0$.

23. $(-8, 0)$

24. $(0, -5)$

25. $(-5, -5)$

26. $(1, -\sqrt{3})$

27. $(9.79, 5.13)$

28. $(-4.26, 31.1)$

In Problems 29–38, use rapid graphing techniques to sketch the graph of each polar equation.

Verify the graphs of Problems 29–38 on a graphing utility.

29. $r = 4\sin\theta$

30. $r = 4\cos\theta$

31. $r = 10\sin 2\theta$

32. $r = 8\cos 2\theta$

33. $r = 5\cos 3\theta$

34. $r = 6\sin 3\theta$

35. $r = 2 + 2\sin\theta$

36. $r = 3 + 3\cos\theta$

37. $r = 2 + 4\sin\theta$

38. $r = 2 + 4\cos\theta$

Problems 39–44 are exploratory problems requiring the use of a graphing utility.

39. Graph each polar equation in its own viewing window: $r = 2 + 2 \sin \theta$, $r = 4 + 2 \sin \theta$, $r = 2 + 4 \sin \theta$

40. Graph each polar equation in its own viewing window: $r = 2 + 2 \cos \theta$, $r = 4 + 2 \cos \theta$, $r = 2 + 4 \cos \theta$

41. (A) Graph each polar equation in its own viewing window: $r = 4 \sin \theta$, $r = 4 \sin 3\theta$, $r = 4 \sin 5\theta$
(B) What would you guess to be the number of leaves for $r = 4 \sin 7\theta$?
(C) What would you guess to be the number of leaves for $r = a \sin n\theta$, $a > 0$ and n odd?

42. (A) Graph each polar equation in its own viewing window: $r = 4 \cos \theta$, $r = 4 \cos 3\theta$, $r = 4 \cos 5\theta$
(B) What would you guess to be the number of leaves for $r = 4 \cos 7\theta$?
(C) What would you guess to be the number of leaves for $r = a \cos n\theta$, $a > 0$ and n odd?

43. (A) Graph each polar equation in its own viewing window: $r = 4 \sin 2\theta$, $r = 4 \sin 4\theta$, $r = 4 \sin 6\theta$
(B) What would you guess to be the number of leaves for $r = 4 \sin 8\theta$?
(C) What would you guess to be the number of leaves for $r = a \sin n\theta$, $a > 0$ and n even?

44. (A) Graph each polar equation in its own viewing window: $r = 4 \cos 2\theta$, $r = 4 \cos 4\theta$, $r = 4 \cos 6\theta$
(B) What would you guess to be the number of leaves for $r = 4 \cos 8\theta$?
(C) What would you guess to be the number of leaves for $r = a \cos n\theta$, $a > 0$ and n even?

In Problems 45–50, change each rectangular equation to polar form. Identify the graph as a line, circle, etc.

45. $x^2 + y^2 = 4$

46. $x + y = 0$

47. $x - \sqrt{3}y = 0$

48. $x^2 + y^2 + 8x = 0$

49. $5y = x^2$

50. $x^2 - y^2 = 1$

In Problems 51–56, change each polar equation to rectangular form. Identify the graph as a line, circle, etc.

51. $r = 3 \cos \theta$

52. $\theta + \pi/3 = 0$

53. $r(4 \sin \theta - \cos \theta) = 1$

54. $r + 5 \sin \theta = 0$

55. $r(2 + \cos \theta) = 1$

56. $r(1 + \cos \theta) = 1$

C

Problems 57–58 are exploratory problems requiring the use of a graphing utility.

57. Graph $r = 1 + 2 \sin(n\theta)$ for various values of n, n a natural number. Describe how n is related to the number of large petals and the number of small petals on the graph and how the large and small petals are related to each other relative to n.

58. Graph $r = 1 + 2 \cos(n\theta)$ for various values of n, n a natural number. Describe how n is related to the number of large petals and the number of small petals on the graph and how the large and small petals are related to each other relative to n.

In Problems 59–62, graph each system of equations on the same set of polar coordinate axes. Then solve the system simultaneously. [Note: Any solution (r_1, θ_1) to the system must satisfy each equation in the system and thus identifies a point of intersection of the two graphs. However, there may be other points of intersection of the two graphs that do not have any coordinates that satisfy both equations. This represents a major difference between the rectangular coordinate system and the polar coordinate system.]

59. $r = 4 \cos \theta$
$r = -4 \sin \theta$
$0 \le \theta \le \pi$

60. $r = 2 \cos \theta$
$r = 2 \sin \theta$
$0 \le \theta \le \pi$

61. $r = 6 \cos \theta$
$r = 6 \sin 2\theta$
$0° \le \theta \le 360°$

62. $r = 8 \sin \theta$
$r = 8 \cos 2\theta$
$0° \le \theta \le 360°$

63. Analytic Geometry. A distance formula for the distance between two points in a polar coordinate system follows directly from the law of cosines:

$$d^2 = r_1^2 + r_2^2 - 2r_1r_2 \cos(\theta_2 - \theta_1)$$

$$d = \sqrt{r_1^2 + r_2^2 - 2r_1r_2 \cos(\theta_2 - \theta_1)}$$

Find the distance (to three decimal places) between the two points $P_1(4, \pi/4)$ and $P_2(1, \pi/2)$.

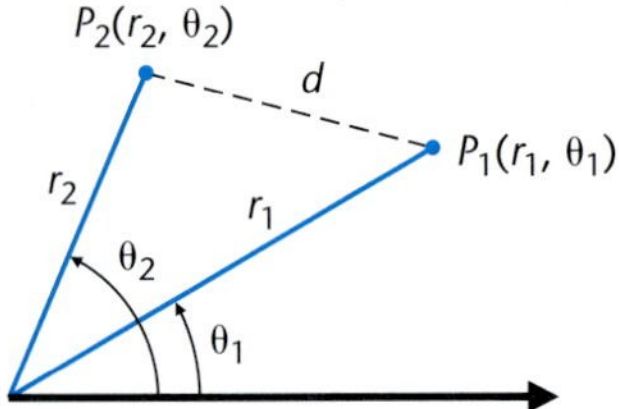

64. Analytic Geometry. Refer to Problem 63. Find the distance (to three decimal places) between the two points $P_1(2, 30°)$ and $P_2(3, 60°)$.

Problems 65–66 refer to the polar diagram in the figure. Polar diagrams of this type are used extensively by serious sailboat racers, and this polar diagram represents speeds in knots of a high-performance sailboat sailing at various angles to a wind blowing at 20 knots.

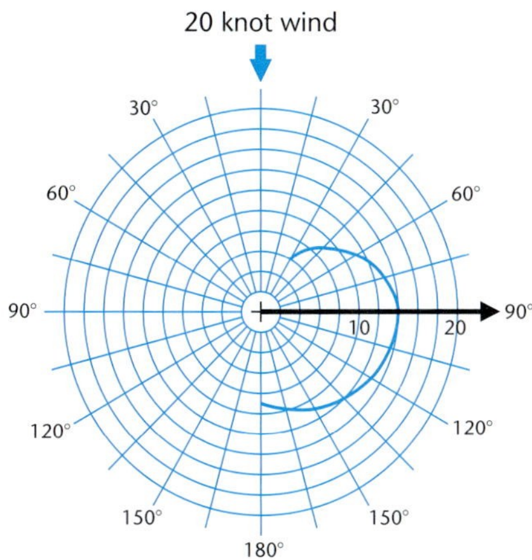

65. Sailboat Racing. Referring to the figure, estimate to the nearest knot the speed of the sailboat sailing at the following angles to the wind: 30°, 75°, 135°, and 180°.

66. Sailboat Racing. Referring to the figure, estimate to the nearest knot the speed of the sailboat sailing at the following angles to the wind: 45°, 90°, 120°, and 150°.

67. Conic Sections. Using a graphing utility, graph the equation

$$r = \frac{8}{1 - e \cos \theta}$$

for the following values of e (called the **eccentricity** of the conic), and identify each curve as a hyperbola, ellipse, or a parabola.

(A) $e = 0.4$ (B) $e = 1$ (C) $e = 1.6$

(It is instructive to explore the graph for other positive values of e.)

68. Conic Sections. Using a graphing utility, graph the equation

$$r = \frac{8}{1 - e \cos \theta}$$

for the following values of e, and identify each curve as a hyperbola, ellipse, or a parabola.

(A) $e = 0.6$ (B) $e = 1$ (C) $e = 2$

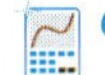

69. Astronomy. (A) The planet Mercury travels around the sun in an elliptical orbit given approximately by

$$r = \frac{3.442 \times 10^7}{1 - 0.206 \cos \theta}$$

where r is measured in miles and the sun is at the pole. Graph the orbit. Use TRACE to find the distance from Mercury to the sun at **aphelion** (greatest distance from the sun) and at **perihelion** (shortest distance from the sun).

(B) Johannes Kepler (1571–1630) showed that a line joining a planet to the sun sweeps out equal areas in space in equal intervals in time (see figure). Use this information to determine whether a planet travels faster or slower at aphelion than at perihelion. Explain your answer.

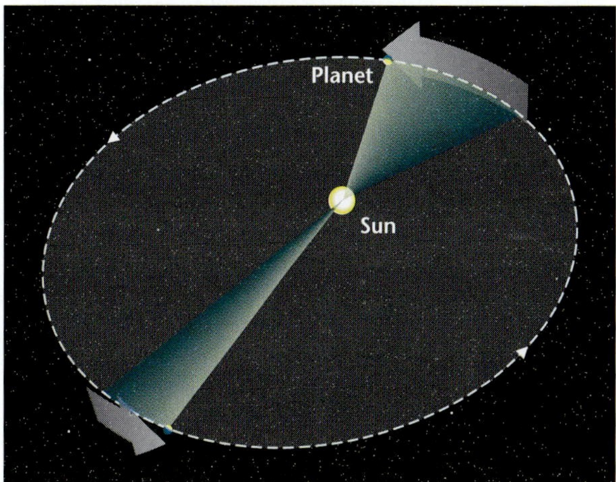

SECTION 7-6 Complex Numbers in Rectangular and Polar Forms

- Rectangular Form
- Polar Form
- Multiplication and Division in Polar Form
- Historical Note

Utilizing polar concepts studied in the last section, we now show how complex numbers can be written in polar form, which can be very useful in many applications. A brief review of Section 1-5 on complex numbers should prove helpful before proceeding further.

• Rectangular Form

Recall from Section 1-5 that a complex number is any number that can be written in the form

$$a + bi$$

where a and b are real numbers and i is the imaginary unit. Thus, associated with each complex number $a + bi$ is a unique ordered pair of real numbers (a, b), and vice versa. For example,

$$3 - 5i \qquad \text{corresponds to} \qquad (3, -5)$$

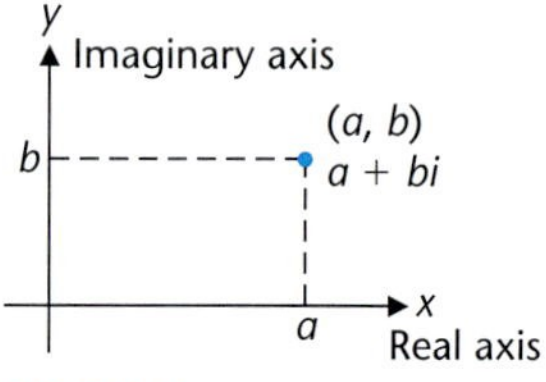

FIGURE 1 Complex plane.

Associating these ordered pairs of real numbers with points in a rectangular coordinate system, we obtain a **complex plane** (see Fig. 1). When complex numbers are associated with points in a rectangular coordinate system, we refer to the x axis as the **real axis** and the y axis as the **imaginary axis.** The complex number $a + bi$ is said to be in **rectangular form.**

EXAMPLE 1 Plotting in the Complex Plane

Plot the following complex numbers in a complex plane:

$$A = 2 + 3i \qquad B = -3 + 5i \qquad C = -4 \qquad D = -3i$$

Solution

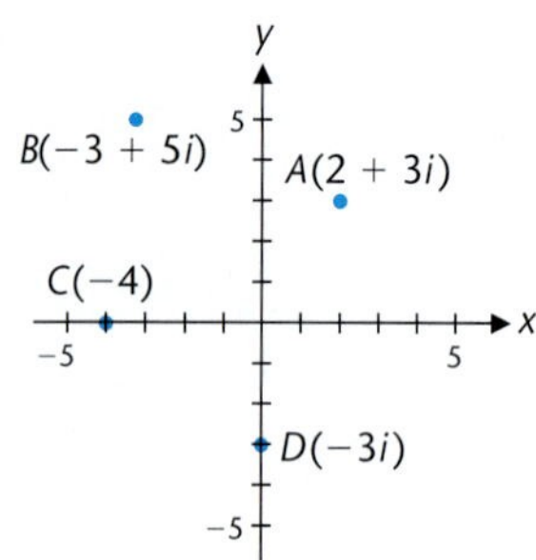

Matched Problem 1 Plot the following complex numbers in a complex plane:

$$A = 4 + 2i \qquad B = 2 - 3i \qquad C = -5 \qquad D = 4i$$

EXPLORE-DISCUSS 1 On a *real number line* there is a one-to-one correspondence between the set of real numbers and the set of points on the line: Each real number is associated with exactly one point on the line, and each point on the line is associated with exactly one real number. Does such a correspondence exist between the set of complex numbers and the set of points in an extended plane? Explain how a one-to-one correspondence can be established.

• Polar Form

Complex numbers also can be written in **polar form.** Using the polar–rectangular relationships from Section 7-5,

$$x = r\cos\theta \qquad \text{and} \qquad y = r\sin\theta$$

we can write the complex number $z = x + iy$ in polar form as follows:

$$z = x + iy = r\cos\theta + ir\sin\theta = r(\cos\theta + i\sin\theta) \tag{1}$$

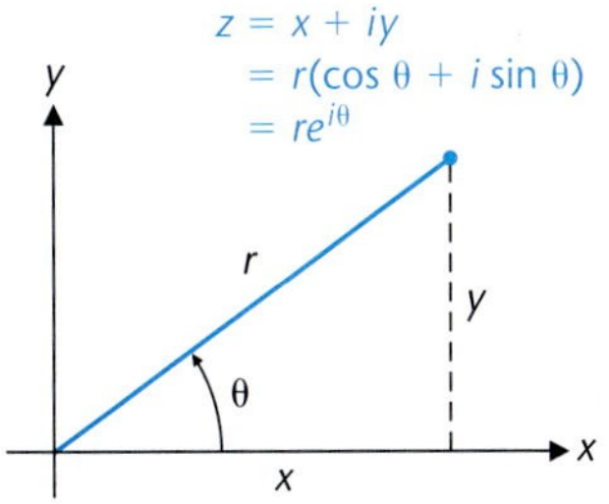

FIGURE 2 Rectangular–polar relationship.

This rectangular–polar relationship is illustrated in Figure 2. In a more advanced treatment of the subject, the following famous equation is established:

$$e^{i\theta} = \cos\theta + i\sin\theta \tag{2}$$

where $e^{i\theta}$ obeys all the basic laws of exponents. Thus, equation (1) takes on the form

$$z = x + yi = r(\cos\theta + i\sin\theta) = re^{i\theta} \tag{3}$$

We will freely use $re^{i\theta}$ as a polar form for a complex number. In fact, some graphing utilities display the polar form of $x + iy$ this way (see Fig. 3, where θ is in radians).

Since $\cos\theta$ and $\sin\theta$ are both periodic with period 2π, we have

$$\left.\begin{aligned}\cos(\theta + 2k\pi) &= \cos\theta\\ \sin(\theta + 2k\pi) &= \sin\theta\end{aligned}\right\} \quad k \text{ any integer}$$

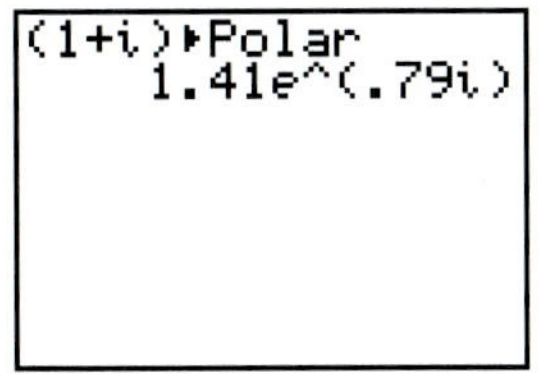

FIGURE 3 $1 + i = 1.41e^{0.79i}$.

Thus, we can write a more general polar form for a complex number $z = x + iy$, as given below, and observe that $re^{i\theta}$ is periodic with period $2k\pi$, k any integer.

DEFINITION 1

General Polar Form of a Complex Number

For k any integer,

$$\begin{aligned} z = x + iy &= r[\cos(\theta + 2k\pi) + i\sin(\theta + 2k\pi)]\\ &= re^{i(\theta+2k\pi)}\end{aligned}$$

The number r is called the **modulus,** or **absolute value,** of z and is denoted by **mod** z, or $|z|$. The polar angle that the line joining z to the origin makes with the polar axis is called the **argument** of z and is denoted by **arg** z. From Figure 2 we see the following relationships:

DEFINITION 2

Modulus and Argument for $z = x + iy$

$$\text{mod } z = r = \sqrt{x^2 + y^2} \qquad \text{Never negative}$$

$$\arg z = \theta + 2k\pi \qquad k \text{ any integer}$$

where $\sin\theta = y/r$ and $\cos\theta = x/r$. The argument θ is usually chosen so that $-180° < \theta \le 180°$ or $-\pi < \theta \le \pi$.

EXAMPLE 2 From Rectangular to Polar Form

Write parts A–C in polar form, θ in radians, $-\pi < \theta \le \pi$. Compute the modulus and arguments for parts A and B exactly; compute the modulus and argument for part C to two decimal places.

(A) $z_1 = 1 - i$ (B) $z_2 = -\sqrt{3} + i$ (C) $z_3 = -5 - 2i$

Solution Locate in a complex plane first; then if x and y are associated with special angles, r and θ can often be determined by inspection.

(A) A sketch shows that z_1 is associated with a special 45° triangle (Fig. 4). Thus, by inspection, $r = \sqrt{2}$, $\theta = -\pi/4$ (not $7\pi/4$), and

$$\begin{aligned} z_1 &= \sqrt{2}\left[\cos\left(-\frac{\pi}{4}\right) + i\sin\left(-\frac{\pi}{4}\right)\right] \\ &= \sqrt{2}e^{(-\pi/4)i} \end{aligned}$$

FIGURE 4

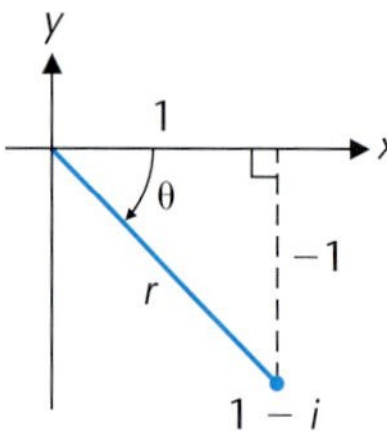

(B) A sketch shows that z_2 is associated with a special 30°–60° triangle (Fig. 5). Thus, by inspection, $r = 2$, $\theta = 5\pi/6$, and

$$\begin{aligned} z_2 &= 2\left(\cos\frac{5\pi}{6} + i\sin\frac{5\pi}{6}\right) \\ &= 2e^{(5\pi/6)i} \end{aligned}$$

FIGURE 5

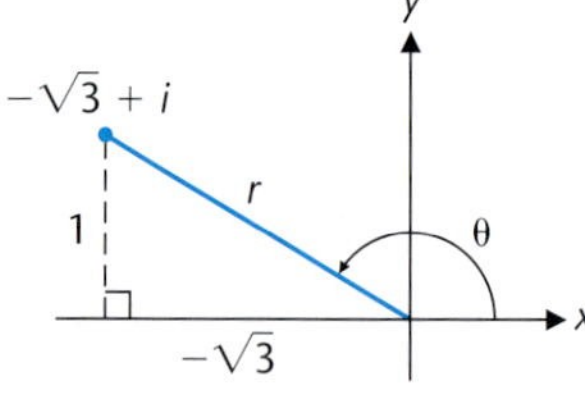

(C) A sketch shows that z_3 is not associated with a special triangle (Fig. 6). So, we proceed as follows:

$$r = \sqrt{(-5)^2 + (-2)^2} = 5.39 \quad \text{To two decimal places}$$

$$\theta = -\pi + \tan^{-1}\tfrac{2}{5} = -2.76 \quad \text{To two decimal places}$$

Thus,

$$z_3 = 5.39\ [\cos(-2.76) + i\ \sin(-2.76)]$$
$$= 5.39e^{(-2.76)i}$$

FIGURE 6

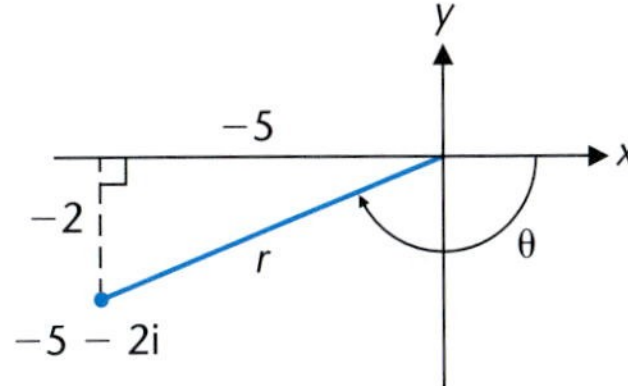

Figure 7 shows the same conversion done by a graphing utility with a built-in conversion routine.

FIGURE 7 $-5 - 2i = 5.39e^{-2.76i}$.

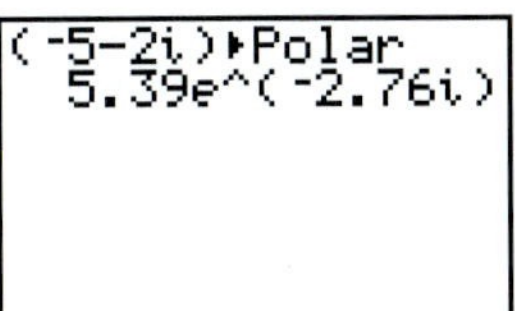

Matched Problem 2 Write parts A–C in polar form, θ in radians, $-\pi < \theta \leq \pi$. Compute the modulus and arguments for parts A and B exactly; compute the modulus and argument for part C to two decimal places.

(A) $-1 + i$ (B) $1 + i\sqrt{3}$ (C) $-3 - 5i$

EXAMPLE 3 From Polar to Rectangular Form

Write parts A–C in rectangular form. Compute the exact values for parts A and B; for part C compute a and b for $a + bi$ to two decimal places.

(A) $z_1 = 2e^{(5\pi/6)i}$ (B) $z_2 = 3e^{(-60°)i}$ (C) $z_3 = 7.19e^{-2.13i}$

Solution (A) $x + iy = 2e^{(5\pi/6)i}$

$$= 2\left(\cos\frac{5\pi}{6} + i\sin\frac{5\pi}{6}\right)$$
$$= 2\left(\frac{-\sqrt{3}}{2}\right) + i2\left(\frac{1}{2}\right)$$
$$= -\sqrt{3} + i$$

(B) $x + iy = 3e^{-60°i}$

$$= 3[\cos(-60°) + i\sin(-60°)]$$

$$= 3\left(\frac{1}{2}\right) + i3\left(\frac{-\sqrt{3}}{2}\right)$$

$$= \frac{3}{2} - \frac{3\sqrt{3}}{2}i$$

```
7.19e^(-2.13i)▸R
ect
        -3.81-6.09i
```

FIGURE 8
$7.19e^{-2.13i} = -3.81 - 6.09i$.

(C) $x + iy = 7.19e^{-2.13i}$

$$= 7.19[\cos(-2.13) + i\sin(-2.13)]$$

$$= -3.81 - 6.09i$$

Figure 8 shows the same conversion done by a graphing utility with a built-in conversion routine.

EXPLORE-DISCUSS 2 If your calculator has a built-in polar-to-rectangular conversion routine, try it on $\sqrt{2}e^{45°i}$ and $\sqrt{2}e^{(\pi/4)i}$, then reverse the process to see if you get back where you started. (For complex numbers in exponential polar form, some calculators require θ to be in radian mode for calculations. Check your user's manual.)

Matched Problem 3 Write parts A–C in rectangular form. Compute the exact values for parts A and B; for part C compute a and b for $a + bi$ to two decimal places.

(A) $z_1 = \sqrt{2}e^{(-\pi/2)i}$ (B) $z_2 = 3e^{120°i}$ (C) $z_3 = 6.49e^{-2.08i}$

EXPLORE-DISCUSS 3 Let $z_1 = \sqrt{3} + i$ and $z_2 = 1 + i\sqrt{3}$

(A) Find z_1z_2 and z_1/z_2 using the rectangular forms of z_1 and z_2.

(B) Find z_1z_2 and z_1/z_2 using the exponential polar forms of z_1 and z_2, θ in degrees. (Assume the product and quotient exponent laws hold for $e^{i\theta}$.)

(C) Convert the results from part B back to rectangular form and compare with the results in part A.

• Multiplication and Division in Polar Form

You will now see a particular advantage of representing complex numbers in polar form: Multiplication and division become very easy. Theorem 1 provides the reason. (The exponential polar form of a complex number obeys the product and quotient rules for exponents: $b^mb^n = b^{m+n}$ and $b^m/b^n = b^{m-n}$.)

Theorem 1 **Products and Quotients in Polar Form**

If $z_1 = r_1e^{i\theta_1}$ and $z_2 = r_2e^{i\theta_2}$, then

1. $z_1z_2 = r_1e^{i\theta_1}r_2e^{i\theta_2} = r_1r_2e^{i(\theta_1+\theta_2)}$

2. $\dfrac{z_1}{z_2} = \dfrac{r_1e^{i\theta_1}}{r_2e^{i\theta_2}} = \dfrac{r_1}{r_2}e^{i(\theta_1-\theta_2)}$

We establish the multiplication property and leave the quotient property for Problem 34 in Exercise 7-6.

$$\begin{aligned}
z_1z_2 &= r_1e^{i\theta_1}r_2e^{i\theta_2} && \text{Write in trigonometric form.}\\
&= r_1r_2(\cos\theta_1 + i\sin\theta_1)(\cos\theta_2 + i\sin\theta_2) && \text{Multiply.}\\
&= r_1r_2(\cos\theta_1\cos\theta_2 + i\cos\theta_1\sin\theta_2 + \\
&\qquad i\sin\theta_1\cos\theta_2 - \sin\theta_1\sin\theta_2)\\
&= r_1r_2[(\cos\theta_1\cos\theta_2 - \sin\theta_1\sin\theta_2) + \\
&\qquad i(\cos\theta_1\sin\theta_2 + \sin\theta_1\cos\theta_2)] && \text{Use sum identities.}\\
&= r_1r_2[\cos(\theta_1+\theta_2) + i\sin(\theta_1+\theta_2)]\\
&= r_1r_2e^{i(\theta_1+\theta_2)} && \text{Write in exponential form.}
\end{aligned}$$

EXAMPLE 4 **Products and Quotients**

If $z_1 = 8e^{45°i}$ and $z_2 = 2e^{30°i}$, find:

(A) z_1z_2 (B) z_1/z_2

Solution (A) $z_1z_2 = 8e^{45°i} \cdot 2e^{30°i}$

$$= 8\cdot 2e^{i(45°+30°)} = 16e^{75°i}$$

(B) $\dfrac{z_1}{z_2} = \dfrac{8e^{45°i}}{2e^{30°i}}$

$$= \tfrac{8}{2}e^{i(45°-30°)} = 4e^{15°i}$$

Matched Problem 4 If $z_1 = 9e^{165°i}$ and $z_2 = 3e^{55°i}$, find:

(A) z_1z_2 (B) z_1/z_2

• Historical Note There is hardly an area in mathematics that does not have some imprint of the famous Swiss mathematician Leonhard Euler (1707–1783), who spent most of his productive life at the New St. Petersburg Academy in Russia and the Prussian Academy in Berlin. One of the most prolific writers in the history of the subject, he is credited with making the following familiar notations standard:

$f(x)$	function notation
e	natural logarithmic base
i	imaginary unit, $\sqrt{-1}$

For our immediate interest, he is also responsible for the extraordinary relationship

$$e^{i\theta} = \cos\theta + i\sin\theta$$

If we let $\theta = \pi$, we obtain an equation that relates five of the most important numbers in the history of mathematics:

$$e^{i\pi} + 1 = 0$$

Answers to Matched Problems

1.

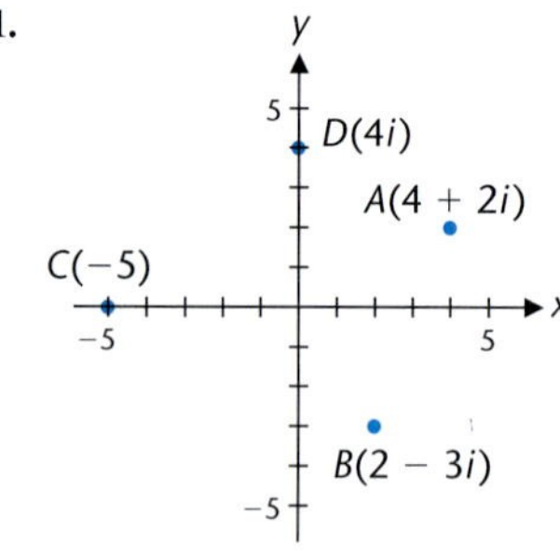

2. (A) $\sqrt{2}[\cos(3\pi/4) + i\sin(3\pi/4)] = \sqrt{2}e^{(3\pi/4)i}$
(B) $2[\cos(\pi/3) + i\sin(\pi/3)] = 2e^{(\pi/3)i}$
(C) $5.83[\cos(-2.11) + i\sin(-2.11)] = 5.83e^{-2.11i}$

3. (A) $-\sqrt{2}i$ (B) $-\frac{3}{2} + \frac{3\sqrt{3}}{2}i$ (C) $-3.16 - 5.67i$

4. (A) $z_1z_2 = 27e^{220°i}$ (B) $z_1/z_2 = 3e^{110°i}$

EXERCISE 7-6

A

In Problems 1–8, plot each set of complex numbers in a complex plane.

1. $A = 3 + 4i, B = -2 - i, C = 2i$
2. $A = 4 + i, B = -3 + 2i, C = -3i$
3. $A = 3 - 3i, B = 4, C = -2 + 3i$
4. $A = -3, B = -2 - i, C = 4 + 4i$
5. $A = 2e^{(\pi/3)i}, B = \sqrt{2}e^{(\pi/4)i}, C = 4e^{(\pi/2)i}$
6. $A = 2e^{(\pi/6)i}, B = 4e^{\pi i}, C = \sqrt{2}e^{(3\pi/4)i}$
7. $A = 4e^{-150°i}, B = 3e^{20°i}, C = 5e^{-90°i}$
8. $A = 2e^{150°i}, B = 3e^{-50°i}, C = 4e^{75°i}$

B

In Problems 9–12, change parts A–C to polar form. For Problems 9 and 10, choose θ in degrees, $-180° < \theta \le 180°$; for Problems 11 and 12, choose θ in radians, $-\pi < \theta \le \pi$. Compute the modulus and arguments for parts A and

B exactly; compute the modulus and argument for part C to two decimal places.

9. (A) $\sqrt{3} + i$ (B) $-1 - i$ (C) $5 - 6i$

10. (A) $-1 + i\sqrt{3}$ (B) $-3i$ (C) $-7 - 4i$

11. (A) $-i\sqrt{3}$ (B) $-\sqrt{3} - i$ (C) $-8 + 5i$

12. (A) $\sqrt{3} - i$ (B) $-2 + 2i$ (C) $6 - 5i$

In Problems 13–16, change parts A–C to rectangular form. Compute the exact values for parts A and B; for part C compute a and b for a + bi to two decimal places.

13. (A) $2e^{(\pi/3)i}$ (B) $\sqrt{2}e^{-45°i}$ (C) $3.08e^{2.44i}$

14. (A) $2e^{30°i}$ (B) $\sqrt{2}e^{(-3\pi/4)i}$ (C) $5.71e^{-0.48i}$

15. (A) $6e^{(\pi/6)i}$ (B) $\sqrt{7}e^{-90°i}$ (C) $4.09e^{-122.88°i}$

16. (A) $\sqrt{3}e^{(-\pi/2)i}$ (B) $\sqrt{2}e^{135°i}$ (C) $6.83e^{-108.82°i}$

In Problems 17–22, find z_1z_2 and z_1/z_2.

17. $z_1 = 7e^{82°i}$, $z_2 = 2e^{31°i}$ **18.** $z_1 = 6e^{132°i}$, $z_2 = 3e^{93°i}$

19. $z_1 = 5e^{52°i}$, $z_2 = 2e^{83°i}$ **20.** $z_1 = 3e^{67°i}$, $z_2 = 2e^{97°i}$

21. $z_1 = 3.05e^{1.76i}$, $z_2 = 11.94e^{2.59i}$

22. $z_1 = 7.11e^{0.79i}$, $z_2 = 2.66e^{1.07i}$

Simplify Problems 23–28 directly and by using polar forms. Write answers in both rectangular and polar forms, θ in degrees.

23. $(1 + i)(2 + 2i)$ **24.** $(\sqrt{3} + i)^2$

25. $(-1 + i)^3$ **26.** $(\sqrt{3} + i\sqrt{3})(1 + i\sqrt{3})$

27. $\dfrac{1 + i}{1 - i}$ **28.** $\dfrac{1 - i\sqrt{3}}{\sqrt{3} + i}$

29. The conjugate of $a + bi$ is $a - bi$. What is the conjugate of $re^{i\theta}$? Explain.

30. How is the product of a complex number z with its conjugate related to the modulus of z? Explain.

C

31. Show that $r^{1/3}e^{\theta/3}$ is a cube root of $re^{i\theta}$.

32. Show that $r^{1/2}e^{\theta/2}$ is a square root of $re^{i\theta}$.

33. If $z = re^{i\theta}$, show that $z^2 = r^2e^{2\theta i}$ and $z^3 = r^3e^{3\theta i}$. What do you think z^n will be for n a natural number?

34. Prove:

$$\frac{z_1}{z_2} = \frac{r_1e^{i\theta_1}}{r_2e^{i\theta_2}} = \frac{r_1}{r_2}e^{i(\theta_1-\theta_2)}$$

APPLICATIONS

35. Forces and Complex Numbers. An object is located at the pole, and two forces **u** and **v** act on the object. Let the forces be vectors going from the pole to the complex numbers $20e^{0°i}$ and $10e^{60°i}$, respectively. Force **u** has a magnitude of 20 pounds in a direction of 0°. Force **v** has a magnitude of 10 pounds in a direction of 60°.
(A) Convert the polar forms of these complex numbers to rectangular form and add.
(B) Convert the sum from part A back to polar form.
(C) The vector going from the pole to the complex number in part B is the resultant of the two original forces. What is its magnitude and direction?

36. Forces and Complex Numbers. Repeat Problem 35 with forces **u** and **v** associated with the complex numbers $8e^{0°i}$ and $6e^{30°i}$, respectively.

SECTION 7-7 De Moivre's Theorem

- De Moivre's Theorem, n a Natural Number
- nth-Roots of z

Abraham De Moivre (1667–1754), of French birth, spent most of his life in London doing private tutoring, writing, and publishing mathematics. He belonged to many prestigious professional societies in England, Germany, and France, and he was a close friend of Isaac Newton.

Using the polar form for a complex number, De Moivre established a theorem that still bears his name for raising complex numbers to natural number powers. More importantly, the theorem is the basis for the *nth-root theorem,* which enables us to find *all n n*th roots of any complex number, real or imaginary.

• De Moivre's Theorem, n a Natural Number

We start with Explore-Discuss 1 and generalize from this exploration.

EXPLORE-DISCUSS 1 By repeated use of the product formula for the exponential polar form $re^{i\theta}$, discussed in the last section, establish the following:

1. $(x + iy)^2 = (re^{\theta i})^2 = r^2e^{2\theta i}$

2. $(x + iy)^3 = (re^{\theta i})^3 = r^3e^{3\theta i}$

3. $(x + iy)^4 = (re^{\theta i})^4 = r^4e^{4\theta i}$

Based on forms 1–3, and for n a natural number, what do you think the polar form of $(x + iy)^n$ would be?

If you guessed $r^ne^{n\theta i}$, you have guessed De Moivre's theorem, which we now state without proof. A full proof of the theorem for all natural numbers n requires a method of proof, called *mathematical induction,* which is discussed in the second section in "Sequence and Series."

Theorem 1 **De Moivre's Theorem**

If $z = x + iy = re^{i\theta}$, and n is a natural number, then

$$z^n = (x + iy)^n = (re^{i\theta})^n = r^ne^{n\theta i}$$

EXAMPLE 1 The Natural Number Power of a Complex Number

Use De Moivre's theorem to find $(1 + i)^{10}$. Write the answer in exact rectangular form.

Solution

$$\begin{aligned}(1 + i)^{10} &= (\sqrt{2}e^{45°i})^{10} && \text{Convert } 1+i \text{ to polar form.}\\ &= (\sqrt{2})^{10}e^{(10\cdot 45°)i} && \text{Use De Moivre's theorem.}\\ &= 32e^{450°i} && \text{Change to rectangular form.}\\ &= 32(\cos 450° + i \sin 450°)\\ &= 32(0 + i)\\ &= 32i && \text{Rectangular form}\end{aligned}$$

Matched Problem 1 Use De Moivre's theorem to find $(1 + i\sqrt{3})^5$. Write the answer in exact polar and rectangular forms.

EXAMPLE 2 The Natural Number Power of a Complex Number

Use De Moivre's theorem to find $(-\sqrt{3} + i)^6$. Write the answer in exact rectangular form.

Solution

$$
\begin{aligned}
(-\sqrt{3} + i)^6 &= (2e^{150°i})^6 && \text{Convert } -\sqrt{3} + i \text{ to polar form.} \\
&= 2^6 e^{(6\cdot 150°)i} && \text{De Moivre's theorem} \\
&= 64e^{900°i} && \text{Change to rectangular form.} \\
&= 64(\cos 900° + i \sin 900°) \\
&= 64(-1 + i0) \\
&= -64 && \text{Rectangular form}
\end{aligned}
$$

[*Note:* $-\sqrt{3} + i$ must be a sixth root of -64, since $(-\sqrt{3} + i)^6 = -64$.]

Matched Problem 2 Use De Moivre's theorem to find $(1 - i\sqrt{3})^4$. Write the answer in exact polar and rectangular forms.

• *n*th Roots of *z*

We now consider roots of complex numbers. We say ***w* is an *n*th root of *z*,** n a natural number, if $w^n = z$. For example, if $w^2 = z$, then w is a square root of z. If $w^3 = z$, then w is a cube root of z. And so on.

EXPLORE-DISCUSS 2 If $z = re^{i\theta}$, then use De Moivre's theorem to show that $r^{1/2}e^{(\theta/2)i}$ is a square root of z and $r^{1/3}e^{(\theta/3)i}$ is a cube root of z.

We can proceed in the same way as in Explore-Discuss 2 to show that $r^{1/n}e^{(\theta/n)i}$ is an nth root of $re^{i\theta}$, n a natural number:

$$
\begin{aligned}
\left[r^{1/n}e^{(\theta/n)i}\right]^n &= (r^{1/n})^n e^{n(\theta/n)i} \\
&= re^{\theta i}
\end{aligned}
$$

But we can do even better than this. The nth-root theorem (Theorem 2) shows us how to find *all* the nth roots of a complex number.

Theorem 2 ***n*th-Root Theorem**

For n a positive integer greater than 1,

$$r^{1/n}e^{(\theta/n + k360°/n)i} \qquad k = 0, 1, \ldots, n-1$$

are the n distinct nth roots of $re^{i\theta}$, and there are no others.

The proof of Theorem 2 is left to Problems 39 and 40 in Exercise 7-7.

EXAMPLE 3 Finding all Sixth Roots of a Complex Number

Find six distinct sixth roots of $-1 + i\sqrt{3}$, and plot them in a complex plane.

Solution First write $-1 + i\sqrt{3}$ in polar form:

$$-1 + i\sqrt{3} = 2e^{120^\circ i}$$

Using the nth-root theorem, all six roots are given by

$$2^{1/6}e^{(120^\circ/6 + k360^\circ/6)i} = 2^{1/6}e^{(20^\circ + k60^\circ)i} \qquad k = 0, 1, 2, 3, 4, 5$$

Thus,

$$w_1 = 2^{1/6}e^{(20^\circ + 0\cdot 60^\circ)i} = 2^{1/6}e^{20^\circ i}$$
$$w_2 = 2^{1/6}e^{(20^\circ + 1\cdot 60^\circ)i} = 2^{1/6}e^{80^\circ i}$$
$$w_3 = 2^{1/6}e^{(20^\circ + 2\cdot 60^\circ)i} = 2^{1/6}e^{140^\circ i}$$
$$w_4 = 2^{1/6}e^{(20^\circ + 3\cdot 60^\circ)i} = 2^{1/6}e^{200^\circ i}$$
$$w_5 = 2^{1/6}e^{(20^\circ + 4\cdot 60^\circ)i} = 2^{1/6}e^{260^\circ i}$$
$$w_6 = 2^{1/6}e^{(20^\circ + 5\cdot 60^\circ)i} = 2^{1/6}e^{320^\circ i}$$

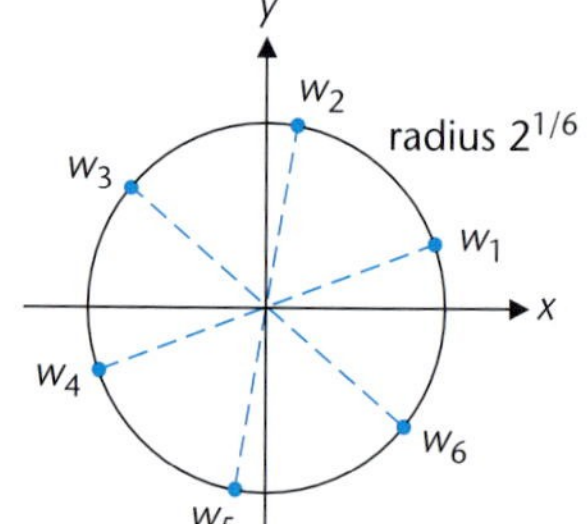

FIGURE 1

All roots are easily graphed in the complex plane after the first root is located. The root points are equally spaced around a circle of radius $2^{1/6}$ at an angular increment of 60° from one root to the next (Fig. 1).

Matched Problem 3 Find five distinct fifth roots of $1 + i$. Leave the answers in polar form, and plot them in a complex plane.

EXAMPLE 4 Solving a Cubic Equation

Solve $x^3 + 1 = 0$. Write final answers in rectangular form, and plot them in a complex plane.

Solution

$$x^3 + 1 = 0$$
$$x^3 = -1$$

We see that x is a cube root of -1, and there are a total of three roots. To find the three roots, we first write -1 in polar form:

$$-1 = 1e^{180^\circ i}$$

Using the nth-root theorem, all three cube roots of -1 are given by

$$1^{1/3}e^{(180°/3+k360°/3)i} = 1e^{(60°+k120°)i} \qquad k = 0, 1, 2$$

Thus,

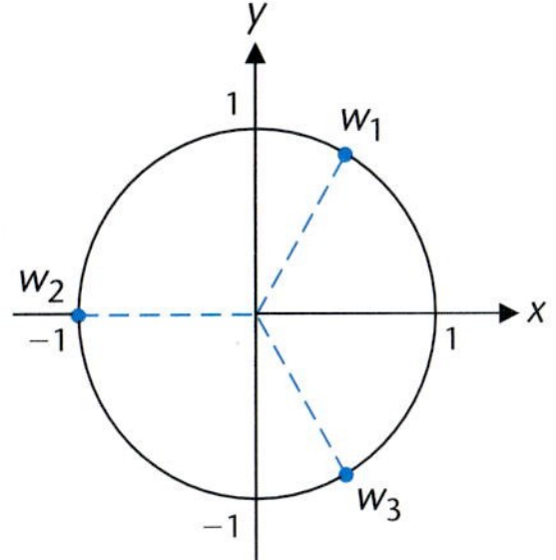

FIGURE 2

$$w_1 = 1e^{60°i} = \cos 60° + i \sin 60° = \frac{1}{2} + i\frac{\sqrt{3}}{2}$$

$$w_2 = 1e^{180°i} = \cos 180° + i \sin 180° = -1$$

$$w_3 = 1e^{300°i} = \cos 300° + i \sin 300° = \frac{1}{2} - i\frac{\sqrt{3}}{2}$$

[*Note:* This problem also can be solved using factoring and the quadratic formula—try it.] The three roots are graphed in Figure 2.

Matched Problem 4 Solve $x^3 - 1 = 0$. Write final answers in rectangular form, and plot them in a complex plane.

Answers to Matched Problems

1. $32e^{300°i} = 16 - i16\sqrt{3}$
2. $16e^{-240°i} = -8 + i8\sqrt{3}$
3. $w_1 = 2^{1/10}e^{9°i}$, $w_2 = 2^{1/10}e^{81°i}$, $w_3 = 2^{1/10}e^{153°i}$, $w_4 = 2^{1/10}e^{225°i}$, $w_5 = 2^{1/10}e^{297°i}$

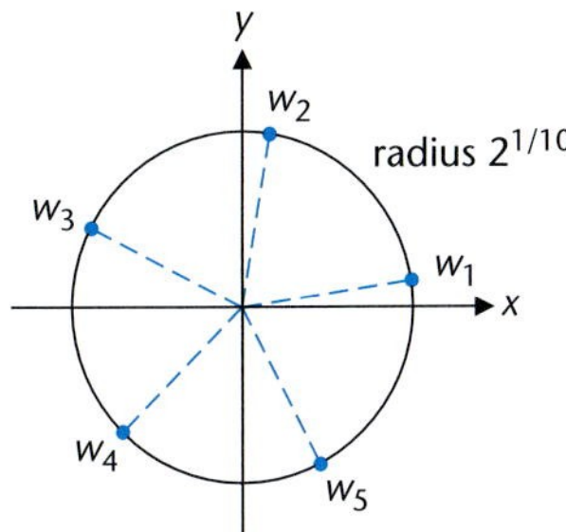

4. $1, -\frac{1}{2} + i\frac{\sqrt{3}}{2}, -\frac{1}{2} - i\frac{\sqrt{3}}{2}$

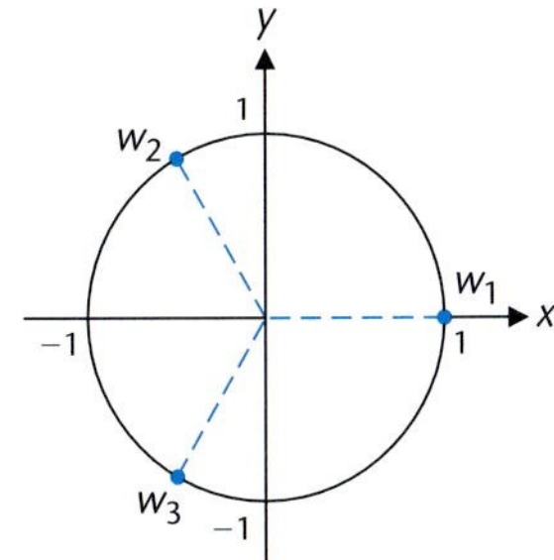

EXERCISE 7-7

A

In Problems 1–6, use De Moivre's theorem to evaluate. Express answers in polar form.

1. $(3e^{40°i})^4$ **2.** $(\sqrt{2}e^{15°i})^5$

3. $(1 - i)^6$ **4.** $(1 + i)^{12}$

5. $\left(\frac{\sqrt{3} - i}{2}\right)^{20}$ **6.** $(1 - i\sqrt{3})^8$

B

In Problems 7–12, find the value of each expression and write the final answer in exact rectangular form. (Verify the results in Problems 7–12 by evaluating each directly on a calculator.)

7. $(-\sqrt{3} - i)^4$ **8.** $(-1 + i)^4$

9. $(1 - i)^8$ **10.** $(-\sqrt{3} + i)^5$

11. $\left(-\frac{1}{2} + \frac{\sqrt{3}}{2}i\right)^3$ **12.** $\left(-\frac{1}{2} - \frac{\sqrt{3}}{2}i\right)^3$

For n and z as indicated in Problems 13–18, find all nth roots of z. Leave answers in polar form.

13. $z = 8e^{30°i}, n = 3$ **14.** $z = 8e^{45°i}, n = 3$

15. $z = 81e^{60°i}, n = 4$ **16.** $z = 16e^{90°i}, n = 4$

17. $z = 1 - i, n = 5$ **18.** $z = -1 + i, n = 3$

For n and z as indicated in Problems 19–24, find all nth roots of z. Write answers in polar form, and plot in a complex plane.

19. $z = 8, n = 3$ **20.** $z = 1, n = 4$

21. $z = -16, n = 4$ **22.** $z = -8, n = 3$

23. $z = i, n = 6$ **24.** $z = -i, n = 5$

In Problems 25–28, determine whether the statement is true or false. If true, explain why. If false, give a counterexample.

25. Every negative real number has two square roots.

26. If w is a cube root of z, then so is the conjugate of w.

27. Every twelfth root of 1 is a fourth root of 1.

28. Every fourth root of 1 is a twelfth root of 1.

29. (A) Show that $1 + i$ is a root of $x^4 + 4 = 0$. How many other roots does the equation have?

(B) The root $1 + i$ is located on a circle of radius $\sqrt{2}$ in the complex plane as indicated in the figure. Locate the other three roots of $x^4 + 4 = 0$ on the figure, and explain geometrically how you found their location.

(C) Verify that each complex number found in part B is a root of $x^4 + 4 = 0$.

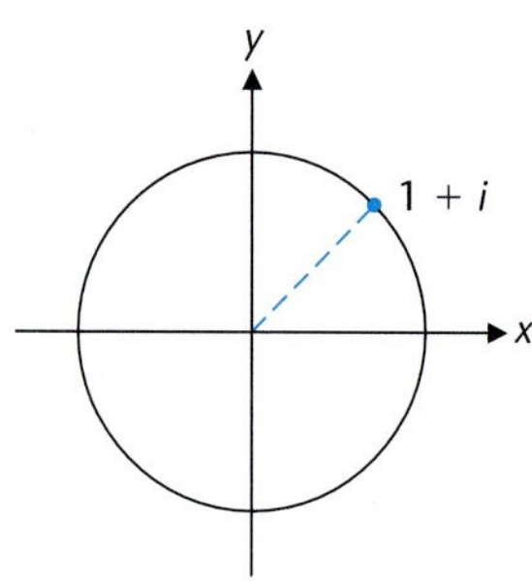

Figure for 29

30. (A) Show that -2 is a root of $x^3 + 8 = 0$. How many other roots does the equation have?

(B) The root -2 is located on a circle of radius 2 in the complex plane as indicated in the figure. Locate the other two roots of $x^3 + 8 - 0$ on the figure, and explain geometrically how you found their location.

(C) Verify that each complex number found in part B is a root of $x^3 + 8 = 0$.

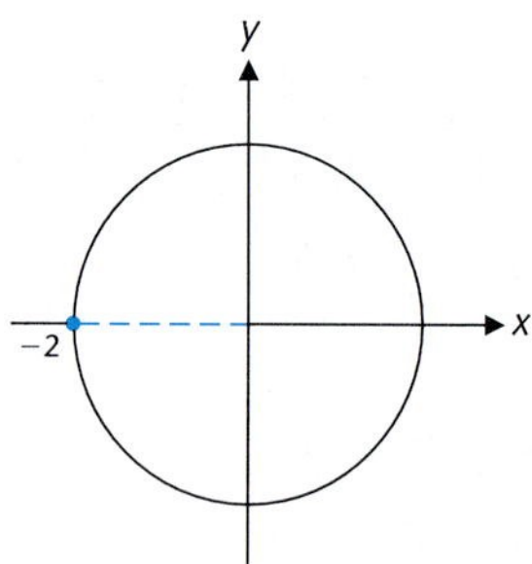

In Problems 31–34, solve each equation for all roots. Write final answers in polar and exact rectangular form.

31. $x^3 + 64 = 0$ **32.** $x^3 - 64 = 0$

33. $x^3 - 27 = 0$ **34.** $x^3 + 27 = 0$

C

For n and z as indicated in Problems 35–38, find all nth roots of z. Write answers in exact rectangular form.

35. $z = -4, n = 2$ **36.** $z = 25i, n = 2$

37. $z = 8i, n = 3$ **38.** $z = -64i, n = 3$

39. Show that

$$[r^{1/n}e^{(\theta/n + k360°/n)i}]^n = re^{i\theta}$$

for any natural number n and any integer k.

40. Show that

$$r^{1/n}e^{(\theta/n+k360°/n)i}$$

is the same number for $k = 0$ and $k = n$.

In Problems 41–44, write answers in polar form.

41. Find all complex zeros for $P(x) = x^5 - 32$.

42. Find all complex zeros for $P(x) = x^6 + 1$.

43. Solve $x^5 + 1 = 0$ in the set of complex numbers.

44. Solve $x^3 - i = 0$ in the set of complex numbers.

In Problems 45 and 46, write answers using exact rectangular forms.

45. Write $P(x) = x^6 + 64$ as a product of linear factors.

46. Write $P(x) = x^6 - 1$ as a product of linear factors.

CHAPTER 7 GROUP ACTIVITY Conic Sections and Planetary Orbits

I Conic Sections in Polar Form

(A) Introduction to Conics. To understand orbits of planets, comets, and other celestial bodies, one must know something of the nature and properties of conic sections. (Conic sections are treated in detail in Chapter 11. Here our treatment will be brief and limited to polar representations.) **Conic sections** get their name because the curves are formed by cutting a complete right circular cone of two nappes with a plane (Fig. 1). Any plane perpendicular to the axis of the cone cuts a section that is a **circle.** Tilt the plane slightly and the section becomes an **ellipse.** If the plane is parallel to one edge of the cone, it will cut only one nappe, and the section will be a **parabola.** Tilt the plane further to the vertical, then it will cut both nappes of the cone and produce a **hyperbola** with two branches. Closed orbits are ellipses or circles. Open (or escape) orbits are parabolas or hyperbolas.

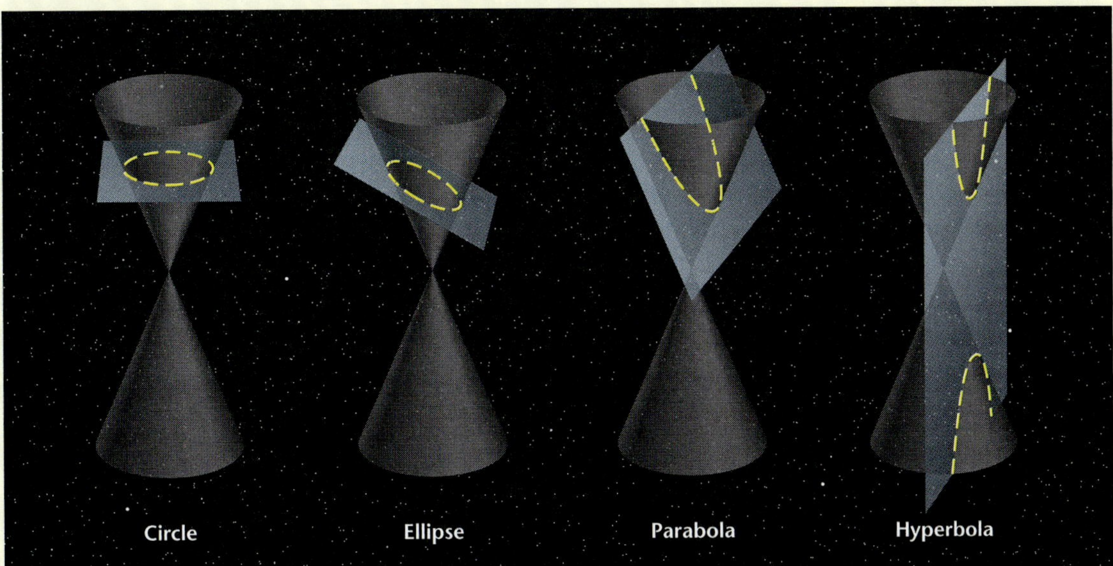

FIGURE 1 Conic sections.

(B) Conics and Eccentricity. Another way of defining conic sections is in terms of their eccentricity. Let F be a fixed point, called the **focus,** and let d be a fixed line, called the **directrix** (see Fig. 2). For positive values of **eccentricity** e, a conic section can be defined as the set of points $\{P\}$ having the property that the ratio of the distance from P to the focus F to the distance from P to the directrix d is the constant e. As we will see, an ellipse, a parabola, or a hyperbola can be obtained by choosing e appropriately.

FIGURE 2 Conic section.

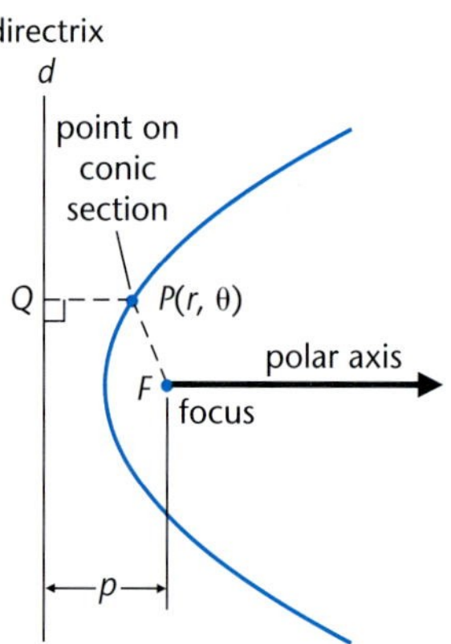

(C) Polar Representation of Conics. A unified treatment of conic sections can be obtained by use of the polar coordinate system. Polar equations of conics are used extensively in celestial mechanics to describe and analyze orbits of planets, comets, satellites, and other celestial bodies.

Problem 1: Polar Equation of a Conic. Use the eccentricity definition of a conic section given in part B to show that the polar equation of a conic is given by

$$r = \frac{ep}{1 - e \cos \theta} \tag{1}$$

where p is the distance between the focus F and the directrix d, the pole of the polar axis is at F, and the polar axis is perpendicular to d and is pointing away from d (see Fig. 2).

Problem 2: Graphing Utility Exploration, $0 < e < 1$. For $0 < e < 1$, use a graphing utility to systematically explore the nature of the changes in the graph of equation (1) as you change the eccentricity e and the distance p. Summarize the results of holding e fixed and changing p and the results of holding p fixed and changing e. For $0 < e < 1$, which conic section is produced?

Problem 3: Graphing Utility Exploration, $e = 1$. For $e = 1$, use a graphing utility to systematically explore the nature of the changes in the graph of equation (1) as you change the distance p. Summarize the results of holding e to 1 and changing p. For $e = 1$, which conic section is produced?

Problem 4: Graphing Utility Exploration, $e > 1$. For $e > 1$, use a graphing utility to systematically explore the nature of the changes in the graph of equation (1) as you change the eccentricity e and the distance p. Summarize the results of holding e fixed and changing p and the results of holding p fixed and changing e. For $e > 1$, which conic section is produced?

II Planetary Orbits

We are now interested in finding polar equations for the orbits of specific planets where the sun is at the pole. Then these equations can be graphed in a graphing utility, and further questions about the orbits can be answered. The material in Table 1, found in the readily available *World Almanac* and rounded to 3 significant digits, gives us enough information to find the polar equation for any planet's orbit.

Problem 5: Polar Equations for the Orbits of Mercury, Earth, and Mars. In all cases the polar axis intersects the planet's orbit at aphelion (the greatest distance from the sun).

(A) Show that Mercury's orbit is given approximately by

$$r = \frac{3.44 \times 10^7}{1 - 0.206 \cos \theta}$$

TABLE 1 The Planets

Planet	Eccentricity	Max. distance from sun (millions of miles)	Min. distance from sun (millions of miles)
Mercury	0.206	43.4	28.6
Venus	0.00677	67.7	66.8
Earth	0.0167	94.6	91.4
Mars	0.0934	155	129
Jupiter	0.0485	507	461
Saturn	0.0555	938	838
Uranus	0.0463	1,860	1,670
Neptune	0.00899	2,820	2,760
Pluto	0.249	4,550	2,760

(B) Show that Earth's orbit is given approximately by

$$r = \frac{9.30 \times 10^7}{1 - 0.0167 \cos \theta}$$

(C) Show that Mars' orbit is given approximately by

$$r = \frac{1.41 \times 10^8}{1 - 0.0934 \cos \theta}$$

Problem 6: Plotting the Orbits for Mercury, Earth, and Mars. Plot all three orbits (Mercury, Earth, and Mars) from the equations in parts A through C in the same viewing window of a graphing utility. Choose the window dimensions so that Mars' orbit fills up most of the window.

Problem 7: Finding Distances and Angles Related to Orbits. Figure 3 represents a schematic drawing showing Earth at two locations during its orbit. Find the straight-line distance between the position at A and the position at B to 3 significant digits. Find the measures of the angles BAO and ABO in degree measure to one decimal place. The Earth's orbit crosses the polar axis at aphelion (the greatest distance from the sun).

FIGURE 3 Earth's orbit.

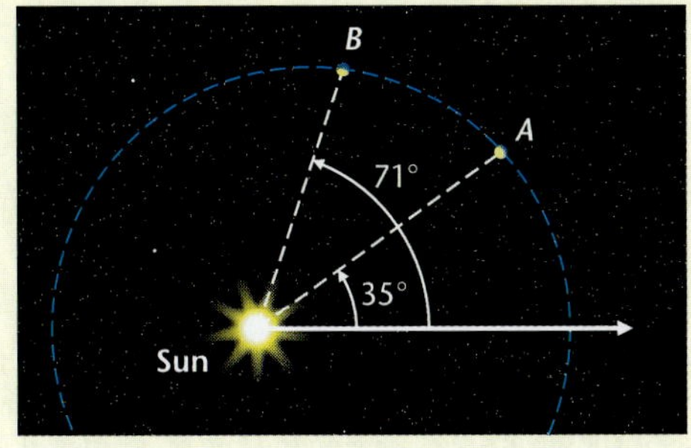

Chapter 7 Review

7-1 LAW OF SINES

An **oblique triangle** is a triangle without a right angle. An oblique triangle is **acute** if all angles are between 0 and 90° and **obtuse** if one angle is between 90 and 180°. The labeling convention shown in these figures is followed in this chapter.

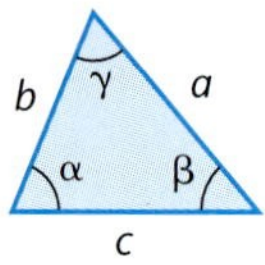

Acute triangle

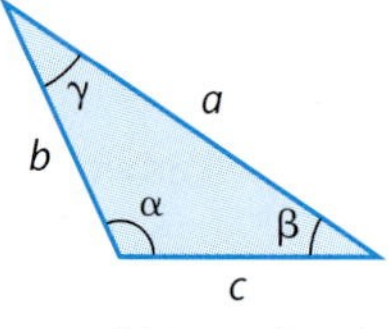

Obtuse triangle

The objective in this and the next section is to solve an oblique triangle given any three of the six quantities indicated in either figure, if a solution exists. The law of sines, discussed in this section, and the law of cosines, discussed in the next section, are used for this purpose. Accuracy in computations is governed by Table 1.

TABLE 1 Triangles and Significant Digits

Angle to nearest	Significant digits for side measure
1°	2
10′ or 0.1°	3
1′ or 0.01°	4
10″ or 0.001°	5

The **law of sines** is given as

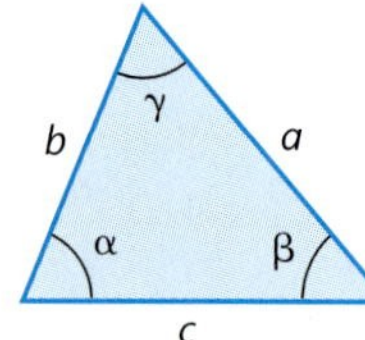

$$\frac{\sin \alpha}{a} = \frac{\sin \beta}{b} = \frac{\sin \gamma}{c}$$

and is generally used to solve the ASA, AAS, and SSA cases for oblique triangles. The AAS case is easily reduced to the ASA case by solving for the third angle first. The SSA case has a number of variations, including the ambiguous case. These variations are summarized in Table 2. Note that the ambiguous case always results in two triangles, one obtuse and one acute.

TABLE 2 SSA Variations

α	a ($h = b \sin \alpha$)	Number of triangles	Figure
Acute	$0 < a < h$	0	
Acute	$a = h$	1	
Acute	$h < a < b$	2	Ambiguous case
Acute	$a \geq b$	1	
Obtuse	$0 < a \leq b$	0	
Obtuse	$a > b$	1	

7-2 LAW OF COSINES

The **law of cosines** is given as

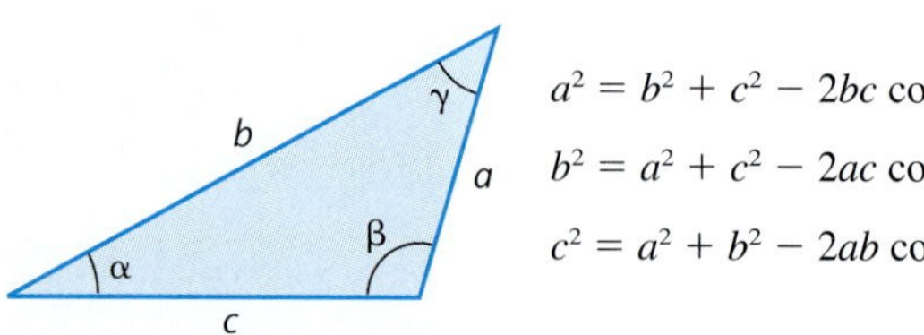

$$a^2 = b^2 + c^2 - 2bc \cos \alpha$$
$$b^2 = a^2 + c^2 - 2ac \cos \beta$$
$$c^2 = a^2 + b^2 - 2ab \cos \gamma$$

and is generally used as the first step in solving the SAS and SSS cases for oblique triangles. After a side or angle is found using the law of cosines, it is usually easier to continue the solving process with the law of sines.

7-3 GEOMETRIC VECTORS

A **scalar** is a real number. A **geometric vector** in a plane is a directed line segment and is represented by an arrow as indicated in the figure.

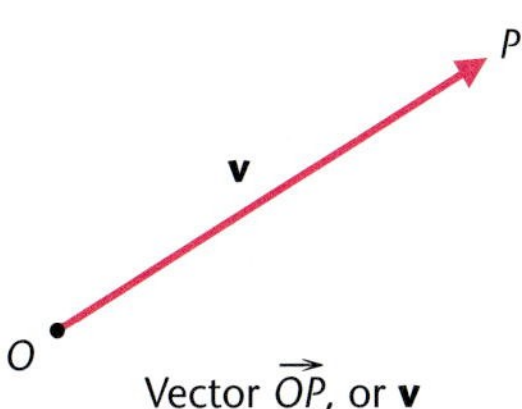

Vector $\overrightarrow{OP}$, or **v**

The point O is called the **initial point,** and the point P is called the **terminal point.**

The **magnitude** of the vector $\overrightarrow{AB}$, denoted by $|\overrightarrow{AB}|$, $|\vec{v}|$, or $|\mathbf{v}|$, is the length of the directed line segment. Two vectors have the **same direction** if they are parallel and point in the same direction. Two vectors have **opposite direction** if they are parallel and point in opposite directions. The **zero vector,** denoted by $\vec{0}$, or **0,** has a magnitude of zero and an arbitrary direction. Two vectors are **equal** if they have the same magnitude and direction. Thus, a vector may be **translated** from one location to another as long as the magnitude and direction do not change.

The **sum of two vectors u and v** can be defined using the **tail-to-tip rule.** The sum of two nonparallel vectors also can be defined using the **parallelogram rule.** Both forms are shown in the following figure:

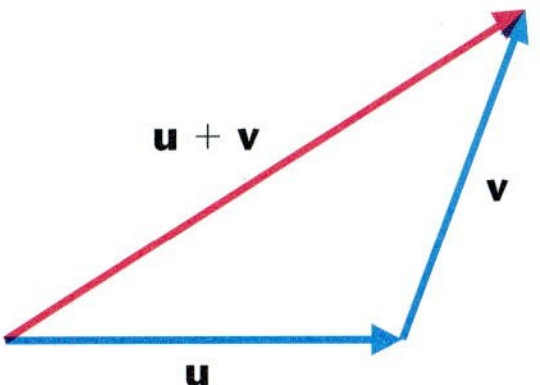

Vector addition: tail-to-tip rule

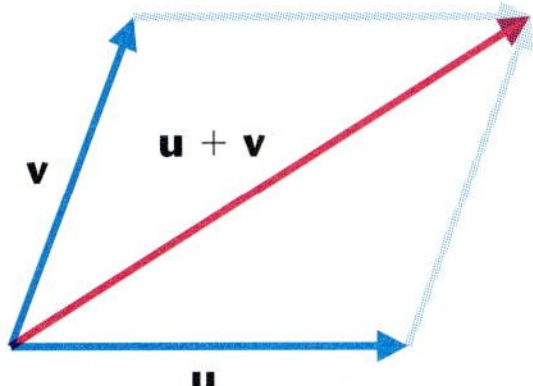

Vector addition: parallelogram rule

The vector **u** + **v** is also called the **resultant** of the two vectors **u** and **v**, and **u** and **v** are called **vector components** of **u** + **v**. Vector addition is **commutative;** that is, **u** + **v** = **v** + **u**.

A vector that represents the direction and speed of an object in motion is called a **velocity vector.** The velocity of an airplane relative to the air is called **apparent velocity,** and the velocity relative to the ground is called the **resultant,** or **actual, velocity.** The resultant velocity is the vector sum of the apparent velocity and wind velocity. Similar statements apply to objects in water subject to currents.

A vector that represents the direction and magnitude of an applied force is called a **force vector.** If an object is subjected to two forces, then the sum of these two forces, the **resultant force,** is a single force acting on the object in the same way as the two original forces taken together.

7-4 ALGEBRAIC VECTORS

A geometric vector $\overrightarrow{AB}$ in a rectangular coordinate system translated so that its initial point is at the origin is said to be in **standard position.** The vector $\overrightarrow{OP}$ such that $\overrightarrow{OP} = \overrightarrow{AB}$ is said to be the **standard vector** for $\overrightarrow{AB}$. This is shown in the following figure.

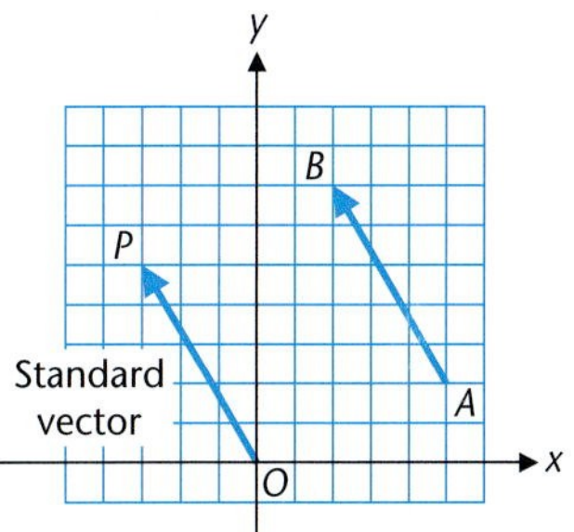

$\overrightarrow{OP}$ is the standard vector for $\overrightarrow{AB}$

Note that the vector $\overrightarrow{OP}$ in the figure is the standard vector for infinitely many vectors—all vectors with the same magnitude and direction as $\overrightarrow{OP}$.

Referring to the above figure, if the coordinates of A are (x_a, y_a) and the coordinates of B are (x_b, y_b), then the coordinates of P are given by

$$(x_p, y_p) = (x_b - x_a, y_b - y_a)$$

Each geometric vector in a coordinate system can be associated with an ordered pair of real numbers, the coordinates of the terminal point of its standard vector. Conversely, every ordered pair of real numbers can be associated with a unique geometric standard vector. This leads to the definition of an **algebraic vector** as an ordered pair of real numbers, denoted by $\langle \boldsymbol{a}, \boldsymbol{b} \rangle$. The real numbers a and b are **scalar components** of the vector $\langle a, b \rangle$.

Two vectors $\mathbf{u} = \langle a, b \rangle$ and $\mathbf{v} = \langle c, d \rangle$ are said to be **equal** if their corresponding components are equal, that is, if $a = c$ and $b = d$. The **zero vector** is denoted by $\mathbf{0} = \langle 0, 0 \rangle$ and has arbitrary direction.

The **magnitude,** or **norm,** of a vector $\mathbf{v} = \langle a, b \rangle$ is denoted by $|\mathbf{v}|$ and is given by

$$|\mathbf{v}| = \sqrt{a^2 + b^2}$$

Geometrically, $\sqrt{a^2 + b^2}$ is the length of the standard geometric vector $\overrightarrow{OP}$ associated with the algebraic vector $\langle a, b \rangle$.

If $\mathbf{u} = \langle a, b \rangle$, $\mathbf{v} = \langle c, d \rangle$, and k is a scalar, then the **sum** of **u** and **v** is given by

$$\mathbf{u} + \mathbf{v} = \langle a + c, b + d \rangle$$

and **scalar multiplication** of **u** by k is given by

$$k\mathbf{u} = k\langle a, b \rangle = \langle ka, kb \rangle$$

If **v** is a nonzero vector, then

$$\mathbf{u} = \frac{1}{|\mathbf{v}|}\mathbf{v}$$

is **a unit vector with the same direction as v.** The **i** and **j** unit vectors are defined as follows:

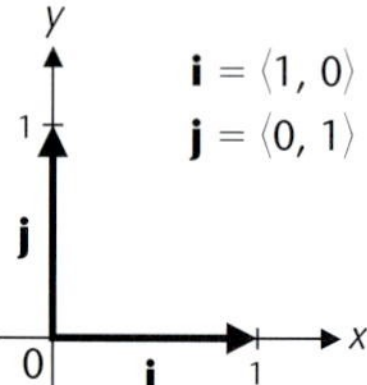

Every algebraic vector can be expressed in terms of the **i** and **j** unit vectors:

$$\mathbf{v} = \langle a, b \rangle = a\mathbf{i} + b\mathbf{j}$$

The following algebraic properties of vector addition and scalar multiplication enable us to manipulate symbols representing vectors and scalars in much the same way we manipulate symbols that represent real numbers in algebra.

Algebraic Properties of Vectors

A. Addition Properties. For all vectors **u**, **v**, and **w**:

1. $\mathbf{u} + \mathbf{v} = \mathbf{v} + \mathbf{u}$ — **Commutative Property**
2. $\mathbf{u} + (\mathbf{v} + \mathbf{w}) = (\mathbf{u} + \mathbf{v}) + \mathbf{w}$ — **Associative Property**
3. $\mathbf{u} + \mathbf{0} = \mathbf{0} + \mathbf{u} = \mathbf{u}$ — **Additive Identity**
4. $\mathbf{u} + (-\mathbf{u}) = (-\mathbf{u}) + \mathbf{u} = \mathbf{0}$ — **Additive Inverse**

B. Scalar Multiplication Properties. For all vectors **u** and **v** and all scalars m and n:

1. $m(n\mathbf{u}) = (mn)\mathbf{u}$ — **Associative Property**
2. $m(\mathbf{u} + \mathbf{v}) = m\mathbf{u} + m\mathbf{v}$ — **Distributive Property**
3. $(m + n)\mathbf{u} = m\mathbf{u} + n\mathbf{u}$ — **Distributive Property**
4. $1\mathbf{u} = \mathbf{u}$ — **Multiplicative Identity**

Certain **static equilibrium** problems can be solved using the material developed in this section. The conditions for static equilibrium are:

1. An object at rest is said to be in **static equilibrium.**
2. For an object located at the origin in a rectangular coordinate system to remain in static equilibrium, at rest, it is necessary that the sum of all the force vectors acting on the object be the zero vector.

7-5 POLAR COORDINATES AND GRAPHS

The following figure illustrates a **polar coordinate system.** The fixed point O is called the **pole** or **origin,** and the horizontal arrow is called the **polar axis.** We have the following **relationships between rectangular coordinates (x, y) and polar coordinates (r, θ):**

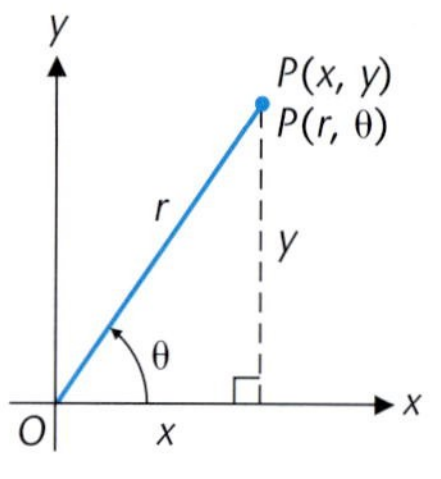

$$r^2 = x^2 + y^2$$

$$\sin\theta = \frac{y}{r} \quad \text{or} \quad y = r\sin\theta$$

$$\cos\theta = \frac{x}{r} \quad \text{or} \quad x = r\cos\theta$$

$$\tan\theta = \frac{y}{x}$$

[*Note:* The signs of x and y determine the quadrant for θ. The angle θ is chosen so that $-\pi < \theta \le \pi$ or $-180° < \theta \le 180°$, unless directed otherwise.]

Polar graphs can be obtained by **point-by-point** plotting much in the same ways graphs in rectangular coordinates are formed. Make a table of values that satisfy the polar equation, plot these points, then join them with a smooth curve.

Graphs can also be obtained by **rapid graphing techniques.** If only a rough sketch of a polar equation involving sin θ or cos θ is desired, we can speed up the point-by-point graphing process by taking advantage of the uniform variation of sin θ and cos θ as θ moves through each set of quadrant values. **Graphing utilities** can produce polar graphs almost instantly.

The following table shows some standard polar curves with their equations:

Standard Polar Graphs

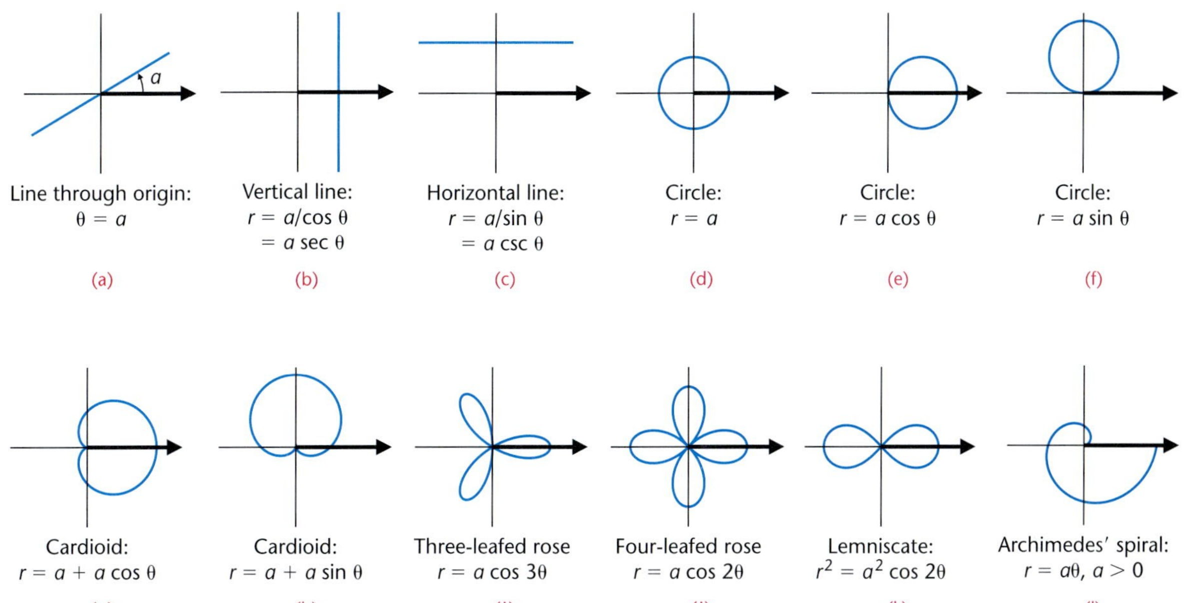

7-6 COMPLEX NUMBERS IN RECTANGULAR AND POLAR FORMS

A **complex number** is a number of the form

$$a + bi$$

where a and b are real numbers and i is the **imaginary unit.** The following figure shows a complex number $a + bi$ plotted in a **complex plane.**

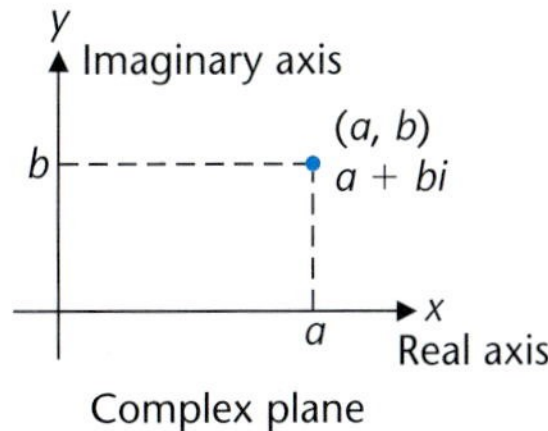

Complex plane

When complex numbers are associated with points in a rectangular coordinate system, we refer to the x axis as the **real axis** and the y axis as the **imaginary axis.** The complex number $a + bi$ is said to be in **rectangular form.**

Complex numbers can also be written in **polar,** or **trigonometric, form** using $x = r \cos \theta$ and $y = r \sin \theta$ as shown in the figure below:

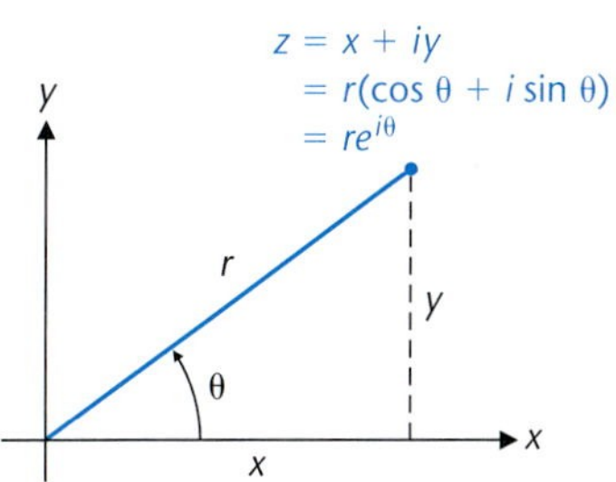

Rectangular–polar relationship

Because of the periodic nature of sine and cosine functions, we have the more **general polar form** for a complex number $z = x + iy$:

$$z = x + iy = r[\cos(\theta + 2k\pi) + i \sin(\theta + 2k\pi)]$$
$$= re^{(\theta + 2k\pi)i}$$

where

$$e^{i\theta} = \cos \theta + i \sin \theta$$

and the quadrant for θ is determined by x and y.

The number r is called the **modulus,** or **absolute value,** of z and is denoted by **mod** z, or $|z|$. The polar angle that the line joining z to the origin makes with the polar axis is called the

argument of z and is denoted by **arg** z. From the figure above we have the following representations of the modulus and argument for $z = x + iy$:

$$\text{mod } z = r = \sqrt{x^2 + y^2} \qquad \text{Never negative}$$

$$\arg z = \theta + 2k\pi \qquad k \text{ any integer}$$

where $\sin \theta = y/r$ and $\cos \theta = x/r$, and θ is usually chosen so that $-\pi < \theta \le \pi$ or $-180° < \theta \le 180°$.

Products and **quotients** of complex numbers in polar form are found as follows: If

$$z_1 = r_1 e^{i\theta_1} \quad \text{and} \quad z_2 = r_2 e^{i\theta_2}$$

then

1. $z_1 z_2 = r_1 e^{i\theta_1} r_2 e^{i\theta_2} = r_1 r_2 e^{i(\theta_1 + \theta_2)}$
2. $\dfrac{z_1}{z_2} = \dfrac{r_1 e^{i\theta_1}}{r_2 e^{i\theta_2}} = \dfrac{r_1}{r_2} e^{i(\theta_1 - \theta_2)}$

7-7 DE MOIVRE'S THEOREM

This section discusses the famous De Moivre theorem and the related nth-root theorem. These theorems make the process of finding natural number powers and all the nth roots of a complex number relatively easy. **De Moivre's theorem** is stated as follows: If

$$z = x + iy = re^{i\theta}$$

and n is a natural number, then

$$z^n = (x + iy)^n = (re^{i\theta})^n = r^n e^{n\theta i}$$

From De Moivre's theorem, we can derive the ***n*th-root theorem:** For n a positive integer greater than 1,

$$r^{1/n} e^{(\theta/n + k360°/n)i} \qquad k = 0, 1, \ldots, n-1$$

are the n distinct nth roots of $re^{i\theta}$, and there are no others.

Chapter 7 Review Exercise

Work through all the problems in this chapter review and check answers in the back of the book. Answers to all review problems are there, and following each answer is a number in italics indicating the section in which that type of problem is discussed. Where weaknesses show up, review appropriate sections in the text.

Problems in this exercise use the following labeling of sides and angles:

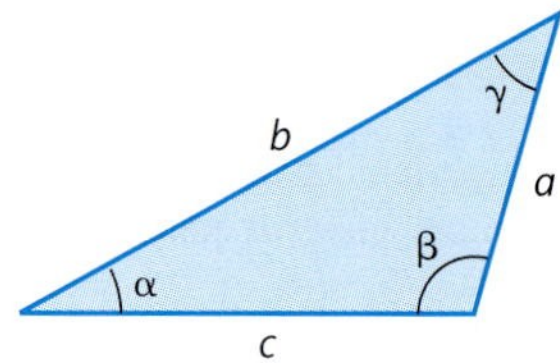

A

In Problems 1–3, determine whether the information in each problem allows you to construct zero, one, or two triangles. Do not solve the triangle.

1. $a = 11$ meters, $b = 3.7$ meters, $\alpha = 67°$
2. $c = 15$ centimeters, $\alpha = 97°$, $\beta = 84°$
3. $a = 18$ feet, $b = 22$ feet, $\alpha = 54°$

4. Referring to the figure at the beginning of the exercise, if $\alpha = 52.6°$, $b = 57.1$ centimeters, and $c = 79.5$ centimeters, which of the two angles, β or γ, can you say for certain is acute and why?

In Problems 5–7, solve each triangle, given the indicated information.

5. $\alpha = 67°$, $\beta = 38°$, and $c = 49$ meters
6. $\alpha = 15°$, $b = 9.1$ feet, and $c = 12$ feet
7. $\gamma = 121°$, $c = 11$ centimeters, and $b = 4.2$ centimeters

8. Given geometric vectors **u** and **v** as indicated in the figure. Find $|\mathbf{u} + \mathbf{v}|$ and θ, given $|\mathbf{u}| = 160$ miles per hour and $|\mathbf{v}| = 55$ miles per hour.

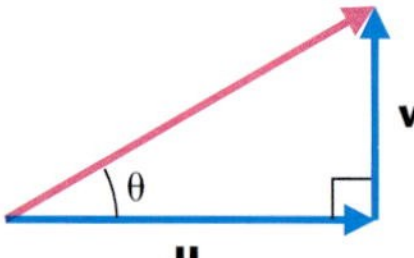

9. Write the algebraic vector $\langle a, b \rangle$ corresponding to the geometric vector $\overrightarrow{AB}$ with endpoints $A(2, 6)$ and $B(5, -1)$.
10. Find the magnitude of the vector $\langle -3, -5 \rangle$.
11. Sketch a graph of $\theta = \pi/6$ in a polar coordinate system.
12. Sketch a graph of $r = 6$ in a polar coordinate system.
13. Plot in a complex plane: $A = 3 + 5i$, $B = -1 - i$, $C = -3i$
14. A point in a polar coordinate system has coordinates $(10, -30°)$. Find all other polar coordinates for the point,

$-360° \le \theta \le 360°$, and verbally describe how the coordinates are associated with the point.

15. Plot in a complex plane: $A = 5e^{30°i}$, $B = 10e^{(-\pi/2)i}$, $C = 7e^{(3\pi/4)i}$

16. (A) Change $1 - i\sqrt{3}$ to polar form, $r \ge 0$, $-180° < \theta \le 180°$.
(B) Change $4e^{(-30°)i}$ to exact rectangular form.

17. (A) Find $[(-1/2) - (\sqrt{3}/2)i]^3$ using De Moivre's theorem. Write the final answer in the exact rectangular form.
(B) Verify the results in part A with a calculator.

18. Find $(2e^{15°i})^4$ using De Moivre's theorem, and write the final answer in exact rectangular form.

B

19. Referring to the figure at the beginning of the exercise, if $a = 434$ meters, $b = 302$ meters, and $c = 197$ meters, then if the triangle has an obtuse angle, which angle must it be and why?

In Problems 20–23, solve each triangle. If a problem does not have a solution, say so. If a triangle has two solutions, say so, and solve the obtuse case.

20. $\beta = 115.4°$, $a = 5.32$ centimeters, $c = 7.05$ centimeters

21. $\alpha = 63.2°$, $a = 179$ millimeters, $b = 205$ millimeters

22. $\alpha = 26.4°$, $a = 52.2$ kilometers, $b = 84.6$ kilometers

23. $a = 19.0$ inches, $b = 27.8$ inches, $c = 26.1$ inches

24. If four nonzero force vectors with different magnitudes and directions are acting on an object at rest, what must the sum of all four vectors be for the object to remain at rest?

25. Given geometric vectors **u** and **v** as indicated in the figure. Find $|\mathbf{u} + \mathbf{v}|$ and α, given $|\mathbf{u}| = 75.2$ kilograms, $|\mathbf{v}| = 34.2$ kilograms, and $\theta = 57.2°$.

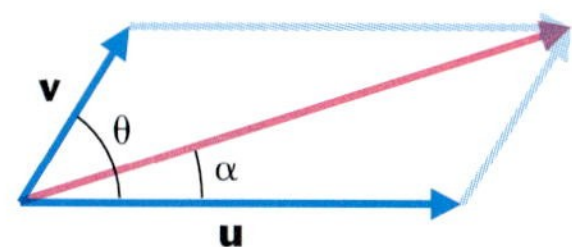

26. Express $\mathbf{u} = \langle 11, -7\rangle$ in terms of **i** and **j** unit vectors.

27. Find $6\mathbf{u} - 3\mathbf{v}$ if $\mathbf{u} = \langle 13, 18\rangle$, and $\mathbf{v} = \langle -9, 21\rangle$.

28. Find a unit vector **u** having the opposite direction as $\mathbf{v} = \langle 8, -6\rangle$.

29. Find a scalar k so that $2\mathbf{j} + k\langle 1, -3\rangle$ is a unit vector.

30. Find scalars k_1 and k_2 so that $k_1\langle 1, -2\rangle + k_2\langle 2, 1\rangle = \mathbf{i}$.

31. Find the vector that has magnitude 3 and the same direction as $5\mathbf{i} - 12\mathbf{j}$.

In Problems 32–35, use rapid sketching techniques to sketch each graph in a polar coordinate system.

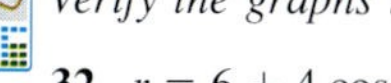

Verify the graphs in Problems 32–35 on a graphing utility.

32. $r = 6 + 4\cos\theta$

33. $r = 8 + 8\sin\theta$

34. $r = 10\cos 2\theta$

35. $r = 8\sin 3\theta$

36. Graph $r = 6\cos(\theta/7)$ for $0 \le \theta \le 7\pi$.

37. Graph $r = 6\cos(\theta/9)$ for $0 \le \theta \le 9\pi$.

38. Graph $r = 8(\sin\theta)^{2n}$ for $n = 1, 2$, and 3. How many leaves do you expect the graph will have for arbitrary n?

39. Graph $r = 3/(1 - e\cos\theta)$ for the following values of e, and identify each curve as an ellipse, a parabola, or a hyperbola.
(A) $e = 0.55$ (B) $e = 1$ (C) $e = 1.7$

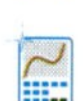

40. Convert $x^2 + y^2 = 6x$ to polar form.

41. Convert $r = 5\cos\theta$ to rectangular form.

42. Change the following complex numbers to polar form, $r \ge 0$, $-180° < \theta \le 180°$: $z_1 = -1 + i$, $z_2 = -1 - i\sqrt{3}$, $z_3 = 5$.

43. Change the following complex numbers to exact rectangular form: $z_1 = \sqrt{2}e^{(\pi/4)i}$, $z_2 = 3e^{210°i}$, $z_3 = 2e^{(-2\pi/3)i}$

44. If $z_1 = 8e^{25°i}$ and $z_2 = 4e^{19°i}$, find:
(A) z_1z_2 (B) z_1/z_2

Leave answers in polar form.

45. (A) Write $(1 + i\sqrt{3})^4$ in exact rectangular form. Use De Moivre's theorem.
(B) Verify part A by evaluating $(1 + i\sqrt{3})^4$ directly on a calculator.

46. Find all cube roots of i. Write final answers in exact rectangular form, and locate the roots on a circle in the complex plane.

47. Find all cube roots of $-4\sqrt{3} + 4i$ exactly. Leave answers in polar form.

48. Show that $4e^{15°i}$ is a square root of $8\sqrt{3} + 8i$.

49. Change the rectangular coordinates $(5.17, -2.53)$ to polar coordinates to two decimal places, $r \ge 0$, $-180° < \theta \le 180°$.

50. Change the polar coordinates $(5.81, -2.72)$ to rectangular coordinates to two decimal places.

51. Change the complex number $-3.18 + 4.19i$ to polar form to two decimal places, $r \ge 0$, $-180° < \theta \le 180°$.

52. Change the complex number $7.63e^{-162.27°i}$ to rectangular form $a + bi$, where a and b are computed to two decimal places.

53. (A) The cube root of a complex number is shown in the figure. Geometrically locate all other cube roots of the number on the figure at the top of the next page, and explain how they were located.

(B) Determine geometrically the other cube roots of the number in exact rectangular form.
(C) Cube each cube root from parts A and B.

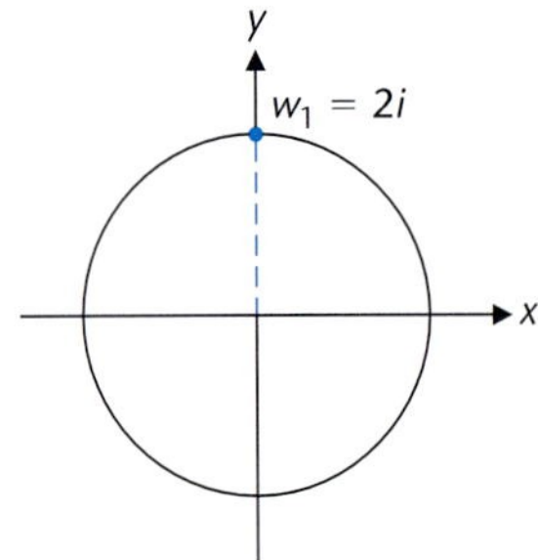

C

54. For an oblique triangle with $\alpha = 23.4°$, $b = 44.6$ millimeters, and a the side opposite angle α, determine a value k so that if $0 < a < k$, there is no solution; if $a = k$, there is one solution; and if $k < a < b$, there are two solutions.

55. Show that for any triangle

$$\frac{a^2 + b^2 + c^2}{2abc} = \frac{\cos\alpha}{a} + \frac{\cos\beta}{b} + \frac{\cos\gamma}{c}$$

56. Let $\mathbf{u} = \langle a, b\rangle$ and $\mathbf{v} = \langle c, d\rangle$ be vectors and m a scalar; prove:
(A) $(\mathbf{u} + \mathbf{v}) = (\mathbf{v} + \mathbf{u})$
(B) $m(\mathbf{u} + \mathbf{v}) = m\mathbf{u} + m\mathbf{v}$

In Problems 57–60, determine whether the statement is true or false. If true, explain why. If false, give a counterexample.

57. If w is a cube root of z, then $|w| \le |z|$.

58. Every tenth root of 1 is a fifth root of 1.

59. If w is both a cube root and a fourth root of z, then $w = 0$ or $w = 1$.

60. The square roots of a real number are also real.

61. Given the polar equation $r = 4 + 4\cos(\theta/2)$,
(A) Sketch a graph of the equation using rapid graphing techniques.
(B) Verify the graph in part A on a graphing utility.

62. (A) Graph $r = -8\sin\theta$ and $r = 8\cos\theta$, $0 \le \theta \le \pi$, in the same viewing window. Use TRACE to determine which intersection point has coordinates that satisfy both equations simultaneously.
(B) Solve the equations simultaneously to verify the results in part B.
(C) Explain why the pole is not a simultaneous solution, even though the two curves intersect at the pole.

63. Find all solutions, real and imaginary, for $x^8 - 1 = 0$. Write roots in exact rectangular form.

64. Write $P(x) = x^3 - 8i$ as a product of linear factors.

APPLICATIONS 

For Problems 65–67, use the navigational compass. Assume directions given in terms of north, east, south, and west are exact.

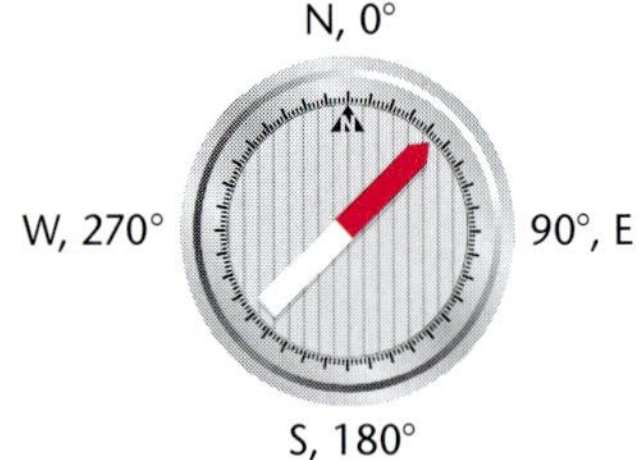

Navigational compass

65. Navigation. An airplane flies east at 256 miles per hour, and another airplane flies southeast at 304 miles per hour. After 2 hours, how far apart are the two planes?

66. Navigation. An airplane flies with an airspeed of 450 miles per hour and a compass heading of 75°. If the wind is blowing at 65 miles per hour out of the north (from north to south), what is the plane's actual direction and speed relative to the ground? Compute direction to the nearest degree and speed to the nearest mile per hour.

★ **67. Navigation.** An airplane that can cruise at 500 miles per hour in still air is to fly due east. If the wind is blowing from the northeast at 50 miles per hour, what compass heading should the pilot choose? What will be the actual speed of the plane relative to the ground? Compute direction to the nearest degree and speed to the nearest mile per hour.

★ **68. Coastal Navigation.** The owner of a pleasure boat cruising along a coast wants to pass a rocky point at a safe distance (see figure). Sightings of the rocky point are made at A and at B, 1.0 mile apart. If the boat continues on the same course, how close will it come to the point? That is, find d in the figure to the nearest tenth of a mile.

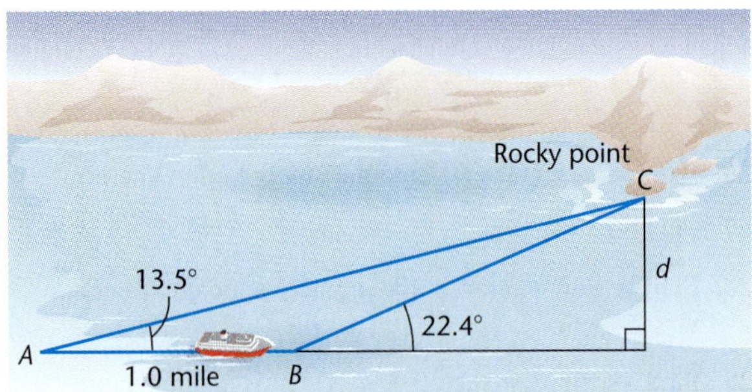

69. Forces. Two forces $\mathbf{u}$ and $\mathbf{v}$ are acting on an object as indicated in the figure. Find the direction and magnitude of the resultant force $\mathbf{u} + \mathbf{v}$ relative to force $\mathbf{v}$.

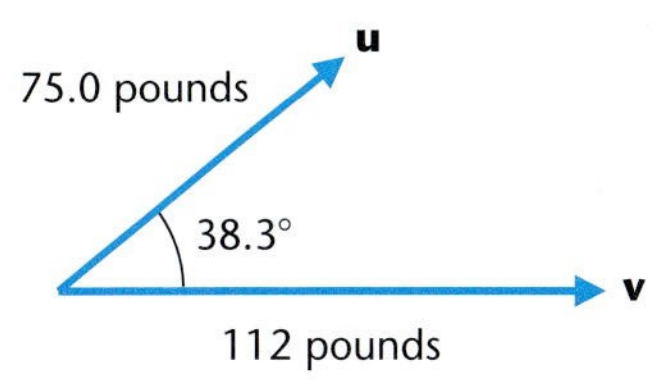

Figure for 69

★ **70 Static Equilibrium.** Two forces **u** and **v** are acting on an object as indicated in the figure. What third force **w** must be added to achieve static equilibrium? Give direction relative to **u**.

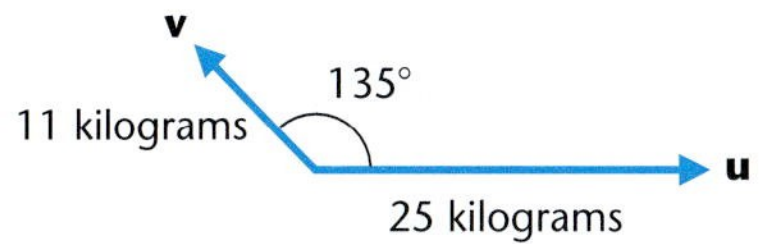

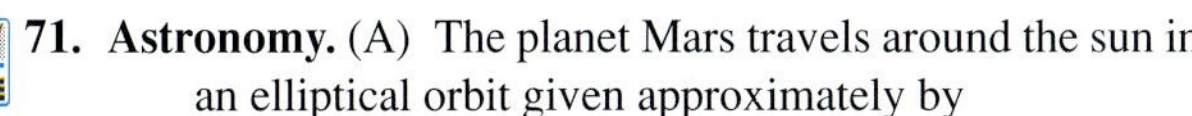

71. Astronomy. (A) The planet Mars travels around the sun in an elliptical orbit given approximately by

$$r = \frac{1.41 \times 10^8}{1 - 0.0934 \cos \theta} \qquad (1)$$

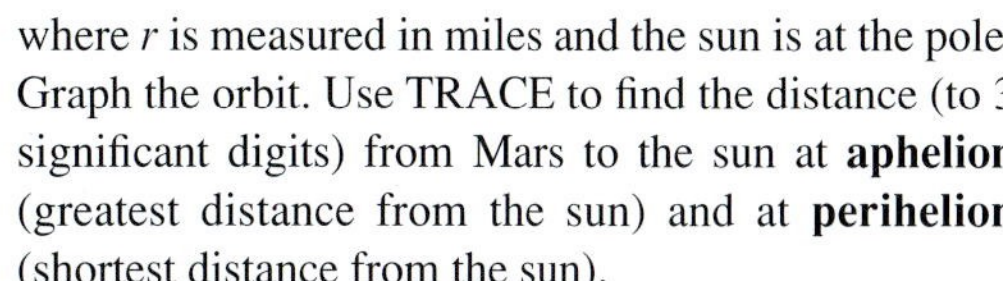

where r is measured in miles and the sun is at the pole. Graph the orbit. Use TRACE to find the distance (to 3 significant digits) from Mars to the sun at **aphelion** (greatest distance from the sun) and at **perihelion** (shortest distance from the sun).

(B) Referring to equation (1), r is maximum when the denominator is minimum, and r is minimum when the denominator is maximum. Use this information to find the distance from Mars to the sun at aphelion and at perihelion.

★ **72 Engineering.** A cable car weighing 1,000 pounds is used to cross a river (see the figure). What is the tension in each half of the cable when the car is located as indicated? Compute the answer to 3 significant digits.

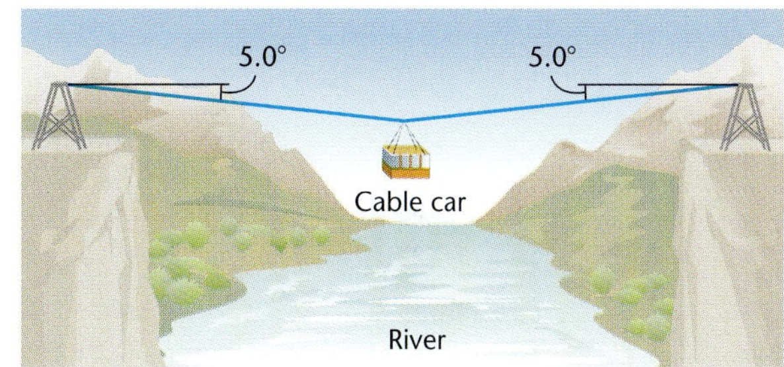

Cumulative Review Exercise Chapters 5–7

Work through all the problems in this cumulative review and check answers in the back of the book. Answers to all review problems, except verifications, are there, and following each answer is a number in italics indicating the section in which that type of problem is discussed. Where weaknesses show up, review appropriate sections in the text.

A

1. In a circle of radius 6 meters, find the length of an arc opposite an angle of 0.31 radian.

2. Solve the triangle:

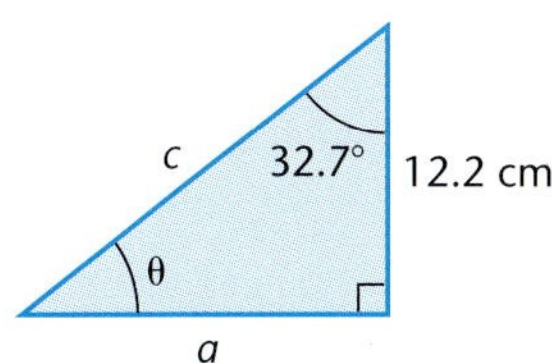

3. In which quadrants is each positive?
(A) $\sin \theta$ (B) $\cos \theta$ (C) $\tan \theta$

4. If $(-3, 4)$ is on the terminal side of an angle θ, find:
(A) $\cos \theta$ (B) $\csc \theta$ (C) $\tan \theta$

5. Find the reference angle associated with each angle θ:
(A) $-3\pi/4$ (B) $245°$ (C) $-30°$

6. Indicate the domain, range, and period of each:
(A) $y = \sin x$ (B) $y = \cos x$ (C) $y = \tan x$

7. Sketch a graph of $y = \cos x$, $-\pi/2 \le x \le 5\pi/2$.

8. Sketch a graph of $y = \tan x$, $-\pi/2 < x < 3\pi/2$.

9. Describe the meaning of a central angle in a circle with radian measure 2.

10. Describe the smallest shift of the graph of $y = \cos x$ to produce the graph of $y = \sin x$.

Verify each identity in Problems 11–14.

11. $\cot \theta \sec \theta = \csc \theta$

12. $\sec x - \cos x = \tan x \sin x$

13. $\sin(x - \pi/2) = -\cos x$

14. $\csc 2x = \frac{1}{2} \csc x \sec x$

15. Use a graphing utility to test whether each of the following is an identity. If an equation appears to be an identity, verify it. If the equation does not appear to be an identity, find

a value of x for which both sides are defined but are not equal.

(A) $\dfrac{\sin^2 x}{\cos x} + \cos x = \csc x$

(B) $\dfrac{\sin^2 x}{\cos x} + \cos x = \sec x$

16. In a triangle, if $a = 32.5$ feet, $c = 77.2$ feet, and $\beta = 61.3°$, without solving the triangle or drawing any pictures, which of the two angles, α or γ, can you say for certain is acute and why?

Solve Problems 17 and 18 to four decimal places.

17. $\sin x = 0.3188$, $0 \le x \le 2\pi$

18. $\tan \theta = -4.076$, $-90° < \theta < 90°$

19. Solve the triangle:

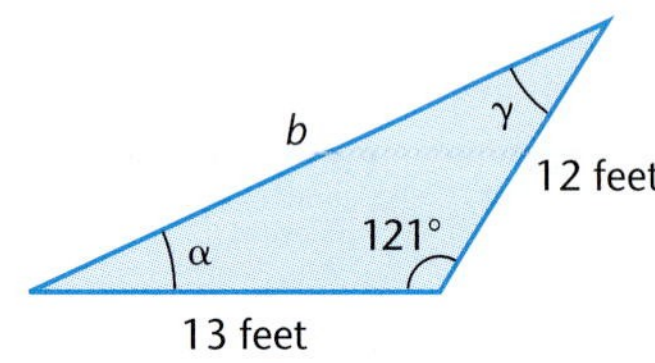

20. Write the algebraic vector $\langle a, b\rangle$ corresponding to the geometric vector $\overrightarrow{AB}$ with endpoints $A(-3, 2)$ and $B(3, -1)$.

21. A point in a polar coordinate system has coordinates $(-5, 150°)$. Find all other polar coordinates for the point, $-360° \le \theta \le 360°$, and verbally describe how the coordinates are associated with the point.

22. Sketch a graph of $r = 6 \cos \theta$ in a polar coordinate system.

23. Plot in a complex plane: $A = -3 + 4i$ and $B = 4e^{60°i}$.

24. Find $(2e^{10°i})^3$. Write the final answer in exact rectangular form.

B

25. Which of the following angles are coterminal with 150°: 30°, $-7\pi/6$, 870°?

26. Change 1.31 radians to decimal degrees to two decimal places.

27. Which of the following have the same value as cos 8?
(A) cos(8 rad) (B) cos 8° (C) $\cos(8 - 4\pi)$

Evaluate Problems 28–37 exactly without a calculator. If the function is not defined at the value, say so.

28. cos 150°
29. $\csc(-\frac{3\pi}{4})$
30. cot 180°
31. $\tan \frac{11\pi}{6}$
32. $\cos^{-1}\frac{1}{2}$
33. $\arctan(-\sqrt{3})$
34. $\arcsin \sqrt{2}$
35. $\sin[\tan^{-1}(-3)]$
36. $\cos\left(\cos^{-1}\dfrac{\sqrt{2}}{3}\right)$
37. $\cos^{-1}[\cos(-\frac{\pi}{4})]$

38. Evaluate to 4 significant digits using a calculator. If a function is not defined, say so.
(A) tan 84°12′55″ (B) sec(−1.8409)
(C) $\tan^{-1}(-84.32)$ (D) $\cos^{-1}(\tan 2.314)$

39. Sketch a graph of $y = 2 - 2\cos(\pi x/2)$, $-1 \le x \le 5$.

40. (A) Find the exact degree measure of $\theta = \cos^{-1}(-\sqrt{3}/2)$ without a calculator.
(B) Find the degree measure of $\theta = \sin^{-1}(-0.338)$ to three decimal places using a calculator.

41. Evaluate $\sin^{-1}(\sin 3)$ with a calculator set in radian mode, and explain why this does or does not illustrate the inverse sine–sine identity.

42. A circular point $P(a, b)$ moves counterclockwise around the circumference of a unit circle starting at (1, 0) and stops after covering a distance of 11.205 units. Explain how you would find the coordinates of point P at its final position and how you would determine which quadrant P is in. Find the coordinates of P to three decimal places and the quadrant for the final position of P.

43. Explain the difference in solving the equation $\tan x = -24.5$ and evaluating $\tan^{-1}(-24.5)$.

44. Find an equation of the form $y = k + a \sin Bx$ that produces the graph:

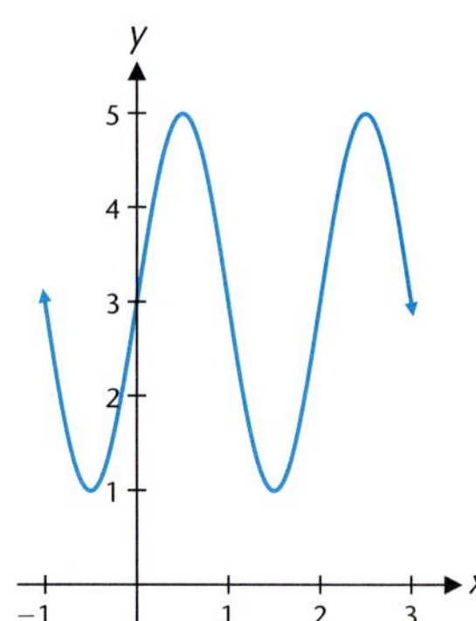

45. Sketch a graph of $y = 3\sin(2x - \pi)$, $-\pi \le x \le 2\pi$. Indicate amplitude A, period P, and phase shift $P.S.$

46. Sketch a graph of $y = 2\tan(\pi x/2 - \pi/2)$, $0 < x < 4$. Indicate the period P and phase shift $P.S.$

47. Sketch a graph of $y = \sin x$ and $y = \csc x$ in the same coordinate system.

48. Describe the smallest left shift and/or reflection that transforms the graph of $y = \cot x$ into the graph of $y = \tan x$.

49. Graph $y = 1/(\cot^2 x + 1)$ in a graphing utility that displays at least two full periods of the graph. Find an equation of the form $y = k + A\sin Bx$ or $y = k + A\cos Bx$ that has the same graph. Graph both equations in the same viewing window, and use TRACE to verify that both graphs are the same.

50. Graph $y = (2 - 2\sin^2 x)/(\sin 2x)$ in a graphing utility that displays at least two full periods of the graph. Find an equation of the form $y = A \tan Bx$ or $y = A \cot Bx$ that has the same graph. Graph both equations in the same viewing window, and use TRACE to verify that both graphs are the same.

51. Given the equation $\sin 2x = 2\sin x$.
(A) Are $x = 0$ and $x = \pi$ solutions?
(B) Is the equation an identity or a conditional equation? Explain.

Verify each identity in Problems 52–57.

52. $\dfrac{\sin u}{1 + \cos u} + \cot u = \csc u$

53. $\sec x + \tan x = \dfrac{\cos x}{1 - \sin x}$

54. $\tan \dfrac{x}{2} = \csc x - \cot x$

55. $\csc^2 \dfrac{x}{2} = 2 \csc x (\csc x + \cot x)$

56. $\dfrac{2}{1 + \cos 2x} = \sec^2 x$

57. $\dfrac{\cos x + \cos y}{\sin x - \sin y} = \cot \dfrac{x - y}{2}$
[*Hint:* Use sum–product identities.]

58. Use a graphing utility to test whether each of the following is an identity. If an equation appears to be an identity, verify it. If the equation does not appear to be an identity, find a value of x for which both sides are defined but are not equal.
(A) $\dfrac{\tan x}{2\tan x - \sin x} = \dfrac{1}{2 + \sin x}$
(B) $\dfrac{\tan x}{2\tan x - \sin x} = \dfrac{1}{2 - \cos x}$

59. Find $\cos(x - y)$ exactly without a calculator, given $\sin x = (-2/\sqrt{5})$, $\cos y = (-2/\sqrt{5})$, x a quadrant IV angle, and y a quadrant III angle.

60. Compute the exact value of $\sin 2x$ and $\cos(x/2)$ without a calculator, given $\sin x = \frac{3}{5}$, $\pi/2 \le x \le \pi$.

Solve Problems 61 and 62 exactly without a calculator, θ in degrees and x real.

61. $2\sin^2 \theta + \sin \theta = 1, 0 \le \theta < 360°$

62. $\sin 2x = \sin x$, all real solutions

63. (A) Solve $\cot x = -2\cos x$ exactly, $0 \le x \le 2\pi$.
(B) Solve $\cot x = -2\cos x$ to three decimal places using a graphing utility, $0 \le x \le 2\pi$.

64. Solve $2\cos x = x - \cos 2x$ to three decimal places for all real solutions using a graphing utility.

In Problems 65–67, solve each triangle labeled as in the figure. If a problem does not have a solution, say so. If a triangle has two solutions, solve the obtuse case.

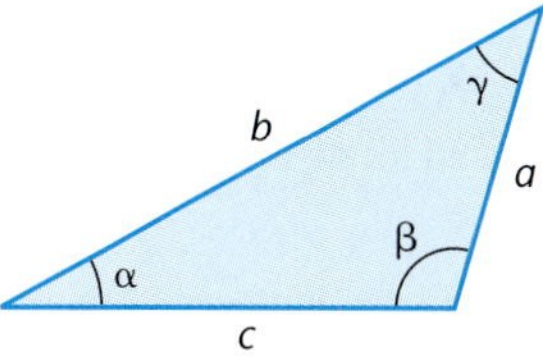

65. $a = 21.3$ meters, $b = 37.4$ meters, $c = 48.2$ meters

66. $\alpha = 125.4°$, $b = 25.4$ millimeters, $a = 20.3$ millimeters

67. $\alpha = 52.9°$, $b = 37.1$ inches, $a = 34.4$ inches

68. Assume in a triangle that γ is acute, $a = 92.5$ centimeters, and $b = 43.3$ centimeters. Which of the angles, α or β, can you say for certain is acute and why?

69. Given geometric vectors as indicated in the figures, find $|\mathbf{u} + \mathbf{v}|$ and α, given $|\mathbf{u}| = 25.3$ pounds, $|\mathbf{v}| = 13.4$ pounds, and $\theta = 48.3°$.

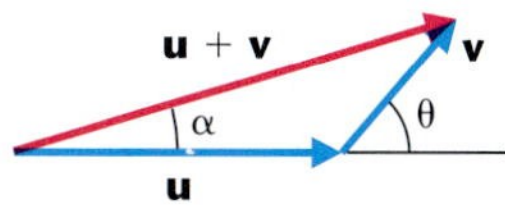

Tail-to-tip rule
(a)

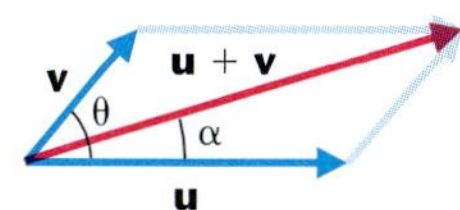

Parallelogram rule
(b)

70. Find $2\mathbf{u} - \mathbf{v} + 3\mathbf{w}$ for:
(A) $\mathbf{u} = \langle -1, 2\rangle$, $\mathbf{v} = \langle 0, -2\rangle$, $\mathbf{w} = \langle 1, -1\rangle$
(B) $\mathbf{u} = 2\mathbf{i} - \mathbf{j}$, $\mathbf{v} = \mathbf{i} + 3\mathbf{j}$, $\mathbf{w} = 2\mathbf{j}$

71. Convert to polar form: $x^2 + y^2 = 8y$

72. Convert $r = -4\cos\theta$ to rectangular form.

Use rapid sketching techniques to graph Problems 73 and 74 in a polar coordinate system.

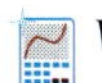

Verify the graphs in Problems 73 and 74 on a graphing utility.

73. $r = 4 + 4\cos\theta$

74. $r = 6\sin 3\theta$

75. Graph $r = 5(\cos 2\theta)^{2n}$, for $n = 1$, 2, and 3. How many leaves do you expect the graph will have for arbitrary n?

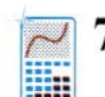

76. Graph $r = e^{\cos\theta} - 2\cos(4\theta)$ using a squared window and 0.05 for a step size for θ. The resulting curve is often referred to as a *butterfly curve*.

77. Change the rectangular coordinates $(-2.78, -3.19)$ to polar coordinates to two decimal places, $r \ge 0$, $-180° < \theta \le 180°$.

78. Change the polar coordinates $(6.22, -4.08)$ to rectangular coordinates to two decimal places.

79. Change $2e^{(-\pi/6)i}$ to exact rectangular form.

80. Change $z = -1 + i\sqrt{3}$ to polar form, θ in degrees.

81. Compute $(1 - i\sqrt{3})^6$ using De Moivre's theorem, and write the final answer in $a + bi$ form.

82. Find all cube roots of $-i$ exactly. Write final answers in the form $a + bi$, and locate the roots on a circle in the complex plane.

83. Change the complex number $-4.88 - 3.17i$ to polar form to two decimal places, $r \geq 0$, $-180° < \theta \leq 180°$.

84. Change the complex number $6.97e^{163.87°i}$ to rectangular form $a + bi$, where a and b are computed to two decimal places.

85. (A) The fourth root of a complex number is shown in the figure. Geometrically locate all other fourth roots of the number on the figure, and explain how they were located.
(B) Determine geometrically the other fourth roots of the number in exact rectangular form.
(C) Raise each fourth root from parts A and B to the fourth power.

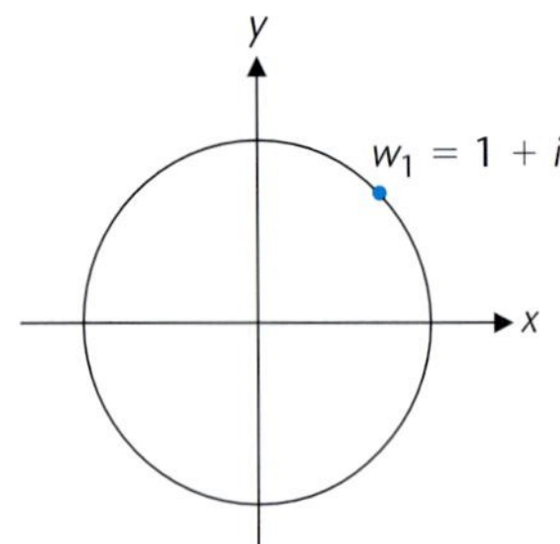

C

86. If, in the figure, the coordinates of A are (1, 0) and arc length s is 1.2 units, find the coordinates of P to 3 significant digits.

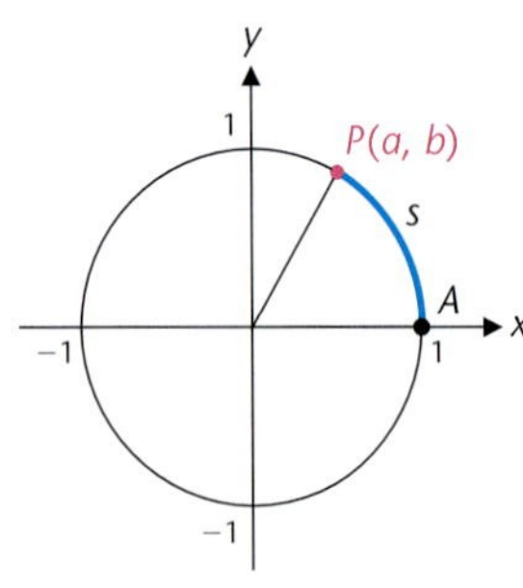

87. Sketch a graph of $y = 1 + \sec x$, $-3\pi/2 < x < 3\pi/2$.

88. The accompanying graph is a graph of an equation of the form $y = A\cos(Bx + C)$, $0 < -C/B < 1$. Find the equation by finding A, B, and C exactly. What is the period, amplitude, and phase shift?

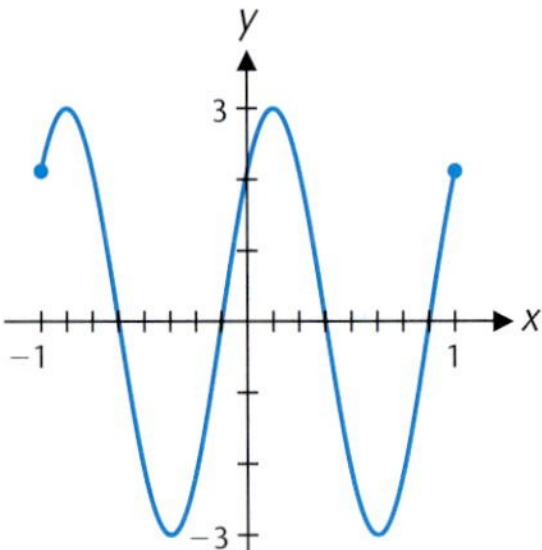

89. Graph $1.6 \sin 2x - 1.2 \cos 2x$ in a graphing utility. (Select the dimensions of a viewing window so that at least two periods are visible.) Find an equation of the form $y = A\sin(Bx + C)$ that has the same graph as the given equation. Find A and B exactly and C to three decimal places. Use the x intercept closest to the origin as the phase shift. To check your results graph both equations in the same viewing window, and use TRACE while shifting back and forth between the two graphs.

90. Write $\csc(\cos^{-1} x)$ as an algebraic expression in x, free of trigonometric or inverse trigonometric functions.

Solve Problems 91 and 92 without a calculator.

91. $\sin(2 \cot^{-1} \frac{3}{4}) = ?$

92. Given $\sec x = -\frac{5}{3}$, $\pi/2 \leq x \leq \pi$, find
(A) $\sin(x/2)$ (B) $\cos 2x$

93. (A) Solve $2 \sin^2 x = 3 \cos x$ exactly for all real solutions, $0 \leq x \leq 2\pi$.

(B) Solve $2 \sin^2 x = 3 \cos x$ to four decimal places using a graphing utility, $0 \leq x \leq 2\pi$.

94. (A) Use rapid sketching techniques to sketch a graph of the polar equation $r^2 = 36 \cos 2\theta$.
(B) Verify the graph in part A using a graphing utility.

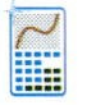

95. (A) Graph $r1 = 2 + 2\cos\theta$ and $r2 = 6\cos\theta$ in the same viewing window, $0 \leq \theta \leq 2\pi$.
(B) Use TRACE to determine how many times the graph of $r2$ crosses the graph of $r1$ as θ goes from 0 to 2π.
(C) Solve the two equations simultaneously to find the exact solutions for $0 \leq \theta \leq 2\pi$.
(D) Explain why the number of solutions found in part C does not agree with the number of times $r1$ crosses $r2$, $0 \leq \theta \leq 2\pi$.

96. Write $P(x) = x^3 + i$ as a product of linear factors.

97. If the three angles of a triangle are known (AAA), can the law of sines and/or the law of cosines be used to solve for the three sides? Explain.

In Problems 98–105, determine whether the statement is true or false. If true, explain why. If false, give a counterexample.

98. If x is a real number for which $\sin^{-1}(\sin x)$ is defined, then $\sin^{-1}(\sin x) = x$.

99. If x is a real number for which $\sin(\sin^{-1} x)$ is defined, then $\sin(\sin^{-1} x) = x$.

100. Every cube root of 1 is a sixth root of 1.

101. Every cube root of -1 is a sixth root of -1.

102. The unit vectors **i** and **j** have opposite directions.

103. If **u** is a positive scalar multiple of **v**, then $|\mathbf{u}| \geq |\mathbf{v}|$.

104. If f is a periodic function and c is a real number, then the graph of f intersects the line $y = c$ infinitely many times.

105. If f is a periodic function, then there exists a real number c such that the graph of f intersects the line $y = c$ infinitely many times.

APPLICATIONS 

106. **Astronomy.** A line from the sun to the Earth sweeps out an angle of how many radians in 5 days?

107. **Meteorology.** A weather balloon is released and rises vertically. Two weather stations C and D, in the same vertical plane as the balloon and 1,000 meters apart, sight the balloon at the same time and record the information given in the figure. At the time of sighting, how high was the balloon to the nearest meter?

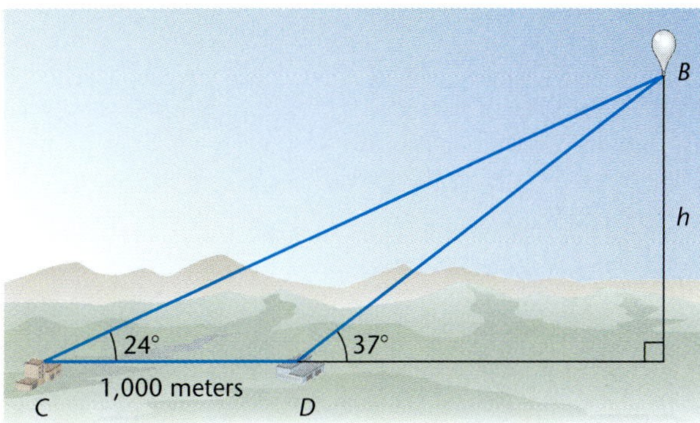

108. **Geometry.** Find the length to two decimal places of one side of a regular pentagon inscribed in a circle with radius 5 inches.

109. **Geometry.** Find $\angle ABC$ to the nearest degree in the rectangular solid shown below.

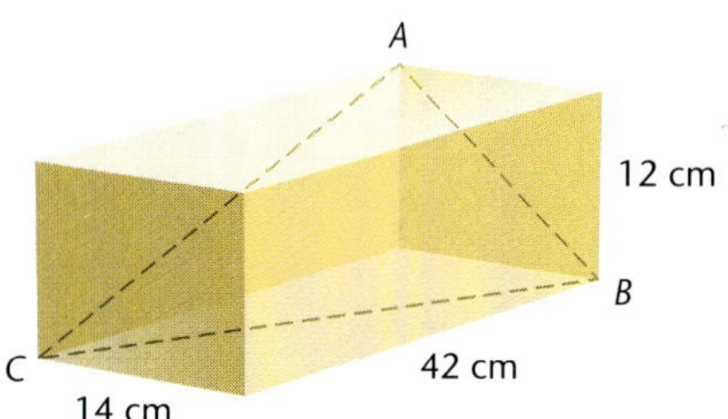

110. **Electric Circuit.** The current I in an alternating electric circuit has an amplitude of 50 amperes and a period of $\frac{1}{110}$ second. If $I = 50$ amperes when $t = 0$, find an equation of the form $I = A \cos Bt$ that gives the current at time $t \geq 0$.

111. **Navigation.** An airplane flies with an airspeed of 260 miles per hour and a compass heading of 110°. If a 36-mile-per-hour wind is blowing out of the north, what is the plane's actual heading and ground speed? Compute direction to the nearest degree and ground speed to the nearest mile per hour.

112. **Engineering.** A 65-pound child glides across a small river on a homemade cable trolley (see figure). What is the tension on each half of the support cable when the child is in the center? Compute answer to nearest pound.

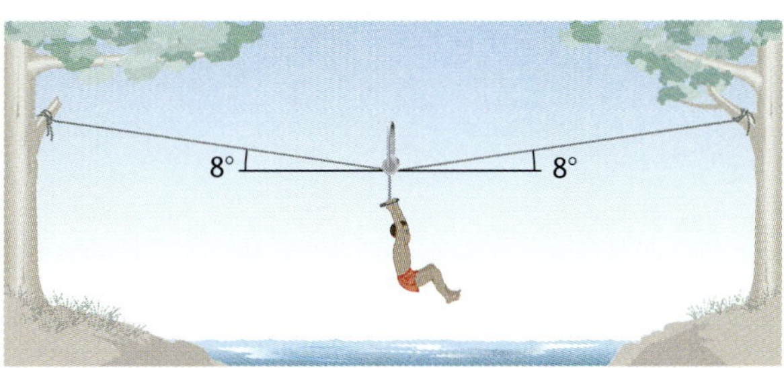

 113. **Geometry.** A circular arc of 10 centimeters has a chord of 8 centimeters as shown in the figure.

(A) Explain how the radius is given by the equation

$$\sin \frac{5}{R} = \frac{4}{R}$$

(B) What difficulties do you encounter in trying to solve the equation in part A exactly using algebraic and trigonometric methods?

(C) Show on a graphing utility how to approximate the radius of the circle R, and find R to three decimal places.

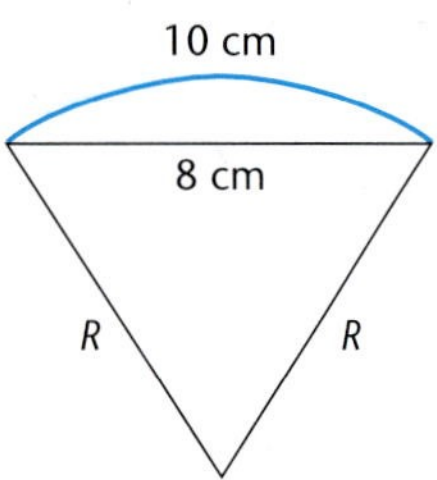

114. Modeling Temperature Variation. The 30-year average monthly temperature, °F, for each month of the year for Washington, D.C., is given in Table 1 *(World Almanac)*.

(A) Using 1 month as the basic unit of time, enter the data for a 2-year period in your graphing utility and produce a scatter plot in the viewing window. Choose $25 \leq y \leq 80$ for the viewing window.

(B) It appears that a sine curve of the form

$$y = k + A \sin(Bx + C)$$

will closely model this data. The constants k, A, and B are easily determined from Table 1. To estimate C, visually estimate to one decimal place the smallest positive phase shift from the plot in part A. After determining A, B, k, and C, write the resulting equation. (Your value of C may differ slightly from the answer book.)

(C) Plot the results of parts A and B in the same viewing window. (An improved fit may result by adjusting your value of C a little.)

TABLE 1 Monthly Average Temperatures, Washington, D.C.

x (mos.)	1	2	3	4	5	6	7	8	9	10	11	12
y (temp.)	31	34	43	53	62	71	76	74	67	55	45	35

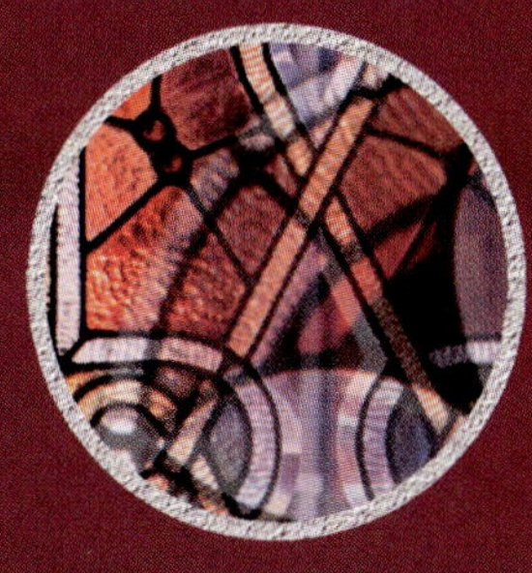

CHAPTER 8

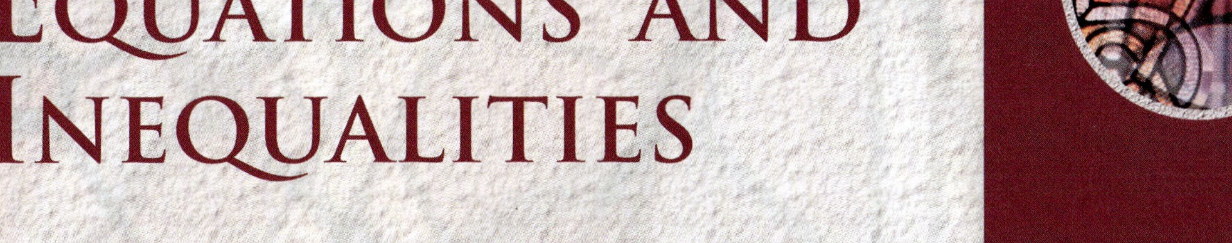

Systems of Equations and Inequalities

In this chapter we move from the standard methods of solving two linear equations with two variables to a method that can be used to solve linear systems with any number of variables and equations. This method is well-suited to computer use in the solution of larger systems. We also discuss systems involving nonlinear equations and systems of linear inequalities. Finally, we introduce a relatively new and powerful mathematical tool called *linear programming.* Many applications in mathematics involve systems of equations or inequalities. The mathematical techniques discussed in this chapter are applied to a variety of interesting and significant applications.

SECTION 8-1 Systems of Linear Equations and Augmented Matrices

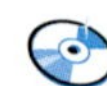

- Substitution: A Brief Review
- Graphing
- Elimination by Addition
- Matrices
- Solving Linear Systems Using Augmented Matrices

In this section we will continue our discussion, began in Section 1-2, of systems involving two linear equations and two variables:

$$\begin{aligned} ax + by &= h \\ cx + dy &= k \end{aligned} \tag{1}$$

where x and y are variables, a, b, c, and d are real numbers called the **coefficients** of x and y, and h and k are real numbers called the **constant terms** in the equations. Recall that a pair of numbers $x = x_0$ and $y = y_0$, also written as an ordered pair (x_0, y_0), is a **solution** of this system if each equation is satisfied by the pair. The set of all such ordered pairs of numbers is called the **solution set** for the system.

In Section 1-2, we used substitution to solve system (1). After briefly reviewing this method, we will explore the relationship between the graph of the equations in a system and the solution of the system and review the standard *elimination-by-addition method.* Then we will introduce augmented matrices to transform the elimination-by-addition method into a solution process that is well-suited for computer use in the solution of linear systems involving large numbers of equations and variables.

To keep the introduction to augmented matrices in this section as simple as possible, we restrict the discussion to two equations with two variables. In the next section, the solution process is generalized and applied to larger linear systems.

• Substitution: A Brief Review

To review some of the basic concepts introduced in Section 1-2, consider the following simple example. At a computer fair, student tickets cost \$2 and general admission tickets cost \$3. If a total of 7 tickets are purchased for a total cost of \$18, how many of each type were purchased?

Let

x = Number of student tickets

y = Number of general admission tickets

Then

$$\begin{aligned} x + y &= 7 \qquad \text{Total number of tickets purchased} \\ 2x + 3y &= 18 \qquad \text{Total purchase cost} \end{aligned}$$

To solve this system by substitution, we solve the first equation for y in terms of x and substitute in the second equation:

$$\begin{aligned} x + y &= 7 & 2x + 3y &= 18 \\ y &= 7 - x & 2x + 3(\mathbf{7 - x}) &= 18 \\ & & -x + 21 &= 18 \\ & & \mathbf{x} &= \mathbf{3} \end{aligned}$$

Now, replace x with 3 in $y = 7 - x$:

$$\begin{aligned} y &= 7 - x \\ &= 7 - \mathbf{3} \\ \mathbf{y} &= \mathbf{4} \end{aligned}$$

Thus the solution is 3 student tickets and 4 general admission tickets. You should check this result in each of the original equations.

• Graphing

Recall that the graph of a linear equation is the line consisting of all ordered pairs that satisfy the equation. If we graph both equations in system (1) in the same coordinate system, then the coordinates of any points that the lines have in common must be solutions to the system. Example 1 illustrates this process for the ticket problem discussed above.

EXAMPLE 1 Solving a System by Graphing

Solve the ticket problem by graphing: $\begin{aligned} x + y &= 7 \\ 2x + 3y &= 18 \end{aligned}$

Solution From Figure 1 we see that:

FIGURE 1

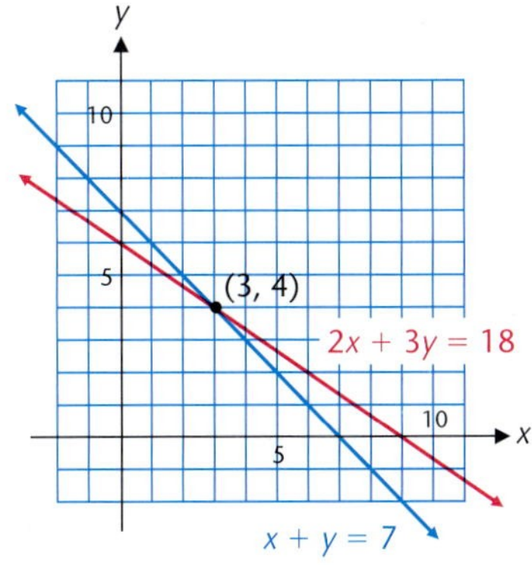

$$\begin{aligned} x &= 3 \qquad \text{Student tickets} \\ y &= 4 \qquad \text{General admission tickets} \end{aligned}$$

Check

$$\begin{aligned} x + y &= 7 & 2x + 3y &= 18 \\ 3 + 4 &\stackrel{?}{=} 7 & 2(3) + 3(4) &\stackrel{?}{=} 18 \\ 7 &\stackrel{\checkmark}{=} 7 & 18 &\stackrel{\checkmark}{=} 18 \end{aligned}$$

Matched Problem 1 Solve by graphing and check:
$$\begin{aligned} x - y &= 3 \\ x + 2y &= -3 \end{aligned}$$

It is clear that the preceding example has exactly one solution, since the lines have exactly one point of intersection. In general, lines in a rectangular coordinate system are related to each other in one of three ways illustrated in the next example.

EXAMPLE 2 Solving Three Important Types of Systems by Graphing

Solve each of the following systems by graphing:

(A) $\begin{aligned} 2x - 3y &= 2 \\ x + 2y &= 8 \end{aligned}$ (B) $\begin{aligned} 4x + 6y &= 12 \\ 2x + 3y &= -6 \end{aligned}$ (C) $\begin{aligned} 2x - 3y &= -6 \\ -x + \tfrac{3}{2}y &= 3 \end{aligned}$

Solutions (A)

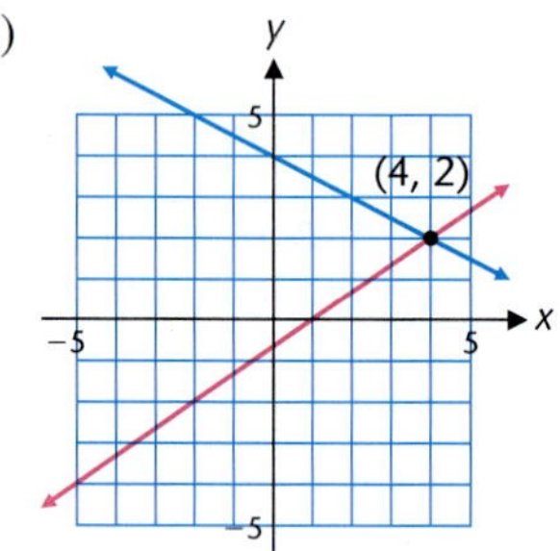

Lines intersect at one point only. Exactly one solution: $x = 4$, $y = 2$

(B)

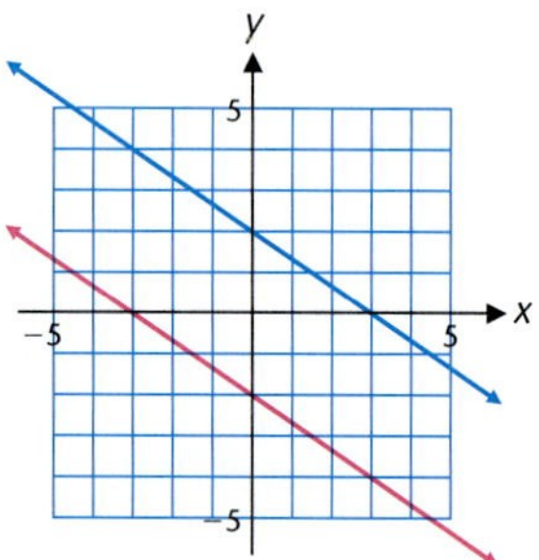

Lines are parallel (each has slope $-\frac{2}{3}$). No solution.

(C)

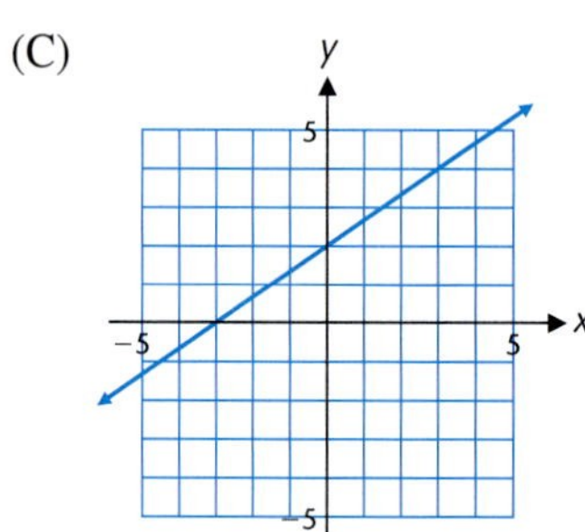

Lines coincide. Infinitely many solutions.

Matched Problem 2 Solve each of the following systems by graphing:

(A) $\begin{aligned} 2x + 3y &= 12 \\ x - 3y &= -3 \end{aligned}$ (B) $\begin{aligned} x - 3y &= -3 \\ -2x + 6y &= 12 \end{aligned}$ (C) $\begin{aligned} 2x - 3y &= 12 \\ -x + \tfrac{3}{2}y &= -6 \end{aligned}$

We now define some terms that can be used to describe the different types of solutions to systems of equations illustrated in Example 2.

Systems of Linear Equations: Basic Terms

A system of linear equations is **consistent** if it has one or more solutions and **inconsistent** if no solutions exist. Furthermore, a consistent system is said to be **independent** if it has exactly one solution (often referred to as the **unique solution**) and **dependent** if it has more than one solution.

Referring to the three systems in Example 2, the system in part A is consistent and independent, with the unique solution $x = 4$ and $y = 2$. The system in part B is inconsistent, with no solution. And the system in part C is consistent and dependent, with an infinite number of solutions: all the points on the two coinciding lines.

EXPLORE-DISCUSS 1

(A) Can a consistent and dependent system have exactly two solutions? Exactly three solutions? Explain.

(B) Solve each of the systems in Example 2 by the substitution method. Based on your results, describe how you might recognize a dependent system or an inconsistent system when using substitution.

By geometrically interpreting a system of two linear equations in two variables, we gain useful information about what to expect in the way of solutions to the system. In general, any two lines in a rectangular coordinate plane must intersect in exactly one point, be parallel, or coincide (have identical graphs). Thus, the systems in Example 2 illustrate the only three possible types of solutions for systems of two linear equations in two variables. These ideas are summarized in Theorem 1.

Theorem 1

Possible Solutions to a Linear System

The linear system

$$ax + by = h$$
$$cx + dy = k$$

must have:

1. Exactly one solution — Consistent and independent

or

2. No solution — Inconsistent

or

3. Infinitely many solutions — Consistent and dependent

There are no other possibilities.

One drawback of finding a solution by graphing is the inaccuracy of hand-drawn graphs. Graphic solutions performed on a graphing utility, however, provide both a useful geometric interpretation and an accurate approximation of the solution to a system of linear equations in two variables.

EXAMPLE 3 Solving a System Using a Graphing Utility

Solve to two decimal places using a graphing utility:

$$\begin{aligned} 5x - 3y &= 13 \\ 2x + 4y &= 15 \end{aligned}$$

Solution First solve each equation for y:

$$\begin{aligned} 5x - 3y &= 13 \\ -3y &= -5x + 13 \\ y &= \tfrac{5}{3}x - \tfrac{13}{3} \end{aligned} \qquad \begin{aligned} 2x + 4y &= 15 \\ 4y &= -2x + 15 \\ y &= -0.5x + 3.75 \end{aligned}$$

Next, enter each equation in a graphing utility [Fig. 2(a)], graph in an appropriate viewing window, and approximate the intersection point [Fig. 2(b)].

FIGURE 2

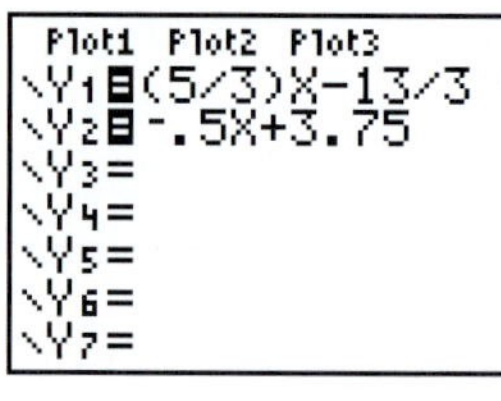

(a) Equation definitions

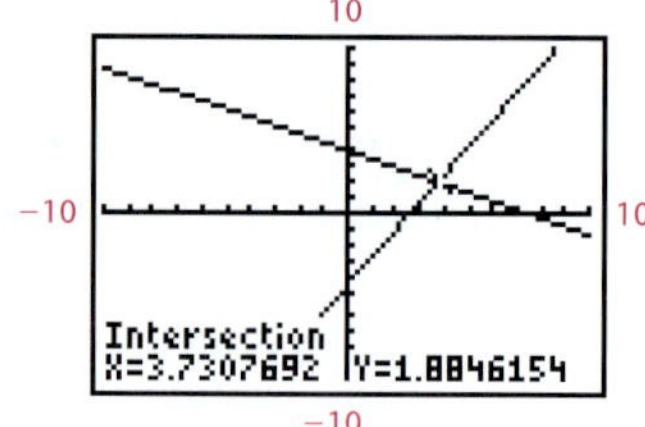

(b) Intersection point

Rounding the values in Figure 2(b) to two decimal places, we see that the solution is

$$x = 3.73 \text{ and } y = 1.88 \qquad \text{or} \qquad (3.73,\ 1.88)$$

Check

$$\begin{aligned} 5x - 3y &= 13 \\ 5(3.73) - 3(1.88) &\stackrel{?}{=} 13 \\ 13.01 &\stackrel{\checkmark}{\approx} 13 \end{aligned} \qquad \begin{aligned} 2x + 4y &= 15 \\ 2(3.73) + 4(1.88) &\stackrel{?}{=} 15 \\ 14.98 &\stackrel{\checkmark}{\approx} 15 \end{aligned}$$

The checks are not exact because the values of x and y are approximations.

Matched Problem 3 Solve to two decimal places using a graphing utility:

$$\begin{aligned} 2x - 5y &= -25 \\ 4x + 3y &= 5 \end{aligned}$$

Graphic methods help us visualize a system and its solutions, frequently reveal relationships that might otherwise be hidden, and, with the assistance of a graphing utility, provide very accurate approximations to solutions.

• Elimination by Addition

Now we turn to **elimination by addition.** This is probably the most important method of solution, since it is readily generalized to higher-order systems. The method involves the replacement of systems of equations with simpler *equivalent systems,* by performing appropriate operations, until we obtain a system with an obvious solution. **Equivalent systems** of equations are, as you would expect, systems that have exactly the same solution set. Theorem 2 lists operations that produce equivalent systems.

Theorem 2 **Elementary Equation Operations Producing Equivalent Systems**

A system of linear equations is transformed into an equivalent system if:

1. Two equations are interchanged.
2. An equation is multiplied by a nonzero constant.
3. A constant multiple of another equation is added to a given equation.

Any one of the three operations in Theorem 2 can be used to produce an equivalent system, but operations 2 and 3 will be of most use to us now. Operation 1 becomes more important later in the section. The use of Theorem 2 is best illustrated by examples.

EXAMPLE 4 Solving a System Using Elimination by Addition

Solve using elimination by addition:
$$\begin{aligned} 3x - 2y &= 8 \\ 2x + 5y &= -1 \end{aligned}$$

Solution We use Theorem 2 to eliminate one of the variables and thus obtain a system with an obvious solution.

$$\begin{aligned} 3x - 2y &= 8 \\ 2x + 5y &= -1 \end{aligned}$$

If we multiply the top equation by 5, the bottom by 2, and then add, we can eliminate y.

$$\begin{aligned} 15x - 10y &= 40 \\ 4x + 10y &= -2 \\ \hline 19x \qquad &= 38 \\ x &= 2 \end{aligned}$$

The equation $x = 2$ paired with either of the two original equations produces an equivalent system. Thus, we can substitute $x = 2$ back into either of the two original equations to solve for y. We choose the second equation.

$$\begin{aligned} 2(2) + 5y &= -1 \\ 5y &= -5 \\ y &= -1 \end{aligned}$$

Solution: $x = 2$, $y = -1$ or $(2, -1)$.

Check

$$\begin{aligned} 3x - 2y &= 8 \\ 3(2) - 2(-1) &\overset{?}{=} 8 \\ 8 &\overset{\checkmark}{=} 8 \end{aligned} \qquad \begin{aligned} 2x + 5y &= -1 \\ 2(2) + 5(-1) &\overset{?}{=} -1 \\ -1 &\overset{\checkmark}{=} -1 \end{aligned}$$

Matched Problem 4 Solve using elimination by addition:

$$\begin{aligned} 6x + 3y &= 3 \\ 5x + 4y &= 7 \end{aligned}$$

Let's see what happens in the elimination process when a system either has no solution or has infinitely many solutions. Consider the following system:

$$\begin{aligned} 2x + 6y &= -3 \\ x + 3y &= 2 \end{aligned}$$

Multiplying the second equation by -2 and adding, we obtain

$$\begin{aligned} 2x + 6y &= -3 \\ -2x - 6y &= -4 \\ \hline 0 &= -7 \end{aligned}$$

We have obtained a contradiction. An assumption that the original system has solutions must be false, otherwise, we have proved that $0 = -7$! Thus, the system has no solution. The graphs of the equations are parallel and the system is inconsistent.

Now consider the system

$$\begin{aligned} x - \tfrac{1}{2}y &= 4 \\ -2x + y &= -8 \end{aligned}$$

If we multiply the top equation by 2 and add the result to the bottom equation, we get

$$\begin{aligned} 2x - y &= 8 \\ -2x + y &= -8 \\ \hline 0 &= 0 \end{aligned}$$

Obtaining $0 = 0$ by addition implies that the two original equations are equivalent. That is, their graphs coincide and the system is dependent. If we let $x = t$, where t is any real number, and solve either equation for y, we obtain $y = 2t - 8$. Thus,

$$(t, 2t - 8) \qquad t \text{ a real number}$$

describes the solution set for the system. The variable t is called a **parameter,** and replacing t with a real number produces a **particular solution** to the system. For example, some particular solutions to this system are

$t = -1$	$t = 2$	$t = 5$	$t = 9.4$
$(-1, -10)$	$(2, -4)$	$(5, 2)$	$(9.4, 10.8)$

Many real-world problems are readily solved by applying two-equation–two-variable methods. Example 5 provides an illustration of an application that leads to such a system.

EXAMPLE 5 **Food Processing**

A food manufacturer produces regular and lite smoked sausages. A regular sausage is 72% pork and 28% turkey, and a lite sausage is 22% pork and 78% turkey. The company has just received a shipment of 2,000 pounds of pork and 2,000 pounds of turkey. How many pounds of each type of sausage should be produced to use all the meat in this shipment?

Solution First we define the relevant variables:

$$x = \text{Number of pounds of regular sausage}$$
$$y = \text{Number of pounds of lite sausage}$$

Next we summarize the given information in Table 1. It is convenient to organize the table so that the quantities represented by variables correspond to columns in the table rather than to rows.

TABLE 1

	Regular sausage	Lite sausage	Total
Pork	72%	22%	2,000
Turkey	28%	78%	2,000

Now we use the information in the table to form equations involving x and y:

$$\begin{pmatrix}\text{Pork in } x \text{ pounds}\\ \text{of regular sausage}\end{pmatrix} + \begin{pmatrix}\text{Pork in } y \text{ pounds}\\ \text{of lite sausage}\end{pmatrix} = \begin{pmatrix}\text{Total}\\ \text{pork}\end{pmatrix}$$
$$0.72x + 0.22y = 2{,}000$$
$$\begin{pmatrix}\text{Turkey in } x \text{ pounds}\\ \text{of regular sausage}\end{pmatrix} + \begin{pmatrix}\text{Turkey in } y \text{ pounds}\\ \text{of lite sausage}\end{pmatrix} = \begin{pmatrix}\text{Total}\\ \text{turkey}\end{pmatrix}$$
$$0.28x + 0.78y = 2{,}000$$

To solve using elimination by addition, we multiply the first equation by 0.78, the second by −0.22, and add:

$$\begin{aligned} 0.5616x + 0.1716y &= 1{,}560 \\ -0.0616x - 0.1716y &= 440 \\ \hline 0.5x \qquad\qquad &= 1{,}120 \\ x &= 2{,}240 \end{aligned} \qquad \begin{aligned} 0.72(2{,}240) + 0.22y &= 2{,}000 \\ 0.22y &= 387.2 \\ y &= 1{,}760 \end{aligned}$$

Producing 2,240 pounds of regular sausage and 1,760 pounds of lite sausage will use all the available pork and turkey.

Check

$$0.72x + 0.22y = 2{,}000 \qquad\qquad 0.28x + 0.78y = 2{,}000$$
$$0.72(2{,}240) + 0.22(1{,}760) \stackrel{?}{=} 2{,}000 \qquad 0.28(2{,}240) + 0.78(1{,}760) \stackrel{?}{=} 2{,}000$$
$$2{,}000 \stackrel{\checkmark}{=} 2{,}000 \qquad\qquad 2{,}000 \stackrel{\checkmark}{=} 2{,}000$$

Matched Problem 5 A food manufacturer produces regular and deluxe rice mixtures by mixing wild rice with long-grain rice. The regular rice mixture is 5% wild rice and 95% long-grain rice, and the deluxe rice mixture is 10% wild rice and 90% long-grain rice. The company has just received a shipment of 120 pounds of wild rice and 1,500 pounds of long-grain rice. How many pounds of each type of rice mixture should be produced to use all the rice in this shipment?

• Matrices

In solving systems of equations using elimination by addition, the coefficients of the variables and the constant terms played a central role. The process can be made more efficient for generalization and computer work by the introduction of a mathematical form called a *matrix.* A **matrix** is a rectangular array of numbers written within brackets. Two examples are

$$A = \begin{bmatrix} 1 & -3 & 7 \\ 5 & 0 & -4 \end{bmatrix} \qquad B = \begin{bmatrix} -5 & 4 & 11 \\ 0 & 1 & 6 \\ -2 & 12 & 8 \\ -3 & 0 & -1 \end{bmatrix} \tag{2}$$

Each number in a matrix is called an **element** of the matrix. Matrix A has six elements arranged in two rows and three columns. Matrix B has 12 elements arranged in four rows and three columns. If a matrix has m rows and n columns, it is called an $\boldsymbol{m \times n}$ **matrix** (read "m by n matrix"). The expression $m \times n$ is called the **size** of the matrix, and the numbers m and n are called the **dimensions** of the matrix. It is important to note that the number of rows is always given first. Referring to (2) above, A is a 2×3 matrix and B is a 4×3 matrix. A matrix with n rows and n columns is called a **square matrix of order $\boldsymbol{n}$.** A matrix with only one column is called a **column matrix,** and a matrix with only one row is called a **row matrix.** These definitions are illustrated by the following:

3×3 | 4×1 | 1×4

$$\begin{bmatrix} 0.5 & 0.2 & 1.0 \\ 0.0 & 0.3 & 0.5 \\ 0.7 & 0.0 & 0.2 \end{bmatrix} \qquad \begin{bmatrix} 3 \\ -2 \\ 1 \\ 0 \end{bmatrix} \qquad \begin{bmatrix} 2 & \frac{1}{2} & 0 & -\frac{2}{3} \end{bmatrix}$$

Square matrix of order 3 | Column matrix | Row matrix

The **position** of an element in a matrix is the row and column containing the element. This is usually denoted using **double subscript notation** a_{ij}, where i is the row and j is the column containing the element a_{ij}, as illustrated below:

$$A = \begin{bmatrix} 1 & 5 & -3 \\ 6 & 0 & -4 \end{bmatrix} \qquad \begin{matrix} a_{11} = 1, a_{12} = 5, a_{13} = -3 \\ a_{21} = 6, a_{22} = 0, a_{23} = -4 \end{matrix}$$

Note that a_{12} is read "a sub one two," not "a sub twelve." The elements $a_{11} = 1$ and $a_{22} = 0$ make up the *principal diagonal* of A. In general, the **principal diagonal** of a matrix A consists of the elements $a_{11}, a_{22}, a_{33}, \ldots$.

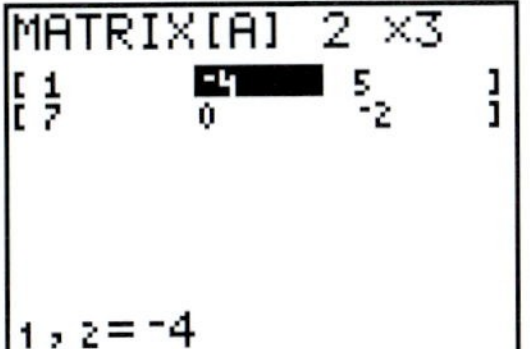

FIGURE 3 Matrix notation on a graphing utility.

Remark. Most graphing utilities are capable of storing and manipulating matrices. Figure 3 shows matrix A displayed in the editing screen of a particular graphing calculator. The size of the matrix is given at the top of the screen, and the position of the currently selected element is given at the bottom. Notice that a comma is used in the notation for the position. This is common practice on graphing utilities but not in mathematical literature.

The coefficients and constant terms in a system of linear equations can be used to form several matrices of interest to our work. Related to the system

$$\begin{aligned} 2x - 3y &= 5 \\ x + 2y &= -3 \end{aligned} \tag{3}$$

are the following matrices:

Coefficient matrix	Constant matrix	Augmented coefficient matrix
$\begin{bmatrix} 2 & -3 \\ 1 & 2 \end{bmatrix}$	$\begin{bmatrix} 5 \\ -3 \end{bmatrix}$	$\left[\begin{array}{cc\|c} 2 & -3 & 5 \\ 1 & 2 & -3 \end{array}\right]$

The augmented coefficient matrix will be used in this section. The other matrices will be used in later sections. The augmented coefficient matrix contains the essential parts of the system—both the coefficients and the constants. The vertical bar is included only as a visual aid to help us separate the coefficients from the constant terms. (Matrices entered and displayed on a graphing calculator or computer will not display this line.)

For ease of generalization to the larger systems in the following sections, we are now going to change the notation for the variables in system (3) to a subscript form (we would soon run out of letters, but we will not run out of subscripts). That is, in place of x and y, we will use x_1 and x_2, respectively, and (3) will be written as

$$\begin{aligned} 2x_1 - 3x_2 &= 5 \\ x_1 + 2x_2 &= -3 \end{aligned}$$

In general, associated with each linear system of the form

$$\begin{aligned} a_{11}x_1 + a_{12}x_2 &= k_1 \\ a_{21}x_1 + a_{22}x_2 &= k_2 \end{aligned} \tag{4}$$

where x_1 and x_2 are variables, is the **augmented matrix** of the system:

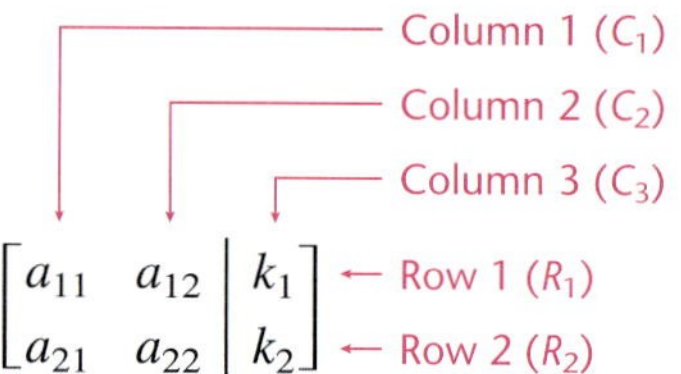

$$\left[\begin{array}{cc|c} a_{11} & a_{12} & k_1 \\ a_{21} & a_{22} & k_2 \end{array}\right]$$

This matrix contains the essential parts of system (4). Our objective is to learn how to manipulate augmented matrices in such a way that a solution to system (4) will result, if a solution exists.

In our earlier discussion of using elimination by addition, we said that two systems were equivalent if they had the same solution. And we used the operations in Theorem 2 to transform a system into an equivalent system. Paralleling this approach, we now say that two augmented matrices are **row-equivalent,** denoted by the symbol $\sim$ between the two matrices, if they are augmented matrices of equivalent systems of equations. And we use the operations listed in Theorem 3 below to transform augmented matrices into row-equivalent matrices. Note that Theorem 3 is a direct consequence of Theorem 2.

Theorem 3 **Elementary Row Operations Producing Row-Equivalent Matrices**

An augmented matrix is transformed into a row-equivalent matrix if any of the following **row operations** is performed:

1. Two rows are interchanged ($R_i \leftrightarrow R_j$).
2. A row is multiplied by a nonzero constant ($kR_i \rightarrow R_i$).
3. A constant multiple of one row is added to another row ($kR_j + R_i \rightarrow R_i$).

[*Note:* The arrow means "replaces."]

• Solving Linear Systems Using Augmented Matrices

The use of Theorem 3 in solving systems in the form of (3) is best illustrated by examples.

EXAMPLE 6 Solving a System Using Augmented Matrix Methods

Solve, using augmented matrix methods:

$$\begin{aligned} 3x_1 + 4x_2 &= 1 \\ x_1 - 2x_2 &= 7 \end{aligned} \qquad (5)$$

Solution We start by writing the augmented matrix corresponding to system (5):

$$\left[\begin{array}{rr|r} 3 & 4 & 1 \\ 1 & -2 & 7 \end{array}\right] \tag{6}$$

Our objective is to use row operations from Theorem 3 to try to transform matrix (6) into the form

$$\left[\begin{array}{rr|r} 1 & 0 & m \\ 0 & 1 & n \end{array}\right] \tag{7}$$

where m and n are real numbers. The solution to system (5) will then be obvious, since matrix (7) will be the augmented matrix of the following system:

$$\begin{aligned} x_1 &= m \qquad x_1 + 0x_2 = m \\ x_2 &= n \qquad 0x_1 + x_2 = n \end{aligned}$$

We now proceed to use row operations to transform (6) into form (7).

Step 1. To get a 1 in the upper left corner, we interchange rows 1 and 2—Theorem 3, part 1:

$$\left[\begin{array}{rr|r} 3 & 4 & 1 \\ 1 & -2 & 7 \end{array}\right] \underset{R_1 \leftrightarrow R_2}{\sim} \left[\begin{array}{rr|r} 1 & -2 & 7 \\ 3 & 4 & 1 \end{array}\right]$$

Now you see why we wanted Theorem 2, part 1.

Step 2. To get a 0 in the lower left corner, we multiply R_1 by -3 and add to R_2—Theorem 3, part 3. This changes R_2 but not R_1. Some people find it useful to write $(-3)R_1$ outside the matrix to help reduce errors in arithmetic:

$$\left[\begin{array}{rr|r} 1 & -2 & 7 \\ 3 & 4 & 1 \end{array}\right] \underset{(-3)R_1 + R_2 \to R_2}{\sim} \left[\begin{array}{rr|r} 1 & -2 & 7 \\ 0 & 10 & -20 \end{array}\right]$$

$$-3 \quad 6 \quad -21$$

Step 3. To get a 1 in the second row, second column, we multiply R_2 by $\frac{1}{10}$—Theorem 3, part 2:

$$\left[\begin{array}{rr|r} 1 & -2 & 7 \\ 0 & 10 & -20 \end{array}\right] \underset{\frac{1}{10}R_2 \to R_2}{\sim} \left[\begin{array}{rr|r} 1 & -2 & 7 \\ 0 & 1 & -2 \end{array}\right]$$

Step 4. To get a 0 in the first row, second column, we multiply R_2 by 2 and add the result to R_1—Theorem 3, part 3. This changes R_1 but not R_2.

$$0 \quad 2 \quad -4$$

$$\left[\begin{array}{rr|r} 1 & -2 & 7 \\ 0 & 1 & -2 \end{array}\right] \underset{2R_2 + R_1 \to R_1}{\sim} \left[\begin{array}{rr|r} 1 & 0 & 3 \\ 0 & 1 & -2 \end{array}\right]$$

We have accomplished our objective! The last matrix is the augmented matrix for the system

$$\begin{aligned} x_1 &= 3 \\ x_2 &= -2 \end{aligned} \tag{8}$$

Since system (8) is equivalent to the original system (5), we have solved system (5). That is, $x_1 = 3$ and $x_2 = -2$.

Check

$$\begin{aligned} 3x_1 + 4x_2 &= 1 \\ 3(3) + 4(-2) &\overset{?}{=} 1 \\ 1 &\overset{\checkmark}{=} 1 \end{aligned} \qquad \begin{aligned} x_1 - 2x_2 &= 7 \\ 3 - 2(-2) &\overset{?}{=} 7 \\ 7 &\overset{\checkmark}{=} 7 \end{aligned}$$

The above process is written more compactly as follows:

Step 1: Need a 1 here

$$\left[\begin{array}{cc|c} 3 & 4 & 1 \\ 1 & -2 & 7 \end{array}\right] \quad R_1 \leftrightarrow R_2$$

Step 2: Need a 0 here

$$\sim \left[\begin{array}{cc|c} 1 & -2 & 7 \\ 3 & 4 & 1 \end{array}\right] \quad (-3)R_1 + R_2 \to R_2$$

$$-3 \quad 6 \quad -21$$

Step 3: Need a 1 here

$$\sim \left[\begin{array}{cc|c} 1 & -2 & 7 \\ 0 & 10 & -20 \end{array}\right] \quad \tfrac{1}{10}R_2 \to R_2$$

$$0 \quad 2 \quad -4$$

Step 4: Need a 0 here

$$\sim \left[\begin{array}{cc|c} 1 & -2 & 7 \\ 0 & 1 & -2 \end{array}\right] \quad 2R_2 + R_1 \to R_1$$

$$\sim \left[\begin{array}{cc|c} 1 & 0 & 3 \\ 0 & 1 & -2 \end{array}\right]$$

Therefore, $x_1 = 3$ and $x_2 = -2$.

Matched Problem 6 Solve, using augmented matrix methods:

$$\begin{aligned} 2x_1 - x_2 &= -7 \\ x_1 + 2x_2 &= 4 \end{aligned}$$

EXPLORE-DISCUSS 2 The summary at the end of Example 6 shows five augmented coefficient matrices. Write the linear system that each matrix represents, solve each system graphically, and discuss the relationship between these solutions.

EXAMPLE 7 Solving a System Using Augmented Matrix Methods

Solve, using augmented matrix methods:

$$\begin{aligned} 2x_1 - 3x_2 &= 7 \\ 3x_1 + 4x_2 &= 2 \end{aligned}$$

Solution

Step 1: Need a 1 here

$$\begin{bmatrix} 2 & -3 & | & 7 \\ 3 & 4 & | & 2 \end{bmatrix} \quad \tfrac{1}{2}R_1 \to R_1$$

Step 2: Need a 0 here

$$\sim \begin{bmatrix} 1 & -\frac{3}{2} & | & \frac{7}{2} \\ 3 & 4 & | & 2 \end{bmatrix} \quad (-3)R_1 + R_2 \to R_2$$

$$-3 \quad \tfrac{9}{2} \quad -\tfrac{21}{2}$$

Step 3: Need a 1 here

$$\sim \begin{bmatrix} 1 & -\frac{3}{2} & | & \frac{7}{2} \\ 0 & \frac{17}{2} & | & -\frac{17}{2} \end{bmatrix} \quad \tfrac{2}{17}R_2 \to R_2$$

$$0 \quad \tfrac{3}{2} \quad -\tfrac{3}{2}$$

Step 4: Need a 0 here

$$\sim \begin{bmatrix} 1 & -\frac{3}{2} & | & \frac{7}{2} \\ 0 & 1 & | & -1 \end{bmatrix} \quad \tfrac{3}{2}R_2 + R_1 \to R_1$$

$$\sim \begin{bmatrix} 1 & 0 & | & 2 \\ 0 & 1 & | & -1 \end{bmatrix}$$

Thus, $x_1 = 2$ and $x_2 = -1$. You should check this solution in the original system.

Matched Problem 7 Solve, using augmented matrix methods:

$$\begin{aligned} 5x_1 - 2x_2 &= 12 \\ 2x_1 + 3x_2 &= 1 \end{aligned}$$

EXAMPLE 8 Solving a System Using Augmented Matrix Methods

Solve, using augmented matrix methods:

$$\begin{aligned} 2x_1 - x_2 &= 4 \\ -6x_1 + 3x_2 &= -12 \end{aligned} \tag{9}$$

Solution

$$\begin{bmatrix} 2 & -1 & | & 4 \\ -6 & 3 & | & -12 \end{bmatrix} \quad \begin{matrix} \tfrac{1}{2}R_1 \to R_1 \text{ (This produces a 1 in the upper left corner.)} \\ \tfrac{1}{3}R_2 \to R_2 \text{ (This simplifies } R_2\text{.)} \end{matrix}$$

$$\sim \begin{bmatrix} 1 & -\frac{1}{2} & | & 2 \\ -2 & 1 & | & -4 \end{bmatrix} \quad 2R_1 + R_2 \to R_2 \text{ (This produces a 0 in the lower left corner.)}$$

$$2 \quad -1 \quad 4$$

$$\sim \begin{bmatrix} 1 & -\frac{1}{2} & | & 2 \\ 0 & 0 & | & 0 \end{bmatrix}$$

The last matrix corresponds to the system

$$\begin{aligned} x_1 - \tfrac{1}{2}x_2 &= 2 \qquad x_1 - \tfrac{1}{2}x_2 = 2 \\ 0 &= 0 \qquad 0x_1 + 0x_2 = 0 \end{aligned}$$

Thus, $x_1 = \frac{1}{2}x_2 + 2$. Hence, for any real number t, if $x_2 = t$, then $x_1 = \frac{1}{2}t + 2$. That is, the solution set is described by

$$(\tfrac{1}{2}t + 2, t) \qquad t \text{ a real number} \tag{10}$$

For example, if $t = 6$, then (5, 6) is a particular solution; if $t = -2$, then (1, −2) is another particular solution; and so on. Geometrically, the graphs of the two original equations coincide and there are infinitely many solutions.

In general, if we end up with a row of 0's in an augmented matrix for a two-equation–two-variable system, the system is dependent and there are infinitely many solutions.

Check The following is a check that (10) provides a solution for system (9) for any real number t:

$$\begin{aligned} 2x_1 - x_2 &= 4 \\ 2(\tfrac{1}{2}t + 2) - t &\overset{?}{=} 4 \\ t + 4 - t &\overset{?}{=} 4 \\ 4 &\overset{\checkmark}{=} 4 \end{aligned} \qquad \begin{aligned} -6x_1 + 3x_2 &= -12 \\ -6(\tfrac{1}{2}t + 2) + 3t &\overset{?}{=} -12 \\ -3t - 12 + 3t &\overset{?}{=} -12 \\ -12 &\overset{\checkmark}{=} -12 \end{aligned}$$

Matched Problem 8 Solve, using augmented matrix methods:

$$\begin{aligned} -2x_1 + 6x_2 &= 6 \\ 3x_1 - 9x_2 &= -9 \end{aligned}$$

EXPLORE-DISCUSS 3 Most graphing utilities can perform row operations. Figure 4 shows the solution to Example 8 on a particular graphing calculator. Consult your manual to see how to perform row operations, and solve Matched Problem 8 on your graphing utility.

FIGURE 4 Performing row operations on a graphing utility.

```
[A]
   [[2  -1 4   ]
    [-6 3   -12]]
*row(1/2,[A],1)→
[A]
   [[1  -.5 2  ]
    [-6 3   -12]]
```

```
*row(1/3,[A],2)→
[A]
   [[1  -.5 2 ]
    [-2 1   -4]]
*row+(2,[A],1,2)
→[A]
   [[1  -.5 2]
    [0  0   0]]
```

EXAMPLE 9 Solving a System Using Augmented Matrix Methods

Solve, using augmented matrix methods:

$$\begin{aligned} 2x_1 + 6x_2 &= -3 \\ x_1 + 3x_2 &= 2 \end{aligned}$$

Solution

$$\left[\begin{array}{cc|c} 2 & 6 & -3 \\ 1 & 3 & 2 \end{array}\right] \quad R_1 \leftrightarrow R_2$$

$$\sim \left[\begin{array}{cc|c} 1 & 3 & 2 \\ 2 & 6 & -3 \end{array}\right] \quad (-2)R_1 + R_2 \to R_2$$

$$\begin{array}{ccc} -2 & -6 & -4 \end{array}$$

$$\sim \left[\begin{array}{cc|c} 1 & 3 & 2 \\ 0 & 0 & -7 \end{array}\right] \quad R_2 \text{ implies the contradiction: } 0 = -7$$

This is the augmented matrix of the system

$$\begin{aligned} x_1 + 3x_2 &= 2 \qquad x_1 + 3x_2 = 2 \\ 0 &= -7 \qquad 0x_1 + 0x_2 = -7 \end{aligned}$$

The second equation is not satisfied by any ordered pair of real numbers. Hence, the original system is inconsistent and has no solution. Otherwise, we have proved that $0 = -7$!

Thus, if we obtain all 0's to the left of the vertical bar and a nonzero number to the right of the bar in a row of an augmented matrix, then the system is inconsistent and there are no solutions.

Matched Problem 9 Solve, using augmented matrix methods:

$$\begin{aligned} 2x_1 - x_2 &= 3 \\ 4x_1 - 2x_2 &= -1 \end{aligned}$$

Summary

For m, n, p real numbers: $p \neq 0$:

Form 1: A Unique Solution (Consistent and Independent)	**Form 2: Infinitely Many Solutions (Consistent and Dependent)**	**Form 3: No Solution (Inconsistent)**
$\left[\begin{array}{cc\|c} 1 & 0 & m \\ 0 & 1 & n \end{array}\right]$	$\left[\begin{array}{cc\|c} 1 & m & n \\ 0 & 0 & 0 \end{array}\right]$	$\left[\begin{array}{cc\|c} 1 & m & n \\ 0 & 0 & p \end{array}\right]$

The process of solving systems of equations described in this section is referred to as *Gauss–Jordan elimination.* We will use this method to solve larger-scale systems in the next section, including systems where the number of equations and the number of variables are not the same.

Answers to Matched Problems

1. 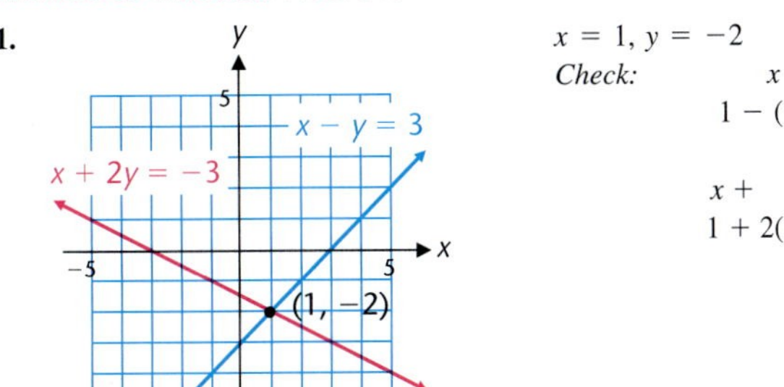

$x = 1$, $y = -2$

Check:

$$\begin{aligned} x - y &= 3 \\ 1 - (-2) &\stackrel{?}{=} 3 \\ 3 &\stackrel{\checkmark}{=} 3 \\ x + 2y &= -3 \\ 1 + 2(-2) &\stackrel{?}{=} -3 \\ -3 &\stackrel{\checkmark}{=} -3 \end{aligned}$$

2. (A) (3, 2), or $x = 3$ and $y = 2$ (B) No solution (C) Infinite number of solutions
3. $(-1.92, 4.23)$, or $x = -1.92$ and $y = 4.23$ **4.** $(-1, 3)$, or $x = -1$ and $y = 3$
5. 840 pounds of regular mix, 780 pounds of deluxe mix
6. $x_1 = -2, x_2 = 3$ **7.** $x_1 = 2, x_2 = -1$
8. The system is dependent. For t any real number, $x_2 = t$, $x_1 = 3t - 3$ is a solution.
9. Inconsistent—no solution

EXERCISE 8-1

A

Match each system in Problems 1–4 with one of the following graphs, and use the graph to solve the system.

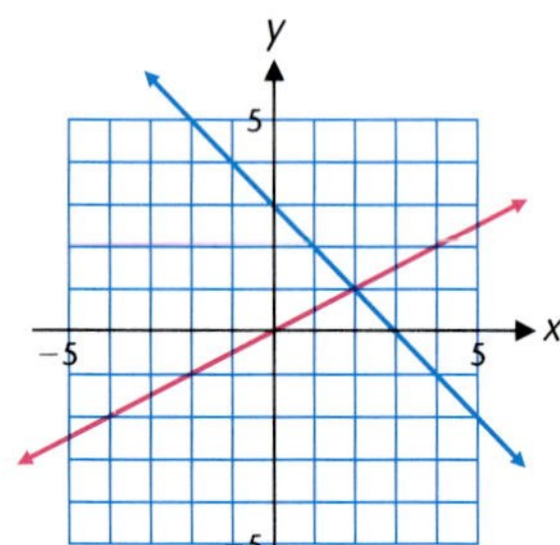

(a)

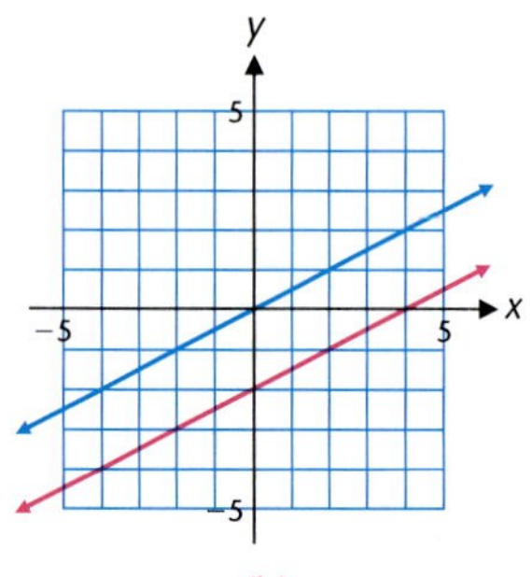

(b)

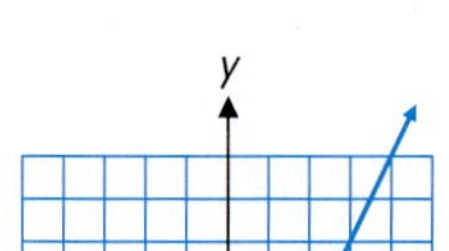

(c)

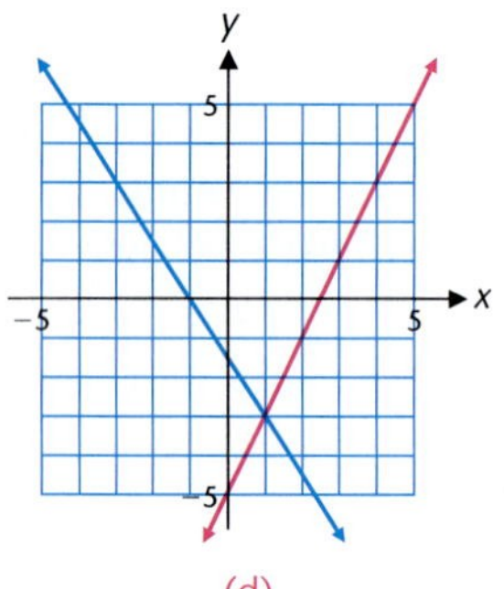

(d)

1. $\begin{aligned} 2x - 4y &= 8 \\ x - 2y &= 0 \end{aligned}$

2. $\begin{aligned} x + y &= 3 \\ x - 2y &= 0 \end{aligned}$

3. $\begin{aligned} 2x - y &= 5 \\ 3x + 2y &= -3 \end{aligned}$

4. $\begin{aligned} 4x - 2y &= 10 \\ 2x - y &= 5 \end{aligned}$

Solve Problems 5–10 by graphing.

5. $\begin{aligned} x + y &= 7 \\ x - y &= 3 \end{aligned}$

6. $\begin{aligned} x - y &= 2 \\ x + y &= 4 \end{aligned}$

7. $\begin{aligned} 3x - 2y &= 12 \\ 7x + 2y &= 8 \end{aligned}$

8. $\begin{aligned} 3x - y &= 2 \\ x + 2y &= 10 \end{aligned}$

9. $\begin{aligned} 3u + 5v &= 15 \\ 6u + 10v &= -30 \end{aligned}$

10. $\begin{aligned} m + 2n &= 4 \\ 2m + 4n &= -8 \end{aligned}$

Solve Problems 11–14 using elimination by addition.

11. $\begin{aligned} 2x + 3y &= 1 \\ 3x - y &= 7 \end{aligned}$

12. $\begin{aligned} 2m - n &= 10 \\ m - 2n &= -4 \end{aligned}$

13. $\begin{aligned} 4x - 3y &= 15 \\ 3x + 4y &= 5 \end{aligned}$

14. $\begin{aligned} 5x + 2y &= 1 \\ 2x - 3y &= -11 \end{aligned}$

Problems 15–24 refer to the following matrices:

$$A = \begin{bmatrix} 3 & -2 & 0 \\ 4 & 1 & -6 \end{bmatrix} \qquad B = \begin{bmatrix} -2 & 8 & 0 \\ -3 & 6 & 9 \\ 4 & 2 & 0 \end{bmatrix}$$

$$C = \begin{bmatrix} 3 & -2 & 0 \end{bmatrix} \qquad D = \begin{bmatrix} -4 \\ 7 \end{bmatrix}$$

15. What is the size of A? Of C?

16. What is the size of B? Of D?

17. Identify all row matrices.

18. Identify all column matrices.

19. Identify all square matrices.

20. How many additional rows would matrix A need to be a square matrix?

21. For matrix A, find a_{12} and a_{23}.

22. For matrix A, find a_{21} and a_{13}.

23. Find the elements on the principal diagonal of matrix B.

24. Find the elements on the principal diagonal of matrix A.

Perform each of the row operations indicated in Problems 25–36 on the following matrix:

$$\left[\begin{array}{rr|r} 1 & -3 & 2 \\ 4 & -6 & -8 \end{array}\right]$$

25. $R_1 \leftrightarrow R_2$ **26.** $\frac{1}{2}R_2 \rightarrow R_2$ **27.** $-4R_1 \rightarrow R_1$

28. $-2R_1 \rightarrow R_1$ **29.** $2R_2 \rightarrow R_2$ **30.** $-1R_2 \rightarrow R_2$

31. $(-4)R_1 + R_2 \rightarrow R_2$ **32.** $(-\frac{1}{2})R_2 + R_1 \rightarrow R_1$

33. $(-2)R_1 + R_2 \rightarrow R_2$ **34.** $(-3)R_1 + R_2 \rightarrow R_2$

35. $(-1)R_1 + R_2 \to R_2$

36. $1R_1 + R_2 \to R_2$

Solve Problems 37 and 38 using augmented matrix methods. Write the linear system represented by each augmented matrix in your solution, and solve each of these systems graphically. Discuss the relationship between the solutions of these systems.

37. $x_1 + x_2 = 7$
$x_1 - x_2 = 1$

38. $x_1 + x_2 = 5$
$x_1 - x_2 = -3$

B

Solve Problems 39–50 using augmented matrix methods:

39. $x_1 - 4x_2 = -2$
$-2x_1 + x_2 = -3$

40. $x_1 - 3x_2 = -5$
$-3x_1 - x_2 = 5$

41. $3x_1 - x_2 = 2$
$x_1 + 2x_2 = 10$

42. $2x_1 + x_2 = 0$
$x_1 - 2x_2 = -5$

43. $x_1 + 2x_2 = 4$
$2x_1 + 4x_2 = -8$

44. $2x_1 - 3x_2 = -2$
$-4x_1 + 6x_2 = 7$

45. $2x_1 + x_2 = 6$
$x_1 - x_2 = -3$

46. $3x_1 - x_2 = -5$
$x_1 + 3x_2 = 5$

47. $3x_1 - 6x_2 = -9$
$-2x_1 + 4x_2 = 6$

48. $2x_1 - 4x_2 = -2$
$-3x_1 + 6x_2 = 3$

49. $4x_1 - 2x_2 = 2$
$-6x_1 + 3x_2 = -3$

50. $-6x_1 + 2x_2 = 4$
$3x_1 - x_2 = -2$

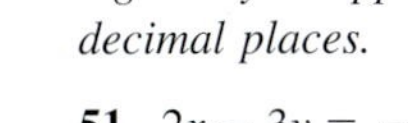

In Problems 51–54 use an intersection routine on a graphing utility to approximate the solution of each system to two decimal places.

51. $2x - 3y = -5$
$3x + 4y = 13$

52. $7x - 3y = 20$
$5x + 2y = 8$

53. $3.5x - 2.4y = 0.1$
$2.6x - 1.7y = -0.2$

54. $5.4x + 4.2y = -12.9$
$3.7x + 6.4y = -4.5$

C

55. The coefficients of the three systems below are very similar. One might guess that the solution sets to the three systems would also be nearly identical. Develop evidence for or against this guess by considering graphs of the systems and solutions obtained using elimination by addition.

(A) $4x + 5y = 4$
$9x + 11y = 4$

(B) $4x + 5y = 4$
$8x + 11y = 4$

(C) $4x + 5y = 4$
$8x + 10y = 4$

56. Repeat Problem 55 for the following systems.

(A) $5x - 6y = -10$
$11x - 13y = -20$

(B) $5x - 6y = -10$
$10x - 13y = -20$

(C) $5x - 6y = -10$
$10x - 12y = -20$

Solve Problems 57–60 using augmented matrix methods. Use a graphing utility to perform the row operations.

57. $0.8x_1 + 2.88x_2 = 4$
$1.25x_1 + 4.34x_2 = 5$

58. $2.7x_1 - 15.12x_2 = 27$
$3.25x_1 - 18.52x_2 = 33$

59. $4.8x_1 - 40.32x_2 = 295.2$
$-3.75x_1 + 28.7x_2 = -211.2$

60. $5.7x_1 - 8.55x_2 = -35.91$
$4.5x_1 + 5.73x_2 = 76.17$

APPLICATIONS 

61. Puzzle. A friend of yours came out of the post office having spent \$19.50 on 32¢ and 23¢ stamps. If she bought 75 stamps in all, how many of each type did she buy?

62. Puzzle. A parking meter contains only nickels and dimes worth \$6.05. If there are 89 coins in all, how many of each type are there?

63. Investments. Bond *A* pays 6% compounded annually and bond *B* pays 9% compounded annually. If a \$200,000 investment in a combination of the two bonds returns \$14,775 annually, how much is invested in each bond?

64. Investments. Past history indicates that mutual fund *A* will earn 14.6% annually and mutual fund *B* will earn 9.8% annually. How should an investment be divided between the two funds to produce an expected return of 11%?

65. Chemistry. A chemist has two solutions of sulfuric acid: a 20% solution and an 80% solution. How much of each should be used to obtain 100 liters of a 62% solution?

66. Chemistry. A chemist has two solutions: one containing 40% alcohol and another containing 70% alcohol. How much of each should be used to obtain 80 liters of a 49% solution?

67. Nutrition. Animals in an experiment are to be kept on a strict diet. Each animal is to receive, among other things, 54 grams of protein and 24 grams of fat. The laboratory technician is able to purchase two food mixes of the following compositions: Mix *A* has 15% protein and 10% fat; mix *B* has 30% protein and 5% fat. How many grams of each mix should be used to obtain the right diet for a single animal?

68. Nutrition—Plants. A fruit grower can use two types of fertilizer in his orange grove, brand *A* and brand *B*. Each bag of brand *A* contains 9 pounds of nitrogen and 5 pounds of phosphoric acid. Each bag of brand *B* contains 8 pounds of nitrogen and 6 pounds of phosphoric acid. Tests indicate that the grove needs 770 pounds of nitrogen and 490 pounds of phosphoric acid. How many bags of each brand should be used to provide the required amounts of nitrogen and phosphoric acid?

69. Delivery Charges. United Express, a nationwide package delivery service, charges a base price for overnight delivery of packages weighing 1 pound or less and a surcharge for each additional pound (or fraction thereof). A customer is billed \$27.75 for shipping a 5-pound package and \$64.50 for shipping a 20-pound package. Find the base price and the surcharge for each additional pound.

70. Delivery Charges. Refer to Problem 69. Federated Shipping, a competing overnight delivery service, informs the customer in Problem 69 that it would ship the 5-pound package for \$29.95 and the 20-pound package for \$59.20.

(A) If Federated Shipping computes its cost in the same manner as United Express, find the base price and the surcharge for Federated Shipping.

(B) Devise a simple rule that the customer can use to choose the cheaper of the two services for each package shipped. Justify your answer.

71. Resource Allocation. A coffee manufacturer uses Colombian and Brazilian coffee beans to produce two blends, robust and mild. A pound of the robust blend requires 12 ounces of Colombian beans and 4 ounces of Brazilian beans. A pound of the mild blend requires 6 ounces of Colombian beans and 10 ounces of Brazilian beans. Coffee is shipped in 132-pound burlap bags. The company has 50 bags of Colombian beans and 40 bags of Brazilian beans on hand. How many pounds of each blend should it produce in order to use all the available beans?

72. Resource Allocation. Refer to Problem 71.

(A) If the company decides to discontinue production of the robust blend and only produce the mild blend, how many pounds of the mild blend can it produce and how many beans of each type will it use? Are there any beans that are not used?

(B) Repeat part A if the company decides to discontinue production of the mild blend and only produce the robust blend.

SECTION 8-2 Gauss–Jordan Elimination

- Reduced Matrices
- Solving Systems by Gauss–Jordan Elimination
- Application

Now that you have had some experience with row operations on simple augmented matrices, we will consider systems involving more than two variables. In addition, we will not require that a system have the same number of equations as variables. It turns out that the results for two-variable–two-equation linear systems, stated in Theorem 1 in Section 8-1, actually hold for linear systems of any size.

Possible Solutions to a Linear System

It can be shown that any linear system must have exactly one solution, no solution, or an infinite number of solutions, regardless of the number of equations or the number of variables in the system. The terms *unique, consistent, inconsistent, dependent,* and *independent* are used to describe these solutions, just as they are for systems with two variables.

• Reduced Matrices

In the last section we used row operations to transform the augmented coefficient matrix for a system of two equations in two variables

$$\left[\begin{array}{cc|c} a_{11} & a_{12} & k_1 \\ a_{21} & a_{22} & k_2 \end{array}\right] \qquad \begin{aligned} a_{11}x_1 + a_{12}x_2 &= k_1 \\ a_{21}x_1 + a_{22}x_2 &= k_2 \end{aligned}$$

into one of the following simplified forms:

$$\begin{array}{ccc} \text{Form 1} & \text{Form 2} & \text{Form 3} \\ \left[\begin{array}{cc|c} 1 & 0 & m \\ 0 & 1 & n \end{array}\right] & \left[\begin{array}{cc|c} 1 & m & n \\ 0 & 0 & 0 \end{array}\right] & \left[\begin{array}{cc|c} 1 & m & n \\ 0 & 0 & p \end{array}\right] \end{array} \tag{1}$$

where m, n, and p are real numbers, $p \neq 0$. Each of these reduced forms represents a system that has a different type of solution set, and no two of these forms are row-equivalent. Thus, we consider each of these to be a different simplified form. Now we want to consider larger systems with more variables and more equations.

EXPLORE-DISCUSS 1 Forms 1, 2, and 3 above represent systems that have, respectively, a unique solution, an infinite number of solutions, and no solution. Discuss the number of solutions for the systems of three equations in three variables represented by the following augmented coefficient matrices.

$$\text{(A)} \left[\begin{array}{ccc|c} 1 & 1 & 1 & 2 \\ 0 & 0 & 0 & 3 \\ 0 & 0 & 0 & 0 \end{array}\right] \qquad \text{(B)} \left[\begin{array}{ccc|c} 1 & 1 & 1 & 2 \\ 0 & 0 & 0 & 0 \\ 0 & 0 & 0 & 0 \end{array}\right] \qquad \text{(C)} \left[\begin{array}{ccc|c} 1 & 0 & 0 & 2 \\ 0 & 1 & 0 & 3 \\ 0 & 0 & 1 & 4 \end{array}\right]$$

Since there is no upper limit on the number of variables or the number of equations in a linear system, it is not feasible to explicitly list all possible "simplified forms" for larger systems, as we did for systems of two equations in two variables. Instead, we state a general definition of a simplified form called a *reduced matrix* that can be applied to all matrices and systems, regardless of size.

DEFINITION 1 **Reduced Matrix**

A matrix is in **reduced form** if:

1. Each row consisting entirely of 0's is below any row having at least one nonzero element.
2. The leftmost nonzero element in each row is 1.
3. The column containing the leftmost 1 of a given row has 0's above and below the 1.
4. The leftmost 1 in any row is to the right of the leftmost 1 in the preceding row.

EXAMPLE 1 **Reduced Forms**

The matrices below are not in reduced form. Indicate which condition in the definition is violated for each matrix. State the row operation(s) required to transform the matrix to reduced form, and find the reduced form.

(A) $\left[\begin{array}{cc|c} 0 & 1 & -2 \\ 1 & 0 & 3 \end{array}\right]$ (B) $\left[\begin{array}{ccc|c} 1 & 2 & -2 & 3 \\ 0 & 0 & 1 & -1 \end{array}\right]$

(C) $\left[\begin{array}{cc|c} 1 & 0 & -3 \\ 0 & 0 & 0 \\ 0 & 1 & -2 \end{array}\right]$ (D) $\left[\begin{array}{ccc|c} 1 & 0 & 0 & -1 \\ 0 & 2 & 0 & 3 \\ 0 & 0 & 1 & -5 \end{array}\right]$

Solutions (A) Condition 4 is violated: The leftmost 1 in row 2 is not to the right of the leftmost 1 in row 1. Perform the row operation $R_1 \leftrightarrow R_2$ to obtain the reduced form:

$$\left[\begin{array}{cc|c} 1 & 0 & 3 \\ 0 & 1 & -2 \end{array}\right]$$

(B) Condition 3 is violated: The column containing the leftmost 1 in row 2 does not have a zero above the 1. Perform the row operation $2R_2 + R_1 \rightarrow R_1$ to obtain the reduced form:

$$\left[\begin{array}{ccc|c} 1 & 2 & 0 & 1 \\ 0 & 0 & 1 & -1 \end{array}\right]$$

(C) Condition 1 is violated: The second row contains all zeros, and it is not below any row having at least one nonzero element. Perform the row operation $R_2 \leftrightarrow R_3$ to obtain the reduced form:

$$\left[\begin{array}{cc|c} 1 & 0 & -3 \\ 0 & 1 & -2 \\ 0 & 0 & 0 \end{array}\right]$$

(D) Condition 2 is violated: The leftmost nonzero element in row 2 is not a 1. Perform the row operation $\frac{1}{2}R_2 \rightarrow R_2$ to obtain the reduced form:

$$\left[\begin{array}{ccc|c} 1 & 0 & 0 & -1 \\ 0 & 1 & 0 & \frac{3}{2} \\ 0 & 0 & 1 & -5 \end{array}\right]$$

Matched Problem 1 The matrices below are not in reduced form. Indicate which condition in the definition is violated for each matrix. State the row operation(s) required to transform the matrix to reduced form and find the reduced form.

(A) $\left[\begin{array}{cc|c} 1 & 0 & 2 \\ 0 & 3 & -6 \end{array}\right]$ (B) $\left[\begin{array}{ccc|c} 1 & 5 & 4 & 3 \\ 0 & 1 & 2 & -1 \\ 0 & 0 & 0 & 0 \end{array}\right]$

(C) $\left[\begin{array}{ccc|c} 0 & 1 & 0 & -3 \\ 1 & 0 & 0 & 0 \\ 0 & 0 & 1 & 2 \end{array}\right]$ (D) $\left[\begin{array}{ccc|c} 1 & 2 & 0 & 3 \\ 0 & 0 & 0 & 0 \\ 0 & 0 & 1 & 4 \end{array}\right]$

• Solving Systems by Gauss–Jordan Elimination

We are now ready to outline the Gauss–Jordan elimination method for solving systems of linear equations. The method systematically transforms an augmented matrix into a reduced form. The system corresponding to a reduced augmented coefficient matrix is called a **reduced system.** As we will see, reduced systems are easy to solve.

The Gauss–Jordan elimination method is named after the German mathematician Carl Friedrich Gauss (1777–1885) and the German geodesist Wilhelm Jordan (1842–1899). Gauss, one of the greatest mathematicians of all time, used a method of solving systems of equations that was later generalized by Jordan to solve problems in large-scale surveying.

EXAMPLE 2 **Solving a System Using Gauss–Jordan Elimination**

Solve by Gauss–Jordan elimination:

$$\begin{aligned} 2x_1 - 2x_2 + x_3 &= 3 \\ 3x_1 + x_2 - x_3 &= 7 \\ x_1 - 3x_2 + 2x_3 &= 0 \end{aligned}$$

Solution Write the augmented matrix and follow the steps indicated at the right to produce a reduced form.

Need a 1 here

$$\left[\begin{array}{ccc|c} 2 & -2 & 1 & 3 \\ 3 & 1 & -1 & 7 \\ 1 & -3 & 2 & 0 \end{array}\right] \quad R_1 \leftrightarrow R_3$$

Step 1: Choose the leftmost nonzero column and get a 1 at the top.

Need 0's here

$$\sim \left[\begin{array}{ccc|c} 1 & -3 & 2 & 0 \\ 3 & 1 & -1 & 7 \\ 2 & -2 & 1 & 3 \end{array}\right] \quad \begin{array}{l} \\ (-3)R_1 + R_2 \to R_2 \\ (-2)R_1 + R_3 \to R_3 \end{array}$$

Step 2: Use multiples of the row containing the 1 from step 1 to get zeros in all remaining places in the column containing this 1.

Need a 1 here

$$\sim \left[\begin{array}{ccc|c} 1 & -3 & 2 & 0 \\ 0 & 10 & -7 & 7 \\ 0 & 4 & -3 & 3 \end{array}\right] \quad 0.1R_2 \to R_2$$

Step 3: Repeat step 1 with the *submatrix* formed by (mentally) deleting the top row.

Need 0's here

$$\sim \left[\begin{array}{ccc|c} 1 & -3 & 2 & 0 \\ 0 & 1 & -0.7 & 0.7 \\ 0 & 4 & -3 & 3 \end{array}\right] \quad \begin{array}{l} 3R_2 + R_1 \to R_1 \\ \\ (-4)R_2 + R_3 \to R_3 \end{array}$$

Step 4: Repeat step 2 with the *entire matrix.*

Need a 1 here

$$\sim \left[\begin{array}{ccc|c} 1 & 0 & -0.1 & 2.1 \\ 0 & 1 & -0.7 & 0.7 \\ 0 & 0 & -0.2 & 0.2 \end{array}\right] \quad (-5)R_3 \to R_3$$

Step 3: Repeat step 1 with the *submatrix* formed by (mentally) deleting the top two rows.

Need 0's here

$$\sim \left[\begin{array}{ccc|c} 1 & 0 & -0.1 & 2.1 \\ 0 & 1 & -0.7 & 0.7 \\ 0 & 0 & 1 & -1 \end{array}\right] \quad \begin{array}{l} 0.1R_3 + R_1 \to R_1 \\ 0.7R_3 + R_2 \to R_2 \\ \end{array}$$

Step 4: Repeat step 2 with the *entire matrix.*

$$\sim\left[\begin{array}{ccc|c} 1 & 0 & 0 & 2 \\ 0 & 1 & 0 & 0 \\ 0 & 0 & 1 & -1 \end{array}\right]$$

The matrix is now in reduced form, and we can proceed to solve the corresponding reduced system.

$$\begin{aligned} x_1 \quad\quad &= 2 \\ x_2 \quad &= 0 \\ x_3 &= -1 \end{aligned}$$

The solution to this system is $x_1 = 2$, $x_2 = 0$, $x_3 = -1$. You should check this solution in the original system.

Gauss–Jordan Elimination

Step 1. Choose the leftmost nonzero column and use appropriate row operations to get a 1 at the top.

Step 2. Use multiples of the row containing the 1 from step 1 to get zeros in all remaining places in the column containing this 1.

Step 3. Repeat step 1 with the **submatrix** formed by (mentally) deleting the row used in step 2 and all rows above this row.

Step 4. Repeat step 2 with the **entire matrix,** including the mentally deleted rows. Continue this process until it is impossible to go further.

[*Note:* If at any point in this process we obtain a row with all zeros to the left of the vertical line and a nonzero number to the right, we can stop, since we will have a contradiction: $0 = n$, $n \neq 0$. We can then conclude that the system has no solution.]

Remarks

1. Even though each matrix has a unique reduced form, the sequence of steps (algorithm) presented here for transforming a matrix into a reduced form is not unique. That is, other sequences of steps (using row operations) can produce a reduced matrix. (For example, it is possible to use row operations in such a way that computations involving fractions are minimized.) But we emphasize again that we are not interested in the most efficient hand methods for transforming small matrices into reduced forms. Our main interest is in giving you a little experience with a method that is suitable for solving large-scale systems on a computer or graphing utility.

2. Most graphing utilities have the ability to find reduced forms, either directly or with some programming. Figure 1 illustrates the solution of Example 2 on a graphing calculator that has a built-in routine for finding reduced forms. Notice that in row 2 and column 4 of the reduced form the graphing calculator has displayed the very small number −3.5E-13 instead of the exact value 0. This is a common occurrence on a graphing calculator and causes no problems. Just replace any very small numbers displayed in scientific notation with 0.

```
A
         [[2 -2 1  3]
          [3 1  -1 7]
          [1 -3 2  0]]
rref A
     [[1 0 0 2        ]
      [0 1 0 -3.5E-13]
      [0 0 1 -1       ]]
```

FIGURE 1 Gauss–Jordan elimination on a graphing calculator.

Matched Problem 2 Solve by Gauss–Jordan elimination:

$$\begin{aligned} 3x_1 + x_2 - 2x_3 &= 2 \\ x_1 - 2x_2 + x_3 &= 3 \\ 2x_1 - x_2 - 3x_3 &= 3 \end{aligned}$$

EXAMPLE 3 Solving a System Using Gauss–Jordan Elimination

Solve by Gauss–Jordan elimination:

$$\begin{aligned} 2x_1 - 4x_2 + x_3 &= -4 \\ 4x_1 - 8x_2 + 7x_3 &= 2 \\ -2x_1 + 4x_2 - 3x_3 &= 5 \end{aligned}$$

Solution

$$\left[\begin{array}{rrr|r} 2 & -4 & 1 & -4 \\ 4 & -8 & 7 & 2 \\ -2 & 4 & -3 & 5 \end{array}\right] \quad 0.5R_1 \to R_1$$

$$\sim \left[\begin{array}{rrr|r} 1 & -2 & 0.5 & -2 \\ 4 & -8 & 7 & 2 \\ -2 & 4 & -3 & 5 \end{array}\right] \quad \begin{array}{l} (-4)R_1 + R_2 \to R_2 \\ 2R_1 + R_3 \to R_3 \end{array}$$

$$\sim \left[\begin{array}{rrr|r} 1 & -2 & 0.5 & -2 \\ 0 & 0 & 5 & 10 \\ 0 & 0 & -2 & 1 \end{array}\right] \quad 0.2R_2 \to R_2$$

Note that column 3 is the leftmost nonzero column in this submatrix.

$$\sim \left[\begin{array}{rrr|r} 1 & -2 & 0.5 & -2 \\ 0 & 0 & 1 & 2 \\ 0 & 0 & -2 & 1 \end{array}\right] \quad \begin{array}{l} (-0.5)R_2 + R_1 \to R_1 \\ \\ 2R_2 + R_3 \to R_3 \end{array}$$

$$\sim \left[\begin{array}{rrr|r} 1 & -2 & 0 & -3 \\ 0 & 0 & 1 & 2 \\ 0 & 0 & 0 & 5 \end{array}\right]$$

We stop the Gauss–Jordan elimination, even though the matrix is not in reduced form, since the last row produces a contradiction.

The system is inconsistent and has no solution.

Matched Problem 3 Solve by Gauss–Jordan elimination:

$$\begin{aligned} 2x_1 - 4x_2 - x_3 &= -8 \\ 4x_1 - 8x_2 + 3x_3 &= 4 \\ -2x_1 + 4x_2 + x_3 &= 11 \end{aligned}$$

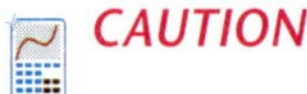

CAUTION Figure 2 shows the solution to Example 3 on a graphing calculator with a built-in reduced form routine. Notice that the graphing calculator does not stop when a contradiction first occurs, as we did in the solution to Example 3, but continues on to find the reduced form. Nevertheless, the last row in the reduced form still produces a contradiction, indicating that the system has no solution.

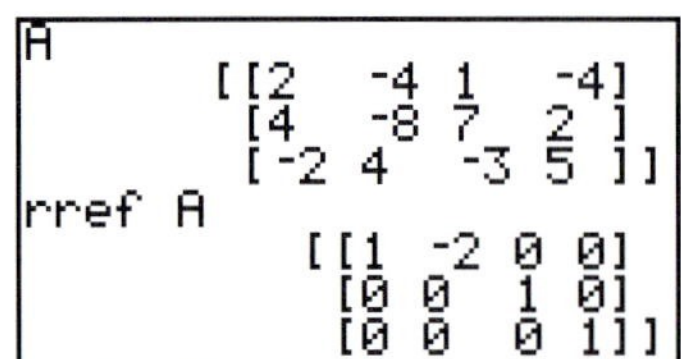

FIGURE 2 Recognizing contradictions on a graphing calculator.

EXAMPLE 4 Solving a System Using Gauss–Jordan Elimination

Solve by Gauss–Jordan elimination:

$$\begin{aligned} 3x_1 + 6x_2 - 9x_3 &= 15 \\ 2x_1 + 4x_2 - 6x_3 &= 10 \\ -2x_1 - 3x_2 + 4x_3 &= -6 \end{aligned}$$

Solution

$$\left[\begin{array}{rrr|r} 3 & 6 & -9 & 15 \\ 2 & 4 & -6 & 10 \\ -2 & -3 & 4 & -6 \end{array}\right] \quad \tfrac{1}{3}R_1 \to R_1$$

$$\sim \left[\begin{array}{rrr|r} 1 & 2 & -3 & 5 \\ 2 & 4 & -6 & 10 \\ -2 & -3 & 4 & -6 \end{array}\right] \quad \begin{array}{l} (-2)R_1 + R_2 \to R_2 \\ 2R_1 + R_3 \to R_3 \end{array}$$

$$\sim \left[\begin{array}{rrr|r} 1 & 2 & -3 & 5 \\ 0 & 0 & 0 & 0 \\ 0 & 1 & -2 & 4 \end{array}\right] \quad R_2 \leftrightarrow R_3$$

Note that we must interchange rows 2 and 3 to obtain a nonzero entry at the top of the second column of this submatrix.

$$\sim \left[\begin{array}{rrr|r} 1 & 2 & -3 & 5 \\ 0 & 1 & -2 & 4 \\ 0 & 0 & 0 & 0 \end{array}\right] \quad (-2)R_2 + R_1 \to R_1$$

$$\sim \left[\begin{array}{rrr|r} 1 & 0 & 1 & -3 \\ 0 & 1 & -2 & 4 \\ 0 & 0 & 0 & 0 \end{array}\right]$$

This matrix is now in reduced form. Write the corresponding reduced system and solve.

$$\begin{aligned} x_1 \quad + x_3 &= -3 \\ x_2 - 2x_3 &= 4 \end{aligned}$$

We discard the equation corresponding to the third (all 0) row in the reduced form, since it is satisfied by all values of x_1, x_2, and x_3.

Note that the leftmost variable in each equation appears in one and only one equation. We solve for the leftmost variables x_1 and x_2 in terms of the remaining variable x_3:

$$x_1 = -x_3 - 3$$
$$x_2 = 2x_3 + 4$$

This dependent system has an infinite number of solutions. We will use a parameter to represent all the solutions. If we let $x_3 = t$, then for any real number t,

$$x_1 = -t - 3$$
$$x_2 = 2t + 4$$
$$x_3 = t$$

You should check that $(-t - 3, 2t + 4, t)$ is a solution of the original system for any real number t. Some particular solutions are

$t = 0$	$t = -2$	$t = 3.5$
$(-3, 4, 0)$	$(-1, 0, -2)$	$(-6.5, 11, 3.5)$

Matched Problem 4 Solve by Gauss–Jordan elimination:

$$\begin{aligned} 2x_1 - 2x_2 - 4x_3 &= -2 \\ 3x_1 - 3x_2 - 6x_3 &= -3 \\ -2x_1 + 3x_2 + x_3 &= 7 \end{aligned}$$

In general,

If the number of leftmost 1's in a reduced augmented coefficient matrix is less than the number of variables in the system and there are no contradictions, then the system is dependent and has infinitely many solutions.

There are many different ways to use the reduced augmented coefficient matrix to describe the infinite number of solutions of a dependent system. We will always proceed as follows: Solve each equation in a reduced system for its leftmost variable and then introduce a different parameter for each remaining variable. As the solution to Example 4 illustrates, this method produces a concise and useful representation of the solutions to a dependent system. Example 5 illustrates a dependent system where two parameters are required to describe the solution.

EXPLORE-DISCUSS 2 Explain why the definition of reduced form ensures that each leftmost variable in a reduced system appears in one and only one equation and no equation contains more than one leftmost variable. Discuss methods for determining if a consistent system is independent or dependent by examining the reduced form.

EXAMPLE 5 Solving a System Using Gauss–Jordan Elimination

Solve by Gauss–Jordan elimination:

$$\begin{aligned} x_1 + 2x_2 + 4x_3 + x_4 - x_5 &= 1 \\ 2x_1 + 4x_2 + 8x_3 + 3x_4 - 4x_5 &= 2 \\ x_1 + 3x_2 + 7x_3 + 3x_5 &= -2 \end{aligned}$$

Solution

$$\left[\begin{array}{rrrrr|r} 1 & 2 & 4 & 1 & -1 & 1 \\ 2 & 4 & 8 & 3 & -4 & 2 \\ 1 & 3 & 7 & 0 & 3 & -2 \end{array}\right] \quad \begin{array}{l} \\ (-2)R_1 + R_2 \to R_2 \\ (-1)R_1 + R_3 \to R_3 \end{array}$$

$$\sim \left[\begin{array}{rrrrr|r} 1 & 2 & 4 & 1 & -1 & 1 \\ 0 & 0 & 0 & 1 & -2 & 0 \\ 0 & 1 & 3 & -1 & 4 & -3 \end{array}\right] \quad R_2 \leftrightarrow R_3$$

$$\sim \left[\begin{array}{rrrrr|r} 1 & 2 & 4 & 1 & -1 & 1 \\ 0 & 1 & 3 & -1 & 4 & -3 \\ 0 & 0 & 0 & 1 & -2 & 0 \end{array}\right] \quad (-2)R_2 + R_1 \to R_1$$

$$\sim \left[\begin{array}{rrrrr|r} 1 & 0 & -2 & 3 & -9 & 7 \\ 0 & 1 & 3 & -1 & 4 & -3 \\ 0 & 0 & 0 & 1 & -2 & 0 \end{array}\right] \quad \begin{array}{l} (-3)R_3 + R_1 \to R_1 \\ R_3 + R_2 \to R_2 \\ \end{array}$$

$$\sim \left[\begin{array}{rrrrr|r} 1 & 0 & -2 & 0 & -3 & 7 \\ 0 & 1 & 3 & 0 & 2 & -3 \\ 0 & 0 & 0 & 1 & -2 & 0 \end{array}\right] \quad \text{Matrix is in reduced form.}$$

$$\begin{aligned} x_1 \qquad - 2x_3 \qquad - 3x_5 &= 7 \\ x_2 + 3x_3 \qquad + 2x_5 &= -3 \\ x_4 - 2x_5 &= 0 \end{aligned}$$

Solve for the leftmost variables x_1, x_2, and x_4 in terms of the remaining variables x_3 and x_5:

$$\begin{aligned} x_1 &= 2x_3 + 3x_5 + 7 \\ x_2 &= -3x_3 - 2x_5 - 3 \\ x_4 &= 2x_5 \end{aligned}$$

If we let $x_3 = s$ and $x_5 = t$, then for any real numbers s and t,

$$\begin{aligned} x_1 &= 2s + 3t + 7 \\ x_2 &= -3s - 2t - 3 \\ x_3 &= s \\ x_4 &= 2t \\ x_5 &= t \end{aligned}$$

is a solution. The check is left for you to perform.

Matched Problem 5 Solve by Gauss–Jordan elimination:

$$\begin{aligned} x_1 - x_2 + 2x_3 - 2x_5 &= 3 \\ -2x_1 + 2x_2 - 4x_3 - x_4 + x_5 &= -5 \\ 3x_1 - 3x_2 + 7x_3 + x_4 - 4x_5 &= 6 \end{aligned}$$

• Application

We now consider an application that involves a dependent system of equations.

EXAMPLE 6 Purchasing

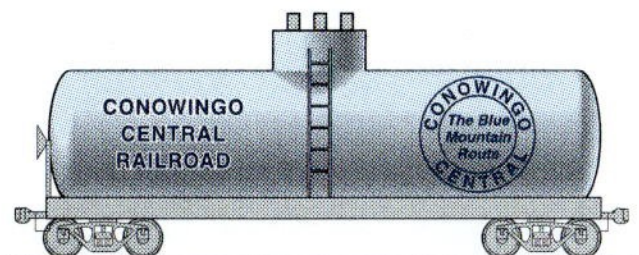

A chemical manufacturer wants to purchase a fleet of 24 railroad tank cars with a combined carrying capacity of 250,000 gallons. Tank cars with three different carrying capacities are available: 6,000 gallons, 8,000 gallons, and 18,000 gallons. How many of each type of tank car should be purchased?

Solution Let

$$\begin{aligned} x_1 &= \text{Number of 6,000-gallon tank cars} \\ x_2 &= \text{Number of 8,000-gallon tank cars} \\ x_3 &= \text{Number of 18,000-gallon tank cars} \end{aligned}$$

Then

$$\begin{aligned} x_1 + x_2 + x_3 &= 24 \qquad \text{Total number of tank cars} \\ 6{,}000x_1 + 8{,}000x_2 + 18{,}000x_3 &= 250{,}000 \qquad \text{Total carrying capacity} \end{aligned}$$

Now we can form the augmented matrix of the system and solve by using Gauss–Jordan elimination:

$$\left[\begin{array}{ccc|c} 1 & 1 & 1 & 24 \\ 6{,}000 & 8{,}000 & 18{,}000 & 250{,}000 \end{array}\right] \quad \tfrac{1}{1{,}000}R_2 \to R_2 \text{ (simplify } R_2\text{)}$$

$$\sim \left[\begin{array}{ccc|c} 1 & 1 & 1 & 24 \\ 6 & 8 & 18 & 250 \end{array}\right] \quad (-6)R_1 + R_2 \to R_2$$

$$\sim \left[\begin{array}{ccc|c} 1 & 1 & 1 & 24 \\ 0 & 2 & 12 & 106 \end{array}\right] \quad \tfrac{1}{2}R_2 \to R_2$$

$$\sim \left[\begin{array}{ccc|c} 1 & 1 & 1 & 24 \\ 0 & 1 & 6 & 53 \end{array}\right] \quad (-1)R_2 + R_1 \to R_1$$

$$\sim \left[\begin{array}{ccc|c} 1 & 0 & -5 & -29 \\ 0 & 1 & 6 & 53 \end{array}\right] \quad \text{Matrix is in reduced form}$$

$$\begin{aligned} x_1 - 5x_3 &= -29 \quad \text{or} \quad x_1 = 5x_3 - 29 \\ x_2 + 6x_3 &= 53 \quad \text{or} \quad x_2 = -6x_3 + 53 \end{aligned}$$

Let $x_3 = t$. Then for t any real number,

$$x_1 = 5t - 29$$
$$x_2 = -6t + 53$$
$$x_3 = t$$

is a solution—or is it? Since the variables in this system represent the number of tank cars purchased, the values of x_1, x_2, and x_3 must be nonnegative integers. Thus, the third equation requires that t must be a nonnegative integer. The first equation requires that $5t - 29 \geq 0$, so t must be at least 6. The middle equation requires that $-6t + 53 \geq 0$, so t can be no larger than 8. Thus, 6, 7, and 8 are the only possible values for t. There are only three possible combinations that meet the company's specifications of 24 tank cars with a total carrying capacity of 250,000 gallons, as shown in Table 1:

TABLE 1

t	6,000-gallon tank cars x_1	8,000-gallon tank cars x_2	18,000-gallon tank cars x_3
6	1	17	6
7	6	11	7
8	11	5	8

The final choice would probably be influenced by other factors. For example, the company might want to minimize the cost of the 24 tank cars.

Matched Problem 6 A commuter airline wants to purchase a fleet of 30 airplanes with a combined carrying capacity of 960 passengers. The three available types of planes carry 18, 24, and 42 passengers, respectively. How many of each type of plane should be purchased?

Answers to Matched Problems

1. (A) Condition 2 is violated: The 3 in row 2 and column 2 should be a 1. Perform the operation $\frac{1}{3}R_2 \rightarrow R_2$ to obtain:

$$\left[\begin{array}{cc|c} 1 & 0 & 2 \\ 0 & 1 & -2 \end{array}\right]$$

(B) Condition 3 is violated: The 5 in row 1 and column 2 should be a 0. Perform the operation $(-5)R_2 + R_1 \rightarrow R_1$ to obtain:

$$\left[\begin{array}{ccc|c} 1 & 0 & -6 & 8 \\ 0 & 1 & 2 & -1 \\ 0 & 0 & 0 & 0 \end{array}\right]$$

(C) Condition 4 is violated: The leftmost 1 in the second row is not to the right of the leftmost 1 in the first row. Perform the operation $R_1 \leftrightarrow R_2$ to obtain:

$$\left[\begin{array}{ccc|c} 1 & 0 & 0 & 0 \\ 0 & 1 & 0 & -3 \\ 0 & 0 & 1 & 2 \end{array}\right]$$

(D) Condition 1 is violated: The all-zero second row should be at the bottom. Perform the operation $R_2 \leftrightarrow R_3$ to obtain:

$$\left[\begin{array}{ccc|c} 1 & 2 & 0 & 3 \\ 0 & 0 & 1 & 4 \\ 0 & 0 & 0 & 0 \end{array}\right]$$

2. $x_1 = 1, x_2 = -1, x_3 = 0$
3. Inconsistent; no solution
4. $x_1 = 5t + 4, x_2 = 3t + 5, x_3 = t$, t any real number
5. $x_1 = s + 7, x_2 = s, x_3 = t - 2, x_4 = -3t - 1, x_5 = t$, s and t any real numbers
6.

t	18-passenger planes x_1	24-passenger planes x_2	42-passenger planes x_3
14	2	14	14
15	5	10	15
16	8	6	16
17	11	2	17

EXERCISE 8-2

A

In Problems 1–8, indicate whether each matrix is in reduced form.

1. $\left[\begin{array}{cc|c} 1 & 0 & -1 \\ 0 & 2 & 6 \end{array}\right]$

2. $\left[\begin{array}{cc|c} 1 & 0 & 5 \\ 0 & 1 & -3 \end{array}\right]$

3. $\left[\begin{array}{ccc|c} 0 & 1 & -2 & 0 \\ 0 & 0 & 0 & 1 \\ 0 & 0 & 0 & 0 \end{array}\right]$

4. $\left[\begin{array}{ccc|c} 1 & -1 & 4 & 0 \\ 0 & 0 & 0 & 0 \\ 0 & 0 & 0 & 1 \end{array}\right]$

5. $\left[\begin{array}{ccc|c} 0 & 0 & 1 & 2 \\ 0 & 1 & 0 & -5 \\ 1 & 0 & 0 & 4 \end{array}\right]$

6. $\left[\begin{array}{ccc|c} 1 & -2 & 4 & 1 \\ 0 & 0 & 1 & -3 \\ 0 & 0 & 0 & 0 \end{array}\right]$

7. $\left[\begin{array}{cccc|c} 0 & 1 & 6 & 0 & -8 \\ 0 & 0 & 0 & 1 & 1 \end{array}\right]$

8. $\left[\begin{array}{ccc|c} 0 & 0 & 1 & 0 \\ 0 & 0 & 0 & 0 \end{array}\right]$

In Problems 9–16, write the linear system corresponding to each reduced augmented matrix and solve.

9. $\left[\begin{array}{ccc|c} 1 & 0 & 0 & -2 \\ 0 & 1 & 0 & 3 \\ 0 & 0 & 1 & 0 \end{array}\right]$

10. $\left[\begin{array}{cccc|c} 1 & 0 & 0 & 0 & -2 \\ 0 & 1 & 0 & 0 & 0 \\ 0 & 0 & 1 & 0 & 1 \\ 0 & 0 & 0 & 1 & 3 \end{array}\right]$

11. $\left[\begin{array}{ccc|c} 1 & 0 & -2 & 3 \\ 0 & 1 & 1 & -5 \\ 0 & 0 & 0 & 0 \end{array}\right]$

12. $\left[\begin{array}{ccc|c} 1 & -2 & 0 & -3 \\ 0 & 0 & 1 & 5 \\ 0 & 0 & 0 & 0 \end{array}\right]$

13. $\left[\begin{array}{cc|c} 1 & 0 & 0 \\ 0 & 1 & 0 \\ 0 & 0 & 1 \end{array}\right]$

14. $\left[\begin{array}{cc|c} 1 & 0 & 5 \\ 0 & 1 & -3 \\ 0 & 0 & 0 \end{array}\right]$

15. $\left[\begin{array}{cccc|c} 1 & -2 & 0 & -3 & -5 \\ 0 & 0 & 1 & 3 & 2 \end{array}\right]$

16. $\left[\begin{array}{cccc|c} 1 & 0 & -2 & 3 & 4 \\ 0 & 1 & -1 & 2 & -1 \end{array}\right]$

B

Use row operations to change each matrix in Problems 17–22 to reduced form.

17. $\left[\begin{array}{cc|c} 1 & 2 & -1 \\ 0 & 1 & 3 \end{array}\right]$

18. $\left[\begin{array}{cc|c} 1 & 3 & 1 \\ 0 & 2 & -4 \end{array}\right]$

19. $\left[\begin{array}{ccc|c} 1 & 0 & -3 & 1 \\ 0 & 1 & 2 & 0 \\ 0 & 0 & 3 & -6 \end{array}\right]$

20. $\left[\begin{array}{ccc|c} 1 & 0 & 4 & 0 \\ 0 & 1 & -3 & -1 \\ 0 & 0 & -2 & 2 \end{array}\right]$

21. $\left[\begin{array}{rrr|r} 1 & 2 & -2 & -1 \\ 0 & 3 & -6 & 1 \\ 0 & -1 & 2 & -\frac{1}{3} \end{array}\right]$

22. $\left[\begin{array}{rrr|r} 0 & -2 & 8 & 1 \\ 2 & -2 & 6 & -4 \\ 0 & -1 & 4 & \frac{1}{2} \end{array}\right]$

Solve Problems 23–42 using Gauss–Jordan elimination.

23. $\begin{aligned} 2x_1 + 4x_2 - 10x_3 &= -2 \\ 3x_1 + 9x_2 - 21x_3 &= 0 \\ x_1 + 5x_2 - 12x_3 &= 1 \end{aligned}$

24. $\begin{aligned} 3x_1 + 5x_2 - x_3 &= -7 \\ x_1 + x_2 + x_3 &= -1 \\ 2x_1 + 11x_3 &= 7 \end{aligned}$

25. $\begin{aligned} 3x_1 + 8x_2 - x_3 &= -18 \\ 2x_1 + x_2 + 5x_3 &= 8 \\ 2x_1 + 4x_2 + 2x_3 &= -4 \end{aligned}$

26. $\begin{aligned} 2x_1 + 7x_2 + 15x_3 &= -12 \\ 4x_1 + 7x_2 + 13x_3 &= -10 \\ 3x_1 + 6x_2 + 12x_3 &= -9 \end{aligned}$

27. $\begin{aligned} 2x_1 - x_2 - 3x_3 &= 8 \\ x_1 - 2x_2 &= 7 \end{aligned}$

28. $\begin{aligned} 2x_1 + 4x_2 - 6x_3 &= 10 \\ 3x_1 + 3x_2 - 3x_3 &= 6 \end{aligned}$

29. $\begin{aligned} 2x_1 - x_2 &= 0 \\ 3x_1 + 2x_2 &= 7 \\ x_1 - x_2 &= -1 \end{aligned}$

30. $\begin{aligned} 2x_1 - x_2 &= 0 \\ 3x_1 + 2x_2 &= 7 \\ x_1 - x_2 &= -2 \end{aligned}$

31. $\begin{aligned} 3x_1 - 4x_2 - x_3 &= 1 \\ 2x_1 - 3x_2 + x_3 &= 1 \\ x_1 - 2x_2 + 3x_3 &= 2 \end{aligned}$

32. $\begin{aligned} 3x_1 + 7x_2 - x_3 &= 11 \\ x_1 + 2x_2 - x_3 &= 3 \\ 2x_1 + 4x_2 - 2x_3 &= 10 \end{aligned}$

33. $\begin{aligned} -2x_1 + x_2 + 3x_3 &= -7 \\ x_1 - 4x_2 + 2x_3 &= 0 \\ x_1 - 3x_2 + x_3 &= 1 \end{aligned}$

34. $\begin{aligned} 2x_1 + 5x_2 + 4x_3 &= -7 \\ -4x_1 - 5x_2 + 2x_3 &= 9 \\ -2x_1 - x_2 + 4x_3 &= 3 \end{aligned}$

35. $\begin{aligned} 2x_1 - 2x_2 - 4x_3 &= -2 \\ -3x_1 + 3x_2 + 6x_3 &= 3 \end{aligned}$

36. $\begin{aligned} 2x_1 + 8x_2 - 6x_3 &= 4 \\ -3x_1 - 12x_2 + 9x_3 &= -6 \end{aligned}$

37. $\begin{aligned} 4x_1 - x_2 + 2x_3 &= 3 \\ -4x_1 + x_2 - 3x_3 &= -10 \\ 8x_1 - 2x_2 + 9x_3 &= -1 \end{aligned}$

38. $\begin{aligned} 4x_1 - 2x_2 + 2x_3 &= 5 \\ -6x_1 + 3x_2 - 3x_3 &= -2 \\ 10x_1 - 5x_2 + 9x_3 &= 4 \end{aligned}$

39. $\begin{aligned} 2x_1 - 5x_2 - 3x_3 &= 7 \\ -4x_1 + 10x_2 + 2x_3 &= 6 \\ 6x_1 - 15x_2 - x_3 &= -19 \end{aligned}$

40. $\begin{aligned} -4x_1 + 8x_2 + 10x_3 &= -6 \\ 6x_1 - 12x_2 - 15x_3 &= 9 \\ -8x_1 + 14x_2 + 19x_3 &= -8 \end{aligned}$

41. $\begin{aligned} 5x_1 - 3x_2 + 2x_3 &= 13 \\ 2x_1 - x_2 - 3x_3 &= 1 \\ 4x_1 - 2x_2 + 4x_3 &= 12 \end{aligned}$

42. $\begin{aligned} 4x_1 - 2x_2 + 3x_3 &= 3 \\ 3x_1 - x_2 - 2x_3 &= -10 \\ 2x_1 + 4x_2 - x_3 &= -1 \end{aligned}$

43. Consider a consistent system of three linear equations in three variables. Discuss the nature of the solution set for the system if the reduced form of the augmented coefficient matrix has
(A) One leftmost 1
(B) Two leftmost 1's
(C) Three leftmost 1's
(D) Four leftmost 1's

44. Consider a system of three linear equations in three variables. Give examples of two reduced forms that are not row equivalent if the system is
(A) Consistent and dependent
(B) Inconsistent

C

Solve Problems 45–50 using Gauss–Jordan elimination.

45. $\begin{aligned} x_1 + 2x_2 - 4x_3 - x_4 &= 7 \\ 2x_1 + 5x_2 - 9x_3 - 4x_4 &= 16 \\ x_1 + 5x_2 - 7x_3 - 7x_4 &= 13 \end{aligned}$

46. $\begin{aligned} 2x_1 + 4x_2 + 5x_3 + 4x_4 &= 8 \\ x_1 + 2x_2 + 2x_3 + x_4 &= 3 \end{aligned}$

47. $\begin{aligned} x_1 - x_2 + 3x_3 - 2x_4 &= 1 \\ -2x_1 + 4x_2 - 3x_3 + x_4 &= 0.5 \\ 3x_1 - x_2 + 10x_3 - 4x_4 &= 2.9 \\ 4x_1 - 3x_2 + 8x_3 - 2x_4 &= 0.6 \end{aligned}$

48. $\begin{aligned} x_1 + x_2 + 4x_3 + x_4 &= 1.3 \\ -x_1 + x_2 - x_3 &= 1.1 \\ 2x_1 + x_3 + 3x_4 &= -4.4 \\ 2x_1 + 5x_2 + 11x_3 + 3x_4 &= 5.6 \end{aligned}$

49. $\begin{aligned} x_1 - 2x_2 + x_3 + x_4 + 2x_5 &= 2 \\ -2x_1 + 4x_2 + 2x_3 + 2x_4 - 2x_5 &= 0 \\ 3x_1 - 6x_2 + x_3 + x_4 + 5x_5 &= 4 \\ -x_1 + 2x_2 + 3x_3 + x_4 + x_5 &= 3 \end{aligned}$

50. $\begin{aligned} x_1 - 3x_2 + x_3 + x_4 + 2x_5 &= 2 \\ -x_1 + 5x_2 + 2x_3 + 2x_4 - 2x_5 &= 0 \\ 2x_1 - 6x_2 + 2x_3 + 2x_4 + 4x_5 &= 4 \\ -x_1 + 3x_2 - x_3 - x_5 &= -3 \end{aligned}$

APPLICATIONS 

Solve Problems 51–72 using Gauss–Jordan elimination.

★ **51. Puzzle.** A friend of yours came out of the post office after spending \$14.00 on 15¢, 20¢, and 35¢ stamps. If she bought 45 stamps in all, how many of each type did she buy?

★ **52. Puzzle.** A parking meter accepts only nickels, dimes, and quarters. If the meter contains 32 coins with a total value of \$6.80, how many of each type are there?

★★ **53. Chemistry.** A chemist can purchase a 10% saline solution in 500 cubic centimeter containers, a 20% saline solution in 500 cubic centimeter containers, and a 50% saline solution in 1,000 cubic centimeter containers. He needs 12,000 cubic centimeters of 30% saline solution. How many containers of each type of solution should he purchase in order to form this solution?

★★ **54. Chemistry.** Repeat Problem 53 if the 50% saline solution is available only in 1,500 cubic centimeter containers.

55. Geometry. Find a, b, and c so that the graph of the parabola with equation $y = a + bx + cx^2$ passes through the points $(-2, 3)$, $(-1, 2)$, and $(1, 6)$.

56. Geometry. Find a, b, and c so that the graph of the parabola with equation $y = a + bx + cx^2$ passes through the points $(1, 3)$, $(2, 2)$, and $(3, 5)$.

57. Geometry. Find a, b, and c so that the graph of the circle with equation $x^2 + y^2 + ax + by + c = 0$ passes through the points $(6, 2)$, $(4, 6)$, and $(-3, -1)$.

58. Geometry. Find a, b, and c so that the graph of the circle with equation $x^2 + y^2 + ax + by + c = 0$ passes through the points $(-4, 1)$, $(-1, 2)$, and $(3, -6)$.

59. Production Scheduling. A small manufacturing plant makes three types of inflatable boats: one-person, two-person, and four-person models. Each boat requires the services of three departments, as listed in the table. The cutting, assembly, and packaging departments have available a maximum of 380, 330, and 120 labor-hours per week, respectively. How many boats of each type must be produced each week for the plant to operate at full capacity?

	One-person boat	Two-person boat	Four-person boat
Cutting department	0.5 h	1.0 h	1.5 h
Assembly department	0.6 h	0.9 h	1.2 h
Packaging department	0.2 h	0.3 h	0.5 h

60. Production Scheduling. Repeat Problem 59 assuming the cutting, assembly, and packaging departments have available a maximum of 350, 330, and 115 labor-hours per week, respectively.

★ **61. Production Scheduling.** Rework Problem 59 assuming the packaging department is no longer used.

★ **62. Production Scheduling.** Rework Problem 60 assuming the packaging department is no longer used.

★ **63. Production Scheduling.** Rework Problem 59 assuming the four-person boat is no longer produced.

★ **64. Production Scheduling.** Rework Problem 60 assuming the four-person boat is no longer produced.

65. Nutrition. A dietitian in a hospital is to arrange a special diet using three basic foods. The diet is to include exactly 340 units of calcium, 180 units of iron, and 220 units of vitamin A. The number of units per ounce of each special ingredient for each of the foods is indicated in the table. How many ounces of each food must be used to meet the diet requirements?

	Units per ounce		
	Food A	**Food B**	**Food C**
Calcium	30	10	20
Iron	10	10	20
Vitamin A	10	30	20

66. Nutrition. Repeat Problem 65 if the diet is to include exactly 400 units of calcium, 160 units of iron, and 240 units of vitamin A.

★ **67. Nutrition.** Solve Problem 65 with the assumption that food C is no longer available.

★ **68. Nutrition.** Solve Problem 66 with the assumption that food C is no longer available.

★ **69. Nutrition.** Solve Problem 65 assuming the vitamin A requirement is deleted.

★ **70. Nutrition.** Solve Problem 66 assuming the vitamin A requirement is deleted.

71. Sociology. Two sociologists have grant money to study school busing in a particular city. They wish to conduct an opinion survey using 600 telephone contacts and 400 house contacts. Survey company A has personnel to do 30 telephone and 10 house contacts per hour; survey company B can handle 20 telephone and 20 house contacts per hour. How many hours should be scheduled for each firm to produce exactly the number of contacts needed?

72. Sociology. Repeat Problem 71 if 650 telephone contacts and 350 house contacts are needed.

SECTION 8-3 Systems Involving Second-Degree Equations

- Solution by Substitution
- Other Solution Methods

If a system of equations contains any equations that are not linear, then the system is called a **nonlinear system.** In this section we investigate nonlinear systems involving second-degree terms such as

$$\begin{array}{lll} x^2 + y^2 = 5 & x^2 - 2y^2 = 2 & x^2 + 3xy + y^2 = 20 \\ 3x + y = 1 & xy = 2 & xy - y^2 = 0 \end{array}$$

It can be shown that such systems have at most four solutions, some of which may be imaginary. Since we are interested in finding both real and imaginary solutions to the systems we consider, we now assume that the replacement set for each variable is the set of complex numbers, rather than the set of real numbers.

• Solution by Substitution

The substitution method used to solve linear systems of two equations in two variables is also an effective method for solving nonlinear systems. This process is best illustrated by examples.

EXAMPLE 1 Solving a Nonlinear System by Substitution

Solve the system: $x^2 + y^2 = 5$
$3x + y = 1$

Solution Solve the second equation for y in terms of x; then substitute for y in the first equation to obtain an equation that involves x alone.

$$3x + y = 1$$

$$y = 1 - 3x \quad \text{Substitute this expression for } y \text{ in the first equation.}$$

$$x^2 + y^2 = 5$$

$$x^2 + (1 - 3x)^2 = 5$$

$$10x^2 - 6x - 4 = 0 \quad \text{Simplify and write in standard quadratic form.}$$

$$5x^2 - 3x - 2 = 0 \quad \text{Divide through by 2 to simplify further.}$$

$$(x - 1)(5x + 2) = 0$$

$$x = 1, -\tfrac{2}{5}$$

If we substitute these values back into the equation $y = 1 - 3x$, we obtain two solutions to the system:

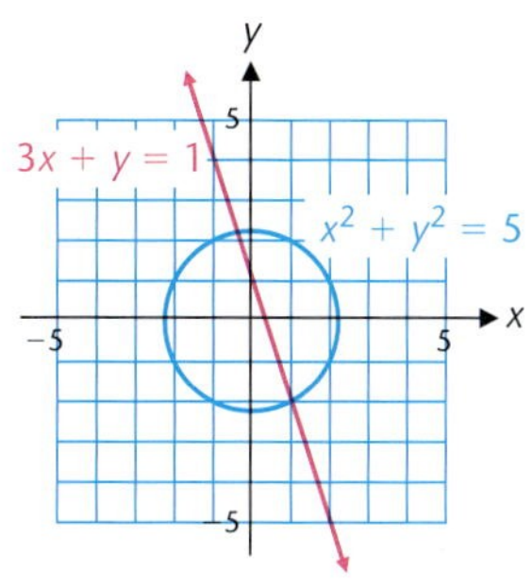

FIGURE 1

$$x = 1 \qquad\qquad x = -\tfrac{2}{5}$$

$$y = 1 - 3(1) = -2 \qquad y = 1 - 3(-\tfrac{2}{5}) = \tfrac{11}{5}$$

A check, which you should provide, verifies that $(1, -2)$ and $(-\frac{2}{5}, \frac{11}{5})$ are both solutions to the system. These solutions are illustrated in Figure 1. However, if we substitute the values of x back into the equation $x^2 + y^2 = 5$, we obtain

$$
\begin{array}{cc}
x = 1 & x = -\frac{2}{5} \\
1^2 + y^2 = 5 & (-\frac{2}{5})^2 + y^2 = 5 \\
y^2 = 4 & y^2 = \frac{121}{25} \\
y = \pm 2 & y = \pm\frac{11}{5}
\end{array}
$$

It appears that we have found two additional solutions, $(1, 2)$ and $(-\frac{2}{5}, -\frac{11}{5})$. But neither of these solutions satisfies the equation $3x + y = 1$, which you should verify. So, neither is a solution of the original system. We have produced two **extraneous roots,** apparent solutions that do not actually satisfy both equations in the system. This is a common occurrence when solving nonlinear systems.

It is always very important to check the solutions of any nonlinear system to ensure that extraneous roots have not been introduced.

Matched Problem 1 Solve the system:
$$
\begin{aligned}
x^2 + y^2 &= 10 \\
2x + y &= 1
\end{aligned}
$$

EXPLORE-DISCUSS 1 In Example 1, we saw that the line $3x + y = 1$ intersected the circle $x^2 + y^2 = 5$ in two points.

(A) Consider the system

$$
\begin{aligned}
x^2 + y^2 &= 5 \\
3x + y &= 10
\end{aligned}
$$

Graph both equations in the same coordinate system. Are there any real solutions to this system? Are there any complex solutions? Find any real or complex solutions.

(B) Consider the family of lines given by

$$3x + y = b \qquad b \text{ any real number}$$

What do all these lines have in common? Illustrate graphically the lines in this family that intersect the circle $x^2 + y^2 = 5$ in exactly one point. How many such lines are there? What are the corresponding value(s) of b? What are the intersection points? How are these lines related to the circle?

EXAMPLE 2 Solving a Nonlinear System by Substitution

Solve: $x^2 - 2y^2 = 2$
$xy = 2$

Solution Solve the second equation for y, substitute in the first equation, and proceed as before.

$$xy = 2$$

$$y = \frac{2}{x}$$

$$x^2 - 2\left(\frac{2}{x}\right)^2 = 2$$

$$x^2 - \frac{8}{x^2} = 2$$

$$x^4 - 2x^2 - 8 = 0 \qquad \text{Multiply both sides by } x^2 \text{ and simplify.}$$

$$u^2 - 2u - 8 = 0 \qquad \text{Substitute } u = x^2 \text{ to transform to quadratic form and solve.}$$

$$(u - 4)(u + 2) = 0$$

$$u = 4, -2$$

Thus,

$$x^2 = 4 \quad \text{or} \quad x^2 = -2$$

$$x = \pm 2 \qquad x = \pm\sqrt{-2} = \pm i\sqrt{2}$$

For $x = 2$, $y = \frac{2}{2} = 1$. For $x = i\sqrt{2}$, $y = \frac{2}{i\sqrt{2}} = -i\sqrt{2}$.

For $x = -2$, $y = \frac{2}{-2} = -1$. For $x = -i\sqrt{2}$, $y = \frac{2}{-i\sqrt{2}} = i\sqrt{2}$.

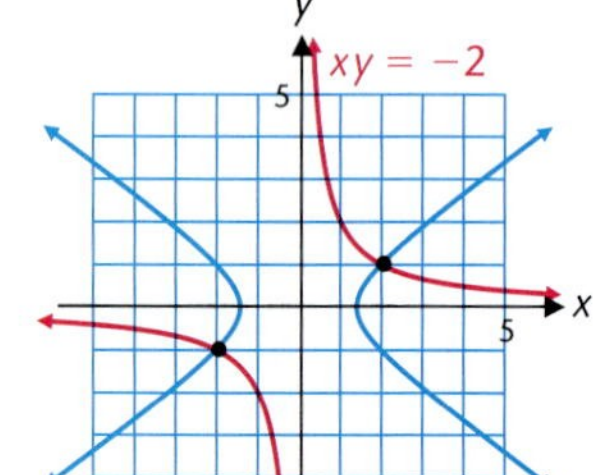

FIGURE 2

Thus, the four solutions to this system are (2, 1), (−2, −1), $(i\sqrt{2}, -i\sqrt{2})$, and $(-i\sqrt{2}, i\sqrt{2})$. Notice that two of the solutions involve imaginary numbers. These imaginary solutions cannot be illustrated graphically (see Fig. 2); however, they do satisfy both equations in the system (verify this).

Matched Problem 2 Solve: $3x^2 - y^2 = 6$
$xy = 3$

EXPLORE-DISCUSS 2 (A) Refer to the system in Example 2. Could a graphing utility be used to find the real solutions of this system? The imaginary solutions?

(B) In general, explain why graphic approximation techniques can be used to approximate the real solutions of a system, but not the complex solutions.

EXAMPLE 3 Design

An engineer is to design a rectangular computer screen with a 19-inch diagonal and a 175-square-inch area. Find the dimensions of the screen to the nearest tenth of an inch.

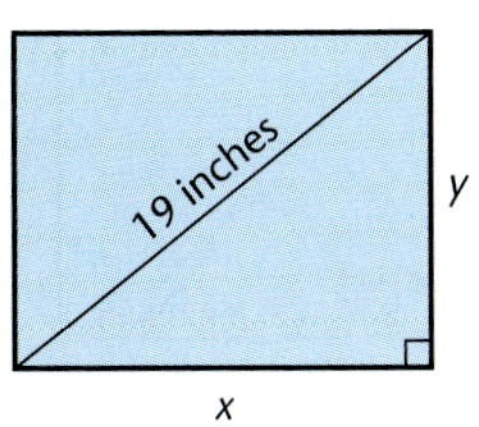

FIGURE 3

Solution Sketch a rectangle letting x be the width and y the height (Fig. 3). We obtain the following system using the Pythagorean theorem and the formula for the area of a rectangle:

$$x^2 + y^2 = 19^2$$
$$xy = 175$$

This system is solved using the procedures outlined in Example 2. However, in this case, we are only interested in real solutions. We start by solving the second equation for y in terms of x and substituting the result into the first equation.

$$y = \frac{175}{x}$$

$$x^2 + \frac{175^2}{x^2} = 19^2$$

$$x^4 + 30{,}625 = 361x^2 \quad \text{Multiply both sides by } x^2 \text{ and simplify}$$

$$x^4 - 361x^2 + 30{,}625 = 0 \quad \text{Quadratic in } x^2$$

Solve the last equation for x^2 using the quadratic formula, then solve for x:

$$x = \sqrt{\frac{361 \pm \sqrt{361^2 - 4(1)(30{,}625)}}{2}}$$
$$= 15.0 \text{ inches or } 11.7 \text{ inches}$$

Substitute each choice of x into $y = 175/x$ to find the corresponding y values:

For $x = 15.0$ inches,

$$y = \frac{175}{15} = 11.7 \text{ inches}$$

For $x = 11.7$ inches,

$$y = \frac{175}{11.7} = 15.0 \text{ inches}$$

Assuming the screen is wider than it is high, the dimensions are 15.0 by 11.7 inches.

Matched Problem 3 An engineer is to design a rectangular television screen with a 21-inch diagonal and a 209-square-inch area. Find the dimensions of the screen to the nearest tenth of an inch.

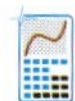

Since Example 3 is only concerned with real solutions, graphic techniques can also be used to approximate the solutions (see Fig. 4). As we saw in Section 2-1, graphing a circle on a graphing utility requires two functions, one for the upper half of the circle and another for the lower half. [Note: since x and y must be nonnegative real numbers, we ignore the intersection points in the third quadrant—see Fig 4(a).]

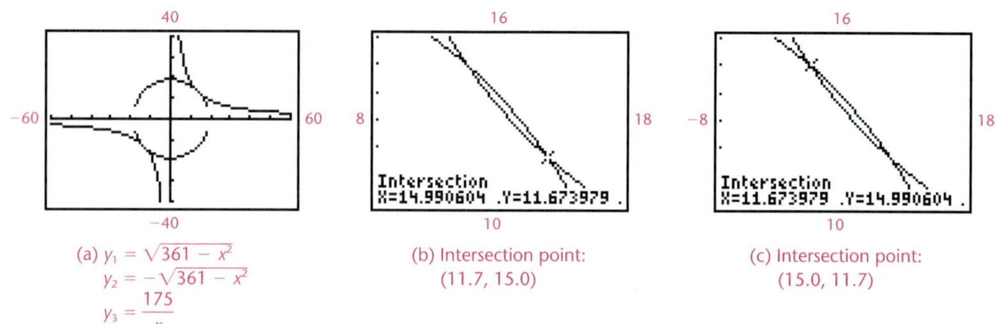

FIGURE 4 Graphic solution of $x^2 + y^2 = 19^2$, $xy = 175$.

(a) $y_1 = \sqrt{361 - x^2}$, $y_2 = -\sqrt{361 - x^2}$, $y_3 = \dfrac{175}{x}$

(b) Intersection point: (11.7, 15.0)

(c) Intersection point: (15.0, 11.7)

• Other Solution Methods

We now look at some other techniques for solving nonlinear systems of equations.

EXAMPLE 4 Solving a Nonlinear System by Elimination

Solve:
$$
\begin{aligned}
x^2 - y^2 &= 5\\
x^2 + 2y^2 &= 17
\end{aligned}
$$

Solution This type of system can be solved using elimination by addition. Multiply the second equation by -1 and add:

$$
\begin{aligned}
x^2 - y^2 &= 5\\
-x^2 - 2y^2 &= -17\\
\hline
-3y^2 &= -12\\
y^2 &= 4\\
y &= \pm 2
\end{aligned}
$$

Now substitute $y = 2$ and $y = -2$ back into either original equation to find x.

For $y = 2$,

$$x^2 - (2)^2 = 5$$
$$x = \pm 3$$

For $y = -2$,

$$x^2 - (-2)^2 = 5$$
$$x = \pm 3$$

Thus, $(3, -2)$, $(3, 2)$, $(-3, -2)$, and $(-3, 2)$, are the four solutions to the system. The check of the solutions is left to you.

Matched Problem 4 Solve:
$$\begin{aligned} 2x^2 - 3y^2 &= 5 \\ 3x^2 + 4y^2 &= 16 \end{aligned}$$

EXAMPLE 5 **Solving a Nonlinear System Using Factoring and Substitution**

Solve:
$$\begin{aligned} x^2 + 3xy + y^2 &= 20 \\ xy - y^2 &= 0 \end{aligned}$$

Solution Factor the left side of the equation that has a 0 constant term:

$$xy - y^2 = 0$$

$$y(x - y) = 0$$

$$y = 0 \qquad \text{or} \qquad y = x$$

Thus, the original system is equivalent to the two systems:

$$\begin{aligned} y &= 0 \\ x^2 + 3xy + y^2 &= 20 \end{aligned} \qquad \text{or} \qquad \begin{aligned} y &= x \\ x^2 + 3xy + y^2 &= 20 \end{aligned}$$

These systems are solved by substitution.

First System

$$\begin{aligned} y &= 0 \\ x^2 + 3xy + y^2 &= 20 \end{aligned}$$

Substitute $y = 0$ in the second equation, and solve for x.

$$\begin{aligned} x^2 + 3x(0) + (0)^2 &= 20 \\ x^2 &= 20 \\ x &= \pm\sqrt{20} = \pm 2\sqrt{5} \end{aligned}$$

Second System

$$\begin{aligned} y &= x \\ x^2 + 3xy + y^2 &= 20 \end{aligned}$$

Substitute $y = x$ in the second equation and solve for x.

$$\begin{aligned} x^2 + 3xx + x^2 &= 20 \\ 5x^2 &= 20 \\ x^2 &= 4 \\ x &= \pm 2 \end{aligned}$$

Substitute these values back into $y = x$ to find y.

For $x = 2$, $y = 2$. For $x = -2$, $y = -2$.

The solutions for the original system are $(2\sqrt{5}, 0)$, $(-2\sqrt{5}, 0)$, $(2, 2)$, and $(-2, -2)$. The check of the solutions is left to you.

Matched Problem 5 Solve:

$$x^2 - xy + y^2 = 9$$
$$2x^2 - xy = 0$$

Example 5 is somewhat specialized. However, it suggests a procedure that is effective for some problems.

EXAMPLE 6 Graphic Approximations of Real Solutions

Use a graphing utility to approximate real solutions to two decimal places:

$$x^2 - 4xy + y^2 = 12$$
$$2x^2 + 2xy + y^2 = 6$$

Solution Before we can enter these equations in our graphing utility, we must solve for y:

$$x^2 - 4xy + y^2 = 12 \qquad 2x^2 + 2xy + y^2 = 6$$
$$y^2 - 4xy + (x^2 - 12) = 0 \qquad y^2 + 2xy + (2x^2 - 6) = 0$$

Applying the quadratic formula to each equation, we have

$$y = \frac{4x \pm \sqrt{16x^2 - 4(x^2 - 12)}}{2} \qquad y = \frac{-2x \pm \sqrt{4x^2 - 4(2x^2 - 6)}}{2}$$
$$= \frac{4x \pm \sqrt{12x^2 + 48}}{2} \qquad = \frac{-2x \pm \sqrt{24 - 4x^2}}{2}$$
$$= 2x \pm \sqrt{3x^2 + 12} \qquad = -x \pm \sqrt{6 - x^2}$$

Since each equation has two solutions, we must enter four functions in the graphing utility, as shown in Figure 5(a). Examining the graph in Figure 5(b), we see that there are four intersection points. Using the built-in intersection routine repeatedly (details omitted), we find that the solutions to two decimal places are $(-2.10, 0.83)$, $(-0.37, 2.79)$, $(0.37, -2.79)$, and $(2.10, -0.83)$.

FIGURE 5

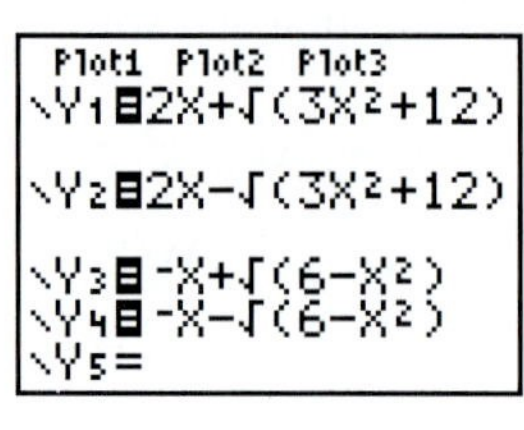

(a)

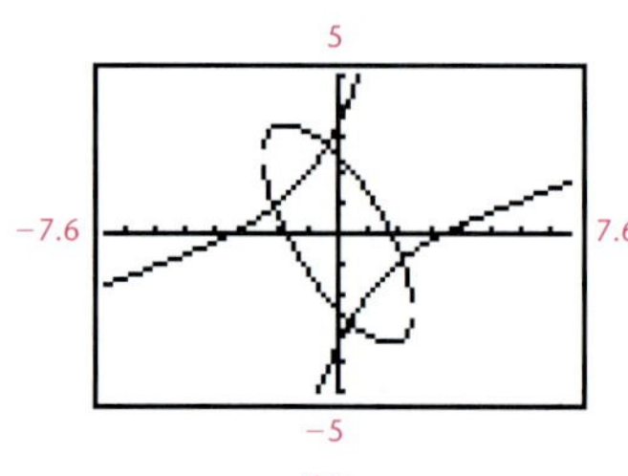

(b)

Matched Problem 6 Use a graphing utility to approximate real solutions to two decimal places:

$$x^2 + 8xy + y^2 = 70$$
$$2x^2 - 2xy + y^2 = 20$$

Answers to Matched Problems

1. $(-1, 3), (\frac{9}{5}, -\frac{13}{5})$ **2.** $(\sqrt{3}, \sqrt{3}), (-\sqrt{3}, -\sqrt{3}), (i, -3i), (-i, 3i)$ **3.** 17.1 by 12.2 in
4. $(2, 1), (2, -1), (-2, 1), (-2, -1)$ **5.** $(0, 3), (0, -3), (\sqrt{3}, 2\sqrt{3}), (-\sqrt{3}, -2\sqrt{3})$
6. $(-3.89, -1.68), (-0.96, -5.32), (0.96, 5.32), (3.89, 1.68)$

EXERCISE 8-3

A

Solve each system in Problems 1–12.

1. $x^2 + y^2 = 169$, $x = -12$

2. $x^2 + y^2 = 25$, $y = -4$

3. $8x^2 - y^2 = 16$, $y = 2x$

4. $y^2 = 2x$, $x = y - \frac{1}{2}$

5. $3x^2 - 2y^2 = 25$, $x + y = 0$

6. $x^2 + 4y^2 = 32$, $x + 2y = 0$

7. $y^2 = x$, $x - 2y = 2$

8. $x^2 = 2y$, $3x = y + 2$

9. $2x^2 + y^2 = 24$, $x^2 - y^2 = -12$

10. $x^2 - y^2 = 3$, $x^2 + y^2 = 5$

11. $x^2 + y^2 = 10$, $16x^2 + y^2 = 25$

12. $x^2 - 2y^2 = 1$, $x^2 + 4y^2 = 25$

B

Solve each system in Problems 13–24.

13. $xy - 4 = 0$, $x - y = 2$

14. $xy - 6 = 0$, $x - y = 4$

15. $x^2 + 2y^2 = 6$, $xy = 2$

16. $2x^2 + y^2 = 18$, $xy = 4$

17. $2x^2 + 3y^2 = -4$, $4x^2 + 2y^2 = 8$

18. $2x^2 - 3y^2 = 10$, $x^2 + 4y^2 = -17$

19. $x^2 - y^2 = 2$, $y^2 = x$

20. $x^2 + y^2 = 20$, $x^2 = y$

21. $x^2 + y^2 = 9$, $x^2 = 9 - 2y$

22. $x^2 + y^2 = 16$, $y^2 = 4 - x$

23. $x^2 - y^2 = 3$, $xy = 2$

24. $y^2 = 5x^2 + 1$, $xy = 2$

An important type of calculus problem is to find the area between the graphs of two functions. To solve some of these problems it is necessary to find the coordinates of the points of intersections of the two graphs. In Problems 25–32, find the coordinates of the points of intersections of the two given equations.

25. $y = 5 - x^2, y = 2 - 2x$

26. $y = 5x - x^2, y = x + 3$

27. $y = x^2 - x, y = 2x$

28. $y = x^2 + 2x, y = 3x$

29. $y = x^2 - 6x + 9, y = 5 - x$

30. $y = x^2 + 2x + 3, y = 2x + 4$

31. $y = 8 + 4x - x^2, y = x^2 - 2x$

32. $y = x^2 - 4x - 10, y = 14 - 2x - x^2$

33. Consider the circle with equation $x^2 + y^2 = 5$ and the family of lines given by $2x - y = b$, where b is any real number.
(A) Illustrate graphically the lines in this family that intersect the circle in exactly one point, and describe the relationship between the circle and these lines.
(B) Find the values of b corresponding to the lines in part A, and find the intersection points of the lines and the circle.
(C) How is the line with equation $x + 2y = 0$ related to this family of lines? How could this line be used to find the intersection points in part B?

34. Consider the circle with equation $x^2 + y^2 = 25$ and the family of lines given by $3x + 4y = b$, where b is any real number.
(A) Illustrate graphically the lines in this family that intersect the circle in exactly one point, and describe the relationship between the circle and these lines.
(B) Find the values of b corresponding to the lines in part A, and find the intersection points of the lines and the circle.

(C) How is the line with equation $4x - 3y = 0$ related to this family of lines? How could this line be used to find the intersection points and the values of b in part B?

C

Solve each system in Problems 35–42.

35. $2x + 5y + 7xy = 8$
$xy - 3 = 0$

36. $2x + 3y + xy = 16$
$xy - 5 = 0$

37. $x^2 - 2xy + y^2 = 1$
$x - 2y = 2$

38. $x^2 + xy - y^2 = -5$
$y - x = 3$

39. $2x^2 - xy + y^2 = 8$
$x^2 - y^2 = 0$

40. $x^2 + 2xy + y^2 = 36$
$x^2 - xy = 0$

41. $x^2 + xy - 3y^2 = 3$
$x^2 + 4xy + 3y^2 = 0$

42. $x^2 - 2xy + 2y^2 = 16$
$x^2 - y^2 = 0$

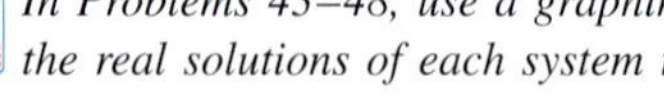

In Problems 43–48, use a graphing utility to approximate the real solutions of each system to two decimal places.

43. $-x^2 + 2xy + y^2 = 1$
$3x^2 - 4xy + y^2 = 2$

44. $-x^2 + 4xy + y^2 = 2$
$8x^2 - 2xy + y^2 = 9$

45. $3x^2 - 4xy - y^2 = 2$
$2x^2 + 2xy + y^2 = 9$

46. $5x^2 + 4xy + y^2 = 4$
$4x^2 - 2xy + y^2 = 16$

47. $2x^2 - 2xy + y^2 = 9$
$4x^2 - 4xy + y^2 + x = 3$

48. $2x^2 + 2xy + y^2 = 12$
$4x^2 - 4xy + y^2 + x + 2y = 9$

APPLICATIONS

49. Numbers. Find two numbers such that their sum is 3 and their product is 1.

50. Numbers. Find two numbers such that their difference is 1 and their product is 1. (Let x be the larger number and y the smaller number.)

51. Geometry. Find the lengths of the legs of a right triangle with an area of 30 square inches if its hypotenuse is 13 inches long.

52. Geometry. Find the dimensions of a rectangle with an area of 32 square meters if its perimeter is 36 meters long.

53. Design. An engineer is designing a small portable television set. According to the design specifications, the set must have a rectangular screen with a 7.5-inch diagonal and an area of 27 square inches. Find the dimensions of the screen.

54. Design. An artist is designing a logo for a business in the shape of a circle with an inscribed rectangle. The diameter of the circle is 6.5 inches, and the area of the rectangle is 15 square inches. Find the dimensions of the rectangle.

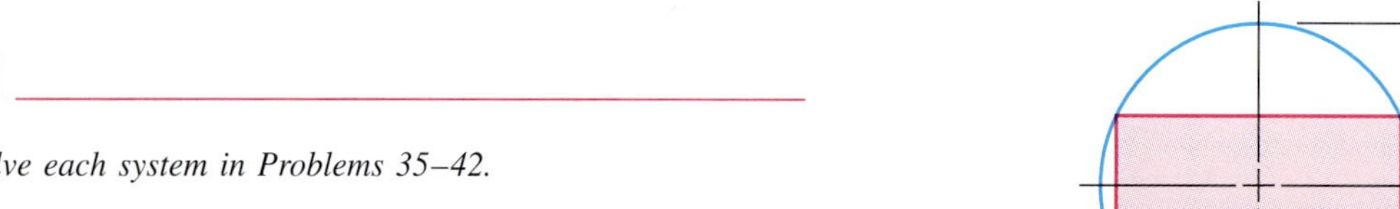

★ **55. Construction.** A rectangular swimming pool with a deck 5 feet wide is enclosed by a fence as shown in the figure. The surface area of the pool is 572 square feet, and the total area enclosed by the fence (including the pool and the deck) is 1,152 square feet. Find the dimensions of the pool.

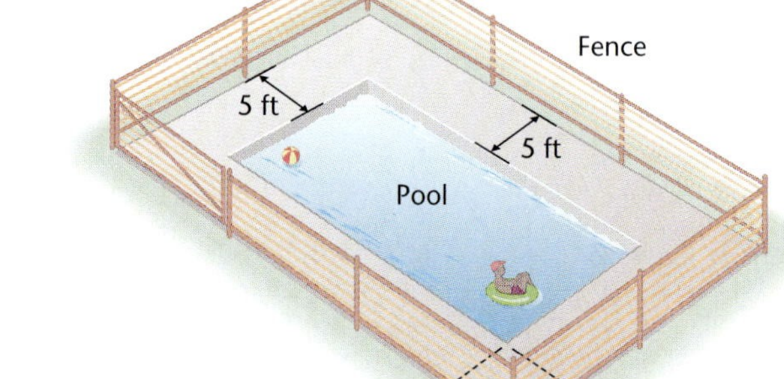

★ **56. Construction.** An open-topped rectangular box is formed by cutting a 6-inch square from each corner of a rectangular piece of cardboard and bending up the ends and sides. The area of the cardboard before the corners are removed is 768 square inches, and the volume of the box is 1,440 cubic inches. Find the dimensions of the original piece of cardboard.

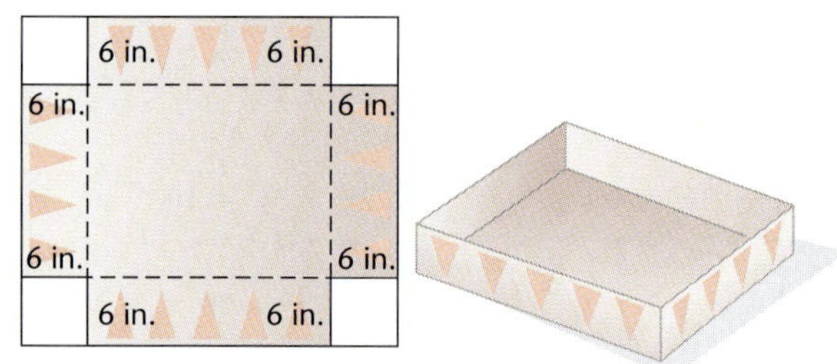

★★ **57. Transportation.** Two boats leave Bournemouth, England, at the same time and follow the same route on the 75-mile trip across the English Channel to Cherbourg, France. The average speed of boat *A* is 5 miles per hour greater than the average speed of boat *B*. Consequently, boat *A* arrives at Cherbourg 30 minutes before boat *B*. Find the average speed of each boat.

★★ **58. Transportation.** Bus *A* leaves Milwaukee at noon and travels west on Interstate 94. Bus *B* leaves Milwaukee 30 minutes later, travels the same route, and overtakes bus *A* at a point 210 miles west of Milwaukee. If the average speed of bus *B* is 10 miles per hour greater than the average speed of bus *A*, at what time did bus *B* overtake bus *A*?

SECTION 8-4 Systems of Linear Inequalities in Two Variables

- Graphing Linear Inequalities in Two Variables
- Solving Systems of Linear Inequalities Graphically
- Application

Many applications of mathematics involve systems of inequalities rather than systems of equations. A graph is often the most convenient way to represent the solutions of a system of inequalities in two variables. In this section, we discuss techniques for graphing both a single linear inequality in two variables and a system of linear inequalities in two variables.

• Graphing Linear Inequalities in Two Variables

We know how to graph first-degree equations such as

$$y = 2x - 3 \qquad \text{and} \qquad 2x - 3y = 5$$

but how do we graph first-degree inequalities such as

$$y \le 2x - 3 \qquad \text{and} \qquad 2x - 3y > 5$$

Actually, graphing these inequalities is almost as easy as graphing the equations. But before we begin, we must discuss some important subsets of a plane in a rectangular coordinate system.

A line divides a plane into two halves called **half-planes.** A vertical line divides a plane into **left** and **right half-planes** [Fig. 1(a)]; a nonvertical line divides a plane into **upper** and **lower half-planes** [Fig. 1(b)].

FIGURE 1 Half-planes.

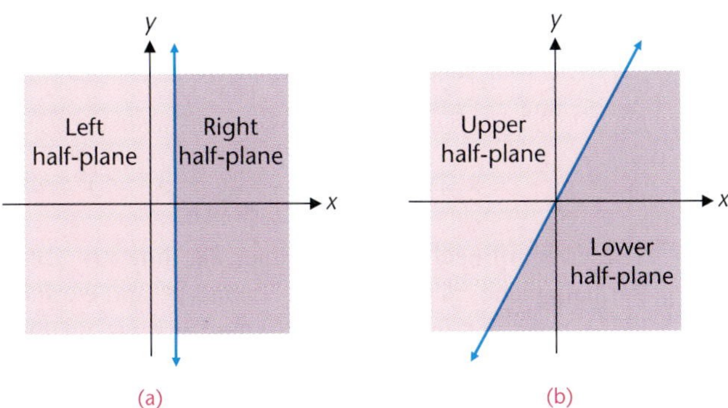

EXPLORE-DISCUSS 1 Consider the following linear equation and related linear inequalities:

(1) $2x - 3y = 12$ (2) $2x - 3y < 12$ (3) $2x - 3y > 12$

(A) Graph the line with equation (1).

(B) Find the point on this line with x coordinate 3 and draw a vertical line through this point. Discuss the relationship between the y coordinates of the points on this line and statements (1), (2), and (3).

(C) Repeat part B for $x = -3$. For $x = 9$.

(D) Based on your observations in parts B and C, write a verbal description of all the points in the plane that satisfy equation (1), those that satisfy inequality (2), and those that satisfy inequality (3).

Now let's investigate the half-planes determined by the linear equation $y = 2x - 3$. We start by graphing $y = 2x - 3$ (Fig. 2). For any given value of x, there is exactly one value for y such that (x, y) lies on the line. For the same x, if the point (x, y) is below the line, then $y < 2x - 3$. Thus, the lower half-plane corresponds to the solution of the inequality $y < 2x - 3$. Similarly, the upper half-plane corresponds to the solution of the inequality $y > 2x - 3$, as shown in Figure 2.

FIGURE 2

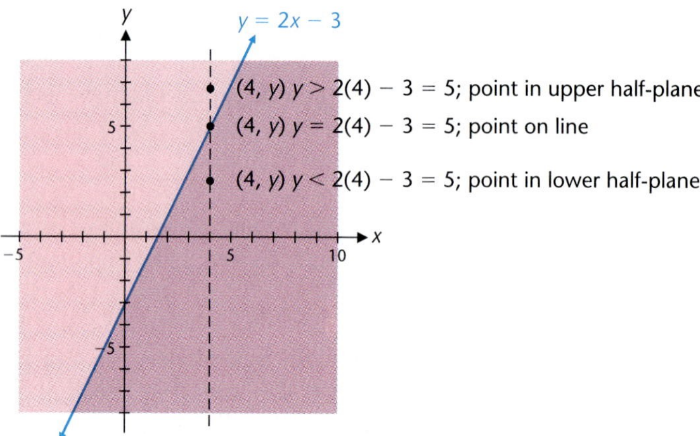

The four inequalities formed from $y = 2x - 3$ by replacing the = sign by $\geq$, $>$, $\leq$, and $<$, respectively, are

$$y \geq 2x - 3 \qquad y > 2x - 3 \qquad y \leq 2x - 3 \qquad y < 2x - 3$$

The graph of each is a half-plane. The line $y = 2x - 3$, called the **boundary line** for the half-plane, is included for $\geq$ and $\leq$ and excluded for $>$ and $<$. In Figure 3, the half-planes are indicated with small arrows on the graph of $y = 2x - 3$ and then graphed as shaded regions. Included boundary lines are shown as solid lines, and excluded boundary lines are shown as dashed lines.

FIGURE 3

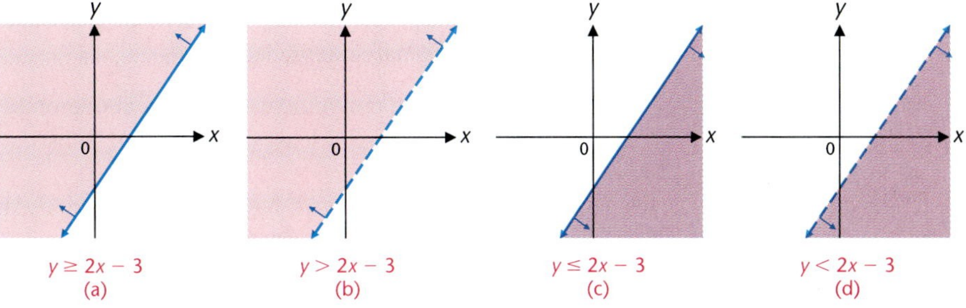

Theorem 1 **Graphs of Linear Inequalities in Two Variables**

The graph of a linear inequality

$$Ax + By < C \qquad \text{or} \qquad Ax + By > C$$

with $B \neq 0$, is either the upper half-plane or the lower half-plane (but not both) determined by the line $Ax + By = C$.

If $B = 0$, then the graph of

$$Ax < C \qquad \text{or} \qquad Ax > C$$

is either the left half-plane or the right half-plane (but not both) determined by the line $Ax = C$.

As a consequence of Theorem 1, we state a simple and fast mechanical procedure for graphing linear inequalities.

Procedure for Graphing Linear Inequalities in Two Variables

Step 1. Graph $Ax + By = C$ as a dashed line if equality is not included in the original statement or as a solid line if equality is included.

Step 2. Choose a test point anywhere in the plane not on the line and substitute the coordinates into the inequality. The origin (0, 0) often requires the least computation.

Step 3. The graph of the original inequality includes the half-plane containing the test point if the inequality is satisfied by that point, or the half-plane not containing that point if the inequality is not satisfied by that point.

EXAMPLE 1 **Graphing a Linear Inequality**

Graph: $3x - 4y \leq 12$

Solution

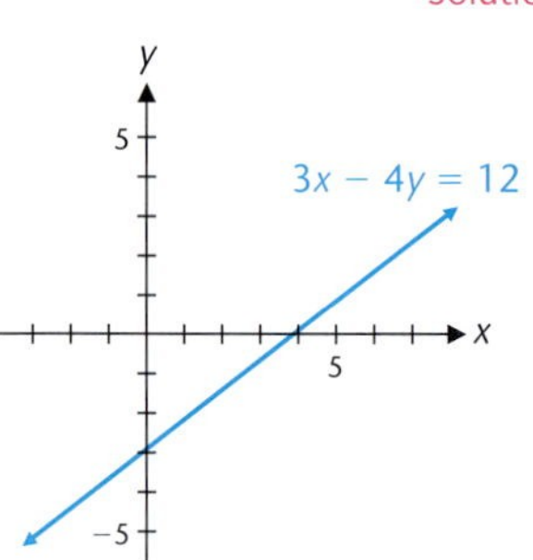

FIGURE 4

Step 1. Graph $3x - 4y = 12$ as a solid line, since equality is included in the original statement (Fig. 4).

Step 2. Pick a convenient test point above or below the line. The origin (0, 0) requires the least computation. Substituting (0, 0) into the inequality

$$3x - 4y \le 12$$

$$3(0) - 4(0) = 0 \le 12$$

produces a true statement; therefore, (0, 0) is in the solution set.

Step 3. The line $3x - 4y = 12$ and the half-plane containing the origin form the graph of $3x - 4y \le 12$ (Fig. 5).

FIGURE 5

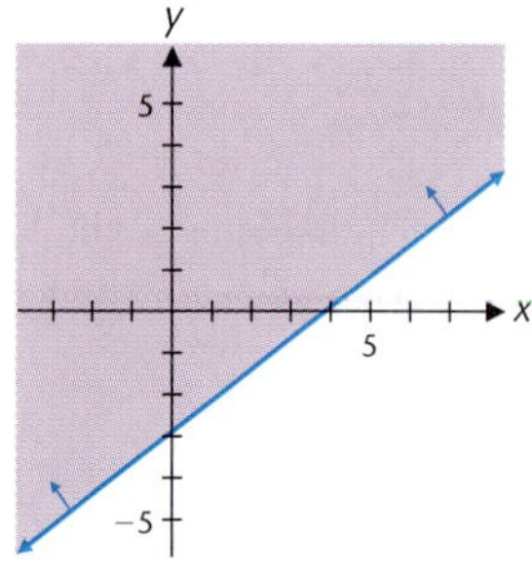

Matched Problem 1 Graph: $2x + 3y < 6$

EXAMPLE 2 Graphing a Linear Inequality

Graph: (A) $y > -3$ (B) $2x \le 5$

Solutions

(A) The graph of $y > -3$ is shown in Figure 6.

(B) The graph of $2x \le 5$ is shown in Figure 7.

FIGURE 6

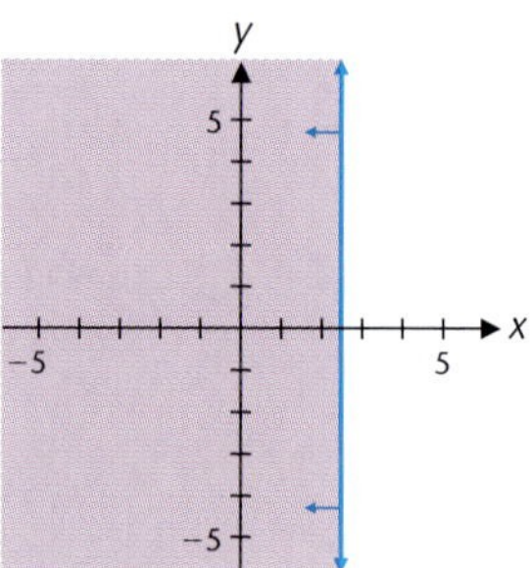

FIGURE 7

Matched Problem 2 Graph: (A) $y \le 2$ (B) $3x > -8$

• Solving Systems of Linear Inequalities Graphically

We now consider systems of linear inequalities such as

$$\begin{aligned} x + y &\geq 6 \\ 2x - y &\geq 0 \end{aligned} \qquad \text{and} \qquad \begin{aligned} 2x + y &\leq 22 \\ x + y &\leq 13 \\ 2x + 5y &\leq 50 \\ x &\geq 0 \\ y &\geq 0 \end{aligned}$$

We wish to **solve** such systems **graphically**—that is, to find the graph of all ordered pairs of real numbers (x, y) that simultaneously satisfy all the inequalities in the system. The graph is called the **solution region** for the system. To find the solution region, we graph each inequality in the system and then take the intersection of all the graphs. To simplify the discussion that follows, **we will consider only systems of linear inequalities where equality is included in each statement in the system.**

EXAMPLE 3 Solving a System of Linear Inequalities Graphically

Solve the following system of linear inequalities graphically:

$$\begin{aligned} x + y &\geq 6 \\ 2x - y &\geq 0 \end{aligned}$$

Solution First, graph the line $x + y = 6$ and shade the region that satisfies the inequality $x + y \geq 6$. This region is shaded in blue in Figure 8(a). Next, graph the line $2x - y = 0$ and shade the region that satisfies the inequality $2x - y \geq 0$. This region is shaded in red in Figure 8(a). The solution region for the system of inequalities is the intersection of these two regions. This is the region shaded in both red and blue in Figure 8(a), which is redrawn in Figure 8(b) with only the solution region shaded for clarity. The coordinates of any point in the shaded region of Figure 8(b) specify a solution to the system. For example, the points (2, 4), (6, 3), and (7.43, 8.56) are three of infinitely many solutions, as can be easily checked. The intersection point (2, 4) can be obtained by solving the equations $x + y = 6$ and $2x - y = 0$ simultaneously.

FIGURE 8

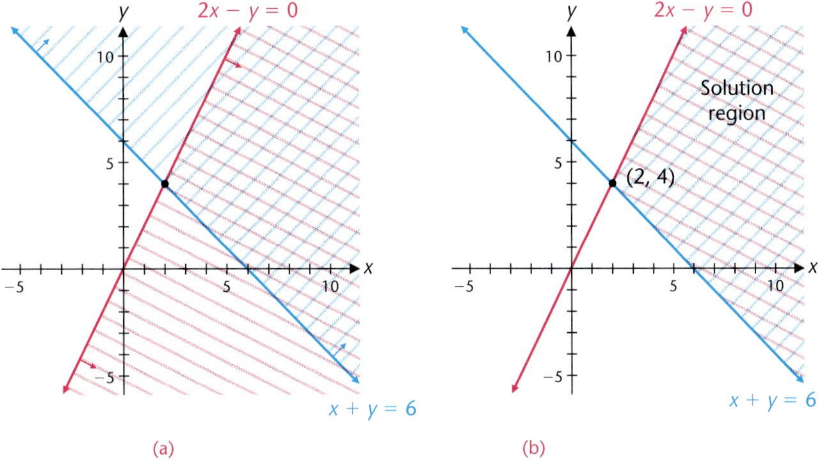

Matched Problem 3 Solve the following system of linear inequalities graphically:

$$\begin{aligned} 3x + y &\le 21 \\ x - 2y &\le 0 \end{aligned}$$

EXPLORE-DISCUSS 2 Refer to Example 3. Graph each boundary line and shade the regions obtained by reversing each inequality. That is, shade the region of the plane that corresponds to the inequality $x + y < 6$ and then shade the region that corresponds to the inequality $2x - y < 0$. What portion of the plane is left unshaded? Compare this method with the one used in the solution to Example 3.

The points of intersection of the lines that form the boundary of a solution region play a fundamental role in the solution of linear programming problems, which are discussed in the next section.

Definition 1

Corner Point

A **corner point** of a solution region is a point in the solution region that is the intersection of two boundary lines.

The point (2, 4) is the only corner point of the solution region in Example 3; see figure (b).

EXAMPLE 4 Solving a System of Linear Inequalities Graphically

Solve the following system of linear inequalities graphically, and find the corner points.

$$\begin{aligned} 2x + y &\le 22 \\ x + y &\le 13 \\ 2x + 5y &\le 50 \\ x &\ge 0 \\ y &\ge 0 \end{aligned}$$

Solution The inequalities $x \ge 0$ and $y \ge 0$, called **nonnegative restrictions,** occur frequently in applications involving systems of inequalities since x and y often represent quantities that can't be negative—number of units produced, number of hours worked, etc. The solution region lies in the first quadrant, and we can restrict our attention to that portion of the plane. First, we graph the lines

$$2x + y = 22$$ Find the x and y intercepts of each line; then sketch the line through these points, as shown in Figure 9.

$$x + y = 13$$

$$2x + 5y = 50$$

Next, choosing (0, 0) as a test point, we see that the graph of each of the first three inequalities in the system consists of its corresponding line and the half-plane lying below the line, as indicated by the arrows in Figure 9. Thus, the solution region of the system consists of the points in the first quadrant that simultaneously lie on or below all three of these lines—see Figure 9.

FIGURE 9

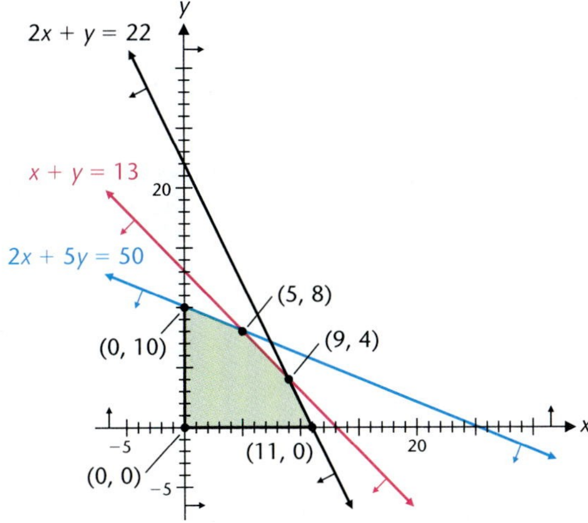

The corner points (0, 0), (0, 10), and (11, 0) can be determined from the graph. The other two corner points are determined as follows:

Solve the system

$$\begin{aligned} 2x + 5y &= 50 \\ y + \quad y &= 13 \end{aligned}$$

to obtain (5, 8).

Solve the system

$$\begin{aligned} 2x + y &= 22 \\ x + y &= 13 \end{aligned}$$

to obtain (9, 4).

Note that the lines $2x + 5y = 50$ and $2x + y = 22$ also intersect, but the intersection point is not part of the solution region, and hence, is not a corner point.

Matched Problem 4 Solve the following system of linear inequalities graphically, and find the corner points:

$$\begin{aligned} 5x + y &\geq 20 \\ x + y &\geq 12 \\ x + 3y &\geq 18 \\ x &\geq 0 \\ y &\geq 0 \end{aligned}$$

As we saw in Section 8-1, a graphing utility is a useful tool for graphing lines and finding intersection points. Most graphing utilities also have some limited capabilities for shading regions that satisfy inequalities. Figure 10(a) shows a solution for Example 3 that could be produced on most graphing utilities. On newer models (for example, the TI-83 and TI-96), it is a simple operation to shade the region above or below a graph. Using this option to shade the points that do not satisfy a given inequality, as discussed in Explore-Discuss 2, clearly displays the solution to a system of inequalities. Figure 10(b) shows the result of applying this method to Example 4.

FIGURE 10

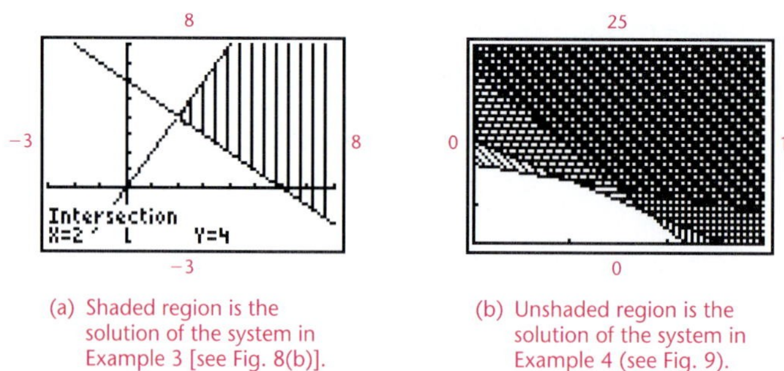

(a) Shaded region is the solution of the system in Example 3 [see Fig. 8(b)].

(b) Unshaded region is the solution of the system in Example 4 (see Fig. 9).

If we compare the solution regions of Examples 3 and 4, we see that there is a fundamental difference between these two regions. We can draw a circle around the solution region in Example 4. However, it is impossible to include all the points in the solution region in Example 3 in any circle, no matter how large we draw it. This leads to the following definition.

DEFINITION 2 **Bounded and Unbounded Solution Regions**

A solution region of a system of linear inequalities is **bounded** if it can be enclosed within a circle. If it cannot be enclosed within a circle, then it is **unbounded.**

Thus, the solution region for Example 4 is bounded and the solution region for Example 3 is unbounded. This definition will be important in the next section.

• Application

EXAMPLE 5 **Production Scheduling**

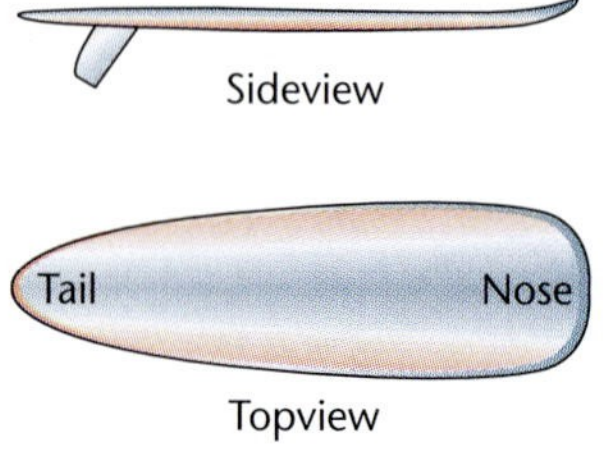

A manufacturer of surfboards makes a standard model and a competition model. Each standard board requires 6 labor-hours for fabricating and 1 labor-hour for finishing. Each competition board requires 8 labor-hours for fabricating and 3 labor-hours for finishing. The maximum labor-hours available per week in the fabricating and finishing departments are 120 and 30, respectively. What combinations of boards can be produced each week so as not to exceed the number of labor-hours available in each department per week?

Solution To clarify relationships, we summarize the information in the following table:

	Standard model (labor-hours per board)	**Competition model (labor-hours per board)**	**Maximum labor-hours available per week**
Fabricating	6	8	120
Finishing	1	3	30

Let

$$x = \text{Number of standard boards produced per week}$$

$$y = \text{Number of competition boards produced per week}$$

These variables are restricted as follows:

Fabricating department restriction:

$$\begin{pmatrix}\text{Weekly fabricating time}\\ \text{for } x \text{ standard boards}\end{pmatrix} + \begin{pmatrix}\text{Weekly fabricating time}\\ \text{for } y \text{ competition boards}\end{pmatrix} \le \begin{pmatrix}\text{Maximum labor-hours}\\ \text{available per week}\end{pmatrix}$$

$$6x \quad + \quad 8y \quad \le \quad 120$$

Finishing department restriction:

$$\begin{pmatrix}\text{Weekly finishing time}\\ \text{for } x \text{ standard boards}\end{pmatrix} + \begin{pmatrix}\text{Weekly finishing time}\\ \text{for } y \text{ competition boards}\end{pmatrix} \le \begin{pmatrix}\text{Maximum labor-hours}\\ \text{available per week}\end{pmatrix}$$

$$1x \quad + \quad 3y \quad \le \quad 30$$

Since it is not possible to manufacture a negative number of boards, x and y also must satisfy the nonnegative restrictions

$$x \ge 0$$

$$y \ge 0$$

Thus, x and y must satisfy the following system of linear inequalities:

$$6x + 8y \le 120 \quad \text{Fabricating department restriction}$$

$$x + 3y \le 30 \quad \text{Finishing department restriction}$$

$$x \ge 0 \quad \text{Nonnegative restriction}$$

$$y \ge 0 \quad \text{Nonnegative restriction}$$

Graphing this system of linear inequalities, we obtain the set of **feasible solutions,** or the **feasible region,** as shown in Figure 11. For problems of this type and for the linear programming problems we consider in the next section, solution regions are often referred to as feasible regions. Any point within the shaded area, including the boundary lines, represents a possible production schedule. Any point outside the shaded area represents an impossible schedule. For example, it would be possible to produce 12 standard boards and 5 competition boards per week, but it would not be possible to produce 12 standard boards and 7 competition boards per week (see the figure).

FIGURE 11

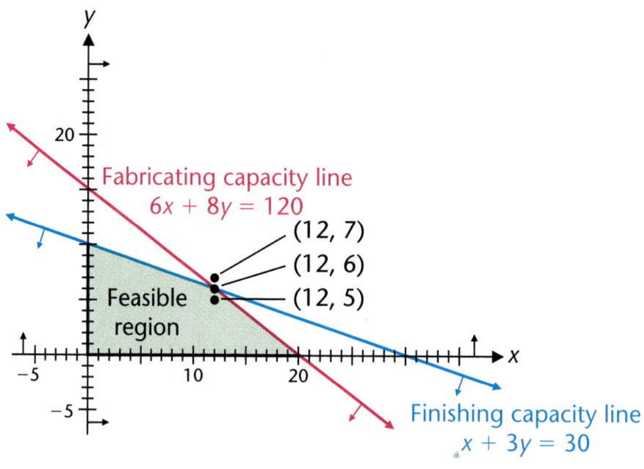

Matched Problem 5 Repeat Example 5 using 5 hours for fabricating a standard board and a maximum of 27 labor-hours for the finishing department.

Remark. Refer to Example 5. How do we interpret a production schedule of 10.5 standard boards and 4.3 competition boards? It is not possible to manufacture a fraction of a board. But it is possible to ***average*** 10.5 standard and 4.3 competition boards per week. In general, we will assume that all points in the feasible region represent acceptable solutions, even though noninteger solutions might require special interpretation.

Answers to Matched Problems

1.

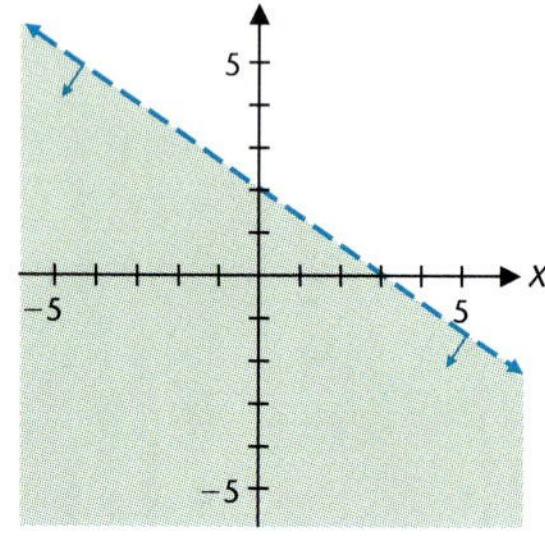

2. (A)

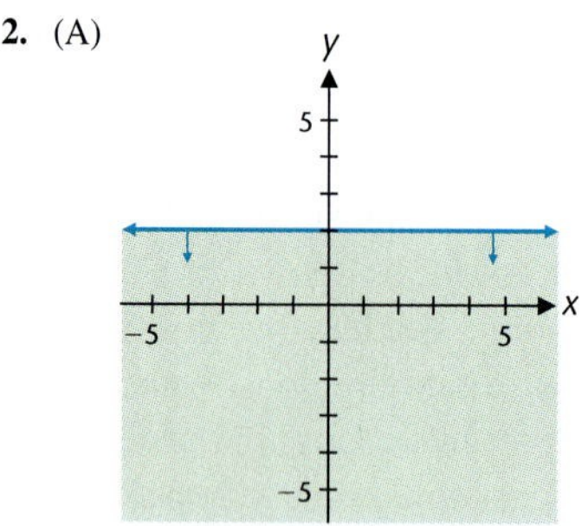

(B)

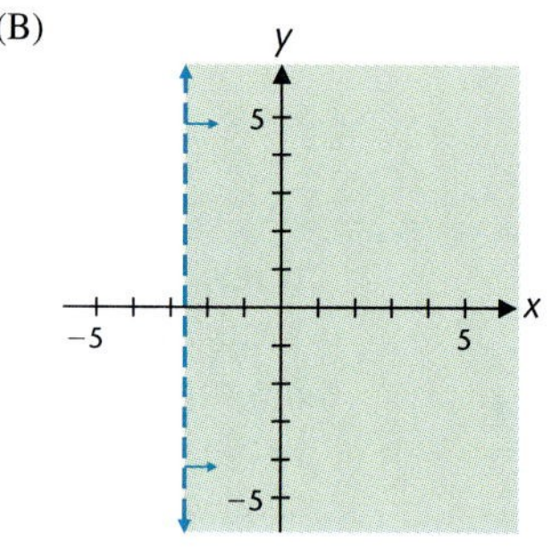

3.

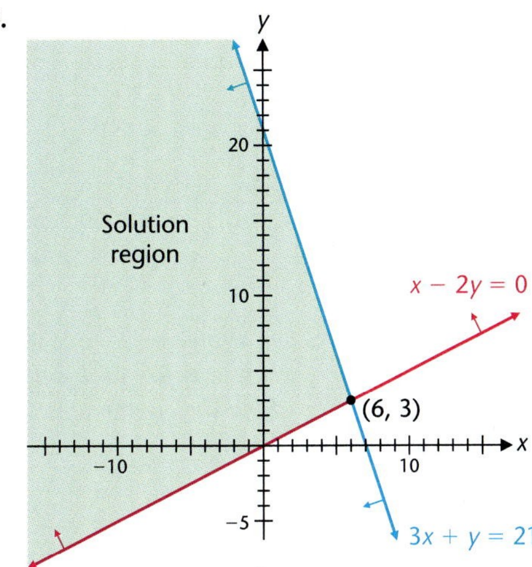

4.

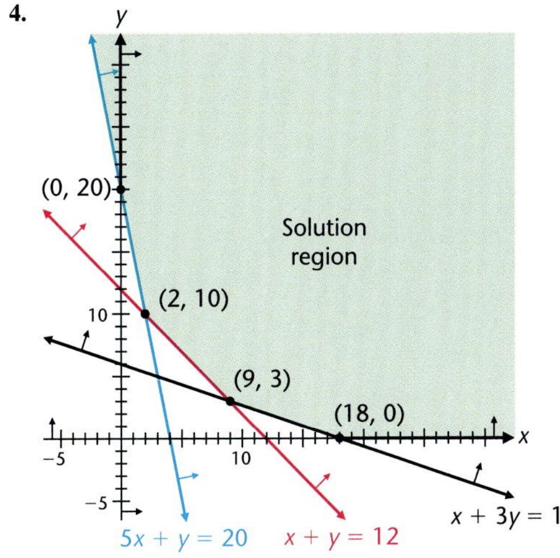

5.

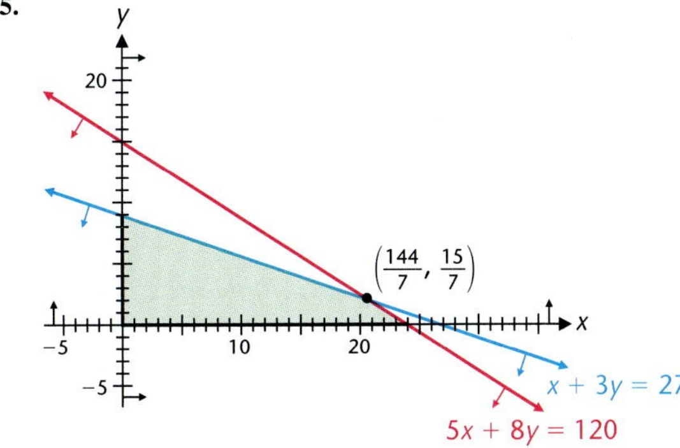

EXERCISE 8-4

A

Graph each inequality in Problems 1–10.

1. $2x - 3y < 6$ **2.** $3x + 4y < 12$ **3.** $3x + 2y \geq 18$

4. $3y - 2x \geq 24$ **5.** $y \leq \frac{2}{3}x + 5$ **6.** $y \geq \frac{1}{3}x - 2$

7. $y < 8$ **8.** $x > -5$ **9.** $-3 \leq y < 2$

10. $-1 < x \leq 3$

In Problems 11–14, match the solution region of each system of linear inequalities with one of the four regions shown in the figure below.

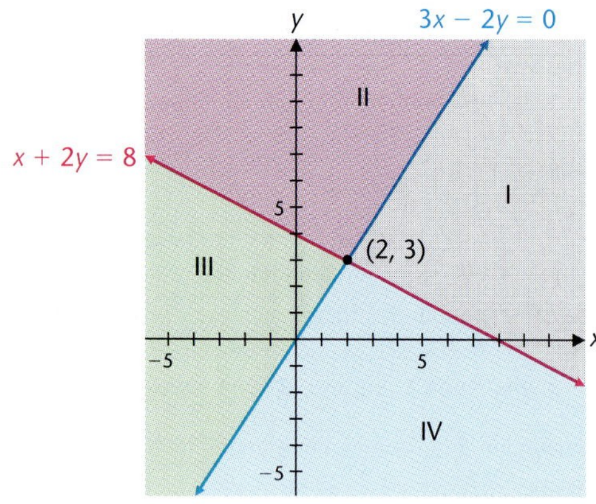

11. $x + 2y \le 8$
$3x - 2y \ge 0$

12. $x + 2y \ge 8$
$3x - 2y \le 0$

13. $x + 2y \ge 8$
$3x - 2y \ge 0$

14. $x + 2y \le 8$
$3x - 2y \le 0$

In Problems 15–20, solve each system of linear inequalities graphically.

15. $x \ge 5$
$y \le 6$

16. $x \le 4$
$y \ge 2$

17. $3x + y \ge 6$
$x \le 4$

18. $3x + 4y \le 12$
$y \ge -3$

19. $x - 2y \le 12$
$2x + y \ge 4$

20. $2x + 5y \le 20$
$x - 5y \le -5$

B

In Problems 21–24, match the solution region of each system of linear inequalities with one of the four regions shown in the figure below. Identify the corner points of each solution region.

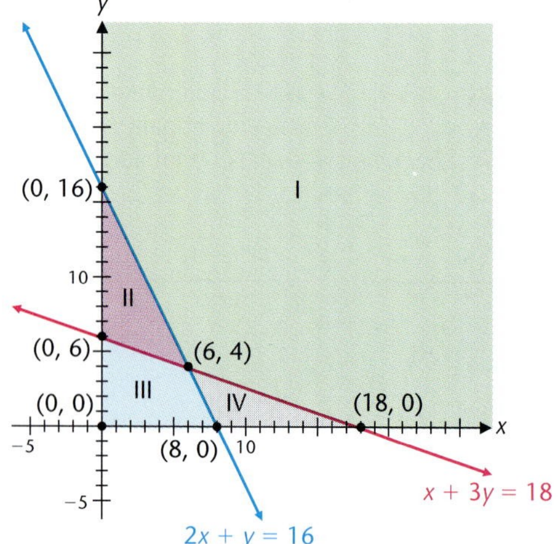

21. $x + 3y \le 18$
$2x + y \ge 16$
$x \ge 0$
$y \ge 0$

22. $x + 3y \le 18$
$2x + y \le 16$
$x \ge 0$
$y \ge 0$

23. $x + 3y \ge 18$
$2x + y \ge 16$
$x \ge 0$
$y \ge 0$

24. $x + 3y \ge 18$
$2x + y \le 16$
$x \ge 0$
$y \ge 0$

In Problems 25–36, solve the systems graphically, and indicate whether each solution region is bounded or unbounded. Find the coordinates of each corner point.

25. $2x + 3y \le 6$
$x \ge 0$
$y \ge 0$

26. $4x + 3y \le 12$
$x \ge 0$
$y \ge 0$

27. $4x + 5y \ge 20$
$x \ge 0$
$y \ge 0$

28. $5x + 6y \ge 30$
$x \ge 0$
$y \ge 0$

29. $2x + y \le 8$
$x + 3y \le 12$
$x \ge 0$
$y \ge 0$

30. $x + 2y \le 10$
$3x + y \le 15$
$x \ge 0$
$y \ge 0$

31. $4x + 3y \ge 24$
$2x + 3y \ge 18$
$x \ge 0$
$y \ge 0$

32. $x + 2y \ge 8$
$2x + y \ge 10$
$x \ge 0$
$y \ge 0$

33. $2x + y \le 12$
$x + y \le 7$
$x + 2y \le 10$
$x \ge 0$
$y \ge 0$

34. $3x + y \le 21$
$x + y \le 9$
$x + 3y \le 21$
$x \ge 0$
$y \ge 0$

35. $x + 2y \ge 16$
$x + y \ge 12$
$2x + y \ge 14$
$x \ge 0$
$y \ge 0$

36. $3x + y \ge 30$
$x + y \ge 16$
$x + 3y \ge 24$
$x \ge 0$
$y \ge 0$

C

In Problems 37–44, solve the systems graphically, and indicate whether each solution region is bounded or unbounded. Find the coordinates of each corner point.

37. $x + y \le 11$
$5x + y \ge 15$
$x + 2y \ge 12$

38. $4x + y \le 32$
$x + 3y \le 30$
$5x + 4y \ge 51$

39. $3x + 2y \ge 24$
$3x + y \le 15$
$x \ge 4$

40. $3x + 4y \le 48$
$x + 2y \ge 24$
$y \le 9$

41. $x + y \le 10$
$3x + 5y \ge 15$
$3x - 2y \le 15$
$-5x + 2y \le 6$

42. $3x - y \ge 1$
$-x + 5y \ge 9$
$x + y \le 9$
$y \le 5$

 **43.** $\begin{aligned} 16x + 13y &\le 119 \\ 12x + 16y &\ge 101 \\ -4x + 3y &\le 11 \end{aligned}$

44. $\begin{aligned} 8x + 4y &\le 41 \\ -15x + 5y &\le 19 \\ 2x + 6y &\ge 37 \end{aligned}$

APPLICATIONS

45. Manufacturing—Resource Allocation. A manufacturing company makes two types of water skis: a trick ski and a slalom ski. The trick ski requires 6 labor-hours for fabricating and 1 labor-hour for finishing. The slalom ski requires 4 labor-hours for fabricating and 1 labor-hour for finishing. The maximum labor-hours available per day for fabricating and finishing are 108 and 24, respectively. If x is the number of trick skis and y is the number of slalom skis produced per day, write a system of inequalities that indicates appropriate restraints on x and y. Find the set of feasible solutions graphically for the number of each type of ski that can be produced.

46. Manufacturing—Resource Allocation. A furniture manufacturing company manufactures dining room tables and chairs. A table requires 8 labor-hours for assembling and 2 labor-hours for finishing. A chair requires 2 labor-hours for assembling and 1 labor-hour for finishing. The maximum labor-hours available per day for assembly and finishing are 400 and 120, respectively. If x is the number of tables and y is the number of chairs produced per day, write a system of inequalities that indicates appropriate restraints on x and y. Find the set of feasible solutions graphically for the number of tables and chairs that can be produced.

★ **47. Manufacturing—Resource Allocation.** Refer to Problem 45. The company makes a profit of \$50 on each trick ski and a profit of \$60 on each slalom ski.

(A) If the company makes 10 trick and 10 slalom skis per day, the daily profit will be \$1,100. Are there other feasible production schedules that will result in a daily profit of \$1,100? How are these schedules related to the graph of the line $50x + 60y = 1{,}100$?

(B) Find a feasible production schedule that will produce a daily profit greater than \$1,100 and repeat part A for this schedule.

(C) Discuss methods for using lines like those in parts A and B to find the largest possible daily profit.

★ **48. Manufacturing—Resource Allocation.** Refer to Problem 46. The company makes a profit of \$50 on each table and a profit of \$15 on each chair.

(A) If the company makes 20 tables and 20 chairs per day, the daily profit will be \$1,300. Are there other feasible production schedules that will result in a daily profit of \$1,300? How are these schedules related to the graph of the line $50x + 15y = 1{,}300$?

(B) Find a feasible production schedule that will produce a daily profit greater than \$1,300 and repeat part A for this schedule.

(C) Discuss methods for using lines like those in parts A and B to find the largest possible daily profit.

49. Nutrition—Plants. A farmer can buy two types of plant food, mix A and mix B. Each cubic yard of mix A contains 20 pounds of phosphoric acid, 30 pounds of nitrogen, and 5 pounds of potash. Each cubic yard of mix B contains 10 pounds of phosphoric acid, 30 pounds of nitrogen, and 10 pounds of potash. The minimum requirements are 460 pounds of phosphoric acid, 960 pounds of nitrogen, and 220 pounds of potash. If x is the number of cubic yards of mix A used and y is the number of cubic yards of mix B used, write a system of inequalities that indicates appropriate restraints on x and y. Find the set of feasible solutions graphically for the amount of mix A and mix B that can be used.

50. Nutrition. A dietitian in a hospital is to arrange a special diet using two foods. Each ounce of food M contains 30 units of calcium, 10 units of iron, and 10 units of vitamin A. Each ounce of food N contains 10 units of calcium, 10 units of iron, and 30 units of vitamin A. The minimum requirements in the diet are 360 units of calcium, 160 units of iron, and 240 units of vitamin A. If x is the number of ounces of food M used and y is the number of ounces of food N used, write a system of linear inequalities that reflects the conditions indicated. Find the set of feasible solutions graphically for the amount of each kind of food that can be used.

51. Sociology. A city council voted to conduct a study on inner-city community problems. A nearby university was contacted to provide sociologists and research assistants. Each sociologist will spend 10 hours per week collecting data in the field and 30 hours per week analyzing data in the research center. Each research assistant will spend 30 hours per week in the field and 10 hours per week in the research center. The minimum weekly labor-hour requirements are 280 hours in the field and 360 hours in the research center. If x is the number of sociologists hired for the study and y is the number of research assistants hired for the study, write a system of linear inequalities that indicates appropriate restrictions on x and y. Find the set of feasible solutions graphically.

52. Psychology. In an experiment on conditioning, a psychologist uses two types of Skinner (conditioning) boxes with mice and rats. Each mouse spends 10 minutes per day in box A and 20 minutes per day in box B. Each rat spends 20 minutes per day in box A and 10 minutes per day in box B. The total maximum time available per day is 800 minutes for box A and 640 minutes for box B. We are interested in the various numbers of mice and rats that can be used in the experiment under the conditions stated. If x is the number of mice used and y is the number of rats used, write a system of linear inequalities that indicates appropriate restrictions on x and y. Find the set of feasible solutions graphically.

SECTION 8-5 Linear Programming

- A Linear Programming Problem
- Linear Programming—A General Description
- Application

Several problems in Section 8-4 are related to the general type of problems called *linear programming problems.* Linear programming is a mathematical process that has been developed to help management in decision making, and it has become one of the most widely used and best-known tools of management science and industrial engineering. We will use an intuitive graphical approach based on the techniques discussed in Section 8-4 to illustrate this process for problems involving two variables.

The American mathematician George B. Dantzig (1914–) formulated the first linear programming problem in 1947 and introduced a solution technique, called the *simplex method,* that does not rely on graphing and is readily adaptable to computer solutions. Today, it is quite common to use a computer to solve applied linear programming problems involving thousands of variables and thousands of inequalities.

• A Linear Programming Problem

We begin our discussion with an example that will lead to a general procedure for solving linear programming problems in two variables.

EXAMPLE 1 Production Scheduling

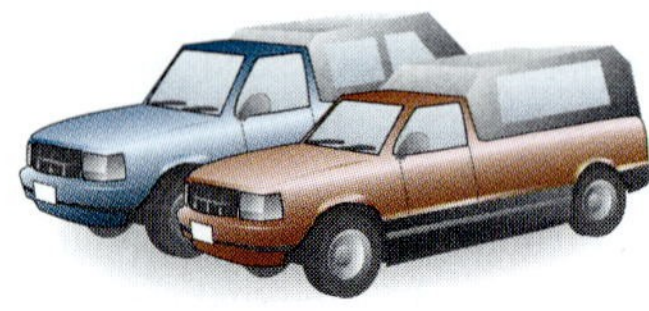

A manufacturer of fiberglass camper tops for pickup trucks makes a compact model and a regular model. Each compact top requires 5 hours from the fabricating department and 2 hours from the finishing department. Each regular top requires 4 hours from the fabricating department and 3 hours from the finishing department. The maximum labor-hours available per week in the fabricating department and the finishing department are 200 and 108, respectively. If the company makes a profit of \$40 on each compact top and \$50 on each regular top, how many tops of each type should be manufactured each week to maximize the total weekly profit, assuming all tops can be sold? What is the maximum profit?

Solution This is an example of a linear programming problem. To see relationships more clearly, we summarize the manufacturing requirements, objectives, and restrictions in the table:

	Compact model (labor-hours per top)	**Regular model (labor-hours per top)**	**Maximum labor-hours available per week**
Fabricating	5	4	200
Finishing	2	3	108
Profit per top	\$40	\$50	

We now proceed to formulate a *mathematical model* for the problem and then to solve it using graphical methods.

Objective Function

The *objective* of management is to *decide* how many of each camper top model should be produced each week in order to *maximize* profit. Let

$$\left.\begin{aligned} x &= \text{Number of compact tops produced per week} \\ y &= \text{Number of regular tops produced per week} \end{aligned}\right\} \quad \text{Decision variables}$$

The following function gives the total profit P for x compact tops and y regular tops manufactured each week:

$$P = 40x + 50y \quad \text{Objective function}$$

Mathematically, management needs to decide on values for the **decision variables** (x and y) that achieve its objective, that is, maximizing the **objective function** (profit) $P = 40x + 50y$. It appears that the profit can be made as large as we like by manufacturing more and more tops—or can it?

Constraints

Any manufacturing company, no matter how large or small, has manufacturing limits imposed by available resources, plant capacity, demand, and so forth. These limits are referred to as **problem constraints.**

Fabricating department constraint:

$$\begin{pmatrix}\text{Weekly fabricating} \\ \text{time for } x \\ \text{compact tops}\end{pmatrix} + \begin{pmatrix}\text{Weekly fabricating} \\ \text{time for } y \\ \text{regular tops}\end{pmatrix} \le \begin{pmatrix}\text{Maximum labor-hours} \\ \text{available per week}\end{pmatrix}$$

$$5x \quad + \quad 4y \quad \le \quad 200$$

Finishing department constraint:

$$\begin{pmatrix}\text{Weekly finishing} \\ \text{time for } x \\ \text{compact tops}\end{pmatrix} + \begin{pmatrix}\text{Weekly finishing} \\ \text{time for } y \\ \text{regular tops}\end{pmatrix} \le \begin{pmatrix}\text{Maximum labor-hours} \\ \text{available per week}\end{pmatrix}$$

$$2x \quad + \quad 3y \quad \le \quad 108$$

Nonnegative constraints: It is not possible to manufacture a negative number of tops; thus, we have the **nonnegative constraints**

$$x \ge 0$$
$$y \ge 0$$

which we usually write in the form

$$x, y \ge 0$$

Mathematical Model

We now have a **mathematical model** for the problem under consideration:

$$\begin{aligned} &\text{Maximize} && P = 40x + 50y && \text{Objective function} \\ &\text{Subject to} && \left.\begin{aligned} 5x + 4y &\le 200 \\ 2x + 3y &\le 108 \end{aligned}\right\} && \text{Problem constraints} \\ & && x, y \ge 0 && \text{Nonnegative constraints} \end{aligned}$$

Graphic Solution

Solving the system of linear inequality constraints **graphically,** as in Section 8-4, we obtain the feasible region for production schedules, as shown in Figure 1.

FIGURE 1

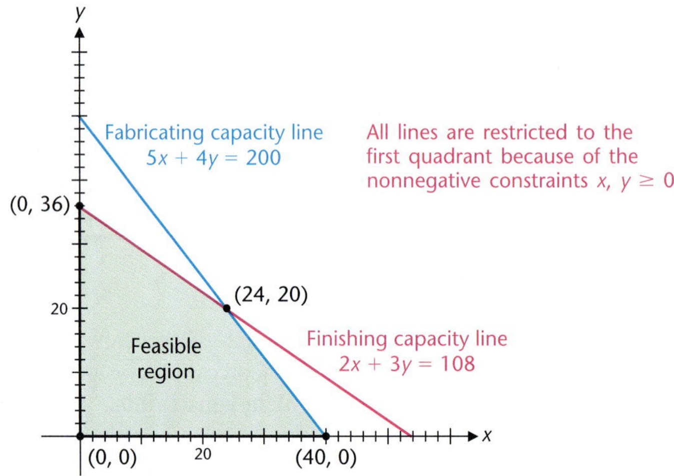

By choosing a production schedule (x, y) from the feasible region, a profit can be determined using the objective function $P = 40x + 50y$. For example, if $x = 24$ and $y = 10$, then the profit for the week is

$$P = 40(24) + 50(10) = \$1{,}460$$

Or if $x = 15$ and $y = 20$, then the profit for the week is

$$P = 40(15) + 50(20) = \$1{,}600$$

The question is, out of all possible production schedules (x, y) from the feasible region, which schedule(s) produces the maximum profit? Such a schedule, if it exists, is called an **optimal solution** to the problem because it produces the maximum value of the objective function and is in the feasible region. It is not practical to use point-by-point checking to find the optimal solution. Even if we consider only points with integer coordinates, there are over 800 such points in the feasible region for this problem. Instead, we use the theory that has been developed to solve linear programming problems. Using advanced techniques, it can be shown that:

If the feasible region is bounded, then one or more of the corner points of the feasible region is an optimal solution to the problem.

The maximum value of the objective function is unique; however, there can be more than one feasible production schedule that will produce this unique value. We will have more to say about this later in this section.

Corner point (x, y)	Objective function $P = 40x + 50y$	
(0, 0)	0	
(0, 36)	1,800	
(24, 20)	1,960	Maximum value of P
(40, 0)	1,600	

Since the feasible region for this problem is bounded, at least one of the corner points, (0, 0), (0, 36), (24, 20), or (40, 0), is an optimal solution. To find which one, we evaluate $P = 40x + 50y$ at each corner point and choose the corner point that produces the largest value of P. It is convenient to organize these calculations in a table, as shown in the margin.

Examining the values in the table, we see that the maximum value of P at a corner point is $P = 1{,}960$ at $x = 24$ and $y = 20$. Since the maximum value of P over the entire feasible region must always occur at a corner point, we conclude that the maximum profit is \$1,960 when 24 compact tops and 20 regular tops are produced each week.

Matched Problem 1 We now convert the surfboard problem discussed in Section 8-4 into a linear programming problem. A manufacturer of surfboards makes a standard model and a competition model. Each standard board requires 6 labor-hours for fabricating and 1 labor-hour for finishing. Each competition board requires 8 labor-hours for fabricating and 3 labor-hours for finishing. The maximum labor-hours available per week in the fabricating and finishing departments are 120 and 30, respectively. If the company makes a profit of \$40 on each standard board and \$75 on each competition board, how many boards of each type should be manufactured each week to maximize the total weekly profit?

(A) Identify the decision variables.
(B) Write the objective function P.
(C) Write the problem constraints and the nonnegative constraints.
(D) Graph the feasible region, identify the corner points, and evaluate P at each corner point.
(E) How many boards of each type should be manufactured each week to maximize the profit? What is the maximum profit?

EXPLORE-DISCUSS 1 Refer to Example 1. If we assign the profit function P in $P = 40x + 50y$ a particular value and plot the resulting equation in the coordinate system shown in Figure 1, we obtain a **constant-profit line (isoprofit line).** Every point in the feasible region on this line represents a production schedule that will produce the same profit. Figure 2 shows the constant-profit lines for $P = \$1{,}000$ and $P = \$1{,}500$.

FIGURE 2

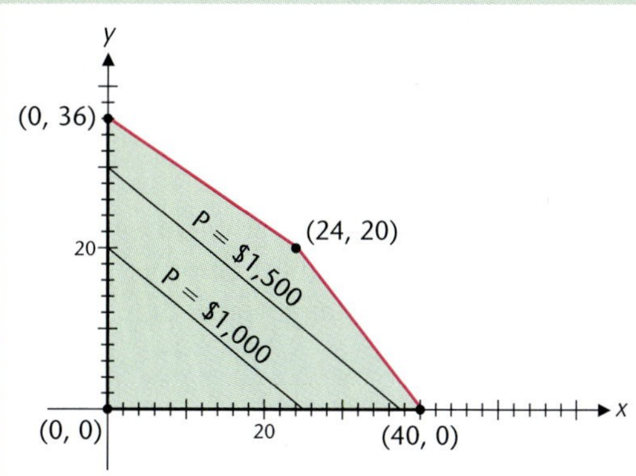

(A) How are all the constant-profit lines related?

(B) Place a straightedge along the constant-profit line for $P = \$1{,}000$ and slide it as far as possible in the direction of increasing profit without changing its slope and without leaving the feasible region. Explain how this process can be used to identify the optimal solution to a linear programming problem.

(C) If P is changed to $P = 25x + 75y$, graph the constant-profit lines for $P = \$1{,}000$ and $P = \$1{,}500$, and use a straightedge to identify the optimal solution. Check your answer by evaluating P at each corner point.

(D) Repeat part C for $P = 75x + 25y$.

• Linear Programming—A General Description

The linear programming problems considered in Example 1 and Matched Problem 1 were *maximization problems* where we wanted to maximize profits. The same technique can be used to solve *minimization problems* where, for example, we may want to minimize costs. Before considering additional examples, we state a few general definitions.

A **linear programming problem** is one that is concerned with finding the **optimal value** (maximum or minimum value) of a linear **objective function** of the form

$$z = ax + by$$

where the **decision variables** x and y are subject to **problem constraints** in the form of linear inequalities and the **nonnegative constraints** $x, y \geq 0$. The set of points satisfying both the problem constraints and the nonnegative constraints is called the **feasible region** for the problem. Any point in the feasible region that produces the optimal value of the objective function over the feasible region is called an **optimal solution.**

Theorem 1 is fundamental to the solving of linear programming problems.

Theorem 1 **Fundamental Theorem of Linear Programming**

Let S be the feasible region for a linear programming problem, and let $z = ax + by$ be the objective function. If S is bounded, then z has both a maximum and a minimum value on S and each of these occurs at a corner point of S. If S is unbounded, then a maximum or minimum value of z on S may not exist. However, if either does exist, then it must occur at a corner point of S.

We will not consider any problems with unbounded feasible regions in this brief introduction. If a feasible region is bounded, then Theorem 1 provides the basis for the following simple procedure for solving the associated linear programming problem:

Solution of Linear Programming Problems

Step 1. Form a mathematical model for the problem:
(A) Introduce decision variables and write a linear objective function.
(B) Write problem constraints in the form of linear inequalities.
(C) Write nonnegative constraints.

Step 2. Graph the feasible region and find the corner points.

Step 3. Evaluate the objective function at each corner point to determine the optimal solution.

Before considering additional applications, we use this procedure to solve a linear programming problem where the model has already been determined.

EXAMPLE 2 Solving a Linear Programming Problem

$$\begin{aligned} &\text{Minimize and maximize} \quad z = 5x + 15y \\ &\text{Subject to} \quad x + 3y \le 60 \\ &\qquad\qquad\quad\; x + y \ge 10 \\ &\qquad\qquad\quad\; x - y \le 0 \\ &\qquad\qquad\qquad x, y \ge 0 \end{aligned}$$

Solution This problem is a combination of two linear programming problems—a minimization problem and a maximization problem. Since the feasible region is the same for both problems, we can solve these problems together. To begin, we graph the feasible region S, as shown in Figure 3, and find the coordinates of each corner point.

FIGURE 3

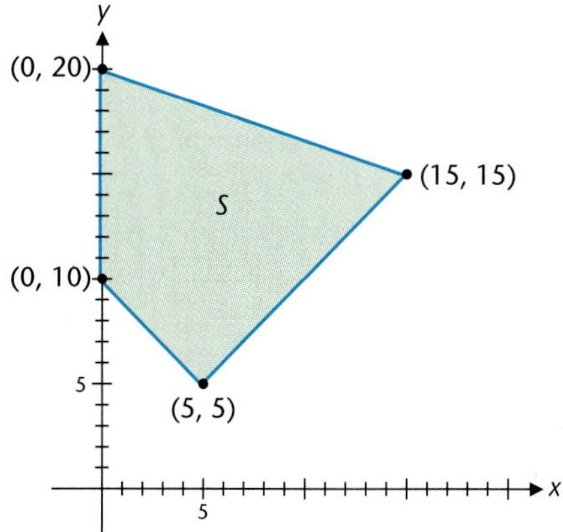

Next, we evaluate the objective function at each corner point, with the results given in the table:

Corner point (x, y)	Objective function $z = 5x + 15y$	
(0, 10)	150	
(0, 20)	300	Maximum value } Multiple optimal solutions
(15, 15)	300	Maximum value } Multiple optimal solutions
(5, 5)	100	Minimum value

Examining the values in the table, we see that the minimum value of z on the feasible region S is 100 at (5, 5). Thus, (5, 5) is the optimal solution to the minimization problem. The maximum value of z on the feasible region S is 300, which occurs at (0, 20) and at (15, 15). Thus, the maximization problem has **multiple optimal solutions.** In general:

If two corner points are both optimal solutions of the same type (both produce the same maximum value or both produce the same minimum value) to a linear programming problem, then any point on the line segment joining the two corner points is also an optimal solution of that type.

It can be shown that this is the only time that an optimal value occurs at more than one point.

Matched Problem 2 Minimize and maximize $z = 10x + 5y$

Subject to

$$
\begin{aligned}
2x + y &\geq 40 \\
3x + y &\leq 150 \\
2x - y &\geq 0 \\
x, y &\geq 0
\end{aligned}
$$

• Application

Now we consider another application where we must first find the mathematical model and then find its solution.

EXAMPLE 3 Agriculture

	Pounds per cubic yard	
	Mix A	**Mix B**
Nitrogen	10	5
Potash	8	24
Phosphoric acid	9	6

A farmer can use two types of plant food, mix A and mix B. The amounts (in pounds) of nitrogen, phosphoric acid, and potash in a cubic yard of each mix are given in the table. Tests performed on the soil in a large field indicate that the field needs at least 840 pounds of potash and at least 350 pounds of nitrogen. The tests also indicate that no more than 630 pounds of phosphoric acid should be added to the field. A cubic yard of mix A costs \$7, and a cubic yard of mix B costs \$9. How many cubic yards of each mix should the farmer add to the field in order to supply the necessary nutrients at minimal cost?

Solution Let

$x =$ Number of cubic yards of mix A added to the field
$y =$ Number of cubic yards of mix B added to the field } Decision variables

We form the linear objective function

$$C = 7x + 9y$$

which gives the cost of adding x cubic yards of mix A and y cubic yards of mix B to the field. Using the data in the table and proceeding as in Example 1, we formulate the mathematical model for the problem:

Minimize	$C = 7x + 9y$	Objective function
Subject to	$10x + 5y \geq 350$	Nitrogen constraint
	$8x + 24y \geq 840$	Potash constraint
	$9x + 6y \leq 630$	Phosphoric acid constraint
	$x, y \geq 0$	Nonnegative constraints

Solving the system of constraint inequalities graphically, we obtain the feasible region S shown in Figure 4, and then we find the coordinates of each corner point.

FIGURE 4

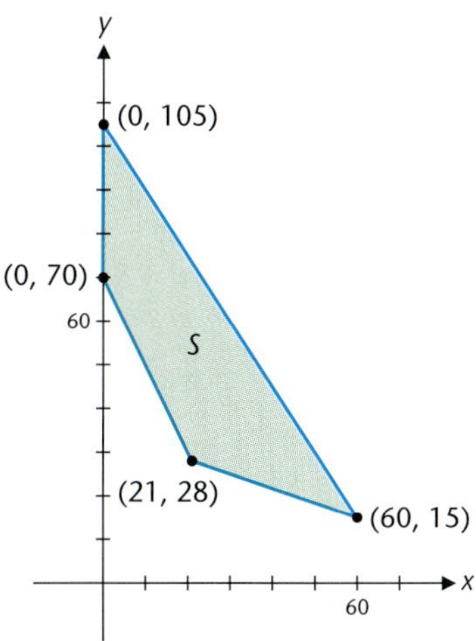

Corner point (x, y)	Objective function $C = 7x + 9y$	
(0, 105)	945	
(0, 70)	630	
(21, 28)	399	Minimum value of C
(60, 15)	555	

Next, we evaluate the objective function at each corner point, as shown in the table in the margin.

The optimal value is $C = 399$ at the corner point (21, 28). Thus, the farmer should add 21 cubic yards of mix A and 28 cubic yards of mix B at a cost of \$399. This will result in adding the following nutrients to the field:

Nitrogen:	$10(21) + 5(28) = 350$	pounds
Potash:	$8(21) + 24(28) = 840$	pounds
Phosphoric acid:	$9(21) + 6(28) = 357$	pounds

All the nutritional requirements are satisfied.

Matched Problem 3 Repeat Example 3 if the tests indicate that the field needs at least 400 pounds of nitrogen with all other conditions remaining the same.

Answers to Matched Problems

1. (A) x = Number of standard boards manufactured each week
   y = Number of competition boards manufactured each week

   (B) $P = 40x + 75y$ (C) $6x + 8y \le 120$ Fabricating constraint
   $x + 3y \le 30$ Finishing constraint
   $x, y \ge 0$ Nonnegative constraints

   (D)

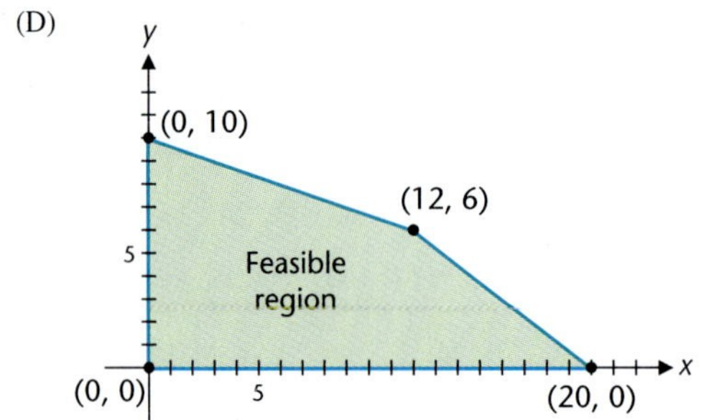

Corner point (x, y)	Objective function $P = 40x + 75y$
(0, 0)	0
(0, 10)	750
(12, 6)	930
(20, 0)	800

   (E) 12 standard boards and 6 competition boards for a maximum profit of \$930
2. Max $z = 600$ at (30, 60); min $z = 200$ at (10, 20) and (20, 0) (multiple optimal solutions)
3. 27 cubic yards of mix A, 26 cubic yards of mix B; min $C = \$423$

EXERCISE 8-5

A

In Problems 1–4, find the maximum value of each objective function over the feasible region S shown in the figure below.

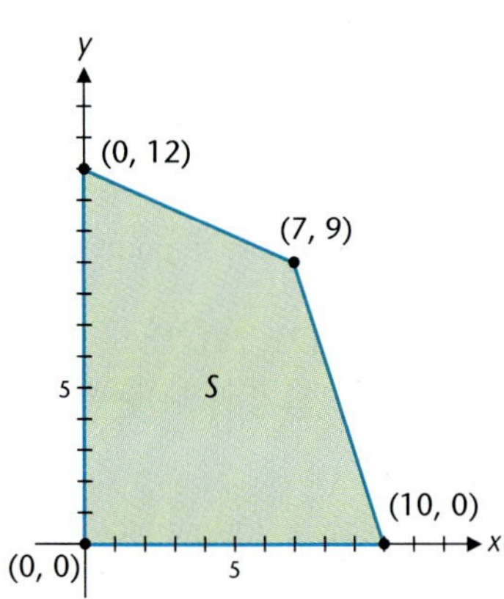

1. $z = x + y$
2. $z = 4x + y$
3. $z = 3x + 7y$
4. $z = 9x + 3y$

In Problems 5–8, find the minimum value of each objective function over the feasible region T shown in the figure below.

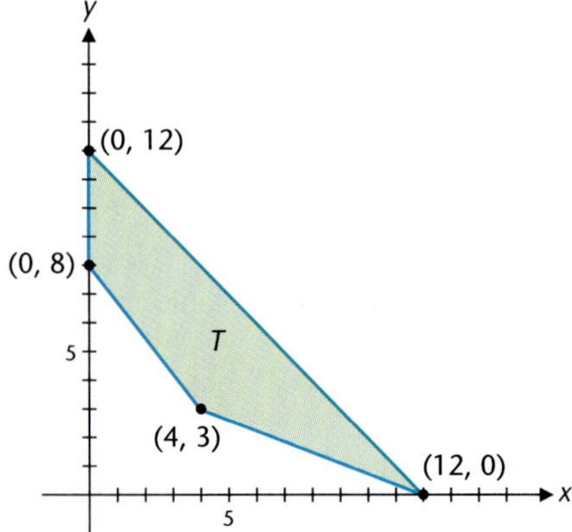

5. $z = 7x + 4y$
6. $z = 7x + 9y$
7. $z = 3x + 8y$
8. $z = 5x + 4y$

B

In Problems 9–22, solve the linear programming problems.

9. Maximize $z = 3x + 2y$
Subject to
$$\begin{aligned} x + 2y &\le 10 \\ 3x + y &\le 15 \\ x, y &\ge 0 \end{aligned}$$

10. Maximize $z = 4x + 5y$
Subject to
$$\begin{aligned} 2x + y &\le 12 \\ x + 3y &\le 21 \\ x, y &\ge 0 \end{aligned}$$

11. Minimize $z = 3x + 4y$
Subject to
$$\begin{aligned} 2x + y &\ge 8 \\ x + 2y &\le 10 \\ x, y &\ge 0 \end{aligned}$$

12. Minimize $z = 2x + y$
Subject to
$$\begin{aligned} 4x + 3y &\ge 24 \\ 4x + y &\le 16 \\ x, y &\ge 0 \end{aligned}$$

13. Maximize $z = 3x + 4y$
Subject to
$$\begin{aligned} x + 2y &\le 24 \\ x + y &\le 14 \\ 2x + y &\le 24 \\ x, y &\ge 0 \end{aligned}$$

14. Maximize $z = 5x + 3y$
Subject to
$$\begin{aligned} 3x + y &\le 24 \\ x + y &\le 10 \\ x + 3y &\le 24 \\ x, y &\ge 0 \end{aligned}$$

15. Minimize $z = 5x + 6y$
Subject to
$$\begin{aligned} x + 4y &\ge 20 \\ 4x + y &\ge 20 \\ x + y &\le 20 \\ x, y &\ge 0 \end{aligned}$$

16. Minimize $z = x + 2y$
Subject to
$$\begin{aligned} 2x + 3y &\ge 30 \\ 3x + 2y &\ge 30 \\ x + y &\le 15 \\ x, y &\ge 0 \end{aligned}$$

17. Minimize and maximize $z = 25x + 50y$
Subject to
$$\begin{aligned} x + 2y &\le 120 \\ x + y &\ge 60 \\ x - 2y &\ge 0 \\ x, y &\ge 0 \end{aligned}$$

18. Minimize and maximize $z = 15x + 30y$
Subject to
$$\begin{aligned} x + 2y &\ge 100 \\ 2x - y &\le 0 \\ 2x + y &\le 200 \\ x, y &\ge 0 \end{aligned}$$

19. Minimize and maximize $z = 25x + 15y$
Subject to
$$\begin{aligned} 4x + 5y &\ge 100 \\ 3x + 4y &\le 240 \\ x &\le 60 \\ y &\le 45 \\ x, y &\ge 0 \end{aligned}$$

20. Minimize and maximize $z = 25x + 30y$
Subject to
$$\begin{aligned} 2x + 3y &\ge 120 \\ 3x + 2y &\le 360 \\ x &\le 80 \\ y &\le 120 \\ x, y &\ge 0 \end{aligned}$$

21. Maximize $P = 525x + 478y$
Subject to
$$\begin{aligned} 275x + 322y &\le 3{,}381 \\ 350x + 340y &\le 3{,}762 \\ 425x + 306y &\le 4{,}114 \\ x, y &\ge 0 \end{aligned}$$

22. Maximize $P = 300x + 460y$
Subject to
$$\begin{aligned} 245x + 452y &\le 4{,}181 \\ 290x + 379y &\le 3{,}888 \\ 390x + 299y &\le 4{,}407 \\ x, y &\ge 0 \end{aligned}$$

C

23. The corner points for the feasible region determined by the problem constraints

$$\begin{aligned} 2x + y &\le 10 \\ x + 3y &\le 15 \\ x, y &\ge 0 \end{aligned}$$

are $O = (0, 0)$, $A = (5, 0)$, $B = (3, 4)$, and $C = (0, 5)$. If $z = ax + by$ and $a, b > 0$, determine conditions on a and b that ensure that the maximum value of z occurs:

(A) Only at A
(B) Only at B
(C) Only at C
(D) At both A and B
(E) At both B and C

24. The corner points for the feasible region determined by the problem constraints

$$\begin{aligned} x + y &\ge 4 \\ x + 2y &\ge 6 \\ 2x + 3y &\le 12 \\ x, y &\ge 0 \end{aligned}$$

are $A = (6, 0)$, $B = (2, 2)$, and $C = (0, 4)$. If $z = ax + by$ and $a, b > 0$, determine conditions on a and b that ensure that the minimum value of z occurs:

(A) Only at A
(B) Only at B
(C) Only at C
(D) At both A and B
(E) At both B and C

APPLICATIONS 

25. Resource Allocation. A manufacturing company makes two types of water skis, a trick ski and a slalom ski. The relevant manufacturing data is given in the table.

(A) If the profit on a trick ski is \$40 and the profit on a slalom ski is \$30, how many of each type of ski should be manufactured each day to realize a maximum profit? What is the maximum profit?

(B) Discuss the effect on the production schedule and the maximum profit if the profit on a slalom ski decreases to \$25 and all other data remains the same.

(C) Discuss the effect on the production schedule and the maximum profit if the profit on a slalom ski increases to \$45 and all other data remains the same.

	Trick ski (labor-hours per ski)	Slalom ski (labor-hours per ski)	Maximum labor-hours available per day
Fabricating department	6	4	108
Finishing department	1	1	24

26. Psychology. In an experiment on conditioning, a psychologist uses two types of Skinner boxes with mice and rats. The amount of time (in minutes) each mouse and each rat spends in each box per day is given in the table. What is the maximum total number of mice and rats that can be used in this experiment? How many mice and how many rats produce this maximum?

	Mice (minutes)	Rats (minutes)	Max. time available per day (minutes)
Skinner box *A*	10	20	800
Skinner box *B*	20	10	640

27. Purchasing. A trucking firm wants to purchase a maximum of 15 new trucks that will provide at least 36 tons of additional shipping capacity. A model *A* truck holds 2 tons and costs \$15,000. A model *B* truck holds 3 tons and costs \$24,000. How many trucks of each model should the company purchase to provide the additional shipping capacity at minimal cost? What is the minimal cost?

28. Transportation. The officers of a high school senior class are planning to rent buses and vans for a class trip. Each bus can transport 40 students, requires 3 chaperones, and costs \$1,200 to rent. Each van can transport 8 students, requires 1 chaperone, and costs \$100 to rent. The officers want to be able to accommodate at least 400 students with no more than 36 chaperones. How many vehicles of each type should they rent in order to minimize the transportation costs? What are the minimal transportation costs?

★ **29. Resource Allocation.** A furniture company manufactures dining room tables and chairs. Each table requires 8 hours from the assembly department and 2 hours from the finishing department and contributes a profit of \$90. Each chair requires 2 hours from the assembly department and 1 hour from the finishing department and contributes a profit of \$25. The maximum labor-hours available each day in the assembly and finishing departments are 400 and 120, respectively.

(A) How many tables and how many chairs should be manufactured each day to maximize the daily profit? What is the maximum daily profit?

(B) Discuss the effect on the production schedule and the maximum profit if the marketing department of the company decides that the number of chairs produced should be at least four times the number of tables produced.

★ **30. Resource Allocation.** An electronics firm manufactures two types of personal computers, a desktop model and a portable model. The production of a desktop computer requires a capital expenditure of \$400 and 40 hours of labor. The production of a portable computer requires a capital expenditure of \$250 and 30 hours of labor. The firm has \$20,000 capital and 2,160 labor-hours available for production of desktop and portable computers.

(A) What is the maximum number of computers the company is capable of producing?

(B) If each desktop computer contributes a profit of \$320 and each portable contributes a profit of \$220, how much profit will the company make by producing the maximum number of computers determined in part A? Is this the maximum profit? If not, what is the maximum profit?

31. Pollution Control. Because of new federal regulations on pollution, a chemical plant introduced a new process to supplement or replace an older process used in the production of a particular chemical. The older process emitted 20 grams of sulfur dioxide and 40 grams of particulate matter into the atmosphere for each gallon of chemical produced. The new process emits 5 grams of sulfur dioxide and 20 grams of particulate matter for each gallon produced. The company makes a profit of 60¢ per gallon and 20¢ per gallon on the old and new processes, respectively.

(A) If the regulations allow the plant to emit no more than 16,000 grams of sulfur dioxide and 30,000 grams of particulate matter daily, how many gallons of the chemical should be produced by each process to maximize daily profit? What is the maximum daily profit?

(B) Discuss the effect on the production schedule and the maximum profit if the regulations restrict emissions of sulfur dioxide to 11,500 grams daily and all other data remains unchanged.

(C) Discuss the effect on the production schedule and the maximum profit if the regulations restrict emissions of sulfur dioxide to 7,200 grams daily and all other data remains unchanged.

★★ **32. Sociology.** A city council voted to conduct a study on inner-city community problems. A nearby university was contacted to provide a maximum of 40 sociologists and research assistants. Allocation of time and cost per week are given in the table.

(A) How many sociologists and research assistants should be hired to meet the weekly labor-hour requirements and minimize the weekly cost? What is the weekly cost?

(B) Discuss the effect on the solution in part A if the council decides that they should not hire more sociologists than research assistants and all other data remains unchanged.

	Sociologist (labor-hours)	Research assistant (labor-hours)	Minimum labor-hours needed per week
Fieldwork	10	30	280
Research center	30	10	360
Cost per week	\$500	\$300	

★★ **33. Plant Nutrition.** A fruit grower can use two types of fertilizer in her orange grove, brand A and brand B. The amounts (in pounds) of nitrogen, phosphoric acid, potash, and chlorine in a bag of each mix are given in the table. Tests indicate that the grove needs at least 480 pounds of phosphoric acid, at least 540 pounds of potash, and at most 620 pounds of chlorine. If the grower always uses a combination of bags of brand A and brand B that will satisfy the constraints for phosphoric acid, potash, and chlorine, discuss the effect that this will have on the amount of nitrogen added to the field.

	Pounds per bag	
	Brand A	**Brand B**
Nitrogen	6	7
Phosphoric acid	2	4
Potash	6	3
Chlorine	3	4

★★ **34. Diet.** A dietitian in a hospital is to arrange a special diet composed of two foods, M and N. Each ounce of food M contains 16 units of calcium, 5 units of iron, 6 units of cholesterol, and 8 units of vitamin A. Each ounce of food N contains 4 units of calcium, 25 units of iron, 4 units of cholesterol, and 4 units of vitamin A. The diet requires at least 320 units of calcium, at least 575 units of iron, and at most 300 units of cholesterol. If the dietitian always selects a combination of foods M and N that will satisfy the constraints for calcium, iron, and cholesterol, discuss the effect that this will have on the amount of vitamin A in the diet.

CHAPTER 8 GROUP ACTIVITY Modeling with Systems of Linear Equations

In this group activity we consider two real-world problems that can be solved using systems of linear equations: heat conduction and traffic flow. Both problems involve using a grid and a basic assumption to construct the model (the system of equations). Gauss–Jordan elimination is then used to solve the model. In the heat conduction problem, the solution of the model is easily interpreted in terms of the original problem. The system in the second problem is dependent, and the solution requires a more careful interpretation.

I Heat Conduction

A metal grid consists of four thin metal bars. The end of each bar of the grid is kept at a constant temperature, as shown in Figure 1. We assume that the temperature at each intersection point in the grid is the average of the temperatures at the four adjacent points in the grid (adjacent points are either other intersection points or ends of bars). Thus, the temperature x_1 at the intersection point in the upper left-hand corner of the grid must satisfy

$$x_1 = \tfrac{1}{4}(\underset{\text{Left}}{40} + \underset{\text{Above}}{0} + \underset{\text{Right}}{x_2} + \underset{\text{Below}}{x_3})$$

Find equations for the temperature at the other three intersection points, and solve the resulting system to find the temperature at each intersection point in the grid.

FIGURE 1

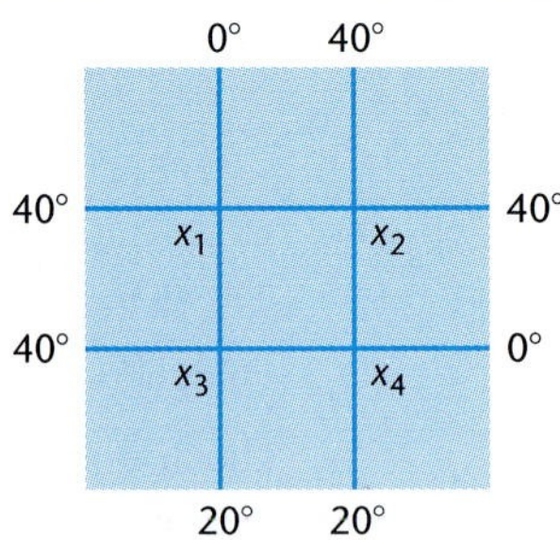

II Traffic Flow

The rush-hour traffic flow for a network of four one-way streets in a city is shown in Figure 2. The numbers next to each street indicate the number of vehicles per hour that enter and leave the network on that street. The variables x_1, x_2, x_3, and x_4 represent the flow of traffic between the four intersections in the network. For a smooth flow of traffic, we assume that the number of vehicles entering each intersection should always equal the number leaving. For example, since 1,500 vehicles enter the intersection of 5th Street and Washington Avenue each hour and $x_1 + x_4$ vehicles leave this intersection, we see that $x_1 + x_4 = 1{,}500$.

FIGURE 2

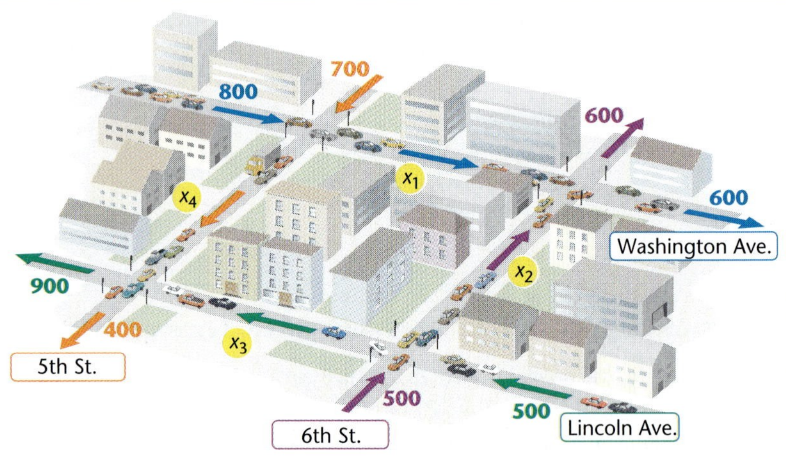

(A) Find the equations determined by the traffic flow at each of the other three intersections.
(B) Find the solution to the system in part A.
(C) What is the maximum number of vehicles that can travel from Washington Avenue to Lincoln Avenue on 5th Street? What is the minimum number?
(D) If traffic lights are adjusted so that 1,000 vehicles per hour travel from Washington Avenue to Lincoln Avenue on 5th Street, determine the flow around the rest of the network.

Chapter 8 Review

8-1 SYSTEMS OF LINEAR EQUATIONS IN TWO VARIABLES; AUGMENTED MATRICES

A system of two linear equations with two variables is a system of the form

$$\begin{aligned} ax + by &= h \\ cx + dy &= k \end{aligned} \qquad (1)$$

where x and y are variables, a, b, c, and d are real numbers called the **coefficients** of x and y, and h and k are real numbers called the **constant terms** in the equations. The ordered pair of numbers (x_0, y_0) is a **solution** to system (1) if each equation is satisfied by the pair. The set of all such ordered pairs of numbers is called the **solution set** for the system. To **solve** a system is to find its solution set.

In general, a system of linear equations has exactly one solution, no solution, or infinitely many solutions. A system of linear equations is **consistent** if it has one or more solutions and **inconsistent** if no solutions exist. A consistent system is said to be **independent** if it has exactly one solution and **dependent** if it has more than one solution.

Two standard methods for solving system (1) were reviewed: **solution by graphing** and **solution using elimination by addition.**

Two systems of equations are **equivalent** if both have the same solution set. A system of linear equations is transformed into an equivalent system if:

1. Two equations are interchanged.
2. An equation is multiplied by a nonzero constant.
3. A constant multiple of another equation is added to a given equation.

These operations form the basis of solution using elimination by addition. The method of solution using elimination by addition can be transformed into a more efficient method for larger-scale systems by the introduction of an *augmented matrix.* A **matrix** is a rectangular array of numbers written within brackets. Each number in a matrix is called an **element** of the matrix. If a matrix has m rows and n columns, it is called an $m \times n$ **matrix** (read "m by n matrix"). The expression $m \times n$ is called the **size** of the matrix, and the numbers m and n are called the **dimensions** of the matrix. A matrix with n rows and n columns is called a **square matrix of order** n. A matrix with only one column is called a **column matrix,** and a matrix with only one row is called a **row matrix.** The **position** of an element in a matrix is the row and column containing the element. This is usually denoted using **double subscript notation** a_{ij}, where i is the row and j is the column containing the element a_{ij}.

For ease of generalization to larger systems, we change the notation for variables and constants in system (1) to a subscript form:

$$\begin{aligned} a_1x_1 + b_1x_2 &= k_1 \\ a_2x_1 + b_2x_2 &= k_2 \end{aligned} \qquad (2)$$

Associated with each linear system of the form (2), where x_1 and x_2 are variables, is the **augmented matrix** of the system:

Column 1 (C_1)
Column 2 (C_2)
Column 3 (C_3)

$$\left[\begin{array}{cc|c} a_1 & b_1 & k_1 \\ a_2 & b_2 & k_2 \end{array}\right] \begin{array}{l} \leftarrow \text{Row 1 } (R_1) \\ \leftarrow \text{Row 2 } (R_2) \end{array} \qquad (3)$$

Two augmented matrices are **row-equivalent,** denoted by the symbol ~ between the two matrices, if they are augmented matrices of equivalent systems of equations. An augmented matrix is transformed into a row-equivalent matrix if any of the following **row operations** is performed:

1. Two rows are interchanged.
2. A row is multiplied by a nonzero constant.
3. A constant multiple of another row is added to a given row.

The following symbols are used to describe these row operations:

1. $R_i \leftrightarrow R_j$ means "interchange row i with row j."
2. $kR_i \to R_i$ means "multiply row i by the constant k."
3. $kR_j + R_i \to R_i$ means "multiply row j by the constant k and add to R_i."

In solving system (2) using row operations, the objective is to transform the augmented matrix (3) into the form

$$\left[\begin{array}{cc|c} 1 & 0 & m \\ 0 & 1 & n \end{array}\right]$$

If this can be done, then (m, n) is the unique solution of system (2). If (3) is transformed into the form

$$\left[\begin{array}{cc|c} 1 & m & n \\ 0 & 0 & 0 \end{array}\right]$$

then system (2) has infinitely many solutions. If (3) is transformed into the form

$$\left[\begin{array}{cc|c} 1 & m & n \\ 0 & 0 & p \end{array}\right] \qquad p \neq 0$$

then system (2) does not have a solution.

8-2 GAUSS–JORDAN ELIMINATION

In the last part of Section 8-1 we were actually using *Gauss–Jordan elimination* to solve a system of two equations with two variables. The method generalizes completely for systems with more than two variables, and the number of variables does not have to be the same as the number of equations.

As before, our objective is to start with the augmented matrix of a linear system and transform it using row operations into a simple form where the solution can be read by inspection. The simple form, called the **reduced form,** is achieved if:

1. Each row consisting entirely of 0's is below any row having at least one nonzero element.
2. The leftmost nonzero element in each row is 1.
3. The column containing the leftmost 1 of a given row has 0's above and below the 1.
4. The leftmost 1 in any row is to the right of the leftmost 1 in the preceding row.

A **reduced system** is a system of linear equations that corresponds to a reduced augmented matrix. When a reduced system has more variables than equations and contains no contradictions, the system is dependent and has infinitely many solutions.

The **Gauss–Jordan elimination** procedure for solving a system of linear equations is given in step-by-step form as follows:

Step 1. Choose the leftmost nonzero column, and use appropriate row operations to get a 1 at the top.

Step 2. Use multiples of the row containing the 1 from step 1 to get zeros in all remaining places in the column containing this 1.

Step 3. Repeat step 1 with the **submatrix** formed by (mentally) deleting the row used in step 2 and all rows above this row.

Step 4. Repeat step 2 with the **entire matrix,** including the mentally deleted rows. Continue this process until it is impossible to go further.

If at any point in the above process we obtain a row with all 0's to the left of the vertical line and a nonzero number n to the right, we can stop, since we have a contradiction: $0 = n, n \neq 0$. We can then conclude that the system has no solution. If this does not happen and we obtain an augmented matrix in reduced form without any contradictions, the solution can be read by inspection.

8-3 SYSTEMS INVOLVING SECOND-DEGREE EQUATIONS

If a system of equations contains any equations that are not linear, then the system is called a **nonlinear system.** In this section we investigated nonlinear systems involving second-degree terms such as

$$\begin{array}{lll} x^2 + y^2 = 5 & x^2 - 2y^2 = 2 & x^2 + 3xy + y^2 = 20 \\ 3x + y = 1 & xy = 2 & xy - y^2 = 0 \end{array}$$

It can be shown that such systems have at most four solutions, some of which may be imaginary.

Several methods were used to solve nonlinear systems of the indicated form: **solution by substitution, solution using elimination by addition,** and **solution using factoring and substitution.** It is always important to check the solutions of any nonlinear system to ensure that extraneous roots have not been introduced.

8-4 SYSTEMS OF LINEAR INEQUALITIES IN TWO VARIABLES

A graph is often the most convenient way to represent the solution of a linear inequality in two variables or of a system of linear inequalities in two variables.

A vertical line divides a plane into **left** and **right half-planes.** A nonvertical line divides a plane into **upper** and **lower half-planes.** Let A, B, and C be real numbers with A and B not both zero, then the **graph of the linear inequality**

$$Ax + By < C \qquad \text{or} \qquad Ax + By > C$$

with $B \neq 0$, is either the upper half-plane or the lower half-plane (but not both) determined by the line $Ax + By = C$. If $B = 0$, then the graph of

$$Ax < C \qquad \text{or} \qquad Ax > C$$

is either the left half-plane or the right half-plane (but not both) determined by the line $Ax = C$. Out of these results follows an easy **step-by-step procedure for graphing a linear inequality in two variables:**

Step 1. Graph $Ax + By = C$ as a broken line if equality is not included in the original statement or as a solid line if equality is included.

Step 2. Choose a test point anywhere in the plane not on the line and substitute the coordinates into the inequality. The origin (0, 0) often requires the least computation.

Step 3. The graph of the original inequality includes the half-plane containing the test point if the inequality is satisfied by that point, or the half-plane not containing that point if the inequality is not satisfied by that point.

We now turn to systems of linear inequalities in two variables. The **solution to a system of linear inequalities in two variables** is the set of all ordered pairs of real numbers that simultaneously satisfy all the inequalities in the system. The graph is called the **solution region.** In many applications the solution region is also referred to as the **feasible region.** To **find the solution region,** we graph each inequality in the system and then take the intersection of all the graphs. A **corner point** of a solution region is a point in the solution region that is the intersection of two boundary lines. A solution region is **bounded** if it can be enclosed within a circle. If it cannot be enclosed within a circle, then it is **unbounded.**

8-5 LINEAR PROGRAMMING

Linear programming is a mathematical process that has been developed to help management in decision making, and it has become one of the most widely used and best-known tools of management science and industrial engineering.

A **linear programming problem** is one that is concerned with finding the **optimal value** (maximum or minimum value) of a linear **objective function** of the form $z = ax + by$, where the **decision variables** x and y are subject to **problem constraints** in the form of linear inequalities and **nonnegative constraints** $x, y \geq 0$. The set of points satisfying both the problem constraints and the nonnegative constraints is called the **feasible region** for the problem. Any point in the feasible region that produces the optimal value of the objective function over the feasible region is called an **optimal solution.** The **fundamental theorem of linear programming** is basic to the solving of linear programming problems: Let S be the feasible region for a linear programming problem, and let $z = ax + by$ be the objective function. If S is bounded, then z has both a maximum and a minimum value on S and each of these occurs at a corner point of S. If S is unbounded, then a maximum or minimum value of z on S may not exist. However, if either does exist, then it must occur at a corner point of S.

Problems with unbounded feasible regions are not considered in this brief introduction. The theorem leads to a simple **step-by-step solution to linear programming problems with a bounded feasible region:**

Step 1. Form a mathematical model for the problem:
(A) Introduce decision variables and write a linear objective function.
(B) Write problem constraints in the form of linear inequalities.
(C) Write nonnegative constraints.

Step 2. Graph the feasible region and find the corner points.

Step 3. Evaluate the objective function at each corner point to determine the optimal solution.

If two corner points are both optimal solutions of the same type (both produce the same maximum value or both produce the same minimum value) to a linear programming problem, then any point on the line segment joining the two corner points is also an optimal solution of that type.

Chapter 8 Review Exercise

Work through all the problems in this chapter review and check answers in the back of the book. Answers to all review problems are there, and following each answer is a number in italics indicating the section in which that type of problem is discussed. Where weaknesses show up, review appropriate sections in the text.

A

Solve Problems 1–6 using elimination by addition.

1. $2x + y = 7$
$3x - 2y = 0$

2. $3x - 6y = 5$
$-2x + 4y = 1$

3. $4x - 3y = -8$
$-2x + \frac{3}{2}y = 4$

4. $y = x^2 - 5x - 3$
$y = -x + 2$

5. $x^2 + y^2 = 2$
$2x - y = 3$

6. $3x^2 - y^2 = -6$
$2x^2 + 3y^2 = 29$

Solve Problems 7–9 by graphing.

7. $3x - 2y = 8$
$x + 3y = -1$

8. $3x - 4y \geq 24$

9. $2x + y \leq 2$
$x + 2y \geq -2$

Perform each of the row operations indicated in Problems 10–12 on the following augmented matrix:

$$\left[\begin{array}{rr|r} 1 & -4 & 5 \\ 3 & -6 & 12 \end{array}\right]$$

10. $R_1 \leftrightarrow R_2$

11. $\frac{1}{3}R_2 \rightarrow R_2$

12. $(-3)R_1 + R_2 \rightarrow R_2$

In Problems 13–15, write the linear system corresponding to each reduced augmented matrix and solve.

13. $\left[\begin{array}{rr|r} 1 & 0 & 4 \\ 0 & 1 & -7 \end{array}\right]$

14. $\left[\begin{array}{rr|r} 1 & -1 & 4 \\ 0 & 0 & 1 \end{array}\right]$

15. $\left[\begin{array}{rr|r} 1 & -1 & 4 \\ 0 & 0 & 0 \end{array}\right]$

16. Find the maximum and minimum values of $z = 5x + 3y$ over the feasible region S shown in the figure.

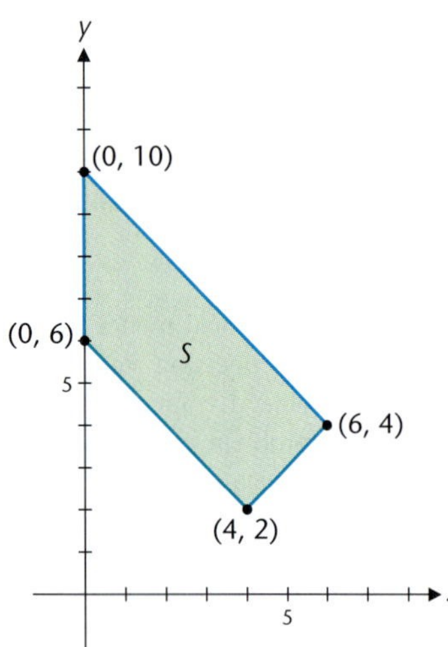

17. Use Gauss–Jordan elimination to solve the system

$$x_1 - x_2 = 4$$
$$2x_1 + x_2 = 2$$

Then write the linear system represented by each augmented matrix in your solution, and solve each of these systems graphically. Discuss the relationship between the solutions of these systems.

18. Use an intersection routine on a graphing utility to approximate the solution of the following system to two decimal places:

$$x + 3y = 9$$
$$-2x + 7y = 10$$

B

Solve Problems 19–24 using Gauss–Jordan elimination.

19. $\begin{aligned} 3x_1 + 2x_2 &= 3 \\ x_1 + 3x_2 &= 8 \end{aligned}$

20. $\begin{aligned} x_1 + x_2 &= 1 \\ x_1 - x_3 &= -2 \\ x_2 + 2x_3 &= 4 \end{aligned}$

21. $\begin{aligned} x_1 + 2x_2 + 3x_3 &= 1 \\ 2x_1 + 3x_2 + 4x_3 &= 3 \\ x_1 + 2x_2 + x_3 &= 3 \end{aligned}$

22. $\begin{aligned} x_1 + 2x_2 - x_3 &= 2 \\ 2x_1 + 3x_2 + x_3 &= -3 \\ 3x_1 + 5x_2 &= -1 \end{aligned}$

23. $\begin{aligned} x_1 - 2x_2 &= 1 \\ 2x_1 - x_2 &= 0 \\ x_1 - 3x_2 &= -2 \end{aligned}$

24. $\begin{aligned} x_1 + 2x_2 - x_3 &= 2 \\ 3x_1 - x_2 + 2x_3 &= -3 \end{aligned}$

Solve Problems 25–27.

25. $\begin{aligned} x^2 - y^2 &= 2 \\ y^2 &= x \end{aligned}$

26. $\begin{aligned} x^2 + 2xy + y^2 &= 1 \\ xy &= -2 \end{aligned}$

27. $\begin{aligned} 2x^2 + xy + y^2 &= 8 \\ x^2 - y^2 &= 0 \end{aligned}$

Solve the systems in Problems 28–30 graphically, and indicate whether each solution region is bounded or unbounded. Find the coordinates of each corner point.

28. $\begin{aligned} 2x + y &\le 8 \\ 2x + 3y &\le 12 \\ x, y &\ge 0 \end{aligned}$

29. $\begin{aligned} 2x + y &\ge 8 \\ x + 3y &\ge 12 \\ x, y &\ge 0 \end{aligned}$

30. $\begin{aligned} x + y &\le 20 \\ x + 4y &\ge 20 \\ x - y &\ge 0 \end{aligned}$

Solve the linear programming problems in Problems 31–33.

31. Maximize $z = 7x + 9y$
Subject to $\begin{aligned} x + 2y &\le 8 \\ 2x + y &\le 10 \\ x, y &\ge 0 \end{aligned}$

32. Minimize $z = 5x + 10y$
Subject to $\begin{aligned} x + y &\le 20 \\ 3x + y &\ge 15 \\ x + 2y &\ge 15 \\ x, y &\ge 0 \end{aligned}$

33. Minimize and maximize $z = 5x + 8y$
Subject to $\begin{aligned} x + 2y &\le 20 \\ 3x + y &\le 15 \\ x + y &\ge 7 \\ x, y &\ge 0 \end{aligned}$

C

34. Solve using Gauss–Jordan elimination:

$$\begin{aligned} x_1 + x_2 + x_3 &= 7{,}000 \\ 0.04x_1 + 0.05x_2 + 0.06x_3 &= 360 \\ 0.04x_1 + 0.05x_2 - 0.06x_3 &= 120 \end{aligned}$$

35. Solve:

$$\begin{aligned} x^2 - xy + y^2 &= 4 \\ x^2 + xy - 2y^2 &= 0 \end{aligned}$$

36. Maximize $z = 30x + 20y$
Subject to $\begin{aligned} 1.2x + 0.6y &\le 960 \\ 0.04x + 0.03y &\le 36 \\ 0.2x + 0.3y &\le 270 \\ x, y &\ge 0 \end{aligned}$

37. Approximate all real solutions to two decimal places:

$$\begin{aligned} x^2 + 4xy + y^2 &= 8 \\ 5x^2 + 2xy + y^2 &= 25 \end{aligned}$$

38. Discuss the number of solutions for the system corresponding to the reduced form shown below if

(A) $m \neq 0$ (B) $m = 0$ and $n \neq 0$
(C) $m = 0$ and $n = 0$

$$\left[\begin{array}{ccc|c} 1 & 0 & -3 & 4 \\ 0 & 1 & 2 & 5 \\ 0 & 0 & m & n \end{array}\right]$$

APPLICATIONS 

39. **Business.** A container holds 120 packages. Some of the packages weigh $\frac{1}{2}$ pound each, and the rest weigh $\frac{1}{3}$ pound each. If the total contents of the container weigh 48 pounds, how many are there of each type of package? Solve using two-equation–two-variable methods.

★ 40. **Geometry.** Find the dimensions of a rectangle with an area of 48 square meters and a perimeter of 28 meters. Solve using two-equation–two-variable methods.

★ 41. **Diet.** A laboratory assistant wishes to obtain a food mix that contains, among other things, 27 grams of protein, 5.4 grams of fat, and 19 grams of moisture. He has available mixes *A*, *B*, and *C* with the compositions listed in the table. How many grams of each mix should be used to get the desired diet mix? Set up a system of equations and solve using Gauss–Jordan elimination.

Mix	Protein (%)	Fat (%)	Moisture (%)
A	30	3	10
B	20	5	20
C	10	4	10

★★ 42. **Puzzle.** A piggy bank contains 30 coins worth \$1.90.
(A) If the bank contains only nickels and dimes, how many coins of each type does it contain?
(B) If the bank contains nickels, dimes, and quarters, how many coins of each type does it contain?

★ 43. **Resource Allocation.** North Star Sail Loft manufactures regular and competition sails. Each regular sail takes 1 labor-hour to cut and 3 labor-hours to sew. Each competition sail takes 2 labor-hours to cut and 4 labor-hours to sew. There are 140 labor-hours available in the cutting department and 360 labor-hours available in the sewing department.
(A) If the loft makes a profit of \$60 on each regular sail and \$100 on each competition sail, how many sails of each type should the company manufacture to maximize its profit? What is the maximum profit?
(B) An increase in the demand for competition sails causes the profit on a competition sail to rise to \$125. Discuss the effect of this change on the number of sails manufactured and on the maximum profit.
(C) A decrease in the demand for competition sails causes the profit on a competition sail to drop to \$75. Discuss the effect of this change on the number of sails manufactured and on the maximum profit.

★ 44. **Nutrition—Animals.** A special diet for laboratory animals is to contain at least 800 units of vitamins, at least 800 units of minerals, and at most 1,300 calories. There are two feed mixes available, mix *A* and mix *B*. A gram of mix *A* contains 5 units of vitamins, 2 units of minerals, and 4 calories. A gram of mix *B* contains 2 units of vitamins, 4 units of minerals, and 4 calories.
(A) If mix *A* costs \$0.07 per gram and mix *B* costs \$0.04 per gram, how many grams of each mix should be used to satisfy the requirements of the diet at minimal cost? What is the minimal cost?
(B) If the price of mix *B* decreases to \$0.02 per gram, discuss the effect of this change on the solution in part A.
(C) If the price of mix *B* increases to \$0.15 per gram, discuss the effect of this change on the solution in part A.

Matrices and Determinants

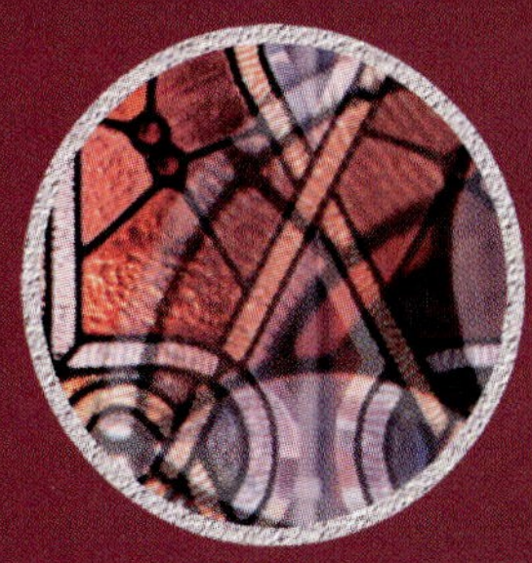

Chapter 9

In this chapter we discuss matrices in more detail. In the first three sections we define and study some algebraic operations on matrices, including addition, multiplication, and inversion. The next three sections deal with the determinant of a matrix.

In the last chapter we used row operations and Gauss–Jordan elimination to solve systems of linear equations. Row operations play a prominent role in the development of several topics in this chapter. One consequence of our discussion will be the development of two additional methods for solving systems of linear equations: one method involves inverse matrices and the other determinants.

Matrices are both a very ancient and a very current mathematical concept. References to matrices and systems of equations can be found in Chinese manuscripts dating back to around 200 B.C. Over the years, mathematicians and scientists have found many applications of matrices. More recently, the advent of personal and large-scale computers has increased the use of matrices in a wide variety of applications. In 1979 Dan Bricklin and Robert Frankston introduced VisiCalc, the first electronic spreadsheet program for personal computers. Simply put, a *spreadsheet* is a computer program that allows the user to enter and manipulate numbers, often using matrix notation and operations. Spreadsheets were initially used by businesses in areas such as budgeting, sales projections, and cost estimation. However, many other applications have begun to appear. For example, a scientist can use a spreadsheet to analyze the results of an experiment, or a teacher can use one to record and average grades. There are even spreadsheets that can be used to help compute an individual's income tax.

SECTION 9-1 Matrices: Basic Operations

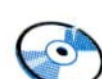

- Addition and Subtraction
- Multiplication of a Matrix by a Number
- Matrix Product

In Section 8-1 we introduced basic matrix terminology and solved systems of equations by performing row operations on augmented coefficient matrices. Matrices have many other useful applications and possess an interesting mathematical structure in their own right. As we will see, matrix addition and multiplication are similar to real number addition and multiplication in many respects, but there are some important differences. To help you understand the similarities and the differences, you should review the basic properties of real number operations discussed in Section A-1.

• Addition and Subtraction

Before we can discuss arithmetic operations for matrices, we have to define equality for matrices. Two matrices are **equal** if they have the same size and their corresponding elements are equal. For example,

$$\overset{2 \times 3}{\begin{bmatrix} a & b & c \\ d & e & f \end{bmatrix}} = \overset{2 \times 3}{\begin{bmatrix} u & v & w \\ x & y & z \end{bmatrix}} \quad \text{if and only if} \quad \begin{matrix} a = u & b = v & c = w \\ d = x & e = y & f = z \end{matrix}$$

The **sum of two matrices** of the same size is a matrix with elements that are the sums of the corresponding elements of the two given matrices.

Addition is not defined for matrices of different sizes.

EXAMPLE 1 **Matrix Addition**

(A) $\begin{bmatrix} a & b \\ c & d \end{bmatrix} + \begin{bmatrix} w & x \\ y & z \end{bmatrix} = \begin{bmatrix} (a+w) & (b+x) \\ (c+y) & (d+z) \end{bmatrix}$

(B) $\begin{bmatrix} 2 & -3 & 0 \\ 1 & 2 & -5 \end{bmatrix} + \begin{bmatrix} 3 & 1 & 2 \\ -3 & 2 & 5 \end{bmatrix} = \begin{bmatrix} 5 & -2 & 2 \\ -2 & 4 & 0 \end{bmatrix}$

Matched Problem 1 Add: $\begin{bmatrix} 3 & 2 \\ -1 & -1 \\ 0 & 3 \end{bmatrix} + \begin{bmatrix} -2 & 3 \\ 1 & -1 \\ 2 & -2 \end{bmatrix}$

```
[A]
     [[2 -3 0 ]
      [1 2   -5]]
[B]
      [[3  1 2]
       [-3 2 5]]
[A]+[B]
     [[5  -2 2]
      [-2 4  0]]
```

FIGURE 1 Addition on a graphing calculator.

Graphing utilities can also be used to solve problems involving matrix operations. Figure 1 illustrates the solution to Example 1B on a graphing calculator.

Because we add two matrices by adding their corresponding elements, it follows from the properties of real numbers that matrices of the same size are commutative and associative relative to addition. That is, if A, B, and C are matrices of the same size, then

$$A + B = B + A \qquad \text{Commutative}$$

$$(A + B) + C = A + (B + C) \qquad \text{Associative}$$

A matrix with elements that are all 0's is called a **zero matrix.** For example, the following are zero matrices of different sizes:

$$\begin{bmatrix} 0 & 0 & 0 \end{bmatrix} \quad \begin{bmatrix} 0 & 0 \\ 0 & 0 \end{bmatrix} \quad \begin{bmatrix} 0 \\ 0 \\ 0 \\ 0 \end{bmatrix} \quad \begin{bmatrix} 0 & 0 & 0 & 0 \\ 0 & 0 & 0 & 0 \\ 0 & 0 & 0 & 0 \end{bmatrix}$$

[*Note:* "0" may be used to denote the zero matrix of any size.]

The **negative of a matrix *M*,** denoted by $-M$, is a matrix with elements that are the negatives of the elements in M. Thus, if

$$M = \begin{bmatrix} a & b \\ c & d \end{bmatrix}$$

then

$$-M = \begin{bmatrix} -a & -b \\ -c & -d \end{bmatrix}$$

Note that $M + (-M) = 0$ (a zero matrix).

If A and B are matrices of the same size, then we define **subtraction** as follows:

$$A - B = A + (-B)$$

Thus, to subtract matrix B from matrix A, we simply subtract corresponding elements.

EXAMPLE 2 **Matrix Subtraction**

$$\begin{bmatrix} 3 & -2 \\ 5 & 0 \end{bmatrix} - \begin{bmatrix} -2 & 2 \\ 3 & 4 \end{bmatrix} = \begin{bmatrix} 3 & -2 \\ 5 & 0 \end{bmatrix} + \begin{bmatrix} 2 & -2 \\ -3 & -4 \end{bmatrix} = \begin{bmatrix} 5 & -4 \\ 2 & -4 \end{bmatrix}$$

Matched Problem 2 Subtract: $[2 \quad -3 \quad 5] - [3 \quad -2 \quad 1]$

• Multiplication of a Matrix by a Number

The **product of a number k and a matrix M,** denoted by kM, is a matrix formed by multiplying each element of M by k.

EXAMPLE 3 **Multiplication of a Matrix by a Number**

$$-2\begin{bmatrix} 3 & -1 & 0 \\ -2 & 1 & 3 \\ 0 & -1 & -2 \end{bmatrix} = \begin{bmatrix} -6 & 2 & 0 \\ 4 & -2 & -6 \\ 0 & 2 & 4 \end{bmatrix}$$

Matched Problem 3 Find: $10\begin{bmatrix} 1.3 \\ 0.2 \\ 3.5 \end{bmatrix}$

EXPLORE-DISCUSS 1 Multiplication of two numbers can be interpreted as repeated addition if one of the numbers is a positive integer. That is,

$$2a = a + a \qquad 3a = a + a + a \qquad 4a = a + a + a + a$$

and so on. Discuss this interpretation for the product of an integer k and a matrix M. Use specific examples to illustrate your remarks.

We now consider an application that uses various matrix operations.

EXAMPLE 4 Sales and Commissions

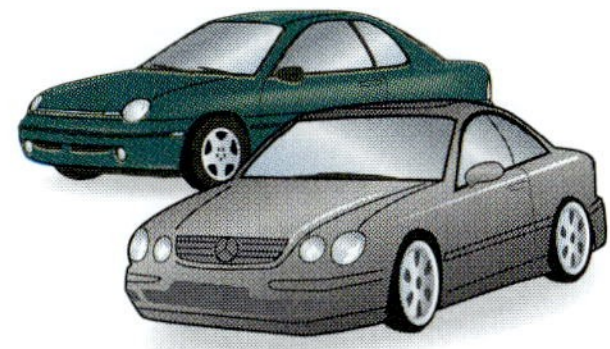

Ms. Fong and Mr. Petris are salespeople for a new car agency that sells only two models. August was the last month for this year's models, and next year's models were introduced in September. Gross dollar sales for each month are given in the following matrices:

$$\begin{array}{l} \\ \text{Fong} \\ \text{Petris} \end{array} \overset{\text{AUGUST SALES}}{\begin{array}{cc} \text{Compact} & \text{Luxury} \end{array}} \begin{bmatrix} \$36{,}000 & \$72{,}000 \\ \$72{,}000 & \$0 \end{bmatrix} = A \qquad \overset{\text{SEPTEMBER SALES}}{\begin{array}{cc} \text{Compact} & \text{Luxury} \end{array}} \begin{bmatrix} \$144{,}000 & \$288{,}000 \\ \$180{,}000 & \$216{,}000 \end{bmatrix} = B$$

For example, Ms. Fong had \$36,000 in compact sales in August and Mr. Petris had \$216,000 in luxury car sales in September.

(A) What were the combined dollar sales in August and September for each salesperson and each model?
(B) What was the increase in dollar sales from August to September?
(C) If both salespeople receive a 3% commission on gross dollar sales, compute the commission for each salesperson for each model sold in September.

Solutions We use matrix addition for part A, matrix subtraction for part B, and multiplication of a matrix by a number for part C.

$$\text{(A)}\quad A + B = \begin{matrix} \text{Compact} \quad \text{Luxury} \\ \begin{bmatrix} \$180{,}000 & \$360{,}000 \\ \$252{,}000 & \$216{,}000 \end{bmatrix} \end{matrix} \begin{array}{l} \text{Fong} \\ \text{Petris} \end{array}$$

$$\text{(B)}\quad B - A = \begin{bmatrix} \$108{,}000 & \$216{,}000 \\ \$108{,}000 & \$216{,}000 \end{bmatrix} \begin{array}{l} \text{Fong} \\ \text{Petris} \end{array}$$

$$\text{(C)}\quad 0.03B = \begin{matrix} \text{Compact} \qquad\qquad \text{Luxury} \\ \begin{bmatrix} (0.03)(\$144{,}000) & (0.03)(\$288{,}000) \\ (0.03)(\$180{,}000) & (0.03)(\$216{,}000) \end{bmatrix} \end{matrix}$$

$$= \begin{bmatrix} \$4{,}320 & \$8{,}640 \\ \$5{,}400 & \$6{,}480 \end{bmatrix} \begin{array}{l} \text{Fong} \\ \text{Petris} \end{array}$$

Matched Problem 4 Repeat Example 4 with

$$A = \begin{bmatrix} \$72{,}000 & \$72{,}000 \\ \$36{,}000 & \$72{,}000 \end{bmatrix} \quad \text{and} \quad B = \begin{bmatrix} \$180{,}000 & \$216{,}000 \\ \$144{,}000 & \$216{,}000 \end{bmatrix}$$

Example 4 involved an agency with only two salespeople and two models. A more realistic problem might involve 20 salespeople and 15 models. Problems of this size are often solved with the aid of a spreadsheet on a personal computer. Figure 2 illustrates a computer spreadsheet solution for Example 4.

	A	B	C	D	E	F	G
1		Compact	Luxury	Compact	Luxury	Compact	Luxury
2		August Sales		September Sales		September Commissions	
3	Fong	\$36,000	\$72,000	\$144,000	\$288,000	\$4,320	\$8,640
4	Petris	\$72,000	\$0	\$180,000	\$216,000	\$5,400	\$6,480
5		Combined Sales		Sales Increases			
6	Fong	\$180,000	\$360,000	\$108,000	\$216,000		
7	Petris	\$252,000	\$216,000	\$108,000	\$216,000		

FIGURE 2

• Matrix Product

Now we are going to introduce a matrix multiplication that may at first seem rather strange. In spite of its apparent strangeness, this operation is well-founded in the general theory of matrices and, as we will see, is extremely useful in many practical problems.

Historically, matrix multiplication was introduced by the English mathematician Arthur Cayley (1821–1895) in studies of linear equations and linear transformations. In Section 9-3, you will see how matrix multiplication is central to the process of expressing systems of equations as matrix equations and to the process of solving matrix equations. Matrix equations and their solutions provide us with an alternate method of solving linear systems with the same number of variables as equations.

We start by defining the product of two special matrices, a row matrix and a column matrix.

DEFINITION 1

Product of a Row Matrix and a Column Matrix

The **product** of a $1 \times n$ row matrix and an $n \times 1$ column matrix is a 1×1 matrix given by

$$\underset{1 \times n}{[a_1 \quad a_2 \quad \cdots \quad a_n]} \underset{n \times 1}{\begin{bmatrix} b_1 \\ b_2 \\ \vdots \\ b_n \end{bmatrix}} = [a_1b_1 + a_2b_2 + \cdots + a_nb_n]$$

Note that the number of elements in the row matrix and in the column matrix must be the same for the product to be defined.

EXAMPLE 5 **Product of a Row Matrix and a Column Matrix**

$$\begin{bmatrix} 2 & -3 & 0 \end{bmatrix} \begin{bmatrix} -5 \\ 2 \\ -2 \end{bmatrix} = [(2)(-5) + (-3)(2) + (0)(-2)]$$

$$= [-10 - 6 + 0] = [-16]$$

Matched Problem 5 $\begin{bmatrix} -1 & 0 & 3 & 2 \end{bmatrix} \begin{bmatrix} 2 \\ 3 \\ 4 \\ -1 \end{bmatrix} = ?$

Refer to Example 5. The distinction between the real number -16 and the 1×1 matrix $[-16]$ is a technical one, and it is common to see 1×1 matrices written as real numbers without brackets. In the work that follows, we will frequently refer to 1×1 matrices as real numbers and omit the brackets whenever it is convenient to do so.

EXAMPLE 6 **Production Scheduling**

A factory produces a slalom water ski that requires 4 labor-hours in the fabricating department and 1 labor-hour in the finishing department. Fabricating personnel receive \$10 per hour, and finishing personnel receive \$8 per hour. Total labor cost per ski is given by the product

$$\begin{bmatrix} 4 & 1 \end{bmatrix} \begin{bmatrix} 10 \\ 8 \end{bmatrix} = [(4)(10) + (1)(8)] = [40 + 8] = [48] \text{ or \$48 per ski}$$

Matched Problem 6 If the factory in Example 6 also produces a trick water ski that requires 6 labor-hours in the fabricating department and 1.5 labor-hours in the finishing department, write a product between appropriate row and column matrices that gives the total labor cost for this ski. Compute the cost.

We now use the product of a $1 \times n$ row matrix and an $n \times 1$ column matrix to extend the definition of matrix product to more general matrices.

DEFINITION 2 **Matrix Product**

If A is an $m \times p$ matrix and B is a $p \times n$ matrix, then the **matrix product** of A and B, denoted AB, is an $m \times n$ matrix whose element in the ith row and jth column is the real number obtained from the product of the ith row of A and the jth column of B. If the number of columns in A does not equal the number of rows in B, then the matrix product AB is **not defined.**

It is important to check sizes before starting the multiplication process. If A is an $a \times b$ matrix and B is a $c \times d$ matrix, then if $b = c$, the product AB will exist and will be an $a \times d$ matrix (see Fig. 3). If $b \neq c$, then the product AB does not exist.

FIGURE 3

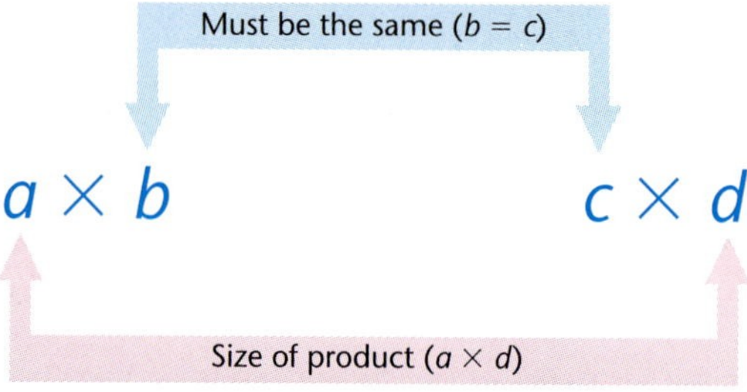

The definition is not as complicated as it might first seem. An example should help clarify the process. For

$$A = \begin{bmatrix} 2 & 3 & -1 \\ -2 & 1 & 2 \end{bmatrix} \quad \text{and} \quad B = \begin{bmatrix} 1 & 3 \\ 2 & 0 \\ -1 & 2 \end{bmatrix}$$

A is 2×3, B is 3×2, and so AB is 2×2. To find the first row of AB, we take the product of the first row of A with every column of B and write each result as a real number, not a 1×1 matrix. The second row of AB is computed in the same manner. The four products of row and column matrices used to produce the four elements in AB are shown in the dashed box below. These products are usually calculated mentally, or with the aid of a calculator, and need not be written out. The shaded portions highlight the steps involved in computing the element in the first row and second column of AB.

$$\underset{2 \times 3}{\begin{bmatrix} 2 & 3 & -1 \\ -2 & 1 & 2 \end{bmatrix}} \underset{3 \times 2}{\begin{bmatrix} 1 & 3 \\ 2 & 0 \\ -1 & 2 \end{bmatrix}} = \begin{bmatrix} [2 \quad 3 \quad -1]\begin{bmatrix} 1 \\ 2 \\ -1 \end{bmatrix} & [2 \quad 3 \quad -1]\begin{bmatrix} 3 \\ 0 \\ 2 \end{bmatrix} \\ [-2 \quad 1 \quad 2]\begin{bmatrix} 1 \\ 2 \\ -1 \end{bmatrix} & [-2 \quad 1 \quad 2]\begin{bmatrix} 3 \\ 0 \\ 2 \end{bmatrix} \end{bmatrix}$$

$$= \underset{2 \times 2}{\begin{bmatrix} 9 & 4 \\ -2 & -2 \end{bmatrix}}$$

EXAMPLE 7 Matrix Product

(A) $\underset{3 \times 2}{\begin{bmatrix} 2 & 1 \\ 1 & 0 \\ -1 & 2 \end{bmatrix}} \underset{2 \times 4}{\begin{bmatrix} 1 & -1 & 0 & 1 \\ 2 & 1 & 2 & 0 \end{bmatrix}} = \underset{3 \times 4}{\begin{bmatrix} 4 & -1 & 2 & 2 \\ 1 & -1 & 0 & 1 \\ 3 & 3 & 4 & -1 \end{bmatrix}}$

(B) $\underset{2 \times 4}{\begin{bmatrix} 1 & -1 & 0 & 1 \\ 2 & 1 & 2 & 0 \end{bmatrix}} \underset{3 \times 2}{\begin{bmatrix} 2 & 1 \\ 1 & 0 \\ -1 & 2 \end{bmatrix}}$ Product is not defined

(C) $\begin{bmatrix} 2 & 6 \\ -1 & -3 \end{bmatrix} \begin{bmatrix} 1 & 2 \\ 3 & 6 \end{bmatrix} = \begin{bmatrix} 20 & 40 \\ -10 & -20 \end{bmatrix}$

(D) $\begin{bmatrix} 1 & 2 \\ 3 & 6 \end{bmatrix} \begin{bmatrix} 2 & 6 \\ -1 & -3 \end{bmatrix} = \begin{bmatrix} 0 & 0 \\ 0 & 0 \end{bmatrix}$

(E) $\begin{bmatrix} 2 & -3 & 0 \end{bmatrix} \begin{bmatrix} -5 \\ 2 \\ -2 \end{bmatrix} = [-16]$

(F) $\begin{bmatrix} -5 \\ 2 \\ -2 \end{bmatrix} \begin{bmatrix} 2 & -3 & 0 \end{bmatrix} = \begin{bmatrix} -10 & 15 & 0 \\ 4 & -6 & 0 \\ -4 & 6 & 0 \end{bmatrix}$

Matched Problem 7

Find each product, if it is defined:

(A) $\begin{bmatrix} -1 & 0 & 3 & -2 \\ 1 & 2 & 2 & 0 \end{bmatrix} \begin{bmatrix} -1 & 1 \\ 2 & 3 \\ 1 & 0 \end{bmatrix}$ (B) $\begin{bmatrix} -1 & 1 \\ 2 & 3 \\ 1 & 0 \end{bmatrix} \begin{bmatrix} -1 & 0 & 3 & -2 \\ 1 & 2 & 2 & 0 \end{bmatrix}$

(C) $\begin{bmatrix} 1 & 2 \\ -1 & -2 \end{bmatrix} \begin{bmatrix} -2 & 4 \\ 1 & -2 \end{bmatrix}$ (D) $\begin{bmatrix} -2 & 4 \\ 1 & -2 \end{bmatrix} \begin{bmatrix} 1 & 2 \\ -1 & -2 \end{bmatrix}$

(E) $\begin{bmatrix} 3 & -2 & 1 \end{bmatrix} \begin{bmatrix} 4 \\ 2 \\ 3 \end{bmatrix}$ (F) $\begin{bmatrix} 4 \\ 2 \\ 3 \end{bmatrix} \begin{bmatrix} 3 & -2 & 1 \end{bmatrix}$

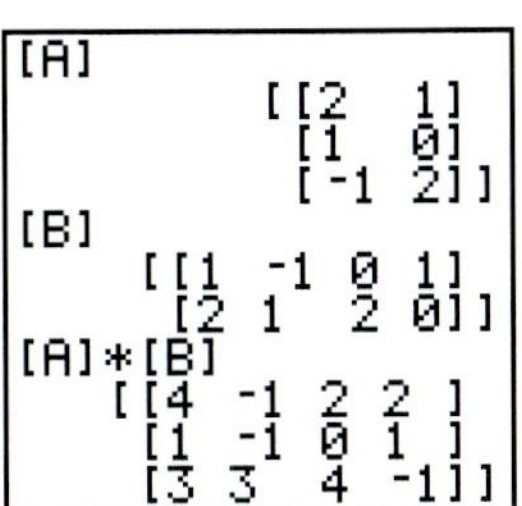

FIGURE 4 Multiplication on a graphing calculator.

Figure 4 illustrates a graphing calculator solution to Example 7A. What would you expect to happen if you tried to solve Example 7B on a graphing calculator?

In the arithmetic of real numbers it does not matter in which order we multiply; for example, $5 \times 7 = 7 \times 5$. In matrix multiplication, however, it does make a difference. That is, AB does not always equal BA, even if both multiplications are defined and both products are the same size (see Examples 7C and 7D). Thus,

Matrix multiplication is not commutative.

Also, AB may be zero with neither A nor B equal to zero (see Example 7D). Thus,

The zero property does not hold for matrix multiplication.

(See Section A-1 for a discussion of the zero property for real numbers.)

Just as we used the familiar algebraic notation AB to represent the product of matrices A and B, we use the notation A^2 for AA, the product of A with itself, A^3 for AAA, and so on.

EXPLORE-DISCUSS 2 In addition to the commutative and zero properties, there are other significant differences between real number multiplication and matrix multiplication.

(A) In real number multiplication, the only real number whose square is 0 is the real number 0 ($0^2 = 0$). Find at least one 2×2 matrix A with all elements nonzero such that $A^2 = 0$, where 0 is the 2×2 zero matrix.

(B) In real number multiplication, the only nonzero real number that is equal to its square is the real number 1 ($1^2 = 1$). Find at least one 2×2 matrix A with all elements nonzero such that $A^2 = A$.

We will continue our discussion of properties of matrix multiplication later in this chapter. Now we consider an application of matrix multiplication.

EXAMPLE 8 Labor Costs

Let us combine the time requirements for slalom and trick water skis discussed in Example 6 and Matched Problem 6 into one matrix:

	Labor-hours per ski	
	Assembly department	Finishing department
Trick ski	6 h	1.5 h
Slalom ski	4 h	1 h

$$\begin{bmatrix} 6\text{ h} & 1.5\text{ h} \\ 4\text{ h} & 1\text{ h} \end{bmatrix} = L$$

Now suppose that the company has two manufacturing plants, X and Y, in different parts of the country and that the hourly rates for each department are given in the following matrix:

	Hourly wages	
	Plant X	Plant Y
Assembly department	\$10	\$12
Finishing department	\$ 8	\$10

$$\begin{bmatrix} \$10 & \$12 \\ \$\ 8 & \$10 \end{bmatrix} = H$$

Since H and L are both 2×2 matrices, we can take the product of H and L in either order and the result will be a 2×2 matrix:

$$HL = \begin{bmatrix} 10 & 12 \\ 8 & 10 \end{bmatrix} \begin{bmatrix} 6 & 1.5 \\ 4 & 1 \end{bmatrix} = \begin{bmatrix} 108 & 27 \\ 88 & 22 \end{bmatrix}$$

$$LH = \begin{bmatrix} 6 & 1.5 \\ 4 & 1 \end{bmatrix} \begin{bmatrix} 10 & 12 \\ 8 & 10 \end{bmatrix} = \begin{bmatrix} 72 & 87 \\ 48 & 58 \end{bmatrix}$$

How can we interpret the elements in these products? Let's begin with the product HL. The element 108 in the first row and first column of HL is the product of the first row matrix of H and the first column matrix of L:

$$\begin{matrix} \text{Plant} & \text{Plant} \\ X & Y \end{matrix}$$
$$\begin{bmatrix} 10 & 12 \end{bmatrix} \begin{bmatrix} 6 \\ 4 \end{bmatrix} \begin{matrix} \text{Trick} \\ \text{Slalom} \end{matrix} = 10(6) + 12(4) = 60 + 48 = 108$$

Notice that \$60 is the labor cost for assembling a trick ski at the California plant and \$48 is the labor cost for assembling a slalom ski at the Wisconsin plant. Although both numbers represent labor costs, it makes no sense to add them together. They do not pertain to the same type of ski or to the same plant. Thus, even though the product HL happens to be defined mathematically, it has no useful interpretation in this problem.

Now let's consider the product LH. The element 72 in the first row and first column of LH is given by the following product:

$$\begin{matrix} \text{Assembly} & \text{Finishing} \end{matrix}$$
$$\begin{bmatrix} 6 & 1.5 \end{bmatrix} \begin{bmatrix} 10 \\ 8 \end{bmatrix} \begin{matrix} \text{Assembly} \\ \text{Finishing} \end{matrix} = 6(10) + 1.5(8)$$
$$= 60 + 12 = 72$$

where \$60 is the labor cost for assembling a trick ski at plant X and \$12 is the labor cost for finishing a trick ski at plant X. Thus, the sum is the total labor cost for producing a trick ski at plant X. The other elements in LH also represent total labor costs, as indicated by the row and column labels shown below:

$$\begin{matrix} \text{Labor costs per ski} \\ \begin{matrix} \text{Plant} & \text{Plant} \\ X & Y \end{matrix} \end{matrix}$$
$$LH = \begin{bmatrix} \$72 & \$87 \\ \$48 & \$58 \end{bmatrix} \begin{matrix} \text{Trick ski} \\ \text{Slalom ski} \end{matrix}$$

Matched Problem 8 Refer to Example 8. The company wants to know how many hours to schedule in each department in order to produce 1,000 trick skis and 2,000 slalom skis. These production requirements can be represented by either of the following matrices:

$$\begin{matrix} & \text{Trick} & \text{Slalom} \\ & \text{skis} & \text{skis} \end{matrix}$$
$$P = \begin{bmatrix} 1{,}000 & 2{,}000 \end{bmatrix} \qquad Q = \begin{bmatrix} 1{,}000 \\ 2{,}000 \end{bmatrix} \begin{matrix} \text{Trick skis} \\ \text{Slalom skis} \end{matrix}$$

Using the labor-hour matrix L from Example 8, find PL or LQ, whichever has a meaningful interpretation for this problem, and label the rows and columns accordingly.

CAUTION Example 8 and Problem 8 illustrate an important point about matrix multiplication. Even if you are using a graphing utility to perform the calculations in a matrix product, it is still necessary for you to know the definition of matrix multiplication so that you can interpret the results correctly.

Answers to Matched Problems

1. $\begin{bmatrix} 1 & 5 \\ 0 & -2 \\ 2 & 1 \end{bmatrix}$ **2.** $[-1 \quad -1 \quad 4]$ **3.** $\begin{bmatrix} 13 \\ 2 \\ 35 \end{bmatrix}$

4. (A) $\begin{bmatrix} \$252{,}000 & \$288{,}000 \\ \$180{,}000 & \$288{,}000 \end{bmatrix}$ (B) $\begin{bmatrix} \$108{,}000 & \$144{,}000 \\ \$108{,}000 & \$144{,}000 \end{bmatrix}$ (C) $\begin{bmatrix} \$5{,}400 & \$6{,}480 \\ \$4{,}320 & \$6{,}480 \end{bmatrix}$

5. $[8]$ **6.** $[6 \quad 1.5]\begin{bmatrix} 10 \\ 8 \end{bmatrix} = [72]$ or \$72

7. (A) Not defined (B) $\begin{bmatrix} 2 & 2 & -1 & 2 \\ 1 & 6 & 12 & -4 \\ -1 & 0 & 3 & -2 \end{bmatrix}$ (C) $\begin{bmatrix} 0 & 0 \\ 0 & 0 \end{bmatrix}$ (D) $\begin{bmatrix} -6 & -12 \\ 3 & 6 \end{bmatrix}$

(E) $[11]$ (F) $\begin{bmatrix} 12 & -8 & 4 \\ 6 & -4 & 2 \\ 9 & -6 & 3 \end{bmatrix}$

8. $PL = \begin{matrix} \text{Assembly} & \text{Finishing} \\ [14{,}000 & 3{,}500] \end{matrix}$ Labor hours

EXERCISE 9-1

A

Perform the indicated operations in Problems 1–18, if possible.

1. $\begin{bmatrix} 5 & -2 \\ 3 & 0 \end{bmatrix} + \begin{bmatrix} -3 & 7 \\ 1 & -6 \end{bmatrix}$

2. $\begin{bmatrix} 0 & 8 \\ 2 & -1 \end{bmatrix} + \begin{bmatrix} 9 & -4 \\ 7 & 5 \end{bmatrix}$

3. $\begin{bmatrix} 4 & 0 \\ -2 & 3 \\ 8 & 1 \end{bmatrix} + \begin{bmatrix} -1 & 2 \\ 0 & 5 \\ 4 & -6 \end{bmatrix}$

4. $\begin{bmatrix} 6 & -2 & 3 \\ 4 & -8 & -7 \end{bmatrix} + \begin{bmatrix} 3 & 9 & -1 \\ 6 & -2 & 4 \end{bmatrix}$

5. $\begin{bmatrix} 4 & 0 \\ -2 & 3 \\ 8 & 1 \end{bmatrix} + \begin{bmatrix} -1 & 0 & 4 \\ 2 & 5 & -6 \end{bmatrix}$

6. $\begin{bmatrix} 6 & -2 & 3 \\ 4 & -8 & -7 \end{bmatrix} + \begin{bmatrix} 3 & 6 \\ 9 & -2 \\ -1 & 4 \end{bmatrix}$

7. $\begin{bmatrix} 5 & -1 & 0 \\ 4 & 6 & 3 \end{bmatrix} - \begin{bmatrix} 2 & 4 & -6 \\ 3 & 5 & -5 \end{bmatrix}$

8. $\begin{bmatrix} 6 & 2 \\ -4 & 1 \\ 3 & 0 \end{bmatrix} - \begin{bmatrix} 0 & 5 \\ -7 & 2 \\ -1 & 0 \end{bmatrix}$

9. $4\begin{bmatrix} 3 & -4 & 7 \\ -2 & 9 & 5 \end{bmatrix}$

10. $5\begin{bmatrix} -7 & 3 & 0 & 9 \\ 4 & -5 & 6 & 2 \end{bmatrix}$

11. $[5 \quad 3]\begin{bmatrix} 4 \\ 7 \end{bmatrix}$

12. $[-2 \quad 4]\begin{bmatrix} 3 \\ -8 \end{bmatrix}$

13. $\begin{bmatrix} -6 & 3 \\ 2 & -5 \end{bmatrix}\begin{bmatrix} 1 \\ 3 \end{bmatrix}$

14. $\begin{bmatrix} 3 & 7 \\ -1 & -9 \end{bmatrix}\begin{bmatrix} 4 \\ -1 \end{bmatrix}$

15. $\begin{bmatrix} 5 & 1 \\ 4 & 6 \end{bmatrix}\begin{bmatrix} 2 & 0 \\ 3 & 8 \end{bmatrix}$

16. $\begin{bmatrix} -2 & 7 \\ 3 & -1 \end{bmatrix}\begin{bmatrix} 4 & -1 \\ 0 & 5 \end{bmatrix}$

17. $\begin{bmatrix} 8 & -3 \\ -5 & 3 \end{bmatrix}\begin{bmatrix} 2 & 0 \\ 0 & 6 \end{bmatrix}$

18. $\begin{bmatrix} 7 & 0 \\ 0 & 3 \end{bmatrix}\begin{bmatrix} 9 & -2 \\ -4 & -1 \end{bmatrix}$

B

Find the products in Problems 19–26.

19. $\begin{bmatrix} 3 & -6 \end{bmatrix}\begin{bmatrix} -2 \\ 5 \end{bmatrix}$

20. $\begin{bmatrix} -4 & 2 \end{bmatrix}\begin{bmatrix} -8 \\ -1 \end{bmatrix}$

21. $\begin{bmatrix} -2 \\ 5 \end{bmatrix}\begin{bmatrix} 3 & -6 \end{bmatrix}$

22. $\begin{bmatrix} -8 \\ -1 \end{bmatrix}\begin{bmatrix} -4 & 2 \end{bmatrix}$

23. $\begin{bmatrix} 5 & 0 & -3 \end{bmatrix}\begin{bmatrix} 1 \\ -2 \\ 6 \end{bmatrix}$

24. $\begin{bmatrix} 6 & -1 & 2 \end{bmatrix}\begin{bmatrix} 4 \\ 0 \\ -4 \end{bmatrix}$

25. $\begin{bmatrix} 1 \\ -2 \\ 6 \end{bmatrix}\begin{bmatrix} 5 & 0 & -3 \end{bmatrix}$

26. $\begin{bmatrix} 4 \\ 0 \\ -4 \end{bmatrix}\begin{bmatrix} 6 & -1 & 2 \end{bmatrix}$

Problems 27–44 refer to the following matrices.

$$A = \begin{bmatrix} 3 & 2 & 0 \\ -1 & 4 & -6 \end{bmatrix} \quad B = \begin{bmatrix} 5 & -2 \\ 1 & 3 \end{bmatrix}$$

$$C = \begin{bmatrix} 1 & 2 & -4 \\ 0 & -2 & 3 \\ 5 & 0 & 4 \end{bmatrix} \quad D = \begin{bmatrix} 2 & 0 \\ -1 & 6 \\ -3 & 7 \end{bmatrix}$$

Perform the indicated operations, if possible.

27. AB **28.** BA **29.** AC

30. CA **31.** A^2 **32.** B^2

33. C^2 **34.** AD **35.** $2CD$

36. $(-5)DB$ **37.** $3AC - 4BD$ **38.** $2BA + CD$

39. $5DA - 6C$ **40.** $3B - 2AD$ **41.** DAC

42. CDB **43.** ADB **44.** BAB

45. Find a, b, c, and d so that

$$\begin{bmatrix} 5 & -2 \\ 8 & 4 \end{bmatrix} + \begin{bmatrix} a & b \\ c & d \end{bmatrix} = \begin{bmatrix} 9 & 3 \\ -1 & 0 \end{bmatrix}$$

46. Find w, x, y, and z so that

$$\begin{bmatrix} w & x \\ y & z \end{bmatrix} + \begin{bmatrix} -3 & 6 \\ -2 & 1 \end{bmatrix} = \begin{bmatrix} -4 & 7 \\ 2 & -5 \end{bmatrix}$$

47. Find x and y so that

$$\begin{bmatrix} 1 & 2x \\ -3x & -1 \end{bmatrix} + \begin{bmatrix} 3 & -3y \\ 6y & 4 \end{bmatrix} = \begin{bmatrix} 4 & 4 \\ -3 & 3 \end{bmatrix}$$

48. Find x and y so that

$$\begin{bmatrix} x & 5 \\ 8 & x \end{bmatrix} + \begin{bmatrix} 2y & 2 \\ 1 & -y \end{bmatrix} = \begin{bmatrix} 4 & 7 \\ 9 & 7 \end{bmatrix}$$

C

49. Find a, b, c, and d so that

$$\begin{bmatrix} 1 & 4 \\ 0 & 1 \end{bmatrix}\begin{bmatrix} a & b \\ c & d \end{bmatrix} = \begin{bmatrix} 1 & 0 \\ 0 & 1 \end{bmatrix}$$

50. Find w, x, y, and z so that

$$\begin{bmatrix} w & x \\ y & z \end{bmatrix}\begin{bmatrix} 1 & 5 \\ 0 & -1 \end{bmatrix} = \begin{bmatrix} 1 & 0 \\ 0 & 1 \end{bmatrix}$$

51. Find a, b, c, and d so that

$$\begin{bmatrix} 2 & 5 \\ 1 & 3 \end{bmatrix}\begin{bmatrix} a & b \\ c & d \end{bmatrix} = \begin{bmatrix} 4 & 7 \\ 3 & 3 \end{bmatrix}$$

52. Find w, x, y, and z so that

$$\begin{bmatrix} w & x \\ y & z \end{bmatrix}\begin{bmatrix} 4 & -3 \\ -3 & 2 \end{bmatrix} = \begin{bmatrix} 8 & 7 \\ 4 & -4 \end{bmatrix}$$

A 2 × 2 ***diagonal matrix*** *is a matrix of the form*

$$A = \begin{bmatrix} a & 0 \\ 0 & d \end{bmatrix}$$

where a and d are any real numbers; if $a = d = 1$, A is called the 2 × 2 ***identity matrix****. In Problems 53–64, determine whether the statement is true or false. If true, explain why. If false, give a counterexample.*

53. If A and B are 2×2 diagonal matrices, then $A + B$ is a 2×2 diagonal matrix.

54. If A and B are 2×2 diagonal matrices, then AB is a 2×2 diagonal matrix.

55. If A and B are 2×2 diagonal matrices, then $AB = BA$.

56. If A and B are 2×2 matrices, then $AB = BA$.

57. If A and B are 2×2 diagonal matrices, then $A + B = B + A$.

58. If A and B are 2×2 matrices, then $A + B = B + A$.

59. The 2×2 zero matrix is a 2×2 diagonal matrix.

60. If A and B are 2×2 diagonal matrices such that $AB = 0$, then $A = 0$ or $B = 0$.

61. If A is the 2×2 identity matrix and B is any 2×2 matrix, then $AB = BA = B$.

62. If A and B are 2×2 diagonal matrices such that $AB = B$ and $B \neq 0$, then A is the 2×2 identity matrix.

63. If A is a 2×2 diagonal matrix such that $A^2 = A$, then A is the 2×2 identity matrix.

64. If A is a 2×2 diagonal matrix such that $A^2 = 0$, then $A = 0$.

APPLICATIONS 

65. Cost Analysis. A company with two different plants manufactures guitars and banjos. Its production costs for each instrument are given in the following matrices:

$$\begin{array}{l} \text{Materials} \\ \text{Labor} \end{array} \overset{\begin{array}{cc} \text{Plant } X \\ \text{Guitar} & \text{Banjo} \end{array}}{\begin{bmatrix} \$30 & \$25 \\ \$60 & \$80 \end{bmatrix}} = A \qquad \overset{\begin{array}{cc} \text{Plant } Y \\ \text{Guitar} & \text{Banjo} \end{array}}{\begin{bmatrix} \$36 & \$27 \\ \$54 & \$74 \end{bmatrix}} = B$$

Find $\frac{1}{2}(A + B)$, the average cost of production for the two plants.

66. Cost Analysis. If both labor and materials at plant X in Problem 65 are increased 20%, find $\frac{1}{2}(1.2A + B)$, the new average cost of production for the two plants.

67. Markup. An import car dealer sells three models of a car. Current dealer invoice price (cost) and the retail price for the basic models and the indicated options are given in the following two matrices (where "Air" means air conditioning):

Dealer invoice price

	Basic car	Air	AM/FM radio	Cruise control	
Model *A*	\$10,400	\$682	\$215	\$182	
Model *B*	\$12,500	\$721	\$295	\$182	$= M$
Model *C*	\$16,400	\$827	\$443	\$192	

Retail price

	Basic car	Air	AM/FM radio	Cruise control	
Model *A*	\$13,900	\$783	\$263	\$215	
Model *B*	\$15,000	\$838	\$395	\$236	$= N$
Model *C*	\$18,300	\$967	\$573	\$248	

We define the markup matrix to be $N - M$ (**markup** is the difference between the retail price and the dealer invoice price). Suppose the value of the dollar has had a sharp decline and the dealer invoice price is to have an across-the-board 15% increase next year. In order to stay competitive with domestic cars, the dealer increases the retail prices only 10%. Calculate a markup matrix for next year's models and the indicated options. (Compute results to the nearest dollar.)

68. Markup. Referring to Problem 67, what is the markup matrix resulting from a 20% increase in dealer invoice prices and an increase in retail prices of 15%? (Compute results to the nearest dollar.)

69. Labor Costs. A company with manufacturing plants located in different parts of the country has labor-hour and wage requirements for the manufacturing of three types of inflatable boats as given in the following two matrices:

Labor-hours per boat

	Cutting department	Assembly department	Packaging department	
	0.6 h	0.6 h	0.2 h	One-person boat
$M =$	1.0 h	0.9 h	0.3 h	Two-person boat
	1.5 h	1.2 h	0.4 h	Four-person boat

Hourly wages

	Plant I	Plant II	
	\$8	\$9	Cutting department
$N =$	\$10	\$12	Assembly department
	\$5	\$6	Packaging department

(A) Find the labor costs for a one-person boat manufactured at plant I.

(B) Find the labor costs for a four-person boat manufactured at plant II.

(C) Discuss possible interpretations of the elements in the matrix products MN and NM.

(D) If either of the products MN or NM has a meaningful interpretation, find the product and label its rows and columns.

70. Inventory Value. A personal computer retail company sells five different computer models through three stores located in a large metropolitan area. The inventory of each model on hand in each store is summarized in matrix M. Wholesale (W) and retail (R) values of each model computer are summarized in matrix N.

Model

	A	*B*	*C*	*D*	*E*	
	4	2	3	7	1	Store 1
$M =$	2	3	5	0	6	Store 2
	10	4	3	4	3	Store 3

	W	*R*	
	\$700	\$840	*A*
	\$1,400	\$1,800	*B*
$N =$	\$1,800	\$2,400	*C*
	\$2,700	\$3,300	*D*
	\$3,500	\$4,900	*E*

(A) What is the retail value of the inventory at store 2?

(B) What is the wholesale value of the inventory at store 3?

(C) Discuss possible interpretations of the elements in the matrix products MN and NM.

(D) If either of the products MN or NM has a meaningful interpretation, find the product and label its rows and columns.

(E) Discuss methods of matrix multiplication that can be used to find the total inventory of each model on hand at all three stores. State the matrices that can be used, and perform the necessary operations.

(F) Discuss methods of matrix multiplication that can be used to find the total inventory of all five models at each store. State the matrices that can be used, and perform the necessary operations.

71. Airfreight. A nationwide airfreight service has connecting flights between five cities, as illustrated in the figure. To represent this schedule in matrix form, we construct a 5×5 **incidence matrix** A, where the rows represent the origins of each flight and the columns represent the destinations. We place a 1 in the ith row and jth column of this matrix if there is a connecting flight from the ith city to the jth city and a 0 otherwise. We also place 0's on the principal diagonal, because a connecting flight with the same origin and destination does not make sense.

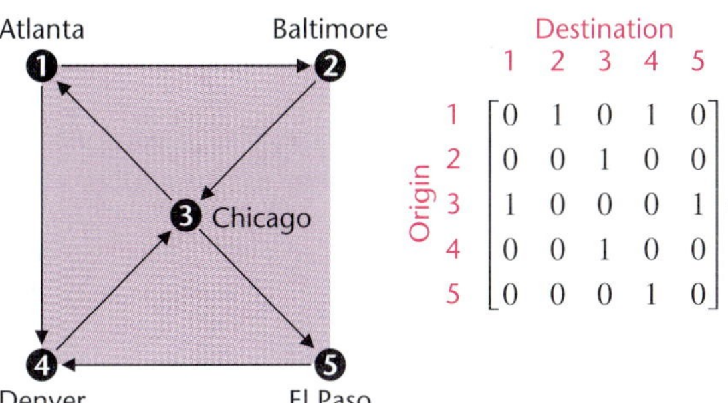

Destination 1 2 3 4 5; Origin 1–5:

$$\begin{matrix} 1 \\ 2 \\ 3 \\ 4 \\ 5 \end{matrix} \begin{bmatrix} 0 & 1 & 0 & 1 & 0 \\ 0 & 0 & 1 & 0 & 0 \\ 1 & 0 & 0 & 0 & 1 \\ 0 & 0 & 1 & 0 & 0 \\ 0 & 0 & 0 & 1 & 0 \end{bmatrix} = A$$

Now that the schedule has been represented in the mathematical form of a matrix, we can perform operations on this matrix to obtain information about the schedule.

(A) Find A^2. What does the 1 in row 2 and column 1 of A^2 indicate about the schedule? What does the 2 in row 1 and column 3 indicate about the schedule? In general, how would you interpret each element off the principal diagonal of A^2? [*Hint:* Examine the diagram for possible connections between the ith city and the jth city.]

(B) Find A^3. What does the 1 in row 4 and column 2 of A^3 indicate about the schedule? What does the 2 in row 1 and column 5 indicate about the schedule? In general, how would you interpret each element off the principal diagonal of A^3?

(C) Compute A, $A + A^2$, $A + A^2 + A^3$, . . . , until you obtain a matrix with no zero elements (except possibly on the principal diagonal), and interpret.

72. Airfreight. Find the incidence matrix A for the flight schedule illustrated in the figure. Compute A, $A + A^2$, $A + A^2 + A^3$, . . . , until you obtain a matrix with no zero elements (except possibly on the principal diagonal), and interpret.

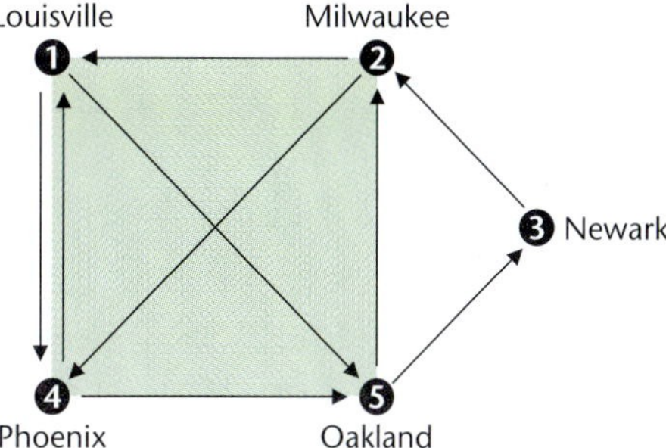

73. Politics. In a local election, a group hired a public relations firm to promote its candidate in three ways: telephone, house calls, and letters. The cost per contact is given in matrix M:

Cost per contact

$$M = \begin{bmatrix} \$0.80 \\ \$1.50 \\ \$0.40 \end{bmatrix} \begin{matrix} \text{Telephone} \\ \text{House call} \\ \text{Letter} \end{matrix}$$

The number of contacts of each type made in two adjacent cities is given in matrix N:

$$\begin{matrix} & \text{Telephone} & \text{House call} & \text{Letter} \end{matrix}$$
$$N = \begin{bmatrix} 1{,}000 & 500 & 5{,}000 \\ 2{,}000 & 800 & 8{,}000 \end{bmatrix} \begin{matrix} \text{Berkeley} \\ \text{Oakland} \end{matrix}$$

(A) Find the total amount spent in Berkeley.

(B) Find the total amount spent in Oakland.

(C) Discuss possible interpretations of the elements in the matrix products MN and NM.

(D) If either of the products MN or NM has a meaningful interpretation, find the product and label its rows and columns.

(E) Discuss methods of matrix multiplication that can be used to find the total number of telephone calls, house calls, and letters. State the matrices that can be used, and perform the necessary operations.

(F) Discuss methods of matrix multiplication that can be used to find the total number of contacts in Berkeley and in Oakland. State the matrices that can be used, and perform the necessary operations.

74. Nutrition. A nutritionist for a cereal company blends two cereals in different mixes. The amounts of protein, carbohydrate, and fat (in grams per ounce) in each cereal are given by matrix M. The amounts of each cereal used in the three mixes are given by matrix N.

$$M = \begin{bmatrix} 4 \text{ g/oz} & 2 \text{ g/oz} \\ 20 \text{ g/oz} & 16 \text{ g/oz} \\ 3 \text{ g/oz} & 1 \text{ g/oz} \end{bmatrix} \begin{matrix} \text{Protein} \\ \text{Carbohydrate} \\ \text{Fat} \end{matrix}$$

(columns: Cereal A, Cereal B)

$$N = \begin{bmatrix} 15 \text{ oz} & 10 \text{ oz} & 5 \text{ oz} \\ 5 \text{ oz} & 10 \text{ oz} & 15 \text{ oz} \end{bmatrix} \begin{matrix} \text{Cereal } A \\ \text{Cereal } B \end{matrix}$$

(columns: Mix X, Mix Y, Mix Z)

(A) Find the amount of protein in mix X.
(B) Find the amount of fat in mix Z.
(C) Discuss possible interpretations of the elements in the matrix products MN and NM.
(D) If either of the products MN or NM has a meaningful interpretation, find the product and label its rows and columns.

SECTION 9-2 Inverse of a Square Matrix

- Identity Matrix for Multiplication
- Inverse of a Square Matrix
- Application: Cryptography

In this section we introduce the identity matrix and the inverse of a square matrix. These matrix forms, along with matrix multiplication, are then used to solve some systems of equations written in matrix form in Section 9-3.

• Identity Matrix for Multiplication

We know that for any real number a

$$(1)a = a(1) = a$$

The number 1 is called the *identity* for real number multiplication. Does the set of all matrices of a given dimension have an identity element for multiplication? That is, if M is an arbitrary $m \times n$ matrix, does M have an identity element I such that $IM = MI = M$? The answer in general is no. However, the set of all **square matrices of order n** (matrices with n rows and n columns) does have an identity.

DEFINITION 1 **Identity Matrix**

The **identity matrix for multiplication** for the set of all square matrices of order n is the square matrix of order n, denoted by I, with 1's along the principal diagonal (from upper left corner to lower right corner) and 0's elsewhere.

For example,

$$\begin{bmatrix} 1 & 0 \\ 0 & 1 \end{bmatrix} \quad \text{and} \quad \begin{bmatrix} 1 & 0 & 0 \\ 0 & 1 & 0 \\ 0 & 0 & 1 \end{bmatrix}$$

are the identity matrices for all square matrices of order 2 and 3, respectively.

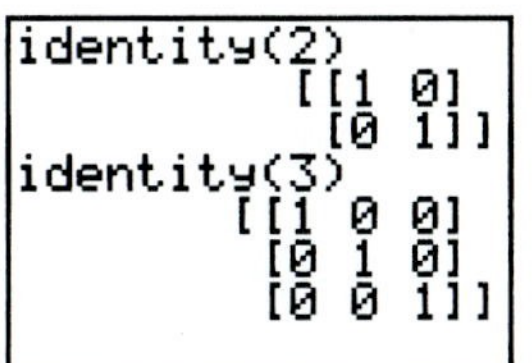

FIGURE 1 Identity matrices.

Most graphing calculators have a built-in command for generating the identity matrix of a given order (see Fig. 1).

EXAMPLE 1 Identity Matrix Multiplication

(A) $\begin{bmatrix} 1 & 0 & 0 \\ 0 & 1 & 0 \\ 0 & 0 & 1 \end{bmatrix}\begin{bmatrix} a & b & c \\ d & e & f \\ g & h & i \end{bmatrix} = \begin{bmatrix} a & b & c \\ d & e & f \\ g & h & i \end{bmatrix}$

(B) $\begin{bmatrix} a & b & c \\ d & e & f \\ g & h & i \end{bmatrix}\begin{bmatrix} 1 & 0 & 0 \\ 0 & 1 & 0 \\ 0 & 0 & 1 \end{bmatrix} = \begin{bmatrix} a & b & c \\ d & e & f \\ g & h & i \end{bmatrix}$

(C) $\begin{bmatrix} 1 & 0 \\ 0 & 1 \end{bmatrix}\begin{bmatrix} a & b & c \\ d & e & f \end{bmatrix} = \begin{bmatrix} a & b & c \\ d & e & f \end{bmatrix}$

(D) $\begin{bmatrix} a & b & c \\ d & e & f \end{bmatrix}\begin{bmatrix} 1 & 0 & 0 \\ 0 & 1 & 0 \\ 0 & 0 & 1 \end{bmatrix} = \begin{bmatrix} a & b & c \\ d & e & f \end{bmatrix}$

Matched Problem 1

Multiply:

(A) $\begin{bmatrix} 1 & 0 \\ 0 & 1 \end{bmatrix}\begin{bmatrix} 3 & -5 \\ 4 & 6 \end{bmatrix}$ and $\begin{bmatrix} 3 & -5 \\ 4 & 6 \end{bmatrix}\begin{bmatrix} 1 & 0 \\ 0 & 1 \end{bmatrix}$

(B) $\begin{bmatrix} 1 & 0 & 0 \\ 0 & 1 & 0 \\ 0 & 0 & 1 \end{bmatrix}\begin{bmatrix} 5 & -7 \\ 2 & 4 \\ 6 & -8 \end{bmatrix}$ and $\begin{bmatrix} 5 & -7 \\ 2 & 4 \\ 6 & -8 \end{bmatrix}\begin{bmatrix} 1 & 0 \\ 0 & 1 \end{bmatrix}$

In general, we can show that if M is a square matrix of order n and I is the identity matrix of order n, then

$$IM = MI = M$$

If M is an $m \times n$ matrix that is not square ($m \neq n$), then it is still possible to multiply M on the left and on the right by an identity matrix, but not with the same-size identity matrix (see Examples 1C and 1D). In order to avoid the complications involved with associating two different identity matrices with each nonsquare matrix, we restrict our attention in this section to square matrices.

EXPLORE-DISCUSS 1 The only real number solutions to the equation $x^2 = 1$ are $x = 1$ and $x = -1$.

(A) Show that $A = \begin{bmatrix} 0 & 1 \\ 1 & 0 \end{bmatrix}$ satisfies $A^2 = I$, where I is the 2×2 identity.

(B) Show that $B = \begin{bmatrix} 0 & -1 \\ -1 & 0 \end{bmatrix}$ satisfies $B^2 = I$.

(C) Find a 2×2 matrix with all elements nonzero whose square is the 2×2 identity matrix.

• Inverse of a Square Matrix

In the set of real numbers, we know that for each real number a, except 0, there exists a real number a^{-1} such that

$$a^{-1}a = 1$$

The number a^{-1} is called the *inverse* of the number a relative to multiplication, or the *multiplicative inverse* of a. For example, 2^{-1} is the multiplicative inverse of 2, since $2^{-1}(2) = 1$. We use this idea to define the *inverse of a square matrix.*

DEFINITION 2

Inverse of a Square Matrix

If M is a square matrix of order n and if there exists a matrix M^{-1} (read "M inverse") such that

$$M^{-1}M = MM^{-1} = I$$

then M^{-1} is called the **multiplicative inverse of M** or, more simply, the **inverse of M.**

The multiplicative inverse of a nonzero real number a also can be written as $1/a$. This notation is not used for matrix inverses.

Let's use Definition 2 to find M^{-1}, if it exists, for

$$M = \begin{bmatrix} 2 & 3 \\ 1 & 2 \end{bmatrix}$$

We are looking for

$$M^{-1} = \begin{bmatrix} a & c \\ b & d \end{bmatrix}$$

such that

$$MM^{-1} = M^{-1}M = I$$

Thus, we write

$$\overset{M}{\begin{bmatrix} 2 & 3 \\ 1 & 2 \end{bmatrix}} \overset{M^{-1}}{\begin{bmatrix} a & c \\ b & d \end{bmatrix}} = \overset{I}{\begin{bmatrix} 1 & 0 \\ 0 & 1 \end{bmatrix}}$$

and try to find a, b, c, and d so that the product of M and M^{-1} is the identity matrix I. Multiplying M and M^{-1} on the left side, we obtain

$$\begin{bmatrix} (2a + 3b) & (2c + 3d) \\ (a + 2b) & (c + 2d) \end{bmatrix} = \begin{bmatrix} 1 & 0 \\ 0 & 1 \end{bmatrix}$$

which is true only if

$$\begin{aligned} 2a + 3b &= 1 \qquad & 2c + 3d &= 0 \\ a + 2b &= 0 \qquad & c + 2d &= 1 \end{aligned}$$

Solving these two systems, we find that $a = 2$, $b = -1$, $c = -3$, and $d = 2$. Thus,

$$M^{-1} = \begin{bmatrix} 2 & -3 \\ -1 & 2 \end{bmatrix}$$

as is easily checked:

$$\overset{M}{\begin{bmatrix} 2 & 3 \\ 1 & 2 \end{bmatrix}} \overset{M^{-1}}{\begin{bmatrix} 2 & -3 \\ -1 & 2 \end{bmatrix}} = \overset{I}{\begin{bmatrix} 1 & 0 \\ 0 & 1 \end{bmatrix}} = \overset{M^{-1}}{\begin{bmatrix} 2 & -3 \\ -1 & 2 \end{bmatrix}} \overset{M}{\begin{bmatrix} 2 & 3 \\ 1 & 2 \end{bmatrix}}$$

Unlike nonzero real numbers, inverses do not always exist for nonzero square matrices. For example, if

$$N = \begin{bmatrix} 2 & 1 \\ 4 & 2 \end{bmatrix}$$

then, proceeding as before, we are led to the systems

$$\begin{aligned} 2a + b &= 1 \qquad & 2c + d &= 0 \\ 4a + 2b &= 0 \qquad & 4c + 2d &= 1 \end{aligned}$$

These systems are both inconsistent and have no solution. Hence, N^{-1} does not exist.

Being able to find inverses, when they exist, leads to direct and simple solutions to many practical problems. In the next section, for example, we will show how inverses can be used to solve systems of linear equations.

The method outlined above for finding the inverse, if it exists, gets very involved for matrices of order larger than 2. Now that we know what we are looking for, we can use augmented matrices as in Section 8-2 to make the process more efficient. Details are illustrated in Example 2.

EXAMPLE 2 Finding an Inverse

Find the inverse, if it exists, of

$$M = \begin{bmatrix} 1 & -1 & 1 \\ 0 & 2 & -1 \\ 2 & 3 & 0 \end{bmatrix}$$

Solution We start as before and write

$$\overset{M}{\begin{bmatrix} 1 & -1 & 1 \\ 0 & 2 & -1 \\ 2 & 3 & 0 \end{bmatrix}} \overset{M^{-1}}{\begin{bmatrix} a & d & g \\ b & e & h \\ c & f & i \end{bmatrix}} = \overset{I}{\begin{bmatrix} 1 & 0 & 0 \\ 0 & 1 & 0 \\ 0 & 0 & 1 \end{bmatrix}}$$

This is true only if

$$\begin{aligned} a - b + c &= 1 & d - e + f &= 0 & g - h + i &= 0 \\ 2b - c &= 0 & 2e - f &= 1 & 2h - i &= 0 \\ 2a + 3b &= 0 & 2d + 3e &= 0 & 2g + 3h &= 1 \end{aligned}$$

Now we write augmented matrices for each of the three systems:

$$\overset{\text{First}}{\left[\begin{array}{rrr|r} 1 & -1 & 1 & 1 \\ 0 & 2 & -1 & 0 \\ 2 & 3 & 0 & 0 \end{array}\right]} \quad \overset{\text{Second}}{\left[\begin{array}{rrr|r} 1 & -1 & 1 & 0 \\ 0 & 2 & -1 & 1 \\ 2 & 3 & 0 & 0 \end{array}\right]} \quad \overset{\text{Third}}{\left[\begin{array}{rrr|r} 1 & -1 & 1 & 0 \\ 0 & 2 & -1 & 0 \\ 2 & 3 & 0 & 1 \end{array}\right]}$$

Since each matrix to the left of the vertical bar is the same, exactly the same row operations can be used on each augmented matrix to transform it into a reduced form. We can speed up the process substantially by combining all three augmented matrices into the single augmented matrix form

$$\left[\begin{array}{rrr|rrr} 1 & -1 & 1 & 1 & 0 & 0 \\ 0 & 2 & -1 & 0 & 1 & 0 \\ 2 & 3 & 0 & 0 & 0 & 1 \end{array}\right] = [M \mid I] \tag{1}$$

We now try to perform row operations on matrix (1) until we obtain a row-equivalent matrix that looks like matrix (2):

$$\left[\begin{array}{ccc|ccc} 1 & 0 & 0 & a & d & g \\ 0 & 1 & 0 & b & e & h \\ 0 & 0 & 1 & c & f & i \end{array}\right] = [I \mid B] \tag{2}$$

(with I labelling the left block and B the right block)

If this can be done, then the new matrix to the right of the vertical bar is M^{-1}! Now let's try to transform (1) into a form like (2). We follow the same sequence of steps as in the solution of linear systems by Gauss–Jordan elimination (see Section 8-2):

$$\overset{\hspace{3em}M\hspace{6em}I}{\left[\begin{array}{rrr|rrr} 1 & -1 & 1 & 1 & 0 & 0 \\ 0 & 2 & -1 & 0 & 1 & 0 \\ 2 & 3 & 0 & 0 & 0 & 1 \end{array}\right]} \quad (-2)R_1 + R_3 \to R_3$$

$$\sim \left[\begin{array}{rrr|rrr} 1 & -1 & 1 & 1 & 0 & 0 \\ 0 & 2 & -1 & 0 & 1 & 0 \\ 0 & 5 & -2 & -2 & 0 & 1 \end{array}\right] \quad \tfrac{1}{2}R_2 \to R_2$$

$$\sim \left[\begin{array}{rrr|rrr} 1 & -1 & 1 & 1 & 0 & 0 \\ 0 & 1 & -\frac{1}{2} & 0 & \frac{1}{2} & 0 \\ 0 & 5 & -2 & -2 & 0 & 1 \end{array}\right] \quad \begin{array}{l} R_2 + R_1 \to R_1 \\ \\ (-5)R_2 + R_3 \to R_3 \end{array}$$

$$\sim \left[\begin{array}{rrr|rrr} 1 & 0 & \frac{1}{2} & 1 & \frac{1}{2} & 0 \\ 0 & 1 & -\frac{1}{2} & 0 & \frac{1}{2} & 0 \\ 0 & 0 & \frac{1}{2} & -2 & -\frac{5}{2} & 1 \end{array}\right] \quad 2R_3 \to R_3$$

$$\sim \left[\begin{array}{rrr|rrr} 1 & 0 & \frac{1}{2} & 1 & \frac{1}{2} & 0 \\ 0 & 1 & -\frac{1}{2} & 0 & \frac{1}{2} & 0 \\ 0 & 0 & 1 & -4 & -5 & 2 \end{array}\right] \quad \begin{array}{l} (-\frac{1}{2})R_3 + R_1 \to R_1 \\ \frac{1}{2}R_3 + R_2 \to R_2 \\ \end{array}$$

$$\sim \left[\begin{array}{rrr|rrr} 1 & 0 & 0 & 3 & 3 & -1 \\ 0 & 1 & 0 & -2 & -2 & 1 \\ 0 & 0 & 1 & -4 & -5 & 2 \end{array}\right] = [I \mid B]$$

Converting back to systems of equations equivalent to our three original systems (we won't have to do this step in practice), we have

$$\begin{array}{lll} a = 3 & d = 3 & g = -1 \\ b = -2 & e = -2 & h = 1 \\ c = -4 & f = -5 & i = 2 \end{array}$$

And these are just the elements of M^{-1} that we are looking for! Hence,

$$M^{-1} = \begin{bmatrix} 3 & 3 & -1 \\ -2 & -2 & 1 \\ -4 & -5 & 2 \end{bmatrix}$$

Note that this is the matrix to the right of the vertical line in the last augmented matrix.

Check Since the definition of matrix inverse requires that

$$M^{-1}M = I \qquad \text{and} \qquad MM^{-1} = I \tag{3}$$

it appears that we must compute both $M^{-1}M$ and MM^{-1} to check our work. However, it can be shown that if one of the equations in (3) is satisfied, then the other is also satisfied. Thus, for checking purposes it is sufficient to compute either $M^{-1}M$ or MM^{-1}—we don't need to do both.

$$M^{-1}M = \begin{bmatrix} 3 & 3 & -1 \\ -2 & -2 & 1 \\ -4 & -5 & 2 \end{bmatrix} \begin{bmatrix} 1 & -1 & 1 \\ 0 & 2 & -1 \\ 2 & 3 & 0 \end{bmatrix} = \begin{bmatrix} 1 & 0 & 0 \\ 0 & 1 & 0 \\ 0 & 0 & 1 \end{bmatrix} = I$$

Matched Problem 2 Let: $M = \begin{bmatrix} 3 & -1 & 1 \\ -1 & 1 & 0 \\ 1 & 0 & 1 \end{bmatrix}$

(A) Form the augmented matrix $[M \mid I]$.
(B) Use row operations to transform $[M \mid I]$ into $[I \mid B]$.
(C) Verify by multiplication that $B = M^{-1}$.

The procedure used in Example 2 can be used to find the inverse of any square matrix, if the inverse exists, and will also indicate when the inverse does not exist. These ideas are summarized in Theorem 1.

Theorem 1 **Inverse of a Square Matrix M**

If $[M \mid I]$ is transformed by row operations into $[I \mid B]$, then the resulting matrix B is M^{-1}. If, however, we obtain all 0's in one or more rows to the left of the vertical line, then M^{-1} does not exist.

EXPLORE-DISCUSS 2 (A) Suppose that the square matrix M has a row of all zeros. Explain why M has no inverse.
(B) Suppose that the square matrix M has a column of all zeros. Explain why M has no inverse.

EXAMPLE 3 Finding a Matrix Inverse

Find M^{-1}, given: $M = \begin{bmatrix} 4 & -1 \\ -6 & 2 \end{bmatrix}$

Solution

$$\left[\begin{array}{cc|cc} 4 & -1 & 1 & 0 \\ -6 & 2 & 0 & 1 \end{array}\right] \quad \tfrac{1}{4}R_1 \rightarrow R_1$$

$$\sim \left[\begin{array}{cc|cc} 1 & -\frac{1}{4} & \frac{1}{4} & 0 \\ -6 & 2 & 0 & 1 \end{array}\right] \quad 6R_1 + R_2 \rightarrow R_2$$

$$\sim \left[\begin{array}{cc|cc} 1 & -\frac{1}{4} & \frac{1}{4} & 0 \\ 0 & \frac{1}{2} & \frac{3}{2} & 1 \end{array}\right] \quad 2R_2 \rightarrow R_2$$

$$\sim \left[\begin{array}{cc|cc} 1 & -\frac{1}{4} & \frac{1}{4} & 0 \\ 0 & 1 & 3 & 2 \end{array}\right] \quad \tfrac{1}{4}R_2 + R_1 \to R_1$$

$$\sim \left[\begin{array}{cc|cc} 1 & 0 & 1 & \frac{1}{2} \\ 0 & 1 & 3 & 2 \end{array}\right]$$

Thus,

$$M^{-1} = \begin{bmatrix} 1 & \frac{1}{2} \\ 3 & 2 \end{bmatrix} \qquad \text{Check by showing } M^{-1}M = I.$$

Matched Problem 3 Find M^{-1}, given: $M = \begin{bmatrix} 2 & -6 \\ 1 & -2 \end{bmatrix}$

EXAMPLE 4 Finding an Inverse

Find M^{-1}, if it exists, given: $M = \begin{bmatrix} 10 & -2 \\ -5 & 1 \end{bmatrix}$

Solution

$$\left[\begin{array}{cc|cc} 10 & -2 & 1 & 0 \\ -5 & 1 & 0 & 1 \end{array}\right] \sim \left[\begin{array}{cc|cc} 1 & -\frac{1}{5} & \frac{1}{10} & 0 \\ -5 & 1 & 0 & 1 \end{array}\right]$$

$$\sim \left[\begin{array}{cc|cc} 1 & -\frac{1}{5} & \frac{1}{10} & 0 \\ 0 & 0 & \frac{1}{2} & 1 \end{array}\right]$$

We have all 0's in the second row to the left of the vertical line. Therefore, M^{-1} does not exist.

Matched Problem 4 Find M^{-1}, if it exists, given: $M = \begin{bmatrix} 6 & -3 \\ -2 & 1 \end{bmatrix}$

Most graphing utilities and computers can compute matrix inverses and can identify those matrices that do not have inverses (see Fig. 2). A matrix that does not have an inverse is often referred to as a **singular matrix.**

FIGURE 2 Finding matrix inverses on a graphing calculator.

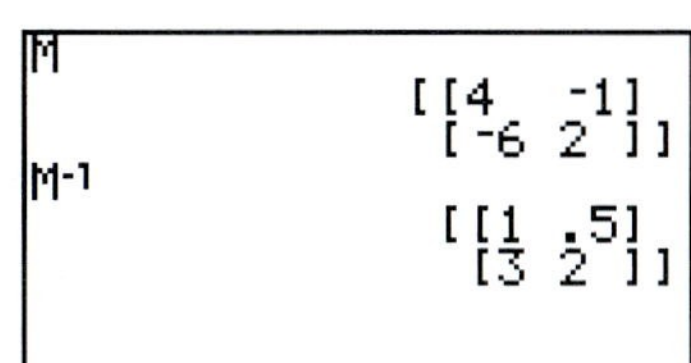

(a) Example 3

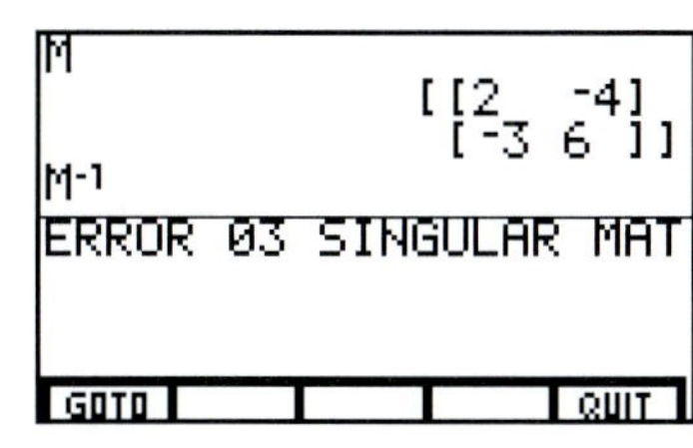

(b) Example 4

• Application: Cryptography

Matrix inverses can be used to provide a simple and effective procedure for encoding and decoding messages. To begin, we assign the numbers 1 to 26 to the letters in the alphabet, as shown below. We also assign the number 27 to a blank to provide for space between words. (A more sophisticated code could include both uppercase and lowercase letters and punctuation symbols.)

A	B	C	D	E	F	G	H	I	J	K	L	M	N
1	2	3	4	5	6	7	8	9	10	11	12	13	14

O	P	Q	R	S	T	U	V	W	X	Y	Z	Blank
15	16	17	18	19	20	21	22	23	24	25	26	27

Thus, the message I LOVE MATH corresponds to the sequence

9 27 12 15 22 5 27 13 1 20 8

Any matrix whose elements are positive integers and whose inverse exists can be used as an **encoding matrix.** For example, to use the 2×2 matrix

$$A = \begin{bmatrix} 4 & 3 \\ 5 & 4 \end{bmatrix}$$

to encode the above message, first we divide the numbers in the sequence into groups of 2 and use these groups as the columns of a matrix with 2 rows. (Notice that we added an extra blank at the end of the message to make the columns come out even.) Then we multiply this matrix on the left by A:

$$\begin{bmatrix} 4 & 3 \\ 5 & 4 \end{bmatrix} \begin{bmatrix} 9 & 12 & 22 & 27 & 1 & 8 \\ 27 & 15 & 5 & 13 & 20 & 27 \end{bmatrix} = \begin{bmatrix} 117 & 93 & 103 & 147 & 64 & 113 \\ 153 & 120 & 130 & 187 & 85 & 148 \end{bmatrix}$$

The coded message is

117 153 93 120 103 130 147 187 64 85 113 148

This message can be decoded simply by putting it back into matrix form and multiplying on the left by the **decoding matrix** A^{-1}. Since A^{-1} is easily determined if A is known, the encoding matrix A is the only key needed to decode messages encoded in this manner. Although simple in concept, codes of this type can be very difficult to crack.

Answers to Matched Problems

1. (A) $\begin{bmatrix} 3 & -5 \\ 4 & 6 \end{bmatrix}$ (B) $\begin{bmatrix} 5 & -7 \\ 2 & 4 \\ 6 & -8 \end{bmatrix}$

2. (A) $\left[\begin{array}{rrr|rrr} 3 & -1 & 1 & 1 & 0 & 0 \\ -1 & 1 & 0 & 0 & 1 & 0 \\ 1 & 0 & 1 & 0 & 0 & 1 \end{array}\right]$ (B) $\left[\begin{array}{rrr|rrr} 1 & 0 & 0 & 1 & 1 & -1 \\ 0 & 1 & 0 & 1 & 2 & -1 \\ 0 & 0 & 1 & -1 & -1 & 2 \end{array}\right]$

(C) $\begin{bmatrix} 1 & 1 & -1 \\ 1 & 2 & -1 \\ -1 & -1 & 2 \end{bmatrix} \begin{bmatrix} 3 & -1 & 1 \\ -1 & 1 & 0 \\ 1 & 0 & 1 \end{bmatrix} = \begin{bmatrix} 1 & 0 & 0 \\ 0 & 1 & 0 \\ 0 & 0 & 1 \end{bmatrix}$

3. $\begin{bmatrix} -1 & 3 \\ -\frac{1}{2} & 1 \end{bmatrix}$ **4.** Does not exist

EXERCISE 9-2

A

Perform the indicated operations in Problems 1–8.

1. $\begin{bmatrix} 6 & -3 \\ 5 & 8 \end{bmatrix}\begin{bmatrix} 1 & 0 \\ 0 & 1 \end{bmatrix}$

2. $\begin{bmatrix} 1 & 0 \\ 0 & 1 \end{bmatrix}\begin{bmatrix} -4 & 9 \\ 7 & -2 \end{bmatrix}$

3. $\begin{bmatrix} 1 & 0 \\ 0 & 1 \end{bmatrix}\begin{bmatrix} 2 & -5 \\ 8 & -4 \end{bmatrix}$

4. $\begin{bmatrix} 2 & 0 \\ -7 & -6 \end{bmatrix}\begin{bmatrix} 1 & 0 \\ 0 & 1 \end{bmatrix}$

5. $\begin{bmatrix} 4 & -2 & 5 \\ 6 & 0 & -3 \\ -1 & 2 & 7 \end{bmatrix}\begin{bmatrix} 1 & 0 & 0 \\ 0 & 1 & 0 \\ 0 & 0 & 1 \end{bmatrix}$

6. $\begin{bmatrix} 1 & 0 & 0 \\ 0 & 1 & 0 \\ 0 & 0 & 1 \end{bmatrix}\begin{bmatrix} 8 & 0 & 3 \\ -2 & -1 & -3 \\ 4 & -6 & 1 \end{bmatrix}$

7. $\begin{bmatrix} 1 & 0 & 0 \\ 0 & 1 & 0 \\ 0 & 0 & 1 \end{bmatrix}\begin{bmatrix} -3 & 2 & -5 \\ 4 & -1 & 6 \\ -3 & 4 & -2 \end{bmatrix}$

8. $\begin{bmatrix} 5 & 1 & 0 \\ 1 & -4 & 1 \\ 0 & 1 & 9 \end{bmatrix}\begin{bmatrix} 1 & 0 & 0 \\ 0 & 1 & 0 \\ 0 & 0 & 1 \end{bmatrix}$

For Problems 9–14, show that the two matrices are inverses of each other by showing that their product is the identity matrix I.

9. $\begin{bmatrix} 1 & 6 \\ 0 & 1 \end{bmatrix}\begin{bmatrix} 1 & -6 \\ 0 & 1 \end{bmatrix}$

10. $\begin{bmatrix} 1 & 0 \\ -3 & 1 \end{bmatrix}\begin{bmatrix} 1 & 0 \\ 3 & 1 \end{bmatrix}$

11. $\begin{bmatrix} 3 & 4 \\ 5 & 7 \end{bmatrix}\begin{bmatrix} 7 & -4 \\ -5 & 3 \end{bmatrix}$

12. $\begin{bmatrix} -6 & 5 \\ -5 & 4 \end{bmatrix}\begin{bmatrix} 4 & -5 \\ 5 & -6 \end{bmatrix}$

13. $\begin{bmatrix} 1 & -3 & 5 \\ 0 & 1 & 4 \\ 0 & 0 & 1 \end{bmatrix}\begin{bmatrix} 1 & 3 & -17 \\ 0 & 1 & -4 \\ 0 & 0 & 1 \end{bmatrix}$

14. $\begin{bmatrix} 1 & 0 & 0 \\ 2 & 1 & 0 \\ -6 & -3 & 1 \end{bmatrix}\begin{bmatrix} 1 & 0 & 0 \\ -2 & 1 & 0 \\ 0 & 3 & 1 \end{bmatrix}$

B

Given M in Problems 15–24, find M^{-1}, and show that $M^{-1}M = I$.

15. $\begin{bmatrix} 1 & 9 \\ 0 & 1 \end{bmatrix}$

16. $\begin{bmatrix} 0 & -1 \\ -1 & 3 \end{bmatrix}$

17. $\begin{bmatrix} -1 & -2 \\ 2 & 5 \end{bmatrix}$

18. $\begin{bmatrix} 3 & -4 \\ -2 & 3 \end{bmatrix}$

19. $\begin{bmatrix} -5 & 7 \\ 2 & -3 \end{bmatrix}$

20. $\begin{bmatrix} 11 & 4 \\ 3 & 1 \end{bmatrix}$

21. $\begin{bmatrix} 1 & -1 & 0 \\ -1 & 1 & -1 \\ 0 & -1 & 1 \end{bmatrix}$

22. $\begin{bmatrix} 2 & -1 & 0 \\ 0 & 1 & 1 \\ 1 & 0 & 1 \end{bmatrix}$

23. $\begin{bmatrix} 1 & 2 & 5 \\ 3 & 5 & 9 \\ 1 & 1 & -2 \end{bmatrix}$

24. $\begin{bmatrix} 1 & -1 & 1 \\ -2 & 3 & 2 \\ 3 & -3 & 2 \end{bmatrix}$

Find the inverse of each matrix in Problems 25–28, if it exists.

25. $\begin{bmatrix} 3 & 9 \\ 2 & 6 \end{bmatrix}$

26. $\begin{bmatrix} 2 & -4 \\ -3 & 6 \end{bmatrix}$

27. $\begin{bmatrix} 2 & 3 \\ 3 & 5 \end{bmatrix}$

28. $\begin{bmatrix} -5 & 4 \\ 4 & -3 \end{bmatrix}$

C

Find the inverse of each matrix in Problems 29–34, if it exists.

29. $\begin{bmatrix} 2 & 2 & -1 \\ 0 & 4 & -1 \\ -1 & -2 & 1 \end{bmatrix}$

30. $\begin{bmatrix} 4 & 2 & -1 \\ 1 & 1 & -1 \\ -3 & -1 & 1 \end{bmatrix}$

31. $\begin{bmatrix} 2 & 1 & 1 \\ 1 & 1 & 0 \\ -1 & -1 & 0 \end{bmatrix}$

32. $\begin{bmatrix} 1 & -1 & 0 \\ 2 & -1 & 1 \\ 0 & 1 & 1 \end{bmatrix}$

33. $\begin{bmatrix} 1 & 5 & 10 \\ 0 & 1 & 4 \\ 1 & 6 & 15 \end{bmatrix}$

34. $\begin{bmatrix} 1 & -5 & -10 \\ 0 & 1 & 6 \\ 1 & -4 & -3 \end{bmatrix}$

35. Show that $(A^{-1})^{-1} = A$ for

$$A = \begin{bmatrix} 3 & 4 \\ 2 & 3 \end{bmatrix}$$

36. Show that $(AB)^{-1} = B^{-1}A^{-1}$ for

$$A = \begin{bmatrix} 3 & 4 \\ 2 & 3 \end{bmatrix} \quad \text{and} \quad B = \begin{bmatrix} 3 & 7 \\ 2 & 5 \end{bmatrix}$$

37. Discuss the existence of M^{-1} for 2×2 diagonal matrices of the form

$$M = \begin{bmatrix} a & 0 \\ 0 & d \end{bmatrix}$$

38. Discuss the existence of M^{-1} for 2×2 upper triangular matrices of the form

$$M = \begin{bmatrix} a & b \\ 0 & d \end{bmatrix}$$

In Problems 39–42, use a graphing utility to find the inverse of each matrix, if it exists.

39. $\begin{bmatrix} 1 & -5 & -2 & -2 \\ 2 & 7 & 0 & -1 \\ 3 & -5 & -2 & 0 \\ 4 & -1 & -2 & 0 \end{bmatrix}$ **40.** $\begin{bmatrix} 1 & -2 & -1 & 0 \\ 2 & 1 & 3 & 5 \\ 1 & 0 & -3 & 2 \\ 0 & 4 & 2 & 6 \end{bmatrix}$

41. $\begin{bmatrix} 1 & 3 & 1 & 6 & 0 \\ 4 & 4 & 2 & 6 & 1 \\ 3 & 4 & 1 & 2 & -2 \\ 3 & -5 & 1 & -4 & 3 \\ -1 & 2 & -1 & -2 & -4 \end{bmatrix}$

42. $\begin{bmatrix} 1 & -1 & -6 & -1 & 3 \\ -2 & 2 & 0 & 2 & 1 \\ 4 & -2 & 0 & -3 & 1 \\ 1 & 3 & 1 & 4 & 2 \\ -1 & 3 & 1 & 1 & 4 \end{bmatrix}$

APPLICATIONS

Problems 43–46 refer to the encoding matrix $A = \begin{bmatrix} 3 & 5 \\ 1 & 2 \end{bmatrix}$

43. Cryptography. Encode the message CAT IN THE HAT with the matrix A given above.

44. Cryptography. Encode the message FOX IN SOCKS with the matrix A given above.

45. Cryptography. The following message was encoded with the matrix A given above. Decode this message.

111 43 40 15 177 68 50 19 116 45 86
29 62 22 121 43 68 27

46. Cryptography. The following message was encoded with the matrix A given above. Decode this message.

99 38 154 58 115 43 121 43 20 7 149
56 86 29 196 73 99 38

Problems 47–50 require the use of a graphing utility. To use the 5 × 5 encoding matrix B given below, form a matrix with 5 rows and as many columns as necessary to accommodate each message.

$$B = \begin{bmatrix} 1 & 0 & 1 & 0 & 1 \\ 0 & 1 & 1 & 0 & 3 \\ 2 & 1 & 1 & 1 & 1 \\ 0 & 0 & 1 & 0 & 2 \\ 1 & 1 & 1 & 2 & 1 \end{bmatrix}$$

47. Cryptography. Encode the message DWIGHT DAVID EISENHOWER with the matrix B given above.

48. Cryptography. Encode the message JOHN FITZGERALD KENNEDY with the matrix B given above.

49. Cryptography. The following message was encoded with the matrix B given above. Decode this message.

41 84 82 44 74 25 56 67 20 54 43 54
89 39 102 44 67 86 44 90 68 135 136
81 149

50. Cryptography. The following message was encoded with the matrix B given above. Decode this message.

22 15 57 5 47 54 58 89 45 84 46 80
87 53 96 51 68 116 39 113 68 135 136
81 149

SECTION 9-3 Matrix Equations and Systems of Linear Equations

- Matrix Equations
- Matrix Equations and Systems of Linear Equations
- Application

The identity matrix and inverse matrix discussed in the last section can be put to immediate use in the solving of certain simple matrix equations. Being able to solve a matrix equation gives us another important method of solving a system of equations having the same number of variables as equations. If the system either has fewer variables than equations or more variables than equations, then we must return to the Gauss–Jordan method of elimination.

• Matrix Equations

Before we discuss the solution of matrix equations, you will probably find it helpful to briefly review the basic properties of real numbers and linear equations discussed in Sections A-1 and 1-1.

EXPLORE-DISCUSS 1 Let a, b, and c be real numbers, with $a \neq 0$. Solve each equation for x.

(A) $ax = b$ (B) $ax + b = c$

Solving simple matrix equations follows very much the same procedures used in solving real number equations. We have, however, less freedom with matrix equations, because matrix multiplication is not commutative. In solving matrix equations, we will be guided by the properties of matrices summarized in Theorem 1.

Theorem 1 **Basic Properties of Matrices**

Assuming all products and sums are defined for the indicated matrices A, B, C, I, and 0, then

Addition Properties

Associative:	$(A + B) + C = A + (B + C)$
Commutative:	$A + B = B + A$
Additive Identity:	$A + 0 = 0 + A = A$
Additive Inverse:	$A + (-A) = (-A) + A = 0$

Multiplication Properties

Associative Property:	$A(BC) = (AB)C$
Multiplicative Identity:	$AI = IA = A$
Multiplicative Inverse:	If A is a square matrix and A^{-1} exists, then $AA^{-1} = A^{-1}A = I$.

Combined Properties

Left Distributive:	$A(B + C) = AB + AC$
Right Distributive:	$(B + C)A = BA + CA$

Equality

Addition:	If $A = B$, then $A + C = B + C$.
Left Multiplication:	If $A = B$, then $CA = CB$.
Right Multiplication:	If $A = B$, then $AC = BC$.

The process of solving certain types of simple matrix equations is best illustrated by an example.

EXAMPLE 1 Solving a Matrix Equation

Given an $n \times n$ matrix A and $n \times 1$ column matrices B and X, solve $AX = B$ for X. Assume all necessary inverses exist.

Solution We are interested in finding a column matrix X that satisfies the matrix equation $AX = B$. To solve this equation, we multiply both sides, on the left, by A^{-1}, assuming it exists, to isolate X on the left side.

$$AX = B$$

$$A^{-1}(AX) = A^{-1}B \quad \text{Use the left multiplication property.}$$

$$(A^{-1}A)X = A^{-1}B \quad \text{Associative property}$$

$$IX = A^{-1}B \quad A^{-1}A = I$$

$$X = A^{-1}B \quad IX = X$$

CAUTION Do not mix the left multiplication property and the right multiplication property. If $AX = B$, then

$$A^{-1}(AX) \neq BA^{-1}$$

Matched Problem 1 Given an $n \times n$ matrix A and $n \times 1$ column matrices B, C, and X, solve $AX + C = B$ for X. Assume all necessary inverses exist.

• Matrix Equations and Systems of Linear Equations

We now show how independent systems of linear equations with the same number of variables as equations can be solved by first converting the system into a matrix equation of the form $AX = B$ and using $X = A^{-1}B$ as obtained in Example 1.

EXAMPLE 2 Using Inverses to Solve Systems of Equations

Use matrix inverse methods to solve the system:

$$\begin{aligned} x_1 - x_2 + x_3 &= 1 \\ 2x_2 - x_3 &= 1 \\ 2x_1 + 3x_2 \quad &= 1 \end{aligned} \tag{1}$$

Solution The inverse of the coefficient matrix

$$A = \begin{bmatrix} 1 & -1 & 1 \\ 0 & 2 & -1 \\ 2 & 3 & 0 \end{bmatrix}$$

provides an efficient method for solving this system. To see how, we convert system (1) into a matrix equation:

$$\overset{A}{\begin{bmatrix} 1 & -1 & 1 \\ 0 & 2 & -1 \\ 2 & 3 & 0 \end{bmatrix}} \overset{X}{\begin{bmatrix} x_1 \\ x_2 \\ x_3 \end{bmatrix}} = \overset{B}{\begin{bmatrix} 1 \\ 1 \\ 1 \end{bmatrix}} \qquad (2)$$

Check that matrix equation (2) is equivalent to system (1) by finding the product of the left side and then equating corresponding elements on the left with those on the right. Now you see another important reason for defining matrix multiplication as we did.

We are interested in finding a column matrix X that satisfies the matrix equation $AX = B$. In Example 1 we found that if $AX = B$ and if A^{-1} exists, then

$$X = A^{-1}B$$

The inverse of A was found in Example 2 in Section 9-2 to be

$$A^{-1} = \begin{bmatrix} 3 & 3 & -1 \\ -2 & -2 & 1 \\ -4 & -5 & 2 \end{bmatrix}$$

Thus,

$$\overset{X}{\begin{bmatrix} x_1 \\ x_2 \\ x_3 \end{bmatrix}} = \overset{A^{-1}}{\begin{bmatrix} 3 & 3 & -1 \\ -2 & -2 & 1 \\ -4 & -5 & 2 \end{bmatrix}} \overset{B}{\begin{bmatrix} 1 \\ 1 \\ 1 \end{bmatrix}} = \begin{bmatrix} 5 \\ -3 \\ -7 \end{bmatrix}$$

and we can conclude that $x_1 = 5$, $x_2 = -3$, and $x_3 = -7$. Check this result in system (1).

Matched Problem 2 Use matrix inverse methods to solve the system:

$$\begin{aligned} 3x_1 - x_2 + x_3 &= 1 \\ -x_1 + x_2 \qquad &= 3 \\ x_1 \qquad + x_3 &= 2 \end{aligned}$$

[*Note:* The inverse of the coefficient matrix was found in Matched Problem 2 in Section 9-2.]

At first glance, using matrix inverse methods seems to require the same amount of effort as using Gauss–Jordan elimination. In either case, row operations must be applied to an augmented matrix involving the coefficients of the system. The advantage of the inverse matrix method becomes readily apparent when solving a number of systems with a common coefficient matrix and different constant terms.

EXAMPLE 3 Using Inverses to Solve Systems of Equations

Use matrix inverse methods to solve each of the following systems:

(A) $x_1 - x_2 + x_3 = 3$
$2x_2 - x_3 = 1$
$2x_1 + 3x_2 = 4$

(B) $x_1 - x_2 + x_3 = -5$
$2x_2 - x_3 = 2$
$2x_1 + 3x_2 = -3$

Solutions Notice that both systems have the same coefficient matrix A as system (1) in Example 2. Only the constant terms have been changed. Thus, we can use A^{-1} to solve these systems just as we did in Example 2.

(A)

$$\overset{X}{\begin{bmatrix} x_1 \\ x_2 \\ x_3 \end{bmatrix}} = \overset{A^{-1}}{\begin{bmatrix} 3 & 3 & -1 \\ -2 & -2 & 1 \\ -4 & -5 & 2 \end{bmatrix}} \overset{B}{\begin{bmatrix} 3 \\ 1 \\ 4 \end{bmatrix}} = \begin{bmatrix} 8 \\ -4 \\ -9 \end{bmatrix}$$

Thus, $x_1 = 8$, $x_2 = -4$, and $x_3 = -9$

(B)

$$\overset{X}{\begin{bmatrix} x_1 \\ x_2 \\ x_3 \end{bmatrix}} = \overset{A^{-1}}{\begin{bmatrix} 3 & 3 & -1 \\ -2 & -2 & 1 \\ -4 & -5 & 2 \end{bmatrix}} \overset{B}{\begin{bmatrix} -5 \\ 2 \\ -3 \end{bmatrix}} = \begin{bmatrix} -6 \\ 3 \\ 4 \end{bmatrix}$$

Thus, $x_1 = -6$, $x_2 = 3$, and $x_3 = 4$

Matched Problem 3 Use matrix inverse methods to solve each of the following systems (see Matched Problem 2):

(A) $3x_1 - x_2 + x_3 = 3$
$-x_1 + x_2 = -3$
$x_1 + x_3 = 2$

(B) $3x_1 - x_2 + x_3 = -5$
$-x_1 + x_2 = 1$
$x_1 + x_3 = -4$

As Examples 2 and 3 illustrate, inverse methods are very convenient for hand calculations because once the inverse is found, it can be used to solve any new system formed by changing only the constant terms. Since most graphing utilities can compute the inverse of a matrix, this method also adapts readily to graphing utility solutions. However, if your graphing utility also has a built-in procedure for finding the reduced form of an augmented coefficient matrix, then it is just as convenient to use Gauss–Jordan elimination. Furthermore, Gauss–Jordan elimination can be used in all cases and, as noted below, matrix inverse methods cannot always be used.

Using Inverse Methods to Solve Systems of Equations

If the number of equations in a system equals the number of variables and the coefficient matrix has an inverse, then the system will always have a unique solution that can be found by using the inverse of the coefficient matrix to solve the corresponding matrix equation.

Matrix equation	Solution
$AX = B$	$X = A^{-1}B$

Remark. What happens if the coefficient matrix does not have an inverse? In this case, it can be shown that the system does not have a unique solution and is either dependent or inconsistent. Gauss–Jordan elimination must be used to determine which is the case. Also, as we mentioned earlier, Gauss–Jordan elimination must always be used if the number of variables is not the same as the number of equations.

• Application

The following application illustrates the usefulness of the inverse method.

EXAMPLE 4 Investment Allocation

An investment adviser currently has two types of investments available for clients: an investment A that pays 10% per year and an investment B of higher risk that pays 20% per year. Clients may divide their investments between the two to achieve any total return desired between 10 and 20%. However, the higher the desired return, the higher the risk. How should each client listed in the table invest to achieve the indicated return?

	Client			
	1	2	3	k
Total investment	\$20,000	\$50,000	\$10,000	k_1
Annual return desired	\$2,400	\$7,500	\$1,300	k_2
	(12%)	(15%)	(13%)	

Solution We first solve the problem for an arbitrary client k using inverses, and then apply the result to the three specific clients.

Let

$$x_1 = \text{Amount invested in } A$$

$$x_2 = \text{Amount invested in } B$$

Then

$$\begin{aligned} x_1 + \quad x_2 &= k_1 \quad \text{Total invested} \\ 0.1x_1 + 0.2x_2 &= k_2 \quad \text{Total annual return} \end{aligned}$$

Write as a matrix equation:

$$\overset{A}{\begin{bmatrix} 1 & 1 \\ 0.1 & 0.2 \end{bmatrix}} \overset{X}{\begin{bmatrix} x_1 \\ x_2 \end{bmatrix}} = \overset{B}{\begin{bmatrix} k_1 \\ k_2 \end{bmatrix}}$$

If A^{-1} exists, then

$$X = A^{-1}B$$

We now find A^{-1} by starting with $[A \mid I]$ and proceeding as discussed in Section 9-2:

$$\left[\begin{array}{cc|cc} 1 & 1 & 1 & 0 \\ 0.1 & 0.2 & 0 & 1 \end{array}\right] \quad 10R_2 \to R_2$$

$$\sim \left[\begin{array}{cc|cc} 1 & 1 & 1 & 0 \\ 1 & 2 & 0 & 10 \end{array}\right] \quad R_2 + (-1)R_1 \to R_2$$

$$\sim \left[\begin{array}{cc|cc} 1 & 1 & 1 & 0 \\ 0 & 1 & -1 & 10 \end{array}\right] \quad R_1 + (-1)R_2 \to R_1$$

$$\sim \left[\begin{array}{cc|cc} 1 & 0 & 2 & -10 \\ 0 & 1 & -1 & 10 \end{array}\right]$$

Thus,

$$A^{-1} = \begin{bmatrix} 2 & -10 \\ -1 & 10 \end{bmatrix} \quad \text{Check} \quad \overset{A^{-1}}{\begin{bmatrix} 2 & -10 \\ -1 & 10 \end{bmatrix}} \overset{A}{\begin{bmatrix} 1 & 1 \\ 0.1 & 0.2 \end{bmatrix}} = \overset{I}{\begin{bmatrix} 1 & 0 \\ 0 & 1 \end{bmatrix}}$$

and

$$\overset{X}{\begin{bmatrix} x_1 \\ x_2 \end{bmatrix}} = \overset{A^{-1}}{\begin{bmatrix} 2 & -10 \\ -1 & 10 \end{bmatrix}} \overset{B}{\begin{bmatrix} k_1 \\ k_2 \end{bmatrix}}$$

To solve each client's investment problem, we replace k_1 and k_2 with appropriate values from the table and multiply by A^{-1}:

Client 1

$$\begin{bmatrix} x_1 \\ x_2 \end{bmatrix} = \begin{bmatrix} 2 & -10 \\ -1 & 10 \end{bmatrix} \begin{bmatrix} 20{,}000 \\ 2{,}400 \end{bmatrix} = \begin{bmatrix} 16{,}000 \\ 4{,}000 \end{bmatrix}$$

Solution: $x_1 = \$16{,}000$ in A, $x_2 = \$4{,}000$ in B

Client 2

$$\begin{bmatrix} x_1 \\ x_2 \end{bmatrix} = \begin{bmatrix} 2 & -10 \\ -1 & 10 \end{bmatrix} \begin{bmatrix} 50{,}000 \\ 7{,}500 \end{bmatrix} = \begin{bmatrix} 25{,}000 \\ 25{,}000 \end{bmatrix}$$

Solution: $x_1 = \$25{,}000$ in A, $x_2 = \$25{,}000$ in B

Client 3

$$\begin{bmatrix} x_1 \\ x_2 \end{bmatrix} = \begin{bmatrix} 2 & -10 \\ -1 & 10 \end{bmatrix} \begin{bmatrix} 10{,}000 \\ 1{,}300 \end{bmatrix} = \begin{bmatrix} 7{,}000 \\ 3{,}000 \end{bmatrix}$$

Solution: x_1 = \$7,000 in A, x_2 = \$3,000 in B

Matched Problem 4 Repeat Example 4 with investment A paying 8% and investment B paying 24%.

Hand solutions of systems involving two equations and two variables can be made more efficient by using a formula for the inverse of a 2×2 matrix.

EXPLORE-DISCUSS 2 The inverse of

$$A = \begin{bmatrix} a & b \\ c & d \end{bmatrix}$$

is

$$A^{-1} = \begin{bmatrix} \dfrac{d}{ad - bc} & \dfrac{-b}{ad - bc} \\ \dfrac{-c}{ad - bc} & \dfrac{a}{ad - bc} \end{bmatrix} = \frac{1}{D}\begin{bmatrix} d & -b \\ -c & a \end{bmatrix}$$

where $D = ad - bc$, provided $D \neq 0$.

(A) Use matrix multiplication to verify this formula. What can you conclude about A^{-1} if $D = 0$? (We will have much more to say about the number D in the next three sections.)

(B) Use this formula to find the inverse of matrix A in Example 4.

Answers to Matched Problems

1.
$$AX + C = B$$
$$(AX + C) - C = B - C$$
$$AX + (C - C) = B - C$$
$$AX + 0 = B - C$$
$$AX = B - C$$
$$A^{-1}(AX) = A^{-1}(B - C)$$
$$(A^{-1}A)X = A^{-1}(B - C)$$
$$IX = A^{-1}(B - C)$$
$$X = A^{-1}(B - C)$$

2. $x_1 = 2, x_2 = 5, x_3 = 0$ **3.** (A) $x_1 = -2, x_2 = -5, x_3 = 4$ (B) $x_1 = 0, x_2 = 1, x_3 = -4$

4. $A^{-1} = \begin{bmatrix} 1.5 & -6.25 \\ -0.5 & 6.25 \end{bmatrix}$; Client 1: \$15,000 in A and \$5,000 in B; Client 2: \$28,125 in A and \$21,875 in B; Client 3: \$6,875 in A and \$3,125 in B

EXERCISE 9-3

A

Write Problems 1–4 as systems of linear equations without matrices.

1. $\begin{bmatrix} 2 & -1 \\ 1 & 3 \end{bmatrix}\begin{bmatrix} x_1 \\ x_2 \end{bmatrix} = \begin{bmatrix} 3 \\ -2 \end{bmatrix}$ **2.** $\begin{bmatrix} -3 & 1 \\ -1 & 2 \end{bmatrix}\begin{bmatrix} x_1 \\ x_2 \end{bmatrix} = \begin{bmatrix} -2 \\ 5 \end{bmatrix}$

3. $\begin{bmatrix} -2 & 0 & 1 \\ 1 & 2 & 1 \\ 0 & 1 & -1 \end{bmatrix}\begin{bmatrix} x_1 \\ x_2 \\ x_3 \end{bmatrix} = \begin{bmatrix} 3 \\ -4 \\ 2 \end{bmatrix}$

4. $\begin{bmatrix} 1 & -2 & 0 \\ -3 & 1 & -1 \\ 2 & 0 & 4 \end{bmatrix}\begin{bmatrix} x_1 \\ x_2 \\ x_3 \end{bmatrix} = \begin{bmatrix} 3 \\ -2 \\ 5 \end{bmatrix}$

Write each system in Problems 5–8 as a matrix equation of the form $AX = B$.

5. $\begin{aligned} 4x_1 - 3x_2 &= 2 \\ x_1 + 2x_2 &= 1 \end{aligned}$ **6.** $\begin{aligned} x_1 - 2x_2 &= 7 \\ -3x_1 + x_2 &= -3 \end{aligned}$

7. $\begin{aligned} x_1 - 2x_2 + x_3 &= -1 \\ -x_1 + x_2 \quad &= 2 \\ 2x_1 + 3x_2 + x_3 &= -3 \end{aligned}$ **8.** $\begin{aligned} 2x_1 \quad + 3x_3 &= 5 \\ x_1 - 2x_2 + x_3 &= -4 \\ -x_1 + 3x_2 \quad &= 2 \end{aligned}$

In Problems 9–12, find x_1 and x_2.

9. $\begin{bmatrix} x_1 \\ x_2 \end{bmatrix} = \begin{bmatrix} 3 & -2 \\ 1 & 4 \end{bmatrix}\begin{bmatrix} -2 \\ 1 \end{bmatrix}$ **10.** $\begin{bmatrix} x_1 \\ x_2 \end{bmatrix} = \begin{bmatrix} -2 & 1 \\ -1 & 2 \end{bmatrix}\begin{bmatrix} 3 \\ -2 \end{bmatrix}$

11. $\begin{bmatrix} x_1 \\ x_2 \end{bmatrix} = \begin{bmatrix} -2 & 3 \\ 2 & -1 \end{bmatrix}\begin{bmatrix} 3 \\ 2 \end{bmatrix}$ **12.** $\begin{bmatrix} x_1 \\ x_2 \end{bmatrix} = \begin{bmatrix} 3 & -1 \\ 0 & 2 \end{bmatrix}\begin{bmatrix} -2 \\ 1 \end{bmatrix}$

B

Write each system in Problems 13–20 as a matrix equation and solve using inverses. [Note: The inverses were found in Problems 17–24 in Exercise 10-2.]

13. $\begin{aligned} -x_1 - 2x_2 &= k_1 \\ 2x_1 + 5x_2 &= k_2 \end{aligned}$
(A) $k_1 = 2, k_2 = 5$
(B) $k_1 = -4, k_2 = 1$
(C) $k_1 = -3, k_2 = -2$

14. $\begin{aligned} 3x_1 - 4x_2 &= k_1 \\ -2x_1 + 3x_2 &= k_2 \end{aligned}$
(A) $k_1 = 3, k_2 = -1$
(B) $k_1 = 6, k_2 = 5$
(C) $k_1 = 0, k_2 = -4$

15. $\begin{aligned} -5x_1 + 7x_2 &= k_1 \\ 2x_1 - 3x_2 &= k_2 \end{aligned}$
(A) $k_1 = -5, k_2 = 1$
(B) $k_1 = 8, k_2 = -4$
(C) $k_1 = 6, k_2 = 0$

16. $\begin{aligned} 11x_1 + 4x_2 &= k_1 \\ 3x_1 + x_2 &= k_2 \end{aligned}$
(A) $k_1 = -2, k_2 = -3$
(B) $k_1 = -1, k_2 = 9$
(C) $k_1 = 4, k_2 = 5$

17. $\begin{aligned} x_1 - x_2 \quad &= k_1 \\ -x_1 + x_2 - x_3 &= k_2 \\ - x_2 + x_3 &= k_3 \end{aligned}$
(A) $k_1 = 1, k_2 = 1, k_3 = 2$
(B) $k_1 = -1, k_2 = 0, k_3 = -4$
(C) $k_1 = 3, k_2 = -2, k_3 = 0$

18. $\begin{aligned} 2x_1 - x_2 \quad &= k_1 \\ x_2 + x_3 &= k_2 \\ x_1 \quad + x_3 &= k_3 \end{aligned}$
(A) $k_1 = -2, k_2 = 4, k_3 = -1$
(B) $k_1 = 2, k_2 = -3, k_3 = 1$
(C) $k_1 = -1, k_2 = 2, k_3 = -5$

19. $\begin{aligned} x_1 + 2x_2 + 5x_3 &= k_1 \\ 3x_1 + 5x_2 + 9x_3 &= k_2 \\ x_1 + x_2 - 2x_3 &= k_3 \end{aligned}$
(A) $k_1 = 0, k_2 = 1, k_3 = 4$
(B) $k_1 = 5, k_2 = -1, k_3 = 0$
(C) $k_1 = -6, k_2 = 0, k_3 = 2$

20. $\begin{aligned} x_1 - x_2 + x_3 &= k_1 \\ -2x_1 + 3x_2 + 2x_3 &= k_2 \\ 3x_1 - 3x_2 + 2x_3 &= k_3 \end{aligned}$
(A) $k_1 = 3, k_2 = -1, k_3 = 0$
(B) $k_1 = 0, k_2 = 4, k_3 = 5$
(C) $k_1 = -2, k_2 = 0, k_3 = 1$

C

For $n \times n$ matrices A and B and $n \times 1$ matrices C, D, and X, solve each matrix equation in Problems 21–26 for X. Assume all necessary inverses exist.

21. $AX = BX + C$ **22.** $AX + BX = C + D$

23. $X = AX + C$ **24.** $X + C = AX - BX$

25. $AX + C = 3X$ **26.** $AX + C = BX - 7X + D$

27. Use matrix inverse methods to solve the following system for the indicated values of k_1 and k_2.

$$\begin{aligned} x_1 + 5.001x_2 &= k_1 \\ x_1 + \quad 5x_2 &= k_2 \end{aligned}$$

(A) $k_1 = 1, k_2 = 1$
(B) $k_1 = 1, k_2 = 0$
(C) $k_1 = 0, k_2 = 1$

Discuss the effect of small changes in the constant terms on the solution set of this system.

28. Repeat Problem 27 for the following system:

$$\begin{aligned} x_1 - 4.001x_2 &= k_1 \\ x_1 - \quad 4x_2 &= k_2 \end{aligned}$$

In Problems 29–32, write each system as a matrix equation and solve by using the inverse coefficient matrix. Use a graphing utility to perform the necessary calculations.

29. $$\begin{aligned} 2x_1 + 4x_2 + 7x_3 &= 26 \\ 5x_1 - 6x_2 + 9x_3 &= -169 \\ 3x_1 + 6x_2 - 5x_3 &= 225 \end{aligned}$$

30. $$\begin{aligned} x_1 + 5x_2 + 4x_3 &= 99 \\ 6x_1 - x_2 + 7x_3 &= 75 \\ 4x_1 + 6x_2 + x_3 &= 125 \end{aligned}$$

31. $$\begin{aligned} x_1 + x_2 - x_3 + 3x_4 &= 75 \\ 2x_1 - x_2 + 3x_3 - x_4 &= 100 \\ 5x_1 + 6x_2 + x_3 + x_4 &= 180 \\ 5x_1 + x_2 + 7x_3 + 2x_4 &= 315 \end{aligned}$$

32. $$\begin{aligned} x_1 + 9x_2 + x_3 + 2x_4 &= 71 \\ 4x_1 + 8x_2 - 5x_3 + x_4 &= 99 \\ 2x_1 + 5x_2 - 6x_3 + 3x_4 &= 113 \\ 6x_1 + 4x_2 + x_3 + 8x_4 &= 81 \end{aligned}$$

APPLICATIONS

Solve using systems of equations and inverses.

33. Resource Allocation. A concert hall has 10,000 seats. If tickets are \$4 and \$8, how many of each type of ticket should be sold (assuming all seats can be sold) to bring in each of the returns indicated in the table? Use decimals in computing the inverse.

	Concert		
	1	2	3
Tickets sold	10,000	10,000	10,000
Return required	\$56,000	\$60,000	\$68,000

34. Production Scheduling. Labor and material costs for manufacturing two guitar models are given in the following table:

Guitar Model	Labor Cost	Material Cost
A	\$30	\$20
B	\$40	\$30

If a total of \$3,000 a week is allowed for labor and material, how many of each model should be produced each week to exactly use each of the allocations of the \$3,000 indicated in the following table? Use decimals in computing the inverse.

	Weekly Allocation		
	1	2	3
Labor	\$1,800	\$1,750	\$1,720
Material	\$1,200	\$1,250	\$1,280

★ **35. Circuit Analysis.** A direct current electric circuit consisting of conductors (wires), resistors, and batteries is diagrammed in the figure.

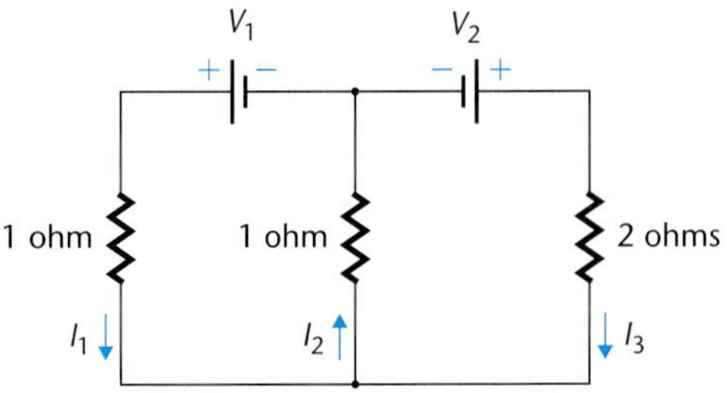

If I_1, I_2, and I_3 are the currents (in amperes) in the three branches of the circuit and V_1 and V_2 are the voltages (in volts) of the two batteries, then *Kirchhoff's* laws* can be used to show that the currents satisfy the following system of equations:

$$\begin{aligned} I_1 - I_2 + I_3 &= 0 \\ I_1 + I_2 &= V_1 \\ I_2 + 2I_3 &= V_2 \end{aligned}$$

Solve this system for:
(A) $V_1 = 10$ volts, $V_2 = 10$ volts
(B) $V_1 = 10$ volts, $V_2 = 15$ volts
(C) $V_1 = 15$ volts, $V_2 = 10$ volts

★ **36. Circuit Analysis.** Repeat Problem 35 for the electric circuit shown in the figure on the next page.

$$\begin{aligned} I_1 - I_2 + I_3 &= 0 \\ I_1 + 2I_2 &= V_1 \\ 2I_2 + 2I_3 &= V_2 \end{aligned}$$

*Gustav Kirchhoff (1824–1887), a German physicist, was among the first to apply theoretical mathematics to physics. He is best-known for his development of certain properties of electric circuits, which are now known as **Kirchhoff's laws.**

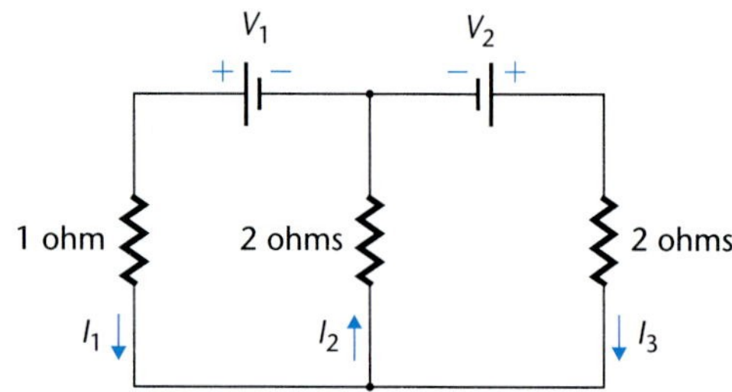

★★ **37. Geometry.** The graph of $f(x) = ax^2 + bx + c$ passes through the points $(1, k_1)$, $(2, k_2)$, and $(3, k_3)$. Determine a, b, and c for:

(A) $k_1 = -2, k_2 = 1, k_3 = 6$
(B) $k_1 = 4, k_2 = 3, k_3 = -2$
(C) $k_1 = 8, k_2 = -5, k_3 = 4$

★★ **38. Geometry.** Repeat Problem 37 if the graph passes through the points $(-1, k_1)$, $(0, k_2)$, and $(1, k_3)$.

Check your answers in Problems 37 and 38 by graphing $y = f(x)$ on a graphing utility and verifying that the graph passes through the indicated points.

39. Diets. A biologist has available two commercial food mixes with the following percentages of protein and fat:

Mix	Protein (%)	Fat (%)
A	20	2
B	10	6

How many ounces of each mix should be used to prepare each of the diets listed in the following table?

	Diet		
	1	2	3
Protein	20 oz	10 oz	10 oz
Fat	6 oz	4 oz	6 oz

SECTION 9-4 Determinants

- Determinants
- Second-Order Determinants
- Third-Order Determinants
- Higher-Order Determinants

• Determinants

In this section we are going to associate with each square matrix a real number, called the **determinant** of the matrix. If A is a square matrix, then the determinant of A is denoted by **det A,** or simply by writing the array of elements in A using vertical lines in place of square brackets. For example,

$$\det\begin{bmatrix} 2 & -3 \\ 5 & 1 \end{bmatrix} = \begin{vmatrix} 2 & -3 \\ 5 & 1 \end{vmatrix} \qquad \det\begin{bmatrix} 1 & -2 & 3 \\ 0 & 5 & -7 \\ -2 & 1 & 6 \end{bmatrix} = \begin{vmatrix} 1 & -2 & 3 \\ 0 & 5 & -7 \\ -2 & 1 & 6 \end{vmatrix}$$

A determinant of **order *n*** is a determinant with n rows and n columns. In this section we concentrate most of our attention on determining the values of determinants of orders 2 and 3. But many of the results and procedures discussed can be generalized completely to determinants of order n.

• Second-Order Determinants

In general, a **second-order determinant** is written as

$$\begin{vmatrix} a_{11} & a_{12} \\ a_{21} & a_{22} \end{vmatrix}$$

and represents a real number as given in Definition 1.

DEFINITION 1 **Value of a Second-Order Determinant**

$$\begin{vmatrix} a_{11} & a_{12} \\ a_{21} & a_{22} \end{vmatrix} = a_{11}a_{22} - a_{21}a_{12} \qquad (1)$$

Formula (1) is easily remembered if you notice that the expression on the right is the product of the **principal diagonal,** from upper left to lower right, minus the product of the **secondary diagonal,** from lower left to upper right.

EXAMPLE 1 **Evaluating a Second-Order Determinant**

$$\begin{vmatrix} -1 & 2 \\ -3 & -4 \end{vmatrix} = (-1)(-4) - (-3)(2) = 4 - (-6) = 10$$

Matched Problem 1 Find: $\begin{vmatrix} 3 & -5 \\ 4 & -2 \end{vmatrix}$

• Third-Order Determinants

A determinant of order 3 is a square array of nine elements and represents a real number given by Definition 2, which is a special case of the general definition of the value of an nth-order determinant. Note that each term in the expansion on the right of equation (2) contains exactly one element from each row and each column.

DEFINITION 2 **Value of a Third-Order Determinant**

$$\begin{vmatrix} a_{11} & a_{12} & a_{13} \\ a_{21} & a_{22} & a_{23} \\ a_{31} & a_{32} & a_{33} \end{vmatrix} = a_{11}a_{22}a_{33} - a_{11}a_{32}a_{23} + a_{21}a_{32}a_{13} - a_{21}a_{12}a_{33} + a_{31}a_{12}a_{23} - a_{31}a_{22}a_{13} \qquad (2)$$

Don't panic! You don't need to memorize formula (2). After we introduce the ideas of *minor* and *cofactor* below, we will state a theorem that can be used to obtain the same result with much less trouble.

The **minor of an element** in a third-order determinant is a second-order determinant obtained by deleting the row and column that contains the element. For example, in the determinant in formula (2),

$$\text{Minor of } a_{23} = \begin{vmatrix} a_{11} & a_{12} \\ a_{31} & a_{32} \end{vmatrix} \qquad \begin{vmatrix} a_{11} & a_{12} & a_{13} \\ a_{21} & a_{22} & a_{23} \\ a_{31} & a_{32} & a_{33} \end{vmatrix}$$

Deletions are usually done mentally.

$$\text{Minor of } a_{32} = \begin{vmatrix} a_{11} & a_{13} \\ a_{21} & a_{23} \end{vmatrix} \qquad \begin{vmatrix} a_{11} & a_{12} & a_{13} \\ a_{21} & a_{22} & a_{23} \\ a_{31} & a_{32} & a_{33} \end{vmatrix}$$

EXPLORE-DISCUSS 1 Write the minors of the other seven elements in the determinant in formula (2).

A quantity closely associated with the minor of an element is the **cofactor of an element** a_{ij} (from the ith row and jth column), which is the product of the minor of a_{ij} and $(-1)^{i+j}$.

DEFINITION 3 **Cofactor**

$$\text{Cofactor of } a_{ij} = (-1)^{i+j}(\text{Minor of } a_{ij})$$

Thus, a cofactor of an element is nothing more than a signed minor. The sign is determined by raising -1 to a power that is the sum of the numbers indicating the row and column in which the element appears. Note that $(-1)^{i+j}$ is 1 if $i + j$ is even and -1 if $i + j$ is odd. Thus, if we are given the determinant

$$\begin{vmatrix} a_{11} & a_{12} & a_{13} \\ a_{21} & a_{22} & a_{23} \\ a_{31} & a_{32} & a_{33} \end{vmatrix}$$

then

$$\text{Cofactor of } a_{23} = (-1)^{2+3}\begin{vmatrix} a_{11} & a_{12} \\ a_{31} & a_{32} \end{vmatrix} = -\begin{vmatrix} a_{11} & a_{12} \\ a_{31} & a_{32} \end{vmatrix}$$

$$\text{Cofactor of } a_{11} = (-1)^{1+1}\begin{vmatrix} a_{22} & a_{23} \\ a_{32} & a_{33} \end{vmatrix} = \begin{vmatrix} a_{22} & a_{23} \\ a_{32} & a_{33} \end{vmatrix}$$

EXAMPLE 2 **Finding Cofactors**

Find the cofactors of -2 and 5 in the determinant

$$\begin{vmatrix} -2 & 0 & 3 \\ 1 & -6 & 5 \\ -1 & 2 & 0 \end{vmatrix}$$

Solution

$$\text{Cofactor of } -2 = (-1)^{1+1}\begin{vmatrix} -6 & 5 \\ 2 & 0 \end{vmatrix} = \begin{vmatrix} -6 & 5 \\ 2 & 0 \end{vmatrix}$$

$$= (-6)(0) - (2)(5) = -10$$

$$\text{Cofactor of } 5 = (-1)^{2+3}\begin{vmatrix} -2 & 0 \\ -1 & 2 \end{vmatrix} = -\begin{vmatrix} -2 & 0 \\ -1 & 2 \end{vmatrix}$$

$$= -[(-2)(2) - (-1)(0)] = 4$$

Matched Problem 2 Find the cofactors of 2 and 3 in the determinant in Example 2.

[*Note:* The sign in front of the minor, $(-1)^{i+j}$, can be determined rather mechanically by using a checkerboard pattern of + and − signs over the determinant, starting with + in the upper left-hand corner:

$$\begin{matrix} + & - & + \\ - & + & - \\ + & - & + \end{matrix}$$

Use either the checkerboard or the exponent method—whichever is easier for you—to determine the sign in front of the minor.]

Now we are ready for the key theorem of this section, Theorem 1. This theorem provides us with an efficient step-by-step procedure, called an algorithm, for evaluating third-order determinants.

Theorem 1 **Value of a Third-Order Determinant**

The value of a determinant of order 3 is the sum of three products obtained by multiplying each element of any one row (or each element of any one column) by its cofactor.

To prove this theorem we must show that the expansions indicated by the theorem for any row or any column (six cases) produce the expression on the right of formula (2). Proofs of special cases of this theorem are left to the C problems in Exercise 9-4.

EXAMPLE 3 **Evaluating a Third-Order Determinant**

Evaluate

$$\begin{vmatrix} 2 & -2 & 0 \\ -3 & 1 & 2 \\ 1 & -3 & -1 \end{vmatrix}$$

by expanding by:

(A) The first row (B) The second column

Solutions (A)

$$\begin{vmatrix} 2 & -2 & 0 \\ -3 & 1 & 2 \\ 1 & -3 & -1 \end{vmatrix} = a_{11}\begin{pmatrix}\text{Cofactor}\\ \text{of } a_{11}\end{pmatrix} + a_{12}\begin{pmatrix}\text{Cofactor}\\ \text{of } a_{12}\end{pmatrix} + a_{13}\begin{pmatrix}\text{Cofactor}\\ \text{of } a_{13}\end{pmatrix}$$

$$= 2\left((-1)^{1+1}\begin{vmatrix} 1 & 2 \\ -3 & -1 \end{vmatrix}\right) + (-2)\left((-1)^{1+2}\begin{vmatrix} -3 & 2 \\ 1 & -1 \end{vmatrix}\right) + 0$$

$$= (2)(1)[(1)(-1) - (-3)(2)] + (-2)(-1)[(-3)(-1) - (1)(2)]$$

$$= (2)(5) + (2)(1) = 12$$

(B)

$$\begin{vmatrix} 2 & -2 & 0 \\ -3 & 1 & 2 \\ 1 & -3 & -1 \end{vmatrix} = a_{12}\begin{pmatrix}\text{Cofactor}\\ \text{of } a_{12}\end{pmatrix} + a_{22}\begin{pmatrix}\text{Cofactor}\\ \text{of } a_{22}\end{pmatrix} + a_{32}\begin{pmatrix}\text{Cofactor}\\ \text{of } a_{32}\end{pmatrix}$$

$$= (-2)\left((-1)^{1+2}\begin{vmatrix} -3 & 2 \\ 1 & -1 \end{vmatrix}\right) + (1)\left((-1)^{2+2}\begin{vmatrix} 2 & 0 \\ 1 & -1 \end{vmatrix}\right)$$

$$+ (-3)\left((-1)^{3+2}\begin{vmatrix} 2 & 0 \\ -3 & 2 \end{vmatrix}\right)$$

$$= (-2)(-1)[(-3)(-1) - (1)(2)] + (1)(1)[(2)(-1) - (1)(0)]$$

$$+ (-3)(-1)[(2)(2) - (-3)(0)]$$

$$= (2)(1) + (1)(-2) + (3)(4) = 12$$

Matched Problem 3 Evaluate

$$\begin{vmatrix} 2 & 1 & -1 \\ -2 & -3 & 0 \\ -1 & 2 & 1 \end{vmatrix}$$

by expanding by:

(A) The first row (B) The third column

```
[A]
     [[2   -2 0 ]
      [-3  1  2 ]
      [1   -3 -1]]
det([A])
              12
```

FIGURE 1

Most graphing utilities will evaluate determinants. Figure 1 shows the evaluation of the determinant in Example 3.

• Higher-Order Determinants

Theorem 1 and the definitions of minor and cofactor generalize completely for determinants of order higher than 3. These concepts are illustrated for a fourth-order determinant in the next example.

EXAMPLE 4 **Evaluating a Fourth-Order Determinant**

Given the fourth-order determinant

$$\begin{vmatrix} 0 & -1 & 0 & 2 \\ -5 & -6 & 0 & -3 \\ 4 & 5 & -2 & 6 \\ 0 & 3 & 0 & -4 \end{vmatrix}$$

(A) Find the minor in determinant form of the element 3.
(B) Find the cofactor in determinant form of the element -5.
(C) Find the value of the fourth-order determinant.

Solutions (A) Minor of 3 $= \begin{vmatrix} 0 & 0 & 2 \\ -5 & 0 & -3 \\ 4 & -2 & 6 \end{vmatrix}$

(B) Cofactor of $-5 = (-1)^{2+1}\begin{vmatrix} -1 & 0 & 2 \\ 5 & -2 & 6 \\ 3 & 0 & -4 \end{vmatrix} = -\begin{vmatrix} -1 & 0 & 2 \\ 5 & -2 & 6 \\ 3 & 0 & -4 \end{vmatrix}$

(C) Generalizing Theorem 1, the value of this fourth-order determinant is the sum of four products obtained by multiplying each element of any one row (or each element of any one column) by its cofactor. The work involved in this evaluation is greatly reduced if we choose the row or column with the greatest number of 0's. Since column 3 has three 0's, we expand along this column:

$$\begin{vmatrix} 0 & -1 & 0 & 2 \\ -5 & -6 & 0 & -3 \\ 4 & 5 & -2 & 6 \\ 0 & 3 & 0 & -4 \end{vmatrix} = 0 + 0 + (-2)(-1)^{3+3}\begin{vmatrix} 0 & -1 & 2 \\ -5 & -6 & -3 \\ 0 & 3 & -4 \end{vmatrix} + 0$$

$$= (-2)\begin{vmatrix} 0 & -1 & 2 \\ -5 & -6 & -3 \\ 0 & 3 & -4 \end{vmatrix}$$ Expand this determinant along the first column.

$$= (-2)\left(0 + (-5)(-1)^{2+1}\begin{vmatrix} -1 & 2 \\ 3 & -4 \end{vmatrix} + 0\right)$$

$$= (-2)(-5)(-1)(-2) = 20$$

Matched Problem 4 Repeat Example 4 for the following fourth-order determinant:

$$\begin{vmatrix} 0 & 4 & -2 & 0 \\ -3 & 3 & -1 & 2 \\ 0 & 6 & 0 & 0 \\ 5 & -6 & -5 & -4 \end{vmatrix}$$

EXPLORE-DISCUSS 2 Write a checkerboard pattern of + and − signs for a fourth-order determinant, and use it to determine the signs of the minors in Example 4.

Remark. Where are determinants used? Many equations and formulas have particularly simple and compact representations in determinant form that are easily remembered. (See Problems 50–54 in Exercise 9-5). Also, in Section 9-6 we will see that the solutions to certain systems of equations can be expressed in terms of determinants. In addition, determinants are involved in theoretical work in advanced mathematics courses. For example, it can be shown that the inverse of a square matrix exists if and only if its determinant is not 0.

Answers to Matched Problems

1. 14 **2.** Cofactor of 2 = 13; cofactor of 3 = −4 **3.** (A) 3 (B) 3

4. (A) $\begin{vmatrix} 0 & -2 & 0 \\ 0 & 0 & 0 \\ 5 & -5 & -4 \end{vmatrix}$ (B) $-\begin{vmatrix} 0 & 4 & 0 \\ -3 & 3 & 2 \\ 0 & 6 & 0 \end{vmatrix}$ (C) −24

EXERCISE 9-4

A

Evaluate each second-order determinant in Problems 1–6.

1. $\begin{vmatrix} 5 & 4 \\ 2 & 3 \end{vmatrix}$

2. $\begin{vmatrix} 8 & -3 \\ 4 & 1 \end{vmatrix}$

3. $\begin{vmatrix} 3 & -7 \\ -5 & 6 \end{vmatrix}$

4. $\begin{vmatrix} 9 & -2 \\ 4 & 0 \end{vmatrix}$

5. $\begin{vmatrix} 4.3 & -1.2 \\ -5.1 & 3.7 \end{vmatrix}$

6. $\begin{vmatrix} -0.7 & -2.3 \\ 1.9 & -4.8 \end{vmatrix}$

Problems 7–14 pertain to the determinant below:

$$\begin{vmatrix} 5 & -1 & -3 \\ 3 & 4 & 6 \\ 0 & -2 & 8 \end{vmatrix}$$

Write the minor of each element given in Problems 7–10. Leave the answer in determinant form.

7. a_{11}

8. a_{33}

9. a_{23}

10. a_{12}

Write the cofactor of each element given in Problems 11–14, and evaluate each.

11. a_{11}

12. a_{33}

13. a_{23}

14. a_{12}

Evaluate Problems 15–20 using cofactors.

Check your answers to Problems 15–20 on a graphing utility.

15. $\begin{vmatrix} 1 & 0 & 0 \\ -2 & 4 & 3 \\ 5 & -2 & 1 \end{vmatrix}$

16. $\begin{vmatrix} 2 & -3 & 5 \\ 0 & -3 & 1 \\ 0 & 6 & 2 \end{vmatrix}$

17. $\begin{vmatrix} 0 & 1 & 5 \\ 3 & -7 & 6 \\ 0 & -2 & -3 \end{vmatrix}$

18. $\begin{vmatrix} 4 & -2 & 0 \\ 9 & 5 & 4 \\ 1 & 2 & 0 \end{vmatrix}$

19. $\begin{vmatrix} -1 & 2 & -3 \\ -2 & 0 & -6 \\ 4 & -3 & 2 \end{vmatrix}$

20. $\begin{vmatrix} 0 & 2 & -1 \\ -6 & 3 & 1 \\ 7 & -9 & -2 \end{vmatrix}$

B

Given the determinant

$$\begin{vmatrix} a_{11} & a_{12} & a_{13} & a_{14} \\ a_{21} & a_{22} & a_{23} & a_{24} \\ a_{31} & a_{32} & a_{33} & a_{34} \\ a_{41} & a_{42} & a_{43} & a_{44} \end{vmatrix}$$

write the cofactor in determinant form of each element in Problems 21–24.

21. a_{11}

22. a_{44}

23. a_{43}

24. a_{23}

Evaluate each determinant in Problems 25–34 using cofactors.

Check your answers to Problems 25–34 on a graphing utility.

25. $\begin{vmatrix} 3 & -2 & -8 \\ -2 & 0 & -3 \\ 1 & 0 & -4 \end{vmatrix}$

26. $\begin{vmatrix} 4 & -4 & 6 \\ 2 & 8 & -3 \\ 0 & -5 & 0 \end{vmatrix}$

27. $\begin{vmatrix} 1 & 4 & 1 \\ 1 & 1 & -2 \\ 2 & 1 & -1 \end{vmatrix}$

28. $\begin{vmatrix} 3 & 2 & 1 \\ -1 & 5 & 1 \\ 2 & 3 & 1 \end{vmatrix}$

29. $\begin{vmatrix} 1 & 4 & 3 \\ 2 & 1 & 6 \\ 3 & -2 & 9 \end{vmatrix}$

30. $\begin{vmatrix} 4 & -6 & 3 \\ -1 & 4 & 1 \\ 5 & -6 & 3 \end{vmatrix}$

31. $\begin{vmatrix} 2 & 6 & 1 & 7 \\ 0 & 3 & 0 & 0 \\ 3 & 4 & 2 & 5 \\ 0 & 9 & 0 & 2 \end{vmatrix}$

32. $\begin{vmatrix} 0 & 1 & 0 & 1 \\ 2 & 4 & 7 & 6 \\ 0 & 3 & 0 & 1 \\ 0 & 6 & 2 & 5 \end{vmatrix}$

33. $\begin{vmatrix} -2 & 0 & 0 & 0 & 0 \\ 9 & -1 & 0 & 0 & 0 \\ 2 & 1 & 3 & 0 & 0 \\ -1 & 4 & 2 & 2 & 0 \\ 7 & -2 & 3 & 5 & 5 \end{vmatrix}$

34. $\begin{vmatrix} 2 & 0 & 0 & 0 & 0 \\ 0 & 3 & 0 & 0 & 0 \\ 0 & 0 & 2 & 0 & 0 \\ 0 & 0 & 0 & 1 & 0 \\ 0 & 0 & 0 & 0 & 4 \end{vmatrix}$

If A is a 3 × 3 matrix, det A can be evaluated by the following ***diagonal expansion.*** *Form a 3 × 5 matrix by augmenting A on the right with its first two columns, and compute the diagonal products $p_1, p_2, \ldots, p_6$ indicated by the arrows:*

$$\begin{bmatrix} a_{11} & a_{12} & a_{13} & a_{11} & a_{12} \\ a_{21} & a_{22} & a_{23} & a_{21} & a_{22} \\ a_{31} & a_{32} & a_{33} & a_{31} & a_{32} \end{bmatrix}$$

$p_4 \quad p_5 \quad p_6 \qquad p_1 \quad p_2 \quad p_3$

Diagonal expansion formula

The determinant of A is given by [compare with formula (2)]

$$\begin{aligned} \det A &= p_1 + p_2 + p_3 - p_4 - p_5 - p_6 \\ &= a_{11}a_{22}a_{33} + a_{12}a_{23}a_{31} + a_{13}a_{21}a_{32} - a_{13}a_{22}a_{31} \\ &\quad - a_{11}a_{23}a_{32} - a_{12}a_{21}a_{33} \end{aligned}$$

[Caution: The diagonal expansion procedure works only for 3 × 3 matrices. Do not apply it to matrices of any other size.]

Use the diagonal expansion formula to evaluate the determinants in Problems 35 and 36.

35. $\begin{vmatrix} 2 & 6 & -1 \\ 5 & 3 & -7 \\ -4 & -2 & 1 \end{vmatrix}$

36. $\begin{vmatrix} 4 & 1 & -5 \\ 1 & 2 & -6 \\ -3 & -1 & 7 \end{vmatrix}$

A square matrix is called an ***upper triangular matrix*** *if all elements below the principal diagonal are zero. In Problems 37–40, determine whether the statement is true or false. If true, explain why. If false, give a counterexample.*

37. If the determinant of an upper triangular matrix is 0, then the elements on the principal diagonal are all 0.

38. If A and B are upper triangular matrices, then $\det(A + B) = \det A + \det B$.

39. The determinant of an upper triangular matrix is the product of the elements on the principal diagonal.

40. If A and B are upper triangular matrices, then $\det(AB) = (\det A)(\det B)$.

C

In Problems 41–46, all the letters represent real numbers. Find an equation that each pair of determinants satisfies, and describe the relationship between the two determinants verbally.

41. $\begin{vmatrix} a & b \\ c & d \end{vmatrix}, \begin{vmatrix} c & d \\ a & b \end{vmatrix}$

42. $\begin{vmatrix} a & b \\ c & d \end{vmatrix}, \begin{vmatrix} b & a \\ d & c \end{vmatrix}$

43. $\begin{vmatrix} a & b \\ c & d \end{vmatrix}, \begin{vmatrix} ka & b \\ kc & d \end{vmatrix}$

44. $\begin{vmatrix} a & b \\ c & d \end{vmatrix}, \begin{vmatrix} a & b \\ kc & kd \end{vmatrix}$

45. $\begin{vmatrix} a & b \\ c & d \end{vmatrix}, \begin{vmatrix} kc + a & kd + b \\ c & d \end{vmatrix}$

46. $\begin{vmatrix} a & b \\ c & d \end{vmatrix}, \begin{vmatrix} a & ka + b \\ c & kc + d \end{vmatrix}$

47. Show that the expansion of the determinant

$$\begin{vmatrix} a_{11} & a_{12} & a_{13} \\ a_{21} & a_{22} & a_{23} \\ a_{31} & a_{32} & a_{33} \end{vmatrix}$$

by the first column is the same as its expansion by the third row.

48. Repeat Problem 47, using the second row and the third column.

49. If

$$A = \begin{bmatrix} 2 & 3 \\ 1 & -2 \end{bmatrix} \quad \text{and} \quad B = \begin{bmatrix} -1 & 3 \\ 2 & 1 \end{bmatrix}$$

show that $\det(AB) = \det A \cdot \det B$.

50. If

$$A = \begin{bmatrix} a & b \\ c & d \end{bmatrix} \quad \text{and} \quad B = \begin{bmatrix} w & x \\ y & z \end{bmatrix}$$

show that $\det(AB) = \det A \cdot \det B$.

If A is an n × n matrix and I is the n × n identity matrix, then the function $f(x) = |xI - A|$ is called the ***characteristic polynomial*** *of A, and the zeros of f(x) are called the* ***eigenvalues*** *of A. Characteristic polynomials and eigenvalues have many important applications that are discussed in more* *advanced treatments of matrices. In Problems 51–54, find the characteristic polynomial and the eigenvalues of each matrix.*

51. $\begin{bmatrix} 5 & -4 \\ 2 & -1 \end{bmatrix}$

52. $\begin{bmatrix} 8 & -6 \\ 3 & -1 \end{bmatrix}$

53. $\begin{bmatrix} 4 & -4 & 0 \\ 2 & -2 & 0 \\ 4 & -8 & -4 \end{bmatrix}$

54. $\begin{bmatrix} -2 & 2 & 0 \\ -1 & 1 & 0 \\ -2 & 4 & 2 \end{bmatrix}$

SECTION 9-5 Properties of Determinants

- Discussion of Determinant Properties
- Summary of Determinant Properties

Determinants have a number of useful properties that can greatly reduce the labor in evaluating determinants of order 3 or greater. These properties and their use are the subject matter for this section.

• Discussion of Determinant Properties

We now state and discuss five general determinant properties in the form of theorems. Because the proofs for the general cases of these theorems are involved and notationally difficult, we will sketch only informal proofs for determinants of order 3. The theorems, however, apply to determinants of any order.

Theorem 1

Multiplying a Row or Column by a Constant

If each element of any row (or column) of a determinant is multiplied by a constant k, the new determinant is k times the original.

Partial Proof

Let C_{ij} be the cofactor of a_{ij}. Then expanding by the first row, we have

$$\begin{vmatrix} ka_{11} & ka_{12} & ka_{13} \\ a_{21} & a_{22} & a_{23} \\ a_{31} & a_{32} & a_{33} \end{vmatrix} = ka_{11}C_{11} + ka_{12}C_{12} + ka_{13}C_{13}$$

$$= k(a_{11}C_{11} + a_{12}C_{12} + a_{13}C_{13})$$

$$= k\begin{vmatrix} a_{11} & a_{12} & a_{13} \\ a_{21} & a_{22} & a_{23} \\ a_{31} & a_{32} & a_{33} \end{vmatrix}$$

Theorem 1 also states that a factor common to all elements of a row (or column) can be taken out as a factor of the determinant.

EXAMPLE 1 **Taking Out a Common Factor of a Column**

$$\begin{vmatrix} 6 & 1 & 3 \\ -2 & 7 & -2 \\ 4 & 5 & 0 \end{vmatrix} = 2\begin{vmatrix} 3 & 1 & 3 \\ -1 & 7 & -2 \\ 2 & 5 & 0 \end{vmatrix}$$

where 2 is a common factor of the first column.

Matched Problem 1 Take out factors common to any row or any column:

$$\begin{vmatrix} 3 & 2 & 1 \\ 6 & 3 & -9 \\ 1 & 0 & -5 \end{vmatrix}$$

EXPLORE-DISCUSS 1 (A) How are $\begin{vmatrix} a & b \\ c & d \end{vmatrix}$ and $\begin{vmatrix} ka & kb \\ kc & kd \end{vmatrix}$ related?

(B) How are $\begin{vmatrix} a & b & c \\ d & e & f \\ g & h & i \end{vmatrix}$ and $\begin{vmatrix} ka & kb & kc \\ kd & ke & kf \\ kg & kh & ki \end{vmatrix}$ related?

Theorem 2 **Row or Column of Zeros**

If every element in a row (or column) is 0, the value of the determinant is 0.

Theorem 2 is an immediate consequence of Theorem 1, and its proof is left as an exercise. It is illustrated in the following example:

$$\begin{vmatrix} 3 & -2 & 5 \\ 0 & 0 & 0 \\ -1 & 4 & 9 \end{vmatrix} = 0$$

Theorem 3 **Interchanging Rows or Columns**

If two rows (or two columns) of a determinant are interchanged, the new determinant is the negative of the original.

A proof of Theorem 3 even for a determinant of order 3 is notationally involved. We suggest that you partially prove the theorem by direct expansion of the determinants before and after the interchange of two rows (or columns). The theorem is illustrated by the following example, where the second and third columns are interchanged:

$$\begin{vmatrix} 1 & 0 & 9 \\ -2 & 1 & 5 \\ 3 & 0 & 7 \end{vmatrix} = -\begin{vmatrix} 1 & 9 & 0 \\ -2 & 5 & 1 \\ 3 & 7 & 0 \end{vmatrix}$$

EXPLORE-DISCUSS 2 (A) What are the cofactors of each element in the first row of the following determinant? What is the value of the determinant?

$$\begin{vmatrix} a & b & c \\ d & e & f \\ d & e & f \end{vmatrix}$$

(B) What are the cofactors of each element in the second column of the following determinant? What is the value of the determinant?

$$\begin{vmatrix} a & b & a \\ d & e & d \\ g & h & g \end{vmatrix}$$

Theorem 4 **Equal Rows or Columns**

If the corresponding elements are equal in two rows (or columns), the value of the determinant is 0.

Proof The general proof of Theorem 4 follows directly from Theorem 3. If we start with a determinant D that has two rows (or columns) equal and we interchange the equal rows (or columns), the new determinant will be the same as the original. But by Theorem 3,

$$D = -D$$

hence,

$$2D = 0$$
$$D = 0$$

Theorem 5 **Addition of Rows or Columns**

If a multiple of any row (or column) of a determinant is added to any other row (or column), the value of the determinant is not changed.

Partial Proof If, in a general third-order determinant, we add a k multiple of the second column to the first and then expand by the first column, we obtain (where C_{ij} is the cofactor of a_{ij} in the original determinant)

$$\begin{vmatrix} a_{11} + ka_{12} & a_{12} & a_{13} \\ a_{21} + ka_{22} & a_{22} & a_{23} \\ a_{31} + ka_{32} & a_{32} & a_{33} \end{vmatrix} = (a_{11} + ka_{12})C_{11} + (a_{21} + ka_{22})C_{21} + (a_{31} + ka_{32})C_{31}$$

$$= (a_{11}C_{11} + a_{21}C_{21} + a_{31}C_{31}) + k(a_{12}C_{11} + a_{22}C_{21} + a_{32}C_{31})$$

$$= \begin{vmatrix} a_{11} & a_{12} & a_{13} \\ a_{21} & a_{22} & a_{23} \\ a_{31} & a_{32} & a_{33} \end{vmatrix} + k\begin{vmatrix} a_{12} & a_{12} & a_{13} \\ a_{22} & a_{22} & a_{23} \\ a_{32} & a_{32} & a_{33} \end{vmatrix} = \begin{vmatrix} a_{11} & a_{12} & a_{13} \\ a_{21} & a_{22} & a_{23} \\ a_{31} & a_{32} & a_{33} \end{vmatrix}$$

The determinant following k is 0 because the first and second columns are equal.

Note the similarity in the process described in Theorem 5 to that used to obtain row-equivalent matrices. We use this theorem to transform a determinant without 0 elements into one that contains a row or column with all elements 0 but one. The transformed determinant can then be easily expanded by this row (or column). An example best illustrates the process.

EXAMPLE 2 Evaluating a Determinant

Evaluate the determinant

$$\begin{vmatrix} 3 & -1 & 2 \\ -2 & 4 & -3 \\ 4 & -2 & 5 \end{vmatrix}$$

Solution We use Theorem 5 to obtain two 0's in the first row, and then expand the determinant by this row. To start, we replace the third column with the sum of it and 2 times the second column to obtain a 0 in the a_{13} position:

$$\begin{vmatrix} 3 & -1 & 2 \\ -2 & 4 & -3 \\ 4 & -2 & 5 \end{vmatrix} = \begin{vmatrix} 3 & -1 & 0 \\ -2 & 4 & 5 \\ 4 & -2 & 1 \end{vmatrix} \qquad 2C_1 + C_3 \to C_3{}^*$$

Next, to obtain a 0 in the a_{11} position, we replace the first column with the sum of it and 3 times the second column:

$$\begin{vmatrix} 3 & -1 & 0 \\ -2 & 4 & 5 \\ 4 & -2 & 1 \end{vmatrix} = \begin{vmatrix} 0 & -1 & 0 \\ 10 & 4 & 5 \\ -2 & -2 & 1 \end{vmatrix} \qquad 3C_2 + C_1 \to C_1$$

Now it is an easy matter to expand this last determinant by the first row to obtain

*C_1, C_2, and C_3 represent columns 1, 2, and 3, respectively.

$$0 + (-1)\left((-1)^{1+2}\begin{vmatrix} 10 & 5 \\ -2 & 1 \end{vmatrix}\right) + 0 = 20$$

Matched Problem 2 Evaluate the following determinant by first using Theorem 5 to obtain 0's in the a_{11} and a_{31} positions, and then expand by the first column.

$$\begin{vmatrix} 3 & 10 & -5 \\ 1 & 6 & -3 \\ 2 & 3 & 4 \end{vmatrix}$$

• Summary of Determinant Properties

We now summarize the five determinant properties discussed above in Table 1 for convenient reference. Even though these properties hold for determinants of any order, for simplicity, we illustrate each property in terms of second-order determinants.

TABLE 1 Summary of Determinant Properties

Property	Examples
1. If each element of any row (or column) of a determinant is multiplied by a constant k, the new determinant is k times the original.	$\begin{vmatrix} 2a & 2b \\ c & d \end{vmatrix} = 2\begin{vmatrix} a & b \\ c & d \end{vmatrix}$ $3\begin{vmatrix} a & b \\ c & d \end{vmatrix} = \begin{vmatrix} 3a & b \\ 3c & d \end{vmatrix}$
2. If every element in a row (or column) is 0, the value of the determinant is 0.	$\begin{vmatrix} a & b \\ 0 & 0 \end{vmatrix} = 0$ $\begin{vmatrix} 0 & b \\ 0 & d \end{vmatrix} = 0$
3. If two rows (or two columns) of a determinant are interchanged, the new determinant is the negative of the original.	$\begin{vmatrix} a & b \\ c & d \end{vmatrix} = -\begin{vmatrix} c & d \\ a & b \end{vmatrix}$ $\begin{vmatrix} a & b \\ c & d \end{vmatrix} = -\begin{vmatrix} b & a \\ d & c \end{vmatrix}$
4. If the corresponding elements are equal in two rows (or columns), the value of the determinant is 0.	$\begin{vmatrix} a & b \\ a & b \end{vmatrix} = 0$ $\begin{vmatrix} a & a \\ c & c \end{vmatrix} = 0$
5. If a multiple of any row (or column) of a determinant is added to any other row (or column), the value of the determinant is not changed.	$\begin{vmatrix} a & b \\ c & d \end{vmatrix} = \begin{vmatrix} a & b \\ c + ka & d + kb \end{vmatrix}$ $\begin{vmatrix} a & b \\ c & d \end{vmatrix} = \begin{vmatrix} a + kb & b \\ c + kd & d \end{vmatrix}$

Answers to Matched Problems

1. $3\begin{vmatrix} 3 & 2 & 1 \\ 2 & 1 & -3 \\ 1 & 0 & -5 \end{vmatrix}$ **2.** 44

EXERCISE 9-5

A

For each statement in Problems 1–10, identify the theorem from this section that justifies it. Do not evaluate.

1. $\begin{vmatrix} 16 & 8 \\ 0 & -1 \end{vmatrix} = 8\begin{vmatrix} 2 & 1 \\ 0 & -1 \end{vmatrix}$

2. $\begin{vmatrix} 1 & -9 \\ 0 & -6 \end{vmatrix} = -3\begin{vmatrix} 1 & 3 \\ 0 & 2 \end{vmatrix}$

3. $-2\begin{vmatrix} 2 & 1 \\ -3 & 4 \end{vmatrix} = \begin{vmatrix} -4 & 1 \\ 6 & 4 \end{vmatrix}$

4. $4\begin{vmatrix} -1 & 3 \\ 2 & 1 \end{vmatrix} = \begin{vmatrix} -4 & 12 \\ 2 & 1 \end{vmatrix}$

5. $\begin{vmatrix} 3 & 0 \\ -2 & 0 \end{vmatrix} = 0$

6. $\begin{vmatrix} 5 & -7 \\ 0 & 0 \end{vmatrix} = 0$

7. $\begin{vmatrix} 5 & -1 \\ 8 & 0 \end{vmatrix} = -\begin{vmatrix} -1 & 5 \\ 0 & 8 \end{vmatrix}$

8. $\begin{vmatrix} 6 & 9 \\ 0 & 1 \end{vmatrix} = -\begin{vmatrix} 0 & 1 \\ 6 & 9 \end{vmatrix}$

9. $\begin{vmatrix} 4 & 3 \\ 1 & 2 \end{vmatrix} = \begin{vmatrix} 4-4 & 3-8 \\ 1 & 2 \end{vmatrix}$

10. $\begin{vmatrix} 3 & 2 \\ 5 & 1 \end{vmatrix} = \begin{vmatrix} 3+4 & 2 \\ 5+2 & 1 \end{vmatrix}$

In Problems 11–14, Theorem 5 was used to transform the determinant on the left to that on the right. Replace each letter x with an appropriate numeral to complete the transformation.

11. $\begin{vmatrix} -1 & 3 \\ 2 & -4 \end{vmatrix} = \begin{vmatrix} -1 & x \\ 2 & 2 \end{vmatrix}$

12. $\begin{vmatrix} -1 & 3 \\ 5 & -2 \end{vmatrix} = \begin{vmatrix} -1 & 3 \\ x & 13 \end{vmatrix}$

13. $\begin{vmatrix} -1 & 2 & 3 \\ 2 & 1 & 4 \\ 1 & 3 & 2 \end{vmatrix} = \begin{vmatrix} -1 & 2 & 0 \\ 2 & 1 & 10 \\ 1 & 3 & x \end{vmatrix}$

14. $\begin{vmatrix} -1 & 2 & 3 \\ 2 & 1 & 4 \\ 1 & 3 & 2 \end{vmatrix} = \begin{vmatrix} -1 & 0 & 3 \\ 2 & x & 4 \\ 1 & 5 & 2 \end{vmatrix}$

Given that

$$\begin{vmatrix} a & b \\ c & d \end{vmatrix} = 10$$

use the properties of determinants discussed in this section to evaluate each determinant in Problems 15–20.

15. $\begin{vmatrix} c & d \\ a & b \end{vmatrix}$

16. $\begin{vmatrix} 2a & 2b \\ c & d \end{vmatrix}$

17. $\begin{vmatrix} a+c & b+d \\ c & d \end{vmatrix}$

18. $\begin{vmatrix} a+b & b \\ c+d & d \end{vmatrix}$

19. $\begin{vmatrix} a & a-b \\ c & c-d \end{vmatrix}$

20. $\begin{vmatrix} a+c & b+d \\ -a & -b \end{vmatrix}$

In Problems 21–24, transform each determinant into one that contains a row (or column) with all elements 0 but one, if possible. Then expand the transformed determinant by this row (or column).

21. $\begin{vmatrix} -1 & 0 & 3 \\ 2 & 5 & 4 \\ 1 & 5 & 2 \end{vmatrix}$

22. $\begin{vmatrix} -1 & 2 & 0 \\ 2 & 1 & 10 \\ 1 & 3 & 5 \end{vmatrix}$

23. $\begin{vmatrix} 3 & 5 & 0 \\ 1 & 1 & -2 \\ 2 & 1 & -1 \end{vmatrix}$

24. $\begin{vmatrix} 2 & 0 & 1 \\ -1 & -3 & 4 \\ 1 & 2 & 3 \end{vmatrix}$

B

For each statement in Problems 25–30, identify the theorem from this section that justifies it.

25. $-2\begin{vmatrix} 1 & 0 & 2 \\ 3 & -2 & 4 \\ 0 & 1 & 1 \end{vmatrix} = \begin{vmatrix} 1 & 0 & 2 \\ -6 & 4 & -8 \\ 0 & 1 & 1 \end{vmatrix}$

26. $\begin{vmatrix} 8 & 0 & 1 \\ 12 & -1 & 0 \\ 4 & 3 & 2 \end{vmatrix} = 4\begin{vmatrix} 2 & 0 & 1 \\ 3 & -1 & 0 \\ 1 & 3 & 2 \end{vmatrix}$

27. $\begin{vmatrix} 1 & 2 & 0 \\ -1 & 3 & 0 \\ 0 & 1 & 0 \end{vmatrix} = 0$

28. $\begin{vmatrix} -2 & 5 & 13 \\ 1 & 7 & 12 \\ 0 & 8 & 15 \end{vmatrix} = -\begin{vmatrix} 5 & -2 & 13 \\ 7 & 1 & 12 \\ 8 & 0 & 15 \end{vmatrix}$

29. $\begin{vmatrix} 4 & 2 & -1 \\ 2 & 0 & 2 \\ -3 & 5 & -2 \end{vmatrix} = \begin{vmatrix} 4-4 & 2 & -1 \\ 2+8 & 0 & 2 \\ -3-8 & 5 & -2 \end{vmatrix}$

30. $\begin{vmatrix} 7 & 7 & 1 \\ -3 & -3 & 11 \\ 2 & 2 & 0 \end{vmatrix} = 0$

In Problems 31–34, Theorem 5 was used to transform the determinant on the left to that on the right. Replace each letter x and y with an appropriate numeral to complete the transformation.

31. $\begin{vmatrix} 2 & 1 & -1 \\ 3 & 4 & 1 \\ 1 & 2 & -2 \end{vmatrix} = \begin{vmatrix} 0 & 0 & -1 \\ x & 5 & 1 \\ -3 & y & -2 \end{vmatrix}$

32. $\begin{vmatrix} 3 & -1 & 1 \\ -2 & 4 & 3 \\ 1 & 5 & 2 \end{vmatrix} = \begin{vmatrix} 0 & -1 & 0 \\ 10 & 4 & 7 \\ x & 5 & y \end{vmatrix}$

33. $\begin{vmatrix} 7 & 9 & 4 \\ 2 & 3 & 1 \\ 3 & 4 & -2 \end{vmatrix} = \begin{vmatrix} -1 & x & 0 \\ 2 & 3 & 1 \\ 7 & y & 0 \end{vmatrix}$

34. $\begin{vmatrix} 5 & 2 & 3 \\ 3 & 1 & 2 \\ -4 & -3 & 5 \end{vmatrix} = \begin{vmatrix} x & 0 & -1 \\ 3 & 1 & 2 \\ 5 & 0 & y \end{vmatrix}$

In Problems 35–42, transform each determinant into one that contains a row (or column) with all elements 0 but one, if possible. Then expand the transformed determinant by this row (or column).

35. $\begin{vmatrix} 1 & 5 & 3 \\ 4 & 2 & 1 \\ 3 & 1 & 2 \end{vmatrix}$

36. $\begin{vmatrix} -1 & 5 & 1 \\ 2 & 3 & 1 \\ 3 & 2 & 1 \end{vmatrix}$

37. $\begin{vmatrix} 5 & 2 & -3 \\ -2 & 4 & 4 \\ 1 & -1 & 3 \end{vmatrix}$

38. $\begin{vmatrix} 5 & 3 & -6 \\ -1 & 1 & 4 \\ 4 & 3 & -6 \end{vmatrix}$

39. $\begin{vmatrix} 3 & -4 & 1 \\ 6 & -1 & 2 \\ 9 & 2 & 3 \end{vmatrix}$

40. $\begin{vmatrix} 2 & 3 & -1 \\ 5 & 4 & 7 \\ -4 & -6 & 2 \end{vmatrix}$

41. $\begin{vmatrix} 0 & 1 & 0 & 1 \\ 1 & -2 & 4 & 3 \\ 2 & 1 & 5 & 4 \\ 1 & 2 & 1 & 2 \end{vmatrix}$

42. $\begin{vmatrix} 2 & 3 & 1 & -1 \\ 3 & 1 & 2 & 1 \\ 0 & 5 & 4 & 0 \\ -1 & 2 & 3 & 0 \end{vmatrix}$

C

Transform each determinant in Problems 43 and 44 into one that contains a row (or column) with all elements 0 but one, if possible. Then expand the transformed determinant by this row (or column).

43. $\begin{vmatrix} 3 & 2 & 3 & 1 \\ 3 & -2 & 8 & 5 \\ 2 & 1 & 3 & 1 \\ 4 & 5 & 4 & -3 \end{vmatrix}$

44. $\begin{vmatrix} -1 & 4 & 2 & 1 \\ 5 & -1 & -3 & -1 \\ 2 & -1 & -2 & 3 \\ -3 & 3 & 3 & 3 \end{vmatrix}$

Problems 45–48 are representative cases of theorems discussed in this section. Use cofactor expansions to verify each statement directly, without reference to the theorem it represents.

45. $\begin{vmatrix} a & b & a \\ d & e & d \\ g & h & g \end{vmatrix} = 0$

46. $\begin{vmatrix} a & b & c \\ kd & ke & kf \\ g & h & i \end{vmatrix} = k\begin{vmatrix} a & b & c \\ d & e & f \\ g & h & i \end{vmatrix}$

47. $\begin{vmatrix} a_1 & b_1 & c_1 \\ a_2 & b_2 & c_2 \\ a_3 & b_3 & c_3 \end{vmatrix} = -\begin{vmatrix} b_1 & a_1 & c_1 \\ b_2 & a_2 & c_2 \\ b_3 & a_3 & c_3 \end{vmatrix}$

48. $\begin{vmatrix} a_1 & b_1 & c_1 \\ a_2 & b_2 & c_2 \\ a_3 & b_3 & c_3 \end{vmatrix} = \begin{vmatrix} a_1 + kc_1 & b_1 & c_1 \\ a_2 + kc_2 & b_2 & c_2 \\ a_3 + kc_3 & b_3 & c_3 \end{vmatrix}$

49. Without expanding, explain why (2, 5) and (−3, 4) satisfy the equation

$$\begin{vmatrix} x & y & 1 \\ 2 & 5 & 1 \\ -3 & 4 & 1 \end{vmatrix} = 0$$

50. Show that

$$\begin{vmatrix} x & y & 1 \\ 2 & 3 & 1 \\ -1 & 2 & 1 \end{vmatrix} = 0$$

is the equation of a line that passes through (2, 3) and (−1, 2).

51. Show that

$$\begin{vmatrix} x & y & 1 \\ x_1 & y_1 & 1 \\ x_2 & y_2 & 1 \end{vmatrix} = 0$$

is the equation of a line that passes through (x_1, y_1) and (x_2, y_2).

52. In analytic geometry it is shown that the area of a triangle with vertices (x_1, y_1), (x_2, y_2), and (x_3, y_3) is the absolute value of

$$\frac{1}{2}\begin{vmatrix} x_1 & y_1 & 1 \\ x_2 & y_2 & 1 \\ x_3 & y_3 & 1 \end{vmatrix}$$

Use this result to find the area of a triangle with vertices (−1, 4), (4, 8), and (1, 1).

53. What can we say about the three points (x_1, y_1), (x_2, y_2), and (x_3, y_3) if the following equation is true?

$$\begin{vmatrix} x_1 & y_1 & 1 \\ x_2 & y_2 & 1 \\ x_3 & y_3 & 1 \end{vmatrix} = 0$$

[*Hint:* See Problem 52.]

54. If the three points (x_1, y_1), (x_2, y_2), and (x_3, y_3) are all on the same line, what can we say about the value of the determinant below?

$$\begin{vmatrix} x_1 & y_1 & 1 \\ x_2 & y_2 & 1 \\ x_3 & y_3 & 1 \end{vmatrix}$$

SECTION 9-6 Cramer's Rule

- Two Equations–Two Variables
- Three Equations–Three Variables

Now let's see how determinants arise rather naturally in the process of solving systems of linear equations. We start by investigating two equations and two variables, and then extend our results to three equations and three variables.

• Two Equations–Two Variables

Instead of thinking of each system of linear equations in two variables as a different problem, let's see what happens when we attempt to solve the general system

$$a_{11}x + a_{12}y = k_1 \tag{1A}$$

$$a_{21}x + a_{22}y = k_2 \tag{1B}$$

once and for all, in terms of the unspecified real constants a_{11}, a_{12}, a_{21}, a_{22}, k_1, and k_2.

We proceed by multiplying equations (1A) and (1B) by suitable constants so that when the resulting equations are added, left side to left side and right side to right side, one of the variables drops out. Suppose we choose to eliminate y. What constant should we use to make the coefficients of y the same except for the signs? Multiply equation (1A) by a_{22} and (1B) by $-a_{12}$; then add:

$$\begin{array}{rr} a_{22}(1A): & a_{11}a_{22}x + a_{12}a_{22}y = k_1a_{22} \\ -a_{12}(1B): & -a_{21}a_{12}x - a_{12}a_{22}y = -k_2a_{12} \\ \hline & a_{11}a_{22}x - a_{21}a_{12}x + 0y = k_1a_{22} - k_2a_{12} \\ & (a_{11}a_{22} - a_{21}a_{12})x = k_1a_{22} - k_2a_{12} \end{array}$$

$$x = \frac{k_1a_{22} - k_2a_{12}}{a_{11}a_{22} - a_{21}a_{12}} \qquad a_{11}a_{22} - a_{21}a_{12} \neq 0$$

What do the numerator and denominator remind you of? From your experience with determinants in the last two sections, you should recognize these expressions as

$$x = \frac{\begin{vmatrix} k_1 & a_{12} \\ k_2 & a_{22} \end{vmatrix}}{\begin{vmatrix} a_{11} & a_{12} \\ a_{21} & a_{22} \end{vmatrix}}$$

Similarly, starting with system (1A) and (1B) and eliminating x (this is left as an exercise), we obtain

$$y = \frac{\begin{vmatrix} a_{11} & k_1 \\ a_{21} & k_2 \end{vmatrix}}{\begin{vmatrix} a_{11} & a_{12} \\ a_{21} & a_{22} \end{vmatrix}}$$

These results are summarized in Theorem 1, **Cramer's rule,** which is named after the Swiss mathematician G. Cramer (1704–1752).

Theorem 1 **Cramer's Rule for Two Equations and Two Variables**

Given the system

$$\begin{aligned} a_{11}x + a_{12}y &= k_1 \\ a_{21}x + a_{22}y &= k_2 \end{aligned} \qquad \text{with} \qquad D = \begin{vmatrix} a_{11} & a_{12} \\ a_{21} & a_{22} \end{vmatrix} \neq 0$$

then

$$x = \frac{\begin{vmatrix} k_1 & a_{12} \\ k_2 & a_{22} \end{vmatrix}}{D} \qquad \text{and} \qquad y = \frac{\begin{vmatrix} a_{11} & k_1 \\ a_{21} & k_2 \end{vmatrix}}{D}$$

The determinant D is called the **coefficient determinant.** If $D \neq 0$, then the system has exactly one solution, which is given by Cramer's rule. If, on the other hand, $D = 0$, then it can be shown that the system is either inconsistent and has no solutions or dependent and has an infinite number of solutions. We must use other methods, such as those discussed in Chapter 8, to determine the exact nature of the solutions when $D = 0$.

EXAMPLE 1 Solving a System with Cramer's Rule

Solve using Cramer's rule:
$$\begin{aligned} 3x - 5y &= 2 \\ -4x + 3y &= -1 \end{aligned}$$

Solution

$$D = \begin{vmatrix} 3 & -5 \\ -4 & 3 \end{vmatrix} = -11$$

$$x = \frac{\begin{vmatrix} 2 & -5 \\ -1 & 3 \end{vmatrix}}{-11} = -\frac{1}{11} \qquad y = \frac{\begin{vmatrix} 3 & 2 \\ -4 & -1 \end{vmatrix}}{-11} = -\frac{5}{11}$$

Matched Problem 1 Solve using Cramer's rule:
$$\begin{aligned} 3x + 2y &= -4 \\ -4x + 3y &= -10 \end{aligned}$$

EXPLORE-DISCUSS 1 Recall that a system of linear equations must have zero, one, or an infinite number of solutions. Discuss the number of solutions for the system

$$ax + 3y = b$$
$$4x + 2y = 8$$

where a and b are real numbers. Use Cramer's rule where appropriate and Gauss–Jordan elimination otherwise.

• Three Equations–Three Variables

Cramer's rule can be generalized completely for any size linear system that has the *same number of variables as equations.* However, it cannot be used to solve systems where the number of variables is not equal to the number of equations. In Theorem 2 we state without proof Cramer's rule for three equations and three variables.

Theorem 2 **Cramer's Rule for Three Equations and Three Variables**

Given the system

$$\begin{aligned} a_{11}x + a_{12}y + a_{13}z &= k_1 \\ a_{21}x + a_{22}y + a_{23}z &= k_2 \\ a_{31}x + a_{32}y + a_{33}z &= k_3 \end{aligned} \qquad \text{with} \qquad D = \begin{vmatrix} a_{11} & a_{12} & a_{13} \\ a_{21} & a_{22} & a_{23} \\ a_{31} & a_{32} & a_{33} \end{vmatrix} \neq 0$$

then

$$x = \frac{\begin{vmatrix} k_1 & a_{12} & a_{13} \\ k_2 & a_{22} & a_{23} \\ k_3 & a_{32} & a_{33} \end{vmatrix}}{D} \qquad y = \frac{\begin{vmatrix} a_{11} & k_1 & a_{13} \\ a_{21} & k_2 & a_{23} \\ a_{31} & k_3 & a_{33} \end{vmatrix}}{D} \qquad z = \frac{\begin{vmatrix} a_{11} & a_{12} & k_1 \\ a_{21} & a_{22} & k_2 \\ a_{31} & a_{32} & k_3 \end{vmatrix}}{D}$$

You can easily remember these determinant formulas for x, y, and z if you observe the following:

1. Determinant D is formed from the coefficients of x, y, and z, keeping the same relative position in the determinant as found in the system of equations.
2. Determinant D appears in the denominators for x, y, and z.
3. The numerator for x can be obtained from D by replacing the coefficients of x (a_{11}, a_{21}, a_{31}) with the constants k_1, k_2, and k_3, respectively. Similar statements can be made for the numerators for y and z.

EXAMPLE 2 Solving a System with Cramer's Rule

Solve using Cramer's rule:

$$\begin{aligned} x + y \phantom{{}- z} &= 2 \\ 3y - z &= -4 \\ x \phantom{{}+3y} + z &= 3 \end{aligned}$$

Solution

$$D = \begin{vmatrix} 1 & 1 & 0 \\ 0 & 3 & -1 \\ 1 & 0 & 1 \end{vmatrix} = 2$$

$$x = \frac{\begin{vmatrix} 2 & 1 & 0 \\ -4 & 3 & -1 \\ 3 & 0 & 1 \end{vmatrix}}{2} = \frac{7}{2} \qquad y = \frac{\begin{vmatrix} 1 & 2 & 0 \\ 0 & -4 & -1 \\ 1 & 3 & 1 \end{vmatrix}}{2} = -\frac{3}{2}$$

$$z = \frac{\begin{vmatrix} 1 & 1 & 2 \\ 0 & 3 & -4 \\ 1 & 0 & 3 \end{vmatrix}}{2} = -\frac{1}{2}$$

Matched Problem 2 Solve using Cramer's rule:

$$\begin{aligned} 3x \phantom{{}-y} - z &= 5 \\ x - y + z &= 0 \\ x + y \phantom{{}+z} &= 1 \end{aligned}$$

In practice, Cramer's rule is rarely used to solve systems of order higher than 2 or 3 by hand, since more efficient methods are available utilizing computer methods. However, Cramer's rule is a valuable tool in more advanced theoretical and applied mathematics.

Answers to Matched Problems

1. $x = \frac{8}{17}, y = -\frac{46}{17}$ **2.** $x = \frac{6}{5}, y = -\frac{1}{5}, z = -\frac{7}{5}$

EXERCISE 9-6

A

Solve Problems 1–8 using Cramer's rule.

1. $x + 2y = 1$; $x + 3y = -1$

2. $x + 2y = 3$; $x + 3y = 5$

3. $2x + y = 1$; $5x + 3y = 2$

4. $x + 3y = 1$; $2x + 8y = 0$

5. $2x - y = -3$; $-x + 3y = 3$

6. $-3x + 2y = 1$; $2x - 3y = -3$

7. $4x - 3y = 4$; $3x + 2y = -2$

8. $5x + 2y = -1$; $2x - 3y = 2$

B

Solve Problems 9–12 to 2 significant digits using Cramer's rule.

9. $0.9925x - 0.9659y = 0$; $0.1219x + 0.2588y = 2{,}500$

10. $0.9877x - 0.9744y = 0$; $0.1564x + 0.2250y = 1{,}900$

11. $0.9954x - 0.9942y = 0$; $0.0958x + 0.1080y = 155$

12. $0.9973x - 0.9957y = 0$; $0.0732x + 0.0924y = 112$

Solve Problems 13–20 using Cramer's rule:

13. $$\begin{aligned} x + y \quad\;\; &= 0 \\ 2y + z &= -5 \\ -x \quad\;\; + z &= -3 \end{aligned}$$

14. $$\begin{aligned} x + y \quad\;\; &= -4 \\ 2y + z &= 0 \\ -x \quad\;\; + z &= 5 \end{aligned}$$

15. $$\begin{aligned} x + y \quad\;\; &= 1 \\ 2y + z &= 0 \\ -y + z &= 1 \end{aligned}$$

16. $$\begin{aligned} x + 3y \quad\;\; &= -3 \\ 2y + z &= 3 \\ -x \quad\;\; + 3z &= 7 \end{aligned}$$

17. $$\begin{aligned} 3y + z &= -1 \\ x \quad\;\; + 2z &= 3 \\ x - 3y \quad\;\; &= -2 \end{aligned}$$

18. $$\begin{aligned} x \quad\;\; - z &= 3 \\ 2x - y \quad\;\; &= -3 \\ x + y + z &= 1 \end{aligned}$$

19. $$\begin{aligned} 2y - z &= -3 \\ x - y - z &= 2 \\ x - y + 2z &= 4 \end{aligned}$$

20. $$\begin{aligned} 2x + y \quad\;\; &= 2 \\ x - y + z &= -1 \\ x + y + z &= 2 \end{aligned}$$

C

In Problems 21 and 22, use Cramer's rule to solve for x only.

21. $$\begin{aligned} 2x - 3y + z &= -3 \\ -4x + 3y + 2z &= -11 \\ x - y - z &= 3 \end{aligned}$$

22. $$\begin{aligned} x + 4y - 3z &= 25 \\ 3x + y - z &= 2 \\ -4x + y + 2z &= 1 \end{aligned}$$

In Problems 23 and 24, use Cramer's rule to solve for y only.

23. $$\begin{aligned} 12x - 14y + 11z &= 5 \\ 15x + 7y - 9z &= -13 \\ 5x - 3y + 2z &= 0 \end{aligned}$$

24. $$\begin{aligned} 2x - y + 4z &= 15 \\ -x + y + 2z &= 5 \\ 3x + 4y - 2z &= 4 \end{aligned}$$

In Problems 25 and 26, use Cramer's rule to solve for z only.

25. $$\begin{aligned} 3x - 4y + 5z &= 18 \\ -9x + 8y + 7z &= -13 \\ 5x - 7y + 10z &= 33 \end{aligned}$$

26. $$\begin{aligned} 13x + 11y + 10z &= 2 \\ 10x + 8y + 7z &= 1 \\ 8x + 5y + 4z &= 4 \end{aligned}$$

It is clear that $x = 0$, $y = 0$, $z = 0$ is a solution to each of the systems given in Problems 27 and 28. Use Cramer's rule to determine whether this solution is unique. [Hint: If $D \neq 0$, what can you conclude? If $D = 0$, what can you conclude?]

27. $$\begin{aligned} x - 4y + 9z &= 0 \\ 4x - y + 6z &= 0 \\ x - y + 3z &= 0 \end{aligned}$$

28. $$\begin{aligned} 3x - y + 3z &= 0 \\ 5x + 5y - 9z &= 0 \\ -2x + y - 3z &= 0 \end{aligned}$$

29. Prove Theorem 1 for y.

30. (Omit this problem if you have not studied trigonometry.) The angles α, β, and γ and the sides a, b, and c of a triangle (see the figure) satisfy

$$\begin{aligned} c &= b\cos\alpha + a\cos\beta \\ b &= c\cos\alpha \qquad\qquad\quad + a\cos\gamma \\ a &= \qquad\qquad c\cos\beta + b\cos\gamma \end{aligned}$$

Use Cramer's rule to express $\cos\alpha$ in terms of a, b, and c, thereby deriving the familiar law of cosines from trigonometry:

$$\cos\alpha = \frac{b^2 + c^2 - a^2}{2bc}$$

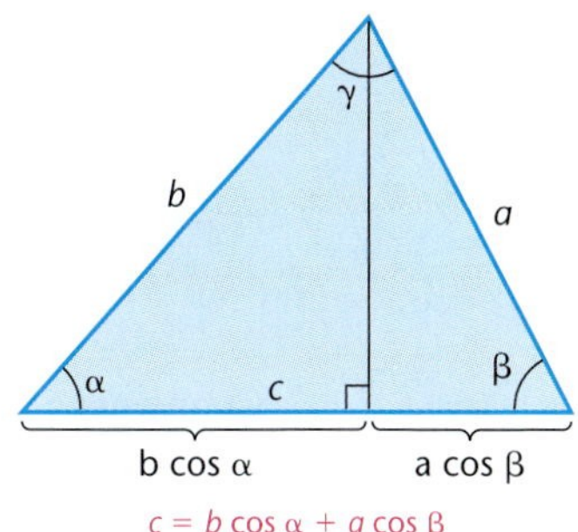

APPLICATIONS 

31. Revenue Analysis. A supermarket sells two brands of coffee: brand A at $\$p$ per pound and brand B at $\$q$ per pound. The daily demand equations for brands A and B are, respectively,

$$\begin{aligned} x &= 200 - 6p + 4q \\ y &= 300 + 2p - 3q \end{aligned} \qquad (1)$$

(both in pounds). The daily revenue R is given by

$$R = xp + yq$$

(A) To analyze the effect of price changes on the daily revenue, an economist wants to express the daily revenue R in terms of p and q only. Use (1) to eliminate x and y in the equation for R, thus expressing the daily revenue in terms of p and q.

(B) To analyze the effect of changes in demand on the daily revenue, the economist now wants to express the daily revenue in terms of x and y only. Use Cramer's rule to solve system (1) for p and q in terms of x and y and then express the daily revenue R in terms of x and y.

32. Revenue Analysis. A company manufactures ten-speed and three-speed bicycles. The weekly demand equations are

$$\begin{aligned} p &= 230 - 10x + 5y \\ q &= 130 + 4x - 4y \end{aligned} \qquad (2)$$

where $\$p$ is the price of a ten-speed bicycle, $\$q$ is the price of a three-speed bicycle, x is the weekly demand for ten-speed bicycles, and y is the weekly demand for three-speed bicycles. The weekly revenue R is given by

$$R = xp + yq$$

(A) Use (2) to express the daily revenue in terms of x and y only.

(B) Use Cramer's rule to solve system (2) for x and y in terms of p and q, and then express the daily revenue R in terms of p and q only.

CHAPTER 9 GROUP ACTIVITY Using Matrices to Find Cost, Revenue, and Profit

A toy distributor purchases model train components from various suppliers and packages these components in three different ready-to-run train sets: the Limited, the Empire, and the Comet. The components used in each set are listed in Table 1. For convenience, the total labor time (in minutes) required to prepare a set for shipping is included as a component.

TABLE 1 Product Components

	Train Sets		
Components	**Limited**	**Empire**	**Comet**
Locomotives	1	1	2
Cars	5	6	8
Track pieces	20	24	32
Track switches	1	2	4
Power pack	1	1	1
Labor (min)	15	18	24

The current costs of the components are given in Table 2, and the distributor's selling prices for the sets are given in Table 3.

TABLE 2 Component Costs

Components	Cost per unit
Locomotive	\$12.52
Car	\$1.43
Track piece	\$0.25
Track switch	\$2.29
Power pack	\$12.54
Labor (per min)	\$0.15

TABLE 3 Selling Prices

Set	Price
Limited	\$54.60
Empire	\$62.28
Comet	\$81.15

The distributor has just received the order shown in Table 4 from a retail toy store.

TABLE 4 Customer Order

Set	Quantity
Limited	48
Empire	24
Comet	12

The distributor wants to store the information in each table in a matrix and use matrix operations to find the following information:

1. The inventory (parts and labor) required to fill the order
2. The cost (parts and labor) of filling the order
3. The revenue (sales) received from the customer
4. The profit realized on the order

(A) Use a single letter to designate the matrix representing each table, and write matrix expressions in terms of these letters that will provide the required information. Discuss the size of the matrix you must use to represent each table so that all the pertinent matrix operations are defined.
(B) Evaluate the matrix expressions in part A.

Shortly after filling the order in Table 4, a supplier informs the distributor that the cars and locomotives used in these train sets are no longer available. The distributor currently has 30 locomotives and 134 cars in stock.

(C) How many train sets of each type can the distributor produce using all the available locomotives and cars? Assume that the distributor has unlimited quantities of the other components used in these sets.
(D) How much profit will the distributor make if all these sets are sold? If there is more than one way to use all the available locomotives and cars, which one will produce the largest profit?

Chapter 9 Review

9-1 MATRICES: BASIC OPERATIONS

Two matrices are **equal** if they are the same size and their corresponding elements are equal. The **sum of two matrices** of the same size is a matrix with elements that are the sums of the corresponding elements of the two given matrices. Matrix addition is **commutative** and **associative.** A matrix with all zero elements is called the **zero matrix.** The **negative of a matrix** M, denoted $-M$, is a matrix with elements that are the negatives of the elements in M. If A and B are matrices of the same size, then we define **subtraction** as follows: $A - B = A + (-B)$. The **product of a number** k **and a matrix** M, denoted by kM, is a matrix formed by multiplying each element of M by k. The **product** of a $1 \times n$ row matrix and an $n \times 1$ column matrix is a 1×1 matrix given by

$$\underset{1 \times n}{\begin{bmatrix} a_1 & a_2 & \cdots & a_n \end{bmatrix}} \underset{n \times 1}{\begin{bmatrix} b_1 \\ b_2 \\ \vdots \\ b_n \end{bmatrix}} = \underset{1 \times 1}{[a_1b_1 + a_2b_2 + \cdots + a_nb_n]}$$

If A is an $m \times p$ matrix and B is a $p \times n$ matrix, then the **matrix**

product of A and B, denoted AB, is an $m \times n$ matrix whose element in the ith row and jth column is the real number obtained from the product of the ith row of A and the jth column of B. If the number of columns in A does not equal the number of rows in B, then the matrix product AB is **not defined. Matrix multiplication is not commutative,** and the **zero property does not hold for matrix multiplication.** That is, for matrices A and B, the matrix product AB can be zero without either A or B being the zero matrix.

9-2 INVERSE OF A SQUARE MATRIX

The **identity matrix for multiplication** for the set of all square matrices of order n is the square matrix of order n, denoted by I, with 1's along the **principal diagonal** (from upper left corner to lower right corner) and 0's elsewhere. If M is a square matrix of order n and I is the identity matrix of order n, then

$$IM = MI = M$$

If M is a square matrix of order n and if there exists a matrix M^{-1} (read "M inverse") such that

$$M^{-1}M = MM^{-1} = I$$

then M^{-1} is called the **multiplicative inverse of M** or, more simply, the **inverse of M.** If the augmented matrix $[M \mid I]$ is transformed by row operations into $[I \mid B]$, then the resulting matrix B is M^{-1}. If, however, we obtain all 0's in one or more rows to the left of the vertical line, then M^{-1} does not exist and M is called a **singular matrix.**

9-3 MATRIX EQUATIONS AND SYSTEMS OF LINEAR EQUATIONS

The following properties of matrices are fundamental to the process of solving matrix equations. Assuming all products and sums are defined for the indicated matrices A, B, C, I, and 0, then:

Addition Properties

Associative:	$(A + B) + C = A + (B + C)$
Commutative:	$A + B = B + A$
Additive Identity:	$A + 0 = 0 + A = A$
Additive Inverse:	$A + (-A) = (-A) + A = 0$

Multiplication Properties

Associative Property:	$A(BC) = (AB)C$
Multiplicative Identity:	$AI = IA = A$
Multiplicative Inverse:	If A is a square matrix and A^{-1} exists, then $AA^{-1} = A^{-1}A = I$.

Combined Properties

Left Distributive:	$A(B + C) = AB + AC$
Right Distributive:	$(B + C)A = BA + CA$

Equality

Addition:	If $A = B$, then $A + C = B + C$.
Left Multiplication:	If $A = B$, then $CA = CB$.
Right Multiplication:	If $A = B$, then $AC = BC$.

A system of linear equations with the same number of variables as equations such as

$$\begin{aligned} a_{11}x_1 + a_{12}x_2 + a_{13}x_3 &= k_1 \\ a_{21}x_1 + a_{22}x_2 + a_{23}x_3 &= k_2 \\ a_{31}x_1 + a_{32}x_2 + a_{33}x_3 &= k_3 \end{aligned}$$

can be written as the matrix equation

$$\overset{A}{\begin{bmatrix} a_{11} & a_{12} & a_{13} \\ a_{21} & a_{22} & a_{23} \\ a_{31} & a_{32} & a_{33} \end{bmatrix}} \overset{X}{\begin{bmatrix} x_1 \\ x_2 \\ x_3 \end{bmatrix}} = \overset{B}{\begin{bmatrix} k_1 \\ k_2 \\ k_3 \end{bmatrix}}$$

If the inverse of A exists, then the matrix equation has a unique solution given by

$$X = A^{-1}B$$

After multiplying B by A^{-1} from the left, it is easy to read the solution to the original system of equations.

9-4 DETERMINANTS

Associated with each square matrix A is a real number called the **determinant** of the matrix. The determinant of A is denoted by **det A,** or simply by writing the array of elements in A using vertical lines in place of square brackets. For example,

$$\det\begin{bmatrix} a_{11} & a_{12} \\ a_{21} & a_{22} \end{bmatrix} = \begin{vmatrix} a_{11} & a_{12} \\ a_{21} & a_{22} \end{vmatrix}$$

A determinant of **order n** is a determinant with n rows and n columns.

The **value of a second-order determinant** is the real number given by

$$\begin{vmatrix} a_{11} & a_{12} \\ a_{21} & a_{22} \end{vmatrix} = a_{11}a_{22} - a_{21}a_{12}$$

The **value of a third-order determinant** is the sum of three products obtained by multiplying each element of any one row (or each element of any one column) by its cofactor. The **cofactor of an element** a_{ij} (from the ith row and jth column) is the product of the minor of a_{ij} and $(-1)^{i+j}$. The **minor of an element** a_{ij} is the determinant remaining after deleting the ith row and jth column. A similar process can be used to evaluate determinants of order higher than 3.

9-5 PROPERTIES OF DETERMINANTS

The use of the following five determinant properties can greatly reduce the effort in evaluating determinants of order 3 or greater:

1. If each element of any row (or column) of a determinant is multiplied by a constant k, the new determinant is k times the original.

$$\begin{vmatrix} 2a & 2b \\ c & d \end{vmatrix} = 2\begin{vmatrix} a & b \\ c & d \end{vmatrix}$$

$$3\begin{vmatrix} a & b \\ c & d \end{vmatrix} = \begin{vmatrix} 3a & b \\ 3c & d \end{vmatrix}$$

2. If every element in a row (or column) is 0, the value of the determinant is 0.

$$\begin{vmatrix} a & b \\ 0 & 0 \end{vmatrix} = 0 \qquad \begin{vmatrix} 0 & b \\ 0 & d \end{vmatrix} = 0$$

3. If two rows (or two columns) of a determinant are interchanged, the new determinant is the negative of the original.

$$\begin{vmatrix} a & b \\ c & d \end{vmatrix} = -\begin{vmatrix} c & d \\ a & b \end{vmatrix} \qquad \begin{vmatrix} a & b \\ c & d \end{vmatrix} = -\begin{vmatrix} b & a \\ d & c \end{vmatrix}$$

4. If the corresponding elements are equal in two rows (or columns), the value of the determinant is 0.

$$\begin{vmatrix} a & b \\ a & b \end{vmatrix} = 0 \qquad \begin{vmatrix} a & a \\ c & c \end{vmatrix} = 0$$

5. If a multiple of any row (or column) of a determinant is added to any other row (or column), the value of the determinant is not changed.

$$\begin{vmatrix} a & b \\ c & d \end{vmatrix} = \begin{vmatrix} a & b \\ c+ka & d+kb \end{vmatrix} \qquad \begin{vmatrix} a & b \\ c & d \end{vmatrix} = \begin{vmatrix} a+kb & b \\ c+kd & d \end{vmatrix}$$

9-6 CRAMER'S RULE

Systems of equations having the same number of variables as equations can also be solved using determinants and Cramer's rule. **Cramer's rule for three equations and three variables** is as follows: Given the system

$$\begin{aligned} a_{11}x + a_{12}y + a_{13}z &= k_1 \\ a_{21}x + a_{22}y + a_{23}z &= k_2 \\ a_{31}x + a_{32}y + a_{33}z &= k_3 \end{aligned} \quad \text{with} \quad D = \begin{vmatrix} a_{11} & a_{12} & a_{13} \\ a_{21} & a_{22} & a_{23} \\ a_{31} & a_{32} & a_{33} \end{vmatrix} \neq 0$$

then

$$x = \frac{\begin{vmatrix} k_1 & a_{12} & a_{13} \\ k_2 & a_{22} & a_{23} \\ k_3 & a_{32} & a_{33} \end{vmatrix}}{D} \qquad y = \frac{\begin{vmatrix} a_{11} & k_1 & a_{13} \\ a_{21} & k_2 & a_{23} \\ a_{31} & k_3 & a_{33} \end{vmatrix}}{D} \qquad z = \frac{\begin{vmatrix} a_{11} & a_{12} & k_1 \\ a_{21} & a_{22} & k_2 \\ a_{31} & a_{32} & k_3 \end{vmatrix}}{D}$$

Cramer's rule can be generalized completely for any size linear system that has the same number of variables as equations. The formulas are easily remembered if you observe the following:

1. Determinant D is formed from the coefficients of x, y, and z, keeping the same relative position in the determinant as found in the system of equations.
2. Determinant D appears in the denominators for x, y, and z.
3. The numerator for x can be obtained from D by replacing the coefficients of x (a_{11}, a_{21}, a_{31}) with the constants k_1, k_2, and k_3, respectively. Similar statements can be made for the numerators for y and z.

Cramer's rule is rarely used to solve systems of order higher than 3 by hand, since more efficient methods are available. Cramer's rule, however, is a valuable tool in more advanced theoretical and applied mathematics.

Chapter 9 Review Exercise

Work through all the problems in this chapter review and check answers in the back of the book. Answers to all review problems are there, and following each answer is a number in italics indicating the section in which that type of problem is discussed. Where weaknesses show up, review appropriate sections in the text.

A

In Problems 1–9, perform the operations that are defined, given the following matrices:

$$A = \begin{bmatrix} 4 & -2 \\ 0 & 3 \end{bmatrix} \quad B = \begin{bmatrix} -1 & 5 \\ -4 & 6 \end{bmatrix} \quad C = \begin{bmatrix} -1 & 4 \end{bmatrix} \quad D = \begin{bmatrix} 3 \\ -2 \end{bmatrix}$$

1. AB
2. CD
3. CB
4. AD
5. $A + B$
6. $C + D$
7. $A + C$
8. $2A - 5B$
9. $CA + C$

10. Find the inverse of

$$A = \begin{bmatrix} 4 & 7 \\ -1 & -2 \end{bmatrix}$$

Show that $A^{-1}A = I$.

11. Write the system

$$\begin{aligned} 3x_1 + 2x_2 &= k_1 \\ 4x_1 + 3x_2 &= k_2 \end{aligned}$$

as a matrix equation, and solve using matrix inverse methods for:

(A) $k_1 = 3, k_2 = 5$ (B) $k_1 = 7, k_2 = 10$
(C) $k_1 = 4, k_2 = 2$

Evaluate the determinants in Problems 12 and 13.

12. $\begin{vmatrix} 2 & -3 \\ -5 & -1 \end{vmatrix}$

13. $\begin{vmatrix} 2 & 3 & -4 \\ 0 & 5 & 0 \\ 1 & -4 & -2 \end{vmatrix}$

14. Solve the system using Cramer's rule:

$$\begin{aligned} 3x - 2y &= 8 \\ x + 3y &= -1 \end{aligned}$$

15. Use properties of determinants to find each of the following, given that

$$\begin{vmatrix} a & b & c \\ d & e & f \\ g & h & i \end{vmatrix} = 2$$

(A) $\begin{vmatrix} g & h & i \\ d & e & f \\ a & b & c \end{vmatrix}$ (B) $\begin{vmatrix} a & 3b & c \\ d & 3e & f \\ g & 3h & i \end{vmatrix}$

(C) $\begin{vmatrix} a & b & a+b+c \\ d & e & d+e+f \\ g & h & g+h+i \end{vmatrix}$

B

In Problems 16–21, perform the operations that are defined, given the following matrices:

$$A = \begin{bmatrix} 1 & 2 \\ 4 & 5 \\ -3 & -1 \end{bmatrix} \quad B = \begin{bmatrix} 6 \\ 0 \\ -4 \end{bmatrix} \quad C = \begin{bmatrix} 2 & 4 & -1 \end{bmatrix}$$

$$D = \begin{bmatrix} 7 & 0 & -5 \\ 0 & 8 & -2 \end{bmatrix} \quad E = \begin{bmatrix} 9 & -3 \\ -6 & 2 \end{bmatrix}$$

16. AD **17.** DA **18.** BC

19. CB **20.** DE **21.** ED

22. Find the inverse of

$$A = \begin{bmatrix} 1 & 0 & 4 \\ -2 & 1 & 0 \\ 4 & -1 & 4 \end{bmatrix}$$

Show that $AA^{-1} = I$.

23. Write the system

$$\begin{aligned} x_1 + 2x_2 + 3x_3 &= k_1 \\ 2x_1 + 3x_2 + 4x_3 &= k_2 \\ x_1 + 2x_2 + x_3 &= k_3 \end{aligned}$$

as a matrix equation, and solve using matrix inverse methods for:
(A) $k_1 = 1, k_2 = 3, k_3 = 3$
(B) $k_1 = 0, k_2 = 0, k_3 = -2$
(C) $k_1 = -3, k_2 = -4, k_3 = 1$

Evaluate the determinants in Problems 24 and 25.

24. $\begin{vmatrix} -\frac{1}{4} & \frac{3}{2} \\ \frac{1}{2} & \frac{2}{3} \end{vmatrix}$

25. $\begin{vmatrix} 2 & -1 & 1 \\ -3 & 5 & 2 \\ 1 & -2 & 4 \end{vmatrix}$

26. Solve for y only using Cramer's rule:

$$\begin{aligned} x - 2y + z &= -6 \\ y - z &= 4 \\ 2x + 2y + z &= 2 \end{aligned}$$

(Find the numerator and denominator first; then reduce.)

27. Discuss the number of solutions for a system of n equations in n variables if the coefficient matrix:
(A) Has an inverse. (B) Does not have an inverse.

28. If A is a nonzero square matrix of order n satisfying $A^2 = 0$, can A^{-1} exist? Explain.

C

29. For $n \times n$ matrices A and C and $n \times 1$ column matrices B and X, solve for X assuming all necessary inverses exist:

$$AX - B = CX.$$

30. Find the inverse of

$$A = \begin{bmatrix} 4 & 5 & 6 \\ 4 & 5 & -6 \\ 1 & 1 & 1 \end{bmatrix}$$

Show that $A^{-1}A = I$.

31. Clear the decimals in the system

$$\begin{aligned} 0.04x_1 + 0.05x_2 + 0.06x_3 &= 360 \\ 0.04x_1 + 0.05x_2 - 0.06x_3 &= 120 \\ x_1 + x_2 + x_3 &= 7{,}000 \end{aligned}$$

by multiplying the first two equations by 100. Then write the resulting system as a matrix equation and solve using the inverse found in Problem 30.

32. $\begin{vmatrix} -1 & 4 & 1 & 1 \\ 5 & -1 & 2 & -1 \\ 2 & -1 & 0 & 3 \\ -3 & 3 & 0 & 3 \end{vmatrix} = ?$

33. Show that

$$\begin{vmatrix} u & v \\ w & x \end{vmatrix} = \begin{vmatrix} u + kv & v \\ w + kx & x \end{vmatrix}$$

34. Explain why the points (1, 2) and (−1, 5) must satisfy the equation

$$\begin{vmatrix} x & y & 1 \\ 1 & 2 & 1 \\ -1 & 5 & 1 \end{vmatrix} = 0$$

Describe the set of all points that satisfy this equation.

APPLICATIONS

35. Resource Allocation. A Colorado mining company operates mines at Big Bend and Saw Pit. The Big Bend mine produces ore that is 5% nickel and 7% copper. The Saw Pit mine produces ore that is 3% nickel and 4% copper. How many tons of ore should be produced at each mine to obtain the amounts of nickel and copper listed in the table? Set up a matrix equation and solve using matrix inverses.

	Nickel	Copper
(A)	3.6 tons	5 tons
(B)	3 tons	4.1 tons
(C)	3.2 tons	4.4 tons

36. Labor Costs. A company with manufacturing plants in North and South Carolina has labor-hour and wage requirements for the manufacturing of computer desks and printer stands as given in matrices L and H:

Labor-hour requirements

	Fabricating department	Assembly department	Packaging department
Desk	1.7 h	2.4 h	0.8 h
Stand	0.9 h	1.8 h	0.6 h

$$L = \begin{bmatrix} 1.7\text{ h} & 2.4\text{ h} & 0.8\text{ h} \\ 0.9\text{ h} & 1.8\text{ h} & 0.6\text{ h} \end{bmatrix}$$

Hourly wages

	North Carolina plant	South Carolina plant
Fabricating department	\$11.50	\$10.00
Assembly department	\$9.50	\$8.50
Packaging department	\$5.00	\$4.50

$$H = \begin{bmatrix} \$11.50 & \$10.00 \\ \$9.50 & \$8.50 \\ \$5.00 & \$4.50 \end{bmatrix}$$

(A) Find the labor cost for producing one printer stand at the South Carolina plant.

(B) Discuss possible interpretations of the elements in the matrix products HL and LH.

(C) If either of the products HL or LH has a meaningful interpretation, find the product and label its rows and columns.

37. Labor Costs. The monthly production of computer desks and printer stands for the company in Problem 36 for the months of January and February are given in matrices J and F:

January production

	North Carolina plant	South Carolina plant
Desks	1,500	1,650
Stands	850	700

$$J = \begin{bmatrix} 1{,}500 & 1{,}650 \\ 850 & 700 \end{bmatrix}$$

February production

	North Carolina plant	South Carolina plant
Desks	1,700	1,810
Stands	930	740

$$F = \begin{bmatrix} 1{,}700 & 1{,}810 \\ 930 & 740 \end{bmatrix}$$

(A) Find the average monthly production for the months of January and February.

(B) Find the increase in production from January to February.

(C) Find $J\begin{bmatrix} 1 \\ 1 \end{bmatrix}$ and interpret.

38. Cryptography. The following message was encoded with the matrix B shown below. Decode the message:

25 8 26 24 25 33 21 41 48 41 30 50
21 32 41 52 52 79

$$B = \begin{bmatrix} 1 & 1 & 0 \\ 1 & 0 & 1 \\ 1 & 1 & 1 \end{bmatrix}$$

Cumulative Review Exercise Chapters 8 and 9

Work through all the problems in this cumulative review and check answers in the back of the book. Answers to all review problems are there, and following each answer is a number in italics indicating the section in which that type of problem is discussed. Where weaknesses show up, review appropriate sections in the text.

A

1. Solve using substitution or elimination by addition:

$$\begin{aligned} 3x - 5y &= 11 \\ 2x + 3y &= 1 \end{aligned}$$

2. Solve by graphing: $\begin{aligned} 2x - y &= -4 \\ 3x + y &= -1 \end{aligned}$

3. Solve by substitution or elimination by addition:

$$\begin{aligned} x^2 + y^2 &= 2 \\ 2x - y &= 1 \end{aligned}$$

4. Solve by graphing: $\begin{aligned} 3x + 5y &\le 15 \\ x, y &\ge 0 \end{aligned}$

5. Find the maximum and minimum value of $z = 2x + 3y$ over the feasible region S:

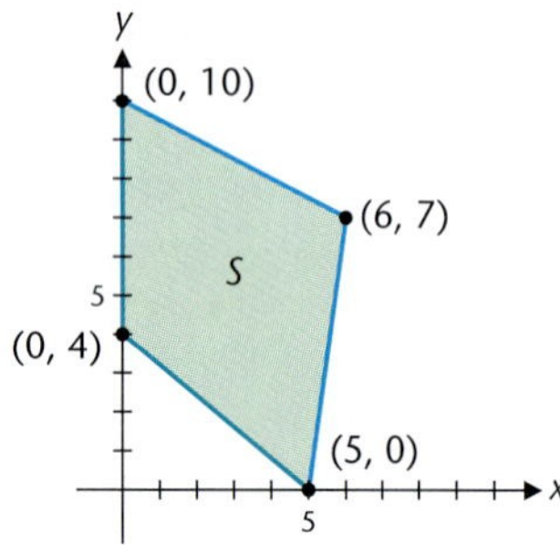

6. Perform the operations that are defined, given the following matrices:

$$M = \begin{bmatrix} 2 & 1 \\ 1 & -3 \end{bmatrix} \quad N = \begin{bmatrix} 1 & 2 \\ -1 & 3 \end{bmatrix}$$

$$P = \begin{bmatrix} 1 & 2 \end{bmatrix} \quad Q = \begin{bmatrix} -1 \\ 2 \end{bmatrix}$$

(A) $M - 2N$ (B) $P + Q$ (C) PQ
(D) MN (E) PN (F) QM

7. Evaluate: $\begin{vmatrix} 0 & 2 & 0 \\ 1 & 3 & 2 \\ -1 & 4 & 3 \end{vmatrix}$

8. Write the linear system corresponding to each augmented matrix and solve:

(A) $\left[\begin{array}{cc|c} 1 & 0 & 3 \\ 0 & 1 & -4 \end{array}\right]$ (B) $\left[\begin{array}{cc|c} 1 & -2 & 3 \\ 0 & 0 & 0 \end{array}\right]$

(C) $\left[\begin{array}{cc|c} 1 & -2 & 3 \\ 0 & 0 & 1 \end{array}\right]$

9. Given the system: $\begin{aligned} x_1 + x_2 &= 3 \\ -x_1 + x_2 &= 5 \end{aligned}$

(A) Write the augmented matrix for the system.
(B) Transform the augmented matrix into reduced form.
(C) Write the solution to the system.

10. Given the system: $\begin{aligned} x_1 - 3x_2 &= k_1 \\ 2x_1 - 5x_2 &= k_2 \end{aligned}$

(A) Write the system as a matrix equation of the form $AX = B$.
(B) Find the inverse of the coefficient matrix A.
(C) Use A^{-1} to find the solution for $k_1 = -2$ and $k_2 = 1$.
(D) Use A^{-1} to find the solution for $k_1 = 1$ and $k_2 = -2$.

11. Given the system: $\begin{aligned} 2x - 3y &= 1 \\ 4x - 5y &= 2 \end{aligned}$

(A) Find the determinant of the coefficient matrix.
(B) Solve the system using Cramer's rule.

12. Use Gauss–Jordan elimination to solve the system

$$\begin{aligned} x_1 + 3x_2 &= 10 \\ 2x_1 - x_2 &= -1 \end{aligned}$$

Then write the linear system represented by each augmented matrix in your solution, and solve each of these systems graphically. Discuss the relationship between the solutions of these systems.

13. Use an intersection routine on a graphing utility to approximate the solution of the following system to two decimal places:

$$\begin{aligned} -2x + 3y &= 7 \\ 3x + 4y &= 18 \end{aligned}$$

B

Solve Problems 14–16 using Gauss–Jordan elimination.

14. $$\begin{aligned} x_1 + 2x_2 - x_3 &= 3 \\ x_2 + x_3 &= -2 \\ 2x_1 + 3x_2 + x_3 &= 0 \end{aligned}$$

15. $$\begin{aligned} x_1 + x_2 - x_3 &= 2 \\ 4x_2 + 6x_3 &= -1 \\ 6x_2 + 9x_3 &= 0 \end{aligned}$$

16. $$\begin{aligned} x_1 - 2x_2 + x_3 &= 1 \\ 3x_1 - 2x_2 - x_3 &= -5 \end{aligned}$$

In Problems 17 and 18, solve each system.

17. $$\begin{aligned} x^2 - 3xy + 3y^2 &= 1 \\ xy &= 1 \end{aligned}$$

18. $$\begin{aligned} x^2 - 3xy + y^2 &= -1 \\ x^2 - xy &= 0 \end{aligned}$$

19. Given $M = \begin{bmatrix} 1 & 2 & -1 \end{bmatrix}$ and $N = \begin{bmatrix} 1 \\ -1 \\ 2 \end{bmatrix}$. Find:

(A) MN (B) NM

20. Given

$$L = \begin{bmatrix} 2 & -1 & 0 \\ 1 & 2 & 1 \end{bmatrix} \quad M = \begin{bmatrix} 1 & 2 \\ -1 & 0 \\ 1 & 1 \end{bmatrix} \quad N = \begin{bmatrix} 2 & 1 \\ -1 & 0 \end{bmatrix}$$

Find, if defined: (A) $LM - 2N$ (B) $ML + N$

21. Solve graphically and indicate whether the solution region is bounded or unbounded. Find the coordinates of each corner point.

$$\begin{aligned} 3x + 2y &\geq 12 \\ x + 2y &\geq 8 \\ x, y &\geq 0 \end{aligned}$$

22. Solve the linear programming problem:

$$\begin{aligned} \text{Maximize} \quad & z = 4x + 9y \\ \text{Subject to} \quad & x + 2y \leq 14 \\ & 2x + y \leq 16 \\ & x, y \geq 0 \end{aligned}$$

23. Given the system: $$\begin{aligned} x_1 + 4x_2 + 2x_3 &= k_1 \\ 2x_1 + 6x_2 + 3x_3 &= k_2 \\ 2x_1 + 5x_2 + 2x_3 &= k_3 \end{aligned}$$

(A) Write the system as a matrix equation in the form $AX = B$.
(B) Find the inverse of the coefficient matrix A.
(C) Use A^{-1} to solve the system when $k_1 = -1$, $k_2 = 2$, and $k_3 = 1$.
(D) Use A^{-1} to solve the system when $k_1 = 2$, $k_2 = 0$, and $k_3 = -1$.

24. Given the system: $$\begin{aligned} x + 2y - z &= 1 \\ 2x + 8y + z &= -2 \\ -x + 3y + 5z &= 2 \end{aligned}$$

(A) Evaluate the coefficient determinant D.
(B) Solve for z using Cramer's rule.

C

25. Use a graphing utility to approximate all real solutions to two decimal places:

$$\begin{aligned} x^2 + 2xy - y^2 &= 1 \\ 9x^2 + 4xy + y^2 &= 15 \end{aligned}$$

26. Discuss the number of solutions for the system corresponding to the reduced form shown below if
(A) $m = 0$ and $n = 0$ (B) $m = 0$ and $n \neq 0$
(C) $m \neq 0$

$$\left[\begin{array}{ccc|c} 1 & 0 & -5 & 2 \\ 0 & 1 & 3 & 6 \\ 0 & 0 & m & n \end{array}\right]$$

27. If a square matrix A satisfies the equation $A^2 = A$, find A. Assume that A^{-1} exists.

28. Which of the following augmented matrices are in reduced form?

$$L = \left[\begin{array}{ccc|c} 1 & 0 & 0 & 2 \\ 0 & 1 & 0 & 0 \\ 0 & 0 & 1 & -1 \end{array}\right] \quad M = \left[\begin{array}{ccc|c} 1 & 0 & 3 & 3 \\ 0 & 1 & -2 & 2 \\ 0 & 0 & 0 & 0 \end{array}\right]$$

$$N = \left[\begin{array}{cc|c} 0 & 0 & 0 \\ 1 & 0 & 2 \\ 0 & 1 & -3 \end{array}\right] \quad P = \left[\begin{array}{cccc|c} 1 & 2 & 0 & 2 & -2 \\ 0 & 0 & 1 & 3 & 1 \end{array}\right]$$

29. Show that

$$k\begin{vmatrix} a & b \\ c & d \end{vmatrix} = \begin{vmatrix} ka & b \\ kc & d \end{vmatrix}$$

30. Show that

$$\begin{vmatrix} a & b \\ c & d \end{vmatrix} = \begin{vmatrix} a & b \\ c + ka & d + kb \end{vmatrix}$$

31. If $M = \begin{vmatrix} a & b \\ c & d \end{vmatrix}$ and $\det M \neq 0$, show that

$$M^{-1} = \frac{1}{\det M}\begin{bmatrix} d & -b \\ -c & a \end{bmatrix}$$

*Recall that a square matrix is called **upper triangular** if all elements below the principal diagonal are zero, and it is called **diagonal** if all elements not on the principal diagonal are zero. A square matrix is called **lower triangular** if all elements above the principal diagonal are zero. In Problems 32–39, determine whether the statement is true or false. If true, explain why. If false, give a counterexample.*

32. The sum of two upper triangular matrices is upper triangular.

33. The product of two lower triangular matrices is lower triangular.

34. The sum of an upper triangular matrix and a lower triangular matrix is a diagonal matrix.

35. The product of an upper triangular matrix and a lower triangular matrix is a diagonal matrix.

36. A matrix that is both upper triangular and lower triangular is a diagonal matrix.

37. If a diagonal matrix has no zero elements on the principal diagonal, then it has an inverse.

38. The determinant of a diagonal matrix is the product of the elements on the principal diagonal.

39. The determinant of a lower triangular matrix is the product of the elements on the principal diagonal.

APPLICATIONS 

40. **Finance.** An investor has $12,000 to invest. If part is invested at 8% and the rest in a higher-risk investment at 14%, how much should be invested at each rate to produce the same yield as if all had been invested at 10%?

41. **Diet.** In an experiment involving mice, a zoologist needs a food mix that contains, among other things, 23 grams of protein, 6.2 grams of fat, and 16 grams of moisture. She has on hand mixes of the following compositions: Mix A contains 20% protein, 2% fat, and 15% moisture; mix B contains 10% protein, 6% fat, and 10% moisture; and mix C contains 15% protein, 5% fat, and 5% moisture. How many grams of each mix should be used to get the desired diet mix?

42 **Purchasing.** A soft-drink distributor has budgeted $300,000 for the purchase of 12 new delivery trucks. If a model A truck costs $18,000, a model B truck costs $22,000, and a model C truck costs $30,000, how many trucks of each model should the distributor purchase to use exactly all the budgeted funds?

43. **Geometry.** Find the dimensions of a rectangle with perimeter 24 meters and area 32 square meters.

44. **Manufacturing.** A manufacturer makes two types of day packs, a standard model and a deluxe model. Each standard model requires 0.5 labor-hour from the fabricating department and 0.3 labor-hour from the sewing department. Each deluxe model requires 0.5 labor-hour from the fabricating department and 0.6 labor-hour from the sewing department. The maximum number of labor-hours available per week in the fabricating department and the sewing department are 300 and 240, respectively.
    (A) If the profit on a standard day pack is $8 and the profit on a deluxe day pack is $12, how many of each type of pack should be manufactured each day to realize a maximum profit? What is the maximum profit?
    (B) Discuss the effect on the production schedule and the maximum profit if the profit on a standard day pack decreases by $3 and the profit on a deluxe day pack increases by $3.
    (C) Discuss the effect on the production schedule and the maximum profit if the profit on a standard day pack increases by $3 and the profit on a deluxe day pack decreases by $3.

45. **Averaging Tests.** A teacher has given four tests to a class of five students and stored the results in the following matrix:

$$\begin{array}{r} \\ \\ \text{Ann} \\ \text{Bob} \\ \text{Carol} \\ \text{Dan} \\ \text{Eric} \end{array} \begin{array}{c} \text{Tests} \\ \begin{array}{cccc} 1 & 2 & 3 & 4 \end{array} \\ \begin{bmatrix} 78 & 84 & 81 & 86 \\ 91 & 65 & 84 & 92 \\ 95 & 90 & 92 & 91 \\ 75 & 82 & 87 & 91 \\ 83 & 88 & 81 & 76 \end{bmatrix} \end{array} = M$$

Discuss methods of matrix multiplication that the teacher can use to obtain the indicated information in parts A–C below. In each case, state the matrices to be used and then perform the necessary multiplications.
    (A) The average on all four tests for each student, assuming that all four tests are given equal weight
    (B) The average on all four tests for each student, assuming that the first three tests are given equal weight and the fourth is given twice this weight
    (C) The class average on each of the four tests

SEQUENCES AND SERIES

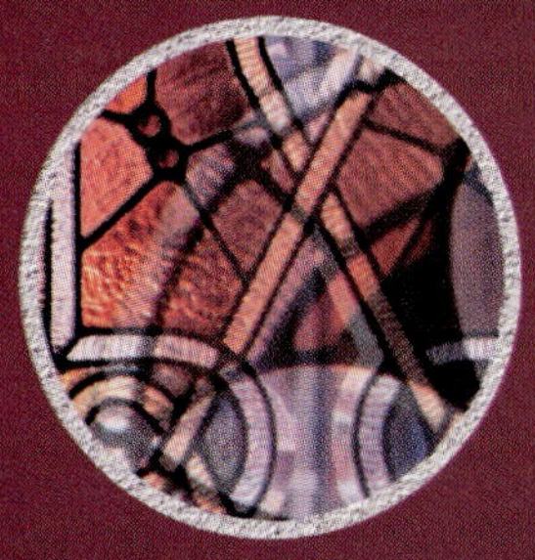

CHAPTER 10

If someone asked you to list all natural numbers that are perfect squares, you might begin by writing

$$1, \quad 4, \quad 9, \quad 16, \quad 25, \quad 36$$

But you would soon realize that it is impossible to actually list all the perfect squares, since there are an infinite number of them. However, this collection of numbers can be represented in several different ways. One common method is to write

$$1, 4, 9, \ldots, n^2, \ldots \qquad n \in N$$

where *N* is the set of natural numbers. A list of numbers such as this is generally called a *sequence.* Sequences and related topics form the subject matter of this chapter. One of the related topics involves a method of proof we have referred to several times earlier in this book—*mathematical induction.* This method enables us to prove conjectures involving infinite sets of successive integers.

SECTION 10-1 Sequences and Series

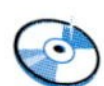

- Sequences
- Series

In this section we introduce special notation and formulas for representing and generating sequences and sums of sequences.

• Sequences

Consider the function f given by

$$f(n) = 2n - 1 \tag{1}$$

where the domain of f is the set of natural numbers N. Note that

$$f(1) = 1, f(2) = 3, f(3) = 5, \ldots$$

The function f is an example of a sequence. A **sequence** is a function with domain a set of successive integers. However, a sequence is hardly ever represented in the form of equation (1). A special notation for sequences has evolved, which we describe here.

To start, the range value $f(n)$ is usually symbolized more compactly with a symbol such as a_n. Thus, in place of equation (1) we write

$$a_n = 2n - 1$$

The domain is understood to be the set of natural numbers N unless stated to the contrary or the context indicates otherwise. The elements in the range are called **terms**

of the sequence: a_1 is the first term, a_2 the second term, and a_n the nth term, or the **general term:**

$$\begin{aligned} a_1 &= 2(1) - 1 = 1 && \text{First term} \\ a_2 &= 2(2) - 1 = 3 && \text{Second term} \\ a_3 &= 2(3) - 1 = 5 && \text{Third term} \\ &\vdots \qquad\quad \vdots \end{aligned}$$

The ordered list of elements

$$1, 3, 5, \ldots, 2n - 1, \ldots$$

in which the terms of a sequence are written in their natural order with respect to the domain values, is often informally referred to as a sequence. A sequence is also represented in the abbreviated form $\{a_n\}$, where a symbol for the nth term is placed between braces. For example, we can refer to the sequence

$$1, 3, 5, \ldots, 2n - 1, \ldots$$

as the sequence $\{2n - 1\}$.

If the domain of a function is a finite set of successive integers, then the sequence is called a **finite sequence.** If the domain is an infinite set of successive integers, then the sequence is called an **infinite sequence.** The sequence $\{2n - 1\}$ above is an example of an infinite sequence.

EXPLORE-DISCUSS 1 The sequence $\{2n - 1\}$ is a function whose domain is the set of natural numbers, and so it may be graphed in the same way as any function whose domain and range are sets of real numbers (see Fig. 1).

FIGURE 1 Graph of $\{2n - 1\}$.

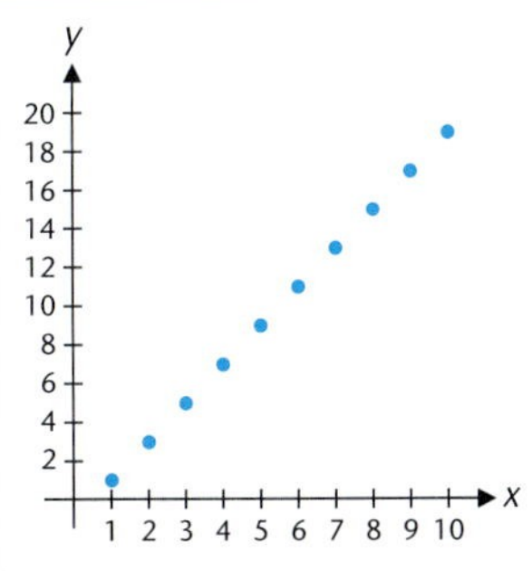

(A) Explain why the graph of the sequence $\{2n - 1\}$ is not continuous.

(B) Explain why the points on the graph of $\{2n - 1\}$ lie on a line. Find an equation for that line.

(C) Graph the sequence $\left\{\dfrac{2n^2 - n + 1}{n}\right\}$. How are the graphs of $\{2n - 1\}$ and $\left\{\dfrac{2n^2 - n + 1}{n}\right\}$ related?

Some sequences are specified by a **recursion formula**—that is, a formula that defines each term in terms of one or more preceding terms. The sequence we have chosen to illustrate a recursion formula is a very famous sequence in the history of mathematics called the **Fibonacci sequence.** It is named after the most celebrated mathematician of the thirteenth century, Leonardo Fibonacci from Italy (1180?–1250?).

EXAMPLE 1 Fibonacci Sequence

List the first six terms of the sequence specified by

$$a_1 = 1$$
$$a_2 = 1$$
$$a_n = a_{n-1} + a_{n-2} \qquad n \geq 3$$

Solution

$$\begin{aligned} a_1 &= 1 \\ a_2 &= 1 \\ a_3 &= a_2 + a_1 = 1 + 1 = 2 \\ a_4 &= a_3 + a_2 = 2 + 1 = 3 \\ a_5 &= a_4 + a_3 = 3 + 2 = 5 \\ a_6 &= a_5 + a_4 = 5 + 3 = 8 \end{aligned}$$

The formula $a_n = a_{n-1} + a_{n-2}$ is a recursion formula that can be used to generate the terms of a sequence in terms of preceding terms. Of course, starting terms a_1 and a_2 must be provided in order to use the formula. Recursion formulas are particularly suitable for use with calculators and computers (see Problems 57 and 58 in Exercise 10-1).

Matched Problem 1 List the first five terms of the sequence specified by

$$a_1 = 4$$
$$a_n = \tfrac{1}{2}a_{n-1} \qquad n \geq 2$$

Now we consider the reverse problem. That is, can a sequence be defined just by listing the first three or four terms of the sequence? And can we then use these initial terms to find a formula for the nth term? In general, without other information, the answer to the first question is no. Many different sequences may start off with the same terms. For example, each of the following sequences starts off with the same three terms:

$$1, 3, 9, \ldots, 3^{n-1}, \ldots$$
$$1, 3, 9, \ldots, 1 + 2(n - 1)^2, \ldots$$
$$1, 3, 9, \ldots, 8n + \frac{12}{n} - 19, \ldots$$

However, these are certainly different sequences. You should verify that these sequences agree for the first three terms and differ in the fourth term by evaluating the general term for each sequence at $n = 1, 2, 3, 4$. Thus, simply listing the first three terms, or any other finite number of terms, does not specify a particular sequence. In fact, it can be shown that given any list of m numbers, there are an infinite number of sequences whose first m terms agree with these given numbers. What about the second question? That is, given a few terms, can we find the general formula for at least one sequence whose first few terms agree with the given terms? The answer to this question is a qualified yes. If we can observe a simple pattern in the given terms, then we may be able to construct a general term that will produce the pattern. The next example illustrates this approach.

EXAMPLE 2 **Finding the General Term of a Sequence**

Find the general term of a sequence whose first four terms are:

(A) $5, 6, 7, 8, \ldots$ (B) $2, -4, 8, -16, \ldots$

Solutions

(A) Since these terms are consecutive integers, one solution is $a_n = n$, $n \geq 5$. If we want the domain of the sequence to be all natural numbers, then another solution is $b_n = n + 4$.

(B) Each of these terms can be written as the product of a power of 2 and a power of -1:

$$2 = (-1)^0 2^1$$
$$-4 = (-1)^1 2^2$$
$$8 = (-1)^2 2^3$$
$$-16 = (-1)^3 2^4$$

If we choose the domain to be all natural numbers, then a solution is

$$a_n = (-1)^{n-1} 2^n$$

Matched Problem 2 Find the general term of a sequence whose first four terms are:

(A) $2, 4, 6, 8, \ldots$ (B) $1, -\frac{1}{2}, \frac{1}{4}, -\frac{1}{8}, \ldots$

In general, there is usually more than one way of representing the nth term of a given sequence. This was seen in the solution of Example 2A. However, unless stated to the contrary, we assume the domain of the sequence is the set of natural numbers N.

EXPLORE-DISCUSS 2 The sequence with general term $b_n = \frac{\sqrt{5}}{5}\left(\frac{1+\sqrt{5}}{2}\right)^n$ is closely related to the Fibonacci sequence. Compute the first 20 terms of both sequences and discuss the relationship. (The values of b_1, b_2, and b_3 are shown in Fig. 2.)

FIGURE 2

```
(1+√(5))/2→A
          1.618033989
√(5)/5*A
          .7236067977
Ans*A
          1.170820393
          1.894427191
```

• Series

If $a_1, a_2, a_3, \ldots, a_n, \ldots$ is a sequence, then the expression $a_1 + a_2 + a_3 + \ldots + a_n + \ldots$ is called a **series.** If the sequence is finite, the corresponding series is a **finite series.** If the sequence is infinite, the corresponding series is an **infinite series.** For example,

$$1, 2, 4, 8, 16 \qquad \text{Finite sequence}$$

$$1 + 2 + 4 + 8 + 16 \qquad \text{Finite series}$$

We restrict our discussion to finite series in this section.

Series are often represented in a compact form called **summation notation** using the symbol $\sum$, which is a stylized version of the Greek letter sigma. Consider the following examples:

$$\sum_{k=1}^{4} a_k = a_1 + a_2 + a_3 + a_4$$

$$\sum_{k=3}^{7} b_k = b_3 + b_4 + b_5 + b_6 + b_7$$

$$\sum_{k=0}^{n} c_k = c_0 + c_1 + c_2 + \cdots + c_n$$

Domain is the set of integers greater than or equal to 0.

The terms on the right are obtained from the expression on the left by successively replacing the **summing index** k with integers, starting with the first number indicated below $\sum$ and ending with the number that appears above $\sum$. Thus, for example, if we are given the sequence

$$\frac{1}{2}, \frac{1}{4}, \frac{1}{8}, \ldots, \frac{1}{2^n}$$

the corresponding series is

$$\frac{1}{2} + \frac{1}{4} + \frac{1}{8} + \cdots + \frac{1}{2^n}$$

or, more compactly,

$$\sum_{k=1}^{n} \frac{1}{2^k}$$

EXAMPLE 3 Writing the Terms of a Series

Write without summation notation: $\sum_{k=1}^{5} \frac{k-1}{k}$

Solution

$$\sum_{k=1}^{5} \frac{k-1}{k} = \frac{1-1}{1} + \frac{2-1}{2} + \frac{3-1}{3} + \frac{4-1}{4} + \frac{5-1}{5}$$

$$= 0 + \frac{1}{2} + \frac{2}{3} + \frac{3}{4} + \frac{4}{5}$$

Matched Problem 3 Write without summation notation: $\sum_{k=0}^{5} \frac{(-1)^k}{2k+1}$

If the terms of a series are alternately positive and negative, it is called an **alternating series.** Example 4 deals with the representation of such a series.

EXAMPLE 4 Writing a Series in Summation Notation

Write the following series using summation notation:

$$1 - \frac{1}{2} + \frac{1}{3} - \frac{1}{4} + \frac{1}{5} - \frac{1}{6}$$

(A) Start the summing index at $k = 1$.
(B) Start the summing index at $k = 0$.

Solutions

(A) $(-1)^{k-1}$ provides the alternation of sign, and $1/k$ provides the other part of each term. Thus, we can write

$$\sum_{k=1}^{6} \frac{(-1)^{k-1}}{k}$$

as can be easily checked.

(B) $(-1)^k$ provides the alternation of sign, and $1/(k + 1)$ provides the other part of each term. Thus, we write

$$\sum_{k=0}^{5} \frac{(-1)^k}{k+1}$$

as can be checked.

Matched Problem 4 Write the following series using summation notation:

$$1 - \frac{2}{3} + \frac{4}{9} - \frac{8}{27} + \frac{16}{81}$$

(A) Start with $k = 1$. (B) Start with $k = 0$.

EXPLORE-DISCUSS 3 (A) Find the smallest number of terms of the infinite series

$$1 + \frac{1}{2} + \frac{1}{3} + \cdots + \frac{1}{n} + \cdots$$

that, when added together, give a number greater than 3.

(B) Find the smallest number of terms of the infinite series

$$\frac{1}{2} + \frac{1}{4} + \cdots + \frac{1}{2^n} + \cdots$$

that, when added together, give a number greater than 0.99. Greater than 0.999. Can the sum ever exceed 1? Explain.

Answers to Matched Problems

1. 4, 2, 1, $\frac{1}{2}$, $\frac{1}{4}$ **2.** (A) $a_n = 2n$ (B) $a_n = (-1)^{n-1}(\frac{1}{2})^{n-1}$ **3.** $1 - \frac{1}{3} + \frac{1}{5} - \frac{1}{7} + \frac{1}{9} - \frac{1}{11}$

4. (A) $\sum_{k=1}^{5} (-1)^{k-1}\left(\frac{2}{3}\right)^{k-1}$ (B) $\sum_{k=0}^{4} (-1)^{k}\left(\frac{2}{3}\right)^{k}$

EXERCISE 10-1

A

Write the first four terms for each sequence in Problems 1–6.

1. $a_n = 2n + 5$

2. $a_n = n^2 + 1$

3. $a_n = \dfrac{3n + 1}{n + 1}$

4. $a_n = (n + 1)^n$

5. $a_n = 3^n + (-2)^n$

6. $a_n = \dfrac{(-1)^n}{n^3}$

7. Write the one-hundredth term of the sequence in Problem 1.

8. Write the fiftieth term of the sequence in Problem 2.

9. Write the two-hundredth term of the sequence in Problem 3.

10. Write the ninety-ninth term of the sequence in Problem 4.

In Problems 11–16, write each series in expanded form without summation notation.

11. $\sum_{k=1}^{5} k$

12. $\sum_{k=1}^{4} k^2$

13. $\sum_{k=1}^{3} \frac{1}{10^k}$

14. $\sum_{k=1}^{5} \left(\frac{1}{3}\right)^k$

15. $\sum_{k=1}^{4} (-1)^k$

16. $\sum_{k=1}^{6} (-1)^{k+1}k$

B

Write the first five terms of each sequence in Problems 17–26.

17. $a_n = (-1)^{n+1}n^2$

18. $a_n = (-1)^{n+1}\left(\frac{1}{2^n}\right)$

19. $a_n = \frac{1}{3}\left(1 - \frac{1}{10^n}\right)$

20. $a_n = n[1 - (-1)^n]$

21. $a_n = (-\frac{1}{2})^{n-1}$

22. $a_n = (-\frac{3}{2})^{n-1}$

23. $a_1 = 7; a_n = a_{n-1} - 4, n \geq 2$

24. $a_1 = a_2 = 1; a_n = a_{n-1} + a_{n-2}, n \geq 3$

25. $a_1 = 4; a_n = \frac{1}{4}a_{n-1}, n \geq 2$

26. $a_1 = 2; a_n = 2a_{n-1}, n \geq 2$

In Problems 27–38, find the general term of a sequence whose first four terms are given.

27. $3, 5, 7, 9, \ldots$

28. $2, -1, -4, -7, \ldots$

29. $4, 9, 16, 25, \ldots$

30. $24, 35, 48, 63, \ldots$

31. $4, 8, 16, 32, \ldots$

32. $1, 4, 27, 256, \ldots$

33. $1, \frac{3}{2}, \frac{5}{3}, \frac{7}{4}, \ldots$

34. $\frac{1}{2}, \frac{3}{4}, \frac{7}{8}, \frac{15}{16}, \ldots$

35. $7, -7, 7, -7, \ldots$

36. $1, -\frac{1}{5}, \frac{1}{25}, -\frac{1}{125}, \ldots$

37. $x, \frac{x^2}{2}, \frac{x^3}{4}, \frac{x^4}{8}, \ldots$

38. $-x, x^4, -x^7, x^{10}, \ldots$

 *Some graphing utilities use special routines to graph sequences (consult your manual). In Problems 39–42, use such routines to graph the first 20 terms of each sequence.*

39. $a_n = 1/n$

40. $a_n = 2 + \pi n$

41. $a_n = (-0.9)^n$

42. $a_1 = -1, a_n = \frac{2}{3}a_{n-1} + \frac{1}{2}$

In Problems 43–48, write each series in expanded form without summation notation.

43. $\sum_{k=1}^{4} \frac{(-2)^{k+1}}{k}$

44. $\sum_{k=1}^{5} (-1)^{k+1}(2k - 1)^2$

45. $\sum_{k=1}^{3} \frac{1}{k}x^{k+1}$

46. $\sum_{k=1}^{5} x^{k-1}$

47. $\sum_{k=1}^{5} \frac{(-1)^{k+1}}{k}x^k$

48. $\sum_{k=0}^{4} \frac{(-1)^k x^{2k+1}}{2k + 1}$

In Problems 49–56, write each series using summation notation with the summing index k starting at k = 1.

49. $1^2 + 2^2 + 3^2 + 4^2$

50. $2 + 3 + 4 + 5 + 6$

51. $\frac{1}{2} + \frac{1}{2^2} + \frac{1}{2^3} + \frac{1}{2^4} + \frac{1}{2^5}$

52. $1 - \frac{1}{2} + \frac{1}{3} - \frac{1}{4}$

53. $1 + \frac{1}{2^2} + \frac{1}{3^2} + \cdots + \frac{1}{n^2}$

54. $2 + \frac{3}{2} + \frac{4}{3} + \cdots + \frac{n + 1}{n}$

55. $1 - 4 + 9 - \cdots + (-1)^{n+1}n^2$

56. $\frac{1}{2} - \frac{1}{4} + \frac{1}{8} - \cdots + \frac{(-1)^{n+1}}{2^n}$

C

The sequence

$$a_n = \frac{a_{n-1}^2 + M}{2a_{n-1}} \qquad n \geq 2, M \text{ a positive real number}$$

can be used to find $\sqrt{M}$ to any decimal-place accuracy desired. To start the sequence, choose a_1 arbitrarily from the positive real numbers. Problems 57 and 58 are related to this sequence.

57. (A) Find the first four terms of the sequence

$$a_1 = 3 \qquad a_n = \frac{a_{n-1}^2 + 2}{2a_{n-1}} \qquad n \geq 2$$

(B) Compare the terms with $\sqrt{2}$ from a calculator.

(C) Repeat parts A and B letting a_1 be any other positive number, say 1.

58. (A) Find the first four terms of the sequence

$$a_1 = 2 \qquad a_n = \frac{a_{n-1}^2 + 5}{2a_{n-1}} \qquad n \geq 2$$

(B) Find $\sqrt{5}$ with a calculator, and compare with the results of part A.

(C) Repeat parts A and B letting a_1 be any other positive number, say 3.

59. Let $\{a_n\}$ denote the Fibonacci sequence and let $\{b_n\}$ denote the sequence defined by $b_1 = 1, b_2 = 3, b_n = b_{n-1} + b_{n-2}$ for $n \geq 3$. Compute 10 terms of the sequence $\{c_n\}$, where $c_n = b_n/a_n$. Describe the terms of $\{c_n\}$ for large values of n.

60. Define sequences $\{u_n\}$ and $\{v_n\}$ by $u_1 = 1, v_1 = 0, u_n = u_{n-1} + v_{n-1}$ and $v_n = u_{n-1}$ for $n \geq 2$. Find the first 10 terms of each sequence, and explain their relationship to the Fibonacci sequence.

 *In calculus, it can be shown that*

$$e^x = \sum_{k=0}^{\infty} \frac{x^k}{k!} \approx 1 + \frac{x}{1!} + \frac{x^2}{2!} + \frac{x^3}{3!} + \ldots + \frac{x^n}{n!}$$

where the larger n is, the better the approximation. Problems 61 and 62 refer to this series. Note that n!, read "n factorial," is defined by 0! = 1 and $n! = 1 \cdot 2 \cdot 3 \cdot \cdots \cdot n$ for $n \in N$.

61. Approximate $e^{0.2}$ using the first five terms of the series. Compare this approximation with your calculator evaluation of $e^{0.2}$.

62. Approximate $e^{-0.5}$ using the first five terms of the series. Compare this approximation with your calculator evaluation of $e^{-0.5}$.

63. Show that: $\sum_{k=1}^{n} ca_k = c \sum_{k=1}^{n} a_k$

64. Show that: $\sum_{k=1}^{n} (a_k + b_k) = \sum_{k=1}^{n} a_k + \sum_{k=1}^{n} b_k$

SECTION 10-2 Mathematical Induction

- Introduction
- Mathematical Induction
- Additional Examples of Mathematical Induction
- Three Famous Problems

• Introduction

In common usage, the word "induction" means the generalization from particular cases or facts. The ability to formulate general hypotheses from a limited number of facts is a distinguishing characteristic of a creative mathematician. The creative process does not stop here, however. These hypotheses must then be proved or disproved. In mathematics, a special method of proof called **mathematical induction** ranks among the most important basic tools in a mathematician's toolbox. In this section mathematical induction will be used to prove a variety of mathematical statements, some new and some that up to now we have just assumed to be true.

We illustrate the process of formulating hypotheses by an example. Suppose we are interested in the sum of the first n consecutive odd integers, where n is a positive integer. We begin by writing the sums for the first few values of n to see if we can observe a pattern:

$$1 = 1 \qquad n = 1$$

$$1 + 3 = 4 \qquad n = 2$$

$$1 + 3 + 5 = 9 \qquad n = 3$$

$$1 + 3 + 5 + 7 = 16 \qquad n = 4$$

$$1 + 3 + 5 + 7 + 9 = 25 \qquad n = 5$$

Is there any pattern to the sums 1, 4, 9, 16, and 25? You no doubt observed that each is a perfect square and, in fact, each is the square of the number of terms in the sum. Thus, the following conjecture seems reasonable:

Conjecture P_n: For each positive integer n,

$$1 + 3 + 5 + \cdots + (2n - 1) = n^2$$

That is, the sum of the first n odd integers is n^2 for each positive integer n.

So far ordinary induction has been used to generalize the pattern observed in the first few cases listed above. But at this point conjecture P_n is simply that—a conjecture. How do we prove that P_n is a true statement? Continuing to list specific cases will never provide a general proof—not in your lifetime or all your descendants' lifetimes! Mathematical induction is the tool we will use to establish the validity of conjecture P_n.

Before discussing this method of proof, let's consider another conjecture:

Conjecture Q_n: For each positive integer n, the number $n^2 - n + 41$ is a prime number.

TABLE 1

n	$n^2 - n + 41$	Prime?
1	41	Yes
2	43	Yes
3	47	Yes
4	53	Yes
5	61	Yes

It is important to recognize that a conjecture can be proved false if it fails for only one case. A single case or example for which a conjecture fails is called a **counterexample.** We check the conjecture for a few particular cases in Table 1. From the table, it certainly appears that conjecture Q_n has a good chance of being true. You may want to check a few more cases. If you persist, you will find that conjecture Q_n is true for n up to 41. What happens at $n = 41$?

$$41^2 - 41 + 41 = 41^2$$

which is not prime. Thus, since $n = 41$ provides a counterexample, conjecture Q_n is false. Here we see the danger of generalizing without proof from a few special cases. This example was discovered by Euler (1707–1783).

EXPLORE-DISCUSS 1 Prove that the following statement is false by finding a counterexample: If $n \geq 2$, then at least one-third of the positive integers less than or equal to n are prime.

• Mathematical Induction

We begin by stating the *principle of mathematical induction,* which forms the basis for all our work in this section.

Theorem 1 **Principle of Mathematical Induction**

Let P_n be a statement associated with each positive integer n, and suppose the following conditions are satisfied:

1. P_1 is true.

2. For any positive integer k, if P_k is true, then P_{k+1} is also true.

Then the statement P_n is true for all positive integers n.

Theorem 1 must be read very carefully. At first glance, it seems to say that if we assume a statement is true, then it is true. But that is not the case at all. If the two conditions in Theorem 1 are satisfied, then we can reason as follows:

P_1 is true.	Condition 1
P_2 is true, because P_1 is true.	Condition 2
P_3 is true, because P_2 is true.	Condition 2
P_4 is true, because P_3 is true.	Condition 2
⋮	⋮

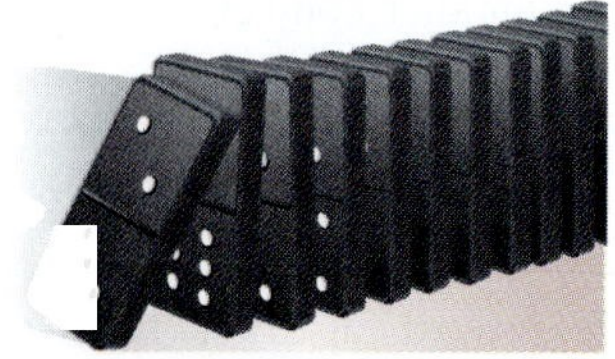

Condition 1: The first domino can be pushed over.
(a)

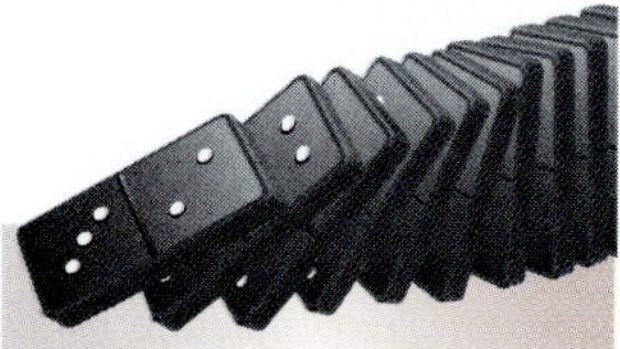

Condition 2: If the kth domino falls, then so does the $(k + 1)$st.
(b)

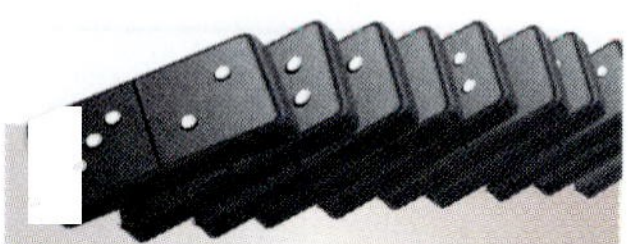

Conclusion: All the dominoes will fall.
(c)

FIGURE 1 Interpreting mathematical induction.

Since this chain of implications never ends, we will eventually reach P_n for any positive integer n.

To help visualize this process, picture a row of dominoes that goes on forever (see Fig. 1) and interpret the conditions in Theorem 1 as follows: Condition 1 says that the first domino can be pushed over. Condition 2 says that if the kth domino falls, then so does the $(k + 1)$st domino. Together, these two conditions imply that all the dominoes must fall.

Now, to illustrate the process of proof by mathematical induction, we return to the conjecture P_n discussed earlier, which we restate below:

$$P_n\text{:}\quad 1 + 3 + 5 + \cdots + (2n - 1) = n^2 \qquad n \text{ any positive integer}$$

We already know that P_1 is a true statement. In fact, we demonstrated that P_1 through P_5 are all true by direct calculation. Thus, condition 1 in Theorem 1 is satisfied. To show that condition 2 is satisfied, we assume that P_k is a true statement:

$$P_k\text{: } 1 + 3 + 5 + \cdots + (2k - 1) = k^2$$

Now we must show that this assumption implies that P_{k+1} is also a true statement:

$$P_{k+1}\text{:}\quad 1 + 3 + 5 + \cdots + (2k - 1) + (2k + 1) = (k + 1)^2$$

Since we have assumed that P_k is true, we can perform operations on this equation. Note that the left side of P_{k+1} is the left side of P_k plus $(2k + 1)$. So we start by adding $(2k + 1)$ to both sides of P_k:

$$1 + 3 + 5 + \cdots + (2k - 1) = k^2 \qquad P_k$$

$$1 + 3 + 5 + \cdots + (2k - 1) + (2k + 1) = k^2 + (2k + 1) \qquad \text{Add } 2k + 1 \text{ to both sides.}$$

Factoring the right side of this equation, we have

$$1 + 3 + 5 + \cdots + (2k - 1) + (2k + 1) = (k + 1)^2 \qquad P_{k+1}$$

But this last equation is P_{k+1}. Thus, we have started with P_k, the statement we assumed true, and performed valid operations to produce P_{k+1}, the statement we want to be true. In other words, we have shown that if P_k is true, then P_{k+1} is also true. Since both conditions in Theorem 1 are satisfied, P_n is true for all positive integers n.

• Additional Examples of Mathematical Induction

Now we will consider some additional examples of proof by induction. The first is another summation formula. Mathematical induction is the primary tool for proving that formulas of this type are true.

EXAMPLE 1 Proving a Summation Formula

Prove that for all positive integers n

$$\frac{1}{2} + \frac{1}{4} + \frac{1}{8} + \cdots + \frac{1}{2^n} = \frac{2^n - 1}{2^n}$$

Proof State the conjecture:

$$P_n\colon\ \frac{1}{2}+\frac{1}{4}+\frac{1}{8}+\cdots+\frac{1}{2^n}=\frac{2^n-1}{2^n}$$

Part 1 Show that P_1 is true.

$$P_1\colon\ \frac{1}{2}=\frac{2^1-1}{2^1}$$
$$=\frac{1}{2}$$

Thus, P_1 is true.

Part 2 Show that if P_k is true, then P_{k+1} is true. It is a good practice to always write out both P_k and P_{k+1} at the beginning of any induction proof to see what is assumed and what must be proved:

$$P_k\colon\ \frac{1}{2}+\frac{1}{4}+\frac{1}{8}+\cdots+\frac{1}{2^k}=\frac{2^k-1}{2^k}$$ We assume P_k is true.

$$P_{k+1}\colon\ \frac{1}{2}+\frac{1}{4}+\frac{1}{8}+\cdots+\frac{1}{2^k}+\frac{1}{2^{k+1}}=\frac{2^{k+1}-1}{2^{k+1}}$$ We must show that P_{k+1} follows from P_k.

We start with the true statement P_k, add $1/2^{k+1}$ to both sides, and simplify the right side:

$$\frac{1}{2}+\frac{1}{4}+\frac{1}{8}+\cdots+\frac{1}{2^k}=\frac{2^k-1}{2^k}$$ P_k

$$\frac{1}{2}+\frac{1}{4}+\frac{1}{8}+\cdots+\frac{1}{2^k}+\mathbf{\frac{1}{2^{k+1}}}=\frac{2^k-1}{2^k}+\mathbf{\frac{1}{2^{k+1}}}$$
$$=\frac{2^k-1}{2^k}\cdot\frac{2}{2}+\frac{1}{2^{k+1}}$$
$$=\frac{2^{k+1}-2+1}{2^{k+1}}$$
$$=\frac{2^{k+1}-1}{2^{k+1}}$$

Thus,

$$\frac{1}{2}+\frac{1}{4}+\frac{1}{8}+\cdots+\frac{1}{2^k}+\frac{1}{2^{k+1}}=\frac{2^{k+1}-1}{2^{k+1}}$$ P_{k+1}

and we have shown that if P_k is true, then P_{k+1} is true.

Conclusion Both conditions in Theorem 1 are satisfied. Thus, P_n is true for all positive integers n.

Matched Problem 1 Prove that for all positive integers n

$$1 + 2 + 3 + \cdots + n = \frac{n(n + 1)}{2}$$

The next example provides a proof of a law of exponents that previously we had to assume was true. First we redefine a^n for n a positive integer, using a recursion formula:

DEFINITION 1

Recursive Definition of a^n

For n a positive integer

$$a^1 = a$$
$$a^{n+1} = a^n a \qquad n \geq 1$$

EXAMPLE 2 **Proving a Law of Exponents**

Prove that $(xy)^n = x^n y^n$ for all positive integers n.

Proof State the conjecture:

$$P_n\text{:} \quad (xy)^n = x^n y^n$$

Part 1 Show that P_1 is true.

$$\begin{aligned} (xy)^1 &= xy && \text{Definition 1} \\ &= x^1 y^1 && \text{Definition 1} \end{aligned}$$

Thus, P_1 is true.

Part 2 Show that if P_k is true, then P_{k+1} is true.

$$\begin{aligned} P_k\text{:} \quad (xy)^k &= x^k y^k && \text{Assume } P_k \text{ is true.} \\ P_{k+1}\text{:} \quad (xy)^{k+1} &= x^{k+1} y^{k+1} && \text{Show that } P_{k+1} \text{ follows from } P_k. \end{aligned}$$

Here we start with the left side of P_{k+1} and use P_k to find the right side of P_{k+1}:

$$\begin{aligned} (xy)^{k+1} &= (xy)^k (xy)^1 && \text{Definition 1} \\ &= x^k y^k xy && \text{Use } P_k\text{: } (xy)^k = x^k y^k \\ &= (x^k x)(y^k y) && \text{Property of real numbers} \\ &= x^{k+1} y^{k+1} && \text{Definition 1} \end{aligned}$$

Thus, $(xy)^{k+1} = x^{k+1}y^{k+1}$, and we have shown that if P_k is true, then P_{k+1} is true.

Conclusion Both conditions in Theorem 1 are satisfied. Thus, P_n is true for all positive integers n.

Matched Problem 2 Prove that $(x/y)^n = x^n/y^n$ for all positive integers n.

Our last example deals with factors of integers. Before we start, recall that an integer p is *divisible* by an integer q if $p = qr$ for some integer r.

EXAMPLE 3 Proving a Divisibility Property

Prove that $4^{2n} - 1$ is divisible by 5 for all positive integers n.

Proof Use the definition of divisibility to state the conjecture as follows:

$$P_n\colon \quad 4^{2n} - 1 = 5r \qquad \text{for some integer } r$$

Part 1 Show that P_1 is true.

$$P_1\colon \quad 4^2 - 1 = 15 = 5 \cdot 3$$

Thus, P_1 is true.

Part 2 Show that if P_k is true, then P_{k+1} is true.

$$P_k\colon \quad 4^{2k} - 1 = 5r \qquad \text{for some integer } r \qquad \text{Assume } P_k \text{ is true.}$$

$$P_{k+1}\colon \quad 4^{2(k+1)} - 1 = 5s \qquad \text{for some integer } s \qquad \text{Show that } P_{k+1} \text{ must follow.}$$

As before, we start with the true statement P_k:

$$\begin{aligned} 4^{2k} - 1 &= 5r && P_k \\ 4^2(4^{2k} - 1) &= 4^2(5r) && \text{Multiply both sides by } 4^2. \\ 4^{2k+2} - 16 &= 80r && \text{Simplify.} \\ 4^{2(k+1)} - 1 &= 80r + 15 && \text{Add 15 to both sides.} \\ &= 5(16r + 3) && \text{Factor out 5.} \end{aligned}$$

Thus,

$$4^{2(k+1)} - 1 = 5s \qquad P_{k+1}$$

where $s = 16r + 3$ is an integer, and we have shown that if P_k is true, then P_{k+1} is true.

Conclusion Both conditions in Theorem 1 are satisfied. Thus, P_n is true for all positive integers n.

Matched Problem 3 Prove that $8^n - 1$ is divisible by 7 for all positive integers n.

In some cases, a conjecture may be true only for $n \geq m$, where m is a positive integer, rather than for all $n \geq 0$. For example, see Problems 49 and 50 in Exercise 10-2. The principle of mathematical induction can be extended to cover cases like this as follows:

Theorem 2 **Extended Principle of Mathematical Induction**

Let m be a positive integer, let P_n be a statement associated with each integer $n \geq m$, and suppose the following conditions are satisfied:

1. P_m is true.
2. For any integer $k \geq m$, if P_k is true, then P_{k+1} is also true.

Then the statement P_n is true for all integers $n \geq m$.

• Three Famous Problems

The problem of determining whether a certain statement about the positive integers is true may be extremely difficult. Proofs may require remarkable insight and ingenuity and the development of techniques far more advanced than mathematical induction. Consider, for example, the famous problems of proving the following statements:

1. **Lagrange's Four Square Theorem, 1772:** Each positive integer can be expressed as the sum of four or fewer squares of positive integers.
2. **Fermat's Last Theorem, 1637:** For $n > 2$, $x^n + y^n = z^n$ does not have solutions in the natural numbers.
3. **Goldbach's Conjecture, 1742:** Every positive even integer greater than 2 is the sum of two prime numbers.

The first statement was considered by the early Greeks and finally proved in 1772 by Lagrange. Fermat's last theorem, defying the best mathematical minds for over 350 years, finally succumbed to a 200-page proof by Prof. Andrew Wiles of Princeton University in 1993. To this date no one has been able to prove or disprove Goldbach's conjecture.

EXPLORE-DISCUSS 2

(A) Explain the difference between a theorem and a conjecture.

(B) Why is "Fermat's last theorem" a misnomer? Suggest more accurate names for the result.

Answers to Matched Problems

1. Sketch of proof. State the conjecture: P_n: $1 + 2 + 3 + \cdots + n = \frac{n(n+1)}{2}$

Part 1. $1 = \frac{1(1+1)}{2}$. P_1 is true.

Part 2. Show that if P_k is true, then P_{k+1} is true.

$$1 + 2 + 3 + \cdots + k = \frac{k(k+1)}{2} \qquad P_k$$

$$1 + 2 + 3 + \ldots + k + (k+1) = \frac{k(k+1)}{2} + (k+1)$$

$$= \frac{(k+1)(k+2)}{2} \qquad P_{k+1}$$

Conclusion: P_n is true.

2. Sketch of proof. State the conjecture: P_n: $\left(\frac{x}{y}\right)^n = \frac{x^n}{y^n}$

Part 1. $\left(\frac{x}{y}\right)^1 = \frac{x}{y} = \frac{x^1}{y^1}$. P_1 is true.

Part 2. Show that if P_k is true, then P_{k+1} is true.

$$\left(\frac{x}{y}\right)^{k+1} = \left(\frac{x}{y}\right)^k\left(\frac{x}{y}\right) = \frac{x^k}{y^k}\left(\frac{x}{y}\right) = \frac{x^k x}{y^k y} = \frac{x^{k+1}}{y^{k+1}}$$

Conclusion: P_n is true.

3. Sketch of proof. State the conjecture: P_n: $8^n - 1 = 7r$ for some integer r

Part 1. $8^1 - 1 = 7 = 7 \cdot 1$. P_1 is true.

Part 2. Show that if P_k is true, then P_{k+1} is true.

$$8^k - 1 = 7r \qquad P_k$$

$$8(8^k - 1) = 8(7r)$$

$$8^{k+1} - 1 = 56r + 7 = 7(8r + 1) = 7s \qquad P_{k+1}$$

Conclusion: P_n is true.

EXERCISE 10-2

A

In Problems 1–4, find the first positive integer n that causes the statement to fail.

1. $3^n + 4^n \geq 5^n$
2. $n^2 - 3n < 100$
3. $17^n - 1$ is divisible by 2^n
4. $n^2 = 5n - 6$

Verify each statement P_n in Problems 5–10 for n = 1, 2, and 3.

5. P_n: $2 + 6 + 10 + \ldots + (4n - 2) = 2n^2$
6. P_n: $4 + 8 + 12 + \ldots + 4n = 2n(n + 1)$
7. P_n: $a^5a^n = a^{5+n}$
8. P_n: $(a^5)^n = a^{5n}$
9. P_n: $9^n - 1$ is divisible by 4
10. P_n: $4^n - 1$ is divisible by 3

Write P_k and P_{k+1} for P_n as indicated in Problems 11–16.

11. P_n in Problem 5
12. P_n in Problem 6
13. P_n in Problem 7
14. P_n in Problem 8
15. P_n in Problem 9
16. P_n in Problem 10

In Problems 17–22, use mathematical induction to prove that each P_n holds for all positive integers n.

17. P_n in Problem 5
18. P_n in Problem 6
19. P_n in Problem 7
20. P_n in Problem 8
21. P_n in Problem 9
22. P_n in Problem 10

B

In Problems 23–26, prove the statement is false by finding a counterexample.

23. If $n > 2$, then any polynomial of degree n has at least one real zero.

24. Any positive integer $n > 7$ can be written as the sum of three or fewer squares of positive integers.

25. If n is a positive integer, then there is at least one prime number p such that $n < p < n + 6$.

26. If a, b, c, d are positive integers such that $a^2 + b^2 = c^2 + d^2$, then $a = c$ or $a = d$.

In Problems 27–42, use mathematical induction to prove each proposition for all positive integers n, unless restricted otherwise.

27. $2 + 2^2 + 2^3 + \cdots + 2^n = 2^{n+1} - 2$

28. $\frac{1}{2} + \frac{1}{4} + \frac{1}{8} + \cdots + \frac{1}{2^n} = 1 - \left(\frac{1}{2}\right)^n$

29. $1^2 + 3^2 + 5^2 + \cdots + (2n - 1)^2 = \frac{1}{3}(4n^3 - n)$

30. $1 + 8 + 16 + \cdots + 8(n - 1) = (2n - 1)^2; n > 1$

31. $1^2 + 2^2 + 3^2 + \cdots + n^2 = \frac{n(n + 1)(2n + 1)}{6}$

32. $1 \cdot 2 + 2 \cdot 3 + 3 \cdot 4 + \cdots + n(n + 1) = \frac{n(n + 1)(n + 2)}{3}$

33. $\frac{a^n}{a^3} = a^{n-3}; n > 3$

34. $\frac{a^5}{a^n} = \frac{1}{a^{n-5}}; n > 5$

35. $a^m a^n = a^{m+n}; m, n \in N$
[*Hint:* Choose m as an arbitrary element of N, and then use induction on n.]

36. $(a^n)^m = a^{mn}; m, n \in N$

37. $x^n - 1$ is divisible by $x - 1; x \neq 1$
[*Hint:* Divisible means that $x^n - 1 = (x - 1)Q(x)$ for some polynomial $Q(x)$.]

38. $x^n - y^n$ is divisible by $x - y; x \neq y$

39. $x^{2n} - 1$ is divisible by $x - 1; x \neq 1$

40. $x^{2n} - 1$ is divisible by $x + 1; x \neq -1$

41. $1^3 + 2^3 + 3^3 + \cdots + n^3 = (1 + 2 + 3 + \cdots + n)^2$
[*Hint:* See Matched Problem 1 following Example 1.]

42. $\frac{1}{1 \cdot 2 \cdot 3} + \frac{1}{2 \cdot 3 \cdot 4} + \frac{1}{3 \cdot 4 \cdot 5} + \cdots + \frac{1}{n(n + 1)(n + 2)} = \frac{n(n + 3)}{4(n + 1)(n + 2)}$

C

In Problems 43–46, suggest a formula for each expression, and prove your hypothesis using mathematical induction, $n \in N$.

43. $2 + 4 + 6 + \cdots + 2n$

44. $\frac{1}{1 \cdot 2} + \frac{1}{2 \cdot 3} + \frac{1}{3 \cdot 4} + \cdots + \frac{1}{n(n + 1)}$

45. The number of lines determined by n points in a plane, no three of which are collinear

46. The number of diagonals in a polygon with n sides

In Problems 47–50, prove the statement is true for all integers n as specified.

47. $a > 1 \Rightarrow a^n > 1; n \in N$

48. $0 < a < 1 \Rightarrow 0 < a^n < 1; n \in N$

49. $n^2 > 2n; n \geq 3$

50. $2^n > n^2; n \geq 5$

51. Prove or disprove the generalization of the following two facts:

$$3^2 + 4^2 = 5^2$$

$$3^3 + 4^3 + 5^3 = 6^3$$

52. Prove or disprove: $n^2 + 21n + 1$ is a prime number for all natural numbers n.

If $\{a_n\}$ and $\{b_n\}$ are two sequences, we write $\{a_n\} = \{b_n\}$ if and only if $a_n = b_n$, $n \in N$. In Problems 53–56, use mathematical induction to show that $\{a_n\} = \{b_n\}$.

53. $a_1 = 1, a_n = a_{n-1} + 2; b_n = 2n - 1$

54. $a_1 = 2, a_n = a_{n-1} + 2; b_n = 2n$

55. $a_1 = 2, a_n = 2^2 a_{n-1}; b_n = 2^{2n-1}$

56. $a_1 = 2, a_n = 3a_{n-1}; b_n = 2 \cdot 3^{n-1}$

SECTION 10-3 Arithmetic and Geometric Sequences

- Arithmetic and Geometric Sequences
- nth-Term Formulas
- Sum Formulas for Finite Arithmetic Series
- Sum Formulas for Finite Geometric Series
- Sum Formula for Infinite Geometric Series

For most sequences it is difficult to sum an arbitrary number of terms of the sequence without adding term by term. But particular types of sequences, *arithmetic sequences* and *geometric sequences*, have certain properties that lead to convenient and useful formulas for the sums of the corresponding *arithmetic series* and *geometric series.*

• Arithmetic and Geometric Sequences

The sequence 5, 7, 9, 11, 13, . . . , $5 + 2(n-1)$, . . . , where each term after the first is obtained by adding 2 to the preceding term, is an example of an arithmetic sequence. The sequence 5, 10, 20, 40, 80, . . . , $5(2)^{n-1}$, . . . , where each term after the first is obtained by multiplying the preceding term by 2, is an example of a geometric sequence.

DEFINITION 1

Arithmetic Sequence

A sequence

$$a_1, a_2, a_3, \ldots, a_n, \ldots$$

is called an **arithmetic sequence,** or **arithmetic progression,** if there exists a constant d, called the **common difference,** such that

$$a_n - a_{n-1} = d$$

That is,

$$a_n = a_{n-1} + d \qquad \text{for every } n > 1$$

DEFINITION 2

Geometric Sequence

A sequence

$$a_1, a_2, a_3, \ldots, a_n, \ldots$$

is called a **geometric sequence,** or **geometric progression,** if there exists a nonzero constant r, called the **common ratio,** such that

$$\frac{a_n}{a_{n-1}} = r$$

That is,

$$a_n = ra_{n-1} \qquad \text{for every } n > 1$$

EXPLORE-DISCUSS 1 (A) Graph the arithmetic sequence 5, 7, 9, ...
Describe the graphs of all arithmetic sequences with common difference 2.

(B) Graph the geometric sequence 5, 10, 20, ...
Describe the graphs of all geometric sequences with common ratio 2.

EXAMPLE 1 Recognizing Arithmetic and Geometric Sequences

Which of the following can be the first four terms of an arithmetic sequence? Of a geometric sequence?

(A) 1, 2, 3, 5, . . . (B) $-1, 3, -9, 27, \ldots$
(C) 3, 3, 3, 3, . . . (D) 10, 8.5, 7, 5.5, . . .

Solution (A) Since $2 - 1 \neq 5 - 3$, there is no common difference, so the sequence is not an arithmetic sequence. Since $2/1 \neq 3/2$, there is no common ratio, so the sequence is not geometric either.

(B) The sequence is geometric with common ratio -3, but it is not arithmetic.

(C) The sequence is arithmetic with common difference 0 and it is also geometric with common ratio 1.

(D) The sequence is arithmetic with common difference -1.5, but it is not geometric.

Matched Problem 1 Which of the following can be the first four terms of an arithmetic sequence? Of a geometric sequence?

(A) 8, 2, 0.5, 0.125, . . .
(B) $-7, -2, 3, 8, \ldots$
(C) 1, 5, 25, 100, . . .

• *n*th-Term Formulas

If $\{a_n\}$ is an arithmetic sequence with common difference d, then

$$a_2 = a_1 + d$$

$$a_3 = a_2 + d = a_1 + 2d$$

$$a_4 = a_3 + d = a_1 + 3d$$

This suggests Theorem 1, which can be proved by mathematical induction (see Problem 63 in Exercise 10-3).

Theorem 1 **The *n*th Term of an Arithmetic Sequence**

$$a_n = a_1 + (n - 1)d \quad \text{for every } n > 1$$

Similarly, if $\{a_n\}$ is a geometric sequence with common ratio r, then

$$a_2 = a_1 r$$
$$a_3 = a_2 r = a_1 r^2$$
$$a_4 = a_3 r = a_1 r^3$$

This suggests Theorem 2, which can also be proved by mathematical induction (see Problem 69 in Exercise 10-3).

Theorem 2 ***n*th Term of a Geometric Sequence**

$$a_n = a_1 r^{n-1} \qquad \text{for every } n > 1$$

EXAMPLE 2 Finding Terms in Arithmetic and Geometric Sequences

(A) If the first and tenth terms of an arithmetic sequence are 3 and 30, respectively, find the fiftieth term of the sequence.
(B) If the first and tenth terms of a geometric sequence are 1 and 4, find the seventeenth term to 3 decimal places.

Solution (A) First use Theorem 1 with $a_1 = 3$ and $a_{10} = 30$ to find d:

$$a_n = a_1 + (n - 1)d$$
$$a_{10} = a_1 + (10 - 1)d$$
$$30 = 3 + 9d$$
$$d = 3$$

Now find a_{50}:

$$a_{50} = a_1 + (50 - 1)3$$
$$= 3 + 49 \cdot 3$$
$$= 150$$

(B) First let $n = 10$, $a_1 = 1$, $a_{10} = 4$ and use Theorem 2 to find r.

$$a_n = a_1 r^{n-1}$$
$$4 = 1r^{10-1}$$
$$r = 4^{1/9}$$

Now use Theorem 2 again, this time with $n = 17$.

$$a_{17} = a_1 r^{16} = 1\,(4^{1/9})^{16} = 4^{16/9} = 11.758$$

Matched Problem 2 (A) If the first and fifteenth terms of an arithmetic sequence are -5 and 23, respectively, find the seventy-third term of the sequence.

(B) Find the eighth term of the geometric sequence $\frac{1}{64}, -\frac{1}{32}, \frac{1}{16}, \ldots$.

• Sum Formulas for Finite Arithmetic Series

If $a_1, a_2, a_3, \ldots, a_n$ is a finite arithmetic sequence, then the corresponding series $a_1 + a_2 + a_3 + \cdots + a_n$ is called an *arithmetic series*. We will derive two simple and very useful formulas for the sum of an arithmetic series. Let d be the common difference of the arithmetic sequence $a_1, a_2, a_3, \ldots, a_n$ and let S_n denote the sum of the series $a_1 + a_2 + a_3 + \cdots + a_n$.

Then

$$S_n = a_1 + (a_1 + d) + \cdots + [a_1 + (n - 2)d] + [a_1 + (n - 1)d]$$

Reversing the order of the sum, we obtain

$$S_n = [a_1 + (n - 1)d] + [a_1 + (n - 2)d] + \cdots + (a_1 + d) + a_1$$

Adding the left sides of these two equations and corresponding elements of the right sides, we see that

$$\begin{aligned} 2S_n &= [2a_1 + (n - 1)d] + [2a_1 + (n - 1)d] + \cdots + [2a_1 + (n - 1)d] \\ &= n[2a_1 + (n - 1)d] \end{aligned}$$

This can be restated as in Theorem 3:

Theorem 3 **Sum of an Arithmetic Series—First Form**

$$S_n = \frac{n}{2}[2a_1 + (n - 1)d]$$

By replacing $a_1 + (n - 1)d$ with a_n, we obtain a second useful formula for the sum:

Theorem 4 **Sum of an Arithmetic Series—Second Form**

$$S_n = \frac{n}{2}(a_1 + a_n)$$

The proof of the first sum formula by mathematical induction is left as an exercise (see Problem 64 in Exercise 10-3).

EXAMPLE 3 **Finding the Sum of an Arithmetic Series**

Find the sum of the first 26 terms of an arithmetic series if the first term is -7 and $d = 3$.

Solution Let $n = 26$, $a_1 = -7$, $d = 3$, and use Theorem 3.

$$S_n = \frac{n}{2}[2a_1 + (n - 1)d]$$

$$S_{26} = \tfrac{26}{2}[2(-7) + (26 - 1)3]$$

$$= 793$$

Matched Problem 3 Find the sum of the first 52 terms of an arithmetic series if the first term is 23 and $d = -2$.

EXAMPLE 4 **Finding the Sum of an Arithmetic Series**

Find the sum of all the odd numbers between 51 and 99, inclusive.

Solution First, use $a_1 = 51$, $a_n = 99$, and Theorem 1 to find n:

$$a_n = a_1 + (n - 1)d$$

$$99 = 51 + (n - 1)2$$

$$n = 25$$

Now use Theorem 4 to find S_{25}:

$$S_n = \frac{n}{2}(a_1 + a_n)$$

$$S_{25} = \tfrac{25}{2}(51 + 99)$$

$$= 1{,}875$$

Matched Problem 4 Find the sum of all the even numbers between -22 and 52, inclusive.

EXAMPLE 5 **Prize Money**

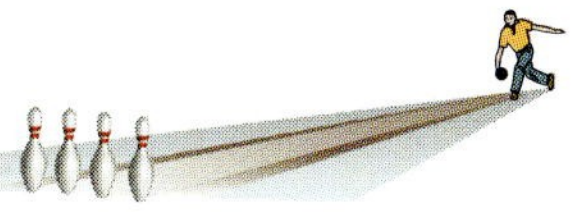

A 16-team bowling league has \$8,000 to be awarded as prize money. If the last-place team is awarded \$275 in prize money and the award increases by the same amount for each successive finishing place, how much will the first-place team receive?

Solution If a_1 is the award for the first-place team, a_2 is the award for the second-place team, and so on, then the prize money awards form an arithmetic sequence with $n = 16$, $a_{16} = 275$, and $S_{16} = 8{,}000$. Use Theorem 4 to find a_1.

$$S_n = \frac{n}{2}(a_1 + a_n)$$

$$8{,}000 = \tfrac{16}{2}(a_1 + 275)$$

$$a_1 = 725$$

Thus, the first-place team receives $725.

Matched Problem 5 Refer to Example 5. How much prize money is awarded to the second-place team?

• Sum Formulas for Finite Geometric Series

If $a_1, a_2, a_3, \ldots, a_n$ is a finite geometric sequence, then the corresponding series $a_1 + a_2 + a_3 + \cdots + a_n$ is called a *geometric series.* As with arithmetic series, we can derive two simple and very useful formulas for the sum of a geometric series. Let r be the common ratio of the geometric sequence $a_1, a_2, a_3, \ldots, a_n$ and let S_n denote the sum of the series $a_1 + a_2 + a_3 + \cdots + a_n$. Then

$$S_n = a_1 + a_1r + a_1r^2 + a_1r^3 + \cdots + a_1r^{n-2} + a_1r^{n-1}$$

Multiply both sides of this equation by r to obtain

$$rS_n = a_1r + a_1r^2 + a_1r^3 + \cdots + a_1r^{n-1} + a_1r^n$$

Now subtract the left side of the second equation from the left side of the first, and the right side of the second equation from the right side of the first to obtain

$$S_n - rS_n = a_1 - a_1r^n$$

$$S_n(1 - r) = a_1 - a_1r^n$$

Thus, solving for S_n, we obtain the following formula for the sum of a geometric series:

Theorem 5 **Sum of a Geometric Series—First Form**

$$S_n = \frac{a_1 - a_1r^n}{1 - r} \qquad r \neq 1$$

Since $a_n = a_1r^{n-1}$, or $ra_n = a_1r^n$, the sum formula also can be written in the following form:

Theorem 6 **Sum of a Geometric Series—Second Form**

$$S_n = \frac{a_1 - ra_n}{1 - r} \qquad r \neq 1$$

The proof of the first sum formula by mathematical induction is left as an exercise (see Problem 70, Exercise 10-3).

If $r = 1$, then

$$S_n = a_1 + a_1(1) + a_1(1^2) + \cdots + a_1(1^{n-1}) = na_1$$

EXAMPLE 6 Finding the Sum of a Geometric Series

Find the sum of the first 20 terms of a geometric series if the first term is 1 and $r = 2$.

Solution Let $n = 20$, $a_1 = 1$, $r = 2$, and use Theorem 5.

$$S_n = \frac{a_1 - a_1r^n}{1 - r}$$

$$= \frac{1 - 1 \cdot 2^{20}}{1 - 2} = 1{,}048{,}575 \quad \text{Calculation using a calculator}$$

Matched Problem 6 Find the sum, to two decimal places, of the first 14 terms of a geometric series if the first term is $\frac{1}{64}$ and $r = -2$.

• Sum Formula for Infinite Geometric Series

Consider a geometric series with $a_1 = 5$ and $r = \frac{1}{2}$. What happens to the sum S_n as n increases? To answer this question, we first write the sum formula in the more convenient form

$$S_n = \frac{a_1 - a_1r^n}{1 - r} = \frac{a_1}{1 - r} - \frac{a_1r^n}{1 - r} \qquad (1)$$

For $a_1 = 5$ and $r = \frac{1}{2}$,

$$S_n = 10 - 10\left(\frac{1}{2}\right)^n$$

Thus,

$$S_2 = 10 - 10\left(\frac{1}{4}\right)$$

$$S_4 = 10 - 10\left(\frac{1}{16}\right)$$

$$S_{10} = 10 - 10\left(\frac{1}{1{,}024}\right)$$

$$S_{20} = 10 - 10\left(\frac{1}{1{,}048{,}576}\right)$$

It appears that $(\frac{1}{2})^n$ becomes smaller and smaller as n increases and that the sum gets closer and closer to 10.

In general, it is possible to show that, if $|r| < 1$, then r^n will get closer and closer to 0 as n increases. Symbolically, $r^n \to 0$ as $n \to \infty$. Thus, the term

$$\frac{a_1 r^n}{1 - r}$$

in equation (1) will tend to 0 as n increases, and S_n will tend to

$$\frac{a_1}{1 - r}$$

In other words, if $|r| < 1$, then S_n can be made as close to

$$\frac{a_1}{1 - r}$$

as we wish by taking n sufficiently large. Thus, we define the **sum of an infinite geometric series** by the following formula:

DEFINITION 3 **Sum of an Infinite Geometric Series**

$$S_\infty = \frac{a_1}{1 - r} \qquad |r| < 1$$

If $|r| \geq 1$, an infinite geometric series has no sum.

EXAMPLE 7 Expressing a Repeating Decimal as a Fraction

Represent the repeating decimal $0.454\,545 \cdots = 0.\overline{45}$ as the quotient of two integers. Recall that a repeating decimal names a rational number and that any rational number can be represented as the quotient of two integers.

Solution

$$0.\overline{45} = 0.45 + 0.0045 + 0.000\,045 + \cdots$$

The right side of the equation is an infinite geometric series with $a_1 = 0.45$ and $r = 0.01$. Thus,

$$S_\infty = \frac{a_1}{1 - r} = \frac{0.45}{1 - 0.01} = \frac{0.45}{0.99} = \frac{5}{11}$$

Hence, $0.\overline{45}$ and $\frac{5}{11}$ name the same rational number. Check the result by dividing 5 by 11.

Matched Problem 7 Repeat Example 7 for $0.818\ 181 \cdots = 0.\overline{81}$.

EXAMPLE 8 **Economy Stimulation**

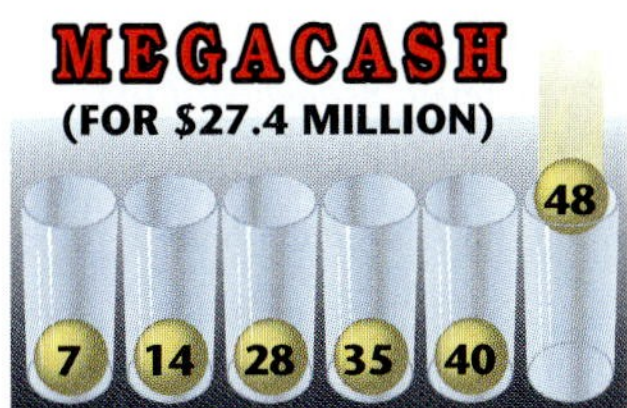

A state government uses proceeds from a lottery to provide a tax rebate for property owners. Suppose an individual receives a \$500 rebate and spends 80% of this, and each of the recipients of the money spent by this individual also spends 80% of what he or she receives, and this process continues without end. According to the **multiplier doctrine** in economics, the effect of the original \$500 tax rebate on the economy is multiplied many times. What is the total amount spent if the process continues as indicated?

Solution The individual receives \$500 and spends 0.8(500) = \$400. The recipients of this \$400 spend 0.8(400) = \$320, the recipients of this \$320 spend 0.8(320) = \$256, and so on. Thus, the total spending generated by the \$500 rebate is

$$400 + 320 + 256 + \cdots = 400 + 0.8(400) + (0.8)^2(400) + \cdots$$

which we recognize as an infinite geometric series with $a_1 = 400$ and $r = 0.8$. Thus, the total amount spent is

$$S_\infty = \frac{a_1}{1 - r} = \frac{400}{1 - 0.8} = \frac{400}{0.2} = \$2{,}000$$

Matched Problem 8 Repeat Example 8 if the tax rebate is \$1,000 and the percentage spent by all recipients is 90%.

EXPLORE-DISCUSS 2

(A) Find an infinite geometric series with $a_1 = 10$ whose sum is 1,000.

(B) Find an infinite geometric series with $a_1 = 10$ whose sum is 6.

(C) Suppose that an infinite geometric series with $a_1 = 10$ has a sum. Explain why that sum must be greater than 5.

Answers to Matched Problems

1. (A) The sequence is geometric with $r = \frac{1}{4}$, but not arithmetic.
(B) The sequence is arithmetic with $d = 5$, but not geometric.
(C) The sequence is neither arithmetic nor geometric.

2. (A) 139 (B) −2 **3.** −1,456 **4.** 570 **5.** \$695

6. −85.33 **7.** $\frac{9}{11}$ **8.** \$9,000

EXERCISE 10-3

A

In Problems 1–2, determine whether the following can be the first three terms of an arithmetic or geometric sequence, and, if so, find the common difference or common ratio and the next two terms of the sequence.

1. (A) $-11, -16, -21, \ldots$ (B) $2, -4, 8, \ldots$
(C) $1, 4, 9, \ldots$ (D) $\frac{1}{2}, \frac{1}{6}, \frac{1}{18}, \ldots$

2. (A) $5, 20, 100, \ldots$ (B) $-5, -5, -5, \ldots$
(C) $7, 6.5, 6, \ldots$ (D) $512, 256, 128, \ldots$

Let $a_1, a_2, a_3, \ldots, a_n, \ldots$ be an arithmetic sequence. In Problems 3–10, find the indicated quantities.

3. $a_1 = 6, d = 5; a_2 = ?, a_3 = ?, a_4 = ?$

4. $a_1 = -11, d = 4; a_2 = ?, a_3 = ?, a_4 = ?$

5. $a_1 = -13, d = -7; a_{21} = ?, S_{21} = ?$

6. $a_1 = 20, d = -6; a_{12} = ?, S_{12} = ?$

7. $a_1 = -1, a_2 = 5; S_{25} = ?$

8. $a_1 = 2, a_2 = 9; a_{30} = ?$

9. $a_1 = 40, a_2 = 95; S_{15} = ?$

10. $a_1 = -8, a_2 = 2; a_{29} = ?$

Let $a_1, a_2, a_3, \ldots, a_n, \ldots$ be a geometric sequence. In Problems 11–16, find each of the indicated quantities.

11. $a_1 = -6, r = -\frac{1}{2}; a_2 = ?, a_3 = ?, a_4 = ?$

12. $a_1 = 12, r = \frac{2}{3}; a_2 = ?, a_3 = ?, a_4 = ?$

13. $a_1 = 81, r = \frac{1}{3}; a_{10} = ?$

14. $a_1 = 64, r = \frac{1}{2}; a_{13} = ?$

15. $a_1 = 3, a_7 = 2{,}187, r = 3; S_7 = ?$

16. $a_1 = 1, a_7 = 729, r = -3; S_7 = ?$

B

Let $a_1, a_2, a_3, \ldots, a_n, \ldots$ be an arithmetic sequence. In Problems 17–24, find the indicated quantities.

17. $a_1 = 5, a_5 = 17; S_{50} = ?$

18. $a_1 = -7, a_8 = 7; S_{100} = ?$

19. $a_1 = 25, a_{15} = 32; a_{64} = ?$

20. $a_1 = -9, a_{12} = -31; a_{45} = ?$

21. $a_4 = 27, a_9 = 62; a_1 = ?$

22. $a_6 = 26, a_{10} = 50; S_{10} = ?$

23. $a_1 = 4, S_{10} = 35; a_{10} = ?, d = ?$

24. $a_1 = -3, S_{12} = 60; a_{12} = ?, d = ?$

Let $a_1, a_2, a_3, \ldots, a_n, \ldots$ be a geometric sequence. Find each of the indicated quantities in Problems 25–30.

25. $a_1 = 1, a_5 = 100; r = ?$

26. $a_1 = 324, a_9 = 4; r = ?$

27. $a_1 = 64{,}000, r = -\frac{1}{2}; S_{10} = ?$

28. $a_1 = 20, r = 2; S_{12} = ?$

29. $a_1 = 40, a_4 = 135; a_2 = ?, a_3 = ?$

30. $a_1 = 625, a_6 = -\frac{32}{5}; a_2 = ?, a_3 = ?, a_4 = ?, a_5 = ?$

31. $S_{51} = \sum_{k=1}^{51} (3k + 3) = ?$ **32.** $S_{40} = \sum_{k=1}^{40} (2k - 3) = ?$

33. $S_7 = \sum_{k=1}^{7} (-3)^{k-1} = ?$ **34.** $S_7 = \sum_{k=1}^{7} 3^k = ?$

35. Find $g(1) + g(2) + g(3) + \cdots + g(51)$ if $g(t) = 5 - t$.

36. Find $f(1) + f(2) + f(3) + \cdots + f(20)$ if $f(x) = 2x - 5$.

37. Find $g(1) + g(2) + \cdots + g(10)$ if $g(x) = (\frac{1}{2})^x$.

38. Find $f(1) + f(2) + \cdots + f(10)$ if $f(x) = 2^x$.

39. Find the sum of all the even integers between 21 and 135.

40. Find the sum of all the odd integers between 100 and 500.

41. Show that the sum of the first n odd natural numbers is n^2, using appropriate formulas from this section.

42. Show that the sum of the first n even natural numbers is $n + n^2$, using appropriate formulas from this section.

43. Find a positive number x so that $-2 + x - 6$ is a three-term geometric series.

44. Find a positive number x so that $6 + x + 8$ is a three-term geometric series.

45. Can a sequence be both arithmetic and geometric? Explain.

46. Given an arbitrary real number x, is there an infinite geometric series with $a_1 = 2$ whose sum is x? Explain.

In Problems 47–50, find the least positive integer n such that $a_n < b_n$ by graphing the sequences $\{a_n\}$ and $\{b_n\}$ with a graphing utility. Check your answer by using a graphing utility to display both sequences in table form.

47. $a_n = 5 + 8n, \; b_n = 1.1^n$

48. $a_n = 96 + 47n, \; b_n = 8(1.5)^n$

49. $a_n = 1{,}000\,(0.99)^n,\ b_n = 2n + 1$

50. $a_n = 500 - n,\ b_n = 1.05^n$

In Problems 51–56, find the sum of each infinite geometric series that has a sum.

51. $3 + 1 + \frac{1}{3} + \cdots$

52. $16 + 4 + 1 + \cdots$

53. $2 + 4 + 8 + \cdots$

54. $4 + 6 + 9 + \cdots$

55. $2 - \frac{1}{2} + \frac{1}{8} - \cdots$

56. $21 - 3 + \frac{3}{7} - \cdots$

In Problems 57–62, represent each repeating decimal as the quotient of two integers.

57. $0.\overline{7} = 0.7777\cdots$

58. $0.\overline{5} = 0.5555\cdots$

59. $0.\overline{54} = 0.545\,454\cdots$

60. $0.\overline{27} = 0.272\,727\cdots$

61. $3.\overline{216} = 3.216\,216\,216\cdots$

62. $5.6\overline{3} = 5.636\,363\cdots$

C

63. Prove, using mathematical induction, that if $\{a_n\}$ is an arithmetic sequence, then

$$a_n = a_1 + (n - 1)d \qquad \text{for every } n > 1$$

64. Prove, using mathematical induction, that if $\{a_n\}$ is an arithmetic sequence, then

$$S_n = \frac{n}{2}[2a_1 + (n - 1)d]$$

65. If in a given sequence, $a_1 = -2$ and $a_n = -3a_{n-1}$, $n > 1$, find a_n in terms of n.

66. For the sequence in Problem 65, find $S_n = \sum_{k=1}^{n} a_k$ in terms of n.

67. Show that $(x^2 + xy + y^2)$, $(z^2 + xz + x^2)$, and $(y^2 + yz + z^2)$ are consecutive terms of an arithmetic progression if x, y, and z form an arithmetic progression. (From U.S.S.R. Mathematical Olympiads, 1955–1956, Grade 9.)

68. Take 121 terms of each arithmetic progression 2, 7, 12, . . . and 2, 5, 8, ... How many numbers will there be in common? (From U.S.S.R. Mathematical Olympiads, 1955–1956, Grade 9.)

69. Prove, using mathematical induction, that if $\{a_n\}$ is a geometric sequence, then

$$a_n = a_1 r^{n-1} \qquad n \in N$$

70. Prove, using mathematical induction, that if $\{a_n\}$ is a geometric sequence, then

$$S_n = \frac{a_1 - a_1 r^n}{1 - r} \qquad n \in N,\ r \neq 1$$

71. Given the system of equations

$$ax + by = c$$
$$dx + ey = f$$

where a, b, c, d, e, f is any arithmetic progression with a nonzero constant difference, show that the system has a unique solution.

72. The sum of the first and fourth terms of an arithmetic sequence is 2, and the sum of their squares is 20. Find the sum of the first eight terms of the sequence.

APPLICATIONS 

73. Business. In investigating different job opportunities, you find that firm A will start you at \$25,000 per year and guarantee you a raise of \$1,200 each year while firm B will start you at \$28,000 per year but will guarantee you a raise of only \$800 each year. Over a period of 15 years, how much would you receive from each firm?

74. Business. In Problem 73, what would be your annual salary at each firm for the tenth year?

75. Economics. The government, through a subsidy program, distributes \$1,000,000. If we assume that each individual or agency spends 0.8 of what is received, and 0.8 of this is spent, and so on, how much total increase in spending results from this government action?

76. Economics. Due to reduced taxes, an individual has an extra \$600 in spendable income. If we assume that the individual spends 70% of this on consumer goods, that the producers of these goods in turn spend 70% of what they receive on consumer goods, and that this process continues indefinitely, what is the total amount spent on consumer goods?

★ **77. Business.** If \$$P$ is invested at $100r\%$ compounded annually, the amount A present after n years forms a geometric progression with a common ratio $1 + r$. Write a formula for the amount present after n years. How long will it take a sum of money P to double if invested at 6% interest compounded annually?

★ **78. Population Growth.** If a population of A_0 people grows at the constant rate of $100r\%$ per year, the population after t years forms a geometric progression with a common ratio $1 + r$. Write a formula for the total population after t years. If the world's population is increasing at the rate of 2% per year, how long will it take to double?

79. Finance. Eleven years ago an investment earned \$7,000 for the year. Last year the investment earned \$14,000. If the earnings from the investment have increased the same amount each year, what is the yearly increase and how much income has accrued from the investment over the past 11 years?

80. Air Temperature. As dry air moves upward, it expands. In so doing, it cools at the rate of about 5°F for each 1,000-foot rise. This is known as the **adiabatic process.**

(A) Temperatures at altitudes that are multiples of 1,000 feet form what kind of a sequence?

(B) If the ground temperature is 80°F, write a formula for the temperature T_n in terms of n, if n is in thousands of feet.

81. Engineering. A rotating flywheel coming to rest rotates 300 revolutions the first minute (see figure). If in each subsequent minute it rotates two-thirds as many times as in the preceding minute, how many revolutions will the wheel make before coming to rest?

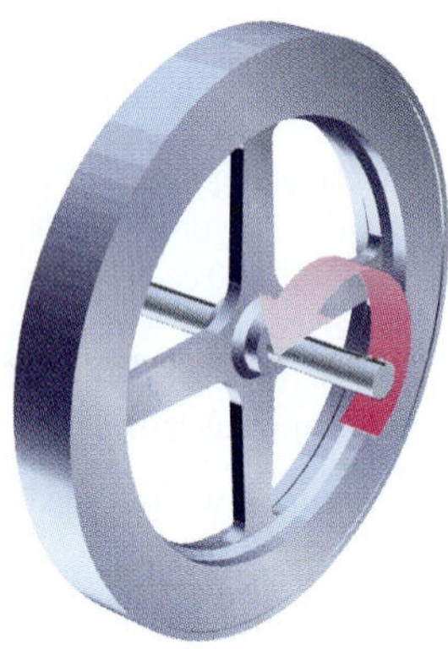

82. Physics. The first swing of a bob on a pendulum is 10 inches. If on each subsequent swing it travels 0.9 as far as on the preceding swing, how far will the bob travel before coming to rest?

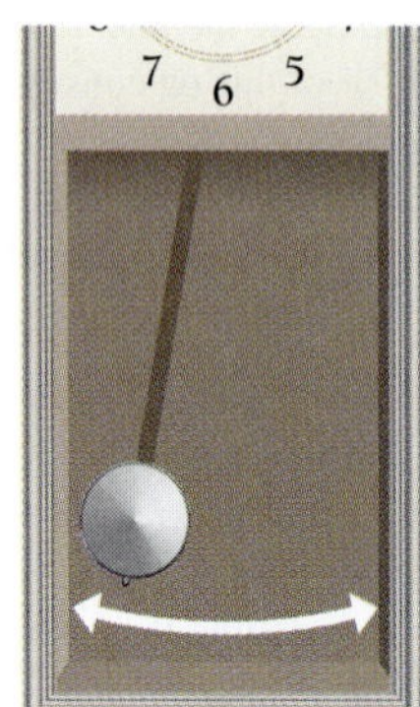

83. Food Chain. A plant is eaten by an insect, an insect by a trout, a trout by a salmon, a salmon by a bear, and the bear is eaten by you. If only 20% of the energy is transformed from one stage to the next, how many calories must be supplied by plant food to provide you with 2,000 calories from the bear meat?

★ **84. Genealogy.** If there are 30 years in a generation, how many direct ancestors did each of us have 600 years ago? By *direct* ancestors we mean parents, grandparents, great-grandparents, and so on.

★ **85. Physics.** An object falling from rest in a vacuum near the surface of the Earth falls 16 feet during the first second, 48 feet during the second second, 80 feet during the third second, and so on.

(A) How far will the object fall during the eleventh second?

(B) How far will the object fall in 11 seconds?

(C) How far will the object fall in t seconds?

★ **86. Physics.** In Problem 85, how far will the object fall during:

(A) The twentieth second? (B) The tth second?

★ **87. Bacteria Growth.** A single cholera bacterium divides every $\frac{1}{2}$ hour to produce two complete cholera bacteria. If we start with a colony of A_0 bacteria, how many bacteria will we have in t hours, assuming adequate food supply?

★ **88. Cell Division.** One leukemic cell injected into a healthy mouse will divide into two cells in about $\frac{1}{2}$ day. At the end of the day these two cells will divide again, with the doubling process continuing each $\frac{1}{2}$ day until there are 1 billion cells, at which time the mouse dies. On which day after the experiment is started does this happen?

★★ **89. Astronomy.** Ever since the time of the Greek astronomer Hipparchus, second century B.C., the brightness of stars has been measured in terms of magnitude. The brightest stars, excluding the sun, are classed as magnitude 1, and the dimmest visible to the eye are classed as magnitude 6. In 1856, the English astronomer N. R. Pogson showed that first-magnitude stars are 100 times brighter than sixth-magnitude stars. If the ratio of brightness between consecutive magnitudes is constant, find this ratio. [*Hint:* If b_n is the brightness of an nth-magnitude star, find r for the geometric progression $b_1, b_2, b_3, \ldots$, given $b_1 = 100b_6$.]

★ **90. Music.** The notes on a piano, as measured in cycles per second, form a geometric progression.

(A) If A is 400 cycles per second and A′, 12 notes higher, is 800 cycles per second, find the constant ratio r.

(B) Find the cycles per second for C, three notes higher than A.

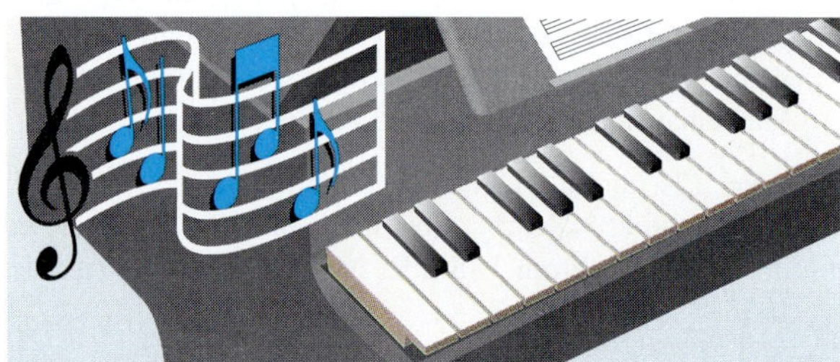

91. Puzzle. If you place 1¢ on the first square of a chessboard, 2¢ on the second square, 4¢ on the third, and so on, continuing to double the amount until all 64 squares are covered, how much money will be on the sixty-fourth square? How much money will there be on the whole board?

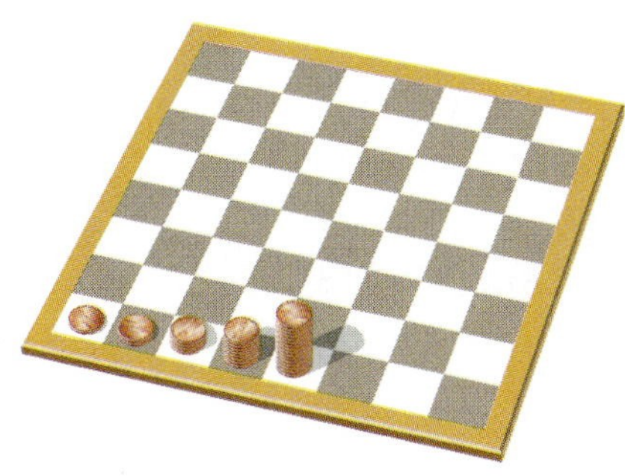

★ **92. Puzzle.** If a sheet of very thin paper 0.001 inch thick is torn in half, and each half is again torn in half, and this process is repeated for a total of 32 times, how high will the stack of paper be if the pieces are placed one on top of the other? Give the answer to the nearest mile.

★ **93. Atmospheric Pressure.** If atmospheric pressure decreases roughly by a factor of 10 for each 10-mile increase in altitude up to 60 miles, and if the pressure is 15 pounds per square inch at sea level, what will the pressure be 40 miles up?

94. Zeno's Paradox. Visualize a hypothetical 440-yard oval racetrack that has tapes stretched across the track at the halfway point and at each point that marks the halfway point of each remaining distance thereafter. A runner running around the track has to break the first tape before the second, the second before the third, and so on. From this point of view it appears that he will never finish the race. This famous paradox is attributed to the Greek philosopher Zeno, 495–435 B.C. If we assume the runner runs at 440 yards per minute, the times between tape breakings form an infinite geometric progression. What is the sum of this progression?

95. Geometry. If the midpoints of the sides of an equilateral triangle are joined by straight lines, the new figure will be an equilateral triangle with a perimeter equal to half the original. If we start with an equilateral triangle with perimeter 1 and form a sequence of "nested" equilateral triangles proceeding as described, what will be the total perimeter of all the triangles that can be formed in this way?

96. Photography. The shutter speeds and f-stops on a camera are given as follows:

$$\text{Shutter speeds:} \quad 1, \tfrac{1}{2}, \tfrac{1}{4}, \tfrac{1}{8}, \tfrac{1}{15}, \tfrac{1}{30}, \tfrac{1}{60}, \tfrac{1}{125}, \tfrac{1}{250}, \tfrac{1}{500}$$

$$\text{f-stops:} \quad 1.4, 2, 2.8, 4, 5.6, 8, 11, 16, 22$$

These are very close to being geometric progressions. Estimate their common ratios.

★★ **97. Geometry.** We know that the sum of the interior angles of a triangle is 180°. Show that the sums of the interior angles of polygons with 3, 4, 5, 6, . . . sides form an arithmetic sequence. Find the sum of the interior angles for a 21-sided polygon.

SECTION 10-4 Binomial Formula

- Factorial
- Binomial Formula

The binomial form

$$(a + b)^n$$

where n is a natural number, appears more frequently than you might expect. The coefficients in the expansion play an important role in probability studies. The *binomial formula,* which we derive below, enables us to expand $(a + b)^n$ directly for n any natural number. Since the formula involves *factorials,* we digress for a moment to introduce this important concept.

• Factorial

For n a natural number, ***n* factorial**—denoted by $n!$—is the product of the first n natural numbers. **Zero factorial** is defined to be 1.

DEFINITION 1 ***n* Factorial**

For n a natural number

$$n! = n(n-1) \cdot \cdots \cdot 2 \cdot 1$$
$$1! = 1$$
$$0! = 1$$

It is also useful to note that:

Theorem 1 **Recursion Formula for *n* Factorial**

$$n! = n \cdot (n-1)!$$

EXAMPLE 1 **Evaluating Factorials**

(A) $4! = 4 \cdot 3! = 4 \cdot 3 \cdot 2! = 4 \cdot 3 \cdot 2 \cdot 1! = 4 \cdot 3 \cdot 2 \cdot 1 = 24$
(B) $5! = 5 \cdot 4 \cdot 3 \cdot 2 \cdot 1 = 120$
(C) $\dfrac{7!}{6!} = \dfrac{7 \cdot \cancel{6!}}{\cancel{6!}} = 7$
(D) $\dfrac{8!}{5!} = \dfrac{8 \cdot 7 \cdot 6 \cdot \cancel{5!}}{\cancel{5!}} = 336$

Matched Problem 1 Find: (A) $6!$ (B) $\dfrac{6!}{5!}$ (C) $\dfrac{9!}{6!}$

CAUTION When reducing fractions involving factorials, don't confuse the single integer n with the symbol $n!$, which represents the product of n consecutive integers.

$$\frac{6!}{3!} \neq 2! \qquad \frac{6!}{3!} = \frac{6 \cdot 5 \cdot 4 \cdot 3!}{3!} = 6 \cdot 5 \cdot 4 = 120$$

Factorials are used in the definition of the important symbol $\dbinom{n}{r}$. This symbol is frequently used in probability studies. It is called the **combinatorial symbol** and is defined for nonnegative r and n, as follows:

DEFINITION 2 **Combinatorial Symbol**

For nonnegative integers r and n, $0 \le r \le n$.

$$\binom{n}{r} = \frac{n!}{r!(n-r)!}$$

$$= \frac{n(n-1)(n-2) \cdot \cdots \cdot (n-r+1)}{r(r-1) \cdot \cdots \cdot 2 \cdot 1}$$

The combinatorial symbol $\binom{n}{r}$ also can be denoted by $C_{n,r}$, ${}_nC_r$, or $C(n, r)$ and read as "n choose r." Many calculators use ${}_nC_r$ to denote the function that evaluates the combinatorial symbol.

EXAMPLE 2 **Evaluating the Combinatorial Symbol**

(A) $\binom{8}{3} = \frac{8!}{3!(8-3)!} = \frac{8!}{3!5!} = \frac{8 \cdot 7 \cdot 6 \cdot 5!}{3 \cdot 2 \cdot 1 \cdot 5!} = 56$

(B) $\binom{7}{0} = \frac{7!}{0!(7-0)!} = \frac{7!}{7!} = 1$ Remember, $0! = 1$.

Matched Problem 2 Find: (A) $\binom{9}{2}$ (B) $\binom{5}{5}$

EXPLORE-DISCUSS 1 (A) Compute the terms of the finite sequence $\binom{8}{0}, \binom{8}{1}, \binom{8}{2}, \ldots, \binom{8}{8}$ (the sequence is graphed in Fig. 1). Is the sequence arithmetic? Geometric? Which term is the largest? The smallest? Find the sum of the corresponding series.

FIGURE 1

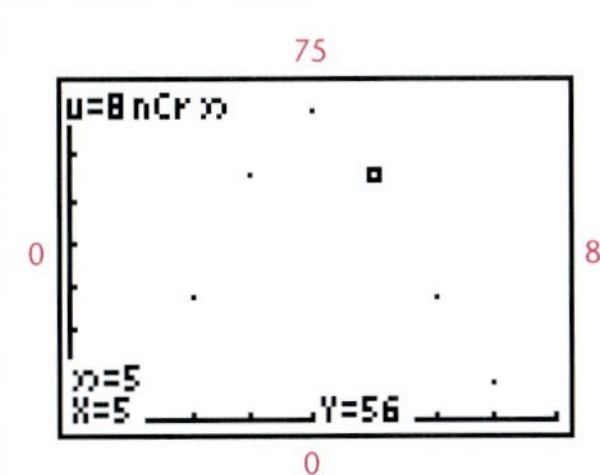

(B) Answer the same questions for the finite sequence $\binom{9}{0}, \binom{9}{1}, \binom{9}{2}, \ldots, \binom{9}{9}$.

• Binomial Formula

We are now ready to try to discover a formula for the expansion of $(a + b)^n$ using ordinary induction; that is, we will look at a few special cases and postulate a general formula from them. We will then try to prove that the formula holds for all natural numbers, using mathematical induction. To start, we calculate directly the first five natural number powers of $(a + b)^n$, arranging the terms in decreasing powers of a:

$$(a + b)^1 = a + b$$

$$(a + b)^2 = a^2 + 2ab + b^2$$

$$(a + b)^3 = a^3 + 3a^2b + 3ab^2 + b^3$$

$$(a + b)^4 = a^4 + 4a^3b + 6a^2b^2 + 4ab^3 + b^4$$

$$(a + b)^5 = a^5 + 5a^4b + 10a^3b^2 + 10a^2b^3 + 5ab^4 + b^5$$

Observations

1. The expansion of $(a + b)^n$ has $n + 1$ terms.
2. The power of a decreases by 1 for each term as we move from left to right.
3. The power of b increases by 1 for each term as we move from left to right.
4. In each term, the sum of the powers of a and b always adds up to n.
5. Starting with a given term, we can get the coefficient of the next term by multiplying the coefficient of the given term by the exponent of a and dividing by the number that represents the position of the term in the series of terms. For example, in the expansion of $(a + b)^4$, the coefficient of the third term is found from the second term by multiplying 4 and 3 and then dividing by 2. Thus, the coefficient of the third term is $(4 \cdot 3)/2 = 6$.

We now postulate the properties for the general case:

$$\begin{aligned}(a + b)^n &= a^n + \frac{n}{1}a^{n-1}b + \frac{n(n-1)}{1 \cdot 2}a^{n-2}b^2 \\ &\qquad + \frac{n(n-1)(n-2)}{1 \cdot 2 \cdot 3}a^{n-3}b^3 + \cdots + b^n \\ &= \frac{n!}{0!(n-0)!}a^n + \frac{n!}{1!(n-1)!}a^{n-1}b + \frac{n!}{2!(n-2)!}a^{n-2}b^2 \\ &\qquad + \frac{n!}{3!(n-3)!}a^{n-3}b^3 + \cdots + \frac{n!}{n!(n-n)!}b^n \\ &= \binom{n}{0}a_n + \binom{n}{1}a^{n-1}b + \binom{n}{2}a^{n-2}b^2 + \binom{n}{3}a^{n-3}b^3 + \cdots + \binom{n}{n}b^n\end{aligned}$$

Thus, we have arrived at the **binomial formula** using ordinary induction:

Theorem 2 **Binomial Formula**

For n a positive integer

$$(a+b)^n = \sum_{k=0}^{n} \binom{n}{k} a^{n-k} b^k$$

We now proceed to prove that the binomial formula holds for all natural numbers n using mathematical induction.

Proof State the conjecture.

$$P_n:\quad (a+b)^n = \sum_{j=0}^{n} \binom{n}{j} a^{n-j} b^j$$

Part 1 Show that P_1 is true.

$$\sum_{j=0}^{1} \binom{1}{j} a^{1-j} b^j = \binom{1}{0} a + \binom{1}{1} b = a + b = (a+b)^1$$

Thus, P_1 is true.

Part 2 Show that if P_k is true, then P_{k+1} is true.

$$P_k:\quad (a+b)^k = \sum_{j=0}^{k} \binom{k}{j} a^{k-j} b^j \qquad \text{Assume } P_k \text{ is true.}$$

$$P_{k+1}:\quad (a+b)^{k+1} = \sum_{j=0}^{k+1} \binom{k+1}{j} a^{k+1-j} b^j \qquad \text{Show } P_{k+1} \text{ is true.}$$

We begin by multiplying both sides of P_k by $(a + b)$:

$$(a+b)^k(a+b) = \left[\sum_{j=0}^{k} \binom{k}{j} a^{k-j} b^j\right](a+b)$$

The left side of this equation is the left side of P_{k+1}. Now we multiply out the right side of the equation and try to obtain the right side of P_{k+1}:

$$\begin{aligned}
(a+b)^{k+1} &= \left[\binom{k}{0}a^k + \binom{k}{1}a^{k-1}b + \binom{k}{2}a^{k-2}b^2 + \cdots + \binom{k}{k}b^k\right](a+b) \\
&= \left[\binom{k}{0}a^{k+1} + \binom{k}{1}a^k b + \binom{k}{2}a^{k-1}b^2 + \cdots + \binom{k}{k}ab^k\right] \\
&\qquad + \left[\binom{k}{0}a^k b + \binom{k}{1}a^{k-1}b^2 + \cdots + \binom{k}{k-1}ab^k + \binom{k}{k}b^{k+1}\right] \\
&= \binom{k}{0}a^{k+1} + \left[\binom{k}{0} + \binom{k}{1}\right]a^k b + \left[\binom{k}{1} + \binom{k}{2}\right]a^{k-1}b^2 + \cdots \\
&\qquad + \left[\binom{k}{k-1} + \binom{k}{k}\right]ab^k + \binom{k}{k}b^{k+1}
\end{aligned}$$

We now use the following facts (the proofs are left as exercises; see Problems 49–51, Exercise 10-4)

$$\binom{k}{r-1} + \binom{k}{r} = \binom{k+1}{r} \qquad \binom{k}{0} = \binom{k+1}{0} \qquad \binom{k}{k} = \binom{k+1}{k+1}$$

to rewrite the right side as

$$\binom{k+1}{0}a^{k+1} + \binom{k+1}{1}a^k b + \binom{k+1}{2}a^{k-1}b^2 + \cdots$$
$$+ \binom{k+1}{k}ab^k + \binom{k+1}{k+1}b^{k+1} = \sum_{j=0}^{k+1}\binom{k+1}{j}a^{k+1-j}b^j$$

Since the right side of the last equation is the right side of P_{k+1}, we have shown that P_{k+1} follows from P_k.

Conclusion P_n is true. That is, the binomial formula holds for all positive integers n.

EXAMPLE 3 **Using the Binomial Formula**

Use the binomial formula to expand $(x + y)^6$.

Solution

$$(x+y)^6 = \sum_{k=0}^{6}\binom{6}{k}x^{6-k}y^k$$
$$= \binom{6}{0}x^6 + \binom{6}{1}x^5y + \binom{6}{2}x^4y^2 + \binom{6}{3}x^3y^3 + \binom{6}{4}x^2y^4 + \binom{6}{5}xy^5 + \binom{6}{6}y^6$$
$$= x^6 + 6x^5y + 15x^4y^2 + 20x^3y^3 + 15x^2y^4 + 6xy^5 + y^6$$

Matched Problem 3 Use the binomial formula to expand $(x + 1)^5$.

EXAMPLE 4 **Using the Binomial Formula**

Use the binomial formula to expand $(3p - 2q)^4$.

Solution

$$(3p - 2q)^4 = [(3p) + (-2q)]^4 \qquad a = 3p,\ b = -2q$$
$$= \sum_{k=0}^{4}\binom{4}{k}(3p)^{4-k}(-2q)^k$$
$$= \binom{4}{0}(3p)^4 + \binom{4}{1}(3p)^3(-2q) + \binom{4}{2}(3p)^2(-2q)^2$$
$$+ \binom{4}{3}(3p)(-2q)^3 + \binom{4}{4}(-2q)^4$$
$$= 81p^4 - 216p^3q + 216p^2q^2 - 96pq^3 + 16q^4$$

Matched Problem 4 Use the binomial formula to expand $(2m - 5n)^3$.

EXPLORE-DISCUSS 2 (A) Compute each term and also the sum of the alternating series $\binom{6}{0} - \binom{6}{1} + \binom{6}{2} - \cdots + \binom{6}{6}$.

(B) What result about an alternating series can be deduced by letting $a = 1$ and $b = -1$ in the binomial formula?

EXAMPLE 5 **Using the Binomial Formula**

Use the binomial formula to find the fourth and sixteenth terms in the expansion of $(x - 2)^{20}$.

Solution In the expansion of $(a + b)^n$, the exponent of b in the rth term is $r - 1$ and the exponent of a is $n - (r - 1)$. Thus,

Fourth term:

$$\binom{20}{3}x^{17}(-2)^3 = \frac{20 \cdot 19 \cdot 18}{3 \cdot 2 \cdot 1}x^{17}(-8) = -9{,}120x^{17}$$

Sixteenth term:

$$\binom{20}{15}x^5(-2)^{15} = \frac{20 \cdot 19 \cdot 18 \cdot 17 \cdot 16}{5 \cdot 4 \cdot 3 \cdot 2 \cdot 1}x^5(-32{,}768) = -508{,}035{,}072x^5$$

Matched Problem 5 Use the binomial formula to find the fifth and twelfth terms in the expansion of $(u - 1)^{18}$.

Answers to Matched Problems

1. (A) 720 (B) 6 (C) 504 **2.** (A) 36 (B) 1
3. $x^5 + 5x^4 + 10x^3 + 10x^2 + 5x + 1$ **4.** $8m^3 - 60m^2n + 150mn^2 - 125n^3$
5. $3{,}060u^{14}$; $-31{,}824u^7$

EXERCISE 10-4

A

Evaluate each expression in Problems 1–12.

1. $7!$ **2.** $5!$ **3.** $\frac{15!}{13!}$ **4.** $\frac{20!}{17!}$ **5.** $4! + 5!$ **6.** $(4 + 5)!$

7. $\frac{8!}{3!5!}$ **8.** $\frac{10!}{2!8!}$ **9.** $\frac{7!}{0!(7 - 0)!}$

10. $\dfrac{12!}{12!(12-12)!}$ **11.** $\dfrac{10!}{7!}$ **12.** $\dfrac{10!}{3!}$

Write each expression in Problems 13–16 as the quotient of two factorials.

13. 9 **14.** 12

15. $6 \cdot 7 \cdot 8$ **16.** $9 \cdot 10 \cdot 11 \cdot 12$

B

Evaluate each expression in Problems 17–22.

17. $\binom{13}{9}$ **18.** $\binom{12}{5}$ **19.** $\binom{14}{7}$

20. $\binom{16}{8}$ **21.** $\binom{100}{97}$ **22.** $\binom{100}{3}$

23. Find the smallest positive integer n such that $n!$ produces an overflow error on your calculator.

24. Find the smallest positive integer n such that $\binom{2n}{n}$ produces an overflow error on your calculator.

Expand Problems 25–30 using the binomial formula.

25. $(2x - 3y)^3$ **26.** $(3u + 2v)^3$ **27.** $(x - 2)^4$

28. $(3p - q)^4$ **29.** $(2x - y)^5$ **30.** $(2x - y)^6$

In Problems 31–38, find the indicated term in each expansion.

31. $(u + v)^{15}$; seventh term **32.** $(a + b)^{12}$; fifth term

33. $(2m + n)^{10}$; ninth term **34.** $(x + 3y)^{13}$; third term

35. $(w - 3)^{20}$; fifth term **36.** $(2x - 5y)^8$; sixth term

37. $(3x - 2y)^8$; sixth term **38.** $(2p - 3q)^7$; fourth term

In Problems 39–42, use a graphing utility to graph each sequence and to display it in table form.

39. Find the number of terms of the sequence

$$\binom{20}{0}, \binom{20}{1}, \binom{20}{2}, \ldots, \binom{20}{20}$$

that are greater than one-half of the largest term.

40. Find the number of terms of the sequence

$$\binom{40}{0}, \binom{40}{1}, \binom{40}{2}, \ldots, \binom{40}{40}$$

that are greater than one-half of the largest term.

41. (A) Find the largest term of the sequence $a_0, a_1, a_2, \ldots, a_{10}$ to three decimal places, where
$$a_k = \binom{10}{k}(0.6)^{10-k}(0.4)^k.$$
(B) According to the binomial formula, what is the sum of the series $a_0 + a_1 + a_2 + \cdots + a_{10}$?

42. (A) Find the largest term of the sequence $a_0, a_1, a_2, \ldots, a_{10}$ to three decimal places, where
$$a_k = \binom{10}{k}(0.3)^{10-k}(0.7)^k.$$
(B) According to the binomial formula, what is the sum of the series $a_0 + a_1 + a_2 + \cdots + a_{10}$?

C

43. Evaluate $(1.01)^{10}$ to four decimal places, using the binomial formula. [*Hint:* Let $1.01 = 1 + 0.01$.]

44. Evaluate $(0.99)^6$ to four decimal places, using the binomial formula.

In Problems 45–48, determine whether the statement is true or false. If true, explain why. If false, give a counterexample.

45. $\binom{n}{r} = \binom{n}{n-r}$

46. If the positive integer n is divisible by a prime p, then $\binom{n}{r}$ is divisible by p for $1 \le r \le n - 1$.

47. If p is a prime, then $\binom{p}{r}$ is divisible by p for $1 \le r \le p - 1$.

48. If p is an odd prime, then $\binom{2p}{p}$ is not divisible by p.

49. Show that: $\binom{k}{r-1} + \binom{k}{r} = \binom{k+1}{r}$

50. Show that: $\binom{k}{0} = \binom{k+1}{0}$

51. Show that: $\binom{k}{k} = \binom{k+1}{k+1}$

52. Show that $\binom{n}{r}$ is given by the recursion formula

$$\binom{n}{r} = \frac{n-r+1}{r}\binom{n}{r-1}$$

where $\binom{n}{0} = 1$.

53. Write $2^n = (1 + 1)^n$ and expand, using the binomial formula to obtain

$$2^n = \binom{n}{0} + \binom{n}{1} + \binom{n}{2} + \cdots + \binom{n}{n}$$

54. Can you guess what the next two rows in **Pascal's triangle,** shown at right, are? Compare the numbers in the triangle with the binomial coefficients obtained with the binomial formula.

1

1 1

1 2 1

1 3 3 1

1 4 6 4 1

SECTION 10-5 Multiplication Principle, Permutations, and Combinations

- Multiplication Principle
- Permutations
- Combinations

We may expand the binomial form $(a + b)^n$ in two steps: first, expand into a sum of 2^n terms, each with coefficient 1; second, group together those terms in which b appears to the same power, obtaining the sum of the $n + 1$ terms of the binomial formula. For example,

$$\begin{aligned}(a + b)^3 &= (a + b)(a + b)^2 = (a + b)(aa + ab + ba + bb) \\ &= aaa + aab + aba + abb + baa + bab + bba + bbb \quad \text{Step 1} \\ &= a^3 + 3a^2b + 3ab^2 + b^3 \quad \text{Step 2}\end{aligned}$$

Consider the term aba of step 1: The first a comes from the first factor of $a + b$, the b comes from the second factor of $a + b$, and the final a from the third factor. Therefore, $\binom{3}{1} = 3$, the coefficient of a^2b in step 2, is the number of ways of choosing b from exactly one of the three factors of $a + b$ in $(a + b)^3$.

In the same way, $\binom{52}{5} = 2{,}598{,}960$ is the number of ways of choosing b from exactly five of the 52 factors of $a + b$ in $(a + b)^{52}$. Analogously, 2,598,960 is the number of 5-card hands which can be chosen from a standard 52-card deck. In this section we study such counting techniques that are related to the sequence $\binom{n}{0}, \binom{n}{1}, \binom{n}{2}, \ldots, \binom{n}{n}$, and we develop important counting tools that form the foundation of probability theory.

• Multiplication Principle

We start with an example.

EXAMPLE 1 Combined Outcomes

Suppose we flip a coin and then throw a single die (see Fig. 1). What are the possible combined outcomes?

Solution To solve this problem, we use a **tree diagram:**

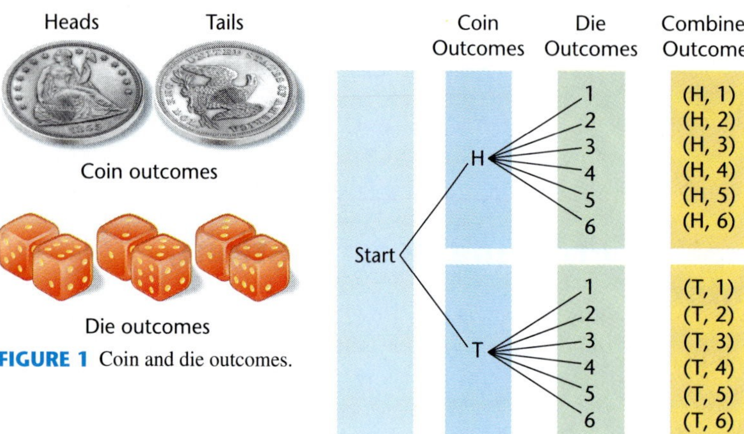

FIGURE 1 Coin and die outcomes.

Thus, there are 12 possible combined outcomes—two ways in which the coin can come up followed by six ways in which the die can come up.

Matched Problem 1 Use a tree diagram to determine the number of possible outcomes of throwing a single die followed by flipping a coin.

Now suppose you are asked, "From the 26 letters in the alphabet, how many ways can 3 letters appear in a row on a license plate if no letter is repeated?" To try to count the possibilities using a tree diagram would be extremely tedious, to say the least. The following **multiplication principle,** also called the **fundamental counting principle,** enables us to solve this problem easily. In addition, it forms the basis for several other counting techniques developed later in this section.

Multiplication Principle

1. If two operations O_1 and O_2 are performed in order, with N_1 possible outcomes for the first operation and N_2 possible outcomes for the second operation, then there are

$$N_1 \cdot N_2$$

possible combined outcomes of the first operation followed by the second.

2. In general, if n operations $O_1, O_2, \ldots, O_n$ are performed in order, with possible number of outcomes $N_1, N_2, \ldots, N_n$, respectively, then there are

$$N_1 \cdot N_2 \cdot \cdots \cdot N_n$$

possible combined outcomes of the operations performed in the given order.

In Example 1, we see that there are two possible outcomes from the first operation of flipping a coin and six possible outcomes from the second operation of throwing a die. Hence, by the multiplication principle, there are $2 \cdot 6 = 12$ possible combined outcomes of flipping a coin followed by throwing a die. Use the multiplication principle to solve Matched Problem 1.

To answer the license plate question, we reason as follows: There are 26 ways the first letter can be chosen. After a first letter is chosen, 25 letters remain; hence there are 25 ways a second letter can be chosen. And after 2 letters are chosen, there are 24 ways a third letter can be chosen. Hence, using the multiplication principle, there are $26 \cdot 25 \cdot 24 = 15{,}600$ possible ways 3 letters can be chosen from the alphabet without allowing any letter to repeat. By not allowing any letter to repeat, earlier selections affect the choice of subsequent selections. If we allow letters to repeat, then earlier selections do not affect the choice in subsequent selections, and there are 26 possible choices for each of the 3 letters. Thus, if we allow letters to repeat, there are $26 \cdot 26 \cdot 26 = 26^3 = 17{,}576$ possible ways the 3 letters can be chosen from the alphabet.

EXAMPLE 2 Computer-Generated Tests

Many universities and colleges are now using computer-assisted testing procedures. Suppose a screening test is to consist of 5 questions, and a computer stores 5 equivalent questions for the first test question, 8 equivalent questions for the second, 6 for the third, 5 for the fourth, and 10 for the fifth. How many different 5-question tests can the computer select? Two tests are considered different if they differ in one or more questions.

Solution

O_1:	Select the first question	N_1:	5 ways
O_2:	Select the second question	N_2:	8 ways
O_3:	Select the third question	N_3:	6 ways
O_4:	Select the fourth question	N_4:	5 ways
O_5:	Select the fifth question	N_5:	10 ways

Thus, the computer can generate

$$5 \cdot 8 \cdot 6 \cdot 5 \cdot 10 = 12{,}000 \text{ different tests}$$

Matched Problem 2 Each question on a multiple-choice test has 5 choices. If there are 5 such questions on a test, how many different response sheets are possible if only 1 choice is marked for each question?

EXAMPLE 3 Counting Code Words

How many 3-letter code words are possible using the first 8 letters of the alphabet if:

(A) No letter can be repeated? (B) Letters can be repeated?
(C) Adjacent letters cannot be alike?

Solutions (A) No letter can be repeated.

O_1: Select first letter	N_1: 8 ways	
O_2: Select second letter	N_2: 7 ways	Because 1 letter has been used
O_3: Select third letter	N_3: 6 ways	Because 2 letters have been used

Thus, there are

$$8 \cdot 7 \cdot 6 = 336 \text{ possible code words}$$

(B) Letters can be repeated.

O_1: Select first letter	N_1: 8 ways	
O_2: Select second letter	N_2: 8 ways	Repeats are allowed.
O_3: Select third letter	N_3: 8 ways	Repeats are allowed.

Thus, there are

$$8 \cdot 8 \cdot 8 = 8^3 = 512 \text{ possible code words}$$

(C) Adjacent letters cannot be alike.

O_1: Select first letter	N_1: 8 ways	
O_2: Select second letter	N_2: 7 ways	Cannot be the same as the first
O_3: Select third letter	N_3: 7 ways	Cannot be the same as the second, but can be the same as the first

Thus, there are

$$8 \cdot 7 \cdot 7 = 392 \text{ possible code words}$$

Matched Problem 3 How many 4-letter code words are possible using the first 10 letters of the alphabet under the three conditions stated in Example 3?

EXPLORE-DISCUSS 1 The postal service of a developing country is choosing a five-character postal code consisting of letters (of the English alphabet) and digits. At least half a million postal codes must be accommodated. Which format would you recommend to make the codes easy to remember?

The multiplication principle can be used to develop two additional methods for counting that are extremely useful in more complicated counting problems. Both of these methods use the factorial function, which was introduced in Section 10-4.

• Permutations

Suppose 4 pictures are to be arranged from left to right on one wall of an art gallery. How many arrangements are possible? Using the multiplication principle, there are 4 ways of selecting the first picture. After the first picture is selected, there are 3 ways of selecting the second picture. After the first 2 pictures are selected, there are 2 ways of selecting the third picture. And after the first 3 pictures are selected, there is only 1 way to select the fourth. Thus, the number of arrangements possible for the 4 pictures is

$$4 \cdot 3 \cdot 2 \cdot 1 = 4! \qquad \text{or} \qquad 24$$

In general, we refer to a particular arrangement, or **ordering,** of n objects without repetition as a **permutation** of the n objects. How many permutations of n objects are there? From the reasoning above, there are n ways in which the first object can be chosen, there are $n - 1$ ways in which the second object can be chosen, and so on. Applying the multiplication principle, we have Theorem 1:

Theorem 1

Permutations of n Objects

The number of permutations of n objects, denoted by $P_{n,n}$, is given by

$$P_{n,n} = n \cdot (n - 1) \cdot \cdots \cdot 1 = n!$$

Now suppose the director of the art gallery decides to use only 2 of the 4 available pictures on the wall, arranged from left to right. How many arrangements of 2 pictures can be formed from the 4? There are 4 ways the first picture can be selected. After selecting the first picture, there are 3 ways the second picture can be selected. Thus, the number of arrangements of 2 pictures from 4 pictures, denoted by $P_{4,2}$, is given by

$$P_{4,2} = 4 \cdot 3 = 12$$

Or, in terms of factorials, multiplying $4 \cdot 3$ by 1 in the form $2!/2!$, we have

$$P_{4,2} = 4 \cdot 3 = \frac{4 \cdot 3 \cdot 2!}{2!} = \frac{4!}{2!}$$

This last form gives $P_{4,2}$ in terms of factorials, which is useful in some cases.

A **permutation of a set of n objects taken r at a time** is an arrangement of the r objects in a specific order. Thus, reasoning in the same way as in the example above, we find that the number of permutations of n objects taken r at a time, $0 \le r \le n$, denoted by $P_{n,r}$, is given by

$$P_{n,r} = n(n - 1)(n - 2) \cdot \cdots \cdot (n - r + 1)$$

Multiplying the right side of this equation by 1 in the form $(n - r)!/(n - r)!$, we obtain a factorial form for $P_{n,r}$:

$$P_{n,r} = n(n - 1)(n - 2) \cdot \cdots \cdot (n - r + 1) \frac{(n - r)!}{(n - r)!}$$

But

$$n(n-1)(n-2)\cdot\cdots\cdot(n-r+1)(n-r)! = n!$$

Hence, we have Theorem 2:

Theorem 2 **Permutation of n Objects Taken r at a Time**

The number of permutations of n objects taken r at a time is given by

$$P_{n,r} = \underbrace{n(n-1)(n-2)\cdot\cdots\cdot(n-r+1)}_{r \text{ factors}}$$

or

$$P_{n,r} = \frac{n!}{(n-r)!} \qquad 0 \le r \le n$$

Note that if $r = n$, then the number of permutations of n objects taken n at a time is

$$P_{n,n} = \frac{n!}{(n-n)!} = \frac{n!}{0!} = n! \qquad \text{Recall, } 0! = 1.$$

which agrees with Theorem 1, as it should.

The permutation symbol $P_{n,r}$ *also can be denoted by* P^n_r, ${}_nP_r$, or $P(n, r)$. Many calculators use ${}_nP_r$ to denote the function that evaluates the permutation symbol.

EXAMPLE 4 **Selecting Officers**

From a committee of 8 people, in how many ways can we choose a chair and a vice-chair, assuming one person cannot hold more than one position?

Solution We are actually asking for the number of permutations of 8 objects taken 2 at a time—that is, $P_{8,2}$:

$$P_{8,2} = \frac{8!}{(8-2)!} = \frac{8!}{6!} = \frac{8\cdot 7\cdot 6!}{6!} = 56$$

Matched Problem 4 From a committee of 10 people, in how many ways can we choose a chair, vice-chair, and secretary, assuming one person cannot hold more than one position?

CAUTION

Remember to use the definition of factorial when simplifying fractions involving factorials.

$$\frac{6!}{3!} \neq 2! \qquad \frac{6!}{3!} = \frac{6 \cdot 5 \cdot 4 \cdot 3!}{3!} = 120$$

EXAMPLE 5 Evaluating $P_{n,r}$

Find the number of permutations of 25 objects taken 8 at a time. Compute the answer to 4 significant digits using a calculator.

Solution

$$P_{25,8} = \frac{25!}{(25-8)!} = \frac{25!}{17!} = 4.361 \times 10^{10} \qquad \text{A very large number}$$

Matched Problem 5 Find the number of permutations of 30 objects taken 4 at a time. Compute the answer exactly using a calculator.

• Combinations

Now suppose that an art museum owns 8 paintings by a given artist and another art museum wishes to borrow 3 of these paintings for a special show. How many ways can 3 paintings be selected for shipment out of the 8 available? Here, the order of the items selected doesn't matter. What we are actually interested in is how many subsets of 3 objects can be formed from a set of 8 objects. We call such a subset a **combination** of 8 objects taken 3 at a time. The total number of combinations is denoted by the symbol

$$C_{8,3} \qquad \text{or} \qquad \binom{8}{3}$$

To find the number of combinations of 8 objects taken 3 at a time, $C_{8,3}$, we make use of the formula for $P_{n,r}$ and the multiplication principle. We know that the number of permutations of 8 objects taken 3 at a time is given by $P_{8,3}$, and we have a formula for computing this quantity. Now suppose we think of $P_{8,3}$ in terms of two operations:

O_1: Select a subset of 3 objects (paintings)

N_1: $C_{8,3}$ ways

O_2: Arrange the subset in a given order

N_2: 3! ways

The combined operation, O_1 followed by O_2, produces a permutation of 8 objects taken 3 at a time. Thus,

$$P_{8,3} = C_{8,3} \cdot 3!$$

To find $C_{8,3}$, we replace $P_{8,3}$ in the above equation with $8!/(8-3)!$ and solve for $C_{8,3}$:

$$\frac{8!}{(8-3)!} = C_{8,3} \cdot 3!$$

$$C_{8,3} = \frac{8!}{3!(8-3)!} = \frac{8 \cdot 7 \cdot 6 \cdot 5!}{3 \cdot 2 \cdot 1 \cdot 5!} = 56$$

Thus, the museum can make 56 different selections of 3 paintings from the 8 available.

A **combination of a set of *n* objects taken *r* at a time** is an r-element subset of the n objects. Reasoning in the same way as in the example, the number of combinations of n objects taken r at a time, $0 \le r \le n$, denoted by $C_{n,r}$, can be obtained by solving for $C_{n,r}$ in the relationship

$$P_{n,r} = C_{n,r} \cdot r!$$

$$C_{n,r} = \frac{P_{n,r}}{r!}$$

$$= \frac{n!}{r!(n-r)!} \qquad P_{n,r} = \frac{n!}{(n-r)!}$$

Theorem 3 **Combination of *n* Objects Taken *r* at a Time**

The number of combinations of n objects taken r at a time is given by

$$C_{n,r} = \binom{n}{r} = \frac{P_{n,r}}{r!} = \frac{n!}{r!(n-r)!} \qquad 0 \le r \le n$$

Note that we used the combination formula in Section 10-4 to represent binomial coefficients.

The combination symbols $C_{n,r}$ and $\binom{n}{r}$ also can be denoted by C_r^n, ${}_nC_r$, or $C(n, r)$.

EXAMPLE 6 Selecting Subcommittees

From a committee of 8 people, in how many ways can we choose a subcommittee of 2 people?

Solution Notice how this example differs from Example 4, where we wanted to know how many ways a chair and a vice-chair can be chosen from a committee of 8 people. In Example 4, ordering matters. In choosing a subcommittee of 2 people, the ordering does not matter. Thus, we are actually asking for the number of combinations of 8 objects taken 2 at a time. The number is given by

$$C_{8,2} = \binom{8}{2} = \frac{8!}{2!(8-2)!} = \frac{8 \cdot 7 \cdot 6!}{2 \cdot 1 \cdot 6!} = 28$$

Matched Problem 6 How many subcommittees of 3 people can be chosen from a committee of 8 people?

EXAMPLE 7 **Evaluating $C_{n,r}$**

Find the number of combinations of 25 objects taken 8 at a time. Compute the answer to 4 significant digits using a calculator.

Solution

$$C_{25,8} = \binom{25}{8} = \frac{25!}{8!(25-8)!} = \frac{25!}{8!17!} = 1.082 \times 10^6$$

Compare this result with that obtained in Example 5.

Matched Problem 7 Find the number of combinations of 30 objects taken 4 at a time. Compute the answer exactly using a calculator.

Remember: In a permutation, order counts. In a combination, order does not count.

To determine whether a permutation or combination is needed, decide whether rearranging the collection or listing makes a difference. If so, use permutations. If not, use combinations.

EXPLORE-DISCUSS 2 Each of the following is a selection without repetition. Would you consider the selection to be a combination? A permutation? Discuss your reasoning.

(A) A student checks out three books from the library.

(B) A baseball manager names his starting lineup.

(C) The newly elected President names his Cabinet members.

(D) The President selects a delegation of three Cabinet members to attend the funeral of a head of state.

(E) An orchestra conductor chooses three pieces of music for a symphony program.

A standard deck of 52 cards involves four suits, hearts, spades, diamonds, and clubs, as shown in Figure 2. Example 8, as well as other examples and exercises in this chapter, refer to this standard deck.

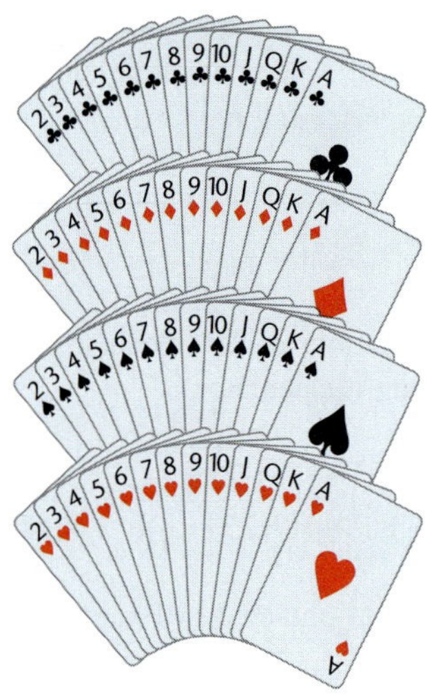

FIGURE 2 A standard deck of cards.

EXAMPLE 8 Counting Card Hands

Out of a standard 52-card deck, how many 5-card hands will have 3 aces and 2 kings?

Solution

O_1: Choose 3 aces out of 4 possible — Order is not important.

N_1: $C_{4,3}$

O_2: Choose 2 kings out of 4 possible — Order is not important.

N_2: $C_{4,2}$

Using the multiplication principle, we have

$$\text{Number of hands} = C_{4,3} \cdot C_{4,2} = 4 \cdot 6 = 24$$

Matched Problem 8 From a standard 52-card deck, how many 5-card hands will have 3 hearts and 2 spades?

EXAMPLE 9 Counting Serial Numbers

Serial numbers for a product are to be made using 2 letters followed by 3 numbers. If the letters are to be taken from the first 8 letters of the alphabet with no repeats and the numbers from the 10 digits 0 through 9 with no repeats, how many serial numbers are possible?

Solution

O_1: Choose 2 letters out of 8 available — Order is important.

N_1: $P_{8,2}$

O_2: Choose 3 numbers out of 10 available — Order is important.

N_2: $P_{10,3}$

Using the multiplication principle, we have

$$\text{Number of serial numbers} = P_{8,2} \cdot P_{10,3} = 40{,}320$$

Matched Problem 9 Repeat Example 9 under the same conditions, except the serial numbers are now to have 3 letters followed by 2 digits with no repeats.

Answers to Matched Problems

1. 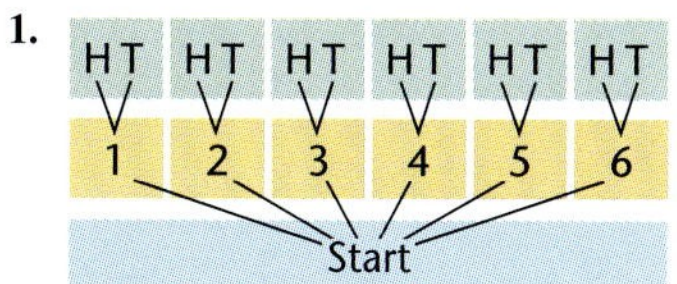

2. 5^5, or 3,125

3. (A) $10 \cdot 9 \cdot 8 \cdot 7 = 5{,}040$ (B) $10 \cdot 10 \cdot 10 \cdot 10 = 10{,}000$ (C) $10 \cdot 9 \cdot 9 \cdot 9 = 7{,}290$

4. $P_{10,3} = \dfrac{10!}{(10-3)!} = 720$ 5. $P_{30,4} = \dfrac{30!}{(30-4)!} = 657{,}720$ 6. $C_{8,3} = \dfrac{8!}{3!(8-3)!} = 56$

7. $C_{30,4} = \dfrac{30!}{4!(30-4)!} = 27{,}405$ 8. $C_{13,3} \cdot C_{13,2} = 22{,}308$ 9. $P_{8,3} \cdot P_{10,2} = 30{,}240$

EXERCISE 10-5

A

Evaluate Problems 1–16.

1. $\dfrac{15!}{12!}$ 2. $\dfrac{20!}{18!}$ 3. $\dfrac{32!}{0!32!}$

4. $\dfrac{25!}{24!1!}$ 5. $\dfrac{9!}{6!3!}$ 6. $\dfrac{7!}{5!2!}$

7. $\dfrac{16!}{4!(16-4)!}$ 8. $\dfrac{18!}{3!(18-3)!}$ 9. $P_{8,5}$

10. $C_{8,5}$ 11. $P_{52,3}$ 12. $P_{13,5}$

13. $C_{13,5}$ 14. $C_{13,4}$ 15. $C_{52,5}$

16. $P_{20,4}$

17. A particular new car model is available with 5 choices of color, 3 choices of transmission, 4 types of interior, and 2 types of engine. How many different variations of this model car are possible?

18. A deli serves sandwiches with the following options: 3 kinds of bread, 5 kinds of meat, and lettuce or sprouts. How many different sandwiches are possible, assuming one item is used out of each category?

19. In a horse race, how many different finishes among the first 3 places are possible for a 10-horse race? Exclude ties.

20. In a long-distance foot race, how many different finishes among the first 5 places are possible for a 50-person race? Exclude ties.

21. How many ways can a subcommittee of 3 people be selected from a committee of 7 people? How many ways can a president, vice president, and secretary be chosen from a committee of 7 people?

22. Suppose 9 cards are numbered with the 9 digits from 1 to 9. A 3-card hand is dealt, 1 card at a time. How many hands are possible where:
(A) Order is taken into consideration?
(B) Order is not taken into consideration?

23. There are 10 teams in a league. If each team is to play every other team exactly once, how many games must be scheduled?

24. Given 7 points, no 3 of which are on a straight line, how many lines can be drawn joining 2 points at a time?

B

25. How many 4-letter code words are possible from the first 6 letters of the alphabet, with no letter repeated? Allowing letters to repeat?

26. A small combination lock on a suitcase has 3 wheels, each labeled with digits from 0 to 9. How many opening combinations of 3 numbers are possible, assuming no digit is repeated? Assuming digits can be repeated?

27. From a standard 52-card deck, how many 5-card hands will have all hearts?

28. From a standard 52-card deck, how many 5-card hands will have all face cards? All face cards, but no kings? Consider only jacks, queens, and kings to be face cards.

29. How many different license plates are possible if each contains 3 letters followed by 3 digits? How many of these license plates contain no repeated letters and no repeated digits?

30. How many 5-digit zip codes are possible? How many of these codes contain no repeated digits?

31. From a standard 52-card deck, how many 7-card hands have exactly 5 spades and 2 hearts?

32. From a standard 52-card deck, how many 5-card hands will have 2 clubs and 3 hearts?

33. A catering service offers 8 appetizers, 10 main courses, and 7 desserts. A banquet chairperson is to select 3 appetizers, 4 main courses, and 2 desserts for a banquet. How many ways can this be done?

34. Three research departments have 12, 15, and 18 members, respectively. If each department is to select a delegate and an alternate to represent the department at a conference, how many ways can this be done?

35. (A) Use a graphing utility to display the sequences $P_{10,0}$, $P_{10,1}, \ldots, P_{10,10}$ and $0!, 1!, \ldots, 10!$ in table form, and show that $P_{10,r} \geq r!$ for $r = 0, 1, \ldots, 10$.
(B) Find all values of r such that $P_{10,r} = r!$
(C) Explain why $P_{n,r} \geq r!$ whenever $0 \leq r \leq n$.

36. (A) How are the sequences $\frac{P_{10,0}}{0!}, \frac{P_{10,1}}{1!}, \ldots, \frac{P_{10,10}}{10!}$ and $C_{10,0}$, $C_{10,1}, \ldots, C_{10,10}$ related?

(B) Use a graphing utility to graph each sequence and confirm the relationship of part A.

C

37. A sporting goods store has 12 pairs of ski gloves of 12 different brands thrown loosely in a bin. The gloves are all the same size. In how many ways can a left-hand glove and a right-hand glove be selected that do not match relative to brand?

38. A sporting goods store has 6 pairs of running shoes of 6 different styles thrown loosely in a basket. The shoes are all the same size. In how many ways can a left shoe and a right shoe be selected that do not match?

39. Eight distinct points are selected on the circumference of a circle.
(A) How many chords can be drawn by joining the points in all possible ways?
(B) How many triangles can be drawn using these 8 points as vertices?
(C) How many quadrilaterals can be drawn using these 8 points as vertices?

40. Five distinct points are selected on the circumference of a circle.
(A) How many chords can be drawn by joining the points in all possible ways?
(B) How many triangles can be drawn using these 5 points as vertices?

41. How many ways can 2 people be seated in a row of 5 chairs? 3 people? 4 people? 5 people?

42. Each of 2 countries sends 5 delegates to a negotiating conference. A rectangular table is used with 5 chairs on each long side. If each country is assigned a long side of the table, how many seating arrangements are possible? [*Hint:* Operation 1 is assigning a long side of the table to each country.]

43. A basketball team has 5 distinct positions. Out of 8 players, how many starting teams are possible if:
(A) The distinct positions are taken into consideration?
(B) The distinct positions are not taken into consideration?
(C) The distinct positions are not taken into consideration, but either Mike or Ken, but not both, must start?

44. How many committees of 4 people are possible from a group of 9 people if:
(A) There are no restrictions?
(B) Both Juan and Mary must be on the committee?
(C) Either Juan or Mary, but not both, must be on the committee?

45. A 5-card hand is dealt from a standard 52-card deck. Which is more likely: the hand contains exactly 1 king or the hand contains no hearts?

46. A 10-card hand is dealt from a standard 52-card deck. Which is more likely: all cards in the hand are red or the hand contains all four aces?

47. A parent is placing an order for five single-dip ice cream cones. If today's flavors are vanilla, chocolate, and strawberry, how many orders are possible? Explain. (*Note:* This type of selection, in which repetition is allowed but order is irrelevant, is neither a combination nor a permutation.)

48. One dozen identical doughnuts are to be distributed among nine students. If each student must receive at least one doughnut, how many distributions are possible? Explain.

CHAPTER 10 GROUP ACTIVITY Sequences Specified by Recursion Formulas

The recursion formula $a_n = 5a_{n-1} - 6a_{n-2}$, together with the initial values $a_1 = 4$, $a_2 = 14$, specifies the sequence $\{a_n\}$ whose first several terms are 4, 14, 46, 146, 454, 1394, ... The sequence $\{a_n\}$ is neither arithmetic nor geometric. Nevertheless, because it satisfies a simple recursion formula, it is possible to obtain an nth-term formula for $\{a_n\}$ that is analogous to the nth-term formulas for arithmetic and geometric sequences. Such an nth-term formula is valuable because it allows us to estimate a term of a sequence without computing all the preceding terms.

If the geometric sequence $\{r^n\}$ satisfies the recursion formula above, then $r^n = 5r^{n-1} - 6r^{n-2}$. Dividing by r^{n-2} leads to the quadratic equation $r^2 - 5r + 6 = 0$, whose solutions are $r = 2$ and $r = 3$. Now it is easy to check that the geometric sequences $\{2^n\} = 2, 4, 8, 16, \ldots$ and $\{3^n\} = 3, 9, 27, 81, \ldots$ satisfy the recursion formula. Therefore, any sequence of the form $\{u2^n + v3^n\}$, where u and v are constants, will satisfy the same recursion formula.

We now find u and v so that the first two terms of $\{u2^n + v3^n\}$ are $a_1 = 4$, $a_2 = 14$. Letting $n = 1$ and $n = 2$ we see that u and v must satisfy the following linear system:

$$2u + 3v = 4$$
$$4u + 9v = 14$$

Solving the system gives $u = -1$, $v = 2$. Therefore, an nth-term formula for the original sequence is $a_n = (-1)2^n + (2)3^n$.

Note that the nth-term formula was obtained by solving a quadratic equation and a system of two linear equations in two variables.

(A) Compute $(-1)2^n + (2)3^n$ for $n = 1, 2, \ldots, 6$, and compare with the terms of $\{a_n\}$.

(B) Estimate the one-hundredth term of $\{a_n\}$.

(C) Show that any sequence of the form $\{u2^n + v3^n\}$, where u and v are constants, satisfies the recursion formula $a_n = 5a_{n-1} - 6a_{n-2}$.

(D) Find an nth-term formula for the sequence $\{b_n\}$ that is specified by $b_1 = 5$, $b_2 = 55$, $b_n = 3b_{n-1} + 4b_{n-2}$.

(E) Find an nth-term formula for the Fibonacci sequence.

(F) Find an nth-term formula for the sequence $\{c_n\}$ that is specified by $c_1 = -3$, $c_2 = 15$, $c_3 = 99$, $c_n = 6c_{n-1} - 3c_{n-2} - 10c_{n-3}$. (Since the recursion formula involves the three terms which precede c_n, our method will involve the solution of a cubic equation and a system of three linear equations in three variables.)

Chapter 10 Review

10-1 SEQUENCES AND SERIES

A **sequence** is a function with the domain a set of successive integers. The symbol a_n, called the ***n*th term**, or **general term**, represents the range value associated with the domain value n. Unless specified otherwise, the domain is understood to be the set of natural numbers. A **finite sequence** has a finite domain, and an **infinite sequence** has an infinite domain. A **recursion formula** defines each term of a sequence in terms of one or more of the preceding terms. For example, the **Fibonacci sequence** is defined by $a_n = a_{n-1} + a_{n-2}$ for $n \geq 3$, where $a_1 = a_2 = 1$. If $a_1, a_2, \ldots, a_n, \ldots$ is a sequence, then the expression $a_1 + a_2 + \cdots + a_n + \cdots$ is called a **series.** A finite sequence produces a **finite series,** and an infinite sequence produces an **infinite series.** Series can be represented using summation notation:

$$\sum_{k=m}^{n} a_k = a_m + a_{m+1} + \cdots + a_n$$

where k is called the **summing index.** If the terms in the series are alternately positive and negative, the series is called an **alternating series.**

10-2 MATHEMATICAL INDUCTION

A wide variety of statements can be proven using the **principle of mathematical induction:** Let P_n be a statement associated with each positive integer n and suppose the following conditions are satisfied:

1. P_1 is true.
2. For any positive integer k, if P_k is true, then P_{k+1} is also true.

Then the statement P_n is true for all positive integers n.

To use mathematical induction to prove statements involving laws of exponents, it is convenient to state a **recursive definition of a^n:**

$$a^1 = a \qquad \text{and} \qquad a^{n+1} = a^n a \qquad \text{for any integer } n \geq 1$$

To deal with conjectures that may be true only for $n \geq m$, where m is a positive integer, we use the **extended principle of mathematical induction:** Let m be a positive integer, let P_n be a statement associated with each integer $n \geq m$, and suppose the following conditions are satisfied:

1. P_m is true.
2. For any integer $k \geq m$, if P_k is true, then P_{k+1} is also true.

Then the statement P_n is true for all integers $n \geq m$.

10-3 ARITHMETIC AND GEOMETRIC SEQUENCES

A sequence is called an **arithmetic sequence,** or **arithmetic progression,** if there exists a constant d, called the **common difference,** such that

$$a_n - a_{n-1} = d \qquad \text{or} \qquad a_n = a_{n-1} + d$$
$$\text{for every } n > 1$$

The following formulas are useful when working with arithmetic sequences and their corresponding series:

$a_n = a_1 + (n - 1)d$ **nth-Term Formula**

$S_n = \frac{n}{2}[2a_1 + (n - 1)d]$ **Sum Formula—First Form**

$S_n = \frac{n}{2}(a_1 + a_n)$ **Sum Formula—Second Form**

A sequence is called a **geometric sequence,** or a **geometric progression,** if there exists a nonzero constant r, called the **common ratio,** such that

$$\frac{a_n}{a_{n-1}} = r \qquad \text{or} \qquad a_n = ra_{n-1} \qquad \text{for every } n > 1$$

The following formulas are useful when working with geometric sequences and their corresponding series:

$a_n = a_1 r^{n-1}$ **nth-Term Formula**

$S_n = \frac{a_1 - a_1 r^n}{1 - r}$ $r \neq 1$ **Sum Formula—First Form**

$S_n = \frac{a_1 - ra_n}{1 - r}$ $r \neq 1$ **Sum Formula—Second Form**

$S_\infty = \frac{a_1}{1 - r}$ $|r| < 1$ **Sum of an Infinite Geometric Series**

10-4 BINOMIAL FORMULA

For n a natural number, ***n* factorial**—denoted $n!$—is defined by

$$n! = n(n - 1) \cdot \dots \cdot 2 \cdot 1 \qquad 1! = 1 \qquad 0! = 1$$

Also, n factorial is given by the **recursion formula**

$$n! = n \cdot (n - 1)!$$

For nonnegative integers r and n, $0 \leq r \leq n$, the **combinatorial symbol** $\binom{n}{r}$ is defined by

$$\binom{n}{r} = \frac{n!}{r!(n - r)!} = \frac{n(n - 1) \cdot \dots \cdot (n - r + 1)}{r(r - 1) \cdot \dots \cdot 2 \cdot 1}$$

For n a positive integer, the **binomial formula** is

$$(a + b)^n = \sum_{k=0}^{n} \binom{n}{k} a^{n-k} b^k$$

10-5 MULTIPLICATION PRINCIPLE, PERMUTATIONS, AND COMBINATIONS

Given a sequence of operations, **tree diagrams** are often used to list all the possible combined outcomes. To count the number of combined outcomes without actually listing them, we use the **multiplication principle:**

1. If operations O_1 and O_2 are performed in order with N_1 possible outcomes for the first operation and N_2 possible outcomes for the second operation, then there are

$$N_1 \cdot N_2$$

possible outcomes of the first operation followed by the second.

2. In general, if n operations $O_1, O_2, \dots, O_n$ are performed in order, with possible number of outcomes $N_1, N_2, \dots, N_n$, respectively, then there are

$$N_1 \cdot N_2 \cdot \dots \cdot N_n$$

possible combined outcomes of the operations performed in the given order.

A particular arrangement or ordering of n objects without repetition is called a **permutation.** The number of permutations of n objects is given by

$$P_{n,n} = n \cdot (n-1) \cdot \cdots \cdot 1 = n!$$

and the number of permutations of n objects taken r at a time is given by

$$P_{n,r} = \frac{n!}{(n-r)!} \qquad 0 \le r \le n$$

A **combination of a set of n elements taken r at a time** is an r-element subset of the n objects. The number of combinations of n objects taken r at a time is given by

$$C_{n,r} = \binom{n}{r} = \frac{P_{n,r}}{r!} = \frac{n!}{r!(n-r)!} \qquad 0 \le r \le n$$

In a permutation, order is important. In a combination, order is not important.

Chapter 10 Review Exercise

Work through all the problems in this chapter review and check answers in the back of the book. Answers to all review problems are there, and following each answer is a number in italics indicating the section in which that type of problem is discussed. Where weaknesses show up, review appropriate sections in the text.

A

1. Determine whether each of the following can be the first three terms of a geometric sequence, an arithmetic sequence, or neither.
(A) $16, -8, 4, \ldots$ (B) $5, 7, 9, \ldots$
(C) $-8, -5, -2, \ldots$ (D) $2, 3, 5, \ldots$
(E) $-1, 2, -4, \ldots$

In Problems 2–5:
(A) Write the first four terms of each sequence.
(B) Find a_{10}. (C) Find S_{10}.

2. $a_n = 2n + 3$ **3.** $a_n = 32(\frac{1}{2})^n$

4. $a_1 = -8;\ a_n = a_{n-1} + 3,\ n \ge 2$

5. $a_1 = -1;\ a_n = (-2)a_{n-1},\ n \ge 2$

6. Find S_∞ in Problem 3.

Evaluate Problems 7–10.

7. $10!$ **8.** $\dfrac{30!}{25!}$

9. $\dfrac{13!}{5!(13-5)!}$ **10.** $P_{8,4}$ and $C_{8,4}$

11. A single die is rolled and a coin is flipped. How many combined outcomes are possible? Solve:
(A) By using a tree diagram
(B) By using the multiplication principle

12. How many seating arrangements are possible with 6 people and 6 chairs in a row? Solve by using the multiplication principle.

13. Solve Problem 12 using permutations or combinations, whichever is applicable.

Verify Problems 14–16 for $n = 1, 2,$ and 3.

14. P_n: $5 + 7 + 9 + \cdots + (2n + 3) = n^2 + 4n$

15. P_n: $2 + 4 + 8 + \cdots + 2^n = 2^{n+1} - 2$

16. P_n: $49^n - 1$ is divisible by 6

In Problems 17–19, write P_k and P_{k+1}.

17. For P_n in Problem 14 **18.** For P_n in Problem 15

19. For P_n in Problem 16

20. Either prove the statement is true or prove it is false by finding a counterexample: If n is a positive integer, then the sum of the series $1 + \frac{1}{2} + \frac{1}{3} + \cdots + \frac{1}{n}$ is less than 4.

B

Write Problems 21 and 22 without summation notation, and find the sum.

21. $S_{10} = \sum_{k=1}^{10} (2k - 8)$

22. $S_7 = \sum_{k=1}^{7} \dfrac{16}{2^k}$

23. $S_\infty = 27 - 18 + 12 + \cdots = ?$

24. Write

$$S_n = \frac{1}{3} - \frac{1}{9} + \frac{1}{27} + \cdots + \frac{(-1)^{n+1}}{3^n}$$

using summation notation, and find S_∞.

25. Six distinct points are selected on the circumference of a circle. How many triangles can be formed using these points as vertices?

26. In an arithmetic sequence, $a_1 = 13$ and $a_7 = 31$. Find the common difference d and the fifth term a_5.

27. The sum of the first ten terms of an arithmetic series is 81. If the sixth term is 10, find the first term a_1 and the common difference d.

28. How many 3-letter code words are possible using the first 8 letters of the alphabet if no letter can be repeated? If letters can be repeated? If adjacent letters cannot be alike?

29. Use the formula for the sum of an infinite geometric series to write $0.727\,272\cdots = 0.\overline{72}$ as the quotient of two integers.

30. Solve the following problems using $P_{n,r}$ or $C_{n,r}$, as appropriate:
(A) How many 3-digit opening combinations are possible on a combination lock with 6 digits if the digits cannot be repeated?
(B) Suppose 5 tennis players have made the finals. If each of the 5 players is to play every other player exactly once, how many games must be scheduled?

Evaluate Problems 31–33.

31. $\dfrac{20!}{18!(20-18)!}$ 32. $\dbinom{16}{12}$ 33. $\dbinom{11}{11}$

34. Which is larger, $\dfrac{987!}{493!}$ or $\dfrac{987!}{(987-493)!}$? Explain.

35. Which is smaller, $\dbinom{1000}{500}$ or $\dbinom{1000}{501}$? Explain.

36. Expand $(x - y)^5$ using the binomial formula.

37. Find the tenth term in the expansion of $(2x - y)^{12}$.

Establish each statement in Problems 38–40 for all natural numbers, using mathematical induction.

38. P_n in Problem 14

39. P_n in Problem 15

40. P_n in Problem 16

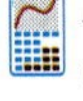
In Problems 41–42, find the smallest positive integer n such that $a_n < b_n$ by graphing the sequences $\{a_n\}$ and $\{b_n\}$ with a graphing utility. Check your answer by using a graphing utility to display both sequences in table form.

41. $a_n = C_{50,n}$, $b_n = 3^n$

42. $a_1 = 100$, $a_n = 0.99a_{n-1} + 5$, $b_n = 9 + 7n$

C

43. How many different families with 5 children are possible, excluding multiple births, where the sex of each child in the order of their birth is taken into consideration? How many families are possible if the order pattern is not taken into account?

44. A free-falling body travels $g/2$ feet in the first second, $3g/2$ feet during the next second, $5g/2$ feet the next, and so on. Find the distance fallen during the twenty-fifth second and the total distance fallen from the start to the end of the twenty-fifth second.

45. How many ways can 2 people be seated in a row of 4 chairs?

46. Expand $(x + i)^6$, where i is the imaginary unit, using the binomial formula.

47. **Transportation.** A distribution center A wishes to distribute its products to 5 different retail stores, B, C, D, E, and F, in a city. How many different route plans can be constructed so that a single truck can start from A, deliver to each store exactly once, and then return to the center?

Prove that each statement in Problems 48–52 holds for all positive integers, using mathematical induction.

48. $\displaystyle\sum_{k=1}^{n} k^3 = \left(\sum_{k=1}^{n} k\right)^2$

49. $x^{2n} - y^{2n}$ is divisible by $x - y$, $x \neq y$

50. $\dfrac{a^n}{a^m} = a^{n-m}$; $n > m$, n, m positive integers

51. $\{a_n\} = \{b_n\}$, where $a_n = a_{n-1} + 2$, $a_1 = -3$, $b_n = -5 + 2n$

52. $(1!)1 + (2!)2 + (3!)3 + \cdots + (n!)n = (n + 1)! - 1$ (From U.S.S.R. Mathematical Olympiads, 1955–1956, Grade 10.)

Problems 53–56 refer to the sequences $\{a_n\}$ and $\{b_n\}$ where $a_n = \dbinom{2n}{n}$ and $b_n = 4^{n-1}$.

53. Find the smallest positive integer n such that $a_n \le b_n$.

54. Are $\{a_n\}$ and $\{b_n\}$ arithmetic sequences? Geometric sequences?

55. Show that $\dfrac{a_{n+1}}{a_n} < 4$ for all positive integers n.

56. Use mathematical induction to show that $a_n < b_n$ for $n \ge 5$.

Additional Topics in Analytic Geometry

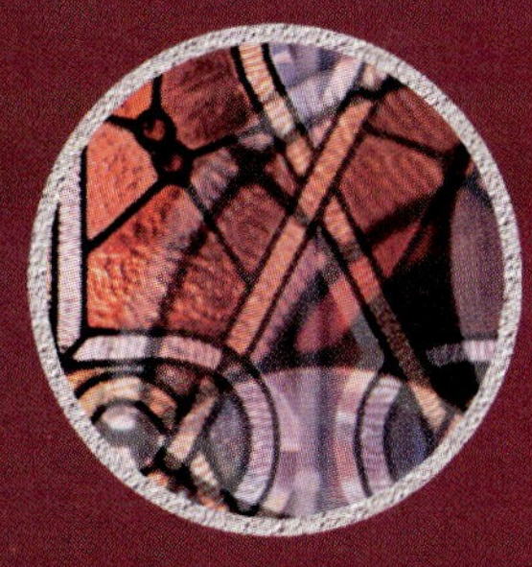

CHAPTER 11

Analytic geometry, a union of geometry and algebra, enables us to analyze certain geometric concepts algebraically and to interpret certain algebraic relationships geometrically. Our two main concerns center around graphing algebraic equations and finding equations of useful geometric figures. We have discussed a number of topics in analytic geometry, such as straight lines and circles, in earlier chapters. In this chapter we discuss additional analytic geometry topics: conic sections and translation of axes.

René Descartes (1596–1650), the French philosopher–mathematician, is generally recognized as the founder of analytic geometry.

SECTION 11-1 Conic Sections; Parabola

- Conic Sections
- Definition of a Parabola
- Drawing a Parabola
- Standard Equations and Their Graphs
- Applications

In this section we introduce the general concept of a conic section and then discuss the particular conic section called a *parabola*. In the next two sections we will discuss two other conic sections called *ellipses* and *hyperbolas*.

• Conic Sections

In Section 2-2 we found that the graph of a first-degree equation in two variables,

$$Ax + By = C \tag{1}$$

where A and B are not both 0, is a straight line, and every straight line in a rectangular coordinate system has an equation of this form. What kind of graph will a second-degree equation in two variables,

$$Ax^2 + Bxy + Cy^2 + Dx + Ey + F = 0 \tag{2}$$

where A, B, and C are not all 0, yield for different sets of values of the coefficients? The graphs of equation (2) for various choices of the coefficients are plane curves obtainable by intersecting a cone* with a plane, as shown in Figure 1. These curves are called **conic sections.**

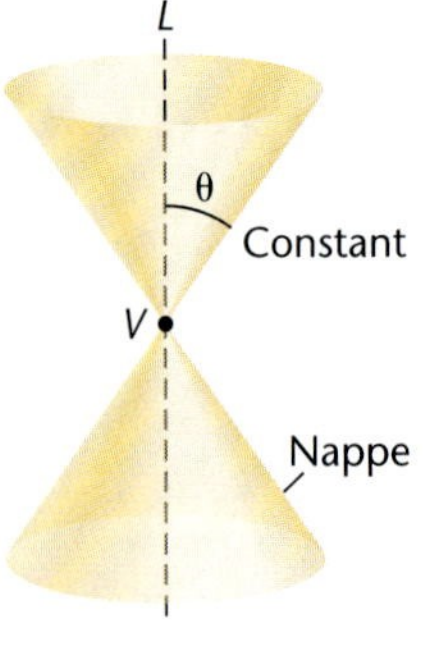

If a plane cuts clear through one nappe, then the intersection curve is called a **circle** if the plane is perpendicular to the axis and an **ellipse** if the plane is not perpendicular to the axis. If a plane cuts only one nappe, but does not cut clear through,

*Starting with a fixed line L and a fixed point V on L, the surface formed by all straight lines through V making a constant angle θ with L is called a **right circular cone.** The fixed line L is called the axis of the cone, and V is its **vertex.** The two parts of the cone separated by the vertex are called **nappes.**

FIGURE 1 Conic sections.

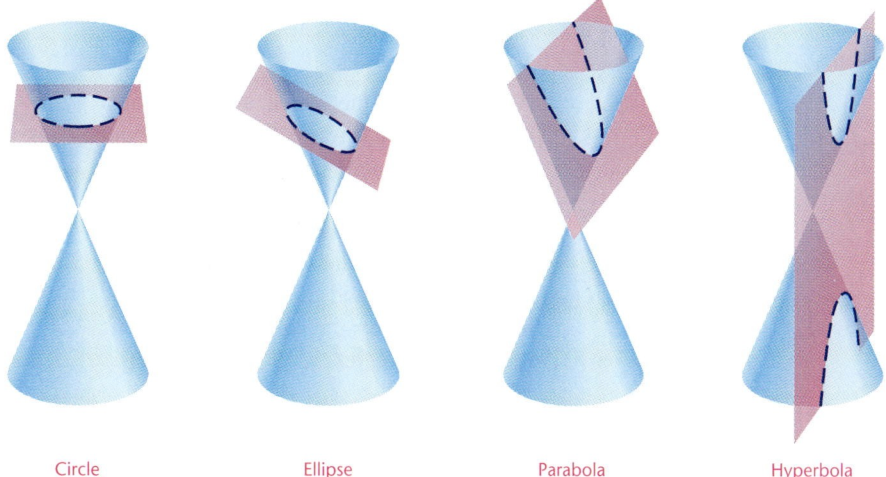

then the intersection curve is called a **parabola.** Finally, if a plane cuts through both nappes, but not through the vertex, the resulting intersection curve is called a **hyperbola.** A plane passing through the vertex of the cone produces a **degenerate conic**—a point, a line, or a pair of lines.

Conic sections are very useful and are readily observed in your immediate surroundings: wheels (circle), the path of water from a garden hose (parabola), some serving platters (ellipses), and the shadow on a wall from a light surrounded by a cylindrical or conical lamp shade (hyperbola) are some examples (see Fig. 2). We will discuss many applications of conics throughout the remainder of this chapter.

FIGURE 2 Examples of conics.

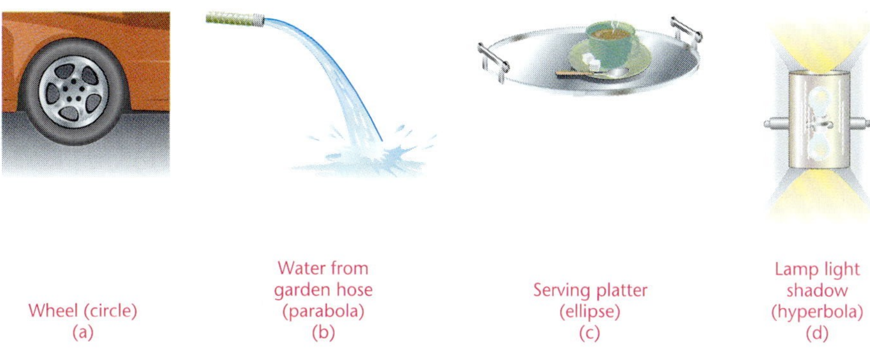

A definition of a conic section that does not depend on the coordinates of points in any coordinate system is called a **coordinate-free definition.** In Section 2-1 we gave a coordinate-free definition of a circle and developed its standard equation in a rectangular coordinate system. In this and the next two sections we will give coordinate-free definitions of a parabola, ellipse, and hyperbola, and we will develop standard equations for each of these conics in a rectangular coordinate system.

• Definition of a Parabola

The following definition of a parabola does not depend on the coordinates of points in any coordinate system:

DEFINITION 1 **Parabola**

A **parabola** is the set of all points in a plane equidistant from a fixed point F and a fixed line L in the plane. The fixed point F is called the **focus,** and the fixed line L is called the **directrix.** A line through the focus perpendicular to the directrix is called the **axis,** and the point on the axis halfway between the directrix and focus is called the **vertex.**

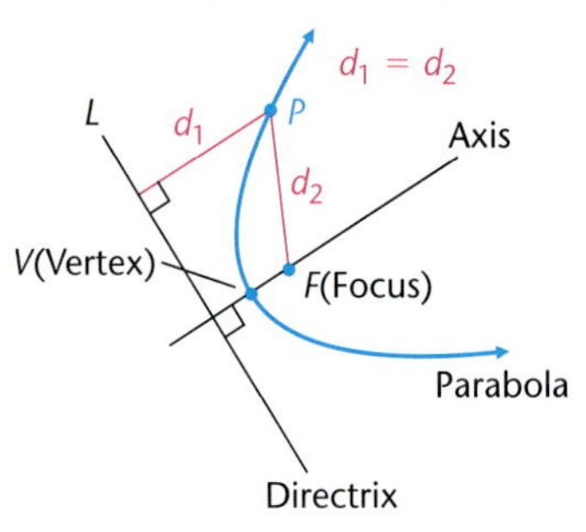

• Drawing a Parabola

Using the definition, we can draw a parabola with fairly simple equipment—a straightedge, a right-angle drawing triangle, a piece of string, a thumbtack, and a pencil. Referring to Figure 3, tape the straightedge along the line AB and place the thumbtack above the line AB. Place one leg of the triangle along the straightedge as indicated, then take a piece of string the same length as the other leg, tie one end to the thumbtack, and fasten the other end with tape at C on the triangle. Now press the string to the edge of the triangle, and keeping the string taut, slide the triangle along the straightedge. Since DE will always equal DF, the resulting curve will be part of a parabola with directrix AB lying along the straightedge and focus F at the thumbtack.

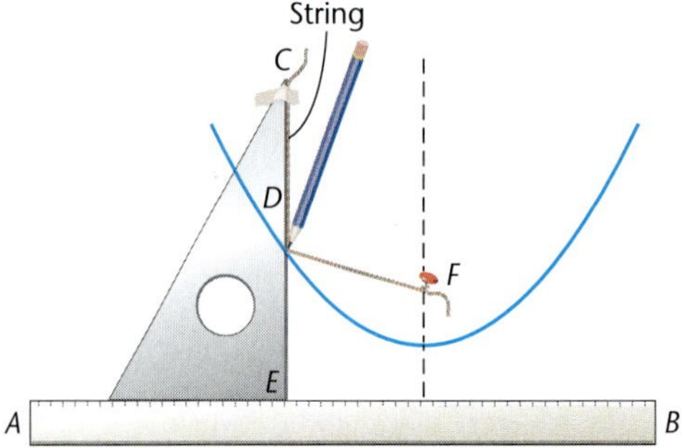

FIGURE 3 Drawing a parabola.

EXPLORE-DISCUSS 1 The line through the focus F that is perpendicular to the axis of a parabola intersects the parabola in two points G and H. Explain why the distance from G to H is twice the distance from F to the directrix of the parabola.

• Standard Equations and Their Graphs

Using the definition of a parabola and the distance-between-two-points formula

$$d = \sqrt{(x_2 - x_1)^2 + (y_2 - y_1)^2} \quad (3)$$

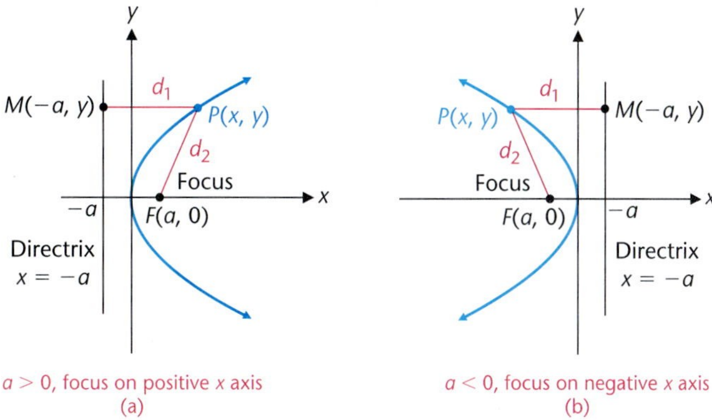

FIGURE 4 Parabola with center at the origin and axis the x axis.

we can derive simple standard equations for a parabola located in a rectangular coordinate system with its vertex at the origin and its axis along a coordinate axis. We start with the axis of the parabola along the x axis and the focus at $F(a, 0)$. We locate the parabola in a coordinate system as in Figure 4 and label key lines and points. This is an important step in finding an equation of a geometric figure in a coordinate system. Note that the parabola opens to the right if $a > 0$ and to the left if $a < 0$. The vertex is at the origin, the directrix is $x = -a$, and the coordinates of M are $(-a, y)$.

The point $P(x, y)$ is a point on the parabola if and only if

$$d_1 = d_2$$

$$d(P, M) = d(P, F)$$

$$\sqrt{(x + a)^2 + (y - y)^2} = \sqrt{(x - a)^2 + (y - 0)^2} \quad \text{Use equation (3).}$$

$$(x + a)^2 = (x - a)^2 + y^2 \quad \text{Square both sides.}$$

$$x^2 + 2ax + a^2 = x^2 - 2ax + a^2 + y^2 \quad \text{Simplify.}$$

$$y^2 = 4ax \tag{4}$$

Equation (4) is the standard equation of a parabola with vertex at the origin, axis the x axis, and focus at $(a, 0)$.

Now we locate the vertex at the origin and focus on the y axis at $(0, a)$. Looking at Figure 5 on the following page, we note that the parabola opens upward if $a > 0$ and downward if $a < 0$. The directrix is $y = -a$, and the coordinates of N are $(x, -a)$. The point $P(x, y)$ is a point on the parabola if and only if

$$d_1 = d_2$$

$$d(P, N) = d(P, F)$$

$$\sqrt{(x - x)^2 + (y + a)^2} = \sqrt{(x - 0)^2 + (y - a)^2} \quad \text{Use equation (3).}$$

$$(y + a)^2 = x^2 + (y - a)^2 \quad \text{Square both sides.}$$

$$y^2 + 2ay + a^2 = x^2 + y^2 - 2ay + a^2 \quad \text{Simplify.}$$

$$x^2 = 4ay \tag{5}$$

Equation (5) is the standard equation of a parabola with vertex at the origin, axis the y axis, and focus at $(0, a)$.

FIGURE 5 Parabola with center at the origin and axis the y axis.

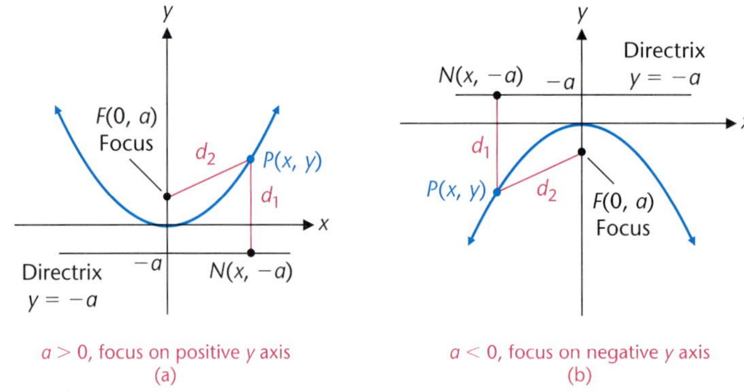

We summarize these results for easy reference in Theorem 1:

Theorem 1 **Standard Equations of a Parabola with Vertex at (0, 0)**

1. $y^2 = 4ax$
Vertex: (0, 0)
Focus: $(a, 0)$
Directrix: $x = -a$
Symmetric with respect to the x axis
Axis the x axis

2. $x^2 = 4ay$
Vertex: (0, 0)
Focus: $(0, a)$
Directrix: $y = -a$
Symmetric with respect to the y axis
Axis the y axis

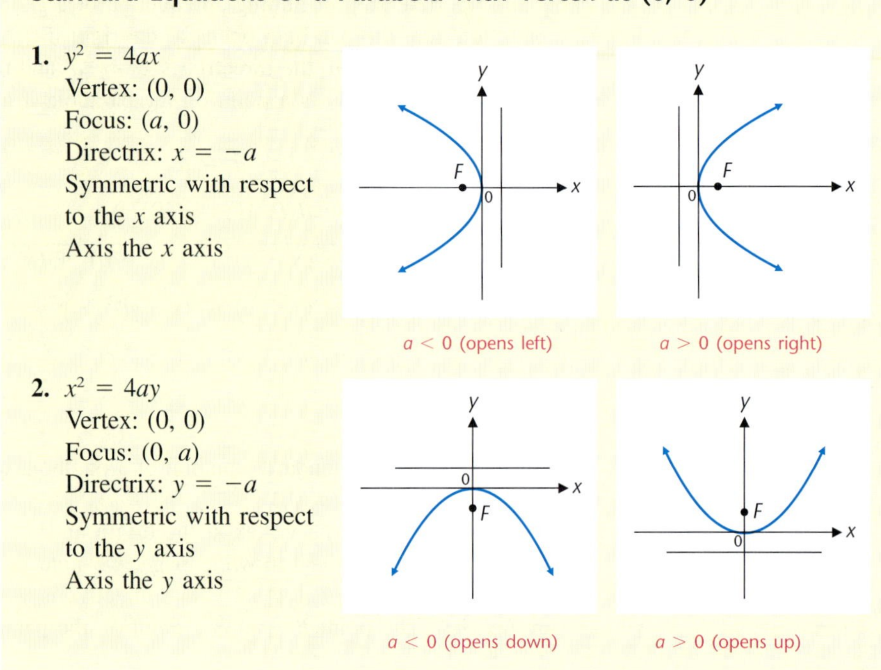

EXAMPLE 1 Graphing $x^2 = 4ay$

Graph $x^2 = -16y$, and locate the focus and directrix.

Solution To graph $x^2 = -16y$, it is convenient to assign y values that make the right side a perfect square, and solve for x. Note that y must be 0 or negative for x to be real. Since the coefficient of y is negative, a must be negative, and the parabola opens downward (Fig. 6).

x	0	±4	±8
y	0	−1	−4

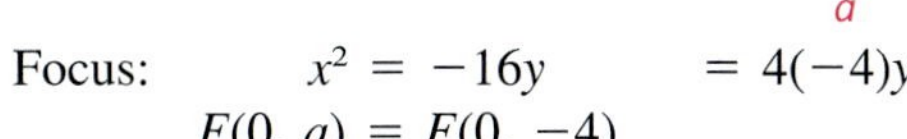

Focus: $x^2 = -16y \qquad = 4(-4)y$

$F(0, a) = F(0, -4)$

Directrix: $y = -a$

$= -(-4) = 4$

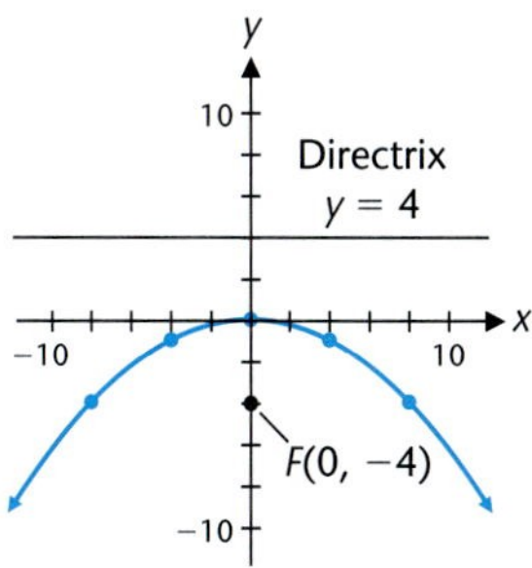

FIGURE 6 $x^2 = -16y$.

Matched Problem 1 Graph $y^2 = -8x$, and locate the focus and directrix.

Remark. To graph the equation $x^2 = -16y$ of Example 1 on a graphing utility, we first solve the equation for y and then graph the function $y = -\frac{1}{16}x^2$. If that same approach is used to graph the equation $y^2 = -8x$ of Matched Problem 1, then $y = \pm\sqrt{-8x}$, and there are two functions to graph. The graph of $y = \sqrt{-8x}$ is the upper half of the parabola, and the graph of $y = -\sqrt{-8x}$ is the lower half (see Fig. 7).

FIGURE 7

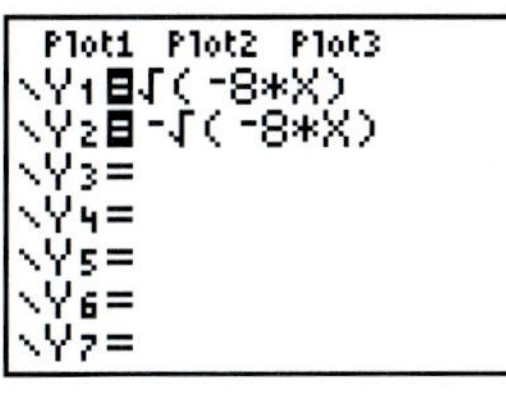

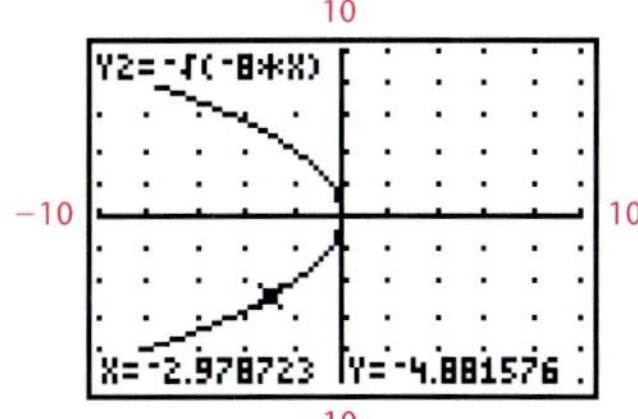

CAUTION A common error in making a quick sketch of $y^2 = 4ax$ or $x^2 = 4ay$ is to sketch the first with the y axis as its axis and the second with the x axis as its axis. The graph of $y^2 = 4ax$ is symmetric with respect to the x axis, and the graph of $x^2 = 4ay$ is symmetric with respect to the y axis, as a quick symmetry check will reveal.

EXAMPLE 2 Finding the Equation of a Parabola

(A) Find the equation of a parabola having the origin as its vertex, the y axis as its axis, and $(-10, -5)$ on its graph.

(B) Find the coordinates of its focus and the equation of its directrix.

Solutions (A) The parabola is opening down and has an equation of the form $x^2 = 4ay$. Since $(-10, -5)$ is on the graph, we have

$$x^2 = 4ay$$
$$(-10)^2 = 4a(-5)$$
$$100 = -20a$$
$$a = -5$$

Thus, the equation of the parabola is

$$x^2 = 4(-5)y$$
$$= -20y$$

(B) Focus: $x^2 = -20y \qquad = 4(\overset{a}{-5})y$

$F(0, a) = F(0, -5)$

Directrix: $y = -a$
$= -(-5)$
$= 5$

Matched Problem 2 (A) Find the equation of a parabola having the origin as its vertex, the x axis as its axis, and $(4, -8)$ on its graph.

(B) Find the coordinates of its focus and the equation of its directrix.

EXPLORE-DISCUSS 2 Consider the graph of an equation in the variables x and y. The equation of its magnification by a factor $k > 0$ is obtained by replacing x and y in the equation by x/k and y/k, respectively. (Of course, a magnification by a factor k between 0 and 1 means an actual reduction in size.)

(A) Show that the magnification by a factor 3 of the circle with equation $x^2 + y^2 = 1$ has equation $x^2 + y^2 = 9$.

(B) Explain why every circle with center at (0, 0) is a magnification of the circle with equation $x^2 + y^2 = 1$.

(C) Find the equation of the magnification by a factor 3 of the parabola with equation $x^2 = y$. Graph both equations.

(D) Explain why every parabola with vertex (0, 0) that opens upward is a magnification of the parabola with equation $x^2 = y$.

• Applications

Parabolic forms are frequently encountered in the physical world. Suspension bridges, arch bridges, microphones, symphony shells, satellite antennas, radio and optical telescopes, radar equipment, solar furnaces, and searchlights are only a few of many items that utilize parabolic forms in their design.

Figure 8(a) illustrates a parabolic reflector used in all reflecting telescopes—from 3- to 6-inch home type to the 200-inch research instrument on Mount Palomar in California. Parallel light rays from distant celestial bodies are reflected to the focus off a parabolic mirror. If the light source is the sun, then the parallel rays are focused at F and we have a solar furnace. Temperatures of over 6,000°C have been achieved by such furnaces. If we locate a light source at F, then the rays in Figure 8(a) reverse, and we have a spotlight or a searchlight. Automobile headlights can use parabolic reflectors with special lenses over the light to diffuse the rays into useful patterns.

Figure 8(b) shows a suspension bridge, such as the Golden Gate Bridge in San Francisco. The suspension cable is a parabola. It is interesting to note that a free-hanging cable, such as a telephone line, does not form a parabola. It forms another curve called a *catenary.*

Figure 8(c) shows a concrete arch bridge. If all the loads on the arch are to be compression loads (concrete works very well under compression), then using physics and advanced mathematics, it can be shown that the arch must be parabolic.

FIGURE 8 Uses of parabolic forms.

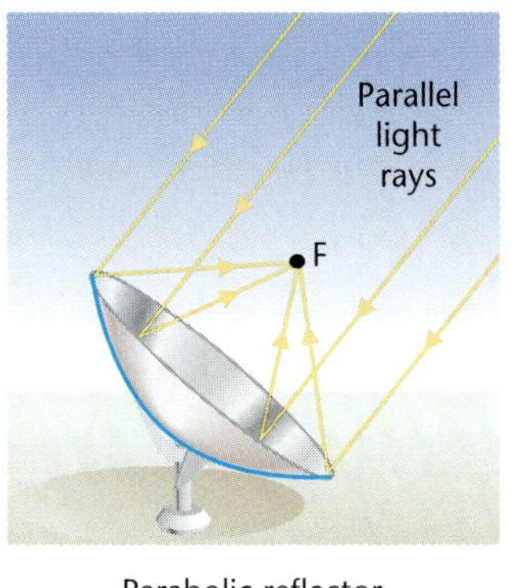

Parabolic reflector

(a)

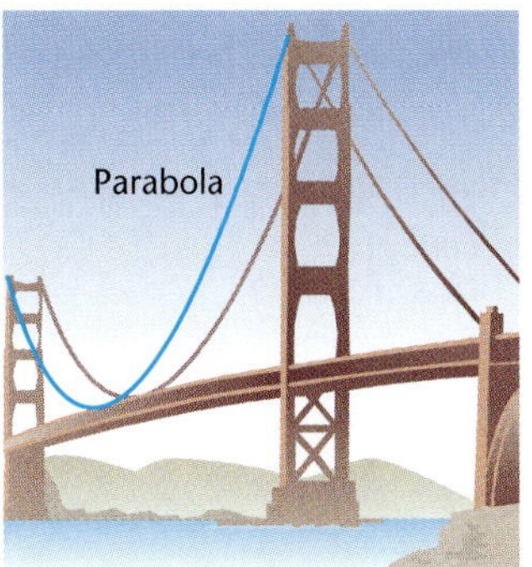

Suspension bridge

(b)

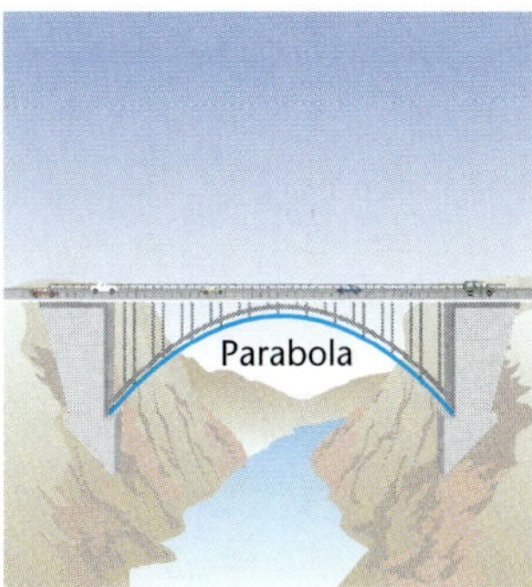

Arch bridge

(c)

EXAMPLE 3 Parabolic Reflector

A **paraboloid** is formed by revolving a parabola about its axis. A spotlight in the form of a paraboloid 5 inches deep has its focus 2 inches from the vertex. Find, to one decimal place, the radius R of the opening of the spotlight.

Solution

Step 1. Locate a parabolic cross section containing the axis in a rectangular coordinate system, and label all known parts and parts to be found. This is a very important step and can be done in infinitely many ways. Since we are in charge, we can make things simpler for ourselves by locating the vertex at the origin and choosing a coordinate axis as the axis. We choose the y axis as the axis of the parabola with the parabola opening upward. See Figure 9 on the following page.

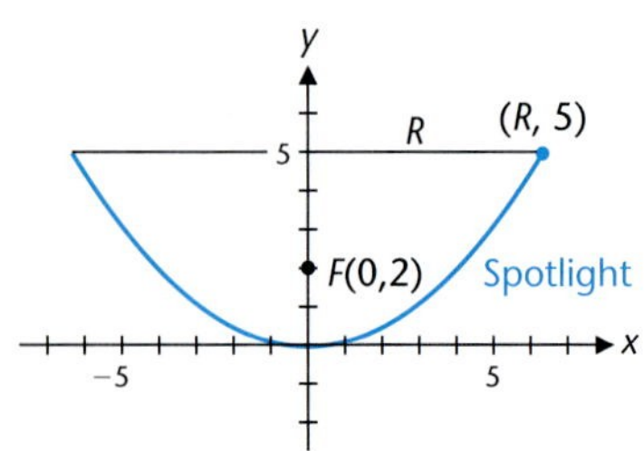

FIGURE 9

Step 2. Find the equation of the parabola in the figure. Since the parabola has the y axis as its axis and the vertex at the origin, the equation is of the form

$$x^2 = 4ay$$

We are given $F(0, a) = F(0, 2)$; thus, $a = 2$, and the equation of the parabola is

$$x^2 = 8y$$

Step 3. Use the equation found in step 2 to find the radius R of the opening. Since $(R, 5)$ is on the parabola, we have

$$R^2 = 8(5)$$
$$R = \sqrt{40} \approx 6.3 \text{ inches}$$

Matched Problem 3 Repeat Example 3 with a paraboloid 12 inches deep and a focus 9 inches from the vertex.

Answers to Matched Problems

1. Focus: $(-2, 0)$
Directrix: $x = 2$

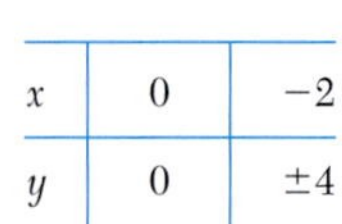

x	0	-2
y	0	± 4

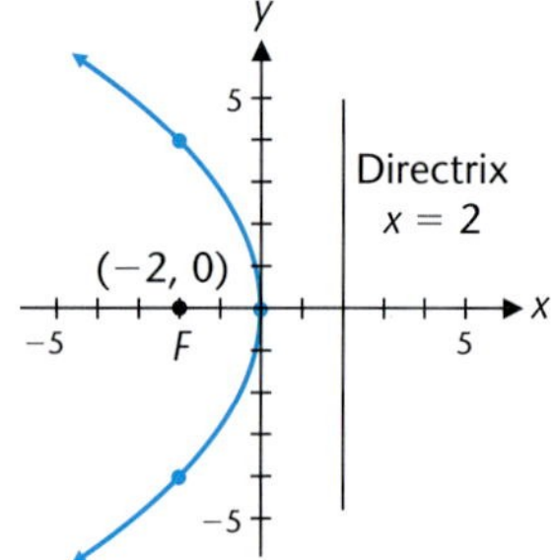

2. (A) $y^2 = 16x$ (B) Focus: $(4, 0)$; Directrix: $x = -4$
3. $R = 20.8$ in.

EXERCISE 11-1

A

In Problems 1–12, graph each equation, and locate the focus and directrix.

1. $y^2 = 4x$ **2.** $y^2 = 8x$ **3.** $x^2 = 8y$
4. $x^2 = 4y$ **5.** $y^2 = -12x$ **6.** $y^2 = -4x$
7. $x^2 = -4y$ **8.** $x^2 = -8y$ **9.** $y^2 = -20x$
10. $x^2 = -24y$ **11.** $x^2 = 10y$ **12.** $y^2 = 6x$

Find the coordinates to two decimal places of the focus for each parabola in Problems 13–18.

13. $y^2 = 39x$ **14.** $x^2 = 58y$ **15.** $x^2 = -105y$
16. $y^2 = -93x$ **17.** $y^2 = -77x$ **18.** $x^2 = -205y$

B

In Problems 19–24, find the equation of a parabola with vertex at the origin, axis the x or y axis, and:

19. Focus $(-6, 0)$

20. Directrix $y = 8$

21. Directrix $y = -5$

22. Focus $(3, 0)$

23. Focus $(0, -\frac{1}{3})$

24. Directrix $y = -\frac{1}{2}$

In Problems 25–30, find the equation of the parabola having its vertex at the origin, its axis as indicated, and passing through the indicated point.

25. x axis; $(-4, -20)$

26. y axis; $(30, -15)$

27. y axis; $(9, -27)$

28. x axis; $(121, 11)$

29. x axis; $(-8, -2)$

30. y axis; $(-\sqrt{2}, 3)$

In Problems 31–34, find the first-quadrant points of intersection for each system of equations to three decimal places.

Check Problems 31–34 with a graphing utility.

31. $x^2 = 4y$
$y^2 = 4x$

32. $y^2 = 3x$
$x^2 = 3y$

33. $y^2 = 6x$
$x^2 = 5y$

34. $x^2 = 7y$
$y^2 = 2x$

35. Consider the parabola with equation $x^2 = 4ay$.
(A) How many lines through $(0, 0)$ intersect the parabola in exactly one point? Find their equations.
(B) Find the coordinates of all points of intersection of the parabola with the line through $(0, 0)$ having slope $m \neq 0$.

36. Find the coordinates of all points of intersection of the parabola with equation $x^2 = 4ay$ and the parabola with equation $y^2 = 4bx$.

37. If a line through the focus contains two points A and B of a parabola, then the line segment AB is called a **focal chord**. Find the coordinates of A and B for the focal chord that is perpendicular to the axis of the parabola $x^2 = 4ay$.

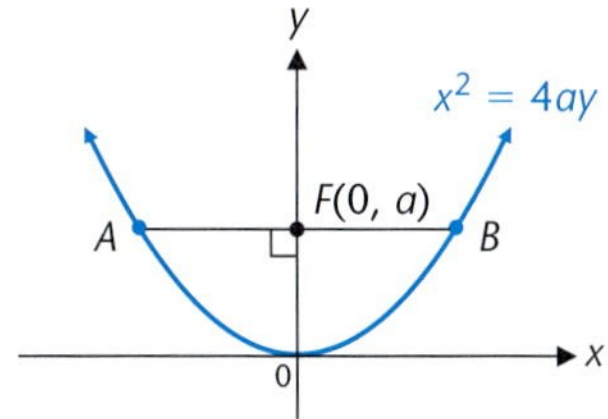

Figure for 37 and 38

38. Find the length of the focal chord AB that is perpendicular to the axis of the parabola $x^2 = 4ay$.

In Problems 39–42, determine whether the statement is true or false. If true, explain why. If false, give a counterexample.

39. If a is real, then the graph of $y^2 = 4ax$ is a parabola.

40. If a is negative, then the graph of $y^2 = 4ax$ is a parabola.

41. Every vertical line intersects the graph of $x^2 = 4y$.

42. Every nonhorizontal line intersects the graph of $x^2 = 4y$.

C

In Problems 43–46, use the definition of a parabola and the distance formula to find the equation of a parabola with:

43. Directrix $y = -4$ and focus $(2, 2)$

44. Directrix $y = 2$ and focus $(-3, 6)$

45. Directrix $x = 2$ and focus $(6, -4)$

46. Directrix $x = -3$ and focus $(1, 4)$

In Problems 47–50, use a graphing utility to find the coordinates of all points of intersection to two decimal places.

47. $x^2 = 8y$, $y = 5x + 4$

48. $x^2 = 3y$, $7x + 4y = 11$

49. $x^2 = -8y$, $y^2 = -5x$

50. $y^2 = 6x$, $2x - 9y = 13$

APPLICATIONS

51. Engineering. The parabolic arch in the concrete bridge in the figure must have a clearance of 50 feet above the water and span a distance of 200 feet. Find the equation of the parabola after inserting a coordinate system with the origin at the vertex of the parabola and the vertical y axis (pointing upward) along the axis of the parabola.

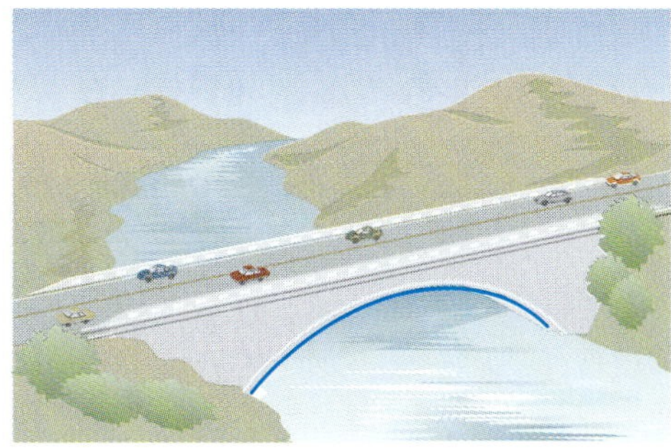

52. Astronomy. The cross section of a parabolic reflector with 6-inch diameter is ground so that its vertex is 0.15 inch below the rim (see the figure).

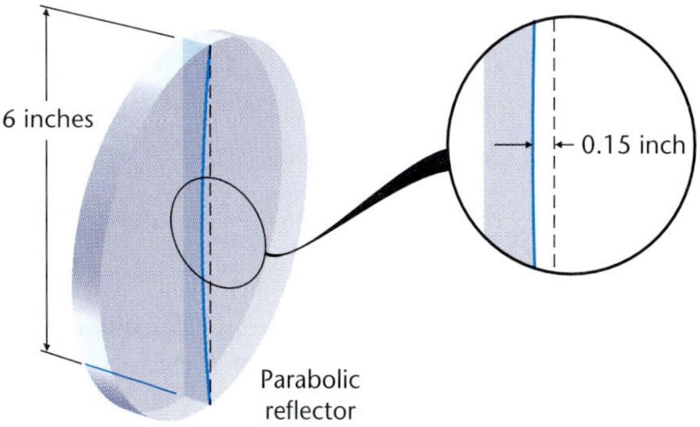

(A) Find the equation of the parabola after inserting an xy coordinate system with the vertex at the origin, the y axis (pointing upward) the axis of the parabola.
(B) How far is the focus from the vertex?

53. Space Science. A designer of a 200-foot-diameter parabolic electromagnetic antenna for tracking space probes wants to place the focus 100 feet above the vertex (see the figure).

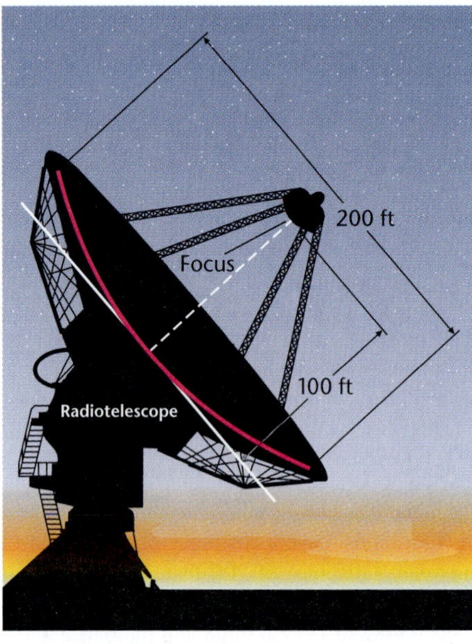

(A) Find the equation of the parabola using the axis of the parabola as the y axis (up positive) and vertex at the origin.

(B) Determine the depth of the parabolic reflector.

54. Signal Light. A signal light on a ship is a spotlight with parallel reflected light rays (see the figure). Suppose the parabolic reflector is 12 inches in diameter and the light source is located at the focus, which is 1.5 inches from the vertex.

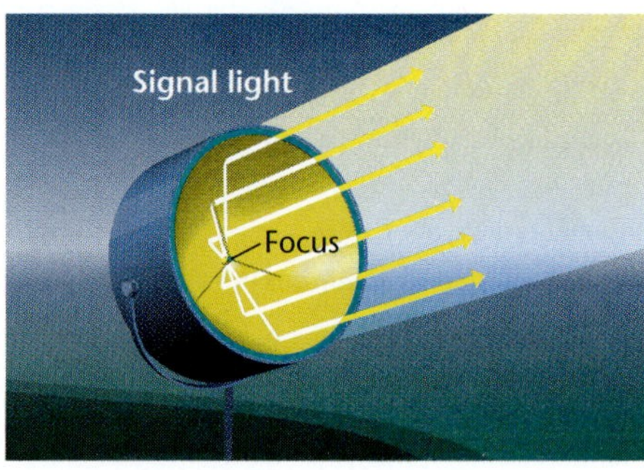

(A) Find the equation of the parabola using the axis of the parabola as the x axis (right positive) and vertex at the origin.

(B) Determine the depth of the parabolic reflector.

SECTION 11-2 Ellipse

- Definition of an Ellipse
- Drawing an Ellipse
- Standard Equations and Their Graphs
- Applications

We start our discussion of the ellipse with a coordinate-free definition. Using this definition, we show how an ellipse can be drawn and we derive standard equations for ellipses specially located in a rectangular coordinate system.

• Definition of an Ellipse

The following is a coordinate-free definition of an ellipse:

DEFINITION 1 **Ellipse**

An **ellipse** is the set of all points P in a plane such that the sum of the distances of P from two fixed points in the plane is constant. Each of the fixed points, F' and F, is called a **focus,** and together they are called **foci.** Referring to the figure, the line segment $V'V$ through the foci is the **major axis.** The perpendicular bisector $B'B$ of the major axis is the **minor axis.** Each end of the major axis,

V' and V, is called a **vertex.** The midpoint of the line segment $F'F$ is called the **center** of the ellipse.

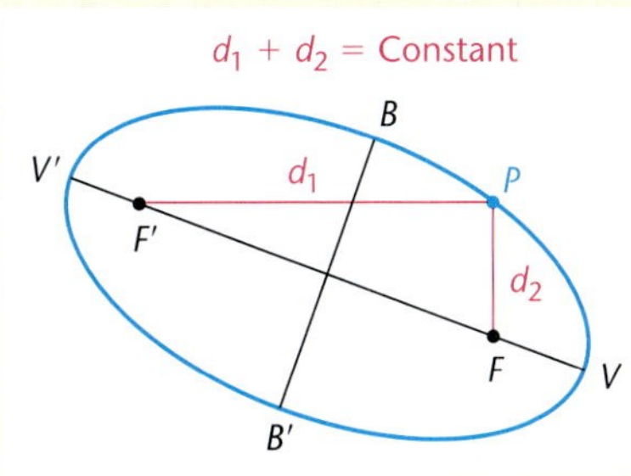

• Drawing an Ellipse

An ellipse is easy to draw. All you need is a piece of string, two thumbtacks, and a pencil or pen (see Fig. 1). Place the two thumbtacks in a piece of cardboard. These form the foci of the ellipse. Take a piece of string longer than the distance between the two thumbtacks—this represents the constant in the definition—and tie each end to a thumbtack. Finally, catch the tip of a pencil under the string and move it while keeping the string taut. The resulting figure is by definition an ellipse. Ellipses of different shapes result, depending on the placement of thumbtacks and the length of the string joining them.

FIGURE 1 Drawing an ellipse.

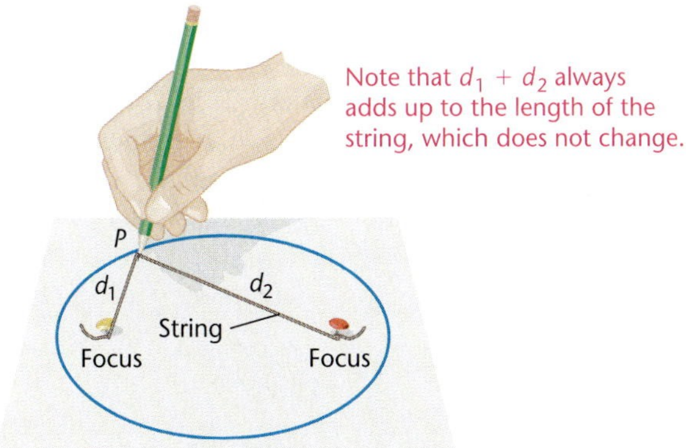

• Standard Equations and Their Graphs

Using the definition of an ellipse and the distance-between-two-points formula, we can derive standard equations for an ellipse located in a rectangular coordinate system. We start by placing an ellipse in the coordinate system with the foci on the x axis equidistant from the origin at $F'(-c, 0)$ and $F(c, 0)$, as in Figure 2.

FIGURE 2 Ellipse with foci on x axis.

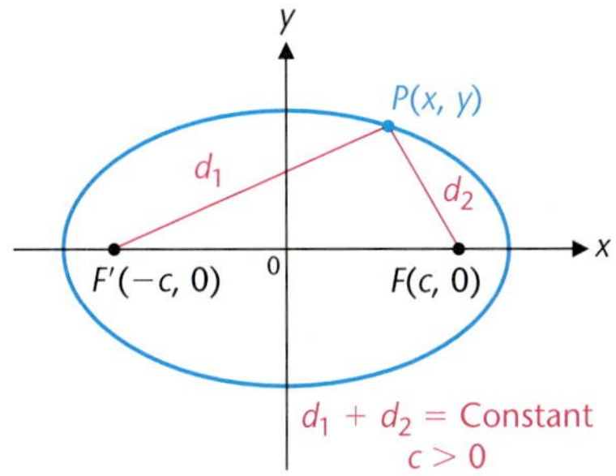

For reasons that will become clear soon, it is convenient to represent the constant sum $d_1 + d_2$ by $2a$, $a > 0$. Also, the geometric fact that the sum of the lengths of any two sides of a triangle must be greater than the third side can be applied to Figure 2 to derive the following useful result:

$$\begin{aligned} d(F', P) + d(P, F) &> d(F', F) \\ d_1 + d_2 &> 2c \\ 2a &> 2c \\ a &> c \end{aligned} \tag{1}$$

We will use this result in the derivation of the equation of an ellipse, which we now begin.

Referring to Figure 2, the point $P(x, y)$ is on the ellipse if and only if

$$\begin{aligned} d_1 + d_2 &= 2a \\ d(P, F') + d(P, F) &= 2a \\ \sqrt{(x + c)^2 + (y - 0)^2} + \sqrt{(x - c)^2 + (y - 0)^2} &= 2a \end{aligned}$$

After eliminating radicals and simplifying, a good exercise for you, we obtain

$$(a^2 - c^2)x^2 + a^2y^2 = a^2(a^2 - c^2) \tag{2}$$

$$\frac{x^2}{a^2} + \frac{y^2}{a^2 - c^2} = 1 \tag{3}$$

Dividing both sides of equation (2) by $a^2(a^2 - c^2)$ is permitted, since neither a^2 nor $a^2 - c^2$ is 0. From equation (1), $a > c$; thus $a^2 > c^2$ and $a^2 - c^2 > 0$. The constant a was chosen positive at the beginning.

To simplify equation (3) further, we let

$$b^2 = a^2 - c^2 \qquad b > 0 \tag{4}$$

to obtain

$$\frac{x^2}{a^2} + \frac{y^2}{b^2} = 1 \tag{5}$$

From equation (5) we see that the x intercepts are $x = \pm a$ and the y intercepts are $y = \pm b$. The x intercepts are also the vertices. Thus,

Major axis length = $2a$

Minor axis length = $2b$

To see that the major axis is longer than the minor axis, we show that $2a > 2b$. Returning to equation (4),

$$\begin{aligned} b^2 &= a^2 - c^2 \qquad a, b, c > 0 \\ b^2 + c^2 &= a^2 \end{aligned}$$

$$b^2 < a^2 \qquad \text{Definition of } <$$

$$b^2 - a^2 < 0$$

$$(b - a)(b + a) < 0$$

$$b - a < 0 \qquad \text{Since } b + a \text{ is positive, } b - a \text{ must be negative.}$$

$$b < a$$

$$2b < 2a$$

$$2a > 2b$$

$$\begin{pmatrix}\text{Length of}\\\text{major axis}\end{pmatrix} > \begin{pmatrix}\text{Length of}\\\text{minor axis}\end{pmatrix}$$

If we start with the foci on the y axis at $F(0, c)$ and $F'(0, -c)$ as in Figure 3, instead of on the x axis as in Figure 2, then, following arguments similar to those used for the first derivation, we obtain

$$\frac{x^2}{b^2} + \frac{y^2}{a^2} = 1 \qquad a > b \tag{6}$$

where the relationship among a, b, and c remains the same as before:

$$b^2 = a^2 - c^2 \tag{7}$$

The center is still at the origin, but the major axis is now along the y axis and the minor axis is along the x axis.

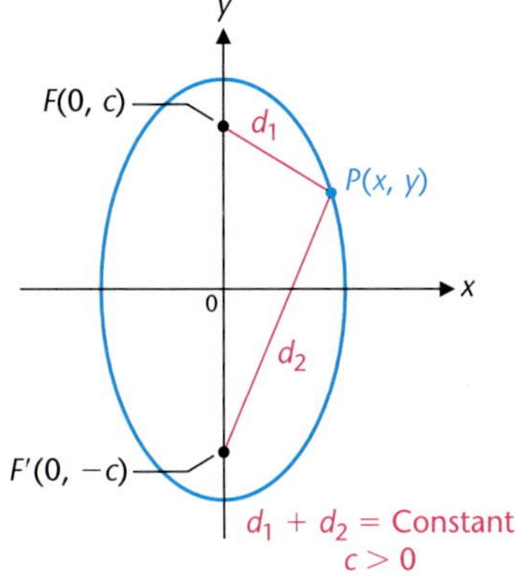

FIGURE 3 Ellipse with foci on y axis.

To sketch graphs of equations of the form (5) or (6) is an easy matter. We find the x and y intercepts and sketch in an appropriate ellipse. Since replacing x with $-x$, or y with $-y$ produces an equivalent equation, we conclude that the graphs are symmetric with respect to the x axis, y axis, and origin. If further accuracy is required, additional points can be found with the aid of a calculator and the use of symmetry properties.

Given an equation of the form (5) or (6), how can we find the coordinates of the foci without memorizing or looking up the relation $b^2 = a^2 - c^2$? There is a simple geometric relationship in an ellipse that enables us to get the same result using the Pythagorean theorem. To see this relationship, refer to Figure 4(a). Then, using the definition of an ellipse and $2a$ for the constant sum, as we did in deriving the standard equations, we see that

$$\begin{aligned} d + d &= 2a \\ 2d &= 2a \\ d &= a \end{aligned}$$

Thus:

The length of the line segment from the end of a minor axis to a focus is the same as half the length of a major axis.

This geometric relationship is illustrated in Figure 4(b). Using the Pythagorean theorem for the triangle in Figure 4(b), we have

$$b^2 + c^2 = a^2$$

or

$$b^2 = a^2 - c^2 \quad \text{Equations (4) and (7)}$$

or

$$c^2 = a^2 - b^2 \quad \text{Useful for finding the foci, given } a \text{ and } b$$

Thus, we can find the foci of an ellipse given the intercepts a and b simply by using the triangle in Figure 4(b) and the Pythagorean theorem.

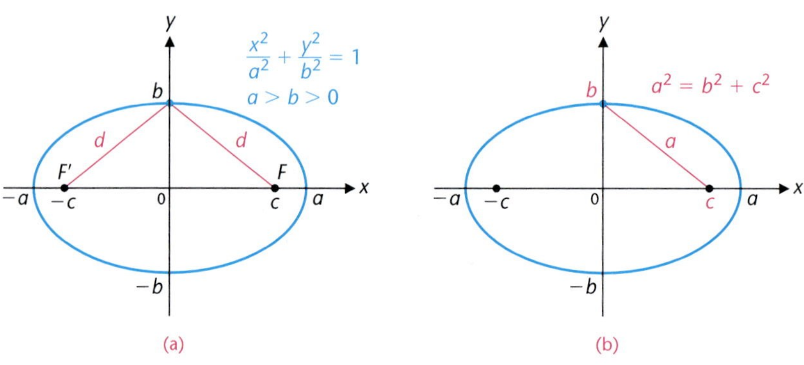

FIGURE 4 Geometric relationships.

We summarize all of these results for convenient reference in Theorem 1.

Theorem 1 **Standard Equations of an Ellipse with Center at (0, 0)**

1. $\frac{x^2}{a^2} + \frac{y^2}{b^2} = 1 \qquad a > b > 0$

x intercepts: $\pm a$ (vertices)
y intercepts: $\pm b$
Foci: $F'(-c, 0)$, $F(c, 0)$

$$c^2 = a^2 - b^2$$

Major axis length $= 2a$
Minor axis length $= 2b$

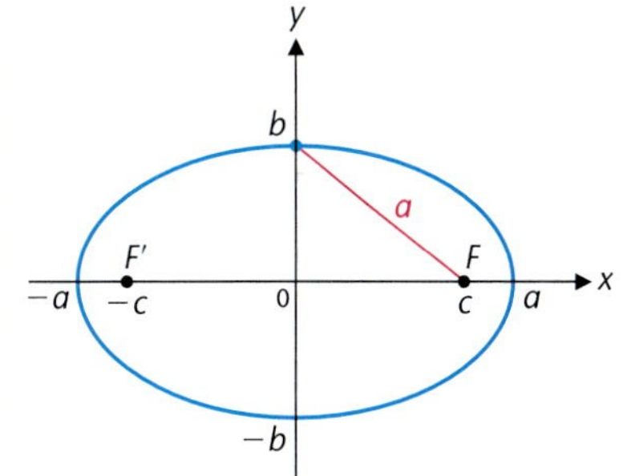

2. $\dfrac{x^2}{b^2} + \dfrac{y^2}{a^2} = 1 \qquad a > b > 0$

x intercepts: $\pm b$
y intercepts: $\pm a$ (vertices)
Foci: $F'(0, -c)$, $F(0, c)$

$$c^2 = a^2 - b^2$$

Major axis length $= 2a$
Minor axis length $= 2b$

[*Note:* Both graphs are symmetric with respect to the x axis, y axis, and origin. Also, the major axis is always longer than the minor axis.]

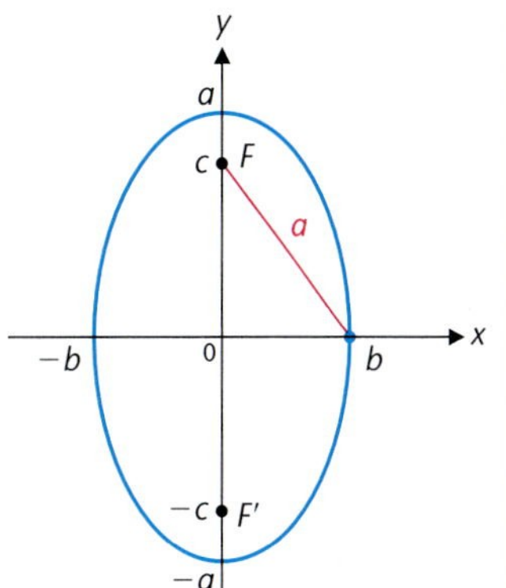

EXPLORE-DISCUSS 1 The line through a focus F of an ellipse that is perpendicular to the major axis intersects the ellipse in two points G and H. For each of the two standard equations of an ellipse with center $(0, 0)$, find an expression in terms of a and b for the distance from G to H.

EXAMPLE 1 Graphing Ellipses

Sketch the graph of each equation, find the coordinates of the foci, and find the lengths of the major and minor axes.

(A) $9x^2 + 16y^2 = 144$ (B) $2x^2 + y^2 = 10$

Solutions (A) First, write the equation in standard form by dividing both sides by 144:

$$9x^2 + 16y^2 = 144$$

$$\frac{9x^2}{144} + \frac{16y^2}{144} = \frac{144}{144}$$

$$\frac{x^2}{16} + \frac{y^2}{9} = 1 \qquad a^2 = 16 \text{ and } b^2 = 9$$

Locate the intercepts:

$$x \text{ intercepts: } \pm 4$$
$$y \text{ intercepts: } \pm 3$$

and sketch in the ellipse, as shown in Figure 5.

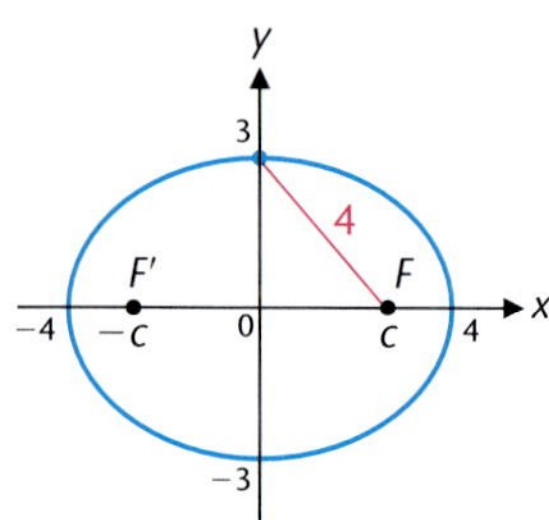

FIGURE 5 $9x^2 + 16y^2 = 144.$

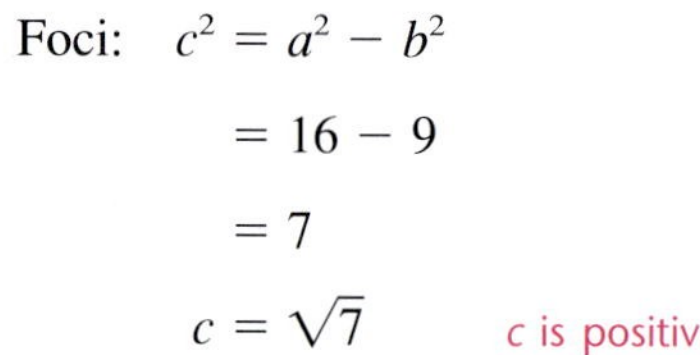

$$\begin{aligned} \text{Foci: } c^2 &= a^2 - b^2 \\ &= 16 - 9 \\ &= 7 \\ c &= \sqrt{7} \qquad \textit{c is positive} \end{aligned}$$

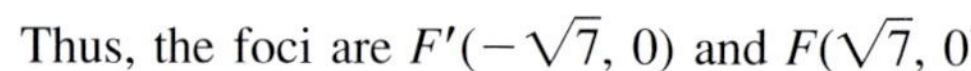

Thus, the foci are $F'(-\sqrt{7}, 0)$ and $F(\sqrt{7}, 0)$.

$$\text{Major axis length} = 2(4) = 8$$
$$\text{Minor axis length} = 2(3) = 6$$

(B) Write the equation in standard form by dividing both sides by 10:

$$2x^2 + y^2 = 10$$

$$\frac{2x^2}{10} + \frac{y^2}{10} = \frac{10}{10}$$

$$\frac{x^2}{5} + \frac{y^2}{10} = 1 \qquad a^2 = 10 \text{ and } b^2 = 5$$

Locate the intercepts:

$$x \text{ intercepts: } \pm\sqrt{5} \approx \pm 2.24$$
$$y \text{ intercepts: } \pm\sqrt{10} \approx \pm 3.16$$

and sketch in the ellipse, as shown in Figure 6.

FIGURE 6 $2x^2 + y^2 = 10.$

$$\begin{aligned} \text{Foci: } c^2 &= a^2 - b^2 \\ &= 10 - 5 \\ &= 5 \\ c &= \sqrt{5} \end{aligned}$$

Thus, the foci are $F'(0, -\sqrt{5})$ and $F(0, \sqrt{5})$.

$$\text{Major axis length} = 2\sqrt{10} \approx 6.32$$
$$\text{Minor axis length} = 2\sqrt{5} \approx 4.47$$

Remark. To graph the equation $9x^2 + 16y^2 = 144$ of Example 1A on a graphing utility we first solve the equation for y, obtaining $y = \pm\sqrt{\dfrac{144 - 9x^2}{16}}$. We then graph each of the two functions. The graph of $y = \sqrt{\dfrac{144 - 9x^2}{16}}$ is the upper half of the ellipse, and the graph of $y = -\sqrt{\dfrac{144 - 9x^2}{16}}$ is the lower half.

Matched Problem 1 Sketch the graph of each equation, find the coordinates of the foci, and find the lengths of the major and minor axes.

(A) $x^2 + 4y^2 = 4$ (B) $3x^2 + y^2 = 18$

EXAMPLE 2 **Finding the Equation of an Ellipse**

Find an equation of an ellipse in the form

$$\frac{x^2}{M} + \frac{y^2}{N} = 1 \qquad M, N > 0$$

if the center is at the origin, the major axis is along the y axis, and:

(A) Length of major axis = 20
Length of minor axis = 12

(B) Length of major axis = 10
Distance of foci from center = 4

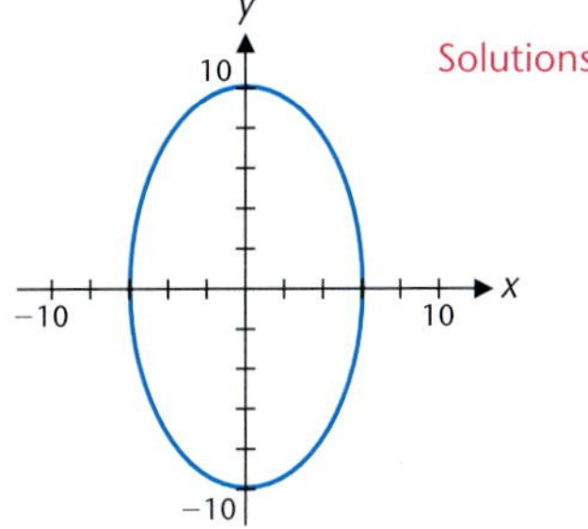

FIGURE 7 $\dfrac{x^2}{36} + \dfrac{y^2}{100} = 1.$

Solutions (A) Compute x and y intercepts and make a rough sketch of the ellipse, as shown in Figure 7.

$$\frac{x^2}{b^2} + \frac{y^2}{a^2} = 1$$

$$a = \frac{20}{2} = 10 \qquad b = \frac{12}{2} = 6$$

$$\frac{x^2}{36} + \frac{y^2}{100} = 1$$

(B) Make a rough sketch of the ellipse, as shown in Figure 8; locate the foci and y intercepts, then determine the x intercepts using the special triangle relationship discussed earlier.

$$\frac{x^2}{b^2} + \frac{y^2}{a^2} = 1$$

$$a = \frac{10}{2} = 5 \qquad b^2 = 5^2 - 4^2 = 25 - 16 = 9$$

$$b = 3$$

$$\frac{x^2}{9} + \frac{y^2}{25} = 1$$

FIGURE 8 $\dfrac{x^2}{9} + \dfrac{y^2}{25} = 1.$

Matched Problem 2 Find an equation of an ellipse in the form

$$\frac{x^2}{M} + \frac{y^2}{N} = 1 \qquad M, N > 0$$

if the center is at the origin, the major axis is along the x axis, and:

(A) Length of major axis = 50
Length of minor axis = 30

(B) Length of minor axis = 16
Distance of foci from center = 6

EXPLORE-DISCUSS 2 Consider the graph of an equation in the variables x and y. The equation of its magnification by a factor $k > 0$ is obtained by replacing x and y in the equation by x/k and y/k, respectively.

(A) Find the equation of the magnification by a factor 3 of the ellipse with equation $(x^2/4) + y^2 = 1$. Graph both equations.

(B) Give an example of an ellipse with center (0, 0) with $a > b$ that is not a magnification of $(x^2/4) + y^2 = 1$.

(C) Find the equations of all ellipses that are magnifications of $(x^2/4) + y^2 = 1$.

• Applications

You are no doubt aware of many occurrences and uses of elliptical forms: orbits of satellites, planets, and comets; shapes of galaxies; gears and cams; some airplane wings, boat keels, and rudders; tabletops; public fountains; and domes in buildings are a few examples (see Fig. 9). A fairly recent application in medicine is the use of elliptical reflectors and ultrasound to break up kidney stones.

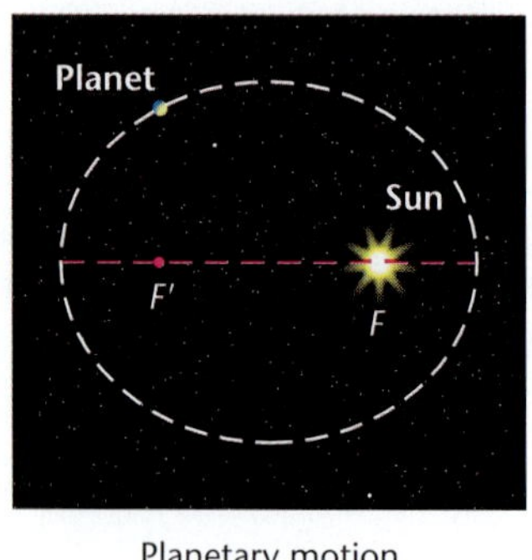

Planetary motion
(a)

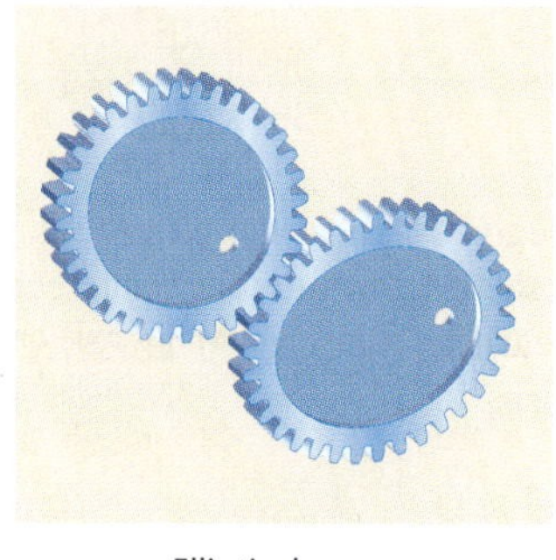
Elliptical gears
(b)

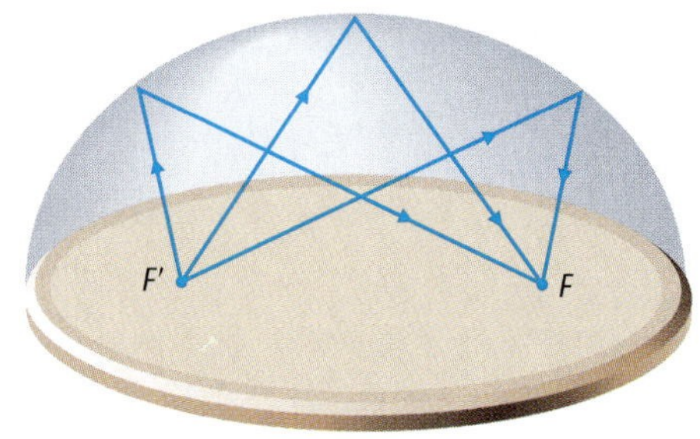

Elliptical dome
(c)

FIGURE 9 Uses of elliptical forms.

Johannes Kepler (1571–1630), a German astronomer, discovered that planets move in elliptical orbits, with the sun at a focus, and not in circular orbits as had been thought before [Fig. 9(a)]. Figure 9(b) shows a pair of elliptical gears with pivot points at foci. Such gears transfer constant rotational speed to variable rotational speed, and vice versa. Figure 9(c) shows an elliptical dome. An interesting property of such a dome is that a sound or light source at one focus will reflect off the dome and pass through the other focus. One of the chambers in the Capitol Building in Washington, D.C., has such a dome, and is referred to as a whispering room because a whispered sound at one focus can be easily heard at the other focus.

Answers to Matched Problems

1. (A)

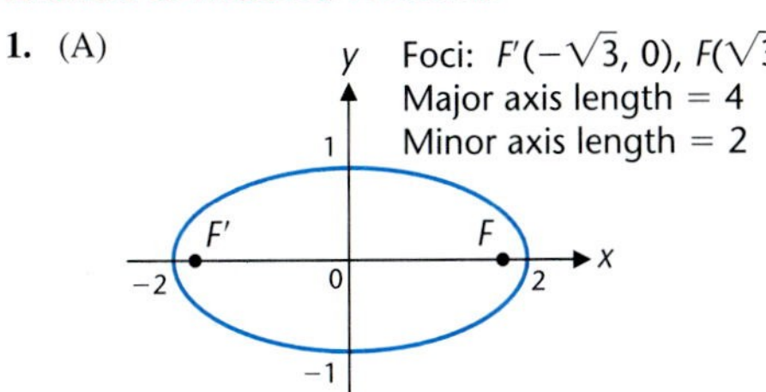

(B)

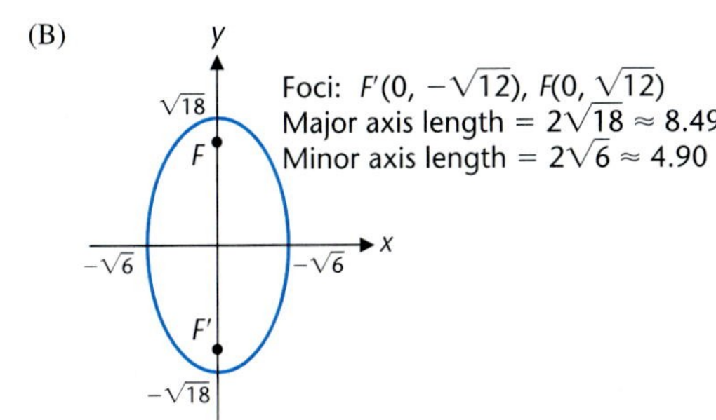

2. (A) $\dfrac{x^2}{625} + \dfrac{y^2}{225} = 1$ (B) $\dfrac{x^2}{100} + \dfrac{y^2}{64} = 1$

EXERCISE 11-2

A

In Problems 1–6, sketch a graph of each equation, find the coordinates of the foci, and find the lengths of the major and minor axes.

1. $\dfrac{x^2}{25} + \dfrac{y^2}{4} = 1$ **2.** $\dfrac{x^2}{9} + \dfrac{y^2}{4} = 1$

3. $\dfrac{x^2}{4} + \dfrac{y^2}{25} = 1$ **4.** $\dfrac{x^2}{4} + \dfrac{y^2}{9} = 1$

5. $x^2 + 9y^2 = 9$ **6.** $4x^2 + y^2 = 4$

B

In Problems 7–12, sketch a graph of each equation, find the coordinates of the foci, and find the lengths of the major and minor axes.

7. $25x^2 + 9y^2 = 225$ **8.** $16x^2 + 25y^2 = 400$

9. $2x^2 + y^2 = 12$ **10.** $4x^2 + 3y^2 = 24$

11. $4x^2 + 7y^2 = 28$ **12.** $3x^2 + 2y^2 = 24$

In Problems 13–18, find an equation of an ellipse in the form

$$\frac{x^2}{M} + \frac{y^2}{N} = 1 \qquad M, N > 0$$

if the center is at the origin, and:

13. Major axis on y axis
Major axis length = 6
Minor axis length = 2

14. Major axis on x axis
Major axis length = 32
Minor axis length = 30

15. Major axis on x axis
Minor axis length = 10
Distance of foci from center = 4

16. Major axis on y axis
Major axis length = 16
Distance of foci from center = 7

17. Major axis on y axis
Major axis length $= 24$
Distance between foci $= 2$

18. Major axis on x axis
Major axis length $= 4$
Distance between foci $= 50$

19. Explain why an equation whose graph is an ellipse does not define a function.

20. Consider all ellipses having $(0, \pm 1)$ as the ends of the minor axis. Describe the connection between the elongation of the ellipse and the distance from a focus to the origin.

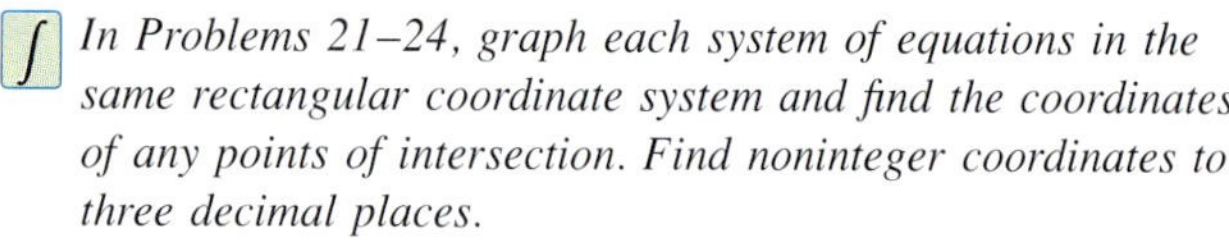

In Problems 21–24, graph each system of equations in the same rectangular coordinate system and find the coordinates of any points of intersection. Find noninteger coordinates to three decimal places.

*Check Problems 21–24 with a graphing utility.**

21. $16x^2 + 25y^2 = 400$
$2x - 5y = 10$

22. $25x^2 + 16y^2 = 400$
$5x + 8y = 20$

23. $25x^2 + 16y^2 = 400$
$25x^2 - 36y = 0$

24. $16x^2 + 25y^2 = 400$
$3x^2 - 20y = 0$

In Problems 25–28, find the first-quadrant points of intersection for each system of equations to three decimal places.

Check Problems 25–28 with a graphing utility.

25. $5x^2 + 2y^2 = 63$
$2x - y = 0$

26. $3x^2 + 4y^2 = 57$
$x - 2y = 0$

27. $2x^2 + 3y^2 = 33$
$x^2 - 8y = 0$

28. $3x^2 + 2y^2 = 43$
$x^2 - 12y = 0$

In Problems 29–32, determine whether the statement is true or false. If true, explain why. If false, give a counterexample.

29. The line segment joining the foci of an ellipse has greater length than the minor axis.

30. There is exactly one ellipse with center $(0, 0)$ and foci $(\pm 1, 0)$.

31. Every line through the center of $x^2 + 4y^2 = 16$ intersects the ellipse in exactly two points.

32. Every nonvertical line through a vertex of $x^2 + 4y^2 = 16$ intersects the ellipse in exactly two points.

C

33. Find an equation of the set of points in a plane, each of whose distance from $(2, 0)$ is one-half its distance from the line $x = 8$. Identify the geometric figure.

34. Find an equation of the set of points in a plane, each of whose distance from $(0, 9)$ is three-fourths its distance from the line $y = 16$. Identify the geometric figure.

*Please note that use of a graphing utility is not required to complete these exercises. Checking them with a g.u. is optional.

In Problems 35–38, use a graphing utility to find the coordinates of all points of intersection to two decimal places.

35. $x^2 + 3y^2 = 20,\ 4x + 5y = 11$

36. $8x^2 + 35y^2 = 3{,}600,\ x^2 = -25y$

37. $50x^2 + 4y^2 = 1{,}025,\ 9x^2 + 2y^2 = 300$

38. $2x^2 + 7y^2 = 95,\ 13x^2 + 6y^2 = 63$

APPLICATIONS

39. Engineering. The semielliptical arch in the concrete bridge in the figure must have a clearance of 12 feet above the water and span a distance of 40 feet. Find the equation of the ellipse after inserting a coordinate system with the center of the ellipse at the origin and the major axis on the x axis. The y axis points up, and the x axis points to the right. How much clearance above the water is there 5 feet from the bank?

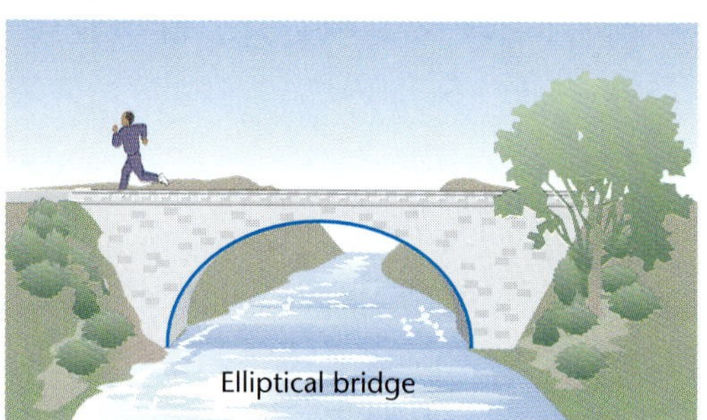

Elliptical bridge

40. Design. A 4×8 foot elliptical tabletop is to be cut out of a 4×8 foot rectangular sheet of teak plywood (see the figure). To draw the ellipse on the plywood, how far should the foci be located from each edge and how long a piece of string must be fastened to each focus to produce the ellipse (see Figure 1 in the text)? Compute the answer to two decimal places.

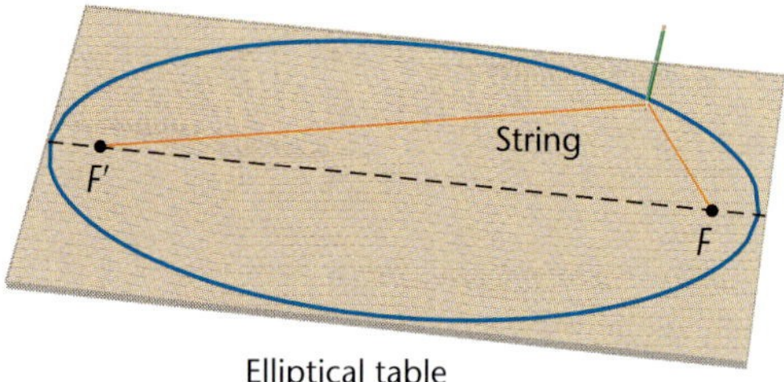

Elliptical table

★ **41. Aeronautical Engineering.** Of all possible wing shapes, it has been determined that the one with the least drag along the trailing edge is an ellipse. The leading edge may be a straight line, as shown in the figure. One of the most famous planes with this design was the World War II British Spitfire. The plane in the figure has a wingspan of 48.0 feet.

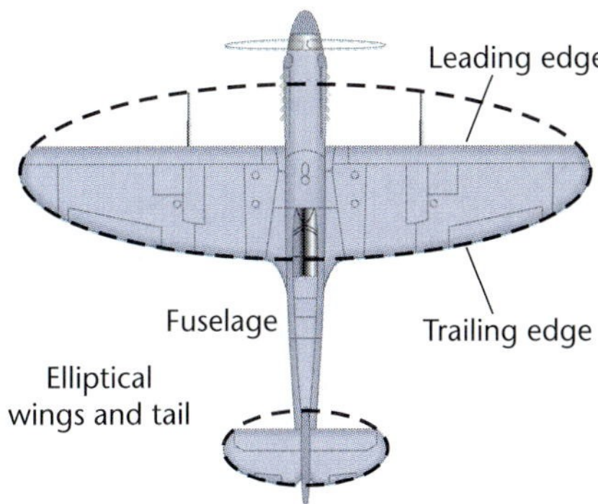

(A) If the straight-line leading edge is parallel to the major axis of the ellipse and is 1.14 feet in front of it, and if the leading edge is 46.0 feet long (including the width of the fuselage), find the equation of the ellipse. Let the x axis lie along the major axis (positive right), and let the y axis lie along the minor axis (positive forward).

(B) How wide is the wing in the center of the fuselage (assuming the wing passes through the fuselage)?

Compute quantities to 3 significant digits.

★ **42. Naval Architecture.** Currently, many high-performance racing sailboats use elliptical keels, rudders, and main sails for the same reasons stated in Problem 41—less drag along the trailing edge. In the accompanying figure, the ellipse containing the keel has a 12.0-foot major axis. The straight-line leading edge is parallel to the major axis of the ellipse and 1.00 foot in front of it. The chord is 1.00 foot shorter than the major axis.

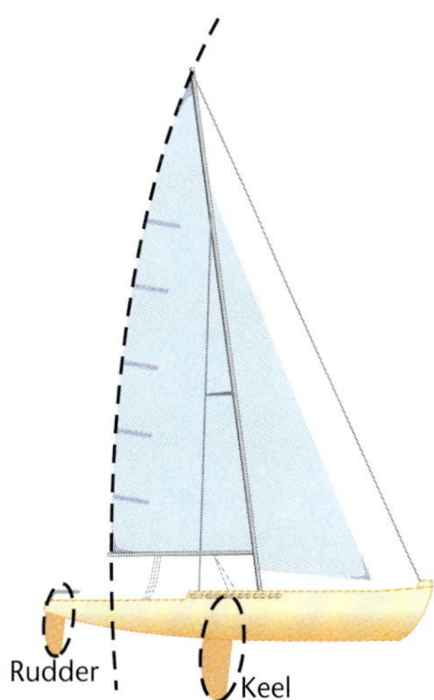

(A) Find the equation of the ellipse. Let the y axis lie along the minor axis of the ellipse, and let the x axis lie along the major axis, both with positive direction upward.

(B) What is the width of the keel, measured perpendicular to the major axis, 1 foot up the major axis from the bottom end of the keel?

Compute quantities to 3 significant digits.

SECTION 11-3 Hyperbola

- Definition of a Hyperbola
- Drawing a Hyperbola
- Standard Equations and Their Graphs
- Applications

As before, we start with a coordinate-free definition of a hyperbola. Using this definition, we show how a hyperbola can be drawn and we derive standard equations for hyperbolas specially located in a rectangular coordinate system.

• Definition of a Hyperbola

The following is a coordinate-free definition of a hyperbola:

DEFINITION 1 **Hyperbola**

A **hyperbola** is the set of all points P in a plane such that the absolute value of the difference of the distances of P to two fixed points in the plane is a positive constant. Each of the fixed points, F' and F, is called a **focus.** The intersection points V' and V of the line through the foci and the two branches of the hyperbola are called **vertices,** and each is called a **vertex.** The line segment $V'V$ is called the **transverse axis.** The midpoint of the transverse axis is the **center** of the hyperbola.

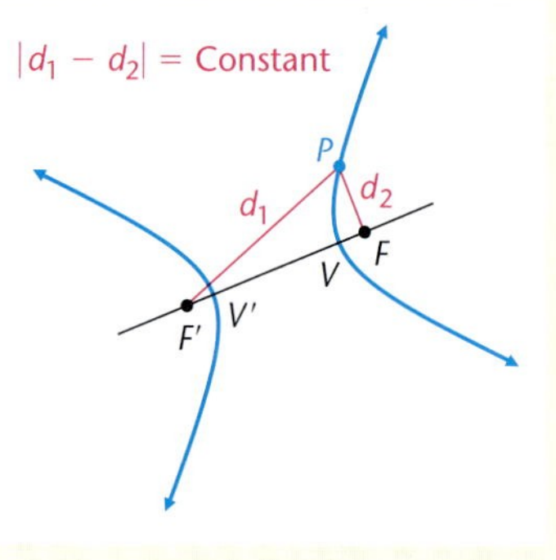

• Drawing a Hyperbola

Thumbtacks, a straightedge, string, and a pencil are all that are needed to draw a hyperbola (see Fig. 1). Place two thumbtacks in a piece of cardboard—these form the foci of the hyperbola. Rest one corner of the straightedge at the focus F' so that it is free to rotate about this point. Cut a piece of string shorter than the length of the straightedge, and fasten one end to the straightedge corner A and the other end to the thumbtack at F. Now push the string with a pencil up against the straightedge at B. Keeping the string taut, rotate the straightedge about F', keeping the corner at F'. The resulting curve will be part of a hyperbola. Other parts of the hyperbola can be drawn by changing the position of the straightedge and string. To see that the resulting curve meets the conditions of the definition, note that the difference of the distances BF' and BF is

$$\begin{aligned} BF' - BF &= BF' + BA - BF - BA \\ &= AF' - (BF + BA) \\ &= \begin{pmatrix}\text{Straightedge}\\ \text{length}\end{pmatrix} - \begin{pmatrix}\text{String}\\ \text{length}\end{pmatrix} \\ &= \text{Constant} \end{aligned}$$

FIGURE 1 Drawing a hyperbola.

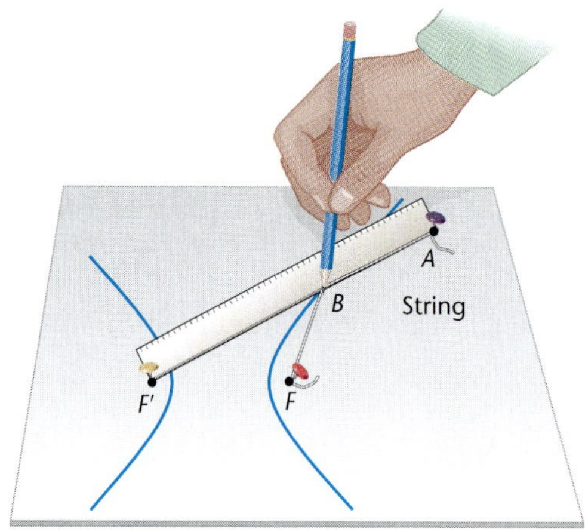

• Standard Equations and Their Graphs

Using the definition of a hyperbola and the distance-between-two-points formula, we can derive the standard equations for a hyperbola located in a rectangular coordinate system. We start by placing a hyperbola in the coordinate system with the foci on the x axis equidistant from the origin at $F'(-c, 0)$ and $F(c, 0)$, $c > 0$, as in Figure 2.

Just as for the ellipse, it is convenient to represent the constant difference by $2a$, $a > 0$. Also, the geometric fact that the difference of two sides of a triangle is always less than the third side can be applied to Figure 2 to derive the following useful result:

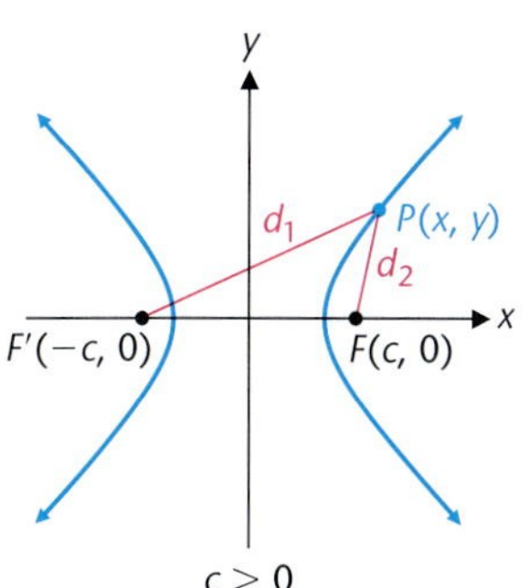

$c > 0$
$|d_1 - d_2|$ = Positive constant

FIGURE 2 Hyperbola with foci on the x axis.

$$|d_1 - d_2| < 2c$$
$$2a < 2c$$
$$a < c \qquad (1)$$

We will use this result in the derivation of the equation of a hyperbola, which we now begin.

Referring to Figure 2, the point $P(x, y)$ is on the hyperbola if and only if

$$|d_1 - d_2| = 2a$$
$$|d(P, F') - d(P, F)| = 2a$$
$$\left|\sqrt{(x + c)^2 + y^2} - \sqrt{(x - c)^2 + y^2}\right| = 2a$$

After eliminating radicals and absolute value signs by appropriate use of squaring and simplifying, another good exercise for you, we have

$$(c^2 - a^2)x^2 - a^2y^2 = a^2(c^2 - a^2) \qquad (2)$$
$$\frac{x^2}{a^2} - \frac{y^2}{c^2 - a^2} = 1 \qquad (3)$$

Dividing both sides of equation (2) by $a^2(c^2 - a^2)$ is permitted, since neither a^2 nor $c^2 - a^2$ is 0. From equation (1), $a < c$; thus, $a^2 < c^2$ and $c^2 - a^2 > 0$. The constant a was chosen positive at the beginning.

To simplify equation (3) further, we let

$$b^2 = c^2 - a^2 \qquad b > 0 \qquad (4)$$

to obtain

$$\frac{x^2}{a^2} - \frac{y^2}{b^2} = 1 \qquad (5)$$

From equation (5) we see that the x intercepts, which are also the vertices, are $x = \pm a$ and there are no y intercepts. To see why there are no y intercepts, let $x = 0$ and solve for y:

$$\frac{0^2}{a^2} - \frac{y^2}{b^2} = 1$$
$$y^2 = -b^2$$
$$y = \pm\sqrt{-b^2} \qquad \text{An imaginary number}$$

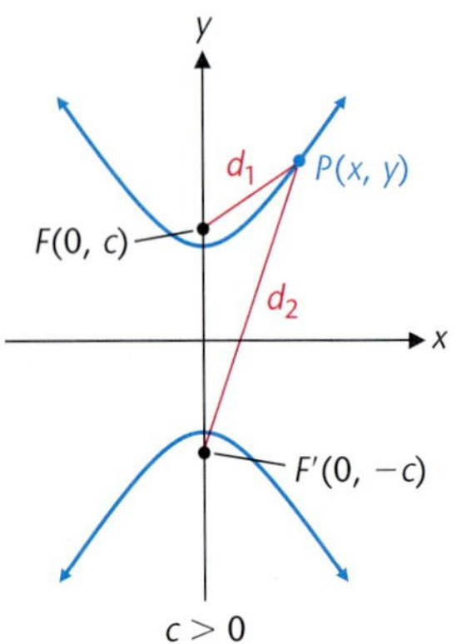

$c > 0$
$|d_1 - d_2|$ = Positive constant

FIGURE 3 Hyperbola with foci on the y axis.

If we start with the foci on the y axis at $F'(0, -c)$ and $F(0, c)$ as in Figure 3, instead of on the x axis as in Figure 2, then, following arguments similar to those used for the first derivation, we obtain

$$\frac{y^2}{a^2} - \frac{x^2}{b^2} = 1 \tag{6}$$

where the relationship among a, b, and c remains the same as before:

$$b^2 = c^2 - a^2 \tag{7}$$

The center is still at the origin, but the transverse axis is now on the y axis.

As an aid to graphing equation (5), we solve the equation for y in terms of x, another good exercise for you, to obtain

$$y = \pm\frac{b}{a}x\sqrt{1 - \frac{a^2}{x^2}} \tag{8}$$

As x changes so that $|x|$ becomes larger, the expression $1 - (a^2/x^2)$ within the radical approaches 1. Hence, for large values of $|x|$, equation (5) behaves very much like the lines

$$y = \pm\frac{b}{a}x \tag{9}$$

These lines are **asymptotes** for the graph of equation (5). The hyperbola approaches these lines as a point $P(x, y)$ on the hyperbola moves away from the origin (see Fig. 4). An easy way to draw the asymptotes is to first draw the rectangle as in Figure 4, then extend the diagonals. We refer to this rectangle as the **asymptote rectangle.**

FIGURE 4 Asymptotes.

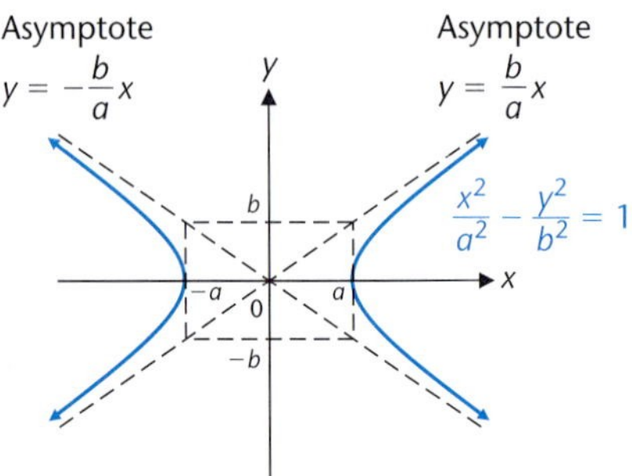

Starting with equation (6) and proceeding as we did for equation (5), we obtain the asymptotes for the graph of equation (6):

$$y = \pm\frac{a}{b}x \tag{10}$$

The perpendicular bisector of the transverse axis, extending from one side of the asymptote rectangle to the other, is called the **conjugate axis** of the hyperbola.

Given an equation of the form (5) or (6), how can we find the coordinates of the foci without memorizing or looking up the relation $b^2 = c^2 - a^2$? Just as with the ellipse, there is a simple geometric relationship in a hyperbola that enables us to get the same result using the Pythagorean theorem. To see this relationship, we rewrite $b^2 = c^2 - a^2$ in the form

$$c^2 = a^2 + b^2 \tag{11}$$

Note in the figures in Theorem 1 below that the distance from the center to a focus is the same as the distance from the center to a corner of the asymptote rectangle. Stated in another way:

A circle, with center at the origin, that passes through all four corners of the asymptote rectangle also passes through all foci of hyperbolas with asymptotes determined by the diagonals of the rectangle.

We summarize all the preceding results in Theorem 1 for convenient reference.

Theorem 1 **Standard Equations of a Hyperbola with Center at (0, 0)**

1. $\dfrac{x^2}{a^2} - \dfrac{y^2}{b^2} = 1$

x intercepts: $\pm a$ (vertices)
y intercepts: none
Foci: $F'(-c, 0)$, $F(c, 0)$

$$c^2 = a^2 + b^2$$

Transverse axis length $= 2a$
Conjugate axis length $= 2b$

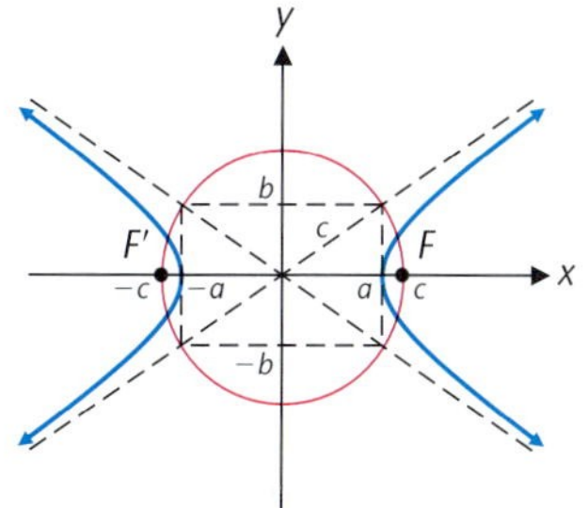

2. $\dfrac{y^2}{a^2} - \dfrac{x^2}{b^2} = 1$

x intercepts: none
y intercepts: $\pm a$ (vertices)
Foci: $F'(0, -c)$, $F(0, c)$

$$c^2 = a^2 + b^2$$

Transverse axis length $= 2a$
Conjugate axis length $= 2b$

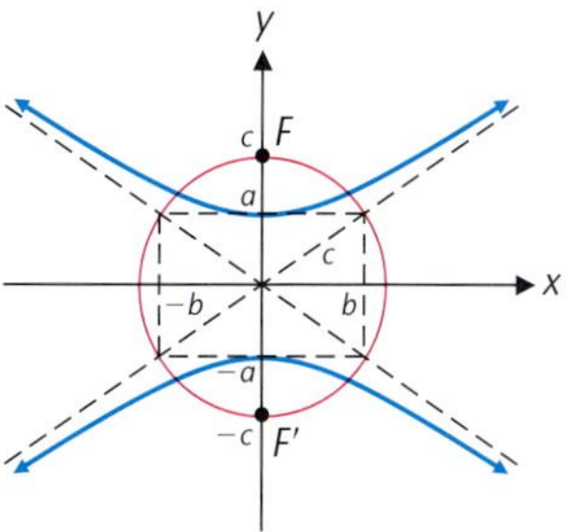

[*Note:* Both graphs are symmetric with respect to the x axis, y axis, and origin.]

EXPLORE-DISCUSS 1 The line through a focus F of a hyperbola that is perpendicular to the transverse axis intersects the hyperbola in two points G and H. For each of the two standard equations of a hyperbola with center (0, 0), find an expression in terms of a and b for the distance from G to H.

EXAMPLE 1 Graphing Hyperbolas

Sketch the graph of each equation, find the coordinates of the foci, and find the lengths of the transverse and conjugate axes.

(A) $9x^2 - 16y^2 = 144$ (B) $16y^2 - 9x^2 = 144$ (C) $2x^2 - y^2 = 10$

Solutions (A) First, write the equation in standard form by dividing both sides by 144:

$$9x^2 - 16y^2 = 144$$

$$\frac{x^2}{16} - \frac{y^2}{9} = 1 \qquad a^2 = 16 \text{ and } b^2 = 9$$

Locate x intercepts, $x = \pm 4$; there are no y intercepts. Sketch the asymptotes using the asymptote rectangle, then sketch in the hyperbola (Fig. 5).

$$\begin{aligned} \text{Foci:} \quad c^2 &= a^2 + b^2 \\ &= 16 + 9 \\ &= 25 \\ c &= 5 \end{aligned}$$

FIGURE 5 $9x^2 - 16y^2 = 144$.

Thus, the foci are $F'(-5, 0)$ and $F(5, 0)$

$$\text{Transverse axis length} = 2(4) = 8$$

$$\text{Conjugate axis length} = 2(3) = 6$$

(B)

$$16y^2 - 9x^2 = 144$$

$$\frac{y^2}{9} - \frac{x^2}{16} = 1 \qquad a^2 = 9 \text{ and } b^2 = 16$$

Locate y intercepts, $y = \pm 3$; there are no x intercepts. Sketch the asymptotes using the asymptote rectangle, then sketch in the hyperbola (Fig. 6). It is important to note that the transverse axis and the foci are on the y axis.

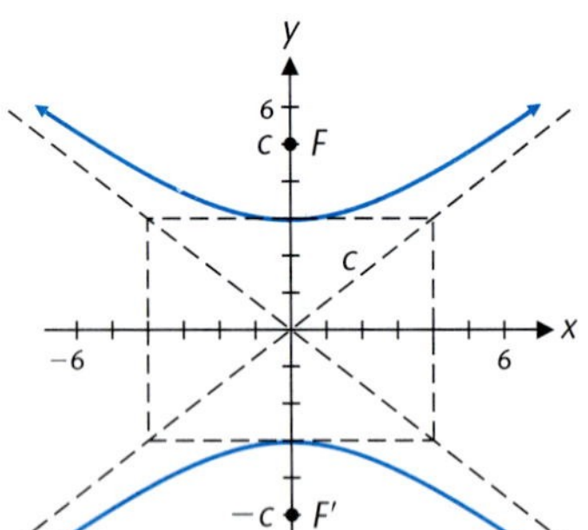

FIGURE 6 $16y^2 - 9x^2 = 144$.

$$\begin{aligned} \text{Foci:} \quad c^2 &= a^2 + b^2 \\ &= 9 + 16 \\ &= 25 \\ c &= 5 \end{aligned}$$

Thus, the foci are $F'(0, -5)$ and $F(0, 5)$.

$$\text{Transverse axis length} = 2(3) = 6$$

$$\text{Conjugate axis length} = 2(4) = 8$$

(C)
$$2x^2 - y^2 = 10$$
$$\frac{x^2}{5} - \frac{y^2}{10} = 1 \qquad a^2 = 5 \text{ and } b^2 = 10$$

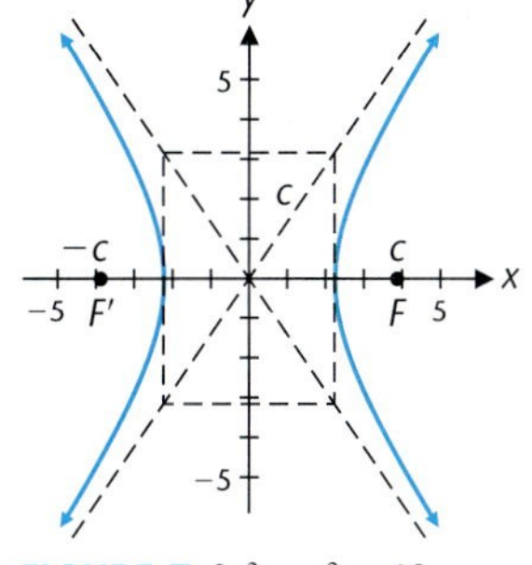

FIGURE 7 $2x^2 - y^2 = 10$.

Locate x intercepts, $x = \pm\sqrt{5}$; there are no y intercepts. Sketch the asymptotes using the asymptote rectangle, then sketch in the hyperbola (Fig. 7).

$$\begin{aligned} \text{Foci:} \quad c^2 &= a^2 + b^2 \\ &= 5 + 10 \\ &= 15 \\ c &= \sqrt{15} \end{aligned}$$

Thus, the foci are $F'(-\sqrt{15}, 0)$ and $F(\sqrt{15}, 0)$.

$$\text{Transverse axis length} = 2\sqrt{5} \approx 4.47$$
$$\text{Conjugate axis length} = 2\sqrt{10} \approx 6.32$$

Remark. To graph the equation $9x^2 - 16y^2 = 144$ of Example 1A on a graphing utility we first solve the equation for y, obtaining $y = \pm\sqrt{\frac{9x^2 - 144}{16}}$. We then graph each of the two functions. The graph of $y = \sqrt{\frac{9x^2 - 144}{16}}$ is the upper half of the hyperbola, and the graph of $y = -\sqrt{\frac{9x^2 - 144}{16}}$ is the lower half.

Matched Problem 1 Sketch the graph of each equation, find the coordinates of the foci, and find the lengths of the transverse and conjugate axes.

(A) $16x^2 - 25y^2 = 400$ (B) $25y^2 - 16x^2 = 400$ (C) $y^2 - 3x^2 = 12$

Hyperbolas of the form

$$\frac{x^2}{M} - \frac{y^2}{N} = 1 \qquad \text{and} \qquad \frac{y^2}{N} - \frac{x^2}{M} = 1 \qquad M, N > 0$$

are called **conjugate hyperbolas.** In Example 1 and Matched Problem 1, the hyperbolas in parts A and B are conjugate hyperbolas—they share the same asymptotes.

CAUTION When making a quick sketch of a hyperbola, it is a common error to have the hyperbola opening up and down when it should open left and right, or vice versa. The mistake can be avoided if you first locate the intercepts accurately.

EXAMPLE 2 Finding the Equation of a Hyperbola

Find an equation of a hyperbola in the form

$$\frac{y^2}{M} - \frac{x^2}{N} = 1 \qquad M, N > 0$$

if the center is at the origin, and:

(A) Length of transverse axis is 12
Length of conjugate axis is 20

(B) Length of transverse axis is 6
Distance of foci from center is 5

Solutions (A) Start with

$$\frac{y^2}{a^2} - \frac{x^2}{b^2} = 1$$

and find a and b:

$$a = \frac{12}{2} = 6 \qquad \text{and} \qquad b = \frac{20}{2} = 10$$

Thus, the equation is

$$\frac{y^2}{36} - \frac{x^2}{100} = 1$$

(B) Start with

$$\frac{y^2}{a^2} - \frac{x^2}{b^2} = 1$$

and find a and b:

$$a = \frac{6}{2} = 3$$

To find b, sketch the asymptote rectangle (Fig. 8), label known parts, and use the Pythagorean theorem:

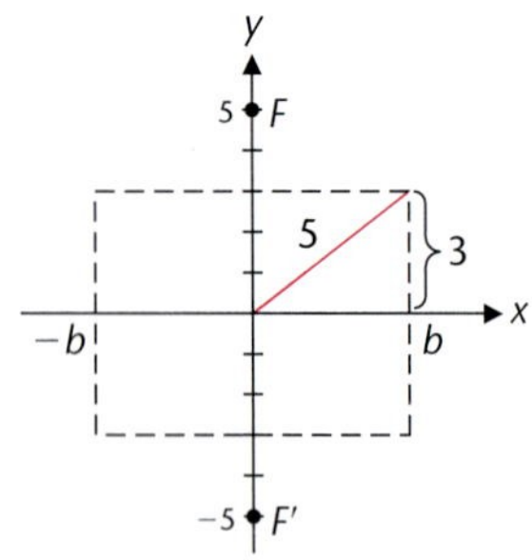

FIGURE 8 Asymptote rectangle.

$$\begin{aligned} b^2 &= 5^2 - 3^2 \\ &= 16 \\ b &= 4 \end{aligned}$$

Thus, the equation is

$$\frac{y^2}{9} - \frac{x^2}{16} = 1$$

Matched Problem 2 Find an equation of a hyperbola in the form

$$\frac{x^2}{M} - \frac{y^2}{N} = 1 \qquad M, N > 0$$

if the center is at the origin, and:

(A) Length of transverse axis is 50
Length of conjugate axis is 30

(B) Length of conjugate axis is 12
Distance of foci from center is 9

EXPLORE-DISCUSS 2

(A) Does the line with equation $y = x$ intersect the hyperbola with equation $x^2 - (y^2/4) = 1$? If so, find the coordinates of all intersection points.

(B) Does the line with equation $y = 3x$ intersect the hyperbola with equation $x^2 - (y^2/4) = 1$? If so, find the coordinates of all intersection points.

(C) For which values of m does the line with equation $y = mx$ intersect the hyperbola $\frac{x^2}{a^2} - \frac{y^2}{b^2} = 1$? Find the coordinates of all intersection points.

• Applications

You may not be aware of the many important uses of hyperbolic forms. They are encountered in the study of comets; the loran system of navigation for pleasure boats, ships, and aircraft; sundials; capillary action; nuclear cooling towers; optical and radiotelescopes; and contemporary architectural structures. The TWA building at Kennedy Airport is a *hyperbolic paraboloid,* and the St. Louis Science Center Planetarium is a *hyperboloid.* (See Fig. 9.)

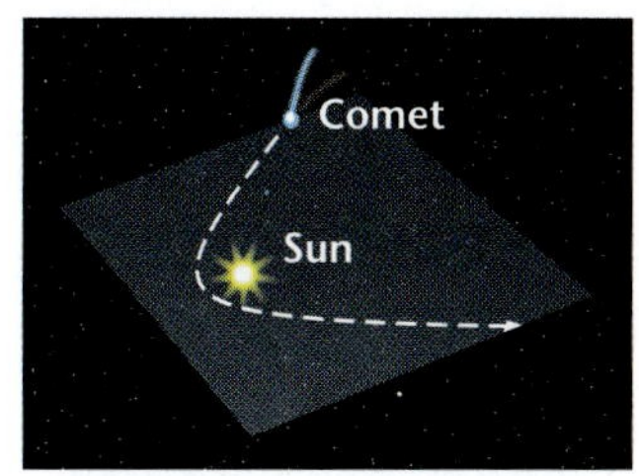

Comet around sun
(a)

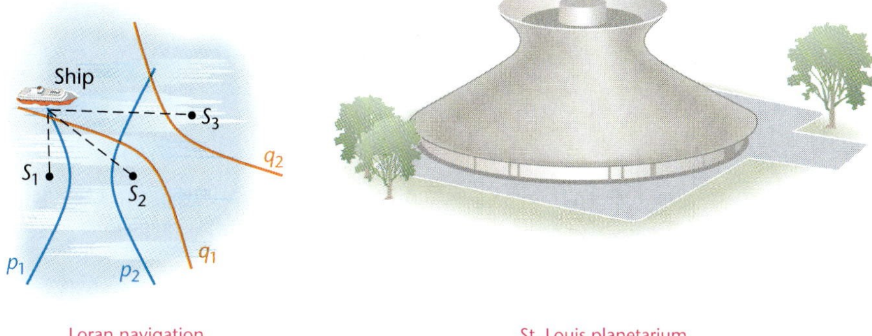

Loran navigation
(b)

St. Louis planetarium
(c)

FIGURE 9 Uses of hyperbolic forms.

Some comets from outer space occasionally enter the sun's gravitational field, follow a hyperbolic path around the sun (with the sun at a focus), and then leave, never to be seen again [Fig. 9(a)]. In the loran system of navigation, transmitting stations in three locations, S_1, S_2, and S_3 [Fig. 9(b)], send out signals simultaneously. A ship with a receiver records the difference in the arrival times of the signals from S_1 and S_2 and also records the difference in arrival times of the signals from S_2 and S_3. The difference in arrival times can be transformed into differences of the distances that the ship is to S_1 and S_2 and to S_2 and S_3. Plotting all points so that these differences in distances remain constant produces two branches, p_1 and p_2, of a hyperbola with foci S_1 and S_2 and two branches, q_1 and q_2, of a hyperbola with foci S_2 and S_3. It is easy to tell which branches the ship is on by noting the arrival times of the signals from each station. The intersection of a branch from each hyperbola locates the ship. Most of these calculations are now done by shipboard computers, and positions in longitude and latitude are given. This system of navigation is widely used for coastal navigation. Inexpensive loran units are now found on many small pleasure boats. Figure 9(c) illustrates a hyperboloid used architecturally. With such structures, thin concrete shells can span large spaces.

Answers to Matched Problems

1. 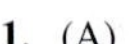
(A)

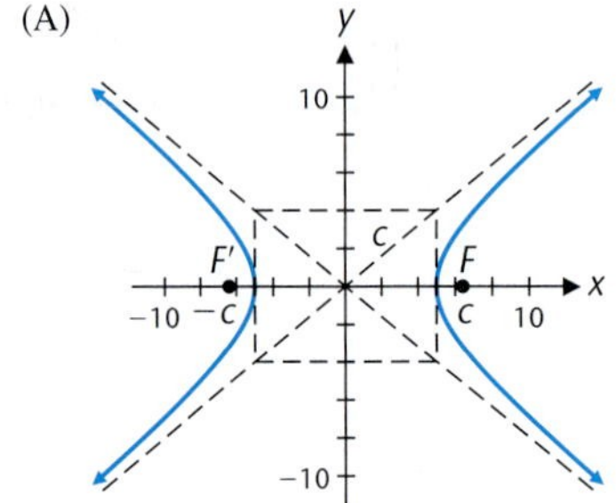

$$\frac{x^2}{25} - \frac{y^2}{16} = 1$$

Foci: $F'(-\sqrt{41}, 0)$, $F(\sqrt{41}, 0)$
Transverse axis length = 10
Conjugate axis length = 8

(B)

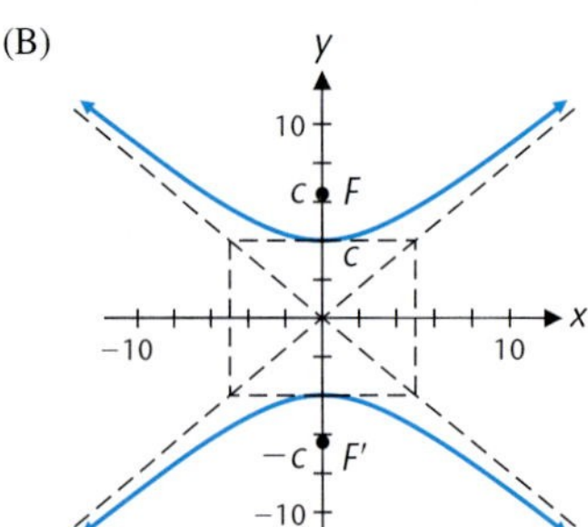

$$\frac{y^2}{16} - \frac{x^2}{25} = 1$$

Foci: $F'(0, -\sqrt{41})$, $F(0, \sqrt{41})$
Transverse axis length = 8
Conjugate axis length = 10

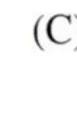
(C)

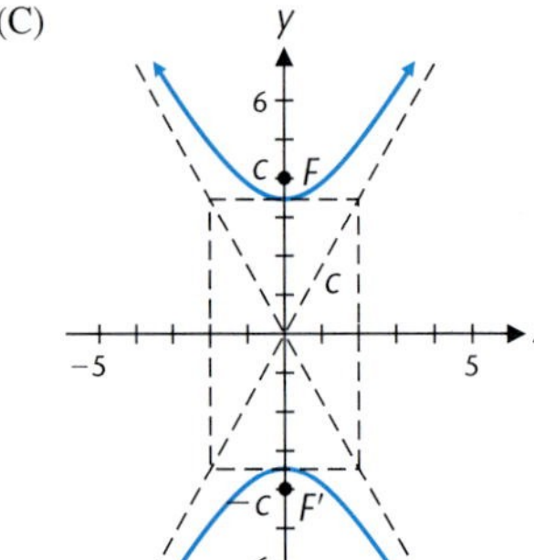

$$\frac{y^2}{12} - \frac{x^2}{4} = 1$$

Foci: $F'(0, -4)$, $F(0, 4)$
Transverse axis length = $2\sqrt{12} \approx 6.93$
Conjugate axis length = 4

2. (A) $\frac{x^2}{625} - \frac{y^2}{225} = 1$ (B) $\frac{x^2}{45} - \frac{y^2}{36} = 1$

EXERCISE 11-3

A

Sketch a graph of each equation in Problems 1–8, find the coordinates of the foci, and find the lengths of the transverse and conjugate axes.

1. $\frac{x^2}{9} - \frac{y^2}{4} = 1$

2. $\frac{x^2}{9} - \frac{y^2}{25} = 1$

3. $\frac{y^2}{4} - \frac{x^2}{9} = 1$

4. $\frac{y^2}{25} - \frac{x^2}{9} = 1$

5. $4x^2 - y^2 = 16$

6. $x^2 - 9y^2 = 9$

7. $9y^2 - 16x^2 = 144$

8. $4y^2 - 25x^2 = 100$

B

Sketch a graph of each equation in Problems 9–12, find the coordinates of the foci, and find the lengths of the transverse and conjugate axes.

9. $3x^2 - 2y^2 = 12$

10. $3x^2 - 4y^2 = 24$

11. $7y^2 - 4x^2 = 28$

12. $3y^2 - 2x^2 = 24$

In Problems 13–18, find an equation of a hyperbola in the form

$$\frac{x^2}{M} - \frac{y^2}{N} = 1 \quad or \quad \frac{y^2}{N} - \frac{x^2}{M} = 1 \quad M, N > 0$$

if the center is at the origin, and:

13. Transverse axis on y axis
Transverse axis length = 10
Conjugate axis length = 18

14. Transverse axis on x axis
Transverse axis length = 22
Conjugate axis length = 2

15. Transverse axis on x axis
Transverse axis length = 14
Distance of foci from center = 9

16. Transverse axis on y axis
Conjugate axis length = 30
Distance of foci from center = 25

17. Conjugate axis on x axis
Conjugate axis length = 12
Distance between foci = $12\sqrt{2}$

18. Conjugate axis on y axis
Transverse axis length = 2
Distance between foci = 48

19. (A) How many hyperbolas have center at (0, 0) and a focus at (1, 0)? Find their equations.
(B) How many ellipses have center at (0, 0) and a focus at (1, 0)? Find their equations.
(C) How many parabolas have vertex at (0, 0) and focus at (1, 0)? Find their equations.

20. How many hyperbolas have the lines $y = \pm 2x$ as asymptotes? Find their equations.

In Problems 21–24, graph each system of equations in the same rectangular coordinate system and find the coordinates of any points of intersection.

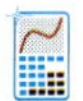

Check Problems 21–24 with a graphing utility.

21. $3y^2 - 4x^2 = 12$
$y^2 + x^2 = 25$

22. $y^2 - x^2 = 3$
$y^2 + x^2 = 5$

23. $2x^2 + y^2 = 24$
$x^2 - y^2 = -12$

24. $2x^2 + y^2 = 17$
$x^2 - y^2 = -5$

In Problems 25–28, find all points of intersection for each system of equations to three decimal places.

Check Problems 25–28 with a graphing utility.

25. $y^2 - x^2 = 9$
$2y - x = 8$
$y \geq 0$

26. $y^2 - x^2 = 4$
$y - x = 6$
$y \geq 0$

27. $y^2 - x^2 = 4$
$y^2 + 2x^2 = 36$

28. $y^2 - x^2 = 1$
$2y^2 + x^2 = 16$

In Problems 29–32, determine whether the statement is true or false. If true, explain why. If false, give a counterexample.

29. The line segment joining the foci of a hyperbola has greater length than the conjugate axis.

30. The line segment joining the foci of a hyperbola has greater length than the transverse axis.

31. Every line through the center of $4x^2 - y^2 = 16$ intersects the hyperbola in exactly two points.

32. Every nonvertical line through a vertex of $4x^2 - y^2 = 16$ intersects the hyperbola in exactly two points.

C

Eccentricity. *The set of points in a plane each of whose distance from a fixed point is e times its distance from a fixed line is a conic section. The positive number e is called the* **eccentricity** *of the conic section. Problems 33 and 34 below and Problems 33 and 34 in Section 12-2 illustrate an approach to defining the conic sections in terms of eccentricity.*

33. Find an equation of the set of points in a plane each of whose distance from (3, 0) is three-halves its distance from the line $x = \frac{4}{3}$. Identify the geometric figure.

34. Find an equation of the set of points in a plane each of whose distance from (0, 4) is four-thirds its distance from the line $y = \frac{9}{4}$. Identify the geometric figure.

In Problems 35–38, use a graphing utility to find the coordinates of all points of intersection to two decimal places.

35. $2x^2 - 3y^2 = 20,\ 7x + 15y = 10$

36. $y^2 - 3x^2 = 8,\ x^2 = -\frac{y}{3}$

37. $24y^2 - 18x^2 = 175,\ 90x^2 + 3y^2 = 200$

38. $8x^2 - 7y^2 = 58,\ 4y^2 - 11x^2 = 45$

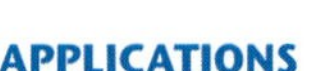

39. **Architecture.** An architect is interested in designing a thin-shelled dome in the shape of a hyperbolic paraboloid, as shown in figure (a). Find the equation of the hyperbola located in a coordinate system [Fig. (b)] satisfying the indicated conditions. How far is the hyperbola above the vertex 6 feet to the right of the vertex? Compute the answer to two decimal places.

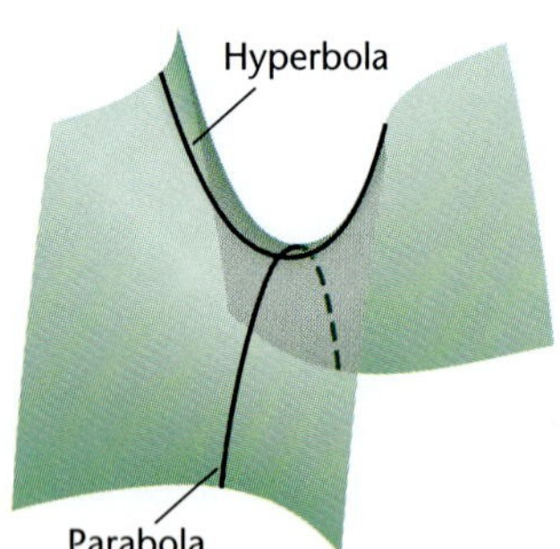

Hyperbolic paraboloid
(a)

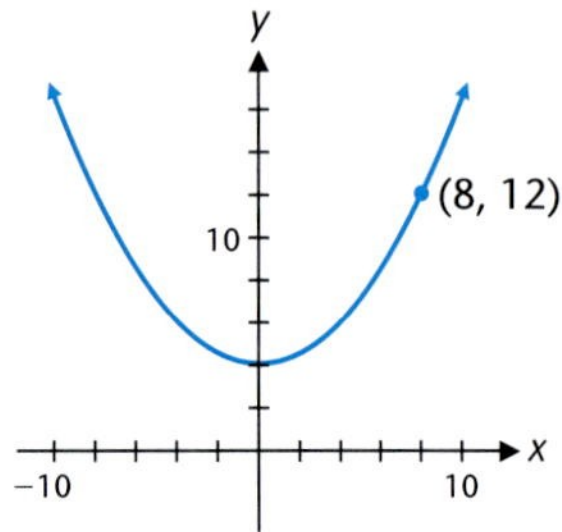

Hyperbola part of dome
(b)

40. **Nuclear Power.** A nuclear cooling tower is a **hyperboloid,** that is, a hyperbola rotated around its conjugate axis, as shown in Figure (a). The equation of the hyperbola in Figure (b) used to generate the hyperboloid is

$$\frac{x^2}{100^2} - \frac{y^2}{150^2} = 1$$

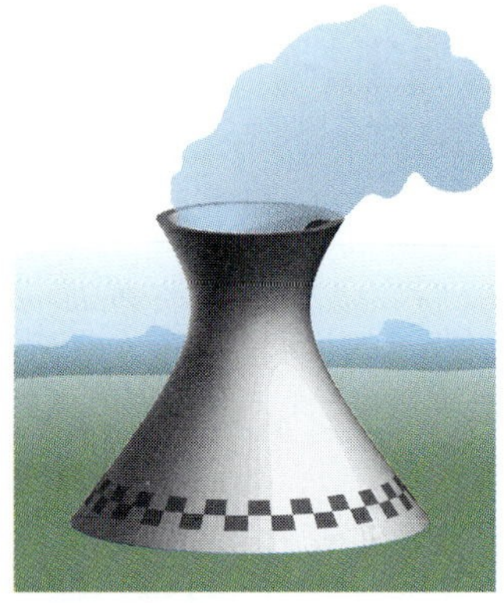

Nuclear cooling tower
(a)

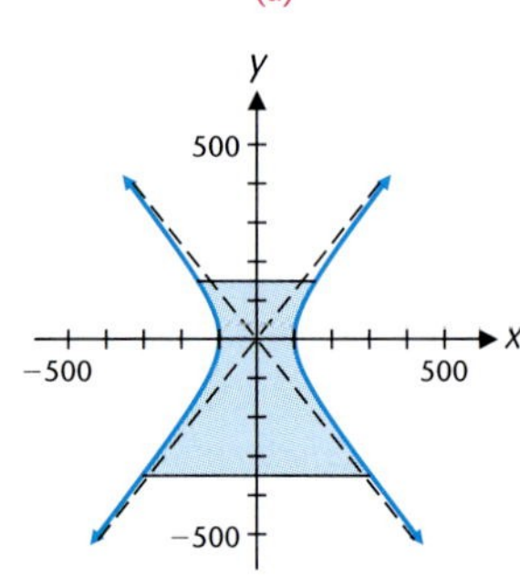

Hyperbola part of dome
(b)

If the tower is 500 feet tall, the top is 150 feet above the center of the hyperbola, and the base is 350 feet below the center, what is the radius of the top and the base? What is the radius of the smallest circular cross section in the tower? Compute answers to 3 significant digits.

41. **Space Science.** In tracking space probes to the outer planets, NASA uses large parabolic reflectors with diameters equal to two-thirds the length of a football field. Needless to say, many design problems are created by the weight of these reflectors. One weight problem is solved by using a hyperbolic reflector sharing the parabola's focus to reflect the incoming electromagnetic waves to the other focus of the hyperbola where receiving equipment is installed (see the figure).

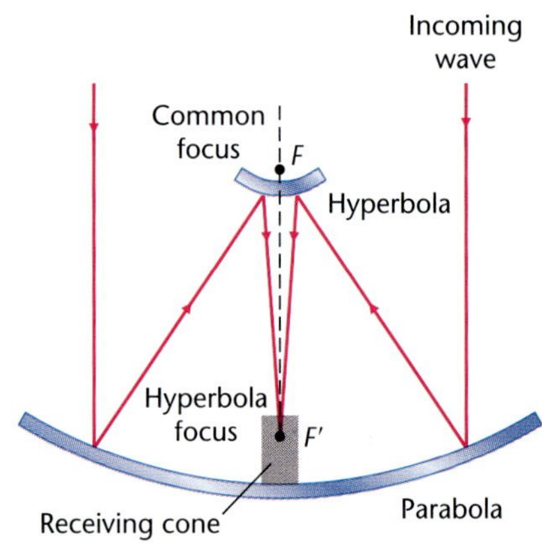

(a)

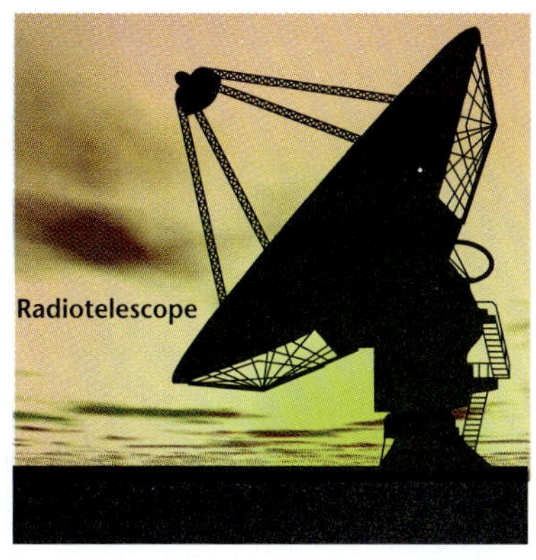

(b)

For the receiving antenna shown in the figure, the common focus F is located 120 feet above the vertex of the parabola, and focus F' (for the hyperbola) is 20 feet above the vertex. The vertex of the reflecting hyperbola is 110 feet above the vertex for the parabola. Introduce a coordinate system by using the axis of the parabola as the y axis (up positive), and let the x axis pass through the center of the hyperbola (right positive). What is the equation of the reflecting hyperbola? Write y in terms of x.

SECTION 11-4 Translation of Axes

- Translation of Axes
- Standard Equations of Translated Conics
- Graphing Equations of the Form $Ax^2 + Cy^2 + Dx + Ey + F = 0$
- Finding Equations of Conics

In the last three sections we found standard equations for parabolas, ellipses, and hyperbolas located with their axes on the coordinate axes and centered relative to the origin. What happens if we move conics away from the origin while keeping their axes parallel to the coordinate axes? We will show that we can obtain new standard equations that are special cases of the equation $Ax^2 + Cy^2 + Dx + Ey + F = 0$, where A and C are not both zero. The basic mathematical tool used in this endeavor is *translation of axes.* The usefulness of translation of axes is not limited to graphing conics, however. Translation of axes can be put to good use in many other graphing situations.

• Translation of Axes

A **translation of coordinate axes** occurs when the new coordinate axes have the same direction as and are parallel to the original coordinate axes. To see how coordinates in the original system are changed when moving to the translated system, and vice versa, refer to Figure 1.

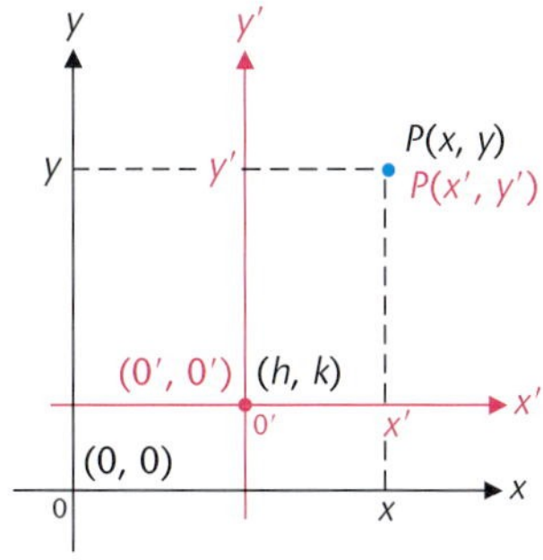

FIGURE 1 Translation of coordinates.

A point P in the plane has two sets of coordinates: (x, y) in the original system and (x', y') in the translated system. If the coordinates of the origin of the translated system are (h, k) relative to the original system, then the old and new coordinates are related as given in Theorem 1.

Theorem 1 **Translation Formulas**

1. $x = x' + h$
   $y = y' + k$

2. $x' = x - h$
   $y' = y - k$

It can be shown that these formulas hold for (h, k) located anywhere in the original coordinate system.

EXAMPLE 1 Equation of a Curve in a Translated System

A curve has the equation

$$(x - 4)^2 + (y + 1)^2 = 36$$

If the origin is translated to $(4, -1)$, find the equation of the curve in the translated system and identify the curve.

Solution Since $(h, k) = (4, -1)$, use translation formulas

$$x' = x - h = x - 4$$

$$y' = y - k = y + 1$$

to obtain, after substitution,

$$x'^2 + y'^2 = 36$$

This is the equation of a circle of radius 6 with center at the new origin. The coordinates of the new origin in the original coordinate system are $(4, -1)$ (Fig. 2). Note that this result agrees with our general treatment of the circle in Section 2-1.

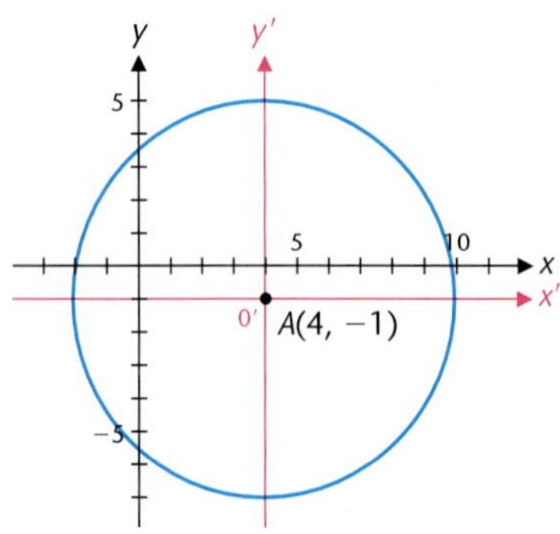

FIGURE 2
$(x - 4)^2 + (y + 1)^2 = 36.$

Matched Problem 1 A curve has the equation $(y + 2)^2 = 8(x - 3)$. If the origin is translated to $(3, -2)$, find an equation of the curve in the translated system and identify the curve.

• Standard Equations of Translated Conics

We now proceed to find standard equations of conics translated away from the origin. We do this by first writing the standard equations found in earlier sections in the $x'y'$ coordinate system with $0'$ at (h, k). We then use translation equations to find the standard forms relative to the original xy coordinate system. The equations of translation in all cases are

$$x' = x - h$$
$$y' = y - k$$

For parabolas we have

$$x'^2 = 4ay' \qquad (x - h)^2 = 4a(y - k)$$
$$y'^2 = 4ax' \qquad (y - k)^2 = 4a(x - h)$$

For circles we have

$$x'^2 + y'^2 = r^2 \qquad (x - h)^2 + (y - k)^2 = r^2$$

For ellipses we have for $a > b > 0$

$$\frac{x'^2}{a^2} + \frac{y'^2}{b^2} = 1 \qquad \frac{(x - h)^2}{a^2} + \frac{(y - k)^2}{b^2} = 1$$
$$\frac{x'^2}{b^2} + \frac{y'^2}{a^2} = 1 \qquad \frac{(x - h)^2}{b^2} + \frac{(y - k)^2}{a^2} = 1$$

For hyperbolas we have

$$\frac{x'^2}{a^2} - \frac{y'^2}{b^2} = 1 \qquad \frac{(x - h)^2}{a^2} - \frac{(y - k)^2}{b^2} = 1$$
$$\frac{y'^2}{a^2} - \frac{x'^2}{b^2} = 1 \qquad \frac{(y - k)^2}{a^2} - \frac{(x - h)^2}{b^2} = 1$$

Table 1 summarizes these results with appropriate figures and some properties discussed earlier.

TABLE 1 Standard Equations for Translated Conics

Parabolas

$(x - h)^2 = 4a(y - k)$

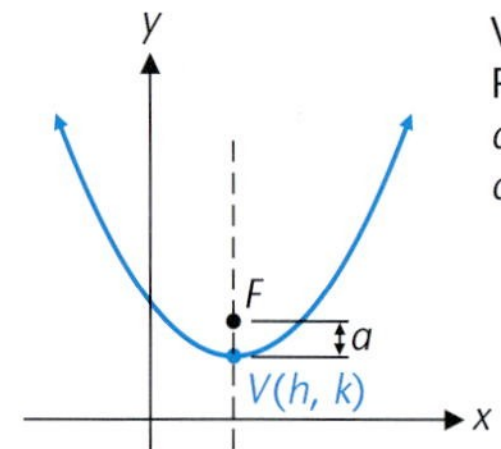

Vertex (h, k)
Focus $(h, k + a)$
$a > 0$ opens up
$a < 0$ opens down

$(y - k)^2 = 4a(x - h)$

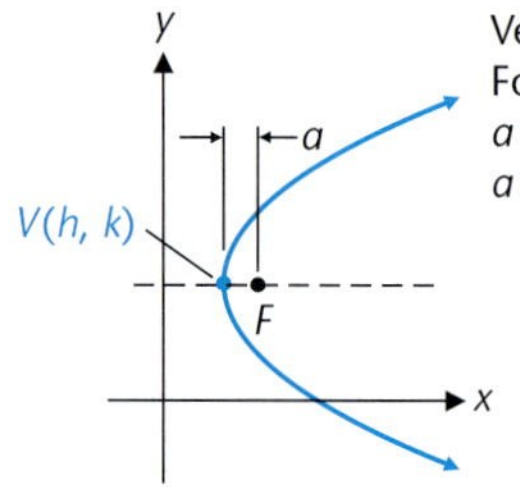

Vertex (h, k)
Focus $(h + a, k)$
$a < 0$ opens left
$a > 0$ opens right

Circles

$(x - h)^2 + (y - k)^2 = r^2$

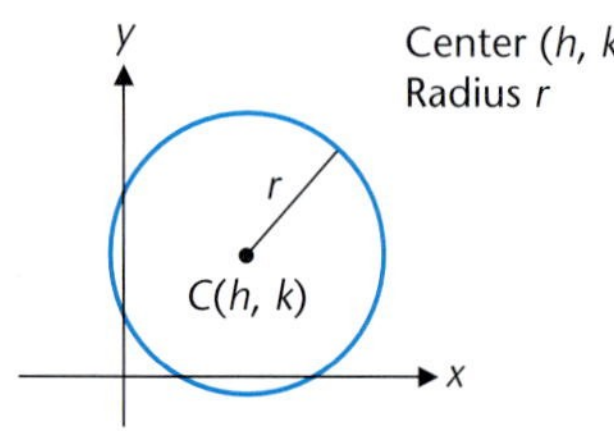

Center (h, k)
Radius r

Ellipses

$$\frac{(x - h)^2}{a^2} + \frac{(y - k)^2}{b^2} = 1 \qquad a > b > 0 \qquad \frac{(x - h)^2}{b^2} + \frac{(y - k)^2}{a^2} = 1$$

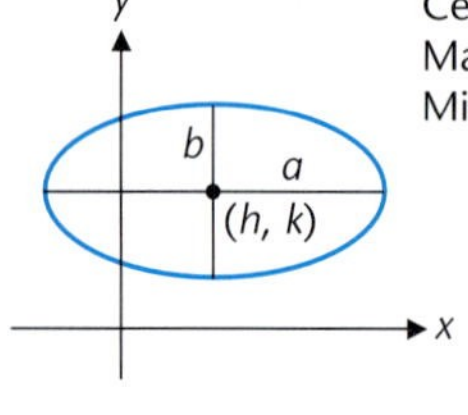

Center (h, k)
Major axis $2a$
Minor axis $2b$

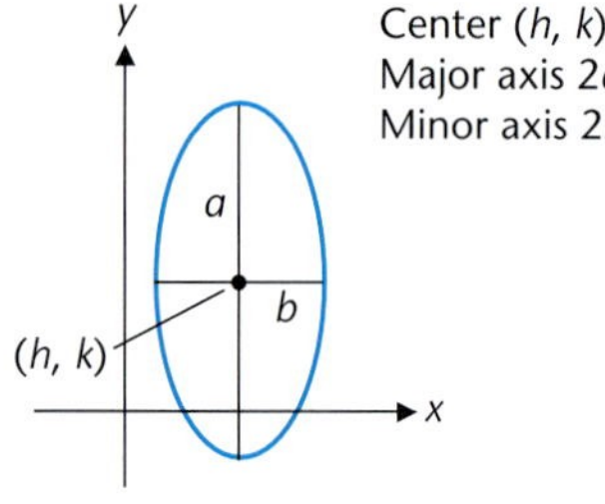

Center (h, k)
Major axis $2a$
Minor axis $2b$

Hyperbolas

$$\frac{(x - h)^2}{a^2} - \frac{(y - k)^2}{b^2} = 1$$

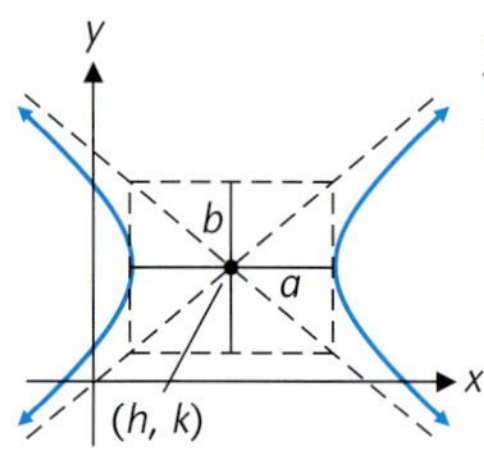

Center (h, k)
Transverse axis $2a$
Conjugate axis $2b$

$$\frac{(y - k)^2}{a^2} - \frac{(x - h)^2}{b^2} = 1$$

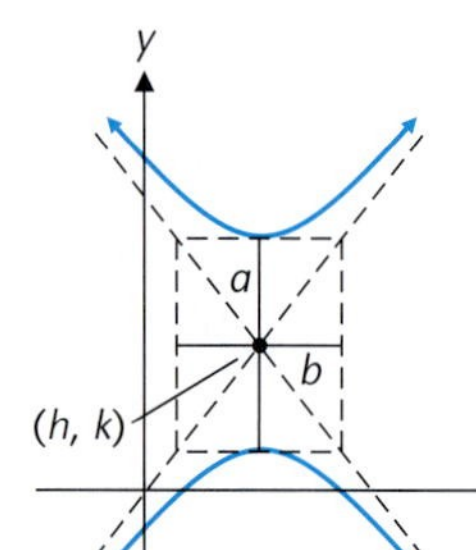

Center (h, k)
Transverse axis $2a$
Conjugate axis $2b$

• Graphing Equations of the Form $Ax^2 + Cy^2 + Dx + Ey + F = 0$

It can be shown that the graph of

$$Ax^2 + Cy^2 + Dx + Ey + F = 0 \tag{1}$$

where A and C are not both zero, is a conic or a degenerate conic or that there is no graph. If we can transform equation (1) into one of the standard forms in Table 1, then we will be able to identify its graph and sketch it rather quickly. The process of completing the square discussed in Section 1-6 will be our primary tool in accomplishing this transformation. A couple of examples should help make the process clear.

EXAMPLE 2 Graphing a Translated Conic

Transform

$$y^2 - 6y - 4x + 1 = 0 \tag{2}$$

into one of the standard forms in Table 1. Identify the conic and graph it.

Solution

Step 1. Complete the square in equation (2) relative to each variable that is squared—in this case y:

$$y^2 - 6y - 4x + 1 = 0$$
$$y^2 - 6y = 4x - 1$$
$$y^2 - 6y + 9 = 4x + 8 \quad \text{Add 9 to both sides to complete the square on the left side.}$$
$$(y - 3)^2 = 4(x + 2) \tag{3}$$

From Table 1 we recognize equation (3) as an equation of a parabola opening to the right with vertex at $(h, k) = (-2, 3)$.

Step 2. Find the equation of the parabola in the translated system with origin $0'$ at $(h, k) = (-2, 3)$. The equations of translation are read directly from equation (3):

$$x' = x + 2$$
$$y' = y - 3$$

Making these substitutions in equation (3) we obtain

$$y'^2 = 4x' \tag{4}$$

the equation of the parabola in the $x'y'$ system.

Step 3. Graph equation (4) in the $x'y'$ system following the process discussed in Section 11-1. The resulting graph is the graph of the original equation relative to the original xy coordinate system (Fig. 3).

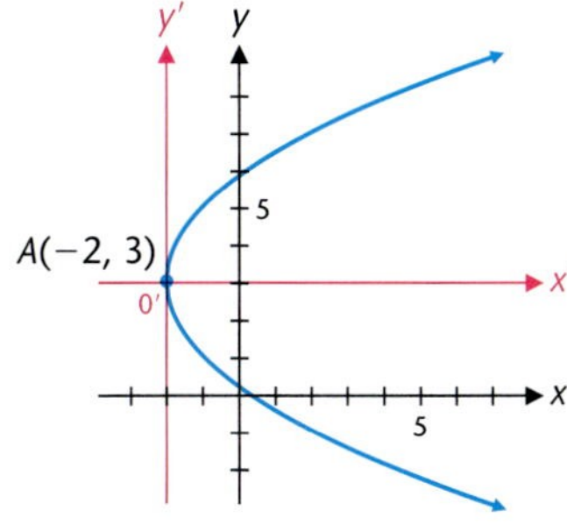

FIGURE 3
$y^2 - 6y - 4x + 1 = 0$.

Matched Problem 2 Transform

$$x^2 + 4x + 4y - 12 = 0$$

into one of the standard forms in Table 1. Identify the conic and graph it.

EXAMPLE 3 Graphing a Translated Conic

Transform

$$9x^2 - 4y^2 - 36x - 24y - 36 = 0$$

into one of the standard forms in Table 1. Identify the conic and graph it. Find the coordinates of any foci relative to the original system.

Solution

Step 1. Complete the square relative to both x and y.

$$\begin{aligned} 9x^2 - 4y^2 - 36x - 24y - 36 &= 0 \\ 9x^2 - 36x \qquad - 4y^2 - 24y \qquad &= 36 \\ 9(x^2 - 4x \qquad) - 4(y^2 + 6y \qquad) &= 36 \\ 9(x^2 - 4x + 4) - 4(y^2 + 6y + 9) &= 36 + 36 - 36 \\ 9(x - 2)^2 - 4(y + 3)^2 &= 36 \\ \frac{(x - 2)^2}{4} - \frac{(y + 3)^2}{9} &= 1 \end{aligned}$$

From Table 1 we recognize the last equation as an equation of a hyperbola opening left and right with center at $(h, k) = (2, -3)$.

Step 2. Find the equation of the hyperbola in the translated system with origin $0'$ at $(h, k) = (2, -3)$. The equations of translation are read directly from the last equation in step 1:

$$\begin{aligned} x' &= x - 2 \\ y' &= y + 3 \end{aligned}$$

Making these substitutions, we obtain

$$\frac{x'^2}{4} - \frac{y'^2}{9} = 1$$

the equation of the hyperbola in the $x'y'$ system.

Step 3. Graph the equation obtained in step 2 in the $x'y'$ system following the process discussed in Section 11-3. The resulting graph is the graph of the original equation relative to the original xy coordinate system (Fig. 4).

FIGURE 4
$9x^2 - 4y^2 - 36x - 24y - 36 = 0.$

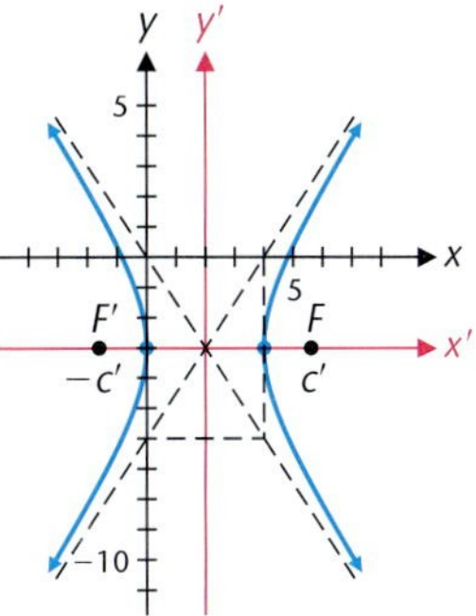

Step 4. Find the coordinates of the foci. To find the coordinates of the foci in the original system, first find the coordinates in the translated system:

$$c'^2 = 2^2 + 3^2 = 13$$
$$c' = \sqrt{13}$$
$$-c' = -\sqrt{13}$$

Thus, the coordinates in the translated system are

$$F'(-\sqrt{13}, 0) \qquad \text{and} \qquad F(\sqrt{13}, 0)$$

Now, use

$$x = x' + h = x' + 2$$
$$y = y' + k = y' - 3$$

to obtain

$$F'(-\sqrt{13} + 2, -3) \qquad \text{and} \qquad F(\sqrt{13} + 2, -3)$$

as the coordinates of the foci in the original system.

Matched Problem 3 Transform

$$9x^2 + 16y^2 + 36x - 32y - 92 = 0$$

into one of the standard forms in Table 1. Identify the conic and graph it. Find the coordinates of any foci relative to the original system.

Remark. A graphing utility provides an alternative approach to graphing equations of the form $Ax^2 + Cy^2 + Dx + Ey + F = 0$. Consider, for example, the equation $9x^2 - 4y^2 - 36x - 24y - 36 = 0$ of Example 3. We write the equation as a quadratic equation in the variable y: $4y^2 + 24y + (-9x^2 + 36x + 36) = 0$. By the quadratic formula, $y = \dfrac{-24 \pm \sqrt{24^2 - 16f(x)}}{8}$, where $f(x) = -9x^2 + 36x + 36$. We then graph each of the two functions in the expression for y. The graph of $y = \dfrac{-24 + \sqrt{24^2 - 16f(x)}}{8}$ is the upper half of the hyperbola, and the graph of $y = \dfrac{-24 - \sqrt{24^2 - 16f(x)}}{8}$ is the lower half.

EXPLORE-DISCUSS 1 If $A \neq 0$ and $C \neq 0$, show that the translation of axes $x' = x + \dfrac{D}{2A}$, $y' = y + \dfrac{E}{2C}$ transforms the equation $Ax^2 + Cy^2 + Dx + Ey + F = 0$ into an equation of the form $Ax'^2 + Cy'^2 = K$.

• Finding Equations of Conics

We now reverse the problem: Given certain information about a conic in a rectangular coordinate system, find its equation.

EXAMPLE 4 Finding the Equation of a Translated Conic

Find the equation of a hyperbola with vertices on the line $x = -4$, conjugate axis on the line $y = 3$, length of the transverse axis $= 4$, and length of the conjugate axis $= 6$.

Solution Locate the vertices, asymptote rectangle, and asymptotes in the original coordinate system [Fig. 5(a)], then sketch the hyperbola and translate the origin to the center of the hyperbola [Fig. 5(b)].

FIGURE 5

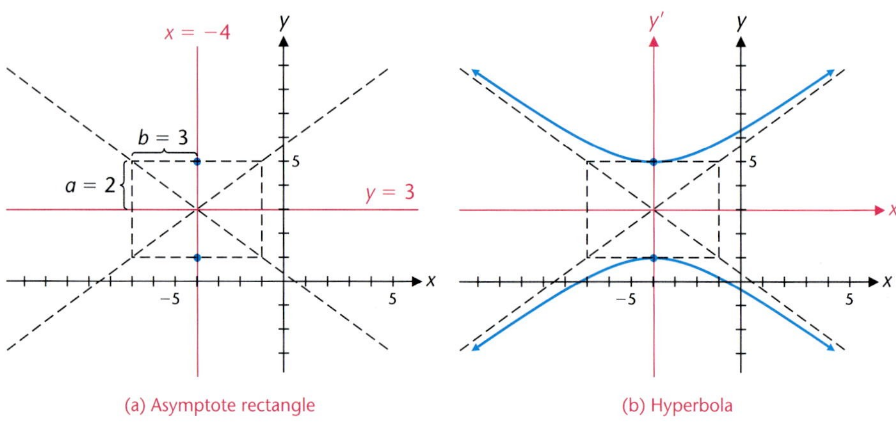

(a) Asymptote rectangle (b) Hyperbola

Next write the equation of the hyperbola in the translated system:

$$\frac{y'^2}{4} - \frac{x'^2}{9} = 1$$

The origin in the translated system is at $(h, k) = (-4, 3)$, and the translation formulas are

$$x' = x - h = x - (-4) = x + 4$$
$$y' = y - k = y - 3$$

Thus, the equation of the hyperbola in the original system is

$$\frac{(y - 3)^2}{4} - \frac{(x + 4)^2}{9} = 1$$

or, after simplifying and writing in the form of equation (1),

$$4x^2 - 9y^2 + 32x + 54y + 19 = 0$$

Matched Problem 4 Find the equation of an ellipse with foci on the line $x = 4$, minor axis on the line $y = -3$, length of the major axis $= 8$, and length of the minor axis $= 4$.

EXPLORE-DISCUSS 2 Use the strategy of completing the square to transform each equation to an equation in an $x'y'$ coordinate system. Note that the equation you obtain is not one of the standard forms in Table 1; instead, it is either the equation of a degenerate conic or the equation has no solution. If the solution set of the equation is not empty, graph it and identify the graph (a point, a line, two parallel lines, or two intersecting lines).

(A) $x^2 + 2y^2 - 2x + 16y + 33 = 0$

(B) $4x^2 - y^2 - 24x - 2y + 35 = 0$

(C) $y^2 - 2y - 15 = 0$

(D) $5x^2 + y^2 + 12y + 40 = 0$

(E) $x^2 - 18x + 81 = 0$

Answers to Matched Problems

1. $y'^2 = 8x'$; a parabola

2. $(x + 2)^2 = -4(y - 4)$; a parabola

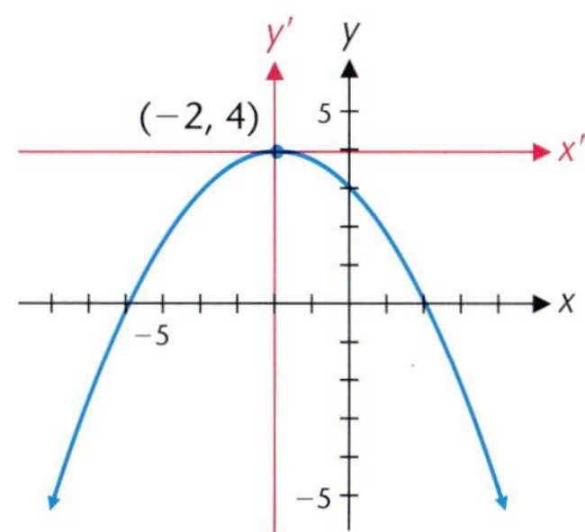

3. $\dfrac{(x+2)^2}{16} + \dfrac{(y-1)^2}{9} = 1$; ellipse   Foci: $F'(-\sqrt{7}-2, 1)$, $F(\sqrt{7}-2, 1)$

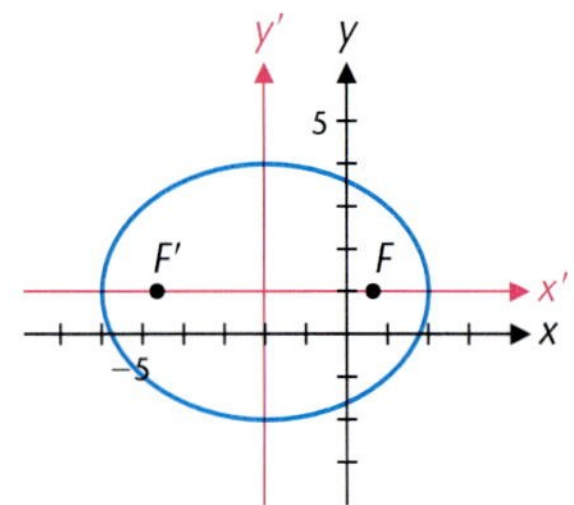

4. $\dfrac{(x-4)^2}{4} + \dfrac{(y+3)^2}{16} = 1$, or $4x^2 + y^2 - 32x + 6y + 57 = 0$

EXERCISE 11-4

A

In Problems 1–8:
(A) Find translation formulas that translate the origin to the indicated point (h, k).
(B) Write the equation of the curve for the translated system.
(C) Identify the curve.

1. $(x-3)^2 + (y-5)^2 = 81$; $(3, 5)$

2. $(x-3)^2 = 8(y+2)$; $(3, -2)$

3. $\dfrac{(x+7)^2}{9} + \dfrac{(y-4)^2}{16} = 1$; $(-7, 4)$

4. $(x+2)^2 + (y+6)^2 = 36$; $(-2, -6)$

5. $(y+9)^2 = 16(x-4)$; $(4, -9)$

6. $\dfrac{(y-9)^2}{10} - \dfrac{(x+5)^2}{6} = 1$; $(-5, 9)$

7. $\dfrac{(x+8)^2}{12} + \dfrac{(y+3)^2}{8} = 1$; $(-8, -3)$

8. $\dfrac{(x+7)^2}{25} - \dfrac{(y-8)^2}{50} = 1$; $(-7, 8)$

In Problems 9–14:
(A) Write each equation in one of the standard forms listed in Table 1.
(B) Identify the curve.

9. $16(x-3)^2 - 9(y+2)^2 = 144$

10. $(y+2)^2 - 12(x-3) = 0$

11. $6(x+5)^2 + 5(y+7)^2 = 30$

12. $12(y-5)^2 - 8(x-3)^2 = 24$

13. $(x+6)^2 + 24(y-4) = 0$

14. $4(x-7)^2 + 7(y-3)^2 = 28$

B

In Problems 15–22, transform each equation into one of the standard forms in Table 1. Identify the curve and graph it.

15. $4x^2 + 9y^2 - 16x - 36y + 16 = 0$

16. $16x^2 + 9y^2 + 64x + 54y + 1 = 0$

17. $x^2 + 8x + 8y = 0$

18. $y^2 + 12x + 4y - 32 = 0$

19. $x^2 + y^2 + 12x + 10y + 45 = 0$

20. $x^2 + y^2 - 8x - 6y = 0$

21. $-9x^2 + 16y^2 - 72x - 96y - 144 = 0$

22. $16x^2 - 25y^2 - 160x = 0$

23. If $A \neq 0$, $C = 0$, and $E \neq 0$, find h and k so that the translation of axes $x = x' + h$, $y = y' + k$ transforms the equation $Ax^2 + Cy^2 + Dx + Ey + F = 0$ into one of the standard forms of Table 1.

24. If $A = 0$, $C \neq 0$, and $D \neq 0$, find h and k so that the translation of axes $x = x' + h$, $y = y' + k$ transforms the equation $Ax^2 + Cy^2 + Dx + Ey + F = 0$ into one of the standard forms of Table 1.

In Problems 25–34, use the given information to find the equation of each conic. Express the answer in the form $Ax^2 + Cy^2 + Dx + Ey + F = 0$ with integer coefficients and $A \geq 0$.

25. A parabola with vertex at $(5, 3)$, and focus at $(5, 11)$.

26. A parabola with focus at $(2, 3)$, and directrix the y axis.

27. An ellipse with vertices $(-3, -2)$ and $(-3, 10)$ and length of minor axis $= 10$.

28. A hyperbola with vertices $(-2, -8)$ and $(4, -8)$ and length of conjugate axis $= 24$.

29. A hyperbola with foci (2, 1) and (6, 1) and vertices (3, 1) and (5, 1).

30. An ellipse with foci $(-3, 0)$ and $(-3, 6)$ and vertices $(-3, -2)$ and $(-3, 8)$.

31. A parabola with axis the y axis and passing through the points (1, 0) and (2, 4).

32. A parabola with vertex at $(-6, 2)$, axis the line $y = 2$, and passing through the point (0, 7).

33. An ellipse with vertices $(1, -1)$, and $(-5, -1)$ that passes through the origin.

34. A hyperbola with vertices at (2, 3), and (2, 5) that passes through the point (4, 0).

C

In Problems 35–40, find the coordinates of any foci relative to the original coordinate system:

35. Problem 15 **36.** Problem 16 **37.** Problem 17

38. Problem 18 **39.** Problem 21 **40.** Problem 22

In Problems 41–44, use a graphing utility to find the coordinates of all points of intersection to two decimal places.

41. $3x^2 - 5y^2 + 7x - 2y + 11 = 0,\ 6x + 4y = 15$

42. $8x^2 + 3y^2 - 14x + 17y - 39 = 0,\ 5x - 11y = 23$

43. $7x^2 - 8x + 5y - 25 = 0,\ x^2 + 4y^2 + 4x - y - 12 = 0$

44. $4x^2 - y^2 - 24x - 2y + 35 = 0,\ 2x^2 + 6y^2 - 3x - 34 = 0$

SECTION 11-5 Parametric Equations

- Parametric Equations and Plane Curves
- Projectile Motion
- Cycloid

• Parametric Equations and Plane Curves

Consider the two equations

$$\begin{aligned} x &= t + 1 \\ y &= t^2 - 2t \end{aligned} \qquad -\infty < t < \infty \tag{1}$$

Each value of t determines a value of x, a value of y, and hence, an ordered pair (x, y). To graph the set of ordered pairs (x, y) determined by letting t assume all real values, we construct Table 1 listing selected values of t and the corresponding values of x and y. Then we plot the ordered pairs (x, y) and connect them with a continuous curve, as shown in Figure 1. The variable t is called a **parameter** and does not appear on the graph. Equations (1) are called **parametric equations** because both x and y are expressed in terms of the parameter t. The graph of the ordered pairs (x, y) is called a **plane curve.**

TABLE 1

t	0	1	2	3	4	−1	−2
x	1	2	3	4	5	0	−1
y	0	−1	0	3	8	3	8

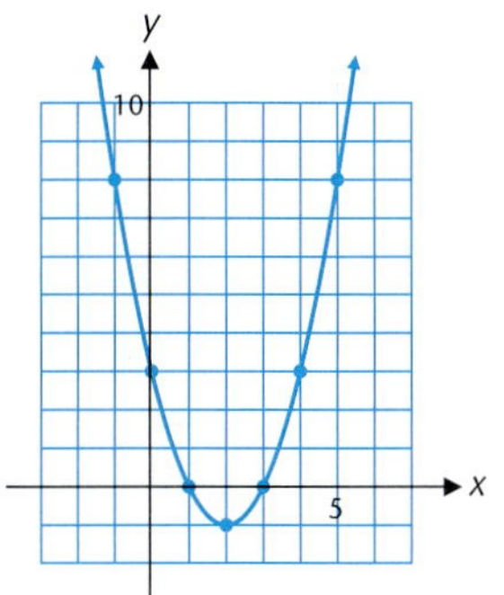

FIGURE 1 Graph of $x = t + 1$, $y = t^2 - 2t$, $-\infty < t < \infty$.

In some cases it is possible to eliminate the parameter by solving one of the equations for t and substituting into the other. In the example just considered, solving the first equation for t in terms of x, we have

$$t = x - 1$$

Then, substituting the result into the second equation, we obtain

$$\begin{aligned} y &= (x - 1)^2 - 2(x - 1) \\ &= x^2 - 4x + 3 \end{aligned}$$

We recognize this as the equation of a parabola, as we would guess from Figure 1.

In other cases, it may not be easy or possible to eliminate the parameter to obtain an equation in just x and y. For example, for

$$\begin{aligned} x &= t + \log t \\ y &= t - e^t \end{aligned} \qquad t > 0$$

you will not find it possible to solve either equation for t in terms of functions we have considered.

Is there more than one parametric representation for a plane curve? The answer is yes. In fact, there is an unlimited number of parametric representations for the same plane curve. The following are two additional representations of the parabola in Figure 1.

$$\begin{aligned} x &= t + 3 \\ y &= t^2 + 2t \end{aligned} \qquad -\infty < t < \infty \qquad (2)$$

$$\begin{aligned} x &= t \\ y &= t^2 - 4t + 3 \end{aligned} \qquad -\infty < t < \infty \qquad (3)$$

The concepts introduced in the preceding discussion are summarized in Definition 1.

DEFINITION 1

Parametric Equations and Plane Curves

A **plane curve** is the set of points (x, y) determined by the **parametric equations**

$$\begin{aligned} x &= f(t) \\ y &= g(t) \end{aligned}$$

where the **parameter** t varies over an interval I and the functions f and g are both defined on the interval I.

Why are we interested in parametric representations of plane curves? It turns out that this approach is more general than using equations with two variables as we have been doing. In addition, the approach generalizes to curves in three- and higher-dimensional spaces. Other important reasons for using parametric representations of plane curves will be brought out in the discussion and examples that follow.

EXAMPLE 1 Graphing Parametric Equations and Eliminating the Parameter

Graph the plane curve given parametrically by

$$x = 8 \cos \theta \qquad -\infty < \theta < \infty \qquad (4)$$
$$y = 4 \sin \theta$$

Identify the curve by eliminating the parameter θ.

Solution Construct a table and graph:

θ	0	$\pi/6$	$\pi/3$	$\pi/2$	$2\pi/3$	$5\pi/6$	π	$7\pi/6$	$4\pi/3$	$3\pi/2$	$5\pi/3$	$11\pi/6$	2π
x	8	$4\sqrt{3}$	4	0	-4	$-4\sqrt{3}$	-8	$-4\sqrt{3}$	-4	0	4	$4\sqrt{3}$	8
y	0	2	$2\sqrt{3}$	4	$2\sqrt{3}$	2	0	-2	$-2\sqrt{3}$	-4	$-2\sqrt{3}$	-2	0

To eliminate the parameter θ, we solve the first equation in (4) for $\cos\theta$, the second for $\sin\theta$, and substitute into the Pythagorean identity $\cos^2\theta + \sin^2\theta = 1$:

$$\cos\theta = \frac{x}{8} \qquad \text{and} \qquad \sin\theta = \frac{y}{4}$$

$$\cos^2\theta + \sin^2\theta = 1$$

$$\left(\frac{x}{8}\right)^2 + \left(\frac{y}{4}\right)^2 = 1$$

$$\frac{x^2}{64} + \frac{y^2}{16} = 1$$

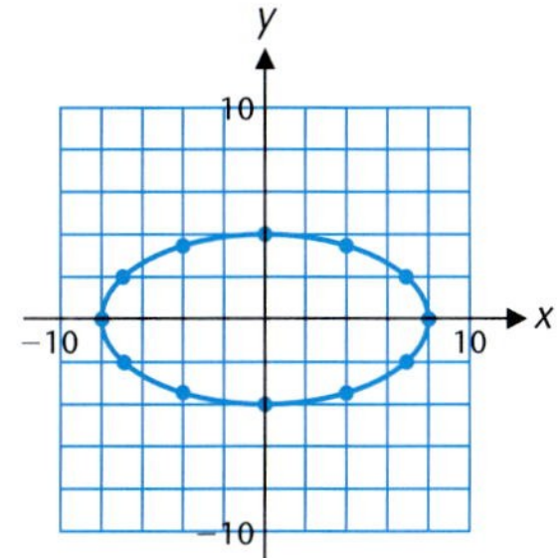

FIGURE 2
Graph of $x = 8 \cos \theta$, $y = 4 \sin \theta$, $-\infty < \theta < \infty$.

The graph is an ellipse (Fig. 2).

Matched Problem 1 Graph the plane curve given parametrically by $x = 4 \cos \theta$, $y = 4 \sin \theta$, $\theta \geq 0$. Identify the curve by eliminating the parameter θ.

EXPLORE-DISCUSS 1 Graph one period ($0 \leq \theta \leq 2\pi$) of each of the three plane curves given parametrically by

$$x_1 = 5 \cos\theta \qquad x_2 = 2 \cos\theta \qquad x_3 = 5 \cos\theta$$
$$y_1 = 5 \sin\theta \qquad y_2 = 2 \sin\theta \qquad y_3 = 2 \sin\theta$$

Identify the curves by eliminating the parameter.

EXAMPLE 2 Parametric Equations for Conic Sections

Find parametric equations for the conic section with the given equation:

(A) $25x^2 + 9y^2 - 100x + 54y - 44 = 0$
(B) $x^2 - 16y^2 - 10x + 32y - 7 = 0$

Solutions (A) By completing the square in x and y we obtain the standard form $\frac{(x-2)^2}{9} + \frac{(y+3)^2}{25} = 1$. So the graph is an ellipse with center $(2, -3)$ and major axis on the line $x = 2$. Since $\cos^2 \theta + \sin^2 \theta = 1$, a parametric representation with parameter θ is obtained by letting $\frac{x-2}{3} = \cos \theta$, $\frac{y+3}{5} = \sin \theta$:

$$x = 2 + 3\cos\theta \qquad -\infty < \theta < \infty$$
$$y = -3 + 5\sin\theta$$

(B) By completing the square in x and y we obtain the standard form $\frac{(x-5)^2}{16} - (y-1)^2 = 1$. So the graph is a hyperbola with center $(5, 1)$ and transverse axis on the line $y = 1$. Since $\sec^2 \theta - \tan^2 \theta = 1$, a parametric representation with parameter θ is obtained by letting $\frac{x-5}{4} = \sec\theta$, $y - 1 = \tan\theta$:

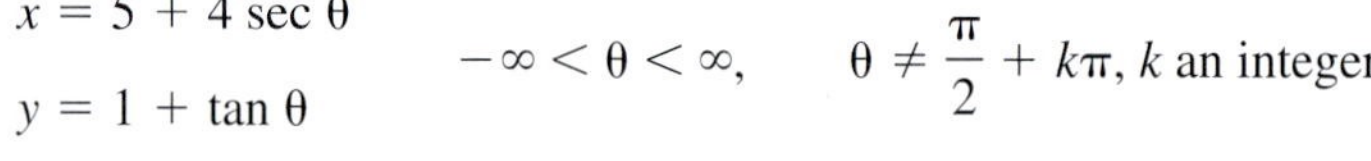

$$x = 5 + 4\sec\theta \qquad -\infty < \theta < \infty, \qquad \theta \neq \frac{\pi}{2} + k\pi,\ k \text{ an integer}$$
$$y = 1 + \tan\theta$$

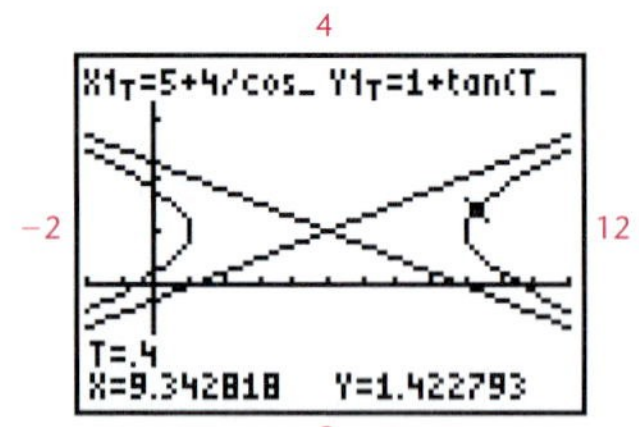

FIGURE 3
$x = 5 + 4\sec\theta$, $y = 1 + \tan\theta$.

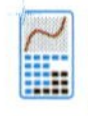
Note that when the parametric equations are graphed using a graphing utility in connected mode, the graph appears to show the asymptotes of the hyperbola (see Fig. 3).

Matched Problem 2

Find parametric equations for the conic section with the given equation:

(A) $36x^2 + 16y^2 + 504x - 96y + 1332 = 0$
(B) $16y^2 - 9x^2 - 36x + 128y + 76 = 0$

• Projectile Motion

Newton's laws and advanced mathematics can be used to determine the path of a projectile. If v_0 is the initial speed of the projectile at an angle α with the horizontal (see Fig. 4) and air resistance is neglected, then the path of the projectile is given by

$$x = (v_0 \cos\alpha)t \qquad 0 \le t \le b \qquad (5)$$
$$y = (v_0 \sin\alpha)t - 4.9t^2$$

The parameter t represents time in seconds, and x and y are distances measured in meters. Solving the first equation in (5) for t in terms of x, substituting into the second equation, and simplifying results in the following equation:

$$y = (\tan \alpha)x - \frac{4.9}{v_0^2 \cos^2 \alpha} x^2 \qquad (6)$$

You should verify this by supplying the omitted details.

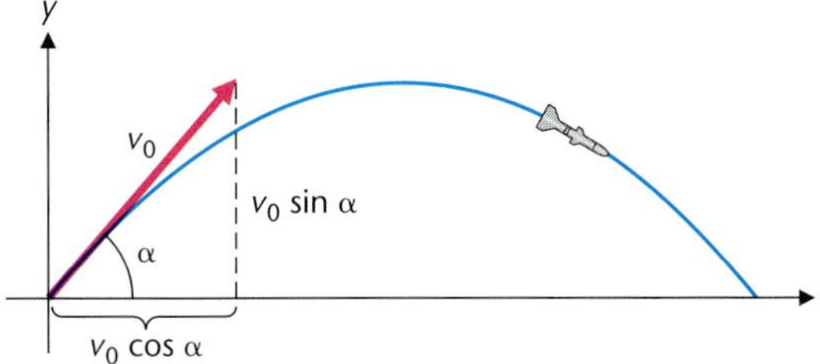

FIGURE 4 Projectile motion.

We recognize equation (6) as a parabola. This equation in x and y describes the path the projectile follows but tells us little else about its flight. On the other hand, the parametric equations (5) not only determine the path of the projectile but also tell us where it is at any time t. Furthermore, using concepts from physics and calculus, the parametric equations can be used to determine the velocity and acceleration of the projectile at any time t. This illustrates another advantage of using parametric representations of plane curves.

The **range of a projectile** is the distance from the point of firing to the point of impact. If we keep the initial speed v_0 of the projectile constant and vary the angle α in Figure 4, we obtain different parabolic paths followed by the projectile and different ranges. The maximum range is obtained when $\alpha = 45°$. Furthermore, assuming that the projectile always stays in the same vertical plane, then there are points in the air and on the ground that the projectile cannot reach, irrespective of the angle α used, $0° \le \alpha \le 180°$. Using more advanced mathematics, it can be shown that the reachable region is separated from the nonreachable region by a parabola called an **envelope** of the other parabolas (see Fig. 5).

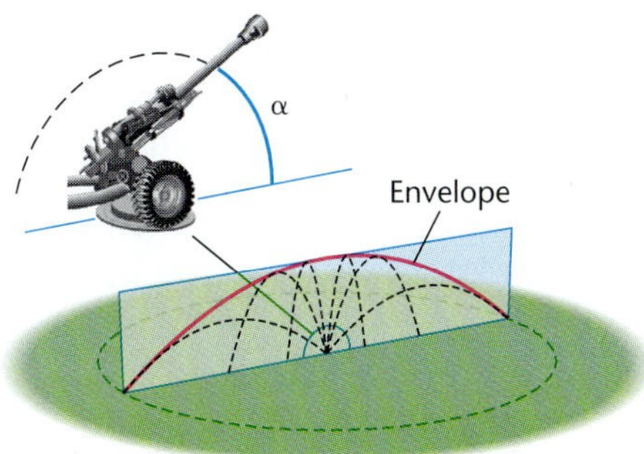

FIGURE 5 Reachable region of a projectile.

• Cycloid

We now consider an unusual curve called a *cycloid,* which has a fairly simple parametric representation and a very complicated representation in terms of x and y only. The path traced by a point on the rim of a circle that rolls along a line is called a **cycloid.** To derive parametric equations for a cycloid we roll a circle of radius a along the x axis with the tracing point P on the rim starting at the origin (see Figure 6).

FIGURE 6 Cycloid.

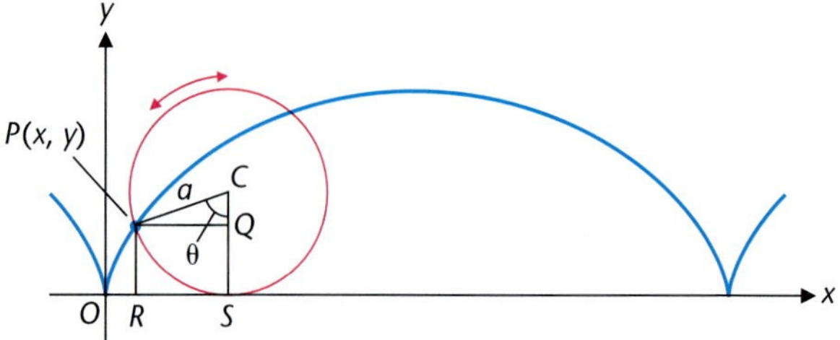

Since the circle rolls along the x axis without slipping (refer to Figure 6), we see that

$$d(O, S) = \text{arc } PS$$
$$= a\theta \qquad \theta \text{ in radians} \tag{7}$$

where S is the point of contact between the circle and the x axis. Referring to triangle CPQ, we see that

$$d(P, Q) = a \sin \theta \qquad 0 \le \theta \le \pi/2 \tag{8}$$
$$d(Q, C) = a \cos \theta \qquad 0 \le \theta \le \pi/2 \tag{9}$$

Using these results, we have

$$x = d(O, R)$$
$$= d(O, S) - d(R, S)$$
$$= (\text{arc } PS) - d(P, Q)$$
$$= a\theta - a \sin \theta$$ Use equations (7) and (8).
$$y = d(R, P)$$
$$= d(S, C) - d(Q, C)$$
$$= a - a \cos \theta$$ Use equation (9) and the fact that $d(S, C) = a$.

Even though θ in equations (8) and (9) was restricted so that $0 \le \theta \le \pi/2$, it can be shown that the derived parametric equations generate the whole cycloid for $-\infty < \theta < \infty$. The graph specifies a periodic function with period $2\pi a$. Thus, in general, we have Theorem 1.

Theorem 1 **Parametric Equations for a Cycloid**

For a circle of radius a rolled along the x axis, the resulting cycloid generated by a point on the rim starting at the origin is given by

$$\begin{aligned} x &= a\theta - a \sin \theta \\ y &= a - a \cos \theta \end{aligned} \qquad -\infty < \theta < \infty$$

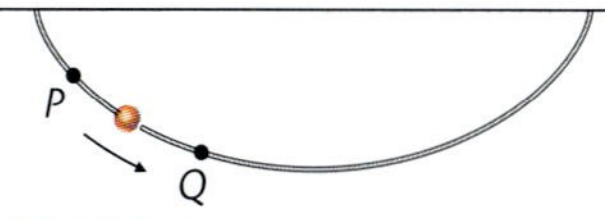

FIGURE 7 Cycloid path.

The cycloid is a good example of a curve that is very difficult to represent without the use of a parameter. A cycloid has a very interesting physical property. An object sliding without friction from a point P to a point Q lower than P, but not on the same vertical line as P, will arrive at Q in a shorter time traveling along a cycloid than on any other path (see Fig. 7).

EXPLORE-DISCUSS 2 (A) Let Q be a point b units from the center of a wheel of radius a, where $0 < b < a$. If the wheel rolls along the x axis with the tracing point Q starting at $(0, a - b)$, explain why parametric equations for the path of Q are given by

$$x = a\theta - b \sin \theta$$

$$y = a - b \cos \theta$$

(B) Use a graphing utility to graph the paths of a point on the rim of a wheel of radius 1, and a point halfway between the rim and center, as the wheel makes two complete revolutions rolling along the x axis.

Answers to Matched Problems

1. $x^2 + y^2 = 16$; circle

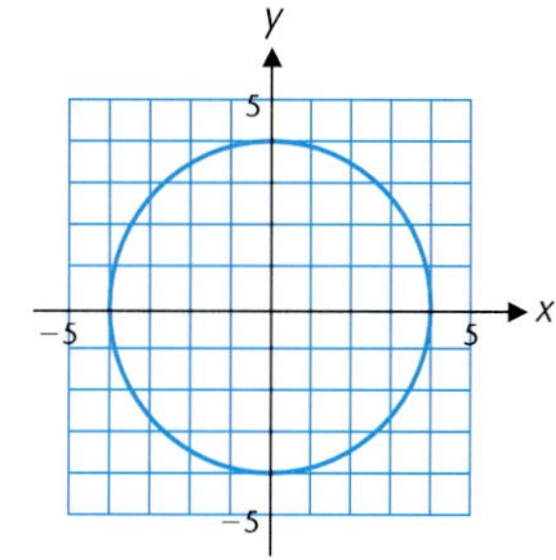

2. (A) $x = -7 + 4 \cos \theta,\ y = 3 + 6 \sin \theta,\ -\infty < \theta < \infty$

(B) $x = -2 + 4 \tan \theta,\ y = -4 + 3 \sec \theta,\ -\infty < \theta < \infty,\ \theta \neq \frac{\pi}{2} + k\pi$, k an integer

EXERCISE 11-5

A

In Problems 1–10, plot each plane curve by use of a table of values (see Example 1). Obtain an equation in x and y by eliminating the parameter, and identify the curve. (In this exercise set, the interval for the parameter is the whole real line, unless stated to the contrary.)

1. $x = -t,\ y = 2t - 2$

2. $x = t,\ y = t + 1$

3. $x = -t^2,\ y = 2t^2 - 2$

4. $x = t^2,\ y = t^2 + 1$

5. $x = 3t,\ y = -2t$

6. $x = 2t,\ y = t$

7. $x = \frac{1}{4}t^2,\ y = t$

8. $x = 2t,\ y = t^2$

9. $x = \frac{1}{4}t^4,\ y = t^2$

10. $x = 2t^2,\ y = t^4$

B

In Problems 11–18, obtain an equation in x and y by eliminating the parameter. Use the simpler of the two forms to plot the curve. Name the curve if it is a curve we have identified.

11. $x = 3 \sin \theta,\ y = 4 \cos \theta$

12. $x = 3 \sin \theta,\ y = 3 \cos \theta$

13. $x = 2 + 2 \sin \theta,\ y = 3 + 2 \cos \theta$

14. $x = 3 + 4 \sin \theta,\ y = 2 + 2 \cos \theta$

15. $x = t - 2,\ y = \dfrac{2}{2 - t};\ t \neq 2$

16. $x = t - 1, y = \dfrac{2}{t-1}; t \neq 1$

17. $x = t - 1, y = \sqrt{t}; t \geq 0$

18. $x = t^3, y = t^2 + 1$

19. If $A \neq 0$, $C = 0$, and $E \neq 0$, find parametric equations for $Ax^2 + Cy^2 + Dx + Ey + F = 0$. Identify the curve.

20. If $A = 0$, $C \neq 0$, and $D \neq 0$, find parametric equations for $Ax^2 + Cy^2 + Dx + Ey + F = 0$. Identify the curve.

C

In Problems 21–26, obtain an equation in x and y by eliminating the parameter. Use the simpler of the two forms to plot the curve. Name the curve if it is a curve we have identified.

21. $x = t^2, y = t^{-2}; t \neq 0$ **22.** $x = e^t, y = e^{-t}$

23. $x = \cos 2\theta, y = 4 \sin \theta$ **24.** $x = 3 \sec^2 \theta, y = 2 \tan^2 \theta$

25. $x = \dfrac{8}{t^2 + 4}, y = \dfrac{4t}{t^2 + 4}$ **26.** $x = \dfrac{4t}{t^2 + 1}, y = \dfrac{4t^2}{t^2 + 1}$

Graph, using a calculator, one period ($0 \leq \theta \leq 2\pi$) of each cycloid in Problems 27 and 28.

27. $x = \theta - \sin \theta, y = 1 - \cos \theta$

28. $x = 2\theta - 2 \sin \theta, y = 2 - 2 \cos \theta$

In Problems 29–32, use a graphing utility to graph the parametric equations. Then eliminate the parameter and find the standard equation for the curve. Name the curve and find its center.

29. $x = 3 + 6 \cos t, y = 2 + 4 \sin t, 0 \leq t \leq 2\pi$

30. $x = 1 + 3 \sec t, y = -2 + 2 \tan t, -\dfrac{\pi}{2} < t < \dfrac{3\pi}{2}, t \neq \dfrac{\pi}{2}$

31. $x = -3 + 2 \tan t, y = -1 + 5 \sec t, -\dfrac{\pi}{2} < t < \dfrac{3\pi}{2}, t \neq \dfrac{\pi}{2}$

32. $x = -4 + 5 \cos t, y = 1 + 8 \sin t, 0 \leq t \leq 2\pi$

33. Find an equation of the form $Ax^2 + Cy^2 + Dx + Ey + F = 0$ that has the same graph as the parametric equations $x = 2 \tan t, y = 5 \tan t, -\dfrac{\pi}{2} < t < \dfrac{\pi}{2}$.

34. Repeat Problem 33 for $x = \cot t$, $y = (3t \cot t)/|t|$, $-\pi < t < \pi$, $t \neq 0$.

In Problems 35–38, find the standard form for each equation. Name the curve and find its center. Use parametric equations to graph the curve on a graphing utility.

35. $25x^2 - 200x - 9y^2 - 18y + 616 = 0$

36. $36x^2 + 360x + 4y^2 - 8y + 760 = 0$

37. $4x^2 - 24x + 49y^2 + 392y + 624 = 0$

38. $16x^2 + 32x - 9y^2 - 36y - 164 = 0$

APPLICATIONS

39. Plane Motion. An object follows a path as given by

$$x = 5 \sin 6\pi t \qquad t \geq 0$$
$$y = 5 \cos 6\pi t$$

where t is time in seconds and x and y are distances in feet.
(A) What are the coordinates of the object when $t = 0.1$ second? Compute answers to one decimal place.
(B) Eliminate the parameter and graph the resulting equation in x and y. Identify the path.

40. Plane Motion. Repeat Problem 39 for

$$x = 4 \sin \pi t \qquad t \geq 0$$
$$y = 2 \cos \pi t$$

41. Projectile Motion. A projectile is fired with an initial speed of 300 meters per second at an angle of 45° to the horizontal. Neglecting air resistance, find:
(A) The time of impact
(B) The horizontal distance covered (range) in meters and kilometers at time of impact
(C) The maximum height in meters of the projectile
Compute all answers to three decimal places using a calculator.

42. Projectile Motion. Repeat Problem 41 if the same projectile is fired at 40° to the horizontal instead of 45°.

CHAPTER 11 GROUP ACTIVITY Focal Chords

Many of the applications of the conic sections are based on their reflective or focal properties. One of the interesting algebraic properties of the conic sections concerns their focal chords.

If a line through a focus F contains two points G and H of a conic section, then the line segment GH is called a **focal chord.** Let $G(x_1, y_1)$ and $H(x_2, y_2)$ be points on the graph of $x^2 = 4ay$ such that GH is a focal chord. Let u denote the length of GF and v the length of FH (see Fig. 1).

FIGURE 1 Focal chord GH of the parabola $x^2 = 4ay$.

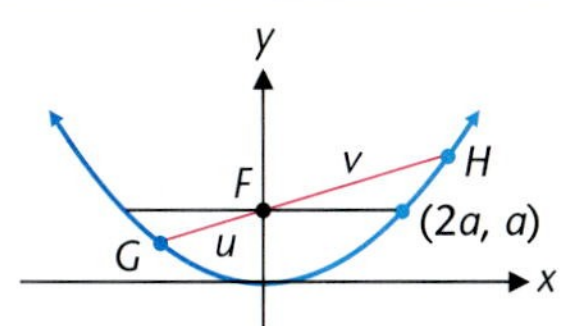

(A) Use the distance formula to show that $u = y_1 + a$.

(B) Show that G and H lie on the line $y - a = mx$, where $m = (y_2 - y_1)/(x_2 - x_1)$.

(C) Solve $y - a = mx$ for x and substitute in $x^2 = 4ay$, obtaining a quadratic equation in y. Explain why $y_1y_2 = a^2$.

(D) Show that $\frac{1}{u} + \frac{1}{v} = \frac{1}{a}$.

(E) Show that $u + v - 4a = \frac{(u - 2a)^2}{u - a}$. Explain why this implies that $u + v \geq 4a$, with equality if and only if $u = v = 2a$.

(F) Which focal chord is the shortest? Is there a longest focal chord?

(G) Is $\frac{1}{u} + \frac{1}{v}$ a constant for focal chords of the ellipse? For focal chords of the hyperbola? Obtain evidence for your answers by considering specific examples.

(H) The conic section with focus at the origin, directrix the line $x = D > 0$, and eccentricity $E > 0$ has the polar equation $r = \frac{DE}{1 + E\cos\theta}$. Explain how this polar equation makes it easy to show that $\frac{1}{u} + \frac{1}{v} = \frac{1}{a}$ for a parabola. Use the polar equation to determine the sum $\frac{1}{u} + \frac{1}{v}$ for a focal chord of an ellipse or hyperbola.

Chapter 11 Review

11-1 CONIC SECTIONS; PARABOLA

The plane curves obtained by intersecting a right circular cone with a plane are called **conic sections.** If the plane cuts clear through one nappe, then the intersection curve is called a **circle** if the plane is perpendicular to the axis and an **ellipse** if the plane is not perpendicular to the axis. If a plane cuts only one nappe, but does not cut clear through, then the intersection curve is called a **parabola.** If a plane cuts through both nappes, but not through the vertex, the resulting intersection curve is called a **hyperbola.** A plane passing through the vertex of the cone produces a **degenerate conic**—a point, a line, or a pair of lines. The figure illustrates the four nondegenerate conics.

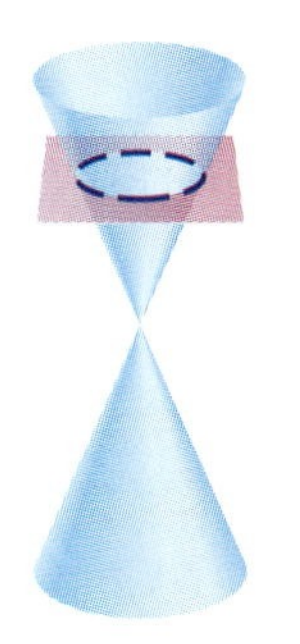
Circle

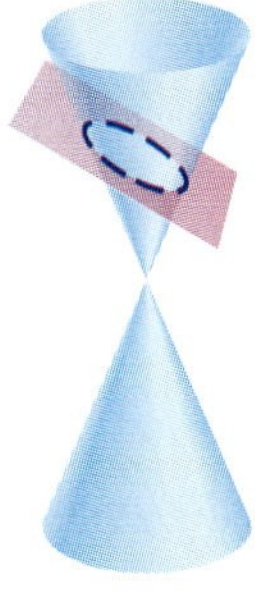
Ellipse

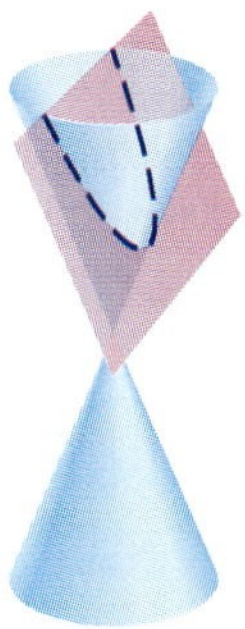
Parabola

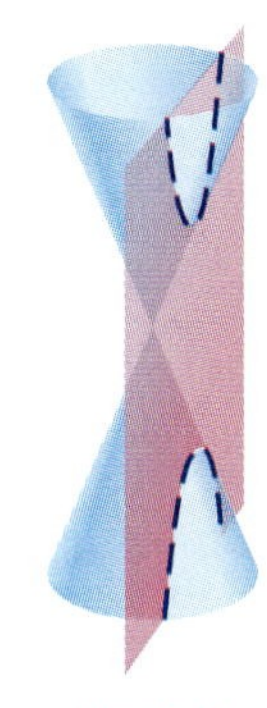
Hyperbola

The graph of

$$Ax^2 + Bxy + Cy^2 + Dx + Ey + F = 0$$

where A, B, and C are not all 0, is a conic.

The following is a coordinate-free definition of a parabola:

Parabola

A **parabola** is the set of all points in a plane equidistant from a fixed point F and a fixed line L in the plane. The fixed point F is called the **focus,** and the fixed line L is called the **directrix.** A line through the focus perpendicular to the directrix is called the **axis,** and the point on the axis halfway between the directrix and focus is called the **vertex.**

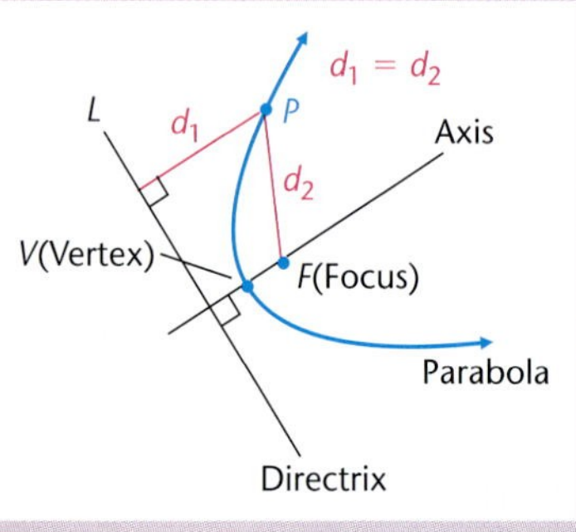

From the definition of a parabola, we can obtain the following standard equations:

Standard Equations of a Parabola with Vertex at (0, 0)

1. $y^2 = 4ax$
Vertex: (0, 0)
Focus: $(a, 0)$
Directrix: $x = -a$
Symmetric with respect to the x axis
Axis the x axis

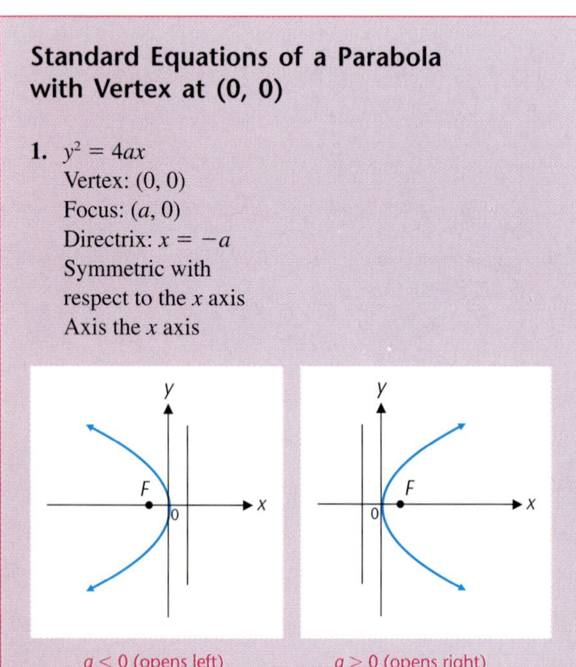

$a < 0$ (opens left) $a > 0$ (opens right)

2. $x^2 = 4ay$
Vertex: (0, 0)
Focus: $(0, a)$
Directrix: $y = -a$
Symmetric with respect to the y axis
Axis the y axis

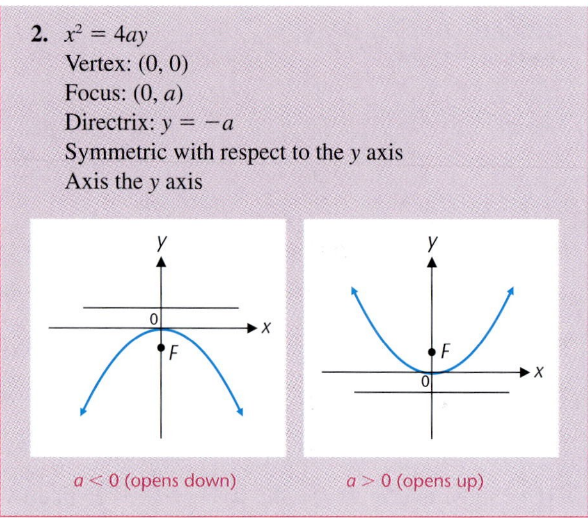

$a < 0$ (opens down) $a > 0$ (opens up)

11-2 ELLIPSE

The following is a coordinate-free definition of an ellipse:

Ellipse

An **ellipse** is the set of all points P in a plane such that the sum of the distances of P from two fixed points in the plane is constant. Each of the fixed points, F' and F, is called a **focus,** and together they are called **foci.** Referring to the figure, the line segment $V'V$ through the foci is the **major axis.** The perpendicular bisector $B'B$ of the major axis is the **minor axis.** Each end of the major axis, V' and V, is called a **vertex.** The midpoint of the line segment $F'F$ is called the **center** of the ellipse.

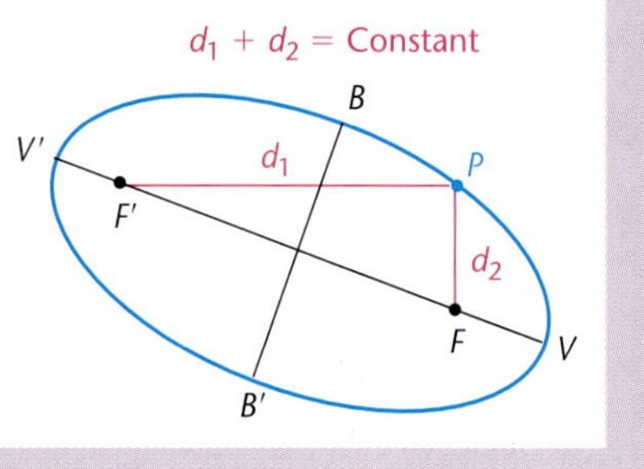

From the definition of an ellipse, we can obtain the following standard equations:

Standard Equations of an Ellipse with Center at (0, 0)

1. $\dfrac{x^2}{a^2} + \dfrac{y^2}{b^2} = 1 \qquad a > b > 0$
   x intercepts: $\pm a$ (vertices)
   y intercepts: $\pm b$
   Foci: $F'(-c, 0)$, $F(c, 0)$

   $$c^2 = a^2 - b^2$$

   Major axis length $= 2a$
   Minor axis length $= 2b$

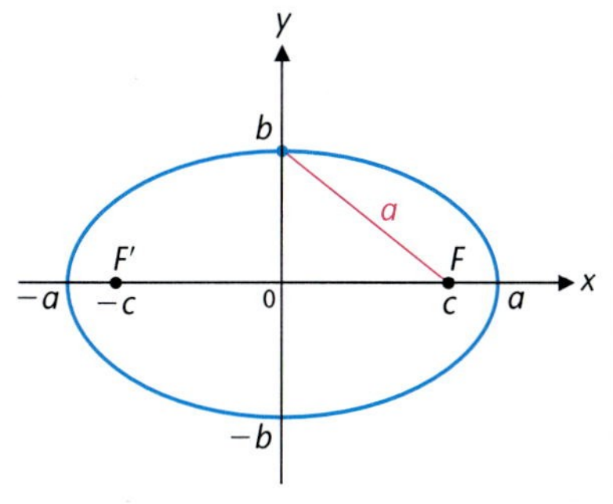

2. $\dfrac{x^2}{b^2} + \dfrac{y^2}{a^2} = 1 \qquad a > b > 0$
   x intercepts: $\pm b$
   y intercepts: $\pm a$ (vertices)
   Foci: $F'(0, -c)$, $F(0, c)$

   $$c^2 = a^2 - b^2$$

   Major axis length $= 2a$
   Minor axis length $= 2b$

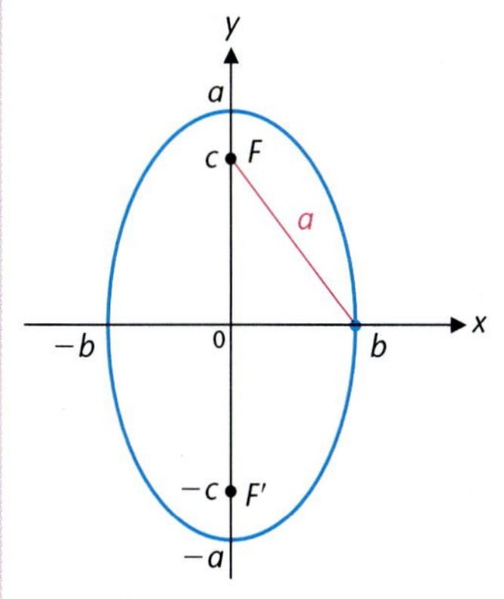

[*Note:* Both graphs are symmetric with respect to the x axis, y axis, and origin. Also, the major axis is always longer than the minor axis.]

11-3 HYPERBOLA

The following is a coordinate-free definition of a hyperbola:

Hyperbola

A **hyperbola** is the set of all points P in a plane such that the absolute value of the difference of the distances of P to two fixed points in the plane is a positive constant. Each of the fixed points, F' and F, is called a **focus.** The intersection points V' and V of the line through the foci and the two branches of the hyperbola are called **vertices,** and each is called a **vertex.** The line segment $V'V$ is called the **transverse axis.** The midpoint of the transverse axis is the **center** of the hyperbola.

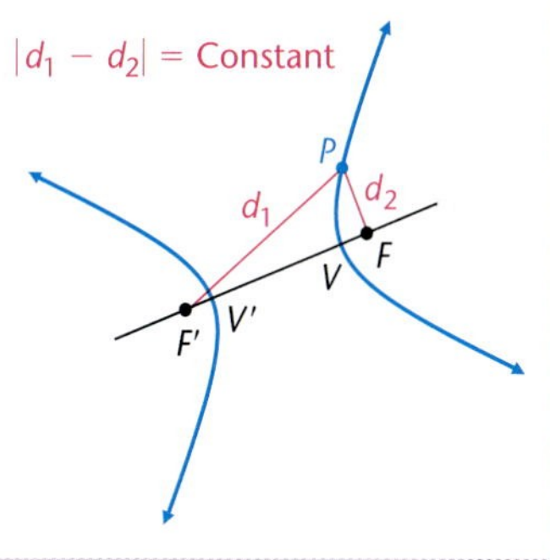

From the definition of a hyperbola, we can obtain the following standard equations:

Standard Equations of a Hyperbola with Center at (0, 0)

1. $\dfrac{x^2}{a^2} - \dfrac{y^2}{b^2} = 1$
   x intercepts: $\pm a$ (vertices)
   y intercepts: none
   Foci: $F'(-c, 0)$, $F(c, 0)$

   $$c^2 = a^2 + b^2$$

   Transverse axis length $= 2a$
   Conjugate axis length $= 2b$

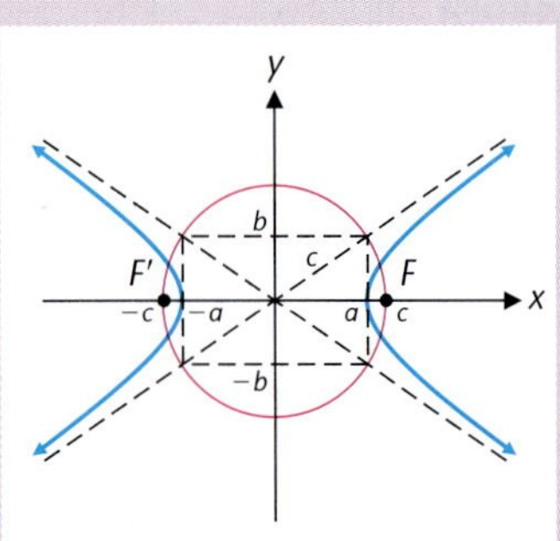

2. $\frac{y^2}{a^2} - \frac{x^2}{b^2} = 1$

x intercepts: none
y intercepts: $\pm a$ (vertices)
Foci: $F'(0, -c)$, $F(0, c)$

$$c^2 = a^2 + b^2$$

Transverse axis length $= 2a$
Conjugate axis length $= 2b$

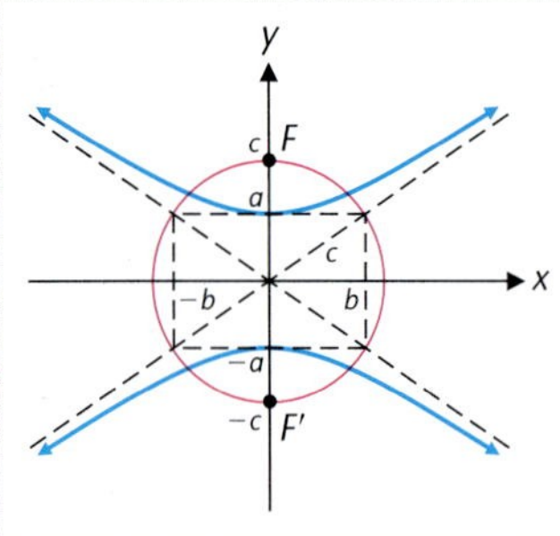

[*Note:* Both graphs are symmetric with respect to the x axis, y axis, and origin.]

11-4 TRANSLATION OF AXES

In the last three sections we found standard equations for parabolas, ellipses, and hyperbolas located with their axes on the coordinate axes and centered relative to the origin. We now move the conics away from the origin while keeping their axes parallel to the coordinate axes. In this process we obtain new standard equations that are special cases of the equation $Ax^2 + Cy^2 + Dx + Ey + F = 0$, where A and C are not both zero. The basic mathematical tool used is *translation of axes.*

A **translation of coordinate axes** occurs when the new coordinate axes have the same direction as and are parallel to the original coordinate axes. **Translation formulas** are as follows:

1. $x = x' + h$
$y = y' + k$

2. $x' = x - h$
$y' = y - k$

where (h, k) are the coordinates of the origin $0'$ relative to the original system.

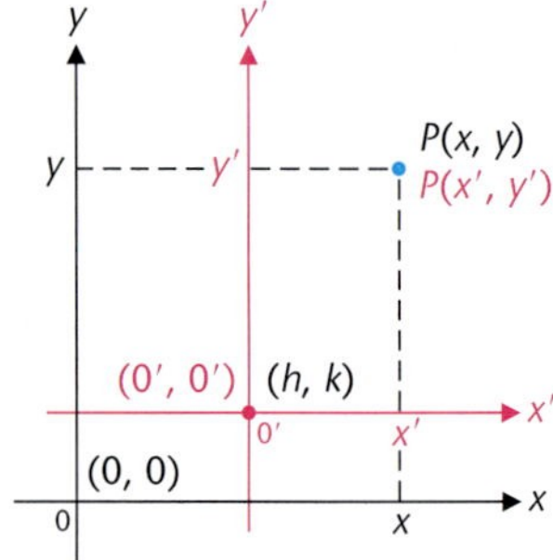

Table 1 lists the standard equations for translated conics.

11-5 PARAMETRIC EQUATIONS

A **plane curve** is the set of points (x, y) given by the **parametric equations**

$$x = f(t) \qquad \text{and} \qquad y = g(t)$$

where the **parameter** t varies over an interval I.

The **path of a projectile** with an initial speed v_0 at an angle α with the horizontal is given by

$$x = (v_0 \cos \alpha)t \qquad \text{and} \qquad y = (v_0 \sin \alpha)t - 4.9t^2, \quad 0 \le t \le b$$

or, after eliminating the parameter t, by

$$y = (\tan \alpha)x - \frac{4.9}{v_0^2 \cos^2 \alpha} x^2$$

where t is time in seconds and x and y are distances in meters. The **range of a projectile** is the distance from the point of firing to the point of impact. If the initial speed v_0 is held constant and the angle α is varied, then the reachable region of the projectile is separated from the nonreachable region by a parabola called an **envelope** of the possible parabolic paths of the projectile.

The path traced by a point on the rim of a circle of radius a that rolls along a straight line is called a **cycloid** and is given by

$$x = a\theta - a \sin \theta \qquad \text{and} \qquad y = a - a \cos \theta, \quad -\infty < \theta < \infty$$

TABLE 1 Standard Equations for Translated Conics

Parabolas

$(x - h)^2 = 4a(y - k)$

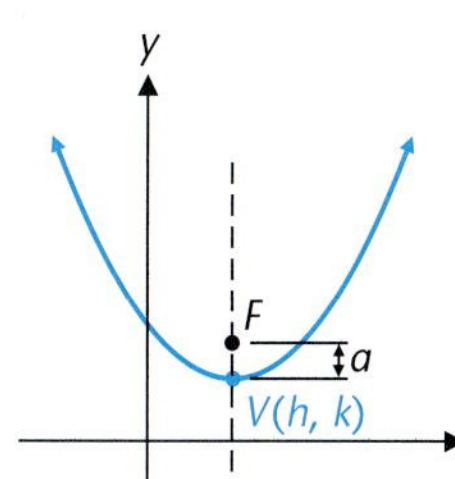

Vertex (h, k)
Focus $(h, k + a)$
$a > 0$ opens up
$a < 0$ opens down

$(y - k)^2 = 4a(x - h)$

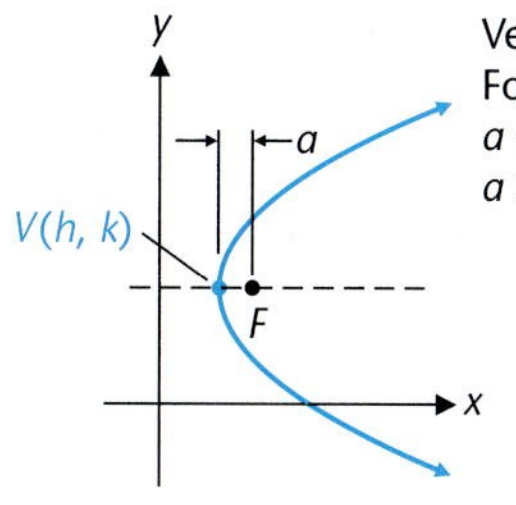

Vertex (h, k)
Focus $(h + a, k)$
$a < 0$ opens left
$a > 0$ opens right

Circles

$(x - h)^2 + (y - k)^2 = r^2$

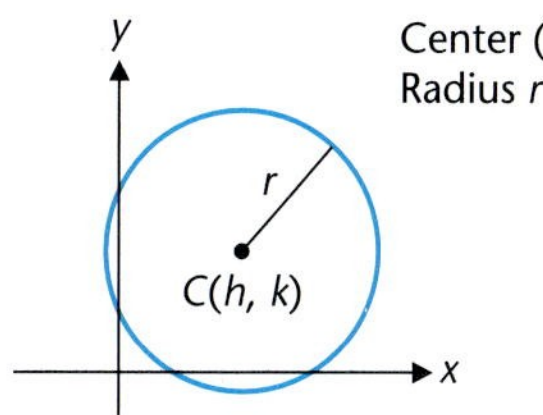

Center (h, k)
Radius r

Ellipses

$$\frac{(x - h)^2}{a^2} + \frac{(y - k)^2}{b^2} = 1 \qquad a > b > 0 \qquad \frac{(x - h)^2}{b^2} + \frac{(y - k)^2}{a^2} = 1$$

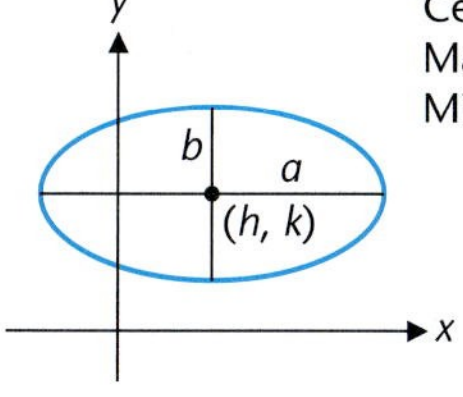

Center (h, k)
Major axis $2a$
Minor axis $2b$

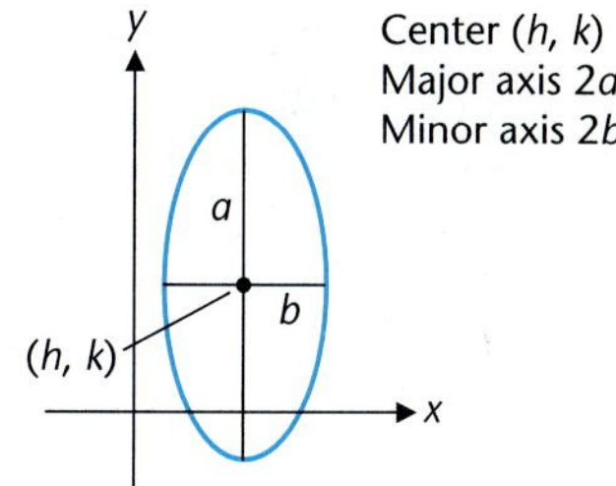

Center (h, k)
Major axis $2a$
Minor axis $2b$

Hyperbolas

$$\frac{(x - h)^2}{a^2} - \frac{(y - k)^2}{b^2} = 1$$

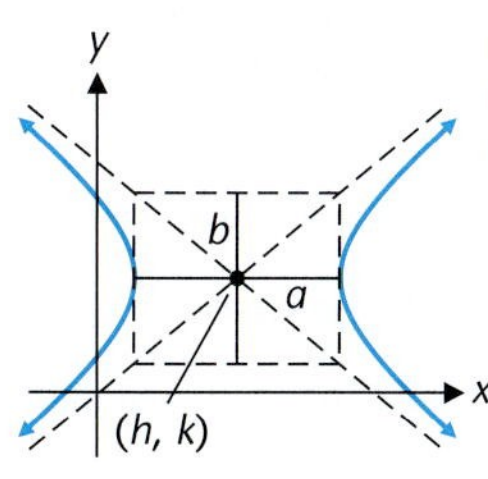

Center (h, k)
Transverse axis $2a$
Conjugate axis $2b$

$$\frac{(y - k)^2}{a^2} - \frac{(x - h)^2}{b^2} = 1$$

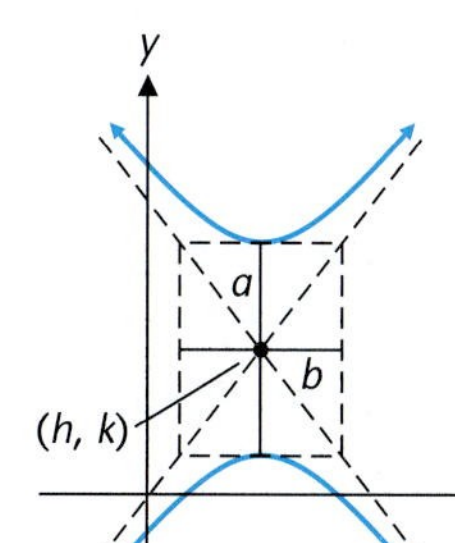

Center (h, k)
Transverse axis $2a$
Conjugate axis $2b$

Chapter 11 Review Exercise

Work through all the problems in this chapter review and check answers in the back of the book. Answers to all review problems are there, and following each answer is a number in italics indicating the section in which that type of problem is discussed. Where weaknesses show up, review appropriate sections in the text.

A

In Problems 1–3, graph each equation and locate foci. Locate the directrix for any parabolas. Find the lengths of major, minor, transverse, and conjugate axes where applicable.

1. $9x^2 + 25y^2 = 225$

2. $x^2 = -12y$

3. $25y^2 - 9x^2 = 225$

In Problems 4–6:
(A) Write each equation in one of the standard forms listed in Table 1 of the review.
(B) Identify the curve.

4. $4(y + 2)^2 - 25(x - 4)^2 = 100$

5. $(x + 5)^2 + 12(y + 4) = 0$

6. $16(x - 6)^2 + 9(y - 4)^2 = 144$

B

7. Find the equation of the parabola having its vertex at the origin, its axis the x axis, and $(-4, 2\sqrt{7})$ on its graph.

8. Find an equation of an ellipse in the form

$$\frac{x^2}{M} + \frac{y^2}{N} = 1 \qquad M, N > 0$$

if the center is at the origin, the major axis is on the y axis, the minor axis has length 4, and the distance between the foci is 62.

9. Find an equation of a hyperbola in the form

$$\frac{x^2}{M} - \frac{y^2}{N} = 1 \qquad M, N > 0$$

if the center is at the origin, the transverse axis has length 10, and the foci are 6 units from the center.

10. Plot the curve given parametrically by

$$x = -t^2$$
$$y = -\tfrac{1}{2}t^2 + 1$$

Obtain an equation in x and y by eliminating the parameter, and identify the curve.

In Problems 11–13, graph each system of equations in the same coordinate system and find the coordinates of any points of intersection.

11. $x^2 + 4y^2 = 32$
$x + 2y = 0$

12. $16x^2 + 25y^2 = 400$
$16x^2 - 45y = 0$

13. $x^2 + y^2 = 10$
$16x^2 + y^2 = 25$

In Problems 14–16, transform each equation into one of the standard forms in Table 1 in the review. Identify the curve and graph it.

14. $16x^2 + 4y^2 + 96x - 16y + 96 = 0$

15. $x^2 - 4x - 8y - 20 = 0$

16. $4x^2 - 9y^2 + 24x - 36y - 36 = 0$

17. Given the parametric equations of a plane curve, $x = -2 + 2 \sin \theta$ and $y = 3 + 4 \cos \theta$, obtain an equation in x and y by eliminating the parameter. Use the simpler of the two forms to plot the curve. Identify the curve.

18. Use a graphing utility to graph $x^2 = y$ and $x^2 = 50y$ in the viewing window $-10 \le x, y \le 10$. Find m so that the graph of $x^2 = y$ in the viewing window $-m \le x, y \le m$, has the same appearance as the graph of $x^2 = 50y$ in $-10 \le x, y \le 10$. Explain.

C

19. Use the definition of a parabola and the distance formula to find the equation of a parabola with directrix $x = 6$ and focus at (2, 4).

20. Find an equation of the set of points in a plane each of whose distance from (4, 0) is twice its distance from the line $x = 1$. Identify the geometric figure.

21. Find an equation of the set of points in a plane each of whose distance from (4, 0) is two-thirds its distance from the line $x = 9$. Identify the geometric figure.

In Problems 22–24, find the coordinates of any foci relative to the original coordinate system.

22. Problem 14 **23.** Problem 15 **24.** Problem 16

25. Given the parametric equations of a plane curve

$$x = 2^t$$
$$y = 2^{-t}$$

obtain an equation in x and y by eliminating the parameter. Use the simpler of the two forms to graph the curve. Identify the curve.

26. Use a graphing utility to find, to two decimal places, the coordinates of all points of intersection of $x^2 - 3y^2 + 9x + 7y - 22 = 0$ and $4x^2 + 5x + 10y - 53 = 0$.

APPLICATIONS

27. Communications. A parabolic satellite television antenna has a diameter of 8 feet and is 1 foot deep. How far is the focus from the vertex?

28. Engineering. An elliptical gear is to have foci 8 centimeters apart and a major axis 10 centimeters long. Letting the x axis lie along the major axis (right positive) and the y axis lie along the minor axis (up positive), write the equation of the ellipse in the standard form

$$\frac{x^2}{a^2} + \frac{y^2}{b^2} = 1$$

29. Space Science. A hyperbolic reflector for a radiotelescope (such as that illustrated in Problem 41, Exercise 11-3) has the equation

$$\frac{y^2}{40^2} - \frac{x^2}{30^2} = 1$$

If the reflector has a diameter of 30 feet, how deep is it? Compute the answer to 3 significant digits.

Cumulative Review Exercise Chapters 10 and 11

Work through all the problems in this cumulative review and check answers in the back of the book. Answers to all review problems are there, and following each answer is a number in italics indicating the section in which that type of problem is discussed. Where weaknesses show up, review appropriate sections in the text.

A

1. Determine whether each of the following can be the first three terms of an arithmetic sequence, a geometric sequence, both, or neither.

(A) 5, 25, 100, . . . (B) 15, 3, −9, . . .
(C) 1, 1, 1, . . . (D) −64, 16, −4, . . .
(E) 17, 119, 833, . . . (F) 1, 3, 6, . . .

In Problems 2–4:
(A) Write the first four terms of each sequence.
(B) Find a_8. (C) Find S_8.

2. $a_n = (-2)^n$

3. $a_n = 6n - 5$

4. $a_1 = -20;\ a_n = a_{n-1} + 4,\ n \ge 2$

5. Evaluate each of the following:

(A) $7!$ (B) $\dfrac{25!}{22!}$ (C) $\dfrac{10!}{(10-4)!\,4!}$

6. Evaluate each of the following:

(A) $\dbinom{12}{6}$ (B) $C_{9,4}$ (C) $P_{8,5}$

In Problems 7–9, graph each equation and locate foci. Locate the directrix for any parabolas. Find the lengths of major, minor, transverse, and conjugate axes where applicable.

7. $25x^2 - 36y^2 = 900$

8. $25x^2 + 36y^2 = 900$

9. $25x^2 - 36y = 0$

10. A coin is flipped three times. How many combined outcomes are possible? Solve:
(A) By using a tree diagram
(B) By using the multiplication principle

11. How many ways can 4 distinct books be arranged on a shelf? Solve:
(A) By using the multiplication principle
(B) By using permutations or combinations, whichever is applicable

12. Plot the curve given parametrically by

$$x = 2t + 3$$
$$y = 4t + 5$$

Obtain an equation in x and y by eliminating the parameter, and identify the curve.

Verify Problems 13 and 14 for n = 1, 2, and 3.

13. P_n: $1 + 5 + 9 + \cdots + (4n - 3) = n(2n - 1)$

14. P_n: $n^2 + n + 2$ is divisible by 2

In Problems 15 and 16, write P_k and P_{k+1}.

15. For P_n in Problem 13

16. For P_n in Problem 14

B

17. Find the equation of the parabola having its vertex at the origin, its axis the y axis, and $(2, -8)$ on its graph.

18. Find an equation of an ellipse in the form

$$\frac{x^2}{M} + \frac{y^2}{N} = 1 \qquad M, N > 0$$

if the center is at the origin, the major axis is the x axis, the major axis length is 10, and the distance of the foci from the center is 3.

19. Find an equation of a hyperbola in the form

$$\frac{x^2}{M} - \frac{y^2}{N} = 1 \qquad M, N > 0$$

if the center is at the origin, the transverse axis length is 16, and the distance of the foci from the center is $\sqrt{89}$.

20. Write $\sum_{k=1}^{5} k^k$ without summation notation and find the sum.

21. Write the series $\frac{2}{2!} - \frac{2^2}{3!} + \frac{2^3}{4!} - \frac{2^4}{5!} + \frac{2^5}{6!} - \frac{2^6}{7!}$ using summation notation with the summation index k starting at $k = 1$.

22. Find S_∞ for the geometric series $108 - 36 + 12 - 4 + \cdots$.

23. How many 4-letter code words are possible using the first 6 letters of the alphabet if no letter can be repeated? If letters can be repeated? If adjacent letters cannot be alike?

24. Let $a_n = 100(0.9)^n$ and $b_n = 10 + 0.03n$. Find the least positive integer n such that $a_n < b_n$ by graphing the sequences $\{a_n\}$ and $\{b_n\}$ with a graphing utility. Check your answer by using a graphing utility to display both sequences in table form.

25. Given the parametric equations of a plane curve

$$x = 2 + 7\cos\theta$$

$$y = -3 + 5\sin\theta$$

obtain an equation in x and y by eliminating the parameter. Use the simpler of the two forms to plot the curve. Identify the curve.

26. Evaluate each of the following:

(A) $P_{25,5}$ (B) $C(25, 5)$ (C) $\binom{25}{20}$

27. Expand $(a + \frac{1}{2}b)^6$ using the binomial formula.

28. Find the fifth and the eighth terms in the expansion of $(3x - y)^{10}$.

Establish each statement in Problems 29 and 30 for all positive integers using mathematical induction.

29. P_n in Problem 13

30. P_n in Problem 14

31. Find the sum of all the odd integers between 50 and 500.

32. Use the formula for the sum of an infinite geometric series to write $2.\overline{45} = 2.454\,545 \cdots$ as the quotient of two integers.

33. Let $a_k = \binom{30}{k}(0.1)^{30-k}(0.9)^k$ for $k = 0, 1, \ldots, 30$. Use a graphing utility to find the largest term of the sequence $\{a_k\}$ and the number of terms that are greater than 0.01.

In Problems 34–36, use a translation of coordinates to transform each equation into a standard equation for a nondegenerate conic. Identify the curve and graph it.

34. $4x + 4y - y^2 + 8 = 0$

35. $x^2 + 2x - 4y^2 - 16y + 1 = 0$

36. $4x^2 - 16x + 9y^2 + 54y + 61 = 0$

37. How many 9-digit zip codes are possible? How many of these have no repeated digits?

38. Use a graphing utility to find, to two decimal places, the coordinates of all points of intersection of $5x^2 + 2y^2 - 7x + 8y - 48 = 0$ and $e^x - e^{-x} - 2y = 0$.

39. Use mathematical induction to prove that the following statement holds for all positive integers:

$$P_n\colon \quad \frac{1}{1\cdot 3} + \frac{1}{3\cdot 5} + \frac{1}{5\cdot 7} + \cdots + \frac{1}{(2n-1)(2n+1)} = \frac{n}{2n+1}$$

C

40. Use the binomial formula to expand $(x - 2i)^6$, where i is the imaginary unit.

41. Use the definition of a parabola and the distance formula to find the equation of a parabola with directrix $y = 3$ and focus $(6, 1)$.

42. An ellipse has vertices $(\pm 4, 0)$ and foci $(\pm 2, 0)$. Find the y intercepts.

43. A hyperbola has vertices $(2, \pm 3)$ and foci $(2, \pm 5)$. Find the length of the conjugate axis.

In Problems 44–47, determine whether the sequence $\{a_n\}$ is arithmetic, geometric, both, or neither.

44. $a_n = \dfrac{2^n}{5^{n+1}}$

45. $a_1 = 3,\ a_{n+1} = a_n + \dfrac{1}{n} \quad n \ge 1$

46. $a_1 = 1,\ a_2 = 1,\ a_{n+2} = a_n \cdot a_{n+1} \quad n \ge 1$

47. $a_n = \dbinom{n}{2} - \dfrac{n^2}{2}$

In Problems 48–56, determine whether the statement is true or false. If true, explain why. If false, give a counterexample.

48. If n is a positive integer, then $n^{10} \ge n!$

49. If n is a positive integer, then $n^n \ge n!$

50. If the integer $n \ge 3$ is divisible by a prime p, then $\dbinom{n}{2}$ is divisible by p.

51. If p is a prime greater than 2, then $\dbinom{p}{2}$ is divisible by p.

52. If P is a point on an ellipse, then the sum of the distances from P to the foci is equal to the length of the major axis.

53. If P is a point on a parabola, then the distance from P to the directrix equals the distance from P to the vertex.

54. If P is a point on a hyperbola, then the absolute value of the differences of the distances from P to the foci is equal to the length of the conjugate axis.

55. The length of the major axis of an ellipse is greater than the length of the minor axis.

56. The length of the transverse axis of a hyperbola is greater than the length of the conjugate axis.

57. How many parabolas have the x axis as axis and pass through the points $(0, -3)$ and $(2, 5)$? Find the equation(s).

58. A hyperbola has its transverse axis on the line $y = -4$ and its conjugate axis on the line $x = 5$, and the transverse axis is twice the length of the conjugate axis. Find the equations of the asymptotes.

59. An ellipse passes through the point $P(8, -3)$, the distance between the foci is 12, and the length of the minor axis is 4. Find the sum of the distances from P to the foci.

60. Seven distinct points are selected on the circumference of a circle. How many triangles can be formed using these 7 points as vertices?

61. Given the parametric equations of a plane curve

$$x = e^{2t} - 4$$
$$y = 1 - e^t$$

obtain an equation in x and y by eliminating the parameter. Use the simpler of the two forms to plot the curve. Identify the curve.

62. Use mathematical induction to prove that $2^n < n!$ for all integers $n > 3$.

63. Use mathematical induction to show that $\{a_n\} = \{b_n\}$, where $a_1 = 3$, $a_n = 2a_{n-1} - 1$ for $n > 1$, and $b_n = 2^n + 1$, $n \ge 1$.

64. Find an equation of the set of points in the plane each of whose distance from $(1, 4)$ is three times its distance from the x axis. Write the equation in the form $Ax^2 + Cy^2 + Dx + Ey + F = 0$, and identify the curve.

APPLICATIONS 

65. Economics. The government, through a subsidy program, distributes \$2,000,000. If we assume that each individual or agency spends 75% of what it receives, and 75% of this is spent, and so on, how much total increase in spending results from this government action?

66. Engineering. An automobile headlight contains a parabolic reflector with a diameter of 8 inches. If the light source is located at the focus, which is 1 inch from the vertex, how deep is the reflector?

67. Architecture. A sound whispered at one focus of a whispering chamber can be easily heard at the other focus. Suppose that a cross section of this chamber is a semielliptical arch which is 80 feet wide and 24 feet high (see the figure). How far is each focus from the center of the arch? How high is the arch above each focus?

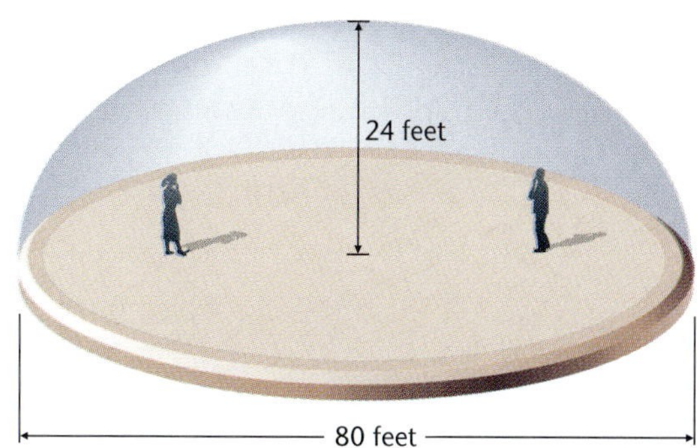

BASIC ALGEBRAIC OPERATIONS

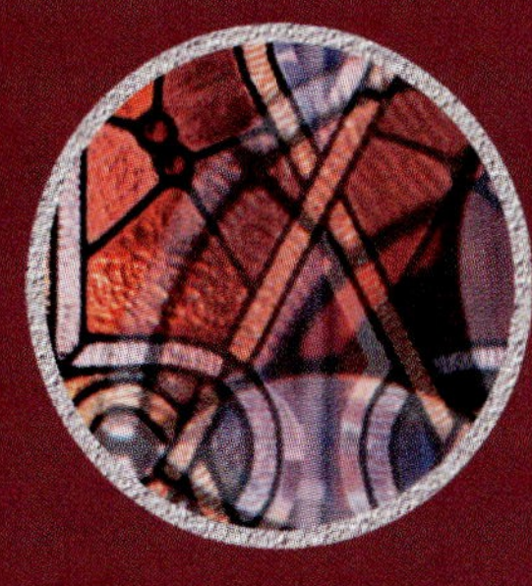

APPENDIX A

Algebra is often referred to as "generalized arithmetic." In arithmetic we deal with the basic arithmetic operations of addition, subtraction, multiplication, and division performed on specific numbers. In algebra we continue to use all that we know in arithmetic, but, in addition, we reason and work with symbols that represent one or more numbers. In this chapter we review some important basic algebraic operations usually studied in earlier courses. The material may be studied systematically before commencing with the rest of the book or reviewed as needed.

SECTION A-1 Algebra and Real Numbers

- Sets
- The Set of Real Numbers
- The Real Number Line
- Basic Real Number Properties
- Further Properties
- Fraction Properties

The rules for manipulating and reasoning with symbols in algebra depend, in large measure, on properties of the real numbers. In this section we look at some of the important properties of this number system. To make our discussions here and elsewhere in the text clearer and more precise, we first introduce a few useful notions about sets.

• Sets

Georg Cantor (1845–1918) developed a theory of sets as an outgrowth of his studies on infinity. His work has become a milestone in the development of mathematics.

Our use of the word "set" will not differ appreciably from the way it is used in everyday language. Words such as "set," "collection," "bunch," and "flock" all convey the same idea. Thus, we think of a **set** as a collection of objects with the important property that we can tell whether any given object is or is not in the set.

Each object in a set is called an **element,** or **member,** of the set. Symbolically,

$a \in A$	means	"a is an element of set A"	$3 \in \{1, 3, 5\}$
$a \notin A$	means	"a is not an element of set A"	$2 \notin \{1, 3, 5\}$

Capital letters are often used to represent sets and lowercase letters to represent elements of a set.

A set is **finite** if the number of elements in the set can be counted and **infinite** if there is no end in counting its elements. A set is **empty** if it contains no elements. The empty set is also called the **null** set and is denoted by $\varnothing$. It is important to observe that the empty set is *not* written as $\{\varnothing\}$.

A set is usually described in one of two ways—by **listing** the elements between braces, { }, or by enclosing within braces a **rule** that determines its elements. For example, if D is the set of all numbers x such that $x^2 = 4$, then using the listing method we write

$$D = \{-2, 2\} \quad \text{Listing method}$$

or, using the rule method we write

$$D = \{x \mid x^2 = 4\} \quad \text{Rule method}$$

Note that in the rule method, the vertical bar | represents "such that," and the entire symbolic form $\{x \mid x^2 = 4\}$ is read, "The set of all x such that $x^2 = 4$."

The letter x introduced in the rule method is a *variable.* In general, a **variable** is a symbol that is used as a placeholder for the elements of a set with two or more elements. This set is called the **replacement set** for the variable. A **constant,** on the other hand, is a symbol that names exactly one object. The symbol "8" is a constant, since it always names the number eight.

If each element of set A is also an element of set B, we say that A is a **subset** of set B, and we write

$$A \subset B \qquad \{1, 5\} \subset \{1, 3, 5\}$$

Note that the definition of a subset allows a set to be a subset of itself.

Since the empty set $\varnothing$ has no elements, every element of $\varnothing$ is also an element of any given set. Thus, the empty set is a subset of every set. For example,

$$\varnothing \subset \{1, 3, 5\} \qquad \text{and} \qquad \varnothing \subset \{2, 4, 6\}$$

If two sets A and B have exactly the same elements, the sets are said to be **equal,** and we write

$$A = B \qquad \{4, 2, 6\} = \{6, 4, 2\}$$

Notice that the order of listing elements in a set does not matter.

We can now begin our discussion of the real number system. Additional set concepts will be introduced as needed.

• The Set of Real Numbers

The real number system is the number system you have used most of your life. Informally, a **real number** is any number that has a decimal representation. Table 1 on the next page describes the set of real numbers and some of its important subsets. Figure 1 illustrates how these sets of numbers are related to each other.

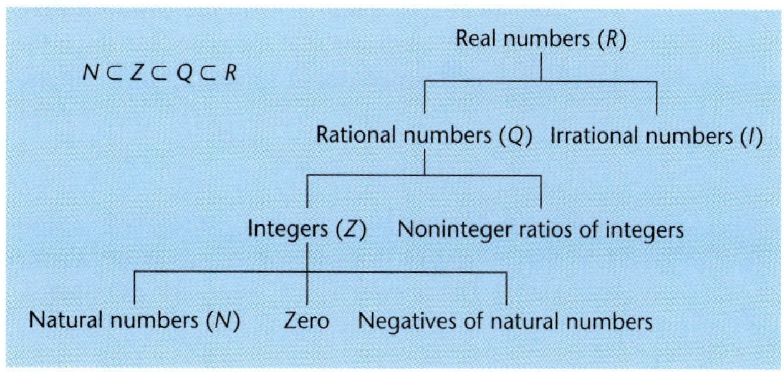

Figure 1 Real numbers and important subsets.

TABLE 1 The Set of Real Numbers

Symbol	Name	Description	Examples
N	Natural numbers	Counting numbers (also called positive integers)	1, 2, 3, . . .
Z	Integers	Natural numbers, their negatives, and 0	. . . , −2, −1, 0, 1, 2, . . .
Q	Rational numbers	Numbers that can be represented as a/b, where a and b are integers and $b \neq 0$; decimal representations are repeating or terminating	$-4, 0, 1, 25, \frac{-3}{5}, \frac{2}{3}, 3.67, -0.33\overline{3}$,* $5.2727\overline{27}$
I	Irrational numbers	Numbers that can be represented as nonrepeating and nonterminating decimal numbers	$\sqrt{2}, \pi, \sqrt[3]{7}$, 1.414213 . . . , 2.71828182 . . .
R	Real numbers	Rational numbers and irrational numbers	

*The overbar indicates that the number (or block of numbers) repeats indefinitely.

• The Real Number Line

A one-to-one correspondence exists between the set of real numbers and the set of points on a line. That is, each real number corresponds to exactly one point, and each point to exactly one real number. A line with a real number associated with each point, and vice versa, as in Figure 2, is called a **real number line,** or simply a **real line.** Each number associated with a point is called the **coordinate** of the point. The point with coordinate 0 is called the **origin.** The arrow on the right end of the line indicates a positive direction. The coordinates of all points to the right of the origin are called **positive real numbers,** and those to the left of the origin are called **negative real numbers.** The real number 0 is neither positive nor negative.

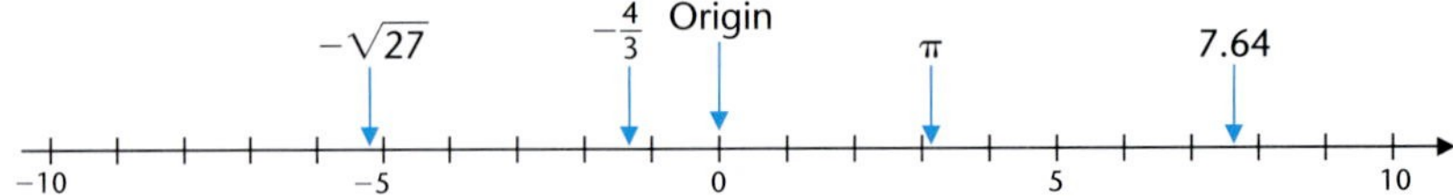

Figure 2 A real number line.

• Basic Real Number Properties

We now take a look at some of the basic properties of real numbers. (See the box on the next page.)

You are already familiar with the **commutative properties** for addition and multiplication. They indicate that the order in which the addition or multiplication of two numbers is performed doesn't matter. For example,

$$4 + 5 = 5 + 4 \qquad \text{and} \qquad 4 \cdot 5 = 5 \cdot 4$$

Is there a commutative property relative to subtraction or division? That is, does $x - y = y - x$ or does $x \div y = y \div x$ for all real numbers x and y (division by 0 excluded)? The answer is no, since, for example,

$$7 - 5 \neq 5 - 7 \qquad \text{and} \qquad 6 \div 3 \neq 3 \div 6$$

Basic Properties of the Set of Real Numbers

Let R be the set of real numbers, and let x, y, and z be arbitrary elements of R.

Addition Properties

Closure: $x + y$ is a unique element in R.

Associative: $(x + y) + z = x + (y + z)$

Commutative: $x + y = y + x$

Identity: 0 is the additive identity; that is, $0 + x = x + 0 = x$ for all x in R, and 0 is the only element in R with this property.

Inverse: For each x in R, $-x$ is its unique additive inverse; that is, $x + (-x) = (-x) + x = 0$, and $-x$ is the only element in R relative to x with this property.

Multiplication Properties

Closure: xy is a unique element in R.

Associative: $(xy)z = x(yz)$

Commutative: $xy = yx$

Identity: 1 is the multiplicative identity; that is, for x in R, $(1)x = x(1) = x$, and 1 is the only element in R with this property.

Inverse: For each x in R, $x \neq 0$, $1/x$ is its unique multiplicative inverse; that is, $x(1/x) = (1/x)x = 1$, and $1/x$ is the only element in R relative to x with this property.

Combined Property

Distributive: $x(y + z) = xy + xz \qquad (x + y)z = xz + yz$

When computing

$$2 + 5 + 3 \qquad \text{or} \qquad 2 \cdot 5 \cdot 3$$

why don't we need parentheses to indicate which two numbers are to be added or multiplied first? The answer is to be found in the **associative properties.** These properties allow us to write

$$(2 + 5) + 3 = 2 + (5 + 3) \qquad \text{and} \qquad (2 \cdot 5) \cdot 3 = 2 \cdot (5 \cdot 3)$$

so it doesn't matter how we group numbers relative to either operation. Is there an associative property for subtraction or division? The answer is no, since, for example,

$$(8 - 4) - 2 \neq 8 - (4 - 2) \qquad \text{and} \qquad (8 \div 4) \div 2 \neq 8 \div (4 \div 2)$$

Evaluate both sides of these equations to see why.

Conclusion

Relative to addition, **commutativity** and **associativity** permit us to change the order of addition at will and insert or remove parentheses as we please. The same is true for multiplication, but not for subtraction and division.

What number added to a given number will give that number back again? What number times a given number will give that number back again? The answers are 0 and 1, respectively. Because of this, 0 and 1 are called the **identity elements** for the real numbers. Hence, for any real numbers x and y,

$$7 + 0 = 7 \qquad 0 + (x + y) = x + y \qquad \text{0 is the additive identity.}$$

$$1 \cdot 6 = 6 \qquad 1(x + y) = x + y \qquad \text{1 is the multiplicative identity.}$$

We now consider **inverses.** For each real number x, there is a unique real number $-x$ such that $x + (-x) = 0$. The number $-x$ is called the **additive inverse** of x, or the **negative** of x. For example, the additive inverse of 4 is -4, since $4 + (-4) = 0$. The additive inverse of -4 is $-(-4) = 4$, since $-4 + [-(-4)] = 0$. It is important to remember:

$-x$ is not necessarily a negative number; it is positive if x is negative and negative if x is positive.

For each nonzero real number x there is a unique real number $1/x$ such that $x(1/x) = 1$. The number $1/x$ is called the **multiplicative inverse** of x, or the **reciprocal** of x. For example, the multiplicative inverse of 7 is $\frac{1}{7}$, since $7(\frac{1}{7}) = 1$. Also note that 7 is the multiplicative inverse of $\frac{1}{7}$. The number 0 has no multiplicative inverse.

We now turn to a real number property that involves both multiplication and addition. Consider the two computations:

$$3(4 + 2) = 3(6) = 18$$

$$3(4) + 3(2) = 12 + 6 = 18$$

Thus,

$$3(4 + 2) = 3(4) + 3(2)$$

and we say that multiplication by 3 *distributes* over the sum $(4 + 2)$. In general, multiplication **distributes** over addition in the real number system. Two more illustrations are given below:

$$2(x + y) = 2x + 2y \qquad (3 + 5)x = 3x + 5x$$

EXAMPLE 1 **Using Real Number Properties**

Which real number property justifies the indicated statement?

Statement	**Property Illustrated**
(A) $(7x)y = 7(xy)$	Associative $(\cdot)$
(B) $a(b + c) = (b + c)a$	Commutative $(\cdot)$
(C) $(2x + 3y) + 5y = 2x + (3y + 5y)$	Associative $(+)$
(D) $(x + y)(a + b) = (x + y)a + (x + y)b$	Distributive
(E) If $a + b = 0$, then $b = -a$.	Inverse $(+)$

Matched Problem 1* Which real number property justifies the indicated statement?

(A) $4 + (2 + x) = (4 + 2) + x$
(B) $(a + b) + c = c + (a + b)$
(C) $3x + 7x = (3 + 7)x$
(D) $(2x + 3y) + 0 = 2x + 3y$
(E) If $ab = 1$, then $b = 1/a$.

• Further Properties

Subtraction and division can be defined in terms of addition and multiplication, respectively:

DEFINITION 1 **Subtraction and Division**

For all real numbers a and b:

Subtraction: $a - b = a + (-b)$ $\quad (-5) - (-3) = (-5) + (3) = -2$

Division: $b\overline{)a} = a \div b = \dfrac{a}{b} = a\left(\dfrac{1}{b}\right) \quad b \neq 0$ $\quad 3 \div 2 = 3\left(\dfrac{1}{2}\right)$

Thus, to subtract b from a, add the negative of b to a. To divide a by b, multiply a by the reciprocal of b. Note that division by 0 is not defined, since 0 does not have a reciprocal. It is important to remember:

Division by 0 is never allowed.

The following properties of negatives can be proved using the preceding properties and definitions.

*Answers to matched problems in a given section are found near the end of the section, before the exercise set.

Theorem 1 **Properties of Negatives**

For all real numbers a and b:

1. $-(-a) = a$
2. $(-a)b = -(ab) = a(-b) = -ab$
3. $(-a)(-b) = ab$
4. $(-1)a = -a$
5. $\frac{-a}{b} = -\frac{a}{b} = \frac{a}{-b} \qquad b \neq 0$
6. $\frac{-a}{-b} = -\frac{-a}{b} = -\frac{a}{-b} = \frac{a}{b} \qquad b \neq 0$

We now state an important theorem involving 0.

Theorem 2 **Zero Properties**

For all real numbers a and b:

1. $a \cdot 0 = 0$
2. $ab = 0$ if and only if $a = 0$ or $b = 0$ or both

EXAMPLE 2 Using Negative and Zero Properties

Which real number property or definition justifies each statement?

Statement	**Property or Definition Illustrated**
(A) $3 - (-2) = 3 + [-(-2)] = 5$	Subtraction (Definition 1 and Theorem 1, part 1)
(B) $-(-2) = 2$	Negatives (Theorem 1, part 1)
(C) $-\frac{-3}{2} = \frac{3}{2}$	Negatives (Theorem 1, part 6)
(D) $\frac{5}{-2} = -\frac{5}{2}$	Negatives (Theorem 1, part 5)
(E) If $(x - 3)(x + 5) = 0$, then either $x - 3 = 0$ or $x + 5 = 0$.	Zero (Theorem 2, part 2)

Matched Problem 2 Which real number property or definition justifies each statement?

(A) $\frac{3}{5} = 3\left(\frac{1}{5}\right)$ (B) $(-5)(2) = -(5 \cdot 2)$ (C) $(-1)3 = -3$

(D) $\frac{-7}{9} = -\frac{7}{9}$ (E) If $x + 5 = 0$, then $(x - 3)(x + 5) = 0$.

EXPLORE-DISCUSS 1 In general, a set of numbers is closed under an operation if performing the operation on numbers in the set always produces another number in the set. For example, the real numbers are closed under addition, multiplication, subtraction, and division, excluding division by 0. Replace each ? in the following tables with T (true) or F (false), and illustrate each false statement with an example. (See Table 1 for the definitions of the sets N, Z, I, Q, and R.)

	Closed under Addition	Closed under Multiplication
N	?	?
Z	?	?
Q	?	?
I	?	?
R	T	T

	Closed under Subtraction	Closed under Division*
N	?	?
Z	?	?
Q	?	?
I	?	?
R	T	T

*Excluding division by 0.

• Fraction Properties

Recall that the quotient $a \div b$, $b \neq 0$, written in the form a/b is called a **fraction.** The quantity a is called the **numerator** and the quantity b is the **denominator.**

Theorem 3 **Fraction Properties**

For all real numbers a, b, c, d, and k (division by 0 excluded):

1. $\frac{a}{b} = \frac{c}{d}$ if and only if $ad = bc$

   $\frac{4}{6} = \frac{6}{9}$ since $4 \cdot 9 = 6 \cdot 6$

2. $\frac{ka}{kb} = \frac{a}{b}$

   $\frac{7 \cdot 3}{7 \cdot 5} = \frac{3}{5}$

3. $\frac{a}{b} \cdot \frac{c}{d} = \frac{ac}{bd}$

   $\frac{3}{5} \cdot \frac{7}{8} = \frac{3 \cdot 7}{5 \cdot 8}$

4. $\frac{a}{b} \div \frac{c}{d} = \frac{a}{b} \cdot \frac{d}{c}$

   $\frac{2}{3} \div \frac{5}{7} = \frac{2}{3} \cdot \frac{7}{5}$

5. $\frac{a}{b} + \frac{c}{b} = \frac{a + c}{b}$

   $\frac{3}{6} + \frac{5}{6} = \frac{3 + 5}{6}$

6. $\frac{a}{b} - \frac{c}{b} = \frac{a - c}{b}$

   $\frac{7}{8} - \frac{3}{8} = \frac{7 - 3}{8}$

7. $\frac{a}{b} + \frac{c}{d} = \frac{ad + bc}{bd}$

   $\frac{2}{3} + \frac{3}{5} = \frac{2 \cdot 5 + 3 \cdot 3}{3 \cdot 5}$

Answers to Matched Problems

1. (A) Associative (+) (B) Commutative (+) (C) Distributive (D) Identity (+)
   (E) Inverse (·)
2. (A) Division (Definition 1) (B) Negatives (Theorem 1, part 2)
   (C) Negatives (Theorem 1, part 4) (D) Negatives (Theorem 1, part 5)
   (E) Zero (Theorem 2, part 1)

EXERCISE A-1

All variables represent real numbers.

A

In Problems 1–8, indicate true (T) or false (F).

1. $4 \in \{3, 4, 5\}$

2. $6 \in \{2, 4, 6\}$

3. $3 \notin \{3, 4, 5\}$

4. $7 \notin \{2, 4, 6\}$

5. $\{1, 2\} \subset \{1, 3, 5\}$

6. $\{2, 6\} \subset \{2, 4, 6\}$

7. $\{7, 3, 5\} \subset \{3, 5, 7\}$

8. $\{7, 3, 5\} = \{3, 5, 7\}$

In Problems 9–14, replace each question mark with an appropriate expression that will illustrate the use of the indicated real number property.

9. Commutative property (+): $x + 7 = ?$

10. Commutative property (·): $uv = ?$

11. Associative property (·): $x(yz) = ?$

12. Associative property (+): $3 + (7 + y) = ?$

13. Identity property (+): $0 + 9m = ?$

14. Identity property (·): $1(u + v) = ?$

In Problems 15–26, each statement illustrates the use of one of the following properties or definitions. Indicate which one.

Commutative (+, ·)	*Subtraction*
Associative (+, ·)	*Division*
Distributive	*Negatives (Theorem 1)*
Identity (+, ·)	*Zero (Theorem 2)*
Inverse (+, ·)	

15. $x + ym = x + my$

16. $7(3m) = (7 \cdot 3)m$

17. $7u + 9u = (7 + 9)u$

18. $-\frac{u}{-v} = \frac{u}{v}$

19. $(-2)(\frac{1}{-2}) = 1$

20. $8 - 12 = 8 + (-12)$

21. $w + (-w) = 0$

22. $5 \div (-6) = 5(\frac{1}{-6})$

23. $3(xy + z) + 0 = 3(xy + z)$

24. $ab(c + d) = abc + abd$

25. $\dfrac{-x}{-y} = \dfrac{x}{y}$

26. $(x + y) \cdot 0 = 0$

B

Write each set in Problems 27–32 using the listing method; that is, list the elements between braces. If the set is empty, write $\varnothing$.

27. $\{x \mid x \text{ is an even integer between } -3 \text{ and } 5\}$

28. $\{x \mid x \text{ is an odd integer between } -4 \text{ and } 6\}$

29. $\{x \mid x \text{ is a letter in "status"}\}$

30. $\{x \mid x \text{ is a letter in "consensus"}\}$

31. $\{x \mid x \text{ is a month starting with B}\}$

32. $\{x \mid x \text{ is a month with 32 days}\}$

33. The set $S_1 = \{a\}$ has only two subsets, S_1 and $\varnothing$. How many subsets does each of the following sets have?
(A) $S_2 = \{a, b\}$
(B) $S_3 = \{a, b, c\}$
(C) $S_4 = \{a, b, c, d\}$

34. Based on the results in Problem 33, how many subsets do you think a set with n elements will have?

In Problems 35–42, each statement illustrates the use of one of the following properties or definitions. Indicate which one.

Commutative $(+, \cdot)$	*Subtraction*
Associative $(+, \cdot)$	*Division*
Distributive	*Negatives (Theorem 1)*
Identity $(+, \cdot)$	*Zero (Theorem 2)*
Inverse $(+, \cdot)$	

35. $-(-y) + 2x = y + 2x$

36. $(ab)(ba) = (ab)(ab)$

37. $(wz)(zw) = w[z(zw)]$

38. $s + (t + 2) = (s + t) + 2$

39. $(n + 2)(m + 3) = n(m + 3) + 2(m + 3)$

40. $p(r - 1) + q(r - 1) = (p + q)(r - 1)$

41. $(2x - 3)(3x + 5) = 0$ if and only if $2x - 3 = 0$ or $3x + 5 = 0$

42. $\dfrac{-y}{-(1 - y)} = \dfrac{y}{1 - y}$

43. If $ab = 0$, does either a or b have to be 0?

44. If $ab = 1$, does either a or b have to be 1?

45. Indicate which of the following are true:
(A) All natural numbers are integers.
(B) All real numbers are irrational.
(C) All rational numbers are real numbers.

46. Indicate which of the following are true:
(A) All integers are natural numbers.
(B) All rational numbers are real numbers.
(C) All natural numbers are rational numbers.

47. Give an example of a rational number that is not an integer.

48. Give an example of a real number that is not a rational number.

In Problems 49 and 50, list the subset of S consisting of (A) natural numbers, (B) integers, and (C) rational numbers.

49. $S = \{-3, -\frac{2}{3}, 0, 1, \sqrt{3}, \frac{9}{5}, 12\}$

50. $S = \{-\sqrt{5}, -1, -\frac{1}{2}, 2, \sqrt{7}, 6, \frac{25}{3}\}$

In Problems 51 and 52, use a calculator to express each number as a decimal fraction. Classify each decimal number as terminating, repeating, or nonrepeating and nonterminating. Identify the pattern of repeated digits in any repeating decimal numbers.

51. (A) $\frac{8}{9}$ (B) $\frac{3}{11}$ (C) $\sqrt{5}$ (D) $\frac{11}{8}$

52. (A) $\frac{13}{6}$ (B) $\sqrt{21}$ (C) $\frac{7}{16}$ (D) $\frac{29}{111}$

53. Indicate true (T) or false (F), and for each false statement find real number replacements for a and b that will provide a counterexample. For all real numbers a and b:
(A) $a + b = b + a$
(B) $a - b = b - a$
(C) $ab = ba$
(D) $a \div b = b \div a$

54. Indicate true (T) or false (F), and for each false statement find real number replacements for a, b, and c that will provide a counterexample. For all real numbers a, b, and c:
(A) $(a + b) + c = a + (b + c)$
(B) $(a - b) - c = a - (b - c)$
(C) $a(bc) = (ab)c$
(D) $(a \div b) \div c = a \div (b \div c)$

C

55. If $A = \{1, 2, 3, 4\}$ and $B = \{2, 4, 6\}$, find:
(A) $\{x \mid x \in A \text{ or } x \in B\}$
(B) $\{x \mid x \in A \text{ and } x \in B\}$

56. If $F = \{-2, 0, 2\}$ and $G = \{-1, 0, 1, 2\}$, find:
(A) $\{x \mid x \in F \text{ or } x \in G\}$
(B) $\{x \mid x \in F \text{ and } x \in G\}$

57. If $c = 0.151515\ldots$, then $100c = 15.1515\ldots$ and

$$100c - c = 15.1515\ldots - 0.151515\ldots$$
$$99c = 15$$
$$c = \tfrac{15}{99} = \tfrac{5}{33}$$

Proceeding similarly, convert the repeating decimal 0.090909 . . . into a fraction. (All repeating decimals are rational numbers, and all rational numbers have repeating decimal representations.)

58. Repeat Problem 57 for 0.181818. . . .

59. To see how the distributive property is behind the mechanics of long multiplication, compute each of the following and compare:

Long Multiplication	Use of the Distributive Property
$\begin{array}{r} 23 \\ \times\ 12 \\ \hline \end{array}$	$\begin{aligned} &23 \cdot 12 \\ &\quad = 23(2 + 10) \\ &\quad = 23 \cdot 2 + 23 \cdot 10 = \end{aligned}$

60. For a and b real numbers, justify each step using a property in this section.

Statement	Reason
1. $(a + b) + (-a) = (-a) + (a + b)$	**1.**
2. $= [(-a) + a] + b$	**2.**
3. $= 0 + b$	**3.**
4. $= b$	**4.**

SECTION A-2 Polynomials: Basic Operations

- Natural Number Exponents
- Polynomials
- Combining Like Terms
- Addition and Subtraction
- Multiplication
- Combined Operations
- Application

In this section we review the basic operations on *polynomials,* a mathematical form encountered frequently throughout mathematics. We start the discussion with a brief review of natural number exponents. Integer and rational exponents and their properties will be discussed in detail in subsequent sections.

• Natural Number Exponents

The definition of a **natural number exponent** is given below:

DEFINITION 1 **Natural Number Exponent**

For n a natural number and a any real number:

$$a^n = \underbrace{a \cdot a \cdot \cdots \cdot a}_{n \text{ factors of } a} \qquad 2^4 = \underbrace{2 \cdot 2 \cdot 2 \cdot 2}_{4 \text{ factors of } 2}$$

Also, the **first property of exponents** is stated as follows:

Theorem 1

First Property of Exponents

For any natural numbers m and n, and any real number a:

$$a^m a^n = a^{m+n} \qquad (3x^5)(2x^7) = (3 \cdot 2)x^{5+7} = 6x^{12}$$

• Polynomials

Algebraic expressions are formed by using constants and variables and the algebraic operations of addition, subtraction, multiplication, division, raising to powers, and taking roots. Some examples are

$$\sqrt[3]{x^3 + 5} \qquad 5x^4 + 2x^2 - 7$$

$$x + y - 7 \qquad (2x - y)^2$$

$$\frac{x - 5}{x^2 + 2x - 5} \qquad 1 + \cfrac{1}{1 + \cfrac{1}{x}}$$

An algebraic expression involving only the operations of addition, subtraction, multiplication, and raising to natural number powers on variables and constants is called a **polynomial.** Some examples are

$$2x - 3 \qquad 4x^2 - 3x + 7$$

$$x - 2y \qquad 5x^3 - 2x^2 - 7x + 9$$

$$5 \qquad x^2 - 3xy + 4y^2$$

$$0 \qquad x^3 - 3x^2y + xy^2 + 2y^7$$

In a polynomial, a variable cannot appear in a denominator, as an exponent, or within a radical. Accordingly, a **polynomial in one variable** x is constructed by adding or subtracting constants and terms of the form ax^n, where a is a real number and n is a natural number. A **polynomial in two variables** x and y is constructed by adding and subtracting constants and terms of the form ax^my^n, where a is a real number and m and n are natural numbers. Polynomials in three or more variables are defined in a similar manner.

Polynomial forms can be classified according to their *degree.* If a term in a polynomial has only one variable as a factor, then the **degree of that term** is the power of the variable. If two or more variables are present in a term as factors, then the **degree of the term** is the sum of the powers of the variables. The **degree of a polynomial** is the degree of the nonzero term with the highest degree in the polynomial. Any nonzero constant is defined to be a **polynomial of degree 0.** The number 0 is also a polynomial but is not assigned a degree.

EXAMPLE 1 Polynomials and Nonpolynomials

(A) Polynomials in one variable:

$$x^2 - 3x + 2 \qquad 6x^3 - \sqrt{2}x - \tfrac{1}{3}$$

(B) Polynomials in several variables:

$$3x^2 - 2xy + y^2 \qquad 4x^3y^2 - \sqrt{3}xy^2z^5$$

(C) Nonpolynomials:

$$\sqrt{2x} - \frac{3}{x} + 5 \qquad \frac{x^2 - 3x + 2}{x - 3} \qquad \sqrt{x^2 - 3x + 1}$$

(D) The degree of the first term in $6x^3 - \sqrt{2}x - \frac{1}{3}$ is 3, the degree of the second term is 1, the degree of the third term is 0, and the degree of the whole polynomial is 3.

(E) The degree of the first term in $4x^3y^2 - \sqrt{3}xy^2$ is 5, the degree of the second term is 3, and the degree of the whole polynomial is 5.

Matched Problem 1 (A) Which of the following are polynomials?

$$3x^2 - 2x + 1 \qquad \sqrt{x - 3} \qquad x^2 - 2xy + y^2 \qquad \frac{x - 1}{x^2 + 2}$$

(B) Given the polynomial $3x^5 - 6x^3 + 5$, what is the degree of the first term? The second term? The whole polynomial?

(C) Given the polynomial $6x^4y^2 - 3xy^3$, what is the degree of the first term? The second term? The whole polynomial?

In addition to classifying polynomials by degree, we also call a single-term polynomial a **monomial,** a two-term polynomial a **binomial,** and a three-term polynomial a **trinomial.**

$\frac{5}{2}x^2y^3$	Monomial
$x^3 + 4.7$	Binomial
$x^4 - \sqrt{2}x^2 + 9$	Trinomial

• Combining Like Terms

We start with a word about *coefficients.* A constant in a term of a polynomial, including the sign that precedes it, is called the **numerical coefficient,** or simply, the **coefficient,** of the term. If a constant doesn't appear, or only a + sign appears, the coefficient is understood to be 1. If only a $-$ sign appears, the coefficient is understood to be -1. Thus, given the polynomial

$$2x^4 - 4x^3 + x^2 - x + 5 \quad 2x^4 + (-4)x^3 + 1x^2 + (-1)x + 5$$

the coefficient of the first term is 2, the coefficient of the second term is -4, the coefficient of the third term is 1, the coefficient of the fourth term is -1, and the coefficient of the last term is 5.

At this point, it is useful to state two additional distributive properties of real numbers that follow from the distributive properties stated in Section A-1.

Additional Distributive Properties

1. $a(b - c) = (b - c)a = ab - ac$

2. $a(b + c + \cdots + f) = ab + ac + \cdots + af$

Two terms in a polynomial are called **like terms** if they have exactly the same variable factors to the same powers. The numerical coefficients may or may not be the same. Since constant terms involve no variables, all constant terms are like terms. If a polynomial contains two or more like terms, these terms can be combined into a single term by making use of distributive properties. Consider the following example:

$$\begin{aligned} 5x^3y - 2xy - x^3y - 2x^3y &= 5x^3y - x^3y - 2x^3y - 2xy \\ &= (5x^3y - x^3y - 2x^3y) - 2xy \\ &= (5 - 1 - 2)x^3y - 2xy \\ &= 2x^3y - 2xy \end{aligned}$$

It should be clear that free use has been made of the real number properties discussed earlier. The steps done in the dashed box are usually done mentally, and the process is quickly mechanized as follows:

Like terms in a polynomial are combined by adding their numerical coefficients.

EXAMPLE 2 Simplifying Polynomials

Remove parentheses and combine like terms:

(A) $2(3x^2 - 2x + 5) + (x^2 + 3x - 7)$

$$\begin{aligned} &= 2(3x^2 - 2x + 5) + 1(x^2 + 3x - 7) \quad \text{Think} \\ &= 6x^2 - 4x + 10 + x^2 + 3x - 7 \\ &= 7x^2 - x + 3 \end{aligned}$$

(B) $(x^3 - 2x - 6) - (2x^3 - x^2 + 2x - 3)$

$$= 1(x^3 - 2x - 6) + (-1)(2x^3 - x^2 + 2x - 3)$$

Think — Be careful with the sign here.

$$= x^3 - 2x - 6 - 2x^3 + x^2 - 2x + 3$$
$$= -x^3 + x^2 - 4x - 3$$

(C) $[3x^2 - (2x + 1)] - (x^2 - 1) = [3x^2 - 2x - 1] - (x^2 - 1)$ Remove inner parentheses first.

$$= 3x^2 - 2x - 1 - x^2 + 1$$
$$= 2x^2 - 2x$$

Matched Problem 2 Remove parentheses and combine like terms:

(A) $3(u^2 - 2v^2) + (u^2 + 5v^2)$

(B) $(m^3 - 3m^2 + m - 1) - (2m^3 - m + 3)$

(C) $(x^3 - 2) - [2x^3 - (3x + 4)]$

• Addition and Subtraction

Addition and subtraction of polynomials can be thought of in terms of removing parentheses and combining like terms, as illustrated in Example 2. Horizontal and vertical arrangements are illustrated in the next two examples. You should be able to work either way, letting the situation dictate the choice.

EXAMPLE 3 Adding Polynomials

Add: $x^4 - 3x^3 + x^2$, $-x^3 - 2x^2 + 3x$, and $3x^2 - 4x - 5$

Solution Add horizontally:

$$(x^4 - 3x^3 + x^2) + (-x^3 - 2x^2 + 3x) + (3x^2 - 4x - 5)$$
$$= x^4 - 3x^3 + x^2 - x^3 - 2x^2 + 3x + 3x^2 - 4x - 5$$
$$= x^4 - 4x^3 + 2x^2 - x - 5$$

Or vertically, by lining up like terms and adding their coefficients:

$$\begin{array}{r} x^4 - 3x^3 + x^2 \qquad\qquad \\ - x^3 - 2x^2 + 3x \qquad \\ 3x^2 - 4x - 5 \\ \hline x^4 - 4x^3 + 2x^2 - x - 5 \end{array}$$

Matched Problem 3 Add horizontally and vertically:

$$3x^4 - 2x^3 - 4x^2, \qquad x^3 - 2x^2 - 5x, \qquad \text{and} \qquad x^2 + 7x - 2$$

EXAMPLE 4 **Subtracting Polynomials**

Subtract: $4x^2 - 3x + 5$ from $x^2 - 8$

Solution

$$\begin{aligned}(x^2 - 8) - (4x^2 - 3x + 5) \\ = x^2 - 8 - 4x^2 + 3x - 5 \\ = -3x^2 + 3x - 13\end{aligned} \qquad \text{or} \qquad \begin{array}{r} x^2 \phantom{{}+3x} - 8 \\ \underline{-4x^2 + 3x - 5} \\ -3x^2 + 3x - 13 \end{array}$$

← Change signs and add.

Matched Problem 4 Subtract: $2x^2 - 5x + 4$ from $5x^2 - 6$

CAUTION

When you use a horizontal arrangement to subtract a polynomial with more than one term, you must enclose the polynomial in parentheses. Thus, to subtract $2x + 5$ from $4x - 11$, you must write

$$4x - 11 - (2x + 5) \qquad \text{and not} \qquad 4x - 11 - 2x + 5$$

• Multiplication

Multiplication of algebraic expressions involves the extensive use of distributive properties for real numbers, as well as other real number properties.

EXAMPLE 5 **Multiplying Polynomials**

Multiply: $(2x - 3)(3x^2 - 2x + 3)$

Solution

$$\begin{aligned}(2x - 3)(3x^2 - 2x + 3) &= 2x(3x^2 - 2x + 3) - 3(3x^2 - 2x + 3) \\ &= 6x^3 - 4x^2 + 6x - 9x^2 + 6x - 9 \\ &= 6x^3 - 13x^2 + 12x - 9\end{aligned}$$

Or, using a vertical arrangement,

$$\begin{array}{rrrr} & 3x^2 & - 2x & + 3 \\ & & 2x & - 3 \\ \hline 6x^3 & - 4x^2 & + 6x & \\ & - 9x^2 & + 6x & - 9 \\ \hline 6x^3 & - 13x^2 & + 12x & - 9 \end{array}$$

Matched Problem 5 Multiply: $(2x - 3)(2x^2 + 3x - 2)$

Thus, to multiply two polynomials, multiply each term of one by each term of the other, and combine like terms.

Products of certain binomial factors occur so frequently that it is useful to develop procedures that will enable us to write down their products by inspection. To find the product $(2x - 1)(3x + 2)$, we will use the popular **FOIL method.** We multiply each term of one factor by each term of the other factor as follows:

F First product ↓ O Outer product ↓ I Inner product ↓ L Last product ↓

$$(2x - 1)(3x + 2) = 6x^2 + 4x - 3x - 2$$

The inner and outer products are like terms and hence combine into one term. Thus,

$$(2x - 1)(3x + 2) = 6x^2 + x - 2$$

To speed up the process, we combine the inner and outer product mentally.

Products of certain binomial factors occur so frequently that it is useful to remember formulas for their products. The following formulas are easily verified by multiplying the factors on the left using the FOIL method:

Special Products

1. $(a - b)(a + b) = a^2 - b^2$
2. $(a + b)^2 = a^2 + 2ab + b^2$
3. $(a - b)^2 = a^2 - 2ab + b^2$

EXPLORE-DISCUSS 1 (A) Explain the relationship between special product formula 1 and the areas of the rectangles in the figures.

$$(a - b)(a + b) \quad = \quad a^2 - b^2$$

(B) Construct similar figures to provide geometric interpretations for special product formulas 2 and 3.

EXAMPLE 6 **Multiplying Binomials**

Multiply:

(A) $(2x - 3y)(5x + 2y) = 10x^2 + 4xy - 15xy - 6y^2 = 10x^2 - 11xy - 6y^2$

(B) $(3a - 2b)(3a + 2b) = (3a)^2 - (2b)^2 = 9a^2 - 4b^2$

(C) $(5x - 3)^2 = (5x)^2 - 2(5x)(3) + 3^2 = 25x^2 - 30x + 9$

(D) $(m + 2n)^2 = m^2 + 4mn + 4n^2$

Matched Problem 6 Multiply:

(A) $(4u - 3v)(2u + v)$ (B) $(2xy + 3)(2xy - 3)$
(C) $(m + 4n)(m - 4n)$ (D) $(2u - 3v)^2$ (E) $(6x + y)^2$

CAUTION Remember to include the sum of the inner and outer terms when using the FOIL method to square a binomial. That is,

$$(x + 3)^2 \neq x^2 + 9 \quad (x + 3)^2 = x^2 + 6x + 9$$

• Combined Operations

We now consider several examples that use all the operations just discussed. Before considering these examples, it is useful to summarize order-of-operation conventions pertaining to exponents, multiplication and division, and addition and subtraction.

Order of Operations

1. Simplify inside the innermost grouping first, then the next innermost, and so on.

$$2[3 - (x - 4)] = 2[3 - x + 4]$$
$$= 2(7 - x) = 14 - 2x$$

2. Unless grouping symbols indicate otherwise, apply exponents before multiplication or division is performed.

$$2(x - 2)^2 = 2(x^2 - 4x + 4) = 2x^2 - 8x + 8$$

3. Unless grouping symbols indicate otherwise, perform multiplication and division before addition and subtraction. In either case, proceed from left to right.

$$5 - 2(x - 3) = 5 - 2x + 6 = 11 - 2x$$

EXAMPLE 7 Combined Operations

Perform the indicated operations and simplify:

(A) $3x - \{5 - 3[x - x(3 - x)]\} = 3x - \{5 - 3[x - 3x + x^2]\}$

$= 3x - \{5 - 3[-2x + x^2]\}$

$= 3x - \{5 + 6x - 3x^2\}$

$= 3x - 5 - 6x + 3x^2$

$= 3x^2 - 3x - 5$

(B) $(x - 2y)(2x + 3y) - (2x + y)^2 = 2x^2 + 3xy - 4xy - 6y^2 - (4x^2 + 4xy + y^2)$

$= 2x^2 - xy - 6y^2 - 4x^2 - 4xy - y^2$

$= -2x^2 - 5xy - 7y^2$

(C) $(2m + 3n)^3 = (2m + 3n)(2m + 3n)^2$

$= (2m + 3n)(4m^2 + 12mn + 9n^2)$

$= 8m^3 + 24m^2n + 18mn^2 + 12m^2n + 36mn^2 + 27n^3$

$= 8m^3 + 36m^2n + 54mn^2 + 27n^3$

Matched Problem 7 Perform the indicated operations and simplify:

(A) $2t - \{7 - 2[t - t(4 + t)]\}$ (B) $(u - 3v)^2 - (2u - v)(2u + v)$
(C) $(4x - y)^3$

• Application

EXAMPLE 8 Volume of a Cylindrical Shell

∫ A plastic water pipe with a hollow center is 100 inches long, 1 inch thick, and has an inner radius of x inches (see the figure on the next page). Write an algebraic expression in terms of x that represents the volume of the plastic used to construct the pipe. Simplify the expression. [*Recall:* The volume V of a right circular cylinder of radius r and height h is given by $V = \pi r^2 h$.]

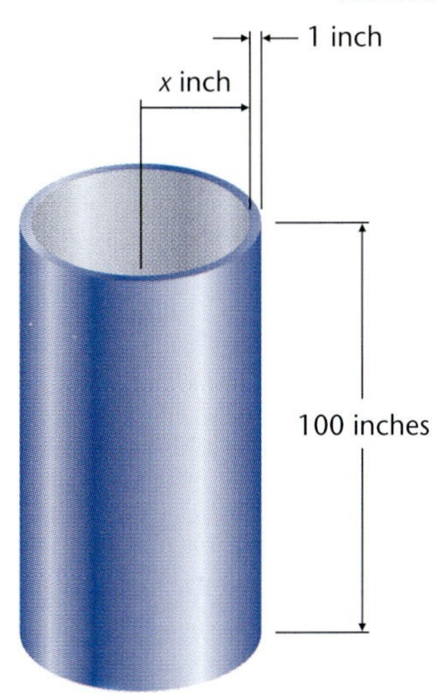

Solution A right circular cylinder with a hollow center is called a **cylindrical shell.** The volume of the shell is equal to the volume of the cylinder minus the volume of the hole. Since the radius of the hole is x inches and the pipe is 1 inch thick, the radius of the cylinder is $x + 1$ inches. Thus, we have

$$\begin{pmatrix}\text{Volume of}\\ \text{shell}\end{pmatrix} = \begin{pmatrix}\text{Volume of}\\ \text{cylinder}\end{pmatrix} - \begin{pmatrix}\text{Volume of}\\ \text{hole}\end{pmatrix}$$

$$\begin{aligned}\text{Volume} &= \pi(x + 1)^2\,100 - \pi x^2 100\\ &= 100\pi(x^2 + 2x + 1) - 100\pi x^2\\ &= 100\pi x^2 + 200\pi x + 100\pi - 100\pi x^2\\ &= 200\pi x + 100\pi\end{aligned}$$

Matched Problem 8 A plastic water pipe is 200 inches long, 2 inches thick, and has an outer radius of x inches. Write an algebraic expression in terms of x that represents the volume of the plastic used to construct the pipe. Simplify the expression.

Answers to Matched Problems

1. (A) $3x^2 - 2x + 1$, $x^2 - 2xy + y^2$ (B) 5, 3, 5 (C) 6, 4, 6
2. (A) $4u^2 - v^2$ (B) $-m^3 - 3m^2 + 2m - 4$ (C) $-x^3 + 3x + 2$
3. $3x^4 - x^3 - 5x^2 + 2x - 2$ **4.** $3x^2 + 5x - 10$ **5.** $4x^3 - 13x + 6$
6. (A) $8u^2 - 2uv - 3v^2$ (B) $4x^2y^2 - 9$ (C) $m^2 - 16n^2$ (D) $4u^2 - 12uv + 9v^2$ (E) $36x^2 + 12xy + y^2$
7. (A) $-2t^2 - 4t - 7$ (B) $-3u^2 - 6uv + 10v^2$ (C) $64x^3 - 48x^2y + 12xy^2 - y^3$
8. Volume $= 200\pi x^2 - 200\pi(x - 2)^2 = 800\pi x - 800\pi$

EXERCISE A-2

A

Problems 1–8 refer to the following polynomials:
(a) $2x^3 - 3x^2 + x + 5$ *(b)* $2x^2 + x - 1$ *(c)* $3x - 2$

1. What is the degree of (a)? **2.** What is the degree of (b)?

3. Add (a) and (b). **4.** Add (b) and (c).

5. Subtract (b) from (a). **6.** Subtract (c) from (b).

7. Multiply (a) and (c). **8.** Multiply (b) and (c).

In Problems 9–28, perform the indicated operations and simplify.

9. $2(x - 1) + 3(2x - 3) - (4x - 5)$

10. $2(u - 1) - (3u + 2) - 2(2u - 3)$

11. $2y - 3y[4 - 2(y - 1)]$

12. $4a - 2a[5 - 3(a + 2)]$

13. $(m - n)(m + n)$

14. $(a + b)(a - b)$

15. $(4t - 3)(t - 2)$

16. $(3x - 5)(2x + 1)$

17. $(3x + 2y)(x - 3y)$

18. $(2x - 3y)(x + 2y)$

19. $(2m - 7)(2m + 7)$

20. $(3y + 2)(3y - 2)$

21. $(6x - 4y)(5x + 3y)$

22. $(3m + 7n)(2m - 5n)$

23. $(3x - 2y)(3x + 2y)$

24. $(4m + 3n)(4m - 3n)$

25. $(4x - y)^2$

26. $(3u + 4v)^2$

27. $(a + b)(a^2 - ab + b^2)$

28. $(a - b)(a^2 + ab + b^2)$

B

In Problems 29–42, perform the indicated operations and simplify.

29. $2x - 3\{x + 2[x - (x + 5)] + 1\}$

30. $m - \{m - [m - (m - 1)]\}$

31. $2\{3[a - 4(1 - a)] - (5 - a)\}$

32. $5b - 3\{-[2 - 4(2b - 1)] + 2(2 - 3b)\}$

33. $(2x^2 - 3x + 1)(x^2 + x - 2)$

34. $(x^2 - 3xy + y^2)(x^2 + 3xy + y^2)$

35. $(x - 2y)^2(x + 2y)^2$

36. $(n^2 + 4nm + m^2)(n^2 - 4nm + m^2)$

37. $(3u - 2v)^2 - (2u - 3v)(2u + 3v)$

38. $(2a - b)^2 - (a + 2b)^2$

39. $(z + 2)(z^2 - 2z + 3) + z - 7$

40. $(y + 3)(y^2 - 3y + 1) + 8y - 1$

41. $(2m - n)^3$

42. $(3a + 2b)^3$

$\int$ *Problems 43–50 are calculus-related. Perform the indicated operations and simplify.*

43. $3(x + h) - 7 - (3x - 7)$

44. $(x + h)^2 - x^2$

45. $2(x + h)^2 - 3(x + h) - (2x^2 - 3x)$

46. $-4(x + h)^2 + 6(x + h) - (-4x^2 + 6x)$

47. $2(x + h)^2 - 4(x + h) - 9 - (2x^2 - 4x - 9)$

48. $3(x + h)^2 + 5(x + h) + 7 - (3x^2 + 5x + 7)$

49. $(x + h)^3 - 2(x + h)^2 - (x^3 - 2x^2)$

50. $(x + h)^3 + 3(x + h) - (x^3 + 3x)$

51. Subtract the sum of the first two polynomials from the sum of the last two: $3m^2 - 2m + 5$, $4m^2 - m$, $3m^2 - 3m - 2$, $m^3 + m^2 + 2$

52. Subtract the sum of the last two polynomials from the sum of the first two: $2x^2 - 4xy + y^2$, $3xy - y^2$, $x^2 - 2xy - y^2$, $-x^2 + 3xy - 2y^2$

C

In Problems 53–56, perform the indicated operations and simplify.

53. $2(x - 2)^3 - (x - 2)^2 - 3(x - 2) - 4$

54. $(2x - 1)^3 - 2(2x - 1)^2 + 3(2x - 1) + 7$

55. $-3x\{x[x - x(2 - x)] - (x + 2)(x^2 - 3)\}$

56. $2\{(x - 3)(x^2 - 2x + 1) - x[3 - x(x - 2)]\}$

57. Show by example that, in general, $(a + b)^2 \neq a^2 + b^2$. Discuss possible conditions on a and b that would make this a valid equation.

58. Show by example that, in general, $(a - b)^2 \neq a^2 - b^2$. Discuss possible conditions on a and b that would make this a valid equation.

59. If you are given two polynomials, one of degree m and the other of degree n, $m > n$, what is the degree of the sum?

60. What is the degree of the product of the two polynomials in Problem 59?

61. How does the answer to Problem 59 change if the two polynomials can have the same degree?

62. How does the answer to Problem 60 change if the two polynomials can have the same degree?

APPLICATIONS 

63. Geometry. The width of a rectangle is 5 centimeters less than its length. If x represents the length, write an algebraic expression in terms of x that represents the perimeter of the rectangle. Simplify the expression.

64. Geometry. The length of a rectangle is 8 meters more than its width. If x represents the width of the rectangle, write an algebraic expression in terms of x that represents its area. Change the expression to a form without parentheses.

★ **65. Coin Problem.** A parking meter contains nickels, dimes, and quarters. There are 5 fewer dimes than nickels, and 2 more quarters than dimes. If x represents the number of nickels, write an algebraic expression in terms of x that represents the value of all the coins in the meter in cents. Simplify the expression.

★ **66. Coin Problem.** A vending machine contains dimes and quarters only. There are 4 more dimes than quarters. If x represents the number of quarters, write an algebraic expression in terms of x that represents the value of all the coins in the vending machine in cents. Simplify the expression.

67. Packaging. A spherical plastic container for designer wristwatches has an inner radius of x centimeters (see the figure on the next page). If the plastic shell is 0.3 centimeters thick, write an algebraic expression in terms of x that represents the volume of the plastic used to construct the container. Simplify the expression. [*Recall:* The volume V of a sphere of radius r is given by $V = \frac{4}{3}\pi r^3$.]

68. Packaging. A cubical container for shipping computer components is formed by coating a metal mold with polystyrene. If the metal mold is a cube with sides x centimeters long and the polystyrene coating is 2 centimeters thick, write an algebraic expression in terms of x that represents the volume of the polystyrene used to construct the container. Simplify the expression. [*Recall:* The volume V of a cube with sides of length t is given by $V = t^3$.]

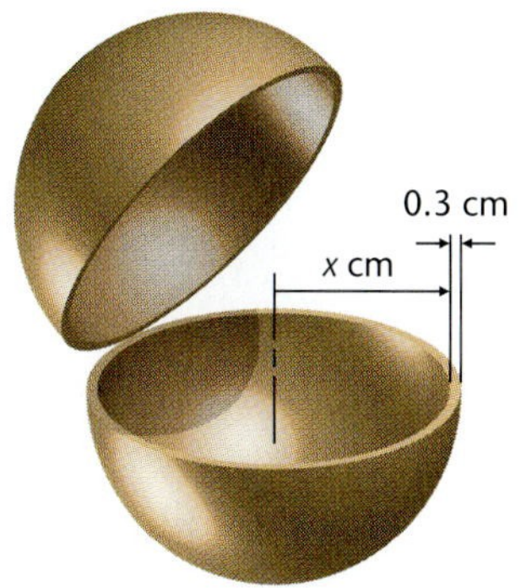

Figure for 67

SECTION A-3 Polynomials: Factoring

- Factoring—What Does It Mean?
- Common Factors and Factoring by Grouping
- Factoring Second-Degree Polynomials
- More Factoring

• Factoring—What Does It Mean?

A **factor of a number** is one of two or more numbers whose product is the given number. Similarly, a **factor of an algebraic expression** is one of two or more algebraic expressions whose product is the given algebraic expression. For example,

$$30 = 2 \cdot 3 \cdot 5 \qquad \text{2, 3, and 5 are each factors of 30.}$$

$$x^2 - 4 = (x - 2)(x + 2) \qquad (x - 2) \text{ and } (x + 2) \text{ are each factors of } x^2 - 4.$$

The process of writing a number or algebraic expression as the product of other numbers or algebraic expressions is called **factoring.** We start our discussion of factoring with the positive integers.

An integer such as 30 can be represented in a factored form in many ways. The products

$$6 \cdot 5 \qquad (\tfrac{1}{2})(10)(6) \qquad 15 \cdot 2 \qquad 2 \cdot 3 \cdot 5$$

all yield 30. A particularly useful way of factoring positive integers greater than 1 is in terms of *prime* numbers.

DEFINITION 1 **Prime and Composite Numbers**

An integer greater than 1 is **prime** if its only positive integer factors are itself and 1. An integer greater than 1 that is not prime is called a **composite number.** The integer 1 is neither prime nor composite.

Examples of prime numbers: 2, 3, 5, 7, 11, 13

Examples of composite numbers: 4, 6, 8, 9, 10, 12

EXPLORE-DISCUSS 1 In the array below, cross out all multiples of 2, except 2 itself. Then cross out all multiples of 3, except 3 itself. Repeat this for each integer in the array that has not yet been crossed out. Describe the set of numbers that remains when this process is completed.

1	2	3	4	5	6	7	8	9	10	11	12	13	14	15	16	17	18	19	20
21	22	23	24	25	26	27	28	29	30	31	32	33	34	35	36	37	38	39	40
41	42	43	44	45	46	47	48	49	50	51	52	53	54	55	56	57	58	59	60
61	62	63	64	65	66	67	68	69	70	71	72	73	74	75	76	77	78	79	80
81	82	83	84	85	86	87	88	89	90	91	92	93	94	95	96	97	98	99	100

This process is referred to as the **sieve of Eratosthenes.** (Eratosthenes was a Greek mathematician and astronomer who was a contemporary of Archimedes, circa 200 B.C.)

A composite number is said to be **factored completely** if it is represented as a product of prime factors. The only factoring of 30 given above that meets this condition is $30 = 2 \cdot 3 \cdot 5$.

EXAMPLE 1 **Factoring a Composite Number**

Write 60 in completely factored form.

Solution

$$60 = 6 \cdot 10 = 2 \cdot 3 \cdot 2 \cdot 5 = 2^2 \cdot 3 \cdot 5$$

or

$$60 = 5 \cdot 12 = 5 \cdot 4 \cdot 3 = 2^2 \cdot 3 \cdot 5$$

or

$$60 = 2 \cdot 30 = 2 \cdot 2 \cdot 15 = 2^2 \cdot 3 \cdot 5$$

Matched Problem 1 Write 180 in completely factored form.

Notice in Example 1 that we end up with the same prime factors for 60 irrespective of how we progress through the factoring process. This illustrates an important property of integers:

Theorem 1 **The Fundamental Theorem of Arithmetic**

Each integer greater than 1 is either prime or can be expressed uniquely, except for the order of factors, as a product of prime factors.

We can also write polynomials in completely factored form. A polynomial such as $2x^2 - x - 6$ can be written in factored form in many ways. The products

$$(2x + 3)(x - 2) \qquad 2(x^2 - \tfrac{1}{2}x - 3) \qquad 2(x + \tfrac{3}{2})(x - 2)$$

all yield $2x^2 - x - 6$. A particularly useful way of factoring polynomials is in terms of prime polynomials.

DEFINITION 2 **Prime Polynomials**

A polynomial of degree greater than 0 is said to be **prime** relative to a given set of numbers if: (*1*) all of its coefficients are from that set of numbers; and (*2*) it cannot be written as a product of two polynomials of positive degree having coefficients from that set of numbers.

Relative to the set of integers:

$x^2 - 2$ is prime
$x^2 - 9$ is not prime, since $x^2 - 9 = (x - 3)(x + 3)$

[*Note:* The set of numbers most frequently used in factoring polynomials is the set of integers.]

A nonprime polynomial is said to be **factored completely relative to a given set of numbers** if it is written as a product of prime polynomials relative to that set of numbers.

Our objective in this section is to review some of the standard factoring techniques for polynomials with integer coefficients. In Chapter 3 we treat in detail the topic of factoring polynomials of higher degree with arbitrary coefficients.

• Common Factors and Factoring by Grouping

The next example illustrates the use of the distributive properties in factoring.

EXAMPLE 2 **Factoring Out Common Factors**

Factor out, relative to the integers, all factors common to all terms:

(A) $2x^3y - 8x^2y^2 - 6xy^3$ (B) $2x(3x - 2) - 7(3x - 2)$

Solutions

(A) $2x^3y - 8x^2y^2 - 6xy^3 = (2xy)x^2 - (2xy)4xy - (2xy)3y^2$

$= 2xy(x^2 - 4xy - 3y^2)$

(B) $2x(3x - 2) - 7(3x - 2) = 2x(3x - 2) - 7(3x - 2)$

$= (2x - 7)(3x - 2)$

Matched Problem 2 Factor out, relative to the integers, all factors common to all terms:

(A) $3x^3y - 6x^2y^2 - 3xy^3$ (B) $3y(2y + 5) + 2(2y + 5)$

EXAMPLE 3 Factoring Out Common Factors

Factor completely relative to the integers:

$$4(2x + 7)(x - 3)^2 + 2(2x + 7)^2(x - 3)$$

Solution

$$\begin{aligned} &4(2x + 7)(x - 3)^2 + 2(2x + 7)^2(x - 3) \\ &= 2(2x + 7)(x - 3)[2(x - 3) + (2x + 7)] \\ &= 2(2x + 7)(x - 3)(2x - 6 + 2x + 7) \\ &= 2(2x + 7)(x - 3)(4x + 1) \end{aligned}$$

Matched Problem 3 Factor completely relative to the integers:

$$4(2x + 5)(3x + 1)^2 + 6(2x + 5)^2(3x + 1)$$

Some polynomials can be factored by first grouping terms in such a way that we obtain an algebraic expression that looks something like Example 2B. We can then complete the factoring by the method used in that example.

EXAMPLE 4 Factoring by Grouping

Factor completely, relative to the integers, by grouping:

(A) $3x^2 - 6x + 4x - 8$ (B) $wy + wz - 2xy - 2xz$
(C) $3ac + bd - 3ad - bc$

Solutions (A) $3x^2 - 6x + 4x - 8$

$= (3x^2 - 6x) + (4x - 8)$ Group the first two and last two terms.

$= 3x(x - 2) + 4(x - 2)$ Remove common factors from each group.

$= (3x + 4)(x - 2)$ Factor out the common factor $(x - 2)$.

(B) $wy + wz - 2xy - 2xz$

$= (wy + wz) - (2xy + 2xz)$ Group the first two and last two terms—be careful of signs.

$= w(y + z) - 2x(y + z)$ Remove common factors from each group.

$= (w - 2x)(y + z)$ Factor out the common factor $(y + z)$.

(C) $3ac + bd - 3ad - bc$

In parts (A) and (B) the polynomials are arranged in such a way that grouping the first two terms and the last two terms leads to common factors. In this problem neither the first two terms nor the last two terms have a common factor. Sometimes rearranging terms will lead to a factoring by grouping. In this case, we interchange the second and fourth terms to obtain a problem comparable to part (B), which can be factored as follows:

$$\begin{aligned} 3ac - bc - 3ad + bd &= (3ac - bc) - (3ad - bd) \\ &= c(3a - b) - d(3a - b) \\ &= (c - d)(3a - b) \end{aligned}$$

Matched Problem 4 Factor completely, relative to the integers, by grouping:

(A) $2x^2 + 6x + 5x + 15$ (B) $2pr + ps - 6qr - 3qs$
(C) $6wy - xz - 2xy + 3wz$

• Factoring Second-Degree Polynomials

We now turn our attention to factoring second-degree polynomials of the form

$$2x^2 - 5x - 3 \qquad \text{and} \qquad 2x^2 + 3xy - 2y^2$$

into the product of two first-degree polynomials with integer coefficients. The following example will illustrate an approach to the problem.

EXAMPLE 5 Factoring Second-Degree Polynomials

Factor each polynomial, if possible, using integer coefficients:

(A) $2x^2 + 3xy - 2y^2$ (B) $x^2 - 3x + 4$ (C) $6x^2 + 5xy - 4y^2$

Solutions (A) $2x^2 + 3xy - 2y^2 = (2x + \underset{?}{\uparrow}\ y)(x - \underset{?}{\uparrow}\ y)$ Put in what we know. Signs must be opposite. (We can reverse this choice if we get $-3xy$ instead of $+3xy$ for the middle term.)

Now, what are the factors of 2 (the coefficient of y^2)?

2	
$1 \cdot 2$	$(2x + y)(x - 2y) = 2x^2 - 3xy - 2y^2$
$2 \cdot 1$	$(2x + 2y)(x - y) = 2x^2 - 2y^2$

The first choice gives us $-3xy$ for the middle term—close, but not there—so we reverse our choice of signs to obtain

$$2x^2 + 3xy - 2y^2 = (2x - y)(x + 2y)$$

(B) $x^2 - 3x + 4 = (x - \quad)(x - \quad)$ Signs must be the same because the third term is positive and must be negative because the middle term is negative.

4	
$2 \cdot 2$	$(x - 2)(x - 2) = x^2 - 4x + 4$
$1 \cdot 4$	$(x - 1)(x - 4) = x^2 - 5x + 4$
$4 \cdot 1$	$(x - 4)(x - 1) = x^2 - 5x + 4$

No choice produces the middle term; hence $x^2 - 3x + 4$ is not factorable using integer coefficients.

(C) $6x^2 + 5xy - 4y^2 = (\ \ x + \ \ y)(\ \ x - \ \ y)$

$\uparrow \quad \uparrow \quad \uparrow \quad \uparrow$
? ? ? ?

The signs must be opposite in the factors, because the third term is negative. We can reverse our choice of signs later if necessary. We now write all factors of 6 and of 4:

6	4
$2 \cdot 3$	$2 \cdot 2$
$3 \cdot 2$	$1 \cdot 4$
$1 \cdot 6$	$4 \cdot 1$
$6 \cdot 1$	

and try each choice on the left with each on the right—a total of 12 combinations that give us the first and last terms in the polynomial $6x^2 + 5xy - 4y^2$. The question is: Does any combination also give us the middle term, $5xy$? After trial and error and, perhaps, some educated guessing among the choices, we find that $3 \cdot 2$ matched with $4 \cdot 1$ gives us the correct middle term. Thus,

$$6x^2 + 5xy - 4y^2 = (3x + 4y)(2x - y)$$

If none of the 24 combinations (including reversing our sign choice) had produced the middle term, then we would conclude that the polynomial is not factorable using integer coefficients.

Matched Problem 5 Factor each polynomial, if possible, using integer coefficients:

(A) $x^2 - 8x + 12$ (B) $x^2 + 2x + 5$
(C) $2x^2 + 7xy - 4y^2$ (D) $4x^2 - 15xy - 4y^2$

• More Factoring

The factoring formulas listed below will enable us to factor certain polynomial forms that occur frequently.

Special Factoring Formulas

1. $u^2 + 2uv + v^2 = (u + v)^2$	**Perfect Square**
2. $u^2 - 2uv + v^2 = (u - v)^2$	**Perfect Square**
3. $u^2 - v^2 = (u - v)(u + v)$	**Difference of Squares**
4. $u^3 - v^3 = (u - v)(u^2 + uv + v^2)$	**Difference of Cubes**
5. $u^3 + v^3 = (u + v)(u^2 - uv + v^2)$	**Sum of Cubes**

The formulas in the box can be established by multiplying the factors on the right.

CAUTION

Note that we did not list a special factoring formula for the sum of two squares. In general,

$$u^2 + v^2 \neq (au + bv)(cu + dv)$$

for any choice of real number coefficients a, b, c, and d. In the first chapter we saw that $u^2 + v^2$ can be factored using complex numbers.

EXAMPLE 6 Using Special Factoring Formulas

Factor completely relative to the integers:

(A) $x^2 + 6xy + 9y^2$ (B) $9x^2 - 4y^2$ (C) $8m^3 - 1$ (D) $x^3 + y^3z^3$

Solutions

(A) $x^2 + 6xy + 9y^2 \boxed{= x^2 + 2(x)(3y) + (3y)^2} = (x + 3y)^2$

(B) $9x^2 - 4y^2 \boxed{= (3x)^2 - (2y)^2} = (3x - 2y)(3x + 2y)$

(C) $8m^3 - 1 \boxed{\begin{aligned}&= (2m)^3 - 1^3\\ &= (2m - 1)[(2m)^2 + (2m)(1) + 1^2]\end{aligned}}$

$= (2m - 1)(4m^2 + 2m + 1)$

(D) $x^3 + y^3z^3 \boxed{= x^3 + (yz)^3}$

$= (x + yz)(x^2 - xyz + y^2z^2)$

Matched Problem 6 Factor completely relative to the integers:

(A) $4m^2 - 12mn + 9n^2$ (B) $x^2 - 16y^2$ (C) $z^3 - 1$ (D) $m^3 + n^3$

EXPLORE-DISCUSS 2

(A) Verify the following factor formulas for $u^4 - v^4$:

$$u^4 - v^4 = (u - v)(u + v)(u^2 + v^2)$$
$$= (u - v)(u^3 + u^2v + uv^2 + v^3)$$

(B) Discuss the pattern in the following formulas:

$$u^2 - v^2 = (u - v)(u + v)$$
$$u^3 - v^3 = (u - v)(u^2 + uv + v^2)$$
$$u^4 - v^4 = (u - v)(u^3 + u^2v + uv^2 + v^3)$$

(C) Use the pattern you discovered in part (B) to write similar formulas for $u^5 - v^5$ and $u^6 - v^6$. Verify your formulas by multiplication.

We complete this section by considering factoring that involves combinations of the preceding techniques as well as a few additional ones. Generally speaking:

When asked to factor a polynomial, we first take out all factors common to all terms, if they are present, and then proceed as above until all factors are prime.

EXAMPLE 7 Combining Factoring Techniques

Factor completely relative to the integers:

(A) $18x^3 - 8x$ (B) $x^2 - 6x + 9 - y^2$ (C) $4m^3n - 2m^2n^2 + 2mn^3$
(D) $2t^4 - 16t$ (E) $2y^4 - 5y^2 - 12$

Solutions

(A) $18x^3 - 8x = 2x(9x^2 - 4)$
$= 2x(3x - 2)(3x + 2)$

(B) $x^2 - 6x + 9 - y^2$

$= (x^2 - 6x + 9) - y^2$ Group the first three terms.

$= (x - 3)^2 - y^2$ Factor $x^2 - 6x + 9$.

$= [(x - 3) - y][(x - 3) + y]$ Difference of squares

$= (x - 3 - y)(x - 3 + y)$

(C) $4m^3n - 2m^2n^2 + 2mn^3 = 2mn(2m^2 - mn + n^2)$

(D) $2t^4 - 16t = 2t(t^3 - 8)$

$= 2t(t - 2)(t^2 + 2t + 4)$

(E) $2y^4 - 5y^2 - 12 = (2y^2 + 3)(y^2 - 4)$

$= (2y^2 + 3)(y - 2)(y + 2)$

Matched Problem 7 Factor completely relative to the integers:

(A) $3x^3 - 48x$ (B) $x^2 - y^2 - 4y - 4$
(C) $3u^4 - 3u^3v - 9u^2v^2$ (D) $3m^4 - 24mn^3$
(E) $3x^4 - 5x^2 + 2$

Answers to Matched Problems

1. $2^2 \cdot 3^2 \cdot 5$ **2.** (A) $3xy(x^2 - 2xy - y^2)$ (B) $(3y + 2)(2y + 5)$
3. $2(2x + 5)(3x + 1)(12x + 17)$
4. (A) $(2x + 5)(x + 3)$ (B) $(p - 3q)(2r + s)$ (C) $(3w - x)(2y + z)$
5. (A) $(x - 2)(x - 6)$ (B) Not factorable using integers (C) $(2x - y)(x + 4y)$
(D) $(4x + y)(x - 4y)$
6. (A) $(2m - 3n)^2$ (B) $(x - 4y)(x + 4y)$ (C) $(z - 1)(z^2 + z + 1)$
(D) $(m + n)(m^2 - mn + n^2)$
7. (A) $3x(x - 4)(x + 4)$ (B) $(x - y - 2)(x + y + 2)$ (C) $3u^2(u^2 - uv - 3v^2)$
(D) $3m(m - 2n)(m^2 + 2mn + 4n^2)$ (E) $(3x^2 - 2)(x - 1)(x + 1)$

EXERCISE A-3

A

In Problems 1–8, factor out, relative to the integers, all factors common to all terms.

1. $6x^4 - 8x^3 - 2x^2$ **2.** $6m^4 - 9m^3 - 3m^2$

3. $10x^3y + 20x^2y^2 - 15xy^3$ **4.** $8u^3v - 6u^2v^2 + 4uv^3$

5. $5x(x + 1) - 3(x + 1)$ **6.** $7m(2m - 3) + 5(2m - 3)$

7. $2w(y - 2z) - x(y - 2z)$ **8.** $a(3c + d) - 4b(3c + d)$

In Problems 9–16, factor completely relative to integers.

9. $x^2 - 2x + 3x - 6$

10. $2y^2 - 6y + 5y - 15$

11. $6m^2 + 10m - 3m - 5$

12. $5x^2 - 40x - x + 8$

13. $2x^2 - 4xy - 3xy + 6y^2$

14. $3a^2 - 12ab - 2ab + 8b^2$

15. $8ac + 3bd - 6bc - 4ad$

16. $3pr - 2qs - qr + 6ps$

In Problems 17–28, factor completely relative to the integers. If a polynomial is prime relative to the integers, say so.

17. $2x^2 + x - 3$ **18.** $3y^2 - 8y - 3$

19. $x^2 + 3x - 8$ **20.** $u^2 + 4uv - 12v^2$

21. $m^2 + m - 6$ **22.** $x^2 + 3xy - 10y^2$

23. $4a^2 - 9b^2$ **24.** $x^2 + 4y^2$

25. $4x^2 - 20x + 25$ **26.** $a^2b^2 - c^2$

27. $a^2b^2 + c^2$ **28.** $9x^2 - 4$

B

In Problems 29–42, factor completely relative to the integers. If a polynomial is prime relative to the integers, say so.

29. $6x^2 + 48x + 72$ **30.** $3z^2 - 28z + 48$

31. $2y^3 - 22y^2 + 48y$

32. $2x^4 - 24x^3 + 40x^2$

33. $16x^2y - 8xy + y$

34. $4xy^2 - 12xy + 9x$

35. $6s^2 + 7st + 3t^2$

36. $6m^2 - mn - 12n^2$

37. $x^3y - 9xy^3$

38. $4u^3v - uv^3$

39. $3m^3 - 6m^2 + 15m$

40. $2x^3 - 2x^2 + 8x$

41. $m^3 + n^3$

42. $r^3 - t^3$

Problems 43–50 are calculus-related. Factor completely relative to the integers.

43. $2x(x + 1)^4 + 4x^2(x + 1)^3$

44. $(x - 1)^3 + 3x(x - 1)^2$

45. $6(3x - 5)(2x - 3)^2 + 4(3x - 5)^2(2x - 3)$

46. $2(x - 3)(4x + 7)^2 + 8(x - 3)^2(4x + 7)$

47. $5x^4(9 - x)^4 - 4x^5(9 - x)^3$

48. $3x^4(x - 7)^2 + 4x^3(x - 7)^3$

49. $2(x + 1)(x^2 - 5)^2 + 4x(x + 1)^2(x^2 - 5)$

50. $4(x - 3)^3(x^2 + 2)^3 + 6x(x - 3)^4(x^2 + 2)^2$

In Problems 51–56, factor completely relative to the integers. In polynomials involving more than three terms, try grouping the terms in various combinations as a first step. If a polynomial is prime relative to the integers, say so.

51. $(a - b)^2 - 4(c - d)^2$

52. $(x + 2)^2 + 9$

53. $2am - 3an + 2bm - 3bn$

54. $15ac - 20ad + 3bc - 4bd$

55. $3x^2 - 2xy - 4y^2$

56. $5u^2 + 4uv - v^2$

C

In Problems 57–72, factor completely relative to the integers. In polynomials involving more than three terms, try grouping the terms in various combinations as a first step. If a polynomial is prime relative to the integers, say so.

57. $x^3 - 3x^2 - 9x + 27$

58. $x^3 - x^2 - x + 1$

59. $a^3 - 2a^2 - a + 2$

60. $t^3 - 2t^2 + t - 2$

61. $4(A + B)^2 - 5(A + B) - 5$

62. $6(x - y)^2 + 23(x - y) - 4$

63. $m^4 - n^4$

64. $y^4 - 3y^2 - 4$

65. $y^4 - 3y^2 - 5$

66. $27a^2 + a^5b^3$

67. $m^2 + 2mn + n^2 - m - n$

68. $y^2 - 2xy + x^2 - y + x$

69. $18a^3 - 8a(x^2 + 8x + 16)$

70. $25(4x^2 - 12xy + 9y^2) - 9a^2b^2$

71. $x^4 + 2x^2 + 1 - x^2$

72. $a^4 + 2a^2b^2 + b^4 - a^2b^2$

APPLICATIONS

73. Construction. A rectangular open-topped box is to be constructed out of 20-inch-square sheets of thin cardboard by cutting x-inch squares out of each corner and bending the sides up as indicated in the figure. Express each of the following quantities as a polynomial in both factored and expanded form.

(A) The area of cardboard after the corners have been removed.

(B) The volume of the box.

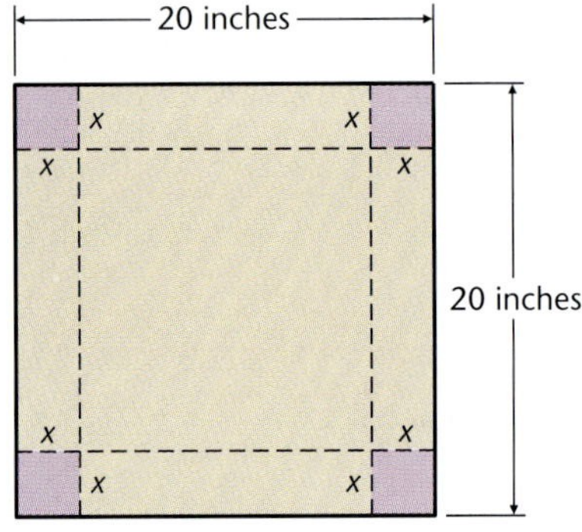

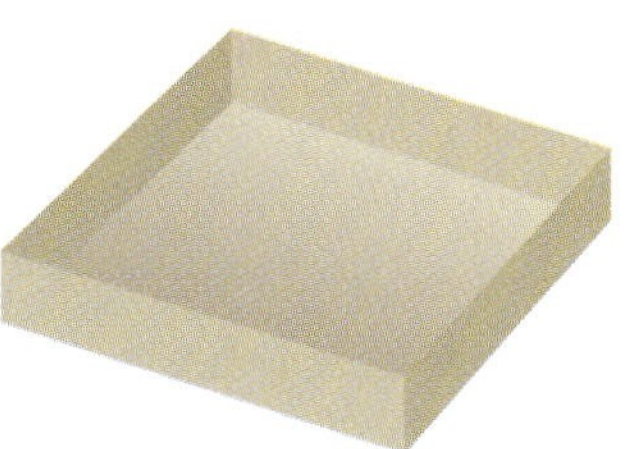

74. Construction. A rectangular open-topped box is to be constructed out of 9- by 16-inch sheets of thin cardboard by cutting x-inch squares out of each corner and bending the sides up. Express each of the following quantities as a polynomial in both factored and expanded form.

(A) The area of cardboard after the corners have been removed.

(B) The volume of the box.

SECTION A-4 Rational Expressions: Basic Operations

- Reducing to Lowest Terms
- Multiplication and Division
- Addition and Subtraction
- Compound Fractions

We now turn our attention to fractional forms. A quotient of two algebraic expressions, division by 0 excluded, is called a **fractional expression.** If both the numerator and denominator of a fractional expression are polynomials, the fractional expression is called a **rational expression.** Some examples of rational expressions are the following (recall, a nonzero constant is a polynomial of degree 0):

$$\frac{x-2}{2x^2-3x+5} \qquad \frac{1}{x^4-1} \qquad \frac{3}{x} \qquad \frac{x^2+3x-5}{1}$$

In this section we discuss basic operations on rational expressions, including multiplication, division, addition, and subtraction.

Since variables represent real numbers in the rational expressions we are going to consider, the properties of real number fractions summarized in Section A-1 play a central role in much of the work that we will do.

Even though not always explicitly stated, we always assume that variables are restricted so that division by 0 is excluded.

• Reducing to Lowest Terms

We start this discussion by restating the **fundamental property of fractions** (from Theorem 3 in Section A-1):

Fundamental Property of Fractions

If a, b, and k are real numbers with $b, k \neq 0$, then

$$\frac{ka}{kb} = \frac{a}{b} \qquad \frac{2 \cdot 3}{2 \cdot 4} = \frac{3}{4} \qquad \frac{(x-3)2}{(x-3)x} = \frac{2}{x}$$

$$x \neq 0,\ x \neq 3$$

Using this property from left to right to eliminate all common factors from the numerator and the denominator of a given fraction is referred to as **reducing a fraction to lowest terms.** We are actually dividing the numerator and denominator by the same nonzero common factor.

Using the property from right to left—that is, multiplying the numerator and the denominator by the same nonzero factor—is referred to as **raising a fraction to higher terms.** We will use the property in both directions in the material that follows.

We say that a rational expression is **reduced to lowest terms** if the numerator and denominator do not have any factors in common. Unless stated to the contrary, factors will be relative to the integers.

EXAMPLE 1 Reducing Rational Expressions

Reduce each rational expression to lowest terms.

(A) $\dfrac{x^2 - 6x + 9}{x^2 - 9} = \dfrac{(x - 3)^2}{(x - 3)(x + 3)}$ Factor numerator and denominator completely. Divide numerator and denominator by $(x - 3)$; this is a valid operation as long as $x \neq 3$ and $x \neq -3$.

$$= \frac{x - 3}{x + 3}$$

(B) $\dfrac{x^3 - 1}{x^2 - 1} = \dfrac{\overset{1}{\cancel{(x - 1)}}(x^2 + x + 1)}{\underset{1}{\cancel{(x - 1)}}(x + 1)}$ Dividing numerator and denominator by $(x - 1)$ can be indicated by drawing lines through both $(x - 1)$'s and writing the resulting quotients, 1's.

$$= \frac{x^2 + x + 1}{x + 1}$$ $x \neq -1$ and $x \neq 1$

Matched Problem 1 Reduce each rational expression to lowest terms.

(A) $\dfrac{6x^2 + x - 2}{2x^2 + x - 1}$ (B) $\dfrac{x^4 - 8x}{3x^3 - 2x^2 - 8x}$

EXAMPLE 2 Reducing a Rational Expression

Reduce the following rational expression to lowest terms.

$$\frac{6x^5(x^2 + 2)^2 - 4x^3(x^2 + 2)^3}{x^8} = \frac{2x^3(x^2 + 2)^2[3x^2 - 2(x^2 + 2)]}{x^8}$$

$$= \frac{2\overset{1}{\cancel{x^3}}(x^2 + 2)^2(x^2 - 4)}{\underset{x^5}{\cancel{x^8}}}$$

$$= \frac{2(x^2 + 2)^2(x - 2)(x + 2)}{x^5}$$

Matched Problem 2 Reduce the following rational expression to lowest terms.

$$\frac{6x^4(x^2 + 1)^2 - 3x^2(x^2 + 1)^3}{x^6}$$

CAUTION

Remember to always factor the numerator and denominator first, then divide out any *common factors.* Do not indiscriminately eliminate *terms* that appear in both the numerator and the denominator. For example,

$$\frac{2x^3 + y^2}{y^2} \neq \frac{2x^3 + \overset{1}{\cancel{y^2}}}{\underset{1}{\cancel{y^2}}} = 2x^3 + 1$$

Since the term y^2 is not a factor of the numerator, it cannot be eliminated. In fact, $(2x^3 + y^2)/y^2$ is already reduced to lowest terms.

• Multiplication and Division

Since we are restricting variable replacements to real numbers, multiplication and division of rational expressions follow the rules for multiplying and dividing real number fractions (Theorem 3 in Section A-1).

Multiplication and Division

If a, b, c, and d are real numbers with $b, d \neq 0$, then:

1. $\frac{a}{b} \cdot \frac{c}{d} = \frac{ac}{bd}$ $\qquad \frac{2}{3} \cdot \frac{x}{x-1} = \frac{2x}{3(x-1)}$

2. $\frac{a}{b} \div \frac{c}{d} = \frac{a}{b} \cdot \frac{d}{c} \quad c \neq 0$ $\qquad \frac{2}{3} \div \frac{x}{x-1} = \frac{2}{3} \cdot \frac{x-1}{x}$

EXPLORE-DISCUSS 1 Write a verbal description of the process of multiplying two fractions. Do the same for the quotient of two fractions.

EXAMPLE 3 Multiplying and Dividing Rational Expressions

Perform the indicated operations and reduce to lowest terms.

(A) $$\frac{10x^3y}{3xy + 9y} \cdot \frac{x^2 - 9}{4x^2 - 12x} = \frac{\overset{5x^2}{\cancel{10x^3y}}}{\underset{3 \cdot 1}{\cancel{3y}(\cancel{x+3})}} \cdot \frac{\overset{1 \cdot 1}{(\cancel{x-3})(\cancel{x+3})}}{\underset{2 \cdot 1}{\cancel{4x}(\cancel{x-3})}}$$

Factor numerators and denominators; then divide any numerator and any denominator with a like common factor.

$$= \frac{5x^2}{6}$$

(B) $$\frac{4 - 2x}{4} \div (x - 2) = \frac{\overset{1}{\cancel{2}}(2 - x)}{\underset{2}{\cancel{4}}} \cdot \frac{1}{x - 2}$$

$x - 2$ is the same as $\frac{x-2}{1}$.

$$= \frac{2 - x}{2(x - 2)} = \frac{-(x - 2)}{2(x - 2)} = -\frac{1}{2}$$

$b - a = -(a - b)$, a useful change in some problems.

(C) $$\frac{2x^3 - 2x^2y + 2xy^2}{x^3y - xy^3} \div \frac{x^3 + y^3}{x^2 + 2xy + y^2}$$

$$= \frac{2x(x^2 - xy + y^2)}{xy(x + y)(x - y)} \cdot \frac{(x + y)^2}{(x + y)(x^2 - xy + y^2)}$$

$$= \frac{2}{y(x - y)}$$

Matched Problem 3 Perform the indicated operations and reduce to lowest terms.

(A) $\dfrac{12x^2y^3}{2xy^2 + 6xy} \cdot \dfrac{y^2 + 6y + 9}{3y^3 + 9y^2}$ (B) $(4 - x) \div \dfrac{x^2 - 16}{5}$

(C) $\dfrac{m^3 + n^3}{2m^2 + mn - n^2} \div \dfrac{m^3n - m^2n^2 + mn^3}{2m^3n^2 - m^2n^3}$

• Addition and Subtraction

Again, because we are restricting variable replacements to real numbers, addition and subtraction of rational expressions follow the rules for adding and subtracting real number fractions (Theorem 3 in Section A-1).

Addition and Subtraction

For a, b, and c real numbers with $b \neq 0$:

1. $\dfrac{a}{b} + \dfrac{c}{b} = \dfrac{a + c}{b}$ $\quad \dfrac{x}{x - 3} + \dfrac{2}{x - 3} = \dfrac{x + 2}{x - 3}$

2. $\dfrac{a}{b} - \dfrac{c}{b} = \dfrac{a - c}{b}$ $\quad \dfrac{x}{2xy^2} - \dfrac{x - 4}{2xy^2} = \dfrac{x - (x - 4)}{2xy^2}$

Thus, we add rational expressions with the same denominators by adding or subtracting their numerators and placing the result over the common denominator. If the denominators are not the same, we raise the fractions to higher terms, using the fundamental property of fractions to obtain common denominators, and then proceed as described.

Even though any common denominator will do, our work will be simplified if the least common denominator (LCD) is used. Often, the LCD is obvious, but if it is not, the steps in the box describe how to find it.

The Least Common Denominator (LCD)

The LCD of two or more rational expressions is found as follows:

1. Factor each denominator completely.
2. Identify each different prime factor from all the denominators.
3. Form a product using each different factor to the highest power that occurs in any one denominator. This product is the LCD.

EXAMPLE 4 Adding and Subtracting Rational Expressions

Combine into a single fraction and reduce to lowest terms.

(A) $\frac{3}{10} + \frac{5}{6} - \frac{11}{45}$ (B) $\frac{4}{9x} - \frac{5x}{6y^2} + 1$

(C) $\frac{x + 3}{x^2 - 6x + 9} - \frac{x + 2}{x^2 - 9} - \frac{5}{3 - x}$

Solutions (A) To find the LCD, factor each denominator completely:

$$\left.\begin{aligned} 10 &= 2 \cdot 5 \\ 6 &= 2 \cdot 3 \\ 45 &= 3^2 \cdot 5 \end{aligned}\right\} \text{LCD} = 2 \cdot 3^2 \cdot 5 = 90$$

Now use the fundamental property of fractions to make each denominator 90:

$$\frac{3}{10} + \frac{5}{6} - \frac{11}{45} = \frac{9 \cdot 3}{9 \cdot 10} + \frac{15 \cdot 5}{15 \cdot 6} - \frac{2 \cdot 11}{2 \cdot 45}$$

$$= \frac{27}{90} + \frac{75}{90} - \frac{22}{90}$$

$$= \frac{27 + 75 - 22}{90} = \frac{80}{90} = \frac{8}{9}$$

(B) $$\left.\begin{aligned} 9x &= 3^2x \\ 6y^2 &= 2 \cdot 3y^2 \end{aligned}\right\} \text{LCD} = 2 \cdot 3^2xy^2 = 18xy^2$$

$$\frac{4}{9x} - \frac{5x}{6y^2} + 1 = \frac{2y^2 \cdot 4}{2y^2 \cdot 9x} - \frac{3x \cdot 5x}{3x \cdot 6y^2} + \frac{18xy^2}{18xy^2}$$

$$= \frac{8y^2 - 15x^2 + 18xy^2}{18xy^2}$$

(C) $$\frac{x+3}{x^2-6x+9} - \frac{x+2}{x^2-9} - \frac{5}{3-x} = \frac{x+3}{(x-3)^2} - \frac{x+2}{(x-3)(x+3)} + \frac{5}{x-3}$$

Note: $$-\frac{5}{3-x} = -\frac{5}{-(x-3)} = \frac{5}{x-3}$$

We have again used the fact that $a - b = -(b - a)$.

The LCD $= (x-3)^2(x+3)$. Thus,

$$\frac{(x+3)^2}{(x-3)^2(x+3)} - \frac{(x-3)(x+2)}{(x-3)^2(x+3)} + \frac{5(x-3)(x+3)}{(x-3)^2(x+3)}$$

$$= \frac{(x^2+6x+9) - (x^2-x-6) + 5(x^2-9)}{(x-3)^2(x+3)}$$

Be careful of sign errors here.

$$= \frac{x^2+6x+9-x^2+x+6+5x^2-45}{(x-3)^2(x+3)}$$

$$= \frac{5x^2+7x-30}{(x-3)^2(x+3)}$$

Matched Problem 4 Combine into a single fraction and reduce to lowest terms.

(A) $\frac{5}{28} - \frac{1}{10} + \frac{6}{35}$ (B) $\frac{1}{4x^2} - \frac{2x+1}{3x^3} + \frac{3}{12x}$

(C) $\frac{y-3}{y^2-4} - \frac{y+2}{y^2-4y+4} - \frac{2}{2-y}$

EXPLORE-DISCUSS 2 What is the value of $\cfrac{16}{\cfrac{4}{2}}$?

What is the result of entering $16 \div 4 \div 2$ on a calculator?

What is the difference between $16 \div (4 \div 2)$ and $(16 \div 4) \div 2$?

How could you use fraction bars to distinguish between these two cases when writing $\cfrac{16}{\cfrac{4}{2}}$?

• Compound Fractions

A fractional expression with fractions in its numerator, denominator, or both is called a **compound fraction.** It is often necessary to represent a compound fraction as a

simple fraction—that is (in all cases we will consider), as the quotient of two polynomials. The process does not involve any new concepts. It is a matter of applying old concepts and processes in the right sequence. We will illustrate two approaches to the problem, each with its own merits, depending on the particular problem under consideration.

EXAMPLE 5 Simplifying Compound Fractions

Express as a simple fraction reduced to lowest terms:

$$\frac{\frac{2}{x} - 1}{\frac{4}{x^2} - 1}$$

Solution *Method 1.* Multiply the numerator and denominator by the LCD of all fractions in the numerator and denominator—in this case, x^2. (We are multiplying by $1 = x^2/x^2$).

$$\frac{x^2\left(\frac{2}{x} - 1\right)}{x^2\left(\frac{4}{x^2} - 1\right)} = \frac{x^2\frac{2}{x} - x^2}{x^2\frac{4}{x^2} - x^2} = \frac{2x - x^2}{4 - x^2} = \frac{x\overset{1}{\cancel{(2 - x)}}}{(2 + x)\underset{1}{\cancel{(2 - x)}}}$$

$$= \frac{x}{2 + x}$$

Method 2. Write the numerator and denominator as single fractions. Then treat as a quotient.

$$\frac{\frac{2}{x} - 1}{\frac{4}{x^2} - 1} = \frac{\frac{2 - x}{x}}{\frac{4 - x^2}{x^2}} = \frac{2 - x}{x} \div \frac{4 - x^2}{x^2} = \frac{\overset{1}{\cancel{2 - x}}}{\underset{1}{\cancel{x}}} \cdot \frac{\overset{x}{\cancel{x^2}}}{\underset{1}{\cancel{(2 - x)}}(2 + x)}$$

$$= \frac{x}{2 + x}$$

Matched Problem 5 Express as a simple fraction reduced to lowest terms. Use the two methods described in Example 5.

$$\frac{1 + \frac{1}{x}}{x - \frac{1}{x}}$$

EXAMPLE 6 Simplifying Compound Fractions

Express as a simple fraction reduced to lowest terms:

$$\frac{\dfrac{y}{x^2} - \dfrac{x}{y^2}}{\dfrac{y}{x} - \dfrac{x}{y}}$$

Solution Using the first method described in Example 5, we have

$$\frac{x^2y^2\left(\dfrac{y}{x^2} - \dfrac{x}{y^2}\right)}{x^2y^2\left(\dfrac{y}{x} - \dfrac{x}{y}\right)} = \frac{x^2y^2\dfrac{y}{x^2} - x^2y^2\dfrac{x}{y^2}}{x^2y^2\dfrac{y}{x} - x^2y^2\dfrac{x}{y}} = \frac{y^3 - x^3}{xy^3 - x^3y} = \frac{\overset{1}{\cancel{(y - x)}}(y^2 + xy + x^2)}{xy\underset{1}{\cancel{(y - x)}}(y + x)}$$

$$= \frac{y^2 + xy + x^2}{xy(y + x)}$$

Matched Problem 6 Express as a simple fraction reduced to lowest terms. Use the first method described in Example 5.

$$\frac{\dfrac{a}{b} - \dfrac{b}{a}}{\dfrac{a}{b} + 2 + \dfrac{b}{a}}$$

Answers to Matched Problems

1. (A) $\dfrac{3x + 2}{x + 1}$ (B) $\dfrac{x^2 + 2x + 4}{3x + 4}$ 2. $\dfrac{3(x^2 + 1)^2(x + 1)(x - 1)}{x^4}$
3. (A) $2x$ (B) $\dfrac{-5}{x + 4}$ (C) mn
4. (A) $\dfrac{1}{4}$ (B) $\dfrac{3x^2 - 5x - 4}{12x^3}$ (C) $\dfrac{2y^2 - 9y - 6}{(y - 2)^2(y + 2)}$
5. $\dfrac{1}{x - 1}$ 6. $\dfrac{a - b}{a + b}$

EXERCISE A-4

A

In Problems 1–20, perform the indicated operations and reduce answers to lowest terms. Represent any compound fractions as simple fractions reduced to lowest terms.

1. $\left(\dfrac{d^5}{3a} \div \dfrac{d^2}{6a^2}\right) \cdot \dfrac{a}{4d^3}$

2. $\dfrac{d^5}{3a} \div \left(\dfrac{d^2}{6a^2} \cdot \dfrac{a}{4d^3}\right)$

3. $\dfrac{2y}{18} - \dfrac{-1}{28} - \dfrac{y}{42}$

4. $\dfrac{x^2}{12} + \dfrac{x}{18} - \dfrac{1}{30}$

5. $\dfrac{3x+8}{4x^2} - \dfrac{2x-1}{x^3} - \dfrac{5}{8x}$

6. $\dfrac{4m-3}{18m^3} + \dfrac{3}{4m} - \dfrac{2m-1}{6m^2}$

7. $\dfrac{2x^2-3x-2}{x^2-4} \div (2x+1)$

8. $\dfrac{3x^2+x-2}{3x^2-2x} \div (3x^2+5x+2)$

9. $\dfrac{2m+3n}{2m^2-5mn-12n^2} \div \dfrac{2m^2-7mn-4n^2}{m^2-8mn+16n^2}$

10. $\dfrac{4y^2-4y+1}{2y^2+5y-3} \div \dfrac{2y^2-3y-2}{2y^2+7y+3}$

11. $\dfrac{2a-b}{a^2-b^2} - \dfrac{2a+3b}{a^2+2ab+b^2}$

12. $\dfrac{x+2}{x^2-1} - \dfrac{x-2}{(x-1)^2}$

13. $m+2-\dfrac{m-2}{m-1}$

14. $\dfrac{x+1}{x-1}+x$

15. $\dfrac{3}{x-2} - \dfrac{2}{2-x}$

16. $\dfrac{1}{a-3} - \dfrac{2}{3-a}$

17. $\dfrac{3}{y+2} + \dfrac{2}{y-2} - \dfrac{4y}{y^2-4}$

18. $\dfrac{4x}{x^2-y^2} + \dfrac{3}{x+y} - \dfrac{2}{x-y}$

19. $\dfrac{\dfrac{x^2}{y^2}-1}{\dfrac{x}{y}+1}$

20. $\dfrac{\dfrac{4}{x}-x}{\dfrac{2}{x}-1}$

B

Problems 21–26 are calculus-related. Reduce each fraction to lowest terms.

21. $\dfrac{6x^3(x^2+2)^2-2x(x^2+2)^3}{x^4}$

22. $\dfrac{4x^4(x^2+3)-3x^2(x^2+3)^2}{x^6}$

23. $\dfrac{2x(1-3x)^3+9x^2(1-3x)^2}{(1-3x)^6}$

24. $\dfrac{2x(2x+3)^4-8x^2(2x+3)^3}{(2x+3)^8}$

25. $\dfrac{-2x(x+4)^3-3(3-x^2)(x+4)^2}{(x+4)^6}$

26. $\dfrac{3x^2(x+1)^3-3(x^3+4)(x+1)^2}{(x+1)^6}$

In Problems 27–40, perform the indicated operations and reduce answers to lowest terms. Represent any compound fractions as simple fractions reduced to lowest terms.

27. $\dfrac{y}{y^2-2y-8} - \dfrac{2}{y^2-5y+4} + \dfrac{1}{y^2+y-2}$

28. $\dfrac{x}{x^2-9x+18} + \dfrac{x-8}{x-6} + \dfrac{x+4}{x-3}$

29. $\dfrac{16-m^2}{m^2+3m-4} \cdot \dfrac{m-1}{m-4}$

30. $\dfrac{x+1}{x(1-x)} \cdot \dfrac{x^2-2x+1}{x^2-1}$

31. $\dfrac{x+7}{ax-bx} + \dfrac{y+9}{by-ay}$

32. $\dfrac{c+2}{5c-5} - \dfrac{c-2}{3c-3} + \dfrac{c}{1-c}$

33. $\dfrac{x^2-16}{2x^2+10x+8} \div \dfrac{x^2-13x+36}{x^3+1}$

34. $\left(\dfrac{x^3-y^3}{y^3} \cdot \dfrac{y}{x-y}\right) \div \dfrac{x^2+xy+y^2}{y^2}$

35. $\dfrac{x^2-xy}{xy+y^2} \div \left(\dfrac{x^2-y^2}{x^2+2xy+y^2} \div \dfrac{x^2-2xy+y^2}{x^2y+xy^2}\right)$

36. $\left(\dfrac{x^2-xy}{xy+y^2} \div \dfrac{x^2-y^2}{x^2+2xy+y^2}\right) \div \dfrac{x^2-2xy+y^2}{x^2y+xy^2}$

37. $\left(\dfrac{x}{x^2-16} - \dfrac{1}{x+4}\right) \div \dfrac{4}{x+4}$

38. $\left(\dfrac{3}{x-2} - \dfrac{1}{x+1}\right) \div \dfrac{x+4}{x-2}$

39. $\dfrac{1+\dfrac{2}{x}-\dfrac{15}{x^2}}{1+\dfrac{4}{x}-\dfrac{5}{x^2}}$

40. $\dfrac{\dfrac{x}{y}-2+\dfrac{y}{x}}{\dfrac{x}{y}-\dfrac{y}{x}}$

Problems 41–44 are calculus-related. Perform the indicated operations and reduce answers to lowest terms. Represent any compound fractions as simple fractions reduced to lowest terms.

41. $\dfrac{\dfrac{1}{x+h}-\dfrac{1}{x}}{h}$

42. $\dfrac{\dfrac{1}{(x+h)^2}-\dfrac{1}{x^2}}{h}$

43. $\dfrac{\dfrac{(x+h)^2}{x+h+2}-\dfrac{x^2}{x+2}}{h}$

44. $\dfrac{\dfrac{2x+2h+3}{x+h}-\dfrac{2x+3}{x}}{h}$

In Problems 45–52, imagine that the indicated "solutions" were given to you by a student whom you were tutoring in this class.

(A) Is the solution correct? If the solution is incorrect, explain what is wrong and how it can be corrected.
(B) Show a correct solution for each incorrect solution.

45. $\dfrac{x^2 + 5x + 4}{x + 4} = \dfrac{x^2 + 5x}{x} = x + 5$

46. $\dfrac{x^2 - 2x - 3}{x - 3} = \dfrac{x^2 - 2x}{x} = x - 2$

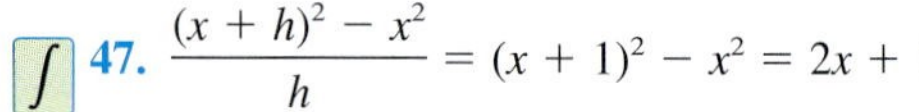

47. $\dfrac{(x + h)^2 - x^2}{h} = (x + 1)^2 - x^2 = 2x + 1$

48. $\dfrac{(x + h)^3 - x^3}{h} = (x + 1)^3 - x^3 = 3x^2 + 3x + 1$

49. $\dfrac{x^2 - 2x}{x^2 - x - 2} + x - 2 = \dfrac{x^2 - 2x + x - 2}{x^2 - x - 2} = 1$

50. $\dfrac{2}{x - 1} - \dfrac{x + 3}{x^2 - 1} = \dfrac{2x + 2 - x - 3}{x^2 - 1} = \dfrac{1}{x + 1}$

51. $\dfrac{2x^2}{x^2 - 4} - \dfrac{x}{x - 2} = \dfrac{2x^2 - x^2 - 2x}{x^2 - 4} = \dfrac{x}{x + 2}$

52. $x + \dfrac{x - 2}{x^2 - 3x + 2} = \dfrac{x + x - 2}{x^2 - 3x + 2} = \dfrac{2}{x - 2}$

C

In Problems 53–56, perform the indicated operations and reduce answers to lowest terms. Represent any compound fractions as simple fractions reduced to lowest terms.

53. $\dfrac{y - \dfrac{y^2}{y - x}}{1 + \dfrac{x^2}{y^2 - x^2}}$

54. $\dfrac{\dfrac{s^2}{s - t} - s}{\dfrac{t^2}{s - t} + t}$

55. $2 - \dfrac{1}{1 - \dfrac{2}{a + 2}}$

56. $1 - \dfrac{1}{1 - \dfrac{1}{1 - \dfrac{1}{x}}}$

In Problems 57 and 58, a, b, c, and d represent real numbers.

57. (A) Prove that d/c is the multiplicative inverse of c/d $(c, d \neq 0)$.
(B) Use part (A) to prove that

$$\frac{a}{b} \div \frac{c}{d} = \frac{a}{b} \cdot \frac{d}{c} \qquad b, c, d \neq 0$$

58. Prove that

$$\frac{a}{b} + \frac{c}{b} = \frac{a + c}{b} \qquad b \neq 0$$

SECTION A-5 Integer Exponents

- Integer Exponents
- Scientific Notation

The French philosopher/mathematician René Descartes (1596–1650) is generally credited with the introduction of the very useful exponent notation "x^n." This notation as well as other improvements in algebra may be found in his *Geometry,* published in 1637.

In Section A-2 we introduced the natural number exponent as a short way of writing a product involving the same factors. In this section we will expand the meaning of exponent to include all integers so that exponential forms of the following types will all have meaning:

$$7^5 \qquad 5^{-4} \qquad 3.14^0$$

• Integer Exponents

Definition 1 generalizes exponent notation to include 0 and negative integer exponents.

DEFINITION 1

a^n, n an integer and a a real number

1. For n a positive integer:

$$a^n = \underbrace{a \cdot a \cdot \cdots \cdot a}_{n \text{ factors of } a} \qquad 3^5 = 3 \cdot 3 \cdot 3 \cdot 3 \cdot 3$$

2. For $n = 0$:

$$a^0 = 1 \qquad a \neq 0 \qquad 132^0 = 1$$

0^0 is not defined

3. For n a negative integer:

$$a^n = \frac{1}{a^{-n}} \qquad a \neq 0 \qquad 7^{-3} = \frac{1}{7^{-(-3)}} = \frac{1}{7^3}$$

Note: In general, it can be shown that for *all* integers n

$$a^{-n} = \frac{1}{a^n} \qquad a^{-5} = \frac{1}{a^5} \qquad a^{-(-3)} = \frac{1}{a^{-3}}$$

EXAMPLE 1 **Using the Definition of Integer Exponents**

Write each part as a decimal fraction or using positive exponents.

(A) $(u^3v^2)^0 = 1 \quad u \neq 0, \quad v \neq 0$

(B) $10^{-3} = \dfrac{1}{10^3} = \dfrac{1}{1{,}000} = 0.001$

(C) $x^{-8} = \dfrac{1}{x^8}$

(D) $\dfrac{x^{-3}}{y^{-5}} = \dfrac{x^{-3}}{1} \cdot \dfrac{1}{y^{-5}} = \dfrac{1}{x^3} \cdot \dfrac{y^5}{1} = \dfrac{y^5}{x^3}$

Matched Problem 1 Write parts (A)–(D) as decimal fractions and parts (E) and (F) with positive exponents.

(A) 636^0 (B) $(x^2)^0 \quad x \neq 0$ (C) 10^{-5}

(D) $\dfrac{1}{10^{-3}}$ (E) $\dfrac{1}{x^{-4}}$ (F) $\dfrac{u^{-7}}{v^{-3}}$

The basic properties of integer exponents are summarized in Theorem 1. The proof of this theorem involves *mathematical induction,* which is discussed in Chapter 10.

Theorem 1 **Properties of Integer Exponents**

For n and m integers and a and b real numbers:

1. $a^m a^n = a^{m+n}$ $\qquad a^5a^{-7} = a^{5+(-7)} = a^{-2}$
2. $(a^n)^m = a^{mn}$ $\qquad (a^3)^{-2} = a^{(-2)3} = a^{-6}$
3. $(ab)^m = a^m b^m$ $\qquad (ab)^3 = a^3b^3$
4. $\left(\dfrac{a}{b}\right)^m = \dfrac{a^m}{b^m} \quad b \neq 0$ $\qquad \left(\dfrac{a}{b}\right)^4 = \dfrac{a^4}{b^4}$
5. $\dfrac{a^m}{a^n} = \begin{cases} a^{m-n} \\ \dfrac{1}{a^{n-m}} \end{cases} \quad a \neq 0$ $\qquad \dfrac{a^3}{a^{-2}} = a^{3-(-2)} = a^5$, $\quad \dfrac{a^3}{a^{-2}} = \dfrac{1}{a^{-2-3}} = \dfrac{1}{a^{-5}}$

EXPLORE-DISCUSS 1 Property 1 in Theorem 1 can be expressed verbally as follows:

To find the product of two exponential forms with the same base, add the exponents and use the same base.

Express the other properties in Theorem 1 verbally. Decide which you find easier to remember, a formula or a verbal description.

EXAMPLE 2 **Using Exponent Properties**

Simplify using exponent properties, and express answers using positive exponents only.*

(A) $(3a^5)(2a^{-3}) = (3 \cdot 2)(a^5a^{-3}) = 6a^2$

(B) $\dfrac{6x^{-2}}{8x^{-5}} = \dfrac{3x^{-2-(-5)}}{4} = \dfrac{3x^3}{4}$

(C) $-4y^3 - (-4y)^3 = -4y^3 - (-4)^3y^3 = -4y^3 - (-64)y^3$

$= -4y^3 + 64y^3 = 60y^3$

*By "simplify" we mean eliminate common factors from numerators and denominators and reduce to a minimum the number of times a given constant or variable appears in an expression. We ask that answers be expressed using positive exponents only in order to have a definite form for an answer. Later in this section we will encounter situations where we will want negative exponents in a final answer.

Matched Problem 2 Simplify using exponent properties, and express answers using positive exponents only.

(A) $(5x^{-3})(3x^4)$ (B) $\dfrac{9y^{-7}}{6y^{-4}}$ (C) $2x^4 - (-2x)^4$

CAUTION Be careful when using the relationship $a^{-n} = \dfrac{1}{a^n}$:

$ab^{-1} \neq \dfrac{1}{ab}$ $\quad ab^{-1} = \dfrac{a}{b}$ and $(ab)^{-1} = \dfrac{1}{ab}$

$\dfrac{1}{a+b} \neq a^{-1} + b^{-1}$ $\quad \dfrac{1}{a+b} = (a+b)^{-1}$ and $\dfrac{1}{a} + \dfrac{1}{b} = a^{-1} + b^{-1}$

Do not confuse properties 1 and 2 in Theorem 1:

$a^3a^4 \neq a^{3\cdot 4}$ $\quad a^3a^4 = a^{3+4} = a^7$ property 1, Theorem 1

$(a^3)^4 \neq a^{3+4}$ $\quad (a^3)^4 = a^{3\cdot 4} = a^{12}$ property 2, Theorem 1

From the definition of negative exponents and the five properties of exponents, we can easily establish the following properties, which are used very frequently when dealing with exponent forms.

Theorem 2 **Further Exponent Properties**

For a and b any real numbers and m, n, and p any integers (division by 0 excluded):

1. $(a^mb^n)^p = a^{pm}b^{pn}$ **2.** $\left(\dfrac{a^m}{b^n}\right)^p = \dfrac{a^{pm}}{b^{pn}}$

3. $\dfrac{a^{-n}}{b^{-m}} = \dfrac{b^m}{a^n}$ **4.** $\left(\dfrac{a}{b}\right)^{-n} = \left(\dfrac{b}{a}\right)^n$

Proof We prove properties 1 and 4 in Theorem 2 and leave the proofs of 2 and 3 to you.

1. $(a^mb^n)^p = (a^m)^p(b^n)^p$ property 3, Theorem 1

$= a^{pm}b^{pn}$ property 2, Theorem 1

$$4.\ \left(\frac{a}{b}\right)^{-n} = \frac{a^{-n}}{b^{-n}} \qquad \text{property 4, Theorem 1}$$

$$= \frac{b^n}{a^n} \qquad \text{property 3, Theorem 2}$$

$$= \left(\frac{b}{a}\right)^n \qquad \text{property 4, Theorem 1}$$

EXAMPLE 3 Using Exponent Properties

Simplify using exponent properties, and express answers using positive exponents only.

(A) $(2a^{-3}b^2)^{-2} = 2^{-2}a^6b^{-4} = \dfrac{a^6}{4b^4}$

(B) $\left(\dfrac{a^3}{b^5}\right)^{-2} = \dfrac{a^{-6}}{b^{-10}} = \dfrac{b^{10}}{a^6}$ or $\left(\dfrac{a^3}{b^5}\right)^{-2} = \left(\dfrac{b^5}{a^3}\right)^2 = \dfrac{b^{10}}{a^6}$

(C) $\dfrac{4x^{-3}y^{-5}}{6x^{-4}y^3} = \dfrac{2x^{-3-(-4)}}{3y^{3-(-5)}} = \dfrac{2x}{3y^8}$

(D) $\left(\dfrac{m^{-3}m^3}{n^{-2}}\right)^{-2} = \left(\dfrac{m^{-3+3}}{n^{-2}}\right)^{-2} = \left(\dfrac{m^0}{n^{-2}}\right)^{-2} = \left(\dfrac{1}{n^{-2}}\right)^{-2} = \dfrac{1}{n^4}$

(E) $(x + y)^{-3} = \dfrac{1}{(x + y)^3}$

Matched Problem 3

Simplify using exponent properties, and express answers using positive exponents only.

(A) $(3x^4y^{-3})^{-2}$ (B) $\left(\dfrac{x^2}{y^4}\right)^{-3}$ (C) $\dfrac{6m^{-2}n^3}{15m^{-1}n^{-2}}$

(D) $\left(\dfrac{x^{-3}}{y^4y^{-4}}\right)^{-3}$ (E) $\dfrac{1}{(a - b)^{-2}}$

In simplifying exponent forms there is often more than one sequence of steps that will lead to the same result (see Example 3B). Use whichever sequence of steps makes sense to you.

EXAMPLE 4 Simplifying a Compound Fraction

Express as a simple fraction reduced to lowest terms:

$$\frac{x^{-2} - y^{-2}}{x^{-1} + y^{-1}} = \frac{\frac{1}{x^2} - \frac{1}{y^2}}{\frac{1}{x} + \frac{1}{y}} = \frac{x^2y^2\left(\frac{1}{x^2} - \frac{1}{y^2}\right)}{x^2y^2\left(\frac{1}{x} + \frac{1}{y}\right)}$$

$$= \frac{y^2 - x^2}{xy^2 + x^2y} = \frac{(y - x)\overset{1}{\cancel{(y + x)}}}{xy\underset{1}{\cancel{(y + x)}}}$$

$$= \frac{y - x}{xy}$$

Matched Problem 4 Express as a simple fraction reduced to lowest terms:

$$\frac{x - x^{-1}}{1 - x^{-2}}$$

• Scientific Notation

Scientific work often involves the use of very large numbers or very small numbers. For example, the average cell contains about 200,000,000,000,000 molecules, and the diameter of an electron is about 0.000 000 000 0004 centimeter. It is generally troublesome to write and work with numbers of this type in standard decimal form. The two numbers written here cannot even be entered into most calculators as they are written. With exponents now defined for all integers, it is possible to express any decimal form as the product of a number between 1 and 10 and an integer power of 10; that is, in the form

$$a \times 10^n \qquad 1 \le a < 10,\ n \text{ an integer, } a \text{ in decimal form}$$

A number expressed in this form is said to be in **scientific notation.**

EXAMPLE 5 **Scientific Notation**

Each number is written in scientific notation:

$$7 = 7 \times 10^0 \qquad 0.5 = 5 \times 10^{-1}$$

$$720 = 7.2 \times 10^2 \qquad 0.08 = 8 \times 10^{-2}$$

$$6{,}430 = 6.43 \times 10^3 \qquad 0.000\ 32 = 3.2 \times 10^{-4}$$

$$5{,}350{,}000 = 5.35 \times 10^6 \qquad 0.000\ 000\ 0738 = 7.38 \times 10^{-8}$$

Can you discover a rule relating the number of decimal places the decimal point is moved to the power of 10 that is used?

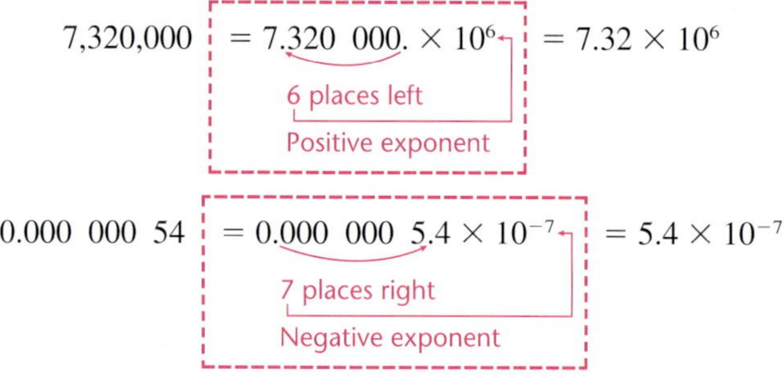

Matched Problem 5 (A) Write each number in scientific notation: 430; 23,000; 345,000,000; 0.3; 0.0031; 0.000 000 683

(B) Write in standard decimal form: 4×10^3; 5.3×10^5; 2.53×10^{-2}; 7.42×10^{-6}

Most calculators express very large and very small numbers in scientific notation. (Elsewhere in the book you will encounter optional exercises that require a graphing calculator. If you have such a calculator, you can certainly use it here. Otherwise, any scientific calculator will be sufficient for the problems in this review.) Consult the manual for your calculator to see how numbers in scientific notation are entered in your calculator. Some common methods for displaying scientific notation on a calculator are shown below.

Number Represented	Typical Scientific Calculator Display	Typical Graphing Calculator Display
$5.427\,493 \times 10^{-17}$	5.427493 -17	5.427493E-17
$2.359\,779 \times 10^{12}$	2.359779 12	2.359779E12

EXAMPLE 6 Using Scientific Notation on a Calculator

Write each number in scientific notation; then carry out the computations using your calculator. (Refer to the user's manual accompanying your calculator for the procedure.) Express the answer to three significant digits* in scientific notation.

$$\frac{325{,}100{,}000{,}000}{0.000\ 000\ 000\ 000\ 0871} = \frac{3.251 \times 10^{11}}{8.71 \times 10^{-14}}$$

$= \boxed{3.732491389\text{E}24}$ Calculator display

$= 3.73 \times 10^{24}$ To three significant digits

*For those not familiar with the meaning of *significant digits,* see Appendix B for a brief discussion of this concept.

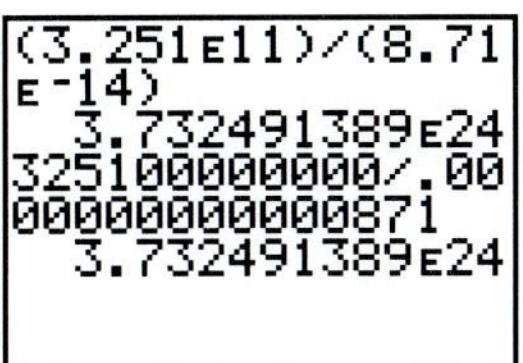

Figure 1

Figure 1 shows two solutions to this problem on a graphing calculator. In the first solution we entered the numbers in scientific notation, and in the second we used standard decimal notation. Although the multiple-line screen display on a graphing calculator allows us to enter very long standard decimals, scientific notation is usually more efficient and less prone to errors in data entry. Furthermore, as Figure 1 shows, the calculator uses scientific notation to display the answer, regardless of the manner in which the numbers are entered.

Matched Problem 6 Repeat Example 6 for:

$$\frac{0.000\ 000\ 006\ 932}{62{,}600{,}000{,}000}$$

EXAMPLE 7 **Measuring Time with an Atomic Clock**

An atomic clock that counts the radioactive emissions of cesium is used to provide a precise definition of a second. One second is defined to be the time it takes cesium to emit 9,192,631,770 cycles of radiation. How many of these cycles will occur in 1 hour? Express the answer to five significant digits in scientific notation.

Solution

$$(9{,}192{,}631{,}770)(60^2) = \boxed{3.309347437\text{E}13}$$

$$= 3.3093 \times 10^{13}$$

Matched Problem 7 Refer to Example 7. How many of these cycles will occur in 1 year? Express the answer to five significant digits in scientific notation.

Answers to Matched Problems

1. (A) 1 (B) 1 (C) 0.000 01 (D) 1,000 (E) x^4 (F) v^3/u^7
2. (A) $15x$ (B) $3/(2y^3)$ (C) $-14x^4$
3. (A) $y^6/(9x^8)$ (B) y^{12}/x^6 (C) $2n^5/(5m)$ (D) x^9 (E) $(a-b)^2$
4. x
5. (A) 4.3×10^2; 2.3×10^4; 3.45×10^8; 3×10^{-1}; 3.1×10^{-3}; 6.83×10^{-7}
(B) 4,000; 530,000; 0.0253; 0.000 007 42
6. 1.11×10^{-19} **7.** 2.8990×10^{17}

EXERCISE A-5

All variables are restricted to prevent division by 0.

A

Simplify Problems 1–16 and write the answers using positive exponents only.

1. $x^{-6}x^6$

2. y^7y^{-7}

3. $(3x^2)(5x^4)(4x^5)$

4. $(2a^3)(7a^4)(3a^9)$

5. $(2x^4z^{-3})^2$

6. $(3u^{-2}v^3)^{-2}$

7. $\left(\dfrac{r^2s^3}{pq^4}\right)^5$

8. $\left(\dfrac{a^3b^2}{2c^5}\right)^2$

9. $\dfrac{10^{19} \cdot 10^{-14}}{10^{-5} \cdot 10^3}$

10. $\dfrac{10^{-9}\cdot 10^{-12}}{10^{-17}\cdot 10^{5}}$ **11.** $\dfrac{6a^{-3}b^{-5}}{2a^{-7}b^{-2}}$

12. $\dfrac{20u^{-4}v^{5}}{4u^{4}v^{-3}}$ **13.** $\left(\dfrac{y^{-2}}{y^{-3}}\right)^{-1}$

14. $\left(\dfrac{w^{-2}}{w^{-4}}\right)^{-2}$ **15.** $\dfrac{12\times 10^{4}}{3\times 10^{-7}}$

16. $\dfrac{15\times 10^{-8}}{5\times 10^{-5}}$

Write the numbers in Problems 17–22 in scientific notation.

17. 45,320,000 **18.** 3,670

19. 0.066 **20.** 0.029

21. 0.000 000 084 **22.** 0.000 497

In Problems 23–28, write each number in standard decimal form.

23. 9×10^{-5} **24.** 3×10^{-3} **25.** 3.48×10^{6}

26. 8.63×10^{8} **27.** 4.2×10^{-9} **28.** 1.6×10^{-7}

B

Simplify Problems 29–44, and write the answers using positive exponents only. Write compound fractions as simple fractions.

29. $\dfrac{27x^{-5}x^{5}}{18y^{-6}y^{2}}$ **30.** $\dfrac{32n^{5}n^{-8}}{24m^{-7}m^{7}}$ **31.** $\left(\dfrac{x^{4}y^{-1}}{x^{-2}y^{3}}\right)^{2}$

32. $\left(\dfrac{m^{-2}n^{3}}{m^{4}n^{-1}}\right)^{2}$ **33.** $\left(\dfrac{2x^{-3}y^{2}}{4xy^{-1}}\right)^{-2}$ **34.** $\left(\dfrac{6mn^{-2}}{3m^{-1}n^{2}}\right)^{-3}$

35. $\left[\left(\dfrac{u^{3}v^{-1}w^{-2}}{u^{-2}v^{-2}w}\right)^{-2}\right]^{2}$ **36.** $\left[\left(\dfrac{x^{-2}y^{3}t}{x^{-3}y^{-2}t^{2}}\right)^{2}\right]^{-1}$

37. $(x+y)^{-2}$ **38.** $(a^{2}-b^{2})^{-1}$ **39.** $\dfrac{1+x^{-1}}{1-x^{-2}}$

40. $\dfrac{1-x}{x^{-1}-1}$ **41.** $\dfrac{x^{-1}-y^{-1}}{x-y}$ **42.** $\dfrac{u+v}{u^{-1}+v^{-1}}$

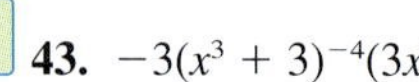

43. $-3(x^{3}+3)^{-4}(3x^{2})$

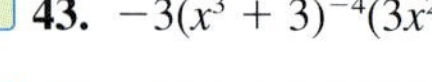

44. $-2(x^{2}+3x)^{-3}(2x+3)$

45. What is the result of entering 2^{3^2} on a calculator?

46. Refer to Problem 45. What is the difference between $2^{(3^2)}$ and $(2^3)^2$? Which agrees with the value of 2^{3^2} obtained with a calculator?

47. If $n = 0$, then property 1 in Theorem 1 implies that $a^m a^0 = a^{m+0} = a^m$. Explain how this helps motivate the definition of a^0.

48. If $m = -n$, then property 1 in Theorem 1 implies that $a^{-n}a^n = a^0 = 1$. Explain how this helps motivate the definition of a^{-n}.

Problems 49–54 are calculus-related. Write each problem in the form $ax^p + bx^q$ or $ax^p + bx^q + cx^r$, where a, b, and c are real numbers and p, q, and r are integers. For example,

$$\frac{2x^4-3x^2+1}{2x^3} = \frac{2x^4}{2x^3}-\frac{3x^2}{2x^3}+\frac{1}{2x^3}$$

$$= x - \frac{3}{2}x^{-1} + \frac{1}{2}x^{-3}$$

49. $\dfrac{4x^{2}-12}{2x}$ **50.** $\dfrac{6x^{3}+9x}{3x^{3}}$

51. $\dfrac{5x^{3}-2}{3x^{2}}$ **52.** $\dfrac{7x^{5}-x^{2}}{4x^{5}}$

53. $\dfrac{2x^{3}-3x^{2}+x}{2x^{2}}$ **54.** $\dfrac{3x^{4}-4x^{2}-1}{4x^{3}}$

Evaluate Problems 55–58, to three significant digits using scientific notation where appropriate and a calculator.

55. $\dfrac{(32.7)(0.000\ 000\ 008\ 42)}{(0.0513)(80{,}700{,}000{,}000)}$

56. $\dfrac{(4{,}320)(0.000\ 000\ 000\ 704)}{(835)(635{,}000{,}000{,}000)}$

57. $\dfrac{(5{,}760{,}000{,}000)}{(527)(0.000\ 007\ 09)}$

58. $\dfrac{0.000\ 000\ 007\ 23}{(0.0933)(43{,}700{,}000{,}000)}$

In Problems 59–64, use a calculator to evaluate each of the following problems to five significant digits. (Read the instruction book accompanying your calculator.)

59. $(23.8)^{8}$ **60.** $(-302)^{7}$

61. $(-302)^{-7}$ **62.** $(23.8)^{-8}$

63. $(9{,}820{,}000{,}000)^{3}$ **64.** $(0.000\ 000\ 000\ 482)^{-4}$

C

Simplify Problems 65–70, and write the answers using positive exponents only. Write compound fractions as simple fractions.

65. $\dfrac{12(a+2b)^{-3}}{6(a+2b)^{-8}}$ **66.** $\dfrac{4(x-3)^{-4}}{8(x-3)^{-2}}$

67. $\dfrac{xy^{-2}-yx^{-2}}{y^{-1}-x^{-1}}$ **68.** $\dfrac{b^{-2}-c^{-2}}{b^{-3}-c^{-3}}$

69. $\left(\dfrac{x^{-1}}{x^{-1}-y^{-1}}\right)^{-1}$ **70.** $\left[\dfrac{u^{-2}-v^{-2}}{(u^{-1}-v^{-1})^{2}}\right]^{-1}$

APPLICATIONS 

71. Earth Science. If the mass of the earth is approximately 6.1×10^{27} grams and each gram is 2.2×10^{-3} pound, what is the mass of the earth in pounds?

72. Biology. In 1929 Vernadsky, a biologist, estimated that all the free oxygen of the earth weighs 1.5×10^{21} grams and that it is produced by life alone. If 1 gram is approximately 2.2×10^{-3} pound, what is the weight of the free oxygen in pounds?

73. Computer Science. If a computer can perform a single operation in 10^{-10} second, how many operations can it perform in 1 second? In 1 minute? Compute answers to three significant digits.

★ **74. Computer Science.** If electricity travels in a computer circuit at the speed of light (1.86×10^5 miles per second), how far will electricity travel in the superconducting computer (see Problem 73) in the time it takes it to perform one operation? (Size of circuits is a critical problem in computer design.) Give the answer in miles, feet, and inches (1 mile = 5,280 feet). Compute answers to three significant digits.

75. Economics. If in the United States in 1999 the national debt was about \$5,680,000,000,000 and the population was about 274,000,000, estimate to three significant digits each individual's share of the national debt. Write your answer in scientific notation and in standard decimal form.

76. Economics. If in the United States in 1999 the gross national product (GNP) was about \$8,870,000,000,000 and the population was about 274,000,000, estimate to three significant digits the GNP per person. Write your answer in scientific notation and in standard decimal form.

SECTION A-6 Rational Exponents

- Roots of Real Numbers
- Rational Exponents

We now know what symbols such as 3^5, 2^{-3}, and 7^0 mean; that is, we have defined a^n, where n is any integer and a is a real number. But what do symbols such as $4^{1/2}$ and $7^{2/3}$ mean? In this section we will extend the definition of exponent to the rational numbers. Before we can do this, however, we need a precise knowledge of what is meant by "a root of a number."

• Roots of Real Numbers

Perhaps you recall that a **square root** of a number b is a number c such that $c^2 = b$, and a **cube root** of a number b is a number d such that $d^3 = b$.

What are the square roots of 9?

$$3 \text{ is a square root of } 9, \text{ since } 3^2 = 9.$$

$$-3 \text{ is a square root of } 9, \text{ since } (-3)^2 = 9.$$

Thus, 9 has two real square roots, one the negative of the other.

What are the cube roots of 8?

$$2 \text{ is a cube root of } 8, \text{ since } 2^3 = 8.$$

And 2 is the only real number with this property. In general:

DEFINITION 1

Definition of an *n*th Root

For a natural number n and a and b real numbers:

a is an nth root of b if $a^n = b$ 3 is a fourth root of 81, since $3^4 = 81$

EXPLORE-DISCUSS 1 Is -4 a cube root of -64?

Is either 8 or -8 a square root of -64?

Can you find any real number b with the property that $b^2 = -64$? [*Hint:* Consider the sign of b^2 for $b > 0$ and $b < 0$.]

How many real square roots of 4 exist? Of 5? Of -9? How many real fourth roots of 5 exist? Of -5? How many real cube roots of 27 are there? Of -27? The following important theorem (which we state without proof) answers these questions.

Theorem 1

Number of Real *n*th Roots of a Real Number *b**

	***n* even**	***n* odd**
b positive	Two real nth roots -3 and 3 are both fourth roots of 81	One real nth root 2 is the only real cube root of 8
b negative	No real nth root -9 has no real square roots	One real nth root -2 is the only real cube root of -8

Thus, 4 and 5 have two real square roots each, and -9 has none. There are two real fourth roots of 5 and none for -5. And 27 and -27 have one real cube root each. What symbols do we use to represent these roots? We turn to this question now.

• Rational Exponents

If all exponent properties are to continue to hold even if some of the exponents are rational numbers, then

$$(5^{1/3})^3 = 5^{3/3} = 5 \qquad \text{and} \qquad (7^{1/2})^2 = 7^{2/2} = 7$$

Since Theorem 1 states that the number 5 has one real cube root, it seems reasonable to use the symbol $5^{1/3}$ to represent this root. On the other hand, Theorem 1 states that 7 has two real square roots. Which real square root of 7 does $7^{1/2}$ represent? We answer this question in the following definition.

*In this section we limit our discussion to real roots of real numbers. After the real numbers are extended to the complex numbers (see Section 1-5), additional roots may be considered. For example, it turns out that 1 has three cube roots: in addition to the real number 1, there are two other cube roots of 1 in the complex number system.

DEFINITION 2

$b^{1/n}$, Principal *n*th Root

For n a natural number and b a real number,

$b^{1/n}$ is the **principal nth root of b**

defined as follows:

1. If n is even and b is positive, then $b^{1/n}$ represents the positive nth root of b.

   $16^{1/2} = 4$ not -4 and 4.
   $-16^{1/2} = -4$ $-16^{1/2}$ and $(-16)^{1/2}$ are not the same.

2. If n is even and b is negative, then $b^{1/n}$ does not represent a real number. (More will be said about this case later.)

   $(-16)^{1/2}$ is not real.

3. If n is odd, then $b^{1/n}$ represents the real nth root of b (there is only one).

   $32^{1/5} = 2$ $(-32)^{1/5} = -2$

4. $0^{1/n} = 0$ $0^{1/9} = 0$ $0^{1/6} = 0$

EXAMPLE 1 **Principal *n*th Roots**

(A) $9^{1/2} = 3$

(B) $-9^{1/2} = -3$ Compare parts (B) and (C).

(C) $(-9)^{1/2}$ is not a real number. (D) $27^{1/3} = 3$

(E) $(-27)^{1/3} = -3$ (F) $0^{1/7} = 0$

Matched Problem 1 Find each of the following:

(A) $4^{1/2}$ (B) $-4^{1/2}$ (C) $(-4)^{1/2}$

(D) $8^{1/3}$ (E) $(-8)^{1/3}$ (F) $0^{1/8}$

How should a symbol such as $7^{2/3}$ be defined? If the properties of exponents are to hold for rational exponents, then $7^{2/3} = (7^{1/3})^2$; that is, $7^{2/3}$ must represent the square of the cube root of 7. This leads to the following general definition:

DEFINITION 3

$b^{m/n}$ and $b^{-m/n}$, Rational Number Exponent

For m and n natural numbers and b any real number (except b cannot be negative when n is even):

$$b^{m/n} = (b^{1/n})^m \quad \text{and} \quad b^{-m/n} = \frac{1}{b^{m/n}}$$

$4^{3/2} = (4^{1/2})^3 = 2^3 = 8$ $\quad 4^{-3/2} = \frac{1}{4^{3/2}} = \frac{1}{8}$ $\quad (-4)^{3/2}$ is not real

$(-32)^{3/5} = [(-32)^{1/5}]^3 = (-2)^3 = -8$

We have now discussed $b^{m/n}$ for all rational numbers m/n and real numbers b. It can be shown, though we will not do so, that all five properties of exponents listed in Theorem 1 in Section A-5 continue to hold for rational exponents as long as we avoid even roots of negative numbers. With the latter restriction in effect, the following useful relationship is an immediate consequence of the exponent properties:

Theorem 2 **Rational Exponent Property**

For m and n natural numbers and b any real number (except b cannot be negative when n is even):

$$b^{m/n} = \begin{cases} (b^{1/n})^m \\ (b^m)^{1/n} \end{cases} \qquad 8^{2/3} = \begin{cases} (8^{1/3})^2 \\ (8^2)^{1/3} \end{cases}$$

EXPLORE-DISCUSS 2 Find the contradiction in the following chain of equations:

$$-1 = (-1)^{2/2} = [(-1)^2]^{1/2} = 1^{1/2} = 1 \qquad (1)$$

Where did we try to use Theorem 2? Why was this not correct?

The three exponential forms in Theorem 2 are equal as long as only real numbers are involved. But if b is negative and n is even, then $b^{1/n}$ is not a real number and Theorem 2 does not necessarily hold, as illustrated in Explore-Discuss 2. One way to avoid this difficulty is to assume that m and n have no common factors.

EXAMPLE 2 **Using Rational Exponents**

Simplify, and express answers using positive exponents only. All letters represent positive real numbers.

(A) $8^{2/3} = (8^{1/3})^2 = 2^2 = 4$ or $8^{2/3} = (8^2)^{1/3} = 64^{1/3} = 4$

(B) $(-8)^{5/3} = [(-8)^{1/3}]^5 = (-2)^5 = -32$

(C) $(3x^{1/3})(2x^{1/2}) = 6x^{1/3+1/2} = 6x^{5/6}$

(D) $\left(\dfrac{4x^{1/3}}{x^{1/2}}\right)^{1/2} = \dfrac{4^{1/2}x^{1/6}}{x^{1/4}} = \dfrac{2}{x^{1/4-1/6}} = \dfrac{2}{x^{1/12}}$

(E) $(u^{1/2} - 2v^{1/2})(3u^{1/2} + v^{1/2}) = 3u - 5u^{1/2}v^{1/2} - 2v$

Matched Problem 2 Simplify, and express answers using positive exponents only. All letters represent positive real numbers.

(A) $9^{3/2}$ (B) $(-27)^{4/3}$ (C) $(5y^{3/4})(2y^{1/3})$ (D) $(2x^{-3/4}y^{1/4})^4$

(E) $\left(\frac{8x^{1/2}}{x^{2/3}}\right)^{1/3}$ (F) $(2x^{1/2} + y^{1/2})(x^{1/2} - 3y^{1/2})$

EXAMPLE 3 Evaluating Rational Exponential Forms with a Calculator

Evaluate to four significant digits using a calculator. (Refer to the instruction book for your particular calculator to see how exponential forms are evaluated.)

(A) $11^{3/4}$ (B) $3.1046^{-2/3}$ (C) $(0.000\ 000\ 008\ 437)^{3/11}$

Solutions (A) First change $\frac{3}{4}$ to the standard decimal form 0.75; then evaluate $11^{0.75}$ using a calculator.

$$11^{3/4} = 6.040$$

(B) $3.1046^{-2/3} = 0.4699$

(C) $$\begin{aligned}(0.000\ 000\ 008\ 437)^{3/11} &= (8.437 \times 10^{-9})^{3/11} \\ &= 0.006\ 281\end{aligned}$$

Matched Problem 3 Evaluate to four significant digits using a calculator.

(A) $2^{3/8}$ (B) $57.28^{-5/6}$ (C) $(83{,}240{,}000{,}000)^{5/3}$

EXAMPLE 4 Simplifying Fractions Involving Rational Exponents

Write the following expression as a simple fraction reduced to lowest terms and without negative exponents:

$$\frac{(1 + x^2)^{1/2}(2x) - x^2(\frac{1}{2})(1 + x^2)^{-1/2}(2x)}{1 + x^2}$$

Solution The negative exponent indicates the presence of a fraction in the numerator. Multiply numerator and denominator by $(1 + x^2)^{1/2}$ to eliminate the negative exponent and simplify.

$$\frac{(1 + x^2)^{1/2}(2x) - x^2(\frac{1}{2})(1 + x^2)^{-1/2}(2x)}{1 + x^2} \cdot \frac{(1 + x^2)^{1/2}}{(1 + x^2)^{1/2}}$$

$$= \frac{2x(1+x^2) - x^3}{(1+x^2)^{3/2}} = \frac{2x + 2x^3 - x^3}{(1+x^2)^{3/2}} = \frac{2x + x^3}{(1+x^2)^{3/2}}$$

$$= \frac{x(2+x^2)}{(1+x^2)^{3/2}}$$

Matched Problem 4 Write the following expression as a simple fraction reduced to lowest terms and without negative exponents:

$$\frac{x^2(\frac{1}{2})(1+x^2)^{-1/2}(2x) - (1+x^2)^{1/2}(2x)}{x^4}$$

Answers to Matched Problems

1. (A) 2 (B) −2 (C) Not real (D) 2 (E) −2 (F) 0
2. (A) 27 (B) 81 (C) $10y^{13/12}$ (D) $16y/x^3$ (E) $2/x^{1/18}$ (F) $2x - 5x^{1/2}y^{1/2} - 3y$
3. (A) 1.297 (B) 0.034 28 (C) 1.587×10^{18}
4. $-(2+x^2)/[x^3(1+x^2)^{1/2}]$

EXERCISE A-6

All variables represent positive real numbers unless otherwise stated.

A

In Problems 1–12, evaluate each expression that results in a rational number.

1. $25^{1/2}$ **2.** $27^{1/3}$ **3.** $9^{3/2}$

4. $8^{2/3}$ **5.** $-16^{1/2}$ **6.** $64^{2/3}$

7. $(-16)^{1/2}$ **8.** $(-64)^{2/3}$ **9.** $(\frac{27}{125})^{2/3}$

10. $(\frac{25}{36})^{3/2}$ **11.** $4^{-5/2}$ **12.** $9^{-5/2}$

Simplify Problems 13–20, and express answers using positive exponents only.

13. $a^{1/3}a^{4/3}$ **14.** $b^{2/5}b^{4/5}$ **15.** $c^{3/5}c^{-1/5}$

16. $d^{1/5}d^{-3/5}$ **17.** $(u^{-5})^{1/10}$ **18.** $(v^{-3/4})^8$

19. $(16x^8y^{-4})^{1/4}$ **20.** $(27x^{-6}y^9)^{1/3}$

B

Simplify Problems 21–30, and express answers using positive exponents only.

21. $\left(\frac{a^{-3}}{b^4}\right)^{1/12}$ **22.** $\left(\frac{m^{-2/3}}{n^{-1/2}}\right)^{-6}$ **23.** $\left(\frac{4x^{-2}}{y^4}\right)^{-1/2}$

24. $\left(\frac{w^4}{9x^{-2}}\right)^{-1/2}$ **25.** $\left(\frac{8a^{-4}b^3}{27a^2b^{-3}}\right)^{1/3}$ **26.** $\left(\frac{25x^5y^{-1}}{16x^{-3}y^{-5}}\right)^{1/2}$

27. $\frac{8x^{-1/3}}{12x^{1/4}}$ **28.** $\frac{6a^{3/4}}{15a^{-1/3}}$ **29.** $\left(\frac{a^{2/3}b^{-1/2}}{a^{1/2}b^{1/2}}\right)^2$

30. $\left(\frac{x^{-1/3}y^{1/2}}{x^{-1/4}y^{1/3}}\right)^6$

In Problems 31–38, multiply, and express answers using positive exponents only.

31. $2m^{1/3}(3m^{2/3} - m^6)$ **32.** $3x^{3/4}(4x^{1/4} - 2x^8)$

33. $(a^{1/2} + 2b^{1/2})(a^{1/2} - 3b^{1/2})$

34. $(3u^{1/2} - v^{1/2})(u^{1/2} - 4v^{1/2})$

35. $(2x^{1/2} - 3y^{1/2})(2x^{1/2} + 3y^{1/2})$

36. $(5m^{1/2} + n^{1/2})(5m^{1/2} - n^{1/2})$

37. $(x^{1/2} + 2y^{1/2})^2$ **38.** $(3x^{1/2} - y^{1/2})^2$

In Problems 39–46, evaluate to four significant digits using a calculator. (Refer to the instruction book for your calculator to see how exponential forms are evaluated.)

39. $15^{5/4}$ **40.** $22^{3/2}$ **41.** $103^{-3/4}$

42. $827^{-3/8}$ **43.** $2.876^{8/5}$ **44.** $37.09^{7/3}$

45. $(0.000\,000\,077\,35)^{-2/7}$ **46.** $(491{,}300{,}000{,}000)^{7/4}$

Problems 47–50 illustrate common errors involving rational exponents. In each case, find numerical examples that show that the left side is not always equal to the right side.

47. $(x + y)^{1/2} \neq x^{1/2} + y^{1/2}$ **48.** $(x^3 + y^3)^{1/3} \neq x + y$

49. $(x + y)^{1/3} \neq \dfrac{1}{(x + y)^3}$ **50.** $(x + y)^{-1/2} \neq \dfrac{1}{(x + y)^2}$

Problems 51–56 are calculus-related. Write each problem in the form $ax^p + bx^q$, where a and b are real numbers and p and q are rational numbers. For example,

$$\frac{2x^{1/3} + 4}{4x} = \frac{2x^{1/3}}{4x} + \frac{4}{4x} = \frac{1}{2}x^{1/3-1} + x^{-1}$$

$$= \frac{1}{2}x^{-2/3} + x^{-1}$$

51. $\dfrac{12x^{1/2} - 3}{4x^{1/2}}$ **52.** $\dfrac{x^{2/3} + 2}{2x^{1/3}}$ **53.** $\dfrac{3x^{2/3} + x^{1/2}}{5x}$

54. $\dfrac{2x^{3/4} + 3x^{1/3}}{3x}$ **55.** $\dfrac{x^2 - 4x^{1/2}}{2x^{1/3}}$ **56.** $\dfrac{2x^{1/3} - x^{1/2}}{4x^{1/2}}$

In Problems 57–60, m and n represent positive integers. Simplify and express answers using positive exponents.

57. $(a^{3/n}b^{3/m})^{1/3}$ **58.** $(a^{n/2}b^{n/3})^{1/n}$

59. $(x^{m/4}y^{n/3})^{-12}$ **60.** $(a^{m/3}b^{n/2})^{-6}$

61. If possible, find a real value of x such that:
(A) $(x^2)^{1/2} \neq x$ (B) $(x^2)^{1/2} = x$ (C) $(x^3)^{1/3} \neq x$

62. If possible, find a real value of x such that:
(A) $(x^2)^{1/2} \neq -x$ (B) $(x^2)^{1/2} = -x$ (C) $(x^3)^{1/3} = -x$

63. If n is even and b is negative, then $b^{1/n}$ is not real. If m is odd, n is even, and b is negative, is $(b^m)^{1/n}$ real?

64. If we assume that m is odd and n is even, is it possible that one of $(b^{1/n})^m$ and $(b^m)^{1/n}$ is real and the other is not?

Problems 65–68 are calculus-related. Simplify by writing each expression as a simple fraction reduced to lowest terms and without negative exponents.

65. $\dfrac{(2x - 1)^{1/2} - (x + 2)(\frac{1}{2})(2x - 1)^{-1/2}(2)}{2x - 1}$

66. $\dfrac{(x - 1)^{1/2} - x(\frac{1}{2})(x - 1)^{-1/2}}{x - 1}$

67. $\dfrac{2(3x - 1)^{1/3} - (2x + 1)(\frac{1}{3})(3x - 1)^{-2/3}(3)}{(3x - 1)^{2/3}}$

68. $\dfrac{(x + 2)^{2/3} - x(\frac{2}{3})(x + 2)^{-1/3}}{(x + 2)^{4/3}}$

APPLICATIONS

69. Economics. The number of units N of a finished product produced from the use of x units of labor and y units of capital for a particular Third World country is approximated by

$$N = 10x^{3/4}y^{1/4} \quad \text{Cobb-Douglas equation}$$

Estimate how many units of a finished product will be produced using 256 units of labor and 81 units of capital.

70. Economics. The number of units N of a finished product produced by a particular automobile company where x units of labor and y units of capital are used is approximated by

$$N = 50x^{1/2}y^{1/2} \quad \text{Cobb-Douglas equation}$$

Estimate how many units will be produced using 256 units of labor and 144 units of capital.

71. Braking Distance. R. A. Moyer of Iowa State College found, in comprehensive tests carried out on 41 wet pavements, that the braking distance d (in feet) for a particular automobile traveling at v miles per hour was given approximately by

$$d = 0.0212v^{7/3}$$

Approximate the braking distance to the nearest foot for the car traveling on wet pavement at 70 miles per hour.

72. Braking Distance. Approximately how many feet would it take the car in Problem 71 to stop on wet pavement if it were traveling at 50 miles per hour? (Compute answer to the nearest foot.)

SECTION A-7 Radicals

- From Rational Exponents to Radicals, and Vice Versa
- Properties of Radicals
- Simplifying Radicals
- Sums and Differences
- Products
- Rationalizing Operations

What do the following algebraic expressions have in common?

$$2^{1/2} \qquad 2x^{2/3} \qquad \frac{1}{x^{1/2} + y^{1/2}}$$

$$\sqrt{2} \qquad 2\sqrt[3]{x^2} \qquad \frac{1}{\sqrt{x} + \sqrt{y}}$$

Each vertical pair represents the same quantity, one in rational exponent form and the other in *radical form.* There are occasions when it is more convenient to work with radicals than with rational exponents, or vice versa. In this section we see how the two forms are related and investigate some basic operations on radicals.

• From Rational Exponents to Radicals, and Vice Versa

We start this discussion by defining an ***n*th-root radical:**

DEFINITION 1

$\sqrt[n]{b}$, *n*th-Root Radical

For n a natural number greater than 1 and b a real number, we define $\sqrt[n]{b}$ to be the **principal *n*th root of *b*** (see Definition 2 in Section A-6); that is,

$$\sqrt[n]{b} = b^{1/n}$$

If $n = 2$, we write $\sqrt{b}$ in place of $\sqrt[2]{b}$

$\sqrt{25} = 25^{1/2} = 5$ $\qquad$ $\sqrt[5]{32} = 32^{1/5} = 2$

$-\sqrt{25} = -25^{1/2} = -5$ $\qquad$ $\sqrt[5]{-32} = (-32)^{1/5} = -2$

$\sqrt{-25}$ is not real $\qquad$ $\sqrt[4]{0} = 0^{1/4} = 0$

The symbol $\sqrt{\ }$ is called a **radical,** n is called the **index,** and b is called the **radicand.**

As stated above, it is often an advantage to be able to shift back and forth between rational exponent forms and radical forms. The following relationships, which are direct consequences of Definition 1 and Theorem 2 in Section A-6, are useful in this regard:

Rational Exponent/Radical Conversions

For m and n positive integers ($n > 1$), and b not negative when n is even,

$$b^{m/n} = \begin{cases} (b^m)^{1/n} = \sqrt[n]{b^m} \\ (b^{1/n})^m = (\sqrt[n]{b})^m \end{cases} \qquad 2^{2/3} = \begin{cases} \sqrt[3]{2^2} \\ (\sqrt[3]{2})^2 \end{cases}$$

Note: Unless stated to the contrary, all variables in the rest of the discussion are restricted so that all quantities involved are real numbers.

EXPLORE-DISCUSS 1 In each of the following, evaluate both radical forms.

$$16^{3/2} = \sqrt{16^3} = (\sqrt{16})^3$$
$$27^{2/3} = \sqrt[3]{27^2} = (\sqrt[3]{27})^2$$

Which radical conversion form is easier to use if you are performing the calculations by hand?

EXAMPLE 1 **Rational Exponents/Radical Conversions**

Change from rational exponent form to radical form.

(A) $x^{1/7} = \sqrt[7]{x}$

(B) $(3u^2v^3)^{3/5} = \sqrt[5]{(3u^2v^3)^3}$ or $(\sqrt[5]{3u^2v^3})^3$ The first is usually preferred.

(C) $y^{-2/3} = \dfrac{1}{y^{2/3}} = \dfrac{1}{\sqrt[3]{y^2}}$ or $\sqrt[3]{y^{-2}}$ or $\sqrt[3]{\dfrac{1}{y^2}}$

Change from radical form to rational exponent form.

(D) $\sqrt[5]{6} = 6^{1/5}$ (E) $-\sqrt[3]{x^2} = -x^{2/3}$ (F) $\sqrt{x^2 + y^2} = (x^2 + y^2)^{1/2}$

Matched Problem 1 Change from rational exponent form to radical form.

(A) $u^{1/5}$ (B) $(6x^2y^5)^{2/9}$ (C) $(3xy)^{-3/5}$

Change from radical form to rational exponent form.

(D) $\sqrt[4]{9u}$ (E) $-\sqrt[7]{(2x)^4}$ (F) $\sqrt[3]{x^3 + y^3}$

• Properties of Radicals

The process of changing and simplifying radical expressions is aided by the introduction of several properties of radicals that follow directly from exponent properties considered earlier.

Theorem 1 **Properties of Radicals**

For n a natural number greater than 1, and x and y positive real numbers:

1. $\sqrt[n]{x^n} = x$ $\quad \sqrt[3]{x^3} = x$
2. $\sqrt[n]{xy} = \sqrt[n]{x}\sqrt[n]{y}$ $\quad \sqrt[5]{xy} = \sqrt[5]{x}\sqrt[5]{y}$
3. $\sqrt[n]{\dfrac{x}{y}} = \dfrac{\sqrt[n]{x}}{\sqrt[n]{y}}$ $\quad \sqrt[4]{\dfrac{x}{y}} = \dfrac{\sqrt[4]{x}}{\sqrt[4]{y}}$

EXAMPLE 2 **Simplifying Radicals**

Simplify:

(A) $\sqrt[5]{(3x^2y)^5} = 3x^2y$

(B) $\sqrt{10}\sqrt{5} = \sqrt{50} = \sqrt{25 \cdot 2} = \sqrt{25}\sqrt{2} = 5\sqrt{2}$

(C) $\sqrt[3]{\dfrac{x}{27}} = \dfrac{\sqrt[3]{x}}{\sqrt[3]{27}} = \dfrac{\sqrt[3]{x}}{3}$ or $\dfrac{1}{3}\sqrt[3]{x}$

Matched Problem 2 Simplify:

(A) $\sqrt[7]{(u^2 + v^2)^7}$ (B) $\sqrt{6}\sqrt{2}$ (C) $\sqrt[3]{\dfrac{x^2}{8}}$

CAUTION In general, properties of radicals can be used to simplify terms raised to powers, not sums of terms raised to powers. Thus, for x and y positive real numbers,

$$\sqrt{x^2 + y^2} \neq \sqrt{x^2} + \sqrt{y^2} = x + y$$

but

$$\sqrt{x^2 + 2xy + y^2} = \sqrt{(x + y)^2} = x + y$$

• Simplifying Radicals

The properties of radicals provide us with the means of changing algebraic expressions containing radicals to a variety of equivalent forms. One form that is often useful is a *simplified form.* An algebraic expression that contains radicals is said to be in **simplified form** if all four of the conditions listed in the following definition are satisfied.

DEFINITION 2 **Simplified (Radical) Form**

1. No radicand (the expression within the radical sign) contains a factor to a power greater than or equal to the index of the radical.
   For example, $\sqrt{x^5}$ violates this condition.
2. No power of the radicand and the index of the radical have a common factor other than 1.
   For example, $\sqrt[6]{x^4}$ violates this condition.
3. No radical appears in a denominator.
   For example, $y/\sqrt{x}$ violates this condition.
4. No fraction appears within a radical.
   For example, $\sqrt{\frac{3}{5}}$ violates this condition.

EXAMPLE 3 **Finding Simplified Form**

Express radicals in simplified form.

(A) $\sqrt{12x^3y^5z^2} = \sqrt{(4x^2y^4z^2)(3xy)}$ Condition 1 is not met.

$= \sqrt{(2xy^2z)^2(3xy)}$ $x^{pm}y^{pn} = (x^my^n)^p$

$= \sqrt{(2xy^2z)^2}\sqrt{3xy}$ $\sqrt[n]{xy} = \sqrt[n]{x}\sqrt[n]{y}$

$= 2xy^2z\sqrt{3xy}$ $\sqrt[n]{x^n} = x$

(B) $\sqrt[3]{6x^2y}\sqrt[3]{4x^5y^2} = \sqrt[3]{(6x^2y)(4x^5y^2)}$ $\sqrt[n]{x}\sqrt[n]{y} = \sqrt[n]{xy}$

$= \sqrt[3]{24x^7y^3}$

$= \sqrt[3]{(8x^6y^3)(3x)}$ Condition 1 is not met.

$= \sqrt[3]{(2x^2y)^3(3x)}$ $x^{pm}y^{pn} = (x^my^n)^p$

$= \sqrt[3]{(2x^2y)^3}\sqrt[3]{3x}$ $\sqrt[n]{xy} = \sqrt[n]{x}\sqrt[n]{y}$

$= 2x^2y\sqrt[3]{3x}$ $\sqrt[n]{x^n} = x$

(C) $\sqrt[6]{16x^4y^2} = [(4x^2y)^2]^{1/6}$ Condition 2 is not met.

$= (4x^2y)^{2/6}$ Note the convenience of using rational exponents.

$= (4x^2y)^{1/3}$

$= \sqrt[3]{4x^2y}$

(D) $\sqrt[3]{\sqrt{27}} = [(3^3)^{1/2}]^{1/3}$

$= (3^3)^{1/6} = 3^{3/6}$ $= 3^{1/2} = \sqrt{3}$

Matched Problem 3 Express radicals in simplified form.

(A) $\sqrt{18x^5y^2z^3}$ (B) $\sqrt[4]{27a^3b^3}\sqrt[4]{3a^5b^3}$ (C) $\sqrt[9]{8x^6y^3}$ (D) $\sqrt{\sqrt[3]{4}}$

• Sums and Differences

Algebraic expressions involving radicals often can be simplified by adding and subtracting terms that contain exactly the same radical expressions. We proceed in essentially the same way as we do when we combine like terms in polynomials. The distributive property of real numbers plays a central role in this process.

EXAMPLE 4 Combining Like Terms

Combine as many terms as possible:

(A) $5\sqrt{3} + 4\sqrt{3} = (5 + 4)\sqrt{3} = 9\sqrt{3}$

(B) $2\sqrt[3]{xy^2} - 7\sqrt[3]{xy^2} = (2 - 7)\sqrt[3]{xy^2} = -5\sqrt[3]{xy^2}$

(C) $3\sqrt{xy} - 2\sqrt[3]{xy} + 4\sqrt{xy} - 7\sqrt[3]{xy} = 3\sqrt{xy} + 4\sqrt{xy} - 2\sqrt[3]{xy} - 7\sqrt[3]{xy}$
$= 7\sqrt{xy} - 9\sqrt[3]{xy}$

Matched Problem 4 Combine as many terms as possible:

(A) $6\sqrt{2} + 2\sqrt{2}$ (B) $3\sqrt[5]{2x^2y^3} - 8\sqrt[5]{2x^2y^3}$

(C) $5\sqrt[3]{mn^2} - 3\sqrt{mn} - 2\sqrt[3]{mn^2} + 7\sqrt{mn}$

• Products

We will now consider several types of special products that involve radicals. The distributive property of real numbers plays a central role in our approach to these problems.

EXAMPLE 5 Multiplication with Radical Forms

Multiply and simplify:

(A) $\sqrt{2}(\sqrt{10} - 3) = \sqrt{2}\sqrt{10} - \sqrt{2} \cdot 3 = \sqrt{20} - 3\sqrt{2} = 2\sqrt{5} - 3\sqrt{2}$

(B) $(\sqrt{2} - 3)(\sqrt{2} + 5) = \sqrt{2}\sqrt{2} - 3\sqrt{2} + 5\sqrt{2} - 15$
$= 2 + 2\sqrt{2} - 15$
$= 2\sqrt{2} - 13$

(C) $(\sqrt{x} - 3)(\sqrt{x} + 5) = \sqrt{x}\sqrt{x} - 3\sqrt{x} + 5\sqrt{x} - 15$
$= x + 2\sqrt{x} - 15$

(D) $(\sqrt[3]{m} + \sqrt[3]{n^2})(\sqrt[3]{m^2} - \sqrt[3]{n}) = \sqrt[3]{m^3} + \sqrt[3]{m^2n^2} - \sqrt[3]{mn} - \sqrt[3]{n^3}$

$= m - \sqrt[3]{mn} + \sqrt[3]{m^2n^2} - n$

Matched Problem 5 Multiply and simplify:

(A) $\sqrt{3}(\sqrt{6} - 4)$ (B) $(\sqrt{3} - 2)(\sqrt{3} + 4)$

(C) $(\sqrt{y} - 2)(\sqrt{y} + 4)$ (D) $(\sqrt[3]{x^2} - \sqrt[3]{y^2})(\sqrt[3]{x} + \sqrt[3]{y})$

• Rationalizing Operations

We now turn to algebraic fractions involving radicals in the denominator. Eliminating a radical from a denominator is referred to as **rationalizing the denominator.** To rationalize the denominator, we multiply the numerator and denominator by a suitable factor that will rationalize the denominator—that is, will leave the denominator free of radicals. This factor is called a **rationalizing factor.** The following special products are of use in finding some rationalizing factors (see Examples 6C, D):

$$(a - b)(a + b) = a^2 - b^2 \tag{1}$$

$$(a - b)(a^2 + ab + b^2) = a^3 - b^3 \tag{2}$$

$$(a + b)(a^2 - ab + b^2) = a^3 + b^3 \tag{3}$$

EXPLORE-DISCUSS 2 Use special products (1) to (3) above to find a rationalizing factor for each of the following:

(A) $\sqrt{a} - \sqrt{b}$ (B) $\sqrt{a} + \sqrt{b}$ (C) $\sqrt[3]{a} - \sqrt[3]{b}$ (D) $\sqrt[3]{a} + \sqrt[3]{b}$

EXAMPLE 6 Rationalizing Denominators

Rationalize denominators.

(A) $\dfrac{3}{\sqrt{5}}$ (B) $\sqrt[3]{\dfrac{2a^2}{3b^2}}$ (C) $\dfrac{\sqrt{x} + \sqrt{y}}{3\sqrt{x} - 2\sqrt{y}}$ (D) $\dfrac{1}{\sqrt[3]{m} + 2}$

Solutions (A) $\sqrt{5}$ is a rationalizing factor for $\sqrt{5}$, since $\sqrt{5}\sqrt{5} = \sqrt{5^2} = 5$. Thus, we multiply the numerator and denominator by $\sqrt{5}$ to rationalize the denominator:

$$\frac{3}{\sqrt{5}} = \frac{3\sqrt{5}}{\sqrt{5}\sqrt{5}} = \frac{3\sqrt{5}}{5}$$

(B) $$\sqrt[3]{\frac{2a^2}{3b^2}} = \frac{\sqrt[3]{2a^2}}{\sqrt[3]{3b^2}} = \frac{\sqrt[3]{2a^2}\sqrt[3]{3^2b}}{\sqrt[3]{3b^2}\sqrt[3]{3^2b}} = \frac{\sqrt[3]{2 \cdot 3^2a^2b}}{\sqrt[3]{3^3b^3}} = \frac{\sqrt[3]{18a^2b}}{3b}$$

(C) Special product (1) suggests that if we multiply the denominator $3\sqrt{x} - 2\sqrt{y}$ by $3\sqrt{x} + 2\sqrt{y}$, we will obtain the difference of two squares and the denominator will be rationalized.

$$\frac{\sqrt{x}+\sqrt{y}}{3\sqrt{x}-2\sqrt{y}} = \frac{(\sqrt{x}+\sqrt{y})(3\sqrt{x}+2\sqrt{y})}{(3\sqrt{x}-2\sqrt{y})(3\sqrt{x}+2\sqrt{y})}$$

$$= \frac{3\sqrt{x^2}+2\sqrt{xy}+3\sqrt{xy}+2\sqrt{y^2}}{(3\sqrt{x})^2-(2\sqrt{y})^2}$$

$$= \frac{3x+5\sqrt{xy}+2y}{9x-4y}$$

(D) Special product (3) above suggests that if we multiply the denominator $\sqrt[3]{m}+2$ by $(\sqrt[3]{m})^2 - 2\sqrt[3]{m} + 2^2$, we will obtain the sum of two cubes and the denominator will be rationalized.

$$\frac{1}{\sqrt[3]{m}+2} = \frac{1[(\sqrt[3]{m})^2 - 2\sqrt[3]{m} + 2^2]}{(\sqrt[3]{m}+2)[(\sqrt[3]{m})^2 - 2\sqrt[3]{m} + 2^2]}$$

$$= \frac{\sqrt[3]{m^2} - 2\sqrt[3]{m} + 4}{(\sqrt[3]{m})^3 + 2^3}$$

$$= \frac{\sqrt[3]{m^2} - 2\sqrt[3]{m} + 4}{m+8}$$

Matched Problem 6 Rationalize denominators.

(A) $\dfrac{6}{\sqrt{2x}}$ (B) $\dfrac{10x^3}{\sqrt[3]{4x}}$ (C) $\dfrac{\sqrt{x}+2}{2\sqrt{x}+3}$ (D) $\dfrac{1}{1-\sqrt[3]{y}}$

Answers to Matched Problems

1. (A) $\sqrt[5]{u}$ (B) $\sqrt[9]{(6x^2y^5)^2}$ or $(\sqrt[9]{6x^2y^5})^2$ (C) $1/\sqrt[5]{(3xy)^3}$ (D) $(9u)^{1/4}$
(E) $-(2x)^{4/7}$ (F) $(x^3+y^3)^{1/3}$

2. (A) u^2+v^2 (B) $2\sqrt{3}$ (C) $(\sqrt[3]{x^2})/2$ or $\frac{1}{2}\sqrt[3]{x^2}$

3. (A) $3x^2yz\sqrt{2xz}$ (B) $3a^2b\sqrt[4]{b^2} = 3a^2b\sqrt{b}$ (C) $\sqrt[3]{2x^2y}$ (D) $\sqrt[3]{2}$

4. (A) $8\sqrt{2}$ (B) $-5\sqrt[5]{2x^2y^3}$ (C) $3\sqrt[3]{mn^2} + 4\sqrt{mn}$

5. (A) $3\sqrt{2}-4\sqrt{3}$ (B) $2\sqrt{3}-5$ (C) $y+2\sqrt{y}-8$ (D) $x+\sqrt[3]{x^2y}-\sqrt[3]{xy^2}-y$

6. (A) $\dfrac{3\sqrt{2x}}{x}$ (B) $5x^2\sqrt[3]{2x^2}$ (C) $\dfrac{2x+\sqrt{x}-6}{4x-9}$ (D) $\dfrac{1+\sqrt[3]{y}+\sqrt[3]{y^2}}{1-y}$

EXERCISE A-7

Unless stated to the contrary, all variables are restricted so that all quantities involved are real numbers.

A

In Problems 1–8, change to radical form. Do not simplify.

1. $m^{2/3}$ **2.** $n^{4/5}$ **3.** $6x^{3/5}$

4. $7y^{2/5}$ **5.** $(4xy^3)^{2/5}$ **6.** $(7x^2y)^{5/7}$

7. $(x + y)^{1/2}$ **8.** $x^{1/2} + y^{1/2}$

In Problems 9–16, change to rational exponent form. Do not simplify.

9. $\sqrt[5]{b}$ **10.** $\sqrt{c}$ **11.** $5\sqrt[4]{x^3}$

12. $7m\sqrt[5]{n^2}$ **13.** $\sqrt[5]{(2x^2y)^3}$ **14.** $\sqrt[9]{(3m^4n)^2}$

15. $\sqrt[3]{x} + \sqrt[3]{y}$ **16.** $\sqrt[3]{x + y}$

In Problems 17–32, write in simplified form.

17. $\sqrt[3]{-8}$ **18.** $\sqrt[3]{-27}$ **19.** $\sqrt{9x^8y^4}$

20. $\sqrt{16m^4y^8}$ **21.** $\sqrt[4]{16m^4n^8}$ **22.** $\sqrt[5]{32a^{15}b^{10}}$

23. $\sqrt{8a^3b^5}$ **24.** $\sqrt{27m^2n^7}$ **25.** $\sqrt[3]{2^4x^4y^7}$

26. $\sqrt[4]{2^4x^5y^8}$ **27.** $\sqrt[4]{m^2}$ **28.** $\sqrt[10]{n^6}$

29. $\sqrt[5]{\sqrt[3]{xy}}$ **30.** $\sqrt{\sqrt[4]{5x}}$ **31.** $\sqrt[3]{9x^2}\sqrt[3]{9x}$

32. $\sqrt{2x}\sqrt{8xy}$

In Problems 33–40, rationalize denominators, and write in simplified form.

33. $\dfrac{1}{\sqrt{5}}$ **34.** $\dfrac{1}{\sqrt{7}}$ **35.** $\dfrac{6x}{\sqrt{3x}}$

36. $\dfrac{12y^2}{\sqrt{6y}}$ **37.** $\dfrac{2}{\sqrt{2} - 1}$ **38.** $\dfrac{4}{\sqrt{6} - 2}$

39. $\dfrac{\sqrt{2}}{\sqrt{6} + 2}$ **40.** $\dfrac{\sqrt{2}}{\sqrt{10} - 2}$

B

In Problems 41–52, write in simplified form.

41. $x\sqrt[5]{3^6x^7y^{11}}$ **42.** $2a\sqrt[3]{8a^8b^{13}}$ **43.** $\dfrac{\sqrt[4]{32m^7n^9}}{2mn}$

44. $\dfrac{\sqrt[5]{32u^{12}v^8}}{uv}$ **45.** $\sqrt[6]{a^4(b - a)^2}$ **46.** $\sqrt[8]{3^6(u + v)^6}$

47. $\sqrt[3]{\sqrt[4]{a^9b^3}}$ **48.** $\sqrt{\sqrt[6]{x^8y^6}}$

49. $\sqrt[3]{2x^2y^4}\sqrt[3]{3x^5y}$ **50.** $\sqrt[4]{4m^5n}\sqrt[4]{6m^3n^4}$

51. $\sqrt[3]{a^3 + b^3}$ **52.** $\sqrt{x^2 + y^2}$

In Problems 53–64, rationalize denominators and write in simplified form.

53. $\dfrac{\sqrt{2m}\sqrt{5}}{\sqrt{20m}}$

54. $\dfrac{\sqrt{6}\sqrt{8c}}{\sqrt{18c}}$

55. $\dfrac{4a^3b^2}{\sqrt[3]{2ab^2}}$

56. $\dfrac{8x^3y^5}{\sqrt[3]{4x^2y}}$

57. $\sqrt[4]{\dfrac{3y^3}{4x^3}}$

58. $\sqrt[5]{\dfrac{4x^2}{16y^3}}$

59. $\dfrac{3\sqrt{y}}{2\sqrt{y} - 3}$

60. $\dfrac{5\sqrt{x}}{3 - 2\sqrt{x}}$

61. $\dfrac{2\sqrt{5} + 3\sqrt{2}}{5\sqrt{5} + 2\sqrt{2}}$

62. $\dfrac{3\sqrt{2} - 2\sqrt{3}}{3\sqrt{3} - 2\sqrt{2}}$

63. $\dfrac{x^2}{\sqrt{x^2 + 9} - 3}$

64. $\dfrac{-y^2}{2 - \sqrt{y^2 + 4}}$

Problems 65–68 are calculus-related. Rationalize the numerators; that is, perform operations on the fractions that eliminate radicals from the numerators. (This is a particularly useful operation in some problems in calculus.)

65. $\dfrac{\sqrt{t} - \sqrt{x}}{t - x}$

66. $\dfrac{\sqrt{x} - \sqrt{y}}{\sqrt{x} + \sqrt{y}}$

67. $\dfrac{\sqrt{x + h} - \sqrt{x}}{h}$

68. $\dfrac{\sqrt{2 + h} + \sqrt{2}}{h}$

In Problems 69–80, evaluate to four significant digits using a calculator. (Read the instruction booklet accompanying your calculator for the process required to evaluate $\sqrt[n]{x}$.)

69. $\sqrt{0.032\ 965}$

70. $\sqrt{419.763}$

71. $\sqrt[3]{45.0218}$

72. $\sqrt[4]{0.098\ 553}$

73. $\sqrt[8]{5.477 \times 10^{-9}}$

74. $\sqrt[7]{4.892 \times 10^{16}}$

75. $\sqrt[5]{9} + \sqrt[5]{9}$

76. $\sqrt[3]{2} + \sqrt[3]{2}$

77. $\sqrt[4]{\sqrt[5]{100}}$ and $\sqrt[20]{100}$

78. $\sqrt[3]{\sqrt[7]{500}}$ and $\sqrt[21]{500}$

79. $\dfrac{1}{\sqrt[3]{9}}$ and $\dfrac{\sqrt[3]{3}}{3}$

80. $\dfrac{1}{\sqrt[3]{2}}$ and $\dfrac{\sqrt[3]{4}}{2}$

C

For what real numbers are Problems 81–84 true?

81. $\sqrt{x^2} = -x$

82. $\sqrt{x^2} = x$

83. $\sqrt[3]{x^3} = x$

84. $\sqrt[3]{x^3} = -x$

In Problems 85 and 86, evaluate each expression on a calculator and determine which pairs have the same value. Verify these results algebraically.

85. (A) $\sqrt{3} + \sqrt{5}$ (B) $\sqrt{2 + \sqrt{3}} + \sqrt{2 - \sqrt{3}}$
(C) $1 + \sqrt{3}$ (D) $\sqrt[3]{10 + 6\sqrt{3}}$
(E) $\sqrt{8 + \sqrt{60}}$ (F) $\sqrt{6}$

86. (A) $2\sqrt[3]{2 + \sqrt{5}}$ (B) $\sqrt{8}$
(C) $\sqrt{3} + \sqrt{7}$ (D) $\sqrt{3 + \sqrt{8}} + \sqrt{3 - \sqrt{8}}$
(E) $\sqrt{10 + \sqrt{84}}$ (F) $1 + \sqrt{5}$

In Problems 87–90, rationalize denominators.

87. $\dfrac{1}{\sqrt[3]{a} - \sqrt[3]{b}}$

88. $\dfrac{1}{\sqrt[3]{m} + \sqrt[3]{n}}$

89. $\dfrac{1}{\sqrt{x} - \sqrt{y} + \sqrt{z}}$

90. $\dfrac{1}{\sqrt{x} + \sqrt{y} - \sqrt{z}}$

[*Hint for Problem 89:* Start by multiplying numerator and denominator by $(\sqrt{x} - \sqrt{y}) - \sqrt{z}$.]

Problems 91 and 92 are calculus-related. Rationalize numerators.

91. $\dfrac{\sqrt[3]{x + h} - \sqrt[3]{x}}{h}$

92. $\dfrac{\sqrt[3]{t} - \sqrt[3]{x}}{t - x}$

93. Show that $\sqrt[kn]{x^{km}} = \sqrt[n]{x^m}$ for k, m, and n natural numbers greater than 1.

94. Show that $\sqrt[m]{\sqrt[n]{x}} = \sqrt[mn]{x}$ for m and n natural numbers greater than 1.

APPLICATIONS 

95. Physics—Relativistic Mass. The mass M of an object moving at a velocity v is given by

$$M = \frac{M_0}{\sqrt{1 - \dfrac{v^2}{c^2}}}$$

where M_0 = mass at rest and c = velocity of light. The mass of an object increases with velocity and tends to infinity as the velocity approaches the speed of light. Show that M can be written in the form

$$M = \frac{M_0 c\sqrt{c^2 - v^2}}{c^2 - v^2}$$

96. Physics—Pendulum. A simple pendulum is formed by hanging a bob of mass M on a string of length L from a fixed support (see the figure). The time it takes the bob to swing from right to left and back again is called the **period** T and is given by

$$T = 2\pi\sqrt{\frac{L}{g}}$$

where g is the gravitational constant. Show that T can be written in the form

$$T = \frac{2\pi\sqrt{gL}}{g}$$

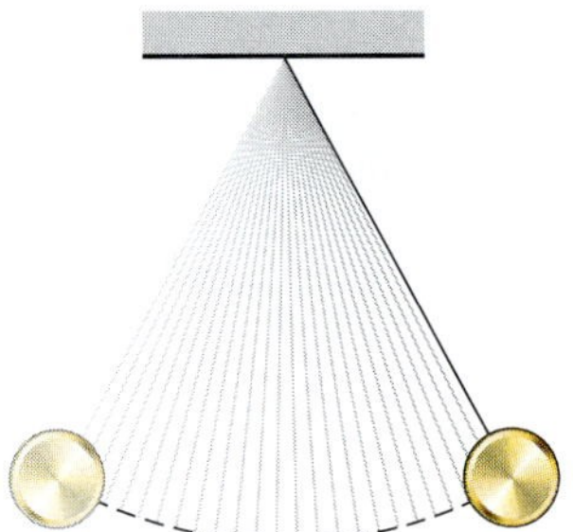

APPENDIX A GROUP ACTIVITY Rational Number Representations

The set of real numbers can be partitioned into two disjoint subsets, the set of rational numbers and the set of irrational numbers. Rational numbers can be represented two ways: as a/b, where a and b are integers and $b \neq 0$, and as terminating or repeating decimal expansions. The irrational numbers can be represented as nonrepeating and nonterminating decimal expansions. In this activity, we want to explore the relationship between the two different methods for representing rational numbers.

Consider the rational number $r = a/b$, where a and b are integers with no common factors and $b \neq 0$.

1. If $b = 10^n$ for a positive integer n, what kind of decimal expansion will r have?
2. If $b = 2^m5^n$ for positive integers m and n, what kind of decimal expansion will r have?
3. If r has a terminating decimal expansion, show that r can be expressed in the form a/b, where $b = 2^m5^n$.
4. Find a and b for the following repeating expansions (the overbar indicates the repeating block):
   (A) $0.\overline{63}$ (B) $0.4\overline{86}$ (C) $0.\overline{846153}$
5. Find a and b for $r = 0.1\overline{9}$ and then find a terminating decimal expansion for r.

Appendix A Review

A-1 ALGEBRA AND REAL NUMBERS

A **set** is a collection of objects called **elements** or **members** of the set. Sets are usually described by **listing** the elements or by stating a **rule** that determines the elements. A set may be **finite** or **infinite.** A set with no elements is called the **empty set** or the **null set** and is denoted $\varnothing$. A **variable** is a symbol that represents unspecified elements from a **replacement set.** A **constant** is a symbol for a single object. If each element of set A is also in set B, we say A is a **subset** of B and write $A \subset B$.

Real numbers:

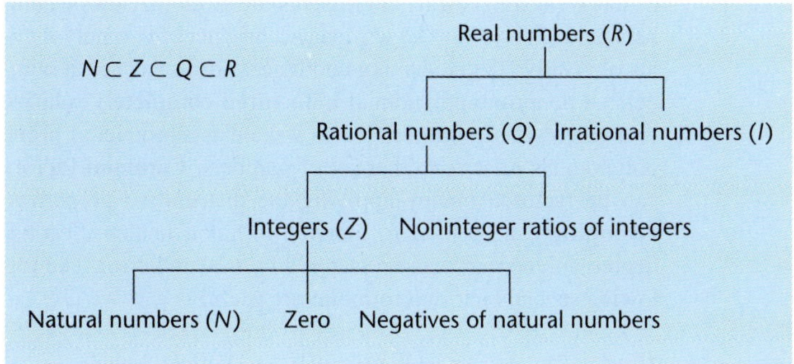

Real number line:

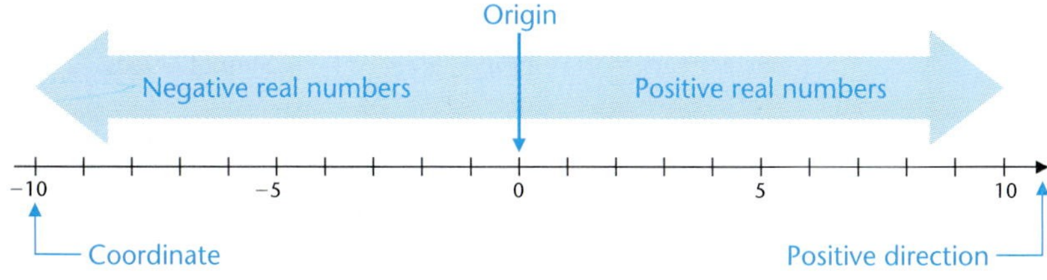

Basic real number properties include **associative properties:** $x + (y + z) = (x + y) + z$ and $x(yz) = (xy)z$; **commutative properties:** $x + y = y + x$ and $xy = yx$; **identities:** $0 + x = x + 0 = x$ and $(1)x = x(1) = x$; **inverses:** $-x$ is the additive inverse or **negative** of x and, if $x \neq 0$, $1/x$ is the multiplicative inverse or **reciprocal** of x; and **distributive property:** $x(y + z) = xy + xz$. **Subtraction** is defined by $a - b = a + (-b)$ and **division** by $a/b = a(1/b)$. Division by 0 is never allowed. Additional properties include **properties of negatives:**

1. $-(-a) = a$
2. $(-a)b = -(ab) = a(-b) = -ab$
3. $(-a)(-b) = ab$
4. $(-1)a = -a$
5. $\dfrac{-a}{b} = -\dfrac{a}{b} = \dfrac{a}{-b} \qquad b \neq 0$
6. $\dfrac{-a}{-b} = -\dfrac{-a}{b} = -\dfrac{a}{-b} = \dfrac{a}{b} \qquad b \neq 0$

zero properties:

1. $a \cdot 0 = 0$
2. $ab = 0$ if and only if $a = 0$ or $b = 0$ or both.

and **fraction properties** (division by 0 excluded):

1. $\dfrac{a}{b} = \dfrac{c}{d}$ if and only if $ad = bc$
2. $\dfrac{ka}{kb} = \dfrac{a}{b}$
3. $\dfrac{a}{b} \cdot \dfrac{c}{d} = \dfrac{ac}{bd}$
4. $\dfrac{a}{b} \div \dfrac{c}{d} = \dfrac{a}{b} \cdot \dfrac{d}{c}$
5. $\dfrac{a}{b} + \dfrac{c}{b} = \dfrac{a + c}{b}$
6. $\dfrac{a}{b} - \dfrac{c}{b} = \dfrac{a - c}{b}$
7. $\dfrac{a}{b} + \dfrac{c}{d} = \dfrac{ad + bc}{bd}$

A-2 POLYNOMIALS: BASIC OPERATIONS

For n and m natural numbers and a any real number:

$$a^n = a \cdot a \cdot \cdots \cdot a \text{ (n factors of a)} \quad \text{and} \quad a^m a^n = a^{m+n}$$

An **algebraic expression** is formed by using constants and variables and the operations of addition, subtraction, multiplication, division, raising to powers, and taking roots. A **polynomial** is an algebraic expression formed by adding and subtracting constants and terms of the form ax^n (one variable), $ax^n y^m$ (two variables), and so on. The **degree of a term** is the sum of the powers of all variables in the term, and the **degree of a polynomial** is the degree of the nonzero term with highest degree in the polynomial. Polynomials with one, two, or three terms are called **monomials, binomials,** and **trinomials,** respectively. **Like terms** have exactly the same variable factors to the same powers and can be combined by adding their **coefficients.** Polynomials can be **added, subtracted,** and **multiplied** by repeatedly applying the distributive property and combining like terms. The **FOIL method** is used to multiply two binomials. **Special products** obtained using FOIL method are:

1. $(a - b)(a + b) = a^2 - b^2$
2. $(a - b)^2 = a^2 - 2ab + b^2$
3. $(a + b)^2 = a^2 + 2ab + b^2$

A-3 POLYNOMIALS: FACTORING

A number or algebraic expression is **factored** if it is expressed as a product of other numbers or algebraic expressions, which are called **factors.** An integer greater than 1 is a **prime number** if its only positive integer factors are itself and 1, and a **composite number** otherwise. Each composite number can be **factored uniquely into a product of prime numbers.** A polynomial is **prime** relative to a given set of numbers (usually the set of integers) if (1) all its coefficients are from that set of numbers, and (2) it cannot be written as a product of two polynomials of positive degree having coefficients from that set of numbers. A nonprime polynomial is **factored completely relative to a given set of numbers** if it is written as a product of prime polynomials relative to that set of numbers. **Common factors** can be factored out by applying the distributive properties. **Grouping** can be used to identify common factors. Second-degree polynomials can be factored by trial and error. The following special factoring formulas are useful:

1. $u^2 + 2uv + v^2 = (u + v)^2$ Perfect Square
2. $u^2 - 2uv + v^2 = (u - v)^2$ Perfect Square
3. $u^2 - v^2 = (u - v)(u + v)$ Difference of squares
4. $u^3 - v^3 = (u - v)(u^2 + uv + v^2)$ Difference of Cubes
5. $u^3 + v^3 = (u + v)(u^2 - uv + v^2)$ Sum of Cubes

There is no factoring formula relative to the real numbers for $u^2 + v^2$.

A-4 RATIONAL EXPRESSIONS: BASIC OPERATIONS

A **fractional expression** is the ratio of two algebraic expressions, and a **rational expression** is the ratio of two polynomials. The rules for adding, subtracting, multiplying, and dividing real number fractions (see Section 1-1 in this review) all extend to fractional expressions with the understanding that **variables are always restricted to exclude division by zero.** Fractions can be **reduced to lowest terms** or **raised to higher terms** by using the fundamental property of fractions:

$$\frac{ka}{kb} = \frac{a}{b} \qquad \text{with } b, k \neq 0$$

A rational expression is **reduced to lowest terms** if the numerator and denominator do not have any factors in common relative to the integers. The **least common denominator** (LCD) is useful for adding and subtracting fractions with different denominators and for reducing **compound fractions** to **simple fractions.**

A-5 INTEGER EXPONENTS

$a^n = a \cdot a \cdot \cdots \cdot a$ (n factors of a) for n a positive integer, $a^0 = 1$ ($a \neq 0$), and $a^n = 1/a^{-n}$ for n a negative integer ($a \neq 0$). 0^0 is not defined.

Properties of integer exponents (division by 0 excluded):

1. $a^m a^n = a^{m+n}$
2. $(a^n)^m = a^{mn}$
3. $(ab)^m = a^m b^m$
4. $\left(\dfrac{a}{b}\right)^m = \dfrac{a^m}{b^m}$
5. $\dfrac{a^m}{a^n} = a^{m-n} = \dfrac{1}{a^{n-m}}$

Further exponent properties (division by 0 excluded):

1. $(a^m b^n)^p = a^{pm} b^{pn}$
2. $\left(\dfrac{a^m}{b^n}\right)^p = \dfrac{a^{pm}}{b^{pn}}$
3. $\dfrac{a^{-n}}{b^{-m}} = \dfrac{b^m}{a^n}$
4. $\left(\dfrac{a}{b}\right)^{-n} = \left(\dfrac{b}{a}\right)^n$

Scientific notation:

$$a \times 10^n \qquad 1 \le a < 10$$

n an integer, a in decimal form.

A-6 RATIONAL EXPONENTS

For n a natural number and a and b real numbers:

$$a \text{ is an } n\text{th root of } b \text{ if } a^n = b$$

The **principal *n*th root** of b is denoted by $b^{1/n}$. If n is odd, b has one real nth root which is the principal nth root. If n is even and $b > 0$, b has two real nth roots and the positive nth root is the principal nth root. If n is even and $b < 0$, b has no real nth roots.

Rational number exponents (even roots of negative numbers excluded):

$$b^{m/n} = (b^{1/n})^m = (b^m)^{1/n} \qquad \text{and} \qquad b^{-m/n} = \frac{1}{b^{m/n}}$$

A-7 RADICALS

An ***n*th root radical** is defined by $\sqrt[n]{b} = b^{1/n}$, where $b^{1/n}$ is the principal nth root of b, $\sqrt{\ }$ is a **radical,** n is the **index,** and b is the **radicand.** Rational exponents and radicals are related by

$$b^{m/n} = (b^m)^{1/n} = \sqrt[n]{b^m} = (b^{1/n})^m = (\sqrt[n]{b})^m$$

Properties of radicals ($x > 0, y > 0$):

1. $\sqrt[n]{x^n} = x$
2. $\sqrt[n]{xy} = \sqrt[n]{x}\sqrt[n]{y}$
3. $\sqrt[n]{\dfrac{x}{y}} = \dfrac{\sqrt[n]{x}}{\sqrt[n]{y}}$

A radical is in **simplified form** if:

1. No radicand contains a factor to a power greater than or equal to the index of the radical.
2. No power of the radicand and the index of the radical have a common factor other than 1.
3. No radical appears in a denominator.
4. No fraction appears within a radical.

Algebraic fractions containing radicals are **rationalized** by multiplying numerator and denominator by a **rationalizing factor** often determined by using a special product formula.

Appendix A Review Exercise

Work through all the problems in this chapter review and check answers in the back of the book. Answers to all review problems are there, and following each answer is a number in italics indicating the section in which that type of problem is discussed. Where weaknesses show up, review appropriate sections in the text.

A

1. For $A = \{1, 2, 3, 4, 5\}$, $B = \{1, 2, 4\}$, and $C = \{4, 1, 2\}$, indicate true (T) or false (F):
(A) $3 \in A$ (B) $5 \notin C$ (C) $B \in A$
(D) $B \subset A$ (E) $B \neq C$ (F) $A \subset B$

2. Replace each question mark with an appropriate expression that will illustrate the use of the indicated real number property:
(A) Commutative $(\cdot)$: $x(y + z) = ?$
(B) Associative $(+)$: $2 + (x + y) = ?$
(C) Distributive: $(2 + 3)x = ?$

Problems 3–7 refer to the following polynomials:
(a) $3x - 4$ *(b)* $x + 2$ *(c)* $3x^2 + x - 8$ *(d)* $x^3 + 8$

3. Add all four.

4. Subtract the sum of (a) and (c) from the sum of (b) and (d).

5. Multiply (c) and (d).

6. What is the degree of (d)?

7. What is the coefficient of the second term in (c)?

In Problems 8–11, perform the indicated operations and simplify.

8. $5x^2 - 3x[4 - 3(x - 2)]$

9. $(3m - 5n)(3m + 5n)$

10. $(2x + y)(3x - 4y)$

11. $(2a - 3b)^2$

In Problems 12–14, write each polynomial in a completely factored form relative to the integers. If the polynomial is prime relative to the integers, say so.

12. $9x^2 - 12x + 4$

13. $t^2 - 4t - 6$

14. $6n^3 - 9n^2 - 15n$

In Problems 15–18, perform the indicated operations and reduce to lowest terms. Represent all compound fractions as simple fractions reduced to lowest terms.

15. $\dfrac{2}{5b} - \dfrac{4}{3a^3} - \dfrac{1}{6a^2b^2}$

16. $\dfrac{3x}{3x^2 - 12x} + \dfrac{1}{6x}$

17. $\dfrac{y - 2}{y^2 - 4y + 4} \div \dfrac{y^2 + 2y}{y^2 + 4y + 4}$

18. $\dfrac{u - \dfrac{1}{u}}{1 - \dfrac{1}{u^2}}$

Simplify Problems 19–24, and write answers using positive exponents only. All variables represent positive real numbers.

19. $6(xy^3)^5$

20. $\dfrac{9u^8v^6}{3u^4v^8}$

21. $(2 \times 10^5)(3 \times 10^{-3})$

22. $(x^{-3}y^2)^{-2}$

23. $u^{5/3}u^{2/3}$

24. $(9a^4b^{-2})^{1/2}$

25. Change to radical form: $3x^{2/5}$

26. Change to rational exponent form: $-3\sqrt[3]{(xy)^2}$

Simplify Problems 27–31, and express answers in simplified form. All variables represent positive real numbers.

27. $3x\sqrt[3]{x^5y^4}$

28. $\sqrt{2x^2y^5}\sqrt{18x^3y^2}$

29. $\dfrac{6ab}{\sqrt{3a}}$

30. $\dfrac{\sqrt{5}}{3 - \sqrt{5}}$

31. $\sqrt[8]{y^6}$

B

32. Write using the listing method:

$\{x \mid x \text{ is an odd integer between } -4 \text{ and } 2\}$

In Problems 33–38, each statement illustrates the use of one of the following real number properties or definitions. Indicate which one.

Commutative $(+, \cdot)$	*Identity* $(+, \cdot)$
Division	*Associative* $(+, \cdot)$
Inverse $(+, \cdot)$	*Zero*
Distributive	*Subtraction*
Negatives	

33. $(-3) - (-2) = (-3) + [-(-2)]$

34. $3y + (2x + 5) = (2x + 5) + 3y$

35. $(2x + 3)(3x + 5) = (2x + 3)3x + (2x + 3)5$

36. $3 \cdot (5x) = (3 \cdot 5)x$

37. $\dfrac{a}{-(b - c)} = -\dfrac{a}{b - c}$

38. $3xy + 0 = 3xy$

39. Indicate true (T) or false (F):
(A) An integer is a rational number and a real number.
(B) An irrational number has a repeating decimal representation.

40. Give an example of an integer that is not a natural number.

41. Given the algebraic expressions:
(a) $2x^2 - 3x + 5$ (b) $x^2 - \sqrt{x - 3}$
(c) $x^{-3} + x^{-2} - 3x^{-1}$ (d) $x^2 - 3xy - y^2$
(A) Identify all second-degree polynomials.
(B) Identify all third-degree polynomials.

In Problems 42–46, perform the indicated operations and simplify.

42. $(2x - y)(2x + y) - (2x - y)^2$

43. $(m^2 + 2mn - n^2)(m^2 - 2mn - n^2)$

44. $5(x + h)^2 - 7(x + h) - (5x^2 - 7x)$

45. $-2x\{(x^2 + 2)(x - 3) - x[x - x(3 - x)]\}$

46. $(x - 2y)^3$

In Problems 47–53, write in a completely factored form relative to the integers.

47. $(4x - y)^2 - 9x^2$

48. $2x^2 + 4xy - 5y^2$

49. $6x^3y + 12x^2y^2 - 15xy^3$

50. $(y - b)^2 - y + b$

51. $3x^3 + 24y^3$

52. $y^3 + 2y^2 - 4y - 8$

53. $2x(x - 4)^3 + 3x^2(x - 4)^2$

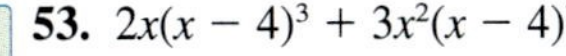

In Problems 54–58, perform the indicated operations and reduce to lowest terms. Represent all compound fractions as simple fractions reduced to lowest terms.

54. $\dfrac{3x^2(x + 2)^2 - 2x(x + 2)^3}{x^4}$

55. $\dfrac{m - 1}{m^2 - 4m + 4} + \dfrac{m + 3}{m^2 - 4} + \dfrac{2}{2 - m}$

56. $\dfrac{y}{x^2} \div \left(\dfrac{x^2 + 3x}{2x^2 + 5x - 3} \div \dfrac{x^3y - x^2y}{2x^2 - 3x + 1}\right)$

57. $\dfrac{1 - \dfrac{1}{1 + \dfrac{x}{y}}}{1 - \dfrac{1}{1 - \dfrac{x}{y}}}$

58. $\dfrac{a^{-1} - b^{-1}}{ab^{-2} - ba^{-2}}$

59. Check the following solution. If it is wrong, explain what is wrong and how it can be corrected, and then show a correct solution.

$$\frac{x^2 + 2x}{x^2 + x - 2} + x + 2 = \frac{x^2 + 3x + 2}{x^2 + x - 2} = \frac{x + 1}{x - 1}$$

In Problems 60–65, perform the indicated operations, simplify and write answers using positive exponents only. All variables represent positive real numbers.

60. $\left(\dfrac{8u^{-1}}{2^2u^2v^0}\right)^{-2}\left(\dfrac{u^{-5}}{u^{-3}}\right)^3$

61. $\dfrac{5^0}{3^2} + \dfrac{3^{-2}}{2^{-2}}$

62. $\left(\dfrac{27x^2y^{-3}}{8x^{-4}y^3}\right)^{1/3}$

63. $(a^{-1/3}b^{1/4})(9a^{1/3}b^{-1/2})^{3/2}$

64. $(x^{1/2} + y^{1/2})^2$

65. $(3x^{1/2} - y^{1/2})(2x^{1/2} + 3y^{1/2})$

66. Convert to scientific notation and simplify:

$$\frac{0.000\ 000\ 000\ 52}{(1{,}300)(0.000\ 002)}$$

Evaluate Problems 67–74 to four significant digits using a calculator.

67. $\dfrac{(20{,}410)(0.000\ 003\ 477)}{0.000\ 000\ 022\ 09}$

68. 0.1347^5

69. $(-60.39)^{-3}$

70. $82.45^{8/3}$

71. $(0.000\ 000\ 419\ 9)^{2/7}$

72. $\sqrt[5]{0.006\ 604}$

73. $\sqrt[3]{3 + \sqrt{2}}$

74. $\dfrac{2^{-1/2} - 3^{-1/2}}{2^{-1/3} + 3^{-1/3}}$

In Problems 75–83, perform the indicated operations and express answers in simplified form. All radicands represent positive real numbers.

75. $-2x\sqrt[5]{3^6x^7y^{11}}$

76. $\dfrac{2x^2}{\sqrt[3]{4x}}$

77. $\sqrt[5]{\dfrac{3y^2}{8x^2}}$

78. $\sqrt[9]{8x^6y^{12}}$

79. $\sqrt{\sqrt[3]{4x^4}}$

80. $(2\sqrt{x} - 5\sqrt{y})(\sqrt{x} + \sqrt{y})$

81. $\dfrac{3\sqrt{x}}{2\sqrt{x} - \sqrt{y}}$

82. $\dfrac{2\sqrt{u} - 3\sqrt{v}}{2\sqrt{u} + 3\sqrt{v}}$

83. $\dfrac{y^2}{\sqrt{y^2 + 4} - 2}$

84. Rationalize the numerator: $\dfrac{\sqrt{t} - \sqrt{5}}{t - 5}$

85. Write in the form $ax^p + bx^q$, where a and b are real numbers and p and q are rational numbers:

$$\frac{4\sqrt{x} - 3}{2\sqrt{x}}$$

C

86. Write the repeating decimal 0.545454 . . . in the form a/b reduced to lowest terms, where a and b are positive integers. Is the number rational or irrational?

87. If $M = \{-4, -3, 2\}$ and $N = \{-3, 0, 2\}$, find:
(A) $\{x \mid x \in M \quad \text{or} \quad x \in N\}$
(B) $\{x \mid x \in M \quad \text{and} \quad x \in N\}$

88. Evaluate $x^2 - 4x + 1$ for $x = 2 - \sqrt{3}$.

89. Simplify: $x(2x - 1)(x + 3) - (x - 1)^3$

90. Factor completely with respect to the integers:

$$4x(a^2 - 4a + 4) - 9x^3$$

91. Evaluate each expression on a calculator and determine which pairs have the same value. Verify these results algebraically.
(A) $\sqrt{3 + \sqrt{5}} + \sqrt{3 - \sqrt{5}}$
(B) $\sqrt{4 + \sqrt{15}} + \sqrt{4 - \sqrt{15}}$
(C) $\sqrt{10}$

In Problems 92–95, simplify and express answers using positive exponents only (m is an integer greater than 1).

92. $\dfrac{8(x-2)^{-3}(x+3)^2}{12(x-2)^{-4}(x+3)^{-2}}$

93. $\left(\dfrac{a^{-2}}{b^{-1}}+\dfrac{b^{-2}}{a^{-1}}\right)^{-1}$

94. $(x^{1/3}-y^{1/3})(x^{2/3}+x^{1/3}y^{1/3}+y^{2/3})$

95. $\left(\dfrac{x^{m^2}}{x^{2m-1}}\right)^{1/(m-1)} \quad m>1$

96. Rationalize the denominator: $\dfrac{1}{1-\sqrt[3]{x}}$

97. Rationalize the numerator: $\dfrac{\sqrt[3]{t}-\sqrt[3]{5}}{t-5}$

98. Write in simplified form: $\sqrt[n+1]{x^{n^2}x^{2n+1}} \quad n>0$

APPLICATIONS 

99. Construction. A circular fountain in a park includes a concrete wall that is 3 feet high and 2 feet thick (see the figure). If the inner radius of the wall is x feet, write an algebraic expression in terms of x that represents the volume of the concrete used to construct the wall. Simplify the expression.

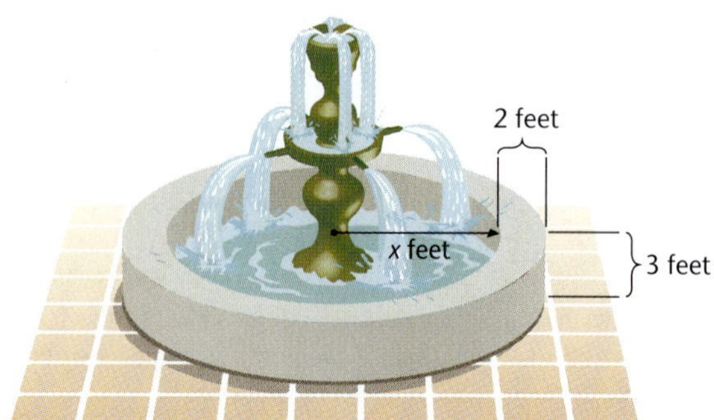

100. Economics. If in the United States in 1999 the total personal income was about $7,770,000,000,000 and the population was about 274,000,000, estimate to three significant digits the average personal income. Write your answer in scientific notation and in standard decimal form.

101. Economics. The number of units N produced by a petroleum company from the use of x units of capital and y units of labor is approximated by

$$N = 20x^{1/2}y^{1/2}$$

(A) Estimate the number of units produced by using 1,600 units of capital and 900 units of labor.

(B) What is the effect on production if the number of units of capital and labor are doubled to 3,200 units and 1,800 units, respectively?

(C) What is the effect on production of doubling the units of labor and capital at any production level?

102. Electric Circuit. If three electric resistors with resistances R_1, R_2, and R_3 are connected in parallel, then the total resistance R for the circuit shown in the figure is given by

$$R = \frac{1}{\dfrac{1}{R_1}+\dfrac{1}{R_2}+\dfrac{1}{R_3}}$$

Represent this compound fraction as a simple fraction.

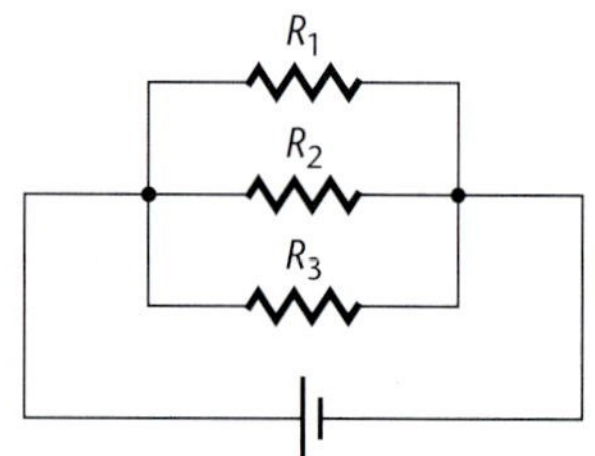

★ **103. Construction.** A box with a hinged lid is to be made out of a piece of cardboard that measures 16 by 30 inches. Six squares, x inches on a side, will be cut from each corner and the middle, and then the ends and sides will be folded up to form the box and its lid (see the figure). Express each of the following quantities as a polynomial in both factored and expanded form.

(A) The area of cardboard after the corners have been removed.

(B) The volume of the box.

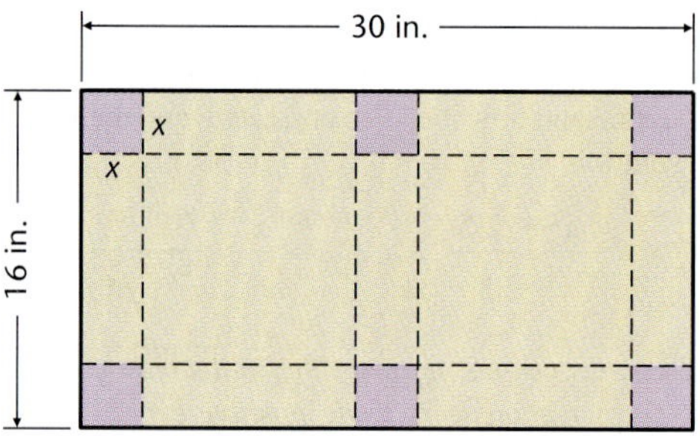

APPENDIX B
Significant Digits

Most calculations involving problems of the real world deal with figures that are only approximate. It therefore seems reasonable to assume that a final answer should not be any more accurate than the least accurate figure used in the calculation. This is an important point, since calculators tend to give the impression that greater accuracy is achieved than is warranted.

Suppose we wish to compute the length of the diagonal of a rectangular field from measurements of its sides of 237.8 meters and 61.3 meters. Using the Pythagorean theorem and a calculator, we find

$$d = \sqrt{237.8^2 + 61.3^2}$$
$$= 245.573\ 878 \cdots$$

d

61.3 meters

237.8 meters

The calculator answer suggests an accuracy that is not justified. What accuracy is justified? To answer this question, we introduce the idea of *significant digits*.

Whenever we write a measurement such as 61.3 meters, we assume that the measurement is accurate to the last digit written. Thus, the measurement 61.3 meters indicates that the measurement was made to the nearest tenth of a meter. That is, the actual width is between 61.25 meters and 61.35 meters. In general, the digits in a number that indicate the accuracy of the number are called **significant digits.** If all the digits in a number are nonzero, then they are all significant. Thus, the measurement 61.3 meters has 3 significant digits, and the measurement 237.8 meters has 4 significant digits.

What are the significant digits in the number 7,800? The accuracy of this number is not clear. It could represent a measurement with any of the following accuracies:

Between 7,750 and 7,850	Correct to the hundreds place
Between 7,795 and 7,805	Correct to the tens place
Between 7,799.5 and 7,800.5	Correct to the units place

In order to give a precise definition of significant digits that resolves this ambiguity, we use scientific notation.

DEFINITION 1

Significant Digits

If a number x is written in scientific notation as

$$x = a \times 10^n \qquad 1 \le a < 10, \ n \text{ an integer}$$

then the number of significant digits in x is the number of digits in a.

Thus,

7.8×10^3	has 2 significant digits
7.80×10^3	has 3 significant digits
7.800×10^3	has 4 significant digits

All three of these measurements have the same decimal representation (7,800), but each represents a different accuracy.

Definition 1 tells us how to write a number so that the number of significant digits is clear, but it does not tell us how to interpret the accuracy of a number that is not written in scientific notation. We will use the following convention for numbers that are written as decimal fractions:

Significant Digits in Decimal Fractions

The number of significant digits in a number with no decimal point is found by counting the digits from left to right, starting with the first digit and ending with the last *nonzero* digit.

The number of significant digits in a number containing a decimal point is found by counting the digits from left to right, starting with the first *nonzero* digit and ending with the last digit.

Applying this rule to the number 7,800, we conclude that this number has 2 significant digits. If we want to indicate that it has 3 or 4 significant digits, we must use scientific notation. The significant digits in the following numbers are underlined:

$$\underline{70{,}007} \qquad \underline{82}{,}000 \qquad \underline{5.600} \qquad 0.000\underline{8} \qquad 0.000\ \underline{830}$$

In calculations involving multiplication, division, powers, and roots, we adopt the following convention:

Rounding Calculated Values

The result of a calculation is rounded to the same number of significant digits as the number used in the calculation that has the least number of significant digits.

Thus, in computing the length of the diagonal of the rectangular field shown earlier, we write the answer rounded to 3 significant digits because the width has 3 significant digits and the length has 4 significant digits:

$$d = 246 \text{ meters}$$ 3 significant digits

One Final Note: In rounding a number that is exactly halfway between a larger and a smaller number, we use the convention of making the final result even.

EXAMPLE 1 **Rounding Numbers**

Round each number to 3 significant digits:

(A) 43.0690 (B) 48.05 (C) 48.15 (D) $8.017\ 632 \times 10^{-3}$

Solutions

(A) 43.1
(B) 48.0
(C) 48.2
Use the convention of making the digit before the 5 even if it is odd, or leaving it alone if it is even. [applies to (B) and (C)]
(D) 8.02×10^{-3}

Matched Problem 1 Round each number to 3 significant digits:

(A) 3.1495 (B) 0.004 135 (C) 32,450 (D) $4.314\ 764\ 09 \times 10^{12}$

Answers to Matched Problem

1. (A) 3.15 (B) 0.004 14 (C) 32,400 (D) 4.31×10^{12}

APPENDIX C
Geometric Formulas

• Similar Triangles

(A) Two triangles are similar if two angles of one triangle have the same measure as two angles of the other.

(B) If two triangles are similar, their corresponding sides are proportional:

$$\frac{a}{a'} = \frac{b}{b'} = \frac{c}{c'}$$

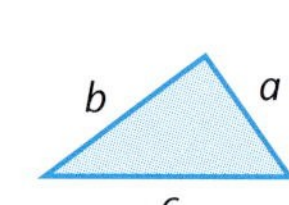

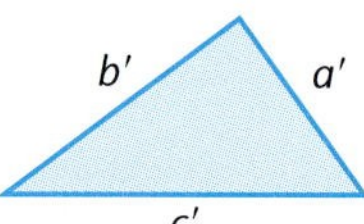

• Pythagorean Theorem

$c^2 = a^2 + b^2$

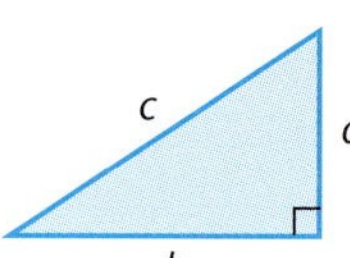

• Rectangle

$A = ab$ Area

$P = 2a + 2b$ Perimeter

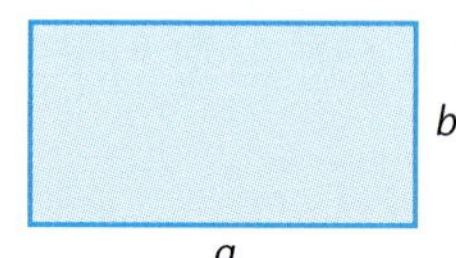

• Parallelogram

$h =$ Height

$A = ah = ab \sin \theta$ Area

$P = 2a + 2b$ Perimeter

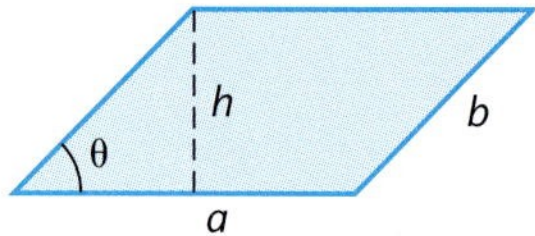

• Triangle

$h =$ Height

$A = \frac{1}{2}hc$ Area

$P = a + b + c$ Perimeter

$s = \frac{1}{2}(a + b + c)$ Semiperimeter

$A = \sqrt{s(s - a)(s - b)(s - c)}$ Area—Heron's formula

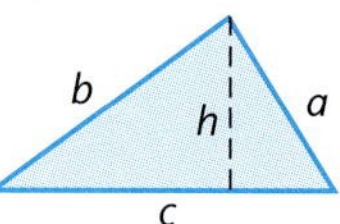

• Trapezoid Base a is parallel to base b.

h = Height

$A = \frac{1}{2}(a + b)h$ Area

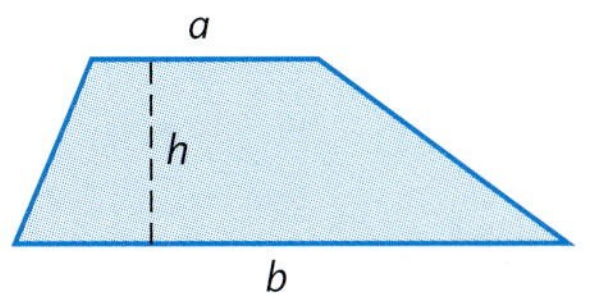

• Circle R = Radius

D = Diameter

$D = 2R$

$A = \pi R^2 = \frac{1}{4}\pi D^2$ Area

$C = 2\pi R = \pi D$ Circumference

$\frac{C}{D} = \pi$ For all circles

$\pi \approx 3.141\ 59$

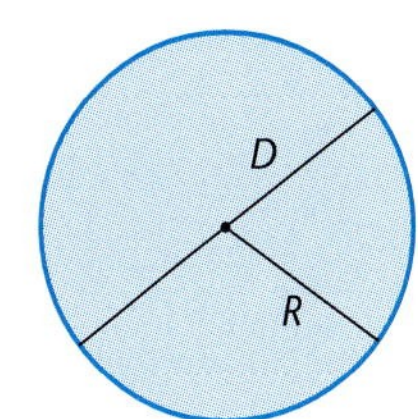

• Rectangular Solid $V = abc$ Volume

$T = 2ab + 2ac + 2bc$ Total surface area

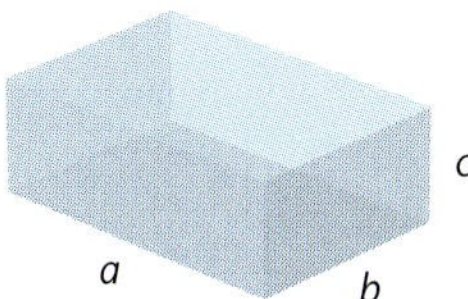

• Right Circular Cylinder R = Radius of base

h = Height

$V = \pi R^2 h$ Volume

$S = 2\pi Rh$ Lateral surface area

$T = 2\pi R(R + h)$ Total surface area

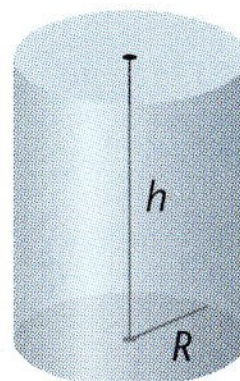

• Right Circular Cone R = Radius of base

h = Height

s = Slant height

$V = \frac{1}{3}\pi R^2 h$ Volume

$S = \pi Rs = \pi R\sqrt{R^2 + h^2}$ Lateral surface area

$T = \pi R(R + s) = \pi R(R + \sqrt{R^2 + h^2})$ Total surface area

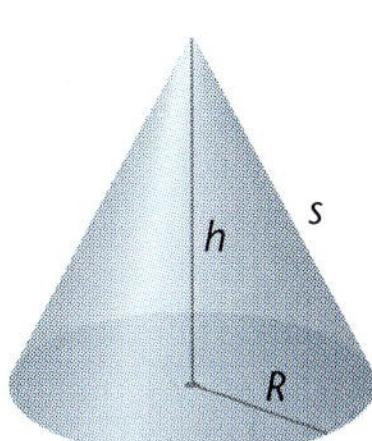

- **Sphere**

R = Radius

D = Diameter

$D = 2R$

$V = \frac{4}{3}\pi R^3 = \frac{1}{6}\pi D^3$ Volume

$S = 4\pi R^2 = \pi D^2$ Surface area

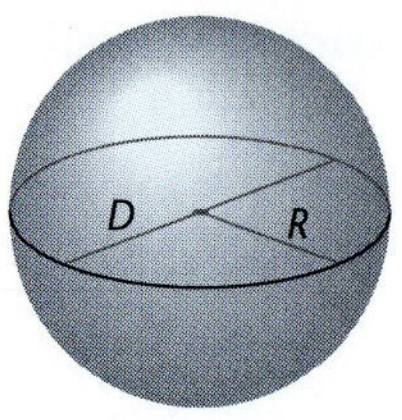

ANSWERS

CHAPTER 1

Exercise 1-1

1. $x = 16$ **3.** No solution **5.** $a = -\frac{23}{4}$ or -5.75 **7.** $x = \frac{16}{5}$ or 3.2 **9.** No solution **11.** $s = -5.93$
13. $y = \frac{16}{9}$ or $1.\overline{7}$ **15.** No solution **17.** $m = \frac{5}{2}$ or 2.5 **19.** No solution **21.** $x = -\frac{15}{2}$ or -7.5 **23.** $a = -\frac{9}{8}$ or -1.125
25. $x = 1.83$ **27.** $x = -8.55$ **29.** $d = (a_n - a_1)/(n - 1)$ **31.** $f = d_1d_2/(d_1 + d_2)$ **33.** $a = (A - 2bc)/(2b + 2c)$
35. $x = (5y + 3)/(2 - 3y)$ **37.** Wrong. There is no solution. **39.** 4 **41.** All real numbers except 0 and 1
43. $x = (by + cy - ac)/(a - y)$ **45.** 24 **47.** 8, 10, 12, 14 **49.** 17 by 10 m **51.** 42 ft **53.** \$90 **55.** \$19,750
57. (A) $T = 30 + 25(x - 3)$ (B) 330°C (C) 13 km **59.** 90 mi **61.** 5,000 trout **63.** 10 gal **65.** 11.25 liters
67. 1.5 h **69.** (A) 216 mi (B) 225 mi **71.** 330 Hz; 396 Hz **73.** 150 cm **75.** 150 ft **77.** $5\frac{5}{11}$ min after 1 P.M.

Exercise 1-2

1. $x = -2, y = -1$ **3.** $x = 3, y = -4$ **5.** $x = -1, y = 2$ **7.** $s = -\frac{5}{2}$ or -2.5, $t = \frac{9}{2}$ or 4.5 **9.** $m = 0, n = 10/3$ or $3.\overline{3}$
11. $x = 10{,}500, y = 57{,}330$ **13.** $u = 0.8, v = 0.3$ **15.** $a = -7/4$ or -1.75, $b = -19/15$ or $-1.2\overline{6}$ **17.** The system has no solutions.
19. $q = x + y - 5, p = 3x + 2y - 12$ **21.** $x = \dfrac{dh - bk}{ad - bc}, y = \dfrac{ak - ch}{ad - bc}, ad - bc \neq 0$ **23.** Airspeed = 330 mph, wind rate = 90 mph
25. 2.475 km **27.** 40mL 50% solution, 60mL 80% solution **29.** 5,200 records **31.** \$7,200 at 10% and \$4,800 at 15%
33. Mexico plant: 75 h; Taiwan plant: 50 h **35.** Mix *A*: 80 g; Mix *B*: 60 g
37. (A) $p = 0.001q + 0.15$ (B) $p = -0.002q + 1.89$ (C) Equilibrium price = \$0.73; Equilibrium quantity = 580 bushels
39. (A) $a = 196, b = -16$ (B) 196 ft (C) 3.5 s **41.** 40 s, 24 s, 120 mi

Exercise 1-3

1. $-8 \le x \le 7$ **3.** $-6 \le x < 6$ **5.** $x \ge -6$
7. $(-2, 6]$ **9.** $(-7, 8)$ **11.** $(-\infty, -2]$
13. $[-7, 2); -7 \le x < 2$ **15.** $(-\infty, 0]; x \le 0$ **17.** $x < 5$ or $(-\infty, 5)$ **19.** $x \ge 3$ or $[3, \infty)$
21. $N < -8$ or $(-\infty, -8)$ **23.** $t > 2$ or $(2, \infty)$ **25.** $m > 3$ or $(3, \infty)$
27. $B \ge -4$ or $[-4, \infty)$ **29.** $-2 < t \le 3$ or $(-2, 3]$ **31.** $(-5, 7]$
33. $(2, 4)$ **35.** $(-\infty, \infty)$ **37.** $(-\infty, -1) \cup [3, 7)$
39. $(1, 5)$ **41.** $(-\infty, 6]$ **43.** $q < -14$ or $(-\infty, -14)$
45. $x \ge 4.5$ or $[4.5, \infty)$ **47.** $-20 \le x \le 20$ or $[-20, 20]$
49. $-30 \le x < 18$ or $[-30, 18)$ **51.** $-8 \le x < -3$ or $[-8, -3)$
53. $-14 < x \le 11$ or $(-14, 11]$ **55.** $x \ge -0.60$ **57.** $-0.255 < x < 0.362$ **59.** $x \le 1$ **61.** $x \ge -\frac{5}{3}$
63. $x > -\frac{3}{2}$ **65.** (A) and (C) $a > 0$ and $b > 0$, or $a < 0$ and $b < 0$ (B) and (D) $a > 0$ and $b < 0$, or $a < 0$ and $b > 0$
67. (A) $>$ (B) $<$ **69.** Positive **71.** (A) F (B) T (C) T **77.** $9.8 \le x \le 13.8$ (from 9.8 to 13.8 km)
79. (A) $x > 40{,}625$ (B) $x = 40{,}625$ **81.** (B) $x > 52{,}000$ (C) Raise wholesale price \$3.50 to \$66.50
83. $2 \le I \le 25$ or $[2, 25]$ **85.** \$2,060 ≤ Benefit reduction ≤ \$3,560

Exercise 1-4

1. $\sqrt{5}$ **3.** 4 **5.** $5 - \sqrt{5}$ **7.** $5 - \sqrt{5}$ **9.** 12 **11.** 12 **13.** 4 **15.** 4 **17.** 9 **19.** $|x - 3| = 4$
21. $|m + 2| = 5$ **23.** $|x - 3| < 5$ **25.** $|p + 2| > 6$ **27.** $|q - 1| \geq 2$

29. x is no more than 7 units from the origin. $-7 \leq x \leq 7$ or $[-7, 7]$

31. x is at least 7 units from the origin. $x \leq -7$ or $x \geq 7$ or $(-\infty, -7] \cup [7, \infty)$

33. y is 3 units from 5. $y = 2, 8$ **35.** y is less than 3 units from 5. $2 < y < 8$ or $(2, 8)$

37. y is more than 3 units from 5. $y < 2$ or $y > 8$ or $(-\infty, 2) \cup (8, \infty)$

39. u is 3 units from -8. $u = -11, -5$

41. u is no more than 3 units from -8. $-11 \leq u \leq -5$ or $[-11, -5]$

43. u is at least 3 units from -8. $u \leq -11$ or $u \geq -5$ or $(-\infty, -11] \cup [-5, \infty)$

45. $1 \leq x \leq \frac{11}{3}$; $[1, \frac{11}{3}]$ **47.** $t < -1$ or $t > 5$; $(-\infty, -1) \cup (5, \infty)$ **49.** $m = -2, -\frac{8}{7}$ **51.** $-2.5 < w < 5.5$; $(-2.5, 5.5)$
53. $u \leq -11$ or $u \geq -6$; $(-\infty, -11] \cup [-6, \infty)$ **55.** $-35 < C < -\frac{5}{9}$ or $(-35, -\frac{5}{9})$ **57.** $-2 < x < 2$ or $(-2, 2)$

59. $-\frac{1}{3} \leq t \leq 1$ or $[-\frac{1}{3}, 1]$ **61.** $t < 0$ or $t > 3$ or $(-\infty, 0) \cup (3, \infty)$ **63.** $(2.9, 3) \cup (3, 3.1)$

65. $(c - d, c) \cup (c, c + d)$ **67.** $x \geq 2$ **69.** $x \leq 1.5$ **71.** $x = -2.2, 1$ **73.** $-3 \leq x \leq 0$

75. $x = 1.4, 5$ **77.** ± 1 **91.** $42.2 < x < 48.6$ **93.** $|P - 500| \leq 20$ **95.** $|A - 12.436| < 0.001$, $(12.435, 12.437)$
97. $|N - 2.37| \leq 0.005$

Exercise 1-5

1. $5 + 9i$ **3.** $3 + 3i$ **5.** $3 + 3i$ **7.** $8 - 10i$ **9.** $7 + 3i$ **11.** -8 **13.** $-12 - 8i$ **15.** $11 + 2i$
17. $13 - i$ **19.** 85 **21.** $0.1 - 0.2i$ **23.** $2 - i$ **25.** $3 - i$ **27.** $7 - 5i$ **29.** $-3 + 2i$ **31.** $8 + 25i$
33. $\frac{5}{7} - \frac{2}{7}i$ **35.** $\frac{2}{13} + \frac{3}{13}i$ **37.** $-\frac{2}{5}i$ or $0 - \frac{2}{5}i$ **39.** $\frac{3}{2} - \frac{1}{2}i$ **41.** $-6i$ or $0 - 6i$ **43.** 0 or $0 + 0i$
45. $i^{18} = -1$, $i^{32} = 1$, $i^{67} = -i$ **47.** $x = 3$, $y = -2$ **49.** $x > 3$ **51.** $x > \frac{2}{3}$ **53.** $33.89 - 20.38i$ **55.** $0.85 - 0.89i$
57. $(a + c) + (b + d)i$ **59.** $a^2 + b^2$ or $(a^2 + b^2) + 0i$ **61.** $(ac - bd) + (ad + bc)i$
63. $i^{4k} = (i^4)^k = (i^2 \cdot i^2)^k = [(-1)(-1)]^k = 1^k = 1$
65. (*1*) Definition of addition; (*2*) Commutative (+) property for *R*; (*3*) Definition of addition

Exercise 1-6

1. $x = 0, 4$ **3.** $t = \frac{3}{2}$ (double root) **5.** $w = -5, \frac{2}{3}$ **7.** $m = \pm 5$ **9.** $c = \pm 3i$ **11.** $y = \pm\frac{3}{2}i$ **13.** $z = \pm 4\sqrt{2}/5$
15. $s = -1 \pm \sqrt{5}$ **17.** $n = 3 \pm 2i$ **19.** $x = 1 \pm \sqrt{2}$ **21.** $x = 1 \pm i\sqrt{2}$ **23.** $t = (3 \pm i\sqrt{7})/2$ **25.** $t = (3 \pm \sqrt{7})/2$
27. $x = 2 \pm \sqrt{5}$ **29.** $r = (-5 \pm \sqrt{3})/2$ **31.** $u = (-2 \pm i\sqrt{11})/2$ **33.** $w = (-2 \pm i\sqrt{5})/3$ **35.** $x = -\frac{5}{4}, \frac{2}{3}$
37. $y = (3 \pm \sqrt{5})/2$ **39.** $x = (3 \pm \sqrt{13})/2$ **41.** $n = -\frac{4}{7}, 0$ **43.** $x = 2 \pm 2i$ **45.** $m = -50, 2$
47. $x = (-5 \pm \sqrt{57})/2$ **49.** $x = (-3 \pm \sqrt{57})/4$ **51.** $u = 1, 2, (-3 \pm \sqrt{17})/2$ **53.** $t = \sqrt{2s/g}$
55. $I = (E + \sqrt{E^2 - 4RP})/2R$ **57.** $x = 1.35, 0.48$ **59.** $x = -1.05, 0.63$
61. If $c < 4$, there are two distinct real roots; if $c = 4$, there is one real double root; and if $c > 4$, there are two distinct imaginary roots.
63. Has real solutions, since discriminant is positive **65.** Has no real solutions, since discriminant is negative
67. $x = \frac{4}{3}\sqrt{6} \pm \frac{2}{3}\sqrt{15}$ or $(4\sqrt{6} \pm 2\sqrt{15})/3$ **69.** $x = \sqrt{2} - i, -\sqrt{2} - i$ **71.** $x = 1, -\frac{1}{2} \pm \frac{1}{2}i\sqrt{3}$
75. $[(-b + \sqrt{b^2 - 4ac})/2a] \times [(-b - \sqrt{b^2 - 4ac})/2a] = [b^2 - (b^2 - 4ac)]/4a^2 = c/a$
77. The $\pm$ in front still yields the same two numbers even if a is negative. **79.** 8, 13 **81.** 12, 14 **83.** 5.12 by 3.12 in.
85. 20% **87.** 100 mph, 240 mph **89.** 13.09 h and 8.09 h **91.** 50 mph
93. 50 ft wide and 300 ft long or 150 ft wide and 100 ft long **95.** 52 mi

Exercise 1-7

1. T **3.** F **5.** F **7.** $x = 14$ **9.** $y = 6$ **11.** $w = -1, 2$ **13.** No solution **15.** $m = \pm\sqrt{3}, \pm i\sqrt{5}$
17. $x = \frac{1}{2}i$ **19.** $y = -64, \frac{27}{8}$ **21.** $m = -1, 3, 1 \pm 2i$ **23.** No solution **25.** $w = -2, 1$ **27.** $z = -1$

29. $x = -\frac{3}{2} + \frac{1}{2}i$ **31.** $y = \frac{1}{3} \pm \frac{i\sqrt{2}}{3}$ **33.** $t = \pm\frac{\sqrt{2}}{2}, \pm\sqrt{2}$ **35.** $z = \frac{3}{2} \pm \frac{3i\sqrt{3}}{2}$ **37.** $m = 0.25$ **39.** $w = 4$

41. $x = -1$ **43.** $x = \pm\sqrt{\frac{5 \pm \sqrt{13}}{6}}$ (four roots) **45.** $x = -4, 39{,}596$ **47.** $x = \left(\frac{4}{5 \pm \sqrt{17}}\right)^5$ **49.** 13.1 in. by 9.1 in.

51. 1.65 ft or 3.65 ft **53.** \$30; 1,600 telephones

Exercise 1-8

1. $-5 < x < 2$ $(-5, 2)$ **3.** $x < 3$ or $x > 7$ $(-\infty, 3) \cup (7, \infty)$ **5.** $0 \le x \le 8$ $[0, 8]$ **7.** $-5 \le x \le 0$ $[-5, 0]$ **9.** $x < -2$ or $x > 2$ $(-\infty, -2) \cup (2, \infty)$

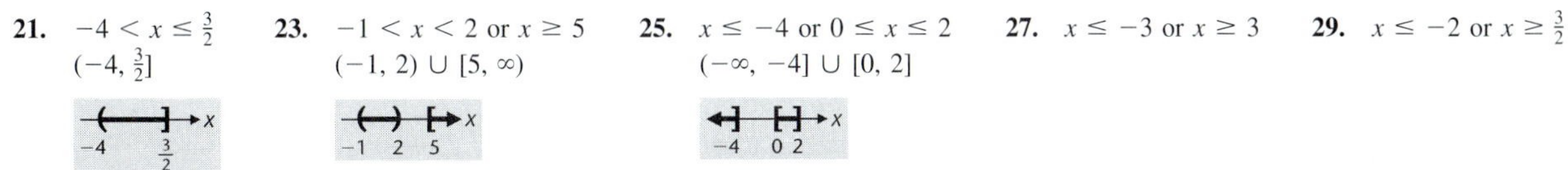

11. $-4 < x \le 2$ $(-4, 2]$ **13.** $x \le -4$ or $x > 1$ $(-\infty, -4] \cup (1, \infty)$ **15.** $-5 \le x \le 0$ or $x > 3$ $[-5, 0] \cup (3, \infty)$ **17.** $-3 < x < 1$ $(-3, 1)$ **19.** $x < 0$ or $x > \frac{1}{4}$ $(-\infty, 0) \cup (\frac{1}{4}, \infty)$

21. $-4 < x \le \frac{3}{2}$ $(-4, \frac{3}{2}]$ **23.** $-1 < x < 2$ or $x \ge 5$ $(-1, 2) \cup [5, \infty)$ **25.** $x \le -4$ or $0 \le x \le 2$ $(-\infty, -4] \cup [0, 2]$ **27.** $x \le -3$ or $x \ge 3$ **29.** $x \le -2$ or $x \ge \frac{3}{2}$

31. $-7 \le x < 3$ **33.** If $a > 0$, the solution set is $(-\infty, r_1) \cup (r_2\ \infty)$. If $a < 0$, the solution set is (r_1, r_2).
35. If $a > 0$, the solution set is R, the set of real numbers. If $a < 0$, the solution set is $\{r\}$. **37.** $x^2 \ge 0$ **39.** No solution; $\varnothing$
41. No solution; $\varnothing$
43. $x \le 2 - \sqrt{5}$ or $x \ge 2 + \sqrt{5}$ $(-\infty, 2 - \sqrt{5}] \cup [2 + \sqrt{5}, \infty)$ **45.** $1 - \sqrt{2} < x < 0$ or $x > 1 + \sqrt{2}$ $(1 - \sqrt{2}, 0) \cup (1 + \sqrt{2}, \infty)$ **47.** $-2 \le x \le -\frac{1}{2}$ or $\frac{1}{2} \le x \le 2$ $[-2, -\frac{1}{2}] \cup [\frac{1}{2}, 2]$ **49.** $-2 \le x \le 2$ $[-2, 2]$

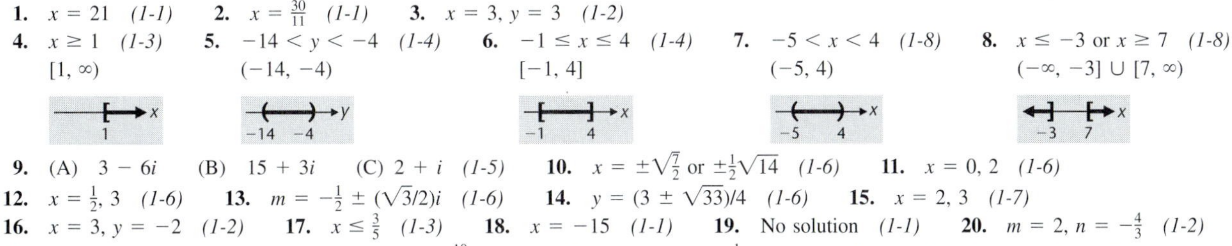

51. (A) Profit: $\$4 < p < \$7$ or (\$4, \$7) (B) Loss: $\$0 \le p < \$4$ or $p > \$7$ or $[\$0, \$4) \cup (\$7, \infty)$ **53.** $2 \le t \le 5$
55. $v > 75$ mph **57.** $5 \le t \le 20$

Chapter 1 Review Exercise*

1. $x = 21$ *(1-1)* **2.** $x = \frac{30}{11}$ *(1-1)* **3.** $x = 3, y = 3$ *(1-2)*
4. $x \ge 1$ *(1-3)* $[1, \infty)$ **5.** $-14 < y < -4$ *(1-4)* $(-14, -4)$ **6.** $-1 \le x \le 4$ *(1-4)* $[-1, 4]$ **7.** $-5 < x < 4$ *(1-8)* $(-5, 4)$ **8.** $x \le -3$ or $x \ge 7$ *(1-8)* $(-\infty, -3] \cup [7, \infty)$

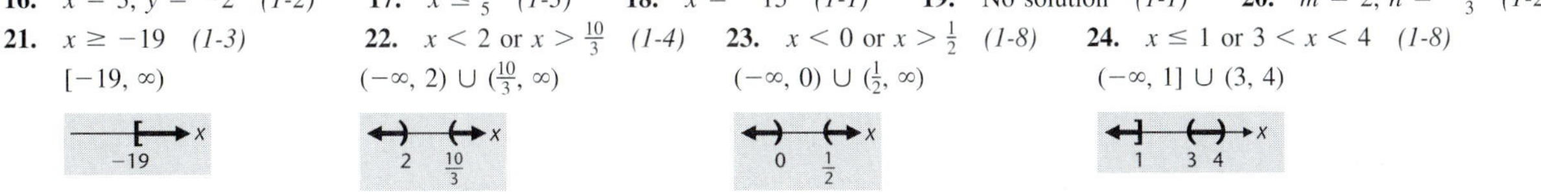

9. (A) $3 - 6i$ (B) $15 + 3i$ (C) $2 + i$ *(1-5)* **10.** $x = \pm\sqrt{\frac{7}{2}}$ or $\pm\frac{1}{2}\sqrt{14}$ *(1-6)* **11.** $x = 0, 2$ *(1-6)*
12. $x = \frac{1}{2}, 3$ *(1-6)* **13.** $m = -\frac{1}{2} \pm (\sqrt{3}/2)i$ *(1-6)* **14.** $y = (3 \pm \sqrt{33})/4$ *(1-6)* **15.** $x = 2, 3$ *(1-7)*
16. $x = 3, y = -2$ *(1-2)* **17.** $x \le \frac{3}{5}$ *(1-3)* **18.** $x = -15$ *(1-1)* **19.** No solution *(1-1)* **20.** $m = 2, n = -\frac{4}{3}$ *(1-2)*
21. $x \ge -19$ *(1-3)* $[-19, \infty)$ **22.** $x < 2$ or $x > \frac{10}{3}$ *(1-4)* $(-\infty, 2) \cup (\frac{10}{3}, \infty)$ **23.** $x < 0$ or $x > \frac{1}{2}$ *(1-8)* $(-\infty, 0) \cup (\frac{1}{2}, \infty)$ **24.** $x \le 1$ or $3 < x < 4$ *(1-8)* $(-\infty, 1] \cup (3, 4)$

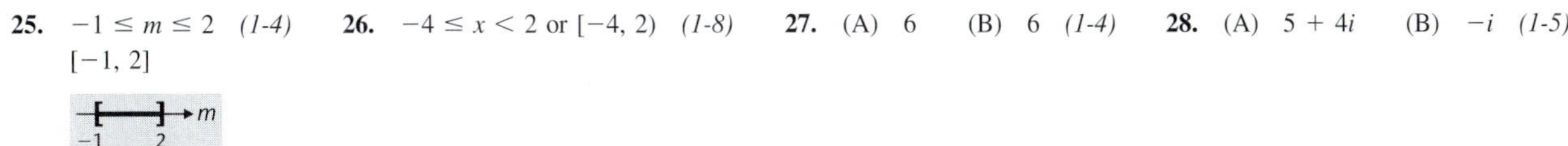

25. $-1 \le m \le 2$ *(1-4)* $[-1, 2]$ **26.** $-4 \le x < 2$ or $[-4, 2)$ *(1-8)* **27.** (A) 6 (B) 6 *(1-4)* **28.** (A) $5 + 4i$ (B) $-i$ *(1-5)*

29. (A) $-1 + i$ (B) $\frac{4}{13} - \frac{7}{13}i$ (C) $\frac{5}{2} - 2i$ *(1-5)* **30.** $u = (-5 \pm \sqrt{5})/2$ *(1-6)* **31.** $u = 1 \pm i\sqrt{2}$ *(1-6)*
32. $x = (1 \pm \sqrt{43})/3$ *(1-6)* **33.** $x = -\frac{27}{8}, 64$ *(1-7)* **34.** $m = \pm 2, \pm 3i$ *(1-7)* **35.** $y = \frac{9}{4}, 3$ *(1-7)* **36.** $x = 0.45$ *(1-1)*
37. $-2.24 \le x \le 1.12$ or $[-2.24, 1.12]$ *(1-3)* **38.** $0.89 - 0.32i$ *(1-5)* **39.** $x = -1.64, 0.89$ *(1-6)*

*The number in parentheses after each answer to a chapter review problem refers to the section in which that type of problem is discussed.

40. $x = 0.94$, $y = 1.02$ *(1-2)* **41.** $M = P/(1 - dt)$ *(1-1)* **42.** $I = (E \pm \sqrt{E^2 - 4PR})/(2R)$ *(1-6)*
43. $y = (5 - x)/(2x - 4)$ *(1-1)* **44.** The correct answer is $x = -1$. *(1-1)*
45. If $c < 9$, there are two distinct real roots; if $c = 9$, there is one real double root; and if $c > 9$, there are two distinct imaginary roots. *(1-6)*
46. All real b and all negative a *(1-3)* **47.** Less than 1 *(1-3)* **48.** $x = 1/(1 - y)$ *(1-1)*
49. $6 - d < x < 6 + d, x \neq 6$ *(1-4)* **50.** $x = \frac{1}{4}\sqrt{3} \pm \frac{1}{4}i$ *(1-6)* **51.** $x = \pm\sqrt{(2 \pm \sqrt{3})/2}$ (four real roots) *(1-7)* **52.** 1 *(1-5)*
$(6 - d, 6) \cup (6, 6 + d)$

6 − d 6 6 + d

53. No solution *(1-8)* **54.** Set of all real numbers *(1-8)*
55. $x \leq -4$ or $-2 \leq x < 0$ or $0 < x \leq 2$ or $x \geq 4$; $(-\infty, -4] \cup [-2, 0) \cup (0, 2] \cup [4, \infty)$ *(1-8)*
56. $u = -31 + 5x - 7y$, $v = 13 - 2x + 3y$ *(1-2)* **57.** (A) Infinite number of solutions (B) No solution *(1-2)*
58. $\frac{5}{3}$ or $-\frac{3}{5}$ *(1-6)* **59.** (A) $H = 0.7(220 - A)$ (B) 140 beats/min (C) 40 years old *(1-1)*
60. 20 mL of 30% solution, 30 mL of 80% solution *(1-1)* **61.** 3 mph *(1-6)*
62. (A) 3 km (B) 16.2 km/h (C) 20.6 min *(1-1, 1-6)* **63.** 85 bags of brand A and 45 bags of brand B *(1-2)*
64. (A) 2,000 and 8,000 (B) 5,000 *(1-6)* **65.** $x = (13 \pm \sqrt{45})/2$ thousand, or approx. 3,146 and 9,854 *(1-6)*
66. $(13 - \sqrt{45})/2 < x < (13 + \sqrt{45})/2$ or approx. $3.146 < x < 9.854$, x in thousands *(1-8)* **67.** $|T - 110| \leq 5$ *(1-4)*
68. 20 cm by 24 cm *(1-6)* **69.** 6.58 ft or 14.58 ft *(1-7)*

CHAPTER 2

Exercise 2-1

1. The y axis **3.** Quadrant III **5.** Quadrant IV **7.** Quadrant II and quadrant IV **9.** Quadrant I and quadrant IV

11.

13. Symmetric with respect to the origin

15. Symmetric with respect to the x axis

17. Symmetric with respect to the x axis, y axis, and origin

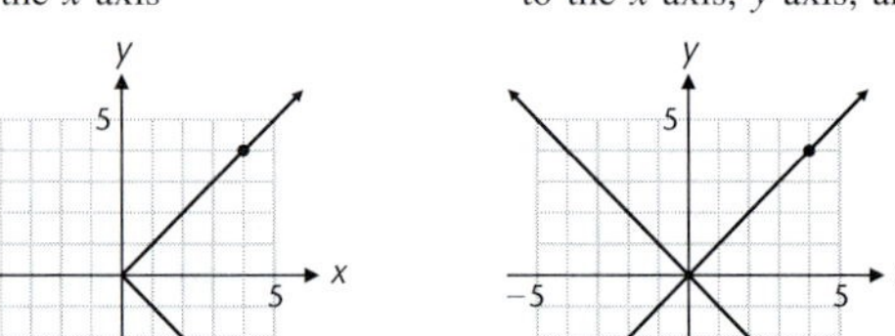

19. $\sqrt{106}$ **21.** $\sqrt{82}$ **23.** $x^2 + y^2 = 16$ **25.** $(x - 3)^2 + (y + 2)^2 = 1$ **27.** $(x - 2)^2 + (y - 6)^2 = 3$
29. (A) 3 (B) -2 (C) $-3, -1, 4$ (D) $-4, 1, 3$

31. (A)

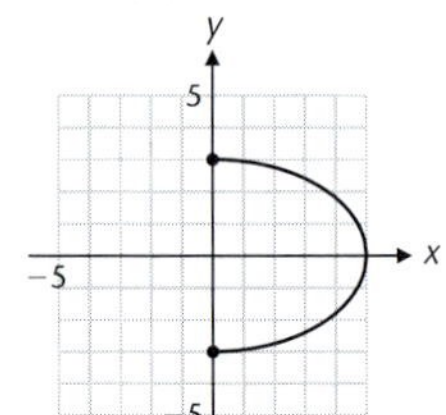

(B)

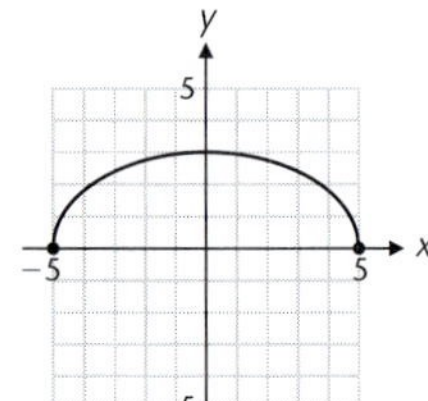

(C)

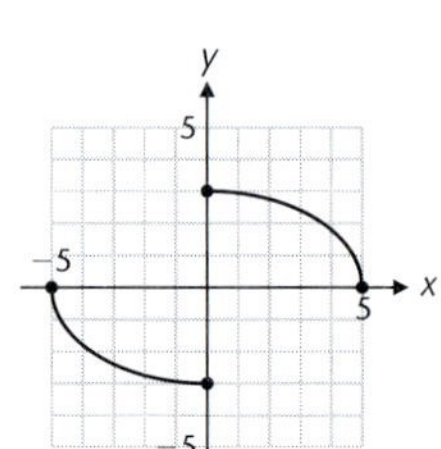

(D)

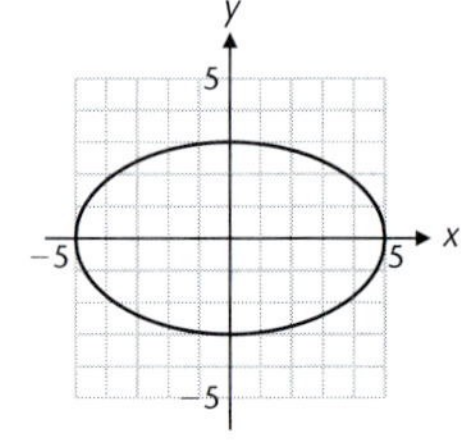

33. Symmetric with respect to the x axis

35. Symmetric with respect to the y axis

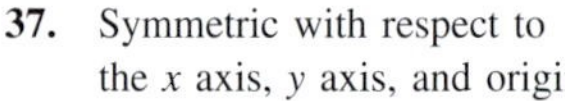

37. Symmetric with respect to the x axis, y axis, and origin

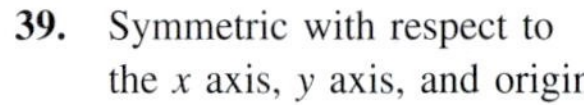

39. Symmetric with respect to the x axis, y axis, and origin

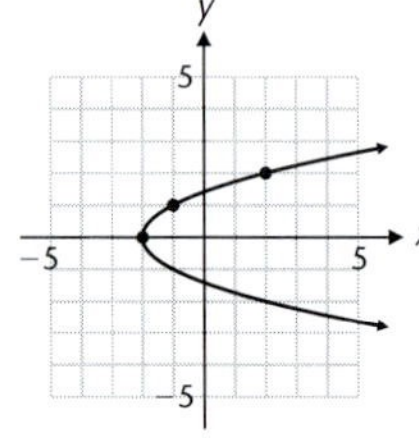

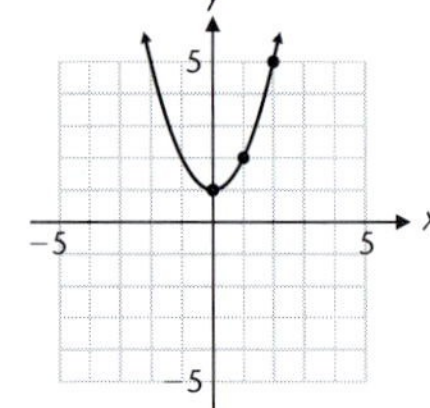

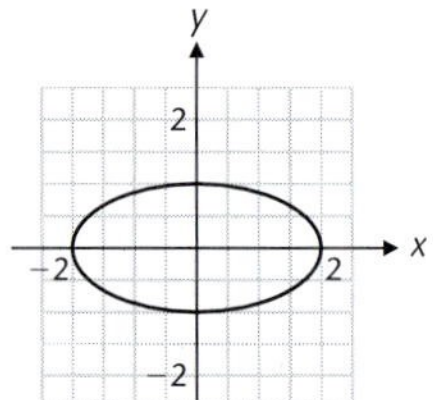

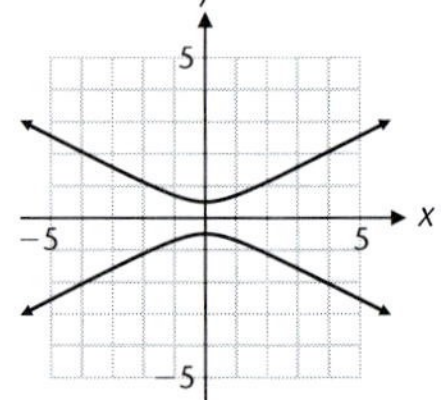

41. Symmetric with respect to the origin

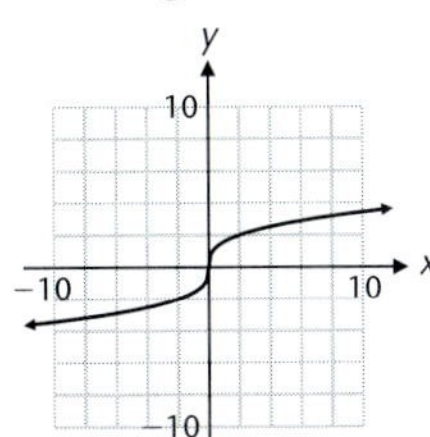

43. Symmetric with respect to the y axis

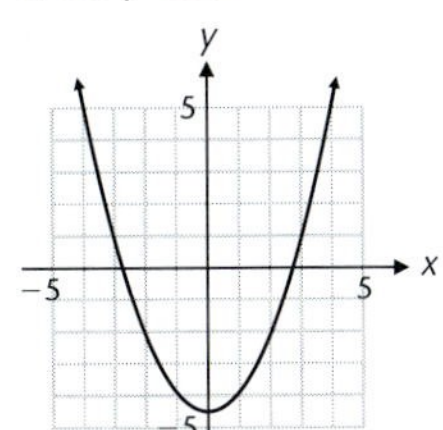

45. Symmetric with respect to the y axis

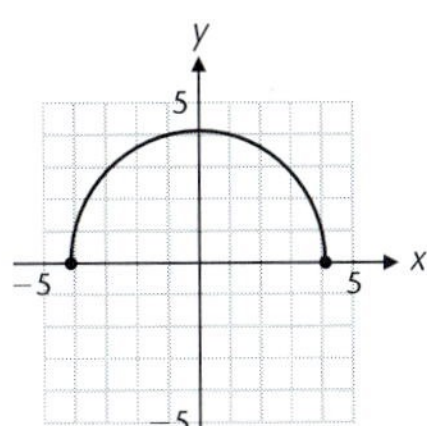

47. Symmetric with respect to the y axis

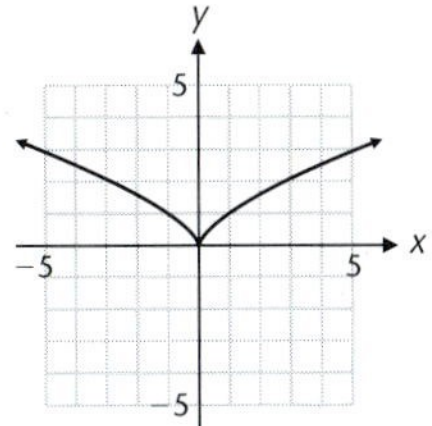

49. Area = 28, perimeter = 26.96 **51.** $x = -12, 4$ **53.** $y = 4$

55. Center: $(-4, 2)$; radius: $\sqrt{7}$ **57.** Center: $(3, 2)$; radius: 7 **59.** Center: $(-4, 3)$; radius: $\sqrt{17}$ **61.**

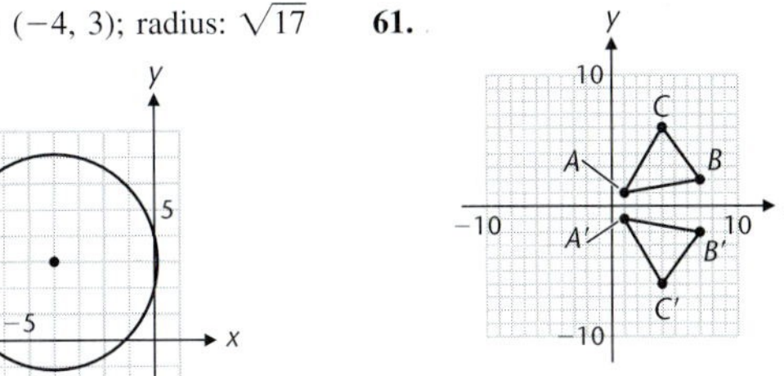

63.

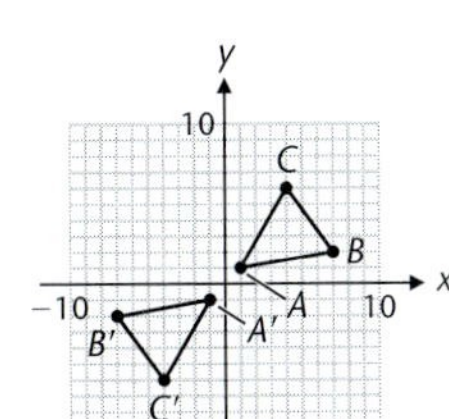

65. $y = \pm\sqrt{3 - x^2}$

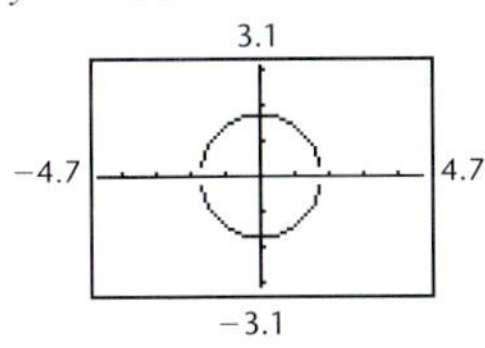

67. $y = -1 \pm \sqrt{2 - (x + 3)^2}$

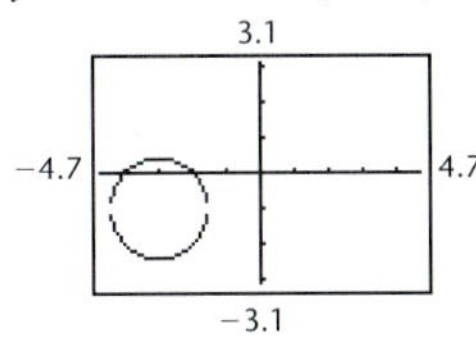

69. Center: $(1, 0)$; radius: 1; $(x - 1)^2 + y^2 = 1$

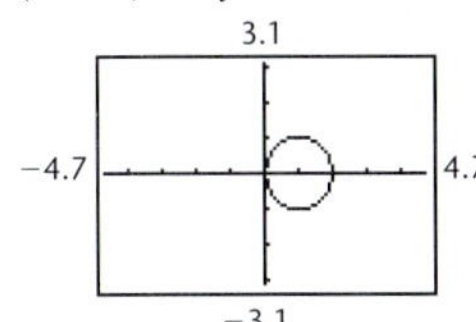

71. Center: $(2, 1)$; radius: 3; $(x - 2)^2 + (y - 1)^2 = 9$

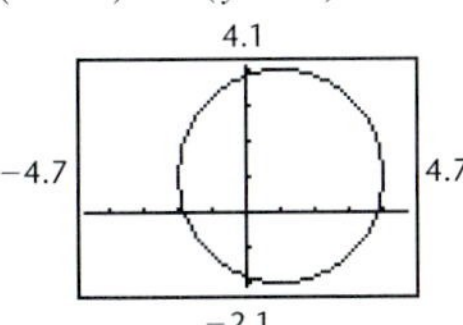

73. Symmetric with respect to the y axis

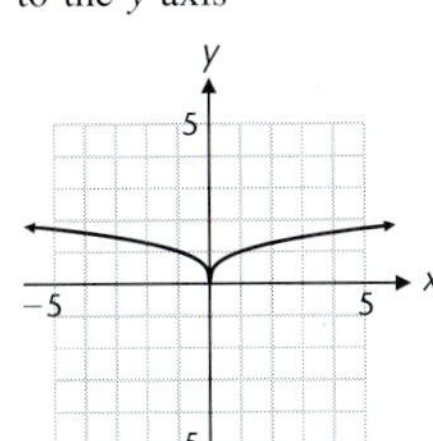

75. Symmetric with respect to the origin

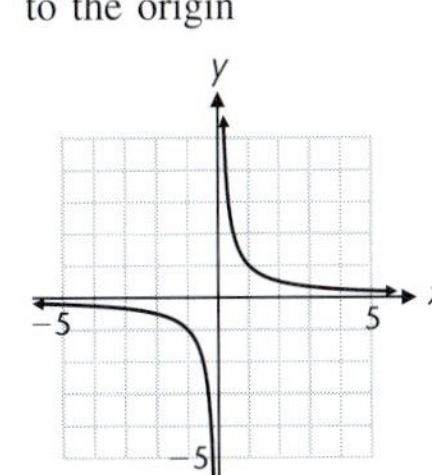

77.

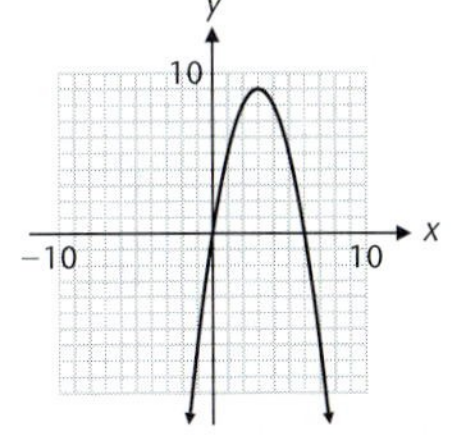

79. $5x + 3y = -2$ **81.** $(x - 4)^2 + (y - 2)^2 = 34$ **83.** $(x - 2)^2 + (y - 2)^2 = 50$ **85.** Yes

87. (A) 3,000 cases (B) Demand decreases by 400 cases (C) Demand increases by 600 cases

89. (A) 53° (B) 68° at 3 P.M. (C) 1 A.M., 7 A.M., 11 P.M.

91.

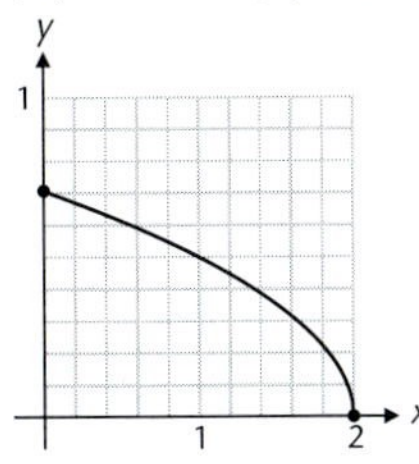

93. 2.5 ft **95.** (A) $(x + 12)^2 + (y + 5)^2 = 26^2$; center: $(-12, -5)$; radius: 26 (B) 13.5 mi

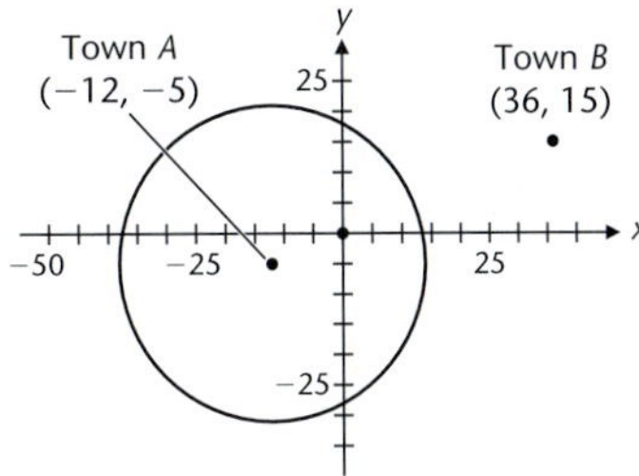

Exercise 2-2

1. x intercept: -2; y intercept: 2; slope: 1; equation: $y = x + 2$ **3.** x intercept: -2; y intercept: -4; slope: -2; equation: $y = -2x - 4$
5. x intercept: 3; y intercept: -1; slope: $\frac{1}{3}$; equation: $y = \frac{1}{3}x - 1$
7. Slope $= -\frac{3}{5}$ **9.** Slope $= -\frac{3}{4}$ **11.** Slope $= \frac{2}{3}$ **13.** Slope $= \frac{4}{5}$

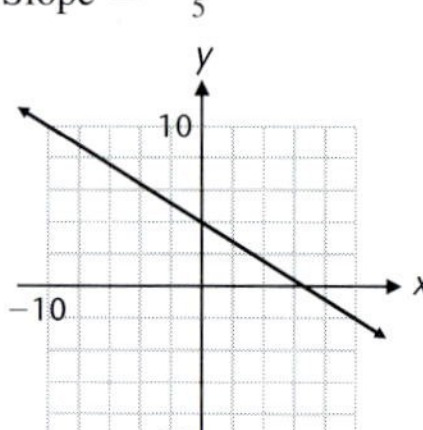
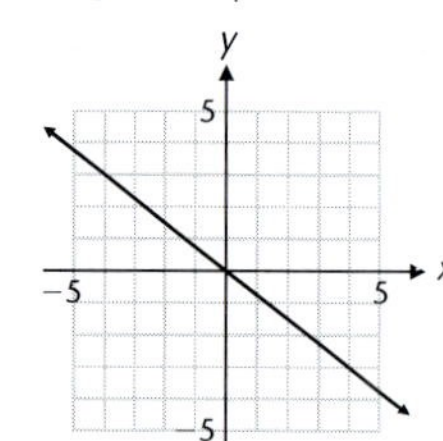
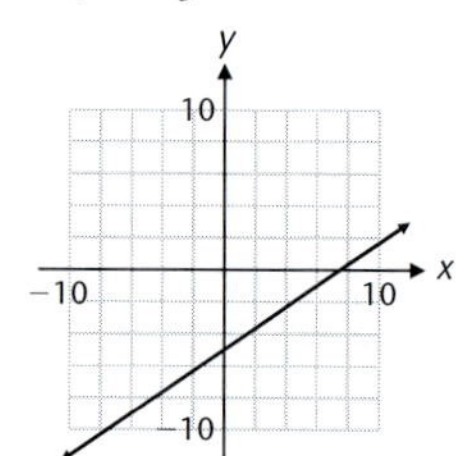

15. Slope $= 2$ **17.** Slope not defined **19.** Slope $= 0$ **21.** $x - y = 0$ **23.** $2x + 3y = -12$

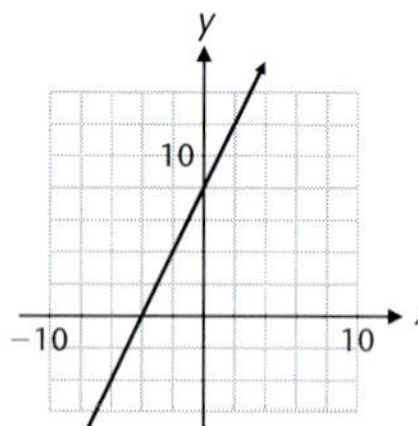
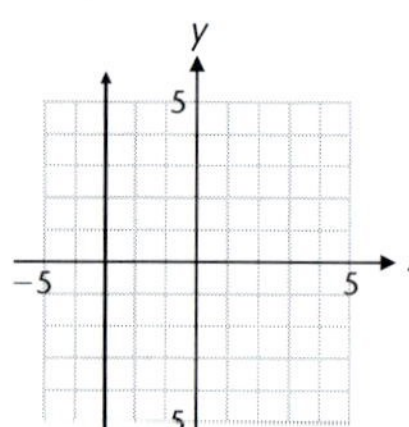
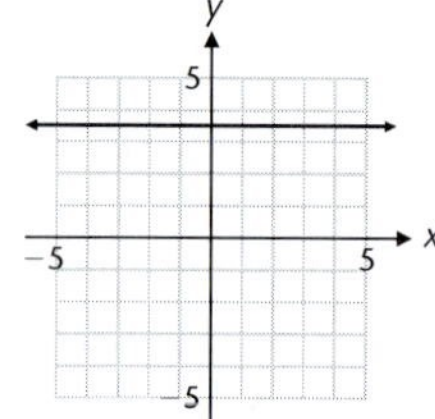

25. $y = -2x + 3$ **27.** $y = \frac{3}{2}x + \frac{23}{2}$ **29.** $y = -4x + 13$ **31.** $x = -3$ **33.** $y = 2$ **35.** $3x + y = 5$
37. $2x + 3y = -12$ **39.** $y = 3$ **41.** $2x + 3y = 23$ **43.** $2x + 5y = 10$ **45.** $x = -2$ **47.** Trapezoid **49.** Rectangle
51. $y = -\frac{6}{7}x - \frac{5}{14}$ **53.** $3x + 4y = 25$ **55.** $x - y = 10$ **57.** $5x - 12y = 232$

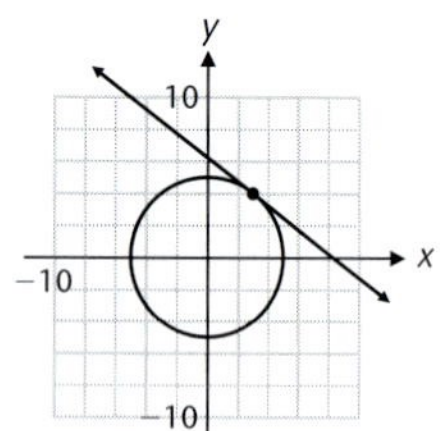
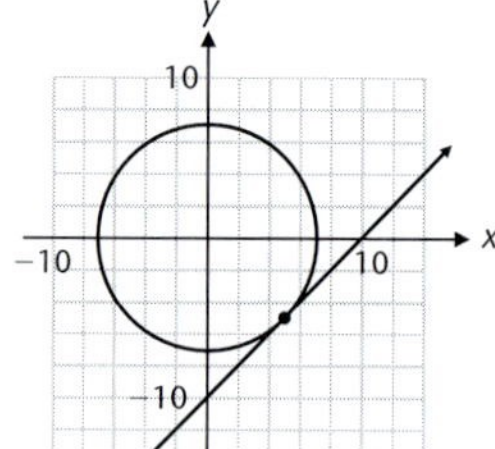
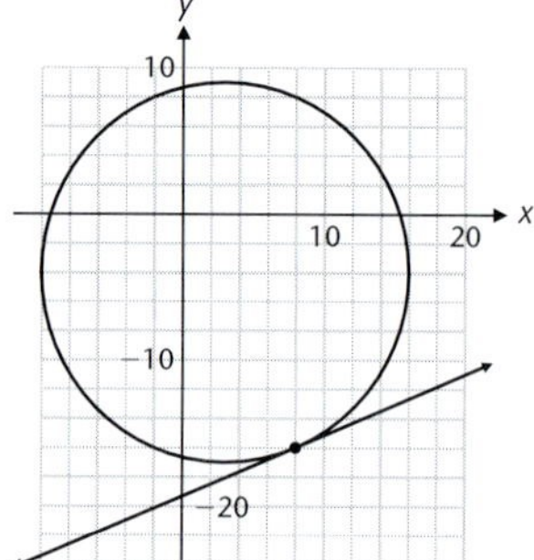

59. (A)

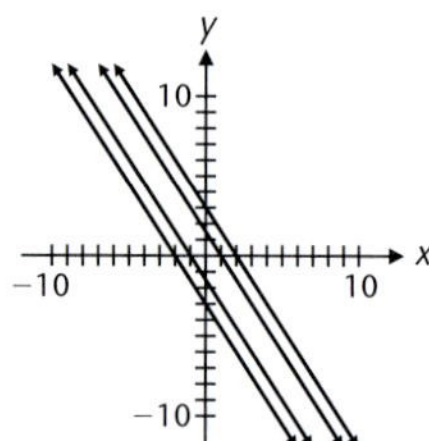

61.

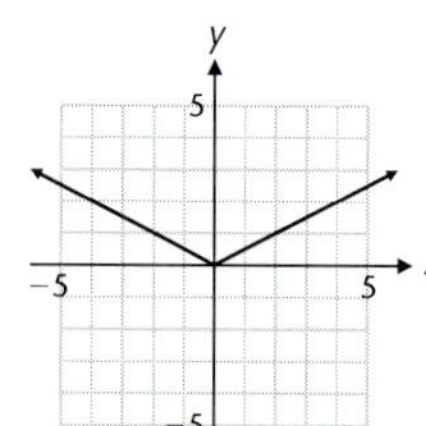

63.

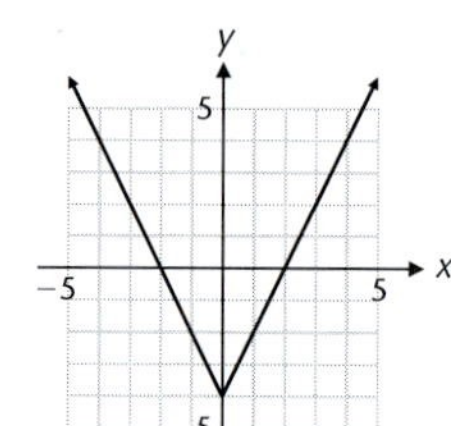

65.

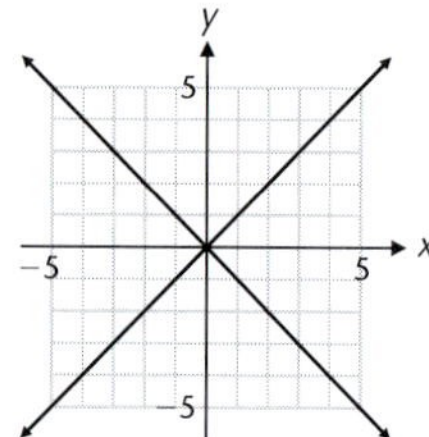

71. $7x - 2y = -14$ **73.** (A)

x	0	5,000	10,000	15,000	20,000	25,000	30,000
B	212	203	194	185	176	167	158

(B) The boiling point drops 9°F for each 5,000-ft increase in altitude.

75. The rental charges are \$25 per day plus \$0.25 per mile driven.

77. (A)

x	0	1	2	3	4
Sales	5.9	6.5	7.7	8.6	9.7
$f(x)$	5.7	6.7	7.7	8.6	9.6

(B)

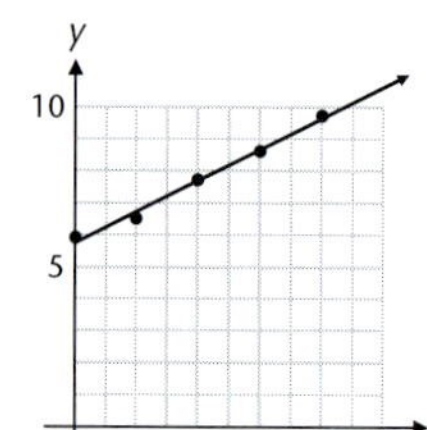

(C) \$10.6 billion, \$17.4 billion

79. (A) $F = \frac{9}{5}C + 32$ (B) 68°F; 30°C (C) $\frac{9}{5}$ **81.** (A) $V = -1{,}600t + 8{,}000, 0 \le t \le 5$ (B) \$3,200 (C) −1,600
83. (A) $T = -5A + 70, A \ge 0$ (B) 14,000 ft (C) −5; the temperature changes −5°F for each 1,000-ft rise in altitude
85. (A) $h = 1.13t + 12.8$ (B) 32.9 h
87. (A) $R = 0.001\,52C - 0.159, C \ge 210$ (B) 0.236
(C) 0.001 52; coronary risk increases 0.001 52 per unit increase in cholesterol above the 210 cholesterol level

Exercise 2-3

1. Function **3.** Not a function **5.** Function **7.** Function; domain = {2, 3, 4, 5}; range = {4, 6, 8, 10} **9.** Not a function
11. Function; domain = {0, 1, 2, 3, 4, 5}; range = {1, 2} **13.** Function **15.** Not a function **17.** Not a function **19.** 4
21. 0 **23.** 10 **25.** 4 **27.** 5.25 **29.** 7 **31.** −1, −5, 6 **33.** A function with domain all real numbers
35. Not a function; for example, when $x = 0, y = \pm 2$ **37.** A function with domain all real numbers
39. Not a function; for example, when $x = 0$, y can be any real number **41.** A function with domain all real numbers except 0
43. Domain: $x \ge 2$ **45.** Domain: all real numbers **47.** Domain: all real numbers except 4
49. Domain: all real numbers except −2 and 1 **51.** Domain: $y \le -1$ or $y \ge 3$ **53.** Domain: all real numbers
55. Domain: $y < -2$ or $y \ge 3$ **57.** $g(x) = 2x^3 - 5$ **59.** $G(x) = 2\sqrt{x} - x^2$
61. Function f multiplies the domain element by 2 and subtracts 3 from the result.
63. Function F multiplies the cube of the domain element by 3 and subtracts twice the square root of the domain element from the result.
65. 3 **67.** $-6 - h$ **69.** $-2h + 11$ **71.** $f(x) = 3x^2 - 5x + 9$ **73.** $m(t) = -2t^2 - 5\sqrt{t} - 2$ **75.** (A) 4 (B) 4
77. (A) $4x + 2h$ (B) $2x + 2a$ **79.** (A) $-8x + 3 - 4h$ (B) $-4x - 4a + 3$
81. (A) $3x^2 - 2 + 3xh + h^2$ (B) $x^2 + ax + a^2 - 2$ **83.** $P(w) = 2w + (128/w), w > 0$ **85.** $h(b) = \sqrt{b^2 + 25}, b > 0$
87. $C(x) = 300 + 1.75x$
89. (A) $s(0) = 0, s(1) = 16, s(2) = 64, s(3) = 144$ (B) $64 + 16h$
(C) Value of expression tends to 64; this number appears to be the speed of the object at the end of 2 s.
91. $V(x) = x(8 - 2x)(12 - 2x)$; domain: $0 < x < 4$
93. $F(x) = 8x + (250/x) - 12$;

x	4	5	6	7
$F(x)$	82.5	78	77.7	79.7

95. $C(x) = 10{,}000(20 - x) + 15{,}000\sqrt{x^2 + 64}$; domain: $0 \le x \le 20$ **97.** $C(v) = 100v + (200{,}000/v)$

Exercise 2-4

1. (A) [−4, 4) (B) [−3, 3) (C) 0 (D) 0 (E) [−4, 4) (F) None (G) None (H) None
3. (A) (−∞, ∞) (B) [−4, ∞) (C) −3, 1 (D) −3 (E) [−1, ∞) (F) (−∞, −1] (G) None (H) None
5. (A) (−∞, 2) ∪ (2, ∞) (B) (−∞, −1) ∪ [1, ∞) (C) None (D) 1 (E) None (F) (−∞, −2], (2, ∞)
(G) [−2, 2) (H) $x = 2$

7. One possible answer:

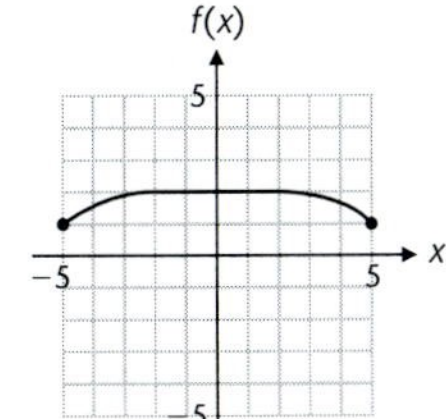

9. One possible answer:

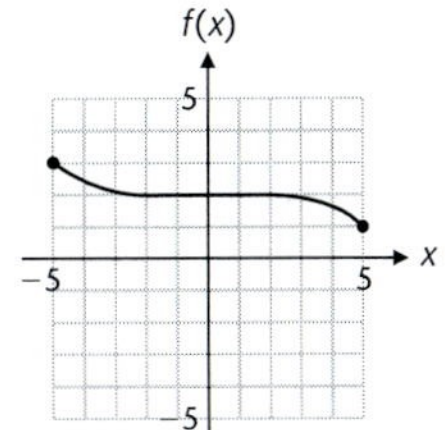

11. One possible answer:

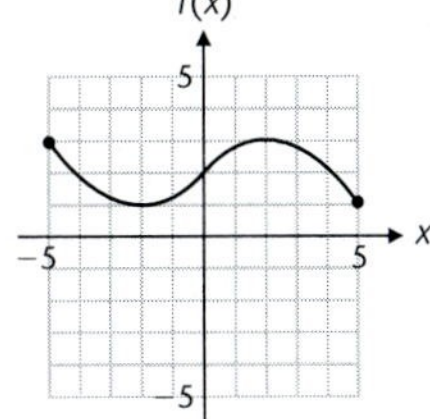

13. Slope = 2, x intercept = −2, y intercept = 4

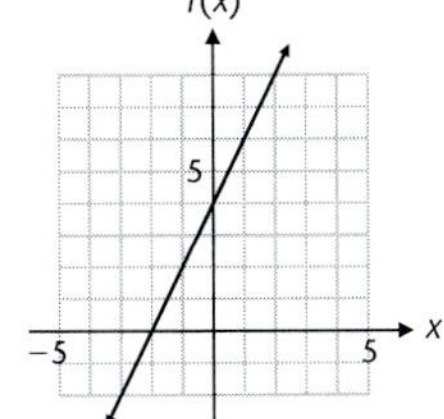

15. Slope = $-\frac{1}{2}$, x intercept = $-\frac{10}{3}$, y intercept = $-\frac{5}{3}$

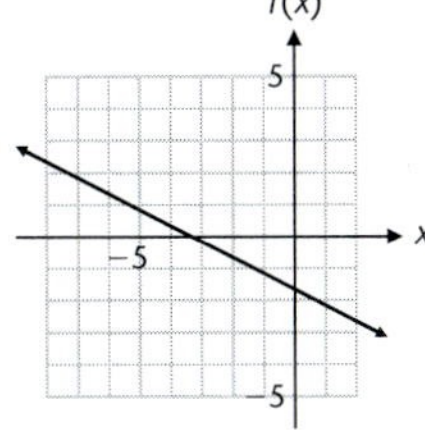

17. $f(x) = -\frac{3}{2}x + 4$

19. Min $f(x) = f(3) = 2$
Range $= [2, \infty)$

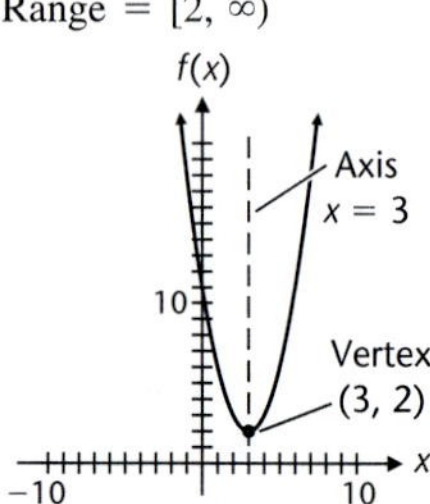

21. Max $f(x) = f(-3) = -2$
Range $= (-\infty, -2]$

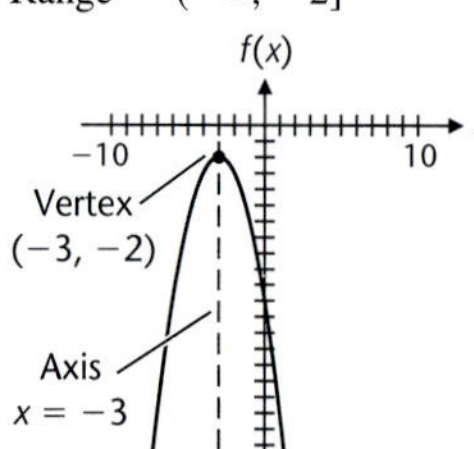

23. x intercepts: $-1, 5$
y intercept $= -5$

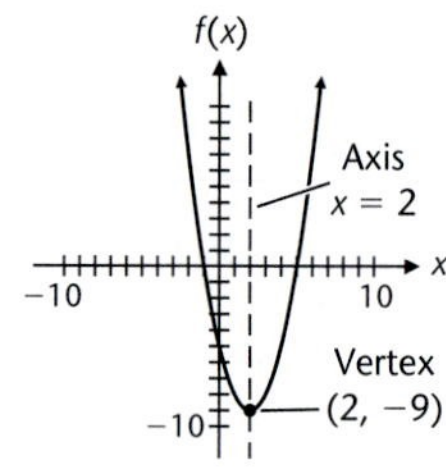

25. x intercepts: 0, 6
y intercept $= 0$

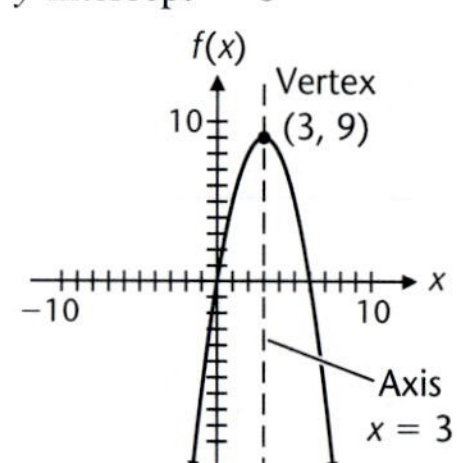

27. Increasing on $[-3, \infty)$
Decreasing on $(-\infty, -3]$

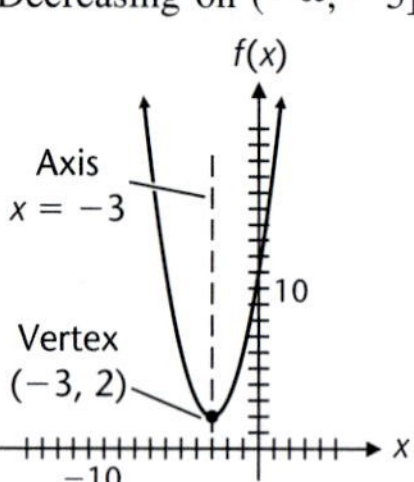

29. Increasing on $(-\infty, 3]$
Decreasing on $[3, \infty)$

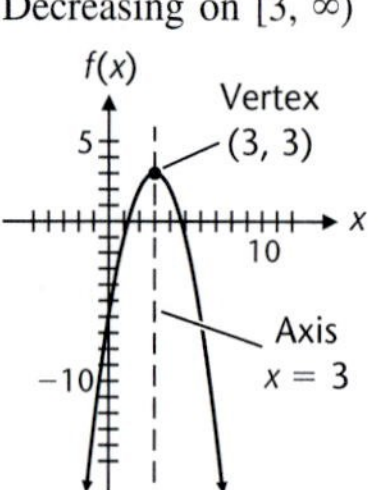

31. Domain: $[-1, 1]$
Range: $[0, 1]$

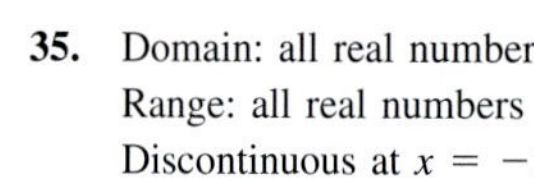

33. Domain: $[-3, -1) \cup (-1, 2]$
Range: $\{-2, 4\}$ (a set, not an interval)
Discontinuous at $x = -1$

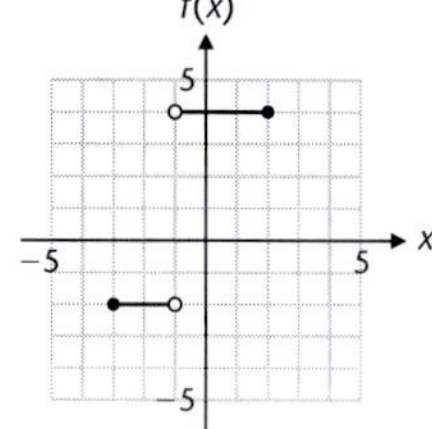

35. Domain: all real numbers
Range: all real numbers
Discontinuous at $x = -1$

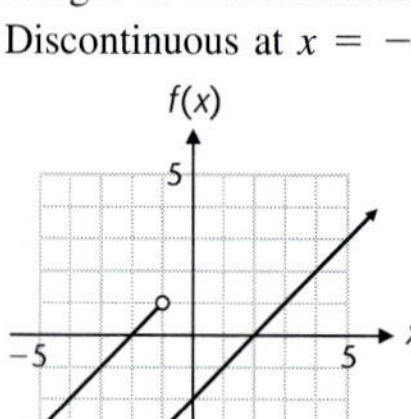

37. Domain: $x \neq 0$, or $(-\infty, 0) \cup (0, \infty)$
Range: $(-\infty, -1) \cup (1, \infty)$
Discontinuous at $x = 0$

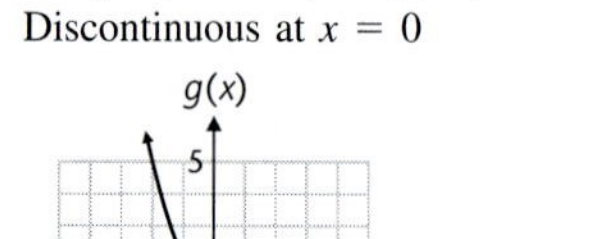

39. Min $f(x) = f(-2) = 1$
Range: $[1, \infty)$
No x intercepts
y intercept $= f(0) = 3$
Increasing on $[-2, \infty)$
Decreasing on $(-\infty, -2]$

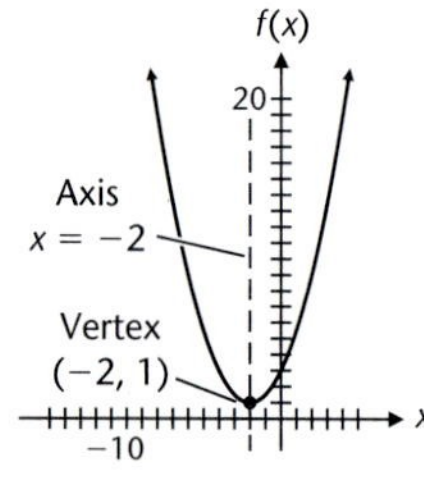

41. Min $f(x) = f(\frac{3}{2}) = 0$
Range: $[0, \infty)$
x intercept $= \frac{3}{2}$
y intercept $= f(0) = 9$
Increasing on $[\frac{3}{2}, \infty)$
Decreasing on $(-\infty, \frac{3}{2}]$

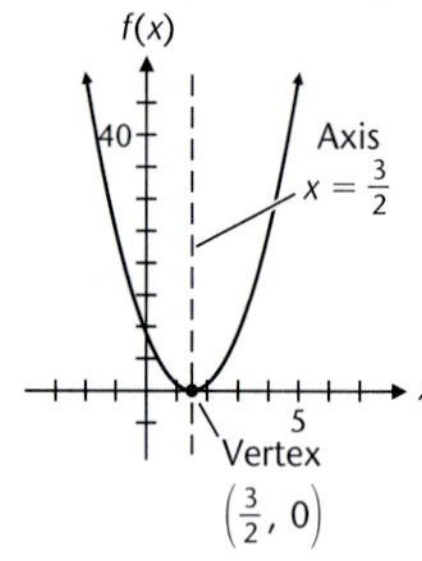

43. Max $f(x) = f(-2) = 6$
Range: $(-\infty, 6]$
x intercepts: $-2 \pm \sqrt{3}$
y intercept $= f(0) = -2$
Increasing on $(-\infty, -2]$
Decreasing on $[-2, \infty)$

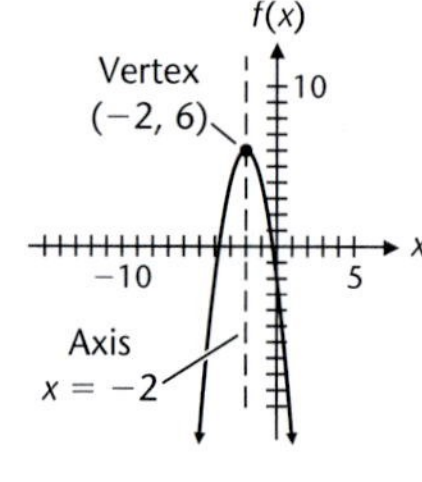

45. $f(x) = \begin{cases} -1 & \text{if } x < 0 \\ 1 & \text{if } x > 0 \end{cases}$

Domain: $x \neq 0$, or $(-\infty, 0) \cup (0, \infty)$
Range: $\{-1, 1\}$ (a set, not an interval)
Discontinuous at $x = 0$

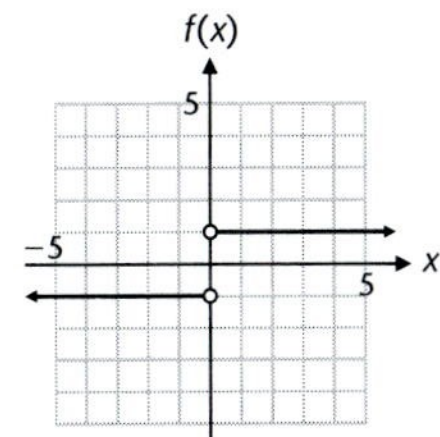

47. $f(x) = \begin{cases} x - 1 & \text{if } x < 1 \\ x + 1 & \text{if } x > 1 \end{cases}$

Domain: $x \neq 1$, or $(-\infty, 1) \cup (1, \infty)$
Range: $(-\infty, 0) \cup (2, \infty)$
Discontinuous at $x = 1$

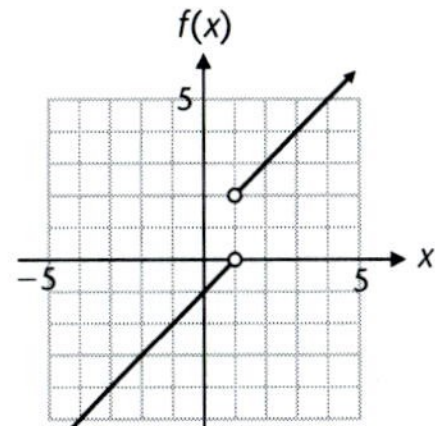

49. $f(x) = \begin{cases} 2 - 2x & \text{if } x < 0 \\ 2 & \text{if } 0 \leq x < 2 \\ -2 + 2x & \text{if } x \geq 2 \end{cases}$

Domain: all real numbers
Range: $[2, \infty)$
No discontinuities

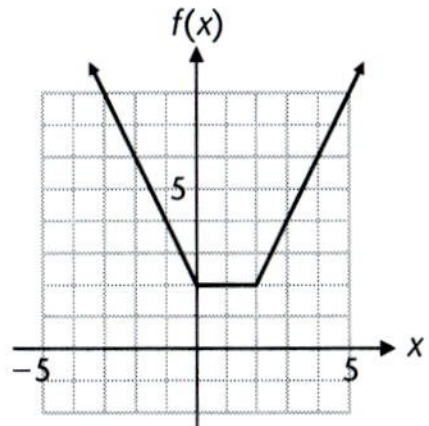

51. Domain: all real numbers
Range: all integers
Discontinuous at the even integers

$$f(x) = \begin{cases} \vdots \\ -2 & \text{if } -4 \leq x < -2 \\ -1 & \text{if } -2 \leq x < 0 \\ 0 & \text{if } 0 \leq x < 2 \\ 1 & \text{if } 2 \leq x < 4 \\ 2 & \text{if } 4 \leq x < 6 \\ \vdots \end{cases}$$

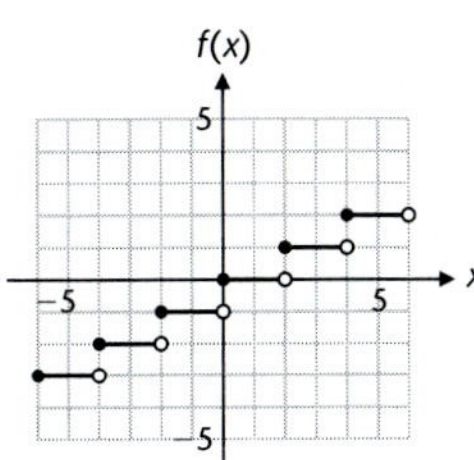

53. Domain: all real numbers
Range: all integers
Discontinuous at rational numbers of the form $k/3$, where k is an integer

$$f(x) = \begin{cases} \vdots \\ -2 & \text{if } -\frac{2}{3} \leq x < -\frac{1}{3} \\ -1 & \text{if } -\frac{1}{3} \leq x < 0 \\ 0 & \text{if } 0 \leq x < \frac{1}{3} \\ 1 & \text{if } \frac{1}{3} \leq x < \frac{2}{3} \\ 2 & \text{if } \frac{2}{3} \leq x < 1 \\ \vdots \end{cases}$$

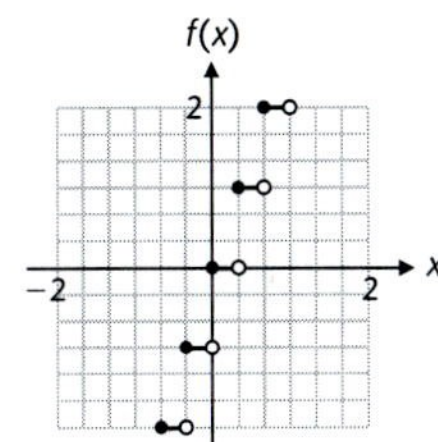

55. Domain: all real numbers
Range: $[0, 1)$
Discontinuous at all integers

$$f(x) = \begin{cases} \vdots \\ x + 2 & \text{if } -2 \leq x < -1 \\ x + 1 & \text{if } -1 \leq x < 0 \\ x & \text{if } 0 \leq x < 1 \\ x - 1 & \text{if } 1 \leq x < 2 \\ x - 2 & \text{if } 2 \leq x < 3 \\ \vdots \end{cases}$$

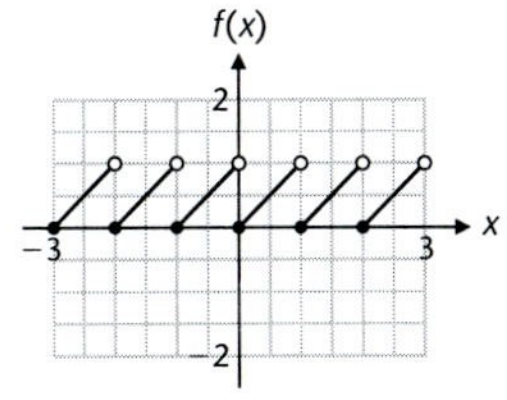

57. Axis: $x = 2$; vertex: $(2, 4)$; range: $[4, \infty)$; no x intercepts

59. (A) One possible answer:

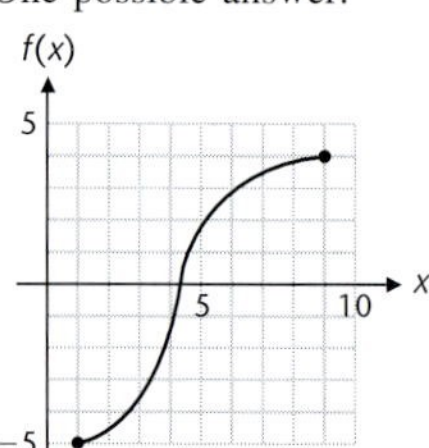

(B) The graph must cross the x axis exactly once.

61. (A) One possible answer:

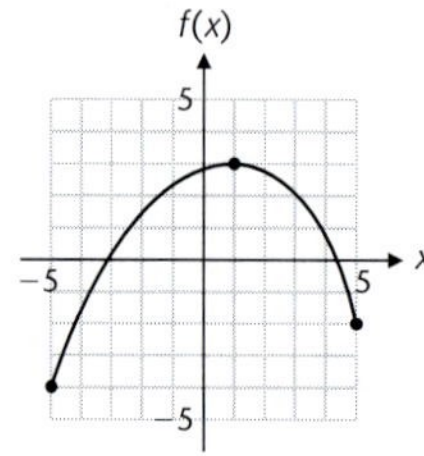

(B) The graph must cross the x axis at least twice. There is no upper limit on the number of times it can cross the x axis.

63. $y = 2x - 1$

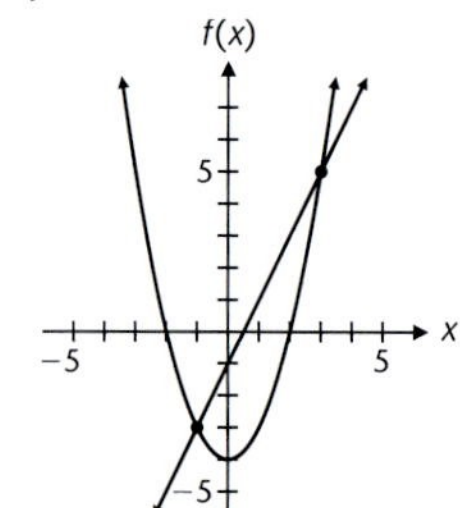

65. (A) $h + 1$ (B)

h	1	0.1	0.01	0.001
Slope	2	1.1	1.01	1.001

; approaching 1

67. Graphs of f and g

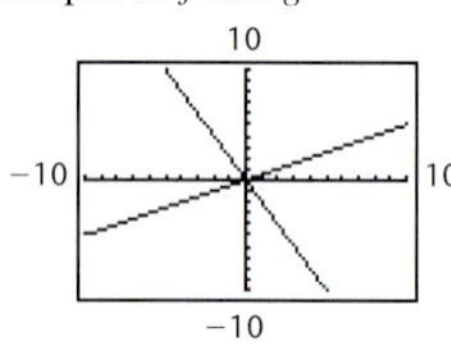

Graph of m

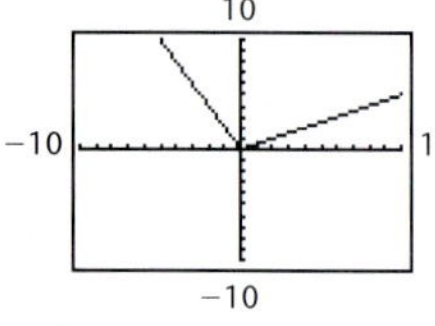

Graph of n

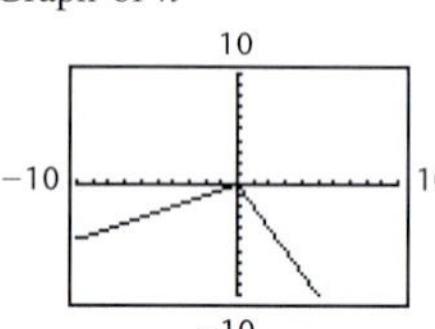

69. Graphs of f and g

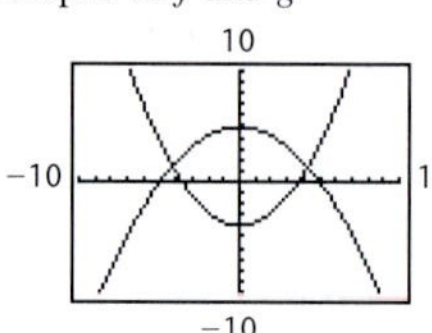

Graph of m

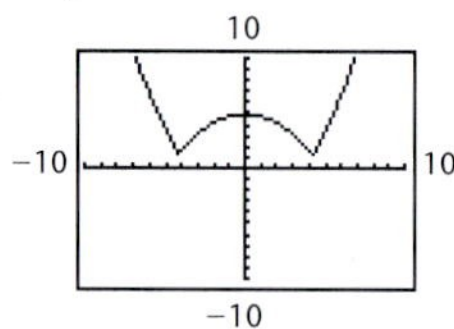

Graph of n

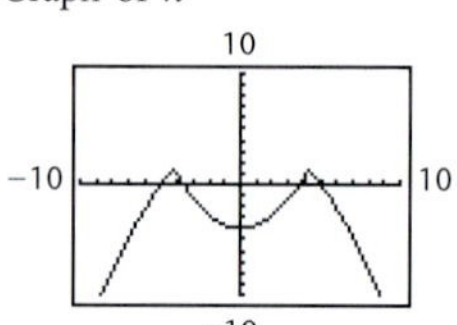

71. Graphs of f and g

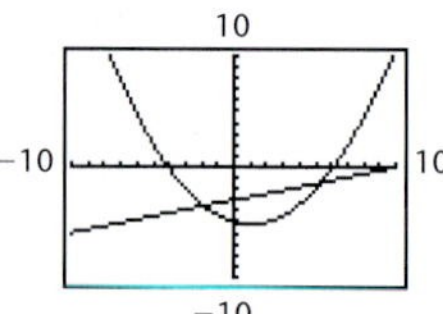

Graph of m

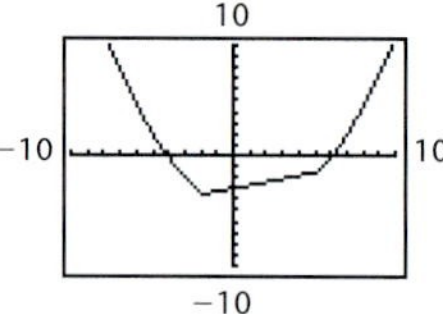

Graph of n

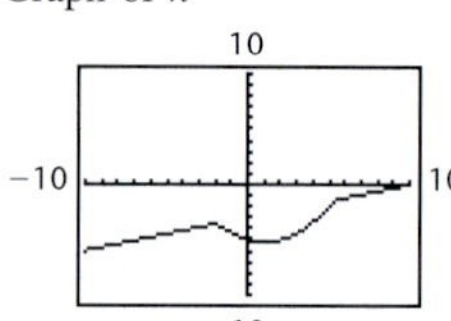

73. $m(x) = \max[f(x), g(x)]$

75. (A)

x	28	30	32	34	36
Mileage	45	52	55	51	47
$f(x)$	45.3	51.8	54.2	52.4	46.5

(B)

(C) $f(31) \approx 53.50$ thousand miles
$f(35) \approx 49.95$ thousand miles

77. (A) $s = f(w) = w/10$
(B) $f(15) = 1.5$ in, $f(30) = 3$ in
(C) Slope $= \frac{1}{10}$
(D)

79. $E(x) = \begin{cases} 200 & \text{if } 0 \le x \le 3{,}000 \\ 80 + 0.04x & \text{if } 3{,}000 < x < 8{,}000 \\ 180 + 0.04x & \text{if } x \ge 8{,}000 \end{cases}$

Discontinuous at $x = 8{,}000$
$E(5{,}750) = \$310$, $E(9{,}200) = \$548$

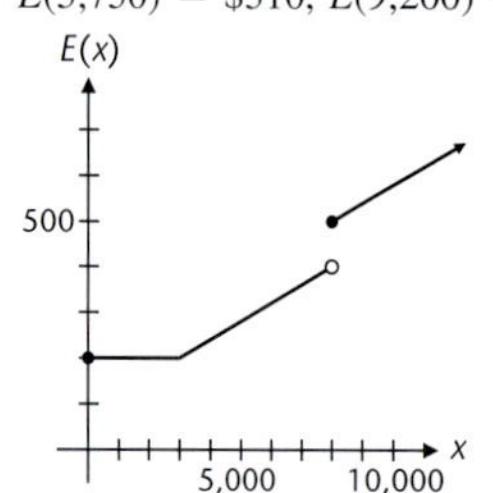

81. (A) $A(x) = 50x - x^2$
(B) Domain: $0 < x < 50$
(C)

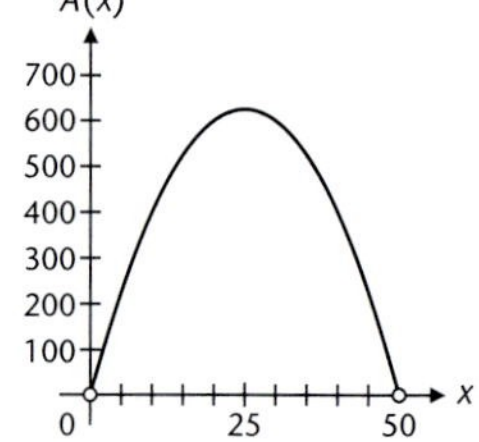

(D) 25 by 25 ft

83.

x	4	-4	6	-6	24	25	247	-243	-245	-246
$f(x)$	0	0	10	-10	20	30	250	-240	-240	-250

; f rounds numbers to the tens place.

85. $f(x) = [\![0.5 + 100x]\!]/100$

87. (A)
$$C(x) = \begin{cases} 15 & 0 < x \le 1 \\ 18 & 1 < x \le 2 \\ 21 & 2 < x \le 3 \\ 24 & 3 < x \le 4 \\ 27 & 4 < x \le 5 \\ 30 & 5 < x \le 6 \end{cases}$$

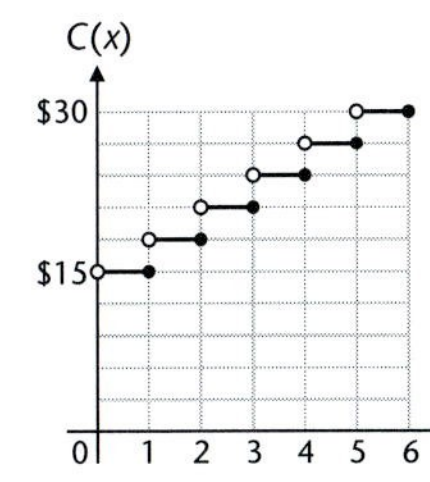

(B) No, since $f(x) \neq C(x)$ at $x = 1, 2, 3, 4, 5$, or 6

89. \$50 per day; maximum income = \$12,500 **91.** (A) $32\sqrt{5} \approx 71.55$ ft/s (B) 15 ft

Exercise 2-5

1. Domain: $[0, \infty)$; Range: $(-\infty, 0]$ **3.** Domain: R; Range: $(-\infty, 0]$ **5.** Domain: R; Range R

7. $(f + g)(x) = 5x + 1$, $(f - g)(x) = 3x - 1$, $(fg)(x) = 4x^2 + 4x$, $(f/g)(x) = 4x/(x + 1)$;
Domain of $f + g$ = Domain of $f - g$ = Domain of $fg = (-\infty, \infty)$, Domain of $f/g = (-\infty, -1) \cup (-1, \infty)$

9. $(f + g)(x) = 3x^2 + 1$, $(f - g)(x) = x^2 - 1$, $(fg)(x) = 2x^4 + 2x^2$, $(f/g)(x) = 2x^2/(x^2 + 1)$; Domain of each function $= (-\infty, \infty)$

11. $(f \circ g)(x) = x^4 - 8x^3 + 16x^2 + 3$, $(g \circ f)(x) = x^4 + 2x^2 - 3$; Domain of $f \circ g$ = Domain of $g \circ f = (-\infty, \infty)$

13. $(f \circ g)(x) = 2(x^3 - 1)^{2/3}$, $(g \circ f)(x) = 8x^2 - 1$; Domain of $f \circ g$ = Domain of $g \circ f = (-\infty, \infty)$

15.

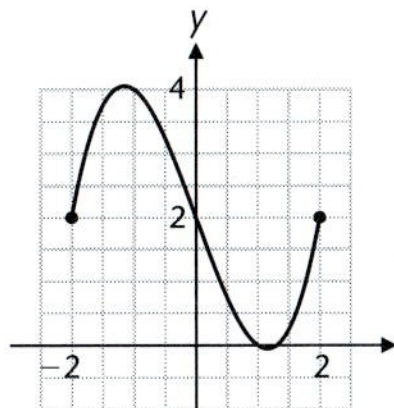

17.

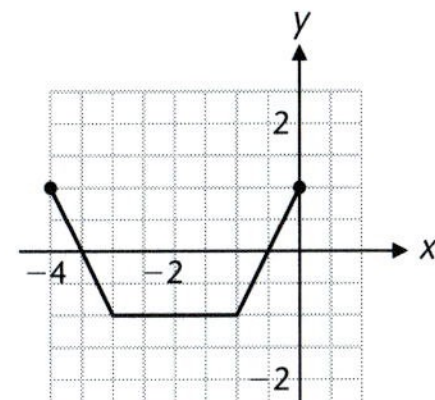

19.

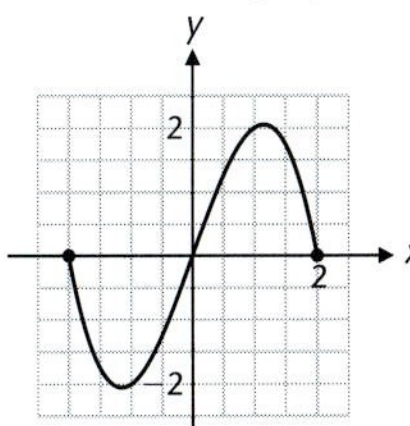

21.

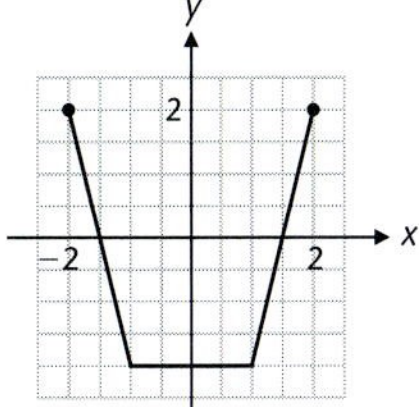

23. The graph of $y = |x|$ is shifted 2 units to the left and reflected in the x axis.

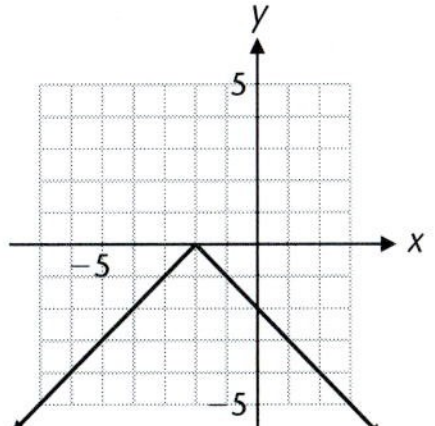

25. The graph of $y = x^2$ is shifted 2 units to the right and 4 units down.

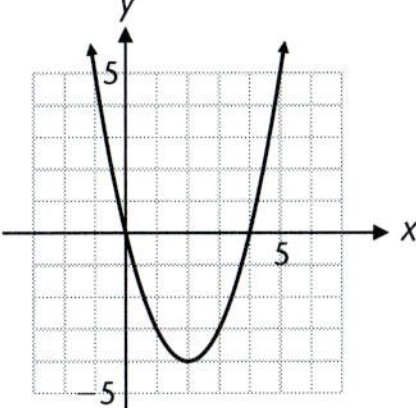

27. The graph of $y = \sqrt{x}$ is vertically expanded by a factor of 2, reflected in the x axis, and shifted 4 units up.

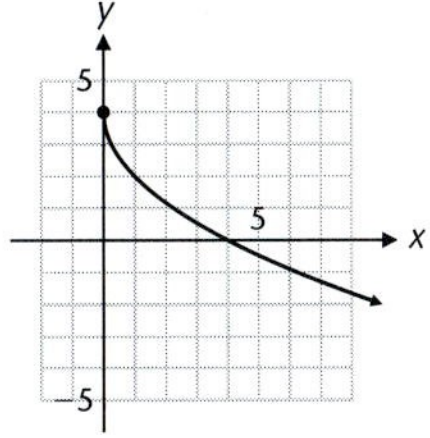

29. $(f + g)(x) = \sqrt{x + 2} + \sqrt{4 - x}$, $(f - g)(x) = \sqrt{x + 2} - \sqrt{4 - x}$, $(fg)(x) = \sqrt{8 + 2x - x^2}$, $(f/g)(x) = \sqrt{(x + 2)/(4 - x)}$;
Domain of $f + g$ = Domain of $f - g$ = Domain of $fg = [-2, 4]$, Domain of $f/g = [-2, 4)$

31. $(f + g)(x) = 8 - 3\sqrt{x}$, $(f - g)(x) = 2 - \sqrt{x}$, $(fg)(x) = 15 - 11\sqrt{x} + 2x$, $(f/g)(x) = (5 - 2\sqrt{x})/(3 - \sqrt{x})$;
Domain of $f + g$ = Domain of $f - g$ = Domain of $fg = [0, \infty)$, Domain of $f/g = [0, 9) \cup (9, \infty)$

33. $(f + g)(x) = \sqrt{x^2 + x - 2} + \sqrt{24 + 2x - x^2}$, $(f - g)(x) = \sqrt{x^2 + x - 2} - \sqrt{24 + 2x - x^2}$,
$(fg)(x) = \sqrt{-x^4 + x^3 + 28x^2 + 20x - 48}$, $(f/g)(x) = \sqrt{(x^2 + x - 2)/(24 + 2x - x^2)}$;
Domain of $f + g$ = Domain of $f - g$ = Domain of $fg = [-4, -2] \cup [1, 6]$, Domain of $f/g = (-4, -2] \cup [1, 6)$

35. $(f \circ g)(x) = \sqrt{4 - x} + 2$, $(g \circ f)(x) = \sqrt{2 - x}$; Domain of $f \circ g = (-\infty, 4]$, Domain of $g \circ f = (-\infty, 2]$

37. $(f \circ g)(x) = (3x - 5)/(x - 2)$, $(g \circ f)(x) = 1/(x + 1)$; Domain of $f \circ g = (-\infty, 2) \cup (2, \infty)$, Domain of $g \circ f = (-\infty, -1) \cup (-1, \infty)$

39. $(f \circ g)(x) = 3|(x - 2)/(x - 3)|$, $(g \circ f)(x) = |x + 2|/(|x + 2| - 3)$; Domain of $f \circ g = (-\infty, 3) \cup (3, \infty)$, Domain of $g \circ f = (-\infty, -5) \cup (-5, 1) \cup (1, \infty)$

41. $y = |x + 1| - 2$ **43.** $y = 3 - \sqrt[3]{x}$ **45.** $y = 1 - (x + 2)^3$

47. $y = \sqrt{x + 2} + 3$

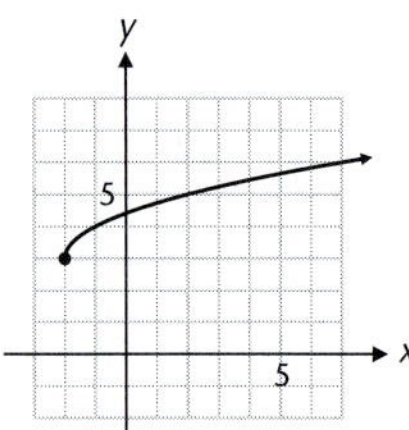

49. $y = -|x - 3|$

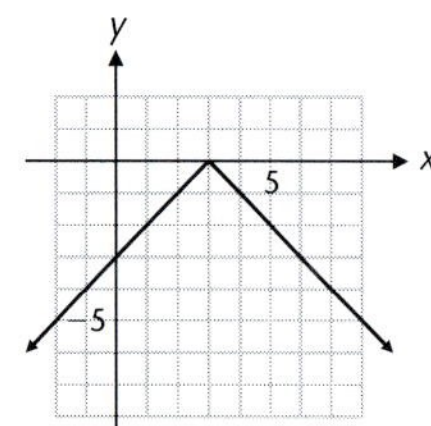

51. $y = -(x + 2)^3 + 1$

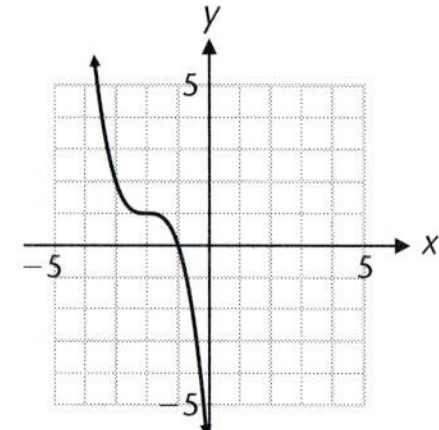

53. Reversing the order does not change the result. **55.** Reversing the order can change the result.

57. Reversing the order does not change the result.

59. $h(x) = (f \circ g)(x)$; $f(x) = x^4$, $g(x) = 2x - 7$ **61.** $h(x) = (f \circ g)(x)$; $f(x) = x^{1/2}$, $g(x) = 4 + 2x$

63. $h(x) = (g \circ f)(x)$; $g(x) = 3x - 5$, $f(x) = x^7$ **65.** $h(x) = (g \circ f)(x)$; $g(x) = 4x + 3$, $f(x) = x^{-1/2}$

67. The graph of $y = |x|$ is reflected in the x axis and vertically expanded by a factor of 3. Equation: $y = -3|x|$.

69. The graph of $y = x^3$ is reflected in the x axis and vertically contracted by a factor of 0.5. Equation: $y = -0.5x^3$.

75. $(f + g)(x) = 2x$, $(f - g)(x) = 2/x$, $(fg)(x) = x^2 - (1/x^2)$, $(f/g)(x) = (x^2 + 1)/(x^2 - 1)$; Domain of $f + g$ = Domain of $f - g$ = Domain of $fg = (-\infty, 0) \cup (0, \infty)$, Domain of $f/g = (-\infty, -1) \cup (-1, 0) \cup (0, 1) \cup (1, \infty)$

77. $(f + g)(x) = 2$, $(f - g)(x) = -2x/|x|$, $(fg)(x) = 0$, $(f/g)(x) = 0$; Domain of $f + g$ = Domain of $f - g$ = Domain of $fg = (-\infty, 0) \cup (0, \infty)$, Domain of $f/g = (0, \infty)$

79. $(f \circ g)(x) = 9 - x$, $(g \circ f)(x) = \sqrt{9 - x^2}$; Domain of $f \circ g = (-\infty, 9]$, Domain of $g \circ f = [-3, 3]$

81. $(f \circ g)(x) = (3x + 1)/(8 - x)$, $(g \circ f)(x) = (4x - 3)/(7 - x)$; Domain of $f \circ g = (-\infty, 3) \cup (3, 8) \cup (8, \infty)$, Domain of $g \circ f = (-\infty, 2) \cup (2, 7) \cup (7, \infty)$

83. $(f \circ g)(x) = \sqrt{x^2 + 1}$, $(g \circ f)(x) = \sqrt{x^2 + 1}$; Domain of $f \circ g = (-\infty, -2] \cup [2, \infty)$, Domain of $g \circ f = (-\infty, \infty)$

85. $P(p) = -70{,}000 + 6{,}000p - 200p^2$

87.

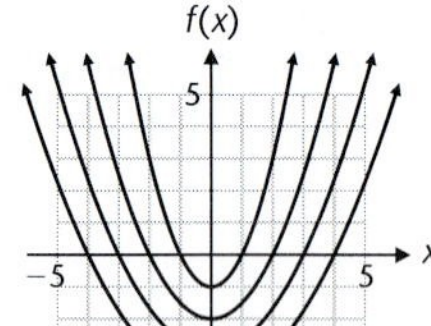

89.

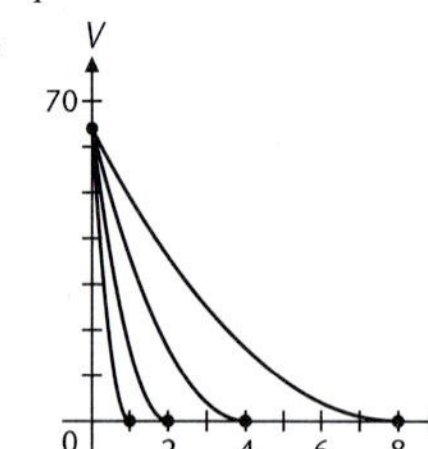

91. (A) $r(h) = \frac{1}{2}h$ (B) $V(h) = \frac{1}{12}\pi h^3$ (C) $V(t) = \frac{0.125}{12}\pi t^{3/2}$

Exercise 2-6

1. One-to-one **3.** Not one-to-one **5.** One-to-one **7.** Not one-to-one **9.** One-to-one **11.** Not one-to-one

13. One-to-one **15.** One-to-one **17.** One-to-one **19.** Not one-to-one **21.** One-to-one

23. One-to-one **25.** Not one-to-one **27.** Not one-to-one **29.** One-to-one

31. Domain of $f^{-1} = [1, 5]$
Range of $f^{-1} = [-4, 4]$

33. Domain of $f^{-1} = [-3, 5]$
Range of $f^{-1} = [-5, 3]$

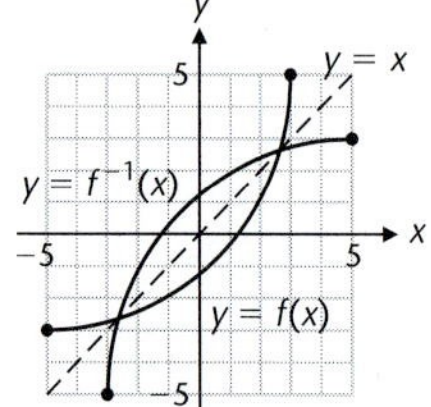

35.

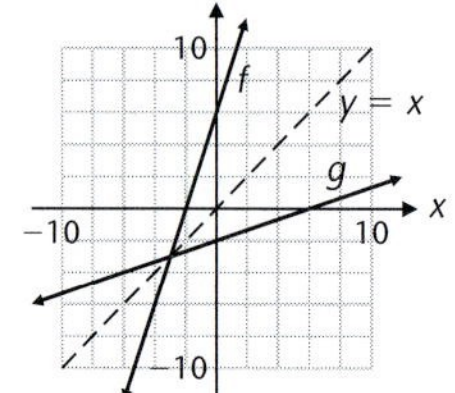

37.

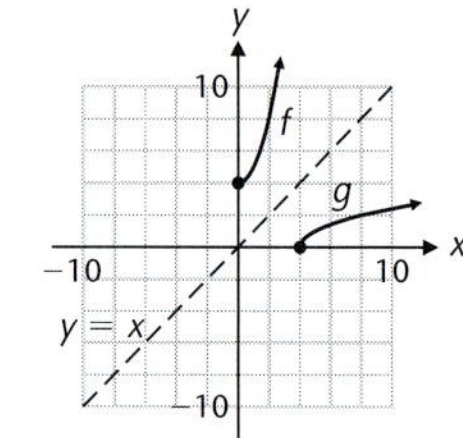

39. 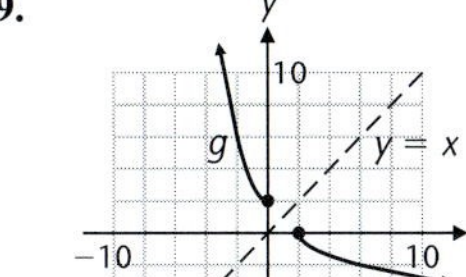

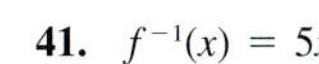

41. $f^{-1}(x) = 5x$ **43.** $f^{-1}(x) = \frac{1}{2}x - \frac{7}{2}$ **45.** $f^{-1}(x) = 5x - 2$ **47.** $f^{-1}(x) = 2/(3 - x)$

49. $f^{-1}(x) = x/(2 - x)$ **51.** $f^{-1}(x) = (5x + 4)/(2 - x)$ **53.** $f^{-1}(x) = 0.5\sqrt[3]{x + 5}$ **55.** $f^{-1}(x) = \frac{1}{3}(x - 2)^5 + \frac{7}{3}$
57. $f^{-1}(x) = 9 - \frac{1}{4}x^2, x \geq 0$ **59.** $f^{-1}(x) = -x^2 + 4x - 1, x \geq 2$
61. The x intercept of f is the y intercept of f^{-1}, and the y intercept of f is the x intercept of f^{-1}.
63. $f^{-1}(x) = 1 + \sqrt{x - 2}$ **65.** $f^{-1}(x) = -1 - \sqrt{x + 3}$

67. $f^{-1}(x) = \sqrt{9 - x^2}$
Domain of $f^{-1} = [-3, 0]$
Range of $f^{-1} = [0, 3]$

69. $f^{-1}(x) = -\sqrt{9 - x^2}$
Domain of $f^{-1} = [0, 3]$
Range of $f^{-1} = [-3, 0]$

71. $f^{-1}(x) = \sqrt{2x - x^2}$
Domain of $f^{-1} = [1, 2]$
Range of $f^{-1} = [0, 1]$

73. $f^{-1}(x) = -\sqrt{2x - x^2}$
Domain of $f^{-1} = [0, 1]$
Range of $f^{-1} = [-1, 0]$

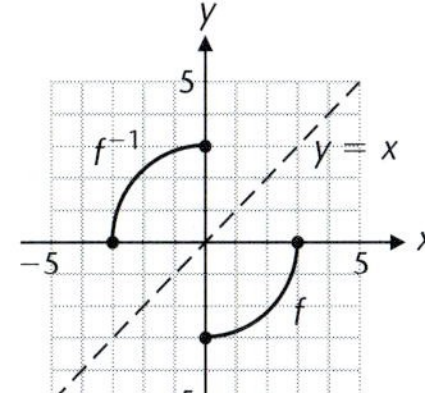

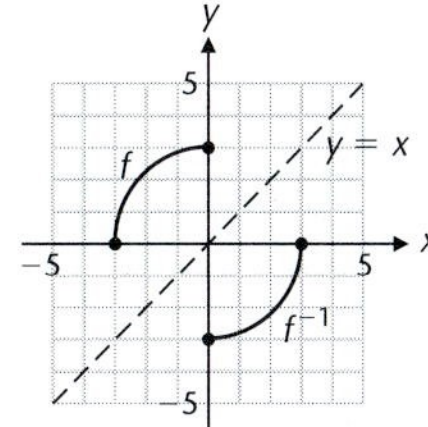

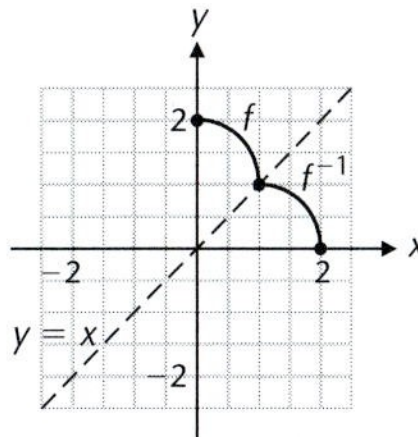

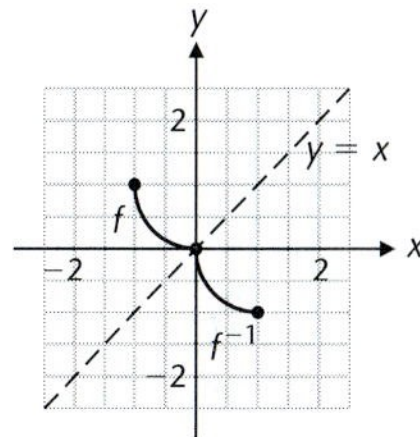

75. $f^{-1}(x) = (x - b)/a$ **77.** $a = 1$ and $b = 0$ or $a = -1$ and b arbitrary **81.** (A) $f^{-1}(x) = 2 - \sqrt{x}$ (B) $f^{-1}(x) = 2 + \sqrt{x}$
83. (A) $f^{-1}(x) = 2 - \sqrt{4 - x^2}, 0 \leq x \leq 2$ (B) $f^{-1}(x) = 2 + \sqrt{4 - x^2}, 0 \leq x \leq 2$

Chapter 2 Review Exercise

1. (A) $\sqrt{45}$ (B) $-\frac{1}{2}$ (C) 2 *(2-1, 2-2)* **2.** (A) $x^2 + y^2 = 7$ (B) $(x - 3)^2 + (y + 2)^2 = 7$ *(2-1)*
3. Center: $(-3, 2)$; radius $= \sqrt{5}$ *(2-1)* **4.** Slope $= -\frac{3}{2}$ *(2-2)* **5.** $2x + 3y = 12$ *(2-2)* **6.** $y = -\frac{2}{3}x + 2$ *(2-2)*

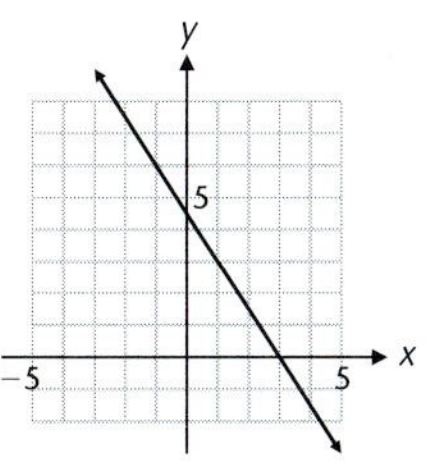

7. Vertical: $x = -3$, slope not defined; horizontal: $y = 4$, slope $= 0$ *(2-2)*
8. (A) Function; domain $= \{1, 2, 3\}$, range $= \{1, 4, 9\}$ (B) Not a function
(C) Function; domain $= \{-2, -1, 0, 1, 2\}$, range $= \{2\}$ *(2-3)*
9. (A) Not a function (B) A function (C) A function (D) Not a function *(2-3)*
10. Parts A and C specify functions. *(2-3)* **11.** 16 *(2-3)* **12.** 1 *(2-3)* **13.** 3 *(2-3)* **14.** $-2a - h$ *(2-3)*
15. $9 + 3x - x^2$ *(2-5)* **16.** $1 + 3x + x^2$ *(2-5)* **17.** $20 + 12x - 5x^2 - 3x^3$ *(2-5)* **18.** $(3x + 5)/(4 - x^2), x \neq \pm 2$ *(2-5)*
19. $17 - 3x^2$ *(2-5)* **20.** $-21 - 30x - 9x^2$ *(2-5)*
21. (A)

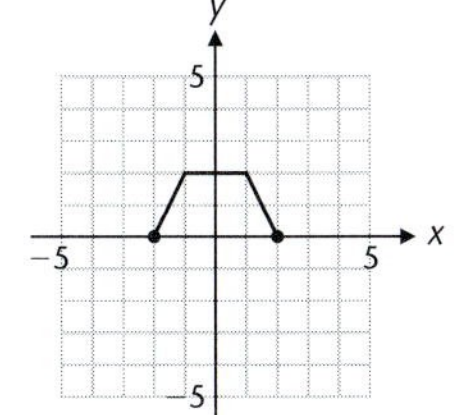

(B)

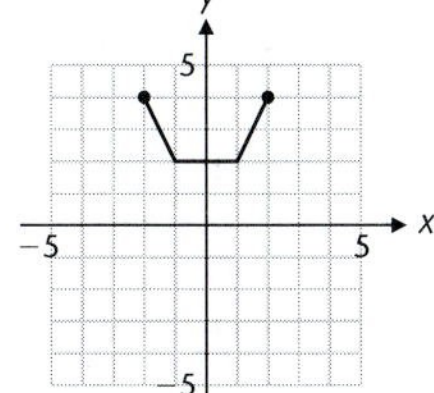

(C)

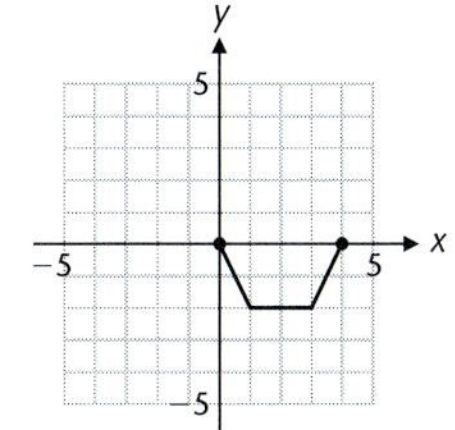

(D)

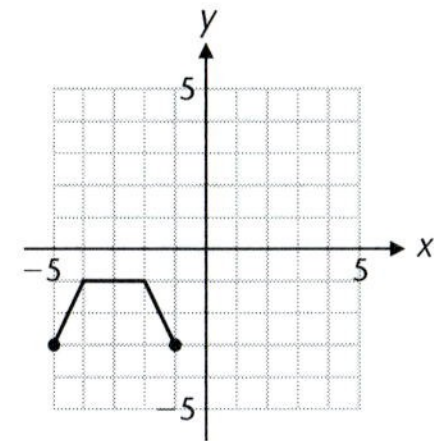

(2-5)

22. (A) g (B) m (C) n (D) f *(2-4, 2-5)*
23. (A) x intercepts: $-4, 0$; y intercept: 0 (B) Vertex: $(-2, -4)$ (C) Minimum: -4 (D) Range: $y \geq -4$ or $[-4, \infty)$
(E) Increasing on $[-2, \infty)$ (F) Decreasing on $(-\infty, -2]$ *(2-4)*
24. Min $f(x) = f(3) = 2$; vertex: $(3, 2)$ *(2-4)*
25. (A) Reflected across x axis (B) Shifted down 3 units (C) Shifted left 3 units *(2-5)*

26. (A) 0 (B) 1 (C) 2 (D) 0 *(2-4)* **27.** (A) $-2, 0$ (B) 1 (C) No solution (D) $x = 3$ and $x < -2$ *(2-4)*
28. Domain $= (-\infty, \infty)$, range $= (-3, \infty)$ *(2-4)* **29.** $[-2, -1], [1, \infty)$ *(2-4)* **30.** $[-1, 1)$ *(2-4)* **31.** $(-\infty, -2)$ *(2-4)*
32. $x = -2, x = 1$ *(2-4)* **33.** $f(x) = 4x^3 - \sqrt{x}$ *(2-3)*
34. The function f multiplies the square of the domain element by 3, adds 4 times the domain element, and then subtracts 6. *(2-3)*
35. (A) $3x + 2y = -6$ (B) $\sqrt{52}$ *(2-1, 2-2)* **36.** (A) $y = -2x - 3$ (B) $y = \frac{1}{2}x + 2$ *(2-2)*
37. It is symmetric with respect to all three. *(2-1)* **38.** $(-\infty, 3)$ *(2-3)*
39. Range $= [-4, \infty)$ *(2-4)* **40.** $[0, 16) \cup (16, \infty)$ *(2-3)*
Intercepts: $x = 1$ and $x = 5$, $y = 5$
Min $f(x) = f(3) = -4$

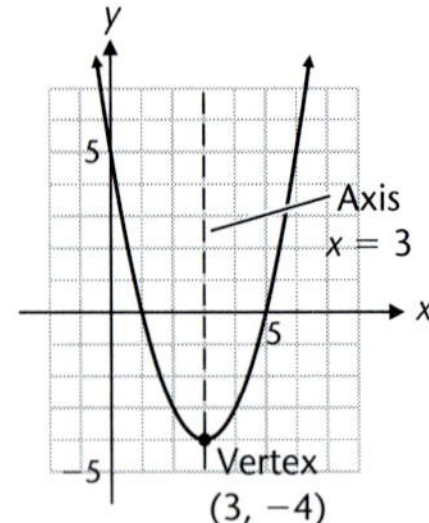

41. (A) $(f \circ g)(x) = \sqrt{|x|} - 8$, $(g \circ f)(x) = |\sqrt{x} - 8|$ (B) Domain of $f \circ g = (-\infty, \infty)$, domain of $g \circ f = [0, \infty)$ *(2-5)*
42. Functions (A), (C), and (D) are one-to-one *(2-6)* **43.** (A) $(x + 7)/3$ (B) 4 (C) x (D) Increasing *(2-6)*
44. Domain $= [-1, 1]$ *(2-4)*
Range $= [0, 1] \cup (2, 3]$
Discontinuous at $x = 0$

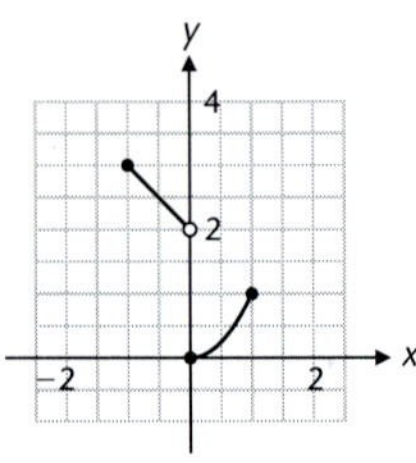

45. The graph of $y = x^2$ is vertically expanded by a factor of 2, reflected in the x axis, and shifted to the left 3 units.
Equation: $y = -2(x + 3)^2$. *(2-5)*
46. $g(x) = 5 - 3|x - 2|$ *(2-5)* **47.** $y = -(x - 4)^2 + 3$ *(2-4, 2-5)*

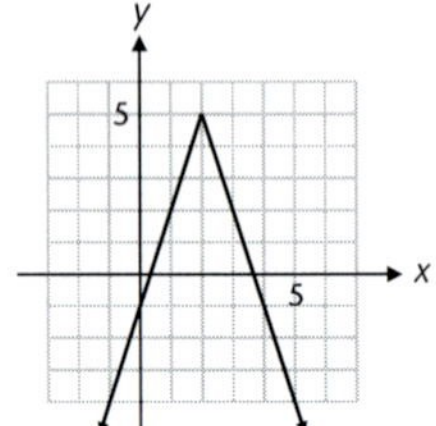

48. (A)

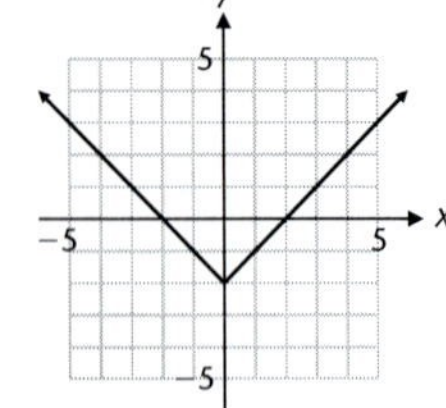

(B)

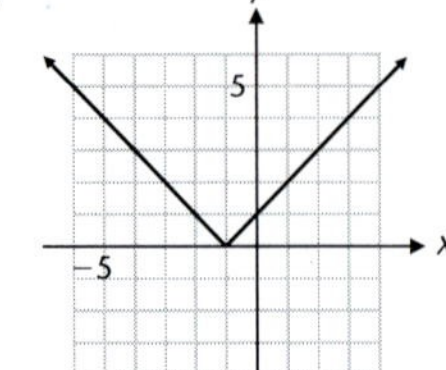

(C)

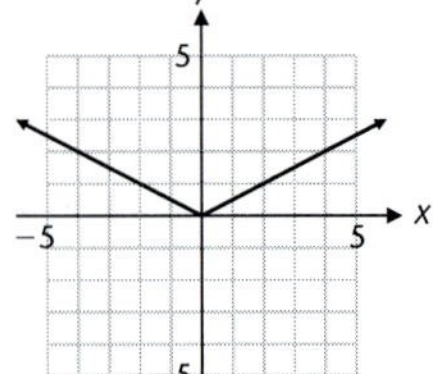

(2-5)

49. (A) $f^{-1}(x) = x^2 + 1$
(B) Domain of $f = [1, \infty) =$ Range of f^{-1}
Range of $f = [0, \infty) =$ Domain of f^{-1}
(C) *(2-6)*

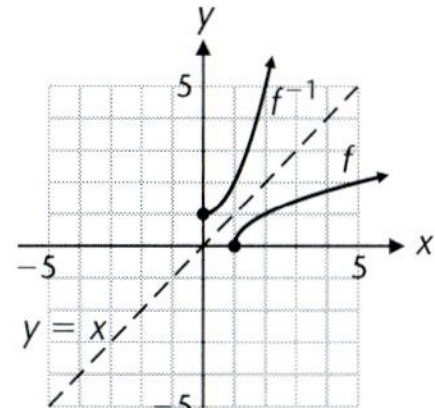

50. $(x - 3)^2 + y^2 = 32$ *(2-1)*

51. Center: $(-2, 3)$, radius $= 4$ *(2-1)*

52. Symmetric with respect to the origin *(2-1)*

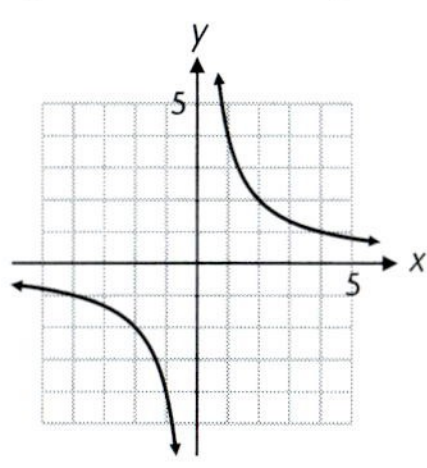

53. Decreasing *(2-2, 2-3)*

54. (A) Domain of $f = [0, \infty) =$ Range of f^{-1}; Range of $f = [-1, \infty) =$ Domain of f^{-1} (B) $\sqrt{x + 1}$ (C) 2 (D) 4 (E) x *(2-6)*

55. The graph of $y = \sqrt[3]{x}$ is vertically expanded by a factor of 2, reflected in the x axis, shifted 1 unit left and 1 unit down. Equation $y = -2\sqrt[3]{x + 1} - 1$ *(2-4)*

56. It is the same as the graph of g shifted to the right 2 units, reflected in the x axis, and shifted down 1 unit. *(2-5)*

57. *(2-5)*

58. $[-5, 5]$ *(1-8, 2-3)*

59. (A) $x^2\sqrt{1 - x}$, domain $= (-\infty, 1]$ (B) $x^2/\sqrt{1 - x}$, domain $= (-\infty, 1)$
(C) $1 - x$, domain $= (-\infty, 1]$ (D) $\sqrt{1 - x^2}$, domain $= [-1, 1]$ *(2-5)*

60. (A) $(3x + 2)/(x - 1)$ (B) $\frac{11}{2}$ (C) x *(2-6)*

61. $f(x) = \begin{cases} -2 & \text{if } x < -1 \\ 2x & \text{if } -1 \le x < 1 \\ 2 & \text{if } x \ge 1 \end{cases}$; domain $= (-\infty, \infty)$, range $= [-2, 2]$ *(2-4)*

62. $x - y = 3$; a line *(2-1, 2-2)*

65. (A)

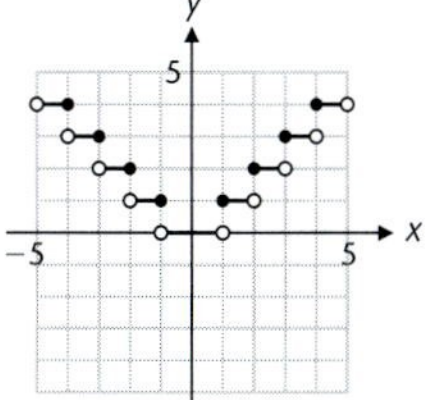

(B) *(2-5)*

66. Domain: All real numbers except $x = 2$,
Range: $y > -3$ or $(-3, \infty)$,
discontinuous at $x = 2$ *(2-4)*

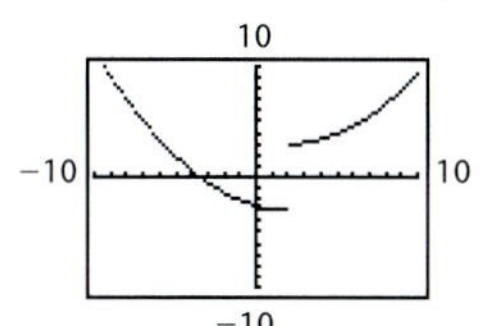

67. (A) (B) *(2-1, 2-5)*

68. (A) The graph must cross the x axis exactly once. (B) The graph can cross the x axis either once or not at all. *(2-4)*

69. (A) $V = -1{,}250t + 12{,}000$ (B) \$5,750 *(2-2)* **70.** (A) $R = 1.6C$ (B) \$168 *(2-2)*

71. $E(x) = \begin{cases} 200 & \text{if } 0 \le x \le 3{,}000 \\ 0.1x - 100 & \text{if } x > 3{,}000 \end{cases}$; $E(2{,}000) = \$200$, $E(5{,}000) = \$400$ *(2-4)*

72. (A)

x	0	5	10	15	20
Consumption	309	276	271	255	233
$f(x)$	303	286	269	252	234

(B)

(C) 217 in 1995, 200 in 2000 (D) Per capita egg consumption is dropping about 17 eggs every 5 years. *(2-4)*

73. (A) $C(x) = \begin{cases} 0.49x & \text{for } 0 \le x < 36 \\ 0.44x & \text{for } 36 \le x < 72 \\ 0.39x & \text{for } 72 \le x \end{cases}$ (B) Discontinuous at $x = 36$ and $x = 72$ *(2-4)*

74. (A) $C = 84{,}000 + 15x$; $R = 50x$ (B) $R = C$ at $x = 2{,}400$ units: $R < C$ for $0 \le x < 2{,}400$; $R > C$ for $x > 2{,}400$ *(2-2)*

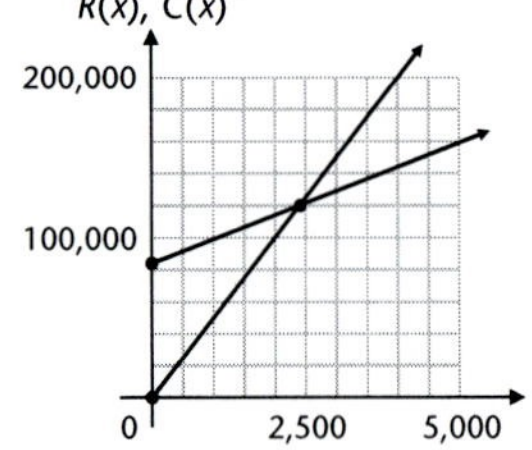

75. $P(p) = -14{,}000 + 700p - 10p^2$ *(2-5)* **76.** 5 ft *(2-1)*

77. (A) $A(x) = 60x - \frac{3}{2}x^2$ (B) $0 < x < 40$ (C) $x = 20$, $y = 15$ *(2-4)*

78. (A) 0 (B) 1 (C) 2 (D) 0 (E) 1 (F) 0 *(2-4)*

Cumulative Review Exercise: Chapters 1 and 2

1. $x = \frac{5}{2}$ *(1-1)* **2.** $x = 1$, $y = -2$ *(1-2)* **3.** $y \ge 5$ *(1-3)* $[5, \infty)$ **4.** $-5 < x < 9$ *(1-4)* $(-5, 9)$ **5.** $x \le -5$ or $x \ge 2$ *(1-8)* $(-\infty, -5] \cup [2, \infty)$

6. (A) $7 - 10i$ (B) $23 + 7i$ (C) $1 - i$ *(1-5)* **7.** $x = -4, 0$ *(1-6)* **8.** $x = -\sqrt{5}, \sqrt{5}$ *(1-6)*

9. $x = 3 \pm \sqrt{7}$ *(1-6)* **10.** $x = 3$ *(1-7)* **11.** $x \ge -\frac{2}{3}$ or $[-\frac{2}{3}, \infty)$ *(1-3)*

12. (A) $2\sqrt{5}$ (B) 2 (C) $-\frac{1}{2}$ *(2-1, 2-2)*

13. (A) $x^2 + y^2 = 2$ (B) $(x + 3)^2 + (y - 1)^2 = 2$ *(2-1)* **14.** Slope: $\frac{2}{3}$; y intercept: -2; x intercept: 3 *(2-2)*

15. (A) Function: domain; $\{1, 2, 3\}$; range: $\{1\}$ (B) Not a function
(C) Function: domain: $\{-2, -1, 0, 1, 2\}$; range: $\{-1, 0, 2\}$ *(2-3)*

16. (A) 20 (B) $x^2 + x + 3$ (C) $9x^2 - 18x + 13$ (D) $2a + h - 2$ *(2-2, 2-5)*

17. (A) Expanded by a factor of 2 (B) Shifted right 2 units (C) Shifted down 2 units *(2-5)*

18. (A) (B) *(2-5)*

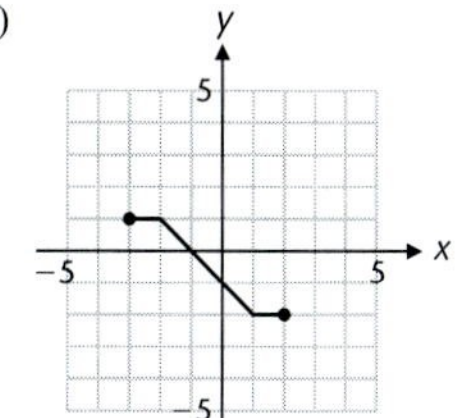

19. No solution *(1-1)* **20.** $x = \frac{1}{2}, 3$ *(1-6)* **21.** $x = 1, \frac{5}{2}$ *(1-7)* **22.** $x = 87/16, y = 5/8$ *(1-2)*

23. $x < \frac{3}{2}$ or $x > 3$ *(1-4)*
$(-\infty, \frac{3}{2}) \cup (3, \infty)$

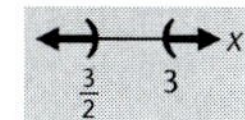

24. $\frac{2}{3} \le m \le 2$ *(1-4)*
$[\frac{2}{3}, 2]$

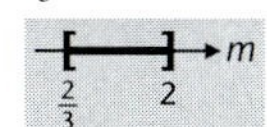

25. $-1 < x < 2$ or $x \ge 5$ *(1-8)*
$(-1, 2) \cup [5, \infty)$

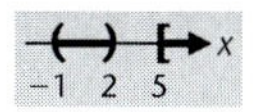

26. $x \ge 2, x \ne 4$ *(1-3)*
$[2, 4) \cup (4, \infty)$

27. (A) 0 (B) $\frac{6}{5}$ (C) $-i$ *(1-5)* **28.** (A) $3 + 18i$ (B) $-2.9 + 10.7i$ (C) $-4 - 6i$ *(1-5)*

29. (A) All real numbers (B) $\{-2\} \cup [1, \infty)$ (C) 1 (D) $[-3, -2]$ and $[2, \infty)$ (E) $-2, 2$ *(2-3, 2-4)*

30. (A) $y = -\frac{3}{2}x - 8$ (B) $y = \frac{2}{3}x + 5$ *(2-2)* **31.** $[-4, \infty)$ *(2-3)* **32.** Range: $[-9, \infty)$ *(2-4)*
Intercepts: $x = -2$ and $x = 4$, $y = -8$
Min $f(x) = f(1) = -9$

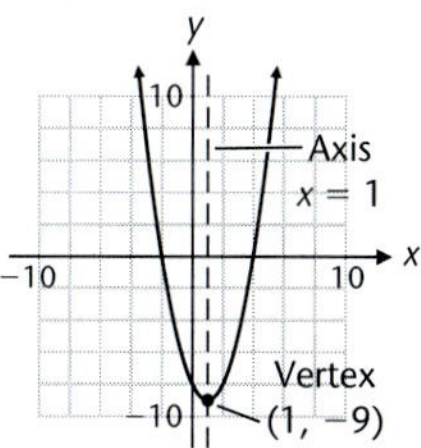

33. $(f \circ g)(x) = \dfrac{x}{3 - x}$;
Domain: $x \ne 0, 3$ *(2-5)*

34. $f^{-1}(x) = \frac{1}{2}x - \frac{5}{2}$ *(2-6)*

35. Domain: all real numbers *(2-4)*
Range: $(-\infty, -1) \cup [1, \infty)$
Discontinuous at $x = 0$

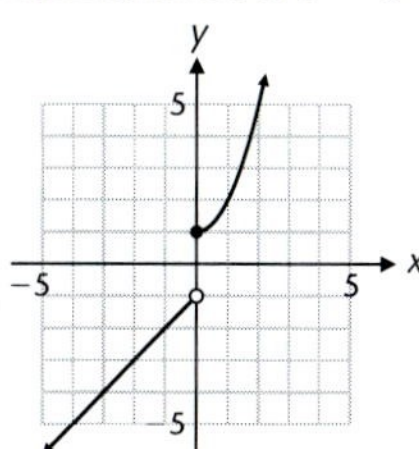

36. (A) (B) *(2-5)*

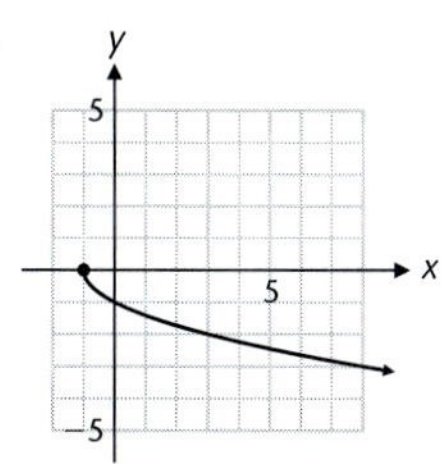

37. The graph of $y = |x|$ is contracted by $\frac{1}{2}$, reflected in the x axis, shifted 2 units to the right and 3 units up; $y = -\frac{1}{2}|x - 2| + 3$. *(2-5)*

38. (A) $f^{-1}(x) = x^2 - 4, x \geq 0$
(B) Domain $f = [-4, \infty) =$ Range f^{-1};
Range $f = [0, \infty) =$ Domain f^{-1}
(C) *(2-6)*

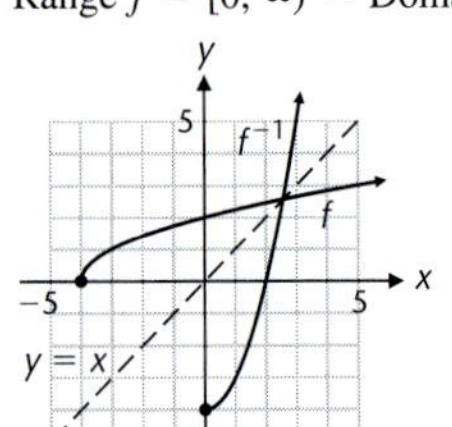

39. Center: $(3, -1)$; radius: $\sqrt{10}$ *(2-1)*

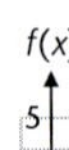

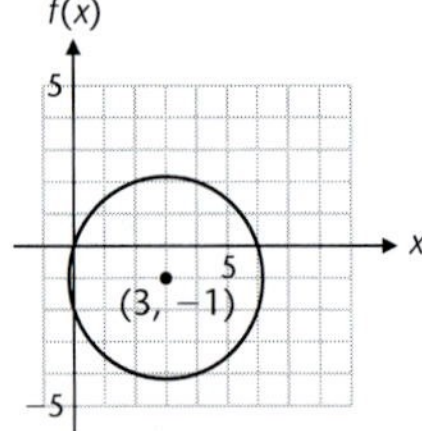

40. Symmetric with respect to the origin *(2-1)*

41. $y = (x + 2)^2 - 3$ *(2-4)* **42.** $y = 3 \pm i\sqrt{5}$ *(1-6)* **43.** $x = -\frac{1}{8}, \frac{27}{8}$ *(1-7)* **44.** $u = \pm\sqrt{3}, \pm 2i$ *(1-7)*
45. $t = \frac{9}{4}$ *(1-7)* **46.** $-18.36 \leq x < 16.09$; $[-18.36, 16.09)$ *(1-3)* **47.** $x = -5.68, 1.23$ *(1-6)*
48. $x = 1.49, y = 0.55$ *(1-2)* **49.** $y = \dfrac{3 - 3x}{x + 4}$ *(1-1)* **50.** $s = 5 + x - 2y, t = -12 - 2x + 5y$ *(1-2)*

51. *(2-5)*

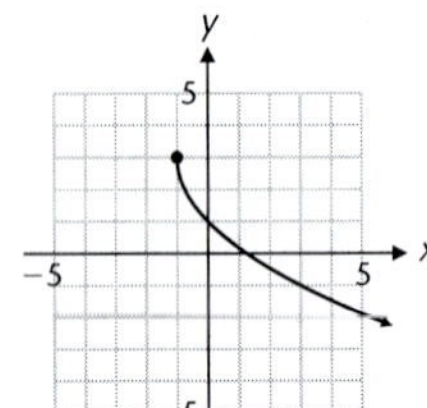

52. 0 *(1-5)* **53.** All a and b such that $a < b$ *(1-3)* **54.** $y = \dfrac{x + x^2}{x - 1}$ *(1-1)*

55. $x = (\sqrt{2} \pm i)/3$ *(1-6)*
56. If $b < -2$ or $b > 2$, there are two distinct real roots; if $b = -2$ or $b = 2$, there is one real double root; and if $-2 < b < 2$, there are two distinct imaginary roots. *(1-6, 1-8)*
57. $x = \pm\sqrt{3\sqrt{2} + 3}$ (two real roots) *(1-7)* **58.** $\dfrac{a^2 - b^2}{a^2 + b^2} + \dfrac{2ab}{a^2 + b^2}i$ *(1-5)*
59. $x < -1$ or $x > 2$; $(-\infty, -1) \cup (2, \infty)$ *(1-8)*
60. $f(x) = \begin{cases} -2x & \text{if } x < -2 \\ 4 & \text{if } -2 \leq x \leq 2 \\ 2x & \text{if } x > 2 \end{cases}$

Domain: all real numbers; Range: $[4, \infty)$ *(2-4)*

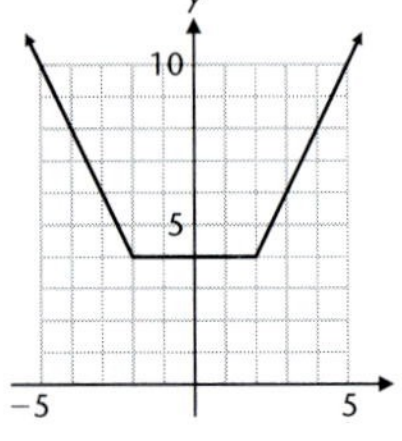

61. (A) Domain g: $[-2, 2]$
(B) $\left(\dfrac{f}{g}\right)(x) = \dfrac{x^2}{\sqrt{4 - x^2}}$; Domain $\left(\dfrac{f}{g}\right)$: $(-2, 2)$
(C) $(f \circ g)(x) = 4 - x^2$; Domain $(f \circ g)$: $[-2, 2]$ *(2-5)*

62. (A) $f^{-1}(x) = 1 + \sqrt{x + 4}$
(B) Domain f^{-1}: $[-4, \infty)$; Range f^{-1}: $[1, \infty)$
(C)

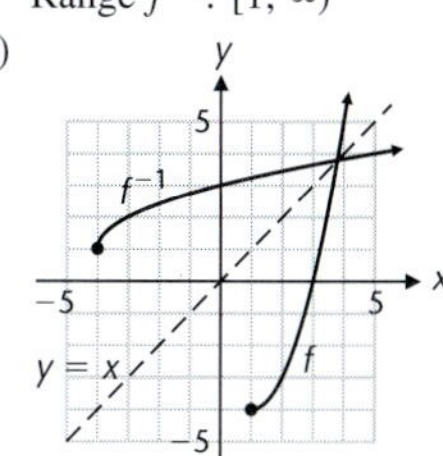

(2-6)

63. $f(x) = \begin{cases} \vdots & \\ 2x + 2 & \text{if } -1 \le x < -\frac{1}{2} \\ 2x + 1 & \text{if } -\frac{1}{2} \le x < 0 \\ 2x & \text{if } 0 \le x < \frac{1}{2} \\ 2x - 1 & \text{if } \frac{1}{2} \le x < 1 \\ 2x - 2 & \text{if } 1 \le x < \frac{3}{2} \\ 2x - 3 & \text{if } \frac{3}{2} \le x < 2 \\ \vdots & \end{cases}$

Domain: all real numbers; Range: $[0, 1)$; discontinuous at $x = k/2$, k an integer *(2-4)*

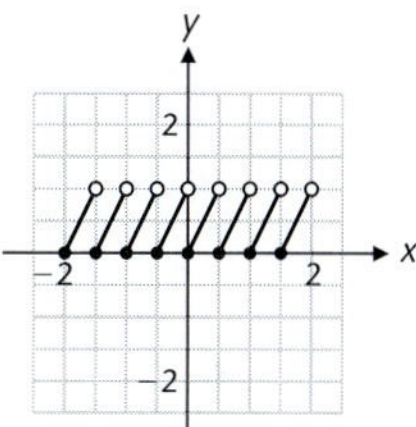

64. 2 or $-\frac{1}{2}$ *(1-6)* **65.** 10.5 min *(1-1)* **66.** 2.5 mph *(1-6)* **67.** 12 gal *(1-1)* **68.** 8,800 books *(1-2)*
69. $|p - 200| \le 10$ *(1-4)* **70.** \$16.50, 5,500 cheese heads *(1-2)*
71. (A) Profit: $\$5.5 < p < \$8$ or (\$5.5, \$8) (B) Loss: $\$0 \le p < 5.5$ or $p > \$8$ or $[\$0, \$5.5) \cup (\$8, \infty)$ *(1-8)*
72. 40 mi from A to B and 75 mi from B to C or 75 mi from A to B and 40 mi from B to C *(1-6)*
73. $x = -900p + 4{,}571$; 1,610 bottles *(2-2)*

74. $C(x) = \begin{cases} 0.06x & \text{if } 0 \le x \le 60 \\ 0.05x + 0.6 & \text{if } 60 < x \le 150 \\ 0.04x + 2.1 & \text{if } 150 < x \le 300 \\ 0.03x + 5.1 & \text{if } 300 < x \end{cases}$

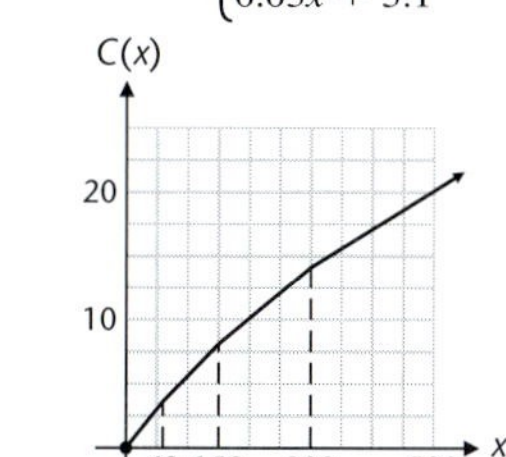

75. (A) $A(x) = 80x - 2x^2$ (B) $0 < x < 40$
(C) 20 × 40 ft *(2-4)*

A(x)
1,000
800
600
400
200
0
25
50
x

76. (A) $f(1) = f(3) = 1, f(2) = f(4) = 0$ (B) $f(n) = \begin{cases} 1 & \text{if } n \text{ is an odd integer} \\ 0 & \text{if } n \text{ is an even integer} \end{cases}$ *(2-4)*
77. (A) $a = 2{,}500$, $b = -16$ (B) 2,500 ft (C) 12.5 s *(1-1, 1-2, 1-6)*

CHAPTER 3

EXERCISE 3-1

1. c **3.** d **5.** h **7.** h, k **9.** $a - 2$; $R = 0$ **11.** $b - 3$; $R = -18$ **13.** $x^2 + 3$; $R = 5$ **15.** $4y^2 - 4y + 1$; $R = 0$
17. $x + 7 + \dfrac{6}{x - 3}$ **19.** $3x - 7 + \dfrac{7}{x + 2}$ **21.** $2x^2 - 3x + 1 - \dfrac{2}{x + 3}$ **23.** 4 **25.** −7 **27.** −2
29. $2x^4 + 2x^3 + 2x^2 - 3x - 3$; $R = 0$ **31.** $x^3 - 4x^2 + 16x - 64$; $R = 240$ **33.** $4x^3 + 3x^2 + x + 1$; $R = -4$
35. $x^5 + 2x^4 - x$; $R = 0$ **37.** $2x^3 + 6x^2 - 4x + 2$; $R = 0$ **39.** $3x^3 + 6x^2 - 3x + 9$; $R = 2$
41. $5x^3 - 3x^2 - 0.6x + 1.88$; $R = -4.624$ **43.** $5x^4 - x^3 + 4.6x^2 + 3.24x - 1.944$; $R = -4.8336$
45. The graph has three x intercepts and two turning points; $P(x) \to \infty$ as $x \to \infty$ and $P(x) \to -\infty$ as $x \to -\infty$

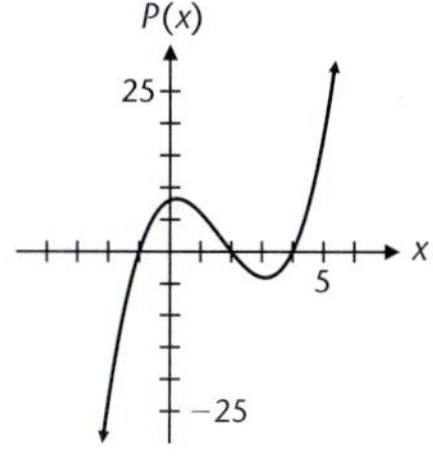

47. The graph has three x intercepts and two turning points; $P(x) \to \infty$ as $x \to \infty$ and $P(x) \to -\infty$ as $x \to -\infty$

P(x)
25
x
−5
−25

49. The graph has one x intercept and two turning points; $P(x) \to -\infty$ as $x \to \infty$ and $P(x) \to \infty$ as $x \to -\infty$

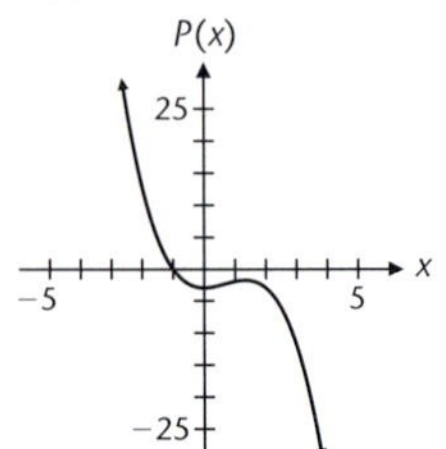

51. The graph has one x intercept and no turning points; $P(x) \to -\infty$ as $x \to \infty$ and $P(x) \to \infty$ as $x \to -\infty$

53. $P(x) = x^3$ **55.** No such polynomial exists. **57.** $x^2 + x + 2$; $R = 0$ **59.** $x^2 - 2x + 1$; $R = x + 3$

61. $x^3 + (2 + i)x^2 + (-3 + 2i)x - 3i$; $R = 0$ **63.** (A) -5 (B) $-40i$ (C) 0 (D) 0

65. The graph has two x intercepts and one turning point; $P(x) \to \infty$ as $x \to \infty$ and as $x \to -\infty$

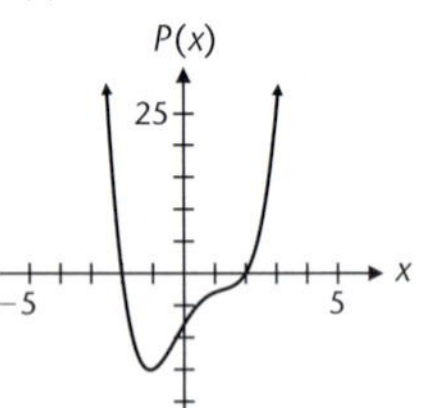

67. The graph has two x intercepts and three turning points; $P(x) \to \infty$ as $x \to \infty$ and as $x \to -\infty$

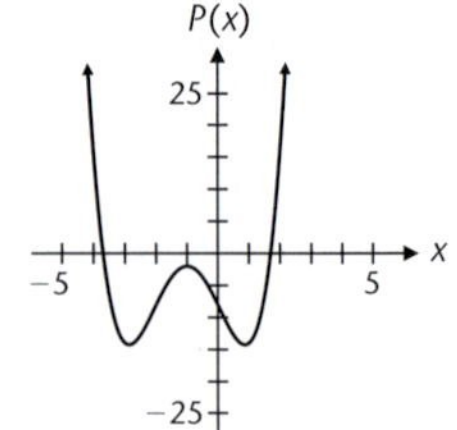
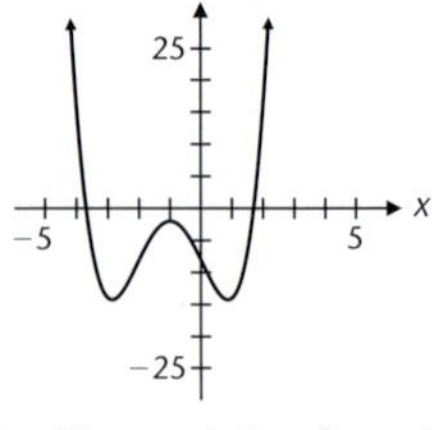

69. The graph has four x intercepts and three turning points; $P(x) \to -\infty$ as $x \to \infty$ and as $x \to -\infty$

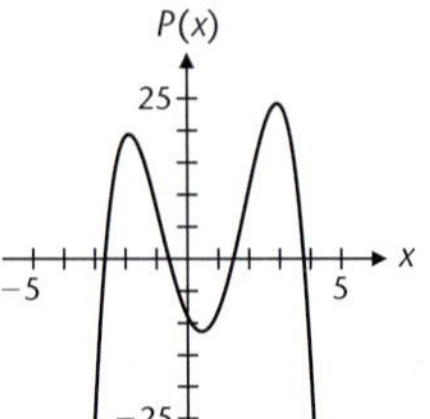

71. The graph has five x intercepts and four turning points; $P(x) \to \infty$ as $x \to \infty$ and $P(x) \to -\infty$ as $x \to -\infty$

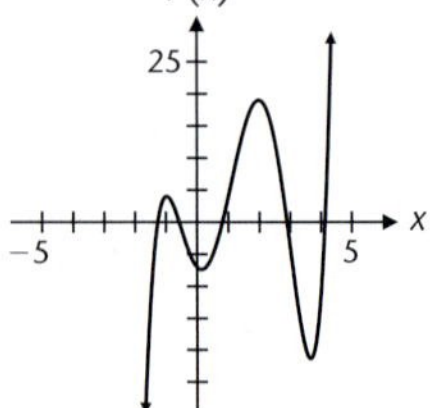

73. (A) In both cases, the coefficient of x is a_2, the constant term is $a_2r + a_1$, and the remainder is $(a_2r + a_1)r + a_0$.
(B) The remainder expanded is $a_2r^2 + a_1r + a_0 = P(r)$.

75. $P(-2) = 81$; $P(1.7) = 6.2$

Exercise 3-2

1. -2 (multiplicity 5), 3 (multiplicity 4); degree of $P(x)$ is 9 **3.** -1, 1 (multiplicity 4), 7 (multiplicity 3); degree of $P(x)$ is 8

5. $P(x) = (x - 2)^3(x + 1)$; degree 4 **7.** $P(x) = x(x - 2)(x + 1 - \sqrt{3})(x + 1 + \sqrt{3})$; degree 4

9. $P(x) = (x + 2)^2(x - 2)^2(x - 3 + i)(x - 3 - i)$; degree 6 **11.** $(x + 2)(x - 1)(x - 3)$, degree 3 **13.** $(x + 2)^2(x - 1)^2$, degree 4

15. $(x + 3)(x + 2)x(x - 1)(x - 2)$, degree 5 **17.** Yes **19.** No **21.** $\pm 1, \pm 2, \pm 5, \pm 10$ **23.** $\pm 1, \pm 5, \pm\frac{1}{2}, \pm\frac{5}{2}$

25. $\pm 1, \pm 5, \pm 25, \pm\frac{1}{2}, \pm\frac{5}{2}, \pm\frac{25}{2}, \pm\frac{1}{3}, \pm\frac{5}{3}, \pm\frac{25}{3}, \pm\frac{1}{6}, \pm\frac{5}{6}, \pm\frac{25}{6}$ **27.** $P(x) = (x + 4)^2(x + 1)$ **29.** $P(x) = (x - 1)(x + 1)(x - i)(x + i)$

31. $P(x) = (2x - 1)[x - (4 + 5i)][(x - (4 - 5i)]$ **33.** $\frac{3}{2}, 1 \pm \sqrt{3}$ **35.** -1, (double root), $\pm\sqrt{3}$ **37.** $\pm 3, -1 \pm \sqrt{2}$

39. $-2, -\frac{1}{2}, 1, 1 \pm i$ **41.** $-4, -3, 2$ **43.** $-0.2, 0, 1.5, 2$ **45.** 3 (double zero), $-2 \pm \sqrt{3}$

47. -1 (double zero), $\frac{2}{3}$, $1 \pm 2i$ **49.** $P(x) = (2x + 3)(3x - 1)(x + 2)$ **51.** $P(x) = (x + 3)[x - (2 + \sqrt{5})][x - (2 - \sqrt{5})]$

53. $P(x) = (x + 2)(2x + 1)(2x - 3)(x - 2)$ **55.** Inequality notation: $2 - \sqrt{3} \le x \le 2 + \sqrt{3}$; interval notation: $[2 - \sqrt{3}, 2 + \sqrt{3}]$

57. Inequality notation: $x \le -1$ or $1 \le x \le 3$; interval notation: $(-\infty, -1] \cup [1, 3]$

59. Inequality notation: $-3 \le x \le \frac{1}{2}$ or $x \ge 2$; interval notation: $[-3, \frac{1}{2}] \cup [2, \infty)$

61. $x^2 - 8x + 41$ **63.** $x^2 - 2ax + a^2 + b^2$ **65.** $1 - 2i, -2$ **67.** $3i, -4$ **69.** $3 + 2i, 1 \pm \sqrt{2}$

71. Inequality notation: $-\frac{5}{2} < x < -1$ or $x > 1$; interval notation: $(-\frac{5}{2}, -1) \cup (1, \infty)$

73. Inequality notation: $x \le -2$ or $-1 < x < 2$ or $3 < x \le 5$; interval notation: $(-\infty, -2] \cup (-1, 2) \cup (3, 5]$

75. $\frac{1}{3}, 6 \pm 2\sqrt{3}$ **77.** $-\frac{5}{2}, \frac{3}{2}, \pm 4i$ **79.** $\frac{3}{2}$ (double zero), $4 \pm \sqrt{6}$ **81.** (A) 3 (B) $-\frac{1}{2} - (\sqrt{3}/2)i, -\frac{1}{2} + (\sqrt{3}/2)i$

83. Max. $= n$, min. $= 1$ **85.** No, since $P(x)$ is not a polynomial with real coefficients (the coefficient of x is the imaginary number $2i$)

87. 2 ft **89.** 0.5 by 0.5 in or 1.59 by 1.59 in

Exercise 3-3

1. There is at least one x intercept in each of the intervals $(-5, -1)$, $(-1, 3)$, and $(5, 8)$
3. There is at least one x intercept in each of the intervals $(-6, -4)$, $(-4, 0)$, $(2, 4)$, and $(4, 7)$ **5.** Zeros in $(0, 1)$, $(3, 4)$, and $(4, 5)$
7. Zeros in $(-3, -2)$, $(-2, -1)$, and $(1, 2)$ **9.** Upper bound: 2; lower bound: -2
11. Upper bound: 3; lower bound: -2 **13.** Upper bound: 2; lower bound: -3
15. (A) Upper bound: 4; lower bound: -2; real zeros in $(-2, -1)$, $(0, 1)$, and $(3, 4)$ (B) 3.2
17. (A) Upper bound: 3; lower bound: -2; real zero in $(-2, -1)$ (B) -1.4
19. (A) Upper bound: 4; lower bound: -3; real zeros in $(-3, -2)$, $(-1, 0)$, $(1, 2)$, and $(3, 4)$ (B) 3.1
21. (A) Upper bound: 3; lower bound: -2; real zeros in $(-2, -1)$ and $(-1, 0)$ (B) -0.5
23. (A) Upper bound: 3; lower bound: -1 (B) 2.25 **25.** (A) Upper bound: 3; lower bound: -4 (B) -3.51, 2.12
27. (A) Upper bound: 2; lower bound: -3 (B) -2.09, 0.75, 1.88 **29.** (A) Upper bound: 1; lower bound: -1 (B) 0.83
31. (A) Upper bound: 5; lower bound: -2; real zeros in $(-2, -1)$, $(1, 2)$, and $(3, 4)$ (B) 3.22
33. (A) Upper bound: 4; lower bound: -4; real zeros in $(-4, -3)$, $(1, 2)$, and $(2, 3)$ (B) 2.92
35. (A) Upper bound: 30; lower bound: -10 (B) -1.29, 0.31, 24.98
37. (A) Upper bound: 30; lower bound: -40 (B) -36.53, -2.33, 2.40, 24.46
39. (A) Upper bound: 20; lower bound: -10 (B) -7.47, 14.03
41. (A) Upper bound: 30; lower bound: -20 (B) -17.66, 2.5 (double zero), 22.66
43. (A) Upper bound: 40; lower bound: -40 (B) -30.45, 9.06, 39.80
45. $x^4 - 3x^2 - 2x + 4 = 0$; $(1, 1)$ and $(1.7, 2.9)$ **47.** $4x^3 - 84x^2 + 432x - 600 = 0$; 2.3 in. or 4.6 in.
49. $x^3 - 15x^2 + 30 = 0$; 1.5 ft

Exercise 3-4

1. $g(x)$ **3.** $h(x)$ **5.** Domain: $(-\infty, -1) \cup (-1, \infty)$; x intercept: 2 **7.** Domain: $(-\infty, -4) \cup (-4, 4) \cup (4, \infty)$; x intercepts: -1, 1
9. Domain: $(-\infty, -3) \cup (-3, 4) \cup (4, \infty)$; x intercepts: -2, 3 **11.** Domain: all real numbers; x intercept: 0
13. Vertical asymptote: $x = 4$; horizontal asymptote: $y = 2$ **15.** Vertical asymptotes: $x = -4$, $x = 4$; horizontal asymptote: $y = \frac{2}{3}$
17. No vertical asymptotes; horizontal asymptote: $y = 0$ **19.** Vertical asymptotes: $x = -1$, $x = \frac{5}{3}$; no horizontal asymptote

21.
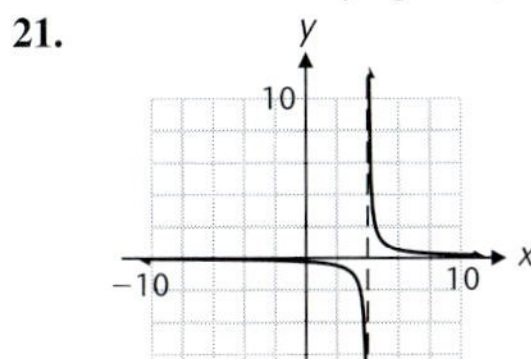

23.
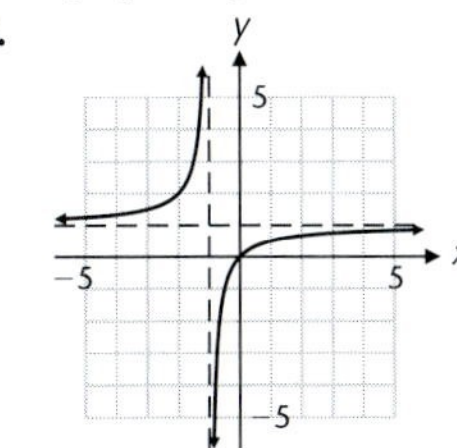

25.
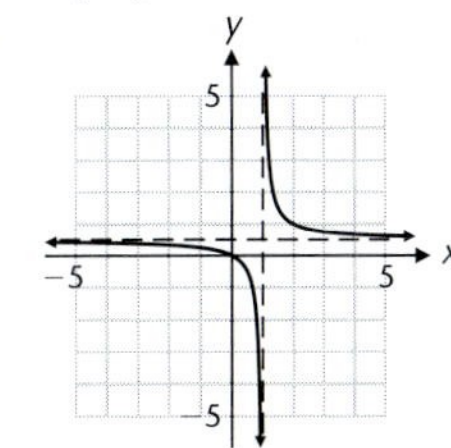

27.
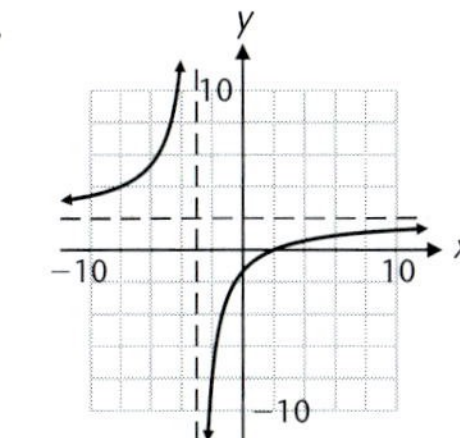

29.
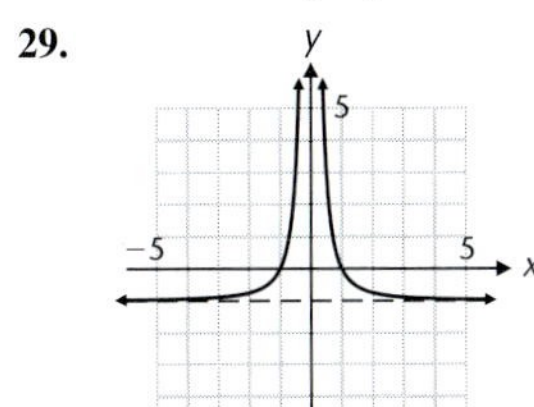

31.
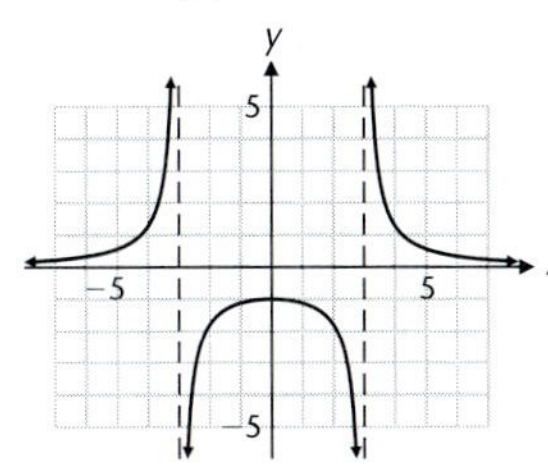

33.
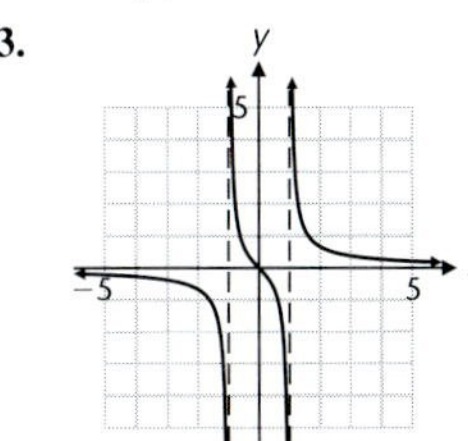

35.
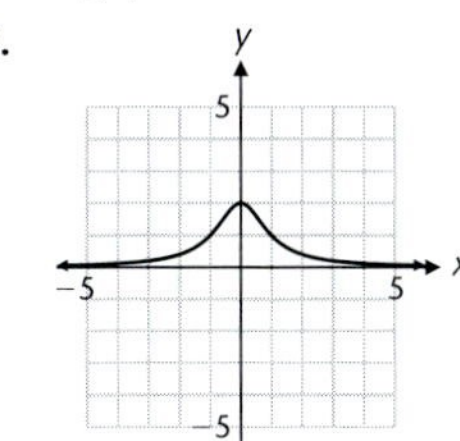

37.
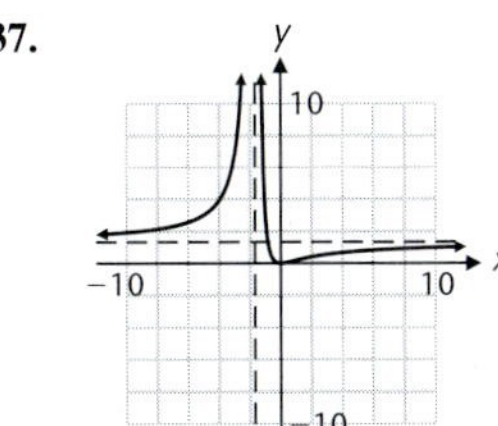

39.
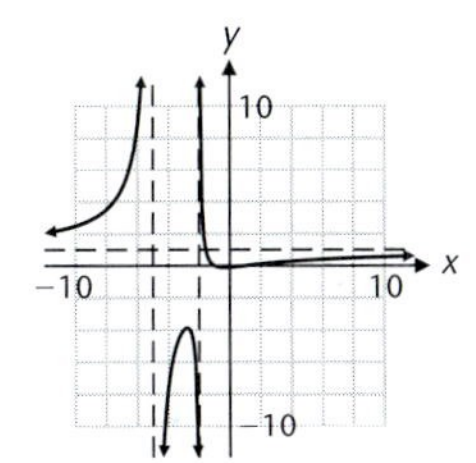

41. The maximum number of x intercepts is 2 and the minimum number is 0. For example, $(x^2 - 1)/x^2$ has two x intercepts and $(x^2 + 1)/x^2$ has none.
43. Vertical asymptote: $x = 1$; oblique asymptote: $y = 2x + 2$ **45.** Oblique asymptote: $y = x$
47. Vertical asymptote: $x = 0$; oblique asymptote: $y = 2x - 3$
49. $f(x) \to 5$ as $x \to \infty$ and $f(x) \to -5$ as $x \to -\infty$; the lines $y = 5$ and $y = -5$ are horizontal asymptotes.
51. $f(x) \to 4$ as $x \to \infty$ and $f(x) \to -4$ as $x \to -\infty$; the lines $y = 4$ and $y = -4$ are horizontal asymptotes.

53.

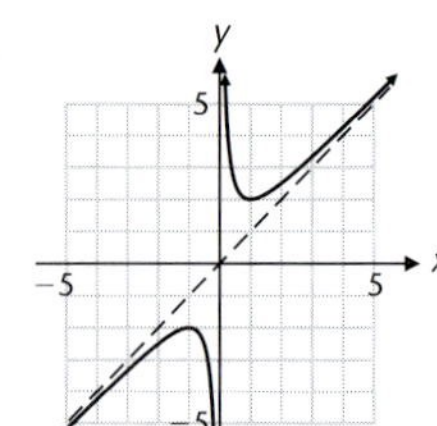

55.

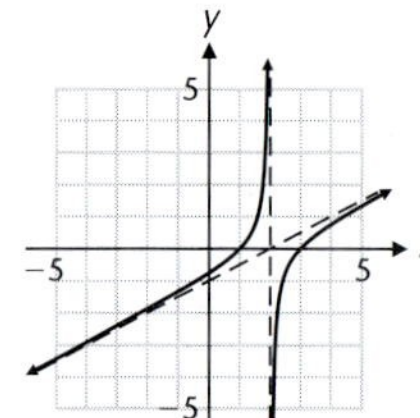

57.

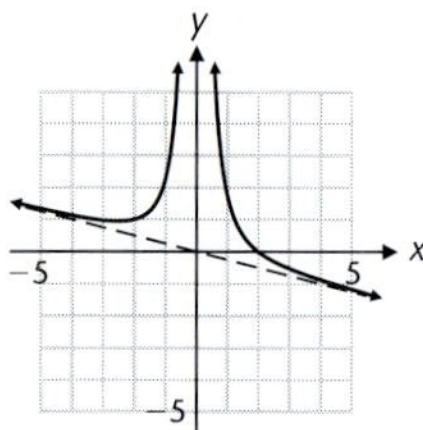

59. $p(x) = x^2 - 1$, $[f(x) - p(x)] \to 0$ as $x \to \pm\infty$ **61.** $p(x) = x^3 + x$, $[f(x) - p(x)] \to 0$ as $x \to \pm\infty$

63. Domain: $x \neq 2$, or $(-\infty, 2) \cup (2, \infty)$; $f(x) = x + 2$

65. Domain: $x \neq 2, -2$, or $(-\infty, -2) \cup (-2, 2) \cup (2, \infty)$; $f(x) = \dfrac{1}{x - 2}$

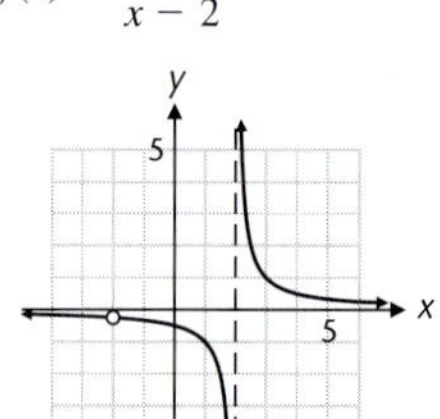

67. $N \to 50$ as $t \to \infty$

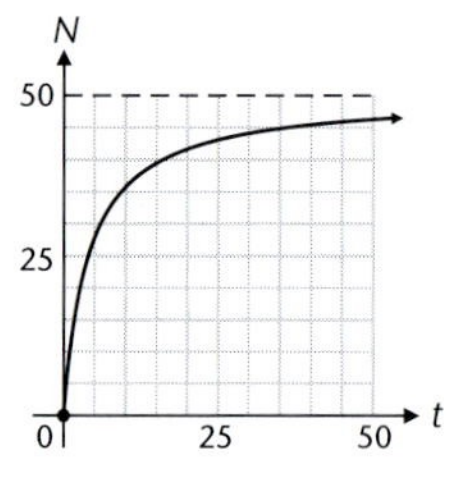

69. $N \to 5$ as $t \to \infty$

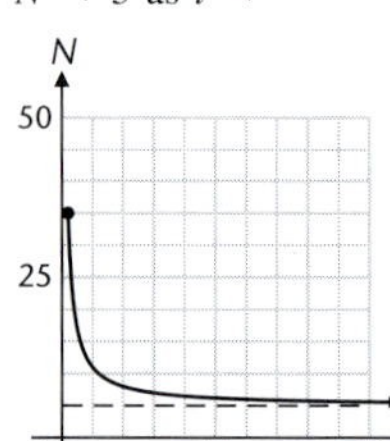

71. (A) $\overline{C}(n) = 25n + 175 + \dfrac{2{,}500}{n}$

(B) 10 yr

(C)

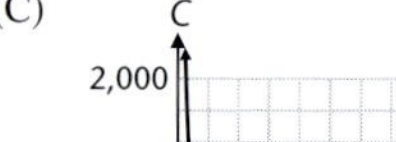

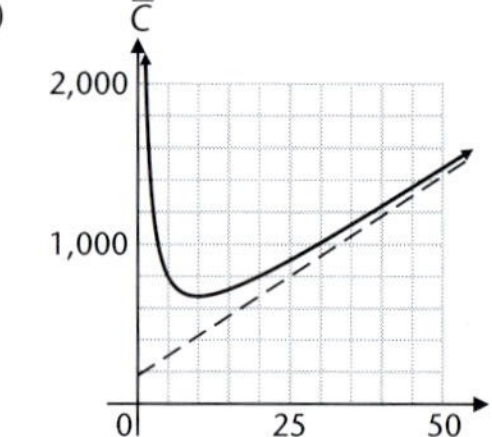

73. (A) $L(x) = 2x + \dfrac{450}{x}$

(B) $(0, \infty)$

(C) 15 ft by 15 ft

(D)

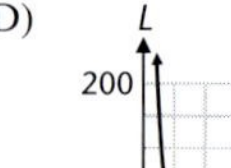

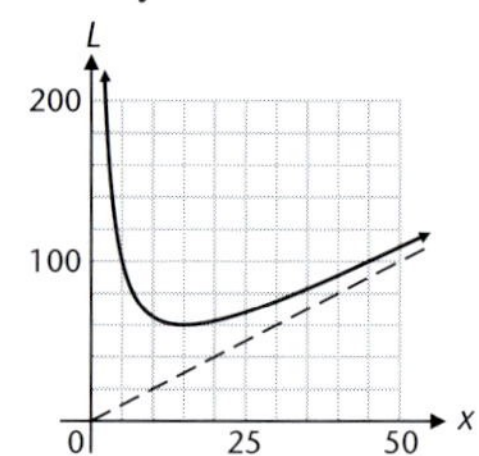

Exercise 3-5

1. $A = 2, B = 3$ **3.** $A = 3, B = -5$ **5.** $A = 2, B = 3, C = -4$ **7.** $A = 3, B = -2, C = 5$

9. $A = 2, B = -1, C = -3, D = 0$ **11.** $\dfrac{-4}{x - 4} + \dfrac{7}{x + 3}$ **13.** $\dfrac{4}{2x + 7} - \dfrac{2}{3x - 2}$ **15.** $\dfrac{4}{x} - \dfrac{5}{2x - 1} - \dfrac{3}{(2x - 1)^2}$

17. $\dfrac{3}{x} + \dfrac{4x + 1}{2x^2 + x + 1}$ **19.** $\dfrac{4x - 5}{x^2 + 1} + \dfrac{2x}{(x^2 + 1)^2}$ **21.** $\dfrac{3}{x + 1} + \dfrac{x - 3}{x^2 - x + 2}$ **23.** $x + 2 + \dfrac{3}{x - 2} + \dfrac{2}{(x - 2)^2} + \dfrac{x + 6}{x^2 + 2x + 4}$

25. $2x + \dfrac{7}{x - 3} + \dfrac{1}{x + 1} - \dfrac{3}{x^2 + 2x + 3}$ **27.** $\dfrac{1}{x + a} - \dfrac{a}{(x + a)^2}$ **29.** $\dfrac{1}{(a - b)(x - a)} - \dfrac{1}{(a - b)(x - b)}$

Chapter 3 Review Exercise

1. $2x^3 + 3x^2 - 1 = (x + 2)(2x^2 - x + 2) - 5$ *(3-1)* **2.** $P(3) = -8$ *(3-1, 3-2)* **3.** $2, -4, -1$ *(3-2)* **4.** $1 - i$ *(3-3)*

5. (A) $P(x) = (x + 2)x(x - 2) = x^3 - 4x$ (B) $P(x) \to \infty$ as $x \to \infty$ and $P(x) \to -\infty$ as $x \to -\infty$ *(3-1)*

6. Lower bound: $-2, -1$; upper bound: 4 *(3-3)* **7.** $P(1) = -5$ and $P(2) = 1$ are of opposite sign. *(3-3)*

8. $\pm 1, \pm 2, \pm 3, \pm 6$ *(3-2)* **9.** $-1, 2, 3$ *(3-2)*

10. (A) Domain: $(-\infty, -4) \cup (-4, \infty)$; x intercept: $\frac{3}{2}$ (B) Domain: $(-\infty, -2) \cup (-2, 3) \cup (3, \infty)$; x intercept: 0 *(3-4)*

11. (A) Horizontal asymptote: $y = 2$; vertical asymptote: $x = -4$
(B) Horizontal asymptote: $y = 0$; vertical asymptotes: $x = -2, x = 3$ *(3-4)*

12. $\dfrac{2}{x - 3} + \dfrac{5}{x + 2}$ *(3-5)*

13. (A) The graph of $P(x)$ has three x intercepts and two turning points; $P(x) \to \infty$ as $x \to \infty$ and $P(x) \to -\infty$ as $x \to -\infty$

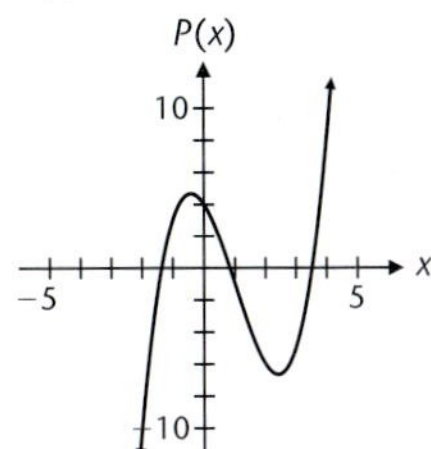

(B) 3.5 *(3-1, 3-3)*

14. $Q(x) = 8x^3 - 12x^2 - 16x - 8$, $R = 5$; $P(\frac{1}{4}) = R = 5$ *(3-1)* **15.** -4 *(3-1)* **16.** $P(x) = [x - (1 + \sqrt{2})][x - (1 - \sqrt{2})]$ *(3-2)*
17. Yes, since $P(-1) = 0$, $x - (-1) = x + 1$ must be a factor. *(3-2)* **18.** $-2, -\frac{1}{2}, 4$ *(3-2)*
19. $P(x) = (x + 2)(2x + 1)(x - 4)$ *(3-2)* **20.** No rational zeros *(3-2)* **21.** $-1, \frac{1}{2}, \dfrac{1 \pm i\sqrt{3}}{2}$ *(3-2)*
22. $(x + 1)(2x - 1)\left(x - \dfrac{1 + i\sqrt{3}}{2}\right)\left(x - \dfrac{1 - i\sqrt{3}}{2}\right)$ *(3-2)*
23. Inequality notation: $x \le -3$ or $-\frac{1}{2} \le x \le 2$; interval notation $(-\infty, -3] \cup [-\frac{1}{2}, 2]$ *(3-2, 3-8)*
24. (A) Upper bound: 7; lower bound: -5 (B) 6.62 (C) $-4.67, 6.62$ *(3-3)*
25. (A) Domain: $(-\infty, -1) \cup (-1, \infty)$; x intercept: 1; y intercept: $-\frac{1}{2}$ (B) Vertical asymptote: $x = -1$; horizontal asymptote: $y = \frac{1}{2}$
(C) *(3-4)*

26. $\dfrac{1}{x} - \dfrac{2}{x - 2} + \dfrac{3}{(x - 2)^2}$ *(3-5)* **27.** $\dfrac{3}{x} + \dfrac{2x - 1}{2x^2 - 3x + 3}$ *(3-5)*
28. $P(x) = [x - (1 + i)][x^2 + (1 + i)x + (3 + 2i)] + (3 + 5i)$ *(3-1)* **29.** $P(x) = (x + \frac{1}{2})^2(x + 3)(x - 1)^3$, degree 6 *(3-2)*
30. $P(x) = (x + 5)[x - (2 - 3i)][x - (2 + 3i)]$, degree 3 *(3-2)* **31.** $\frac{1}{2}, \pm 2, 1 \pm \sqrt{2}$ *(3-2)*
32. $(x - 2)(x + 2)(2x - 1)[x - (1 - \sqrt{2})][x - (1 + \sqrt{2})]$ *(3-2)*
33. Inequality notation: $-3 < x \le -\frac{3}{2}$ or $-\frac{1}{2} < x \le \frac{1}{2}$ or $x > 2$; interval notation: $(-3, -\frac{3}{2}] \cup (-\frac{1}{2}, \frac{1}{2}] \cup (2, \infty)$ *(3-2, 3-8)*
34. Since $P(x)$ changes sign three times, the minimal degree is 3. *(3-3)*
35. $P(x) = a(x - r)(x^2 - 2x + 5)$, and since the constant term, $-5ar$, must be an integer, r must be a rational number. *(3-2)*
36. (A) 3 (B) $-\dfrac{3}{2} \pm \dfrac{3i\sqrt{3}}{2}$ *(3-2)* **37.** (A) Upper bound: 30; lower bound: -30 (B) $-23.54, 21.57$ *(3-3)*
38. *(3-4)*

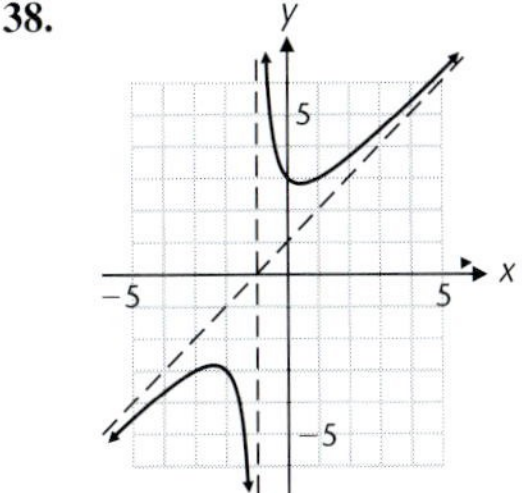

39. $y = 2$ and $y = -2$ *(3-4)* **40.** $\dfrac{2}{x - 3} - \dfrac{3}{x} + \dfrac{x - 1}{x^2 + 1}$ *(3-5)* **41.** $2x^3 - 32x + 48 = 0$, 4×12 ft or 5.2×9.2 ft *(3-2)*
42. $x^3 + 27x^2 - 729 = 0$, 4.8 ft *(3-3)* **43.** $4x^3 - 70x^2 + 300x - 300$, 1.4 in. or 4.5 in. *(3-3)*
44. $x^4 - 7x^2 - 2x + 8$, $(-2, 4)$, $(-1.6, 2.6)$, $(1, 1)$, $(2.6, 6.8)$ *(3-2)*

CHAPTER 4

Exercise 4-1

1.

x	y
−3	0.04
−2	0.11
−1	0.33
0	1
1	3
2	9
3	27

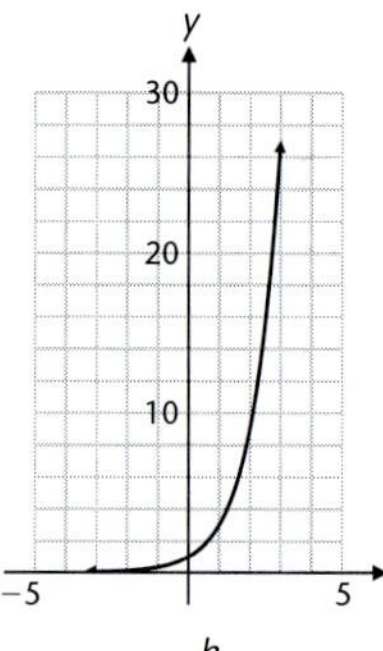

3.

x	y
−3	27
−2	9
−1	3
0	1
1	0.33
2	0.11
3	0.04

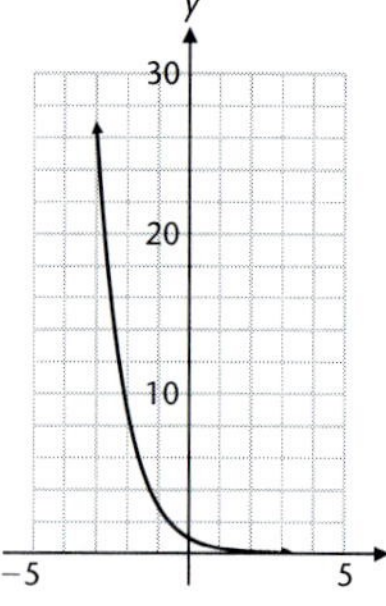

5.

x	$g(x)$
−3	−27
−2	−9
−1	−3
0	−1
1	−0.33
2	−0.11
3	−0.04

7.

x	$h(x)$
−3	0.19
−2	0.56
−1	1.67
0	5
1	15
2	45
3	135

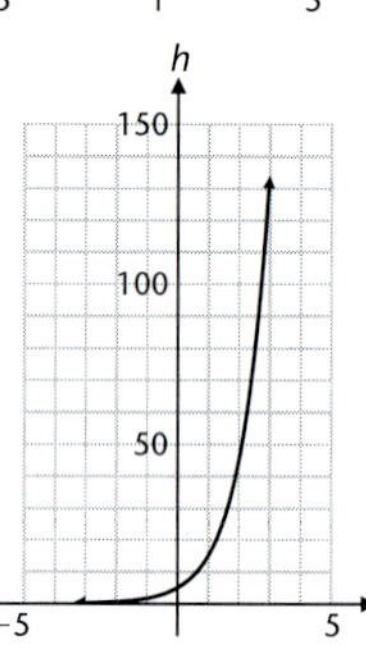

9.

x	$h(x)$
−6	−4.96
−5	−4.89
−4	−4.67
−3	−4
−2	−2
−1	4
0	22

11. 2^{3x+4} **13.** 3^{-4x+7y} **15.** 4^{3x}

17. 10^{4x} **19.** $\frac{5^{3x}}{4^{3y}}$ **21.** $a^2b^2c^4$ **23.** $x = -3.5$ **25.** $x = 0, 2$ **27.** $x = 0.5$ **29.** $x = 6$ **31.** $x = 2$

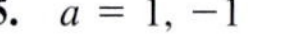

33. $x = 1, 1.5$ **35.** $a = 1, -1$ **37.**

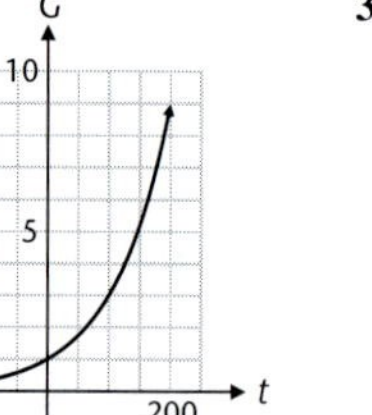

39.

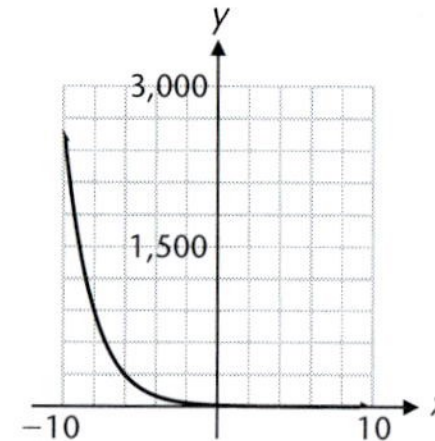

41.

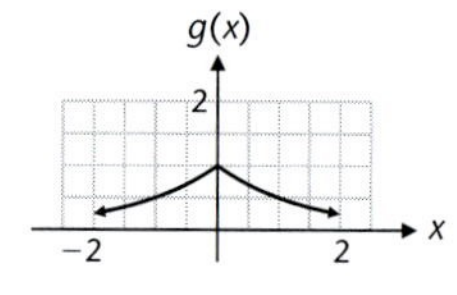

43.

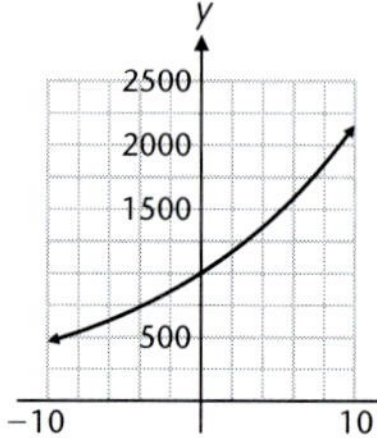

45.

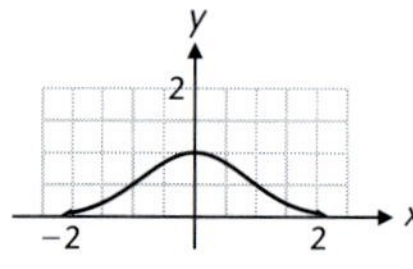

47. $6^{2x} - 6^{-2x}$ **49.** 4 **51.**

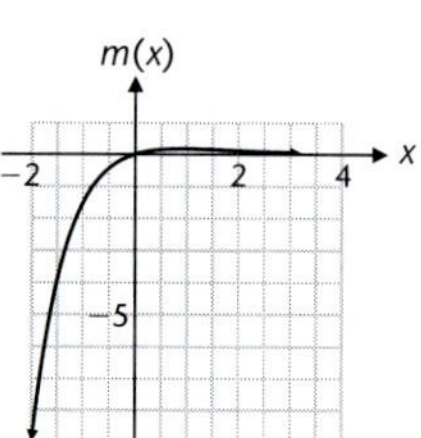

53.

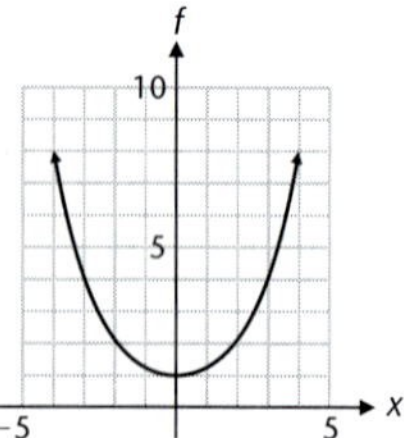

55. (A) 1.46 (B) $f(x) \to \infty$ as $x \to \infty$; $f(x) \to -5$ as $x \to -\infty$; $y = -5$

57. (A) −1.08 (B) $f(x) \to \infty$ as $x \to \infty$; $f(x) \to -\infty$ as $x \to -\infty$; none

59.

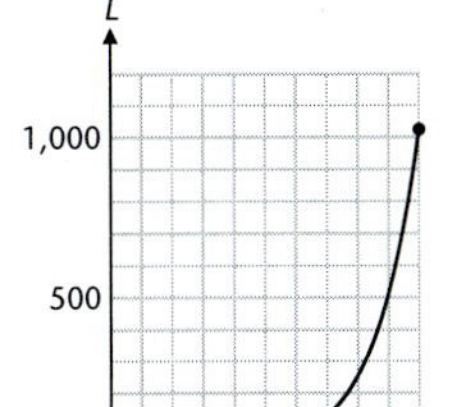

61. (A) 76 flies (B) 570 flies **63.** (A) 19 lb (B) 7.9 lb

65. (A) $4,225.92 (B) $12,002.71 **67.** $9,841 **69.** No

Exercise 4-2

1.

x	y
−3	−0.05
−2	−0.14
−1	−0.37
0	−1
1	−2.72
2	−7.39
3	−20.09

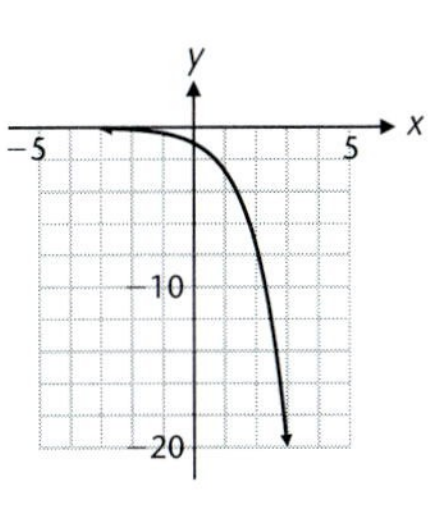

3.

x	y
−5	3.68
−4	4.49
−3	5.49
−2	6.7
−1	8.19
0	10
1	12.21
2	14.92
3	18.22
4	22.26
5	27.18

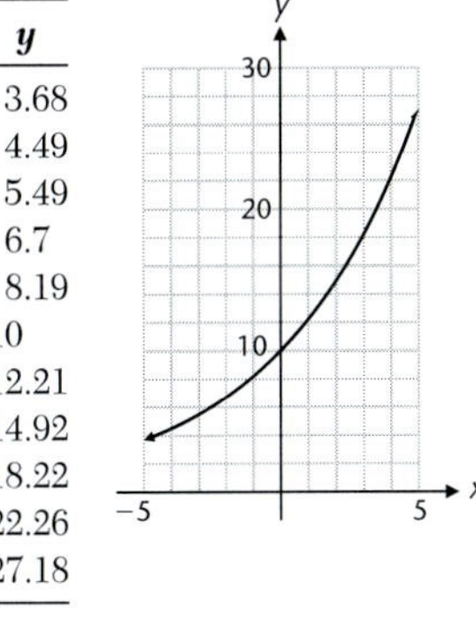

5.

t	$f(t)$
−5	164.87
−4	149.18
−3	134.99
−2	122.14
−1	110.52
0	100
1	90.48
2	81.87
3	74.08
4	67.03
5	60.65

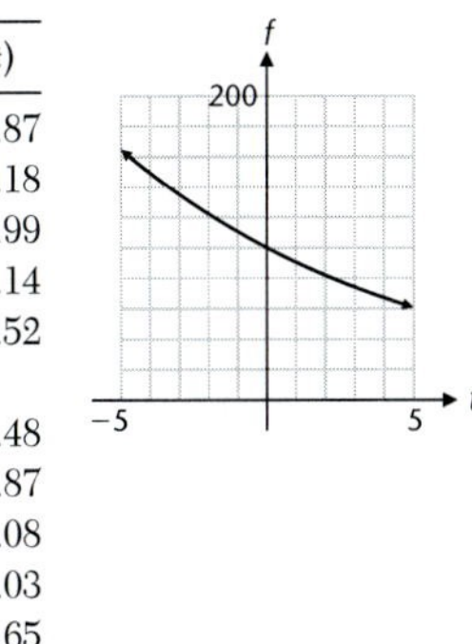

7. e^{-x} **9.** e^{3x} **11.** e^{3x-1} 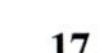 **13.** (A) $1 + 1/m$ is not equal to 1 (B) e 

15.

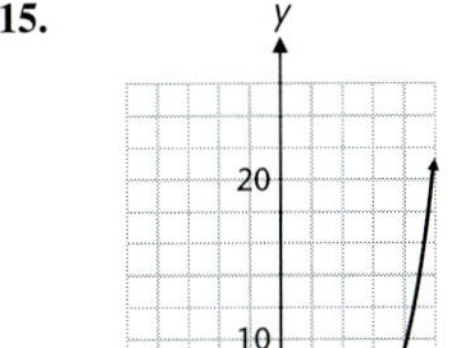

17.

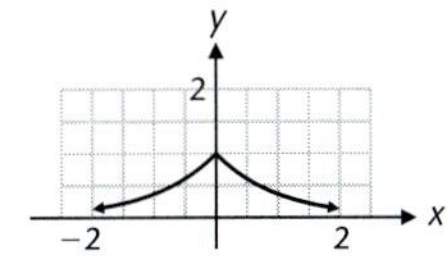

19.

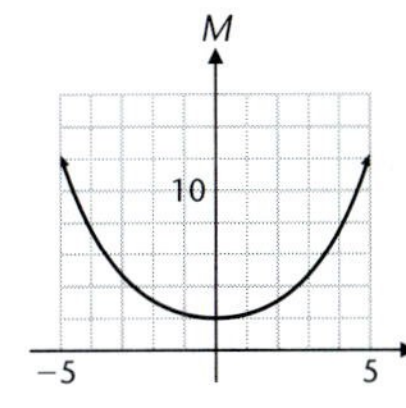

21.

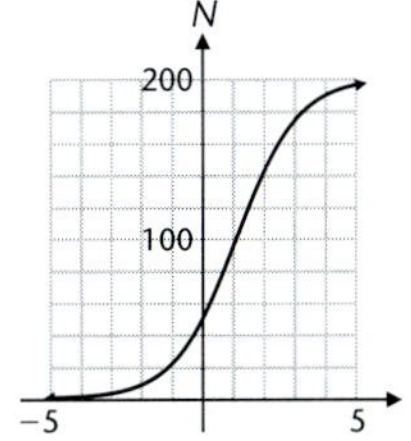

23. $\dfrac{e^{-2x}(-2x - 3)}{x^4}$ **25.** $2e^{2x} + 2e^{-2x}$ **27.** $2e^{2x}$ **29.** $x = 0$ **31.** $x = 0, 5$ **33.**

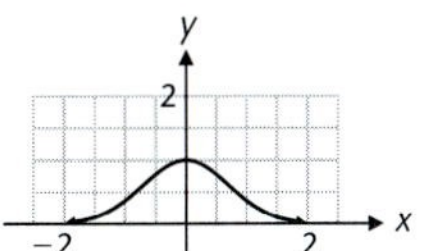

35. (A)

s	$f(s)$
−0.5	4.0000
−0.2	3.0518
−0.1	2.8680
−0.01	2.7320
−0.001	2.7196
−0.0001	2.7184

s	$f(s)$
0.5	2.2500
0.2	2.4883
0.1	2.5937
0.01	2.7048
0.001	2.7169
0.0001	2.7181

(B) $2.718 \cdots \approx e$ **37.**

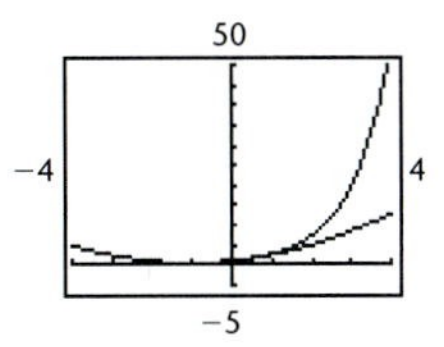

39.

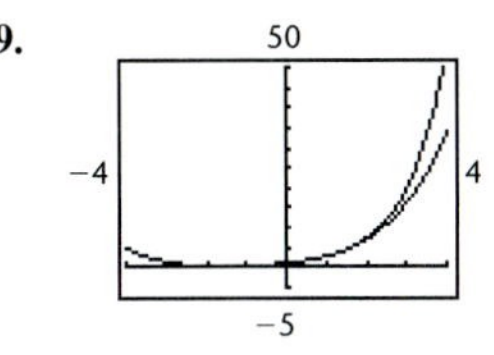

41. As $x \to \infty$, $f_n(x) \to 0$; as $x \to -\infty$, $f_n(x) \to \infty$ if n is even, and $f_n(x) \to -\infty$ if n is odd; $y = 0$ is a horizontal asymptote.
43. 7.1 billion **45.** 2006

47.

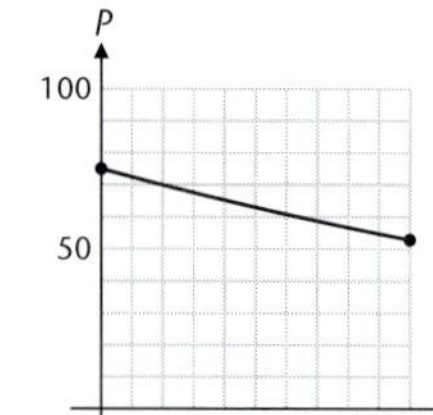

49. (A) 62% (B) 39% **51.** (A) \$10,691.81 (B) \$36,336.69

53. Gill Savings: \$1,230.60; Richardson S&L: \$1,231.00; USA Savings: \$1,229.03 **55.** \$12,197.09

57. (A) 15 million (B) 30 million

59. 40 boards

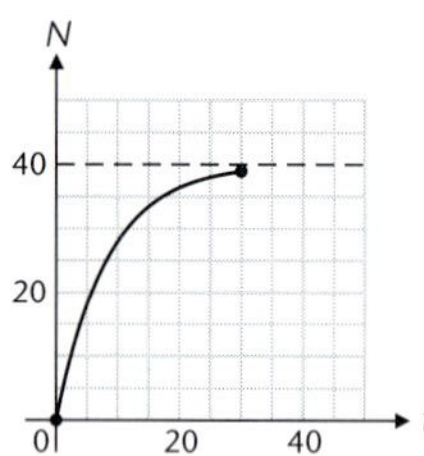

61. 50°F **63.** 0.0009 coulomb

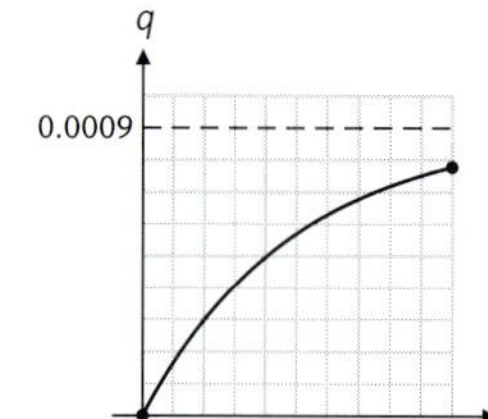

65. 100 deer

67.

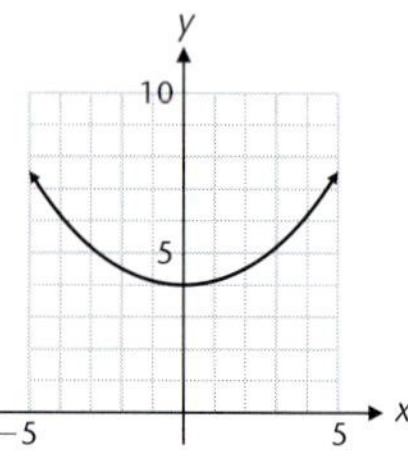

Exercise 4-3

1. $2^6 = 64$ **3.** $10^5 = 100{,}000$ **5.** $9^{1/2} = 3$ **7.** $\left(\frac{1}{4}\right)^{-3} = 64$ **9.** $\log_{10} 0.001 = -3$ **11.** $\log_8 4 = \frac{2}{3}$

13. $\log_{81}\left(\frac{1}{3}\right) = -\frac{1}{4}$ **15.** $\log_{125} 5 = \frac{1}{3}$ **17.** 1 **19.** 0 **21.** 3 **23.** $\frac{1}{2}$ **25.** -3 **27.** x^2 **29.** $(x - 1)^3$

31. 8 **33.** $\frac{1}{25}$ **35.** 4 **37.** 64 **39.** -3 **41.** 4 **43.** 100 **45.** $6 \log_b x + 9 \log_b y$

47. $\frac{3}{2} \log_b u - \frac{5}{3} \log_b v$ **49.** $\log_b m + \log_b n - \log_b p - \log_b q$ **51.** $-4 \log_b a$ **53.** $\frac{1}{2} \log_b (c^2 + d^2)$

55. $\frac{1}{2} \log_b u - \log_b v - 2 \log_b w$ **57.** $\frac{2}{3} \log_b x + \frac{1}{6} \log_b y - \log_b z$ **59.** $\log_b (x^2/y)$ **61.** $\log_b (w/xy)$

63. $\log_b (x^3y^2/\sqrt[4]{z})$ **65.** $\log_b (\sqrt{u}/v^2)^5$ **67.** $\log_b \sqrt[5]{x^2y^3}$ **69.** $5 \log_b (x + 3) + 2 \log_b (2x - 7)$

71. $7 \log_b (x + 10) - 2 \log_b (1 + 10x)$ **73.** $2 \log_b x - \frac{1}{2} \log_b (x + 1)$ **75.** $2 \log_b x + \log_b (x + 5) + \log_b (x - 4)$ **77.** $x = 4$

79. $x = \frac{1}{3}$ **81.** $x = \frac{8}{7}$ **83.** $x = 2$ **85.** $x = 2$ **87.** 3.40 **89.** -0.92 **91.** 3.30 **93.** 0.23 **95.** -0.05

97.

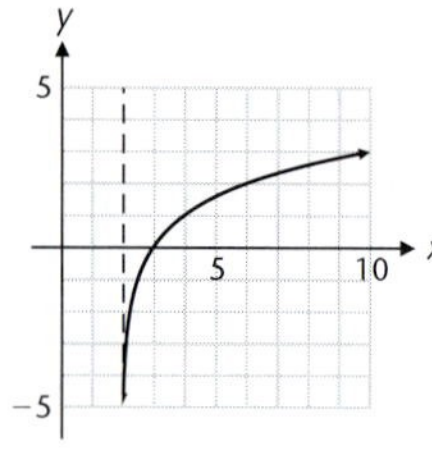

99.

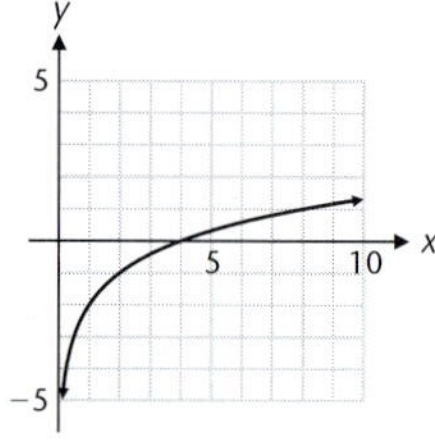

101. (A)

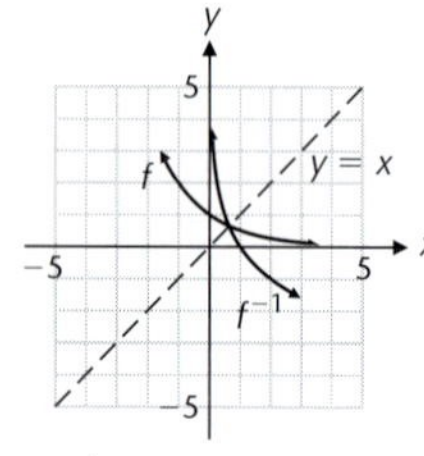

(B) Domain $f = (-\infty, \infty) =$ Range f^{-1}
Range $f = (0, \infty) =$ Domain f^{-1}

(C) $f^{-1}(x) = \log_{1/2} x = -\log_2 x$

103. $f^{-1}(x) = \frac{1}{3}[1 + \log_5 (x - 4)]$ **105.** $g^{-1}(x) = \frac{1}{5}(e^{x/3} + 2)$ **107.** The reflection is not a function, since $y = 3^{x^2}$ is not one-to-one.

109. $x = 100e^{-0.08t}$

Exercise 4-4

1. 4.4408 **3.** -2.3644 **5.** -7.3324 **7.** 6.1242 **9.** 32.45 **11.** 0.1039 **13.** 0.055 68 **15.** 3,407 **17.** 1.238 **19.** 2.320 **21.** -51.083 **23.** 5.192 **25.** 35.779 **27.** -12.169 **29.** 4.6505×10^{21} **31.** 1.4925×10^{-5}

33.

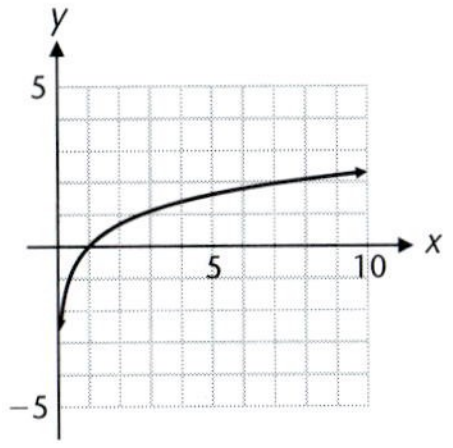

35.

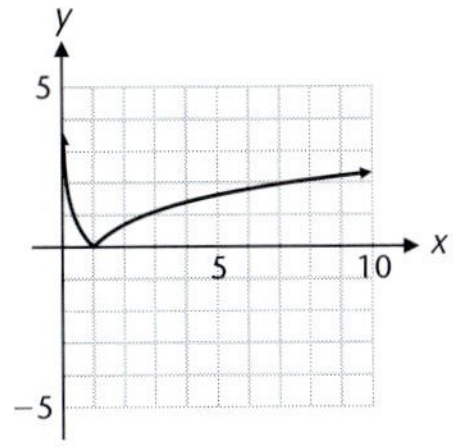

37.

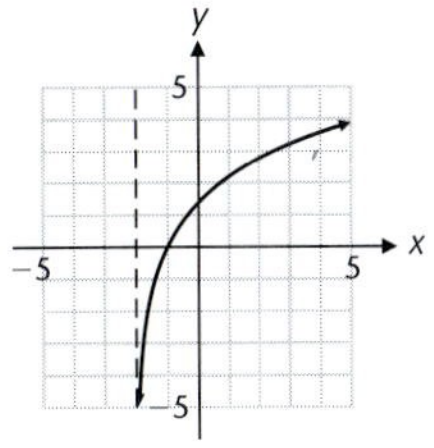

39.

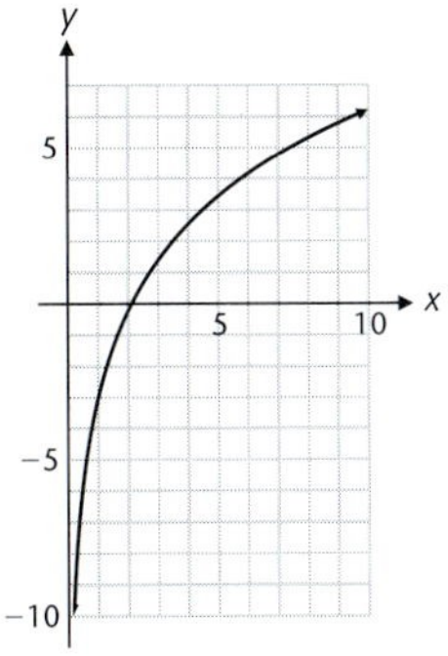

41. The inequality sign in the last step reverses because $\log \frac{1}{3}$ is negative. **43.** (B) Domain = $(1, \infty)$; range = $(-\infty, \infty)$

45. (0.90, −0.11), (38.51, 3.65) **47.** (6.41, 1.86), (93.35, 4.54) **49.**

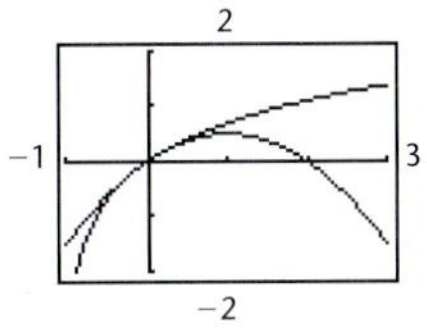

51.

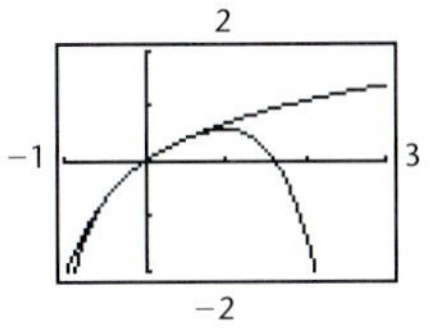

53. (A) 0 decibel (B) 120 decibels **55.** 30 decibels more **57.** 8.6 **59.** 1,000 times more powerful **61.** 7.67 km/s **63.** (A) 8.3, basic (B) 3.0, acid **65.** 6.3×10^{-6} mole/liter

Exercise 4-5

1. 1.44 **3.** 5.88 **5.** -1.68 **7.** 0.0967 **9.** 3.24 **11.** -3.43 **13.** $\frac{5}{2}$ **15.** $8/e^2$ **17.** $x = 10$ **19.** 86.7 **21.** -921 **23.** 5.18 **25.** ± 4.01 **27.** $x = 5$ **29.** $x = 2 + \sqrt{3}$ **31.** $x = \frac{1}{4}(1 + \sqrt{89})$ **33.** $x = 1, e^2, e^{-2}$ **35.** $x = e^e$ **37.** $x = 0.1, 100$ **39.** (B) 2 **41.** (B) $-1.252, 1.707$ **43.** 3.6776 **45.** -1.6094 **47.** -1.7372

49. $r = \dfrac{\ln (A/P)}{t}$ **51.** $I = I_0(10^{D/10})$ **53.** $I = I_0[10^{(6-M)/2.5}]$ **55.** $t = \dfrac{-L}{R}\ln\left(1 - \dfrac{RI}{E}\right)$ **57.** $x = \ln(y \pm \sqrt{y^2 - 1})$

59. $x = \dfrac{1}{2}\ln\dfrac{1 + y}{1 - y}$ **61.**

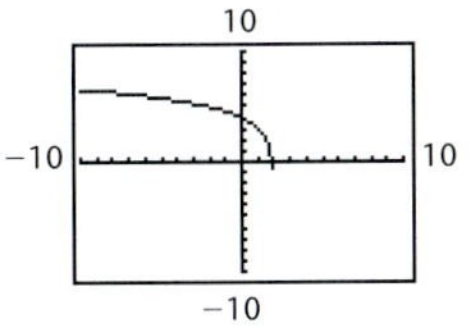

63.

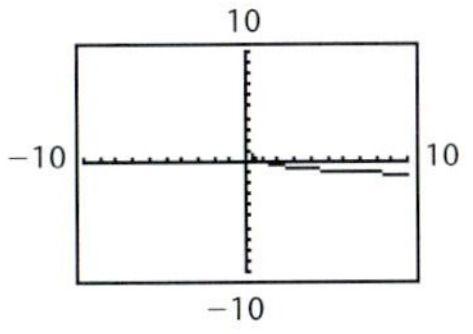

65. 0.38 **67.** 0.55 **69.** 0.57

71. 0.85 **73.** 0.43 **75.** 0.27 **77.** Approx. 5 years **79.** 9.16% **81.** (A) 6 (B) 100 times brighter **83.** Approx. 35 years **85.** 18,600 years old **87.** 7.52 s **89.** $k = 0.40$; $t = 2.9$ h **91.** 10 years

Chapter 4 Review Exercise

1. $\log m = n$ *(4-3)* **2.** $\ln x = y$ *(4-3)* **3.** $x = 10^y$ *(4-3)* **4.** $y = e^x$ *(4-3)* **5.** 7^{2x} *(4-1)* **6.** e^{2x^2} *(4-1)* **7.** $x = 8$ *(4-3)* **8.** $x = 5$ *(4-3)* **9.** $x = 3$ *(4-3)* **10.** $x = 1.24$ *(4-3)* **11.** $x = 11.9$ *(4-3)* **12.** $x = 0.984$ *(4-3)* **13.** $x = 103$ *(4-3)* **14.** $x = 4$ *(4-3)* **15.** $x = 2$ *(4-3)* **16.** $x = -1, 3$ *(4-2)* **17.** $x = 1$ *(4-1)* **18.** $x = \pm 3$ *(4-2)* **19.** $x = -2$ *(4-3)* **20.** $x = \frac{1}{3}$ *(4-3)* **21.** $x = 64$ *(4-3)* **22.** $x = e$ *(4-3)* **23.** $x = 33$ *(4-3)* **24.** $x = 1$ *(4-3)* **25.** 1.145 *(4-3)* **26.** Not defined *(4-3)* **27.** 2.211 *(4-3)* **28.** 11.59 *(4-3)* **29.** $x = 41.8$ *(4-1)* **30.** $x = 1.95$ *(4-3)* **31.** $x = 0.0400$ *(4-3)* **32.** $x = -6.67$ *(4-3)* **33.** $x = 1.66$ *(4-3)* **34.** $x = 2.32$ *(4-5)* **35.** $x = 3.92$ *(4-5)* **36.** $x = 92.1$ *(4-5)* **37.** $x = 2.11$ *(4-5)* **38.** $x = 0.881$ *(4-5)* **39.** $x = 300$ *(4-5)* **40.** $x = 2$ *(4-5)* **41.** $x = 1$ *(4-5)* **42.** $x = \frac{1}{2}(3 + \sqrt{13})$ *(4-5)* **43.** $x = 1, 10^3, 10^{-3}$ *(4-5)* **44.** $x = 10^e$ *(4-5)* **45.** $e^{-x} - 1$ *(4-2)* **46.** $2 - 2e^{-2x}$ *(4-2)*

47. (4-1) 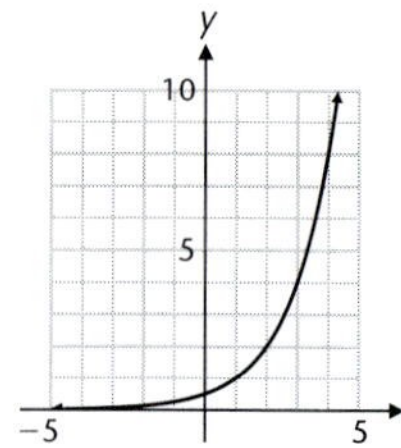

48. (4-2) 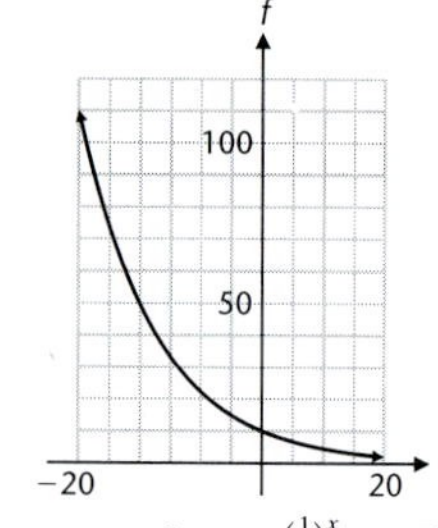

49. (4-3) 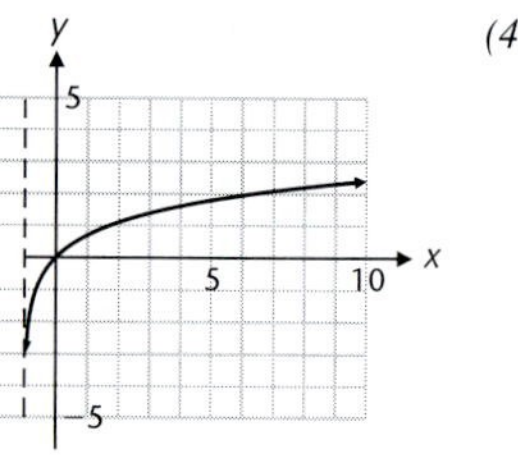

50. (4-2) 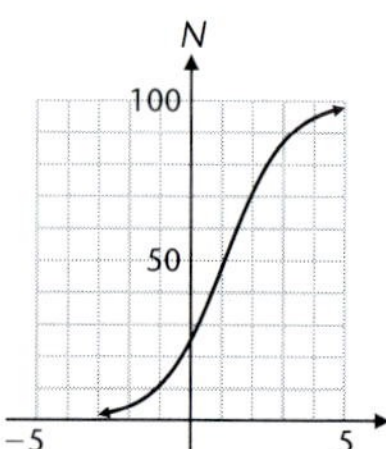

51. $y = -e^x$; $y = (\frac{1}{e})^x = e^{-x}$ *(4-3)*

52. (A) $y = e^{-x/3}$ is decreasing, while $y = 4 \ln (x + 1)$ is increasing without bound. (B) 0.258 *(4-5)* **53.** 0.018, 2.187 *(4-3)*
54. (1.003, 0.010), (3.653, 4.502) *(4-4)* **55.** $I = I_0(10^{D/10})$ *(4-5)* **56.** $x = \pm\sqrt{-2 \ln (\sqrt{2\pi}y)}$ *(4-5)* **57.** $I = I_0(e^{-kx})$ *(4-5)*
58. $n = -\dfrac{\ln [1 - (Pi/r)]}{\ln (1 + i)}$ *(4-5)* **59.** $f^{-1}(x) = e^{x/2} + 1$ *(4-5, 1-7)* **60.** $f^{-1}(x) = \ln (x + \sqrt{x^2 + 1})$ *(4-5, 2-6)*
61. $y = ce^{-5t}$ *(4-3, 4-5)*
62. Domain $f = (0, \infty) =$ Range f^{-1}
Range $f = (-\infty, \infty) =$ Domain f^{-1} *(4-3)*

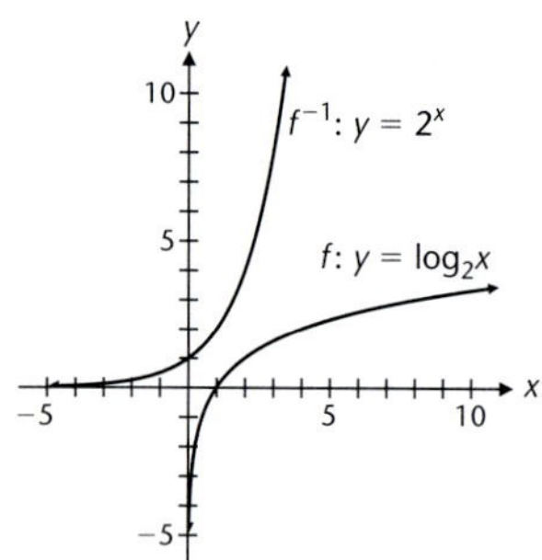

63. If $\log_1 x = y$, then we have $1^y = x$; that is, $1 = x$ for arbitrary positive x, which is impossible. *(4-3)* **65.** 23.4 years *(4-5)*
66. 23.1 years *(4-5)* **67.** 37,100 years *(4-5)* **68.** (A) $N = 2^{2t}$ or $N = 4^t$ (B) 15 days *(4-5)* **69.** $\$1.1 \times 10^{26}$ *(4-2)*
70. (A)

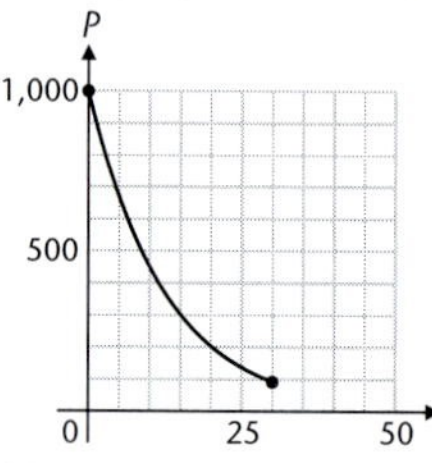

(B) 0 *(4-2)* **71.** 6.6 *(4-4)* **72.** 7.08×10^{16} joules *(4-4)*

73. 50 decibels more *(4-4)* **74.** $k = 0.009\ 42$; 489 ft *(4-2)* **75.** 3 years *(4-5)*

Cumulative Review Exercise: Chapters 3 and 4

1. (A) $P(x) = (x + 1)^2(x - 1)(x - 2)$ (B) $P(x) \to \infty$ as $x \to \infty$ and as $x \to -\infty$ *(3-1)*
2. $3x^3 + 5x^2 - 18x - 3 = (x + 3)(3x^2 - 4x - 6) + 15$ *(3-1)* **3.** $-2, 3, 5$ *(3-2)*
4. $P(1) = -5$ and $P(2) = 5$ are of opposite sign *(3-3)* **5.** 1, 2, -4 *(3-2)*
6. $\dfrac{3}{x + 1} + \dfrac{2}{x - 2}$ *(3-5)* **7.** (A) $x = \log y$ (B) $x = e^y$ *(4-4)* **8.** (A) $8e^{3x}$ (B) e^{5x} *(4-2)*
9. (A) 9 (B) 4 (C) $\frac{1}{2}$ *(4-3)* **10.** (A) 0.371 (B) 11.4 (C) 0.0562 (D) 15.6 *(4-4)*
11. $f(x) = 3 \ln x - \sqrt{x}$ *(4-3)*
12. The function f multiplies the base e raised to a power $\frac{1}{2}$ the domain element by 100 and then subtracts 50. *(4-2)*
13. $P(\frac{1}{2}) = \frac{5}{2}$ *(3-2)* **14.** Part B *(3-1)*

15. (A) The graph of $P(x)$ has four x intercepts and three turning points; $P(x) \to \infty$ as $x \to \infty$ and as $x \to -\infty$ (B) 2.8 *(3-1, 3-3)*

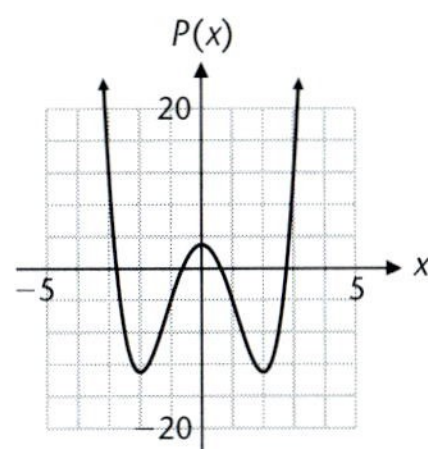

16. (A) Upper bound: 4; lower bound: −6 (B) 3.80 (C) −5.68, 3.80 *(3-3)*

17. $Q(x) = 2x^3 - 5x^2 + 1$, $R = -2$, $P(2) = -2$ *(3-2)* **18.** $3, 1 \pm \frac{1}{2}i$ *(3-2)*

19. $-4, -1, \pm\sqrt{3}$; $P(x) = (x + 4)(x + 1)(x - \sqrt{3})(x + \sqrt{3})$ *(3-2)* **20.** $x \le -2$ or $3 \le x \le 6$; $(-\infty, -2] \cup [3, 6]$ *(3-2)*

21. 2.23 *(3-3)* **22.** $\frac{1}{x} + \frac{2}{x+1} - \frac{5}{(x+1)^2}$ *(3-5)* **23.** $-\frac{2}{x} + \frac{3x - 1}{x^2 - x + 1}$ *(3-5)*

24. (A) Domain: $x \ne -2$; x intercept: −4; y intercept: 4
(B) Vertical asymptote: $x = -2$
Horizontal asymptote: $y = 2$
(C) *(3-4)*

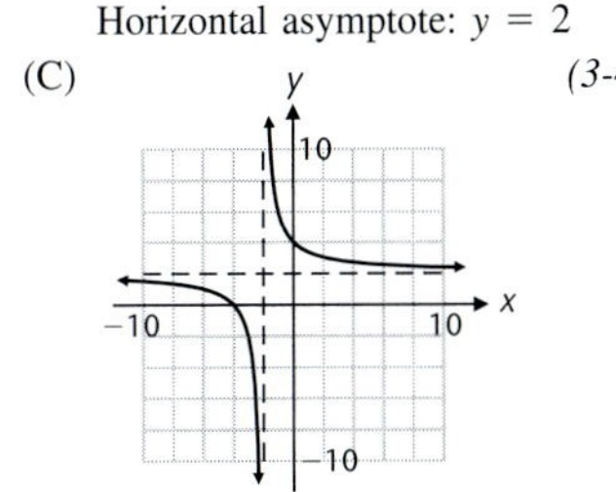

25. $x = -2, 4$ *(4-1)* **26.** $x = -1, \frac{1}{2}$ *(4-2)* **27.** $x = 2.5$ *(4-3)* **28.** $x = 10$ *(4-3)* **29.** $x = \frac{1}{27}$ *(4-3)*

30. $x = 5$ *(4-5)* **31.** $x = 7$ *(4-5)* **32.** $x = 5$ *(4-5)* **33.** $x = e^{0.1}$ *(4-4)* **34.** $x = 1, e^{0.5}$ *(4-5)*

35. $x = 3.38$ *(4-5)* **36.** $x = 4.26$ *(4-4)* **37.** $x = 2.32$ *(4-4)* **38.** $x = 3.67$ *(4-5)* **39.** $x = 0.549$ *(4-5)*

40. *(4-1)*

41. *(4-4)*

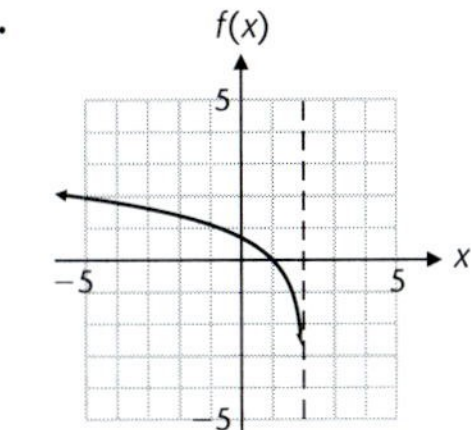

42. *(4-2)*

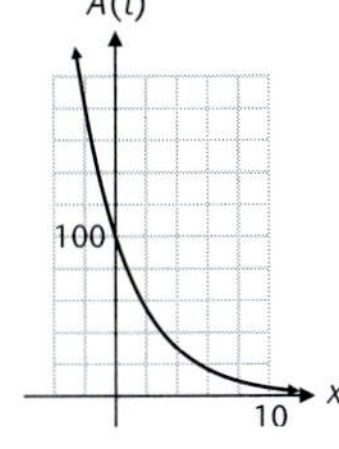

43. *(4-2)*

44. A reflection in the x axis transforms the graph of $y = \ln x$ into the graph of $y = -\ln x$. A reflection in the y axis transforms the graph of $y = \ln x$ into the graph of $y = \ln(-x)$. *(4-3)*

45. (A) For $x > 0$, $y = e^{-x}$ decreases from 1 to 0, while $\ln x$ increases from $-\infty$ to ∞. Consequently, the graphs can intersect at exactly one point. (B) 1.31 *(4-3)*

46. Yes, for example: $P(x) = (x + i)(x - i)(x + \sqrt{2})(x - \sqrt{2}) = x^4 - x^2 - 2$ *(3-2)*

47. (A) Upper bound: 20; lower bound: −30 (B) −26.69, −6.22, 7.23, 16.67 *(3-3)*

48. Vertical asymptote: $x = -2$
Oblique asymptote: $y = x + 2$ *(3-4)*

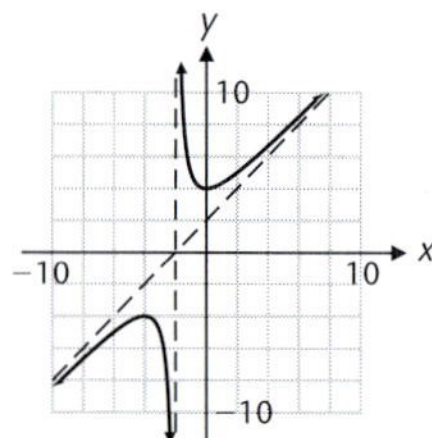

49. $P(x) = (x + 1)^2 x^3 (x - 3 - 5i)(x - 3 + 5i)$; degree 7 *(3-2)*

50. -1 (double root), 2, $2 \pm i\sqrt{2}$; $P(x) = (x + 1)^2(x - 2)(x - 2 - i\sqrt{2})(x - 2 + i\sqrt{2})$ *(3-2)*

51. -2 (double root), -1.88, 0.35, 1.53 *(3-3)* **52.** $\dfrac{-2}{(x-1)} + \dfrac{2}{(x-1)^2} + \dfrac{2x+3}{x^2+x+2}$ *(3-5)*

53. (A) $f^{-1}(x) = e^{x/3} + 2$
(B) Domain $f = (2, \infty) =$ Range f^{-1}
Range $f =$ Domain $f^{-1} = (-\infty, \infty)$
(C) *(4-5)*

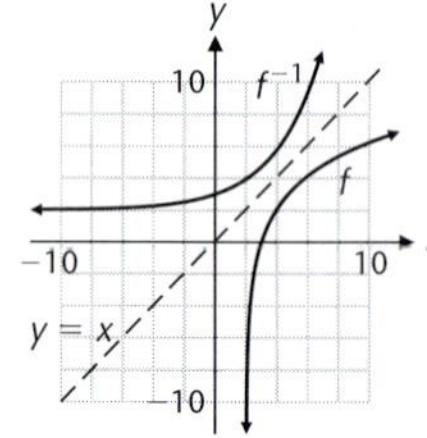

54. $n = \dfrac{\ln(1 + Ai/P)}{\ln(1 + i)}$ *(4-5)*

55. $y = Ae^{5x}$ *(4-5)* **56.** $x = \ln(y + \sqrt{y^2 + 2})$ *(4-5)* **57.** $x = 2$ ft and $y = 2$ ft or $x = 1.3$ ft and $y = 4.8$ ft *(3-2)*
58. 1.8 by 3.3 ft *(3-3)* **59.** (A) 46.8 million (B) 103 million *(4-1)* **60.** 10.2 years *(4-5)* **61.** 9.90 years *(4-5)*
62. 63.1 times more powerful *(4-4)* **63.** 6.31×10^{-4} W/m^2 *(4-4)*

CHAPTER 5

Exercise 5-1

1. $(-1, 0)$ **3.** $(1, 0)$ **5.** $(-1, 0)$ **7.** $(0, -1)$ **9.** $(0, -1)$ **11.** $(0, -1)$ **13.** $(1/2, \sqrt{3}/2)$
15. $(1/\sqrt{2}, -1/\sqrt{2})$ **17.** $(-\sqrt{3}/2, -1/2)$ **19.** $(-1/\sqrt{2}, -1/\sqrt{2})$ **21.** $(0, -1)$ **23.** $(\sqrt{3}/2, -1/2)$
25. a, −; b, − **27.** a, +; b, + **29.** a, +; b, + **31.** a, +; b, − **33.** a, −; b, − **35.** 0; $2k\pi$, k any integer
37. $3\pi/4$; $3\pi/4 + 2k\pi$, k any integer
39. $W(x)$ is the point on a unit circle that is $|x|$ units from (1, 0), in a counterclockwise direction if x is positive and in a clockwise direction if x is negative. $W(x + 4\pi)$ has the same coordinates as $W(x)$, since we return to the same point every time we go around the unit circle any integer multiple of 2π units (the circumference of the circle) in either direction.
41. T **43.** F **45.** T **47.** $-7\pi/4$, $\pi/4$ **49.** $-4\pi/3$, $2\pi/3$ **51.** $-5\pi/6$, $7\pi/6$ **53.** $x = \pi/4 + 2k\pi$, k any integer.

Exercise 5-2

1. (A) a (B) $1/b$ (C) a/b (D) $1/a$ (E) b/a (F) b **3.** 1 **5.** $\frac{1}{2}$ **7.** 1 **9.** $\sqrt{3}$ **11.** Not defined
13. 1 **15.** $\sqrt{2}$ **17.** 1 **19.** Not defined **21.** Quadrant II or III **23.** Quadrant I or II **25.** Quadrant II or IV
27. -64.05 **29.** 12.24 **31.** 0.4043 **33.** Not defined **35.** -1 **37.** $\frac{1}{2}$ **39.** $\sqrt{2}$ **41.** $-\frac{1}{2}$ **43.** -1
45. $\sqrt{3}$ **47.** Not defined **49.** (A) $\sin 0.4 = 0.4$ (B) $\cos 0.4 = 0.9$ (C) $\tan 0.4 = 0.4$
51. (A) $\sec 2.2 = -2$ (B) $\tan 5.9 = -0.4$ (C) $\cot 3.8 = 1$
53. $\sin x < 0$ in quadrants III and IV; $\cot x < 0$ in quadrants II and IV; therefore, both are true in quadrant IV.
55. $\cos x < 0$ in quadrants II and III; $\sec x > 0$ in quadrants I and IV; therefore, it is not possible to have both true for the same value of x.
57. None **59.** $\pi/2$, $3\pi/2$ **61.** $\pi/2$, $3\pi/2$ **63.** (A) 0 to 1 (B) 1 to 0 (C) 0 to -1 (D) -1 to 0
65. 0.8138 **67.** 0.5290 **69.** $\frac{1}{3}$ **71.** $\sqrt{5}$ **73.** -25
75. $\sin x = -\sqrt{3}/2$, $\tan x = -\sqrt{3}$, $\cot x = -1/\sqrt{3}$, $\csc x = -2/\sqrt{3}$, $\sec x = 2$
77. $\cos x = -1/\sqrt{2}$, $\tan x = 1$, $\cot x = 1$, $\csc x = -\sqrt{2}$, $\sec x = -\sqrt{2}$
79. $\cot x = 1/\sqrt{3}$, $\sin x = -\sqrt{3}/2$, $\cos x = -\frac{1}{2}$, $\csc x = -2/\sqrt{3}$, $\sec x = -2$ **81.** π **83.** $5\pi/6$ **85.** $5\pi/6$
87. (A) Identity (5) (B) Identity (9) (C) Identity (1) **89.** 75 m^2 **91.** $12\sqrt{3} \approx 20.78$ in^2
93. $a_1 = 0.5$, $a_2 = 1.377\ 583$, $a_3 = 1.569\ 596$, $a_4 = 1.570\ 796$, $a_5 = 1.570\ 796$; $\pi/2 = 1.570\ 796$

Exercise 5-3

1. 40° **3.** 270° **5.** 6 **7.** 2.5 **9.** $\pi/4$ **11.** $3\pi/2$ **13.** $\pi/6, \pi/3, \pi/2, 2\pi/3, 5\pi/6, \pi$ **15.** $-\pi/4, -\pi/2, -3\pi/4, -\pi$
17. 60°, 120°, 180°, 240°, 300°, 360° **19.** −90°, −180°, −270°, −360° **21.** False **23.** False **25.** False **27.** 5.859°
29. 354.141° **31.** 3°2′31″ **33.** 403°13′23″ **35.** 0.314 **37.** 0.415 **39.** 87.09° **41.** −47.56° **43.** Quadrant II
45. Quadrant IV **47.** Quadrant IV **49.** Quadrantal angle **51.** Quadrant III **53.** Quadrant III **55.** Coterminal
57. Not coterminal **59.** Coterminal **61.** Not coterminal **63.** Coterminal **65.** Coterminal **67.** 24,000 mi
69. The 7.5° angle and θ have a common side. (An extended vertical pole in Alexandria will pass through the center of the earth.) The sun's rays are essentially parallel when they arrive at the earth. Thus, the other two sides of the angles are parallel, since a sun ray to the bottom of the well, when extended, will pass through the center of the earth. From geometry we know that the alternate interior angles made by a line intersecting two parallel lines are equal. Therefore, $\theta = 7.5°$.
71. $7\pi/4$ rad **73.** 200 rad **75.** $\pi/26 \approx 0.12$ rad **77.** 12 **79.** 865,000 mi **81.** 33 ft

Exercise 5-4

1. $\sin\theta = 4/5$, $\cos\theta = 3/5$, $\tan\theta = 4/3$, $\csc\theta = 5/4$, $\sec\theta = 5/3$, $\cot\theta = 3/4$
3. $\sin\theta = \sqrt{3}/2$, $\cos\theta = -1/2$, $\tan\theta = -\sqrt{3}$, $\csc\theta = 2/\sqrt{3}$, $\sec\theta = -2$, $\cot\theta = -1/\sqrt{3}$
5. 0.9272 **7.** −0.2958 **9.** 0.2038 **11.** 9.761 **13.** 108.6 **15.** 0 **17.** $\sqrt{3}$ **19.** $1/\sqrt{2}$ **21.** $\sqrt{2}$
23. Not defined **25.** Not defined **27.** 60° **29.** $\pi/6$ **31.** $\pi/3$ **33.** −1 **35.** −1/2 **37.** $-2/\sqrt{3}$ **39.** −1
41. −2 **43.** $\sqrt{3}/2$ **45.** $-\sqrt{3}$ **47.** $\sqrt{2}$ **49.** Defined for all θ, since $\cos\theta = a/r$ and r is never zero.
51. 90° and 270°, since $\tan\theta = b/a$ and $a = 0$ at $\theta = 90°, 270°$ **53.** 0° and 180°, since $\csc\theta = r/b$ and $b = 0$ at $\theta = 0°$ and 180°
55. 120° or $2\pi/3$ rad **57.** 210° or $7\pi/6$ rad **59.** 240° or $4\pi/3$ rad
61. $\cos\theta = -4/5$, $\tan\theta = -3/4$, $\csc\theta = 5/3$, $\sec\theta = -5/4$, $\cot\theta = -4/3$
63. $\sin\theta = -2/3$, $\tan\theta = 2/\sqrt{5}$, $\csc\theta = -3/2$, $\sec\theta = -3/\sqrt{5}$, $\cot\theta = \sqrt{5}/2$
65. Tangent and secant, since $\tan\theta = b/a$ and $\sec\theta = r/a$ and $a = 0$ if $P(a, b)$ is on the vertical axis (division by zero is not defined)
67. 150°, 210° **69.** $\pi/4, 5\pi/4$ **71.** (A) 1.75 rad (B) (−0.713, 3.936) **73.** 2π units
75. k, $0.866k$, $0.5k$ **79.** (A) 3.31371, 3.14263, 3.14160, 3.14159 (B) $\pi = 3.1315926\cdots$
81. (A) 44.07; −0.32 (B) $y = -0.93x + 1.27$

Exercise 5-5

1. b/c **3.** c/b **5.** b/a **7.** $\cot(90° - \theta)$ **9.** $\sin(90° - \theta)$ **11.** $\cos(90° - \theta)$ **13.** 67.56° **15.** 84.01°
17. 42.06° **19.** $\alpha = 72.2°$, $a = 3.28$, $b = 1.05$ **21.** $\alpha = 46°40'$, $b = 116$, $c = 169$ **23.** $\beta = 67°0'$, $b = 127$, $c = 138$
25. $\beta = 36.79°$, $a = 31.85$, $c = 39.77$ **27.** $\alpha = 35°20'$, $\beta = 54°40'$, $c = 10.4$ **29.** $\alpha = 37°30'$, $\beta = 52°30'$, $a = 7.67$
31. (A) $\cos\theta = OA/1 = OA$ (B) Angle $OED = \theta$; $\cot\theta = DE/1 = DE$ (C) $\sec\theta = OC/1 = OC$
33. (A) As θ approaches 90°, $OA = \cos\theta$ approaches 0. (B) As θ approaches 90°, $DE = \cot\theta$ approaches 0.
(C) As θ approaches 90°, $OC = \sec\theta$ increases without bound.
35. (A) As θ approaches 0°, $AD = \sin\theta$ approaches 0. (B) As θ approaches 0°, $CD = \tan\theta$ approaches 0.
(C) As θ approaches 0°, $OE = \csc\theta$ increases without bound.
39. 228 ft **41.** 127.5 ft **43.** 2,225 mi **45.** 44° **47.** 9.8 m/s^2 **49.** (B)

θ	$C(\theta)$
10°	$368,222
20°	$363,435
30°	$360,622
40°	$360,146
50°	$363,050

51. 0.77 m

Exercise 5-6

1. $2\pi, \pi, 2\pi$ **3.** (A) 1 unit (B) Indefinitely far (C) Indefinitely far
5. (A) $-2\pi, -\pi, 0, \pi, 2\pi$ (B) $-3\pi/2, -\pi/2, \pi/2, 3\pi/2$ (C) No x intercepts
7. $y = \cot x$, $y = \csc x$
9. (A) No vertical asymptotes (B) $-3\pi/2, -\pi/2, \pi/2, 3\pi/2$ (C) $-2\pi, -\pi, 0, \pi, 2\pi$

11. (A) $y = \cos x$

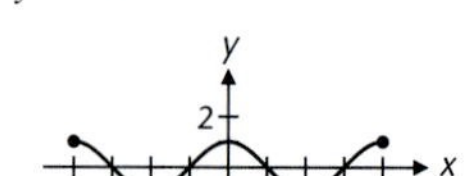

(B) $y = \tan x$

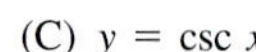

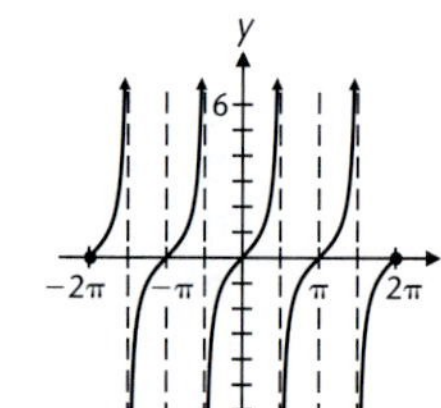

(C) $y = \csc x$

13. (A) A shift of $\pi/2$ to the left will transform the cosecant graph into the secant graph. (The answer is not unique—see part B.)
(B) The graph of $y = -\csc(x - \pi/2)$ is a $\pi/2$ shift to the right and a reflection in the x axis of the graph of $y = \csc x$. The result is the graph of $y = \sec x$.

15. (A)

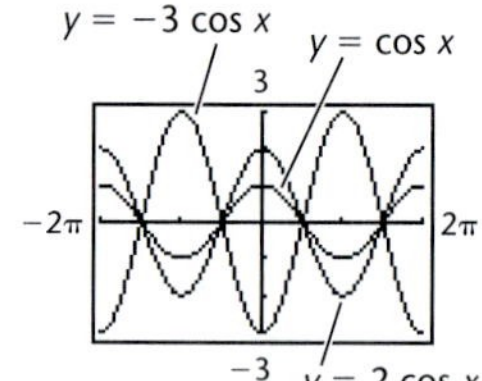

(B) No (C) 1 unit; 2 units; 3 units
(D) The deviation of the graph from the x axis is changed by changing A. The deviation appears to be $|A|$.

17. (A)

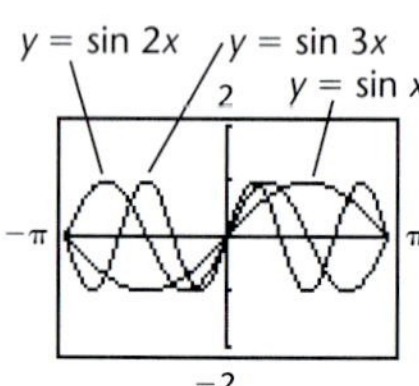

(B) 1; 2; 3 (C) n

19. (A)

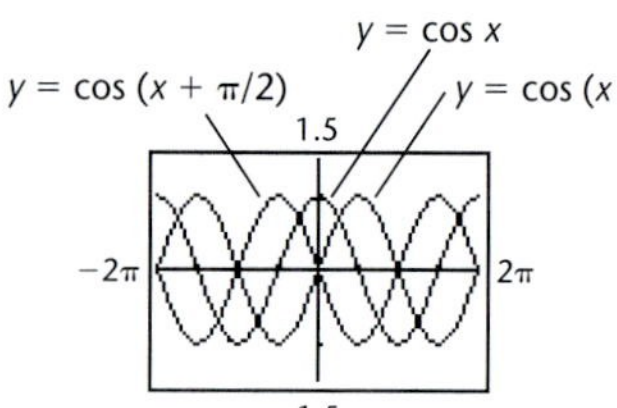

(B) The graph of $y = \cos x$ is shifted $|C|$ units to the right if $C < 0$ and $|C|$ units to the left if $C > 0$.

21. For each case, the number is not in the domain of the function, and an error message of some type will appear.

25. (A) The graphs are almost indistinguishable the closer x is to the origin.
(B)

x	−0.3	−0.2	−0.1	0.0	0.1	0.2	0.3
$\sin x$	−0.296	−0.199	−0.100	0.000	0.100	0.199	0.296

27. True **29.** True **31.** True **33.** False

Exercise 5-7

1. $A = 3, P = 2\pi$

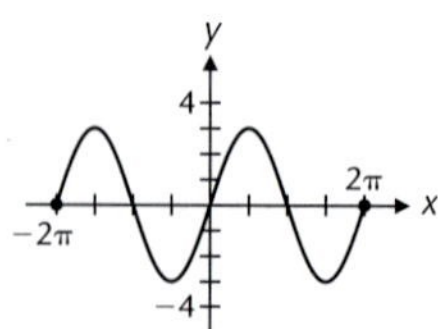

3. $A = \frac{1}{2}, P = 2\pi$

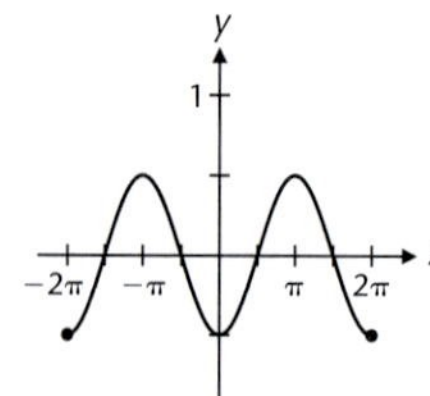

5. $A = 1, P = 2\pi/3$

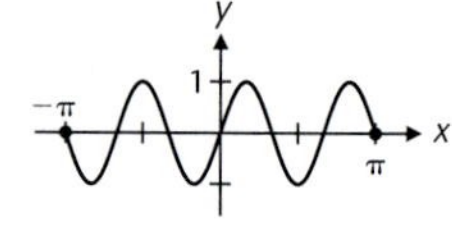

7. $A = 1, P = 4\pi$

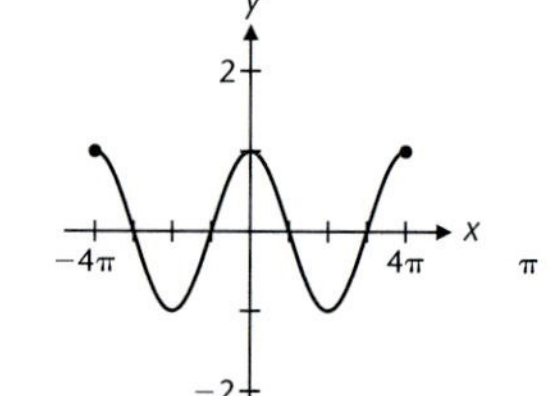

9. $A = 1, P = 2$

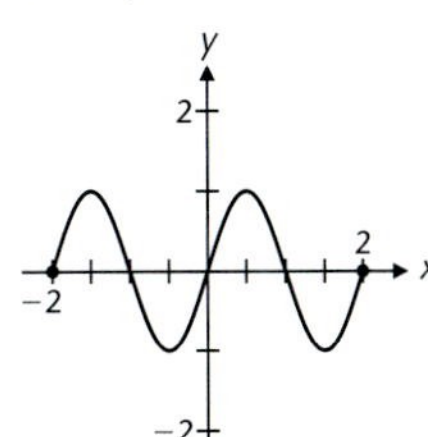

11. $A = 3, P = \pi$

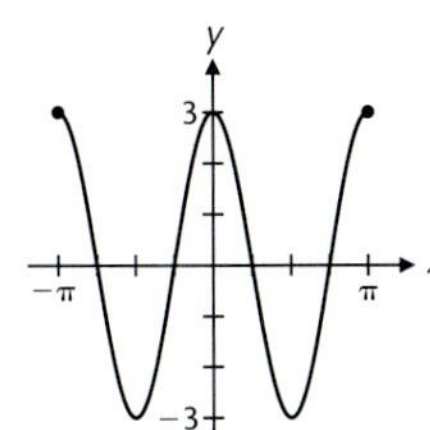

13. $A = \frac{1}{2}, P = 1$

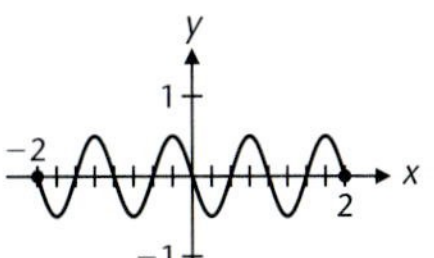

15. $A = 3, P = 4\pi$

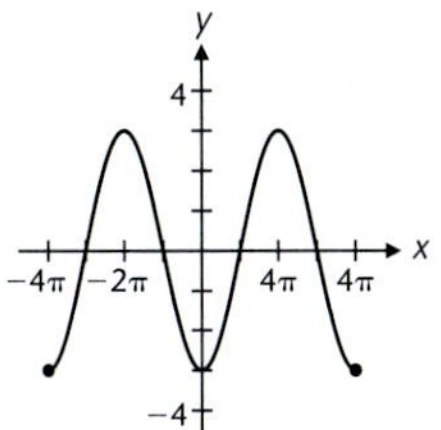

17. $A = 2, P = 4$

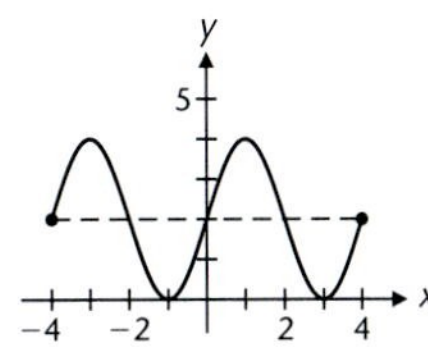

19. $A = 2, P = 4\pi$

21. $y = 3 \sin 4x, -\pi/4 \leq x \leq \pi/2$

23. $y = -10 \sin \pi x, -1 \leq x \leq 2$ **25.** $y = 5 \cos (x/4), -4\pi \leq x \leq 8\pi$ **27.** $y = -0.5 \cos (\pi x/4), -4 \leq x \leq 8$

29. $y = \cos 2x$ **31.** $y = 1 - \cos 2x$

33. $A = 1, P = 2\pi$
Phase shift $= -\pi$

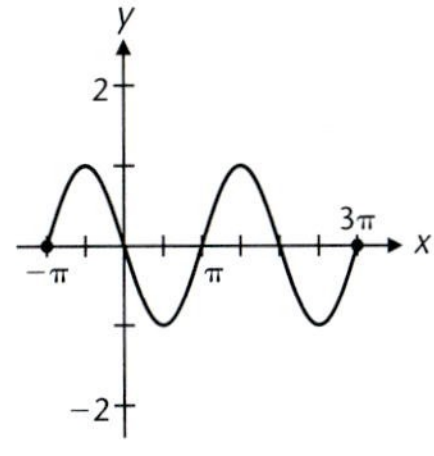

35. $A = \frac{1}{2}, P = 2\pi$
Phase shift $= \pi/4$

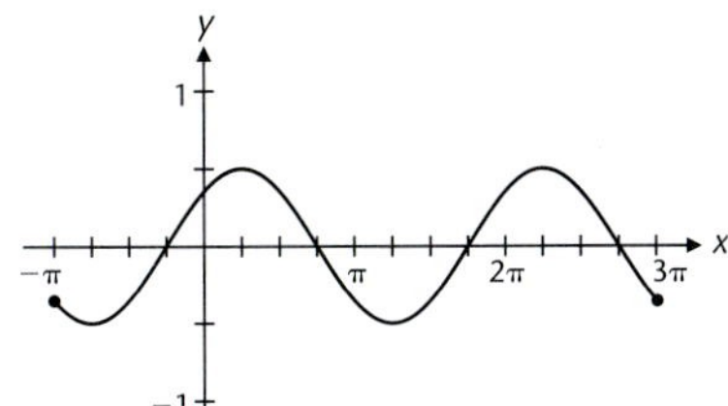

37. $A = 1, P = 2$
Phase shift $= 1$

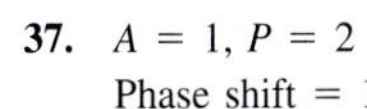

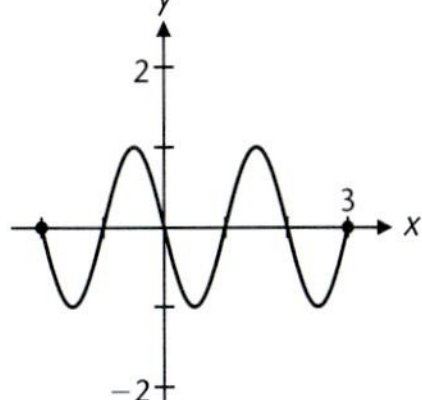

39. $A = 3, P = 2$
Phase shift $= -\frac{1}{2}$

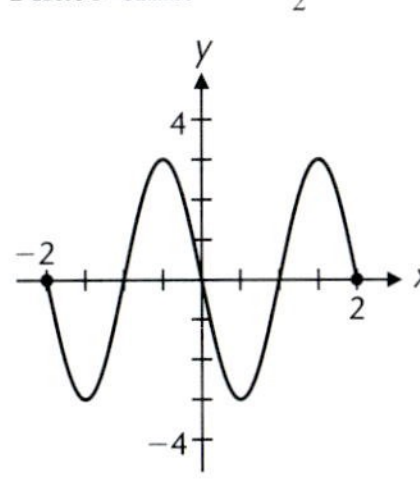

41.

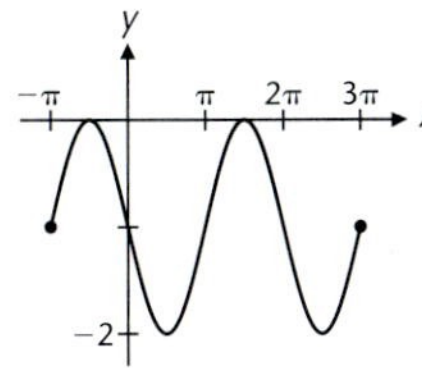

43.

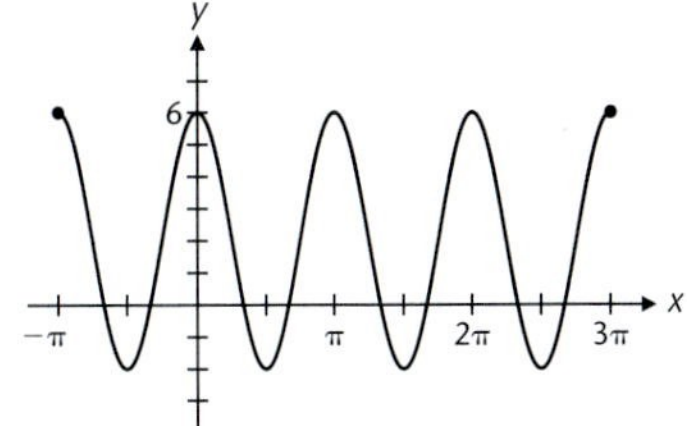

45. $y = -4 \sin (\pi x/2 - \pi/2)$ **47.** $y = \frac{1}{2} \cos (x/4 - 3\pi/4)$ **49.** 2; $y = 4 \sin (2x + 2\pi/3)$, $y = -4 \sin (2x + 2\pi/3)$

51. $A = 3.5, P = 4$
Phase shift $= -0.5$

53. $A = 50, P = 1$
Phase shift $= 0.25$

55. $y = 2 \sin (x + 0.785)$ **57.** $y = 2 \sin (x - 0.524)$

59. $y = 5 \sin (2x - 0.284)$ **61.** True **63.** True

65. The amplitude is decreasing with time. This is often referred to as a **damped sine wave.** Examples are a car's vertical motion, which is damped by the suspension system after the car goes over a bump, and the slowing down of a pendulum that is released away from the vertical line of suspension (air resistance and friction).

67. 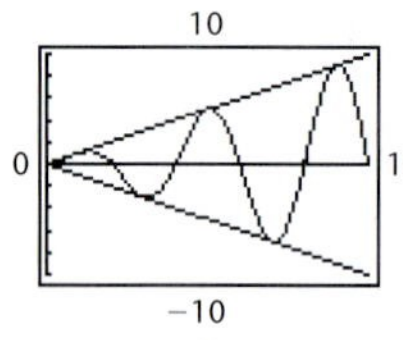

The amplitude is increasing with time. In physical and electric systems this is referred to as **resonance.** Some examples are the swinging of a bridge during high winds and the movement of tall buildings during an earthquake. Some bridges and buildings are destroyed when the resonance reaches the elastic limits of the structure.

69. 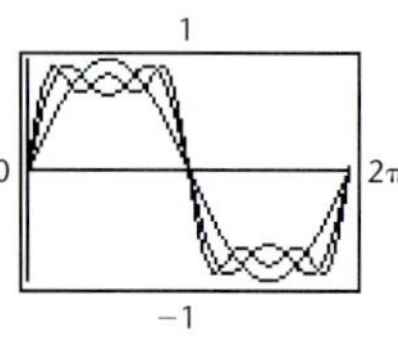

71. $A = \frac{1}{3}$, $P = \pi/4$

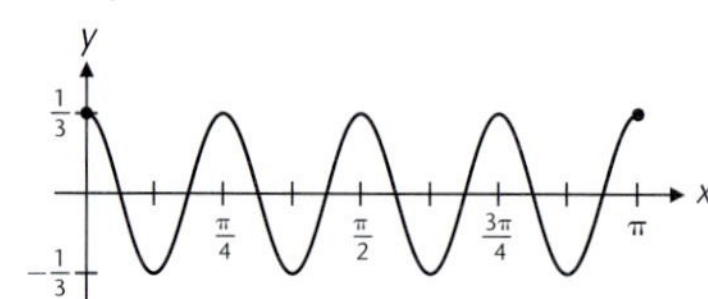

73. $y = -8 \cos 4\pi t$

75. 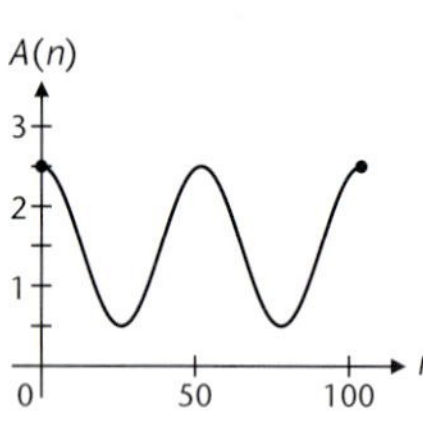

The graph shows the seasonal changes of sulfur dioxide pollutant in the atmosphere; more is produced during winter months because of increased heating.

77. $A = 15$, $P = \frac{1}{60}$
Phase shift $= -\frac{1}{240}$

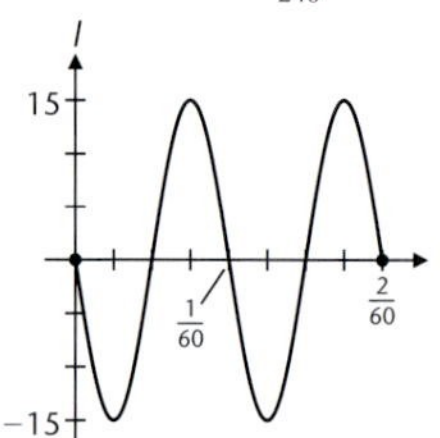

79. $A = 3$, $P = \frac{1}{3}$

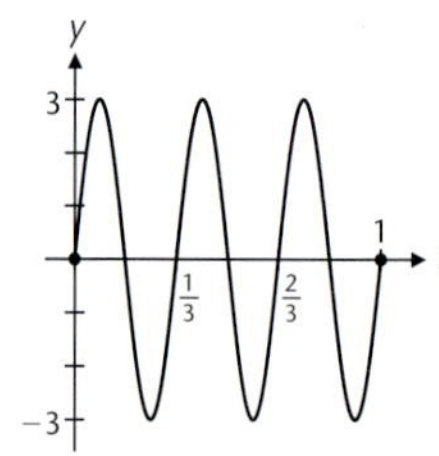

81. (A) 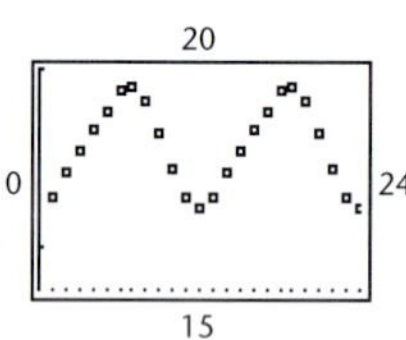

(B) $y = 18.22 + 1.37 \sin(\pi x/6 - 1.75)$

(C) 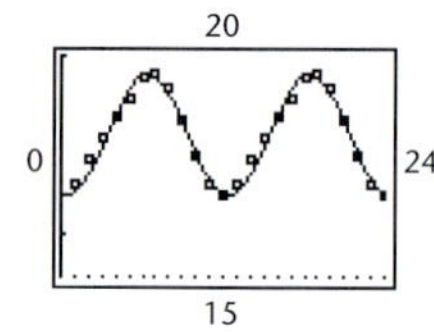

Exercise 5-8

1. Period $= \dfrac{\pi}{4}$

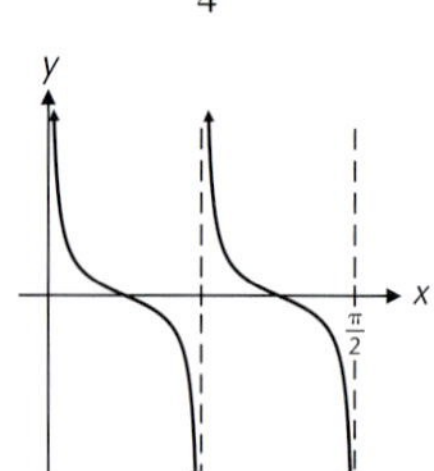

3. Period $= \frac{1}{8}$

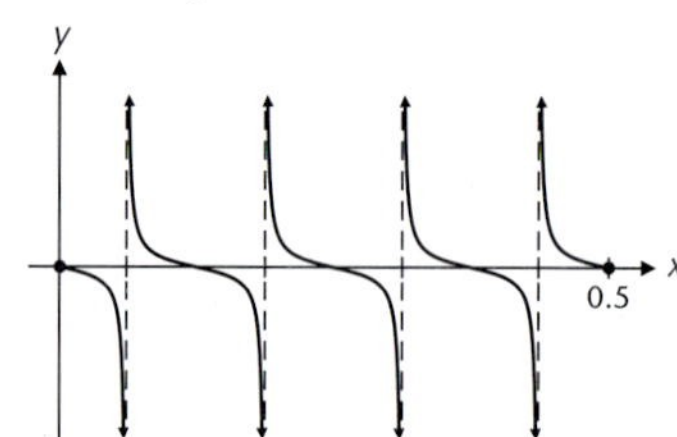

5. Period $= 4\pi$

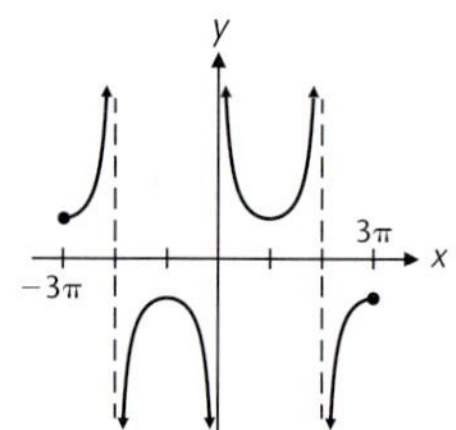

7. Period = 2

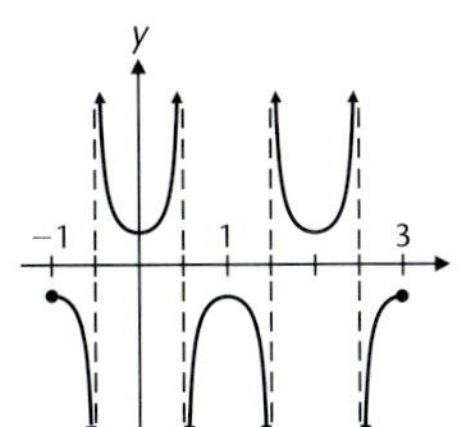

9. Period = π

Phase shift = $-\dfrac{\pi}{2}$

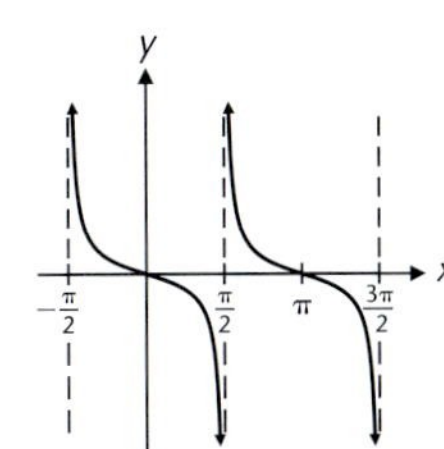

11. Period = $\pi/2$

Phase shift = $-\pi/2$

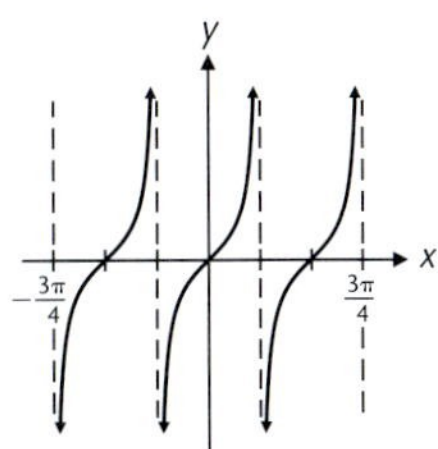

13. Period = 2

Phase shift = $-\frac{1}{2}$

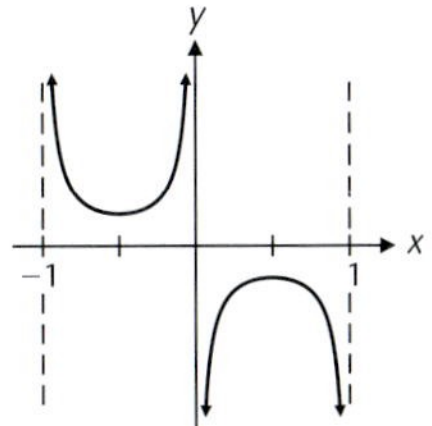

15. False **17.** False **19.** $y = 2 \cot 2x$

21. $y = \cot (x/2)$

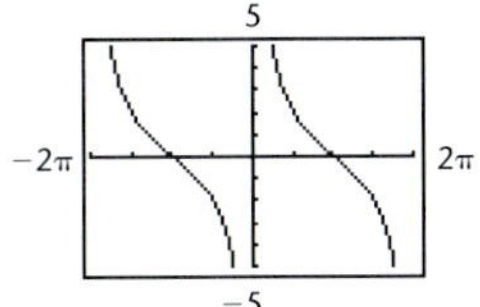

23. Period = 4
Phase shift = 1

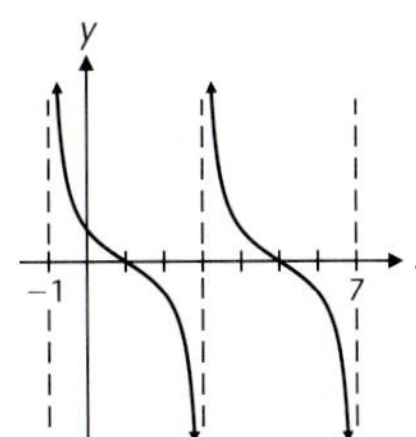

25. Period = 4
Phase shift = −1

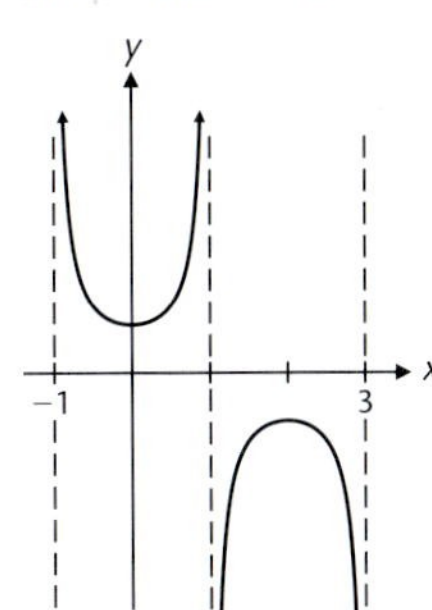

27. $y = \csc 3x$

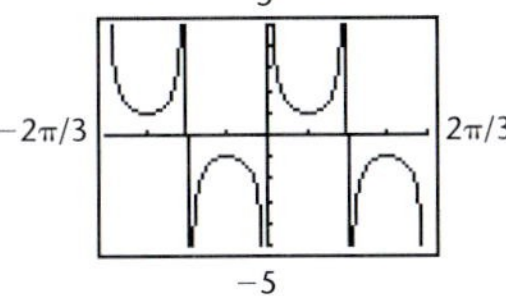

29. $y = \tan 2x$

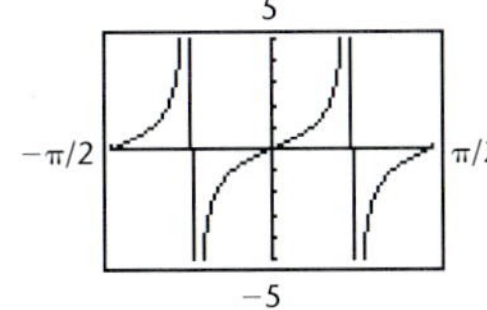

31. (A) $c = 20 \sec (\pi t/2)$, [0, 1). (B)

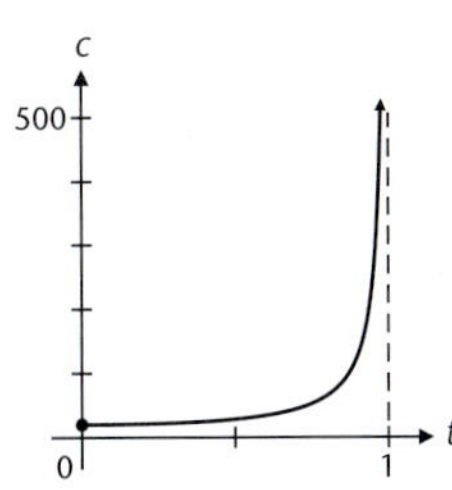

(C) The length of the light beam starts at 20 ft and increases slowly at first, then increases rapidly without end.

Exercise 5-9

1. $\pi/2$ **3.** $\pi/3$ **5.** $\pi/3$ **7.** $\pi/4$ **9.** 0 **11.** $\pi/6$ **13.** 0.6064 **15.** 1.563 **17.** Not defined **19.** $5\pi/6$
21. $-\pi/4$ **23.** $-\pi/6$ **25.** 25 **27.** 1/2 **29.** −0.9810 **31.** 2.645 **33.** −45° **35.** −60° **37.** 180°
39. 43.51° **41.** 21.48°
43. $\sin^{-1} (\sin 2) = 1.1416 \neq 2$ For the identity $\sin^{-1} (\sin x) = x$ to hold, x must be in the restricted domain of the sine function; that is, $-\pi/2 \le x \le \pi/2$. The number 2 is not in the restricted domain.

45.

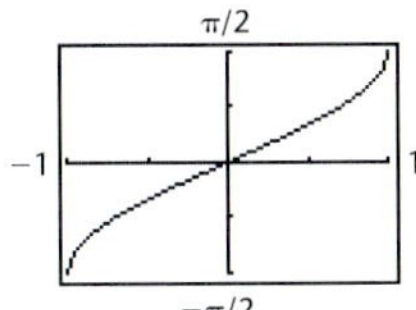

47.

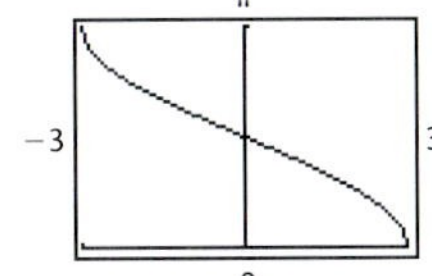

49.

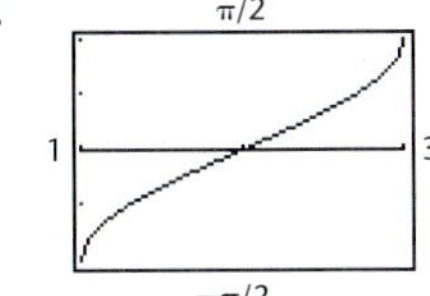

51.

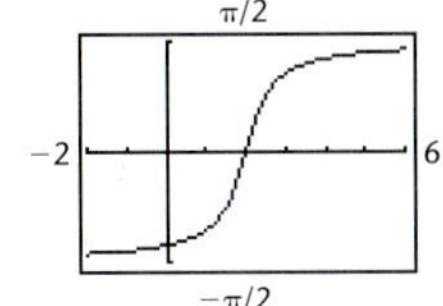

53. (A)

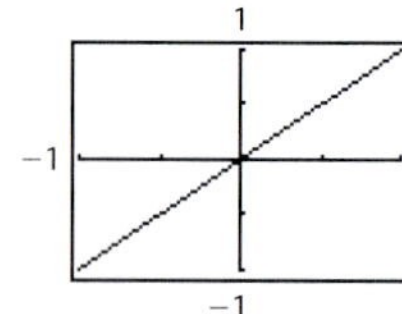

(B) The domain of $\cos^{-1}$ is restricted to $-1 \le x \le 1$; hence no graph will appear for other x.

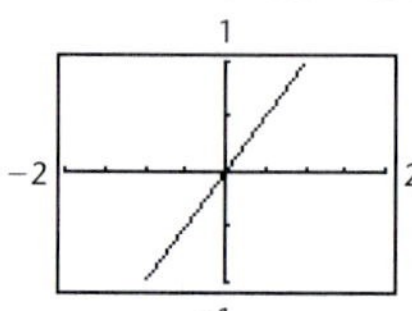

55. $\sqrt{2}/2$ **57.** 1/2 **59.** $\sqrt{1 - x^2}$ **61.** $1/\sqrt{1 + x^2}$ **63.** $f^{-1}(x) = 3 + \cos^{-1}[(x - 4)/2]$, $2 \le x \le 6$

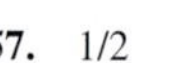

65. (A)

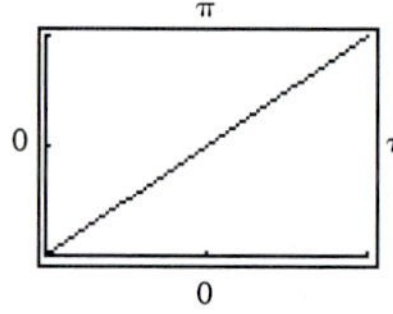

(B) The domain for $\cos x$ is $(-\infty, \infty)$, and the range is $[-1, 1]$, which is the domain for $\cos^{-1} x$. Thus, $y = \cos^{-1}(\cos x)$ has a graph over the interval $(-\infty, \infty)$, but $\cos^{-1}(\cos x) = x$ only on the restricted domain of $\cos x$, $[0, \pi]$.

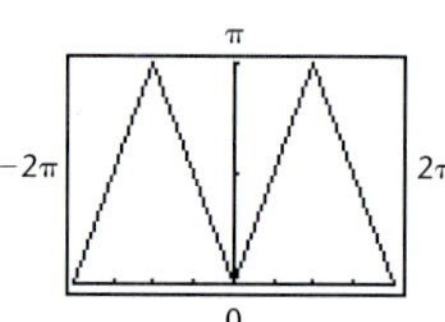

67. 75.38°; 24.41°

69. (A)

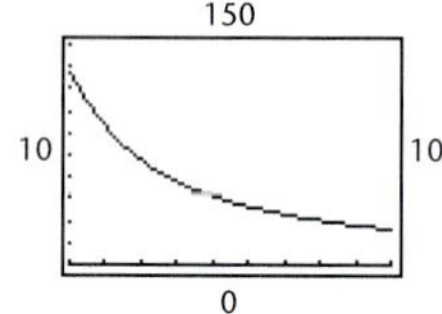

(B) 59.44 mm

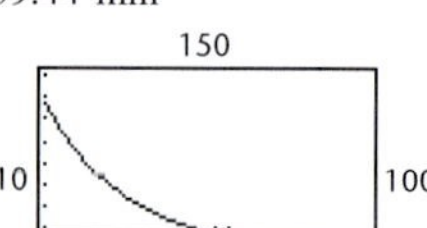

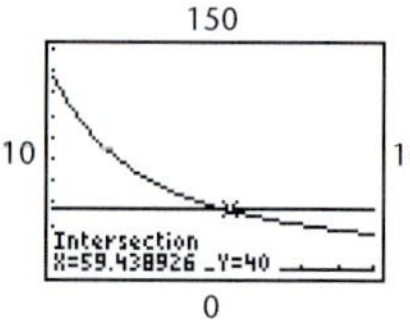

71. 21.59 in.

73. (A)

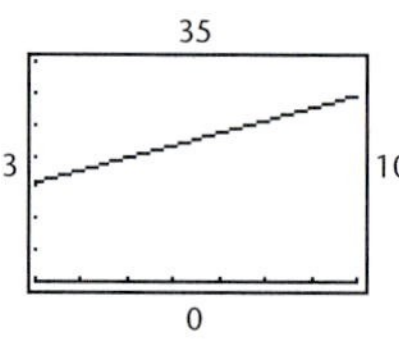

(B) 7.22 in.

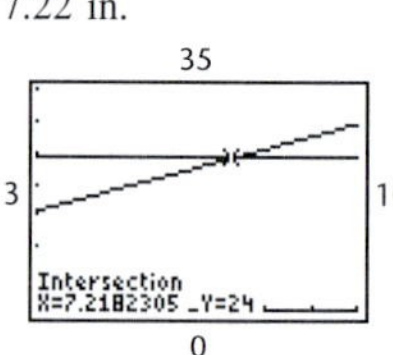

75. (B) 76.10 ft

Chapter 5 Review Exercise

1. 2.5 radians *(5-1)* **2.** 7.5 cm *(5-1)* **3.** $\alpha = 54.8°$, $a = 16.5$ ft, $b = 11.6$ ft *(5-2)*

4. (A) $\pi/3$ (B) 60° (C) $\pi/6$ (D) 30° *(5-2, 5-4)* **5.** (A) III, IV (B) II, III (C) II, IV *(5-3)*

6. (A) $-\frac{3}{5}$ (B) $\frac{5}{4}$ (C) $-\frac{4}{3}$ *(5-3)*

7.

θ°	θ rad	sin θ	cos θ	tan θ	csc θ	sec θ	cot θ
0	0	0	1	0	ND*	1	ND
30	$\pi/6$	1/2	$\sqrt{3}/2$	$1/\sqrt{3}$	2	$2/\sqrt{3}$	$\sqrt{3}$
45	$\pi/4$	$1/\sqrt{2}$	$1/\sqrt{2}$	1	$\sqrt{2}$	$\sqrt{2}$	1
60	$\pi/3$	$\sqrt{3}/2$	1/2	$\sqrt{3}$	$2/\sqrt{3}$	2	$1/\sqrt{3}$
90	$\pi/2$	1	0	ND	1	ND	0
180	π	0	−1	0	ND	−1	ND
270	$3\pi/2$	−1	0	ND	−1	ND	0
360	2π	0	1	0	ND	1	ND

*ND = not defined.

(5-4) **8.** (A) 2π (B) 2π (C) π *(5-6)*

9. (A) Domain = $(-\infty, \infty)$, range = $[-1, 1]$

(B) Domain is the set of all real numbers except $x = \frac{2k + 1}{2}\pi$, k an integer, range is the set of all real numbers *(5-6)*

10. *(5-6)*

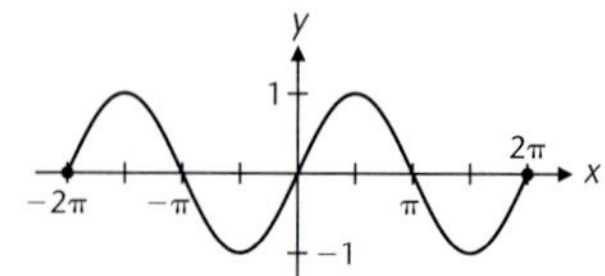

11. *(5-6)*

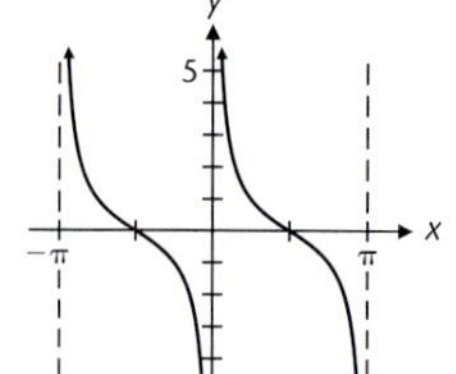

12. The central angle in a circle subtended by an arc of half the length of the radius. *(5-1)*
13. If the graph of $y = \sin x$ is shifted $\pi/2$ units to the left, the result will be the graph of $y = \cos x$. *(5-6, 5-7)* **14.** 78.50° *(5-1)*
15. $\alpha = 49.7°$, $\beta = 40.3°$, $c = 20.6$ cm *(5-2)* **16.** (A) II (B) Quadrantal (C) III *(5-1)* **17.** (A) and (C) *(5-1)*
18. (B) and (C) *(5-3, 5-5)* **19.** (A) $\pi/2, 3\pi/2$ (B) $0, \pi$ (C) $0, \pi$ *(5-3, 5-5)*
20. Since the coordinates of a point on a unit circle are given by $P(a, b) = P(\cos x, \sin x)$, we evaluate $P(\cos(-8.305), \sin(-8.305))$—using a calculator set in radian mode—to obtain $P(-0.436, -0.900)$. Note that $x = -8.305$, since P is moving clockwise. The quadrant in which $P(a, b)$ lies can be determined by the signs of a and b. In this case P is in the third quadrant, since a is negative and b is negative. *(5-5)*
21. $-\frac{1}{2}$ *(5-4)* **22.** $1/\sqrt{2}$ or $\sqrt{2}/2$ *(5-4)* **23.** -1 *(5-4)* **24.** $1/\sqrt{3}$ or $\sqrt{3}/3$ *(5-4)* **25.** $\pi/2$ *(5-4, 5-9)*
26. $-\pi/2$ *(5-4, 5-9)* **27.** $\pi/4$ *(5-4, 5-9)* **28.** $-\pi/3$ *(5-4, 5-9)* **29.** Not defined *(5-4, 5-9)* **30.** $2\pi/3$ *(5-4, 5-9)*
31. $\pi/3$ *(5-4, 5-9)* **32.** Not defined *(5-4, 5-9)* **33.** $\frac{2}{3}$ *(5-4, 5-9)* **34.** $5\pi/6$ *(5-4, 5-9)* **35.** $\sqrt{5}$ *(5-4, 5-9)*
36. $\sqrt{7}/4$ *(5-4, 5-9)* **37.** 0.4431 *(5-3)* **38.** -15.17 *(5-3)* **39.** -2.077 *(5-3, 5-5)* **40.** -0.9750 *(5-2, 5-9)*
41. Not defined *(5-2, 5-9)* **42.** 1.557 *(5-2, 5-9)* **43.** 1.095 *(5-9)* **44.** Not defined *(5-9)*
45. (A) $\theta = -30°$ (B) $\theta = 120°$ *(5-9)* **46.** (A) $\theta = 151.20°$ (B) $\theta = 82.28°$ *(5-9)*
47. $\cos^{-1}[\cos(-2)] = 2$ For the identity $\cos^{-1}(\cos x) = x$ to hold, x must be in the restricted domain of the cosine function; that is, $0 \le x \le \pi$. The number -2 is not in the restricted domain. *(5-9)*
48. $A = 2, P = 2$ *(5-7)* **49.** *(5-7)* **50.** $y = 6\cos 2x, -\pi/2 \le x \le \pi$ *(5-7)*

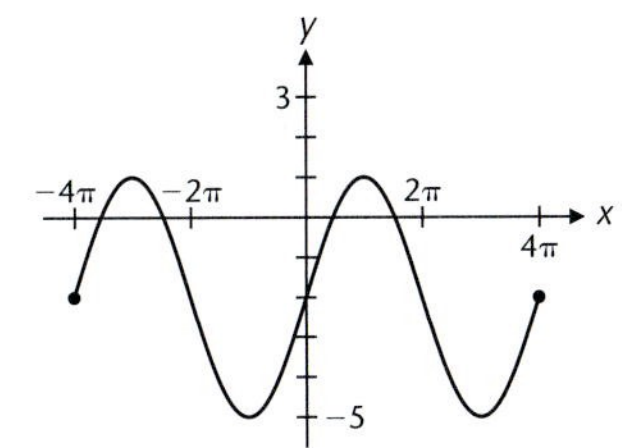

51. $y = -0.5 \sin \pi x, -1 \le x \le 2$ *(5-7)*
52. If the graph of $y = \tan x$ is shifted $\pi/2$ units to the right and reflected in the x axis, the result will be the graph of $y = \cot x$. *(5-6, 5-7)*
53. (A) $\cos x$ (B) $\tan^2 x$ *(5-5)* **54.** *(5-7)* **55.** $A = 2, P = 4$, phase shift $= \frac{1}{2}$ *(5-7)*

56. $y = \cos^{-1} x = \arccos x$ *(5-9)*
Domain $= [-1, 1]$
Range $= [0, \pi]$

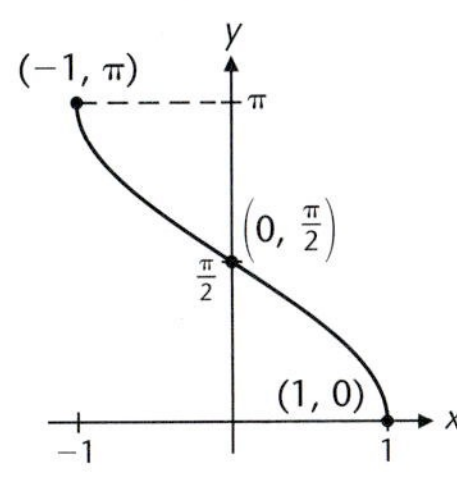

57. $y = \frac{1}{2} + \frac{1}{2}\cos 2x$ *(5-7)*

58. (A) $y = \tan x$ (B) $y = \cot x$ *(5-8)*

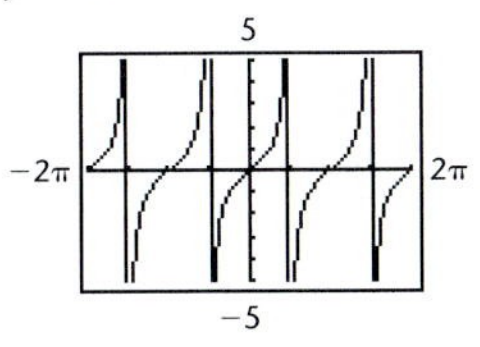

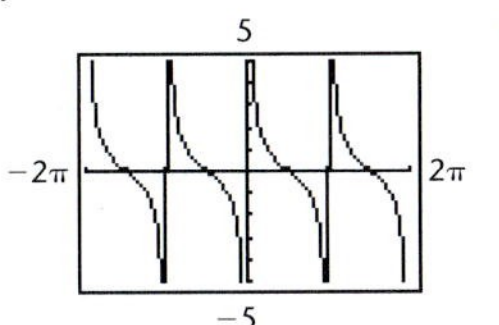

59. (A) 2.5 rad (B) $(-6.41, 4.79)$ *(5-1, 5-3)* **60.** (A) $2\pi/3$ (B) $5\pi/4$ *(5-5)*

61. *(5-6)*

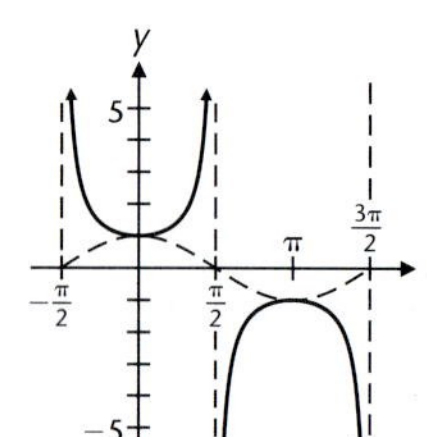

62. $y = \tan^{-1} x = \arctan x$ *(5-9)*
Domain $= (-\infty, \infty)$
Range $= (-\pi/2, \pi/2)$

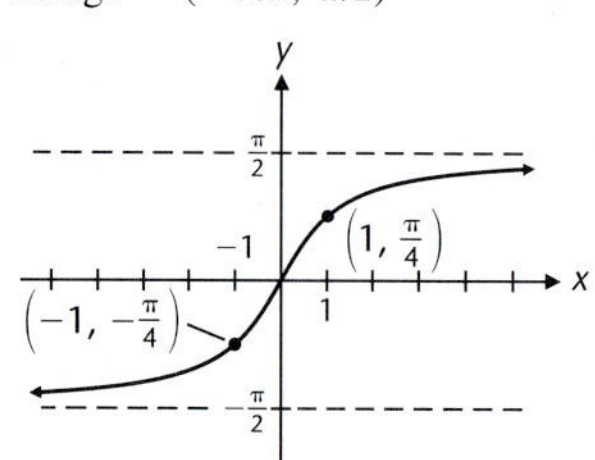

63. $P = 1$; phase shift $= -\frac{1}{2}$ *(5-8)*

64. Period $= 4\pi$; phase shift $= \pi/2$ *(5-8)* **65.** (A) Origin (B) y axis (C) Origin *(5-6)* **66.** $1/\sqrt{1 - x^2}$ *(5-9)*
67. For each case, the number is not in the domain of the function and an error message of some type will appear. *(5-5, 5-9)*
68. $y = 2 \sin (\pi x + \pi/4)$ *(5-7)* **69.** True *(5-3)* **70.** True *(5-4)* **71.** False *(5-4)* **72.** False *(5-6)*
73. True *(5-6)* **74.** False *(5-9)* **75.** False *(5-7)* **76.** True *(5-7)* **77.** Yes *(5-1, 5-6)*
83. $y = 2 \sin (2x + 0.928)$ *(5-7)*

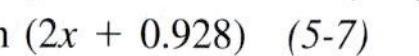

84. (A)

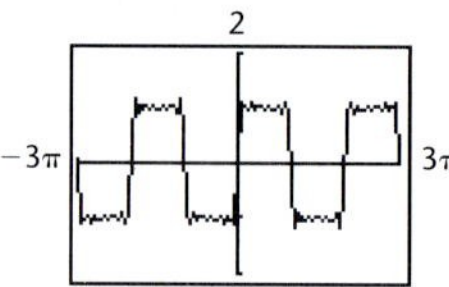

(B) *(5-7)*

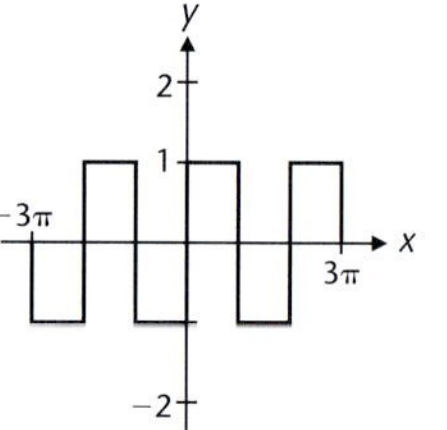

85. $2\pi/5$ rad *(5-1)* **86.** 28.3 cm *(5-2)* **87.** $I = 30 \cos 120\pi t$ *(5-7)*
88. (A) $L = 10 \csc \theta + 15 \sec \theta,\ 0 < \theta < \pi/2$
(B) Length of longest log that can make the corner is 35 ft.

TABLE 2

θ (rad)	0.4	0.5	0.6	0.7	0.8	0.9	1.0
L (ft)	42.0	38.0	35.9	35.1	35.5	36.9	39.6

(C) Length of longest log that can make the corner is 35.1 ft.

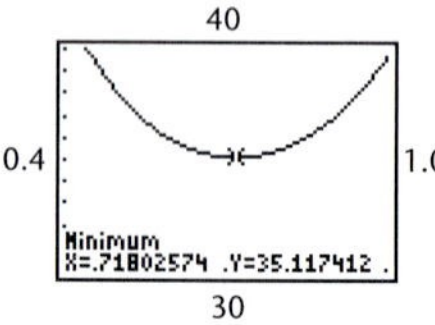

(D) The length L increases without end. *(5-2, 5-5)*
89. (A) $y = 4 - 3 \cos (\pi t/6)$
(B) The graph shows the seasonal changes in soft-drink consumption. Most is consumed in August and the least in February. *(5-7)*
90. (A)

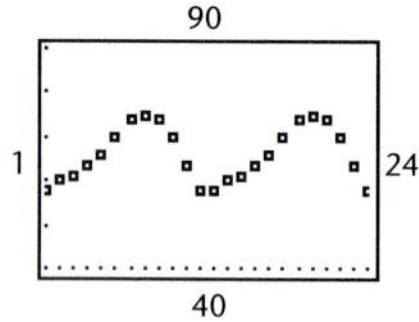

(B) $y = 66.5 + 8.5 \sin [(\pi x/6) - 2.4]$ (C) *(5-7)*

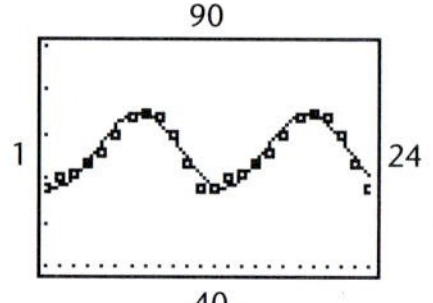

CHAPTER 6

Exercise 6-1

25.

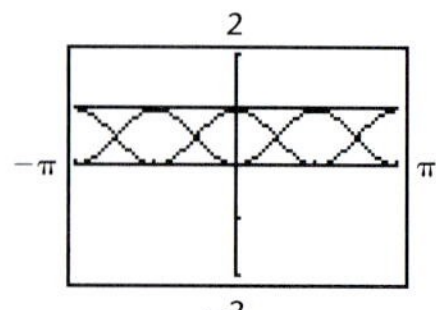

27.

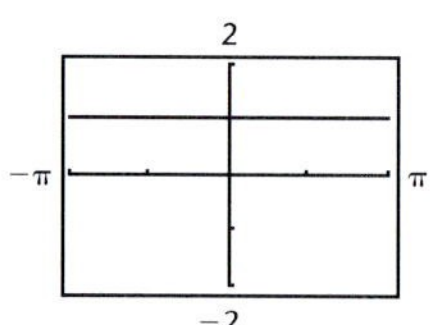
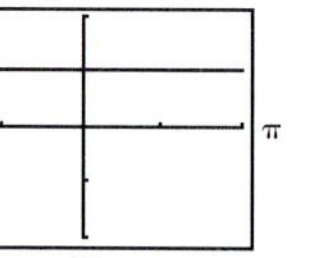

29. No **31.** Yes **33.** No **35.** No

65. Not an identity **67.** An identity **69.** Not an identity **71.** An identity **73.** An identity **75.** Not an identity

83. $g(x) = \cot x$

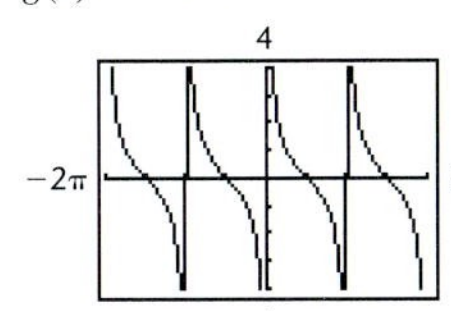

85. $g(x) = -1 + \csc x$

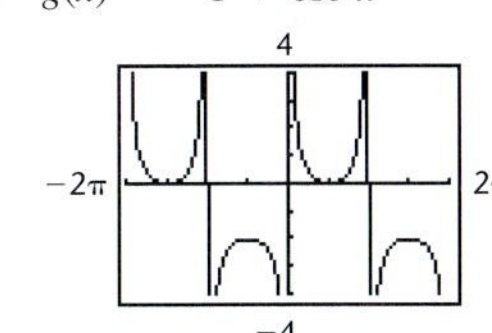

87. $g(x) = 3 \cos x$

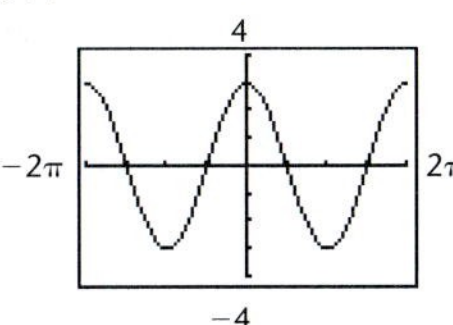

89. III, IV **91.** I, II **93.** All quadrants **95.** I, IV **97.** $a \cos x$ **99.** $a \sec x$

Exercise 6-2

1. Yes **3.** No **5.** Yes **7.** No **9.** Yes **15.** $\frac{\sqrt{2}}{2}(\cos x - \sin x)$ **17.** $\frac{\sqrt{3} + \tan x}{1 - \sqrt{3} \tan x}$ **19.** $-\cos x$ **21.** $\frac{\sqrt{2}}{2}$

23. $\frac{\sqrt{3}}{3}$ **25.** $\frac{\sqrt{2}}{4}(\sqrt{3} - 1)$ **27.** $-\frac{\sqrt{2}}{4}(1 + \sqrt{3})$ **29.** $\sin(x - y) = \frac{-3 - 4\sqrt{8}}{15}$, $\tan(x + y) = \frac{4\sqrt{8} - 3}{4 + 3\sqrt{8}}$

31. $\sin(x - y) = \frac{-2}{\sqrt{5}}$, $\tan(x + y) = \frac{2}{11}$ **47.** $-0.3685, -0.3685; 0.9771, 0.9771$ **49.** $-0.4429, -0.4429; -2.682, -2.682$

51. Evaluate each side for a particular set of values of x and y for which each side is defined. If the left side is not equal to the right side, then the equation is not an identity. For example, for $x = 2$ and $y = 1$, both sides are defined but are not equal.

53. $y_1 = \sin(x + \pi/6)$

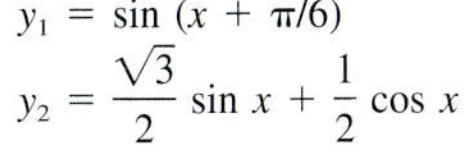

$y_2 = \frac{\sqrt{3}}{2} \sin x + \frac{1}{2} \cos x$

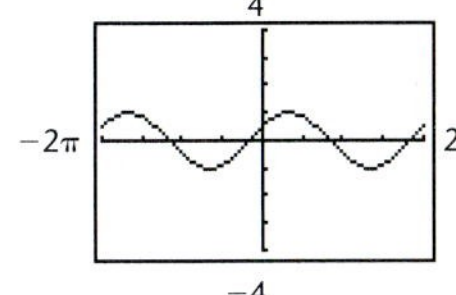

55. $y_1 = \cos(x - 3\pi/4)$

$y_2 = -\frac{\sqrt{2}}{2} \cos x + \frac{\sqrt{2}}{2} \sin x$

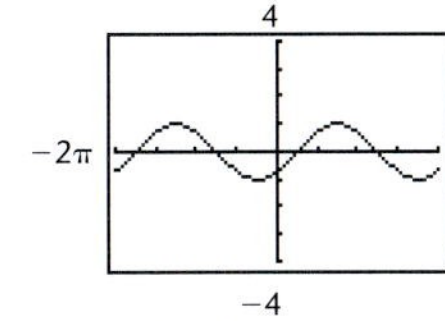

57. $y_1 = \tan(x + 2\pi/3)$

$y_2 = \frac{\tan x - \sqrt{3}}{1 + \sqrt{3} \tan x}$

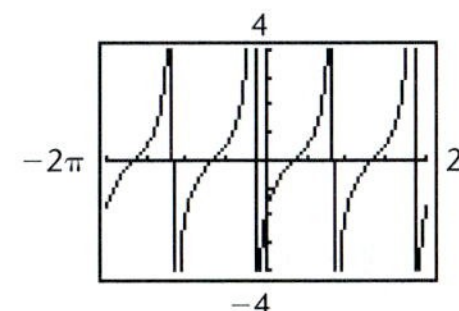

59. $\frac{24}{25}$ **61.** $-\frac{1}{2}$ **63.** $xy + (\sqrt{1 - x^2})(\sqrt{1 - y^2})$

67. $y_1 = \cos 1.2x \cos 0.8x - \sin 1.2x \sin 0.8x$ **73.** (C) 3,510 ft

$y_2 = \cos 2x$

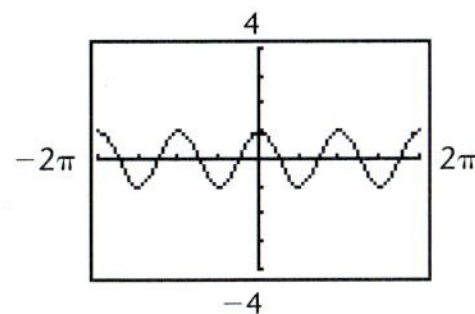

Exercise 6-3

1. $\frac{1}{2} = \frac{1}{2}$ **3.** $-\sqrt{3} = -\sqrt{3}$ **5.** $1 = 1$ **7.** $2 - \sqrt{3}$ **9.** $-\frac{\sqrt{2 - \sqrt{2}}}{2}$

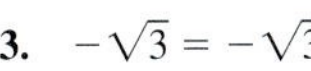

11.

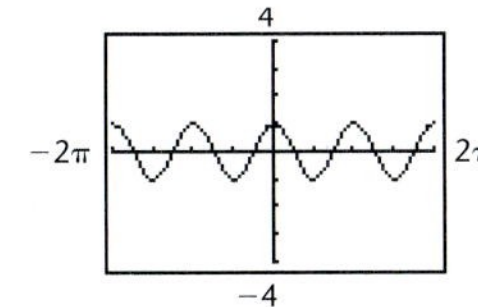

13.

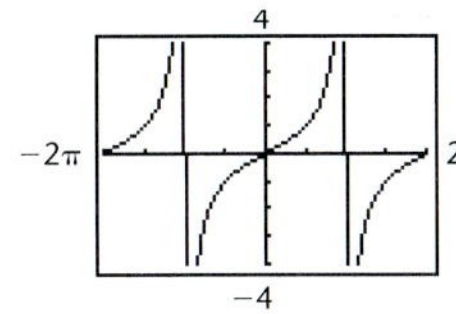

29. No **31.** Yes **33.** No

35. $\sin 2x = -\frac{24}{25}$, $\cos 2x = \frac{7}{25}$, $\tan 2x = -\frac{24}{7}$ **37.** $\sin 2x = -\frac{120}{169}$, $\cos 2x = \frac{119}{169}$, $\tan 2x = -\frac{120}{119}$

39. $\sin \frac{x}{2} = \sqrt{\frac{3 + 2\sqrt{2}}{6}}$, $\cos \frac{x}{2} = -\sqrt{\frac{3 - 2\sqrt{2}}{6}}$, $\tan \frac{x}{2} = -3 - 2\sqrt{2}$ **41.** $\sin \frac{x}{2} = -\frac{2\sqrt{5}}{5}$, $\cos \frac{x}{2} = \frac{\sqrt{5}}{5}$, $\tan \frac{x}{2} = -2$

43. (A) 2θ is a second-quadrant angle, since θ is a first-quadrant angle and $\tan 2\theta$ is negative for 2θ in the second quadrant and not for 2θ in the first.

(B) Construct a reference triangle for 2θ in the second quadrant with $(a, b) = (-3, 4)$. Use the Pythagorean theorem to find $r = 5$. Thus, $\sin 2\theta = \frac{4}{5}$ and $\cos 2\theta = -\frac{3}{5}$.

(C) The double-angle identities $\cos 2\theta = 1 - 2 \sin^2 \theta$ and $\cos 2\theta = 2 \cos^2 \theta - 1$.

(D) Use the identities in part C in the form $\sin \theta = \sqrt{\frac{1 - \cos 2\theta}{2}}$ and $\cos \theta = \sqrt{\frac{1 + \cos 2\theta}{2}}$

The positive radicals are used because θ is in quadrant I.

(E) $\sin \theta = 2\sqrt{5}/5$; $\cos \theta = \sqrt{5}/5$

45. (A) $-0.723\ 35 = -0.723\ 35$ (B) $-0.588\ 21 = -0.588\ 21$ **47.** (A) $-3.2518 = -3.2518$ (B) $0.892\ 79 = 0.892\ 79$

49. $y_1 = y_2$ for $[-\pi\ \pi]$

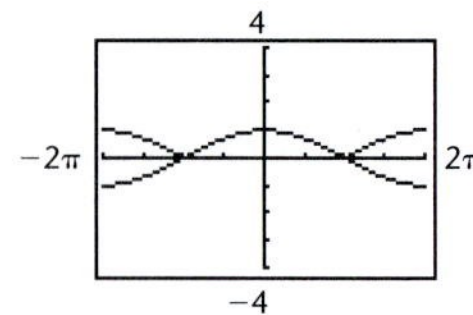

51. $y_1 = y_2$ for $[-2\pi, 0]$

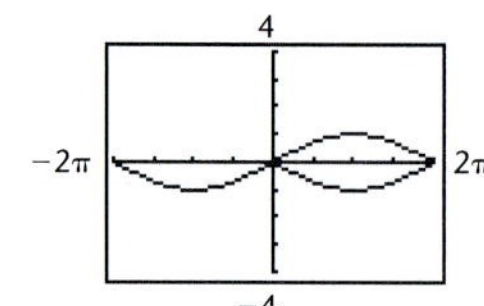

57. $-\frac{7}{25}$ **59.** $-\frac{24}{7}$ **61.** $\sqrt{5}/5$

63. $\tan(x/2)$

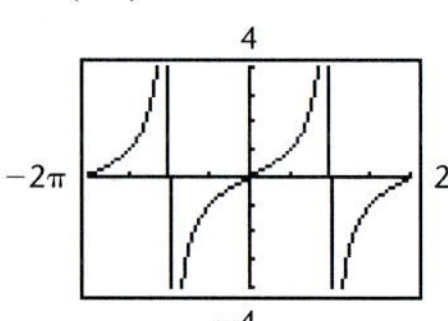

65. $1 + 2\sin x$

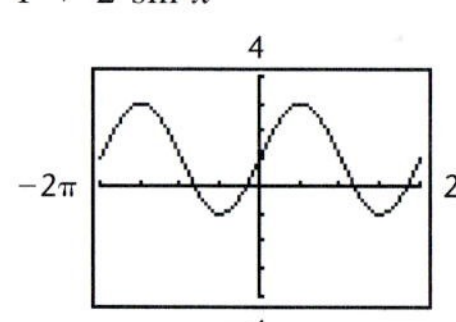

67. $\sec 2x$

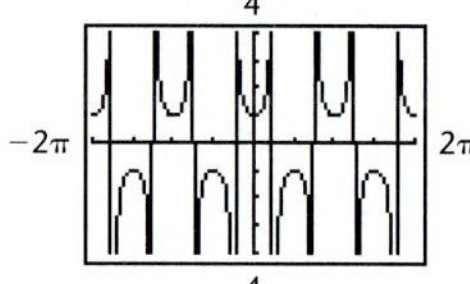

69. $x = \frac{224}{17} \approx 13.176$ m; $\theta = 28.955°$ **71.** (A) $d = \dfrac{v_0^2 \sin 2\theta}{32 \text{ ft/s}^2}$ (B) $\theta = 45°$

73. (B)

TABLE 1

n	10	100	1,000	10,000
A_n	2.93893	3.13953	3.14157	3.14159

(C) A_n appears to approach π, the area of the circle with radius 1.

(D) A_n will not exactly equal the area of the circumscribing circle for any n no matter how large n is chosen; however, A_n can be made as close to the area of the circumscribing circle as we like by making n sufficiently large.

Exercise 6-4

1. $\frac{1}{2}\sin 4m + \frac{1}{2}\sin 2m$ **3.** $\frac{1}{2}\cos 2u - \frac{1}{2}\cos 4u$ **5.** $2\sin 2t\cos t$ **7.** $2\sin 7w\sin 2w$ **9.** $\frac{1}{4}$ **11.** $\dfrac{1+\sqrt{2}}{4}$

13. $-\dfrac{\sqrt{6}}{2}$ **15.** $-\dfrac{\sqrt{2}}{2}$

19. Let $x = u + v$ and $y = u - v$, and solve the resulting system for u and v in terms of x and y, then substitute the results into the first identity. The second identity will result after a small amount of algebraic manipulation. **25.** Yes **27.** No **29.** No

31. (A) $-0.34207 = -0.34207$ (B) $-0.05311 = -0.05311$ **33.** (A) $-0.19115 = -0.19115$ (B) $-0.46541 = -0.46541$

35. $y_2 = 2\sin\dfrac{3x}{2}\cos\dfrac{x}{2}$

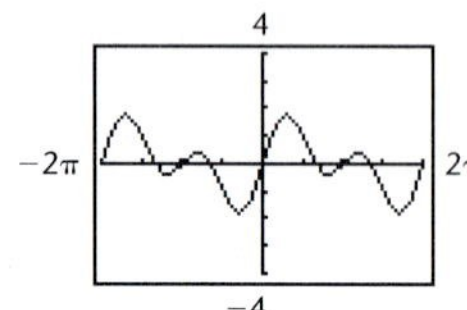

37. $y_2 = -2\sin x\sin 0.7x$

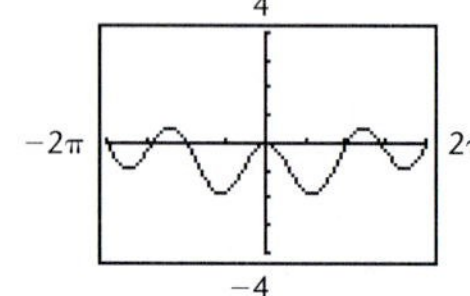

39. $y_2 = \frac{1}{2}(\sin 4x + \sin 2x)$

4

−2π 2π

−4

41. $y_2 = \frac{1}{2}(\cos 1.6x - \cos 3x)$

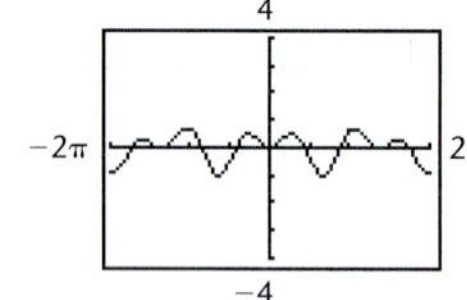

45. (A)

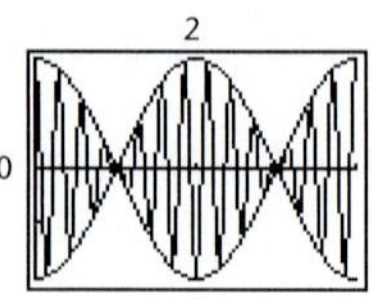

(B) $y_1 = \cos(30\pi x) + \cos(26\pi x)$
Graph same as part A

47. (A)

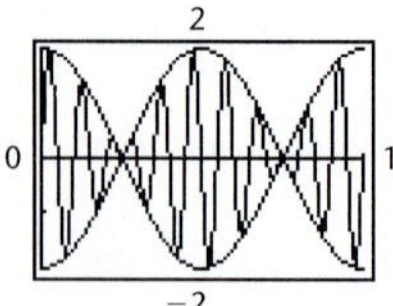

(B) $y_1 = \sin(22\pi x) + \sin(18\pi x)$
Graph same as part A

49. (B)

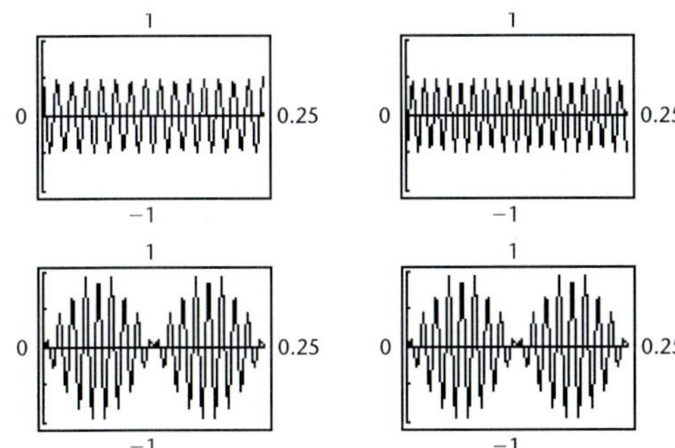

Exercise 6-5

1. $7\pi/6$, $11\pi/6$ **3.** $7\pi/6 + 2k\pi$, $11\pi/6 + 2k\pi$, k any integer **5.** $2\pi/3$ **7.** $2\pi/3 + k\pi$, k any integer **9.** 30°, 330° **11.** 1.1279, 5.1553 **13.** 74.0546° **15.** $3.5075 + 2k\pi$, $5.9172 + 2k\pi$, k any integer **17.** 0.3376 **19.** 2.7642 **21.** $k(180°)$, $135° + k(180°)$, k any integer **23.** 0, $2\pi/3$, π, $4\pi/3$ **25.** 210°, 330° **27.** 60°, 180°, 300° **29.** $\pi/3$, π, $5\pi/3$ **31.** 41.81° **33.** 1.911 **35.** 0.3747, 2.767 **37.** Infinitely many **39.** One **41.** Infinitely many **43.** 0.3747, 2.7669 **45.** $0.3747 + 2k\pi$, $2.7669 + 2k\pi$, k any integer **47.** $(-1.1530, 1.1530)$ **49.** $[3.5424, 5.3778]$, $[5.9227, \infty)$ **51.** 1.8183

53. $\tan^{-1}(-5.377)$ has exactly one value, -1.387; the equation $\tan x = -5.377$ has infinitely many solutions, which are found by adding $k\pi$, k any integer, to each solution in one period of $\tan x$.

55. 0, $3\pi/2$ **57.** π **59.** 0.1204, 0.1384

61. (A) The largest zero for f is 0.3183. As x increases without bound, $1/x$ tends to 0 through positive numbers, and $\sin(1/x)$ tends to 0 through positive numbers. $y = 0$ is a horizontal asymptote for the graph of f.
(B) Infinitely many zeros exist between 0 and b, for any b, however small. The exploration graphs suggest this conclusion, which is reinforced by the following reasoning: Note that for each interval $(0, b]$, however small, as x tends to zero through positive numbers, $1/x$ increases without bound, and as $1/x$ increases without bound, $\sin(1/x)$ will cross the x axis an unlimited number of times. The function f does not have a smallest zero, because, between 0 and b, no matter how small b is, there is always an unlimited number of zeros.

63. 0.009235 s **65.** 50.77° **67.** 123° **69.** 2.267 rad **71.** (A) 12.4575 mm (B) 2.6496 mm **73.** $(r, \theta) = (0, 0), (0, 180°), (0, 360°)$ **75.** $\theta = 67°$

Chapter 6 Review Exercises

5. $\frac{1}{2}(\cos 5\alpha + \cos \alpha)$ *(6-4)* **6.** $2 \cos 7x \sin 2x$ *(6-4)* **7.** $\cos x$ *(6-2)* **8.** $135° + k(360°)$, $225° + k(360°)$, k any integer *(6-5)* **9.** $k\pi$, $\pi/4 + k\pi$, k any integer *(6-5)* **10.** $\pm 34.7648° + k(360°)$, k any integer *(6-5)* **11.** -0.0065 *(6-5)* **12.** $0.5943 + 2k\pi$, $2.5473 + 2k\pi$, k any integer *(6-5)* **13.** 3.1855 *(6-5)* **14.** (A) Not an identity (B) An identity *(6-1)* **24.** $\dfrac{-2 - \sqrt{3}}{4}$ *(6-4)* **25.** $-\dfrac{\sqrt{6}}{2}$ *(6-4)* **26.** No *(6-1)* **27.** Yes *(6-3)* **28.** Yes *(6-2)* **29.** No *(6-2)* **30.** $\pi/3$, $2\pi/3$, $4\pi/3$, $5\pi/3$ *(6-5)* **31.** 0°, 120° *(6-5)* **32.** $k\pi$, $\pi/6 + 2k\pi$, $5\pi/6 + 2k\pi$, k any integer *(6-5)* **33.** $k\pi$, $\pi/6 + 2k\pi$, $11\pi/6 + 2k\pi$, k any integer *(6-5)* **34.** $120° + k(360°)$, $240° + k(360°)$, k any integer *(6-5)* **35.** $14.34° + k(180°)$, k any integer *(6-5)* **36.** $0.6259 + 2k\pi$, $2.516 + 2k\pi$, k any integer *(6-5)* **37.** 1.178, 2.749 *(6-5)* **38.** Two *(6-5)* **39.** Infinitely many *(6-5)* **40.** None *(6-5)* **41.** Infinitely many *(6-5)* **42.** 1.4903 *(6-5)* **43.** $x < 1.4903$ *(6-5)* **44.** -0.6716, 0.6716 *(6-5)* **45.** $[-0.6716, 0.6716]$ *(6-5)*

46. (A) Yes (B) Conditional equation, since the equation is false for $x = 1$ and $y = 1$, for example, and both sides are defined at $x = 1$ and $y = 1$. *(6-1)*

47. $\sin^{-1} 0.3351$ has exactly one value, and the equation $\sin x = 0.3351$ has infinitely many solutions. *(5-9, 6-5)*

48. (A) Not an identity (B) An identity *(6-1)*

49. $y_2 = \frac{1}{2}\cos x + \dfrac{\sqrt{3}}{2}\sin x$ *(6-2)*

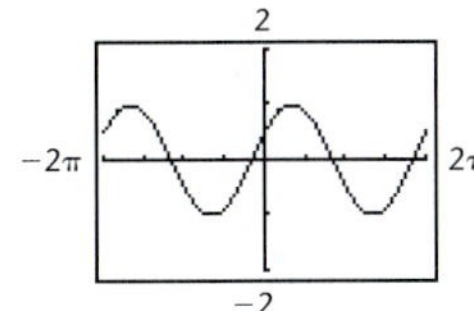

50. (A) 0, $2\pi/3$, $4\pi/3$ (B) 0, 2.0944, 4.1888 *(6-5)* **51.** -2.233, 0.149 *(6-5)* **52.** (A) $3/\sqrt{10}$ or $3\sqrt{10}/10$ (B) 7/25 *(6-3)* **53.** $-24/25$ *(6-3)* **54.** 24/25 *(6-2)*

55. (A) $0, \pi/3, 2\pi/3$ (B) 0, 1.0472, 2.0944 *(6-5)*

56. (A) 0.6817, 1.3183

(B) As x increases without bound, $1/(x - 1)$ tends to 0 through positive numbers, and $\sin [1/(x - 1)]$ tends to 0 through positive numbers. $y = 0$ is a horizontal asymptote for the graph of f.

(C) There are infinitely many zeros in any interval containing $x = 1$. The number $x = 1$ is not a zero because $\sin [1/(x - 1)]$ is not defined at $x = 1$. *(6-5)*

57. $x = \sqrt{27}$; $x = 5.196$ cm, $\theta = 30.000°$ *(6-3)* **58.** 0.00346 s *(6-5)*

59. $y = 0.6 \cos (184\pi t)$ $y = -0.6 \cos (208\pi t)$

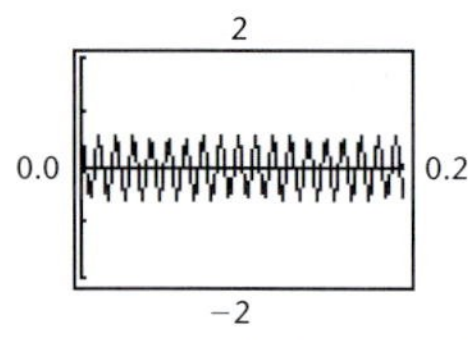

2
0.0 0.2
−2

$y = 0.6 \cos (184\pi t) - 0.6 \cos (208\pi t)$ $y = 1.2 \sin (12\pi t) \sin (196\pi t)$

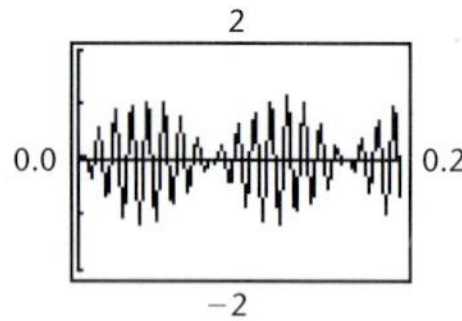

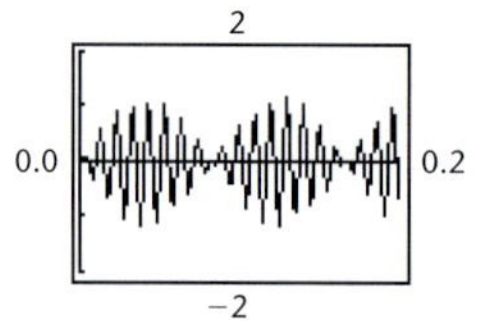

(6-4)

60. Height = 7.057 ft, radius = 21.668 ft

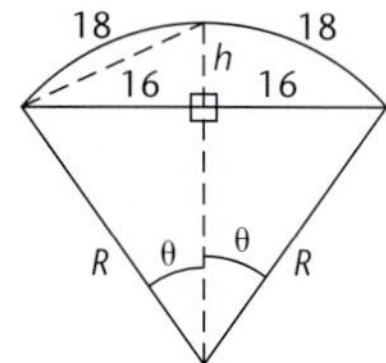

From the figure, $R\theta = 18$ and $\sin \theta = 16/R$. From these two equations, solving each for R in terms of θ and setting the results equal to each other, we obtain the desired trigonometric equation. *(6-5)*

CHAPTER 7

Exercise 7-1

1. $\gamma = 79°$, $a = 41$ ft, $b = 20$ ft **3.** $\beta = 40°$, $a = 16$ km, $c = 5.8$ km **5.** $\alpha = 49°$, $a = 53$ yd, $b = 66$ yd
7. $\beta = 81°$, $b = 16$ cm, $c = 12$ cm **9.** Two triangles; case (c) **11.** One triangle; case (b) **13.** Zero triangles; case (a)
15. Zero triangles; case (e) **17.** One triangle; case (d) **19.** One triangle; none of the cases
21. $\beta = 49.5°$, $a = 20.0$ ft, $c = 4.81$ ft **23.** $\gamma = 58.1°$, $a = 140$ m, $c = 129$ m **25.** No solution
27. Triangle 1: $\beta = 63.4°$, $\gamma = 77.7°$, $c = 46.7$ in.; Triangle 2: $\beta = 116.6°$, $\gamma = 24.5°$, $c = 19.8$ in. **29.** No solution
31. $\alpha = 22°10'$, $\gamma = 128°20'$, $c = 89.8$ mm **33.** $k = 25.2 \sin 42.3° = 16.9599$ **35.** Left side: 16.204; right side: 16.073
37. 4.06 mi, 2.47 mi **39.** 353 ft **41.** 5.8 in., 3.1 in. **43.** 4.42×10^7 km, 2.39×10^8 km **45.** 159 ft
47. $R = 7.76$ mm, $s = 13.4$ mm

Exercise 7-2

1. Angle γ is acute. A triangle can have at most one obtuse angle. Since α is acute, then, if the triangle has an obtuse angle, it must be the angle opposite the longer of the two sides, b and c. Thus, γ, the angle opposite the shorter of the two sides, c, must be acute.
3. $a = 6.03$ yd, $\beta = 56.6°$, $\gamma = 52.2°$ **5.** $c = 14.0$ mm, $\alpha = 20°40'$, $\beta = 39°0'$
7. If the triangle has an obtuse angle, then it must be the angle opposite the longest side; in this case, β.
9. $\alpha = 23.0°$, $\beta = 94.9°$, $\gamma = 62.1°$ **11.** $\alpha = 67.3°$, $\beta = 54.6°$, $\gamma = 58.1°$ **13.** No solution
15. $b = 23.5$ inches, $\alpha = 28.3°$, $\gamma = 25.5°$ **17.** No solution **19.** $\alpha = 30.7°$, $\gamma = 110.9°$, $c = 21.0$ in.
21. $\alpha = 49.1°$, $\beta = 102.9°$, $\gamma = 28.0°$
23. Triangle 1: $\beta = 70.3°$, $\alpha = 51.3°$, $a = 5.99$ m; Triangle 2: $\beta = 109.7°$, $\alpha = 11.9°$, $a = 1.58$ m **25.** No solution **33.** 120 yd
35. 5.81 ft **37.** 121 mi **39.** 74.1 m **41.** 0.284 rad **43.** $\alpha = 31°50'$, $\beta = 50°10'$, $\gamma = 98°0'$ **45.** $\angle CAB = 33°$
47. 24,800 mi

Exercise 7-3

1. $|\mathbf{u} + \mathbf{v}| = 78$ mph, $\theta = 67°$ **3.** $|\mathbf{u} + \mathbf{v}| = 41$ kg, $\theta = 45°$ **5.** $|\mathbf{u}| = 12$ lb, $|\mathbf{v}| = 21$ lb **7.** $|\mathbf{u}| = 388$ mph, $|\mathbf{v}| = 41$ mph
9. $|\mathbf{u} + \mathbf{v}| = 77$g, $\alpha = 15°$ **11.** $|\mathbf{u} + \mathbf{v}| = 23$ knots, $\alpha = 6°$ **13.** $|\mathbf{u}| = 12$ kg, $|\mathbf{v}| = 6.0$ kg

15. $|\mathbf{u}| = 109$ mph, $|\mathbf{v}| = 160$ mph **17.** True **19.** False **21.** True **23.** False **25.** 260 mph at 282°
27. 288°, 7.6 knots **29.** 3,900 lb at 72° **31.** (A) 388 lb (B) 4,030 lb **33.** To the right

Exercise 7-4

1. $\langle 7, 2\rangle$ **3.** $\langle -4, 8\rangle$ **5.** $\langle -2, 9\rangle$ **7.** 15 **9.** 75 **11.** 493
13. (A) $\langle 1, 4\rangle$ (B) $\langle 3, -2\rangle$ (C) $\langle 14, -1\rangle$ **15.** (A) $\langle -2, 1\rangle$ (B) $\langle -6, -3\rangle$ (C) $\langle -10, -1\rangle$
17. $\mathbf{v} = -8\mathbf{i}$ **19.** $\mathbf{v} = 6\mathbf{i} - 12\mathbf{j}$ **21.** $\mathbf{v} = -5\mathbf{i} - 2\mathbf{j}$ **23.** $5\mathbf{i} + 2\mathbf{j}$ **25.** $-16\mathbf{j}$ **27.** $-8\mathbf{j}$ **29.** $\mathbf{u} = \left\langle -\frac{1}{\sqrt{2}}, \frac{1}{\sqrt{2}} \right\rangle$
31. $\mathbf{u} = \langle -\frac{12}{13}, \frac{5}{13}\rangle$ **33.** False **35.** False **45.** Left side: 760 lb; right side: 761 lb **47.** Left cable: 897 lb; right cable: 732 lb
49. Member *AB* has a compression force of 231 lb; member *CB* has a tension force of 462 lb
51. For *AB*: compression: 2,360 lb; for *CB*: tension: 2,000 lb

Exercise 7-5

1.

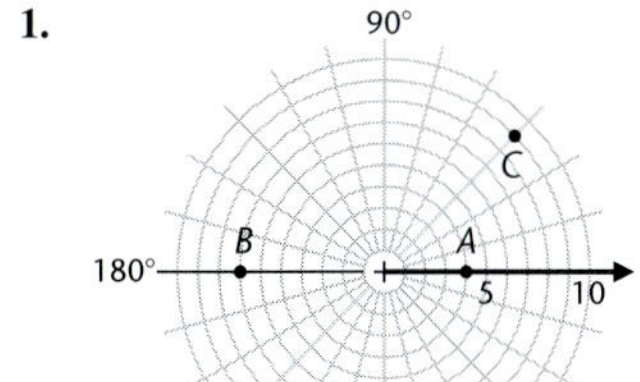

3.

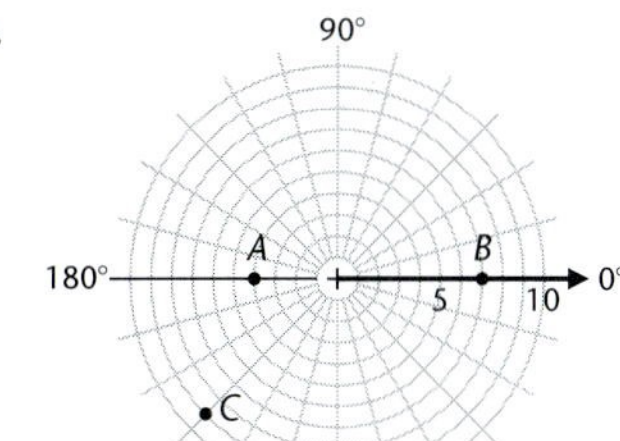

5.

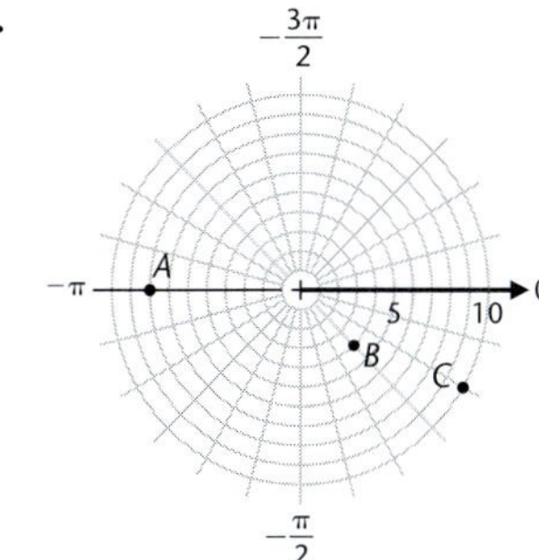

7.

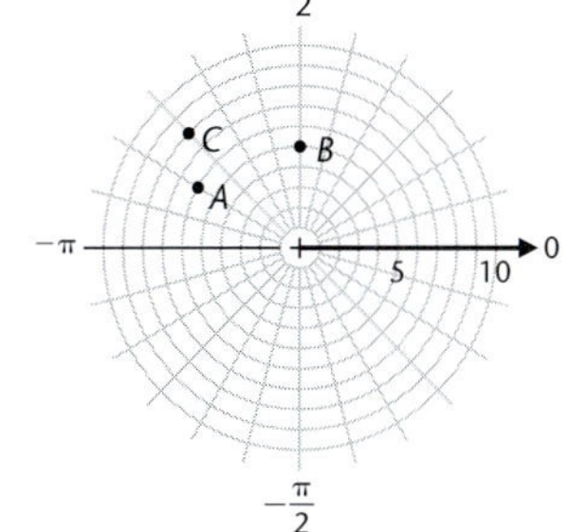

9. $(5, -\pi/4)$: The polar axis is rotated $\pi/4$ rad clockwise (negative direction), and the point is located 5 units from the pole along the positive polar axis. $(5, 7\pi/4)$: The polar axis is rotated $7\pi/4$ rad counterclockwise (positive direction), and the point is located 5 units from the pole along the positive polar axis. $(-5, -5\pi/4)$: The polar axis is rotated $5\pi/4$ rad clockwise (negative direction), and the point is located 5 units from the pole along the negative polar axis.

11.

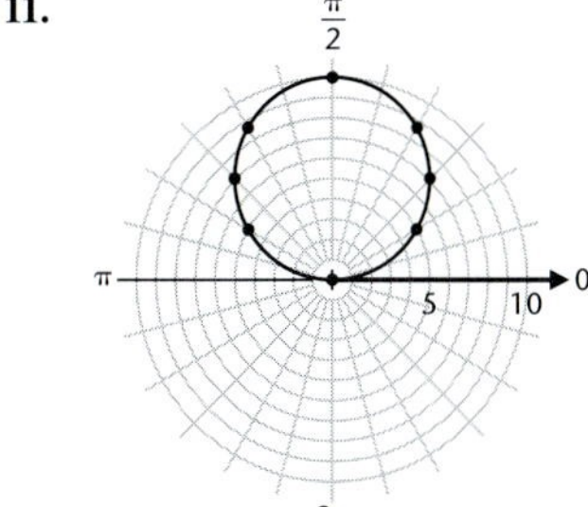

13.

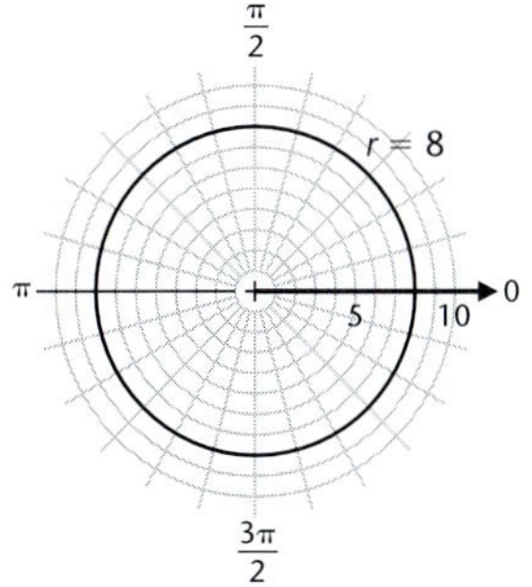

15.

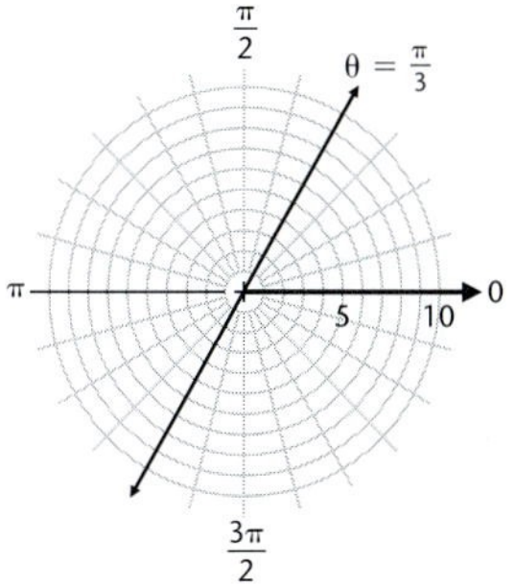

17. (5.196, 3.000) **19.** (1.848, −0.765) **21.** (2.078, 3.688) **23.** (8, 180°) **25.** $(5\sqrt{2}, -135°)$ **27.** (11.05, 27.7°)

29.

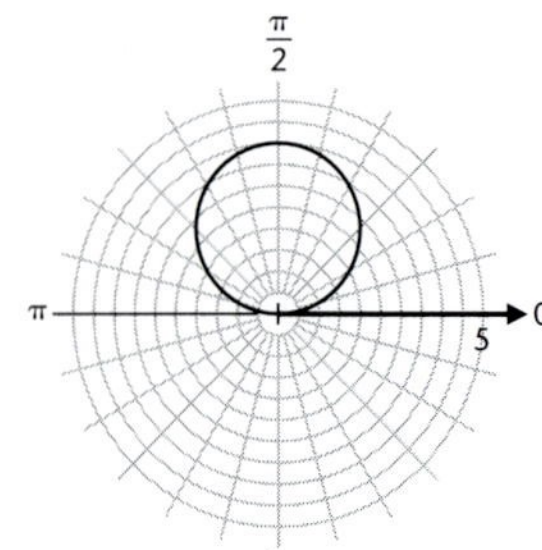

31.

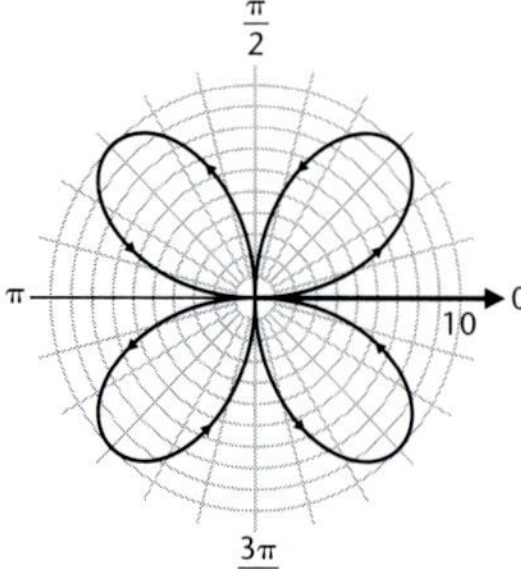

33.

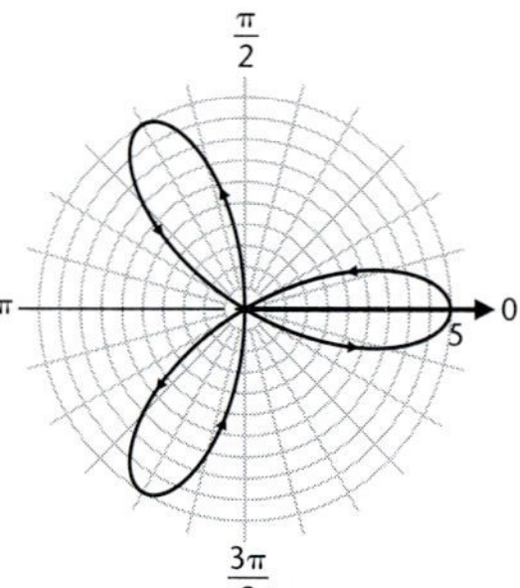

35.

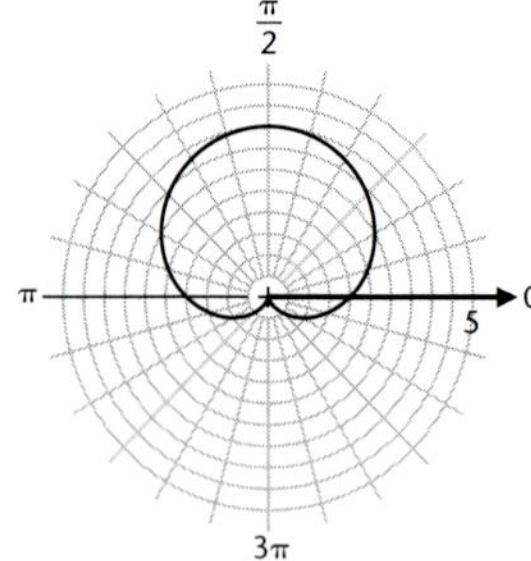

37.

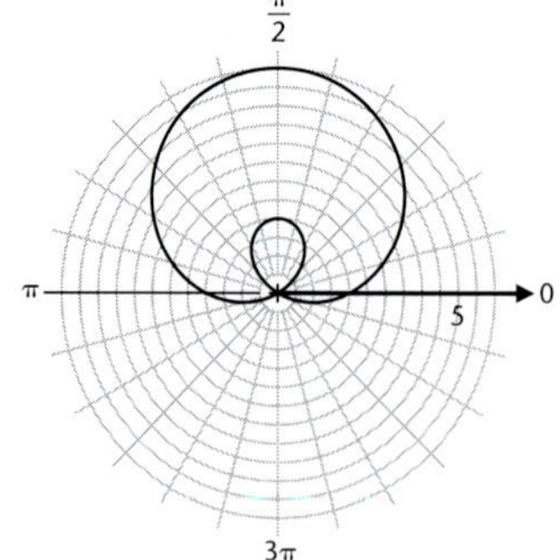

39.

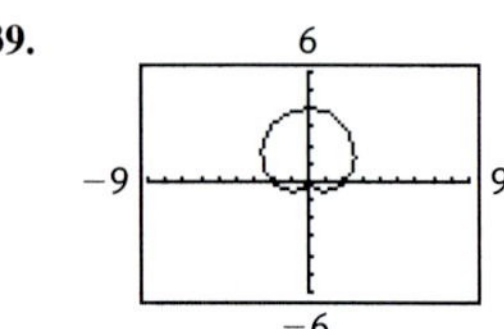

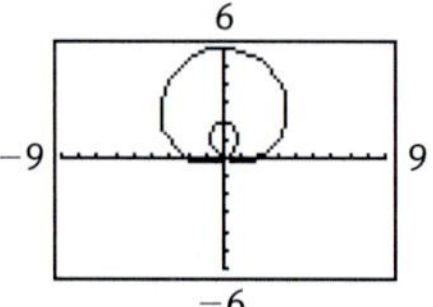

41. (A)

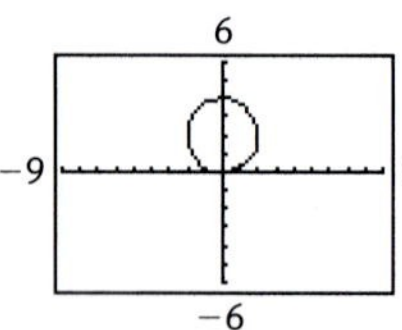

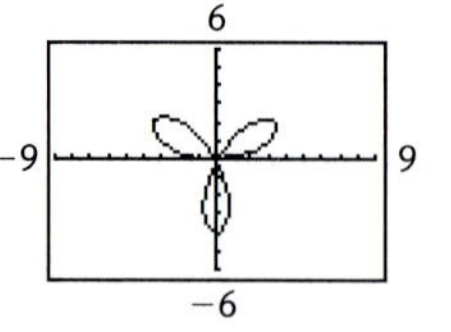

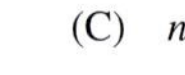

(B) 7 (C) n

43. (A)

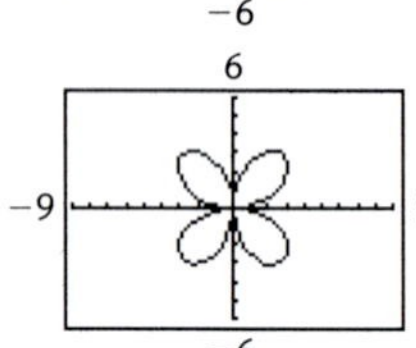

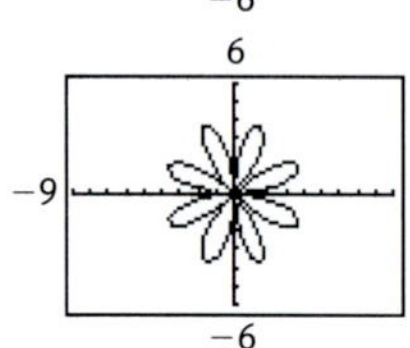

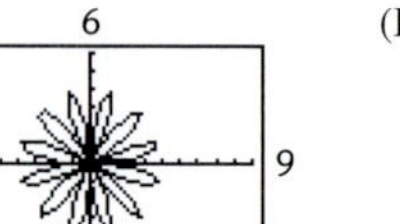

(B) 16 (C) $2n$

45. $r = 2$; circle **47.** $\tan\theta = 1/\sqrt{3}$ or $\theta = \pi/6$; line **49.** $r = 5\tan\theta\sec\theta$; parabola **51.** $x^2 + y^2 = 3x$; circle

53. $4y - x = 1$; line **55.** $3x^2 + 4y^2 = 1 - 2x$; ellipse

57. For each n, there are n large petals and n small petals. For n odd, the small petals are within the large petals; for n even, the small petals are between the large petals.

59.

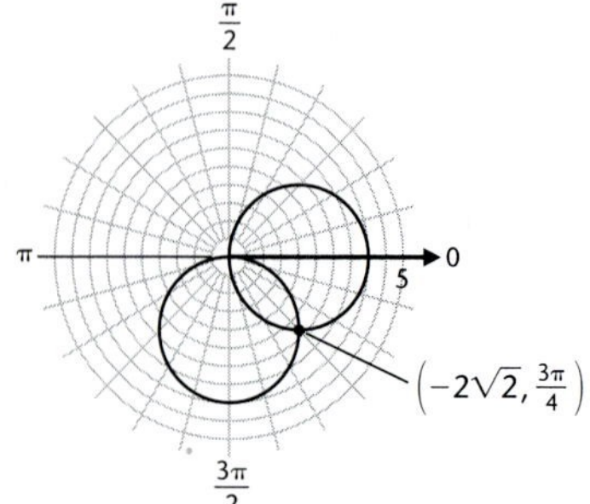

$(r, \theta) = (-2\sqrt{2}, 3\pi/4)$ [*Note:* (0, 0) is not a solution of the system even though the graphs cross at the origin.]

61.

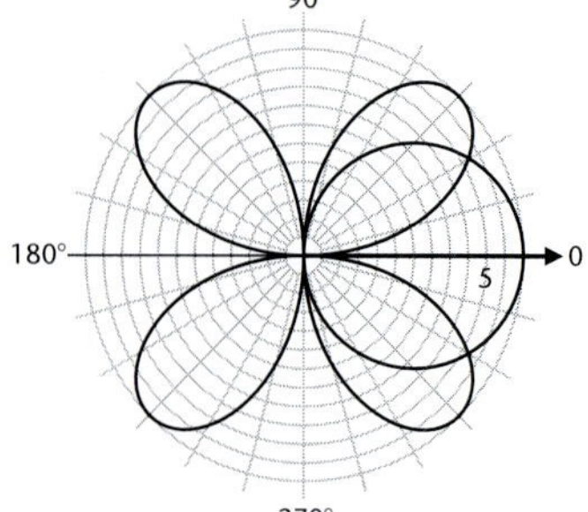

$(r, \theta) = (0, 90°), (0, 270°), (3\sqrt{3}, 30°), (-3\sqrt{3}, 150°)$ [*Note:* (0, 0) is not a solution of the system even though the graphs cross at the origin.]

63. 3.368 units **65.** 6 k, 13 k, 12 k, 9 k

67. (A) Ellipse (B) Parabola (C) Hyperbola

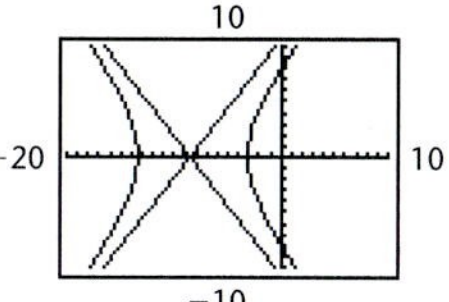

69. (A) Aphelion: 4.34×10^7 mi; Perihelion: 2.85×10^7 mi

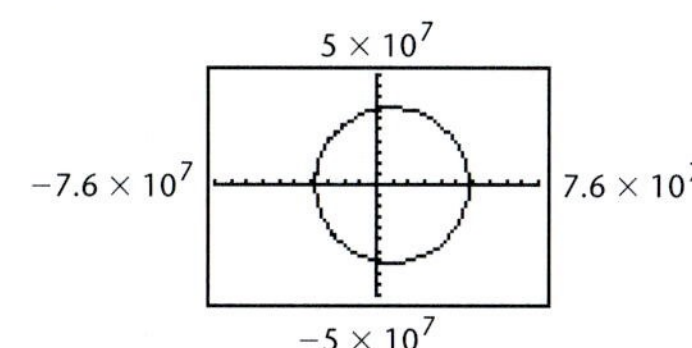

(B) Faster at perihelion. Since the distance from the sun to Mercury is less at perihelion than at aphelion, the planet must move faster near perihelion in order for the line joining Mercury to the sun to sweep out equal areas in equal intervals of time.

Exercise 7-6

1. **3.** **5.** **7.**

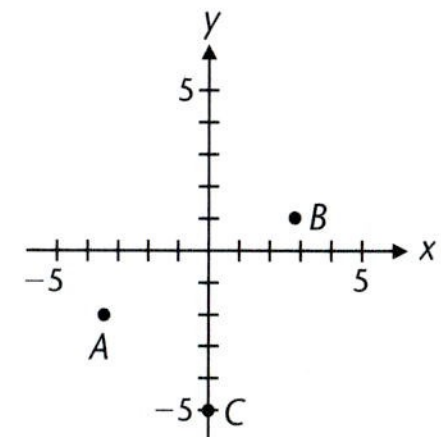

9. (A) $2e^{30°i}$ (B) $\sqrt{2}e^{(-135°)i}$ (C) $7.81e^{(-50.19°)i}$ **11.** (A) $\sqrt{3}e^{(-\pi/2)i}$ (B) $2e^{(-5\pi/6)i}$ (C) $9.43e^{2.58i}$

13. (A) $1 + i\sqrt{3}$ (B) $1 - i$ (C) $-2.35 + 1.99i$ **15.** (A) $3\sqrt{3} + 3i$ (B) $-i\sqrt{7}$ (C) $-2.22 - 3.43i$

17. $14e^{113°i}$; $3.5e^{51°i}$ **19.** $10e^{135°i}$; $2.5e^{(-31°)i}$ **21.** $36.42e^{4.35i}$; $0.26e^{(-0.83i)}$ **23.** $4i$; $4e^{90°i}$ **25.** $2 + 2i$; $2\sqrt{2}e^{45°i}$

27. i; $e^{90°i}$ **29.** $re^{i(-\theta)} = re^{-i\theta}$ **33.** $z^n = r^n e^{n\theta i}$

35. (A) $(20 + 0i) + (5 + i5\sqrt{3}) = 25 + i5\sqrt{3}$ (B) $26.5e^{19.1°i}$ (C) 26.5 lb at an angle of 19.1°

Exercise 7-7

1. $81e^{160°i}$ **3.** $8e^{90°i}$ **5.** $e^{120°i}$ **7.** $-8 + 8\sqrt{3}i$ **9.** 16 **11.** 1

13. $w_1 = 2e^{10°i}$, $w_2 = 2e^{130°i}$, $w_3 = 2e^{250°i}$ **15.** $w_1 = 3e^{15°i}$, $w_2 = 3e^{105°i}$, $w_3 = 3e^{195°i}$, $w_4 = 3e^{285°i}$

17. $w_1 = 2^{1/10}e^{(-9°)i}$, $w_2 = 2^{1/10}e^{63°i}$, $w_3 = 2^{1/10}e^{135°i}$, $w_4 = 2^{1/10}e^{207°i}$, $w_5 = 2^{1/10}e^{279°i}$

19. $w_1 = 2e^{0°i}$, $w_2 = 2e^{120°i}$, $w_3 = 2e^{240°i}$ **21.** $w_1 = 2e^{45°i}$, $w_2 = 2e^{135°i}$, $w_3 = 2e^{225°i}$, $w_4 = 2e^{315°i}$

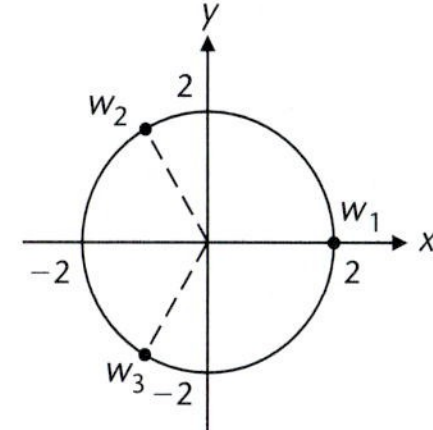

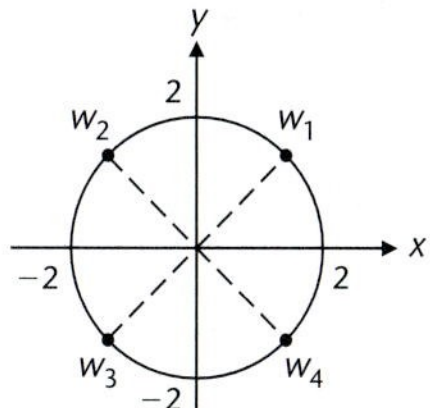

23. $w_1 = 1e^{15°i}$, $w_2 = 1e^{75°i}$, $w_3 = 1e^{135°i}$, $w_4 = 1e^{195°i}$, $w_5 = 1e^{255°i}$, $w_6 = 1e^{315°i}$

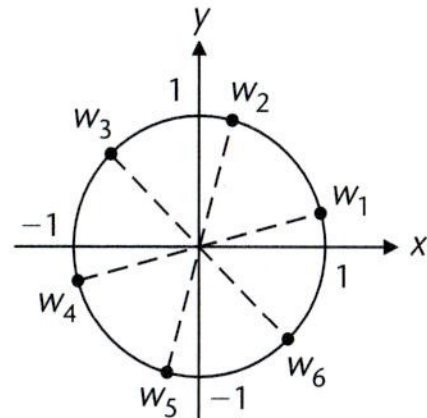

25. True **27.** False

29. (A) $(1 + i)^4 + 4 = -4 + 4 = 0$; there are three other roots.

(B) The four roots are equally spaced around the circle. Since there are four roots, the angle between successive roots on the circle is $360°/4 = 90°$.

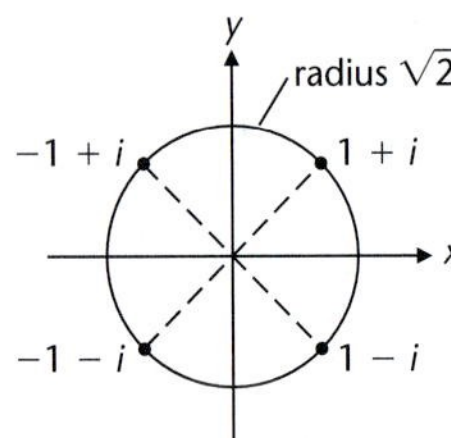

(C) $(-1 + i)^4 + 4 = -4 + 4 = 0$; $(-1 - i)^4 + 4 = -4 + 4 = 0$; $(1 - i)^4 + 4 = -4 + 4 = 0$

31. $x_1 = 4e^{60°i} = 2 + 2\sqrt{3}i$, $x_2 = 4e^{180°i} = -4$, $x_3 = 4e^{300°i} = 2 - 2\sqrt{3}i$

33. $x_1 = 3e^{0°i} = 3$, $x_2 = 3e^{120°i} = -\frac{3}{2} + \frac{3\sqrt{3}}{2}i$, $x_3 = 3e^{240°i} = -\frac{3}{2} - \frac{3\sqrt{3}}{2}i$ **35.** $w_1 = 2i$, $w_2 = -2i$

37. $w_1 = \sqrt{3} + i$, $w_2 = -\sqrt{3} + i$, $w_3 = -2i$ **41.** $x_1 = 2e^{0°i}$, $x_2 = 2e^{72°i}$, $x_3 = 2e^{144°i}$, $x_4 = 2e^{216°i}$, $x_5 = 2e^{288°i}$

43. $x_1 = e^{36°i}$, $x_2 = e^{108°i}$, $x_3 = e^{180°i}$, $x_4 = e^{252°i}$, $x_5 = e^{324°i}$

45. $P(x) = (x - 2i)(x + 2i)[x - (-\sqrt{3} + i)][x - (-\sqrt{3} - i)][x - (\sqrt{3} + i)][x - (\sqrt{3} - i)]$

Chapter 7 Review Exercise

1. 1 *(7-1)* **2.** 0 *(7-1)* **3.** 2 *(7-1)*

4. Angle β is acute. A triangle can have at most one obtuse angle. Since α is acute, then, if the triangle has an obtuse angle, it must be the angle opposite the longer of the two sides, b and c. Thus, β, the angle opposite the shorter of the two sides, b, must be acute. *(7-2)*

5. $\gamma = 75°$, $a = 47$ m, $b = 31$ m *(7-1)* **6.** $a = 4.0$ ft, $\beta = 36°$, $\gamma = 129°$ *(7-1, 7-2)* **7.** $\alpha = 40°$, $\beta = 19°$, $a = 8.2$ cm *(7-1)*

8. $\theta = 19°$, $|\mathbf{u} + \mathbf{v}| = 170$ mi/h *(7-3)* **9.** $\langle 3, -7 \rangle$ *(7-4)* **10.** $\sqrt{34}$ *(7-4)*

11. *(7-5)* **12.** *(7-5)* **13.** *(7-6)*

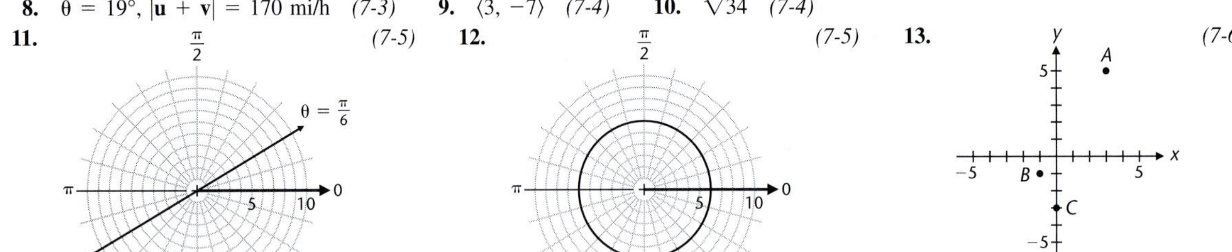

14. $(-10, -210°)$: The polar axis is rotated 210° clockwise (negative direction), and the point is located 10 units from the pole along the negative polar axis. $(-10, 150°)$: The polar axis is rotated 150° counterclockwise (positive direction), and the point is located 10 units from the pole along the negative polar axis. $(10, 330°)$: The polar axis is rotated 330° counterclockwise, and the point is located 10 units from the pole along the positive polar axis. *(7-5)*

15. *(7-6)* **16.** (A) $2e^{(-60°)i}$ (B) $2\sqrt{3} - 2i$ *(7-6)* **17.** (A) 1 (B) 1 *(7-7)*

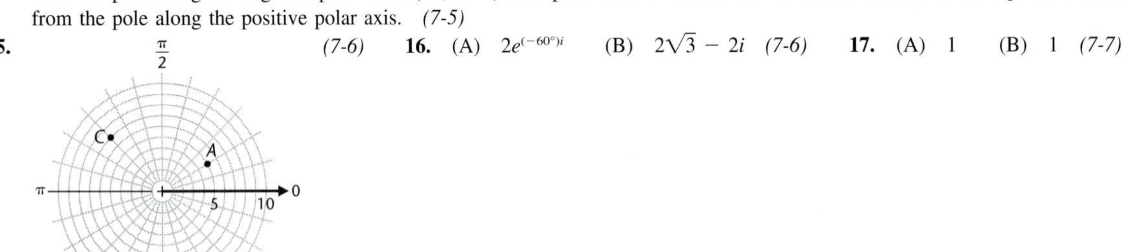

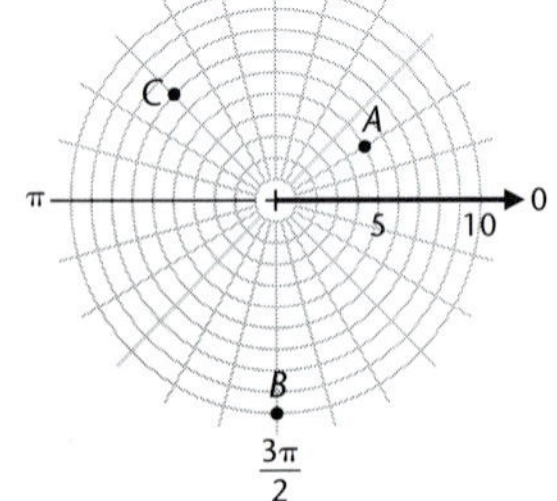

18. $8 + i8\sqrt{3}$ *(7-7)* **19.** If the triangle has an obtuse angle, then it must be the angle opposite the longest side; in this case, α. *(7-2)*

20. $b = 10.5$ cm, $\alpha = 27.2°$, $\gamma = 37.4°$ *(7-2)* **21.** No solution *(7-1)*

22. Two solutions. Obtuse case: $\beta = 133.9°$, $\gamma = 19.7°$, $c = 39.6$ km *(7-1)* **23.** $\alpha = 41.1°$, $\beta = 74.1°$, $\gamma = 64.8°$ *(7-1, 7-2)*

24. The sum of all of the force vectors must be the zero vector for the object to remain at rest. *(7-4)*

25. $|\mathbf{u} + \mathbf{v}| = 98.0$ kg, $\alpha = 17.1°$ *(7-3)* **26.** $\mathbf{u} = 11\mathbf{i} - 7\mathbf{j}$ *(7-4)*

27. $\langle 105, 45 \rangle$ *(7-4)* **28.** $u = \left\langle -\frac{4}{5}, \frac{3}{5} \right\rangle$ *(7-4)* **29.** $k = \frac{6 \pm \sqrt{6}}{10}$ *(7-4)* **30.** $k_1 = \frac{1}{5}, k_2 = \frac{2}{5}$ *(7-4)* **31.** $\left\langle \frac{15}{13}, -\frac{36}{13} \right\rangle$ *(7-4)*

32. *(7-5)* **33.** *(7-5)* **34.** *(7-5)*

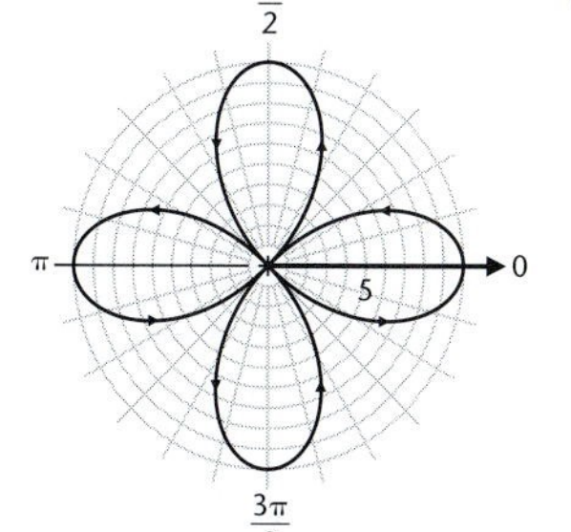

35. *(7-5)* **36.** *(7-5)* **37.** *(7-5)*

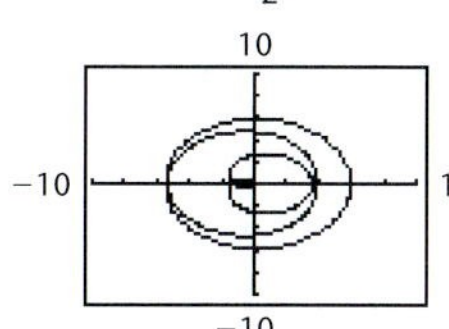

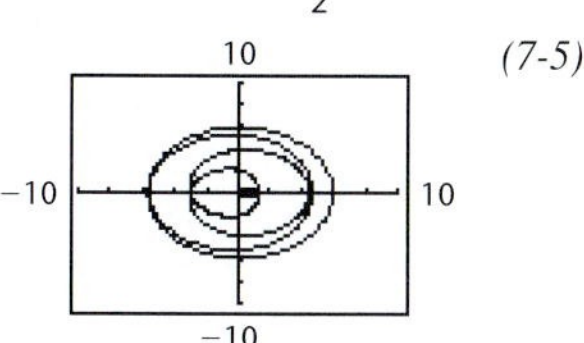

38. $n = 1$ $n = 2$ $n = 3$ Two leaves for all n *(7-5)*

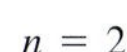

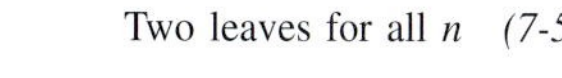

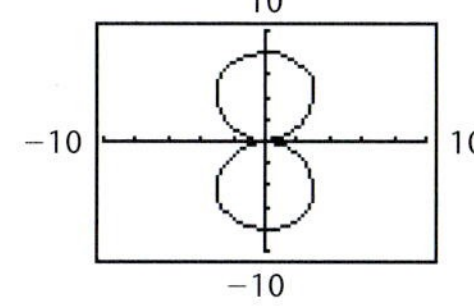

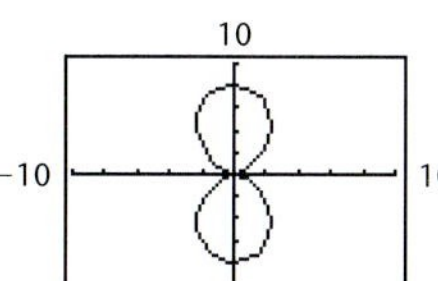

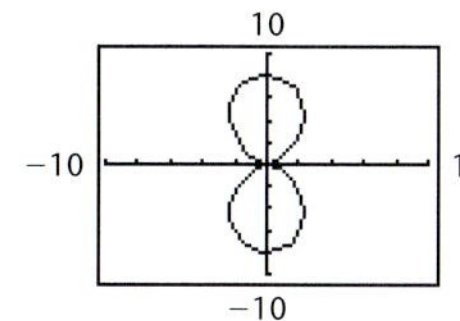

39. (A) Ellipse (B) Parabola (C) Hyperbola *(7-5)*

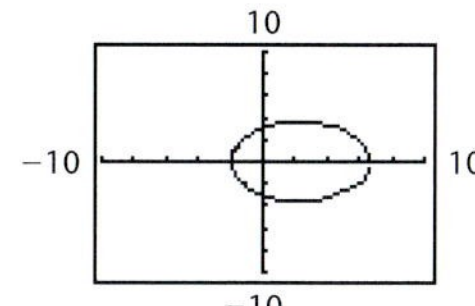

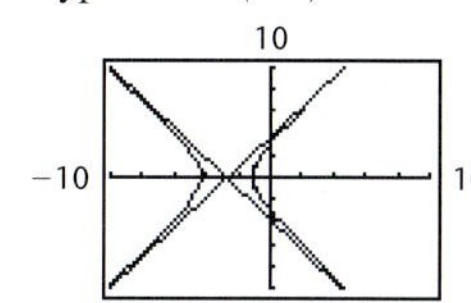

40. $r^2 = 6r \cos \theta$ or $r = 6 \cos \theta$ *(7-5)* **41.** $x^2 + y^2 = 5x$ *(7-6)*

42. $z_1 = \sqrt{2}e^{135°i}$, $z_2 = 2e^{(-120°)i}$, $z_3 = 5e^{0°i}$ *(7-6)* **43.** $z_1 = 1 + i$, $z_2 = (-3\sqrt{3}/2) - \frac{3}{2}i$, $z_3 = -1 - i\sqrt{3}$ *(7-6)*

44. (A) $32e^{44°i}$ (B) $2e^{6°i}$ *(7-6)* **45.** (A) $-8 - 8\sqrt{3}i$ (B) $-8 - 13.86i$ *(7-7)*

46. $w_1 = \frac{\sqrt{3}}{2} + \frac{1}{2}i$ *(7-7)*

$w_2 = -\frac{\sqrt{3}}{2} + \frac{1}{2}i$

$w_3 = -i$

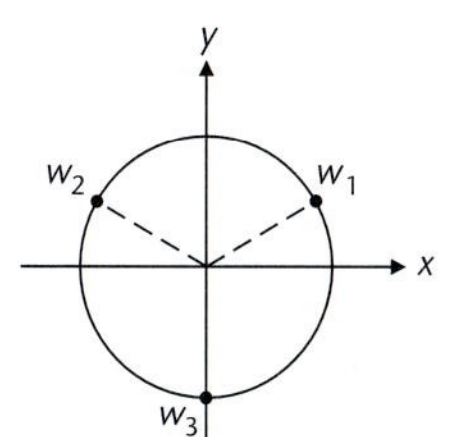

47. $2e^{50°i}, 2e^{170°i}, 2e^{290°i}$ *(7-7)* **48.** $(4e^{15°i})^2 = 16e^{30°i} = 8\sqrt{3} + 8i$ *(7-7)*
49. $(5.76, -26.08°)$ *(7-5)* **50.** $(-5.30, -2.38)$ *(7-5)* **51.** $5.26e^{127.20°i}$ *(7-6)* **52.** $-7.27 - 2.32i$ *(7-6)*
53. (A) There are a total of three cube roots, and they are spaced equally around a circle of radius 2.

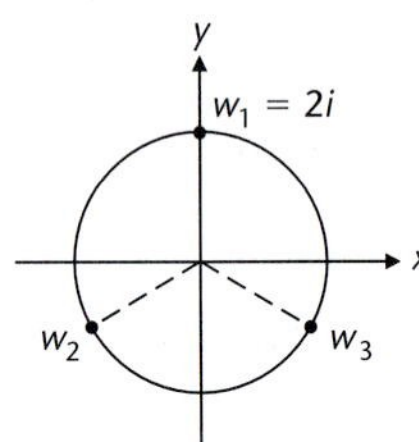

(B) $w_2 = -\sqrt{3} - i$, $w_3 = \sqrt{3} - i$ (C) The cube of each cube root is $-8i$. *(7-7)*
54. $k = 44.6 \sin 23.4°$ *(7-1)* **57.** False *(7-7)* **58.** False *(7-7)* **59.** True *(7-7)* **60.** False *(7-7)*
61. (A) (B) *(7-5)*

62. (A) The coordinates of P represent a simultaneous solution.

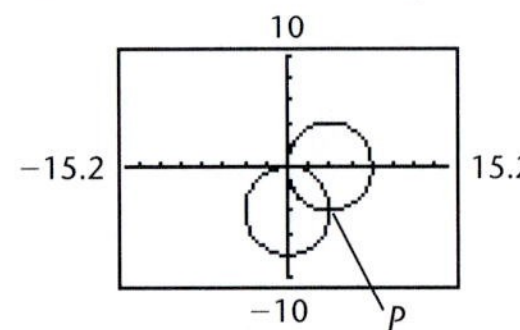

(B) $r = -4\sqrt{2}$, $\theta = 3\pi/4$ (C) The two graphs go through the pole at different values of θ. *(7-5)*
63. $1, -1, i, -i, \sqrt{2}/2 + i\sqrt{2}/2, \sqrt{2}/2 - i\sqrt{2}/2, -\sqrt{2}/2 + i\sqrt{2}/2, -\sqrt{2}/2 - i\sqrt{2}/2$ *(7-7)*
64. $P(x) = (x + 2i)[x - (-\sqrt{3} + i)][x - (\sqrt{3} + i)]$ *(7-7)* **65.** 438 mi *(7-3)* **66.** 438 mph at 83° *(7-3)*
67. 86°, 464 mph *(7-3)* **68.** 0.6 mi *(7-1)* **69.** 177 lb at 15.2° relative to **v** *(7-3)* **70.** 19 kg at 204° relative to **v** *(7-4)*
71. (A) Distance at aphelion: 1.56×10^8 mi
Distance at perihelion: 1.29×10^8 mi
(B) Distance at aphelion: 1.56×10^8 mi
Distance at perihelion: 1.29×10^8 mi *(7-5)*

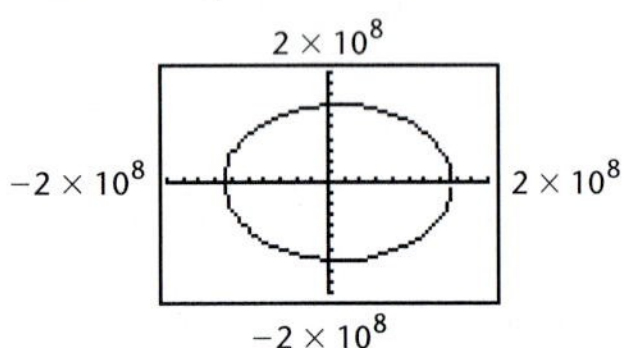

72. 5,740 lb *(7-4)*

Cumulative Review Exercise Chapters 5–7

1. 1.86 m *(5-3)* **2.** $\theta = 57.3°$, 14.5 cm, 7.83 cm *(5-5)* **3.** (A) I, II (B) I, IV (C) I, III *(5-4)*
4. (A) $-\frac{3}{5}$ (B) $\frac{5}{4}$ (C) $-\frac{4}{3}$ *(5-4)* **5.** (A) $\pi/4$ (B) 65° (C) 30° *(5-4)*
6. (A) Domain: all real numbers; Range: $-1 \le y \le 1$; Period: 2π (B) Domain: all real numbers; Range: $-1 \le y \le 1$; Period: 2π
(C) Domain: all real numbers except $x = \pi/2 + k\pi$, k an integer
Range: all real numbers; Period: π *(5-6)*

7. *(5-6)*

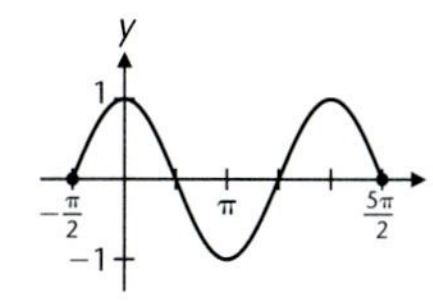

8. *(5-6)*

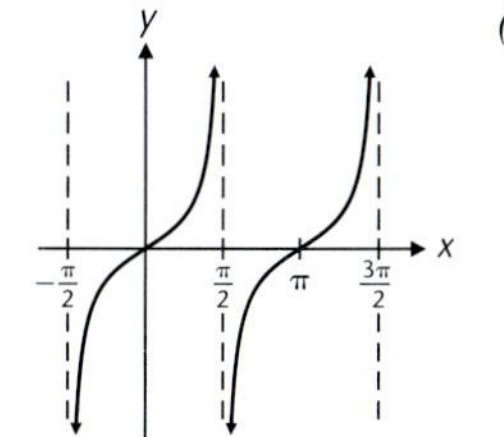

9. The central angle of a circle subtended by an arc of twice the length of the radius. *(5-3)*

10. If the graph of $y = \cos x$ is shifted $\pi/2$ units to the right, the result will be the graph of $y = \sin x$ *(5-6, 5-7)*

15. (A) Not an identity (B) An identity *(6-1)*

16. Angle α is acute. A triangle can have at most one obtuse angle. Since β is acute, then, if the triangle has an obtuse angle, it must be the angle opposite the longer of the two sides, a and c. Thus, α, the angle opposite the shorter of the two sides, a, must be acute. *(7-2)*

17. 0.3245, 2.8171 *(6-5)* **18.** $-76.2154°$ *(6-5)* **19.** $b = 22$ ft, $\alpha = 28°$, $\gamma = 31°$ *(7-2, 7-1)* **20.** $(6, -3)$ *(7-4)*

21. $(5, -30°)$: The polar axis is rotated 30° clockwise (negative direction), and the point is located 5 units from the pole along the positive polar axis. $(-5, -210°)$: The polar axis is rotated 210° clockwise (negative direction), and the point is located 5 units from the pole along the negative polar axis. $(5, 330°)$: The polar axis is rotated 330° counterclockwise (positive direction), and the point is located 5 units from the pole along the positive polar axis. *(7-5)*

22. *(7-6)*

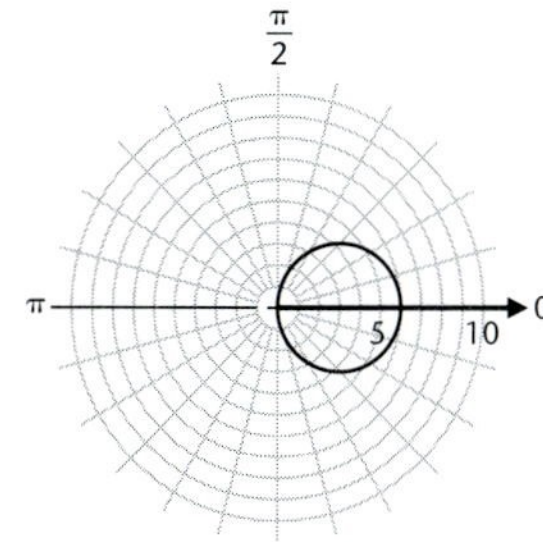

23. *(7-6)*

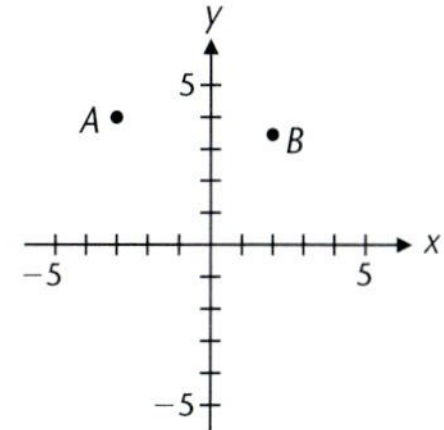

24. $4\sqrt{3} + 4i$ *(7-7)*

25. $-7\pi/6$, 870° *(5-3)* **26.** 75.06° *(5-3)* **27.** (A) and (C) *(5-4)* **28.** $-\sqrt{3}/2$ *(5-4)* **29.** $-\sqrt{2}$ *(5-2)*

30. Not defined *(5-4)* **31.** $-\sqrt{3}/3$ *(5-2)* **32.** $\frac{\pi}{3}$ *(5-9)* **33.** $-\frac{\pi}{3}$ *(5-9)* **34.** Not defined *(5-9)*

35. $-\dfrac{3\sqrt{10}}{10}$ *(5-9)* **36.** $\sqrt{2}/3$ *(5-9)* **37.** $\frac{\pi}{4}$ *(5-9)*

38. (A) 9.871 (B) -3.748 (C) -1.559 (D) Not defined *(5-4)*

39. *(5-7)*

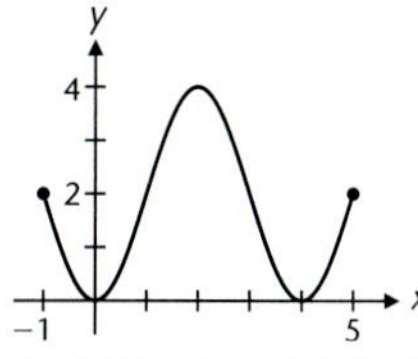

40. (A) 150° (B) $-19.755°$ *(5-9)*

41. $\sin^{-1}(\sin 3) = 0.142$. For the identity $\sin^{-1}(\sin x) = x$ to hold, x must be in the restricted domain of the sine function; that is, $-\pi/2 \le x \le \pi/2$. The number 3 is not in the restricted domain. *(5-9)*

42. Since the coordinates of a point on a unit circle are given by $P(a, b) = P(\cos x, \sin x)$, we evaluate $P(\cos(11.205), \sin(11.205))$—using a calculator set in radian mode—to obtain $P(0.208, -0.978)$. The quadrant in which $P(a, b)$ lies can be determined by the signs of a and b. In this case P is in the fourth quadrant, since a is positive and b is negative. *(5-2)*

43. The equation has infinitely many solutions [$x = \tan^{-1}(-24.5) + k\pi$, k any integer]; $\tan^{-1}(-24.5)$ has a unique value (-1.530 to three decimal places). *(5-9, 6-5)*

44. $y = 3 + 2 \sin \pi x$ *(5-7)*

45. $A = 3$; $P = \pi$; P.S. $= \pi/2$ *(5-7)* **46.** $P = 2$; P.S. $= 1$ *(5-8)*

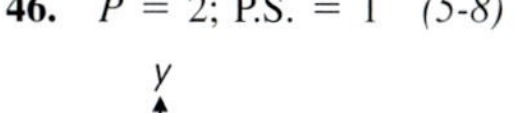

47. *(5-6)*

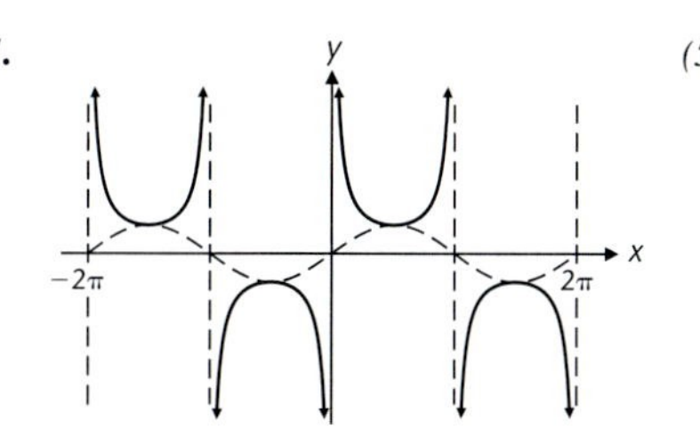

48. If the graph of $y = \cot x$ is shifted to the left $\pi/2$ units and reflected in the x axis, the result will be the graph of $y = \tan x$. *(5-6, 5-7)*

49. $y = \frac{1}{2} - \frac{1}{2}\cos 2x$ *(5-7)*

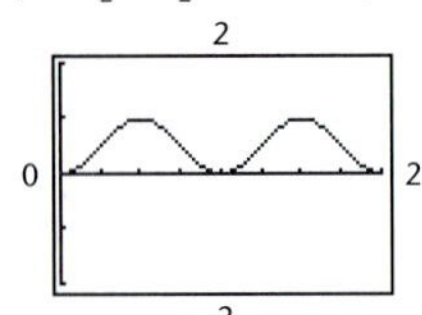

50. $y = \cot x$ *(5-7, 5-8)*

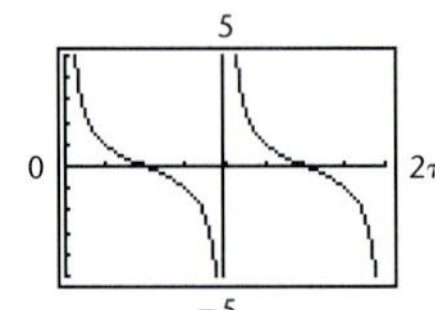

51. (A) Yes (B) Conditional, since both sides are defined at $x = \pi/2$, for example, but $\pi/2$ is not a solution. *(6-1)*

58. (A) Not an identity (B) An identity *(6-1)* **59.** 0 *(6-2)*

60. $\sin 2x = -\frac{24}{25}$, $\cos(x/2) = \sqrt{\frac{1}{10}}$ or $\sqrt{10}/10$ *(5-4, 6-3)* **61.** 30°, 150°, 270° *(6-5)*

62. $k\pi$, $\pi/3 + 2k\pi$, and $-\pi/3 + 2k\pi$, all for k any integer *(6-5)*

63. (A) $\pi/2$, $3\pi/2$, $7\pi/6$, $11\pi/6$ (B) 1.571, 3.665, 4.712, 5.760 *(6-5)* **64.** $x = 0.926$ *(6-5)*

65. $\alpha = 25.0°$, $\beta = 47.8°$, $\gamma = 107.2°$ *(7-1, 7-2)* **66.** No solution *(7-1)* **67.** $\beta = 120.7°$, $\gamma = 6.4°$, $c = 4.81$ in *(7-1)*

68. β must be acute. A triangle can have at most one obtuse angle, and since γ is acute, the obtuse angle, if present, must be opposite the longer of the two sides a and b. *(7-2)*

69. $|\mathbf{u} + \mathbf{v}| = 35.6$ lb, $\alpha = 16.3°$ *(7-3, 7-1, 7-2)* **70.** (A) $\langle 1, 3 \rangle$ (B) $3\mathbf{i} + \mathbf{j}$ *(7-4)* **71.** $r = 8 \sin \theta$ *(7-5)*

72. $x^2 + y^2 = -4x$ *(7-5)*

73. *(7-5)*

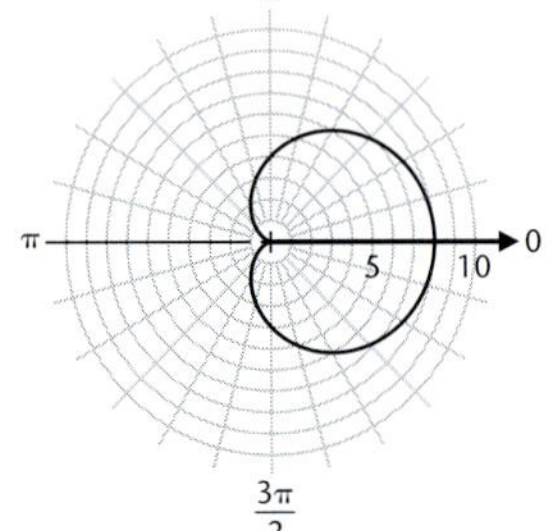

74. *(7-5)*

π/2, π, 0, 5, 10, 3π/2

75. $n = 1$

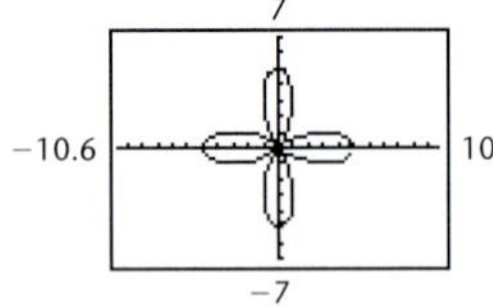

$n = 2$

7, −10.6, 10.6, −7

$n = 3$

7, −10.6, 10.6, −7

four leaves for all n *(7-5)*

76. *(7-5)*

4, −6.06, 6.06, −4

77. (4.23, −131.07°) *(7-5)* **78.** (−3.68, 5.02) *(7-5)* **79.** $\sqrt{3} - i$ *(7-6)*

80. $z = 2e^{120°i}$ *(7-6)* **81.** $64 + 0i = 64$ *(7-7)* **82.** $w_1 = \sqrt{3}/2 - \frac{1}{2}i$, $w_2 = i$, $w_3 = -\sqrt{3}/2 - \frac{1}{2}i$, *(7-8)*

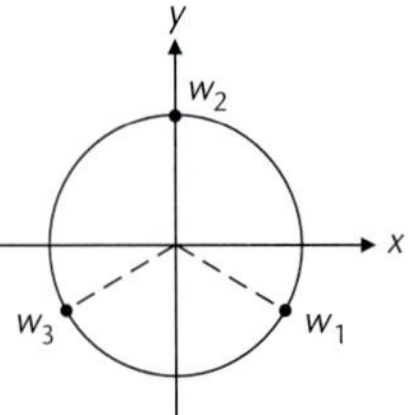

83. $5.82e^{(-146.99°)i}$ *(7-6)* **84.** $-6.70 + 1.94i$ *(7-6)*

85. (A) There are a total of four fourth roots, and they are spaced equally around a circle of radius $\sqrt{2}$.

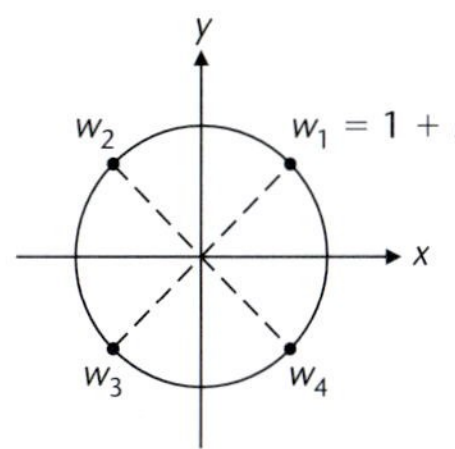

(B) $w_2 = -1 + i$, $w_3 = -1 - i$, $w_4 = 1 - i$ (C) The fourth power of each fourth root is -4. *(7-7)*

86. $a = \cos 1.2 = 0.362$, $b = \sin 1.2 = 0.932$ *(5-2)* **87.** *(5-8)*

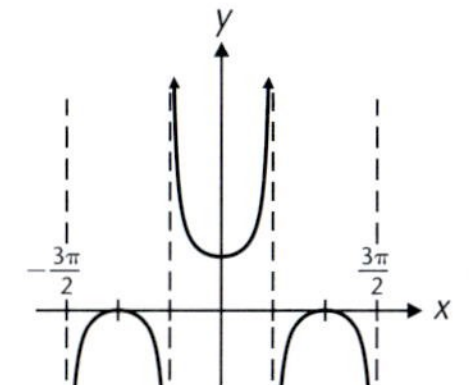

88. $y = 3 \cos (2\pi x - \pi/4)$; amplitude $= 3$, period $= 1$, P.S. $= \frac{1}{8}$ *(5-7)*

89. $y = 2 \sin (2x - 0.644)$ *(5-7)* **90.** $1/\sqrt{1 - x^2}$ *(5-9)* **91.** $\frac{24}{25}$ *(6-3, 5-9)* **92.** (A) $\sqrt{\frac{8}{10}} = 2\sqrt{5}/5$ (B) $-\frac{7}{25}$ *(6-3)*

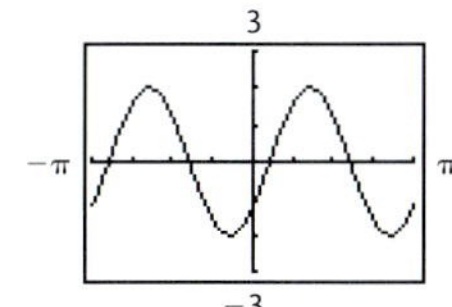

93. (A) $\pi/3$, $5\pi/3$ (B) 1.0472, 5.2360 *(6-5)*

94. (A)

$\frac{\pi}{2}$

π 10 0

$\frac{3\pi}{2}$

(B) *(7-5)*

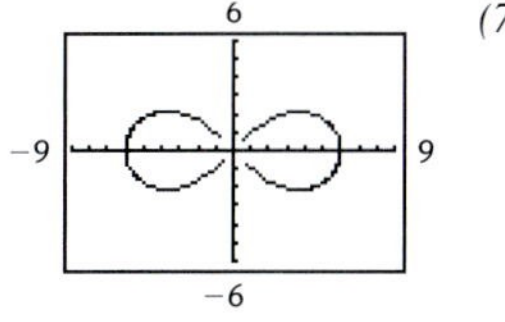

95. (A)

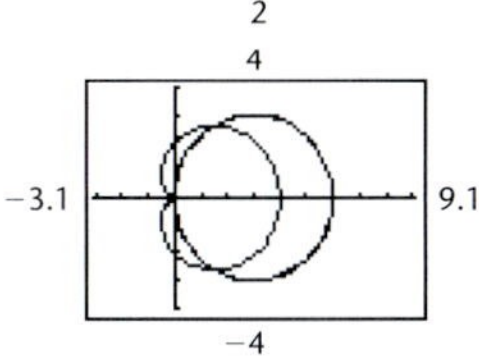

(B) 6 (C) $(3, \pi/3)$, $(3, 5\pi/3)$

(D) The points on $r2$ and $r1$ arrive at the intersection points for different values of θ, except for the two found in part C. *(7-7)*

96. $P(x) = (x - i)[x - (\sqrt{3}/2 - i/2)][x - (-\sqrt{3}/2 - i/2)]$ *(7-7)* **97.** No *(7-1, 7-2)* **98.** False *(5-9)* **99.** True *(5-9)*
100. True *(7-7)* **101.** False *(7-7)* **102.** False *(7-3, 7-4)* **103.** False *(7-4)* **104.** False *(5-6)* **105.** True *(5-6)*
106. $2\pi/73$ rad *(5-3)* **107.** 1,088 m *(5-5)* **108.** 5.88 in. *(5-5 or 7-2)* **109.** 76° *(7-2)* **110.** $I = 50 \cos 220\pi t$ *(5-7)*
111. 274 mph at 117° *(7-3)* **112.** Both have a tension of 234 lb *(7-4)*

113. (A) Add the perpendicular bisector of the chord as shown in the figure. Then, $\sin\theta = 4/R$ and $\theta = 5/R$. Substituting the second into the first, we obtain $\sin 5/R = 4/R$.

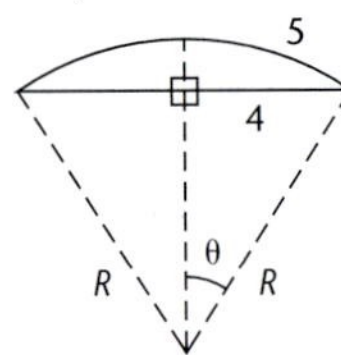

(B) R cannot be isolated on one side of the equation.

(C) Plot $y_1 = \sin 5/R$ and $y_2 = 4/R$ in the same viewing window and solve for R at the point of intersection using a built-in routine (see figure). $R = 4.420$ cm.

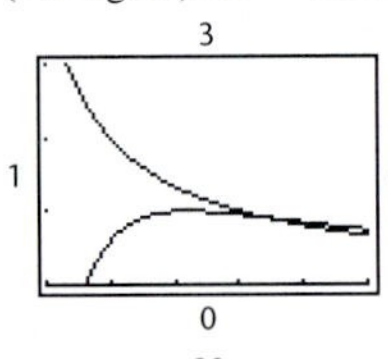

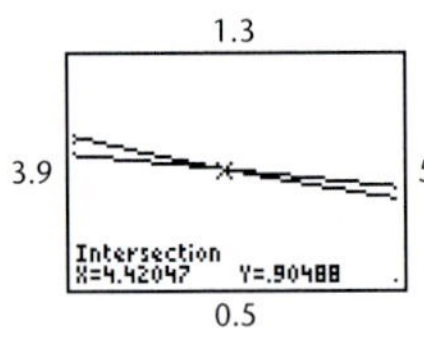

(6-5)

114. (A)

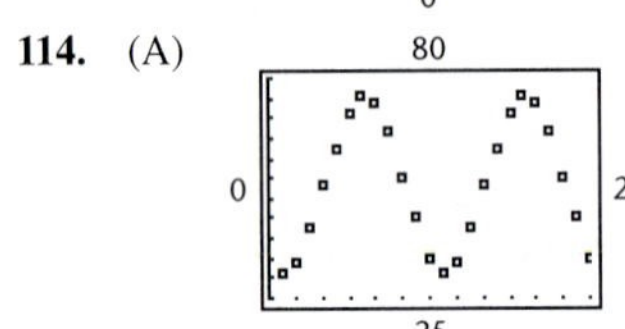

(B) $y = 53.5 + 22.5 \sin(\pi x/6 - 2.1)$ (C)

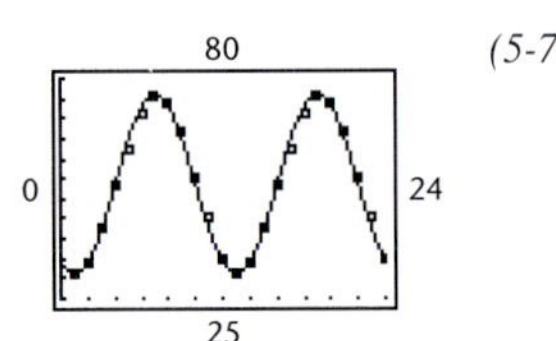

(5-7)

CHAPTER 8

Exercise 8-1

1. B, no solution **3.** D, $(1, -3)$ **5.** $(5, 2)$ **7.** $(2, -3)$ **9.** No solution (parallel lines) **11.** $(2, -1)$ **13.** $(3, -1)$

15. 2×3, 1×3 **17.** C **19.** B **21.** $-2, -6$ **23.** $-2, 6, 0$ **25.** $\left[\begin{array}{cc|c} 4 & -6 & -8 \\ 1 & -3 & 2 \end{array}\right]$ **27.** $\left[\begin{array}{cc|c} -4 & 12 & -8 \\ 4 & -6 & -8 \end{array}\right]$

29. $\left[\begin{array}{cc|c} 1 & -3 & 2 \\ 8 & -12 & -16 \end{array}\right]$ **31.** $\left[\begin{array}{cc|c} 1 & -3 & 2 \\ 0 & 6 & -16 \end{array}\right]$ **33.** $\left[\begin{array}{cc|c} 1 & -3 & 2 \\ 2 & 0 & -12 \end{array}\right]$ **35.** $\left[\begin{array}{cc|c} 1 & -3 & 2 \\ 3 & -3 & -10 \end{array}\right]$

37. $x_1 = 4$, $x_2 = 3$; each pair of lines has the same intersection point.

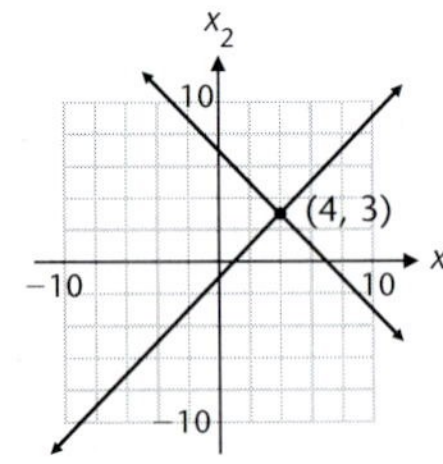

$x_1 + x_2 = 7$
$x_1 - x_2 = 1$

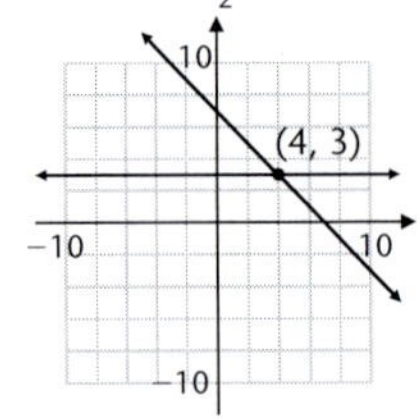

$x_1 + x_2 = 7$
$-2x_2 = -6$

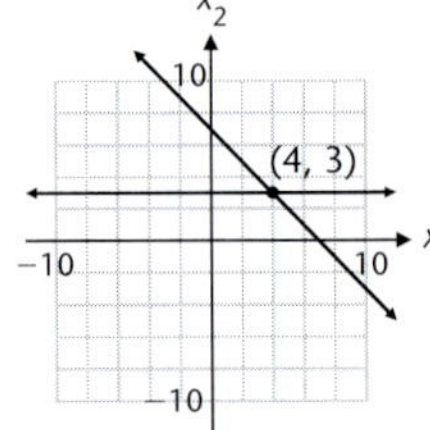

$x_1 + x_2 = 7$
$x_2 = 3$

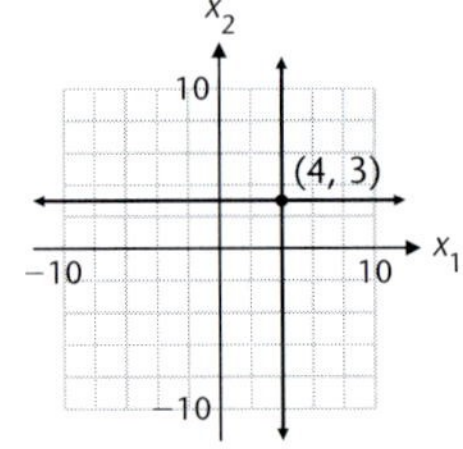

$x_1 = 4$
$x_2 = 3$

39. $x_1 = 2$ and $x_2 = 1$ **41.** $x_1 = 2$ and $x_2 = 4$ **43.** No solution **45.** $x_1 = 1$ and $x_2 = 4$

47. Infinitely many solutions; for any real number s, $x_2 = s$, $x_1 = 2s - 3$

49. Infinitely many solutions; for any real number s, $x_2 = s$, $x_1 = \frac{1}{2}s + \frac{1}{2}$

51. $(1.12, 2.41)$ **53.** $(-2.24, -3.31)$ **55.** (A) $(-24, 20)$ (B) $(6, -4)$ (C) No solution **57.** $(-23.125, 7.8125)$

59. $(3.225, -6.9375)$ **61.** 25 32¢ stamps, 50 23¢ stamps **63.** \$107,500 in bond A and \$92,500 in bond B

65. 30 liters of 20% solution and 70 liters of 80% solution **67.** 200 g of mix A and 80 g of mix B

69. Base price = \$17.95, surcharge = \$2.45/lb **71.** 5,720 lb of the robust blend and 6,160 lb of the mild blend.

Exercise 8-2

1. No **3.** Yes **5.** No **7.** Yes **9.** $x_1 = -2$, $x_2 = 3$, $x_3 = 0$ **11.** $x_1 = 2t + 3$, $x_2 = -t - 5$, $x_3 = t$, t any real number

13. No solution **15.** $x_1 = 2s + 3t - 5$, $x_2 = s$, $x_3 = -3t + 2$, $x_4 = t$, s and t any real numbers **17.** $\left[\begin{array}{cc|c} 1 & 0 & -7 \\ 0 & 1 & 3 \end{array}\right]$

19. $\left[\begin{array}{ccc|c} 1 & 0 & 0 & -5 \\ 0 & 1 & 0 & 4 \\ 0 & 0 & 1 & -2 \end{array}\right]$ **21.** $\left[\begin{array}{ccc|c} 1 & 0 & 2 & -\frac{5}{3} \\ 0 & 1 & -2 & \frac{1}{3} \\ 0 & 0 & 0 & 0 \end{array}\right]$ **23.** $x_1 = -2$, $x_2 = 3$, $x_3 = 1$ **25.** $x_1 = 0$, $x_2 = -2$, $x_3 = 2$

27. $x_1 = 2t + 3, x_2 = t - 2, x_3 = t$, t any real number **29.** $x_1 = 1, x_2 = 2$ **31.** No solution
33. $x_1 = 2t + 4, x_2 = t + 1, x_3 = t$, t any real number **35.** $x_1 = s + 2t - 1, x_2 = s, x_3 = t$, s and t any real numbers
37. No solution **39.** $x_1 = 2.5t - 4, x_2 = t, x_3 = -5$ for t any real number **41.** $x_1 = 1, x_2 = -2, x_3 = 1$
43. (A) Dependent with two parameters (B) Dependent with one parameter (C) Independent (D) Impossible
45. $x_1 = 2s - 3t + 3, x_2 = s + 2t + 2, x_3 = s, x_4 = t$ for s and t any real numbers **47.** $x_1 = -0.5, x_2 = 0.2, x_3 = 0.3, x_4 = -0.4$
49. $x_1 = 2s - 1.5t + 1, x_2 = s, x_3 = -t + 1.5, x_4 = 0.5t - 0.5, x_5 = t$ for s and t any real numbers
51. 15¢ stamps: $3t - 100$, 20¢ stamps: $145 - 4t$, 35¢ stamps: t, where $t = 34, 35$, or 36
53. 10% containers: $6t - 24$, 20% containers: $48 - 8t$, 50% containers: t, where $t = 4, 5$, or 6
55. $a = 3, b = 2, c = 1$ **57.** $a = -2, b = -4, c = -20$ **59.** 20 one-person boats, 220 two-person boats, 100 four-person boats
61. One-person boats: $t - 80$, two-person boats: $-2t + 420$, four-person boats: t, $80 \le t \le 210$, t an integer
63. No solution; no production schedule will use all the labor-hours in all departments. **65.** 8 oz food A, 2 oz food B, 4 oz food C
67. No solution **69.** 8 oz food A, $-2t + 10$ oz food B, t oz food C, $0 \le t \le 5$ **71.** Company A: 10 h, company B: 15 h

Exercise 8-3

1. $(-12, 5), (-12, -5)$ **3.** $(2, 4), (-2, -4)$ **5.** $(5, -5), (-5, 5)$ **7.** $(4 + 2\sqrt{3}, 1 + \sqrt{3}), (4 - 2\sqrt{3}, 1 - \sqrt{3})$
9. $(2, 4), (2, -4), (-2, 4), (-2, -4)$ **11.** $(1, 3), (1, -3), (-1, 3), (-1, -3)$ **13.** $(1 + \sqrt{5}, -1 + \sqrt{5}), (1 - \sqrt{5}, -1 - \sqrt{5})$
15. $(\sqrt{2}, \sqrt{2}), (-\sqrt{2}, -\sqrt{2}), (2, 1), (-2, -1)$ **17.** $(2, 2i), (2, -2i), (-2, 2i), (-2, -2i)$ **19.** $(2, \sqrt{2}), (2, -\sqrt{2}), (-1, i), (-1, -i)$
21. $(3, 0), (-3, 0), (\sqrt{5}, 2), (-\sqrt{5}, 2)$ **23.** $(2, 1), (-2, -1), (i, -2i), (-i, 2i)$ **25.** $(-1, 4), (3, -4)$ **27.** $(0, 0), (3, 6)$
29. $(1, 4), (4, 1)$ **31.** $(-1, 3), (4, 8)$
33. (A) The lines are tangent to the circle.

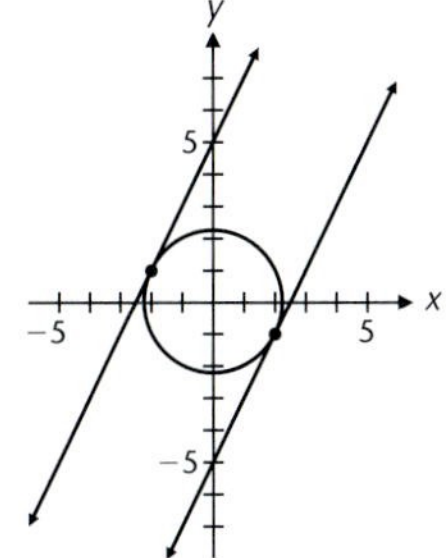

(B) $b = 5$, intersection point is $(2, -1)$; $b = -5$, intersection point is $(-2, 1)$
(C) The line $x + 2y = 0$ is perpendicular to all the lines in the family and intersects the circle at the intersection points found in part B. Solving the system $x^2 + y^2 = 5$, $x + 2y = 0$ would determine the intersection points.

35. $(-5, -\frac{3}{5}), (-\frac{3}{2}, -2)$ **37.** $(0, -1), (-4, -3)$ **39.** $(2, 2), (-2, -2), (\sqrt{2}, -\sqrt{2}), (-\sqrt{2}, \sqrt{2})$
41. $(-3, 1), (3, -1), (-i, i), (i, -i)$ **43.** $(-1.41, -0.82), (-0.13, 1.15), (0.13, -1.15), (1.41, 0.82)$
45. $(-1.66, -0.84), (-0.91, 3.77), (0.91, -3.77), (1.66, 0.84)$ **47.** $(-2.96, -3.47), (-0.89, -3.76), (1.39, 4.05), (2.46, 4.18)$
49. $\frac{1}{2}(3 - \sqrt{5}), \frac{1}{2}(3 + \sqrt{5})$ **51.** 5 in. and 12 in. **53.** 6 by 4.5 in. **55.** 22 by 26 ft **57.** Boat A: 30 mph; boat B: 25 mph

Exercise 8-4

1.

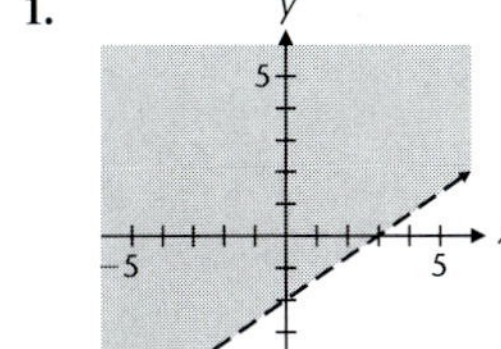

3.

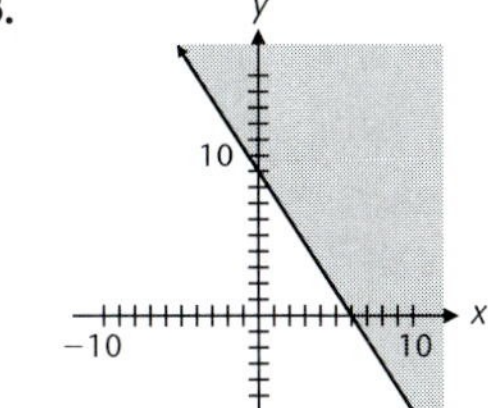

5.

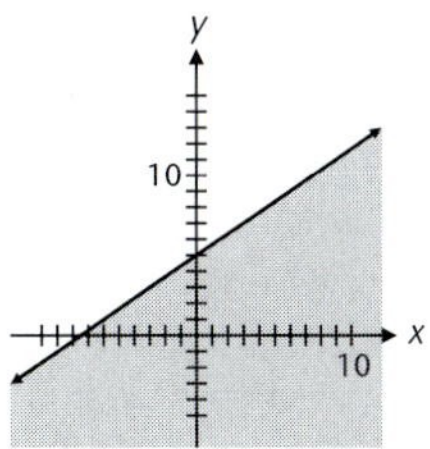

7.

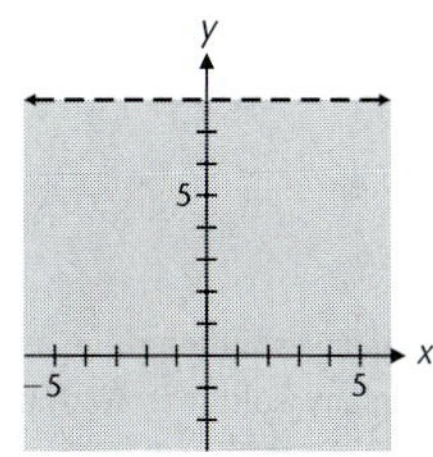

9.

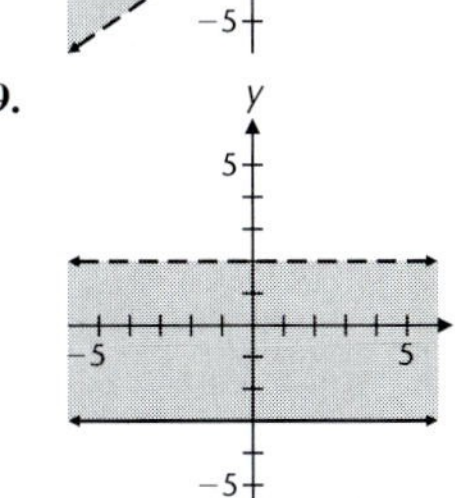

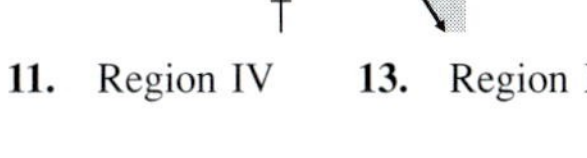
11. Region IV **13.** Region I

15.

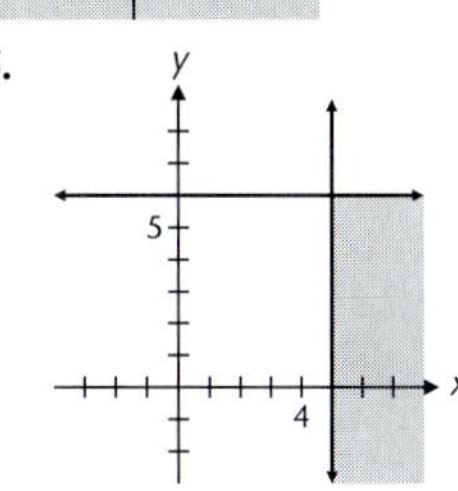

17.

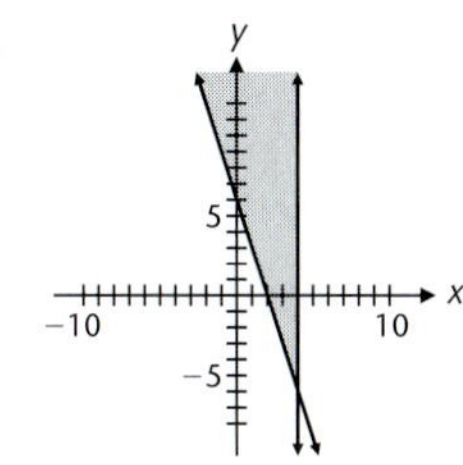

19.

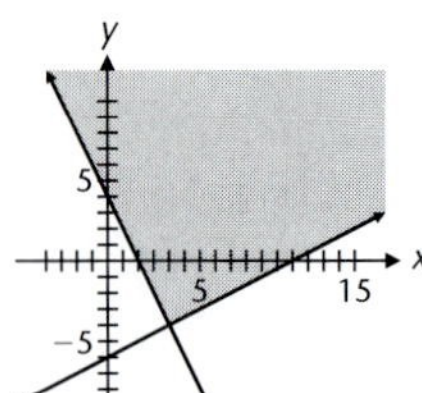
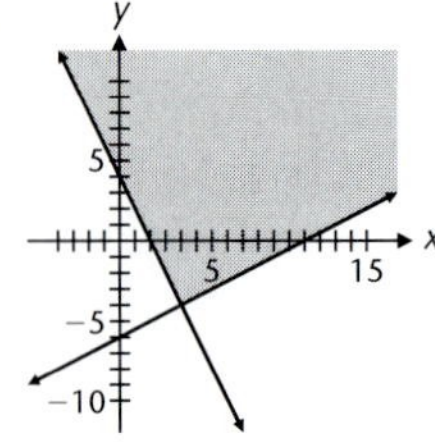

21. Region IV; corner points (6, 4), (8, 0), (18, 0)

23. Region I, corner points: (0, 16), (6, 4), (18, 0)

25. Corner points: (0, 0), (3, 0), (0, 2)
Bounded

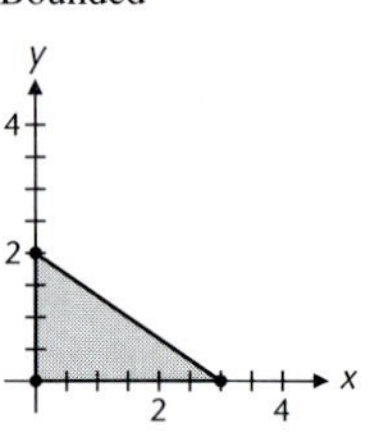

27. Corner points: (5, 0), (0, 4)
Unbounded

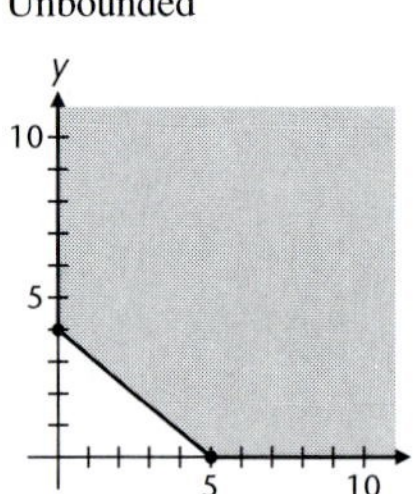

29. Corner points: (0, 0), (0, 4), $(\frac{12}{5}, \frac{16}{5})$, (4, 0)
Bounded

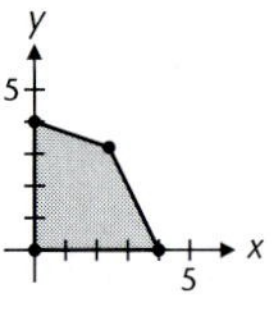

31. Corner points: (0, 8), (3, 4), (9, 0)
Unbounded

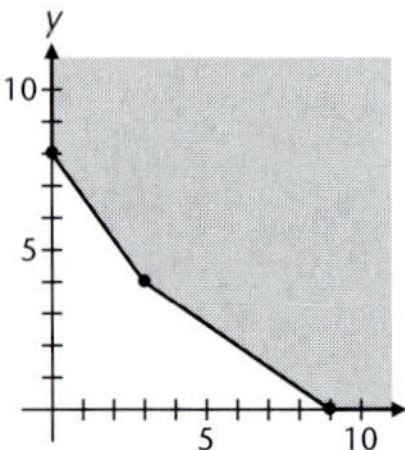

33. Corner points: (0, 0), (0, 5), (4, 3), (5, 2), (6, 0)
Bounded

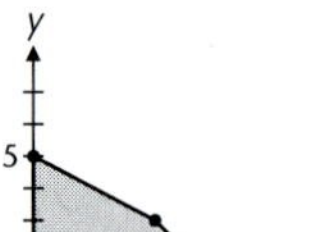

35. Corner points: (0, 14), (2, 10), (8, 4), (16, 0)
Unbounded

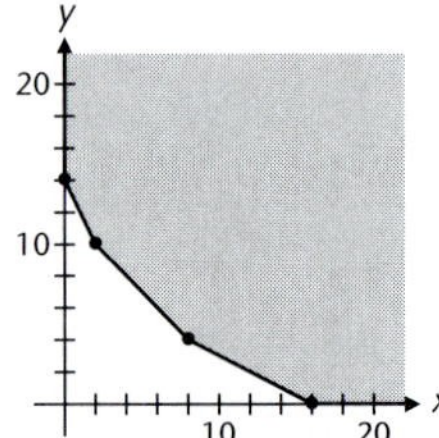

37. Corner points: (2, 5), (10, 1), (1, 10)
Bounded

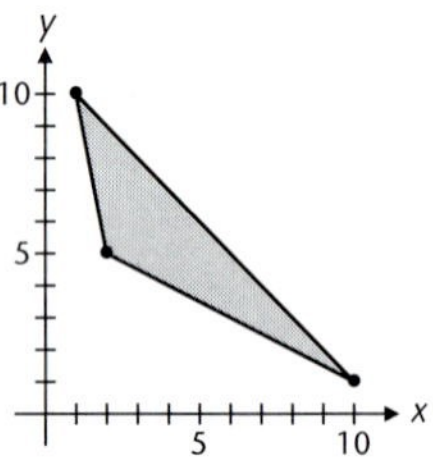

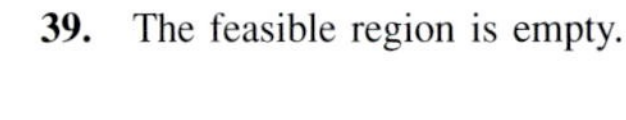

39. The feasible region is empty.

41. Corner points: (0, 3), (5, 0), (7, 3), (2, 8)
Bounded

43. Corner points: (1.27, 5.36), (2.14, 6.52), (5.91, 1.88) Bounded

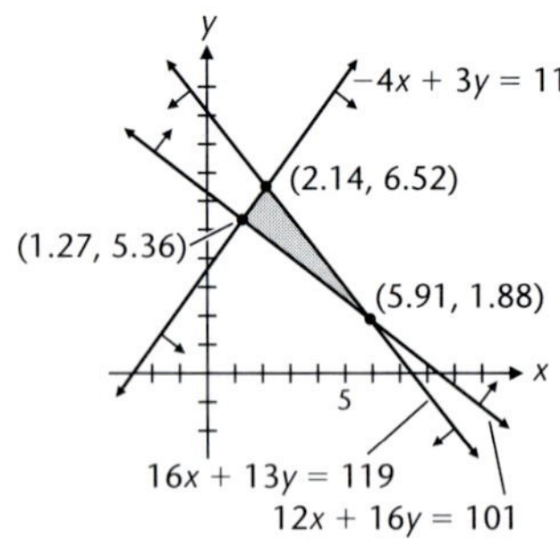

45.
$$\begin{aligned} 6x + 4y &\le 108 \\ x + y &\le 24 \\ x &\ge 0 \\ y &\ge 0 \end{aligned}$$

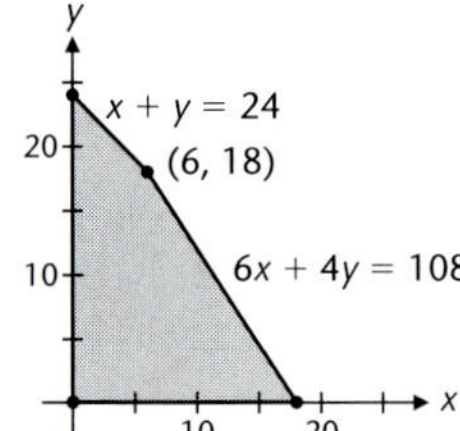

47. (A) All production schedules in the feasible region that are on the graph of $50x + 60y = 1{,}100$ will result in a profit of \$1,100.

(B) There are many possible choices. For example, producing 5 trick and 15 slalom skis will produce a profit of \$1,150. The graph of the line $50x + 60y = 1{,}150$ includes all the production schedules in the feasible region that result in a profit of \$1,150.

49. $20x + 10y \geq 460$
$30x + 30y \geq 960$
$5x + 10y \geq 220$
$x \geq 0$
$y \geq 0$

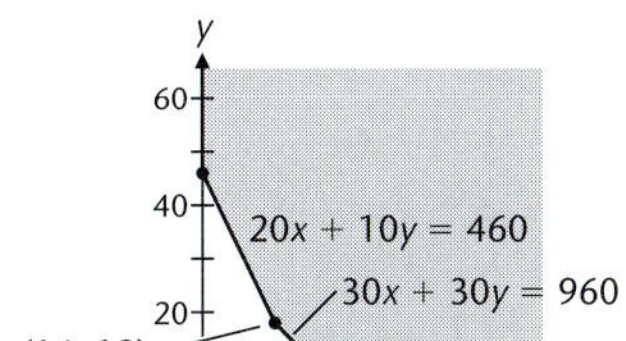

51. $10x + 30y \geq 280$
$30x + 10y \geq 360$
$x \geq 0$
$y \geq 0$

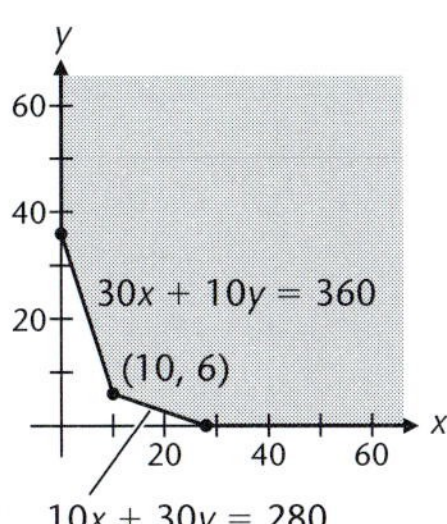

Exercise 8-5

1. Max $z = 16$ at (7, 9) **3.** Max $z = 84$ at (0, 12) and (7, 9) (multiple optimal solutions) **5.** Min $z = 32$ at (0, 8)
7. Min $z = 36$ at (12, 0) and (4, 3) (multiple optimal solutions) **9.** Max $z = 18$ at (4, 3) **11.** Min $z = 12$ at (4, 0)
13. Max $z = 52$ at (4, 10) **15.** Min $z = 44$ at (4, 4)
17. Min $z = 1{,}500$ at (60, 0); max $z = 3{,}000$ at (60, 30) and (120, 0) (multiple optimal solutions)
19. Min $z = 300$ at (0, 20); max $z = 1{,}725$ at (60, 15) **21.** Max $P = 5{,}507$ at $x = 6.62$ and $y = 4.25$
23. (A) $a > 2b$ (B) $\frac{1}{3}b < a < 2b$ (C) $b > 3a$ (D) $a = 2b$ (E) $b = 3a$
25. (A) 6 trick skis, 18 slalom skis; \$780 (B) The maximum profit decreases to \$720 when 18 trick and no slalom skis are produced. (C) The maximum profit increases to \$1,080 when no trick and 24 slalom skis are produced.
27. 9 model *A* trucks, 6 model *B* trucks, \$279,000
29. (A) 40 tables, 40 chairs; \$4,600 (B) The maximum profit decreases to \$3,800 when 20 tables and 80 chairs are produced.
31. (A) Max $P = \$450$ when 750 gal is produced using the old process exclusively.
(B) The maximum profit decreases to \$380 when 400 gal is produced using the old process and 700 gal using the new process.
(C) The maximum profit decreases to \$288 when 1,440 gal is produced using the new process exclusively.
33. The nitrogen will range from a minimum of 940 lb when 40 bags of brand *A* and 100 bags of brand *B* are used to a maximum of 1,190 lb when 140 bags of brand *A* and 50 bags of brand *B* are used.

Chapter 8 Review Exercise

1. (2, 3) *(8-1)* **2.** No solution (inconsistent) *(8-1)* **3.** Infinitely many solutions $(t, (4t + 8)/3)$, for any real number t *(8-1)*
4. $(-1, 3), (5, -3)$ *(8-3)* **5.** $(1, -1), (\frac{7}{5}, -\frac{1}{5})$ *(8-3)* **6.** $(1, 3), (1, -3), (-1, 3), (-1, -3)$ *(8-3)*

7. *(8-1)*

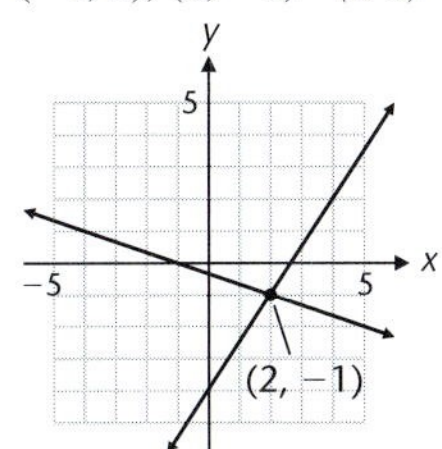

8. *(8-4)*

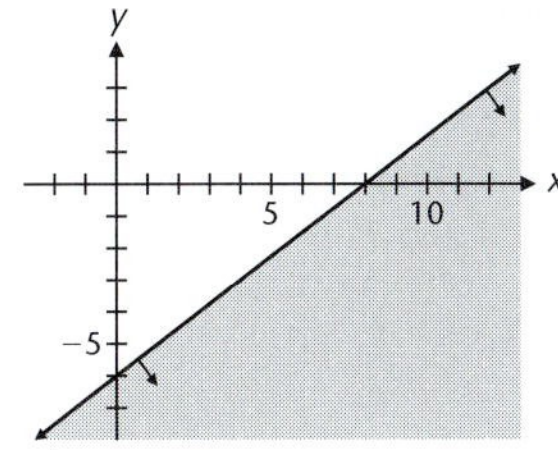

9. *(8-4)*

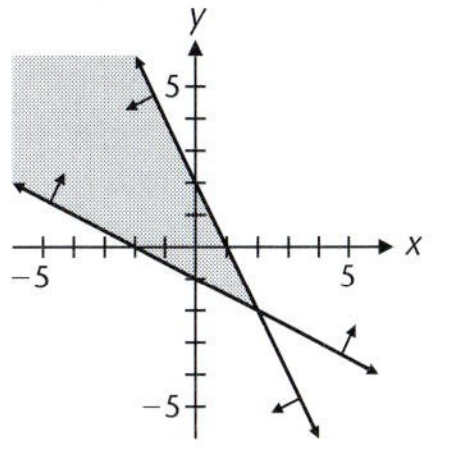

10. $\left[\begin{array}{rr|r} 3 & -6 & 12 \\ 1 & -4 & 5 \end{array}\right]$ *(8-1)* **11.** $\left[\begin{array}{rr|r} 1 & -4 & 5 \\ 1 & -2 & 4 \end{array}\right]$ *(8-1)* **12.** $\left[\begin{array}{rr|r} 1 & -4 & 5 \\ 0 & 6 & -3 \end{array}\right]$ *(8-1)* **13.** $x_1 = 4$ *(8-2)*
$x_2 = -7$
$(4, -7)$

14. $x_1 - x_2 = 4$ *(8-2)*
$0 = 1$
No solution

15. $x_1 - x_2 = 4$ *(8-2)*
$x_1 = t + 4$, $x_2 = t$, t any real number

16. Min $z = 18$ at (0, 6); max $z = 42$ at (6, 4) *(8-5)*

17. $x_1 = 2, x_2 = -2$; each pair of lines has the same intersection point. *(8-1)*

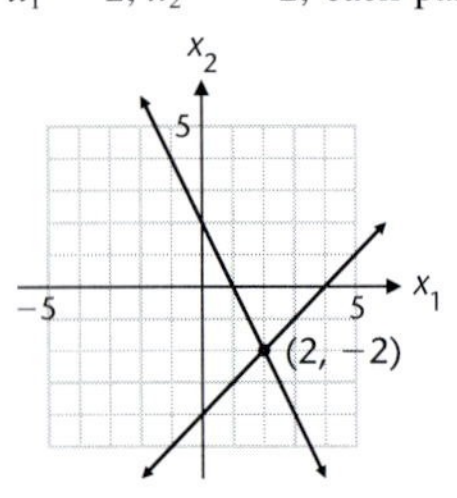

$x_1 - x_2 = 4$
$2x_1 + x_2 = 2$

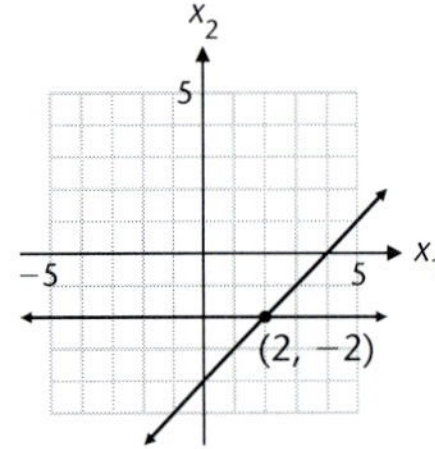

$x_1 - x_2 = 4$
$3x_2 = -6$

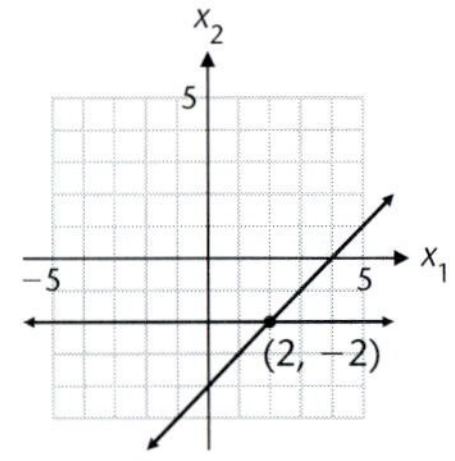

$x_1 - x_2 = 4$
$x_2 = -2$

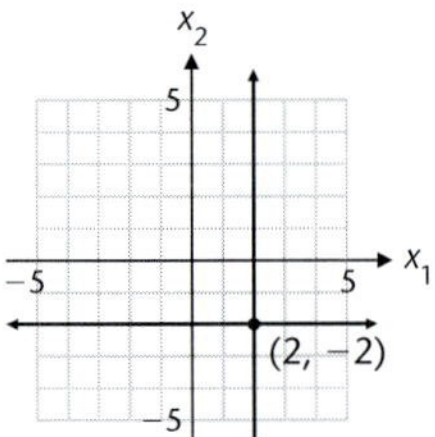

$x_1 = 2$
$x_2 = -2$

18. (2.54, 2.15) *(8-1)* **19.** $x_1 = -1, x_2 = 3$ *(8-2)* **20.** $x_1 = -1, x_2 = 2, x_3 = 1$ *(8-2)* **21.** $x_1 = 2, x_2 = 1, x_3 = -1$ *(8-2)*
22. Infinitely many solutions: $x_1 = -5t - 12, x_2 = 3t + 7, x_3 = t$, t any real number *(8-2)* **23.** No solution *(8-2)*
24. Infinitely many solutions: $x_1 = -\frac{3}{7}t - \frac{4}{7}, x_2 = \frac{5}{7}t + \frac{9}{7}, x_3 = t$, t any real number *(8-2)*
25. $(2, \sqrt{2}), (2, -\sqrt{2}), (-1, i), (-1, -i)$ *(8-3)* **26.** $(1, -2), (-1, 2), (2, -1), (-2, 1)$ *(8-3)*
27. $(2, -2), (-2, 2), (\sqrt{2}, \sqrt{2}), (-\sqrt{2}, -\sqrt{2})$ *(8-3)*

28. Corner points: (0, 0), (0, 4), (3, 2), (4, 0) *(8-4)*
Bounded

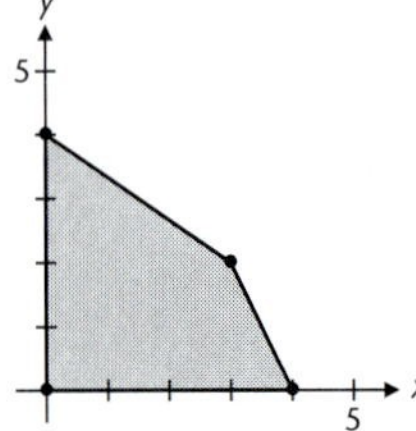

29. Corner points: (0, 8), $(\frac{12}{5}, \frac{16}{5})$, (12, 0) *(8-4)*
Unbounded

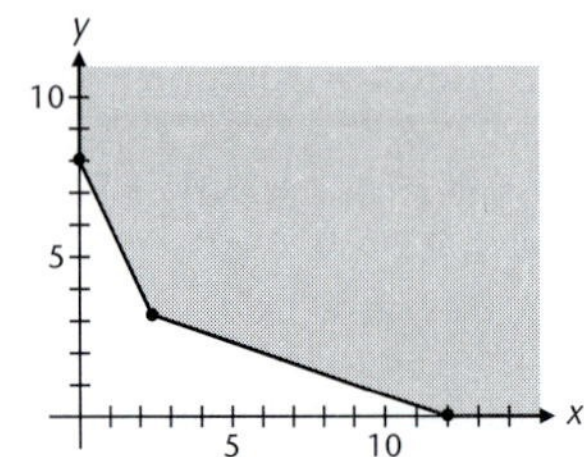

30. Corner points: (4, 4), (10, 10), (20, 0) *(8-4)*
Bounded

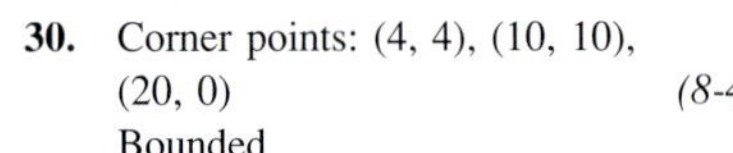

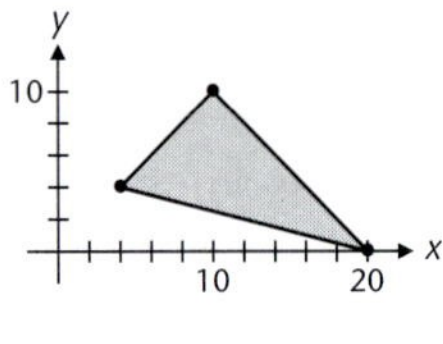

31. Max $z = 46$ at (4, 2) *(8-5)* **32.** Min $z = 75$ at (3, 6) and (15, 0) (multiple optimal solutions) *(8-5)*
33. Min $z = 44$ at (4, 3); max $z = 82$ at (2, 9) *(8-5)* **34.** $x_1 = 1{,}000, x_2 = 4{,}000, x_3 = 2{,}000$ *(8-2)*
35. $(2, 2), (-2, -2), (\frac{4}{7}\sqrt{7}, -\frac{2}{7}\sqrt{7}), (-\frac{4}{7}\sqrt{7}, \frac{2}{7}\sqrt{7})$ *(8-3)* **36.** Max $z = 26{,}000$ at (600, 400) *(8-5)*
37. $(-2.16, -0.37), (-1.09, 5.59), (1.09, -5.59), (2.16, 0.37)$ *(8-3)*
38. (A) A unique solution (B) No solution (C) An infinite number of solutions *(8-2)*
39. $\frac{1}{2}$-lb packages: 48, $\frac{1}{3}$-lb packages: 72 *(8-1, 8-2)* **40.** 6 by 8 m *(8-3)* **41.** 40 g mix A, 60 g mix B, 30 g mix C *(8-2)*
42. (A) 22 nickels, 8 dimes (B) $22 + 3t$ nickels, $8 - 4t$ dimes, and t quarters, $t = 0, 1, 2$ *(8-1, 8-2)*
43. (A) Maximum profit is $P = \$7{,}800$ when 80 regular and 30 competition sails are produced.
(B) The maximum profit increases to $8,750 when 70 competition and no regular sails are produced.
(C) The maximum profit decreases to $7,200 when no competition and 120 regular sails are produced. *(8-5)*
44. (A) The minimum cost is $C = \$13$ when 100 g of mix A and 150 g of mix B are used.
(B) The minimum cost decreases to $9 when 50 g of mix A and 275 g of mix B are used.
(C) The minimum cost increases to $28.75 when 250 g of mix A and 75 g of mix B are used. *(8-5)*

CHAPTER 9

Exercise 9-1

1. $\begin{bmatrix} 2 & 5 \\ 4 & -6 \end{bmatrix}$ **3.** $\begin{bmatrix} 3 & 2 \\ -2 & 8 \\ 12 & -5 \end{bmatrix}$ **5.** Not defined **7.** $\begin{bmatrix} 3 & -5 & 6 \\ 1 & 1 & 8 \end{bmatrix}$ **9.** $\begin{bmatrix} 12 & -16 & 28 \\ -8 & 36 & 20 \end{bmatrix}$ **11.** $[41]$ **13.** $\begin{bmatrix} 3 \\ -13 \end{bmatrix}$

15. $\begin{bmatrix} 13 & 8 \\ 26 & 48 \end{bmatrix}$ **17.** $\begin{bmatrix} 16 & -18 \\ -10 & 18 \end{bmatrix}$ **19.** $[-36]$ **21.** $\begin{bmatrix} -6 & 12 \\ 15 & -30 \end{bmatrix}$ **23.** $[-13]$ **25.** $\begin{bmatrix} 5 & 0 & -3 \\ -10 & 0 & 6 \\ 30 & 0 & -18 \end{bmatrix}$ **27.** Not defined

29. $\begin{bmatrix} 3 & 2 & -6 \\ -31 & -10 & -8 \end{bmatrix}$ **31.** Not defined **33.** $\begin{bmatrix} -19 & -2 & -14 \\ 15 & 4 & 6 \\ 25 & 10 & -4 \end{bmatrix}$ **35.** $\begin{bmatrix} 24 & -32 \\ -14 & 18 \\ -4 & 56 \end{bmatrix}$ **37.** Not defined

39. $\begin{bmatrix} 24 & 8 & 24 \\ -45 & 122 & -198 \\ -110 & 110 & -234 \end{bmatrix}$ **41.** $\begin{bmatrix} 6 & 4 & -12 \\ -189 & -62 & -42 \\ -226 & -76 & -38 \end{bmatrix}$ **43.** $\begin{bmatrix} 32 & 28 \\ 42 & -78 \end{bmatrix}$ **45.** $a = 4, b = 5, c = -9, d = -4$

47. $x = 5, y = 2$ **49.** $a = 1, b = -4, c = 0, d = 1$ **51.** $a = -3, b = 6, c = 2, d = -1$ **53.** True **55.** True

57. True **59.** True **61.** True **63.** False **65.**

	Guitar	Banjo
Materials	\$33	\$26
Labor	\$57	\$77

67.

	Basic car	Air	Markup AM/FM radio	Cruise control
Model *A*	\$3,330	\$77	\$42	\$27
Model *B*	\$2,125	\$93	\$95	\$50
Model *C*	\$1,270	\$113	\$121	\$52

69. (A) \$11.80 (B) \$30.30 (C) *MN* gives the labor costs per boat at each plant
(D)

	Plant I	Plant II
One-person boat	\$11.80	\$13.80
Two-person boat	\$18.50	\$21.60
Four-person boat	\$26.00	\$30.30

(MN = the matrix above)

71. (A) $A^2 = \begin{bmatrix} 0 & 0 & 2 & 0 & 0 \\ 1 & 0 & 0 & 0 & 1 \\ 0 & 1 & 0 & 2 & 0 \\ 1 & 0 & 0 & 0 & 1 \\ 0 & 0 & 1 & 0 & 0 \end{bmatrix}$; There is one way to travel from Baltimore to Atlanta with one intermediate connection; there are two ways to travel from Atlanta to Chicago with one intermediate connection. In general, the elements in A^2 indicate the number of different ways to travel from the *i*th city to the *j*th city with one intermediate connection.

(B) $A^3 = \begin{bmatrix} 2 & 0 & 0 & 0 & 2 \\ 0 & 1 & 0 & 2 & 0 \\ 0 & 0 & 3 & 0 & 0 \\ 0 & 1 & 0 & 2 & 0 \\ 1 & 0 & 0 & 0 & 1 \end{bmatrix}$; There is one way to travel from Denver to Baltimore with two intermediate connections; there are two ways to travel from Atlanta to El Paso with two intermediate connections. In general, the elements in A^3 indicate the number of different ways to travel from the *i*th city to the *j*th city with two intermediate connections.

(C) $A + A^2 + A^3 + A^4 = \begin{bmatrix} 2 & 3 & 2 & 5 & 2 \\ 1 & 1 & 4 & 2 & 1 \\ 4 & 1 & 3 & 2 & 4 \\ 1 & 1 & 4 & 2 & 1 \\ 1 & 1 & 1 & 3 & 1 \end{bmatrix}$; It is possible to travel from any origin to any destination with at most 3 intermediate connections.

73. (A) \$3,550 (B) \$6,000 (C) *NM* gives the total cost per town.

(D)

	Cost per town
Berkeley	\$3,550
Oakland	\$6,000

(NM = the matrix above)

(E)

	Telephone call	House call	Letter
$[1\ \ 1]N =$	3,000	1,300	13,000

(F) $N\begin{bmatrix} 1 \\ 1 \\ 1 \end{bmatrix} = \begin{bmatrix} 6{,}500 \\ 10{,}800 \end{bmatrix}$ Berkeley / Oakland (Total contacts)

Exercise 9-2

1. $\begin{bmatrix} 6 & -3 \\ 5 & 8 \end{bmatrix}$ **3.** $\begin{bmatrix} 2 & -5 \\ 8 & -4 \end{bmatrix}$ **5.** $\begin{bmatrix} 4 & -2 & 5 \\ 6 & 0 & -3 \\ -1 & 2 & 7 \end{bmatrix}$ **7.** $\begin{bmatrix} -3 & 2 & -5 \\ 4 & -1 & 6 \\ -3 & 4 & -2 \end{bmatrix}$ **15.** $\begin{bmatrix} 1 & -9 \\ 0 & 1 \end{bmatrix}$ **17.** $\begin{bmatrix} -5 & -2 \\ 2 & 1 \end{bmatrix}$

19. $\begin{bmatrix} -3 & -7 \\ -2 & -5 \end{bmatrix}$ **21.** $\begin{bmatrix} 0 & -1 & -1 \\ -1 & -1 & -1 \\ -1 & -1 & 0 \end{bmatrix}$ **23.** $\begin{bmatrix} -19 & 9 & -7 \\ 15 & -7 & 6 \\ -2 & 1 & -1 \end{bmatrix}$ **25.** Does not exist **27.** $\begin{bmatrix} 5 & -3 \\ -3 & 2 \end{bmatrix}$ **29.** $\begin{bmatrix} 1 & 0 & 1 \\ \frac{1}{2} & \frac{1}{2} & 1 \\ 2 & 1 & 4 \end{bmatrix}$

31. Does not exist **33.** $\begin{bmatrix} -9 & -15 & 10 \\ 4 & 5 & -4 \\ -1 & -1 & 1 \end{bmatrix}$ **37.** M^{-1} exists if and only if all the elements on the main diagonal are nonzero.

39. $\begin{bmatrix} -0.4 & 0.8 & 1.8 & -1.4 \\ 0.1 & -0.2 & -0.7 & 0.6 \\ -0.85 & 1.7 & 3.95 & -3.6 \\ -0.1 & -0.8 & -1.3 & 1.4 \end{bmatrix}$ **41.** $\begin{bmatrix} -0.75 & 3.75 & -4 & 0.75 & 3.5 \\ -0.5 & -0.5 & 1 & -0.5 & -1 \\ 2.5 & -10.5 & 11.5 & -1.5 & -9.5 \\ 0.125 & 1.375 & -1.75 & 0.375 & 1.5 \\ -0.75 & 0.75 & -0.5 & -0.25 & 0 \end{bmatrix}$

43. 14 5 195 74 97 37 181 67 49 18 121 43 103 41 **45.** GREEN EGGS AND HAM
47. 21 56 55 25 58 46 97 94 48 75 45 58 63 45 59 48 64 80 44 69 68 104 123 72 127
49. LYNDON BAINES JOHNSON

Exercise 9-3

1. $\begin{aligned} 2x_1 - x_2 &= 3 \\ x_1 + 3x_2 &= -2 \end{aligned}$ **3.** $\begin{aligned} -2x_1 + x_3 &= 3 \\ x_1 + 2x_2 + x_3 &= -4 \\ x_2 - x_3 &= 2 \end{aligned}$ **5.** $\begin{bmatrix} 4 & -3 \\ 1 & 2 \end{bmatrix}\begin{bmatrix} x_1 \\ x_2 \end{bmatrix} = \begin{bmatrix} 2 \\ 1 \end{bmatrix}$ **7.** $\begin{bmatrix} 1 & -2 & 1 \\ -1 & 1 & 0 \\ 2 & 3 & 1 \end{bmatrix}\begin{bmatrix} x_1 \\ x_2 \\ x_3 \end{bmatrix} = \begin{bmatrix} -1 \\ 2 \\ -3 \end{bmatrix}$

9. $x_1 = -8, x_2 = 2$ **11.** $x_1 = 0, x_2 = 4$ **13.** (A) $x_1 = -20, x_2 = 9$ (B) $x_1 = 18, x_2 = -7$ (C) $x_1 = 19, x_2 = -8$

15. (A) $x_1 = 8, x_2 = 5$ (B) $x_1 = 4, x_2 = 4$ (C) $x_1 = -18, x_2 = -12$

17. (A) $x_1 = -3, x_2 = -4, x_3 = -2$ (B) $x_1 = 4, x_2 = 5, x_3 = 1$ (C) $x_1 = 2, x_2 = -1, x_3 = -1$

19. (A) $x_1 = -19, x_2 = 17, x_3 = -3$ (B) $x_1 = -104, x_2 = 82, x_3 = -11$ (C) $x_1 = 100, x_2 = -78, x_3 = 10$

21. $X = (A - B)^{-1}C$ **23.** $X = (I - A)^{-1}C$ **25.** $X = (3I - A)^{-1}C$

27. (A) $x_1 = 1, x_2 = 0$ (B) $x_1 = -5{,}000, x_2 = 1{,}000$ (C) $x_1 = 5{,}001, x_2 = -1{,}000$

29. $x_1 = 13, x_2 = 21, x_3 = -12$ **31.** $x_1 = 35, x_2 = -5, x_3 = 15, x_4 = 20$

33. Concert 1: 6,000 \$4 tickets and 4,000 \$8 tickets; Concert 2: 5,000 \$4 tickets and 5,000 \$8 tickets; Concert 3: 3,000 \$4 tickets and 7,000 \$8 tickets

35. (A) $I_1 = 4, I_2 = 6, I_3 = 2$ (B) $I_1 = 3, I_2 = 7, I_3 = 4$ (C) $I_1 = 7, I_2 = 8, I_3 = 1$

37. (A) $a = 1, b = 0, c = -3$ (B) $a = -2, b = 5, c = 1$ (C) $a = 11, b = -46, c = 43$

39. Diet 1: 60 oz mix *A* and 80 oz mix *B*; Diet 2: 20 oz mix *A* and 60 oz mix *B*; Diet 3: 0 oz mix *A* and 100 oz mix *B*

Exercise 9-4

1. 7 **3.** −17 **5.** 9.79 **7.** $\begin{vmatrix} 4 & 6 \\ -2 & 8 \end{vmatrix}$ **9.** $\begin{vmatrix} 5 & -1 \\ 0 & -2 \end{vmatrix}$ **11.** $(-1)^{1+1}\begin{vmatrix} 4 & 6 \\ -2 & 8 \end{vmatrix} = 44$ **13.** $(-1)^{2+3}\begin{vmatrix} 5 & -1 \\ 0 & -2 \end{vmatrix} = 10$

15. 10 **17.** −21 **19.** −40 **21.** $(-1)^{1+1}\begin{vmatrix} a_{22} & a_{23} & a_{24} \\ a_{32} & a_{33} & a_{34} \\ a_{42} & a_{43} & a_{44} \end{vmatrix}$ **23.** $(-1)^{4+3}\begin{vmatrix} a_{11} & a_{12} & a_{14} \\ a_{21} & a_{22} & a_{24} \\ a_{31} & a_{32} & a_{34} \end{vmatrix}$ **25.** 22 **27.** −12 **29.** 0

31. 6 **33.** 60 **35.** 114 **37.** False **39.** True

41. $\begin{vmatrix} a & b \\ c & d \end{vmatrix} = -\begin{vmatrix} c & d \\ a & b \end{vmatrix}$; interchanging the rows of this determinant changes the sign.

43. $\begin{vmatrix} ka & b \\ kc & d \end{vmatrix} = k\begin{vmatrix} a & b \\ c & d \end{vmatrix}$; multiplying a column of this determinant by a number *k* multiplies the value of the determinant by *k*.

45. $\begin{vmatrix} kc + a & kd + b \\ c & d \end{vmatrix} = \begin{vmatrix} a & b \\ c & d \end{vmatrix}$; adding a multiple of one row to the other row does not change the value of the determinant.

49. $49 = (-7)(-7)$ **51.** $f(x) = x^2 - 4x + 3$; 1, 3 **53.** $f(x) = x^3 + 2x^2 - 8x$; −4, 0, 2

Exercise 9-5

1. Theorem 1 **3.** Theorem 1 **5.** Theorem 2 **7.** Theorem 3 **9.** Theorem 5 **11.** $x = 0$ **13.** $x = 5$ **15.** −10

17. 10 **19.** −10 **21.** 25 **23.** −12 **25.** Theorem 1 **27.** Theorem 2 **29.** Theorem 5 **31.** $x = 5, y = 0$

33. $x = -3, y = 10$ **35.** −28 **37.** 106 **39.** 0 **41.** 6 **43.** 14

45. Expand the left side of the equation using minors. **47.** Expand both sides of the equation and compare.

49. This follows from Theorem 4.

51. Expand the determinant about the first row to obtain $(y_1 - y_2)x - (x_1 - x_2)y + (x_1y_2 - x_2y_1) = 0$. Then show that the two points satisfy this linear equation.

53. If the determinant is 0, then the area of the triangle formed by the three points is 0. The only way this can happen is if the three points are on the same line—that is, the points are collinear.

Exercise 9-6

1. $x = 5, y = -2$ **3.** $x = 1, y = -1$ **5.** $x = -\frac{6}{5}, y = \frac{3}{5}$ **7.** $x = \frac{2}{17}, y = -\frac{20}{17}$ **9.** $x = 6{,}400, y = 6{,}600$

11. $x = 760, y = 760$ **13.** $x = 2, y = -2, z = -1$ **15.** $x = \frac{4}{3}, y = -\frac{1}{3}, z = \frac{2}{3}$ **17.** $x = -9, y = -\frac{7}{3}, z = 6$

19. $x = \frac{3}{2}, y = -\frac{7}{6}, z = \frac{2}{3}$ **21.** $x = 4$ **23.** $y = 2$ **25.** $z = \frac{5}{2}$

27. Since $D = 0$, the system has either no solution or infinitely many. Since $x = 0, y = 0, z = 0$ is a solution, the second case must hold.

31. (A) $R = 200p + 300q - 6p^2 + 6pq - 3q^2$
(B) $p = -0.3x - 0.4y + 180, q = -0.2x - 0.6y + 220, R = 180x + 220y - 0.3x^2 - 0.6xy - 0.6y^2$

Chapter 9 Review Exercise

1. $\begin{bmatrix} 4 & 8 \\ -12 & 18 \end{bmatrix}$ *(9-1)* **2.** $[-11]$ *(9-1)* **3.** $[-15 \quad 19]$ *(9-1)* **4.** $\begin{bmatrix} 16 \\ -6 \end{bmatrix}$ *(9-1)* **5.** $\begin{bmatrix} 3 & 3 \\ -4 & 9 \end{bmatrix}$ *(9-1)*

6. Not defined *(9-1)* **7.** Not defined *(9-1)* **8.** $\begin{bmatrix} 13 & -29 \\ 20 & -24 \end{bmatrix}$ *(9-1)* **9.** $[-15 \quad 18]$ *(9-1)* **10.** $\begin{bmatrix} 2 & 7 \\ -1 & -4 \end{bmatrix}$ *(9-2)*

11. (A) $x_1 = -1, x_2 = 3$ (B) $x_1 = 1, x_2 = 2$ (C) $x_1 = 8, x_2 = -10$ *(9-3)* **12.** -17 *(9-4)* **13.** 0 *(9-4, 9-5)*

14. $x = 2, y = -1$ *(9-6)* **15.** (A) -2 (B) 6 (C) 2 *(9-5)* **16.** $\begin{bmatrix} 7 & 16 & -9 \\ 28 & 40 & -30 \\ -21 & -8 & 17 \end{bmatrix}$ *(9-1)*

17. $\begin{bmatrix} 22 & 19 \\ 38 & 42 \end{bmatrix}$ *(9-1)* **18.** $\begin{bmatrix} 12 & 24 & -6 \\ 0 & 0 & 0 \\ -8 & -16 & 4 \end{bmatrix}$ *(9-1)* **19.** $[16]$ *(9-1)* **20.** Not defined *(9-1)*

21. $\begin{bmatrix} 63 & -24 & -39 \\ -42 & 16 & 26 \end{bmatrix}$ *(9-1)* **22.** $\begin{bmatrix} -1 & 1 & 1 \\ -2 & 3 & 2 \\ \frac{1}{2} & -\frac{1}{4} & -\frac{1}{4} \end{bmatrix}$ *(9-2)*

23. (A) $x_1 = 2, x_2 = 1, x_3 = -1$ (B) $x_1 = 1, x_2 = -2, x_3 = 1$ (C) $x_1 = -1, x_2 = 2, x_3 = -2$ *(9-3)*

24. $-\frac{11}{12}$ *(9-4)* **25.** 35 *(9-4, 9-5)* **26.** $y = \frac{10}{5} = 2$ *(9-6)*

27. (A) A unique solution (B) Either no solution or an infinite number *(9-3)*

28. No. *(9-3)* **29.** $X = (A - C)^{-1}B$ *(9-3)* **30.** $\begin{bmatrix} -\frac{11}{12} & -\frac{1}{12} & 5 \\ \frac{10}{12} & \frac{2}{12} & -4 \\ \frac{1}{12} & -\frac{1}{12} & 0 \end{bmatrix}$ or $\frac{1}{12}\begin{bmatrix} -11 & -1 & 60 \\ 10 & 2 & -48 \\ 1 & -1 & 0 \end{bmatrix}$ *(9-2)*

31. $x_1 = 1{,}000, x_2 = 4{,}000, x_3 = 2{,}000$ *(9-3)* **32.** 42 *(9-5)*

33. $\begin{vmatrix} u + kv & v \\ w + kx & x \end{vmatrix} = (u + kv)x - (w + kx)v = ux + kvx - wv - kvx = ux - wv = \begin{vmatrix} u & v \\ w & x \end{vmatrix}$ *(9-5)*

34. Theorem 4 in Section 9-5 implies that both points satisfy the equation. All other points on the line through the given points will also satisfy the equation. *(9-5)*

35. (A) 60 tons at Big Bend, 20 tons at Saw Pit (B) 30 tons at Big Bend, 50 tons at Saw Pit
(C) 40 tons at Big Bend, 40 tons at Saw Pit *(9-3)*

36. (A) \$27 (B) Elements in LH give the total cost of manufacturing each product at each plant.
(C) *(9-1)*

	North Carolina	South Carolina	
$LH =$	\$46.35	\$41.00	Desks
	\$30.45	\$27.00	Stands

37. (A) $\begin{bmatrix} 1{,}600 & 1{,}730 \\ 890 & 720 \end{bmatrix}$ (B) $\begin{bmatrix} 200 & 160 \\ 80 & 40 \end{bmatrix}$ (C) $\begin{bmatrix} 3{,}150 \\ 1{,}550 \end{bmatrix}$ Desks / Stands
Total production of each item in January *(9-1)*

38. GRAPHING UTILITY *(9-2)*

Cumulative Review Exercise: Chapters 8 and 9

1. $x = 2, y = -1$ *(8-1)* **2.** $(-1, 2)$ *(8-1)* **3.** $(-\frac{1}{5}, -\frac{7}{5}), (1, 1)$ *(8-3)* **4.** *(8-4)*

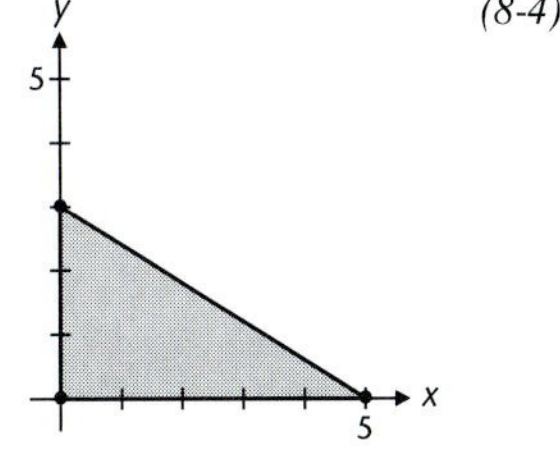

5. Maximum: 33; Minimum: 10 *(8-5)*

6. (A) $\begin{bmatrix} 0 & -3 \\ 3 & -9 \end{bmatrix}$ (B) Not defined (C) $[3]$ (D) $\begin{bmatrix} 1 & 7 \\ 4 & -7 \end{bmatrix}$ (E) $[-1, 8]$ (F) Not defined *(9-1)*

7. -10 *(9-4)* **8.** (A) $x_1 = 3, x_2 = -4$ (B) $x_1 = 2t + 3, x_2 = t$, t any real number. (C) No solution *(8-1)*

9. (A) $\left[\begin{array}{cc|c} 1 & 1 & 3 \\ -1 & 1 & 5 \end{array}\right]$ (B) $\left[\begin{array}{cc|c} 1 & 0 & -1 \\ 0 & 1 & 4 \end{array}\right]$ (C) $x_1 = -1, x_2 = 4$ *(8-1, 8-2)*

10. (A) $\begin{bmatrix} 1 & -3 \\ 2 & -5 \end{bmatrix}\begin{bmatrix} x_1 \\ x_2 \end{bmatrix} = \begin{bmatrix} k_1 \\ k_2 \end{bmatrix}$ (B) $A^{-1} = \begin{bmatrix} -5 & 3 \\ -2 & 1 \end{bmatrix}$ (C) $x_1 = 13, x_2 = 5$ (D) $x_1 = -11, x_2 = -4$ *(9-3)*

11. (A) 2 (B) $x = \frac{1}{2}$, $y = 0$ *(9-6)*

12. $x_1 = 1$, $x_2 = 3$; each pair of lines has the same intersection point. *(8-1)*

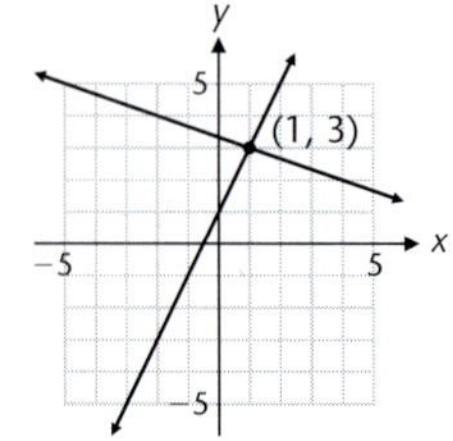

$x_1 + 3x_2 = 10$
$2x_1 - x_2 = -1$

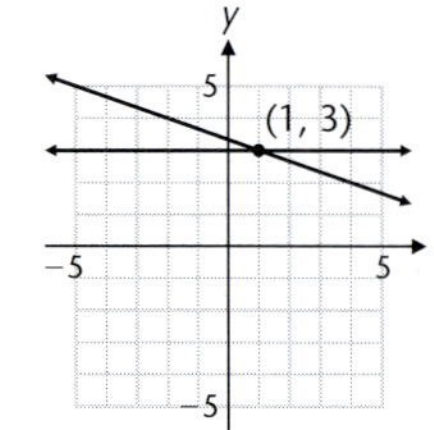

$x_1 + 3x_2 = 10$
$-7x_2 = -21$

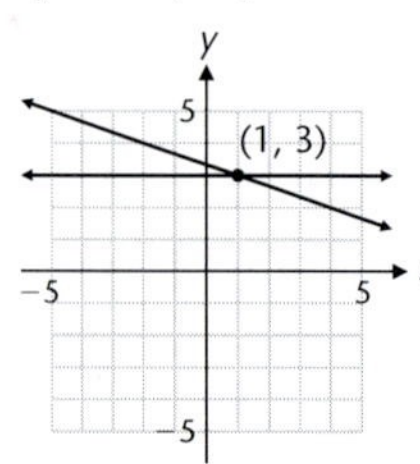

$x_1 + 3x_2 = 10$
$x_2 = 3$

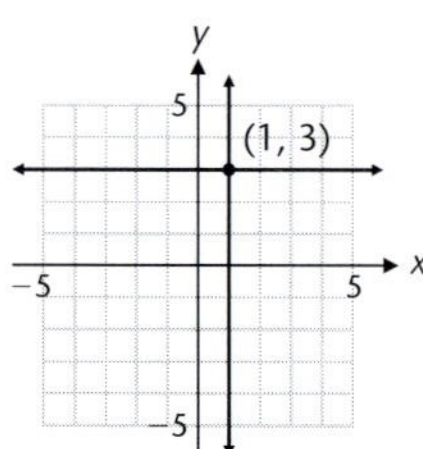

$x_1 = 1$
$x_2 = 3$

13. (1.53, 3.35) *(8-1)* **14.** (1, 0, −2) *(8-2)* **15.** No solution *(8-2)* **16.** $(t - 3, t - 2, t)$ t any real number *(8-2)*

17. (1, 1), (−1, −1), $(\sqrt{3}, \sqrt{3}/3)$, $(-\sqrt{3}, -\sqrt{3}/3)$ *(8-3)* **18.** $(0, i)$, $(0, -i)$, (1, 1), (−1, −1) *(8-3)*

19. (A) [−3] (B) $\begin{bmatrix} 1 & 2 & -1 \\ -1 & -2 & 1 \\ 2 & 4 & -2 \end{bmatrix}$ *(9-1)* **20.** (A) $\begin{bmatrix} -1 & 2 \\ 2 & 3 \end{bmatrix}$ (B) Not defined *(9-1)*

21. *(8-4)* **22.** 63 *(8-5)*

y
10
(0, 6)
5
(2, 3)
(8, 0)
x
5
10

23. (A) $\begin{bmatrix} 1 & 4 & 2 \\ 2 & 6 & 3 \\ 2 & 5 & 2 \end{bmatrix}\begin{bmatrix} x_1 \\ x_2 \\ x_3 \end{bmatrix} = \begin{bmatrix} k_1 \\ k_2 \\ k_3 \end{bmatrix}$ (B) $A^{-1} = \begin{bmatrix} -3 & 2 & 0 \\ 2 & -2 & 1 \\ -2 & 3 & -2 \end{bmatrix}$ (C) (7, −5, 6) (D) (−6, 3, −2) *(9-3)*

24. (A) $D = 1$ (B) $z = 32$ *(9-5, 9-6)* **25.** (−1.35, 0.28), (−0.87, −1.60), (0.87, 1.60), (1.35, −0.28) *(8-3)*

26. (A) Infinite number of solutions (B) No solution (C) Unique solution *(8-2)*

27. $A = I$, the $n \times n$ identity *(9-3)* **28.** L, M, and P *(8-2)* **32.** True *(9-1)* **33.** True *(9-1)* **34.** False *(9-1)*

35. False *(9-1)* **36.** True *(9-2)* **37.** True *(9-2)* **38.** True *(9-4)* **39.** True *(9-4)*

40. $8,000 at 8% and $4,000 at 14% *(8-1, 8-2)* **41.** 60-g mix A, 50-g mix B, 40-g mix C *(8-2)*

42. 1 model *A* truck, 6 model *B* trucks, and 5 model *C* trucks; or 3 model *A* trucks, 3 model *B* trucks, and 6 model *C* trucks; or 5 model *A* trucks and 7 model *C* trucks. *(8-2)*

43. 8 by 4 m *(8-3)*

44. (A) Manufacturing 400 standard and 200 deluxe day packs produces a maximum weekly profit of $5,600.
(B) The maximum weekly profit increases to $6,000 when 0 standard and 400 deluxe day packs are manufactured.
(C) The maximum weekly profit increases to $6,600 when 600 standard and 0 deluxe day packs are manufactured. *(8-5)*

45. (A) $M\begin{bmatrix} 0.25 \\ 0.25 \\ 0.25 \\ 0.25 \end{bmatrix} = \begin{bmatrix} 82.25 \\ 83 \\ 92 \\ 83.75 \\ 82 \end{bmatrix}\begin{matrix} \text{Ann} \\ \text{Bob} \\ \text{Carol} \\ \text{Dan} \\ \text{Eric} \end{matrix}$ (B) $M\begin{bmatrix} 0.2 \\ 0.2 \\ 0.2 \\ 0.4 \end{bmatrix} = \begin{bmatrix} 83 \\ 84.8 \\ 91.8 \\ 85.2 \\ 80.8 \end{bmatrix}\begin{matrix} \text{Ann} \\ \text{Bob} \\ \text{Carol} \\ \text{Dan} \\ \text{Eric} \end{matrix}$

(C) Class averages

	Test 1	Test 2	Test 3	Test 4
[0.2 0.2 0.2 0.2 0.2]M =	[84.4	81.8	85	87.2]

(9-1)

CHAPTER 10

Exercise 10-1

1. 7, 9, 11, 13, . . . **3.** $2, \frac{7}{3}, \frac{5}{2}, \frac{13}{5}, \ldots$ **5.** 1, 13, 19, 97, . . . **7.** 205 **9.** $\frac{601}{201}$ **11.** 1 + 2 + 3 + 4 + 5

13. $\frac{1}{10} + \frac{1}{100} + \frac{1}{1{,}000}$ **15.** −1 + 1 −1 + 1 **17.** 1, −4, 9, −16, 25 **19.** 0.3, 0.33, 0.333, 0.3333, 0.333 33

21. $1, -\frac{1}{2}, \frac{1}{4}, -\frac{1}{8}, \frac{1}{16}$ **23.** 7, 3, −1, −5, −9 **25.** $4, 1, \frac{1}{4}, \frac{1}{16}, \frac{1}{64}$ **27.** $a_n = 1 + 2n$ **29.** $a_n = (n + 1)^2$ **31.** $a_n = 2^{n+1}$

33. $a_n = (2n - 1)/n$ **35.** $a_n = 7(-1)^{n+1}$ **37.** $a_n = x^n/2^{n-1}$

39.

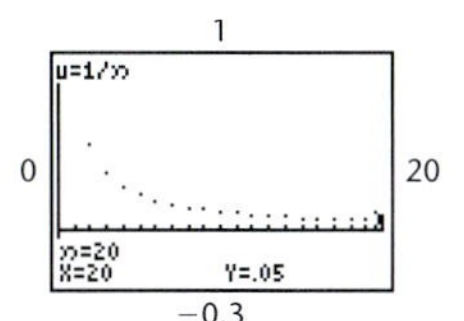

41.

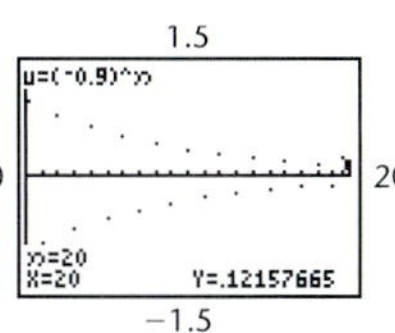

43. $\frac{4}{1} - \frac{8}{2} + \frac{16}{3} - \frac{32}{4}$ **45.** $x^2 + \frac{x^3}{2} + \frac{x^4}{3}$ **47.** $x - \frac{x^2}{2} + \frac{x^3}{3} - \frac{x^4}{4} + \frac{x^5}{5}$ **49.** $\sum_{k=1}^{4} k^2$ **51.** $\sum_{k=1}^{5} \frac{1}{2^k}$ **53.** $\sum_{k=1}^{n} \frac{1}{k^2}$ **55.** $\sum_{k=1}^{n} (-1)^{k+1} k^2$

57. (A) 3, 1.83, 1.46, 1.415 (B) $\sqrt{2} \approx 1.4142$ (C) For $a_1 = 1$: 1, 1.5, 1.417, 1.414

59. The values of c_n are approximately 2.236 (i.e., $\sqrt{5}$) for large values of n.

61. Series approx. of $e^{0.2} = 1.221\ 400\ 0$; calculator value of $e^{0.2} = 1.221\ 402\ 8$

Exercise 10-2

1. 3 **3.** 5 **5.** P_1: $2 = 2 \cdot 1^2$; P_2: $2 + 6 = 2 \cdot 2^2$; P_3: $2 + 6 + 10 = 2 \cdot 3^2$

7. P_1: $a^5a = a^{5+1}$; P_2: $a^5a^2 = a^5(a^1a) = (a^5a)a = a^6a = a^7 = a^{5+2}$; P_3: $a^5a^3 = a^5(a^2a) = a^5(a^1a)a = [(a^5a)a]a = a^8 = a^{5+3}$

9. P_1: $9^1 - 1 = 8$ is divisible by 4; P_2: $9^2 - 1 = 80$ is divisible by 4; P_3: $9^3 - 1 = 728$ is divisible by 4

11. P_k: $2 + 6 + 10 + \cdots + (4k - 2) = 2k^2$; P_{k+1}: $2 + 6 + 10 + \cdots + (4k - 2) + (4k + 2) = 2(k + 1)^2$

13. P_k: $a^5a^k = a^{5+k}$; P_{k+1}: $a^5a^{k+1} = a^{5+k+1}$ **15.** P_k: $9^k - 1 = 4r$; P_{k+1}: $9^{k+1} - 1 = 4s$; $r, s \in N$

23. $n = 4$, $p(x) = x^4 + 1$ **25.** $n = 23$ **43.** $2 + 4 + 6 + \cdots + 2n = n(n + 1)$

45. $1 + 2 + 3 + \cdots + (n - 1) = n(n - 1)/2$, $n \ge 2$ **51.** $3^4 + 4^4 + 5^4 + 6^4 \ne 7^4$

Exercise 10-3

1. (A) Arithmetic with $d = -5$; -26, -31 (B) Geometric with $r = -2$; -16, 32 (C) Neither
(D) Geometric with $r = \frac{1}{3}$; $\frac{1}{54}$, $\frac{1}{162}$

3. $a_2 = 11$, $a_3 = 16$, $a_4 = 21$ **5.** $a_{21} = -153$, $S_{21} = -1{,}743$ **7.** $S_{25} = 1{,}775$ **9.** $S_{15} = 6{,}375$

11. $a_2 = 3$, $a_3 = -\frac{3}{2}$, $a_4 = \frac{3}{4}$ **13.** $a_{10} = \frac{1}{243}$ **15.** $S_7 = 3{,}279$ **17.** $S_{50} = 3{,}925$ **19.** $a_{64} = 56.5$ **21.** $a_1 = 6$

23. $a_{10} = 3$, $d = -\frac{1}{9}$ **25.** $r = \pm\sqrt{10}$ **27.** $S_{10} = 42{,}625$ **29.** $a_2 = 60$, $a_3 = 90$ **31.** $S_{51} = 4{,}131$ **33.** $S_7 = 547$

35. $-1{,}071$ **37.** $\frac{1{,}023}{1{,}024}$ **39.** 4,446 **43.** $x = 2\sqrt{3}$ **45.** Yes **47.** 66 **49.** 133 **51.** $S_\infty = \frac{9}{2}$ **53.** No sum

55. $S_\infty = \frac{8}{5}$ **57.** $\frac{7}{9}$ **59.** $\frac{6}{11}$ **61.** $3\frac{8}{37}$ or $\frac{119}{37}$ **65.** $a_n = (-2)(-3)^{n-1}$ **67.** *Hint:* $y = x + d$, $z = x + 2d$

71. $x = -1$, $y = 2$ **73.** Firm *A*: \$501,000; firm *B*: \$504,000 **75.** \$4,000,000 **77.** $A = P(1 + r)^n$; approx. 12 yr

79. \$700 per year; \$115,500 **81.** 900 **83.** 1,250,000 **85.** (A) 336 ft (B) 1,936 ft (C) $16t^2$ **87.** $A = A_0 2^{2t}$

89. $r = 10^{-0.4} = 0.398$ **91.** $\$9.223 \times 10^{16}$; $\$1.845 \times 10^{17}$ **93.** 0.0015 psi **95.** 2 **97.** 3,420°

Exercise 10-4

1. 5,040 **3.** 210 **5.** 144 **7.** 56 **9.** 1 **11.** 720 **13.** 9!/8! **15.** 8!/5! **17.** 715 **19.** 3,432

21. 161,700 **25.** $8x^3 - 36x^2y + 54xy^2 - 27y^3$ **27.** $x^4 - 8x^3 + 24x^2 - 32x + 16$

29. $32x^5 - 80x^4y + 80x^3y^2 - 40x^2y^3 + 10xy^4 - y^5$ **31.** $5{,}005u^9v^6$ **33.** $180m^2n^8$ **35.** $392{,}445w^{16}$ **37.** $-48{,}384x^3y^5$

39. 5 **41.** (A) $a_4 = 0.251$ (B) 1 **43.** 1.1046 **45.** True **47.** True

Exercise 10-5

1. 2,730 **3.** 1 **5.** 84 **7.** 1,820 **9.** 6,720 **11.** 132,600 **13.** 1,287 **15.** 2,598,960 **17.** $5 \cdot 3 \cdot 4 \cdot 2 = 120$

19. $P_{10,3} = 10 \cdot 9 \cdot 8 = 720$ **21.** $C_{7,3} = 35$ subcommittees; $P_{7,3} = 210$ **23.** $C_{10,2} = 45$

25. No repeats: $6 \cdot 5 \cdot 4 \cdot 3 = 360$; with repeats: $6 \cdot 6 \cdot 6 \cdot 6 = 1{,}296$ **27.** $C_{13,5} = 1{,}287$

29. $26 \cdot 26 \cdot 26 \cdot 10 \cdot 10 \cdot 10 = 17{,}576{,}000$ possible license plates; no repeats: $26 \cdot 25 \cdot 24 \cdot 10 \cdot 9 \cdot 8 = 11{,}232{,}000$

31. $C_{13,5}C_{13,2} = 100{,}386$ **33.** $C_{8,3}C_{10,4}C_{7,2} = 246{,}960$

35. (B) $r = 0$, 10 (C) Each is the product of r consecutive integers, the largest of which is n for $P_{n,r}$, and r for $r!$

37. $12 \cdot 11 = 132$ **39.** (A) $C_{8,2} = 28$ (B) $C_{8,3} = 56$ (C) $C_{8,4} = 70$

41. 2 people: $P_{5,2} = 20$; 3 people: $P_{5,3} = 60$; 4 people: $P_{5,4} = 120$; 5 people: $P_{5,5} = 120$

43. (A) $P_{8,5} = 6{,}720$ (B) $C_{8,5} = 56$ (C) $2 \cdot C_{6,4} = 30$

45. There are $C_{4,1} \cdot C_{48,4} = 778{,}320$ hands that contain exactly one king, and $C_{39,5} = 575{,}757$ hands that contain no hearts, so the former is more likely.

47. 21

Chapter 10 Review Exercise

1. (A) Geometric (B) Arithmetic (C) Arithmetic (D) Neither (E) Geometric *(10-1, 10-3)*
2. (A) 5, 7, 9, 11 (B) $a_{10} = 23$ (C) $S_{10} = 140$ *(10-1, 10-3)*
3. (A) 16, 8, 4, 2 (B) $a_{10} = \frac{1}{32}$ (C) $S_{10} = 31\frac{31}{32}$ *(10-1, 10-3)*
4. (A) $-8, -5, -2, 1$ (B) $a_{10} = 19$ (C) $S_{10} = 55$ *(10-1, 10-3)*
5. (A) $-1, 2, -4, 8$ (B) $a_{10} = 512$ (C) $S_{10} = 341$ *(10-1, 10-3)* **6.** $S_\infty = 32$ *(10-3)* **7.** 3,628,800 *(10-4)*
8. 17,100,720 *(10-4)* **9.** 1,287 *(10-4)* **10.** $P_{8,4} = 1{,}680$; $C_{8,4} = 70$ *(10-5)*
11. (A) 12 combined outcomes: (B) $6 \cdot 2 = 12$ *(10-5)* **12.** $6 \cdot 5 \cdot 4 \cdot 3 \cdot 2 \cdot 1 = 720$ *(10-5)* **13.** $P_{6,6} = 6! = 720$ *(10-5)*

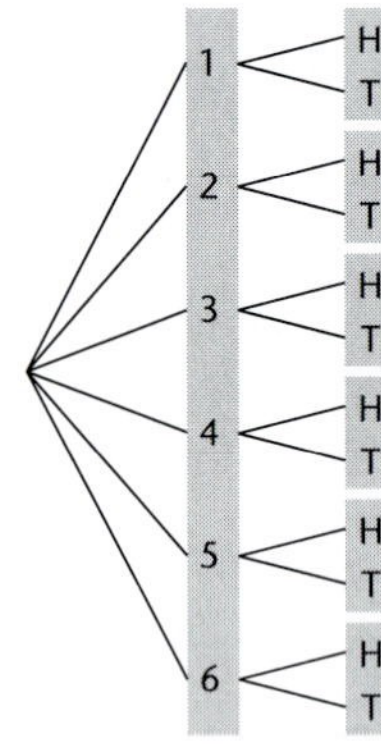

14. P_1: $5 = 1^2 + 4 \cdot 1$; P_2: $5 + 7 = 2^2 + 4 \cdot 2$; P_3: $5 + 7 + 9 = 3^2 + 4 \cdot 3$ *(10-2)*
15. P_1: $2 = 2^{1+1} - 2$; P_2: $2 + 4 = 2^{2+1} - 2$; P_3: $2 + 4 + 8 = 2^{3+1} - 2$ *(10-2)*
16. P_1: $49^1 - 1 = 48$ is divisible by 6; P_2: $49^2 - 1 = 2{,}400$ is divisible by 6; P_3: $49^3 - 1 = 117{,}648$ is divisible by 6 *(10-2)*
17. P_k: $5 + 7 + 9 + \cdots + (2k + 3) = k^2 + 4k$; P_{k+1}: $5 + 7 + 9 + \cdots + (2k + 3) + (2k + 5) = (k + 1)^2 + 4(k + 1)$ *(10-2)*
18. P_k: $2 + 4 + 8 + \cdots + 2^k = 2^{k+1} - 2$; P_{k+1}: $2 + 4 + 8 + \cdots + 2^k + 2^{k+1} = 2^{k+2} - 2$ *(10-2)*
19. P_k: $49^k - 1 = 6r$ for some integer r; P_{k+1}: $49^{k+1} - 1 = 6s$ for some integer s *(10-2)* **20.** $n = 31$ is a counterexample *(10-2)*
21. $S_{10} = -6 - 4 - 2 + 0 + 2 + 4 + 6 + 8 + 10 + 12 = 30$ *(10-3)* **22.** $S_7 = 8 + 4 + 2 + 1 + \frac{1}{2} + \frac{1}{4} + \frac{1}{8} = 15\frac{7}{8}$ *(10-3)*
23. $S_\infty = \frac{81}{5}$ *(10-3)* **24.** $S_n = \sum_{k=1}^{n} \frac{(-1)^{k+1}}{3^k}$; $S_\infty = \frac{1}{4}$ *(10-3)* **25.** $C_{6,3} = 20$ *(10-5)* **26.** $d = 3$, $a_5 = 25$ *(10-3)*
27. $a_1 = -9$, $d = \frac{19}{5}$ *(10-3)* **28.** 336; 512; 392 *(10-5)* **29.** $\frac{8}{11}$ *(10-3)* **30.** (A) $P_{6,3} = 120$ (B) $C_{5,2} = 10$ *(10-5)*
31. 190 *(10-4)* **32.** 1,820 *(10-4)* **33.** 1 *(10-4)* **34.** $\frac{987!}{493!}$ *(10-4)* **35.** $\binom{1000}{501}$ *(10-5)*
36. $x^5 - 5x^4y + 10x^3y^2 - 10x^2y^3 + 5xy^4 - y^5$ *(10-4)* **37.** $-1{,}760x^3y^9$ *(10-4)* **41.** 29 *(10-4)* **42.** 26 *(10-1)*
43. $2^5 = 32$; 6 *(10-5)* **44.** 49 g/2 ft; 625 g/2 ft *(10-3)* **45.** 12 *(10-5)*
46. $x^6 + 6ix^5 - 15x^4 - 20ix^3 + 15x^2 + 6ix - 1$ *(10-4)* **47.** $P_{5,5} = 120$ *(10-5)* **53.** $n = 5$ *(10-1)*
54. $\{a_n\}$ is neither arithmetic nor geometric; $\{b_n\}$ is geometric *(10-3)*

CHAPTER 11

Exercise 11-1

1.

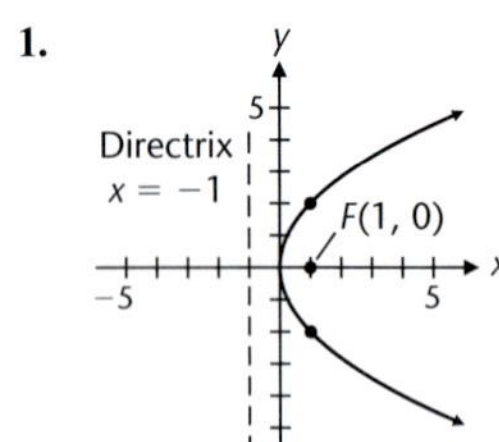

3.

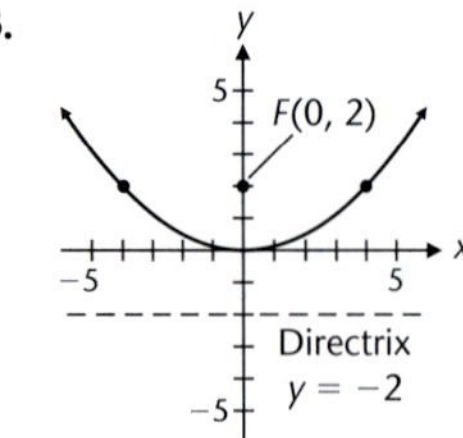

5.

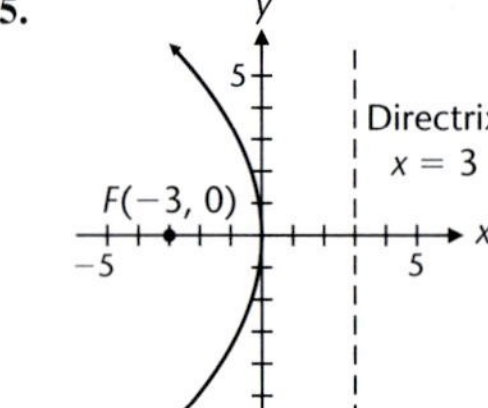

7.

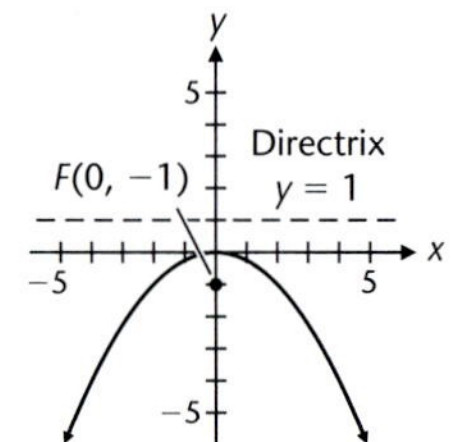

9.

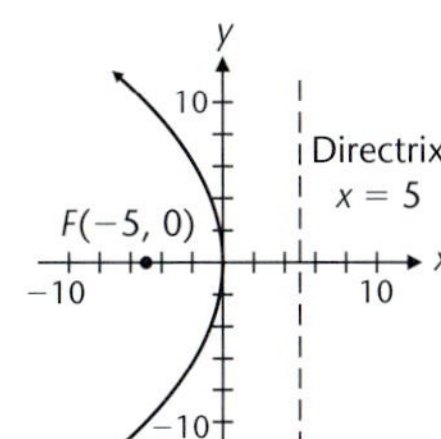

11.

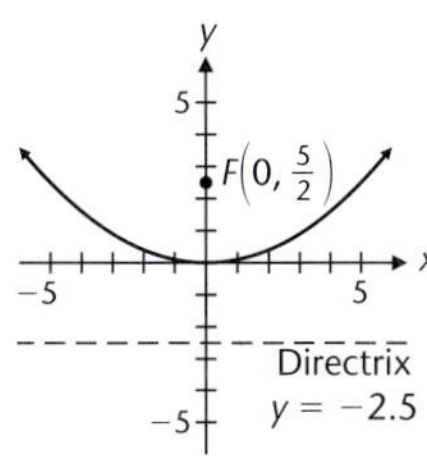
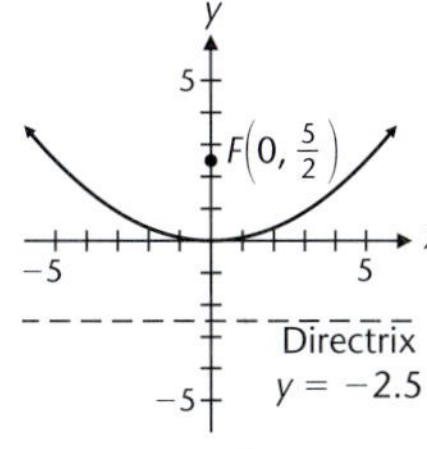

13. (9.75, 0) **15.** (0, −26.25) **17.** (−19.25, 0)

19. $y^2 = -24x$ **21.** $x^2 = 20y$ **23.** $x^2 = -\frac{4}{3}y$ **25.** $y^2 = -100x$ **27.** $x^2 = -3y$ **29.** $y^2 = -\frac{1}{2}x$

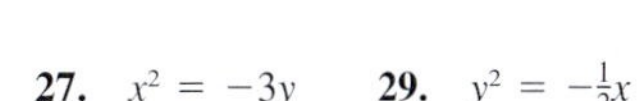

31.

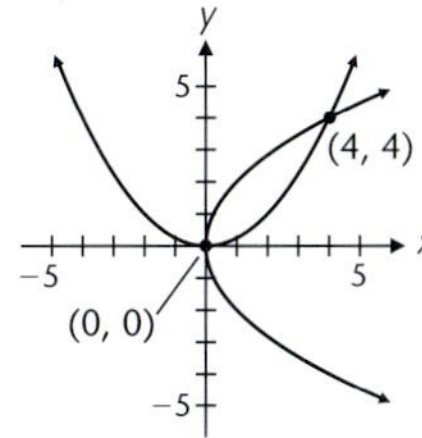

33. (5.313, 5.646); (0, 0)

35. (A) 2; $x = 0$ and $y = 0$ (B) (0, 0), $(4am, 4am^2)$

37. $(\pm 2a, a)$ **39.** False **41.** True **43.** $x^2 - 4x - 12y - 8 = 0$ **45.** $y^2 + 8x - 8x + 48 = 0$

47. (−0.78, 0.08), (40.78, 207.92) **49.** (−6.84, −5.85), (0, 0) **51.** $x^2 = -200y$

53. (A) $y = 0.0025x^2, -100 \le x \le 100$ (B) 25 ft

Exercise 11-2

1. Foci: $F'(-\sqrt{21}, 0)$, $F(\sqrt{21}, 0)$
Major axis length = 10
Minor axis length = 4

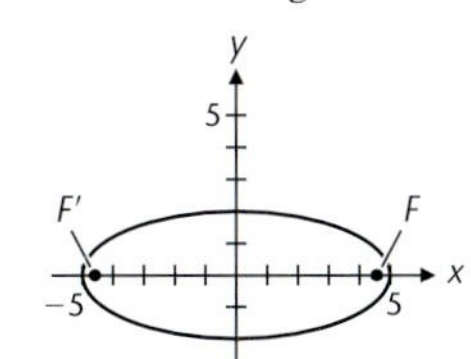

3. Foci: $F'(0, -\sqrt{21})$, $F(0, \sqrt{21})$
Major axis length = 10
Minor axis length = 4

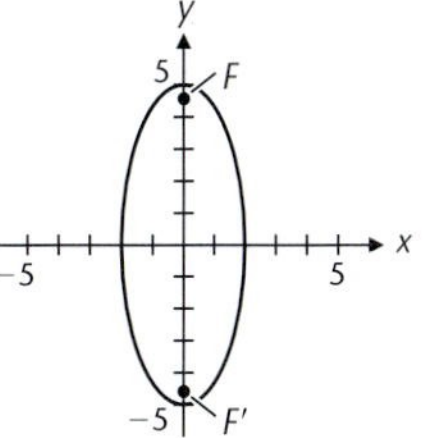

5. Foci: $F'(-\sqrt{8}, 0)$, $F(\sqrt{8}, 0)$
Major axis length = 6
Minor axis length = 2

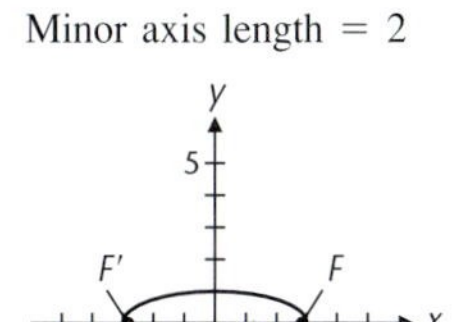
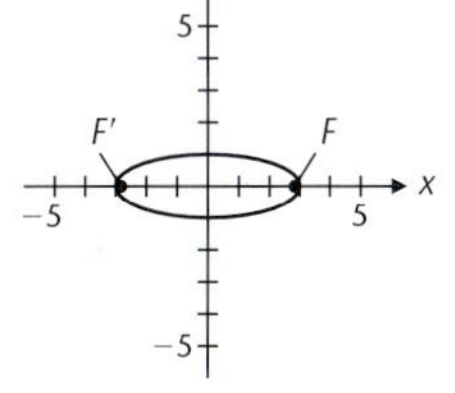

7. Foci: $F'(0, -4)$, $F(0, 4)$
Major axis length = 10
Minor axis length = 6

9. Foci: $F'(0, -\sqrt{6})$, $F(0, \sqrt{6})$
Major axis length = $2\sqrt{12} \approx 6.93$
Minor axis length = $2\sqrt{6} \approx 4.90$

11. Foci: $F'(-\sqrt{3}, 0)$, $F(\sqrt{3}, 0)$
Major axis length = $2\sqrt{7} \approx 5.29$
Minor axis length = 4

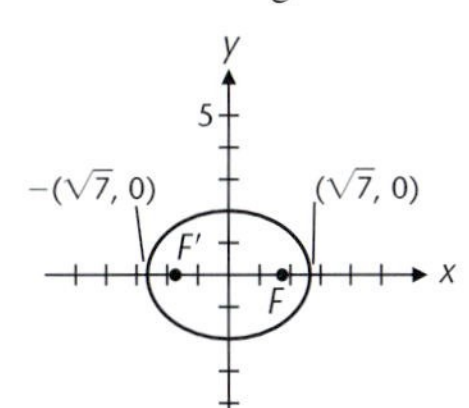

13. $x^2 + \dfrac{y^2}{9} = 1$ **15.** $\dfrac{x^2}{41} + \dfrac{y^2}{25} = 1$ **17.** $\dfrac{x^2}{143} + \dfrac{y^2}{144} = 1$ **19.** It does not pass the vertical line test.

21.

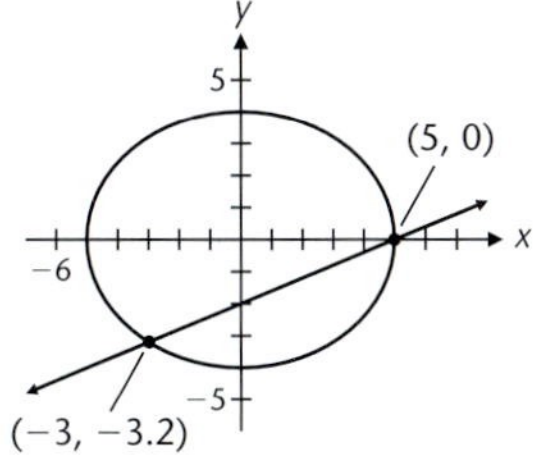

23.

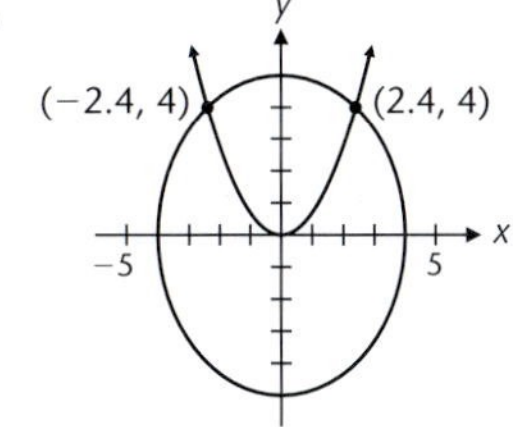

25. (2.201, 4.403)

27. (3.565, 1.589) **29.** False **31.** True **33.** $\dfrac{x^2}{16} + \dfrac{y^2}{12} = 1$; ellipse

35. $(-0.46, 2.57), (4.08, -1.06)$ **37.** $(\pm 3.64, \pm 9.50)$ **39.** $\dfrac{x^2}{400} + \dfrac{y^2}{144} = 1$; 7.94 ft **41.** (A) $\dfrac{x^2}{576} + \dfrac{y^2}{15.9} = 1$ (B) 5.13 ft

Exercise 11-3

1. Foci: $F'(-\sqrt{13}, 0), F(\sqrt{13}, 0)$
Transverse axis length = 6
Conjugate axis length = 4

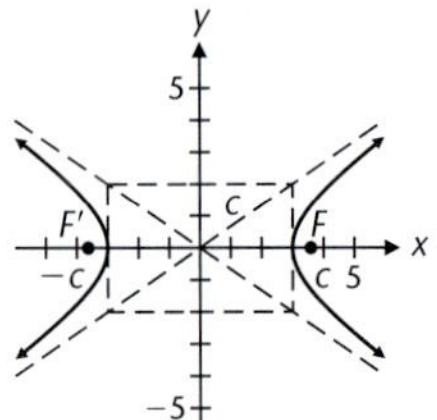

3. Foci: $F'(0, -\sqrt{13}), F(0, \sqrt{13})$
Transverse axis length = 4
Conjugate axis length = 6

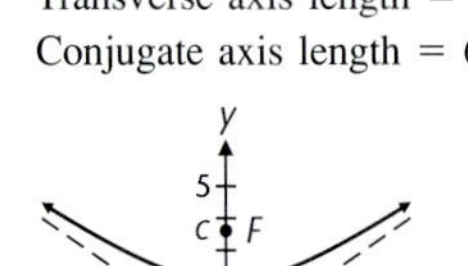
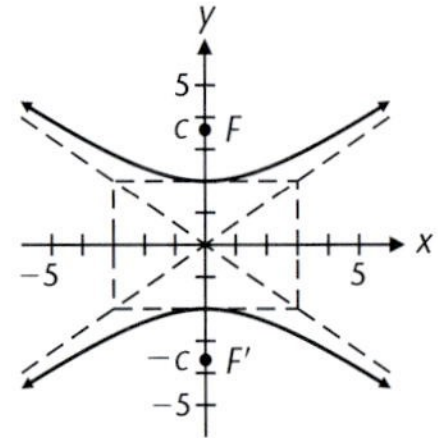

5. Foci: $F'(-\sqrt{20}, 0), F(\sqrt{20}, 0)$
Transverse axis length = 4
Conjugate axis length = 8

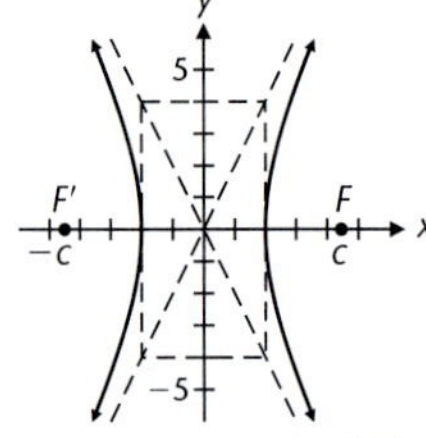

7. Foci: $F'(0, -5), F(0, 5)$
Transverse axis length = 8
Conjugate axis length = 6

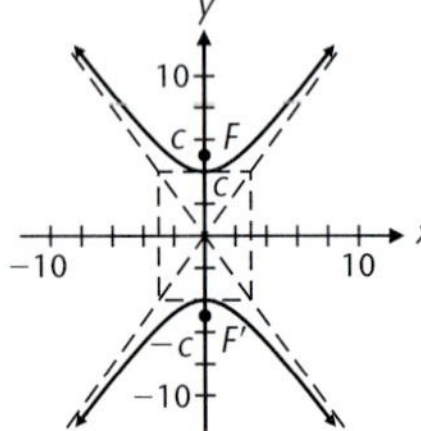
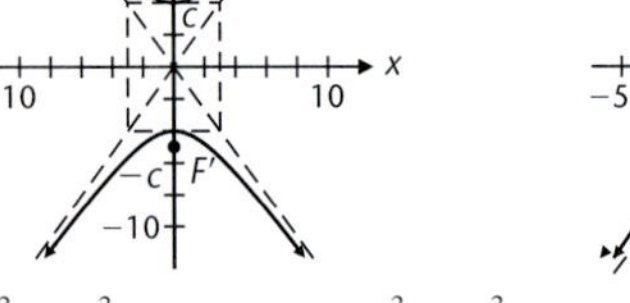

9. Foci: $F'(-\sqrt{10}, 0), F(\sqrt{10}, 0)$
Transverse axis length = 4
Conjugate axis length = $2\sqrt{6} \approx 4.90$

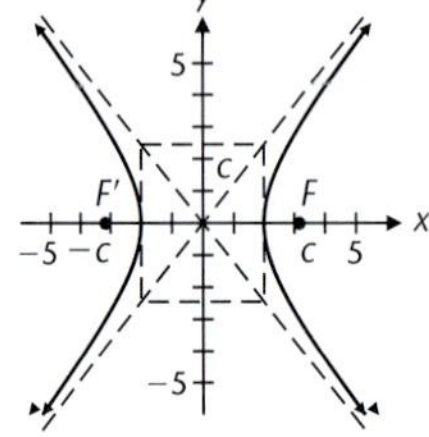

11. Foci: $F'(0, -\sqrt{11}), F(0, \sqrt{11})$
Transverse axis length = 4
Conjugate axis length = $2\sqrt{7} \approx 5.29$

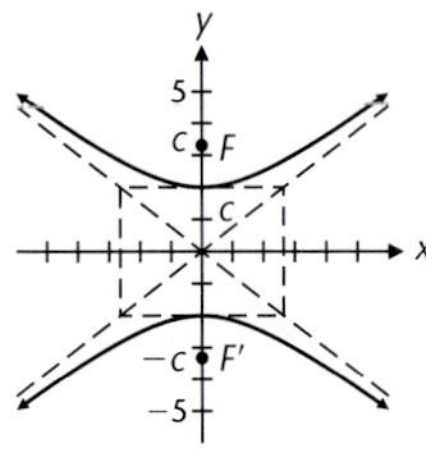

13. $\dfrac{y^2}{25} - \dfrac{x^2}{81} = 1$ **15.** $\dfrac{x^2}{49} - \dfrac{y^2}{32} = 1$ **17.** $\dfrac{y^2}{36} - \dfrac{x^2}{36} = 1$

19. (A) Infinitely many; $\dfrac{x^2}{a^2} - \dfrac{y^2}{1 - a^2} = 1$ $(0 < a < 1)$ (B) Infinitely many; $\dfrac{x^2}{a^2} + \dfrac{y^2}{a^2 - 1} = 1$ $(a > 1)$ (C) One; $y^2 = 4x$

21.

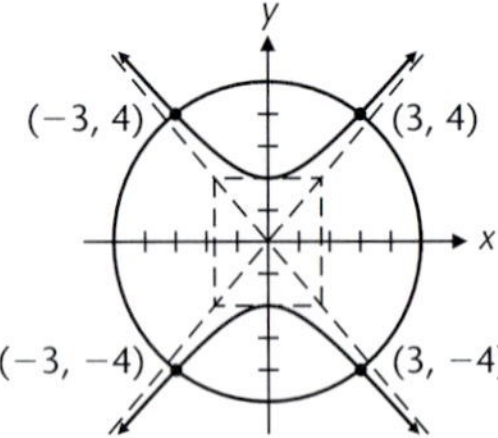

23.

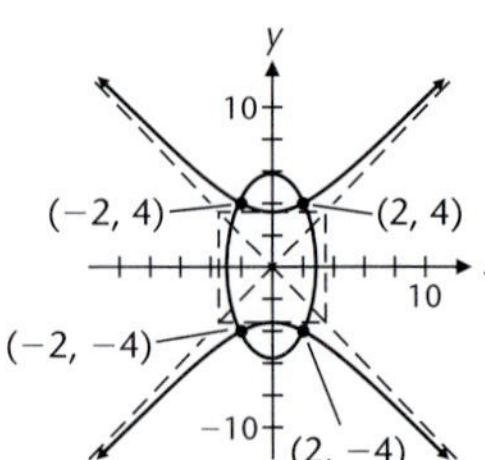

25. $(-1.389, 3.306), (6.722, 7.361)$

27. $(\pm 3.266, \pm 3.830)$ **29.** True **31.** False **33.** $\dfrac{x^2}{4} - \dfrac{y^2}{5} = 1$; hyperbola **35.** $(-4.73, 2.88), (3.35, -0.90)$

37. $(\pm 1.39, \pm 2.96)$ **39.** $\dfrac{y^2}{16} - \dfrac{x^2}{8} = 1$; 5.38 ft above vertex **41.** $y = \frac{4}{3}\sqrt{x^2 + 30^2}$

Exercise 11-4

1. (A) $x' = x - 3, y' = y - 5$ (B) $x'^2 + y'^2 = 81$ (C) Circle

3. (A) $x' = x + 7, y' = y - 4$ (B) $\dfrac{x'^2}{9} + \dfrac{y'^2}{16} = 1$ (C) Ellipse

5. (A) $x' = x - 4, y' = y + 9$ (B) $y'^2 = 16x'$ (C) Parabola

7. (A) $x' = x + 8, y' = y + 3$ (B) $\dfrac{x'^2}{12} + \dfrac{y'^2}{8} = 1$ (C) Ellipse **9.** (A) $\dfrac{(x - 3)^2}{9} - \dfrac{(y + 2)^2}{16} = 1$ (B) Hyperbola

11. (A) $\dfrac{(x + 5)^2}{5} + \dfrac{(y + 7)^2}{6} = 1$ (B) Ellipse **13.** (A) $(x + 6)^2 = -24(y - 4)$ (B) Parabola

15. $\frac{(x-2)^2}{9} + \frac{(y-2)^2}{4} = 1$; ellipse

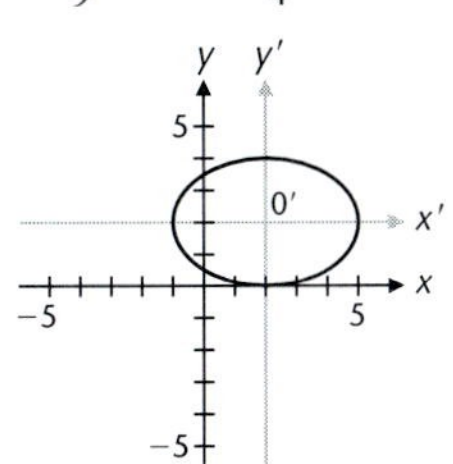

17. $(x+4)^2 = -8(y-2)$; parabola

19. $(x+6)^2 + (y+5)^2 = 16$; circle

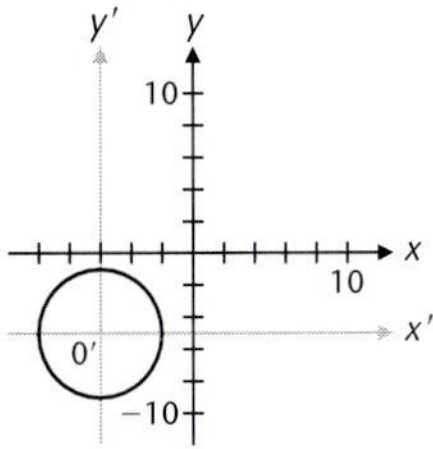

21. $\frac{(y-3)^2}{9} - \frac{(x+4)^2}{16} = 1$; hyperbola

23. $h = \frac{-D}{2A}$, $k = \frac{D^2 - 4AF}{4AE}$

25. $x^2 - 10x - 32y + 121 = 0$ **27.** $36x^2 + 25y^2 + 216x - 200y - 176 = 0$ **29.** $3x^2 - y^2 - 24x + 2y + 44 = 0$
31. $4x^2 - 3y - 4 = 0$ **33.** $x^2 + 5y^2 + 4x + 10y = 0$ **35.** $F'(-\sqrt{5} + 2, 2)$, $F(\sqrt{5} + 2, 2)$ **37.** $F(-4, 0)$
39. $F'(-4, -2)$, $F(-4, 8)$ **41.** (1.18, 1.98), (6.85, −6.52) **43.** (−1.72, −1.87), (−0.99, 2.06)

Exercise 11-5

1. $y = -2x - 2$;
straight line

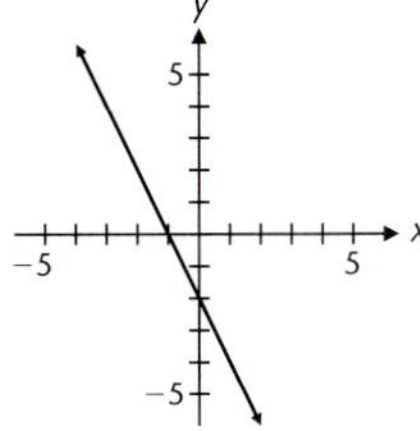

3. $y = -2x - 2$, $x \le 0$;
a ray (part of a straight line)

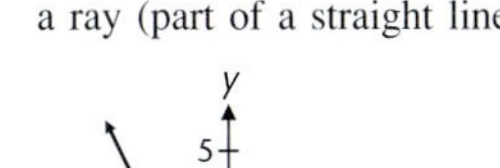

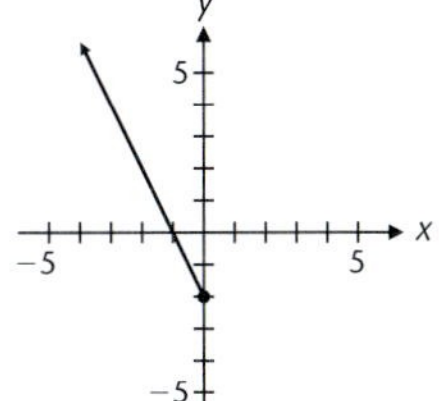

5. $y = -\frac{2}{3}x$;
straight line

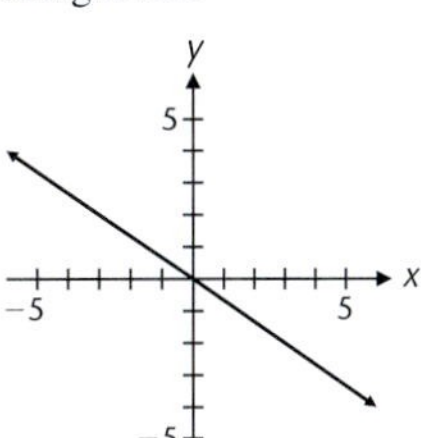

7. $y^2 = 4x$;
parabola

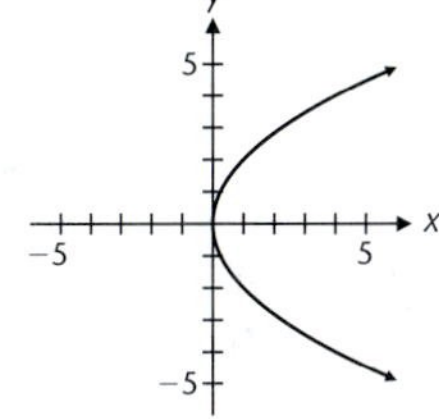

9. $y^2 = 4x$, $y \ge 0$;
parabola (upper half)

11. $\frac{x^2}{9} + \frac{y^2}{16} = 1$;
ellipse

13. $(x-2)^2 + (y-3)^2 = 4$;
circle

15. $y = -\frac{2}{x}$;
hyperbola

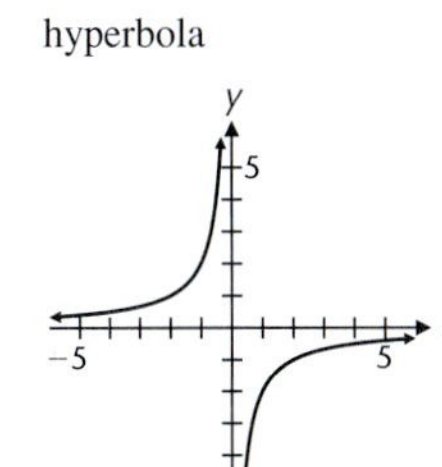

17. $y^2 = x + 1$, $y \ge 0$, $x \ge -1$;
parabola (upper half)

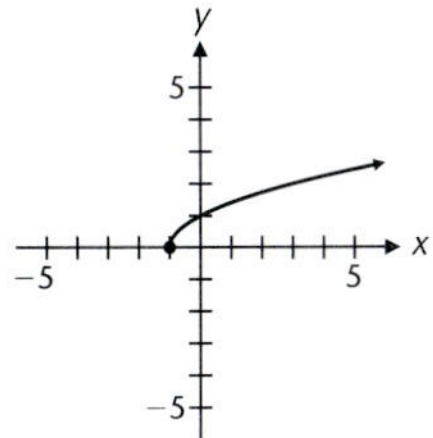

19. $x = t$, $y = \frac{At^2 + Dt + F}{-E}$, $-\infty < t < \infty$; parabola

21. $y = 1/x$, $x > 0$;
hyperbola (one branch)

23. $y^2 = -8(x - 1), -1 \le x \le 1$; part of a parabola

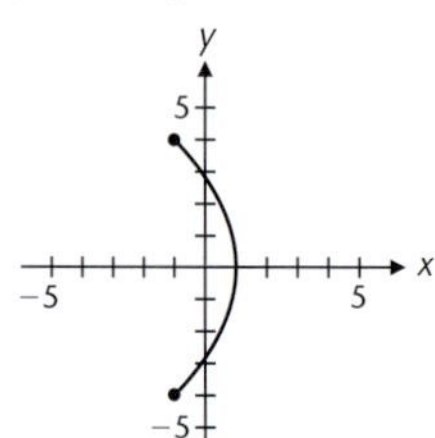

25. $x^2 + y^2 = 2x, x \neq 0$ or $(x - 1)^2 + y^2 = 1, x \neq 0$; circle (note hole at origin)

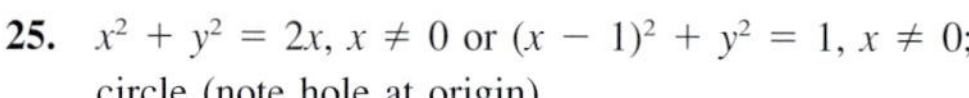

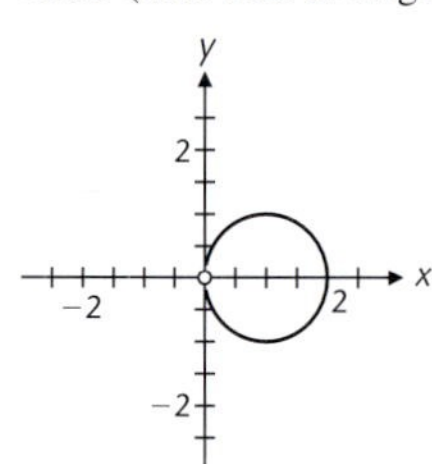

27.

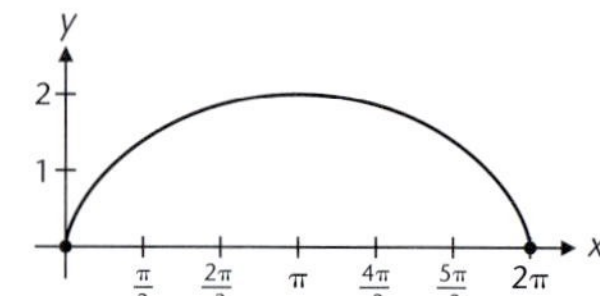

29. $\dfrac{(x-3)^2}{36} + \dfrac{(y-2)^2}{16} = 1$; ellipse with center (3, 2)

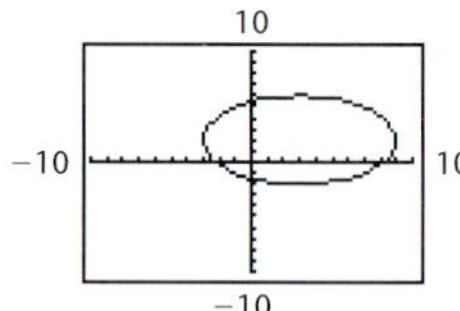

31. $\dfrac{(y+1)^2}{25} - \dfrac{(x+3)^2}{4} = 1$; hyperbola with center $(-3, -1)$

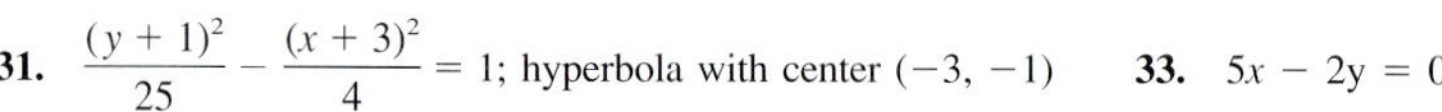

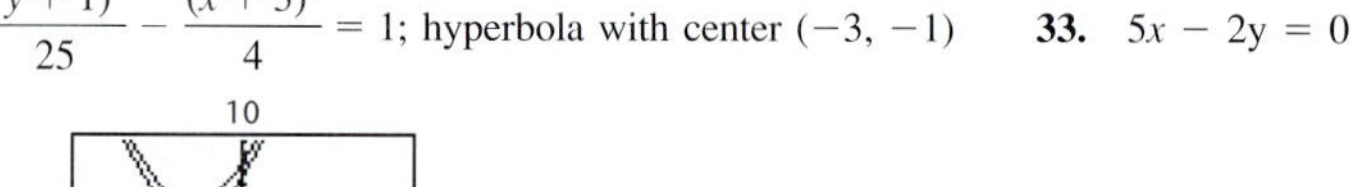

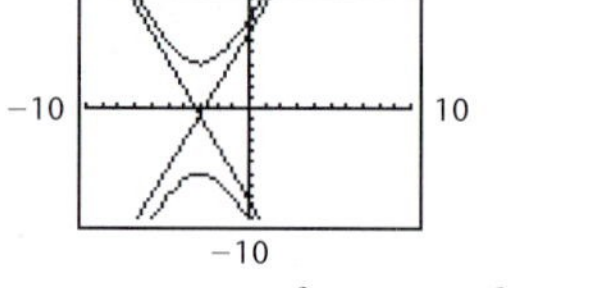

33. $5x - 2y = 0$

35. $\dfrac{(y+1)^2}{25} - \dfrac{(x-4)^2}{9} = 1$; hyperbola with center $(4, -1)$;

$x = 4 + 3 \tan t,\ y = -1 + 5 \sec t,\ -\dfrac{\pi}{2} < t < \dfrac{3\pi}{2},\ t \neq \dfrac{\pi}{2}$

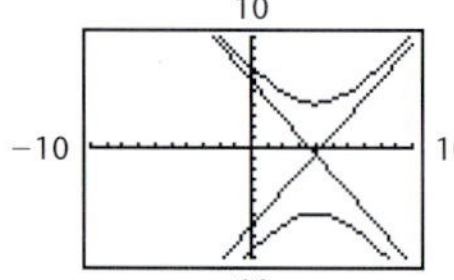

37. $\dfrac{(x-3)^2}{49} + \dfrac{(y+4)^2}{4} = 1$; ellipse with center $(3, -4)$;

$x = 3 + 7 \cos t,\ y = -4 + 2 \sin t,\ 0 \le t \le 2\pi$

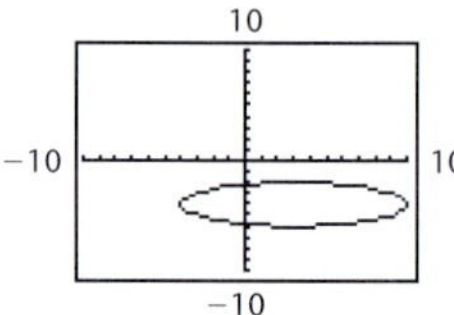

39. (A) $(4.8, -1.5)$
(B) $x^2 + y^2 = 25$; circle

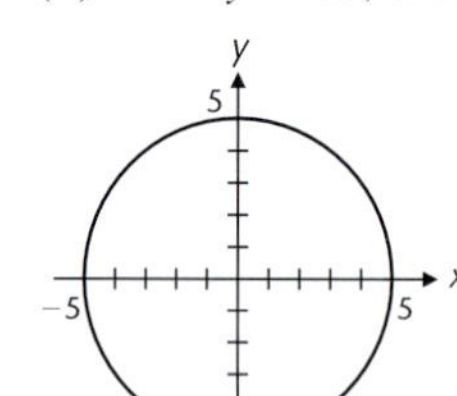

41. (A) 43.292 s (B) 9,183.619 m; 9.184 km (C) 2,295.918 m

Chapter 11 Review Exercise

1. Foci: $F'(-4, 0), F(4, 0)$ *(11-2)*
Major axis length = 10
Minor axis length = 6

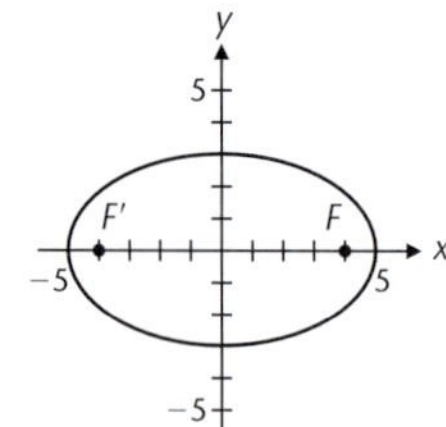

2. *(11-1)*

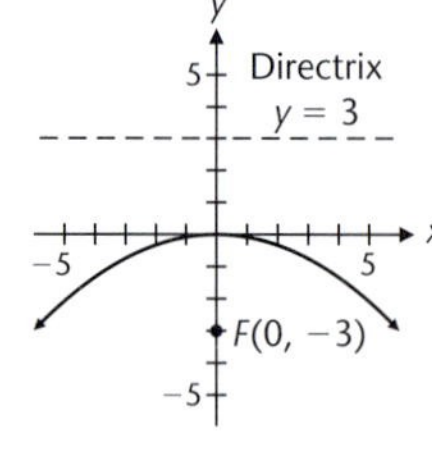

3. Foci: $F'(0, -\sqrt{34}), F(0, \sqrt{34})$ *(11-3)*
Transverse axis length = 6
Conjugate axis length = 10

4. (A) $\dfrac{(y+2)^2}{25} - \dfrac{(x-4)^2}{4} = 1$ (B) Hyperbola *(11-4)*

5. (A) $(x + 5)^2 = -12(y + 4)$ (B) Parabola *(11-4)*

6. (A) $\dfrac{(x-6)^2}{9} + \dfrac{(y-4)^2}{16} = 1$ (B) Ellipse *(11-4)*

7. $y^2 = -7x$ *(11-1)*

8. $\dfrac{x^2}{4} + \dfrac{y^2}{965} = 1$ *(11-2)*

9. $\dfrac{x^2}{25} - \dfrac{y^2}{11} = 1$ *(11-3)*

10. $y = \frac{1}{2}x + 1$, $x \leq 0$; a ray (part of a straight line) *(11-5)*

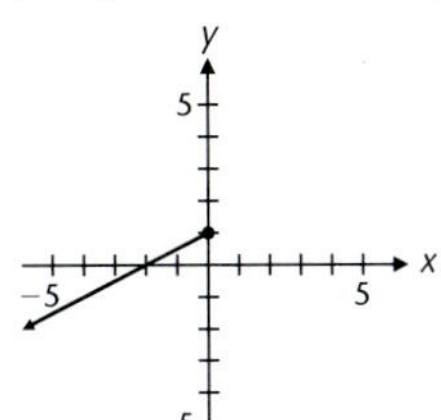

11. *(11-2, 8-3)*

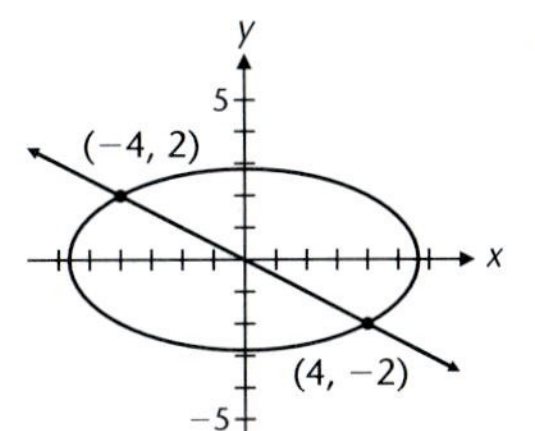

12. *(11-1, 11-2, 8-3)*

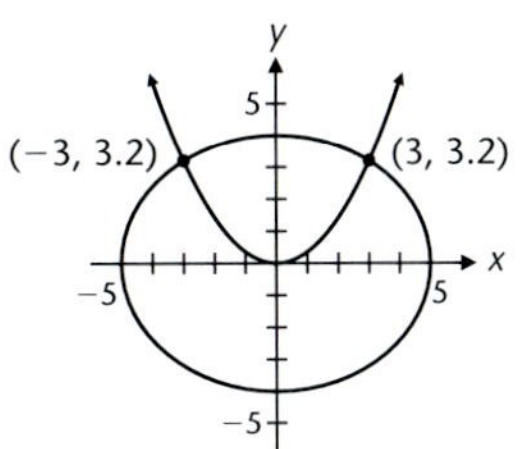

13. *(11-2, 8-3)*

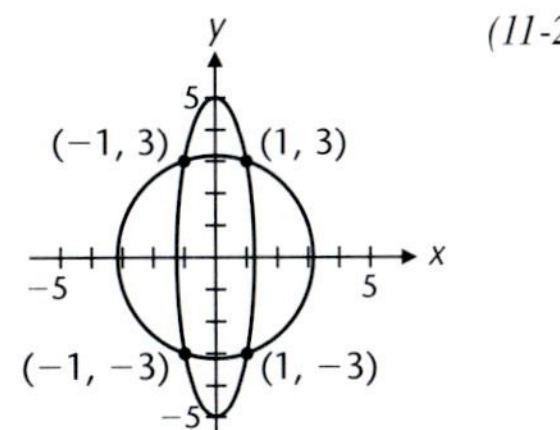

14. $\frac{(x+3)^2}{4} + \frac{(y-2)^2}{16} = 1$; ellipse *(11-4)*

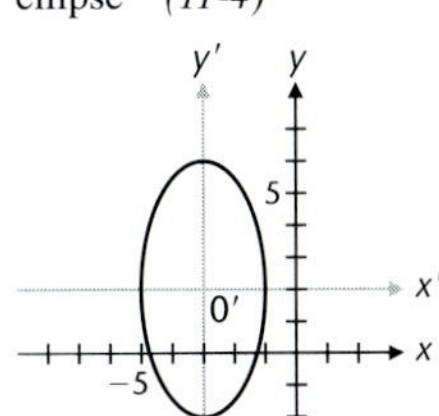

15. $(x-2)^2 = 8(y+3)$; parabola *(11-4)*

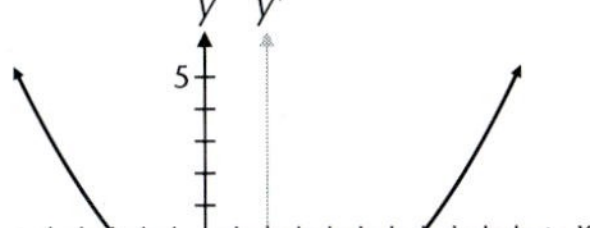

16. $\frac{(x+3)^2}{9} - \frac{(y+2)^2}{4} = 1$; hyperbola *(11-4)*

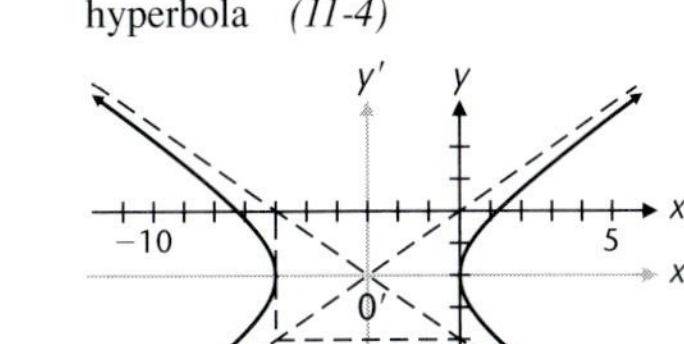

17. $\frac{(x+2)^2}{4} + \frac{(y-3)^2}{16} = 1$; ellipse *(11-5)*

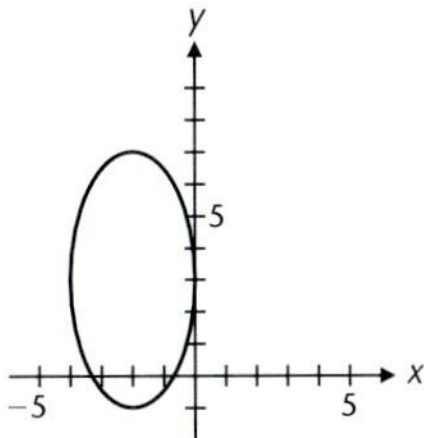

18. $m = 0.2$; $x^2 = 50y$ is a magnification by a factor 50 of $x^2 = y$ *(11-1)*

19. $(y-4)^2 = -8(x-4)$, or $y^2 - 8y + 8x - 16 = 0$ *(11-1)*

20. $\frac{x^2}{4} - \frac{y^2}{12} = 1$; hyperbola *(11-3)*

21. $\frac{x^2}{36} + \frac{y^2}{20} = 1$; ellipse *(11-2)*

22. $F'(-3, -\sqrt{12} + 2)$, $F(-3, \sqrt{12} + 2)$ *(11-4)*

23. $F(2, -1)$ *(11-4)*

24. $F'(-\sqrt{13} - 3, -2)$, $F(\sqrt{13} - 3, -2)$ *(11-4)*

25. $y = \frac{1}{x}$, $x > 0$; hyperbola (one branch) *(11-5)*

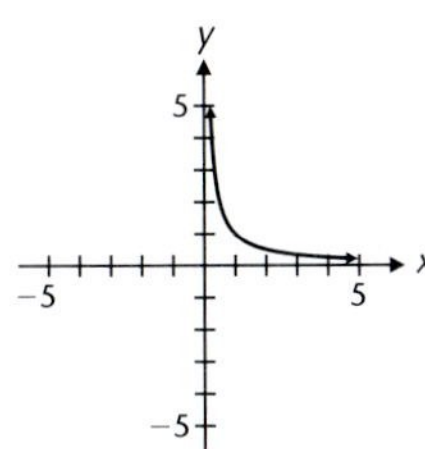

26. (2.09, 2.50), (3.67, −1.92) *(11-4)*

27. 4 ft *(11-1)*

28. $\frac{x^2}{5^2} + \frac{y^2}{3^2} = 1$ *(11-2)*

29. 4.72 ft deep *(11-3)*

Cumulative Review Exercise: Chapters 10 and 11

1. (A) Neither (B) Arithmetic (C) Both (D) Geometric (E) Geometric (F) Neither *(10-3)*
2. (A) −2, 4, −8, 16 (B) $a_8 = 256$ (C) $S_8 = 170$ *(10-3)*
3. (A) 1, 7, 13, 19 (B) $a_8 = 43$ (C) $S_8 = 176$ *(10-3)*
4. (A) −20, −16, −12, −8 (B) $a_8 = 8$ (C) $S_8 = -48$ *(10-3)*
5. (A) 5,040 (B) 13,800 (C) 210 *(10-4)* **6.** (A) 924 (B) 126 (C) 6,720 *(10-4, 10-5)*
7. Foci: $F'(-\sqrt{61}, 0)$, $F(\sqrt{61}, 0)$ *(11-3)*
Transverse axis length = 12
Conjugate axis length = 10

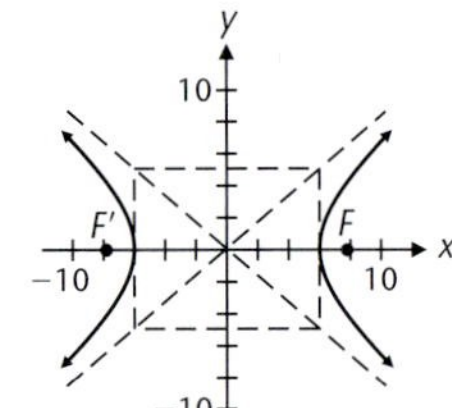

8. Foci: $F'(-\sqrt{11}, 0)$, $F(\sqrt{11}, 0)$ *(11-2)*
Major axis length = 12
Minor axis length = 10

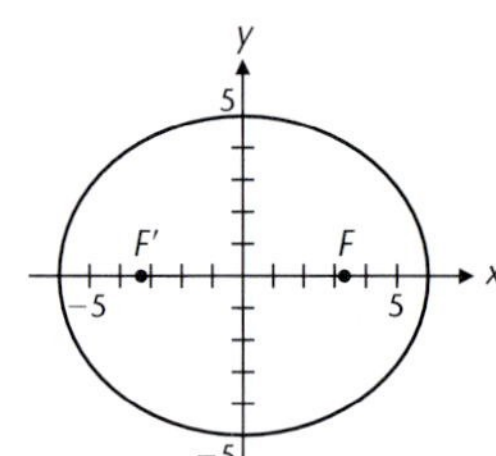

9. *(11-1)*

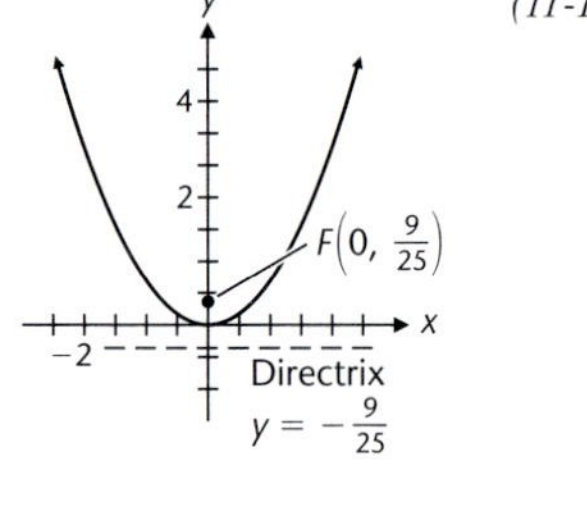

10. 8 combined outcomes: (B) $2 \cdot 2 \cdot 2 = 8$ *(10-5)* **11.** (A) $4 \cdot 3 \cdot 2 \cdot 1 = 24$ (B) $P_{4,4} = 4! = 24$ *(10-5)*

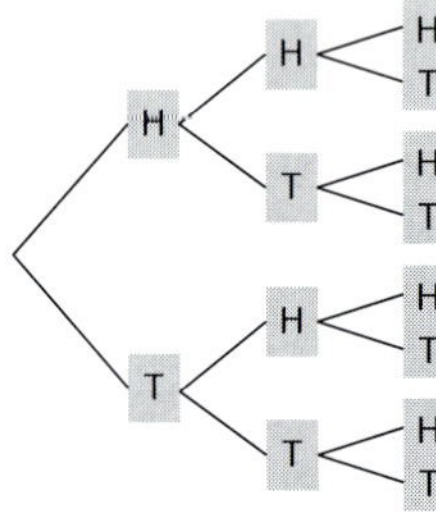

12. $y = 2x - 1$; straight line *(11-5)*

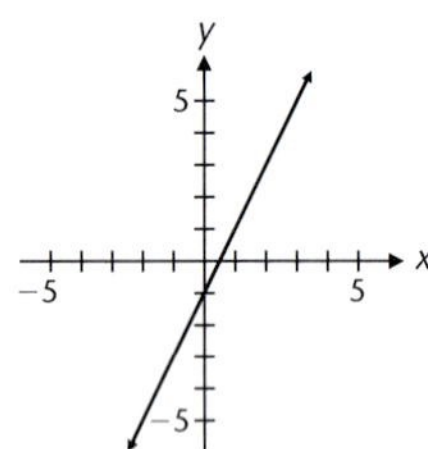

13. P_1: $1 = 1(1)$; P_2: $1 + 5 = 2(3)$
P_3: $1 + 5 + 9 = 3(5)$ *(10-2)*
14. P_1: $1^2 + 1 + 2 = 4$ is divisible by 2
P_2: $2^2 + 2 + 2 = 8$ is divisible by 2
P_3: $3^2 + 3 + 2 = 14$ is divisible by 2 *(10-2)*
15. P_k: $1 + 5 + 9 + \cdots + (4k - 3) = k(2k - 1)$
P_{k+1}: $1 + 5 + 9 + \cdots + (4k - 3) + (4k + 1) = (k + 1)(2k + 1)$ *(10-2)*
16. P_k: $k^2 + k + 2 = 2r$ for some integer r
P_{k+1}: $(k + 1)^2 + (k + 1) + 2 = 2s$ for some integer s *(10-2)*
17. $y = -2x^2$ *(11-1)* **18.** $\frac{x^2}{25} + \frac{y^2}{16} = 1$ *(11-2)* **19.** $\frac{x^2}{64} - \frac{y^2}{25} = 1$ *(11-3)* **20.** $1 + 4 + 27 + 256 + 3{,}125 = 3{,}413$ *(10-1)*
21. $\sum_{k=1}^{6} \frac{(-1)^{k+1}2^k}{(k+1)!}$ *(10-1)* **22.** 81 *(10-3)* **23.** 360; 1,296; 750 *(10-5)* **24.** $n = 22$ *(10-3)*

25. $\frac{(x-2)^2}{49} + \frac{(y+3)^2}{25} = 1$; ellipse *(11-5)* **26.** (A) 6,375,600 (B) 53,130 (C) 53,130 *(10-4, 10-5)*

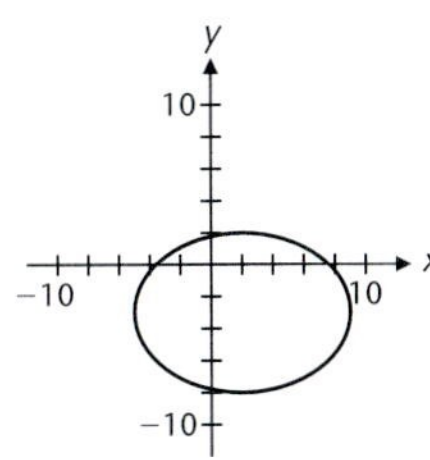

27. $a^6 + 3a^5b + \frac{15}{4}a^4b^2 + \frac{5}{2}a^3b^3 + \frac{15}{16}a^2b^4 + \frac{3}{16}ab^5 + \frac{1}{64}b^6$ *(10-4)* **28.** $153{,}090x^6y^4$; $-3{,}240x^3y^7$ *(10-4)* **31.** 61,875 *(10-3)*
32. $\frac{27}{11}$ *(10-3)* **33.** $a_{22} = 0.236$; 8 terms *(10-4)*

34. $(y-2)^2 = 4(x+3)$; parabola *(11-1)*

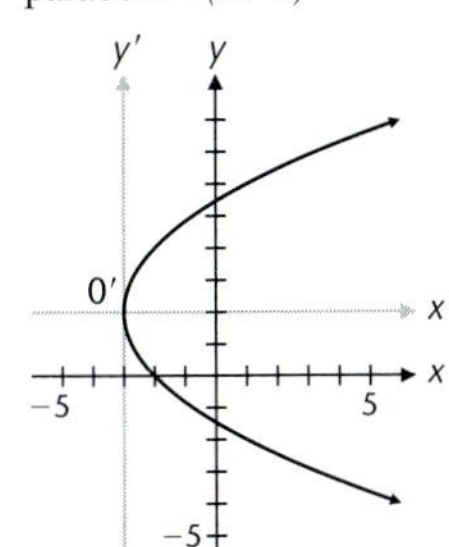

35. $\frac{(y+2)^2}{4} - \frac{(x+1)^2}{16} = 1$; hyperbola *(11-3)*

36. $\frac{(x-2)^2}{9} + \frac{(y+3)^2}{4} = 1$; ellipse *(11-2)*

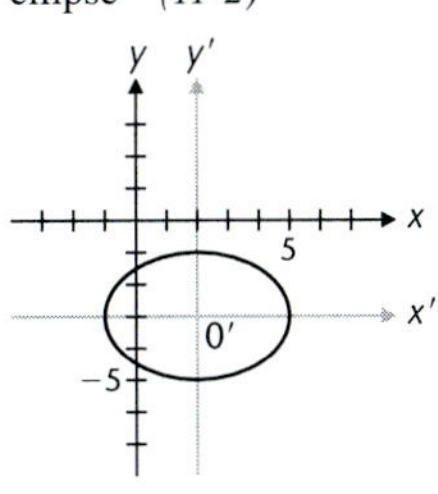

37. $10^9 = 1{,}000{,}000{,}000$; 3,628,800 *(10-5)* **38.** (−2.26, −4.72), (1.85, 3.09) *(11-4)*
40. $x^6 - 12ix^5 - 60x^4 + 160ix^3 + 240x^2 - 192ix - 64$ *(10-4)* **41.** $(x-6)^2 = -4(y-2)$ or $x^2 - 12x + 4y + 28 = 0$ *(11-1)*
42. $\pm 2\sqrt{3}$ *(11-2)* **43.** 8 *(11-3)* **44.** Geometric *(10-3, 10-4)* **45.** Neither *(10-3)* **46.** Both *(10-3)*
47. Arithmetic *(10-3)* **48.** False *(10-2, 10-4)* **49.** True *(10-2, 10-4)* **50.** False *(10-2, 10-4)* **51.** True *(10-2, 10-4)*
52. True *(11-2)* **53.** False *(11-1)* **54.** False *(11-3)* **55.** True *(11-2)* **56.** False *(11-3)*
57. One parabola; $y^2 = 8x + 9$ *(11-4)* **58.** $y + 4 = \pm\frac{1}{2}(x - 5)$ *(11-3, 11-4)* **59.** $4\sqrt{10}$ *(11-2)* **60.** $C_{7,3} = 35$ *(10-5)*
61. $x + 4 = (y-1)^2$, $y < 1$; lower half of a parabola (excluding the vertex) *(11-5)*

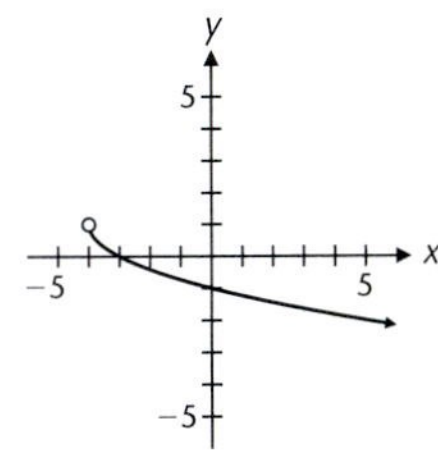

64. $x^2 - 8y^2 - 2x - 8y + 17 = 0$; hyperbola *(11-3)* **65.** \$6,000,000 *(10-3)* **66.** 4 in. *(11-1)* **67.** 32 ft, 14.4 ft *(11-2)*

APPENDIX A

Exercise A-1

1. T **3.** F **5.** F **7.** T **9.** $7 + x$ **11.** $(xy)z$ **13.** $9m$ **15.** Commutative (·) **17.** Distributive
19. Inverse (·) **21.** Inverse (+) **23.** Identity (+) **25.** Negatives **27.** {−2, 0, 2, 4} **29.** {a, s, t, u} **31.** ∅
33. (A) 4 (B) 8 (C) 16 **35.** Negatives **37.** Associative (·) **39.** Distributive **41.** Zero **43.** Yes
45. (A) T (B) F (C) T **47.** $\frac{3}{5}$ and −1.43 are two examples of infinitely many
49. (A) $S = \{1, 12\}$ (B) $S = \{-3, 0, 1, 12\}$ (C) $S = \{-3, -\frac{2}{3}, 0, 1, \frac{9}{5}, 12\}$
51. (A) 0.888 888 . . . ; repeating; repeated digit: 8 (B) 0.272 727 . . . ; repeating; repeated digits: 27
(C) 2.236 067 977 . . . ; nonrepeating and nonterminating (D) 1.375; terminating
53. (B) is false, since, for example, $5 - 3 \neq 3 - 5$ (D) is false, since, for example, $9 \div 3 \neq 3 \div 9$
55. (A) {1, 2, 3, 4, 6} (B) {2, 4} **57.** $\frac{1}{11}$ **59.**

$$\begin{array}{r} 23 \\ \times 12 \\ \hline 46 \\ 230 \\ \hline 276 \end{array} \qquad \begin{aligned} 23 \cdot 12 &= 23(2 + 10) \\ &= 23 \cdot 2 + 23 \cdot 10 \\ &= 46 + 230 \\ &= 276 \end{aligned}$$

Exercise A-2

1. 3 **3.** $2x^3 - x^2 + 2x + 4$ **5.** $2x^3 - 5x^2 + 6$ **7.** $6x^4 - 13x^3 + 9x^2 + 13x - 10$ **9.** $4x - 6$ **11.** $6y^2 - 16y$
13. $m^2 - n^2$ **15.** $4t^2 - 11t + 6$ **17.** $3x^2 - 7xy - 6y^2$ **19.** $4m^2 - 49$ **21.** $30x^2 - 2xy - 12y^2$ **23.** $9x^2 - 4y^2$
25. $16x^2 - 8xy + y^2$ **27.** $a^3 + b^3$ **29.** $-x + 27$ **31.** $32a - 34$ **33.** $2x^4 - x^3 - 6x^2 + 7x - 2$ **35.** $x^4 - 8x^2y^2 + 16y^4$
37. $5u^2 - 12uv + 13v^2$ **39.** $z^3 - 1$ **41.** $8m^3 - 12m^2n + 6mn^2 - n^3$ **43.** $3h$ **45.** $4hx - 3h + 2h^2$ **47.** $4hx - 4h + 2h^2$
49. $3hx^2 - 4hx + 3h^2x - 2h^2 + h^3$ **51.** $m^3 - 3m^2 - 5$ **53.** $2x^3 - 13x^2 + 25x - 18$ **55.** $9x^3 - 9x^2 - 18x$
57. $(1 + 1)^2 \neq 1^2 + 1^2$; either a or b must be zero. **59.** m **61.** Now the degree is less than or equal to m.
63. Perimeter $= 2x + 2(x - 5) = 4x - 10$ **65.** Value $= 5x + 10(x - 5) + 25(x - 3) = 40x - 125$
67. Volume $= \frac{4}{3}\pi(x + 0.3)^3 - \frac{4}{3}\pi x^3 = 1.2\pi x^2 + 0.36\pi x + 0.036\pi$

Exercise A-3

1. $2x^2(3x^2 - 4x - 1)$ **3.** $5xy(2x^2 + 4xy - 3y^2)$ **5.** $(5x - 3)(x + 1)$ **7.** $(2w - x)(y - 2z)$ **9.** $(x + 3)(x - 2)$
11. $(2m - 1)(3m + 5)$ **13.** $(2x - 3y)(x - 2y)$ **15.** $(4a - 3b)(2c - d)$ **17.** $(2x + 3)(x - 1)$ **19.** Prime
21. $(m - 2)(m + 3)$ **23.** $(2a + 3b)(2a - 3b)$ **25.** $(2x - 5)^2$ **27.** Prime **29.** $6(x + 2)(x + 6)$ **31.** $2y(y - 3)(y - 8)$
33. $y(4x - 1)^2$ **35.** Prime **37.** $xy(x - 3y)(x + 3y)$ **39.** $3m(m^2 - 2m + 5)$ **41.** $(m + n)(m^2 - mn + n^2)$
43. $2x(3x + 1)(x + 1)^3$ **45.** $2(3x - 5)(2x - 3)(12x - 19)$ **47.** $9x^4(9 - x)^3(5 - x)$ **49.** $2(x + 1)(x^2 - 5)(3x + 5)(x - 1)$
51. $[(a - b) - 2(c - d)][(a - b) + 2(c - d)]$ **53.** $(2m - 3n)(a + b)$ **55.** Prime **57.** $(x + 3)(x - 3)^2$
59. $(a - 2)(a + 1)(a - 1)$ **61.** Prime **63.** $(m - n)(m + n)(m^2 + n^2)$ **65.** Prime **67.** $(m + n)(m + n - 1)$
69. $2a[3a - 2(x + 4)][3a + 2(x + 4)]$ **71.** $(x^2 - x + 1)(x^2 + x + 1)$
73. (A) $4(10 - x)(10 + x) = 400 - 4x^2$ (B) $4x(10 - x)^2 = 400x - 80x^2 + 4x^3$

Exercise A-4

1. $\frac{a^2}{2}$ **3.** $\frac{22y + 9}{252}$ **5.** $\frac{x^2 + 8}{8x^3}$ **7.** $\frac{1}{x + 2}$ **9.** $\frac{1}{2m + n}$ **11.** $\frac{2b^2}{(a + b)^2(a - b)}$ **13.** $\frac{m^2}{m - 1}$ **15.** $\frac{5}{x - 2}$
17. $\frac{1}{y + 2}$ **19.** $\frac{x - y}{y}$ **21.** $\frac{4(x + 1)(x - 1)(x^2 + 2)^2}{x^3}$ **23.** $\frac{x(2 + 3x)}{(1 - 3x)^4}$ **25.** $\frac{(x + 1)(x - 9)}{(x + 4)^4}$ **27.** $\frac{1}{y - 1}$ **29.** -1
31. $\frac{7y - 9x}{xy(a - b)}$ **33.** $\frac{x^2 - x + 1}{2(x - 9)}$ **35.** $\frac{(x - y)^2}{y^2(x + y)}$ **37.** $\frac{1}{x - 4}$ **39.** $\frac{x - 3}{x - 1}$ **41.** $\frac{-1}{x(x + h)}$ **43.** $\frac{x^2 + xh + 4x + 2h}{(x + h + 2)(x + 2)}$
45. (A) Incorrect (B) $x + 1$ **47.** (A) Incorrect (B) $2x + h$ **49.** (A) Incorrect (B) $\frac{x^2 - 2}{x + 1}$ **51.** (A) Correct
53. $\frac{-x(x + y)}{y}$ **55.** $\frac{a - 2}{a}$

Exercise A-5

1. 1 **3.** $60x^{11}$ **5.** $4x^8/z^6$ **7.** $r^{10}s^{15}/(p^5q^{20})$ **9.** 10^7 **11.** $3a^4/b^3$ **13.** $1/y$ **15.** 4×10^{11} **17.** 4.532×10^7
19. 6.6×10^{-2} **21.** 8.4×10^{-8} **23.** 0.000 09 **25.** 3,480,000 **27.** 0.000 000 004 2 **29.** $3y^4/2$ **31.** x^{12}/y^8
33. $4x^8/y^6$ **35.** $w^{12}/(u^{20}v^4)$ **37.** $1/(x + y)^2$ **39.** $x/(x - 1)$ **41.** $-1/(xy)$ **43.** $-9x^2/(x^3 + 3)^4$ **45.** 64
49. $2x - 6x^{-1}$ **51.** $\frac{5}{3}x - \frac{2}{3}x^{-2}$ **53.** $x - \frac{3}{2} + \frac{1}{2}x^{-1}$ **55.** 6.65×10^{-17} **57.** 1.54×10^{12} **59.** 1.0295×10^{11}
61. -4.3647×10^{-18} **63.** 9.4697×10^{29} **65.** $2(a + 2b)^5$ **67.** $(x^2 + xy + y^2)/(xy)$ **69.** $(y - x)/y$ **71.** 1.3×10^{25} lb
73. 10^{10} or 10 billion; 6×10^{11} or 600 billion **75.** 2.07×10^4 dollars per person; \$20,700 per person

Exercise A-6

1. 5 **3.** 27 **5.** -4 **7.** Not a real number **9.** $\frac{9}{25}$ **11.** $\frac{1}{32}$ **13.** $a^{5/3}$ **15.** $c^{2/5}$ **17.** $1/u^{1/2}$ **19.** $2x^2/y$
21. $1/(a^{1/4}b^{1/3})$ **23.** $xy^2/2$ **25.** $2b^2/(3a^2)$ **27.** $2/(3x^{7/12})$ **29.** $a^{1/3}/b^2$ **31.** $6m - 2m^{19/3}$ **33.** $a - a^{1/2}b^{1/2} - 6b$
35. $4x - 9y$ **37.** $x + 4x^{1/2}y^{1/2} + 4y$ **39.** 29.52 **41.** 0.030 93 **43.** 5.421 **45.** 107.6
47. $x = y = 1$ is one of many choices. **49.** $x = y = 1$ is one of many choices. **51.** $3 - \frac{3}{4}x^{-1/2}$ **53.** $\frac{3}{5}x^{-1/3} + \frac{1}{5}x^{-1/2}$
55. $\frac{1}{2}x^{5/3} - 2x^{1/6}$ **57.** $a^{1/n}b^{1/m}$ **59.** $1/(x^{3m}y^{4n})$ **61.** (A) $x = -2$, for example (B) $x = 2$, for example (C) Not possible
63. No **65.** $(x - 3)/(2x - 1)^{3/2}$ **67.** $(4x - 3)/(3x - 1)^{4/3}$ **69.** 1,920 units **71.** 428 ft

Exercise A-7

1. $\sqrt[3]{m^2}$ or $(\sqrt[3]{m})^2$ (first preferred) **3.** $6\sqrt[5]{x^3}$ (not $\sqrt[5]{6x^3}$) **5.** $\sqrt[5]{(4xy^3)^2}$ **7.** $\sqrt{x + y}$ **9.** $b^{1/5}$ **11.** $5x^{3/4}$ **13.** $(2x^2y)^{3/5}$
15. $x^{1/3} + y^{1/3}$ **17.** -2 **19.** $3x^4y^2$ **21.** $2mn^2$ **23.** $2ab^2\sqrt{2ab}$ **25.** $2xy^2\sqrt[3]{2xy}$ **27.** $\sqrt{m}$ **29.** $\sqrt[15]{xy}$
31. $3x\sqrt[3]{3}$ **33.** $\sqrt{5}/5$ **35.** $2\sqrt{3x}$ **37.** $2\sqrt{2} + 2$ **39.** $\sqrt{3} - \sqrt{2}$ **41.** $3x^2y^2\sqrt[5]{3x^2y}$ **43.** $n\sqrt[4]{2m^3n}$
45. $\sqrt[3]{a^2(b - a)}$ **47.** $\sqrt[4]{a^3b}$ **49.** $x^2y\sqrt[3]{6xy^2}$ **51.** In simplified form **53.** $\sqrt{2}/2$ or $\frac{1}{2}\sqrt{2}$ **55.** $2a^2b\sqrt[3]{4a^2b}$
57. $\sqrt[4]{12xy^3}/(2x)$ or $[1/(2x)]\sqrt[4]{12xy^3}$ **59.** $(6y + 9\sqrt{y})/(4y - 9)$ **61.** $(38 + 11\sqrt{10})/117$ **63.** $\sqrt{x^2 + 9} + 3$ **65.** $1/(\sqrt{t} + \sqrt{x})$

67. $1/(\sqrt{x+h}+\sqrt{x})$ **69.** 0.1816 **71.** 3.557 **73.** 0.092 75 **75.** 1.602 **77.** Both are 1.259 **79.** Both are 0.4807

81. $x \le 0$ **83.** All real numbers **85.** A and E, B and F, C and D **87.** $\dfrac{\sqrt[3]{a^2}+\sqrt[3]{ab}+\sqrt[3]{b^2}}{a-b}$

89. $\dfrac{(\sqrt{x}-\sqrt{y}-\sqrt{z})[(x+y-z)+2\sqrt{xy}]}{(x+y-z)^2-4xy}$ **91.** $\dfrac{1}{\sqrt[3]{(x+h)^2}+\sqrt[3]{x(x+h)}+\sqrt[3]{x^2}}$

93. $\sqrt[kn]{x^{km}} = (x^{km})^{1/kn} = x^{km/kn} = x^{m/n} = \sqrt[n]{x^m}$

Appendix A Review Exercise

1. (A) T (B) T (C) F (D) T (E) F (F) F *(A-1)* **2.** (A) $(y+z)x$ (B) $(2+x)+y$ (C) $2x+3x$ *(A-1)*
3. x^3+3x^2+5x-2 *(A-2)* **4.** $x^3-3x^2-3x+22$ *(A-2)* **5.** $3x^5+x^4-8x^3+24x^2+8x-64$ *(A-2)* **6.** 3 *(A-2)*
7. 1 *(A-2)* **8.** $14x^2-30x$ *(A-2)* **9.** $9m^2-25n^2$ *(A-2)* **10.** $6x^2-5xy-4y^2$ *(A-2)* **11.** $4a^2-12ab+9b^2$ *(A-2)*
12. $(3x-2)^2$ *(A-3)* **13.** Prime *(A-3)* **14.** $3n(2n-5)(n+1)$ *(A-3)* **15.** $(12a^3b-40b^2-5a)/(30a^3b^2)$ *(A-4)*
16. $(7x-4)/[6x(x-4)]$ *(A-4)* **17.** $(y+2)/[y(y-2)]$ *(A-4)* **18.** u *(A-4)* **19.** $6x^5y^{15}$ *(A-5)* **20.** $3u^4/v^2$ *(A-5)*
21. 6×10^2 *(A-5)* **22.** x^6/y^4 *(A-5)* **23.** $u^{7/3}$ *(A-6)* **24.** $3a^2/b$ *(A-6)* **25.** $3\sqrt[5]{x^2}$ *(A-7)* **26.** $-3(xy)^{2/3}$ *(A-7)*
27. $3x^2y\sqrt[3]{x^2y}$ *(A-7)* **28.** $6x^2y^3\sqrt{xy}$ *(A-7)* **29.** $2b\sqrt{3a}$ *(A-7)* **30.** $(3\sqrt{5}+5)/4$ *(A-7)* **31.** $\sqrt[4]{y^3}$ *(A-7)*
32. $\{-3,-1,1\}$ *(A-1)* **33.** Subtraction *(A-1)* **34.** Commutative (+) *(A-1)* **35.** Distributive *(A-1)*
36. Associative (·) *(A-1)* **37.** Negatives *(A-1)* **38.** Identity (+) *(A-1)* **39.** (A) T (B) F *(A-1)*
40. 0 and −3 are two examples of infinitely many. *(A-1)* **41.** (A) (a) and (d) (B) None *(A-2)* **42.** $4xy-2y^2$ *(A-2)*
43. $m^4-6m^2n^2+n^4$ *(A-2)* **44.** $10xh+5h^2-7h$ *(A-2)* **45.** $2x^3-4x^2+12x$ *(A-2)* **46.** $x^3-6x^2y+12xy^2-8y^3$ *(A-2)*
47. $(x-y)(7x-y)$ *(A-3)* **48.** Prime *(A-3)* **49.** $3xy(2x^2+4xy-5y^2)$ *(A-3)* **50.** $(y-b)(y-b-1)$ *(A-3)*
51. $3(x+2y)(x^2-2xy+4y^2)$ *(A-3)* **52.** $(y-2)(y+2)^2$ *(A-3)* **53.** $x(x-4)^2(5x-8)$ *(A-3)* **54.** $\dfrac{(x-4)(x+2)^2}{x^3}$ *(A-4)*
55. $2m/[(m+2)(m-2)^2]$ *(A-4)* **56.** y^2/x *(A-4)* **57.** $(x-y)/(x+y)$ *(A-4)* **58.** $-ab/(a^2+ab+b^2)$ *(A-4)*
59. Incorrect; correct final form is $(x^2+2x-2)/(x-1)$ *(A-4)* **60.** $\frac{1}{4}$ *(A-5)* **61.** $\frac{5}{9}$ *(A-5)* **62.** $3x^2/(2y^2)$ *(A-6)*
63. $27a^{1/6}/b^{1/2}$ *(A-6)* **64.** $x+2x^{1/2}y^{1/2}+y$ *(A-6)* **65.** $6x+7x^{1/2}y^{1/2}-3y$ *(A-6)* **66.** 2×10^{-7} *(A-5)*
67. 3.213×10^6 *(A-5)* **68.** 4.434×10^{-5} *(A-5)* **69.** -4.541×10^{-6} *(A-5)* **70.** 128,800 *(A-6)* **71.** 0.01507 *(A-6)*
72. 0.3664 *(A-7)* **73.** 1.640 *(A-7)* **74.** 0.08726 *(A-6)* **75.** $-6x^2y^2\sqrt[5]{3x^2y}$ *(A-7)* **76.** $x\sqrt[3]{2x^2}$ *(A-7)*
77. $\sqrt[5]{12x^3y^2}/(2x)$ *(A-7)* **78.** $y\sqrt[3]{2x^2y}$ *(A-7)* **79.** $\sqrt[3]{2x^2}$ *(A-7)* **80.** $2x-3\sqrt{xy}-5y$ *(A-7)*
81. $(6x+3\sqrt{xy})/(4x-y)$ *(A-7)* **82.** $(4u-12\sqrt{uv}+9v)/(4u-9v)$ *(A-7)* **83.** $\sqrt{y^2+4}+2$ *(A-7)* **84.** $1/(\sqrt{t}+\sqrt{5})$ *(A-7)*
85. $2-\frac{3}{2}x^{-1/2}$ *(A-7)* **86.** $\frac{6}{11}$; rational *(A-1)* **87.** (A) $\{-4,-3,0,2\}$ (B) $\{-3,2\}$ *(A-1)* **88.** 0 *(A-7)*
89. x^3+8x^2-6x+1 *(A-2)* **90.** $x(2a+3x-4)(2a-3x-4)$ *(A-3)* **91.** All three have the same value. *(A-7)*
92. $\frac{2}{3}(x-2)(x+3)^4$ *(A-5)* **93.** $a^2b^2/(a^3+b^3)$ *(A-5)* **94.** $x-y$ *(A-6)* **95.** x^{m-1} *(A-6)*
96. $(1+\sqrt[3]{x}+\sqrt[3]{x^2})/(1-x)$ *(A-7)* **97.** $1/(\sqrt[3]{t^2}+\sqrt[3]{5t}+\sqrt[3]{25})$ *(A-7)* **98.** x^{n+1} *(A-7)*
99. Volume $= 3\pi(x+2)^2-3\pi x^2 = 12\pi x+12\pi$ ft³ *(A-2)* **100.** 2.84×10^4 dollars per person; \$28,400 per person *(A-5)*
101. (A) 24,000 units (B) Production doubles to 48,000 units
(C) At any production level, doubling the units of capital and labor doubles production. *(A-6)*

102. $R = \dfrac{R_1R_2R_3}{R_2R_3+R_1R_3+R_1R_2}$ *(A-4)*

103. (A) $A = 480-6x^2 = 6(80-x^2)$ (B) $V = x(16-2x)(15-1.5x) = 240x-54x^2+3x^3$ *(A-3)*

APPLICATIONS INDEX

SUBJECT INDEX